11th European Space Power Conference (ESPC 2016)

E3S Web of Conferences Volume 16 (2017)

Thessaloniki, Greece
3 - 7 October 2016

Editors:

Arturo Fernandez

ISBN: 978-1-5108-4119-2

Printed from e-media with permission by:

Curran Associates, Inc.
57 Morehouse Lane
Red Hook, NY 12571

proceedings.com

Some format issues inherent in the e-media version may also appear in this print version.

Printed by Curran Associates, Inc. (2017)

For additional information, please contact EDP Sciences – Web of Conferences at the address below.

EDP Sciences – Web of Conferences
17, Avenue du Hoggar
Parc d'Activité de Courtabœuf
BP 112
F-91944 Les Ulis Cedex A
France

Phone: +33 (0) 1 69 18 75 75
Fax: +33 (0) 1 69 28 84 91

contact-edps@webofconferences.org

Additional copies of this publication are available from:

Curran Associates, Inc.
57 Morehouse Lane
Red Hook, NY 12571 USA
Phone: 845-758-0400
Fax: 845-758-2633
Email: curran@proceedings.com
Web: www.proceedings.com

TABLE OF CONTENTS

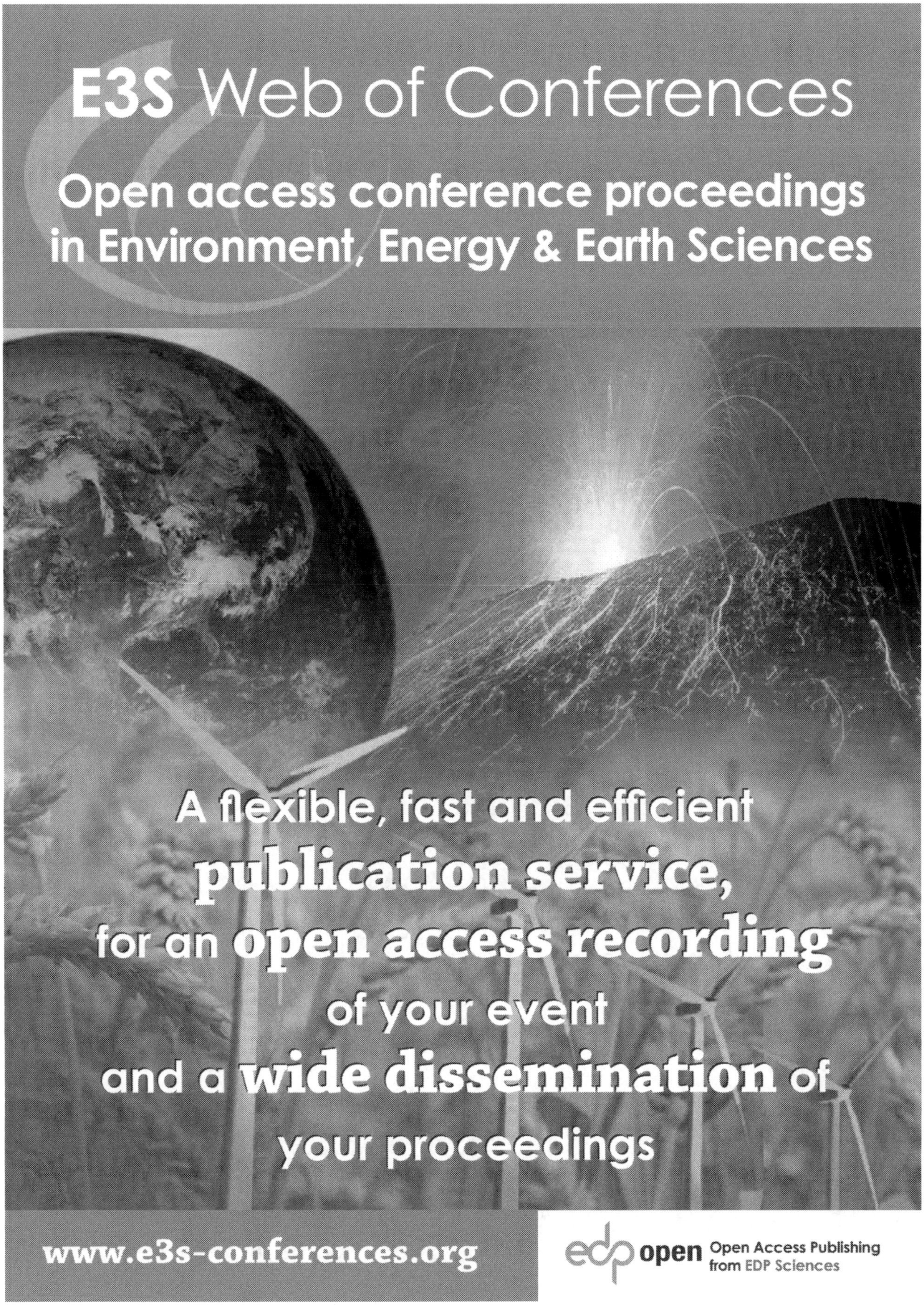
E3S Web of Conferences
Open access conference proceedings
in Environment, Energy & Earth Sciences
A flexible, fast and efficient
publication service,
for an open access recording
of your event
and a wide dissemination of
your proceedings
www.e3s-conferences.org
edp open Open Access Publishing
from EDP Sciences

edp sciences

E3S Web of Conferences

Statement of Peer review

In submitting conference proceedings to E3S Web of Conferences, I certify to the Publisher that I adhere to the **Policy on Publishing Integrity** of the journal in order to safeguard good scientific practice in publishing.

1. All articles have been subjected to peer review administered by the proceedings editors.
2. Reviews have been conducted by expert referees, who have been requested to provide unbiased and constructive comments aimed, whenever possible, at improving the work.
3. Proceedings editors have taken all reasonable steps to ensure the quality of the materials they publish and their decision to accept or reject a paper for publication has been based only on the merits of the work and the relevance to the journal.

Title, date and place of the conference: 11th EUROPEAN SPACE POWER CONFERENCE ESPC 2016, 3-7 October 2016, PORTO PALACE THESSALONIKI, GREECE

Proceedings editor(s):
Visa: ARTURO FERNANDEZ
On behalf of H. Barde
Head of TEC-EP Division

20-04-2017

Date and editor's signature
Visa: ARTURO FERNANDEZ
On behalf of H. Barde
Head of TEC-EP Division

17, avenue du Hoggar - PA de Courtabœuf – BP 112 - 91944 Les Ulis Cedex A (France)
Tél. : 33 (0)1 69 18 75 75 - Fax : 33(0)1 69 07 45 17 – www.edpsciences.org

ADVANCED PHOTOVOLTAIC POWER SYSTEM DEVELOPMENT AT THE U.S. AIR FORCE RESEARCH LABORATORY

John Merrill[(1)], David Wilt[(1)], David Chapman[(1)], Geoff Bradshaw[(1)], Kyle Montgomery[(1)], Nathan Gapp[(1)] and Bernie Carpenter[(2)]

[(1)] US Air Force Research Laboratory, 3550 Aberdeen Ave SE, KAFB, NM 87117, Email: AFRL.VSSVOrgMailbox@us.af.mil
(2) Aerospace Corporation, Email: Bernie.F.Carpenter@aero.org

ABSTRACT

Photovoltaics continue to be the primary source of electrical power for most near-Sun space missions. The desire to enhance or enable new space missions through higher efficiency, increased specific power (W/kg), increased volumetric power density (W/m3) and improved radiation resistance, along with decreased costs, continues to push the development of novel solar cell and array technologies. To meet present and future space power requirements, advanced multijunction solar cells and novel cell technologies are being pursued. These efforts have resulted in a continual advancement in performance, but new paradigms will be required to continue that performance trend. Similarly, new array technologies are being investigated and developed to meet the ever increasing power system performance requirements.

1. INTRODUCTION

Space photovoltaic (PV) development has made continual and remarkable progress, averaging roughly 0.5% per year improvement in absolute efficiency over many decades (fig. 1). Early solar cell development activities were driven solely to meet spacecraft performance needs, as PV materials costs were considered to be too high for terrestrial applications. Beginning in the early 1970's, investment in space PV development began to wane as the conventional wisdom of the day asserted that PV generation capability was limited to approximately 1 kW of spacecraft power and beyond that nuclear power systems would be utilized. In addition, financial forecasts indicated that photovoltaics would always be too expensive for terrestrial application. Fortunately, these apparent constraints turned out to be inaccurate, and PV development and applications have flourished for both terrestrial and spacecraft needs.

A wide variety of PV technologies have been investigated and developed in order to attain the continual efficiency improvements demonstrated, with current 1-Sun space solar cells projected to attain >35% efficiency in the near future. History has shown that advancements in cell efficiency, as long as the new technologies have a comparable or better environmental durability than current state of the art, are quickly adopted for widespread use in space for both commercial and government spacecraft. This speaks to the leveraging impact increases in photovoltaic system performance has on the rest of the spacecraft design.

This paper focuses on photovoltaic power system development efforts within the Space Vehicles Directorate of the Air Force Research Laboratory (AFRL) specifically focusing on advanced solar cell, blanket and array technologies. AFRL has interest and efforts in energy storage and power management technologies however this presentation will focus on the energy generation component of the power system.

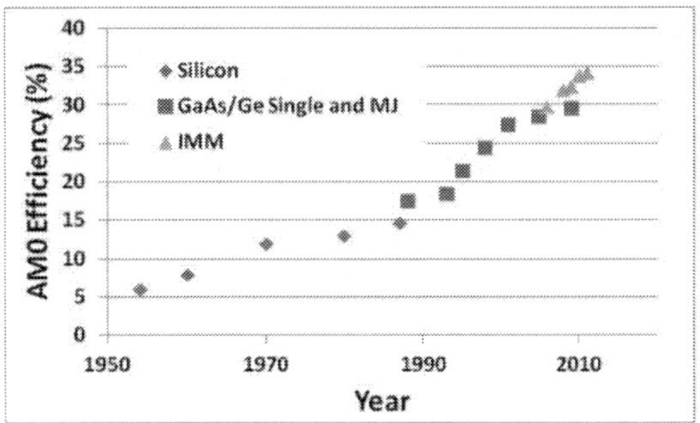

Figure 1. Historical performance trend for space solar cells.

2. PHOTOVOLTAIC DEVICE DEVELOPMENT

Conventional lattice matched germanium (Ge) based triple junction solar cells have largely reached their practical efficiency limit of ~30%. Moving to higher efficiencies has driven solar cell developers to investigate a range of alternative cell technologies, including metamorphic structures, novel materials (e.g. dilute nitride) and novel structures (e.g. quantum enhanced devices). Within the AFRL research portfolio, the emphasis in next generation cells has focused primarily on Inverted Metamorphic Multijunction (IMM) devices [1]. IMM devices consist of conventional III-V materials, but are grown inverted such that the critical high bandgap cells are produced lattice matched to the epitaxial substrate, thus preserving their high performance. The lower bandgap cells in the multjunction stack are then grown lattice mismatched in order to enable the optimum bandgap combinations to be achieved. This approach has resulted in air mass zero (AM0) beginning of life (BOL) efficiencies ranging from 33% to 37%. AFRL/MANTECH is nearing the completion of an IMM manufacturing development program with an IMMX+ cell expected to begin AIAA S-111 qualification in the near future [2]. The IMMX+ cell offers ~33% BOL efficiency and perhaps more importantly, offers a 10% performance benefit at end-of-life (EOL) and at operating temperature, compared to conventional 30% class triple junction solar cells.

In addition to the performance advantage, the IMMX+ is being developed to be delivered to blanket and array manufacturers as a pre-integrated string of solar cells, known as Flex String Arrays (FSAs) (fig. 2). FSAs will be available in custom cell length strings and custom configurations and should offer significantly lower blanket and panel integration costs, compared to conventional discrete cells. A full cell repair/replacement process for the FSAs has been developed and has passed preliminary qualification confidence testing.

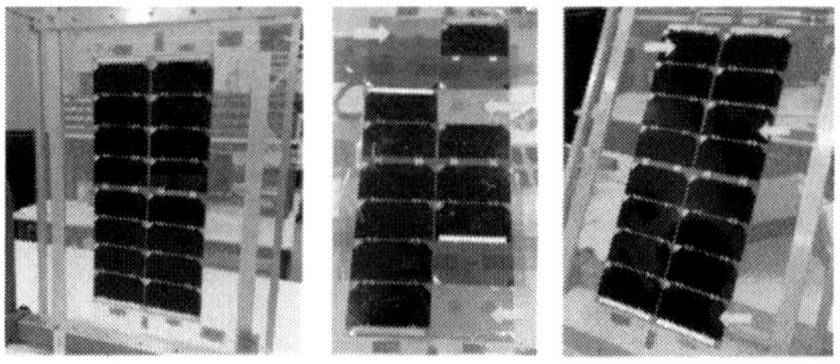

Figure 2. Four 4-cell IMM FSAs assembled into 16-cell string (l), three cells being replaced (center) and string following re-work (r) [2].

AFRL is also investigating a range of other cell technologies which offers high performance. These include dilute nitride based multijunction solar cells [3], GaAs based quantum dot photovoltaics [4], mechanically bonded multijunction solar cells [5] and solar concentrator solar cells [6]. The mechanically bonded and concentrator cells are intended for use in space solar concentrator solar arrays that will be described in the array section.

3. CELL AND BLANKET TECHNOLOGIES

In addition to new solar cells, AFRL is also developing ancillary technologies of benefit to a wide range of cell and array options. The first is a solar cell coverglass replacement technology known as pseudomorphic glass (PMG) [7]. PMG consists of small glass beads (e.g. fused silica) imbedded in space qualified silicone adhesive and offers a flexible, fully encapsulation alternative to conventional coverglass. PMG is currently produced in thin sheets via a doctor blade casting process, coated with a UV rejection coating (Optical Coating Solutions) and then bonded to panels once the cell strings have been integrated. This technology has been validated in space and is being used on multiple commercial LEO satellites.

With the interest of both space and terrestrial solar cell developers in thinner crystalline cells, in order to reduce mass, cost and/or increase performance, the potential for cells to fracture and lose performance may be increased. To address this potential issue, AFRL has developed an advanced solar cell metal matrix composite (MMC) metallization consisting of low-cost, low-purity, multi-wall carbon nanotubes imbedded within the bulk silver metallization [8]. This technology has demonstrated the ability to electrically bridge >40 micron fractures in the underlying semiconductor material and was recently integrated with triple junction space solar cells. That work demonstrated that the Jsc of the cells with MMC metallization was maintained even after the cells were intentionally fractured, whereas cells with conventional metallization demonstrated ~50% reduction in Jsc under similar fracturing tests.

4. SOLAR ARRAY TECHNOLOGIES

In contrast to the widespread adoption of new solar cell technologies, the transition of new solar array technologies has historically been very difficult. Recently however, several new array technologies have been developed which are bucking that trend and moving towards commercial acceptance.

The MOSAIC solar array, under development by Vanguard Space Technologies [9], is based on a modular (e.g. Lego-like) architecture whereby modules containing a single string of solar cells are mechanically attached to a pre-wired frame to form completed panels (fig 3). The MOSAIC frames contain internal mechanisms that enable synchronised deployment with a simple single actuator. The MOSAIC system was designed to be compatible with smaller spacecraft (e.g. ESPA class) utilizing 28Vdc direct energy transfer bus topology and a total power of approximately 1.5 kW. MOSAIC arrays offer the potential for lower cost (reduced non-reoccurring engineering and custom array fabrication costs), faster development (all components are modular and can be warehoused) and the ability to test and fly new cell and panel technologies as part of the operational array. This testing feature is a new capability that is being utilized in two upcoming AFRL sponsored MOSAIC flight experiments (SurreySat and STPSAT IV).

 DOI: 10.1051/e3sconf/20171601001

Figure 3. MOSAIC module (l) and pre-wired MOSAIC panel frame capable of holding six modules (r).

In the area of small arrays, AFRL is also working with Deployable Space Systems (DSS) on the Aladdin solar array (fig 4). This array design leverages recent advancements in strained composite technology and has the potential to offer very low cost and high reliability given the very low part count. The array consists of a strained composite structure (similar to tape measure but with a wider and flatter mid-section). In the folded configuration, cells are placed in patches separated by areas which enable the structure to fold. When the structure is flattened, it is able to be folded and stowed for launch. Once on-orbit, the array deploys using the stored strain energy in the composite structure.

The largest solar array currently being commercialized is the DSS ROSA solar array [10]. ROSA (Roll Out Solar Array) uses two strained composite slit-tube booms to deploy and tension a flexible solar array blanket. ROSA is currently under qualification development testing by Space Systems Loral (SSL) for their next generation of GEO spacecraft. In parallel with the SSL activity, AFRL is working with NASA and DSS to perform a flight demonstration of ROSA aboard the ISS (fig 5). The ISS flight experiment is primarily a structural validation of the ROSA (deployment, stiffness, etc.) although the experiment will include multiple active solar cell strings.

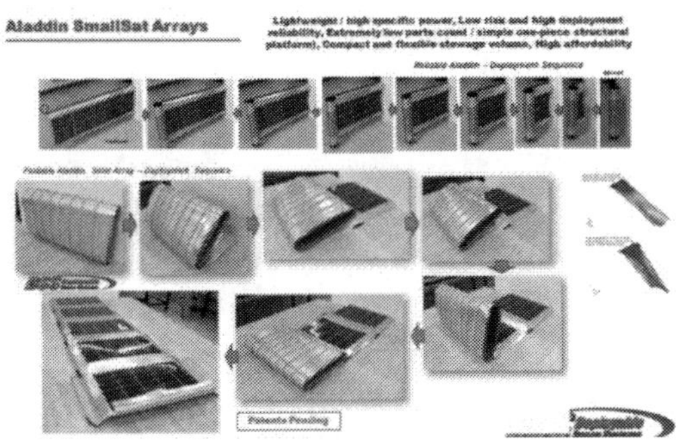

Figure 4. The DSS Aladdin solar array shown in a rolled configuration (top) and a fold-out configuration (bottom).

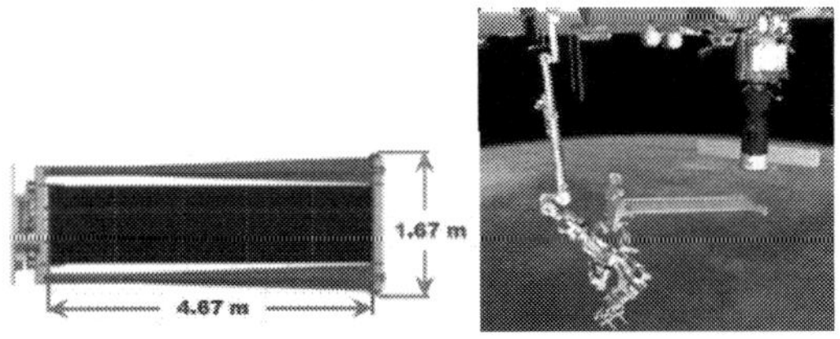

Figure 5. The DSS ROSA solar array ISS experiment (l) and shown during testing aboard the ISS at the end of the robotic arm (r).

The ROSA ISS hardware has been delivered to NASA for integration and is expected to be launched to ISS in the trunk of the DRAGON in early 2017. Once on-orbit, ROSA will be grappled by the robotic arm, deployed and characterized for approximately 7-days. The base on the ROSA experiment includes mechanical actuators to induce loads to the deployed ROSA hardware. The response of ROSA to these disturbances will be assessed using a variety of on-board sensors and photogrammetry tools. Results of the on-orbit testing will be used to validate ground models. The active solar cell strings will also be characterized via I-V testing. At the conclusion of the experiment, ROSA will be retracted and re-stowed in the DRAGON trunk which will burn up upon re-entry.

AFRL is also active in the development of solar concentrator technologies as they offer potential for cost, performance and environmental durability benefits. Current efforts are primarily considering refractive concentrator technology, given their benefits related to shape error tolerance and soiling degradation. In addition, AFRL is investigating the potential to modify high performance terrestrial concentrator technologies (Semprius) for use in space. These systems have well developed manufacturing process and utilize very small (~600 micron) cells enabling very thin packaging, high environmental shielding and potentially low cost. Given the past challenges of space solar concentrators and their requirements for precise pointing, these systems will have to show considerable benefit in order to transition to operational use. Fortunately there have been a number of successful uses and demonstrations of space solar concentrators, thus there is hope that this technology may be adopted in the near future.

5. CONCLUSIONS

Improvements in solar cell and array technology continue to transition to operational use given the tremendous benefit these technologies offer to the overall spacecraft. Rather than diminishing, the number of solar cell technologies that have the potential to provide >30% efficiency is expanding. Given the limited size of the space marketplace, it may be challenging for all of the potential technologies to be commercially viable.

6. REFERENCES

1. Experimental results from performance improvement and radiation hardening of inverted metamorphic multi-junction solar cells, Pravin Patel; Daniel Aiken; Andreea Boca; Benjamin Cho; Daniel Chumney; Brad Clevenger; Arthur Cornfeld; Navid Fatemi; Yong Lin; James Mccarty; Fred Newman; Paul Sharps; John Spann; Mark Stan; Jeff Steinfeldt; Tansen Varghese, Photovoltaic Specialists Conference (PVSC), 2011 37th IEEE.

2. Manufacturing Improvement of Inverted Metamorphic Multijunction (IMM) Solar Cells and Flex String Array (FSA) Confidence Testing on the AFRL Mantech Program, Benjamin Cho, Daniel Derkacs , Kip Hazlett, Christopher Kerestes, Chelsea Mackos, Nathaniel Miller, Bed Pantha, Pravin Patel, Paul Sharps , Boyd Shaw, Steve Whipple, Lei Yang, Proc. 43rd IEEE PVSC (2016)

3. Effects of in situ annealing on GaInNAs solar cells Sarah Kurtz; Richard King; Daniel Law; Aaron Ptak; John Geisz; Nasser Karam 2013 IEEE 39th Photovoltaic Specialists Conference (PVSC) Pages: 2095 – 2099

4. Intermediate band solar cell design using InAs quantum dots in AlAsSb cladding, Seth M. Hubbard; Staffan Hellstroem; Zachary S. Bittner; Ramesh B. Laghumavarapu; Diana Huffaker, Photovoltaic Specialist Conference (PVSC), 2015 IEEE 42nd

5. 35.8% space and 38.8% terrestrial 5J direct bonded cells, P. T. Chiu; D. C Law; R. L. Woo; S. B. Singer; D. Bhusari; W. D. Hong; A. Zakaria; J. Boisvert; S. Mesropian; R. R. King; N. H. Karam, 2014 IEEE 40th Photovoltaic Specialist Conference (PVSC)

6. Six-Junction (6J) Microscale Concentrating Photovoltaics (CPV) Concepts for Space Applications, Matthew Lumb, Brent Fisher, Ken Schmieder, Matthew Meitl, Proc. 43rd IEEE PVSC (2016)

7. Novel flexible solar cell coverglass for space photovoltaic devices, David Wilt; Neil Snyder; Phillip Jenkins; Andy Gray, 2013 IEEE 39th Photovoltaic Specialists Conference (PVSC)

8. Silver–Carbon-Nanotube Metal Matrix Composites for Metal Contacts on Space Photovoltaic Cells, Omar K. Abudayyeh; Nathan D. Gapp; Cayla Nelson; David M. Wilt; Sang M. Han, IEEE Journal of Photovoltaics, Year: 2016, Volume: 6, Issue: 1, Pages: 337 - 342

9. Increasing the TRL level of new PV technologies using modular solar panels, N. Walmsley; T. Stern; D. Wilt, Photovoltaic Specialists Conference (PVSC), 2012 38th IEEE

10. Commercialization of Deployable Space Systems' roll-out solar array (ROSA) technology for Space Systems Loral (SSL) solar arrays, Bao Hoang; Steve White; Brian Spence; Steven Kiefer, 2016 IEEE Aerospace Conference

IN FLIGHT DETECTION AND LOCALISATION OF SHORT CIRCUITS ON SOLAR GENERATORS

Henri Barde [(1)], **Emmanuelle Carla** [(2)], **Nicolas Neugnot** [(2)], **Etienne Rapp** [(3)]

[(1)] *European Space Agency (ESA), henri.barde@esa.int*
[(2)] *Airbus DS France, 31 rue des Cosmonautes 31402 Toulouse France, Email:emmanuelle.carla@airbus.com*
[(3)] *CNES, 2 place Maurice Quentin 75039 Paris Cedex 01 France, Email:Etienne.Rapp@cnes.fr*

ABSTRACT

In the last decades a number of failures in which a solar cell is short-circuited to the panel substrate have been reported or suspected. In the meantime protective measures, like single string connection and use of bleed resistors in the grounding circuit have been implemented. However it is still not always clear whether such an anomaly occurred and the in-flight instrumentation is usually not sufficient to provide unambiguous evidence. This paper describes a simple way to identify both the occurrence of such a cell to substrate short circuit and the panel on which it is located.
The paper is then completed with the main results from the CNES study performed in 2012 to demonstrate the performance, analyse the reliability and assess the industrial feasibility of such a system.

1. INTRODUCTION

Detailed information on failure occurring in orbiting spacecraft is seldom released. However, when looking at the history of spacecraft anomalies, it appears that solar array failures represent a very significant fraction of the insurance claims [1]. Even if not all of them are of electrical nature (deployment failures are a significant contributor) it is known, even if not published, that cell to substrate short circuits are far from being uncommon. Experience also shows that identifying the exact cause of an anomaly and especially if the suspected short circuit is the result of an external cause, like a micro meteorite impact, or a consequence of another electrical event is challenging. The usual spacecraft instrumentation is designed to support nominal and anticipated contingency operations, but in general not to help a detailed investigation. The present paper is a contribution to this issue.

2. DESCRIPTION OF THE PROBLEM

The earlier designs of solar generators generally made use of a direct connection of the panels structure to the spacecraft electrical reference point, for the purpose of avoiding large differential voltages to build up. As shown in a simplified way in Figure 1, the consequence of this direct connection was that a short-circuit of a cell to the substrate could result into a loss of power, potentially significant if the cells were connected both in series and parallel (in a so-called matrix configuration). Some spacecraft built in the 1980s experienced a significant amount of such power reduction, which, in some cases, were synchronous of

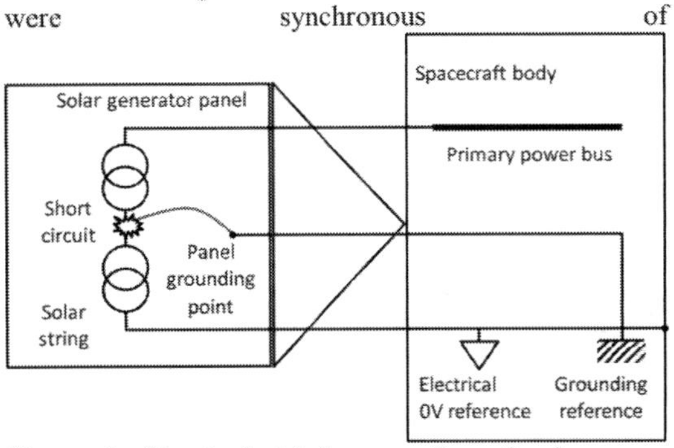

Figure 1: Direct electrical connection of solar array structure to spacecraft body

an attitude disturbance, suggesting that an impact could have been at the origin of this power loss. In other cases, this could also be interpreted as an ejection of mass, as the result of a component blow-up.
The relatively frequent occurrence of such phenomena led to the implementation of preventive measures, namely:

- The "single string" layout, where each string of series connected cells is paralleled to the other ones via a blocking diode, limiting any potential loss to this of the affected string;
- The electrical isolation of the structure panels between them and from the spacecraft electrical reference point;
- And the introduction of bleed resistors of a few 10 kΩ, to provide a path for charges to flow and so ensure a limited voltage gradient between both.

This is summarised in Figure 2.
With these protections, a cell to substrate short circuit occurring on a panel can only produce a negligible stray current and at least two such events have to occur on the same panel to induce the loss of one string.
However, even with these precautions taken, some failures are still reported and raise the question of the actual "chicken and egg" situation mentioned above: are impacts frequent, but not visible as long as they do not affect several times the same panel, or are the occasional disturbances seen on the attitude the result of energetic

events generated on-board? Being able to detect such short circuits even when they do not produce observable effects would contribute to a better understanding of the problem.

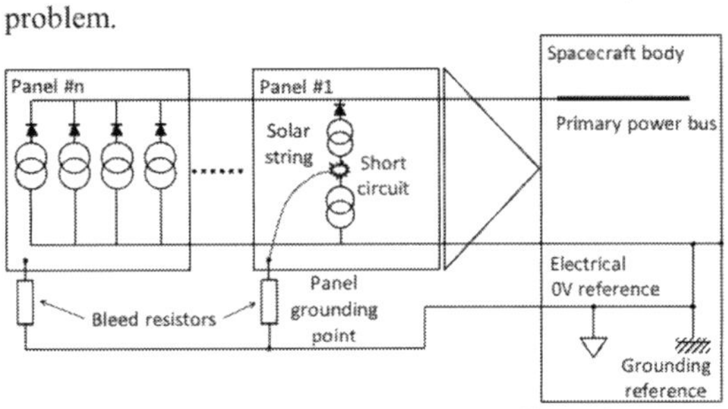

Figure 2: Protective measures against cell to substrate short circuit

3. PROPOSED SOLUTION

The first step is the detection of the short-circuit. To this end the network of bleed resistors usually in place can be slightly modified to provide this information, as shown in Figure 3, by the introduction of an additional sense resistor between the panels grounding line and the spacecraft electrical reference.

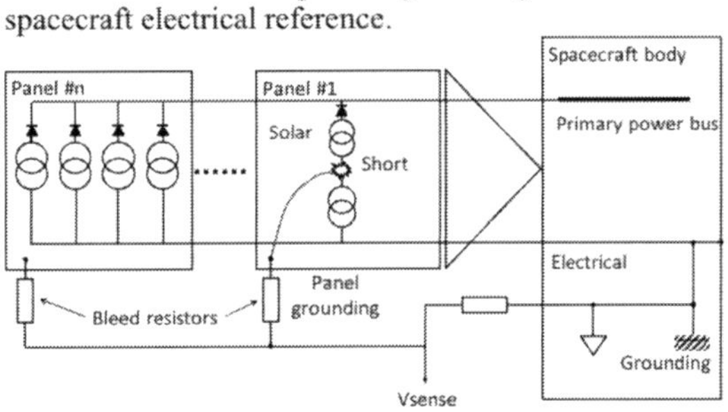

Figure 3: Protective measures against cell to substrate short circuit

In the absence of any short-circuit, the voltage Vsense across this resistor is essentially null, being the product of its resistance by the charge draining current, in the range of nanoamperes. Now, when a cell becomes short-circuited to a panel, the voltage across the fraction of the string between the zero volt and this cell is applied to the resistive network made of the bleed resistors and of the sense resistor. Knowing their respective values allows to determine the voltage at the short circuit, i.e. the position of the failed cell inside its string.

The selection of the ohmic value of the sense resistor depends on the maximum voltage that can be reached by a string, the number of short-circuits that may exist and the input voltage range of the analogue telemetry. As an example, for a 3 panels array with an upper cell shorted on each, a maximum voltage of 100V (beginning of life, cold conditions, full illumination) and a 5 V telemetry range, the sense resistor value should not be larger than 5/300 of the common value of the bleed resistors. In this case a short-circuit on a single panel at half the string in normal operating conditions would be around a few 10% of the full range, which is acceptable.

More information can be extracted from this very simple circuit, though, considering that, in its practical implementation, the grounding network is redundant. By properly selecting the values of the prime and redundant resistors, different divider ratio can be set for each solar panel, making possible the identification of the panel on which the short circuit has occurred. This is illustrated in a simplified form by Figure 4, where a three panels array is considered.

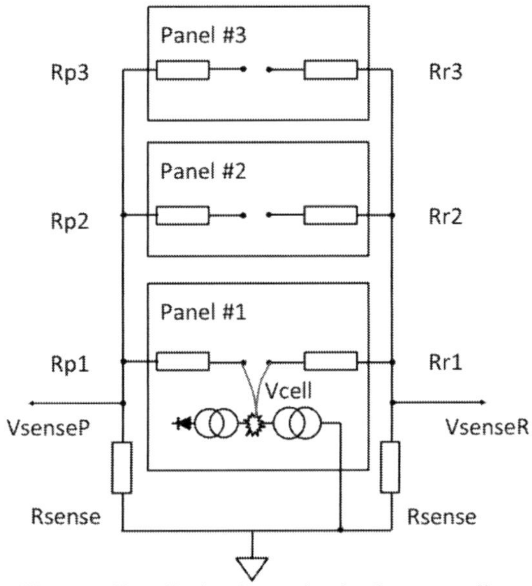

Figure 4: Use of both prime and redundant grounding networks for short circuit localisation

The ratio between VsenseP and VsenseR is given by:

$$\frac{VsenseP}{VsenseR} = \frac{R1+Rsense}{R2+Rsense} \qquad (1)$$

where

$$R1 = \frac{R.Rr1}{Rp1+Rr1+R} \qquad (2)$$

$$R2 = Rp1.\frac{R}{Rp1+Rr1+R} \qquad (3)$$

and

$$R = \frac{(Rp2+Rr2).(Rp3+Rr3)}{Rp2+Rr2+Rp3+Rr3} \qquad (4)$$

The details of the quite simple calculation are provided in [2]. Therefore if all ratios on

$$VsensePVsenseR = \frac{R1+Rsense}{R2+Rsense} \qquad (1)$$

right side are different, which is easy to ensure by a

proper selection of the bleed resistors values, then the identification of the panel is unambiguous. For instance a series of increasing ohmic values can be attributed to the prime side and the same series in opposite order on the redundant one.

At this stage the benefit is already significant, since the identification of the panel provides an indication of the distance from the spacecraft centre of mass at which the impact or the ejection occurred. In addition, in the relatively frequent situation where the strings are laid down parallel to the spacecraft body (as illustrated by Figure 5) the identification of the shorted cell together with the characteristics of the attitude disturbance allow to know on which side of the panel the event occurred. This is important in making the distinction between e.g. a micro-meteorite impinging on a cell or the ejection of a blown component from the rear side.

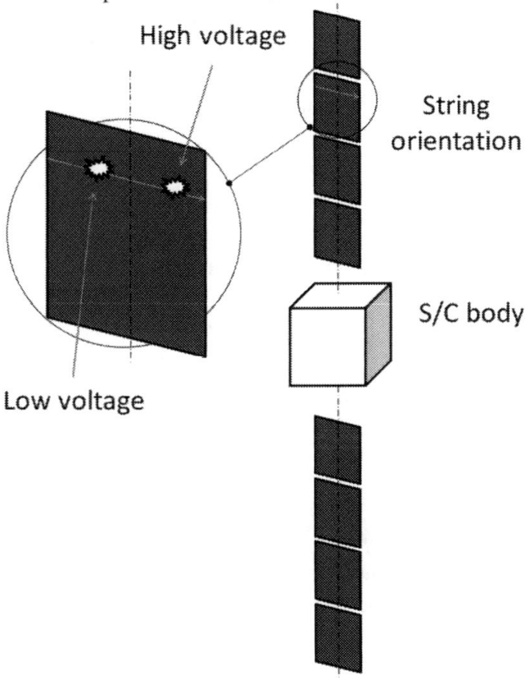

Figure 5: Knowledge of panel and rank in the string of short-circuited cell allows approximate location of the defect

4. OTHER ASPECTS OF THE CIRCUIT

4.1 Measurement noise and filtering

The normal operation of the solar array regulator generates an AC component of the current delivered by one or more sections of the solar array. A S3R [3], for instance, will produce a square wave, the current of the toggling section varying from its operating to short-circuit value. The capacitive coupling that exists between the cells and the substrates means that this AC component will be picked up by the circuit described above. It is therefore advisable to filter out this noise, e.g. by connecting a capacitor in parallel to the sense resistor.

4.2 Selection of bleed resistors

Implementing resistors of different values on each panel has the drawback of introducing a slight complication in the design and procurement process. A way to reduce this – modest – added burden is to achieve the different ohmic values by adequate series/parallel arrangement of identical resistors. For instance, the circuit of Figure 4 could be made of resistors of 2R, R and R/2 value by using respectively two in series, a single one or two in parallel. The corresponding ratios between VsenseP and VsenseR would then be in the order of 4, 1 and 0.25, values that can be easily discriminated. As a side – positive – effect, having two different values for the same information allows to process the highest one to achieve the best resolution.

4.3 Multiple events

It is important to assess if, in case of multiple short-circuits, on one or several panels, it is still possible to identify, for each of them, both the applied voltage and the panel location.

This is rather straightforward for multiple events on the same panel: whatever the characteristics of the source voltage, resulting from the combination of the various strings involved, the ratio between the two sensed voltages will stay the same. It may be more difficult to understand the exact configuration but, at least, further events would be detected, as long as they modify the source voltage.

In case a second event occurs on a panel different from the one hosting the initial short-circuit, the situation is more complex: indeed different combinations of panel and applied voltage may produce the same VsenseP/VsenseR ratio. This is illustrated by Figure 6 where a second failure, generating a panel voltage between zero and twice the first panel one, is applied either to panel 2 or panel 3. One can see that a range of ratio values are common to both panels. In this case, the initial ratio, after the first failure, is close to 0.26, i.e. a short-circuit close to the negative side of a string would probably not be detected. In most of the cases, though, a new short would modify the sensed voltages and, for a range of values, would allow identification of the newly affected panel. In the example shown, any value of the ratio above 0.64 would be indicative of a short circuit on panel #3.

In summary:

- Any short-circuit occurring on a first panel would be detected; the panel and applied voltage would be identified;
- Any further short-circuit on the same panel would be detected but the exact event would be difficult to characterise;
- Any short-circuit occurring on another panel would

be detected, provided that it concerns a sizeable fraction of the string; the panel identification may or may not be possible, depending of the applied voltage.

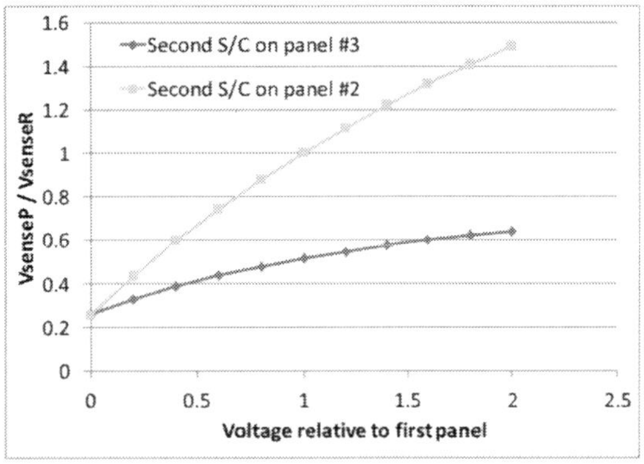

Figure 6: VsenseP/VsenseR ratio as a function of the second short–circuit voltage, the panel as a parameter

5. RELIABILITY ANALYSIS

The objective of reliability analysis is to consolidate the previous design but also to check that the operation of the satellite is not degraded even in the event of a failure. For that, several analyses have been performed in the frame of the CNES study and are presented hereafter:

- ESD: The ESD analysis allowed to check that the detection system is immune to the phenomena of electrostatic discharge on the one hand and that it is not likely to trigger an electrostatic discharge on the other hand, even in the event of a failure. This is justified by the fact that there are no metallic parts floating since they are always connected through a bleed resistance to the spacecraft structure, and because no new materials are included (same harness and same bleed resistors used).

- FMECA: Failure Mode, Effects and Criticality Analysis proved that no failure can degrade the operation of the satellite. In fact, in event of failure only the localization and detection function can be lost but no propagation to the satellite.

- Stress analysis: During a failure, system components will not exceed their nominal operating conditions.

6. INDUSTRIALIZATION PLAN

The conclusion of the CNES study shows that for a solar array configuration that is to say between 3 and 5 panels, it is possible to detect the defective panel when the default voltage exceeds 10V for a 50V bus and 20V for a 100V bus, considering the worst case tolerances. Moreover, it is possible to discriminate the cell where the default occurs with a precision of +/- 1 cell for a 50V bus and +/- 2 cells for a 100V bus. Noise due to power supply regulator switching can add temporary uncertainties (few seconds) on the localization performances approximately of +/- 1 cell. This is due to a power variation on the bus (and therefore a modification of duty cycle or commutation from section).

It seems definitely a very interesting idea that should be applied without much effort. To do so, an industrialization plan has been defined for Airbus DS telecom platform Eurostar 3000 (E3000).

The list of materials has been gathered and all the components required for this application are already qualified and used on the current platform. No additional and expensive qualification campaigns are therefore required.

The location of the additional resistors, the type of wire and the connector including the filter have been analysed without any showstopper identified so far.

The industrialization on Eurostar 3000 platform is therefore demonstrated and currently waiting for a potential customer willing to apply it. Other programs can also take advantage of such technology like the Astrobus platform, Metop-NG…

7. CONCLUSION

The circuit described in this paper allows, at a very moderate expense, to detect almost all short-circuits that could occur between solar cells and substrate in a solar generator and in many cases to identify the affected panel and the voltage at the short-circuit.

Its implementation on spacecraft would allow to get a very valuable statistic of such events and, in some specific failure cases, to bring an important contribution to the understanding of the causes.

8. REFERENCES

1. Rodiek, J.A. & Brandhorst H.W.Jr. (2008) Solar array reliability in satellite operations, *Photovoltaic Specialists Conference, 2008. PVSC '08. 33rd IEEE.*

2. Carla E., Barde H., System and method for detecting and locating an insulation flaw in a solar generator on a space vehicle, US patent US 2014/0232416

3. Weinberg A.H., O'Sullivan D.M., Limit cycling regulator apparatus for plural parallel power sources, US patent US4186336 A

DOI: 10.1051/e3sconf/20171601003

NOVEL DEPLOYMENT MECHANISM FOR CONVENTIONAL SOLAR ARRAY ENHANCEMENT

Paul A. Hodgetts [1], Jean-Paul Collette [2], Don de Wilde [3]

[1] *ESA/ESTEC, Keplerlaan 1, 2200 AG Noordwijk, The Netherlands, Email: Paul.Hodgetts@esa.int*
[2] *WalOpt, Avenue E. Ysaye 44, 4053 Embourg, Belgium, Email: jpcollette@walopt.com*
[3] *ESA/ESTEC, Keplerlaan 1, 2200 AG Noordwijk, The Netherlands, Email: Don.de.Wilde@esa.int*

ABSTRACT

A novel mechanism is described, by which flexible blankets could be deployed from existing solar panel designs. These blankets could be covered with flexible cells, or they could be reflective films to form a concentrator array. Either way, the performance of an existing array design could be enhanced.

1. INTRODUCTION

A conventional solar array comprises a backbone of rigid panels and a yoke, fixed in deployed configuration by lockable hinges. The performance of such an array (W/kg) could be enhanced in one of two ways:

- Lightweight blankets could be extended from either side of each rigid panel, equipped with flexible cells. When deployed, these blankets would lie in the same plane as the rigid panels. The additional cell area would increase the power output of the array, while not significantly increasing its total mass.

- Reflective membranes could be deployed from either side of each rigid panel, inclined at 60 ° with respect to the plane of the respective panel. In this way, a reflective trough would be formed, thereby increasing the incident light on the existing cells of the rigid panels.

In either case, in stowed configuration, these blankets or foils would have to be wrapped around a cylindrical drum, positioned at either side of each panel.

These would have to be released and unwound in orbit, in such a way that their deployment would not inhibit the deployment of the rigid panel assembly.

A deployment mechanism, which fulfils these requirements, has been invented by the author (Patent No. WO 2015/062650 A1) the description of which is the subject of this paper.

2. MECHANISM DESCRIPTION

The principle of the mechanism is based on a crossed tether, wrapped around two wheels, forming a figure of eight. Attached to the tether are slender rods which are forced out laterally, when the wheels rotate and the tether is drawn.

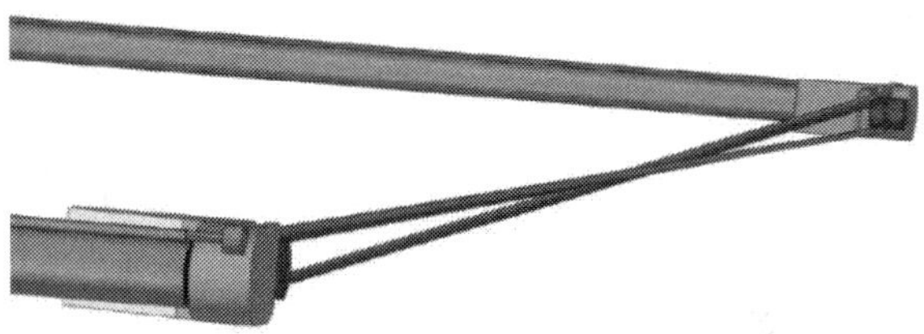

Figure 1. Core of Deployment Mechanism Showing Crossed Tether and Slender Rods

At the opposite end of the panel, an identical mechanism exists with a crossed tether and a pair of slender rods.

Attached to the ends of these slender rods are spreader bars, running the full length of the panel, one on either panel side. To each spreader bar, a blanket or reflective film is attached, which is wrapped around a cylindrical drum when stowed, with the spreader bars positioned adjacently.

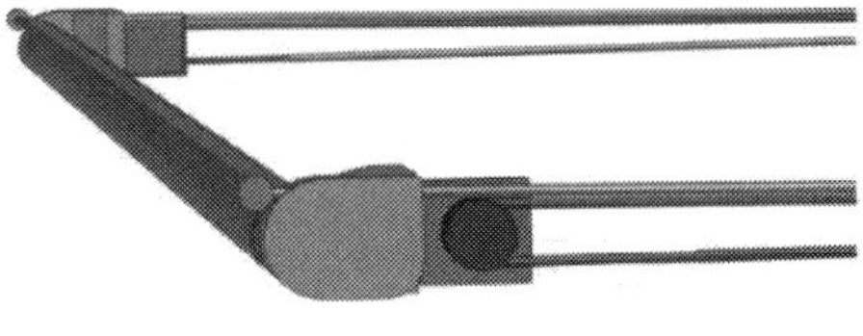

Figure 2. Slender Rods, Spreader Bar and Blanket/Reflective Foil in Stowed Configuration

DOI: 10.1051/e3sconf/20171601003

The complete assembly of two tethers, four slender rods, two spreader bars and two cylindrical drums, forms a rectangular shape.

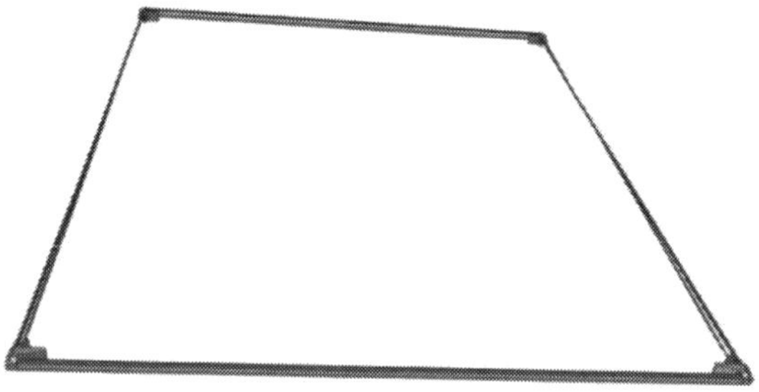

Figure 3. Deployment Mechanism and Blanket/Foil Assembly

The assembly fits around a conventional panel: the rods are accommodated in the edge member profile of the panel substrates, to which the panel hinges are attached.

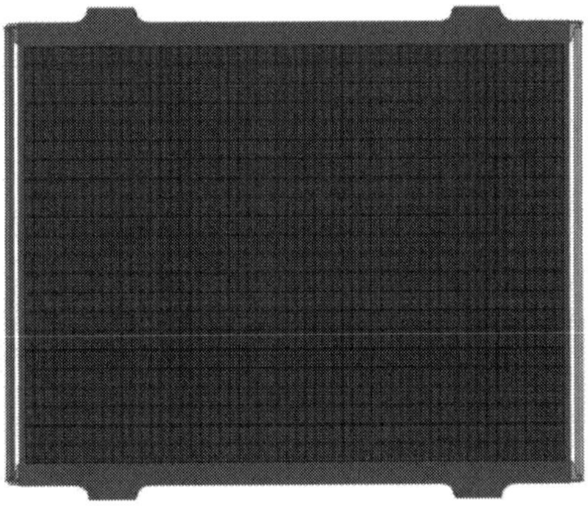

Figure 4. Accommodation of Deployment Mechanism and Blanket/Foil Assembly

The motion of the tethers is suppressed, until all panel hinges, on either side of a given panel, latch in orbit.

With the release of the tether, the rods are drawn by rotational springs connected to the pulley wheels. The motion of each mechanism, on either side of a given panel, is synchronised by the blanket/foil, as it is unwound from the cylindrical drums.

The slender rods are guided by a slot in the cylindrical drum support housing, which also provides a hinge location for the wheels, around which the tether is wound.

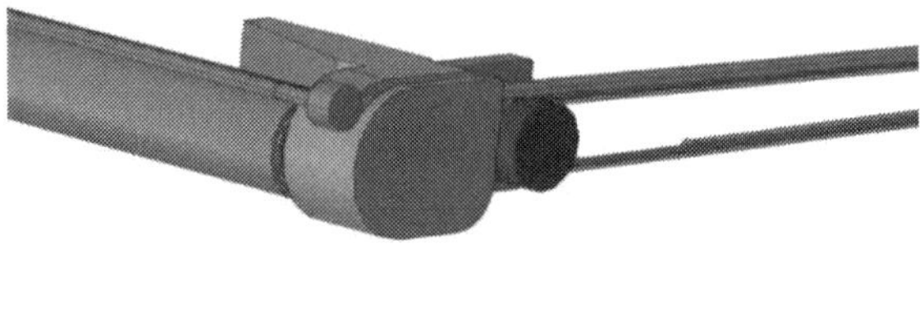

Figure 5. Cylindrical Drum Support Housing and Guidance Slot

At the end of deployment, the ends of these rods are forced into a recess, where they lodge.

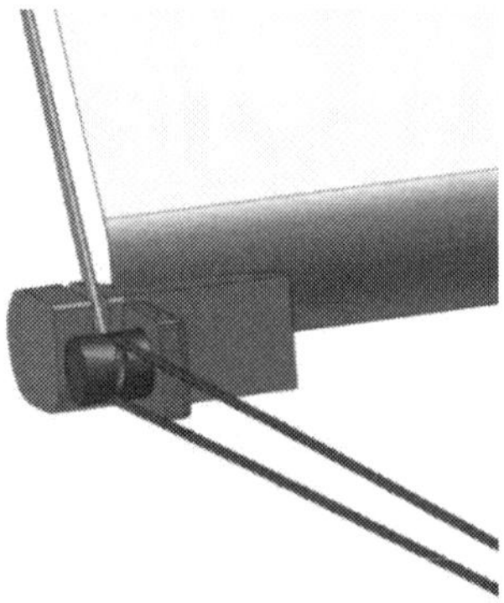

Figure 6. Slender Rod, Lodged in Recess

Figure 7. Reflective Foils in Fully Deployed Configuration

The dimensions of the recess are such that, in the case of a reflective foil, these become inclined at 60 degrees with respect to the plane of the solar panel (Figure 7) thereby forming a reflective trough.

In the case of cell covered flexible blankets, the recess is such that the blankets would end up in a near flat orientation, with respect to the rigid panels.

3. MECHANISM KINEMATICS

In the case of the mechanism deploying a reflective film as part of a concentrator array, as the end of the slender rod rotates around the tether wheel (Figure 8) it subtends an angle θ with respect to its initial position, referred to as the mechanism deployment angle. The resulting inclination of the slender rod ψ is the foil inclination angle.

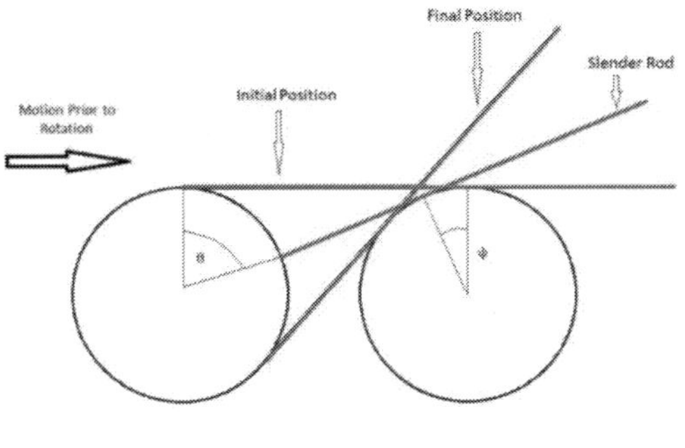

Figure 8. Schematic of Mechanism Showing Initial and Final Positions of Slender Rod

The kinematic relationship between these angles has been calculated (Figure 9). As can be seen, the foil inclination angle increases most rapidly with respect to mechanism deployment angle during mid-deployment.

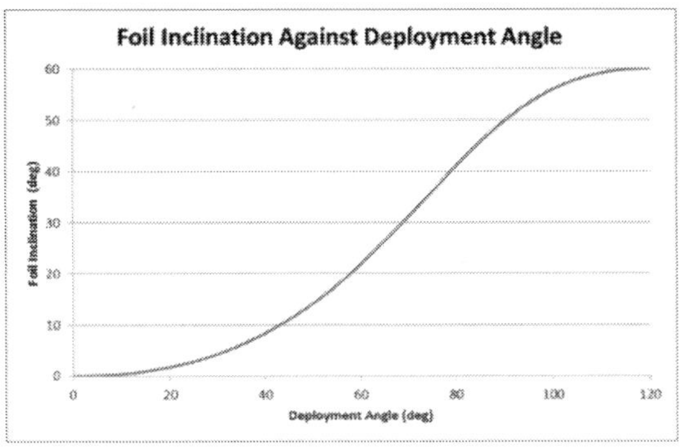

Figure 9. Kinematic Relationship between Foil Inclination Angle and Mechanism Deployment Angle

At the start and end of deployment, the slope is zero.

4. DEPLOYMENT CHARACTERISTICS

Since the mechanism is released the moment the panel hinges of a given panel latch, in the case of a multi-panel array, blanket/foil deployment may take place while other panels are still in motion. Any jolts imposed on the wing, due to the blanket foil deployment, therefore need to be kept to a minimum.

As the mechanism involves two slender rod/spreader bar assemblies, which move simultaneously in opposite lateral directions, their inertial loads counteract one another, thereby generating a near zero in-plane impulse.

In the case of the reflective foil application, where significant out-of-plane motion exists, the kinematics show that as the slender rods start to rotate, much of their translational kinetic energy will be transferred into rotational kinetic energy. This will result in a rapid translational deceleration of the slender rod/spreader bar assemblies.

Precisely how the rods and reflective foils reach their final 60° orientation will depend on frictional losses in the mechanism and the extent to which elastic energy is stored in the rods as they flex. While this is beyond the scope of the present study, expectations are that a soft end latch is achievable by means of tuning the frictional parameters via latch design and material selection.

5. FUTURE DEVELOPMENT

This deployment scheme is currently being further investigated by WalOpt, based in Liège (Belgium) as part of a feasibility study of concentrator solar arrays.

Multi-body dynamic analyses of the mechanism are being performed, based on panel dimensions typical of a telecom satellite array. In addition, a small scale breadboard model is to be manufactured, to assess its functionality. Particular attention will be paid to the suppression of wrinkles in the reflective film, since these are known to generate a non-uniform illumination of the solar cells.

6. REFERENCES

[1] "Small Satellite REIMEI for Auroral Observations", H. Saito, et al., Proceedings of the 61st International Astronautica Congress, 2011.

ARA MK4 SOLAR ARRAY DEVELOPMENT

Ron van der Ven [(1)], Ed Bongers [(2)]

[(1)] *Airbus Defence and Space Netherlands, Mendelweg 30, 2333 CS LEIDEN, The Netherlands, Email: r.vd.ven@airbusDS.nl*
[(2)] *Airbus Defence and Space Netherlands, Mendelweg 30, 2333 CS LEIDEN, The Netherlands, Email: e.bongers@airbusDS.nl*

ABSTRACT

The ARA (Advanced Rigid Array) Mk4 solar array development program is conducted under an ESA-Artes contract.
In order to enter new markets with recurring potential such as the market for small to medium range Comsats and moreover to remain competitive with its Solar Arrays on existing markets, Airbus DS Netherlands needed to evolve its current product line of ARA Mk3 and FRED to a new generation Solar Array, the ARA Mk4. The main goals of the ARA Mk4 development program are solar array cost and risk reduction, robustness, performance increase and readiness for growth capability.
The requirements for the ARA Mk4 solar array are enveloping the needs for a number of intended applications such as EDRS-C, GMP-T, MetOp-SG and Geostar-3. Furthermore the requirements are based on launch with today's launchers. The EDRS-C program has been the first customer for the ARA Mk4 solar array.
The ARA Mk4 solar array has been divided into a number of units such as PVA, Panel Substrates, Yoke, Holddown and Deployment mechanisms and Inter-Panel Harness. These units are developed and qualified before verification at wing assembly level. Wing level tests, including sine vibration and acoustic noise testing, has been performed successfully.
In addition to hardware development also the analysis tools, integration approach, qualification status control and documentation generation for recurring projects have been upgraded.
The outline of the ARA Mk4 development program, the obtained results and the achieved status with regard to the set goals are presented in this paper.

1. DEVELOPMENT PROGRAM DEFINITION

As a first step an inventory is made for all potential improvements and lessons learned from the previous solar array programs. These improvements have been categorized and trade-offs have been performed to set the goals for the ARA Mk4 development. The potential applications for ARA Mk4 have been used to determine an enveloping key requirements baseline. The potential solar array applications include EDRS-C, GMP-T, MetOp-SG and Geostar-3.
The ARA Mk4 solar array has been divided into a number of units such as PVA (Photo Voltaic Assembly), Panel Substrates, Yoke, Holddown and Release System, Deployment Mechanisms and Interpanel Harness. For production and GSE (Ground Support Equipment) development the maximum panel dimensions are set at 2.5 m by 3.5 m.
The ARA Mk4 solar array range has been set from 2 panel configurations being the low end, and to 5 panel configurations being the high end of the range.
The low and high end configurations have been used to derive the requirements for the individual units. The units have been designed against the so derived unit level requirements.

A breadboard program has enabled establishing the ultimate properties by testing flight representative test samples. The ultimate performance limits have been used to determine the qualification levels. For thermal qualification a robustness margin of 5°C is selected on top of the qualification margin of 10°C and an uncertainty of 10°C to minimize the risk of thermal delta qualification needs in future programs.

2. DEVELOPMENT GOALS

The ARA Mk4 Solar Array development goals are:

Cost reduction of 25%, by:

- Upgrade of solar array subsystem design
- Selection of cost effective materials
- Selection of cost effective processes
- Selection of competitive suppliers
- Thorough manufacturing engineering
- Increased efficiency of Engineering and AIT (Assembly Integration & Test) processes
- Reduction of the throughput time
- Standardization and streamlining of documentation
- Improvement of tools (design & managerial)

Risk reduction of 50%, by:

- Higher margins (thermal, mechanical, electrical), both for the production process and for in-orbit operation

 DOI: 10.1051/e3sconf/20171601004

- Double sourcing through suppliers and materials/processes selection (obsolescence risk)
- Increased robustness of production processes

Performance increase:

- Lower mass (target is 6%)
- Higher (local) stiffness (target is an increase of the deployed wing frequency by 5%)
- Adjustable electrical interface for SA (Solar Array) Release Mechanisms

Readiness for growth capability:

- PEF implementation capability
- Implementation of thin film high efficiency solar cell technology

3. UNIT DEVELOPMENT

3.1 Panel substrates

Until ARA Mk4 Airbus DS NL always subcontracted the production of panel substrates Built-to-Print. For ARA Mk4 it has been decided to define besides the dimensions and interface positions, only the design influencing parameters such as facesheet skin ply orientation, carbon fibres, honeycomb core type and height. The selection of all other materials (e.g. resin for the facesheet skin, adhesives) and processes is the responsibility of the supplier. The advantage is that the supplier can determine the most cost effective approach for manufacturing.

Airbus DS NL has selected three suppliers for the ARA Mk4 panel substrates: two European suppliers (Airborne Aerospace and Airbus DS Spain) to have redundant suppliers also for ESA programs and one US supplier (ATK) for US solar array opportunities.

First the general panel substrate baseline has been defined in terms of materials, processes and production sequence. After that breadboard samples have been made to verify the approach, followed by engineering samples (including 1 full size panel by each supplier) and qualification samples. Testing included structural, thermal and electrical tests. The achieved thermal qualification range is -188°C to +162°C.

The unit level tests have demonstrated a significant strength and stiffness improvement for the panel hinge connections.

For the ARA Mk4 qualification wing Airborne Aerospace has provided the panel substrates since Airborne Aerospace is the supplier for the first application program, EDRS-C.

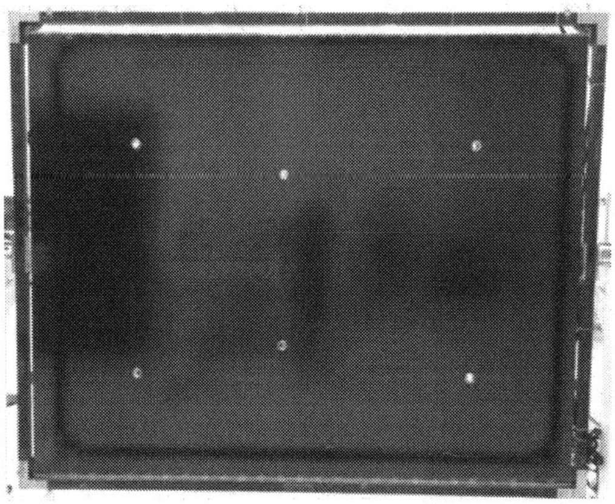

Figure 1. Solar panel substrate.

3.2 Yoke

The yoke connects the solar panels to the spacecraft and ensures that the solar panels are positioned out of the S/C shadow. The yoke design is in general different for every solar array application due to mission requirements and S/C designs. The non-recurring effort to come to a new yoke design has always been quite extensive and the ARA Mk4 goal has been to develop a simple and straightforward approach for project specific yoke designs.

A trade-off on several yoke design concepts including yoke panel, Y-shaped yoke, I-shaped yoke and V-shaped yoke has resulted in the V-shaped yoke as the ARA Mk4 baseline. Main reasons were S/C radiator field of view, yoke mass and yoke stiffness.

The ARA Mk4 V-shaped yoke design comprises an ARA Mk4 root hinge with a standard interface to a root bracket creating the required V-angle, square CFRP (Carbon Fibre Reinforced Plastic) yoke tubes, tip brackets and a flexprint panel. The flexprint panel is a CFRP sandwich panel, based on the panel substrate technology above and serves to collect the electrical harness from the solar panels and to route this harness towards the yoke tubes.

For a project specific yoke design only the yoke tube length and the bracket angles need to be adapted. The root - and tip brackets are made of Titanium using ALM (Additive Layer Manufacturing); this manufacturing method enables a quick and low cost switch for project specific parts.

Yoke samples have been tested to verify the structural and thermal properties of the bracket to tube and bracket to flexprint panel connections. At yoke level proofloading, stiffness measurements and thermal vacuum cycling tests have been performed.

The achieved yoke deployed out-of-plane stiffness is 8380 Nm/rad which is a factor of 1.5 above the previous

ARA design. The achieved thermal qualification range is -100°C to +120°C.

Figure 2. Yoke Assembly.

3.3 Hold Down and Release Mechanism

In a parallel program a multi-functional holddown and release mechanism called NELS (Non Explosive Low Shock) is under development. This NELS concept is compatible with the ARA Mk4 solar array design.

The NELS holddown and release system will provide an increased pretension capability for ARA Mk4 solar arrays of 15 kN which is more than a factor 2 higher than for the previous ARA design, while the release shock has been reduced to a level below 10g.

Another improvement of the NELS holddown and release system is that the operational voltage for the release mechanism can be tuned within a range of 20-50V to one of the available interface voltages of the S/C.

The details of the NELS holddown and release mechanism have been presented in [1].

Figure 3. NELS HDRS.

3.4 Deployment Mechanisms

The deployment mechanisms for ARA Mk4 solar arrays include a root hinge, panel hinges, a partial deployment mechanism, a deployment damper and the synchronisation system. Several improvements have been implemented for the deployment mechanisms, the most important ones are listed below.

The electrical wiring transfer at the root hinge has been changed into a flat cable loom design. This concept enables a simple and effective design that complies to the latest ECSS derating requirements. The root hinge has a standard interface to the yoke structure; the root hinge design can be used for the complete ARA Mk4 range.

The latch and stop-bolt configuration of the panel hinge has been optimised, which results in an improved deployed stiffness and a higher strength allowable compared to the previous ARA baseline.

A new partial deployment mechanism concept has been introduced that allows partial deployment of the outer panel also for 2 panel wings such as EDRS-C. The partial deployment mechanism is derived from the Eurostar 3000 partial deployment mechanism design [2] with ARA Mk4 interface adaptations.

Figure 4. Partial Deployment Hinge.

Furthermore, the assembly, integration and test aspects have been improved for the deployment mechanisms which have resulted in shorter integration and test durations and less non-conformances.

3.5 Interpanel Harness

The ARA Mk4 interpanel harness design uses flexprint assemblies with 12 tracks of flat copper with a derating equivalent for AWG20 wires. The flat copper tracks are laminated in between 2 Kapton® foils with a thickness of 25 micron. The connection to the panel level power and signal harness wires is done inside a PES (Poly Ether Sulfone) connector that is mounted on the edge of the solar panel.

Compared to the old ARA baseline, double insulation

DOI: 10.1051/e3sconf/20171601004

has been implemented to increase the robustness against ESD (Electro Static Discharge) type phenomena in orbit. Furthermore, the supply chain and production aspects have been improved.

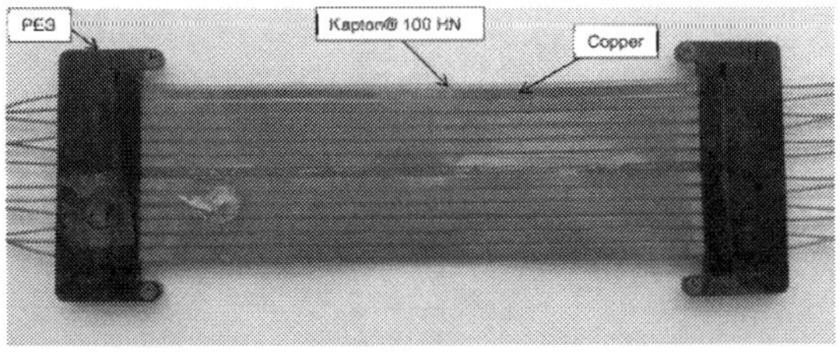

Figure 5. Flexprint Assembly.

3.6 PVA

The objective has been to demonstrate compatibility of the ARA Mk4 panel substrate design with state-of-the-art PVA. For this purpose a number of DVT (Design Verification Test) coupons have been made. The following PVA/solar cell technologies have been tested:

- AZUR 3G30C advanced with integral diode (8x4 cm²) and 100 micron coverglass CMX 100 AR applied by Airbus DS Ottobrunn
- AZUR 3G30C advanced with integral diode (6x12 cm²) and 100 micron coverglass CMX 100 AR applied by Airbus DS Ottobrunn
- XTJ (triple junction GaAs 73.52cm² net cell area) with an P3 external shunt diode and 100 micron coverglass CMG 100 AR applied by Spectrolab
- ZTJ (triple junction GaAs) 85x85mm² (65cm² net cell area) with an external shunt diode and 100 micron coverglass CMX 100 AR applied by SolAero

Figure 6. 3G30C 8x4 DVT coupon.

4. WING LEVEL VERIFICATION

After successful unit level development completion, a qualification wing has been built and tested to verify the wing level performance.

The ARA Mk4 qualification wing design is identical to the EDRS-C flight solar array and consists of a yoke plus 2 panels and has partial deployment capability for the outboard solar panel.

The wing level analyses included structural analysis with NASTRAN, ACTRAN for fatigue analysis, ADAMS for deployment analysis, ESATAN for thermal analysis and the Airbus DS Ottobrunn in-house electrical power calculation tool.

The wing level performance tests included wing mass, thermal knife released deployment (partial and full), alignment, stiffness, ELM for solar cell inspection (Electro Luminescence Method) and electrical power measurements. These tests have been performed at the Airbus DS Leiden facilities. The deployed wing stiffness test confirmed the unit level achieved stiffness improvements; the ARA Mk4 QM wing deployed frequency is 0.35 Hz.

At IABG Munich the stowed wing has been subjected to a 3-axis sine vibration test and an acoustic noise test. With a minimum out-of-plane notch of 3.7g the ARA Mk4 target of >2g has been met with ample margin. The acoustic noise spectrum envelopes the levels of Ariane 5 and Falcon 9.

The wing finite element model (FEM) has been correlated with the results of the sine vibration test. The correlated FEM provides a sound basis for structural analysis on follow-on ARA Mk4 solar array applications.

Figure 7. ARA Mk4 QM wing.

5. REALIZED GOALS

During the course of the ARA Mk4 development program the status with regard to the set goals has been monitored as key performance indicators. In the figures below the evolution during the program is shown.

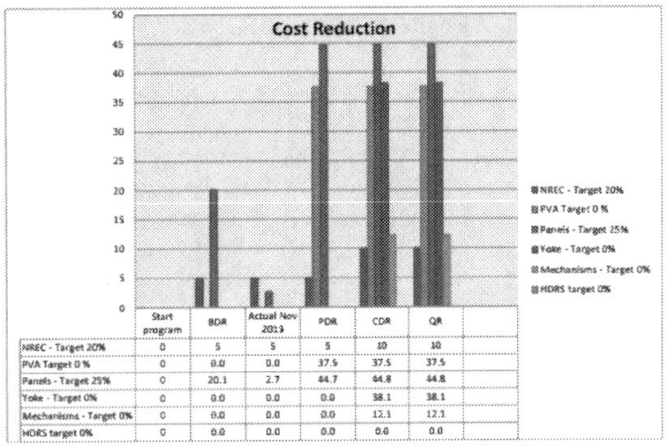

Figure 8. Cost Reduction (goal overall 25%).

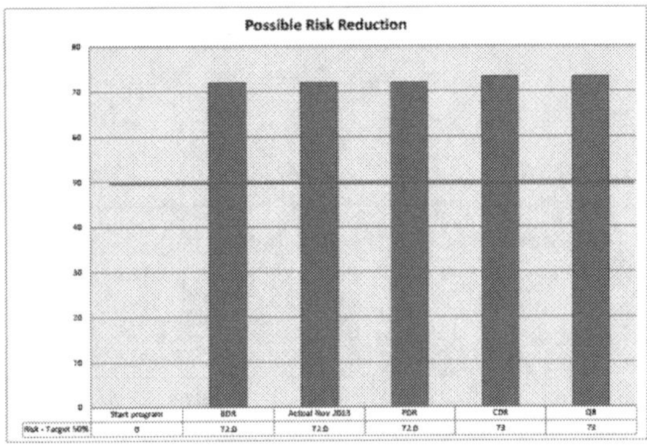

Figure 9. Risk Reduction (goal 50%).

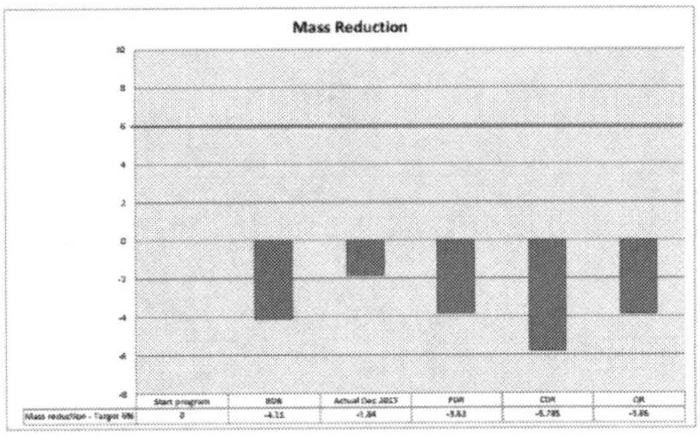

Figure 10. Mass Reduction (goal 6%).

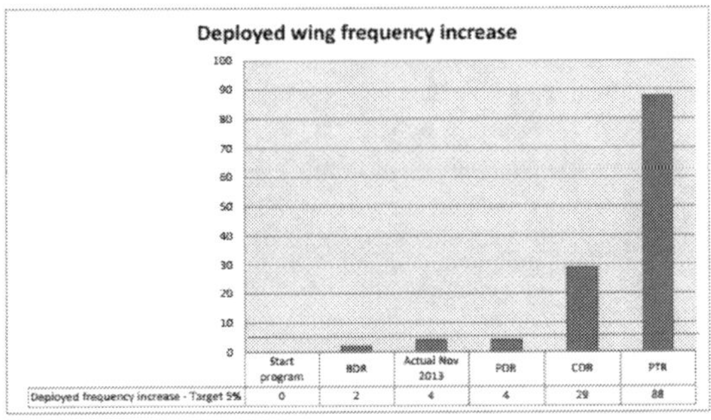

Figure 11. Deployed Wing Frequency Increase (goal 5%).

The figures above show that all development goals have been met with the exception of the mass reduction goal. Several mass reduction options were identified (e.g. thin ply CFRP for the panel substrates) but the majority resulted in increased cost or increased risks. In these cases it has been decided to not introduce the mass reduction option. For the Partial Deployment Mechanism a significant mass reduction of about 300 grams per wing has been realized.

6. NEW ARA MK4 APPLICATIONS

The successful completion of the ARA Mk4 development program has put Airbus DS Netherlands in the position to offer competitive solar array proposals.

After the EDRS-C first application the new ARA Mk4 solar array applications are:

- Exomars
- MetOp-SG
- Quantum
- JUICE

The Exomars, MetOp-SG and JUICE programs have a large number of special solar array features such as for instance increased honeycomb core height for Exomars, skewed solar array deployment angels for MetOp-SG and lateral deployable panels plus a very wide temperature range for JUICE. These features will further extend the ARA Mk4 range.

7. REFERENCES

1. J. Augustijn, E. Bongers, T. Konink, J. Koning (2015). Development Of Non Explosive Low Shock (NELS) Holddown And Release System. 16th European Space Mechanisms & Tribology Symposium ESMATS 2015 Bilbao, Spain

2. Qualification Status List PD Hinge - E3SG-ASO-QSL-ES3000-001, issue 6

 DOI: 10.1051/e3sconf/20171601005

DEVELOPMENT OF INNOVATIVE MECHANICAL FLEXIBLE SOLAR ARRAY ARCHITECTURE

Bernard BOULANGER [(1)], Yannick BAUDASSÉ [(1)], François GUINOT [(1)], Laurent D'ABRIGEON [(1)], Mikaël THIBAUDEAU [(1)], Marina HEIM [(2)], Fabrice BUFFE [(2)], Etienne RAPP [(2)]

(1) THALES ALENIA SPACE France, 5 allée des Gabians BP 99,06156 Cannes La Bocca Cedex (France) Email: Bernard.Boulanger@thalesaleniaspace.com

(2) CNES France, 18 avenue Edouard Belin, 31401 Toulouse Cedex 9 (FRANCE) Email: Marina.Heim@cnes.fr

ABSTRACT

Flexible solar array architecture is proposed instead of rigid one, for high power solar array used from space tug mission to GEO satellite one, but also for specific scientific mission. THALES ALENIA SPACE has developed with Centre National d'Etudes Spatiales (CNES) support, innovative mechanical flexible solar array architecture, using 22 m long carbon tape-springs.
The first part of the study was to design, predict, verify carbon tape-springs mechanical behaviour during deployment and coiling phases at room temperature. THALES ALENIA SPACE built mechanical flexible solar array mock-up full scale width / half scale length and performed deployment and rolling-up tests.
The second part was to design, predict carbon tape-springs mechanical behaviour with large temperature range due to solar array missions. THALES ALENIA SPACE used again flexible solar array mock-up, performed deployment/coiling tests with innovative tape springs.

This 11th ESPC paper describes results of these development activities and the THALES ALENIA SPACE forecast on this flexible solar array market.

1. INTRODUCTION

The double facing deployed tape-springs concept (see Fig. 1) is a real technological gap and has been patented by THALES ALENIA SPACE. This concept permits to have a very light structure with high stiffness and stability in its deployed configuration.
Functional tests realized on a full scale flexible solar-array demonstrator and correlation with mathematical models reinforce the confidence in the reliability of the system.

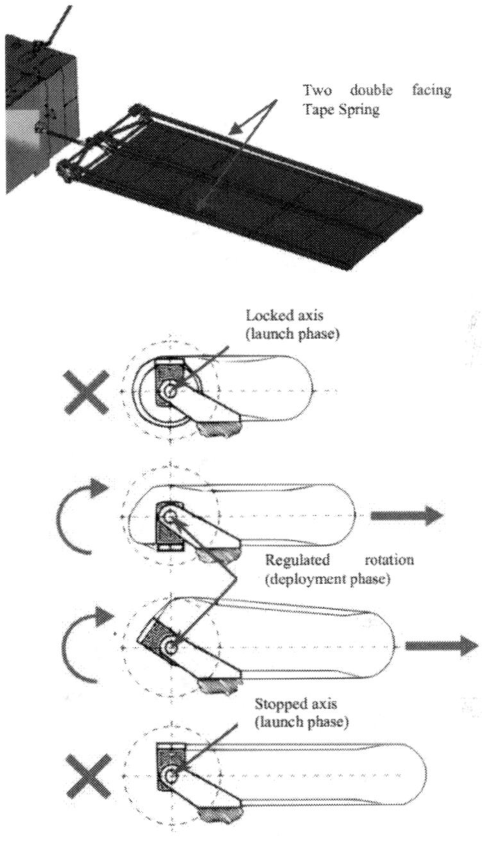

Figure 1. Solar wing with double facing deployed tape-springs concept.

The deployment is made reliable thanks to a gear motor control system in parallel of tape springs actuation:

- deployment control (kinematics, speed regulation)
- in case of "harsh point", possibility to briefly over-motorize

Coiling the substrate offers major advantages :

- High modularity/accommodation,
- Very high power range till 35-50KW.
- Optimized weight/power ratio
- Optimized stowed volume under the fairing
- Rolling/Unrolling on demand

DOI: 10.1051/e3sconf/20171601005

This innovative concept has been developed for solar array but could be also used for additional applications such as thermal shield baffles for telescopes.

2. CARBON TAPE SPRING DEVELOPMENT

2.1. Stiffness considerations

The rollable solar array mechanical structure (THALES ALENIA SPACE patent) is based on tape-springs coiling/uncoiling. The global concept consists of tape-springs bent into two. One tape-springs extremity is fixed while the other one is allowed to rotate around a mandrel. On this mechanical structure, a flexible blanket supporting the flexible solar cells and the electrical routing is attached.

The tape-spring design (geometric properties and material choice) has been done to be in agreement with the classic rigid solar array specifications: the frequency of the first vibration mode when deployed must be ≥ 0.1 Hz.

The mechanical design optimization has been done using a finite element software (MSC Marc). First a model validation has been done thanks to experimental comparisons on tape-springs samples.

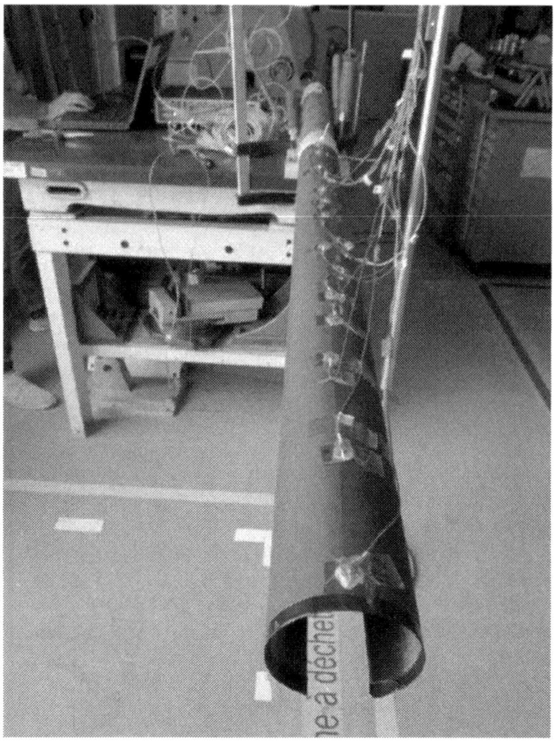

Figure 2. Experimental setup for modal analysis.

Figure 2 shows the experimental set-up that has been used to identify the modal behavior of a tape-spring sample to be compared to simulation results. Accelerometers have been regularly placed along the tape-spring in order to measure the response of a shock transmitted through an impact hammer.

Figure 3. Modal shapes, simulation results

Mode	Theoretical Expected Frequency (Hz)	Frequency obtained by test (Hz)
1	-	15
2	28	29
3	41	43
4	67	64

Table 1. Frequencies comparison.

Table 1 shows that the simulations perfectly predict the modal behavior of tape-springs. It is now assumed that this modeling procedure can be used to identify the frequencies of the global flexible solar array (comprising flexible blanket, stiffeners), and to validate the geometric and material choices.

The full scale global finite element model consists in two tape-springs made of carbon fibers lay-up which are modeled as shell, so is the flexible blanket (considered as a stacking of different materials such as copper and solar cells). The stiffeners are modeled as beams.

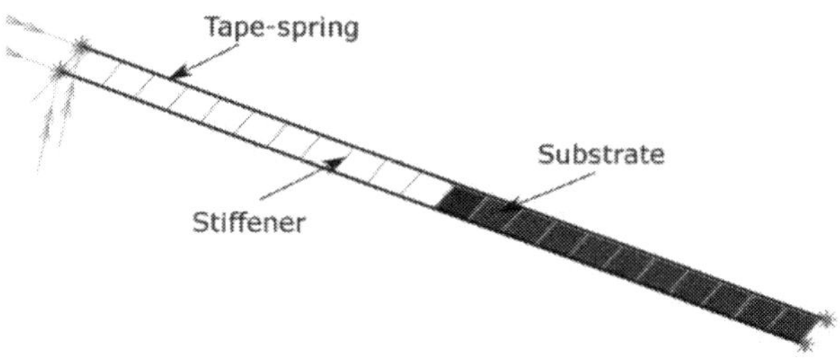

Figure 4. Full scale global finite element model.

 DOI: 10.1051/e3sconf/20171601005

The model described in figure 4 is used to perform a highly nonlinear analysis to set the flexible solar array in its deployed configuration. The modal analysis is then carried out around this stable position.

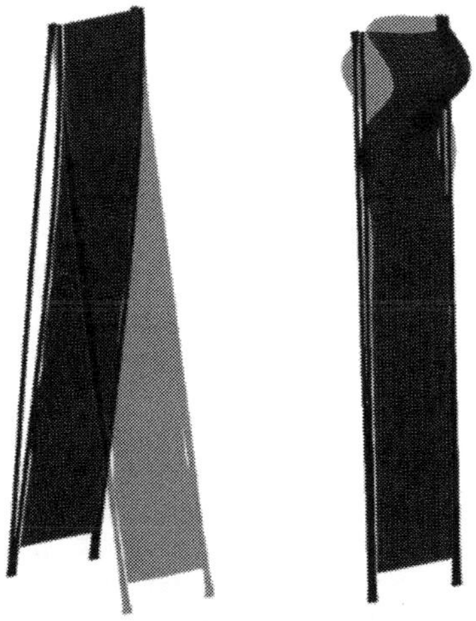

Figure 5. First 2 modal shapes

Several simulations have been run in order to identify the tape-springs characteristics that give a 0.1 Hz for the 1st mode frequency. These geometrical properties and material characteristics lead to a mass of 28 kg for the 2 tape-springs, whereas the blanket weights 74 kg.

This tape-springs design have been used to produce several 22m long tape-springs in order to build and test functional mockups as it can be seen in figures 6 and 7.

Figure 6. Deployment test and torque measurement

Figure 7. Flexible equivalent substrate mounted on the tape-springs structure.

Several mechanical tests have been carried out on the tape-springs themselves and on the mock-up. First the production quality has been checked and validated thanks to bending tests on tape-springs samples.

Figure 8. Nonlinear bending test performed on a 3m long tape-spring sample.

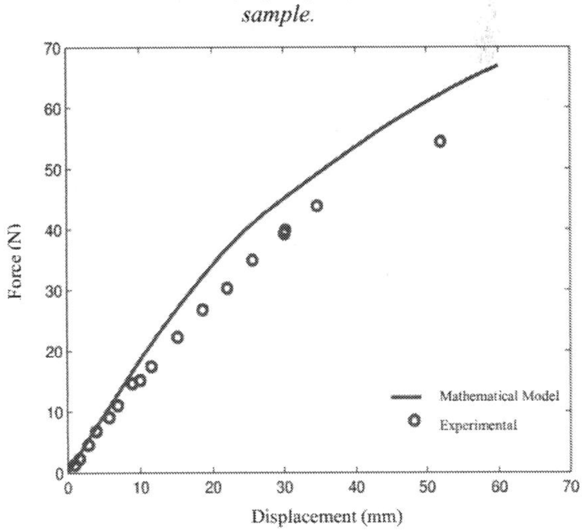

Figure 9. Force-displacement relationship for the sample bending test.

A good correlation between simulation and experimental tests can be seen on figure 9. Motorization tests have also been run thanks to the mockup measurement setup. The motorization torque value during the deployment phase has been compared to a numerical simulation results. The model used for coiling/uncoiling phase is an analytical model that has been developed by THALES ALENIA SPACE and CNRS these last years [1]. This beam model with a flexible cross section is able to simulate the highly nonlinear and dynamic behavior of tape-springs. In this case, this fast and accurate modeling strategy is more suitable than the classical finite element shell models.

Figure 10. Compact configuration

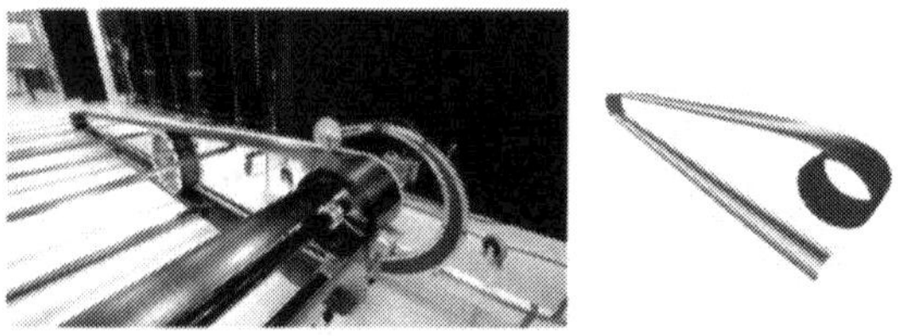

Figure 11. Deployed configuration.

Figures 10 and 11 illustrate two different steps of the deployment simulation performed with the THALES ALENIA SPACE analytical model. The computed motorization torque is compared to the experimental values. Figure 12 shows that the model is able to predict the tape-spring motorization torque with an excellent accuracy.

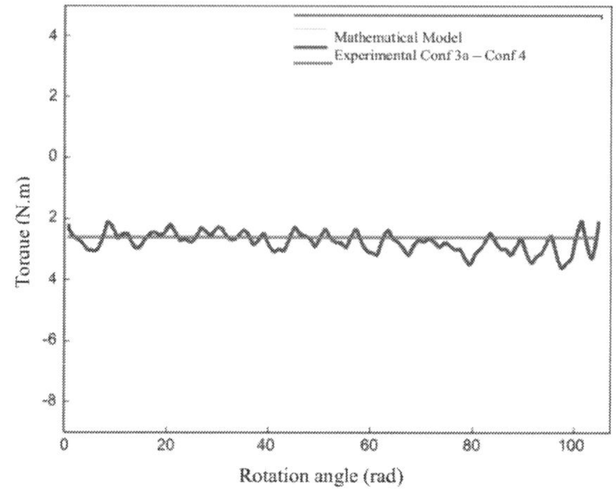

Figure 12. Torque vs. rotation relationship during deployment phase.

In addition to stiffness specifications, the tape-springs thermoelastic behavior is a crucial point in the flexible solar array design. Solar Array are long structures submitted to high thermal variation typically from -200°C to +100°C. Stability of structure is requested for sun pointing but also in order not to disturb satellite AOCS.

2.2. Thermoelastic and other considerations

Classic thin composite shells are usually submitted to twisting phenomena when temperature varies ; coiling is then impossible. Figure 13 shows an example of a carbon tape-spring made of a usual anti-symmetric layup. A new lay-up design has been modelled and patented by THALES ALENIA SPACE in order to solve this critical issue. Figure 14 shows the evolution of the layup design, from a highly twisted tape-spring to a very stable one.

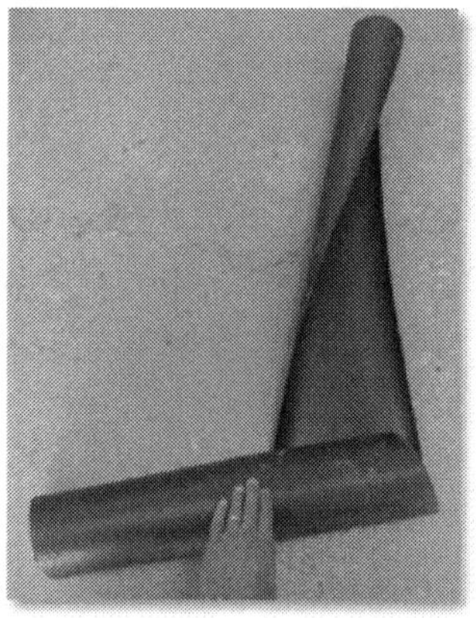

Figure 13. Twisting after thermal excursion.

 DOI: 10.1051/e3sconf/20171601005

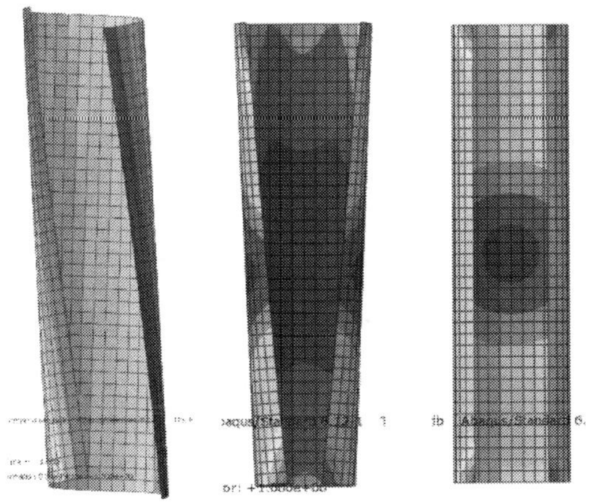

Figure 14. Simulations of 3 different layup designs in order to get rid of the twisting phenomena under 150°C temperature variation.

2.3. Comparison with another concept

The THALES ALENIA SPACE "bent into 2" tape-springs concept gives the flexible solar array structure a similar stiffness to the usual rigid panels, which is crucial for satellite attitude control. Also this solution brings a symmetrical strength to the wing, easier to manage than straight solution having a buckling limitation on one direction.

One can consider another solution using straight tape-springs. In this concept the flexible blanket is deployed from a canister located at the extremity of the solar array. So the tape-spring mass is divided by approximately 2, whereas a heavy mass is deployed. Modal analysis have been performed for comparison purpose (Table 2).

	Estimated mass per wing	Computed frequency of 1st mode
THALES ALENIA SPACE concept	Blanket : 74 kg Tape-springs : 29 kg Total : 103 kg	0.1 Hz
Straight concept	Blanket : 70 kg Tape-springs : 12 kg Canister : 5 kg Total : 87 kg	0.019 Hz

Table 2. Frequencies comparison for both concepts. Considered electrical power 23.6kW with 2 wings

Those simulations show that the straight concept, despite its 20% lower mass, has a 5 times lower frequency.

The additional tape-springs mass (THALES ALENIA SPACE concept) increases considerably the global stiffness as the two extremities are attached to a rigid part linked to the satellite frame, contrary to the straight concept that has a free extremity and a distant mass.

Figure 15. Modal analysis of a straight tape-spring solar array concept.

2.4. Development conclusions

The developments performed today for flexible mechanical architecture have been initiated with university labs (several thesis about the mechanical behaviour of tape-springs have been performed) and has been carried out with several contracts with agencies.

The development has bought and consolidated several advantages of the THALES ALENIA SPACE solution at different levels:

a/ Mechanical Architecture:

- Great and symmetrical stiffness and strength confirmed
- No deported mass at wing tip
- Thermo-elastic stability (no bending effect out of plane)
- One fixed point to avoid rotary electrical connection of network
- Stability versus buckling
- Bi-stable design to get an automatic stowage of TS
- New Solar Array possibility of wing rewinding / unwinding during flight

b/ Mathematical modelling:

- Non-linear FEM are now correlated with test on tape springs authorizing

DOI: 10.1051/e3sconf/20171601005

mathematical early optimization of the structure
- Bi-stable composite sizing capability demonstrated

c/ Manufacturability of long tape springs:

- Manufacturing processes:
  - Thin ply Molding
  - CFRP Pull-Press Technology Successful for full scale models

d/ Mechanical architecture mock-up:

- Deployment feasibility under 1g demonstrated in horizontal direction
- Functional behaviour of Tape Spring characterization
- Rolling capability (bi-stable)
- Synchronization of deployment
- Raising of interface to PVA blanket needs (lateral struts)

3. THALES ALENIA SPACE FORECAST ON FLEXIBLE SOLAR ARRAY MARKET

The Flexible solar array is expected to bring at the same time the following key advantages with regard to rigid array:

- Enhanced capability of power supply for equivalent mass and stowage volume, up to 50 kW on classical GEO platform.
- Mass reduction of about 35% for the same power supply and network cell thickness
- Stowed bulk reduction for the same power supply.
- Recurring cost reduction about 20% compared to classical rigid architecture thanks to simplified mechanical architecture and AIT.
- Stowage in orbit capability

Therefore, the flexible solar arrays will have interest for numerous markets including:

- GEO Telecom application having rising power demand with limited bulk under fairing, strong mass constraint and still requesting better competitiveness.
- Constellations Applications are good candidates for flexible array use while this solution is a great competitiveness lever in addition of any parallel development for network and cells cost saving. Also the stowed volume reduction can be a great advantage to increase the amount of satellites in a coupled launch.
- Missions with small satellite and high power request where the flexible compact design is a major advantage or a mandatory performance like for Space Tug mission.
- Specific missions requesting in orbit stowage: The flexible array can authorize rewinding / unwinding, for example to allow temporary high orbital loads (carrier, desorbitation, landing...), to operate next to other space vehicles (desorbiter, supply modules...) or to cross Van Allen belt.

For GEO Market the trend identified by THALES ALENIA SPACE is a progressive introduction of the flexible arrays from 2018~2020 till 2025, first for for high power payloads and later with a progressive dissemination to lower powerfull missions. The maturation of this technology will be obtained in parallel to a progressive decrease of the classic rigid architecture.

For some specific scientific missions, THALES ALENIA SPACE already identified that the feasibility of such missions relies on the use of flexible solar array due to the high power request with very limited stowage volume and mass budget.

The trend toward flexible array seems also to be confirmed by THALES ALENIA SPACE suppliers and competitors since numerous flexible array solutions are currently being developed and qualified by competitors.

4. SUMMARY

The work performed by THALES ALENIA SPACE with CNES support has raised the flexible architecture maturity level and demonstrated the feasibility of the patented concept with its main advantage being the simplicity of the mechanism and structure.

Also the parallel work on theoretical approach for large nonlinear structure finite element modeling and resolution combined with experimental activities has clearly demonstrated the capability of prediction by mathematical models and will authorize faster development and design performance adaptation of the tape springs.

The manufacturing of the tape spring with the requested composite layout has been demonstrated on mock-up with full scale model and so with manufacturing processes directly reusable for future flight models. This activity has then matured the manufacturing process of this "unusual" long composite structure of up to 44m. In addition, this manufacturing process is not limited by the length of the tape spring. The suppliers partnership has been fruitful and should naturally lead to next qualification models manufacturing.

The next step for the development of flexible solar array will consist in developping the interface of the electrical blanket with the structure and in maturing the PVA

 DOI: 10.1051/e3sconf/20171601005

blanket . This will lead to the manufacturing of a first EQM combining the mechanical architecture with the electrical network and afterwards the start of environmental testing.

ACKNOWLEDGMENT
The work reported here was possible due to CNES support and ESA for support of Tape Spring technologies.

5. REFERENCES

Useful references to past works concerning tape spring and flexible solar array architecture are in the following list.

1. A Rod Model With Flexible Thin-Walled Cross Sections. Application To The Folding Of Tape Springs. F. Guinot, S. Bourgeois, B. Cochelin, L. Blanchard, International Journal Of Solids And Structures, Volume 49, Issue 1, January 2012, Pages 73-86.
2. Modélisation De Structures Multi-Rubans Pour Applications Spatiales. P. Marone-Hitz, B. Cochelin, S. Bourgeois, F. Guinot. Actes Du 11eme Colloque National En Calcul Des Structures, 13-17 Mai 2013, Giens.

DEVELOPMENT OF A NEW, HIGH-POWER SOLAR ARRAY FOR TELECOMMUNICATION SATELLITES

C. G. Zimmermann, A. Bals, A. Übner, F. Schlerka and A. Schindler

Airbus DS GmbH, 81663 Munich, Germany, email: claus.zimmermann@airbus.com

ABSTRACT

Airbus is currently developing the Next Generation Solar Array (NGSA) for telecommunication satellites. It is based on a hybrid array concept which combines a conventional rigid panel array with lightweight, semi-rigid lateral panels. The main figures of merit power/mass and power/volume can be doubled through this concept. Mechanically, the semi-rigid panels are the key new element. Through acoustic testing as well as sine vibration testing in air and in vacuum it was verified that these panels are suitable as cell support in stowed configuration. With the help of finite element modelling it is demonstrated that the semi-rigid panels are compatible with a free deployment. Electrically, the new array is to be equipped with a new generation of 4 junction solar cells with efficiencies above 30%. The increased radiation dose due to electric orbit raising has to be taken into account to arrive at the optimum shielding while still minimizing the array mass. By adjusting the ratio of rigid to semi-rigid panels and through the choice of solar cell type and mass, the NGSA can be tailored in a wide range to needs of a given platform. This is illustrated for the solar array to be flown on the new Airbus platform Eurostar Neo.

1. INTRODUCTION

Airbus is currently developing the new tele-communication platform Eurostar Neo, which is optimized to an even higher degree than its predecessor Eurostar 3000 for the payload. This payload centric design imposes new, much more stringent requirements onto the solar array. Most notably, the stowage volume available to the solar array on the satellite sidewall is reduced, both in terms of surface area as well as in stack height. At the same time, the power demand has increased by 50%, compared to previous Eurostar solar arrays. Due to the focus on electric orbit raising, the radiation fluences are also increased up to five times.

To address these challenges, we rely on the concept of a hybrid solar array, which was already suggested several years ago[1]. There, a backbone, resembling a conventional rigid panel array is combined with lightweight, semi-rigid lateral panels. During launch, these panels are folded in between the rigid panels and are thus protected. This Next Generation Solar Array (NGSA) is developed in the frame of the ESA ARTES funding lines.

OVERALL SOLAR ARRAY DESIGN – THE HYBRID CONCEPT

These key features of the NGSA are illustrated in Fig. 1 in deployed and in Fig. 2 in stowed configuration.

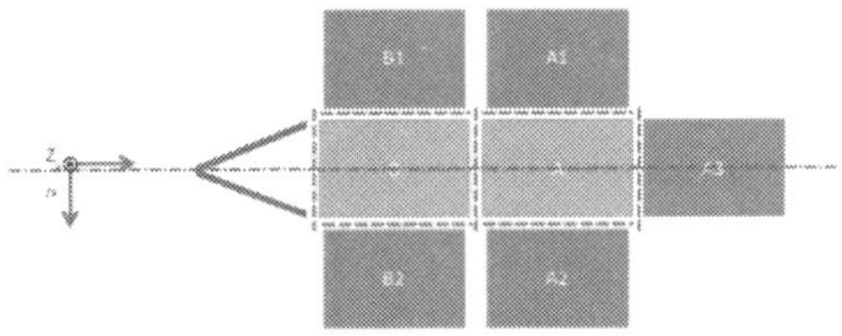

Figure 1: Schematic illustration of the solar array concept. Conventional rigid panels are indicated in light green and the new semi-rigid lateral panels are dark green.

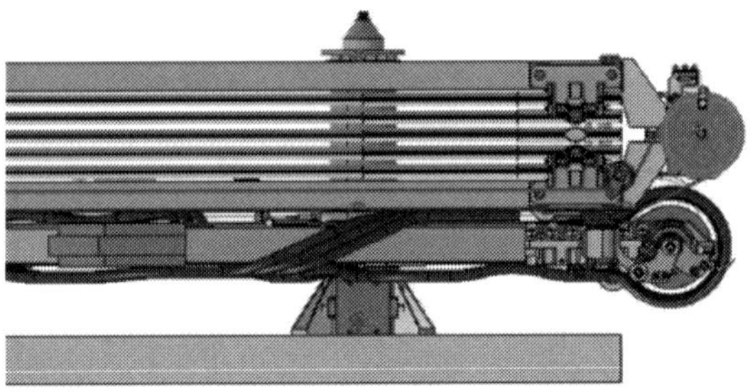

Figure 2: Cross sectional view through the solar array stack in stowed configuration.

In Fig. 2 the most important advantage of the hybrid array concept is directly evident, the semi rigid panels are much thinner than the conventional honeycomb panels and in addition can be stacked much closer than in conventional arrays. Therefore the stowage volume is greatly improved. In addition, their aerial mass is 5 times lower compared to the rigid panel and combined with the use of lightweight cells, the power to mass ratio of the array is also improved significantly. Based on the maximum configuration with 5 semi-rigid panels sandwiched between two rigid panels, illustrated in Figs. 1 and 2, a whole array family can be constructed based on the needs of a given platform.

In Table 1, the performance figures of the NGSA concept are summarized for a 2 rigid / 5 semi-rigid panel version consisting of Eurostar 3000 panel sizes, which equate to a surface area of 9 m² per panel. Eurostar 3000 is Airbus's current solar array for telecommunication missions, consisting of a conventional rigid in-line panel configuration. As reference, the performance of the largest version, consisting of 5 rigid panels, is included as well. In terms of radiation dose, a classical 15 year GEO mission with purely chemical transfer orbit is assumed and the end-of-life (EOL) summer solstice (SS) power figures are computed. Both mass and stowage volume figures are "real" figures in the sense that they include yoke, yoke hinge as well as all hinges and hold-down and release units. The stack height is calculated from the satellite sidewall up to the outermost panel.

Table 1: Main performance figures of the NGSA in a standard GEO mission scenario.

version	E3000 5L	NGSA 2+5	NGSA 2+5	NGSA 2+5
cell type	3J 30%	3J 30%	3J 30%	4J 33%
cell thickness [µm]	140	140	20	20
coverglass [µm]	100	100	50	50
mass [kg]	302	284	205	205
power EOL SS [kW]	19.8	26.9	26.5	29.5
power/mass [W/kg]	65	95	129	146
power/ volume [kW/m3]	4.2	7.2	7.2	8.0

According to Table 1, the NGSA concept is capable of approximately doubling the main figures of merit, power/mass and power/volume, compared to the conventional rigid panel arrays in use today. Table 1 also highlights the main contributing factors. The hybrid array concept by itself improves power/mass by 44% and the stowage volume by 72%, as evident from the third column in Table 1. The reduction in solar cell assembly (SCA) mass from a classical 140 µm cell/100 µm coverglass configuration of 3.5 g for a 30 cm² SCA to a lightweight 20 µm cell and 50 µm coverglass configuration of 1 g in column 4 improves power/mass by an additional 37%. Finally using a lightweight 4J cell with a 10% increased EOL efficiency of 29% at 10^{15} 1 MeV e- /cm² results in an identical 10% increase in power/mass and power/volume again, as shown in column 5.

2. CONCEPT VERIFICATION

This increase in power/mass and power/volume by a factor of two is especially attractive due to the fact that it is achieved by a solar array design that relies in large parts on components with extensive space heritage. What has to be ensured, however, is that the solar cells on the semi-rigid panels can actually survive the mechanical loads occurring during launch. Although the hybrid solar array concept has been suggested several years ago, the actual feasibility of the concept has not been experimentally verified up to now.

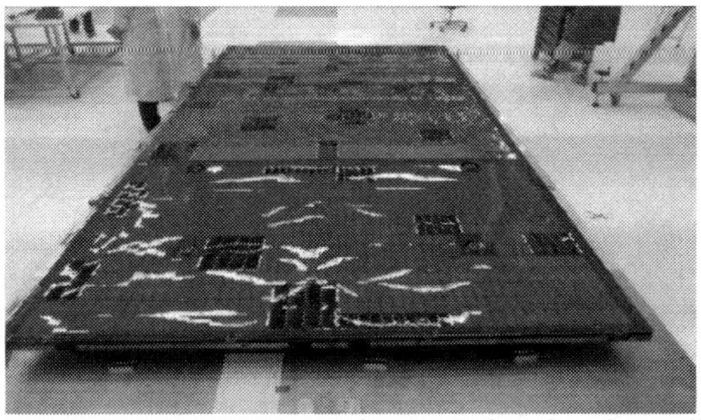

Figure 3: Semi-rigid panel equipped with solar cells and SCA mass simulators.

For the experimental verification, two semi-rigid panels, 3.9x 2.3 m in size were built and equipped with 30% efficient triple junction cells in representative areas, as illustrated in Fig. 3. The cell thickness varied from 20 – 80 µm and 50 µm and 100 µm coverglasses were used in order to cover a range of SCA masses. The two semi-rigid panels were sandwiched between two rigid panels, and the entire stack was subjected to three axes sine vibration testing, with 4 g input out-of-plane, notched for the first resonance frequency of the rigid panels, and 6 g for the two in-plane tests.

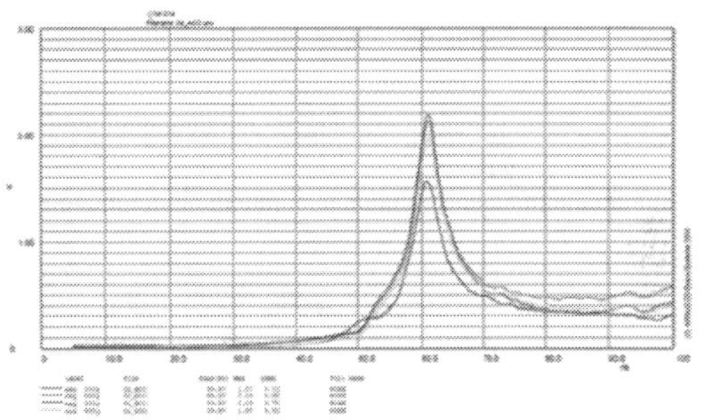

Figure 4: Accelerations measured during the in-plane sine vibration testing on the 2 rigid and 2 semi-rigid panels at identical position.

All sine vibration tests were carried out successfully. The cell integrity was not affected. Both rigid as well as semi-rigid panels were equipped with accelerometers during test. In Fig. 4 the accelerations measured on 4 sensors located at the same position on each panel during the in plane test are shown as an example. All accelerometers show the same resonance frequency around 60 Hz. In particular no resonance frequency of the semi-rigid panels, expected in the lower frequency range is observed. A more detailed analysis confirmed that the enclosed air in the stack resulted in significant damping as well as an in-phase movement of the semi-

 DOI: 10.1051/e3sconf/20171601006

rigid panels with the rigid panels.

In addition, standard acoustic noise testing was carried out on the same stack with a maximum sound pressure level of 145 dB for 70 s and with 155 dB for 10 s. Again the test was completed successfully, with no cracked solar cells as well as no damage to the rigid or semi-rigid substrates observed.

These sine vibration and acoustic tests confirmed the basic feasibility of the hybrid solar array concept. They however also highlighted the important effect air damping has on the stack. Since during an actual launch the evacuation of the fairing proceeds much faster than the acceleration loads are reduced, two important conclusions result: i) for the semi-rigid panels it is not possible to rely on the effect of air damping and ii) the stacked configuration has to be subjected to a sine vibration test under vacuum.

3. SEMI-RIGID PANELS

The central element of the NGSA concept are the semi-rigid panels. Their design is illustrated schematically in Fig. 5. The panel consists of a thin, Aramid fiber re-enforced sheet. It is thus closely based on the Eurostar 3000 AOCS flap, which already has an extensive space heritage. Carbon black included in the resin serves as protection of the Aramid fiber against ultraviolet (UV) light. For the use as cell support in the NGSA application it is strengthened through additional corrugations in a frame like configuration, which provide the required stiffness in deployed configuration. In addition the corrugations act as interface point for deployment hinges. As illustrated in Fig. 5, these corrugations divide the panel in three approximately equal areas.

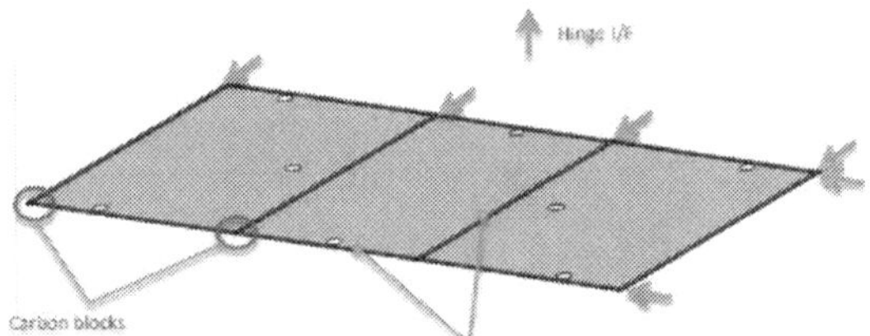

Figure 5: Schematic illustration of the semi-rigid panel.

The damping system, which was developed to substitute the air damping, consists of two main elements. The first one is focused onto the corrugations and prevents the in plane movement of the semi-rigid panels within the stack. The second part of the damping system limits the amplitude the sheet can perform in the area between the corrugations.

As pointed out before, for a complete verification, also the part of the launch with virtually no atmosphere has to be simulated on ground. Rather than attempting to house a conventional shaker in a vacuum vessel, a unique, lightweight vacuum enclosure was developed that can be mounted on a standard shaker. To balance the atmospheric pressure while still keeping the weight down, plates in sandwich configurations were devised. In addition, the size of the test specimen was limited to one third of a semi-rigid panel. This greatly helped to reduce the loads the sandwich plates have to withstand due to the atmospheric pressure. Nevertheless, the testing of a 1/3 panel is still fully representative, since the boundary conditions at the position of the corrugations are the same everywhere on the panel. The vacuum setup mounted on a shaker is shown in Fig. 6. Its resonance frequency is ≈ 100 Hz and thus outside the range of interest for the actual sine vibration test and the base pressure that can be routinely achieved is 15 mbar.

Figure 6: Setup for sine vibration testing under vacuum.

Several 1/3 panel semi-rigid flaps were built and equipped with mass dummies as well as 80 SCAs consisting of 80µm solar cells and 100 µm cover glasses. Sine vibration testing with one up to three semi-rigid panels sandwiched between two rigid panels were carried out under vacuum. The input levels for the three axes were identical to the testing performed in air. In all cases the cell integrity during the dynamic loading could be experimentally verified. Thus the feasibility of the hybrid array concept is validated under all environmental conditions.

In Fig. 7 the readings of one accelerometer on the semi-rigid panel are plotted during an out of plane sine vibration test in a stack consisting of one semi-rigid and two rigid panels. Two main resonances around 23 Hz and around 45 Hz are visible. The first one can be attributed to the semi-rigid panel itself, whereas the second one belongs to the rigid panels, as confirmed by accelerometer measurements on the rigid panels themselves. The appearance of this second peak is an

DOI: 10.1051/e3sconf/20171601006

indirect consequence of the damping system that was incorporated into the design. While the semi-rigid panels can still oscillate freely, as shown by the presence of their resonance frequency, they are nevertheless coupled to the rigid panels and thus the rigid resonance frequency is visible in the acceleration response of the semi-rigid panels as well.

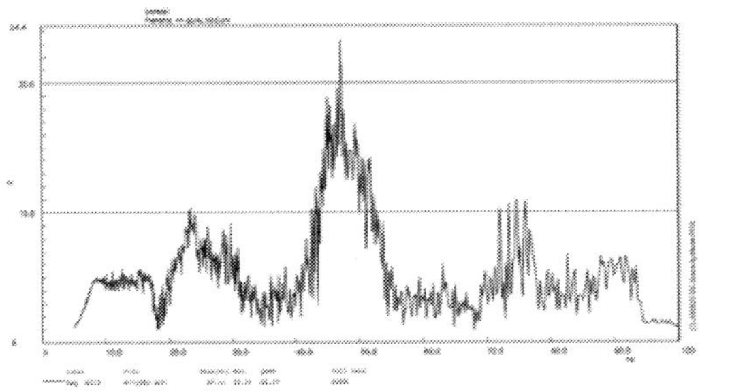

Figure 7: Measured accelerations on a semi-rigid panel during a sine vibration test in vacuum.

The second main consideration for the semi-rigid panels is their strength and stiffness in deployed configuration and, since this is going to be the limiting case, during deployment. The NGSA design uses C-Spring hinges, one at every one of the four corrugations, due to the low mass and the simplified wing integration which is possible with this solution. In order to not overly complicate the overall deployment mechanism, the lateral semi-rigid panels are to be deployed freely. The loads introduced into the corrugations, especially during latch-up, have to remain within acceptable limits.

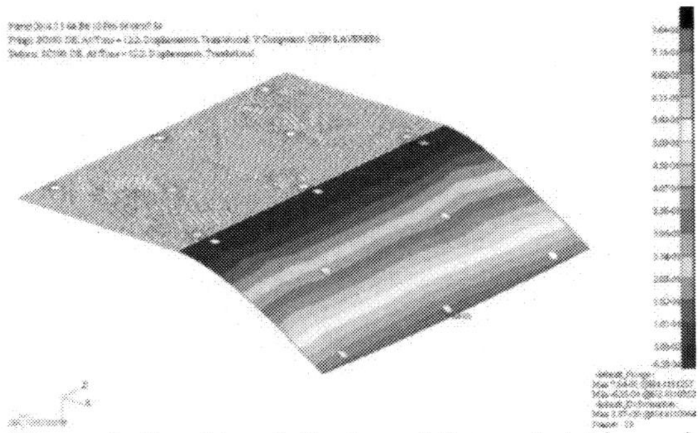

Figure 8: Resulting deflection of the semi-rigid panel during free deployment according to FEM modelling.

To address this topic, a non-linear FEM model was setup and correlated based on the bending curve of the corrugations alone in a 1 g environment. The resulting deformation of a lateral semi-rigid flap during free deployment is illustrated in Fig. 8. The deformation of the outermost edge of the ≈ 2 m wide flap is ≈ 0.8 m, which results in a fully acceptable radius of curvature.

In addition, the maximum bending moment in the corrugations is far below the critical moment. This demonstrates that the planned free deployment of the lateral panels is feasible.

4. PVA CONSIDERATIONS

While from a mechanical viewpoint the semi-rigid panel is the most important element of the NGSA, from the electrical side it is the solar cell assembly.

The NGSA solar array is ideally suited to make maximum use of the next generation of III-V solar cells, which will reach efficiencies above 30% through the addition of subcells and a current matched design. Several of the new cell concepts will automatically result in thin and thus lightweight solar cells. Alternatively cells relying on a Ge bottom cell can be thinned to the desired weight. Within Europe the 4G32 solar cell is under development, with a target efficiency of 28.5 % at 10^{15} 1 MeV e-/cm^2.

As pointed out in Table 1, the SCA mass has a significant impact on the power/mass ratio. The need to have an as lightweight SCA as possible, however, has to be balanced with the radiation shielding requirements. Electric orbit raising (EOR) is a major trend for GEO satellites. Due to the significant mass savings on platform level and the reduce launch cost resulting from it the majority of GEO satellites will employ electric propulsion.

For electric orbit raising missions the radiation dose is significantly increased, due to the additional non-ionizing dose inflicted predominantly by protons in the inner Van Allen belt. This inner belt is crossed multiple times in electric orbit raising trajectories in comparison to only once in a chemical transfer. The expected radiation dose will vary significantly in actual missions, depending on the launcher chosen. One challenge for the PVA layout will be to find an optimal solution for the bulk of the EOR missions, while still ensuring sufficient voltage for the worst case missions. This necessitates an electrical layout that makes use of the non-ideal fill factor of a section by placing the fixed bus voltage in the "knee" of the I-V curve. In this region, voltage can be essentially traded for current.

The expected 1 MeV electron fluence for a typical EOR +GEO radiation environment is plotted in Fig. 9 as a function of frontside shielding. An infinite backside shielding was assumed in this calculation. Therefore the contributions of both sides have to be added to calculate the total dose. As indicated by the blue lines in Fig. 9, the dose shielding curve can be approximated linearly with two slopes. Both lines cross at approximately 130 µm equivalent Si shielding. This represents the

DOI: 10.1051/e3sconf/20171601006

minimum amount of shielding that should be present both on the front as well as on the rearside. Additional shielding then can be distributed on front or rearside as seen fit. For the NGSA SCA, a 100 µm coverglass has been chosen, which comes fairly close to this minimum shielding requirement. For a Ge based current matched cell, the question arises how much Ge can be assumed as "inactive" and thus as part of the rearside shielding. Assuming an active Ge thickness of 20 µm, the shielding of the inactive Ge, rearside metallization adhesive and semi-rigid substrate amounts to 175 µm equivalent Si shielding thickness for a 50 µm Ge based cell. For such a cell the total equivalent 1 MeV e-fluence amounts to 1.8×10^{15} e-/cm^2. For typical III-V multijunction cells, such a fluence is already in the steep part of the cell degradation curve. For this reason, an 80 µm equivalent Ge thickness was chosen as baseline cell for the NGSA in EOR missions. The mass of such a cell is still compatible structurally with the semi-rigid support, while at the same time the additional 30 µm in rearside shielding on the semi-rigid flaps result in an additional ≈1 % in absolute cell efficiency at end of life.

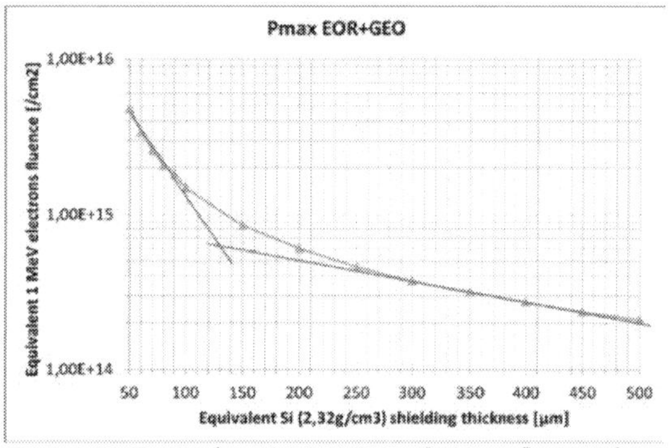

Figure 9: Equivalent 1 MeV P_{max} electron fluence for a combined typical electric orbit raising (EOR) and GEO radiation environment as a function of frontside shielding thickness. Infinite backside shielding was assumed.

5. NGSA IMPLEMENTATION ON EUROSTAR NEO

For Eurostar NEO the NGSA concept will be implemented for the first time, based on a family of 4 rigid and up to 4 semi-rigid panels. The largest version is illustrated in Fig. 10. In this configuration a maximum of 2 semi-rigid panels are sandwiched between two rigid panels each. A double yoke is included to prevent antenna shadowing completely, whereas the innermost rigid panel without side panels facilitates a wider viewing angle of the star sensor. In the baseline configuration, all rigid panels will be equipped with 6x12 cm^2, 140 µm thick triple junction cells and all semi-rigid panels with 80 µm thick 4x8 cm^2 four junction cells. The resulting performance figures are summarized in Table 2.

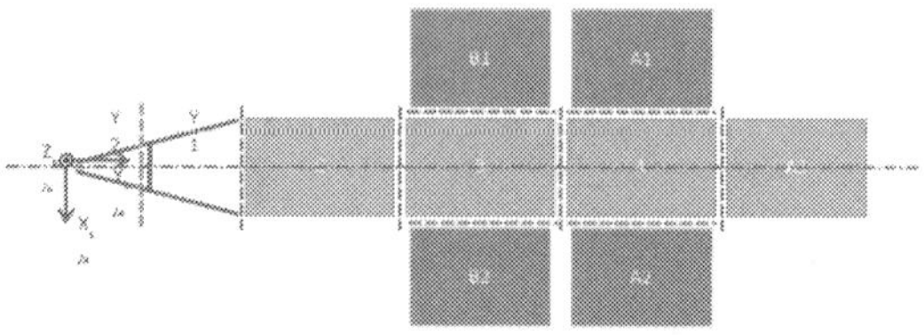

Figure 10: NGSA configuration for the Eurostar Neo platform.

In comparison with Table 1 it is evident that the Eurostar Neo array exploits to a great deal the lower stowage volume provided by the NGSA concept. In addition in a high power version, where the entire array is equipped with 4G32 cells, the maximum end-of life power is increased by an additional 2 kW to ≈ 27 kW and the power/mass ratio is increased to 88 W/kg, a 35% improvement compared to a Eurostar 3000 array.

Table 2: Eurostar Neo implementation of the NGSA.

version	E NEO R4S4 LARS / 4G32	E NEO R4S4 4G32
cell	6x12 cm^2 LARS 26.5%, 4x8 cm^2 4G32 28.5 %	4x8 cm^2 4G32 28.5 %
cell (Ge) thickness	140 µm & 80 µm	80 µm
coverglass	100 µm	100 µm
equivalent fluence (Pmax)	1.5×10^{15}	1.5×10^{15}
power [kW]	24.6	26.6
mass [kg]	314	302
stack height [mm]	264	264
power/mass [W/kg]	79	88
power/volume [kW/m^3]	6.4	7

6. CONCLUSION

The NGSA technology provides an attractive solar array solution for even the highest power GEO telecommunication spacecraft, both in terms of mass as well as stowage volume. It is ideally suited to exploit the benefits of the next generation of lightweight four junction solar cells. Compared to conventional in-line rigid arrays power/mass and power/volume can be doubled. This improvement is achieved through an evolutionary concept with extensive space heritage in all major components like the rigid backbone or the semi-rigid substrate.

Through changing the rigid/semi-rigid panel ratio and through the cell selection in terms of weight and form factor it is possible to tailor the array to the needs of a particular platform to find the optimum solution in terms of power, mass, stowage volume and cost. This was demonstrated for the Eurostar Neo platform. The associated solar array will be the first NGSA array to be launched. It will be among the first arrays to fly > 30% four junction cell technology for telecommunication applications.

7. ACKNOWLEDGEMENTS

This work was funded in part through the ESA ARTES5.2 and 3-4 lines.

8. REFERENCES

1 V. G. Baghdasarian, "Hybrid solar panel array", patent US5785280 A, 1998

CASOLBA: BALLOON CALIBRATION OF SOLAR CELLS

Etienne Rapp [1], Valérie Pichetto [1], Christian Elisabelar[1], Carsten Baur[2], Denis Bausch[1], Maxime Jeunet[1]

(1) CNES, 18 avenue Edouard Belin 31401 Toulouse Cedex 9, France
etienne.rapp@cnes.fr; valerie.pichetto@cnes.fr; christian.elisabelar@cnes.fr; denis.bausch@cnes.fr

(2) European Space Agency (ESA), Keplerlaan 1, 2201 AZ Noordwijk, The Netherlands
carsten.baur@esa.int

ABSTRACT

In order to make accurate measurements, sun simulators used for solar cell characterisation need to be calibrated with standard solar cells. These so-called secondary working standards are derived from primary reference cells and are used for every cell measurement.
To generate primary reference cells, balloon calibration campaigns are a good compromise between accuracy, flight opportunity and cost.
CNES calibrates primary standards with stratospheric balloon flights since 1976. Safety rules limited the flight opportunities, so it was not possible to propose the calibration service since 2005.
A new balloon facility is developed in CNES with the help of ESA in order to calibrate space solar cells with high accuracy. The first flight is planned in April 2017 in Alice Springs, Australia.

1. INTRODUCTION

The methods for AM0 calibration of solar cells fall into two broad categories; extraterrestrial and synthetic. Extraterrestrial methods attempt to measure the solar cell near the AM0 condition by measuring cells at high altitude (low air mass) or in space. Examples of this method include high altitude airplane, International Space Station and high altitude balloon. In the ideal case, these methods require no knowledge of the solar spectrum. High altitude balloons have long been recognized as the industry standard of extraterrestrial, AM0 solar cell calibration [1].

A recent study performed by ESA in collaboration with AZUR SPACE, Fraunhofer ISE and PTB (Physikalisch-Technische Bundesanstalt – German national metrological institute) demonstrated that both methods - synthetic method used by PTB and CNES balloon - were in perfect agreement within the measurement uncertainties claimed (≤ 1%), at least for the top and the middle cell results. Only for the bottom cell there was a higher deviation in the order of 2.5% [2].
Synthetic methods rely on the knowledge of the AM0 spectrum. Ultimately, regardless of how much care is taken measuring the cell performance, the community is still reluctant to adopt the synthetic method as an alternative option that is fully equivalent to balloon or Lear Jet calibrations. Therefore, CNES, upon ESA request, decided to reinitiate a balloon calibration capability by developing a complete new payload to be launched on a new nacelle.

2. HISTORY AND BACKGROUND

Figure 1: 1976-1996 flight configuration

The first flight occurred in 1976. The principles of the experiment were:

- Open stratospheric balloon (100 000m^3,

altitude between 34km and 40km)

- Dedicated sun-pointed nacelle (110kg) under the balloon
- Sunshield (collimator, to avoid the parasitic light coming from the earth limbo and from the sun reflexion on the balloon envelope)
- Passive thermal control: the cells are cooled during the ascent then the temperature increases after sun pointing. The temperature is recorded continuously during the flight.
- Measurement of the open circuit voltage and near short circuit current: the voltmeter is switched from open circuit to a calibrated resistor (shunt) by means of electromechanical relays.
- Coefficients of correction: ks (sun-earth distance) and kn (residual atmosphere for low air mass (0,005<AM<0,014).

Figure 2: 1998-2001 flight configuration (collimator removed)

For the 1998 campaign, the nacelle is extended to carry more TTC (tracking, telemetry and control) and localisation equipment.
The measuring electronics is improved: it allows the calibration of 30 cells by performing measurements of the I-V characteristics from short circuit to open circuit at each temperature. The mass of the nacelle is now 185kg. The analogic 2-axes orientation system was used since the first flight.
In June 2001, some cells were degraded by the drop of the ballast (lead balls), so a second flight was planned in October. However, the nacelle was destroyed at landing.

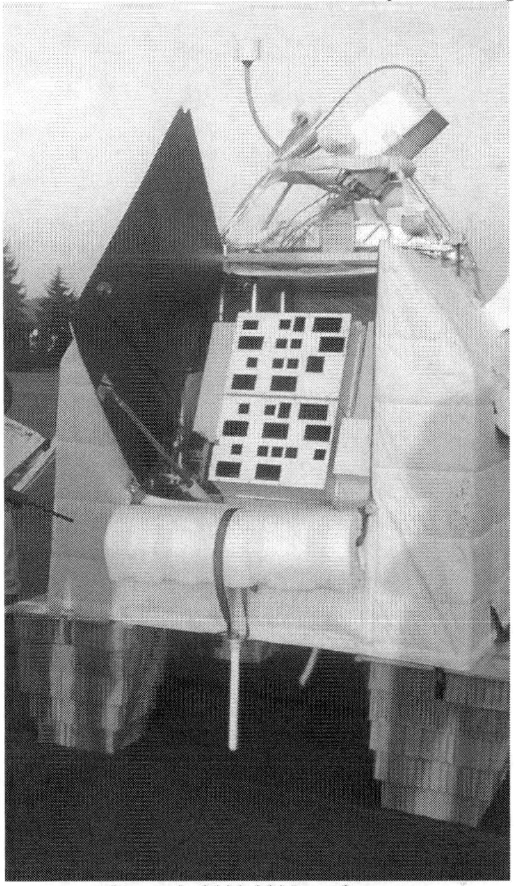

Figure 3: 2003-2005 configuration

In 2003, a multi-purpose pointed nacelle is used, with better performances. The total mass of the nacelle reaches 380kg.
After this flight, it was no more possible to fly in France.
In 2005, the balloon is launched from Teresina base in Brazil.
After 2005, no further campaign was organised due to either a lack of customer requests, i.e. not enough cells to calibrate, or the unavailability of the nacelle. The possibility to design a "mini-Casolba" nacelle, 50kg, able to fly 6 to 8 cells was also investigated. But flying a small nacelle is as difficult as a big one.

In 2014, ESA asked CNES to look at the possibility to restart the activity. A programme has been presented at the 17th AM0 Workshop on Space Solar Cell Calibration in April 2014. It was received with enthusiasm by the

 DOI: 10.1051/e3sconf/20171602001

community and CNES decided to start the project, with the help of ESA.

3. BALLOON AND NACELLE

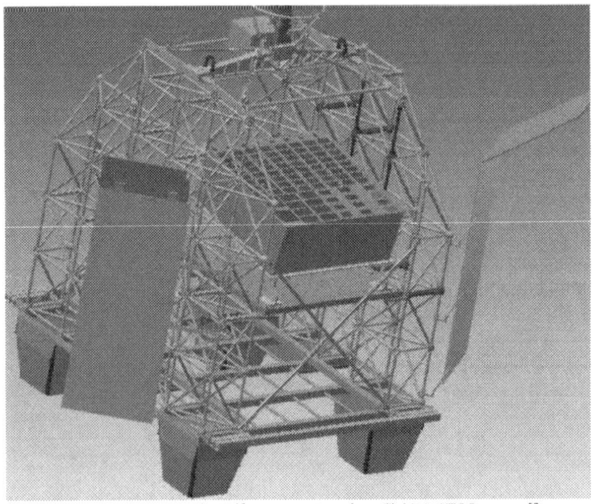

Figure 4: plate and collimator in the CARMEN nacelle

The CARMEN nacelle is a heavy multi-mission facility which provides all facilities: accurate pointing, energy, TM/TC (telemetry and telecommand) and data storage. The sun pointing accuracy is 1° in azimuth and 0.5° in elevation.
The maximum weight is 1100kg. Up to 4 scientific missions can fly simultaneously.

The open stratospheric balloon has a volume of 400 000m^3 and allows a flight altitude of 36km for a 750kg load.

4. MEASUREMENT SYSTEM

The measurement system is based on CNES know-how, with a passive thermal control (the temperature of the cells decreases during the ascent and during non-pointed periods). The current-voltage curves (100 points from Voc to Isc – and 100 points from Isc to Voc) of the cells are measured continuously during the sun-pointed periods. The temperature of the cell is measured at the beginning and at the end of the I-V curve measurement. The temperature range is [10°C; 40°C] or more.
The measurement electronics is shared into 3 MCCS (Module de Calibration de Cellules Solaires) boxes. Each box is able to measure successively 32 solar cells. The electronics is able to measure the cell voltage up to 5.7V and cell current up to 2.8A.
The measurement time is one second per cell. One measuring session consists of the measurement of the I-V characteristics of all the cells and the complete measuring sequence comprises all the sessions of a sun-pointed period. Up to 3 measuring sequences can be performed during the flight at different sun elevation or different altitudes in order to have a better view on the effect of residual atmosphere and to improve the accuracy of the corrections. In between the measuring sequences, the nacelle is pointed at 90° of the sun direction and the cells are cooled down.

Cells are mounted on specific holders. The sun-oriented plate is able to support 63 mid-size holders (for cells up to 42*85 mm^2), and 25 large-size holders (for cells up to 71*148mm^2). Double-size holders can be designed if needed. Each holder carries a temperature sensor (AD590M) and a protection by-pass diode. It is possible to mount NSCAP (NRL-NASA-AFRL Near Space Characterization of Advanced Photovoltaics) or also other type of cell holders on the plate by means of appropriate adaptors.

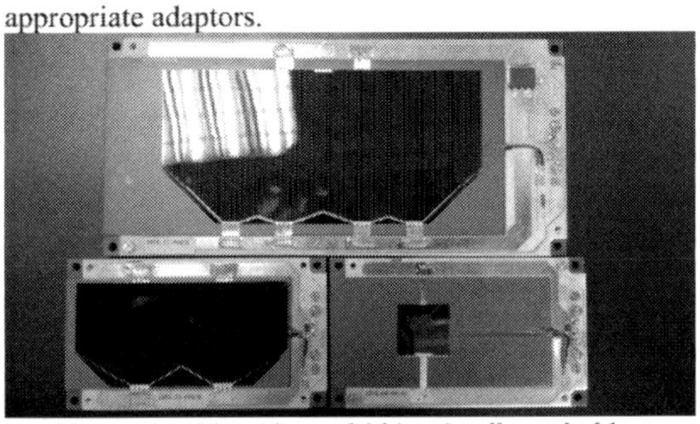

*Figure 5: 12*6, 8*4 and 2*2cm^2 cells on holders*

A collimator is placed in front of the plate in order to avoid any parasitic light and control the temperature range. The front mask of the collimator reproduces the shape of the cells and is redesigned for each flight.

All the flight data will be provided to the customers: rough current-voltage curves, temperature and current in the temperature sensor, but also corrections of the sun-earth distance and atmospheric spectral absorption (air mass), altitude and localisation.

5. UNCERTAINTY ANALYSIS

The specification states that the uncertainties shall be lower than 1%.
A list of the different uncertainties has been made from the previous flights: 8 of them are due to the environment, 6 come from the method used, 1 from the operator, 24 from the measuring instruments and finally 11 uncertainties come from the measurand.

Once every uncertainty was analysed we concluded that the main uncertainties come from the coefficient kn, the MCCS and the temperature.

The temperature is measured with a probe. The datasheet states that the probe is accurate to less than 0.5°C. Thus, it is possible to calculate the uncertainties

DOI: 10.1051/e3sconf/20171602001

related to the temperature and take this uncertainty into account for the calculation of the currents and kn. For more accurate verification, the measurement of the current in the probe will be provided.
The coefficient kn is defined as below:

$$kn = \frac{\int R_s(\lambda).S_{AM0}(\lambda).d\lambda}{\int R_s(\lambda).S_{AMX}(\lambda).d\lambda} \quad (1)$$

So in order to calculate kn, one needs the spectral response of the solar cells (R_s), the solar spectrum at AM0 (S_{AM0}) and finally the solar spectrum at the altitude of the solar cells (S_{AMX}).

In order to calculate the uncertainties related to kn, we have to study the coefficient AM, the spectral responses and the solar spectrum. This leads to a Type B uncertainty. With the data from previous flight one finds the Type A uncertainty. Finally one obtains the following equations:

$$U_{cA}(kn) = 3{,}4738.10^{-4} \quad (2)$$
$$U_{cB}(kn) = 1{,}1611.10^{-4} \quad (3)$$

$$U_c(kn) = \sqrt{U_{cA}^2(kn) + U_{cB}^2(kn)} = 3{,}6627.10^{-4} \quad (4)$$

Due to the components there is a difference between the empirical and theoretical voltages at the output of the analogic-digital converter. So, in order to correct this error a calibration is required. A first calibration was made but the range of the temperatures being wide; this calibration did not correct the voltages precisely. Indeed the components of the MCCS are sensitive to the temperature. Hence, different transfer functions are used depending on the temperature. The first one is used when the cells' temperatures are between -10 and 25°C. The second transfer function is taken between 25°C and 40°C. Finally, the last transfer function is used as soon as the temperature reaches 40°C. The uncertainty will probably be reduced after the flight, taking into account the actual MCCS temperature range.

The idea of one transfer function, which depends of the temperature, was considered. However, once implemented, we saw that the errors did not decrease significantly.

The study performed on the MCCS gave the following uncertainties. They are expressed in ampere.

Range 2.8 A

$$U_c(I_{sc\_gross}) = 1{,}7953.10^{-4}.I_{sc\_gross} + 5{,}3258.10^{-5} \quad (5)$$

Range 250 mA

$$U_c(I_{sc\_gross}) = 1{,}3008.10^{-4}.I_{sc\_gross} + 5{,}6639.10^{-6} \quad (6)$$

Using the following formula we can determine the uncertainty of the corrected current.

$$U^2{}_c = \sum_{i=1}^{N} \left[\frac{\partial f}{\partial x_i}\right]^2 . u^2(x_i) \quad (7)$$

$$I_{sc\_corrected} = I_{sc\_gross}.kn.ks \quad (8)$$

The uncertainties of the corrected current of each cell will be calculated after the flight.

6. REPRODUCIBILITY

Three cells are measured frequently in order to check the reproducibility, during the period 1999-2005.

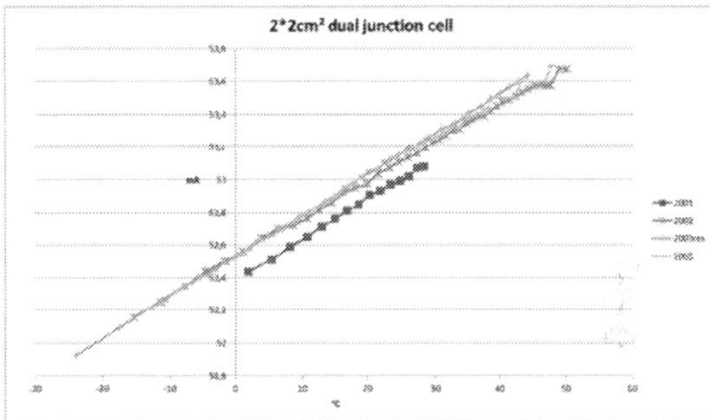

Figure 6:DJ cell reproducibility

For a dual-junction cell shown in Fig. 6, the maximum difference between measurements is 0.24% at 25°C; the standard deviation is 0.14% (including a weighting factor equal to 1.7 for 4 samples).

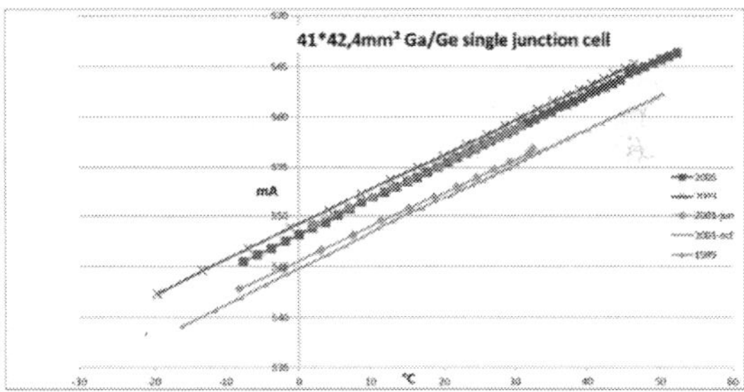

Figure 7: GaAs cell reproducibility

For GaAs single-junction cell shown in Fig. 7, the maximum difference between measurements is 0.64% at 25°C; the standard deviation is 0.49% (including a weighting factor equal to 1.4 for 5 samples).

DOI: 10.1051/e3sconf/20171602001

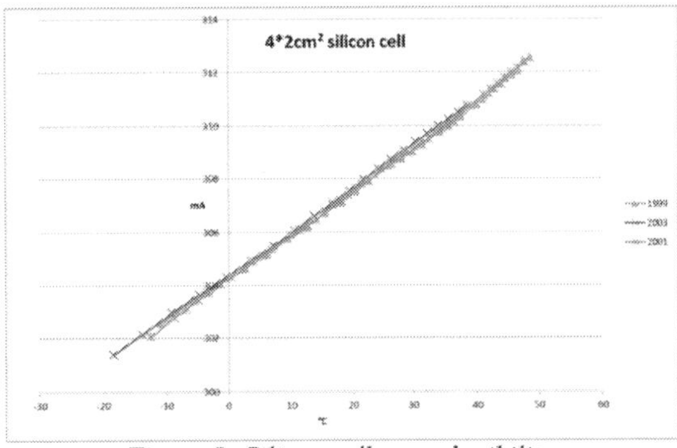

Figure 8: Silicon cell reproducibility

For the Silicon cell given in Fig. 8, the maximum difference between measurements is 0.06% at 25°C; the standard deviation is 0.16% (including a weighting factor equal to 2.3 for 3 samples).

These cells will be re-calibrated during the next flight, using the new configuration and pieces of equipment.

7. REFERENCES

1. P. Jenkins, D. Wilt, D. Scheiman, B. Anspaugh, M. Pisczcor, D. Snyder, D. Ball, J. Juneau, H. Franco, D. Orr, S. Cannon, J. Lorentzen, R. Walters, M. Krasowski and B. Reed. A new air mass zero calibration facility using the high altitude Balloon method. In *Proceedings of the 9th European Space Power Conference*, St. Raphaël, France, 2011.
2. C. Baur, G. Siefer, R. Kern, S.Winter. Primary solar cell standards - comparison of extraterrestrial and synthetic calibration. In *Proceedings of the 9th European Space Power Conference*, St. Raphaël, France, 2011.

RECENT ADVANCES IN SYNTHETIC CALIBRATIONS OF MULTI JUNCTION SOLAR CELLS AND THEIR CORRESPONDING COMPONENT CELLS

Florian Witt[(1)], Ingo Kröger[(1)], Stefan Winter[(1)]

[(1)] *Physikalisch-Technische Bundesanstalt Braunschweig, Bundesallee 100, 38114 Braunschweig, Germany*
Florian.Witt@ptb.de

ABSTRACT

Precise calibrations of primary reference standards for Multi Junction Solar Cells are crucial for the exact characterization of photovoltaic devices for space applications. To complement the established but rarely performed atmospheric calibration campaigns via balloon flights a reliable laboratory based, synthetic, calibration method is desirable. Differential spectral responsivity measurements are an established calibration method for solar cells in terrestrial application. Here it is shown that such synthetic calibrations can also be performed with lowest measurement uncertainties on Component Cells but also on whole Multi Junction Solar cell devices.

1. INTRODUCTION

The quality control in the field of Multi Junction Solar Cells (MJSC) for space applications relies until now on primary reference standards which are directly measured close to AM0 like conditions at high altitudes [1]. A disadvantage of such calibration campaigns is the long time between the flights, for Europe no mission was performed up to now since 2005. Another drawback are the high costs of such a single flight as well as the risks of failure and loss of samples. These points can be addressed by the application of a laboratory based, synthetic, calibration method. For solar cells used in terrestrial applications differential spectral responsivity (DSR) calibrations have been established as method with lowest measurement uncertainties for the short circuit current under terrestrial (1000 W/m^2 irradiance with an AM1.5 spectrum at 25C°) standard test conditions obtainable [2]. This method has also been applied to determine the short circuit current under space test condtions (I_{STC}) of Component Cells (a reference solar cell where only a single junction of a MJSC device is electrically active) but with higher measurement uncertainties as for silicon reference solar cells [3]. The space-like test conditions were defined as follows: ~1360 W/m^2 irradiance with an AM0 spectrum (according to Wehrli [4]) at 25C°. Within the European Metrology Research Program (EMRP) project SolCell methods and devices to improve DSR calibrations of Component as well as MJSC solar cells are developed. In this work recently obtained results are presented.

2. THE DIFFERENTIAL SPECTRAL RESPONSIVITY CALIBRATION METHOD

The short circuit current of any solar cell is determined by the cell's absolute spectral responsivity ($s(\lambda)$ with λ as wavelength) and the spectral irradiance ($E_{AMx}(\lambda)$ with x=0 or x=1.5 respectively) at which the cell is measured (equation 1). Therefore, to determine precisely the I_{STC} value of a solar cell under a given reference spectrum a precise determination of its spectral responsivity is necessary.

$$I_{STC} = \int s(\lambda)\cdot E_{AMx}(\lambda)\mathrm{d}\lambda. \qquad (1)$$

The absolute spectral responsivity is determined in the following way. A monochromatic modulated probe beam generates a small AC current in the device under test (DUT) which is kept close to short circuit conditions. The wavelength of the beam is varied and the current answer in dependence of the wavelength is determined via a trans-impedance/ lock-in amplifier combination. In parallel, the DUT is illuminated by a constant DC bias light field. To consider the inhomogeneity of the DUT as well as the influence of the sample's housings, all measurements are performed with both light fields over illuminating the sample. Due to the modulation of the monochromatic irradiance, the measured quantity is the differential spectral $\tilde{s}(\lambda, I_{bias})$.Therefore it is necessary to perform the measurements at different irradiance levels below and above the targeted test conditions (either 1000 W/m^2 for AM 1.5 like conditions or the integrated irradiance value (~1360 W/m^2) of the applied AM0 spectrum). Here I_{bias} corresponds to the DC current the solar cell is generating at the given DC bias light field [5].
The absolute irradiance of the monochromatic light field is determined by a primary calibrated irradiance standard traced to a cryogenic radiometer. This reference is measured subsequently to the DUT. Due to the monitor based measurement design these independent measurements can be directly compared. To determine the absolute spectral responsivity from the differential method at first a differential (non-spectral) AM0 weighted responsivity $\tilde{s}(I_{bias})$ is needed. It is

 DOI: 10.1051/e3sconf/20171602002

obtained by equation 2.

$$\tilde{s}\left(I_{\text{bias}}\right)=\frac{\int_0^\infty \tilde{s}\left(\lambda,I_{\text{bias}}\right)\cdot E_{\lambda,\text{AM0}}(\lambda)\,\mathrm{d}\lambda}{\int_0^\infty E_{\lambda,\text{AM0}}(\lambda)\,\mathrm{d}\lambda} \qquad (2)$$

If the investigated solar cells would be linear, the short circuit current under given test conditions could be simply calculated from a single $\tilde{s}\left(I_{bias}\right)$. Due to the in reality occurring non-linear behavior, it becomes necessary to solve equation 3 to obtain the I_{STC} value of the DUT.

$$E=\int_0^{I_{\text{STC}}}\frac{1}{\tilde{s}\left(I_{\text{bias}}\right)}\mathrm{d}\,I_{\text{bias}} \qquad (3)$$

From this equation it becomes apparent that the I_{STC} is defined implicitly as integration limit. Equation 3 is solved for I_{STC} therefore iteratively by increasing the I_{STC} value until E equals the absolute irradiance of the given reference spectrum.

$$s\left(\lambda,I_{\text{STC}}\right)=\frac{I_{\text{STC}}}{\int_0^{I_{\text{STC}}}\frac{1}{\tilde{s}\left(\lambda,I_{\text{bias}}\right)}\mathrm{d}\,I_{\text{bias}}} \qquad (4)$$

When the I_{STC} value is obtained, also the absolute spectral responsivity at the given test conditions can be subsequently calculated from equation 4.

3. THE LASER DSR-FACILITY

For the measurements presented in this work, a tunable laser source was used as monochromatic radiation source (figure 1). In contrast to standard DSR setups basing on classical light sources combined with monochromators, this setup possesses a higher spectral irradiance. This is used on the one hand to improve the signal to noise ratio of the modulated signal and on the other hand to increase the homogeneity and size of the monochromatic light field. As shown in [6] inhomogeneity as well as wavelength deviations of the monochromatic light field dominate the overall measurement uncertainty of $s(\lambda)$ and I_{STC}. With the use of multiple diffusors and a wavelength dependent positioning of the imaging optics the inhomogeneity is kept below 1% for all wavelengths in the area of the sample. For further reducing the influence of the light's field inhomogeneity the reference scans the light field for every wavelength in the whole area of the DUT. To address the issue of wavelength deviations a calibrated array-spectroradiometer is used for live monitoring the monochromatic signal.

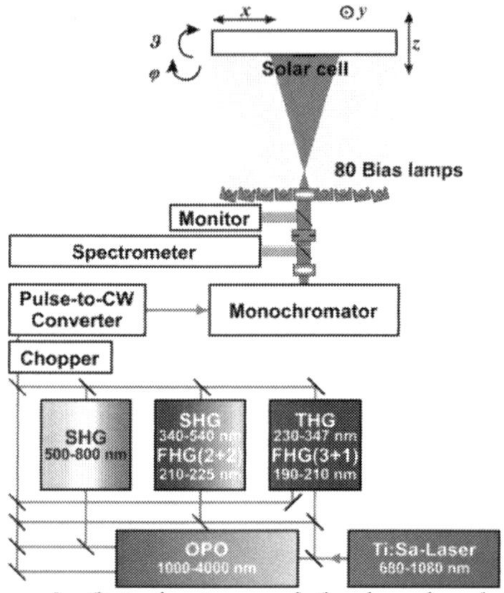

Figure 1: Optical setup used for laser-based DSR measurements. A tunable Ti:Sa-Laser in combination with multiple frequency multipliers as well as an optic parametric oscillator allow an accessible spectral range from 190nm to 4000nm.

The temperature control of the samples is realized with Peltier elements which can directly cool or heat the DUT under illumination. The stability in temperature is found commonly in the range of millikelvins. The temperature of the DUT can be chosen freely and is commonly set to 25°C according to terrestrial standard test conditions. The temperature coefficient of I_{STC} is determined additionally by DSR measurements at three different temperatures.

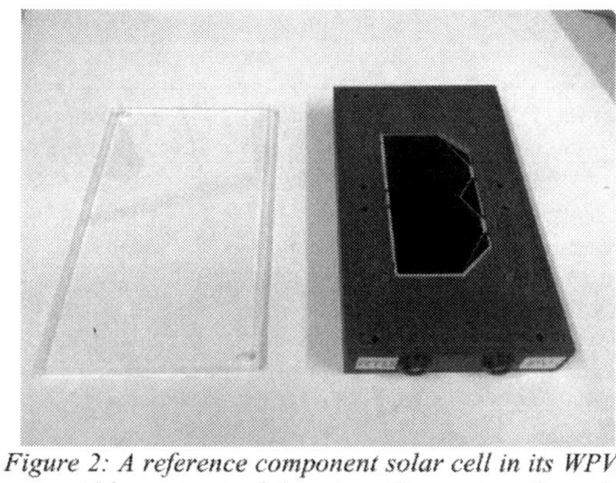

Figure 2: A reference component solar cell in its WPVS compatible customized housing. Connectors for solar cell and temperature sensor positioned at the bottom side. Cover shield for transportation and storage on the left.

DOI: 10.1051/e3sconf/20171602002

4. SYNTHETIC CALIBRATION OF COMPONENT SOLAR CELLS

The described procedure was applied to standard triple junction component solar cells composed of a InGaP top cell, an InGaAs middle Cell and a Germanium bottom cell. In each of the three investigated devices only one of the three sub junctions were electrically active. The cells were built into WPVS compatible housings (see figure 2). Close to the solar cell a PT 100 temperature sensor was attached. In figure 3 differential spectral responsivity measurement results of each DUT is shown for selected bias light levels. A first observation here is that the devices $\tilde{s}(\lambda, I_{bias})$ vary only little with the DC irradiance.

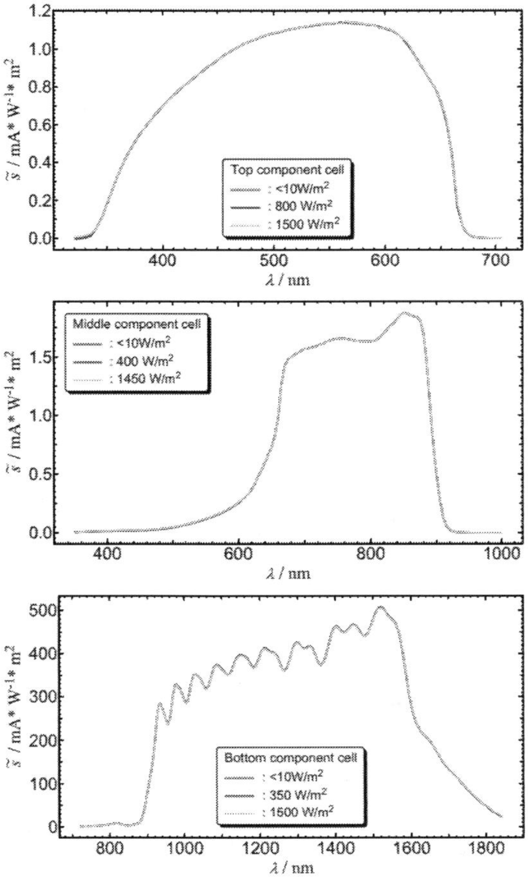

Figure 3: The differential spectral responsivity of each investigated component cell is plotted at three different DC light bias irradiance levels.

This is further supported by the plot of the AM0 weighted differential responsivity (calculated from equation 2) of all measured bias light levels in figure 4.

According to equation 3 and 4 the I_{STC} and the absolute spectral responsivity $s(\lambda)$ is calculated (latter shown in figure 5). With a Monte Carlo simulation (for details see [6]) the measurement uncertainties of all calibration results are determined. For the top and middle component cells a relative measurement uncertainty (MU) of the I_{STC} of 0.5 % (k=2) was achieved. In the case of the bottom junction component solar cell 1% relative measurement uncertainty (k=2) for the I_{STC} was achieved. Here the calibration uncertainty of the used germanium irradiance standard was found as one of the main limiting factors.

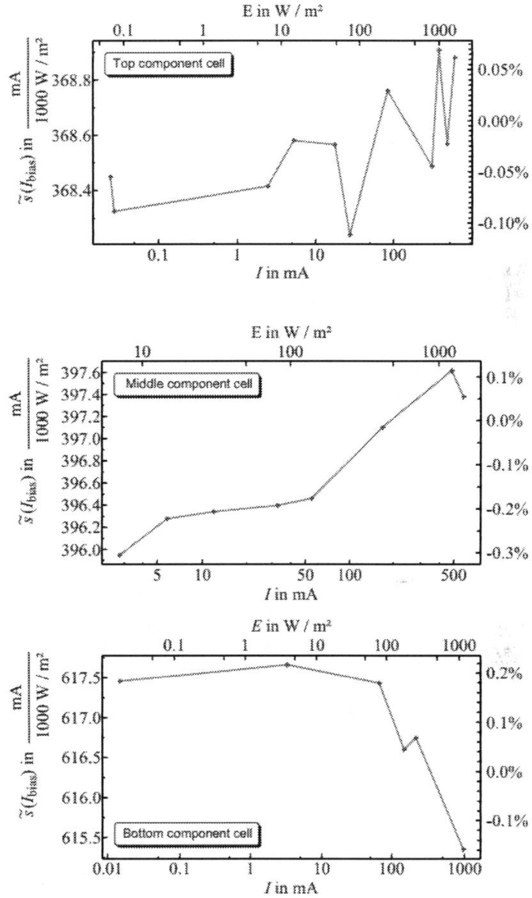

Figure 4:AM0 weighted DSR $\tilde{s}(I_{bias})$ calculated from the AM0 reference spectrum and the $\tilde{s}(\lambda, I_{bias})$ measurements for all investigated component cells. Here I corresponds to the measured DC current of the DUT which is later converted to the corresponding irradiance level E. The integral non-linearity of all samples is below 0.5%.

This important result states that it is now possible to achieve a MU on the same order of magnitude as

 DOI: 10.1051/e3sconf/20171602002

obtainable by balloon flight calibrations.

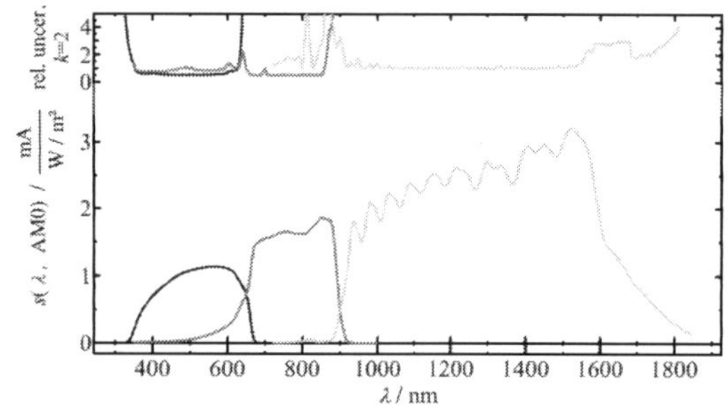

Figure 5: Spectral Responsivity at AM0 test conditions of Top Comp. Cell (black), Mid Comp. Cell (red) and Bottom Comp. Cell (green) and the associated relative measurement uncertainty.

5. SYNTHETIC CALIBRATION OF MULTI-JUNCTION SOLAR CELLS

The application of the DSR method on MJSC is a non-trivial experiment. Due to the serial connection of the sub junctions it is impossible to generate an AC current by exciting a single sub junction monochromatically. To solve this issue and to determine the relative DSR from a MJSC device it is established to apply polychromatic bias lights which are chosen so that for each sub junction of the device an own DC excitation source is available [7]. These sources are adjusted so that currents in the sub junctions not under test are generated in excess compared to the junction under test. In this way a modulated current generated in the targeted sub junction can also be measured over the whole device. Similar to the investigations on the component cells different dc irradiation levels were defined, so that the junction under test was operated below and above 1366W/m^2. In figure 6 the application of such polychromatic light sources is shown. Here high power LEDs (470 nm LEDs for the excitation of the top junction and 780nm LEDs for the excitation of the middle junction are used. The bottom junction is excited by halogen lamps). Additionally, it is necessary to apply a constant DC offset voltage since by the polychromatic illumination the junction under test is shifted to negative operation voltages.

Results from the measurements regarding the top and middle sub junction of the DUT are shown in figure 7. Here the voltage dependence of the DSR at a single polychromatic bias light combination is shown.

Figure 6: MJSC under polychromatic bias light illumination during laser-DSR calibration.

As can be seen the voltage has only a slight influence on the DSR. To take the slight voltage dependence into account the $\tilde{s}(\lambda, I_{bias})$ measurements were performed at three different voltage levels and subsequently averaged before $\tilde{s}(I_{bias})$ was calculated. In figure 8 $\tilde{s}(\lambda, I_{bias})$ as well as the calculated $s(\lambda)$ are shown.

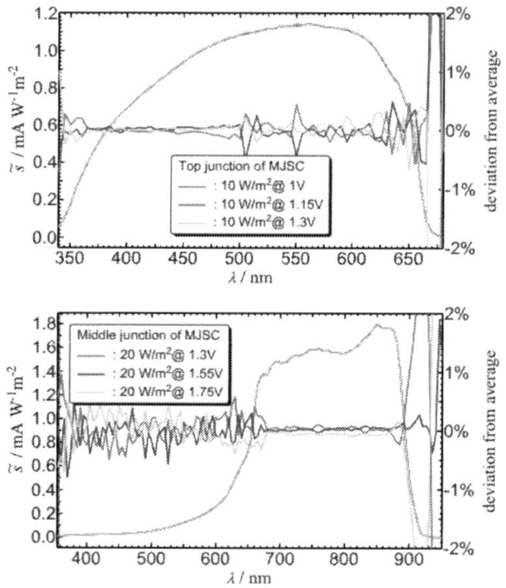

Figure 7: Voltage dependence of Top and Middle junction DSR of MJSC at fixed bias light level. The annotated irradiance level corresponds to the irradiance from the current limiting light source (For the top junction the 470 nm LEDs, for the middle junction the 780 nm LEDs).

As observed with the component cells before, the sub junctions show a linear behaviour. The I_{STC} value of the top sub junction was obtained with a relative uncertainty of about 1%. For the middle junction similar results are obtained.

DOI: 10.1051/e3sconf/20171602002

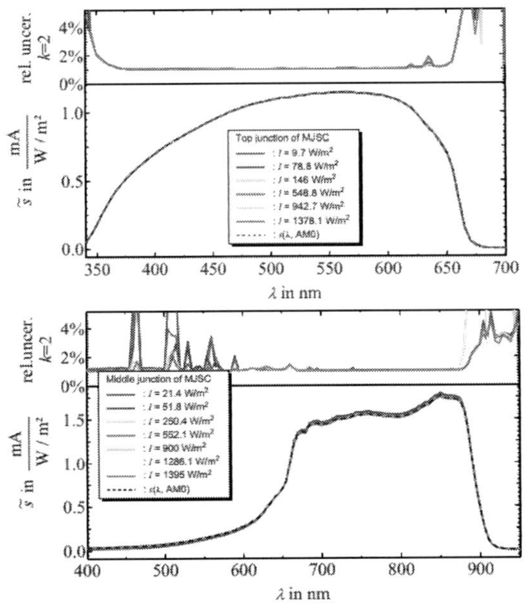

Figure 8: $\tilde{s}(\lambda, I_{bias})$ *of top and middle junction of MJSC as well* $s(\lambda)$. *Additionally, the calculated relative MU is shown.*

During measurements of the bottom junction issues related of a parasitic parallel resistance was observed as reported in literature. E.g. an artefact of the middle sub junction of the device was found in the bottom sub junction signal and the DSR of the bottom junction was suppressed (upper graph in figure 9). As expected it was not possible to suppress the artefact by the variation of the dc bias voltage.

To correct the measured data a method published by Siefer et al in [8] was applied to the obtained DSR results. In the lower graph of figure 9 the corrected data are presented. It is notable that almost no non-linearity at different bias light levels is observed anymore, as expected from experiments on bottom component cells. Experimentally due to insufficient irradiance from the polychromatic bias light it was only possible to measure the DSR of the bottom sub junction at an estimated absolute irradiance of about 500W/m^2. A preliminary evaluation results in an I_{STC} value comparable to the currents generated in bottom junction component cells. Here further work is necessary to verify these findings and to achieve DC bias light irradiance levels sufficient to generate currents above the expected I_{STC} in the bottom junction of the MJSC device.

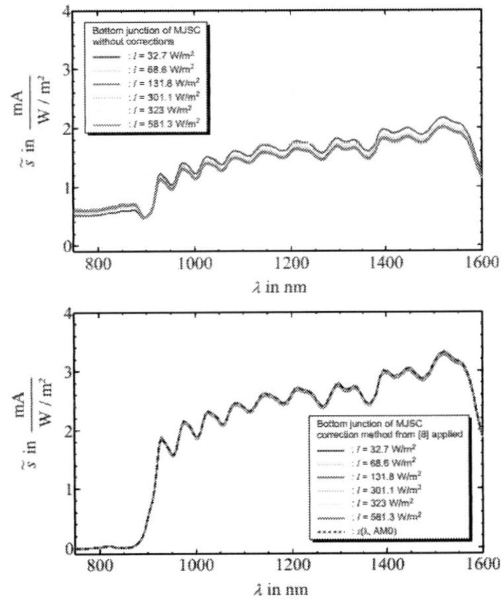

Figure 9 Top: DSR results from the polychromatic measurement of the bottom sub junction of the MJSC. An artefact of the middle junction and a large non-linearity in irradiance of the sub junction is observed. Bottom: After applying the correction procedure from [8] all measurements, show the expected linear behaviour.

6. CONCLUSION

In this work we presented the results from a calibration campaign which states that synthetic calibration of MJSC and their corresponding component cells can be performed and achieve measurement uncertainties comparable to balloon flight calibrations. With these tools in future, a reliable and less expensive method for Si-traceble primary calibrations of component cell references in the field of space photovoltaics is available. To give further prove of the suitability of this method for the quality control of component cells used as primary working standards a direct comparison to balloon flights need to be addressed in future.

7. ACKNOWLEDGEMENTS

We thank AZUR SPACE for the supply with samples. The research work leading to this article was carried out within the EMRP ENG51 project SolCell. The EMRP is jointly funded by the EMRP participating countries within EURAMET and the European Union.

DOI: 10.1051/e3sconf/20171602002

8. REFERENCES

1. Pichetto, V. (2002). CASOLBA Calibration of Solar Cells–Measurement and Calibration Procedures. In*Proc. 29'" IEEE Photovoltaics Specialists Conference., New Orleans* (p. 1010)

2. Winter, S., Metzdorf, J., Emery, K., Fabero, F., Hishikawa, Y., Hund, B., ... & Warta, W. (2005). The results of the second world photovoltaic scale recalibration. In *Photovoltaic Specialists Conference, 2005. Conference Record of the Thirty-first IEEE* (pp. 1011-1014). IEEE.

3. Baur, C., Siefer, G., Kern, R., & Winter, S. (2011). Primary Solar Cell Standards-Comparison Of Extraterrestrial And Synthetic Calibration. In *ESA Special Publication* (Vol. 690, p. 12).

4. Wehrli, C. (1985) "Extraterrestrial Solar Spectrum", Publication no. 615, Physikalisch-Meteorologisches Observatorium + World Radiation Center (PMO/WRC) Davos Dorf, Switzerland.

5. S. Winter, T. Wittchen, J. Metzdorf (2000) "Primary Reference Cell Calibration at the PTB Based on an Improved DSR Facility"; in "Proc. 16th European Photovoltaic Solar Energy Conf.",(p.4).

6. Winter, S., Fey, T., Kröger, I., Friedrich, D., Ladner, K., Ortel, B., ... & Witt, F. (2014). Design, realization and uncertainty analysis of a laser-based primary calibration facility for solar cells at PTB. *Measurement, 51*, (pp. 457-463).

7. Meusel, M., Baur, C., Létay, G., Bett, A. W., Warta, W., & Fernandez, E. (2003). Spectral response measurements of monolithic GaInP/Ga (In) As/Ge triple-junction solar cells: Measurement artifacts and their explanation. *Progress in Photovoltaics: Research and Applications, 11*(8), (pp.499-514).

8. Siefer, G., Baur, C., & Bett, A. W. (2010). External quantum efficiency measurements of germanium bottom subcells: Measurement artifacts and correction procedures. In *Photovoltaic Specialists Conference (PVSC), 2010 35th IEEE* (pp. 000704-000707). IEEE.

DIRECT COMPARISON OF GROUND AND IN-FLIGHT MEASUREMENTS OF NEW MULTI-JUNCTION SOLAR CELL TECHNOLOGIES

Phillip P. Jenkins[(1)], Benjamin Cho[(2)], David Wilt[(3)], James H. Ermer[(4)], Eric Rehder[(5)], Daniel Law[(6)], Michael J. Krasowski[(7)], Lawrence Greer[(8)] and Joseph Flatico[(9)]

[(1)] Naval Research Laboratory, 4555 Overlook Ave. SW, Washington, DC, USA, nrl_pv_research@nrl.navy.mil
[(2)] SolAero Technologies, 10420 Research Rd. SE Albuquerque, NM, USA, email, Benjamin_Cho@solaerotech.com
[(3)] Air Force Research Laboratory, 3550 Aberdeen Ave Kirtland AFB, NM, USA, AFRL.VSSVOrgMailbox@us.af.mil
[(4)] Boeing-Spectrolab, 12500 Gladstone Ave, Sylmar, CA, USA, email, james.h.ermer@boeing.com
[(5)] Boeing-Spectrolab, 12500 Gladstone Ave, Sylmar, CA, USA, email, eric.m.rehder@boeing.com
[(6)] Boeing-Spectrolab, 12500 Gladstone Ave, Sylmar, CA, USA, email, daniel.c.law@boeing.com
[(7)] NASA Glenn Research Center, 21000 Brookpark Rd. Cleveland, OH, USA, email, Michael.J.Krasowski@nasa.gov
[(8)] NASA Glenn Research Center, 21000 Brookpark Rd. Cleveland, OH, USA, email, lawrence.greer@nasa.gov
[(9)] Ohio Aerospace Institute, 22800 Cedar Point Rd. Cleveland, OH, USA, email, joseph.m.flatico@nasa.gov

ABSTRACT

The pace of development for new photovoltaic structures such as the inverted meta-morphic (IMM) III-V cells, exceeds the pace at which spectrally matched, calibrated, solar cells can be created due to the infrequent opportunities of balloon or space based calibration. As such, laboratories must rely on previous generation AM0 calibrated cells and make appropriate corrections. Equally challenging is measuring the same cells in space without a controlled laboratory environment. It is within this context, the 3rd Forward Technology Solar Cell Experiment (FTSCE III), measured three, four and five junction advanced technology solar cells in space. This report compares the beginning-of-life in-flight measurements with the laboratory measurements made before flight.

1. INTRODUCTION

Historically, the technique for AM0 calibration of solar cells was to measure cells under near AM0 conditions on a high altitude balloon (~35km) or aircraft. Unfortunately for the world-wide solar cell community, there has not been a high-altitude balloon flight since 2005 by either NASA or CNES, the two traditional providers of these flights. Since that time, there has been an entire generation of GaInP/GaAs/Ge multi-junction flight qualified cells that did not benefit from the use of AM0 calibrated reference cells. Instead, laboratories used previous generation calibrated cells, which have very similar spectral response, and applied either a spectral correction factor, or relied on high fidelity solar simulators to minimize the difference between the previous generation and current generation cells. These practices have proven to be relatively successful substantially due the fact that all cells grown on germanium substrates had very similar spectral response, and beginning-of-life (BOL) cells behave as two-junction cells with respect to current matching, because the Ge cell has a large amount of excess current. Not to understate the challenge of measuring cells without an AM0 standard, but for the most part, BOL cells grown on germanium behave as if the current is controlled by at most two junctions.

However, the next generation of the highest efficiency solar cells under development no longer have a current rich germanium bottom cell and all junctions in these new cells are much closer to current matched. Each sub cell has a narrower spectral response, making it more difficult to rely on calibration standards of previous generation cell types. It is within this context a space solar cell experiment was flown on the International Space Station (ISS) with three and four junction IMM cells as well as five-junction semiconductor bonded cells. This paper compares the in-flight BOL and pre-flight measurements to examine the efficacy of ground-based measurements of IMM III-V cells where no calibrated standard cell exists.

The 3rd Forward Technology Solar Cell Experiment (FTSCE III) was deployed during the Space Shuttle STS-134 mission in May 2011, and was part of a lager mission known as the 8th Materials on the International Space Station Experiment, MISSE-8. This experiment recorded current vs. voltage (I-V) curves and environmental conditions during each orbit. The experiment ran for 26 months and carried advanced technology cells from a variety of US suppliers. The Air Force Research Laboratory, Space Vehicles Directorate (AFRL/RVSV) and Lockheed Martin Corporation sponsored samples of laboratory solar cells such as three and four-junction IMM cells and strings from SolAero and Spectrolab, and semiconductor bonded technology (SBT) cells from Spectrolab. The particular cells highlighted in this paper were very stable over time.

DOI: 10.1051/e3sconf/20171602003

2. CATALOG OF CELLS

Tab.1 shows the matrix of multi-junction cells where high fidelity ground (I-V) measurements were available. Also shown is the bias condition cells were held at when not being measured. Fig. 1 shows the locations of the active solar cell samples.

2.1 SolAero Cells

The majority of SolAero cells were mounted on a honeycomb substrate instrumented with two temperature sensors. Most cells were encapsulated between two cover glasses (DCG), front and back, although one string was mounted using a rigid substrate CTE matched to GaAs. Two of the large area 3J-IMM cells (7, 8) were mounted to a sparse grid of synthetic fibers used on ATK's UltraFlex™ solar array designs. One temperature sensor was mounted to the back of one of these cells. Details of the SolAero (formerly Emcore) FTSCE III coupons can be found in [1]. Details of the 3J-IMM and 4J-IMM cell designs (including spectral response data) are found in [2, 3].

2.2 Spectrolab Cells

The Spectrolab cells were mounted on two honeycomb substrates and fully grouted with a space-grade RTV compound. The 3J-IMM and 4J-IMM designs are similar to those in reference [4]. The SBT cells were five junction stacks, with the top three cells grown lattice matched to GaAs and the bottom two junctions matched to InP [5]. During experiment integration, one of the large area 3J-IMM cells experienced a cabling issue, which prevented any on-orbit measurements being collected on that cell.

3. SPACE-BASED MEASUREMENTS

The measurement electronics were developed at the NASA Glenn Research Center, for the first FTSCE, and a "build to print" with minor modifications was built for FTSCE III. A comprehensive description of the circuitry is found in [6]. The electronics were attached to a

Table 1. Listing of the III-V cells made by SolAero and Spectrolab on FTSCE III.

Vendor	Type	Size	Bias Condition	Photo Ref. Number
SolAero	3J-IMM	3-cell string	Pmax	1, 2
SolAero	3J-IMM	26.3 cm^2	Pmax	3, 4
SolAero	4J-IMM	4 cm^2	Isc	5, 6
SolAero/ATK	3J-IMM	26.3 cm^2	Pmax	7, 8
Spectrolab	3J-IMM	3-cell string	Pmax	9,10
Spectrolab	3J-IMM	20 cm^2	Pmax	11,12
Spectrolab	4J-IMM	4 cm^2	Isc	16,17
Spectrolab	SBT	1 cm^2	Isc	18,19
Spectrolab	UTJ	26.3 cm^2	Isc	20, 21

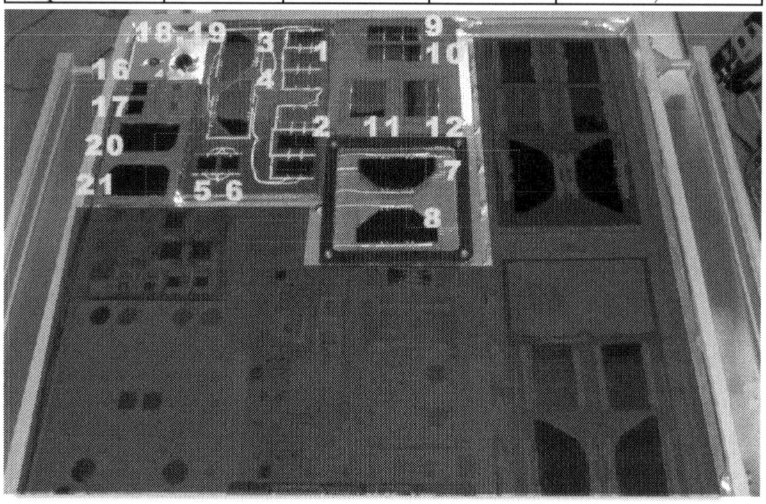

Figure 1. Photograph of the MISSE-8 zenith face highlighting Spectrolab and SolAero on FTSCE III.

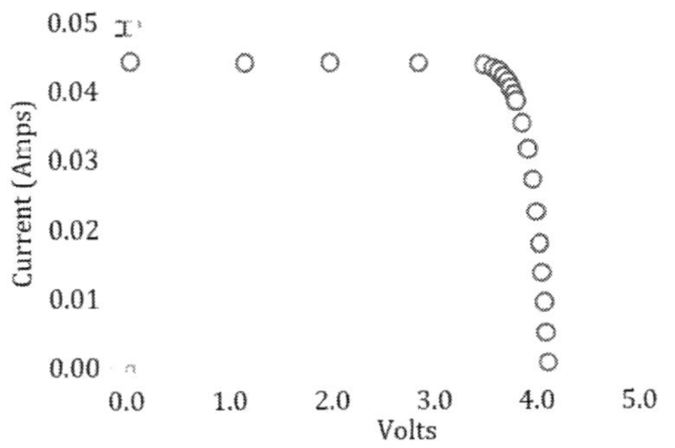

Figure 2. Typical I-V curve obtained from FTSCE III.

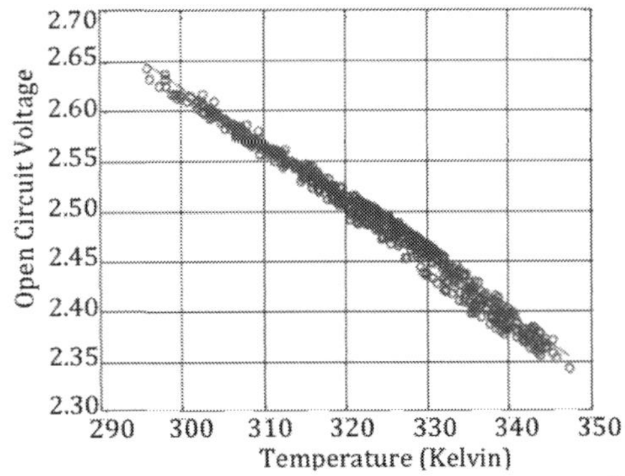

Figure 3 Typical temperature vs. open circuit voltage for a three junction cell.

separate deck and stacked behind the solar panel deck (seen in Fig. 1). The electronics consisted of a central motherboard communicating with six daughter boards and relaying data to the central MISSE-8 communication board [7]. FTSCE III did not track the sun, and the best sun angle possible on each orbit was limited by the solar beta angle, which varied from +75 to −75 degree s with zero crossings roughly once a month. The FTSCE III motherboard would query the six daughter boards every 30 seconds for temperature and sun angle data and record these data points every 10 minutes into a rotating buffer, where the previous orbit's worth of environmental data was kept. Using the record of the previous orbit, the motherboard used a *relative* set point of temperature or sun angle to acquire I-V curves.

4. DATA ANALYSIS METHODOLOGY

Fig. 2 is a typical I-V curve on a multi-junction cell collected by FTSCE III. Only those I-V curves with a recorded sun angle between normal incidence and 45° are considered. At angles larger than 45°, the space station structure can create shadows or reflections. Even with this restricted angle, there are spurious data points from ISS operations such as vehicle docking or extra vehicular activities (EVAs). Only in a few cases did we attempt to remove these spurious data, as they do not significantly affect the results.

The I-V curves are reduced to the performance parameters short circuit current (Isc), open circuit voltage (Voc), current at maximum power (Imp), voltage at maximum power (Vmp), and maximum power (Pmp). These data were corrected to the same reporting condition as the pre-flight characterization by correcting for sun angle, sun-earth distance, and temperature. The voltages, Voc and Vmp, were also adjusted based on the current contribution between the measured Isc when data were collected, and the nominal AM0 current.

Temperature measurement of these cells was particularly challenging. While the on-orbit temperature coefficient measurements of open circuit voltage show the expected linear behaviour as in Fig. 3, the *absolute* temperature of the cells was difficult to determine since a single temperature sensor was employed to monitor the temperature of all cells on a substrate. Furthermore, the ability of the temperature sensor to measure the temperature of the cell is problematic due to the non-metallic face sheets the cells were attached to on the honeycomb substrate. While cell voltage is a strong function of temperature, the short circuit current (Isc) and maximum power current (Imp) are only weakly temperature dependent.

The ability to extract a high quality current temperature coefficient is made more complicated by the nature of the test conditions on-orbit. Unlike a laboratory environment, where one would maintain a constant intensity and vary the cell temperature, on-orbit the cell temperature is varied by the changing intensity due to off-pointing of the sun. Therefore, the current temperature coefficient depends on accurately correcting for cosine loss, which has its own uncertainty. Fig. 4 shows Imp vs. temperature corrected for intensity for the 3J-IMM marked "2" string in Fig.1.

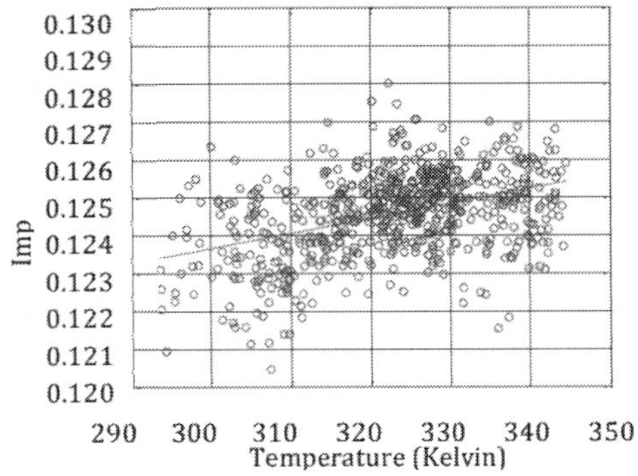

Figure 4. Imp vs. temperature for a 3J-IMM string.

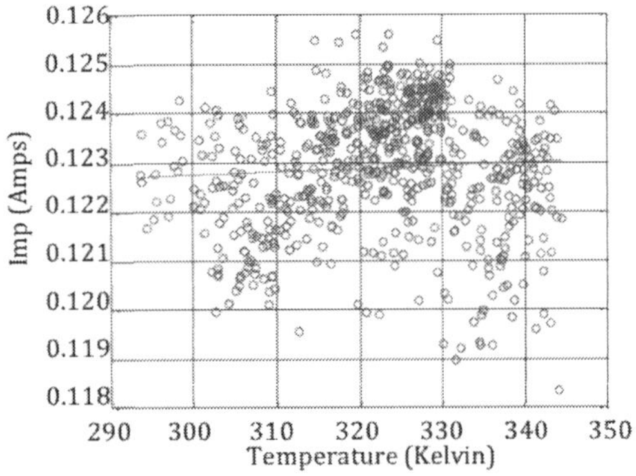

Figure 5. Imp vs temperature for another 3J-IMM string.

Fig. 5 shows the Imp vs. temperature plot for the other SolAero 3J-IMM string (number "1" in Fig. 1) showing considerably less temperature dependence than the similar string (Fig. 4). The slope of the fitted lines in the graphs of figures 3, 4 and 5 correspond to the temperature coefficients used to correct these data to standard operating conditions, and are presented explicitly later in the paper.

5. FLIGHT VS. GROUND MEASUREMENTS

Because the flight test articles are mounted on honeycomb substrates, large area pulse solar simulators (LAPSS) are used in laboratories to illuminate the cells for I-V measurements. The short pulse (~1-2 milliseconds) of the LAPSS prevents the cells from heating during the measurements and the temperature of the panels can be maintained. LAPSS systems do present their own measurement challenges though.

Typically, these simulators have uniform spectral content and intensity for only one or two milliseconds. With few exceptions, laboratories do not have proper instrumentation to characterize the intensity and full spectral content of the flash pulse. This type of characterization is commonplace for continuous solar simulators, but for LAPSS systems the instrumentation is often prohibitively expensive. Without full understanding of the LAPSS spectral content, making a spectral mismatch correction is not possible [8]. Thus, it is usually the practice, to choose the closest spectrally matched calibrated cells, and hope the fidelity of the solar simulator is a good enough match to AM0 that the cell I-V curve will be accurately reproduced. A ground based calibrated isotype sub cells would be the best alternative, but are often not available for developmental cells where cell-to-cell differences can be relatively large.

Table 2. On-orbit performance parameters (BOL) and temperature coefficients compared to laboratory pre-flight ground measurements (GRD) for SolAero cells.

SolAero Cell Type/Parameter	4J-IMM	4J-IMM	3J-IMM string (rigid)	3J-IMM string (DCG)	3J-IMM Large area on honeycomb	3J-IMM Large area on honeycomb	3J-IMM Large area on UltraFlex™	3J-IMM Large area on UltraFlex™
BOL Pwr Watts)	0.165	0.171	0.965	0.957	1.049	1.100	1.091	1.102
GRD Pwr (Watts)	0.179	0.179	0.997	0.996	1.062	1.112	1.113	1.106
BOL Voc (Volts)	3.09	3.13	8.88	8.87	2.98	2.95	2.95	2.98
GRD Voc (Volts)	3.237	3.229	9.048	9.045	3.000	3.014	3.018	3.027
BOL Isc (Amps)	0.066	0.066	0.132	0.131	0.448	0.449	0.445	0.447
GRD Isc (Amps)	0.0654	0.0673	0.131	0.132	0.441	0.436	0.445	0.443
BOL Imp (Amps)	0.062	0.063	0.124	0.123	0.405	0.428	0.426	0.424
GRD Imp (Amps)	0.0633	0.0634	0.124	0.124	0.402	0.419	0.422	0.421
BOL Vmp (Volts)	2.668	2.719	7.779	7.782	2.591	2.571	2.560	2.598
GRD Vmp (Volts)	2.829	2.829	8.03	8.022	2.645	2.652	2.638	2.63
Percent difference of the ground based measurement compared to the on-orbit measurement (minus sign, red) = lower than on-orbit								
% diff Power	8.2	4.7	3.4	4.0	1.2	1.0	2.0	0.4
% diff Voc	4.8	3.2	1.9	1.9	0.6	2.1	2.5	1.5
% diff Isc	-1.0	2.0	-0.8	0.8	-1.5	-3.0	0.0	-1.1
% diff Imp	2.1	0.6	0.2	0.9	-0.8	-2.1	-1.0	-0.8
% diff Vmp	6.0	4.1	3.2	3.1	2.1	3.2	3.0	1.2
On-orbit measured temperature coefficients								
Voc (mV/°C)	-6.71	-6.58	-6.51	-5.37	-6.17	-5.57	-4.86	-6.00
Vmp (mV/°C)	-7.19	-7.16	-7.07	-5.89	-6.42	-6.37	-6.58	-7.24
Jsc (μA/cm²/°C)	5.51	0.88	10.37	6.72	10.16	3.20	22.40	18.89
Jmp (μA/cm²/°C)	0.69	-4.45	4.93	0.79	9.22	-0.38	6.63	9.66

5.1 SolAero Cells Measurement Comparison

Tab. 2 shows BOL on-orbit temperature coefficients, and the I-V performance measurements compared to the pre-flight laboratory measurements made at SolAero corrected to AM0 and 28°C. Across all cells, the laboratory measurements show higher Voc and Vmp values. This difference, between the on-orbit Voc and Vmp is likely due to systemic errors in determining the temperature of individual cells on-orbit. In the laboratory, temperature is well controlled and cells are at a uniform temperature. In space, where the thermal environment is driven by the changing view to the sun, the Earth, and deep space, the temperature of a cell maybe considerably different than the temperature sensor itself. This observation is consistent with the location of the temperature sensor being underneath the substrate face sheet. The front side of the face sheet with the cells being illuminated by the sun, will naturally be warmer than backside of the face sheet where the temperature sensor is located.

Further evidence for the systemic errors of the on-orbit temperature measurement, is the fact that the 4J-IMM cells show the greatest Voc difference implying the 4J-IMM cells are running hotter than other cells on the same panel. The 4J-IMM cells were the only cells not biased near their maximum power point (Pmax), but at Isc. This would cause the 4J-IMM cells to run hotter than those biased at Pmax, since ~25% more of the solar energy would be converted to heat compared to other cells on the plate where the bias load would shed energy from the cell.

It is not unreasonable to argue that the laboratory Voc measurements and the on-orbit Voc should be very close in value if they were measured at the same temperature. If we assume that the laboratory measurements of temperature are more reliable than the on-orbit temperature measurements, we can use the temperature coefficients to calculate the implied temperature difference and adjust the I-V parameters accordingly.

Doing so, reduces the percentage difference between the lab and on-orbit *power* measurements to roughly the percentage differences in Imp, as the Vmp values are then in good agreement in most instances. That leaves approximately a ±2% band around the on-orbit measurements, with the 3J-IMM large area cells consistently measured lower current values in the lab than on-orbit, and the 4J-IMM and the 3J-IMM strings measured as higher in current in the lab than the on-orbit data would suggest. Given the uncertainty in the spectral content of the LAPSS and the lack of spectrally matched calibrated reference cells, the laboratory and on-orbit measurements are remarkably close.

5.2 Spectrolab Cells Measurement Comparison

The Spectrolab coupons were measured in much the same way in the laboratory as the SolAero panels. That is, with calibrated GaInP/GaAs/Ge isotype cells,

Table 3. On-orbit performance parameters (BOL) and temperature coefficients compared to laboratory pre-flight ground measurements (GRD) for Spectrolab cells.

BOL at 22°C & 28°C, GRD at 22°C & 28°C	22°C SBT	22°C SBT	22°C 4J-IMM	22°C 4J-IMM	22°C UTJ	22°C UTJ	28°C 3J-IMM string	28°C 3J-IMM string	28°C Large Area 3J-IMM
BOL Pwr Watts)	0.0372	0.0384	0.153	0.155	0.982	1.010	0.571	0.541	0.736
GRD Pwr (Watts)	0.0368	0.0352	0.155	0.156	0.996	1.001	0.588	0.596	0.765
BOL Voc (Volts)	4.785	4.673	4.354	4.326	2.644	2.643	8.657	8.721	2.872
GRD Voc (Volts)	4.680	4.660	4.300	4.320	2.700	2.670	8.88	8.83	2.93
BOL Isc (Amps)	0.0097	0.0097	0.0409	0.0415	0.4633	0.4575	0.0843	0.0827	0.337
GRD Isc (Amps)	0.0097	0.0096	0.0420	0.0420	0.4600	0.4540	0.083	0.082	0.33
BOL Imp (Amps)	0.0090	0.0094	0.0396	0.0402	0.4254	0.4353	0.0757	0.0725	0.305
GRD Imp (Amps)	0.0090	0.0090	0.0400	0.0400	0.4240	0.4240	0.076	0.077	0.31
BOL Vmp (Volts)	4.157	4.085	3.875	3.859	2.309	2.321	7.547	7.458	2.415
GRD Vmp (Volts)	4.090	3.910	3.870	3.890	2.350	2.360	7.73	7.74	2.50
Percent difference of the ground based measurement compared to the on-orbit measurement (minus sign, red) = lower than on-orbit									
% diff Power	-1.2	-8.5	0.9	0.3	1.4	-1.0	2.9	10.2	3.9
% diff Voc	-2.2	-0.3	-1.2	-0.1	2.1	1.0	2.6	1.2	2.0
% diff Isc	0.3	-1.3	2.6	1.3	-0.7	-0.8	-1.5	-0.8	-1.3
% diff Imp	0.5	-4.4	1.0	-0.5	-0.3	-2.6	0.4	6.2	0.3
% diff Vmp	-1.6	-4.3	-0.1	0.8	1.8	1.7	2.4	3.8	3.5
On-orbit measured temperature coefficients									
Voc (mV/°C)	-10.66	-10.55	-8.08	-8.25	-6.04	-5.66	-5.94	-5.38	-5.33
Jsc (µA/cm2/°C)	1.75	1.98	7.75	4.20	8.19	12.76	2.24	6.40	13.36
Jmp (µA/cm2/°C)	8.49	2.88	7.47	4.04	7.03	5.72	1.45	5.19	8.10
Vmp (mV/°C)	-10.02	-9.88	-8.22	-8.46	-6.55	-6.46	-6.43	-5.31	-5.63

without using a spectral correction between the cells under test and the calibrated standards. As before, this type of measurement relies on having a high fidelity solar simulator to make these measurements meaningful. The laboratory measurements in this case were made at two different temperatures, with most cells measured at 22°C and several measured at 28°C. For the results presented, in Tab. 3, the on-orbit data was corrected to the temperature of the laboratory results. There are several interesting observations to make regarding the comparative results.

As a whole, the laboratory measurements are again, remarkably close to the on-orbit measurements with just two outliers. Unlike the SolAero cells where the laboratory Voc values were all higher than the on-orbit measurements, the Spectrolab cells show an even mix of higher and lower Voc values. It is still likely though, that the on-orbit temperature sensor does not faithfully represent the temperature of each cell as well as in the laboratory.

The outliers, one of the SBT cells, is significantly lower than on-orbit, and one of the 3J-IMM strings significantly higher, are interesting because the Vmp values are further off than the Voc values, so one cannot immediately attribute the difference to temperature differences between the lab and on-orbit. Likewise, the Imp values are significantly further off than the Isc values. Since both Imp and Vmp are off, and temperature seems an unlikely cause, it suggests that spectral current matching maybe more of an issue for these two cells. If this is so, it makes the case for obtaining calibrated isotype cells for the cell under test or fully characterizing the spectral content of the LAPSS for performing a spectral mismatch correction.

6. SUMMARY

The FTSCE III on-orbit data of IMM and other advanced multi-junction technologies have been analysed and compared to the pre-flight measurements. Given the lack of calibration standards for these cell types, laboratory measurements were successful at predicting the BOL performance of these cells. There were however errors attributed to temperature discrepancies between on-orbit and the laboratory, and potentially a lack of spectral matching between LAPSS systems and AM0.

Acknowledgement: This work was supported by the United States Air Force Research Laboratory, Space Vehicles Directorate.

7. REFERENCES

1. Cho, B., Lutz, R., Pappan, J., Downard, E., Cornfeld, A., Fatemi, N., Stan, M.A., Sharps, P., Su, C., Billets, S., Gasner, S., Howard, A., "IMM experimentation in the next frontier: Emcore's participation in the MISSE-8 program," *35th IEEE Photovoltaic Specialists Conference* pp.110-112, 20-25 June 2010

2. A. B. Cornfeld, M. Stan, T. Varghese, J. Diaz, A. V. Ley, B. Cho, A. Korostyshevsky, D. J. Aiken and P. R. Sharps, "Development of a large area inverted metamorphic multi-junction (IMM) highly efficient AM0 solar cell," *33rd IEEE Photovoltaic Specialists Conference,* 11-16 May 2008

3. A. B. Cornfeld, D. Aiken, B. Cho, A. Vance Ley, P. Sharps, M. Stan, T. Varghese, "Development of a four sub-cell inverted metamorphic multi-junction (IMM) highly efficient AM0 solar cell," *35th IEEE Photovoltaic Specialists Conference* pp.105-109, 20-25 June 2010

4. D. C. Law, X. Q. Liu, J. C. Boisvert, E. M. Redher, C. M. Fetzer, S. Mesropian, R. R. King, K. M. Edmondson, B. Jun, R. L. Woo, D. D. Krut, P. T. Chiu, D. M. Bhusari, S. K. Sharma, N. H. Karam, "Recent progress of Spectrolab high-efficiency space solar cells," *38th IEEE Photovoltaic Specialists Conference* pp. 3146-3149, 3-8 June 2012

5. P. T. Chiu, D. C. Law, R. L. Woo, S. B. Singer, D. Bhusari, W. D. Hong and A. Zakaria, J. Boisvert, S. Mesropian, R. R. King, N. H. Karam, "35.8% space and 38.8% terrestrial 5J direct bonded cells," *40th IEEE Photovoltaic Specialists Conference* pp. 11-13, 8-13 June 2014

6. M. Krasowski, L. Greer, J. Flatico, P. Jenkins, D. Spina, "A Hardware and Software Perspective of the Fifth Materials on the International Space Station Experiment (MISSE-5)," NASA Technical Memorandum, NASA/TM-2005-213840, August, 2005.

7. Michael J. Krasowski, Norman F. Prokop, Joseph M. Flatico, Lawrence C. Greer, Phillip P. Jenkins, Philip G. Neudeck, Liangyu Chen, and Danny C. Spina, "CIB: An Improved Communication Architecture for Real-Time Monitoring of Aerospace Materials, Instruments, and Sensors on the ISS," The Scientific World Journal, vol. 2013, Article ID 185769, 12 pages, 2013. doi:10.1155/2013/185769.

8. Clay H. Seaman, "Calibration of solar cells by the reference cell method—The spectral mismatch problem," Solar Energy, Volume 29, Issue 4, 1982, Pages 291-298, ISSN 0038-092X,

GAAS DISPLACEMENT DAMAGE DOSIMETER BASED ON DIODE DARK CURRENTS

Jeffrey H. Warner[1], Scott R. Messenger[2], Cory D. Cress[1], Nicolas J-H Roche[3], Michael Yakes[1], Kenneth Clarke[1], Mitchell F. Bennett[4], Phillip P. Jenkins[1], Robert J. Walters[1]

[(1)]US Naval Research Laboratory, Washington, DC 20375 (USA), Email: dd-effects@nrl.navy.mil

[(2)]Northrup Grumman, Baltimore, MD 21240

[(3)] The George Washington University, Washington, DC 20375

[(4)] Sotera Defense Solutions, Annapolis Junction, MD 20701

ABSTRACT

GaAs diode dark currents are correlated over a very large proton energy range as a function of displacement damage dose (DDD). The linearity of the dark current increase with DDD over a wide range of applied voltage bias deems this device an excellent candidate for a displacement damage dosimeter. Additional proton testing performed in situ enabled error estimate determination to within 10% for simulated space use.

1. INTRODUCTION

The concepts of nonionizing energy loss (NIEL) and displacement damage dose (DDD) have been shown to be useful for correlating displacement damage produced by different particles in irradiated semiconductor [1-9]. This research has culminated over the past few years in the context of solar cell degradation in the production of a MATLAB-based software package called Solar Cell Radiation Environment Analysis Models (SCREAM) [10, 11]. SCREAM has been very successful in providing valid explanations involving several high profile space programs such as GPS [12] and TacSat4 [13]. In GPS spacecraft, there has been an anomalous degradation of the solar arrays since the beginning of the program, not agreeing with those results as predicted using the existing environment models [4]. SCREAM was used to calculate the expected damage on shielded silicon arrays using on-orbit dosimetry provided by BDD detectors from Los Alamos National Laboratories. The BDD data were used as input to provide a time dependent accumulation of the DDD behind the coverglass and adhesive shielding and the associated solar cell degradation was then determined. In this case the on-orbit solar cell degradation and model calculations remained to disagree which effectively eliminated the cell degradation as the issue. For the TacSat4 spacecraft [13], the solar array output was degrading at a much higher rate than expected which prompted a reanalysis of the environment and damage assessments. In this case, there were two payloads on-board the spacecraft to directly use on the analyses: 1) the Compact Environment Anomaly Sensor (CEASE) [14] and 2) a solar cell experiment yielding full current-voltage (IV) data using similar solar cells and shielding conditions. Again, SCREAM was used successfully to analyze the CEASE data and predict the cell degradation, this time agreeing with the solar cell experiment data. Therefore, in this case, it was the environment which was much different than expected and two parallel payloads were used to confirm the higher rate of degradation experienced by the power panels.

It has become evident from these two examples that on-orbit *displacement damage* dosimetry is necessary to resolve anomalies and largely has been ignored in past space programs. There are several detector packages that can be used to generate displacement damage data such as CEASE [14] and CARMEN-2 [15]. Each of these packages are very powerful and can quantify several radiation hazards such as TID, spacecraft charging and SEE. However, each of these are very costly to integrate on a satellite and none directly yield displacement damage information, but are obtained through secondary means. For example, the CARMEN-2 package contains an optically stimulated luminescent (OSL) detector from which the light output of a calibrated LED is used to quantify displacement damage.

The detectors and radiation effects section of NRL has recently been awarded a new-start program to investigate a detector that can easily be used to quantify displacement damage in real time onboard a spacecraft with low size, weight and power (SWaP). The Realtime Radiation Displacement Damage Dosimeter (R2D3) program is designed to accomplish this in 4 years and culminates in the production of a space qualified sensor for space use. There are several variations of the design which can accommodate varying degrees of sensitivity for both ground and space use. Different shielding schemes are being considered as well to encompass a wide energy range of the radiation environment. This work with be presented in a future publication.

The purpose of this paper is to demonstrate that when operating a GaAs diode in forward bias voltage mode, the measured dark current can be used as a metric for measuring the amount of absorbed displacement damage over a large proton energy range. In the present device, the forward biased dark current is dominated by generation-recombination in the junction. This simplifies the sensor's ground based calibration, improves accuracy, and also allows the device to be operated over a larger range of voltages thereby making this device an ideal candidate for a displacement damage dosimeter. Other advantages can be elucidated by different diode growth designs and gives some flexibility to the overall R2D3 system. The basic concepts of operation will be presented in this paper.

2. EXPERIMENTAL DETAILS

The samples used in this study were $p^{+}n$ GaAs photodiodes grown on GaAs substrates by molecular beam epitaxy. A detailed description of the photodiodes is given in [9]. The wafers were grown and processed at NRL. A sample consisted of three photodiodes (2.4 mm diameter) and four solid diodes (1 mm in diameter). Current-voltage measurements were performed in the dark (DIV) on each sample both before and after irradiation. The DIV measurements were performed using a Keithley 2635A source measure unit using a four point probe technique both at NRL and at the radiation facilities when needed.

Two sets of irradiation campaigns were performed. The first set of irradiations were performed using 2 and 225 MeV protons. The 2 MeV proton irradiations were performed at the NRL Pelletron facility up to a maximum fluence of 10^{13} h^{+}/cm^{2}. The 225 MeV proton irradiations were performed at the Francis H. Burr Proton Therapy Center to a maximum fluence of $3x10^{14}$ h^{+}/cm^{2}. Different samples were irradiated at open circuit at several fluence levels and brought back to NRL for electrical characterization. Dosimetry was accomplished using Faraday cups in both facilities and all irradiations were performed at room temperature with low enough beam currents to avoid any significant increase in sample temperature. Typical flux values were $\sim 10^{9}$ $h^{+}/cm^{2}/sec$. The 2 and 225 MeV irradiations were performed in vacuum and air, respectively.

The second set of irradiations were performed to simulate the on-orbit response. The samples were exposed to 50 MeV protons at the TRIUMF radiation facility located in Vancouver, BC. The samples were mounted in printed circuit boards and were wire bonded using 1 mil (25.4 micrometer) gold wires to enable in situ current measurements (Figure 1). Each device was connected to enable four point probe measurements. Two irradiation experiments were performed. For the first one, full DIV data were taken prior to and following incremental fluences to develop full characteristic degradation data (calibration curve). For the second one, full DIV data were taken in situ while the beam was on (during proton exposure). These the

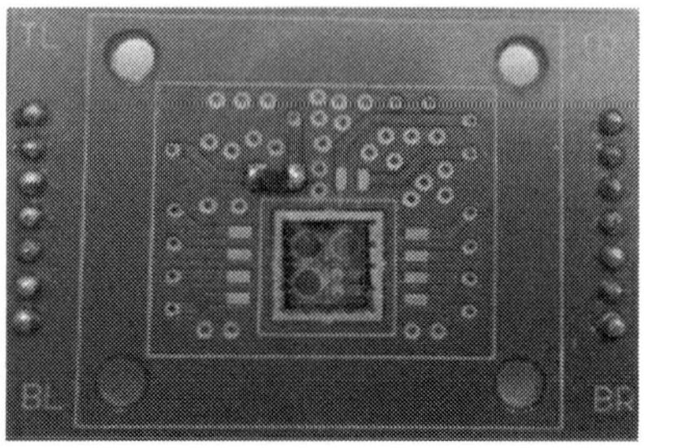

Figure 1. Image showing a sample (5 devices) mounted and wire bonded in a printed circuit board for in situ characterization.

measured current values were then used to predict corresponding fluence values (via DDD) using the calibration data. The typical beam flux at TRIUMF was 10^{9} $h^{+}/cm^{2}/sec$ but was reduced to 10^{8} $h^{+}/cm^{2}/sec$ during in situ characterization to eliminate proton-induced beam current effects. These irradiations were also performed at room temperature and in air. Dosimetry was accomplished using an ion chamber.

3. EXPERIMENTAL RESULTS

3.1 Characteristic Degradation Curve Development

The response of the GaAs diode dark current as a function of forward bias voltage both prior to and after 2 MeV and 225 MeV proton irradiation with proton fluences up to $2x10^{14}$ cm^{-2} are shown in Figure 2 and Figure 3, respectively. The dashed line represents the pre-radiation DIV response while the solid lines correspond to DIV response after proton exposure with increasing fluence levels. In all cases, the dark current increases with increasing proton fluence over the entire voltage range. Figure 4 shows the same data in Figure 2 where the measured dark current values at fixed voltage bias points 0.2V – 0.8V) are plotted as a function of 2 MeV proton fluence. There is a noted linear dependence in current as a function of the fluence for all bias voltage points shown. The 2 and 225 MeV proton dark current data are combined and plotted as a function of DDD in Figure 5 using total NIEL data for GaAs [2]. The NIEL values for 2 and 225 MeV protons on GaAs is $2.89x10^{-2}$ and $3.21x10^{-3}$ MeV cm2/g. The linearity of the data with DDD is seen to extend ~1 order of magnitude by adding the 225 MeV data to the plot.

The data in Figure 5 can now be classified as characteristic degradation data, or calibration data, representing the dark current increase for these particular GaAs p-n junction diodes. The effects of proton irradiation on this device can be determined using the calibration data in conjunction with the proton NIEL. The fact that these data are linear over a large proton energy range and at a wide voltage bias range gives us good confidence that this device structure is an excellent candidate for a displacement damage

DOI: 10.1051/e3sconf/20171602004

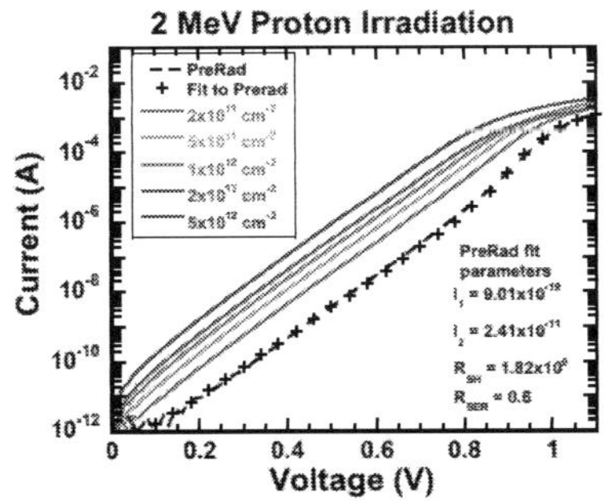

Figure 2. Dark current-voltage response of a p$^+$n GaAs photodiode following 2 MeV proton irradiation up to a fluence level of 5x10^{12} cm^{-2}. The pre-irradiation curve is represented by the dashed line, and the + symbols represent a fit using equation 1. The other lines correspond to increasing fluence levels and are not fits. Data is plotted as lines to clearly illustrate trends. The data resolution is 10 mV.

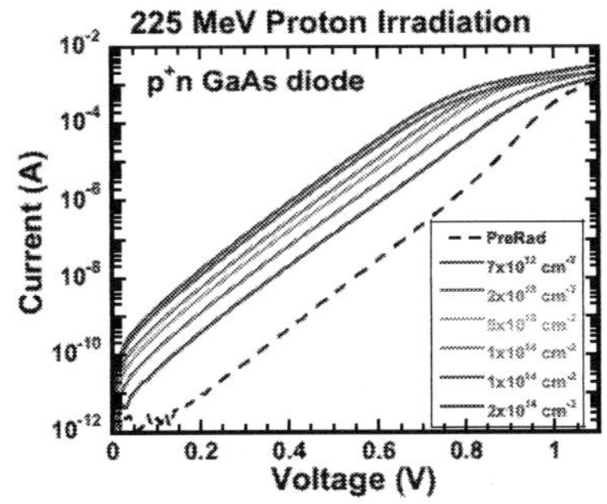

Figure 3. Dark current-voltage data of a p$^+$n GaAs photodiode following 225 MeV proton irradiation up to a fluence level of 2x10^{14} cm^{-2}. The pre-irradiation curve is represented by the dashed line. The other lines correspond to increasing fluence levels and are not fits. Data is plotted as lines to clearly illustrate trends. The data resolution is 10 mV.

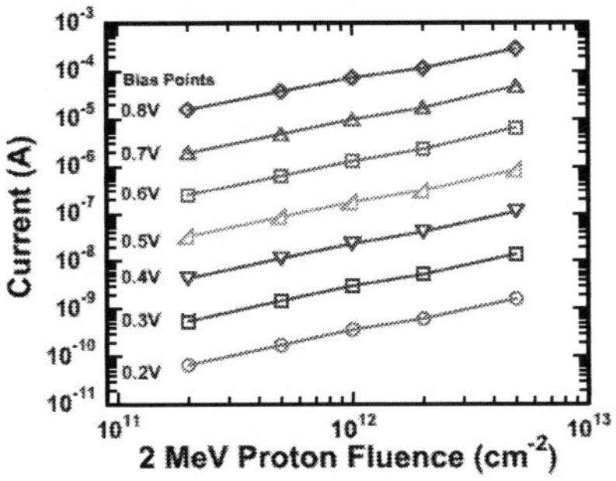

Figure 4. Measured current values at fixed voltages (0.2-0.8 V) as a function of 2 MeV proton fluence for p$^+$n GaAs photodiodes. These data were extracted from Figure 2.

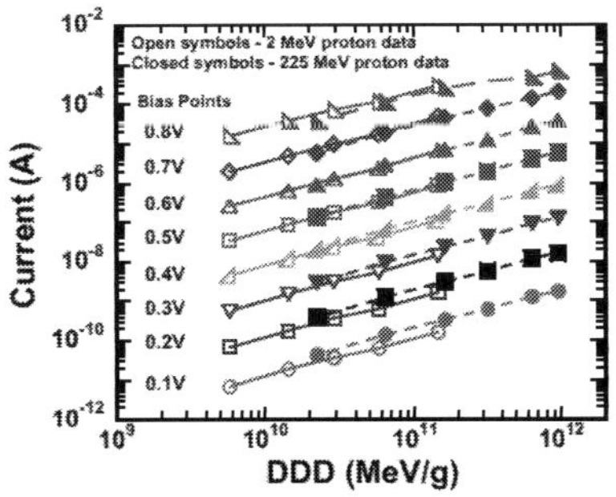

Figure 5. Measured current values at fixed voltages as a function of displacement damage dose for p$^+$n GaAs photodiodes. Data were extracted from Figure 2 and Figure 3. The open and closed symbols correspond to the 2 MeV and 225 MeV proton data, respectively. Each curve corresponds to a different bias voltage.

dosimeter for either monoenergetic or spectral proton energies. The data at low and high bias values do show some effects of nonlinearity and can be explained by the presence of shunt and series resistance effects, respectively. Nonetheless, the acceptable voltage range of linearity is ~0.4-0.7V which gives an excellent trade space in absolute current measurement capabilities. That is, different bias values can be applied to this structure to bring the current into a desired measurement range in the final circuit of the R2D3. A full voltage sweep would provide a lot of data to work with in practice.

3.2 Simulation of On-Orbit Performance

To simulate the on-orbit response and check the general feasibility that this system will work in space, additional proton irradiations were performed on GaAs devices. The devices were packaged (see Figure 1) so the diode dark currents could be measured *in situ* during proton exposure. The irradiations were performed using 50 MeV protons at TRIUMF. Two sets of radiation experiments were performed in this effort. The first set of experiments set out to define the characteristic dark current degradation data (similar to the data in Figure 5) of which the results are shown in Figure 6. The fluence values were converted to DDD using a 50 MeV proton NIEL value of 4.11x10^{-3} MeV cm^2/g.

The current data shown in Figure 6 has three functional dependencies (regions) with displacement damage dose in contrast to the data of Figure 5 where only a linear dependence is observed. For the 50 MeV proton data set, the radiation experiments started at much lower fluence values in order to find the onset of measurable dark current increase. In region 1, there is a minimal change in dark current (mechanism unknown) until the DDD reaches ~7x10^7 MeV/g. Thereafter, there

 DOI: 10.1051/e3sconf/20171602004

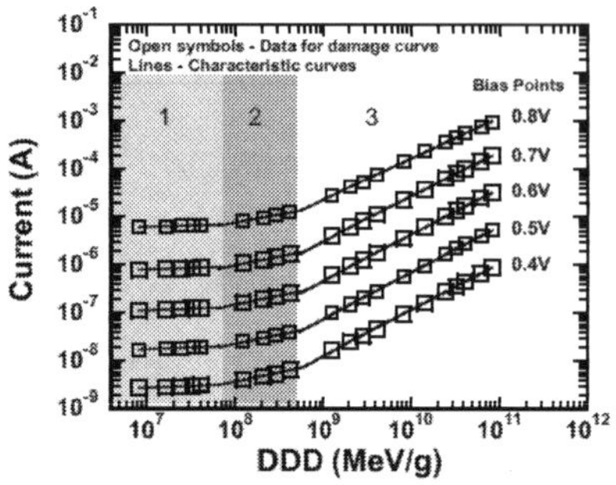

Figure 6. Dark current as a function of displacement damage dose for 50 MeV proton irradiations at Triumf. The data points correspond to the characteristic damage curves at each bias voltage and lines are coplotted at various slopes to guide the eye.

is an increase in slope as defined by region 2 until ~6x10^8 MeV/g. Region 3 is then defined showing the linear behavior with the dark current and DDD. Region 3 is consistent with the dark current being dominated by recombination processes also presented in Figure 5. The slope of the curve in region 2 is consistent with the dark current being dominated by diffusion processes. This was previously not observed in earlier datasets because experiments were not carried out at these lower DDD values. The lines in Figure 6 for region 2 and 3 are the characteristic damage curves at each bias point determined from fits using equations (4) and (5), which are presented below.

The second set of experiments taken at TRIUMF aimed to test how well the measured dark current can be used to predict the particle fluence (DDD) using the characteristic degradation data generated from the first set of experiments. The beam fluence at the facility is measured in terms of monitor counts. Full dark current voltage sweeps were measured with the 50 MeV proton beam on *in situ* at random increasing monitor counts. The monitor counts were converted to a known fluence level. These results for several discrete forward bias voltages are shown in Figure 7 where the lines represent the characteristic degradation data (first experiment) and the data points represent the *in situ* measured dark current. The agreement is excellent over a large DDD and voltage bias range. An estimate of errors can be obtained through these experiments. The predicted DDD as a function of actual DDD is plotted in Figure 8. The different symbol types correspond to current values measured at different forward bias voltages. The predicted DDD was calculated using the characteristic degradation data curves (first experiment) and the dark current that was measured in situ (second experiment). The solid line represents the actual fluence values. Data above and below this line correspond to an overestimate and underestimate, respectively.

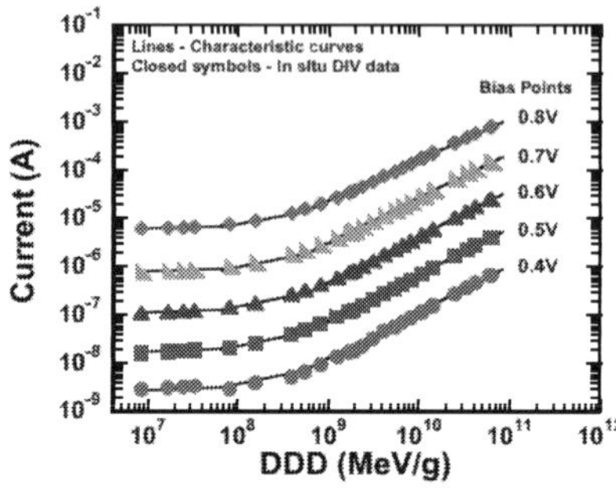

Figure 7. Dark current as a function of DDD. The lines correspond to the characteristic damage curves at each bias voltage and the symbols represent the dark current data measured in situ during 50 MeV proton exposures.

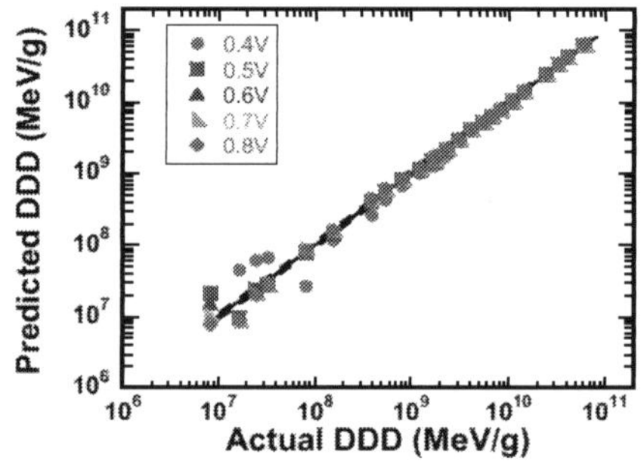

Figure 8. Predicted DDD as a function of actual DDD. The solid line represents the actual DDD. Data points below the line correspond to underestimate of the DDD and points above are overestimates. The two dashed lines are representative of ±10%.

To quantitatively access the uncertainty in using the dark current to predict the DDD on-orbit, a plot showing the percent error is shown in Figure 9. The horizontal dashed lines represent ±10% error. For the most part, the data are bound by these limits with the exception of DDD values below ~ 3x10^7 MeV/g and the 0.4V dataset. The percent error actually decreases for increasing DDD values.

The reason for the large deviations at low DDD values is evident from Figure 6. For DDD values below 3x10^7 MeV/g, a small change in measured current leads to a large uncertainty in the predicted DDD value because of the very steep slope of the characteristic degradation data evident when plotted as DDD vs dark current. When the current degradation mechanism changes from being diffusion current dominated to recombination current dominated, there is a decrease in the slope of the characteristic damage curve with a corresponding reduction in uncertainty in the predicted DDD. Although there is a large percent error for DDD values below 3x10^7 MeV/g, this would not be so

 DOI: 10.1051/e3sconf/20171602004

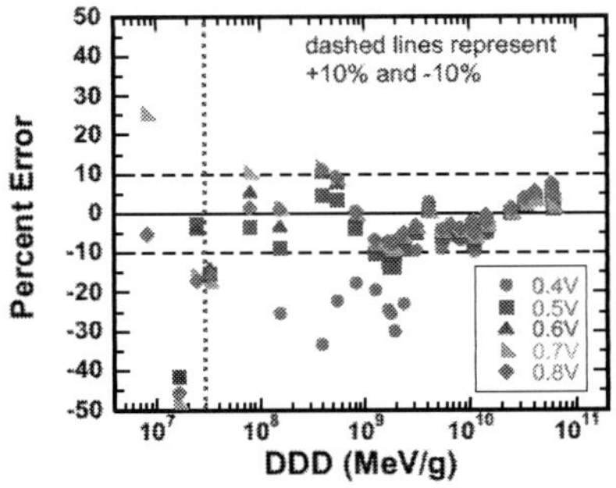

Figure 9. Calculated percent error between the predicted and actual DDD. The horizontal dashed lines represent ±10%. The data points correspond to data measured at fixed voltage values.

important for most space missions. For example, consider the GPS orbit. A dose value of ~3x10^7 MeV/g (indicated by vertical line in Figure 9), which is ~ the sensitivity limit of the current sensor design, would be reached in ~0.5 days in orbit if this device had 1 mil of aluminum shielding. This dose value would be reached much sooner for device with no shielding. Therefore, this sensor can be used for real-time DDD monitoring. Since this dose value is achieved fairly early in most missions, the large uncertainties observed at the low dose values are not important for space applications. Also, the same DDD value can be predicted at different forward bias voltages so the accuracy could be improved by averaging these DDD values.

4. DISCUSSION

The results presented in Section 3 can be qualitatively explained using conventional p-n junction theory. The forward-biased current in the dark of a junction is given by [16]:

$$I(V) = I_1\left[e\left(\frac{V - IR_{SER}}{k_bT}\right) - 1\right] + I_2\left[e\left(\frac{V-IR_{SER}}{2k_bT}\right) - 1\right] + \frac{V-IR_{SER}}{R_{SH}} \quad (1)$$

where I_1 is the diffusion current constant in the quasi-neutral regions of the device, I_2 is the recombination-generation current constant in the space charge region (SCR), R_{SER} is the series resistance, R_{SH} is the shunt resistance, k_B is the Boltzmann constant and T is the temperature. The prefactors I_1 *and* I_2 can be expressed as

$$I_1 = \frac{qn_i^2\sqrt{D_n}}{N_A\sqrt{\tau_n}} + \frac{qn_i^2\sqrt{D_p}}{N_D\sqrt{\tau_p}} \quad (2)$$

$$I_2 = \frac{n_ikT}{\sqrt{\tau_h\tau_n}}\frac{W_0}{\sqrt{V_b}}\frac{\pi}{2} \quad (3)$$

$D_{n,p}$ are the electron and hole diffusion coefficients, respectively, n_i is the intrinsic carrier concentration, $\tau_{n,p}$ are the minority carrier life time in the n- and p-regions, respectively, $N_{A,D}$ are the acceptor and donor doping concentrations on each side of the junction, V_b is the built-in voltage and W_0 is the width of the SCR with no applied bias. The diffusion and recombination current constants can be simplified and related to particle fluence with the following substitutions: $L^2 = D\tau$ and for $\tau = (N_T\sigma v_{th})^{-1}$ with the defect concentration given by $N_T = k\varphi$ (assuming linear trap introduction rates), (2) and (3) can be re-written as:

$$I_1 = I_1{}^*\varphi^{1/2} \quad (4)$$

$$I_2 = I_2{}^*\varphi \quad (5)$$

where I_1^* and I_2^* are defined in [16]. Equations (4) and (5) indicates that the diffusion dark current constant has a square root dependence with particle fluence while the recombination dark current constant is linear relationship with fluence (or displacement damage dose when fluence is converted to DDD using NIEL).

Each dark current dataset was fit to equation 1 using a nonlinear least squares fitting routine to determine the fitting parameters I_1 and I_2, R_{SH}, and R_{SER}. An example of such a fit is shown in Figure 2 represented by the + symbols. The values for the fitting constants for the pre-radiation data are given in the figure. The values for the fitting constants to the pre-radiation data are shown in the figure. Depending on the device technology, structure, and measurement voltage range, the forward bias dark current can be characterized by 4 regions: 1) diffusion, 2) recombination, 3) shunt resistance and 4) series resistance. The dark current due to recombination in the depletion region usually dominates at low voltage bias and the dark current associated with diffusion in the quasi-neutral region dominates at higher voltage bias regions. The magnitude where each mechanism contributes to the forward bias current depends on varied parameters such as the material type, doping and layer thicknesses. For the GaAs diodes in the present work, these regions are illustrated in the pre-radiation data. For V<0.05V, shunt resistance dominates the current. From 0.05<V<0.75 the recombination current (n_2=2) dominates. Thereafter, there is an increase in the slope when the diffusion current (n_1=1) begins to dominate, and eventually series resistance dominates the current flow causing a "roll-over" or flattening effect which is more evident in radiation data than the pre-radiation data set for this data set. However, after irradiation, the slope is fairly constant over a large bias range (0.1 V to 0.8 V) indicating that the current is dominated by one of the mechanisms discussed above.

Prior to irradiation, the diffusion current constant I_1 was already ~ 8 orders of magnitude smaller than the recombination component I_2. After irradiation, the recombination current continued to dominate the forward bias current. The dark current in these GaAs

 DOI: 10.1051/e3sconf/20171602004

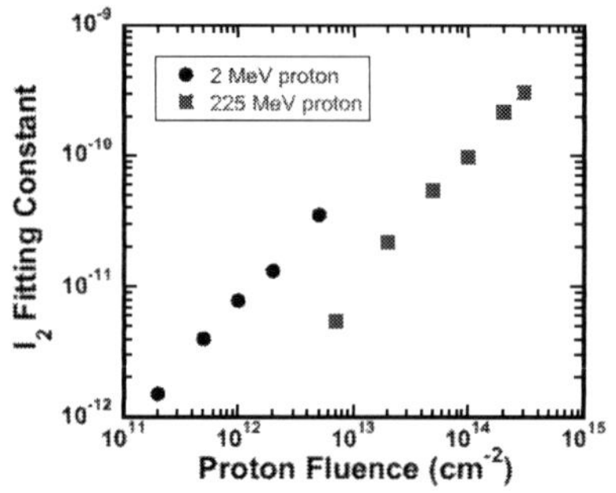

Figure 10. I_2 fitting term as a function of proton fluence for the 2 MeV and 225 MeV proton datasets. Each dataset was fit using equation 1 to extract the I2 fitting term.

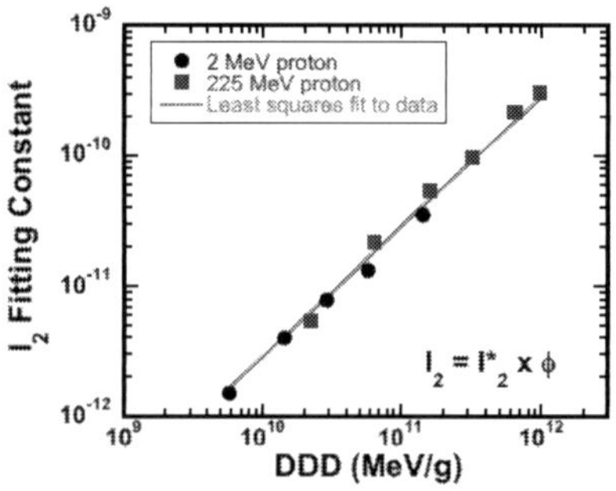

Figure 11. I_2 fitting term as a function of displacement damage dose. The data collapse and can be represented by a single characteristic curve defined by equation 5 (fluence in this equation must be converted to displacement damage dose).

diodes over this fluence range is thereby dominated by Shockley-Read-Hall recombination (SRH) in the SCR. The fitting results for I_2 as a function of proton fluence is shown in Figure 10 for both the 2 MeV and 225 MeV proton data. It is clear that the DIV response depends on the both the proton energy and fluence. These same data have been re-plotted in Figure 11 as a function of DDD by multiplying the proton fluence by the corresponding NIEL value for protons in GaAs (2.89×10^{-2} and 3.21×10^{-3} MeVcm^2/g for 2 and 225 MeV protons, respectively). When plotted as a function of DDD, the data collapse and can be represented by a single characteristic curve (solid line in Figure 11). The rate at which defects are introduced and their effect on the forward bias dark current is linearly proportional to the total NIEL.

The basic mechanisms and dependence of the diffusion and recombination current constants with particle fluence can be qualitatively explained by fitting the data in Figure 11 to equations. (4) or (5). The data in were fit using equation. (5) and the slope was determined to be 1.013, which confirms the dark current of this diode over this DDD range is dominated by recombination in the depletion region. However, the 50 MeV proton data set in Figure 6 shows that the current increase is dominated be either diffusion (region 2) or recombination (region 3) processes depending on the DDD value.

5. R2D3 DEVELOPMENT

The *in situ* electrical characterization has given confidence that the device can be used in a space environment to predict DDD. There does not appear to be any current injection annealing and GaAs devices are also well-known to be insensitive to thermal annealing until ~200°C . The R2D3 development of the prototype is currently underway. It is desired to have data collection, storage and telemetry as well as temperature measurement contained in the R2D3 package. For the measurement electronics, different levels of processing complexity in the implementation are being explored. At the high end of complexity, we are looking at a radiation-hardened version of the 8051 microcontroller. This is the same microcontroller used on the Microelectronics and Photonics Test Bed (MPTB) experiment payload. An alternative approach to this would be a rad-hard Field Programmable Gate Array (FPGA) that implements a simple state machine to make the necessary analog measurements and send the data to the host spacecraft. In both cases, a rad-hard high-resolution analog-to-digital converter will make the necessary measurements.

6. CONCLUSIONS

A characteristic degradation curve of the dark current in irradiated GaAs p-n junction diodes was obtained as a function of DDD following proton irradiations from 2 to 225 MeV. The linearity of the characteristic damage curve gives credence to a novel DD dosimeter for space applications. Furthermore, the wide voltage bias range over which the linearity is obtained simplifies the on-orbit measurement requirements. Additional proton testing at 50 MeV was performed to simulate measurements in space and yielded error estimates within 10% of expected values. The R2D3 system will contain the GaAs device with the associated measurement electronics for data collection, storage and telemetry and offers a low SWaP and low cost option for DD dosimetry for space and ground applications. A full voltage sweep in orbit will yield a large amount of data to improve error estimates. Future research will involve electron testing and temperature effects.

7. ACKNOWLEDGEMENTS

The authors would like to sincerely thank TRIUMF

radiation facility for the proton beam time in order to carry out the 50 MeV proton radiation experiments.

8. REFERENCES

[1] G. P. Summers, E. A. Burke, P. Shapiro, S. R. Messenger, and R. J. Walters, "Damage correlations in semiconductors exposed to gamma, electron and proton radiations," *IEEE Trans. Nucl. Sci.*, vol. 40, no. 6, pp. 1372-1379, 1993.

[2] J. Insoo, M. A. Xapsos, S. R. Messenger, E. A. Burke, R. J. Walters, G. P. Summers, and T. Jordan, "Proton nonionizing energy loss (NIEL) for device applications," *IEEE Trans. Nucl. Sci.*, vol. 50, no. 6, pp. 1924-1928, 2003.

[3] J. Insoo, K. Wousik, and R. Evans, "Electron Nonionizing Energy Loss for Device Applications," *IEEE Trans. Nucl. Sci.*, vol. 56, no. 6, pp. 3229-3235, 2009.

[4] G. P. Summers, E. A. Burke, and M. A. Xapsos, "Displacement damage analogs to ionizing radiation effects," *Radiation Measurements*, vol. 24, no. 1, pp. 1-8, 1995.

[5] C. Inguimbert, and S. Messenger, "Equivalent Displacement Damage Dose for On-Orbit Space Applications," *IEEE Trans. Nucl. Sci.*, vol. 59, no. 6, pp. 3117-3125, 2012.

[6] S. R. Messenger, G. P. Summers, E. A. Burke, R. J. Walters, and M. A. Xapsos, "Modeling solar cell degradation in space: A comparison of the NRL displacement damage dose and the JPL equivalent fluence approaches," *Progress in Photovoltaics: Research and Applications*, vol. 9, no. 2, pp. 103-121, 2001.

[7] J. H. Warner, S. R. Messenger, R. J. Walters, and G. P. Summers, "Displacement damage correlation of proton and silicon ion radiation in GaAs," *IEEE Trans. Nucl. Sci.*, vol. 52, no. 6, pp. 2678-2682, 2005.

[8] J. H. Warner, S. R. Messenger, R. J. Walters, G. P. Summers, J. R. Lorentzen, D. M. Wilt, and M. A. Smith, "Correlation of Electron Radiation Induced-Damage in GaAs Solar Cells," *IEEE Trans. Nucl. Sci.*, vol. 53, no. 4, pp. 1988-1994, 2006.

[9] J. H. Warner, R. J. Walters, S. R. Messenger, G. P. Summers, S. M. Khanna, D. Estan, L. S. Erhardt, and A. Houdayer, "High-energy proton irradiation effects in GaAs devices," *IEEE Trans. Nucl. Sci.*, vol. 51, no. 5, pp. 2887-2895, 2004.

[10] S. R. Messenger, E. M. Jackson, J. H. Warner, and R. J. Walters, "Scream: A new code for solar cell degradation prediction using the displacement damage dose approach," *IEEE 2010 35th Photovoltaic Specialists Conference (PVSC)*. pp. 001106-001111, 20-25 June 2010.

[11] S. R. Messenger, E. A. Jackson, J. H. Warner, and R. J. Walters, "Advancements to SCREAM: Multiple spectrum input and ShielDDDose options," *IEEE 2012 38th Photovoltaic Specialists Conference (PVSC)*. pp. 003281-003286, 3-8 June 2012.

[12] S. R. Messenger, E. M. Jackson, J. H. Warner, R. J. Walters, T. E. Cayton, C. Yue, R. W. Friedel, R. M. Kippen, and B. Reed, "Correlation of Telemetered Solar Array Data With Particle Detector Data On GPS Spacecraft," *IEEE Trans. Nucl. Sci.*, vol. 58, no. 6, pp. 3118-3125, 2011.

[13] P. P. Jenkins, D. C. Bentz, J. Barnds, C. R. Binz, M. S.R., W. J.H., M. J. Krasowski, N. F. Prokop, D. C. Spina, M. O'Neill, M. Eskenazi, H. W. Brandhorst, E. Downard, and K. C. Crist, "TACSAT-4 Solar Cell Experiment: Two Years in Orbit," *Proceedings of 10th European Space Power Conference*, 13–17 April 2014

[14] B. K. Dichter, J. O. McGarity, M. R. Oberhardt, V. T. Jordanov, D. J. Sperry, A. C. Huber, J. A. Pantazis, E. G. Mullen, G. Ginet, and M. S. Gussenhoven, "Compact environmental anomaly sensor (CEASE): a novel spacecraft instrument for in situ measurements of environmental conditions," *IEEE Trans. Nucl. Sci.*, vol. 45, no. 6, pp. 2758-2764, 1998.

[15] C. Deneau, J. R. Vaille, L. Dusseau, J. Mekki, P. Garcia, F. Bezerra, E. Lorfevre, and R. Ecoffet, "First In-Flight Data Analysis of Displacement Damage on the OSL Sensor On-Board CARMEN-2," *IEEE Trans. Nucl. Sci.*, vol. 58, no. 3, pp. 939-944, 2011.

[16] J. C. Bourgoin, and M. Zazoui, "Irradiation-induced degradation in solar cell: characterization of recombination centres," *Semiconductor Science and Technology*, vol. 17, no. 5, pp. 453, 2002.

DOI: 10.1051/e3sconf/20171602005

SUBCELL LIGHT CURRENT-VOLTAGE CHARACTERIZATION OF IRRADIATED MULTIJUNCTION SOLAR CELL

Don Walker, John Nocerino, Yao Yue, Colin J. Mann, and Simon H. Liu

The Aerospace Corporation, 2310 E. El Segundo Blvd., El Segundo, CA 90245, U.S.A, Email: don.walker@aero.org

ABSTRACT

The degradation of individual subcell J-V parameters, such as short circuit current, open circuit voltage, fill factor, and power of a GaInP/GaInAs/Ge triple junction solar cell by 1 MeV electrons were derived utilizing the spectral reciprocity relation between electroluminescence and external quantum efficiency. After exposure to a fluence of 1 x 10^{15} 1 MeV electrons, it was observed that up to 67% of the voltage loss is from the middle, GaInAs subcell. Also, the dark saturation current of the Ge and GaInAs subcells increased but a simultaneous decrease in ideality factor caused a reduction of the open circuit voltage. The reduced ideality factor further indicates a change in the primary recombination mechanism.

1. INTRODUCTION

Understanding the effects that influence the current-voltage (I-V) characteristics of space photovoltaics is critical to predicting their on-orbit performance[1-3]. By investigating the effects of space radiation on the underlying device parameters we can better understand and predict the performance of on-orbit solar arrays. Current state-of-the-art solar cells are made from direct bandgap III-V materials because of their high efficiency and radiation hardness. Multijunction solar cells are more radiation hard than previous generation single junction gallium arsenide (GaAs) solar cells due to the radiation hardness of the current limiting indium gallium phosphide based top cell[4]. While current degradation can be tuned for and characterized by quantum efficiency measurements, it has only been recently discovered that the open circuit voltage of each subcell of a monolithic GaInP/GaInAs/Ge can be derived along with the partial dark and light current-voltage parameters. By utilizing the reciprocity theorem between electroluminescence and external quantum efficiency [5], the subcell dark and light current-voltage curves can be derived before and after irradiation [6-8]. From the subcell dark and light current-vottage curves we can determine the radiation induced degradation of each subcell's short circuit current, open circuit voltage, fill factor, power, ideality, and dark saturation current by 1 MeV electrons at a fluence of 1 x 10^{15} electrons/cm^2. The ability to measure all the degradation current-voltage parameters of each subcell can lead to better degradation modelling. Understanding how each subcell degrades in voltage adds a new dimension of potentially tuning for less voltage degradation after particle irradiation

2. EXPERIMENT

Triple-junction ATJ solar cells of GaInP/GaInAs/Ge were purchased from SolAero Technologies®. The solar cells were 2cm x 2cm and did not have a coverglass. Light current-voltage, quantum efficiency, and electroluminescence measurements were performed before and after electron radiation at The Aerospace Corporation's Photovoltaic Evaluation And Research Laboratory (PEARL). The short circuit current density (J_{sc}), open circuit voltage (V_{oc}), fill factor, and efficiency were determined under a calibrated AM0 solar simulator on a temperature-controlled stage at 301.15 K (28°C). The simulator was calibrated using ATJ primary, JPL balloon flown standards. Quantum efficiency measurements were conducted using a Newport monochromator that was calibrated using a silicon photodiode and InGaAs diode that are National Institute of Standards and Technology traceable. Electroluminescence measurements were also performed at 301.15 K using a Horiba monochromator with a 2d, thermoelectrically cooled silicon CCD, and a temperature controlled, liquid nitrogen cooled extended InGaAs linear array. Electron radiation was conducted at the National Institute of Standards and Technology radiation facility. The triple junction cells were exposed to electrons at an energy of 1 MeV and a fluence of 1 x 10^{15} electrons/cm^2.

3. RESULTS AND DISCUSSION

The spectral reciprocity relation as defined by Eq. 1 relates the external luminescence (ϕ_{em}) and quantum efficiency (Q_e) to the voltage (V) of a solar cell, where ϕ_{bb} is the black body photon flux, q is the electron charge, k is the Boltzmann constant, and T is temperature. By taking electroluminescence spectra of each junction in a multijunction solar cell at different injection currents you can calculate the voltage of each subcell at each injection current. The open circuit voltage of each subcell is derived from the voltage calculated using Eq. 1 at an injection current equal to the short circuit current of the individual subcell.

$$\phi_{em} = Q_e \phi_{bb} \left[e^{\frac{qV}{kT}} - 1 \right] \tag{1}$$

The electroluminescence of each junction was taken before and after irradiating the triple junction cell with a

DOI: 10.1051/e3sconf/20171602005

fluence of 1 x 10^{15} 1 MeV electrons/cm^2 as can be seen in Fig. 1. The electroluminescence spectrum of each junction was taken at injection current densities logarithmically ranging over approximately 25 points from 25 μA/cm^2 to 35 mA/cm^2. After exposure, to electron radiation, it is observed in Fig. 1a that the electroluminescence of GaInP and GaInAs is reduced by over two orders of magnitude. Also, the relative peak intensity between GaInP and GaInAs changes such that GaInP is brighter than GaInAs. The electroluminescence of the Ge subcell also decreases, losing about half of its electroluminescence. Also, the luminescence below 1400nm is quenched after electron radiation. The luminescence below 1400 is believed to be from direct transfer states in the Ge.

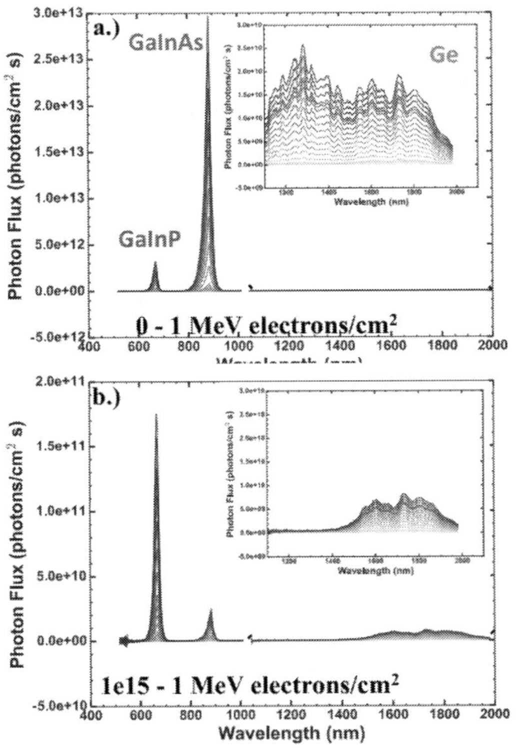

Figure 1. Electroluminescence of GaInP, GaInAs, and Ge subcells in a triple junction solar cell before irradiation with 1 MeV electrons (a.) and after a fluence of 1 x 10^{15} 1 MeV electrons/cm^2 (b.).

By using the reciprocity relation we can calculate the voltage of each subcell for each injection current and derive subcell dark current-voltage characteristics for each subcell before and after electron radiation as seen in Fig 2. The most obvious observed change is an increase in the dark current for each subcell at any given voltage. Less obvious, is the change in ideality of each subcell as well as the change in the saturation current density. By fitting the subcell dark current-voltage curves, we can see that ideality becomes more ideal, closer to 1 from the Ge and GaInAs subcell (Tab. 1).

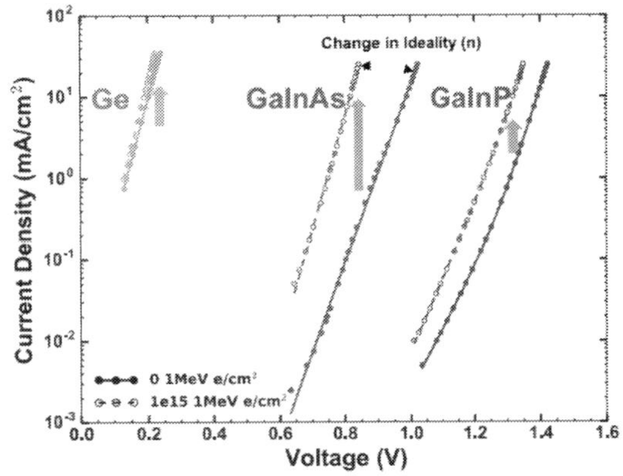

Figure 2. Dark current-voltage properties before and after irradiation with 1 x 10^{15} 1 MeV electrons/cm^2. The dark current-voltage curves were derived by using the spectral reciprocity relationship. The GaInAs subcell showed the greatest increase in dark current. Also, the GaInAs and Ge subcells had a decrease in their respective ideality factor.

The decrease in ideality indicates that the subcells are becoming more dominated by radiative recombination after irradiation, albeit there is much less radiative recombination as indicated by the electroluminescence spectra.

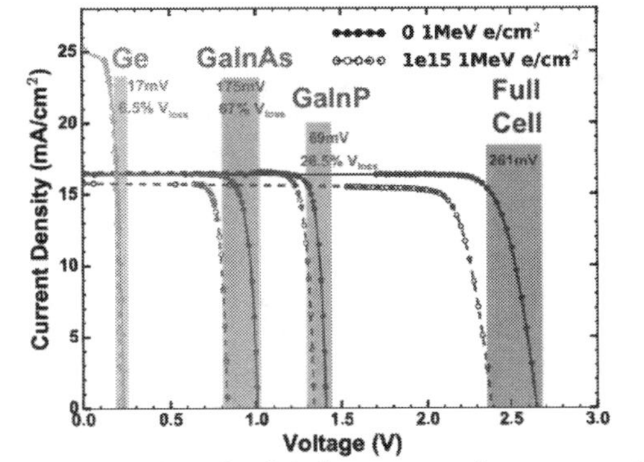

Figure 3. The subcell light current-voltage properties are derived by using superposition. The photocurrent derived from the respective quantum efficiency spectrum and the AM0 spectra were used to derive the subcell light current-voltage properties pre and post irradiation with 1 x 10^{15} 1 MeV electrons/cm^2.

Finally, by offsetting the dark current-voltage curves using the derived photocurrent from integrating the AM0 spectrum with the quantum efficiency spectrum

DOI: 10.1051/e3sconf/20171602005

Table 1

1 MeV e^-	Cell	J_{sc} (mA/cm^2)	ff	P_{max} (W)	V_{oc} (volts)	V_{loss} (%)	J_o (mA/cm^2)	n
0	Full	16.42	0.84	0.147	2.65			
	J_1	17.30	0.88	0.086	1.41		3.75E-19	1.20
	J_2	17.26	0.83	0.058	1.01		1.50E-10	1.52
	J_3	26.44	0.64*	0.016*	0.23		8.88E-03	1.11
1×10^{15}	Full	15.78	0.83	0.124	2.39			
	J_1	17.16	0.87	0.080	1.34	26.45	2.11E-17	1.26
	J_2	16.34	0.85	0.046	0.83	67.06	4.44E-11	1.21
	J_3	26.35	0.65*	0.015*	0.21	6.49	6.85E-03	1.00

*Table 1. The measured and derived light current-voltage properties of a multijunction solar cell (Full) its subcells(J_1, J_2, and J_3) before and after irritation with 1×10^{15} 1 MeV electrons/cm^2. The dark current-voltage parameters for each subcell are also presented. (*These values were calculated and presented for completion, but have error associated with not being able to capture the full 'knee' of the subcell current-voltage curve.)*

we can produce light current-voltage curves for each subcell. As can be seen in Fig. 3 and Tab. 1, the subcell fill factors, Voc, Jsc, and power. Previously, this has not been accomplished as the lower injection currents needed to obtain the knee of the current-voltage curve had not been obtained [7, 8]. Utilizing the derived subcell voltages we can now determine where the greatest loss in voltage comes from in the multijunction solar cell by evaluating the $V_{i_{loss}}$ factor as described in Eq. 2.

$$V_{i_{loss}} = \frac{\Delta V_i}{\Delta V_{Total\ Loss}} = \frac{V_{i_0} - V_{i_{1e15}}}{V_0 - V_{1e15}} \quad (2)$$

In Eq. 2, $V_{i_{loss}}$ is the normalized voltage loss for each junction (i). $V_{i_{loss}}$ is simply a way to determine what subcell contributes to the greatest loss in voltage of the multijunction solar cell. In this case the GaInAs contributes 67.06% of the voltage loss, GaInP contributes 26.45%, and Ge contributes 6.49% of the loss.

4. CONCLUSION

By deriving the subcell current-voltage properties and characterizing their radiation induced degradation, we can now identify where most of the voltage is coming from, radiation effects on fill factor, as well as the root cause for voltage degradation in multijunction solar cells. The ability to measure the degradation of subcell current-voltage properties has implications in better predicting radiation degradation of space photovoltaics, as well as developing more radiation hard solar cells.

5. REFERENCES

1. H. Y. Tada, J. R. Carter, B. E. Anspaugh, and R. G. Downing, "Solar cell radiation handbook," Jet Propulsion Laboratory, California Institute of Technology 82-69, 1982.
2. B. E. Anspaugh, "GaAs solar cell radiation handbook," Jet Propulsion Laboratory, California Institute of Technology 96-9, 1996.
3. K. C. Reinhardt, C. S. Mayberry, B. P. Lewis, and T. L. Kreifels, "Multijunction solar cell iso-junction dark current study," in *Photovoltaic Specialists Conference, 2000. Conference Record of the Twenty-Eighth IEEE*, 2000, pp. 1118-1121.
4. K. C. Reinhardt, Y. K. Yeo, P. H. Ostdiek, and R. L. Hengehold, "Junction characteristics of electron-irradiated $Ga_{0.5}In_{0.5}P$ n+-p diodes and solar cells," *Journal of Applied Physics*, vol. 81, pp. 3700-3706, 1997.
5. T. Kirchartz and U. Rau, "Detailed balance and reciprocity in solar cells," *Physica Status Solidi a-Applications and Materials Science*, vol. 205, pp. 2737-2751, Dec 2008.
6. T. Kirchartz, U. Rau, M. Hermle, A. W. Bett, A. Helbig, and J. H. Werner, "Internal voltages in GaInP/GaInAs/Ge multijunction solar cells determined by electroluminescence measurements," *Applied Physics Letters*, vol. 92, Mar 2008.
7. S. Roensch, R. Hoheisel, F. Dimroth, and A. W. Bett, "Subcell I-V characteristic analysis of gainp/gainas/ge solar cells using electroluminescence measurements," *Applied Physics Letters*, vol. 98, Jun 2011.
8. R. Hoheisel, F. Dimroth, A. W. Bett, S. R. Messenger, P. P. Jenkins, and R. J. Walters, "Electroluminescence analysis of irradiated GaInP/GaInAs/Ge space solar cells," *Solar Energy Materials and Solar Cells*, vol. 108, pp. 235-240, Jan 2013.

EXTERNAL QUANTUM EFFICIENCY AND FIRST RESULTS OF ELECTRIC PERFORMANCE MEASUREMENTS ON A QUADRUPLE JUNCTION SPACE SOLAR CELL

Grás, A. (1), Jüngst, G. (2), Campesato, R. (3), Gori, G. (3), Greco, E. (3)

(1) INTA-SPASOLAB, Carretera de Ajalvir km. 4, 28850 Torrejón de Ardoz (Spain), Email: grassa@inta.es
(2) ISDEFE Consulting, C/ Beatriz de Bobadilla 3, 28040 Madrid (Spain), Email: jungstg.pers_externo@inta.es
(3) CESI S.p.A., Via Rubattino 134, 20134 Milano (Italy), Email: roberta.campesato@cesi.it

ABSTRACT

External Quantum Efficiency (EQE) and Current-Voltage (I-V) measurements of a monolithic quadruple junction (4J) space solar cell AlGaInP (J1)/ AlInGaAs (J2) / InGaAs (J3) / Ge (J4) manufactured by CESI (Milano, Italy) have been performed at INTA-Spasolab. In order to determine electrical cell performance of this 4J solar cell type under AM0 irradiance and cell temperature standard conditions, the external quantum efficiency (EQE) measurement system and steady-state solar simulators at the cell testing laboratory INTA-Spasolab are to be upgraded. The purpose of this work is to present both the upgrading of INTA-Spasolab facilities and the results of the characterization of a 4J solar cell under development.

1. INTRODUCTION

Electrical performance characterization of new coming space multi-junction solar cells have required solar simulators to evolve from basic, single source systems to multi-source systems capable of providing test illumination conditions much closer to the reference AM0 spectrum. For accurate I–V measurements of monolithic multi-junction (MJ) solar cells, the spectrum of the solar simulator has to be adjusted such that each sub-cell of the MJ cell generates the same photocurrent as under the reference spectrum. The procedures to adjust the multi-source systems [1-4] usually involve the calculation of the so-called spectral mismatch factor for each junction for which EQE measurement of the sub-cells are necessary.

Manufacturers of space solar cells are working on increasing the number of junctions of III-V semiconductor based solar cells by a combination of semiconductors with different band-gaps to further improve the already high efficiencies of InGaP/InGaAs/Ge triple junction solar cells (above 30% in AM0 conditions) widely used in space applications due to their radiation resistance. Solar simulators are to be conditioning by adding and/or adapting spectrally their light sources in order to provide an accurate AM0 spectral match according to specific devices.

CESI is investigating the design of a monolithic quadruple junction (4J) space solar cell AlGaInP/ AlInGaAs/ (In)GaAs/ Ge that can be grown entirely by MOCVD toward the optimization of a 32% 4J lattice matched solar cell solution for space applications [5]. The approach to optimise the final four junction device is an iterative one: the solar cell structure is designed and simulated by the code, and then the solar cells are manufactured and tested. The experimental results from characterisation provide the feedback needed to improve the input parameters to the code and then to restart optimisation. The InGaAs junction was found to be the limiting sub-cell in the first four junction devices samples that have been manufactured and tested. This sub-cell determined a low radiation resistance in such devices. A 4J solar cell with modification of the former InGaAs structure [6] has been characterized at INTA-Spasolab.

2. TEST EQUIPMENT

2.1 EQE Measurement System

The EQE measurement system of INTA-Spasolab as well as the electrical characterisation method of multi-junction solar cells is described elsewhere in [2, 7]. As depicted in Fig. 1, bias light illumination can be easily adapted by optical bandpass filter change, so that measurements can be switched quickly from three junctions to four junctions III-V cells.

Figure 1. Optical Bandpass Filter Mount.

DOI: 10.1051/e3sconf/20171602006

Long (LP) and short (SP) bandpass filters used on bias light sources for EQE measurements on 4J sub-cells are listed in Tab. 1.

Table 1. Bandpass filters for EQE 4J cell measurements

Sub-cell	Bandpass Filter
J1	LP700 nm
J2	SP540 nm + LP830 nm
J3	SP600 nm
J4	Coldglass KG5

2.2 4J Solar Simulator

A commercial steady-state solar simulator based on a short-arc xenon lamp (Xe) with a light set up so far used for measurements of state-of-art triple junction solar cells (3J) GaInP/ InGaAs/Ge has been upgraded. As depicted in Fig. 2, spectrally filtered halogen light sources can be placed and combined on an external ring as required to meet finally effective irradiance unity at each sub-cell of the 4J cell. In total, three sets of assembled halogen light sources (Ha1, Ha2 and Ha3) have been utilised. The incident angle of this supplementary-light illumination is fixed within 30° to maintain cosine signal response of solar cells.

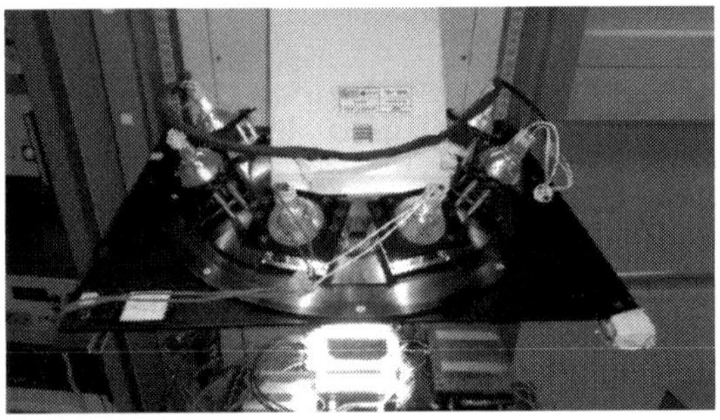

Figure 2. Additional halogen light sources for a single simulator solar simulator.

In order to achieve AM0 equivalent conditions of illumination with a four- light source simulator, a set of single-junction reference solar cells (SCRC) is used to adjust the irradiance level of each light source on the test plane in a way that the AM0 calibration value of each reference cell is achieved. The selected 4J SCRC set represent a spectral response range similar to the 4J cell to be tested, which is shown in Fig. 3. This set is comprised by the following:

- CESI model CTJ-4J, 80x40mm² with 30.15cm² active area: AlInGaP/Ge (J1), AlInGaAs/Ge (J2) and InGaAs/Ge (J3). All of them with active Germanium sub-cell.
- AZUR SPACE 3G28, 80x40mm²: Ge (J4)

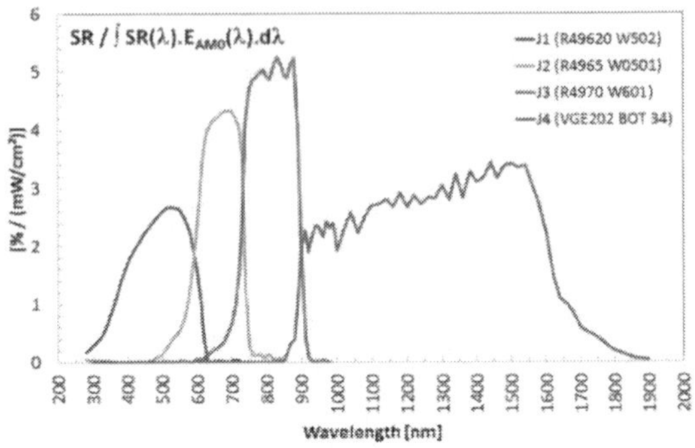

Figure 3. Spectral Response curves of the 4J single component reference cells (SCRC) set.

The spectral distribution curves of each light source of the 4J solar simulator at AM0 conditions are depicted in Fig. 4.

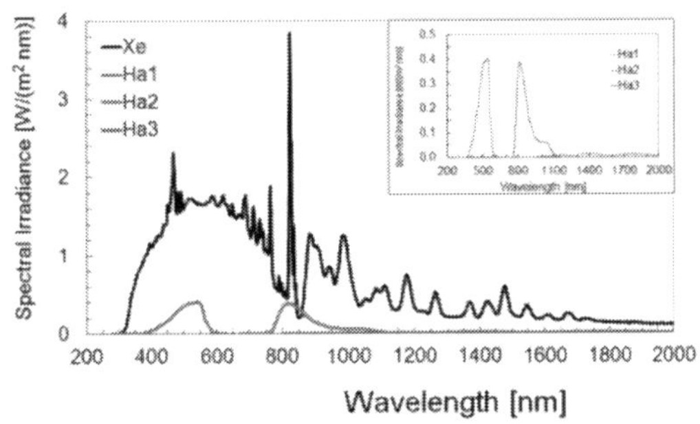

Figure 4. Spectral distribution of light source sets

The contribution of the different light sources of the 4J solar simulator to the short circuit current of each SCRC at AM0 equivalent illumination is indicated in Tab. 2

Table 2. Contribution of each light source to the short-circuit current of the SCRC at AM0 equivalent illumination

Light Source	J1	J2	J3	J4
Xe	90%	99%	83%	87%
Ha1	10%	1%	0%	0%
Ha2	0%	0%	17%	3%
Ha3	0%	0%	0%	10%

The nonconformity in the spatial uniformity (UF) of irradiance in the working plane for each junction and corresponding relevant class of solar simulator according to [8] is in Tab. 3.

Table 3. Nonconformity of irradiance uniformity

4J Solar Simulator	J1	J2	J3	J4
UF Nonconformity	±5%	±2%	±6%	±3%
Class	B	A	C	B

DOI: 10.1051/e3sconf/20171602006

3. TEST RESULTS

3.1 External Quantum Efficiency

EQE measurement results of the aforementioned 4J solar cell obtained using the EQE system previously described are presented in Fig. 5.

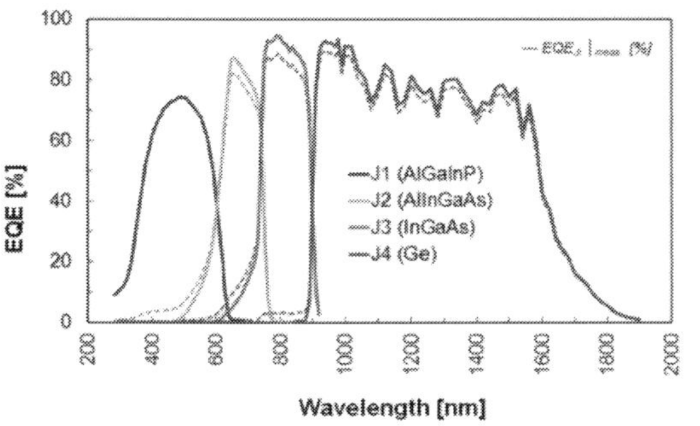

Figure 5. EQE results of the 4J solar cell

Measured EQE curves for the higher bandgap junctions show artefacts in the spectral range of the lower bandgap junctions. The presence of non-desired parasitic EQE when performing EQE measurement of a sub-cell could be related either to low shunts resistance of the sub-cell under test [9] or to luminescence coupling [10]. EQE measured values are then to be corrected to remove such artefacts. Although the physics involved are different, the mathematical procedure in [9, 10] to correct such artefacts is similar. The measured EQE values of the higher band junctions have been then corrected to get the actual EQE values. The results in Tab. 4 show the contribution of actual EQE values to the measured EQE ones.

Table 4. Contribution of actual EQE values to the measured EQE for each sub-cell of the 4J cell

$EQE_{Ji}\|_{meas}$	EQE_{J1}	EQE_{J2}	EQE_{J3}	EQE_{J4}
J1	**100%**	-	-	-
J2	5.9%	**94.1%**	-	-
J3	0.6%	10.4%	**89.0%**	-
J4	0.0%	0.15%	4.35%	**95.5%**

3.2 Current-Voltage Curve

The testing procedure to establish AM0 equivalent illumination conditions for multi-junction solar cells applied at INTA-Spasolab is described in [2, 4]. This procedure requires the measurement of the relative spectral irradiance of each light source at different lamp intensities to compute the corresponding spectral mismatch correction factor for each junction using relative spectral measurements of both, SCRC and related junction of the 4J cell, from the previously measured EQE in Fig. 3 and Fig. 4. The spectral mismatch parameter for each junction was found to be the same for the full operating range of each source.

The spectral mismatch factors to be applied when adjusting the 4J solar simulator to AM0 equivalent illumination conditions for the 4J solar cell are given in Tab. 5.

Table 5. Applied Spectral Mismatch Factors (M)

Light Source	J1	J2	J3	J4
Xe	1.005	0.999	1.000	1.026
Ha1	0.970	0.782	-	-
Ha2	-	0.183	1.003	0.932
Ha3	-	-	-	1.004

During the 4J solar simulator setting process, the AM0 effective irradiance for each sub-cell junction is computed by using the measured short-circuit current values of the 4J SCRC set and applying the corresponding spectral mismatch factor. Effective irradiance results are given in Tab. 6, where the junction used to set the different light sources in bold is highlighted.

Table 6. AM0 effective irradiance for each sub-cell

AM0 effective irradiance	J1	J2	J3	J4
Xe	0.903	0.993	0.829	0.880
Xe+**Ha1**	0.999	1.000	0.824	0.888
Xe+Ha1+**Ha2**	0.999	1.000	1.004	0.919
Xe+Ha1+Ha2+**Ha3**[1]	**0.999**	**1.000**	**1.004**	**1.020**

[1] 4J Solar Simulator adjusted to Isc-AM0

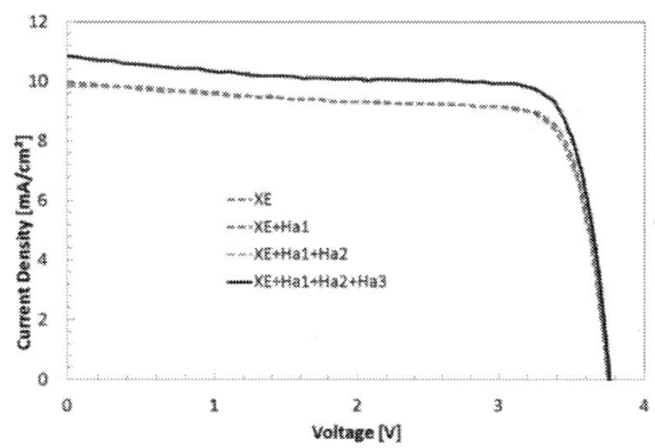

Figure 6. Current-voltage curves of 4J solar cell during the 4J solar simulator setting process.

Current-voltage curves of the 4J solar cell measured during the 4J solar simulator adjustment process are depicted in Fig. 6. These curves indicate that the InGaAs junction (J3) was the limited sub-cell under the Xe light source, because, when adding the second light source Ha1, no major impact in the I-V curve shape of the 4J cell is noticed. Nevertheless, adding the third light source Ha2 to achieve eventually AM0 equivalent illumination conditions in this sub-cell gives rise to an increased short-circuit current of the 4J solar cell. This

DOI: 10.1051/e3sconf/20171602006

is sustained as expected, when adding the fourth light source Ha3, since Ha3 only contributes to increase the irradiance level on the Ge sub-cell (J4). The limiting cell at AM0 illumination conditions as derived from the I-V curves of 4J SCRC set in Fig. 7 is AlInGaAs (J2). This has been also confirmed by further increasing the illumination level of Ha2 light source and noticing no relevant impact on the I-V curve shape.

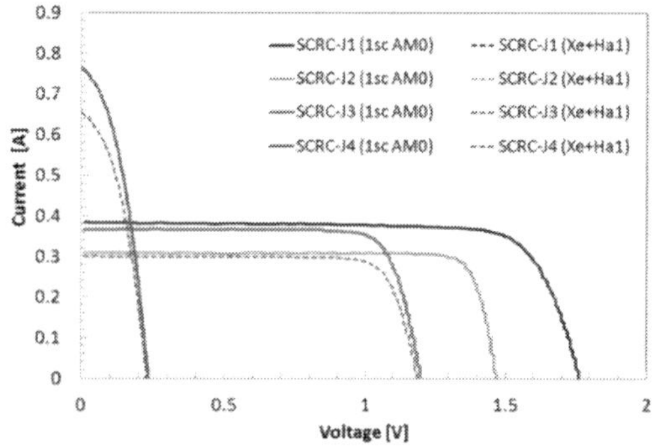

Figure 7. Current-voltage curves of 4J SCRC set during the 4J solar simulator setting process.

4. CONCLUSIONS

In this work external quantum efficiency (EQE) results of a 4J solar cell by CESI have been determined at INTA-Spasolab. The applied measurement method allows the EQE measurement system to be configured quickly, by changing optical bandpass filters, in order to switch from 3J to 4J solar cells. The EQE curves of higher bandgap junctions are reported including corrections of the parasitic EQE.

After adapting a solar simulator utilised so far for electric performance measurements on triple junction solar cells, first results of EP measurements of a 4J solar cell at 25°C have been obtained. The results confirm that the modified CESI 4J solar cell with the new proposed InGaAs structure [6] is not the limiting sub-cell under AM0 illumination conditions.

INTA-Spasolab future works are addressed to consolidate accurate effective irradiance adjustments. Spectrometric characterization as performed in [11] are planned to determine measurement uncertainty of electrical performance measurement at AM0 standard test conditions.

5. REFERENCES

1. Emery, K. et al. (2000). "Procedures for evaluating multi-junction concentrators". In *Proc. 28th IEEE Photovoltaic Specialists Conference*, pp1126-1130.
2. Gras, A. et al. (2002). "Analysis for multi-junction solar cell measurements at Spasolab". In *Proc. 6th European Space Power Conference*, ESA SP-502, pp577-580.
3. Meusel, M. et al. (2006). "Characterization of monolithic III–V multi-junction solar cells-challenges and application". *Solar Energy Materials & Solar Cells* 90, pp3268–3275.
4. Gras, A. et al. (2008). "Multi-Source solar simulator characterization for establishing AM0 equivalent conditions", In *Proc. 8th European Space Power Conference*. ESA SP-661 (CD-ROM).
5. Campesato, R. et al. (2013). "Electron and proton irradiation on high efficiency III-V solar cells based on three and four junctions". In *Proc. 28th European Photovoltaic Solar Energy Conference*, pp531-535
6. Campesato, R., Gori, G., Casale & M., Gabetta, G. (2014). "High Efficiency Four junction lattice matched solar cells for space applications: analysis of radiation hardness against 1 MeV electrons". In *Proc. 29th European Photovoltaic Solar Energy Conference*, pp2031-2035.
7. Blanco Moriana, G., Fernández-Marín, J., Jüngst, G. & Taylor, A. (2010). "Testing Methods to Quantify the Effects of Irradiation of Various Energies Protons and Electrons on the Performance of Triple Junction Solar Cells". In *Proc. 25th European Photovoltaic Solar Energy Conference*, pp872-876.
8. ECSS-E-ST-20-08C Rev. 1 (18 July 2012). "Space Engineering - Photovoltaic assemblies and components". ECSS Secretariat, ESA-ESTEC Requirements & Standards Division, Noordwijk, The Netherlands.
9. Siefer, G., Baur, C. & Bett, A. (2010). "External quantum efficiency measurements of germanium bottom sub-cells: measurement artefact and its solutions". In *Proc. 35th IEEE Photovoltaic Specialists Conference*, pp704 –707.
10. Steiner et al. (2013). "Measuring IV Curves and Sub-cell Photocurrents in the Presence of Luminescent Coupling". *IEEE J. Photovolt., vol. 3, no. 2*, pp879-887.
11. Jüngst, G., Gras, A. & Baur, C. (2013). "Measurement uncertainty for electric performance calibration of triple junction solar cells at Spasolab". In *Proc. 28th European Photovoltaic Solar Energy Conference*, pp540-543.

SIMULATING LARGE AREA, HIGH INTENSITY AM0 ILLUMINATION – TEST RESULTS FROM BEPICOLOMBO AND SOLAR ORBITER QUALIFICATION

C. Oberhüttinger [(1)], H. Nesswetter [(1)], D. Quabis [(1)], C. G. Zimmermann [(1)]

[(1)] Airbus Defence and Space, 81663 Munich, Germany, Email: carola.oberhuettinger@airbus.com

ABSTRACT

Two ESA high temperature missions, BepiColombo and Solar Orbiter, are planned to start in 2018. Severe environmental conditions with sun intensities up to 10.6 resp. 12.8 solar constants will be encountered during both missions.

Different tests have been performed to qualify the main component for the solar array, the solar cell. The key qualification test for these high temperature, high intensity missions is a combined test which exposes a large number of cells simultaneously to the complete AM0 spectrum at the required irradiance and temperature.

During the last European Space Power Conference in 2014, the development and performance of the appropriate test setup in the VTC1.5 chamber located at ESTEC was presented. In the meanwhile, the BepiColombo 4000 hour qualification test is successfully finished. This paper summarizes the results of the qualification test and highlights the main findings.

Specifically, the following topics will be treated. Different methods like electrical performance, thermo-optical and external quantum efficiency measurements have been used to characterize the behaviour of the solar cells after illumination under these conditions. A special focus has been put on the electrical performance. A comparison to other solar cell qualification tests under solely UV radiation has been undertaken. The results have also been compared to a theoretical model. However, the paper will not cover only characterization results but will also give some insight in challenges experienced during the test execution itself. Deviating from other solar cell qualification tests, a representatively equipped photovoltaic assembly on carbon fibre reinforced cyanate has also been included. On these coupon segments, solar cell assemblies connected to shunt diodes and placed next to optical surface reflectors have been exposed to AM0 illumination to qualify the solar cells including their surroundings which therefore covers also contamination effects. Last but not least, first results from the Solar Orbiter qualification are presented. This test with additional 1000 hours and increased intensity has been completed recently.

1. INTRODUCTION

Two ESA high temperature missions, BepiColombo (BC) and Solar Orbiter (SolO), are planned to start in 2018. The severe environmental conditions with sun intensities up to 10.6 resp. 12.8 solar constants (SCs) results in predicted solar cell operating temperatures up to 205°C. A specific cell design was therefore developed for these missions [1, 2]. Different tests have been performed to qualify the solar array according to the ECSS standards [3]. For one of the main components, the solar cell, the key qualification test for these high temperature, high intensity missions was a combined test which exposes a large number of cells simultaneously to the complete AM0 spectrum at the required irradiance and temperature. In the last European Space Power Conference (ESPC), the development and performance of the corresponding test setup was presented [4]. This paper now covers the results of the completed qualification test.

2. QUALIFICATION TEST DESCRIPTION

The BepiColombo AM0 Long Duration Illumination Qualification Test is a 4000 h test which was conducted from November 2013 to September 2015 in the VTC1.5 chamber located at ESTEC. The illumination intensity was set to 10 SCs +/- 2.5 SCs AM0. Solar cell test temperature was 190°C resp. 205°C (600 h). Both solar cell assemblies (SCAs) and representatively equipped photovoltaic assembly (PVA) on carbon fibre facesheets have been investigated. They were individually connected to an external shunt resistor to operate them close to the maximum power point. The samples have been mounted on a specifically developed mechanical ground support equipment (MGSE) which provides the possibility to test the samples in the light beam under an angle of 70°.

For monitoring reasons, different additional equipment has been used. On the one hand, molecular organic contamination control samples have been used to determine quantitatively and qualitatively the occurring contamination. On the other hand, an infrared (IR) camera was placed inside the chamber to examine the arising solar cell temperatures. This camera was providing live pictures during the complete test duration and was accessible via internet. Furthermore, thermocouples on some witness SCAs as well as the current and voltage of the individual SCAs and coupon segments under qualification were continuously recorded and checked by remote access.

Due to lamp restrictions, the complete test duration had to be split into sub-cycles with envisaged duration of about 400 hours for each cycle. The test breaks have

been used for detailed solar cell characterizations at ambient conditions.
As the Solar Orbiter mission submits the solar array to even higher sun intensities, the SolO representative solar cells were tested for further 1000 hours at the required increased intensity of 10.5 SCs +/- 2.3 SCs and 190°C using the same setup. This additional test was performed from May to August 2016.

3. BEPICOLOMBO SOLAR CELL QUALIFICATION TEST RESULTS

As qualification samples, 16 40x40mm² solar cells and 17 80x40mm² solar cells with the specific design for high temperature, high intensity missions and equipped with interconnectors and cover glasses have been placed inside the VTC1.5 chamber for the test on SCA level. These SCAs have been bonded to aluminium holder plates for handling reasons. The following inspections have been used to characterize the samples: visual inspection, electroluminescence measurements (ELM), electrical performance measurements, dark IV, thermo-optical properties and external quantum efficiency (EQE). In the following subsections, the main findings of certain measurement techniques are summarized.

3.1 Live Data Analysis of the Cell Parameters

The individual cell currents and voltages were recorded with a sampling rate of 300 s. Due to the lamp movement and degradation, which is described in section 5.1, the intensity on the samples was varying. This also lead to a slightly changing temperature on the SCAs and because of the temperature/intensity dependence of the cell parameters the current and voltage values were also varying. These effects overweigh the minor degradation of the cells which is presented below.
However, with the continuous live data it is possible to detect major changes in cell performance. This was used to determine the occurrence time for a cell crack. For the corresponding cell a crack was detected during ELM inspection after test cycle 4. Fig. 1 shows for the test cycles 2 – 5 the voltage of that cell related to the neighboring cells, to exclude lamp variations as good as possible. The moving average (black curve) was calculated to reduce the remaining fluctuations even more. During cycles 2 – 4 the affected cell had a voltage of about 98% of the cells next to it and in cycle 5 it was only 96%. A cell crack reduces the maximum power point voltage because of increased surface recombination, what confirms that the crack occurred between lamp switch-off for cycle 4 and the ELM inspection. It could therefore be concluded that the cell crack did not occur during the exposure test itself.

3.2 Electrical Performance

The main focus of the test was on the evolution of the

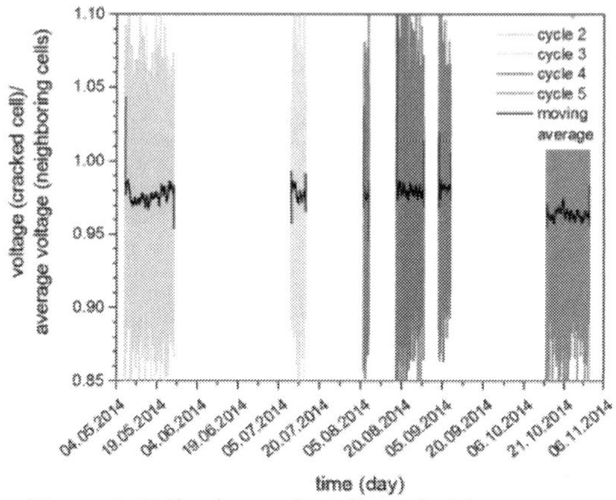

Figure 1. Cell voltage of a cell cracked between test cycle 4 and 5 related to the neighboring cells.

electrical characteristics of the solar cells at the different test steps. In average there is no difference for the 40x40 mm² and 80x40 mm² solar cells. Therefore, Tab. 1 shows the average degradation in maximum power (P_{max}), voltage V_{pmax} at P_{max}, current I_{pmax} at P_{max}, short circuit current (I_{sc}), open circuit voltage (V_{oc}), and fill factor FF of all SCAs after 4000 h test.

ΔP_{max} (%)	ΔV_{pmax} (%)	ΔI_{pmax} (%)	ΔI_{sc} (%)	ΔV_{oc} (%)	ΔFF (%)
-3.0	-0.2	-2.8	-2.3	-0.3	-0.4

Table 1. Average degradation of all SCAs after 4000 h test duration.

The main degradation is related to the current parameters and consequently a bit more pronounced to P_{max}. This gives a clear indication that the degradation is mainly caused by transmission losses, i.e. cover glass adhesive darkening and contamination.
In Fig. 2, the average degradation in I_{sc} of all samples is plotted over the test hours. It was of special interest if the observed behaviour is comparable to results obtained in further solar cell qualification tests under solely UV illumination. Therefore, test data from corresponding UV tests for BepiColombo have been included as well. In theory, the UV degradation D over time t of photovoltaic modules can be calculated according to Eq. (1) [5]:

$$D(t) = 1 - n \cdot \ln(1 + a \cdot t) \qquad (1)$$

with parameters n and a derived from a fit to the UV test results. The theoretical degradation describes very well the measured cell degradation both for the UV and the AM0 test. The slightly higher degradation in short circuit current for the AM0 test samples compared to the theoretical curve results from contamination effects.

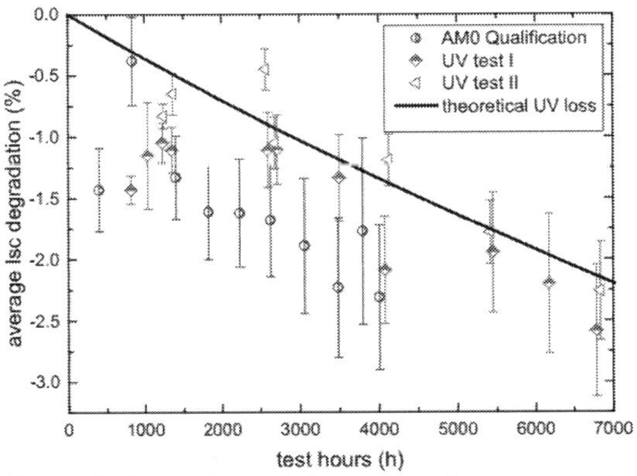

Figure 2. Comparison of electrical performance results for AM0 and UV tests.

3.3 External Quantum Efficiency

At begin of test (BOT) and end of test (EOT), the external quantum efficiency of the SCAs has been measured. With this measurement technique, essential changes in the solar cell behaviour can be observed. Exemplarily for the SCA with the highest I_{sc} degradation of -3.5% the external quantum efficiency is shown in Fig. 3.

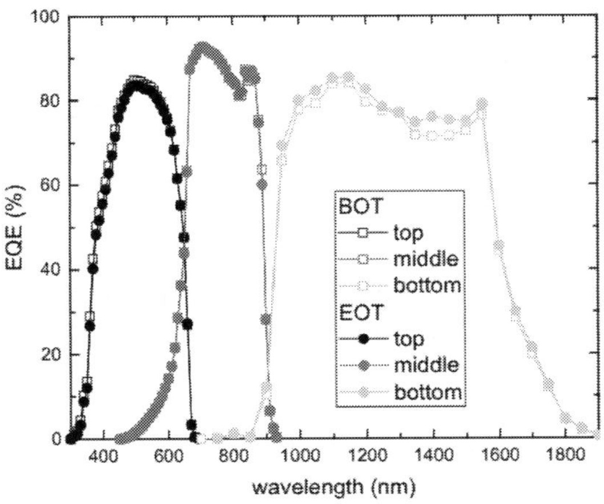

Figure 3. External quantum efficiency for the SCA with highest I_{sc} degradation.

No evidence for drastic changes in the solar cell characteristics has been determined. For the same cell the ratio of the remaining factor for the EQE of the top and middle cells is shown in Fig. 4 to see more details. The bottom cell is not considered because of higher uncertainties in that measurement. Furthermore, the bottom cell will not limit the triple-junction solar cell current under any mission point and condition.

It is clearly visible that the degradation takes mainly place in the short UV wavelength range. For the first point up to 350 nm it has to be mentioned that the

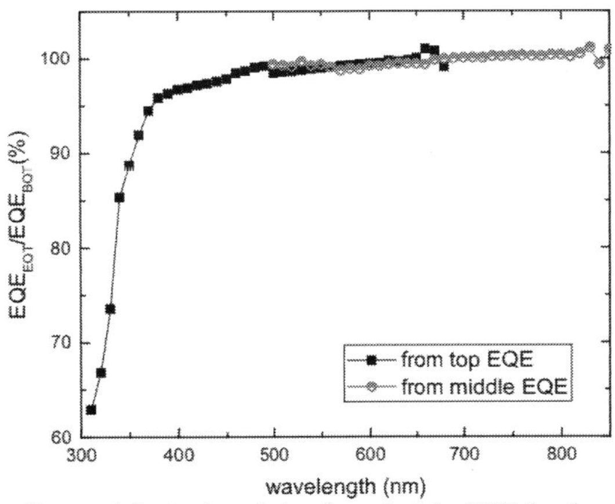

Figure 4. Ratio (remaining factor) of the EQE for the top and middle cells.

absolute values of the EQE are pretty small which leads to higher measurement uncertainties.

The EQE can be converted to spectral response (SR) values in units of A/W and together with the AM0 sun spectrum the short circuit current I_{sc} can be calculated:

$$I_{sc} = \int SR_{top}(\lambda) \cdot AM0(\lambda) d\lambda \tag{2}$$

By using the top cell SR, a top cell limitation is assumed, which is correct for not particle irradiated cells. Calculating the I_{sc} degradation of the cell from Fig. 3 and Fig. 4 according to Eq. (2) leads to theoretical degradation under AM0 of -1.4%. This value is lower than the measured one because of two reasons: the accuracy of the EQE measurement itself and the spectrum of the used flasher system for the electrical performance measurements is of cause not identical to the real AM0 sun spectrum.

3.4 Thermo-Optical Measurements

Another important topic is the change of thermo-optical properties of the solar cells with time. A degradation in absorption or emissivity could strongly influence the whole solar array performance as it would lead to higher end of life cell temperatures. The AM0 Long Duration Illumination Test was therefore used to confirm the stability of the thermo-optical properties. Only minor changes within measurement accuracy have been determined. The range of the solar absorptance α_s, the IR normal thermal emittance ε_N as well as the corresponding ratio α_s/ε_N is shown in Tab. 2.

	α_s (-)	ε_N (-)	α_s/ε_N (-)
Min	0.84	0.82	0.97
Max	0.90	0.88	1.04

Table 2. Range of the thermo-optical parameters for the SCAs without difference for BOT and EOT.

DOI: 10.1051/e3sconf/20171602007

4. COUPON SEGMENTS

In addition to the single qualification samples described above, also six PVA representative coupon segments have been part of the AM0 Test. Each coupon segment has been equipped with six 40x40 mm² solar cells connected to external shunt diodes and placed next to optical surface reflectors (OSRs). A picture including all six coupon segments is depicted in Fig. 5. Due to late availability of one key component, the coupons were not tested from the very beginning and a total test duration of 3166 h was reached. The aim of the coupon segments was to qualify the performance of the SCAs including their surroundings, i.e. OSRs, adhesives and facesheet, which therefore covers also contamination issues.

Figure 5. Picture of six coupon segments including solar cells, shunt diodes and OSRs.

The electrical characteristics of the solar cells interconnected to short strings was also measured at BOT, at intermediate inspection steps and EOT. The average degradation of all relevant electrical parameters is shown in Tab. 3.

ΔP_{max} (%)	ΔV_{pmax} (%)	ΔI_{pmax} (%)	ΔI_{sc} (%)	ΔV_{oc} (%)	ΔFF (%)
-2.4	-0.3	-2.1	-1.4	-0.3	-0.7

Table 3. Average degradation of the six coupons after 3166 h test duration.

The degradation is like for the SCAs mainly for the current parameters. The lower degradation values are caused by the shorter test duration. Additionally, the configuration with strings instead of single solar cells can lead to lower degradation in short circuit current, if the limiting cell in the string has a comparatively low degradation. The overall electrical performance results of the coupon segments confirm the suitability of the whole PVA ensemble for the BC and SolO mission needs.

With this coupon segments it was not only possible to characterize the behaviour of the solar cells themselves, but also to measure the characteristics of the utilised shunt diodes. The individual shunt diodes have been measured at ambient conditions before assembly with the solar cells. After the long duration illumination test, the shunt diodes were electrically separated from the solar cells by cutting the interconnectors. The measurement of the EOT shunt diode characteristics was performed by using a contact technique as shown in Fig. 6.

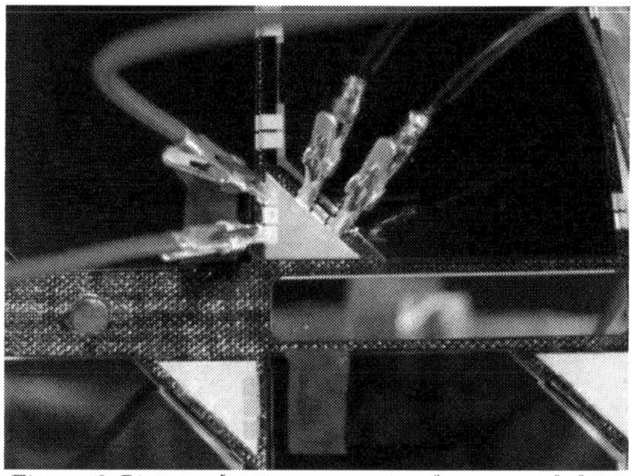

Figure 6. Picture showing contacting for external shunt diode characterization.

The average results for BOT and EOT are shown in Tab. 4.

	V_{forw} (@3 A) (V)	I_{rev} (@-4 V) (μA)
BOT	0.842	0.004
EOT	0.805	0.070

Table 4. Average shunt diode parameters BOT and EOT at ambient conditions.

The forward voltage at a current level of 3 A was slightly reduced by about 40 mV over the test duration. This is even an improvement/advantage, because the voltage loss is reduced in case of a forward biased shunt diode. The reverse current at -4 V was slightly increased, but is in average and also for each individual diode still well below the BOT acceptance criteria of <2 μA.

5. CHALLENGES DURING TEST EXECUTION

The setup for this very specific test was new developed. Besides expected measures to compensate the lamp degradation, also some unexpected challenges occurred during test execution. Some examples are explained in more detail below.

5.1 Lamp movement and degradation

The absolute beam intensity at the location of the

sample plates was determined with a lamp mapping in three planes BOT and EOT for each test cycle. The 25 kW Xenon lamps used showed on the one hand an unpredictable movement of the beam as no optics was present in the beam, which resulted in intensity resp. temperature variations. This was compensated as good as possible by applying an appropriate thermal washer configuration below the samples [4]. On the other hand, the lamp intensity suffered from certain intensity degradation with minor differences between the individual lamps. This was taken into account by a stepwise increase of lamp current. To decide on the point in time and the amount of lamp current increase, the temperature readings of the IR camera and the thermocouples on the witness cells as well as the individual cell currents from the live data acquisition were used. The average solar cell current values for all samples on one plate over 24 h were used for this evaluation. By doing this, the movement of the lamp, which has also influence on the individual cell current, was excluded as much as possible.

5.2 Test Artefact

As described above, an IR camera inside the chamber was used to monitor the arising solar cell temperatures. After starting the first exposure cycle, so called "red corners" have been observed for the large qualification samples only (see Fig. 7). This could be explained by a edge delamination of the solar cell from the aluminium holder plate induced by the different coefficient of thermal expansion (CTE) of both materials. As the shear stress in the bond line for the large cells is twice as big as for the small cells, this phenomenon was reduced to the 80x40 mm² qualification samples. It could be proven both by analysis and test that this was a test artefact only. This phenomenon is observed at high intensities/temperatures and additionally under vacuum conditions as the weaker thermal contact between cell and holder plate causes a bad temperature distribution. In the flight hardware configuration with panel substrates out of carbon fibre instead of aluminium plates for the test sample holders, the CTE difference is much smaller and this effect cannot occur.

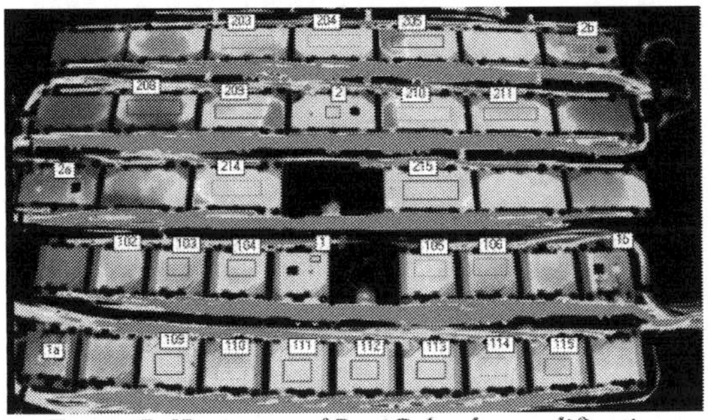

Figure 7. IR picture of BepiColombo qualification samples.

5.3 Wire Darkening

During the exposure cycles, the qualification samples had to be tested under operating conditions. Therefore, cabling of the individual samples had to be present which led to unrepresentative high cable bundles exposed to the illumination. This setup is in contrast to the solar array flight configuration as the cables are routed on the panel rear side. Due to the height, there was a bad thermal contact of the cable bundles to the aluminium sample plates. The wires therefore have been exposed to extremely high temperatures which caused darkening of the wires and also certain contamination. To allow continuation of the test, cable shields have been introduced to protect the cabling from further degradation. These aluminium shields are shown in Fig. 8.

Figure 8. Picture of aluminium shields for wire protection.

6. FIRST SOLAR ORBITER QUALIFICATION TEST RESULTS

The SolO representative 16 40x40 mm² single qualification cells have been tested further in the SolO AM0 Qualification Test for 1000 hours under the required higher intensity of up to 12.8 SCs leading to an overall test duration of 5000 hours for these samples. This additional test with 1000 h was split in three test runs. To optimize the available test setup, the qualification samples were re-arranged to center them in the region with the highest illumination homogeneity. These new sample locations are depicted in green in Fig. 9 (a). The other positions were equipped with dummy cells to have the same thermo-optical properties like in the BepiColombo test. Therefore, it was not necessary to change the thermal washer concept compared to the previous test [4]. A picture from the test samples on the MGSE is shown in Fig. 9 (b).
The main focus of the characterization was again the electrical performance measurement of the individual SCAs after each test cycle. Prior to the SolO test start a new BOT measurement was performed, with slightly higher values compared to EOT from the BC AM0 test. The difference is within the measurement accuracy, but could be an indication for a small annealing during the storage period between the two qualification tests. The average degradation of the 16 present SCAs in the SolO

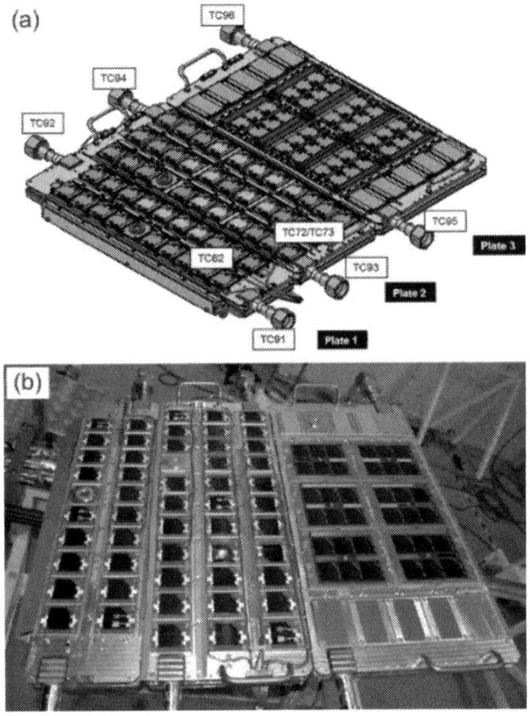

Figure 9. (a) Overview of sample locations for SolO Qualification Test (shown in green). (b) Picture of the samples on the MGSE.

Qualification Test compared to BOT SolO (1000 h) and BOT BC (5000 h) are listed in Tab. 5.

Δ to	ΔP_{max} (%)	ΔV_{pmax} (%)	ΔI_{pmax} (%)	ΔI_{sc} (%)	ΔV_{oc} (%)	ΔFF (%)
BOT SolO	-0.8	-0.5	-0.3	-0.3	-0.4	-0.1
BOT BC	-3.1	-0.5	-2.6	-2.5	-0.5	-0.1

Table 5. Average degradation of SolO qualification SCAs for the additional 1000 h test duration and compared to BOT BC with 5000 h in summary.

For this additional long duration illumination test with 1000 h only a small additional degradation in the current parameters was observed, which is a consistent continuation of the BC AM0 test. The variations in the voltage parameters are mainly caused by measurement accuracy. This can be seen in the V_{pmax} parameter, which was in average identical for BOT BC and BOT SolO. The updated graph from Fig. 2 with the results from the SolO qualification test is shown in Fig. 10.
The offset between the theoretical UV degradation and the real measurements is relatively stable after short test duration. This means the root cause for the higher degradation, which is most likely contamination, occurs mainly at the beginning and saturates later. This is an

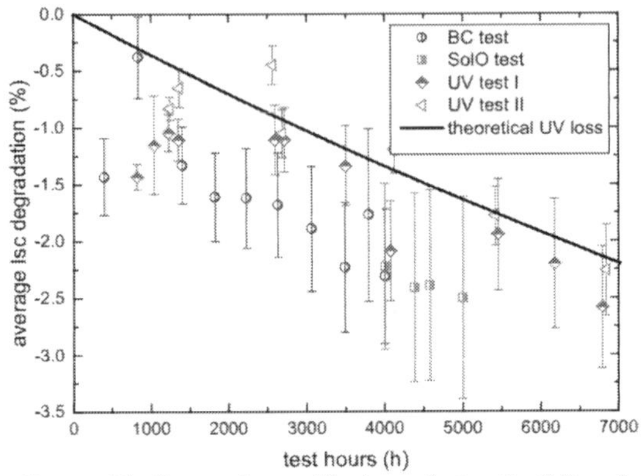

Figure 10. Comparison of I_{sc} degradation for BC and SolO AM0 and UV tests.

expected behaviour for the outgassing of adhesive, for example.

7. CONCLUSIONS

The BepiColombo and Solar Orbiter AM0 Long Duration Illumination Qualification Tests have been successfully completed. The validity of the solar cell design for these high temperature, high intensity missions needs could be confirmed. The results with respect to solar cell behaviour and further characteristics have been presented. In addition, some insights have been given in challenges which occurred during test execution.

8. ACKNOWLEDGMENTS

The work on the BepiColombo and Solar Orbiter solar array including this tests was carried out under a programme of and funded by the European Space Agency. The view expressed herein can in no way be taken to reflect the official opinion of the European Space Agency.

9. REFERENCES

1. Zimmermann, C.G., Nömayr, C., Köstler, W., Caon, A., Fernandez E., Baur, C. & Fiebrich H.K. (2011). Photovoltaic Technology Development for the BepiColombo Mission. In *Proc. 9th European Space Power Conference ESPC 2011*, 2103064.

2. Zimmermann, C.G., Nömayr, C., Kolb, M. & Caon, A. (2011). A solar cell design for the BepiColombo high intensity – high temperature mission. In *Proc. IEEE Photovoltaic Specialists Conference PVSC 2011*, pp. 003713-003718.

3. Andreev, T., Zimmermann, C., Löhberg, A., de Jong, S., Schuhmacher, U., Schneider, J., Köstler, W., Parketta, P., Zäh, E., Holzbauer, R., Fink, M., Hall, J., Fernandez Lisbona, E., Lundmark, K.,

DOI: 10.1051/e3sconf/20171602007

Baur, C., Taylor, S., Panin, F., Caon, A. (2015). Qualifying Solar Array Components for High Intensity, High Temperature Environments – The Test Approach for the Bepi Colombo Solar Arrays, In *Proc. 10th European Space Power Conference ESPC 2014*, ESA SP-719 (CD-ROM).

4. Oberhüttinger, C., Quabis, D. & Zimmermann, C.G. (2014). Simulating large area, high intensity AM0 illumination on earth - representative testing at elevated temperatures for the BepiColombo and SolO missions. In *Proc. 10th European Space Power Conference ESPC 2014*, ESA SP-719 (CD-ROM).

5. Zimmermann, C.G. (2008). Time dependent degradation of photovoltaic modules by ultraviolet light. *Appl. Phys. Lett.* 92, 241110.

METAL MATRIX COMPOSITE SOLAR CELL METALLIZATION

David M. Wilt[(1)], Geoffrey Bradshaw[(1)], Lt. Nathan Gap[(1)], Omar K. Abudayyeh[(2)], Cayla Nelson[(2)], Sang Han[(2)], Nathanael Cox[(3)], Aaron Rape[(3)], Brian Landi[(3)] and Steve Whipple[(4)]

[(1)]Space Vehicles Directorate, Air Force Research Laboratory
3550 Aberdeen Ave SE, Kirtland Air Force Base, NM 87111, USA, AFRL.VSSVOrgMailbox@us.af.mil
[(2)]The University of New Mexico, Albuquerque, NM USA
[(3)]Rochester Institute of Technology, Rochester, NY, USA
[(4)]SolAero Technologies Corp, Albuquerque, NM, USA

ABSTRACT

Advanced solar cells are moving to ever thinner formats in order to save mass and in some cases improve performance. As cells are thinned, the possibility that they may fracture or cleave due to mechanical stresses is increased. Fractures of the cell can degrade the overall device performance if the fracture propagates through the contact metallization, which frequently occurs. To address this problem, a novel semiconductor metallization system based on multi-walled carbon nanotube (CNT) reinforcement, termed metal matrix composite (MMC) metallization is under investigation. Electro-mechanical characterization of MMC films demonstrate their ability to provide electrical conductivity over >40 micron wide cracks in the underlying semiconductor, with the carbon nanotubes bridging the gap. In addition, these materials show a "self-healing" behaviour, electrically reconnecting at ~30 microns when strained past failure. Triple junction (TJ) space cells with MMC metallization demonstrated no loss in J_{sc} after intentional fracture, whereas TJ cells with conventional metallization suffer up to 50% J_{sc} loss.

1. INTRODUCTION

Advanced crystalline solar cells used on-earth and in-space are rapidly moving towards thinner devices. For terrestrial applications, the emphasis on thinner devices is driven primarily to reduce material costs, although there can be device performance benefits. In space, thinner triple junction cells are becoming more common due to their mass savings and inverted metamorphic multijunction (IMM) solar cells, consisting of 10 to 15 microns of epitaxy, are nearing market entry. Despite the advantages of thinner crystalline solar cells, they could be more likely to fracture and cleave in response to mechanical stress. Frequently, fractures that remain within the semiconductor layers do not severely impact the overall device performance. Unfortunately, high stresses at the crack tips generally lead to fracturing of the metallization and subsequent loss of electrical performance. Recently, Kajari-Schröderet al. [1] have characterized microcracks in crystalline-silicon-based photovoltaic cells, using electroluminescence after artificial aging and snow damage (Fig. 1). These microcracks can electrically disconnect areas of the cells and lead to substantial power loss [2]. To mitigate the power loss and increase the lifetime of solar cells, we have developed and investigated silver (Ag) metal films imbedded with single and multi-walled carbon nanotubes (CNTs), also known as metal matrix composites (MMCs), for the reinforced mechanical strength against stress-induced cracks [3, 4, 5]. Numerous techniques exist today to deposit bulk MMCs, including powder metallurgy, melting and solidification, and electrochemical routes [6].

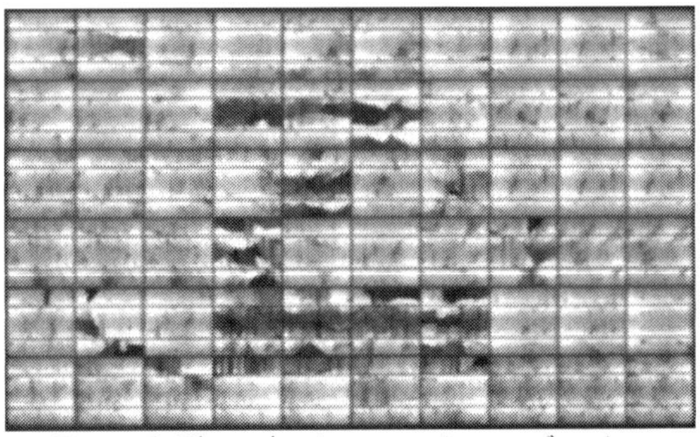

Figure 1. Electroluminescence image of a micro-cracked PV module. Dark regions are electrically inactive areas [1].

In this investigation, MMC films have been deposited by either metal electroplating combined with CNT spray coating or by metal evaporation combined with CNT film transfer, although some preliminary investigation of screen printing MMC films has also been conducted [7]. Both single wall and multi-wall CNT materials have been examined and the MMC structure has primarily consisted of a layer-by-layer design (~2 µm Ag base layer / CNT film / ~2 µm Ag overcoat). A range of CNT film loadings have been examined to assess the impact on MMC gapping capability (distance that the MMC films are able to electrically bridge).

DOI: 10.1051/e3sconf/20171603001

MMC electro-mechanical crack bridging capability was examined using a specifically designed test apparatus, affectionately called the RACK (Resistance Across Cleaves and cracKs). For this testing, MMC grid lines were formed on a semi-insulating indium phosphide (InP) substrate. InP was chosen as it is electrically insulating and more importantly, particularly easy to cleave. The sample is then bonded across two circuit boards and then mounted to a micro-positioning stage. The grid lines are then connected to the data acquisition system allowing ~30mA of current to be passed through each gridline and their resistance to be measured. A single cleave is then formed in the sample, perpendicular to the grid lines. The conductivity of the grid lines is then measured as the circuit boards are pulled apart in micron increments (and brought back together). The fracture spacing (gap width) is measured via optical microscope. Figure 2 shows a drawing of the overall RACK configuration and a photo of a sample under test.

Figure 2. (l) drawing of RACK configuration (note: two voltage measurements were actually made for each of the four (4) gridlines), (r) photo of sample during RACK testing.

2. MMC Electro-Mechanical Behavior

Figure 3 shows the RACK data from pure Ag metallization and a single-walled CNT MMC (SW-MMC) sample [5]. All of the Ag samples and many of the SW-MMC samples failed electrically at the initial crack. This is partly due to the 2 µm to > 6µm initial crack that formed when the sample was cleaved. The shaded region indicates the initial crack length. The Ag-only electrode fails electrically upon cracking (crack width ~3.7 µm), while the SW-MMC maintains electrical connection after initial fracture (crack width ~2.8 µm), and fails electrically after an additional extension of 2.7 µm, with a total gap width of ~5.5 µm. This is representative of the successful SW-MMC samples, which generally failed below 6 µm.

Figure 4 shows the maximum and average gaps before the loss of electrical connection for multi-walled CNT MMC samples as a function of CNT loading [3]. The pull tests reveal that the electrical connection is maintained across larger gaps with higher CNT surface coverage, reaching a maximum of 42-µm-wide gap. As the CNT surface coverage increases from 35% to 96%, we observe a monotonic increase in the average gap. Electroplated Ag lines without CNTs are also analyzed as controls, using RACK test. Unlike MMC films, the Ag gridlines did not withstand the initial crack generation (~4-µm-wide), and the electrical connection is immediately and irrecoverably lost.

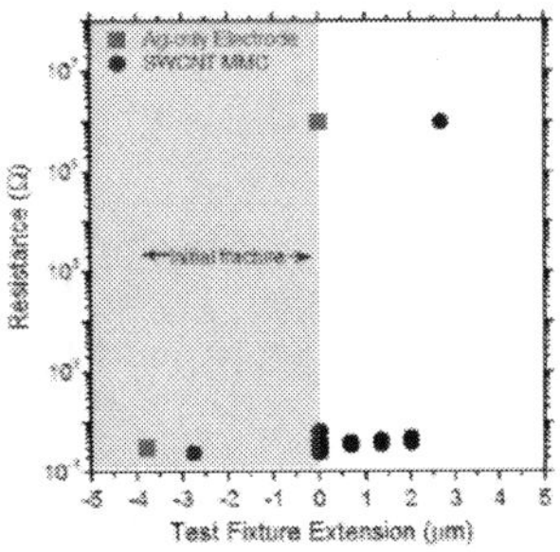

Figure 3 - Electrical resistance versus gap for a Ag-only electrode compared to a single walled MMC electrode. The shaded region indicates the initial crack length.

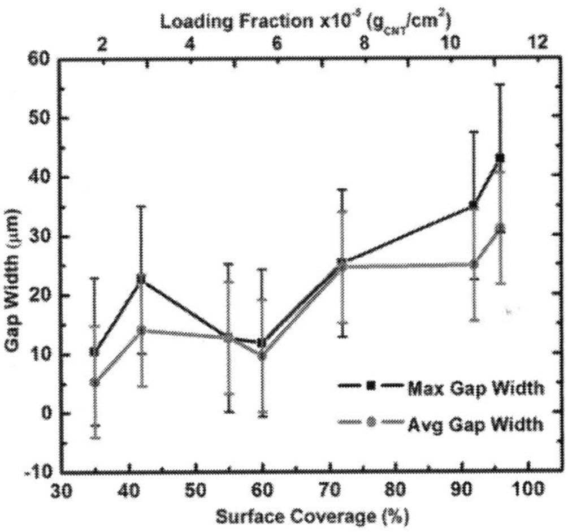

Figure 4 - Maximum and average gap widths achieved before lost connection as a function of CNT surface coverage and loading fraction.

SEM micrograph images of fractured MMC samples show how the multi-walled MMC metallization is able to maintain electrical connection across significant gaps (>40 microns). Figure 5 shows the multi-walled CNT bridging a 9 micron wide fracture. The CNTs are anchored in the Ag matrix, indicating a good adhesion between the functionalized CNTs and the Ag matrix. Once fractured, the total current (30 mA) is conducted through only a limited number of CNTs at the fracture point, as shown in Fig. 5. We have not determined the maximum current density that MMCs can tolerate after fracture, but have pushed them to 150 mA (2500 A/cm^2 equivalent) without conductivity loss.

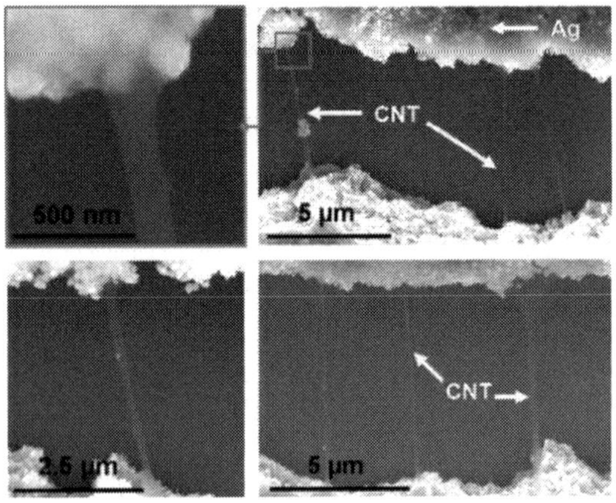

Figure 5 - SEM images of CNTs anchored in Ag layers clearly showing CNTs bridging cracks up to 9-µm-wide.

As a test, MMC samples were pulled apart until they lost electrical contact and then were slowly brought back together while checking for electrical connection. Surprisingly, the MMC samples re-established electrical connection well before the sample edges were brought back into contact. Figure 6 shows the log of resistance as a function of physical gap for samples with different MWCNT loading as the gap is closed. The sample with the highest CNT loading showed electrical reconnection at >40 microns with the lower loaded samples reconnecting at <10 microns. It is believed that strong electric fields at the ends of the dangling CNT fibers is likely responsible for assisting them in reconnecting before the surfaces are brought in contact.

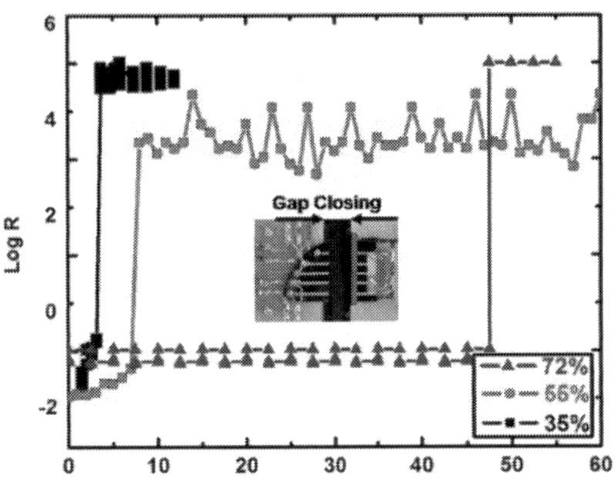

Figure 6 - Log plot of resistance vs. gap across fractured MMC gridlines with three different MWCNT surface coverages.

In order to test to see if this "self-healing" behaviour was repeatable, one sample was tested numerous times (pulled to electrical isolation, brought back together and then pulled apart again, ~10x). The data (figure 7) shows that for the second and third pulls, the maximum gap distance was reduced from ~40 microns to ~25 microns, but was stable at that point for all of the successive pulls.

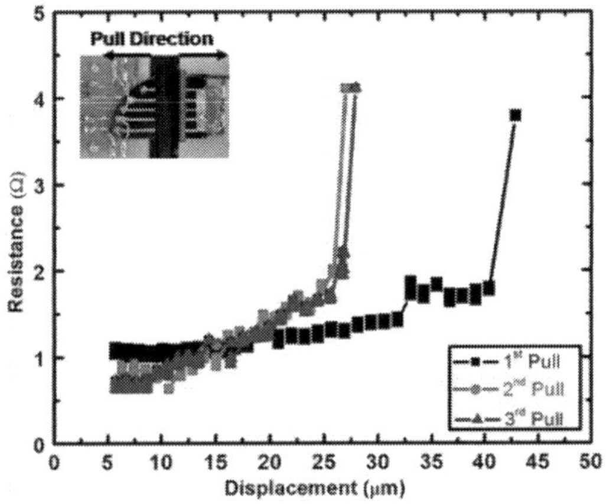

Figure 7 - Strain failure test of a MMC gridline (96% CNT surface coverage using drop cast method). The maximum gap,at which the electrical connection is lost, is reduced by ~34% after the 2nd pull test but remained constant after subsequent pull tests

3. MMC Cell Integration

Representative InGaP/GaAs/Ge space triple junction cells were fabricated, some with conventional evaporated metallization (controls) and some with only thin (~1000A) evaporated seed layers for the grid metallization. The cells with the seed layers had their metallization built up as either electroplated Ag or MMC consisting of electroplated Ag and spray deposited multi-wall CNT. I-V measurements of the devices demonstrated relatively comparable initial electrical performance (~27% AM0 efficiency) for all of the devices.

The 2x2 cm cells were then intentionally cracked by wrapping them around a 6 cm radius cylinder (back of the cell to the cylinder). Figure 8 shows the light I-V (LIV) measurements as well as the electroluminescence images of the control cells pre and post cracking. The top two cells have evaporated Ag metallization whereas the bottom two cells have electroplated Ag metallization. The EL and I-V data show that fracturing can sever the metallization, leading to a reduction in cell collecting area and frequently a significant (~50%) reduction in J_{sc}. This data also suggests that there isn't a

 DOI: 10.1051/e3sconf/20171603001

significant difference in the response of the cells to cracking based on the metal deposition process (evaporation vs. plating).

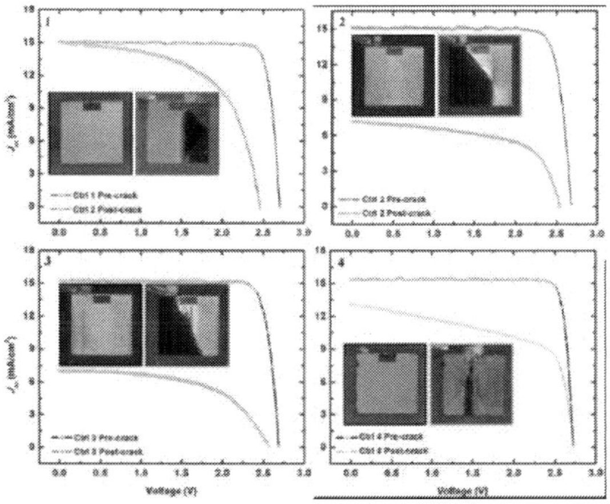

Figure 8 – LIV characteristics and electroluminescence response of TJ control samples pre- and post-cracking. Graphs (1) & (2) represent control samples with electron beam evaporated Ag gridlines. Graphs (3) & (4) represent control samples with electrochemically deposited Ag gridlines.

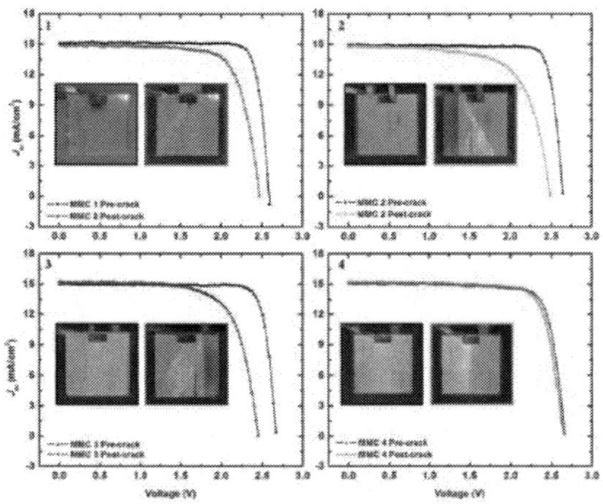

Figure 9 – LIV characteristics and electroluminescence response of TJ cell with MMC gridlines.

Figure 9 shows the LIV and electroluminescence data for the cells with MMC metallization. The striking feature of this data set is the lack of J_{sc} degradation due to cracking when MMC metallization is used. The cells do show variable V_{oc} and FF degradation, but the current is essentially unchanged. This is particularly important as cells are connected in series, thus current loss in a single cell can limit the overall string. This data represents the first integration of MMC metallization with multijunction solar cells and further analysis and optimization is currently underway.

4. CONCLUSIONS

In this study, we have developed and characterized Ag-CNT metal matrix composite (MMC) films that maintain electrical conductivity upon mechanical fracture of the substrate. Our composite lines are capable of bridging cracks in the underlying semiconductor substrates >40 µm. In addition to maintaining the electrical conductivity over the stress-induced cracks, the composite films can re-establish the electrical connection when the cracks close up. This MMC technology has been integrated with Ge based triple junction cells and has demonstrated a clear benefit in terms of J_{sc} maintained following cell cracking. Additional work is underway to optimize and fully characterize this technology.

5. REFERENCES

1. S. Kajari-Schröder, Kunze I, U. Eitner, and M. Köntges, "Spatial and orientational distribution of cracks in crystalline photovoltaic modules generated by mechanical load tests," Solar Energy Mater Solar Cells, vol. 95, pp. 3054-3059, 2011.
2. M. Köntges, I. Kunze, S. Kajari-Schröder, X. Breitenmoser, and B. Bjrneklett, "The risk of power loss in crystalline silicon based photovoltaic modules due to microcracks," Solar Energy Mater Solar Cells, vol.95, pp. 1131-1137, 2011.
3. O. Abudayyeh, N. Gapp, C. Nelson, D. Wilt, and S. Han, "Silver-Carbon-Nanotube Metal Matrix Composites for Metal Contacts on Space Photovoltaic Cells," IEEE J. of Photovoltaics, 2015
4. O. Abudayyeh, D. Wilt, G. Bradshaw and S. Han, "Crack-Tolerant Silver-Carbon-Nanotubes Metal Matrix Composites as Photovoltaic Gridlines," IEEE J. of Photovoltaics, 2016 (in review)
5. N. Cox, M. Pham, N. Gapp, A. Rape, J. Rossi, D. Wilt, B. Landi, "Electrical Characterization of Carbon Nanotube Metal Matrix Composite Solar Cell Electrodes Under Mechanical Stress," Proc. 43rd IEEE Photovoltaic Specialist Conference, 2016.
6. S. R. Bakshi, D. Lahiri, and A. Agarwal, "Carbon nanotube reinforced metal matrix composites - a review," International Materials Reviews, vol. 55, pp. 41-64, 2010.
7. A. Manoussakis, O. Abudayyeh, N. Gapp and D. Wilt, "Screen Printed Metal Matrix Composite Contacts for Crack Tolerant Solar Cells," Proc. 43rd IEEE Photovoltaic Specialist Conference, 2016.

NEXT GENERATION RADIATION HARD IMM SPACE SOLAR CELLS

P. Sharps, D. Aiken, B. Cho, S. Cruz, D. Derkacs, N. Fatemi, A. Haas, C. Kerestes, N. Miller, B. Pantha, P. Patel, M. Stan, A. Stavrides, J. Steinfeldt, C. Struempel, and S. Whipple

SolAero Technologies Corporation
10420 Research Rd. SE
Albuquerque, NM 87123 USA
Paul_Sharps@solaerotech.com

ABSTRACT

High efficiency space solar cells based on the $GaInP_2$/InGaAs/Ge triple junction lattice matched device with 1 sun, AM0 efficiencies of approximately 30% are commercially available from several companies. Multiple approaches have been suggested for significantly surpassing the 30% efficiency level, but achieving this in a commercially viable cell has not been easy. We review those approaches, pointing out the advantages and challenges of each. SolAero has been focused on developing the inverted metamorphic multi-junction (IMM) solar cell, and has recently demonstrated a radiation hard version. We present the most recent performance data, including the response to particle radiation. The IMM cell can be used in a number of rigid or flexible configurations, and considerable effort has been focused on cell packaging and panel integration. The IMM device is ready to start qualification to the ECSS and AIAA standards.

1. INTRODUCTION

The $GaInP_2$/InGaAs/Ge triple junction lattice matched solar cell is the primary source of space satellite power. Several vendors provide commercially available cells, with performance of approximately 30% under 1 sun, AM0 illumination [1]. While silicon solar cells were initially used for space power, they were replaced by III-V based devices, first by the GaAs single junction cell [2] and then eventually by the previously mentioned triple junction cell. While the III-V devices were more expensive ($/Watt) than silicon on a bare cell level, at the array and panel level the III-V devices were a lower cost option [3, 4]. This was due to the III-V device higher efficiency and superior radiation hardness compared to silicon, which led to smaller area arrays, with fewer cells.

The current situation is in some ways similar to the transition from silicon to III-Vs. Surpassing the 30% efficiency level for space cells requires a new cell architecture. Higher efficiency cells require either additional junctions or a better match of the junctions to the solar spectrum, or a combination of both. Added growth and processing increases the cost at the bare cell level. The question is how much additional bare cell cost is acceptable. If the higher cell performance is mission enabling, or if the $/W at the spacecraft level are equal or lower than current technologies, then a higher bare cell cost will be acceptable. The decision needs to be made at the mission level.

2. HIGHER EFFICIENCY CELL OPTIONS

Options for higher efficiency cells have been discussed previously [5], with pros and cons for each approach. Since that time progress has been made with several of the approaches, notably wafer bonding, dilute nitrides, upright metamorphic, and IMM. While the wafer bonding approach has achieved very high terrestrial efficiencies [6], it will not be considered due to the very high expense of growths on two substrates (one being very expensive InP) and subsequent processing, which includes chemi-mechanical polishing as well as chemical treatments to ensure good high quality bonding with no voids. It is simply too expensive for 1 sun space applications. Updates on the other approaches will be provided in the next few sections.

2.1 Dilute Nitrides

The dilute nitride approach involves inserting a lattice matched GaInAsN 1.0 eV junction between the second junction and the bottom Ge junction (Figs. 1). The approach was first presented in 1998 [7] and created considerable interest as a way to create a four junction, lattice matched higher efficiency device.

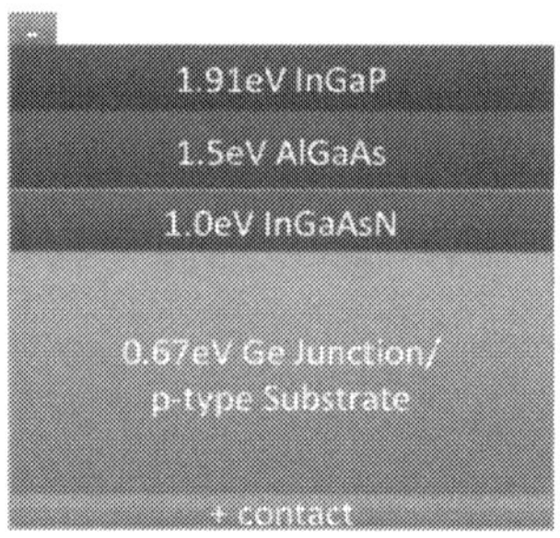

Figure 1. Cross section schematic of a four junction device incorporating a GaInAsN 1 eV junction.

 DOI: 10.1051/e3sconf/20171603002

Unfortunately, poor material quality due to short minority carrier diffusion lengths and high carbon background for MOCVD grown GaInAsN [8] led to poorly performing devices, and interest in the dilute nitride approach waned. Several years after this MBE grown GaInAsSbN demonstrated lower background carbon doping which allowed for p-i-n junctions with a high internal field to be used to counter the still poor minority carrier diffusion lengths and low voltages. Triple junction $GaInP_2$/InGaAs/GaInAsSbN devices were demonstrated. Efforts are currently being made to commercialize both triple and quadruple junction devices [9]. While the MBE grown dilute nitride junction is improved over a MOCVD grown junction, there are still issues with poor minority carrier diffusion lengths. The dilute nitride junctions have yet to achieve the voltages expected for the corresponding band gaps. Another challenge is that while MBE grown dilute nitride material is better quality than MOCVD grown material, MOCVD is the better option for growing high quality phosphide materials such as $GaInP_2$. Growing a three or four junction device with optimized performance would require two separate growths, since neither growth option is best for the complete structure. Another issue with MOCVD grown dilute nitrides is the high quantity of nitrogen precursor required and the associated high precursor costs [8]. The growth/performance/cost issue needs to be resolved.

2.2 Upright Metamorphic

The upright metamorphic approach (Fig. 2) has also been studied since 1997 [10]. The challenge has been growth of high quality, lattice mismatched material with good minority carrier properties. All of the epitaxial junctions, which provide a high percentage of the power conversion, have threading dislocations and recombination centers, which increases the junction dark current. For 1 sun applications the metamorphic devices have never surpassed lattice matched devices. In addition, the metamorphic grades add growth time, and the mechanical strain due to the lattice mismatch requires a thick germanium substrate, which adds weight as well as further cost. Weight reduction requires thinning the substrate, and ensuring that the final solar cell assembly is attached to a rigid coverglass or carrier.

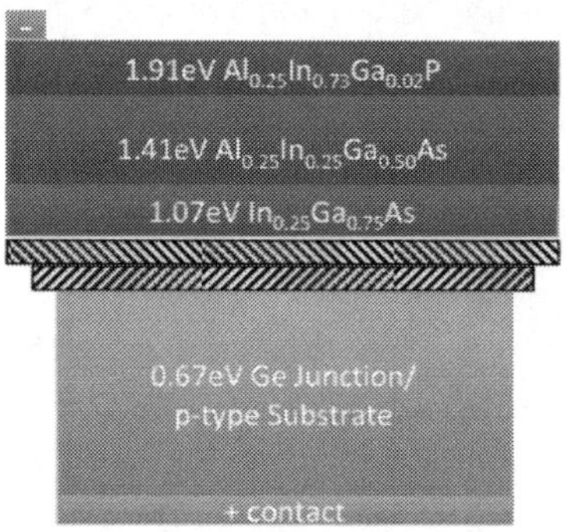

Figure 2. Cross sectional schematic of an upright metamorphic four junction device. All epitaxial junctions are grown lattice mismatched with respect to the Ge junction.

2.3 Inverted Metamorphic

The IMM space cell (Fig. 3) has been studied since about 2007 [11, 12], and devices with up to 6 junctions have been demonstrated [13]. The device is grown inverted, with lattice matched high band gap junctions grown first, followed by two lattice mismatched junctions. Subsequently the structure is inverted, mounted on a carrier, and the growth substrate removed. While readily surpassing the 30% efficiency level, the challenge has been the added costs associated with growth and processing. In addition, there have been some questions about the radiation hardness of the lower band gap junctions. These two issues have been overcome, and in the following sections we address first the performance concern and then secondly the cost and manufacturability concern.

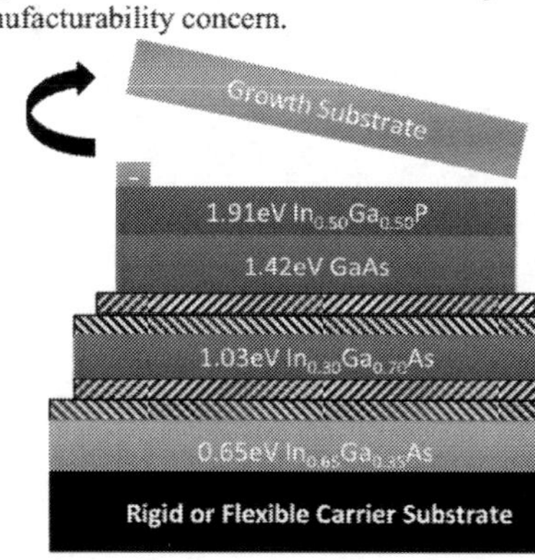

Figure 3. Cross sectional schematic of the IMM four junction device.

3. IMM PERFORMANCE

Recent development efforts at SolAero on the IMM cell have resulted in very high radiation hardness. While some beginning of life (BOL) performance has been sacrificed, the end of life (EOL) performance has been much improved. Fig. 4 shows a comparison between SolAero's $GaInP_2$/InGaAs/Ge ZTJ efficiency and the IMM efficiency as a function of exposure to 1 MeV electrons. The IMM shows a marked improvement over the ZTJ cell over a wide fluence range. In addition, a curve for an "annealed" or "healed" IMM cell is also shown. Radiation damage in a solar cell can be

 DOI: 10.1051/e3sconf/20171603002

annealed due to a number of factors, including thermal load, light soaking, and reverse bias exposure [14]. For thermal healing, the amount of efficiency restoration is dependent on the time at a given temperature. In Fig. 4 the improvement was based on the ECSS post radiation exposure testing procedure. Fig. 5 shows the relative improvement in IMM efficiency for both the AIAA S-111 and ECSS post radiation testing procedures, after exposure to 1 MeV electrons to a fluence of 1 x 10^{15} e/cm^2. With the AIAA standard there is an absolute 0.5% efficiency improvement, and with the ECSS standard the improvement is 0.8% absolute. With radiation damage annealing in the IMM accounted for, the efficiency is 28.3% for 1 sun (135.3 mW/cm^2), 28°C, AM0 conditions after exposure to 1 MeV electrons to a fluence of 1 x 10^{15} e/cm^2. This is a 12% relative improvement over the ZTJ cell, and 87% EOL/BOL for these conditions.

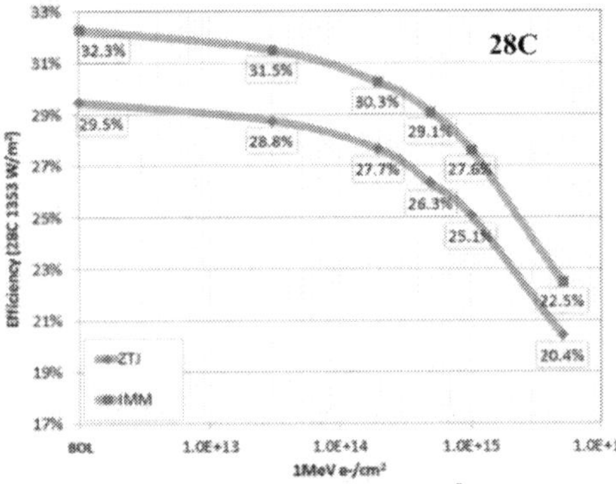

Figure 4. Efficiency (135.3 mW/cm^2, 28°C) as a function of 1 MeV fluence for the ZTJ cell and the IMM cell, post radiation and healed post radiation.

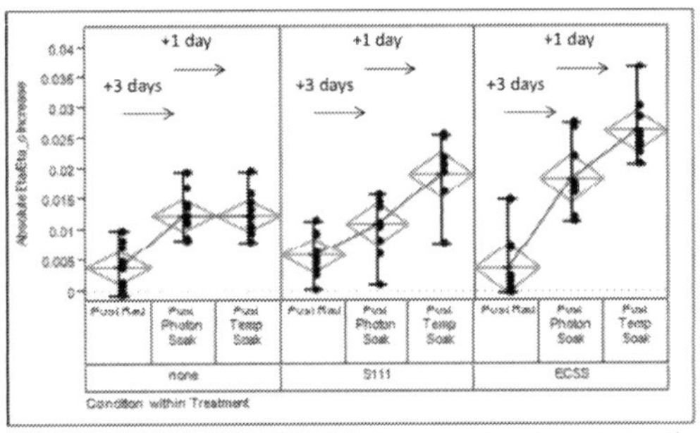

Figure 5. Improvement in efficiency for the IMM cell after exposure to 1 MeV electrons to a fluence of 1 x 10^{15} e/cm^2, for the AIAA S111 and ECSS standards post radiation test procedure.

4. COST AND MANUFACTURABILITY

A major step forward in reducing IMM costs has been the reduction in MOCVD growth time and the incorporation of a number of automated steps in the IMM process flow. The reduction in growth time was achieved through a simplification of the growth structure, particularly the metamorphic buffer layers. Automation, an example of which is shown in the wet bench in Fig. 6, increases throughput and yield, and reduces labor costs.

Figure 6. Automated etching bench that includes robotics for cassette handling of wafers.

A step by step analysis of the IMM growth and process steps have indicated the major cost drivers. We have worked on a number of our process steps as well as with a number of vendors to determine lower cost process and materials alternatives that will still provide the same final solar cell performance. In one case the process/material costs were reduced by 70%, and in another by 55%, all without compromising performance. We are working through a process step cost pareto to ensure that we are working on areas that will provide the best return for our efforts.

Cell size has a large impact on panel integration costs through reduction in piece part count and touch labor on a per Watt basis [15]. Large area IMM devices on 4" wafers of ~60 cm^2 have been demonstrated. In addition, IMM manufacturing on 6" wafers to achieve even larger cell form factors has begun at SolAero. A comparison of ~27.5 cm^2, ~60 cm^2 (4" substrate), and 73.5 cm^2 (6" substrate) form factors is shown in Fig. 7.

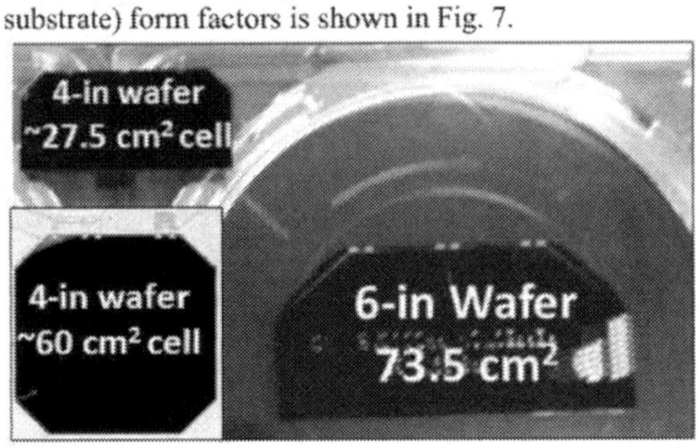

Figure 7. A comparison of cell form factors for 6 inch and 4 inch wafers.

DOI: 10.1051/e3sconf/20171603002

Based on the IMM cell performance and manufacturing status it is ready to start qualification to the AIAA S-111 and ECSS-E-ST-20-08 cell standards.

5. INTEGRATION

Integration of the IMM cell offers more possibilities than the standard upright triple junction cell. The carrier substrate in the schematic in Fig. 3 can be rigid or flexible. Preliminary work has been done to verify the ability of the IMM cell to survive space conditions for some of these integrated configurations. Fig. 8 shows an IMM cell string on a rigid panel before and after 12,000 LEO thermal cycles. The cells are forward biased into electroluminescence, and there are no indications of damage to the cells. The coupon pre/post measurements showed no change in performance. An additional 6 coupons have been tested to a similar level with no performance degradation seen. While a full qualification is needed, both the radiation performance and the survivability to thermal cycling indicate that the IMM cell can be used as a "drop in replacement" for the traditional multi-junction cell.

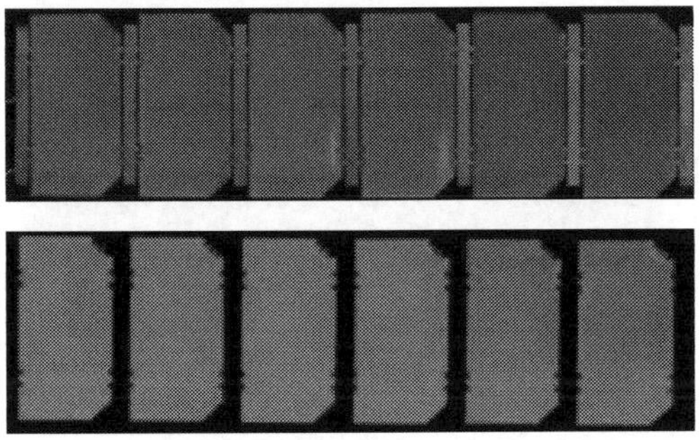

Figure 8. Pre (top string)/post (bottom string) luminescent pictures of an IMM string on a rigid substrate which underwent 12,000 LEO thermal cycles. Performance testing indicated no change in efficiency.

In addition to a rigid panel configuration, the IMM cell can be used on a flexible panel. Fig. 9 shows two 8 cell IMM strings on kapton, which are mounted on flexible mesh scrim. This configuration can then be rolled or folded into a more compact array.

Figure 9. Two 8 IMM cell strings connected and mounted on mesh scrim.

Initial confidence testing has been done on the flexible mounted strings, and consisted of 20x roll and unroll at room temperature, 5 day soaks at ± 60 °C, multi-axis vibration testing intended to mimic stowed flight conditions, thermal vacuum exposure (80 °C for a minimum of 24 hours in a vacuum of <10-5 torr, 8 thermal vacuum cycles from -137 to 120 °C), and finally hot GEO thermal cycling (2000 cycles from -175 to +140 °C). The mounted strings exhibited essentially no loss of output power as measured by large area pulsed solar simulator (LAPSS) measurements during and after the confidence test suite (Fig. 10), indicating excellent survivability of these string assemblies.

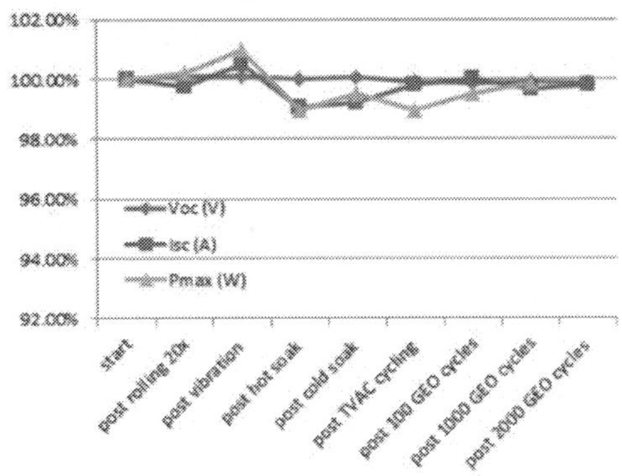

Figure 10. % remaining factors through confidence testing, as measured by LAPSS

6. CONCLUSIONS

The IMM cell developed at SolAero demonstrates superior performance at BOL and EOL compared to current state of the art triple junction technologies as well as other proposed next generation technologies. The cell is capable of being made in a high volume. There is a roadmap to cost parity ($/W) with heritage triple junction cells at the panel level. The cell is ready to start qualification to the AIAA S-111 and the ECSS-E-ST-20-08 cell standards.

7. REFERENCES

1. http://solaerotech.com/products/space-solar-cells-coverglass-interconnected-cells-cic/; http://www.spectrolab.com/space.htm; http://www.azurspace.com/index.php/en/products/products-space

2. G. C. Datum and S. A. Billets, "Gallium Arsenide Solar Arrays – A Mature Technology", in *22nd IEEE Photovoltaics Specialist Conference*, 1991, p. 1422

3. Navid S. Fatemi, Howard E. Pollard, Hong Q. Hou, and Paul R. Sharps, "Solar Array Trades Bewteen Very High-Efficiency Multi-Junction and Si Space Solar Cells", *Proc. 28th PVSC*, Anchorage, AK, 2000, 1083-1086.

4. D. M. Murphy, M. I. Eskenazi, S. F. White, and B. R. Spence, "Thin-Film and Crystalline Solar Cell Array System performance Comparisons", *Proc. 29th IEEE PVSC*, New Orleans, LA, 2002, 782-787.

5. P. R. Sharps, A. Cornfeld, M. Stan, A. Korostyshevsky, V. Ley, B. Cho, T. Varghese, J. Diaz, and D. Aiken, "The Future of High Efficiency, Multi-Junction Solar Cells", *Proc. 33rd IEEE PVSC*, San Diego, CA, 2008.

6. F. Dimroth, M. Grave, P. Beutel, U. Fiedler, C. Karcher, T. Tibbits, E. Oliva, G. Siefer, M. Schachtner, A. Wekkeli, A. Bett, R. Krause, M. Piccin, N. Blanc, C. Drazek, E. Guiot, B. Ghyselen, T. Salvetat, A. Tauzin, T. Signamarcheix, A. Dobrich, T. Hannappel, and K. Schwarzburg, "Wafer Bonded Four-junction GaInP/GaAs//GaInAsP/GaInAs Concentrator Cells With 44.7% Efficiency", *Progress in Photovoltaics*, **22** (3) 2014, 277-282

7. H. Hou, K. Reinhardt, S. Kurtz, J. Gee, A. Allerman, B. Hammons, P. Chang, and E. Jones, "Novel InGaAsN pn Junction for High Efficiency Multi-Junction Solar Cells", *2nd World Conf. and Exhibition on Photovoltaic Solar Energy Conversion*, Vienna, Austria, 1998, 3600-3603.

8. N. Li, P. Sharps. M. Stan, F. Newman, J. Hills, and H. Hou, "Development of 1.25 eV InGaAsN for Triple Junction Solar Cells", *Proc. 28th IEEE PVSC,* Anchorage, AK, 2000.

9. http://www.sj-solar.com/about_us/space.php

10. R. Hoffman, N. Fatemi, M. Stan, P. Jenkins, V. Weizer, D. Scheiman, and D. Brinker, "High Efficiency InGaAs-on-GaAs Devices for Monolithic Multi-junction Solar Cell Applications", *2nd World Conf. and Exhibition on Photovoltaic Solar Energy Conversion*, Vienna, Austria, 1998, 3604-3608.

11. A. Cornfeld, M. Stan, T. Varghese, J. Diaz, V. Ley, B. Cho, A. Korostyshevsky, D. Aiken, and P. Sharps, "Development of a Large Area Inverted metamorphic Multi-junction (IMM) Highly Efficient Solar Cell", *Proc. 33rd IEEE PVSC*, San Diego, CA, 2008.

12. A. Cornfeld, D. Aiken, B. Cho, V. Ley, P. Sharps, M. Stan, and T. Varghese, "Development of a Four Sub-cell Inverted Metamorphic Multi-junction (IMM) Highly Efficient Solar Cell", *Proc. 35th IEEE PVSC*, Honolulu, HI, 2010.

13. P. Patel, D. Aiken, A. Boca, B. Cho, D. Chumney, M. Clevenger, A. Cornfeld, N. Fatemi, Y. Lin, J. McCarty, F. Newman, P. Sharps, J. Spann, M. Stan, J. Steinfeldt, C. Strautin, and T. Varghese, "Experimental Results on Performance Improvement and Radiation Hardness of Inverted Metamorphic Multi-junction Solar Cells", *Proc. 37th IEEE PVSC*, Seattle, WA, 2011.

14. SolAero Technologies Corp report EWRP0458G, "ZTJ AIAA Characterization and Qualification Report".

15. B. Cho, M. Winter, I. Aeby, P. Patel, N. Fatemi, and P. Sharps, " The One-per-wafer ZTJ Solar Cell From Emcore- Confidence Testing and Volume Manufacturing Performance," in *38th IEEE Photovoltaic Specialist Conference*, 2012, p. 25.

31% EUROPEAN INGAP/GAAS/INGANAS SOLAR CELLS FOR SPACE APPLICATION

Roberta Campesato[1] Antti Tukiainen [2], Arto Aho[2], Gabriele Gori [1], Riku Isoaho[2], Erminio Greco [1], Mircea Guina [2]

(1) CESI S.p.A., Via Rubattino 54, 20134 Milan, Italy: mail: campesato@cesi.it

(2) Tampere University of Technology, Optoelectronics Research Centre, PO Box 692, FI-33101 Tampere, Finland

ABSTRACT

We report a triple junction InGaP/GaAs/InGaNAs solar cell with efficiency of ~31% at AM0, 25 °C fabricated using a combined molecular beam epitaxy (MBE) and metal-organic chemical vapour deposition (MOCVD) processes. The prototype cells comprise of InGaNAs (Indium Gallium Nitride Arsenide) bottom junction grown on a GaAs (Gallium Arsenide) substrate by MBE and middle and top junctions deposited by MOCVD. Repeatable cell characteristics and uniform efficiency pattern over 4-inch wafers were obtained. Combining the advantages offered by MBE and MOCVD opens a new perspective for fabrication of high-efficiency space tandem solar cells with three or more junctions. Results of radiation resistance of the sub-cells are also presented and critically evaluated to achieve high efficiency in EOL conditions.

1. INTRODUCTION

Space applications require more and more efficient solar cells. From the standard InGaP/InGaAs/Ge triple junction solar cells with AM0 efficiency >29%, research is addressing the development of more than three junction devices and the substitution of germanium with 1 eV materials in order to overcome 31% efficient solar cells. The highest efficiency, i.e. 35.8% at AM0, has been demonstrated for a 5-junction cell fabricated by direct semiconductor bonding technique, i.e. not a monolithic process [1]. Using MBE, monolithic lattice-matched solar cells with efficiencies as high as 44% measured in concentrated sunlight have been demonstrated [2]. The increase in efficiency compared to standard approaches is made possible owing to MBE's capability to synthesize high-quality dilute nitride alloys, such as InGaNAsSb, which can be grown lattice matched onto GaAs or Ge substrates and exhibit a bandgap ranging from 0.8 eV to 1.43 eV. InGaNAsSb can also be grown by MOCVD but the efficiency increase has been hindered by incorporation of impurities and high background carrier concentrations that are specific to MOCVD processes [3]. In contrast, using MBE, low background carrier concentrations and high quantum efficiency and current production have been demonstrated with high open-circuit voltage (V_{oc}) by MBE-grown 1 eV materials [4]. Even though MBE has proven to be compatible with mass production of III-V semiconductor devices, its use for volume fabrication of solar cells has not yet been widely adopted, allegedly due to higher throughput offered by MOCVD.

In this paper, we report results obtained from III-V solar high-efficiency multi-junction solar cells grown by the MOCVD and MBE techniques. The technique presented in [5] combines two active solar cell junctions grown by different methods into single monolithic multi-junction structure. This opens a new perspective for fabrication of space tandem solar cells with three or more junctions.

2. THEORETICAL MODEL

The elevated complexity of a III-V multi-junction solar cell, in terms of number of layers, opto-electronic material properties and physical phenomena involved, makes a pure experimental optimization of the device impracticable. A reliable model is therefore necessary to boost the optimization process.

CESI developed a proprietary code to simulate the optical and the electrical behaviour of a multi-junction solar cell.

As concern the electrical model, each sub-cell is separately simulated taking into account the diffusion and the recombination dark currents relative to each junction according to a two diode model (Hovel).

Once the current voltage (*I–V*) curve of each sub-cell is determined, the *I–V* curve of the multi-junction device is numerically calculated as the series of the n sub-cells that constitute the device. The effect of series and shunt resistances are also included in the model.

All of the electrical parameters of each semiconductor material used in the cell structure and necessary for the calculations (e.g. carrier mobility, electron and hole lifetimes, surface recombination velocity at interfaces etc.) are loaded in the material database. The database is continuously updated both from literature and from experimental results performed by CESI.

Of particular importance for a reliable simulation result are the optical interference effects not negligible for a structure where the thickness of the layers is comparable with the optical wavelengths.

The optical model implemented in CESI code is based on the transfer matrix formalism where each layer of the device is described by a 2×2 matrix and the stack of layers can then be represented by a matrix which is the product of the individual layer matrices [6]. This method allows calculation of the reflectivity and the

DOI: 10.1051/e3sconf/20171603003

incident optical spectrum on each sub-cell taking into account the interference effects. As the electrical parameters of each material, the optical ones (refractive index and extinction coefficient) are entered in the material database.
The best methodology in order to optimize the device performances is an iterative approach: the solar cell structure is designed, simulated, manufactured and tested. The experimental results obtained from different characterization techniques (e.g. Spectral response, reflectivity, photoluminescence, current-voltage curve, Hall measurements etc.) provide the feedback needed to improve the input parameters to the code and then to restart the optimization.
The CESI model, updated for the optimization of a Triple junction InGaP/ GaAs/ InGaNAs solar cells, has been very useful in determination of the optimal band gaps and the thickness of the bases of the two top junctions (InGaP/GaAs).
The optimization of a solar cell manufactured using a combination of molecular beam epitaxy (MBE) and metal-organic chemical vapor deposition (MOCVD) processes is very challenging because it is also necessary to consider the effect of MOCVD overgrowth on the InGaNAs sub-cell. A typical effect due to MOCVD overgrowth is the shift of the external quantum efficiency (EQE) of dilute nitride junction towards shorter wavelengths. This effect is rather typical for dilute nitride heterostructures experiencing thermal annealing.
Therefore, the blue-shift of the band edge in InGaNAs must be included in the model in order to properly balance the photocurrent generated in each sub-cell and thus to optimize the device efficiency.
Using the code a theoretical efficiency of 32% is obtained for InGaP/InGaAs/InGaNAs solar cells at AM0, 25 C.

3. MANUFACTURING

The process for production of TJ solar cells using combined MBE-MOCVD technique consists in:

- Growth of InGaNAs junction on GaAs substrate by MBE;
- Transfer of wafers from MBE reactor to MOCVD reactor;
- Growth of GaAs and InGaP junction by MOCVD.

The InGaNAs junction was grown using a Veeco GEN20 MBE system. The MBE process was started by thermal desorption of the native oxides from an epiready 4" p-GaAs(100) substrate. A p-GaAs buffer layer followed by a lattice-matched p-InGaP back surface field layer were then grown. The p-i-n diode, consisting of a p-GaAs base layer, a nominally undoped InGaNAs layer and an n-GaAs emitter was grown next. The structure was finalized by growing window and contact layers that were used as transfer interface for the MOCVD process. Additional details of the epitaxy of the InGaNAs sub-cells can be found elsewhere [4],[6–7].
After the MBE growth the substrates were transferred to the MOCVD system using a packaging procedure properly defined to reduce the possibility for contamination of the overgrowth interface.
InGaP and GaAs junctions were grown using a VEECO E450 MOCVD reactor. The nucleation layer was optimized to maintain a high quality interface between the MBE grown structure and MOCVD grown structure. An (Al)GaAs-based p^{++}/n^{++} tunnel junction was deposited after the nucleation layer. Then, the GaAs middle junction, the top tunnel junction and the InGaP top junction were deposited. The GaAs middle junction and the InGaP top junction had to be made thinner than those of the standard CESI InGaP/GaInAs/Ge (CTJ30) solar cell structure [8]. To determine the proper thickness for the GaAs junction, single n-GaAs overlayers with different thicknesses and thus with different growth and annealing times were deposited by MOCVD. Based on the effect of the the thickness of the n-GaAs overlayer on the J_{sc} of the InGaNAs junction, the triple-junction solar cell structures were designed and fabricated using the combined MBE-MOCVD process.
The wafers were characterized using light microscopy (Leica DFC280 and Nikon L200) and PL (Accent RPM2000). Electrical characteristics of 2×2 cm^2 triple-junction solar cells were measured by *I-V* and EQE. EQEs were measured using a self-made system comprising of a 250 W QTH lamp, a DK480 monochromator, a chopper, an infinity-corrected 50× Olympus objective lens, a NIST-calibrated Ge detector as a reference, electrical and optical biasing circuits for multi-junction cells, and a lock-in amplifier. *I–V* measurements were taken using a dual source WACOM simulator that was calibrated at AM0 illumination using a Secondary Working Standard (SWS) set for the InGaP/GaAs/Ge triple-junction cell calibrated at Spasolab.

4. EXPERIMENTAL RESULTS

a) b)

Figure 1. Surface micrographs taken from epiwafers a) with single-junction dilute nitride solar cell before MOCVD-regrowth

b) with triple-junction solar cell after the MOCVD process

Morphology of MBE-grown dilute nitride junctions showed only few surface defects. Mainly small defects with circular shape and typical oval defects (Fig. 1a) were detected. The number of surface defects, mainly conical ones, was slightly increased after MOCVD growth but the surface morphology was of high quality (Fig. 1b).

The behaviour of the short-circuit current density (J_{sc}) of the InGaNAs junction as a function of the thickness of the MOCVD-grown n-GaAs over layer is shown in Fig. 2. When the n-GaAs layer thickness is larger than 2μm, the short circuit current of the dilute nitride junction falls below 18 mA/cm^2.

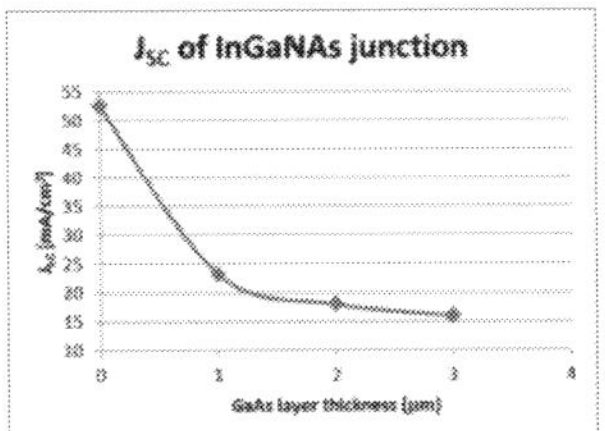

Figure 2. J_{sc} of the InGaNAs junction as a function of the thickness of the MOCVD-grown n-GaAs overlayer

The Jsc decrease after MOCVD growth is linked to the light absorption of the over layers and to the properties of dilute nitride junction that are altered by the annealing suffered during re-growth process as discussed in [5].

In fact, the energy gap of InGaNAs slightly increases as demonstrated by Photoluminescence and Spectral Response measurement reported in figure 3. As reported in [5] during the MOCVD process the emission peak corresponding to the dilute nitride junction is slightly shifted towards shorter wavelengths and this explains at least in part the reduced J_{sc}. Such behavior is rather typical for dilute nitride heterostructures experiencing thermal annealing [9–13].

The EQEs of dilute nitride junction before and after MOCVD overgrowth are shown in Fig. 3. Here the blueshift of the long wavelength edge of the EQE curve is visible as well and is related to blueshift of the band edge. The EQE is also increased due to MOCVD overgrowth for the shorter wavelengths.

As the MOCVD processes are typically run at higher temperature than the corresponding MBE processes and contain contaminants like hydrogen and carbon, it was expected that the MOCVD overgrowth could have an adverse effect on the properties of the InGaNAs junction either by thermal load or by introduction of impurities. Nevertheless, PL and EQE results indicate that the characteristics of the dilute nitride sub-cell were not degraded during overgrowth except for a slight blue shift of the absorption edge. An additional sign of preserving the high quality of dilute-nitride material was that the PL peak width did not get wider after overgrowth.

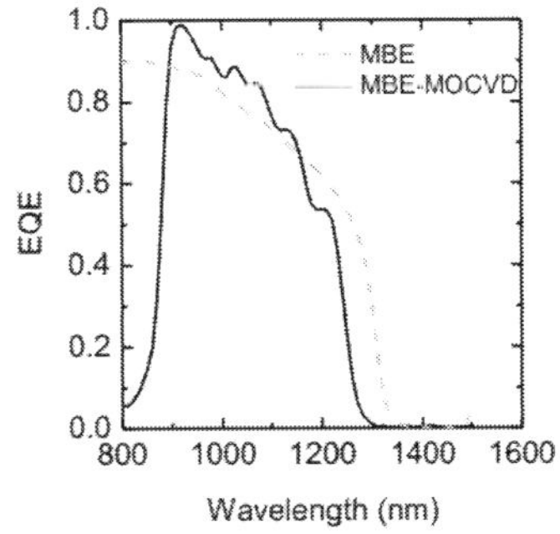

Figure 3. EQE measurements for a typical dilute nitride single-junction solar cell before the MOCVD process (solid black line) and for a dilute nitride sub-cell in a triple-junction solar cell after the MOCVD overgrowth process (dashed green line)

GaAs and InGaP junction were designed and their growth parameters optimized to produce TJ solar cells by the combined MBE-MOCVD technique. Design and optimization was made keeping into account the effects of MOCVD growth on the properties of the InGaNAs junction and improving the InGaP growth parameters.

After the optimization, 2×2 cm^2 solar cells with 30.8% AM0 efficiency at Beginning-Of-Life (BOL) were produced (Fig. 4).

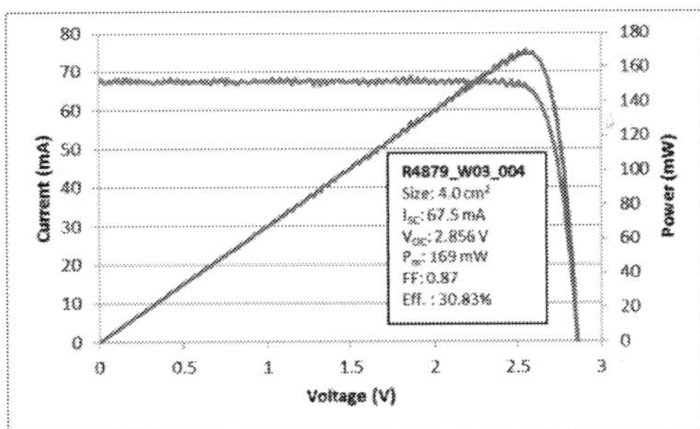

Figure 4. Electrical performances of InGaP/GaAs/InGaNAs triple-junction solar cells measured under dual source WACOM solar simulator at 1 sun, 25 °C

The main difference between experimental and theoretical expectation was related to the short circuit current (I_{sc}), mainly linked to InGaP junction that needs further improvements.

For optimization of End-Of-Life (EOL) efficiency of TJ solar cells, electron irradiation was performed on InGaP, GaAs and InGaNAs isotype solar cells. 2×2 cm^2 solar cells were irradiated with 1 MeV electrons at 2E15 e$^-$/cm^2 and 5E15 e$^-$/cm^2 fluences. The irradiation was executed at ILU-6 accelerator at Centre for Radiation Research and Technology of Institute of Nuclear Chemistry and Technology (INCT) in Warsaw (Poland).

Remaining factors of isotype solar cells after irradiation are reported in Tab. 1 and Tab. 2.

Table 1: Remaining factors at 2E15 e^-/cm^2 fluence

Solar cell structure	I_{sc} RF	V_{oc} RF	P_m RF
InGaP	0.99	0.97	0.95
GaAs	0.85	0.89	0.74
InGaNAs	0.69	0.90	0.59

Table 2: Remaining factors at 5E15 e^-/cm^2 fluence

Solar cell structure	I_{sc} RF	V_{oc} RF	P_m RF
InGaP	0.97	0.95	0.89
GaAs	0.74	0.86	0.61
InGaNAs	0.56	0.82	0.40

The analysis of the short circuit currents of the three junctions to determine the limiting sub-cell and the EOL behaviour (table 3) is very intriguing: at BOL the limiting junction is the top one whereas after irradiation of 2e15 e^-/cm^2, the limiting sub-cell is the middle.

Table 3: Short circuit currents of sub-cells after 1 MeV electrons (cell area 2×2 cm²)

I_{sc} (mA)	BOL	2e15 e/cm²	5e15 e/cm²
InGaP	68	66	65
GaAs	70	59	52
InGaNAs	98	72	52

To improve the EOL efficiency it is important to increase the current of the GaAs sub-cell. Following the model explained in chapter 3, the structure of the GaAs junction will be properly modified to increase its short circuit current without penalising the thermal load on the InGaNAs.

5. CONCLUSIONS

InGaP/GaAs/InGaNAs solar cells have been manufactured by a monolithic approach using the combination of MBE and MOCVD techniques.

The results indicate that efficiency near 31% in AM0, 25 °C is reachable. Comparison between the theoretical and experimental results also suggests that further improvement in the BOL efficiency can be obtained by optimization of the InGaP sub-cell.

EOL performance of the InGaP/GaAs/ InGaNAs solar cells is dominated by the behaviour of the middle junction. Therefore, to increase the EOL performance of the TJ solar cells the short circuit current of the GaAs sub-cell needs to be increased.

6. Acknowledgements

This work was performed under a programme of and funded by the European Space Agency (contracts N.: 4000108058/13/NL/FE, 4000112560/14/NL/FE). The view expressed herein can in no way be taken to reflect the official opinion of the European Space Agency.

7. REFERENCES

1. Chiu P. T., Law D. C., Woo R.L., Singer S. B., Bhusari D., Hong W. D., Zakaria A., Boisvert J., Mesropian S., King R. R., Karam N. H., Proceedings of the 40th IEEE PVSC, 2014, DOI: 10.1109/PVSC.2014.6924957.

2. Green M. A., Emery K., Hishikawa Y., Warta W., Dunlop E. D., Solar cell efficiency tables (version 41), Progress in Photovoltaics: Research and Applications 2013; 21: 1, DOI: 10.1002/pip.2352

3. Friedman D. J., Geisz J. F., Kurtz S. R., Olson J. M., 1-eV solar cells with GaInNAs active layer, Journal of Crystal Growth 1998; 195: 409, DOI: 10.1016/S0022-0248(98)00561-2.

4. Aho A., Tukiainen A., Polojärvi V., Korpijärvi V.–M., Gubanov A., Salmi J., Guina M., Laukkanen P., Lattice Matched Dilute Nitride Materials for III-V High-Efficiency Multi-Junction Solar Cells: Growth Parameter Optimization in Molecular Beam Epitaxy, Proceedings of the 26th EU PVSEC, 2011, 58, DOI:10.4229/26thEUPVSEC2011-1AO.8.3.

5. A. Tukiainen, A. Aho, G. Gori, V. Polojärvi, M. Casale, E. Greco, R. Isoaho, T. Aho, M. Raappana, R. Campesato and M. Guina, High-efficiency GaInP/GaAs/GaInNAs solar cells grown by combined MBE-MOCVD technique, Progress in Photovoltaics: Research and Applications, Accepted for publication, 2016.

6. Aho A., Tukiainen A., Korpijärvi V.–M., Polojärvi V., Salmi J., Guina M., Comparison of GaInNAs and InGaNAsSb solar cells grown by plasma-assisted molecular beam epitaxy, AIP Conference Proceedings 2012: 49, DOI:10.1063/1.4753831.

7. Aho A., Polojärvi V., Korpijärvi V.–M., Salmi J., Tukiainen A., Laukkanen P., Guina M., Composition dependent growth dynamics in molecular beam epitaxy of GaInNAs solar cells, Solar Energy Materials and Solar Cells. 2014; 124: 150, DOI: 10.1016/j.solmat.2014.01.044.

8. Gori G., Campesato R.: Photovoltaic Cell Having a high Conversion Efficiency, PCT I09111-WO 2009

DOI: 10.1051/e3sconf/20171603003

9. Klar P. J., Grüning H., Koch J., Schäfer S., Volz K., Stoltz W., Heimbrodt W., (Ga, In)(N, As)-fine structure of the band gap due to nearest-neighbor configurations of the isovalent nitrogen, Physical Review B 2001; 64: 121203, DOI: 10.1103/PhysRevB.64.121203.

10. Pavelescu E.–M., Jouhti T., Dumitrescu M., Klar P.J., Karirinne S., Fedorenko Y., Pessa M., Growth-temperature-dependent (self-)annealing-induced blueshift of photoluminescence from 1.3 μm GaInNAs/GaAs quantum wells, Applied Physics Letters 2003; 83: 1497, DOI: 10.1063/1.1601309.

11. Kudrawiec R., Pavelescu E.–M., Andrzejewski J., Misiewicz J., Gheorghiu A., Jouhti T., Pessa M., The energy-fine structure of GaInNAs/GaAs multiple quantum wells grown at different temperatures and postgrown annealed, Journal of Applied Physics 2004; 96: 2909, DOI: 10.1063/1.1774258.

12. Karirinne S., Pavelescu E.–M., Konttinen J., Jouhti T., Pessa M., The behaviour of optical and structural properties of GaInNAs/GaAs quantum wells upon annealing, New Journal of Physics 2004; 6: 192, DOI: 10.1088/1367-2630/6/1/192.

13. Pavelescu E.–M., Wagner J., Komsa H.–P., Rantala T.T., Dumitrescu M., Pessa M., Nitrogen incorporation into GaInNAs lattice-matched to GaAs: The effects of growth temperature and thermal annealing, Journal of Applied Physics

E3S Web of Conferences **16**, 03004 (2017) DOI: 10.1051/e3sconf/20171603004
ESPC 2016

LIMITING EFFICIENCIES OF NOVEL SOLAR CELL CONCEPTS IN SPACE

A. Martí[(1)], A.Datas [(1)], J.R.González[(2)], C. Baur [(2)]

[(1)] *Instituto de Energía Solar, Universidad Politécnica de Madrid, Email:amarti@etsit.upm.es*
[(2)] *European Space Agency Email:Jose.Ramon.Gonzalez@esa.int*

ABSTRACT

In this work we study the limiting efficiency in space environmental conditions of three novel solar cell concepts (hot carrier solar cells, multiple exciton generation solar cells and intermediate band solar cells) and how this limiting efficiency is impacted by the temperature and degradation due to radiation. Comparisons are made with state of the art triple-junction solar cells whose performance is taken as reference. In the last section of the work we briefly review the status related to the experimental achievements of these cells to date.

1. NOVEL SOLAR CELL CONCEPTS OVERVIEW

In order to exceed the limiting efficiency of single gap solar cells, three novel solar cell approaches have been proposed: the hot carrier solar cell (HCSC), the multiple exciton generation solar cell (MEGSC) and the intermediate band solar cell (IBSC).

In the HCSC (Fig. 1a) photons (*hv*) create electron-hole-pairs. However, the excited electrons and holes are not allowed to interact with the semiconductor lattice so that they do not cool down and, therefore, they do not lose their energy in thermalization processes with the semiconductor lattice. It is from this property that the HCSC heritages its efficiency advantage over single gap solar cells. The extraction of these hot electrons and holes from the solar cell demands, however, the use of special contacts (named energy selective contacts) in order to make possible that electrons cool down reversibly to the contact temperature at the time they increase their electro-chemical potential. The HCSC was proposed by Ross and Nozik [1] and reviewed by Wuerfel [2, 3] who realised of the necessity of the energy selective contacts.

In the multiple exciton generation solar cell (MEGSC, Fig. 1b) high energy photons (*hv*) create high energy excited electron-hole pairs (or excitons). This pairs do not ideally lose their energy through thermalization processes but, instead, create one or more additional electron-hole pairs (e_1-h_1; e_2-h_2, e_3-h_3) depending on how many times their energy exceeds the energy of the gap. The MEGSC, proposed by Nozik [4], is an evolution of the impact ionization solar cell (IISC) proposed by Werner, Brendel and Queisser [5]. The difference relies on the fact that, while the IISC was thought to be implemented in bulk semiconductors, the MEGSC is envisaged to be implemented using quantum dots. By using quantum dots, the probability of one photon creating more than one electron-hole pair is increased because the momentum conservation selection rule is not required.

The intermediate band solar cell (IBSC) is based on the idea of implementing a semiconductor like material that, instead of exhibiting one single gap, it would exhibit two bandgaps (Fig. 1c). The two bandgaps appear when an "intermediate band" (IB) is created inside a conventional high bandgap semiconductor host. This IB allows the absorption of below bandgap energy photons that, in conventional solar cells are wasted. This absorption occurs through the successive promotion of electrons from the valence band (VB) to the IB and from the IB to the conduction band (CB). Thanks to the formation of the additional bandgap, the IBSC is capable of performing, when properly optimized, with an efficiency close to that of a triple junction solar cell [6].

The limiting efficiency of these novel concepts has been calculated in other works for operation on Earth surface and often, for operation for the sun assumed as a black body at 6000 and maximum concentration (46050 suns). However, this limiting efficiency has never been calculated systematically for AM0 [7] to our knowledge. We think this is necessary as a first step to guide the possible application of these novel concepts in space. In addition, when assuming operation on Earth, the solar cell is usually assumed to operate at 300 K. We think that, with operation in space conditions in mind, evaluating the impact of temperature on cell efficiency is also of relevance. Finally, the limiting efficiency of these novel concepts is usually calculated assuming the so-called radiative limit, a concept by which electron-hole pairs can only disappear by emitting one photon (radiative recombination). However, in space, solar cell degradation by radiation plays an important role and, in the next sections, we will also explore the impact of this degradation on the performance of these cells.

 DOI: 10.1051/e3sconf/20171603004

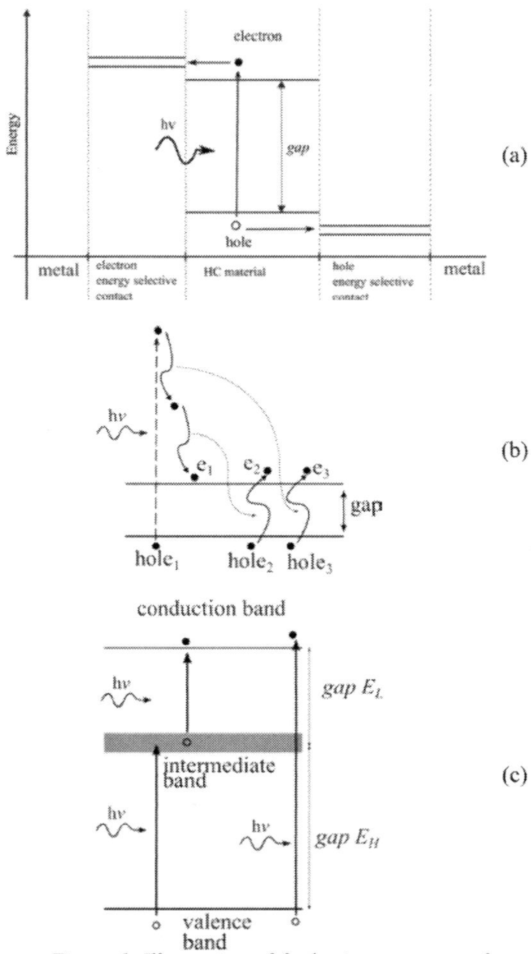

Figure 1. Illustration of the basic structure and operation of (a) a hot carrier solar cell; (b) a multiple exciton generation solar cell and (c) an intermediate band solar cell.

2. HOT CARRIER SOLAR CELL

Figure 2 plots the limiting efficiency of the HCSC for AM0 illumination as a function of the semiconductor bandgap. The physical model used to calculate this limiting efficiency is known as "energy conservation model" adapted for AM0 spectral conditions [2]. The impact of the temperature, in the range from 103 K to 413 K (minimum temperature has been considered for a standard Geostationary Orbit -GEO- mission; maximum temperature has been considered for a Low-Earth Orbit -LEO- mission) is also illustrated. As it can be seen, the optimum bandgap for operation at 300 K is located at around 0.9 eV (with a 51.8 % limiting efficiency) and shifts towards higher values (1 eV) when the temperature of operation is increased to 413 K and towards lower values (0.6 eV) when the temperature is decreased to 103 K. The limiting efficiency decreases from 51.8 % to 41.7 % when the temperature is increased to 413 K and increases to 76.8 % when the temperature is decreased from 300 K to 103 K.

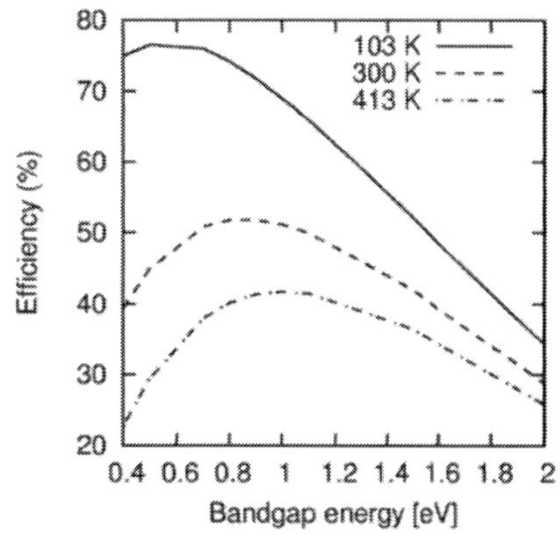

Figure 2. Limiting efficiency of the hot carrier solar cell for AM0 illumination conditions as a function of the semiconductor bandgap and the temperature of operation of the cell.

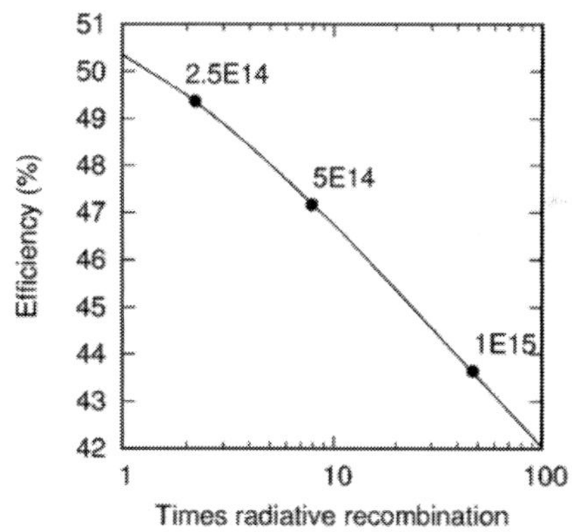

Figure 3. Impact of non-radiative recombination (measured as times the radiative recombination) on the limiting efficiency of the hot carrier solar cell. The temperature for the case illustrated has been assumed 300 K and the bandgap of the cell 0.9 eV which corresponds to the optimum gap for operation at this temperature.

Figure 3 illustrates the impact of the degradation of the cell due to the appearance of non-radiative recombination measured in times the radiative recombination. In this respect, a factor 5, for example, implies that radiation has introduced non-radiative recombination in the cell that is equivalent to 5 times the radiative recombination that is calculated according to detailed balance arguments. As it can be observed, efficiency degrades from 51.8 % to 42.0 % when non-radiative recombination is 100 times the radiative recombination. In order to have a useful meaning in space environment, these factors should be correlated with radiation fluence. However, this is impossible to do, at least while the HCSC remains mostly a theoretical concept since, for example, the materials to implement the concept with have not been clearly identified yet. In this respect, the references to fluence that appear in the Figure refer to fluences that have been estimated to introduce the indicated non-radiative recombination in conventional triple junction solar cells according to the explanations that will be given in Section 5.

3. MULTIPLE EXCITON GENERATION SOLAR CELL

Figure 4 plots the limiting efficiency of the MEGSC for AM0 illumination as a function of the semiconductor bandgap. The physical model used to calculate this limiting efficiency is the one described in [5] adapted for the AM0 spectral conditions. The optimum bandgap for operation at 300 K is found to be 0.73 eV (with a 43.1 % limiting efficiency), 0.89 eV for 413 K (with a 33.9%) and 0.32 eV for 103 K (and a limiting efficiency of 68.1 %).

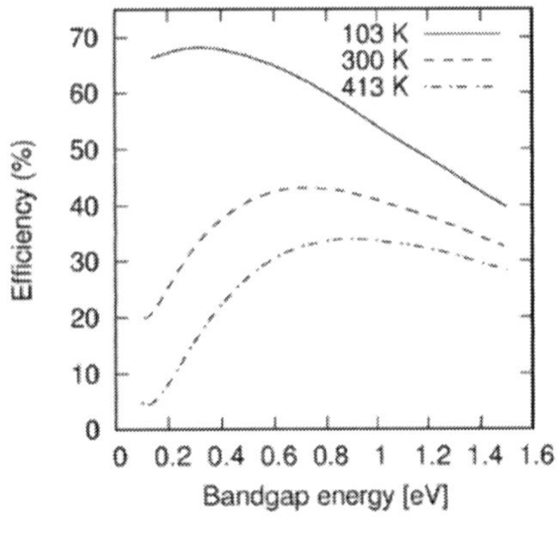

Figure 4. Limiting efficiency of the multiple exciton generation solar cell for AM0 illumination conditions as a function of the semiconductor bandgap and the temperature of operation of the cell.

As we did for the HCSC, Figure 5 illustrates now the impact of the appearance of non-radiative recombination in the efficiency of the cell. Fluences producing the same non-radiative recombination in 3J-MJSC are indicated for reference. As it can be observed, non-radiative recombination seems to have a greater impact on the MEGSC than in the HCSC since, for example, its efficiency degrades below 38 % for a non-radiative recombination 10 times the radiative one while the limiting efficiency of the HCSC remains above 46 % for the same non-radiative recombination.

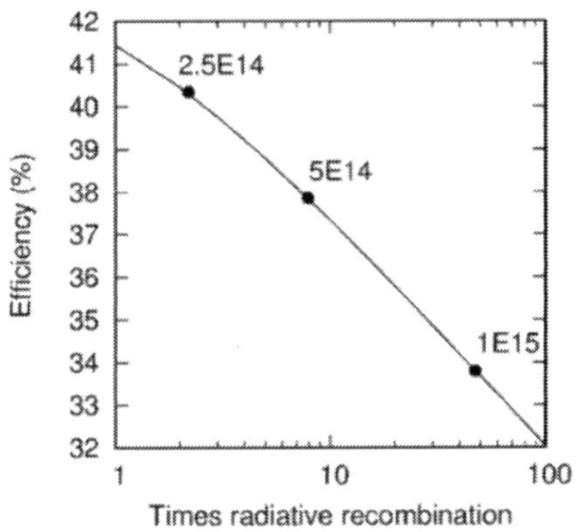

Figure 5. Impact of non-radiative recombination (measured as times the radiative recombination) on the limiting efficiency of the multiple exciton generation cell. The temperature for the case is illustrated has been assumed 300 K and the bandgap of the cell 0.73 eV which corresponds to the optimum for operation at this temperature.

4. INTERMEDIATE BAND SOLAR CELL

As introduced in section 1, the IBSC is characterized by two bandgaps E_L and E_H. Figure 6 plots the limiting efficiency of the IBSC for AM0 illumination as a function of the lowest bandgap E_L at the time the value of E_H is optimized. The physical model used to calculate this limiting efficiency is the one described in [6] adapted for the AM0 spectral conditions. The optimum bandgap for operation at 300 K is found to be E_L=0.85 eV (with a 45.8 % limiting efficiency and E_H=1.38 eV). The limiting efficiency degrades to 39 % when the cell operates at 413 K and increases to 61 % when the temperature decreases down to 103 K.

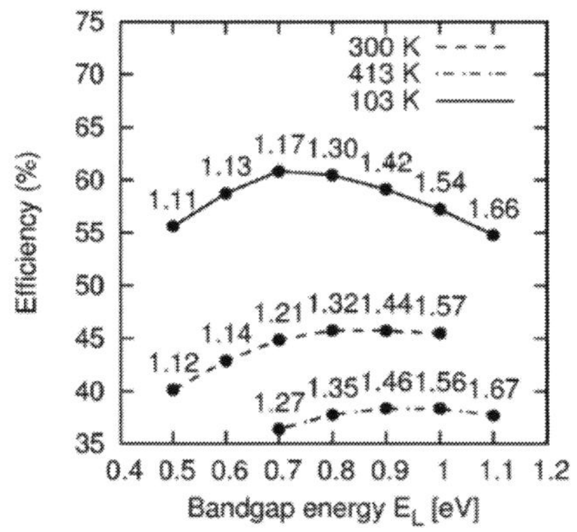

Figure 6. Limiting efficiency of the intermediate band solar for AM0 illumination conditions as a function of the semiconductor bandgap E_L and the temperature of operation of the cell. Labels indicate the optimum value for E_H.

Figure 7 shows the impact on the efficiency of the optimum cell (E_L=0.85 eV and E_H=1.38 eV) for operation at 300 K of non-radiative recombination. As it can be seen, efficiency would degrade down to 38.5 % when non-radiative recombination becomes one hundred times the radiative one. We indicate in the figure, for reference, the fluences that in the 3J-MJSC are estimated to introduce the same non-radiative recombination

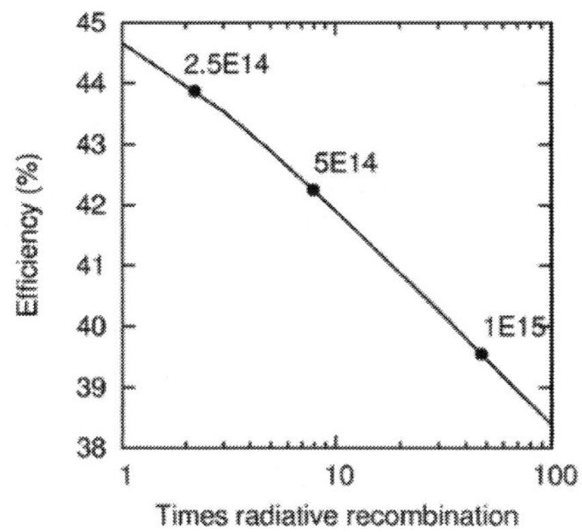

Figure 7. Impact of non-radiative recombination (measured as times the radiative recombination) on the limiting efficiency of the intermediate band solar cell. The temperature for the case is illustrated has been assumed 300 K for the bandgaps E_L=0.85 eV and E_H=1.38 eV which corresponds to the optimum gaps for operation at this temperature.

5. 3J-MULTIJUNCTION SOLAR CELL

It will be illustrative to undertake, for the solar cell commonly used today in space (the GaInP/InGaAs/Ge lattice matched tandem solar cell), the same study we have carried out in the previous sections for the novel solar cell concepts.

In this respect, Figure 8 plots the limiting efficiency of this 3J-MJSC as a function of the temperature. The limiting efficiency has been calculated using detailed balance arguments [8, 9] assuming 1.88 eV, 1,41 eV and 0,66 eV gaps for the GaInP, InGaAs and Ge respectively. Since due to the lattice matching constrains these gaps are not the ideal ones for a tandem configuration of three solar cells connected in series, a better current matching and efficiency is obtained if the GaInP cell is assumed not to absorb all the sunlight above its bandgap but, instead, 86 % of it while it allows the remaining light to reach the InGaAs middle cell. In contrast, total light absorption has been assumed for the InGaAs and Ge solar cells. No radiative coupling between the cells has been assumed in these calculations.

In the same figure we also plot, for comparison, the dependence of the limiting efficiency with the temperature of the HCSC, MEGSC and IBSC which efficiency was optimum at 300 K. This dependence shows to be linear with the temperature and Table 1 collects this slope for each of the cells. In this respect, the 3J-MJSC shows the lowest dependence (0.055 efficiency points of decrease per degree) while the HCSC shows the highest (almost 0.1 points of efficiency per degree).

Table 1 Decrease in efficiency points per K degree for the HCSC, MEGSC and IBSC which operation is optimum at 300 K.

	3J-MJSC	HCSC	MEGSC	IBSC
K^{-1}	-0.055	-0.099	-0.093	-0.069

With respect to the degradation induced by non-radiative recombination, Figure 9 plots the limiting efficiency of the 3J-MJSC as a function of the non-radiative recombination. The correlation between "fluence" and the "times radiative recombination" factor advanced in previous sections has been estimated as follows. From public data published by AzurSpace [10] for their commercial triple junction solar cell 3G30C-Advanced it has been possible to correlate that 1 MeV [e/cm2] fluences of 2.5×10^{14}, 5×10^{14} and 10^{15} degrade the cell beginning of life (BOL) efficiency by 3.1, 5.8 and 10.2 % respectively. Assuming this degradation percentages in the limiting efficiency of the ideal GaInP/InGaAs/Ge solar cell (Figure 9) leads to non-radiative factors of 2.2, 7.9 and 47.3 respectively.

These are the factors that we also indicated for reference in the efficiency plots of Figures 3, 5 and 7.

Table 2 summarises the impact on the limiting efficiency of the different novel concepts studied in this work of this levels of non-radiative recombination. As it can be observed, the MEGSC seems to exhibit the highest dependence to non-radiative recombination followed by the HCSC and IBSC.

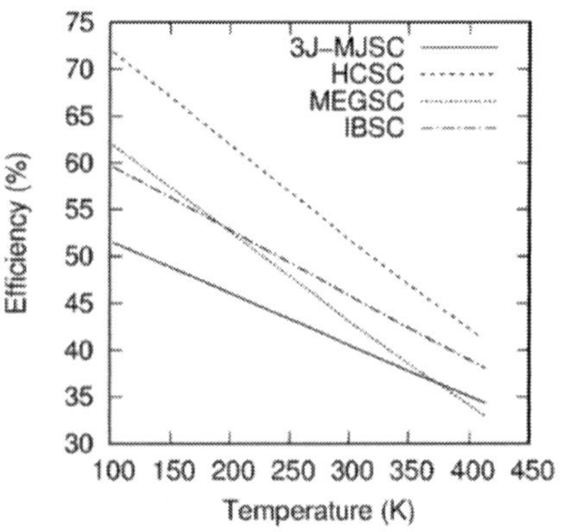

Figure 8. Impact of temperature on the limiting efficiency of the 3J-MJSC (InGaP/InGaAs/Ge lattice matched tandem solar cell), HCSC (0.9 eV), MEGSC (0,73 eV) and IBSC (E_L=0.8 eV and E_H=1.32 eV). The value of the indicated gaps were found to be optimum for photovoltaic conversion of these cells at 300 K.

Table 2. Limiting efficiency of several novel concepts for several that 1 MeV [e/cm2] fluences. For the calculation we have assumed that the given fluence introduced in the novel concept the same average non-radiative recombination that it would introduce in a 3J-MJSC. For the 3J-MJSC we have assumed a InGaP/InGaAs/Ge lattice matched solar cell. The increment Δ refer to the variation of the efficiency between the 1×10^{15} fluence and the BOL efficiency.

	BOL	2.5×10^{14}	5×10^{14}	1×10^{15}	Δ
HCSC	51.8	49.4	71.2	43.6	-15.8 %
MEGSC	43.1	40.3	37.9	33.8	-21.6 %
IBSC	45.8	43.9	42.3	39.6	-13.5 %
3J-MJSC	41.8	40.5	39.4	37.5	-10.3 %

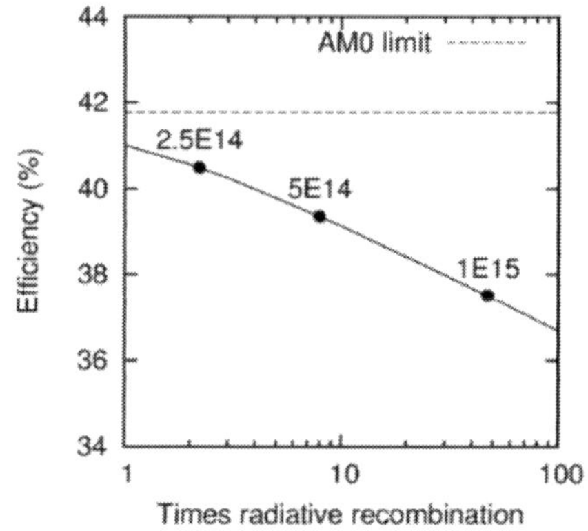

Figure 9. Impact of non-radiative recombination in the limiting efficiency of a 3J-MJSC InGaP/InGaAs/Ge lattice matched tandem solar cell. Dots indicate the 1 MeV [e/cm2] fluence that is estimated to induce the corresponding non-radiative recombination. (3 gaps in series: 0.7 eV, 1.1 eV and 1.7 eV).

6. BRIEF REVIEW OF THE STATUS OF EMPIRICAL RESEARCH ON NOVEL CONCEPTS

The results above show the limiting efficiency of the HCSC, MEGSC and IBSC for space operation conditions as well as the impact of the temperature and degradation by non-radiative recombination. However, looking forward the real use of these cells, it must be taken into account that, so far, none of these cells have demonstrated the promised efficiencies although, to more or less extent the physical principles involved in their operation have been demonstrated.

Hence, for the case of the IBSC, the absorption of two below bandgap energy photons to create one electron-hole pair has been demonstrated, for example, in several implementations that use III-V semiconductor quantum dots [11-15]. In addition, it has also been proved that the presence of the intermediate band does not limit the output voltage of the cell [16, 17].

In the MEGSC context, Semonin et al. [18] have demonstrated a solar cell based on colloidal quantum dots that exhibited an external quantum efficiency higher than one.

Research on HCSC focuses, on one side, on finding a suitable hot carrier material and, on the other side, on finding suitable energy selective contacts. As for the hot carrier material concerns, Conibeer et al. have proposed

[19] materials such as InN, HfN and SnSi. Yao and Köning have proposed BSb [20]. As for the energy selective contacts, the most studied option is based on the exploitation of different variations of the resonant tunnelling using nanostructures [21]. Dimmock et al. [22] have proposed and investigated a HCSC implemented on a metal-semiconductor junction.

7. CONCLUSIONS

HCSC, MEGSC and IBSC promise efficiencies of 51.8 %, 43.1% and 45.8 % in ideal conditions for AM0 illumination and 300 K. Regarding the dependence of these limiting efficiencies with temperature, the HCSC has revealed the highest dependence and the IBSC the lowest (see Table I).

As for the impact of non-radiative recombination in this efficiency, assuming this would introduce the same non-radiative recombination factor than in a 3J-MJSC, the MEGSC has exhibited the highest sensibility to non-radiative recombination while the IBSC has exhibited the lowest (see Table II).

In the laboratory, however, no cell has demonstrated efficiencies exceeding the efficiency of single gap solar cells. Therefore, their practical application in space will still need to wait a few years more. In spite of this, all the concepts have provided strong experimental support demonstrating that the physical principles they are based on are correct. It is also noticeable that the three concepts demand the use of nanostructures, in a way or another, for their practical implementation.

8. ACKNOWLEDGMENTS

This work has been funded by the European Space Agency through the Contract No. 4000115109/15/NL/FE

9. REFERENCES

1. Ross, R.T., and Nozik, A.J. (1982). Efficiency of hot-carrier solar energy converters. Journal of Applied Physics *53*, 3813-3818.

2. Würfel, P., Brown, A.S., Humphrey, T.E., and Green, M.A. (2005). Particle conservation in the hot-carrier solar cell. Progress in Photovoltaics: Research and Applications *13*, 277-285.

3. Wurfel, P. (1997). Solar energy conversion with hot electrons from impact ionisation. Solar Energy Materials and Solar Cells *46*, 43-52.

4. Nozik, A.J. (2002). Quantum dot solar cells Physica E *14*, 115-120.

5. Jiirgen H. Werner, Rolf Brendel, and Queisser, H.J. (1994). New Upper Efficiency Limits for semiconductor solar cells. Proceedings of the First World Conference on Photovoltaic Energy Conversion.

6. Luque, A., and Martí, A. (1997). Increasing the efficiency of ideal solar cells by photon induced transitions at intermediate levels. Physical Review Letters *78*, 5014–5017.

7. 2000 ASTM Standard Extraterrestrial Spectrum Reference E-490-00.

8. Shockley, W., and Queisser, H.J. (1961). Detailed Balance Limit of Efficiency of p-n Junction Solar Cells. Journal of Applied Physics *32*, 510-519.

9. Martí, A., and Araújo, G.L. (1996). Limiting efficiencies for photovoltaic energy conversion in multigap systems Solar Energy Materials and Solar cells *43*, 203-222.

10. AzurSpace URL:http://www.azurspace.com/index.php/en/products/products-space/space-solar-cells. Accessed: 2016-06-02. (Archived by WebCite® at http://www.webcitation.org/6hxzSFIDP).

11. Martí, A., Antolin, E., Stanley, C.R., Farmer, C.D., Lopez, N., Diaz, P., Canovas, E., Linares, P.G., and Luque, A. (2006). Production of Photocurrent due to Intermediate-to-Conduction-Band Transitions: A Demonstration of a Key Operating Principle of the Intermediate-Band Solar Cell. Physical Review Letters *97*, 247701-247704.

12. López, E., Datas, A., Ramiro, I., Linares, P.G., Antolín, E., Artacho, I., Martí, A., Luque, A., Shoji, Y., Sogabe, T., et al. (2016). Demonstration of the operation principles of intermediate band solar cells at room temperature. Solar Energy Materials and Solar Cells *149*, 15-18.

13. Datas, A., López, E., Ramiro, I., Antolín, E., Martí, A., Luque, A., Tamaki, R., Shoji, Y., Sogabe, T., and Okada, Y. (2015). Intermediate Band Solar Cell with Extreme Broadband Spectrum Quantum Efficiency. Physical Review Letters *114*, 157701.

14. Tamaki, R., Shoji, Y., Okada, Y., and Miyano, K. (2014). Spectrally resolved intraband transitions on two-step photon absorption in InGaAs/GaAs quantum dot solar cell. Applied Physics Letters *105*, -.

15. Iñigo Ramiro , Elisa Antolín, Jinyoung Hwang, Alan Teran, Andy Martin, Joanna Millunchick, Jamie Phillips, Martí, A., and Luque, A. (2016). Three-Bandgap Absolute Quantum Efficiency in Intermediate Band Solar Cells. To appear published at the 43 IEEE PVSC.

16. Linares, P.G., Martí, A., Antolín, E., Farmer, C.D., Ramiro, I., Stanley, C.R., and Luque, A. (2012). Voltage recovery in intermediate band solar cells. Solar Energy Materials and Solar cells *98*, 240-244.

17. Ramiro, I., Antolin, E., Linares, P.G., Lopez, E., Artacho, I., Datas, A., Marti, A., Luque, A., Steer, M.J., and Stanley, C.R. (2014). Two-photon photocurrent and voltage up-conversion in a quantum dot intermediate band solar cell. In Photovoltaic Specialist Conference (PVSC), 2014 IEEE 40th. pp. 3251-3253.

18. Semonin, O.E., Luther, J.M., Choi, S., Chen, H.-Y., Gao, J., Nozik, A.J., and Beard, M.C. (2011). Peak External Photocurrent Quantum Efficiency Exceeding 100% via MEG in a Quantum Dot Solar Cell. Science *334*, 1530-1533.

19. Conibeer, G., Shrestha, S., Huang, S.J., Patterson, R., Xia, H.Z., Feng, Y., Zhang, P.F., Gupta, N., Tayebjee, M., Smyth, S., et al. (2015). Hot carrier solar cell absorber prerequisites and candidate material systems. Solar Energy Materials and Solar Cells *135*, 124-129.

20. Yao, Y., and König, D. (2015). Comparison of bulk material candidates for hot carrier absorber. Solar Energy Materials and Solar Cells *140*, 422-427.

21. Dimmock, J.A.R., Day, S., Kauer, M., Smith, K., and Heffernan, J. (2014). Demonstration of a hot-carrier photovoltaic cell. Prog. Photovoltaics *22*, 151-160.

22. Dimmock, J.A.R., Kauer, M., Stavrinou, P.N., and Ekins-Daukes, N.J. (2015). A metallic hot carrier photovoltaic cell. In Physics Simulation and Photonic Engineering of Photovoltaic Devices Iv, Volume 9358, A. Freundlich, J.F. Guillemoles and M. Sugiyama, eds.

SPACE SOLAR CELLS – 3G30 AND NEXT GENERATION RADIATION HARD PRODUCTS

W. Guter, F. Dunzer, L. Ebel, K. Hillerich, W. Köstler, T. Kubera, M. Meusel, B. Postels, C. Wächter

AZUR SPACE Solar Power GmbH, Theresienstr. 2, 74072 Heilbronn, Germany, wolfgang.guter@azurspace.com

ABSTRACT

The 3G30-Advanced, AZUR SPACE's latest qualified solar cell product, provides highest end-of-life efficiencies in space. The cell reaches 27.8% at a fluence of 5 E14 cm^{-2} and 26.5% at a fluence of 1 E15 cm^{-2} 1 MeV electrons.

The cell mass can be reduced to a minimum by substrate thinning, the cell cost can be reduced by implementation of large area configurations and even higher radiation hardness can be achieved by using AZUR's proprietary 3G30-1E16+ design. Various configurations are currently in production.

The increasing demand for cells suited for LEO applications, made AZUR to develop a novel upright metamorphic triple junction solar cell with a BOL efficiency of 31% designed for a fluence of 1 E14 cm^{-2} 1 MeV electrons. This cell design is already in production.

AZUR's next generation product 4G32 comprises an upright metamorphic 4-junction device with 28.5% EOL (1 E15 cm^{-2} 1 MeV electrons) efficiency. Hence, the 4G32 even surpasses the EOL efficiency of the lattice-matched 3-junction cell 3G30-Advanced. It utilizes the excess current of the Ge subcell by a metamorphic cell concept and a fourth junction added to the stack. This cell will be qualified by mid-2017.

This paper summarizes the results and achievements for various 3G and 4G solar cell products from AZUR SPACE, including radiation hardness and cell formats.

1. INTRODUCTION

Future telecommunication satellites continue to require more power for technologies, such as 3D TV and ultra HD TV. The electrical power supplied by solar generators increases correspondingly. High-performance, lightweight and cost effective solar generators are required in order to decrease cost. For the solar cell, this translates into high end-of-life [EOL] efficiency, low weight and low production costs.

Besides this, the demand for cells with even higher radiation hardness also increases. Solar electric orbit raising or orbits with high particle irradiation intensity are the main reasons.

On the other hand, the demand for cells suitable for LEO missions also increases. This requires cells with high beginning-of-life [BOL] efficiencies.

AZUR SPACE is active in all three market segments.

AZUR's latest best-in-class solar cell product in the market "3G30C-Advanced" provides a high end-of-life efficiency of 27.8% [1]. Low production costs have been accomplished by a lean cell design and by optimized production processes. In order to reduce the mass of a panel to a minimum, the 3G30C-Advanced may be thinned down to about 80 µm. Customers can also select from various formats. Starting from the classical 4 x 8 cm^2 size and leading to enhanced cell shapes, such as AZUR's "halfpipe design" wafer utilization and packing density on the panel can be increased while production cost is further reduced.

The move toward solar electric propulsion even increases the demand on radiation-hardness of the solar cell compared to rapid conventional chemical orbit raising. Since the cell collects an increased dose of high-energy particles during multiple traverses of the Van Allen Belt, the target equivalent fluence of 1 MeV electrons increases from about 4 E15 cm^{-2} up to 5 E16 cm^{-2} and more. AZUR has developed a new product segment to meet this demand. The advancement of the radiation-hardening technology from the 3G30C yields a new class of solar cells called 3G30C-1E16+ for increased EOL requirements. This cell yields an efficiency of 21.2% at a fluence of 1 E16 cm^{-2} 1 MeV electrons.

The move toward constellation programs and other LEO missions requires cells with high BOL efficiency. In order to meet these demands, AZUR has developed the 3G31C solar cell design. This cell comprises a metamorphic triple-junction solar cell adapted for a fluence of 1 E14 cm^{-2} 1 MeV electrons and yields BOL efficiencies of up to 31%.

Next generation solar cells will exceed the EOL efficiencies of the 3G30C-Advanced and utilize the excess current of the Ge subcell. Metamorphic solar cell concept [2] have demonstrated considerable potential. Upright and inverted devices lattice-matched or metamorphic all target approximately the same bandgap combination, although achieved with different materials [3]. AZUR SPACE proposed a 4-junction solar cell design [4] in order to reach EOL efficiencies of 28.5% at 1 E15 cm^{-2} 1 MeV electrons. The upright cell concepts keep epitaxy costs low and are highly compatible with standard cell processing and well-established lay-down technology.

DOI: 10.1051/e3sconf/20171603005

All these developments follow AZUR's roadmap for LEO, GEO and MEO missions, as shown in Fig. 1.

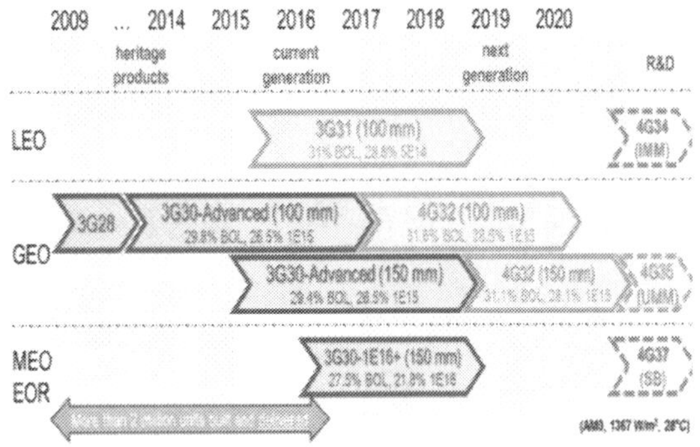

Figure 1. Product roadmap of AZUR for three market segments. Colours indicate cell technology: lattice-matched (blue), UMM (orange), inverted metamorphic (IMM, green) or semiconductor bonded (SB, green). Today more than 2 million cells have been delivered.

2. 3G30 SOLAR CELL PRODUCTS

All 3G30 solar cell products are based on a triple-junction design with subcells made from InGaP, InGaAs and Ge, all lattice-matched to the Ge growth substrate.

2.1 Radiation Hardness

AZUR's proprietary and fully patent protected end-of-life concept bases upon the selection of radiation-hard semiconductor materials, such as InGaP and Ge, an advanced cell design for increased minority carrier diffusion length in the InGaAs cell and a distributed Bragg reflector (DBR) for a reduced thickness of the InGaAs cell (Fig. 2).

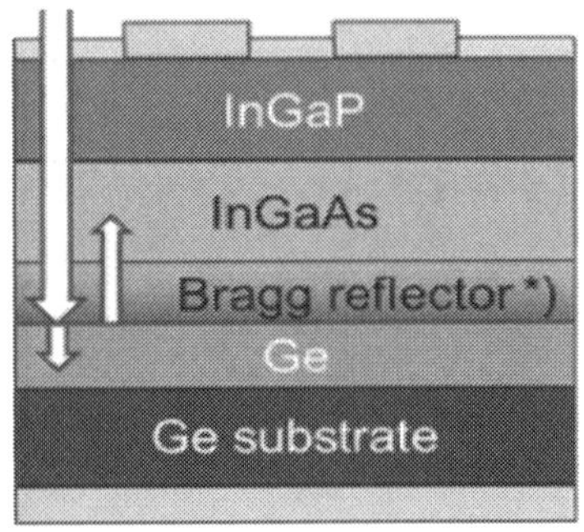

Figure 2. Structure of the 3G30C solar cell with a bandgap sequence of 1.9, 1.4 and 0.66 eV and AZUR's proprietary and patent protected EOL design including a distributed Bragg reflector (*).

High-energy particles indeed do impair the InGaP material. But temperatures around 25°C are sufficient to reduce the displacement damage by annealing to a great extent.

The Ge material is an indirect-gap semiconductor used for a relatively thick Ge subcell (>> 20 µm). The carriers generated by phonon-induced absorption hardly contribute at BOL. Hence, the reduction of diffusion length hardly affects the cell performance at EOL. Furthermore, the Ge subcell generates substantial excess current (~40%) compared to the other two subcells. A minor reduction of current density does not affect the triple-junction solar cell performance.

However, high-energy particles significantly affect the InGaAs material [5]. Annealing of the displacement damage requires temperatures well above 100°C [6]. This is usually undesired for geostationary satellites. Continuous and intensive research by AZUR SPACE and partners lead to a unique radiation-hardening approach. On the one hand, the layers are designed to increase minority carrier diffusion lengths and to reduce surface recombination. On the other hand, a distributed Bragg reflector is implemented to reduce the cell thickness and thus lower the minimally required diffusion lengths while maintaining the cell's optical thickness for light absorption.

AZUR's radiation-hardening concept yields the highest available EOL performances as summarized in Tab. 1.

3G30	BOL	5E14	1E15	3E15	1E16
V_{OC} [mV]	2700	2564	2522	2424	2324
I_{SC} [mA]	520	514	502	461	397
V_{mp} [mV]	2411	2290	2246	2174	2085
η [%]	29.5	27.8	26.5	23.4	18.7

Table 1. Average efficiencies of AZUR's 3G30C at different fluences of 1 MeV electrons. The cell is designed for the 8 E14 cm^{-2} range (1367 W/m^2).

2.2 Cell Types and Assemblies

In order to comply with existing module designs the 3G30C is offered in the well-established 4 x 8 cm² design with cropped corners.

In order to optimize production cost, AZUR transferred production from 100 mm wafer technology to 150 mm technology. This allows cell sizes of up to 6 x 12 cm² in the cropped corner format. AZUR's latest enhanced "halfpipe design" with an area of up to 82 cm² further optimizes wafer utilization as well as packing density on the module. This leads to a doubled cost reduction on panel level. Fig. 3 shows some cell formats.

In order to reduce cell weight, a combined grinding and etching process is used to thin down the Ge substrates to 80 µm. This represents a cost effective trade-off between weight reduction and yield in production and handling. Other cell thicknesses are also feasible.

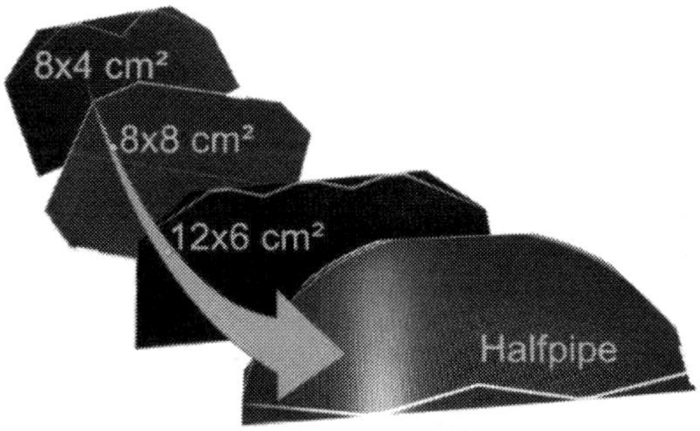

Figure 3. Different 3G30C solar cell formats ranging from 30 cm² to 70 cm² active area.

AZUR SPACE has advanced its bare cell technology to a higher integration step. Typical solar cell assemblies (SCAs) feature multiple configurations of integral and external bypass diodes, coverglass thickness down to 50 µm and multiple toe interconnectors for welding reliability. Fig. 4 shows two examples.

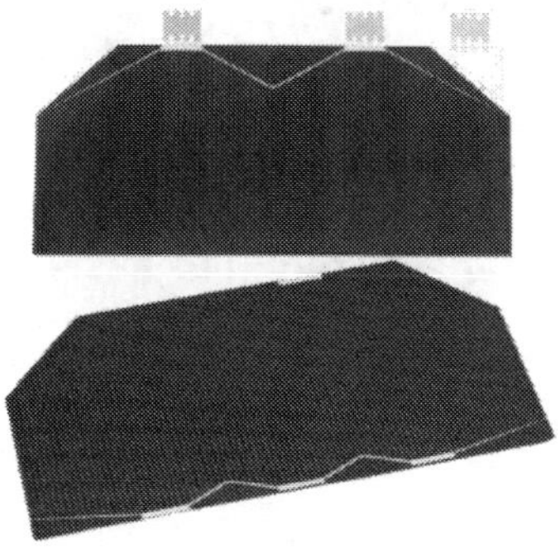

Figure 4. 3G30A assembly with external Si bypass diode (top) and a 3G30C with integral bypass diode (bottom).

2.3 3G30-1E16+

Solar electric orbit raising makes completely new demands on the solar generator and the solar cell. Power demand rises, as the solar generator is also used to power the ion thrusters during raising of the satellite. During the journey, the spacecraft traverses the Van Allen Belts multiple times and hence is exposed to a higher dose of high-energy particles than during conventional chemical orbit raising.

AZUR SPACE has used its unique skills in radiation hardening to develop an unequalled EOL design for a solar cell. The 3G30C-1E16+ is optimized for 1 MeV electron fluences above 1 E16 cm^{-2}. Tab. 2 summarizes some results measured on this device. At high fluences this cell clearly exceeds the already excellent performance of the 3G30C-Advanced. At 1 E16 cm^{-2} an average efficiency of 21.2% has been reached. This outshines all past EOL solar cell designs. Fig. 5 compares the two exceptional devices vs. fluence.

3G30-1E16+	BOL	1E16	2.5E16
V_{OC} [mV]	2700	2352	2256
I_{SC} [mA]	470	457	414
V_{mp} [mV]	2457	2070	1993
η [%]	27.4	21.2	17.9

Table 2. Average efficiencies of AZUR's 3G30C-1E16+ at different fluences of 1 MeV electrons (1367 W/m²).

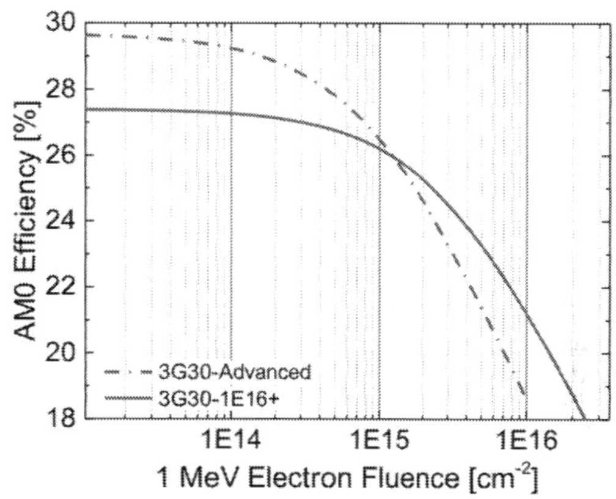

Figure 5. Comparison of AZUR's 3G30C and 3G30C-1E16+ depending on 1 MeV electron fluences.

3. 3G31 SOLAR CELLS

The 3G31C solar cell is based on a triple-junction design with subcells made from InGaP, InGaAs and Ge. This is the same material combination as used for the 3G30C. However, the alloy composition of the arsenides and phosphides features an about 5%(abs.) higher In content. As a result, the bandgaps are shifted to longer wavelengths (Fig. 6) and the upper subcells exhibit a lattice-mismatch of less than 0.5% to the Ge growth substrate. This cell concept is called upright metamorphic (UMM).

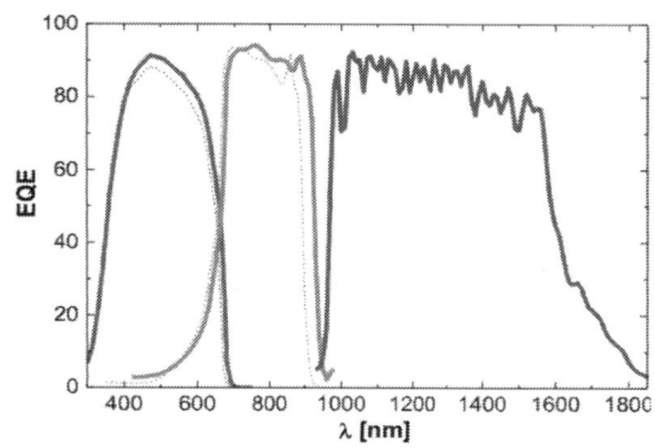

Figure 6. Quantum efficiency of AZUR's 3G31C (solid) vs. the top and middle cell of 3G30C (dashed).

DOI: 10.1051/e3sconf/20171603005

The 3G31C features a step graded buffer structure with three steps and overshoot layer made from InGaAs (Fig. 7). This buffer structure generates misfit dislocation to dissipate strain and simultaneously avoids the penetration of threading dislocations into the photoactive layers. This technology has been developed for many years by AZUR SPACE and partners and has been introduced into the 3C44 [7], AZUR's current metamorphic triple-junction solar cell for CPV applications. Now, this technology also found its use for space solar cells. In addition to the metamorphic buffer, the structure features a distributed Bragg reflector, also shown in Fig. 7.

In production, this structure yields BOL efficiencies of 30.7% (Tab. 3). It still features upright processing and may be combined with a thinning process to reduce thickness from 145 µm to 80 µm.

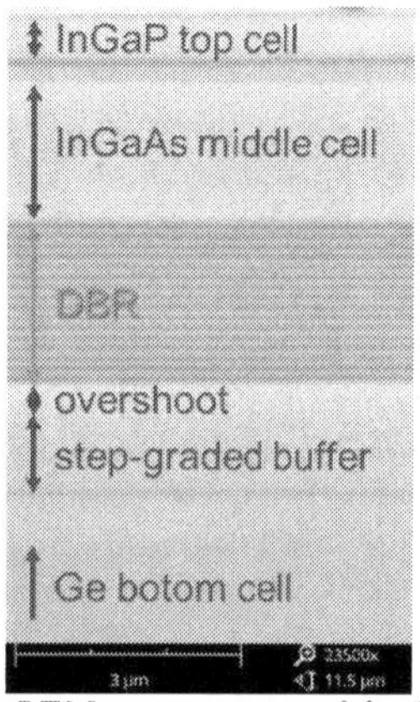

Figure 7: REM cross-section of the 3G31C. A distributed Bragg reflector (DBR) and a step-graded metamorphic buffer with overshoot complement the three subcells.

3G31	BOL	1E14
V_{OC} [mV]	2605	2549
I_{SC} [mA]	565	552
V_{mp} [mV]	2330	2272
η [%]	30.7	29.5

Table 3. Efficiencies of AZUR's 3G31C at BOL and at a fluence of 1 E14 cm^{-2} 1 MeV electrons (1367 W/m^2).

4. 4G32 SOLAR CELL PRODUCTS

The 4G32 solar cell is based on a four-junction design with subcells made from AlInGaP, AlInGaAs, InGaAs and Ge. The upper subcells exhibit a large lattice-mismatch to the Ge growth substrate. Hence, again AZUR has chosen an upright metamorphic (UMM) approach for this solar cell structure.

4.1 Mechanical Design

The about 1.5% lattice-mismatch between the Ge substrate and the InGaAs subcell is compensated by a step-graded metamorphic buffer structure similar to the 3G31C. Fig. 8 illustrates the layer structure. Since the lattice-mismatch for this device is much stronger than for a 3G31C, it is essential that the misfit dislocations dissipate 100% of the strain and that the threading dislocation density stays below 10^6 cm^{-2}. AZUR SPACE currently implements metamorphic buffers in three products: the 3C44 (for CPV), the 3G31 (for LEO) and the 4G32 (for GEO).

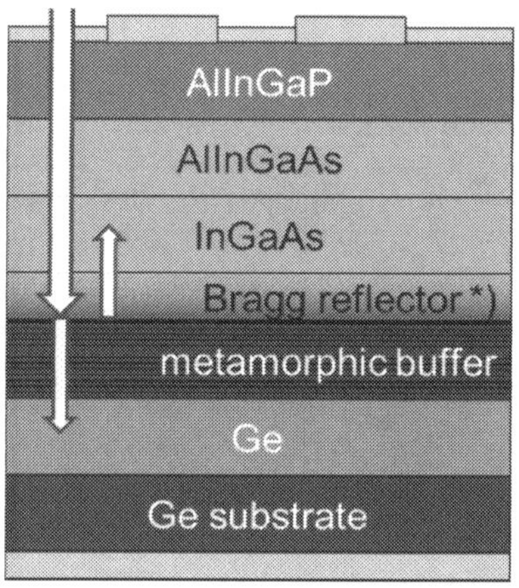

Figure 8. Structure of the 4G32C solar cell with a bandgap sequence of 1.9, 1.4, 1.1 and 0.66 eV and AZUR's proprietary and patent protected EOL design.

4.2 Radiation Hardness

The 4G32C achieves an almost optimal bandgap sequence of 1.9, 1.4, 1.1 and 0.66 eV (Fig. 9) with subcells made from AlInGaP, AlInGaAs, InGaAs and Ge, respectively (Fig. 8). The external quantum efficiency from Fig. 9 depicts the high quality of each subcell and the well-balanced current generation in all four subcells.

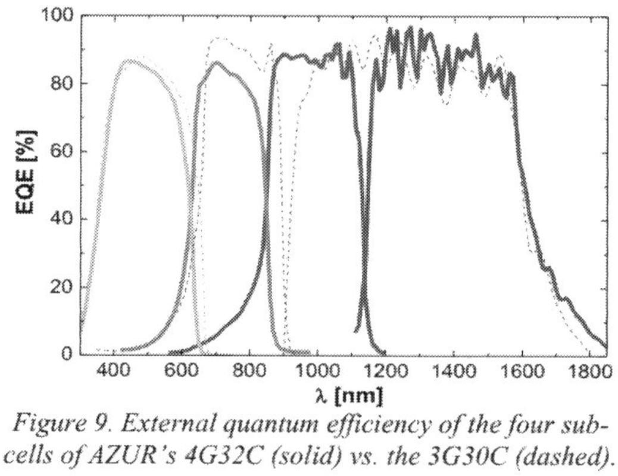

Figure 9. External quantum efficiency of the four sub-cells of AZUR's 4G32C (solid) vs. the 3G30C (dashed).

 DOI: 10.1051/e3sconf/20171603005

Exceptional radiation hardness of the cell is achieved by the implementation and adaption of AZUR's proprietary EOL design.
The 4G32C employs the radiation hard semiconductor materials AlInGaP and Ge. This is comparable to the 3G30C, where InGaP and Ge are used.
The InGaAs cell benefits from a distributed Bragg reflector similar to the radiation-hard 3G30C structure. This allows reducing the cell's thickness while keeping its optical thickness for light absorption constant (Fig. 8). However, the reflectivity range is adapted to the 4J design and the materials fit the lattice of the metamorphic buffer. The external quantum efficiency of the InGaAs subcell (Fig. 9, around 1000 nm) also shows the reflectivity of the DBR by an increased signal in the longer wavelength region from 1000 to 1100 nm.
A thinner AlInGaAs cell reduces the impact of a deteriorated minority carrier diffusion length caused by high-particle irradiation in space.
A total remaining factor of 95% in V_{OC} after 1 E15 cm^{-2} 1 MeV electrons is feasible. This outperforms the 3G30C's remaining factor by about 1% and facilitates efficiencies of up to 28.5% EOL (1 E15 cm^{-2} 1 MeV electrons). Fig. 10 shows plotted IV characteristics of the 4G32C measured at BOL and at EOL with efficiencies of 30.7% and 27.8%, respectively. Tab. 4 summarizes the relevant IV parameters at different 1 MeV electron fluences.
The production of the qualification lot for the 4G32C with a thickness of 80 μm is scheduled for the end of 2016, in order to facilitate a full qualification according to ECSS by mid-2017 [8].
The advancement program for the 4G32C is scheduled for the next three years and includes further improvement of the Al-compound subcells regarding material quality, further enhancement of radiation-hardness as well as the shift to 150 mm wafer technology. The rightmost column of Tab. 4 shows the target EOL performance for the 4G32C-Advanced.

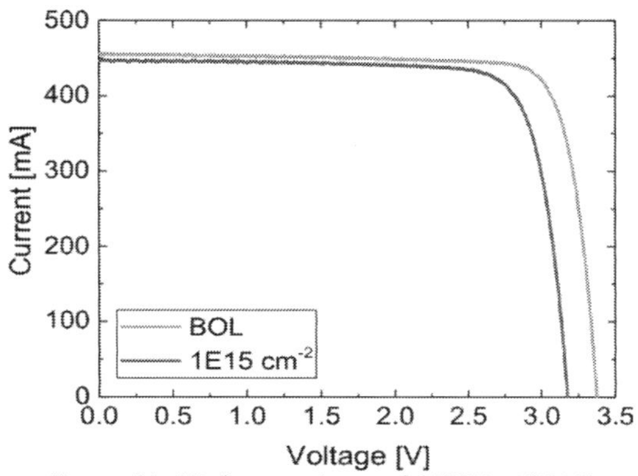

Figure 10. IV characteristics of AZUR's 4G32C (1367 W/m²) at beginning-of-life and at a 1 MeV electron fluence of 1E15 cm⁻².

4G32	BOL	1E15	5E15	1E15 target
V_{OC} [V]	3.37	3.18	3.01	3.25
I_{SC} [mA]	454	445	390	445
V_{mp} [V]	2.92	2.73	2.62	2.79
η [%]	30.7	27.8	22.7	28.5

Table 4. Performance of the 4G32C after irradiation with 1 MeV electrons (1367 W/m²) and target value for 4G32C-Advanced.

5. FURTHER CONCEPTS

Alternative approaches to surpass the 30% efficiency level at a fluence of 1 E15 cm^{-2} 1 MeV electrons use other technology building blocks, also investigated by AZUR and partners.
Inverted metamorphic four-junction solar cells developed in cooperation with Fraunhofer ISE can also reduce the dissipation of excess current in the bottom cell. The bottleneck with this concept is the low radiation-hardness of the three (In)GaAs-based subcells. In order to bring this concept to a success, we are working on advanced cell designs with several distributed Bragg reflectors for improved radiation-hardness. A scalable metal-bonding process has already been established.
Another cell approach assembles the optimal bandgap combination from subcells grown on different substrates. Thus, we follow an approach of direct semiconductor bonding with transparent layers. Fig. 11 shows a scanning acoustic microscopy image (SAM) of two solar cell wafers based on GaAs and InP substrates bonded at Fraunhofer ISE with a very homogeneous bonding area and only a very small number of voids. These voids can be correlated to surface defects on the bonding interfaces as shown in Fig. 11 on the right.

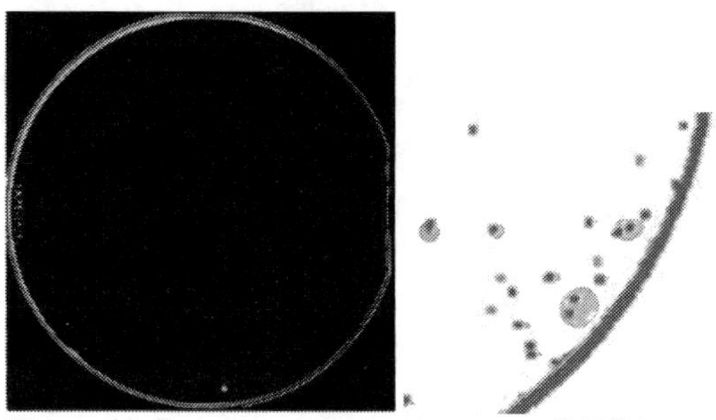

Figure 11. Scanning acoustic microscopy (SAM) image of a cells based on InP and GaAs wafers successfully semiconductor bonded (left). Voids (grey areas) arise from larger surface defects (right, dots).

6. SUMMARY

AZUR's space qualified 3G30C-Advanced provides the world's highest EOL efficiencies and is available in several sizes and formats up to 82 cm^2. The cells may be thinned down to 80 µm or less in order to reduce weight.
Solar electric orbit raising requires radiation-hard designs up to an equivalent 1 MeV electron fluence of 5 E16 cm^{-2}. This demand can be met by the enhanced cell design of the 3G30C-1E16+, yielding efficiencies of 21.2% at 1 E16 cm^{-2} 1 MeV electrons.
The third market segment of LEO missions requires high BOL efficiencies. AZUR SPACE has developed a 3G31C cell design to meet these demands.
The 4G32C is AZUR's next generation product, succeeding the 3G30C. It bases upon a four-junction upright metamorphic solar cell structure and implements AZUR's proprietary EOL design. The cell achieves extraordinary EOL efficiencies of 27.8% at 1 E15 cm^{-2} 1 MeV electrons. Qualification will be completed in 2017. The 4G32C-Advanced will reach 28.5% at 1 E15 cm^{-2} 1 MeV electrons

7. ACKNOWLEDGEMENT

The authors would like to thank the colleagues from AZUR SPACE and Fraunhofer ISE for their support. Part of the R&D work presented in this paper was supported by the European Space Agency (ESA contracts No. 4000112560/14/NLFE and 4000118033/16/NL/CLP) and the German Aerospace Agency DLR (Contract No. 50RN1501)

8. REFERENCES

1. Guter, W., Ebel, L., Fuhrmann, D., Köstler, W. & Meusel, M. (2014). Development, Qualification and Production of Space Solar Cells with 30% EOL Efficiency, Proc. of the European Space Power Conference, Noordwijkerhout.

2. Geisz, J., Kurtz, S., Wanlass, M., Ward, J., Duda, A., Friedman, D., Olson, J., McMahon, W., Moriaty, T. & Kiehl, J. (2007). High-efficiency GaInP/GaAs/InGaAs triple-junction solar cells grown inverted with a metamorphic bottom junction, APL, vol. 91, p. 23502/1.

3. Meusel, M., Bensch, W., Fuhrmann, D., Guter, W., Kern, R., Köstler, W, Löckenhoff, R., Strobl, G.F.X., Klinger, V., Siefer, G., Welser, E., Dimroth, F. & Bett, A. (2010). III-V Multijunction Solar Cells - From Current Space and Terrestrial Products, Proc. of the 25th EUPVSEC, Valencia, p. 18.

4. Strobl, G.F.X., Fuhrmann, D., Guter, W., Khorenko, V., Köstler, W. & Meusel, M. (2012). About AZUR's "3G30-advanced" Space Solar Cell and Next Generation Product, Proc. of the 27th EUPVSEC, Frankfurt, p. 1BO10.3.

5. Zazoui, M., Warner, J.H., Summers, G.P., Messenger, S.R. & Lorentzen, J.R. (2005). Radiation response mechanisms in multifunction III/V space solar cells, Conf. Rec. 35th IEEE Photovoltaic Spec. Conf., 2005, pp. 542–547.

6. Brandt, C., Baur, C., Caon, A., Müller-Buschbaum, P., Zimmermann, C. & Andreev, T. (2013). The Influence of High Temperatures on Radiation Damage of $GaInP_2$/GaAs/Ge Triple Junction Cells, IEEE Journal of Photovoltaics, Vol. 3, No. 2.

7. Khorenko, V., Strobl, G.F.X., Guter, W., Fuhrmann, D., Ebel, L., Kern, R., Meusel, M. & Köstler, W. (2014). Roadmap for the next generations of European Space Solar Cells, Proc. of the 29th EU PVSEC, Amsterdam, 4DO.13.1

8. ESA Contract No. 4000118033/16/NL/CLP, Next Generation Upright Metamorphic 4-Junction Space Solar cells.

DOI: 10.1051/e3sconf/20171603006

ADVANCES IN DILUTE NITRIDE MULTI-JUNCTION SOLAR CELLS FOR SPACE POWER APPLICATIONS

F. Suarez [(1)], T. Liu [(1)], A. Sukiasyan [(1)], J. Lang [(1)], E. Pickett [(1)], E. Lucow [(1)], T. Bilir [(1)], S. Chary [(1)], R. Roucka[(1)], I. Aeby [(1)], L. Zhang [(1)], S. Siala [(1)]

[(1)] Solar Junction Corp., 401 Charcot Avenue, San Jose, CA 95131 U.S.A, Email: fsuarez@sj-solar.com

ABSTRACT

A sub-cell with bandgap of around 1 eV is required to improve the efficiency of multi-junction solar cells beyond what is possible with legacy triple-junction architectures [1]. Solar Junction Corporation has been focused since 2007 on the development and commercialization of dilute nitride materials to be used as the 1eV sub-cell in a fully lattice matched multi-junction solar cell. Initial focus on the terrestrial concentrating photovoltaics (CPV) market led to Solar Junction Corp.'s achievement of multiple world records in multi-junction solar cell efficiency with its triple-junction cells on GaAs [2], [3]. These solar cells have been available as commercial products since 2010. Solar Junction Corp. has leveraged its high-quality, manufacturable dilute nitride material to develop and introduce an entirely new class of space solar cells capable of reaching over 33% AM0 conversion efficiency in a four-junction (4J) configuration lattice matched to active Ge substrates, with a clear line of sight to 36% AM0 efficiency in five- or six-junction devices that can be manufactured more cost-effectively than devices relying on metamorphic technologies.

In this paper, we review the latest performance and qualification results of Solar Junction Corp.'s lattice matched 4J-on-Ge space solar cells and CIC (Cell-Interconnect-Coverglass) products incorporating GaInNAsSb dilute nitride material. We also report on the production readiness of these advanced space solar cells manufactured using an optimized hybrid Molecular Beam Epitaxy (MBE) / Metal Organic Vapor Phase Epitaxy (MOVPE) growth process.

1. INTRODUCTION

Within the multi-junction solar cell industry, it has long been known that to improve efficiency beyond what is possible with existing triple-junction solar cells, there needs to be a material that can provide a 1eV bandgap. GaInNAsSb, a material variant historically referred to as 'dilute nitride,' is the only known material that absorbs in the IR and is lattice-matched to other materials used in multi-junction solar cells [4], [5]. By adjusting the amount of nitrogen and indium in GaInNAsSb, the lattice constant remains matched to GaAs and germanium, thereby preserving the existing reliability paradigm of lattice-matched compound semiconductor structures while enabling the bandgap to be tuned from 0.8 eV to 1.4 eV, as shown in Fig 1.

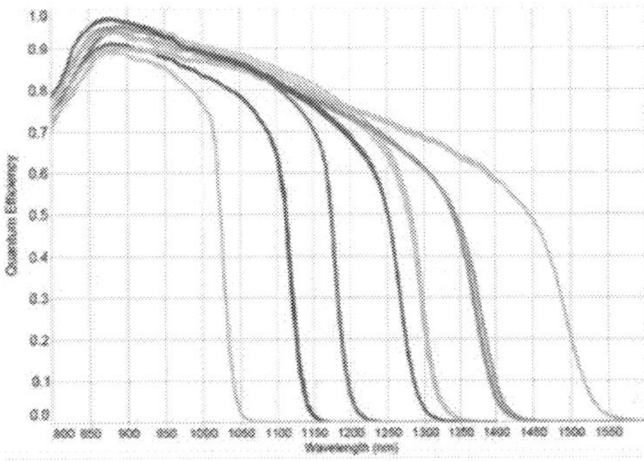

Figure 1. Band gap tunability characteristics of Solar Junction's Dilute Nitride material.

Developing high quality dilute nitride material for a 1eV sub-cell proved to be challenging despite substantial investment and development efforts by numerous industrial and research groups over many years. Solar Junction Corp. has taken a leadership role in improving the dilute nitride material (e.g. the use of Sb as a surfactant). These efforts have resulted in record efficiencies achieved in GaInNAsSb-containing triple junction (3J) solar cells [2], [3] and enabled Solar Junction Corp. to develop an extensive U.S. and international patent portfolio covering dilute nitride materials for solar cells.

Initially developed to meet the needs of terrestrial CPV systems, Solar Junction Corp.'s high-reliability lattice-matched solar cell technology has now been adapted for use in the AM0 illumination condition of space. This technology forms the foundation for Solar Junction Corp.'s multi-generational efficiency roadmap, depicted in Fig. 2, which is expected to enable the development and production of devices exceeding 36% AM0 conversion efficiency.

Lattice matched 3J space cells on GaAs with 31% conversion efficiency have already been developed [6]. Solar Junction Corp.'s 3J dilute nitride cells have been operating in low earth orbit on two picosats launched in June 2014 by The Aerospace Corporation, and have shown excellent on-orbit performance, with equivalent or superior degradation factors when compared to conventional space qualified InGaP/InGaAs/Ge 3J cells.

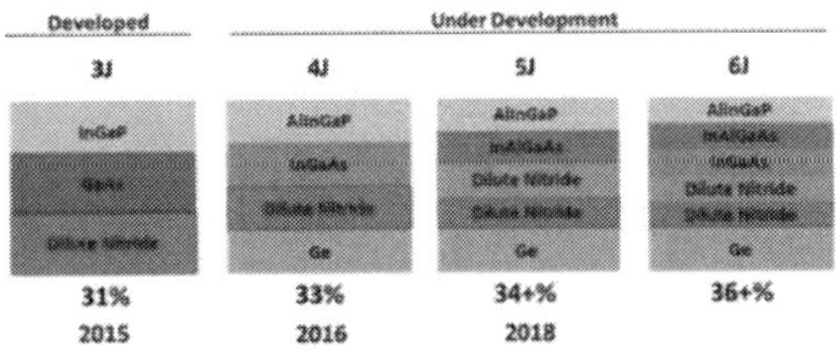

Figure 2. Solar Junction Corp. multi-junction space solar cell roadmap.

Fig. 3 shows the baseline and on-orbit Current vs. Voltage of these 3J dilute nitride solar cells [7].

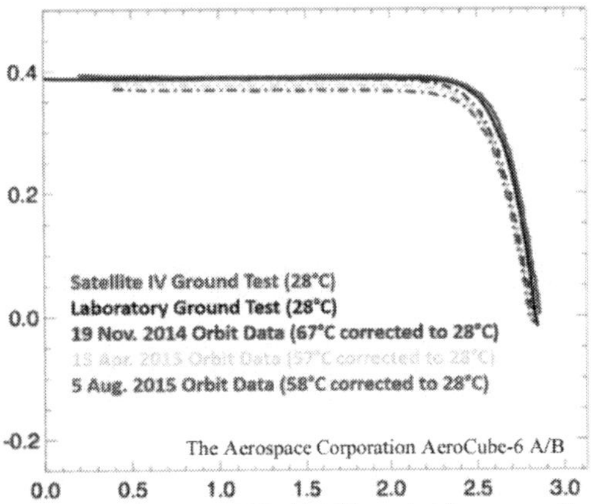

Figure 3. Change in I-V of 3J dilute nitride space cells in low-earth orbit since June 2014 [7] (Courtesy of The Aerospace Corporation)

2. 4J Ge SPACE SOLAR CELLS

A 4J space solar cell product incorporating an upright growth, lattice matched dilute nitride structure on heritage Ge substrates is currently under development with a target conversion efficiency of 33% AM0 by the end of this year. The structure of the 4J device is shown (inset) in Fig. 4.

Recently, Solar Junction Corp. successfully developed and demonstrated prototype 4J cells with 33% AM0 conversion efficiency. Electrical performance characteristics of the prototype 4J dilute nitride solar cell are shown in Fig. 3. To the best of our knowledge, this is the highest AM0 efficiency ever recorded in a lattice-matched solar cell. Current density versus voltage of a prototype cell with AM0 conversion efficiency of 33% is shown in Fig. 4(b) and its corresponding quantum efficiency is as shown in Fig. 5.

Preliminary stress testing of 4J lattice matched solar cells has confirmed the reliability and robustness of this design. Cells subjected to humidity exposure at 45°C, 95% RH, for 1000 hours remained stable.

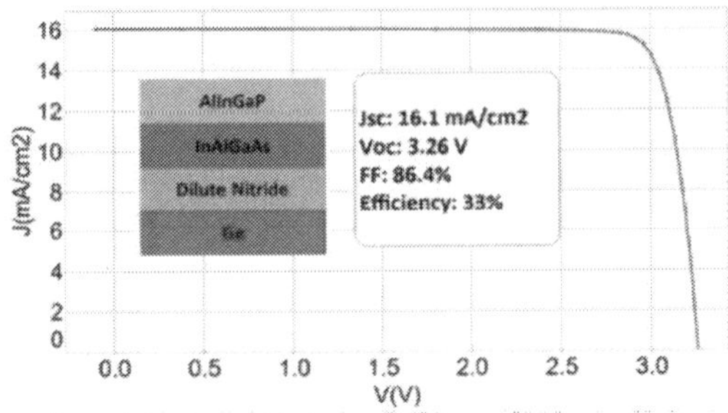

Figure 4. Current Density vs. Voltage of (inset) lattice matched 4J on Ge space cell.

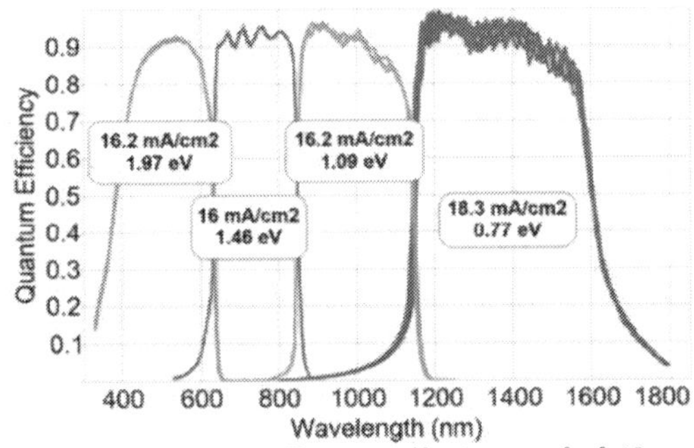

Figure 5. Quantum efficiency of lattice matched 4J on Ge incorporating dilute nitride junction.

Additionally, these 4J cells have performed well under irradiation and temperature annealing tests per ECSS-E-ST-20-08C. Proton irradiation effects were evaluated on sample 4J cells and the results show similar degradation to present generation space qualified InGaP/InGaAs/Ge triple junction solar cells. The results of these proton radiation effects tests are shown in Fig. 6 and Fig. 7. An end-of-life (EOL) design is planned to further enhance the radiation performance of these 4J cells.

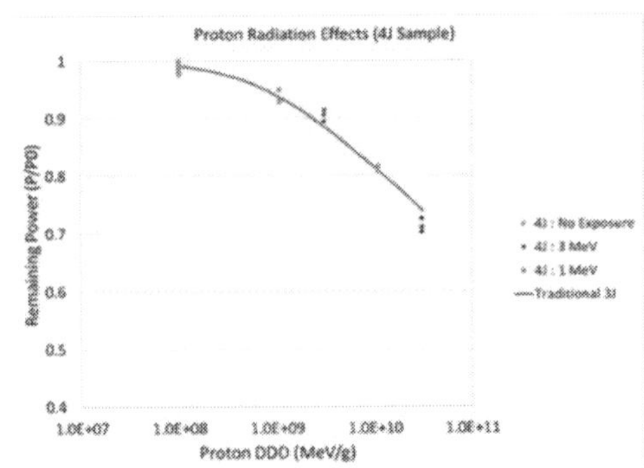

Figure 6. Remaining Power (P/Po) vs. Proton Displacement Damage of lattice matched 4J on Ge cell.

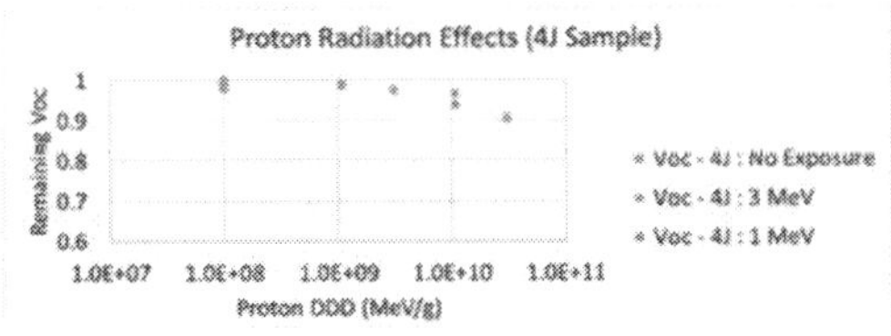

(a)

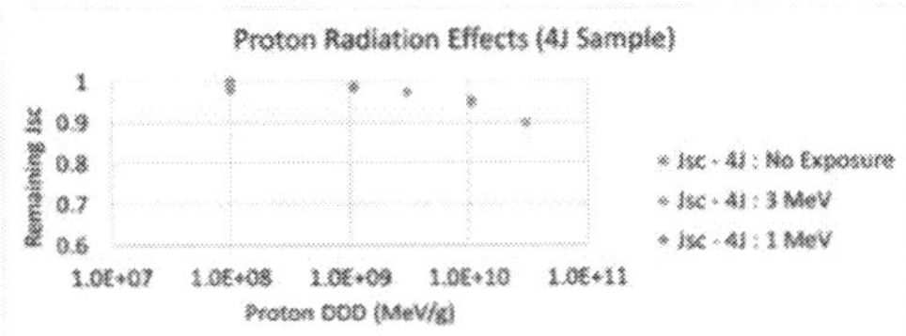

(b)

Figure 7. Remaining Voc (a) and Jsc (b) vs. Displacement Damage from 1 and 3 MeV protons for lattice matched 4J on Ge space cell.

3. MANUFACTURING OF LATTICE MATCHED 4J SPACE CELLS

The improved solar cell efficiency provided by the lattice matched GaInNAsSb material system is realized using a proven, cost effective, and relatively simple epitaxy manufacturing process that is consistent with the production of present generation space solar cells. Incorporating Solar Junction Corp.'s lattice matched GaInNAsSb material does not require increased thickness of the material stack to accommodate defect suppression layers, nor does it require additional capital assets for complex operations such as epitaxial release and substrate re-bonding. The hybrid MBE/MOVPE growth employed in creating the 4J epitaxial structure leverages the advantages of both growth techniques and allows for optimal cost and performance for the epitaxial wafers. Additionally, uniform epitaxy is achieved consistently, enabling high performance large area solar cells such as the legacy 30 cm^2 '2-per' (see Fig. 8) and 59.5 cm^2 '1-per' form factors produced on 100 mm diameter wafers.

Solar Junction Corp.'s lattice matched 4J on Ge space solar cells provide a drop-in replacement for heritage 3J cells with similar mechanical and thermal characteristics, making it possible to leverage existing integration equipment and heritage materials and processes for the manufacture of the cell and the integration of the cell into a CIC device.

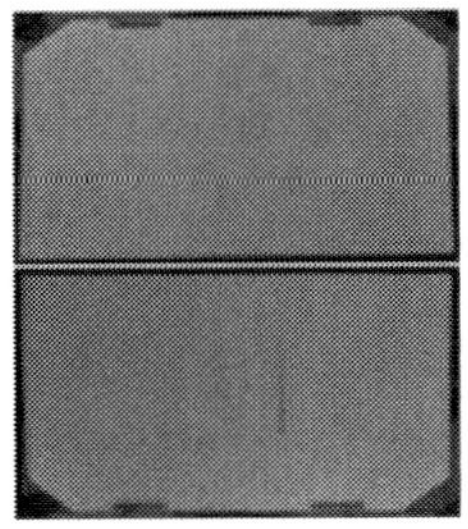

Figure 8. Electroluminescence of 2-per 4J space cells.

4. SUMMARY

Building on Solar Junction Corp.'s breakthrough development of 3J-on-GaAs solar cells incorporating lattice matched GaInNAsSb dilute nitride material for the CPV market, we now report the development and demonstration of lattice matched, dilute nitride-based 4J-on-Ge space solar cells achieving 33% AM0 conversion efficiency. Solar Junction Corp. is currently developing its CIC assembly capabilities and performing space qualification testing for its 4J product, with commercial production of this new generation of space solar cell planned for 2017.

5. REFERENCES

1. Friedman, D.J., Geisz, J.F., Kurtz, S.R. & Olson, J.M. (1998). 1-eV Solar Cells with GaInNAs Active Layer. J. of Crystal Growth 195, (1-4), pp409-415.
2. Wiemer, M, Sabnis, V & Yuen, H. (2011). 43.5% Efficient Lattice Matched Solar Cells. in Proc. of SPIE, 810804-5.
3. Jones-Albertus, R., et al. (2013). Pushing Solar Cell Efficiencies Even Higher. CLEO - Applications and Technology, San Jose, California, USA.
4. Kondow, M, et al. (1996) Jpn. J. Appl. Phys. (35), pp11273-75
5. Kurtz, S.R., Allerman, A.A., Jones, E.D., Gee, J.M. Banas, J.J. & Hammons, B.E. (1999) "InGaAsN Solar Cells with 1.0 eV Band Gap, Lattice Matched to GaAs", Appl. Phys. Lett. 74 (5), pp729-731.
6. Suarez, F. et al. (2015). Presentation at the 33rd Annual Space Power Workshop, Manhattan Beach, California, USA.
7. Liu, S., et al. (2015). Presentation at the 33rd Annual Space Power Workshop, Manhattan Beach, California, USA.

NOVEL CONCEPTS FOR HIGH-EFFICIENCY LIGHTWEIGHT SPACE SOLAR CELLS

F. Cappelluti[(1)], G. Ghione[(1)], M. Gioannini[(1)], G. Bauhuis[(2)], P. Mulder[(2)], J. Schermer [(2)], M. Cimino[(3)], G. Gervasio[(3)], G. Bissels[(4)], E. Katsia[(4)], T. Aho[(5)], T. Niemi[(5)], M. Guina[(5)], D. Kim [(6)], J. Wu [(6)], H. Liu [(6)]

[(1)] Department of Electronics and Telecommunications, Politecnico di Torino, Cso Duca degli Abruzzi 24, 10129 Torino, Italy, Email: federica.cappelluti@polito.it
[(2)] Institute for Molecules and Materials, Radboud University, Heyendaalseweg 135, 6525 AJ Nijmegen, The Netherlands, Email:g.bauhuis@science.ru.nl, p.mulder@science.ru.nl, j.schermer@science.ru.nl
[(3)] Thales Alenia Space, Strada Antica di Collegno 253, 10146 Torino (Italy), Email: Giuseppe.Gervasio@thalesaleniaspace.com
[(4)] tf2 devices B.V., Heyendaalseweg 135, 6525 AJ Nijmegen, The Netherlands, Email:gunther.bissels@tf2devices.com, eleni.katsia@tf2devices.com
[(5)] Optoelectronics Research Centre,Tampere University of Technology, Korkeakoulunkatu3,33720,Tampere, P.O.Box692, 33101Tampere, Finland, Email: mircea.guina@tut.fi
[(6)] Department of Electronic and Electrical Engineering, University College London, Torrington Place, London WC1E 7JE, United Kingdom, Email: huiyun.liu@ucl.ac.uk

ABSTRACT

One of the key issues in the design and development of a satellite Photovoltaic Assembly (PVA) is the trade-off to be made between the available volume located to the PVA, its mass and the total amount of power that the solar panels have to guarantee to the spacecraft. The development of high-efficiency, flexible, lightweight solar cells is therefore instrumental to the design of future satellites providing enhanced missions and services. Based on the consolidated development of GaAs-based single junction and lattice matched triple-junction solar cells, several research efforts are being pursued worldwide to further increase the efficiency and reduce mass. Promising approaches include thin-film technologies such as Inverted Metamorphic and Epitaxial Lift-Off (ELO), and the use of nanostructures or highly mismatched alloys grown by MBE. We propose here an alternative path towards the development of lightweight GaAs-based solar cells with the potential to exceed the Shockley-Queisser (SQ) limit of single junction cells. Our approach is based on the synergistic combination of thin-film design, quantum dots (QDs) absorption, and photonic nanostructures. Challenges and opportunities offered by the use of QDs are discussed. A cost-effective and scalable fabrication process including ELO technology and nanoimprint lithography is outlined. Finally, a proof-of-concept design, based on rigorous electromagnetic and physics-based simulations, is presented. Efficiency higher than 30% and weight reduction close to 90% - owing to the substrate removal - makes the proposed device to rank record power-to-weight ratio, with the potential to become a cost-effective, attractive option for next generation space solar cells.

1. INTRODUCTION

Today's industry standard for space solar arrays is provided by triple-junction III-V solar cells on a GaAs or Ge substrate having efficiency of 30%. In these cells the typical substrate thickness is about 150-200 μm (notice that the useful active device thickness for photovoltaic conversion is less than 5 μm), resulting in an average weight of the bare solar cell close to 90 mg/cm^2 and a power–to weight ratio of about 400 W/kg. Besides their large weight, such cells are not flexible. Indeed, the development of flexible high efficiency solar cells will enable the design of new solar arrays architectures such as rollable solar arrays or inflatable ones with the possibility to be used in satellites (commercial and scientific) or moon bases. At the satellite level, increased power-to-weight ratio and flexibility may allow for knock-on advantages in terms of launch cost savings or payload enhancement or higher performances, depending on the demands of the specific application. Thus, there is a strong effort worldwide to develop new III-V cell technologies aiming at higher efficiency and lower mass. In these approaches, the wafers only serve as a production template and are removed to obtain genuine thin-film solar cells.

Besides remarkable mass and cost reduction (if wafer reuse is possible) the thin-film approach is theoretically superior since it allows for a substantial increase of efficiency in radiative-limited cells, like the GaAs-based one, through photon recycling. Thin-film configurations with back-side mirrors have been the key to the most recent efficiency record of 28.8% [1]. By further optimization of the back reflector, an efficiency higher than 30% is within reach. Nano-engineering of the semiconductor material might yield even higher

 DOI: 10.1051/e3sconf/20171603007

efficiency with a less complex technology than multi-junction cells. Quantum dot cells have been extensively investigated, motivated by the idea of extending the photon harvesting at longer wavelengths and improve the short circuit current taking advantage of sub-bandgap optical transitions enabled by QDs. However, one of the main shortcomings of this approach is the small QD optical cross-section, that has prevented up to now the demonstration of a significant efficiency increase with respect to single-junction cells. In this respect, the thin-film geometry is a prerequisite to the implementation of light trapping techniques, that can allow for a large enhancement of the QD short circuit current. In particular, nanostructuring the bottom or top wide-gap window layers, combined with suitable reflectors and Anti-Reflection Coating (ARC) layers, may provide efficient light trapping within the absorbing layers. Nanostructuring of the thin-film cell may be realized through a cost-effective technology such as Nanoimprint Lithography (NIL) that allows to pattern even subwavelength periodic gratings over large areas.

2. TECHNOLOGY

2.1 THIN-FILM FLEXIBLE CELLS BY EPITAXIAL LIFT-OFF

For the current generation of InGaP/GaAs/Ge triple junction cells, the Ge wafer on which they are produced is an integral part of the cell structure. However, for the next generation solar cells the Ge (or GaAs) wafers might only serve as a production template that can be removed to obtain genuine thin-film solar cells. The enabling technology for this approach will be the Epitaxial Lift-Off (ELO). ELO allows removal of the III-V photo-sensitive layers in such a way that the wafers can be reused.

For this purpose, a typically 10 nm thick $Al_xGa_{1-x}As$ ($x > 0.6$) release layer is deposited before the actual III-V cell structure. This sacrificial layer allows for the separation of the device structure from its native semi-conductor wafer by selective wet etching using an aqueous HF solution [2]. The thin, single crystal cell structures obtained by the ELO technology can be mounted on low-cost foreign carriers, like metal or plastic foils, to achieve stability during further processing. The fact that these carriers can be selected on the basis of their material properties rather than of crystal growth demands, opens the way for the development of innovative device structures. Owing to the large selectivity ($>10^6$) of the HF solution for etching of Al(Ga)As over GaAs and Ge, the expensive semi-conductor wafer is not affected, and can be reused as a production platform for subsequent thin-film cells [3].

ELO involves lateral etching of the release layer. Nevertheless, increasing the size of the structures subjected to the process does not impose fundamental problems, as long as the released part of the thin-film structure is bent away during the process, leaving the etch front easily accessible for the etch solution [2]. In this way, the process was up-scaled from (typically) 1 cm^2 samples to complete 2" diameter and further to 4" diameter wafers (see Fig. 1). It was furthermore demonstrated that the technology can be applied to III-V cells from any supplier of epi-structures if the thin-film structure on the wafer is properly prepared to enable the ELO process step.

Besides reusing the semiconductor wafers, the ELO technology also enables the production of unique thin-film device geometries that allow for a substantial increase in cell efficiency by photon recycling. Since GaAs solar cells are limited by radiative recombination, the application of a back-contact/mirror thin-film cell geometry resulted in thin-film GaAs cells with a world record efficiency of 26.1% in 2009 [4]. Subsequently, theoretical studies indicated that additional improvements in the back-side mirror reflectivity beyond 95% will increase the cell efficiency super-linearly [5], resulting in the current 28.8% efficiency record reported by Alta Devices [1]. Further improvements in this respect have the potential to yield single-junction thin-film GaAs cells with an efficiency exceeding 30%.

Figure 1. Two high-efficiency, thin-film III-V cells on a 2" diameter metal foil acting as mirror and back contact produced at Radboud University (left) and a large number of small thin-film III-V cells for concentrator system applications produced on a 4" diameter foil at tf2-devices (right).

2.2 QUANTUM DOT MATERIAL

For the implementation of high-efficiency QDSCs, it is essential to achieve a substantial improvement in the optical absorption of QD arrays by increasing the number of QDs. To this end, *Sugaya et al.* have reported ultra-high stacked InGaAs/GaAs QDSCs [6]. A 400-stack QD structure was grown without using a strain balancing technique, which led to the total QD density of 2×10^{13} cm^{-2}. Also, for high in-plane QD density, we have previously demonstrated Sb-mediated InAs QD growth [7]. In-plane QD density of 1×10^{11} cm^{-2} was achieved by depositing Sb before the QD growth, which led to an increase in the photon absorption, and hence an improvement in the short-circuit current.

In this work, we have grown a high-stack (100-stack)

 DOI: 10.1051/e3sconf/20171603007

InAs/GaAs QD structure with high in-plane QD density, as shown in Fig. 2. High in-plane QD density (8×10^{10} cm^{-2}) is achieved by applying the Sb-mediated QD growth technique, as previously demonstrated [7]. The AFM images in Fig 3 show that the QD structure has a good surface morphology and high in-plane QD density even after 100 stacks of QD layers.

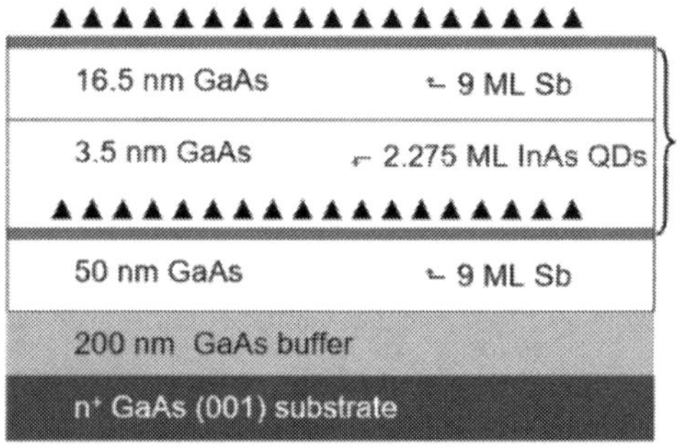

Figure 2. Structure of the high-stack (100-stack) InAs/GaAs QD structure with high in-plane QD density using Sb-mediated growth technique.

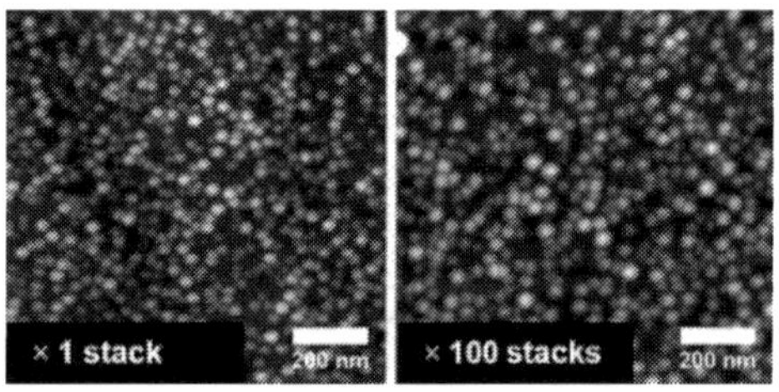

Figure 3. AFM images of InAs QDs (Sb-mediated growth) for 1-stack (left) and 100-stack (right) InAs/GaAs QD structures.

2.3 Nanostructures for photonic management

Nanophotonic gratings will be realized implementing a process already applied for demonstrating moth-eye broad-band antireflection coatings (ARC) [8].

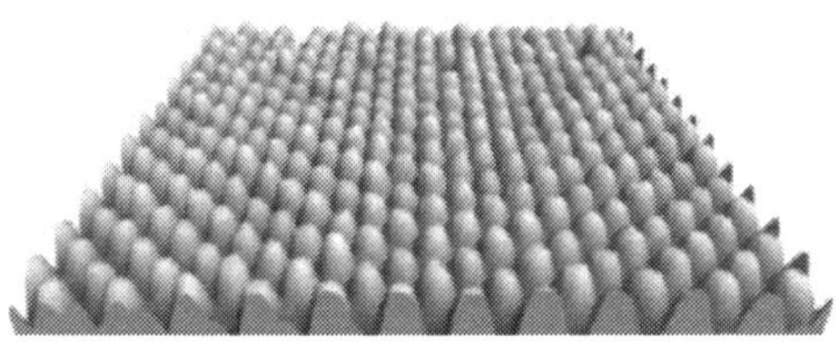

Figure 4. Nanostructured AlInP acting as a broad-band ARC for a GaAs-based solar cell.

In particular, we will use a fabrication process based on replication by Nanoimprint Lithography (NIL) to pattern even subwavelength period gratings over large areas. When employing the NIL technique, the main component of the overall process cost, besides labour and capital expenses, is the cost of the wafer-masters, that is around 2000-3000 €/master. Since the master can be replicated up to ~100 times, mass production enabled by NIL can bring the cost of nano-structuring down to few euros per wafer. Using NIL, wafers with size up to 6" can be patterned in single step. Moreover, since NIL can be integrated as a part of the normal process flow to define contact-grids, the NIL process is deemed to be significantly cheaper (also in terms of scalability) than the standard deposition of dielectric films.

When it comes to defining light-trapping by microstructuring the back contact (which will also act as a reflector) we will pay special attention to the choice of metals forming the contact layer and to the microstructuring strategy, to ensure high reflection without sacrificing the electrical performance of the contact. A preliminary experiment showing the role of the metals forming the back contact reflector is shown in Fig. 5 [9].

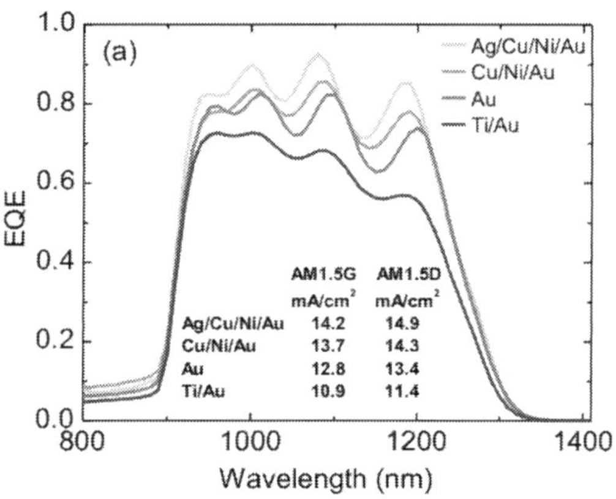

Figure 5 External quantum efficiency for 1-eV solar cell with a back contact reflector formed of different metal combinations. Inset shows the equivalent current generation.

3. PROOF-OF-CONCEPT DESIGN

In this section we discuss, as a proof-of-concept, the design of a realistic configuration for a thin-film InAs/GaAs QD cell with light-trapping and photon-recycling, that makes use of the technological approaches described so far. The design is carried out through rigorous electromagnetic simulations, based on Finite-Difference Time-Domain (FDTD) and Rigorous Coupled Wave Analysis (RCWA), and QD-aware-transport simulations [11].

3.1 Electrical structure and model parameters

To focus the ideas we consider a deep junction design with a thick lightly doped emitter and a thin base layer.

DOI: 10.1051/e3sconf/20171603007

In high quality materials, the deep junction design has been shown to achieve higher open circuit voltage and is particularly suited for taking advantage of the photon recycling effect [13]. A sketch of the device cross-section and the associated energy band diagram at short-circuit condition is shown in Fig. 6. A stack of 20 InAs/GaAs QD layers - embedded in a 420 nm thick intrinsic region - is placed at the bottom of the lightly doped emitter region, next to the junction, in order to maximize carrier collection efficiency at short-circuit.

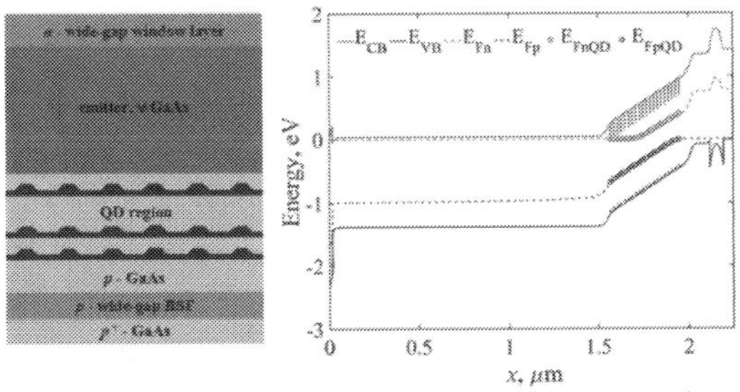

Figure 6. *Left: Cross-section of the analysed QD-based solar cell. An n^+ GaAs cap layer (not shown) is used for the realization of the top metal contact. Right: Calculated energy band diagram at short-circuit condition under AM1.5G illumination.*

Electrical simulations are carried out by including in a classical drift-diffusion-Poisson formulation the detailed description of charge carrier transitions in the QD states. To this aim, QDs are modelled as a three-level system, including electron/hole ground state (GS), excited state (ES) and the two dimensional wetting-layer state (WL). Carrier exchange among QD states and with bulk states occurs through optical and thermal excitation and relaxation processes. The characteristic scattering times and the optical absorption properties assumed for the QD material are summarised in Table 1 and Fig. 7, respectively. As seen in Fig. 7, QDs sub-bandgap states extend the harvested wavelength range beyond the GaAs band edge. However, due to the low absorption of these states, some light management scheme need to be employed to reach a significant enhancement of the short-circuit current, as discussed in Sec. 3.2.

The model also includes barrier radiative recombination, with radiative coefficient of 2.0×10^{-10} cm^3s^{-1}, and defect induced non-radiative recombination modelled according to Shockley Read Hall (SRH) theory, with characteristic lifetime of 0.5 μs. In high quality GaAs crystal, SRH lifetimes as high as 1 μs are achievable [10]. All the presented simulations are done at ambient temperature, T=300 K. Photovoltaic characteristics are computed under 1 sun AM1.5G illumination.

Further details on the modelling approach may be found in [11], [12].

Table 1. *QD parameters*

QD density, N_{QD}, [cm^{-2}]	6×10^{10}
State degeneracy, $g_{ES}^{e,h}$, $g_{GS}^{e,h}$	4, 2
WL Density of States, $N_{WL}^{e,h}$, [cm^{-2}]	2.4×10^{12}
$\Delta E_{B\text{-}WL}^{e}$, $\Delta E_{WL\text{-}ES}^{e}$, $\Delta E_{ES\text{-}GS}^{e}$, [meV]	40, 162, 70
$\Delta E_{B\text{-}WL}^{h}$, $\Delta E_{WL\text{-}ES}^{h}$, $\Delta E_{ES\text{-}GS}^{h}$, [meV]	4, 40, 16
$\tau_{CAP,\,B\to WL}^{e}$, $\tau_{CAP,\,WL\to ES}^{e}$, $\tau_{CAP,\,ES\to GS}^{e}$, [ps]	0.3, 1.2, 1.2
$\tau_{CAP,\,B\to WL}^{h}$, $\tau_{CAP,\,WL\to ES}^{h}$, $\tau_{CAP,\,ES\to GS}^{h}$, [ps]	0.1, 0.1, 0.1
τ_r^{WL}, τ_r^{ES}, τ_r^{GS}, [ns]	1, 1, 1

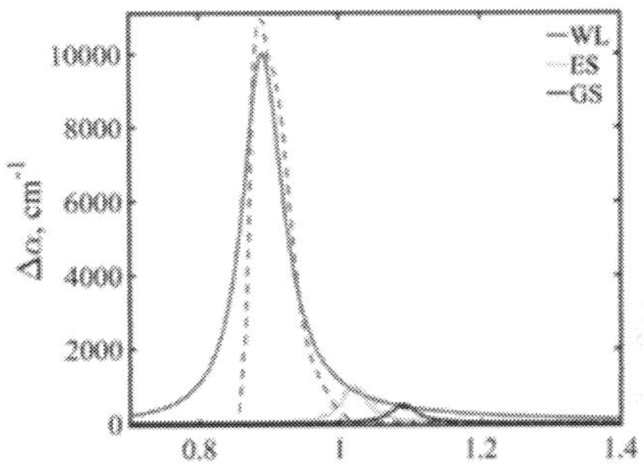

Figure 7. *Wavelength dependence of the absorption coefficient of the QD material as estimated by literature data (solid lines) and approximated by Lorentz oscillators for the sake of electromagnetic simulations (dashed lines).*

3.2 Photonic structures design

For the sake of illustrating the impact of photon management on the QD cell performance, we describe a couple of representative designs: i) a thin-film device combining a nanostructured periodic grating on the top surface with a planar reflector at the rear side; ii) a thin-film device with planar antireflection coating and a periodic grating on the rear side. The first design aims at reaching very low reflectivity over the whole GaAs and QD wavelength range and doubling photocurrent at sub-bandgap wavelengths. The second design aims at investigating the achievable QD photocurrent enhancement through light-trapping schemes.

A sketch of the first structure is reported in Fig.8 (a): the top-surface periodic grating acts as broadband antireflection coating (ARC), with very low reflectivity in the 400-1200 nm range. The top-surface grating may also provide some light-trapping enhancement in the wavelength range of about 800-1100 nm by exciting higher order diffraction modes. The presented results refer to the ARC-optimized geometry realized in [8], with nanocone size features of 80/300/440 nm (top diameter /base diameter /height). With such dimensions, reflectivity well below 5% is achieved in a broadband range covering the GaAs band edge and the QD interband transitions [8].

We estimated the absorption enhancement (Fig. 3 and Table 2) in the GaAs/QD active region through FDTD simulations. Fig. 8 compares the absorbed photon for various device configurations: i) a conventional, wafer-based single-pass cell with planar ARC; ii) a wafer-based cell with nanostructured ARC (the nanocone); iii) a thin-film cell with nanostructured ARC and ideal reflector (modelled as perfect electrical conductor).

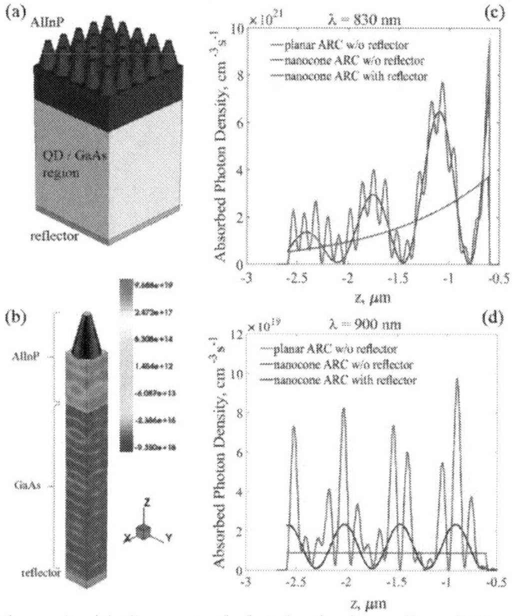

Figure 8. *(a) Conceptual sketch of solar cell exploiting nanostructured ARC and reflector. (b) 3D distribution of the absorbed photon density (cm^{-3}) at λ=900 nm for the device with nanocone ARC and planar reflector. (c), (d) Cutline across the z-axis of the absorbed photon density at λ=830 nm and λ=900 nm, respectively, for the various structures discussed in the text.*

Table 2. *Estimated Gain in the Total Absorbed Photon density of the solar cells exploiting the nanostructured ARC with and w/o reflector with respect to a conventional cell (with planar ARC and no reflector), for wavelengths below (830 nm) and above (900 nm) the GaAs band edge.*

Structure	Normalized Integrated Absorbed Photon Density	
	830 nm	900 nm
Nanocone ARC w/o reflector	1.2	1.3
Nanocone ARC with reflector	1.4	2.7

The conventional, exponentially decreasing photo-generation in i) is replaced by significant interference patterns in configurations ii) and iii). As highlighted in Table II, both configurations ii) and iii) provide gain in the total absorbed photon density. At 900 nm (where the GaAs absorption length is ~ 0.1 cm), configuration iii) allows for a factor 2 increase of photogeneration when compared to configuration ii).

The basic structure of the second photonic design, the light-trapping enhanced cell, is shown in Fig. 9. The topmost layer is an ARC which can be realized by a conventional planar thin-film (e.g. silicon nitride SiNx) or by the nanostructured "moth eye" layer previously discussed. On the bottom of the cell we have a periodic micro- or nano- structured AlInP layer acting as diffraction grating and a metal mirror (Ag in the present simulations) to reflect light as efficiently as possible. By proper optimization of the grating geometrical parameters, light is coupled into high order diffraction modes propagating outside of the escape cone, realizing a strong enhancement of the absorbed photon density. An example of the solar cell absorbance (computed through RCWA simulations) for different thin-film configurations, including both lamellar (2D) triangular gratings and pyramidal (3D) gratings is shown in Fig. 10. The absorbance spectra shows a marked increase around the GaAs band-edge and in the QD range.

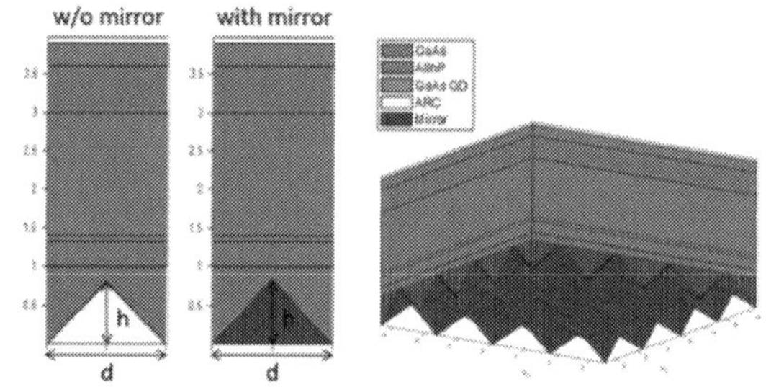

Figure 9. *Structure of the thin-film cell with rear –side pyramidal grating.*

From the absorbance spectra in Fig. 10 we can estimate the photogenerated current in a specific wavelength range under a specific incident solar irradiance. The grating geometry is optimized to maximize the photocurrent in the QD range (895 nm - 1200 nm). 2D and 3D diffraction gratings with optimized geometry provide a QD photocurrent of about 1.47 mA/cm^2 and 4 mA/cm^2, respectively, against 0.3 mA/cm^2 and 0.49 mA/cm^2 provided by the single-pass configuration and the thin-film one with planar mirror. Thus, an increase of 5 (2D grating) and 13 (3D grating) times is achieved by the light-trapping enhanced thin-film cells with respect to the conventional wafer-based cell.

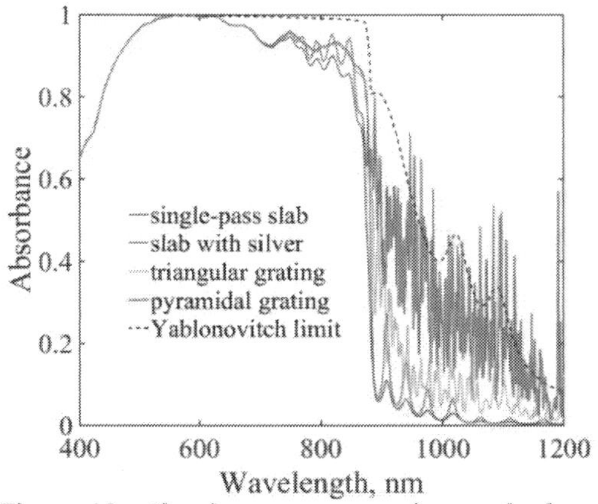

Figure 10. *Absorbance spectra of several photonic configurations: wafer-based (single-pass), thin-film with planar silver mirror, thin-film with 2D rear-side triangular grating, and thin-film with rear-side pyramidal grating. The black dashed line indicates the classical Lambertian limit [14].*

3.3 Device-level simulations

Based on the results in Sec. 3.2, we study here the performance of the QD cell structure described in Sec. 3.1 considering a conventional wafer-based configuration (single-pass), a thin-film configuration exploiting a planar mirror (double-pass) and a thin-film configuration with textured rear grating (i.e. implementing efficient light-trapping). The last one is studied in the so called Lambertian limit [15] (full LT), corresponding in the 1D electrical simulation presented here to a maximum photogeneration enhancement of about $2n^2=25$, n being the GaAs refractive index, in the limit of very weak absorption. Details on the numerical approach may be found in [16]. Residual reflection and shadowing loss are accounted for by a wavelength independent attenuation factor of 5% of the incident solar irradiance.

The performance of the QD cells are compared to those of a single-junction GaAs cell with the same epitaxial structure, except for the QD stack region which is replaced by a lightly n-doped GaAs region of equal thickness.

The External Quantum Efficiency (EQE) of the regular GaAs cells and QD cells in the various configurations is analysed in Fig. 11. The enhancement of photo-conversion in the thin-film configurations is well visible at the GaAs band edge and QD wavelengths. For the sake of model validation, the measured EQE of the first fabricated prototypes - wafer configuration - is shown in Fig. 12, pointing out the good qualitative agreement between simulations and measurements in terms of QD behaviour. On the other hand, in the above gap region, the measured EQE is substantially lower than the predicted one owing to the absence of the ARC and to non-optimal material quality.

Despite the significant increase of short-circuit current achieved by the QD cell with light-trapping thin-film

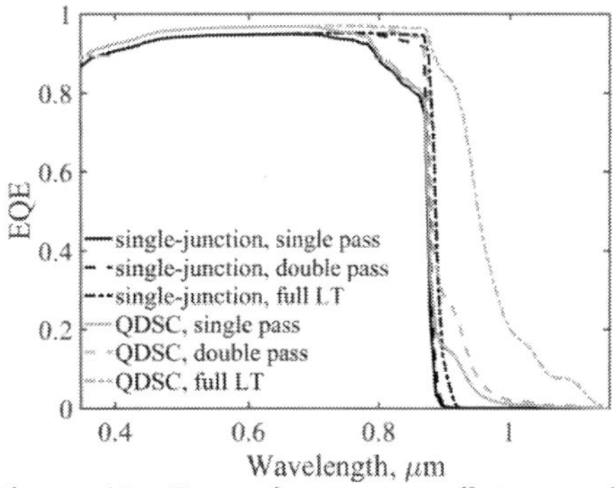

Figure 11. *External quantum efficiency of single-junction and QD thin film cells, for single-pass, double-pass, and full light-trapping (LT) configurations.*

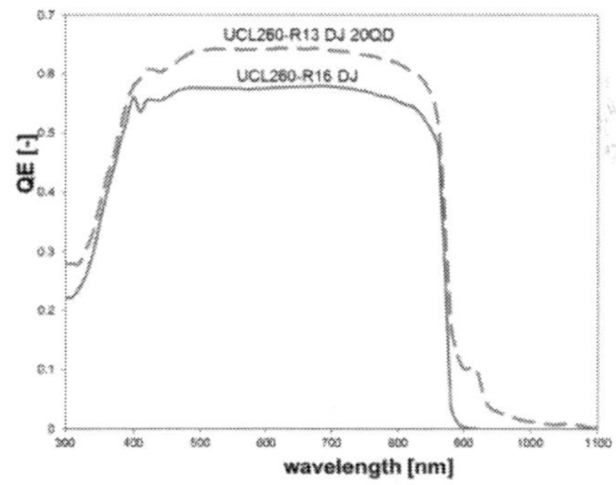

Figure 12. *Measured EQE for the first fabricated wafer-based prototypes: single-junction (solid line) and 20 QD layers QDSC (dashed line).*

configuration, the predicted efficiency (about 24.4 %) remains substantially lower than the efficiency (26.2 %) of the wafer-based regular cell. The reason must be sought in the significant degradation of the open circuit voltage suffered by the QD cell. This is a very well known – and somewhat controversial – issue of QD cells. One of the main causes for such penalty relies on the inherent nature of QDs as radiative recombination centres under high injection conditions [11]. In fact, at open circuit condition, electrically injected carriers undergo relaxation and radiative recombination through the WL and QD confined states making larger the recombination currents across the cell. One possible strategy to overcome such limitation is by exploiting doping in order to fill the QD states and block relaxation and recombination processes. Indeed, by QD selective doping, substantial mitigation of the open circuit voltage has been experimentally observed in [17], [18], and theoretically explained in view of the suppression of radiative recombination through QDs [12]. An overview of the impact of QD doping (implemented through QD

DOI: 10.1051/e3sconf/20171603007

direct doping [17]) and of the various photonic configurations on the achievable efficiency of the QD cells is reported in Fig. 13. For the sake of comparison, the efficiency of the corresponding regular GaAs cells is also shown. A significant efficiency increase is achieved in both single-junction and QD thin-film cells with respect to their wafer counterpart. Moreover, in the light-trapping configuration, an absolute improvement of about 0.7% is achieved by the thin-film QD cell with optimum doping level (around 14 e/dot) with respect to the thin-film single junction cell. On the other hand, even though promising, the predicted efficiency values still fall below the target efficiency of 30%. As already mentioned, the recent efficiency record in GaAs solar cells has been achieved by photonic enhancement of the open circuit voltage [5]. In fact, for high quality material, radiative recombination becomes the dominant loss mechanism in the cell and its effective suppression through the so called photon-recycling effect is required to approach the SQ limit. To make significant the re-absorption of the radiative emission and the consequent open circuit voltage enhancement, the emitted photons must be trapped within the cell active region. Owing to the isotropic nature of radiative emission, surface texturing is not necessary, but very high reflectivity at the back-surface mirror is needed to make single-junction GaAs cells to reach, and possibly overcome, 30% efficiency [5]. The impact of photon-recycling may be expected to be even more dramatic in high-quality QD solar cells wherein, as previously discussed, the ultimate limit to the attainable photovoltage is set by radiative recombination through the QDs themselves.

Thus, having established a baseline design for the regular and QD cells, we investigate to what extent photon recycling may enhance their maximum achievable efficiency. To this aim, we follow the approach in [10], introducing in the electrical model a radiative recombination rate (both for the bulk and QD recombination rate) that scales inversely with the Photon Recycling (PR) factor, defined as the ratio between the amount of re-absorbed photons and the radiatively emitted photons in the cell active region.

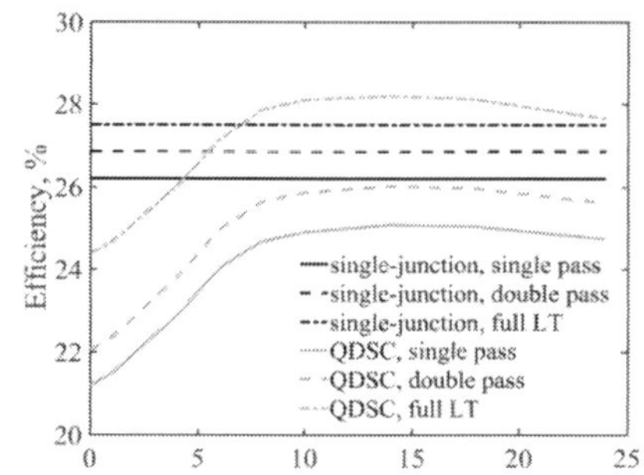

Figure 13. *Efficiency of the QD cell vs. QD doping density for the different optical configurations. The horizontal lines show the efficiency of the benchmark single-junction cell.*

Clearly, the PR factor depends on the detailed photonic configuration of the cell. Just to focus the ideas, one could notice that some amount of photon recycling is always present even in wafer-based cells and strongly enhanced by placing a high reflective mirror in thin-film devices. Calculations in [10] for a simple 2 μm thick GaAs cell report a 80% PR factor for the wafer-based configuration, and higher than 95% PR factor for a thin-film configuration with Au mirror.

Fig. 14 shows the influence of the PR mechanism on the short-circuit current, open circuit voltage, and efficiency of regular GaAs cells and QD cells in the double-pass and light-trapping configuration (including a highly reflective mirror on the backside of the grating, see Fig. 9). As expected, photon-recycling mostly affects the open circuit voltage, with a superlinear increase at high PR factor (markedly stronger in the QD cells with respect to the regular ones). There is also a noticeable effect on the short-circuit current which can be explained in view of slightly improved internal quantum efficiency in the GaAs absorption region. Fill Factor (not shown here) shows instead a negligible increase as the PR factor increases. Similar results for single-GaAs cells with planar optics – i.e. the double-pass configuration – have been already reported by others [5], [13], [19], [20]. Finally, from Fig. 14 we may

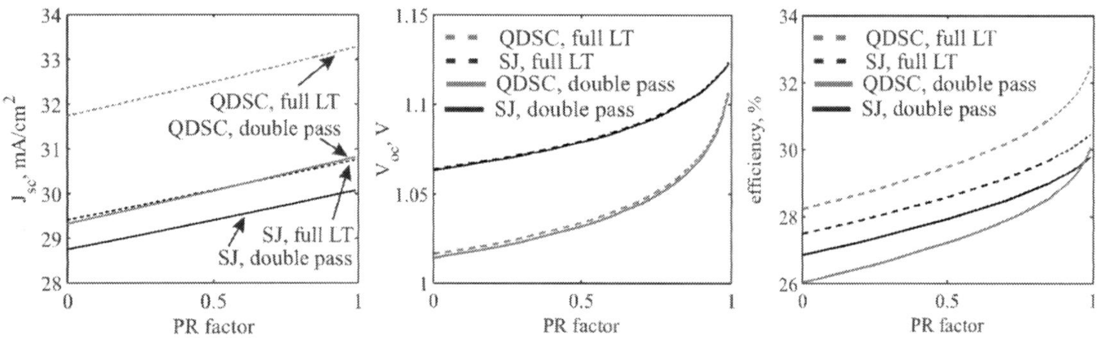

Figure 14. *Short circuit current (top), open circuit voltage (middle) and efficiency (bottom) as a function of the photon recycling factor (PR factor) for the single-junction and 20-layers QDSC cells in the double-pass and light-trapping thin-film configurations.*

observe that the QD double-pass cell overcomes the single-junction one only for PR factor higher than 97%. On the other end, in the full light-trapping scheme, the QD cell shows higher efficiency than the single-junction one regardless of the PR factor, and overcomes the 30% target already for PR factor of 70%. Thus, by optimizing material quality, light-trapping scheme and exploiting high reflective mirrors, QD solar cells with efficiency higher than 30% are feasible.

4. CONCLUSIONS

We have presented a novel approach for the realization of high-efficiency, lightweight and flexible space solar cells, based on the integration of promising technologies such as thin-film fabrication via Epitaxial Lift-Off, light-trapping by means of photonic gratings, and nanostructured semiconductors. Efficient photon management within the thin-film device enables to enhance the sun harvesting by quantum dots to such an extent that the quantum dot cell efficiency will be larger than the efficiency of single-junction bulk cells. The engineering of the quantum dot material and of the photonic management structures are discussed. Based on accurate electromagnetic and physics-based simulation simulations, the two processes are considered together in evaluating the performances of the proposed cell through a proof-of-concept design targeting higher than 30% efficiency.

5. ACKNOWLEDGEMENTS

This project has received funding from the European Union's Horizon 2020 research and innovation programme under grant agreement No 687253.

6. REFERENCES

[1] M.A. Green, K. Emery, Y. Hishikawa, W. Warta, E.D. Dunlop, "Solar cell efficiency tables (version 47)", *Prog. Photovolt: Res. Appl.* 24, 3 (2016).

[2] G.J. Bauhuis, P. Mulder, J.J. Schermer: "Thin-film III-V solar cells using epitaxial lift-off" Chapter 21 in: "High-efficiency solar cells: physics, materials and devices" X. Wang, Z.M. Wang (eds.), Springer Int. Pub. (2013).

[3] G.J. Bauhuis, P. Mulder, E.J. Haverkamp, J.J. Schermer, E. Bongers, G. Oomen, W. Köstler, G. Strobl, "Wafer reuse for repeated growth of III-V solar cells", *Prog. Photovolt: Res. Appl.* 18, 155 (2010).

[4] G.J. Bauhuis, P. Mulder, E.J. Haverkamp, J.C.C.M. Huijben, J.J. Schermer, "26.1% thin-film GaAs solar cells using epitaxial lift-off ", *Sol. Energy Mat. Sol. Cells,* 93, 1488 (2009).

[5] O.D. Miller, E. Yablonovitch, S.R. Kurtz, "Strong internal and external luminescence as solar cells approach the Shockley-Queisser limit", *IEEE J. Photovolt.* 2, 303 (2012).

[6] T. Sugaya, O. Numakami, R. Oshima, S. Furue, H. Komaki, T. Amano, K. Matsubara, Y. Okano, and S. Niki, "Ultra-high stacks of InGaAs/GaAs quantum dots for high efficiency solar cells", *Energy Environ. Sci.*, 5 (3), pp. 6233–6237 (2012).

[7] F. K. Tutu, J. Wu, P. Lam, M. Tang, N. Miyashita, Y. Okada, J. Wilson, R. Allison, and H. Liu, "Antimony mediated growth of high-density InAs quantum dots for photovoltaic cells", *Appl. Phys. Lett.*, 103 (4), p. 043901, (2013).

[8] J. Tommila et al., *Sol. Energy Mat. Sol. Cells*, 94 (10), pp. 1845-1848 (2010).

[9] T. Aho, A. Aho, A. Tukiainen, V. Polojärvi, M. Raappana, and M. Guina, "Optically enhanced GaInNAs solar cell", EUPVSEC 2016 (2016).

[10] M.P. Lumb, M.A. Steiner, J.F. Geisz, and R.J. Walters, "Incorporating photon recycling into the analytical drift-diffusion model of high efficiency solar cells," *J. Appl. Phys.*, 116(19), p.194504 (2014).

[11] M. Gioannini et al., IEEE J. Photovoltaics, vol. 3, no. 4, pp. 1271–1278 (2013).

[12] F. Cappelluti, M. Gioannini, A. Khalili. "Impact of doping on InAs/GaAs quantum-dot solar cells: A numerical study on photovoltaic and photoluminescence behavior," *Sol. Energy Mat. Sol. Cells*, 157, 209-220 (2016).

[13] M. A. Steiner, J. F. Geisz, I. Garcia, D. J. Friedman, A. Duda, 22 and S. R. Kurtz, *J. Appl. Phys.* 113, 123109 (2013).

[14] E. Yablonovitch and G. D. Cody, "Intensity enhancement in textured optical sheets for solar cells", *IEEE Trans. Electron Dev.* 29(2), 300 (1982).

[15] Mokkapati, S. and Catchpole, K. R., "Nanophotonic light trapping in solar cells", *J. Appl. Phys.*, 112, 101101 (2012).

[16] F. Cappelluti, et al., "Numerical Study of Thin-Film Quantum-Dot Solar Cells Combining Selective Doping and Light-Trapping Approaches," 43rd IEEE PVSC, Portland, Oregon, (2016).

[17] P. Lam et al., "Voltage recovery in charged InAs/GaAs quantum dot solar cells," *Nano Energy*, vol. 6, pp. 159 – 166, (2014).

[18] S. Polly, D. Forbes, K. Driscoll, S. Hellstrom, and S. Hubbard, "Delta-doping effects on quantum-dot solar cells," *IEEE J. Photovoltaics*, vol. 4, no. 4, pp. 1079–10 857 (2014).

[19] Wang X, Khan MR, Gray JL, Alam MA, Lundstrom MS. "Design of GaAs solar cells operating close to the Shockley–Queisser Limit," *IEEE J. Photovoltaics*, Apr;3(2):737-44 (2013).

[20] Walker, A. W., et al. "Impact of Photon Recycling on GaAs Solar Cell Designs," *IEEE J. Photovoltaics,* 5.6 636-1645 (2015).

PERFORMACE OF DILUTE NITRIDE TRIPLE JUNCTION SPACE SOLAR CELL GROWN BY MBE

Arto Aho, Riku Isoaho, Antti Tukiainen, Ville Polojärvi, Marianna Raappana, Timo Aho, Mircea Guina

Tampere University of Technology, Optoelectronics Research Centre, Korkeakoulunkatu 3, 33720 Tampere, Finland, Email:arto.j.aho@tut.fi

ABSTRACT

Dilute nitride arsenide antimonide compounds offer widely tailorable band-gaps, ranging from 0.8 eV to 1.4 eV, for the development of lattice-matched multijunction solar cells with three or more junctions. Here we report on the performance of GaInP/GaAs/GaInNAsSb solar cell grown by molecular beam epitaxy. An efficiency of 27% under AM0 conditions is demonstrated. In addition, the cell was measured at different temperatures. The short circuit current density exhibited a temperature coefficient of 0.006 mA/cm^2/°C while the corresponding slope for the open circuit voltage was -6.8 mV/°C. Further efficiency improvement, up to 32%, is projected by better current balancing and structural optimization.

1. INTRODUCTION

Commercial space solar cells have reached 30% beginning of life (BOL) efficiencies [1]. For even higher efficiencies and more efficient spectral utilization, the development of new materials is needed. In lattice matched GaInP/GaInAs/Ge cell the spectral range from 0.7 eV to 1.4 eV is covered by the Ge sub-junction. In this case, there is a high overall quantum defect between photon energies and the band-gap energy of Ge leading to inefficient energy conversion. Consequently, Ge junction is generating approximately only 0.2 V. Lattice matched 1 eV band-gap dilute nitrides, such as GaInNAsSb alloys, offer lower quantum defect and higher voltage generation. The best band-gap voltage offset value demonstrated for dilute nitride junction is 0.49 V [2]. Therefore, with current balanced ~1 eV dilute nitride junction, up to 0.3 V higher operation voltage could be produced than with Ge junction. Projective calculations based on diode modelling and the measured performance of dilute nitride cells have shown that for realistic GaInP/GaAs/GaInNAsSb design, the BOL efficiency could increase approximately 3 percentage points when compared to a Ge-based cell [3]. Moreover, the incorporation of dilute nitrides in a four junction cell with Ge bottom junction could give additional 2 percentage points higher efficiency, yielding ultimately 35% BOL efficiency [3]. Here we demonstrate a triple junction GaInP/GaAs/GaInNAsSb solar cell grown by molecular beam epitaxy (MBE), which sets a roadmap towards demonstrating higher efficiencies and solar cells incorporating higher number of junctions.

2. EXPERIMENTS

The GaInP/GaAs/GaInNAsSb solar cell structure was grown monolithically on p-GaAs substrate using a Veeco GEN20 MBE-system. Detailed description for the fabrication process can be found elsewhere [4, 5]. Subsequent to the growth, the wafer was processed into 4x4 mm^2 solar cells. TiAu and NiAu alloys were used for p- and n-side contacts, respectively. For anti-reflection coating, we used a double-layer TiO_x/SiO_x thin film [6]. For the measurements, we used a spectrally adjustable in-house built three-band solar simulator, which was set to AM0 spectral conditions with separately calibrated single junction GaInP, GaAs and GaInNAsSb cells. ASTM E490 AM0 spectrum (1366 W/m^2) [7] was used for reference. External quantum efficiency (EQE) measurements were performed using an in-house built, monochromator based, measurement system. In addition, for different spectral regions, proper set of band pass filters were used, which ensures monochromatic probe beam. The EQE system response was calibrated with Si and Ge detectors. For the performance analysis and performance projections we used diode modelling employing the parametrization described earlier [3, 8].

3. RESULTS AND DISCUSSION

The current-voltage (*I-V*) performance of the GaInP/GaAs/GaInNAsSb cell illuminated with AM0 three band solar simulator is shown in Fig. 1a. The short circuit current density (J_{sc}) was 17.2 mA/cm^2. Our analysis concluded that the cell was current limited by the GaInP top cell. The open circuit voltage (V_{oc}) and fill factor (*FF*) were 2.67 V and 80%, respectively. The band-gaps for GaInP/GaAs/GaInNAsSb cell are 1.9 eV, 1.4 eV and 0.98 eV, respectively. These values are revealed by the EQE measurement performed at 22°C, which is shown in Fig. 1b. In addition, Fig. 1b reveals the EQE measurement at 80°C. The J_{sc} for the sub-junctions calculated using the EQE measurements are presented in the Tab. 1. For the GaAs sub-junction, the limiting current was estimated from both EQE and using a three band solar simulator measurements to give realistic values. For GaInP top cell and for GaInNAsSb bottom cell the values were directly calculated from the EQE measurements and AM0 spectrum.

 DOI: 10.1051/e3sconf/20171603008

Table 1. Analysis of the EQE measurements.

Sub-cell	Band-gap$_{22°C}$ (eV)	$J_{sc22°C}$ (mA/cm^2)	Band-gap$_{80°C}$ (eV)	$\Delta J_{sc(25\rightarrow 80°C)}$ (mA/cm^2)
GaInP	1.88	17.1	1.85	+0.4
GaAs	1.41	17.7*	1.37	+0.4
GaInNAsSb	0.98	18.5	0.96	-0.2

*Limited EQE measurement performance for the GaAs sub-junction, J_{sc}-values are estimated for GaAs cell based on the EQE and solar simulator measurements.

For the absolute comparison and projective analysis of GaInP/GaAs/GaInNAsSb cell performance, diode modelled *I-V*-characteristics of a current matched cell are also presented in Fig. 1a. The comparison reveals that with further current matching and band-gap tuning, 4 percentage points higher efficiencies are projected for the optimized GaInP/GaAs/GaInNAsSb cell. For better current matching, structural optimization is needed for GaInP and GaAs sub-cells, which we have started with very promising results. Therefore, we believe that the J_{sc} can be increased at least to 17.7 mA/cm^2. The triple junction cell has 0.18 V lower V_{oc} than the simulated cell. This voltage drop can be reduced with band-gap engineering and material improvements.

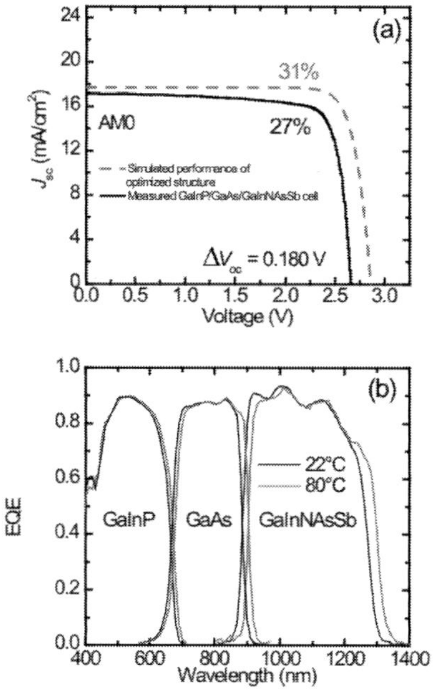

Figure 1. (a) AM0 biased IV-characteristics of GaInP/GaAs/GaInNAsSb solar cell and the simulated performance of GaInP/GaAs/GaInNAsSb cell. (b) EQE characteristics of GaInP/GaAs/GaInNAsSb solar cell measured at 22°C and 80°C.

Approximately 0.1 V higher V_{oc} can be obtained by using higher band-gap for the GaInNAsSb junction. For the remaining voltage gap of 0.08 V, structural and material quality optimizations are the key factors. Ultimately, additional 0.05-0.09 mV improvement for the GaInNAsSb sub-junction might be expected with optimization and improved design, yielding efficiency as high as 32% for the triple junction cell.

Moreover, the temperature coefficients for the cell were defined by the *I-V* measurements performed in the temperature range of 25-90°C. The relative change of the J_{sc} value for the GaInP/GaAs/GaInNAsSb cell is presented in Fig. 2a, while Fig. 2b shows the dependence on the cell V_{oc} as the function of the cell temperature. The J_{sc} temperature coefficient was 0.006 mA/cm^2/°C while the corresponding value for V_{oc} was -6.8 mV/°C. The J_{sc} temperature coefficient calculated from EQE was also 0.006 mA/cm^2.

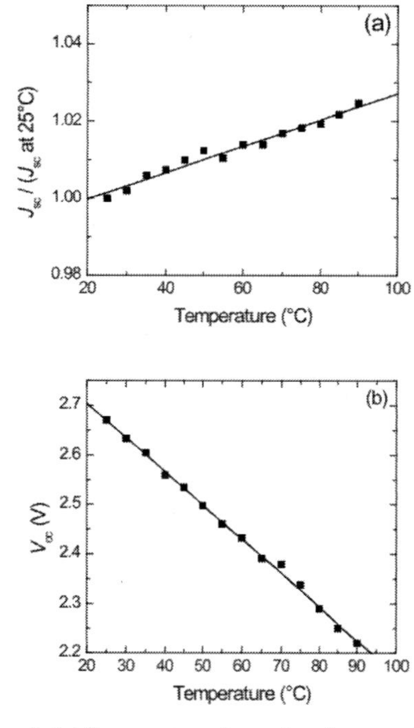

Figure 2. (a) Temperature dependent J_{sc} normalized to J_{sc} at 25°C. (b) Temperature dependence of V_{oc} for GaInP/GaAs/GaInNAsSb cell.

4. CONCLUCION AND FUTURE WORK

We have demonstrated a high performance GaInP/GaAs/GaInNAsSb space solar cell with an efficiency of 27% at AM0 conditions. Using realistic diode model simulation, we show that the efficiency of GaInP/GaAs/GaInNAsSb cell could reach 31%, following current matching and band-gap optimization of the dilute nitride bottom cell. Ultimately, with further optimization of the triple junction cell interfaces and material quality, an BOL efficiency as high as 32% is feasible. We also determined the temperature coefficients for J_{sc} and V_{oc}, which were 0.006 mA/cm^2/°C and -6.8 mV/°C, respectively.

5. ACKNOWLEDGEMENTS

The authors acknowledge the financial support provided by the Finnish Funding Agency for Technology and Innovation, Tekes (projects 'Solar III-V' #40120/09 and 'Nextsolar' #40239/12). This work was also partly performed under a programme of and funded by the European Space Agency (contract N.: 4000108058/13/NL/FE). The view expressed herein can in no way be taken to reflect the official opinion of the European Space Agency.

6. REFERENCES

1. A. W. Bett, F. Dimroth, W. Guter, R. Hoheisel, E. Oliva, S. P. Philipps, J. Schöne, G. Siefer, M. Steiner and A. Wekkeli (2009). Highest efficiency multi-junction solar cell for terrestrial and space applications. 24th European Photovoltaic Solar Energy Conference. 21-25 September 2009, Hamburg, Germany, pp1-6.

2. J. Tommila, A. Aho, A. Tukiainen, V. Polojärvi, J. Salmi, T. Niemi and M. Guina. Moth-eye antireflection coating fabricated by nanoimprint lithography on 1 eV dilute nitride solar cell. Prog. Photovoltaics Res. Appl. vol. 21. pp1158-1162.

3. A. Aho, A. Tukiainen, V. Polojärvi and M. Guina, "Dilute Nitride Space Solar Cells: Towards 4 Junctions," 10th European Space Power Conference ESPC 2014, 13-17 April, 2014, Noordwijkerhout, the Netherlands, vol. 719, pp1-3, 2014.

4. A. Aho, R. Isoaho, A. Tukiainen, V. Polojärvi, T. Aho, M. Raappana and M. Guina. (2015). Temperature coefficients for GaInP/GaAs/GaInNAsSb solar cells. AIP Conference Proceedings. vol. 1679.

5. A. Aho, V. Polojärvi, V. -M. Korpijärvi, J. Salmi, A. Tukiainen, P. Laukkanen and M. Guina. (2014). Composition dependent growth dynamics in molecular beam epitaxy of GaInNAs solar cells. Solar Energy Mater. Solar Cells. vol. 124, pp150-158.

6. A. Aho, J. Tommila, A. Tukiainen, V. Polojärvi, T. Niemi and M. Guina. (2014). Moth eye antireflection coated GaInP/GaAs/GaInNAs solar cell. AIP Conference Proceedings 1616. pp33-36. DOI: http://dx.doi.org/10.1063/1.4897022.

7. ASTM E490: Standard tables for AM0 reference solar spectral irradiances.

8. A. Aho, A. Tukiainen, V. Polojärvi and M. Guina, Performance assessment of multijunction solar cells incorporating GaInNAsSb. Nanoscale Res. Lett. vol. 9, pp1-7.

 DOI: 10.1051/e3sconf/20171603009

STATUS OF FOUR-JUNCTION CELL DEVELOPMENT AT FRAUNHOFER ISE

D. Lackner[(1,*)], O. Höhn[(1)], A.W. Walker[(1)], M. Niemeyer[(1)], P. Beutel[(1)], G. Siefer[(1)], M. Schachtner[(1)] V. Klinger[(1)], F. Oliva[(1)], K. Hillerich[(2)], T. Kubera[(2)], W. Guter[(2)], A.W. Bett[(1)], F. Dimroth[(1)]

[(1)] *Fraunhofer Institute for Solar Energy Systems ISE, Heidenhofstrasse 2, 79211 Freiburg (Germany),*
**Email: david.lackner@ise.fraunhofer.de*
[(2)] *AZUR SPACE Solar Power GmbH, Theresienstrasse. 2, 74072 Heilbronn, (Germany*

ABSTRACT

Four-junction solar cells are being developed for space applications as they promise higher efficiencies compared to the present GaInP/GaInAs/Ge triple-junction industry standard. There are multiple technological routes to achieve four-junction cells with the ideal bandgap combination of 1.9 eV, 1.4 eV 1.05 eV and 0.7 eV. This includes metamorphic growth concepts and direct semiconductor wafer bonded technology. All cell designs have their specific advantages and challenges. Therefore, at Fraunhofer ISE a plurality of different four-junction cell concepts is under investigation. The current status of the development and a discussion of so far achieved characteristics are presented in this work.

1. INTRODUCTION

Four-junction cells are being developed as the next generation product for space applications to replace today's triple-junction cells where $Ga_{0.49}In_{0.51}P$ (1.9 eV) and $Ga_{0.99}In_{0.01}As$ (1.4 eV) are grown on Ge (0.7 eV) [1-3]. The triple-junction cells are well developed and e.g. the 3G30 product provided by AZUR SPACE reaches 29.5% begin-of-life and 26.5% conversion efficiency under end-of-life conditions ($1*10^{15}$ e^-/cm^2 1 MeV electron irradiation) [4]. Slightly lower end-of-life results are achieved by Spectrolab [5] and SolAero [6]. However, there is a strong pull for more efficient, lighter and more radiation hard cell designs. The route to higher efficiencies clearly demands a higher number of junctions. Calculations of the ideal bandgaps for four-junctions solar cells shows that for the third sub-cell a bandgap energy of around 1 eV is imperative. From *Figure 1* it can be easily seen that, aside from the diluted nitride material, there is no choice for a lattice matched material to GaAs or Ge with the desired 1 eV bandgap. The $Ga_{0.93}In_{0.07}N_{0.03}As_{0.97}$ material has been intensively investigated as a lattice matched 1 eV junction [7-9]. However, due to a low material quality for metal-organic vapour phase epitaxy (MOVPE) grown materials [10], such four-junction cells have failed to surpass the triple-junction performance[11]. Thus, new cell concepts have evolved where the lattice matching constraint has been dropped.

They contain either one or more metamorphic buffer layers where the lattice constant is increased, so that GaInAs junctions with different Indium contents are used as sub-cells [1, 3, 12]. For example, the upright metamorphic four-junction solar cell architecture developed by AZUR SPACE (4G32) is introduced to space qualification by the end of 2016 [1]. This cell architecture is an evolutionary development of an upright metamorphic triple-junction cell previously developed at Fraunhofer ISE for terrestrial concentrator applications [13]. Recently a four-junction inverted metamorphic cell has been introduced as a product by SolAero where 32.7% efficiency is reported at begin-of-live while 26.4% remain after irradiation with 1 MeV electrons with a fluence of $1*10^{15}$ e/cm^2 according to the data sheet [6]. This corresponds to a remaining factor in maximum power of 81%

Another possibility to realise four-junction solar cells is using the direct semiconductor bonding technology. Here a set of sub-cells has been grown on different substrates and thus on different lattice constants and are bonded afterwards [2, 14]. Both approaches, the metamorphic growth concept and the wafer bonding technology can also be combined.

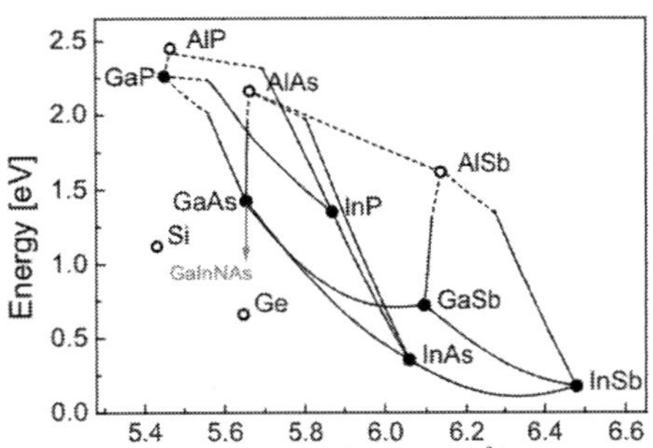

Figure 1. III-V semiconductors: The bandgap is plotted as a function of lattice constant. The filled (open) circles represent (in-)direct semiconductors. The solid (dashed) lines represent the according (in-) direct ternaries.

Figure 2 shows four-junction structures that are being developed at Fraunhofer ISE for the use in space or terrestrial concentrator photovoltaic applications.

 DOI: 10.1051/e3sconf/20171603009

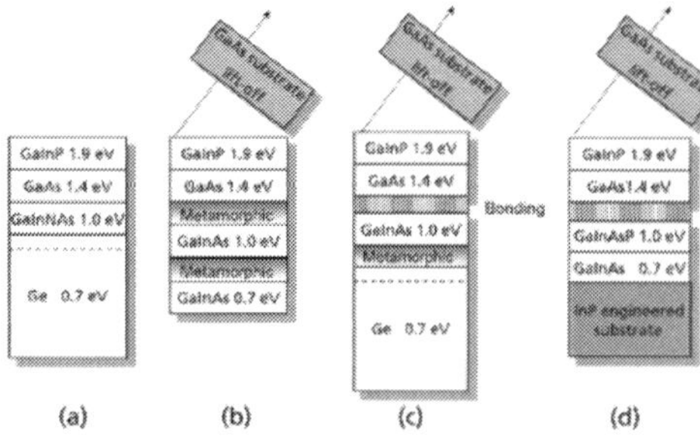

Figure 2. Absorber and substrate materials for four-junction solar cells: (a) dilute nitride-based, (b) inverted metamorphic, (c) bonded & metamorphic on Ge and (d) bonded InP based.

The four-junction cell architecture consisting of GaInP/AlGaAs//GaInAsP/GaInAs has been developed by Fraunhofer ISE, SOITEC and CEA-LETI for the use in terrestrial concentrator systems. This solar cell holds the actual record conversion efficiency of 46.0% at 508 suns under the AM1.5d spectrum [15, 16]. The cell is built from two separate dual-junction cells grown on GaAs and InP respectively. Each epitaxy process contains only lattice matched materials and therefore achieves high material quality. The two dual-junction solar cells are bonded to form a four-junction solar cell. Here it is noteworthy to mention that an adapted cell design can be also benefical for space applications. Spectrolab has also worked on wafer-bonded cell architectures and reported a 5-junction device with an efficiency of 35.8% at AM0 conditions [2]. This also proves that more junctions lead to better performances. Fraunhofer ISE and AZUR SPACE are currently developing more radiation hard designs based on the direct wafer bonding approach.

2. EXPERIMENTAL METHODS

All epitaxial layers were deposited by metal-organic vapour phase epitaxy with a multi-wafer AIX2800G4-TM reactor on 4" Ge, GaAs or InP substrates using standard precursors and process conditions.

The four-junction solar cell made on germanium and containing a metamorphic GaInAs solar cell (see concept *Figure 2c*) used semiconductor bonding technology. At Fraunhofer ISE we use an Ayumi Industries Co. Ltd bonding chamber (SAB-100) with a fast atom beam surface activated process. Here, an argon beam in a high vacuum environment ($<3*10^{-6}$ Pa) is used to clean and deoxidize the surfaces of both wafers just before bringing the surfaces in contact.

Photolithography, wet chemical etching and evaporation of anti-reflection coatings and metal contacts are used to finalize the solar cells.

The transfer matrix method [17] has been used to identify the best bandgaps and absorber thicknesses for an optimized begin-of-life performance. Here, the absorption and transmission of every single layer within the solar cell is calculated. Moreover, assuming that every absorbed photon in the active cell layers creates an electron-hole pair (i.e. an internal quantum efficiency of unity), the sub-cell short circuit current density (J_{sc}) for every junction is calculated for the AM0 spectrum. This calculated value is used as a benchmark value during our development. In order to provide a benchmark for the V_{OC} of each four-junction device, dark currents as function of the bandgap and material were analysed. Eventually the dark currents were empirically fitted from the best single-junction data available in literature and Fraunhofer ISE. For such idealized cells, a fill factor > 88% is calculated. However, as a one diode model with an ideality-factor of 1 was used for modelling the sub-cells and distributed resistance effects are neglected, this fill factor depicts an upper limit. In a realistic device we expect to achieve fill factors close to 86%

3. TECHNOLOGICAL BUILDING BLOCKS

3.1 Metamorphic Growth

Metamorphic buffers are used to change the lattice constant during the growth which will inevitably introduce crystal defects. The requirements of such a metamorphic buffer include the following: change the in-plane lattice constant to the desired target (pseudo substrate) while keeping a low surface roughness, contain the defects in the buffer to allow for high quality material to be grown above the buffer, and lastly be transparent to the transmitted light. For increasing the lattice constant starting from Germanium or GaAs (see Figure 1) this is typically realized by growing a number of layers with an increased Indium composition for every layer. (step-graded buffers). During the continuous growth, the critical thickness is passed and the layers will relax by forming misfit segments which change the in-plane lattice constant.

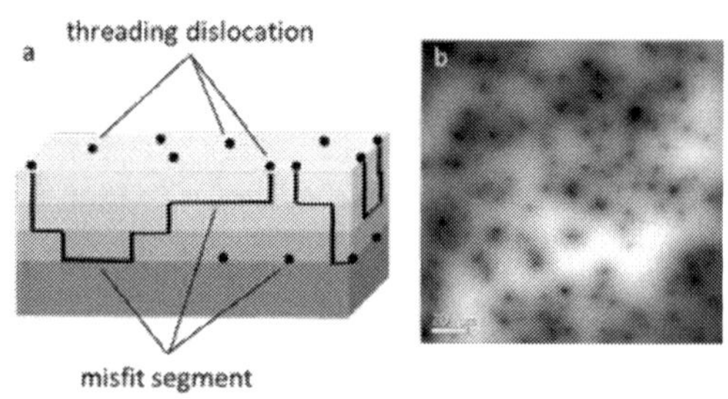

Figure 3. a) Schematic of a step-graded buffer showing misfit segments and threading dislocations. b) Cathodoluminescence image of a GaInAs double heterostructure above a step-graded buffer on GaAs with a target Indium composition of 20%. The threading dislocation density is below $1*10^{6}\ cm^{-2}$.

DOI: 10.1051/e3sconf/20171603009

However, every misfit segment typically begins and ends with a threading dislocation (*Figure 3*a).

The threading dislocations continue in the subsequent growth, where they form non-radiative recombination centres in the active region of the solar cell. The density of the threading dislocation depends on the buffer design and growth conditions including temperature, misfit grading rate, growth rate, V-III ratio, surfactants, possible wafer miscut orientations, material choice and target lattice constant. Thus a careful engineering of the process is necessary. A cathodoluminescence image can visualize the threading dislocation density (TDD), as the areas of high non–radiative recombination around a threading dislocation appear dark (*Figure 3*b). For high performance sub-cells, TDDs below $1{*}10^6$ cm^2 are typically desired. We achieved such low TDDs in GaInAs buffers up to a misfit of 1.3 % [18].

3.2 Direct Semiconductor Bonding

The requirements of a semiconductor bond within a multi-junction cell are low electrical resistance, no optical absorption or reflection and high mechanical stability. This is readily fulfilled for most direct semiconductor bonds as the two semiconductors are in direct contact and form covalent bonds after a short annealing treatment at moderate temperatures. When the doping of the two joined layers is high enough, typical bond resistances lead to a voltage loss of less than 1 mV at current densities below 17 mA/cm^2 (see *Figure 4)*. Bonding allows the combination of a wide range of different materials. Low resistance bonds between various materials have already been demonstrated including Si//GaAs, InP//GaAs and GaSb//GaInAs [19-21]

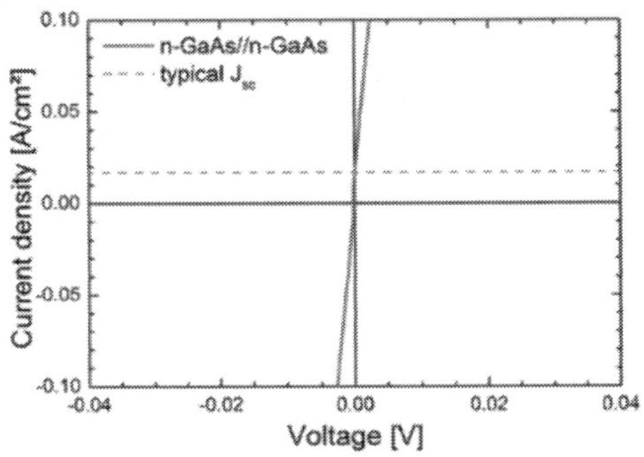

Figure 4.Typical I-V curve of a n-GaAs/n-GaAs direct bond after annealing at 400°C for 60 s. The dashed grey line symbolizes the typical short-current density of four-junction cells at AM0. The voltage loss due to bond resistance at 17 mA/cm^2 is below 1 mV.

However a critical condition for direct semiconductor bonding is the surface roughness which typically needs to be below 1 nm (RMS). Furthermore, all particles present at the surface will cause voids in the bond or even wafer breakage. Thus, the wafers are currently treated after epitaxy before the bonding with a chemical mechanical polishing process to ensure a clean surface and a low surface roughness.

With this process, full 4" wafers have been bonded with no or very few voids (*Figure 5*) suggesting that this method is viable for large area devices such as space solar cells.

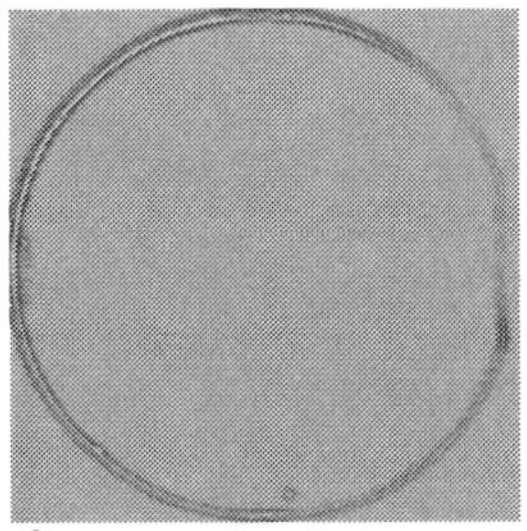

Figure 5. Surface acoustic microscopy image of a 4" InP//GaAs direct bond performed at Fraunhofer ISE. The pair of wafers is fully bonded except one void at the bottom (bright spot) corner most likely caused by a particle.

4. FOUR-JUNCTION CELLS

In the following, the actual status of three four-junction designs that have been realized at Fraunhofer ISE will be presented (*Figure 2a, b, c).* Note that some designs have only been explored for terrestrial applications so far and thus are not optimized, i.e. current matched under the AM0 spectrum. However, all four-junction concepts have calculated realistic efficiency potentials above 36% when optimized for begin of-life performance. The calculated benchmark efficiencies including V_{OC} and J_{SC} values for the begin-of-life optimized designs (see *Figure 2)* are displayed in *Table 1*).

Table 1: Calculated benchmark efficiencies for the begin-of-life optimized four junction structures

	J_{sc} [mA/cm^2]	V_{oc} [V]	η [%]
dilute nitride-based (a)	16.4	3.46	36.9
inverted metamorphic (b)	16.4	3.62	38.4
bonded & metamorphic on Ge (c)	16.4	3.46	36.9
bonded InP based (d)	16.4	3.56	37.9

4.1 Dilute nitride based four-junction

Dilute nitride $Ga_{0.93}In_{0.07}N_{0.03}As_{0.97}$ seems a perfect choice to progress from the standard GaInP/GaAs/Ge triple-junction to a four-junction cell (*Figure 2a)* as the material can reach a bandgap of 1 eV while being lattice matched to Ge. However, up to now minority carrier lifetimes in MOVPE grown material are well below 1 ns which is not sufficient to achieve high performances. It should be noted though, that MBE growth has already led to excellent dilute nitride devices [22] which shows that these current problems challenges may be overcome in the future. However, for MOVPE-grown

 DOI: 10.1051/e3sconf/20171603009

materials so far background doping levels above $5*10^{16}$ cm^{-3} have been found due to a strong carbon incorporation. This is one reason why only diffusion lengths in the order of 100 nm are measured. As a consequence, internal quantum efficiencies below 60% and V_{OC} values below 500 mV for 1 eV materials are typical today. When incorporating such a sub-cell into the triple-junction, the four-junction J_{SC} is currently significantly reduced [11] (*Figure 6*). This resulted in a low efficiency of only 13% for this device.

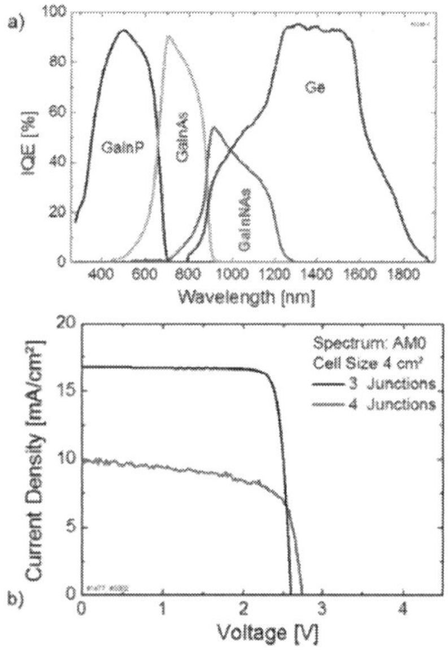

Figure 6. Internal quantum efficiency (IQE)(a) and I-V characteristics (b) of a GaInP/GaAs/GaInNAs/Ge four-junction cell under AM0. The cell current is clearly limited by the low performance of the GaInNAs junction. The I-V curves under AM0 at the bottom shows the comparison between the four-junction (red) and the according triple-junction cell (black). While the J_{SC} is significantly reduced, only a small amount in V_{OC} could be gained (Data taken from [11]).

4.2 Inverted metamorphic four-junction

The inverted metamorphic approach also requires only one growth process on one wafer. This comes at the cost of two metamorphic buffers which need to be graded from the lattice constant of GaAs to a new larger lattice constant with a misfit of 3.8% in total. Considering that high quality buffers with a low TDD are typically grown with a grading rate of 0.5 - 1 µm/% misfit, another 5 to 10 µm of extra epitaxial layers are added for the buffer. After the growth, the substrate needs to be removed and the epitaxial layers need to be transferred to a support substrate that can be a Si wafer or a metal film. At Fraunhofer ISE, such inverted metamorphic cells structures consisting of GaInP (1.9 eV), GaAs (1.42 eV), $Ga_{0.73}In_{0.27}As$ (1.05 eV) and $Ga_{0.47}In_{0.53}As$ (0.75 eV) sub-cells have been realized and have been processed into solar cells by AZUR SPACE. The I-V curve of the best present cell is displayed in *Figure 7*. Currently it reaches an efficiency of 29.4% under AM0 conditions. Compared with the simulated benchmark (see *Table 1*) the biggest improvement potential is in V_{OC}. This is partly due to the fact that the lowest sub-cell still has a TDD of $6*10^6$ cm^{-2} which can be improved further. However, a TDD value below $2*10^6$ cm^{-2} has already been reached in a standalone buffer sample, and thus should result in an improved voltage of the GaInAs sub-cells. Improved devices are currently being prepared.

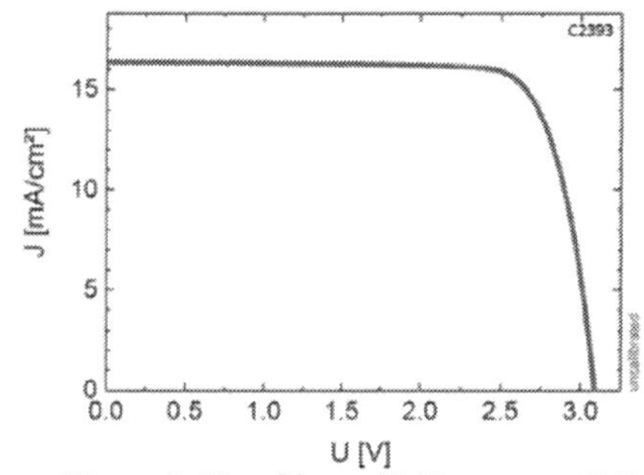

Figure 7. Uncalibrated I-V curve at AM0 for an inverted metamorphic four-junction cell with a 4 cm² cell area.

The high begin-of-life efficiency potential of this cell design has recently been proven by France *et al.* who reported an efficiency value of 35.3% under AM0 [12]. One downside of this approach for space use is, however, that the GaAs and GaInAs junctions show weak radiation hardness compared to GaInP or Ge [23].

4.3 Bonded and metamorphic on Ge

The principles of this four-junction cell concept has been developed at Fraunhofer ISE [18]. Currently Fraunhofer ISE and AZUR SPACE continue optimizing the performance for terrestrial usage within the European funded project CPVMatch. This four-junction cell is built from two separate growths: the first is an inverted dual-junction (GaInP 1.9eV and $Al_{0.03}Ga_{0.97}As$ 1.47 eV) on GaAs while the second consists of an in-diffused Ge cell (0.66 eV) with a metamorphic buffer structure and a $Ga_{0.81}In_{0.19}As$ cell (1.15 eV). Those two wafers are joined by a direct wafer bond and the GaAs substrate is removed before photolithography and metal evaporation. So far this cell has only been optimized for the use in terrestrial concentrator systems and thus the choosen bandgaps and current matching is not ideal for the AM0 spectrum. The metamorphic buffer for this GaInAs cell has been optimized to have a TDD < $1*10^6$ cm^{-2}. This allows for the growth of a high quality GaInAs sub-cell. At the bandgap of 1.15 eV a sub-cell V_{OC} of 0.754 V has already been reached. This is of comparable material quality to a GaAs junction grown on a GaAs substrate.

The low maximum external quantum efficiency (EQE) in *Figure 8* of every sub-cell is caused by the higher grid shading of the concentrator cell design. The calculated sub-cell currents that would be generated under the AM0 spectrum are also shown in the image. A redistribution of the current for perfect current matching would lead to 15.8 mA/cm^2.

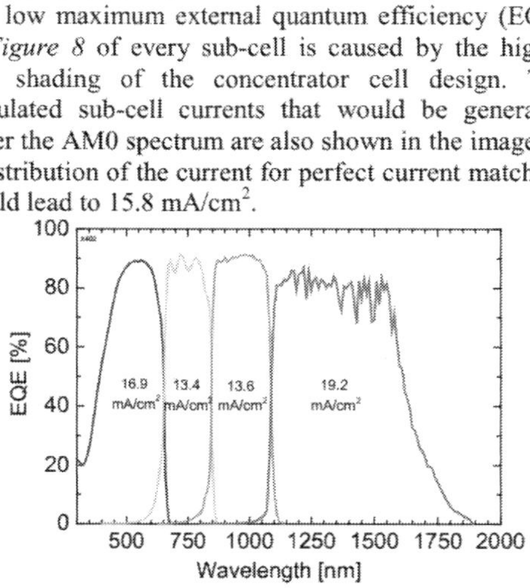

Figure 8. External Quantum Efficiency (EQE) of the metamorphic four-junction cell on Ge. The calculated subcell currents under AM0 are included in the figure. Since the cell was optimized for terrestrial applications, a high current mismatch is observed under the AM0 spectrum.

The measured V_{OC} is 3.32 V at 1 sun under the AM1.5d spectrum. For an ideal bandgap combination for AM0 conditions of 1.9 eV, 1.42 eV, 1.05 eV and 0.66 eV, the two middle cells would have to be reduced in terms of bandgap. Taking this into account, a V_{OC} of 3.17 eV would be expected assuming the same material quality. This is close to the simulated value of 3.46 V showing the maturity of the cell development and high expected performance of such a design.

5. CONCLUSION

To surpass the performance of existing triple-junction space solar cells, four or five-junctions are necessary. At Fraunhofer ISE, different four-junction cell architectures are being developed. All of these structures have been simulated based on realistic material data and the calculated realistic efficiency is above 36% AM0 when optimized for begin-of-life conditions. For the monolithic direct growth on Ge the challenge is in integrating a 1.eV junction. Because the lattice matched material GaInNAs is of too low quality so far, when being deposited by MOVPE. Thus either metamorphic buffer structures and/or wafer bonding have been introduced. The development status of three four-junction cells has been presented: "dilute nitride based", "inverted metamorphic", "metamorphic and bonded on Ge". Still more R&D is needed to achieve the benchmark efficiencies beyond 36 %.

6. ACKNOWLEDGEMENTS

The authors like to acknowledge device processing and characterization by M. Scheer, K. Wagner, S. Stättner, R. Freitas, R. Koch, E. Fehrenbacher, A. Wekkeli, S.K. Reichmuth and E. Schäffer. This work was supported the European Space Agency ESA-ESTEC (contract No. 4000112560/14/NL/FE, 4000104852 and RFQ 3-11296/07/NL/GLC), the German Aerospace DLR (contract No. 50RN1301 and 50JR0641) and the European Union (contract No. 283798). M. Niemeyer acknowledges funding of his PhD work by the German Federal Environmental Foundation (DBU).

7. REFERENCES

1. Guter, W., Dunzer, F., Ebel, L., Hillerich, K., Köstler, W., Kubera, T., . . . & Wächter, C. (2016). Space Solar Cells – 3G30 And Next Generation Radiation Hard Products. in Proc. European Space Power Conference, Thessaloniki, Greece - to be published.
2. Chiu, P.T., Law, D.C., Woo, R.L., Singer, S.B., Bhusari, D., Hong, W.D., . . . & Karam, N.H. (2014) 35.8% space and 38.8% terrestrial 5J direct bonded cells. *2014 IEEE* 40th Photovoltaic Specialists Conference (PVSC). 0011-3.
3. Patel, P., Aiken, D., Boca, A., Cho, A.Y., Chumney, D., Clevenger, M.B., . . . & Varghese, T. (2012) Experimental results from performance improvement and radiation hardening of inverted metamorphic multijunction solar cells IEEE Journal of Photovoltaics. 2(3), 377-81.
4. AZUR SPACE Solar Power GmbH, http://www.azurspace.com/index.php/en/products/products-space/space-solar-cells
5. Spectrolab, http://www.spectrolab.com/solarcells.htm
6. SolAero, http://solaerotech.com/products/space-solar-cells-coverglass-interconnected-cells-cic/
7. Volz, K., Lackner, D., Nemeth, I., Kunert, B., Stolz, W., Baur, C., . . . & Bett, A.W. (2008) Optimization of annealing conditions of (GaIn)(NAs) for solar cell applications. Journal of Crystal Growth. 310(7-9), 2222-8.
8. Geisz, J.F. & Friedman, D.J. (2002) III-N-V semiconductors for solar photovoltaic applications. Semiconductor Science and Technology. 17(8), 769-77.
9. Garrod, T.J., Kirch, J., Dudley, P., Kim, S., Mawst, L.J. & Kuech, T.F. (2011) Narrow band gap GaInNAsSb material grown by metal organic vapor phase epitaxy (MOVPE) for solar cell applications Journal of Crystal Growth. 315(1), 68-73.

DOI: 10.1051/e3sconf/20171603009

10. Baur, C., Bett, A.W., Dimroth, F., van Riesen, S., Kunert, B., Traversa, M., . . . & Stolz, W. (2003). Development of a 1.0 eV (GaIn)(NAs) Solar Cell. in Proc. Technical Digest of the 3rd World Conference on Photovoltaic Energy Conversion, Osaka, Japan. pp.677-680
11. Essig, S., Stämmler, E., Rönsch, S., Oliva, E., Schachtner, M., Siefer, G., . . . & Dimroth, F. (2011). Dilute nitrides for 4- and 6- junction space solar cells. in Proc. 9th European Space Power Conference, St.-Raphael, France.
12. France, R.M., Geisz, J.F., Garcia, I., Steiner, M.A., McMahon, W.E., Friedman, D.J., . . . & Olavarria, W.J. (2015) Design Flexibility of Ultrahigh Efficiency Four-Junction Inverted Metamorphic Solar Cells. IEEE Journal of Photovoltaics. PP(99), 1-6.
13. Guter, W., Schöne, J., Philipps, S.P., Steiner, M., Siefer, G., Wekkeli, A., . . . & Dimroth, F. (2009) Current-matched triple-junction solar cell reaching 41.1% conversion efficiency under concentrated sunlight. Applied Physics Letters 94(22), 223504-6.
14. Dimroth, F., Tibbits, T.N.D., Beutel, P., Karcher, C., Oliva, E., Siefer, G., . . . & Hannappel, T. (2014) Development of high efficiency wafer bonded 4-junction solar cells for concentrator photovoltaic applications. 40th IEEE Photovoltaic Specialists Conference. 0006-10
15. Tibbits, T.N.D., Beutel, P., Grave, M., Karcher, C., Oliva, E., Siefer, G., . . . & Dobrich, A. (2014). New Efficiency Frontiers With Wafer-Bonded Multi-Junction Solar Cells. in Proc. 29th European Photovoltaic Solar Energy Conference, Amsterdam, Netherlands. pp 1975-8
16. Green, M.A., Emery, K., Hishikawa, Y., Warta, W. & Dunlop, E.D. (2016) Solar cell efficiency tables (version 47). Progress in Photovoltaics: Research and Applications. 24(1), 3-11.
17. Létay, G., Breselge, M. & Bett, A.W. (2003). Calculating the Generation Function of III-V Solar Cells. in Proc. Proceedings of the 3rd World Conference on Photovoltaic Energy Conversion, Osaka, Japan. pp 741-4
18. Dimroth, F., Tibbits, T.N.D., Niemeyer, M., Predan, F., Beutel, P., Karcher, C., . . . & Signamarcheix, T. (2015). Four-junction wafer bonded concentrator solar cells. in Proc. Photovoltaic Specialist Conference (PVSC), 2015 IEEE 42nd, pp 0006-10
19. Essig, S. & Dimroth, F. (2013) Fast atom beam activated wafer bonds between n-Si and n-GaAs with low resistance. ECS Journal of Solid State Science and Technology. 2(9), Q178-81.
20. Predan, F., Reinwald, D., Klinger, V. & Dimroth, F. (2015) Transparent and electrically conductive GaSb/Si direct wafer bonding at low temperatures by argon-beam surface activation. Applied Surface Science. 353(1203 - 1207.
21. Tauzin, A., Lagoutte, E., Salvetat, T., Guelfucci, J., Bogumilowicz, Y., Imbert, B., . . . & Dimroth, F. (2016) InP-based composite substrates for four junction concentrator solar cells. AIP Conference Proceedings. 1679(1), pp 040009.
22. Sabnis, V., Yuen, H. & Wiemer, M. (2012). High-efficiency multijunction solar cells employing dilute nitrides. in Proc. 8th International Conference on Concentrating Photovoltaic Systems, Toledo, Spain.
23. Yamaguchi, M. (2001) Radiation-resistant solar cells for space use. Solar Energy Materials and Solar Cells. 68(1), 31-53.

DOI: 10.1051/e3sconf/20171603010

THE NEXT GENERATION OF GERMANIUM SUBSTRATES: EXPOGER™

Pieter Arickx [(1)], Rufi Kurstjens [(1)], Wim Geens [(1)], Kristof Dessein [(1)]

[(1)] Umicore Electro-Optic Materials, Watertorenstraat 33 – B2250 Olen (Belgium),
Email: pieter.arickx@eu.umicore.com

ABSTRACT

Umicore has been the leading Germanium substrate manufacturer over the last 20 years. We are only able to maintain that position by constantly improving the quality of our product. In this work, Umicore wants to present how it will continue this tradition by adding even more value to its product.

In the beginning of 2014 we proposed to our customers to help them in achieving their yield improvement goals. Two main concerns were expressed: the breakage rate in their production line and non-uniform epi-growth. After two years of development, Umicore has bundled these two improvements into a new product. The ExpoGer™ wafer combines a better mechanical yield with a uniform, epi-ready surface enabling a narrower distribution of the electrical performance.

The breakage rate of the customers is related to the wafers that suffer from early failure. The percentage of early failures is related to its thickness. With the trend towards even thinner substrates, it is important to add features that reduce the percentage of early failures. Two features are crucial: the shape of the substrate edge and the stress relief etch, which influences the wafer strength.

Just as for the strength, two distinctive aspects define the surface quality. On the one hand, there are the number and size of surface defects, which have an important influence on the metal-organic chemical vapour deposition (MOCVD) growth performed at Umicore's customers. On the other hand, drying marks, which are invisible to the naked eye, may lead to a local, non-uniform growth and they can be a precursor to time dependent haze issues.

Umicore is able to demonstrate the impact of the ExpoGer™ processes on the strength and surface quality of its wafers and thus on the production yield of its customers.

1. INTRODUCTION

Germanium substrates are mainly used as the epitaxial growth substrate of high-efficiency, triple junction solar cells for space applications. Staying true to Umicore's value proposition, we developed specific improvements that increase the Ge substrate quality significantly. By focusing on our customer's wishes, we were able to create a new product: the ExpoGer™ wafer.

When our customers perform the MOCVD growth to create the triple junction solar cell, any deviation from a perfect surface will be a nucleation point for a defect or a cause of a non-uniform epi-stack. These imperfections can be caused by surface roughness, crystal originated pits (COPs), particles or non-uniform layers of native oxide. The area and density of these imperfections will determine the impact on the growth and the final solar cell performance.

The improvements focus on avoiding the larger area and higher density imperfections. Through adjusting the cleaning and drying, the unwanted surface roughness, irregular oxide thickness, particles and drying marks could be avoided. Additionally, the adjusted clean also reduces the impact of COPs. The formation of COPs during crystal pulling was not addressed in this work.

Wafer breakage occurs due to brittle fracture. The stressed material does not deform plastically and it fails in the elastic regime with very little absorption of energy. Small chips or cracks can be the nucleation site for a fracture if enough stress is applied on that specific location. To avoid these defects, two processes have been updated. First, the edge now has a more rounded profile, making it more robust towards impact and the process itself is inherently less prone towards chipping. Secondly, these microscopic cracks on the etched surface are avoided by using a newly developed etch in a very controlled environment. The roughness of the ExpoGer™ wafer is significantly lower than our current standard wafers and thus has less nucleation sites for a fracture.

This paper will show our first, internal results for 100 mm substrates. Umicore is currently transferring these developments to 150 mm as well.

2. MECHANICAL STRENGTH

2.1 Wafer Edge

When the edge of a wafer has sharp corners, stress is concentrated in these points. Geometric discontinuities cause an object to experience a local increase in the intensity of a stress field. Examples of shapes that cause these concentrations are cracks, sharp corners, holes, and changes in the cross-sectional area of the object. High local stresses can cause objects to fail more

DOI: 10.1051/e3sconf/20171603010

quickly, so engineers must design the geometry to minimize stress concentrations. During processing, upon impact with the edge, chipping or wafer fracture will occur more easily for sharp edges. To avoid this stress concentration, the wafer edge can be bevelled or round.

Chen et al [1,2] have compared 4 types of edge profile, schematically represented at the bottom of Figure 1. They have tested the breakage energy necessary to induce wafer breakage by designing a drop test where a 200 g weight is dropped from some height on the edge of the wafer. The recorded breakage energy is seen to increase as the edge is more and more rounded.

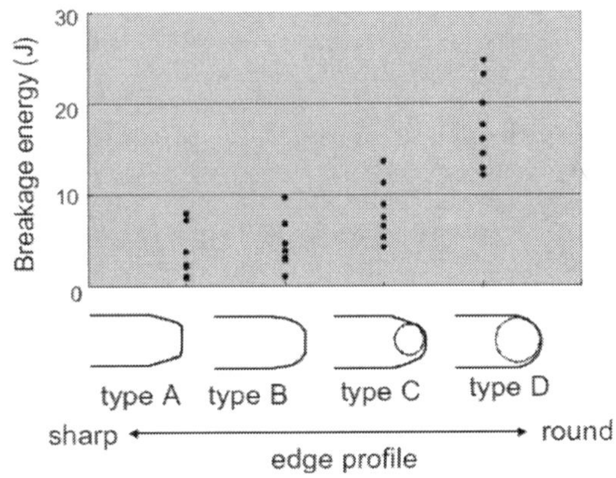

Figure 1. The results of a mechanical drop test for four kinds wafer edge profile. [1,2]

Chen et al then compared these experimental results to wafer breakage ratios for the four types of edges from a production environment. The breakage ratio of 10000 wafers processed with each type of edge design is shown in Figure 2. It can be seen that the breakage ratio drops from 3400 ppm to roughly 1200 ppm when going from the edge with sharp corners to the round edge.

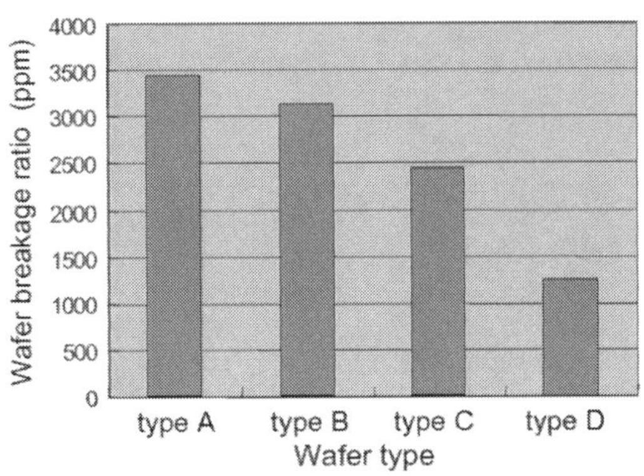

Figure 2. The wafer breakage ratio of type A to type D in a production line tested on 10000 wafers. [1,2]

Guided by the above, we were able at Umicore to obtain the rounded edge profile for our wafers. Not only was the edge grinding wheel modified to fit these new requirements, the entire process flow needed to be carefully reviewed as not all steps are perfectly symmetrical, i.e. the removal and the front side of the wafer is not necessarily equal to the removal on the back side (e.g. only the front side is polished).

In Figure 3 the profile of the rounded edge can be seen.

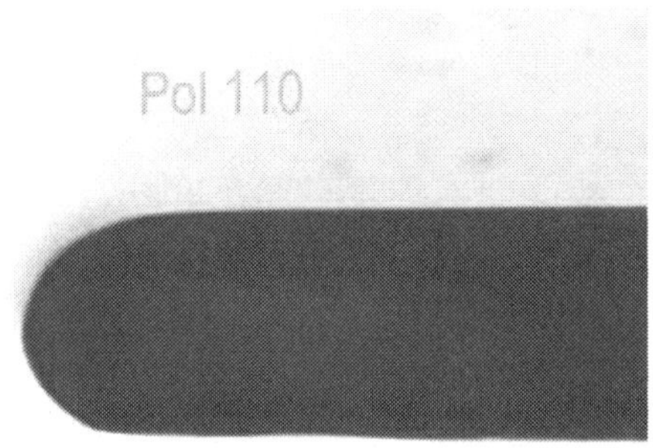

Figure 3. Round edge profile of a polished wafer along the (110) crystal direction.

2.2 Wafer Surface

The stress relief etch immediately follows the mechanical processing steps of the wafering process, as this step needs to remove all the mechanically induced damage from the previous steps, such as wafer sawing and surface grinding. The subsurface damage (SSD) is inversely related to the breakage strength, as can be seen in Eq. 1:

$$\sigma_f = \frac{K_{Ic}}{\sqrt{\pi a}} \tag{1}$$

where K_{Ic} is the stress intensity factor (in MPa.m$^{1/2}$), σ_f is the fracture stress and a is the crack length.

As the breakage strength is related to the crack length, it is clear that the SSD needs to be removed to obtain the strongest substrate possible, while maintaining the geometrical characteristics of the wafer. Furthermore, the surface needs to remain smooth, as uncontrolled etching would induce deeper defects which could serve as crack nuclei. The etchant thus needs to be designed to meet these requirements.

DOI: 10.1051/e3sconf/20171603010

The fracture strength of the wafer is measured by the ball-on-ring method (Instrom uniaxial set-up) where the wafer rests on a ring with an internal diameter smaller than the wafer itself. This technique withholds edge defects such as chipping or small cracks and only tests the structural strength of the surface. When the ball (with a diameter of 5 mm) brings down a certain force on the surface, the opposite side experiences a strain force. The unpolished backside experiences this strain force as the wafers are always placed with the polished side up. The ball is gradually lowered (0.01 inch/min) until the strain breaks the wafer. The breaking load is usually expressed as pounds-force per square inch (lbf). Typically, breaking loads are distributed according to a Weibull distribution.

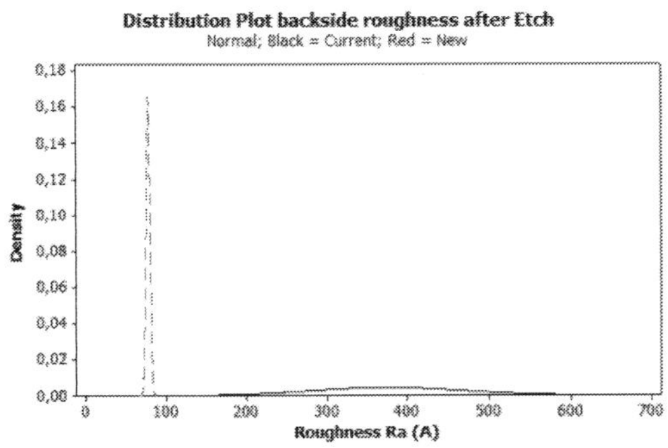

Figure 4. The roughness of the etched (back) side of the wafer of the current (black) and the new (red) process.

Umicore has replaced the current, stress relief etch process, by a newly developed etch process that improves the surface related strength of the wafer. Due to a highly controlled process, the roughness stays well below the roughness of our current substrates (Figure 4) and the average strength of 175 µm (final thickness) substrates almost doubles, as clearly demonstrated in Figure 5.

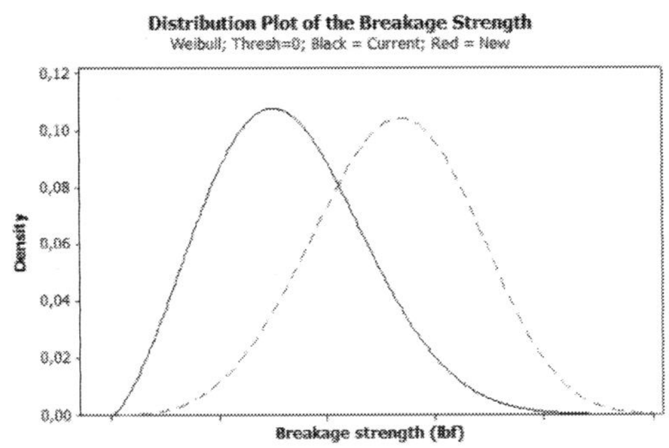

Figure 5. The breakage strength the wafer etched with the current (black) and the new (red) process.

Both improvements act as an enabler for the storage and transportation of thin 100 mm substrates in multi-wafer shippers (MWS).

3. Surface Quality

3.1 An Alternative Clean

The final clean process has an impact on the size of the defects present on the wafer surface. The purpose of the new clean was to drastically alter the behaviour with respect to the defects and uniformity of the wafer process while maintaining the cleaning efficiency. If the defects or uniformity variations on the wafer surface can stay below a certain threshold, they will not act as initiation sites for MOCVD growth defects.

The Umicore final clean holds two major parts: a particle clean to remove any leftover abrasive from the polishing step and a metal clean to remove trace metals from the wafer surface.

The particle clean is evaluated via the particle removal efficiency. We have developed a method to intentionally contaminate our substrates with the particles used as the abrasive in the polishing process. We measure before and after deliberate contamination and after the clean. Figure 6 is a demonstration of this method, measured by a KLA-Tencor Candela, a surface inspection system.

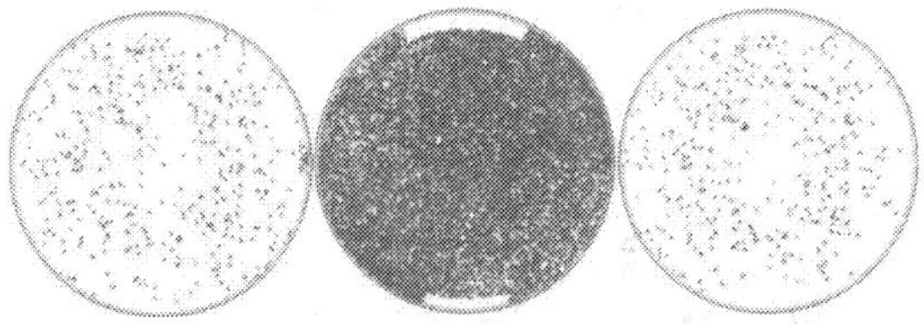

Figure 6. A clean, contaminated and a re-cleaned wafer (from left to right) measured by a KLA-Tencor Candela.

The particle removal efficiency (PRE) can be calculated as shown in Eq. 2.

$$PRE = \left(1 - \frac{Particles_{AftClean} - Particles_{BefCont}}{Particles_{AftCont} - Particles_{BefCont}} \right) \cdot 100 \qquad (2)$$

As shown in Figure 7, the PRE of the new clean is 100%.

DOI: 10.1051/e3sconf/20171603010

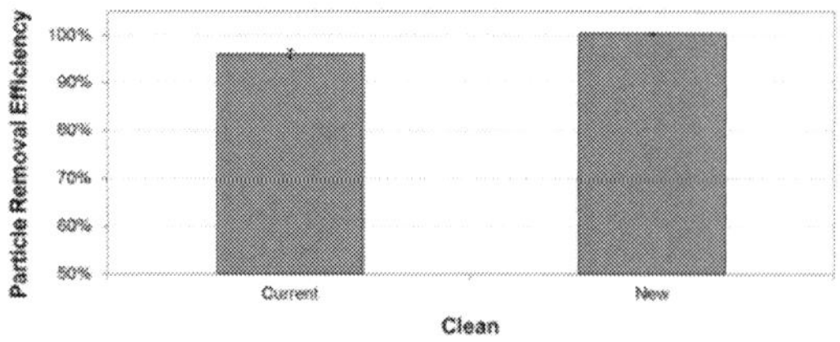

Figure 7. The particle removal efficiency of the current and the new clean process.

The main feature of the new clean is of course its neutral behaviour towards the crystallographic properties and uniformity of the wafer. Crystallographic features on the substrate surface will not be altered. Figure 8 shows that a substrate processed via the standard process shows more surface defects. Both wafers were sampled adjacent to each other from the crystal. The standard clean causes a larger degradation of the final surface made "visible" by the KLA-Tencor Candela tool.

Figure 8. The occurrence of surface anomalies on the substrate's surface depending on the final clean binned as a function of the anomaly size with smallest elements in bin 1 and the largest elements in bin 5.

Trace metals are detrimental for triple junction solar cells and should thus be avoided at all time. A clean was developed with the same metal removal capabilities, while having less impact on the quality of the polished surface. Figure 9 demonstrates that for a myriad of metals, the new clean is similar or even better than the current clean. These measurements were done by TXRF, where the metals were etched by HF from the substrate surface.

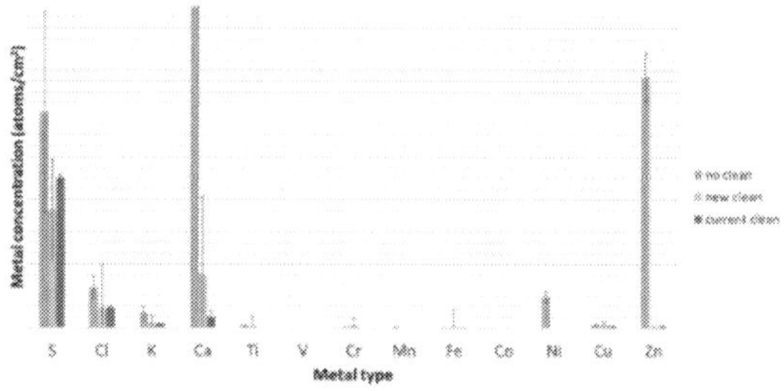

Figure 9. The occurrence of metals on the substrate's surface as a function of the final clean.

3.2 New Drying Technology

As described in the introduction, the occurrence of drying marks will lead to growing defects and in a further stage can even support the formation of haze on the wafer surface. The marks itself are created when water is allowed to evaporate on the wafer surface. The normal drying process uses the Marangoni effect to prevent this from happening. To further improve the drying of wafers, Umicore recently acquired a new tool where the entire process was redesigned, albeit still being based on the Marangoni effect.

In the current process, the contact area between the wafers and the supporting combs of the dryer is still significant. When the alignment of the wafer into the comb is not perfect, water can be trapped between the comb and the wafer and evaporate on the surface, leaving a drying mark, as shown in Figure 10.

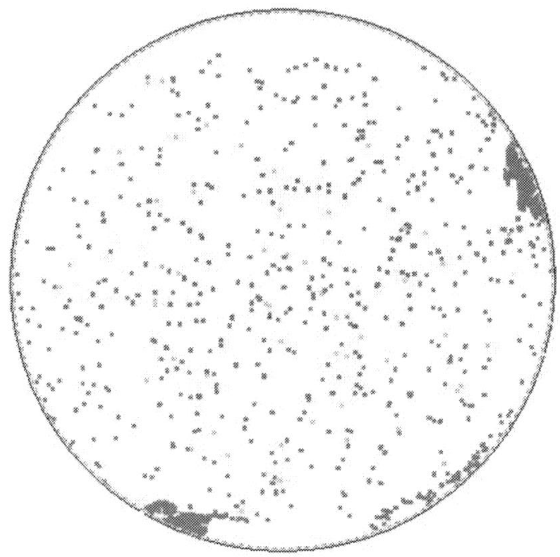

Figure 10. The grey areas at the sides of the wafer indicate drying marks.

The new process has a minimal contact area between the wafer and the positioning comb and special features are in place to avoid misalignment and minimize the chance of water being trapped between comb and wafer. This allowed Umicore to reduce the occurrence and the impact of drying by evaporation, instead of the Marangoni effect drying. Figure 11 shows a KLA-Tencor SurfScan image, another surface inspection method, where the drying marks of the new tool are indicated. It is clear that these are minimal compared to the current process.

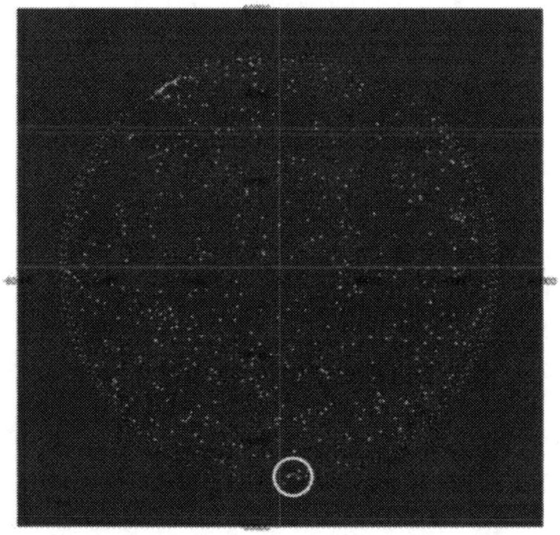

Figure 11. The circled area at the side of the wafer indicates a drying mark.

4. CONCLUSIONS

It has been demonstrated that the new generation of Umicore Germanium substrates are both stronger and show a promising improvement of the surface quality. With the ExpoGer™ wafer, Umicore addresses two main concerns of its customers.

The stronger substrate was obtained by redesigning the edge grinding process and the stress relief etch, for the best combination of an impact-resistant edge with a smooth and strong substrate back side. These improvements enable transportation in MWS, which in turn enable automatic handling of wafers.

The quality of the surface was improved by developing new particle and metal cleans, while at the same time improving the drying process at the end of the cleaning.

5. ACKNOWLEDGEMENTS

The authors would like to thank the European Space Agency (ESA) for the funding of the project with Contract No 4000110920/14/NL/CBi/fk (ESA-Imager III).

6. REFERENCES

1. P.-Y. Chen, M.-H. Tsai, W.-K. Yeh, M.-H. Jing, & Y. Chang. (2010). Relationship between wafer edge design and its ultimate mechanical strength. Microelectronic Engineering, vol. 87, no. 11, pp. 2065-2070.

2. P.-Y. Chen, M.-H. Tsai, W.-K. Yeh, M.-H. Jing, & Y. Chang. (2010). Relationship between wafer fracture reduction and controlling during the edge manufacturing process. Microelectronic Engineering, vol. 87, no. 10, pp. 1809-1815.

BOL AND EOL CHARACTERIZATION OF AZUR 3G LILT SOLAR CELLS FOR ESA JUICE MISSION

Victor Khorenko[(1)], Carsten Baur[(2)], Gerald Siefer[(3)], Michael Schachtner[(3)], Seonyong Park[(4)], Bruno Boizot[(4)], Jacques C. Bourgoin[(4)], Mariacristina Casale[(5)], Roberta Campesato[(5)]

[(1)] *AZUR SPACE Solar Power GmbH, Theresienstr.2, 74072 Heilbronn, Germany, Email:victor.khorenko@azurspace.com*
[(2)] *European Space Agency, Keplerlaan 1, 2201AZ Noordwijk, The Netherland, Email: carsten.baur@esa.int*
[(3)] *Fraunhofer Institute for Solar Energy Systems, Freiburg, 79110, Germany, Email: gerald.siefer@ise.fraunhofer.de*
[(4)] *Laboratoire des Solides Irradiés, CNRS-UMR 7642, CEA-DRF-IRAMIS, Ecole Polytechnique, Université Paris-Saclay, 91128 Palaiseau Cedex, France, Email:seonyong.park@polytechnique.edu*
[(5)] *CESI S.p.A., Milan, 20134, Italy, Email: mariacristina.casale@cesi.it*

ABSTRACT

In the present paper, we describe the results of electrical characterization of AZUR SPACE triple-junction solar cells at a sun light intensity of 3.7% AM0 and temperatures down to -150°C. At these conditions, which are relevant for the anticipated ESA JUICE mission, the cell efficiency reaches 33.5 % at BOL. Special attention has been paid to the establishing of an in-situ characterization procedure for defining EOL cell characteristics after electron and proton irradiation at low temperature low intensity condition. It was shown that solar cells irradiated at low temperature exhibit a strong recovery effect within short time after stopping the irradiation whereas the absolute value of the recovery depends on the irradiation fluence and particle type. Further on, it was demonstrated that the degradation of the maximum power, Pmp, is much stronger than the degradation of Isc and Voc values. Experimentally defined remaining factors for electron and proton irradiation and the quantification of the observed recovery effects allow a realistic prediction of the solar cell performance at JUICE mission conditions and are essential for the planned solar cell qualification activities.

1. INTRODUCTION

The feasibility of solar cells for powering a spacecraft at a distance of up to five astronomic units from the sun has been successfully demonstrated with Si cells for the Rosetta mission and with triple-junction GaAs-based cells for the Juno mission. Notwithstanding, such deep space missions that combine low solar intensities with very low temperatures, which are often referred to as low intensity low temperature (LILT), likely pose the ultimate limit for photovoltaics technology in space.

The anticipated ESA JUpiter ICy moons Explorer (JUICE) mission will continuously operate at a light intensity and a temperature of 50W/m^2 and -130°C, respectively. It has been shown that at LILT conditions solar cells become very sensitive to electrical shunts and may suffer from the so-called "flat spot" (or "broken knee") effect that may unpredictably appear and cause a drop of fill factor at temperatures below -80°C. Despite the increase of the cell voltage at low temperatures, both effects (shunts and flat spot) significantly lower the power output of the cell that is reduced due to low light intensity as well.

Shunts are caused by imperfections or defects in p-n junctions within the solar cell structure or on the cell mesa edges. They are observable in dark IV measurements at room temperature and can be in principle significantly diminished by improving the overall manufacturing process. The appearance of the flat spot effect has been attributed to the formation of local Metal-Semiconductor-Like (MSL) junctions at the interface between the front side metallization and top sub-cell [1]. It was assumed that the formation of a MSL-junction occurs preferentially on local microscopic defects where crystal quality of the semiconductor material is disturbed. A general improvement of the cell epitaxy and manufacturing process and, as an additional measure, a suitable choice of the front grid metallization, have allowed for avoiding the flat spot effect completely [2, 3].

Another challenge consists in a reliable characterization of LILT solar cells, especially at End-of-Life (EOL) conditions. In contrast to the cells powering spacecraft in Earth orbit, the particle irradiation of LILT cells takes place at continuously lower temperatures. Thus, the determination of the EOL performance shall be preferable done at low temperatures in order to avoid any temperature annealing. This annealing will improve cell characteristics and lead to an overestimation of the EOL performance, which would be critical for designing of the spacecraft power sub-system.

In this paper, we present the latest results of the beginning-of-life (BOL) and end-of-life (EOL) electrical characterization of triple-junction AZUR solar cells and solar cell assemblies (SCAs) under LILT

 DOI: 10.1051/e3sconf/20171603011

conditions. We paid special attention to the investigation of post-irradiation annealing und the quantification of annealing factors as a function of the irradiation fluence.

2. MANUFACTURING AND SELECTION OF LILT SOLAR CELLS

The solar cells were manufactured using AZUR 3G28 lattice matched triple junction solar cell structure based on the InGaP/InGaAs/Ge material combination. For BOL and EOL characterization after particle irradiation at room temperature, 4×8cm^2 solar cells and solar cell assemblies (SCAs) were used, whereas the evaluation of the in-situ EOL performance and annealing effect study were done on 2×2cm^2 bare cells fabricated using the same manufacturing processes. For all device designs, the cell thickness was chosen to be 80 μm.

Besides AZUR's standard quality inspection, all cells were subjected to additional acceptance tests which included room temperature electrical characterization at low light intensity and in the dark. The objective of these tests was to exclude solar cells with shunts from the later characterization; thus, the cells with fill factor FF<0.82 and I_{fw}>1e-2mA at V_{fw}=1.3V (in the dark IV characteristic) were sorted out.

3. TEST FACILITIES

The pre-selection and room temperature characterization of solar cells and SCAs was performed by AZUR SPACE. BOL and EOL measurements of 4×8 cm^2 cells after room temperature irradiation were done at CESI, whereas BOL and EOL characterization of 2×2 cm^2 cells including EQE measurements at room and low temperatures was carried out by Fraunhofer ISE.

For the investigation of EOL performance at LILT conditions, the experimental setup of the Laboratoire des Solides Irradiés (LSI) of École Polytechnique in Palaiseau was used (Fig.1). The samples were placed inside a cryostat that was mounted onto a beam line of an irradiation facility. After finishing the irradiation, the cell holder is rotated towards a quartz window for illuminating the cell und recording its IV characteristic. To maintain the correct temperature during particle irradiation and subsequent measurements, the setup was preliminarily calibrated by comparing temperature records from the thermocouple built in the cell holder and a thermocouple glued to the surface of a test cell by a heat conductive adhesive.

Particle irradiation was performed with electrons and protons using SIRIUS irradiation facility of LSI and ARAMIS irradiation facility of CSNSM, respectively.

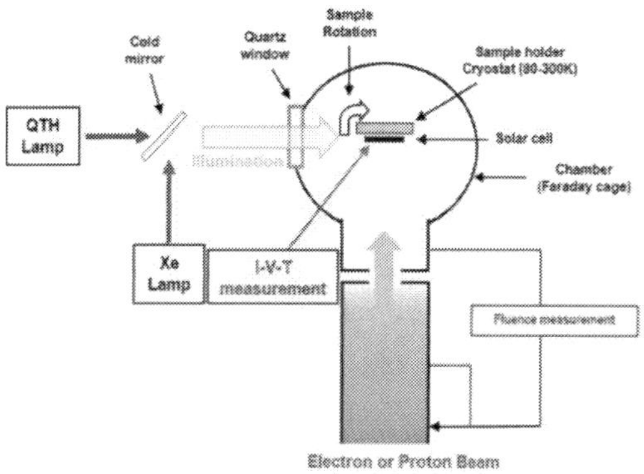

Figure 1. Experimental set-up for in-situ EOL characterisation at LILT conditions

4. ELECTRICAL PERFORMANCE

4.1 Begin-of-Life Performance

Electrical measurements performed at 3.7%AM0 and temperatures down to -150°C show an almost perfect shape of the IV characteristics with no signs of a flat spot effect as shown in Fig. 2.

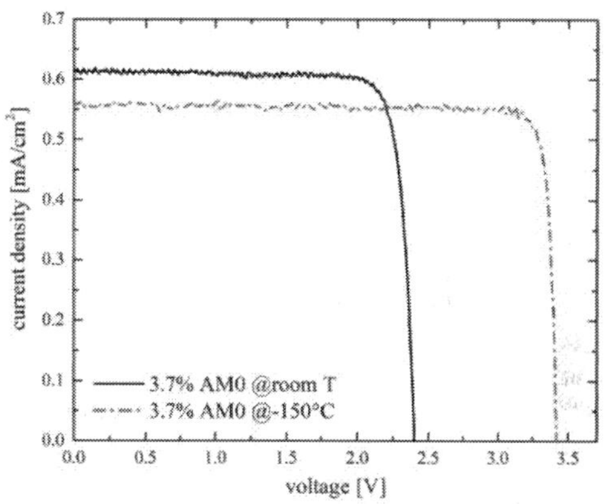

Figure 2. Low intensity (3.7%AM0) IV characteristics of LILT solar cells at room temperature and -150°C

In terms of overall performance, low illumination intensity results in a BOL efficiency of approx. 25% at room temperature, which increases up to 33.5% at -150°C. Due to temperature effects, at LILT the cell voltage increases up to 3.41V. Solar cell characteristics averaged over 100 pcs of 4×8 cm^2 cells are given in Table 1. For comparison, standard AM0 data of the same cell group are presented, too. It is to note that the spread on BOL cell parameters at LILT was found to be the same as the one at standard AM0 conditions.

Table 1. BOL performance of AZUR LILT solar cells at standard AM0 and 3.7% AM0 room and low temperature conditions

	@AM0 28°C	@LIRT	@LILT
Jsc [mA/cm²]	16.65	0.61	0.55
Voc [V]	2.69	2.40	3.41
Pmp [mW]	1178	38	51
FF	0.87	0.86	0.9
Eta [%]	28.5	25.1	33.5

As expected and suggested by the theory, the fill factor starts to increase linearly with lowering temperature. However, at temperatures below -100°C this increase starts to saturate and at -150°C, typical fill factor values are in the range 0.89 - 0.91 as shown in Fig. 3. This effect and its scatter is related to tunnelling effects that become the dominating contributors of the dark current at low temperatures [4].

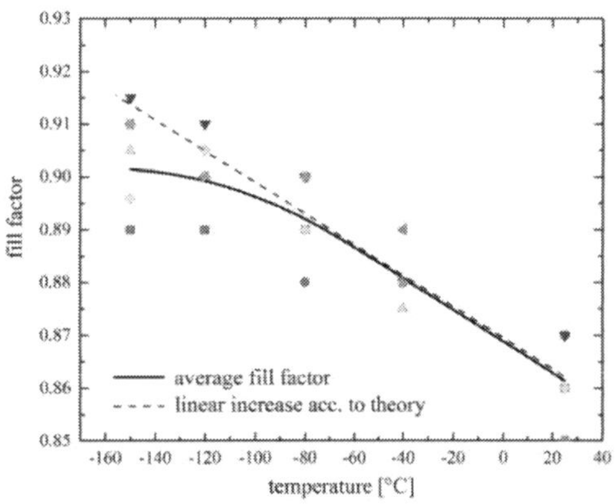

Figure 3. Temperature dependence of the fill factor at 3.7AM0.

4.2 End-Of-Life (EOL) Performance

As mentioned above, the main issue to be taken into account for the correct determination of the EOL performance is an appropriate test procedure that would allow for identifying the influence of temperature effects on the cell characteristics (annealing). The applied test procedure consisted in the irradiation of solar cells inside a cryostat at -150°C and subsequent repetitive electrical measurements at -150°C, 28°C and -150°C again. At each temperature, a series of IV curves was recorded to allow for observing time dependent changes of cell characteristics. Fig. 4 shows BOL and EOL IV characteristics of a triple-junction LILT solar cell irradiated with 3e15 cm^{-2} 1 MeV electrons at low temperature and measured in-situ at LILT conditions directly after stopping irradiation, after keeping the cell at these conditions for 30 minutes and after room temperature annealing. As can be seen, primary EOL cell characteristics (s. EOL1 curve in Fig 4) exhibits a noticeably recovery (EOL2 curve) within the first few minutes after irradiation, whereas keeping the cell at the same conditions longer than 10 minutes does not result in any further improvement (curves EOL3 and EOL4). However, after the cell annealing at higher temperatures, further recovery is observed (EOL-RA curve). Repetitive cell annealing at the same temperature does not change the cell characteristics anymore.

Thus, it can be concluded that a correct determination of the EOL solar cell performance at LILT conditions shall be based on in-situ measurements followed by the extraction of corresponding annealing coefficients. Knowing these coefficients would allow for avoiding very costly and time consuming low temperature irradiation testing during the solar cell qualification phase.

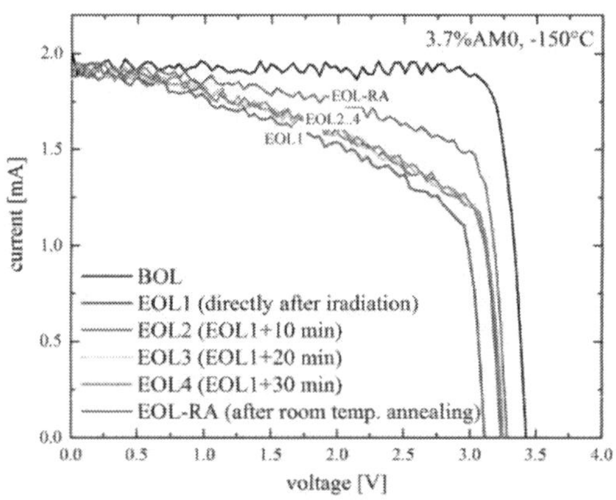

Figure 4. LILT BOL and EOL IV characteristics recorded directly after irradiation and after keeping the cell at LILT conditions for different time.

The second observation related to the behaviour of EOL solar cells at LILT conditions is that the most significant degradation occurs for Pmp compared to Isc and Voc values. A physical explanation of this "shunt"-like effect is given in [5].

The absolute value of the cell parameters recovery at LILT conditions depends on the irradiation fluence as illustrated in Fig. 5. After the highest applied 1 MeV electron fluence of 3e15 cm^{-2}, the primary recovery (stabilization at low temperature) of Pmp consisted of approx. 6 % and only approx. 2% after a fluence of 7.5e14 cm^{-2}. It is, however, to note that the recovery seems to be fast enough and, for the selected time scale, fluence independent: the stabilization was reached within 10 minutes in all measurements.

A similar dependence is observed for solar cells after room temperature annealing: up to 12% of improvement in Pmp after a fluence of 3e15 cm^{-2} (compared to avg. Pmp value after stabilization) and approx. 2% after a fluence of 7.5e15 cm^{-2}.

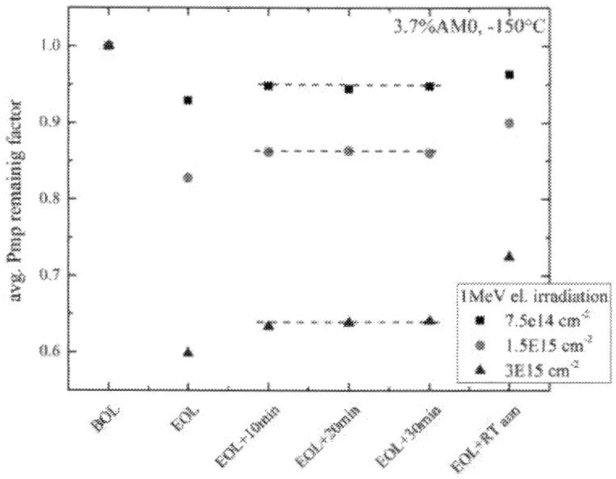

Figure 5. Recovery of LILT Pmp values after 1 MeV electron irradiation with different fluences

For the practical application, the most importance have EOL LILT characteristics after stabilization as the particle irradiation in space has a flux of several orders of magnitude lower than the one used in tests and the generation of radiation defects in solar cells is in equilibrium with their possible recovery. Tab. 2 summarizes LILT remaining factors of solar cells measured after 1 MeV electron irradiation and stabilization at low temperature. Similar data were also obtained for low temperature proton irradiation and are summarized in Tab. 3.

Table 2. Average LILT remaining factors measured at 3.7% AM0, -150°C after 1MeV electron irradiation and stabilization at LILT.

	5E14cm⁻²	1E15cm⁻²	2E15cm⁻²
RF Isc	0.99	0.96	0.89
RF Voc	0.95	0.93	0.91
RF Pmp	0.89	0.76	0.66
RF Eta	0.89	0.76	0.66

In order to verify the remaining factors obtained in the in-situ measurements, BOL and EOL characterization at LILT conditions was also performed at Fraunhofer ISE. As the EOL cells measured at ISE did receive unavoidable room temperature annealing during the transfer from the irradiation facility, their EOL data were corrected using annealing coefficients extracted from in-situ data.

Table 3. Average LILT remaining factors measured at 3.7% AM0, -150°C after 1MeV proton irradiation and stabilization at LILT.

	1e11 cm⁻²	2e11 cm⁻²	4e11 cm⁻²
RF Isc	0.97	0.95	0.86
RF Voc	0.94	0.92	0.91
RF Pmp	0.86	0.82	0.73
RF Eta	0.86	0.85	0.73

One issue is not completely understood, yet. Whereas the distribution of LILT cell performance at BOL is relatively narrow and corresponds to this one at standard AM0 conditions, EOL cell parameters both, after stabilization and room temperature annealing, exhibit a much wider spread that magnitude correlates with the irradiation fluence (Fig. 6). Furthermore, the observed dependence of this spread on the kind of particles (electrons or protons) suggests a dedicated role of particle scattering mechanisms and distribution of irradiation defects in the cell material.

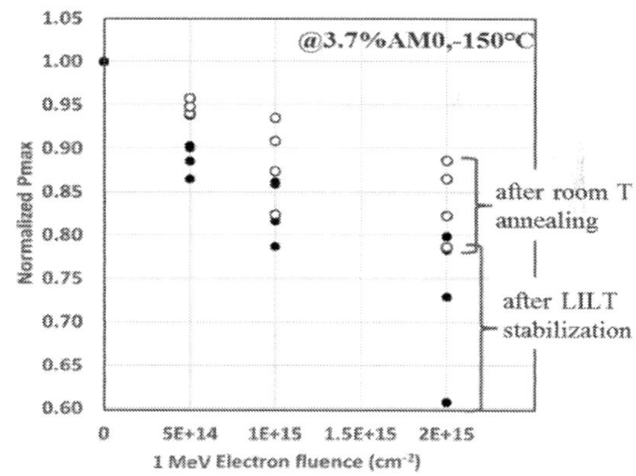

Figure 6. Spread of Pmp values at LILT after 1 MeV electron irradiation with different fluences

5. CONCLUSION

In conclusion, we performed extensive BOL and EOL characterization of 3G28 triple-junction solar cells at LILT conditions relevant for a mission to planet Jupiter. At 3.7% AM0 and -150°C, AZUR LILT solar cells show a superior efficiency of 33.5%. The need in correct determination of the EOL performance under these specific conditions led to the establishing of a sophisticated test procedure to avoid temperature related annealing effects. Using in-situ IV measurements, it was demonstrated that the degradation of Pmp of solar cells irradiated at low temperature is much stronger than the degradation of Isc and Voc values. It was also shown that solar cells irradiated at low temperatures exhibit a strong recovery effect within short time after the irradiation. The absolute value of the recovery depends on the irradiation fluence and particle type. For a fluence of 3e15cm^{-2} 1 MeV electrons, the recovery of Pmp at low temperature consisted of ~ 6% and the room temperature annealing – of 12%. Experimentally defined remaining factors for electron and proton irradiation allow for a realistic prediction of the solar cell performance at JUICE mission conditions whereas the quantification of annealing effects is essential for the planned cell qualification activity.

6. ACKNOWLEDGMENTS

Present work was supported by the European Space Agency under contract no. 4000109645.

7. REFERENCES

1. V. G. Weizer and J. D. Broder, "On the cause of the flat-spot phenomenon observed in silicon solar cells at low temperatures and low intensities", Journal of Applied Physics 53(8), pp. 5926-5930 (1982)
2. V. Khorenko, G.F.X. Strobl, R. Hoheisel, F. Dimroth, R. Campesato, M. Casale, C. Baur, "Triple-junction solar cell design for low intensity low temperature space applications", Proc. of the 9th European Space Power Conference ESPC, ISBN 978-92-9092-257-5, France (2011)
3. V. Khorenko, C. Baur, G. Siefer , J.C. Bourgoin R. Campesato, M. Casale, "Performance analysis of AZUR 3G28 triple-junction solar cells optimized for operation in Jupiter environment". Proc. of the 10th European Space Power Conference ESPC, the Netherlands (2014)
4. R. Hoheisel, R. J. Walters, and A. W. Bett, "Low temperature effects in photovoltaic devices for deep space missions," in Proceedings of the 42nd IEEE Photovoltaic Specialist Conference, 2015, pp. 1–5.
5. S. Park, J. C. Bourgoin, O. Cavani, V. Khorenko, C. Baur, and B. Boizot, "Origin of the degradation of triple junction solar cells at low temperature," *at this conference*.

QUALIFICATION TEST RESULTS OF IMM TRIPLE-JUNCTION SOLAR CELLS, SPACE SOLAR SHEETS, AND LIGHTWEIGHT&COMPACT SOLAR PADDLE

Mitsuru Imaizumi [1], Tatsuya Takamoto [2], Naoyuki Kaneko [3], Yukishige Nozaki [4], Takeshi Ohshima [5]

[1] Japan Aerospace Exploration Agency, 2-1-1 Sengen, Tsukuba, Ibaraki, 305-8505, Japan, Email: imaizumi.mitsuru@jaxa.jp
[2] SHARP, 492 Minosho-cho, Yamatokoriyama-shi, Nara, 639-1186, Japan, E-mail: takamoto.tatsuya@sharp.co.jp
[3] NEC, 1 Nisshin-cho, Fuchu, Tokyo 183-8551, Japan, Email: n-kaneko@ps.jp.nec.com
[4] NEC Space Technologies, 1 Nisshin-cho, Fuchu, Tokyo 183-8551, Japan, Email: y-nozaki@bu.jp.nec.com
[5] National Institutes for Quantum and Radiological Science and Technology, 1233Watanuki, Takasaki, Gunma 370-1292, Japan, Email: ohshima.takeshi@qst.go.jp

ABSTRACT

Development of a high specific power, low stowage volume and lightweight solar paddle utilizing high-efficiency, thin-film, InGaP/GaAs/InGaAs inverted metamorphic triple-junction solar cell has been completed. The cell exhibits the efficiency of as high as 32% with sufficient radiation resistance. Owing to the high efficiency of the cell, the paddle achieved the specific power of greater than 150 W/kg. Qualification and reliability tests for the cell and paddle were carried out. The results confirmed that both the cell and paddle have enough capability to use them in space. Space demonstration of the paddle is now in preparation.

1. INTRODUCTION

Solar paddles with high specific power (W/kg) and compact stowage volume have been demanded to meet the requirements of high electrical output power for spacecraft within limited launch volume [1-4]. We started the development of such solar paddles about a decade ago, and completed the development last year. To realize high specific power, high-efficiency, thin-film, InGaP/GaAs/InGaAs inverted metamorphic triple-junction (IMM-3J) solar cells have been developed to apply to the solar panels of the paddle. The cells are formed in a thin and flexible cell array structure which is called space solar sheet (SSS) [5]. We completed the development of SSS with InGaP/GaAs dual-junction thin-film (TF-2J) solar cell in 2011 [6]. However, since the efficiency of the TF-2J cell was as low as around 25%, the target specific power of 150 W/kg at paddle level was not realized.

In this paper, we introduce the performance of the high specific power solar paddle as well as the IMM-3J solar cells utilized in the paddles. The results of qualification and reliability tests for the cells and paddles are presented. Also, space demonstration test results on the SSS, which are proving high environmental resistance, are shown. In addition, we now have a plan to demonstrate the paddle in space. The details of the plan are explained.

2. IMM TRIPLE-JUNCTION SOLAR CELL

The IMM-3J thin film solar cells are fabricated by inverted growth of InGaP top-cell, GaAs middle-cell and InGaAs bottom-cell by metal-organic vapour phase epitaxy (MOVPE). The cell layers were mounted on a film with the thickness of 10-20 μm after the removal of GaAs substrates, and then the cell fabrication process was carried out. Figure 1 exhibits a photograph of the typical figure of completed IMM-3J bare cells.

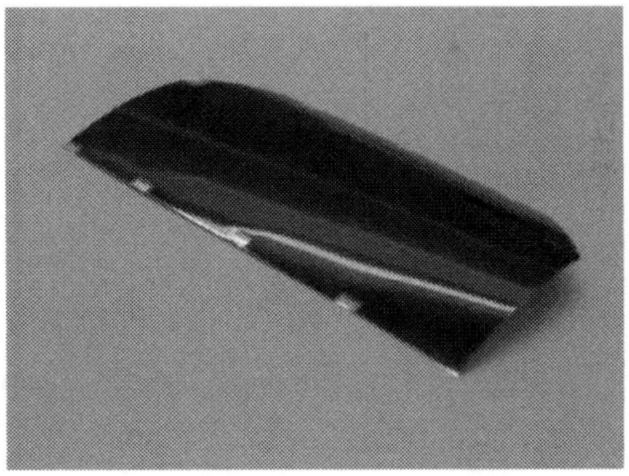

Figure 1. Typical figure of bare IMM-3J solar cells after completion of the fabrication process.

Two types of IMM-3J space solar cells with different design have been developed. One is Type A which has rather radiation hard design in order the remaining factor of maximum power (*P*max) to be greater than 85% after irradiation of 1 MeV electrons with a fluence of 1×10^{15} e^-/cm^2. Meanwhile, the other, Type B, is designed for missions in which high radiation tolerance is not required. Therefore, beginning-of-life (BOL) efficiency of Type B cell is higher than that of Type A

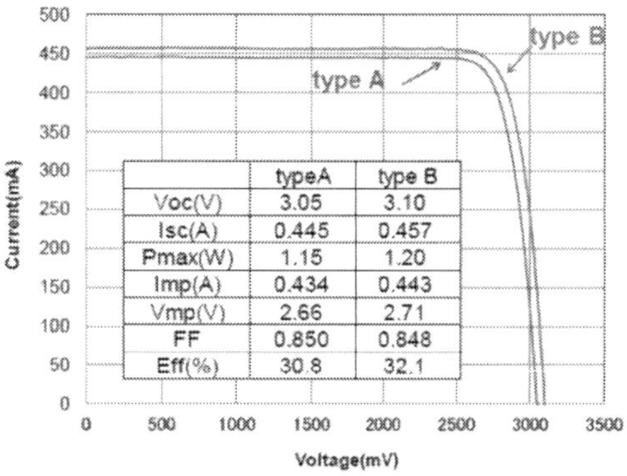

	typeA	type B
Voc(V)	3.05	3.10
Isc(A)	0.445	0.457
Pmax(W)	1.15	1.20
Imp(A)	0.434	0.443
Vmp(V)	2.66	2.71
FF	0.850	0.848
Eff(%)	30.8	32.1

Figure 2. Light I-V characteristics of two types of IMM-3J space solar cells (27.4cm^2, AM0, 25°C).

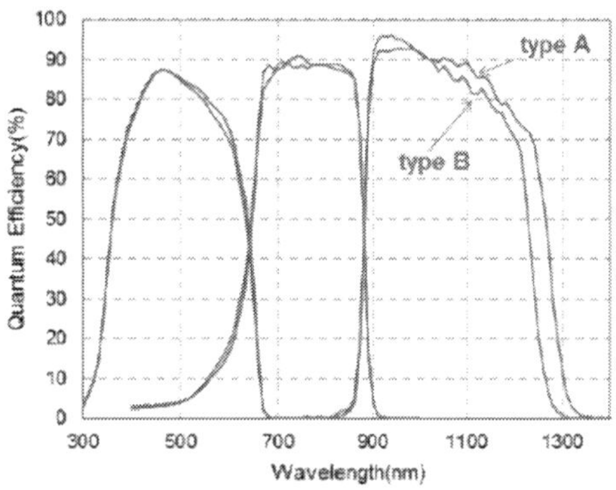

Figure 3. Comparison of external quantum efficiency (EQE) of Type A and B IMM-3J solar cells.

one.

Fig. 2 presents light current-voltage (I-V) characteristics of the two types of IMM-3J cells. The cell area is 27.4 cm^2. The best efficiency of Type A is 30.8% and that of Type B is 32.1% under AM0, 1 sun (136.7 mW/cm^2) condition.

The external quantum efficiency (EQE) of the two types of IMM-3J cell is exhibited in Fig. 3. The generation current ratio of Type A cell was designed with taking account of the radiation condition (1 MeV electrons, fluence of 1×10^{15} e$^-$/cm^2). The deterioration of EQE of the InGaAs bottom cell due to the electron irradiation is found to be greater than that of the InGaP top and GaAs middle cells [7].

For Type B cell, we can adopt a current matching design since the current margin for radiation degradation is not required. Thus, Type B cell has an InGaAs bottom cell with wider band gap (lower indium content) compared to Type A cell. As a result, open-circuit voltage (*V*oc) of Type B cell is about 50 mV higher due to the wider bandgap of InGaAs. In addition,

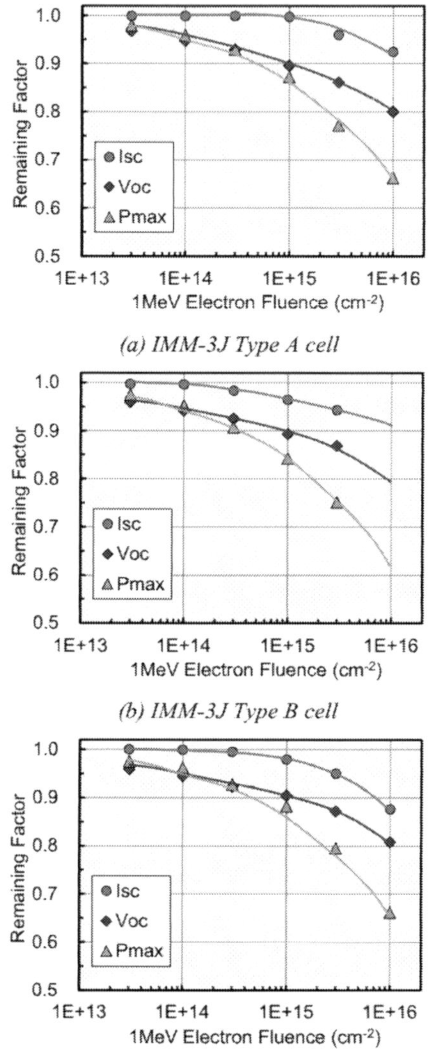

(a) IMM-3J Type A cell

(b) IMM-3J Type B cell

(c) InGaP/GaAs/Ge space cell (SHARP 502)

Figure 4. Degradation characteristics of the IMM-3J Type A and B cells and a current InGaP/GaAs/Ge 3J cell obtained from 1 MeV electron irradiation.

the thickness of the InGaP top cell can be increased because of less current margin for the GaAs middle cell. This results in slightly higher EQE of InGaP top cell of Type B than that of Type A cell. Consequently, short-circuit current (*I*sc) of Type B cell increased by 2.7% from that of Type A cell.

The radiation resistance of Type A and B IMM-3J cells against 1 MeV electrons are compared in Fig. 4. For reference, the irradiation data of a current InGaP/GaAs/Ge 3J space solar cell (SHARP #502) are posted in the figure. All the radiation tests were carried

DOI: 10.1051/e3sconf/20171603012

out at the Takasaki Advanced Radiation Research Institute of National Institute for Quantum and Radiological Science and Technology (QST). From the comparison with #502 cell, equivalent radiation resistance is confirmed for Type A IMM-3J cell. Note that Type B IMM-3J cell also indicates sufficient radiation resistant properties.

3. QUALIFICATION OF SPACE SOLAR SHEET

Fig. 5 illustrates a schematic of cross-sectional view of the Space Solar Sheet (SSS). The SSS utilizes thin silicon bypass diodes and flat-type interconnectors, both of which were specially designed for SSS. An in-plane stress relief structure is applied to the interconnectors for thermal-cycle reliability. The thin-film IMM-3J cells described above are embedded in transparent silicone adhesive, and the cells and the bas-bar tubs as terminals at the ends of SSS are connected by the interconnectors, forming thin-film solar cell arrays (typically five or six cells in series). The thin silicon bypass diodes are located at the cropped corner of the IMM-3J cells, as seen in the current 3J cell panels, and connected by also specially designed flat-type interconnectors. The surfaces of the arrays are laminated with front and back sheet materials.

Two types of pairs of shielding materials are employed for SSSs. One uses transparent polymer films for both sides (film-type SSS), and the other uses thin coverglass (typically 50 μm thick) for the front side and a reinforced carbon fiber sheet for the back side (glass-type SSS). Both are shown in Fig. 6. Total thickness of the both SSSs is less than 0.2 mm and the specific power is approximately 0.6 W/g for both types of SSSs.

Reliability tests were performed for the SSSs to prove the space environment tolerance. The specimens for the reliability tests were sheets with a single string consisting of five IMM-3J cells connected in series. Three specimens were prepared for the respective tests. Tab. 1 depicts the test items and their conditions. The differences in the test condition for film-type and glass-type are based on difference of potential application in space. The acceptance criterion of all the tests is that the change or degradation of power output characteristic parameters, namely, *I*sc, *V*oc, maximum power (*P*max), fill factor (*FF*), voltage at maximum power (*V*mp), and current at maximum power (*I*mp), is within ±2.5%. The test results are also shown in the column next to the test condition in Tab. 1.

As indicated in Tab.1, no significant change in the parameters was observed. In addition no change in appearance, such as discoloration, deformation or crack propagation was confirmed. Therefore, the reliability tests results clearly confirmed that the both types of SSS have sufficient tolerance for space environment. Furthermore, we should mention that the dark lines observed in the electroluminescence (EL) image of the IMM-3J cells which were originally existed before the tests did not influence on output performance of the SSSs throughout the tests.

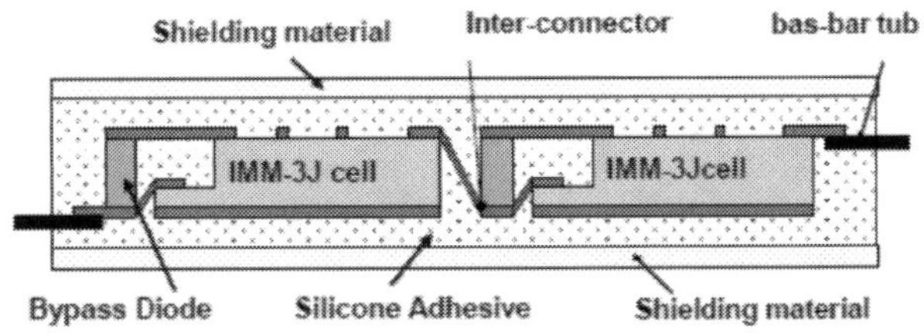

Figure 5. A schematic of cross-sectional view of Space Solar Sheet.

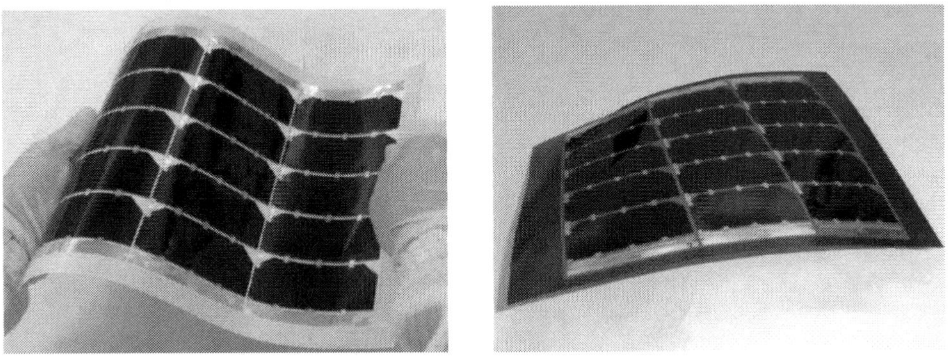

(a) Film type *(b) Glass type*

Figure 6. Photograph of two types of Space Solar Sheet.

Table 1. Summary of reliability tests for two types of Space Solar Sheets

Test item	Film type sheet		Glass type sheet	
	condition	Result characteristic change	condition	Result characteristic change
Thermal cycling	temp. :-178 °C⇔+161 ° C 1200 cycles	< ±1%	temp. : -180 °C⇔+120 °C 1000 cycles	< ±1%
Humidity and temperature	temp.: 65 °C, humidity: 90% duration : 720 h	< ±1%	temp.: 65 °C, humidity: 90% duration :720 h	< ±1%
High temperature vacuum	pressure: <133.3×10^{-5} Pa temp.: 160 °C, duration : 168 h	< ±1%	pressure: <133.3×10^{-5} Pa temp.: 160 °C, duration : 168 h	< ±1%
High temperature	temp.: 161 °C duration :1000 h	< ±1%	temp.: 150 °C duration :1000 h	< ±1%
Reverse current	Reverse current : 500 mA temp. :161 °C, duration :1000 h	< ±1%	Reverse current : 500 mA temp.: 150 °C, duration :1000 h	< ±1%

Table 2. Summary of tests for lightweight and compact solar paddle

Test item	Target/condition	Result
Vibration	In plane: (swp rate: 2oct/min) 0-25 Hz: 15g 25-50 Hz: 12g 50-100 Hz: 8g Out of plane: (swp rate: 2oct/min) 0-50 Hz: 12g 50-100 Hz: 8g	**OK**
Deployment	Vertical (main) panels deployed Lateral (sub) panels deployed	**OK**
Acoustic	145.8 dB	**OK**
Thermal vacuum (coupon)	Temp. range: −125 °C ↔ +110 °C Cycle: 8	**OK**
Thermal cycle (coupon)	Temp. range: −173 °C ↔ +128 °C Cycle: 1370	**OK**

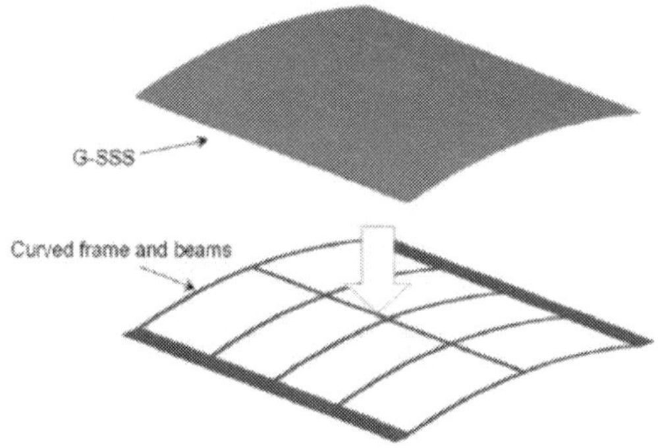

Figure 7. Schematic drawing of the lightweight panel structure utilizing the space solar sheets.

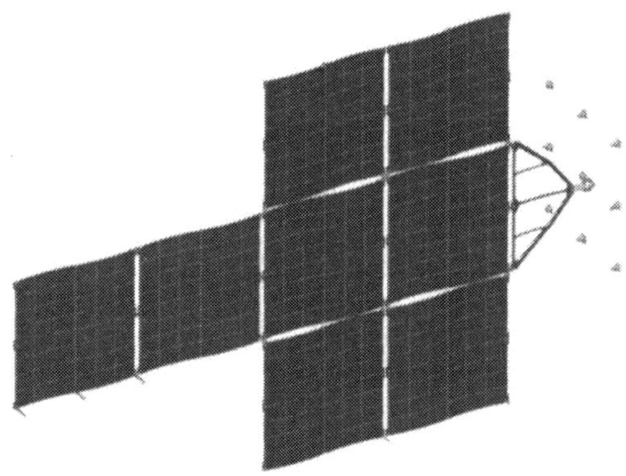

Figure 8. Schematic drawing of the engineering model of the lightweight and compact solar paddle.

 DOI: 10.1051/e3sconf/20171603012

4. LIGHTWEGHT AND COMPACT SOLAR PADDLE

The development of a new lightweight and compact solar paddle using lightweight solar panel with the glass-type SSSs has been completed [8] in 2015.

The lightweight solar panel employs a frame-type structure as shown in Fig. 7. The SSS is a thin sheet with thin coverglass, typically 50 μm thick, and they are attached on the frame structure like paper. We call this configuration "SHOJI structure" which is a Japanese traditional partition between rooms. The frame is made of a standard aluminium honeycomb but the thickness is as thin as 11 mm. To increase the out-of-plane stiffness of the thin panels, a curved structure is applied, and it prevents SSSs on the panels from cracks generated by vibration or other mechanical shocks. The SSSs are attached to the curved frame by Velcro fasteners. The influence of the curved shape on output power due to oblique incidence on the SSS surface is calculated to be 2 % decrease or less.

The engineering model (EM) of the lightweight and compact solar paddle was manufactured for environmental testing. Fig. 8 depicts a schematic of the EM solar paddle. It consists of eight panels, and the size of the panel is 1.75 m × 1.65 m. The weight of the paddle is 40.63 kg. The estimated output power calculated using a cell efficiency of 30% is about 6.5 kW, which results in the specific power of the paddle of about 160 W/kg. The architecture is inherently modular and scalable. For increased power requirements, just the number of the unit panel can be increased. New lightweight hinges, holding/releasing devices and synchronization mechanisms for deployment have been designed, developed and integrated in the EM solar paddle.

5. TEST RESULTS OF LIGHTWEGHT AND COMPACT SOLAR PADDLE

Various tests were performed for the EM solar paddle to qualify the new design and devices. Most of the panel surface was occupied with the glass-type SSSs without solar cells (dummy sheets), while the real ones with healthy IMM-3J cells were attached for some important areas of the panels. Tab. 2 summarizes the tests, their conditions and the obtained results. The EM solar paddle survived all the tests and the new structure and configuration have been qualified successfully. Fig. 9 posts a photograph of the setting for the vibration tests. Note the compactness of the folded eight panels. The height is around 2/3 compared with the current folding/stowing technology with eight of 25 mm-thick panels.

Figure 9. Setting of the EM solar paddle for the vibration tests.

6. FLIGHT DEMONSTRATIONS

Currently, the electrical output characteristics (*I*sc and *V*oc) of three prototype glass-type IMM-3J SSSs has been collected in orbit as the world–first demonstration of such thin solar panel structure. This demonstration is being performed on the scientific telescope satellite "HISAKI" launched in September, 2013. HISAKI was injected in an elliptical orbit at an altitude of 950×1150 km, at an inclination angle of 31°, and for an orbit period of 106 minutes. Two IMM-3J cells are connected in series and covered with 50 μm-thick coverglass. The trend data of remaining factors of the *I*sc and *V*oc are exhibited in Fig. 10. So far the SSSs show no degradation on *I*sc and about 2% degradation in *V*oc

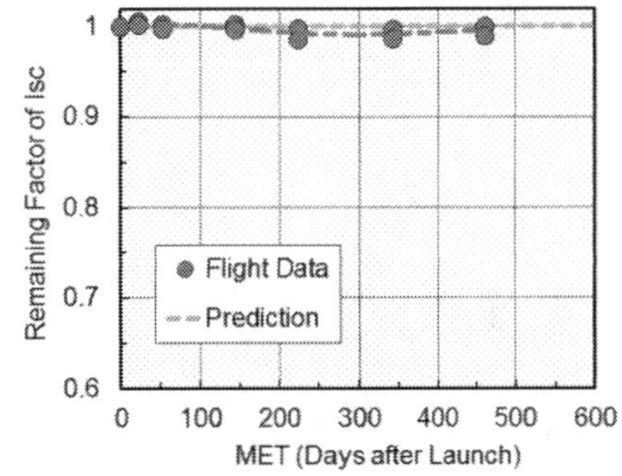

(a) Short-circuit current

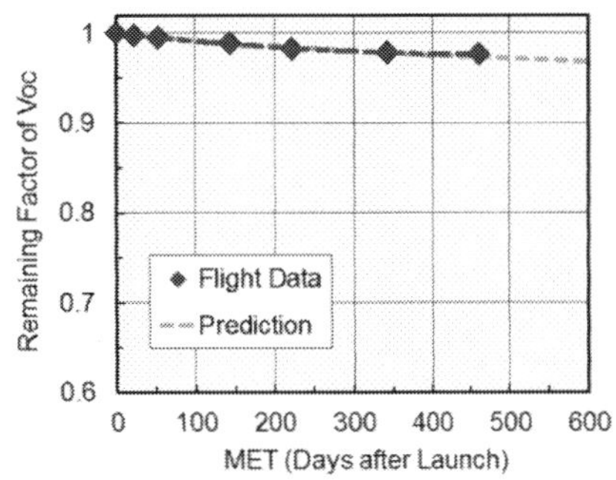

(b) Open-circuit voltage

Figure 10. Trend of remaining factors of (a) short-circuit current and (b) open-circuit voltage of the glass-type SSSs with IMM-3J cells on HISAKI.

which meets our radiation degradation prediction, and no anomaly has been observed. These flight data prove good environmental durability and reliability of glass-type SSSs.

For film-type SSS, a space demonstration is now ready. A demonstration panel which consists of six film-type SSSs has been developed. For electrical connection between SSSs, the conductive thermo-compression bonding tape produced by Dexerials Corp. was employed [9]. The panel is mounted on JAXA's ISS transfer vehicle KONOTORI (HTV) No. 6 which is waiting to be launched in late 2016. The primary purpose of this demonstration is to confirm durability of film-type SSSs in launch/mechanical environment. Environmental tests for the flight panel shown in Fig. 11 have already completed successfully.

Furthermore, a demonstration test for the lightweight and compact solar paddle is now planned using a small, innovative technology demonstration satellite. A small-size solar paddle composed of five panels will be manufactured with using the same deployment mechanisms with practical size paddles. Thus, the prior purpose of this demonstration is to examine the function of the developed three-dimensional deployment system.

Figure 11. Lightweight solar panel flight demonstration unit "SFINKS".

7. SUMMARY

Development of a high specific power, low stowage volume and lightweight solar paddle utilizing high-efficiency, thin-film, InGaP/GaAs/InGaAs inverted metamorphic triple-junction solar cell has been completed. The cell exhibits the efficiency of as high as 32% with sufficient radiation resistance. Owing to the high efficiency of the cell, the paddle achieved specific power of greater than 150 W/kg. Qualification and reliability tests for the cell, SSSs, and paddle were carried out. The results confirmed that both the cell and paddle have enough capability to apply them in space. Space demonstration of the new paddle is now underway.

8. REFERENCES

1. R. Stribling, (2006) 'Boeing High Power Thin Film Solar Array', 4th International Energy Conversion Engineering Conference and Exhibit, AIAA No. 2006-4013.

2. M. Eskenazi, S. White, B. Spence, M. Douglas, M. Glick, A. Pavlick, D. Murphy, M. O'Neill, A. J. McDanal, and M. Piszczor, (2005) 'Promising Results from Three NASA SBIR Solar Array Technology Development Programs', NASA/CP-2005-213431.

3. K. Seifart, T. Schmidt, W. Göhler, and S. Langlois, (2005) 'Structural Design of Flexible Solar Generators', In Proc. 7th European Space Power Conference, ESA SP-589 (CD-ROM), European Space Agency, Noordwijk, The Netherlands.

4. Y. Combet, (2013) 'State of the Art and Future Challenges for Space Telecommunication Satellites Solar Generators' In Proc. 28th European Photovoltaic Solar Energy Conference and Exhibition, 1BO.12.1, Paris, France, 30 Sep - 4 Oct.

5. T. Takamoto, T. Agui, A. Yoshida, K. Nakaido, H. Juso, K. Sasaki, K. Nakamura, H. Yamaguchi, T. Kodama, H. Washio, M. Imaizumi, and M. Takahashi, (2010) 'World's Highest Efficiency Triple-junction Solar Cells Fabricated by Inverted Layers Transfer Process', In Proc. 35th IEEE Photovoltaic Specialists Conference, Honolulu, Hawaii USA, pp.412-417.

6. M. Imaizumi, , K. Shimazaki, , Y. Kobayashi, M. Takahashi, K. Nakamura, T. Takamoto, S. Sato, T. Ohshima, (2011) 'Properties of a Qualified Space Solar Sheet with InGaP/GaAs Dual-Junction Solar Cells', In Proc. 9th European Space Power Conference, ESA SP-690 (CD-ROM), European Space Agency, Noordwijk, The Netherlands.

7. T. Takamoto, H. Washio, and H. Juso, (2014) 'Application of InGaP/GaAs/InGaAs triple junction solar cells to space use and concentrator photovoltaic', In Proc. 40th. IEEE Photovoltaic Specialist Conference, Tampa, Florida, USA.

8. K. Shimazaki, M. Takahashi, M. Imaizumi, T. Takamoto, T. Ito, Y. Nozaki, H. Kusawake, (2014) 'Lightweight solar paddle with high specific power of 150 W/kg', In Proc. 10th European Space Power Conference, ESA SP-719 (CD-ROM), European Space Agency, Noordwijk, The Netherlands.

9. See http://www.dexerials.jp/en/

Multijunction Solar Cell Performance in Mars Orbiter Mission (MOM) conditions

Uma BR †, M. Sankaran, and , Suresh E. Puthanveettil

Power System Group, ISRO Satellite Centre (ISAC), Bangalore, India

† ***Email ID: uma@isac.gov.in, Tel. no: +91 8025083531***

1. Introduction

Mars Orbiter Mission (MOM) is the maiden interplanetary mission of ISRO. The mission was conceived to be a technology demonstrator and includes five payloads which are designed for the observation of the physical features of Mars and its outer atmosphere. The MOM scientific payloads are supported by mainframe systems like power, thermal, communication system, control, digital, and propulsion systems. The electrical power system of MOM comprises of a single power bus tied to a Li-ion battery and power generation is through photovoltaic solar array using state-of-art multi junction solar cells (MJSCs).

This paper discusses about the array operating requirements, array configuration, and predicted array performance versus the observed in-orbit performance at MOM orbit conditions. Also presents about the MOM array performance prediction and array observation models.

Key Words: Photovoltaic, Multi junction solar cells (MJSC), Low intensity and low temperature (LILT), solar radiation pressure (SRP).

2. Solar Array Design Requirements

Photovoltaic solar array (PVA) is a prime source of power for satellites not only for LEO, MEO, and GEO missions but also for various near-earth interplanetary missions and rover missions to Moon and Mars. Advances in solar cell technology in the last three to four decades, starting from the ~ 10 % efficient single-crystal silicon solar cell to the state-of-art ~ 28 % efficient multijunction solar cells have made it possible to cater high power requirements of spacecraft and as well as it can operate in critical environment conditions. Hence, solar array has chosen to be the prime power generation source for MOM spacecraft.

MOM solar array was required to provide up to 850 W of spacecraft load power out of which the maximum consumption of power was of communication and an additional 375W of battery charging power in all the mission phases.

Mass and hence the solar array area being at a premium for the MOM mission, use of ~28% efficiency (1AM0) multijunction solar cells was the prime choice. . The dilemma of MOM solar array design was whether to design the array for the operating conditions of Martian orbit, Earth orbit or both. Our expertise suggested that the solar array be designed for Martian orbit. This would entail accounting for the off-optimal operation loss for the array during the near Earth orbit phase but optimal operation at Martian orbit. Every additional watt of power generated by the solar array, in the Martian phase means, longer periods of communication from MOM to Earth and longer mission life.

Hence, the MOM solar array was designed for the Martian phase and during the Earth bound phases, the solar array was planned to be off-set away from the sun to control the solar cell operating temperature. The mission designers were confident of ensuring this because of the following reasons:

- Battery tied configuration of the power bus
- Amount of autonomy built into MOM
- Any mission safe mode is also power safe

Such a design approach to off-set the solar array to ensure reasonable temperature of the solar cells was in well agreement with the need to minimize the Solar Radiation Pressure (SRP) torque during the cruise phase. This operation met the mission requirements throughout the Earth bound and cruise phases.

The design of the MOM mission orbit transfer strategy resulted in the spacecraft orbiting Earth several times with the low perigee altitude of near 200km, gaining orbital velocity with each burn of the liquid engine aboard the spacecraft. During the design phase, a worst case estimate of 88 transitions through the Van Allen radiation belt was envisaged. Hence, the MOM solar array design accounted for the damage due to the radiation flux through the anticipated 88 transitions through the radiation belts in addition to the predicted solar flares. In actual the MOM solar array experienced total 40 Van Allen belt crossing. The total dose of particulate radiation after each burn was estimated by considering actual orbit parameters using generic radiation models. The array performance was predicted by applying radiation remaining factors also.

With the change in solar intensity, the solar cell temperature also changes. Considering GEO spacecraft as reference, for spacecraft at large distances from the sun, the solar intensity and solar cell temperature are both low, while for spacecraft near the sun they both are high. This aggravates the difficulties of operating solar arrays both far from and near to the sun.

The MOM spacecraft during near Earth bound phase experienced 1353W/m^2 and temperature of +55°C similar to GEO spacecraft. In Cruise phase the spacecraft has undergone different sun intensity starting from 1400W/m^2 to 700W/ m^2 and temperature from +55°C to -2 °C .The MOM orbit around Mars features low solar irradiance (493 W/m^2 ≈ 0.38 AM0) and low temperature (-20°C), this conditions known as Low Intensity and Low Temperature (LILT) affects the solar cell performance. In all the above orbit conditions the MOM solar array performed as per the design and prediction.

3. MOM Solar Array Configuration

MOM solar array is a rigid, deployable single wing array with three solar panels having total area of 7.52 m^2. The panels were populated by ~ 2,000 space qualified multijunction solar cell assemblies of ~28 % BOL efficiency at RT, 1AM0. S. These solar cells are arranged into 16 switchable strings controlled by a Taper Charge Regulator (TCR) and connected to battery tied single bus. Fig.1 shows the three flight panels of the MOM mission assembly.

Fig1. MOM Solar Array

4. MOM Array Performance prediction

A solar array is Ns x Np matrix of solar cells. Prediction of solar array power in space depends on a number of factors like type of solar cell, its characteristics, operating temperature of array, particulate radiation environment, array orientation angle, seasonal variation of sun declination angle, solar intensity and various other degradation factors. The voltage and current of the solar cell for a given solar intensity (X) and temperature (T) is defined by equations 1 to 3.

$$I_{sc}(X,T) = X.I_{sc}(1,T) \quad (1)$$

$$V_{oc}(X,T) \equiv V_{oc}(1,T) + C.\ln(X) \quad (2)$$

$$V_L(X,T) \equiv V_L(1,T) + C.\ln(X) \quad (3)$$

Where C is a constant. The solar cell parameters at different intensity and temperature throughout the mission were calculated using equations (1-3).

The generated electrical power of solar array is obtained by the following equation (4) and the solar array power prediction/ estimation model is shown in Fig 2.

$$P = I*V*i*\cos(\delta)*\cos(\psi) \quad (4)$$

Where
P = Power generated
V= Bus Voltage
I = Solar array Generated Current
i = Intensity Factor
δ = Sun declination angle
ψ= Sun off set angle

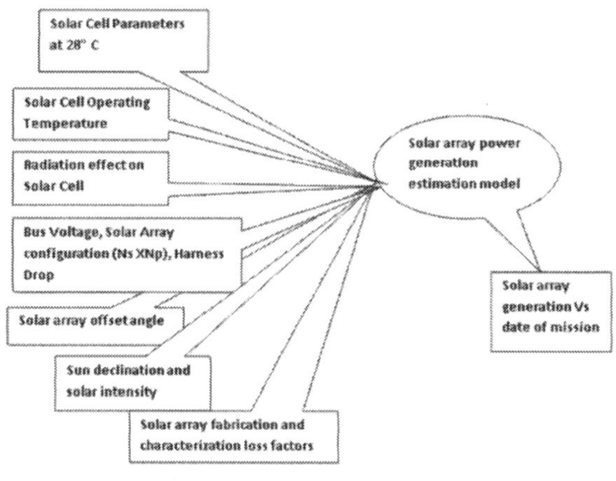

Fig2. Solar array power generation estimation model

 DOI: 10.1051/e3sconf/20171604001

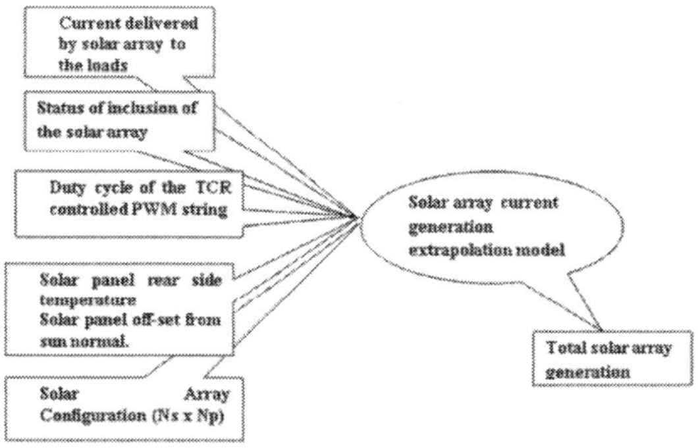

Fig3. On board solar array generation estimation model

5. In-Orbit Performance of MOM array

MOM spacecraft was launched on November 05, 2013 by Polar Satellite Launch Vehicle (PSLV) from SHAR, Sriharikota, India. The solar array was deployed after 45 minutes of vehicle liftoff. The solar array supported the spacecraft power requirement and battery charging requirement except during eclipse and liquid engine burn maneuvers during orbit raising operations.

The observability available on-board MOM spacecraft for performance verification of the solar array are the following:

- Current delivered by solar array to the loads/ Load Currents
- Status of inclusion/exclusion of the solar array strings
- Duty cycle of the TCR controlled PWM string
- Solar panel rear side temperature
- Solar panel off-set from sun normal.

Using these data available in telemetry, a method was established for extrapolating the total power generation capacity of MOM solar array. Such extrapolations were carried out at various points of the MOM mission. The on board array generation model is shown in fig 3. This in-orbit telemetry derived generation estimate was compared against the theoretical predictions throughout the mission.

The solar array performance was observed for various operational conditions at different orbit phases i.e

- Different sun aspect angles of the array
- Different solar panel temperatures
- Effect of Van Allen belts transition
- Effect of LILT etc.

i. Different sun aspect angles and operating temperatures

In order to meet the design requirement of operating the solar array in Earth bound orbit phase, to control the operating temperature and also to minimize the solar radiation pressure, the solar array was operated at different offset angles from Sun and at different operating temperatures. The array performance was monitored during this phase and the array performance was matched with the prediction and is shown in Fig 4.

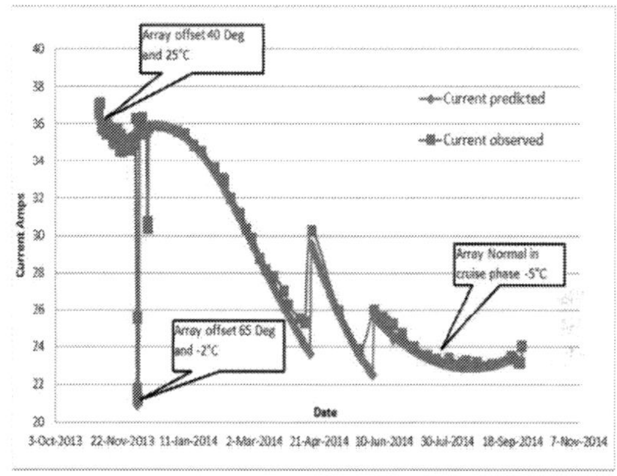

Fig4. Solar array predicted Vs observed/ telemetered performance at different offset angles and temperature in near earth orbit and cruise phases.

ii. Effect of Van Allen belts transition

Due to orbit transfer strategy to raise the MOM orbit, solar array crossed 40 Van Allen radiation belt during this period. The total dose of particulate radiation was estimated by using generic radiation models. The array performance was predicted for 32 Van Allen belt crossing by applying radiation remaining factors. The solar array performance was compared with observed data on 5th November, 2013 immediately after launch and on 21st November, 2013 after several / ~32 Van Allen radiation belt crossing (after five orbit maneuvers), and datas have matched within the accuracy of ±1% and are shown in fig 5 and fig 6.

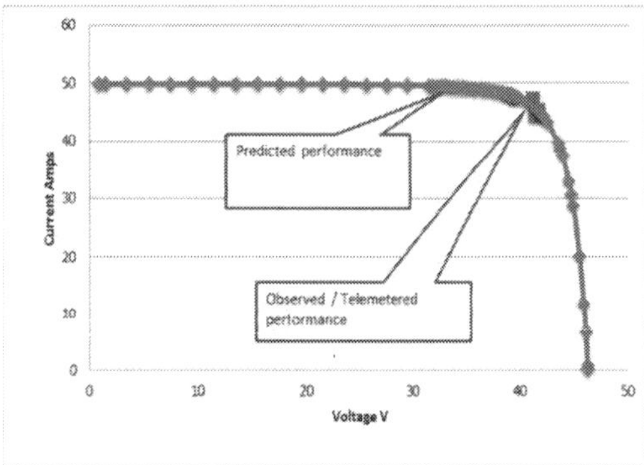

Fig5. Solar array predicted Vs observed/ telemetered performance on launch day (Nov 5, 2013)

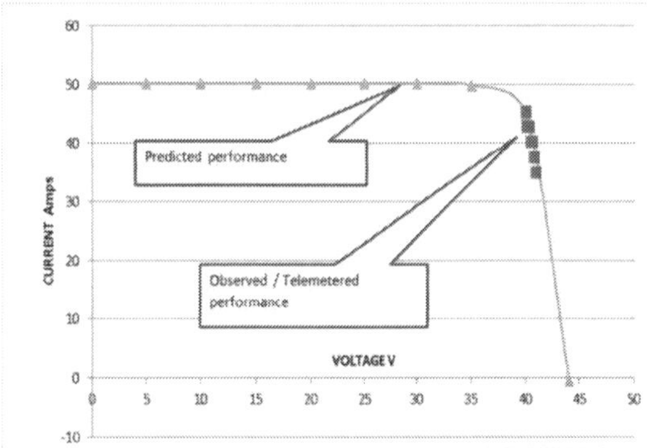

Fig6. Solar array predicted Vs observed/ telemetered performance after 32 Van Allen belt crossing

iii. Effect of LILT

Due to orbital condition of mission, MOM array has experienced low solar irradiance from 700 W/m^2 to 493 W/m^2 ≈ 0.37 AM0 and temperature from -2 °C to as low as -22°C conditions known as Low Intensity and Low Temperature (LILT). The observed solar solar cell performance is well matched with designed prediction within the telemetered accuracy of 2% as can be seen in fig 7.

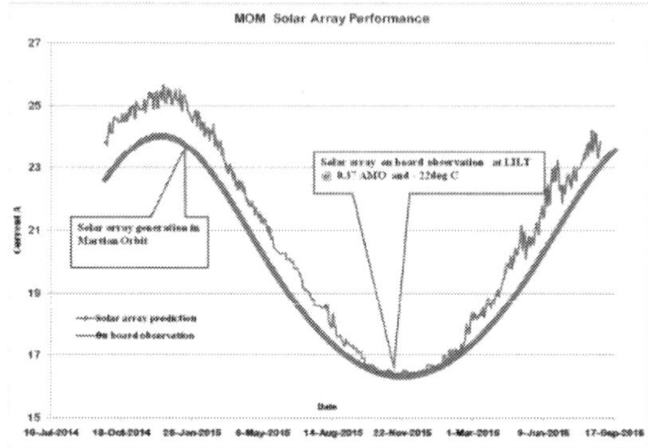

Fig7. Solar array predicted Vs observed/ performance in Martian orbit

6. Conclusion

A brief summary of MOM solar array design, design requirements, methodology for array performance prediction and in-orbit performance analyses at different orbit phases of MOM spacecraft were presented. The performance of the MOM solar array under all operating conditions is as per design and prediction as can be seen from the figure 4, 5, 6 and 7 throughout the mission.

7. Acknowledgement

The authors are grateful to Director, ISAC and Deputy Director, CPA for their guidance and support for this work. Also we wish to thank all individuals of solar panel division for their contribution in realizing the solar array for MOM.

8. References

1. "*Solar Array Design handbook*", RauschenbachH.S.,Van Nostrand Reinhold Co.,New York, USA,1980.
2. "*Solar Cell Radiation Handbook*", H.Y.Tada, J.R.Carter, Jr., B.E. Anspaugh, R.G. Downing, Third Edition, JPL,NASA, California, USA, 1982
3. "*Design and performance of the MER(Mars Exploration Rovers) solar arrays*",Paul M.Stella, Richard C.Ewell and Julie J.Hoskin, 2005 IEEE.
4. *"In Orbit performance of Ultra Triple junction solar arrays",* L. Anikst,Z. Maassarani et.al. 2013 IEEE

Photovoltaic Assemblies For The Power Generation Of The Exomars Missions

Emanuele Ferrando[(1)], Pietro Zanella[(1)], Stefano Riva[(1)], Giulia Damonte[(1)], Romano Romani[(1)], Luigi Ferrante[(1)]

[(1)] Leonardo SpA, Viale Europa snc, Nerviano (MI),
Email: emanuele.ferrando@leonardocompany.com, pietro.zanella@leonardocompany.com, stefano.riva01@leonardocompany.com, giulia.damonte@leonardocompany.com, romano.romani@leonardocompany.com, luigi.ferrante@leonardocompany.com

ABSTRACT

This paper provides a detailed description of the Photovoltaic Assembly (PVA) and, in particular, how the extreme environmental and operative requirements of the Exomars missions are met. "Exomars" is a program established by the European Space Agency to investigate the Martian environment and to validate new technologies in view of future Mars exploration missions. The first mission will be launched in 2016 and will carry a spacecraft composed of the Trace Gas Orbiter (TGO) and the Entry, descent and landing Demonstrator Module (EDM). The second mission is scheduled on year 2018 and will carry a rover vehicle. Leonardo PVA contribution is on either the TGO, Carrier and Rover vehicle solar arrays. Moreover we were deeply involved in the design of the power conditioning and distribution electronics of the TGO leveraging on our Power Management and Distribution heritage for scientific platforms.

As far as Exomars SA's (Solar Array) are concerned, they are significantly different: the TGO is based on two deployable wing system, whose attitude is controlled by the SADM. Each wing comprehends two panels.

The Rover vehicle solar array assembly (SAA) is made of a fixed panel and four deployable (primary and secondary) panels. This SAA is unique with respect to standard deployable systems because of the motor deployment control and trimming possibilities during Martian soil exploration.

Each of these missions imposed major design challenges: for the TGO the aero braking maneuvers will cause a significant dissipation of the kinetic energy leading to a solar array temperature increase up to +175°C on the rear face, while +135°C on the front (active) face. On the other hand, during the cruise phase the same PVA will experience extreme cold temperatures down to -209°C.

For the rover SAA the biggest challenge is related to surviving Mars planet environment (i.e. dust, wind and charge accumulation) and the associated mechanical and electrical constraints. We duly considered the experience gained by NASA-JPL on previous Mars lander missions and the selected technical solutions, all taken from the European heritage, represent the evolution to what already successfully flown.

Finally the bio burden and planetary protection requirements are also not a standard in SA field. A special manufacturing, assembly and test sequence will be implemented to capitalize our previous experience on the Mars drill tool development. For both the arrays European PVA technology was exclusively used: more specifically high efficiency III-V compounds solar cells and a new glass grounding network.

1. PVA technology for MARS missions

The Photovoltaic Assembly is part of the solar array, which design is driven by the complexity of the various missions to Mars. Different solutions has been implemented depending on the specific missions.

1.1 Exomars TGO PVA

On the Exomars TGO the aero braking sequence will be the most demanding phase. The aero braking manoeuvre will be performed near Mars. It is aimed to reduce the orbital period via about several passes through the upper level of the Mars atmosphere. The kinetic energy of the spacecraft that will be dissipated to get the necessary delta-V will cause a solar array temperature increase up to +175°C on the rear face, so mainly affecting discrete EEE parts and wiring, while +135°C on the front face, so for the solar cells and bypass diodes.

Figure 1. EXM TGO artistic view (courtesy of ESA).

SA is composed of two deployable wings, each one made of two identical panels and one Solar Array Drive Mechanism (SADM). Each panel has an electrical network consisting of 10 electrical sections with 6/7 strings each, for a global 65 parallel x 22 series configuration. Each string is protected by 2 blocking diodes grouped at section level into a single diode board. As distance between panel is very low during

 DOI: 10.1051/e3sconf/20171604002

stowed phase (8 mm), the rear side layout has been routed in a specific way to avoid any overlapping between elements while keeping a common unique configuration.
The AZUR space 3G28 is the selected solar cell baseline in combination with an external Si planar bypass diode (SIBPD). These two items followed an almost independent manufacturing process flow and are integrated one to each other at string formation level.

Figure 2. EXM TGO satellite during vibrational test (courtesy of ESA).

Table 1. EXM TGO PVA main characteristics.

Solar cell Assembly	AZUR Space 3G 28% GaAs/Ge solar cell Dow Corning DC93500 adhesive CMG QioptiQ Coverglasses
Protection diode	AZUR Space SIBD silicon diode Dow Corning DC93500 adhesive QioptiQ CMG Coverglasses
PVA layout	2 wings of 2 panels each 1 panel with 10 section of 6/7 strings each 65 strings (22 cells each) per panel (5720 solar cells)
Installed Power	- Begin Of Life - 1 solar constant, AM0, 28°C - Max available Power @ SA: 6360 W

1.2 Exomars Rover PVA

Exomars Rover Vehicle solar array is a unique equipment, different from anything else developed before in Europe. No off the shelf solution are readily available on the market. The biggest challenge in terms of development and qualification efforts is related to surviving Mars planet environment (i.e. dust and wind) and the associated mechanical and electrical constraints.
Due to the above, the experience gained by the Americans on previous Mars lander missions was duly considered and the selected technical solutions are in continuity with respect to what already successfully flown.
The solar array will be released and deployed following landing, prior to the commencement of surface operations.
For maintaining mission operation flexibility, the solar array may be partially deployed (primary panels only) or fully deployed (primary & secondary panels) whilst on the Lander Platform.
Prior to egress, the Rover remains connected to the Lander Platform via electrical umbilical's. Solar array deployment is completed after the Rover has egressed onto the surface (if not fully deployed on the Lander Platform).

Figure 3. EXM Rover vehicle (courtesy of ESA).

During egress and mobile operations the deployed solar array will be latched into its nominal deployed configuration and exposed to mechanical loading arising from the rough terrain. The Rover will only drive during daylight and will be stationary at the end and beginning of each sol. During these stationary periods, the angles of the deployable panels will be adjusted by rotation about the primary hinges to increase the incident solar energy.
The need for the mechanisms to operate in the Mars atmosphere imposes extra requirements with respect to typical space projects, where operation is normally in vacuum. In particular, the presence of dust in the Mars atmosphere could contaminate the mechanisms affecting their functionality. These constraints could affect the SAA mechanisms, although no tests in such Martian environment are foreseen for the SAA mechanisms.
The solar array assembly comprises the following parts:

- Fixed SA panel with PVA and wiring harness
- Primary deployable SA panels with PVA and wiring harness

- Secondary deployable SA panels with PVA and wiring harness.

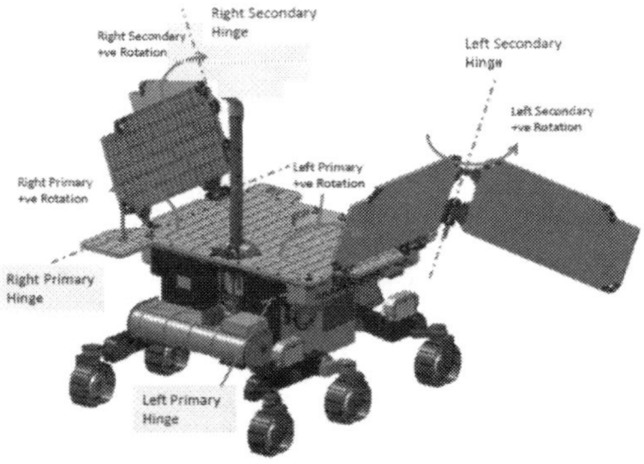

Figure 4. EXM Rover vehicle deployment (courtesy of Airbus).

For what concerns the electrical parts (PVA and wiring harness) the following parts are considered:
- Solar cell assemblies, string blocking diodes, thermistors and bleed resistors
- Electrical harness to bulkhead connectors
- Electrical harness to ADE for deployment lines

The selected solar cell is the AZUR Space 3G30 with an external Si by pass diode. A specific requirement on the glass grounding network.
It is mandatory to create a common ground network for all the glass exposed surfaces. This means the use of Indium Tin Oxide (ITO) coating on top of the glass surface and a stable connection of this coating to the PVA network.

To do so a specific glass was selected with anti-reflective coating on top of ITO coating. Moreover the glass layout is such that the cell and the by-pass diode are covered by the same glass. With the above solution the grounding of the ITO is done via the by-pass diode top contact, partly exposed. A drop of Silicone conductive adhesive is the connection in between the glass edge conductive coating and the reference network point.

Table 2. EXM Rover PVA main characteristics.

Solar cell Assembly (including protection diode)	AZUR Space 3G 30% GaAs/Ge solar cell AZUR Space SIBD silicon diode Dow Corning DC93500 adhesive QioptiQ CMG ITO unique Coverglasses Nusil CV3-2640 conductive adhesive
PVA layout	40 strings of 18 cells each. Divided into - 4 sections with 6 strings each (deployable panels) - 2 sections with 8 strings each (central panel) (720 solar cells in total)
Installed Power	- Begin Of Life - 1 solar constant, AM0, 28°C - Max available Power @ SA: 828 W

1.3 Exomars Carrier PVA

The Exomars Carrier is by sure the less demanding mission amongst the ones here described. Based on that the proposed PVA is the evolution of EXM TGO solutions.

Figure 5. EXM Rover vehicle deployment (courtesy of Airbus).

SA is composed by 8 base panels (body mounted), surrounded by 6 deployable panels. The rear side of the panels is thermally shielded with a Multi Layer Insulator (MLI).
The most important characteristic of this solar array is the relatively complex panel layout and distribution and the extremely limited number of fatigue cycles to whom the PVA will be subjected.

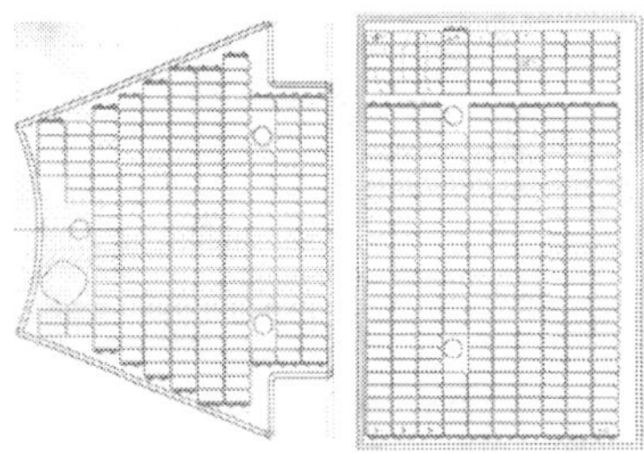

Figure 6. EXM Carrier panel.
Base panel layout (left), deployable panel (rigth)

The presence of MLI impose the double insulation on each of the rear side electrical components. This is performed using conformal coating on exposed surfaces and kapton insulation patches.
Some improvements has been selected regarding the rear side integration technology. One of them is related to the usage of bi-adhesive layer between for fiberglass board of EEE parts, instead of RTV adhesive.

The selected solar cell is the Azur space 3G30 in combination to the Azur SIBD silicon protection diode.

Table 3. EXM TGO PVA main characteristics.

Solar cell Assembly	AZUR Space 3G 30% GaAs/Ge solar cell Dow Corning DC93500 adhesive CMG QioptiQ Coverglasses
Protection diode	AZUR Space SIBD silicon diode Dow Corning DC93500 adhesive QioptiQ CMG Coverglasses
PVA layout	6 base panels with 9 strings 6 base "modified" panels with 8 strings 6 deployable panels with 11 strings All the strings with 26 cells each 1 section per panel (3536 solar cells)
Installed Power	- Begin Of Life - 1 solar constant, AM0, 28°C - Max available Power @ SA: 4084 W

2. Qualification

As a general approach the PVA qualification have been achieved at two levels: solar cell assembly and coupon assembly. Several tests have been already performed at solar cell assembly level so the final adaptation to the specific project requirements is accomplished by means of a delta Type Approval Test (TAT) according to European Standard guidelines while the second by means of Design Verification Test (DVT) coupons.

2.1 Solar Cell Assembly qualification

Leonardo heritage comprehends a full qualification performed in accordance to relevant ECSS for solar cell assembly base on Azur 3G28 triple junction (EXM TGO configuration) and more recently Azur 3G30.
Delta TAT has been defined in order to cover the different project environment involving assemblies based on 3G30 either on EXM Carrier either on EXM Rover.
The Delta TAT are focused on the differences with respect to standard PVA with a consistent qualification and flight heritage. More precisely great attention was given to:

- ESD grounding network and conductive glass stability
- External Si diode assembly
- Characterization of the solar cells against different Mars light conditions
- Long term storage at SCA level

2.2 DVT qualification

The purpose of DVT qualification is to demonstrate that the whole PVA technology is suitable to withstand the whole stresses induced by the mission environment, including qualification and design margins.
The same coupon dimension (420x440 mm2) has been adopted for all projects.

Table 4. EXM TGO PVA main characteristics.

EXM TGO	2 strings with 22 cells each
EXM Carrier	2 strings with 22 cells each
EXM Rover	2 strings with 18 cells each

The DVTs are manufactured using the same materials, mechanical parts, components and processes that are used for the FM panels.

Figure 7. EXM TGO DVT coupon

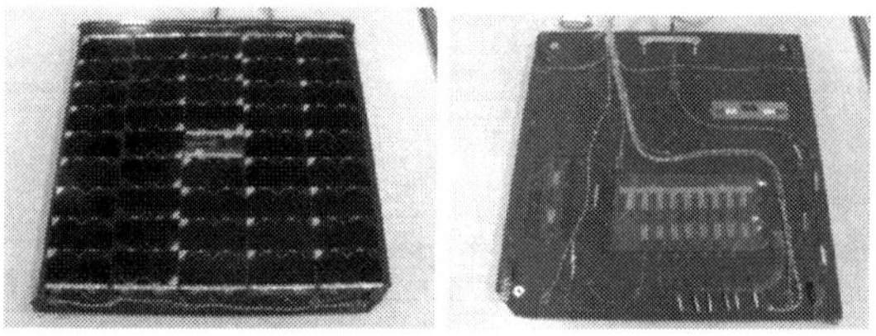

Figure 7. EXM Carrier DVT coupon

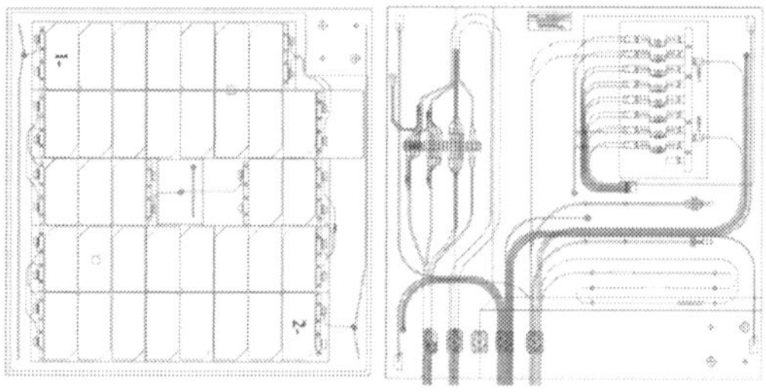

Figure 7. EXM Carrier DVT coupon

The qualification coupons will be submitted to a complete environmental test campaign consisting in:

- Thermal vacuum baking and cycling
- Accelerated ambient pressure fatigue cycles.

Moreover for the EXM Rover sample, specific test are foreseen

- Corona test
- Exposure to Mars wind and dust environment

2.3 STM/EM/QM models

No additional qualification models are foreseen except for the Rover that will foresee an STM, an EM and a QM models.
The STM will be structurally and thermally representative of the equipment in terms of mass distribution, stiffness of primary load parts, thermal control and mechanical and thermal interfaces. Actuator functionality is not required though it shall be manually reconfigurable to the stowed and deployed conditions.
The STM shall be subject to qualification level mechanical test. No thermal tests are foreseen at SAA level and they can be performed after the integration together with the rest of the Rover module.
The PVA STM will be integrated on panel substrates. The result of the integration will be one set of STM panels (fixed panel, primary deployable left panel, secondary deployable left panel, primary deployable right panel, secondary deployable right panel), populated with mass dummies.
Front side thermo-optical representativeness will be achieved using black anodized Al plates of approximately the same size of solar cell assemblies.
The Mechanisms STM will be structurally and thermally representative of the equipment in terms of mass distribution, stiffness of primary load paths, thermal control and mechanical and thermal interfaces. No actuator functionality will be provided, but it will be possible to manually configure the model to the stowed and deployed conditions. Before delivery, the model will be submitted to vibration and thermal tests, in order to reproduce the specified mechanical environment at qualification level and correlate the thermal model. At SAA level, the model will allow verification of the structural and thermal characteristics of the proposed design.
A QM model shall be subject to the full equipment level qualification test sequence and environment, including adequate qualification levels, duration, and margins as agreed by the customer. This shall include life test when it is required to be demonstrated.
Following its successful qualification test sequence, a QM model shall not be acceptable for flight, unless otherwise agreed.
Nevertheless it is Selex opinion that the QM panels, properly refurbished can be considered as FM spares in case of needs.
All mechanisms shall be qualified using a QM. The Statement of Work requires a QM of the full Solar Array Assembly.
Its two sub-assemblies, namely the PVA and the mechanisms will undergo a qualification campaign on their respective QM. Then, integrated tests at QM level will also be performed.
A proper splitting between stand-alone and integrated tests is proposed, described in the test plan, in order to optimise the effectiveness of the test campaign and the associated schedule.
The QM will be submitted to the same test sequence as per FM with the only difference of the testing levels that in this case will be aligned to the wider qualification range rather to be reduced to the acceptance levels.
The Mechanisms QM will be built to the full flight standard. Before testing, it will be submitted to TBD (one-to-three) cycles of dry heat microbial reduction (DHMR), in order to prove compatibility with this process. It will be subject then to the specified qualification test sequence and environment. After the environmental testing, the model (or the relevant parts of it: the mechanisms) will be subject to life testing in order to complete its verification. Then, the model will be integrated into the SAA QM for assembly level qualification activities.
The SAA QM will be subject to the qualification campaign. At the end, a bake-out in T/V chamber is foreseen which covers also DHMR aspects.
The QM will comply with planetary Protection and cleanliness requirements and procedures. The QM is to be used to trial run PP practices in readiness for FM campaign. A bio-burden budget will be assigned to the FM based on the exposed and encapsulated surface area of the subsystem.

3. Qualification results

Qualification at component level has been successfully achieved at solar cell assembly level, finalizing the 3G28 and 3G30 full coverage as per PVA ECSS of the Exomars mission.
Specific test are foreseen to cover the ITO coating network that are to be completed.
DVT campaign is successfully completed for EXM TGO, while it is running on EXM carrier and to be started on EXM rover.
Expected end of qualification for EXM carrier is foreseen by end of 2016, while during 2017 for EXM Rover.

4. Conclusion

Leonardo is proud of the PVA technology developed so far for the various Exomars programs in which there was our direct involvement.
Several solutions implemented for complying with program specific requirements can be proposed on a variety of space missions to near Earth or Interplanetary orbit.
Leonardo would like to thank all the companies that cooperate as customers and/or supplier in the frame of the above projects that are: TAS France - Cannes, Airbus D&S UK, SENER and Airbus DS NL.
We also wish to thank the European and Italian Space Agencies for the continuous support in the technology development and AZUR Space for the open minded technical and programmatic collaboration in the frame of the majority of the past and running programs.

DOI: 10.1051/e3sconf/20171604003

ELECTRO STATIC DISCHARGE TESTING ON METEOSAT 3RD GENERATION PHOTOVOLTAIC ASSEMBLY

Giulia Damonte [1], Giorgio Saltanocchi [1], Pietro Zanella [1], Emanuele Ferrando [1], Stefano Riva [1], Gianfranco Meniconi [2], Damiano Pagano [2], Fabrizio Scortecci [2]

[1] Leonardo Company, Viale Europa s.n.c., 20014 Nerviano (Mi), Italy, Email: giulia.damonte@leonardocompany.com
[2] Aerospazio Tecnologie, Strada di Ficaiole, 53040 Rapolano Terme (Si), Italy, Email: fscortecci@aerospazio.com

ABSTRACT

The electrostatic discharge testing on Photovoltaic Assemblies has been thoroughly investigated and performed during the last decade to ensure durability of solar arrays, especially in cases of high power satellites. *Leonardo Company* together with *Aerospazio Tecnologie* performed the test for the first time in the framework of Meteosat Third Generation (MTG, a Geostationary satellite of 2.5kW, mounting 3G solar cells) with two goals: the first was to qualify MTG PVA technology against possible ESD occurrences during mission and the second was to characterize the PVA design limits against self-sustained arcs.
This paper illustrates the ESD analysis, the test set up preparation and the test results obtained in the framework of MTG ESD test campaign.

1 UNDERSTANDING OF THE ESD PHENOMENA

Space environment substorms can generate Inverted Voltage Gradient (IVG) on solar arrays: the satellite body and solar arrays substrate charges up to several kilovolts compared to the space environment, while dielectric parts (mainly solar cells coverglasses) build up a differential charge of the order of hundreds of Volts by the effect of a different rate of emissions of secondary and photo electrons.
When the electric field generated by the different surface potentials exceeds the threshold value, the satellite capacitance is discharged (blow off), leading also to the release of the charge stored in the solar panels dielectric surfaces (flashover), i.e. solar cells coverglasses and insulation sheet. Since the capacitance of spacecraft (on the order of one hundred pF) is not as great as the capacitance of the dielectric surfaces (of the order of µF), the flashover discharge is the major current source of electrostatic discharge. Flashover discharges depends on the plasma velocity and propagation area [2]; in the ESD testing it is simulated through an external capacitance, dimensioned in order to provide the same energy stored on cover-glasses during mission [4].

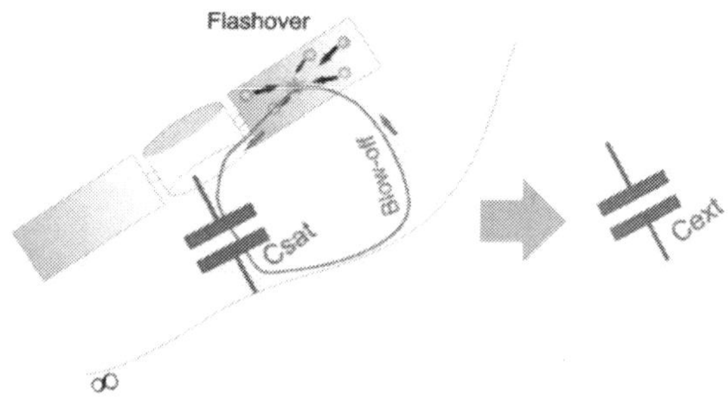

Figure 1: Blow off and flashover impression

The primary discharge generates local plasma environment by vaporization of metal materials at the discharge site, creating a temporary conductive path that can lead to secondary arcs, powered by polarized parts, as can be solar array strings. In case of sufficiently high voltage / current values (typically > 50V / 2A), the secondary arcs in the inter-string area can be self-sustained by the Solar Array power, collecting a string current or even a section current, if no blocking diodes are adopted at string level. Self-sustained arcs can seriously damage the PVA and its insulation layer at the arc site.

2 ESD HAZARD FOR MTG MISSION

MTG PVA shows a strong design against ESD phenomena for the relatively low levels of operating current and voltages, for strings and sections disposition on the panels and for the adopted materials:

- All the metallic parts on the substrate that are not part of the power generation net are grounded to the substrate
- All metallic parts that are part of the Solar panel structure and deployment system are grounded to the satellite through a wires grounding net
- The adopted wires have antistatic external jacket, grounded throughout the substrate surface.
- Blocking diodes protect each single string
- Strings voltage and current levels are quite low (less than 110 V and 600mA during the whole mission in nominal and failure cases)
- Strings and sections are disposed on the panels in a way that maximizes the inter-section gaps (> 5mm) and the inter-strings gaps (> 0.5mm) and that minimizes the voltage differences between adjacent

cells in all nominal (< 50V) and failure cases (< 110V).

This said, the main interest of the present test campaign is in the sustained arc threshold characterization phase more than in MTG ESD qualification phase, which was quite certain to be completed in a successful way since the beginning.

The MTG qualification phase test parameters were chosen between the maximum voltage / current combination encountered during mission in all nominal and failure cases (see Table 1). It is important to notify that the failure cases listed in Table 1 reflect voltage / current combination that occur only in string interruption cases, which, in practice, could hardly lead to sustained arcs. The choice of this test parameters followed a conservative approach and provides an ulterior margin to the results exposed in paragraph 5.

Table 1: Mission Voltage / Current for MTG ESD qualification

Case	Maximum Inter-cell Voltage gap	String biasing
Nominal hot case	27 V	523 mA
Failure 1 cold case	55 V	436 mA
Failure 2 cold case	107 V	466 mA

As mentioned before, even in failure cases, the MTG voltage / current (string current, since blocking diodes are adopted at each string) levels are quite low if compared with threshold limits included in ECSS and literature data [1], [3], [5].

3 ESD TEST COUPONS AND TEST PLAN

The test was conducted on two ESD coupons (SN010 and SN040), fully representative of MTG front side PVA and of MTG solar panels substrate.

Coupons front side mounts two strings of three Azur Space solar cells, fully functional, 3G30 advanced type.

Coupons rear side mounts MTG representative wires and boards and additional capacitances to simulate the MTG complete string missing cells, as it is addressed in paragraph 3.1.

The two coupons are identical except for what concerns the inter-string gap:

- 0.5 mm gap (worst case manufacturing gap) on Coupon SN010
- 1mm gap (nominal gap) on Coupon SN040

SN010 coupon was planned to qualify MTG PVA, therefore to be tested at the voltage / current levels reported in Table 1; SN040 was planned to characterize the PVA limits against sustained arcs occurrences, therefore to be tested at increasing voltage and current levels: from 60V / 550mA to 110V / 2A and 123V / 1.5A. Complete lists of tests are included in the result Table 2 and Table 3.

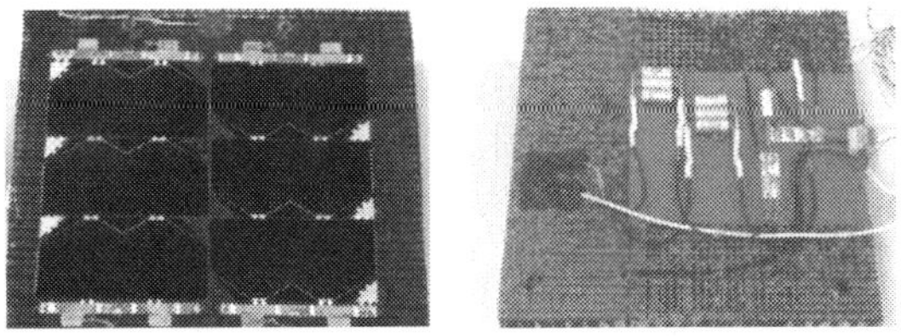

Figure 2: Testing coupon picture. Front and rear side

Higher gap coupons were also available for testing in case threshold limits for sustained arc could be reached with SN040 coupon.

3.1 String equivalent capacitances

Since the energy stored in string capacitances plays an important role during secondary arc occurrences, the missing cells of a complete flight model string (27 cells in series) are considered using an equivalent circuit that accounts for differential and common mode effects [1].

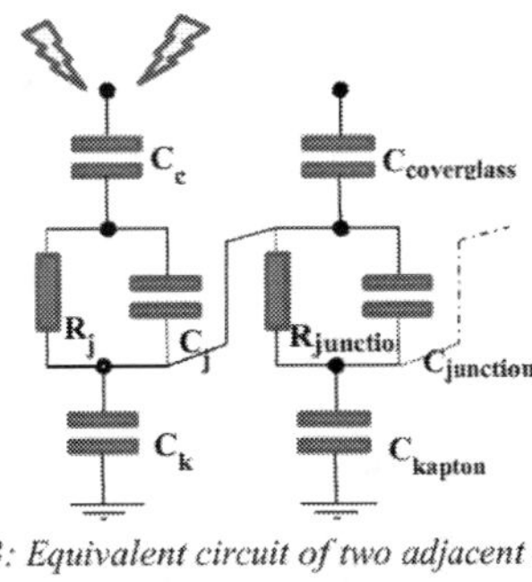

Figure 3: Equivalent circuit of two adjacent in line cells

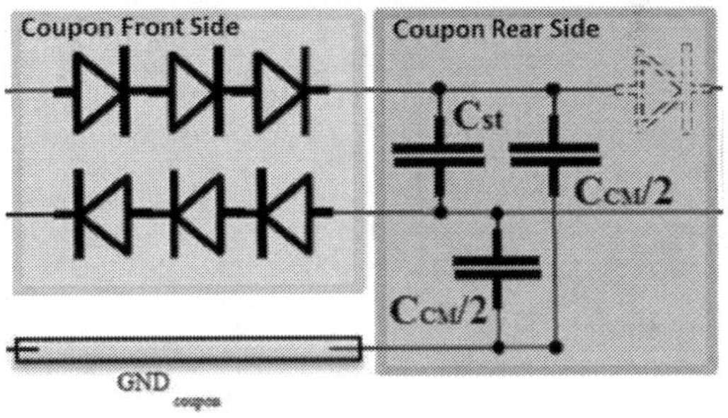

Figure 4: Equivalent capacitances circuit on testing coupon

Figure 4 shows the equivalent capacitances circuit mounted on coupons SN010 and SN040 rear side, C_{st} being the capacitance of the missing cells chain, C_{CM} being the capacitance through the Kapton for the missing cells.

 DOI: 10.1051/e3sconf/20171604003

$$C_{st} = \frac{1}{2} \frac{\sqrt{C_k C_j}}{th\left(\frac{N_S}{2}\sqrt{\frac{C_k}{C_j}}\right)} = 63\text{nF} \tag{1}$$

Where C_K is the capacitance through the Kapton of one cell as in Eq. 2 (ε_r provided by insulation layer manufacturer)

$$C_k = \frac{A}{d}\varepsilon_0\varepsilon_r \tag{2}$$

C_J is the unction capacitance for one solar cell, as measured during MTG solar cells qualification.

$$C_{CM} = 638 \text{ nF} \tag{3}$$

Inductive effects are already present in the test cabling wire length.

4 ESD TEST SETUP

The ESD campaign was conducted in the facility of Aerospazio Tecnologie, company highly qualified for electric propulsion and thermal-vacuum testing and for space environment simulation.

ESD test was performed at high vacuum level (pressure lower than $1x10^{-6}$ Torr), in IVG environment, generated by polarizing the coupon at -5kV and flushing it with high energy electrons generated by an electron gun. The electrons were flushed over the complete surface of the coupon through a raster and they were accelerated by the electron gun in order to reach the coupon substrate with an energy as close as possible to the coverglass secondary electrons emission peak (E_e < 1kV). The electron gun parameters (energy, focalization, current etc.) were changed by an e-gun controller while beam deflection was set using two external signal generators.

A full characterization of the plume over a surface area equal to the coupon one was performed at 80 cm from the electron gun, were the coupon was then positioned.

The coupon suspension system was reduced to the minimum, through the use of lacing chord, in order to avoid any possible interference with the test.

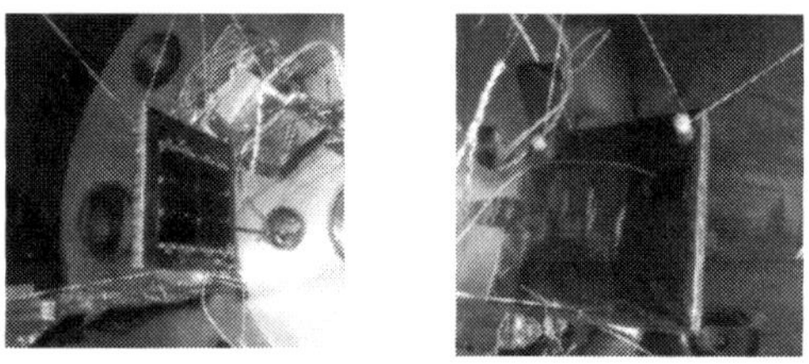

Figure 5 Coupon suspended in the vacuum chamber through lacing chord

The complete electrical test set up is illustrated in Figure 6.

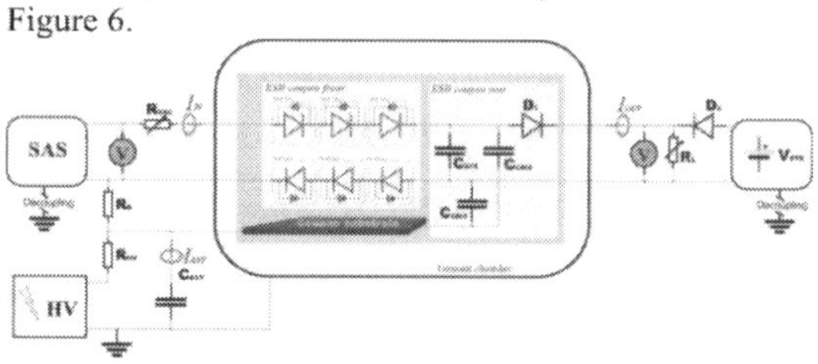

Figure 6: ESD test setup

The coupon strings are connected to a Solar Array Simulator device (SAS): one string is biased to a positive potential with respect to the conductive substrate, simulating the positive end of solar array, the other string is grounded through a resistance to the conductive substrate, simulating the negative end of solar array. Since it is of primary importance for the test to have fast transient to short circuit, the dynamic responses of SAS alone and SAS + testing cables (twisted ones) were measured though the use of a mercury switch, between 1A and 3A [1].

SAS alone showed a response of 1-2 μs, SAS + testing cable (twisted) showed a response of 4 μs; these values being similar to the illuminated solar array response and well below the duration of a typical primary discharge.

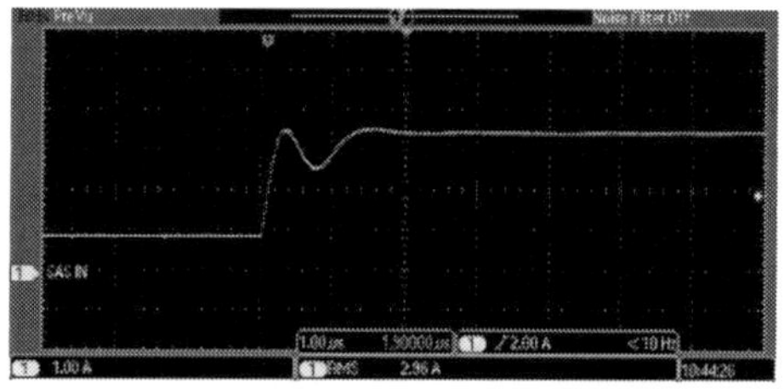

Figure 7: SAS dynamic response from 1A to 3A.

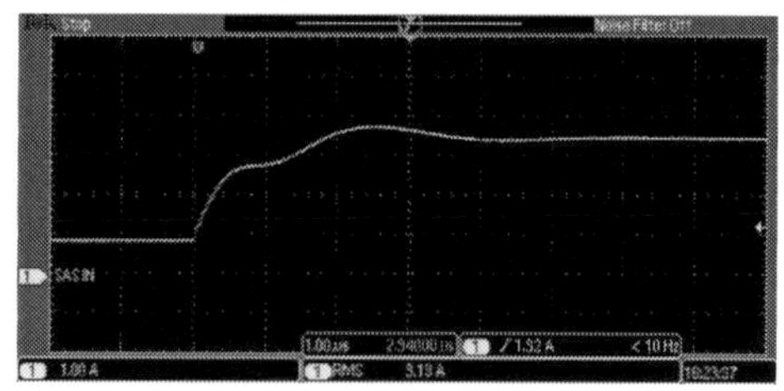

Figure 8: SAS + testing cables dynamic response from 1A to 3A

A variable resistance R_L allows the SAS to impose different voltages and currents to the strings.

An additional power supply (V_{STR} with its protection diode D_2) and an additional blocking diode D_1 allow to maintain the desired voltage gap between the two mini-strings and not to lose part of the arc current in the R_L during secondary arc occurrence, operating as in a diode switch [6]. In a secondary arc case, the voltage at the string ends drops, diode D1 is reversed and SAS makes

its limitation current immediately available.
For Coupon SN010 testing, the SAS biased the strings at the different mission cases current (523mA for test 1, 438mA for test 2 and 446mA for test 3) and current limitation was set to values slightly higher than the biasing one. During Coupon SN040 testing, instead, the biasing current was kept constant (450mA), while the SAS current limitation was increased at each test step (550mA, 1A, 1.5A, 2A): this to protect solar cells and to limit solar cells heating [(1)].

The coupon substrate and the strings are kept at a potential of -5kV by the High Voltage power supply in the left bottom corner of Figure 6.
The capacitance in parallel to the high voltage power supply is discharged during primary discharges occurrences on the testing coupon, simulating satellite structure (blow off) and flashover discharges in space, being the first one negligible compared to the second one as already mentioned in paragraph 1.
C_{EXT} capacitance, therefore, is dimensioned in order to simulate flashover discharge, i.e. to provide the same energy stored on cover-glasses during mission. Calculation exposed in Eq. 4 provides a capacitance value in the range of 3 to 15nF.

$$\frac{1}{2}C_{EXT}V_{test}^2 = \frac{1}{2}C_{CVG}\left(V_{CVG}-V_{SUB}\right)^2 \tag{4}$$

Where V_{test} = -5kV, typical values of ($V_{CVG} - V_{SUB}$) are 200V - 400V [4], C_{CVG} is the capacitance of the coverglasses of one MTG panel.
The external capacitance used for MTG test is 5nF.

The setup, showed in the picture included Figure 9, was mounted as near as possible to the vacuum chamber and the length of the wire was optimized in order to have the minimum inductance.

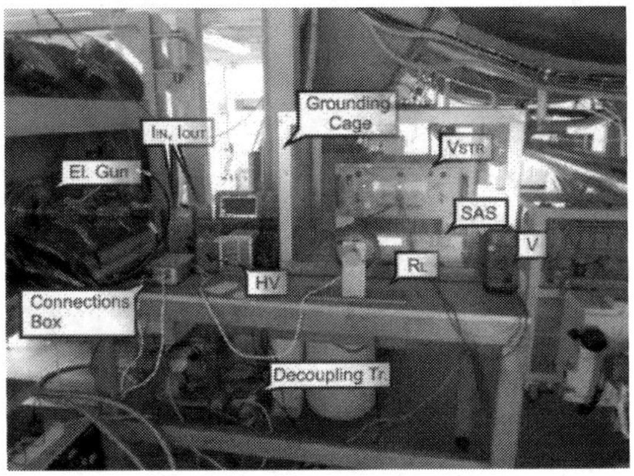

Figure 9: Set up picture

Discharges were recorded with a video camera positioned in front of the vacuum chamber window and discharges signals were registered by an oscilloscope connected with the three current probes indicated in green in Figure 6.

Figure 10: Recording camera

5 TEST RESULTS

The tests performed on the two coupons, together with their results are listed in Table 2 and Table 3.

For each test step, in order to have a sufficient statistics for ESD impact evaluation, at least ten primary discharges were recorded in the inter-string gap before passing to the following test step. The total number of discharges recorded for each test step was about 3 times higher, including also discharges within the same string and on coupon edges.

The IV curve of each coupon was measured before and after its ESD test campaign; after each test step of the same test campaign, the coupon functionality was checked inside the chamber through continuity and insulation test.

As already mentioned in the previous paragraphs, Coupon SN010 (with 0.5mm inter-string gap) was submitted to the MTG mission voltages and currents levels, according to the first three rows of Table 2.
None of the MTG test conditions applied to Coupon SN010 (Test 1 to 3) caused sustained secondary arcs. After test the coupon did not record any electrical performance degradation in terms of IV curve performances and PVA insulation towards the substrate.
It is however important to record that the discharges in some cases generated recombination spot on solar cells with no effect on the string electrical performance (see Figure 11).

[1] The measured coupon heating with 534mA biasing was of 25°C

DOI: 10.1051/e3sconf/20171604003

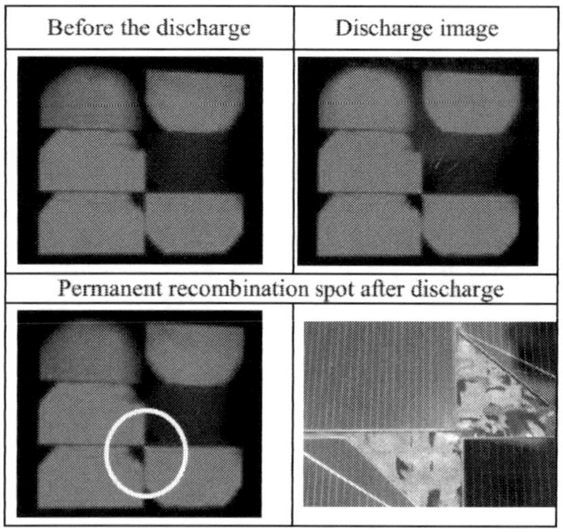

Figure 11: Recombination spot damages on solar cells

Coupon SN040 (with 1mm inter-string gap), was planned to be used for the voltage / current threshold limits determination, with the test conditions reported in Table 3. Strings biasing was kept at 450mA for the whole test phase, while the current immediately available through SAS was increased from test step 4 to 6. Since no sustained secondary arcs occurred during test 4 to 6, Coupon SN040 was also tested in a more severe environment (test 7), up to SAS limits: V=123V and I_{lim} = 1.5A.

Since also test 7 did not show the occurrence of any sustained arc, the same test conditions, were applied on Coupon SN010 (test 8 in Table 2).

Table 2: MTG ESD Qualification phase on SN010 coupon (inter-string gap = 0.5mm).

Test n°	V Gap [V]	ISAS [mA]	Test Condition MTG Reference	Results
Test 1	27	523	MTG mission Nominal case	Passed
Test 2	55	438	MTG mission Failure 1 case	Passed
Test 3	107	446	MTG mission Failure 2 case	Passed
Test 8	123	1500 (limitation)	ESD Characterization	Failed

Table 3: ESD Characterization phase on SN040 coupon (inter-string gap = 1mm)

Test n°	V Gap [V]	I_{SAS} lim [mA]	Results
Test 4	90, 110	550	Passed
Test 5	90, 110	1000	Passed
Test 6	60, 90, 110	2000	Passed
Test 7	123	1500	Passed

During Test 8, after 9 primary arcs recording on Coupon SN010 in the inter-string gap, a sustained arc occurred, determining a permanent solar cells and insulation layer damage.

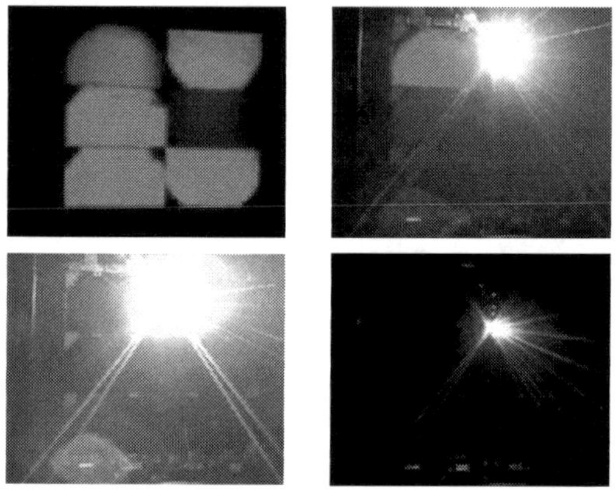

Figure 12: Sustained arc recording on Coupon SN010 (gap = 0.5mm, V=123V, I_{lim} = 1.5A)

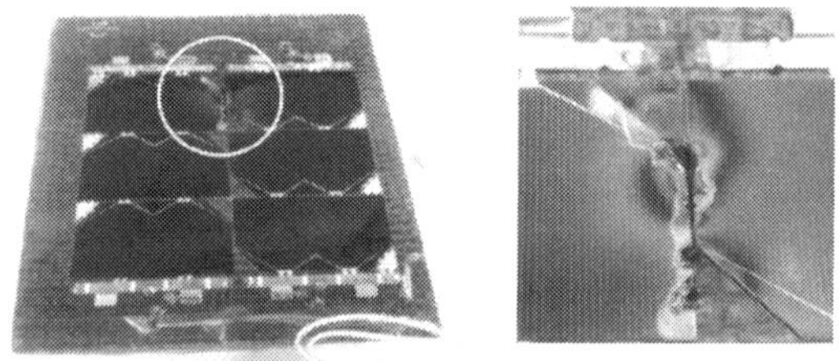

Figure 13: Damaged coupon after sustained arc occurrence

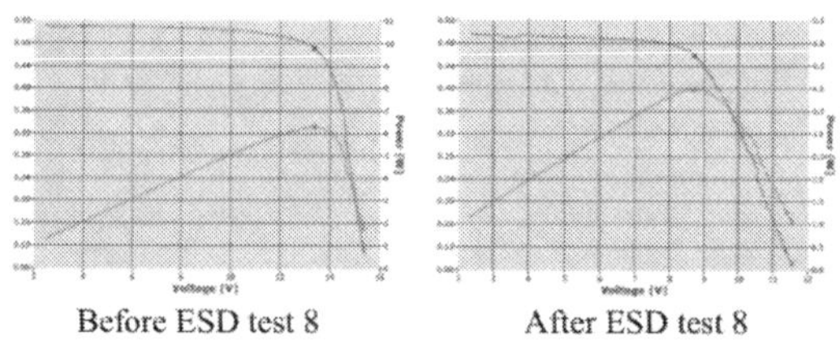

Figure 14: Coupon 010 IV curve before and after sustained arc occurrence

Typical discharge pictures are showed in Figure 15 and typical discharge signals are included in Figure 16 to Figure 19.

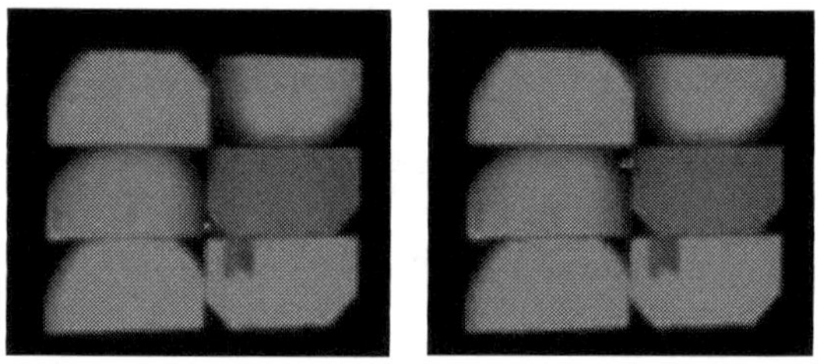

Figure 15: Typical discharges pictures (Coupon SN040)

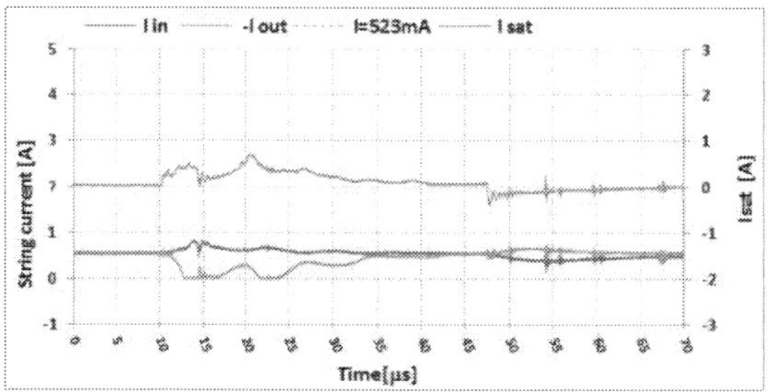

Figure 16: Typical discharge for 27V inter-strings voltage gap (523mA biasing, Coupon SN010)

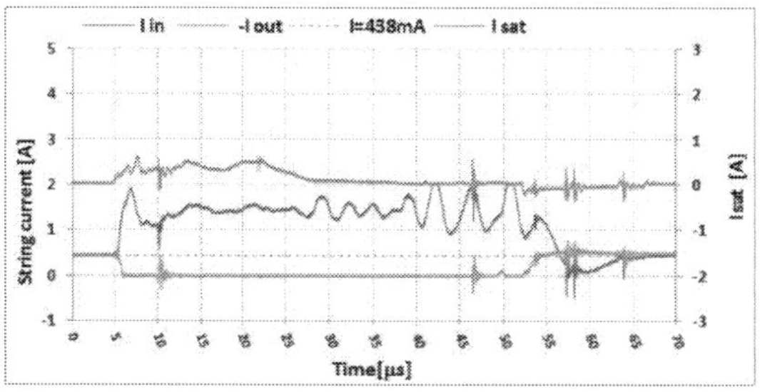

Figure 17: The strongest discharge recorded at 54V inter-strings voltage gap (438mA biasing, Coupon SN010)

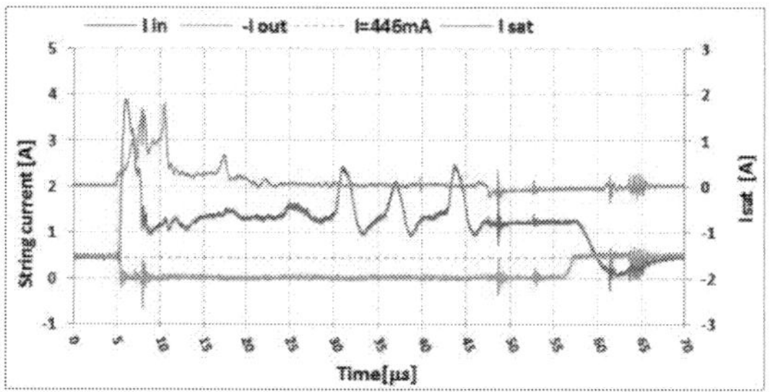

Figure 18: Typical discharge signal at 107V inter-string gap (446mA biasing, Coupon SN010)

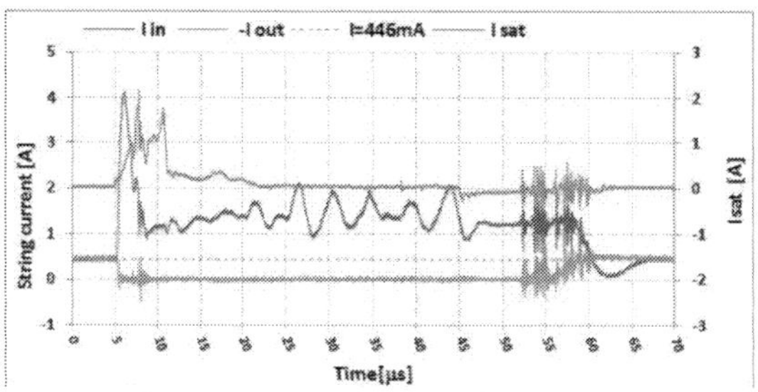

Figure 19: Typical discharge signal at 123V inter-string voltage gap (446mA biasing, SAS current limitation at 1.5A, Coupon SN010)

6 CONCLUSIONS AND FURTHER DEVELOPMENTS

The here presented test campaign showed MTG panel design (0.5 mm gap as a very worst case) to have wide voltage and current margins to the threshold limits for sustained arcs occurrences: 107V / 446mA against a limit of 123V / 1.5A.

The campaign was also a great chance for Leonardo Company and Aerospazio Tecnologie to develop competences in ESD testing on PVA.

That said, some test aspects could be addressed only partially and are worthy further investigations:

- Coupon temperature role (during present test the coupon temperature at different coupon biasing, was indirectly measured before and after each test step)
- The effects of electron beam current density and impact spot on the coupon
- The effect of testing time (electron flushing) on discharge rate.
- The discharges direction along the strings and in the inter-string gaps

The test setup itself has room for improvement for what concerns the discharges recording system (that shall be triggered to the discharges signals at the oscilloscope), the coupon surface voltage recorder (not available at this stage) and the instruments power limitations (maximum reached 123V / 1.5A).

Finally, the results here exposed shall be integrated with a more extended test campaign including higher voltages and string biasing levels and a wider set of inter-string gaps in order to have a more exhausting characterization of the limits for permanent damages occurrences on PVA due to ESD phenomena.

7 REFERENCES

1. Payan, D., Schwander, D., (2005). Solar array test set-up proposed by CNES in the frame of European standardization. Prevention of secondary arching induced by electrostatic discharge. In 9th Spacecraft charging technology conference, Tsukuba, Japan.
2. Okumura, T., Nitta, K., Takahashi, M., Suzuki, T., Toyoda, K., (2010). Flashover plasma characteristics on 5m^2 solar array panels in a simulated plasma environment of geostationary orbit and low Earth orbit. In 48th AIAA, Orlando, Florida.
3. ECSS-E-ST-20-06C
4. Cho, M., Shikata, Y., Hosoda, S., Toyoda, K., Amorim, E., Goka, T., Levy, L., Reulet, R., Payan, D., (2003). Issues concerning the international standard of ESD ground test for GEO satellite solar array. In 8th Spacecraft Charging Technology Conference, Huntsville, Alabama.

5. Levy, L., (2002). Extensive data on ESD induced secondary arcs on solar arrays. In Space Power, Proceedings of the Sixth European Conference, Porto, Portugal.
6. Payan, D., Schwander, D., Catani, J.P., (2002). Solar Array Dynamic Simulator Prevention of Solar Short-Circuits due to Electrostatic Discharges. In Space Power, Proceedings of the Sixth European Conference, Porto, Portugal ESA SP-502 (CD-ROM), European Space Agency, Noordwijk, The Netherlands.

E3S Web of Conferences **16**, 04004 (2017) DOI: 10.1051/e3sconf/20171604004
ESPC 2016

ORIGIN OF THE DEGRADATION OF TRIPLE JUNCTION SOLAR CELLS AT LOW TEMPERATURE

Seonyong Park(1), Jacques C. Bourgoin(1), Olivier Cavani(1), Victor Khorenko(2), Carsten Baur(3), Bruno Boizot(1)

(1) Laboratoire des Solides Irradiés, CNRS-UMR 7642, CEA-DRF-IRAMIS, Ecole Polytechnique, Université Paris-Saclay, 91128 Palaiseau Cedex, France, Email:seonyong.park@polytechnique.edu
(2) AZUR SPACE Solar Power GmbH, Theresienstr.2, 74072 Heilbronn, Germany, Email: victor.khorenko@azurspace.com
(3) European Space Agency, Keplerlaan 1, 2201AZ Noordwijk, The Netherlands, Email: carsten.baur@esa.int

ABSTRACT

The degradation of solar cells under irradiation by high energy particles (electrons, protons) is the consequence of the introduction of defects trapping minority carriers, which are then not collected by the junction. However, at low temperature, defects located in the space charge region can also induce a tunneling current that results in an apparent decreases of the maximum power. The degradation produced by this tunneling current can depend on temperature, since the concentration of defects created by an irradiation is usually temperature dependent, and can be larger than the degradation associated with carrier recombination. For instance, as we shall see below, an irradiation with 1 MeV electrons at 120 K with a fluence of 3.0 x 10^{15} /cm^2 induces a decrease of less than 10 % in the short-circuit current (I_{SC}) and open-circuit voltage (V_{OC}) of triple junction (TJ) cells, but a decrease of about 40 % in the maximum power (P_{MAX}), which implies that more than half of the total degradation of P_{MAX} should be assigned to another loss mechanism, tunneling in this case. In this work, we demonstrate that this additional degradation must indeed be ascribed to a tunneling process and we investigate the variation of the tunneling current versus fluence induced by electron irradiation in TJ cells, in order to tentatively ascribe the tunneling components to specific sub-cells.

1. INTRODUCTION

Gallium-Arsenide (GaAs) based TJ solar cells, made of three sub-cells: Gallium-Indium-Phosphide (GaInP), Gallium-Arsenide (GaAs) and Germanium (Ge), are state-of-the-art solar cells for powering satellites [1]. These cells are also selected for deep space missions [2-4]. In this respect, it is necessary to determine the temperature dependence of their electrical characteristics, in particular following electron and proton irradiations. This study deals with TJ solar cells developed by AZUR SPACE Solar Power GmbH for the planned Jupiter ICy moons Explorer (JUICE) mission of the European Space Agency (ESA). To investigate the performances of these cells, the corresponding environmental conditions: temperature around 120 K, weak sun light intensity (3.7 % AM0), high energy electron and proton irradiations, have to be reproduced in the laboratory. The work consists of simulating these specific conditions and to monitor the changes induced by irradiation of key parameters, such as short-circuit current (I_{SC}), open-circuit voltage (V_{OC}), maximum power (P_{MAX}), fill factor (FF), and conversion efficiency (η).

2. EXPERIMENTAL CONDITIONS

In order to reproduce the irradiation conditions of a mission to Jupiter, the cells are placed in a liquid nitrogen cryostat whose chamber is connected to the accelerator beam through a thin stainless steel foil. The solar simulator satisfying the specific solar spectrum (equivalent to 3.7 % AM0) has been realized by a combination of a Xenon (Xe) and Quartz Tungsten Halogen (QTH) lamps separated by a cold filter. The adjustment of the simulator has been realized by secondary working standards which have been derived from reference standards assuming a linear dependence of I_{SC} and light intensity. TJ solar cells have a size of 4 cm^2 and a thickness of 80 or 140 μm. The electron irradiation has been carried out at 1 MeV using the SIRIUS facility of LSI (Laboratoire des Solides Irradiés) in Ecole Polytechnique. The cell temperature during irradiation was set at 123 K and three fluences were used: 7.5 x 10^{14}, 1.5 x 10^{15}, and 3.0 x 10^{15} cm^{-2}.

3. ELECTRON IRRADIATION OF TRIPLE JUNCTION SOLAR CELLS

In the course of irradiation, we noticed at low temperature the appearance of an apparent "shunt" effect decreasing P_{MAX} significantly, compared to the decrease expected from the changes of V_{OC} and I_{SC} values. Fig. 1 shows the relative variations of I_{SC}, V_{OC}, and P_{MAX} values versus electron fluence. As expected I_{SC}, V_{OC}, and P_{MAX} decrease as the fluence increases. After a fluence of 3.0 x 10^{15} cm^{-2}, P_{MAX} has been degraded up to about 60 % of the original value. However, I_{SC} and V_{OC} have decreased to 92 % and 95 %, respectively. The amount of degradation observed for P_{MAX} is therefore much larger than expected from the degradation of I_{SC} and V_{OC}.

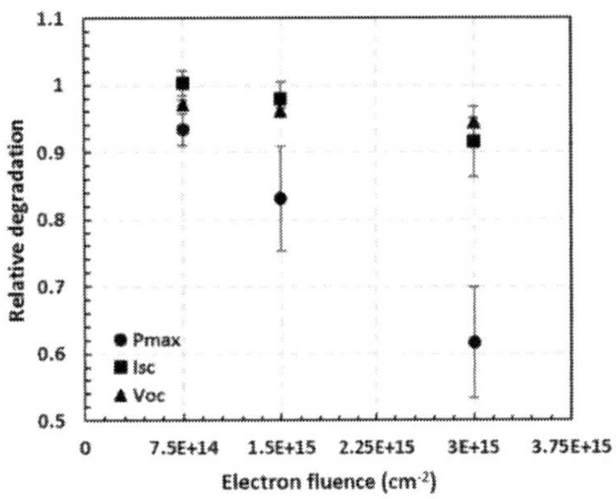

Figure 1. Relative degradation of I_{SC}, V_{OC}, and P_{MAX} measured at 123 K directly after 1 MeV electron irradiation. All values have been divided by BOL values and shown with standard deviations.

As illustrated in Fig. 2, this large degradation correlates with a decrease of the current under light illumination as the voltage increases. The Beginning of Life (BOL) current-voltage (I-V) characteristics under illumination exhibits a fill factor FF of about 90 %. However, after electron irradiation, the End Of Life (EOL) I-V characteristics drops clearly starting from 1 V, resulting in a strong decrease of FF down to 65 %.

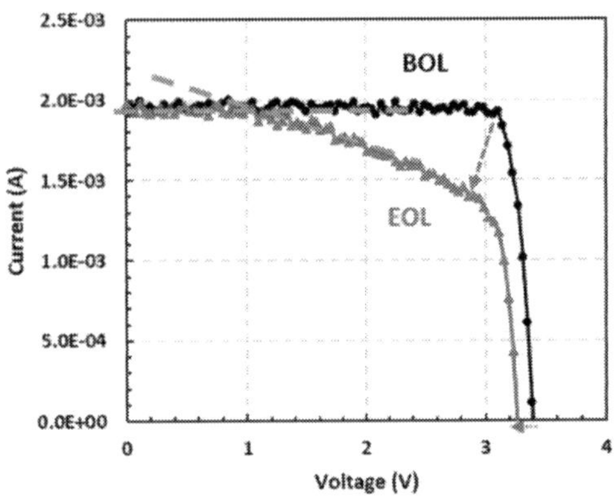

Figure 2. Light I-V characteristics of a TJ solar cell at 123 K before and after 1 MeV electron irradiation (1.5 x 10^{15} /cm²).

Fig. 2 also shows that EOL light I-V characteristics are not linear. The slope of I-V curve increases at 1 V resulting in a significant loss of power generation. This decrease of current as voltage increases cannot be ascribed to a normal shunt effect. As shown in Fig. 3, the subtraction of the EOL light currents from BOL light currents also shows a non-linearity. This demonstrates that the change of I-V characteristics does not come from a shunt resistance. In the dark I-V characteristics (see Fig. 4), it is clearly observed that, after irradiation, the dark current increases abruptly from 0.5 V. This increase can be correlated with the drop of the photo current in the light I-V characteristics starting at 1 V. This correlation between the degradation of P_{MAX} and the increase of the dark current (I_{Dark}), which we label now excess current, after the irradiation has been quantitatively verified by plotting P_{MAX} versus the value of I_{Dark}, at 2 V.

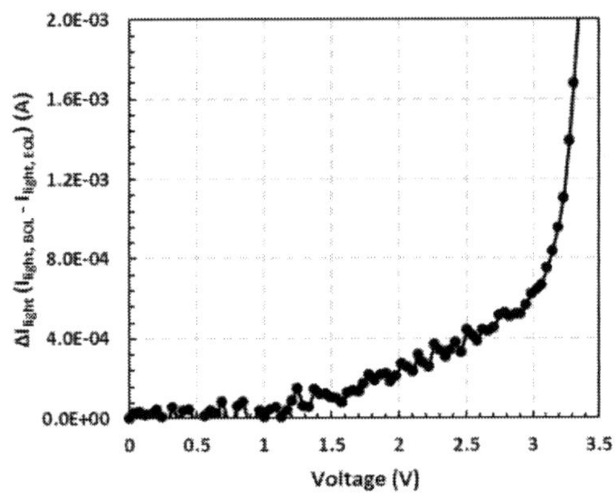

Figure 3. Difference of currents under the illumination: subtraction of EOL light I-V values from BOL light I-V values (linear scale) of a TJ solar cell at 123 K before and after 1 MeV electron irradiation (1.5 x 10^{15} /cm²).

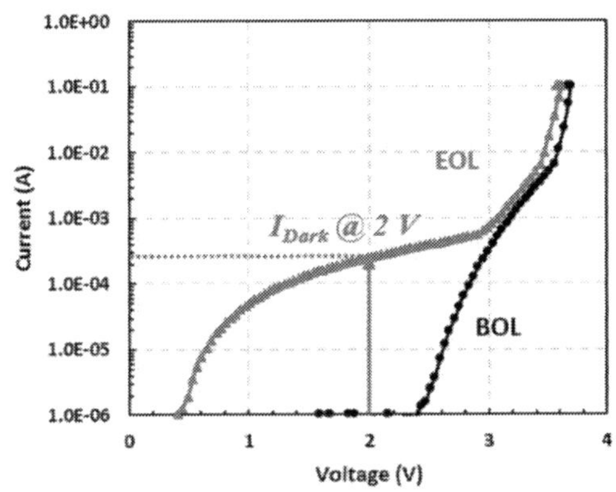

Figure 4. Dark I-V characteristics (log scale) of a TJ solar cell at 123 K before and after 1 MeV electron irradiation (1.5 x 10^{15} /cm²).

Fig. 5 shows indeed that there is a decreasing tendency of P_{MAX} as I_{Dark} increases. The data in the figure are labelled in order to illustrate their dependence on the fluence: as the fluence increases, I_{Dark} increases, which in turns leads to higher P_{MAX} degradation. It remains to understand the origin of this excess current at low voltage in the dark I-

 DOI: 10.1051/e3sconf/20171604004

V characteristics. For this, we studied the dependence of this current versus temperature and fluence.

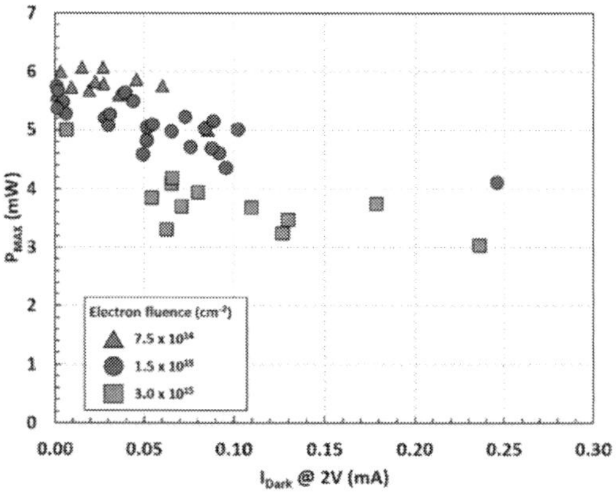

Figure 5. Relationship between P_{MAX} and I_{Dark} at 2 V.

4. VARIATION OF THE DARK CURRENT VERSUS FLUENCE AND TEMPERATURE

In order to reveal the relationship between the decrease of P_{MAX} and the dark current, we have irradiated a TJ cell at low temperature with increasing fluences. As the fluence, i.e. the defect concentration, increases, the threshold voltage at which the excess dark current starts to appear decreases. (Fig. 6)

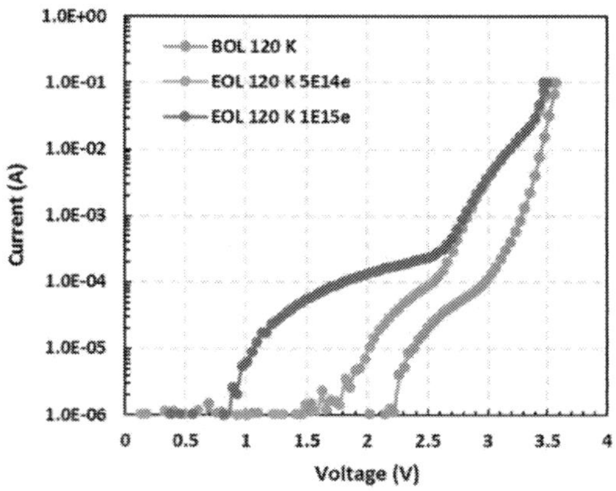

Figure 6. Dark I-V characteristics of a TJ cell after two accumulative fluences.

Thus, the existence of this excess current is directly related to the presence of defects. Then, we tried to determine which part of the cell is responsible for this excess current. For this, we investigated component cells, i.e. cell structures in which only one of photovoltaic pn junctions (top, middle and bottom) is active, in order to determine from which subcell the excess current originates.

The results are shown in Fig. 7: clearly, before irradiation, the excess current comes from the Bottom (Ge) subcell. Following irradiations, top and bottom cells exhibit excess current while the middle cell does not.

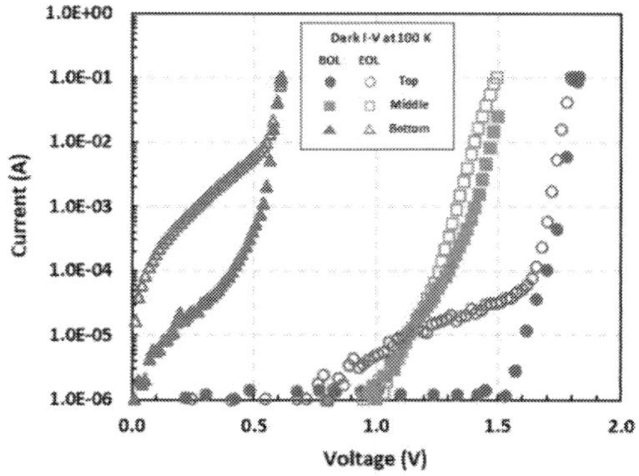

Figure 7. Dark I-V characteristics of top, middle and bottom component cells at 80 K before irradiation and after 1 MeV electron irradiation (3.0 x 10^{15} /cm²).

We finally investigated the temperature dependence of the excess current in the bottom cell. The result described in Fig. 8 shows that the excess current is temperature independent. The dark current increases faster as the temperature increases, due to the temperature dependence of the barrier height of the junction. In fact, the current which changes depending on the temperature is the thermal current regime; the other part of the current, the excess current, (see the red dotted line in Fig. 8) is not modified by changing the temperature.

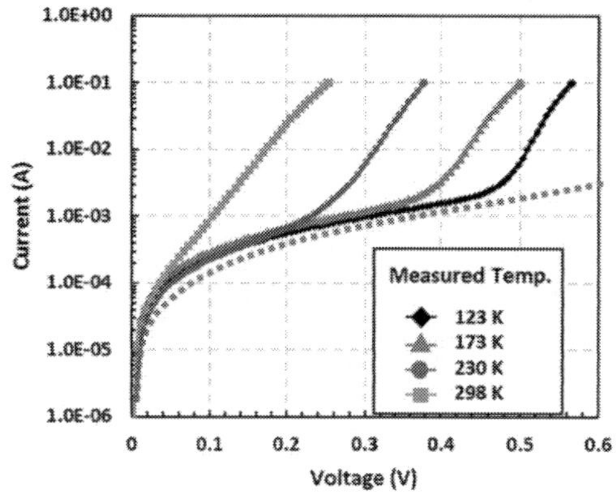

Figure 8. Dark I-V characteristics of a Ge bottom cell at various temperatures after 1 MeV electron irradiation and annealing at 123 K.

5. ORIGIN OF THE EXCESS CURRENT

Because the excess current is not temperature dependent and because its amplitude increases (to some extent) with the defect concentration, we ascribe it to a tunneling effect [4-5]. Fig. 1 in [5] presents a typical dark I-V characteristics of a Ge tunnel diode. Depending on the range of applied bias, and on the concentration of defects, two dark currents mechanism can appear. At low voltage, the current flows by direct band to band tunneling. As the voltage increases, one starts to see the excess current induced by tunneling through defects in the forbidden band. At higher voltage, the dominating current is due to band to band thermal excitation. In our case, the current we observe is the excess current induced by the defects created by the irradiation. The dark current of a Ge bottom cell increases as the temperature increases. This increase is due to the temperature dependence of the barrier height of the junction. In fact, in Fig. 8, the step-like increase of the current results from thermoionic emission (thermal regime) which is the function of temperature, whereas the red dotted line is temperature-independent which is the sign of tunneling mechanism [6]. Thus the variation of the dark current versus fluence, its temperature independence and the shape of the I_{Dark} (V) suggest that the excess current at low voltages in the TJ cells is related to a tunneling regime. It corresponds to carrier tunneling from a band state to a defect state located in the gap before decaying to the other band on the other side of the junction.

6. CONCLUSION

We irradiated TJ solar cells with electrons to study the influence of irradiation on the performances of the solar cell at low temperature in conditions corresponding to the Jupiter environment. After irradiation, the degradation of P_{MAX} was significantly higher than what was expected from the degradation of I_{sc} and V_{oc}. The reason is the existence of an apparent "shunt effect". We demonstrate that this phenomenon is directly related to an increase of the dark current after the irradiation, resulting in the decrease of P_{MAX}, induced by a tunneling current taking place in the top and bottom sub-cells. In conclusion, at low temperature a large fraction of the degradation of P_{MAX} originates from carriers tunneling in some of the junctions of TJ cells.

7. REFERENCES

1. Yamaguchi, M., Takamoto, T., Araki. K. & Ekins-Daukes, N. (2003). III–V compound multi-junction solar cells: present and future. Solar Energy Materials and Solar Cells. 75(1), 261-269.
2. Bett, A.W., Dimroth, F., Guter, W., Hoheisel, R., Oliva, E., Philipps, S.P., Schone, J., Siefer, G., Steiner, M., Wekkeli, A., Welser, E., Meusel, M., Kostler, W. & Strobl, G. (2009). Highest Efficiency Multi-Junction Solar Cell for Terrestrial and Space Applications. 24th European Photovoltaic Solar Energy Conference and Exhibition (PVSEC), Hamburg, Germany, pp1-6.
3. Baur, C., Khorenko, V., Siefer, G., Bourgoin, J.C., Casale, M., Campesato, R., Duzellier, V. & Inguimbert, V. (2013). Development Status of Triple-junction Solar Cells Optimized for Low Intensity Low Temperature Applications. 39th IEEE Photovoltaic Specialists Conferences (PVSC), TAMPA, United States, pp3237-3242.
4. Hoheisel, R., Walters, R.J., & Bett, A.W. (2015). Low Temperature Effects in Photovoltaic Devices for Deep Space Missions. 42nd IEEE Photovoltaic Specialists Conferences (PVSC), New Orleans, LA, pp1-5.
5. Meyerhofer, D., Brown, G.A. & Sommers, Jr. H.S. (1962). Degenerate Germanium. I. Tunnel, Excess, and Thermal Current in Tunnel Diodes. Physical Review. 126(4), 1329-1341.
6. Chynoweth, A.G., Feldmann, W.L. & Logan R.A. (1961). Excess Tunnel Current in Silicon Esaki Junctions. Physical Review. 121(3), 684-694.

STATUS OF SOLAR GENERATOR RELATED TECHNOLOGY DEVELOPMENT ACTIVITIES SUPPORTING THE JUICE MISSION

Carsten Baur[(1)], Victor Khorenko[(2)], Gerald Siefer[(3)], Virginie Inguimbert[(4)], Seonyong Park[(5)], Bruno Boizot[(5)], Jacques C. Bourgoin[(5)], Mariacristina Casale[(6)], Roberta Campesato[(6)], Hans-Georg Schnell[(7)], Andreas Gerhard[(7)], Pietro Zanella[(8)], Emanuele Ferrando[(8)], Xavier Reutenauer[(9)], Ed Bongers[(10)], Ana Gras[(11)]

[(1)] European Space Agency, Keplerlaan 1, 2201 AZ Noordwijk (The Netherlands), Email: carsten.baur@esa.int
[(2)] AZUR SPACE Solar Power GmbH, Theresienstr.2, 74072 Heilbronn (Germany), Email: victor.khorenko@azurspace.com
[(3)] Fraunhofer Institute for Solar Energy Systems, 79110 Freiburg (Germany), Email: gerald.siefer@ise.fraunhofer.de
[(4)] ONERA/DESP, 2 avenue Edouard Belin, 31055 Toulouse Cédex 4 (France), Email: virginie.inguimbert@onera.fr
[(5)] Laboratoire des Solides Irradiés, CNRS-UMR 7642, CEA-DRF-IRAMIS, Ecole Polytechnique, Université Paris-Saclay, 91128 Palaiseau Cedex (France), Email: bruno.boizot@polytechnique.edu
[(6)] CESI S.p.A., Via Rubattino, 54, 20134 Milan (Italy), Email: mariacristina.casale@cesi.it
[(7)] Airbus DS GmbH, Ludwig-Bölkow-Allee, 81663 Munich (Germany), Email: hans-georg.schnell@airbus.com
[(8)] Leonardo - Finmeccanica, Viale Europa, 20014 Nerviano (Mi) (Italy), Email: emanuele.ferrando@leonardocompany.com
[(9)] Thales Alenia Space France, 26 avenue J.F. Champollion, 31037 Toulouse Cedex 1 (France), Email: xavier.reutenauer@thalesaleniaspace.com
[(10)] Airbus DS NL, Mendelweg 30, 2333 CS Leiden (The Netherlands), Email: E.Bongers@airbusDS.nl
[(11)] INTA (SPASOLAB), Ctra. Ajalvir km.4, 28850 Torrejon de Ardoz (Madrid) (Spain), Email: grassa@inta.es

ABSTRACT

The paper provides an overview of the current status of several technical development activities initiated by the European Space Agency (ESA) to support the JUICE mission to the Jovian system. First of all, the qualification status of the solar cells to be used in the JUICE mission will be reported. Then, the conclusions from a dedicated activity aiming at assessing the potential degradation of triple-junction solar cells upon primary discharges will be discussed. Finally, the results on the coupon tests currently running at ESA will be presented. The coupons consist of representative solar cell assemblies including coverglasses with a conductive Indium Tin Oxide (ITO) layer. Dedicated coverglass grounding technologies are tested on the coupons which connect the conductive coverglass surfaces to the panel ground. It will be shown how the resistivity of the materials used in the coupons evolves upon submission to extreme thermal cycles.

1. INTRODUCTION

The European JUpiter ICy moon Explorer (JUICE) mission to the Jovian system poses major challenges to the solar generator subsystem. Since the satellite on its cruise to Jupiter will make use of a Venus gravity assist, the satellite will be exposed to both, hot temperatures and high sun illumination and very cold temperatures and low sun illumination. Temperatures will thus range, between +140-150°C for closest sun approach and extremely cold temperatures of <-220°C reached around Jupiter when the satellite is in eclipse. Sun intensities received by the solar array are ranging between 2.4 suns (around 3300W/m^2) and very low light intensities of down to 3.7% air mass "zero" (AM0) which are in the order of 50W/m^2. Finally, the JUICE mission is also characterized by an extremely harsh irradiation environment (in terms of high energetic charged particles such as electrons and protons) which leads to high degradation on solar cell level due to particle irradiation. But also other effects like detrimental primary discharges as a consequence of high surface charging have to be taken into account. Dedicated technology development activities (TDAs) have been initiated by ESA in order to demonstrate the capability of a certain technology to sustain those extreme conditions. In the solar generator domain three activities are currently still ongoing or have been just finished:

1. JUICE Pre-qualification of Integrated LILT solar cells
2. Assessment of performance degradation of solar cells during the JUICE mission due to primary discharges
3. JUICE Coupon Test

This paper will give an overview on the status and/or the conclusions of those three activities.

DOI: 10.1051/e3sconf/20171604005

2. ACTIVITY DESCRIPTIONS

2.1 JUICE Pre-qualification of Integrated LILT solar cells

The objective of this activity was to run a full qualification test program in line with ECSS-E-ST-20-08 [1] on bare cell and solar cell assembly (SCA) level, with a reduced number of test samples per subgroup, but including additional cell variants as a backup in the test program. In total 5 different solar cell variants have been produced by AZUR SPACE Solar Power GmbH (AZUR) including thin configurations (80µm) and configurations of standard thickness (140µm). An overview of the cell variants is given in Tab. 1. The difference in cell design ("Standard" vs. "LILT") is mainly related to a modification of the front side metallisation aiming at mitigating the so-called "flat spot effect"[2–4]. However, as it was demonstrated in a precursor activity also the cells with standard design did not show any flat spot effect [5]. Thus, initially, Cell A with standard cell design and a thickness of 80µm was baselined for JUICE. The other combinations served as backup. A thin cell is preferred due to the mass criticality of the JUICE mission.

Table 1. Solar cell variants produced by AZUR.

Cell structure	Cell Design	Ge substrate thickness
A	Standard	80µm
A	Standard	140µm
A	LILT	80µm
A	LILT	140µm
B	Standard	80µm

SCAs of all 5 cell variants have been manufactured by Airbus Defence and Space GmbH (ADS) and Leonardo Company (Leonardo). All cell and SCA variants have been submitted as a minimum to a screening test as described in [5] and a full beginning of life (BOL) characterisation covering illumination and temperature conditions around Venus and Jupiter. The screening of the cells and SCAs is required to sort out shunted cells which will clearly underperform under low intensity conditions.

A significant part of this activity was also dedicated to a quantification of the annealing effects that have been observed in precursor activities [5].

2.2 Assessment of performance degradation of solar cells during the JUICE mission due to primary discharges

The purpose of this activity was to test the vulnerability of the SCAs against potential degradation as a consequence of primary discharges when a non-conductive coverglass is used. The operation of solar cells under low intensity low temperature (LILT) conditions implies that shunt resistances created by primary discharges have a much stronger impact on the solar cell performance compared to missions where the cells are operated at higher sun intensities. Therefore, a total number of 20 samples have been submitted to 100 primary discharges each, in order to identify whether the cell performance would be affected and if "yes" to quantify a degradation factor.

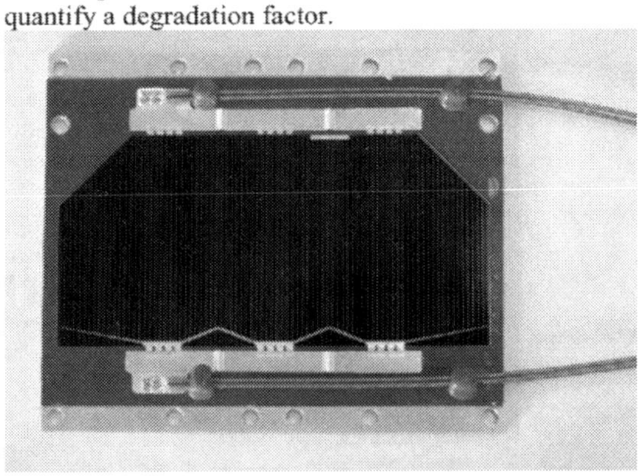

Figure 1. Solar cell "coupons" submitted to primary discharge tests. The coupon consisted of SCAs that were mounted on an Al plate with Kapton.

The samples that were tested were single-cell coupons as shown in Fig. 1. These "coupons" are SCAs mounted on Al plates covered with Kapton. This configuration was chosen in order to allow for an easier beginning of test (BOT) and end of test (EOT) characterisation that was performed at the Fraunhofer ISE. Since the BOT and EOT characterisation had to be performed also at LILT conditions the samples had to be mounted inside a cryostat. The Al plate with holes at the edges allowed for screwing the samples to the holder which was needed for arriving at a good thermal contact.

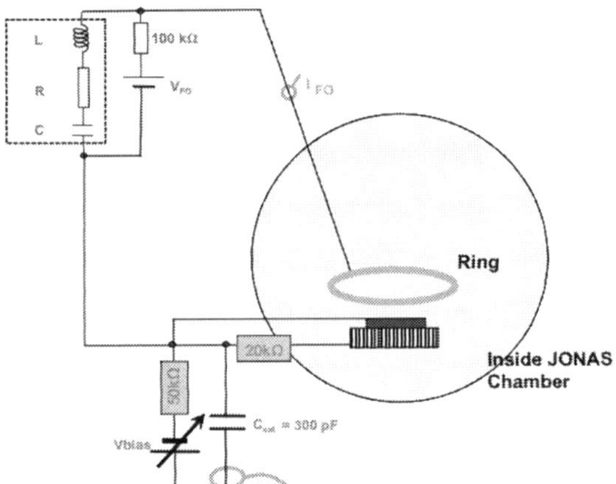

Figure 2. Experimental setup for the ESD accumulation test.

The electrostatic discharge tests have been carried out at

DOI: 10.1051/e3sconf/20171604005

ONERA at their JONAS facility using a flash-over simulator to represent the missing coverglasses on a real panel and assuming the discharge of a full panel as a worst case.

A schematic of the test setup is shown in Fig. 2 and is described in more detail in [6].

2.3 JUICE coupon test

The extreme thermal environment of the JUICE mission required a dedicated test program that aimed at demonstrating that existing substrate and PVA (photovoltaic assembly) technology could be used. All European solar generator and PVA suppliers were contacted to provide dedicated samples to be tested in the Mechanical Systems Laboratory (MSL) of ESA-ESTEC to be submitted to the test conditions and the test sequence as provided in Tab. 2.

Table 2. Thermal cycling test sequence for JUICE solar array qualification, substrate samples and full coupons.

Step	No. of cycles	Conditions
1a	2 thermal cycles	+160°C to -140°C
1b	2 thermal cycles	+80°C to -160°C
2	6 TV cycles	-100°C to -240°C
3a	3 thermal cycle	+160°C to -140°C
3b	3 thermal cycle	+80°C to -160°C
4	90 TV cycles	-100°C to -240°C
5	45 TV cycles	-100°C to -210°C

An overview of the tested samples are given in Tab. 3.

Table 3. Overview of samples submitted to the thermal cycling test sequence in ESTEC facilities.

No	Sample description
1	Kapton peel test samples
2	Flatwise tensile test samples
3	Full coupon with photovoltaic assembly (PVA)

3. RESULTS

3.1 JUICE Pre-qualification of Integrated LILT solar cells

The first major achievement in the framework of the pre-qualification activity was the identification of the solar cell baseline. BOL measurements revealed that cell A (cp. Tab. 1) suffers from a feature known as "majority carrier barrier effect" [7,8]. This effect is a fully reversible drop in cell performance, once the solar cell is operated below a certain threshold temperature. Fig. 3 shows the light I-V characteristic of two solar cells under LILT conditions. At -120°C both cells provide almost identical power. However, once the cells are measured at -150°C, cell B clearly outperforms cell A. Cell A suffers from the majority carrier barrier effect which is obvious from the characteristic signature close to the open circuit voltage. It should be mentioned that this is true for all 4 variants of Cell A given in Tab. 1.

Since during the JUICE mission operating temperatures are dropping also below -120°C, it was decided to change to Cell B for the JUICE mission in an 80μm thick configuration as a baseline.

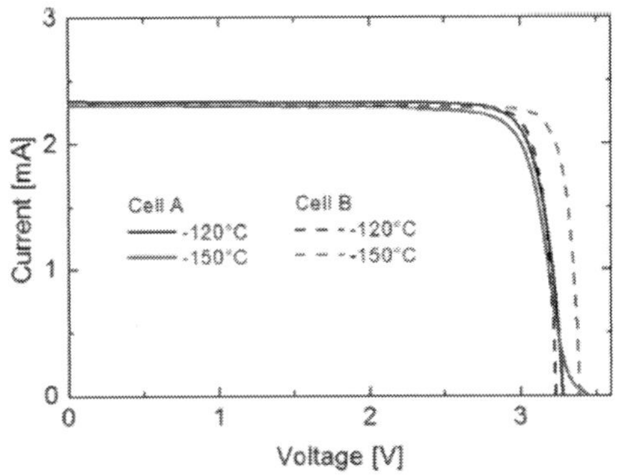

Figure 3. Light I-V curves of two solar cell variants measured under LILT conditions at -120°C and -150°C.

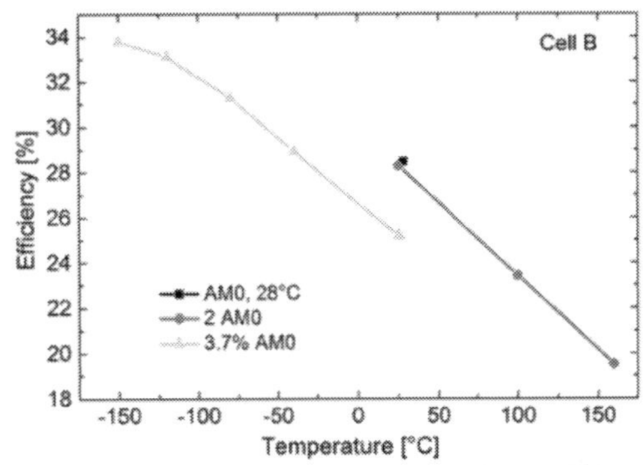

Figure 4. Evolution of the solar cell efficiency of cell B as a function of temperature for different illumination conditions.

Fig. 4 shows the evolution of the efficiency of cell B against temperature and for different illumination conditions. The black dot is the initial measurement under standard testing conditions (AM0, 28°C) performed at AZUR. The red points are derived from measurements performed at INTA Spasolab under high illumination conditions (up to 160°C, 2 suns) to understand the performance of the solar cells during the Venus gravity assist. LILT characterisation of the cells were carried out at CESI (down to -150°C, 3.7% AM0). For the high temperature part and the low temperature part the temperature coefficient follows nicely a straight line except for the very cold temperature regime where a saturation is reached. This is a known phenomenon and can be explained by tunnelling currents which at low temperatures start to become a dominant contributor [9]. Nevertheless, efficiencies at LILT conditions reach very high values of 34% (at -150°C,

 DOI: 10.1051/e3sconf/20171604005

3.7% AM0) with no sign of a flat spot effect.

Electron and proton irradiation tests have been performed at room temperature on cells and SCAs, respectively, with subsequent characterisation under LILT conditions. In addition, low temperature irradiation tests with in-situ LILT measurements have been performed by Ecole Polytechnique to assess the annealing factor that has been observed in precursor activities. The results of those tests are discussed in dedicated publications presented at this conference [10,11]. Other test activities such as UV, thermal cycling on cell and SCA level as well as the life test are not finished, yet.

3.2 Assessment of performance degradation of solar cells during the JUICE mission due to primary discharges

Bare cells of cell structure A and B of the two different thicknesses have been processed into SCAs and single-cell coupons (cp. Fig. 1) by ADS and Leonardo (applied coverglasses had no conductive layer). Tab. 4 lists the cell variants together with the number of samples per type that have been submitted to the ESD test. Numbers in brackets give the number of cells that showed degradation. Cell structure A* is an 80µm cell where the same cutting process is applied as for the 140µm cell. The standard process for cutting 80µm cells differs slightly from the one applied for 140µm cells. Cell A* was manufactured after the results of the first 10 samples were analysed, since it was initially assumed that the cutting process could have been responsible for the high number of degraded samples

Table 4. Coupon variants submitted to the ESD test. The number in brackets in the last column indicates the number of samples showing degradation.

Cell structure	Thickness [µm]	SCA manufacturer	No. of samples
A	80	ADS	5 (4)
A	140	Leonardo	5 (1)
A*	80	ADS	1 (1)
B	80	ADS	2 (0)
B	140	ADS	2 (0)
B	80	Leonardo	2 (0)
B	140	Leonardo	3 (2)

To track any potential degradation during the test sequence inside the test chamber, dark I-V measurements have been performed after the occurrence of every 10 ESD events (cp. Fig. 5). Furthermore, the location of the ESDs has been recorded by a video camera. Before starting the test and after the 100 ESDs electroluminescence, pictures were taken to detect potential shunts and their location (cp. Fig. 6). Fig. 7 shows the evolution of the dark current at a fixed voltage of 2V as a function of the number of ESDs for all cells. When a degradation is observed (increasing dark current) this is mostly related to one single detrimental ESD event after which the dark current remains stable for the rest of the test program. Some cells also showed accumulative degradation as a consequence of multiple detrimental ESDs.

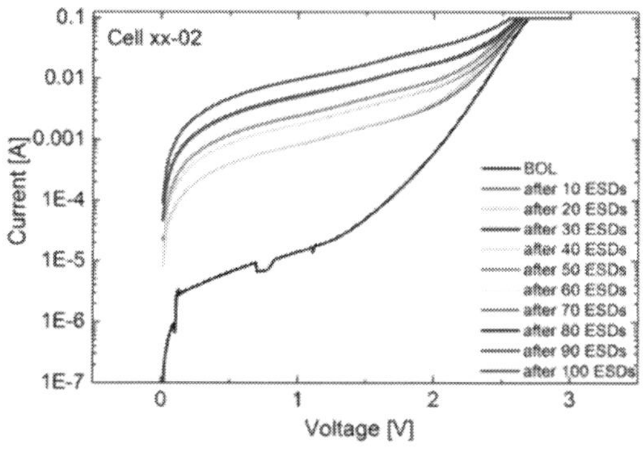

Figure 5. Dark current measurements of cell xx-02 after every 10 ESDs performed in situ at the JONAS facility of ONERA.

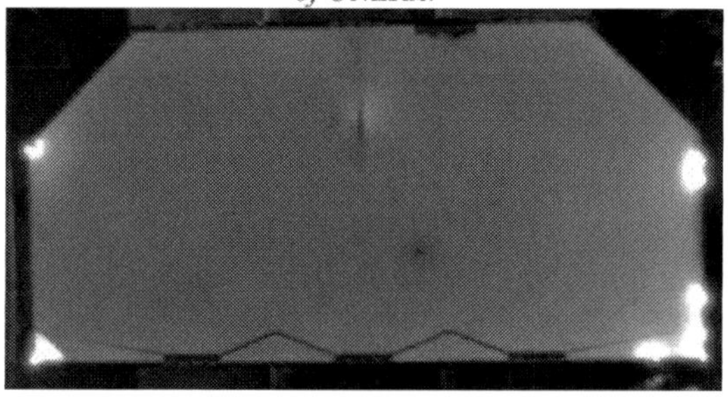

Figure 6. Electroluminescence picture of cell xx-02 after 100 ESDs. In this case, several ESD events have been introduced shunts (bright spots).

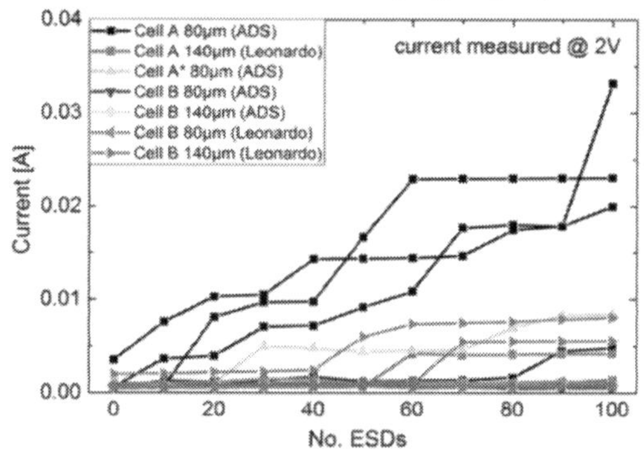

Figure 7. Dark current at a fixed voltage of 2V as a function of number of ESDs.

Initially, only the first 10 coupons of Tab. 4 have been submitted to the ESD test campaign: while almost all the 80µm coupons by ADS showed some degradation, the opposite was true for the 140µm coupons from

Leonardo where only one coupon showed some degradation. And this degradation could even not be correlated to an ESD event. Thus, the first assumption was that the thickness of the cell could be an important parameter for the vulnerability of the cell towards primary discharges. Since the cutting process of 80µm cells differs slightly from the 140µm cells, a dedicated 80µm cell was manufactured where the same cutting process as for the 140µm cell variant (cell A^*) was applied. However, also this cell showed a clear degradation. Thus, this theory was discarded. In parallel, other coupons were manufactured using cell structure B (which had become the baseline for JUICE at that time). As can be seen in Fig. 7, in case of cell B only 2 out of 9 samples show some degradation and for 4 cells with 80µm thickness even no degradation is observed. However, since cell A and cell B only differ slightly in their epitaxial structure, the cell structure itself is not considered to play a role in this test. Thus, it was concluded that basically all cell variants can potentially be affected by primary discharges.

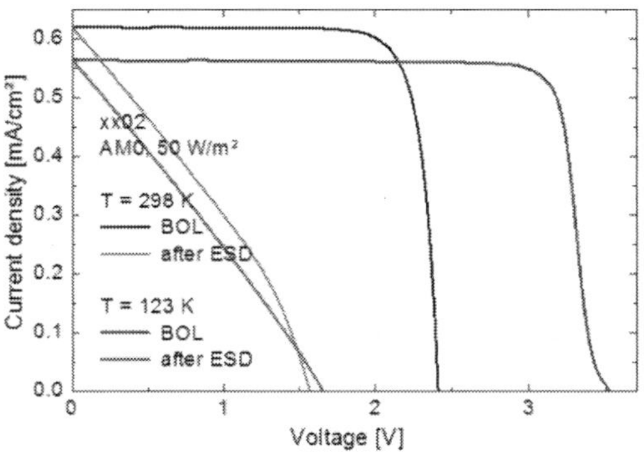

Figure 8. Light I-V curves of a coupon before and after ESD testing under low intensity conditions at room and low temperature.

The impact of the degradation was subsequently quantified by I-V measurements under LILT conditions at Fraunhofer ISE. An example of a coupon before and after the ESD test is shown in Fig. 8. There, light I-V measurements are shown under low intensity conditions at room temperature and -150°C. As mentioned before, the cumulative ESD events introduced one or more shunts at the cell edges leading to a significant drop in cell performance under those conditions. It should be highlighted that under full AM0 this drop in cell performance is much less. While under LILT conditions the drop in efficiency is almost 85%, the same cell shown in Fig. 8 would only drop by 10% under AM0 (room temperature) conditions.

However, as a conclusion of these results, it is of major importance to protect the solar cells against primary discharges for the JUICE mission. Due to the fact that the solar generator is requested to meet a very stringent equipotential requirement of 1V, the coverglasses are anyway covered with a conductive Indium Tin Oxide (ITO) layer. This together with an appropriate coverglass grounding network should protect the solar cells against primary discharge events to happen.

3.3 JUICE coupon test

As explained in 2.3 all European solar generator and PVA suppliers were invited to provide their existing technology for tests under JUICE representative conditions. In a first step, only Kapton peel and flatwise tensile test samples were submitted to the test sequence. Samples were provided by Thales Alenia Space (TAS) (with Patria), Airbus Defence and Space NL (ARA Mk4 samples from Airborne Composites and ATK) as well as by ADS (Eurostar 3000 substrate). All samples have been successfully submitted to the thermal cycling test sequence without significant drop of bond strength when compared to uncycled reference samples. Subsequently, two test campaigns with full coupons (test article No. 3 in Tab. 2) were launched which have just recently been finished. In total, 4 coupons have been tested: 2 front side and 2 back side coupons. The separation into back and front side coupon was necessary to allow for a good thermal contact with the "cryo box" side wall that was used in the test facility where the cryogenic cycles were performed (and thus a significant reduction of test time). The two front side coupons are shown in Fig. 5. The dimension of the coupons was in all cases $410x260mm^2$ due to the limitations of the cryo box. Leonardo laid down 2 strings of 8 and 16 cells in series, respectively, on a substrate provided by TAS (manufactured by Patria). ADS did 4 strings of 5 cells in series on a Eurostar 3000 substrate. Both substrate technologies have significant space heritage of several ESA and commercial missions. One back side coupon per company was equipped with the typical back side elements such as blocking diodes, bleeder resistors and thermal sensors. The substrate used by Leonardo for the back side coupon was again provided by TAS/Patria while the substrate for the back side coupon from ADS was of Ara Mk4 type provided by ADS NL (Airborne Composites). As explained above, there is a stringent requirement to meet an equipotential on the solar generator of 1V. This made the application of ITO coated coverglasses mandatory. Special attention was therefore paid on a possible degradation of the resistivity of the ITO layers and/or different coverglass grounding concepts applied on the test coupons.

In Fig. 9, it can be seen that both companies used patches to interconnect the conductive surfaces of the coverglasses. While Airbus DS GmbH was using one specific coverglass grounding technology, Leonardo Company decided to test several combinations of patches and adhesives. Also in one string, a solution was applied where the coverglasses were directly connected to the shunt diode. However, this solution has to be discarded for the JUICE mission since it does not allow to meet the 1V equipotential requirement. On the contrary, each coverglass will have the potential of the cell it is connected to, leading to a quite inhomogeneous potential distribution.

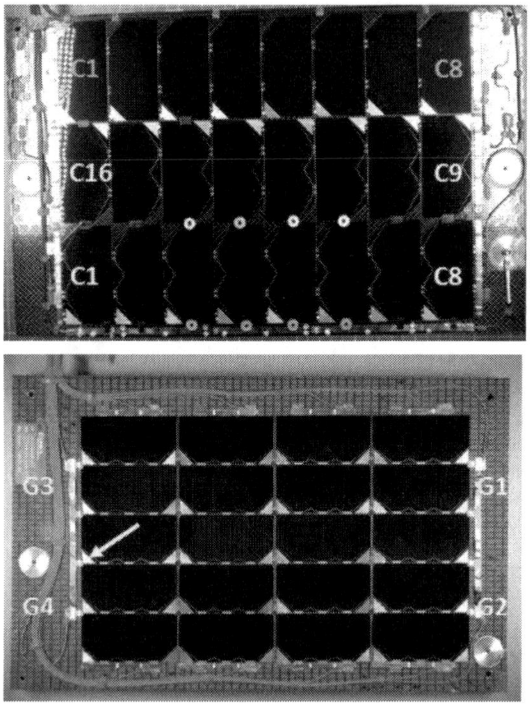

Figure 9. Front side coupons of Leonardo company (top) and ADS (bottom).

Before and after each test step (cp. Tab. 2), the coupons were investigated by visual inspection and an electrical health check was performed (including flasher test and electroluminescence measurements). During the cycling, the electrical elements were checked for continuity and an insulation check (cells vs. substrate) was performed during two cycles (beginning and end of each test sequence). Overall, the test demonstrates a significant confidence in most of the electrical elements. However, a few points of concern have been found which are currently subject to a more in-depth investigation:

1. On the Leonardo coupon, one string showed a slight degradation at P_{max} of 4% after bake-out step 1b of Tab. 2 which then remained stable during the test campaign. All other strings on the Leonardo and ADS coupons did not show any degradation at all. Therefore, it could be concluded that the solar cells do not suffer from the cycling profile.
2. A coverglass crack was noticed on one of the ADS SCAs after step 4 which affected the coverglass grounding path (see below).
3. Some components on the back side coupons of both companies lost contact to the substrate (adhesive failure) without loss of functionality of the component itself.

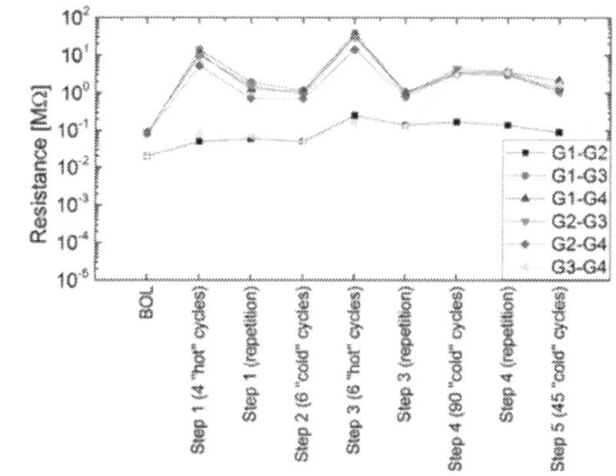

Figure 10. Resistance evolution of coverglass grounding network on ADS front side coupon.

Apart from those points, the major interest of the test campaign was focussed on the performance of the coverglass grounding network solutions. Fig. 10 shows the evolution of the coverglass grounding network resistivity as a function of the cycling steps. In order to have more information about the resistivity as a function of the current path length, the coverglass network grounding bars on each side of the coupon were cut in the middle and resistance measurements were performed between the 4 points G1 to G4 shown in Fig. 9 (bottom). The resistance was derived by applying 50 V between any two points and measuring the current. Obviously, the current paths with the shortest distance – G1-G2 and G3-G4 – have the lowest resistivity. For those points mainly the resistivity of the ITO itself is measured. For all the other measurements, the resistivity values are quite similar which could be expected since the current has to pass over the same minimum number of patches (3). While before any thermal cycling the resistance was in all cases below 100kΩ, a strong increase was observed after the first step with values in the order of 10MΩ. However, repeating this measurement after a few days showed a significant recovery with values being now in the range of 1-2MΩ. The cold cycles in step 2 did not lead to any change in the resistivity. However, additional "hot" cycles in step 3 again lead to very high values which again recovered after a few days. The last cycles consisting of two sequences of "cold" cycles did not lead to a further increase of the resistance resulting in a final resistivity (for the long current paths) of 1-2 MΩ. The sudden increase of the resistivity between G3-G4 can be explained by a crack in the coverglass of cell 3 in string 4 (numbering the strings from right to left in Fig. 9 bottom) exactly at the position of the grounding bar connection apparently interrupting the ITO coating (arrow in Fig 9 bottom). This results in a current path equivalent to the others (where the current has to flow over at least 3 patches). Overall, it can be concluded that the resistivity of the coverglass grounding network increased by roughly one order of magnitude with respect to the BOL measurement. However, the maximum values in terms of resistivity would still allow for meeting the 1V equipotential requirement.

On the coupon from Leonardo, the interesting measurements with respect to the grounding network solutions were done on string 2 (consisting of 16 cells in series) since the solution used on string 1 (coverglasses connected to cell) cannot be used for the JUICE mission as explained above. Patches are applied between the short edges of adjacent cells and from the short edges to the grounding bar (only on one side). 3 different coverglass grounding solutions were applied labelled solution A, B and C. An overview of which coverglass grounding solution has been used between the different cells of string 2 of the Leonardo coupon is given in Tab. 4.

Table 4. Overview on which coverglass grounding solution has been used between the different cells of string 2 of the Leonardo coupon.

Cells in string 2	CG network solution
1, 16	A
2, 15	A
3, 14	B
4, 13	B
5, 12	C
6, 11	C
7, 10	A
8, 9	A

Figure 11. Application of a measurement probe on top of the outer edge of the coverglass. The other electrical connection is done via the grounding bar on the left side (not visible in this picture).

Measurements of the resistance was carried out by using on one side the connection of the grounding bar and by probing on the other side directly on top of the coverglass (see Fig. 11). By this means, the current had to flow over the length of two cells and over two grounding patches per cell pair (one patch between the cells + 1 patch from cell to grounding bar). As for the ADS coupon, resistances were derived from applying 50 V between the two contact points and measuring the current.

The evolution of the resistance of the coverglass grounding network is given in Fig. 12 showing that all solutions exhibit a resistance of clearly less than 100kΩ at the end of the test. For solutions A and B there is at least one path that shows a significant change after one of the test steps. Only solution C remains very stable. However, it has to be pointed out that the statistics are very poor.

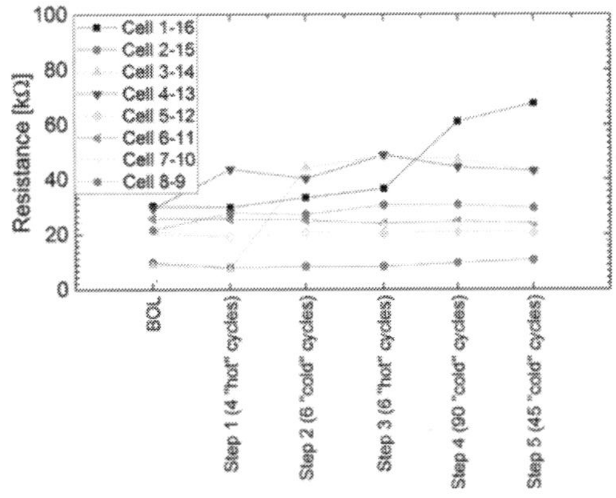

Figure 12. Resistance evolution of coverglass grounding network on Leonardo front side coupon.

Additional measurements have been performed probing each individual coverglass separately (on both sides a probe was placed directly onto the coverglass) in order to check the performance of the ITO layer itself. In average over the 24 cells the resistivity changed from 14.2 kΩ before the test to 11.5 kΩ after the test. These values are well in line with those quoted by Qioptiq in the order of 5kΩ/sq. Taking into account that the way of measuring gives rise to some variability in the results it can be concluded that the ITO resistivity remains very stable during the cycling test.

Additional coupons of both companies are currently being manufactured and are planned to be tested in the upcoming months. On one side, these additional tests are aimed at improving the coverglass grounding network solutions taking into account what has been learned from the present tests and on the other side shall increase statistics.

4. CONCLUSION

The status of running or finished TDAs initiated by ESA to support the JUICE mission has been provided. In the pre-qualification activity the major finding was the detection of the majority carrier barrier effect in one cell structure that led to a reorientation of the cell baseline. Test activities are ongoing with UV, thermal cycling and life test still to be performed. A significant amount of effort went into the electron and proton irradiation campaigns which are discussed in separate contributions to this conference [10,11].

The activity related to the evaluation of potential performance losses upon primary discharges has been finished. It was concluded that there is a potential risk that one ESD event can cause an unacceptable degradation of the solar cell performance under LILT conditions. Therefore, a protection against primary discharges became mandatory for the JUICE mission. Since an ITO coating together with a coverglass grounding network was meanwhile already baselined due to the 1V equipotential requirement on the solar array, this conclusion did not lead to a design change. It was mentioned that under AM0 conditions the observed degradations were much less compared to LILT conditions. However, the findings within this activity are concerning enough to justify the implementation of a dedicated activity to investigate further whether primary discharges could cause detrimental performance losses also for other more standard missions (operating under AM0 and above).

Finally, the coupon test campaign with 4 coupons (two front and two back side coupons) has been successfully completed with a few observations that still need to be investigated. However, it can be concluded that the PVA basically survives the JUICE thermal environment and that there are also coverglass grounding network solutions that survive the cycling profile and meet the stringent equipotential requirement.

5. ACKNOWLEDGEMENT

The authors would like to thank a number of people that contributed significantly to the generation of the huge amount of data that served as a basis for the analysis provided in this paper: Elvira Fehrenbacher, Michael Schachtner, Alexander Wekkeli from Fraunhofer ISE, Bernd Lehmann, Stephane Roure, Carl Hall, Bob Witteveen, Natascha Klein from ESA, Gerald Jüngst, Ainhoa Martínez de Olcoz, Alberto Vegas at INTA and Jean-Michel Siguier from ONERA. Furthermore, C. Baur likes to thank his colleagues from TEC-EPG: Emilio Fernandez, Paul Hodgetts, Antonio Caon, José Ramón González and Gianfelice D'Accolti for fruitful discussions on testing and interpretation of test results. Last but not least, C. Baur likes to thank the JUICE project team Sam Verstaen, Robert Furnell and Fabio Panin from ESA and Thomas Schirmann and Didier Loche from Airbus DS Toulouse for the good collaboration and support in the execution of the TDAs.

6. REFERENCES

[1] ECSS-E-ST-20-08C, Rev. 1, "Space engineering - Photovoltaic assemblies and components." 2012.

[2] V. G. Weizer and J. D. Broder, "On the cause of the flat-spot phenomenon observed in silicon solar cells at low temperatures and low intensities," *J. Appl. Phys.*, vol. 53, no. 8, pp. 5926–5930, 1982.

[3] C. J. Gelderloos, K. B. Miller, R. J. Walters, G. P. Summers, and S. R. Messenger, "Low intensity low temperature performance of advanced solar cells," in *Proceedings of the 29th IEEE Photovoltaics Specialists Conference*, New Orleans, Louisiana, USA, 2002, pp. 804–807.

[4] R. Campesato, C. Flores, G. Damonte, and S. Taylor, "Modelling and analysis of multi junction solar cells at low temperature/low intensity conditions," in *Proceedings of the 22nd European Photovoltaic Solar Energy Conference*, Milano, Italy, 2007, pp. 680–683.

[5] C. Baur, V. Khorenko, G. Siefer, J. C. Bourgoin, M. Casale, R. Campesato, S. Duzellier, and V. Inguimbert, "Development status of triple-junction solar cells optimized for low intensity low temperature applications," in *39th IEEE Photovoltaic Specialists Conference*, 2013, pp. 3237–3242.

[6] V. Inguimbert, P. Sarrailh, J.-C. Mateo-Velez, J.-M. Siguier, C. Baur, B. Boulanger, A. Gerhard, P. Pelissou, M. Sevoz, and D. Payan, "Measurements of the Flashover Expansion on a Real-Solar Panel-Preliminary Results of EMAGS3 Project," *IEEE Trans. Plasma Sci.*, vol. 41, no. 12, pp. 3370–3379, Dec. 2013.

[7] R. Hoheisel, R. J. Walters, and A. W. Bett, "Low temperature effects in photovoltaic devices for deep space missions," in *Proceedings of the 42nd IEEE Photovoltaic Specialist Conference*, 2015, pp. 1–5.

[8] R. Hoheisel and A. W. Bett, "Experimental Analysis of Majority Carrier Transport Processes at Heterointerfaces in Photovoltaic Devices," *IEEE J. Photovolt.*, vol. 2, no. 3, pp. 398–402, Jul. 2012.

[9] J. A. Del Alamo and R. M. Swanson, "Forward-bias tunneling: a limitation to bipolar device scaling," *IEEE Electron Device Lett.*, vol. EDL-7, no. 11, pp. 629–631, 1986.

[10] V. Khorenko, C. Baur, G. Siefer, M. Schachtner, S. Park, B. Boizot, J. C. Bourgoin, M. Casale, and R. Campesato, "BOL and EOL characterization of AZUR 3G LILT solar cells for ESA JUICE mission," *at this conference*.

[11] S. Park, J. C. Bourgoin, O. Cavani, V. Khorenko, C. Baur, and B. Boizot, "Origin of the degradation of triple junction solar cells at low temperature," *at this conference*.

 DOI: 10.1051/e3sconf/20171604006

THE BEPICOLOMBO MERCURY PLANETARY ORBITER (MPO) SOLAR ARRAY DESIGN, MAJOR DEVELOPMENTS AND QUALIFICATION

A. Loehberg [1], C. Gruenwald [1], J. Birkel [1], A. Brandl [1], N. Kuebler [1], H. Perrin [1], T. Andreev [1], U. Schuhmacher[2], S. Taylor [3]
[1] Airbus DS GmbH, 81663 Munich, Germany
[2] Airbus DS GmbH, 88039 Friedrichshafen, Germany
[3] ESA/ESTEC, Keplerlaan 1, 2200AG Noordwijk, The Netherlands

ABSTRACT

The BepiColombo Mercury Planetary Orbiter (MPO) solar array has been designed to guarantee the operational requirements in a very severe environment, mainly characterized by high temperatures and high light intensity. Furthermore, the qualification of the MPO solar array requires ca. 5500 thermal cycles.
The final configuration of the solar array is the result of a very complex system analysis aiming at reducing the solar array operative temperatures. The outcome of this was an extensive development and test effort on components which provide sufficiently high temperature resistance and a wing with unique design characteristics.
The MPO solar generator is composed of one wing consisting of three panels and provides an average power output up to 1800W during the nominal 1 Earth year mission around Mercury. The wing design is characterised by temperature reduction measures. The flight wing has already passed the majority of the environmental test program.

1 INTRODUCTION

The BepiColombo mission to planet Mercury is a joint project between the European Space agency (ESA) and JAXA of Japan. Two scientific orbiters, the Mercury Planetary Orbiter (MPO) and the Mercury Magnetospheric Orbiter (MMO) are dedicated to the investigation of the planet.
In Figure 1 the Mercury Composite Spacecraft (MCS) configuration with the Mercury Transfer Module (MTM), MPO and MMO is shown in launch configuration (solar arrays are stowed).
ESA is responsible for the complete MCS system and MTM, MPO plus MOSIF which acts as a sun shield and an interface to the MMO. The MMO satellite design and development is managed by JAXA.

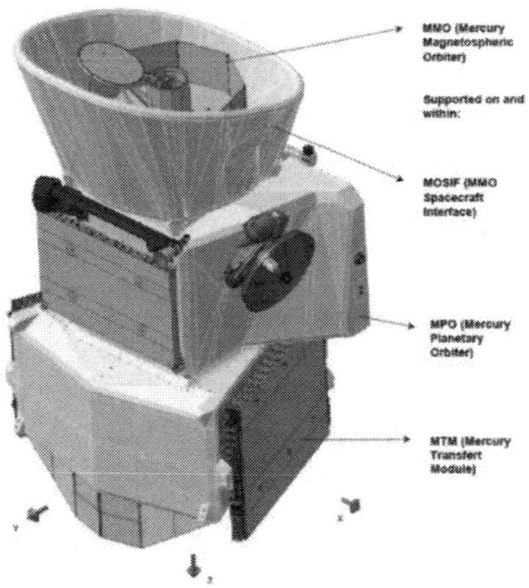

Figure 1. Mercury Composite Spacecraft in stacked configuration – launch configuration

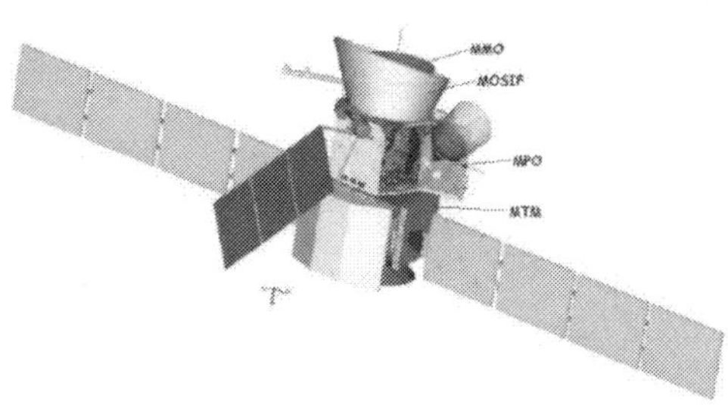

Figure 2. Mercury Composite Spacecraft – cruise configuration

The MTM is used for the transfer to Mercury and hosts an electrical propulsion system with ion thrusters that significantly drive the MTM solar array power requirements. At the end of a 6 years cruise phase to Mercury the MTM is separated from the MCS. After an orbit capture and lowering MMO and MPO are finally separated before the MPO descends to its final orbit.

As a consequence of the vicinity of planet Mercury to the Sun, the two satellites and the MTM are exposed to extreme light intensities up to 11 solar constants which corresponds to 15.4kW/m². This leads to severe operational conditions of the solar arrays. The MPO solar array is additionally exposed to infrared radiation (between 13.6kW/m² at perihelion and at the subsolar point and 6W/m² at the night side) and the reflection of sun light (albedo 12% in average but varies considerably) from Mercury's surface. Those boundary conditions result in very high solar array temperatures.

Since 2005 the Airbus DS GmbH Solar Array division has been involved in the development activities related to solar arrays for the MPO and MTM. Concerning the ESA BepiColombo solar arrays, Airbus DS GmbH Ottobrunn (Munich) is the selected as solar array supplier for the MPO and MTM solar arrays; for the MTM solar array Airbus DS B.V NL has been involved as sub-contractor who is responsible for the MTM overall wing design and assembly. The solar arrays are being procured via Airbus DS GmbH Friedrichshafen who is in the role of the industrial prime for the BepiColombo mission.
In the paper from the ESPC 2011 [1] the evolution of the solar array requirements as a result of technology constraints and major development activities related to high temperature solar arrays are outlined.
The qualification approach for the developed electrical components has been presented in ESPC 2013 [2].
This paper is describing the MPO wing design and qualification activities.

2 MPO WING DESIGN CHARACTERISTICS

In the following the MPO wing design characteristics are described.

2.1 Solar-Mercury Environment and related Orbits as Design Drivers

Figure 3 shows the Mercury orbit around the Sun and the MPO orbit around Mercury.

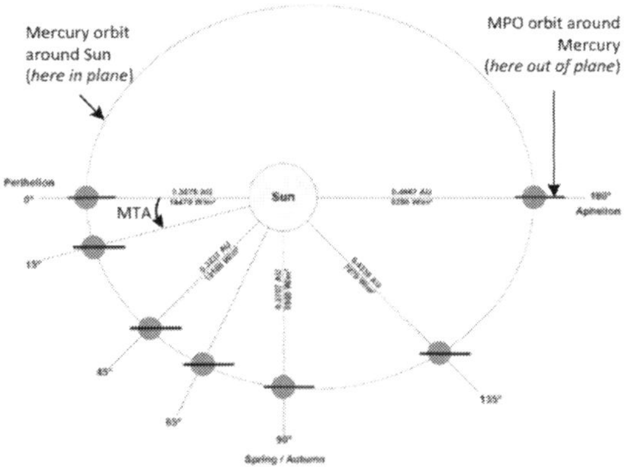

Figure 3. Orbits

In the following graphic the spacecraft (SC) positions in relation to the planet Mercury and the corresponding Sun direction are visualized for three Mercury true anomalies (MTAs).

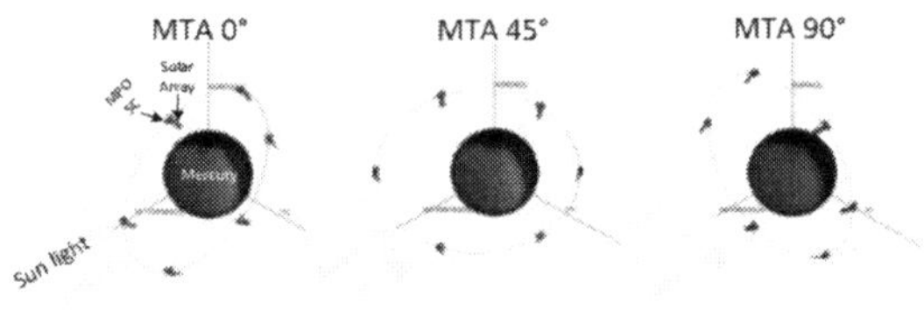

Figure 4. Sun direction

The MPO spacecraft (SC) design and the orbits lead to the fact that during the science mission the Sun light, the Mercury infrared heat and it's albedo hit the MPO solar array from many different directions. In addition to this the spacecraft sidewall reflects sun light and emits infrared heat in direction of the solar array. This is visualized by the following graphic.

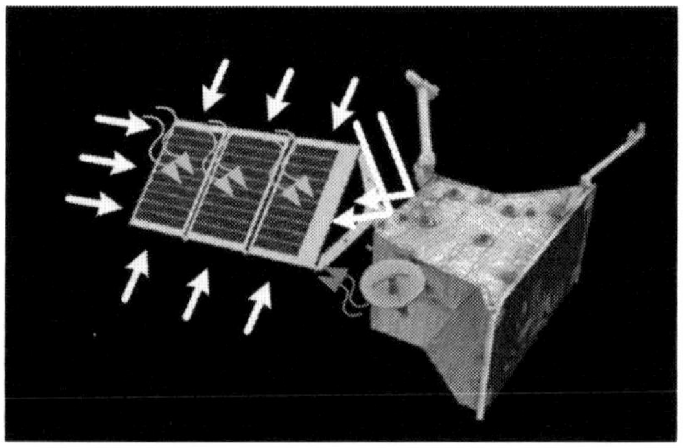

Figure 5. Five "heat sources" from varying directions (graphic copied from [7] and modified)

The variation of intensity and direction of these five "heat sources" during the mission lead to the need that heat protections have to be implemented on all sides of the solar wing.
Figure 5 revealed that solar array operation will be performed under high inclination angles in order to reduce the effective impinging sun light onto the solar array surface. Detailed studies have been performed in order to assess the angle performance of the array [3,4]. This is in particular important for precise performance predication under all circumstances.

As also visible in Figure 4 the SC passes several times the Mercury eclipse which leads to the necessity of robustness of the solar array against ca. 5500 thermal-cycles down to ca. -170°C.

2.2 OSRs and Kapton-VDA on Front Side

In between the solar cell strings optical solar reflectors (OSRs) are applied.
The Y-shaped yoke and the first inner quarter of the first panel are fully covered by OSRs instead of solar cells, as due to the hot SC sidewall these areas will become too hot for the chosen and qualified solar cell technology.

At areas with geometrical inhomogeneities on the panel surface Kapton-VDA is used as second surface mirror instead of OSRs.

The following photo shows the wing front side respectively the solar cell and OSR pattern.

Figure 6. MPO solar array front side

For the skins of the sandwich panels ultra-high thermal conducting carbon fibres are chosen to enable a good heat distribution and an efficient cooling of the solar cells by the OSRs. Also for the carbon honeycomb core high conductive carbon fibres are selected to use the panel's rear side as radiator cooler.

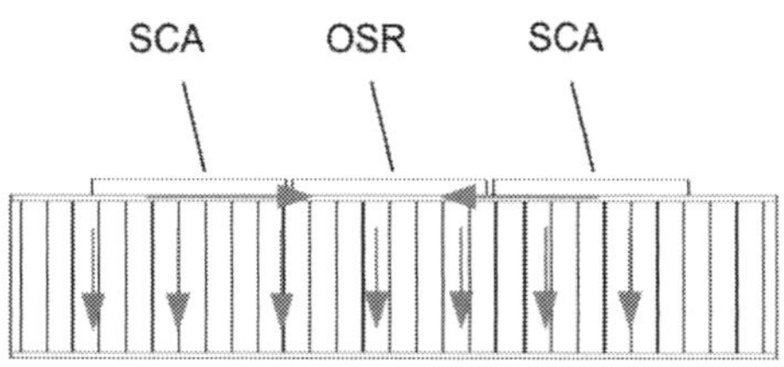

Figure 7. Heat distribution via high conductive fibres

2.3 OSRs and Kapton-VDA on Rear Side

Due to reflections of the SC sidewall even the rear side of the yoke and the inner panel are equipped with OSRs. At areas where wires are routed Kapton-VDA is used instead.

The following photo shows the wing rear side.

Figure 8. MPO solar array rear side

The next photo shows the rear side of the inner panel in more detail.

Figure 9. Rear side of inboard panel

2.4 Increase of Solar Aspect Angle and Edge Shields

Despite the usage of the OSRs the temperature of the panels has to be decreased further in order to not exceed the solar cell and substrate qualification temperatures. For this reason the solar panel surface is tilted outwards the Sun. By this the solar aspect angle (SAA) is increased.

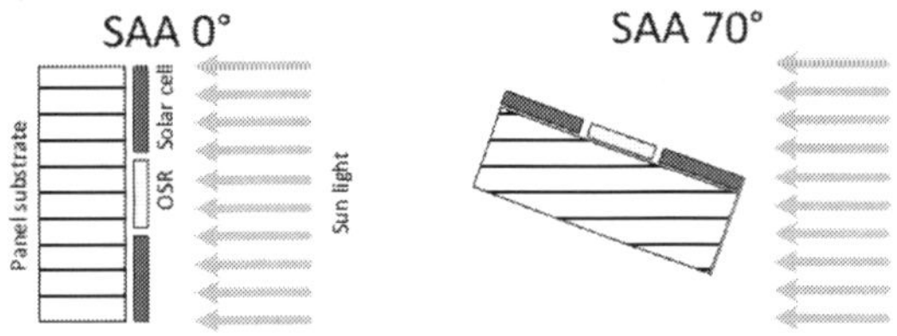

Figure 10. SAA

Drawbacks caused by this are a reduction of the generated electrical power and the need of panel edge shields.

Figure 11. White edge shields

On the Sun averted side the IR emissivity is kept small. Because of the different thermal expansion of the panel substrates and the edge shields, the shields require a bearing with one fix point and one or more sliding/floating points. The predicted temperature of these shields is about +450°C.

2.5 Wiring

Also the electrical wiring turned out to be a very ambitious challenge.

In addition to the five heat sources as shown in Figure 4 the self-heating increases the wire temperatures further. Especially in the vicinity of the hot SC sidewall many measures which decrease the temperatures for one orbit point (e.g. usage of a Sun shield) lead to the opposite effect in another orbit point (e.g. by prevention of IR emittance into deep space).

The solution of this problem was found in a combination of routing the wires on the yoke rear side, usage of sun shields made from white coated Titanium, usage of glass fabrics (Nextel) as heat protection and development of a high temperature electrical insulation (+315°C) for the wires.

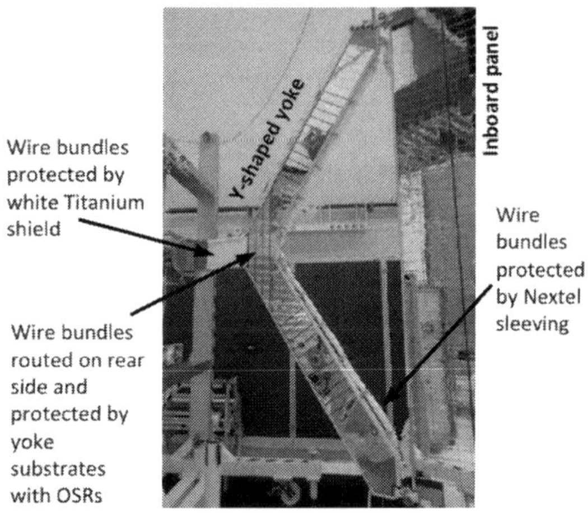

Figure 12. Wire routing at yoke

2.6 High Temperature Structures

For the carbon fibre reinforced sandwich panels and yoke substrates the typically used epoxy resins and adhesives were substituted by cyanate based materials which are cured at high temperature. The determination of the correct process parameters which lead to a reliable quality and reproducible mechanical strength was one of the most demanding issues of the project.

Like for most other space structures also for the MPO solar panels is the launch the sizing mechanical load case. Therefore the highest stresses occur at ambient temperatures.

But as the MCS separation events (MTM, MPO, MOSIF and MMO) will occur near Mercury/Sun the structures have to provide a notable mechanical strength also at high temperatures. The solar arrays are already deployed during these events which lead to significant bending moments when the modules are separated. The upper qualification temperature for the substrates is about +280°C. The combination of such high temperatures with the mechanical loading was tested on sample level only, not on wing level.

For areas where this temperature is exceeded and for mechanisms Titanium was chosen as base material.

2.7 Solar Cell Assemblies

The solar cell assemblies (SCAs) which provide sufficient robustness against high intensity solar light (ca. 11 SC) in combination with high operating temperatures (up to +215°C) are one of the key technologies for the BepiColombo mission. Details of this aspect of the qualification have been reported [2,5,6].

 DOI: 10.1051/e3sconf/20171604006

Also other electrical components like diodes and resistors were partly modified and tested for the specific mission needs.

2.8 Coatings and other thermo-optical Surfaces

For the qualification testing of white coatings, OSRs and Kapton-VDA and Nextel under high temperature conditions in combination with UV/VUV see [5].

2.9 Reduction of Heat Flux into the Spacecraft

Obvious design characteristics of the MPO solar array are the measures to reduce the thermal flux into the spacecraft at the interface areas between the solar array and the SC. Both, the yoke-hinge (root-hinge) and the hold down and release mechanisms (HDRMs), are covered by MLI or shields.

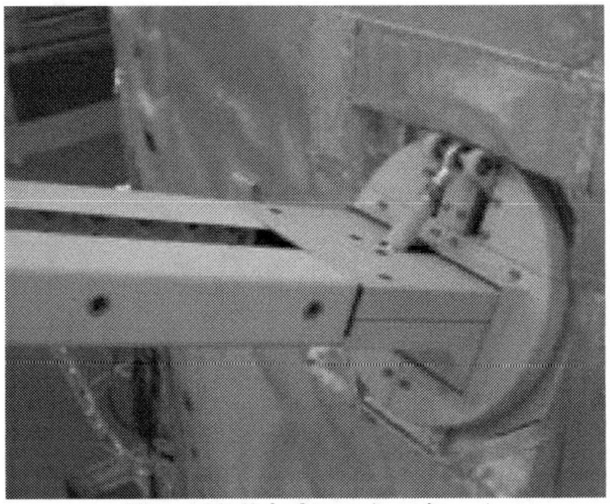

Figure 13. Yoke hinge (root hinge)

Figure 14. HDRM on SC sidewall under MLI

3 MPO WING QUALIFICATION

Beside the continuation of the component qualifications and preparation and testing of qualification-samples and DVT-coupons the wing assembly and testing have been the main tasks for the last months.

3.1 Thermal vacuum testing

One of the major wing tests starts already before wing assembly. It is common practice that the thermal vacuum (TV) testing is done on panel level (respectively yoke level). Therefor the panels and the yoke, which are fully equipped with solar cells and OSRs, have been tested for 10 TV cycles from -170°C to +215°C. This testing was combined with a bake out to reduce the outgassing of the synthetic materials during the mission.

Figure 15. Panel 1 and 2 suspended for integration in the TV chamber

3.2 Wing deployment tests

During the launch the wing is folded on the SC side wall. The wing deployment is tested before and after the environmental test program which shall simulate launch conditions (e.g. shaker tests).

DOI: 10.1051/e3sconf/20171604006

Figure 16. Wing folded to SC sidewall simulator

Figure 17. Wing deployed

3.3 Shaker tests

By the shaker tests the sine base excitations which are acting on the solar array during launch are simulated. This test is done in 3 axes (1x out-of-plane, 2x in-plane). The structural integrity of the panels and yoke was monitored by ca. 60 accelerometers, 81 strain gauges and 5 load cells.

Figure 18. Shaker test / out-of-plane

Figure 19. Shaker test / in-plane

Acoustic tests were performed only on single panel level (EQM panel) and will be performed on system level.

3.4 Electrical performance and health checks

The standard electrical characterisations were done before and after environmental testing.

- Power performance (flasher test)
- Electroluminescence (ELM)
- Electrical health checks (insulation, grounding, continuity, etc.)

4 SUMMARY

Fundamental technology developments were needed for the realisation of the BepiColombo MPO solar array. Technology challenges like UV and thermal cycling degradation under extreme temperature conditions were investigated and a new cell design based on the Azur 3G28 triple junction solar cell was validated under relevant environmental conditions. Due to technology limitations extensive system level analyses have led to design and operations modifications on system level and the MPO solar array. Even with the modifications the environment still imposes quite a challenge on the PVA, solar array structures and other components like high temperature diodes. Dedicated developments were performed in the frame of the project and have been incorporated into the MPO solar array design.

5 ACKNOWLEDGEMENT

The work on the Bepi Colombo solar array was carried out under a programme of and funded by the European Space Agency. The view expressed herein can in no way be taken to reflect the official opinion of the European Space Agency.

Besides the authors of this paper a large team has worked on the development of the BepiColombo solar arrays. Contributions of the following colleagues are highly appreciated: C. Zimmermann and C. Noemayr (PVA technologies); J. Müller (design MPO S/A); R. Sebek (design MTM S/A); F. Ustamuijc (structural developments).

The authors express their thanks to AZUR Space Power GmbH, XPERION Aerospace GmbH, W. L. Gore & Associates GmbH, CNM D+T Microelectronica, GKN Aerospace GmbH heavily involved into the technology development.

6 REFERENCES

1. Petry D., Grünwald. C., Löhberg, A., Brandl, A., Oxynos-Lauschke, C., Andreev, T., van der Ven, R., Schuhmacher, U., Fugger, S., Caon, A., Fiebrich, H.K., (2011). The BepiColombo Mercury Transfer Module and Mercury Planetary Orbiter Solar Array Design and Development. ESPC 2011 - European Space Power Conference.
2. Andreev, T., Zimmermann, C., Löhberg, A., de Jong, S., Schuhmacher, U., Schneider, J., Köstler, W., Parketta, P., Zäh, E., Holzbauer, R., Fink, M., Hall, J., Fernandez Lisbona, E., Lundmark, K., Baur, C., Taylor, S., Panin, F., Caon, A. (2015). Qualifying Solar Array Components for High Intensity, High Temperature Environments – The Test Approach for the Bepi Colombo Solar Arrays, In Proc. 10th European Space Power Conference ESPC 2014, ESA SP-719 (CD-ROM).
3. T. Hülsheger, C. Brandt, A. Caon, H. K. Fiebrich and T. Andreev, "The Angular Performance Behaviour of Triple Junction Solar Cells", In 9th European Space Power Conference, ESA SP-2102638 (CD-ROM), Saint-Raphaël, France, 2011.
4. Brandt, C., Hülsheger, T., Baur, C., Caon A., Andreev, T., "The Influence of High Light Inclination Angles on the Performance of $GaInP_2/GaAs/Ge$ Triple Junction Solar Cells", In: 38th Photovoltaic Specialist Conference (IEEE), Austin, Texas, USA, pp. 1572-1578, June 2011.
5. Nömayr, C., Zimmermann, C., Premysl, J., "Degradation of Solar Array Components in a Combined UV/VUV High Temperature Test Environment, In 11th Space Power Conference ESPC 2016, 3rd to 7th of October 2016, Thessaloniki, Greece
6. Oberhüttinger, C. Quabis, D., Nesswetter, H., Zimmermann, C., "Simulating Large Area, High Intensity AM0 illumination - Test Results from BepiColombo and Solar Orbiter Qualification", In 11th Space Power Conference ESPC 2016, 3rd to 7th of October 2016, Thessaloniki, Greece
7. http://sci.esa.int/bepicolombo/55747-artistic-impression-of-bepicolombo/

DEGRADATION OF SOLAR ARRAY COMPONENTS IN A COMBINED UV/VUV HIGH TEMPERATURE TEST ENVIRONMENT

Christel Nömayr [(1)], **Claus Zimmermann** [(1)], **Premysl Janik**[(2)], **Christopher Semprimoschnig**[(2)]

[(1)] *Airbus Defence and Space, 81663 Munich, Germany, Email: christel.anke.noemayr@airbus.com*
[(2)] *ESA/ESTEC, Keplerlaan 1, 2200AG Noordwijk, The Netherlands*

ABSTRACT

BepiColombo is the joint mission of the European Space Agency (ESA) and the Japanese Aerospace Exploration Agency (JAXA) to explore the planet mercury. The European contributions, namely the mercury transfer module (MTM) and the mercury planetary orbiter (MPO), are both powered by deployable solar arrays. Many materials and technologies are at their limit under the harsh high-intensity, high-temperature (HIHT) conditions of the mission. Synergistic effects like photo fixation and photo enhanced contamination by ultra violet and vacuum ultra violet radiation (UV/VUV) on sunlit surfaces are considered to play an important role in the HIHT environment of the BepiColombo mission.
A design verification test under UV/VUV conditions of sun exposed materials and technologies on component level is presented which forms part of the overall verification and qualification of the solar array design of the MTM and MPO. The test concentrates on the self-contamination aspects and the resulting performance losses of the solar array under high intensity and elevated temperature environment representative for the photovoltaic assembly (PVA).

1. INTRODUCTION

A classical approach to predict contamination levels is based on the surface temperature in comparison to the temperature of the outgassing source. Especially on very hot surfaces contamination would then not be an issue since any outgassing molecule would not stick to the surface at all or would not have a sufficiently long residence time for the build-up of a contamination layer. However, laboratory as well as in-orbit experiments [1] have shown that when sunlit, warmer surfaces become more degraded by contamination than they would have been expected by a pure assessment of their surface temperature. Also the observed degradation of a GPS solar array might be explained to be caused partially by photo enhanced contamination effects [2].
The exact mechanisms that play a role for contamination and degradation of UV/VUV irradiated surfaces are complicated. One explanation for the formation of non-volatile contamination layers is that the adsorbed molecules polymerize to species with longer residence times (= photo fixation). Another mechanism discussed is the photochemical modification of a molecule immediately before it is sticking to the surface leading to a different sticking behaviour (= photo enhanced deposition). The exact photochemistry and the interplay between outgassing molecule arrival rate and photonic density are complicating the phenomenon when setting up a model to calculate expected contamination levels.
Besides the formation of the contamination layer itself, the photochemical development of the contamination film over time, typically characterized by a darkening, has to be considered. One important consequence of this behaviour for solar arrays is the decrease of transmission of the cover glasses which translates into solar cell performance loss. Another important consequence is the higher absorbance of thermal control materials e.g. OSRs which has an impact on the overall solar array temperature. The photochemistry of the representative contamination mixture would have to be understood in detail in order to analyse the degradation by modelling.
All the aforementioned aspects and difficulties made it necessary to identify an alternative experimental approach to assess the impact of combined outgassing and UV/VUV under relevant mission conditions on the BepiColombo solar array. The idea was to test the UV/VUV effect in combination with outgassing as close as possible to the final application. The test plan therefore tried to mimic not only the environmental conditions as far as possible in an on-ground based simulation, but also the most relevant material constellations and solar array outgassing sources.

2. DESIGN OF TEST COUPON

The focus of the test was on the materials and technologies of the PVA [3]: solar cell with cover glass, OSRs, Kapton VDA (vapour deposited aluminium), panel edge shield coatings, and cable bundle sleeves. All of the items have a direct or indirect impact on the solar array performance by power generation and thermal control. In parallel the main outgassing contributors which can be considered for the outgassing environment of these items were included in the sample design: Cyanate ester face skins and film adhesives, RTV (room temperature vulcanizing) silicones, and Kapton foil material.

 DOI: 10.1051/e3sconf/20171604007

There were several constraints to be considered for the design of the test coupon. Due to the test conditions and the uniqueness of facilities that can create them, the available test space was limited to 200mm x 200mm.

Another difficulty was to ensure thermal control during the test. Typical solar array coupon designs contain representative to the flight design, a piece of CFC (carbon fibre composite) faceskin / honeycomb core sandwich as a substrate. For a reliable temperature control in a vacuum test environment, however, the intimate contact of the samples to the facility base plate (which provides cooling and heating) is required. This is even more delicate with the heat input by UV/VUV illumination from top. The solution was to bond the solar cell assemblies (SCA), optical solar reflectors (OSRs), and Kapton VDA as well as a cable bundle sample to representative cyanate ester face skin material. The face skin was bonded directly onto Invar plates being screwed directly onto the sample plate. This is shown in the cross-sectional setup of an SCA sample in Fig. 1. The sample plate was then mounted to the facility plate which allowed the sample temperature to be controlled.

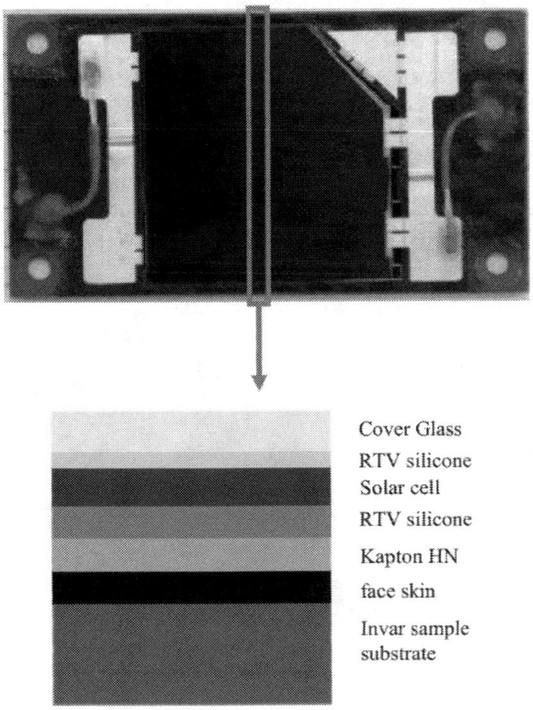

Figure 1. SCA sample with schematic cross-section

In order to mimic the situation on the solar array panel, the test specimen were arranged as representative as possible on the sample base plate as shown in Fig. 2. It was not possible to implement all the view factors e.g. the 90° orientation of the panel edge shields. Yet the tested condition represented part of the combination which can be assumed to represent an envelope in terms of degradation.

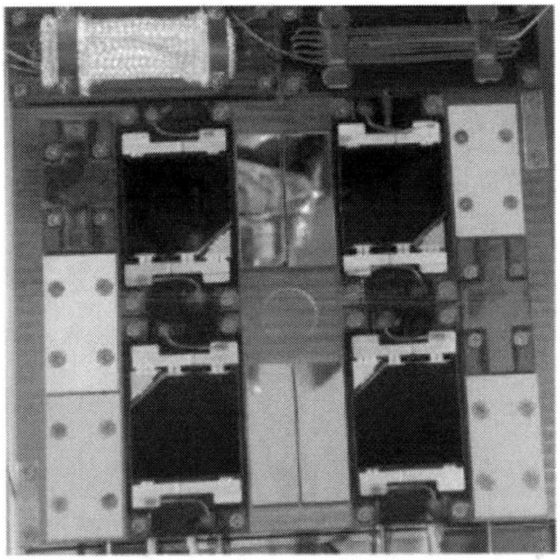

Figure 2. Test coupon arrangement on sample plate

In order to simulate the significantly higher temperature for the cable bundle with Nextel sleeve, which is part of the transfer harness as illustrated in Fig. 3, it was necessary to thermally isolate it by washers to the rest of the samples and induce additional heating by current injection through the cable bundle.

Figure 3. Cable bundle with Nextel sleeve

A second cable bundle without sleeve was kept at solar cell temperature, again with the help of current injection.

In order to test the required setting for the test temperatures a set of pre-test samples was used with embedded thermocouples to read out the obtained temperatures with a master thermocouple used as facility control reference during the test.

3. EXPERIMENTAL

The test was conducted in the CROSS1 vacuum facility at ESTEC. The facility offers exposure to ultraviolet (up to 12 SC) as well as vacuum ultraviolet. The base plate of the facility can be heated and cooled to control the

 DOI: 10.1051/e3sconf/20171604007

sample temperature.
The UV intensity varied between 7 and 12 solar constants (SC) on the sample plate. The VUV intensity during each test phase varied due to the contamination of the lamp which is inside the chamber. Because of this fact, intermediate cleaning steps were undertaken. The measured VUV intensity itself was lower compared to the UV intensity. This was predominantly related to the aspect angle of 45° of the VUV illumination. The measured values ranged from <1 to 9 SC.
The test temperatures for the SCA samples were set to 200°C for 2400 test hours (cycle 1-4) and 215°C (cycle 5) for the last 600 test hours. The Nextel cable bundle was set to 275°C for 2400 test hours followed by a stepwise temperature increase over the last 600h from 275°C up to 294°C. The higher temperature and the temperature profile were made possible by using current heating of the cable bundle. The temperature of the cable bundles were controlled during the whole test by embedded thermocouples.
Lower temperatures, especially during the transfer phase to mercury, were neglected in the test in favour to simulate the combination of outgassing and UV/VUV in the operational phase of the MPO. The samples underwent a bake out before test with temperatures up to 200°C which is also performed on the flight hardware.
The test temperatures of OSRs, Kapton VDA, and white shield coatings were not controlled. Pre-test results showed temperatures for cycle 1-4 of 183°C for Kapton VDA, 179°C for OSRs, and 185 to 189°C for the white coating samples. In the pre-test for cycle 5 the Kapton VDA samples showed temperatures around 203°C, the OSR temperature increased to 197°C, and the white coatings sample temperature ranged between 204 to 207°C. The temperature spread of the samples was also monitored during the test with an IR camera. A snapshot during test is shown in Fig. 4.

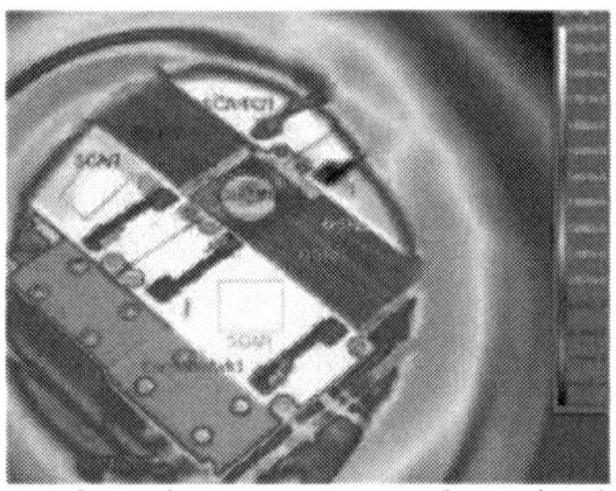

Figure 4. Infra-red camera picture of samples during test exposure

The samples were illuminated with a UV solar aspect angle (SAA) of 0° and VUV SAA of 45°, whereas in operational flight conditions the MPO solar panel surface will be inclined between a SAA of 71.7° and 83.8° [4]. This approximation of UV and VUV irradiation conditions was accepted, however, in order to collect as many equivalent sun hours as possible in a reasonable test time.
The contamination during the test was monitored by witness samples on the sample plate as well as on the facility cold shroud.
The test was split into in total five test runs (cycles) with intermediate inspections after 200, 600, 1400, and 2400 hours. The full set of inspections was performed before test and after 3000 test hours. The characterizations of the samples included electrical performance measurement of SCA samples under different angles, and thermo-optical measurements under normal incidence as well as for the thermal control materials under different angles (only OSRs, VDA, and white coatings before test and after 3000 hours). All samples were visually inspected after each test cycle.

4. RESULTS

4.1. Visual

The visual inspections showed especially on the white coating materials a significantly darkening as can be seen in Fig. 5a. The Kapton VDA is shown in Fig. 5 b and a darkening was also obvious from visual inspection. The SCA samples showed a darkening of the wire insulations and adhesive potter points; however the solar cell and its cover glass respectively did not show an obvious contamination. An example for the SCA samples is depicted in Fig. 5 c. Also on the OSR no obvious contamination layer was observed as can be seen in Fig. 5 d. The cable bundle with Nextel sleeve in Fig. 5 e showed also a pronounced darkening.

a
b
c
d
e

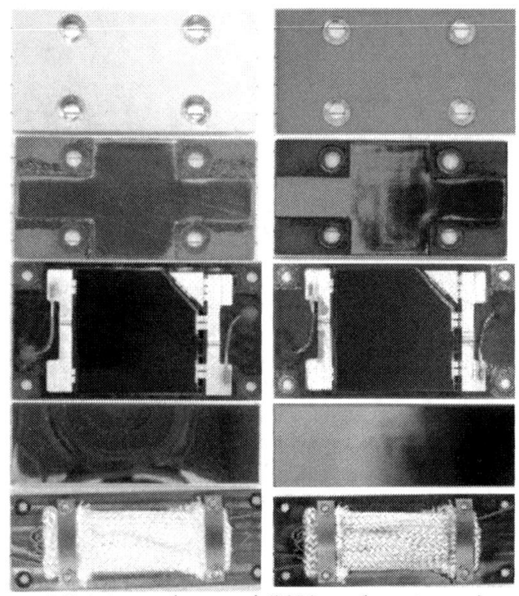

Figure 5.Pristine and exposed (3000 test hours) samples

DOI: 10.1051/e3sconf/20171604007

4.2. Solar Cell Performance

The solar cell current showed a higher degradation in the UV/VUV tests than in UV tests on solar cell assemblies and a test conducted at AM0 [4]. In Figure 6 the resulting difference in short circuit current to begin of test is plotted. The test hours of the UV/VUV test were scaled with a factor of 3.45 to account for the difference in SAA illumination in the other tests with UV and AM0. There a SAA of 70° was tested while in the UV/VUV the samples were irradiated by UV with a SAA of 0° resulting in a higher acceleration.
The lifetime test duration was defined as 6700 hours. The UV tests yielded a degradation of around -2.5%. In comparison to this value, the interpolated percentage value from the UV/VUV degradation lies at -6.1% after this test time. The difference of roughly 4% can directly be attributed to additional contaminants in combination with the VUV radiation: The VUV radiation is completely absorbed on the SCA surface and does not penetrate deeper, for example to the cover glass adhesive. Therefore there is no reason to attribute the degradation observed to additional degradation in these layers. During the UV test, these layers were exposed to the same UV dose as in this test. In the AM0 test, on the other hand, a similar contamination environment was present, but not the enhancing effect of VUV radiation.

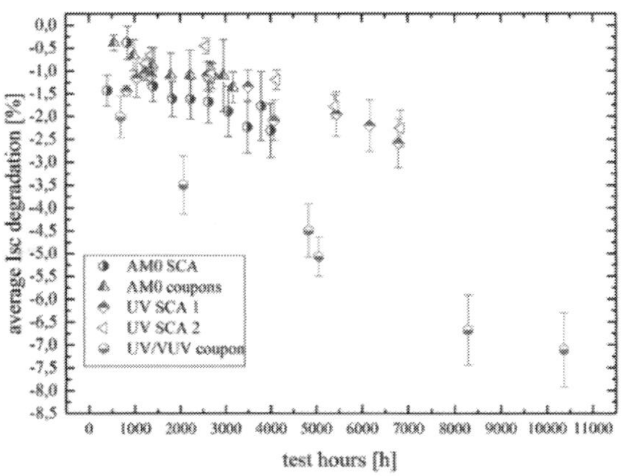

Figure 6. Comparison of Isc degradation in different tests (note: UV/VUV test hours scaled to equivalent UV exposure time under SAA 70°)

It is out of the scope of this work to discuss the effect of VUV for photo enhanced contamination and degradation of the contamination layer. In any case the high energetic electromagnetic radiation will cause further photochemical reactions on the molecules that are produced in the outgassing environment.
A consequence of the test results is that an additional contamination loss factor has to be considered in the performance prediction of the solar array for the BepiColombo mission.

4.3. Thermo-optical Properties

The SCA samples were also characterized in their thermo-optical properties over test time. However, neither the SCA absorbance nor the emittance values showed an increase.
In general all samples showed a stable emittance over high temperature exposure and UV/VUV irradiation.
The solar absorption values degraded for the thermal control materials over exposure time. The Kapton VDA samples showed an increase in absorption, which is after 3000 test hours still in a linear regime as can be seen in Fig. 7. The increase in solar absorption is most probably a combination of intrinsic degradation of polyimide, which is directly facing the UV/VUV radiation, and a contamination effect. The degradation in the polyimide layer after irradiation is also observed as a darkening on the wire insulation. A quantification of the contribution of the two mechanisms for the increase in alpha is however not feasible, since a comparison with results without VUV and contaminants is not available.

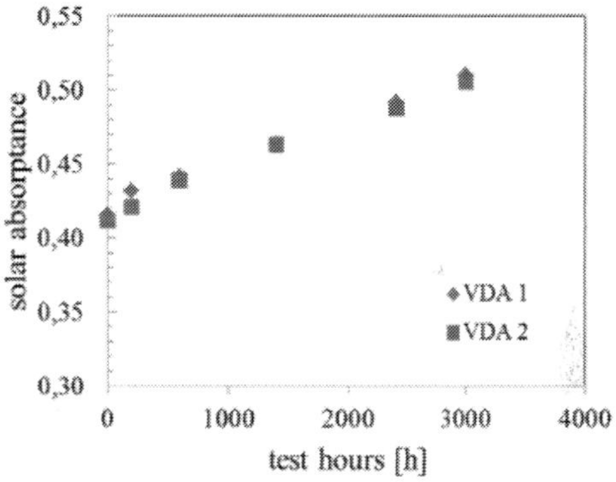

Figure 7. Solar absorptance values of Kapton VDA over test duration

Several white coatings were tested: White Keplacoat and white Keronite (both plasma oxide coatings), and PSZ (plasma sprayed zirconia). The coatings are applied on aluminium. In Fig. 8 the increase in solar absorption of white Keronite is shown. The observed increase is again in line with the visual observation. There seems to be already a trend towards saturation, however, a stable value is not yet achieved after 3000h.

DOI: 10.1051/e3sconf/20171604007

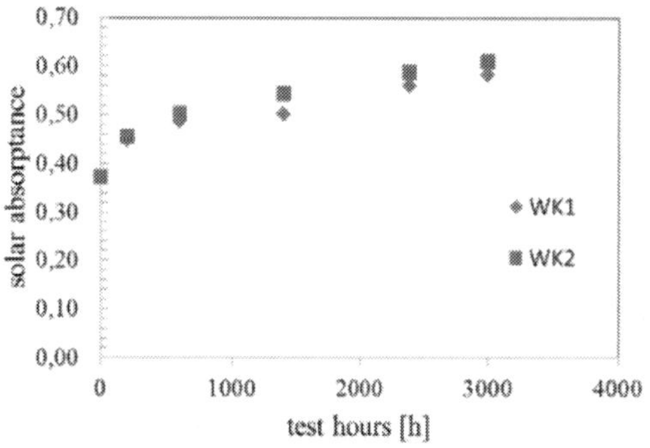

Figure 8. Solar absorptance values of white Keronite samples over test duration

The Nextel sleeve wrapped around a cable bundle darkened visually after exposure. The measurement of the thermo-optical properties yielded no degradation of the emissivity. The solar absorptivity was 0.38 on the exposed sample after 3000 test hours. A pristine sample measured in parallel had a value of 0.05. A second exposed sample which was used for the pre-test with total duration of approximately 340 hours had a value of 0.21. In this case it was not feasible to measure the thermo-optical properties at further intermediate test steps because the Nextel sleeve had to be removed from the sample and cut in order to get a flat surface for measurement.

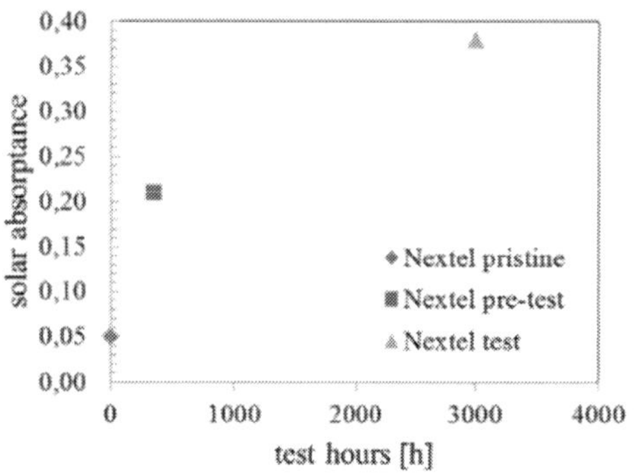

Figure 9. Solar absorptance values of Nextel samples over test duration.

The solar absorption values of the OSR samples show a significant increase over test time. After 3000 test hours the increase has still a linear behaviour. A long term linear increase in absorption was also observed for OSRs during 11 years of the SOHO mission [5]. The increase was suspected to stem from contamination induced degradation.

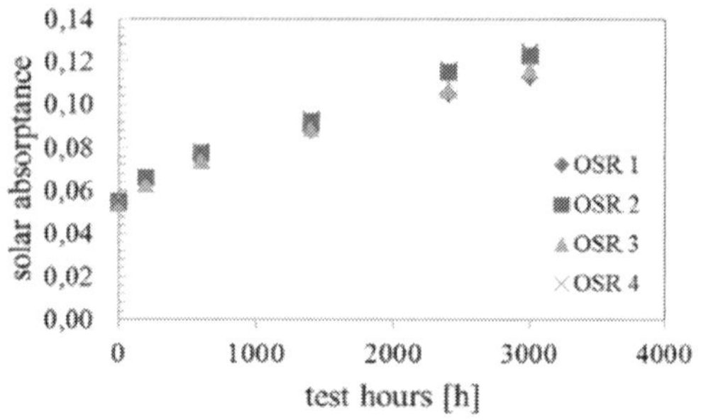

Figure 10. Solar absorptance values of OSRs over test duration

5. SUMMARY

A novel test approach was included in the qualification of solar array materials for the high temperature and high intensity mission BepiColombo. The aim of the test was to mimic the contamination environment in combination with the harsh environmental conditions during mission to gain insight into the impact of synergistic effects on the performance of the solar arrays of MPO and MTM.

The results show that significantly different degradation behaviour can be observed in terms of solar cell current degradation. The other cell parameters showed a similar behaviour than in other tests. This can be attributed to a contamination effect that will reduce the light input and thus impacts the cell current only. In comparison to the optical properties the solar cell current is more sensitive to the contamination.

The degradation of the thermo-optical properties can be used to validate input values for thermal analysis. A difficulty that arises in this case is that the alpha values did not saturate within the test time of 3000 hours. For cases where a lifetime is fairly close to the timeframe covered in the test a conservative but ultimately unphysical approach of a linear extrapolation can be used to identify the limitations for the overall thermal analysis. In some cases this extrapolation would lead to very critical consequences. Therefore an extension of the test with part of the samples has been started.

6. ACKNOWLEDGEMENT

This work was carried out under a program of and funded by the European Space Agency. The view expressed herein can in no way be taken to reflect the official opinion of the European Space Agency.

The constructive and helpful contributions of Cristina Borrero del Pino, Andreas Löhberg, Yuriy Butenko, Adrian Tighe, Jose Ramon Gonzalez, and Emilio Fernandez Lisbona are gratefully acknowledged.

7. REFERENCES

1. Hall, D.F. et al, Photo-Enhanced Spacecraft Contamination Deposition, *AIAA 20th Thermophysics Conference*, AIAA-85-0953, 1985.

2. Stewart, T.B. et al, Photochemical Spacecraft Self-Contamination: Laboratory Results and Systems Impacts, *AIAA Thermophysics, Plasmadynamics and Lasers Conference*, AIAA-88-2728, 1988.

3. Löhberg, A. et al, The BepiColombo mercury planetary orbiter (MPO) solar array design, major developments and qualification, this conference

4. Oberhüttinger, C. et al, Simulating large area, high intensity AM0 illumination - test results from BepiColombo and Solar Orbiter qualification, this conference

5. Poinas P., SOHO OSR degradation 11 years of flight data, report TEC-MCT/2008/3635/ln/PP

EXPLORATORY RESEARCH ON RADIOISOTOPE THERMOELECTRIC GENERATORS FOR DEEP SPACE MISSIONS

D. Freis [(1)], **J.F. Vigier** [(1)], **K. Popa** [(1)], **T. Wiss** [(1)], **J.-C. Griveau** [(1)], **E. D'Agata** [(2)], **J. Somers** [(1)]

[(1)] *European Commission, Joint Research Centre, Institute for Transuranium Elements, PO Box 2340, 76125 Karlsruhe, Germany, Email:Daniel.Freis@ec.europa.eu*
[(2)] *European Commission, Joint Research Centre, Institute for Energy and Transport, Petten, The Netherlands*

ABSTRACT

To support the development of a Radioisotope Power Source (RPS) for exploration missions into deep space based on ^{241}Am, the European Commission's Joint Research Centre (JRC) has recently launched an exploratory research project, scheduled for a period of two years. It aims at finding innovative solutions for higher performance, safety and reliability of such a device. In addition, it will serve as a start-up platform for this field of research and enable identification of research areas, where JRC can significantly complement and support the already existing activities in Europe. The project is divided into several subtasks and will investigate

- safety related properties of americium (oxide) as a heat source,
- compatibility of americium compounds with cladding and structural materials,
- properties of alternative americium compounds,
- properties of advanced thermoelectric materials.

The new exploratory research project will be introduced together with an overview on the available facilities and capabilities of JRC in this domain. Alternative americium forms with potential improved stability versus the oxides are discussed and innovative thermoelectric materials based on actinides are introduced.

1. INTRODUCTION

The European Space Agency (ESA) has decided to develop a Radioisotope Heating Unit (RHU) and a Radioisotope Power Source (RPS) for future exploration missions into deep space and contracted several research organisations, mostly in UK, with the implementation [1]. Unlike past ESA missions, which utilized solar panels as energy source, the new device will be powered by ^{241}Am, a transuranic isotope, which generates heat by decay. The heat will be either used directly to keep sensitive equipment at a required temperature or converted to electricity by means of thermoelectric converters (thermocouples). Future concepts might also include more efficient conversion mechanisms, such as Stirling motors [1], [2].

Most of the historical nuclear power sources for exploration missions were either deployed by the United States or the former Soviet Union and were based on ^{238}Pu [2], [3], a radionuclide with a high isotopic power of 0.567 W/g and a half-life of 87.7 years [4]. Unfortunately, there is a global shortage of this isotope and the costs associated with its production are high. Therefore, ESA has decided to alternatively use ^{241}Am, which has a half-life of 432.2 years. ^{241}Am is a decay product of ^{241}Pu ($T_{1/2}$ = 14.35 y) and can be extracted chemically from existing plutonium stocks in Europe. ^{241}Am, however, has some disadvantages compared to ^{238}Pu, such as a relatively low power density of 0.114 W/g, chemical instability at high temperatures and higher radiation levels.

In order to support the development of a European RPS the European Commission's Joint Research Centre (JRC) has launched an exploratory research project. Within the frame of this project JRC will perform an assessment of the safety and reliability relevant properties of americium oxide, which is the currently considered chemical form for application in space batteries, (e.g. measurement of thermal properties, investigation of helium release behaviour, swelling behaviour and oxygen release kinetics, ageing studies, examination of self-irradiation damage) and its compatibility to cladding and structural materials. In addition, alternative americium compounds will be investigated for their thermal and mechanical properties as well as their chemical stability under space mission conditions. Innovative candidate materials for thermo-electric conversion containing actinides will be assessed at high temperatures, utilizing the unique facilities and competences at JRC-ITU and JRC-IET.

2. ^{241}AM AS POWER PROVIDING ISOTOPE

^{241}Am is a transuranic element which is created in nuclear reactors by neutron capture and subsequent decay. Despite the half-life of "only" 432.2 years ^{241}Am is one of the main contributors to the long-term radiotoxicity of spent nuclear fuel, due to its decay product ^{237}Np, which has a half-life of 2.144 million years. It is, therefore, a key isotope for research on partitioning and transmutation with the aim to destroy the long-lived radioactive waste in dedicated irradiation facilities [5]. However, up to now ^{241}Am was never utilized as power providing isotope for space applications (to our knowledge), and some of its

DOI: 10.1051/e3sconf/20171605001

characteristics are different to ^{238}Pu and need to be considered.

Radiation: ^{241}Am decays by alpha decay to ^{237}Np with a specific activity of $1.271 \cdot 10^{11}$ Bq/g and it emits significant gamma radiation (Tab. 1). Therefore, it has to be handled under strict radiation protection measures, and special equipment, such as shielding and remote operation tools, is required. Due to the low energy of the emitted gamma rays (main line at 59 keV) it can be shielded relatively easy with few millimetres of lead (or comparable), but in opposite to ^{238}Pu at least some shielding has to be provided. Therefore, additional payload for a RPS powered by ^{241}Am might be necessary.

Table 1. ^{241}Am Gamma Emission [4].

Type	Energy(keV)	Emission Probability
	JEFF-3.1	**JEFF-3.1**
γ	59.5412 (± 0.0001)	3.600e-1 (± 1.0e-3)
γ	26.3448 (± 0.0002)	2.430e-2 (± 3.2e-4)
γ	27.040 (± 0.060)	6.26e-3 (± 9.5e-4)
γ	33.1964 (± 0.0003)	1.217e-3 (± 3.9e-5)
γ	43.420 (± 0.020)	6.59e-4 (± 2.3e-5)
γ	42.720 (± 0.040)	3.9e-4 (± 1.1e-4)
γ	98.970 (± 0.020)	2.23e-4 (± 1.1e-5)
γ	102.960 (± 0.020)	2.09e-4 (± 1.1e-5)
γ	55.550 (± 0.040)	1.84e-4 (± 2.2e-5)
γ	125.310 (± 0.020)	4.10e-5 (± 1.2e-6)
γ	69.760 (± 0.020)	2.88e-5 (± 3.0e-6)
γ	28.510 (± 0.050)	2.77e-5 (± 3.0e-6)
γ	123.000 (± 0.040)	1.080e-5 (± 3.9e-7)

Chemical stability: In order to avoid burnup and potential dispersion into the atmosphere during launch abort or re-entry accidents a chemical and mechanical stable form of americium under oxidizing conditions is desired. Therefore, the currently existing RPS designs include dioxide pellets encapsulated in an iridium cladding [2]. Unfortunately, americium exists in different oxidation states and the dioxide is prone to phase changes at elevated temperatures and in vacuum [6]. Fig. 1 illustrates the phase change in vacuum at 1200°C from the dioxide (AmO_2) to the sesquioxide (Am_2O_3) as well as the associated phase change from cubic (alpha) to hexagonal structure.

Because of this behaviour, the oxide form might be not the optimal solution with respect to long term stability of the pellets in the vacuum of space. Alternative compounds, such as $AmAlO_3$, could show a more stable behaviour at high temperatures under both, oxidizing and reducing conditions. Within the frame of the presented exploratory research project JRC will investigate a number of candidate americium compounds for their chemical stability and resistance towards self-irradiation effects.

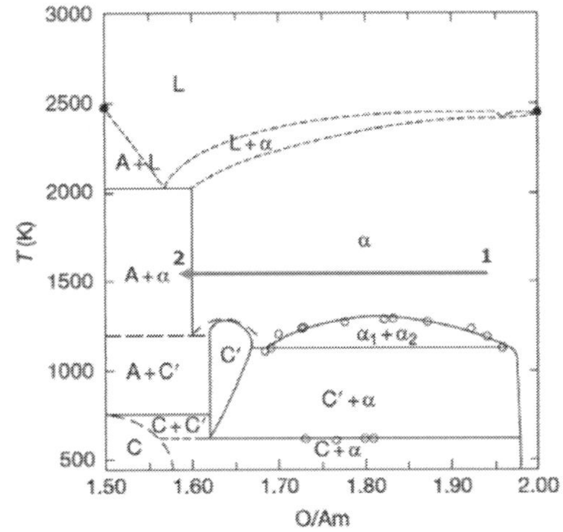

Figure 1. Americium-Oxygen Phase Diagram [6].

Thermal power: Tab. 2 shows a comparison between the thermal properties of the oxides of ^{238}Pu and ^{241}Am. $^{241}AmO_2$ has a power density of 0.101 W/g, which is about one fifth of $^{238}PuO_2$ (0.493 W/g). As a consequence it is necessary to use much more material in order to achieve the same power if americium is chosen. Therefore, it would be very interesting for an americium driven RPS to find possibilities to partly compensate the low power density of ^{241}Am by increased efficiency of the electricity conversion. This could be f.i. achieved by higher operating temperatures, thereby increasing the Carnot efficiency. Therefore, the exploratory research project will look into alternative americium forms, which are long-term resistant to high temperatures, but also candidate thermos-electric materials with high temperature resistance based on actinides.

Table 2. Comparison of ^{238}Pu and ^{241}Am.

	Pu-238	**Am-241**
Chemical form:	PuO_2	AmO_2
Half life:	87.7 y	432.2 y
Specific decay power (oxide):	0.493 W/g	0.101 W/g
Specific weight (oxide at 95% TD):	10.89 g/cm³	11.10 g/cm³
Thermal conductivity at 1500 K [7]:	~ 2.8 W/m K	~ 2.8 W/m K (~ 1 W/m K for Am_2O_3)

The disadvantage of ^{241}Am low power density, compared to ^{238}Pu, is accompanied by the advantage of

 DOI: 10.1051/e3sconf/20171605001

longer half-life, which makes it the ideal power providing isotope for very long lasting missions into the outer region of the solar system or even into the interstellar space and the Oort cloud. Fig. 2 shows the decrease of thermal power of ^{241}Am compared to ^{238}Pu over an period of 200 years. While the plutonium has decayed to circa 20% of the power at BOM, the americium source is still producing about 80% of the initial power. However, it should be noted that a large fraction of the power decrease of an RPS comes from the degeneration of the employed thermoelectric material. Therefore, the exploratory research project is looking into very long-term stability of potential americium compounds as well as alternative actinide based thermo-electric materials.

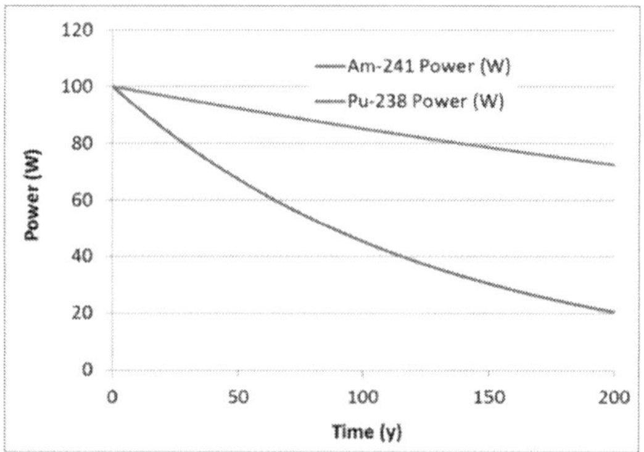

Figure 2. Decrease of ^{241}Am and ^{238}Pu power over 200 y

Thermal conductivity: The thermal conductivity of americium dioxide was found to be slightly lower compared to plutonium dioxide [7]. However, it should be noted that it decreases dramatically with deviation from stoichiometry and the sesquioxide shows a very low thermal conductivity (~ 1 W/m K at 1500 K, see Tab. 2). Since surface temperature, power density and thermal conductivity determine the central pellet temperature during operation, a low thermal conductivity can represent a limiting factor for a desired hot leg temperature of the thermocouples. In addition, a significantly increased thermal conductivity can increase the design flexibility of the RPS and enable relevant space and weight savings. Therefore, we will also investigate alternative americium compounds, and composites, with respect to their thermal properties (including thermal conductivity) at high temperatures.

3. THE MINOR ACTINIDE LABORATORY AT JRC-ITU

Due to strong gamma radiation, working with ^{241}Am can result in high dose rates for the operating personnel. Therefore, remote operated and shielded equipment is advantageous or even necessary in order to prepare americium based pellets for RPS. JRC-ITU possesses a unique infrastructure for handling of highly radiative actinide materials, the so called Minor Actinide Laboratory (MA-Lab) [8]. It is of high importance for safety research on Transmutation fuel in Europe, as it is one of the only dedicated facilities for the synthesis of minor actinide containing materials, either for property measurements or for the preparation of test pins for irradiation experiments.

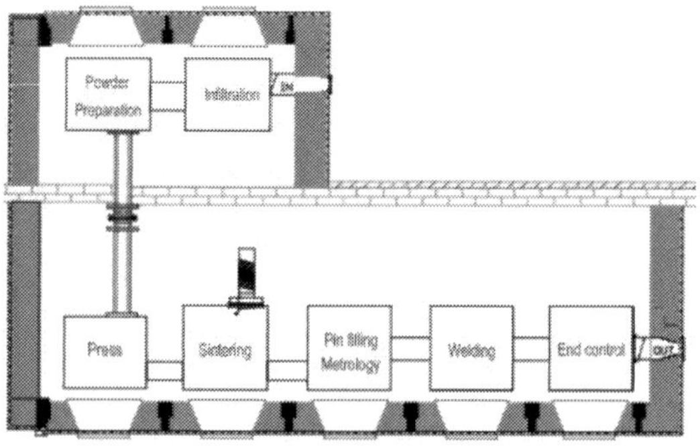

Figure 3. Minor Actinide Laboratory at JRC-ITU [8].

The MA-lab consists of seven glove-boxes with protection walls forming two separate chains. A schematic lay-out of the Ma-Lab is shown in Fig. 3. The glove boxes are shielded by 50 cm water and 5 cm of lead. Based on the thickness of the water and lead wall, the mass limits have been calculated to 150 g of ^{241}Am or 5 g of ^{244}Cm. The glove boxes can be accessed from behind manually if radiation levels are low, for experiment preparation or maintenance. In addition, tele-manipulators and remote operated automated equipment can be used for operation at high dose rates.

The glove boxes of the minor actinide laboratory are configured as complete preparation chain for minor actinide containing samples from the base material to the fully encapsulated sample, and the MA-Lab represents an ideal infrastructure for preparation of highly radiating americium pellets and fully qualified fuel pins.

The synthesis of the base material (powder) is performed in the glove box named "infiltration". The process is dust-free, based on the so-called gel supported precipitation [9] and the porous bead infiltration technique [10]. This process is highly flexible and easily adapted to the requirements and specifications of new sample compositions. The next glove box contains a calcination furnace and other equipment for powder preparation. The prepared powders are dust free, the individual beads typically show heterogeneous size distributions between 30 μm to 120 μm and are ideal for pressing pellets. The ready to press powder can be transferred via an automated channel to the next glove box, where it can be pressed to pellets. After sintering in reducing or oxidizing atmosphere (glove box "Sintering") the pellets are fully characterized and inserted into a cladding. Finally, pin welding and non-destructive weld examination are performed in the two last, alpha free glove boxes.

4. CONCLUSIONS

The European Commission's Joint Research Centre has launched a new exploratory research project in order to find innovative solutions and to improve safety and long-term reliability of ^{241}Am based power sources for future ESA missions into deep space. The project utilizes the unique facilities of JRC for preparing, handling and characterising highly radiating americium compounds. It investigates particular aspects of americium oxide, such as its chemical and self-irradiation stability and looks into promising alternative candidate compounds for americium and new thermo-electric materials containing actinides.

5. REFERENCES

1. ESA (2011). Robotic Exploration Technology Plan, Programme of Work 2009-2014.
2. Schmidt, George, Tom Sutliff, and Leonard Dudzinski (2008). "Radioisotope power: A key technology for Deep Space Exploration." (2008).
3. O'Brien, R. C. et al. (2008). "Safe radioisotope thermoelectric generators and heat sources for space applications." Journal of Nuclear Materials 377.3: 506-521.
4. Magill, J., and N. F. Magill (2010). "Nucleonica: A Platform for Organisational Knowledge Management in the Nuclear Domain." ENC.
5. Magill, J., et al. (2003). "Impact limits of partitioning and transmutation scenarios on the radiotoxicity of actinides in radioactive waste." Nuclear Energy-London 42.5: 263-278.
6. Konings, Rudy (2011). Comprehensive Nuclear Materials: Five volume Set. Newnes, Chapter 2.02.
7. Nishi, Tsuyoshi, et al. (2011). "Heat capacities and thermal conductivities of AmO 2 and AmO 1.5." Journal of Nuclear Materials 414.2: 109-113.
8. Fernandez, A., J. McGinley, and J. Somers (2008). "Minor Actinide Laboratory at JRC-ITU: Fuel Fabrication Facility." Proceedings of the Atalante 2008 Conference "Nuclear Fuel Cycles for a Sustainable Future,'Nimes.
9. J. SOMERS, A. FERNANDEZ, (2005). J. Amer. Ceram. Soc., 88, 827.
10. A. FERNANDEZ, D. HAAS, R. (2002). KONINGS, J. SOMERS, J. Amer. Ceram. Soc., 85, 694 (2002).

RECENT JOINT STUDIES RELATED TO THE DEVELOPMENT OF SPACE RADIOISOTOPE POWER SYSTEMS

Daniel P. Kramer [(1)], Richard Ambrosi [(2)], Mark Sarsfield [(3)], Emily Jane Watkinson[(2)], Ramy Mesalam [(2)], Hugo Williams [(2)], Chadwick Barklay [(1)], Tim Tinsley [(3)], Steve Goodrich [(1)], Timothy Pierson [(1)], Christofer Whiting [(1)]

[(1)] University of Dayton, Dayton, Ohio, USA, Email:daniel.kramer@udri.udayton.edu
[(2)] University of Leicester, Leicestershire, UK, Email: rma8@leicester.ac.uk
[3)] National Nuclear Laboratory, Sellafield, Cumbria, UK, Email: mark.sarsfield@nnl.co.uk

ABSTRACT

Over the last several years there has been a mutually beneficial ongoing technical interchange between the U.K and the U.S. related to various aspects of space radioisotope power systems (RPS). While this interchange has been primarily focused on materials based activities, it has also included some aspects related to safety, environmental, and lessons learned during the application of RPSs by the U.S. during the last fifty years. Recent joint technical RPS endeavors have centered on the development of a possible "cold" ceramic surrogate for $^{238}PuO_2$ and $^{241}AmO_x$ and the irradiation of thermoelectrics and other materials at expected RPS related neutron fluences. As the U.S. continues to deploy and Europe develops RPS capability, on-going joint RPS technical interfaces will continue to enhance each entities' endeavors in this nuclear based power technology critical for deep space exploration.

1. BACKGROUND

Over the last several decades over twenty-five U.S. Radioisotope Power Systems (RPS) have been successfully employed on a number of deep space missions including Galileo (launched 1989 to Jupiter), Ulysses (launched 1990 as a Solar orbital), Cassini (launched 1997 to Saturn), and New Horizons (launched 2006 - fly-by Pluto 2015); and Mars Science Laboratory's rover Curiosity (landed on Mars in 2012).

U.S. RPSs utilize the heat released from the decay of the radioisotope plutonium-238 via thermoelectrics to provide all of the electrical power needed for the spacecraft. Plutonium-238 is employed since it is a strong alpha emitter with a half-life of ~87.7 years. This provides the needed thermal energy and lifetime which allows the completion of long duration missions.

The employed fuel is an oxide, $^{238}PuO_2$, in the form of a ceramic pellet. The individual pellets are fabricated using various classical ceramic processing operations such as ball milling, screening, and hot pressing. Each pellet is ~150gms and produces ~62.5W_{th}. Current U.S. RPSs utilize a lead telluride based thermoelectric material known as TAGS-85 which converts ~5 to 6% of the available thermal energy provided by the radioactive fuel into electricity available for powering a spacecraft.

In Europe over the last several years there has been interest in the development of a European RPS capability for possible future ESA supported space exploratory missions. These efforts are currently based on the application of americium-241 in oxide form as the radioisotope heat source. The radioisotope americium-241 will be separated from used nuclear fuel at the National Nuclear Laboratory in the U.K. ^{241}Am has a half-life of ~432 years and the RPS fuel would be $^{241}AmO_x$ also in the form of ceramic pellets likely fabricated again using ceramic processing operations.

By their inherent application of a radioactive fuel, RPSs need to successfully operate in a radiation environment. The total radiation field which the spacecraft and the RPS itself is exposed to will come from the radioactive fuel and, depending on the specific mission profile, the space environment such as is encountered in missions close to Saturn or Jupiter. The current RPS design being investigated at the University of Leicester will utilize bismuth telluride based thermoelectrics to convert some of the decay heat from the $^{241}AmO_x$ fuel form into electrical power for a spacecraft.

2. RPS FUEL CeO_2 SURROGATE PROCESSING

2.1 Cold Press + Sintering Studies

$^{241}AmO_x$ and $^{238}PuO_2$ are both radioactive making it difficult to readily employ in experiments related to the development and the understanding of how various physical/chemical properties (such as particle size, particle size distribution, etc.) impact their ceramic processing characteristics. Therefore, the application of a non-radioactive surrogate material for performing various endeavours in the U.K. and the U.S. in support of our respective space nuclear power efforts would result in reduced personnel exposure, and additionally in reduced associated costs.

One surrogate material currently under investigation in both the U.K. and the U.S. is cerium dioxide (CeO_2). Cerium dioxide has a number of chemical and physical

 DOI: 10.1051/e3sconf/20171605002

characteristics which makes it a potentially attractive cold surrogate for $^{241}AmO_x$ and $^{238}PuO_2$. It can also be processed into pellets utilizing various ceramic processing techniques which can be employed in experiments. One interesting chemical characteristic of the two actinide oxide based fuels are that they may release oxygen atoms under various reducing atmospheres which results in the formation of sub-stoichiometric oxides. This tendency is similar to CeO_2 which also tends to release oxygen atoms under various reducing conditions forming sub-oxides with different crystal structures [1-3]. The reduction of CeO_2 to Ce_2O_3 with hydrogen also has been shown to result in the formation of over thirty intermediate sub-oxide compositions [4].

Several joint U.K./U.S. studies have been performed between the University of Leicester (U of L), National Nuclear Laboratory (NNL), and the University of Dayton (UD) investigating some of the ceramic processing characteristics of CeO_2. Experiments have been performed using both commercially available CeO_2 powder of different specific particle sizes, and internally (NNL and U of L) produced powder some of which has also been supplied to UD for additional joint investigations.

In one study led by NNL, CeO_2 powder was produced using an "oxalate" technique similar to what is employed in the processing of $^{238}PuO_2$ and likely $^{241}AmO_x$ powder which typically consists of a precipitation and then the calcination of the precipitate. In this particular study on the chemical preparation of CeO_2 powder, two different precipitation temperatures were used followed by one of three different calcination temperatures. Several gram quantities of each of the six powders were also sent to UD where their individual ceramic processing characteristics were studied.

Small ceramic discs have been fabricated at UD using NNL produced CeO_2 powders employing standard ceramic cold pressing techniques (hydraulic pressing of CeO_2 powder loaded in metal dies). The obtained discs have measured as-pressed green densities which range from ~55% to ~67% theoretical density using a 100% theoretical density for CeO_2 of 7.2 gm/cm^3. After cold pressing, some of the green discs were next sintered at 1400°C for one hour in an air atmosphere furnace. Figure 1 shows an example of one of the sintered CeO_2 pellets. After sintering, the theoretical density of the pellets in this study were measured and ranged from ~77%TD to ~85%TD. Since plutonium-238 is an alpha emitter, it is desirable to not fully densify the $^{238}PuO_2$ fuel pellets above ~90% of theoretical density. This facilitates helium removal under various environmental conditions. It is expected that future $^{241}AmO_x$ pellets will also be fabricated in the same general range of percent theoretical density.

Figure 1. Relative size of a CeO_2 pellet sintered at 1400°C for one hour.

After sintering, several of the obtained pellets were mounted and polished to examine their microstructures as a function of processing parameters and to obtain microhardness measurements. Figure 2 shows two of the ~0.8cm in diameter pellets fabricated in this study after mounting and polishing. The microhardness results showed some differences as a function of precipitation and calcination temperatures as the Vickers Hardness values varied from a mean of ~565 to ~648. The hardness values were obtained employing ASTM C1327-08 using a load of 300gms and a dwell time of 15 seconds.

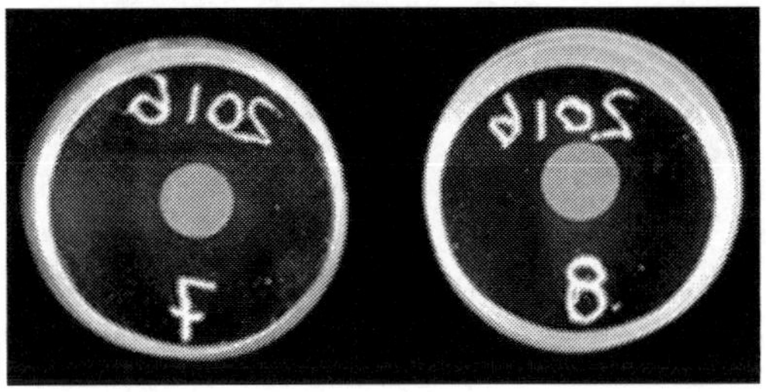

Figure 2. Sintered ~0.8 cm in diameter CeO_2 ceramic discs after mounting and polishing prior to the microhardness tests.

2.2 Hot Pressing Processing Studies

In addition to the CeO_2 NNL/U of L/UD processing study discussed above, a number of other ceramic processing studies have been performed using differently processed CeO_2 powders and other high temperature ceramic sintering techniques such as spark plasma sintering, and specifically at UD hot pressing which is discussed below.

Hot pressing is one of the ceramic processes currently being employed in the fabrication of $^{238}PuO_2$ fuel pellets. Hot pressing employs temperature + pressure during a specified time-temperature-pressure (t-T-P) cycle and has been used in the fabrication of CeO_2 pellets [5]. While hot pressing of ceramic components is generally more complicated compared to cold press +

 DOI: 10.1051/e3sconf/20171605002

sinter, the additional application of pressure can greatly enhance sintering/densification.

Figure 3 is a photograph taken at UD during a CeO_2 hot pressing experiment showing the pressure ram (top) while it is applying a load to the top die punch. This hot press utilizes an induction furnace (coil structure in figure) which heats the graphite die which indirectly heats and sinters the CeO_2 powder within the die. After the completion of the hot press process during an experiment, the sintered CeO_2 pellet was then removed from the die for further investigations.

Figure 3. During the hot pressing of a CeO_2 pellet.

A number of hot press experiments using various t-T-P processing parameters have been performed in the fabrication of CeO_2 hot pressed pellets. Examination of the microstructure of a ceramic pellet can greatly enhance the understanding of the time, temperature, and pressure processing parameters. Several of the obtained CeO_2 hot pressed processed pellets have been cut, mounted, and polished for microstructural analysis. One example of an obtained hot pressed CeO_2 pellet's microstructure is shown in Figure 4.

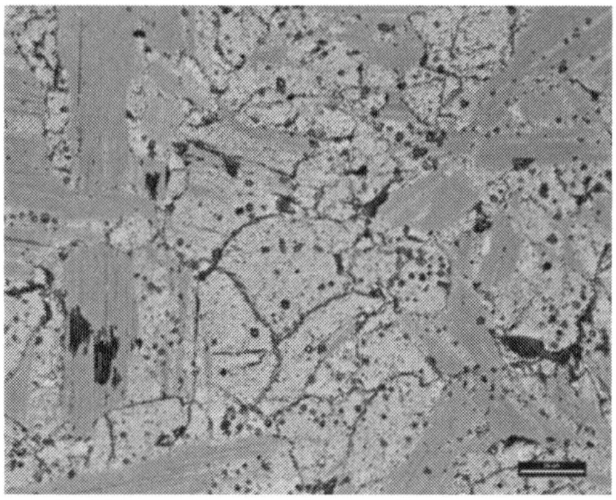

Figure 4. Example of a hot pressed CeO_2 pellet microstructure.

The hot pressed CeO_2 pellet microstructure shown in Figure 4 exhibits three distinct phases. The black areas are porosity which is desirable in RPS fuel pellets. The other two phases, one columnar and one nodular, have previously been determined to be different general sub-oxides of CeO_{2-x} where x is either ~0.1 or ~0.25 [5]. This reduction in the oxidation state of the CeO_2 when subjected to reducing processing conditions is analogous to what is expected to happen to $^{238}PuO_2$ and $^{241}AmO_x$ under similar conditions. However, the magnitude of the reduction is a function of the particular powder composition being processed.

2.3 NEUTRON RADIATION STUDIES ON RPS THERMOELECTRIC MATERIALS

Over the years the U.S. has utilized several different thermoelectric materials (including SiGe and TAGS-85) to convert some of the radioisotope decay heat into electrical power. In addition, new thermoelectric materials are currently being investigated for possible application in future RPSs. For example, UD has assisted the U.S. Department of Energy and NASA (Jet Propulsion Laboratory) in studying the effects of neutron irradiation on a potential new generation of skutterudite based RPS thermoelectric materials.

The RPS design being investigated by the University of Leicester would utilize bismuth telluride based thermoelectrics to convert some of the decay heat from the $^{241}AmO_x$ fuel form into electrical power for a spacecraft. In September 2015, the University of Leicester and the University of Dayton signed an Implementing Arrangement under the auspices of a Memorandum of Understanding between the U.K. Department of Energy and Climate Change and the U.S. Department of Energy. This Arrangement greatly facilities joint space nuclear power related projects to occur between the two universities. Following the signing of the Arrangement, UD and U of L discussed and arranged for neutron irradiation experiments on U of L supplied BiTe based thermoelectrics. Some of these experiments were performed during November 2015 at a university based reactor facility close to UD in which UD has recently performed similar types of neutron irradiation experiments.

In preparation for the radiation experiments, U of L thoroughly characterized various electrical properties of four BiTe thermoelectric modules which were then shipped to UD. Figure 5 shows four U of L supplied thermoelectric modules after being positioned on a test frame prior to the reactor run. The positioning of the modules within the test frame was carefully performed to ensure that all four of the modules were subjected to the same neutron fluence.

DOI: 10.1051/e3sconf/20171605002

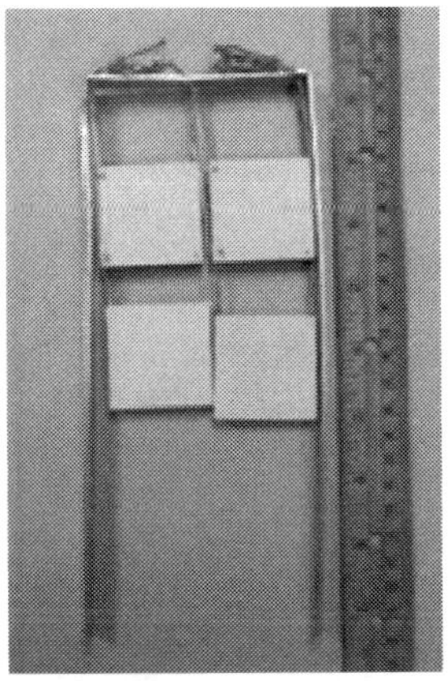

Figure 5. Four bismuth telluride based thermoelectric modules just prior to being loaded into the nuclear reactor.

Figure 6 shows the experimental rig being loaded into a tube located next to the core of the reactor prior to the experimental run. The reactor power used during the neutron radiation experiment was selected to correspond to neutron fluence levels anticipated during a future space mission. After completion of this particular experiment, the thermoelectric modules activated (as expected) and were stored at the reactor facility for a period of time. The modules were later shipped to UD where they were eventually packaged for shipment to U of L following all international guidelines. U of L is currently performing post-performance electrical and other tests on the irradiated modules and these results will be compared to the pre-irradiation test data.

Figure 6. Experimental rig with the four bismuth telleride thermoelectric modules being loaded into the reactor pool.

3. SUMMARY

Over the last several years there has been an ongoing technical interchange between the University of Leicester and the National Nuclear Laboratory in the U.K. and the University of Dayton in the U.S. related to various aspects of space radioisotope power systems (RPS). This technical relationship has been mutually beneficial in several materials based technical areas such as the ceramic processing of CeO_2 surrogate fuel pellets, and in the study of the effects of neutron radiation on potential RPS materials including thermoelectric modules. It is anticipated that these relationships will continue to develop and grow, which will help to ensure that internationally RPS technology continues to play a major role in deep space exploration of the solar system and beyond.

4. ACKNOWLEDGEMENTS

The University of Dayton conducted this research under U.S. Department of Energy/Office of Space & Defense Power Systems contract No. DE-NE0000422, their continuing support is greatly appreciated.

The University of Leicester and the National Nuclear Laboratory also greatly appreciate the support of the European Space Agency for funding the space nuclear power technology development studies.

5. REFERENCES

[1] Pattern: 03-065-5923 (CeO_2), Inorganic Crystal Structure Database (ICSD), National Institute of Standards and Technology (NIST), Gaithersburg MD.

[2] Pattern: 03-065-7999 (Ce_4O_7), Inorganic Crystal Structure Database (ICSD), National Institute of Standards and Technology (NIST), Gaithersburg MD.

[3] Pattern: 00-023-1048 (Ce_2O_3), Inorganic Crystal Structure Database (ICSD), National Institute of Standards and Technology (NIST), Gaithersburg MD.

[4] Bevan, D.J.M., "Ordered Intermediate Phases in the System CeO_2-Ce_2O_3," J. Inorganic and Nuclear Chemistry, v. 1, Pergamon Press Ltd., London, UK, page 49-59, 1955.

[5] Kramer, D.P., Pierson, T.M., Sjoblom, C.O., Grant, D.W., Goodrich, S.M., Barlay, C.D., and Whiting, C.E., "Hot Pressing of CeO_2 Ceramic Pellets," In Proc. 2015 Nuclear and Emerging Technologies for Space - NETS, Albuquerque, N.M.

THE SEPARATION OF ^{241}AM FROM AGED PLUTONIUM DIOXIDE FOR USE IN RADIOISOTOPE POWER SYSTEMS

M.J. Sarsfield [(1)], C. Campbell [(1)], C. Carrigan [(1)], M.J. Carrott [(1)], J-Y. Colle [(2)], D. Freis [(2)], C. Gregson [(1)], T. Griffiths [(1)], J. Holt [(1)], P. Lajarge [(2)], C. J. Maher [(1)], D. Manara [(2)], B. McLuckie [(1)], C. Mason [(1)], M. Naji [(2)], R.J. Taylor[(1)], T. Tinsley[(1)], J. Somers [(2)], K. Stephenson [(3)], J-F. Vigier [(2)]

[(1)]National Nuclear Laboratory, Sellafield, Seascale, Cumbria, UK, Email:mark.sarsfield@nnl.co.uk
[(2)]European Commission, Joint Research Centre, P.O. Box 2340, 76125 Karlsruhe, Germany.
[(3)]European Space Agency, ESTEC TEC-EPS, PO Box 299 - 2200 AG Noordwijk, The Netherlands.

ABSTRACT

Electrical power sources used in outer planet missions are a key enabling technology for data acquisition and communications. State–of-the-art power sources generate electricity from alpha decay of ^{238}Pu via thermoelectric conversion. However, production of ^{238}Pu requires specialist facilities including a nuclear reactor, a source of ^{237}Np for target irradiation and hotcells to chemically separate neptunium and plutonium within the irradiated targets. These specialist facilities are expensive to build and operate, so naturally, a more economical alternative is attractive to the industry. Within Europe ^{241}Am is considered a promising alternative heat source for radioisotope thermoelectric generators (RTGs) and radioisotope heating units (RHUs). As a daughter product of ^{241}Pu decay, ^{241}Am exists in 1000 kgs quantities within the UK civil plutonium stockpile.

A chemical separation process is required to extract the ^{241}Am in a pure form and this paper describes the AMPPEX process (Americium and Plutonium Purification by Extraction), successfully developed over the past five years to isolate ^{241}Am in high yield (> 99%) and to a high purity (> 99%).

The process starts by dissolving plutonium dioxide in nitric acid with the aid of a silver(II) catalyst, which is generated electrochemically. The solution is then conditioned and fed to a PUREX type solvent extraction process, where the plutonium is separated from the americium and silver. The plutonium is converted back to plutonium dioxide and the americium is fed forward to a second solvent extraction step. Here the americium is selectively extracted leaving the silver in the aqueous phase. The americium is stripped from the solvent and recovered from solution as americium oxalate, which is calcined to give americium dioxide as the final product.

This paper will describe the development of the separation process over a series of six solvent extraction separation trials using centrifugal contactors. The material produced (~ 4g ^{241}Am) was used to make ceramic pellets to establish the behaviour of americium oxide material under high temperature (1450°C) sintering conditions.

The chemical separation process is now demonstrated at concentrations expected on the full scale facility taking this process to TRL 4-5.

1. INTRODUCTION

Over the last few years there has been an interest by the European Space Agency (ESA) to develop power systems for space missions within Europe [1] and a study was initiated to explore the development of radioisotope powered systems suitable for producing heat and electrical power. A wide range of isotopes were considered and while ^{238}Pu was the most attractive from a performance perspective, the most accessible and economically viable isotope in Europe proved to 241-americium (^{241}Am) with a half-life of 432.7 years and a specific thermal power output of 0.1146 W_{th}/g.

The radioisotope ^{241}Am is a by-product of the civil nuclear energy industry. Over the last few decades France and the UK have provided a service to reprocess spent nuclear fuel from nuclear reactor sites, separating the uranium and plutonium and storing these products as the oxides ready for reuse in new fuel. Some of the plutonium has been stored for many decades and one of the isotopes (^{241}Pu), present within the plutonium, has beta decayed to the isotopically pure ^{241}Am. A programme of work was initiated by ESA in 2009 to develop a process to isolate ^{241}Am from plutonium dioxide (PuO_2) [2, 3], while separate parallel studies established how the isotope would be incorporated in to power systems [4]. This paper provides an overview of the radioisotope production aspects of the programme.

2. CHEMICAL SEPARATION

The outline process to separate americium from plutonium dioxide powder is shown in Fig. 1. The process involves dissolving the plutonium dioxide using a silver catalyst and then separating the plutonium from americium and silver by solvent extraction. Next the silver is removed in a second solvent extraction process and the purified americium nitrate solution is treated with oxalic acid to precipitate americium oxalate. This oxalate solid is heated and decomposed converting to an americium dioxide product ready for making ceramic pellets. The ultimate aim of the radioisotope production project is to demonstrate a feasible method of producing a pure product that can be made into a ceramic heat source.

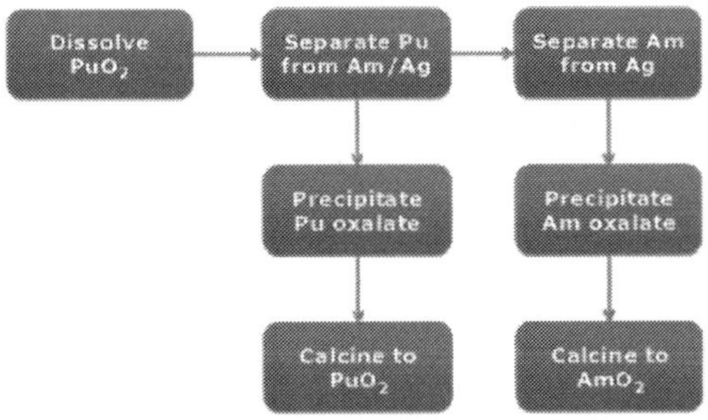

Figure 1. Outline of the chemical separation process.

3. DEVELOPMENT OF THE SEPARATION PROCESS

The development work required to underpin the concept design of the plant focused on the dissolution, solvent extraction and americium finishing processes as the most technically immature aspects that needed to be demonstrated. This demonstration was performed in three experimental runs. Run 1 demonstrated the proof of concept at low americium concentrations with relatively small total amounts of plutonium dioxide. Runs 2 and 3 demonstrated that the process could work at normal operating concentrations (Run 2) as well as beyond the bounding case concentrations (Run 3) of americium whilst maintaining the product quality and minimising wastes. In all of our solvent extraction development runs a 32 stage centrifugal contactor rig was used with counter-current solvent and aqueous feeds into banks of 2 or 4 contactors with 1cm rotors the operational details of which can be found elsewhere [5]. The development results are discussed in more detail in the following sections.

3.1. PLUTONIUM DIOXIDE DISSOLUTION

This process allowed the very insoluble plutonium dioxide powder to be dissolved ready for the next stage in the process.

In Run 1 of the development work a 100 mL electrochemical cell (Fig. 2) based on the CEPOD design [6] was used producing a range of plutonium concentrations from 5 – 200 gPu/L all within a 2 hour reaction period. The cell consisted of an outer platinum anode and titanium or tantalum cathode separated by a porous alumina membrane. Variables such as the current, anode surface area, anode design, stirrer speed, silver nitrate and nitric acid concentration were explored. The current efficiency was generally between 70–80%. In Runs 2 and 3 a 500 mL cell (Fig. 2) was employed that contained similar anode and cathode design but also had space for a UV/vis fibre optic probe that allowed progress of the reaction to be measured in-situ, thus reducing the manual operations required.

Figure 2. The two sizes of dissolution cell used; left) 100 mL cell; right 500 mL cell with UV/vis dip probe.

The 500 mL cell performed less efficiently (60-70% current utilised) but 80 g of plutonium dioxide was completely dissolved within 140 minutes using 0.1 mol/L silver in nitric acid. During dissolution the UV/vis signal for silver(II) (600 nm; shoulder) is reduced while PuO_2 is dissolving and the Pu(VI) peak at 830 nm grows until the dissolution is complete. Once complete, the silver(II) concentration rapidly increases to a steady state providing an excellent indication of a completed reaction. Next NOx gas is passed through the liquor at 50°C for around 30 minutes and the UV/vis spectrum is monitored until the Pu(VI) signal at 830 nm is replaced by the characteristic Pu(IV) spectrum (Fig 3). Once conditioned, the solution is filtered and analysed for nitric acid and plutonium concentrations. The liquor is then made up to ~100 gPu/L and 4.5 mol/L nitric acid ready to be fed to the solvent extraction stage. It was at this point additional uranium and neptunium was added to increase their concentrations to that representative of fully aged plutonium dioxide. Neptunium is generated from the radioactive decay of americium and uranium from the decay of plutonium isotopes.

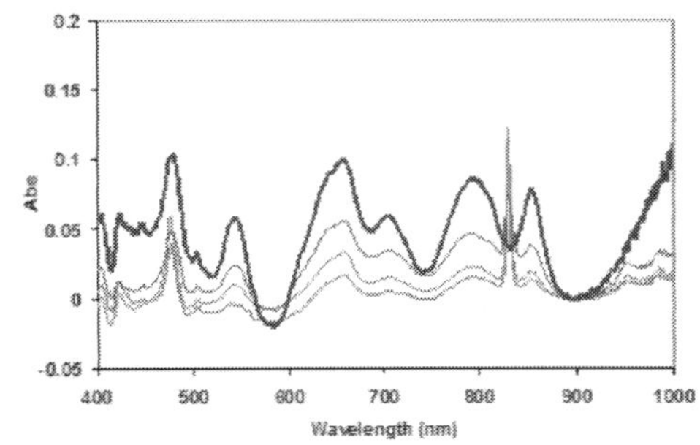

Figure 3. The UV/vis spectrum of the plutonium nitrate solution during NOx conditioning as it changes from Pu(VI) to Pu(IV).

3.2 SEPARATION OF PLUTONIUM FROM

AMERICIUM

In this part of the process a flowsheet was designed based on well established PUREX technology. In Run 1 of the development work the objective was to assess how all of the main elements behaved during extraction into 30% TBP/OK solvent followed by stripping of the plutonium using a hydroxylamine nitrate solution to reduce the extractable Pu(IV) to the poorly extracted Pu(III). At this stage it is important to ensure as much of the plutonium, neptunium and uranium as possible is removed from the americium stream. This is to ensure a pure product because the next solvent extraction stage does not provide such a good decontamination factor for these elements. Another concern was whether the acidity carried over within the solvent would slow the kinetics of plutonium reduction [7].

Once proof of concept was established the amounts of americium in the feed were increased (Tab. 1). The americium concentration in Run 3 was increased to 5.2 g/L by adding the americium produced in Run 2 (~1.7 gAm) to the dissolved aged plutonium dioxide in Run 3.

Table 1. Feed concentrations to the first solvent extraction process

	Full scale process	Run 1	Run 2	Run 3
	g/L	g/L	g/L	g/L
Np	0.27	0.23	0.10	0.26
U	0.39	0.33	0.58	0.42
Pu	95	89	98	97.7
Am	4.80	0.92	2.94	5.20
Ag	10.79	10.79	7.90	9.58
Volume (mL)	3000	500	633	727
Nitric acid (mol/L)	3.5	3.5	4.5	4.2

Plutonium was effectively extracted in Run 1 with only 0.008 gPu/L following the americium product (Tab. 2). The stripping was successfully demonstrated at ambient glovebox temperatures (20-30°C) with > 99.9% of the plutonium recovered. After modelling the plutonium and nitric acid behaviour in the extract and stripping stages of the flowsheet, the feed nitric acid concentration was increased (Tab. 1) to aid extraction and increase the acidity of the americium/silver product. These improvements gave a reduction in the amount of impurities in the americium product (Tab. 2) and ultimately reduced the amount of radioactivity in the waste streams. With an increase in feed acidity the amount of plutonium that followed the americium dropped by nearly two orders of magnitude by %w/w (Tab. 2). The plutonium was sufficiently free of silver to meet the product specifications required, providing a fresh plutonium product free of americium. Most of the uranium was extracted, 90% of which remained in the solvent during the stripping stage. As expected [8], the neptunium was split between the plutonium product (55%) and the americium product (41%) with the remainder left in the solvent.

Table 2. Relative concentrations of key elements in the americium/silver product.

	Run 1 %w/w	Run 2 %w/w	Run 3 %w/w
Np	9.50%	1.04%	2.13%
U	0.10%	0.03%	0.03%
Pu	1.70%	0.05%	0.02%
Am	89%	98.90%	97.8%

The main impurity in the americium product, other than silver (~5g/L), is neptunium. The amount of neptunium extracted with the plutonium could be increased further by increasing the acidity and temperature and controlling the nitrous acid concentration to generate more extractable Np(VI) [8], however it was decided that the decontamination factors downstream were sufficient to not warrant further development in this area.

3.3 SEPARATION OF AMERICIUM FROM SILVER

The strategy used for separating americium from silver was solvent extraction using a chelating nitrogen donor ligand TODGA (Fig. 4) that is known to extract americium very well but tends to form a third phase quite readily in hydrocarbon diluents [9]. Third phase formation can cause difficulties during the extraction process and needs to be avoided. An additive, DHOA (Fig. 4), was used to mitigate this third phase effect and allow sufficient americium loading of the solvent. The solvent formulation was optimised to provide good extraction at 2-3 mol/L nitric acid (D_{Am} ~ 50) while allowing efficient stripping at low (~ 0.01 mol/L) nitric acid concentration ($D_{Am} < 0.1$).

During the development of this part of the flowsheet one objective was to minimise the volume of solvent used and increase the concentration of americium in the final product, from 0.4 gAm/L in Run 1 to 6 gAm/L in Run 3. This enabled the product americium nitrate solution could be used directly in the oxalate precipitation stage, avoiding an evaporation step and reducing process costs.

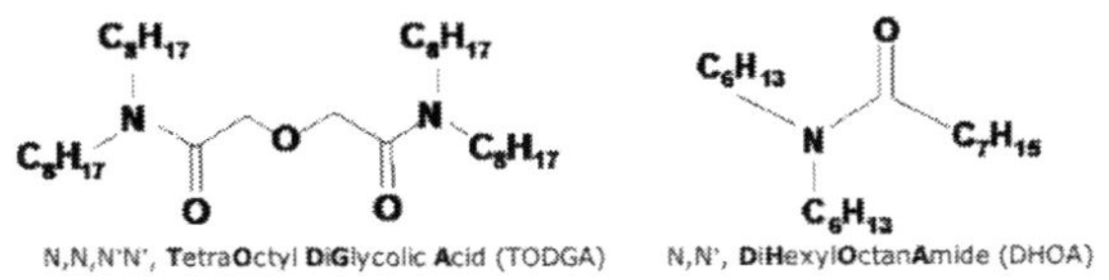

Figure 4. Reagents used in the solvent formulation used for the americium silver separation

Analysis of the products from this second solvent extraction stage demonstrated that extraction of americium is excellent and the amount that is lost to the silver stream is low, typically 1 x 10^{-6} g/L, which is approaching the practical levels of unavoidable cross contamination within our centrifugal contactor rig. The decontamination of silver from the americium product was excellent with typically less than 0.1 mg/L in the americium product (Tab. 3). The main contaminant is again neptunium but this is significantly reduced because most of what is extracted remains within the solvent. Only around 15% of the original neptunium with a decontamination factor of 6.7 to 11.6 times less neptunium in the americium product compared to the feed solution. The final purity of the americium as a percentage of the key elements analysed is > 99% (Tab. 3).

Table 3. Relative concentrations of key elements in the americium product stream after the americium/silver separation

	Run 1 %w/w	**Run 2** %w/w	**Run 3** %w/w
Np	1.14%	0.21%	0.34%
U	0.10%	0.02%	0.03%
Pu	0.53%	0.11%	0.04%
Am	98.21%	99.63%	99.59%
Ag	0.02%	0.03%	0.001%

4. AMERICIUM OXALATE PRECIPITATION

In order to obtain an americium oxide powder the americium nitrate solution in Tab. 3 needs to be converted to a suitable. Americium oxalate is a good precursor to the formation of an americium oxide because the powder formed is easily filtered and readily decomposes to the oxide. A system was developed for continuous precipitation using a vessel with a spout that overflowed onto a simple filtration bed (Fig. 5). The americium nitrate and oxalic acid stocks were pumped into the precipitator at a set flowrate that gave a steady state average residence time within the vessel of around 14 mins. This design helped to minimise the extremity dose to operators with very little manual intervention required. Around 3.5 g of americium were recovered in one run and the pale yellow precipitate (Fig. 6) was easily collected from the filter membrane after washing the solid free of oxalic and nitric acids with demineralised water. The americium nitrate feed contains 0.25 – 0.3 mol/L nitric acid and the amount of americium lost to the filtrate is typically 1 mgAm/L giving an americium recovery of > 99.9%.

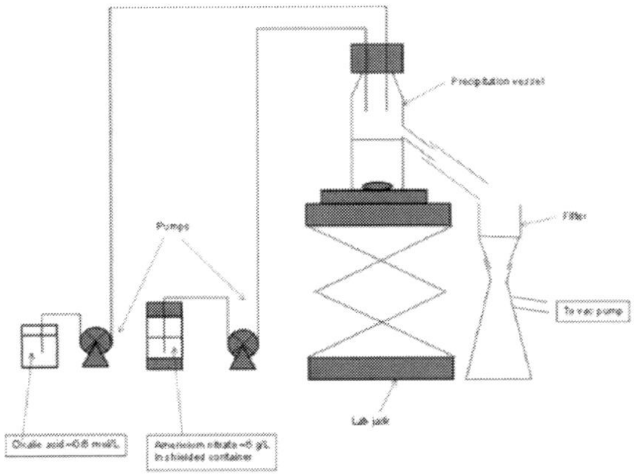

Figure 5. Experimental setup for the precipitation of americium oxalate.

Another benefit of the oxalate precipitation step is an additional decontamination factor for neptunium of around 5. The solubility of americium in the filtrate is consistent with published literature values [10] at ambient temperatures (23-25°C), the final nitric acid (0.23 mol/L) and oxalic acid (0.1 mol/L) concentrations. Increasing the precipitation temperature to 60°C leads to an increase in americium losses to around 10 mg/L, which is also acceptable from a recovery viewpoint and provides the option of influencing the americium oxalate particle size distribution and morphology. It is expected, from the behaviour of plutonium oxalate, that the size and morphology of the americium oxide will be similar to the parent oxalate, which may be important when making ceramic heat source pellets.

5. AMERICIUM OXIDE FORMATION

The americium oxalate needs to be converted from the americium oxalate to the oxide, which will be the feed material for making the ceramic heat source. A comprehensive study on the behaviour of cerium oxalate formation and conversion to the oxide was performed to establish how the change in powder properties changed as the precipitation and calcination conditions were altered. These surrogate oxide powders were then pressed into pellets and sintered at temperatures up to 1600°C. The objective of this study was to establish whether by altering the precipitation and calcination conditions optimal powder properties could be established that would provide suitable ceramic material without the need for milling of the powder. The details of this study will be published elsewhere; suffice to say that the optimal conditions decided were a precipitation at 25°C and decomposition of the oxalate at 650°C. This turned the oxalate from pale yellow to a black oxide powder (Fig. 6).

Figure 6. Americium oxalate before (left) and after (right) decomposing to the oxide.

For each of the development runs a sample of the oxide was analysed to establish the concentration of the key elements as impurities in the sample. An example of the final purity of the americium with respect to the other key elements analysed is compared against the americium nitrate product for development Run 3 (Tab. 4).

The analytical results demonstrate that the relative amount of neptunium has decreased, consistent with the analysis of the filtrate liquors from the americium oxalate precipitation experiments. The purity of the americium appears to be very good (> 99%), however, these results are preliminary and do not establish whether there are any volatile species such as carbon remaining within the material.

Table 4. Concentrations of key elements in the americium nitrate solution and americium dioxide for development Run 3.

	Americium nitrate	Americium dioxide
	%w/w	%w/w
Np	0.34%	0.07%
U	0.03%	0.03%
Pu	0.04%	0.02%
Am	99.59%	99.88%
Ag	0.001%	0.001%

After heat treating this material to 1450°C in a reducing atmosphere (5%H_2/Argon) for 4 hours followed by oxidising at 800°C for 4 hours the resulting AmO_2 material was tested for americium assay by calorimetry to determine the power output per gram of material. The results showed an assay of 100.4% ±0.43 % confirming the high purity of the sample.

6. AMERICIUM OXIDE PELLET FORMATION

The americium oxide powder was then used in a small number of experiments to produce a sintered ceramic pellet. The powder morphology is consistent with particle size and shape of surrogate cerium dioxide material produced using identical oxalate precipitation methods. In our studies on cerium dioxide, made from the decomposition of cerium oxalate, particle morphology is retained upon decomposition to the oxide. The similarity of the particle shapes (Fig. 7) would suggest that this also occurs when americium oxalate is decomposed to americium dioxide, suggesting that cerium oxalate and oxide is a good surrogate for americium oxide powders.

Figure 7. Comparison of cerium oxide (left) and americium oxide (right) powder morphology. Powders produced under the same oxalate precipitation conditions of continuous precipitation at 25°C.

When making ceramic pellets the chemical behaviour of americium oxides needs consideration. It is well documented that americium dioxide loses oxygen when heated to beyond 800°C to form an oxide with a lower oxygen to americium ratio AmO_{2-x} (where x = 0 - 0.5) and can occur at even lower temperatures in a reducing atmosphere [11, 12]. This is an important safety consideration because in launch accident scenarios it is possible for the heat source material to reach high temperatures (< 1800°C). The loss of oxygen at relatively low temperatures means the cladding that encapsulates the heat source needs to be designed to deal with such gas losses. If it is not possible to manage such gas releases safely then the use of AmO_2 may be excluded. It is then also important to understand the conditions under which ceramic Am_2O_3 can be made and stored.

There is also the potential for americium to volatilise either as AmO_2(g), AmO(g) or Am(g) and so tests were performed to establish how the oxide stoichiometry changes under different sintering atmospheres and whether significant mass loss of material occurs.

In order to explore some of these expected difficulties a series of pelleting trials were performed within shielded hot cells where the powder was manipulated using remotely operated mechanical arms to minimise the radiation dose to the operators. Around 400 mg of as received americium dioxide powder was pressed into a "green pellet" that and then sintered in a high temperature furnace. The pellets were sintered at 1450°C under varying atmospheres (Tab. 5) to give a sintered pellet. The sintered pellets produced were analysed for geometric dimensions and mass to calculate the density

 DOI: 10.1051/e3sconf/20171605003

of the pellets. The densities and % of theoretical Density (TD) are shown in Tab. 6.

No detectable levels of americium were lost from any of the sintering trials all of which were performed at 1450°C, consistent with previous literature reports [13, 14].

Table 5 Sintering atmospheres and pressing pressure of the americium oxide pellets made.

Pellet No.	Sintering Atmosphere	Die diameter (mm)	Press. (MPa)
1	Ar/H_2/2000 ppm H_2O	5.40	480
2	Ar/H_2/2000 ppm H_2O	5.40	480
3	Ar/H_2	5.40	480
4	Air	5.40	480
5	Argon	5.40	480
6	Ar/H_2	5.40	480
7	Ar/H_2	5.25	508
8	Ar/H_2	5.25	323

In most cases, where a reducing atmosphere is employed, sufficiently high pellet densities were obtained (~ 90% TD), although cracking of the pellet around the circumference (Fig. 8) occurred (with the exception of pellet 3). While it is not clear why cracking occurs there are a number of possible contributing factors:

- The mechanism may be end-capping where the pellet expands as it comes out of the die as the stress is relieved there is radial expansion. There are many factors that influence the likelihood of end-capping [15] such as powder characteristics, pressing pressure, tapering of the die, material of fabrication of the die and lubrication of the die. Given the small size of the discs this mechanism seems less likely.
- The loss of oxygen upon sintering of the sample could cause high porosity leading to weakened pellets and linked pores leading to cracks.
- The changes in phases from face centred cubic (fcc) AmO_2 phase to body centred cubic (bcc) Am_2O_{3+y} phase to hexagonal Am_2O_3 phase may result in stresses within the material as the pellet heats, loses oxygen and cools down [12].
- Thermal expansion of the hexagonal Am_2O_3 phase is higher than fcc AmO_2 phase and may result in stresses during thermal cycling.
- The volatile impurity found within the material could also be causing high porosity and cracks as it is released during pre-treatment.

Table 6. Density and theoretical density of the americium oxide pellets made before (green) and after (sintered) sintering at 1450°C.

Pellet No.	Green Pellet		Sintered Pellet		Am-O form
	Density (g/cm^3)	TD* (%)	Density (g/cm^3)	TD* (%)	
1	6.61	56.2	9.36	80	Am_2O_3
2	6.85	58.2	8.5	72	Am_2O_3
3	6.78	57.6	10.71	91	Am_2O_3
4	6.69	57.3	9.58	82	AmO_2
5	6.62	56.2	N/A	N/A	Am_2O_3
6	6.56	55.7	N/A	N/A	Am_2O_3
7	6.63	56.3	10.71	91	Am_2O_3
8	5.89	50.0	10.57	90	Am_2O_3

*TD for AmO_2 = 11.68 g/cm^3; for Am_2O_3 = 11.77 g/cm^3 [16].

Pre-treatment of the green pellet performed at 1050°C for 10 hours under the sintering atmosphere to drive off any impurities did not prevent this cracking.

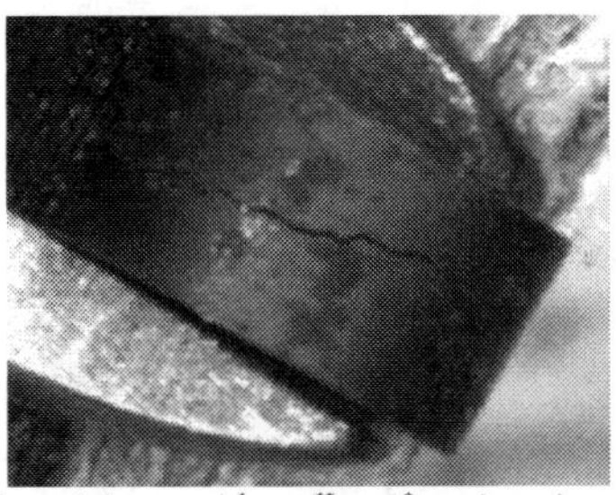

Figure 8. Americium oxide pellet after sintering showing cracking around the circumference (Pellet 7).

When forming pellets 1-6 a small sample of americium oxide powder was sintered by the side of the pellet and then analysed to determine the crystal form of the material post sintering. The powders were analysed using Raman spectroscopy and powder X-Ray diffraction (PXRD).

The reduced form of americium oxide tends to form and this is identified by the A-type hexagonal Am_2O_3 seen in the PXRD diffractogram (Fig. 9) and the characteristic Raman spectrum (Fig. 10). Even without a reducing atmosphere (Argon, only pellet 5) complete reduction occurs.

An AmO_2 pellet could only be produced if sintering is performed in air and the temperature is held at 800°C for

 DOI: 10.1051/e3sconf/20171605003

a period to ensure a stoichiometry of $AmO_{2.00}$.

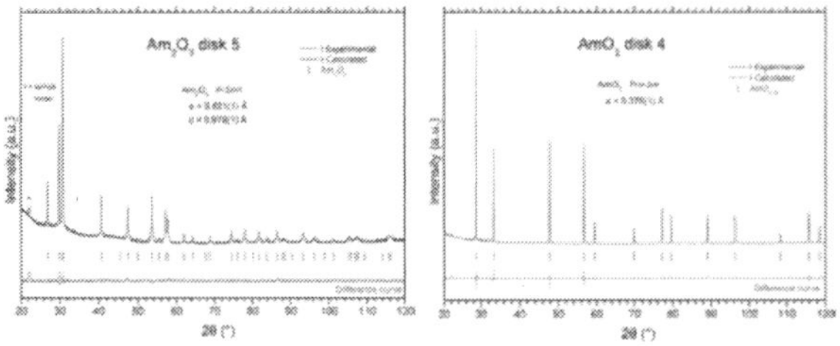

Figure 9. Examples of the PXRD results for a reduced Am_2O_3 (left) and oxidised AmO_2 (right) pellet.

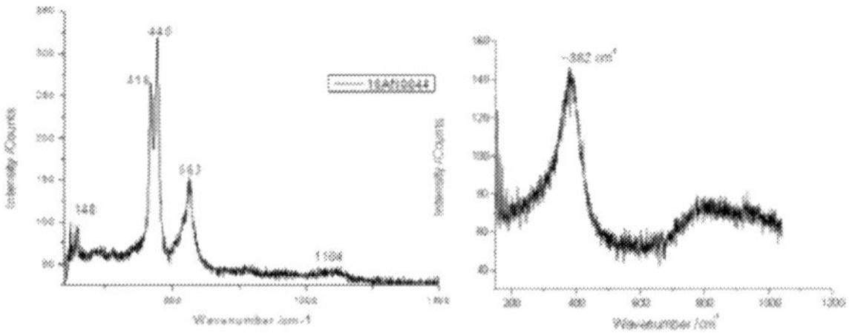

Figure 10. Examples of the Raman results for a reduced Am_2O_3 (left) and oxidised AmO_2 (right) pellet.

Note the Raman spectrum of the hexagonal A-type Am_2O_3 is the first time a clean spectrum with low contamination from other phases has been reported [17].

The purity of the Am_2O_3 PXRD samples and lack of re-oxidation in the manipulation of the sample between analytical facilities demonstrates that the reduced form of americium oxide can be handled albeit in environments that are nominally inert atmospheres (< 1% O_2). This provides confidence that this material can be controlled to avoid oxidation using standard inertion systems if this was to be the final chemical form.

7. CONCLUSIONS

The separation process to extract americium from aged plutonium dioxide is mature and demonstrated at actual americium concentration levels with no detriment to the recovery (> 99%) or purity (> 99%) of the americium produced. This takes the TRL level from 3 to 4-5. The requirements for building a full scale facility and dealing with the radioactive waste streams generated are now fully understood. Investment is now required to build the full scale facility to establish a throughput high enough to support the ESA mission.

From these preliminary pelleting trials it is clear that the reduced Am_2O_3 form is readily generated even in the absence of a reducing hydrogen atmosphere, consistent with reported thermodynamic data. Pellets containing Am_2O_3 are sufficiently densified but further development work on the sintering conditions and how they relate to the particle morphology remains. If Am_2O_3 is the preferred chemical form then the next development stage will also need to assess chemical interactions with potential cladding materials. Further studies are aimed at producing a Radioisotope Heat Source prototype using 10s g of americium oxide material.

8. ACKNOWLEDGEMENTS

The authors would like to thank the European Space Agency for financial support and the Nuclear Decommissioning Authority and Sellafield Ltd. for providing the aged plutonium dioxide.

9. REFERENCES

1. Summerer, L. & Stephenson, K. (2011). Proc. IMechE G: J. Aero. Eng. 225, 129–143.
2. Cordingley, L. Rice, T. Sarsfield, M. J. Stephenson, K. & Tinsley T. (2011). Paper s27_4cordi, Proc. '9th European Space Power Conference', Saint Raphaël, France, 6–10 June, (ESA SP-690, October, 2011).
3. Baker, S.R. Bell, K.J. Brown, J. Carrigan, C. Carrott, M.J. Gregson, C. Clough, M. Maher, C.J. Mason, C. Rhodes, C.J. Rice, T.G. Sarsfield, M.J. Stephenson, K. Taylor, R.J. Tinsley, T.P. Woodhead, D.A. & Wiss, T. (2014). Paper nps5_sarsfield, Proc. '10th European Space Power Conference', Noordwijkerhout, The Netherlands, 15–17 April, (ESA SP-719, May 2014).
4. Williams, H. R. Ambrosi, R. M. Bannister, N. P. Samara-Ratna, & P. Sykes, (2012). J. Int. J. Energy Res., 36, 1192.
5. Carrott, M.J. Bell, K. Brown, J. Geist, A. Gregson, C.R. Heres, X. Maher, C.J. Malmbeck, R. Mason, C. Modolo, G. Muellich, U. Sarsfield, M.J. Wilden, A. & Taylor R.J., (2014). Solv. Extr. Ion Exch. 32, 447.
6. Ryan, J.L. Bray, L.A. & Boldt, A.L. (1987). "Dissolution of PuO_2 or NpO_2 using electrolytically regenerated reagents", United States Patent 4686019.
7. Koltunov, V.S.& Zhuravleva, G.I. (1978). Radiokhim. 20, 94.
8. Taylor, R.J. Gregson, C.R. Carrott, M.J. Mason, C. & Sarsfield, M.J. (2013). Solv. Extr. Ion Exch. 31, 442.
9. Sasaki, Y. Sugo, Y. Suzuki, S. & Kimura, T. (2005). Anal. Chim. Acta, 543, 31.
10. Burney, G.A. & Porter, J.A. (1967). Inorg. Nucl. Chem. Letts., **3**, 79.

11. Chikalla, T.D. & Eyring, L. (1968). J. Inorg. Nucl. Chem., 30, 133.

12. Lebreton, F. Belin, R.C. Delahaye, T. & Blanchart, P. (2012). J. Solid State Chem., 196, 217.

13. Nishi, T. Takano, M. Itoh, A. Akabori, M. Arai, Y. Minato, K. & Numata, M. (2008). J. Nucl. Mats., 373, 295.

14. Nishi, T. Itoh, A. Ichise, K. & Arai, Y. (2011). J. Nucl. Mats., 414, 109.

15. Balakrishna, P. Chakraborthy, K.P. & Singh, A. (1996). Ind. J. Engin. Mats. Sci., 3, 196.

16. D.R. Lide, "Handbook of Chemistry and Physics", 81st Ed., pp 4-40, (2000-2001).

17. Horlait, D. Caraballo, R. Lebreton, F. Jégou, C. Roussel, P. & Delahaye, T. (2014). J. Solid State Chem., 217, 159.

INVESTIGATION ON THE USE OF AMERICIUM OXIDE FOR SPACE POWER SOURCES: RADIATION DAMAGE STUDIES

T. Wiss[(1)], D. Freis[(1)], J.-C. Griveau[(1)], K. Popa[(1)], J.-F. Vigier[(1)], J. Somers[(1)], E. D'Agata[(2)], R.J.M. Konings[(1)], O. Benes[(1)], J.-Y. Colle[(1)], O. Dieste[(1)]

[(1)]European Commission, Joint Research Centre, Directorate G – Nuclear Safety & Security, P.O. box 2340, 76125 Karlsruhe, Germany, Email:thierry.wiss@ec.europa.eu
[(2)] European Commission, Joint Research Centre, Directorate G – Nuclear Safety & Security, P.O. Box 2, NL-1755 ZG Petten, The Netherlands, Email:elio.dagata@ec.europa.eu

ABSTRACT

Within Europe ^{241}Am is a feasible alternative to ^{238}Pu that can provide a heat source for radioisotope thermoelectric generators (RTGs) and radioisotope heating units (RHUs) to be used in Electrical power sources used in outer planet missions. In the EPSO-SPACE* project developed as an exploratory research at JRC, a new type of nuclear power source, to provide heat and electricity for deep space missions is considered. It will consist of a ^{241}Am based compound with aerogel thermal insulation and thermocouples in close contact with the heat source. The assembly is inside an iridium encapsulation to withstand all relevant accident scenarios. The design will be modular and combinable for required electrical output.
Several candidate Americium compounds will be investigated for chemical stability at high temperature and for self-irradiation damage. New thermo-electric converter materials containing actinides will be assessed, and a robust encapsulation designed. Safety analyses will be performed including launch explosion and re-entry accidents. The research will conclude in a conceptual design of a prototype power source. In the first part of this study, americium dioxide will be considered from the point of view of its chemical durability and of its behavior against radiation damage and helium formation, two aspects to be carefully investigated due to the high alpha-activity of the americium. Transmission electron Microscopy (TEM) and helium thermal desorption spectrometry (TDS) experiments will be described and results on aged (more than 30 years) AmO_2 reported. Some comparison with $^{238}PuO_2$ based RTG's will be discussed.

1. INTRODUCTION

In the space conquest century, several satellites are to be sent to the outer space. In order to keep instruments running over several decades, with no direct physical contact to Earth, these space crafts need long-time batteries [1, 2]. ^{238}Pu has been the isotope of choice for five decades. O'Brien et al. [3] have made an detailed review on the different radioisotopes to be used in as alternatives to ^{238}Pu that is traditionally used in radioisotope thermoelectric generators and heating units. The main problem of ^{238}Pu is the availability and price. In addition, having a half-life of 87.5y, after several years of operation, the power of the plutonium-based batteries diminishes. Hence, the electronics need to be far more complicated than usual.
As alternative, it has been proposed to use ^{241}Am. To its credit, the reserves of this actinide at the moment are large, and more can be produced.
One possibility of obtaining ^{241}Am is by reprocessing spent fuel, which contains about 12% ^{241}Pu. (from the total Pu inventory at discharge). Because it spontaneously converts to ^{241}Am, ^{241}Pu can be extracted and may be used to generate further ^{241}Am. However, this process is slow: Half of the original amount of ^{241}Pu decays to ^{241}Am in about 15 years, and the ^{241}Am amount reaches a maximum after 70y.
As ^{241}Am has a significantly longer half-life than ^{238}Pu (432y vs. 87y), the power of the batteries would be much more stable, simplifying the design of the electronics. As withdraw, the peak-power would be lower (114.7mW/g vs. 390mW/g), and the weight needed much higher than usual batteries, but the lower price on the precursor material makes it more suitable.
To be able to design suitable space batteries, more studies on $^{241}AmO_2$ need to be made. To this matter, a comparison between its microstructure evolution to the one of $^{238}PuO_2$ can be of use.
The chemical and physical consequences of self-radiation in actinide compounds are rather complex. It is often difficult to deduce the true physicochemical properties of substances when superimposed by radioactive decays and their effects.
Alpha-decays in oxides typically results in the formation of points defects, mainly from the recoil nucleus and to the formation of a helium atom when the alpha-particle comes to rest. The point defects can also precipitate into larger extended defects (e.g. dislocation loops) and the helium atoms into bubbles. To observe these types of microstructural features the ideal probe is a Transmission Electron Microscope that enables with modern instrument to easily visualize objects in the nanometer (even lower) scale. In addition to conventional imaging, elemental (EDX) and spectroscopic analyses (EELS) proved to be of additional help to characterize the change in

microstructure due to alpha-decay damage [4].
An effect of self-radiation and radioactive decay is the change in impurity content of the compound, due to the ingrowing daughter(s). This change influences the electrical resistivity as well as the lattice parameters. Other effects, which can be regarded as primary effects of the decay process are: the ionization, electronic excitation and atomic displacements. The relative magnitude of each of these phenomena depends on the type and energy of the particle and on the substances.
To our Knowledge, the effect of ^{241}Am self-irradiation has been previously studied for AmO_2, however, the majority of the work conducted has been limited to the effect of self-irradiation on the lattice parameter [5, 6] and thermal conductivity [7]. Much more work was performed on $^{238}PuO_2$ since this compounds has already been used in RTG's including batteries for space probes and even in pacemakers [8, 9].

2. EXPERIMENTAL

The original $^{238}PuO_2$ from a ^{242}Cm RTG fabricated in 1968 with 450 mg of ^{242}Cm-oxyde powder delivering a total thermal power of 45 Watts was retrieved. The curium from the RTG was isolated from additional americium and cerium content by successive selective dissolution of the actinides and precipitation of an oxidized-americium carbonate. After 4.5 half-lives ($T_{1/2}$ = 162 d) the ^{238}Pu was isolated and converted into an oxalate before being calcined into an oxide powder. The isotopic analysis has revealed enrichment by 98.936 % of the isotope-238.
Two AmO_2 samples were retrieved from the JRC – ITU archives after several years of storage [10]. No particular precautions regarding the storage were taken-samples were stored in air and at room temperature.
The investigations have been performed on a FEI Tecnai G^2 TEM, equipped with a GATAN Tridiem GIF camera, an EDAX energy dispersive X-ray analysis system, a tomography suite and a HAADF (High-Angle Annular Dark Field) detector for the STEM (Scanning Transmission Electron Microscopy) imaging. The TEM has been adapted for the examination of highly active or irradiated nuclear materials thanks to a flange that has been inserted in the octagon hosting the objective lenses, and a glove box mounted on this flange around the compustage [11]. Thus, the sample can be moved from the glove box where it was prepared to the microscope making use of a La Calhène DPTE® system.

3. TEM INVESTIGATION

Fig. 1 shows a bright field transmission electron microscope (TEM) image of a sample of AmO_2 having cumulated a dose of $1.4x10^{20}$ $\alpha \cdot g^{-1}$ after 36 years of storage. A population of very small intragranular nanometer sized bubbles was observed. In this area the bubbles are homogeneously distributed and their concentration amounts around $5x10^{22}$ m^{-3}. In some other areas as reported in [4] some larger bubbles of several nanometers were founds but with a rather low concentration. Very few extended defects were found even at the damage level of 36 dpa which highlights an excellent resistance to alpha-damage build-up as also shown in [10].

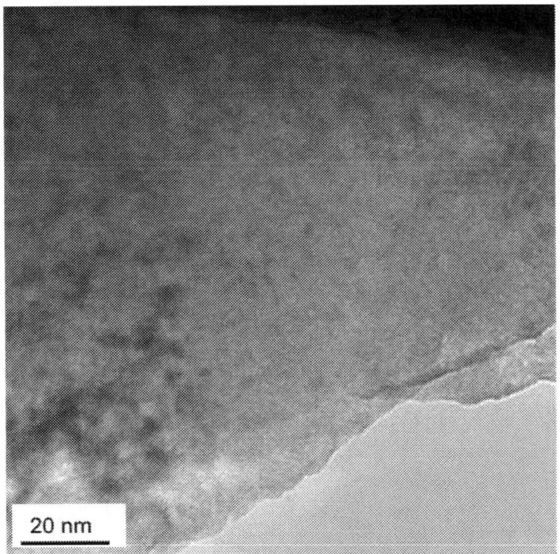

Figure 1. TEM bright field image of $^{241}AmO_2$ after cumulating $1.4x10^{20}$ $\alpha.g^{-1}$ and showing a high concentration of ~ $5x10^{22}$ m^{-3} sub-nanometric helium bubbles (tinny white spots).

The microstructure of the $^{238}PuO_2$ has also been investigate as shown in the TEM bright field micrograph of Fig. 2. Helium bubbles of less than a nanometer were observed. Their concentration was measured to be $5x10^{23}$ m^{-3} hence one order of magnitude higher than for the $^{241}AmO_2$ sample for a cumulated dose 3 times higher than in the later sample. However, the size at this resolution is extremely difficult to assess very precisely and in addition to an increase in concentration it needs to e considered that also the size increases with helium formation. The AmO2 samples have also shown larger bubbles and even if not at a high concentration potentially containing a non-negligible part of the total inventory. More work needs to be performed to assess the retained fraction of helium by thermal desorption spectrometry as performed in the $^{238}PuO_2$ sample for example [12].

 DOI: 10.1051/e3sconf/20171605004

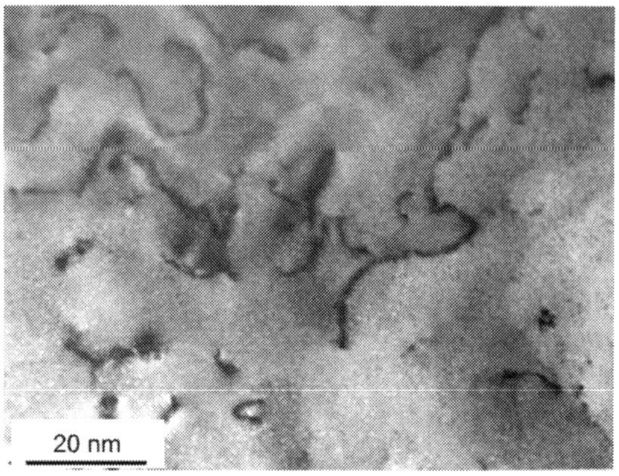

Figure 2. TEM bright field image of $^{238}PuO_2$ after cumulating $3.67x10^{20}\alpha.g^{-1}$ and showing a high concentration of ~ $5x10^{23}$ m^{-3} nanometric helium bubbles.

4. XRD ANALYSES

In the actinide oxides there is a homogeneous distribution of alpha-damage leading to a rapid saturation of the lattice parameter [13-15]. The graph in Fig. 3 shows that the saturation of the lattice parameter for the lattice parameter for the $^{238}PuO_2$ and the $^{241}AmO_2$ is just below 0.3 %. The self-radiation effects on the lattice parameter of PuO_x have been studied on $^{238}PuO_x$ samples [13, 16]. Lattice parameter increase was determined for samples stored at ambient temperature in oxygen or in vacuum, for samples stored at liquid nitrogen temperature and for samples containing 5% Fe (III) and kept under pure oxygen at room temperature.

The ingrowth of damage is frequently described by a simple exponential function. The lattice parameter change $\Delta a/a_0$ can be described [16] by:

$$\Delta a/a_0 = A(1-e^{-B\lambda t}) \quad (1)$$

Where

λ: decay constant
t: time
A: value at saturation
B: rate constant for simultaneous annealing of defects.

The value A can be influenced by the history of the specimen used [15, 17]. The effects due to atomic displacements are rather complex functions describing the relative sink strengths of a given specimen for interstitial and vacancies. The formation of extended defects like dislocation loops [18] or of helium bubbles [19, 20] can provide additional sinks. Local maximum and minimum values observed for the lattice parameter as a function of accumulated damage before reaching a saturation level have been attributed to the trapping of interstitials into (helium) sinks. At saturation, the recombination rate is equal to the production rate of Frenkel defects. In the experiments of Chikalla and Turcotte [14] saturation is observed for lattice expansion of ~0.3 %. This value is in substantial agreement with the measurements of Noe and Fuger [13] in various systems where a lattice expansion at saturation of 0.28 % was measured.

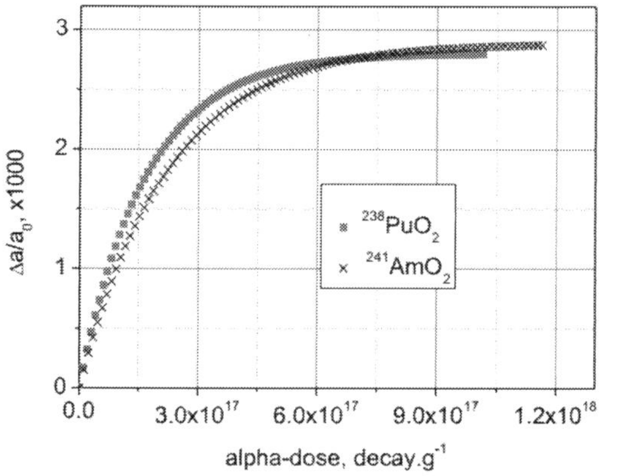

Figure 3. relative lattice parameter evolution as function of accumulated alpha-damage $^{241}AmO_2$ stored in vacuum from ref. [5] and $^{238}PuO_2$ stored in vacuum from ref.[14] respectively.

It is therefore to be expected that a lattice swelling of about 2% can be reach at saturation, hence after 100 days for the $^{241}AmO_2$ and after less than a month for the $^{238}PuO_2$.

Larger swelling, however, has to be accounted for due to the radiogenic helium formation. From the helium bubble distribution in the $^{238}PuO_2$ sample analysed by TEM it was calculated that a volumetric swelling of 3% could be reached [12].

5. CONCLUSION

Despite the high α-irradiation dose in AmO_2, the samples studied remained crystalline. The microstructure examination of this material revealed that only few extended defects are formed. Given the very high alpha-dose the material has experienced it is indicating a rather good resistance to radiation damage build-up. On the other side there is a background of sub-nanometric bubbles that can be observed in this material. The lattice parameter evolution shows saturation at a value of 0.28 % but the formation of bubbles at very large cumulated dose should be accounted for when determining the total swelling of such compounds. Such parameters are of importance when designing batteries envelops for example. In comparison with $^{238}PuO_2$ which was substantially more studied, $^{241}AmO_2$ shows a good behaviour against damage build-up which makes it a good candidate for RTG's from this perspective. Work is in progress to determine the helium behaviour in the AmO_2 and especially to determine the helium retention capacity this fluorite type material. The design and particularly the porosity should account for such parameters.

6. REFERENCES

1. Prelas, M. Weaver, C. Watermann, M. Lukosi, E. Schott, R., Wisniewski, D. (2014). A review of nuclear batteries. Progress in Nuclear Energy. 75, 117-149.

2. Lange, R., Carroll, W. (2008). Review of recent advances of radioisotope power to explore the solar system, Energy Conv. Manag. 43(3), 393-401.

3. O'Brien, R.C. Ambrosi, R.M. Bannister, N.P. Howe, S.D. and Atkinson, H.V. (2008). Safe radioisotope thermoelectric generators and heat sources for space applications. J. Nucl. Mater. 377, 506-521.

4. Wiss, T., Dieste Blanco, O., Tacu, A., Janssen, A., Talip, Z., Colle, J. Y., Martin, P., Konings, R. J. M. (2015). TEM Study of Alpha-damaged Plutonium and Americium Dioxides. J. Mater. Res. 30, 1544-1554.

5. Hurtgen, C., Fuger, J. (1977), Self-irradiation Effects in Americium Oxides, Inorg. Nucl. Chem. Lett. 13, 179-188

6. Nellis, W.J. (1977). The Effect of Self-Radiation on Crystal Volume, Inorganic and Nuclear Chemistry Letters 13, 393-398.

7. Schmidt, H.E., Richter, J., Matzke, H., Van Geel, J., (1994). The Effect of Self-Irradiation on the Thermal Conductivity of Plutonium and Americium Oxides, in: T.W. Tong (Ed.) Thermal Conductivity 22, Technoic Publ. Co., Lancaster, PA, 920-925.

8. Boucher, R., Quéré, Y. (1981). Sources d'énergie au plutonium pour stimulateurs cardiaques, J. Nucl. Mater. 100, 132-136.

9. Alais, M., Berger, R., Boucher, R., Gasper, K.A., Laurens, P. (1974). Nucl. Tech. 26, 307-319.

10. Prieur, D. Vigier, J.-F. Wiss, T. Janssen, A. Rothe, J. Cambriani, A. Somers. (2014). J. Structural investigation of self-irradiation damaged AmO_2, Journal of Solid State Chemistry 212, 7-12.

11. Wiss, T., Thiele, H., Janssen, A., Papaioannou, D., Rondinella, V.V., Konings. R.J.M. (2013). Recent results of Microstructural Characterization of Irradiated Light Water Reactor Fuels using Scanning and Transmission Electron Microscopy, JOM 64, 1390-1395

12. Wiss, T., Hiernaut, J. P., Roudil, D., Colle, J. Y., Maugeri, E., Talip, Z.,Janssen, A., Rondinella, V. V., Konings, R. J. M., Matzke, H., Weber, W. (2014) Evolution of spent fuel in dry storage conditions for millennia and beyond. J. Nucl. Mater. 451, 198-206.

13. Noe, M. Fuger, J. (1974). Self-radiation effects on the lattice parameter of $^{238}PuO_2$, Inorganic and Nuclear Chemistry Letters 10, 7-19.

14. Chikalla, T. D., Turcotte, R. P. (1973). Radiation Effects 19, 93-98.

15. Fuger, J. Matzke, H. (1991). Handbook on the Physics and Chemistry of the Actinides (eds. Freeman, A. J. & Keller, C.), Elsevier Science Publishers B.V., 641 -684.

16. Weber, W. (1984). Alpha-irradiation damage in CeO_2 and PuO_2. Radiation Effects 83, 145-156.

17. Turcotte, R., Chikalla, T. (1973). Annealing of self-radiation damage in $^{238}PuO_2$, Radiation Effects 19, 99-108.

18. Jonnet, J. Van Uffelen, P., Wiss, T., Staicu, D., Rémy, B., Rest, J. (2008). Growth mechanisms of interstitial loops in α-doped UO_2 samples, Nuclear Instruments and Methods in Physics Research, Section B: Beam Interactions with Materials and Atoms. 266(12-13), 3008-3012.

19. Maugeri, E., Wiss, T., Hiernaut, J. P., Desai, K., Thiriet, C., Rondinella, V. V., Colle, J. Y., Konings, R. J. M. (2009). Helium solubility and behaviour in uranium dioxide. J. Nucl. Mater. 385, 461-466.

20. Rondinella, V. V., Wiss, T., Matzke, H., Hiernaut, J. P., Fromknecht, R. (2000). in Mass and Charge Transport in Inorganic Materials, Adv. in Science and Technology 499-506.

BATTERY FOR EXTENDED TEMPERATURE RANGE EXOMARS ROVER MISSION

Steve AMOS[(1)], Paul BROCHARD[(2)]

[(1)] *AIRBUS Defence and Space, Gunnels Wood Road, Stevenage – United Kingdom, Email: stephen.amos@airbus.com*

[(2)] *Saft, Space and Defence Division, BP 1039, 86060 Poitiers – France, Email: paul.brochard@saftbatteries.com*

ABSTRACT

The robotic exploration of Mars plans to deliver a European rover on the surface of Mars during the RSP (Rover & Surface Platform) mission. Exomars is a broad cooperation between ESA and Roscosmos, with important contribution from NASA.

NASA already sent several rovers on Mars using RTG (Radio isotopic Thermoelectric Generator) to provide electricity on-board. RTGs are not used by ESA, therefore an electrochemical storage system is developed for this mission.

Nevertheless, the rover is equipped with Radio-isotopic Heater Units (RHU) supplied by the Russians, which aim to prevent the internal temperature of the rover dropping below -40°C, whereas an external Rover temperature as low as -125°C, can be expected on the Martian surface. The drawback is that during the rover encapsulation and the cruise between the Earth and the red planet, the internal Rover temperature will reach +40°C to +50°C during months.

The rechargeable battery designed for the rover shall then withstand the high temperatures exposure followed by charge / discharge cycles at low temperatures when on surface of Mars.

Another point to highlight is that the mission objective is to try to find well-preserved organic material, particularly from the very early history of the planet, from the subsurface.
Consequently, the contamination of the rover and all its equipment must be controlled similar to medical tools to avoid depositing spores that could pollute the samples during the on surface phase of the mission.

This paper is presenting the innovative battery designed to withstand these extended operating temperatures while being able to power the rover during night and days (even during local sand storm).
The thermal control of the battery was of the utmost criticality to answer to this wide temperature range.
The rover design was well advanced when the battery development started, therefore adding challenging schedule and mass targets!

1. THE EXOMARS MISSION

The mission phases with the associated battery environment are summed up in the following table.
A sol in a Martian "day" (1 sol = 24.66 hours on Earth).

Table 1: Exomars mission phases

Phase	Duration	Temperature
Storage	15 months	+22°C
Prelaunch encapsulated (inside the rover)	9 months	+50°C
Cruise (Earth → Mars)	9 months	+40°C
Mars on-surface	220 sols	-40°C < T<+50°C

The temperatures stated in the above table are those of the mounting plane below the battery (conductive coupling). The rover bathtub structure internal temperature can drop to -115°C adding a radiative coupling with the external surfaces of the battery.

The battery is maintained during the storage phase up to the arrival on the Martian surface. This maintenance aims to keep a minimum state of charge of the battery and to balance it in order to avoid spread in cells' voltage. The battery will experience maximum of 140 cycles to maintain the optimum battery State of Charge (SOC) i.e. 50% ± 10% for minimum fade and also for 3 health checks during the cruise of the rover.

When on-surface, the battery is charged during daylight with the solar panels after preheating to avoid too low charge temperature. It is delivering energy for peak demands during the day, and during the night when the solar panels are not illuminated. A typical 3-days electrical profile is given on Fig. 1.
It is supplying the scientific instruments and the locomotion subsystem to allow the rover travelling autonomously across the Martian surface.

 DOI: 10.1051/e3sconf/20171606001

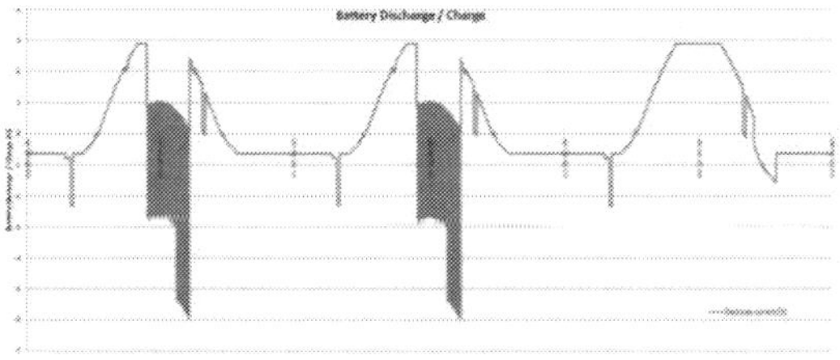

Figure 1. Battery typical 3-days electrical profile

2. THE BATTERY DESIGN

The battery contains the following hardware to fulfil its tasks:

- Electro-chemical cells
- Cabling & inter-connections
- Structural parts
- Passive & active thermal controls
- An isolation switch
- An embedded balancing system

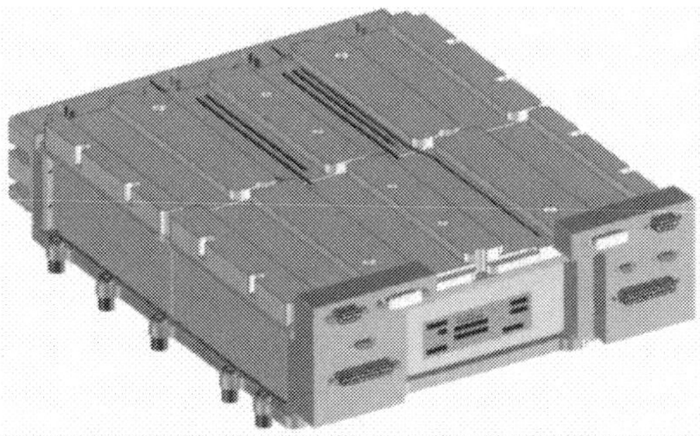

Figure 2. Exomars Rover battery

Electro-chemical Li-ion cells: MP176065 Xtd

Saft selected the MP176065Xtd cell (Xtd standing for Extended temperatures) to cope with the wide climatic range of the Exomars mission.

Manufacturer	Saft
Nameplate capacity	5.6 Ah
Nominal voltage	3.65 V
Operating voltage range	2.5V / 4.2V

Key benefits:

- Excellent operating life in calendar, cycling and floating conditions.
- Unrivalled operating temperature range

A balancing system is embedded inside the battery to compensate the spread in the cells' voltage created by the gradients of the wide operating temperature range of the battery.

The balancing is activated only during few maintenance cycles that take place during the long term storage phases (On-ground and Cruise).
As soon as the battery operates on a daily basis on the Martian surface, the battery is autonomously balanced and does not require any external action.

The isolation switch is a latching type allowing several ON/OFF cycles.
It disconnects the battery from the rover when it is not needed -or only from time to time- especially during the storage periods. It will be finally closed prior to the descent to the Martian surface and will remain closed for the full duration of the on-surface mission.

The battery main performances are sump up in the table below.

Table 2: Battery Fact Sheet

Parameter	*Value*
Technology	Li-ion
Nameplate energy	1140 Wh
Operating Voltage	21.0 V to 29.4 V
Mass	< 10.5 kg
Dimensions	300 x 375 x 100 mm
Charge current	Up to 3.0 A @ T < 0°C Up to 5.2 A @ T > 0°C
Discharge current	Up to 15 A

The battery volume requirement was very challenging due to limited and tortuous space within the Rover bathtub.

The mass requirement was also very ambitious due to late development of the battery compared to the other equipments.

The battery is using mechanical parts made of Aluminium Beryllium alloy to keep the mass as low as possible.

3. ELECTRICAL PERFORMANCES

The electrical performances of the battery rely mainly on the type of cell which was selected.

The choice was based on the following criteria:

- A limited degradation during the high temperature phases.
- A good energy restitution when charged and discharge at low temperature on the Martian surface.
- A good ability to cycle to achieve the maximum duration on surface.
- A good energy density to limit the mass of the battery to the lowest achievable figure.

The acceptance tests carried out on the EM battery (Engineering Model) gave the following performances:

Table 3: EM Battery performances

Performance	Temperature		
	-20°C	+25°C	+55°C
Energy (Wh)	909	1145	1128
Internal resistance(mΩ)	124	46	46

Because of the limited time for the development of the battery, and in order to assess the performances of the battery along the mission; it was decided to carry out three life-tests on cells, one for each critical phase of the mission (pre-launch encapsulated/cruise/ on-surface).

These life-tests include a margin of 1.5 times (on the duration for the calendar tests and on the number of charge/discharge cycles for the cycling tests).

The figures hereafter are showing the irreversible energy losses due to the ageing of the cells during the life-tests.

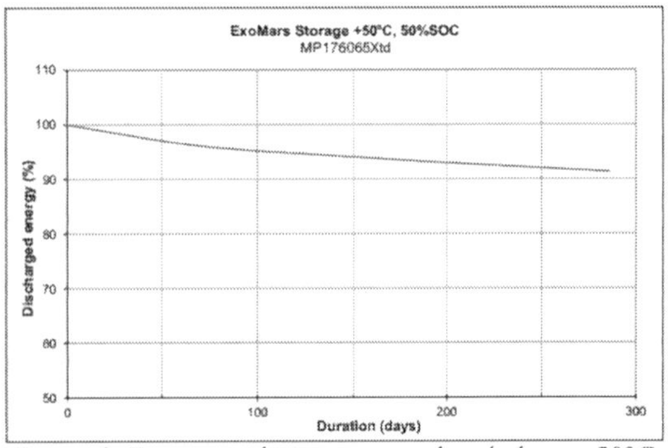

Figure 3. Energy evolution, encapsulated phase +50°C.

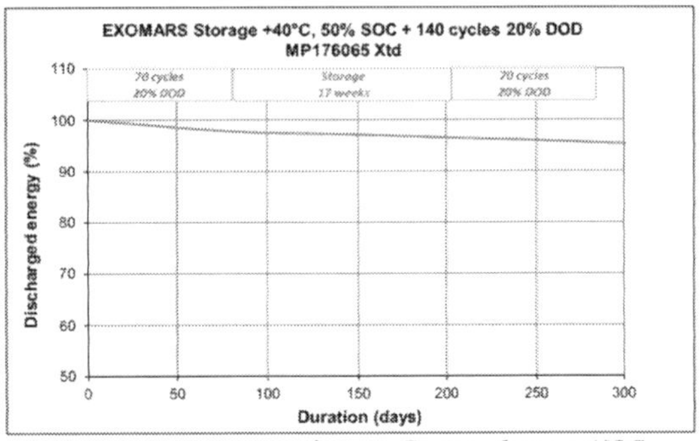

Figure 4. Energy evolution, Cruise phase +40°C.

The cell state of charge is 50% during these two phases.

The figures emphasis that the degradation of performances between the cruise phase (mix of calendar & cycling) at +40°C is half of those of the encapsulated phase at +50°C; showing the criticality of the elevated temperature in the ageing of the cells.

The life-tests were also able to estimate the cell self-discharge during the storage phases.

The results show that the higher temperature, the higher self-discharge:
1.5 mV/day (roughly 1.1% per month) at +40°C;
0.2 mV/day (roughly 1.7 % per month) at +50°C.

The on Martian surface life test exhibit an outstanding performance as it exhibits no loss after the entire phase when charging the battery at -10°C and discharging at -20°C as shown on Fig. 5 below.

We can also notice that the recovered energy is sensitive with the temperature.

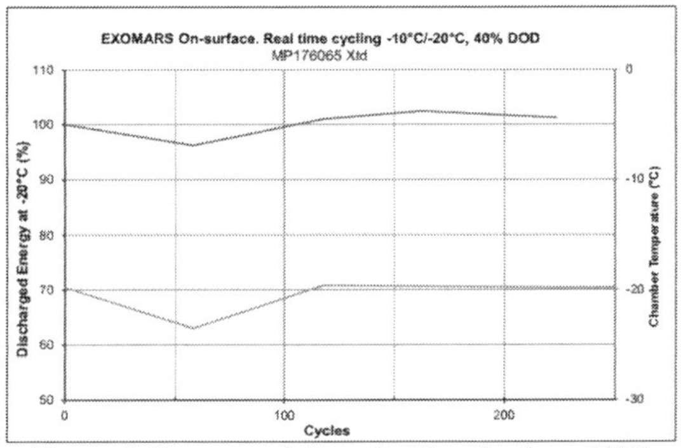

Figure 5. Energy evolution, On Mars Surface -20°C.

The end-of-life performances are assessed based on the above life-tests results.
The performances are excellent and in-line with the requirements of the need.

Table 4: Energy versus life phase

Phase	*Energy*	*Conditions*
Beginning of Life	1140 Wh	+50°C, 50% SOC
End of Prelaunch, encapsulated	1020 Wh	+40°C, 50% SOC
End of Cruise	980 Wh	+40°C, 50% SOC
End of On Mars Surface / End of Life	790 Wh	-20°C, 40% DOD cycling

DOI: 10.1051/e3sconf/20171606001

4. THERMO-MECHANICAL DESIGN

There was almost no flexibility on the mechanical and thermal interface of the battery, as already said, due to the late development of the equipment in the timeline of the rover program. For example, the type and location of the feet was frozen. This is additional challenges for the thermal and mechanical performances of the battery.

The thermal design is critical as the temperatures surrounding the battery are swinging in a wide range.

Passive and active thermal controls are necessary to keep the battery temperature compatible with the operating range of the cells.

The passive thermal control uses feet equipped with thermal spacers that limit the conductive thermal flux between the battery and its mounting plane.

The external surfaces of the battery are covered with a low emissivity metallized tape to reduce the radiative coupling with the cold walls of the Rover.

The active thermal control is almost traditional with two types of heaters and thermal sensors.

One heater keeps the battery minimum temperature over a given threshold (the "survival heater") whereas the second one ("the warm-up heater") aims to pre-heat the cells to allow the charge process. This way, it is limiting the degradations during the cycling of the battery.

The following figures are giving a view of the cells temperatures in cold and hot cases on the Martian surface phase of the mission.

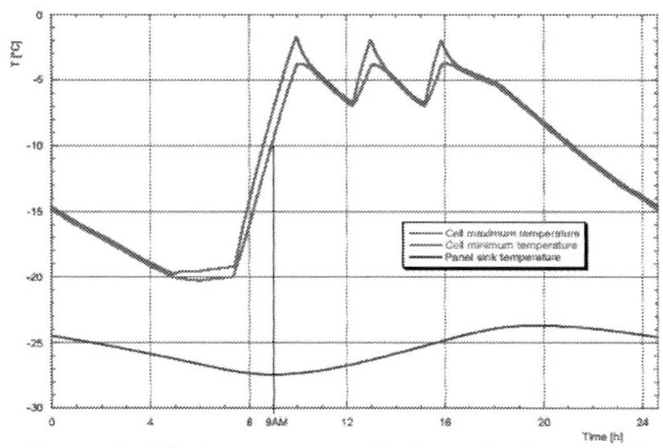

Figure 6. Cells' temperature, On Mars Surface, 1 sol, Cold case (-27°C).

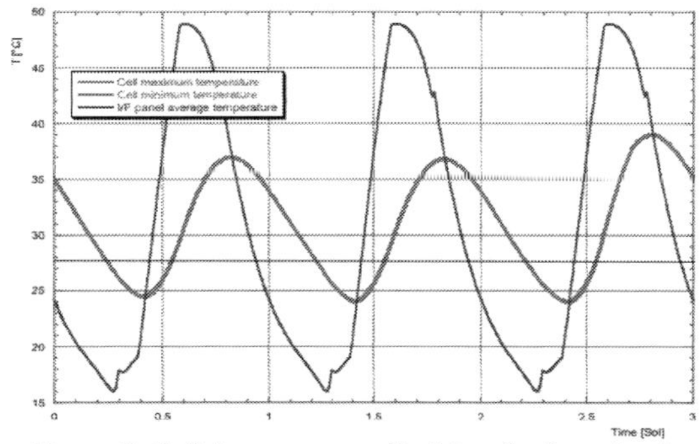

Figure 7. Cells'temperature, On Mars Surface, 3 sols, Hot case.

Another point to highlight is the fact that the launch temperature will be +50°C. Then the harsh vibrations and pyro-shocks will induce stresses in the materials when exposed to this non-standard temperature for a spacecraft battery at the time of a launch.

Besides these environmental induced stresses during the launch, some lower level vibrations and shocks will be seen by the battery during the separation from the descent module and the touch down on the Martian surface.

Moreover, the temperature gradients through the thermal spacers between the mounting plane and battery structure are generating additional thermo-elastic loads which had to be carefully looked at during the development.

5. PLANETARY PROTECTION

As the main scientific objective of the mission is to search for traces of past and present life, it is of the utmost importance to have a sterile battery; even if it is not directly in contact with the samples to be analysed.

It is quite a new and unusual requirement for a battery. Then it was a great challenge to manufacture the Exomars Rover battery in such sterile conditions.

The planetary protection requirement can be met by the following means:

- Battery design
- Manufacturing facilities
- Test facilities
- Packaging design
- Process control with assays and bioburden measurement

The battery is designed to ease the cleaning by means of smooth surfaces and easy access to surfaces to be cleaned. The material selection was also critical and non-metallic ones were screened before use inside the battery.

DOI: 10.1051/e3sconf/20171606001

The battery is assembled in a cabinet with biocontrol similar to those used in medical labs (see picture hereafter). This cabinet is further located in an ISO 8 class clean room.

Figure 8. Sterile cabinet in ISO 8 clean room.

Prior to the assembly, all the parts are decontaminated with sterile IPA (IsoPropyl Alcohol) or other technics and the personnel are wearing sterile garments (gown, gloves, face mask...) similar to conditions as in a hospital operation room.

The contamination level is controlled and assays are performed on the battery with swabs. The bioburden load is assessed according to the ESA ECSS-Q-ST-70-55C.

The main requirement applicable to the battery is a bioburden less than 1000 spores / m^2. The first assays done on the EM battery manufactured without all the specific dispositions shows around 500 spores /m^2.

6. REFERENCES

1. Stephane Zimmermann (2009). The ExoMars Rover Power System. NASA battery workshop, 2009
2. Goeff Dudley (2010). ExoMars Rover Battery Modelling & Life Tests. NASA battery workshop, 2010
3. Rich Ewell (2010). Mars Exploration Rovers: 1 Year and Counting. NASA battery workshop.

 DOI: 10.1051/e3sconf/20171606002

MARS EXPRESS LITHIUM ION BATTERIES PERFORMANCE ANALYSIS

G. Dudley [1], R. Blake[2] & L. Lucas[3]

[1] ESA/ESTEC Keplerlaan 1, 200 AG Noordwijk, the Netherlands (retired), Email: geoff.dudley@planet.nl
[2] SCISYS UK, Methuen Park, Chippenham, SN14 0GB United Kingdom, Email: rick.blake@scisys.co.uk
[3] LSE Space GmbH, Robert-Bosch-Straße 7, 64293 Darmstadt, Germany, Email: luke.lucas@esa.int

ABSTRACT

Now more than 12 years in orbit, Mars Express battery telemetry during some of the deepest discharge cycles has been analysed with the help of the ESTEC lithium ion cell model. The best-fitting model parameter sets were then used to predict the energy that is expected to be available before the battery voltage drops below the minimum value that can support the power bus. This allows mission planners to determine what future power profiles could be supported without risk of entering safe mode. It also gives some more insights into the ageing properties of these batteries.

1. INTRODUCTION

After Proba-1, Mars Express, launched in June 2003 for an initial 4-year mission, is the spacecraft with the longest duration in orbit that relies upon lithium ion batteries to provide main-bus energy storage. Because of an error in the power subsystem discovered soon after launch, the maximum available energy available from the solar arrays was reduced to 72% and power management became particularly critical in order to permit planned science operations without the risk of entering safe mode.

The spacecraft carries 3 identical batteries each with a 6s16p configuration, (16 parallel-connected strings of 6 Sony 18650HC cells in series). Each battery is connected to a separate battery charge-discharge regulator, controlled such as to ensure equal current sharing between batteries. Telemetry channels provide battery voltage, temperature, charge current and discharge current. The high precision of the measurements after calibration is shown in Tab. 1.

Table 1. Battery telemetry details

Parameter	Calibration Resolution	Frequency
Voltage	0.0075 V	1/32 seconds
Current	0.0026 A	1/32 seconds
Temperature	0.0267 °C	1/32 seconds

The batteries cannot be discharged individually to determine their useful capacity so the problem is to predict how much energy would be available under a given power and ambient battery temperature profile using telemetry date from discharge-charge cycles with relatively shallow depths of discharge.

Several battery models of varying complexity have been used to assess battery state of health using telemetry data from the longest eclipses [1]. Comparison of ground test battery performance data with early in-flight eclipse data using the lithium ion battery model developed at ESTEC [2] was reported at the 2006 NASA Aerospace Battery Workshop [3]. Despite some difficulties, which will be described later, it was possible to arrive at a reasonable estimate of battery energy capacity loss. Now, 10 years later and after improvements in the model and the availability of ground test data on aged batteries this analysis is re-visited and extended to include recent telemetry.

2. MARS EXPRESS MISSION

After a six-month cruise plus orbit insertion phase, Mars Express entered into an elliptical polar orbit around Mars in December 2003. There are approximately 3.5 orbits per day and the spacecraft is subjected to two eclipse seasons each Mars year (a season roughly every 10 months). The number of eclipses and maximum eclipse duration varies from one season to another as can be seen from Fig. 1. During eclipses the spacecraft bus has to be maintained by the batteries whilst at other times the spacecraft is in continuous sunlight but the batteries are sometimes required to supplement power from the solar arrays.

At the time of writing Mars Express has just entered the 2016/2017 (16th) eclipse season. At the end of 2015, the total number of discharge cycles was approximately 33605, of which the majority (26409) were less than 0.25 Wh. There were 4168 discharges in the range 25 – 100Wh/battery and 3022 discharges in the range 100 – 250 Wh – more than the manufacturer's qualification limits for the nominal mission of 2397 and 1531 respectively, but the batteries are still well below the 510 limit for discharges above 250 Wh.

The mission is currently approved until end 2018, with a decision on a further extension to 2019/20 expected in autumn 2016.

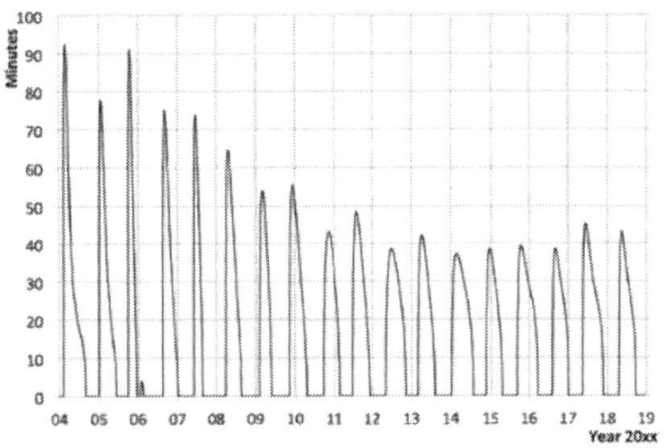

Figure 1. Eclipse duration versus date 2004-2018

The performance of the batteries is a key part of any long term planning particularly with regard to the ability of the spacecraft to operate and survive future eclipse seasons. Preliminary analysis of the next two eclipse seasons in 2016 and 2017 is already underway and while the 2016 season appears relatively benign in terms of eclipse duration and power availability (season coincides with perihelion) the 2017 eclipses presents significant challenges as the they are longer, power availability will be lower and more heater power will be required as Mars passes through aphelion.

From analyses of the longest eclipses it was realised that deeper discharges would be needed to get more reliable battery state of health information. To this end six special deep discharge tests (DDTs) were carried out between 2008 and 2012. Outside of eclipse seasons, the solar arrays were deliberately pointed away from the sun in order to discharge the batteries until the first reached 21V or a time limit expired after which the arrays were automatically restored to sun-pointing. Since 2012 these have not been continued for fear of increasing the stress on the batteries.

3. APPLICATION OF CELL MODEL TO BATTERY TELEMETRY

In the ESTEC model for the ABSL 18650 HC cell [2] the cell voltage is determined by:

1. The EMFs of the positive and negative electrodes as a function of state of charge. The cell EMF is the difference between these two.
2. The variation in cell EMF with temperature as a function of state of charge
3. The cell internal resistance (due to electrolyte, current- collectors and electrode-electrolyte interfacial resistances), independent of state of charge but a strong function of temperature
4. Diffusion polarisation in the electrodes, independent of state of charge but a strong function of temperature
5. EMF hysteresis (the EMF versus state of charge curve differs slightly between charge and discharge), a strong function of state of charge but independent of temperature

Tests on cycled cells [4] show that 1, 2 and 5 are not affected significantly by ageing until very close to end of life. The cell capacity decreases but the EMFs of the electrodes are simply 'compressed' in proportion to their diminishing capacity. The extent to which the two electrodes lose capacity is however different and the relative state of charge between them at the reference state of 4.2 V does change with ageing. Item 3 increases strongly with ageing whereas the item 4 increase is modest. It was found that ageing could be incorporated into the model simply by modifying the values of 6 parameters:

1. Internal resistance [Res_factor] – amount by which the beginning of life (BoL) resistance has to be multiplied
2. Diffusion resistance [Res_factord] – as above for diffusion resistance
3. Positive electrode capacity factor [Cfp] – amount by which BoL positive electrode capacity has to be multiplied
4. Negative electrode capacity factor [Cfn] - as above for the negative electrode
5. Positive state of charge (SoC) at a cell voltage of 4.2 [Ahpinit] – deviation from BoL value of 0
6. Negative SoC at a cell voltage of 4.2 [Ahninit] – as above for negative electrode

Characterising a cycled cell therefore reduces to the determination of the values of the 6 above parameters. Because of the lack of characterisation data for aged cells at temperatures around -5 to +5°C that the Mars Express batteries experience, it has had to be assumed that at any given point in time the changes to the parameters due to ageing are the same at all temperatures.

3.1 Model update

Best-fit values of Res_factord found in earlier reported MEX telemetry analyses [3] where the batteries are cold were in the region of 0.5 rather than 1, whereas in ground tests at 20°C, expected values near to 1 were found. The explanation for this only became evident following recent and precise current interruption resistance measurements on other lithium ion cell types [5]. These showed that while the interfacial resistance (activation polarisation) contribution to the cell internal resistance R_i is negligible at 20°C and above, it can become significant at lower temperatures. Moreover it typically takes several hundred milliseconds to approach a steady value because of the large double-layer capacity. At the time of the development of the

18650HC cell model (2004) the current interruption measurements used recorded the resistance after 70 to 100 mS. It is now realised that this would have been too short a time to capture the full interfacial resistance contribution at 0°C and thus would have lead to an under-estimation of R_i at low temperatures. The diffusion resistance R_d was calculated from the total steady state polarisation resistance R_{tot} under constant current from the relation:

$$R_d = 1.5\ (R_{tot} - R_i) \qquad (1)$$

In consequence it would be too large and result in the observed low values of Res_factord. The model was updated by fitting R_d to step cycle test data [2] at 0 and 20°C and deriving values at other temperatures from the Arrhenius slope defined by these two. R_i is then obtained from Eq. 1 instead of the current interruption resistance measurements. Although a rather rough fix, dictated by the absence of better resistance measurements on the 18650 HC cell, the updated model results in a reduction of the root mean square (RMS) fit errors with the Mars Express telemetry by some 5–20%, and Res-factord parameter values much closer to 1, showing that it is a definite improvement.

4. BATTERY TELEMETRY ANALYSIS

Since the cell model calculates the cell potential under a given imposed current profile, the simplest way of fitting the model parameters to the telemetry data is by imposing the battery telemetry current and temperature profiles on the model of a single cell scaled to the battery configuration, and comparing the resulting voltage prediction to the voltage telemetry. This implies that all cells in a battery have the same characteristics and state of charge. Since individual cell voltages are not monitored there is no other option, but ground tests have confirmed that cells in a battery do remain well matched until very near end of life [4]. The harness resistance from the battery to the point of battery voltage measurement was set to 0.0343 ohms.

The Berkeley Madonna application [6] used to implement the model includes a curve-fitting algorithm that can find a best-fit solution by varying an arbitrary number of parameters and running the model repeatedly until a minimum root-mean-square (RMS) error between measured and modelled battery voltage is reached. The algorithm (which is not publically documented) requires upper and lower limits to be set for each parameter together with two different initial guesses. However the best-fit parameter values that result differ depending on the limits and guesses that are chosen. This is because the algorithm only finds a 'local' minimum in the 6-dimensional parameter space. What is really wanted is the 'deepest' or 'global' minimum. Like the 'travelling salesman' problem this is a so-called 'NP-hard' problem for which there is no known way of finding with certainty this global minimum. Consequently it is necessary to run the curve-fitting algorithm repeatedly with different sets of limits and guesses and to assume that the ones with the smallest RMS errors are close to the global minimum. This is very processor – intensive since each current and temperature dataset consists of several thousand time – value pairs and the curve fit routine for each set of guesses and limits runs the model typically 300 – 500 times, taking several minutes on an average desktop computer. An auto step size integration (between 0.036 and 3600 seconds) was chosen (5th order Runge-Kutta) with a tolerance of 1E-7. and the curve-fitting tolerance set to 1E-6. An 8-electrode slice model was used which is a good compromise between precision and compute speed [2].

Limits were always set well outside of plausible values, to avoid influencing the results, for example Cfp and Cfn upper limits were always > 1.1 and Res_factor lower limits to 0.9 or less. Should any curve fit return a parameter value very close to a limit it was re-run using a wider limit. Occasionally implausible parameter vales were obtained such as Cfn >1 or C_0> 1.5 and these fits were also not used. First and second guesses were varied randomly below and above expected parameter values. Small changes often produced a very different combination of parameter values. At other times they would result in an identical set of parameters.

5. RESULTS

Fig. 2 shows a typical model fit to the DDT6 telemetry. The RMS value of 0.03V obtained corresponds to 5 mV at cell level which is considered a good fit considering the number of factors that could degrade it amongst which are:

- Uncertainty and variability of actual cell internal temperatures.
- Cells no longer perfectly matched
- Difference between telemetry timestamps and instant of acquisition (important near major current changes)
- Possible current offset errors

To investigate the possible magnitude of the last item some test curve fits were performed with a current offset error as a 7th parameter but this did not enhance the fits nor result in any consistent current error value so was dropped from subsequent analyses.

Fig. 3 shows the results for all 3 batteries from DDT 6. Parameter values from the local minimum found for each curve fit are plotted against RMS error in V (battery level). It can be seen that there are a number of different combinations of parameters that give a similar quality of fit though there tends to be a cluster of points close to the best fit. These most likely correspond to minor minima in the vicinity of the global minimum caused by 'noise' in the data.

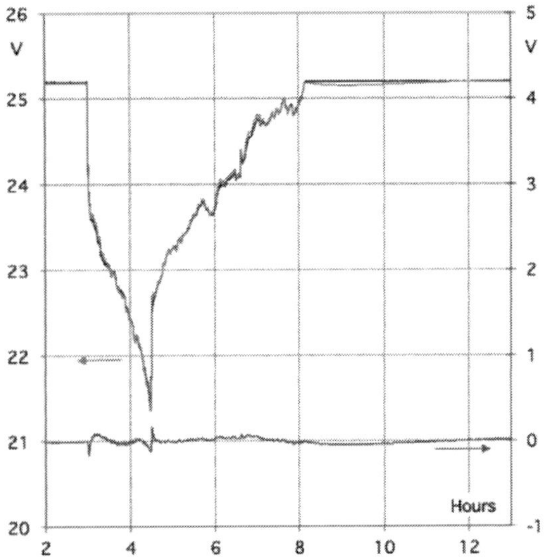

Figure 2. Telemetry (black), model (red), error (blue)

A still better fit that has not been found cannot be excluded but seems very unlikely. The general distribution of points for each parameter is very similar for all three batteries.

The battery cell capacity in Ah at zero current (C_0) is approximated by:

$$C_0 = 1.487\ Cfn + Ahninit - Ahpinit \qquad (2)$$

1.487 Ah is the BoL capacity based on the model electrode EMF data. (C_0 is approximate because it counts the cell as charged when the positive electrode reaches the BoL EMF corresponding to a cell at 4.25 V and fully discharged when the negative electrode reaches the BoL EMF corresponding to a cell at 2.5 V. In both cases the slope of EMF versus SoC of the other electrode is fairly flat so the error will be small).

Whilst the batteries appear to remain well matched to each other, battery 3 shows a slightly lower C_0 value compared to 1 and 2.

Tab. 2 shows details of the 9 cycles for battery 1 from different times in the mission that have been analysed with the aim of identifying how the battery has aged.

Table 2. Analysed Cycles

Cycle date	Ah discharged (cell)	RMS fit mV (battery)	Comments
27/02/04	0.599	18	Eclipse season 1
04/11/05	0.717	22	Eclipse season 3*
10/09/06	0.438	27	Eclipse season 4
22/01/08	0.713	30	DDT 1
26/06/09	0.667	25	DDT 3
23/07/10	0.678	25	DDT 5
07/05/12	0.633	33	DDT 6
20/10/15	0.345	25	Eclipse season 15
04/12/15	0.351	24	Eclipse season 15

**Note: eclipse + shadowing by MARSIS payload boom.*

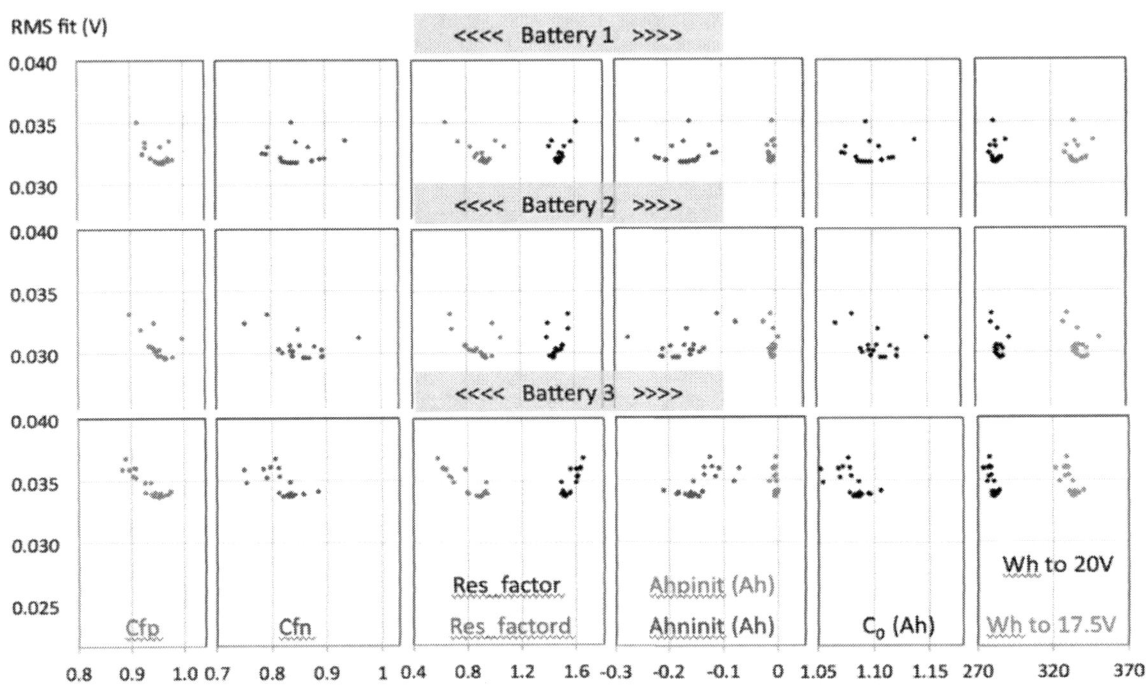

Figure 3. Curve fit results for DDT 6 (all batteries)

 DOI: 10.1051/e3sconf/20171606002

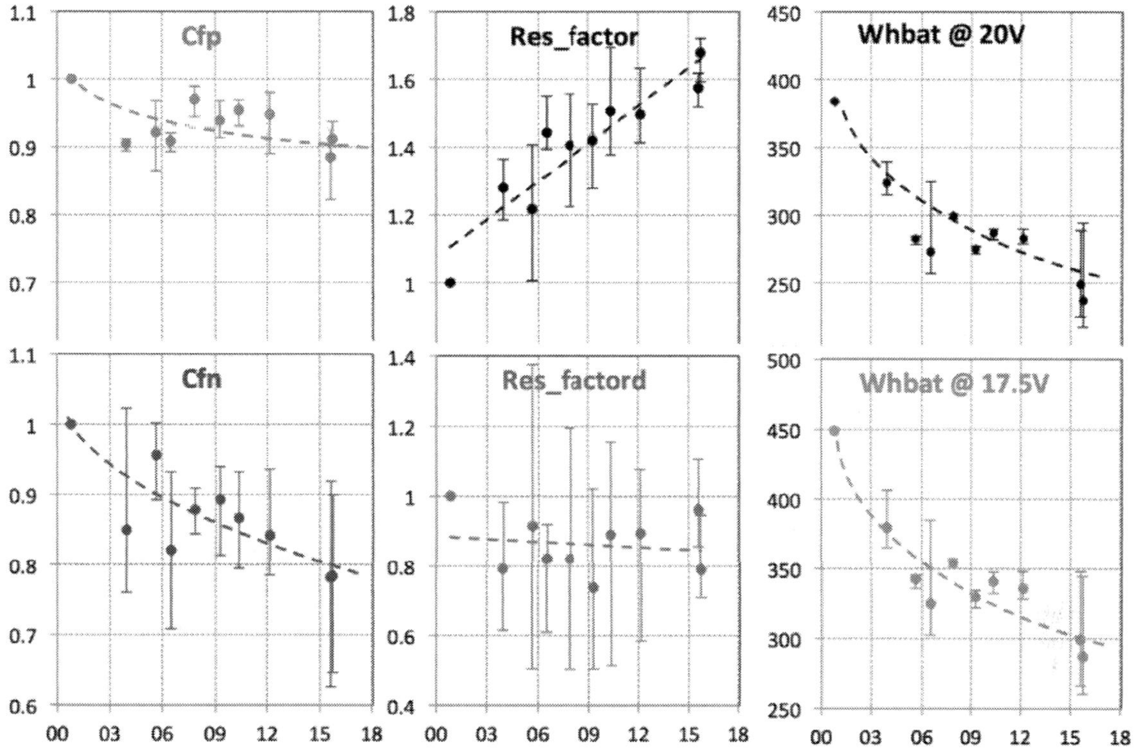

Figure 4. Average parameter values for battery 1as a function of date (20xx)

Fig. 4 shows the parameter values averaged over all curve fits (except those excluded for reasons described in section 4) plotted against date (last two digits of the year). BoL values are also shown positioned at 2001, the approximate time of manufacture of the cells. The error bars encompass the scatter in parameter values from the curve fits. The DDT results show a much smaller scatter than those from eclipse data because they were from deeper discharges. No curve fit from any DDT cycle gave rise to implausible parameter values but even so the scatter in C_0 and Wh values in Fig.3 shows the danger of relying on the results of a single curve fit as was often done in the past. In contrast, curve fits from the shallower eclipse cycles gave much larger parameter value scatter. This is a direct result of the shape of cell electrode EMFs as a function of state of charge, shown in Fig. 5. EMF curves for a negative electrode which has lost 14% of its capacity and another which has BoL capacity but has lost 0.1 Ah of lithium are also shown. The EMF curves for these two possibilities only diverge significantly when the cell is more than 0.6 Ah discharged.

In curve fits from shallow cycles impossible parameter

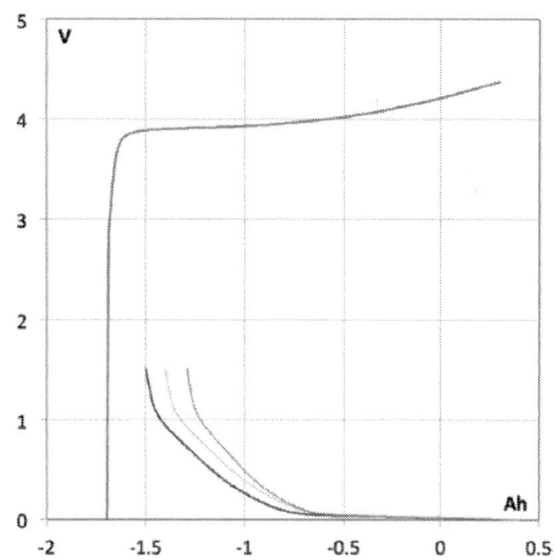

Figure 5. Sony 18650 HC electrode EMFs. Red: positive electrode; blue: negative electrode; green: negative electrode with 0.1 Ah lost lithium; violet: negative electrode with 14% capacity loss

 DOI: 10.1051/e3sconf/20171606002

values now sometimes appear – for example in the last eclipse cycle the analyses with the smallest RMS error show values of Cfn well over 1 and values of C_0 well in excess of 1.47. In the case of cycle 10/09/06, only one third of the curve fits did not have such parameters and these were associated with smaller than expected C_0 values. Others, rejected for producing Cfn values in the region of 1.01 to 1.3 produced C_0 values in the expected region of 1.13 to 1.18 leading to a likely underestimate of available capacity.

From Fig. 5 it may appear surprising that the scatter in estimated capacity is not even larger since no discharge exceeded 0.713 Ah. That this is not the case is due to the combination of capacity and state of charge loss of the negative electrode. In addition, the state of charge of the part of the electrode nearest to the electrolyte (which contributes to the cell voltage) is lower than the average during discharge due to the lithium diffusion concentration gradient. Fig. 6 shows the equivalent Ah discharged of the electrode slices nearest to the electrolyte from the model fitted to the DDT 6 telemetry.

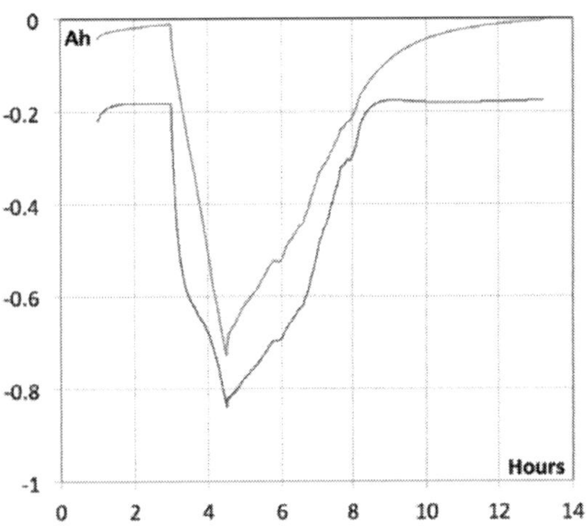

Figure 6. Ah discharged of electrode slices next to electrolyte during DDT 6. Red: positive electrode; blue: negative electrode;

Whereas the discharge during DDT 6 was -0.633 Ah (at cell level), the part of the negative electrode next to the electrolyte has discharged to -0.82 Ah (75% of aged cell capacity). Consequently the part of the EMF curve where the negative EMF starts to change rapidly is actually surpassed.

Res_factor and Res_factord show much scatter but high values for one correlate with low values for the other since the total steady-state resistance must be the same as shown in Eq. 1. Cfn and Ahninit also vary in a coupled fashion for the reasons explained earlier in connection with Fig. 5. As found in a ground tests [4], Res_factor increases greatly with ageing due to the build up of reaction layers on the surface of the electrode active material particles. The activation and surface layer polarisation contributions also make an increasingly large contribution at low temperatures [5] and an increase with ageing is expected but has not been studied by the authors. In contrast Res_factord has not increased much, behaviour also found in ground tests [4]. This might be expected since the parameter is determined by the rate of diffusion of lithium ions in the electrolyte in the pores of the electrodes and would be less affected by surface effects. The overall average value of around 0.85 rather than 1 for a BoL at 20°C is more likely due to the updated model still generating a slightly high diffusion resistance value at 0°C than indicating a real drop with time. Res_factor values correspondingly appear to be about 0.1 too high relative to BoL.

The results suggest that the positive electrode has lost capacity pre-launch and during transfer to Mars but not much since in orbit. It is less than was observed in ground tests at 20°C [5] so this may be explained by the relatively low temperatures in orbit. The slightly higher values found from the DDT analyses compared to the eclipse analyses is interesting but are thought more likely to be a shortcoming of the model rather than real. The negative electrode is down to around 80 to 90% of its BoL capacity, similar to ground tests. The results also indicate that this electrode has lost around 0.15 to 0.2 Ah of lithium when fully charged due to reactions with the electrolyte, also similar to ground tests [5]. These two effects combine to reduce C_0 down to 0.9 Ah or 60% of BoL for the last cycle analysed.

6. BATTERY PERFORMANCE PREDICTION

C_0 takes no account of the 60% increase in internal resistance that also has a strong effect on the usable capacity down to a specified cut-off voltage – the quantity that is much more relevant to spacecraft operations. To calculate this, parameter sets for each retained curve fit were fed into the model. A simple singe RC battery thermal model was used based on thermal capacity and resistance values that reproduced quite well the battery temperature telemetry. The starting temperature was set to -2°C, a typical value for the cycles analysed. A one- hour open-circuit followed by a 6A constant current discharge was simulated. (this provides a margin over the average recent discharge currents of 4.48 A). The initial electrode state of charge was adjusted so that the starting open-circuit voltage was 25.2. The Wh discharged to 20V and to 17.5V were extracted and are also plotted in Fig. 4. 20 V is the nominal battery discharge alarm surveillance (BDAS) limit and 17.5 V is the minimum settable value. The spread in Wh values between curve fits is less than that

DOI: 10.1051/e3sconf/20171606002

for C_0 (not shown) and varies from as little as 1% for the best DDT to over 30% for the most recent eclipse cycles. Differences between successive DDTs which show increases in capacity as well as drops greatly exceed the smallest scatter values and it is not clear whether they are real and, for example, show possible recovery effects from lowering the voltage that batteries were held at between certain eclipses [7] or are due to shortcomings of the model.

Battery performance is enhanced at higher temperatures, so it is of interest to investigate what increase in useful Wh could be obtained if a battery deep discharge is started at say +5°C instead of -2°C. In fact with a 6A discharge there is only a gain of 6Wh at 20V cut-off and 4.5Wh at a 17.5V cut-off. This because by the time these voltages have been reached the battery has warmed by self-heating to over 16°C even when starting at -2°C.

7. CONCLUSIONS AND FURTHER WORK

It has proved possible to fit the Mars Express battery 1 telemetry to the ESTEC battery model with RMS errors between 15 and 33 mV (2.5 and 5.5 mV at cell level), as good as one could expect and comparable to that obtained when fitting ground test data. Better cell resistance measurements on BoL cells at 0°C and -10°C, from which implementation of the temperature and SoC dependence of activation polarisation (5) could be added, could further improve the model accuracy.

This approach should be applicable to other spacecraft that use batteries containing lithium ion cells for which the ESTEC model parameters are available provided they deliver battery current, voltage and temperature telemetry of sufficient precision.

Despite the good fits there is a large scatter in parameter values depending on the initial settings of curve fits. Discharges of more than 0.6 Ah at cell level are required to obtain reliable capacity values. This due to flat EMF versus Ah of the negative electrode which is the electrode that determines the cell capacity. The situation will be different for cells that use other than than hard carbon negative electrodes.

If this type of analysis is to be used routinely, there is a strong need for further automating the choosing of curve fit starting parameters and retrieval of the parameter values. With the absence of scripting capability in Berkeley Madonna, it is currently necessary to do this manually, which is very time-consuming even though each curve fit is quite fast. This is why Mars Express Batteries 2 and 3 have so far only been analysed for one cycle (DDT 6). An automated approach coupling Matlab with the EcosimPro implementation of the model has already been investigated [8].

Although fairly complex, the big advantage of this model approach is that the behaviour of the spacecraft battery under any given current / power / temperature profile can be determined in order to tell in advance whether a proposed payload usage or failure scenario could be supported without entering safe mode.

8. REFERENCES

1. Ferretti, R Denis, M Moorhouse, A Dudley, G Ecale, E Rabenau, E Page, O Schulster, J & Ormston, T (2006). Mars Express Power Subsystem In-flight Experience. SpaceOps 2006 Conference, Rome Italy AIAA-2006-5857

2. Dudley, G. De Roche, J Tonicello, F, & Thwaite. C. (2005). Electrical/Thermal Model of a Sony 18650HC Li-Ion Cell, *Proc. 7th European Space Power Conference* (Ed. A. Wilson), ESA SP-589 (CD-ROM), European Space Agency, Noordwijk, The Netherlands.

3. Dudley, G. Thwaite, C. & Walpole, A. (2006). Use of Electrical/Thermal Cell Model for Mars Express Battery Health Assessment. *NASA Aerospace Battery Workshop*, Huntsville USA

4. Dudley, G. Buckle, R. Hendel, B. Mattle, T. & Spurrett, R. (2008). Ageing of Sony 18650 HC Cells in LEO Life Tests. *Proc. 8th European Space Power Conference*, (Eds. H. Lacoste & L. Ouwehand) ESA SP-661 (CD-ROM), European Space Agency, Noordwijk, The Netherlands.

5. Dudley, G. (2014). Extension of Lithium Ion Cell Model to include Transient and Low-temperature behavior. *Proc. 10th European Space Power Conference*, (Ed. L. Ouwehand) ESA SP-719, European Space Agency, Noordwijk, The Netherlands.

6. http://www.berkeleymadonna.com

7. Sousa, B. & van der Pols, CL. Breath in, breath out how healthy are the batteries on Mars and Venus Express? . *http://www.spaceops2012.org/proceedings/documents/id1254584-Paper-001.pdf* SpaceOps 2012 Conference, Stockholm, Sweden

8. Ringbeck, F. Marongiu, A. & Becker, J. (2016). Report on Battery health status estimation based on modeling and advanced data processing techniques. Final report of ESTEC contract 4000114068/15/NL/HK with RWTH Aachen University

 DOI: 10.1051/e3sconf/20171606003

LI-ION COTS CELLS FOR LOW TEMPERATURE MARS LANDERS

Rachel Buckle [(1)], Carl Thwaite [(1)], Sanjay Vijendran [(2)], Max Schautz [(2)]

[(1)] ABSL Space Products, F4 Culham Science Centre, Abingdon, UK. Email: rachel.buckle@absl.enersys.com
[(2)] ESTEC, Keplerplaan, 1, 2201AZ Noordwijk, The Netherlands

ABSTRACT

This digest describes the work undertaken to develop a secondary Li-ion battery for use in European space exploration missions from Commercial off the Shelf (COTS) cells. The battery was required to operate at temperatures below -20°C and survive a freeze/thaw cycle.

A previous paper [1] described the search for a suitable cell, and the initial data from the module life test. As the results from this testing were extremely promising, the project was extended to determine the module performance up to 720 Sols at -20°C.

Very low temperature survivability tests were also performed on the selected cells with good results. Further work has therefore been carried out to determine the longer term effects following a freeze/thaw cycle.

The final part of the project was to produce a performance model of the cells. Results from the simulations have been compared to test data for the selected cell type.

1 INTRODUCTION

Operation of rovers/landers on the Martian surface requires overcoming several operational barriers for the power system. The solar power is limited due to the distance from the sun and the prevalence of dust storms. Temperatures range from -80°C on a summer night and -130°C on a winter night, up to +30°C at the equator in summer at noon [2]. These factors mean that power could be extremely limited and little thermal control available at times. Several means of generating electrical power, along with maintaining a suitable thermal environment, are available. These include solar arrays, nuclear systems and fuel cells.

As part of the Mars Robotic Exploration Preparation (MREP) programme, ESA is funding development of a low temperature Li-ion battery based on commercial cell technologies. The main objective of the project was to identify a COTS cell that would operate (charge and discharge) at a maximum temperature of -20°C. In addition, the cells needed to survive an emergency case with temperatures reaching as low as -100°C for four weeks.

The previous paper presented an assessment of the operability of COTS Li-ion cells at low temperature for use on a Martian lander. Initially a market survey was conducted to down-select a number of cells for detailed testing. Four cell types were then tested to determine the cell discharge energy and voltage at the required temperature, the charge capability and life performance. From these data, the best performing cell-type was selected for module life testing.

The initial test data demonstrated that the selected cell type had low fade under the required charge/discharge profile up to 180 Sols, and was robust to very low temperature storage.

This paper presents further results from this low temperature COTS project. The module life test was extended to 720 Sols. In addition, the longer term effects of very low temperature storage at -100°C were determined. An electrical model has also been constructed to predict the performance of this cell type under low temperature cycling. Comparative results from this work will be presented.

2 BATTERY REQUIREMENTS

The initial battery requirements for this project are based on a realistic mission specification from the European Space Agency (ESA); they are summarised in Table 1.

Table 1. Project battery requirements [3]

Energy at EoL	200 Wh
Energy density @ -20°C	125 Wh/kg
Average discharge voltage @ -20°C	28V
Lowest temperature for charging	-20°C
Cycle requirement	180 Sols

Data from the end of the 180 Sols showed that the module had lost only 6% of the BoL capacity when measured at -20°C. The project was therefore extended with an updated goal of reaching 720 Sols.

3 MODULE LIFE TEST

3.1 Test Description

The selected cell type was built into an 8s8p module and tested at low temperature (minimum -20°C cell

DOI: 10.1051/e3sconf/20171606003

temperature) and realistic rates. The module was sized using the information on charge/discharge rates, voltages and degradation determined in the cell selection testing [1].

The test profile is shown in Figure 1. Capacity checks at a chamber temperature of -20°C were carried out before the cycle testing, at the end of the original 180 Sol requirement, and after the extended goal of 720 Sols was reached.

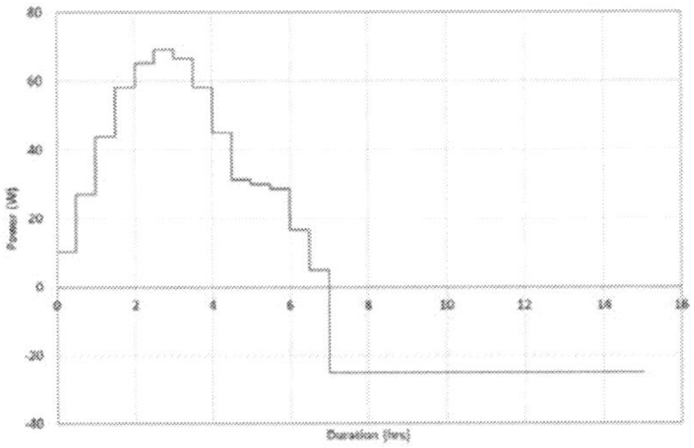

Figure 1. Power profile for module test

Unfortunately, between the end of the 180 Sols and the beginning of the life test extension, the thermal chamber failed, causing the temperature of the battery to rise above 0°C. Although no water was observed on the module or in the connectors, several cells were found to be less than 3.0V (Figure 2). The resulting investigation determined that Cell 8 in String 7 had been shorted.

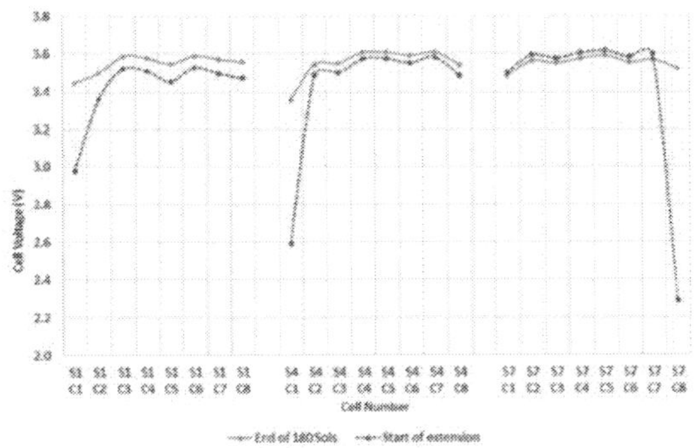

Figure 2. Mod#1 cell open circuit voltages

As the cells are sensitive to low temperature, any failure in this module (Mod#1) could be attributed to the test conditions, or the cell short. Therefore, to be able to confidently determine whether this cell type can perform 720 Sols at -20°C, a second module was built (Mod#2) and cycled under the same regime. Therefore, results from both modules will be presented in this paper.

3.2 Life Test Results

The results from the capacity measurements are shown in Table 2. After cycling at -20°C for 720 Sols, both modules have less than 15% fade. Mod#1 has higher fade and a higher maximum dispersion at the end of discharge (EoD) than Mod#2. This difference is likely due to the module short circuit. The measured data shows Mod#1 lost 3% capacity during the period between the end of 180 Sols and the start of the extension. In addition, the EoD voltage of the Mod#1 at end of life (EoL) was higher than for Mod#2, which restricted the measured capacity.

Table 2. Capacity measurement data

	Mod#1	Mod#2
Capacity loss	15 %	10 %
Energy loss	14 %	11%
EoD dispersion String1	256 mV	672 mV
EoD dispersion String4	1110 mV	664 mV
EoD dispersion String7	364 mV	580 mV

The discharge voltage during Mod#1 capacity measurements at -20°C are shown in Figure 3. At beginning of life (BoL), the cells were more closely balanced than for later measurements, so all cells remained above 2.5V during the discharge to 24V (3.0V/cell on average). As the cells aged, differences in the cells and the temperatures experienced meant that their resistance and/or capacity changed at different rates. In order to prevent cell over-discharge, the module discharge was limited by cell voltage. Therefore, as the dispersion increased, the module EoD voltage also increased, restricting the measured capacity.

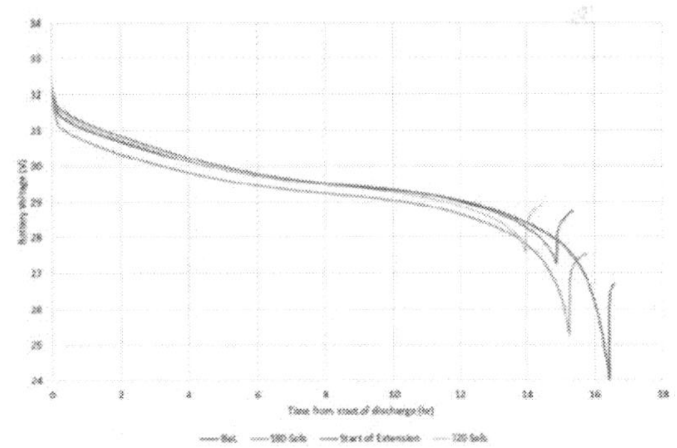

Figure 3. Mod#1 discharge voltage curves (-20°C)

For comparison, the Mod#2 capacity measurements at -20°C are shown in Figure 4. This module shows different behaviour to Mod#1, in that the capacity at EoL was not limited by cell dispersion. When the module was discharged to 24V, all cells in the

DOI: 10.1051/e3sconf/20171606003

monitored strings remained above 2.5V.

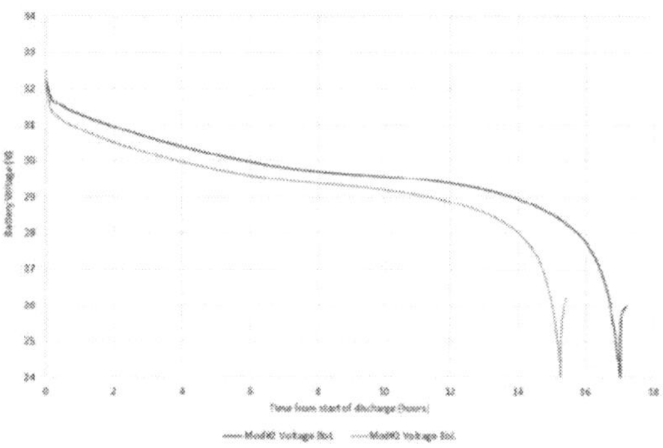

Figure 4. Mod#2 discharge voltage curves (-20°C)

The EoD voltages for both modules throughout the full life test are shown in Figure 5. The 720 Sol requirement was met by both modules, with EoD voltages well above the 24V minimum.

At the start of the test, Mod#1 shows a significantly lower voltage than Mod#2. This can be attributed to the temperature of the tests; Mod#1 was around 4°C colder (on average) than Mod#2 during the first 180 Sols (Figure 9). For the remaining cycles, the voltages of both modules are very similar. The rate of decrease of the EoDV is comparable for both modules, with the actual voltage more affected by the test temperature than by the fade. Considering a cell in Mod#1 was shorted twice, these results show the resilience of the cells, as well as the low level of fade and resistance increase of this cell type when cycled at low temperatures.

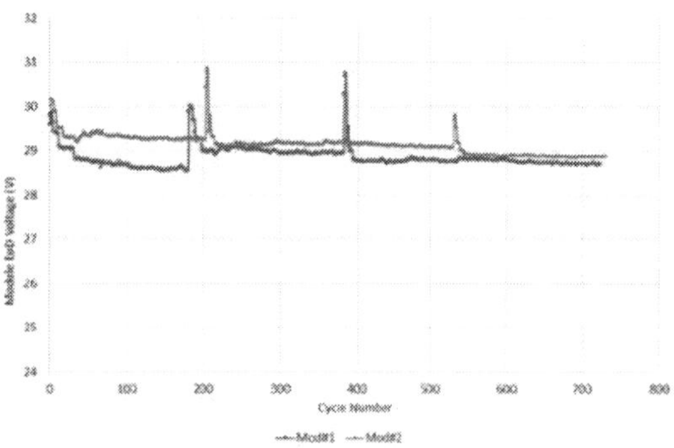

Figure 5. EoD voltages of Mod#1 and Mod#2

During the cycling there are several points at which the EoD voltages of both modules show sharp increases. For Mod#1, the increase at 180 Sols corresponds with the end of the initial cycling requirement, and the restart of the test.

The next voltage spike, at 380 cycles for Mod#1, occurred at the same time as the increase seen after 210 cycles for Mod#2. The reason for these is another failure of the thermal chamber. The temperature again rose past 0°C, with the consequent risk of condensation forming on the module. Results from Mod#1 show that when the temperature started to rise, the cell dispersion initially reduced (Figure 6). This is expected, as the cell resistance will decrease. However after another two Sols, the dispersion in String 7 increased dramatically throughout the following cycle. When this increase occurred, the battery temperature was around 12°C, so any ice in the chamber would be melting and a short circuit due to condensation was possible.

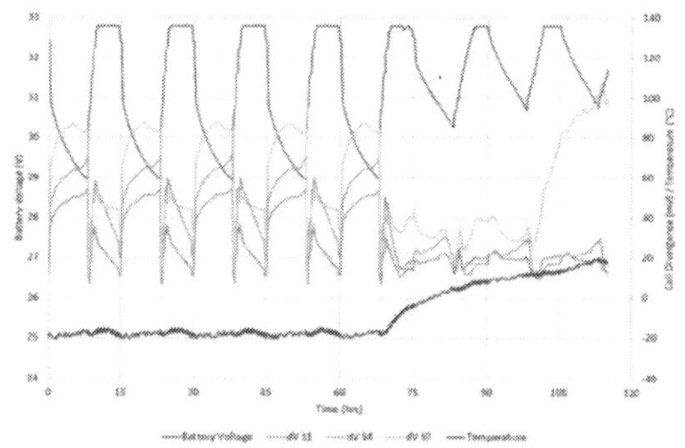

Figure 6. Mod#1 battery temperature and cell dispersion during chamber failure

Unlike Mod#1, no evidence of cell shorting could be seen on Mod#2. There was no dramatic increase in the dispersion for any of the strings during cycling (Figure 7). This was expected, as Mod#2 was equipped with IP67 rated connectors to prevent moisture ingress.

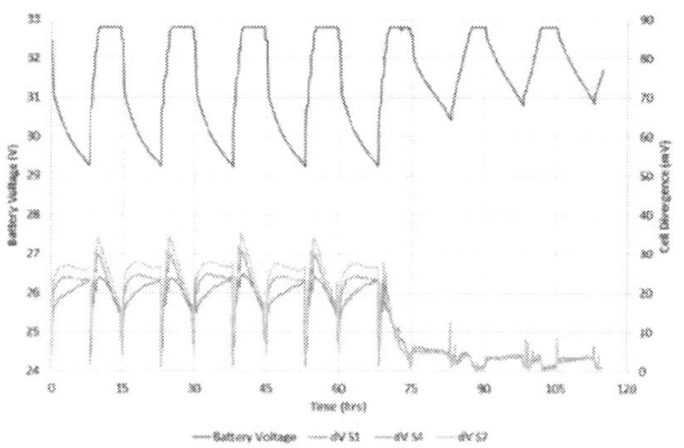

Figure 7. Mod#2 battery voltage and cell dispersion during chamber failure

Due to the lack of balancing electronics on ABSL batteries, and the possibility of Li-plating on the cells if the EoC voltage increases sufficiently, the cell dispersion is an important factor. Figure 8 shows a comparison of the maximum EoC voltage dispersion for each monitored string in the two modules. Apart from

DOI: 10.1051/e3sconf/20171606003

String 7 on Mod#1 (which contains the shorted cell), the remaining strings on both modules showed similar levels of dispersion increase, with the actual dispersion in the strings remaining below 50mV at EoL. Although the dispersion of the shorted string was 150mV, the maximum cell voltage remained less than 4.2V, and the dispersion remained stable. Even this relatively high level of dispersion had no obvious effect on the battery level results.

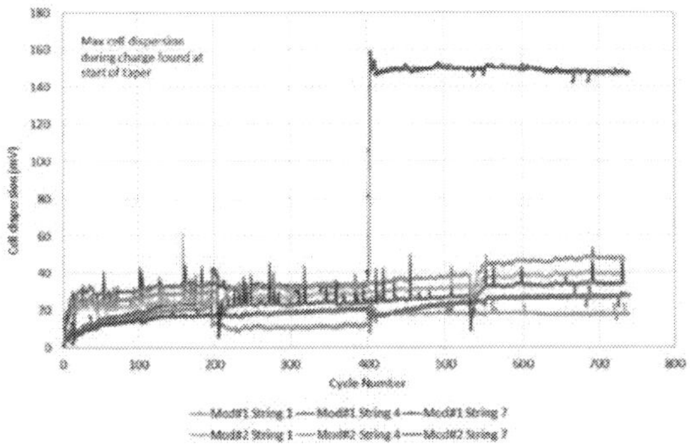

Figure 8. Cell voltage dispersion in the two modules

The average temperature of the cells in both modules at EoC is shown in Figure 9. During the life test, the average chamber temperature was around -30°C; the cells were warmer as the module was insulated and so retained any dissipated energy.

The average cell temperature can be seen to be between -15°C and -20°C. This is an expected result, as the test temperature was fixed so that the coldest cell was at -20°C at end of charge (EoC). However, the data show that the EoC temperature for all cells in Mod#2 was above -20°C for the majority of the test (Figure 10). This occurred because both modules were contained in the same thermal chamber. Mod#1 was colder, with a minimum cell temperature just below -20°C. Therefore, the cell temperatures of Mod#1 limited the chamber temperature, in order to ensure there was no lithium plating in either module.

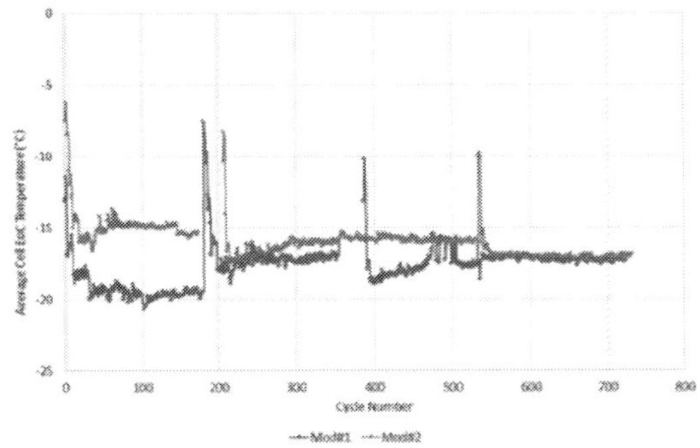

Figure 9. Comparison of module EoC temperatures

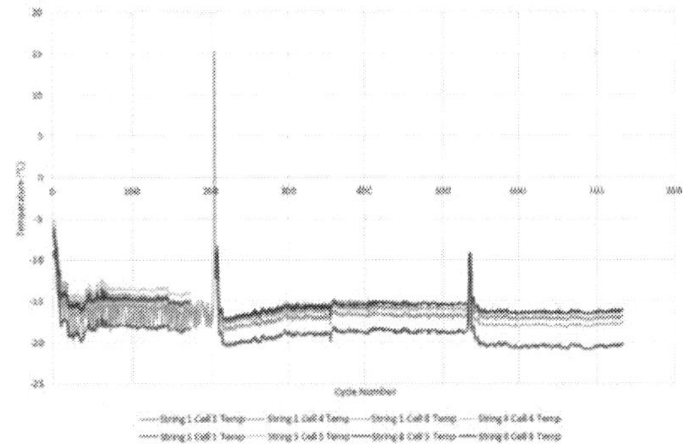

Figure 10. Mod#2 monitored cell EoC temperatures

A comparison of the project statement of work (SoW) requirements with the results from both modules is shown in Table 3. The data shows that the modules met, or exceeded, all the requirements of this activity.

Table 3. Comparison with project requirements (-20°C)

	SoW Req.	Mod#1	Mod#2
EoL Energy	200 Wh	432 Wh	467 Wh
EoL Energy density	125 Wh/kg	140 Wh/kg	152 Wh/kg
Average discharge voltage	28 V	28.7 V	29.6 V
Lowest temperature during charge	-20°C	-20°C	-20°C
Test duration	720 Sols	720 Sols	720 Sols

In summary, both modules completed the 720 Sol extended requirement, with 10-15% fade. The EoD voltages showed no rapid changes, which would be indicative of severe lithium plating occurring. Although Mod#1 had a high voltage dispersion within a string due to the short circuit, the cells remained stable during subsequent cycling. The cells in Mod#2 also stayed balanced during the 720 Sol test, with similar changes in dispersion to the non-shorted strings in Mod#1.

4 SURVIVABILITY TESTS

In addition to the cycling element of the project, the battery was also required to survive an emergency case whereby the cells would be at very low temperature. This was defined as -100°C, to ensure the cell electrolyte was frozen.

In the previous paper, the survivability and recoverability of the cells after storage at -100°C was discussed. Cells were stored at a range of states of charge (SoC) between 10 and 100% for four weeks. Testing showed these conditions had no significant impact on the cells. However, these data only showed the effect on the capacity, resistance, and 10 representative cycles. No data were available on the long term cycling performance.

 DOI: 10.1051/e3sconf/20171606003

The activity was therefore extended to determine the longer term effects. The post-storage cycling requirement was increased from 10 to 180 Sols to match the original module cycling requirement. This paper describes these results.

The cycling regime was based on the module life test, with the power and voltage levels scaled to individual cell level. During the charge, the post-storage cells were charged in one stage using an average power, rather than the 14 stage regime used for the module.

The EoD voltages for the cells during the post-storage cycling is shown in Figure 11. All cells were tested individually. The results show that the EoD voltage dropped gradually throughout the test as the cells degraded. However, the 180 Sols was achieved with a large margin above the minimum voltage of 3.0V. The voltage difference between the cells at EoD increased during the cycling from 50mV at the start of the test (post freeze) to 120mV after the 180 cycles had been completed. This increase in the range will be due to differences in fade and/or resistance change of the cell. No link between the low temperature storage SoC and the post-storage cell cycling performance was observed.

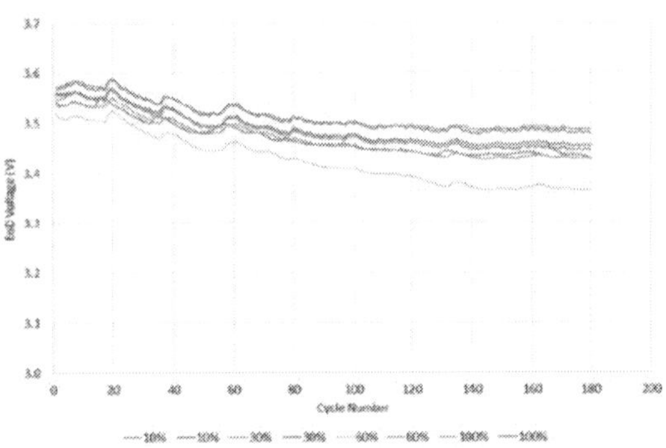

Figure 11. Cell EoDV during extended cycling at -20°C (SoC refers to that during storage at -100°C)

The average cell EoC voltage and the voltage range during the extended cycling test are shown in Figure 12. These values are the open circuit voltages of the cells, 30 seconds into the rest step following the charge. The voltages give an indication of how well the cells are likely to stay balanced after a freeze/thaw cycle. The data show that the EoC voltage range increased from 20mV (once the test was in equilibrium) up to approximately 38mV. This result is in line with the results from the 8s8p module test at -20°C which showed an increase in the dispersion of 15-20mV over the 180 cycles (Figure 8).

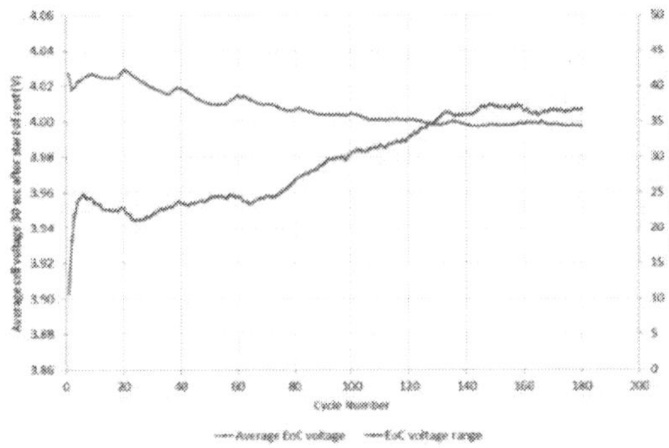

Figure 12. Average cell EoC voltage and range during extended cycling at -20°C

As the objective of the cycling was to determine the effect of the freeze/thaw cycling on the long term performance, it is pertinent to compare the post-storage cell voltages with the module test data. Figure 13 shows the average end of discharge voltage for the cells in the module, compared to the average value for the post-storage cells. Initially, the EoD voltage of the module is higher than that of the post-storage cells, because the module temperature was higher. At the start of the tests both thermal chambers are at -20°C, but because the module was insulated the cell temperature was higher than this. However, after 35 cycles the module cell temperature was at -20°C, so the voltages can be directly compared to the frozen/thawed cells. The difference in voltages was still maintained, implying that the cells stored at -100°C may have increased in resistance during the freeze/thaw cycle. The long term decrease in the voltage also appears to be slightly higher for the cells that were stored at low temperature. However, the tests are not directly comparable as there were differences in the cycling conditions between the module and the frozen cells. The main difference was that the module used a 14 stage charge regime, whereas the cells were charged in one stage using the average power.

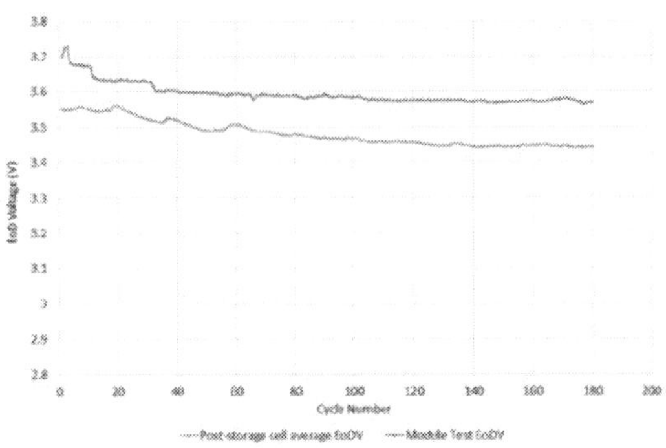

Figure 13. Comparison of EoDV during cycling after low temp storage and during module test

 DOI: 10.1051/e3sconf/20171606003

Another measure of the effect of the long term storage and subsequent cycling is the change in capacity. The capacity was measured at -20°C, as this is representative of the capacity that would be available to a lander. The plot in Figure 14 shows that the cells faded by between 6-9% over the duration of the 180 Sol test. This result is again in-line with the results from Mod#1, which lost 7% capacity over 180 cycles.

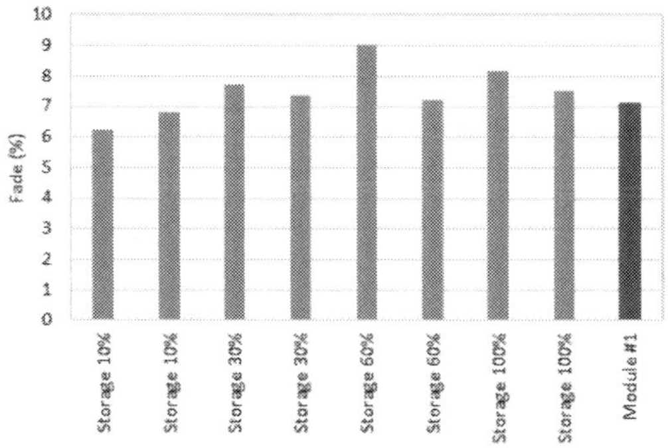

Figure 14. Capacity fade of post-storage cells after cycling at -20°C (SoC refers to that during -100°C storage)

In summary, cells that had been stored at -100°C for 4 weeks were cycled for 180 Sols under a similar regime to the module test. The data showed that the cells were able to complete the cycling requirement, with an end of discharge voltage well above the minimum value, and that the fade and EoC voltage range were in line with the module results.

5 BATTERY MODELLING

An additional part of the low temperature project was the development of an electrical model to be used for general performance simulation. This has been developed for the selected cell type by performing a suite of characterisation tests. The outcome of this work has not previously been reported.

The electrical model uses a distributed nodal network that represents segments of electrode material with inter-nodal resistive links. The chosen approach allows modelling of lithium ion concentration gradients through the thickness of the electrodes. This is important for the calculation of diffusion resistance, especially under the severe rate limiting conditions seen as the temperature decreases towards -20°C.

In order to characterise the cell for the electrical model, testing at multiple temperatures was carried out to determine the following parameters:

- **EMF/SoC**
  Characteristic cell open circuit voltage curve.
- **Internal Resistance**
  Electrical resistance from tags, foils etc, as well as ionic (diffusion) resistances.
- **Thermoneutral potential**
  Effect of temperature on the open circuit voltage.

The effect of temperature on the 'EMF' curve is shown in Figure 15. The data show large differences between the voltage curves at ambient, and those at low temperatures. These differences are more than can be explained by the thermoneutral potential. It is thought that the diffusion rates of ions through the electrodes is extremely slow, such that the current used (C/200) was too fast.

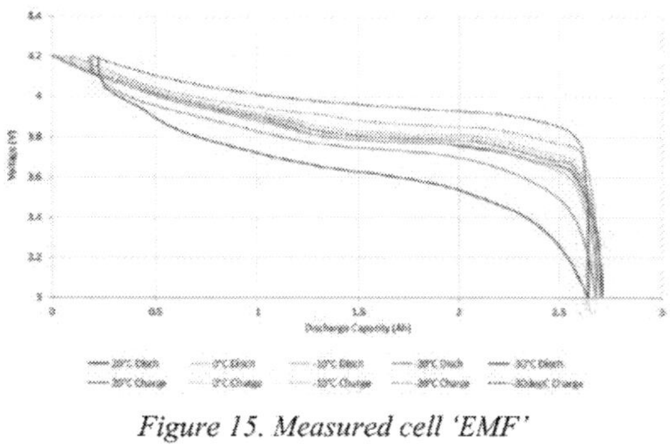

Figure 15. Measured cell 'EMF'

The plots in Figure 17 and Figure 17 show comparisons of the cell characterisation test results with the electrical model at 0°C and -20°C, respectively. The model matches the experimental data well at 0°C. As the temperature is decreased to -20°C, the slow diffusion within the electrodes limits the rate at which the cells can charge/discharge, and the modelling becomes more challenging.

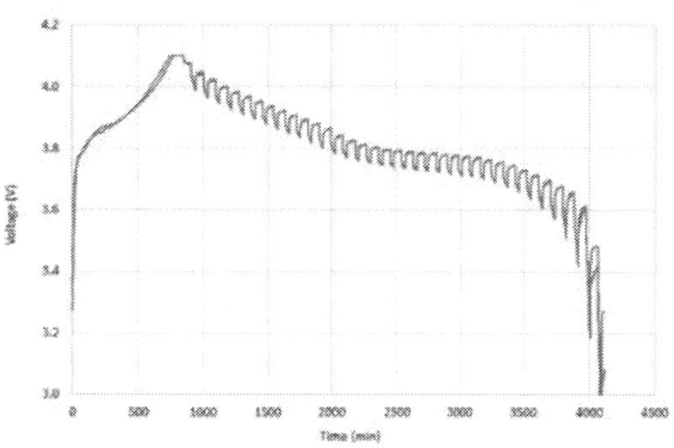

Figure 16. Comparison of test and model results at 0°C

DOI: 10.1051/e3sconf/20171606003

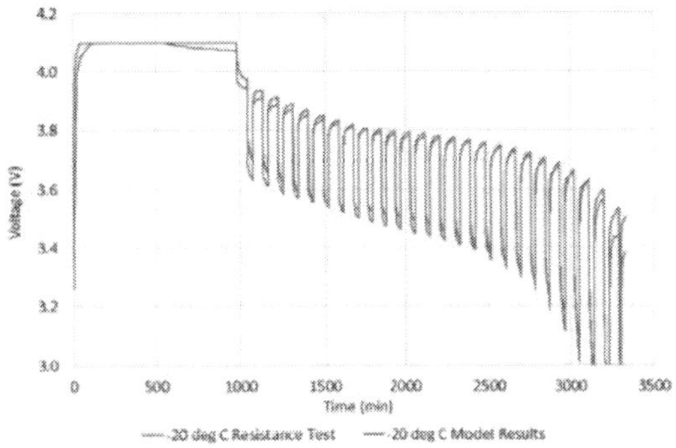

Figure 17. Comparison of test and model results at -20°C

A comparison between the electrical model, and the measured data from Mod#1 (once equilibrium was gained), are shown in Figure 18. Although the EoD voltages are a reasonable match, the model does not follow the voltage during the taper charge very well. The life test module remains at the maximum taper voltage for all charge steps except for the last one, whereas the model shows the voltage dropping almost from the start of the charge. This is likely to be due to different diffusion behaviour within the cells that the model was based on (single cells), and those within an insulated module. In addition, the electrical model cannot take account of the complex thermal behaviour within the module, for example cell-to-cell gradients.

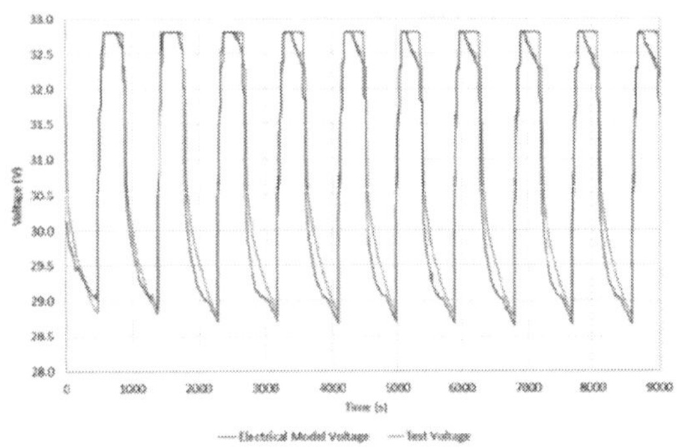

Figure 18. Comparison of model voltage with module life test results

In summary, the low temperature conditions utilised for this project were challenging to model due to low diffusion rates within the electrode having a large effect on the 'resistance' and available capacity. Although there are clearly limitations on the accuracy of the model, the simulation is on the edge of the feasible cell operational window, where the cells' rate capability displays extremely high sensitivity to temperature. However, the simulation results have managed to capture the overall behaviour reasonably.

6 DEVELOPMENT PLAN

Following this testing, it is the view of ABSL that the technology (i.e. the selected cell type for use in low temperature space applications) is currently at TRL4, approaching TRL5. A plan has therefore been developed to increase the TRL to 7.

An overview of the proposed route from TRL4 to TRL7 is shown in Figure 5. The green squares represent complete or partially complete activities, the blue squares represent tests that are ongoing, but may need to be repeated depending on updated power profiles.

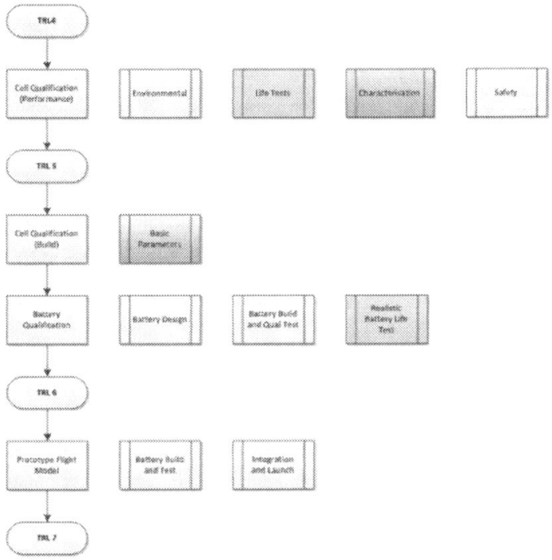

Figure 19. Development Plan

To move from the current position to TRL5, the initial stages of cell qualification should be completed, such that the cell performance in all areas can be confirmed. For example, the cell survival in typical space conditions (vacuum, radiation) and to launch (vibration testing) would need to be confirmed.

To achieve TRL6, further qualification work would be required, so that a fully representative prototype battery can be designed and manufactured. This would include testing to determine spot weld settings, and to obtain data for piece part manufacture.

Finally, to achieve TRL7, a prototype would be manufactured and tested using a standard flight battery process and levels. This would then be delivered for integration into the Martian lander for system level testing prior to launch.

7 CONCLUSIONS AND FURTHER WORK

Work has been undertaken to develop a secondary Li-ion battery from Commercial-off-the-Shelf (COTS) cells than operate at temperatures below -20°C. A previous paper [1] described the search for a suitable cell, and showed the data from the initial module test.

This paper presented the results from the extended module life test. An 8s8p module of COTS cell has now successfully completed 720 Sols under a representative Martian lander power profile.

In addition, the long term effects of cycling after very low temperature storage were presented. Even after these extreme storage conditions, the cell completed the required 180 Sols.

The final part of the project was to produce a performance model of the cells.

A plan has been developed for developing the TRL of this technology to a level suitable for space applications.

8 ACKNOWLEDGEMENTS

This work was performed within the project *Development of a Low Temperature Lithium Ion Battery and Survivability Tests*. The authors want to thank the European Space Agency for funding this project through ESA's Mars Robotic Exploration Preparation (MREP) Programme.

9 REFERENCES

1. Buckle, R., Thwaite, C., Vijendran, S., Schautz, M.; A Low Temperature Li-ion Battery for Mars Landers based on COTS Cells; 10th European Space Power Conference, Noordwijkerhout, The Netherlands, 13-17 April 2014

2. Future Power Systems for Space Exploration: Executive Summary; ESTEC/Contract No. 14565/00/NL/WL

3. ESA; SOW: Development of a Low Temperature Lithium Ion Battery and Survivability Tests, Issue 1, 01/09/2010

SAFT VES16 SOLUTION FOR SMALL GEO

Emmanuel Bonneau [(1)], **Stéphane Remy** [(1)]

[(1)] *Saft, Space and Defence Division, Rue Georges Leclanché 86060 Poitiers – France,*
Email: emmanuel.bonneau@saftbatteries.com, stephane.remy@saftbatteries.com

ABSTRACT

At the end of 2008, Saft has designed the VES16 cell [1], a new Li-ion cell aiming to address Low Earth Orbit (LEO) satellite market.

In the frame of ESA GSTP 5.2 activity, Saft has then developed and successfully qualified [2] a range of batteries based on the VES16 cell and its autonomous Simplified Balancing System (SBS), targeting both LEO and Geostationary Earth Orbit (GEO) missions

The very good performance of the VES16 cell demonstrated in GEO mission cycling configuration allows now Saft to propose it on small geostationary spacecraft with typical power lower than 10kW, offering a simpler, autonomous and competitive solution compared to Saft high capacity cells solutions with VES180 qualified on several telecommunication spacecraft.

This paper will present the design of the VES16, its performance for GEO mission, the configuration of batteries based on this cell, with their benefit and drawbacks, and will detail two current GEO applications using VES16 batteries.

1. VES16 LI-ION CELL DESIGN

The VES16 Li-ion cell has been designed by Saft especially for space applications.

The development was conducted with the support and partnership of CNES French agency, until the qualification successfully completed in July 2011.

Figure 1. VES16 Li-ion cell.

The design of the VES16 cell is based on Nickel Cobalt Aluminium (NCA) electrochemistry. The VES16 cell has then a NCA positive electrode and a graphite negative electrode, and is filled with a $LiPF_6$ based electrolyte.

It combines benefits and heritage of low capacity MPS Li-ion cells in term of size and mechanics, and advantages of Nickel oxide base Li-ion chemistry of VES100-140-180 cells.

Interfaced in a D-size, the cell has a can in stainless steel connected to the negative polarity whereas the positive polarity is on the top of the cell.

Regarding safety aspects, a circuit breaker is included in the cell design as part of the positive terminal for protection against abusive conditions. Only the internal cell pressure is used to detect failures and to trigger the cell open circuit. This activation is non-reversible. This embedded device is very useful to architect the battery in SP configuration.

As a second barrier, each cell is equipped with a weakened cover that prevents the cell to burst (leak before burst design), operating when the internal pressure rises above the average 27 bar absolute pressure.

Its main characteristics are as follows:

Dimensions (Ø x H)	33 x 60 mm (D-size)
Weight	≤ 114.4 g
Volume	0.051 dm^3
Voltage range	[2.7 ; 4.1] V
Nominal capacity (C)	4.5 Ah on 4.1-2.7V @ C/2, 20°C
Nominal energy	16 Wh on 4.1-2.7V @ C/2, 20°C
Specific energy	> 140 Wh/kg
Internal resistance	≤ 35 mΩ @ 20% DoD
Operating temperature	[+10 ; +30] °C
Mech. design margins	EWR & ECSS compliant

Table 1. VES16 Li-ion main characteristics.

 DOI: 10.1051/e3sconf/20171606004

2. VES16 PERFORMANCE FOR GEO MISSIONS

2.1. Calendar losses

The design of the VES16 cell is based on similar electrochemistry as VES100-140-180 cells, which follows the same calendar ageing laws induced by lithium corrosion at the negative electrode.

Fig. 2 below presents the capacity fading of the VES16 cell versus time and for different temperatures. The average data obtained from results of the calendar life-tests are represented with dots.

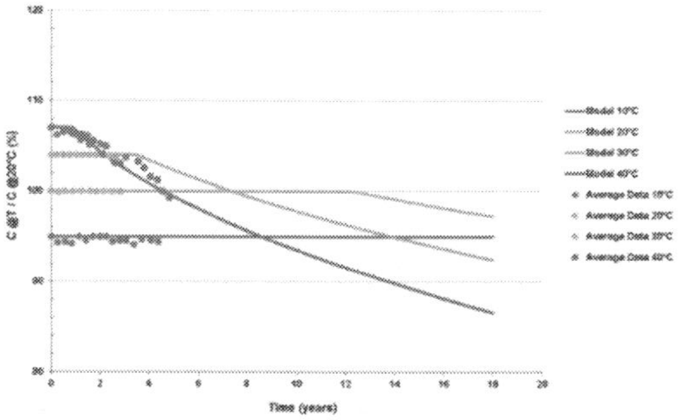

Figure 2. Capacity fading versus time & temperature.

For the VES16 cell, the capacity degradation at +20°C appears only after more than 12 years, once that the complete Lithium reserve included in the negative electrode has been consumed.

This makes it particularly suitable to extended lifetimes involving longer on-ground storage periods and in-orbit Geostationary Earth Orbit (GEO) mission durations.

2.2. GEO mission cycling performance

Accelerated GEO life-tests performed at 60%, 70% and 80% of Depth of Discharge (DoD) have demonstrated the very good performance of the VES16 cell for GEO applications.

As an example, accelerated GEO life-test without solstice (worst case compared to real-time or semi-accelerated life-test) at 80% DoD and End-Of-Charge Voltage (EOCV) of 4.05V shows about only 2% of energy losses after 30 seasons (15 years of GEO mission) and less than 4% after 45 seasons completed, as shown on Fig. 3 below.

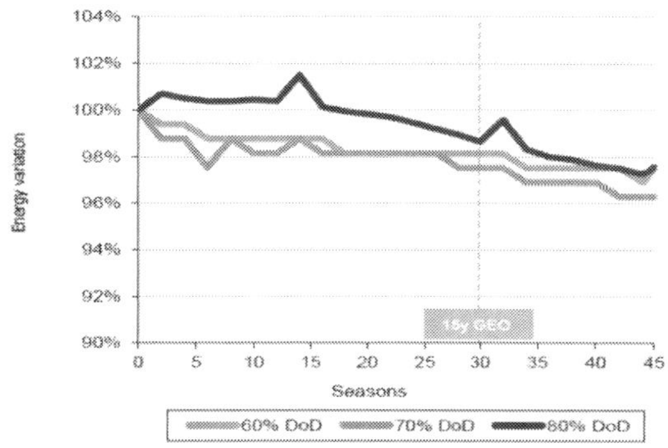

Figure 3. Variation of energy in check-ups.

Moreover the VES16 offers a very good stability of the End-Of-Discharge Voltage (EODV) during the 23rd discharge (72 min) of eclipse seasons, as shown on Fig. 4, Fig. 5 and Fig. 6 below, respectively during GEO cycling at 60%, 70% and 80% DoD.

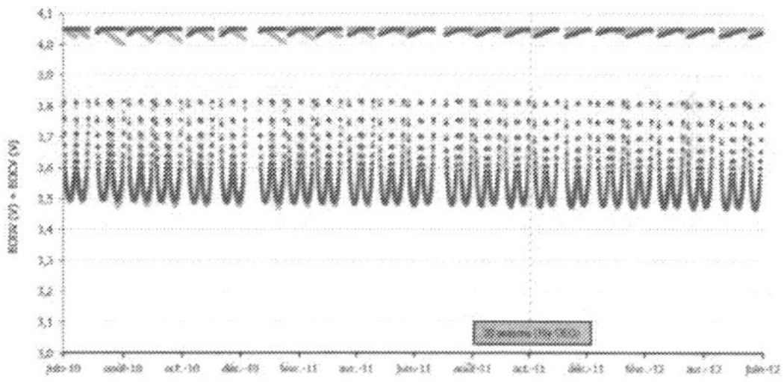

Figure 4. EODV evolution during VES16 accelerated GEO cycling at 60% DoD.

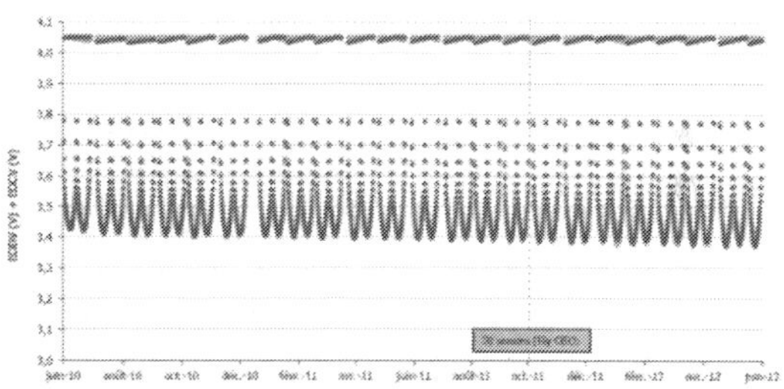

Figure 5. EODV evolution during VES16 accelerated GEO cycling at 70% DoD.

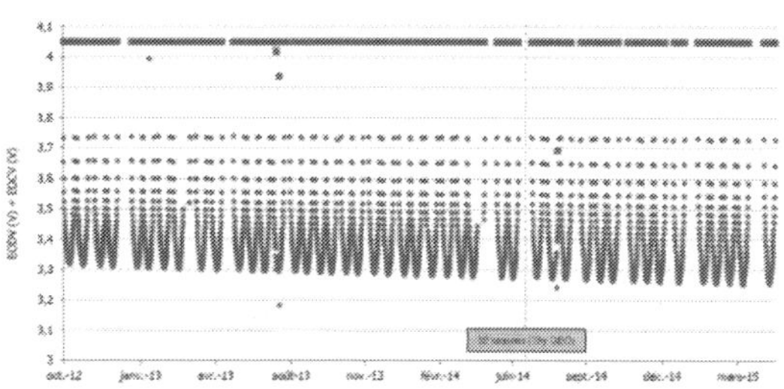

Figure 6. EODV evolution during VES16 accelerated GEO cycling at 80% DoD.

Semi-accelerated life-test has also recently started for analysing the performance of VES16 cells for GEO cycling with additional Plasmic Propulsion System (PPS) cycles and demonstrating its ability for full GEO electric propulsion spacecraft [3].

It consists in two additional PPS cycles at 10% @ 20°C during equinox periods (two GEO cycles per day at 80% DoD) and six PPS cycles at 20% DoD per day during solstice periods.

After 14 seasons, the results show that the effect of PPS discharge on the GEO cycling tests are negligible in term of energy loss (see Fig. 7), EoDV drop and internal resistance increase, perfectly line with predictions.
This confirms the excellent ability of VES16 cell to sustain GEO PPS missions.

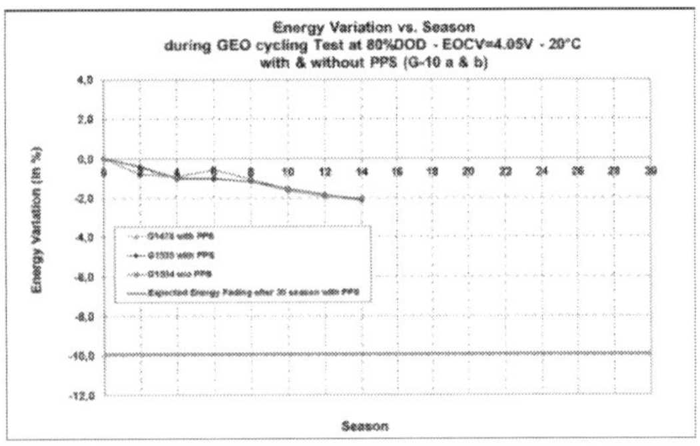

Figure 7. Energy variation in GEO cycling with & without PPS.

3. VES16 BATTERIES CONFIGURATION

The VES16 batteries design is based on Serial-Parallel (S-P) architecture, compared to the Parallel-Serial (P-S) architecture of VES100-140-180 batteries design.

Cells are first connected in series to constitute strings and obtain the required battery voltage.

Strings are then connected in parallel to reach the capacity needed for mission.

This architecture, interesting for small cells capacity, is especially adapted to configurations with P >> S and compatible of unregulated bus.

But this also implies to have safety devices to put in open circuit the string in failure in order to prevent the risks associated to abusive conditions.

Three levels are implemented at cell level, comprising:

- separator shut down effect,
- a circuit breaker inside the cell on the positive connexion,
- two weakenings on the cover for the venting system.

Due to the S-P architecture, a cell in failure leads in the worst case to loss the associated string and so a part of the battery capacity, compared to the P-S architecture where this will lead to a reducing of the battery voltage. It has been shown by test [4] during the unlikely event of a soft short, that the behaviour of the battery can be either a stabilization of the overcharged and over-discharged cells in a sort of steady state, or in the cell opening leading to the loss of one string.

Regarding electronics, a Simplified Balancing System (SBS), implemented across each cell, is used to balance cell voltages within each string at the end of charge and to reduce the cell degradation by limiting excursion beyond 4.1 Volts.

Balancing is made by draining a portion of the charge current above a voltage threshold.

Fig. 8 below gives an example of balancing done on the GSTP-5 10S5P VES16 Battery during GEO cycling after a string was voluntarily unbalanced (voltage spread within the string was at least 50 mV). Cell voltages convergence and stabilization (15mV maximum unbalance per string) is reached at half season.

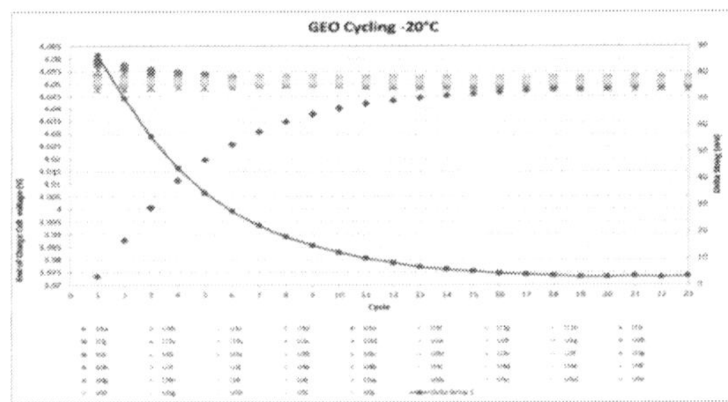

Figure 8. GEO balancing test performed on 10S5P VES16 Battery with string unbalanced.

This balancing system is fully autonomous and requires no command or telemetries.

The Power Conditioning and Distribution Unit (PCDU) is thus simplified as compared to a solution with high capacity cells such as VES180SA cells (voltage of each packs of cells in parallel shall be monitored, command is necessary for managing by-pass, balancing system is implemented in the PCDU if not integrated in the battery).

4. GEO APPLICATIONS WITH VES16 CELLS

Saft is currently working on two GEO applications where the VES16 cell has been selected (Tab. 2), and probably more in the near future.

Program	Orbit	Life	Battery configuration	DoD
Program X	GEO	8.5 yrs	11SxP (x>20)	50-70%
Program Y	GEO	15 yrs	11SyP (y>50)	50-70%

Table 2. VES16 Li-ion Batteries for GEO applications.

These programs are roughly presented in sections 4.1 and 4.2. Section 4.3 will give a trade-off between a VES16 battery and a VES180SA battery that we could propose for a small GEO application (study case).

4.1. Program X

For this program, the choice of VES16 cells has been driven by the atypical mission profile especially due to a long term period storage of more than 10 years. Considering the acceptance and integration activities, the on-ground lifetime reaches 16 years.

At the End of Life (EoL), i.e. after the on-ground period and the in-orbit period of 8.5 years, the 11S24P battery sized to fulfil this mission (DoD > 50%) is predicted to have less than 7% of energy losses EoL.

Regarding the design, the 11SxP battery is composed of four modules, connected in parallel. A junction box is fitted on one of the modules for proposing a single electrical interface. Interconnection harness is included in the battery.

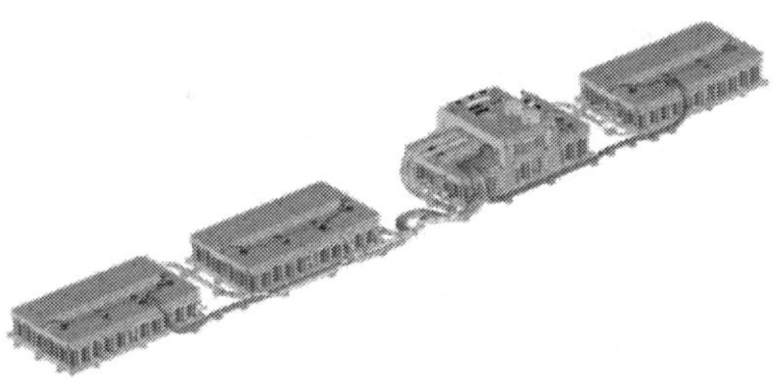

Figure 9. Illustration of the 11SxP battery.

Dimensions of one module		Weight of battery
Length	408 mm	42.0 kg
Width	240 mm	
Height without Junction Box	98 mm	
Height with Junction Box	160 mm	

Table 3. Dimensions and weight of the 11SxP VES16 battery.

4.2. Program Y

For this GEO mission, Saft is currently designing an 11SyP battery, able to provide up to 7 kW (DoD of 50-70%), for a life-time of 15 years (30 seasons).

The Battery is composed of four identical battery packs (Fig. 10) to be connected in parallel.

Figure 10. Illustration of the battery pack.

Dimensions		Weight
Length	454 mm	24.8 kg
Width	338 mm	
Height	163 mm	

Table 4. Dimensions and weight of the battery pack.

Both batteries requires no specific commands, they are interfaced with power connectors and telemetry connectors (voltage, temperature, heaters if any).

For the voltage, intermediate measurement like half string voltages is implemented to assess battery health along life on ground.

The combination of available calendar tests and cycling tests are very helpful to accurately design life in orbit either in solstice or equinox periods with a VES16 battery.

4.3. Trade-off design

As an example, we could imagine a small GEO mission which would be defined as follows:

- AIT and storage: 3 years @ 20°C
- Mission: 10 years @ 20°C during eclipses seasons and 10°C during solstices periods
- Discharge power during eclipses: 1500 W

DOI: 10.1051/e3sconf/20171606004

- Eclipse seasons: 45 eclipses, the longest one lasting 72 minutes
- Battery voltage range: [38.4 V – 49.2 V]
- Maximum DoD: 80%

With a VES16 solution, a 12S12P battery (Fig. 11) allows to fulfil the proposed mission.

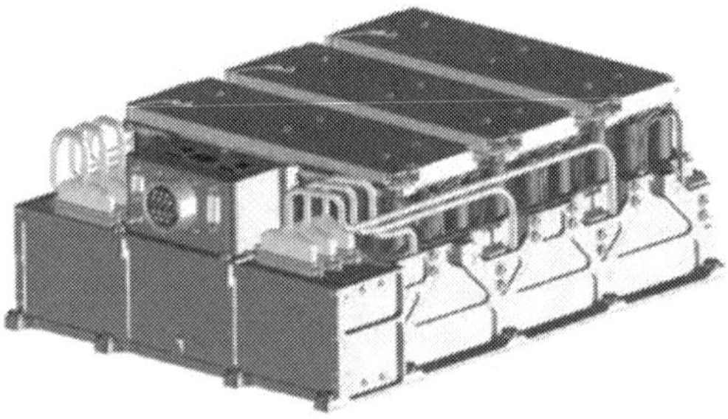

Figure 11. Illustration of the 12S12P VES16 battery.

The maximum DoD is 79%. The energy losses at EoL are less than 3%.

Physical characteristics of the 12S12P VES16 battery are as follows:

Dimensions		Weight
Length	454 mm	24.8 kg
Width	338 mm	
Height	163 mm	

Table 5. Dimensions and weight of the 12S12P VES16 battery.

With a VES180SA solution, the sizing shows that a 2P12S battery (Fig. 12) is necessary.

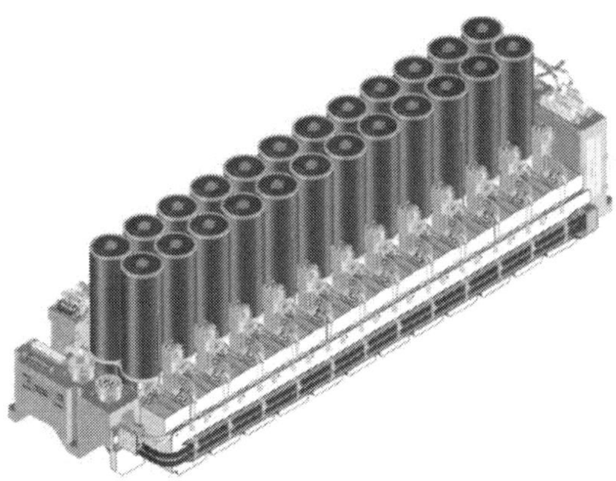

Figure 12. Illustration of the 2P12SVES180SA battery.

The maximum DoD is 44%. The energy losses at EoL are about 1%.

Physical characteristics of the 2P12S VES180SA battery are the followings:

Dimensions		Weight
Length	872 mm	43.7 kg
Width	270 mm	
Height	269 mm	

Table 6. Dimensions and weight of the 2P12S VES180SA battery.

With this simple study case, we can see that a VES16 solution can be preferable for a small GEO mission, offering a smaller and lighter battery, requiring nor telemetry or command.

Of course, this trade-off is dependent on platform level features. The result of the trade-off is also depending on the capacity needed by the battery sizing, i.e. if you are far or not from a multiple of large capacity cell. Assuming that a VES180 is equivalent to 10 VES16 in parallel, you can figure out which capacity points are more interesting in VES16 or VES180.

Based on the modularity of VES16 batteries, there is no technical limit to achieve a very large configuration as for example 30 kW, but however, we consider that 10 kW virtual limit is relevant to question the use of VES16 battery in GEO spacecraft.

It can be as well identified some interesting product lines commonalities, between GEO and LEO batteries solutions at prime level. Said differently: same product, 2 types of missions, GEO and LEO.

5. REFERENCES

1. Remy, S., Prevot, D., Reulier, D. & Vigier, F. (2011). VES16 Li-ion cell for satellites. 9th European Space Power Conference.
2. Remy, S., Mosset, E., Lefeuvre, S. & Nestoridi, M. (2014). VES16 Li-ion batteries range qualification. 10th European Space Power Conference.
3. Prevot, D., Borthomieu, Y., Malet, F., Rebuffel, C., Carre, A. & Cenac-Morthe, C. (2016). Saft solutions for full GEO electric propulsion. 11th European Space Power Conference.
4. Nestoridi, M., Mourra, O., Mosset, E., Lefeuvre, S. (2014). Effect of an induced cell soft-short at VES16 battery module. 11th European Space Power Conference.

PERFORMANCES OF SAFT LITHIUM-ION CELLS IN LEO CYCLING

D. Prevot [(1)], Y. Borthomieu [(1)], E. Ligneel [(1)], R. Hague [(1)], J-P. Peres [(1)], C. Cenac-Morthe [(2)]

[(1)] *Saft Space and Defence Division BP 1039, 86060 Poitiers, France, Yannick.Borthomieu@saftbatteries.com, Didier.Prevot@saftbatteries.com, Eric.Ligneel@saftbatteries.com, Richard.Hague@saftbatteries.com, Jean-paul.Peres@saftbatteries.com*

[(2)] *CNES, 18 avenue Edouard Belin, 31401 Toulouse Cedex9, France, Celine.Cenac-Morthe@cnes.fr*

ABSTRACT

Based on good LEO lifetime tests results, Saft proposes currently VES16 Lithium-ion for 12 years mission in Low Earth Orbit (LEO) up to 30%DoD or more within a temperature range from +10°C to +30°C.
The first flight experimentation started at the end of 2015, confirming its ability to withstand the environmental stresses induced by a space launch and the in-flight constraints generated by a LEO mission.
The upcoming launches scheduled from 2016 will significantly increase the flight heritage of this cell.

From the beginning of VES16 Lithium-ion cell development, a wide lifetime tests program has been implemented in order to evaluate the different degradation modes due to the conditions applied in LEO cycling, as presented in Table 1.
This program has allowed assessing the influence of different parameters such as EOCV (End Of Charge Voltage), DoD (Depth of Discharge), charge current and temperature.

New MP176065 XTD Lithium-ion cell, developed & qualified by Saft, with an operating range extended from -40°C to +85°C, has already been selected for EXOMARS Rover mission.
Promising results in accelerated LEO cycling conditions at 30%DOD has been observed and compared to VES16 lithium-ion cell results. Manufactured on a mass production industrial line, this cell is well suited for low cost missions (eg. Mega Constellations…).

The article will thus present the whole LEO cycling results available for the two cells, and will provide afterwards the correlation status of Saft Li-ion Model (SLIM) with all the experimental data acquired.

Keywords: DOD: Depth of Discharge; **OCV**: Open Circuit Voltage; **EOCV**: End Of Charge Voltage; **EODV**: End Of Discharge Voltage; **GEO**: Geosynchronous Orbit; **IR**: Internal Resistance; **LEO**: Low Earth Orbit; **MEO**: Medium Earth Orbit; **SLIM**: Saft Li-Ion Model; **SoC**: State of Charge.

1. VES16 LITHIUM-ION CELLS LEO LIFE-TIME TESTS RESULTS

This chapter focuses on the analysis of the life-test results performed in accelerated (A) or in real time (RT) mode according to VES16 lithium-ion cells LEO life-tests program described in Tab.1.

Life-Test Mode	Tests Conditions			Cycle Number done
	EOCV	DOD	TEMP	
Acceler.	4.05V	20%	20°C	37 400
Acceler.	4.05 V	30%	20°C	33 720
Real Time	4.05V	20%	20°C	27 000
Real Time	4.05V	20%	30°C	25 000
Real Time	3.9V	20%	20°C	27 000
Real Time	4.05V	30%	20°C	27 000
Real Time	4.05V	30%	30°C	15 500
Real Time	4.05V	10% 15% 20%	10°C	15 000
Real Time	4.05V	20%	20°C	15 000
Real Time	4.05 V	40%	20°C	15 500
Real Time	C/3 4.05V	20%	5°C	3 500
Real Time	4.075V	20%	20°C	15 500
Real Time	4.1 V	20%	20°C	15 500
Real Time	C/5, C/3, C/2.5, C/2 4.05V	20%	20°C	12 500
Real Time	4.05V	50%	20°C	3 500

Table 1: VES16 Li-ion cells On-Going LEO Life-Tests Program

 DOI: 10.1051/e3sconf/20171606005

These LEO Life-Test results have permitted to assess the influence of the cycling parameters such as DOD/charge current, EOCV and temperature on the evolution of the cell electrical characteristics such as EODV, Capacity, Internal Resistance and Energy, as presented in the following paragraphs.

The evolution of these cell electrical characteristics is measured through a check-up sequence, involving a residual discharge and a reference test at 4.1Volts @20°C, performed at 500 cycles, then every 1000 cycles.

This reference test is based on charge up to 4.1V, then a discharge down to 2.7V under C/2 discharge rate @20°C with pulses 1.5C-30seconds at 20& 60%DOD.

It allows also demonstrating the correlation of the LEO life-test results with the SLIM V3 ageing prediction model.

1.1 EOC Voltage Influence

The evaluation of the EOCV impact on the VES16 cell behaviour is performed from the results obtained in LEO cycling at 20%DOD - @20°C.

That is done on QM2/QM3 cells in the EOC Voltage range from 3.9 to 4.05Volts, then on FM3 cells in the EOC Voltage range from 4.05 to 4.1Volts.

After respectively 25000 cycles on QM3 cells & 27000 cycles on QM2 cells in LEO cycling at 20%DOD, no significant effect of EOCV between 3.9 & 4.05Volts on EOD Voltage variation slope and IR0%DOD increase, as presented in Figure 1 and 2.

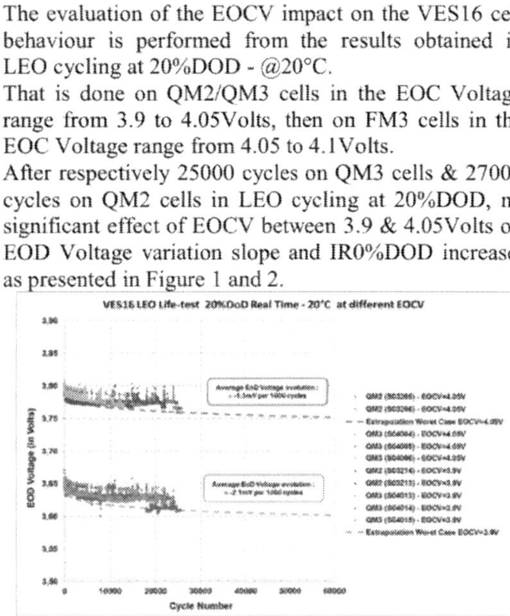

Figure 1. EOD Voltage evolution vs. EOCV & Cycle Number

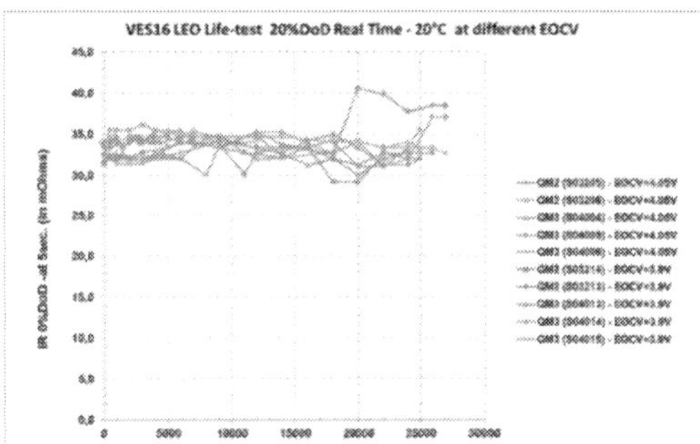

Figure 2. Internal Resistance vs. EOCV & Cycle Number

We can just note that capacity loss is reduced by 4% (i.e. 6% after extrapolation at 60000 cycles) when EOC Voltage is limited to 3.9Volts, refer to Fig. 3. However, the available capacity is reduced by 9% due to the limitation of SOC (≈15%) between EOCV = 4.05 and 3.9Volts.

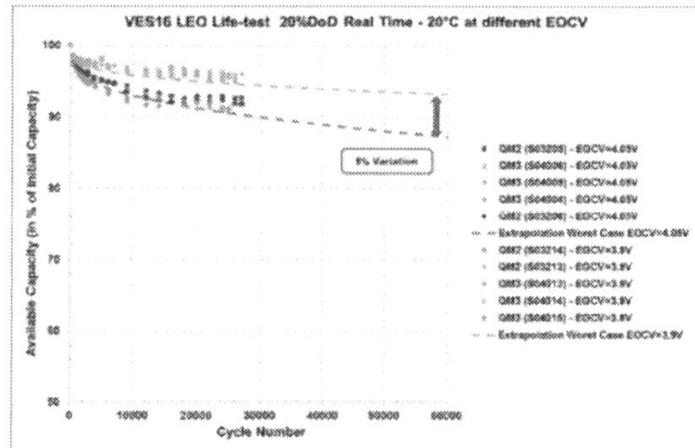

Figure 3. Available capacity vs. EOCV & Cycle Number

After 15500 cycles in LEO cycling at 20%DOD on FM3 cells, no significant effect of EOCV between 4.05 & 4.1Volts has been observed on EOD Voltage variation slope, as presented in Figure 4.

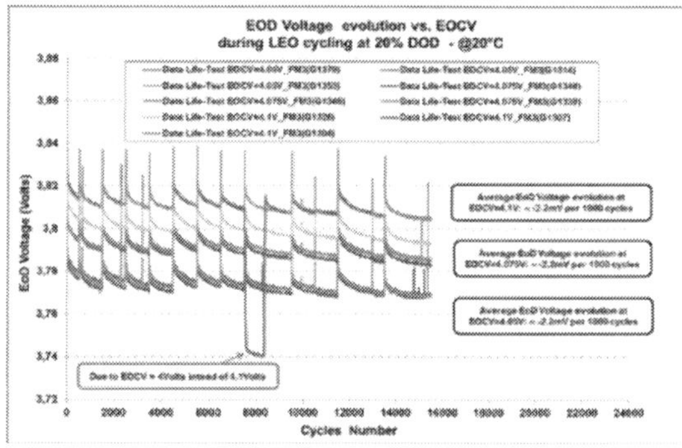

Figure 4. EOD Voltage evolution vs. EOCV & Cycle Number

Negligible variation observed on Capacity/Energy losses (less than 1%) and on IR at 20% & 60%DOD increase, as presented in Fig. 5, 6, 7 and 8.

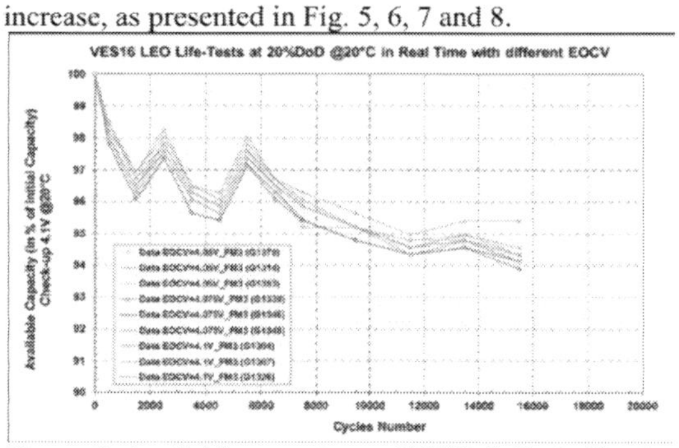

Figure 5. Available capacity vs. EOCV & Cycle Number

 DOI: 10.1051/e3sconf/20171606005

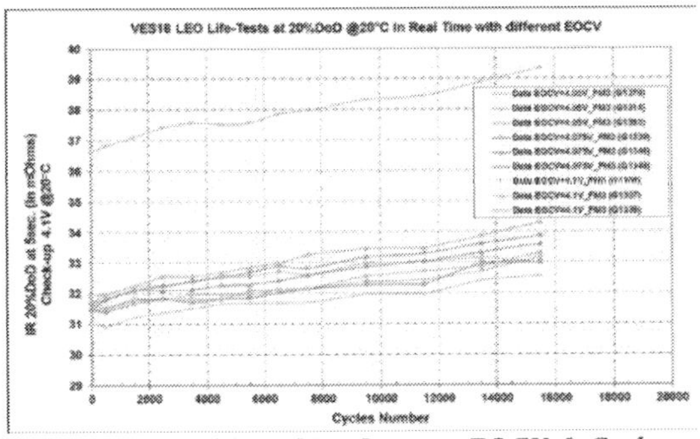

Figure 6. IR 20%DOD - 5sec. vs. EOCV & Cycle Number

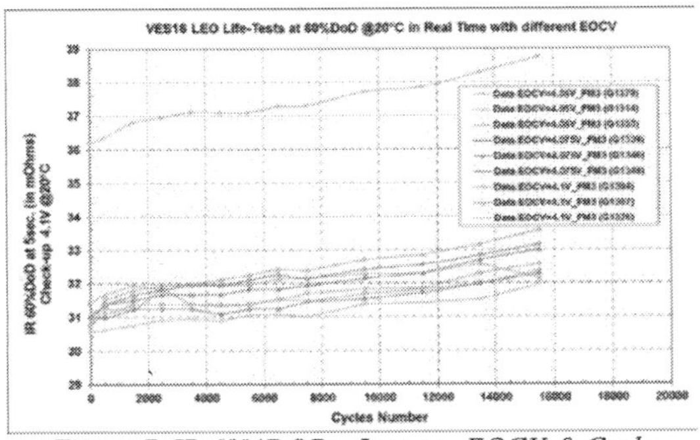

Figure 7. IR 60%DOD - 5sec. vs. EOCV & Cycle Number

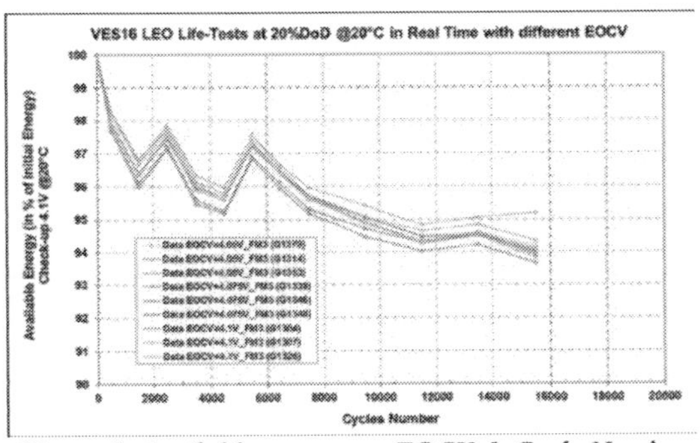

Figure 8. Available energy vs. EOCV & Cycle Number

1.2 DOD & Charge Current Influence

The evaluation of DOD & charge current impact on the VES16 cell behaviour is performed from the results obtained in LEO cycling at EOCV=4.05Volts - @20°C.

That is done on QM2/QM3 cells within DOD range from 20 to 30%, then on FM cells within DOD range from 20% to 50%.

In LEO cycling conditions, DOD & charge current change together. That comes from the fact that the charge duration is limited and requires to compensate DOD increase by a higher charge current, as presented in Table 2.

DOD (%)	Current (Amps)	Maximum Duration (minutes)
20	C/5	65
30	C/3	65
40	C/2.5	65
50	C/2	65

Table 2: Charge conditions vs. DoD

After respectively 27000 cycles on QM2 and 25000 cycles on QM3 cells in LEO cycling at EOCV=4.05Volts @20°C, a very limited EOD voltage evolution slope increase is observed from -1.5mV to -1.6mV per 1000 cycles between 20% & 30%DoD, as presented in Fig. 9.

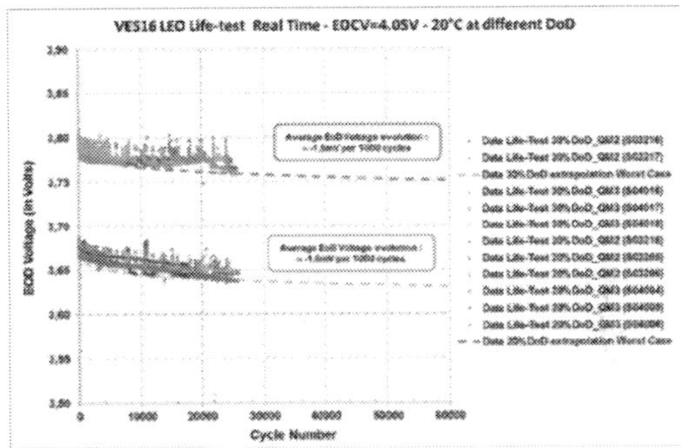

Figure 9. EOD Voltage evolution vs. DOD & Cycle

Higher capacity loss is observed when DoD increases, ≈1.5% between 20% & 30%DoD after respectively 27000 cycles on QM2 cells & 25000cycles on QM3 cells, and ≈2% expected at 60000 cycles, as presented in Fig. 10.

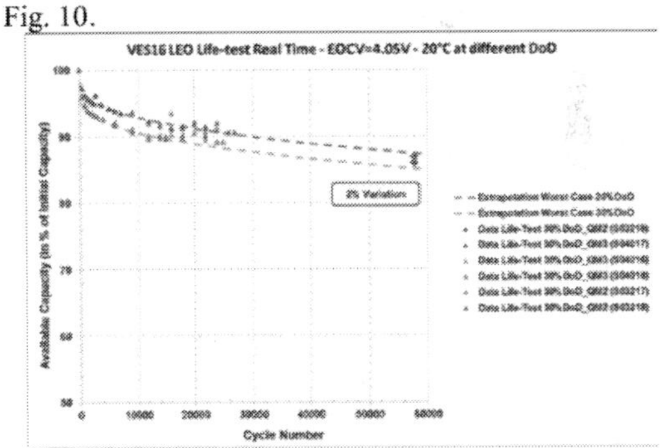

Figure 10. Available capacity vs. DOD & Cycle Number

No visible effect of DOD & charge current increase on the QM2/QM3 cells internal resistance, as presented in Fig.11.

DOI: 10.1051/e3sconf/20171606005

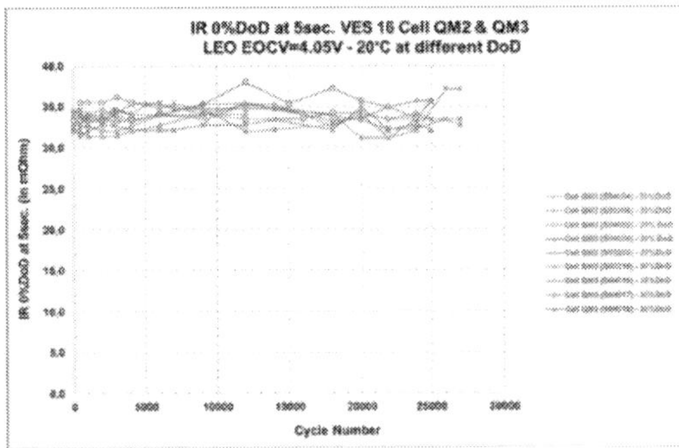

Figure 11. Internal Resistance vs. DOD & Cycle Number

After respectively 15500 cycles on FM cells in LEO cycling at EOCV=4.05Volts @20°C, significant EOD voltage evolution slope increase from -2.2 to -3.5mV per 1000 cycles between 20% & 40%DOD, as presented in Fig. 12.

Promising results are observed on FM cells in LEO cycling at 50%DoD with EOD voltage evolution slope which stabilizes from 1500 cycles, as shown in Fig. 12.

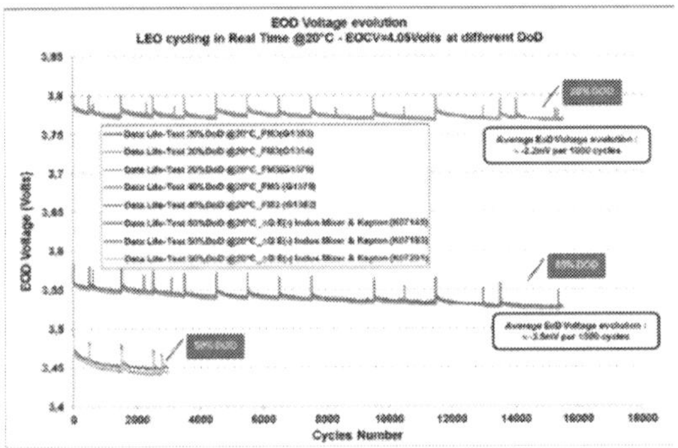

Figure 12. EOD Voltage evolution vs. DOD & Cycle

Higher energy/capacity losses are observed when DOD increases, ≈4% between 20% & 40%DOD after 15500 cycles and ≈2% between 40% & 50%DOD after 3500 cycles, as presented in Fig. 13 and 14.

At 60000 cycles, these energy/capacity losses should be less than about 9% between 20% & 40%DoD and less than about 7% between 40% & 50%DoD.

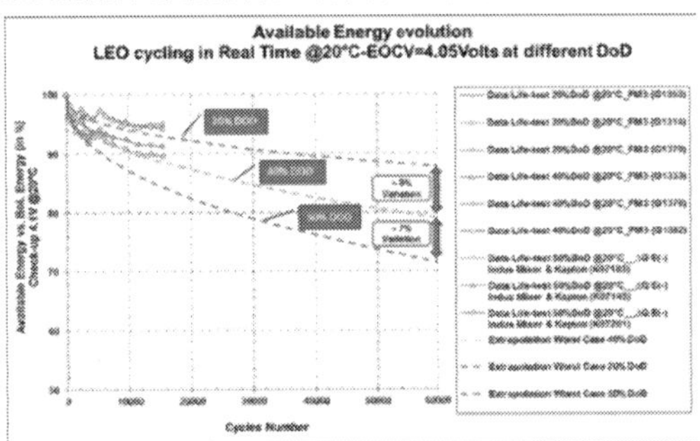

Figure 13. Available energy vs. DOD & Cycle Number

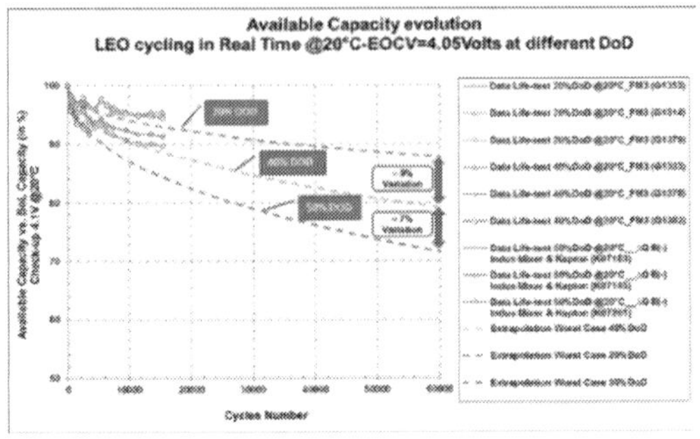

Figure 14. Available capacity vs. DOD & Cycle Number

No visible effect of DOD & charge current increase on FM cells internal resistance evolution, as presented in Fig. 15 and 16.

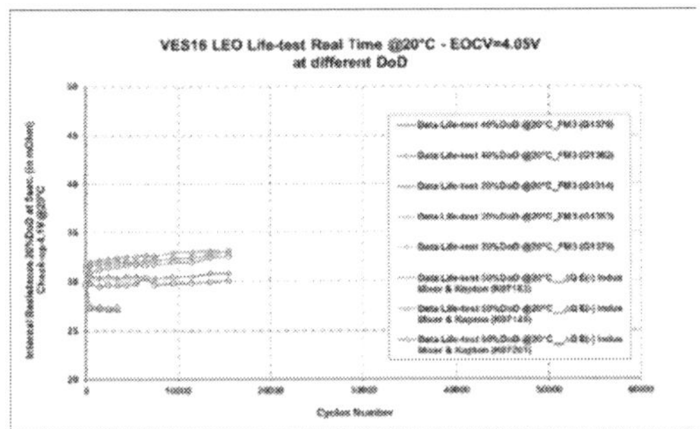

Figure 15. IR 20%DOD vs. DOD & Cycle Number

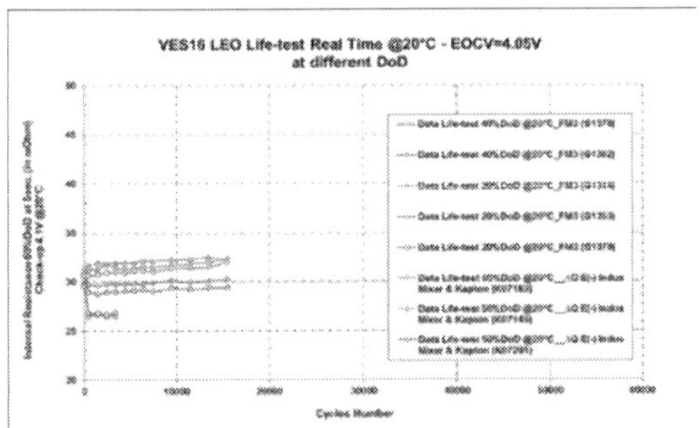

Figure 16. IR 60%DOD vs. DOD & Cycle Number

1.3 Temperature Influence

The evaluation of temperature impact on the VES16 cell behaviour is performed from the results obtained in LEO cycling at EOCV=4.05Volts - 20%DO.

That is done on QM2/QM3 cells within the temperature range from 20 to 30°C, then on FM cells within the temperature range from 5 to 20°C.

After respectively 20000 cycles on QM2 cells & 25000 cycles on QM3 cells in LEO cycling at EOCV=4.05Volts – 20%DOD, weak EOD voltage evolution slope increase is observed from -1.5mV to -

DOI: 10.1051/e3sconf/20171606005

2.3mV per 1000 cycles between 20°C & 30°C, as presented in Fig. 17.

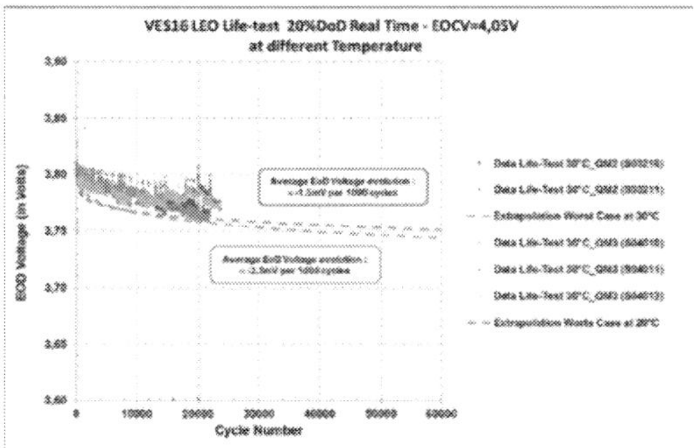

Figure 17. EOD Voltage evolution vs. Temp. & Cycle Number

When the temperature increases from 20°C to 30°C, higher capacity loss is observed, ≈4% after respectively 20000 cycles on QM2 cells & 25000cycles on QM3 cells, and ≈7% expected at 60000 cycles, as presented in Fig. 18.

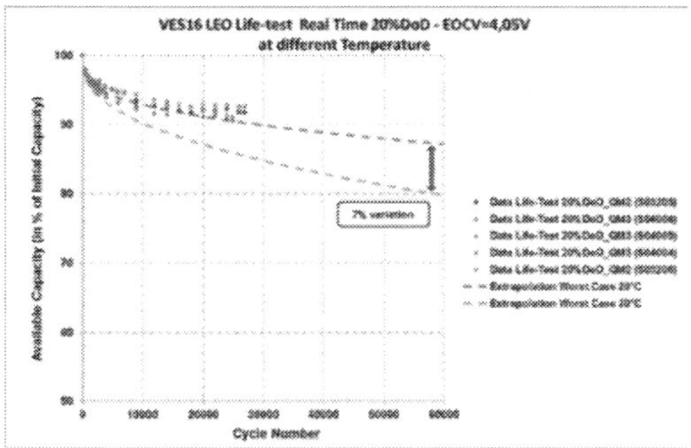

Figure 18. Available capacity vs. Temp. & Cycle Number

No visible effect of DOD & charge current increase on QM2/QM3 cells internal resistance, as presented in Fig. 19.

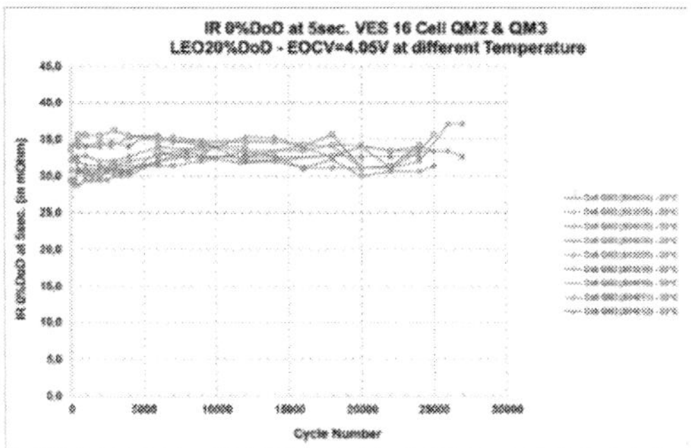

Figure 19. Internal Resistance vs. Temp. & Cycle Number

After 15000 cycles on FM3 cells in LEO cycling at EOCV=4.05Volts - 20%DOD, the same EOD voltage evolution slope is observed at 10°C & 20°C, about -2.4mV per 1000 cycles, as presented in Fig. 20.

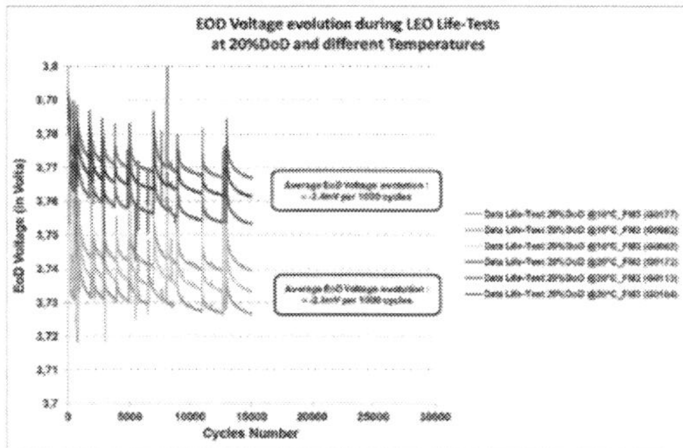

Figure 20. EOD Voltage evolution vs. Temp. & Cycle Number

No significant capacity/energy losses are also measured within the temperature range from 10°C to 20°C, as presented in Fig. 21 and 22.

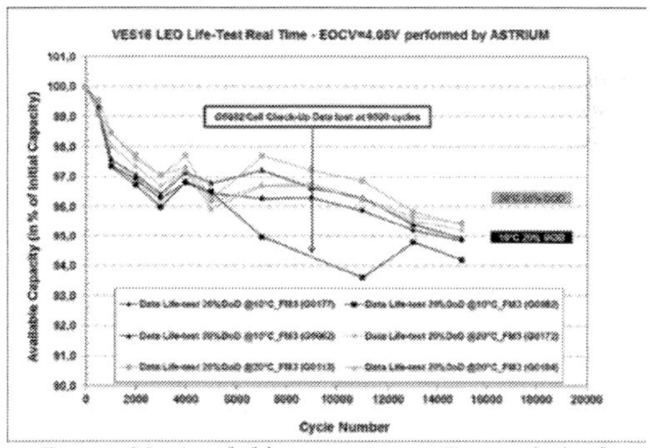

Figure 21. Available capacity vs. Temp. & Cycle Number

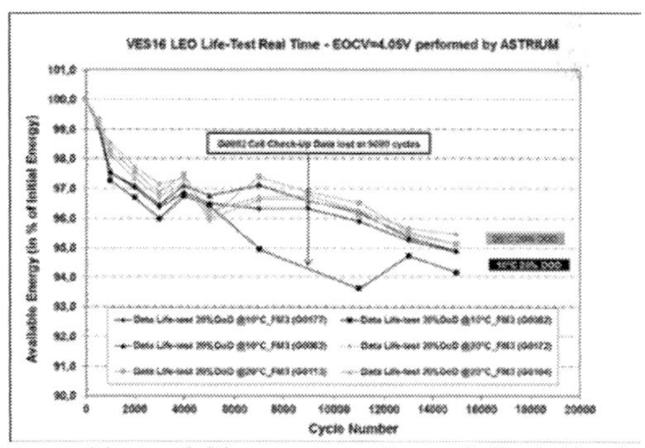

Figure 22. Available energy vs. Temp. & Cycle Number

No more visible effect of the temperature between 10°C & 20°C on the FM3 cells internal resistance evolution, as presented in Fig. 23 and 24.

DOI: 10.1051/e3sconf/20171606005

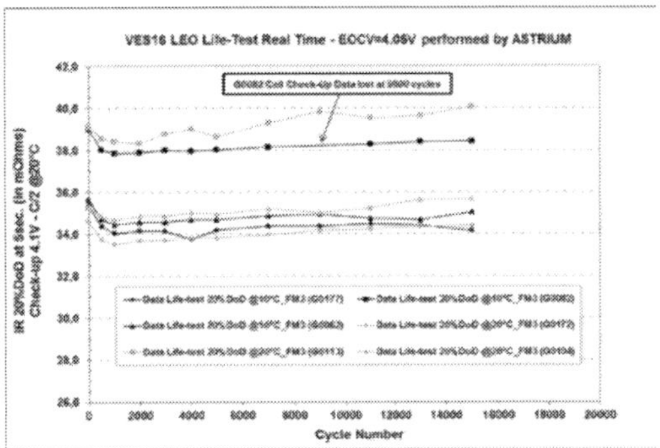

Figure 23. IR 20%DOD vs. Temp. & Cycle Number

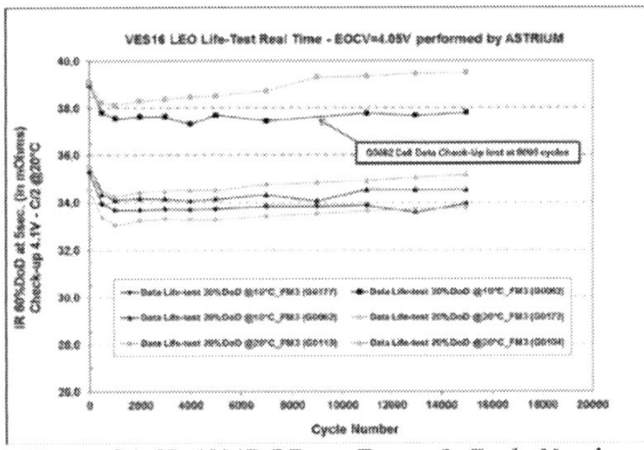

Figure 24. IR 60%DOD vs. Temp. & Cycle Number

After 3500 cycles on FM cells in LEO cycling at EOCV=4.05Volts - 20%DOD with C/3 charge current, the EOD voltage evolution slopes observed at +5°C & +20°C are very similar, respectively -5.4 and -5.1 mV per 1000 cycles, as presented in Fig. 25.

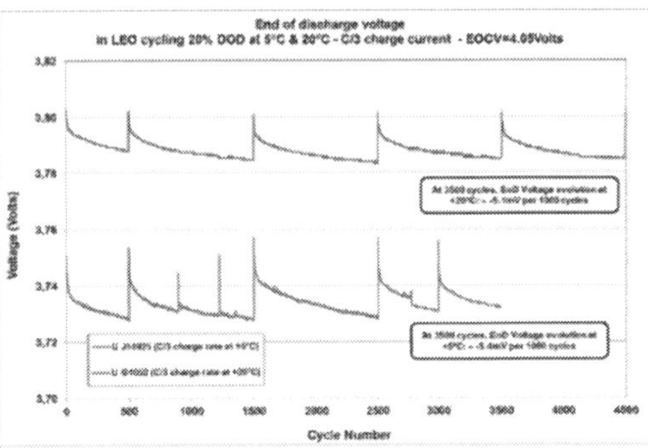

Figure 25. EOD Voltage evolution vs. Temp. & Cycle Number

No significant capacity/energy losses are also measured within the temperature range from +5°C to +20°C, as presented in Fig. 26 and 27.

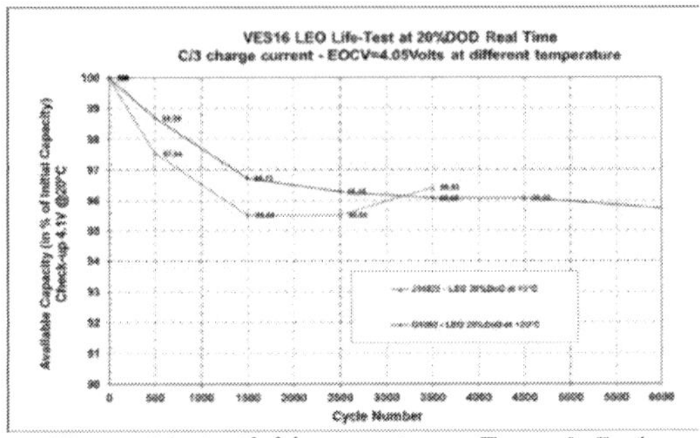

Figure 26. Available capacity vs. Temp & Cycle Number

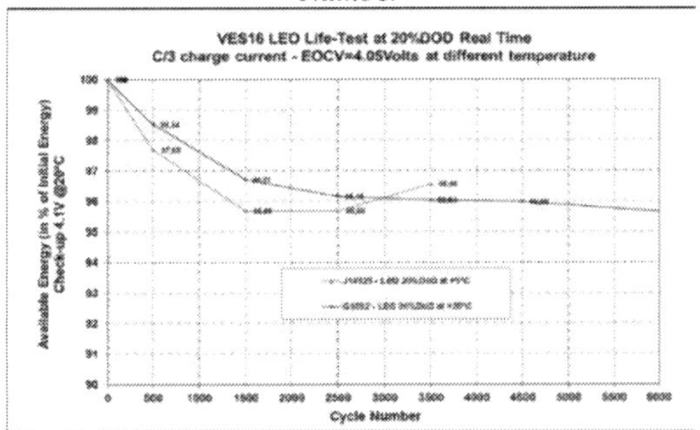

Figure 27. Available energy vs. Temp & Cycle Number

No more visible effect of the temperature between +5°C & +20°C on the FM cells internal resistance evolution, as presented in Fig. 28 and 29.

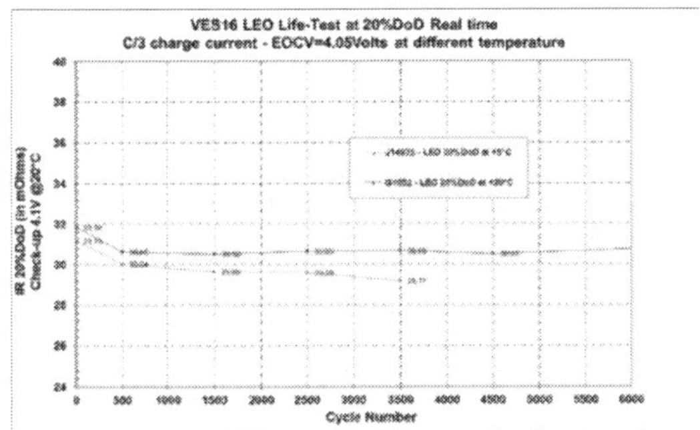

Figure 28. IR 20%DOD vs. Temp. & Cycle Number

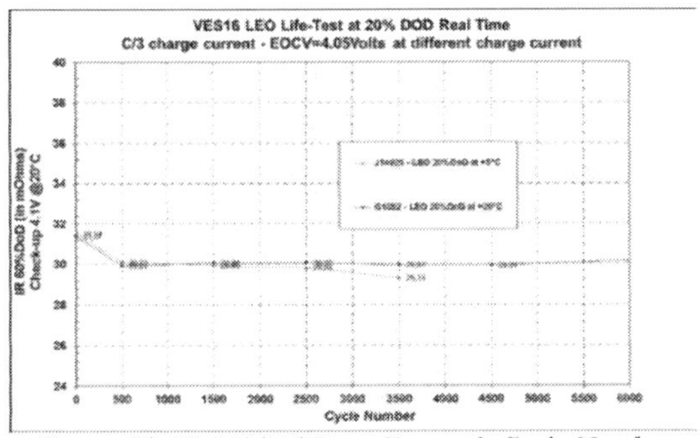

Figure 29. IR 60%DOD vs. Temp. & Cycle Number

DOI: 10.1051/e3sconf/20171606005

1.4 SLIM Correlation based on LEO Life-Test Results

SLIM is a macroscopic electrochemical model based on energy. Its main purpose is to predict the VES and MPS cell performances during the life for GEO, MEO and LEO missions.

This model is based on electrochemical characteristics such as Energy, Capacity, EMF, Internal resistance, EOCV. It uses fading and calendar law effects on energy and internal impedance vs. time, temperature, End of Charge voltage. The degradation mechanisms at electrode levels have been set up based on the Destructive Physical Analyses that were focused on the electrochemical changes. The model is also able to deliver the battery performances using mission figures and profiles.

SLIM V3 correlation has been done from 30%DOD LEO cycling Life-test conditions involving:
- Charge at C/3 constant current up to 4.05Volts, then tapering up to I = 20mA or maximum duration = 65 minutes.
- Discharge at 2.31 Amps during 35 minutes equivalent to 1.35Ah discharged.

As shown in Fig. 30, the EOD voltages measured or extrapolated are much higher than the predicted ones, around 15mV after 25000 cycles on QM3, 25 mV after 27000 cycles on QM2, and 50mV at 50000 cycles.

As presented in Fig.31, the available capacity calculated by SLIM V3 fits at less 6% with the worst case 30%DoD LEO life-tests data after 50000 cycles.

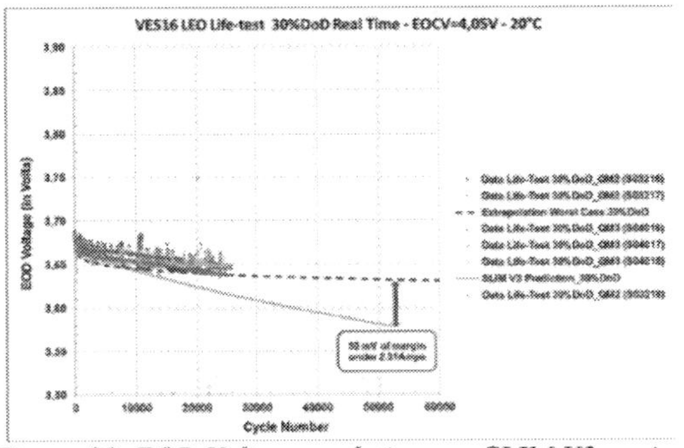

Figure 30. EOD Voltage evolution vs. SLIM V3 ageing model

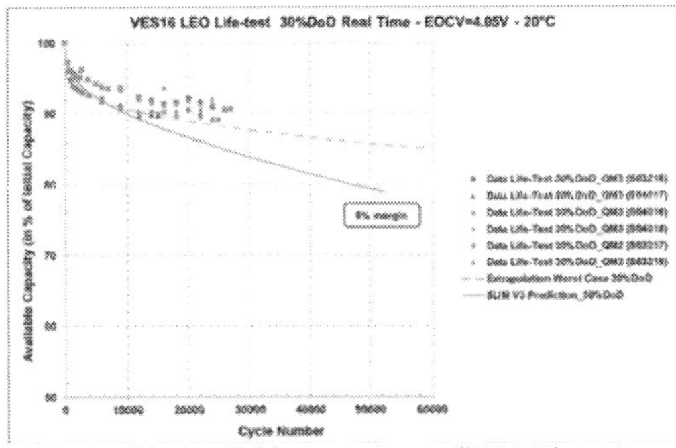

Figure 31. Available capacity vs. SLIM V3 ageing model

1.5 Conclusion

VES16 Li-ion cell electrical performances evolution observed in LEO Life-tests is excellent in the whole range of DOD up to 50%, with EOC voltage in the range from 3.9 to 4.1V and with a temperature in the range from 5°C to 30°C.
That can be explained by the very good ability of the negative electrode design to sustain LEO cycling conditions with larger charge current up to 2.57Amps and lower temperature down to 0°C.
Up to 50000 cycles, the error between the available capacity/EODV predicted by SLIM V3 ageing model and the ones extrapolated from the LEO life-tests data at 30%DOD are respectively less than 50mV & 6%.

2. MP176065 XTD LITHIUM-ION CELLS LEO LIFE-TIME TESTS RESULTS

New MP176065 XTD Lithium-ion cell presents promising results in accelerated LEO cycling conditions at 30%DOD, comparable to VES16 lithium-ion cell ones. We can even observe a lower capacity loss of ≈4% after 6000 cycles, as shown in Fig.32.

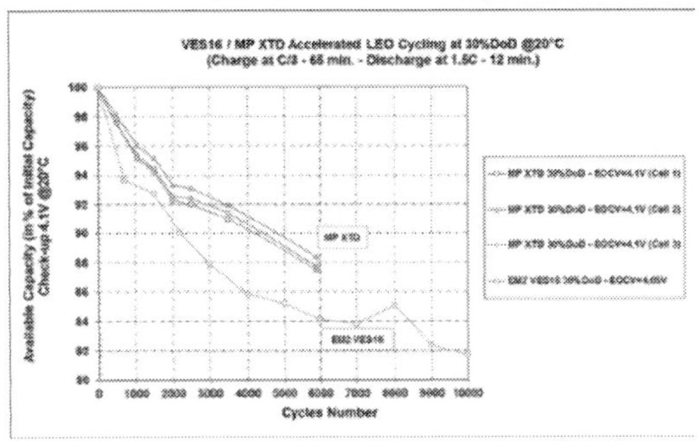

Figure 32. MP176065 XTD / VES16 available capacity evolution in accelerated LEO cycling at 30%DOD

DOI: 10.1051/e3sconf/20171606005

3. CONCLUSION

The excellent results, observed on VES16 Li-ion cell in LEO Lifetime tests, allow to foreseen its use for LEO mission up to 12 years or more, depending on DOD.

New MP176065 XTD Lithium-ion cell presents promising results in accelerated LEO cycling conditions at 30%DOD. However, its lifetime database shall be strengthened to assess and to validate its space use in LEO missions.

4. REFERENCES

[1]9th ESPC 2011, Article 2103465 / Session 3: Cells and Batteries Design , "VES16 Li ion cell for satellite" - S. Rémy, D. Prevot, D. Reulier & Dr. F. Vigier

[2]9th ESPC 2011, Article 2103465 / Session 23: Battery Management - Modelling, "SLIM Battery Modelling Features" – Y. Borthomieu, D. Prevot

[3]10th ESPC 2014, Article S14-1 / Session 14: Battery Experience and modelling, "VES16 Lithium-ion cell lifetime tests results & SLIM correlation" - D. Prevot, Y. Borthomieu & C. Cenac-Morthe

ROSETTA LANDER BATTERIES EXPERIENCE DURING ALL OPERATION PHASES

Céline Cénac-Morthé [(1)], Laurence Mélac [(1)], Stephane Fredon [(1)], Laurène Gillot [(1)], Philippe Gaudon [(1)], Michael Maibaum [(2)], Barbara Cozzoni [(2)], Valentina Lommatsch [(2)], Koen Geurts [(2)], Paul Brochard [(3)], Alex Clark [(4)]

[(1)] Centre National d'Etudes Spatiales, Toulouse, France, Celine.Cenac-Morthe@cnes.fr
[(2)] Deutsches Zentrum für Luft- und Raumfahrt, Cologne, Germany, Michael.Maibaum@dlr.de
[(3)] Saft, Poitiers, France, Paul.BROCHARD@saftbatteries.com
[(4)] ABSL, Abingdon, United Kingdom, alex.clarke@absl.enersys.com

ABSTRACT

Rosetta is an ambitious ESA mission, launched in March 2004 from Kourou and which performed a rendezvous with comet 67/P Churyumov-Gerasimenko. Its lander, Philae, achieved landing on comet soil on the 12th November 2014 and performed 64 hours of science activities on its batteries before going into hibernation due to lack of solar energy.

Philae is operated by the Lander Control Centre (LCC) at DLR Cologne Germany and the Science Operations and Navigation Centre (SONC) at CNES Toulouse France. The Lander system was provided by a European consortium (Germany, France, Italy, Hungary, Finland, UK, Ireland, Switzerland and Austria) and supports a scientific payload of 10 instruments.

The Philae battery system was provided by CNES, it is composed of a Saft primary battery (1518Wh) and an ABSL secondary battery (151Wh). The primary is made of non-rechargeable LSH20 ($LiSOCl_2$) Saft cells and the secondary of rechargeable ABSL Li-ion 18650HC.

For the Philae mission, the energetic constraint was very important. Indeed, before launch, the operations had to be planned considering variability of several parameters (descent duration, communication slots, comet temperature, solar power availability, etc.). Since Rosetta launch, cells and batteries have been stored and specific ground test plans have been identified in order to follow the battery ageing and to validate the final Philae operation schedule. From ground test results, an electrical model of the batteries was developed to help the operations scheduling.

During cruise, the operations consisted of secondary batteries monitoring and tests and primary battery conditioning. During separation and on-comet operations, the behaviour of the batteries system was checked and electric simulations helped with activities scheduling.

Firstly, this paper will describe the Philae mission. In a second part, the batteries system will be presented. The ground strategy will be detailed. Finally, the operations of Philae batteries system will be described.

1. ROSETTA LANDER: PHILAE

Rosetta is an ESA mission launched in March 2004, with the objective to study the comet 67P/Churyumov-Gerasimenko. The space probe, Rosetta, is composed of several payloads, particularly, Philae, the lander, [1]. After 10 years of cruise phase, the last three of them in hibernation, Philae was switched on again in March 2014. The Philae's Separation, Descent and Landing (SDL) on the comet occurred in November 2014. During three days, different science measurements were realised, exploiting the energy stored in the batteries. This operational phase is called First Science Sequence (FSS). After this phase, a Long Term Science phase (LTS) should have started. Due to landing conditions with respect to illumination, it happened only after 7 months of hibernation, [2].

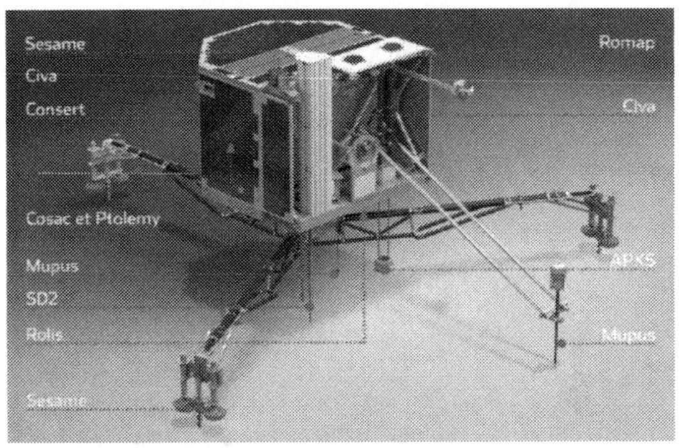

Figure 1. View of Philae and instruments on board

The Philae lander was designed by a European consortium under the leadership of DLR Cologne (German aerospace centre). It weighs about 100kg and includes ten instruments. Philae power was supplied through an umbilical line to the Rosetta probe during the cruise phase and by solar panel with Maximum Power Point Trackers (MPPT) and the batteries system during SDL and on-comet phases. The batteries system provided by CNES will be detailed.

2. BATTERIES SYSTEM

The batteries system is composed of a primary battery and a secondary battery (Figure 2).

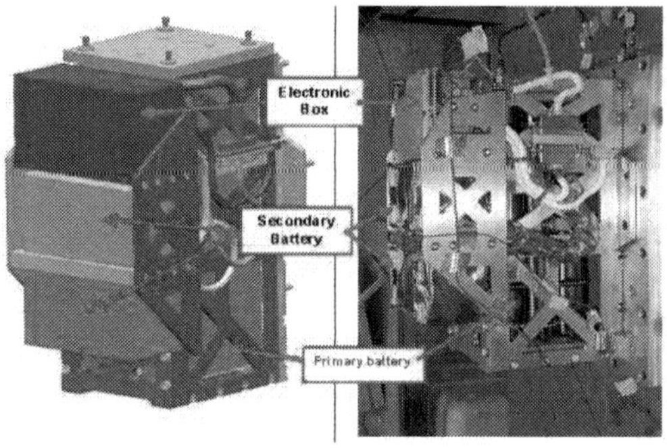

Figure 2. Batteries system

2.1. The primary battery

The primary battery is equipped with LSH20 cells manufactured by Saft (LiSOCl2). The configuration of the battery is 4 strings with 8 cells in series (8s4p). The nominal capacity at BOL (Beginning Of Life) is 52 Ah (on 240Ω/15mA, at 20°C for 2V cut-off voltage). The open circuit voltage is 29.36 V at 20°C. The range of storage temperatures is [-60°C; +85°C]. The range of operational temperatures is [-30°C; +60°C]. A depassivation circuit is implemented in order to impose discharge current pulses and to break the passivation layer that was formed during storage. Moreover, it shows the voltage of each string.

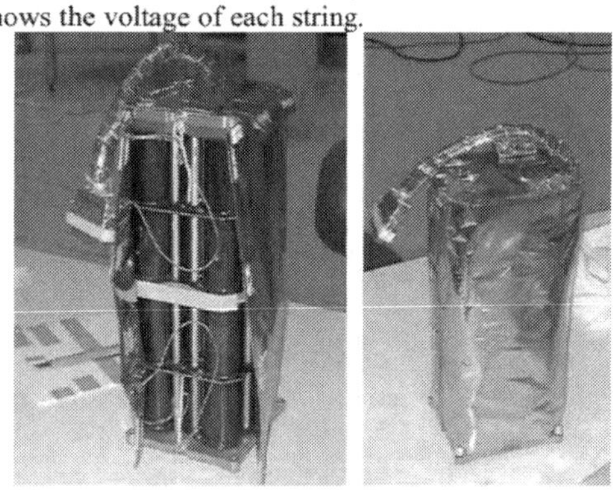

Figure 3. Philae primary battery (Saft)

2.2. The secondary battery

The secondary battery is manufactured by ABSL with Sony 18650 HC cells. There are two blocks of these. The configuration of the first block is 2p7s and of the second block is 7s2p. The second block had to be added later in the design phase, therefore its configuration is different. A monitoring circuit is connected to block 1. It ensures the monitoring of cell-pair voltages, the high and low voltage detections and the cell balancing.
Each battery is equipped with thermistors and heaters to ensure thermal control.

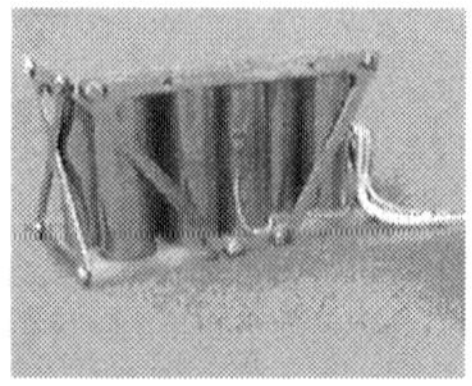

Figure 4. One block of Philae secondary battery (ABSL)

3. GROUND STRATEGY

Different ground activities were realised during the mission cruise phase in order to:

- Check that the primary battery cells are in good health, which can only be realized on ground.
- Prepare the operations during SDL and on-comet phase [3].

3.1. Ground tests

Concerning the primary battery, a storage strategy was organised. The storage conditions on ground change as a function of flight conditions (temperature, durations) to be the most representative on ground of the flight cell life.
Every two years, a few stored cells were discharged to verify their health. These tests consisted in a first depassivation followed by a capacity measurement on six cells. The results showed that after fifteen discharge pulses, the cell voltages reached their nominal value (between 3.2V and 3.3V) for the 0.85A discharge current at 20°C. Moreover, the measured capacity is around 9Ah (under the discharge current: 0.2A; temperature: 20°C; End Of Charge Voltage: 2V). Since the beginning of the test (6 years), the loss of capacity has been inferior to 1%.

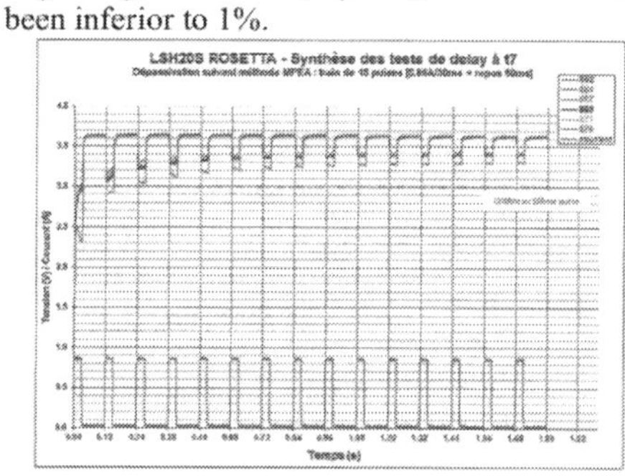

Figure 5. Depassivation test on several stored LSH20 cells – Saft data

Moreover, in order to define the maximum power discharge pulse allowed during landing without turning off the lander, different power pulses were imposed and cell voltages were measured. Finally, before landing, a typical discharge profile of the mission was imposed on a spare battery to ensure that the power peak during landing would not lead to a too low bus voltage (condition of lander switch off) and to evaluate the

DOI: 10.1051/e3sconf/20171606006

available energy.

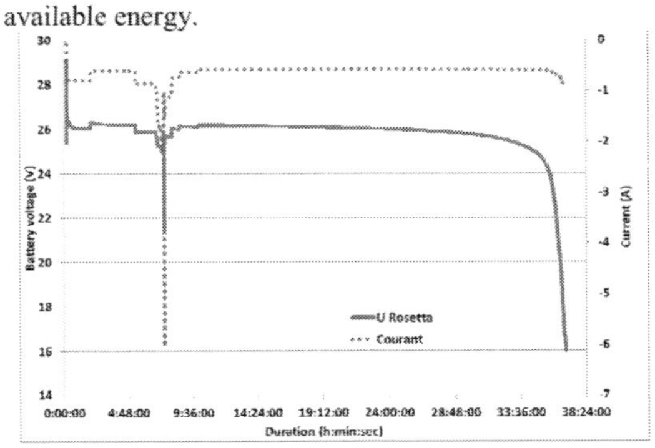

Figure 6. Typical discharge profile test on 8s3p battery

The tests on secondary cells allowed to define their behaviour in severe temperature conditions [-40°C; +40°C] and discharge conditions [C/10;C/2.5].

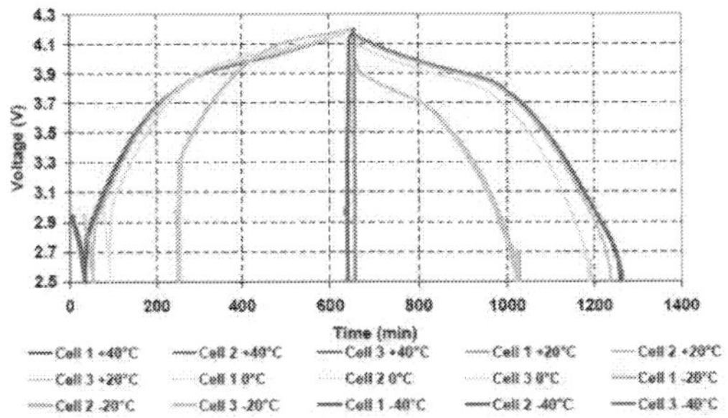

Figure 7. Capacity measurements on 18650HC – ABSL data

3.2. Model

From the different test results, the batteries system was modelled to define the available energy at separation and the bus voltage to prepare the operations [7].

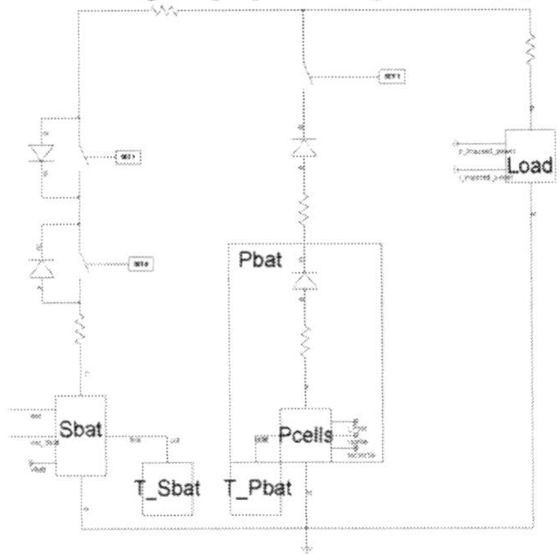

Figure 8. Batteries system model

Each battery was modelled from an equivalent electrical circuit:

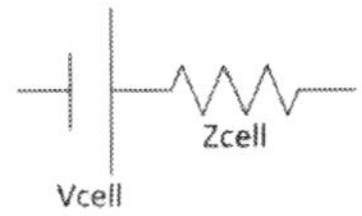

Figure 9. Equivalent electrical circuit used for each battery model

The different components were modelled from the data of battery manufacturers. They are expressed in function of temperature, state of charge and/or open circuit voltage.

Initially, each battery was modelled and validated separately. For the primary battery, the different ground tests indicated previously were used to realize the validation of cells and battery models. For example, the typical discharge presented on *Figure 6*, was simulated with the primary battery model. The results were shown on *Figure 10*. The voltage errors, expressed in percent, between simulation and test results were equal to -1.3% during 15 W discharge power and +0.12% during 140W discharge power peak. The capacity error, expressed in percent, between simulation and test results were equal to 1%.

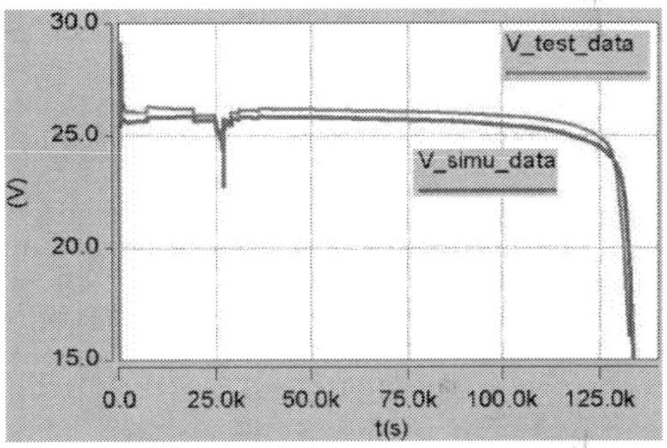

Figure 10. Comparison of primary battery simulation results with the test results of typical discharge profile

The model of secondary battery was always used for CNES Myriade satellites and validated from telemetries, as described in [5].

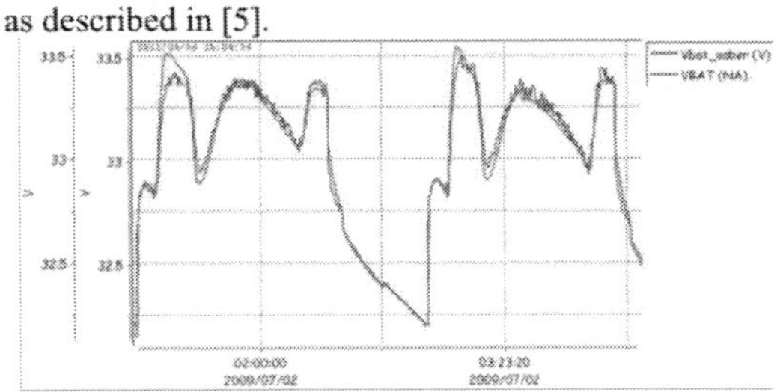

No ground or flight tests allowed validating the batteries system model. In the model, the values of harness resistances were defined from the cable types and from the length of harness. The ones of diode resistances and diode voltage drops were defined from manufacturers'

datasheet.
Thanks to the batteries system model, a dedicated operational loop was prepared as illustrated in *Figure 11*.

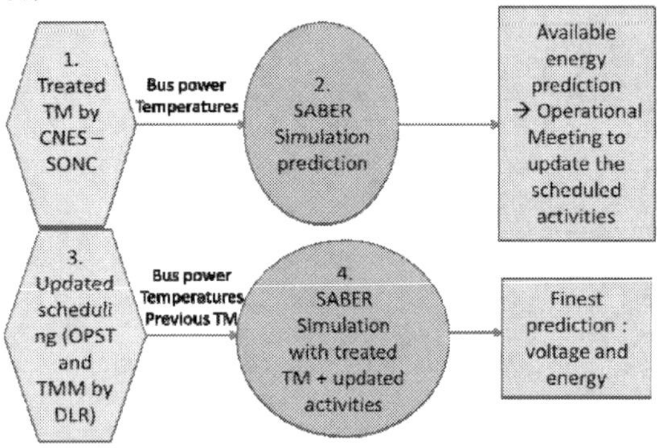

Figure 11. Operational loop used for the prediction of available energy

The objective was to integrate telemetry in the model and provisional power consumption in order to predict in real time the energy available in the batteries and support the scheduling of activities during on-comet operations. The telemetries were downloaded and treated in order to be used on the batteries system model. A first prediction of available energy was given to lander team. The lander team (DLR/SONC) in accordance with the scientists, updated the scheduling of activities and gave us the new power profile. This last one was integrated with telemetries to give a finest prediction of available energy and bus voltage. The data of bus voltage was used to define the switching off of the lander.

4. OPERATIONS

As presented previously, the Philae mission was divided into several phases: cruise and post-hibernation, SDL, FSS and LTS. For each phase, the batteries operations will be detailed.

4.1. Cruise

During cruise, the state of charge of the secondary battery was monitored regularly thanks to the battery voltage measurement. An unexpected decrease in state of charge was detected. So, the secondary battery was charged regularly in order to avoid under-discharge, particularly during the hibernation phase, see *Figure 12*. In addition to this, some tests of the power system were realised, like the charge of batteries through Philae power subsystem.
The operations of conditioning of the primary battery began at Philae wake-up. It consisted in several depassivations to ensure a sufficient voltage of the primary battery at the beginning of SDL. After ten years without any discharge, the first voltage at cell level is around 2.43 V (19.419V/8) at low temperature (3°C), see *Table 1*. After 6 months without any discharge, the first voltage at cell level is around 3.12 V (24.970/8) at 10°C.

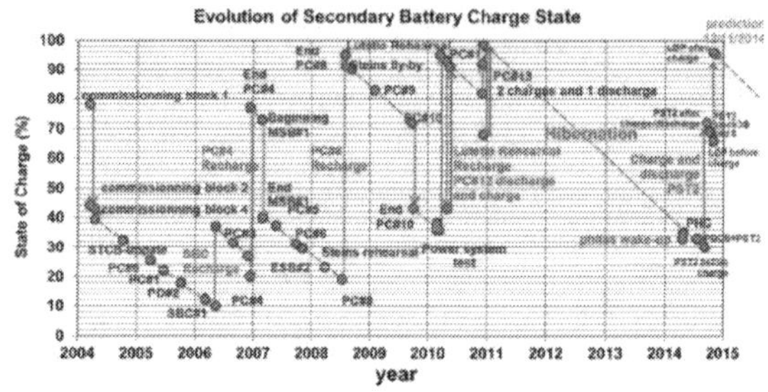

Figure 12. Evolution of secondary state of charge state during cruise

	March 2004	April 2014	October 2014
Storage condition	< -30°C	< -40°C	< -40°C
Test temperature	10°C	3°C	11.5°C
1st measurement : string 1 voltage after 15ms depassivation	21.3V	19.419V	24.970V
string 4 voltage after 485ms of depassivation	25.9V	25.657V	26.152V
Last measurement : string 4 after 4785ms	NA	25.84V	26.170V
Max voltage difference between strings on the 5th cycle	0.23V	0.549V	0.449V
Max voltage difference between strings on the 40th cycle	NA	0.503V	0.485V

Table 1. Synthesis of the different depassivations realised during cruise

4.2. SDL/FSS

SDL and the first part of FSS were scheduled on an automatic block. Some simulations were performed to ensure that the power peak during touch-down would not induce a decrease of voltage too great and trigger the lander switch-off, see *Figure 13*.

 DOI: 10.1051/e3sconf/20171606006

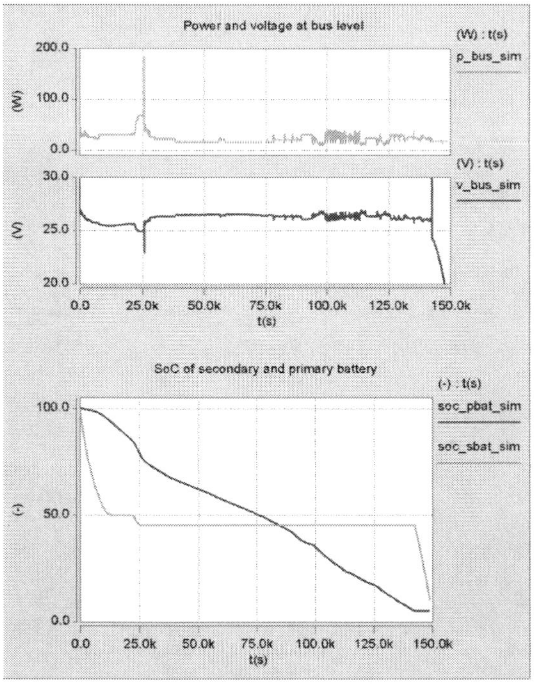

Figure 13. Simulation results dedicated to SDL/FSS first operations

The telemetry received during the first Rosetta-Philae visibility slot after separation indicated that the batteries system behaviour was nominal. Indeed, bus voltage was around 27.20V just after connection. Both primary and secondary batteries are giving current to the bus (I_PBATT = 429,45mA and I_SBATT = 487,6 mA). The temperatures were in the expected range: 26°C for the primary battery and 16°C for the secondary battery. The solar arrays provided energy during all descent phase which permitted to have a lower discharge current on the battery.

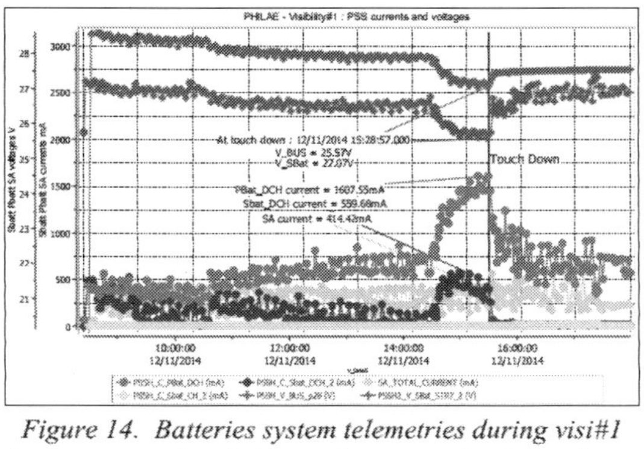

Figure 14. Batteries system telemetries during visi#1 (green : secondary battery voltage, purple : bus voltage, red : primary battery current, blue : secondary battery discharge current, yellow: total solar arrays current)

Through the solar array currents, the spin rotation of the lander and its stabilisation were observed, indicating the non-nominal touch-down of Philae in the target landing site Agilkia and a final resting landing in Abydos.

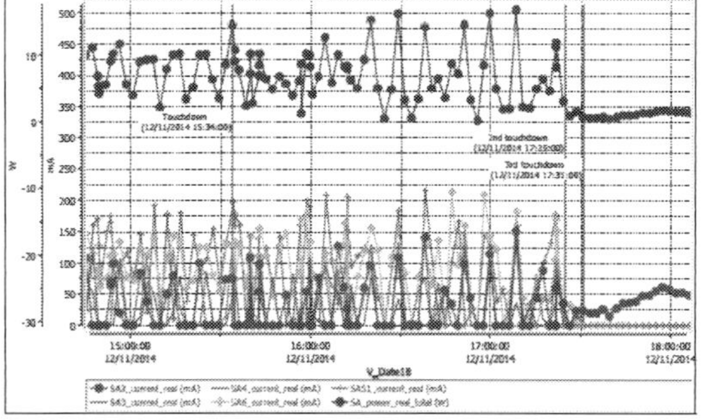

Figure 15. Solar arrays currents and power during SDL (green: current of SA n°2, purple: currents of SA n°1 and 5, red: current of SA n°3, blue: SA total power, yellow: current of SA n°6, brown: current of SA n°4)

In order to understand the situation, the foreseen activities (movement/deployment) in the second visibility were postponed and the execution of science measurements not requiring any mechanical activities was chosen. So, the power consumption was reduced. As expected thanks to simulation, the lander energy was mainly provided by the primary battery, see *Figure 18*. After the second visibility, some simulations were realised using telemetry and estimated power consumption, see *Figure 16*. The results indicated that the lander should go into standby mode two hours before the start of the fifth visibility.

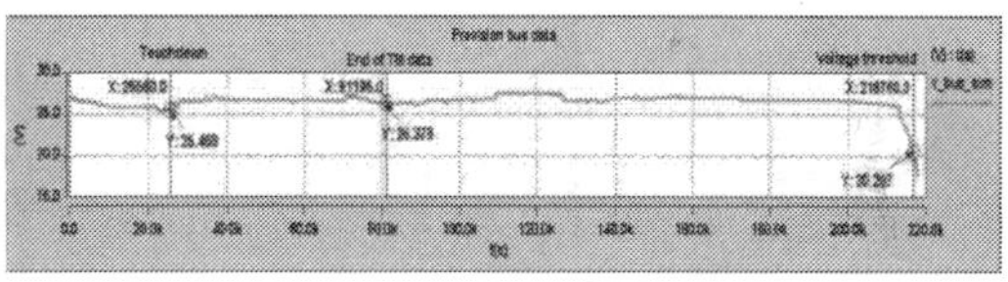

Figure 16. Prediction after the second visibility. Lander standby mode at 2014/11/14 20:37 UTC while VISI#5 is at 22:30 UTC

So, for the third and fourth visibilities, it was decided to perform the mechanical deployments of MUPUS and APX and the SD2 drilling before a complete discharge of the batteries. COSAC sample analysis was preferred to Ptolemy due to lower power consumption. This reduction of power consumption should allow to have enough energy to download the science results.

 DOI: 10.1051/e3sconf/20171606006

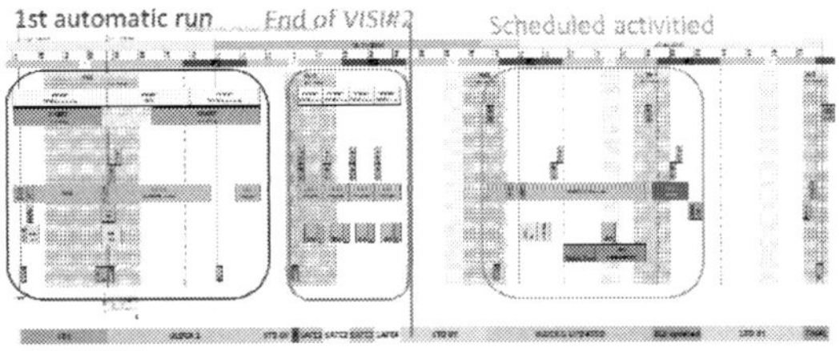

Figure 17. Scheduling of activities displayed in SONC's tool MOST (Science Operation and Navigation Centre – CNES)[4]

During the third and fourth visibilities, all science activities were successfully realised. After including last telemetry and a power consumption prediction, simulations confirmed that the lander could be supplied by the batteries system until the fifth (last) visibility.

Indeed, during the fifth and last visibility, the bus voltage reached 20.3V which corresponds to the minimum bus voltage before switch-off. Nevertheless, all science activities were realised during SDL/FSS. The batteries system operated as expected despite low temperatures and power peaks.

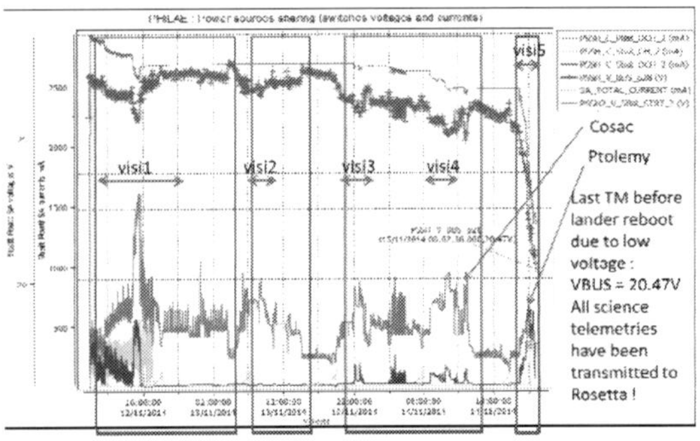

Figure 18. Batteries system telemetry from the beginning to the end of SDL/FSS (green : secondary battery voltage, purple : bus voltage, red : primary battery current, blue : secondary battery discharge current, yellow: total solar arrays current)

Telemetry and final simulations were compared in order to evaluate the model. Some discrepancies can be seen as illustrated in Figure 19. Some hypotheses were considered:

- Difficulty to evaluate the real capacity of secondary battery in flight
- Degradations of primary battery on flight slightly more important than the ageing defined on ground
- Approximation of harness resistances, voltage drop of diodes between batteries and bus bar

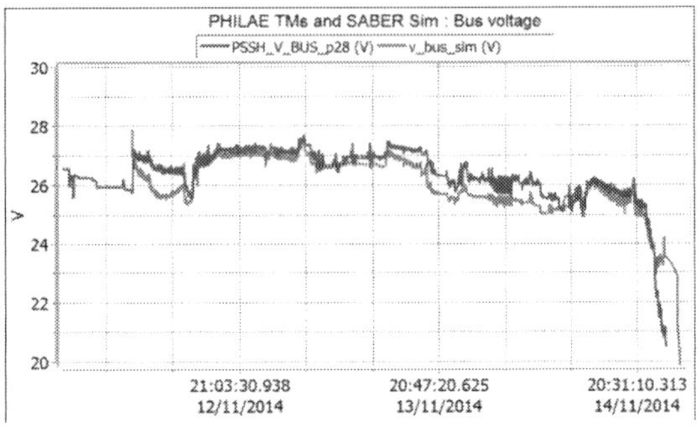

Figure 19. Telemetries and simulation results comparison on bus voltage

4.3. LTS

After the SDL and FSS activities, Philae switched to wake-up mode. Some thermal and power conditions had to be fulfilled in order for Philae to wake up.
In mid-June 2015, a first contact between Rosetta and Philae occurred. The telemetry of the secondary battery voltage indicated that the battery was charged (SoC around 86%) despite the cold temperature of charge (from 0°C). From the extrapolations of charge current telemetry, the secondary battery energy was estimated, by simulation, at 75 Wh. According to these elements, the secondary battery seemed in good health. Nevertheless, due to the following sporadic RF contacts and the uncertainties in lander health, it was not possible to re-activate any payloads during LTS.

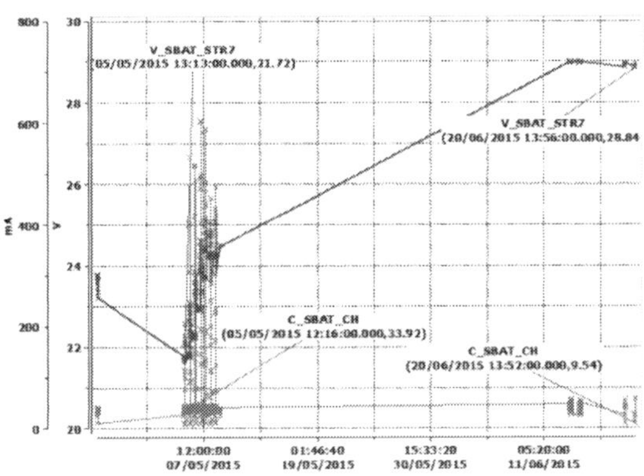

Figure 20. Last telemetry of secondary battery voltage

5. CONCLUSION

Philae achieved its objectives. All the science activities were realised. The design of the batteries system allowed an efficient monitoring and good performance. The storage strategy and the different ground tests have proven useful for preparation of operations. The developed model was integrated into the operational loop. It helped with the operations strategy and

DOI: 10.1051/e3sconf/20171606006

decisions, particularly during SDL and FSS. The batteries system ensured this mission.

6. REFERENCES

1. Ulamec S., Balazs A., Biele J., Espinasse S., Fantinati C., Fischer H.-H., Fischer H., Flamini E., Fronton J.-F., Gaudon P., Geurts K., Hilchenbach M., Jansen S., Kletzkine P., Krause C., Küchemann O., Maibaum M., Melac L., Pätz B., Peragin E., Roll R., Sangiovanni G., Sous S., Tropolle E., Warmbold G., Wiillnecker R., Rosetta Lander – System Status after five years in Space,

2. S.Ulamec et al , Rosetta Lander – Landing and operations on comet 67P/Churyumov-Gerasimenko, Acta Astronautica, j.actaastro.2015.12.050

3. Moussi A., Fronton J.-P., Gaudon P., Delmas C., Lafaille V., Jurado E., Durand J., Hallouard J., Mangeret M., Charpentier A., Ulamec S., Fintinati S., Geurts K., Salatti K., Bibring J.-P., Boehnhardt H. (2015), The Philae lander: science planning and operations, Acta Astronautica, j.actaastro.2015.12.050.

4. Moussi A., Fronton J.-P., Gaudon P., Delmas C., Lafaille V., Jurado E., Durand J., Hallouard J., Mangeret M., Charpentier A., Ulamec S., Fintinati S., Geurts K., Salatti K., Bibring J.-P., Boehnhardt H. (2015), Philae: science scheduling and unknown context, lessons learned, 66th International Astronautical Congress

5. Fredon S., Melac L., Simmons N, Long-Duration Leo Mission Experience With ABSL Batteries Onboard The Myriade SmallSAT Platform, Proceedings of the 9th European Space Power Conference, held 6-10 June 2011 at Saint Raphael,

ACCELERATING RATE CALORIMETRY TESTS OF LITHIUM-ION CELLS BEFORE AND AFTER STORAGE DEGRADATION AT HIGH TEMPERATURE

Omar Samuel Mendoza-Hernandez [1, 2], **Shuichi Taniguchi** [2], **Hiroaki Ishikawa** [3], **Kohei Tanaka** [4], **Seisuke Fukuda** [1, 4], **Yoshitsugu Sone** [1, 4], **Minoru Umeda** [1, 2]

[1] *Japan Aerospace Exploration Agency, 3-1-1 Yoshinodai, Chuo-ku, Sagamihara, Kanagawa, 252-5210(Japan), Email: sone.yoshitsugu@jaxa.jp*

[2] *Nagaoka University of Technology, 1603-1 Kamitomioka, Nagaoka, Niigata, 940-2188 (Japan), Email: mumeda@vos.nagaokaut.ac.jp*

[3] *Industrial Technology Institute of Ibaraki Prefecture, Ibaraki, Japan*

[4] *The Graduate University of Advanced Studies, SOKENDAI, 3-1-1 Yoshinodai, Chuo-ku, Sagamihara, Kanagawa, 252-5210 (Japan) Email: sone.yoshitsugu@jaxa.jp*

ABSTRACT

Understanding the behavior of Li-ion cells during thermal runaway is critical to evaluate the safety of these energy storage devices under outstanding conditions. Li-ion cells possess a high energy density and are used to store and supply energy to many aerospace applications. Incidents related to the overheating or thermal runaway of these cells can cause catastrophic damages that could end up costly space missions; therefore, thermal studies of Li-ion cells are very important for ensuring safety and reliability of space missions. This work evaluates the thermal behavior of Li-ion cells before and after storage degradation at high temperature using accelerating rate calorimeter (ARC) equipment to analyze the thermal behavior of Li-ion cells under adiabatic conditions. Onset temperature points of self-heating and thermal runaway reactions are obtained. The onset points are used to identify non-self-heating, self-heating and thermal runaway regions as a function of state of charge. The results obtained can be useful to develop accurate thermo-electrochemical models of Li-ion cells.

1. INTRODUCTION

Li-ion cells are an essential part of the power systems for satellites, rockets and spacecraft, they serve as energy storage devices and provide power during eclipse time. Aerospace applications demand compact and lightweight battery systems, this implies the use of devices with a high energy density which due to their high energetic components are not inherently safe. Li-ion cells possess a high energy density but can cause catastrophic incidents that could end up costly space missions, those incidents are mainly related to the overheating or thermal runaway of Li-ion cells which can lead to fire and explosion.

The thermal runaway behavior of a Li-ion cell is dominated by the exothermic reactions between its electrolyte and electrode materials [1-3]. Thermal runaway occurs when the exothermic reactions go out of control, thus the self-heating rate of the cell increases to the point that it begins to generate more heat than what can be dissipated [4,5]. Understanding the behavior of Li-ion cells during thermal runaway is very important for ensuring their safety and reliability.

The use of accelerating rate calorimeter (ARC) equipment allows the thermal analysis of Li-ion cells under adiabatic conditions [2, 3, 5-7]. ARC measurements are useful to obtain onset temperature points of self-heating and thermal runaway reactions. Recently, safety concerns related to use of damaged or degraded Li-ion cells have increased, mainly focusing on the thermal behavior of degraded cells, however, there are a few investigations on this matter. In this work, the thermal behavior of non-degraded and degraded Li-ion cells is analyzed by carrying out ARC measurements. Degradation tests were conducted by storing fully charged Li-ion cells at 80 °C. Thermal mapping plots of degraded and non-degraded cells are made, these plots are used to identify non-self-heating, self-heating and thermal runaway regions as a function of state of charge. The results obtained can be used to develop accurate thermo-electrochemical models of Li-ion cells.

2. EXPERIMENTAL METHODOLOGY

Commercial 18650 Li-ion cells with a nominal capacity of 2550 mAh were used in this work. The cathode and anode electrode materials are $LiCoO_2$ and graphite.

2.1 Storage degradation

Li-ion cells were fully charged (4.2 V) and stored in a thermostatic chamber at 80 °C for 168 hours.

2.2 Accelerating rate calorimetry tests

Calorimetry tests were carried out at different state of charges utilizing an accelerating rate calorimeter (ARC 2000™, Columbia Scientific Industries). Once the cells were set to the desired state of charge, they were place inside the ARC cavity. A thermocouple was placed on the surface of the cell to record its temperature. The ARC tests were conducted following a Heat-Wait-Search method. During the heating step, the temperature of ARC's cavity is increased in periods of 5 °C follow by a waiting step of 30 min and a searching step of 5 min. If during the searching period a heating rate larger than 0.05 °C min^{-1} is detected, it is considered that a self-heating (exothermic) reaction has started, and then the ARC shifts to adiabatic mode and tracks the cell's temperature until the end of the exothermic reaction. If no exothermic reaction is detected, the next heating step is initiated. The ARC tests were carried out at 0, 25, 50, 75 and 100% SOC.

3. RESULTS AND DISCUSSION

Fig. 1 shows the ARC measurements of non-degraded Li-ion cells at 0, 25, 50, 75 and 100% SOC. A temperature drop is observed at some SOCs, this likely due to the volatilization of the electrolyte. From the ARC measurements, onset temperature points of self-heating and thermal runaway reactions are obtained by tracking the heating rate of the cells when an exothermic reaction has been detected. We consider that a self-heating reaction has started when the cell heating rate is larger than 0.05 °C min^{-1}, and thermal runaway reaction starts when the cell heating rate is larger than 1 °C min^{-1}.

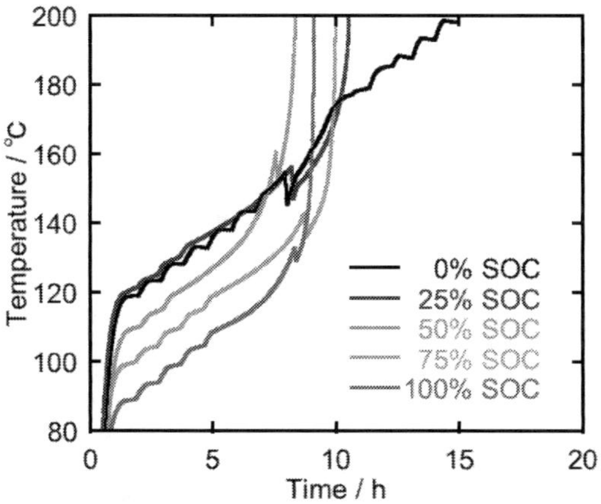

Figure 1. ARC measurements of non-degraded Li-ion cells at 0, 25, 50, 75 and 100% SOC.

Tab. 1 shows the onset temperature points of self-heating and thermal runaway reactions for non-degraded Li-ion cells. It can be noticed that onset temperatures of the exothermic reactions tend to decrease by increasing the state of charge. This reflects the influence of the state of charge on the thermal stability of the electrode materials, where highly delithiated electrode materials become more reactive [8].

Table 1. Onset temperature point of self-heating and thermal runaway of non-degraded Li-ion cells.

SOC (%)	Onset temperature of self-heating (°C)	Onset temperature of Thermal runaway (°C)
0	152.1	-
25	130.2	173.4
50	110.7	172.1
75	121.2	165.3
100	100.0	144.9

Fig. 2 shows a thermal mapping plot of non-degraded cells, this plot summarizes the ARC results. Non-self-heating, self-heating and thermal runaway regions are identified as a function of SOC. In the non-self-heating (blue bars) region, no exothermic reactions take place. In the self-heating region (yellow bars) exothermic reactions with a heating rate larger than 0.05 °C min^{-1} occur. In the thermal runaway region (red bars) the heating rate of the exothermic reactions is larger than 1 °C min^{-1}.

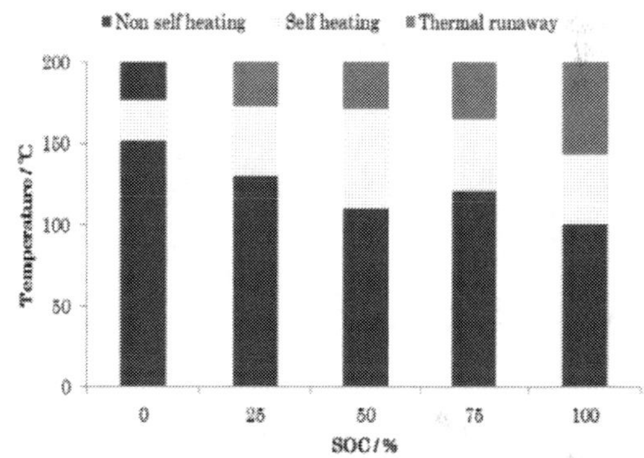

Figure 2. Thermal mapping of non-degraded cells as a function of SOC.

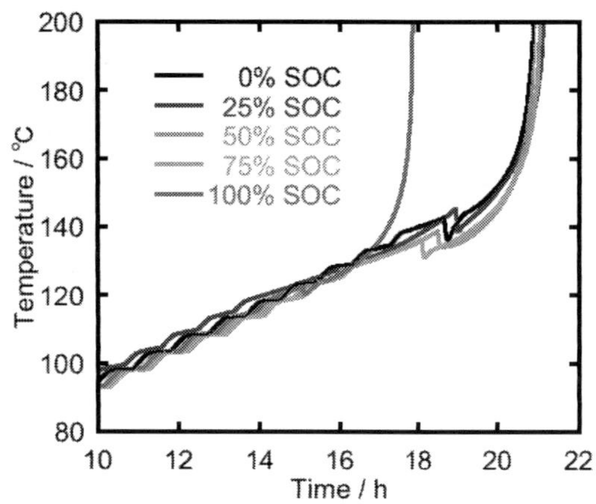

Figure 3. ARC measurements of degraded Li-ion cells at 0, 25, 50, 75 and 100% SOC.

Fig. 3 shows the thermal behavior of degraded Li-ion cells during the ARC measurements at 0, 25, 50, 75 and 100% SOC. Tab. 2 summarises the onset temperature points of self-heating and thermal runaway reactions for the degraded cells. In the case of a degraded Li-ion cell, a thermal runaway reaction was observed at 0% SOC.

Table 2. Onset temperature point of self-heating and thermal runaway of degraded Li-ion cells.

SOC (%)	Onset temperature of self-heating (°C)	Onset temperature of Thermal runaway (°C)
0	140.5	168.2
25	120.3	173.9
50	129.8	163.8
75	125.3	167.6
100	115.1	153.4

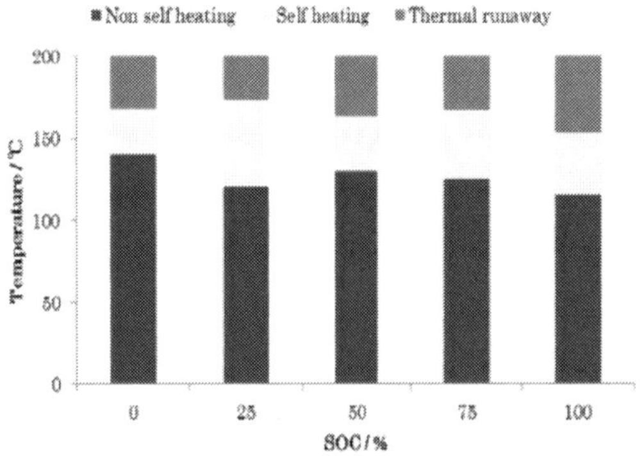

Figure 4. Thermal mapping of degraded Li-ion cells as a function of SOC.

Fig. 4 shows a thermal mapping plot of degraded Li-ion cells. At 0% SOC a thermal runaway region is identified, while in the case of a non-degraded Li-ion cell any thermal runaway region is observed at 0% SOC. A thermal runaway reaction takes place when the exothermic reactions go out of control, the exothermic reactions usually start when the solid electrolyte interface (SEI), generated at the anode electrode material, reacts with the electrolyte and electrode materials [9]. Usually when a Li-ion cell is degraded, the SEI tends to grow, and this increment in the SEI might be accelerating the exothermic reactions until thermal runaway.

4. CONCLUSIONS

The thermal behavior of non-degraded and degraded Li-ion cells was analyzed by carrying out accelerating rate measurements. The ARC results were used to make thermal mapping plots, where non-self-heating, self-heating and thermal runaway regions are identified as a function of state of charge. The results suggest that degraded Li-ion cells tend to become thermally unstable at low state of charges. The results obtained can be used to develop a thermo-electrochemical model of Li-ion cells, which can be very useful to analyze the impact of the state of degradation of a Li-ion cell on its thermal stability.

5. ACKNOWLEGMENT

This research is partially supported by the 'International Joint Research Program for Innovative Energy Technology' of the Ministry of Economy, Trade and Industry (METI).

6. REFERENCES

1. T. D. Hatchard, D. D. MacNeil, A. Basu, J. R. Dahn. (2001). Thermal Model of Cylindrical and Prismatic Lithium-ion Cells. J. Electrochem. Soc. 148, A755-A761.
2. I. Uchida, H. Ishikawa, M. Mohamedi, M. Umeda. (2003). AC-Impedance Measurements during Thermal Runaway Process in Several Lithium/polymer Batteries. J. Power Sources 119–121, 821-825.
3. J.S. Gnanaraj, E. Zinigrad, L. Asraf, H.E. Gottlieb, M. Sprecher, D. Aurbach, M. Schmidt. (2003). The Use of Accelerating Rate Calorimetry (ARC) for the Study of the Thermal Reactions of Li-ion Battery Electrolyte Solutions J. Power Sources 119–121, 794-798.
4. R.M. Spotnitz, J. Weaver, G. Yeduvaka, D.H. Doughty, E.P. Roth. (2007). Simulation of Abuse Tolerance of Lithium-ion Battery Packs. J. Power Sources 163, 1080–1086.
5. B.K. Mandal, A.K. Padhi, Z. Shi, S. Chakraborty, R. Filler. (2006). Thermal Runaway Inhibitors for Lithium Battery Electrolytes. J. Power Sources 161, 1341–1345.
6. H. Ishikawa, O. Mendoza, Y. Sone, M. Umeda. (2012). Study of Thermal Deterioration of Lithium-ion Secondary Cell Using an Accelerating Rate Calorimeter (ARC) and AC Impedance. J. Power Sources 198, 236-242.
7. O. S. Mendoza-Hernandez, H. Ishikawa, Y. Nishikawa, Y. Maruyama, M. Umeda. (2015). Cathode Material Comparison of Thermal Runaway Behavior of Li-ion Cells at Different State of Charges including Over Charge. J. Power Sources 280, 499-504.
8. Y. Furushima, C. Yanagisawa, T. Nakagawa, Y. Aoki, N. Muraki. (2011). Thermal Stability and Kinetics of Delithiated $LiCoO_2$. J. Power Sources 196, 2260-2263.
9. P. Roder, B. Stiaszny, J. C. Ziegler, N. Baba, P. Lagaly, H. Wiemhofer. The Impact of Calendar Aging on the Thermal Stability of a $LiMn_2O_4$-$Li(Ni_{1/3}Mn_{1/3}Co_{1/3})O_2$ / graphite Lithium-ion Cell (2014). J. Power Sources 268, 315-325.

E3S Web of Conferences **16**, 07002 (2017) DOI: 10.1051/e3sconf/20171607002
ESPC 2016

BATTERY SIMULATION TOOL FOR WORST CASE ANALYSIS AND MISSION EVALUATIONS

Stéphane Lefeuvre [(1)], **Abdossamad Benkhalfia**[(2)], **Edouard Mosset** [(1)], **Maria Nestoridi** [(3)], **Olivier Mourra** [(3)]

[(1)] *Saft, Space and Defence Division, rue Georges Leclanché 86060 Poitiers - France,*
Email:stephane.lefeuvre@Saftbatteries.com, Edouard.Mosset@Saftbatteries.com
[(2)] *Saft (external), rue Georges Leclanché 86060 Poitiers - France,*
Email: Abdossamad.Benkhalfia@Saftbatteries.com
[(3)] *ESA/ESTEC, Keplerlaan 1, Postbox 299, 2200 AG Noordwijk - The Netherland,*
Email: Olivier.Mourra@esa.int, Maria.Nestoridi@esa.int

ABSTRACT

Saft recently designed, developed and qualified in the frame of Research and Development contract (GSTP-5) with the European Space Agency (ESA) three new VES16 batteries: 8S4P (4 parallel strings of 8 cells in series), 10S5P, 10S16P [1]. In order to guarantee the cells balancing (to avoid cell overcharge or over discharge during battery charge/discharge cycles), Saft introduced in parallel of each cell a simple electronic circuit called the Simplified Balancing System (SBS), which absorbs some current when a cell is close to its end of charge voltage.

The complete batteries (cells and SBS) were analysed with a proper Worst Case Analysis (WCA) to demonstrate the battery performance at the EOL of the targeted missions, including all possible effects that can influence its behaviour (cells and SBS components initial tolerances, radiation degradations, ageing and temperature effects).

Saft performed this analysis in PSpice and developed VES16 cell model and SBS models to run Monte Carlo Analysis (MCA).

The first part of this paper presents the PSpice models including their respective variable parameters at SBS and cell level. Then the second part of the paper introduces to the reader the model parameters that were chosen and identified to perform Monte Carlo Analysis simulations. The third part reflects some MCA results for a VES16 battery module. Finally the reader will see some other simulations that were performed by re-using the battery model for an another Saft battery cell type (MP XTD) for a specific space application, at high temperature.

1. CELL & SBS PSPICE MODELS

In order to perform Monte Carlo Analysis, Part Stress Analysis (in nominal or fault operation) and worst case analysis, it appeared very quickly that the use of Ppsice would be an asset for a battery containing electronics.

The Pspice models have been developed for the VES16 cells and SBS, based on basic Pspice models already available in the native Ppsice library (resistors, conductance blocks, Laplace blocks (for transfer functions), table blocks (for linearized functions), etc.).

1.1 Cell model

The first model that was developed is the VES16 cell model. This model is presented on Fig. 1. It takes into consideration:

- the Open Circuit Voltage (Voc) as a function of the Depth of Discharge (DoD) of a VES16 cell,
- the individual cell capacity,
- the cell internal resistance,
- the cell self-discharge rate,
- the cell ageing factor.

The cell is a modelled by a voltage source (generated by the E generator in block 5, representing the cell open voltage, Voc) with a current source in parallel (conductance G in block 3, representing the self-discharge of the cell) and a resistance in series (Ri in block 4, representing the internal cell resistance).

The open cell voltage source is generated by using a table (block 5), in which the transfer function Open circuit Voltage versus the cell Depth of Discharge (Voc vs DOD) is modelised.

The DOD of the cell is obtained by taking into account the current flowing through the cell (H sensor in block 6), the intrinsic capacity of the cell (block 2), and the initial DOD of the cell (conductance G of block 2).

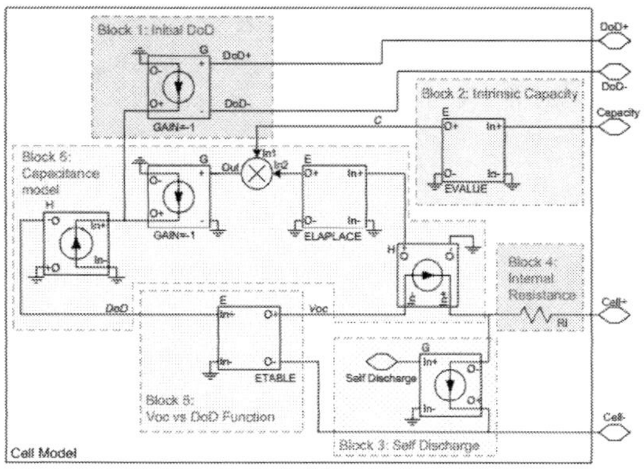

Figure 1. VES16 cell model

 DOI: 10.1051/e3sconf/20171607002

On Fig. 2, the reader can see the cell linearized function Open circuit Voltage versus the cell Depth of Discharge (Voc vs DOD) that was modelled in the block 5 for the VES16 cell model. The model allows over-discharge down to the voltage reversal value and overcharge up to 5V for fault battery operation.

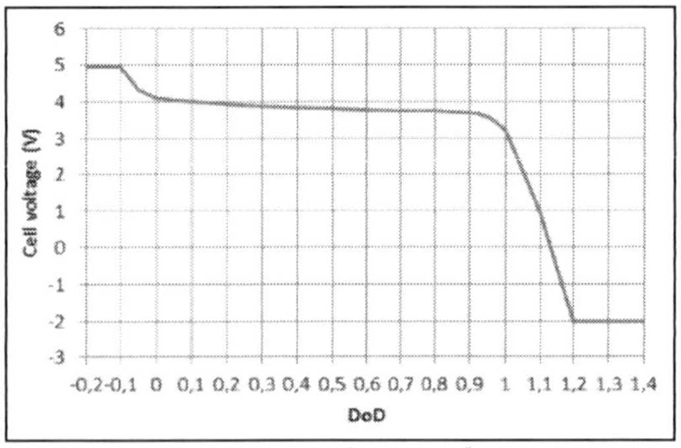

Figure 2. VOC vs DOD Transfer Function

1.2 SBS model

The second model that was developed is the SBS model. As previously said each battery cell is connected in parallel to a SBS.

The SBS is a simple circuitry: above a certain cell voltage value U_0 (called trigger voltage threshold), the SBS absorbs some current directly proportional to the cell voltage (like a resistor called later R_s). Below the trigger voltage threshold, the SBS consumes a reduced leakage current. The SBS I/V characteristic is presented on Fig. 3.

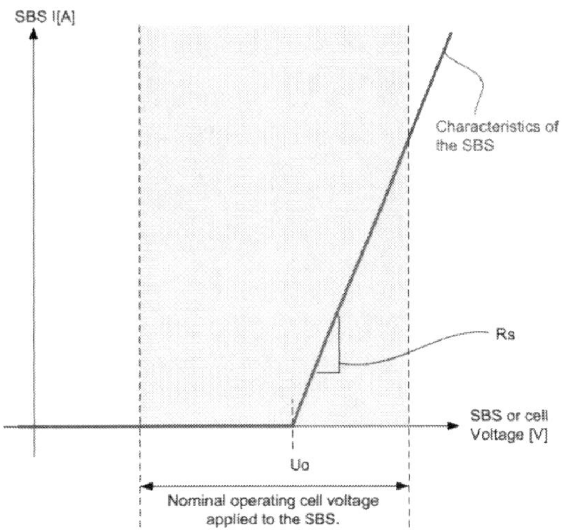

Figure 3. SBS I/V characteristics

The SBS model is presented on Fig. 4. It includes the following parameters:

- trigger voltage threshold, U_0 (computed in block 2, and then activated by block 3, when the cell voltage is higher than U_0)
- R_s (block 5)
- leakage resistor (block 1)

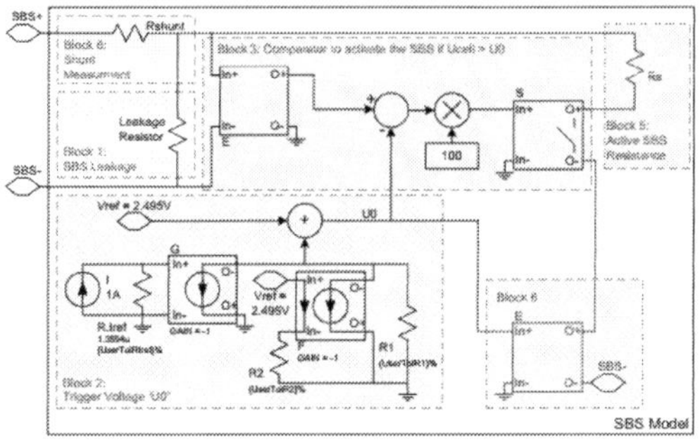

Figure 4. SBS model

2. MODEL PARAMETERS FOR WORST CASE ANALYSIS OF VES16 BATTERY MODULES

2.1 Statistical data of the models parameters

Before to run the Monte Carlo Analysis simulations in PSpice, it was essential to select which parameters will be randomly sampled in a chosen range. The following parameters were selected:

- cell capacity,
- cell internal resistance,
- self-discharge rate,
- SBS trigger voltage threshold, U_0, as a combination of resistors values and internal IC parameters,
- SBS Leakage resistor,
- Internal SBS resistor (R_s) when the SBS is active.

It must be also noted the study assumed the typical LEO mission parameters defined in the Tab. 1.

Parameter	Value
Mission Duration	15 years
TID	15 krad
Min/Max Temperature	-20°C / 60°C
Orbit duration	87 minutes
Eclipse Duration	27 minutes
Discharge rate	C/3
Charge rate	C/5

Table 1. Mission parameters

2.2 EOL Statistical data of the cell model

In order to choose the cell parameters variation ranges, Saft used its internal database of VES16 Cells Acceptance Tests results (used at BOL for individual and batch acceptance), and some cells cycling tests results (tests performed to assess the ageing degradation on the cells).

For instance Fig. 5 presents the record of cell capacity measurements made during VES 16 Cell Acceptance Test.

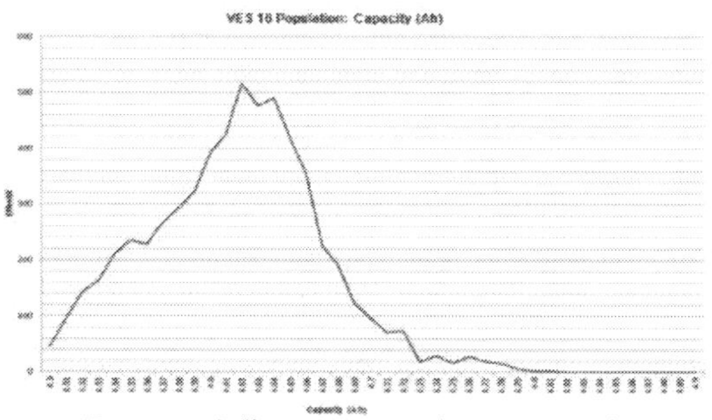

Figure 5. Cell capacity production records

Similarly, Fig. 6 and Fig. 7 present the internal resistance and the self-discharge rate of the cell measured during Cell Acceptance Tests.

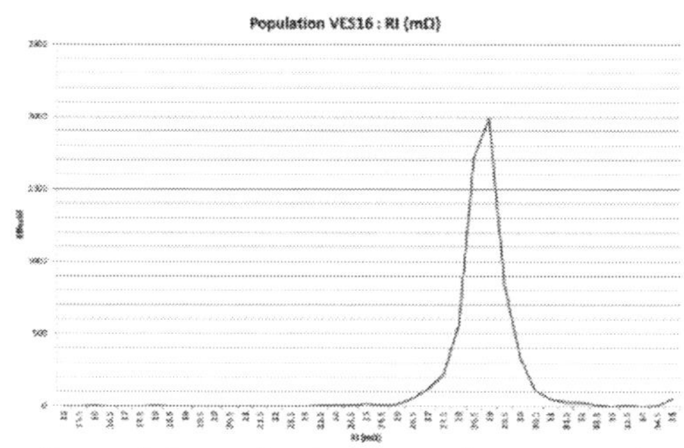

Figure 6. Internal resistance production records

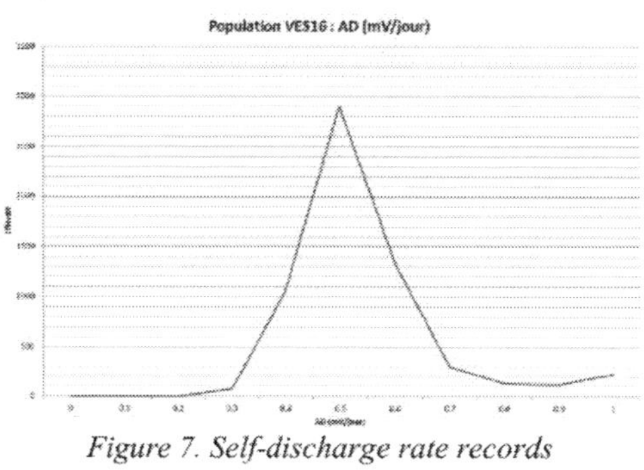

Figure 7. Self-discharge rate records

From these data, the inputs for the BoL model were computed. The distributions were approximated by a normal law. Characteristics of the distribution are given in Tab. 2.

	Capacity	Internal resistance	Self-discharge rate
Min	4.50 Ah	27.7 mΩ	0 mV/day
Max	4.78 Ah	35.0 mΩ	1 mV/day
σ	0.06 Ah	0.47 mΩ	0.1 mV/day

Table 2. Cell statistical parameters from manufacturing data

One can highlight that to minimise the parameters variations on the battery performances, Saft decided to select the cells to be placed in the same string. This allows minimising the differences of capacity, internal resistance and self-discharge rate of cells placed in the same string. Tab. 3 presents these differences for cells in the same strings, after cells selection.

	Capacity	Internal resistance	Self-discharge rate
Δ String	0.07 Ah	5%	0.1 mV/day

Table 3. String constraints

At battery model level, this matching is taken into account. For each string of the battery model, a mean value for capacity, internal resistance and self-discharge rate are randomly chosen (Gauss distribution) as per Tab. 2 and added to an individual cell specific complementary value, randomly chosen as per the values given in Tab. 3.

In addition to these BoL statistic laws, the model includes a correction factor, to be set at the beginning of the simulation, in order to modify the cell capacity and the cell internal resistance to find the EOL variations. The correction factor is extracted from the Saft data measured at cell level (Fig. 8 and Fig. 9).

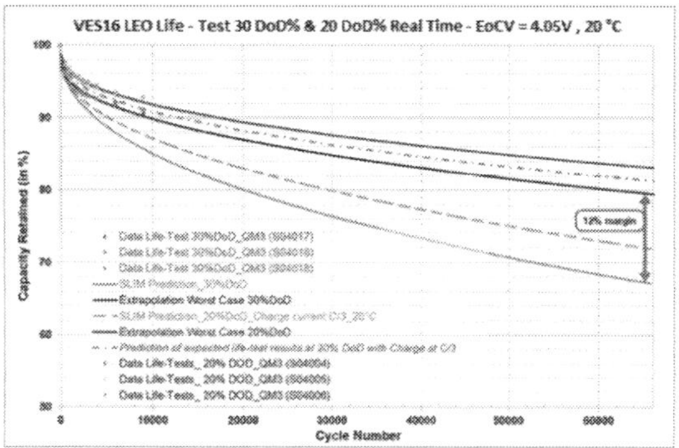

Figure 8. Capacity evolution vs number of cycles

DOI: 10.1051/e3sconf/20171607002

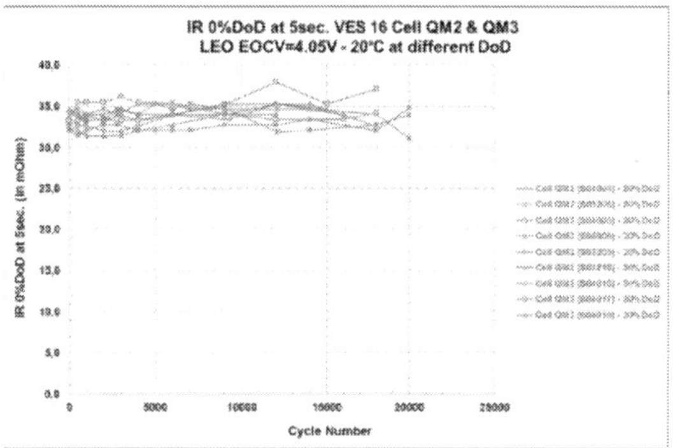

Figure 9. Cell Internal Resistance evolution vs number of cycles, at 20% DoD and 30% DoD

It must be noted that it was also observed and measured that the self-discharge rate remains constant between BOL and EOL.

2.3 EOL Statistical data of the SBS model

For the SBS, the SBS components manufacturer datasheet, some total dose radiation tests results and the SBS screening tests results (post assembly) helped to find the boundaries of the SBS parameters range.
It must be noted that normal distributions were chosen for the random sampling of all parameters.

For physical resistors in the SBS schematic (R_s and resistors used to set the BoL U_0 value), influence of temperature and ageing were computed as per ECSS-Q-TM-30-12A. As an example, Tab. 5 presents the results for a 1%/100ppm resistor.

	Tolerance (as σ value)
Initial	0.34%
Due to temperature	0.149%
Due to ageing	0.085%
Due to radiation	---
Total	**0.38%**

Table 4. Resistors statistical parameters

The leakage current induced by the SBS in stand-by mode (when inactive) has been computed from data gathered during PCBA acceptance test. Characteristic of the distribution is given in Tab. 4 and was implemented as a resistor in parallel of the SBS as described on Fig. 4.

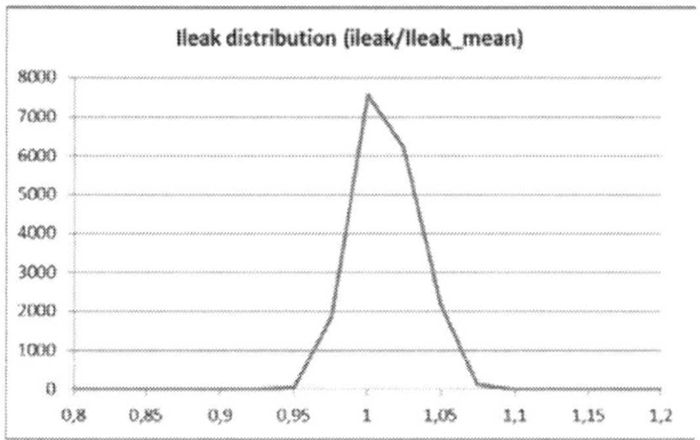

Figure 10. SBS leakage current manufacturing data

	I leakage/Ileak mean
Mean	1
σ	2,1%

Table 5. SBS leakage current statistical parameters from manufacturing data

Similarly to the cell characteristics, the U_0 characteristics of the SBS are matched by string. In the model, the computation of U_0 is based on acceptance data presented in Fig. 11 and Tab. 6.

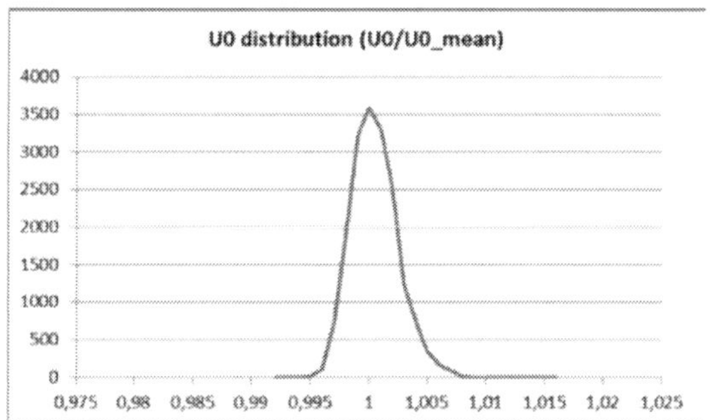

Figure 11. U_0 threshold manufacturing data

	U_0/U_0 mean
Mean	1
σ	0,2%
Δ String	0,3%

Table 6. U_0 threshold statistical parameters from manufacturing data

DOI: 10.1051/e3sconf/20171607002

Radiations are the main contributors to the SBS ageing by modifying key internal variables that directly affect the level at which the SBS becomes active. Based on results from TID tests performed on the core component of the SBS, the impact of the radiation have been evaluated as a drift and a spread value.

The SBS model has been upgraded afterward to take into account these effects by adding resistors. For example, to take into account these variations, block 2 presented in §2.2 has been changed to the schematic presented in Fig. 12, highlighted in red.

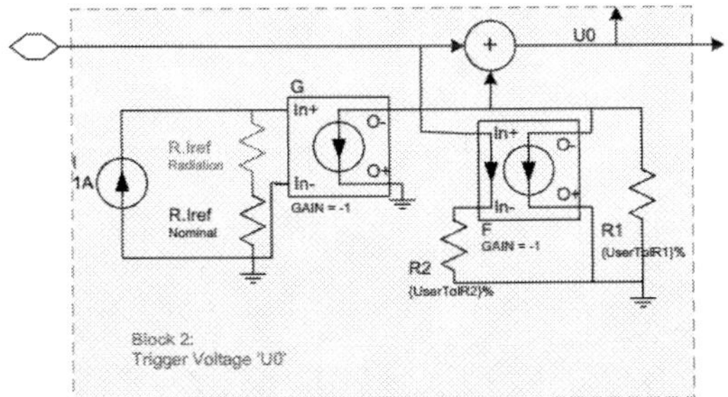

Figure 12. Modified model, with effects of radiations

3. MONTE CARLO ANALYSIS

Before performing the full battery model MCA, static MCA were performed to verify the variation of the SBS characteristics and verification of its specific worst case analysis.

3.1 MCA at battery cell level

Fig. 13 presents the statistical MCA simulation results (BOL) of the capacity performed to verify the cell model versus the Tab. 2 and Tab. 3, previously described.

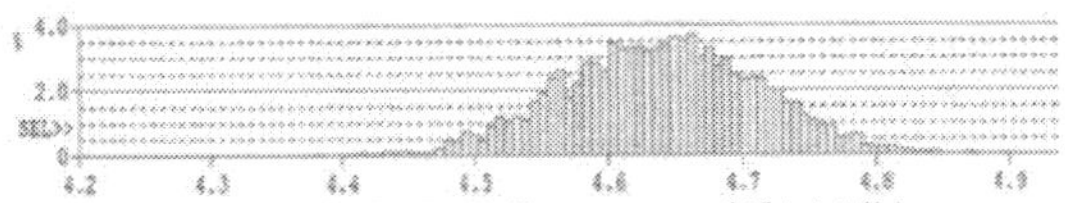

Figure 13. BoL Cell capacity (Ah) MCA

Cell Internal Resistance and self-discharge rate were verified similarly.

3.2 MCA at SBS level

In Fig. 14, the MCA simulation gives the SBS characteristic $I=F(V)$ using the BoL model presented in §3.3 The impact on U0 can clearly be seen, while the effect of Rs change seems lighter.

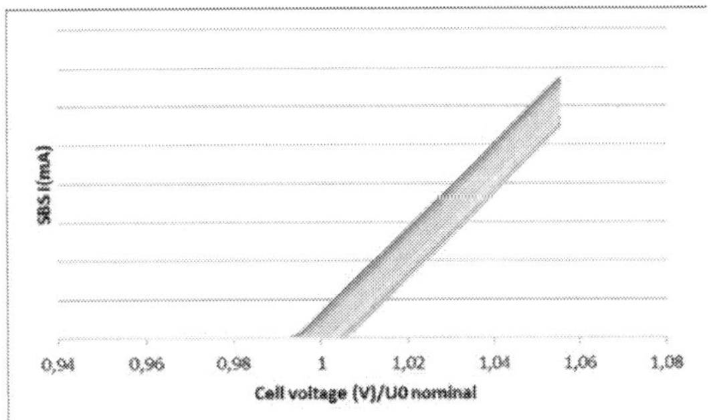

Figure 14. SBS I=f(V) statistical BOL characteristics

3.3 MCA at battery level

Several MCA were performed in transient to analyse all the behaviours of the cells and SBS for several batteries, several LEO missions charge/discharge profiles and different satellites life times.

As an example, Fig. 15 introduces to the reader the WCA results obtained with the MCA of 800 runs, of the maximum voltage un-balance between different cells of the same battery at the BOL, for the mission having the parameters summarised in Tab. 1.

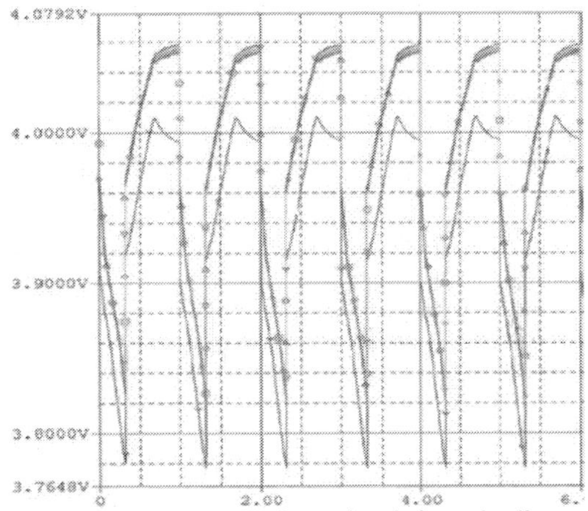

Figure 15: First Cycles with unbalanced cells

The simulation was continued until several hundred of battery cycles (charge and discharge) were reached. Fig.16 presents the maximum voltage un-balance between different cells of the same battery after 450 cycles.

DOI: 10.1051/e3sconf/20171607002

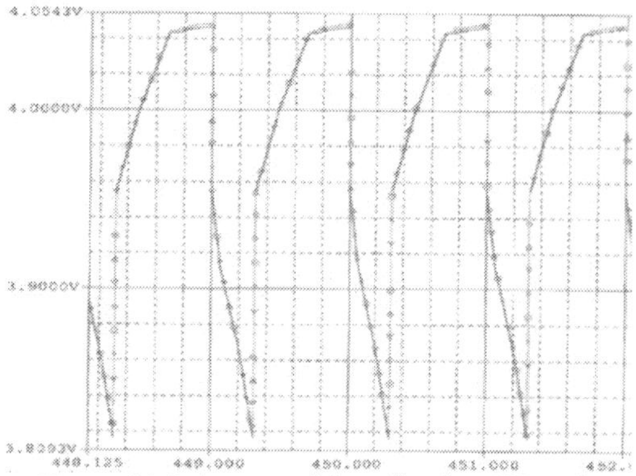

Figure 16:Balancing of the cells after 450 cycles

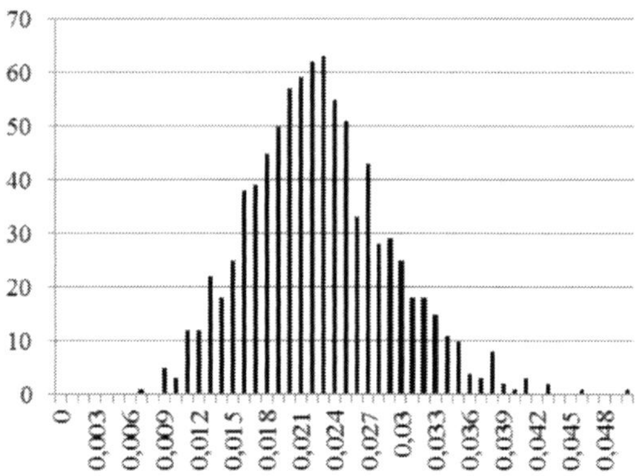

Figure 17. WCA result example of 8S4P(800 runs) – Maximum voltage string un-balance between different cells of the same battery at the EOL

From an unbalance of 50mV in the strings, following the 800 runs performed, the cell voltages dispersions at the End of Life (Fig. 17) were analysed.

These two graphs (Fig.16 and Fig. 17) confirm the correct operation of the SBS (no cells are overcharged or over-discharged during the mission) from an unbalanced state illustrated in Fig. 15 at the start of the simulation to a balanced state illustrated in Fig. 16 and Fig. 17.

4. PSPICE MODEL FOR OTHER ANALYSIS

After the development and use of the VES16 model for LEO satellite applications, it appeared that the tool could also used to define a Saft MP XTD battery maintenance charge profile for a rover during ground and cruise phases of a mission. The profile is presented in Fig. 18.

Due to particular long cruise duration and the specific high temperature, it was necessary to define a charge sequence to perform the maintenance of the battery. The sequence needed to be simple (few actions from the on-board computer), to prevent cell overcharges (over 4.2V for MP XTD cells) while allowing the SBS to balance the cells.

The model, including a new MP XTD cell model, and the simulations allowed defining the most suitable charge sequence taking into account battery and user constraints. This sequence is currently being tested on a prototype battery at Saft.

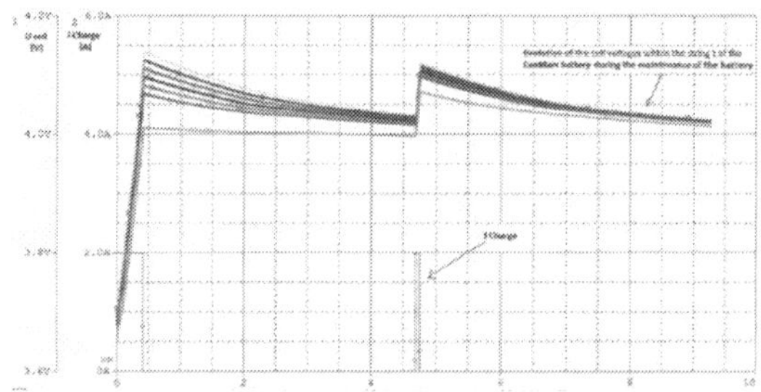

Figure 18. Maintenance profile during mission

5. CONCLUSION AND FURTHER WORK

Several PSpice models, including VES16 cell model and Simplified Balancing System were defined in order to model completely Saft VES16 batteries with Simplified Balancing System. First goal was to perform Monte Carlo Analysis in transient with these models to demonstrate the correct operation of the SBS at EoL.
Models proved to be adaptable enough to perform mission analysis not related to WCA, with a different cell, in this case the Saft MP XTD.

The tool is very useful to simulate different on ground scenarios (battery maintenance frequencies, recharge protocols…) as well as on orbit battery management scenarios.
The tool is currently used by the Electrical Engineer team and the next step is to improve the ease of use allowing direct use by Saft Batteries Engineers.

6. REFERENCES

1. Remy, S., Mosset, E., Lefeuvre, S., Nestoridi, M. (2014) VES16 Li-Ion batteries range qualification. 10th European Space Power Conference.

2. PSpice manual, available on : http://www.orcad.com/resources/orcad-downloads

LOT ACCEPTANCE, ABUSE AND LIFE TESTING OF VARTA LITHIUM POLYMER POUCH CELLS

Amy Anderson [(1)], Alan Kane [(1)], Thomas Parry [(1)], Alexander Wright [(1)]

(1) Clyde Space, Unit 5B, Skypark 5, 45 Finnieston St, Glasgow, G3 8JU, United Kingdom
Email: amy.anderson@clyde.space, alan.kane@clyde.space, thomas.parry@clyde.space, alexander.wright@clyde.space

ABSTRACT

The aim of this paper is to describe the qualification, characterisation and acceptance testing performed on Varta Lithium Polymer cells in order to assess their suitability for LEO cube-sat and small satellite applications.

Verification of battery operational performance and safety is a multi-layered process that involves continuous assessment of cell parameters at cell, battery and system level from initial cell selection through to final battery configuration.

The tests performed involved assessing individual cell performance relating to capacity under a variety of environmental conditions as well as establishing cell safety via abuse testing for small satellite systems.

1. INTRODUCTION

The Varta cell is a COTS lithium ion polymer cell format which offers high volumetric energy densities in a rectangular form factor. The rectangular footprint makes the cell ideal for mounting in stacked PCB configurations and gives a flexible starting point for modular battery configurations ranging from 10Whr packs to 150Whr and beyond.

Clyde Space has extensive experience in designing and testing batteries based on lithium ion polymer pouches for LEO applications [1].

Consistency of cell build quality and performance was assessed via the lot acceptance process. Beyond this the cell responses to external and internal fault conditions as well as capacity retention with continuous cycling was characterised.

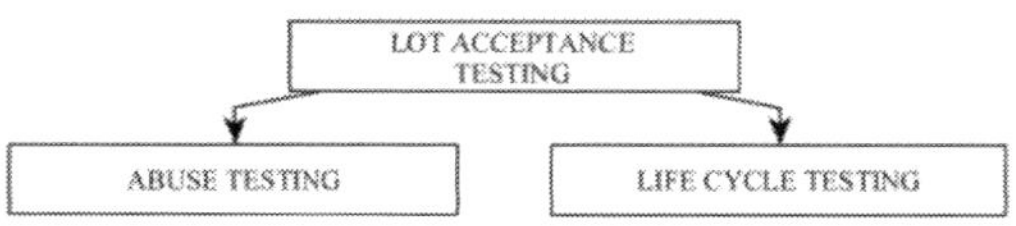

Figure 1. Test Overview for Cell Acceptance and Characterisation

2. LOT ACCEPTANCE TESTING

The lot acceptance testing was performed in accordance with Clyde Space's previously established test flow for cell lot acceptance procedures which was co-developed with ESA support. This testing comprises tests to characterise cell performance as well ensure build quality and consistency across the entire batch.

A total of 36 cells were randomly allocated from the lot for acceptance testing.

2.1 Physical Characteristics

The randomly selected cells were photographed and had their physical characteristics recorded and were visually inspected in order to flag any defects or build inconsistencies prior to further test. No concerns or abnormalities were identified at this stage and all initially selected cells progressed to active cycle testing.

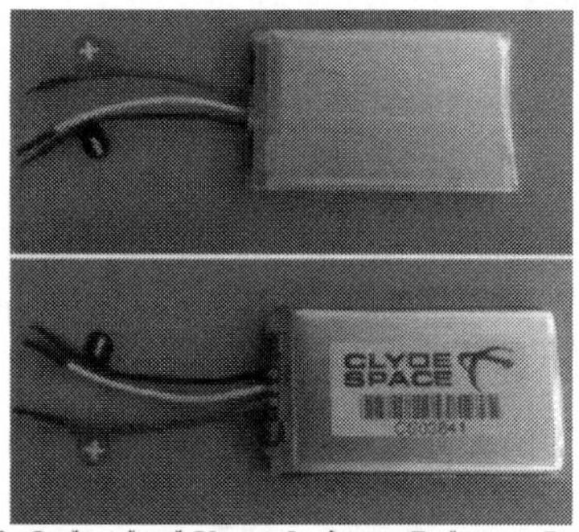

Figure 2. Individual Varta Lithium Polymer Pouch Cell

2.2 Cell Discharge Capacity Assessment

Prior to further testing each individual cell was cycled at C/5 under ambient conditions in order to provide a representation of the expected capacity distribution for the entire lot. Each cell was charged via constant current to at C/5 to 4.2V and then charged at constant voltage until meeting a 10mA cut off. Discharge was performed by applying a constant current load of C/5 until reaching a programmed cut-off voltage of 3.0V.

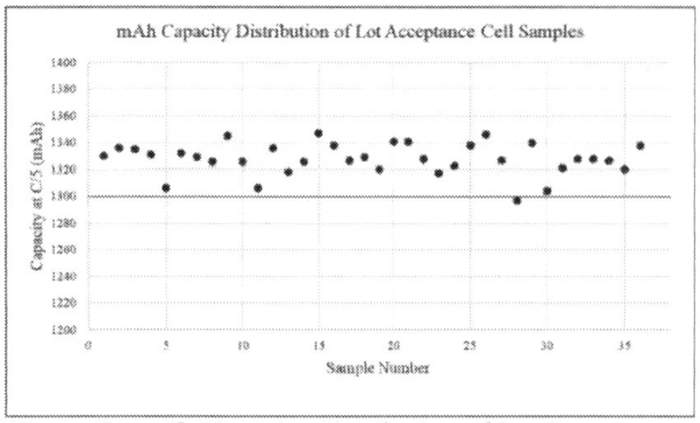

Figure 3. mAh Capacity Distribution of Lot Acceptance Cell Samples

The manufacturer's rated nominal capacity for the cells was 1300mAh at C/5, with a nameplate minimum capacity of 1250mAh. Any cells found to be outside these rated margins would be discarded and not used as part of the lot acceptance testing.

The results for the allocated lot acceptance samples gave a good grouping of recorded discharge capacity with all cells performing within the expected range.

2.3 Discharge Capacity Assessment versus Operational Temperature

Sixteen cells were allocated for testing of extractable cell capacity versus various environmental temperatures. For each of the test points given below the cell under test was subjected to three capacity cycles under a fixed discharge rate and temperature as described in Tab 1.

Test No.	Test Conditions (Discharge Rate, Temperature)	Measured Capacity Cycle 1 (mAh)	Measured Capacity Cycle 2 (mAh)	Measured Capacity Cycle 3 (mAh)
1	C/2, +40 °C	1303	1306	1298
2	C/2, +22 °C	1265	1257	1255
3	C/2, +0 °C	1196	1196	1194
4	C/2, -20 °C	865	832	796
5	C/5, +40 °C	1344	1338	1340
6	C/5, +22 °C	1329	1338	1335
7	C/5, +0 °C	1286	1292	1290
8	C/5, -20 °C	1135	1132	1128
9	C/10, +40 °C	1385	1383	1382
10	C/10, +22 °C	1342	1342	1342
11	C/10, +0 °C	1295	1294	1293
12	C/10, -20 °C	1183	1193	1193
13	C/15, +40 °C	1342	1340	1338
14	C/15, +22 °C	1349	1348	1345
15	C/15, +0 °C	1309	1309	1305
16	C/15, -20 °C	1208	1210	1217

Table 1. Measured Capacity versus Fixed Discharge Rate and Temperature

A graphical representation of the cell output voltage versus state of charge is given in Fig 5, Fig 6, Fig 7 and Fig 8.

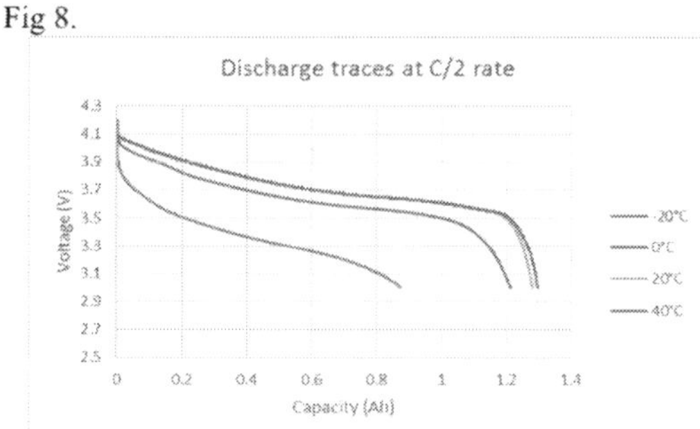

Figure 4. Discharge Slopes at -20degC, 0degC, 20degC and 40degC for a C/2 Discharge Rate

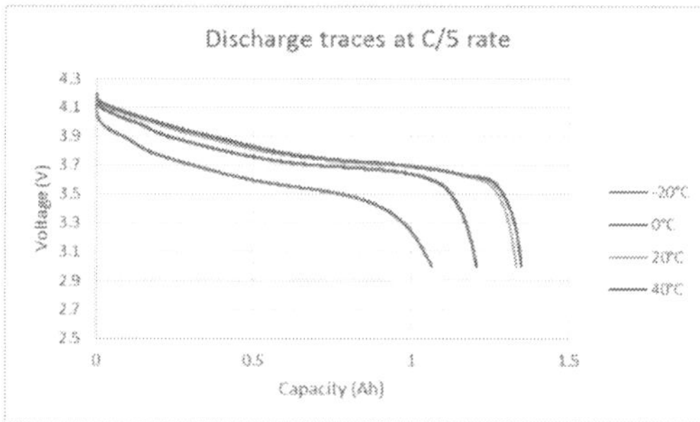

Figure 5. Discharge Slopes at -20degC, 0degC, 20degC and 40degC for a C/5 Discharge Rate

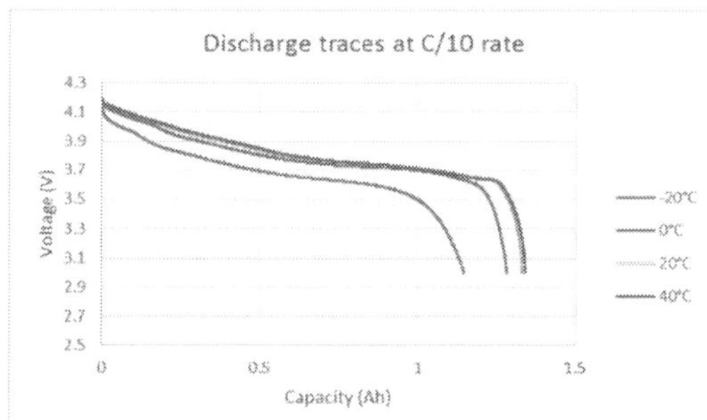

Figure 6. Discharge Slopes at -20degC, 0degC, 20degC and 40degC for a C/10 Discharge Rate

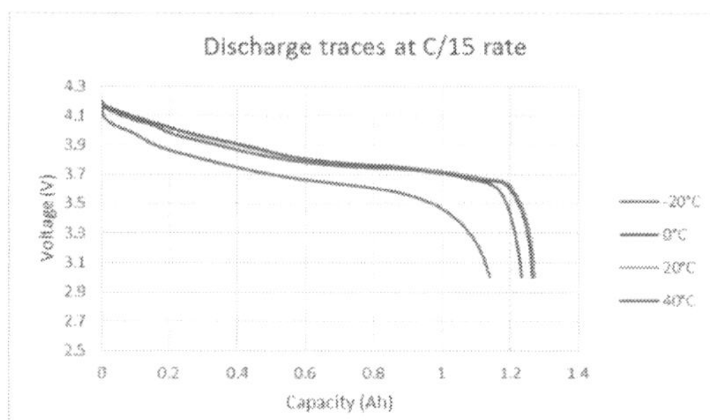

Figure 7. Discharge Slopes at -20degC, 0degC, 20degC and 40degC for a C/15 Discharge Rate

It can be seen that the cells exhibited very consistent

DOI: 10.1051/e3sconf/20171607003

output voltages for a given C rate across the range of temperatures under test.

Of particular interest are the experimental results for the cells cycled at -20°C conditions. This is known to be an operational region where cell capacity is poor for lithium ion cells due to several rate-limiting factors [2]. For cells tested at a discharge rate of C/5 or lower at -20°C the Ahr capacity was over 85% the nominal capacity available under ambient conditions.

For the cell cycled at C/2, -20 °C there is a perceptible reduction in cell capacity per additional discharge cycle even for the low amount of cycles performed.

Despite the reduced impact on usable capacity per cycle at higher discharge rates, due consideration to permanent cell lifetime will still have be given when discharging cells at high C rates – this has yet to be characterised.

3. ABUSE TESTING

The abuse testing performed was in order to support qualification of the cell to levels suitable for stowage on the ISS. This comprised tests of the cell-based electronic protection circuitry as well as forcing physical fault conditions within the cell itself. Fifteen cells were selected from the same lot as LAT testing.

Test Name	Description	Cell Allocation
External Short	Application of 10 mΩ load to cell for three hours to confirm short circuit protection.	3
Overcharge	Application of overcharge fault to cell to confirm safe activation of protection circuit.	3
Overdischarge	Overdischarge of cell at 1C to 0V to confirm safe overdischarge capability.	3
Simulated Internal Short Test	Physical crushing of cell along longitudal axis to generate internal short.	2
Heat-To-Vent	Cell temperature increased to 150°C and periodically ramped until thermal runaway.	2

Table 2. Cell Allocation for Abuse Testing

3.1 External Short, Overcharge and Overdischarge Tests

These tests were performed in accordance with NASA work instruction EP-WI-032 [3] with no catastrophic events. In all cases for the samples tested the electronic protection circuit cleanly activated when a fault condition was applied.

3.2 Simulated Internal Short Test

Each of the cells allocated for this test were compressed along the main body of the pouch between two flat plates in 2KN increments until a fault condition could be observed. For applied forces of up to 70KN both cells remained intact with no deviation in output voltage or pouch rupture.

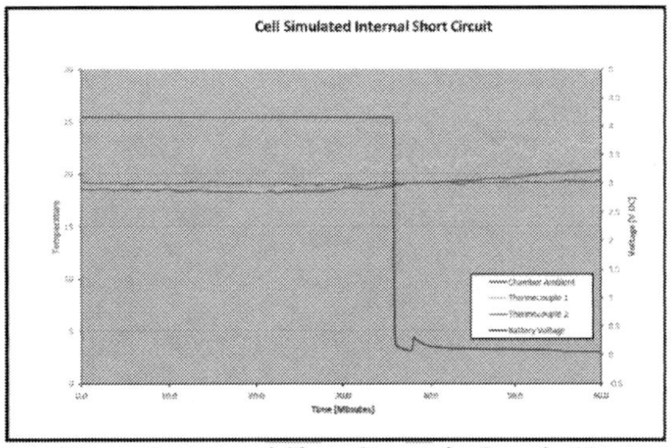

Figure 8. Internal Short Circuit for Sample 10

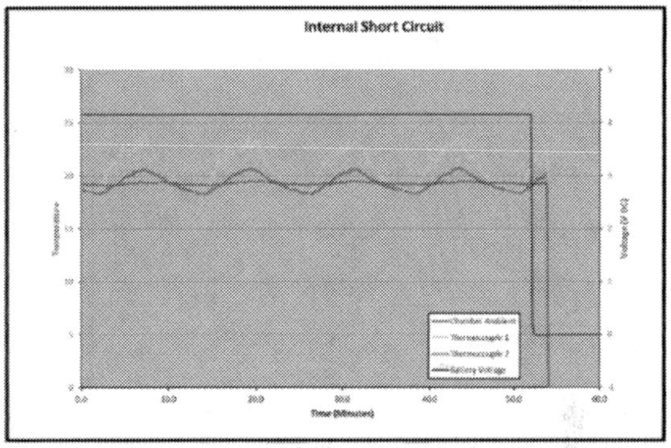

Figure 9. Internal Short Circuit for Sample 11

3.3 Heat-to-Vent

The allocated cells were placed within a thermal chamber with the chamber temperature adjusted every 30 minutes increasing the temperature increments from 66°C to 70°C then in steps of 5°C to 100 °C, the steps were then changed to 10°C steps until the cell under test went into thermal runaway (approx. 150°C) with resultant rupture and venting.

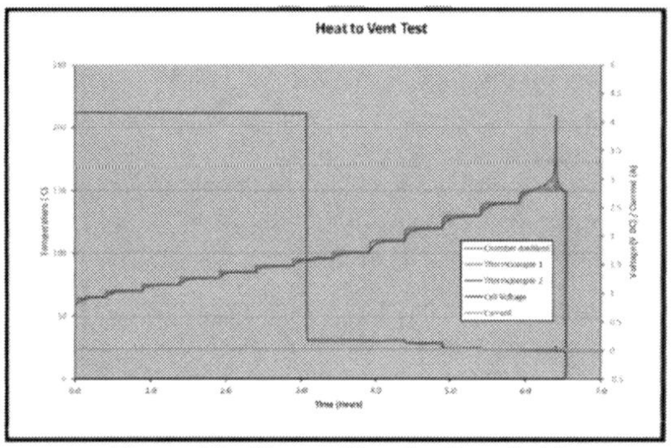

Figure 10. Heat to Vent for Sample 12

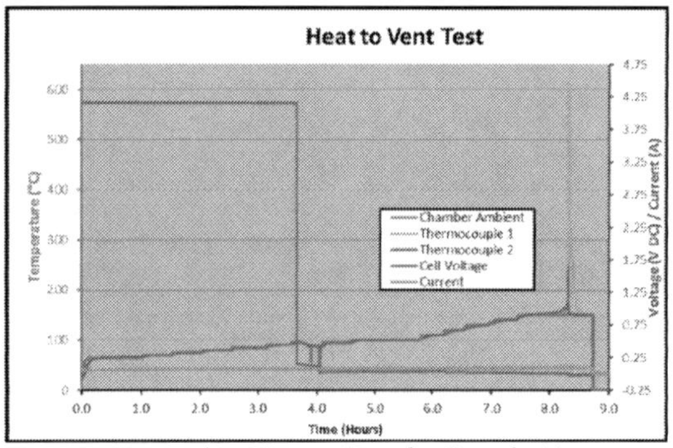

Figure 11. Heat to Vent for Sample 13

Both samples tested exhibited similar responses to the over temperature conditions. Not graphed above is the drop-out of the cell voltage from 4.2V to 0V for both samples at approximately 90°C. From the physical position of the cell fuse with respect to the cell tabs and cell voltage sense points this is believed to be caused by the low-side fuse melting and disconnecting the negative terminal at high temperatures.

Further physical analysis on the samples will be performed to confirm this.

4. LIFE CYCLING

Life cycling of the Varta cell is currently ongoing internally at Clyde Space in order to assess permanent capacity fade of the cells under continuous cycling conditions representative of typical depths of discharge, frequency and ambient temperatures for cubesat applications.

This is comprised of two parts; partial depth of discharge cycling representative of LEO conditions and full depth of discharge cycling to baseline the cells performance with respect to the manufacturer's datasheet parameters.

4.1. Full depth of Discharge Cycling

The cycle protocol consists of an alternating charge/discharge cycle operating between 0% and 100% depth of discharge at C/5. 100% depth of discharge is defined as the point at which the cell voltage reaches the lower 3V threshold.

For this test four cells were selected with varying initial capacities in order to assess if rate of capacity fade was tied to this parameter.

Cell	Initial Discharge Capacity (Ahr)	Initial Discharge Capacity (Whr)	Initial Charge Capacity (Whr)
CS105021	1.424	5.410	5.251
CS105025	1.384	5.184	5.314
CS105027	1.219	4.520	5.303
CS105059	1.272	4.903	5.228

Table 3. Cell Allocation for Full Depth of Discharge Cycling

At present the continuous full depth of discharge cycling has reached cycle 90. Fig. 12 shows the measured relationship between cell discharge capacity and cycle number; the results that widely deviate from the main trends are due to interruptions in the data collection.

This has been plotted versus the predicted capacity fade for this cell type.

The predicted capacity fade has been generated from the manufacturers datasheet which states an expected discharge capacity of >70% of C capacity after 500 cycles at C/5 and an expected discharge capacity of >80% of C after 300 cycles.

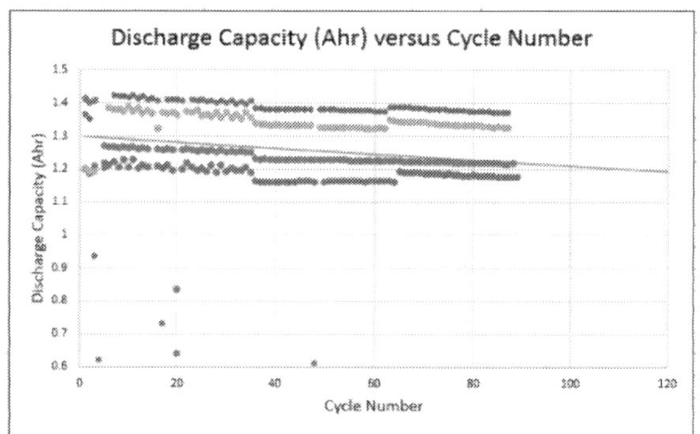

Figure 12. Discharge Capacity (Ahr) versus Cycle Number

 DOI: 10.1051/e3sconf/20171607003

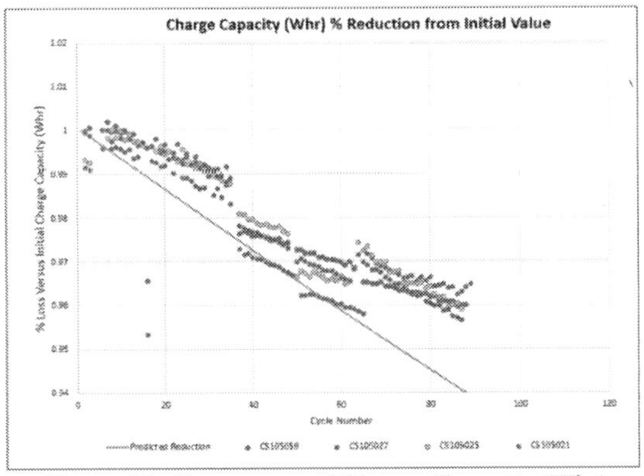

Figure 13. Percentage of Charge Capacity (Whr) Reduction versus Cycle Number

Fig. 13 shows the measured Whr charge capacity per cycle as a percentage of the initial Whr charge capacity for each cell. This is to compensate for the initial variation in cell capacities and assess the rate of change independent of initial capacity.

The "stepping" observed between cycle 40 and cycle 60 is due to an alteration in test setup which resulted in an additional voltage drop in the sense points used to calculate cell capacity. This did not alter the rate of capacity fade however did decrease the measured capacity with respect to the original value.

Outside of this the behaviour of the cells with respect to capacity fade was consistent despite the +/-10% variation in initial capacity for the cells tested. For the first 40 cycles and final 20 cycles the percentage loss per cycle is almost independent of the actual cell capacity.

Fig. 14 shows the change in voltage profile for CS105021 after 87 complete cycles to 100% DoD. The discharge curve for the final cycle with respect to the first exhibits an additional ohmic drop for the entire duration of the discharge period but otherwise the cut-off voltage of 3.0V is reached in a similar amount of time. This indicates that the measured capacity loss is mostly due to an increase in the internal cell resistance with depletion of active materials within the cell not yet playing a major part to capacity loss.

The cycle profile used for this test is to a much deeper depth of discharge than typical applications for these cells. Typically the target design maximum depth of discharge is 20% - the rate of capacity fade is therefore higher than for the cells in the full depth of discharge testing than is expected in a system designed with appropriate depth of discharge limitations.

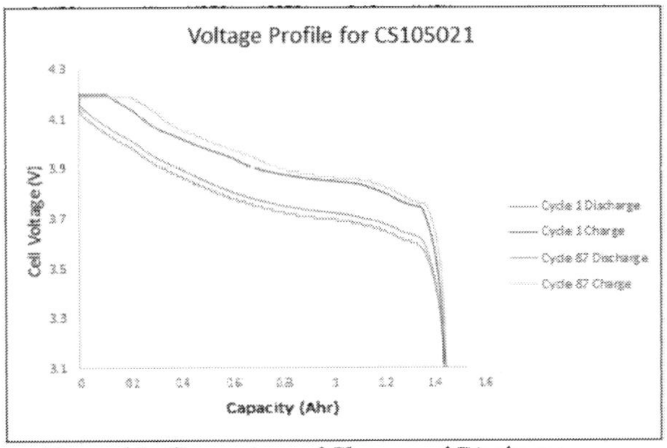

Figure 14. Comparison of Charge and Discharge Voltage Profile for CS105021 for Cycle 1 and Cycle 87

4.2. Partial Depth of Discharge Cycling

The partial depth of discharge testing is yet to commence but the method is included for reference purposes.

The cycle protocol consists of an alternating charge/discharge cycle operating between 0% and 20% depth of discharge at C/5. The 20% depth of discharge point is initially set by the nominal watt hour rating of the cell and will be fixed through the life cycle testing. Of particular interest is the drift with end of discharge voltage with continuous cycling.

Every 250 cycles a full charge and discharge cycle is performed to 100% DoD in addition to a resistance check performed via the current interruption method in order to assess the impact on overall cell capacity and quantify the cell internal resistance.

To provide a representative sample six cells will be cycled under identical conditions.

Item	Mechanical Configuration	Depth of Discharge Cut-Off	Discharge Rate	Ambient Temperature
1	Unconstrained 1s1p cell	20%	C/5	20°C
2	Unconstrained 1s1p cell	20%	C/5	20°C
3	Unconstrained 1s1p cell	20%	C/5	20°C
4	Unconstrained 1s1p cell	20%	C/5	20°C
5	Unconstrained 1s1p cell	20%	C/5	20°C
6	Unconstrained 1s1p cell	20%	C/5	20°C

Table 4. Cell Allocation for Partial Depth of Discharge Cycling

5. CONCLUSION

Characterisation of the Varta cells capacity fade with response to continuous cycling is still ongoing at this time however a body of experimental data for cell performance as well as response to external and internal fault conditions has been generated.

The lot acceptance testing has shown good cell build consistency with respect to manufacturer's parameters as well as promising cell performance at low ambient temperatures.

The abuse testing shows that the cell-level protection circuit can reliably inhibit versus electrical fault conditions; this is of paramount importance for applications involving manned flight. Response to over temperature conditions show that the temperature required to trigger a thermal runaway and cell rupture scenario is well in excess of the cells operational maximum temperature of +50°C.

At present the individual cells have been utilised in a range of battery configurations from bespoke designs to modular cubesat variants.

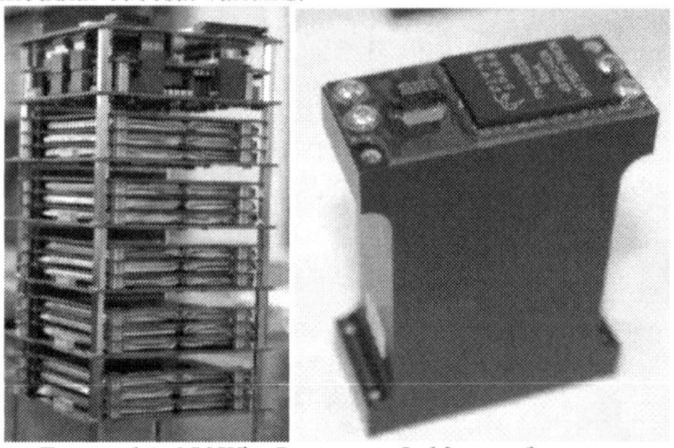

Figure 14. 150Whr Battery in 3s10p configuration (left). 20Whr Module in 4s1p configuration (right).

6. REFERENCES

1. Clark, C. & Simon, E. (2007). Evaluation of Lithium Polymer Technology for Small Satellite Applications. 21st Annual AIAA/USU Conference on Small Satellites.

2. Ji, Y., Zhang, Y. & Wang, C. (2013). Li-Ion Cell Operation at Low Temperatures. J. Electrochem. Soc. 2013, Volume 160, Issue 4, Pages A636-A649.

3. Jeevarajan, J. (2014). EP-WI-032. Engineering Evaluation, Qualification and Flight Acceptance Tests for Lithium-ion Cells and Battery Packs for Small Satellite Systems, National Aeronautics and Space Administration, pp5–15.

A NOVEL IN-FLIGHT SPACE BATTERY HEALTH ASSESSMENT SYSTEM

Brandon Buergler [(1)], **François Bausier** [(1)]

[(1)] *ESA-ESTEC, Keplerlaan 1, 2200 AG Noordwijk, NL, Email: brandon.buergler@esa.int, francois.bausier@esa.int*

ABSTRACT

The assessment of an in-flight battery health status is of great interest. However, it is very difficult to perform with sufficient accuracy because the battery is constantly operated and no full discharge during a space mission can be done. Currently, a method for performing a health assessment is via the correlation of battery capacity with internal resistance, which can be more easily measured. Another method is to fit the on board battery voltage using the telemetry of current and temperature as inputs by advanced modelling techniques and to derive the ageing parameters of the battery. This is done by optimising the parameters so that the fitted voltage curve is as close as possible to the actual measured voltage telemetry curve. Then an assessment of the battery capacity and energy can be done by simulating a battery discharge at the desired conditions with the required voltage limits. However, these two methods often require extensive battery life tests on ground to estimate the ageing behaviour for a particular mission.

In this paper a novel health assessment system is proposed by which a more straight forward and accurate assessment of the battery health in terms of capacity and energy can be performed. The proposed system consists to use one or some battery string(s), which can be safely disconnected via the PCDU (Power Conditioning and Distribution Unit) from the battery for a dedicated capacity measurement while the main part of the battery continues to operate normally. In this paper, the viability of the proposed system is demonstrated and the implications on the battery and the power system are discussed.

1. INTRODUCTION

It is difficult to assess the health status of a spacecraft battery in operation. Due to design, operational and safety constraints, it is essentially impossible to fully discharge the battery in order to accurately measure the currently available capacity. The battery state of health in terms of capacity can be crucial when a mission is to be extended beyond the initially planned duration. In contrast, it is easier to assess the internal resistance of the battery. Every time an increasing current step occurs on the battery bus, the voltage across the battery terminals drops. The internal resistance of the battery can be calculated from this voltage drop and the difference of current before and after the step. Due to limited telemetry sampling rate in the range of one measurement every few seconds, the data has to be statistically evaluated. Only current steps with a certain threshold can be included in the evaluation. A prediction of the capacity fade of a spacecraft battery has then to be based on the relationship between the internal resistance and remaining capacity, which is probably a complex one. The basis for this approach is the availability of experimental data for the cells in similar conditions to the ones encountered in the spacecraft of which the remaining battery capacity is to be determined. This calls for the conduction of long duration life tests on single cells of representative batteries. In practice there is always a difference between the experimental test conditions and the real conditions of a satellite in operation. So the predicted capacity will almost certainly deviate from the real one. The subject is currently under discussion among the space power community worldwide and an adequate solution to this issue would be of great interest. Currently an ISO standard is being prepared on this topic (ISO/TR 20891 Space systems - Space batteries - Guidelines for easing in-flight health assessment). The methods that are currently being used are mostly based on fitting the telemetry data of the in-flight battery by adapting the ageing parameters of the battery models. The derived values for capacity fade and resistance increase are then validated towards other telemetry data before an assessment of capacity and energy are done. Examples of this approach are given in references [1-4]. However, in practice it is difficult to achieve an accuracy better than 10-20% with these methods. If it were possible to assess the battery capacity with improved accuracy, it would be possible to better:

- know the battery health status in order to optimise the spacecraft operations, in particular in case of mission extension beyond the initially planned duration
- keep the spacecraft battery at a sufficiently high state of charge for nominal operations and recovery while mitigating the battery capacity fading
- get in-flight data about battery capacity fading in order to improve knowledge and optimise future space power systems
- investigate the cause of potential power system degradation or failure cases

2. SYSTEM DESCRIPTION

A new system for accurate battery health assessment is proposed here, in which the battery capacity is directly deduced from on board measurements. A scheme of the proposed system is shown in Figure 1. Within the battery, one or more "measurement" strings are connected separately via dedicated terminals leading to the PCDU or another electronic unit. In the following the solution is described assuming one measurement string, keeping in mind that it could be similarly implemented with two or more strings in parallel. This measurement string can be disconnected from the battery main part via the PCDU (via a switch) when the battery is fully charged (e.g. just before an eclipse or with a dedicated charge circuit). Then it can be fully discharged to determine the remaining capacity of the measurement string. A procedure for determining the internal resistance of the string(s) can also be implemented as well as a state of charge (SOC) determination from the measurement of the disconnected string electromotive force (EMF) at zero current. The advantage of this approach is that the satellite can remain in the nominal state but with one battery string less. No risky satellite manoeuvres nor solar array depointing nor battery deep discharge need to be done.

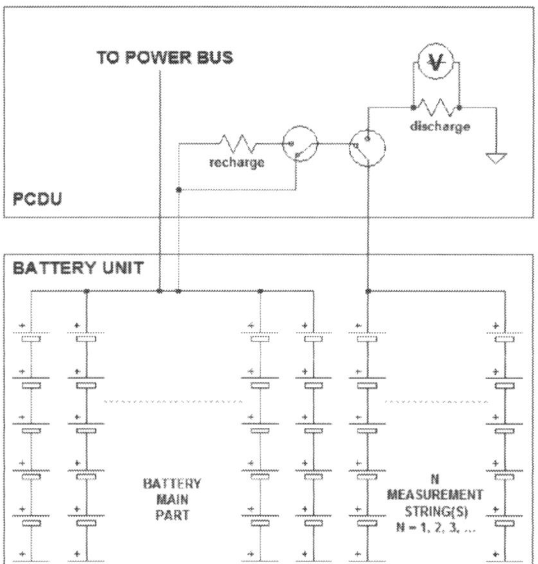

Figure 1 – Proposed battery and PCDU configuration for determining the capacity of an in-flight battery with one or more measurement string(s) connected via the PCDU to the main battery bus.

After recharging the measurement string to the bus voltage (e.g. via a resistor), the switch can be closed again and the string is reconnected to the other battery strings. This approach requires only small changes to the PCDU and to the batteries with a limited cost and weight impact. The telemetry system will have to include the current and voltage values of the individual string. From these telemetry values the actual battery capacity can be calculated. The discharge can be simply done by a resistor (in which case the current can be deduced with accuracy from the voltage telemetry) or by a current source to achieve constant current discharge.

Note that Figure 1 features two switches in order to cope with the accidental connection of the measurement string to the battery main part while the voltages are not equalized, which could stress the string. Some precautions or protection may also be foreseen to mitigate the risk to overdischarge cells in the measurement string.

Physically, the measurement string(s) could be located among the other strings at any chosen location in the battery in order to be exposed to representative environmental conditions.

3. ASSESSMENT BY SIMULATION

Some simulations have been done using the Power Energy Platform Simulation PEPS developed at ESTEC [5], in particular to check the influence of the harness and switch resistance on the cell current and voltage of the measurement string.

Figure 2 shows a circuit where one measurement string would be connected inside the PCDU after the battery harness during normal operation of the battery (i.e. no health check being done).

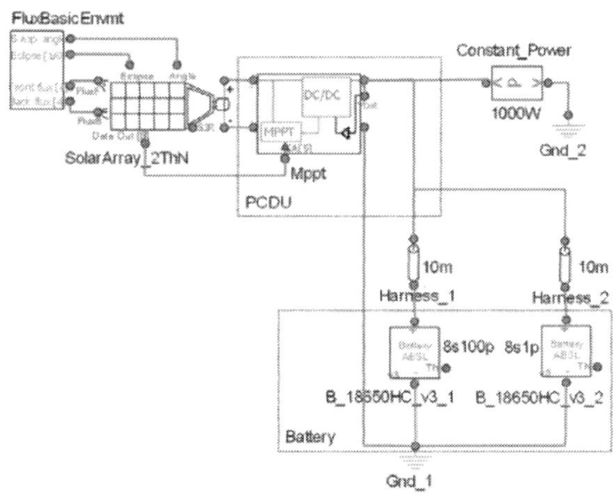

Figure 2 – Simulation diagram (measurement string connected after battery harness; no discharge circuit represented in this diagram).

DOI: 10.1051/e3sconf/20171607004

In this case the current and voltage of the string are not representative of the main battery strings (Figures 3 and 4). This is due to the difference of equivalent series resistance of the main battery strings (R_harness_main × number of strings in main battery) and of the measurement string (R_harness_measurement_string × 1). The single string is connected to the PCDU with a lower resistance as compared to a string in the main battery, which leads to a more severe cycling. An evaluation of battery capacity on the basis of such a system would lead to an overestimation of the battery capacity fading.

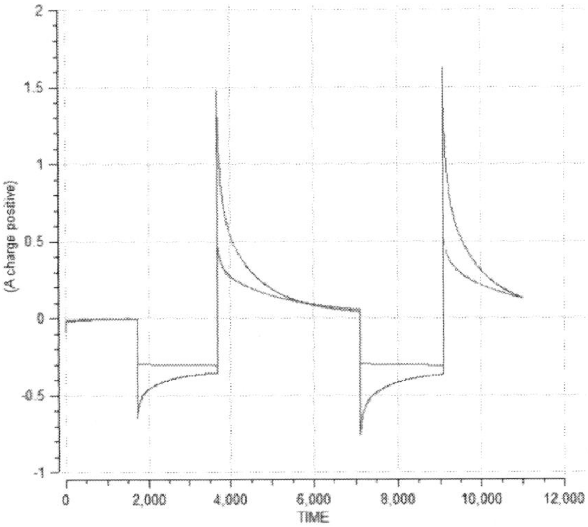

Figure 3 – Comparison of cell current during LEO eclipse cycles for diagram Fig 2 (red=battery main part, blue=measurement string). The measurement string is cycled with higher current than the strings in the battery.

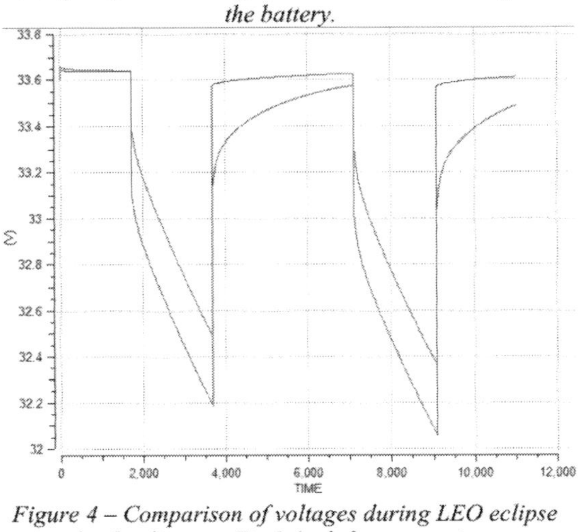

Figure 4 – Comparison of voltages during LEO eclipse cycles for diagram Fig 2 (red=battery main part, blue=measurement string). The measurement string is cycled with a larger voltage range than the strings in the battery.

A way to solve this issue would be to adjust R_harness_measurement_string equal to R_harness_main x number of strings in main battery. However, in practice this solution is difficult to achieve with good accuracy.

A solution to mitigate the influence of the main harness and connection is to connect the measurement string on the battery internal bus bar as shown in Figure 5.

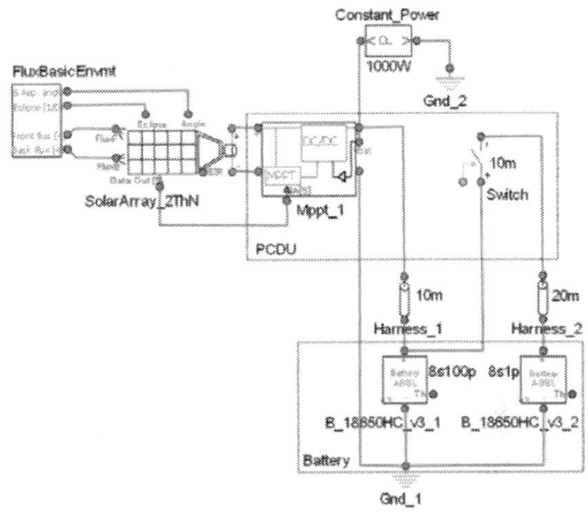

Figure 5 – Simulation diagram (measurement string connected before battery harness; no discharge circuit represented in this diagram).

The simulations performed on the basis of this improved configuration are shown in Figure 6 and Figure 7. In this case the currents and voltages of the cells in the measurements string are very similar to the ones in the battery main part.

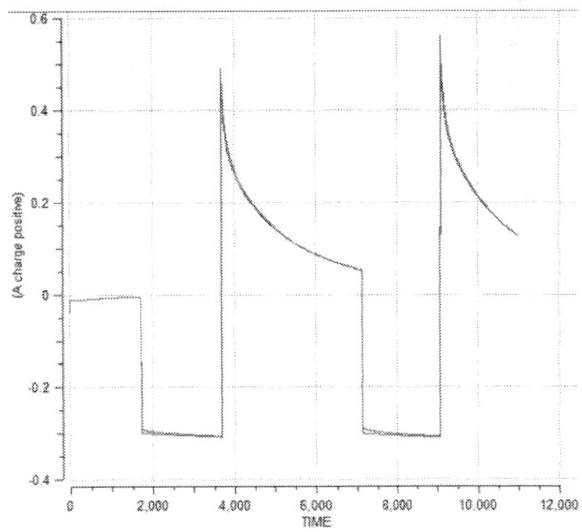

Figure 6 – Comparison cell current during LEO eclipse cycles for diagram Fig 5 (red=battery main part, blue=measurement string).

DOI: 10.1051/e3sconf/20171607004

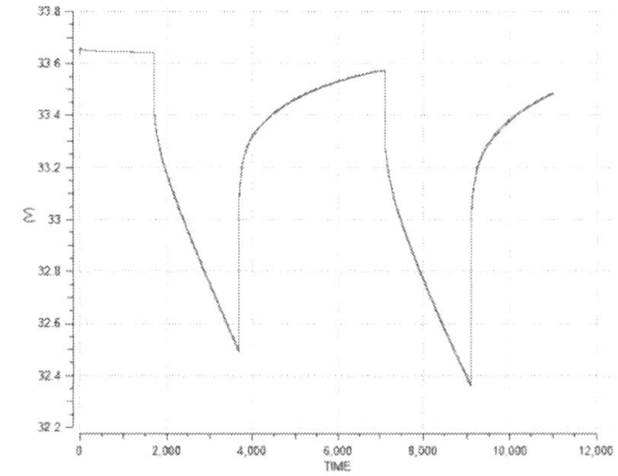

Figure 7 – Comparison of voltages during LEO eclipse cycles for diagram Fig 5 (red=battery main part, blue=measurement string)

The total connection resistance of the measurement string has to be as low as possible to achieve an acceptable match between this string and the strings in the main battery (same current and same voltage during cycling). The possibility to practically achieve a low enough resistance (much lower than the string resistance) and the influence of all parasitic resistances (including the switch resistances) have to be considered to obtain the desired accuracy. An additional resistance of 3% (TBC) for the measurement string as compared to the main battery strings seems to be a reasonable aim. A way to minimise the effect of this resistance is to connect several measurement strings in parallel inside the battery instead of only one. Finally the complete system including the discharge circuit has been simulated (Figure 8).

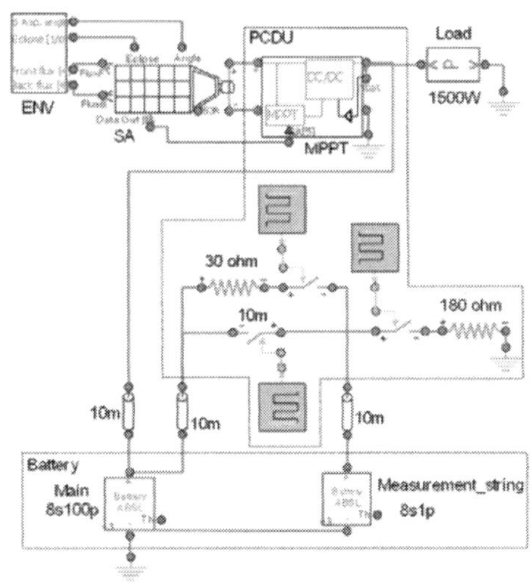

Figure 8 – Simulation diagram of complete system with one measurement string (battery capacity fading of 30%).

In this simulation diagram 3% additional harness resistance for the measurement string have been taken into account. The results are shown in Figures 9 to 11.

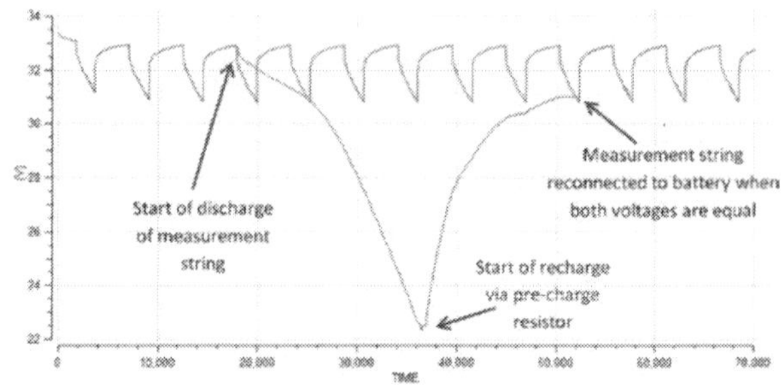

Figure 9 – Battery voltage vs. measurement string voltage during LEO eclipse cycles.

Figure 9 shows that after some cycles the measurement string is disconnected from the main battery after having been fully charged (about 33V battery voltage or 4.1V/cell). The measurement string is then connected to the 180 ohm resistor and is discharged at a current of about C/10 (150mA) down to the minimum voltage. The maximum power dissipated in this resistor is around 6 W at 33 V. Ideally the discharge could be done down to 20V (2.5V/cell) plus some safety margin. However, due to simulation model constraints it has been stopped at 22.5V (2.8V/cell) in this simulation.

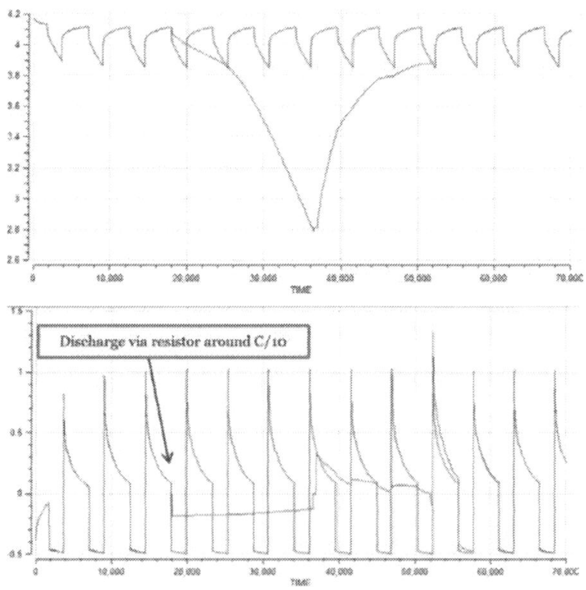

Figure 10 – Cell voltage (top, [V]) and cell current (bottom, [A]) (red=battery main part, blue=measurement string).

Then the measurement string is disconnected from the 180 ohm resistor and connected to the pre-charge resistor (30 ohm) until it reaches again the same voltage as the main battery. At this time it is reconnected to the main battery.

Based on the measurement string voltage and current telemetries it is possible to determine the remaining capacity (and internal resistance) of the measurement string and therefore of the main battery: in this case 70% of the BOL (Beginning of Life) capacity. Considering a telemetry accuracy better than 1% EOL (End of Life) e.g. based on high accuracy 0.1% resistors, an accuracy of the capacity/energy measurement better than a few percent can be expected. Figure 10 shows that the measurement string voltage and current satisfactorily mimic the main battery string voltage and current both before and after the full discharge/charge cycle of the measurement string. Only one or two eclipse cycles are necessary to recover the perfect matching.

In addition the SoC (State of Charge) is shown in Figure 11. Both measurement string cells and main battery cells behave the same way before and after the discharge/charge cycle.

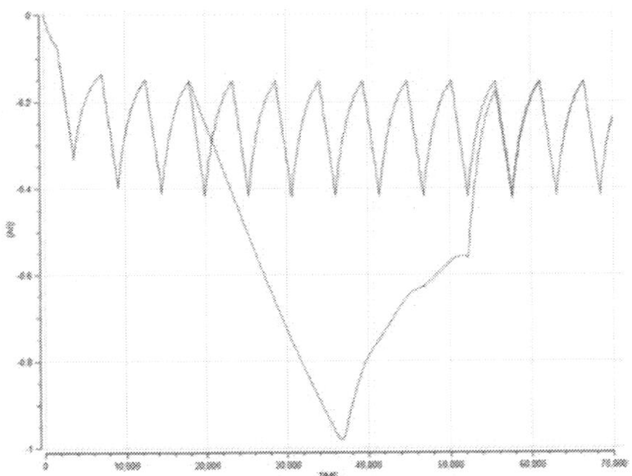

Figure 11 – Cell state of charge in Ah extracted from the cells (red=battery main part, blue=measurement string).

4. CRITICAL ANALYSIS

The last system configuration shown in Figure 8 is promising. An advantage of the system is that it is possible to fully discharge a representative part of the battery for obtaining the overall battery capacity. Capacity assessments on the basis of partial discharges only, have uncertainties concerning the full capacity. Open circuit voltage (OCV) measurements (i.e. battery voltage measurements at zero current) could also be done if needed. Some capacity determination methods are based on OCV measurements, which are not available on satellite batteries without the proposed system. However the following points are worth to be noted to optimise the system:

- The level of degradation of the measurement string due to the full discharge tests has to be considered. Obviously such full discharge test should not be performed too often; otherwise the measurement string would degrade more rapidly than the main battery ones (especially at BOL). For example a frequency of once a year would not degrade the measurement string significantly.
- In the measurement string there are already several cells and the discharged capacity of this string will already account for cells with higher and lower than average capacities. Thus, the measurement of one string will represent an average cell in the battery, however, without information about the variability from cell to cell. Having several measurement strings instead of one would allow for an even more representative capacity figure that could account for the locally different environment in a larger battery system including such with more than one module.
- The optimal number of measurement strings in parallel has to be determined on a case by case basis, also taking into account the series resistance achieved by the connection circuit via the PCDU, the level of fidelity desired and the cost.
- The number of switches needed in the connection/discharge circuit (via PCDU) may vary depending on the desired robustness against faulty battery connection and/or faulty discharge, which could potentially stress the measurement string(s) during AIT activities or in flight. Two switches in series for the battery connection and two switches in series for the discharge would offer a better robustness against a single failure or a wrong command.
- The way the switches are commanded may also vary according to the desired robustness. Two level commands (e.g. arm and fire) would offer a better robustness against a single failure or a wrong command.
- In case it is not convenient to fully charge the battery before the measurement, a partial discharge can also be performed. The remaining energy starting from the actual end of charge voltage during the mission (e.g. at 80% SOC) to the given battery end of discharge voltage seems to be a suitable reference energy for assessing the state of health as well.
- This new system was initially based on ABSL18650HC cell in sp (series-parallel) configuration where a large number of strings are in parallel. Nevertheless the same principle seems valid for other batteries for example based on Saft VES16 or MPS (sp configuration) or even larger Saft cells (ps configuration) with balancing electronics. However, dedicated balancing electronics need to be foreseen for the measurement string(s). Note that this string contribute to the nominal battery performance, except during the string full discharge

test where some operational measures can be taken if necessary (in case the measurement string represents a significant part of the whole battery).

- The detailed impacts on the overall power system in terms of FMECA (Failure Modes, Effects and Criticality Analysis), hardware needed, passivation, representativeness of the selected string, battery bus variation during recharge of separated string and so forth need yet to be determined in detail. Finally tests will be needed to verify the expected performances.

5. IMPACT ON BATTERY AND PCDU

For the battery itself and the PCDU the following adaptions would have to be made:

Battery:
Internally the battery design could be the same but one (or more) string(s) would have to be taken from the main battery internal bus on the positive side.
In order to connect the measurement string to the PCDU and back to the main battery positive bus, two extra pins (or more depending on the resistance of the connection) are needed. These pins could be provided on the already existing battery power connector or it could be placed on a dedicated connector.

PCDU:
As shown on Figure 1 the PCDU would have to include:

- Switches (e.g. relays) including commands to connect the measurement string to the main battery, to the discharge circuit or to the pre-charge resistor (with redundancy for operational/safety reasons).
- A discharge circuit: the simplest is a resistor (with voltage telemetry allowing to measure both voltage and current with high accuracy) but a current source could also be implemented to achieve constant current discharge.
- A pre-charge resistor (to recharge the measurement string to the main battery voltage before to connect to it).
- Two extra pins (or more) to connect the measurement string.

6. CONCLUSIONS

A new system for determining with accuracy the health status of an in-flight space battery has been described with an expected accuracy of better than a few percent. No modelling or simulations, which induce inaccuracies are needed for the capacity determination. Expensive long duration life tests on cell or battery level can also be avoided to some extent. The measurement string current during the capacity assessment can be controlled independently of the satellite operations.
There is currently an increasing interest in the space power community to get such an accurate battery health assessment solution.
Preliminary assessments and analyses of this new system have been conducted and the outcome is promising.

7. REFERENCES

[1] D. Genc, C. Thwaite, "Proba-1 and Mars Express: An ABSL Lithium-Ion Legacy", European Space Power Conference, (2011).
[2] D. Curzon, C. Thwaite, R. Buckle, "Lithium-Ion Battery Performance on board geostationary and lunar orbiter NASA missions - SDO and LRO", European Space Power Conference, (2011).
[3] Y. Borthomieu, D. Gambini, "Saft Li-Ion VES140S Battery Flight Experience Return on TAS Space Bus Platforms", 10th European Space Power Conference, (2014).
[4] Final report TRP activity "Battery capacity estimation based on modelling and advanced data processing techniques" , ESA Contract Nr. 4000114068.
[5] A. Fernandez, G. D'Accolti, B. Buergler, B. Garcia, "PEPS. A Tool for Power System Simulation", 10th European Space Power Conference, ESA, Noordwijkerhout, The Netherlands, 2014

E3S Web of Conferences 16, 07005 (2017) DOI: 10.1051/e3sconf/20171607005
ESPC 2016

INTERNAL IMPEDANCE OF THE LITHIUM-ION SECONDARY CELLS USED FOR REIMEI SATELLITE AFTER THE ELEVEN YEARS OPERATION IN SPACE

Yoshitsugu Sone [1, 2], **Hiromi Watanabe** [3], **Kohei Tanaka** [2], **Omar Samuel Mendoza-Hernandez** [1], **Seisuke Fukuda** [1, 2], **Masayuki Itagaki** [4], **Keita Ogawa** [5], **Kazushi Asamura** [1, 2], **Atsushi Yamazaki** [1], **Hiroyuki Nagamatsu** [1], **Yosuke Fukushima** [1, 2], **Hirofumi Saito** [1, 3]

[1] *Japan Aerospace Exploration Agency, 3-1-1 Yoshinodai Chuo, Sagamihara, Kanagawa, 252-5210 Japan, Email:sone.yoshitsugu@jaxa.jp*
[2] *The Graduate University of Advanced Studies (SOKENDAI), 3-1-1 Yoshinodai, Chuo, Sagamihara, Kanagawa, 252-5210 Japan, Email:fukuda.seisuke@jaxa.jp*
[3] *The University of Tokyo, Saito Lab., 3-1-1 Yoshinodai Chuo-ku, Sagamihara, Kanagawa, 252-5210 Japan, Email:saito.hirofumi@jaxa.jp*
[4] *Tokyo University of Science, 2641 Yamazaki, Noda, Chiba, 278-8510 Japan, Email:itagaki@rs.noda.tus.ac.jp*
[5] *Advanced Engineering Services Co. Ltd., Mitsui Building 7F, 1-6-1, Takezono, Tsukuba, Ibaraki, 305-0032 Japan, Email:k_ogawa@aes.co.jp*

ABSTRACT

The lithium-ion secondary batteries have been widely used for the space programs, today. Among them, REIMEI was one of the first satellites using lithium-ion secondary battery. In 2005, the satellite was launched, and injected into the low earth polar orbit. Eleven years has passed since the launch and over 60,000 cycles of charge and discharge was experienced in space.
The lithium-ion secondary cell of the REIMEI battery was designed using spinel manganese oxide type material for the positive electrode, and the graphitized type carbon for the negative electrode. The cell case was made of aluminium laminated film and the structure was reinforced by the epoxy resin and aluminium housing. After the operation of eleven years, the cells still maintain the appropriate uniform balance and operative. In order to identify the internal condition of the battery/cell, we calculated the ac impedance by the pulse duration to the on-board battery.

1. INTRODUCTION

Today, lithium-ion secondary battery is widely used for the spacecraft and satellites. When the lithium-ion secondary cells started to be used for the space programs, 'off-the-shelf' lithium-ion secondary cells received much attention. The European piggy-back satellite 'PROBA' uses lithium-ion batteries produced by AEA using lithium-ion secondary cells from SONY. The same technology was also applied to European planetary missions such as the satellite 'ROSETTA' and the 'MARS-EXPRESS' [1].
The Japanese satellite 'REIMEI' was also one of the examples where the off-the-shelf lithium-ion secondary cells were used [2-6]. The satellite was an engineering demonstrator for the advanced space technologies. [2, 3] Initially it also had a scientific mission to observe aurora around the polar region using three cameras with different filters for the wave length of the light. The science mission was terminated in eight years after the launch. Today, the satellite is still maintained to observe the degradation mechanisms of the battery in space.
Recently, the 'passivation' of the satellite in the end-of-life was focused [7, 8], and the estimation of the health condition like impedance [9] of the batteries/cells in space was also discussed.
In this paper, we discuss the ac and dc impedance conditions of the batteries/cells after the long term operation of REIMEI satellite in space.

Figure 1. Photograph of the 'REIMEI' satellite (Code name before the launch was INDEX) nderdevelopment.

Table 1. The outline of the satellite 'REIMEI'.

Designation Code	2005-031B
Launch Date	06:10, August 24, 2005 (JST)
Location	Baikonur Cosmodrome
Size	60 cm x 60 cm x 70 cm
Mass	60 kg
Orbiter	Quasi-circular orbit
Altitude	610 - 654 km
Inclination	97.8 degrees
Period	97 minutes
Attitude control	Three-axis stabilization

Table 2 The specifications of the lithium-ion secondary cell for REIMEI battery.

Electrode	Spinel $LiMn_2O_4$ (positive), Graphite type carbon (negative)
Name plate capacity	3 Ah
Mass	75 g
Volume	145 mm x 80 mm x 4 mm
Energy density	158 Wh/kg, 340 Wh/L
Charge voltage	4.1 V (nominal), 4.2 V (temporal)
Discharge limitation	3.0 V

2. OUTLINE OF REIMEI AND CELL/BATTERY

'REIMEI' satellite before launch is shown Fig. 1. The satellite was launched in August, 2015 [2, 3]. It was an engineering demonstrator for the cutting-edge satellite technology. It also carried onboard instruments for scientific aurora observations which were suitable for a small satellite. The outline of the satellite is shown in Tab. 1.

For the main bus battery, the off-the-shelf type lithium-ion secondary cells were used for the REIMEI battery [4, 5]. The specifications of the cell are listed in Tab. 2. It was a lithium-ion secondary cell, whose positive material was spinel manganese oxide and the negative was graphitized carbon. The upper limitation for the charge voltage was 4.2 V/cell. We decided that the nominal charge voltage should be up to 4.1 V/cell to realize the longer operational period of the satellite.

For a battery, seven cells were connected in series in order to realize the 28 V-class-bus-voltage. Two batteries were connected in parallel and installed into the aluminium case. The cell case was aluminium laminated file, and the cells were installed inside the rigid aluminium case and the case was filled with epoxy resin. The cell and battery are shown in Fig. 2.

Fig. 3 shows the block diagram of the electrical power subsystem for REIMEI satellite [5]. Each voltage of individual battery could be monitored. The range for the monitoring of the output voltage was between 0 and 35 V, and the data was acquired by 8 bit data recorder. Thus, the telemetry data for the battery voltage had the accuracy of ±0.136 V [5].

The current for the charge and the discharge was also monitored individually for each battery between -3.0 A

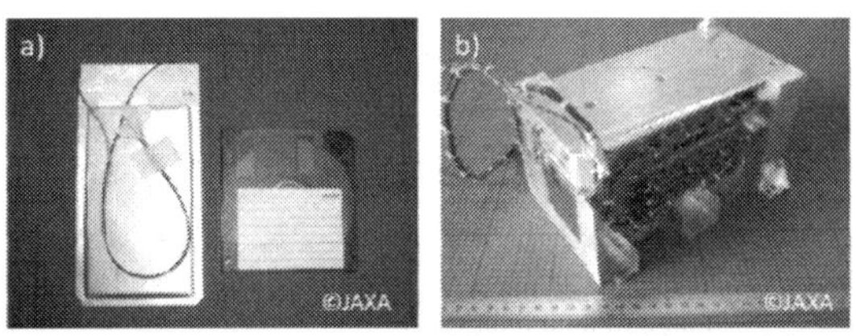

Figure 2. The photographs of the cell (a) and battery (b) for stellite 'REIMEI'.

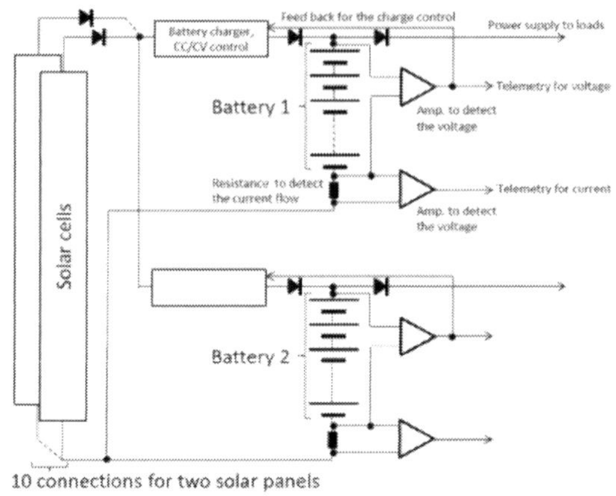

Figure 3. The Block diagram of the electrical power subsystem on-boarded to the satellite 'REIMEI'.

(for the discharge) and +3.0 A (for the charge). The current was also acquired by 8 bit-data-recorder, and had the accuracy of ±23 mA [5].

3. ORDINAL OPERATION OF THE BATTERY

REIMEI was launched in August, 2005. It was injected into the low-earth-polar orbit with the 600 km altitude. Then, the satellite continuously experiences daytime for 35 minutes and eclipse for 62 minutes during one revolution around the Earth.

For the charge, the 'constant current/constant voltage (CC-CV) control' has been used. The battery was

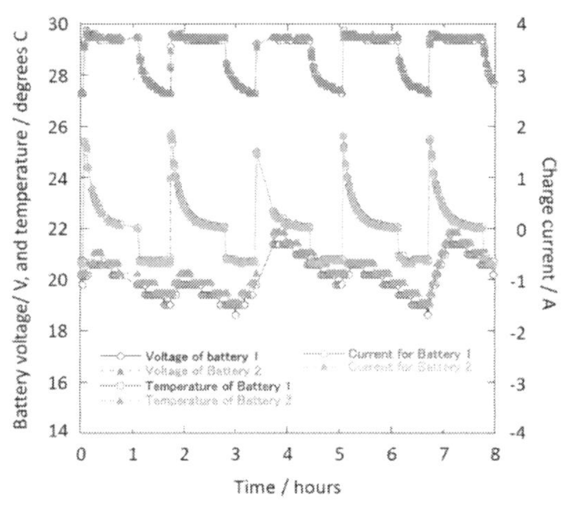

Figure 4. Typical in-flight data of the battery during charge and discharge for 8 hours on September 25st, 2012. CC-CV control was applied. For the CV control, V1 level (4.1 V/cell) was used for the charge control.

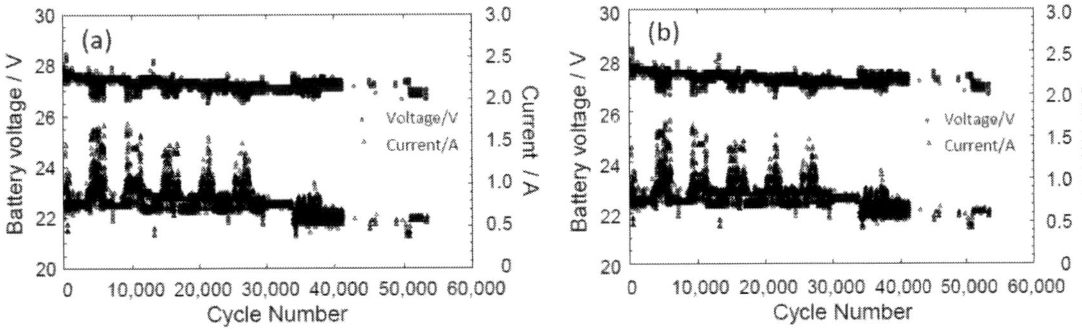

Figure 5. The trend of the voltage and current in the 'end of discharge'.
The responses from the 'Battery 1'(a), and 'Battery 2'(b) are shown.
*(Ref. [5] Sone, Y. et al., (2015) Electrochemistry, **84(1)**, 12-16.)*

initially charged using 1.5 A constant current. When the battery voltage reached up to 28.7 V (4.1 V/cell), the power control unit maintained the constant voltage to the end of the daytime. The charge voltage could be occasionally raised up to 29.4 V (4.2 V/cell) to keep the safer operation during the malfunctions.

The discharge current was ca. 1.6 ~ 1.8 A for the nominal operation of the satellite. The main scientific mission of the satellite was the observation of aurora around the polar region using three cameras with different filters for the wave length of the light. The most sever operation was the observation of aurora in the end of the night time. Each aurora camera requested 0.5 A extra electricity, and the maximum current requested for the operation of REMEI was 3.3 A in total. Fig. 4 shows the typical charge and discharge curves during the operation of REIMEI. The dc impedance of the battery was large enough, and the battery voltage soon arrived to the CV control region when the charge started. The temperature was well controlled between 19 and 22 degrees C by the heater control.

The trend of the 'end of discharge voltage' of the battery is shown in Fig. 5 [5]. By monitoring the drop of the discharge voltage by the changed of the load and the discharge current, we maintained the end of discharge voltage above the 'under-voltage-control (UVC)' level. We have several blank of data due to the natural disaster like earthquake and typhoon, but the operation was continuously maintained.

Even after the operation for eleven years, the two different batteries have been supplying almost same current individually to the satellite loads in the orbit.

4. IN-ORBIT IMPEDANCE ANALYSIS

In 2013, the scientific mission like aurora observation was dismissed. After the termination of the scientific missions, we decided to operate the satellite to confirm the operability of the lithium-ion secondary cells in space.

It is well known that the impedance of the cell is increasing with cycles, and the precise understanding of electrochemistry inside the cell is very important for the prolonged operation of the battery. For that purpose, we tried to understand the internal impedance of the cells and batteries.

Fig. 6 shows the trend of the dc impedance obtained from the REIMEI battery. When the charge of the battery starts, the large increase in voltage is always observed. Furthermore, the large drop of discharge voltage can also be observed, when the discharge of the battery starts. From the change, we calculated the dc impedance of the battery.

The trend shown in Fig. 6 depicted that the calculated dc impedance did not change even after the operation for years. It is well known that the repeated use of the

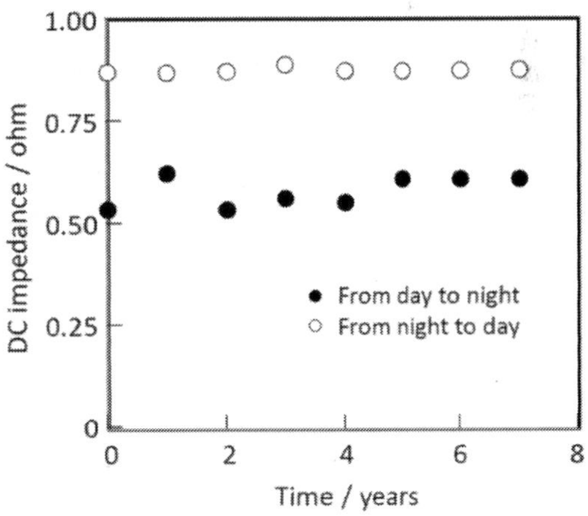

Figure 6. The trend of dc impedance calculated from the REIMEI battery.

The impedance was calculated from the change in voltage and current at the end and beginning of

DOI: 10.1051/e3sconf/20171607005

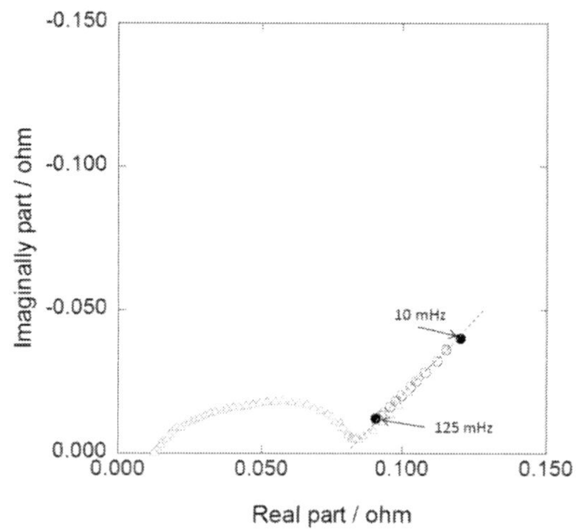

Figure 7. Nyquist plot calculated from the pulse duration to the REIMEI battery.

The result calculated from the in-orbit data (○), and the simulation based the terrestrial experiment (△) are shown.

(Ref. [6] Sone, Y. et al., (2015) 1M06, Proc.83rd Meeting of the Electrochem. Soc. of Japan.)

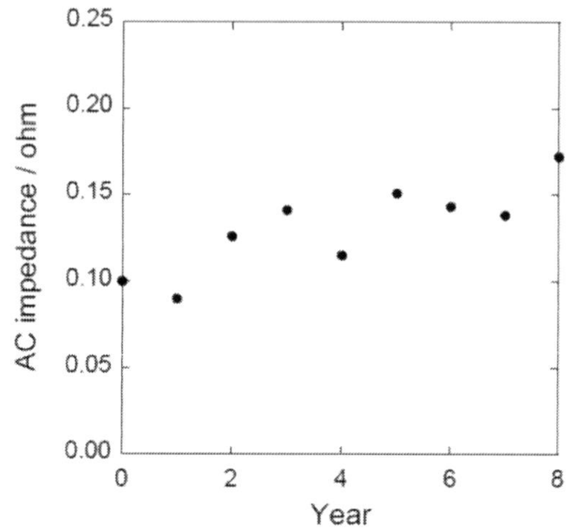

Figure 8. The trend of AC impedance of the lithium-ion secondary cell calculated from the pulse duration to the REIMEI battery.

lithium-ion secondary cells for charge and discharge causes the increasing thickness of the SEI, which finally causes the decrease in the discharge voltage and capacity. The results we obtained through the dc impedance suggested that the information of the dc impedance did not reflect the internal conditions of the battery/cell. We imagined that the impedance loss by the leads were the major part calculated in this case.

Then, we decided to calculate the ac impedance of the battery/cell through the in-orbit operation. When the pulse duration was induced to the batteries, the drop of discharge voltage could be observed. Based on the change in potential, we used Fourier transformation, and tried to obtain the information about ac impedance.

However, we had a problem. The REIMEI battery uses 7 cells to realize the 28 V-class-bus-voltage. To detect the bus voltage, the range for the output voltage between 0 and 35 V was divided by 8 bit, thus the accuracy of the voltage is ±0.136 V. Furthermore, the data is acquired every 8 seconds. Thus, the ac impedance with low frequency region could be obtained through the mathematical method.

However, most information about the degradation of the cells is usually obtained as the charge transfer resistance. For the consideration, Nyquist plots are commonly used. Fig. 7 shows the Nyquist plot obtained from the on-boarded REIMEI battery. The ac impedance less than 100 mHz were obtained [6].

In Fig. 7, semi circler region is also shown. It was very fortunate that the cells which were obtained at almost the same time as the flight ones were still stored for the terrestrial measurements. By using the cells, the basic shape of the Nyquist plot of the cell could be obtained. Based on the information, we simulated the Nyquist plot whose lower frequency region could overlap each other. We considered that the diffusion area for the original Nyquist plot was observed by the calculation based on the in-orbit data. We also believed that the sum of the resistances among the solvent and the charge transfer regions could be obtained by the linear extrapolation of the calculated ac impedance.

Today, we are trying to observe the change in ac impedance inside the batteries/cells. Fig. 8 shows the trend of the increase in the ac impedance. Now, the constant increase in impedance is obtained, which reflects the performance of the battery taking place after the long term operation of REIMEI satellite. It is well known that the increase in ac impedance proceeds due to the degradation of capacity and formation of Solid Electrolyte Interface (SEI).

5. SUMMARY

The satellite 'REIMEI' was designed using off-the-shelf lithium-ion secondary cells for the main bus battery. The spinel manganese oxide was used for the positive and the graphitized carbon was used for the negative. Since the satellite was injected to the low earth orbit which spends 97 minute for one revolution, the battery must be charge and discharged very rapidly.

Even after the eleven years operation, the REIMEI satellite is still operative. Two batteries are used for the satellite, and the distributed current to each battery was almost equal, which shows that the tow batteries for the satellite still maintain good balance.

Direct current impedance and alternated current impedance was calculated to analyse the condition of the batteries. Through the ac impedance, constant increase in impedance was observed, which suggest the ordinal degradation proceeds inside the cell.

6. ACKNOWLEDGEMENT

This research is partially supported by 'International Joint Research Program for Innovative Energy Technology' of Ministry of Economy, Trade and Industry (METI).

7. REFERENCES

1. Spurrett, R., Thwaite, C., Slimm, M. and Lizius, D. (2002) Lihium-Ion Batteries for Space, *Proc. 6th Euro. Space Power Conf.*, 477-482.
2. Saito, H., Masumoto, Y., Mizuno, T., Miura, A., Hashimoto, M., Ogawa, H., Tachikawa, S., Oshima, T., Choki, A., Fukuda, H., Hirahara, M., Okano, S. (2001). INDEX: Piggy-back Satellite for Aurora Observation and Technology Demonstration", Acta Astronautica, 48, 723-735.
3. Saito, H., Hirahara, M., Mizuno T., Fukuda S., Fukushima, Y., Asamura, K., Nagamatsu, H., Tanaka, K., Sone, Y., Okuizumi, N., Mita, M., Uno, M., Yanagawa, Y., Takahara, T., Kaneda, R., Honma, T., Sakanoi, T., Miura, A., Ikenaga, T., Ogawa, K., Masumoto, Y. (2011). Small Satellite REIMEI for Auroral Observations, Acta Astronautica, 69, 499-513.
4. Uno, M., Ogawa, K., Takeda, Y., Sone, Y., Tanaka, K., Mita, M., Saito, H. (2011). "Development and On-orbit Operation of Lithium-ion Pouch Battery for Small Scientific Satellite "REIMEI"", J. Power Sources, 196, 8755-8763.
5. Sone, Y., Watanabe, H., Tanaka, K., Fukuda, S., Ogawa K., Asamura, K., Yamazaki, A., Nagamatsu, H., Fukushima, Y., and Saito, H. (2015). Long Term Operability of Li-ion Battery under Micro-gravity Condition Demonstrated by the Satellite "REIMEI", Electrochemistry, 84(1), 12-16.
6. Sone, Y., Watanabe, H., Tanaka, K., Fukuda, S., Itagaki, M., Ogawa K., Asamura, K., Yamazaki, A., Nagamatsu, H., Fukushima, Y., and Saito, H. (2015). Analysis of the Battery Condition of REIMEI Satellite after the Long-Term Operation in Space, 1M05, Proceedings of the 83rd Meeting of the Electrochemical Society of Japan, The Electrochemical Society of Japan, Suita, Japan.
7. Bausier, F., Tonicello, F., and Soares, T. (2014). Passivation of Spacecraft Electrical power System at End of Mission, Proceedings of the 10th European Space Power Conference, ESA SP-719 (CD-ROM), European Space Agency, Noordwijk, The Netherlands.
8. Alcindor, P., Kimber, R., Remy, S., and Prevot, D. (2014). Passivating Li-ion Batteries in Orbit at the End of the Spacecraft Life, Proceedings of the 10th European Space Power Conference, ESA SP-719 (CD-ROM), European Space Agency, Noordwijk, The Netherlands.
9. Dudley, G. Extension of Lithium Ion Cell model to Include Transient and Low Temperature, Proceedings of the 10th European Space Power Conference, ESA SP-719 (CD-ROM), European Space Agency, Noordwijk, The Netherlands.

 DOI: 10.1051/e3sconf/20171607006

NON-INTRUSIVE BATTERY HEALTH MONITORING

Laurent GAJEWSKI [(1)], Celine CENAC-MORTHE [(2)], Aurore CARRE [(3)], Patrice SIMON [(4)], Pierre-Louis TABERNA[(4)]

[(1)] *Airbus Defence and Space, 31 rue des cosmonautes - Z.I. du Palays - 31402 Toulouse - France, Email: laurent.gajewski@airbus.com*
[(2)] *CNES, 18 avenue Edouard Belin – 31400 Toulouse - FRANCE, Email: Celine.Cenac-Morthe@cnes.fr*
[(3)] *ESTEC, Keplerlaan 1, PO Box 299, 2200 AG Noordwijk, The Netherlands, Email: Aurore.Carre@esa.int*
[(4)] *CIRIMAT, Université Paul Sabatier - 118, Route de Narbonne - 31062 Toulouse – France, Email: simon@chimie.ups-tlse.fr, taberna@chimie.ups-tlse.fr*

ABSTRACT

The "Non-intrusive battery health monitoring", developed by Airbus Defence and Space (ADS) in cooperation with the CIRIMAT-CNRS laboratory and supported by CNES, aims at providing a diagnosis of the battery ageing in flight, called State of Health (SOH), using only the post-treatment of the battery telemetries. The battery current and voltage telemetries are used by a signal processing tool on ground to characterize and to model the battery at low frequencies which allows monitoring the evolution of its degradation with great accuracy. The frequential behaviour estimation is based on inherent disturbances on the current during the nominal functioning of the battery. For instance, on-board thermal control or equipment consumption generates random disturbances on battery current around an average current. The battery voltage response to these current random disturbances enables to model the low frequency impedance of the battery by a signal processing tool. The re-created impedance is then compared with the evolution model of the low frequencies impedance as a function of the battery ageing to estimate accurately battery degradation. Hence, this method could be applied to satellites which are already in orbit and whose battery telemetries acquisition system fulfils the constraints determined in the study. This innovative method is an improvement of present state-of-the-art and is important to have a more accurate in-flight knowledge of battery ageing which is crucial for mission and operation planning and also for possible satellite mission extension or deorbitation. This method is patented by Airbus Defence and Space and CNES.

1. BATTERY MONITORING CONTEXT

For a satellite, the battery ensures a critical function: store and provide the electrical energy during the entire lifetime of the spacecraft without any interruption. The battery is an electrochemical system which undergoes degradations along its operational life. Typically, its degradation results from the calendar effect and the cycling effect depending on the type of mission. For Geostationary Earth Orbit (GEO) missions, the baseline is presently 15 years in orbit with 1350 battery cycles. For Low Earth Orbit (LEO) missions, the battery has to comply with 55 000 cycles during 10 years. The degradations of a battery are characterized by a loss of capacity linked to an increase of the internal resistance. The optimisation of the battery management can be achieved by monitoring its in-flight behaviour and by quantifying its degradation. The battery monitoring can be declined itself in 3 different ways. First way is to check the coherency between predicted degradations and the real ones. Second way is linked to the battery failure and anomaly detection with eventually mission adaptation for recovery. Then, the last is related to the mission extension and deorbitation decision and is in line with the current issue of space debris law.

2. STATE OF THE ART

On-ground, the estimation of the battery SOH is relatively easy and consists to measure the remaining capacity and the DC internal resistance during a complete discharge from a full State of Charge (SoC). However, this method takes a long time (generally several hours) during satellite integration phase. In addition, the transposition on-board is also impossible as a complete discharge in flight means the loss of the satellite. It is also difficult to measure the internal resistance because such a measure needs clean battery current transition.

For this moment, the in-flight battery ageing assessment consists in correlation of the battery voltage evolution with existing steady state battery model. The model is tuned thanks to a fading coefficient in order to match the battery model voltage with the real voltage related to the real current profile. The obtained fading coefficient corresponds to the ageing of the battery. However, there is still large uncertainty on the battery fading because of inherent uncertainty of these models at end of life. The

DOI: 10.1051/e3sconf/20171607006

voltage observed in flight corresponds to a small visibility window on the discharge (10-20% for LEO satellite) and do not allow to know accurately the voltage fall due to no remaining energy in the battery, as shown on the next figure.

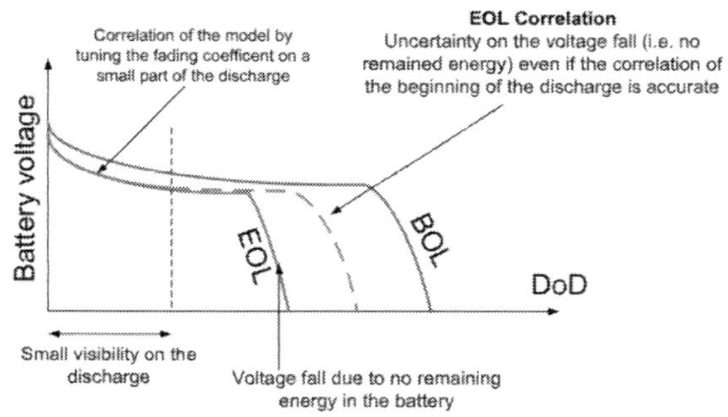

Figure 1. Standard method for battery diagnosis

3. MONITORING BASED ON IMPEDANCE SPECTROSCOPY

Previous R&D studies led by ADS and CIRIMAT (Chemical laboratory in Toulouse) in 2007-2008 pointed out that the analysis of the low frequencies impedance (between 0,1Hz – 1Hz) allows characterizing the Li-ion battery ageing. As represented in Fig 2, the impedance of the Li-ion cells increases with the ageing.

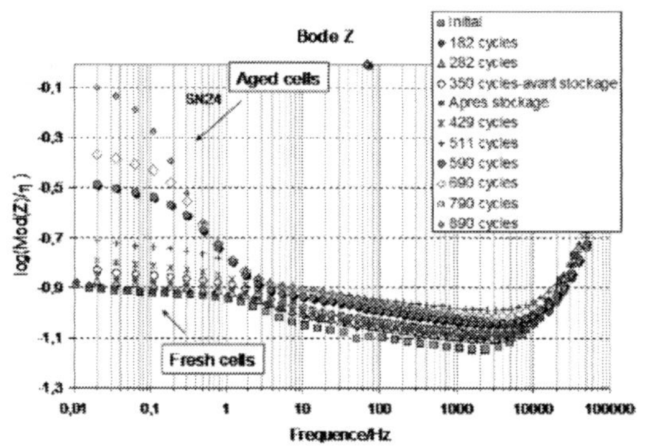

Figure 2. Low frequencies battery impedance evolution due to ageing

A first CNES study has been performed by ADS. Its main purpose was to define a monitoring method of the Low Frequencies (LF) battery impedance, applicable in-flight, to determine the ageing. The method, based on impedance spectroscopy, consists in injecting disturbances (low frequencies sinus). As described on Fig. 3, a dedicated module injects the disturbances and acquires the current and voltage of the battery. The recorded TM's are then transmitted to the ground. A post-treatment enables to assess the battery impedance. In parallel, on ground, a reference test has been performed on the same Li-ion accumulator (accumulators, battery string or complete battery) in order to characterise the impedance evolution (impedance measurements) related to the ageing (test of standard capacity and internal resistance). Note that the impedance measurement takes advantage of typical life tests which are nominally performed in frame of satellite programs (to follow the battery performances along the lifetime). The comparison between the measured impedance from in flight data and the impedance measured during the reference test leads to the in-flight battery ageing.

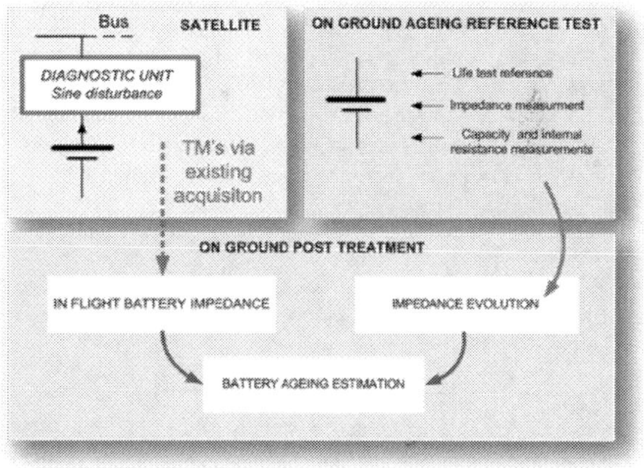

Figure 3. Embedded system for battery monitoring

The method has been proved on a typical LEO cycling life test where an impedance spectroscopy measurement has been regularly performed on 8S1P assembly.

The following Nyquist diagram shows the high resolution of ageing and the stability of the measure along the time. Indeed, the evolution of the LF battery impedance as function of the ageing over 10 weeks is visible and corresponds to a really weak ageing (around 1% of degradation observed on 10 weeks).

 DOI: 10.1051/e3sconf/20171607006

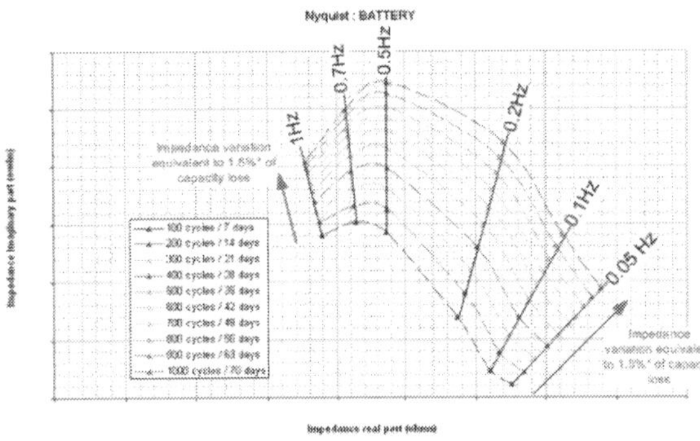

Figure 4. Fine resolution of the ageing thanks to impedance spectroscopy

The battery ageing assessment based on the impedance spectroscopy takes the advantage of the high accuracy of this method. However, for ADS, it is not credible to start the development of a dedicated unit for battery diagnosis which implies cost, time, mass, qualification phase for a non-critical function. Moreover, this system generates disturbances at power bus level for diagnosis which is not convenient at satellite system level.

4. NON INTRUSIVE BATTERY HEALTH MONITORING PRINCIPLE

The NiBHM was developed to obtain a diagnosis of the Battery SOH, keeping the accuracy of the impedance spectroscopy method without any on-board unit to perform the diagnosis. It is inspired by the methods of complex systems modelling (bridge, buildings, fans). Indeed, they cannot be easily modelled by physical insights and their characterization from the injection of canonic disturbances (sine, square waves) is generally impossible. The solution consists in using the random natural disturbances applied at the input of the system. First, as shown in Fig 5, the natural disturbances and the system response are measured in time domain. Then, the natural input disturbances are injected into a model, represented by a transfer function. An optimization algorithm enables to select the best model structure (best transfer function form) which is composed by a vector of n optimized parameters. These parameters are optimized so that the difference between the model response and the real system response to real disturbances is minimized.

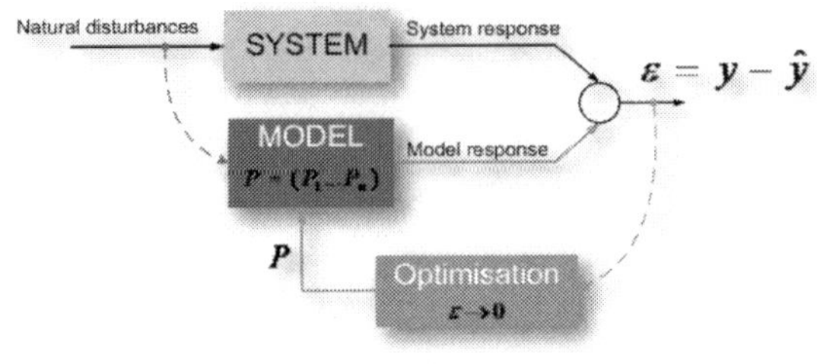

Figure 5. Process of identification based on natural random disturbances

The same principle can be applied to battery. On a spacecraft, the battery current has natural random disturbances, due to the switch on/off of the units and the active thermal control. Hence, from raw battery current and voltage telemetries, the low frequencies impedance can be estimated by post-treatment without any embedded system. Then, if the evolution of the battery impedance as a function of the ageing is known, it is possible to deduce the battery ageing.

The method has been proven thanks to the typical LEO life test previously described. Every 100 LEO cycles (equivalent to one week) an impedance measurement is made between 0.05Hz and 1Hz, thanks to the injection of sine disturbances. Then random disturbances profile is injected on the battery current (this profile is representative of a realistic profile observable on satellite). The battery voltage response is acquired. Finally, the battery impedance is estimated only using the disturbances on the battery current. The method is applied only in discharge so that no additional disturbances due to charge regulation mislead the impedance restitution.

Fig 6 shows the matching between the reference impedance measured on 10 weeks with sine disturbances (in thick lines) and the restitution thanks to the NiBHM process (in large lines). The error of ageing prediction in this demonstration test corresponds to few weeks, which is really accurate (for recall, the ageing on 10 weeks is quantified to 1% of capacity loss).

DOI: 10.1051/e3sconf/20171607006

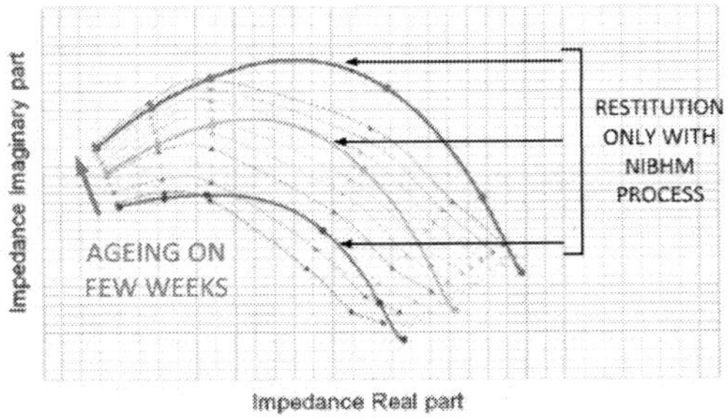

Figure 6. Restitution with NiBHM on a life test

Hence, the restitution of the impedance based on the NiBHM process allows estimating the ageing by a comparison with the set of impedances measurements which are scaled with the capacity loss.

ADS demonstrates also that the TM acquisition systems on present in-flight ADS satellites can be compatible with this method (sampling frequency, Analog-Digital Converter resolution). The real characteristics of several acquisition systems have been emulated on the life test to see the impact on impedance restitution. A dedicated ESA study has been led to quantify the error estimation of the ageing due to uncertainties introduced by acquisition system. It shows also the limitation of oldest ones which are presently in flight.

5. IN-FLIGHT APPLICATION OF THE NIBHM PROCESS

The NiBHM process, in flight, is described as follows.

(1) <u>On ground reference test</u>

The on-ground reference test is necessary to know the LF battery impedance as a function of the ageing. It consists in a life test where regular impedance measurement occurs coupled with capacity and internal resistance measurement.

(2) <u>Raw battery TM's acquisitions</u>

The raw battery current and voltage telemetries have to be acquired with an optimum sampling rate during a discharge phase on few minutes. The ADC resolution implies some uncertainties which have to be taken into account on the ageing estimation during the post treatment.

(3) <u>Battery ageing estimation</u>

The post treatment phase is performed on-ground on any computer with NiBHM tool. The tool is able to provide the impedance estimation after filtering, impedance identification and validation. Then, it performs the comparison with evolution law of the impedance as a function of the ageing. The estimation of the ageing is then provided. The post treatment process is really fast: less than 10 seconds to treat a raw data collection for one measure.

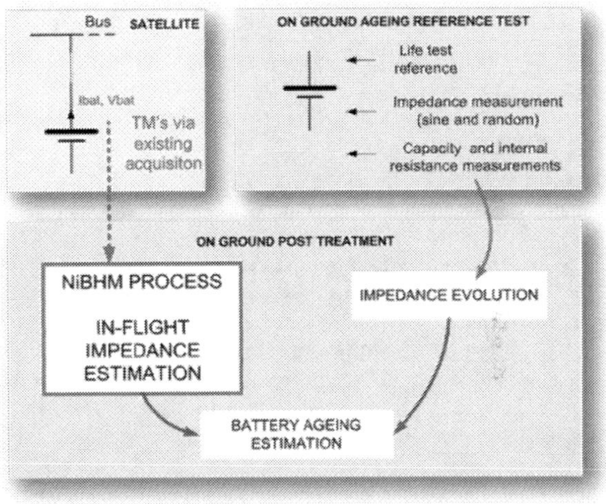

Figure 7. In-flight application of the NiBHM

If no on ground reference test is performed, it is not possible to give an absolute figure of ageing. However, it is possible to monitor the evolution of the impedance which gives a trend for battery degradation. It is also possible to perform the treatment on satellites fleet, to know at the present moment the SOH of every battery.

6. IN-FLIGHT APPLICATION ON TRIAL

The NiBHM process is presently on trial on 2 CNES satellites which have a common technology of battery with the same level of ageing.
Every 6 months, a set of 6 measures (battery current and voltage during several minutes) is performed on each satellite.
Each measure is integrated to NiBHM tool. As no on-ground reference test exists on this type of battery (because the related life tests were performed 10 years ago), the tool is used for a comparative analysis between the 2 satellites.
Thanks to correlation method (presented in §2), the observed level of degradation of both batteries is similar. Consequently, the expected result is to provide practically the same battery impedance with NiBHM process.

As shown, on the next figure, the impedance restitution for each measure is really close from the battery of satellite 1 and the one of satellite 2. The obtained Nyquist diagram allows the restitution of the order of the internal DC resistance of the batteries.

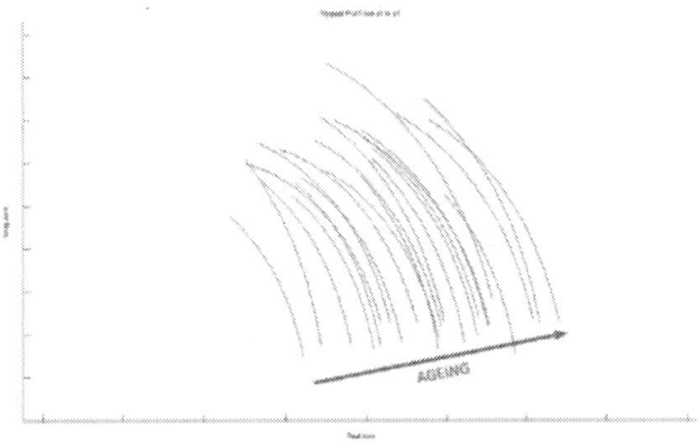

Figure 8. NiBHM on trial on CNES satellites – Restitution of Nyquist diagrams

A present CNES study aims at improving the filtering of raw telemetries in order to reach a better accuracy of the method. The final target for NiBHM process is to estimate the battery ageing with +/-1% of accuracy.

7. CONCLUSION

In conclusion, based on practically ten years of background, the CNES, ESA and internal ADS studies demonstrates that it is possible to monitor the battery ageing thanks to impedance spectroscopy. The patended NiBHM process developed by ADS from 2011 allows providing a diagnosis based on the accuracy of the impedance spectroscopy, without any modification on present space segment (if the acquisition system on-board fulfils the NiBHM requirements).
On the basis of this promising method, the NiBHM tool can be enriched to enlarge its applicability area by covering a larger range of satellites bus and a larger range of battery temperature for example.

8. REFERENCES

1. Gajewski, L., Simon, P. & Taberna, P.L. (2012) Patent (FR1202266), Procédé d'estimation du viellissement d'une batterie

2. Gajewski, L., (2014), Non-Intrusive Battery Health Monitoring, Space Power Workshop Conference

3. Gajewski, L., Boussouf, L., (2016), Diagnostic batterie non intrusif, CCT CNES conference

RHOMBOHEDRAL IRON TRIFLUORIDE WITH A HIERARCHIZED MACROPOROUS/MESOPOROUS TEXTURE FROM GASEOUS FLUORINATION OF IRON DISILICIDE

Diane Delbègue [(1)(2)], Katia Guérin [(1)(2)], Barbara Laik [(3)], Jean-Pierre Pereira-Ramos [(3)], Moulay-Tahar Sougrati [(4)], Céline Cénac-Morthe [(5)]

[(1)] Université Clermont Auvergne, Université Blaise Pascal, Institut de Chimie de Clermont-Ferrand, BP 10448, F-63000 CLERMONT-FERRAND, France, Diane.Delbegue@univ-bpclermont.fr
[(2)] CNRS, UMR 6296, ICCF, F-63171 Aubière, France
[(3)] Université Paris Est Créteil, Institut de Chimie et des Matériaux Paris-Est, UMR CNRS 7182 Thiais, France
[(4)] Université Montpellier II, Institut Charles Gerhardt de Montpellier, UMR CNRS 5253 Montpellier, France
[(5)] Centre National d'Etudes Spatiales, Toulouse, France, Celine.Cenac-Morthe@cnes.fr

ABSTRACT

Stable low temperature rhombohedral iron trifluoride has been obtained by the fluorination under the pure fluorine gas of iron disilicide. The combination of both unusual fluorination process and precursor avoids to get unhydrated crystalline FeF_3 particles and allows the formation of hierarchized channels of mesoporous/macroporous texture favorable for lithium diffusion. The fluorination mechanism proceeds by temperature steps from the formation, for a fluorination temperature below 200 °C, of an amorphous phase and an intermediate iron difluoride identified mainly by ^{57}Fe Mössbauer spectroscopy before getting, as soon as a fluorination temperature of 260 °C is reached, the rhombohedral FeF_3. Both amorphous and crystallized samples display good ability for electrochemical process when used as cathode in lithium-ion battery. The low diameter of rhombohedral structure channels is balanced by an appropriate mesoporous texture and a capacity of 225 $mAh.g^{-1}$ after 5 cycles for a discharge cut-off of 2.5 V vs. Li^+/Li at a current density of C/20 has been obtained and stabilized at 95 $mAh.g^{-1}$ after 116 cycles.

1. INTRODUCTION

Recently, Li-storage batteries have drawn tremendous attention due to their large potential market for electric vehicles and electronic consumables and devices. To realize satisfactory power and energy densities, new cathode materials must be explored [1-3]. Conversion reactions which imply the full reduction by lithium of a compound containing a metal at a high oxidation state into metal are promising reactions for designing high energy density batteries because of the large electron exchange during the electrochemical process. In order to insure high output voltage, only metal halides, for which the ionicity of the bonding is important, are good candidates and among the metal halides, only the metal fluorides do not dissolve into the electrolyte [4, 5]. Most of the researches on metal fluorides/Li batteries were devoted to surpass the insulator behavior of fluorides. Many attempts have been then dedicated to decreasing the particle size of such materials and to increase their conductivity by high energy ball-milling them with conductive carbon [6] or focusing on hydrated phases, because of the presence of water favoring the ionic conductivity [7].

Iron trifluoride has attracted substantial interest due to its electromotive force value of 3.2 V and a large theoretical capacity of 711 $mAh.g^{-1}$ for a redox process of three exchanged electrons [8], [9]. This capacity value is larger from the practical capacity obtained for the operational cathode materials of secondary batteries made of $LiFePO_4$ or $LiCoO_2$. Three electron process of metal fluorides results from two types of redox mechanism: a first one which proceeds at high potential and is due to an insertion mechanism, the second one which proceeds at lower potential and is due to a conversion mechanism. Looking at the only insertion mechanism into iron trifluorides leads to a theoretical capacity of 237 $mAh.g^{-1}$ larger than the practical 170 $mAh.g^{-1}$ of today's cathode materials.

Iron trifluoride crystallizes into three anhydrous polymorphs and two amorphous phases [10]. The three polymorphs known are the rhombohedral form which is the stable high temperature phase denoted r-FeF_3, the orthorhombic hexagonal tungsten bronze phase HTB-FeF_3 and the cubic pyrochlore structure Pyr-FeF_3 [11]. Many hydrated phases also exist, such as HTB-$FeF_3{\cdot}0.33H_2O$ and tetragonal $FeF_3{\cdot}3H_2O$ [12],[13]. Most of the electrochemical properties in lithium batteries of FeF_3 have been obtained for such hydrated phases because iron fluorides are mainly synthesized through aqueous solution-based methods [14-17]. For dehydrated phase, the highest performance reported thus far has been obtained by Myung et al. [18] with a capacity as high as 224 $mAh.g^{-1}$ (1.5-4.5V) at C/10 rate with capacity retention of 71 % after 100 cycles for a r-FeF_3 obtained via the evaporation method at 400°C under argon of $FeF_3{\cdot}3H_2O$. In spite of their better performances, dehydrated phases are little studied, maybe because of the high temperature needed to get them. Indeed, for example, to obtain rhombohedral phase, temperatures higher than 400°C are needed [19] and lead to huge particle sizes not favorable for electrochemical purpose.

 DOI: 10.1051/e3sconf/20171608001

In this work, the fluorination under pure gaseous fluorine at low temperature has been applied on $FeSi_2$. The resulting materials have been characterized depending on their structure by X-ray diffraction, Raman and ^{57}Fe Mössbauer spectroscopies and their texture by Scanning Electron Microscopy SEM and nitrogen adsorption isotherms. Finally, a preliminary electrochemical study has been performed in lithium batteries.

2. RESULTS

$FeSi_2$ ·has been fluorinated under a flux of pure gaseous fluorine at various temperatures up to 350 °C. $FeSi_2$ present two phases: α-$FeSi_2$ and β-$FeSi_2$. X-ray diffraction, Raman and ^{57}Fe Mössbauer spectroscopies show that the synthesis involves three steps for the formation of r-FeF_3. For a fluorination temperature of 150 °C, α-$FeSi_2$ is preserved and β-$FeSi_2$ is converted into FeF_2. At 200 °C, α-$FeSi_2$ and FeF_2 are totally converted into amorphous FeF_3. Finally for a temperature higher than 260 °C, this amorphous FeF_3 turns into r-FeF_3. The Fe-F bonding and their structural organization are key points for electrochemical purpose, but the texture is another strategic point which is often not clearly characterized. We have chosen to characterize in depth the texture evolution upon increase of the fluorination temperature. Indeed, morphology and texture as well as physical-chemical properties including the surface behavior of crystalline FeF_3 can be assumed to be different from that of amorphous FeF_3 obtained from the low temperature fluorination.
According to IUPAC classifications, the nitrogen adsorption/desorption measurements show type IV isotherms, indicating mesoporous morphology of fluoride compounds with the presence of a hysteresis loop. The shape of the hysteresis loop and the steep desorption branch classify it as of the H_2 type for the lowest fluorination temperatures. This occurs for assemblies of tubular capillaries open at both ends [20]. The dimensions responsible for the adsorption branch of the isotherm are heterogeneously distributed and the dimensions responsible for desorption are of equal size. For a fluorination temperature of 350 °C, H_2 type turns into an H_3 type with an adsorption branch steeper at saturation pressure. This occurs for assemblies of capillaries with slit shapes or with very wide bodies and a narrow, short neck. It can be surmised that tubular capillaries have been closed at one end with increasing fluorination temperature.
The SSA, calculated by the standard BET method, increases with fluorination temperature up to about 48 $m^2.g^{-1}$ for the amorphous compound (Tab. 1). It then decreases for a fluorination temperature of 350 °C. Such high value of SSA is unusual for fluoride compounds (SSA of crystalline FeF_3 lower than 7 $m^2.g^{-1}$ [21]) and is only obtained by solution nanostructuration [7]. The origin of the value of SSA is then totally different in our case. Here, SSA is due to pore formation into a micrometric texture, whereas for classical nanostructured fluorides, SSA is due to surface development of dense nanoparticles. Such differences on texture should act on lithium diffusion process during electrochemistry. In the case of nanostructures materials, lithium will act directly with each nanoparticle surface, whereas with porous micrometric particles, lithium will have to diffuse into the porosity.
Pore size distribution is the highest at about 3.5 nm for the compound fluorinated at 150 °C. The size is practically the same for the sample synthesized at 200 °C but the mesoporous porous volume is twice as much. Two mesopore size distributions are present for the sample fluorinated at 350 °C: the one centered at 3.5 nm and other larger centered at 15 nm which consequently increases the pore mesopore volume up to 0.106 $cm^3.g^{-1}$. The mesoporous texture is connected to a macroporous texture of lower pore volume. The macropore volume increases with the fluorination temperature because of the pore collapse with improvement of crystallization.
It is to note that the acetylene black –fluorinated sample mixture, ground for electrochemical purpose has also been analyzed by nitrogen adsorption isotherm. The SSA obtained is the weighted average of the SSA of each component of the mixture. The conditions applied for grinding can be concluded not to modify the texture of the sample and only favor a good dispersion of conductive carbon into insulating fluoride.

Fluorination temperature	150 °C	200 °C	300 °C
SSA ($m^2.g^{-1}$)	13.3	48.0	32.1
V_{micro} ($cm^3.g^{-1}$)	0.000	0.002	0.001
V_{meso} ($cm^3.g^{-1}$)	0.030	0.060	0.106
V_{macro} ($cm^3.g^{-1}$)	0.009	0.014	0.032
Medium pore size (nm)	3.5	3.5	3.5 & 15

Table 1. Adsorptive properties of $FeSi_2$ fluorinated at different temperatures.

Fig. 1 shows the SEM micrograph of the raw material which appears as a dense material. At low fluorination temperature of 150°C, this material is recovered by a foam. With fluorination temperature increase, some numerous well-defined slit pores are formed. SEM micrograph of the amorphous material presents a porous material with a high total pore volume in agreement with its unusually high SSA of 48 $m^2.g^{-1}$. During the reaction, the formation and escape of a gas (SiF_4) from the precursor enlarged the surface area of FeF_3. The pore size determined by SEM is in between 150 and 200 nm, i.e. in the macropore range. This pore size does not clearly evolve as soon as fluoride is formed for temperature fluorination of 200 °C. Macroporosity identified by SEM must be connected to a deeper mesoporosity characterized only by nitrogen isotherms, not visible by SEM because of the resolution of such a technique. This texture leads to predict a favorable lithium diffusion process.

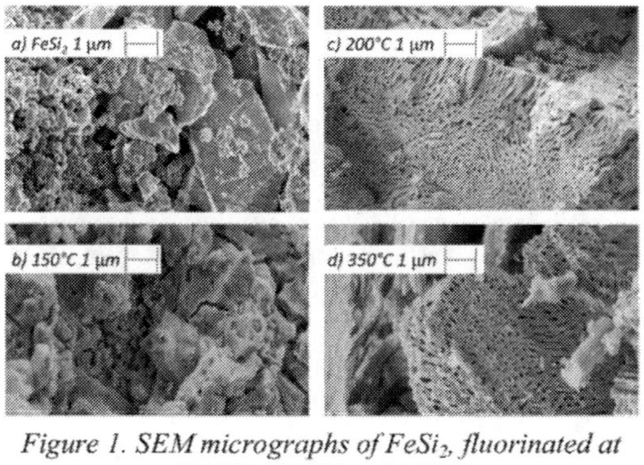

Figure 1. SEM micrographs of $FeSi_2$, fluorinated at 150°C, 200°C and 350°C

The r-FeF_3 phase used as cathode in secondary lithium batteries stands out from the other trifluoride iron phases by both its high and flat operating voltage [6]. Our electrochemical measurements confirm these facts. During the first discharge, a long flat plateau at 3.35 V is clearly observed, followed by an ill-defined process centered at about 3.0 V. The electrochemical properties were measured by using galvanostatic discharge-charge experiments. To get the best electrochemical fingerprint both in terms of capacity and working potential, galvanostatic experiments were performed at 50 and 60 °C in order to facilitate conduction and diffusion processes. This implies to use an electrolyte stable at such high temperature of cycling in our operating conditions. $LiClO_4$ PC 1M appears as a suitable electrolyte. The profiles of galvanostatic discharge-charge were then measured in the range of 2.5-4.3 V. As in this voltage range carbon has no contribution in the specific capacity), the observed charge/discharge capacities can be directly and solely assigned to the electroactive iron fluoride. [22]

The specific capacity was about 250 $mAh.g^{-1}$, very close to the theoretical value of ≈ 240 $mAh.g^{-1}$ when considering the transfer of one electron as specified in eq. 1:

$FeF_3 + e^- + Li^+ \rightarrow LiFeF_3$ Eq. 1

Besides, each of the two contributions involved roughly 0.5 electron. This behavior is in agreement with a transformation reported in literature operating initially through a two-phase region related with the 3.35 V voltage plateau up to $Li_{0.5}FeF_3$ and then through a solid-solution reaction centered at 3 V that involves insertion of lithium in $Li_{0.5+x}FeF_3$ up to $LiFeF_3$ [9],[23] . The reverse scan depicts a dissymmetric charge curve with two main electrochemical steps and leads to a coulombic efficiency of 88 %. Subsequent cycles kept the same shape even if a decrease of the capacity is observed. The step at higher potential is more impacted in particular during the first ten cycles, while the second one remained quite stable. From cycle 10 to cycle 20, a stabilization of the capacity is achieved and interesting capacities about 100 $mAh.g^{-1}$ are still available.

Fig. 2 and Fig. 3 evidenced the influence of the temperature on the electrochemical behavior of the r-FeF_3 material. Presented cycles are recorded at 60 °C. The first discharge kept the same previous profile observed at 50 °C with the well-marked two-step insertion process. The major difference appears at lower potentials. Indeed, instead of a rapid potential decrease, characteristic of the end of the insertion step, a third process begins below 2.7 V highlighting the involvement of a partial conversion according to equation 2:

$LiFeF_3 + 2e^- + 2Li^+ \rightarrow Fe + 3\ LiF$ Eq. 2

The restored capacity of 317 $mAh.g^{-1}$ confirmed this statement as about 80 $mAh.g^{-1}$ more were obtained with respect to the expected theoretical capacity for a Li uptake of 1 Li per mole of FeF_3. The temperature increase seems to favor conversion that happens at higher potential while no effect on insertion has been noticed.

Electrochemical cycling was performed on crystallized FeF_3 over 100 cycles. Fig. 2 shows the voltage versus the capacity for selected 100 cycles. After 100 cycles, r-FeF_3 shows 70 % capacity loss and the discharge capacity was still as large as 95 $mAh.g^{-1}$. There are few examples of electrochemical cycling in a short operating voltage (2.5 -4.3 V) in the literature on r-FeF_3, but for example, recently J. H. Tan et al. have shown a first discharge capacity of about 115 $mAh.g^{-1}$ and after 100 cycles a capacity of 70 $mAh.g^{-1}$ [24]. The loss of capacity is higher in our case, but for the first discharge and for the 100th cycle, the delivered capacity were obviously higher. This large difference of capacity on a same compound can be explained by the difference in the porosity, induced by the liberation of SiF_4 during fluorination, which can support the lithium diffusion. Indeed, some

 DOI: 10.1051/e3sconf/20171608001

recent works have studied the impact of the specific surface area of FeF_3 on its electrochemical properties. For example, R. Ma et al. have embedded a r-FeF_3 of specific surface area of 6.9 $m^2.g^{-1}$ into a carbon replica and a consecutive surface area of 204.7 $m^2.g^{-1}$ of the mixture: an increase of more than 90 % of both discharge and charge capacities were registered. It was explained as a buffering effect of the carbon matrix which plays an important role in interconnecting the FeF_3 particles as well as accommodating the volume change during the electrochemical reaction [21]. These experimental data together with ours seem to indicate that the high specific surface area and the porous structure contribute to facilitate the lithium insertion and improve the electrochemical properties.

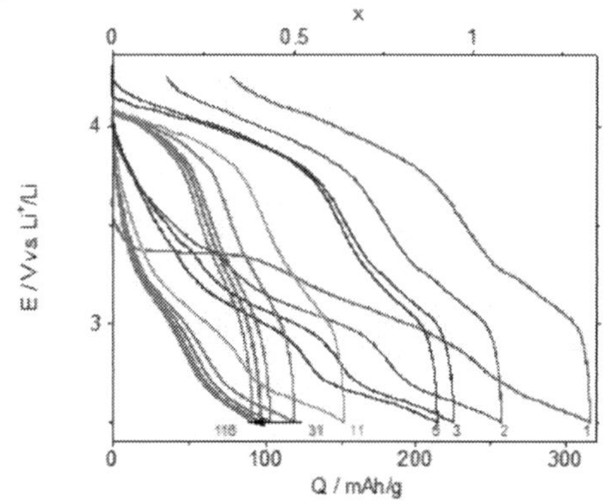

Figure 2. Charge/discharge cycles of a Li/crystallized r-FeF_3 cell obtained at 60 °C

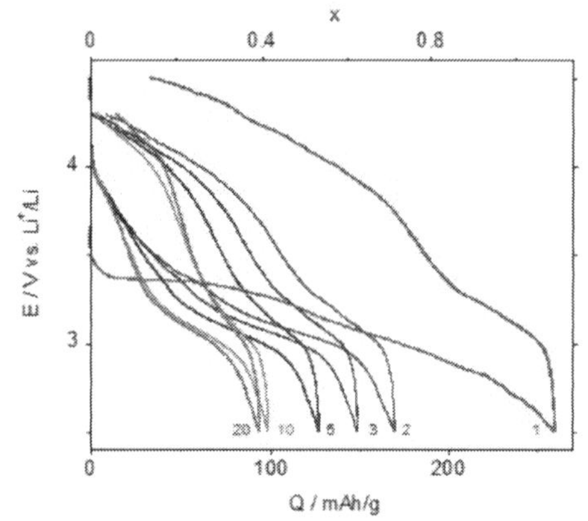

Figure 3. Charge/discharge cycles of a Li/crystallized r-FeF_3 cell obtained at 50 °C

3. CONCLUSION

Whereas most of the electrochemical studies on FeF_3 as cathode material in secondary lithium battery focus on HTB hydrated phase, this work focuses on the potentiality of the rhombohedral phase. The known low performances of this phase result from its high temperature synthesis, which leads to huge crystalline insulating particles. A new synthesis process has been developed in order to form the rhombohedral phase at low temperature, i.e. from 260 °C by applying pure fluorine gas on iron disilicide. Iron disilicide has also been chosen for its release during fluorination of SiF_4 which is a known pore-forming gas. In this way, a mesoporous texture centered at 3.5 nm connected with a macroporous one nearby 150 nm is formed as soon as the fluorination process starts. Rhombohedral phases have good ability as cathode material in lithium battery with a lower hysteresis and a lower capacity loss upon cycle for amorphous phase and a higher capacity for rhombohedral one.
This work opens the way to the use of many other precursors usable under pure fluorine gas to get other pore size distribution. The nature of the carbon (graphene for example) used during mecanochemistry can also be tuned in order to improve the electrochemical performances of the present materials.

4. REFERENCES

1. Tarascon, J.M., & Armand, M. (2008). Building Better Batteries. Nature 451, 652-657.

2. Bruce, P.G., Scrosati, B. & Tarascon, J.-M. (2008). Nanomaterials for Rechargeable Lithium Batteries. Angewandte Chemie International Edition 47(16), 2930-2946.

3. Goodenough, J.B. & Kim., Y. (2010). Challenges for rechargeable Li batteries. Chem. Mater. 22 587-603.

4. Arai, H., Okada, S., Sakurai, Y. & Yamaki, J.-i. (1997). Cathode performance and voltage estimation of metal trihalides. J. Power Sources 68(2), 716-719.

5. Bervas, M., Mansour, A.N., Yoon, W.-S., Al-Sharab, J.F., Badway, F., Cosandey, F., Klein, L.C. & Amatucci, G.G. (2006). Investigation of the Lithiation and Delithiation Conversion Mechanisms of Bismuth Fluoride Nanocomposites. J. Electrochem. Soc. 153(4), A799-A808.

6. Badway, F., Pereira, N., Cosandey, F. & Amatucci, G.G. (2003). Carbon-Metal Fluoride Nanocomposites: Structure and Electrochemistry

of FeF 3 : C. J. Electrochem. Soc. 150(9), A1209-A1218.

7. Li, C., Gu, L., Tong, J., Tsukimoto, S. & Maier, J. (2011). A Mesoporous Iron-Based Fluoride Cathode of Tunnel Structure for Rechargeable Lithium Batteries. Adv. Functional Mater. 21(8), 1391-1397.

8. Doe, R.E., Persson, K.A., Meng, Y.S .& Ceder, G. (2008). First-Principles Investigation of the Li−Fe−F Phase Diagram and Equilibrium and Nonequilibrium Conversion Reactions of Iron Fluorides with Lithium. Chem. Mater. 20(16), 5274-5283.

9. Conte, D. & Pinna, N. (2014). A review on the application of iron(III) fluorides as positive electrodes for secondary cells. Mater. Renew. Sustain. Energy 3(4), 1-22.

10. Ferey, G., Leblanc, M., De Pape, R. & Pannetier, J., (1985). Competing Spin Interactions and Frustration Effects in Fluorides, in Inorganic Solid Fluorides, P. Hagenmuller, Editor Academic Press, pp 395-414.

11. Louvain, N., Fakhry, A., Bonnet, P., El-Ghozzi, M., Guerin, K., Sougrati, M.-T., Jumas, J.-C. & Willmann, P. (2013). One-shot versus stepwise gas-solid synthesis of iron trifluoride: investigation of pure molecular F_2 fluorination of chloride precursors. Cryst. Eng. Comm. 15(18), 3664-3671.

12. Leblanc, M., Ferey, G., Chevallier, P., Calage, Y. & De Pape, R. (1983). Hexagonal tungsten bronze-type FeIII fluoride: $(H_2O)0.33FeF_3$; crystal structure, magnetic properties, dehydration to a new form of iron trifluoride. J. Solid State Chem. 47(1), 53-58.

13. Ferey, G., Leblanc, M., De Pape, R., Passaret, M. & Bothorel-Razazi, M.P. (1975). Cristallisation par voie hydrothermale des fluorures FeF_3; FeF_3, H_2O; FeF_3, $3H_2O$; et NH_4FeF_4. J. Crystal Growth 29(2), 209-211.

14. Kim, S.-W., Seo, D.-H., Gwon, H., Kim, J. & Kang, K. (2010). Fabrication of FeF_3 Nanoflowers on CNT Branches and Their Application to High Power Lithium Rechargeable Batteries. Adv. Mater. 22(46), 5260-5264.

15. Li, C., Gu, L., Tong, J. & Maier, J. (2011). Carbon Nanotube Wiring of Electrodes for High-Rate Lithium Batteries Using an Imidazolium-Based Ionic Liquid Precursor as Dispersant and Binder: A Case Study on Iron Fluoride Nanoparticles. ACS Nano 5(4), 2930-2938.

16. Li, C., Yin, C., Mu, X. & Maier, J. (2013). Top-Down Synthesis of Open Framework Fluoride for Lithium and Sodium Batteries. Chem. Mater. 25(6), 962-969.

17. Tan, J., Liu, L., Hu, H., Yang, Z., Guo, H., Wei, Q., Yi, X., Yan, Z., Zhou, Q., Huang, Z., Shu, H., Yang, X. & Wang, X. (2014). Iron fluoride with excellent cycle performance synthesized by solvothermal method as cathodes for lithium ion batteries. J. Power Sources 251, 75-84.

18. Myung, S.-T., Sakurada, S., Yashiro, H. & Sun, Y.-K. (2013). Iron trifluoride synthesized via evaporation method and its application to rechargeable lithium batteries. J. Power Sources 223, 1-8.

19. Greneche, J.M., Varret, F., Leblanc, M. & Ferey, G. (1987). Mössbauer spectroscopy study of crystallization of amorphous iron (III) fluorides: Influence of experimental conditions. Solid State Comm. 61(12), 813-816.

20. Leofanti, G., Padovan, M., Tozzola, G. & Venturelli, B. (1998). Surface area and pore texture of catalysts. Catalysis Today 41, 207-219.

21. Ma, R., Wang, M., Tao, P., Wang, Y., Cao, C., Shan, G., Yang, S., Xi, L., Chung, J.C.Y. & Lu, Z. (2013). Fabrication of FeF_3 nanocrystals dispersed into a porous carbon matrix as a high performance cathode material for lithium ion batteries. J. Mater. Chem. A 1(47), 15060-15067.

22. Collins, J., Gourdin, G., Foster, M. & Qu, D. (2015). Carbon surface functionalities and SEI formation during Li intercalation. Carbon 92, 193-244.

23. Yamakawa, N., Jiang, M., Key, B. & Grey, C.P. (2009). Identifying the Local Structures Formed during Lithiation of the Conversion Material, Iron Fluoride, in a Li Ion Battery: A Solid-State NMR, X-ray Diffraction, and Pair Distribution Function Analysis Study. J. Am. Chem. Soc. 131(30), 10525-10536.

24. Tan, H.J., Smith, H.L., Kim, L., Harding, T.K., Jones, S.C. & Fultz, B. (2014). Electrochemical Cycling and Lithium Insertion in Nanostructured FeF_3 Cathodes. J. Electrochem. Soc. 161(3), A445-A449.

ENHANCED POWER SOURCES COMPATIBLE WITH EXTENDED THERMAL ENVIRONMENT

Yvan Reynier (1), Florence Fusalba (1), Julien Labbé (2), Valery Gineste (2), Brandon Bürgler (3)

(1) CEA-LITEN (French Atomic and Renewable Energy Commission), 17 rue des Martyrs, 38054 Grenoble (France) yvan.reynier@cea.fr
(2) Airbus Defence and Space, 51-61, route de Verneuil, 78130 Les Mureaux, France valery.gineste@astrium.eads.net
(3) European Space Agency, ESTEC, Keplerlaan 1, Postbox 299, 2200 AG Noordwijk, The Netherlands brandon.buergler@esa.int

ABSTRACT

Today the race to lower the cost to orbit is stronger than ever, with the arrival of new competitors like Space-X. A way to lighten European next generation launchers (Ariane 6) is to use novel primary and secondary batteries for powering avionics, pyro systems, guidance and control electronics, as well as communication systems.
Such application requires unrestricted operations at low temperature (-20°C) and safety behaviour at high temperature (70°C), in addition to a low self-discharge rate because most of the life time of the power source goes by under storage prior to launch. Several battery technologies could be used for this application. While the major challenge for Lithium primary cells is the power capability, regarding secondary Li-ion batteries it may be self-discharge. Both technologies will need a good low temperature power rate capability.
CEA developed 18650 Li-ion cells and AA Li-CF_x cells in order to tackle those particular challenges.

1. INTRODUCTION

Electrical power generation has not been a big issue for the Ariane Launch Vehicle family since the beginning, because the range of power needed to supply electrical systems was very low, mainly since Thrust Vector Control (TVC) systems was essentially hydraulic and range of energy needed was also very low (mission duration, essentially Geostationary Transfer Orbits (GTO) are very short).
Consequently, power source technologies remained unchanged for a long time: Ag-Zn batteries for avionics and Ni-Cd batteries for pyros.

This is going to be strongly modified in the frame of the next European launcher Ariane 6 that will be "more electrical" than current European Launch Vehicles, and could be an "all electrical vehicle".
From the launcher system point of view, it seems appropriate to focus on the functional and the High Power TVC need as they represent critical functions with respect to on board equipment weight and impact on launcher performance.
The pyrotechnics and safeguard functions could be assessed later, in light of the results obtained during the study. More precisely, the electrochemical technologies that have been chosen to be prototyped and tested could be assessed for these two applications in a second loop that could take place after the first test campaign.
Rechargeable batteries are especially needed for the TVC function where high specific energy and specific power are requested with high currents charge capability.
For functional power supply (FPS), primary batteries are preferred for highest energy density, even if high energy secondary batteries may be appropriate. Cycle life requirements of these missions are minimal.
CEA strategy to meet the high power supply (HPS) TVC target was to select a « power sized » chemistry to embed more useful capacity or avoid battery oversizing while better operating at low temperature range (-20°C) in a power mode. Two technologies were selected and optimized in 18650 format: NMC-LTO with high power sizing (~20C rate) and NMC-Si for medium power (5C) high energy sizing. Silicon offers very high specific capacity at the expense of low cyclability due to volume expansion during lithium insertion [1]. However since launcher is not recovered a few cycles is enough to validate performance prior to launch and fulfil the mission.
For the FPS application, CF_x primary lithium battery was selected for its very high energy density (>275Wh/kg possible at cell level). Blending with other materials (MnO_2, V_2O_5) was studied but did not show any improvement regarding low temperature performance. After electrode and design optimizing to maximize power capability, prototypes developed in AA 14500 format were also tested.

2. EXPERIMENTAL

Pack sizing calculation made for the TVC application lead to a cell capacity best matched by the 18650 format. This cylindrical cell is a Li-ion standard first produced by Sony Corporation in the 1990's, who was the first producer of Li-ion cells. It was later adopted by all the industry in consumer market electronics, and is even used today for automotive applications (Tesla

 DOI: 10.1051/e3sconf/20171608002

model S) and low earth orbit satellites.

The format is defined as follows (Figure 1):

- Height : 65 mm
- Diameter : 18 mm
- Hard casing (Steel) with crimped positive cap

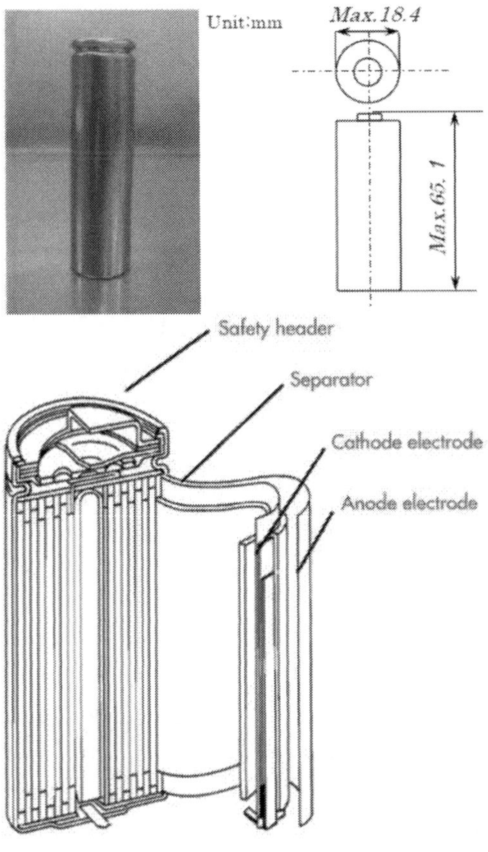

Figure 1. Picture of CEA 18650 cell and internal design

The casing is made of steel and ordered from Hohsen Corporation. The cap includes a safety vent. No PTC device was included to minimize internal resistance.

- For the NMC/Si secondary system, an electrode length of 760 mm was calculated, resulting in 2.4 Ah capacity at 3.4 V average voltage
- Due to the thinner electrodes, the NMC/LTO system positive is 1080 mm long, with a capacity of 0.6 Ah at 2.3 V average voltage.

Primary cells were defined to be manufactured using a stainless steel AA 14500 laser welded cylindrical format (Figure 2).

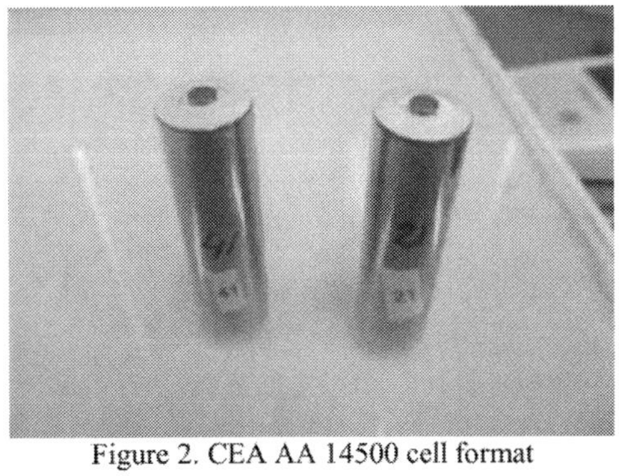

Figure 2. CEA AA 14500 cell format

The format is defined as follows:

- Height : 50 mm
- Diameter : 14 mm
- Glass to metal seals, laser welded cap

Thanks to the glass to metal seal and all laser welded design, this cell can withstand the high vacuum levels of space.
For this primary system, an electrode length of 656 mm was calculated, resulting in 1.9 Ah capacity at 2.4 V average voltage.

Mission profiles and test plans were designed by Airbus DS and prototype cells tested at their facilities in representative conditions.

3. RESULTS

Performance results in standard conditions as well as in representative mission profile for both technologies were collected and compared.

3.1. Li-ion secondary cells

Cells were assembled in 18650 format and their performance compared.
Thanks to silicon's very high specific capacity (>1000mAh/g) compared to graphite (350mAh/g), high energy electrodes can be designed with relatively low thickness. It enable fast Li ion diffusion and high resulting power. Electrolyte was also specifically tuned for low temperature and silicon (ethylene carbonate free composition).
18650 cells using nickel manganese cobalt oxide cathode and silicon anode (NMC/Si) showed a relatively high energy density of 200 Wh/kg for power-sized cell and very low impedance of 16 mΩ at 1 kHz. The rate capability curve at room temperature is shown in Fig. 3. From this curves the continuous power can be extracted and is presented in Tab 1. The cells keep steady performance in the C/3 to 5C current range.

DOI: 10.1051/e3sconf/20171608002

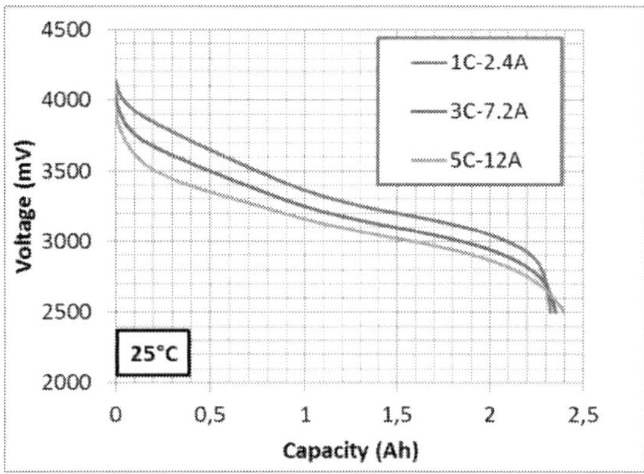

Figure 3. Rate capability of 18650 NMC-Si prototypes

Table 1. Continuous power capability at 25°C for NMC-Si 18650 cells

C-rate	Wh/kg	W/kg (90% SoC)	W/kg (50%SoC)
0,33	198	74	65
1	186	218	192
3	181	625	555
5	178	992	893

Lithium titanium oxide (LTO) is known for its high power capability and fast charge tolerance.

NMC/LTO cells were developed and designed for very high power. They delivered 65% capacity at 20C (Fig. 4). However, low temperature performance was not good enough to meet the target, probably due to further LTO material conductivity improvement need (for example by tuning particle size).

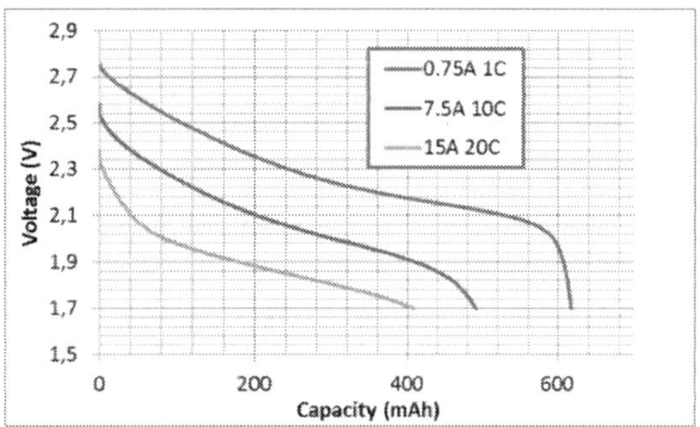

Figure 4. Rate capability of 18650 NMC-LTO cells

After a first selection of the most promising chemistry, namely (NMC/Si), it was further tested at low and high temperature by Airbus defence and space (DS).

NMC/Si 18650 cells passed all the TVC launchers mission profiles designed by Airbus DS to validate the technologies, ranging from -20 to +70°C, as summarized in Tab 2. (Example in Fig. 5). Vibration testing was also successful: no artefact on voltage was noted during vibration test and cells were fully operational after vibration.

Table 2: 18650 cells test plan summary

18650 : NMC – Si	
Sequence High Power Supply (HPS)	**Status**
Mission cycling: HPS boosts at **-20°C**	Passed √
Mission cycling: HPS boosts at **-10°C**	Passed √
Mission cycling: HPS boosts at **0°C**	Passed √
Mission cycling: HPS boosts at **20°C**	Passed √
Mission cycling: HPS boosts at **70°C**	Passed √
Mission cycling: FPS GTO/GTO+ at **-20 °C**	Passed √
Calendar test: 6 months at 20°C	Passed √
Mechanical tests : - Critical frequency search test #1 - Vibration : Sine + Random - Critical frequency search test #2	Passed √

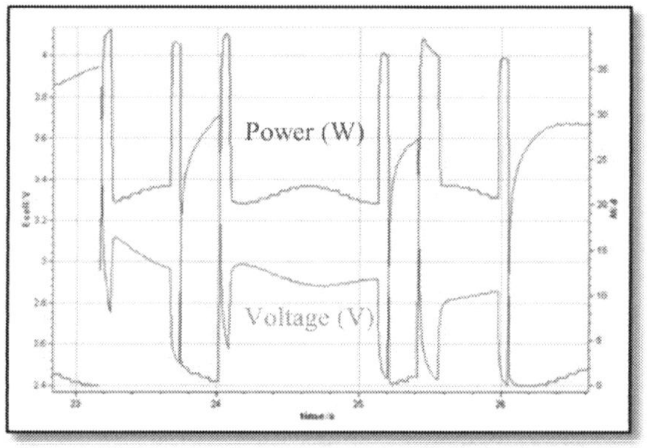

Figure 5. Detail of HPS boost mission at -20°C (40W pulses) for TVC launchers application

Calendar aging results showed very good promise with less than 20% self-discharge after a year at 45°C. After 6 months at room temperature self-discharge is less than 3% (Fig. 6) which can be extrapolated to less than 6% self-discharge after the target 15 months of storage before being used.

18650 secondary cells were also tested on a FPS mission (preliminary devoted to primary cells) at -20°C and passed the profile, which is very encouraging for minimizing the number of different storage systems in the launcher (one cell technology for all applications).

The final specifications of 18650 secondary cells are presented in Tab 3. A high power of 950 W/kg even at -20°C allows the required 40W of power needed for TVC to be obtained at all tested temperatures.

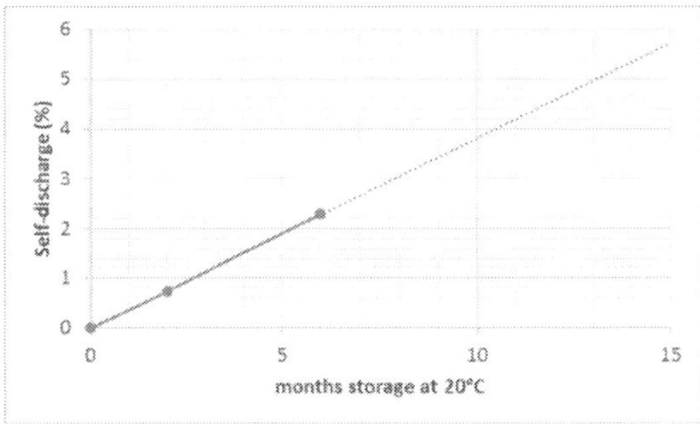

Figure 6: 18650 Self discharge at room temperature

This performance is to be compared to state of the art 18650 cells: for instance the Sony US18650VTC5 [2] considered to be one of the best high power cell on the market. It exhibits a resistance of 150 mΩ at 10A discharge and -20°C. In the same conditions, the developed CEA cells have less than 120 mΩ resistance (Fig.7).

Table 3. NMC/Si 18650 cell specifications

18650 NMC/Si cells specification	
Nominal voltage:	3.4 V
Nominal capacity:	2.4 Ah
Energy:	8.2 Wh
Internal resistance:	16 mOhm (1kHz) 31 mOhm (5s 100% SoC)
Max continuous current:	12.5 A
Energy density:	**200** Wh/kg (C/3) 180 Wh/kg (5C)
Power density:	**950** W/kg from -20 to +70°C (40W, 90% SoC)

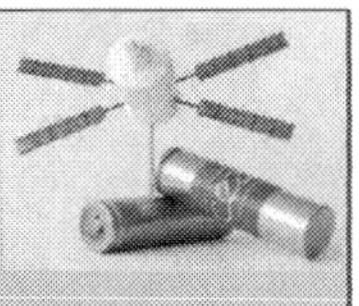

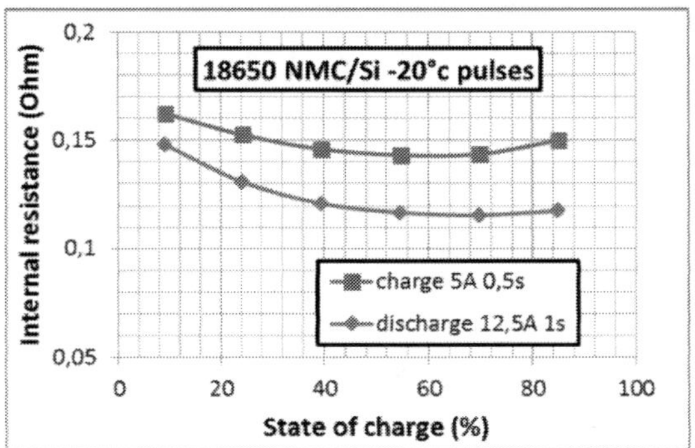

Figure 7. Pulse power capability at -20°C (NMC/Si)

3.2. Lithium primary cells

LASER welded AA cells demonstrated exceptionally high energy density (295 Wh/kg) at the very high rate of 800 mA (ca. 0.4C) for a primary battery (Fig 8). This rate was chosen because it is the average rate of the FPS mission. Cells passed launchers FPS mission profiles at +20°C and -20°C (Fig 9), as well as vibration testing, as shown in Tab. 4.

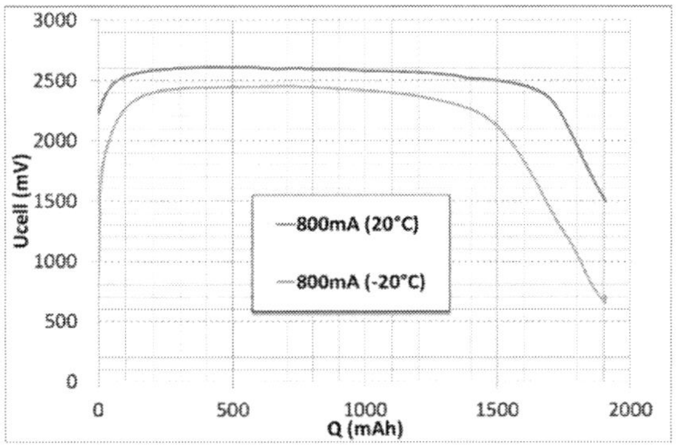

Figure 8. AA CF_x cells discharge at 800mA (C/2.5)

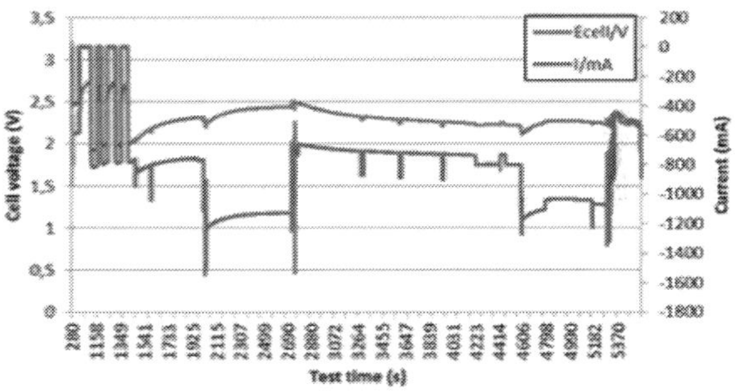

Figure 9. FPS GTO DDO (geostationary transfer orbit, direct deorbitation) profile at -20°C after vibration test

Table 4 : AA CFx cells test plan summary

AA: CFx-Li	
Sequence Functional Power Supply (FPS)	**Status**
Reference test: Total capacity and internal resistance measurement at room temperature	Passed √
Mission cycling: FPS GTO DDO at **-20°C**	Passed √
Mission cycling: FPS GTO DDO at **20°C**	Passed √
Calendar test: 3 months at 20°C	Passed √
Mechanical tests: - Critical frequency search test #1 - Vibration : Sine + Random - Critical frequency search test #2	Passed √

Calendar ageing results after 2 months showed no measurable capacity loss.

It must be noted that several cells showed abnormal behaviour with noisy voltage which was attributed to the low reproducibility of manual assembly with metallic lithium. However thanks to the 3 cells par test used, there was always at least one cell operating as expected to evaluate performance.

Final specifications of AA primary CF_x cells are shown in Tab 5: 260 W/kg is achieved at -20°C and a constant current of 1.6A. This discharge rate is especially high for the CF_x technology known for its usually poor power performance [3, 4].

Table 5. CF_x/Li AA cell specifications

AA CF_x/Li cells specifications

Nominal voltage:	2.5 V
Nominal capacity:	1.9 Ah
Energy:	4,75 Wh
Internal resistance:	130 mOhm (1s)
Max continuous current:	0.8 A
Max pulse current:	1.6 A
Energy density:	**295** Wh/kg (0.8A, RT) 220 Wh/kg (0.8A, -20°C) 300 Wh/kg (1.6A, -20°C)
Power density	**260** W/kg (1,6A)

4. CELLS COMPLIANCE EVALUATION

The cell level requirements for the different mission profiles are summarized in Tab 6. They were calculated with a 1.4 cell to battery integration factor and derived from the mission specifications and Airbus DS targets regarding modules weight.

18650 cells measurements (as confirmed by mission profile testing) show it is compliant regarding both energy and power density for the TVC missions. It is close to FPS GTO DDO compliance although it might be slightly too low energy taking into account self-discharge before mission. Also the voltage profile of the technology being not flat enough it is not compliant versus Airbus DS target voltage range.

Table 6 : Cell level requirements for the different missions

	Energy density (Wh/kg)	Power density (W/kg)
FPS GTO DDO	200	260
FPS GTO PD	273	80
TVC 1 boost mission	16	700
TVC 2 boost mission	14.5	560

AA 14500 cells were designed to fulfil both FPS GTO DDO and the longer GTO perihelia decrease (PD) missions, taking into account the maximum power and energy density requirement of both profiles. It passed GTO DDO mission confirming the compliance with the specification, but should also be tested against GTO PD mission.

5. SYSTEM LEVEL TRADE OFF

Following the test plan results, Airbus DS evaluated both technologies at system level. Final compliance matrix is presented in Tab 7. The NMC/Si was considered as fully compliant for high power supply one or two boosts missions to operate TVC. Primary CF_x cells were deemed partially compliant for FPS mission due to low voltage reached at -20°C (-15% vs. target at system level) at the beginning of the profile. However proper conditioning should solve this issue. Indeed preliminary test showed that initial current pulses just prior to mission can heat up the cell to avoid the voltage drop.

Table 7 : Compliance matrix for the different configurations

Cell type	Mission	Cell assembly (module)	Modules total mass	Target	Comments
14500 CFx-Li	GTO DDO (FPS)	22s13p	6,4 kg	< 7kg	Compliant but not at performance level (cold temp)
14500 CFx-Li	GTO PD (FPS)	22s13p x4	25,6 kg	< 28 kg	Compliant (TBC @ cold temp)
18650 NMC-Si	GTO DDO (FPS)	15s23p	20,3 kg	< 7 kg	uncompliant
18650 NMC-Si	1 boost (TVC)	41s7p	16,9 kg	< 20 kg	compliant
18650 NMC-Si	2 boost (TVC)	41s7p	16,9 kg	< 25 kg	compliant

As expected, 18650 cells did not comply with the FPS GTO DDO mission although the energy and power density requirement were met. Indeed the very small voltage window of this application ([53V-63V]) lead to largely oversizing the module which became too heavy.

Airbus DS evaluated the final technology readiness level (TRL) of primary cells at 4 and secondary cells at 5.

6. FURTHER ACTIVITIES

Despite several unexpected issues during cells development on both primary and secondary technologies, the final deliverable passed all the planned tests. 18650 secondary cells even passed the High energy FPS GTO/GTO+ profile, proving it can be used at least for two applications.

A total of 99 18650 and 58 AA cells were assembled during the project.

Considering the success of the first test plan, several activities could be pursued:

First, the already good performance of the secondary NMC/Si and primary CF_x technologies could be further improved by working on materials and processing.

The chemistries could also be put into optimized cell formats qualified for space application, for instance D-size cells with more capacity to minimize the number of cells in parallel in modules and increase the energy

DOI: 10.1051/e3sconf/20171608002

density.

Mission profile testing also showed that thermal management is critical to evaluate performance. Small modules could be assembled to be representative of the true thermal environments of the cells during launch. Then the cell technologies could be qualified for application on future launchers like Ariane 6.

Finally system management may be adapted to the short length of the target mission in order to save weight on the module

7. ACKNOWLEDGEMENTS

This work was conducted under Programme of and funded by the European Space Agency.

8. REFERENCES

1. S. D. Beattie D. Larcher M. Morcrette B. Simon and J.-M. Tarascon, Si Electrodes for Li-Ion Batteries-A New Way to Look at an Old Problem, *J. Electrochem. Soc.*, 155 2 A158 (2008)

2. Sony US18650VTC5 cell datasheet https://www.powerstream.com/p/us18650vtc5-vtc5.pdf

3. Ganesan Nagasubramanian, Fabrication and Testing Capabilities for 18650 Li/(CFx)n Cells, *Int. J. Electrochem. Sci.*, 2 913 – 922 (2007)

4. Brochard, P.; Godillot, G.; Peres, J.P.; Corbin, J.; Espinosa, A., Very high specific energy, medium power Li/CFx primary battery for launchers and space probes *10th European Space Power Conference. Proceedings*, 59 (2014)

HIGH SPECIFIC ENERGY LITHIUM CELLS FOR SPACE EXPLORATION

F. Farmakis [(1)], N. Georgoulas [(1)], I. Karafyllidis [(1)], I. Amoiridis [(1)], C. Elmasides [(1)], S. Balomenou [(2)], D. Tsiplakides [(2)], M. Nestoridi [(3)]

[(1)] *Democritus University of Thrace, Kimmeria Campus 67100 Xanthi, Greece, Email: farmakis@ee.duth.gr*
[(2)] *CERTH, 6th km. Charilaou-Thermi Rd., GR570 01, Thermi, Thessaloniki, Greece, Email: stellab@cperi.certh.gr*
[(3)] *ESA/ESTEC, PO Box 299, 2200 AG Noordwijk, The Netherlands, Email: Maria.Nestoridi@esa.int*

ABSTRACT

The paper discusses development under an ESA TRP activity (Contract No. 4000109879/13/NL/LvH) with a target of high specific energy Lithium-ion cells, capable of operating under low temperature conditions, i.e. -40 °C. Such cells may be encountered in future exploration missions, which do not consider the use of Radioisotope Heater Units. During the activity, ≥1 Ah silicon-based high energy density prototype cells, following components characterization and optimization, were designed, developed, manufactured and tested under room and sub-zero temperature conditions down to -40 °C.

The developed and tested prototype cells exhibited energy density of around 208 Wh/Kg at room temperature under C/10 charge-discharge rate within voltage range of 2.8 V and 4.1 V. Moreover, the prototype cells could retain and deliver more than 75% of their capacity at room temperature upon cycling at -40 °C, demonstrating an energy density of 140 Wh/kg.

1. LITERATURE SURVEY FOR CELL COMPONENTS

A thorough literature and patent survey was conducted regarding materials (anodes, cathodes, electrolytes) for low temperature Li-ion batteries. Currently, the state-of-the-art space qualified batteries by European manufactures can reach up to 155 Wh/kg (at C/2) at 25 °C [1], and provide satisfactory performance in cycling only down to 0 °C. Below 0 °C the charging of these Li-ion batteries is the main limitation. The current solution in order to charge the battery is the use of external power and heaters to bring the cell to relatively high temperatures since there is no electrochemical system that can efficiently charge at low temperature. However, the use of external heaters decreases system efficiency and robustness while increasing its complexity and weight.

1.1 Anodes

The most important rival to the graphite anode is silicon (or silicon-based materials), thanks to its excellent theoretical specific capacity (around 3800 mAh/g at room temperature, RT). However, even though silicon delivers a high galvanostatic capacity, it suffers from capacity decrease during cycling (cycle fading). Until now, it has been shown that this capacity fading against cycling is due to the volume change during lithiation and delithiation that can exceed 300% of its initial volume. Various technological approaches have been attempted that have as goal to enhance the mechanical integrity of silicon [2]. Nanostructured silicon, silicon-carbon composites, binders and adhesion layers are some of the attempts that are presented. Some of the proposed solutions are promising, but they appear complex and difficult to implement in a production line. Our approach was to develop amorphous silicon film in micro-grain structure.

1.2 Cathodes

Today's technologically important cathodes fall into two broad categories: metal oxides and polyanionic compounds [3]. In general, it is accepted that the cathode kinetics (lithium intercalation) dominate the generally poor rate capability of Li-ion cells at low temperatures. Therefore cathodes with high specific capacity and rate capability at subzero temperatures should be identified.

1.3 Electrolytes

Most compositions of lithium electrolytes are based on solutions of one or more lithium salts in mixtures of two or more solvents. The rationale behind this mixed solvent formulation is that the diverse and often contradicting requirements of battery applications can hardly be met by any individual compound. Low temperature performance represents a significant challenge for Li-ion battery electrolyte solutions. To improve the temperature range of operation of lithium-ion cells, modification of the electrolyte formulation is practiced through the use of (a) optimized solvents blends, (b) use of novel co-solvents, (c) the use of electrolyte additives, and (d) use of novel electrolyte salts, to achieve improved cell resilience to warm temperature exposures.

In view of these, we focused on the development of electrolytes that are capable to operate under sub-zero temperatures. Formulations were based on $LiPF_6$ solutions of carbonates with the addition of a low freezing point agent and a SEI former.

2. ELECTRODES AND LOW TEMPERATURE ELECTROLYTES CHARACTERIZATION

The development of the silicon anode and the low temperature electrolyte was conducted by preparing and testing small Swagelok® T-type half and full cells.

2.1 Graphite and Silicon-based anodes

Two types of anodes have been evaluated: (i) "standard" commercial graphite anode and (ii) silicon ones. The physicochemical characteristics of the produced Si were assessed by profilometry, conductivity measurements and Scanning Electron Microscopy (SEM). A number of techniques were applied in order to evaluate the performance and electrochemical characteristics of the fabricated materials and cells (in the form of Swagelok® T-type half and full cells), including galvanostatic cycling, cyclic voltammetry and Electrochemical Impedance Spectroscopy (EIS).

2.2 Cathode electrodes

Metal-based oxide cathodes were procured, from different Suppliers, and were deployed in Swagelok® T-type half and full cells. The main conclusion from the electrochemical characterization of cathodes was that electrodes received from *Supplier 1* exhibited high specific capacity values at room temperature (190 mAh/g) with more than 2 mAh/cm^2. When half-cells were cycled at subzero temperatures excellent performance was obtained even at -40 °C with a specific capacity of more than 50 mAh/g (or 0.7 mAh/cm^2). Those results were highly promising and therefore the above mentioned electrodes (from *Supplier 1*) were used for the construction of full cells.

2.3 Development and characterization of low temperature Electrolytes

A series of electrolyte formulations for low temperature (LT) operation were prepared. As reference, a commercial carbonate-based electrolyte was used. The conductivities of the electrolytes were measured at the temperature range of -40 °C to 25 °C. Fig. 1 summarizes the conductivities for all electrolyte samples at the two extreme (25 °C and -40 °C) and one intermediate (-10 °C) temperature. The main conclusions form these data are the following:

- The conductivity of all samples at 25 °C is comparable to the standard solution.
- The conductivity of standard (reference) solution, without the addition of any anti-freezing agent, dropped to practically zero value for temperatures below -20 °C, suggesting that the solution was frozen.
- All modified electrolytes showed better performance. The highest conductivity at -40 °C was performed by sample 32, implying a major advantage of this composition for low temperature applications.

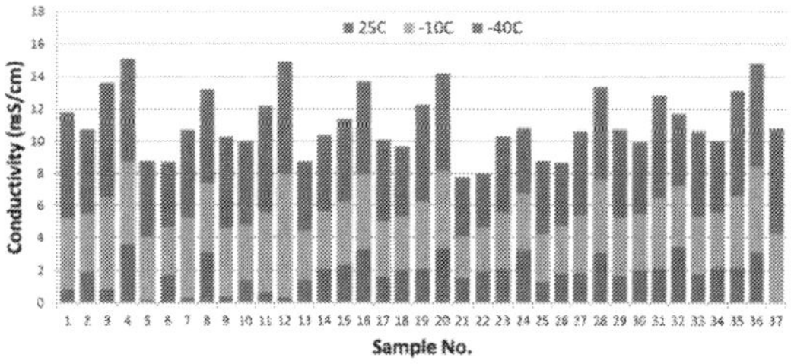

Figure 1. Conductivities of all modified electrolytes at 25 °C, -10 °C and -40 °C.

2.4 Full Cells Characterization

Based on the results and analysis for single elements (anodes, cathode, LT electrolytes), full cells with Si anode and selected cathode with best performing electrolytes (designated as 32* and 20*) were built and all relevant electrochemical and performance tests were conducted under RT to subzero temperatures.

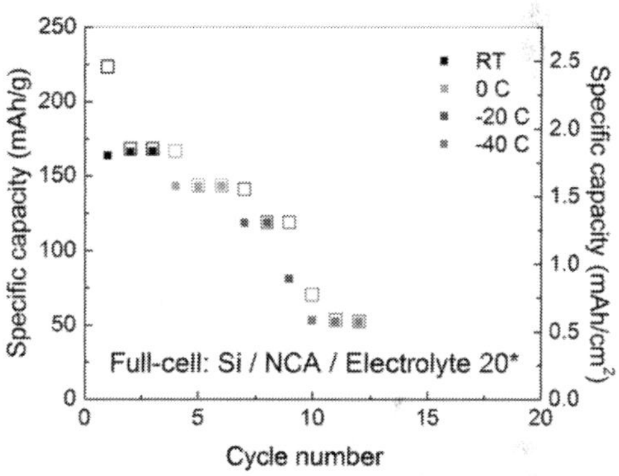

Figure 2. Specific capacity of Swagelok® full cells with electrolyte 20 at various temperatures. Symbols: open-charging, closed-discharging*

After activating the cells and performing 3 cycles at RT at C/10, they were subjected to various low temperatures (0, -20, -40 °C) for three galvanostatic cycles with the same current than the RT. It is shown (Fig. 2) that the electrochemical system with 20* LT electrolyte provides excellent results at low temperatures. More specifically, the cells are able to charge and discharge even at -40 °C delivering around 1/3 of the capacity at RT.

As verification for the absence of Li plating during lithiation, the voltage between the silicon anode and a lithium reference electrode was monitored. Fig. 3 shows the voltage between the anode and the Li reference electrode during charging and discharging at various temperatures. It is evident that this voltage is always above zero, when the full cell voltage attains

DOI: 10.1051/e3sconf/20171608003

4.1 V, and thus we can assume that no Li plating occurs in this voltage.

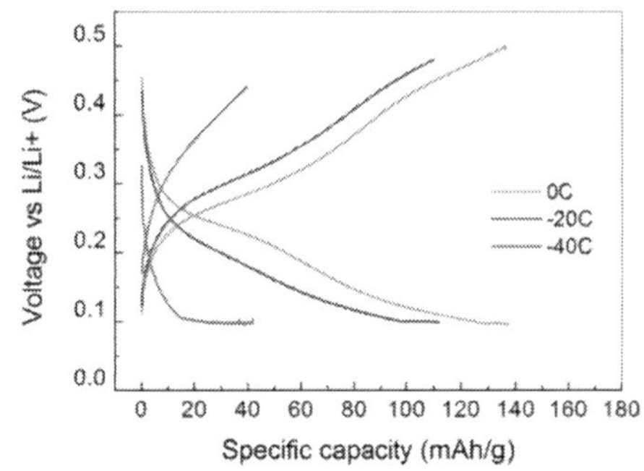

Figure 3. Voltage between Si anode and Li reference electrode in a full cell at various temperatures during charging and discharging.

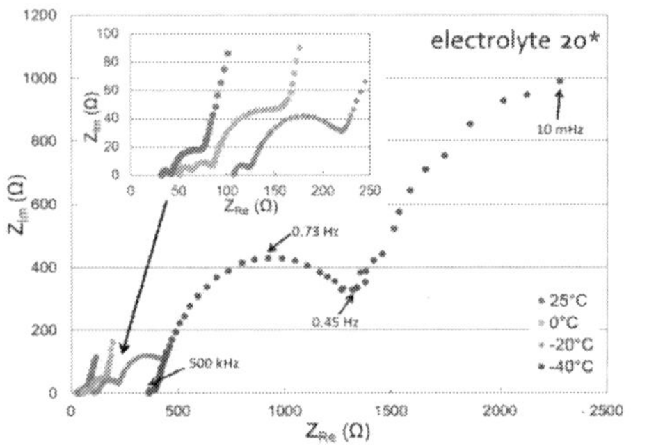

Figure 4. Electrochemical Impedance for full cell with electrolyte 20 at various temperatures.*

EIS performed at full cell at various temperatures shows three semicircles for T<0 °C (Fig. 4).

The small semicircle in HF region was ascribed to the migration of lithium-ion through SEI film as well as contact problems, the semicircle in the MF region was attributed to the electronic properties of the material, and the semicircle in the LF region was related to charge transfer process, while the inclined line in the LF region (frequency<0.45 Hz) represents, as in all Nyquist plots, the solid state diffusion.

3. DEVELOPMENT OF PROTOTYPE CELLS AND EVALUATION

Two versions of prototype cells (both 0.4 Ah and 1 Ah) were developed and tested: Prototypes V1 and V2. Materials were chosen according to the results of materials' performance and electrochemical characterizations on T-type cells. At first a prototype cell was manufactured and tested and finally a modification was designed, manufactured and tested in order to further improve the assembly procedure, enhance leakage control, and lead to smaller and lighter packaging for the cells. The basic design of the prototype cell is illustrated in Fig. 5. The cell case is made from polypropylene, and the poles from copper and aluminum for anode and cathode, respectively. A finite element analysis was conducted in order to provide an improved geometry model.

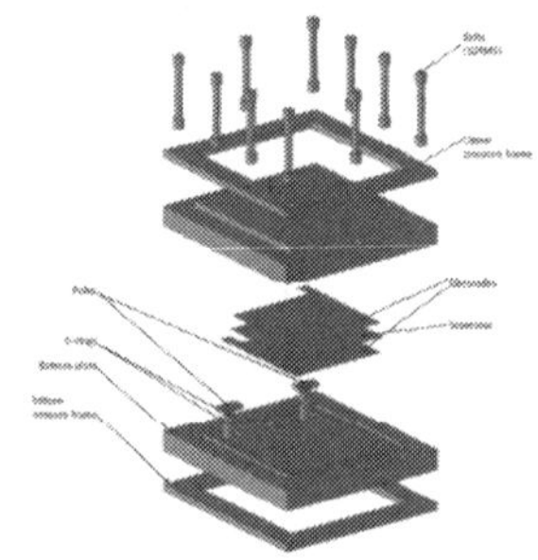

Figure 5. Breakdown of the 3D drawing of the prototype cell.

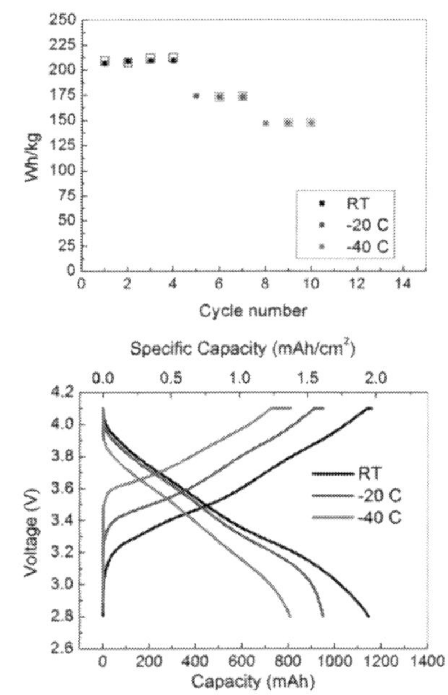

Figure 6. Specific energy as a function of cycling at RT, -20 °C and -40 °C (top) and the relevant voltage vs capacity profiles for the 3rd cycle (bottom). Symbols: open-charging, closed-discharging

Galvanostatic cycling at room temperature was conducted to activated cells between 4.1 V and 2.8 V at C/10 and C/15 discharging and charging rate, respectively. Following, charge/discharge cycles were performed at low temperatures, down to -40 °C. As shown in Fig. 6, the 1 Ah cell demonstrate capacity of more than 1 Ah and more than 200 Wh/Kg of specific energy. In addition, the cell can effectively provide more than 1.0 mAh/cm^2 at low temperatures down to -40 °C

 DOI: 10.1051/e3sconf/20171608003

and they retain more than 70% of the RT capacity.
Further to this direction, two (2) new V2 prototypes (0.4 Ah and 1.0 Ah of nominal capacity) were tested in cycling tests. The performance of the 1.1 Ah cell at BoL at RT and -40 °C is shown in Fig. 7. The cells achieved an energy density of 207 Wh/kg and 160 Wh/kg, respectively. The cycling performance of the cell at -40 °C is demonstrated in Fig. 8. Up to 45 cycles, the capacity remains stable and up to 75% of its initial value at RT. When cycling further progresses, the cell's capacity rapidly fade down to zero. There are several possible explanations, such as Si anode or cathode damage, short circuit due to Li dendrites, electrolyte decomposition, packaging failure etc. The latter is evident even after simple visual inspection of the cells following cycling testing (post mortem analysis).

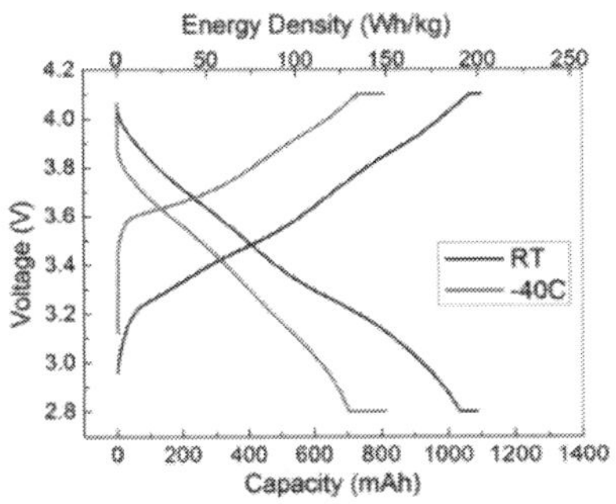

Figure 7. Voltage-capacity profile at BoL at C/10 at RT and -40 °C.

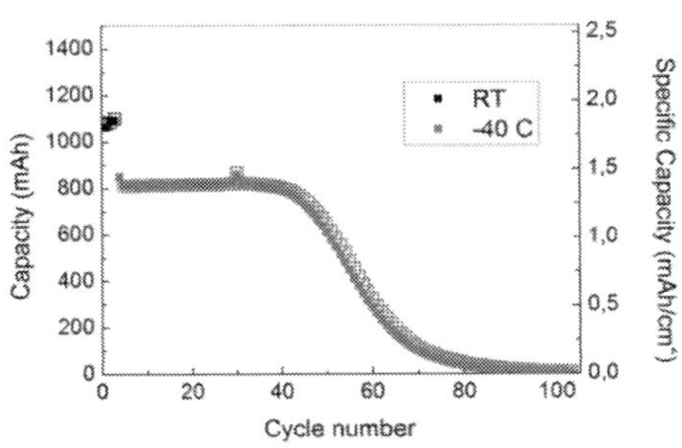

Figure 8. Capacity as a function of galvanostatic cycling at -40 °C. Symbols: open-charging, closed-discharging

During cycling, the internal resistance (IR drop) of the cells was determined by the current interruption (or current-off) method. The results are summarized in Fig. 9.
The main conclusions are the following:

1. The ohmic resistance exhibits a stable performance (IR depends weakly on % DoD) for the first 40-50 cycles. After that, the behaviour changes and the ohmic resistance at low (0%) and high (100%) DoD is significantly higher than that at intermediate DoD values.
2. The increase in ohmic resistance is in accordance with the performance (cycling capacity test) of the cell and concomitant capacity fades down.

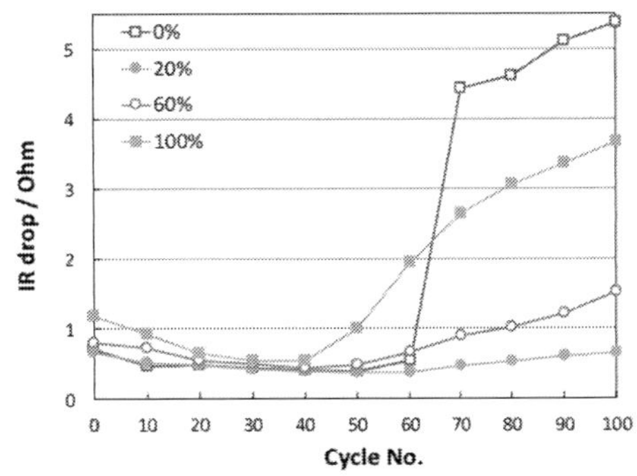

Figure 9. Ohmic resistance evolution during cycling at -40 °C and at different % DoD.

4. CONCLUSIONS

During this activity, silicon-based high energy density prototype cells with 1 Ah nominal capacity, following components characterization and optimization, were designed, developed, manufactured and tested under room and subzero temperature conditions down to -40 °C. The cells exhibit energy density of around 208 Wh/Kg (cell chemistry is considered) at room temperature under C/10 charge-discharge rate within voltage range of 2.8 V and 4.1 V. Moreover, the prototype cells could deliver more than 75% of their capacity at room temperature when cycled at -40 °C (C/15 charge and C/10 discharge rate), demonstrating an energy density of 140 Wh/kg.
To our knowledge, these results are the best reported in terms of capacity retention at -40 °C at high energy cells (>200 Wh/kg).

5. REFERENCES

1. Ratnakumar, B.V. & Smart, M.C. (2010). Lithium Plating Behavior in Lithium-ion Cells. ECS Transactions, 25(36), 241-252.
2. Liang, B., Liu, Y. & Xu, Y. (2014). Silicon-based materials as high capacity anodes for next generation lithium ion batteries. J. Power Sources, 267, 469-490.
3. Doeff, M. M. (2012). Encyclopedia of Sustainability Science and Technology. Springer, New York.

E3S Web of Conferences **16**, 08004 (2017) DOI: 10.1051/e3sconf/20171608004
ESPC 2016

VL51ES (Generation 6) LI-ION CELL FOR SATELLITES

Defer M.[1], Borthomieu Y.[1], Ligneel E.[1], Badet S.[2], Chocinski D.[1]

[1] *Saft, Space and Defence Division, BP 1039, 86060 Poitiers – France, Email: david.chocinski@saftbatteries.com;*

mathieu.defer@saftbatteries.com; yannick.borthomieu@saftbatteries.com; eric.ligneel@saftbatteries.com

[2] *Saft, 111 Bd A Daney, 33074 Bordeaux, Email: sebastien.badet@saftbatteries.com*

ABSTRACT

This paper presents the design of Saft's VL51ES (Generation 6) Li-Ion cell, the main challenges in the course of the development, the main BOL characteristics and performances achieved during the development program. Finally, it also describes how this cell fits in Saft's battery range and the benefits of it.

1. INTRODUCTION

Considering the long cell development and qualification cycle for space application, Saft decided in 2011 to start at its own expenses the development of G6 cells (VL51ES) with GEO application as primary target in order to take over G5 cells (VES180SA). The CDR was held in June 2015 and the cell qualification started in first trimester 2016.

While keeping cell mechanical, electrical and thermal interfaces as close as possible to G5 in order to ensure compatibility with current platforms, the main development axes taken into account were as follows:

- Upgrade to industrial and cost-effective solutions (tab less, radial welds, can, metallic winding mandrel, terminals, axial holding...);
- Improvement of performances compare to G5;
- Enlargement of specifications to lower scraps where it makes sense.

The challenge was to reuse as much as possible the materials and parts, the designs and processes used for industrial applications often bringing cost reductions while seeking to ensure equivalent level of product quality, performance and robustness as current space G5 cells.

The qualification of two2 manufacturing plants (one in France and one in US) was included in the scope for capacity back-up and to address markets locally. They will fulfill ESA ECSS normative referential with same level of demonstration as for G5 cell qualification in the frame of Artes8 program.

In addition, another main topic was to select appropriate design for adequate tightness of the cell life long and validate it by representative wear out tests.

Finally, in parallel to VL51ES cell development, the validation of the associated modules designs and processes in head up and head down configurations is conducted to assess feasibility of battery designs with the new cell, and conduct a qualification in module configuration mechanically and thermally representative of the final application.

2. MECHANICAL ASPECTS

2.1 General Design

VL51ES space cell design combines Saft space reliability and a maximum of regular industrial processes.

In order to maintain as far as possible current satellite battery design, the VL51ES cell dimensions (Figure 1) are close to the one of the current VES180SA space cell.

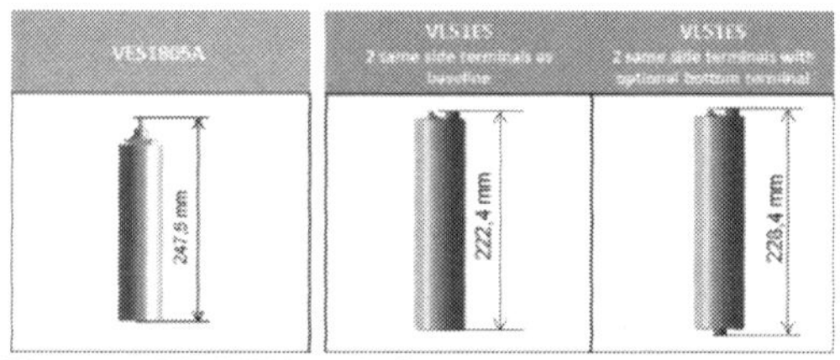

Figure 1. VL51ES cell dimensions

VL51ES cell can have the power connections on the same side (top side) or on opposite side by the

DOI: 10.1051/e3sconf/20171608004

addition of a specific positive terminal on the bottom of the cell (Figure 2).

Figure 2. VL51ES Li-ion cell mechanical envelop

2.2 Internal design: Tabless and Jelly roll holding designs

As mentioned previously, in order to be close to the industrial cell road map, VL51ES cell design applies industrial cell heritage for some design key points: The tab less and the jelly roll holding designs.

The advantages of the tabless design are:

- More robust and reproducible
- More easy to produce
- Less scraps
- More reliable

In VES180SA cells the jelly roll is hold in position, axially, by only screwing it in the positive cover.

For VL51ES cell, jelly roll and electrode collectors are axially held in position, by means of shims on both positive and negative sides.
Therefore, contrary to VES180SA cell, jelly roll adjustement radially in diameter is not required due to axial holding.

The pros and con of the industrial jelly roll holding design vs. current VES180SA design are the following (Table 1):

VL51ES cell	VES180SA cell
Pros	**Pros**
Design close to industrial cell	Jelly roll diameter is maximized
Low stress through internal electrical connection due to design	Proven concept regarding mechanical environments
Cons	**Cons**
Chain of dimension to be mastered axially	Manual labor spent for jelly roll dimensional adjustement on cell can (just slipping process)

Table 1. Jelly roll holding advantages and disadvantages

2.3 Mechanical validation

The challenge behind was to study and validate the use of such solutions for space applications.

Proof of mechanical concept was performed on preliminary Engineering Model (EM) prototypes including both vibrations and shocks tests campaigns.

Then EM and Qualification Model (QM) prototype cells were mechanically tested in battery structures. The tested cells included specific worst cases representative of design and process limits. They were selected or manufactured depending on:

- The gap between the jelly roll (which is not adjusted) and the can, and also depending on the axial holding
- The axial compression due to the shims and the jelly roll edge folding induced by the tabless design.

QM cells were already succesfully submitted to the following tests:

- Standard Vibration test

For sine vibration, the following profile was applied with cell EOCV between 3.85 V and 3.9 V.

Sweep rate: 2 oct/min per axis	Frequencies (Hz)	Levels
// to the mounting plane (X-Y) ⊥ to the mounting plane (Z - cell axis)	5 - 22.3 22.3 - 55 55 - 70 70 - 100	±10mm 20 g 25 g 20 g

Table 2. Sine vibration profile

For random vibration, the following profile was applied with cell EOCV between 3.85 V and 3.9 V.

Random vibe			
Duration: 3 minutes per axis			
Oz (perpendicular; cell axis)		OxOy (parallel to mounting plane)	
Frequency (Hz)	Level (g^2/Hz)	Frequency (Hz)	Level (g^2/Hz)
20 65-400 2000	0.092 0.3 0.06	20 80-1000 2000	0.025 0.1 0.05
Global = 17.39 gRMS		Global = 12.85 gRMS	

Table 3. Random vibration profile

- Shock test

The following shock levels were applied depending on the cella axis. Cell state of charge is at 40%.

3 times per axis along a unique direction	Frequencies (Hz)	Levels (g)
⊥ to the mounting plane (Z)	100	60
	1000	1000
	2700	2150
	10000	2150
3 times per axis along a unique direction	Frequencies (Hz)	Levels (g)
// to the mounting plane (X-Y)	100	60
	1000	1000
	2000	2000
	10000	2000

Table 4. Shock test levels for each axis

- Constant acceleration

The success criteria of these tests are established by taking care of ECSS requirements and customers product specifications.

In parallel to these tests, mechanical margins are being evaluated by finite element analysis with material properties and load cases defined from theoretical studies and/or by tests taking into account the internal pressure built-up and jelly roll swelling life long linked to ageing, cycling and irradiation.

Figure 3. 3S4P VL51ES module for mechanical tests

3. CELL PERFORMANCES

3.1 Generality

VL51ES electrochemistry was chosen by considering several aspects:

3.1.1 Supplying aspect

As far as possible, the electrode components must be available during all the cell life manufacturing and in synergy with industrial cell roadmap. A small material change in the cell definition will lead to time consuming cell delta qualification. A preliminary study of the supplier robustness was performed and gave orientation to the active materials to use in VL51ES cell.

3.1.2 Heritage aspect

VL51ES electrochemical design cell was established according to the heritage acquired on the current industrial cell. The electrochemistry is based on a NCA/Graphite technology which is flightworthy and has already demonstrated good cycling and calendar life characteristics.

3.1.3 Specific ernergy improvement vs. VES180SA cells

A lot of work was done to increase the cell specific energy by working on the optimization of the cell electrode design with a specific focus on the electrode porosity, electrode loading, the thickness of the separator and of the foils. In that we were able to reach an average specific energy of ~170 Wh/kg at 4.1V, charge C/10, discharge C/2 compared to 155 Wh/kg for VES180SA cell. This kind of modification induced a lot of impacts that were handled at product and process level.

3.2 QM cells characteristics

The main contributors to the cell specific energy increase are the thicknesses of the foils and of the separator.
The chosen electrode design on QM cells is a compromise between initial performances, supplier materials availability and expected cycling durability. The chosen active materials were processed in order to be similar in term of raw energy to VES180SA cell characteristics.

 DOI: 10.1051/e3sconf/20171608004

3.2.1 Standard characteristics

The first electrical tests on QM VL51ES cell give the following data (Table 2):

	VES180SA	VL51ES (QM)
Average capacity (Ah) @ 4,1 V	49,4	51
Average energy (Wh) @4,1 V	178	182
Average mass (g)	1128	1076
Average specific energy (Wh/kg)	158	169
Voltage range	2,7 V - 4,1 V	
Internal resistance (mΩ)	< 2,5	
Temperature	+10°C - +60°C	

Table 5. VES180SA and VL51ES cell characteristics

These results are in line with the VL51ES cell specification.

3.2.2 Cell energy for various EOCV

Cells energy was evaluated for various SOC. The data are reported on Figure 4. The behavior is typical for a NCA based positive active material.

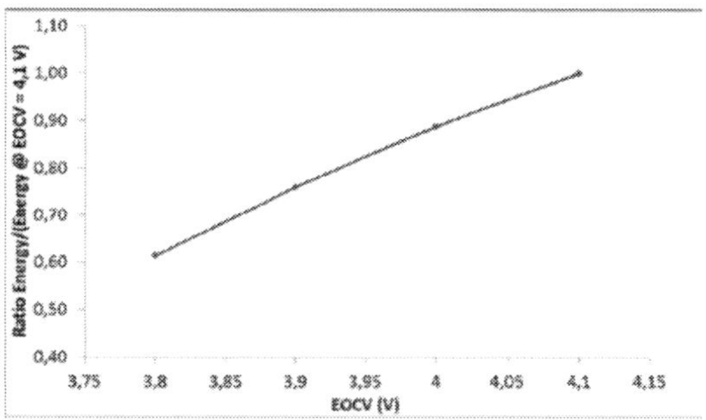

Figure 4. Cell energy ratio depending on EOCV

3.2.3 Cell energy for various temperatures

Cells were characterized at : +0°C, +10°C, +20°C, +30°C, +40°C, and +55°C at D/2. The data are reported on Figure 5.

At +0°C, the cell can delivered 85% of the reference energy at +20°C. Discharge energy is gained by using the cell up to +40°C. Above +40°C, the gain is not significative.

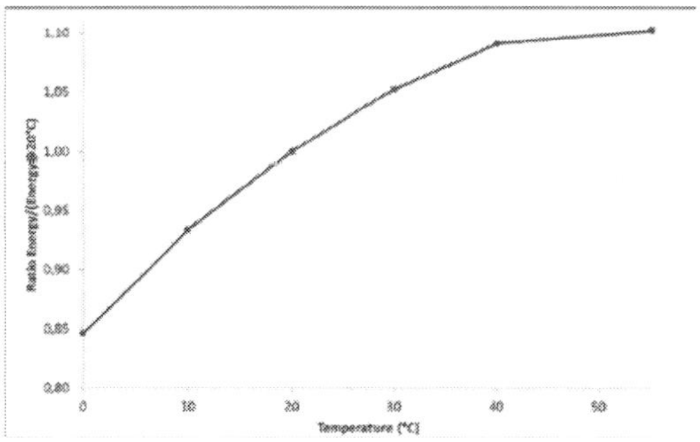

Figure 5. Cell energy ratio depending on temperature

3.2.4 Cell energy for various discharge powers

At +20°C, cells were characterized for the following discharge powers: 45 W, 90 W, 120 W, and 180 W.

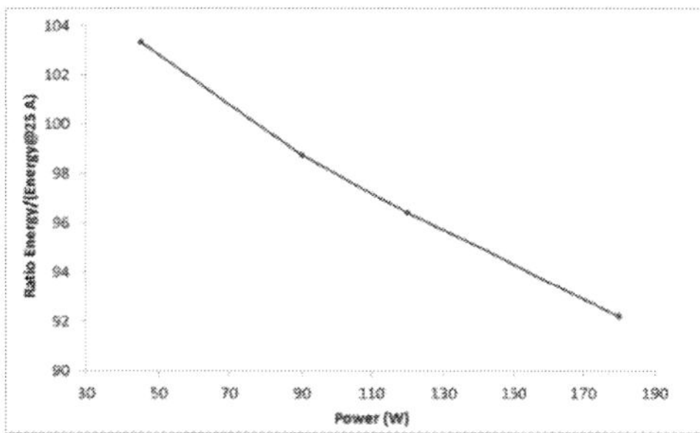

Figure 6. Cell energy ratio depending on discharge power

As usual, the delivered energy decrease when we increase the discharge power. At 1C, the cell delivers 92% of the discharge energy measured at D/2.

3.2.5 Self discharge characterization

At +20°C, cell self discharge was mesasured over 3 months for an open charge voltage of 3.9 V. The voltage data are reported on Figure 7.

 DOI: 10.1051/e3sconf/20171608004

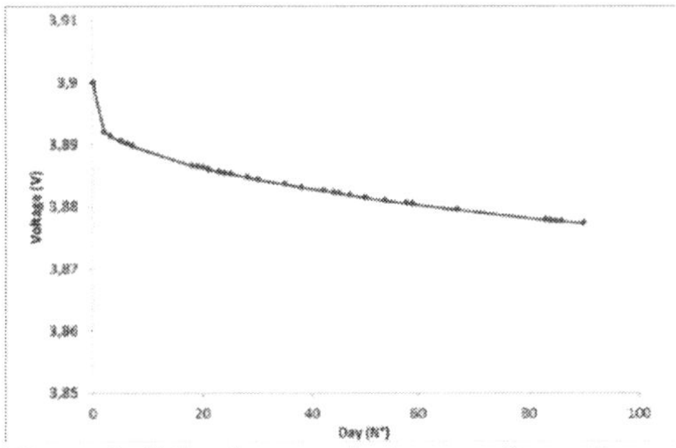

Figure 7. Cell voltage evolution over 3 months

Over 90 days the voltage drop is less than 0.2 mV/day. This represent less than 5% energy drop over this period.

3.3 Cell durability

3.3.1 Nominal cell cycling performances

The validation tests campaign is in progress on EM and QM cells to evaluate the specific energy End of Life. It includes, GEO accelerated testing with and without solstice, calendar life tests at different temperatures with pressure sensors, as well as irradiation with pressure sensors. The results will be compared to results obtained on VES180SA cell.

The expected cell cycling performance are the following:

- Cell energy at end of life over 160 Wh after 19 years GEO cycling at 80% DoD. EM and QM results are currently compliant after 20 seasons on EM and 8 seasons on QM.

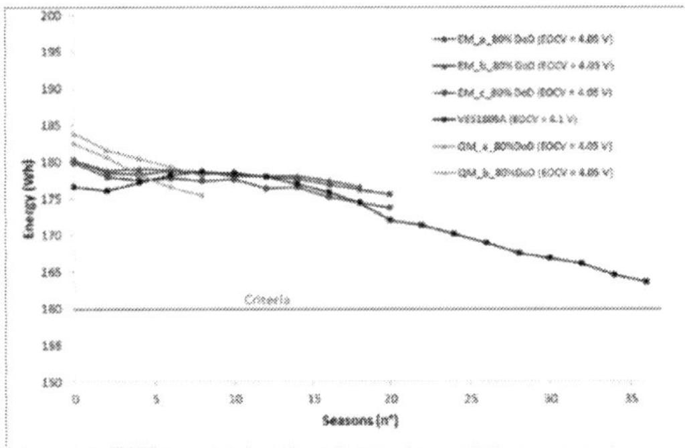

Figure 8. VES180SA and VL51ES cell energies in accelerated GEO cycling at 80%DoD without solstice

The corresponding accelerated GEO cycling profile without solstice is reported on Figure 9.

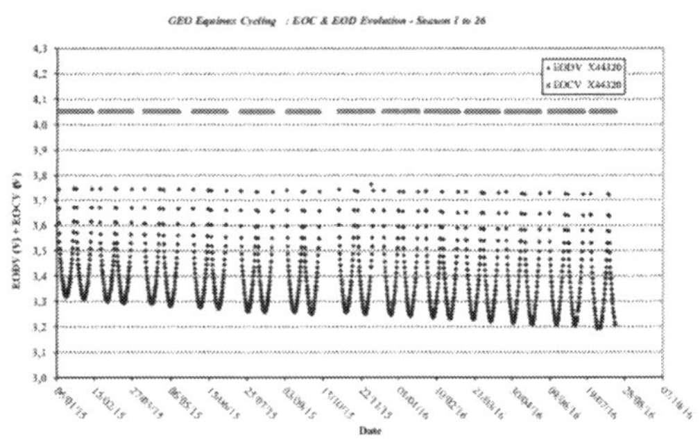

Figure 9. Accelerated GEO cycle profile without solstice (maximum 80% DoD)

For EM cells, End Of Discharge Voltages for the deepest discharge per season are reported on Figure 10. We have an identical behavior between VES180SA and VL51ES. The criteria which is EODV ≥ 3.14 V at 32 seasons is currently fulfilled.

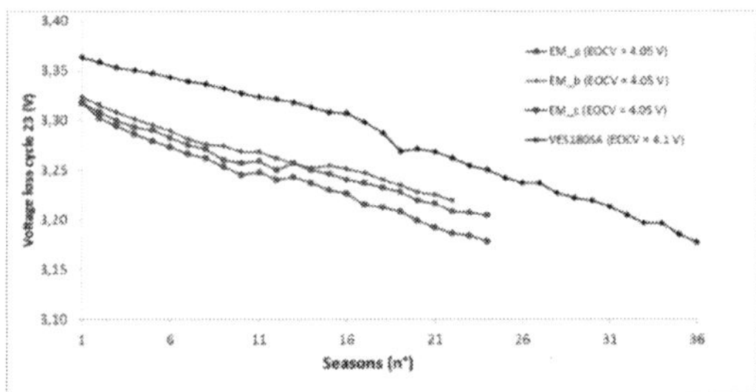

Figure 10. EODV at 80% DoD depending on cycle number during accelerated GEO cycling without solstice

- Cell pressure below 6.89 bars at the cell end of life (not pressure vessel) with an irradiation dose compatible with space applications. A calendar test at 60°C at OCV = 4.1 V on a cell equipped of pressure sensor is currently in progress. The results will be correlated with the cell pressure model. In parallel some cells equipped of pressure sensor are irradiated up to 8 Mrad in order to established the cell internal pressure increase depending on the irradiation rate.

In order to fit with new generation full electric satellites, QM VL51ES cell are tested by taking care of solicitations due to the satellite plasmic propulsion systems. The cycling test profils have been adapted depending on the customer specifications.

3.3.2 Cell process validation

The cell electrode manufacturing processes and treatments have slight differences between manufacturing plants and present some tolerances on parameters (electrode porosity, loading and thickness).
Therefore, sensitivity and limit studies are conducted to optimize these processes.
Specific EM & QM cells were manufactured and tested to cover worst cases. Moreover different types of filling, impregnation and formation processes were evaluated to achieve the best compromise in term of cell performance and cell processability.

4. CELL IMPLEMENTATION

4.1 Cell pack characteristics

VL51ES cell interface was developed in order to be plug & play in the current Saft cell pack structure in head down or head up configurations.

Compare to a standard VES180SA cell pack, VL51ES cell pack will present:

- Lower weight
- Lower height

In addition, preliminary thermal tests and modelling did not show significant different thermal characteristics between a VES180SA cell pack and a VL51ES cell pack.

4.2 Preliminary cell pack validation

At QM level, the cell in battery configuration will be thermally tested by thermal vaccum tests confirming the thermal characteristic of a cell pack. Currently in progress, the results are expected in second trimester 2016.

4.3 Battery range

Based on the EM and QM cell design, a battery range started to be evaluated and compared to the current VES180SA battery range.

On the table, we reported the benefit in mass and in energy to use VL51ES (G6) cells instead of VES180SA (G5) cells in (20S to 24S)-XP batteries.

	Battery Power Benefit G6 vs G5 (kW)	Battery Mass Reduction G6 vs G5 (kg)
	Average	Average
2P Battery	0.3	-2.5
3P Battery	0.5	-3.7
4P Battery	0.7	-4.9
5P Battery	0.8	-6.2
2*3P Dual Battery	1.0	-7.4
2*4P Dual Battery	1.3	-9.9
2*5P Dual Battery	1.7	-12.3

Table 6. Benefit of VL51ES (G6) cell batteries vs. VES180SA (G5) batteries (20S to 24S structures)

The benefits are significant for large batteries.

A battery development range is started for NEOSAT program with VL51ES cells.

5. CONCLUSIONS AND FUTURE WORK

The electrochemical and mechanical designs of Saft's VL51ES Li-Ion cell were finalized in third trimester 2015. QM cells were built and have started tests. The first electrical and mechanical results are satisfactory. Beginning of life Qualification Review is expected in late October 2016.

6. REFERENCES

[1] Y. Wang, J. Jiang, J.R. Dahn, Electrochemistry Comm., 9, 2534 (2007)
[2] K. Shizuka, C. Kiyohare, K. Shima, Y. takeda, J. Power Sources, 166, 233 (2007)
[3] Julien Bréger, thèse "Combined NMR and PDF Studies of Positive Electrode Materials for Rechargeable Li-ion Batteries" (2006)
[4] Z. Lu, L.Y. Beaulieu, R.A. Donaberger, C.L. Thomas, J.R. Dahn, J. Electrochemical Soc., 149 [6], A778 (2002)
[5] Z. Lu, J.R. Dahn, J. Electrochemical Soc., 149 [7], A815 (2002)
[6] J.-H. Kim, Y.-K. Sun, J. Power Sources, 119-121, 166 (2003)
[7] J. Jiang, J.R. Dahn, Electrochimica Acta, 50, 4778 (2005)

DOI: 10.1051/e3sconf/20171608005

BEYOND LITHIUM-ION: LITHIUM- SULPHUR BATTERIES FOR SPACE?

Maria Nestoridi [1], Henri Barde [1]

*[1] ESA/ESTEC, Keplerlaan 1, 2200 AG Noordwijk - The Netherlands,
Maria.Nestoridi@esa.int, Henri.Barde@esa.int*

ABSTRACT

Lithium-ion (Li-ion) batteries are established as the state of the art [1] rechargeable batteries for terrestrial and space applications today since the launch of Proba 1 satellite in 2001. [2]
At the moment there is strong interest by all stakeholders related or influenced by the battery markets on two systems: The rechargeable Li-air (Li-O_2) and Li-Sulfur (Li-S) batteries. There have been many studies on both technologies during the past decades but since major challenges are still to be overcome, none of the two technologies has been yet commercialized.
Li-S is believed to reach mass commercialization towards the end of the decade whereas Li-O_2 is expected to be available after 2030. Therefore, discussion to follow hereby will focus on Li-S.
Li-S cells are regarded as one of the most promising systems for next generation batteries due to their high theoretical capacity, the abundant and low cost sulfur resources and lithium-ion comparable cathode production techniques. [RD3] If Li-S batteries were to be successfully developed and reach their theoretical maximum, batteries over six times lighter than the conventional lithium-ion ones, would be available. [RD4]
Sion Power in the US and OXIS Energy Ltd. in Europe are the major companies producing Li-S cells.
Prototype Cells were procured from Oxis Energy, UK and characterisation tests were performed at ESA-ESTEC Battery Life Test Facility in Noordwijk, Netherlands. The results are presented here, mainly in order to enhance basic understanding on existing technology in Europe and show relevant trends.
Consequences at power system level, if this technology was to be adopted for satellite applications, are also addressed in this paper.

1. OXIS SPECIFICATIONS FOR ULTRA-LIGHT CELLS

The OXIS Cell Specifications as per OXIS technical manual for the Ultra-light prototype cells are provided in Tab. 1. [5]
Six Ultralight prototype cells were procured and tested at ESA-ESTEC Battery Life Test Facility.

Part Number	POA0084
Nominal Voltage	2.1 V
Cell Capacity, discharged at: 0.2C & 20°C to 1.9V	6.5 Ah typical
Cell Dimensions • Length • Width • Height	 146mm 76mm 7mm
Tab Dimensions • Length • Width	 25mm 20mm
Cell Mass	55g
Specific Energy	**248Wh/kg**
Volumetric Energy	180Wh/L without tabs
Operating Temperature • Discharge • Charge • Storage	 -20°C to +30°C Not tested, recommend 20°C Not tested, recommend 10-30°C
Charging Method	Constant current only, with time limit and voltage limit cut offs
Charging Current	650mA (C/10)
Charging Time Limit	10 hours from 0% State of Charge
Charge Voltage Limit	2.6V
Discharge Rates • Constant Current	 4.9A (0.75C) is maximum tested
Discharge Voltage Limit	1.9 V
Cycle Life, at 0.1C charge, 0.2C discharge and 20°C	39 cycles to 80% Beginning of Life

Table 1. OXIS Ultra-Light Cells Specifications.[5]

DOI: 10.1051/e3sconf/20171608005

An important remark at this stage is that due to the limited number of cells more than one test was performed with each cell. Therefore considering that those cells are prototypes and that each cell has a considerable testing history that would naturally influence aging and capacity fading , the actual numbers of capacity are not to be considered, unless it's stated otherwise. The general trends and characteristics of this technology is where the reader's attention is meant to be drawn in this paper.

2. TEST SET UP at ESTEC

The cells were first placed inside an abuse chamber due to their prototype nature. The abuse chamber itself was flushed with nitrogen for safety reasons and was sited inside an environmental chamber so that the temperature could be adjusted as needed.

Figure 1. Test Set-up at ESA-ESTEC Battery Life Test Facility

3. LI-ION VS LI-S

The OXIS Lithium Sulphur battery cells tested typically operate between 2.6 and 1.9 V in contrast to Lithium-ion cells that operate between 4.2 and 2.7 V.

In Fig.2 the typical voltage curves and temperature evolution for common Li-ion cells are presented for comparison with relevant data from OXIS Li-S cells tests. The major difference in the voltage curves is that the Li-S discharge curve comprises of two plateaus. This is a known characteristic of the Li-S cell which is believed to be attributed to a series of structural and morphological changes [6] during the charge/discharge process involving the formation of soluble lithium polysulfides Li_2S_x (1st plateau) and insoluble sulfides Li_2S_2/Li_2S (2nd plateau) in the liquid electrolyte.

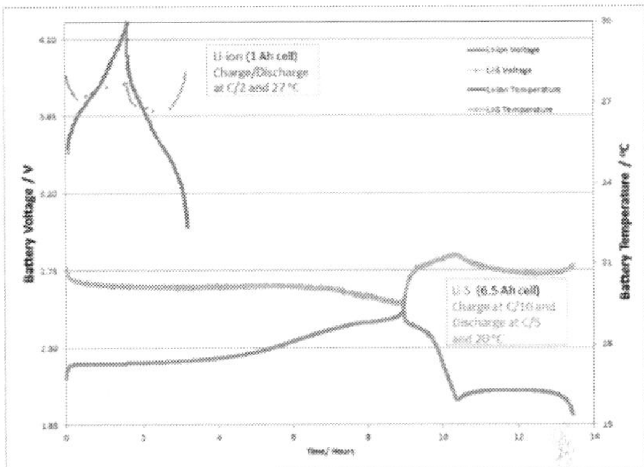

Figure 2. Li-ion vs Li-S Voltage Curves and Temperature during charge and discharge

4. TEST RESULTS

The results from the tests conducted with OXIS Ultra-Light Li-S cells at the ESA-ESTEC Battery Life Test Facility are presented in this section.

In Fig.3 the voltage curve and relevant capacity obtained with a fresh cell during charge at C/10 and 20°C are presented. Slightly lower capacity value than the one defined at OXIS datasheet is obtained. This is attributed to the fact that data in current OXIS datasheet correspond to tests performed at 30 °C whereas tests by ESA were conducted at 20 °C.

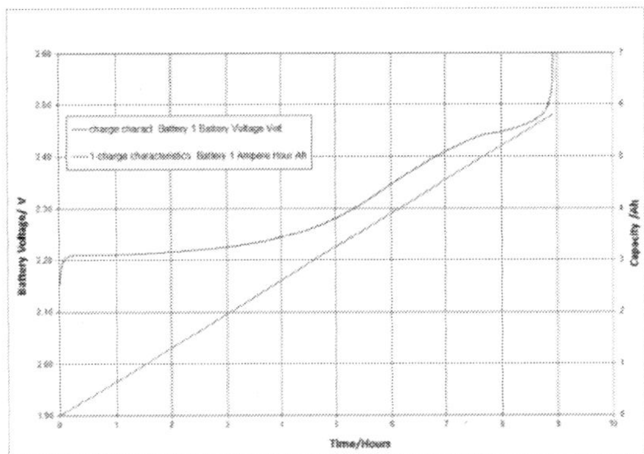

Figure 3. Charge Characteristics of Oxis Ultra-Light Li-S cell at C/10 and 20°C

DOI: 10.1051/e3sconf/20171608005

In Fig.4 the voltage curves vs capacity for discharge from 2.6 V to 1.9 V and 20 °C are presented for different C-rates, ranging from C to 0.1C. It can be seen that for C, 0.25 C and 0.5 C only the first voltage plateau is exhibited during discharge. For 0.3C, 0.2C and 0.1 C the second voltage plateau can also be seen and the capacity obtained is significantly greater than for the higher rates.

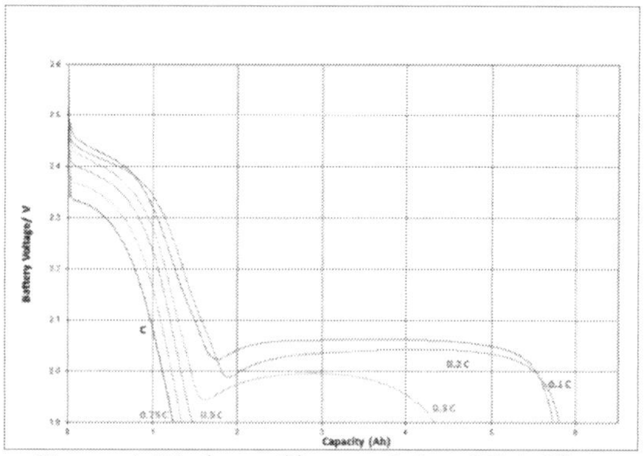

Figure 4. Discharge Characteristics of Oxis Ultra-Light Li-S cells at 20°C and different C-rates

In Fig.5 the voltage curves for discharges at C/5 for temperature values ranging from -20 °C to + 20 °C are shown. It is evident that the End of Discharge cut-off voltage for tests conducted at temperatures below 20 °C is lower than the standard 1.9 V value. This was done in order to compensate for the low temperature slow kinetics and allow the full discharge reaction to occur in order to maximise the cell capacity obtained. For -20 °C no capacity is virtually obtained. The cell can operate at temperatures from -10 °C, however the performance is significantly decreased below 0 °C.

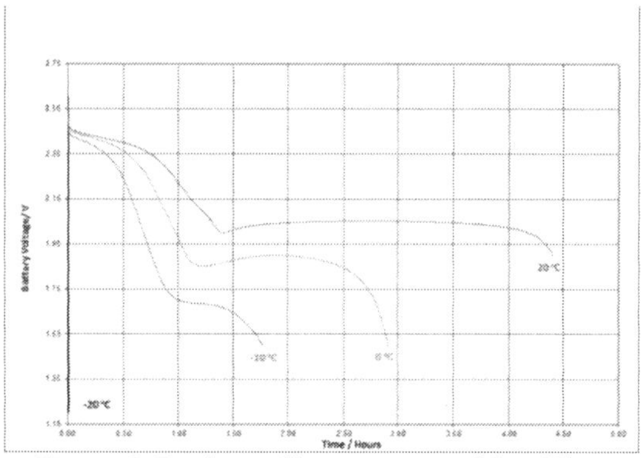

Figure 5. Discharge Characteristics of Oxis Ultra-Light Li-S cells at C/5 and different temperatures

In Fig.6 the Internal Resistance of OXIS Ultra-Light Li-S cells at different temperatures and different DoDs is presented. From these data, a significant increase in internal resistance is shown to occur, for all DoDs, for temperature values lower than 0 °C.

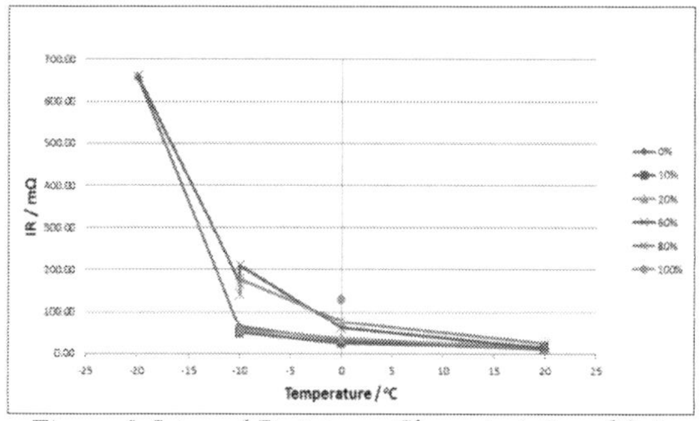

Figure 6. Internal Resistance Characteristics of Oxis Ultra-Light Li-S cells at different Temperatures for different DoDs

In Fig.7 internal resistance data for different C-rates and DoDs are presented. The C-rate does not seem to have a significant influence in internal resistance for a specific DoD value. The internal resistance seems to exhibit minimum values mostly at 10% or 20% DoDs and it reaches a maximum value at higher DoDs.

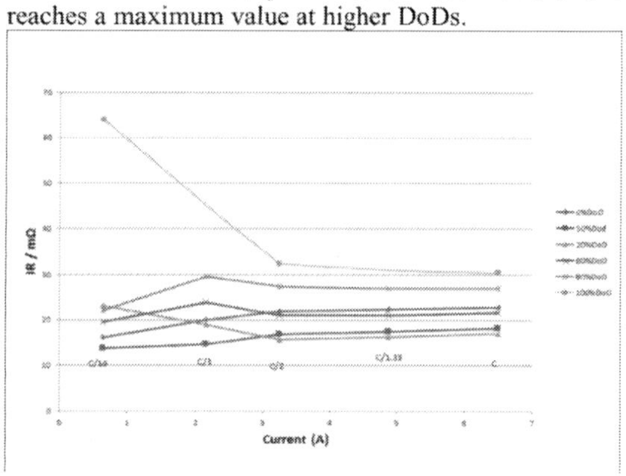

Figure 7. Impact of C-rate in IR Internal Resistance Characteristics of Oxis Ultra-Light Li-S cells at different Temperatures for different DoDs

In Fig.8 the Internal resistance values, for a discharge at C/5 at 20°C, are marked along with the relevant DoDs on the voltage curve of an Oxis Ultra-Light Li-S cell. As shown from data in Fig.7 previously, here the internal resistance at the beginning of the discharge reaches a minimum value, i.e. at the first voltage plateau, and a maximum one at the end of the discharge.

 DOI: 10.1051/e3sconf/20171608005

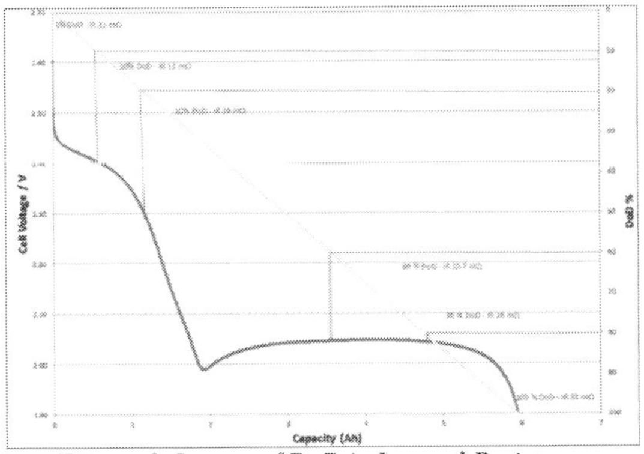

Figure 8. Impact of DoD in Internal Resistance Characteristics of Oxis Ultra-Light Li-S cells at 20°C

In Fig.9 specific energy data vs cycle number are presented for a number of cycling regimes carried out with OXIS Lithium Sulphur cells. The initial specific energy and its relevant progression with cycling is significantly higher at 20 °C compared to 0°C when cycling at 100 % DoD and C/10 charge and C/5 discharge rates. Cycling at the latter C-rates but at lower DoD proved to improve cell cycle life. Cycling at 20 % DoD and 20 °C seems to result in better cyclability, compared to higher DoDs, even when cycling was conducted at faster rates, i.e. C/3 charge and C/2 discharge. Cycling that was conducted at the latter rates and 20% DoD but at 0 °C, despite stability in cycling, exhibited lower specific energy values than the same test conducted at 20 °C.

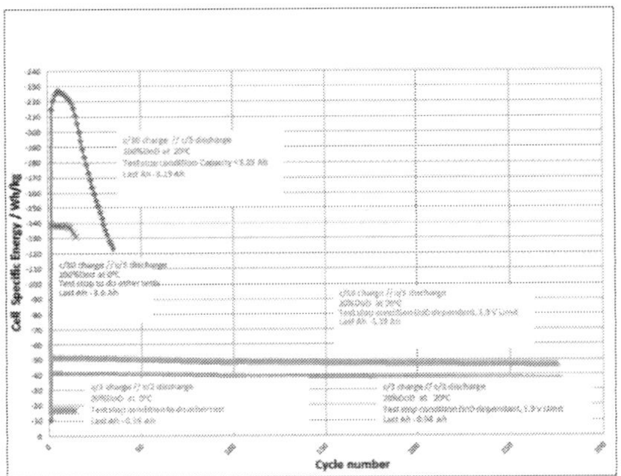

Figure 9. Cycling Tests Synthesis

5. ASSESMENT vs SPACE REQUIREMENTS

All test data presented in this paper show that the DoD plays a significant role in the Li-S cell cyclability. According to existing data, Li-S cells could be considered as a potential energy storage for satellites, for LEO missions, where the spacecraft battery is discharged at a low DoD. In contrast, further work, aiming to improve stability of the cells' performance with cycling at high DoDs, is required in order to make Li-S battery technology a good candidate for GEO missions.

On the rate capability side, when the cells are discharged at rates equal or higher to 0.5 C the second discharge voltage plateau is not reached and therefore the obtained capacity is limited, implying that not all discharge reactions are occurring at high rates. For LEO and GEO applications, the battery needs to be capable to perform discharge at ~ C/2 rates. Therefore, if Li-S battery technology was to be considered for satellite applications the relevant discharge rate capability should be improved.

The data presented in this paper show that the temperature has also a great impact on the specific energy obtained from these Li-S cells. Poor specific energy values were obtained when the cell was cycled at 0 °C. Nowadays the operational temperature window for Li-ion satellite batteries is commonly between 0 and 30 °C. It may be that this window should be pushed upward, provided that a higher operating temperature does not adversely affect the ageing.

6. IMPACT ON EPS DESIGN

It may appear premature to try anticipating what could be the impact of the use of Li-S batteries onto spacecraft Electrical Power System (EPS) but such a reflection, though, is not worthless, even if the technology is still at an early stage of development.

The first impact would be on the battery itself: the available capacities, which, in a context of spin-in, could range from very small (portable appliances) to quite large (Electric vehicle of power grid levelling), will have an influence on the series/parallel assembly, provided that the electrical properties allow paralleling at cell level. The need for balancing is also an important aspect. This depends on the capability of the technology to accept a moderate level of overcharging without degradation of performance, as was the case for the alkaline NiCd and NiH_2. In this case, a small excess capacity return after a discharge allows bringing back all individual cells to their maximum State of Charge. A dedicated test has been performed on one of the cells available at ESA at C/100 overcharging rate for a duration of 27 hours, followed by a 11 hours open circuit and a discharge attempt.

The Fig.10 shows the cell voltage and temperature as a function of time.

The very first outcome is that, following this 0.27 C overcharging, the cell was no longer able to deliver any energy. The second result, of little use in view of the previous one, though, is that even at a charging rate as low as C/100, no stabilisation is achieved, i.e. the simple balancing approach mentioned above cannot be relied on and a dedicated functionality would have to be implemented. On top of this, the End of Charge will have to be managed properly to avoid damaging the cell. As mentioned above, the current results are subject to the cell test history and consequent aging; further tests are therefore required to characterise Li-S cells throughout their lifetime for the missions of interest.

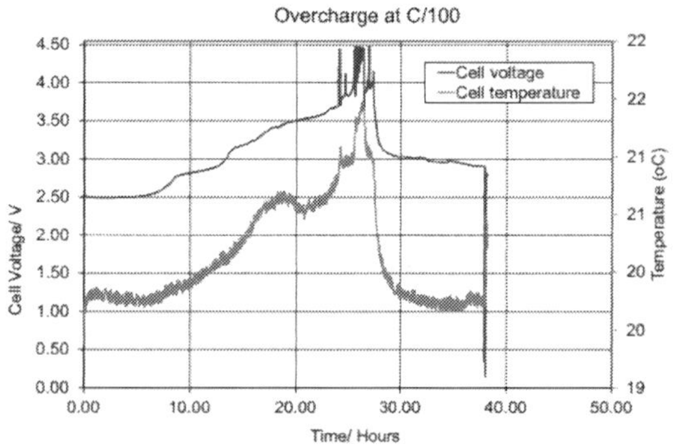

Figure 10. Cell voltage and temperature during low rate overcharge test

At EPS level, the first striking feature of Li-S is the fact that the voltage versus discharged capacity curve is not monotonic. The immediate consequence is that the battery voltage will probably not be usable as a capacity indicator, requiring some sort of Ampere.hour meter to estimate the actual State of Charge (SoC). This has been routinely done in the past: the battery management of the METOP and ENVISAT ESA spacecraft used that approach but their NiCd technology accepted easily a full discharge. Things are different in case the battery needs to be disconnected to avoid irrecoverable damages. This may result from a situation where the Ah meter, usually a software process, is no longer operating. Therefore, a voltage threshold appears safer but it needs to be selected carefully. Looking at Fig.4, at a typical discharge rate of C/3, the voltage difference between the end of the first plateau and the full discharge is quite small, introducing the risk of an unwanted disconnection.

A positive feature, on the opposite, is the charge management technique, which calls for a complete stop of the charge under constant (or available) current upon reaching the cut-off voltage. The absence of taper means that, for Low Earth Orbit (LEO) missions, the solar array would be better used, since there is no need to shunt available power to limit the battery current. This might have to be mitigated by the coulombic round trip efficiency, though, which has been difficult to assess during the various tests, the relatively quick decay of available capacity introducing a bias difficult to evaluate. It does not seem, though, that it is very far from unity. The round trip energetic efficiency was found close to 90 %, which is a reasonable value, given similar uncertainty. Altogether it can be expected that the introduction of Li-S in EPS would involve significant changes, possibly of a larger magnitude than these met when transitioning from alkaline technologies to Li-ion. This would be the price to pay for accessing what would be a significantly better performing technology

7. CONCLUSION

The maturity of Li-S batteries, as of today, is not yet sufficient to consider this technology other than a potential satellite energy storage mean. Nevertheless, taking into account the present status and the theoretical capabilities of Li-S electrochemical system, potentially yielding specific energy performances six times better than Li-ion, Li-S is undoubtedly an interesting battery technology to work on, as a preparation for the future energy storage on-board satellites.

8. REFERENCES

1. Defer, M., Du Peyrat, D., Prévot, D., Borthomieu, Y. (2008) Qualification of high specific energy li-ion cell VES180SA, 8th European Space Power Conference, ESA-SP Vol. 661
2. Genc, Z. D., Thwaite. C. (2011) Proba-1 and Mars Express: An ABSL Lithium-ion Legacy, 9th European Space Power Conference, ESA SP Vol. 690
3. Hagen, M., Dörfler, S., Fanz, P., Berger, T. ., Speck, R., Tübke, J., Althues, H., Hoffmann, M.J., Scherr, C., Kaskel. S. (2013) Development and costs calculation of lithium–sulfur cells with high sulfur load and binder free electrodes, Journal of Power Sources 224, 260–268.
4. Bruce, P.G., Freunberger, S.A.,. Hardwick L., Tarascon. J. M. (2012) Li–O_2 and Li–S batteries with high energy storage, Nature Materials 11, 19-29.
5. Oxis Technical Manual, Ultra-Light cells 2015.
6. Manthiram, A., Fu, Y., Su., Y. S. (2012) Challenges and Prospects of Lithium–Sulfur Batteries. Accounts of Chemical Research, 2013 May 21;46(5):1125-34

DOI: 10.1051/e3sconf/20171608006

High specific energy Lithium Sulfur cell for space application

Bruno Samaniego[(1)], **Emmanuelle Carla**[(1)], **Laura O'Neill**[(2)], **Maria Nestoridi**[(3)]

(1) Airbus DS France, 31 rue des Cosmonautes 31402 Toulouse France, Email: emmanuelle.carla@airbus.com; bruno.samaniego@airbus.com
(2) OXIS Energy England, E1 Culham Science Centre OX14 3DB Abingdon, Oxfordshire UK, Email:Laura.ONeill@oxisenergy.com
(3) ESA-ESTEC, Keplerlaan 1, 2200AG Noordwijk, NL, E-mail: Maria.Nestoridi@esa.int

ABSTRACT

The battery energy density remains a key parameter accounting for the satellite mass budget. As illustration to this, the rechargeable battery still represents 100 to 200 of kilograms for a typical Eurostar 3000 satellite, which can represent up to 5% of the total mass, and about 100 kilograms for the next meteorological satellite program MetOp-SG. Any reduction in weight in these applications has therefore significant financial benefits, considering that the launch cost for such a satellite can be around 10k€/kg.

Lithium-ion technology represented a revolution in terms of specific energy compared to Ni-Cd and is currently the most used and well suited for spacecraft. But it has also many drawbacks like price, some safety issues and its toxicity. Lithium-Sulfur (Li-S) cells are likely to become the next generation of energy storage to replace them.

One of the reasons is that sulfur is an abundant element so it's more affordable than cobalt used in Li-Ion cells. On top of that, Li-S cells are safer and more environmentally friendly. But the main advantage of this technology is the high energy density: around 5 times higher than Li-Ion cells.

The major obstacle for application is due to dissolved polysulfide shuttling between anode and cathode. This phenomeno leads to permanent loss of active mass from the cathode into the electrolyte and onto the Li metal anode (passivating the Li anode with insoluble Li_2S), severe self-discharge, low efficiency and fast capacity decay.

Airbus DS has been testing and characterizing prototype Li-S cells manufactured by OXIS Energy Ltd. since 2014, demonstrating the potential and fast evolution of the cells performance. This paper presents the last test results on a set of different batches provided by OXIS and performed at Airbus DS premises in the frame of an ESA Innovation Triangle Initiative (ITI).

1. INTRODUCTION

This paper will first present a brief state of the art on Lithium-Sulfur technology, followed by the description of OXIS' cells tested during this study. After that, the test plan is described and the test results summarized and analysed to illustrate the current state of Li-S technology and its astonishing evolution.

2. STATE OF THE ART

A Lithium-Sulfur Battery (LSB) is a secondary battery composed of lithium metal anode and sulfur-based cathode.

The trend for the secondary battery requirements is to be cheaper, safer, environmentally friendly and especially to have a high energy density.

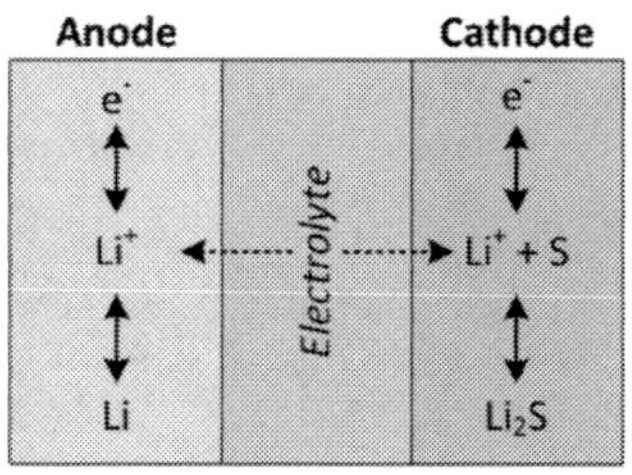

Figure 1: Lithium- Sulfur electrochemical process

The Li-S battery works on the basis of redox reactions between lithium metal anode and sulfur cathode. The reaction in Li-S battery is a reversible conversion reaction. The sulfur cathode offers a theoretical capacity of 1675 mAh/g and a theoretical energy density of 2600 Wh/kg with the full utilization of sulfur in the following process:

$$16\,Li^+ + S_8 + 16\,e^- \leftrightarrow 8\,Li_2S \quad (1)$$

Lithium-sulfur secondary battery technology is currently the only rechargeable battery technology that can reach 400 Wh/kg energy density and 1350 cycles [RD1]. A better understanding and further research about the problems of this new technology could allow reaching even higher energy density in the future.

2.1 Cathode materials

Elemental sulfur is an attractive choice as a cathode material for high-specific-energy rechargeable lithium cells because of its high theoretical specific capacity and the fact that it is very low cost and non-toxic. Sulfur can react with lithium ions between 1.5 and 3.0 V (vs.

 DOI: 10.1051/e3sconf/20171608006

Li/Li^+) producing soluble and insoluble lithium polysulfides. A major drawback of sulfur is its low electronic conductivity (5×10^{-30} S/cm at 25 °C) compared to the cathode materials used for Li-Ion batteries. In comparison to Sulphur, carbon based cathodes have several orders of magnitude higher conductivity values. For example, the conductivity of MCMB (MesoCarbon MicroBeads) is around 10^3 S/cm.

2.2 Anode materials

In the Li-S cells, dendrite deposition of Li metal is not as severe as observed in other lithium cell systems. This is attributed to the dissolved polysulfides (PS), which react with the anode and prevent the growth of Li dendrites.

Actually, while facilitating cell performance, the dissolved PS causes severe redox shuttle between the sulfur cathode and Li anode, which results in low columbic efficiency for charging and a fast self-discharge rate for storage [RD2].

2.3 Electrolyte

PS anions and anionic PS radicals are extremely reactive, they participate in many types of reactions such as basic, nucleophilic, redox, and radical reactions. PS is known to react with most of the common electrolyte solvents, such as esters, carbonates, and phosphates. It appears that the suitable solvents for the Li-S cell electrolytes are limited within the linear and cyclic ethers, such as dimethyl ether (DME) and 1, 3-dioxolane (DOL).

So far, Lithium trifluoromethanesulfonate (or lithium triflate) $LiSO_3CF_3$ and lithium bis(trifluoromethanesulfony)amide (or LiTFSA) $LiN(SO_2CF_3)_2$ have been reported to be the most suitable salt for the electrolyte of Li-S cells, of which $LiN(SO_2CF_3)_2$ is superior in providing higher ionic conductivity and less corrosion to the Al substrate [RD30].

2.4 Electrochemical behaviour

Based on the phase change of sulfur species, the discharge process can be divided into four reduction regions (see *Figure 2*):

•Region I: a solid / liquid two-phase reduction from elemental sulfur to Li_2S_8, which shows the first upper voltage plateau at 2.2 - 2.3 V. In this region, the formed Li_2S_8 dissolves into liquid electrolyte to become a liquid cathode. This process leaves numerous voids in the cathode.

•Region II: a liquid / liquid single-phase reduction from the dissolved Li_2S_8 to low- order PS, during which the cell voltage steeply declines and the solution's viscosity gradually increases with a decrease in the length of S-S chain and an increase in the number (concentration) of PS anions. The solution's viscosity reaches a maximum value in the end of the discharge region.

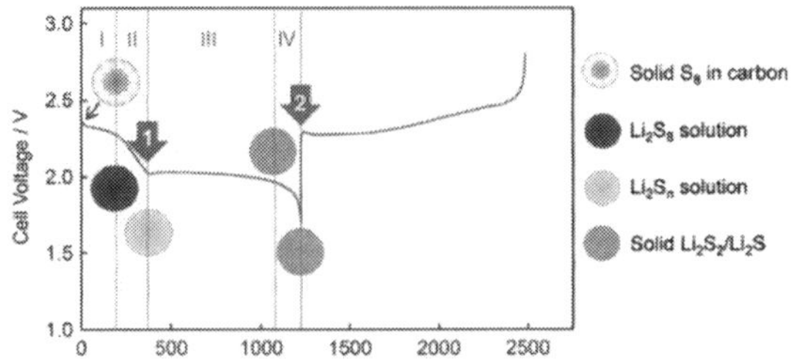

Figure 2: Typical discharge and charge profile of the first cycle of Li-S cells [RD3]

•Region III: a liquid / solid two-phase reduction from the dissolved low-order PS to insoluble Li_2S_2 or Li_2S, during which the following equations are competing with each other. This region forms the second lower voltage plateau at 1.9 - 2.1 V, which contributes to the major capacity of a Li-S cell.

•Region IV: A solid / solid reduction from insoluble Li_2S_2 to Li_2S. This process is kinetically slow and generally suffers from high polarization due to the non-conductive and insoluble natures of Li_2S_2 and Li_2S.

Among the four regions above, Regions I and II show the highest redox shuttle, during which the cell suffers from the highest self-discharge rate and the cell's theoretical capacity can be seldom obtained. Region III contributes to the major capacity of a Li-S cell. Region IV becomes very short or even vanishes.

2.5 Safety

Li-S secondary batteries have demonstrated safe performance consistent with the technology's current state of development. Failures of early prototypes have been experienced, but they are of a magnitude and type that will yield to improvements in cell chemistry, packaging and electronic controls for cell charge / discharge management. It has to be noted that no fire or explosion occurred during the tests performed in the frame of the study.

The research and development areas specifically targeted to enhance Li-S battery safety are:

•Electrode stabilization,

•Electrolyte composition,

•Electronic controls and packaging designs internal and external to the cell.

 DOI: 10.1051/e3sconf/20171608006

3. OXIS LITHIUM-SULFUR CELLS

OXIS provides different cells to be tested at Airbus DS facilities in order to analyse the evolution of the performances. Different batches of 5-6 cells have been specified:

- Batch 1a: Ultra-light cells of 6.5A.h with an energy density of 250Wh/kg;
- Batch 1b: Long-life cells of 10A.h with an energy density of 150Wh/kg;
- Batch 2: Ultra-light cells of 12A.h with an energy density around 300Wh/kg;
- Batch 3: Long-life cells of 20A.h with an energy density of 190Wh/kg.

The cells currently under test are described hereafter.

3.1 OXIS POA0084 6.5Ah Ultra-Light Pouch Cells

The main characteristics of the POA0084 Ultra-Light cells are listed in the following table.

OXIS POA0084 6.5Ah Ultra-Light Pouch Cells	
Nominal Voltage	2.1V
Cell capacity	6.5 Ah (0.2C, 20°C to 1.9V)
Cell dimensions	146 x 76 x 7 mm
Cell mass	55g
Specific energy	248 Wh/kg
Operating temperature	5°C to +30°C

Table 1: OXIS POA0084 6.5Ah Ultra-Light Pouch Cells characteristics

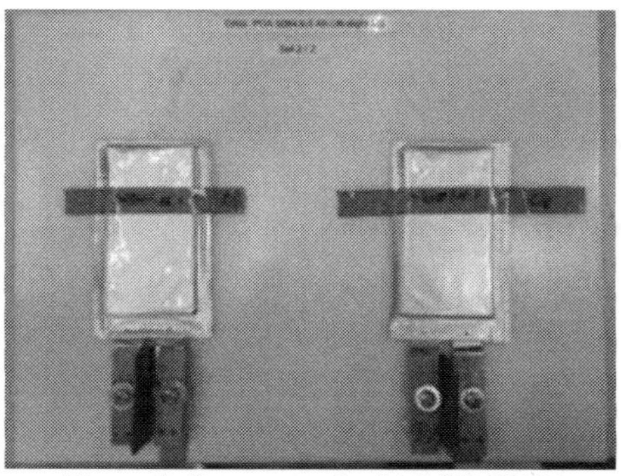

Figure 3: Test set-up for Li-S POA0084

3.2 OXIS POA0122 10Ah Long Life Cells

The main characteristics of the POA0122 Long Life cells are listed in the following table.

OXIS POA0122 10Ah Long Life Cells	
Nominal Voltage	2.05V
Cell capacity	10 Ah (2.3A, 30°C to 1.5V)
Cell dimensions	174 x 112 x 7.3 mm
Cell mass	135g
Specific energy	152 Wh/kg
Operating temperature	5°C to +60°C

Table 2: OXIS POA0122 10Ah Long Life Cells characteristics

Figure 4: Test set-up for Li-S POA0122

3.3 OXIS POA0217 12Ah Ultra-Light Pouch Cells

The main characteristics of the POA0217 Ultra-Light cells are listed in the following table.

OXIS POA0217 12Ah Ultra-Light Pouch Cells	
Nominal Voltage	2.1V
Cell capacity	12 Ah (0.2C, 30°C to 1.9V)
Cell dimensions	174 x 112 x 7.3 mm
Cell mass	90g
Specific energy	300 Wh/kg at 0.2C and 30°C
Operating temperature	5°C to +30°C

Table 3: OXIS POA0217 12Ah Ultra-Light Pouch Cells characteristics

 DOI: 10.1051/e3sconf/20171608006

Figure 5: Test set-up for Li-S POA0217

4. TEST PLAN

The test plan is designed to evaluate the interest of this technology for space systems. For each cell batch delivery, the following tests will be performed.

4.1 Initial Reference test

The objective of this test is to confirm the characteristics shown on OXIS' datasheet. It is performed at the temperature indicated on the datasheet for the capacity measurement and performed at C/5 with internal resistance measurements at 0, 10, 20, 60, 80 and 100% of Depth of discharge (DoD).
In order to assess the self-discharge, two reference tests will be performed: a first one with the discharge starting just after the end of the charge and a second one with one hour pause between the two phases.

4.2 LEO cycling test

If Li-S is meant to be used for satellite applications, it will need to face a high number of cycles representative of any LEO mission. For this test, a discharge at C/3 down to 20% of DoD has been selected with a charge rate of C/5. A hysteresis method is selected for charge management, so the voltage of the cell is always higher than end of charge voltage (EOCV) – 100mV.

4.3 GEO cycling test

GEO missions based on Li-Ion technology use high discharge currents but such high currents are currently one of the weak points of Li-S. A test of cycling at 80% DoD and another one following a real GEO season profile are performed. The impact of EoCV is also assessed by performing one test at 80% DoD with the maximum voltage as EoCV and another one with a reduced EoCV, expecting to increase the life of the cell, even if slightly degrading the total capacity.
After testing the first batch of cells (POA0084) at C/1.5, it was observed that the cell did not achieve the required capacity. It was then decided to characterize the discharge currents and perform a GEO cycling at C/5.

4.4 Launcher test

The wide temperature range expected for Li-S batteries make them very attractive for launcher applications since they often require starting at very low temperatures. A power profile coming from an Ariane 5 GTO DDO mission is run on one cell.

4.5 Cells characterisation tests

The last type of tests includes the characterisation of the cells at different temperatures: +20°C for comparison and minimum and maximum temperatures included in the datasheet. The discharge is performed at C/5 with internal resistance measurements as per the reference test.
Current rates tests are done at Reference test temperature and at C/5, C/2 and C rates for discharge.

5. TEST RESULTS

5.1 Initial Reference test

At the time of writing, the tests results available are those on POA0084 Ultra-Light cells and POA0122 Long-Life cells. Partial results are available for the POA0217 Ultra-light cellS.

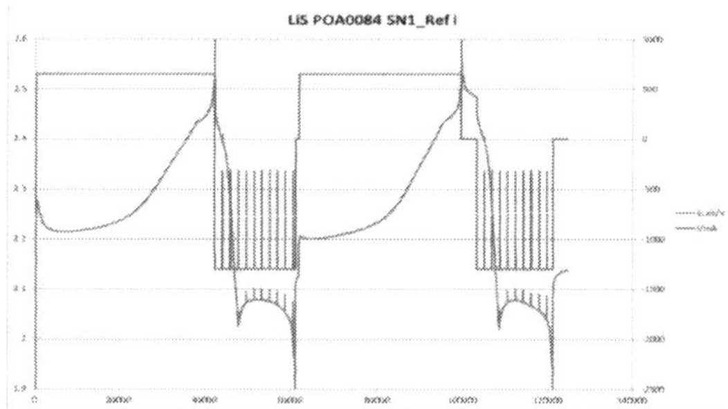

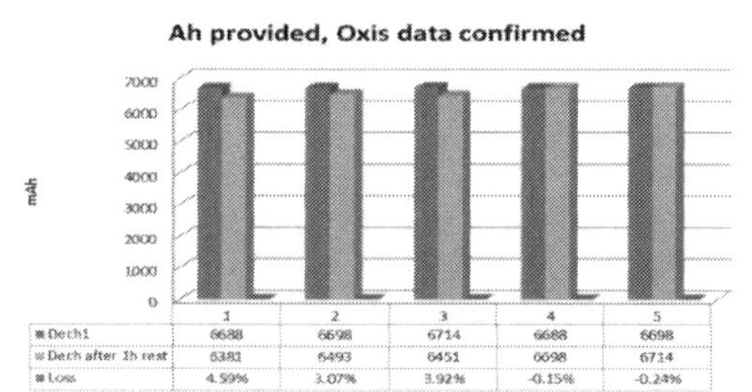

Figure 6: Initial Reference Test on Li-S POA0084

As it can be observed, the POA0084 capacity of 6.5Ah is confirmed by all the cells. The self-discharge after one hour in open-circuit varies from 4.6% to 0%.

 DOI: 10.1051/e3sconf/20171608006

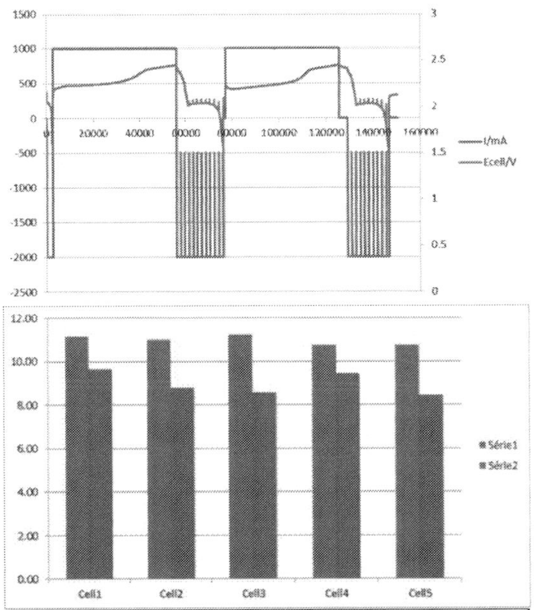

	Cell1	Cell2	Cell3	Cell4	Cell5
Déch 1 (Ah)	11.16	11.00	11.21	10.73	10.74
Déch 2 (Ah) after 1h open c	9.65	8.78	8.56	9.42	8.43
loss (%)	13.55	20.19	23.67	12.21	21.54

Figure 7: Initial Reference Test on Li-S POA0122

As it can be observed, the cell capacity shows always values higher than 10 A.h. However, in this type of cell the self-discharge is more important: between 12% and 24% of loss capacity in just one hour of open-circuit.

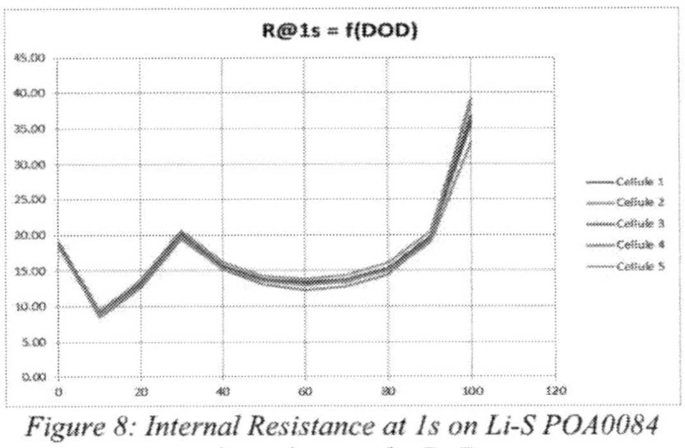

Figure 8: Internal Resistance at 1s on Li-S POA0084 depending on the DoD

Internal resistance measurements, with the Ultra Light Cells, show values between 10mΩ and 40mΩ (at 1s) depending on the DoD. All the five cells show the same trend. Measurements at 30s can go up to 80 mΩ.

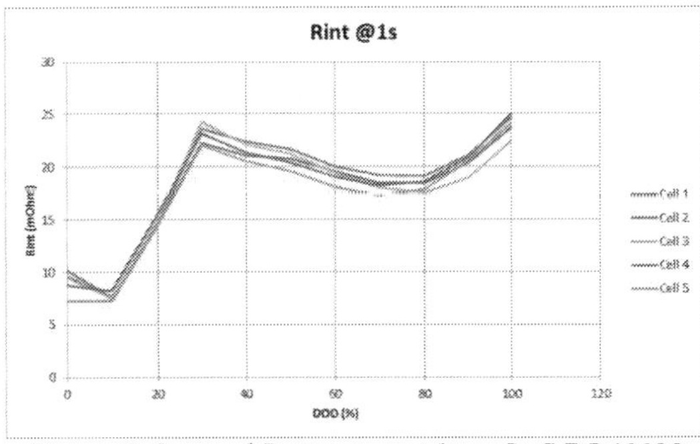

Figure 9: Internal Resistance at 1s on Li-S POA0122 depending on the DoD

Internal resistance measurements, with the long life cells, show values between 7mΩ and 25mΩ (at 1s) depending on the DoD (much lower than the Ultra-light cells). All the five cells show the same trend.

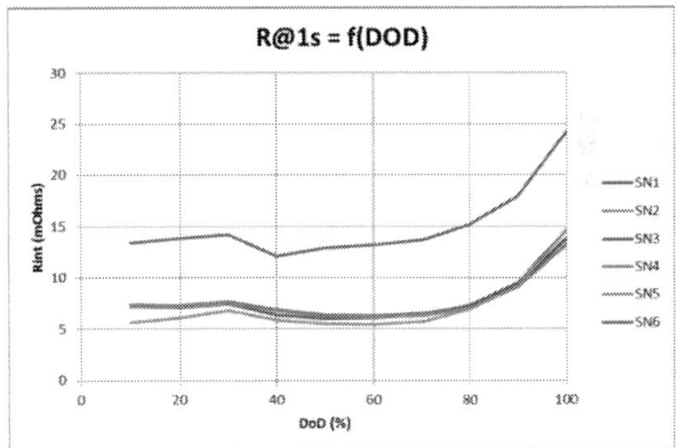

Figure 10: Internal Resistance at 1s on Li-S POA0217 depending on the DoD

A new batch of Ultra-light cells shows a lower internal resistance that the first one. The resistance values decreased from a range of 10 to 40mΩ (at 1s) down to 5 to 15mΩ (at 1s) for five out of six samples.

5.2 LEO cycling test

In figure 11 the hysteresis method for charge management can be observed: once the EOCV (2.6V) is reached, the charge ends, the cell voltage goes down progressively and reaches 2.5V (EOCV-100mV) just when the discharge phase starts.

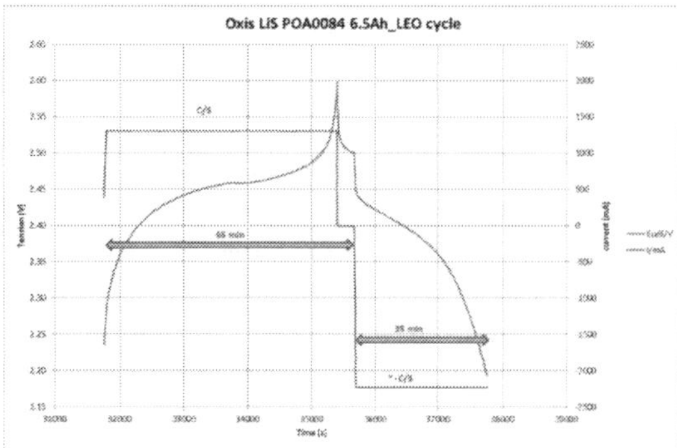

Figure 11: LEO cycling Test on Li-S POA0084

The evolution of the End Of Discharge voltage shows the progressive loss of capacity.

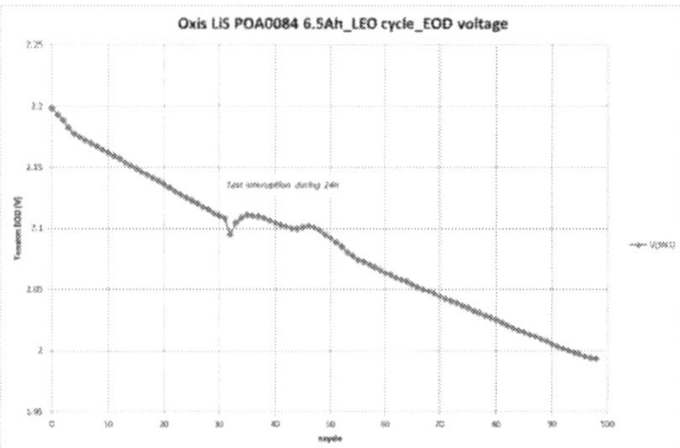

Figure 12: EODV evolution during LEO cycling Test on Li-S POA0084

At the time of writing, the POA0217 Ultra-Light cell has completed 55 cycles and the test is still running. The next table summarizes the performance of the other cells.

	20% DoD cycles
Ultra-light POA0084 6.5Ah	98
Long-life POA0122 10Ah	100
Ultra-light POA0217 12Ah	>55 so far

Table 4: Number of LEO cycles at 20% DoD performed

5.3 GEO cycling test

GEO cycling was initially foreseen with a discharge rate of C/1.5 but as it can be observed, the value of Ah expected (5.2Ah) was higher than the capacity that could be actually provided (3.97Ah).

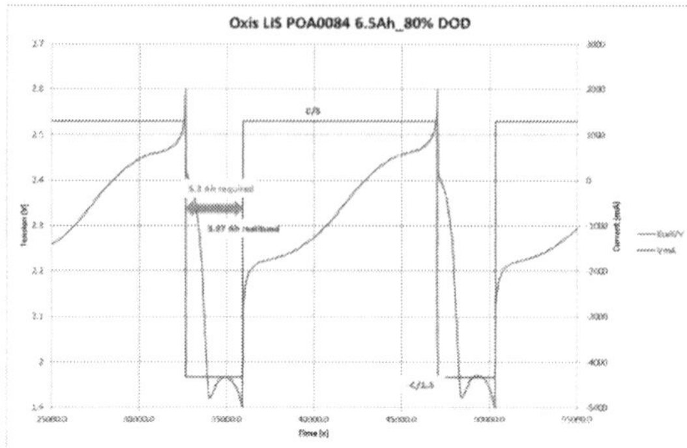

Figure 13: GEO cycling Test on Li-S POA0084

Unlike LEO cycling, the cell cannot provide enough capacity. The test was stopped when the fading reached 83%.

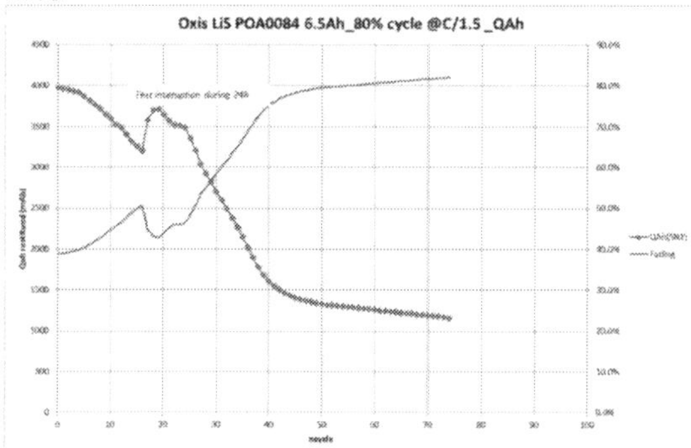

Figure 14: Fading and capacity provided during GEO cycling Test on Li-S POA0084

The cell following a real GEO eclipse season could provide enough capacity until eclipse number 8.

It was then decided to assess higher levels of DoD at lower discharge currents: C/5 instead of C/1.5.

At the time of writing, the POA0217 Ultra-Light cell has completed 25 cycles and the test is still running. The next table summarizes the performance of the other cells.

	80% DoD cycles
Ultra-light POA0084 6.5Ah	73
Long-life POA0122 10Ah	48
Ultra-light POA0217 12Ah	>25 so far

Table 5: Number of cycles at 80% DoD performed

5.4 Launcher test

As explained, a power profile coming from an Ariane 5 GTO DDO mission has been run on one cell per batch.

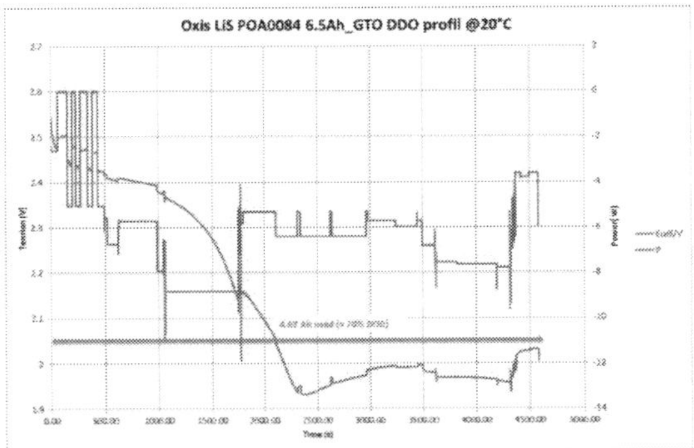

Figure 15: Launcher profile Test on Li-S POA0084

The launcher profile test showed that the cell could withstand without any problem this type of mission. The DoD reached at the end of the mission was 70%, letting enough margin for its application.

The two other cells passed without any problem the adapted launcher profile.

5.5 Cells characterisation tests

Characterisation tests include discharge at different current rates and temperature values. As it has already been demonstrated by the GEO cycling, the discharge current clearly impacts the capacity that can be provided by the cell. C/5 guarantees the nominal capacity, but when it is increased up to C/2 this is reduced by 22%. At 0.75C, the capacity is reduced by 75%.

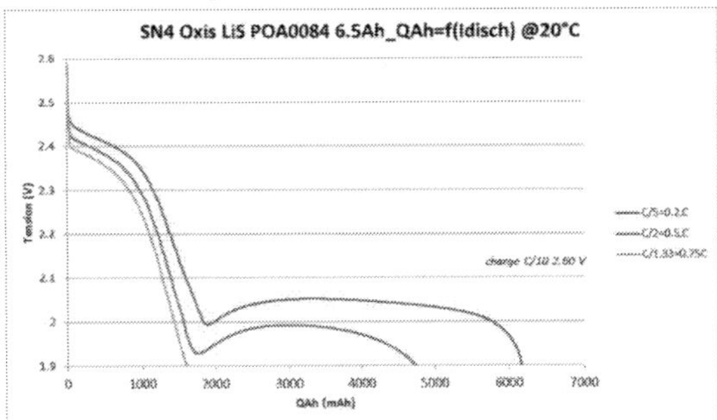

Figure 16: C-rate characterisation Test on Li-S POA0084

The tolerance to high C-rates has been considerably enhanced with the new batch of Ultra-light cells, reducing the loss from 75% down to 27% as it can be observed on the next figure.

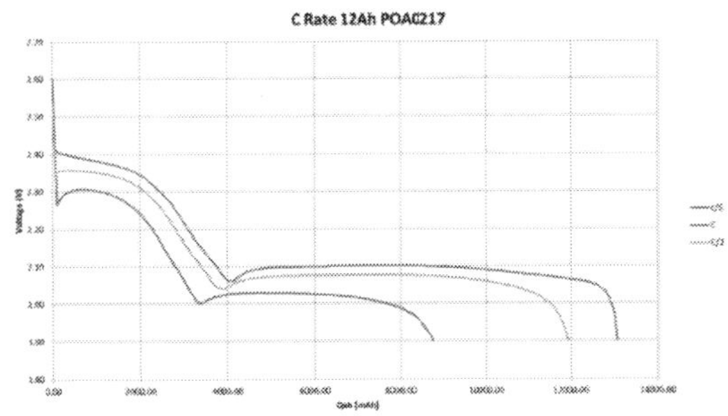

Figure 17: C-rate characterisation Test on Li-S POA0217

The following figures show that the impact at capacity and internal resistance level increases when the temperature decreases. No dramatic differences are observed for +20°C and +30°C but a huge capacity loss appears at +5°C.

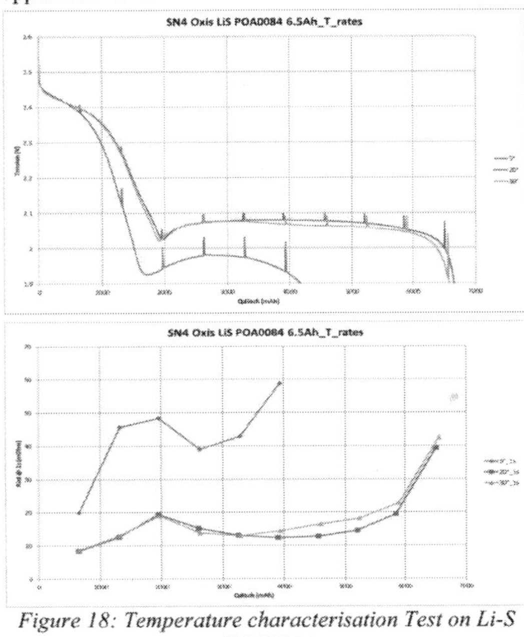

Figure 18: Temperature characterisation Test on Li-S POA0084

Again, this issue has been assessed for the last batch of Ultra-Light cells, reducing the sensitivity to the temperature. Again, as it can be observed on the figure below, no major difference appears between +20°C and +30°C, but the difference with regard to the +5°C case has been considerably reduced.

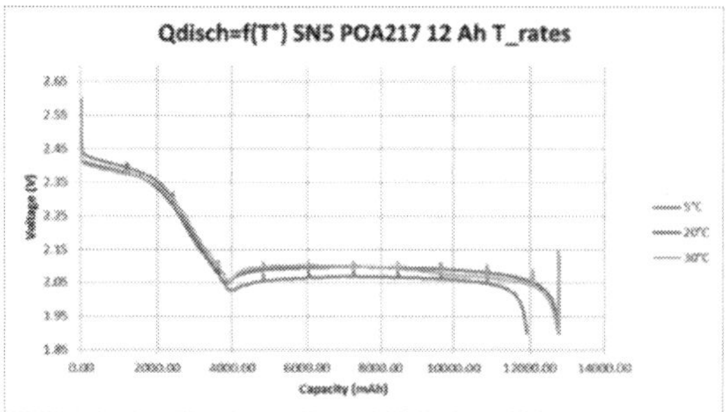

Figure 19: Temperature characterisation Test on Li-S POA0217

POA0122, can operate at a higher maximum temperature (+60°C) and provides a similar capacity compared to +30°C.

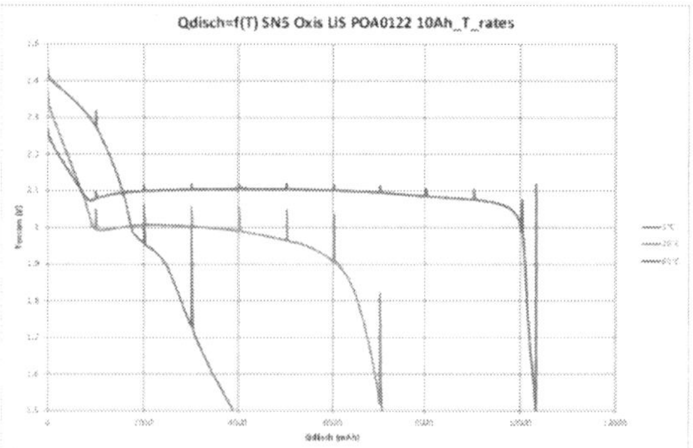

Figure 20: Temperature characterisation Test on Li-S POA0122

6. CONCLUSION

At the time of writing, only two cells have been fully tested and the newer versions of them can show only partial results. However, some conclusions can be drawn.

- It is demonstrated by testing that Li-S is a technology with a specific energy superior to that of Li-Ion, i.e 306 Wh/kg delivered by a POA0217 Ultra-light cell at Airbus DS premises.
- The major drawback of Li-S batteries for spacecraft applications continues to be the cycling capability. Up to now, only 100 cycles have been achieved at a 20% DoD. An additional cell has been provided by OXIS for batches 2 and 3 in order to study the impact of EoCV but there are no significant results at the time of writing. However, background from previous tests performed by OXIS and Airbus DS has shown a much higher cycling capability on other Li-S cells by reducing the EoCV (up to 1400 cycles at 20% DoD with a capacity loss of 15%) by reducing for instance the voltage from 2.6V down to 2.45V. Similar results are expected to be obtained with the additional cells under test for batch 2 and 3. If the cycling capability proves to indeed be significantly improved by a 150 mV reduction in EoCV also for this latest Li-S technology, it would be worth considering for certain mission profiles. Thanks to the significant leap in specific energy for Li-S batteries compared to the Li-ion ones, despite the for-mentioned decrease of the EoCV, Li-S batteries would still deliver a higher specific energy compared to Li-ion.
- Two of the main drawbacks up to now such as the dependency of the capacity on the temperature and the current rate have been addressed.

Despite the limitations to this moment, Li-S battery technology shows a great potential for many space applications like launchers or LEO satellite missions. Performances regarding cycling, rate capability and temperature need to be however further improved.
Li-S technology is just at the beginning and already shows great potential and constant evolution. The results of the tests performed during this ITI are encouraging and suggest that Li-S will in fact become the next generation of energy storage.

7. REFERENCES

RD 1: Min-Kyu Song, Yuegang Zhang and Elton J. Cairns "A Long-Life, High-Rate Lithium/Sulfur Cell: A Multifaceted Approach to Enhancing Cell Performance" - Nano letters, 2013 - ACS Publications

RD 2: Y.V. Mikhaylik and al., "High energy rechargeable Li-S cells for EV application: status, remaining problems and solutions", ECS Transactions, vol.25, n°35, pp. 23-34, 2010

RD 3: S. S. Zhang, "liquid electrolyte lithium / sulfur battery: fundamental chemistry, problems, and solutions", Journal of Power Sources 231 (2013) 153 – 162

RD 4: Test plan for Li-S cells, 2125.SP.CE.16. 2177.ASTR, Issue 02

E3S Web of Conferences **16**, 09001 (2017) DOI: 10.1051/e3sconf/20171609001
ESPC 2016

High Specific Li-ion cells based on high voltage materials

E. LIGNEEL[(1)], M. DEFER[(1)], F. FISCHER[(2)], E. DUMONT[(2)], F. FUSALBA[(3)], M. CHAMI[(3)], V. GINESTE[(4)], M. NESTORIDI[(5)]

(1) SAFT Space and Defense Division, Poitiers, France, Email: mathieu.defer@saftbatteries.com, eric.ligneel@saftbatteries.com
(2) SAFT Research Department, Bordeaux, France, Email: florent.fischer@saftbatteries.com, erwan.dumont@saftbatteries.com
(3) CEA-Liten, Grenoble, France, Email: florence.fusalba@cea.fr, marianne.chami@cea.fr
(4) Airbus Defence & Space, Toulouse, France, Email: valery.gineste@airbus.com
(5) ESA/ESTEC, Keplerlaan 1, Postbox 299, 2200 AG Noordwijk - The Netherland, Email: Maria.Nestoridi@esa.int

ABSTRACT

This paper presents the results obtain in the frame of the ESA TRP: High specific Li-ion cells based on high voltage materials.
As there is a need of high energy cells for Space application, a low TRL study was performed by Saft, CEA and ADS in order to developed high energy cells prototypes with a space proven cell mechanical design. The program was divided into 3 main parts.

1) Bibliographic study on the most promising active materials, 2nd a laboratory
2) Laboratory scale study of the selected active materials
3) Realization and electrochemical evaluation of format D prototype cells

1. INTRODUCTION

ESA initiates this TRP activity in order that all European space missions would benefit by having more lightweight energy storage. By increasing the energy density of the lithium ion cells the weight of the spacecraft battery, and consequently the satellite weight will be reduced. In space applications, in Europe, lithium ion technology has been established as the state-of-the-art battery technology, since the launch of Proba 1 satellite in 2001. Lithium-ion cells provided a great improvement over their preceded technologies by reducing the mass and volume of the batteries and subsequently the cost of satellite launch, since that is directly related to the weight of the satellite. Further development towards higher energy density and therefore lighter lithium-ion cells will also enable the increase of functional payload weight and mission capabilities for space applications.

2. OBJECTIVES

The objective of this activity is to develop, manufacture and evaluate Lithium-ion prototype cells with high specific energy (**≥250 Wh/kg**). This work will be done by SAFT in collaboration with CEA and ADS. It consists in choosing potential active materials, to test them separately in half coin cell versus Li°, and then to integrate them in full prototype cells. The prototype cell design is a space proven Format D industrial design used by SAFT.
Prototype cells must present cyclability with a fading of less than 20% after 300 cycles in Geostationary Earth Orbit (GEO) and Low Earth Orbit (LEO) conditions between -10°C and +50°C.

3. ACHIEVEMENTS

3.1 Part 1: Bibliographic study and Active material selection

3.1.1 Positive active material

As presented on Figure 1, hight mass capacity and high potential can be reach by using NCA ($Li(Ni_xCo_yAl_z)O_2$) and HE-NMC ($Li(Li_aNi_{1-x-y-a}Mn_xCo_y)O_2$) materials.

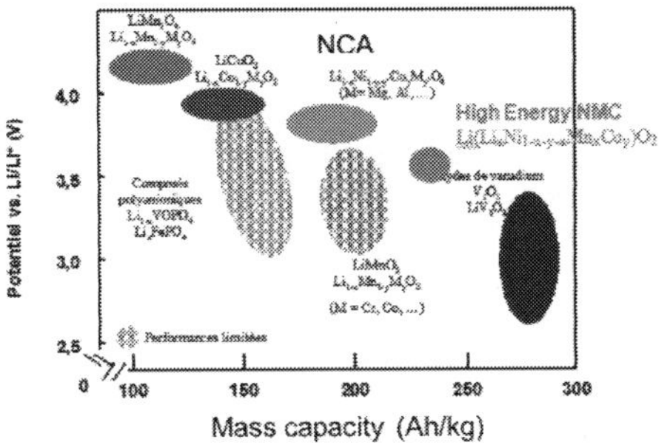

Figure 1: Comparison of several types of cathode materials depending on their specific capacity and their potential of insertion / deinsertion of lithium

It has to be noted that NCA can be competitive versus HE-NMC if it is charged over its standard potential (4.2 V vs. Li°). This can accelerate its degradation during cycling if the grade is not coated.

For the next step of the study, HENMC and NCA materials will be tested in half coin cell.

3.1.2 Negative active material

On negative side, the purpose is to use an active material which has a low potential and a specific capacity close to Li°.

The target can be reached by using Si. However the use of pure Si is not recommended due it important volume change when lithiated (up to 300%). This behavior doesn't permit to obtain a negative electrode with a stable SEI and conductive network during cycling.

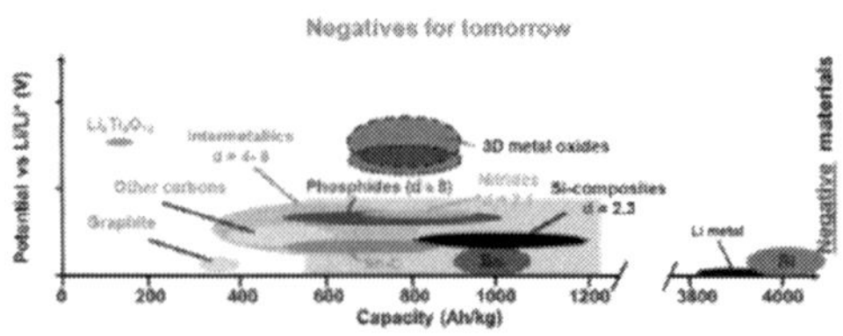

Figure 2: Potential use vs. Specific capacity of Negative active materials

A compromise is to use a mix of Si with Graphite or other non-metal compound.
For the next step, SAFT test Silicon carbon alloys and a Silicon non-metal compound alloy.
CEA adopt the same strategy by testing a Silicon carbon alloy.

3.2 Half coin cells cyclings

3.2.1 Positive electrode

NCA1 is non coated active material whereas NCA2 is a coated active material with the same stoichiometry. Despite a higher charge voltage and a coating, NCA active materials present a higher fading and specific capacity than grades of HENMC materials (HENMC 1 & 2a) (Figure 3).
3 grades of HENMC were tested. Bad performances are obtained on grade named HENMC2b. The others present a good stability.

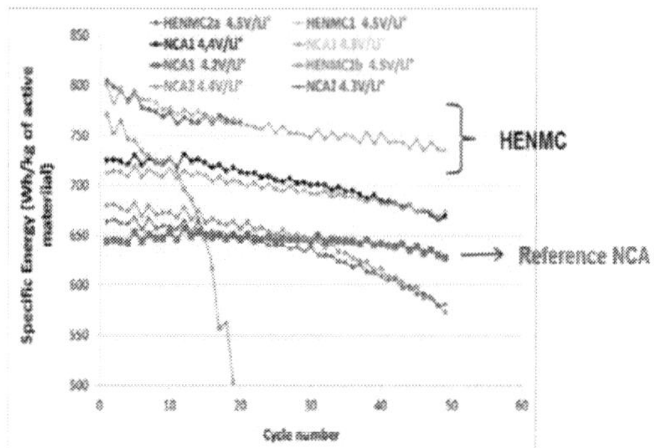

Figure 3: Comparisons of HENMC & NCA, Specific Energy during Cycling at C/5

SAFT proposed to use **HENMC 2a** in prototype cells.

3.2.2 Negative electrode

3.2.2.1 *SAFT electrode*

A comparison between a Si carbon alloys mixed with additional graphite (SiC2/C) and a non-metal Si compound alloy (SiX) was done (Figure 4). After 20 cycles, both materials present the same fading.

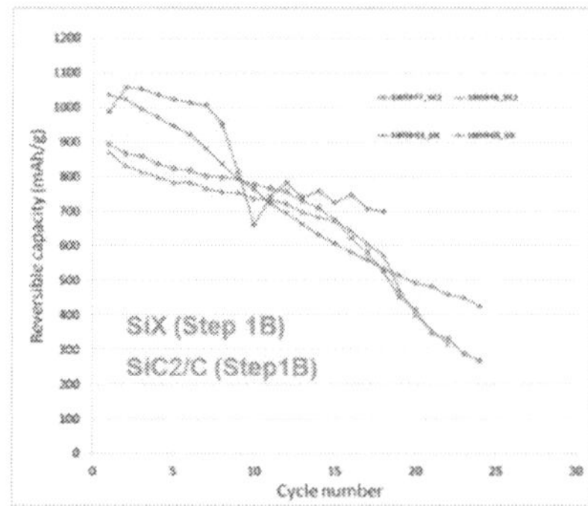

Figure 4: SAFT Negative materials, Cycling tests at C/5

This comparison was realized with a standard electrolyte. Material SiC2/C was tested (Figure 5) with an optimized electrolyte rich in VC and FEC.

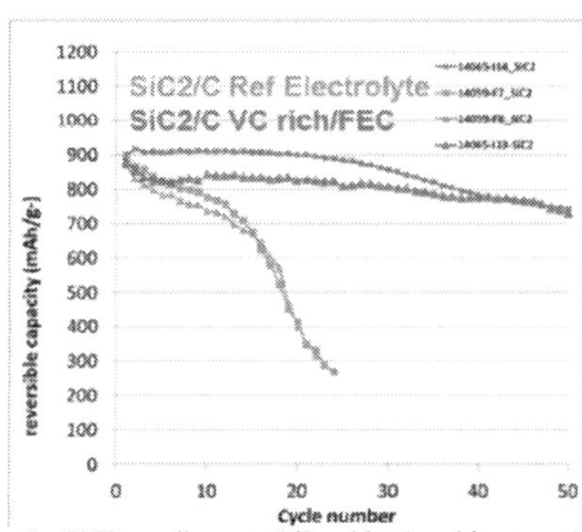

Figure 5: SiC2 cycling at C/5 with & without optimized electrolyte

The cycling results show the strong impact of electrolyte formulation on the cycling performances

DOI: 10.1051/e3sconf/20171609001

of a Si based electrode. It plays on the nature (chemical and mechanical) of SEI negative electrode which is able to manage active material volume change during cycling.

SiC2/C was implemented in prototype cells.

3.2.2.2 CEA electrode

CEA proposes a Si carbon alloy electrode.

In order to limit its volume change during cycling, the electrode has to be cycled with a limited charged capacity. A good stability over 100 cycles can be obtained by cycling it with charge capacity of 750 mAh/g (Formulation D).

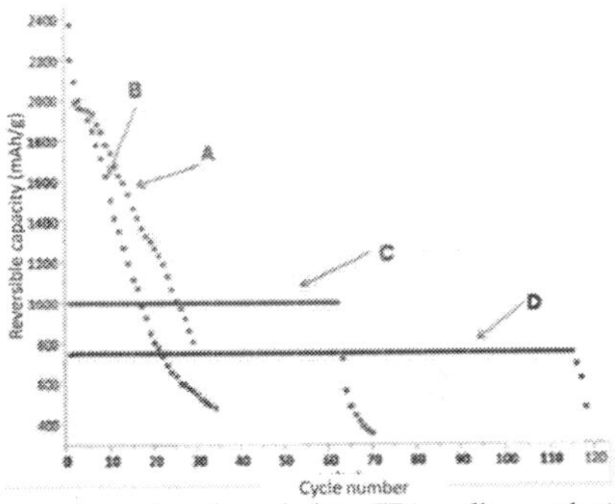

Figure 6: Negatives electrode from CEA cycling results (cycling at C/5)

3.3 Format D Prototype cells

3.3.1 Cell design characteristics

Format D mechanical design is usually use for space cell VES16.

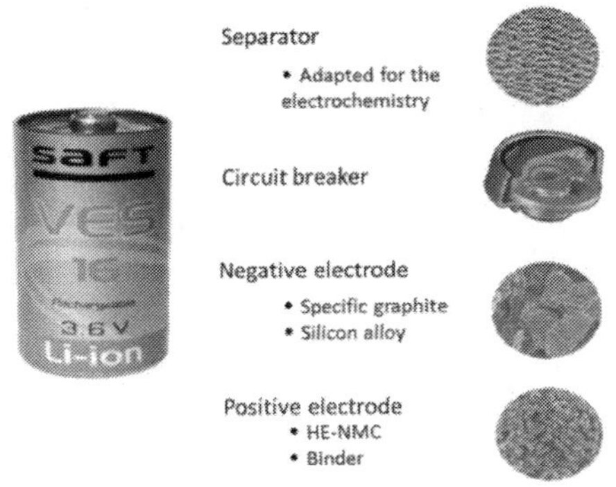

Figure 7: General overview of the cell

The main risks are identified and mitigated by specific countermeasure.

The gas generation during formation of the used electrochemistry is not well known. In order to not open the circuit breaker during cycling, the internal pressure was released after formation.

Despite silicon anode composition adjustments, there is still a possibility of high jelly roll swelling leading to separator puncture and tearing. This was mitigated by using a specific separator and manufacturing low jelly roll diameters.

2 series were manufactured.

- **Serie 1**: Cells manufactured with SAFT HE-NMC2a based positive electrode design and SAFT SiC2/C based negative electrode design.
- **Serie 2**: Cells manufactured SAFT HE-NMC2a based positive electrode design and CEA Si based negative electrode design (Formulation D).

In order to favor cell cyclability, specific energy lower than 220 Wh/kg at D/10 are proposed (Table 1).

	SAFT (-) electrode	CEA (-) electrode
NominalCapacity	7.25 Ah	6.87 Ah
Nominal Energy	24 Wh	22.3 Wh
Specific Energy	220 Wh/kg	200 Wh/kg

Table 1: Nominal characteristics at (C/10 & D/10) at 25°C

At D/2, the main unknown parameter is the electrode behavior at high current rate in prototype D cell configuration. The estimation is based on SAFT heritage (Table 2).

	SAFT (-) electrode	CEA (-) electrode
NominalCapacity	6.2 Ah	6.65 Ah
Nominal Energy	20.8 Wh	21.6 Wh
Specific Energy	190 Wh/kg	194 Wh/kg

Table 2: Nominal characteristics at (C/10 & D/2) at 25°C

3.3.2 Characterisation results

3.3.2.1 Characterization at D/10 and D/2

Initially, the voltage range was [2V – 4.4V]. But due to a high polarization of the electrodes, the obtained average specific energies are lower than design calculation (Figure 9) in this voltage range.

		Discharge (D/10)		
		Ah	Wh	Wh/kg
Serie 1 (Saft)	Average	6,43	21,0	**190**
Serie 2 (Saft/CEA)	Average	5,92	19,0	**171**

Table 3: Serie 1 & Serie 2 cell characterization at D/10 for EOCV = 4.4V

		Discharge (D/2)		
		Ah	Wh	Wh/kg
Serie 1 (Saft)	Average	6,12	19,1	**173**
Serie 2 (Saft/CEA)	Average	5,58	17,3	**155**

Table 4: Serie 1 & Serie 2 cell characterization at D/2 for EOCV = 4.4V

For Serie 1, another characterization was performed at D/10 & D/2 with a higher cell end of charge voltage (EOCV) defined at 4.5 V. Not enough cells were available for this kind of characterization on Serie 2.

		Discharge (D/10)		
		Ah	Wh	Wh/kg
Serie 1 EOCV = 4.4 V	Average	6,43	21,0	**190**
Serie 1 EOCV = 4.5 V	Average	6,56	21,5	**195**

Table 5: Serie 1 cell characterization at D/10 for EOCV = 4.4 V & 4.5 V

		Discharge (D/2)		
		Ah	Wh	Wh/kg
Serie ~~1~~ EODV = 4.5 V	Average	6,12	19,1	**173**
Serie ~~1~~ EODV = 4.5 V	Average	6,47	20,5	**186**

Table 6: Serie 2 cell characterization at D/2 for EOCV = 4.4 V & 4.5 V

As the cell polarization is more important at high current rate, the gain in capacity and energy is more important at D/2 than at D/10. For this last rate, the gain is negligible.

At D/2, the obtained specific energy is close to what was expected by design (186 Wh/kg vs. 190 Wh/kg).

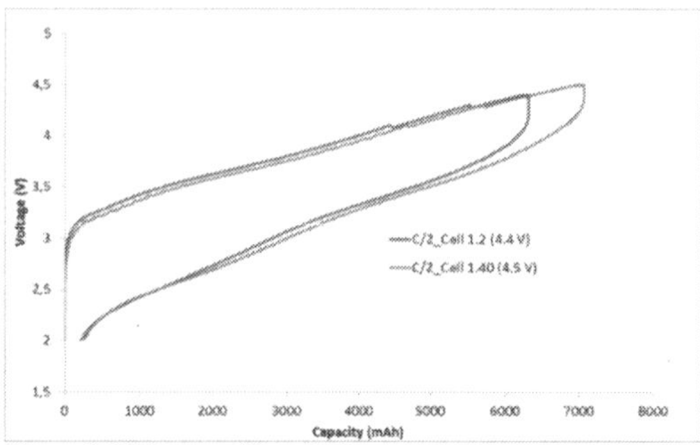

Figure 8: Voltage profiles of cells respectively charged at 4.4 V & 4.5 V

At D/2, voltage profiles of two cells from Serie 1 respectively charged at 4.4 V and 4.5 V were compared. The cells present the same irreversible capacity after the charge. This implies that the charge procedure up to 4.5 V (current rate at C/50) has not significantly degraded the cell.

3.3.2.2 Rate capability characterization

For Serie 1, characterizations were performed for various current rates: C/10, C/3, C/2, C/1.3, C and 1.37C. Not enough cells were available for this kind of characterization on Serie 2.

Characterization at C and 1.37C were not able to be done due to the high self-heating of the cell and reached temperatures higher than the safety limit (+45°C).

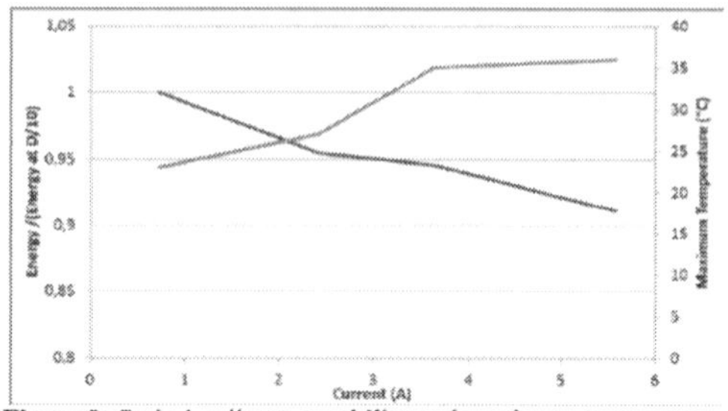

Figure 9: Serie 1 cell rate capability and maximum temperature

The energy drop is limited when the discharge current increase. The important temperature increase of the cell should have favorised the high current rate behavior.

3.3.2.3 Characterization at various temperatures

Cells were characterized at: -10°C, +20°C and +50°C at D/2 with EOCV = 4.4 V.

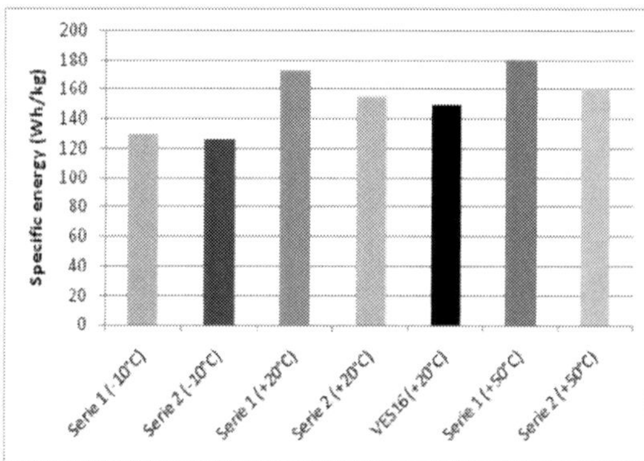

Figure 10: Prototype cell characterization results at various temperatures

For the characterization at -10°C, the cells still present good specific energies. Cells from Serie 2 present a lower energy drop (-18% for Serie 2 & -25% for Serie 1).

At +50°C, the cells did not gain significant energies (~ 4-5%) due to an increase of the cell irreversible capacity and internal resistance.

3.3.3 Self-discharge at 3.5 V

The capacity losses were around 0.12% per day for both Series.

Cell	Self Discharge	
N°	mV/day	%capacity loss/day
Serie 1	1,5	0,12
Serie 2	1,4	0,12

Table 7: Serie 1 & Serie 2 self-discharge results

The cells are far lower the TRP requirements of 5% capacity loss/day.

3.3.4 Prototype cells cycling performances

Cells were tested in GEO (80% DoD, Discharge at D/2) (Figure 10) and LEO (20% DoD, Discharge at D/2) (Figure 11) cycling conditions.

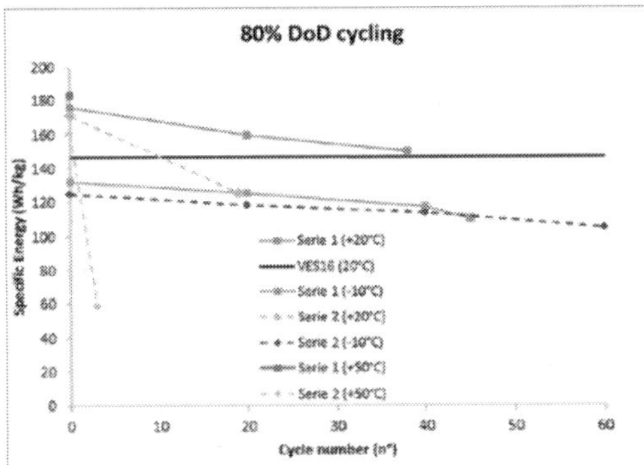

Figure 11: Prototype cell GEO cycling tests results

At +20°C and -10°C, the constant cell fading is probably due to the instability of the negative electrode of both series. This is mostly due to the high volume changes of the electrode during cycling.
At +50°C, cells from Serie 1 present sudden death before the 1st check-up and cells from Serie 2 only performed 3 cycles. These poor results could be due to internal mechanical stresses at electrodes level in the cells at such temperature.

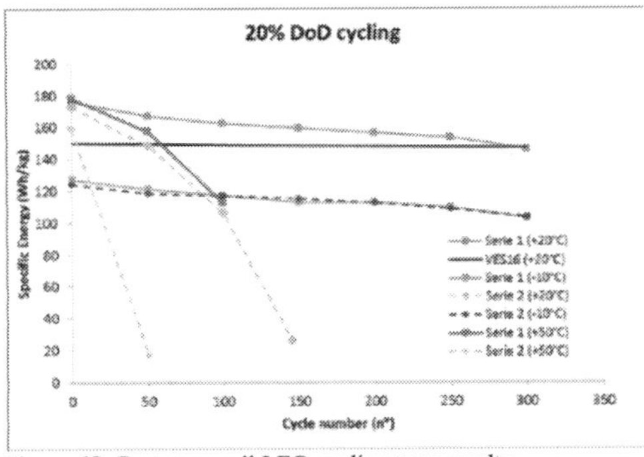

Figure 12: Prototype cell LEO cycling tests results

At +20°C and -10°C, cells from Serie 1 were able to cycle 300 cycles with energy losses below 20%. The low DoD cycling has limited the volume change of the negative electrode and improve its stability during cycling.
At -10°C, cells from Serie 2 were able to cycle 300 cycles with energy losses below 20%.
At +20°C, before the 1st check-up (20 cycles), Serie 2 electrode presents a similar fading as Serie 1 confirming the better stability of Si based electrode at low DoD cycling. Then a higher degradation rate was observed. Post mortem analyses confirm that the cells encountered a mechanical issue due to high stress on the electrode (high volume changes).
At +50°C, cells from Serie 2 only performed 50 cycles. Cells from Serie 1 present sudden death after 100 cycles. Again, these poor results could be due to internal mechanical stresses at electrodes level in the cells at such temperature.

4. CONCLUSION

4.1 Main achievements

HENMC and Si carbon alloys were selected for evaluation and implementation in Format prototype cells.
Preliminary results on coin cell permit to design full prototype cells.
Prototypes cells were successfully assembled and formed. However they present lower specific energy then design prediction. This is mainly due to the high polarization of positive and negative electrode.
The cycling stability in LEO cycling conditions at 20°C and -10°C are conformed to the objectives. At +50°C, the cells seem to encountered mechanical issue leading to their sudden death.
In GEO cycling conditions, the high volume change of the electrodes may induce a high cell degradation which did not permit to reach the objectives.

4.2 Way forward

Work can be done to Improve (-) and (+) electrode polarization at high current by working on material coating and/or electrode formulation. An additional test at higher end of charge voltage has shown that the specific energy design prediction can be reached if the electrode polarization is lower.
Prototype D cell design was not adapted to electrode volume change during cycling. Mechanical issues were observed on cells from Serie 2 and probably on cells from Serie 1 cycled at +50°C. The solutions could be:

- To develop prototype cell with a mechanical design generating lower stress on the electrode: Pouch cell with stack electrodes can be a solution.
- To introduce less SiC in negative electrode in order to limit volume change during cycling.

Using another cell design can be the opportunity to gain as well specific energy.

CATHODE MATERIALS FOR HIGH ENERGY DENSITY LITHIUM BATTERIES

G. Lefèvre [(1)], J.B. Ducros [(1)], M. Nestoridi [(2)], F. Renard [(3)], J.F. Colin [(1)], D. Peralta [(1)], M. Chakir [(4)], M. Chapuis [(1)] S. Martinet [(1)]

[(1)] *Commissariat à l'Energie Atomique-LITEN, 17 Rue des Martyrs 38054 Grenoble Cedex 9, France, Emails:Guillaume.Lefevre2@cea.fr, Jean-Baptiste.Ducros@cea.fr, Jean-Francois.Colin@cea.fr, David.Peralta@cea.fr, Marlene.Chapuis@cea.fr, Sebastien.Martinet@cea.fr*

[(2)] *ESA-ESTEC, Keplerlaan 1, 2200 AG Noordwijk, The Netherlands, Email: Maria.Nestoridi@esa.int*

[(3)] *PRAYON S.A - Siège de Engis - Rue Joseph Wauters, 144 B-4480 ENGIS, Belgium, Email: Frenard@prayon.be*

[(4)] *RENAULT, Technocentre, 1 avenue du Golf, 78288 Guyancourt, France, Email: Mohamed.chakir@renault.com*

ABSTRACT

Two classes of cathode materials have been developed for high energy density applications. The Li-rich layered oxide material with the general formula $Li_{1+x}M_{1-x}O_2$ (M = Ni, Mn, Co) and the lithium manganese silicate Li_2MnSiO_4. Both materials have theoretical capacities higher than commercialized ones, which may give rise to higher energy density batteries.

Li-rich materials have been prepared by solid state and co-precipitation routes. Transmission Electron Microscopy (TEM) characterization showed an irreversible evolution of the structure through a spinel phase during the first charge. Electron Energy Loss Spectroscopy (EELS) also showed a continuous cation migration during cycling of the material leading to charge/discharge voltage decay. The redox process has been studied by X-Ray Diffraction (XRD) in synchrotron facilities (ESRF, Grenoble, France). Ni/Mn ratio has been identified to have a great role on capacity fading of the material. Finally, a Li-rich optimized composition has been prepared and stable reversible capacity of 250 $mAh.g^{-1}$ has been obtained.

Li_2MnSiO_4 has a large theoretical specific capacity (333 mAh/g) through exchange of 2 lithium ions per formula unit. The thermal stability due to strong Si-O bonds makes Li_2MnSiO_4 a very promising material for future energy storage in space applications. Preparation in inert atmosphere showed beneficial improvements of LMSO's electrochemical properties. Nano-sizing and carbon coating have been effective ways to improve electronic conductivity and therefore electrochemical performance. Up to 1.66 Li per formula unit can be re-inserted in the 1st cycle. XRD analysis showed complete amorphization of Li_2MnSiO_4 after the 1st charge at 4.8 V with complete modification of the charge/discharge curves in the next cycles. Increasing the carbon coating ratio limits capacity loss during cycling but did not avoid amorphization. Finally influence of voltage window on structure stability was investigated. Careful choice of upper limit voltage has been showed to stabilize Li_2MnSiO_4 structure but for now is still limited to low Li^+ insertion/extraction from the host material.

1. INTRODUCTION

Driven by recent deployment of electric vehicles, researches on new lithium-ion battery technologies are constantly increasing. To meet demands of automotive/aerospatial sector, new cathode materials targeting high energy density, low cost, eco-friendliness and safety are investigated. Aerospace sector requests state-of-the-art lithium-ion batteries to supply electrical power when solar energy is not available, for emergency cases and during high-power pulsed payloads. Both terrestrial and space transportation are benefiting to this R&D rising.

Layered lithiated transition metal oxide material such as $LiMO_2$ (M =Ni, Mn, Co) represents the most commercialized class of cathode for Li-ion secondary batteries. Since few years, a polyanionic compound - $LiFePO_4$– gains market shares because of its high structural stability leading to higher safety properties. These materials are all limited in term of specific capacity in their domain of stability (150-200 mAh/g). However, aerospace applications require batteries with more and more power and embedded energy, at lower cost. In this frame, recent developments on Li-rich layered oxides and upstream research on polyanionic lithium manganese silicate (Li_2MnSiO_4) have been carried out by CEA.

Li-rich materials have very high practical energy densities and very high specific capacities (>250 mAh/g). It is economically attractive due to its composition mainly rich in manganese which is cheaper than cobalt. From a structural point of view, the lithium-rich layered materials can be regarded as a composite material built of $LiMO_2$ and Li_2MO_3 domains (M = Mn, Ni) [1]. The high specific capacity (> 250 $mAh.g^{-1}$) of Li-rich materials is attributed to the reversible capacity coming from the redox activity of cations in the $LiMO_2$ component and to an additional capacity corresponding to the electrochemical activation of the Li_2MO_3 component above 4.4 V [2]. This activation process involves oxidation of oxygen from the structure but the process is not really understood, especially the reversibility is not clear.

Since 2005 [3] researchers work on silicate cathodes. Li_2MnSiO_4 shows a theoretical high specific capacity of 333 mAh/g through exchange of 2 lithium ions by formula unit. 1st and 2nd lithium extraction has been computed to occur at high voltage of 4.1 and 4.5V respectively [4] leading to high energy density. Covalent Si-O bonds avoid oxygen release in excessive conditions [5], which improves battery safety. Absence of Ni and Co enables low cost and eco-friendliness. In addition to these characteristics, well spread-out raw materials occurrence is a key-factor for strategic sectors like space. As for polyanionic phosphates [6] low electronic conductivity is counterbalanced by size decrease and coating with conductive agent like amorphous carbon. Effect of particle size, atmosphere preparation, carbon coating and voltage window on electrochemical performance and structural stability will be discussed.

2. RESULTS AND DISCUSSION

2.1. Li-rich materials; recent developments

Li-rich materials with general formula $Li_{1+x}M_{1-x}O_2$ (M = Ni, Mn, Co) were investigated in CEA in 2007. In early developments, Li-rich materials were prepared by solid state route. Lately, the use of co-precipitation synthesis allowed higher performances (stable 250mAh/g capacity and higher tapped density) and a better control of its morphology (Fig.1). Such properties are desirable for the realization of high energy density electrodes. The synthesis has been scaled-up in a pilot co-precipitation equipment, with 1 kg batch size.

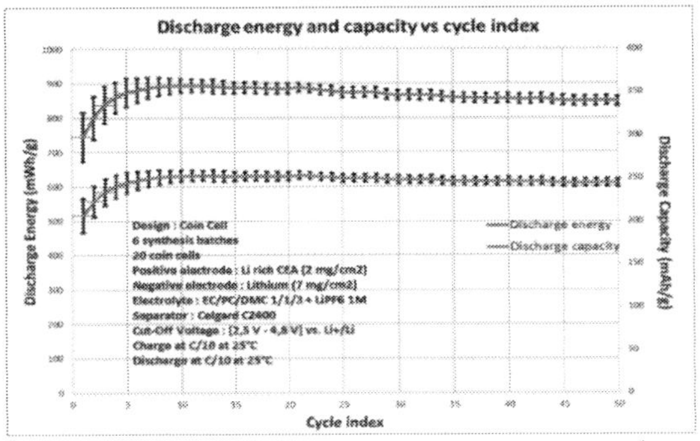

Figure 1. Evolution of the discharge energy and capacity of Li-rich material obtained in half cells (vs. Li) using [2.5 – 4.8] V vs. Li^+/Li cut-off voltages.

Despite very interesting high capacities and attractive price due to its high content in cheap manganese, some challenges still have to be overcome (a large irreversible capacity during the first cycle and the fade of the average cell voltage during the cycle life)

Our group worked simultaneously on Li-rich synthesis development and on its structure evolution understanding. The results from TEM (Fig. 2), Scanning Electron Microscopy (SEM), in-situ synchrotron diffraction measurements (ESRF, Grenoble, France) and electrochemical measurements have been analyzed in parallel to get better understanding of the voltage decay and capacity fading mechanisms.

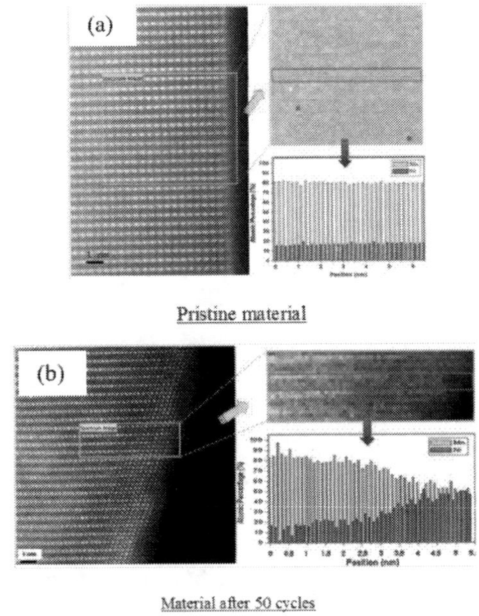

Figure 2. TEM image and chemical composition of Li-rich material (a) before cycling and (b) after 50 cycles.

First, *in situ* X-Ray diffraction study combined with a transmission electronic microscopy study led to the identification of an irreversible structural change during first charge, accompanied with the formation of a secondary phase with spinel structure at the surface of the particle, which can be the origin of the irreversible capacity observed in the first cycle [7].

According to the coupled EELS/TEM study, the voltage decay is due probably to the continuous cationic migration during cycling [8] (Fig.2 and 3).

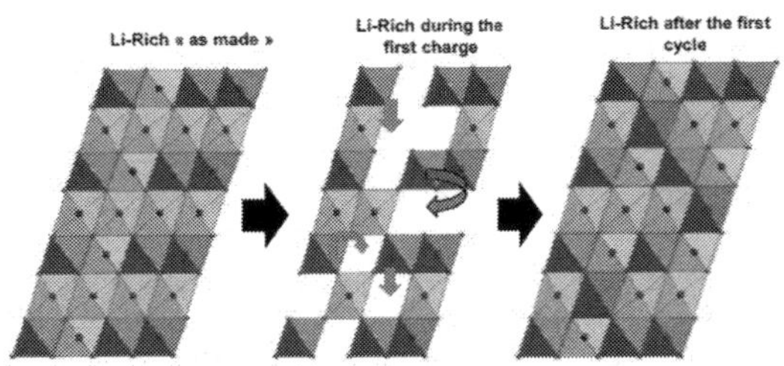

Figure 3. Illustration of the structural instability of the Lithium-rich layered materials.

DOI: 10.1051/e3sconf/20171609002

Finally an *in situ* X-ray Absorption Spectroscopy (XAS) study at ESRF coupled with fine electrochemical analyses gave the identification of the different electrochemical processes and of the link between high and low potential processes in the material [9]. These conclusions were completed and confirmed by a composition study with different Mn/Ni ratios in the aim to understand the Mn role on the voltage decay [10].

Solve this structural instability leading to voltage fading during cycle life is the priority.

As a result, CEA strategy is now to work on a modified Li-rich structure stopping voltage decay. The first results are very promising (Fig.4). Synthesis optimizations are ongoing.

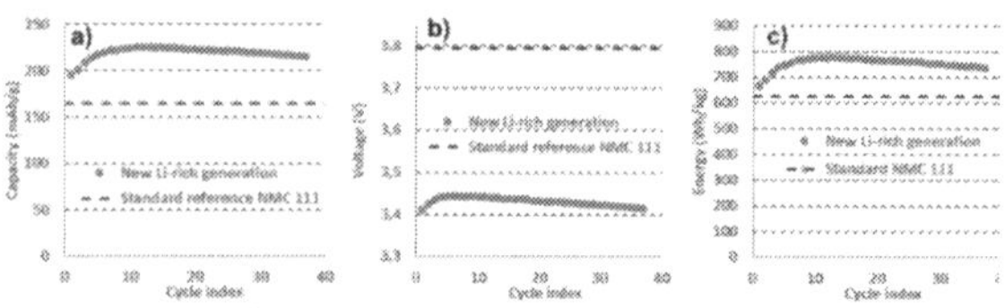

Figure 4: Electrochemical performances of the new generation of Li-rich cathode developed by CEA. Capacity (a), average potential (b) and specific energy (c) as a function of the cycle number.

2.2. Li_2MnSiO_4 material study

Four polymorphs have been reported in literature [11]. The two firsts, $Pmn2_1$ and Pmnb, are usually obtained at lower temperature and have orthorhombic lattice related to β-Li_3PO_4. The two others, $P2_1/n$ and Pn are monoclinic and are formed at higher temperature.

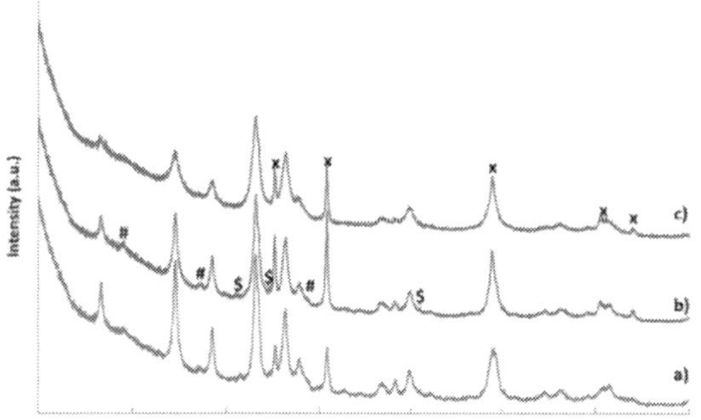

Figure 5. XRD patterns of Li_2MnSiO_4 (a), Li_2MnSiO_4/C with 3wt.% of C (b) and Li_2MnSiO_4/C with 8wt.% of C (c) ; x: MnO ; #: Li_2SiO_3 ; \$: Mn_2SiO_4.

Li_2MnSiO_4 samples were prepared by sol-gel process. Carbon coating (C-coating) was obtained in a second step, by pyrolysis of the polymer (P123) at 650°C for 30 minutes under Ar. The carbon coating ratio was controlled by the quantity of polymer added in solution. Three samples have been prepared having 0; 3 and 8 wt.% C-coating ratio, as determined by ThermoGravimetric Analysis (TGA). Fig. 5. shows the powder XRD of the three samples.

All XRD patterns correspond to Li_2MnSiO_4 material with orthorhombic phase with $Pmn2_1$ space group. However, cubic MnO impurity was also observed for all synthesis. Furthermore, Li_2SiO_3 and Mn_2SiO_4 impurities have been found for two samples (0 and 3wt.% carbon content). These impurities have been regularly reported in different papers [12, 13]. Carbon-coated materials have broader diffraction peaks than uncoated one. It shows that the coherence length is smaller in the C-coated samples. This is probably due to the action of P123 copolymer in solution, which inhibits growth and crystallization of particles [14]. However, SEM picture showed similar particle size in samples with and without C-coating (Fig.6). The primary particles were spherical with a mean diameter of 50 nm. The primary particles are agglomerated together to form micrometric size secondary particles.

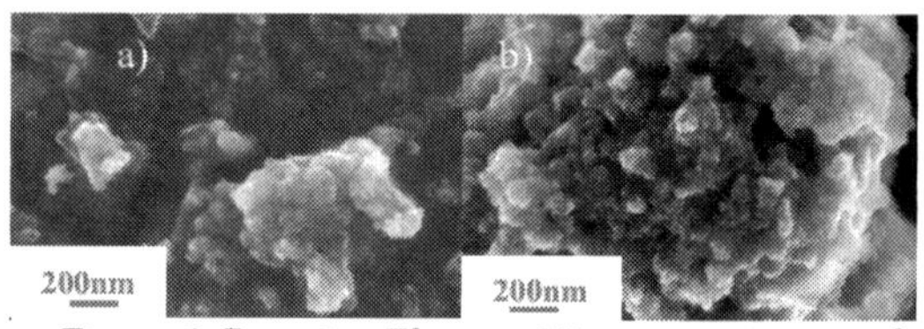

Figure 6. Scanning Electron Microscopy images of Li_2MnSiO_4 (a) and Li_2MnSiO_4/C with 8wt.% of C (b).

Electrochemical performance have been studied in coin cells vs. Li counter electrode, in galvanostatic mode, between 1.5 and 4.8 V.

Fig. 7 shows the 1st cycle obtained at C/20 (1C = 333 mA/g) for Li_2MSiO_4 without C-coating, after a 48h exposition to air or full preparation in Ar.

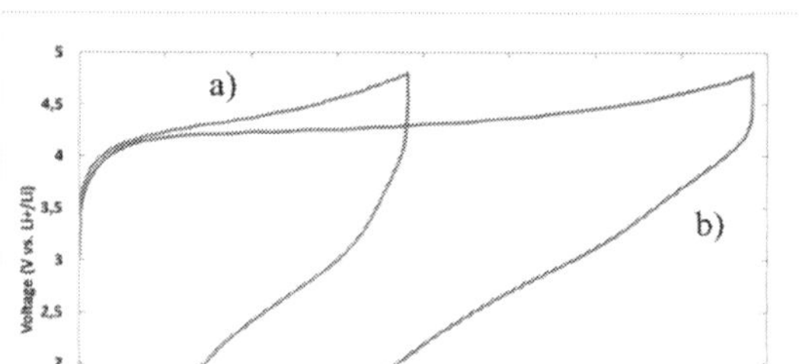

Figure 7. Galvanostatic charge and discharge between 1.5 and 4.8V of Li_2MnSiO_4 which was exposed to air 48h (a) and prepared in argon (b).

 DOI: 10.1051/e3sconf/20171609002

Charge capacity was two times higher when storage of Li_2MnSiO_4 took place in Ar atmosphere. Actually, charge capacity increased from 76 mAh/g in air to 156 mAh/g in Ar. Discharge capacities were respectively 56 mAh/g in air and 97 mAh/g in Ar. It has been shown by XRD than Li_2MnSiO_4 reacts with air to form Li_2CO_3.

Formation of Li_2CO_3 is most probably due to combined oxidation of Mn^{2+} and lithium extraction at the surface, followed by reaction of lithium with CO_2 from air to form Li_2CO_3 [15]. The presence of such isolating material explains lower performance of Li_2MnSiO_4 electrodes exposed to air. We also showed by XRD that the presence of carbon-coating slows down the reactivity of Li_2MnSiO_4 in air. However, the carbon layer does not completely protect the material.

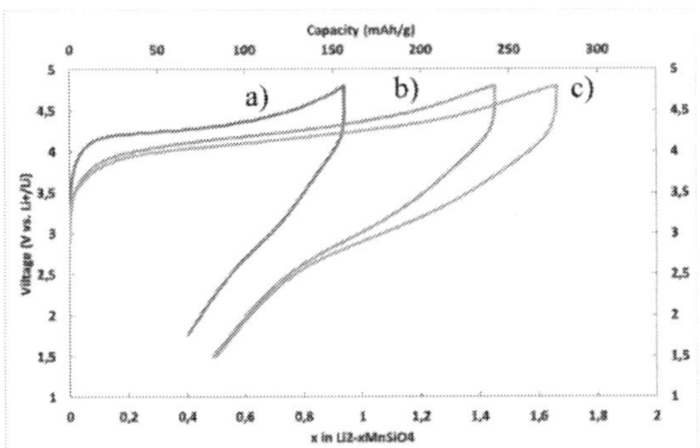

Figure 8. Galvanostatic curves between 1.5 and 4.8V of Li_2MnSiO_4/C prepared in argon with 0wt.% (a), 3wt.% (b) and 8wt.% of carbon (c).

In the next part of this study, all materials from pyrolysis to coin cells were prepared in inert atmosphere to avoid lithium extraction and insulating Li_2CO_3 at surface of particles. Fig. 8 shows the influence of carbon-coating (0; 3 and 8 wt.% of carbon) on the 1st cycle. Charge capacities reached 277, 242 and 77 mAh/g; discharge capacities reached 197 mAh/g, 161 mAh/g and 97 mAh/g for samples with respectively 8, 3 and 0 wt.% of carbon. Higher C-coating ratio clearly increased kinetics of the composite Li_2MnSiO_4/C materials. Extraction and insertion of lithium superior to 1 per formula unit is achieved for 8wt.% of carbon.

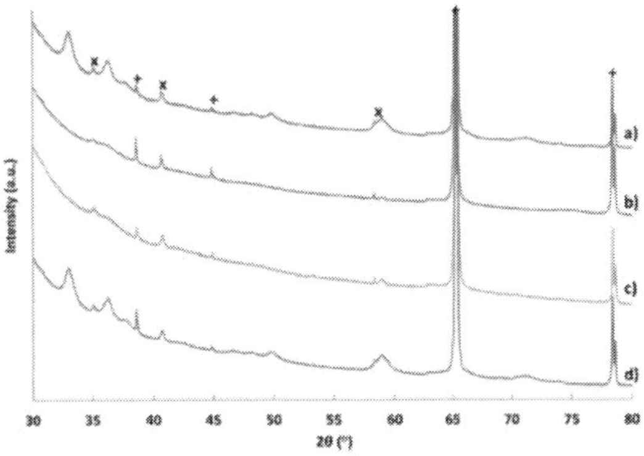

Figure 9. XRD patterns of Li_2MnSiO_4/C electrodes, after 20 cycles at C/20 1.5-4.0V (a), after 20 cycles at C/20 1.5-4.8V (b), after 1 charge at C/20 until 4.8V (c) and pristine (d) ; x: MnO ; +: Al current collector.

1st and 2nd galvanostatic cycle of Li_2MnSiO_4/C with 8wt.% of C between 1.5 and 4.8V are plotted on fig. 10. Shape of charge and discharge of cycle 2 were completely different from cycle 1. A steeple polarization is observed during 2nd charge supposing a probable major change in the Li_2MnSiO_4/C composite material during the first galvanostatic cycle. XRD measurements on Li_2MnSiO_4/C electrode before and after the first oxidation are shown in Fig.9c and d. After the first oxidation, all the diffraction peaks from Li_2MnSiO_4 vanished. Thus, a complete amorphization of the material occurred during 1st charge. Amorphization might limiting lithium diffusion. Capacity loss due to this phenomenon has been reported in the literature [16].

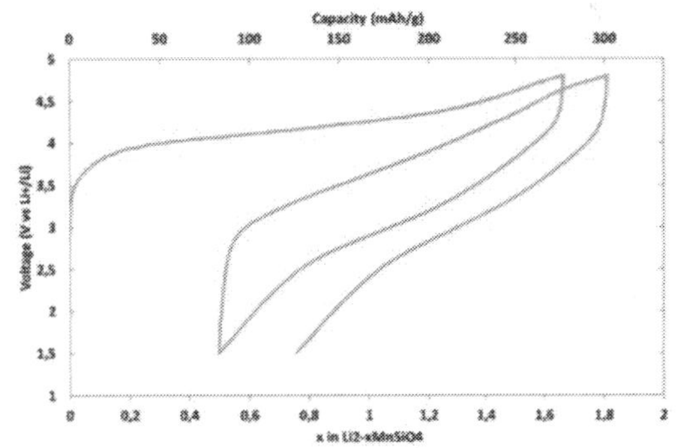

Figure 10. First and second galvanostatic cycles of Li_2MnSiO_4/C (8wt.% of C) with a voltage window of 1.5-4.8V.

 DOI: 10.1051/e3sconf/20171609002

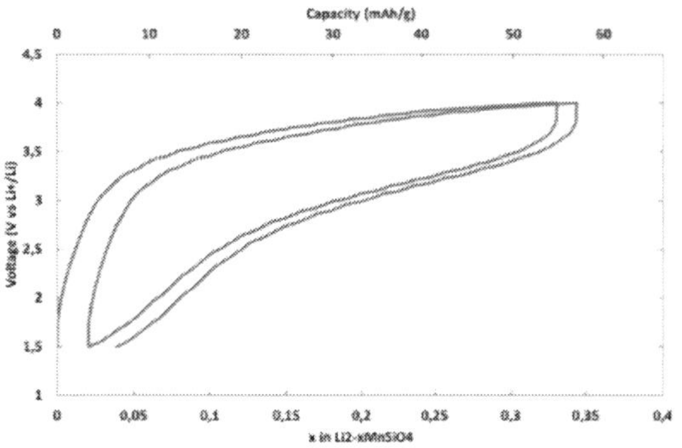

Figure 11. First and second galvanostatic cycles of Li_2MnSiO_4/C (8wt.% of C) with a voltage window of 1.5-4.0V.

One major hypothesis is the instability of Mn at oxidation state (III) and (IV) obtained during lithium extraction, which causes major distortion and finally loss of long range order into structure [17]. Actually, deep charge i.e. deep lithium extraction irreversibly damaged Li_2MnSiO_4 structure and led to gradual capacity loss. On the opposite, the shape of the charge/discharge curves was really similar in the two first cycles when the upper cut-off voltage was limited to 4.0 V vs. Li^+/Li (fig. 11). As a consequence, the capacity was stable upon cycling, as shown in Fig. 12d. However only 0.3 lithium per formula unit was exchanged, greatly limiting the capacity of the material.

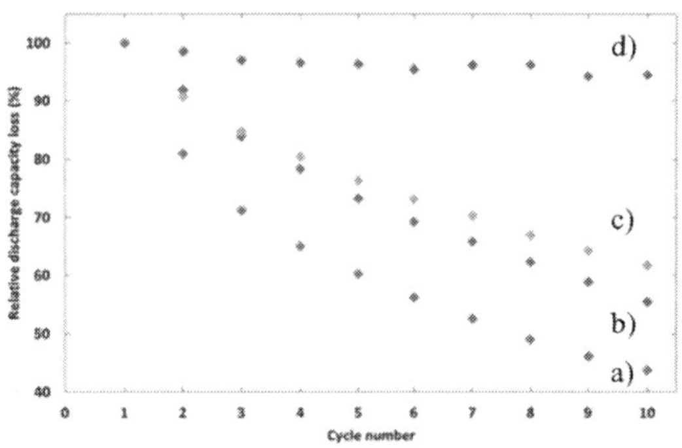

Figure 12. Relative discharge capacity loss versus cycle number of Li_2MnSiO_4 (1.5-4.8V) (d); Li_2MnSiO_4/C (3wt.% of C, 1.5-4.8V) (c); Li_2MnSiO_4/C (8wt.% of C, 1.5-4.8V) (b); Li_2MnSiO_4/C (8wt.% of C, 1.5-4.0V)(a).

XRD showed that the structure was maintained in this narrower voltage window, even after 20 cycles (Fig. 9a). It is a clue that structure is maintained. However only 0.3 lithium per formula unit is exchanged. Nevertheless XRD on electrodes after 20 cycles of charge-discharge (10 cycles at C/20 followed by 10 cycles at C/10) is performed on both voltage window to assess structure preservation. No peaks of Li_2MnSiO_4 remained in electrode cycled until 4.8V, which is supported yet by XRD pattern after one charge (Fig.9b). To the contrary structure of Li_2MnSiO_4 is kept in electrode cycled until 4.0V. It is source of good cyclability for silicates.

3. CONCLUSION

Li-rich and Lithium manganese silicate have been studied as cathode material for lithium-ion batteries.

Spherical particles of Li-rich have been synthesized by co-precipitation route. TEM and in-situ diffraction measurements analyses have been performed to understand structural modifications of the structure during cycling. The formation of an irreversible spinel phase and cation migrations upon cycling are probably responsible for performance loss of the material. Control of the Mn/Ni ratio lead to stable and reversible discharge capacity of 250 mAh/g.

Li_2MnSiO_4 was prepared by sol-gel synthesis. Careful preparation in inert atmosphere improved its electrochemical performance avoiding insulating layer of lithium carbonate around particles and lowering polarization. Up to 1.66 Li^+ ions can be reinserted in 8% carbon coated Li_2MnSiO_4 composite during the 1st discharge. Amorphization during its first charge was probably responsible for capacity loss upon cycling. Stabilization of the structure was observed using appropriate voltage window. However, only 0.3 Li^+ ions can be reversibly extracted from the structure. A new path of investigation is open and will lead to better understanding of lithium manganese silicate and maybe will provide solutions to counterbalance its structural instability.

4. ACKNOWLEDGEMENTS

CEA acknowledges financial support from Renault S.A. on the Li-Rich study. PhD work on silicates cathode is supported by ESA and Prayon.

5. REFERENCES

[1] Boulineau A., Simonin L., Colin J.-F., Canevet E., Daniel L. & Patoux S. (2012). *Chem. Mater.* 24, 3558-3566.

[2] Thackeray M.M., Kang S.-H., Johnson C.S., Vaughey J.T., Benedek R. & Hackney S.A. (2007). *J. Mater. Chem.* 17, 3112-3125.

[5] Nyten A., Abouimrane A., Armand M., Gustafsson T. & Thomas J. O. (2005). *Electrochem. Commun.* **7**(2), 156-160.

[4] Arroyo-de Dompablo, M. E.; Armand M.; Tarascon J. M. & Amador U. (2006). *Electrochem. Commun.* **8**(8), 1292-1298.

[5] Wei Q., Zheng H. & Huang Y. (2013). *ECS Transactions.* Batteries and Energy Technology (General)-221st ECS Meeting, 2012, **45**(29), 219-226.

[6] Yuan, L-X. ; Wang, Z-H; Zhang, W-X.; Hu, X-L.; Chen, J-T.; Huang, Y-H. & Goodenough, J.B.

DOI: 10.1051/e3sconf/20171609002

(2011). *Energ. Environ. Sci.* **4**(2), 269-284.
[7] Simonin L., Colin J.-F., Ranieri V., Canévet E., Martin J-F., Bourbon C., Baehtz C., Strobel P., Daniel L. & Patoux S. 2012. *J. Mater. Chem.* 22, 11316-11322.
[8] Boulineau A., Simonin L., Colin J.-F., Canévet E., Daniel L. & Patoux S. 2012 *Chem. Mater.* 24 (18), 3558–3566
[9] Boulineau A., Simonin L., Colin J.-F., Bourbon C. & Patoux S. (2013). *Nano Lett.* **13** (8), 3857–3863.
[10] Peralta D., Colin J-F., Boulineau A., Simonin L., Fabre F., Bouvet J., Feydi P., Chakir M., Chapuis M. & Patoux S. (2015). *J. Power Sources*. 280, 687-694
[11] Gummow R. J. & He Y. (2014). *J. Power Sources*. 253, 315-331.
[12] Dominko R., Bele M., Gaberscek M., Meden A., Remskar M. & Jamnik J. (2006). *Electrochem. Commun.* **8**(2), 217–222.
[13] Mali G., Meden A. & Dominko R. (2010). *Chem. Commun.* 46, 3306–3308.
[14] Zhu H., He H., Xin X., Ma X., Zan L. & Zhang Y. (2015). *Electrochim. Acta.* 155, 116-124.
[15] Matsumoto K., Kuzuo R., Takeya K. & Yamanaka A. (1999). *J Power Sources.* 81-82, 558-561.
[16] Dominko R., Bele M., Kokalj A., Gaberscek M. & Jamnik J. (2007). *J Power Sources*. 174, 457–461.
[17] Saracibar A., Wang Z., Carroll K. J., Meng Y. S. & Dompablo M. E. Arroyo-de. (2015). *J. Mater. Chem. A.* **3**(11), 6004-6011.

CARBON TOLERANT FUEL ELECTRODES FOR REVERSIBLE SOFC OPERATING ON CARBON DIOXIDE

Kalliopi Maria Papazisi [(1)], Dimitrios Tsiplakides [(1)], Stella Balomenou [(1)]
Ivar Wærnhus [(2)], Crina S. Ilea [(2)], Arild Vik [(2)], Max Schautz [(3)]

[(1)] CPERI/CERTH, 6th klm Charilaou-Thermi Rd., GR 57001 Thessaloniki, Greece, Email: papazisi@cperi.certh.gr
[(2)] CMR Prototech, Fantoftvegen 38, 5072 Bergen, Norway, Email: ivar.warnhus@prototech.no
[(3)] European Space Research and Technology Centre (ESTEC), The Netherlands, Email: Max.Schautz@esa.int

ABSTRACT

A challenging barrier for the broad, successful implementation of Reversible Solid Oxide Fuel Cell (RSOFC) technology for Mars application utilizing CO_2 from the Martian atmosphere as primary reactant, remains the long term stability by the effective control and minimization of degradation resulting from carbon built up. The perovskitic type oxide material $La_{0.75}Sr_{0.25}Cr_{0.9}Fe_{0.1}O_{3-\delta}$ (LSCF) has been developed and studied for its performance and tolerance to carbon deposition, employed as bi-functional fuel electrode in a Reversible SOFC operating on the CO_2 cycle (Solid Oxide Electrolysis Cell/SOEC: CO_2 electrolysis, Solid Oxide Fuel Cell/SOFC: power generation through the electrochemical reaction of CO and oxygen). A commercial state-of-the-art NiO-YSZ (8% mol Y_2O_3 stabilized ZrO_2) cermet was used as reference material. CO_2 electrolysis and fuel cell operation in 70% CO/CO_2 were studied in the temperature range of 900-1000°C. YSZ was used as electrolyte while LSM-YSZ/LSM ($La_{0.2}Sr_{0.8}MnO_3$) as oxygen electrode. Results showed that LSCF had high and stable performance under RSOFC operation.

1. INTRODUCTION

Solid Oxide Fuel Cells (SOFC) and Electrolyzers (SOEC) hold great potential for the security of energy sector, providing solutions for efficient and environmentally-friendly electricity generation, energy storage, carbon capture and chemical synthesis. The anticipated progress that can pave the way to practical application of SOFC/SOEC technology will emerge from specific scientific and engineering advances addressing cost reduction and improvement of performance and stability of cells and units [1].

A key priority for research and development is to fully exploit and improve the fuel flexibility advantage of SOFC/SOEC able to operate on a wide range of fuels and particularly carbon-containing fuels such as mixture of carbon monoxide and carbon dioxide. Ni containing cermets based on YSZ (8% mol Y_2O_3 stabilized ZrO_2), ScSZ (Sc_2O_3 stabilized ZrO_2) and/or rare earth or yttrium doped ceria are the predominant anode materials, however, upon prolonged operation under practical feeds and conditions, degradation occurs caused by (i) redox instability of the cermet when exposed to oxidizing atmosphere (volume change and coarsening of the Ni-phase) (ii) agglomeration of Ni particles due to either thermal or overpotential sintering, (iii) unfavorable deposition of carbon onto nickel through.

Remedies for the anode tolerance enhancement focus on the development and deployment of alternative materials, such as perovskite structures and ceramic composites [2]. The introduction of $CeO_{2-\delta}$ based additives and layers has been considered as one of the most promising approaches in anode development, and Gadolinium Doped Ceria ($CeO_2(Gd_2O_3)$, GDC) is frequently used as a substitute of YSZ. The advantage of ceria compared to YSZ is associated with its high catalytic activity for combustion reactions, e.g. CO and hydrocarbon oxidation [2, 3].

The role of SOFCs in the energy sector goes beyond their conventional operation and utilization for electricity generation; thus the development and realization of new features and functionalities for SOFC/SOEC devices are ranked high in the research agenda. In this context the concept of reversible or regenerative SOFC (RSOFC) has attracted great interest as a means for sustainable energy storage. A reversible solid oxide fuel cell is a device that can operate in both fuel cell mode, i.e. for generating electricity through the electrochemical combination of a fuel (e.g. H_2 or CO) with oxygen, and in electrolysis mode, i.e. for the production of fuel (e.g. hydrogen from water, CO from CO_2, or syngas from mixtures of water and CO_2) when the needed energy is provided by renewable sources. Given the reversible operation of a single solid state device, the need for efficient performance and stability, as well as reversibility of electrodes operation is crucial. A RSOFC operating on the CO_2 cycle (SOEC mode: electrolysis of CO_2, SOFC mode: electrochemical reaction of CO and oxygen for power generation), faces similar problems regarding degradation and carbon built up, as mentioned above for the C-containing fueled SOFCs. In this study, under current ESA activity in collaboration with Prototech CMR (4000108849/13/NL/EK), LSCF ($La_{0.75}Sr_{0.25}Cr_{0.9}Fe_{0.1}O_{3-\delta}$) has been developed and tested under RSOFC operation on the CO_2 cycle in comparison with the state-of-art Ni-YSZ fuel electrode. Testing under C-forming conditions, showed promising stable performance of the perovskitic electrode compared to a commercial Ni-YSZ.

2. EXPERIMENTAL

2.1 Button Cells Preparation

All experiments were carried out on 8YSZ electrolyte-supported cells. The membranes had a thickness of either 1.5 ± 0.5 mm or 300 μm and a diameter of 20 mm. The surface of both anode and cathode electrodes was ~1.76 cm^2 (diameter=15 mm). All oxygen electrodes, LSM-YSZ/LSM (anode under SOEC and cathode under SOFC operation), were prepared with the use of commercial inks (Fuel Cell Materials). Commercial ink was also used in the case of the state-of-the-art NiO-YSZ fuel electrode, while laboratory synthesized powders were used for the perovskitic LSCF fuel electrodes. Application of the powders was achieved by means of an ink prepared using a terpineol-based ink vehicle. In all cases screen printing was applied for electrodes' development. The ink used consisted of the powder material well-mixed (ball milling) with the ink vehicle in a 60-70%wt total solids content.

A standard Ni-YSZ//YSZ//LSM-YSZ/LSM cell was used as a performance reference and in order to standardize procedures (NIYSZ-I) commercial ink, composition: 66% NiO - 36% $(Y_2O_3)_{0.08}(ZrO_2)_{0.92}$, solids loading 73-77% wt., was employed.

The as-prepared perovskite powders LSCF were developed under previous ESA activity (Contract No. 21767/08/NL/LvH) and used as fuel electrodes. In the case of perovskite electrodes, a thin gadolinium-doped ceria (GDC20-TC, $Gd_{0.20}Ce_{0.80}O_{2-x}$) interlayer was applied between the fuel electrode and the YSZ electrolyte in order to avoid formation of $La_2Zr_2O_7$ (a non-conducting phase), thus ensuring high cell efficiency and prolonged lifetime.

The procedure followed for electrodes' development is:

1. Screen printing of the fuel electrode

Typically, up to 5 layers with intermediate drying steps at 105°C for about 15 min were applied in all fuel electrodes in order to achieve a loading of 10-15 mg cm^{-2}.

2. Firing/sintering of the fuel electrode

Screen printing was followed by an appropriate firing/sintering step depending on the material in order to have a well performing electrode. In the case of the cells were CGO interlayer was applied, this interlayer was screen printed and sintered before the perovskitic material. An additional screen printed layer of platinum was applied above the perovskitic material to ensure good current collection. The sintering procedure followed for the different electrodes was:

- NiO-YSZ: 500°C for 2 h with a heating rate of 2°C min^{-1} and then at 1350°C for 2 h with the same heating rate.
- CGO: 500°C for 2 h with a heating rate of 2°C min^{-1} and then at 1300°C for 5 h with the same heating rate.
- LSCF: 500°C for 2 h with a heating rate of 2°C min^{-1} and then at 1200°C for 3 h with the same heating rate.
- Pt: 1000°C for 1 h with a heating rate of 1°C min^{-1}.

3. Screen printing of the oxygen electrode

Next step is the development of the oxygen electrode which consists of the cermet electrode LSM-YSZ and LSM as current collector. For this purpose, the first step is the application of LSM-YSZ layers followed by the application of LSM layers with intermediate drying at 105°C for 15 min.

4. Firing/sintering of the oxygen electrode

The final step is firing/sintering of the cathode (and consequently the full cell) according to the following procedure: 500°C for 2 h with a heating rate of 2°C min^{-1} and then at 1150°C for 3 h.

2.2 Button Cells Testing-Electrochemical Measurements

The electrochemical characterization and button cells' performance evaluation under both CO_2 electrolysis and $CO:CO_2/O_2$ fuel cell mode was realized in a Probostat™ (NorECs) test cell using standard two electrode setup using platinum meshes as current collectors. Gases were fed in anode and cathode chambers directly onto corresponding electrodes, through mass flowmeters (Bronkhorst). More specifically, during electrolysis mode CO_2 and He were fed in the cathode and anode chamber, respectively, while under fuel cell mode a mixture of $CO:CO_2$ was supplied in the anode while O_2 was fed in the cathode. An Autolab PGSTAT302N (Metrohm), equipped with a frequency response analyzer (FRA32M), was used for the electrochemical measurements and for the performance of potentiostatic impedance measurements over a frequency range of 100 kHz to 0.01 Hz and 10 mV ac amplitude. The inner and the outer compartment of the cell were connected to mass flowmeters for the supply of the appropriate gasses. During fuel cell operation H_2 or CO/CO_2 mixture (70/30 % vol.) was fed to the anode (fuel electrode) and oxygen (100%) to the cathode (oxygen electrode), while during electrolysis mode CO_2 was fed to the cathode (fuel electrode) and He to the anode in order to remove the produced oxygen.

3. RESULTS AND DISCUSSION

3.1 Electrochemical measurements obtained for button cells with the thick 1.5 mm YSZ electrolyte

All following performances refer to button cells developed using the thick 1.5 mm YSZ electrolyte. Fig. 1 presents the steady state I-V curves obtained for the button cell Ni-YSZ//YSZ//LSM-YSZ/LSM under CO_2 electrolysis using commercial state-of-the art materials. The enhanced performance by increasing temperature is obvious, reaching a current density of 293 mA cm^{-2} at 1.1 V operation potential. Fig. 2, which presents the Nyquist plots obtained under 1.1 V dc potential shows that resistance series is decreasing by increasing operation temperature. The deviating EIS at 950°C is probably attributed to deterioration of the cell performance which could be related to partial nickel re-oxidation due to CO_2 presence.

Fig. 3 shows the I-V curves obtained under CO-CO_2

 DOI: 10.1051/e3sconf/20171609003

(70-30%vol)/O_2 fuel cell operation at the same temperature range. The same trend of enhanced performance by increasing temperature is observed. At 1000°C the current density obtained is 140 mA cm^{-2} at 0.7 V which is much lower compared to H_2/O_2 fuel cell (295 mA cm^{-2}) at the same conditions. Apart from this, the button cell suffered from severe degradation resulting in more than 50% loss of its initial performance (indicated as fresh) after only 2 h of operation at 900°C, as shown in Fig. 4. This effect was expected as Ni cermet electrodes suffer from carbon deposition. This is also obvious by the Nyquist plots shown in Fig. 5 where a significant increase of the polarization resistance after 2 h of operation under CO-CO_2 mixture to the fuel electrode is observed.

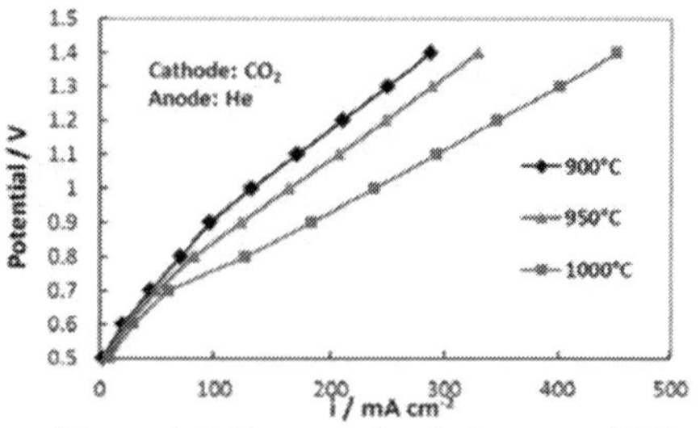

Figure 1. I-V curves for the button cell Ni-YSZ//YSZ//LSM-YSZ/LSM prepared using commercial state-of-the art materials at 900, 950 and 1000°C under CO_2 electrolysis operation.

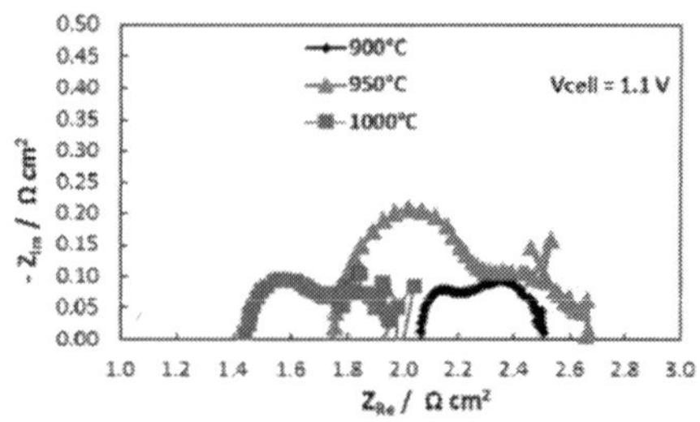

Figure 2. Nyquist plots for the button cell Ni-YSZ//YSZ//LSM-YSZ/LSM prepared using commercial state-of-the art materials at 900, 950 and 1000°C under 0.7 V dc potential under CO_2 electrolysis operation.

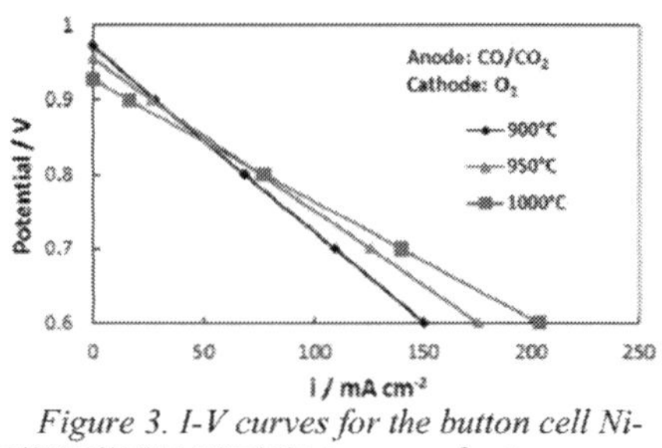

Figure 3. I-V curves for the button cell Ni-YSZ//YSZ//LSM-YSZ/LSM prepared using commercial state-of-the art materials at 900, 950 and 1000°C under CO-CO_2/O_2 fuel cell operation.

The corresponding IV curves and Nyquist plots for the button cell employing LSCF, i.e. of the type LSCF/CGO//YSZ//LSM-YSZ/LSM, at the temperature range 900-1000°C under CO_2 electrolysis are presented in Figs. 6 and 7, respectively. At 1000°C and 1.1 V the current density is 305 mA cm^{-2} which decreases to 161 mA cm^{-2} at 900°C. According to Nyquist plots (Fig. 7), series resistance decreased from 2.18 to 1.42 Ω cm^2 and polarization from 0.73 to 0.40 Ω cm^2 by increasing temperature from 900 to 1000°C.

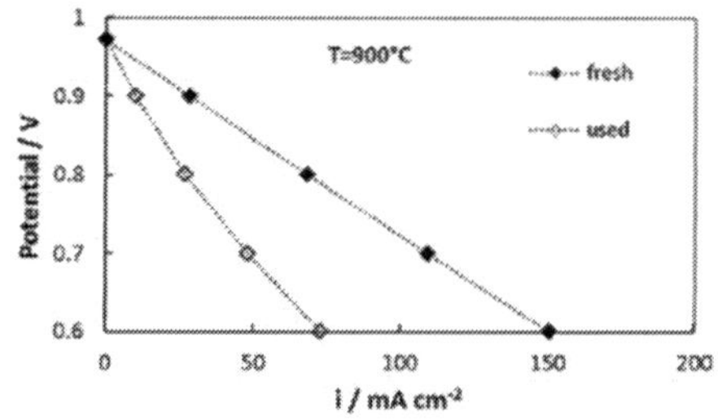

Figure 4. Comparison of the current density-voltage curves obtained for the button cell Ni-YSZ//YSZ//LSM-YSZ/LSM prepared using commercial state-of-the art materials at 900°C under CO-CO_2/O_2 fuel cell operation on its fresh and after 2 h of operation (used) state.

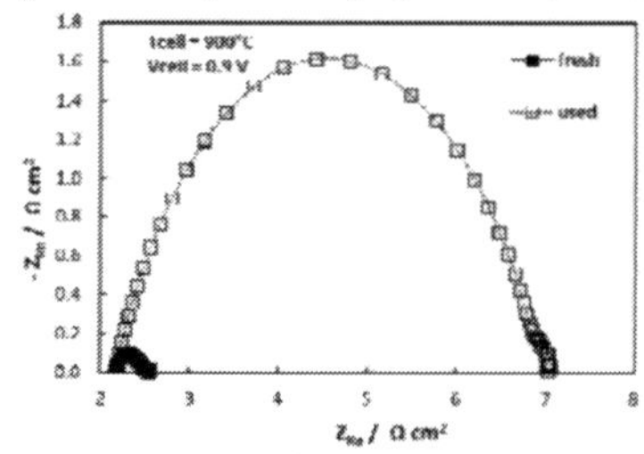

Figure 5. Nyquist plots for the button cell Ni-YSZ//YSZ//LSM-YSZ/LSM prepared using commercial state-of-the art materials at 900oC under 0.9 V dc potential under CO-CO_2/O_2 fuel cell operation on its fresh and after 2 h of operation (used) state.

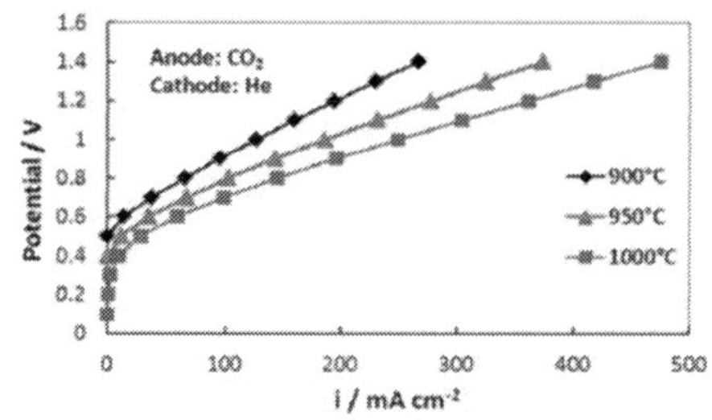

Figure 6. I-V curves for the button cell LSCF/CGO//YSZ//LSM-YSZ/LSM at 900, 950 and 1000°C under CO_2 electrolysis operation.

The current density obtained at 0.7 V under CO-CO_2 fueled SOFC with the perovskitic anode LSCF is 128 mA cm^{-2} at 1000°C as shown in Fig. 8. This performance is similar to that obtained under H_2/O_2 fuel cell operation

DOI: 10.1051/e3sconf/20171609003

(123 mA cm^{-2}, not presented here) indicating that this material performs the same under H_2 and carbon containing fuels, which in turn indicates not only high electrocatalytic efficiency for H_2 and CO oxidation but also sufficient tolerance to carbon deposition and potential for reversibility. The Nyquist plots presented in Fig. 9, show that both series and polarization resistance increase by decreasing temperature. It is worth noting that compared to Ni-YSZ button cell, LSCF presents lower series resistance and significant higher stability during all modes of operation and fuels used.

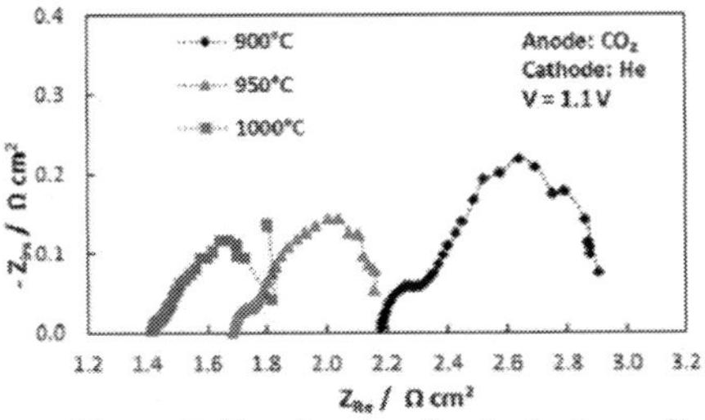

Figure 7. Nyquist plots for the button cell LSCF/CGO//YSZ//LSM-YSZ/LSM at 900, 950 and 1000°C under 0.7 V dc potential under CO_2 electrolysis operation.

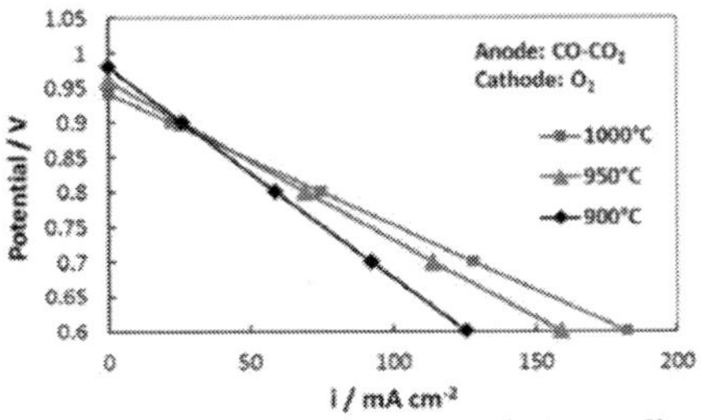

Figure 8. I-V curves for the button cell LSCF/CGO//YSZ//LSM-YSZ/LSM at 900, 950 and 1000°C under CO-CO_2 fuel cell operation.

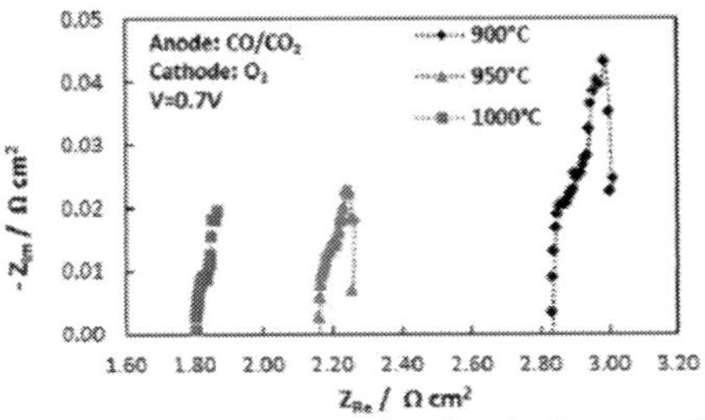

Figure 9. Nyquist plots for the button cell LSCF/CGO//YSZ//LSM-YSZ/LSM at 900, 950 and 1000°C under 0.7 V dc potential under CO-CO_2 fuel cell operation.

3.2 Electrochemical measurements obtained for the button cell LSCF/CGO//YSZ//LSM-YSZ/LSM with 300 μm YSZ electrolyte

As already discussed, the best performed fuel electrode materials in terms of current density and stability was LSCF. In order to further evaluate its performance in cells with thinner electrolytes, cells with YSZ of 300 μm thickness were developed.

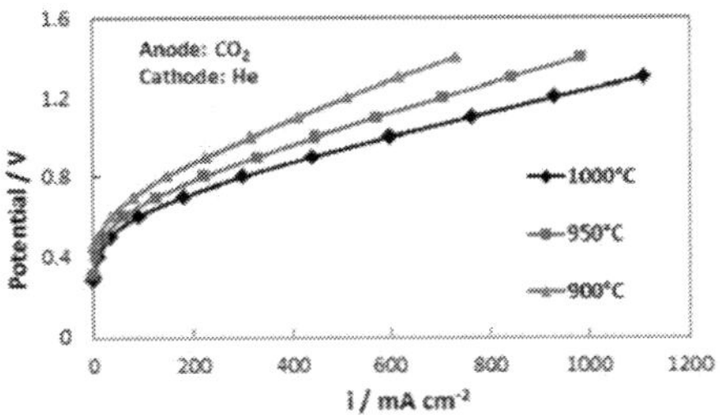

Figure 10. I-V curves for the button cell LSCF/CGO//YSZ//LSM-YSZ/LSM (YSZ=300μm) under CO_2 electrolysis operation.

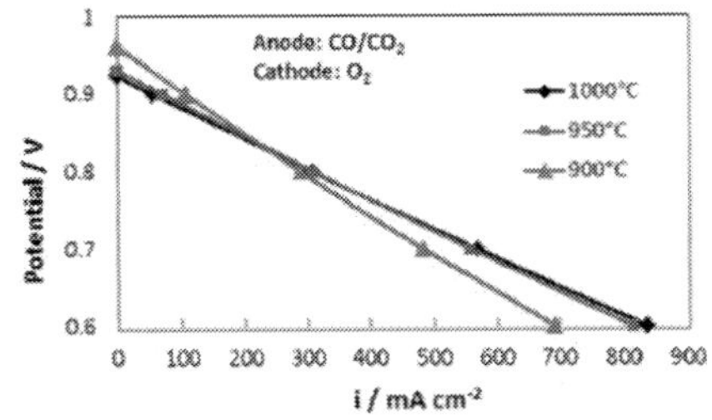

Figure 11. I-V curves for LSCF/CGO//YSZ//LSM-YSZ/LSM (YSZ=300μm) under CO-CO_2 fuel cell operation.

The performance of such cells is presented in Fig. 10 and 11, for the electrolysis and fuel cell mode, respectively. At 1000°C, under CO_2 electrolysis, current density was 762 mA cm^{-2} at 1.1 V, while under FC operation it was 378 mA cm^{-2} at 0.7 V. The degradation rate under constant potential application (1.1 V) was 0.2 mA cm^{-2} min^{-1} during electrolysis operation while no degradation was measured under fuel cell operation, indicating a very stable material for reversible operation in an RSOFC fueled with CO_2.

An accelerated stability test under RSOFC operation was conducted at 1000°C for a button cell of the same type, LSCF/CGO//YSZ//LSM-YSZ/LSM. In that case, a constant current was applied under both operations (CO_2 electrolysis and CO-CO_2/O_2 fuel cell mode) and the voltage versus time was recorded. The procedure (accelerated test profile) followed was: **(a)** Electrolysis operation by feeding CO_2 and He at the fuel and oxygen electrode, respectively, under constant current operation, 0.5 A cm^{-2}, for 20 min and **(b)** Fuel cell operation by feeding CO/CO_2 (70/30) and O_2 at the fuel and oxygen electrode, respectively, under constant current operation, 0.34 A cm^{-2}, for 30 min.

The operation time and current density for each mode were selected in order to produce (under electrolysis) and consume (under fuel cell) the same amount of gases.

Totally, 32 reversible cycles were recorded. Fig. 12a presents the performance of the cell under CO_2 electrolysis while Fig. 12b under CO-CO_2 fuel cell

DOI: 10.1051/e3sconf/20171609003

operation. The cell exhibits high initial performance in both CO_2 SOEC (762 mA/cm^2 at 1.1V) and CO/CO_2 SOFC (378 mA/cm^2 at 0.7V) operation. At the same time, it exhibits remarkable stability, as negligible degradation is observed during the whole duration of normal tests as well as the accelerated test profile of 32 cycles. Therefore, cell's reversibility under cycling between the two operation modes is clearly manifested.

4. CONCLUSIONS

During this study it is concluded that although the state-of-the-art Ni-YSZ electrode presents high activity for both CO_2 electrolysis and CO/CO_2 vs. O_2 fuel cell operation it suffers from significant degradation initiated during electrolysis of CO_2, and resulting in >50% performance loss within the first complete cycle of measurements. This severe degradation is a combined result of the instability of Ni–YSZ in CO–CO_2 environment as well as of the evolvement of carbon deposition, in the form of Ni carbonyls. The later species are highly volatile, making it important to choose operating conditions for which carbonyl formation is less favorable. Moreover, Ni–YSZ composites are severely damaged by re-oxidation caused by pure CO_2. In a practical terrestrial system this problem would be tackled by addition of CO or H_2, to the electrolyzer feed in order to ensure reducing gas composition for the fuel electrode. On the contrary the laboratory synthesized LSCF perovskite had a more stable performance. The steady state IV curves and AC impedance measurements of the prepared cells under both CO_2 splitting and CO oxidation revealed the superior performance. Cells with LSCF showed high initial performance in CO_2 SOEC and CO/CO_2 SOFC operation and exhibited remarkable stability and reversibility under cycling between operation modes.

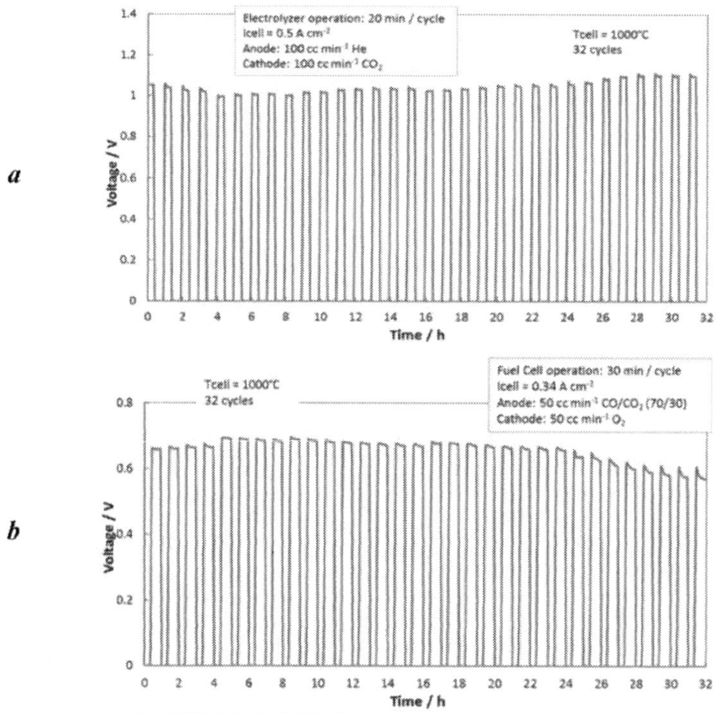

Figure 12. Voltage recorded for LSCF/CGO//YSZ//LSM-YSZ/LSM (YSZ=300μm) cell a. under CO_2 electrolysis operation (I_{cell}=0.5 A cm^{-2}) and b. under CO-CO_2 fuel cell operation (I_{cell}=0.34 A cm^{-2}).

5. REFERENCES

1. Nicholas, J.D. (2013). Highlights from the 2013 National Science Foundation Solid Oxide Fuel Cell Promise, Progress, and Priorities (SOFC-PPP) Workshop. *Interface* 22(4), 49-54.

2. Papazisi, K.M., Balomenou, S., Tsiplakides, D. (2010). Synthesis and characterization of $La_{0.75}Sr_{0.25}Cr_{0.9}M_{0.1}O_3$ perovskites as anodes for CO-fuelled solid oxide fuel cells. *J. Appl. Electrochem.* 40, 1875-1881.

3. Sapountzi, F.M., Brosda, S., Papazisi, K.M., Balomenou, S.P., Tsiplakides, D. (2012). Electrochemical performance of $La_{0.75}Sr_{0.25}Cr_{0.9}M_{0.1}O_3$ perovskites as SOFC anodes in CO/CO_2 mixtures. *J. Appl. Electrochem.* 42, 727-735.

INVESTIGATION OF ADVANCED COMPONENTS IN A HIGH PRESSURE SINGLE-CELL ELECTROLYSER FOR THE DEVELOPMENT OF A HP-PEM-ELY STACK AS PART OF A REGENERATIVE FUEL CELL SYSTEM

Dimitrios K. Niakolas (1), Stylianos Neophytides (1), Constantinos G. Vayenas(2), Alexandros Katsaounis(2), Nikolaos Athanasopoulos(3), Stella Balomenou (4), Kalliopi-Maria Papazisi (4), Dimitrios Tsiplakides (4), Max Schautz (5)

(1) Foundation for Research and Technology, Hellas (FORTH), Greece, Email: niakolas@iceht.forth.gr
(2) Department of Chemical Engineering, University of Patras, Greece, Email: cgvayenas@upatras.gr
(3) Advent Technologies SA, Greece, Email: nikos.athanasop@gmail.com
(4) Centre for Research and Technology - Hellas (CERTH), Greece, Email: stellab@cperi.certh.gr
(5) European Space Research and Technology Centre (ESTEC), The Netherlands, Email: max.schautz@esa.int

ABSTRACT

The objective of the presented work, done under current ESA activity (Contract No. 4000109578/13/NL/SC), is the performance and tolerance evaluation of selected components and materials for the development of a High Pressure, Polymer Electrolyte Membrane (PEM) Electrolyser (HP-PEM-ELY) Stack, aiming to operate at 80 bar with a performance output of 0.3 A/cm^2 at 1.6 V. An extensive study was performed on a single-cell high pressure PEM electrolyser manifold, leading to a list of materials with suitable properties and engineering solutions towards operation in space environment. This investigation provided the necessary feedback for the design of a HP-PEM-ELY stack, which is also discussed. The ultimate target of the current ESA activity is to implement research findings, develop and operate a complete regenerative fuel cell system, comprising of a High Temperature Fuel Cell Stack and the HP-PEM-ELY stack. System aspects are briefly discussed.

1. INTRODUCTION

Regenerative fuel cells offer significant advantages over rechargeable batteries as energy storage devices, especially in applications where high specific energy density is required as in the case of telecommunication satellites. A closed loop regenerative PEM fuel cell system (RPEMFCS) operates as a high pressure PEM electrolyser when excess power is available from the solar array (charging mode/reactant regeneration) and as a PEM fuel cell when the solar array does not generate power i.e. when the satellite is in eclipse (discharge mode). It has to be mentioned that operation of an electrolyser at high pressure is advantageous due to (i) reduction of internal cell resistance, (ii) reduction of volume of gas bubbles, which facilitates the transportation of liquid water and (iii) production of compressed H_2 and O_2, which makes the electrolyser to perform at the same time as a gas compressor.

Regenerative fuel cells, of a discrete or a unitized design. have been considered as potential energy storage solution for future space missions [1]. The ultimate objective of the current ESA activity is the research and development of a regenerative PEM fuel cell (RPEMFC) system for space applications, targeting at (a) large telecommunication platforms in geostationary orbit, GEO (Commercial application) and (b) planetary exploration missions (Science application). Based on the materials research and development, which has been carried out up to now, two stacks (a high-pressure PEM electrolyser and a high-temperature PEM fuel cell) are under manufacturing and preliminary testing, aiming to be incorporated in a regenerative fuel cell test bench for further evaluation [2]. Complete testing of the system will be performed under a predefined load profile corresponding to the needs of a telecom platform in GEO orbit.

2. MATERIALS TESTING AND ELECTROCHEMICAL MEASUREMENTS ON A SINGLE HP-PEM-ELY

The single-cell HP-PEM-ELY experiments employed commercial, low temperature, 5x5 cm^2 reinforced Catalyst Coated Membranes (CCMs - FumaTech) with Pt-based (cathode) and IrOx-based (anode) electrodes. It has to be mentioned that several materials have been examined as materials or components, including various types of current collectors, sealing materials, bipolar plates and CCMs of various thicknesses. Furthermore, a complete study on efficient and stable Ir-based electrocatalysts for the Oxygen Evolution Reaction (OER), which has been performed under the current and previous ESA activities (Contracts 21113/07/NL/LvH and 22329/09/NL/CBI) is behind the targeted development. In this study, the results mainly deal with those components that have been finally selected to be applied in the development of the stack including: (i) properly treated Titanium plates as O_2 current collector, (ii) carbon paper GDL (Freudenberg) as H_2 current collector, (iii) sealing materials based on FKM200

DOI: 10.1051/e3sconf/20171609004

(Freudenberg), PTFE, etc. and (iv) properly treated Titanium film for the cathode (H_2) side bipolar plate.
Fig. 1 shows the single-cell HP-PEM-ELY manifold, operated in a lab-bench top facility with pressure equilibration for anode and cathode chambers, so that all measurements there performed under no differential pressure. The selected O_2 current collector is placed inside a carbon bipolar plate.

Figure 1. Single-cell High Pressure ELY reactor.

The first noteworthy remark from the single cell measurements came out by using under preliminary tests carbon paper (with GDL) as current collector in both electrodes. During the electrochemical measurements, depicted in Fig. 2, there was a continuous degradation of the anode (O_2) current collector, accompanied with the observation of a black-colored outlet feed from the O_2 side. The degradation of the O_2 current collector is attributed to the oxidation of carbon by the evolved O_2, towards the formation of CO_2, which further induces the mechanical disintegration of the current collector. Moreover, it explains, up to a certain extent, the fact that by increasing the operating pressure, apart from the overpotential increase in the ohmic region of the I-V curves, there seems to be a shift of the slope in the activation overpotential region.

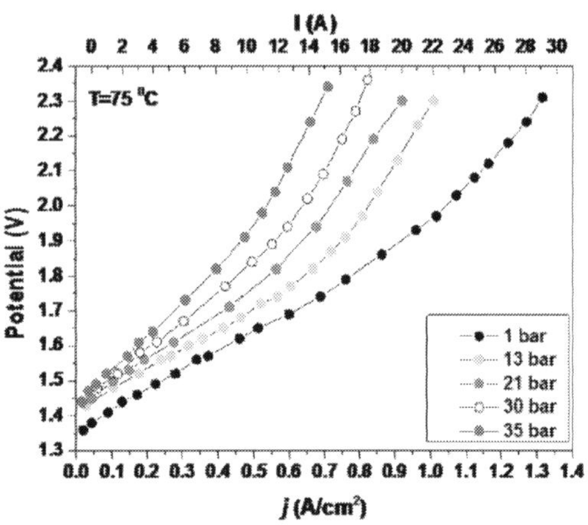

Figure 2. Comparative I-V plots of an MEA which comprises carbon paper as O_2 and H_2 current collector.

In terms of performance, the results in Fig. 2 show that above 13 bar of operating pressure the specific membrane electrode assembly (MEA) cannot reach the desired performance, (0.3 A/cm^2 at 1.6 V), and this is primarily attributed to the degradation of the O_2 current collector, whereas the thickness of the membrane (250 μm) is as additional factor.
The best performing MEAs under high pressure resulted by applying a combination of: (i) new type of sealing materials (e.g. PTFE and other), (ii) properly treated porous Titanium plate as O_2 current collector and (iii) carbon paper as H_2 current collector. Fig. 3 shows the comparison in performance of two different MEAs at P=1 bar and T=73 ^{0}C. Their main difference lies on the fact that one MEA has carbon paper as H_2 current collector, while the other comprises carbon paper and properly treated Ti foil. The thickness of both MEAs is approximately 150 μm and the sealing has been made through a combination of PTFE gaskets. Moreover, both MEAs have been assembled in such a way, so that they have the same thickness overall.
I-V plots in Fig. 3 show that both MEAs have similar performance and it is obvious that the presence of treated Ti foil on the H_2 side does not have a negative effect. On the contrary, it seems that it facilitates the overall performance, especially at high current densities. This is most probably attributed to the fact that it ensures better contact of the carbon paper with the electrode and facilitates the current collection. Thus, treated Ti foil as an additional current collector on the H_2 side seems as a promising solution for the HP-PEM-ELY stack.

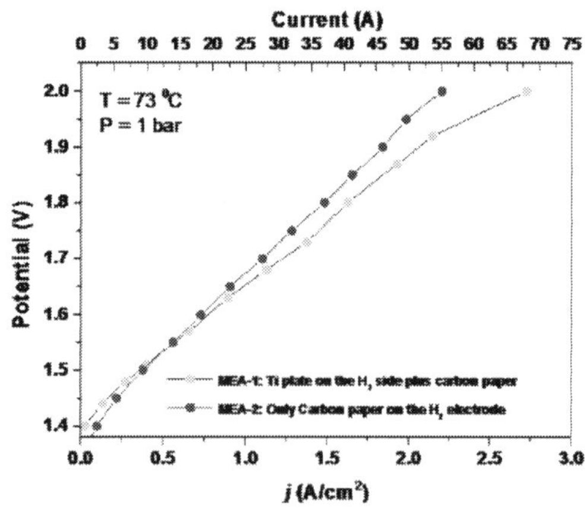

Figure 3. I-V curves of two different MEAs that comprise treated sintered Ti plate as O_2 current collector. MEA-1 comprises carbon paper and treated Ti foil as H_2 current collector, while MEA-2 has only carbon paper with GDL.

The same MEA-1 in Fig. 3, with the treated Ti foil on the H_2 side, was also examined under high pressure. In addition, it was left inactive for approximately 1 month

 DOI: 10.1051/e3sconf/20171609004

in order to find out whether it would need a second re-activation stage or if the 1st activation process, which always takes place prior its operation, is sufficient. Fig. 4 shows the performance before and after one (1) month of idleness, including the operation at high pressure.

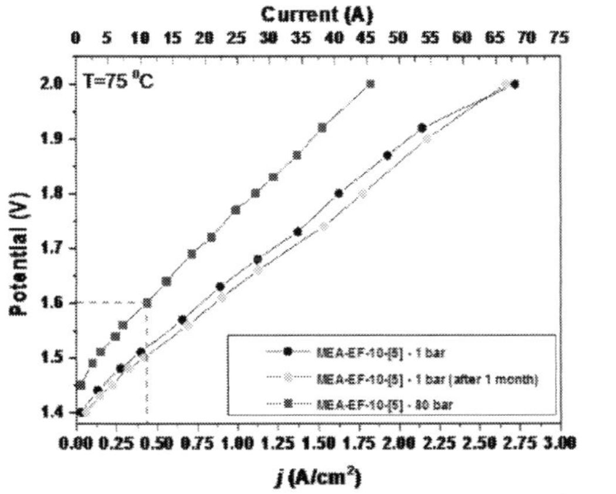

Figure 4. I-V curves of MEA-1, which comprises treated sintered Ti plate as O_2 current collector and treated Ti foil plus carbon paper as H_2 current collectors.

The first interesting remark from Fig. 4, is that the 1st activation step is enough and despite the fact that the electrolyzer was idle for approximately one month this did not affect its performance, which was quite similar at P= 1 bar and T=75 °C. The second important remark is that it managed to operate on a stable basis at P=80 bar with a current density of 0.44 A/cm^2 at 1.6 V, which is higher than the required (0.30 A/cm^2 at 1.6 V), yielding thus an efficiency of 89%. Moreover, the whole I-V plot shows low overpotential at 80 bar. Specifically, there is a rational increase of the Open Circuit Potential, due to the high operating pressure value (80 bar), but the activation, ohmic and concentration overpotentials have not significantly increased. There seems to be a degradation, which can be attributed to ohmic losses that could be related to gradual loosening of the reactor at P = 80 bar, which however did not result in important gas leakage.

The above thorough measurements have provided necessary feedback and a selected list of materials/components and solutions for the design and manufacture of the HP-PEM-ELY stack.

3. DESIGN OF THE HIGH PRESSURE ELECTROLYSER STACK

A preliminary design of the High Pressure Electrolyser (HP-PEM-ELY) stack is presented in the following, Figs. 5 and 6. It has to be stressed that the maximum pressure of 80 bar does not allow the use of large area MEAs, due to the high developed axial and bending forces at the bolts and the end plates. Gaskets, O-rings and bipolar plates should withstand hydrogen embrittlement effects at temperatures up to 82°C. Regarding structural parts, the most significant properties are high specific strength and stiffness. In particular, materials and geometry of the various peripheral parts were studied thoroughly before being selected. The most suitable materials have been identified in order to fulfil the main design objectives for the following parts:

- **End plates** (Cell structured Aluminium)
- **Force regulators and base** (Aluminium & Stainless steel)
- **Bolts** (Aluminium)
- **Piping system** (Aluminium)
- **Gaskets** (PTFE-type)
- **Springs** (high strength SS)
- **Treated Ti plates** current collector (O_2 side)
- **Carbon paper with GDL** current collector (H_2 side)
- **Treated Ti foil** bipolar plates

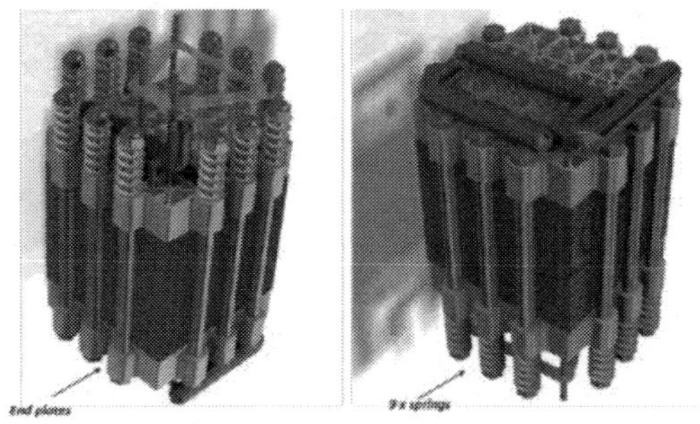

Figure 5. Two different iso-views of the design for HP-PEM-ELY stack and the various parts.

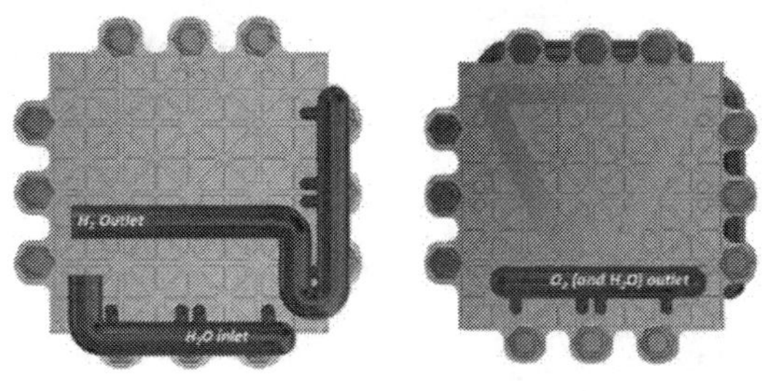

Figure 6. Plan views of the end-plates of the design for HP-PEM-ELY stack

4. DESIGN OF THE REGENERATIVE FUEL CELL SYSTEM

The complete regenerative fuel cell system, comprising of a High Temperature Fuel Cell Stack and the HP-PEM-ELY stack, will be based on the existing R-PEMFC test bench at the premises of CERTH, developed and operated under previous projects on REGENERATIVE PEM FUEL CELLS (Contracts 21113/07/NL/LvH and 22329/09/NL/CBI). Besides the test items (PEM fuel cell and electrolyzer stacks), the reactants' storage and feed subsystems, the thermal management system and the electronic peripherals, functional and performance requirements have been defined and described in detail. The main requirements of the regenerative PEM fuel cell system are:

- *Continuous Operation*. The system will operate in a continuous cycle between power production and reactant regeneration. Following a specific test load profile, the system will operate under fuel cell (FC) mode in order to provide constant power and consecutively in electrolyser (ELY) mode for complete regeneration of the reactants.
- *Closed loop operation*. Reactants (hydrogen and oxygen) produced during electrolyser (ELY) mode are used for power generation during fuel cell (FC) mode. Water produced during FC mode shall be circulated to produce the reactants during ELY mode.

The requirements of the pressurised H_2 and O_2 gas handling/storage system need careful selection of materials and safety specifications to be used. In order to meet these requirements, custom-made products are employed or developed, ensuring proper and safe operation in the long term testing period. The sizing of the gas tanks was based on the stoichiometry of water splitting reaction (H_2/O_2:2/1); therefore, in order to have a balanced electrolyser pressure (no pressure difference between anode and cathode compartments) the H_2 tank has double the volume of the O_2 tank. The power provisions for the system are at the level of 5kW for the fuel cell stack and 3 kW for the electrolyzer stack.

Based on these requirements, and more specifically, reactants storage at elevated pressure (70-100 bar) and avoidance of moving parts (e.g. compressors, pumps), the complete closed loop system has been designed and is under realization (refurbishment of existing infrastructure).

Regarding test planning. the system shall be cycled in a profile representative for a GEO telecommunication satellite, in a 1-year accelerated test profile representative of 15 years of field operation, covering both eclipse seasons (2 x 45 eclipses per year with a duration of 4.9 min to 72 min) as well as the sunlight season (need for electric propulsion (thrust) for orbit correction once per week).

5. CONCLUSIONS AND FUTURE WORK

The key remarks and future planning for development under current ESA activity on regenerative PEM fuel cells (4000109578/13/NL/SC), are summarized here:

- High pressure single-cell measurements operated on a stable basis up to P=80 bar at T= 75 ^{0}C with a performance above the desired threshold of j=0.3 A/cm^2 at V=1.6 V.
- There is low but gradual decrease of the current density by increasing the operating pressure.
- The treated Ti plate on the O_2 side is an effective current collector under high pressure conditions.
- The treated Ti foil on the H_2 side in combination with carbon paper as current collector, seem as proper bipolar and current collecting materials, under high pressure conditions.
- A combination of PTFE-type gaskets as sealing materials, shows to be effective for stable operation at high pressure values (80 bar).
- The so far obtained results from the single-cell HP-PEM-ELY setup have provided the necessary feedback on the proper materials and components for the design of a 1.5 kW light-weight HP-PEM-ELY stack, which is currently under development.
- Upon completion of the closed loop regenerative system (refurbishment of existing infrastructure) preliminary tests for qualifying the functionality of the system and test items (FC and ELY stacks) will commence.

6. REFERENCES

1. ESA Electrochemical Energy Storage Roadmap, 2011.
2. Executive Summary Report from activity GTF2065: "Regenerative PEM Fuel Cells" (ESA Contract 22329/09/NL/CBI)

POLYMERIC ELECTROLYTE MEMBRANE PHOTOELECTROCHEMICAL (PEM-PEC) CELL WITH A WEB OF TITANIA NANOTUBE ARRAYS AS PHOTOANODE AND GASEOUS REACTANTS

M.N. Tsampas, T. Stoll, G. Zafeiropoulos

FOM institute DIFFER - Dutch Institute For Fundamental Energy Research,

De Zaale 20, 5612 AJ, Eindhoven, The Netherlands, Email: m.tsampas@differ.nl

ABSTRACT

A novel photoelectrochemical (PEC) cell design is proposed and investigated for H_2 production with gaseous reactants. The core of the cell is a membrane electrode assembly (MEA) that consists of a TiO_2 photoanode of nanotube arrays, a Pt/C counter and reference electrodes and a polymeric electrolyte membrane (PEM) with proton conductivity, which serves both as compact reactor for water splitting and as gas separator. The design was inspired by PEM electrolysis technology and modified appropriately for allowing illumination and it is also equipped with a third compartment which enables the use of a hydrogen reference electrode.

Photoanodes of titania nanotube arrays, TNTAs, were developed, for the first time, on a Ti-web of microfiber substrates, by electrochemical anodization. The performance of TNTAs/Ti-web photoanodes were evaluated in both gaseous and liquid reactants. Due to the presence of reliable reference electrode in gas phase direct comparison of the results was possible. Gas phase operation with He or Air as carrier gases and only 2.5% of water content exhibits very promising photoefficiency in comparison with conventional PEC cells.

1. INTRODUCTION

Renewable energy sources are highly desirable in this era of dwindling petroleum reserves and increasing environmental concerns. Solar energy can cover a substantial share of global energy needs, but due to the intermittency and dilute nature of sunlight, storage in energy-dense media, such as hydrogen, is essential. Solar hydrogen can be produced using an electrolyzer powered by photovoltaics (PV) [1,2].

Photoelectrochemical (PEC) cells present an alternative to the photovoltaic PV-electrolyzer combination by integration of photoabsorption and photocatalysis in a single unit. The potential advantages of PEC cells over the coupled PV-electrolysis system are that they can enable the use of earth abundant, cheap electrode materials and provide a more compact and integrated reactor design [1,2].

A PEC cell consists of photoactive materials, such as semiconductors, that absorb light and produce charge carriers. Most of the PEC cells, described so far in the literature, are designed to operate in liquid phases, which have a dual role, serving simultaneously both as the electrolyte and as the oxidation/reduction substrate [1,2]. Inspired by the reactor design of polymeric electrolyte membrane (PEM) electrolyzers and fuel cells [3], few research groups [2,4] have attempted to separate the two electrodes of PEC cells (where the two electrochemical half-reactions take place) with an ionically conductive polymeric membrane. In such systems the membrane acts both as electrolyte and as support for the photoelectrodes. This novel strategy seems very promising since it allows the direct separation of the reaction products, while also the polymeric membrane electrolyte does not need to be replenished during long term operation as in the case of liquid electrolytes, which is essential for scaling up PEC devices [1,2].

The viability of PEC cells with PEM electrolysers designs (PEM-PEC) has been demonstrated in acidic, alkaline and neutral conditions, but examples with gaseous feedstock for the anode remains rare [2,4,5]. Nevertheless the use of solid-state electrolytes in conjunction with gaseous reactants have a series of advantages over liquid-phase reactors such as operation at elevated temperatures and pressures (for improved kinetics), direct production of compressed H_2 and hindering of gas bubbles formation which can impede catalytic reactions [3,4]. The latter is very important especially for space applications where bubble formation leads to more severe performance degradation due to microgravity environment [6].

Another appealing idea where the PEM-PEC water splitting devices can find application is the production of hydrogen from the water contained in ambient air [7]. Air based operation in PEM-PEC cells can decrease maintenance costs, as the possibility of corrosion and poisoning is reduced. In addition, liquid pumping systems are not needed since natural convection of air can feed the water vapor, and the risk of freezing is minimized. Furthermore, capturing water from the air implies that virtually no liquid water is needed for

 DOI: 10.1051/e3sconf/20171609005

operation, making it a water-neutral process [7].

It is thus obvious that gas phase operation in PEM-PEC cells is a promising research approach for photoelectrochemistry which deserves further investigation. The transition from aqueous to solid state electrolytes is a challenging process due to the different reactor configurations and electrode requirements. In conventional PEC cells with aqueous electrolytes the photoelectrodes generally consist of semiconductor modified conductive glass substrates that are immersed in the liquid phase [1,2]. This configuration cannot be directly transferred to PEM-PEC cells, where high porosity of the substrate is required in order to optimize the reactants supply to the electrode/electrolyte interfaces [3].

The aim of our efforts is to develop and an integrated PEM-PEC system with gaseous reactants for unassisted water splitting operation based on the Z-scheme approach [1], where nanostructured n-type and a p-type photocatalyst drives for the oxygen and hydrogen evolution reaction respectively. Our present work is focused on the following aspects:

(i) development of a novel PEM-PEC cell reactor equipped with reference electrode compartment for allowing comparison between PEC studies,

(ii) fabrication of a photoanode composed of TiO_2 nanotube arrays on a high surface area Ti-web of microfibers porous substrate and investigation of the correlation between anodization time and nanotube length,

(iii) evaluation and comparison, for the first time, of the fabricated photoanode performance in liquid and gas phase (with He and Air as carrier gases) operation.

Our ultimate goal is to load those TNTAs with visible light sensitive photocatalysts and to examine the effect of nanostructuring on the photoelectrode performance.

2. EXPERIMENTAL

Prior to anodization, Ti-webs (gas diffusion layer, Bekinit, 20μm microfibers, 80% porosity, Figure2) were cleaned by sonication in acetone and in ethanol for 20 min, respectively and rinsed with deionized water. All the experiments were performed in the electrolyte containing ethylene glycol, 0.3 wt% ammonium fluoride and 2% vol% deionized water.

In Figure 1 is illustrated the anodization setup, i.e. typical two-electrode configuration with Ti-web as the counter electrode, along with SEM images of the Ti-web before and after the anodization process. Titania nanotube arrays TNTAs were generated by potentiostatic anodization under a constant applied potential of 30V for 1h. After the anodization, the samples were properly rinsed with deionized water to remove the occluded ions and then were calcined at 450°C for 5h under air to increase the crystallinity

Surface morphologies of the TNTAs were characterized with a Scanning electron microscope (FEI Quanta 3D FEG instrument). The crystal phases were analysed by X-ray diffraction (PanAlytical X'pert PRO MRD) using a Cu K_{α} tube.

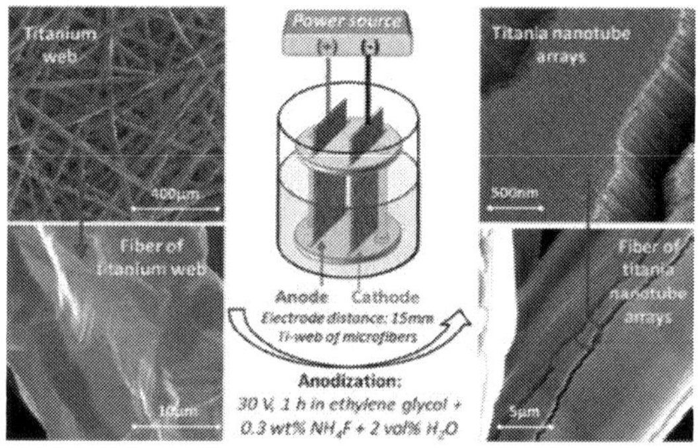

Figure 1. Basic setup and fabrication process of titania nanotube arrays on a Ti-web of microfibers, by potentiostatic electrochemical anodization.

Ti-web photoanode was placed on the top side of a Nafion® perfluorinated membrane (180μm thick, Alfa Aesar) while on the bottom side of the membrane a Pt counter-electrode (supported on carbon cloth, with 0.5mg/cm^2 loading) was placed. The reference electrode which is also Pt/C electrode (with the same characteristics) is deposited on the top of the membrane as presented in Figure 2. MEA was pressed between two graphitic flow-fields plates.

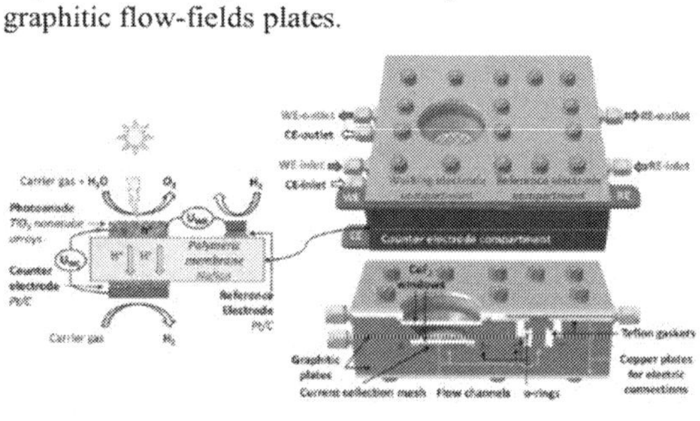

Figure 2. (Left) Configuration and operation of the membrane (photo)electrode assembly. (Right) Novel PEM-PEC reactor design with three compartments for accommodating anode, cathode and reference electrodes.

The three different compartments (sealed with rubber rings) of the novel PEM-PEC cells are illustrated in Figure 2. Teflon gaskets are used to electrically isolate anodic and reference current collection which are in the same side of the cell and are presented with the brown plates. Two Ca_2F windows are used for allowing illumination and ensuring sealing in the anodic compartment and an Au mesh for the current collection.

DOI: 10.1051/e3sconf/20171609005

Each compartment was fed with gaseous streams, made by bubbling He or Air or H_2 through thermostated gas saturators containing H_2O. The cathode was always fed with 2.5 vol% H_2O in He (100 ml/min) and reference with 2.5 vol% H_2O in H_2 (30 ml/min).

The photoelectrochemical studies were carried out using an electrochemical work station (Vertex, Ivium) under the irradiation of a UV-LED lamp (M365LP1, 365±15nm, ThorLabs, 10mW/cm^2 at the photoanode) equipped with pulse modulator (DC2200, ThorLabs) to perform the chopped irradiation linear sweep voltammetry.

3. RESULTS AND DISCUSSION

3.1. PEM-PEC electrode design

The electrodes used in PEM devices typically consist of carbon supports loaded with the electrocatalytically active phase (usually in powder form) which is also in contact with the polymeric membrane, as depicted in Figure 3a [3].

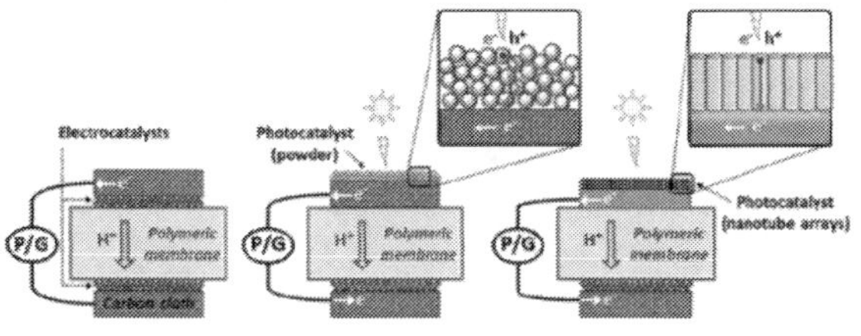

Figure 3. Schematic diagram of the PEM cell (a), PEM-PEC with power photocatalyst (b) and nanotube arrays photocatalyst (c). In all cases the cell is connected to a potentiostat-galvanostat, P/G.

The current gas phase PEM-PEC studies are focused on a very similar design where the electrocatalyst is replaced by a photocatalyst (e.g. TiO_2, WO_3), which however is deposited on the opposite side of the substrate (Figure 1), for allowing light absorption [4,5]. This solution presents two main disadvantages: (i) it provides many recombination pathways of photogenerated species at the powder grain boundaries and (ii) long term stability issues occur, since the photocatalytic powder eventually falls away from the substrate after prolonged operation [3].

To overcome the aforementioned issues we hereby propose in this study an alternative design for PEM-PEC photoelectrodes, which is investigated for the first time in conjunction with gaseous reactants. This design is based on a the fabrication of titania nanotubes arrays (TNTAs) photoanode obtained by electrochemical anodization of a titanium porous substrate (composed of a web of Ti-microfibers). This substrate was recently proposed in PEM-electrolysis studies as alternative material due to the vulnerability of carbon to electro-oxidation [8].

This strategy allows to obtain an unidimensional ordered nanotube architecture that offers a wide specific surface and an excellent electrical channel for charge transfer facilitating this way the separations of the photogenerated electron-hole pairs (Figure 1c).

Although titania nanotubes can be formed by other routes, the electrochemical anodization method is regarded as one of the relatively simple techniques to synthesize TiO_2 nanotubes with large surface area [9]. Moreover, the one-dimensional and highly ordered nanotube architecture offers an excellent electrical channel for vectorial charge transfer so that photoinduced electron-holes pairs can be effectively separated, resulting in an obvious improvement in the photoelectochemical performance. Anodization of Ti-web of microfibers has been very recently reported as a template for loading Pd nanoparticles for alcohol electro-reforming [10], however to our knowledge is the first time that this material is evaluated as photoanode. The unique characteristics of this structure, i.e. high surface area and porosity, make them ideally compatible with membrane electrolytes.

3.2. Reference electrode considerations

While the use of reference electrodes in aqueous phase electrochemistry research is well established, there are only few reports on their application to PEM systems [11]. Among the different possibilities we decided to create a third compartment in our cell which interface a Pt/C electrode deposited on an uncoated area of the polymeric membrane as depicted on the Figure 2a. This compartment is fed by (humidified) hydrogen and thus a reversible hydrogen electrode, RHE is established.

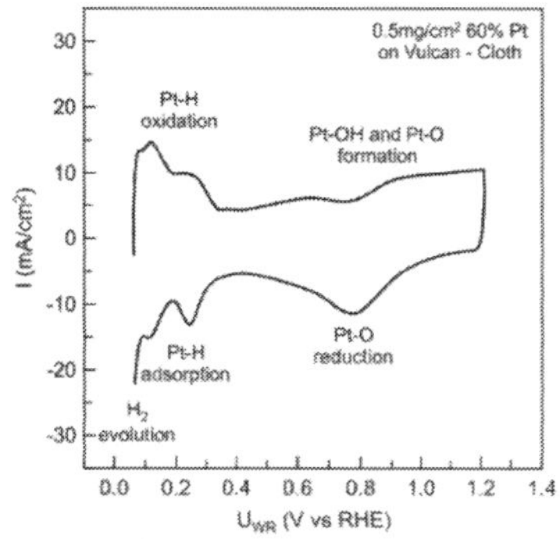

Figure 4. Cyclic voltammetry curves with scan rate of 50mV/s of commercial Pt on Vulcan electrocatalyst obtained in our PEM-PEC device with feed of 2.5% H_2O in He.

In order to validate the reliability of the reference electrode we performed cyclic voltammetry (CV) measurements in commercial carbon supported Pt electrocatalyst. Figure 4 shows that all the features of CV on Pt were observed, i.e. the well-known hydrogen adsorption and desorption peaks in the potential region

DOI: 10.1051/e3sconf/20171609005

of 0 to 0.3V vs RHE, as well as the Pt oxide formation and its counterpart reduction peak at ~0.75V. These results are in good agreement with literature reports obtained in aqueous phase [12], indicating that the PEM-PEC developed in this study is equipped with a reliable reference electrode.

3.3. Photoelectrochemical characterization

The use of gaseous reactants as feed for future PEC installations could be a way to reduce complexity and cost because water supply management can be very simple. Additionally, air based PEM-PEC systems could be used to capture water molecules from ambient air for H_2 production, in this case humidified air supply could be achieved by natural convection [7]. Therefore, in order to investigate this possibility in our study, air was also used as a carrier gas apart from He. The present studies are focused, for the first time, on the comparison of a conventional PEC with aqueous electrolyte (0.1M H_2SO_4) and our gas phase PEM-PEC system operating at 2.5% H_2O vapour pressure which corresponds to 22°C with 100% relative humidity.

Figure 5a and b depicts the linear sweep voltammetry (LSV) measurement of the TNTAs electrodes for water oxidation in a conventional PEC cell with and in our home made PEM-PEC reactor, respectively. Using the PEM-PEC reactor, apart from the gas phase experiments (with He and Air as carrier gas for H_2O steam), water was also poured as droplets on the anode in order to resemble operation with liquid reactants. As one can see by comparing Figure 5a (for 1h anodization) with the dashed line in Figure 5b, the obtained performance with liquid H_2O in the PEM-PEC (apart from the dark current) is almost identical.

Figure 5b shows also that the effect of carrier gas is important only below 0.3V while above 0.4V the photoanode performance is only 8-10% lower when Air is used as carrier gas instead of He. In the low applied potential region the presence of oxygen molecules (when Air is used as carrier gas) act as mediators for recombination [11] and thus cause back reactions which deteriorate photoefficiency. Oxygen-related side reactions are initiated by electron capture by molecular oxygen. To minimize the detrimental impact of oxygen, electrons should be rapidly extracted from the semiconductor to a conducting back contact. The fact that for applied potential higher than 0.4V the presence of oxygen has low impact on the photoefficiency implies that our TNTAs are capable of fast electron injection inside the TiO_2 valence band that limits the possible recombination pathway due their nanostructure. The photocurrent in gas phase is 0.43 mA and 0.40mA at 1.23V when He and Air are used as carrier gas respectively. These values correspond to 50% and 47% of the current obtained at the same potential in liquid phase, which is very interesting taking into account that only 2.5% of water in the gas phase can give half of the photocurrent.

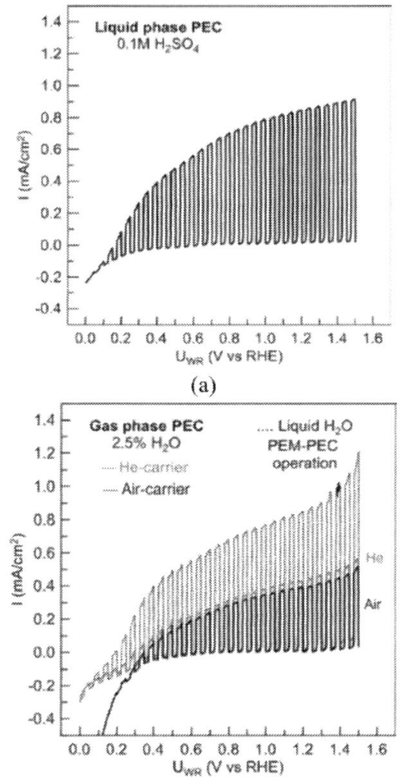

Figure 5. Linear voltammetry sweeps (2.5mV/s) on the photoanode (anodized for 1h) under chopped illumination for water splitting (a) in aqueous electrolytes pH=1 and (b) in polymeric electrolyte with He (top)and Air (bottom) carrier gases for 2.5% H_2O. The dashed line corresponds to operation when the photoanode was poured with droplets of water before assembling the PEM-PEC cell.

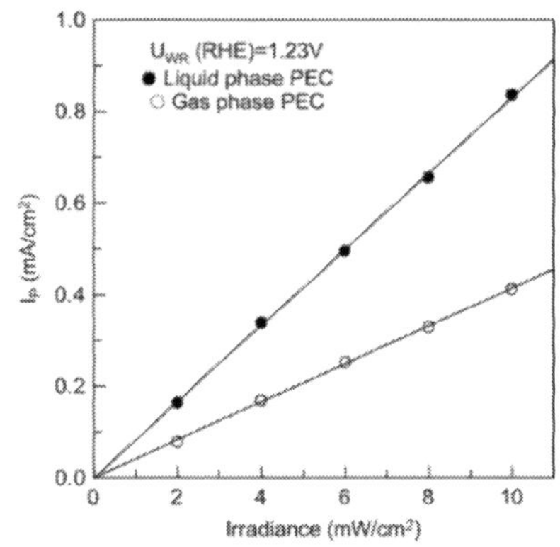

Figure 6. Effect of irradiation on the photocurrent in both aqueous and gas phase PEC.

These results give, for the first time, a comparison of the photoactivity of a photoelectrode in an air-based PEM-PEC vs a conventional PEC cell. The shape of the LSVs in both environments are very similar which illustrates that our proposed PEM-PEC cell design is equipped with a reliable reference electrode, that allows comparison of photoelectrodes in liquid or gas operation.

Figure 6 presents the effect of irradiation on the photocurrent in liquid and gas phase PEC operation. In both cases a linear relationship is observed, showing that photocurrent level reflects the amount of photogenerated charge carriers, limited by the number of photons and efficiency of the photocatalyst. Taking into account the AM1.5 Global and the AM0 which are the standard solar irradiance spectra that are used for terrestrial and space applications, the irradiance in the UV region is 3.9 and 9.1mW/cm^2 respectively [1]. Therefore the expected photocurrent at 1.23V during gas phase operation will be around 0.34mA/cm^2 for terrestrial and 0.16mA/cm^2 for space applications.

4. CONCLUSIONS

TiO_2 nanotubes arrays photoanodes were fabricated on a Ti-web of microfibers by electrochemical anodization.. These novel photoanodes were first evaluated for water splitting and ethanol reforming in conventional PEC cells with liquid electrolytes and then in our novel PEM-PEC cell with gaseous reactants. This solid state electrolyte cell is equipped with a reliable reference electrode that made possible to compare, for the first time, the performance of the fabricated TNTAs/Ti-web in liquid and gas phase operation. It is demonstrated that gas phase PEC operation with only 2.5 % H_2O results to photocurrent close to the 50% in comparison with the liquid operation. Taking into account the cost reduction of air based PEM-PEC cell operation arising from the lack of a liquid pumping system (which is essential for conventional PEC systems), one can conclude that this performance is very promising. The one-dimensional (1D) nanostructures, which provide high electron mobility, specific surface area and mechanical strength, in addition with the porous nature of the Ti-web substrate, makes the fabricated photoanodes ideal for interfacing polymeric membrane electrolytes. Furthermore, this architecture plays a key role in the performance of the TNTAs during gas phase operation since similar results were obtained when either Air or He was used as a carrier gas for water splitting.

5. REFERENCES

1. Van de Krol, R. & Grtzel, M. (2012). Photoelectrochemical Hydrogen Production, Springer, New York.
2. Bosserez, T., Rongé, J., van Humbeeck, J., Haussener, S. & Martens, J. (2015). Design of compact photoelectrochemical cells for water splitting. Oil & Gas Science and Technology – Rev. IFPEN. 70(5), 877-889.
3. Carmo, M., Fritz, D.L., Mergel & J., Stolten D. (2013). A comprehensive review on PEM water electrolysis. Int. J. Hydrogen Energy. 38, 4901-4934.
4. Iwu, K.O., Galeckas, A., Kuznetsov, A.Y. & Norby, T. (2013). Solid-state photoelectro-chemical H_2 generation with gaseous reactants. Electrochim. Acta. 97, 320-325.
5. Iwu, K.O., Galeckas, A., Diplas, S., Seland, F., Kuznetsov, A.Y. & Norby, T. (2014). Effects of temperature, triazole and hot-pressing on the performance of TiO2 photoanode in a solid-state photoelectrochemical cell. Electrochim. Acta. 115, 66-74.
6. Matsushima H., Nishida T., Konishi Y., Fukunaka Y., Ito Y. & Kuribayashi K., (2003). Water electrolysis under microgravity: Part 1. Experimental technique, Electrochim. Acta. 48, 4119–4125.
7. Ronge J., Deng S., Pulinthanathu Sree S., Bosserez T., Verbruggen S.W., Kumar Singh N., Dendooven J., Roeffaers M.B.J., Taulelle F., De Volder M., Detavernier C. & Martens J.A., (2014). Air-based phototelectrochemical cell capturing water molecules from ambient air for hydrogen production, RSC Adv. 4, 29286–29290.
8. S.A. Grigoriev, P. Millet, S. S. Volobuev, V. N. Fateev, (2009). Optimization of porous current collectors for PEM water electrolysers, Int. J. hydrogen Energy. 34, 4968-4973.
9. Grimes, C.A. & Mor, G.K. (2009). TiO2 Nanotube Arrays: Synthesis, Properties and Applications, Springer, Berlin
10. Chen Y.X., Lavacchi A., Miller H.A., Bevilacqua M., Filippi J., Innocenti M., Marchionni A., Oberhauser W., Wang L. & Vizza F., (2014). Nanotechnology makes biomass electrolysis more energy efficient than water electrolysis, Nat. Commun., 5, 4036, 1-6.
11. Brightman E., Dodwell J., van Dijk N. & Hinds G., (2015) In situ characterisation of PEM water electrolysers using a novel reference electrode, Electrochem. Commun. 52, 1–4
12. Ji W., Qi W., Tang S., Peng H. & Li S., (2015) Hydrothermal Synthesis of Ultrasmall Pt Nanoparticles as Highly Active Electrocatalysts for Methanol Oxidation, Nanomaterials. 5(4), 2203-2211.

 DOI: 10.1051/e3sconf/20171609006

GRAPHENE-BASED SYSTEMS FOR ENHANCED ENERGY STORAGE

Aris Amplianitis(1),(3), Athanasios Masouras (3), Giorgos Gkikas (2), Zampia Kalogridi (1),(2), Athanasios Baltopoulos (2), Antonios Vavouliotis (1),(3), Stavros Tsantzalis (3), Vassilis Kostopoulos (3) and Ugo Lafont (4)

(1) PLEIONE ENERGY S.A., www.pleione-energy.com (Greece), Email: amplianitis@pleione-energy.com
(2) ADAMANT COMPOSITES Ltd., www.adamant-composites.gr (GREECE), Email: gkikas@adamant-composites.gr
(3) APPLIED MECHANICS & VIBRATIONS LABORATORY, Dept. Mechanical and Aeronautics Engineering, Polytechnical School, University of Patras, www.aml.mech.upatras.gr (GREECE), Email: kostopoulos@mech.upatras.gr
(4) EUROPEAN SPACE AGENCY (ESA), www.esa.int (THE NETHERLANDS), Email: ugo.lafont@esa.int

ABSTRACT

Extensive global research efforts have focused on the exploitation of graphene for enhanced energy storage. Novel graphene-based composite material electrodes have been developed, in many cases with reports of outstanding performance. However, the development of these composites involve extremely complex and costly procedures/methods whose scalability and eventual commercial exploitation is extremely hard [1]. Within the present activity the use of graphene nanotechnology is exploited to manufacture electrodes for supercapacitors. The goal however is to achieve electrodes with increased specific energy density (compared to the currently commercially available products) using proven and simple manufacturing procedures that can easily be scaled-up and offer competitive products. The roadmap was developed under the framework of European Space Agency highlighting the main advantages brought up from this technology. The activity is separated in three parallel routes; the development and test planning of small–scale production of graphene based materials via the tape casting technology, the establishment of a reliable and low cost industrial production process (scale-up) for these materials and the development and testing of an energy storage demonstrator that shall incorporate the novel electrodes and exhibit their favorable characteristics in energy storage applications for use in space.

1. INTRODUCTION

Graphene is an atomically thin, planar membrane of carbon atoms tightly packed into a two-dimensional (2D) honeycomb lattice with exceptional properties, particularly electronic and electrochemical ones.

As a kind of carbon material, graphene sheet (GS) has attracted increasing attention in a variety of fields because of its high theoretical specific surface area (2600 m^2g^{-1}), good flexibility, superior chemical/thermal stability, and extraordinary electrical, thermal, and mechanical properties. Its unique structure and associated outstanding properties make graphene highly promising for a wide range of applications in electronics, sensors, and energy storage/conversion.

In the case of supercapacitors, the desire for using carbon materials is based on the mechanism of double-layer capacitance. They store the charges electrostatically using reversible adsorption of ions of the electrolyte onto active materials that are electrochemically stable and have a high accessible surface area. Although graphene has a large theoretical surface area, this area will not result in high electrode capacitance if it is not accessible to ions. Therefore, it should be noted that the capacitance is largely dependent on the layering and stacking configuration of graphene, and much research has been conducted to create graphene structures that display large ion-accessible surface areas [1]. The agglomeration and restacking of graphene nano-sheets during the drying process decrease the electrolyte-ion-accessible surface area, and thus limits the electrochemical performance. To take full advantage of graphene in supercapacitors, dispersing carbon materials, metal oxides/hydroxides, and conducting polymer onto graphene to form graphene-based composites is considered an effective and practical method. These materials will not only prevent the agglomeration and restacking of graphene during the drying process but also contribute to the specific capacitance themselves thus improving the whole supercapacitor performance.

More specifically, transition metal oxides such as MnO_2 and RuO_2 can exhibit high specific capacitance (pseudocapacitance) via the redox reaction at the interface of the electrode material and the electrolyte [2], [3]. Conducting polymers can offer capacitive behaviors because of the presence of a delocalized π-electron conjugation system, which can enable fast and reversible redox reactions [4][5]. In addition, conducting polymers also have the advantages of cost effectiveness, environmental benignity, high conductivity in a doped state, and adjustable redox activity through chemical modifications [6]. Thus, conducting polymers are considered as good electrode materials for supercapacitors. However, the limitation of using conducting polymers as standalone materials for supercapacitors is that they can rapidly degrade as a result of swelling and shrinkage, resulting in poor cycle stability [1]. The introduction of conducting polymers into graphene not only can prevent the agglomeration and restacking of graphene during the drying process to enhance the EDL capacitance of graphene but also can improve the mechanical and electrochemical cycling stability of the polymers. Thus, a combination of graphene and electrically conducting polymers is beneficial for improving the electrochemical performance. A variety of polymers such as polypyrrole (PPy), polyaniline (PANI), poly-3,4ethylenedioxythiophene (PEDOT) and polythiphene (PT) grown on the graphene sheets have been investigated for electrode materials [5].

In the present work three types of graphene based composite materials are developed and tested for incorporation in supercapacitors: Graphene Nanoplatelets (GNPs) with PVDF binder composite, GNPs with PVDF and transition metal oxide MnO_2 and GNPs with PVDF and conductive polymer PANI

On a later stage of the activity further investigation will be performed for the favorable characteristics of these materials and how could efficiently be scaled-up to produce large films for industrial incorporation in energy storage devices.

2. MATERIALS DEVELOPMENT AND TESTING

Since the discovery of graphene, many routes for its synthesis have been reported. Two main routes are widely available: either exfoliation of natural or synthetic graphite using, mechanical, chemical, liquid-phase or electrochemical exfoliation techniques, or direct formation methods from organic molecules such as methane and other hydrocarbon sources [1]. While the second method can produce high-quality graphene films, it is not suitable for large scale economical production of graphene mainly due to it high cost [7]. Chemical exfoliation, on the other hand, has the advantages of scalability and low cost although it is difficult to obtain single-layer high-quality graphene because of the defect creation during exfoliation [8]. In the present work, low cost chemically reduced Graphene Nanoplatelets (GNPs) where employed with an overall thickness of approximately 3-10 nanometers (approx. 4 layers thick). As the main goal of the activity is to employ cheap and scalable production of composite films, the tape casting method was chosen. Slurry preparation starts with magnetic stirring of the polymeric binder (PVDF) with NMP solvent, followed by nanoparticle dispersion and thorough sonication of the mixture. From then on, the process involves the feeding of slurry in front of a doctor blade and the formation of a film with the required thickness. A drying stage follows where the solvent evaporates. While the followed procedure for tape casting electrodes from a slurry is not innovative, additional production steps were applied during the development to improve the structural integrity of the films. Annealing of the materials after evaporation of the solvent was performed at a temperature closer to the melting point of PVDF, at this stage the crystallization of PVDF produces an electroactive phase of crystallinity (β-phase) which is necessary for energy applications [9].

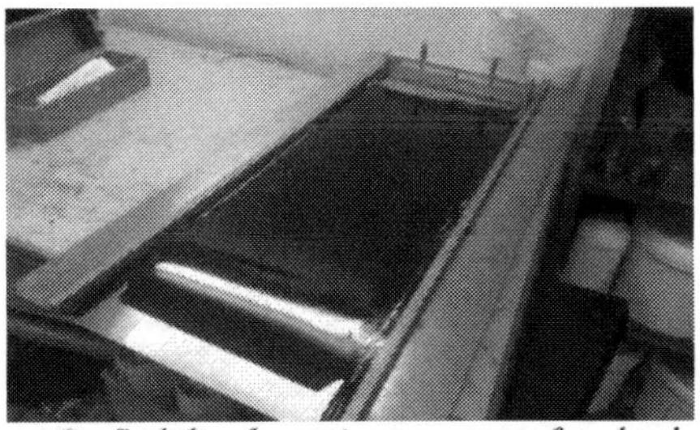

Figure 1. Stabilized casting process for buckyfilms production

The developed composites and their respective compositions are:

- **GNP/PVDF**: 80 wt% Graphene Nano-platelets (GNPs) with 20 wt% PVDF Binder
- **GNP/PANI**: 76 wt% GNPs, 19 wt% PVDF and 5 wt% PANI
- **GNP/MnO₂**: 73 wt% GNPs, 18 wt% PVDF and 9 wt% MnO_2

The properties evaluation of the produced materials includes physical characterization, electrical resistivity measurements and electrochemical characterization which is necessary for materials intended for energy storage applications.

Morphology texture of the produced composite was investigated by Scanning Electron Microscopy (SEM, LEO SUPRA 35VP).

The surface and through the thickness electrical resistance of the produced materials was investigated via four probe measurement methods (surface and through thickness).

Electrochemical performance (electrode capacitance) of the materials was investigated by Cyclic Voltammetry in a 3-electrode configuration set-up using an AUTOLAB potentiostat/galvanostat analyzer. Tests were performed on a sample with active area of 0.7 cm^2 and active material mass of 15 mg. The electrolyte was KOH solutions of 0,5 and 2M and the scan rates were adjusted at 10, 20, 30, 40, 50, 100, 200 mV/s. Ag/AgCl reference electrode was used. The specific capacitance

DOI: 10.1051/e3sconf/20171609006

of the electrodes was calculated using Eq.1.

$$C = \int_{E_1}^{E_2} i(E)dE \,/\, 2(E_2 - E_1)mv \qquad \text{Eq.1}$$

where ∫i(E)dE is the total voltammetric charge obtained by integration of positive and negative sweep in cyclic voltammetry graphs, (E2 – E1) is the potential window width, m is the mass, v is the scan rate.

In the first phase of the activity, various film production parameters were tested in order to optimize both the procedure and the quality of the produced films. Optimum active material contents were also investigated. In later stage (results not presented here) the finalized electrodes shall be incorporated in actual supercapacitors cells (together with appropriate electrolyte) and Galvanostatic charge and discharge cyclic experiments are scheduled to be performed.

For the aforementioned purposes, an appropriate laboratory cell enclosure has been manufactured, with the ability to accommodate both 2 and 3-electrode set-ups, as shown in Figure 2.

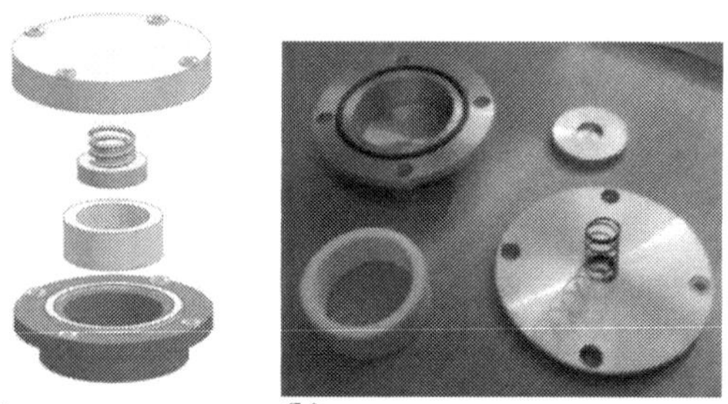

(a) **(b)**

Figure 2. (a) Blueprint of the cell configuration and (b)Two-electrode cell

3. RESULTS AND DISCUSION

3.1. Self - Standing attributes of the produced materials

One of the targets of this activity is the production of self–standing electrodes for energy storage applications. The advanced mechanical properties of GNPs and their high surface area are promising characteristics to achieve the production of self-standing GNP films with structural integrity and energy storage performance (specific capacitance and capacity). The first results proved that it is feasible to produce self-standing electrodes with 80% wt. content of GNPs (Figure 3):

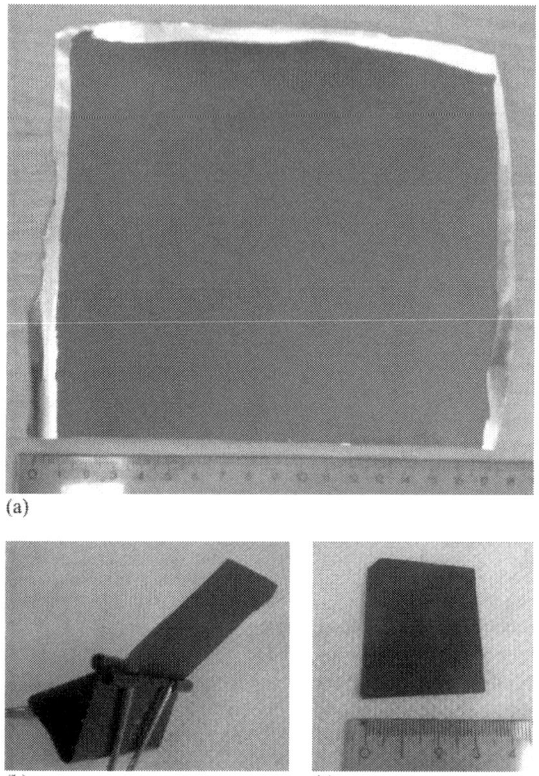

(a) (b) (c)

Figure 3. (a) GNP film after manufacturing on aluminum substrate. (b), (c) Self-standing GNP-film (electrode)

3.2. Microstructural Analysis

Microstructural analysis of the produced materials was performed by SEM. Several micrographs captured in different magnifications in order to investigate the internal structure of the produced films.

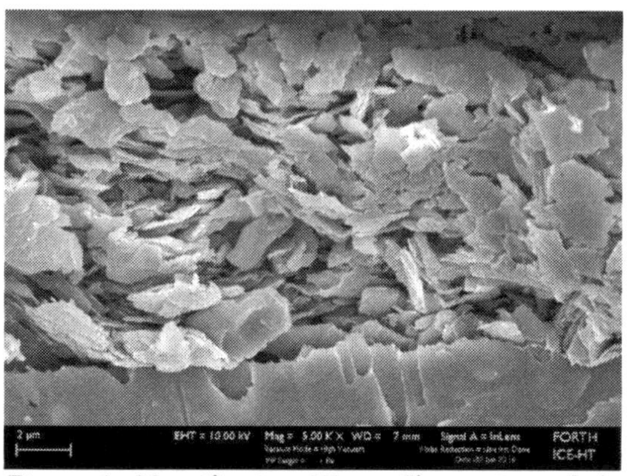

Figure 4. Magnification 5Kx – Closer view of the internal microstructure of GNP/PVDF composite

 DOI: 10.1051/e3sconf/20171609006

Figure 5. Magnification 3Kx – Close view of the internal microstructure of the GNP/PANI composite.

Figure 6. Magnification 2.5Kx – Closer view of the internal microstructure of GNP/MnO₂ composite.

In general, the microstructure of the films that were produced is uniform and the GNPs are well distributed in the structure exhibiting an inter-particular porosity. This intrinsic porosity is beneficial in that it will allow better electrode/electrolyte contact.

3.3. Electrical Resistivity Results

Electrical resistance of the samples was measured for the evaluation of their electrical behaviour. Electrical resistance of electrodes must be low to avoid high internal resistance of the future supercapacitor cell. Figure 7. Through the thickness electrical resistance of the produced electrodes (Ω).Figure 7 presents the results of electrical resistance through the thickness of the produced composites.

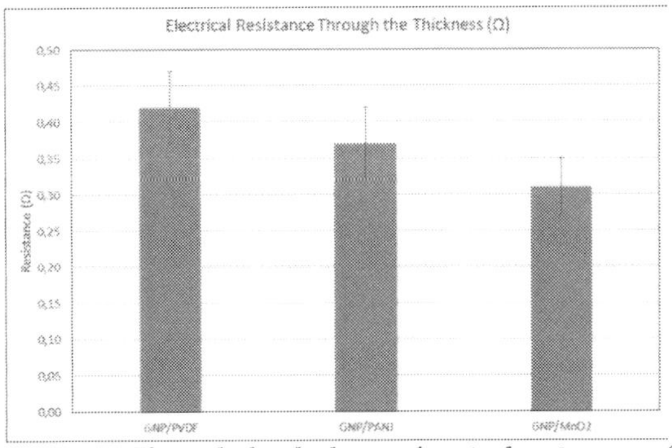

Figure 7. Through the thickness electrical resistance of the produced electrodes (Ω).

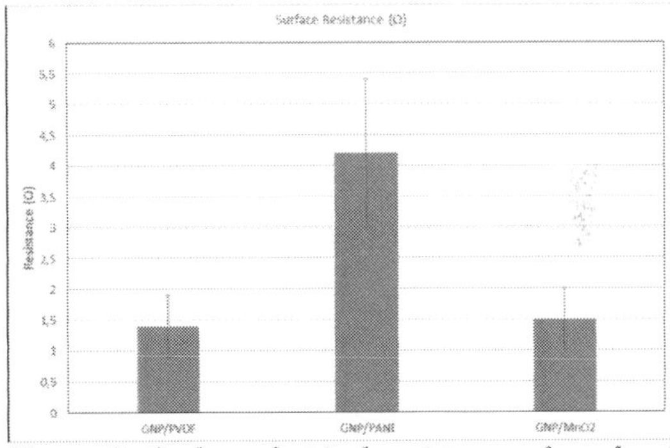

Figure 8. Surface electrical resistance of produced electrodes (Ω).

3.4. Cyclic Voltammetry Results

During this test campaign, it was achieved to extract one set of clear measurements for the three different materials developed.

The cyclic voltammetry tests on the GNP/PVDF composite were performed on samples with active area of 0.7 cm^2 and active material mass 15 mg. The electrolyte was 2 M KOH solution and the scan rates was adjusted at 10, 20, 30, 40, 50, 100, 200 mV/s. Ag/AgCl reference electrode was used. The results are presented in Figure 9 and Figure 10.

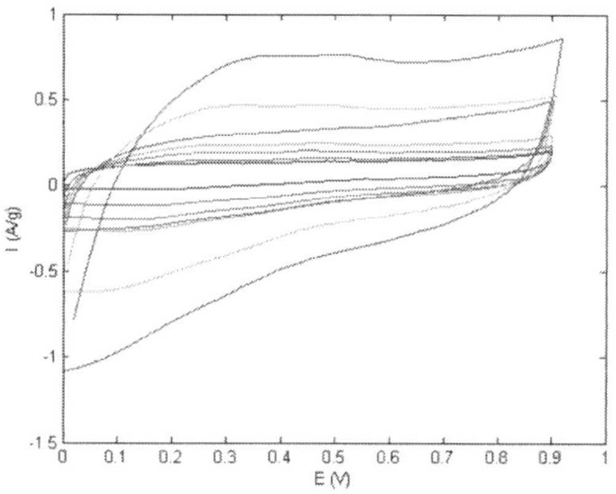

Figure 9. Working Electrode Current vs Reference Electrode Potential for the GNP/PVDF composite.
Scan Rates: 10(—), 20(—), 30(—), 40(—), 50(—), 100(—), 200(—) mV/s

In Figure 9. it is observed that up until the scan rate of 100 mV/s, the CVs retain a nearly symmetrical rectangular shape within the range of scan rates employed, which is characteristic of electro-chemical double-layer capacitive behaviour [10]. Furthermore, no great peaks appear which indicates that the super-capacitive behaviour is free from redox reactions or is purely based on the electrostatic mechanism. On the contrary the GNP/PANI and GNP/MnO_2 electrodes exhibit increasing peaks with increased scan rate and departure for the symmetrical rectangular shape due to the presence of pseudo-capacitive activity of the PANI and metal oxide.

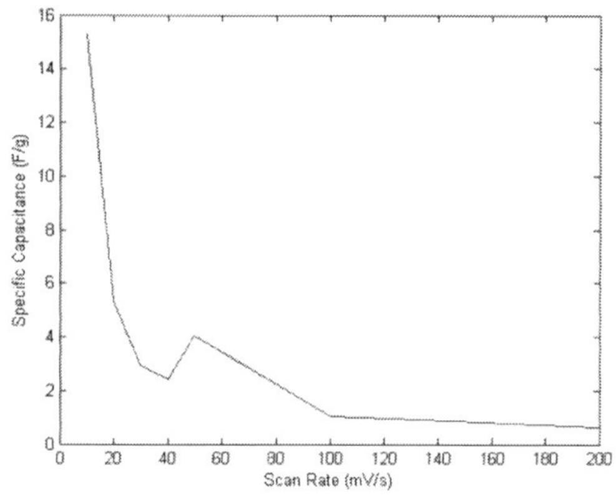

Figure 10. Evolution of the specific capacitance as function of the scan rate for the GNP/PVDF composite electrode

Results of the cyclic voltammetry tests performed on the GNP/PANI composite are presented in Figure 11 and Figure 12.

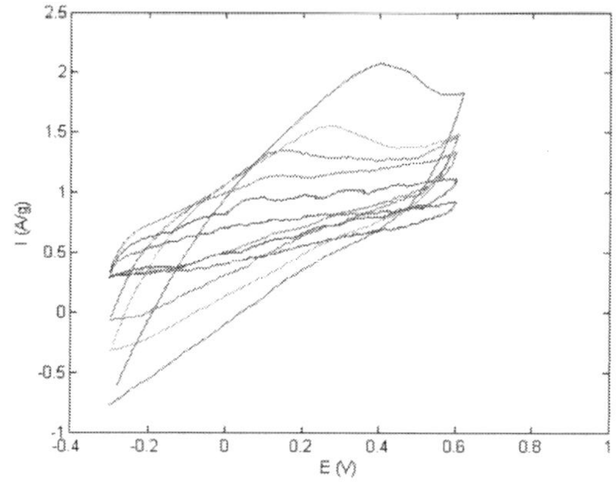

Figure 11. Working Electrode Current vs Reference Electrode Potential for the GNP/PANI composite.
Scan Rates: 10(—), 20(—), 30(—), 50(—), 100(—), 200(—) mV/s

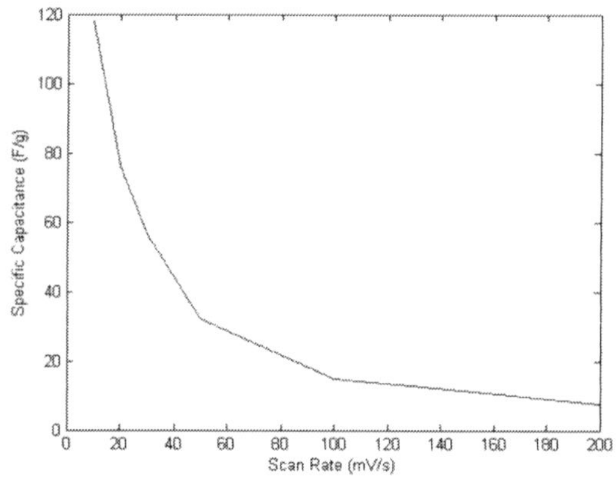

Figure 12. Evolution of the specific capacitance as function of the scan rate for the GNP/PANI composite electrode.

Finally, similar cyclic voltammetry tests were also performed on the GNP/MnO_2 composite. The results are presented in Figure 13 and Figure 14.

DOI: 10.1051/e3sconf/20171609006

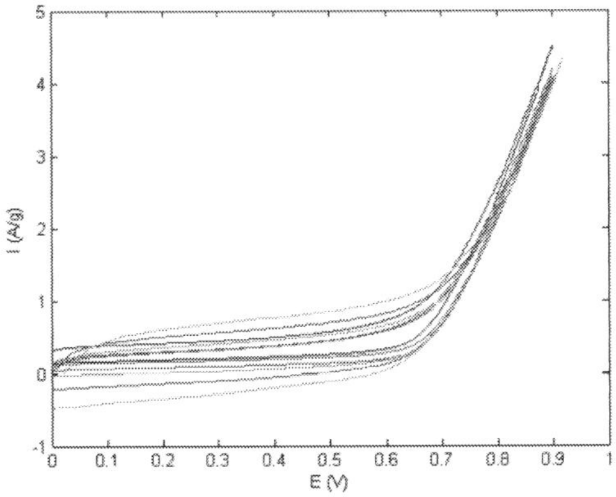

Figure 13. Working Electrode Current vs Reference Electrode Potential for the GNP/MnO_2 composite.
Scan Rates: 10(—), 20(—), 30(—), 50(—), 100(—), 200(—) mV/s

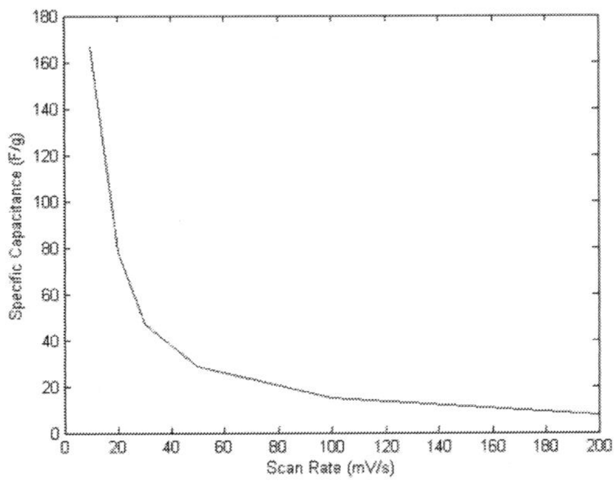

Figure 14. Evolution of the specific capacitance as function of the scan rate for GNP/MnO_2 composite electrode

Figures 10,12 and 14 present the exhibited decrease of specific capacitance with increasing CV scan rate. This decrease has to do with reaction kinetics between the electrode and the electrolyte and therefore becomes more apparent in the electrodes that involve slower faradic reactions such as the GNP/MnO_2. These faradic reactions however are the reason these electrodes exhibit high pseudocapacitance.

Comparison of the specific capacitance results are presented below (at scan rate: 10 mv/s) in Figure 15. Incorporation of PANI and MnO_2 improve the specific capacitance of the electrode from the base value of the reference electrode only with GNPs. Specific capacitance enhanced over an order of magnitude by the presence of PANI and MnO_2 but still further improvement is needed.

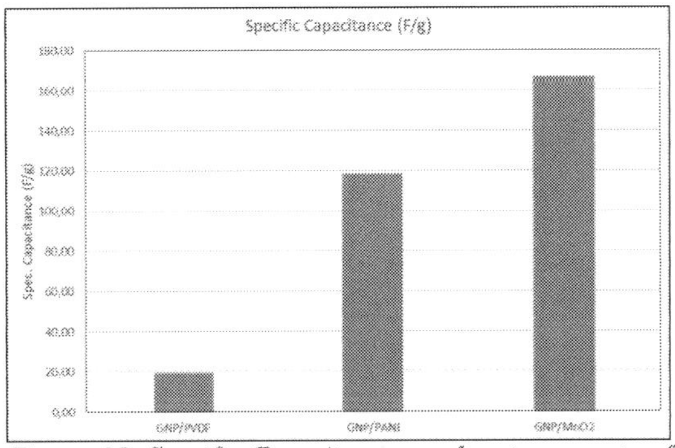

Figure 15. Specific Capacitance results at scan rate of 10 mv/s.

4. PROCESS INDUSTRIALIZATION

Next essential part of the project is the investigation and development of a scale-up manufacturing procedure for the selected graphene based composite materials. Scaling up the production while at the same time maintaining the favorable performance of these materials is the ultimate goal towards a commercialization plan that involves the incorporation of these materials in high energy density supercapacitors. To achieve that, parameters such as auto-feeding of the slurry on the doctor blade, velocity of the blade for the filming, on site drying without removing the final films from the casting substrate were investigated and a small semi-automated production line was designed and manufactured. Currently the automated procedure is being fine-tuned and the first electrodes sheets that are being produced are being tested for their performance. A blueprint of the final film applicator is shown in *Figure 16.*

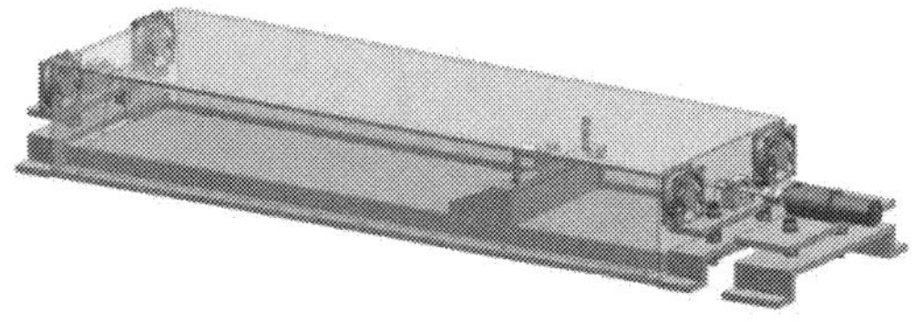

Figure 16. Blueprint of the film applicator device

5. CONCLUSION

Within the present activity the production of graphene based composite electrodes using a proven and economical tape casting method is investigated in order to later establish a scale-up manufacturing procedure for supercapacitor electrodes. Initial 3-electrode CV measurements of the laboratory scale samples exhibit the favorable characteristics of the graphene based electrodes with a maximum specific capacitance of 165 Fg^{-1} for the Graphene/MnO_2 composite. Furthermore, the produced electrodes are free-standing and flexible. In relevant works, specific capacitances of Graphene/MnO_2 composite electrodes are reported at 130 Fg^{-1} [10], 180 Fg^{-1} [2] and 216 Fg^{-1} [11]. However the capacitance of a supercapacitor is strongly dependent on the cell configuration used for the electrochemical measurement and it is always significantly higher when using a three-electrode system[12]. Therefore, in order to have a clearer image of their performance, the produced electrodes need to be integrated in 2-electrode cells and measured during Galvanostatic charging and discharging.

Efforts will now focus on further improvements to achieve higher specific capacitance for the produced materials while keeping the production cost low. First steps towards the establishment of a reliable and low cost industrial production process (scale-up) for these materials has been achieved and the first electrodes sheets that are being produced will need to be tested for their performance before the scale up can be considered successful. These results will be reported in future work when the entire project finalizes and concludes.

6. AKNOWLEDGMENTS

- Cyclic voltammetry testing of the produced composites was done at the Laboratory of Chemical & Electrochemical Processes of the Department of Chemical Engineering, University of Patras.
- The work reported in the present document pertains to the ESA commissioned activity "Graphene-based systems for enhanced energy storage" (ESA-GADGET-4000112857/14/NL /PA-CCN1).

7. REFERENCES

[1] A. Rashid bin Mohd Yusoff, *Graphene-Based Energy Devices*. 2015.

[2] J. Yan, Z. Fan, T. Wei, W. Qian, M. Zhang, and F. Wei, "Fast and reversible surface redox reaction of graphene–MnO2 composites as supercapacitor electrodes," *Carbon N. Y.*, vol. 48, no. 13, pp. 3825–3833, Nov. 2010.

[3] G. Yu, L. Hu, N. Liu, H. Wang, M. Vosgueritchian, Y. Yang, Y. Cui, and Z. Bao, "Enhancing the supercapacitor performance of graphene/MnO2 nanostructured electrodes by conductive wrapping," *Nano Lett.*, vol. 11, no. 10, pp. 4438–4442, 2011.

[4] M. Liang and L. Zhi, "Graphene-based electrode materials for rechargeable lithium batteries," *J. Mater. Chem.*, vol. 19, no. 33, p. 5871, 2009.

[5] H. Gómez, M. K. Ram, F. Alvi, P. Villalba, E. Stefanakos, and A. Kumar, "Graphene-conducting polymer nanocomposite as novel electrode for supercapacitors," *J. Power Sources*, vol. 196, no. 8, pp. 4102–4108, 2011.

[6] Z. Song, T. Xu, M. L. Gordin, Y. B. Jiang, I. T. Bae, Q. Xiao, H. Zhan, J. Liu, and D. Wang, "Polymer-graphene nanocomposites as ultrafast-charge and -discharge cathodes for rechargeable lithium batteries," *Nano Lett.*, vol. 12, no. 5, pp. 2205–2211, 2012.

[7] L. Zhaoping and Z. Xufeng, *Graphene, Energy storage and conversion Application*. CRC Press Taylor & Francis Group.

[8] M. D. Stoller, S. Park, Y. Zhu, J. An, and R. S. Ruoff, "Graphene-Based Ultracapacitors," pp. 6–10, 2008.

[9] P. Martins, A. C. Lopes, and S. Lanceros-Mendez, "Electroactive phases of poly(vinylidene fluoride): Determination, processing and applications," *Prog. Polym. Sci.*, vol. 39, no. 4, pp. 683–706, 2014.

[10] Y. He, W. Chen, X. Li, Z. Zhang, J. Fu, C. Zhao, and E. Xie, "Freestanding three-dimensional graphene/Mno2 composite networks as ultralight and flexible supercapacitor electrodes," *ACS Nano*, vol. 7, no. 1, pp. 174–182, 2013.

[11] S. Chen, J. Zhu, X. Wu, Q. Han, and X. Wang, "Graphene oxide-Mno2 nanocomposites for supercapacitors," *ACS Nano*, vol. 4, no. 5, pp. 2822–2830, 2010.

[12] Q. Cheng, J. Tang, J. Ma, H. Zhang, N. Shinya, and L.-C. Qin, "Graphene and carbon nanotube composite electrodes for supercapacitors with ultra-high energy density.," *Phys. Chem. Chem. Phys.*, vol. 13, no. 39, pp. 17615–24, 2011.

CORROSION OF CARBON SUPPORTS, INDUCED BY HIGH-TEMPERATURE PEMFC OPERATING CONDITIONS

Maria Rau [(1)], **Florina Jung** [(1)], **Carsten Cremers** [(1)], **Jens Tübke** [(1)]

(1) *Fraunhofer Institute for Chemical Technology ICT, Joseph-von-Fraunhofer-Str. 7, 76327 Pfinztal, Germany,*
E-mail: jens.tübke@ict.fraunhofer.de

ABSTRACT

In high-temperature polymer electrolyte membrane fuel cells (HT-PEMFC) one of the detrimental factors that lead to loss of performance is the corrosion of the electrocatalyst support. In both anode and cathode, platinum (Pt) supported on carbon is generally the selected electrocatalyst. Platinum significantly increases the speed of the anode and cathode reactions, but also causes carbon oxidation (corrosion of the support) and the consequent degradation of the electrode materials. In a HT-PEMFC, corrosion is influenced by temperature. For this reason, this paper will present a detailed evaluation of the carbon support degradation under HT-PEMFC conditions. The carbon oxidation of two types of carbon supports will be compared.

1. INTRODUCTION

High-temperature polymer electrolyte membrane fuel cells (HT-PEMFCs) are considered very promising electrochemical devices for the production of energy from hydrogen. Due to the range of operating temperatures (between 120°C and 180°C) the electroadsorption of CO in the active surface is significantly reduced in this type of fuel cell, which reduces the negative effects of CO in the electrodes [1]. Another advantage of HT-PEMFCs is that they allow the co-generation of cooling, heat and power (CHP) for auxiliary power unit applications [2]. However, a high efficiency of these electrochemical devices requires a higher stability of both electrodes (anode and cathode).
One of the damaging factors which play a role in the decay of electrocatalytic activity of the Pt supported on carbon (Pt/C) electrodes used in fuel cells is the corrosion of the carbon support. The influence of several factors, including the electrode fabrication parameters (Teflon content and sintering temperature) has been determined [3]. Electrochemical measurements carried out at 160°C for phosphoric acid fuel cells (PAFC) in the potential range between 0.6 V (vs. RHE) and 1.0V (vs. RHE) indicate that the corrosion rate increases with an increase in Pt content, and is most noticeable at 0.6V. These results show the strong influence of the presence of the supported metal on the extent of carbon corrosion under real conditions [4]. Moreover the influence of Pt activation on the corrosion of carbon in gas diffusion electrodes was investigated using differential electrochemical mass spectroscopy [5].
In a HT-PEMFC a phosphoric-acid-doped polybenzylimidazole (PBI) membrane for the ionic exchange of protons has been demonstrated to be the most successful electrolyte system for high-temperature operation [6]. Considering this, the evaluation of the corrosion of Pt-based electrocatalysts in two different supports under HT-PEMFC conditions in the potential window between 0.0 V (vs. RHE) and 1.0 V (vs. RHE) will be evaluated in this paper. A three electrode cell connected to a mass spectrometer [7] was used, which allows the study of a gas diffusion electrode.

2. EXPERIMENTAL PROCEDURE

Measurements were carried out in a three electrode cell working under HT-PEMFC conditions and connected to a mass spectrometer [7]. The working electrodes tested consisted of homemade Pt supported on carbon nanotubes (Pt/CNT) or carbon vulcan (Pt/CV) and deposited on a GDL H2315 (FREUDENBERG FCCT SE & CO.KG9). The electrodes had a geometric electrode area of 3 cm^2 and platinum loadings of 1.0 ± 0.1 mg_{Pt} cm^{-2}. H_3PO_4 85% (Carl Roth) was used as electrolyte. A Gaskatel HydroFlex® reversible hydrogen electrode (RHE) in the same medium was used as a reference electrode, and a large area Pt electrode as counterelectrode. Phosphoric-acid-doped polybenzimidazole (PBI) membranes were used as the polymer electrolyte membrane. Prior to the evaluation of the electrocatalytic corrosion, the cell was purged with a pure nitrogen flow, which was continued during the entire measurement procedure in order to maintain the electrode in an inert atmosphere. Under these conditions both electrodes were characterized through cyclic voltammetry at a potentiodynamic sweep of 10 mV / s in H_3PO_4 85% at 160°. The electrodes were cycled fifteen times until a stable profile was obtained. The last measurements corresponding to each electrode are presented in Figures 1 and 2.
A sequence of potential steps with a 5 min. duration pro step (time required to obtain the corresponding steady state current) in the range of $0.0 \leq E / V$ (vs. RHE) ≤ 1.0 was then applied to the electrodes (Figure 3). With this procedure, the corrosion of the electrodes operating as an anode (under 0.2V vs. RHE) or a cathode (over 0.8 V

 DOI: 10.1051/e3sconf/20171610001

vs. RHE) under inert conditions was evaluated.

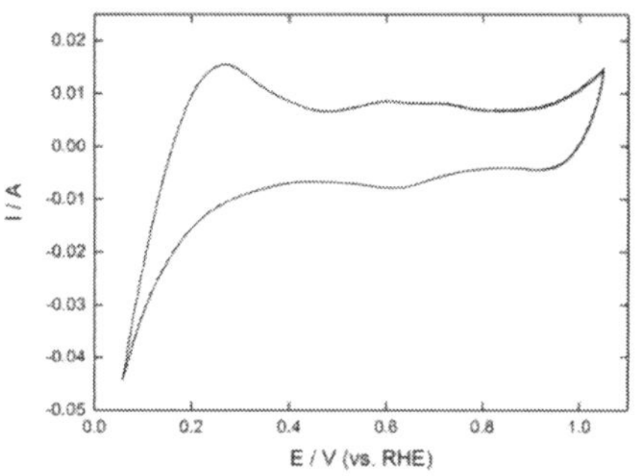

Figure 1. Cyclic voltammetry of Pt/CNT at 10 mV / s in H_3PO_4 85% at 160°C

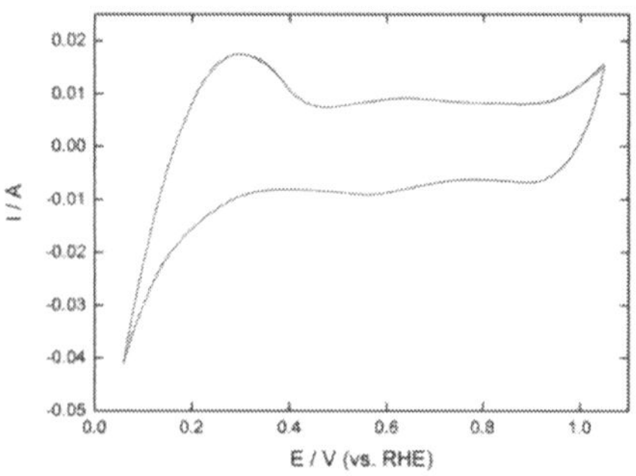

Figure 2. Cyclic voltammetry of Pt/CV at 10 mV / s in H_3PO_4 85% at 160°C

Figure 4 shows the dependence of the current on the potential obtained from the evaluation of the applied potential and the measured current. Simultaneously, measurements of the ionic mass 44, which is related to CO_2 formation and carbon corrosion, were obtained with the mass spectrometer (MS) connected on line. The mass spectrometer is a Pfeiffer vacuum system with a Prisma™ quadrupole analyzer.

Finally, the CVs of the electrodes in N_2 were recorded in order to demonstrate the stability of the materials.

3. RESULTS AND DISCUSSION

Figure 3 shows the potential program applied and the corresponding current obtained for a Pt/CNT (black) and Pt/CV (red) electrode with pure N_2. Under these conditions, the oxidation of the carbon supports can be attributed to independent processes: hydrogen evolution under 0.2 V vs. RHE and oxygen evolution over 0.6 V vs. RHE. It can be observed that for the potential steps under 0.6 V vs. RHE the currents measured reached a plateau, while for higher potentials a constant decay was noted. This could be due to a massive degradation of the electrocatalyst at these higher potentials, which did not allow the stabilization of the measured current during the acquisition time. From this measurement, the dependence of the current on the potential was obtained (Figure 4). At lower potentials (under 0.2 V vs. RHE) the measured currents are generated by the evolution of hydrogen. By contrast, at higher potentials (over 0.6 V vs. RHE) the current measured is due to oxygen evolution. From the comparison of the current dependences illustrated in Figure 4, the same dependence of the current on the potential was observed but for Pt/CV the currents for potentials over 0.5 V vs. RHE were slightly higher than for Pt/CNT. A possible cause could be that a larger area of activated Pt was exposed on the electrode supported on carbon vulcan.

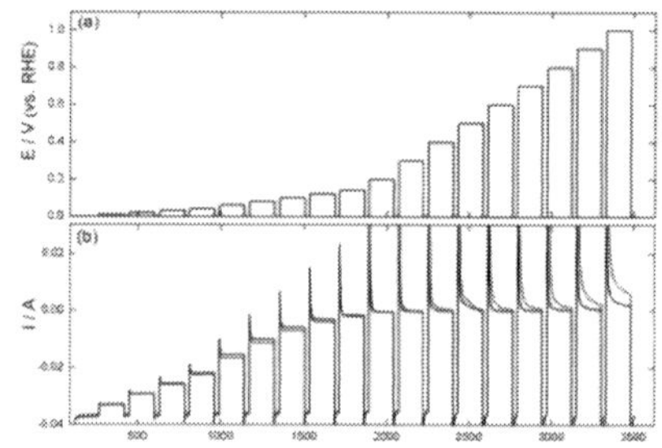

Figure 3. (a) Sequence of potential steps (a) applied for the evaluation of the carbon corrosion, (b) measured currents for Pt/CNT (black) and Pt/CV (red).

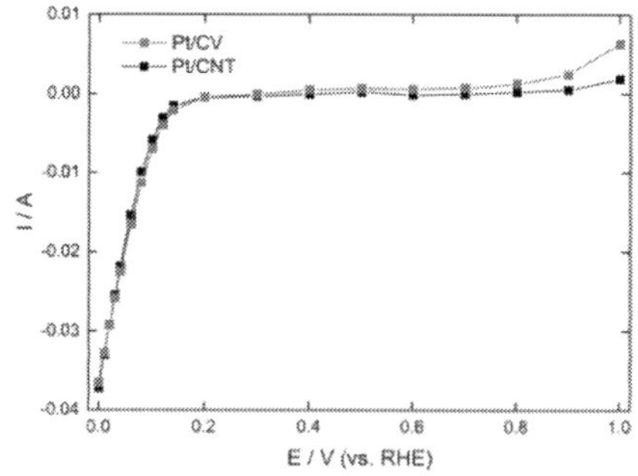

Figure 4. Dependence of the current on the potential for Pt/CNT (black) and Pt/CV (red).

In parallel to the electrochemical measurements (Figures 3 and 4), the ionic mass current for mass 44 (associated with CO_2) was recorded by a mass spectrometer connected to the three electrode cell. The dependence of the ionic mass currents for mass 44 on the potential, measured during the application of the potential steps (Figure 3), is illustrated in Figure 5. Here it can be noted that the ionic currents measured for

Pt/CV are four times greater than for Pt/CNT. These results are in agreement with the experimental investigation on carbon support corrosion rates in lower temperature PEM fuel cells carried out at the Fraunhofer Institute for Chemical Technology ICT under real operating conditions.

In Figure 5 two oxidation domains, represented by a peak in the oxidation ionic current, were determined. One small peak in the carbon corrosion is observed at 0.4V vs. RHE for Pt/CV and at 0.5V vs. RHE for Pt/CNT. A second oxidation peak was measured for higher potentials: at 0.9V vs. RHE for Pt/CV and at 1.0V vs. RHE for Pt/CNT. It is worth noting that even though these results are in good agreement with the previous observations made by Passalacqua *et al.* [4], a small difference in the measured potentials of higher corrosion could be related to the use of the PBI membrane, which could generate a displacement in the recorded values.

Figure 5 shows that in the range of potentials lower than 0.2 V vs. RHE, where the hydrogen oxidation takes place, no significant corrosion of the support (for both materials) was detected under HT-PEMFC conditions. By contrast, at higher potentials over 0.8 V vs. RHE a higher corrosion was determined. In this context, the higher corrosion of HT-PEMFC cathodes, which operates at higher potentials, is illustrated.

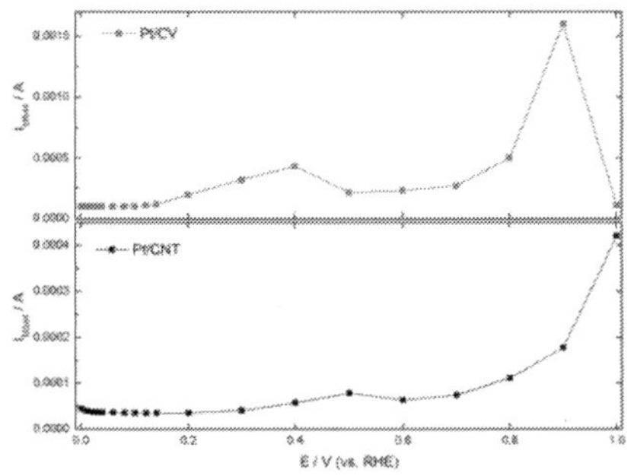

Figure 5. Dependence of the ionic mass current for mass 44 on the potential for Pt/CNT and Pt/CV.

In order to determine the intrinsic corrosion of the carbon support in the absence of platinum, the dependence of the ionic mass current for mass 44 on the potential for a carbon vulcan electrode (without addition of Pt) was obtained following the procedure already described for the Pt-containing electrodes. This evaluation is shown in Figure 6. Here it can be noted that for the carbon vulcan (CV) electrode (represented in blue) there is only one peak in the carbon dioxide formation at 0.9V vs. RHE and that the measured currents of this peak are in the same order as those of Pt/CV (represented in red). These results agree with previous research [4] that proved the massive corrosion observed at potentials higher than 0.9 V vs. RHE.

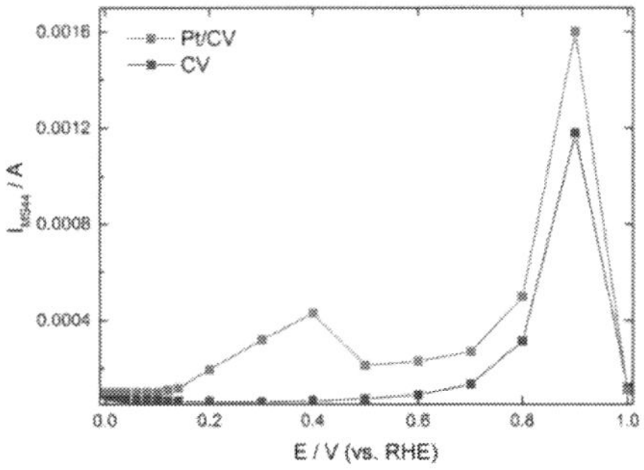

Figure 6. Dependence of the ionic mass current for mass 44 on the potential for Pt/CV (red) and CV (blue).

With this evidence it is possible to confirm that the presence of Pt in the HT-PEMFC has an important impact on the corrosion of the electrocatalyst support. This is despite of the fact that at potentials over 0.8 V vs. RHE the corrosion of the carbon support was identified as a common process resulting from the nature of the support and the highly corrosive working conditions of a HT-PEMFC.

In regard to the peak of corrosion observed at 0.4V vs. RHE for Pt/CV and at 0.5V vs. RHE for Pt/CNT, it can be noted that, as the active electrocatalyst is Pt in both cases, and as this Pt has the same intrinsic electrocatalytic activity for both electrodes, the difference in the corrosion peak is directly related to the support and the electrode morphology.

4. CONCLUSION

In the investigations described, different carbon supports were tested as electrodes used under HT-PEMFC operating conditions. It was found that the carbon corrosion currents are significantly smaller for the Pt/CNT electrode than for Pt/CV. Moreover a shift of about 0.1V to more oxidative potentials was observed for the corrosion peaks of Pt/CNT but only for lower potentials under 0.2 V vs. RHE. Not significant corrosion of the electrodes was noted. By contrast, for the higher potential range a significant oxidation of the carbon support for both studied materials was observed. These results provide an overview of the support oxidation taking place under inert conditions, where no electrode reaction takes place. In this context the impact of the potentials applied to the electrodes was determined. This provides valuable information while the electrode process is taking place. Moreover, information concerning the stability of the electrode material during the hydrogen evolution can be useful for the further development of Pt-based electrodes for water electrolysis.

5. ACKNOWLEDGEMENT

This work was financially supported by the German Federal Ministry for Economics and by the European Commission through the ERANET+ Electromobility.

6. REFERENCES

1. T. J. Schmidt, J. Baurmeister, ECS Trans., 3 (2006) 861 - 869.
2. G. Giacoppo, O. Barbera, A. Carbone, I. Gatto, A. Saccà, R. Pedicini, E. Passalacqua, Int. J. Hydrogen Energy, 38 (2013) 11619 - 11627.
3. N. Giordano, E. Passalacqua, V. Recupero, M. Vivaldi, E. J. Taylor, G. Wilenski, Elecrrochim. Aacta, 35 (1990) 1411 -1421.
4. E. Passalacqua, . L. Antonuci, M. Vivaldi, A. Patti, V. Antonucci, N. Giordano, K. Kinoshita, Electrochim. Acta, 37 (1992) 2725 - 2730.
5. J. Willsau, J. Heitbaum, J. Electroanal. Chem., 161 (1984) 93-101.
6. R. Kerr, H. R. García, M. T. Romero, C. Terkelsen, H. A. Hjuler, T. Steenberg, ECS Transactions, 64 (2014) 973 - 982.
7. C. Niether, M. S. Rau, C. Cremers, K. Pinkwart, J. Tübke, J. of Electroanal. Chem., 747 (2015) 97 - 103.

HIGH TEMPERATURE PEM FUEL CELL STACKS WITH ADVENT TPS MEAS

Stylianos Neophytides[(1)], Maria K. Daletou[(1)], Nikolaos Athanasopoulos[(2)], Nora Gourdoupi[(2)],

Emory De Castro[(2)], Max Schautz[(3)]

[(1)] Foundation of Research and Technology-Hellas, Institute of Chemical Engineering Sciences FORTH/ICE-HT, Patras 26504, Greece, Email: neoph@iceht.forth.gr
[(2)] Advent Technologies S.A., Patras Science Park, Stadiou Street, Patras26504, Greece, Email: ngourdoupi@advent-energy.com
[(3)] European Space Research and Technology Centre (ESTEC), The Netherlands, Email: max.schautz@esa.int

ABSTRACT

High power/high energy applications are expected to greatly benefit from high temperature Polymer Electrolyte Membrane Fuel Cells (PEMFCs). In this work, a combinatorial approach is presented, in which separately developed and evaluated MEAs, design and engineering are employed to result in reliable and effective stacks operating above 180°C and having the characteristics well matched to applications including auxiliary power, micro combined heat and power, and telecommunication satellites.

1. INTRODUCTION

The high temperature polymer electrolyte membrane fuel cells (HT PEMFCs) offer significant flexibility when used as energy converters in stationary, mobile and portable power devices. When compared to low temperature PEM fuel cells, subsystems for membrane humidification and carbon monoxide clean up can be eliminated. In addition to simplifying fuel cell systems, when used in electrolysis mode, dilute, contaminated hydrogen can be concentrated and purified.

Fuel cell components such as the electrocatalyst and electrode, microporous layer, membrane, gaskets, compression assembly elements, bipolar plates, end plates etc. are of great importance.

The Membrane Electrode Assembly (MEA) is the core component of the stack and should have specific properties to withstand the strong conditions during the fuel cell operation. The state of the art MEA technology is based on H_3PO_4 imbibed polymer electrolytes and Pt based catalysts. The most widely used and studied polymer for HT PEMFC is polybenzimidazole (PBI) and its derivatives imbibed with phosphoric acid since this class has excellent mechanical and thermal properties and high ionic conductivity under anhydrous conditions. A stability of >18,000 hrs of continuous operation has been recorded [1]. Alternative polymers that have gained ground in recent years are pyridine based aromatic polyethers [2]. The acid–base interactions between these types of polymers and the phosphoric acid results in ionic materials that are highly conductive and have shown very promising results during the fuel cell operation for 4,000 hrs [3].

In addition to the MEA, the performance of a PEM fuel cell stack is affected by many internal and external factors, such as fuel cell design and assembly, degradation of materials, operational conditions, and impurities or contaminants. Furthermore, the design and the concept of a HT PEMFC stack are mainly determined by the application and the environmental conditions, and in most cases cost and weight limitations need to be considered.

For example in space applications, where regenerative fuel cells could become the superior choice as an alternative to rechargeable (secondary) batteries, a lightweight system is the most critical requirement.

This work focuses on the design and manufacturing of:

-HT PEM fuel cells stacks of 1 kW electrical and 1 kW thermal power for APU and CHP applications using graphite resin bipolar plates and Advent's TPS® MEAs.

-Design of HT PEM fuel cell of 3 kW electric power using metal bipolar plates and Advent TPS MEAs for regenerative PEM fuel cell system for telecommunication satellites.

2. RESULTS AND DISCUSSION

2.1 HTPEM FC stacks with graphite plates for APU and CHP applications.

Based on our previous experience the active area of choice for a HT PEMFC of 1 kW electric power output is 177 cm^2 for APU and CHP applications. The first step was to build an air cooled stack in order to study the bipolar plate design, the gaskets and the operation of the stack at 1 kW. For its construction, 36 Advent TPS MEAs of 177 cm^2 active area (GDE area 191 cm^2) were used.

The individual MEA parts are shown in Figure 1.

The 36-MEAs stack system -which was used as a "proof of concept"- was manufactured utilizing materials such as stainless steel end plates, graphite composite bipolar plates with fins, gold plated current collectors, cartridge heater for start up, which were not optimized in terms of design, weight and volume.

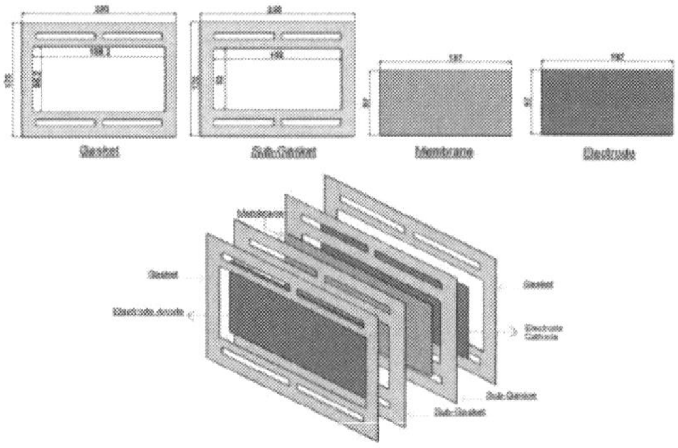

Figure 1: Advent TPS MEA dimensions.

The material of the bipolar plate was a composite of graphite and polymer resin and the flow field was based on a design of prior knowledge. This has been modified accordingly at the inlets and outlets of the gases in order to cope with possible internal leaks. As shown in Fig.2, the fins of the plates were used for the air cooling approach.

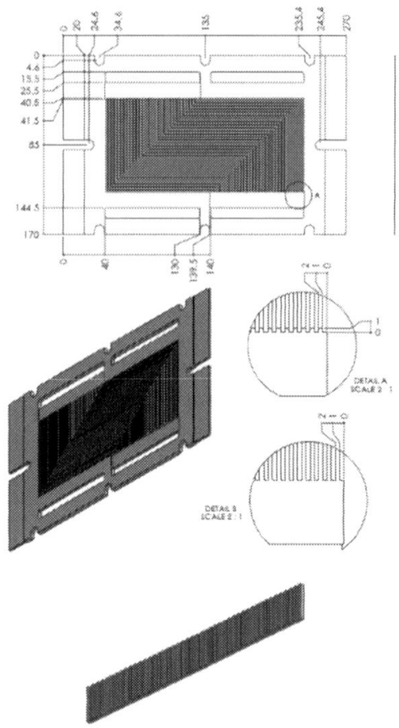

Figure 2: View of the top side of the bipolar plate and block for inlet and outlet of the gases.

Upon assembly, the stack (Picture 1) was checked and found free of external (from the stack core out) and internal leakages (from one compartment to another). The stack was consequently coupled with a steam reforming LPG/LNG fuel processor. For the start up of the stack, cartridges were placed on the bipolar plates edges (cartridges' temperature was 200°C).

The MEA performance was consistent and the power output of 1 kW was reached as shown in Fig. 3 using reformate gas (80.4% H_2, 18% CO_2, 1.25% CH_4, 0.35 % CO) and air. The reformate and air flows were running in counter-flow and the maximum temperature reached was 183°C, recorded at the centre of the stack.

Picture 1: Side view of the 1kW air cooled stack.

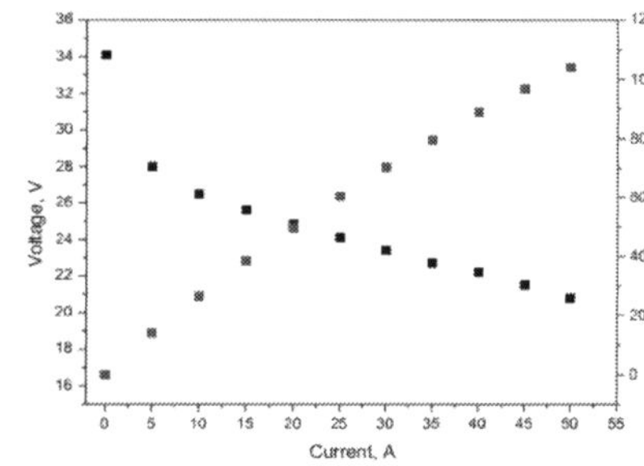

Figure 3. I-V curve of 1 kW air cooled stack, Temperature at the center of the stack (highest): 183°C, Feed: Reformate gas (1 m^3/h) / air (3.9 m^3/hr).

A new concept of external oil cooling has been recently developed as a continuation and optimization of the air cooled stack, which can be applied over a wider range of applications, including CHP, and can be integrated more efficiently in the thermal balance of the Balance of Plant (BoP) of the fuel processor. The external heating/cooling design approach was successfully developed according to the temperature uniformity and heating/cooling rate limitations. Apart from the cooling design, additional changes have been made on the peripheral structure of the current stack in order to reduce the weight of the final system. Weight and volume optimization of the stack has been achieved through the extensive use of advanced parametric CAD/CAE software (Fig.4).

DOI: 10.1051/e3sconf/20171610002

Figure 4: 3D iso view of the liquid cooled stack

Detailed mechanical and thermo-mechanical models have been developed for each part separately in order to predict the stress and strain fields and reduce the weight of the stack. Special attention was given for the uniform distribution of the compression forces over the MEAs through the compression of the specially designed endplates. Moreover the cost development of the final prototype was reduced due to the usage of the aforementioned numerical modeling studies. Advanced alloys and plastics with high specific mechanical properties have been selected, while the selection of the materials for the various parts was based strictly on the optimization of the weight and the final cost.

Based on theoretical calculations, a 41-MEAs stack has been assembled, using MEAs of 177 cm^2 area that comprise an optimized cross linked acid-filled membrane. It should be noted that the performance of this type of MEA is lower than those used in the air cooled stack, but were chosen due to their improved tolerance to mechanical stresses and higher operational temperatures. The bipolar plates were of the same material, same active area and design as in the air cooled stack but with fins excluded.

In Picture 2, the oil cooled stack is depicted.

Picture 2: Side view of the 1kW liquid cooled stack.

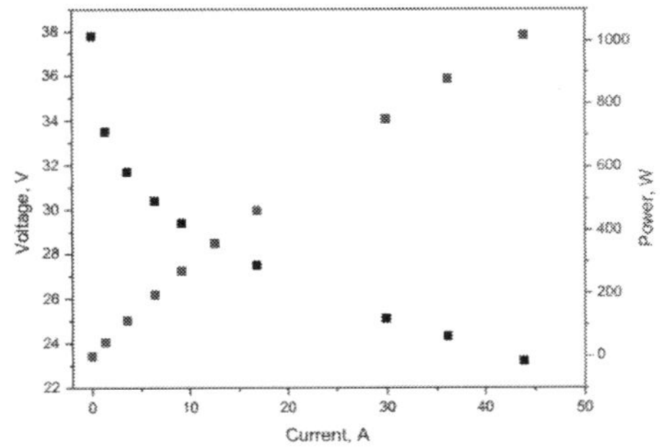

Figure 5. I-V curve of 1 kW liquid cooled stack, Temperature: 193°C, Feed: H_2 (0.84 m^3/h) / air (4 m^3/hr).

The MEA performance was consistent and the power output of 1 kW was reached with hydrogen and air, as shown in Fig. 5.

2.2 HT PEMFC stacks with metal plates for telecom satellites

As already mentioned, RFCs (Regenerative fuel cells) are considered a promising alternative for applications requiring either high power levels or missions with long eclipse duration, hence high energy demand, due to their potentially high energy density compared to batteries. Within the framework of ESA Contract No. 4000109578/13/NL/SC, among others, a 3 kW liquid cooled HT PEMFC stack has been designed. To date, a thorough and extended study has been performed, leading to the selection of the materials that will be used for the realization of the final device. The selected metal and peripheral parts have specific properties in order to achieve a lightweight and reliable structure for space environment according to the specifications set by ESA. Based on our calculations, the dimensions of the bipolar plate are (105.1 mm x 216.6 mm), with active area equal to 100 cm^2 as depicted in Fig.6.

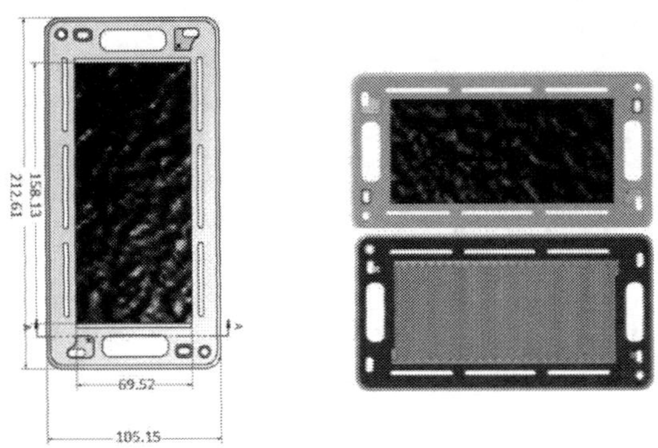

Figure 6. MEA dimensions and bipolar plate cross section.

 DOI: 10.1051/e3sconf/20171610002

The material of bipolar plates was chosen to be Tantalum (99.9%) coated Titanium alloy. The iso-view drawings of the HT PEMFC stack are presented in Fig. 7.

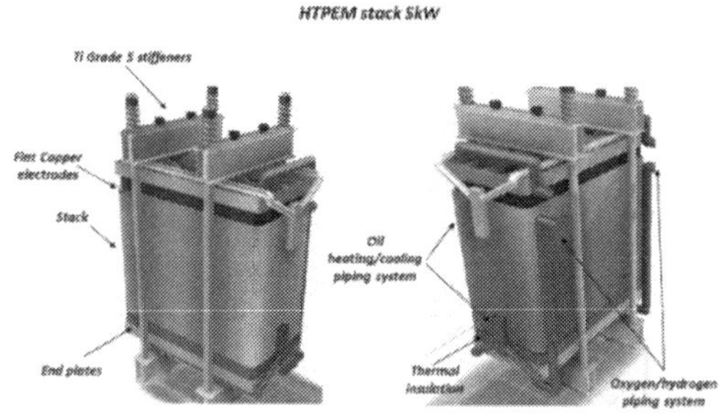

Figure 7. Two iso-views of the HT PEMFC stack and the various parts.

For the metal stack design, the heating/cooling approach will be different compared to the graphitic HT PEMFC stack presented in section 2.1. The coolant/heating oil will flow inside the rectangular channels of all the bipolar plates, as can be observed in the detail drawing of Fig. 8.

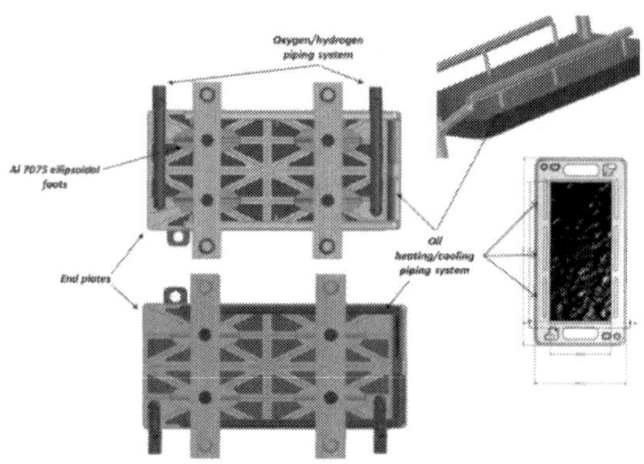

Figure 8. Two different plan-views of the 3 kW HT PEMFC stack where internal heating/cooling is represented.

Different design approaches have been considered in order to simulate the final power density of the stack using various thicknesses of the bipolar plate and Tantalum coating. According to calculations and designs, the maximum power density of a 5 kW stack is 1.2 kg/kW and can be achieved if 14.5 μm Tantalum coating and Titanium bipolar plate of 0.1 mm thickness are used. This power density is within that needed for viable satellite applications. Further progress is currently underway.

Acknowledgements
Funding from the EU's Seventh Framework Program (FP7/2007-2013) for the FCH JTI under the grant agreements 325358 IRMFC and 325368 DeMStack, as well as from the European Space Agency under the Contract No. 4000109578/13/NL/SC.

References

[1]Yu, S., Xiao L. & Benicewicz B.C. (2008). Durability Studies of PBI-based High Temperature PEMFC. Fuel Cells 8(3-4), 165–174.
[2] Kallitsis, J.K., Andreopoulou, A.K., M.K. Daletou & Neophytides, S. (2016). Pyridine Containing Aromatic Polyether Membranes, Ch. 5 in "High Temperature Polymer Electrolyte Fuel Cells- Approaches, Status and Perspectives", J.O. Jensen, D. Aili, H.A. Hjuler, Q. Li (Eds.), 91-126.
[3]http://fuelcell.com/wp-content/uploads/2013/10/Advent-MEA-Product-Sheet.pdf

OPTIMUM ENERGY MANAGEMENT OF PEM FUEL CELL SYSTEMS BASED ON MODEL PREDICTIVE CONTROL

Chrysovalantou Ziogou[(1)], Michael C. Georgiadis[(2)], Spyros Voutetakis[(1)], Simira Papadopoulou[(3)]

[(1)] *Chemical Process and Energy Resources Institute (CPERI), Centre for Research and Technology Hellas (CERTH), Thessaloniki, Greece, Email: cziogou@cperi.certh.gr, paris@cperi.certh.gr*
[(2)]*Department of Chemical Engineering, Aristotle University of Thessaloniki, Thessaloniki, Greece, Email:mgeorg@auth.gr*
[(3)] *Department of Automation Engineering, Alexander Technological Educational Institute of Thessaloniki, Thessaloniki, Greece, Email: shmira@autom.teithe.gr*

ABSTRACT

This work presents an optimum energy management framework, which is developed for integrated Polymer Electrolyte Membrane (PEM) fuel cell systems. The objective is to address in a centralized manner the control issues that arise during the operation of the fuel cell (FC) system and to monitor and evaluate the system's performance at real time. More specifically the operation objectives are to deliver the demanded power while operating at a safe region, avoiding starvation, and concurrently minimize the fuel consumption at stable temperature conditions. To achieve these objectives a novel Model Predictive Control (MPC) strategy is developed and demonstrated. A semi-empirical experimentally validated model is used which is able to capture the dynamic behaviour of the PEMFC. Furthermore, the MPC strategy was integrated in an industrial-grade automation system to demonstrate its applicability in realistic environment. The proposed framework relies on a novel nonlinear MPC (NMPC) formulation that uses a dynamic optimization method that recasts the multivariable control problem into a nonlinear programming problem using a warm-start initialization method and a search space reduction technique which is based on a piecewise affine approximation of the variable's feasible space.

The behaviour of the MPC framework is experimentally verified through the online deployment to a small-scale fully automated PEMFC unit. During the experimental scenarios the PEMFC system demonstrated excellent response in terms of computational effort and accuracy with respect to the control objectives.

1. INTRODUCTION AND MOTIVATION

The shift towards a low carbon, efficient and secure economy requires targeted deployment of innovative technologies and increased exploitation of renewable energy sources. Furthermore, it is recognized that a technological shift and the development of new clean technologies are vital for a successful transition to a decarbonized and sustainable future economy. Although a number of diverse technologies exist that aim at the same target, the synergy between the increased use of renewable energy sources, renewable hydrogen and electricity from fuel cells represent one of the promising ways to realize sustainable energy. Fuel cell and hydrogen technologies have the potential to contribute to the ambitious energy and climate objectives of the European Union for 2020 which are the reduction of the greenhouse gas emissions by 20%, the increase of the share of renewable energy to 20% and the improvement of the energy efficiency by 20% [1].

1.1 Importance of Control Strategies to FC systems

In order to take advantage of FC's great potential a number of technological challenges must be addressed that require significant efforts of research and development. Continuous research efforts are imperative for the realization of FC system-level goals for performance, predictability, stability, and other properties through appropriate analysis, design and implementation. In this context, the optimum energy management through appropriate control methods can function as a catalyst that transforms technological innovation to systems engineering novelties.

The impact of control technology is evident in a wide range of application areas, including fuel cells, as it is the necessary facilitator for achieving desired objectives and fulfilling application-specific goals [2]. Fuel cell systems exhibit fast dynamics, nonlinearities and uncertainties that constitute challenges requiring appropriate control in order to be confronted effectively. The use of efficient control strategies would not only increase the performance of these systems, but would increase the number of operational hours as their lifetime is preserved by operating at optimal levels and also reduce the cost per produced kilowatt-hour. Overall control can be considered as an enabling technology for the optimum energy management of fuel cell systems. Based on this motivation initially the experimental system is described along with the basic modelling approach (Section 2). Subsequently the advanced control (Section 3) and structure of the online energy management framework is discussed (Section 4) which is deployed to the PEMFC into consideration.

 DOI: 10.1051/e3sconf/20171610003

2. PEM FUEL CELL EXPERIMENTAL SETUP AND MODELING

In this work a small-scale fully automated fuel cell unit was used which was designed and constructed at CPERI/CERTH. Fig. 1 presents the simplified process and instrumentation (P&ID) diagram of the experimental PEMFC along with the measured variables and the control actions that will be determined by the the advanced control scheme. The variables that constitute the Input Output (I/O) field of the unit are acquired by a Supervisory control and Data Acquisition (SCADA) system. The PEM fuel cell is by Electrochem®, has a Nafion 1135 (89μm thickness) membrane electrode assembly (MEA) with an active area of 25cm^2, while the power is drawn from the fuel cell through a DC load (KIKUSUI®). The air and the hydrogen are supplied from pressurized cylinders by two mass flow controllers (MFCs). Hydrators and heated lines are used to provide the proper humidity of the inlet gases. The temperature is maintained by an air cooling subsystem and an electrical resistance.

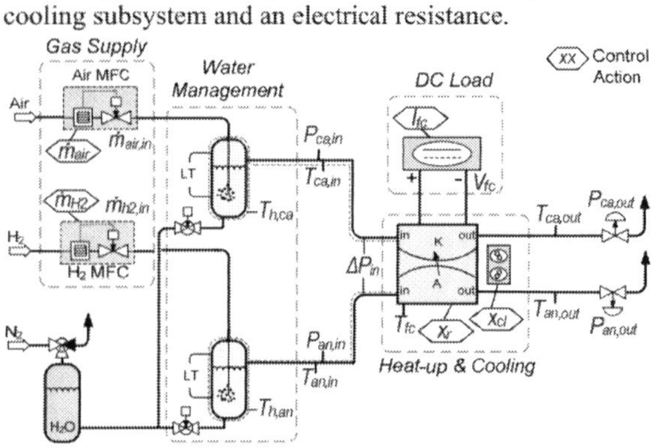

Figure 1. Simplified P&ID of CERTH/CPERI's PEMFC unit

The objective is to manage the produced energy by proper manipulation of the four distinct, yet interacting, subsystems which are related to the power production, the gas supply, the temperature and the water management. Each subsystem has its own control objective and a number of sensors and actuators are used to monitor the system's behavior.

The inlet gasses, the FC temperature and the power are controlled by the proposed model-based framework while the hydrator's temperature and the heated lines are controlled by conventional PIs due to their slow dynamic behaviour. All other system components (pumps, heaters, valves, etc) are controlled by the SCADA system.

2.1 Nonlinear Dynamic Model

The fuel cell unit (Fig. 1) was modelled by an experimentally validated semi-empirical nonlinear dynamic model [3] which considers the mass dynamics in the gas flow channels, the gas diffusion layers (GDL) and the membrane. Also the respective electrochemical equations are used for the calculation of the voltage and an energy balance differential equation is integrated to capture the temperature evolution. An illustrative representation of the involved variables and main components of the PEMFC model is show at Fig. 2.

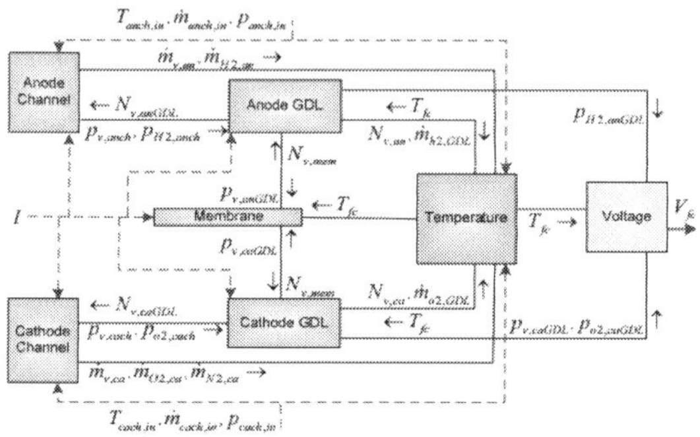

Figure 2. PEMFC mathematical model structure

The developed model couples both the theoretical validity (mechanistic equations) and the inherent simplicity of the empirical application (semi-empirical equations). The described dynamic model is not system dependent as it can be adjusted to describe any other PEM fuel cell system by performing a number of sequentially executed actions. These actions are determination of

- the physical characteristics of the PEM fuel cell,
- the unit specific parameters
- the empirical parameters utilized by the electrical subsystem of the model.

The value of the empirical parameters are determined by a systematic parameter estimation procedure as long as experimental data or data from the manufacturer are provided.

3. ADVANCED MODEL-BASED PREDICTIVE CONTROL STRATEGIES

The objective of advanced control strategies is to achieve a set of predefined conditions for the process and maintain the operation at the desired or optimal values [4]. Model predictive control (MPC) also known as receding horizon control (RHC), is becoming a preferred control strategy for a large number of industrial processes [5]. The main reasons for the increased popularity include the ability to handle state and input constraints, dynamic nonlinearities of the process into consideration and simultaneous satisfaction of economical and operating constraints under an integrated optimization methodology. Moreover recent advances in optimization enable the move towards direct online optimizing control. These characteristics in combination with the demanding operation of a PEMFC, signify that the application of NMPC to such

DOI: 10.1051/e3sconf/20171610003

systems is a suitable approach. In the current work an MPC formulation is selected to effectively handle the interrelated control objectives of a PEMFC.

In general MPC refers to a methodology which makes explicit use of a process model to optimize the future predicted behavior of a process. The main objective is to obtain control actions that minimize a cost function related to selected objectives or performance indices of the system. At each sampling time an optimal control problem is solved using measurements acquired from the system and it yields the appropriate control inputs for the system.

MPC computes online a finite-time constrained optimization problem over a prediction horizon (T_p), using the current state of the process as the initial state. The optimization yields an optimal control sequence ($u_k..u_{k+Nc}$) over a control horizon (T_c), which is partitioned into N_c intervals and only the first control action (u_k) for the current time is applied to the system. We consider the following formulation of the MPC problem [6]:

$$\min_u J = \sum_{j=1}^{Np} (\hat{y}_{k+j} - y_{sp,k+j})^T QR(\hat{y}_{k+j} - y_{sp,k+j}) + \sum_{l=0}^{Nc-1} \Delta u_{k+l}^T R1 \Delta u_{k+l} \quad (1)$$

$$\text{s.t.:} \quad \dot{x} = f_d(x,u), \quad y = g(x,u) \quad (1a)$$

$$e_k = (y_{meas} - y_{pred})_k \quad (1b)$$

$$\hat{y}_{k+j} = y_{pred,k+j} + e_k \quad (1c)$$

where u,y,x are the manipulated, the controlled and the state variables, $y_{meas}, y_{pred}, y_{sp}$ are the predicted, the measured variables and the desired set-points and Q, R are the output tracking and the input move weights. The minimization of functional J (Eq(1)) is subject to constraints of u and y.

In this work, two MPC-based strategies are combined in a unified control framework in order to take advantage of their synergistic benefits [7]. The first methodology is an online Nonlinear Model Predictive control (NMPC) strategy, which is very appealing due to its ability to handle dynamic nonlinearities of the process under consideration [6]. The second methodology is an explicit or multi-parametric Model Predictive Control (mpMPC) strategy. The mpMPC can provide the optimal solution in real-time, as the solution is computed offline and can be implemented online by simple look-up functions [8]. Both approaches handle state and input constraints and satisfy operating objectives. These characteristics in combination with the tight operation of a PEMFC, signify that an advanced control framework is a suitable approach.

4. INTEGRATED ONLINE ENERGY MANAGEMENT FRAMEWORK

Overall the online energy management framework used in this work, has a number of interacting entities that form a computer-aided platform for the monitoring and control of the PEMFC unit [9]. The information flow, from the I/O field to the end user, is shown at Fig. 3. Furthermore, Fig. 3 illustrates the interconnection of the control system's entities and the I/O field including the conceptual flow of information from the signal acquisition to the final presentation at the user level through the development of a graphical interface.

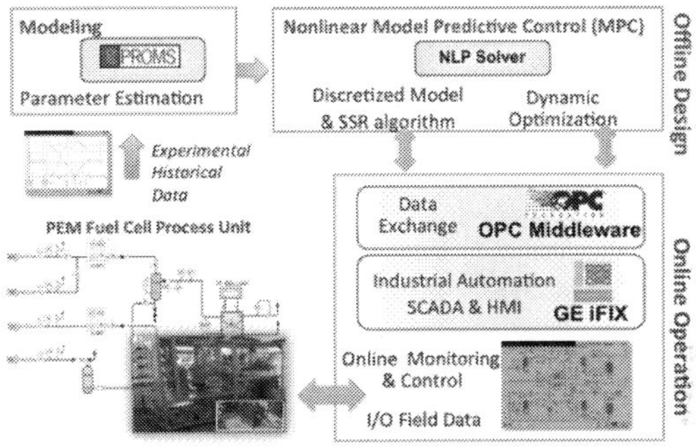

Figure 3. Information flow of the Energy Management Framework of CERTH/CPERI's PEMFC unit

The MPC framework is deployed online to the fuel cell unit using a custom made OPC-based interface that was developed for the communication between the optimizer and the SCADA system [8]. The developed framework has the following desired features:

- Fast calculation of the optimal control actions while taking into account the physical and operating constraints.
- Flexibility to adapt to changing fuel cell response under the influence of disturbances or during start-up and shutdown.
- Incorporation of a multitude of performance criteria under strict computational time demands.
- Easily deployable and maintainable control and optimization solutions.

As far as the MPC related parameters are concerned, the prediction horizon is selected to be sufficiently long with respect to the dynamics of the fuel cell (T_p=5sec) divided into N_p intervals. The control horizon (T_c) was set to be equal to the sampling time of the SCADA system (500ms) and it is divided into N_c. Finally the selected optimization method is the direct transcription using a reduced gradient NLP solver. The nonlinear fuel cell model is discretized based on OCFE. More specifically there are 10 finite elements (NE) with 4 collocation points (N_{cop}) each.

5. PEM FUEL CELL EXPERIMENTAL RESULTS

Overall the objectives for the control system are to effectively address the issue of power generation in an optimum manner. In this context the optimality is defined by the following three terms:

- Operation at a safe region regardless of the load fluctuations.
- Minimization of the fuel consumption and air supply.
- Maintenance of stable temperature conditions ensuring proper gas humidification.

The safe operation is maintained by controlling the reactants at a certain excess ratio level in order to avoid starvation caused by sub-stoichiometric reaction conditions at the cathode and the anode. The safe operating region for the cathode and the anode is defined by two unmeasured variables, the oxygen and hydrogen excess ratios (λ_{O2}, λ_{H2}), expressed as the ratios of the input flow of each gas to the consumed quantities per unit time due to the reaction [10]:

$$\lambda_{O2} = \frac{\dot{m}_{O2,cach,in}}{\dot{m}_{O2,caGDL}} \quad (2)$$

$$\lambda_{H2} = \frac{\dot{m}_{H2,anch,in}}{\dot{m}_{H2,anGDL}} \quad (3)$$

where $\dot{m}_{O2,cach,in}, \dot{m}_{H2,anch,in}$ are the oxygen and hydrogen input flows at the channels while $\dot{m}_{O2,caGDL}, \dot{m}_{H2,anGDL}$ are the respective reacted quantities. In order to reach the required excess ratio set-point the air and hydrogen flows ($\dot{m}_{air}, \dot{m}_{H2}$) are used as manipulated variables. The safety of the operation is ensured by maintaining the excess ratios above one ($\lambda_{O2,SP}>1$, $\lambda_{H2,SP}>1$).

The power generation and the starvation avoidance objective can be achieved by control actions that aim at an accurate set-point tracking of P_{SP}, $\lambda_{O2,SP}$, $\lambda_{H2,SP}$. On the other hand the temperature control ($T_{fc,SP}$) involves two mutually exclusive subsystems, one for the heat-up and another for the cooling

5.1 Online PEMFC Operation

An indicative result form the online operation of the PEMFC is presented. The response of the MPC strategy to various power demands was studied at specific operating conditions of temperature and pressure (T=338K, Pt=1bar) is shown in this section. The power modifications were within a range of 1W to 5W, which covers the operational range of the system. The power response to the corresponding control actions of the manipulated variables (Fig. 4,5) as the power demand changes while the oxygen and hydrogen excess ratio remains at a constant set-point ($\lambda_{O2,sp}$ = 3, $\lambda_{H2,sp}$ = 2.5).

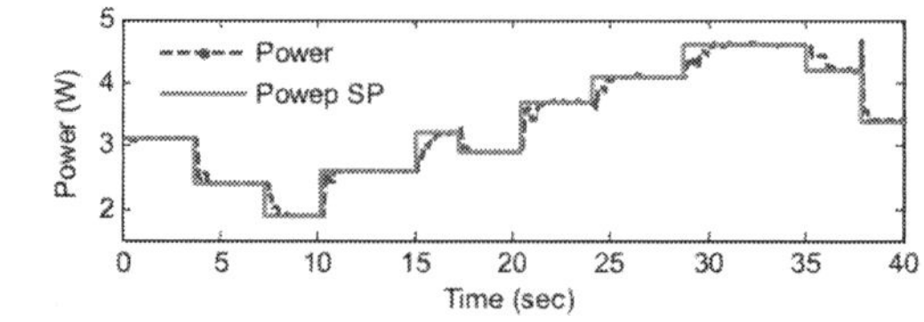

Figure 4. Power demand changes and delivered power by the NMPC controller

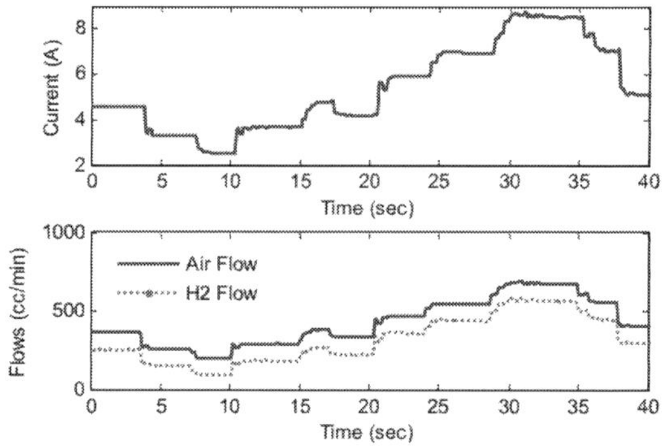

Figure 5. Control actions for the manipulated variables of (a) Current and (b) air and hydrogen mass flows derived by the NMPC controller

The NMPC controller is able to control the fuel cell power to any admissible set-point while maintaining the required excess ratio level.

The mean squared error (MSE) for each objective and some performance metrics are presented at Tab 1.

Table 1. Performance of the Online NMPC framework

MSE from Set-point	Performance Metrics
Power: 20mW	Max. opt. time: 419ms
Temperature: 0.6°C	Average opt. time (SP change): 320ms
λ_{O2}: 1.5*10^{-3}	Average opt. time (steady state): 275ms
λ_{H2}: 2.1*10^{-3}	

The NMPC controller that was deployed to the PEMFC system can efficiently address the issues of power generation in a safe and controlled manner. Overall, it has been clearly illustrated that the NMPC framework is able to deal with uncertainties and achieve trajectory tracking in a satisfactory manner. Also, the NMPC controller has computational requirements that satisfy the time sampling interval without convergence failures while satisfying realistic constraints imposed by the nature of the PEM fuel cell system.

The presented initial results demonstrate that the proposed control method can be applied online at realistic time frame and enable the achievement of multiple, often conflicting, objectives in a dynamically changing environment. Furthermore, due to the nonlinear nature of the model the NMPC approach is capable to adjust to the fuel cell degradation that will shift the operation of the system out of the initial area of operation in time.

6. CONCLUSIONS

In this work an integrated energy management approach that relies on a MPC strategy is presented using an advanced MPC strategy to address the operation objectives of a PEMFC unit. A synergy that reveals the potential of exploiting at best the intrinsic features of two model-based control methods (explicit and nonlinear MPC) and the handling of an interesting control problem that involves the optimum operation of an experimental PEM fuel cell unit is demonstrated by the proposed framework. A challenging multivariable nonlinear control problem with measured and unmeasured variables that involves concurrently four operation objectives for a PEM fuel cell system is addressed.

The proposed framework is deployed to the experimental fuel cell and it is validated online. The performance of the MPC strategy shows promising behavior in terms of computational demands and at the same time it guarantees that the PEMFC system operates at an optimum manner.

7. REFERENCES

1. European Commission (2011). A roadmap for moving to a competitive low carbon economy in 2050, COM(2011) 112, Brussels.

2. CSS (2011). The Impact of Control Technology. T. Samad and A.M. Annaswamy (eds.), IEEE Control Systems Society, www.ieeecss.org (Accessed March 2016).

3. Ziogou C., Voutetakis S., Papadopoulou S., Georgiadis M.C. (2011). Modeling, simulation and experimental validation of a PEM fuel cell system, Computers and Chemical Engineering, 35 (9), 1886-1900.

4. Qin, S. and Badgwell, T. (2003). A survey of industrial model predictive control technology. *Control Engineering Practice*, 11(7), 733-764.

5. Bauer, M., and Craig, I.K. (2008). Economic assessment of advanced process control – A survey and framework. *Journal of Process Control*, 18 (1), 2-18.

6. Allgöwer F., Findeisen R., Nagy Z. K. (2004). Nonlinear Model Predictive Control: From Theory to Application, J. Chin. Inst. Chem. Engrs., 35 (3), 299-315.

7. Ziogou C., Pistikopoulos E. N., Georgiadis M. C., Voutetakis S., Papadopoulou S., (2013). Empowering the performance of advanced NMPC by multi-parametric programming – An application to a PEM fuel cell system, Industrial Engineering and Chemistry Research, 52 (13), 4863–4873.

8. Bemporad, A., Morari, M., Dua, V., Pistikopoulos, E. N. (2002). The explicit linear quadratic regulator for constrained systems. Automatica, 38, 3-20.

9. Ziogou C., Papadopoulou S., Georgiadis M. C., Voutetakis S. (2013). On-line nonlinear model predictive control of a PEM fuel cell system, Journal of Process Control, 23(4), 483-492.

10. Pukrushpan, J. T., Peng H., and Stefanopoulou, A.G. (2004). Control-oriented modeling and analysis for automotive fuel cell systems. Journal of Dynamic Systems Measurement and Control-Transactions of the ASME, 126(1), 14-25.

OPTIMIZED HIGH TEMPERATURE PEM FUEL CELL & HIGH PRESSURE PEM ELECTROLYSER FOR REGENERATIVE FUEL CELL SYSTEMS IN GEO TELECOMMUNICATION SATELLITES

Jarle Farnes [1], Dmitry Bokach [2], Sander ten Hoopen[3]

Kim Skåtun[4], Max Schautz [5], Xavier Geneste[6], Arild Vik [7],

[1] Researcher at CMR Prototech, Fantoftvegen 38 PO Box 60345892 Bergen Norway, Jarle.farnes@prototech.no
[2]Researcher at CMR Prototech, Fantoftvegen 38 PO Box 60345892 Bergen Norway, Dmitry.bokach@prototech.no
[3]CEO at Ideevolutie, European Space Innovation Centre (ESIC) Kapteynstraat 12201 BB Noordwijk, The Netherlands, sander.tenhoopen@ideevolutie.nl
[4]Project engineer at CMR Prototech, Fantoftvegen 38 PO Box 60345892 Bergen Norway, kim.skatun@prototech.no
[5] European Space Agency, ESTEC - Keplerlaan 1, 2201 AZ Noordwijk Zh, The Netherlands, max.schautz@esa.int,
[6] European Space Agency, ESTEC - Keplerlaan 1, 2201 AZ Noordwijk Zh, The Netherlands, Xavier.geneste@esa.int
[7] Technology Director at CMR Prototech, Fantoftvegen 38 PO Box 6034 5892 Bergen Norway, arild.vik@prototech.no

ABSTRACT

Next generation telecommunication satellites will demand increasingly more power. Power levels up to 50 kW are foreseen for the next decades. Battery technology that can sustain up to 50 kW for eclipse lengths of up to 72 minutes will represent a major impact on the total mass of the satellite, even with new Li-ion battery technologies. Regenerative fuel cell systems (RFCS) were identified years ago as a possible alternative to rechargeable batteries. CMR Prototech has investigated this technology in a series of projects initiated by ESA focusing on both the essential fuel cell technology, demonstration of cycle performance of a RFCS, corresponding to 15 years in orbit, as well as the very important reactants storage systems. In the last two years the development has been focused towards optimising the key elements of the RFCS; the HTPEM fuel cell and the High Pressure PEM electrolyser. In these ESA activities the main target has been to optimise the design by reducing the mass and at the same time improve the performance, thus increasing the specific energy. This paper will present the latest development, including the main results, showing that significant steps have been taken to increase TRL on these key components.

1. INTRODUCTION

A Regenerative Fuel Cell System (RFCS) based on hydrogen and oxygen consists of a storage system for reactants (H_2, O_2 and H_2O), a fuel cell, and an electrolyser. During charging, the electrolyser converts water to hydrogen and oxygen by supply of photovoltaic power. During discharge the fuel cell converts hydrogen and oxygen to water and generates electrical power and waste heat. Within the ESA Technology Research Programme (TRP) and ARTES 5 programs activities have been undertaken to develop and test a Regenerative Fuel Cell System to replace batteries on GEO telecommunication satellites in the long term. A first order comparison has been performed of a fuel cell system with Li-ion batteries [8] and it showed a significant mass advantage for large platforms with power levels approaching 20 kW. Next generation telecommunication satellites will put increasingly higher requirements on the power supply system. Power levels of up to 50 kW are foreseen for the next 15 years. Today, rechargeable batteries serve as secondary power, but battery systems that can sustain >30 kW for eclipse lengths of up to 72 minutes will represent a major impact on the total mass of the satellite, even with new Li-Ion battery technologies, thus increasing launch cost correspondingly.

CMR Prototech has conducted a series of projects (studies and hardware tests) for ESA which have been relevant in the development of an energy storage system for satellites based on fuel cells (FC). They have been more thoroughly described in the previous paper submitted to ESPC2014 [10] and can be summarized as: *Hydrogen Storage Technologies* [1], *Regenerative H2/O2 Fuel Cell* [3], *Innovative Gas Storage on Satellites* [4], *Advanced Energy Storage System* [5]. *Demonstration of a closed loop H2/O2 RFCS* [6], *Metal Hydride Hydrogen & Heat Storage System (ongoing until 2018)* [7]. CMR Prototech also participated as subcontractor in a TRP activity; Fuel Cells for Telecom Systems – System Study [2], where Thales Alenia Space was the main contractor. The study concluded that it is expected that the RFCS will be more attractive than batteries on a performance point of view, for telecom satellites with bigger payloads. Based on the results from this system study and the recent developments of the RFCS, an ongoing (2015-16) system study "Fuel cells for extra-large telecommunication satellites" [9] funded by ESA under the ARTES1 program was initiated. Prime contractor is Airbus D&S Toulouse, and CMR Prototech is subcontractor to Airbus D&S. The main purpose of the activity is to design a RFCS based power subsystem for large telecommunication satellites (25 kW and beyond), and compare it to a conventional one based on Li-ion batteries in order to identify the power level at which the RFCS becomes clearly the preferred choice from the full system point of view.

A contract change notice (CCN) to the ESA project "Demonstration of a closed loop H2/O2 RFCS", [6], was initiated in December 2013 to optimise and improve the design of the HT PEM fuel cell and high pressure PEM electrolyser. The aim of this project is to develop light weight technologies for use in telecommunication satellites. The target is to operate the fuel cell at 10 bars and temperatures up to 200°C. The electrolyser would be designed for operation up to 100 bars. The project includes testing of short stacks in standalone configuration and a 1 kW full stack in a closed loop configuration. The results from these activities are expected to give a sound basis for further development of a complete RFCS elegant breadboard.
This paper is mainly focused on the results from the optimisation process of the HT PEM fuel cell and High pressure PEM electrolyser.

2. RFCS System description

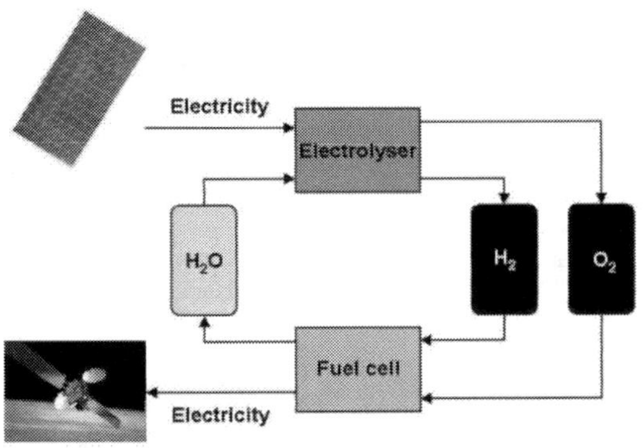

Figure 1 Principle of a Regenerative Fuel Cell System

The principle of a Regenerative Fuel Cell System (RFCS) is to decouple the energy storage from the electrochemical electrodes in an electrochemical energy storage system and thus store energy as element with low mass such as hydrogen and oxygen instead of Lithium and transition metal based oxides. While Lithium batteries has a maximum storage capacity of about 250 Wh/kg on cell level, a RFCS may store up to 1000 Wh/kg for system with long discharging cycles. However, when used in relevant applications, both the Li-ion batteries and the RFCS will meet limitations which reduce their effective storage capacity.
For telecom satellites, the longest discharging period is 72 minutes, which is too short to obtain the maximum energy density of a RFCS. However, the studies have shown a potential for significantly reduced system mass by replacing Li-ion batteries with the RFC.

The main components of CMR Prototech's RFCS are the High Temperature PEM (HTPEM) fuel cell, the High Pressure PEM electrolyser, the storage tanks and the water removal units. In addition, there is a Balance of Plant (BoP) system thermal hardware to dissipate excess heat from the fuel cell.

The HT-PEM (High Temperature Proton Exchange Membrane) fuel cell is a fuel cell technology utilizing phosphoric acid-impregnated membrane and operating from 150°C to 230°C. The higher temperature means that the excess heat can be used to release hydrogen from a metal hydride based storage system. The HT-PEM is also advantageous compared to the lower temperature fuel cells in telecommunication satellites, mainly because of the thermal management becoming much easier at higher temperatures, since radiations from a black body increases with fourth square of the absolute temperature (~*T4*). This will save mass on the thermal hardware compared to lower temperature fuel cells.

Figure 2 HTPEM FC stack used in testing of the closed loop RFCS breadboard demonstrator in 2012 [6] (here without thermal insulation)

The concept of combining the fuel cell with a metal hydride storage tank is particularly interesting. Such a system is being developed in another ESA project [7] [11], where Prototech are working together with Fotec in Austria to combine the HTPEM fuel cell with a metal hydride storage tank which can store both hydrogen and excess heat from the fuel cell. This will allow the thermal hardware which dissipates the excess heat to be reduced significantly in terms of mass, or even removed completely.

The High pressure PEM electrolyser will operate between each fuel cell cycle and produce the required oxygen and hydrogen. Current state of development allows generating hydrogen and oxygen directly by the electrolyser at 100-200 bar pressure. It makes it possible to exclude heavy and power-hungry gas compressors from the system, which is beneficial for the overall system mass and reliability.

3. Development activities

In order to increase the Technology Readiness Level (TRL) of the RFCS, several development activities have been undertaken in the last two years (2014-2016). The main focus is towards the key components, the fuel cell and electrolyser, in order to reach the mass targets and performance.

3.1. HTPEM Fuel Cell

To realize the RFCS within the calculated mass targets, several key issues in the technology have to be addressed. The most important is to improve the FC performance, both in respect to efficiency, specific energy and increased operating temperature. Several development steps have been undertaken to bring the fuel cell technology towards use in telecom satellites. Target development steps in this project:

- Operation of FC on pure oxygen and hydrogen
- Operation of FC at higher pressure, up to 10 bar-a
- Incorporation of liquid cooling
- Increase of FC specific energy, target 1 kW/kg @ nominal power
- Obtain cell voltage of 0,8 V/cell @ nominal current (0,33 A/cm^2) and stack efficiency of 65% LHV Beginning of Life (BoL)
- Increase operating temperature, >200 °C

Fuel cell development: In order to meet the target specific power density, a **new concept** of the bipolar plate (BP) was needed. Thick carbon composite plates with machined gas and cooling channels simply could not meet the requirement due to intrinsic properties of the material. The new concept is based on thin metallic bipolar plates with liquid cooling in each plate. Each bipolar plate is made of two thin (0.1-0.2 mm) corrugated metal sheets welded together. (See example in fig.3).

Figure 3 Example of metallic bipolar plates from supplier showing two bipolar plates welded together, incorporating liquid cooling channels in between the plates

Stack concept development (plate shape): The way the bipolar plate is designed greatly affects the maximum possible stack specific power density, as well as most of the other stack parameters such as pressure drop for all reactants, reactants distribution, temperature difference across the plate etc. Reactants manifolds plays a very important role as well. Several design concepts were analysed to identify which concept will allow the highest specific power density. The approach we have chosen was the following. We would develop different BP designs were the active MEA area is kept constant for all design, and then introduce them into an analysis tool containing parameters such as plate thickness, distance of various elements to the edge, size and position of elements etc. Based on that, we would calculate the plate weight and power-to-weight ratios.

Table 1 Calculated specific power density for different BP concepts at nominal power

	Square shape	Circle shape	External O_2 manifolds	
Specific power density	1.09	0.87	1.13	kW /kg
MEA Area/Total Area	45.6 %	35.1 %	58.0 %	%

As one can see in Tab. 1 the concept with external O2 manifold has the highest power density to the other concepts, providing a specific power of **>1 kW/kg**, which is above the targeted value. External manifolds are beneficial when operating at higher pressure levels and any small external leakages from the cells are contained within the system. Thus this concept was chosen for the further design.

A sensitivity analysis was also performed. By changing each design parameter in the range of ± 10%, we would then calculate the impact of this change on the plate mass. By doing so, we were able to identify several parameters that have the most impact on weight. The sensitivity analysis showed that there are only a few parameters which are really important, like the plate thickness, BP material (density) and MEA specific power density. The analyses also show that the side length of the BP plays a very important role. This has to do with the relative size of the active area of the BP and the area required for gas/liquid cooling inlets/outlets-

Material choice & protective coating: The important parameters for the bipolar plates bulk (core) material are electrical conductivity, corrosion resistance in the HTPEM FC environment, mechanical strength and also the density, which determines the total stack mass and its specific power density. An assessment of new materials for the fuel cell stack has been performed for the bipolar plates base metal, coating for bipolar plates

DOI: 10.1051/e3sconf/20171610004

and sealing material for the stack. It is known that materials commonly used to manufacture metallic bipolar plates (various grades of stainless steel) cannot be used as such in their pristine form: they either start corroding, or develop even thicker passive surface layer, which greatly increases the electrical resistance of the plate. A common way to counter these effects is by using coating, which serves both for corrosion protection and for improving the electrical conductivity. In order to identify the best combinations of base materials and protective coatings we have tested several combinations in real and simulated fuel cell environment. The combinations were chosen based on a set of requirement. For the bipolar plate we considered mechanical properties, low density, compatibility to coating material (adhesion), availability and non-prohibitive cost, and for the coating we looked at adhesion properties to the base metal, stability in the HTPEM environment (Phosphoric acid, H_2 & O_2 & 200 ^{0}C) and good electrical properties. A short stack, which consisted of potential bipolar plate materials with coating, was operated at 180 °C in a steady-state fuel cell operation (between 0,7 and 0,8 V/cell) from 09.02.2015 to 22.02.2015 continuously for 310 hours using pure oxygen and hydrogen as reagents. After the testing, the stack was disassembled, and the plates were analysed for signs of coating delamination or other defects. Based on the results a superalloy was chosen as the base metal for the novel fuel cell bipolar plates. It should be noted that lighter base materials, i.e. titanium potentially can provide an increased specific power compared to the superalloy, but will require an improved coating.

Test results of short stack

A full size short stack consisting of three cells has been manufactured and tested under relevant operating conditions. This test program has had the goal to verify the functional performance of the concept and design with respect to pressure/temperature capabilities, leak tightness and compatibility to the harsh environment in the fuel cell, especially related to the presence of oxygen and phosphoric acid (in the electrolyte). Another goal has been to evaluate the fuel cell performance with particular focus on performance at different pressure levels. At the time of writing the test program is not completed and only the main preliminary results are presented.

Figure 4 Full size short stack installed into the test stand and ready for testing

Test conditions =>

The short stack was integrated into an automated test stand including flow, pressure, temperature control and load control.

- 180 °C stack temperature, limited by membrane specifications
- Pressure: 1-7 bar-a on H^2, O^2 and cooling loop
- Reactants: Hydrogen and oxygen
- Temperature control: Internal liquid cooling
- Constant Current operation: 0 - 0,5 A/cm^2
- Membranes were fresh and unused before testing

Results =>

The short stack was pressure tested with nitrogen to verify leak tightness before fuel cell operation. The stack showed very good leak tightness, both against internal and external leakages, which also was expected to improve further under increased temperature.

The short stack has been operated on a steady 180 °C for several days of testing, including approximately 35 hours of operation and 7 thermal cycles from room temperature. The stack materials have shown good performance, accept from a gasket material on the external casing. This material showed not to have adequate long term performance at the operating temperature, probably also as a result of exposure to oxygen, and need to be replaced by another compatible material.

The short stack has been operated at different pressure levels to see how the performance changes. Previous tests performed in an earlier ESA project [6] on a single cell HTPEM showed a significant increase of cell voltage with increased operating pressure (25-27% increased cell voltage at 10 bar). The fuel cell was initially operated at 1 bar-a and the pressure was stepwise increased up to 5 bar-a. Polarization curves were logged at each step. Figure 5 (a) shows the polarization curves measured for each pressure level, and Figure 5 (b) shows the relative voltage gain for each

 DOI: 10.1051/e3sconf/20171610004

pressure level compared to 1 bar-a.

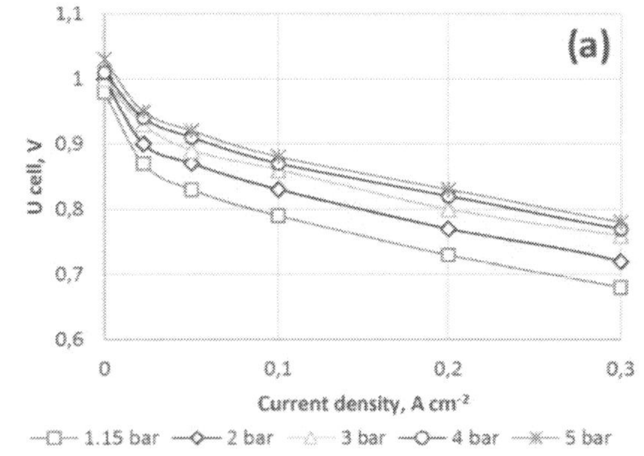

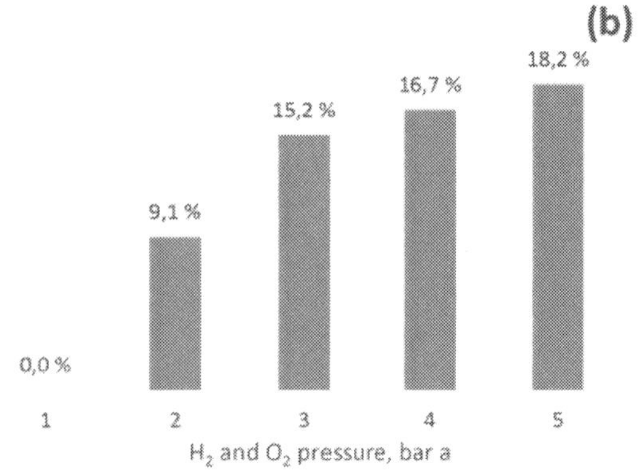

Figure 5 (a) polarization curves measured during fuel cell operation at different pressure levels; (b) relative voltage gain for each pressure level compared to fuel cell operation at 1 bar-a pressure

The test clearly showed the positive trend by increasing the pressure on the anode and cathode. The largest voltage increase is gained when the pressure is increased to 3 bar-a. A 15 % increase is measured. From 3 to 5 bar-a the gain is 3 %, which is still significant in this context. These results correlate well with the single cell test that was performed in the previous mentioned ESA project [6]. Based on those results, and assuming that the relative increase will be the same from 5 to 10 bar-a, another 9 % increase of voltage can be expected. This corresponds to an estimated cell voltage of 0,84 V at 0,3 A/cm^2. If we compare this value to the target we set for the project, which was 0,8 V at 0,33 A/cm^2, we see that the target can be reached.
Also knowing that this short stack is a prototype with potential for further improvements, we expect the performance to be further increased.

The short stack was operated at different pressure levels to verify stable operation at elevated pressure. The target pressure was 10 bar-a, however, the test stand was limited by external components in the cooling loop to 7 bar-a for this test. Modification of the cooling loop hardware is needed to go all the way to 10 bar-a. At 7 bar-a on the cathode, anode and cooling system the short stack operated stable throughout the test, showing no signs to external or internal leakages in the stack. The next step in the test program is to modify the test stand and perform a 10 bar-a test.

The materials' compatibility to the harsh environment has been analysed after the first set of tests by disassembling the short stack and inspecting the individual components. Of particular interest is to evaluate the condition of the bipolar plates after they have been assembled and operated in realistic conditions. This inspection provided some first indications of the materials' durability. Further analyses are planned for after the test program is completed. The main results from the inspection showed that:

- The coating on the bipolar plates was intact (visual microscope inspection), and has good adhesion to the bulk material
- The bipolar plates withstood the compression forces during assembly
- The contact resistance of the bipolar plates has not changed.

Due to these results the bipolar plates were reused in the continuation of the test program.

The results from the short stack testing have shown that the concept works as intended, and the performance is according to the target values we defined for the project. This allows us to continue to the next step of the test program. A 1kW single stack is being assembled at the time of writing, and will shortly undergo a test program in standalone operation, equivalent to the short stack, and in closed loop, together with the high pressure electrolyser. A CAD model of the assembled stack is shown in (Fig. 3).

Figure 6 CAD model of the optimised HTPEM fuel cell stack

 DOI: 10.1051/e3sconf/20171610004

The goal of the full stack testing is to get more long term data of the stack operation, which will provide us more insight about the material performances and knowledge for further design optimisations.

3.2. High Pressure PEM Electrolyser

Another important element is to develop an efficient high pressure electrolyser (>= 100 bar). The oxygen has to be stored at high pressures, and high pressure hydrogen is advantageous in order to remove water impurities ahead of a metal hydride hydrogen storage tank. A high pressure electrolyser eliminates the need of a heavy compressor. The degradation rates of the electrolyser also need to be significantly lower than those of the fuel cell. While the fuel cell operation is estimated to around 2000 hours, the electrolyser shall operate up to 20 000 hours. Like the fuel cell, the electrolyser needs to be optimised for use in telecom satellites. The optimisation includes a redesign in order to reduce the mass, and the use of state of the art PEM (Proton Exchange Membrane) MEAs customized for high pressure operation.

Electrolyser development: The development of the high pressure electrolyser within these activities has been done in collaboration with Ideevolutie (The Netherlands). The main challenge in engineering a high performance, light weight and intrinsically safe high pressure electrolyser cell and stack is to find principal solutions to effectively contain the high pressure hydrogen and high pressure oxygen. Key development steps which have been addressed to adapt the technology for space, and to improve system performance include the following:

- High pressure operation 100 bar and above
- Equal pressure on anode (oxygen) and cathode (hydrogen)
- Increased specific power, 1 kW/kg @ nominal power input of 1 A/cm^2
- Increased stack efficiency, 90% @ BoL

In order to realise these specifications, the main activities of the project were focused on:

- Design of the HP ELY bipolar plates and other stack components
- Weight optimisation. The high pressure electrolyser must achieve the desired specific power density. The elements with the greatest mass reduction potential are identified: bipolar plates and compression plate, manifold plate and tie bolt assembly.
- Performance optimisation. High pressure operation poses additional challenges towards process performance. Advanced MEA technology is to be applied in the electrolyser.

Cell stack design: To fulfil the objective to optimise the specific power to 1 kw/kg, the design of the optimised electrolyser has been divided in two parts; optimisation of the cell assembly (bipolar plates, sealing MEA) and the compression assembly (compression plates, tie bolts, current collectors, etc.).

Table 2 Design specifications of a high pressure electrolyser for large telecom satellites

Number of cells	34	
Current Density	1,0	A/cm^2
Active area	70	cm^2
Cell voltage @ current density	1,75	V
Stack Current	70	A
Stack voltage	59,5	V
Stack power (theoretical)	4,2	kW
Stack power (practical)	4,9	kW

In order to calculate the gravimetric (kWe/kg) and volumetric power density (kWe/L) of the electrolyser stack design, the operating parameters need to be defined. A volumetric hydrogen flow of 1 Nm^3/h is utilized for the definition of stack size, current density, cell voltage and stack power (Tab. 2). A cell potential of 1,75 V is specified at a current density of 1 A/cm2.

A first prototype stack has been developed. This stack features the low mass electrolyser cell design, but relied on the proven 'heavy duty' compression plates. This design is used for validation tests of cell design.

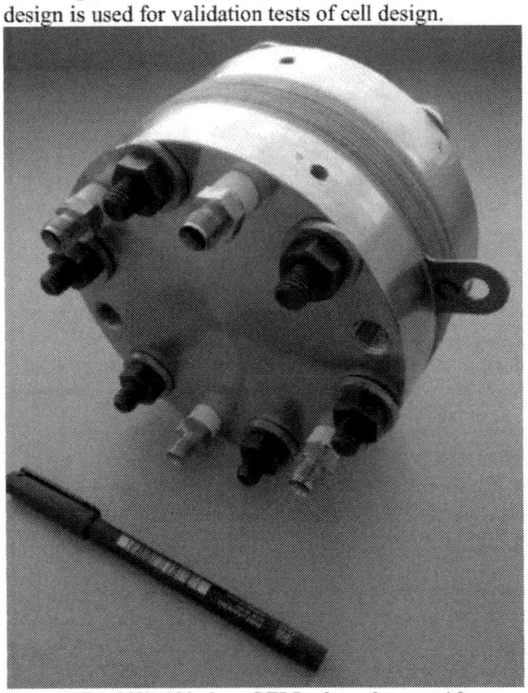

Figure 7 1kW 100 bar PEM electrolyser with mass optimised cell plates, but with non-optimised endplates

DOI: 10.1051/e3sconf/20171610004

Compression assembly: The main design strategy was centred on a mass optimisation of the compression assembly. Novel principal solutions were introduced with respect to a lightweight compression plate in combination with fitting inserts.

Figure 8 First iteration of a mass optimised compression assembly

The mass of the stack is reduced to 4,9 kg and 1,83 l for a 1 Nm^3/hr high pressure electrolyser. This translates to 0,85 kW/kg, taking into account that 4,2 kW power shall be supplied to generate the hydrogen. On the stack level, the cells contribute to 58% of the total mass, while the compression plates contribute to 22%.
We see further potential to reach mass targets by reducing the cell footprint and apply a new concept for the compression assembly.

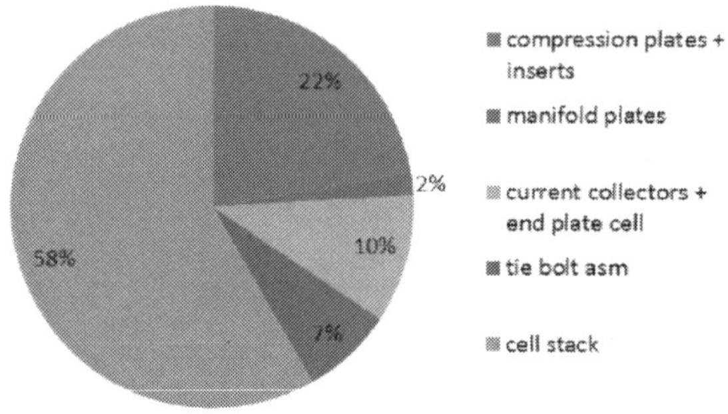

Figure 9 Graphical representation of weight distribution for

Test results of a 1kW high pressure PEM electrolyser stack

A full size 1kW electrolyser stack consisting of seven cells has been manufactured and tested under relevant operating conditions. The goal of the test program has been to verify the functional performance of the concept and design with respect to pressure capabilities, leak tightness and operational performance. The test program consists of both standalone testing and in closed loop configuration together with the fuel cell. At the time of writing the stack has been through the first functional tests in standalone configuration, and the plan is to incorporate the stack into the existing closed loop system when the 1kW fuel cell stack has been assembled (as described in section 3.1).

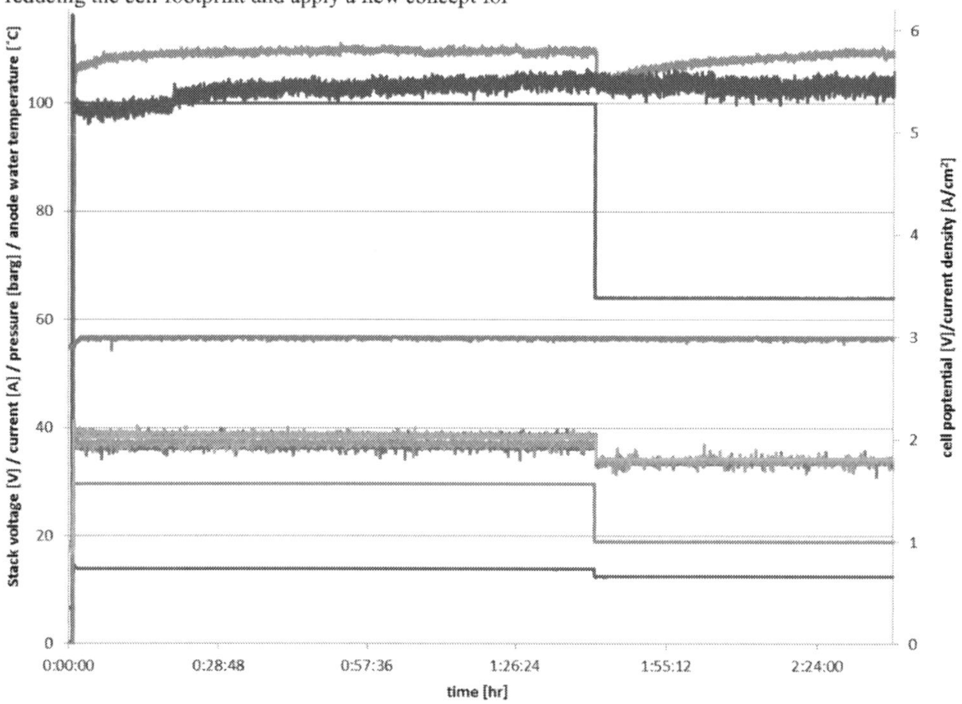

Figure 10: High equal pressure operation: red=stack voltage, green=cathode pressure, purple=anode pressure, blue=water temperature, orange=current density, all curves centred around 1,7V = 7 individual cell voltages

In standalone configuration the stack has been subjected to > 50 hrs of high pressure testing (100 bar on anode and cathode). Figure 10 shows a test where the electrolyser is operated at 100 bars on the anode and cathode. After a phase were the current density is 1,5 A/cm^2 it is reduced to 1,0 A/cm^2. The stack shows good performance by having cell voltages centred around 1,7 V/cell at 1,0 A/cm^2. The stack is at the time of writing ready to go through start/stop cycling performance tests, and later incorporated into the closed loop system for closed loop testing.

4. Conclusion and future work

CMR Prototech in partnership with other R&D companies and Primes have conducted a series of projects for ESA related to the development of a RFCS for telecommunication satellites.
The experimental tests from 2012 [6] have shown that RFCS has the potential to last for the entire lifetime of 15 years in orbit for a satellite, with a satisfactory degradation below 0,01 U/cycle. In the ongoing optimisation activities of the HTPEM fuel cell and High Pressure electrolyser it has been shown that the specific power of 1 kW/kg can be reached on stack level.
The functional performance of the developed HTPEM fuel cell has been tested through a number of tests, including pressure, leak tightness, material compatibility to the harsh fuel cell environment, and performance at different pressure levels. At 5 bar-a a cell voltage of 0,77 V at 0,3 A/cm^2 has been measured. The estimated voltage at 10 bar-a corresponds to 0,84 V. If we compare this value to the project target, which was 0,8 V at 0,33 A/cm^2, we see that the target can be reached. A 1kW single stack is being assembled at the time of writing, and will be tested in standalone operation and in closed loop, together with the high pressure electrolyser.
The High Pressure Electrolyser has been subjected to > 50 hrs of high pressure testing in standalone configuration (100 bar on anode and cathode). The stack shows good performance by having cell voltages centred around 1,7 V/cell at 1,0 A/cm^2, and is ready to be tested for start/stop cycling performance, and closed loop operation.

5. Acknowledgement

The work presented here was performed within projects supported by the European Space Agency. The authors want to thank the European Space Agency and all the individual partners in the projects.

6. REFERENCES

[1] J. Byrknes et.al, *"Summary Report"*, ESA project: "Hydrogen Storage Technologies", ESA contract no. 17925/04/NL7CP, December 2005
Prime contractor: CMR Prototech
Sub-contractor: IFE

[2] I. Nann, H. Weydahl, *"Technical Note 4 – System Design and System Study"*, ESA project: Fuel Cells for Telecom Systems – System Study, ESA contract (AO/1-5525/07/NL/LvH), September 2008
Prime contractor: Thales Alenia Space (Cannes)
Sub-contractor: CMR Prototech

[3] J. Byrknes, J. W. Sørensen, *"Final Report"*, ESA project: "Regenerative H2/O2 Fuel Cell", ESA contract no. 18312/04/NL/JD, August 2005
Prime contractor: CMR Prototech
Sub-contractor: New Energy Systems

[4] Ø. Torvanger, *"Summary Report"*, ESA project: "Innovative Gas Storage on Satellites", ESA contract no. 19399/05/ NL/PM, November 2007
Prime contractor: CMR Prototech,Sub contractor: IFE

[5] H. Weydahl, *"Summary Report"*, ESA project: "Advanced Energy Storage System", CCN 1 to ESA contract no 19399/05/ NL/PM, November 2007
Prime contractor: CMR Prototech

[6] D. Bokach, J. Farnes, *"Final Report"*, ESA project: "Demonstration of a closed loop H2/O2 RFCS", ESA contract no. 22556/09/NL/LvH
Prime contractor: CMR Prototech
Sub-contractor: Astrium (Friedrichshafen)

[7] ESA project, *"Metal Hydride Hydrogen & Heat Storage System"*, ESA contract no. 4000105330/12/NL/CLP, Ongoing project
Prime contractor: CMR Prototech,
Sub-contractor: FOTEC/FHWN (Austria)

[8] I. Nann, *"Technical note 5 – Trade-off Li-ion-RFCS"*, ESA project: Fuel Cells for Telecom Systems – System Study, ESA contract no. 21349/09/NL/LVH, November 2008, Prime contractor: Thales Alenia Space (Cannes), Sub contractor: CMR Prototech

[9] ESA ARTES 1 project, *"Fuel cell for extra-large telecommunication satellites"*, ESA tender AO/1-7843/14/NL/NR, Ongoing project
Prime contractor: Airbus D&S Toulouse
Sub-contractor: CMR Prototech

[10] J. Farnes, *"Recent developments of regenerative fuel cell systems for satellites"*, *conference 10th European Space Power Conference 2014*

[11] A. Reissner, *"Development of a hydrogen and heat storage system for satellite applications"*, conference 10th European Space Power Conference 2014

REGENERATIVE ENERGY STORAGE SYSTEM FOR SPACE EXPLORATION MISSIONS

Ivar Wærnhus [(1)], Crina S. Ilea [(1)], Arild Vik [(1)], Dimitrios Tsiplakides[(2)], Stella Balomenou[(2)], Kalliopi Papazisi[(2)], Max Schautz[(3)]

(1) CMR Prototech AS, Fantoftvegen 38, 5072 Bergen, NORWAY, Email: ivar.warnhus@prototech.no
(2) Centre for Research and Technology - Hellas (CERTH, 6th km Charilaou-Thermi Rd., GR-57001 Thessaloniki, GREECE, Email: dtsiplak@cperi.certh.gr
(3) European Space Agency, ESTEC -Keplerlaan 1, 2201 AZ Noordwijk Zh, The Netherlands, Email: max.schautz@esa.int

ABSTRACT

This paper describes the development and testing of a 1 kW reversible solid oxide fuel cell intended for energy storage on space exploration missions, particularly for long term Mars exploration. The energy is stored as H_2 or CO produced by electrolysis of H_2O or CO_2. The reactants are then converted back to its original composition by producing electricity.

The breadboard was operated for 1250 hours alternating between electrolyser mode and fuel cell mode with H_2/H_2O as reactants. During the tests, as long as the mechanical integrity of the system was maintained, no degradation effect was observed. At the end of the test period, the fuel cell was operated for three full cycles (approx. 50 hours) with CO/CO_2 as reactants. The performance on CO/CO_2 was lower than for hydrogen, but sufficient to be used in a compact energy storage system for Mars exploration.

1. INTRODUCTION

Future exploration missions, including human missions to the Moon and Mars, are expected to have increasingly demanding operational requirements. Generating electrical power, as well as maintaining a specific thermal environment, are both critical capabilities for any mission. However, in the case of exploration, both a wide range of mission types (robotic, human, ISRU etc.) and a variety of environments exist; from interplanetary space to the attenuated and redshifted lighting on the Martian surface. Different technologies already exist to provide power in different operating conditions, and in different types of space missions, including Batteries, Solar Arrays, Nuclear Systems and Fuel Cell Technologies.

The present research focuses on a regenerative solid oxide fuel cell system (RSOFCS) that use H_2O or CO_2 as main medium. These cells when charging absorb electric energy and electrolyse H_2O and/or CO_2 into H_2 and/or CO and O_2. These two reactant gasses are then stored. When the cells discharge, H_2/CO and O_2 are recombined back into H_2O/CO_2 with production of electric energy. CO_2 is available directly from the Martian atmosphere, which means that no reactants need to be carried by the space craft, representing a mass reduction of about 50%.

Previous papers have shown the feasibility of operating with CO as fuel [1] and the reversibility of the RSOFC electrodes on single cells and short stacks [2, 3]. CO is in general an unstable molecule that may decompose to solid carbon blocking the active surface or transport channels. It was therefore decided to demonstrate the closed loop system with H_2/H_2O as reactants. Due to the high cooling flow needed for a standard SOFC, it was not possible to replace air with pure oxygen, which is certainly needed for space operation, during the long term testing. Thus, for the tests presented here, the breadboard is operated with air under atmospheric conditions and no storage of oxygen.

Figure 1. RSOFC stack box

DOI: 10.1051/e3sconf/20171610005

This paper describes the construction and test results from a 1 kW RSOFC operated for >1000 hours on H_2/H_2O, and for >50 hours on CO/CO_2 carried out under ESA contract no 4000105059. In the ongoing activity (4000108849/13/NL/EK) these shortcomings will be addressed and a test program on $CO/CO_2/O_2$ at elevated pressures representing a full Martian winter season (180 sols) is prepared.

2. EXPERIMENTAL

The system includes the following main components:

- A SOFC fuel cell inside a heated chamber,
- Electrical heaters + gas-gas and gas-water heat exchangers for thermal control and water evaporation,
- Compressors and gas tanks for gas storage up to 6 bar,
- Water separators and steam generator, and
- Circulation fan for cooling.

The RSOFC consists of two 30 layer CFY-stacks from Plansee/IKTS [4] numbered stack 2, stack 3 and stack 4. The stacks were mounted on top of each other with a common fuel manifold and an external air manifold designed for both stacks (Fig. 1). The compressors were diaphragm pumps from Air Dimensions suitable for both H_2, and CO. The storage tanks were two ordinary steel tanks (2 x 500 litres /10 bars) able to store ca 9 000 normal litres (Nl) of H_2, sufficient for 7 kWh of electricity. Due to the large surface area of the stack (130 mm x 150 mm), a high cooling flow was required, thus the stack was operated on air, with high excess air flow, both in fuel cell and electrolyzer mode, respectively. The focus for this test was to demonstrate the reversibility of the materials on a real size application, i.e. 1 kW stack.

The following test program and defined success targets were applied: i) Fuel cell mode: 0.5 kW for 5 hours followed by 1 kW for 4 hours; ii) Electrolyzer mode: Fuel for 7 kWh of electricity produced in 7 hours; iii) Continuous testing for 1000 hours with less than 10% degradation, and iv) Two full electrolysis / fuel cell cycles starting with CO2 and producing CO as fuel.

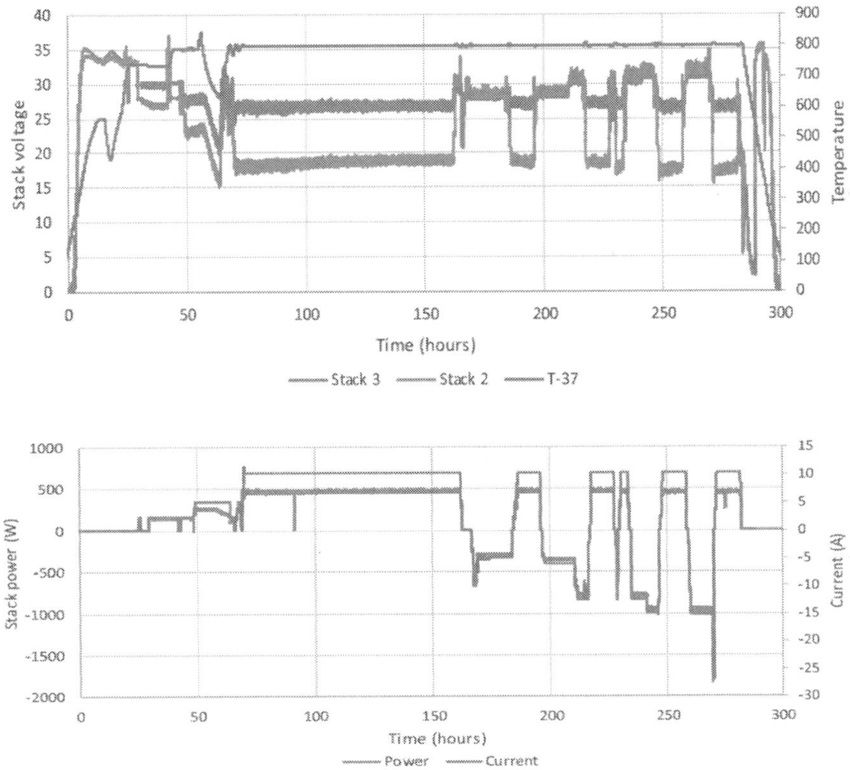

Figure 2. Stack voltage, current, power and temperature during first 300 hours.

DOI: 10.1051/e3sconf/20171610005

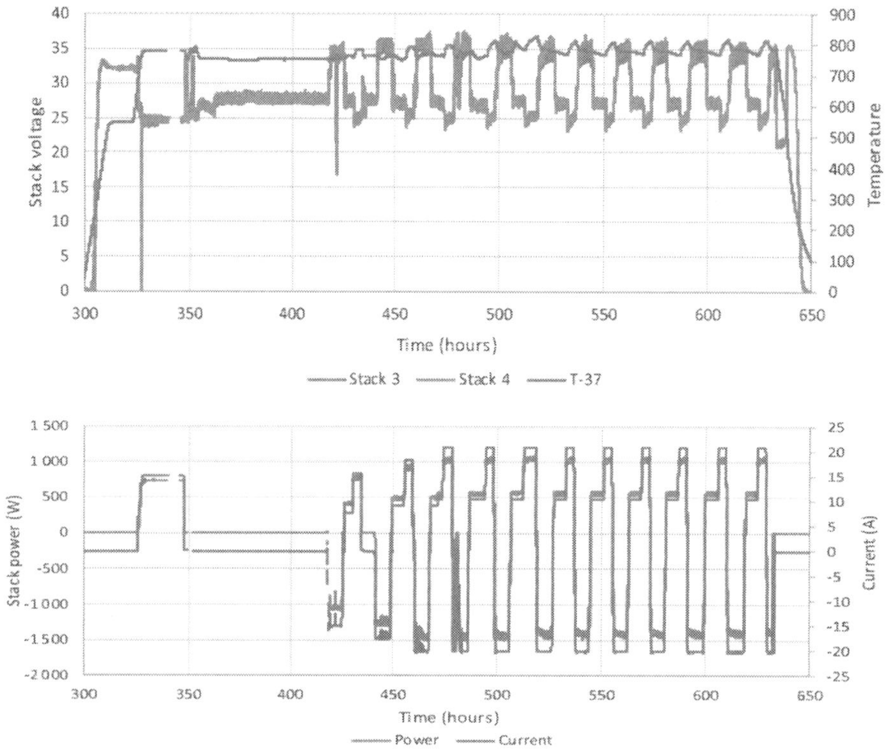

Figure 3. Stack voltage, current, power and temperature from hour 300 to 650.

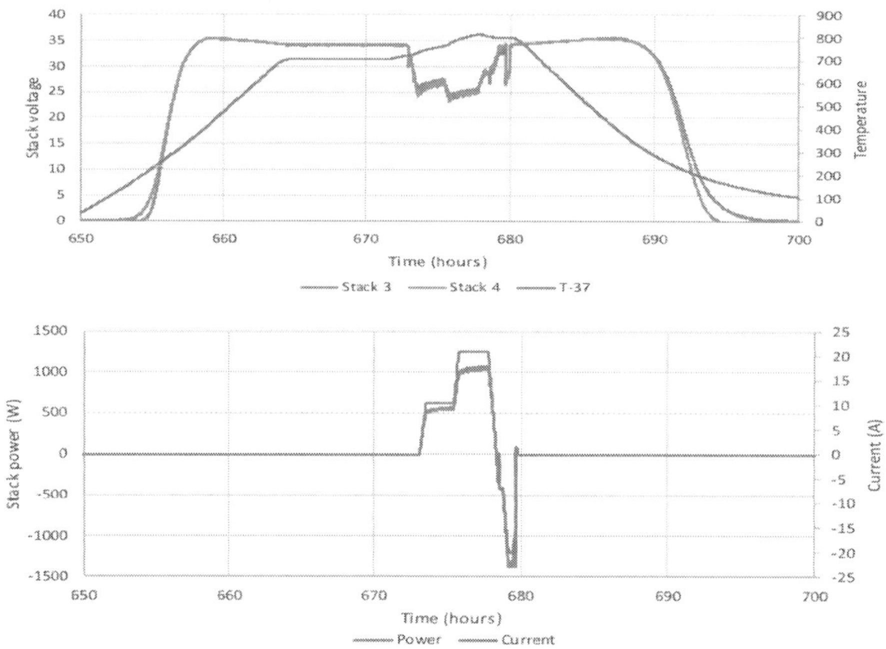

Figure 4. Stack voltage, current, power and temperature hour 650 to 700.

 DOI: 10.1051/e3sconf/20171610005

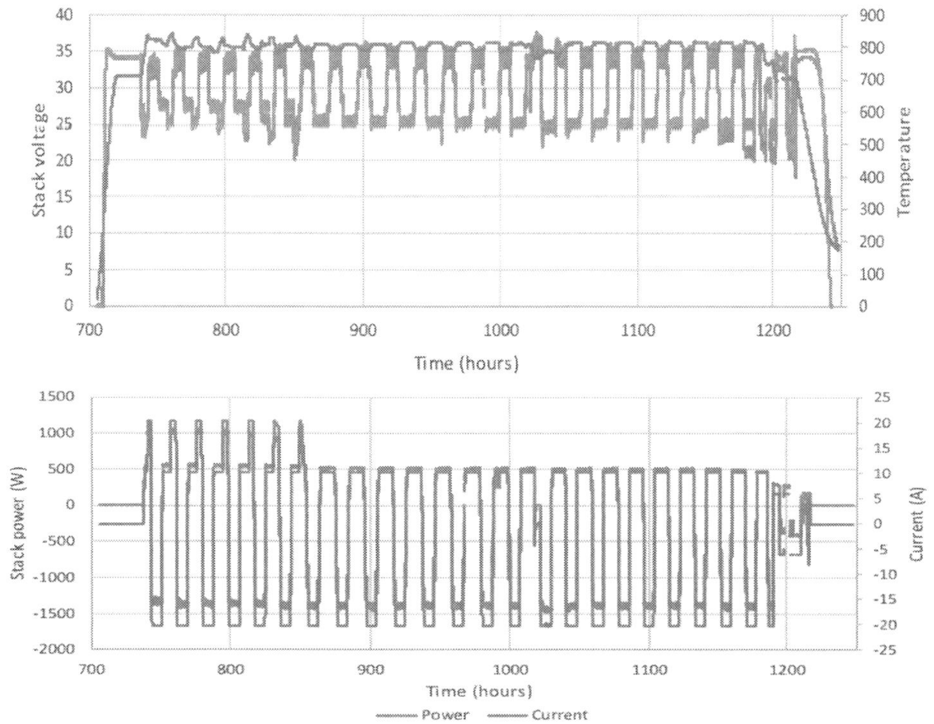

Figure 5. Stack voltage, current, power and temperature from hour 700 to 1250.

3. RESULTS AND DISCUSSIONS

3.1. Fuel cell performance on H_2/H_2O

Figure 2 to figure 5 shows the observed stack voltages (stack 2, stack 3 and stack 4), temperatures, electric power and stack current during the long term H_2/H_2O test cycles. The data is separated into four different data sets, each including one thermal cycle. The first 300 hours were also used to tune the automatic control system. It was also observed that stack 2 did not show the expected performance and it was decided to run down the SOFC and replace the stack with stack no 4. The two next thermal cycles, after 650 and 700 hours, were due to some technical issues with the external power supply powering the stack during electrolysis. When these issues were solved, the system could operate uninterrupted until 1250 hours.

At 830 hours (Fig. 5), the cell voltage at 20 A dropped significantly compared to the previous cycle, indicating an initial failure of the stacks. To avoid further damage, the rest of test was limited to 500W and 10 A in fuel cell mode.

Typical performance of two full cycles is shown in Fig. 6. The voltage measurements are averaged for one minute for reducing the noise level of the data. Here, the system is starting in fuel cell mode, first at 10 A for 6 hours producing 565 W before the current was increased to 20 A for 4 hours, providing 1065 W. Then the current was reduced to zero and reversed to -20 A producing H_2 for 7 hours, with power consumption of 1390 W. Thus, each cycle produces 7.65 kWh in fuel cell mode and consumes 10.36 kWh in electrolysis mode, giving a round trip efficiency of 74 %.

To calculate the degradation, the recorded data at identical currents were compared. Fig. 7 shows the performance of each stack at 10.4 A and 20.9 A (fuel cell mode) and -20.2 A (electrolysis mode). The data from both fuel cell operation modes (at 10 A and 20 A) show essentially no degradation up to 830 hours. However, after 830 hours some degradation is evident. On the other hand, the electrolysis mode seemed to be unaffected by the 830 hour incident, no degradation was observed during electrolysis mode. The difference in performance at the beginning and at the end of the experiments reveals a total degradation of 8.5 %, measured at constant current of 10.4 A, which is within the project target of <10 %. However, for the first 800

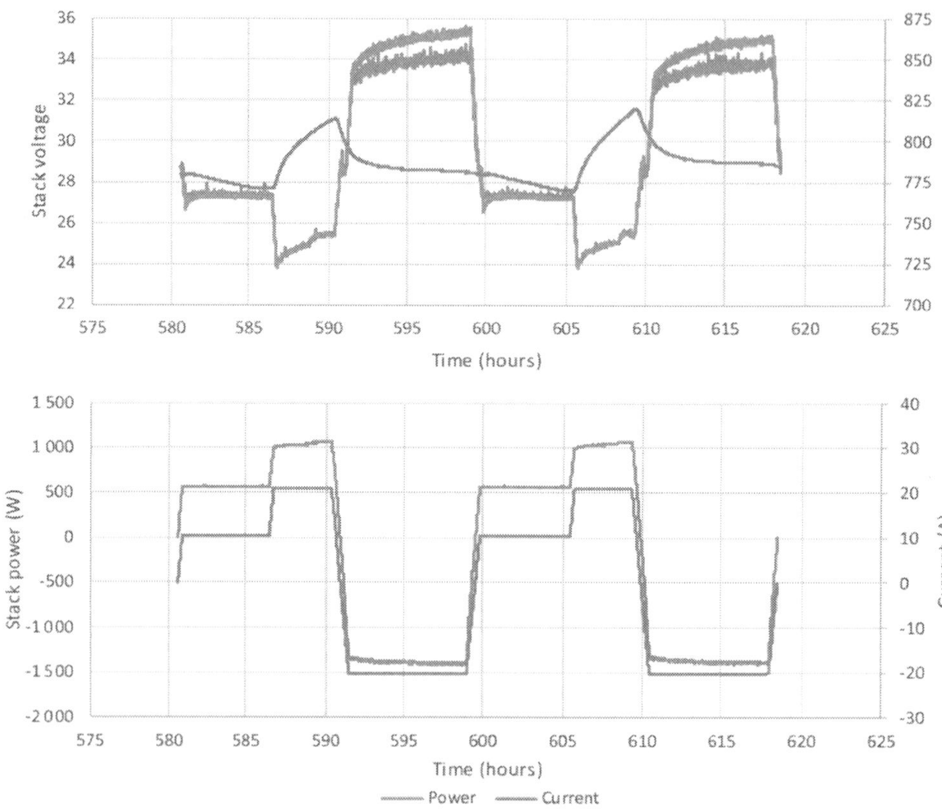

Figure 6. Stack voltage, current, power and temperature during two full cycles

hours, there is essentially no degradation, or at least less than 1 % per 1000 hours, also including three thermal cycles. It is expected that degradation rates <1 %/khour can be reached for systems operating also for longer periods when the thermal and mechanical system is sufficiently robust.

3.2 Fuel cell performance on CO/CO_2

After 1250 hours successfully operation on H_2/H_2O, the system was tested with CO_2 replacing H_2O as the starting gas for electrolysis. Previously, CO from bottles has been successfully tested as SOFC fuel [1], but this experiment was the first demonstration of an energy storage system using CO_2 as starting reactant. Data for stack no 4 (voltage, current and temperature) is given in Fig. 8. The electrolysis voltage has increased from 34.7 V at -20 A to 36.5 V at 17.5 A and the fuel cell voltage drops from 27.5 V at 10 A to 23.3 V at 7.0 A when H_2O is replaced with CO_2. The lower performance is a result partly of slower reaction rates for CO than for H_2, partly of reduce fuel quality since unconverted CO_2 in the produced fuel could not be separated from the CO such as H_2O can be separated from the hydrogen. Still, the round trip efficiency at cell level is close to 64%, which is an acceptable level.

Using the observed OCV of 28.8 V for both gas compositions, the area specific resistance (ASR) of stack no 4 during electrolysis increased from 1.25 $\Omega\cdot cm^2$ to 1.86 $\Omega\cdot cm^2$, an increase of 0.6 $\Omega\cdot cm^2$. Calculations of ASR from in fuel cell mode using the observed OCV (30.2 V for H_2/H_2O while only 27.8 V for CO/CO_2) give us a value of 2.0 $\Omega\cdot cm^2$ for H_2 operation and 2.6 $\Omega\cdot cm^2$ for CO operation, also here a difference of 0.6 $\Omega\cdot cm^2$. This is as expected since the exchange of reactants only affect the electrode kinetics which according to theory should be identical in both directions. The performance of a CO/CO_2 reversible fuel cell system is somewhat lower than of a H_2/H_2O system as a consequence of both higher ASR and lower OCV, however, this is partly retrieved on the system level since there is no need for energy for steam production.

 DOI: 10.1051/e3sconf/20171610005

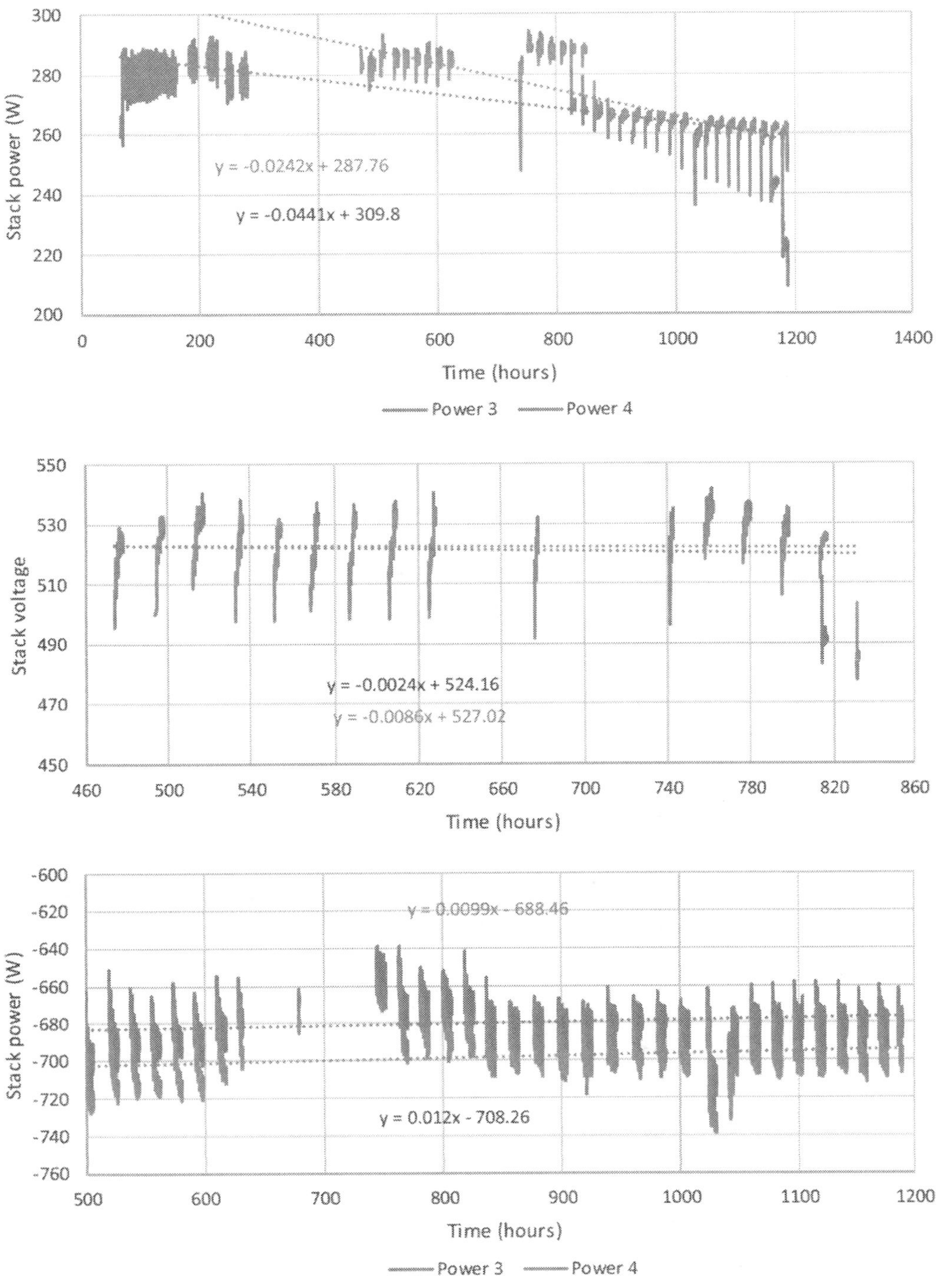

Figure 7. Power vs. time at 10.4 A (top), 20.9 A (middle) and -20.2 A (bottom)

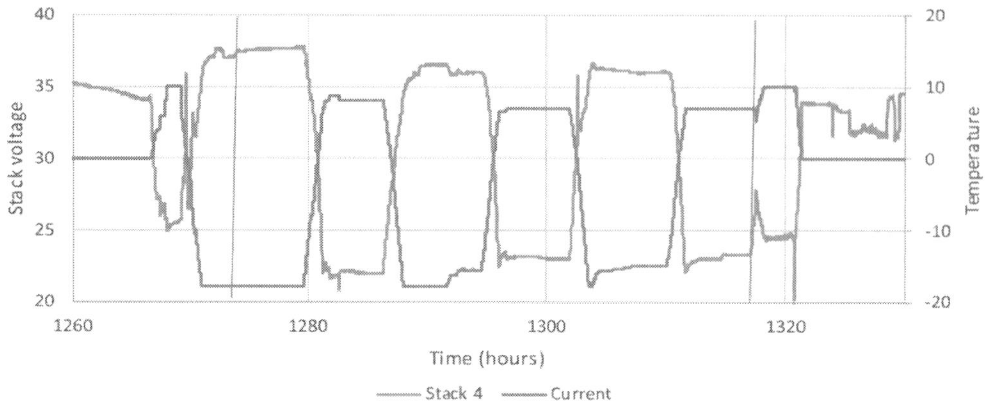

Figure 8. Voltage and current on operation with CO/CO₂

4. CONCLUSIONS

The following results and conclusions have been obtained in the project:

- A functional test bench has been manufactured suitable for further development of the technology.
- Sufficient performance and stability for reversible operation can be achieved with the standard SOFC materials used in terrestrial application.
- As long as the mechanical integrity of the stacks is preserved, degradation is not traceable within 1000 hours operation.
- The SOFC can be operated with both H_2/H_2O and CO/CO_2 as reactants. Lower performance with CO/CO_2 can be partly compensated by increased temperature.

As a final remark, material development, design, manufacture and testing of the 1 kW RSOFC system have provided invaluable experience and knowledge for the further development of a regenerative energy storage system for space exploration missions. The results also show that all the project success targets have been achieved and no major technological obstacles have been revealed. The development continues within the ongoing ESA project *Regenerative fuel cells for Mars exploration*, with the Centre for Research & Technology-Hellas (CERTH) as a partner, on sorbent materials for thermal compression of CO_2 at relevant temperature and pressure range (standard mechanical compressors are not viable in space) and tolerant electrocatalyst materials for CO and CO_2 SOFC/SOEC reversible operation [6].

5. REFERENCES

1. Wærnhus, I., Vik, A., Strand, A., Tsiplakides, D., Balomenou, S., & Papazisi, K. (2011). High temperature fuel cells for future mars missions. 9th European Space Power Conference, Saint Raphael, France.

2. Heggland, O.R., Wærnhus, I., Holst, B. & Ilea, C.S. Nanopowders for reversible oxygen electrodes in SOFC and SOEC (2012). 10th European SOFC Forum, Lucern, Switzerland. (B0711)

3. Wærnhus, I., Ilea, C.S., Vik, A., Heggland, O.R. & Holst, B. (2014). Regenerative energy storage system for space exploration missions. 10th European Space Power Conference, Noordwijkerhout, The Netherlands

4. Megel, S., Kusnezoff, M. et al. (2011). High Efficiency CFY-Stack for High Power Applications. ECS Transactions, 35(1). 269-277

5. Megel, S., Kusnezoff, M. et al. (2012). CFY-Stack: from electrolyte supported cells to high efficiency SOFC stacks. 10th European SOFC Forum (A1203)

6. Papazisi, K. M., Tsiplakides D., Balomenou S., Wærnhus, I., Vik, A., Schautz, M., (2016). Carbon tolerant fuel electrodes for reversible SOFC operating on carbon dioxide. 11th European Space Power Conference, Thessaloniki, Greece

FAILURE RATE MEASUREMENT ON SILICON DIODES REVERSE POLARIZED AT HIGH TEMPERATURE

D. Osorno, E. Sanchis-Kilders, E. Maset, D. Gilabert, A. Ferreres, J. Jordán, V. Esteve, J.L. Gasent-Blesa

Dpt. IE / ETSE, Universidad de Valencia, Avd. de la Universitat s/n, 46100 Burjassot, Spain, daniel.osorno@uv.es

ABSTRACT

This paper calculates the failure rate on reversed polarized silicon diodes with the aim to justify, experimentally, the rules of the European Space Agency (ESA) which are referred to the component life's extension, the reliability increase and the end of life performance enhancement, by using oversized devices (derating rules).
In order to verify the derating rules, 80 silicon diodes are used, which are reverse polarized in a high temperature environment. The diodes are divided in 4 groups of 20 diodes, applying a different voltage to each group, in order to relate the failure rate to the applied derating rule. The experiment described in this paper is developed using a temperature accelerated test to check the leakage current in reverse polarization (HTRB - High Temperature Reverse Bias), with the purpose of obtaining results applying an acceleration factor in order to reduce the test duration.
By using a thermal model of the whole system and the equations that describe the reverse polarized diode behaviour, it is possible to stress the 80 diodes up to very high temperature avoiding the runaway effect.
Finally, the failure rate is calculated and a revision of the derating rules are proposed by using the experimental result obtained.

1. INTRODUCTION

ESA (European Space Agency) uses rules to extend the life of the devices, thus guarantee the proper operation for the duration of the mission [1]. This practice, which oversizes the devices and makes them work below their maximum ratings, is known as "derating".
But these rules, mandatory in space missions, increase the economic cost, the weight and the size of the selected devices to use for the design. These rules also reduces the list of a available devices for design. Thus complicating the suitable selection of these devices (which are not always available for space).

M. de Economía y Competitividad of Gobierno de España and FEDER; Ref. ESP2013-47349-C6-5-R y ESP2104-56169-C6-4-R.
Conselleria d'Educació, Formació i Ocupació, Generalitat Valenciana; Ref. PROMETEO/2012/044.

In the end, the selection of the needed devices becomes a difficult task and increases the cost of the development. Of course this assures a higher reliability.
For example, the derating rules established by ESA for a Schottky diode define maximum values not to be surpassed and are as follows:

- 75 % of maximum forward current,
- 75 % of maximum reverse voltage,
- 50 % of maximum dissipated power and
- a maximum junction temperature of 110 °C or 40 °C below the manufacturer's maximum rating, whichever is lower.

Thus, in the case of the reverse voltage, the maximum value permitted under derating conditions would be 75 % of the maximum reverse voltage specified in the datasheet of the manufacturer.
Additionally, NASA (National Aeronautics and Space Administration) also proposes derating rules for devices, although they are different, in some cases, from the rules proposed by ESA.
For example, the rules from NASA of 2003 [2], for the same Schotkky diode of the last example, establishes the following deratings:

- 50 % of maximum forward current,
- 70 % of peak inverse voltage (PIV),
- 50 % of maximum peak current,
- 80 % of maximum junction temperature (does not exceed 125 °C or 40 °C below the manufacturer's maximum rating, whichever is lower).

And some years before, in 1991, NASA had published derating rules [3] that resembled those currently used by ESA, but different to the used nowadays. In the example of the Schottky diode, the former compulsory derating asked by NASA then was:

- 75 % of maximum reverse voltage,
- 50 % of maximum dissipated power and
- a maximum junction temperature of 110 °C.

Other references [4] also propose that the failure rate is related to the stress and follows the 5th power law, which says, for example, that an increase of a 15 % of the applied voltage to the device, implies doubling the failure rate.
Summarising, the derating rules proposed by ESA and NASA, and other laws also related to derating, applied to electronic devices, are all based on mathematical models and also in the practical experience. However, it is possible to consider that a 10 year long mission should have a different requirement than a mission that is only 6 months long, thus, the derating rules should be different. Also, when both rules (ESA's and NASA's) are compared, some differences are detected that could mean that the reliability of electronic devices is still not fully understood.
A more in depth research is therefore needed, to better understand the reliability of semiconductor devices, together with experimental results to verify this research. The following question should be answered: are derating the rules applicable to any mission? and, could these rules be updated with new results and in order to optimise cost, mass and volume of space missions? [5] [6]
The experimental results obtained that correspond to the failure rate as a function of the applied derating can be used to justify and/or update the existing derating rules applied in space missions.

2. EXPERIMENT DEVELOPMENT

This paper follows these steps:

A) Selection of the device under test.

B) Selection of the variable to analyse.

C) Experimental setup.

D) Auxiliary components.

E) Proposed thermal model.

F) Analysis of the acceleration factor.

The first task to do consist in selecting the device and the variable to test. The next task is to propose an experimental setup. Then all additionally needed components of the setup will be identified. Finally, the thermal model and the acceleration factor will be calculated in order to predict the results of the experiment.
These steps should guarantee positive experimental results showing the failure rate of the devices.

2.1. Selection of the device under test

In order to simplify the experiment, a diode is considered as a device under test (DUT), because other different devices, like transistors imply additional complexity. The final selection of the DUT is based on the diode list found in the EPPL (European Preferred Parts List) [7].
A commercial equivalent for each diode of this list is identified, and then, a selection is made keeping in mind, availability, electric characteristics, package and price of this equivalent diode (see Tab. 1).

Table 1. Diode selection

EPPL	Commercial	Package	Price (€)
1N6642U	1N6642D2A	DLCC2	24
BYW81-200	DPG15I200PA	TO-220AC	1.23
BYW81-200	BYV79E-20	TO-220AC	0.98
BYW52-200	BYW51-200G	TO-220AB	0.76
BYW52-200	NO		
1N5811US	1N5811	PTH	
1N5806US	1N5806	DO-204AP	4.45
1N5416US	NO		
1N5417US	1N5417	PTH	
1N5418US	1N5418	PTH	
1N5420US	1N5420	PTH	
1N54189UR-1	NO		
STPS1045	STPS1045B-TR	D-PAK	0.47
STPS6045	STPS6045CW	TO-247	3.4
STPS40100	NO	TO-220AB	0.76
1N5822U	1N5822	PTH	
1N5615	NO		
1N5617	1N5617	PTH	
1N5619	1N5619	PTH	
1N5623	NO		
STPS20100	NO		
1N5416	NO		
1N5417	NO		
1N5418	NO		
1N5420	NO		
1N5806U	1N5806D2A	TO-220AB	14.76
1N5819U	1N5819HW-7-F	TO-220AB	0.24
1N5614	NO		
1N5616	NO		
1N5618	1N5618	PTH	
1N5550	NO		
1N5552	NO		
1N5554	NO		
BAS70	BAS70-05	SOT-23	0.03

After discarding the diodes which have no commercial equivalent or have a different package than SMD (in order to solder the device on a PCB to a copper plane), the diode STPS1045 is selected as the best option for this test, due to its size, package and price. Additionally, this diode is an European low voltage Schottky, which is commonly used in space missions in output rectifiers of 3.3 V and 5 V.
This diode has a maximum reverse voltage of 45 V, and a reverse leakage current of 1.8 mA at 100 °C (45 V).

2.2. Selection of the variable to analyse

ESA's diode derating rules consider: direct current, reverse voltage, dissipated power and junction temperature.
In the first analysis, the dissipated power and the junction temperature won't be studied, because these characteristics can be calculated as a function of the direct current and/or reverse voltage.

Figure 1. Setup

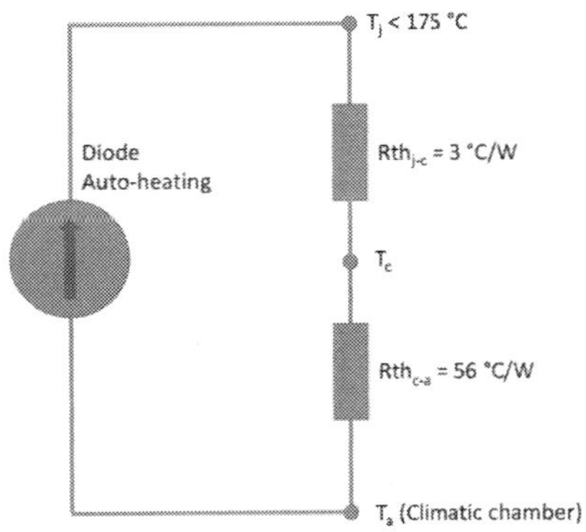

Figure 2. Thermal diagram

Finally, the reverse voltage (V_R) is chosen as a characteristic to analyse. The reverse current will be measured using a shunt resistor connected in series to the DUT and independently from the other diodes. Additionally, the influence of the self-heating effect is less important in reverse polarization than in direct polarization.

2.3. Experimental setup

Once the diode and the characteristic to analyse is chosen, it is proposed to carry out an accelerated test increasing the temperature [8], [9] y [10], also known as an HTRB (High Temperature Reverse Bias) test. The derating rules will be applied as reverse biased diode, limiting the maximum reverse voltage and it will be studied how this extends life of the devices.
In order to guarantee the thermal uniformity of all DUT, they must be soldered to the copper plane of the PCB as close to each other as possible, and placed inside a climatic chamber, which will control the ambient temperature surrounding the devices and all its setup (see Fig. 1).
During the test continuous reverse voltage stress will be applied to the diodes reverse and temperature will be kept constant.
The total population of diodes used in this experiment is 80, which will be divided in 4 groups of 20 diodes and each group is reverse biased with a different voltage (V_R).
In order to obtain results, that could show which is the most convenient derating to apply, the following reverse voltages (V_R) are proposed:

- $V_R = 100\,\%$ (45 V)
- $V_R \approx 80\,\%$ (36 V)
- $V_R \approx 50\,\%$ (24 V)
- $V_R \approx 30\,\%$ (12 V)

2.4. Auxiliary components

To measure the reverse leakage current (I_R) that appears when the diode is reverse polarized, which is the variable to analyse, a shunt resistors will be used in series with each diode.
The value of this shunt resistor must maximize the measurement resolution and, at the same time, the voltage drop should not interfere in the experiment ($V_{R_{shunt}} < 2\,\%$ of the voltage V_R), then finally, the shunt resistor's value selected are:

- V_R=45 V $\Rightarrow$ 75 Ω
- V_R=35 V $\Rightarrow$ 108.33 Ω
- V_R =24 V $\Rightarrow$ 390 Ω
- V_R =12 V $\Rightarrow$ 390 Ω

2.5. Proposed thermal model

With the aim of guaranteeing that the DUT does not reach the maximum junction temperature and/or thermal runaway due to self-heating during the experiment, the operating temperature for the HTRB must be chosen using an appropriate thermal model of the setup.
To do this, the thermal resistance from junction-to-case ($Rth_{j-c} = 3\,°C/W$) and from case-to-ambient ($Rth_{c-a} = 56\,°C/W$) is obtained from the datasheet of the DUT. This last parameter depends on the copper area which the DUT is soldered to.
These constants are included into the thermal diagram of the used setup, obtaining thus the thermal model (see Fig. 2).
On the other hand, from the I_R vs. V_R curve of the datasheet of the DUT (see Fig. 3), the I_R values can be obtained for different points of T_j (25 °C, 50 °C, 75 °C, 100 °C, 125 °C and 150 °C) for the four used values of V_R (45 V, 35 V, 24 V and 12 V) (see Fig. 4). Next, with these values, it is possible to obtain the equations of the regression curves given by any curve fitting software, by using 5th order (see Fig. 4 and equations 1, 2, 3 y 4), for each of the four reverse voltages of this test.
With these four equations, being $V_R = 45\,V$ the voltage with more self-heating influence in the junction temperature, a recursive simulation is carried out, using

DOI: 10.1051/e3sconf/20171611001

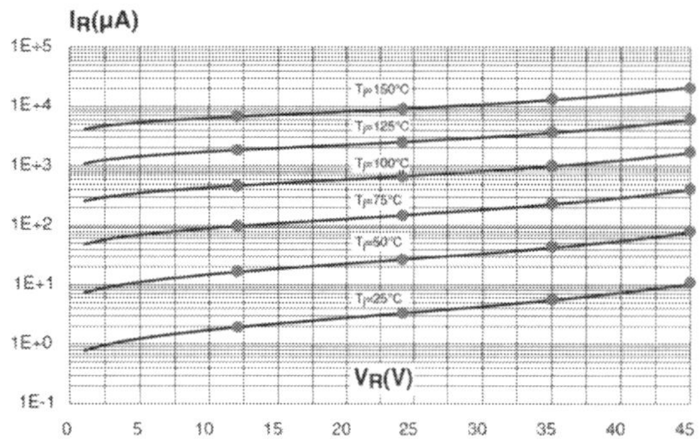

Figure 3. Relationship I_R vs. V_R found in the DUT's datasheet.

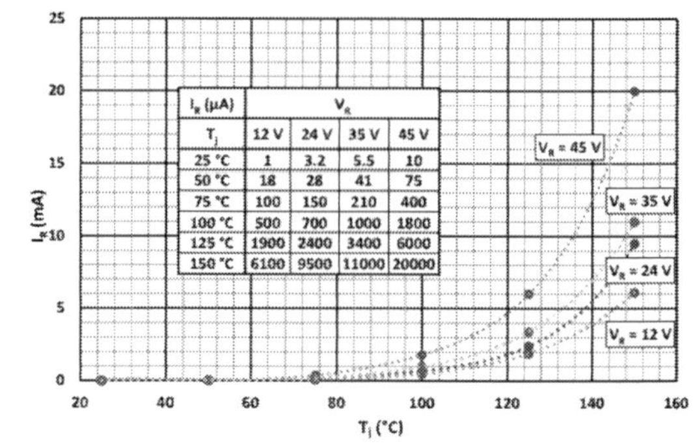

Figure 4. Values extraction of I_R as a function of T_j and V_R, and adjustment to 5th polynomial order.

the equation of the dissipated power of the diode (Eq. 5) and the expression that relates T_j with this dissipated power (Eq. 6) in order to verify the diode does not thermally runaway. In this case we can determine the maximum operating temperature to stress the diodes, before thermal runaway occurs.

$$\begin{aligned} V_R = 45\,\text{V} \Rightarrow I_R(\text{mA}) &= \\ = 3.7248 \cdot 10^{-06} \cdot {T_j}^5 - 1.2997 \cdot 10^{-03} \cdot {T_j}^4 &+ \\ + 1.8231 \cdot 10^{-01} \cdot {T_j}^3 - 1.2068 \cdot 10^{+01} \cdot {T_j}^2 &+ \\ + 3.6960 \cdot 10^{+02} \cdot T_j - 4.06500 \cdot 10^{+03} & \end{aligned} \tag{1}$$

$$\begin{aligned} V_R = 35\,\text{V} \Rightarrow I_R(\text{mA}) &= \\ = 1.7916 \cdot 10^{-06} {T_j}^5 - 6.1835 \cdot 10^{-04} \cdot {T_j}^4 &+ \\ + 8.7004 \cdot 10^{-02} \cdot {T_j}^3 - 5.8015 \cdot 10^{+00} \cdot {T_j}^2 &+ \\ + 1.7912 \cdot 10^{+02} \cdot T_j - 1.9820 \cdot 10^{+03} & \end{aligned} \tag{2}$$

$$\begin{aligned} V_R = 24\,\text{V} \Rightarrow I_R(\text{mA}) &= \\ = 2.6767 \cdot 10^{-06} \cdot {T_j}^5 - 9.6205 \cdot 10^{-04} \cdot {T_j}^4 &+ \\ + 1.3530 \cdot 10^{-01} \cdot {T_j}^3 - 8.9491 \cdot 10^{+00} \cdot {T_j}^2 &+ \\ + 2.7331 \cdot 10^{+02} \cdot T_j - 3.0008 \cdot 10^{+03} & \end{aligned} \tag{3}$$

$$\begin{aligned} V_R = 12\,\text{V} \Rightarrow I_R(\text{mA}) &= \\ = 5.8795 \cdot 10^{-07} \cdot {T_j}^5 - 1.7472 \cdot 10^{-04} \cdot {T_j}^4 &+ \\ + 2.2493 \cdot 10^{-02} \cdot {T_j}^3 - 1.4188 \cdot 10^{+00} \cdot {T_j}^2 &+ \\ + 4.2512 \cdot 10^{+01} \cdot T_j - 4.6400 \cdot 10^{+02} & \end{aligned} \tag{4}$$

$$P_D(W) = I_R \cdot V_R - R_{shunt} \cdot {I_R}^2 \tag{5}$$

where,

- P_D is the dissipated power in W
- I_R is the reverse leakage current in A
- V_R is the reverse voltage in V
- R_{shunt} is the value of the shunt resistor in Ω

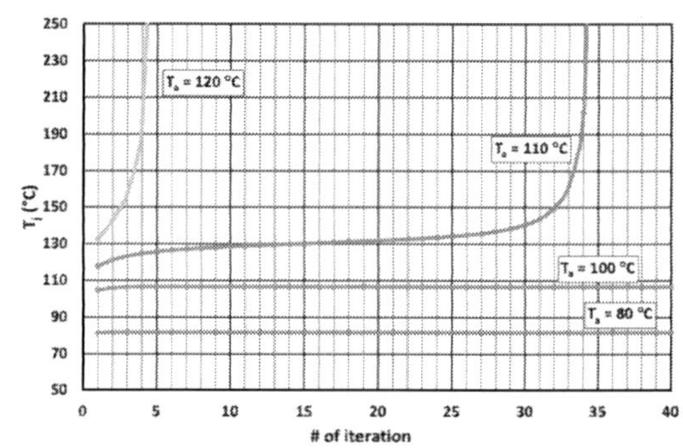

Figure 5. Results of the recursive simulation where a thermal runaway is observed for T_j = 110 °C and T_j = 120 °C

$$T_j(°C) = T_a + (Rth_{j-c} + Rth_{c-a}) \cdot P_D \tag{6}$$

where,

- T_j is the junction temperatures expressed in °C
- T_a is the operating temperature expressed in °C
- Rth_{j-c} is the thermal resistance from junction-to-case expressed in W/°C
- Rth_{c-a} is the thermal resistance from case-to-ambient expressed in W/°C
- P_D is the dissipated power in W.

The result given by the recursive simulation (see Fig. 5) is that, with an operating temperature of about 110 °C, diodes blocking V_R = 45 V reach the thermal runaway (T_j > 150 °C) destroying the silicon junction of the semiconductor. Therefore, the temperature of the experiment is reduced by 10 °C and so fixed at 100 °C, programming the climatic chamber used.

2.6. Analysis of the acceleration factor

As said before, the experiment shown in this paper stressed the diodes after derating their limits [1], which makes more difficult to obtain a failure. To observe a failure an accelerated test will be done.
By using the Arrhenius' equation (Eq. 7) in order to calculate this acceleration factor, customized by NASA (Eq. 8), where the constant of activation energy is (E_A = 0.2656 eV) [11] and knowing the operating temperature of the experiment (T_a = 100 °C), the obtained acceleration factor is equal to 8 units. Por example, 1000 hours of this test are equivalent to 8000 hours of operation of the DUT (at 25 °C).

$$A_T = \frac{\lambda_{T1}}{\lambda_{T2}} = e^{\left[\frac{E_A}{k}\left(\frac{1}{T_1} - \frac{1}{T_2}\right)\right]} \quad (7)$$

where,

- A_T acceleration factor
- T_1 y T_2 is the temperature 1 and 2 respectively in °C
- λ_{T1} and λ_{T2} is the failure rate at temperature 1 and 2 respectively
- E_A is the activation energy in eV
- k is the Boltzmann constant in eV/°C

$$\pi_T = \frac{\lambda_{T_j}}{\lambda_{25°C}} = e^{\left[-3091\left(\frac{1}{T_j - 273} - \frac{1}{298}\right)\right]} \quad (8)$$

where,

- π_T is the acceleration factor
- λ_{T_j} is the failure rate at the operating temperature
- $\lambda_{25\,°C}$ is the failure rate at 25 °C
- T_j is the junction temperature of the semiconductor in K

3. RESULTS

Until now, the results (see Fig. 6) show a quite stable behaviour of the reverse leakage current for the four derated diode groups. In a first transient stage, an increase of the reverse leakage current is observed until it reaches the steady-state after the adjustment of temperature of the climatic chamber up to 100 °C.
The standard calculation method to obtain the failure rate is shown in Eq. 9.

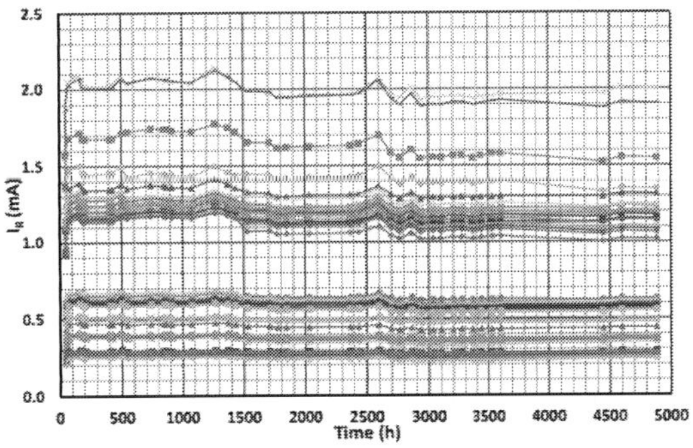

Figure 6. Reverse leakage current evolution

$$\lambda = \frac{x}{TDH \cdot AF} \quad (9)$$

where,

- λ is the failure rate in number of failures / 10^6 hours
- x is the number of failures
- TDH is the summation of the number of units in operation multiplied by the time of operation.
- AF is the acceleration factor

And, additionally, the mean time to failure is defined as:

$$MTTF = \frac{1}{\lambda} \quad (10)$$

where,

- $MTTF$ is the mean time to failure (time to fail of 63% of the population).
- λ is the failure rate.

Although, degradation has not been detected and no broken diodes appeared after 4887 hours of test, and the exact value of the failure rate cannot be obtained. Anyhow, is possible to obtain comply with predicted maximum value of the failure rate in case one diode brakes down just one hour after measuring the devices.
Finally, the failure rate can be calculated by using x=1, TDH = 80 devices multiplied by 4887 hours, and AF = 8 (as previously calculated) in Eq. 9:

$$\lambda < \frac{1}{80 \cdot 4887 \cdot 8} = 0.320/10^6 \quad (11)$$

The mean time to failure (MTTF), in this case, using the equation 10 is:

$$MTTF > \frac{1}{\lambda} = 3\,127\,680\,\mathrm{h} \quad (12)$$

4. CONCLUSION

This paper studies the effect of the derating rules, applied to reversed polarized Schottky diodes, to the failure rate with the aim of comparing the derating rules published by ESA or NASA with the experimental results.
A high temperature reverse bias experiment is developed using a population of 80 diodes divided in 4 groups (4 different derating values) of 20 diodes each, with the aim of measuring the leakage current of every single diode reverse polarized.
With 4887 h under test at 100 °C, which is equivalent to more than 39 000 h at 25 °C, no degradation has been detected and no broken diode happened, obtaining a maximum value of the failure rate equal to 0.320 failures / 10^6 h.
The absence of failures in the diodes, after 4887 h under test at 100 °C with a derating between 100 % and 30 %, means that in short term (taking into account the scope of this work with only 4887 h under test) the derating has not reduced the failure rate.
This fact points out the derating rules should be reviewed in order to take account the duration of the mission, for instance, the increase of the derating values at least in 6-months missions.

5. REFERENCES

[1] ESA - ESTEC, Noordwijk, The Netherlands. *Space product assurance: Derating - EEE components*, ECSS-Q-ST-30-11C Rev 1, Oct. 4 2011.

[2] NASA - Goddard Space Flight Center, Greenbelt, Maryland. *Instructions for EEE Parts Selection, Screening, Qualification, and Deratings*, EEE-INST-002, May 2003.

[3] NASA, Washington DC. *Reliability preferred practices for design and test*, Memorandum 4322, Sept. 1991.

[4] Richard Denning. *Applied R&M manual for defence system*. SARS, MoD, Abbey Wood, May 2012.

[5] Zhaofeng Huang. De-rating design and analysis for safety and reliability. In *Reliability and Maintainability Symposium (RAMS), 2014 Annual*, pages 1–7, Jan 2014.

[6] Kwok-Wai Ma. Highly reliable dc/dc power module for demanding applications-from design to field tracking. In *Applied Power Electronics Conference and Exposition, 1997. APEC '97 Conference Proceedings 1997., Twelfth Annual*, volume 2, pages 647–654 vol.2, Feb 1997.

[7] ESA. *European preferred parts list*, ISSUE 30, Dec. 2015.

[8] G.C. Sikora and L.E. Miller. Application of power step stress techniques to transistor life predictions. In *Physics of Failure in Electronics, 1964. Third Annual Symposium on the*, pages 30–42, Sept 1964.

[9] N. F. Bechtold and C. L. Hanks. Failure-rate studies on silicon rectifiers. *American Institute of Electrical Engineers, Part I: Communication and Electronics, Transactions of the*, 77(1):49–56, March 1958.

[10] J.A. McLinn. Creating a sensible derating system. In *Reliability and Maintainability Symposium, 2006. RAMS '06. Annual*, pages 362–367, Jan 2006.

[11] Department of defense, Washington DC. *Military handbook: Reliability prediction of electronic equipment*, MIL-HDBK-217F, Dec. 2, 1991.

MINIATURISED INTEGRATED CIRCUIT CORE ELEMENT FOR POINT OF LOAD (POL) CONVERSION

S. Pappalardo [1], A. Alessandro [2], C. Santagati [3], C. Ardizzone [4], C. Ribellino [5], I. Mirabella [6], O. Principato [7], F. Tonicello [8]

[1] *STMicroelectronics S.p.A., Stradale Primosole n.50, 95121 – Catania, Italy. Email: salvo.pappalardo@st.com*
[2] *STMicroelectronics S.p.A., Stradale Primosole n.50, 95121 – Catania, Italy. Email: agatino.alessandro@st.com*
[3] *STMicroelectronics S.p.A., Stradale Primosole n.50, 95121 – Catania, Italy. Email: carmelo.santagati@st.com*
[4] *STMicroelectronics S.p.A., Stradale Primosole n.50, 95121 – Catania, Italy. Email: carmelo.ardizzone@st.com*
[5] *STMicroelectronics S.p.A., Stradale Primosole n.50, 95121 – Catania, Italy. Email: lillo.ribellino@st.com*
[6] *STMicroelectronics S.p.A., Stradale Primosole n.50, 95121 – Catania, Italy. Email: ignazio.mirabella@st.com*
[7] *STMicroelectronics S.p.A., Stradale Primosole n.50, 95121 – Catania, Italy. Email: orazio.principato@st.com*
[8] *European Space Agency, European Space Technology Centre, Keplerlaan 1, P. O. Box 299, 2200 AG, NOORDWIJK, The Netherlands. E-Mail: Ferdinando.Tonicello@esa.int*

ABSTRACT

In the last years Point of Load converters (PoL) appeared as a clear bottleneck for the realisation of digital applications for space (to supply FPGA's, microprocessors, memories, specialised digital circuits, etc.). Many preparatory activities have been conducted by ESA to define a target specification, taking into account promising European technologies and user's requirements. This paper provides a brief technical description of a PoL IC device designed and manufactured by STMicroelectronics and shows the main test results and performances in order to demonstrate that this integrated solution embeds the most important features that a Point of Load (PoL) device should have. The device is the outcome of a design activity executed in relation to the TRP activity entitled "Miniaturised core element for Point of Load (PoL) conversion" (T703-305EP of the TRP work plan 2011-2013). The realization of this product enables STMicroelectronics to compete in the global world market, since this is a prerequisite for the future availability of PoL products, for their commercial success and their sustainability.

An export licence-free, performing and advanced PoL ASIC (Application Specific Integrated Circuit) would be a major achievement.

It would encounter interest from users and would resolve the problem of supplying digital loads with the required accuracy and dynamical performances.

1. INTRODUCTION

Typical switching Point of Load ASICs, used in distributed power conversion terrestrial systems and in those applications requiring low power dissipation, are generally sensitive to radiation effects, both Total Dose and Single Event Effects. Today a variety of commercial switching DC-DC converters are available from many American & European suppliers (including STMicroelectronics), but European manufacturers and suppliers are mainly oriented in developing products towards industrial, automotive, computer peripheral and consumer applications rather than for space applications.

At the moment mainly non-European suppliers are providing RH or Rad-tolerant PoL converters for space applications. Some of these devices are QML qualified and are procurable according to an electrically screened SMD. This represents a potential risk to the European space industry, as these MIL qualified devices are subject to US export licenses and/or US Department of Commerce rules. Obviously, under the given policy constraints and commercial conditions associated with the U.S. source, the development of a European supplier alternative is expected to result in significant cost and schedule advantages.

Project's objective from STMicroelectronics was to design, develop and manufacture a European HI-Rel, Rad-Hard Point of Load (PoL) core and is supported by ESA and National Space Agency. The Point of Load ASIC (hereafter called RHFPOL01) is suitable for an extensive range of Space applications (meteo, communication, earth observation satellites) including those ones with mission durations in the range of 15-18 years. STMicroelectronics is committed to get a product for the worldwide space customers, which implies primarily QML-V qualification and the insertion to the European Preferred Part List. ST will be open to discuss also the additional ESCC Evaluation and Qualification providing specific support.

2. RHFPOL01 ASIC

The RHFPOL01 device is a radiation hardened, high efficiency, monolithic, synchronous step-down switching regulator with high precision internal voltage reference and integrated power MOSFETs for synchronous conversion. The device is developed using ST built-in SOI (Silicon On Insulator) technology that offers excellent performance against SEL (Single Event Latch-up) effects. The regulator converts the input

 DOI: 10.1051/e3sconf/20171611002

voltage V_{IN} in the range [3.0V ÷ 12V] to output voltage V_{OUT} from 0.8V up to 0.85xV_{IN} and is able to provide up to 7A output current. In Fig.1 a typical application is shown in which the RHFPOL01 device is embedded.

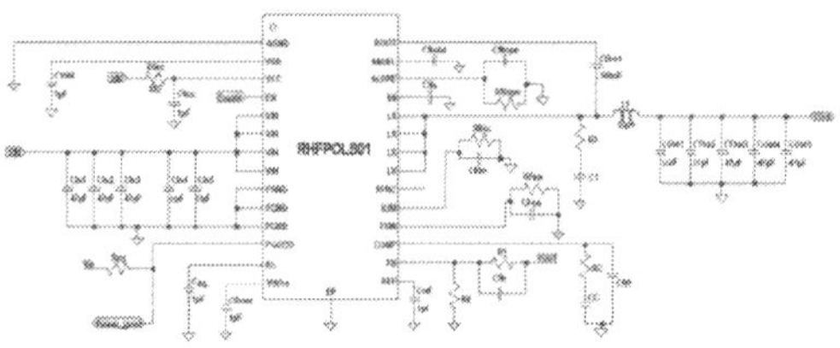

Figure 1 - RHFPOL01 Application Circuit

The major application fields for this device are:

- Point of load regulation for space applications
- FPGA, DSP, CPU and ASICS supply for space applications
- Low-Voltage, High-Density Distributed Power Systems

2.1 RHFPOL01 Block Diagram & Features

The block diagram of RHFPOL01 is shown in Fig. 2. The controller is based on peak current mode control loop architecture which, with its embedded error amplifier and external compensation, enables a stable operation with a wide range of output filter configurations (including Multi-Layer Ceramic Capacitor – MLCC- solutions) ensuring fast response to load transient and very stable switching frequency. The fault management consists of output over-voltage protection (not-latched) and latched/auto-recovery over-current and thermal protections.

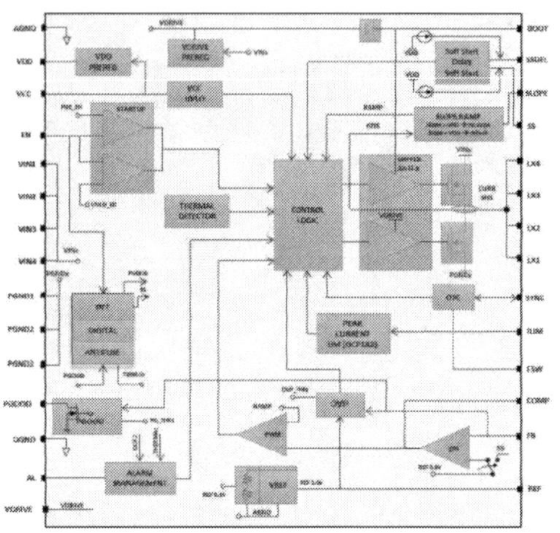

Figure 2 - RHFPOL01 Block Diagram

Low R_{DS_ON} N-channel MOSFETs are used for both HS (high-side) and LS (low-side) switches and also the bootstrap MOSFET driver diode is embedded, for minimum external components provision.

2.2 Start-up and soft-start

The dedicated enable (EN), programmable soft-start (SS) and soft-start delay pins (SSDEL) offer an easy control on the power sequencing and inrush current. Once V_{CC} voltage is above the V_{CC_UVLO} rising threshold, the device waits for enable pin (EN) assertion and then begins the soft-start. In order to add flexibility to turn-on management, the enable pin (EN) has two thresholds: if EN < 0.7V, the IC is fully OFF and the current consumption is typically 400μA. If EN > 1.24V the regulator starts switching, after the SSDEL time elapses. When EN pin is kept between the two thresholds, the main blocks of the regulator are turned ON and the RHFPOL01 device is ready to start regulation. It should be noted that the external filtering capacitors connected between function programming pins (FSW, ILIM and SLOPE) and GND (as shown in Fig. 1), have a direct impact on the IC wake-up timing. In order to be sure that the soft-start sequence starts only when the functions programming pin voltage is stable, the soft-start delay (SSDEL) has been added. The typical start-up sequence is depicted in Fig. 3.

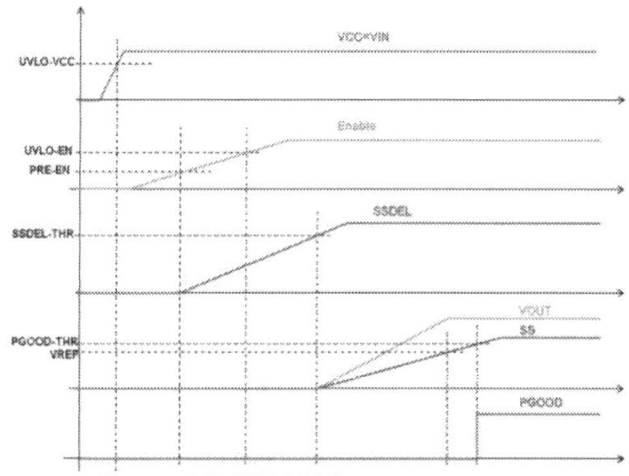

Figure 3 RHFPOL01 start-up sequence

During start-up phase, in order to prevent in-rush current when the V_{OUT} voltage is too low, two switching frequency changes are implemented. This feature is clearly shown in the Fig. 4 (assuming a final programmed frequency of 500 KHz).

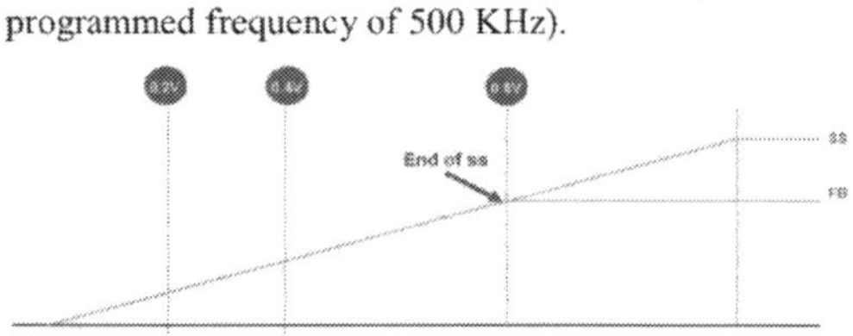

Figure 4 - Switching frequency changes during start-up

As long as voltage on pin SS is < 0.2V, the switching frequency is ¼ of programmed frequency. When SS

DOI: 10.1051/e3sconf/20171611002

value is between 0.2V and 0.4V the switching frequency is ½ of the programmed frequency. After SS reaches 0.4V, the switching frequency nominally reaches the programmed one.

2.3 Control loop

The simplified block diagram of RHFPOL01 control loop is shown in Fig.5. It consists in a constant-frequency peak current-mode control and employs two control loops: an external loop to control the output voltage and an inner loop to control the peak current which flows through the coil. The external loop (voltage loop) compares the feedback voltage V_{FB} with the internally generated precise 0.8V reference voltage through the error amplifier EA.

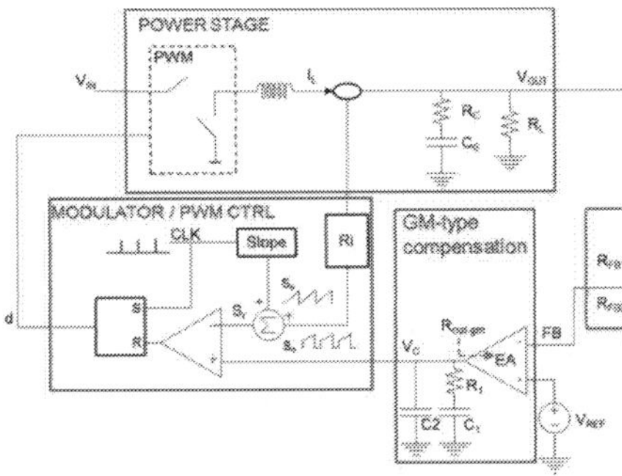

Figure 5 - RHFPOL01 Control Loop

The error amplifier EA is a trans-conductance amplifier (OTA) that, starting from the difference between reference and feedback voltages, adapts the voltage level on node V_C (pin COMP), where external RC compensation network is connected, in order to provide the proper regulation. In the inner loop (i.e. current loop), the current sense circuit converts the current flowing through the coil into a sense voltage Sn according to the gain factor R_i. In addition, the voltage ramp Se, programmable by the external R_{SLOPE} resistor, is added to the Sn signal to get stable operation in case of duty cycle greater than 50%. The voltages V_C and the resulting ramp Sr (see Fig. 5) are compared by means of a PWM comparator and a Pulse Width Modulated (PWM) signal is generated, defining the duty cycle d. The power stage can be seen as a controlled current generator, which provides the current I_L to the power stage output capacitor and load.

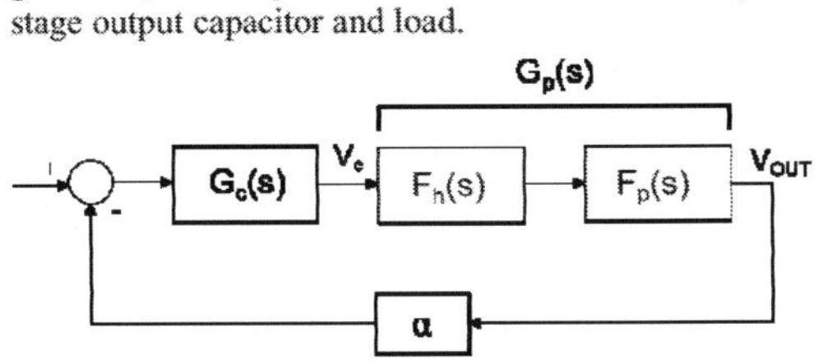

Figure 6: RHFPOLS01 control loop function

Complete small signal model of the converter is obtanined according with PWM switch model, proposed in [3], and small signal model of the peak current mode, proposed in [4]. By combining the open loop transfer function with the compensator and the feedback network, the closed loop transfer function is obtanied:

$$G_{loop}(s) = \alpha(s)\cdot G_p(s)\cdot G_c(s) \qquad (1)$$

Where:

$$G_p(s)=G_p(0)\cdot\frac{\left(1+\frac{s}{\omega_z}\right)}{\left(1+\frac{s}{\omega_p}\right)}\cdot\frac{1}{1+\frac{s}{\omega_n Q_p}+\frac{s^2}{\omega_n^2}} \qquad (2)$$

$$G_C(s)=g_m\cdot R_{out\_gm}\cdot\frac{\left(1+\frac{s}{\omega_{zc}}\right)}{\left(1+\frac{s}{\omega_{pc1}}\right)\cdot\left(1+\frac{s}{\omega_{pc2}}\right)} \qquad (3)$$

$$\alpha(s)=\frac{R_{FB2}}{R_{FB2}+R_{FB1}}\cdot\frac{\left(1+\frac{s}{\omega_{zcfb}}\right)}{\left(1+\frac{s}{\omega_{pcfb}}\right)} \qquad (4)$$

Equations of first order pole and zero of the output filter are:

$$\omega_Z=\frac{1}{R_C\cdot C_O}; \qquad (5)$$

$$\omega_P=\frac{1}{(R_L+R_C)\cdot C_O}+\frac{T_S}{L\cdot C_O}\left(m_C\cdot(1-D)-\frac{1}{2}\right) \qquad (6)$$

The second order term in the equation (2) gives double pole at half of switching frequency. The equation (3) represents the contribution of GM-type compensation network. The relative zero and two poles are below:

$$\omega_{zc}=\frac{1}{R_1\cdot C_1};\quad \omega_{pc1}\cong\frac{1}{R_{out\_gm}\cdot C_1};\quad \omega_{pc2}\cong\frac{1}{R_1\cdot C_2} \qquad (7)$$

Finally, the feed-forward capacitor add pole-zero factor α spaced:

$$\omega_{zcfb}=\frac{1}{R_{fb1}\cdot C_{fb}};\quad \omega_{pcfb}=\frac{1}{C_{fb}\cdot\frac{R_{FB1}\cdot R_{FB2}}{R_{FB1}+R_{FB2}}}=\frac{\omega_{zcfb}}{\alpha} \qquad (8)$$

Generally, R1 and C1 are chosen to compensate the pole due to the output capacitor.

To improve the response of the converter up to 75% of step load, the EA is designed to have not-linear g_m for large input offset. Here below measurement result of the gm versus the feedback voltage is shown.

 DOI: 10.1051/e3sconf/20171611002

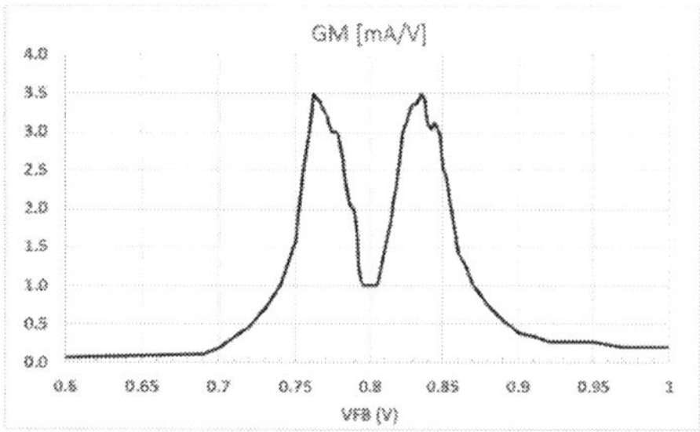

Figure 7: Error amplifier trans-conductance.

2.3.1 Stability, phase and gain margin vs model

To evaluate in detail the stability of the converter, a proper estimation of poles and zeros must be obtained according with internal circuitry tolerances, output capacitor variation and compensation network. Dedicated model is provided by ST in worksheet format to help the final user on choosing proper compensation elements. For instance, a comparison between test bench measurements and the model is shown in Fig. 8 and Fig. 9.

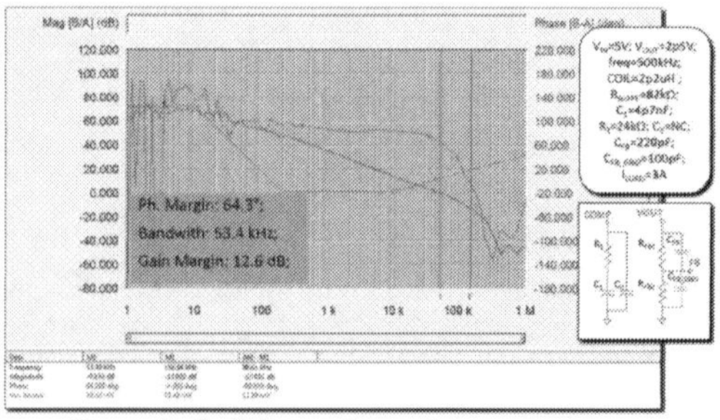

Figure 8: Stability and open loop measurement

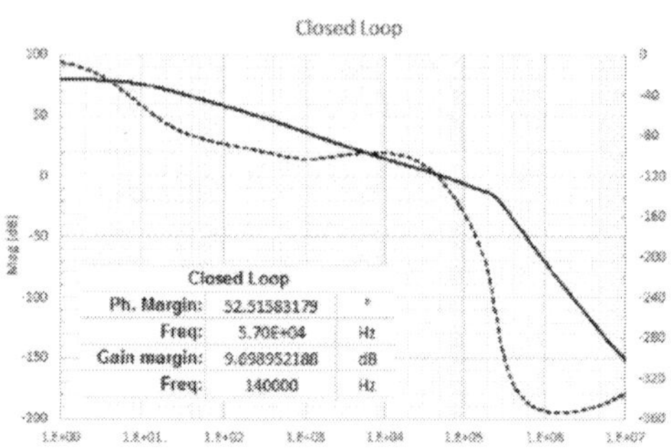

Figure 9: Stability and open loop worksheet result

2.4 Fast load transient response

When fast load transient is applied to the output, the device is able to react very quickly in order to limit output voltage overshoot or undershoot. This feature is implemented by the combination of a fast error amplifier response and a proper control logic scheme that is driven by the use of two fast comparators, called HL (High Load) and LL (Low Load). When the voltage FB goes below the HL reference point REF_HL (see Fig. 10), the HL comparator forces ON the HS power switch. Typically, REF_HL level is -5% below the nominal voltage loop reference of 800mV. The HL comparator output remains latched to high level voltage until the programmed current limit is reached or the output voltage goes back into the linear control regulation voltage level. So, a proper voltage hysteresis is implemented: the comparator output goes low when the feedback goes back again to -2.5% of the regulation threshold (see Fig.10). If during the ON phase current limit is reached, minimum OFF time period around 200ns is applied.

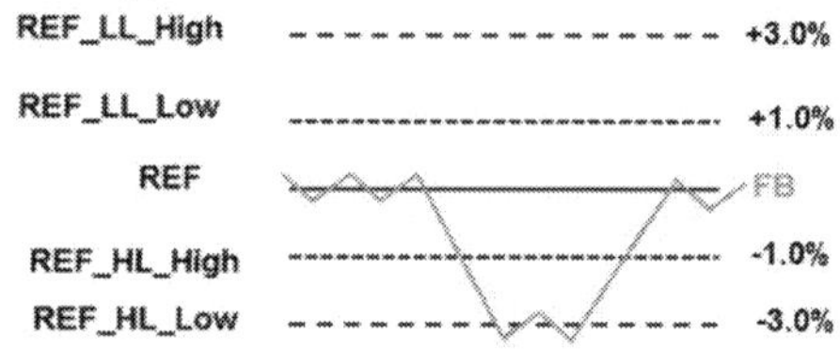

Figure 10 - Fast Load Transient Response

In opposite way, when the load current quickly decreases, the FB voltage can show an overshoot. The LL comparator output goes high when the FB goes above the LL reference point voltage REF_LL (typically +7% above the nominal 800mV reference) and an OFF time is forced. The OFF time is kept till either the FB goes back to the hysteresis threshold (typically +2.5%) or a negative inductor current threshold is reached. In fact, the device naturally works in full PWM mode, but when fast negative step load occurs, the current on the coil can quickly go below zero. So, a zero crossing comparator, with a typical threshold of -1.5A, is implemented avoiding the inductor coil goes deeply below zero. When the zero crossing threshold is reached, switching activity is stopped and the power stage is forced in open loop condition (point LX is in high impedance). Of course, the two comparator themselves are not enough for a smooth operation but, as mentioned above, they also need the concurrent action of the error amplifier. A fast error amplifier with dynamic bias has been designed in order to improve the load-transient response. In practice the error amplifier must be able to quickly charge or discharge the compensation capacitor (on pin COMP) changing its output in non-linear shape at a step change in the input of the error amplifier itself. So, when one of the two comparators drives ON or OFF time, in parallel the bandwidth and the large signal current capability (including the large signal current gain) of the error amplifier are increased. Finally, this solution allows a fast load transient response with lower output capacitance and smaller output voltage drop.

2.5 Synchronization & Interleaving

Two RHFPOL01 ASICs can be configured in Master and Slave mode. The clock synchronization is realized configuring one IC as Master (with a proper switching frequency) and the second IC as Slave, shorting its pin FSW to GND and by connecting together the pin SYNC of each device. When the device is the MASTER its pin SYNC is configured as clock output and gives out the internal switching clock frequency with a 180° phase shifting. If the device is a SLAVE its pin SYNC is configured as clock input, and the device uses the clock received on pin SYNC to synchronize its internal switching clock.

In synchronized mode, each device will have its own output and compensation network (see Fig. 11a), while in the interleaved configuration the two devices share the same output and compensation network (see Fig. 11b). The main benefits of interleaved configuration are a doubled capability of output current and the reduction of RMS current absorption from the input filter, preventing beating frequency noise and therefore allowing a reduction in the size and cost of the input filter, at equal ripple conditions.

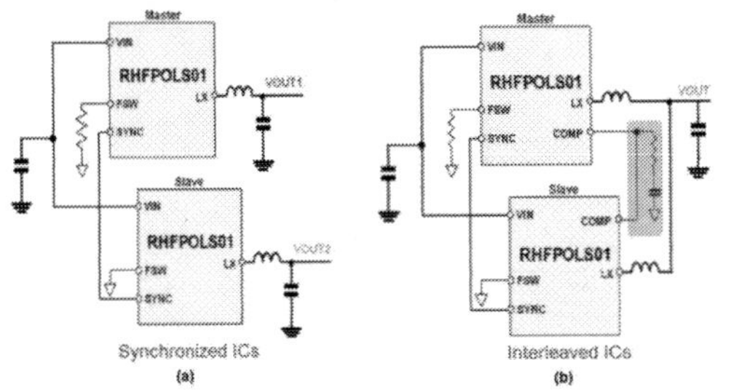

Figure 11 - RHFPOL01: Synchronized & Interleaved ICs

The Fig.12 shows the experimental result of two RHFPOL01 ICs that are working in interleaved configuration providing up to 14A output current.

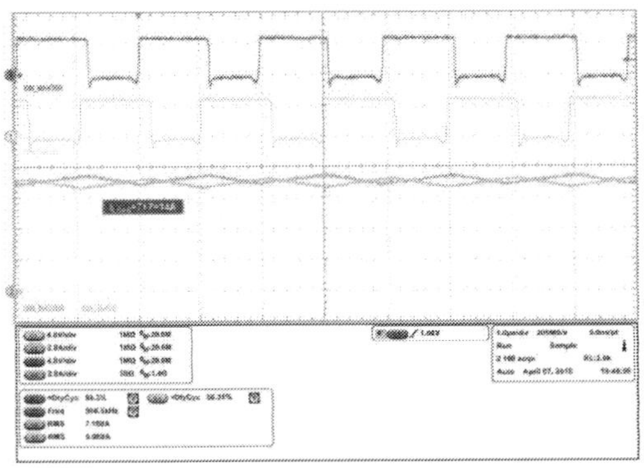

Figure 12: Measure on two interleaved ICs

To guarantee uniform heat distribution and balancing of the electrical stress between the two devices, the total output current I_{LOAD} is shared inside each device (I_{OUT1}, I_{OUT2}) with an accuracy of 10%. It means that:

$$I_{OUT1,2} < I_{LOAD}/2 \pm 10\% \qquad (9)$$

A brief description of the approach used to guarantee the load sharing accuracy is shown in Fig. 13. Shorting together the nodes COMP_M (master) and COMP_S (slave) and also trimming both internal current sensor and slope circuitries, a good accuracy will be ensured.

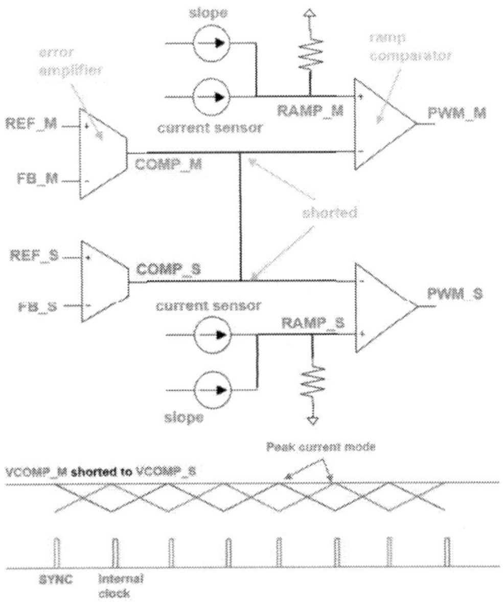

Figure 13: Block diagram and signals for load sharing

It can be interesting to highlight that a strength of this approach is that, slightly changing the value of the external R_{SLOPE} resistor, a fine tuning of final accuracy is possible. In this way, possible asymmetries (due to coil, board resistive paths ...) can be recovered and the accuracy can be improved.

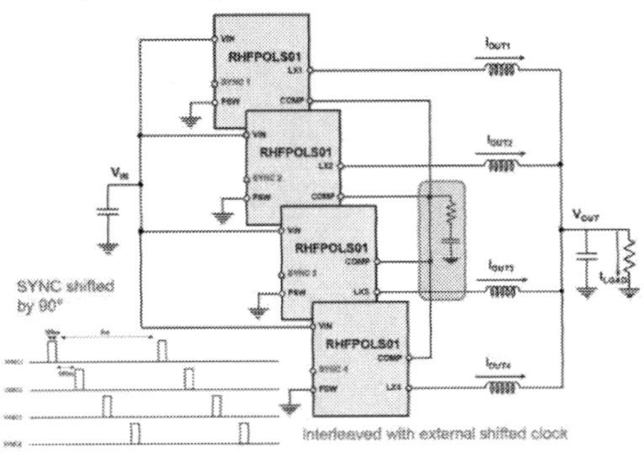

Figure 14: four Slave Interleaved ICs at 500 kHz

To increase even more the output current capability, it is

possible to connect more than two devices ("*n*" in general) in interleaved configuration. In this case, in order to guarantee a proper phase shifting (360°/*n*), all the devices are configured as slave and then an external clock signal must be provided to each device.

In Fig.14 is shown a configuration with four slave devices connected in interleaving mode able to provide up to 28A (here phase shifting is 90°, of course).

Also in this case the load sharing accuracy is guaranteed:

$$I_{OUT1,2,3,4} < I_{LOAD}/4 \pm 10\% \qquad (10)$$

To perform this measurement in our lab, two dedicated boards for double load sharing were connected in parallel, as shown in the figure below.

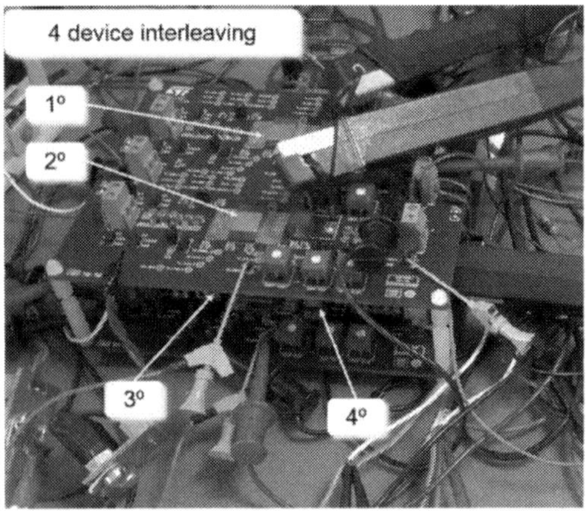

Figure 15: Board equipment of four Interleaved ICs

Even if the resistive paths have not been optimized, the final accuracy was in spec, as can be seen in the measurements highlighted in Fig.16:

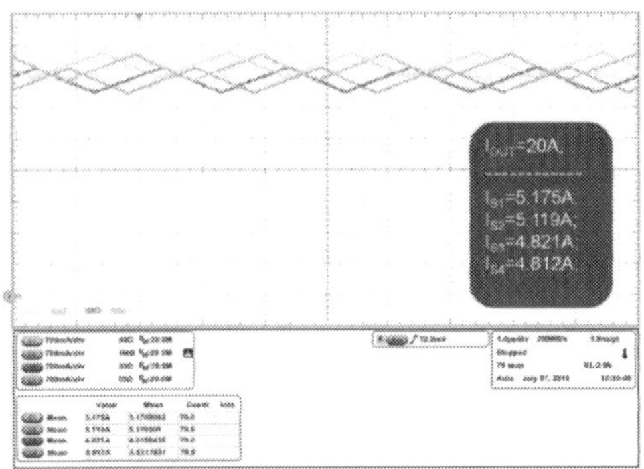

Figure 16: Measure of four Interleaved ICs

2.6 RH_OTP and Run Trimming

The RHFPOL01 embeds a radiation hardened OTP (One Time Programmable circuit, realized with anti-fuse cells) that gives the final user the possibility of tuning the accuracy of some key parameters (such as V_{OUT}, Freq, Ilim...) directly on the application board and during the device regulation, practically like a "run trimming". In this way, the final accuracy will be higher because also the effect of the external passive elements (FB resistor, external resistor to set frequency...) will be compensated during measurement and trimming operation.

The "Run Trimming" consists of two phases: the simulation and the burning phases. During the simulation phase, the user can choose the best trimming value by using the dedicated software and hardware that implement i2c (Inter Integrated Circuit) communication protocol. After trimming the pattern is verified by simulation phase and the final user can proceed with the final burning phase. The figure below depicts the Run Trimming for V_{OUT} parameter performed during the normal switching activity of the device.

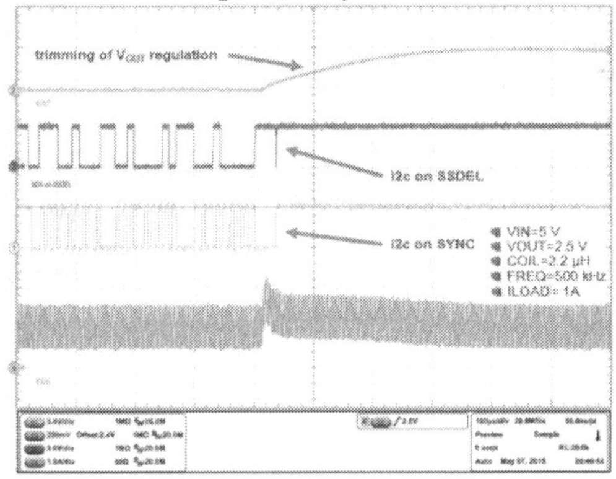

Figure 17 - RHFPOL01: V_{OUT} Run Trimming

2.7 Protection and Alarms

The RHFPOL01 device features a full set of protections and output voltage monitoring:

- PGOOD (power good function) realized by means an open drain output at pin PGOOD, which is externally pulled-up to VIN if the output voltage is in ±10% of regulation window. During start-up, turn-off and fault detection the PGOOD pin is forced low.
- OVP - Over Voltage Protection: The output overvoltage protection is triggered as soon as the output regulated voltage exceeds +25% of the nominal value.
- ILIM1: 1st level of over current protection.
- ILIM2: 2nd level of over current protection.
- OVTP - Over Temperature Protection. The strategy used for this kind of protection took into account that it is required (by ECSS derating rules) that Si devices never exceed the maximum allowable junction temperature of 110°C (in steady state operation), or up to the rated limit for transients. The RHFPOL01 implements the following thermal shutdown feature:

when the junction temperature reaches 155°C the device turns OFF both internal MOSFETs and then pin PGOOD is forced low. When the junction temperature drops again to around 135°C, a delay time on pin AL elapses before the device restarts with a new soft-start sequence (see Fig.18).

The OVP and ILIM1 alarms are non-latching alarms. They put the device in T_{OFF} (ILIM1) or in high impedance conditions (OVP) but the regulation continues normally when the alarm condition elapses.
The ILIM2 and OVTP alarms are managed in a more complex way. A programmable waiting time (cooling window) and a proportional watchdog window are implemented according with the value of the external capacitor (C_{AL}) connected to pin AL. In this way the user has the possibility to manage the cooling of the chip (through the waiting time) and to distinguish if there is a persistent alarm condition or not (through the watching window). For both ILIM2 and OVTP protections, the RHFPOL01 can be configured in "Latched" or "Hiccup" operation mode in case of persistent events. For thermal alarms, if Latched or Hiccup mode is desired, the value of C_{AL} must be carefully chosen to select a watching window duration respectively longer or shorter than the time required by the device to trigger the thermal protection after the cooling time is elapsed. In the figure below is shown the Latched OVTP protection in case of 16 consecutive thermal events: note that in this case the over temperature condition can only appear as a transient one and cannot be indefinite, respecting the above mentioned ECSS derating rules.

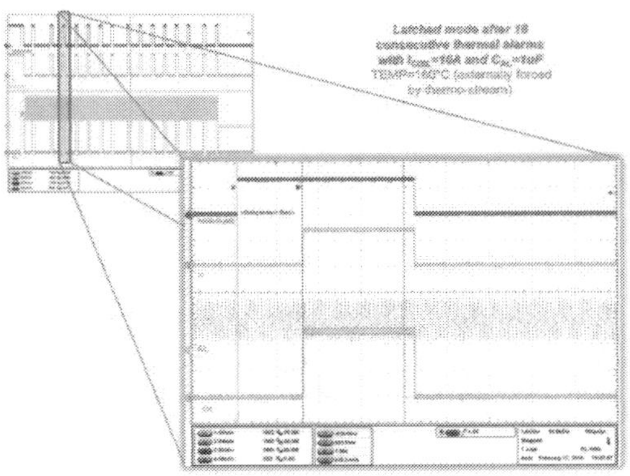

Figure 18 - RHFPOL01: Latched mode OVTP

2.8 RHFPOL01 Efficiency and Current Capability

Several efficiency measures have been performed on the silicon at different V_{IN}/V_{OUT}, switching frequency and inductor values, matching the expected target. In the figures below some experimental results are reported:

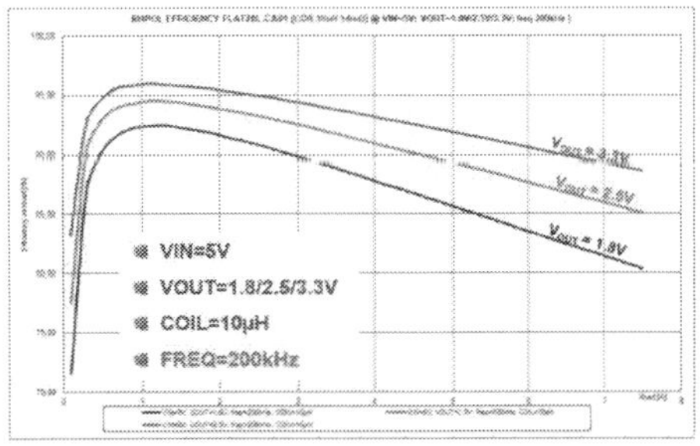

Figure 19 - RHFPOL01 Efficiency @ different V_{OUT}

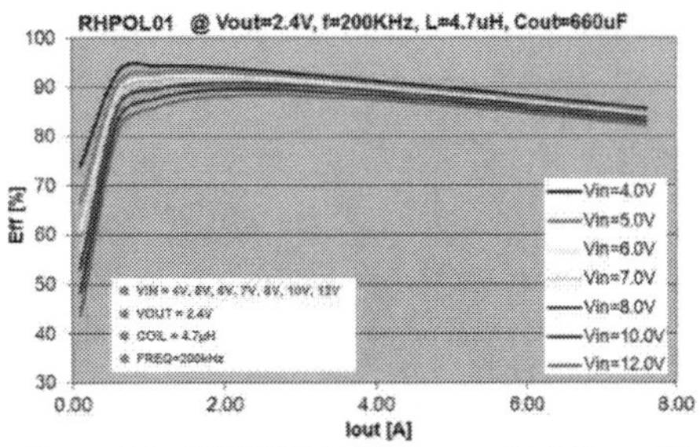

Figure 20 - RHFPOL01 Efficiency @ different V

Test performed in our lab showed that the RHFPOL01, mounted in the official FLAT28L hermetic ceramic package, is able to provide a current capability well above its project target of 7A. In fig.21 the result of an experimental test is shown during which the device provides an output current of 10A. By means of a thermal camera, in this condition we were able to measure a Pack temperature value of 110°C.

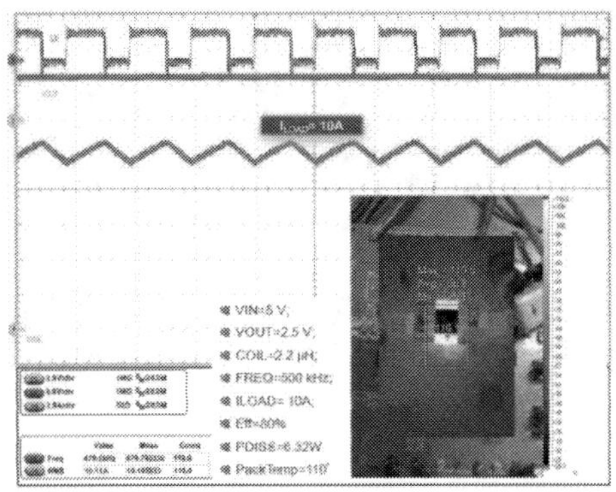

Figure 21 - RHFPOL01: Output Current Capability

2.9 RHFPOL01 Application Boards

In order to evaluate all the technical features of RHFPOL01 device, a very versatile Evaluation Board (shown in Fig.22) has been realized. Of course, this board has to be used for debug and evaluation purposes only and then it was not optimized in terms of

 DOI: 10.1051/e3sconf/20171611002

dimensions and count of external components. In addition (see Fig. 16), a dedicated application board for the current sharing feature has been also manufactured, in order to optimize the symmetry that is required for this kind of application test.

Figure 22 - RHFPOL01: Evaluation Board

Finally, a dedicated demo-board is under manufacturing (at the time of this paper) with the aim to exalt the key technical features of RHFPOL01 device (like the transient response and the system efficiency) through the choice of a proper setting and by minimizing the count of external components and PCB dimensions.
The top view and the real dimension of this demo-board are depicted in Fig. 23.

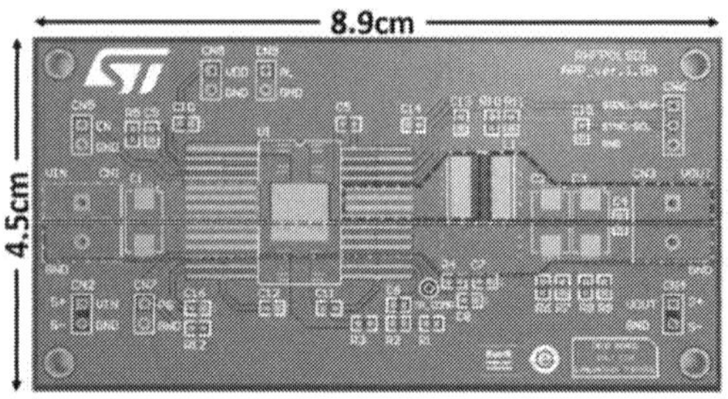

Figure 23 - RHFPOL01: Demo Board

The resulting pin-out of RHFPOL01 device was thus the best trade-off between the minimization of parasitic effects of FLAT28L package and the need to avoid complications in the layout of application boards. In Fig.23 are highlighted the paths involved in the power distribution.

3. PSPICE model

To allow a better customization of final application the PSPICE macro-model including all the main features of RHFPOL01 is already available. Comparison tests between the silicon and PSPICE model have been performed with good results as the one shown in Fig. 24 describing the behaviour of the device at start-up.

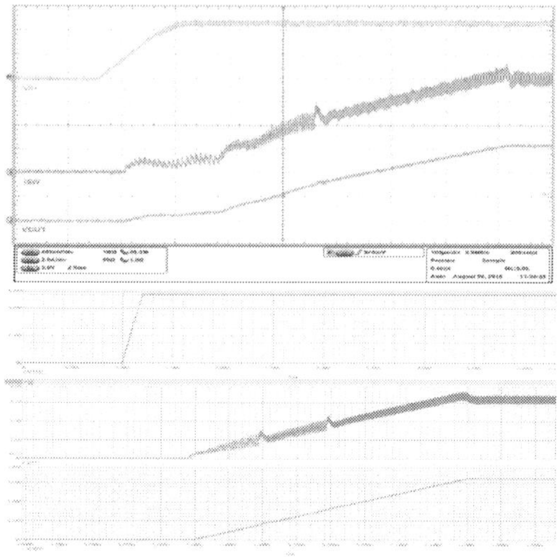

Figure 24: Simulation comparison Silicon vs PSPICE

In this way, the final user can explore the behaviour of its system before to realize the final application board.

4. RADIATION TARGET

The final product must be able to survive in harsh conditions that are different for each orbit assigned to satellites. The key radiation targets are:

- Operate in the Mil-Temperature range (-55°C to +125°C);
- To be robust in radiation (total ionizing dose) up to 100 krad (Si) at standard exposure rate (i.e. 50rad/sec) and low dose rate (i.e. 0.01 rad/sec);
- To be robust in radiation (SEE) up to 70 MeV/mg/cm2 LET, SEU and SEL immune, not affected by SEGR or SEBO and not suffer from appreciable performance degradation to SET.

Preliminary TID radiation tests (HDR & LDR) have already been successfully performed on RHFPOL01 prototype up to 300 krad(Si) at high dose rate (11.4 krad(Si)/h) and up to 100 krad(Si) at low dose rate (0.1 krad(Si)/h) using gamma rays from Cobalt-60 source.

5. REFERENCES

1. ESA Statement of Work (2012) – ESTEC Ref: TEC-EPM/2012/645.
2. M. Triggianese: Power Distribution by Point of Load Conversion – ESTEC Ref: TEC-EPC/2010/528/FT
3. V. Volperian, "Simplified analysis of PWM converters using model of PWM switch part I: continuous conduction mode", IEEE Trans. On Power Electronics, vol. 26, no.3 pages 490-496.
4. Raymond B. Ridley: A new small-signal model for current-mode control, Ph. D. Dissertation, Virginia Polytechnic and State University, Nov. 27th 1990.

PLANAR MAGNETICS FOR OPTIMISED MANUFACTURING, PERFORMANCE AND LIMITATIONS

Erich Strixner (1), **Dr. Frank Herty** (2), **Vittorio Pascucci** (3)

(1) *Airbus Defence and Space (Airbus DS GmbH), Robert-Koch-Str. 1, 82024 Taufkirchen, Germany, Email: Erich.Strixner@airbus.com*
(2) *Airbus Defence and Space (Airbus DS GmbH), Claude-Dornier-Str.88090 Immenstaad, Germany, Email: Frank.Herty@airbus.com*
(3) *Former Airbus Defence and Space (Airbus DS GmbH), Robert-Koch-Str. 1, 82024 Taufkirchen, Germany, Email: VittoPascu@gmail.com*

ABSTRACT

The LION Navigator, as a standard product line of Airbus Defence and Space, has started a design for manufacturing initiative for switched mode power supplies in the year 2010.

An analysis of the overall production effort of power supplies reveals that a high proportion is related to hand soldering of through-hole parts, integration of the PCB into the frame and conventional magnetics which are manufactured with low automatization. A concept study was originated to evaluate the technical prospects for an improved design for manufacturing strategy.
The technical approach is related to substitution of remaining through-hole parts by SMD devices if it is economic, to cool Power-SMDs via PCB to structure and, last but not least, to replace conventional magnetics with planar ones, as main part of the study. The windings of planar magnetics are integrated into the PCB with a highly automated process. They provide advantages due to good cooling capability and a low component height but advanced calculation methods are necessary to succeed in the design. As a follow up to the study the development of the power converter for the LION Navigator was started in order to detail the hardware related to an improved design for manufacturing.

1. INTRODUCTION

The surface mount technology was developed by IBM in the 1960s and became widely used in the late 1980s. Hardware gets much smaller due to reduced part sizes and by a placement on both sides of the PCB. Planar Magnetics integrated into a PCB came up in the 1990s to achieve a flat design of magnetics combined with optimised manufacturing capability. Power applications were limited that time because PCB technology was not advanced as it is today. Multilayer PCB Technology is a standard now with sufficient copper density for space applications as well. This offers the opportunity to integrate magnetics into the PCB. The manufacturing line gains in a reduced manufacturing effort compared to conventional magnetics where enamelled copper wires are wounded on U-, E-, Pot - or Toroid Cores. Planar magnetics have a low height but need more PCB area compared to wire wound standard magnetics. The proximity losses are lower due to the different winding geometry achieving advantages for high current applications. E-Planar cores allow the development of traditional transformers and chokes as well as integrated planar magnetics to achieve optimized core utilization. The development is more oriented towards a prediction of performance with a mathematical approach in order to limit the development time.

2. PRODUCTION EFFORT FOR POWER CONVERTERS

2.1 Through-Hole Parts

In the last decades the manufacturing technology for space hardware has changed. The traditional approach of using parts in through-hole technology [THT] is depicted in Figure 1. There are a lot of different manufacturing steps necessary to finalise a power converter for space applications. Manual work by population of THT parts, preparing wires, hand soldering, screwing of power semiconductors to the frame especially is time consuming and limits the production output of hardware.

2.2 Magnetics

The most common approach for magnetics is the usage of coated toroid ring cores as depicted in Figure 1 where enamelled copper wire is directly applied. E- Cores, RM-Cores and pot cores with coil former, needs less manufacturing effort because of simple and well known automatization techniques. Additionally, a set of tests are necessary to prove their workmanship and their quality level. This is needed because they consist of a combination of a core or core half's, enamelled copper wire, in some cases a coil former and of the coating to resist against humidity and particles. In conclusion, magnetics are complex and time consuming parts that have a high contribution to the overall parts cost.

 DOI: 10.1051/e3sconf/20171611003

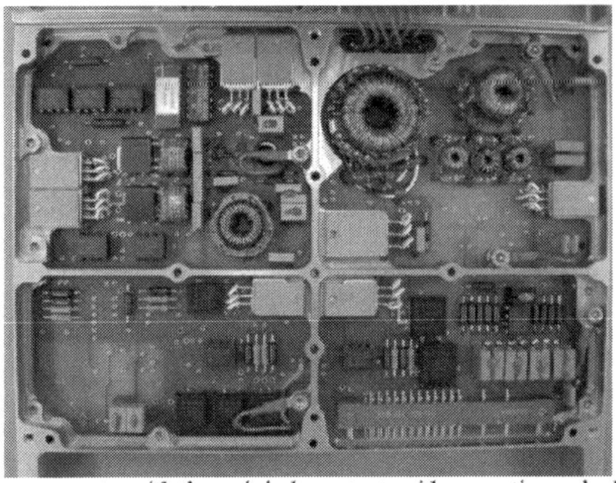

Figure 1: Traditional power converter with through-hole parts, toroid magnetics and a lot of manual mountings.

2.3 Integration of the PCB into the frame

The integration of a populated PCB containing power converter hardware consists of a number of manual tasks.

Screwing of TO packages to the frame and connecting the pins to the PCB via flex or wire and, in some cases crimped connectors are used where every pin has to be connected to a single wire. These steps are necessary for this kind of technology to achieve a reliable operation under high vibration loads as a standard for space hardware.

3. OPTIMISED MANUFACTURING

The concept study done at Airbus DS in 2010 was initiated to develop solutions for manufacturing of improved power converters. The manual or less automated production steps as described in the chapter 2 shall be optimised as much as possible reaching performance data similar to conventional power converters.

3.1 SMD power devices

It is nowadays a standard to use nearly 100% SMD parts for all small signal components. There are a few exceptions where no SMD parts are available on the space market or if THT parts are high runners with economic preference.

Most SMD power parts are small in size and they guarantee an efficient automated mounting. The drawback is that it is difficult to cool them when they are mounted on the PCB. A SMD 0.5 device for example has a limited dissipation of about 500mW. Higher power capabilities without losing standard mounting are possible by cooling of SMDs directly through PCB or substrate to structure.

3.2 Integration of magnetics into the PCB

The windings of planar magnetics are integrated into the PCB with a highly automated process shown in Figure 2.

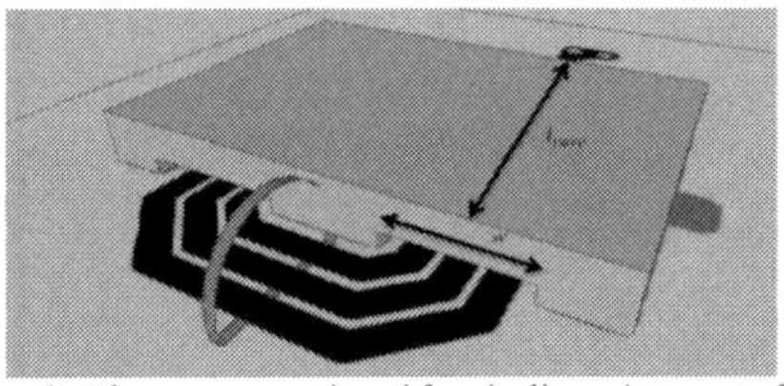

Figure 2: Planar magnetic with windings integrated into the PCB.

There are 4 different kinds of mechanical constructions suited for planar magnetics.

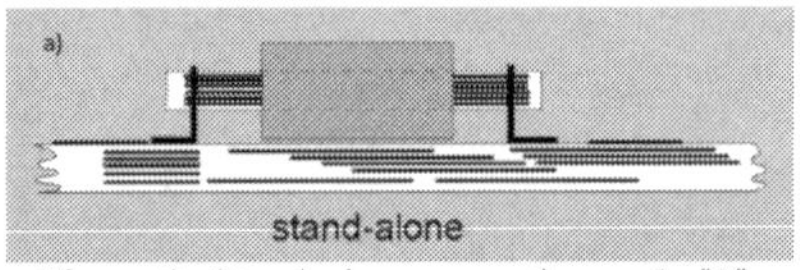

Figure 3: Stand-alone mounting as in [1].

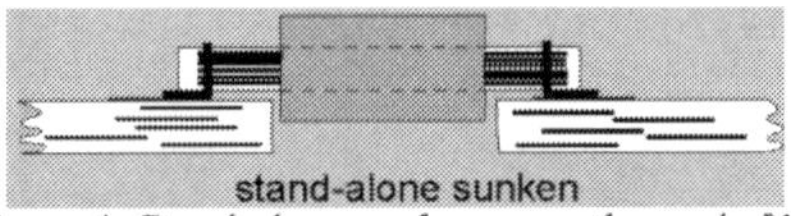

Figure 4: Stand-alone sunken mounting as in [1].

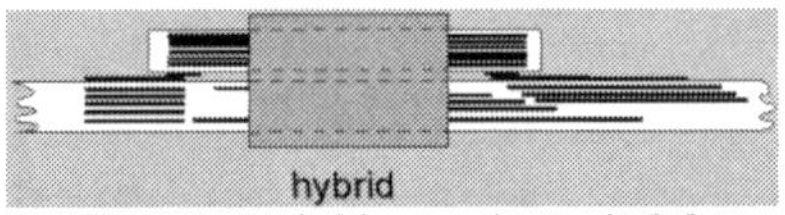

Figure 5: Hybrid mounting as in [1].

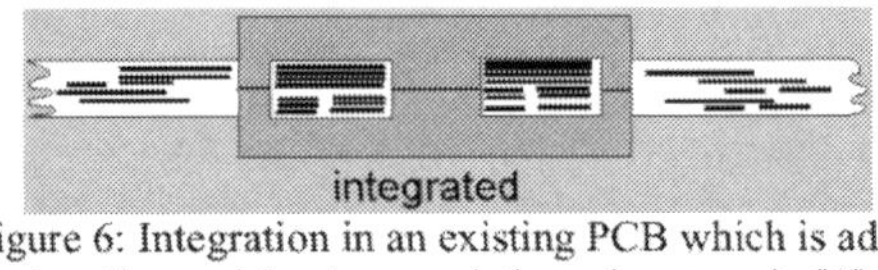

Figure 6: Integration in an existing PCB which is additionally used for the population of parts as in [1].

The hybrid and integrated mounting techniques are mechanically stable and provide a high copper utilization as well. They are best suited for the high environmental requirements related to space applications.

4. ENVIRONMENTAL CONDITIONS FOR QUALIFICATION

4.1 Vibration Loads

4.1.1 Sine Vibration

Airbus DS GmbH has already finalised an ESA Qualification for planar magnetics. Only mature mechanical constructions withstand the high vibration levels related to space requirements. The used vibration levels are shown in Table 1.

Table 1: Sine Vibration Qualification Levels

Frequency [Hz]	Load	Sweep Rate
5 - 20	max. shaker ampl.	2 oct/min
20 - 100	20 g	2 oct/min
100 - 200	15 g	1 oct/min

4.1.2 Random Vibration

The random vibration levels are shown in Table 2.

Table 2: Random Vibration Qualification Levels

Axis	Frequency [Hz]	Qualification Level
Parallel to mounting plane (X, Y axis)	20 – 80	+ 3 dB/oct.
	80 – 400	0.1 g^2/Hz
	400- 2000	-3 dB/oct.
	G rms	10.0
Perpendicular to mounting plane (Z axis)	20- 80	+ 3 dB/oct.
	80 – 400	0.5 g^2/Hz
	400-200	-6 dB/oct.
	G rms	18.4

4.1.3 Shock

The shock levels are shown in Table 3

Table 3: Shock Qualification Levels

Frequency	100 Hz	2kHz	10kHz
Shock response spectrum (Q=10) at unit interface	20g	2000g	2000g

4.2 Thermal Vaccum Test

The thermal vacuum test data are:
Duration: 7days
Temperature: 80°C under
Pressure: 1 x 10-5 Pa

4.3 Thermal cycling

Number of thermal cycles: 200
Temperature ranges: -55°C (-5°C/+0°C) to +110°C(+5°C/-0°C)
Ramp rate: 10±2°C
Dwell time at temperature extreme: 15 minutes
The duration of each cycle amounts to 1 hour.

5. MECHANICAL CONSTRUCTION OF PLANAR MAGNETICS

5.1 Requirements on the PCB

The usage of a PCB layer stack with high copper utilization is preferred. This allows an adequate power density resulting in a competitive size of planar magnetics. A copper thickness of 70µm, 105µm, 210µm up to 400µm ensures a high power density especially in high current applications. A copper thickness up to 210µm supports high frequency operation with planar magnetics without exceeding acceptable proximity losses in most of the applications.

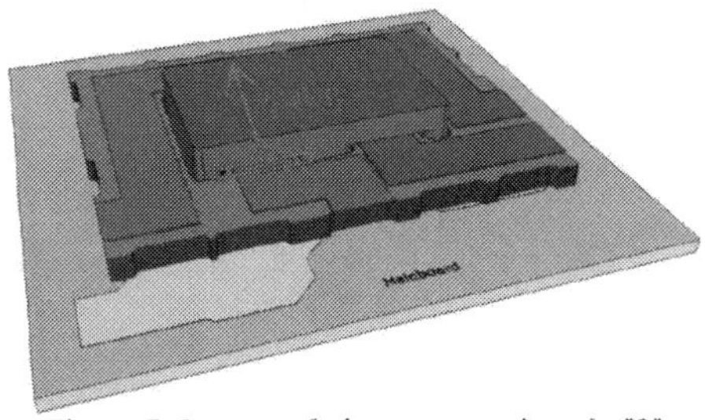

Figure 7: Integrated planar magnetic as in [3]

5.2 Requirements on the planar mounting

Planar magnetics are used for commercial electronics since more than 2 decades. This mounting approach is considered as far as possible, due to its strong heritage. This ensures a low risk for the qualification campaign.

5.3 Requirements on the coating

There are different coatings commonly used for space applications. Some of them are Mapsil, Nusil and Parylene C. The first two are silicone conformal coatings while the last one is a polymer film type. Populated PCBs for space have to withstand hard environmental conditions such as vibrations and thermal cycling as described in chapter 4. The combination of PCB, part construction, mounting technique and coating forms a system that has to perform well under these conditions. EI- or EE-planar magnetics as shown in Figure 2 and Figure 7 are glued to the Motherboard on the bottom side. On the topside of the PCB there are two cavities left for mechanical margin as, in the construction of planar magnetics (e.g.: EE, EI, RM cores), the thermal expansion coefficient of coatings commonly used in space is different to that of ferrite materials.

DOI: 10.1051/e3sconf/20171611003

Table 4: Property assessment of well-known conformal coatings

Silicone conformal coatings: Mapsil, Nusil	**Polymer film conformal coating: Parylene C**
Silicone is roughly equivalent to a very soft rubber.	Nonporous and transparent polymer film
The liquid can be applied via spraying, dipping or brushing.	Applied by vapour based deposition. It offers the best possible coverage of any conformal coating material
Thick coat is needed to be effective. Per IPC Standards, between 76um and 203um	Forms a thin coating between 10um and 15um (50um max).
The reproducibility depends on the process automatization.	High reproducibility in thickness due to deposited as a vapour in a vacuum chamber.
It tends to fill cavities of parts	It does not fill cavities of parts due its thin thickness and the applied deposition technique.

A way to avoid mechanical stress around the cavities as shown in Figure 7 is to use a thin coating like Parylene-C which guarantees good coverage on nearly inaccessible areas as well. As a result the cavities are not filled with coating avoiding pressure to the ferrite core. Such a mismatch in expansion coefficient in combination with filled cavities can lead to fail in qualification by cracks in the ferrite material or by a lift-off of the core half on component side.

6. OVERVIEW OF APPLICATIONS

The first application which was optimised for manufacturing is the power supply for the LION Navigator followed by a mass memory hardware related to high output current up to 30A. The range was finally completed by a high voltage application in the kV range.

6.1 Lion Navigator

The power converter for the LION Navigator as part of the standard product line of Airbus DS GmbH is depicted in Figure 8. It converts the primary power bus into 3 insulated output voltages with an output power of 30W. Customer specific requirements demand an operation on the unregulated 28V and the regulated 50V bus with minor adaptations of small signal parts only. Therefor the converter has a wide input voltage range from 16.5V to 55V. The PCB features 9 planar magnetics with different functions. The input filter, the power and the current transformer, the secondary common mode filters, a small signal pulse transformer and a 1MHz oscillator. In this first application only one magnetic was implemented in a conventional way. This was an initial learning curve which led to an improvement of calculation methods of magnetics. The modelling of leakage inductance and, more important, capacitive coupling within the winding arrangement helps to reduce iteration loops in the design and to treat issues before the first hardware is built. An estimation of the leakage inductance is needed to calculate the losses in snubbers and clamping networks. Knowledge of capacitive coupling is valuable to determine common-mode and differential-mode coupling in filters and transformers.

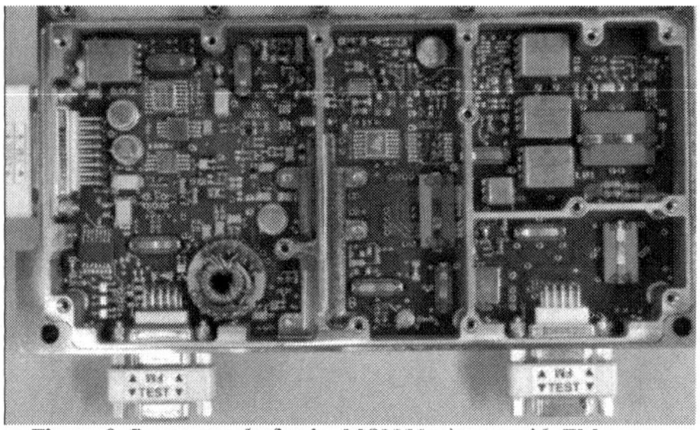

Figure 8: Power supply for the LION Navigator with FM status, Outputs= 3.3V, 5.0V, 9.2V; Sum 30W; Input: 16.5V to 55V

6.2 Mass Memory

The PSO Mass Memory Converter depicted in Figure 9 operates on a 28V power bus and supplies 3.3V/30A to the digital load. It is optimised for the production line by the use of planar magnetics, cooling of SMDs via PCB to structure. The mounting of the populated PCB into the frame is simple because all parts are directly mounted on the PCB. All windings are integrated into a 10 layer PCB. No temperature cycling of the magnetics is necessary to prove workmanship of the winding arrangement. Parylene-C coating guarantees to apply no stress to the magnetic cores and is applied in one step

DOI: 10.1051/e3sconf/20171611003

for all the other populated parts. A verification of the inductance is performed as an in-circuit measurement on the populated board.

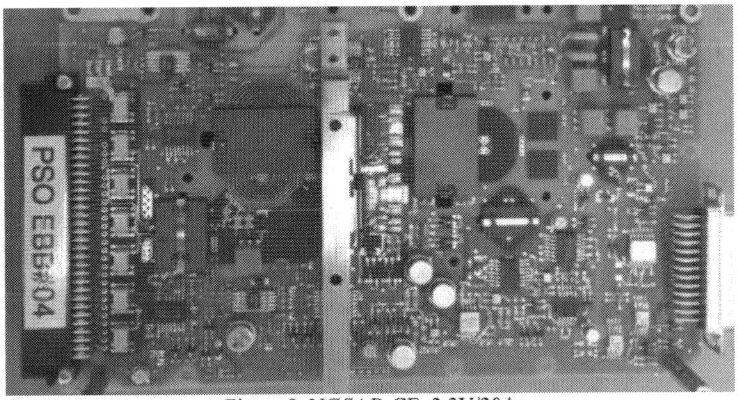

Figure 9: NGSAR-CE, 3.3V/30A

6.3 HVHP technology development

Airbus DS GmbH completed research activities in order to extend the field of applications to high voltage planar magnetics. Thanks to the DLR technology development program called "HVHP" (FKZ.: 50 RN 1302) a 2kV high voltage planar transformer prototype was developed and tested successfully. The prototype is shown in Figure 10. While this kind of transformer has almost the same performances in terms of mass and dissipation, the expected cost savings for manufacturing can be reduced to roughly 50% compared to the toroidal version of the transformer. The reasons for this are the simpler geometry and hence, the simpler potting process and reduced effort in terms of in-production testing.

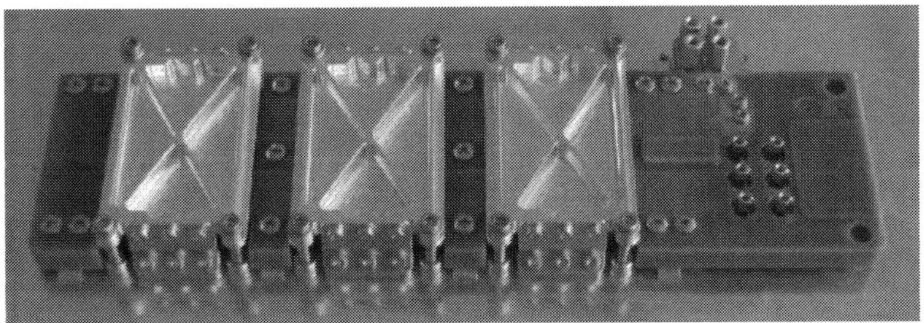
Figure 10: Planar Magnetic as part of the Flat Top Resonance Converter, 2kV, 2500W

7. PERFORMANCE AND LIMITATIONS OF THE TECHNOLOGY

Planar magnetics have many advantages for future designs. The most dominant is the simple manufacturing process of magnetics due to integration of the windings in a multilayer PCB. They are capable to resist high vibration loads if the mechanical construction is mature. Prototype manufacturing takes longer compared to wire wound magnetics because a complex multi-layer PCB needs to be designed and manufactured. Due to this there is a demand of improved prediction methods to get the development more and more into the direction of succeeding in the first run. The parameters of interest are a more precise determination of the power dissipation of magnetics, a prediction of the leakage inductance and the capacitive coupling between the layers of the PCB. A well selected PCB stack and improved cooling concepts are an important part of the development. This knowledge helps to reduce the development time.

7.1 PCB and Forced Cooling of magnetics

The ECSS-Q-ST-70-12C limits the copper thickness within a multilayer PCBs to 70um. This is necessary to fulfil the high reliability requirements on PCBs used for space applications. Improved future technologies are necessary to gain in higher power density.

The E-Planar-Cores occupy more PCB area compared to wire wound magnetics and gain from a low component height. The heat dissipation can be spread over a large area of the PCB reducing operational temperatures. There plain geometry opens new effective ways of forced cooling as depicted in Figure 10. The cores are surrounded by aluminium elements transferring the power dissipation directly to the structure. Compared to Toroid or Pot cores a very low thermal resistance can be reached to cool the PCB and the ferrite core as well. This advantage makes that core geometry well suited for high power applications. The preferred properties of PCBs are a high thermal conductivity and low εr.

7.2 Power Dissipation of Magnetics

PCBs for power converters are complex due to the co-existence of high power and small signal components on the same PCB. Additionally the parts generate heat which has to be transferred to the mounting plate to keep the operational temperature within an acceptable range. A good calculation of losses is important to reduce the iteration loops of a design. A planar magnetic generates power dissipation by the AC-flux density in the ferrite core. The windings are affected by I^2R current losses. They can be generated by DC- or by AC-current. The latter one is affected by the proximity effect which generates additionally losses by current displacement. A combination of the two effects can occur as well depending on the application. Gaped cores generate heat in the winding arrangement due to leakage flux.

Loss calculations are possible by advanced mathematical formulas or by finite element analysis. The mathematical approach is complex but offers the advantage that the different losses can be isolated in DC-I^2R losses, AC-I^2R losses per winding layer and gap losses. An understanding of the loss generation by the different effects combined helps to do a first optimisation. The calculation by finite element method requires mostly a high data processing power which exceeds that one of a standard desktop computer. It provides a detailed view on the results. Modelling takes longer but is more accurate in the end.

7.3 Prediction of leakage inductance and capacitive coupling

Leakage inductance is a common parasitic component in multiple-winding magnetics. It is due to the stray magnetic flux leaking out of the core, between adjacent windings, thus not linked to all the windings. Coupling capacitance is mainly generated at the interface between turns (of the same winding or between different windings). It is effective in case of a non-zero relative voltage swing.

Leakage inductance generates losses in RC-Snubbers and RCD-Clamping networks while coupling capacitance reduces the resonance frequency of a magnetic. In most applications these components are undesired and shall therefore be minimized.

Planar technology is inherently prone to high coupling capacitance due to the flat copper tracks and the small distance between layers. On the other hand, the same characteristics typically tend to decrease the leakage inductance.

Table 5 outlines how low values of leakage inductance and parasitic capacitance are contrary properties of a magnetic. A typical optimization scenario consists in reducing the effective parasitic capacitance while keeping the leakage inductance under a given value. Approaches 3 and 4, in particular, have the advantage of reducing the effective coupling capacitance without significantly affecting the leakage inductance, they should therefore be preferred.

Table 5: Common magnetic optimization techniques

N		C_{eff}	L_{Leak}
1	Winding Interleaving	↑	↓
2	Reduced thickness of layer stack by less insulation	↑	↓
3	Air (vacuum) spacers between windings	↓	-
4	Voltage swing matching within the layer-stack	↓	-

The properties of planar magnetics are very reproducible compared to wire wound types. This guarantees stable performance parameters and simplifies the modelling. The Gyrator-Capacitor model in [2] offers good flexibility and accuracy to analyse and compare different possible designs solutions. Figure 11 shows an example of application of the Gyrator-Capacitor model where the windings coupled to the core are represented by gyrators while the magnetics core is modelled by a capacitor. The magnetization admittance ($1/\mathcal{R}_{core}$) becomes a capacitor connected in series with the magnetic side of all the gyrators. Similarly, leakage admittance paths between adjacent windings are modelled by capacitors connected in between windings which are represented by gyrators. Capacitive coupling within the layer stack is modelled by connecting capacitors between the centre of the windings on the electric side. Thus, a multiple of two Gyrators is necessary to represent each winding, in order to make the point of average voltage electrically accessible.

 DOI: 10.1051/e3sconf/20171611003

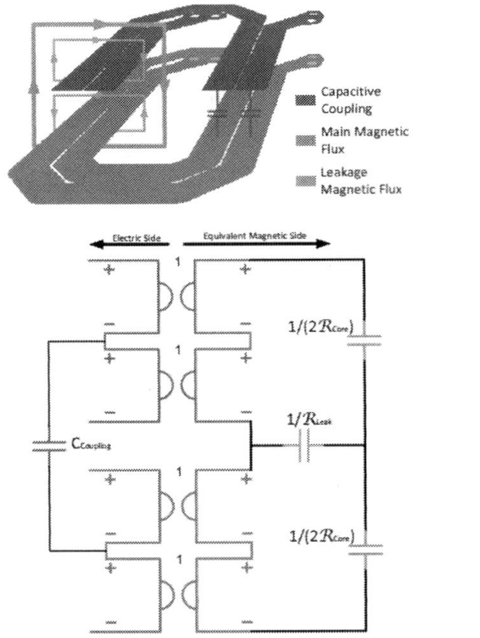

Figure 11: Example of Coupling Capacitance and Leakage Inductance with Gyrator-Capacitor Model as in [2]

The parasitic components (Coupling Capacitance and Leakage Reluctance) in the model can be estimated with the following formulas:

$C_{Coupling} \approx \varepsilon_0 \cdot \varepsilon_r \cdot S_{Ovlp} / d, \ \varepsilon_0 = 8.85 \cdot 10^{-12}\,F/m$

$\mathcal{R}_{Leak} \approx w/(l_{Avg} \cdot \mu_0), \qquad \mu_0 = 4\pi \cdot 10^{-7}\,N/A^2$

Where ε_0 is the electrical field constant, ε_r the permittivity of the dielectric, S_{Ovlp} the total overlapping surface, d the vertical distance between the windings, w the winding width and l_{Avg} the average turn length.

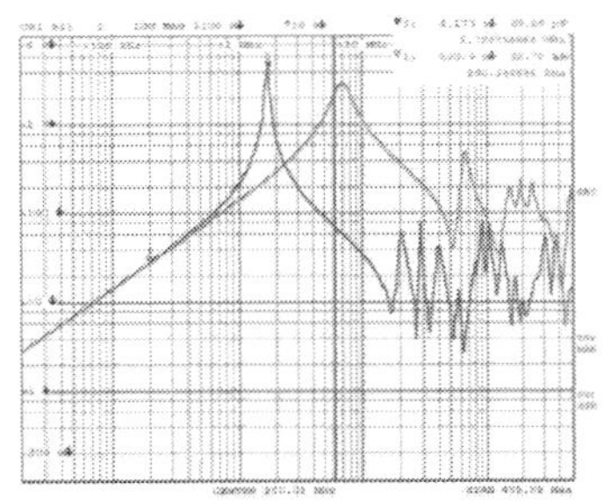

Figure 12: Impedance plot of a coupled magnetic without optimisation in blue colour and with optimisation in green.

7.4 Planar magnetics with limited performance

Planar chokes using Pot- and E-cores have usually a single airgap to provide energy storage capability. Applications with a high operational AC switching current generate high AC magnetic fields around the airgap and, as a consequence, generate high copper losses on adjacent PCB tracks. Applications for reference are the Phase Shift Converter with its resonance choke or a Flyback in discontinuous mode. Toroid cores with their distributed airgap are mostly the best choice for this kind of applications.

8. REFERENCES

1. Application Note of Philips Magnetic Products Planar E Core (7/1997)
2. David C. Hamill (1994). Gyrator-Capacitor-Modeling A Better Way of Understanding Magnetic Components, Department of Electronic and Electrical Engineering, University of Surry, Guildford GU2 5XH, Unoited Kingdom, IEEE 0-7803-1456-5/94
3. Strixner E., Godzik, S., Dr. Drechsler E. (2011). Optimization of an Integrated Planar - Magnetic for active Antenna Panels. In Proc. of the 9th European Space Power Conference. Saint-Raphaël. France

HIGH POWER SELF-ALIGNED, TRENCH-IMPLANTED 4H-SiC JFETs

K. Vamvoukakis[(1)], A. Stavrinidis[(1)], D. Stefanakis[(1)], G. Konstantinidis[(1)], M. Kayambaki[(1)], K. Zekentes[(1,2)]

[(1)] *IESL/ FORTH, Vassilika Vouton, PO Box 1385 Heraklion, Greece*
Email: vamvk@hotmail.com
[(2)]*Grenoble-INP, IMEP-LAHC, F-38000 Grenoble, France Email: konstantinos.zekentes@imep.grenoble-inp.fr*

ABSTRACT

The process technology for the fabrication of 4H-SiC trenched-implanted-gate 4H–SiC vertical-channel JFET (TI-VJFET) has been developed. The optimized TI-VJFETs have been fabricated with self-aligned nickel silicide source and gate contacts using a process sequence that greatly reduces process complexity as it includes only four lithography steps. A source-pillars sidewall oxidation and subsequent removal of the metallization from the top of the sidewall oxide ensured isolation between gate and source. Optimum planarization of the source pillars top has been performed by cyclotene spin coating and etch back. The effect of the channel geometry on the electrical characteristics has been studied by varying its length (0.3 and 1.2μm) and its width (1.5-5μm). The voltage blocking exhibits a triode shape, which is typical for a static-induction transistor (SIT) operation. The transistors exhibited high ON current handling capabilities (Direct Current density >1kA/cm^2) and values of R_{ON} ranging from 6 - 12 mΩ·cm^2 depending on the channel length. Maximum voltage blocking was 800V limited by the edge termination. The maximum voltage gain was 51. Most transistors were normally-on. Normally-off operation has been observed for transistors lower than 2μm channel width (mask level) and deep implantation.

1. INTRODUCTION

Power transistors based on wide band gap semiconductors are subject of intensive research the last decade. The high current and high voltage capabilities of these devices, largely exceeding that of Si devices, represent a challenge on various aspects related to the device fabrication, characterization and packaging.

Currently, SiC and GaN are the wide band gap semiconductors of choice for fabricating high power devices. SiC is more appropriate for high voltage (>1000V) applications and more mature from the technology point of view as many devices are commercially available. For instance SiC devices exhibit 6 times higher current density ate 600 V [1].

4H-SiC-based Metal-Oxide Semiconductor Field-Effect Transistor (MOSFETs), Bipolar Junction Transistors (BJTs) and JFETs have been considered as candidate devices since their technology is more mature than other transistor types.

All things considered, the MOSFET is the most promising device in the SiC field [2]. Its technology is more mature, it is a normally-off device and with little demand on the drive circuits. However, reliability issues (oxide and threshold voltage stability) were the main obstacles for their commercial release despite the long and important effort devoted towards their optimization. Reliability is a minor issue in the commercial devices but it becomes a major one when these devices are considered for space applications.

The SiC BJT main advantages as power switch are its low conduction loss combined with fast switching. Moreover, BJTs are generally very robust devices meeting in many cases space requirements. The most crucial parameter for SiC BJT optimization was the current gain at application temperature. Current gain values lower than 30 were obtained for a long time by various research groups. This issue has been addressed successfully by performing special surface passivation based mainly on oxides [3]. Again, there is a high risk that this surface passivation will not be effective under space conditions.

JFETs have the lowest overall losses of switching devices and their fabrication process is quite straight forward. The main drawback with VJFETs is that they are usually normally-on (depletion mode), which is considered unsafe in power applications. Moreover, JFETs are very robust devices with a channel not seriously affected from surface effects. There is no a major issue up to the moment limiting their application in space conditions. This was the reason for developing SiC JFETs at FORTH.

Once the main device type has been chosen, the next question is to choose the appropriate JFET geometry The usual 4H-SiC JFETs are either purely vertical (VJFETs with Static Induction Transistors-SITs operation) or Semi-lateral (SLJFET). Optimum SLJFET fabrication requires two-step implantation and re-epitaxial growth after material etching on specific locations. Trenched VJFET requires only one implantation step and the less lithography steps. The only difficulty is the tight control of the dimensions, which are of the order of 1-2 μm for both source pillars and gate pitch arising, thus, an issue of tight registry control. For the above reasons, the trenched and implanted TI-VJFET has been chosen as optimum device.

 DOI: 10.1051/e3sconf/20171612001

The 4H-SiC VJFETs/SITs have been developed during the 90's and early 00's [4] for microwave applications and are incorporated in radar systems. High power 4H-SiC SITs with excellent high power performance have been demonstrated later [5, 6] and an excellent review is given in [7].

The purpose of the present study was to thoroughly investigate all 4H-SiC SITs process steps with the ultimate goal to simplify as much as possible their fabrication. Towards this aim different lithographic approaches, metallization schemes, isolation dielectrics and gate implantation conditions have been tested. This effort resulted in the fabrication of SiC TI-VJFET with self-aligned nickel silicide source and gate contacts including only four to six lithography steps depending on the edge termination scheme.

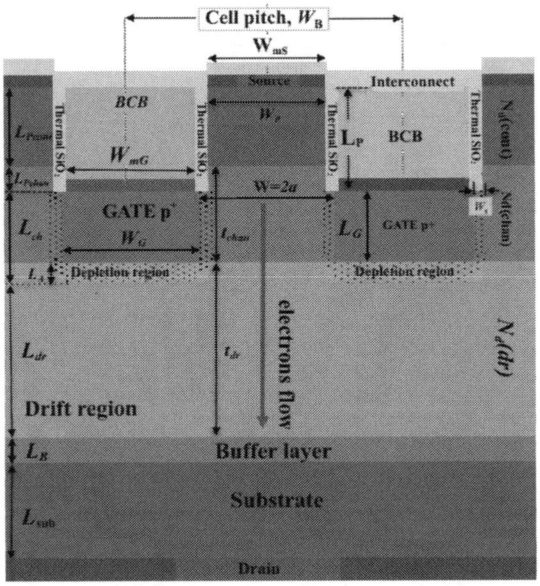

Figure 1. Schematic of the device cross-section.

2. DEVICE PROCESSING

The cross-section of the optimized 4H-SiC TI-VJFETs is shown in Fig. 1. The current flows vertically through an array of the source "pillars", which is controlled from the P-type implanted gate trenched areas between each pillar.

A series of optimization runs have been performed in order to choose this structure. Different epitaxial structures and gate implantation geometries were used. The epitaxial structure corresponding to Fig. 1, was composed by a 8 μm (L_{dr}) blocking thick n-layer with $N_D=5\times10^{15}$ cm^{-3}, an 1.2μm ($L_{ch}+L_{Pchan}$) channel layer n-layer with $N_d=1\times10^{16}$ cm^{-3} and a top 1 μm (L_{Pcont}) thick n+ layer with $N_d=1\times10^{19}$cm^{-3}. Gate regions were formed by Al ion implantation and post-implantation annealing (see details below) resulting in an implanted layer with p-type doping of 1×10^{19}cm^{-3}. Two different implantation depths (L_G=0.3 and 1.1μm), have been employed. The two transistor types will be distinguished by mentioning them as the "short" and "long" channel VJFET.

The technology of 4H-SiC SITs are simpler than that of MOSFETs or BJTs and the main difficulty lies with the small width (~1μm) of the source pillars requiring the use of advanced photolithography equipment (steppers, e-beam lithography) or self-aligned approach in the case of contact lithography. Indeed, our initial tests with conventional contact lithography showed its limitation not allowing the fabrication of structures with small channel widths a necessary condition for high voltage blocking in this type of devices.

The main issue in the case of the self-aligned process is related to the metal tears and wings, which should be avoided in order to have a large gate-source breakdown. In the case of Fig. 2(a) these features are not excessive but a study, relative to the use of different photoresists, allowed eliminating them completely (Fig. 2c).

Source fingers and gate trenches were formed by reactive ion etching (RIE) by using the first lithographic step.

Gate regions were formed by ion implantation followed by a thermal annealing at 1600-1800°C. The same multilayer metal mask was used to perform the above RIE etching and form gate regions by ion implantation. Aluminium was implanted at temperature of 425 °C with normally incident ion beam (sample's surface and ion beam are perpendicular). Individual implantation

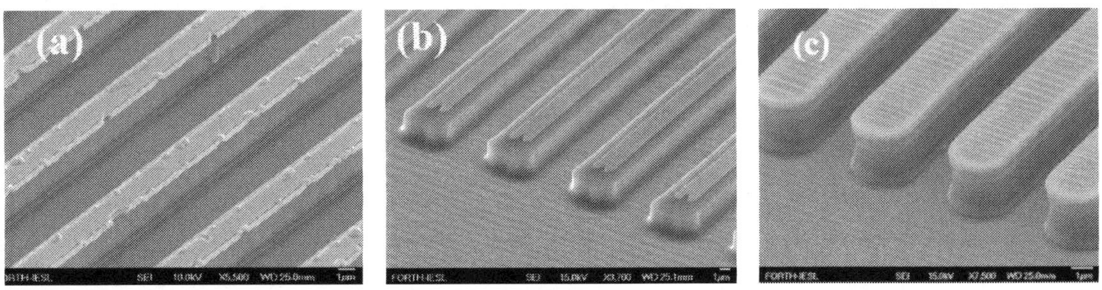

Figure 2. Source pillars after metallization deposition in the case of non-optimized self-aligned (a) conventional-contact-lithography (b) and optimized self-aligned (c) processes.

energies and fluencies were calculated using the SRIM 2D simulation program to form the Al profile. The gate profile has been checked by Secondary Ion Mass Spectroscopy (SIMS) performed on as-implanted and as-post-implantation-annealed witness samples.

The effect on the surface morphology of the SiC-surface-coverage by a C-cap during the post-implantation annealing, is obvious from Fig. 2 as a step-bunching is observed in the case without C-cap (Fig. 2b&c) despite the fact that an optimized process based on SiH_4 overpressure has been used. Indeed, trying to use SiH_4 overpressure to avoid dopant redistribution and surface step bunching is a tricky process because it can result easily to a Si deposition on top of SiC material. Therefore, the use of C-cap is the optimum solution.

Following, a 50 nm thick Ni contact is deposited by e-beam evaporation and patterned (second photolithography step) to remove the metal from the edge termination and device separation regions. Nickel has been chosen as metallization on both source and gates as it results to lowest contact resistance in the case of n-type SiC. Al-based metallizations result in lowest contact resistance in the case of p-type material. However, obtaining a low contact resistance on source n-type material is crucial for low device losses and for this reason Ni was our choice.

Nickel is deposited in a self-aligned way on both source pillar tops and gate trenches (Fig.2). In this way, Ni is deposited on the edge of the pillar tops i.e. on top of the thermal oxide and quite probably on the pillar sidewalls.

(a) (b) (c)

Figure 3. SEM images taken (a) after RTA at 1000°C, (b) after BCB deposition and etch-back from the source fingers and (c) after 2nd etch-back on the gate pad.

A study on optimum gate-source isolation showed the necessity of a dielectric layer on the sidewalls of the source pillars. So, the samples are oxidized to form a 35 nm thick oxide on the Si-face and a 240 nm thick oxide on mesa-structure sidewalls. Note that, the thick oxide film grown on the sidewalls consumes unintentionally Al-doped source-finger-sidewalls and prevents

This metal coverage, on the edge and sidewalls, is believed to be one the reasons for the observed gate-source-leakage in our previous devices [8]. So, a new approach namely a salicide-like process has been included in our process. The salicide process is widely used in Si-technology for self-aligning purposes and involves the reaction of a thin transition metal film with

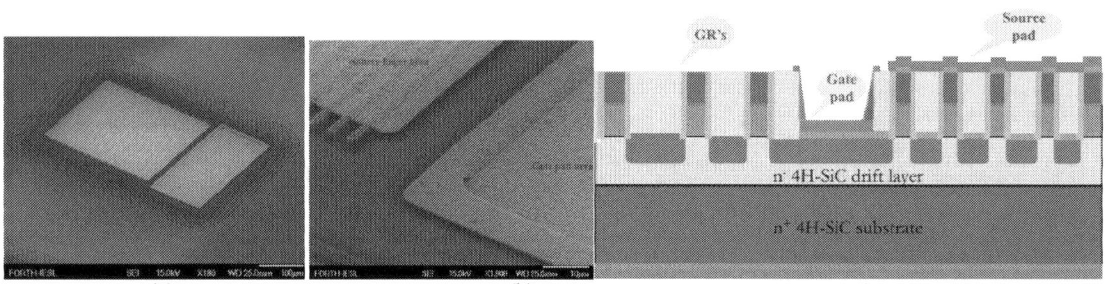

(a) (b) (c)

Figure 4. (a) and (b) SEM photos of the completed device. (c) Device cross-section schematic (Dark to light green: high to lower doping n- SiC, red: p- SiC, yellow: cyclotene, grey: SiO_2, light orange: Ni contact, dark orange: Au overlay.

formation of a leaky gate-source p^+-n^+ SiC junction. The horizontal oxide is removed subsequently by RIE.

silicon in the active regions of the device, ultimately forming a metal silicide contact through a series of annealing and/or etch processes. However, in the

DOI: 10.1051/e3sconf/20171612001

present case the salicide process has been used successfully for gate-source isolation purposes as the gate-source leakage has been greatly reduced.
The silicide process starts by a thermal annealing of Ni contacts at a temperature of 750°C. This anneal temperature is selected because it is enough for a nickel silicide phase formation between the Nickel/SiC physical contact, but it is not enough for a Nickel/SiO_2 reaction and there is an obvious distinction between "pure" Nickel and Nickel silicide. After that, a diluted standard SPM chemical solution (H_2SO_4:H_2O_2) is used to selectively remove the non-reacted Nickel metal from all the unwanted areas (pillar sidewalls and source pillar edges). Following, the device is annealed a second time at 1000°C leading to the best source-gate ohmic contact performance (Fig.3a).

A 100 nm thick Ni layer is also deposited on the wafer back side and then, the samples are annealed at 1000°C in vacuum to form ohmic contacts.

The morphology of the annealed front contacts is shown in Fig.3a.

The next process step is the application of a trench filling dielectric its curing and etching-back to expose the mesa-structure tops with source contacts. The usefulness of this step is two fold. First, to planarize the source pillar tops for interconnecting them by a metal deposition. Second, to avoid any short between the source and the gate. Polyimide and benzocyclobutene (BCB or simply cyclotene) have been tested as the dielectric for filling the gate trenches and in both cases a planarization and an etch back process are required. It has been found that the polyimide is not very suitable for planarization purposes while the BCB-cyclotene has excellent dielectric and planarization properties. The BCB film is deposited all-over the surface by spin coating and it is much thicker than the height of the source pillars. It is etched-back down to the source finger tops (Fig. 3b) with pure SF_6 reactive-ion-etching (RIE).

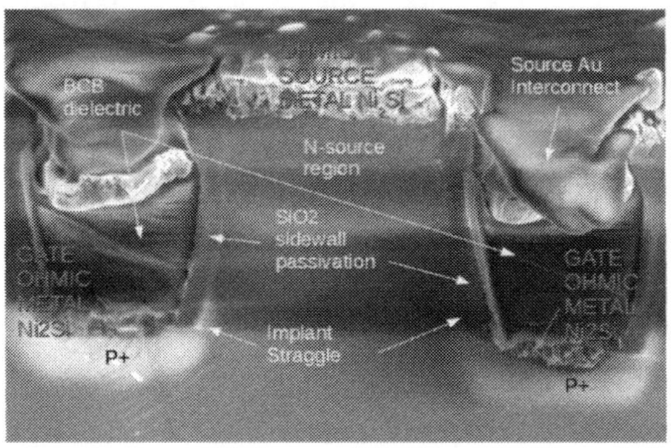

Figure 5. SEM micrograph of a shorth device cross-section.

The etching back of the BCB is happen in two steps. The planarization of the BCB with the source pillar tops is performed firstly. Then, a lithography is performed (the third lithography step) followed by the RIE etching to open the gate pad contacts (Fig. 3c).

Finally, a final fourth lithographic step was employed for depositing and patterning the common source and gate pad Ni/Au metal overlay (Fig. 4a&b). Edge termination is provided by a self-aligned multiple floating-guard-ring (GRs) structure (Fig.4c).

The cross-section of a fabricated TI-VJFET is shown in Fig. 5.
Devices with different channel widths (1.5, 2, 2.5, 3, 4 and 5 μm) have been fabricated and allowed better understanding of transistors operation.

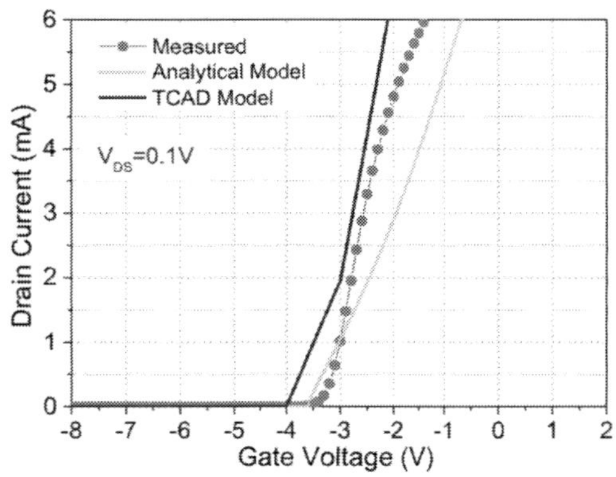

Figure 6. Experimental and simulated transfer characteristics for V_{DS}=0.1V of the long channel device with 2μm of channel width.

3. DEVICE MEASUREMENTS

On-wafer I-V measurements were performed with a Keithley 4200 parametric analyser up to 200V. On-state I-V characteristics were also performed in pulsed mode to avoid self-heating of the devices and the gate has been biased below its 2.7-V built-in potential value to maintain voltage-control capability (high I_D/I_G gain). Measurements at higher voltages have been performed with a Tekronix 370 curve tracer while the samples were covered with a liquid of high dielectric strength (Fluorinert©).

The measured characteristics were modeled with an analytical model and a finite element (TCAD) model. The analytical approach is based on typical semiconductor theory [9] for extracting output ON and transfer I-V as well as C-V characteristics, while a combination of previously published models [10] is involved for the simulation of forward blocking I-Vs. An iterative approach has been adopted to take into account the voltage drop in the drain region and thus, connect extrinsic and intrinsic drain voltage. Both electric-field independent and dependent mobility models have been considered to evaluate their suitability for the JFET modelling. Finally, the model describes in a unified way the linear and saturation operating regions.

The TCAD SILVACO[c] tool has been also employed for a finite element simulation. A simplified structure

Figure 6 shows typical experimental transfer I-V characteristic for the long devices. Both models agree on the value of the threshold voltage when taking into account the correct channel width, which often differs from the that of the mask due mainly to lateral straggling of the gate implantation. The long channel devices have a higher, in real value, threshold voltage as the lateral straggling reduces the effective channel width.

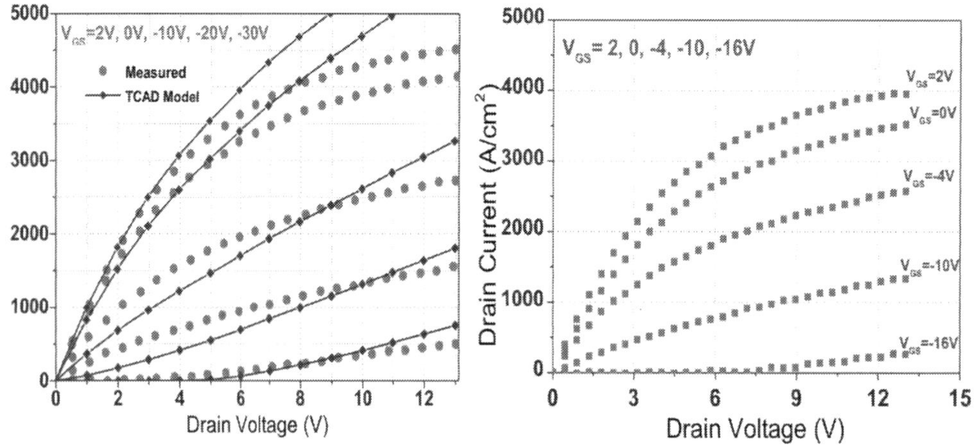

Figure 7. Pulsed measurements and simulated output characteristics of the long channel device with 3µm of channel width (right) and of the short channel device with again 3µm of channel width (left).

similar to that of the figure 1 has been simulated without taking into account edge termination and Gaussian tails of the implanted atoms space profiles. The default SILVACO values for material parameters have been used as a first approach before any refining to better describe experimental results.

Figure 7 shows the output ON I-V simulated and experimental characteristics for the two device types. The experimental curves correspond to pulsed measurements with long OFF times in order to approach isothermal measurements conditions. This allows to not consider device self-heating in the simulation models.

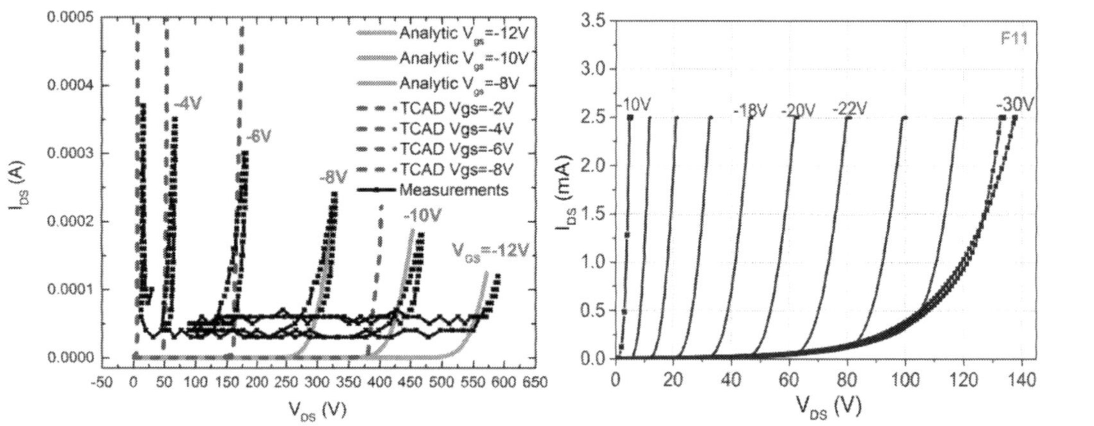

Figure 8. Experimental and simulated blocking characteristics of the long channel device with 1.5 µm of channel width (left) and of the short channel device with again 1.5 µm of channel width (right).

The shorter length device seems difficult to be correctly modelled with the analytical model as the drain-bias induced effect has to be taken into account even for the ON characteristics in order to correctly model the JFET operation. Long channel devices result in lower current densities due to higher R_{ON} values (around 6 and 12 $m\Omega \cdot cm^2$ for the short and long devices respectively).

The transistor blocking (Fig. 8) exhibits a triode shape, which is typical for a static-induction transistor operation. The short channel device presents very low blocking voltages (and corresponding voltage gain values <10) not exceeding 250 V (best case). The long channel devices exhibited blocking voltages up to 800V and voltage gain values up to 50. This difference between the two devices is expected from the SIT mode of operation [6]. Nevertheless, the breakdown voltage of the long channel devices is limited from the fact that the tops of the guard rings (Fig. 4) are highly doped.

course, this drop of the C_{OSS} value does not appear if we limit the graph in low V_{DS} values.

4. CONCLUSIONS

This paper presents the design consideration, fabrication and electrical characterization of 4H-SiC TI-VJFETs with different channel lengths and widths.

Generally speaking, the 4H-SiC TI-VJFETs are characterized by a rather simple process technology in comparison to the transistors like MOSFETs and BJTs. The major fabrication issues deal with the demanding lithography resolution and the isolation of the gate-source junction. The former has been addressed by a self-aligned approach offering in addition the advantage of a reduced number of photolithography steps. The latter has been addressed by incorporating in the process

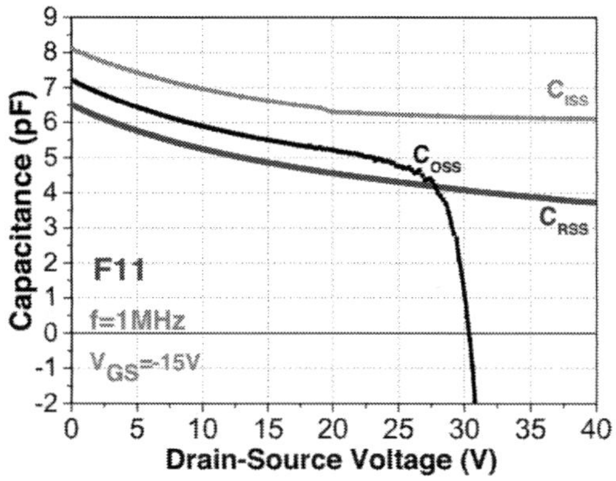

(a)

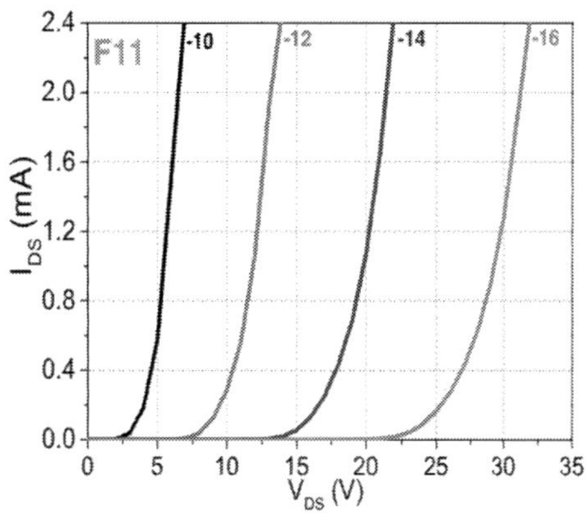

(b)

Fig. 9: (a) C_{ISS}, C_{OSS}, C_{RSS} measurements at f=1MHz, (b) Blocking characteristics of the same short channel device.

In figure 9a, we can see the C_{iss}, C_{oss}, C_{rss} from the short channel devices with the smallest channel width (1.5μm nominal). The gate bias was equal to -15V, during the measurements with V_{DS} sweep, to ensure the off state of the device. All the curves exhibit a descending tense with the V_{DS} rise. The C_{OSS} decrease very abruptly reaching 0 at around V_{DS}=30V and for higher bias values continue to negative capacitance values. Negative capacitance is usually a sign for strong inductive contributions to the impedance. Capacitive and inductive contributions are exclusive. You can only see one or the other, at least at fixed frequency. The increase of inductance can be explained from the fact that the channel becomes conductive above a V_{DS} value (SIT effect). Indeed, by looking the corresponding blocking I-V characteristic (Fig. 9b) the channel becomes conducting for this combination of V_{GS} and V_{DS} values. This conduction in the channel leads to negative values of C_{DS} and consequently of C_{OSS}. Of

a series of steps like a sidewall oxidation and a salicide-like approach for the Ni metallization on top of source fingers. The beneficial effect to the SiC surface, of a C-cap formation prior to post-implantation annealing has been demonstrated. Cyclotene seems to be more suitable for planarization and gate-to-source isolation purposes.

The electrical characteristics showed that the deeper gate transistors exhibit higher on-resistance values and lower leakage currents as expected. The blocking characteristics and consequently the rated voltage class are determined by a static induction effect and therefore rather from the channel geometry than from the drift layer doping and thickness. The advantage of the hereby proposed technology is its simplicity as only four photolithography steps have been employed for the edge termination of self-aligned guard rings.

5. Acknowledgments

The work has been supported by GSRT (Project Kripis – PROENYL).

6. REFERENCES

1. Quadrennial Technology Review 2015, US Dept. of Energy http://energy.gov/under-secretary-science-and-energy/quadrennial-technology-review-2015
2. T. Kimoto, J. A. Cooper, "*Fundamentals of Silicon Carbide Technology: Growth, Characterization, Devices and Applications*" Wiley-IEEE Press, 2014
3. H.-S. Lee, M. Domeij, C.M. Zetterling, M. Östling, F. Allerstam, and E.Ö. Sveinbjörnsson, IEEE Electron Device Lett, Vol. 28, No. 11, p 1007 (2007)
4. G. D. De Salvo, "*SiC Materials and Devices*", Vol. 1, World Scientific, Singapore, 2006, p. 293
5. D. Y. Li, P. Alexandrov, J. Zhao, *IEEE Trans. Electron Dev.*, Vol. 55 (8), p. 1880, (2008).
6. D. Sheridan C., A. Ritenour, V. Bondarenko, J. B. Casady, R. Kelley, *Mat. Sci. Forum*, 679-80, p. 583, 2010.
7. V. Veliadis in "*Advanced Semiconductor Materials and Devices Research: III-Nitrides and SiC*", 2009: p. 407
8. A. Stavrinidis, G. Konstantinidis, M. Kayambaki, F. Cayrel, D. Alquier, Z. Gao and K. Zekentes, Mat. Sci. Forum. *717-720* (2012) pp. 1049-1052
9. B. J. Baliga, "*Modern Power Devices*", J. Wiley, 1992, Chapter 4.
10. C. Bulucea, A. Rusu, *Sol. St. Electron.*, vol. 30, p. 1227, 1987 and A. G. M. Strollo, P. Spirito, *IEEE Trans. Electron Dev*, Vol. 38, No. 8, 1991, p. 1943

DOI: 10.1051/e3sconf/20171612002

GAN TRANSISTOR IN SPACE EQUIPMENTS: FROM SELECTION TO APPLICATIONS

Philippe Fayt & Marc Fossion, Paul Maynadier & Michaël Notarianni

Thales Alenia Space Belgium & France
Emails: firstname.name@thalesaleniaspace.com

ABSTRACT

Two years ago, at ESPC-2014 results of ESA activity on GaN transistor survey has been presented. The main outcome was the existence of an unknown phenomena in silicon MOSFET: the dynamic Rdson.

Today we are the time of studying introduction of this technology into products. At least two suppliers (EPC-co & GaN Systems) are proposing stabilized GaN transistors components which are worth spending some time to try to implement in next generation product prototypes.

Space applications aren't very different from terrestrial ones when we speak about benefits of using GaN technology into dc-dc converters: lower losses & more compact designs thanks to higher switching frequencies. Next to reliability analyses proposed by the component vendors to be able to address the automotive markets, there is of course quite a list "space" specific questions that need also to be answered to raise the confidence that GaN will one day be put is orbit: hermetic packaging & assembly on PCBs, radiation harness, reliability impact of radiations, cooling in the absence of air flow, thermal cycling, drivers, deratings to apply…

In this article we presents some prototyping results & shares some feedback & concerns about the introduction of GaN into space qualified dc-dc products.

1. CONTEXT

GaN based transistors are promising for power applications because of their higher performances (low losses & high speed) and low size (from transistor size itself up to higher switching frequency ability)
Their introduction is progressive on other terrestrial applications, thanks to their progressive cost decrease. But their introduction for space application requests also at least :

- radiation hardness (both total dose and heavy ions)
- robustness & reliability
- thermal draining (no convection)

2. SELECTION

A market survey of commercially available GaN transistors and diodes has been made by mid-2013, under ARTES 5.1 (Advanced Research in Telecommunications Systems) programme element funded by the European Space Agency, and entitled "Demonstration of High Power/High Efficiency Converter Using Commercially Available GaN" (contract 4F.067).

This survey is peculiar to a telecom application being a flight DC/DC converter, fed by the nowadays common 100V satellite bus and with a power level of 600W. Small converters like POL (Point of Load) are clearly out of scope. This lead to make a survey on at least 200V rated GaN components with a high current capability.

The contract encompasses the market survey and the procurement of the relevant parts [4], followed by a breadboard manufacturing using the best commercial GaN. The breadboard will be compared in performance with "classical" Si power MOSFET and diode within the selected DC/DC converter: a SMART topology i.e. a buck followed by a push-pull converter (see §5.1) further.

Even for GaN transistors having to working at 100V bus or above, the Ron resistance becomes now comparable or better than Si MOSFET for power converters. However one important parameter has appeared in GaN HEMT components datasheets: dynamic Rdson. The phenomena seems now understood [6], measurements methods are now available. There is no more bad surprises & we almost don't have to work with data sheets not mentioning dynamic Rdson. As a rough order of magnitude, one has to consider a factor x1.6 more conduction losses when using HEMT normally on devices in dc-dc switching conditions. Some experimental devices are even appearing where no dynamic Rdson are claimed [5].

Nevertheless, the interest of switching GaN is obvious thanks to their very low capacitances, allowing parallel operation without degrading significantly the switching losses (contrary to Si). When the manufacturers will solve or at least improve the dynamic Ron performance, we will take full benefit of this promising component even at 100V bus.

Another important point also linked to the 100V bus voltage is the fact that the very well-known Figure of Merit (FOM) RDSon*QG is not relevant. For the targeted power converter of 600W, the QG losses represent only 4% of the switching losses against 75% due to Coss. We suggest for high power output and voltage above 50V a FOM = DynamicRDSon*Coss.

Finally, we see that within 2..3 years packaging has strongly evolved. All GaN components using the good old TO-xx family of packages have passed over. Indeed high current transients requires low inductance packaging. Furthermore GaN HEMT transistors are now a 4 pin device as its almost mandatory to use Kelvin source to avoid parasitic turn-off.

For diodes, it seems that SiC devices have better performances than GaN in both forward voltage Vd(Id) and reverse current Ir. The main interest to work with those WBG diodes is their dv/dt capability. Indeed, it is widely known that Si Schottky diodes have a limitation at 10V/ns whereas GaN and SiC have no such rating. Although SiC diodes are available their fail radiation testing. GaN diodes are still lagging behind in terms of performance. So if GaN transistors are to be used in classical buck or boost configurations, it is mandatory to resort to proceed to synchronous rectification.

For dc-dc applications, besides some specific operating conditions [7] dc-dc converters are hard to build using normally-on devices. For terrestrial applications, several component suppliers are moving forward in the cascade direction. For space applications this makes much less sense since a radiation hardened silicon MOSFET is needed to create the normally off feature on top of the GaN device which is used to manage high voltage.

The selection was done with the following criteria :

- Normally Off
- No leaded parts

When prospection was initiated mid 2013 were several parts coming from different manufacturers: EPC-co, Microsemi, GaN Systems, RFMD, Avogy, Transphorm, Panasonic, Fujitsu semiconductor, IR, MicroGaN, ... The world is changing extremely fast. There is no 3 months quarters happening without a new player coming in, some other disappearing & other strategic alliances. The survey turned-out to be a permanent task to be stay up to date.

ESA itself has launched the development (project ESA A/O1-7688/13/NL/RA) of radhard components: "GaN devices for space based DCDC power conversion application» contract between ESA and OnSemi. Thales Alenia Space is of involved in the project as end-user of these components.

While waiting for true radhard space qualified components, in order to proceed with investigations on GaN applicability into products, a pre-selection has to be made and Efficient Power Conversion Corp. & Gan System Inc. are the candidates retained for current investigations in prototypes within Thales Alenia Space.

3. EXPERIMENTAL RESULTS

After component tests at parts level, implementation exercises started on several low, medium & high applications.

3.1 Multi-purpose low power dc-dc

As part of the roadmap for product evolutions of the low power dc-dc, GaN transistors have been tested. In this type of low power applications, the use of GaN transistor could be viewed as completely overkill: current levels are significantly below the real capabilities of the transistors. Furthermore using advanced technology as GaN (which is still far away for being completely trusted) in a very low power allow to use it with considerable derating hence leading to a higher confidence in the reliability of the devices used in the context of this application. Such low power application also allow designers to get familiar with controlling, driving and electrically interconnecting GaN HEMT devices without having to take care on power dissipation aspects yet!

Figure 1. GaN-Systems devices under test

The interest of using GaN transistors in such low power equipment (20...40W) mainly lies in the availability of high voltage transistors capable of managing high input voltage ranges: one hand unregulated bus (22V to 37V) and on the other hand telecommunication satellite power bus at 100V. Today managing this large range requires two different hardware variants: one for high voltage with "high voltage" MOSFET transistors and another hardware variant for equipped with "low voltage" MOSFET. Using GaN allow to handle both voltages without compromising dc-dc efficiency. Next to that, as expected, size & weight of the converter can be improved by using higher switching frequencies.

3.2 EPC (Electrical Power Converter) for TWTA

Electrical Power Conditioning unit for high voltage Travelling Wave Tube Amplifier is classified as a mid-range power product with powers in the order of 200 to 600Watts. They are present in satellite with rather high quantities: with 20 to 70 tubes per satellites. A recent trend in this product family in the assembly of so-called "dual EPC" where one module deliver high voltages for 2 TWTA resulting in a significant size, weight & cost reduction at satellite level. For an equipment design perspective delivering twice more power from a unit is not that a simple tasks.

In the context of an ARTES 5.1 activity (see §2), GaN normally-on HEMT components have been tested as replacement parts for the usual silicon MOSFET transistors in the primary side of the dc-dc converter.

Figure 2. electrical prototype of GaN dc-dc EPC (Electrical Power Converter) *for TWTA*

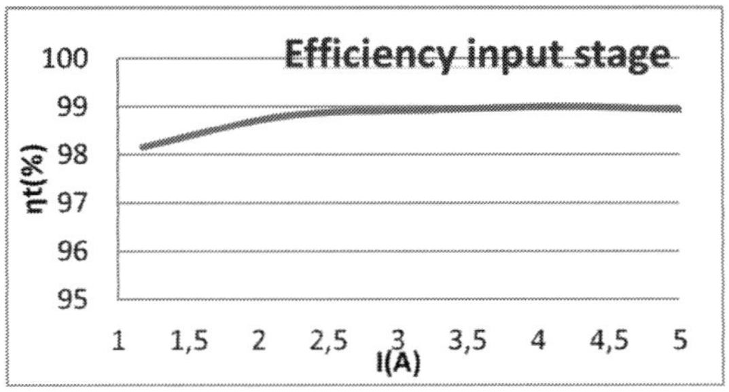

Figure 3. measured efficiency for 100V input bus.

Efficiency of 99.0% at 250KHz achieved for the input stage. After several PCB layout iterations perfect turn-on, turn-off waveform were achieved. The interest of using GaN lies (as usual) in the frequency increase: practically it means it is possible to transfer power level of dual EPC in a HV transformer of a single EPC at same efficiency.

Lessons learned from these prototypes are that thermal management of GaN is much more complex than with classical silicon MOSFET & hence to be considered as a drawback of the GaN technology. Although GaN devices are dissipating much less than silicon MOSFET, there physical side being significantly lower makes the heat dissipation more complex. Not surprisingly also, the speed of this transistors request careful PCB layout to prevent oscillations, spikes & gate overstress.

3.3 PPU

Power supply system for electrical propulsion is the family of product where real high power are being considered with a target of 5kW per module. Clearly in these type of products we are coming close to component ratings & thermal challenges are very complex to manage. The introduction of GaN devices with their new packaging options (flip chip for Efficient Power Corp. & the innovative packaging of GaNSystems) into high power space product will require significant efforts to create new mechanical structure & assembly process to be able to evacuate high power dissipation without air flow while still managing high currents i.e. keeping low inductances everywhere.

3.4 Low power RF units

As described in [2], telecommunication receivers are widely used in geostationary satellites. On this low power application, the power supply delivers around 10W through several output voltages, with strong requirements on output conducted ripple. The switching frequency has been set at 1MHz and the switching speed of the device was up to 30V/ns. The prototype was 50% smaller but owing an additional tool, it was integrated in a RF unit. The RF output clearly "saw" the new switching frequency and additional simple LC filter suppress it. The prototype worked 1000 hours at 25°C without failure.

Figure 4. Low power DCDC in RF unit

DOI: 10.1051/e3sconf/20171612002

3.5 EPC (Electrical Power Converter) for SSPA

Then the application on solid state power amplifier was investigated. Usually such power applications uses bridge legs and a specific work has been done on the optimization of legs commutation [1] : through phase opposition method measurements and dedicated software, a prototype has been built and its characteristics was checked in different configurations. Finally, the efficiency was now driven by the losses in the magnetics devices.

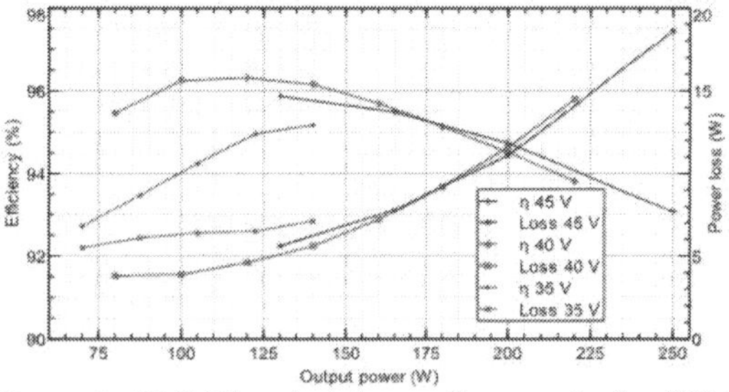

Figure 5. EPC (Electrical Power Converter) *for SSPA core efficiency*

3.6 Low voltage application

GaN transistors were also evaluated to supply digital application needing low voltage (<3V) high current (>10A) applications. Despite their very low Ron, the conductions losses are still "I²" dependant and parallelisation of several devices has been done owing to intrinsic properties of GaN device (positive temperature coefficient, …).

4. FEEDBACK & WAY-FORWARD

All the above mentioned prototypes allow to get better confidence on the electrical & thermal aspects of implementing GaN devices in dc-dc converters. On top the central reliability question, there is of course quite a list "space" specific questions that need also to be answered prior to an introduction of GaN transistors into usual space units.

Hermetic packaging is still needed for most space missions. Hence as specific hermetic packaging needs to be developed for space applications. Currently solutions are converging to SMD05 packages with 4 accesses. The design of this package is a complex exercise to get the heat out of a very small die while minimizing the added parasitic: too high inductance of the connections would completely kill the possibility to increase the switching frequency & hence kill one of the main interest of using GaN devices. Furthermore this specific process of encapsulating die following standardized ECSS processes will significantly increase the recurrent cost of the component. Today the main factor for selecting a component remains part cost!

Reliability & Robustness: using components in Space applications requires a complete understanding of the physics of failures. Indeed this allow to design circuits with enough margins (but not too much also) to reach high reliability figures. Over the experiments conducted within Thales several component failures have been observed for a variety of reasons. First one is the high sensitivity of the gate of the GaN. Optimal & safe gate drive of GaN is an art on itself.

This raise the question of margins & deratings to be applied: the low threshold is very low & can easily be crossed during transients. The performance of the GaN is only achieved for high gate drive. While doing this the gate voltage come close the max ratings. And there is not a lot of headroom for margins. Crossing the max rating lead to component destruction or highly accelerated aging.

Thales experiments also showed unexpected increases of the gate current. While gate current is negligible on fresh components, silent increase of the leakage current of the gate has been observed both during accelerated aging tests and also during irradiation tests. Of course this change of behavior creates some worries & we aren't yet able to declare if this would have any impact on the long reliability of the component. These observations all converge to same point: put a lot of attention on the design of gate drivers!

Radiation tests have been conducted also. As published in NASA reports [8], GaN devices are very robust to total ionizing dose. Threshold drift is can almost be considered negligible for most of standard space missions. The breakdown voltage of requires of course requires to define a safe operating area for each transistor when exposed to heavy ions. Finally gate current leakage has been observed and the phenomena is still not yet explained.

5. CONCLUSION

Today we are the time of pre-introduction of GaN HEMT into products. This introduction in products is going its way & at least isn't completely blocked. Whereas for example SiC components are still reported to dramatically fail at heavy ion testing.

Promises of higher density thanks to increased switching frequencies are met. Product introduction means also investigating all aspects such as procurement, long term reliability, cooling, radiation robustness. Of course questions & worries are raising & they are not fully answered yet, but perspectives are very attractive.

Thales thanks ESA & CNES teams for their support & advises during these GaN investigation projects.

6. ABBREVIATIONS AND ACRONYMS

PPU : Power Propulsion Unit
EPC : Electronic Power Conditioner
EPC-co : Efficient Power Conversion Corp.

7. REFERENCES

1. Delamare, G. (2015). GaN HEMT for space power supplies. PhD INP Toulouse.
2. Delamare, G. & Maynadier, P. (2014). GaN HFET-based DCDC converters for low-power RF satellite equipment, ESPC 2014.
3. Svalgaard, L. & LeSager, P. (2003). What Helio-sesimology Teaches us About the Sun. In Proc. 4th. ISCS 'Solar Variability as an Input to the Earth's Environment' (Eds. H. Lacoste & H. Xerxes), ESA SP-535 (CD-ROM), European Space Agency, Noordwijk, The Netherlands.
4. GaN survey (Philippe Fayt, Vincent Przybylski, & Christophe Delepaut) *7th Wide Band Gap Semiconductors and Components* Workshop. 11-12 September, 2014. Frascati (Rome), Italy
5. Characterization and Application of 600 V Normally off GaN Transistors in Hard Switching DC/DC Converters (Thomas Heckel and Lothar Frey Stefan Zeltner)
6. Mechanisms responsible for dynamic ON-resistance in GaN high-voltage HEMTs Proceedings of the 2012 24th International Symposium on Power Semiconductor Devices and ICs 3-7 June 2012 - Bruges, Belgium Donghyun Jin and Jesús A. del Alamo
7. A very high frequency step-down DC/DC converter for spaceborne envelope-tracking SSPA (Fabien Leroy · Nicolas Le Gallou · Christophe Delepaut · Olivier Deblecker · Fortunato Dualibe) ESPC-2014
8. Status of the Wide Bandgap Working Group:Gallium Nitride Leif Scheick Jet Propulsion Laboratory, 5th NASA Electronic Parts and Packaging (NEPP) Program Electronic Technology Workshop June 23-26, 2014, NASA GSFC, Greenbelt, MD.

DESIGN AND FABRICATION OF SILICON-ON-SILICON-CARBIDE SUBSTRATES AND POWER DEVICES FOR SPACE APPLICATIONS

P.M. Gammon [(1)], C.W. Chan [(1)], F. Gity [(2)], T. Trajkovic [(3)], V. Kilchytska [(4)], L. Fan [(1)], V. Pathirana [(3)], G. Camuso [(3)], K. Ben Ali [(4)], D. Flandre [(4)], P.A. Mawby [(1)], and J.W. Gardner [(1)]
[(1)] *School of Engineering, University of Warwick, Coventry, CV4 7AL, UK, Email: P.M.Gammon@warwick.ac.uk*
[(2)] *Tyndall National Institute at National University of Ireland, Cork, Prospect Row, Ireland*
[(3)] *Cambridge Microelectronics Limited, Cambridge, United Kingdom*
[(4)] *Universite Catholique de Louvain, Louvain-la-Neuve, Belgium*

ABSTRACT

A new generation of power electronic semiconductor devices are being developed for the benefit of space and terrestrial harsh-environment applications. 200-600 V lateral transistors and diodes are being fabricated in a thin layer of silicon (Si) wafer bonded to silicon carbide (SiC). This novel silicon-on-silicon-carbide (Si/SiC) substrate solution promises to combine the benefits of silicon-on-insulator (SOI) technology (i.e device confinement, radiation tolerance, high and low temperature performance) with that of SiC (i.e. high thermal conductivity, radiation hardness, high temperature performance). Details of a process are given that produces thin films of silicon 1, 2 and 5 μm thick on semi-insulating 4H-SiC. Simulations of the hybrid Si/SiC substrate show that the high thermal conductivity of the SiC offers a junction-to-case temperature *ca.* 4× less that an equivalent SOI device; reducing the effects of self-heating, and allowing much greater power density. Extensive electrical simulations are used to optimise a 600 V laterally diffused metal-oxide-semiconductor field-effect transistor (LDMOSFET) implemented entirely within the silicon thin film, and highlight the differences between Si/SiC and SOI solutions.

1. INTRODUCTION

Space is an expanding and changing market for power electronics. This is in part due to the rise of all electric propulsion systems for fine orbit control, as demonstrated for ESA mission GOCE, where a 600W/1.5kV QinetiQ T5 ion thruster [1] has been employed, and more widely for the orbit control of Geostationary Earth Orbit (GEO) telecommunications satellites. Furthermore, the big brother of the T5, the 5kW T6, has been developed to provide the impulse during the cruise phase of ESA's BepiColumbo mission to Mercury [2]. Another driver for space power electronics is the potential to move to high voltage transmission, beyond the 120 V maximum used today. This would allow harness weight, which can be up to 10% of satellite weight, to be greatly reduced given the reduction in I^2R losses [3]. Furthermore, 300 V solar cell output could open up the possibility of a direct drive power unit, the solar cells connected directly to hall thrusters without the need for an intermediate PPU, so saving significant weight and boosting efficiency.

A 2007 report entitled "*Extreme Environments Technologies for Future Space Science Missions*" by NASA's Jet Propulsion Laboratory [4] reviewed the state of space electronics and produced a roadmap for technology development if we are to explore some of the solar system's harshest environments. This described the benefit that could be gained from producing electronics that could survive and work reliably outside of the temperature controlled chamber; thus exposing the electronics to extreme highs, lows, and cycles of temperature. The potential system benefit of this approach would improve the efficiency of the cooling system, and increase the overall lifetime, reliability, and science capability of the mission.

The roadmap [4] cited wide bandgap semiconductors as the future for high temperature missions predicting performance up to 500°C, while low-voltage technology will continue to be dominated by silicon-on insulator (SOI) technology up to a maximum temperature of 300°C. Nearly ten years on from the publishing of this roadmap, silicon (Si) and silicon-on insulator (SOI) solutions remain almost entirely dominant in space power applications [5]. By confining the active area of the device, SOI devices achieve low leakage at high temperature [6]. SOI is often the rad-hard choice at low voltage too, the small volume of active material leading to excellent single event immunity [7]. However, a thick buried-oxide, which is required to support high voltage devices, make SOI devices much more sensitive to total ionizing dose (TID) effects.

Nearly ten years on from [4], wide bandgap semiconductors such as silicon carbide (SiC) and gallium nitride [8] are yet to make a significant impact on the space power market, though SiC Schottky diodes are being integrated into the solar inverter solution on BepiColombo [9]. SiC, in particular, has the potential to produce power systems that are smaller and lighter with improved efficiency, and high radiation and thermal tolerances. However, integration of full SiC solutions that include switching devices such as metal-oxide-semiconductor field-effect transistors (MOSFETs) are still held back by problems with long term reliability (particularly concerning the oxide), stability (threshold voltage and on-resistance drift) and packaging still hold back the continued integration of SiC both in space and terrestrially. Alternative solutions using JFETs [10]

DOI: 10.1051/e3sconf/20171612003

have been suggested, but not yet widely adopted.

In this paper, we introduce a new generation of lateral power electronic semiconductor devices that are being developed for the benefit of both space and terrestrial harsh-environment applications. The silicon-on-silicon carbide (Si/SiC) semiconductor substrate will attempt to improve upon current SOI technology by removing the buried oxide (BOX) layer that prevents efficient heat removal and causes poor TID tolerance. In its place will be a semi-insulating SiC substrate, with its high thermal conductivity and radiation hardness.

We will first introduce this concept showing the potential thermal benefits of the solution. We will then briefly summarise the results of simulations showing the advantages of this solution over conventional SOI. Furthermore, we will show for the first time the method used to fabricate our wafer bonded Si/SiC solution and the resulting wafers. At the conference, we will present the results of currently on-going device fabrication trials that are producing Si/SiC power devices in this substrate, specifically for space applications. The devices will include a 600 V LDMOSFET power transistor, and Schottky and PiN diodes.

2. THE Si/SiC CONCEPT

Lateral power devices including laterally diffused MOSFETs (LDMOS), lateral insulated gate bipolar transistors (LIGBT), and Schottky and PiN diodes, rated from 50 to 600 V will be implemented within a Si thin film 2 μm thick. The thin Si film sits directly on the surface of a high resistivity, semi-insulating 4H-SiC substrate. This can be pictured in Figure 1 alongside a proposed LDMOS Si/SiC transistor.

The Si/SiC solution benefits from the advantages of both materials, where thermally it is most akin to the highly thermally conductive SiC substrate, but

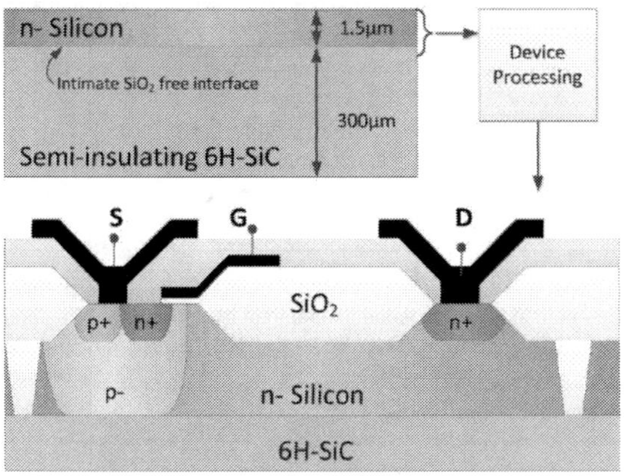

Figure 1: Top, the Si/SiC material system that has been developed. Bottom, proposed Si/SiC LD-MOS transistor.

electrically these devices are Si, or SOI devices. This can be summarised as follows:

SiC-like Thermal Performance: The thermal performance of the substrate is almost equivalent to SiC, resulting in a much lower temperature difference (ΔT) between the Si/SiC device's internal junction temperature and the local ambient temperature [11], compared to bulk Si or SOI. This can be exploited in one of three ways. In comparison to bulk Si or SOI, a Si/SiC device can operate either:

- in exactly the same ambient temperature, dissipating the same power, but with improved performance due to a reduction in the effects of self-heating;
- at a much higher temperature (as much as 100°C higher) for the same performance;
- or at the same temperature but at much higher power (as much as 4×), for the same performance.

Si Electronics. Electrically devices implemented in a Si/SiC substrate are akin to those in SOI or much closer, silicon-on-sapphire. Trapped charge at the SiC-oxide interface results in a MOSFET channel resistance that becomes limiting at voltage ratings of 600 V or below [6, 8]. The Si/SiC device will not suffer this problem, its channel being at a conventional Si/SiO_2 interface. Similarly, diodes and LIGBTs will benefit from the low built in potential at a Si p-n junction (~0.7 V) compared to SiC (~3 V). Furthermore, the processing of Si rather than SiC means that it is compatible with more conventional cleanroom equipment and procedures.

Radiation Immunity. The Si/SiC substrate is expected to offer high radiation hardness. Immunity to single event effects is expected for the same reason as SOI, due to the small volume of active material. Compared to SOI, Si/SiC devices are expected to have a higher TID tolerance due to the absence of a thick BOX layer under the active Si film and thus a lesser effect from radiation-induced oxide charge trapping on device performance. Furthermore, the SiC substrate itself has a high Si-C bond energy (large bandgap energy of 3.2 eV) and a high atomic displacement threshold in SiC of 21.8 eV [12], which is significantly higher than in Si.

Previous studies on Si/SiC electronics have concentrated on demonstrating MOSFET behaviour [13, 14] and the implementation of RF devices [15, 16]. Both showed that the effects of self-heating on the forward characteristics can be suppressed, unlike equivalent SOI devices power devices. These studies have shown experimentally [14] that the channel mobility of a Si/SiC MOSFET at 300°C and hence its channel resistance was just 10% worse than it was at room temperature, compared to an 83% reduction for an equivalent Si bulk device. Experimental results from RF devices implemented in a complicated Si/poly-Si/poly-SiC substrate [16], show that the negative-resistance effect of self-heating (whereby device resistance increases with voltage due to a rise in internal temperature) is minimised in the Si/SiC device, unlike an equivalent SOI device. However, breakdown voltage,

DOI: 10.1051/e3sconf/20171612003

leakage current and maximum oscillation frequency were all shown to worsen. Other groups have considered SOI where the BOX and Si top layer is transferred onto a SiC substrate [17]. However, this is a methodology that our own modelling [11] proves to only fractionally reduce the self-heating problems of conventional SOI.

3. FABRICATING Si/SiC SUBSTRATES

100 mm Si/SiC wafers were fabricated by wafer bonding Norstel semi-insulating ($\geq 1.10^7$ Ω-cm) on-axis 4H-SiC to Icemos Technology Ltd SOI with a buried oxide 5 μm thick, and a lightly n- doped device layer (5 - 45 Ω-cm). Wafers have been produced with a device layer thicknesses 1, 2 and 5 μm thick. The full process used to develop the wafers can be seen in Figure 2.

Initial trials directly bonding the two virgin wafers via a number of methods (hydrophilic, hydrophobic, with and without a deposited interfacial oxide, and with and without a pre bonding anneal to limit outgassing) all resulted in either the handle wafer shearing under the strain, or in a high density of voids, visible in the hydrophobic bonded wafers the inset to Figure 3.

The problem of the voids was suspected to be the result of outgassing while the layers were undergoing a long high temperature anneal to form a permanent bond between the wafers. To overcome this problem, the first step taken in the bonding process (Figure 2) was to etch a grid of trenches 2 μm deep into the SiC surface prior to bonding, giving any gas an escape route during annealing. Next, after a proprietary surface plasma treatment, a hydrophobic bonding process was performed to form a temporary bond between the wafers.

A 2 hour, 1200°C anneal then formed a permanent bond between the wafers. However, the lattice mismatch between the materials resulted in a visible bow, the Si side of the wafer being convex. This still allowed, in the fourth step, for the Si handle wafer to be ground away down to the oxide later, which is simply removed with hydrofluoric acid.

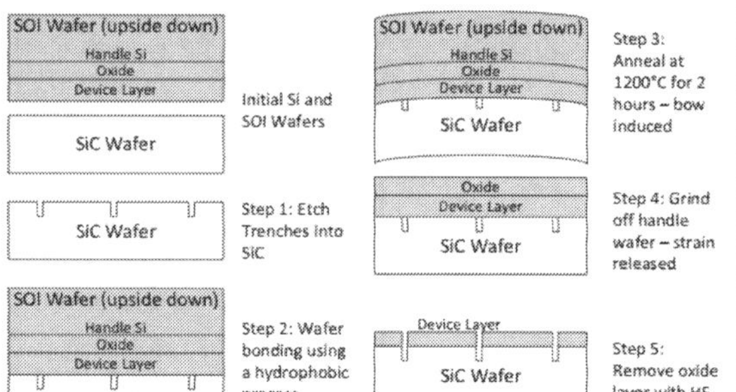

Figure 2: The Si/SiC substrate fabrication process.

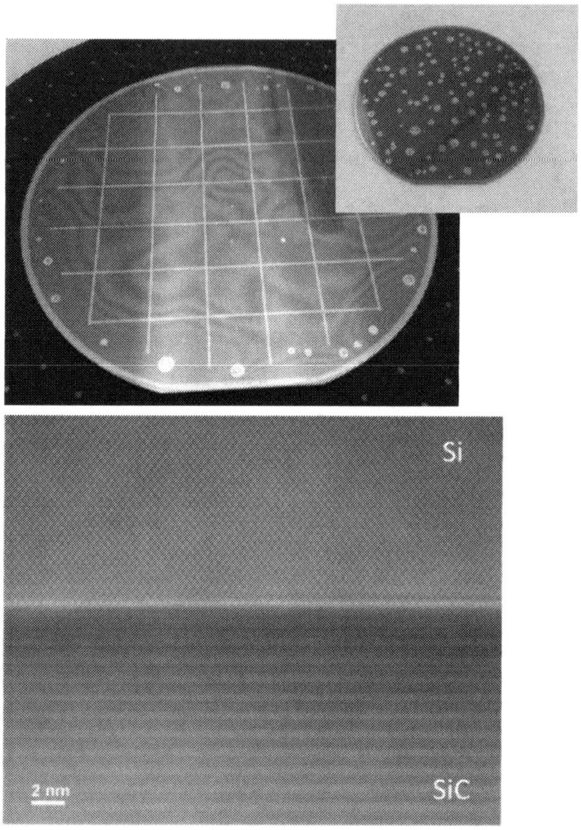

Figure 3: Above, a 100 mm wafer bonded Si/SiC wafer with a 1 μm Si device layer. Inset, the results of bonding without the grid etching. Below, a TEM image of the Si/SiC interface showing no sign of a SiO_2 interfacial layer.

A 2 hour, 1200°C anneal then formed a permanent bond between the wafers. However, the lattice mismatch between the materials resulted in a visible bow, the Si side of the wafer being convex. This still allowed, in the fourth step, for the Si handle wafer to be ground away down to the oxide later, which is simply removed with hydrofluoric acid.

The resulting 100 mm Si/SiC wafer can be seen in Figure 3. This particular wafer had a 1 μm Si device layer. 2 and 5 μm Si layers have also be successfully transferred. Transmission electron microscopy was used to analyse the interface between the wafers. This found that some voids were present at the interface, which may have an effect on current transport at this interface. Also, while the image in Figure 3 shows a clean oxide-free interface, some areas viewed did have a thin amorphous layer up to 2 nm thick which is highly likely to be silicon dioxide. Efforts will be made in the future to remove this by high temperature annealing as has been successful elsewhere [18]. X-ray diffraction analysis revealed no strain in the resulting device layer.

DOI: 10.1051/e3sconf/20171612003

4. SIMULATING Si/SiC DEVICES

Recent work published by the authors [19] revealed the ability of devices implemented on Si/SiC substrates to dissipate effectively heat, and therefore operate at greater efficiency. 600 V LDMOS devices were implemented in the Si/SiC substrate and compared with very similarly implemented devices in SOI. The two layouts are shown in Figure 4. The physical structure of the two active regions of the devices were initially identical, each 42 µm-long, 0.2 µm-thick and with the same "IOS" layout that sees the gate contact extend over a thick field oxide over the drift region. These are based on SOI devices originally developed by Philips in the 1990s [20].

Despite their identical origins, for both devices to support the same blocking voltage (600 V) the linear doping used in the Si/SiC had to be half that of the SOI doping. This is because the drift region in the SOI solution is effectively narrower than that in the Si/SiC, the comparatively thin BOX having sufficient capacitance to form a depletion region in the drift region above.

The results of the simulation study [19] showed that both these devices would break down above 640 V at both room temperature and at 300°C. I-V results from the forward DC characteristics of the devices are shown in Figure 5. Comparing for each device the difference between simulations that ignore the effects of temperature (isothermal) and those that take it into account (non-isothermal), reveals the impact of removing the buried oxide. The isothermal and non-isothermal simulations for the Si/SiC devices are nearly identical with minimal junction temperature, the waste

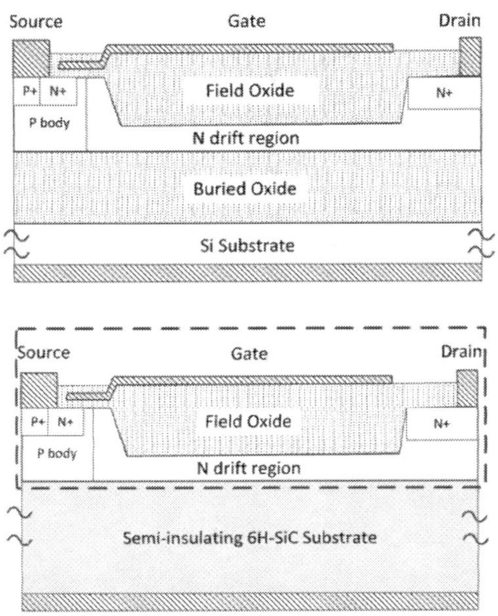

Figure 4: The layout of the simulated SOI (above) and Si/SiC (below) LDMOS transistors.

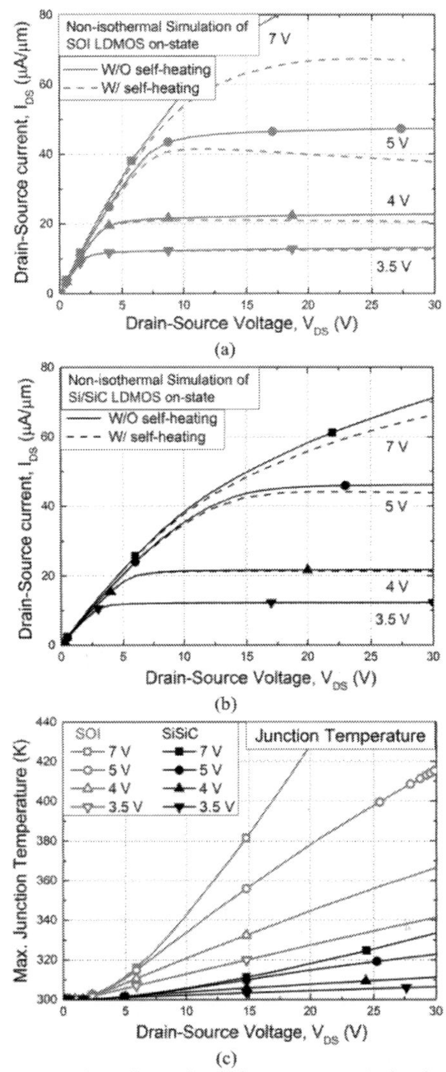

Figure 5: The effect of self-heating on (a) the forward characteristics of the SOI and (b) the Si/SiC LDMOS, as well as (c) their junction temperature. (a) and (b) show both isothermal results ignoring the effects of self-heating (solid) and non-isothermal results including the effects of heating (dashed).

heat generated in the layer having been efficiently channelled through the SiC substrate. However, the SOI can be seen to suffer so-called negative resistance, with the devices failing to remove the heat from the device effectively, so resulting in significant decreased current at a given gate and drain voltage, and significant internal heating.

Further detailed simulations are shown in [19].

5. FABRICATING AND TESTING Si/SiC DEVICES FOR SPACE APPLICATIONS.

At the conference, the results of on-going fabrication trials will be presented. Power devices including Schottky and PiN diodes, and LDMOS devices akin to those in Figure 4 have been fabricated, leading to some very interesting, and unexpected results.

We will present the results of these entirely new devices at the conference for the first time.

6. CONCLUSION

The results presented detail an on-going project to fabricate the first lateral power devices in a Si/SiC substrate. The novel combination of materials offers significant thermal benefits over current state of the art SOI power electronics, reducing self-heating effects, which will improve maximum operating temperature, and efficiency at any given temperature. These devices will be of particular interest to the space sector; the low volume of active material, the absence of a BOX and the wide bandgap substrate all adding to the radiation tolerance of the final solution. Novel Si/SiC wafer bonding has led to the development of 1, 2 and 5 µm thin films of silicon directly bonded to semi-insulating 4H-SiC. Efforts to produce the first power diodes and transistors in this substrate are on-going.

7. ACKNOWLEDGMENTS

The work presented in this paper has been carried out as part of the EU funded SaSHa Project (Si on SiC for the Harsh Environment of Space): www.sashaproject.eu/. We gratefully acknowledge the support of the European commission for this funding.

Dr. Peter Gammon would also like to acknowledge support from the Royal Academy of Engineering, and from the EPSRC (Project EP/N00647X/1).

8. REFERENCES

[1] M. Coletti, *et al.*, "Solar Electric Propulsion Subsystem Architecture for an All Electric Spacecraft," *Advances in Spacecraft Technologies*, p. 123, 2011.

[2] S. D. Clark, *et al.*, "BepiColombo electric propulsion thruster and high power electronics coupling test performances," in *Proc. 33rd Int. Electr. Propuls. Conf*, 2013.

[3] D. J. Hoffman, *et al.*, "Concept design of high power solar electric propulsion vehicles for human exploration," 2011.

[4] E. Kolawa, "Extreme Environments Technologies for Future Space Science Missions," NASA JPL (2007).

[5] J. R. Schwank, *et al.*, "Radiation effects in SOI technologies," *IEEE Transactions on Nuclear Science*, vol. 50, pp. 522-538, 2003.

[6] P. G. Neudeck, *et al.*, "High-temperature electronics - a role for wide bandgap semiconductors?," *Proceedings of the IEEE*, vol. 90, pp. 1065-1076, 2002.

[7] P. M. Shea, "Lateral Power MOSFETs Hardened Against Single Event Radiation Effects," University of Central Florida Orlando, Florida, 2011.

[8] P. M. Gammon, "Silicon and the wide bandgap semiconductors, shaping the future power electronic device market," in *Ultimate Integration on Silicon (ULIS), 2013 14th International Conference on*, 2013, pp. 9-13.

[9] E. Maset, *et al.*, "Accelerated Life Test for SiC Schottky Blocking Diodes in High-Temperature Environment," *Device and Materials Reliability, IEEE Transactions on*, vol. 9, pp. 557-562, 2009.

[10] B. Reese, *et al.*, "Silicon carbide power processing unit for Hall effect thrusters," in *Aerospace Conference, 2012 IEEE*, 2012, pp. 1-6.

[11] C. Chan, *et al.*, "Simulations of a lateral PiN diode on SiSiC substrate for high temperature applications," *Materials Research Forum*, 2014.

[12] A. L. Barry, *et al.*, "Energy dependence of electron damage and displacement threshold energy in 6H silicon carbide," *Nuclear Science, IEEE Transactions on*, vol. 38, pp. 1111-1115, 1991.

[13] Y. Sasada, *et al.*, "Junction Formation via Direct Bonding of Si and 6H-SiC," *Materials Science Forum*, vol. 778-780, p. 714, 2014.

[14] H. Shinohara, *et al.*, "Si metal-oxide-semiconductor field-effect transistor on Si-on-SiC directly bonded wafers with high thermal conductance," *Applied Physics Letters*, vol. 93, p. 122110, 2008.

[15] L. G. Li, *et al.*, "Dynamics of SiO2 Buried Layer Removal from Si-SiO2-Si and Si-SiO2-SiC Bonded Substrates by Annealing in Ar," *Journal of Electronic Materials*, vol. 43, pp. 541-547, 2014/02/01 2014.

[16] S. Lotfi, *et al.*, "LDMOS-transistors on semi-insulating silicon-on-polycrystalline-silicon carbide substrates for improved RF and thermal properties," *Solid-State Electronics*, vol. 70, pp. 14-19, 2012.

[17] S. G. Whipple, *et al.*, "Demonstration of Hybrid Silicon-on-Silicon Carbide Wafers and Electrical Test Structures with Improved Thermal Performance," *MRS Online Proceedings Library*, vol. 911, pp. null-null, 2006.

[18] L. G. Li, *et al.*, "Oxygen out-diffusion from buried layers in SOI and SiC–SOI substrates," *Solid-State Electronics*, vol. 54, pp. 153-157, 2010.

[19] C. W. Chan, *et al.*, "Analysis of Linear-Doped Si/SiC Power LDMOSFETs Based on Device Simulation," *IEEE Transactions on Electron Devices* p. (In Press), 2016

[20] E. Arnold, *et al.*, "High-temperature performance of SOI and bulk-silicon RESURF LDMOS transistors," in *Power Semiconductor Devices and ICs, 1996. ISPSD '96 Proceedings., 8th International Symposium on*, 1996, pp. 93-96.

SILICON CARBIDE AND MAGNETORRESISTIVE TECHNOLOGIES FOR SOLID STATE POWER CONTROLLERS

A. Garrigós[1], J. M. Blanes[1], R. Gutiérrez[1], J. Borrell[1], E. Maset[2], D. Ramirez[2], S. Cardoso[3]

[1] *Industrial Electronics Group, UMH, Spain, e-mail: augarsir@umh.es*
[2] *Instrumentation and Industrial Electronics Laboratory, UV, Spain*
[3] *INESC Microsystems and Nanotechnologies and Physics Dept, Instituto Superior Tecnico, UL, Portugal*

ABSTRACT

The aim of this paper is to make an introduction, review and preliminary investigation tasks of the application of Silicon Carbide (SiC) power semiconductors and magnetoresistive (MR) current sensors for Solid State Power Controllers (SSPCs) and controlled switches, especially for high temperature environment and/or high voltage applications.

1. INTRODUCTION

Wide BandGap (WBG) power semiconductors are nowadays being developed for many power conversion applications. In particular, Silicon Carbide (SiC) very attractive for high-voltage, high-power and high temperature conditions where traditional silicon semiconductors find serious limitations. During the last European Space Power Conference (ESPC 2014), a complete session was dedicated to WBG devices and applications, including a comprehensive paper of potential applications for the space sector [1]. In section 4.2 of that work appears an interesting application which actually is the staring point of this research "*3. high voltage transistors can also be used as output power switches in ion-propulsion to improve the redundancy scheme and allow to connect the nominal or redundant converter to the nominal or redundant thruster*".

The most relevant properties and prized properties of SiC are high critical electrical field, eight times higher than Si, good thermal conductivity, three time higher and higher electron saturation velocity, two times higher Those characteristics impact directly on higher voltage, higher temperature and higher frequency of operation.

Second technology of interest is magnetoresistive (MR) current sensors. Shunt current sensing is probably the most used technique nowadays. Despite its simplicity and well-established technique for space applications they have some drawbacks which some of them could be exacerbated in high voltage applications. First, and the most obvious, it is an invasive method that consumes power and requires certain level of redundancy to avoid open-circuit failure of sensing resistor, but also, in high voltage applications, requires high voltage transistors (current mirror and output current source) to stand with operating voltages and achieve high CMRR. Galvanic isolation is also an interesting topic of discussion.

MR current sensing is becoming a promising alternative to shunt current sensing since overcomes some of the problems abovementioned. In addition, MR technology is widely spread in several space sensor applications (position, magnetometers...) and current sensing has also been investigated recently [2].

2. SiC BASED SOLID STATE PROTECTION CIRCUITS: LITERATURE REVIEW

Focusing on Solid State Power Controller (SSPC) and Solid State Circuit Breaker (SSCB), first found references that uses commercial devices date back to 2007 and 2008, when is suggested to conceive an SSPC using SiC JFET, [3,4]. The target was 270Vdc aircraft power distribution and the rationale behind was Si-MOSFET voltage limitation and the substitution of traditional electromagnetic circuit breakers. Experimental validation of the proposed circuits was carried out using Semisouth and SiCED power semiconductors, currently these companies do not exist anymore. Further, in [3] it is suggested the use of paralleled power transistors and parallel multimodule approach.

In chronological order, next interesting reference [5] corresponds to a patent filled in may 2010. It describes power distribution system using bidirectional SSPCs based on two anti-series SiC JFET and anti-parallel diodes.

Three more works date from 2011. In [6] the authors detail the design and the experimental validation of an SSPC based on 1200V/20A SiC MOSFET. Unlike JFET, MOSFET a normally-off device and, therefore, circuit topology is notably different and more similar to what is commonly found with Si devices. Again, main target is aircraft power distribution at 270Vdc. The same group presents in [7] the design and the experimental validation of a Solid State Circuit Breaker (SSCB) capable to operate at high current and high temperature. Finally, in [8] is described a bidirectional 600V/60A SSCB using two SiC normally-on JFET with a common source and single driving circuit for both devices.

In parallel, from 2010 and 2014 a slightly different SSCB approach is developed using SiC Static Induction Transistors (SIT), which have similar structure to normally-on JFET but smaller channel length, offering less conduction resistance and higher blocking voltage capabilities [9]. 400Vdc power distribution for data centres was the main application.

During 2015 two related works were also published. In [10,11] is presented a review of the state of art of SSCB using WBG, as well as a complete design using a normally-on SiC JFET, UJN1205K from USCi. SSCB is also validated at 400V and 180A and the work discusses the benefits of having normally-on devices.

In spite of several application areas are described in those works, none of them is dedicated exclusively to space power electronics. Space high voltage systems, in the range of few kV, are obvious potential candidates to make use of this technology, but also typical space power distribution systems, which are often limited to 100Vdc in high power satellites and are dominated by Si power MOSFETs could be benefited due to high temperature operation.

To conclude this section and as an anecdote it is worth to mention that Si JFET where proposed as high voltage solid state circuits in the early seventies [12].

3. BACKGROUND AND INITIAL RESULTS

Before to initiate the design of a solid state power switch as a basic building block of a SSPC, some considerations and previous experimental results will be described. Those results, which came from other research activities, mainly focus on DC and transient characterization of SiC power devices.

3.1. SiC JFET: Infineon IJW120R100T1

The IJW120R100T1 is a commercially available normally-on SiC JFET from Infineon. It is 1200V device with an specified maximum continuous drain source current of 10A (V_{GS}=0V, T_C=150°C, R_{thJC}=$R_{thJC,max}$).

Normally-on vs. normally-off power devices as solid-state power controllers is an interesting question that should be addressed. At first sight, the traditional approach using normally-off devices seems to be the preferred option; however, recent studies [11,13] also propose the use of normally-on devices.

Anyway, normally-off behaviour can be achieved by using the cascode configuration with a low voltage MOSFET, please refer to figure 1. At initial conditions, the low voltage MOSFET is in off-state and sets the source of the JFET to positive potential relative to its gate, hence JFET remains in off-state. When low voltage MOSFET turns on, gate to source voltage of the JFET goes to zero, keeping it in on-state.

One major drawback for this option is the absence of direct control of the JFET, so different approaches using negative driving signal and always on-state MOSFET are suggested [14].

Another interesting point for cascode configurations is the possibility to create high voltage devices with lower voltage, but faster and lower on-resistence normally-on SiC JFETs, please refer to figure 1. This structure, known as super cascode, has been recently proposed [15] and it could be of interest in very high voltage applications or to standardize a complete family with different voltages by using the same basic building block.

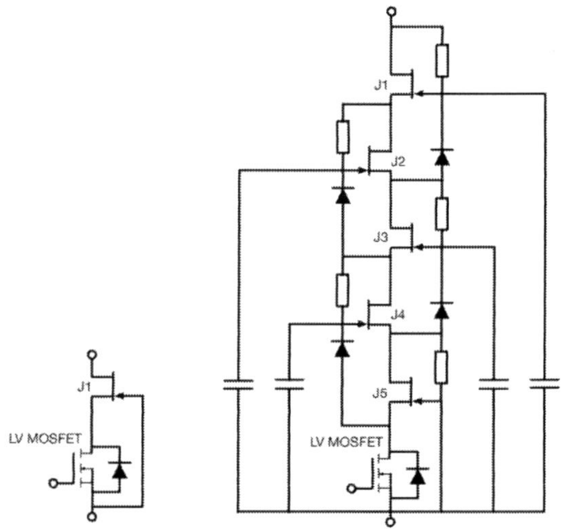

Figure 1. Left side: cascode configuration, right side: super cascode configuration

Next oscilloscope screenshots show the transient response of the IJW120R100T1 working at 25°C and 175°C. Being a protection device, both DC and transient characteristics are extremely important. On-resistance temperature dependence will determine device losses during nominal operation and dynamic characterization including temperature effects is of great importance for driver design and determining turn-off times. On top of this, thermal stress is high in case of operation.

Figure 2 shows switching-on-and-off test with two initial current conditions. First measurement turns on at 0A and turns off at 19A. Second measurement turns on at 19A and turns off at 28A. These measurements are performed at 25°C.

 DOI: 10.1051/e3sconf/20171612004

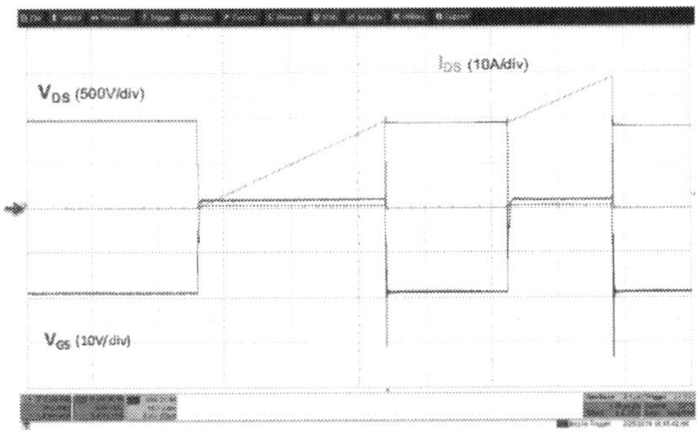

Figure 2. IJW120R100T1 switching characterization. Operating conditions: Vds=0.8·Vds(max); T=25°C, Vgs=-19V..2V; Rg(on)=20 ohm; Rg(off)=2.4 ohm.

The detail of the hard-switching transition at 19A are shown in figure 3 for turn-off and figure 4 for turn-on.

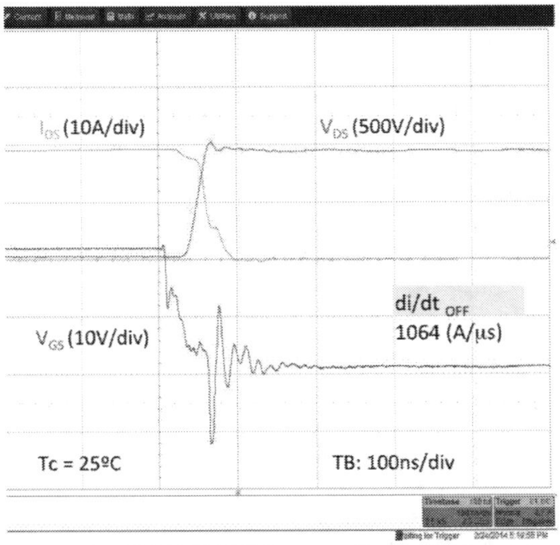

Figure 3. IJW120R100T1 turn-off detail. Operating conditions:
Vds=0.8·Vds(max); Id=0,75·Id(max)=19.5A, T=25°C, Vgs=-19V..2V; Rg(on)=20 ohm; Rg(off)=2.4 ohm.

Switching behaviour at 25°C and 175°C is shown in figures 5 and 6. One may note that turn-on performance is very similar, dissipated energy in turn-off is at 175°C lesser than at 25°C, also, gate signals remain very similar, indicating that input capacitances do not vary significantly with temperature.

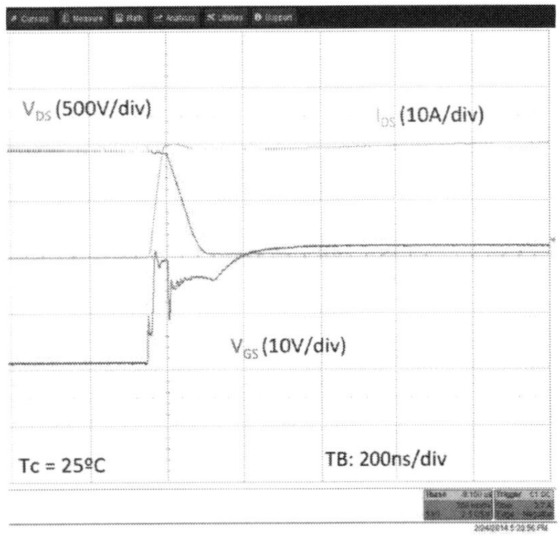

Figure 4. IJW120R100T1 turn-on detail. Operating conditions:
Vds=0.8·Vds(max); Id=0,75·Id(max)=19.5A, T=25°C, Vgs=-19V..2V; Rg(on)=20 ohm; Rg(off)=2.4 ohm.

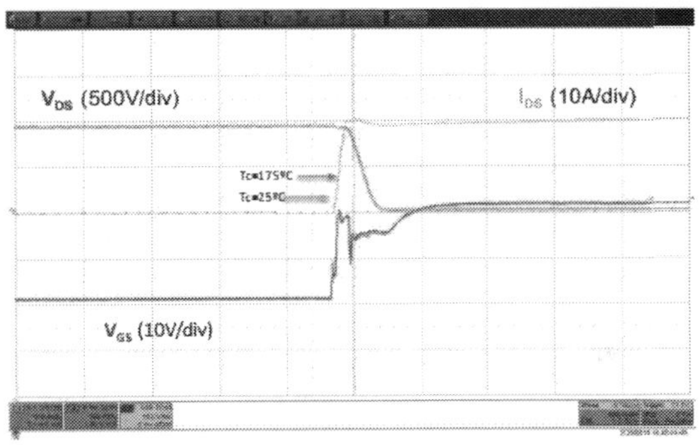

Figure 5. IJW120R100T1 turn-on detail at 25°C and 175°C. Operating conditions:
Vds=0.8·Vds(max); Id=0,75·Id(max)=19.5A, Vgs=-19V..2V; Rg(on)=20 ohm; Rg(off)=2.4 ohm.

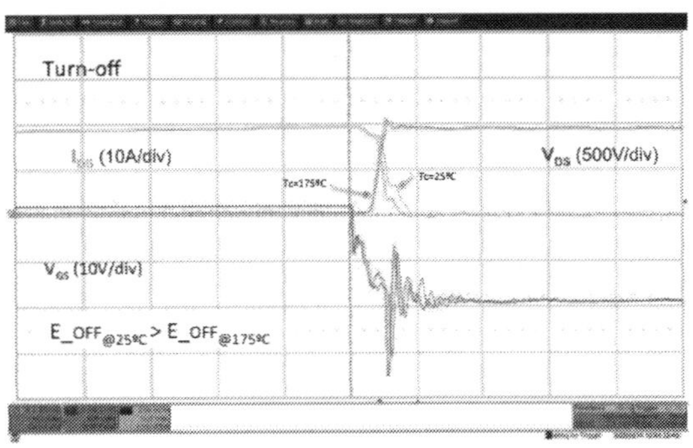

Figure 6. IJW120R100T1 turn-off detail at 25°C and

175°C. Operating conditions: Vds=0.8·Vds(max); Id=0,75·Id(max)=19.5A, Vgs=-19V..2V; Rg(on)=20 ohm; Rg(off)=2.4 ohm.

3.2. SiC JFET: USCi UJC1210K

The UJC1210K is a commercially available JFET SiC integrated cascode in TO-247 package. An SPICE model is also provided by the manufacturer which allows initial circuit design.

Figure 7 shows the simulation schematic of a double pulse circuit, similar to the experimental results presented in the previous section. Simulation results are also included in figures 8 and 9, showing, respectively, switching off and on details.

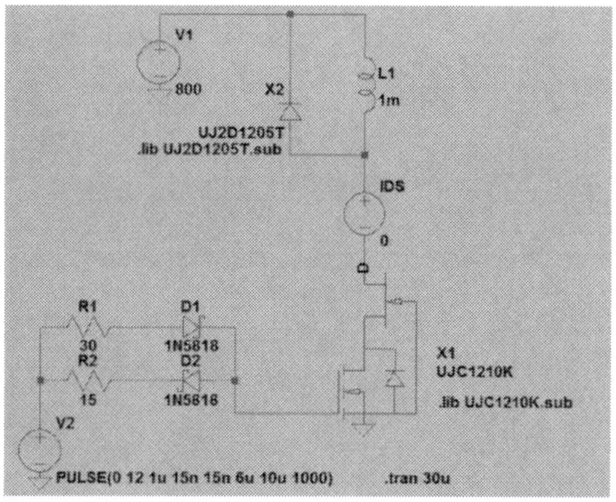

Figure 7. UJC1210K double-pulse LTSpice simulation schematic.

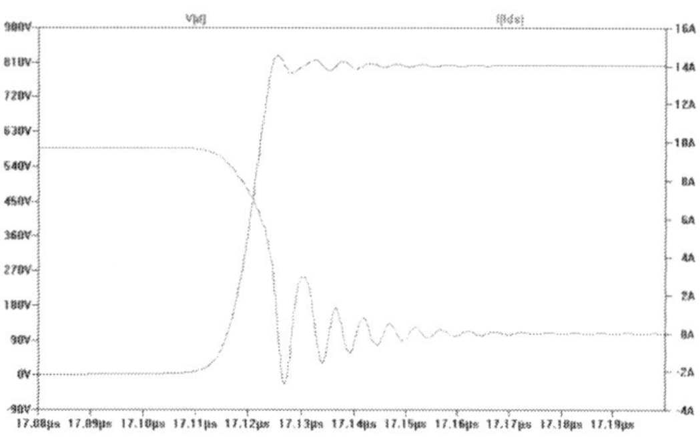

Figure 8. UJC1210K switching off detail. Operating conditions: Vds=800V; Id=10A; T=25°C; Vgs=0V.. 12V; Rgon=30 ohm; Rgoff=15 ohm.

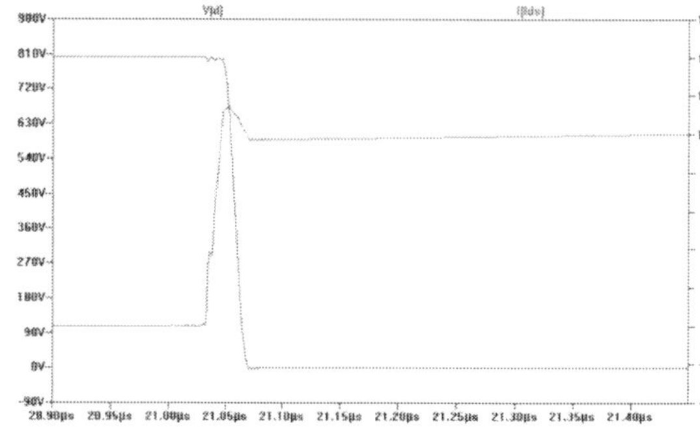

Figure 9. UJC1210K switching on detail. Operating conditions: Vds=800V; Id=10A; T=25°C; Vgs=0V.. 12V; Rgon=30 ohm; Rgoff=15 ohm.

3.3 MR current sensors

MR current sensors have been used extensively in industrial, scientific and/or biomedical applications. These sensors are based on a Wheatstone topology resistive bridge that properly biased will bring high voltage sensitivity and linearity (Figure 10). Based on their sensing principle (Biot-Savart law) MR current sensors provide inherent galvanic isolation and no-dissipative self heating features [16]. Ultimate MR sensors based on the tunnel MR effect (TMR) provide low offset and hysteresis values with high input impedance and low power supply consumption. Additionally, they offer interesting dynamic response. Figure 11 shows the 0 to +7.5 A step response of a TMR based current sensor compared with a comercial Hall current probe. In the frequency domain, different MR current sensor technologies were studied and modelled [16, 17]. Stability against radiation is a key factor to consider when designing electronic system for spacecraft applications. In this topic various results were obtained on TMR current sensors showing their feasibility to be used in on-board electronic converters, [18]. All the above mentioned features are now being analyzed to be applied in a future SiC-MR based SSPC development.

4. CONCLUSIONS

This paper presents the planned activities and some previous results which form the background for a new research project in SiC based solid state power controllers and the use of magnetoresistive current sensors in space applications. These activities are financially supported by a Spanish national research project which lasts from 2016 to 2018. First designs and electrical simulations are planned to be carried out during 2016. The field of application is oriented to high voltage systems.

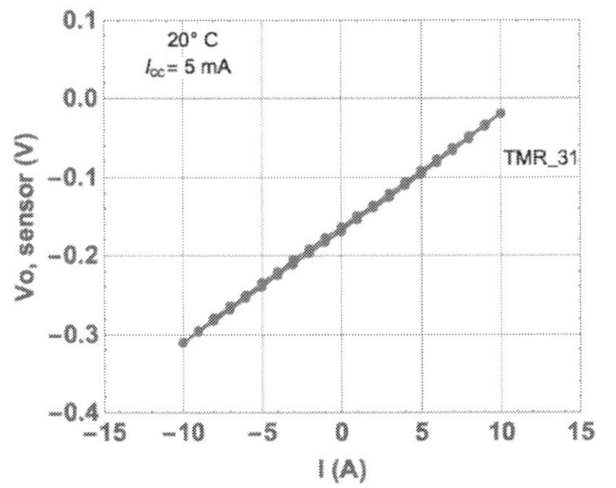

Figure 10. Output voltage of a TMR current sensor under an input current sweep from -10A to +10A.

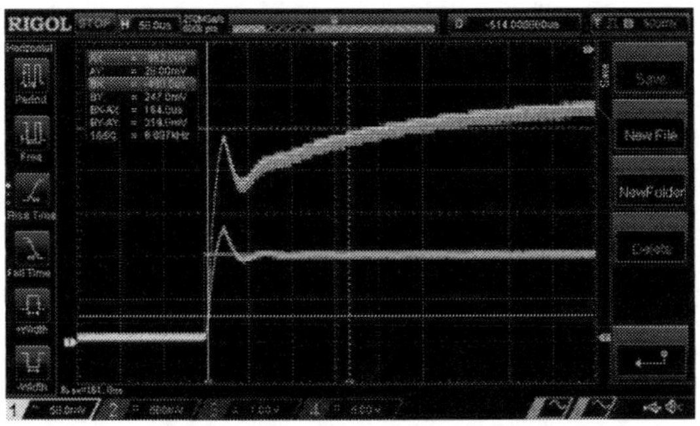

Figure 11. Output voltage of a TMR current sensor under an 7.5A input current step. CH1: TMR output voltage, CH2: Agilent TCP202 Hall current probe output response, CH4: Step control voltage.

5. ACKNOWLEDGMENTS

The authors would like to thank the Spanish Ministry of Economy and Competitiveness which has supported these research activities under projects ESP2015-68117-C2-1-R and ESP2015-68117-C2-2-R.

6. REFERENCES

1. Bausier, F., Massetti, S. & Tonicello, F. (2014). Silicon Carbide for space power applications, 10th ESPC, Noordwijkerhout, The Netherlands.

2. Slatter, R. & Goffin, B. (2014). Magnetoresistive current sensors for high accuracy, high bandwidth current measurement in spacecraft power electronics, 10th ESPC, Noordwijkerhout, The Netherlands.

3. Feng, X. (2007). SiC based Solid State Power Controller, University of Kentucky Doctoral Dissertations. Paper 543.

4. Feng, X. & Radun, A. V. (2008). SiC based Solid State Power Controller, IEEE APEC.

5. Rozman, G. I. & Swenson, J. C. (2013). Bus-tie SSPCs for DC power distribution system, US8,344,544 B2.

6. Guo, Y. B., Bhat, K. P., Aravamudhan, A., Hopkins D. C. & Hazelmeyer, D. R. (2011). High current and thermal transient design of a SiC SSPC for aircraft application, IEEE APEC.

7. Bhat, K. P., Guo, Y. B., Yang, Hazelmeyer, D. R. & Hopkins D. C. (2012). Results for an Al/AIN composite 350ºC SiC Solid-State Circuit Breaker module, IEEE APEC.

8. Urcioli, D. P., Veladis, V., Ha, H. C. & Lubormirsky, V. (2011). Demonstration of a 600-V, 60-A, bidirectional silicon carbide solid-state circuit breaker, IEEE APEC.

9. Sato, Y., Tanaka, Y., Fukui, A., Yamasaki, M., & Ohasi, H. (2014). SiC-SIT circuit breakers with controllable interruption voltage for 400-V DC distribution systems, IEEE Trans. Power Electron. 29(5), 2597-2605.

10. Shen, Z. J., Sabui, G., Miao, Z. & Shuai, Z. (2015) Wide Bandgap Solid State Circuit Breakers for DC power systems: device and circuit considerations, IEEE Trans. Electron Devices 62(2), 294-300.

11. Miao, Z., *et al* (2015) A self-powered ultra-fast DC solid state circuit breaker using a normally-on SiC JFET, IEEE APEC.

12. Mitchell, R. R., Ciccanti, A. D. & Bartlemay J. M. (1970). High voltage solid state switches for space applications, IEEE PESC.

13. Miao, Z., *et al* (2015). A self-powered bidirectional DC solid state circuit breaker using two normally-on SiC JFETs, IEEE ECCE.

14. -. (2014). CoolSiC 1200V Power Transistor IJW120R100T1 datasheet.

15. Bendel, J. & Li, X. (2015). Using "normally on" JFETs in power systems, Bodo's Power Systems.

16. Sánchez, J., Ramírez, D., Amaral, S., Cardoso, S., & Freitas, P. P. (2012). Electrical ammeter based on spin-valve sensor, Review of Scientific Instruments, 83, 105-113.

17. Ravelo, S. I., Ramírez, D., Sánchez, J., Cardoso, S., Ferreira, R. & Freitas, P. J. P. (2013). Fractional modelling of the AC large-signal frequency

DOI: 10.1051/e3sconf/20171612004

response in magnetoresistive current sensors, Sensors, 13(12), 17516-17533.

18. Ravelo, S. I., Ramírez, D., Sánchez, J., Cardoso, S., Ferreira, R. & Freitas, P. J. P. (2015). Total ionizing dose (TID) evaluation of magnetic tunnel junction (MTJ) current sensors, Sensors and Actuators A: Physical, 225, 119-127.

PROTOTYPING AND CHARACTERIZATION OF 1.2KV SIC SCHOTTKY DIODES FOR TWTA APPLICATION: THE CHALLENGE TO MEET THE USER SPECIFICATION

E. Maset [(1)], E.Sanchis-Kilders[(1)], S. Massetti [(2)], J. Montserrat[(3)], P. Godignon [(3)], J.Moreno [(4)], E. Cordero[(4)], J. Bevan[(5)]

[(1)] *University of Valencia, LEII, 46100 Burjassot (Spain), Email: Enrique.maset@uv.es*
[(2)] *European Space Agency (ESTEC), Kepleerlaan 1 PObox 299,2200AG Noordwijk, The Netherlands, Email: silvia.massetti@esa.int*
[(3)] *Centro Nacional de Microelectronica (CNM-CSIC), Campus UAB 08193 Barcelona (Spain), Email:philippe.godignon@cnm.es*

[(4)] *Alter Technology, Calle de la Majada, Tres Cantos, (Spain), Email: juan.moreno@altertechnology.com*

[(5)] *Micross 2-4 Hellesdon Park Road Drayton High Road Norwich NR6 5DR, Email: juan.bevan@micross.com*

1. INTRODUCTION

In Space applications the interests in SiC devices is growing due the benefits in terms of saving mass, increasing power, and for extreme environments (high temperature and high radiations levels).

Benefits based on the main features of SiC base material (high energy gap, high electric field breakdown in combination with reasonably high electron mobility and high thermal conductivity - 3 times higher than Si -) led to the following expected and in some cases already proven capabilities for power application: low on-state voltage (100 to 200 times lower on resistance x total gate charge with respect to Si), low leakage currents (order of magnitude lower than in Si devices), low recovery charge, fast turn-on and turn-off, high blocking voltage (> 2 KV), high power density [1-2]. Higher reliable operating junction temperature have been also demonstrated above 300°C [3-4].

The high thermal conductivity combined with the large band gap and high critical field provides SiC devices a great advantage. More specifically, the large bandgap and the breakdown field allow very thin and highly doped thicknesses in the stall region and therefore with very high conductivity. This implies that have the ability to switch high frequency power with less loss.

The EPC (Electronic Power Conditioning) for TWT (Travelling Wave Tube) is a converter that provides high voltage (up to about 8 kV) and average power (up to about 300W), and therefore is an application where SiC devices can be particularly attractive. An EPC feeds one or two traveling wave tubes, forming a high power transmitter. A telecommunications satellite can embark from a few tens to over 100 of these sets. Therefore any topological simplification and reduction of mass per set is essential. The EPC typically operate under a baseplate temperature about 65°C, however traveling wave tubes are exposed to baseplate temperatures around 95°C. Increased operating temperature for EPC, using SiC semiconductors, allow simplification of the satellite relative its thermal design, a reduction of the distance between the EPC and the TWT and therefore reduced size and mass, increased system reliability.

In the paper a summary of the technological issues encountered during the design of the SiC diode, in order to comply the optimized performances needed for the rectifier module in a next generation EPC for TWTA working at temperature higher than 65°C is reported. The static, switching and endurance electrical test characterization is specifically addressed.

The present study makes part of Work Plan of the Basic Technology Research Programme (TRP) funded by ESA falling in the category of Technology push for a new generation of space electronics units.

2. CRITICAL ANALYSYS

For the high voltage rectifier module, inside the EPC, the diodes should have the following specification to represent an improve respect to the silicon diodes use actually. Tab. 1 shows the SiC diodes requirements. Regarding the switching performances, the new diodes requires a near zero recovery current and near zero reverse recovery time, in comparative with the silicon ones.

Table 1.- Electrical SiC diodes requirements.

SiC Diodes parameter	Target specifications
$V_{breakdown}$	1200V goal 1500V
I_f max at T_{op} max (case) ≥150°C	2A
I_r max @ V_r =1300V	≤10uA@25°C ≤50uA@130°C
V_f @ 0.1A	≤0.95V @ 25°C ≤0.70V@100°C
I_{max} non repetitive surge	100A (t_{pulse}=10us)

SiC Diodes parameter	Target specifications
current	
I_{max} repetitive surge current	30A
C_{in} @400V, 1MHz	<30 pF
Qc @400V, 1MHz	≤20nC @ 25°C <30nC @ T_j max

The key parameter for the devices was the forward voltage drop at 0.1A and 100°C. However, we also have to consider a low leakage current at 1200V (<10uA). We also have to deal with a low input capacitance and a short switching turn-off time. Both these last requirements need to minimize the Schottky diode area, which do not benefit the forward voltage drop and the surge current capability.

On top of this a specific hermetic package compatible with a 6 Lead Leadless Carrier Chip (6L-LCC) package need to be considered to avoid the requalification of the potting process of the HV section of EPC. Then, a compromise has to be found in the design.

For these specifications, we can consider two types of devices, a pure Schottky or a Junction Barrier Schottky diode [5]. The cross section of both devices is shown in Fig. 1. The difference is in the P+ rings included in the active part of the JBS. The ratio P/N between the width of the P+ ring (L_P) and the distance between P+ ring (L_N) is a key parameter in the design of the JBS.

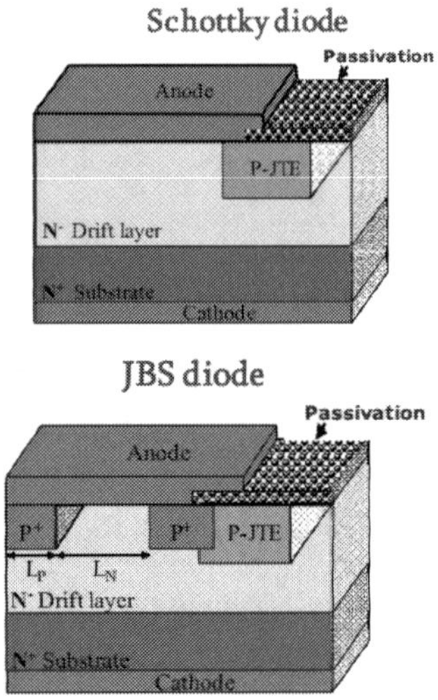

Figure 1: Schematic cross section of a Schottky diode and a JBS diode, including high voltage termination (JTE).

In the Tab. 2, we select the 3 main design/technology parameters, which will influence the trade-off between forward voltage drop, reverse leakage current and switching time.

We have designed JBS diodes with two Schottky barrier height and we have seen that with the high barrier configuration we do not fill the forward voltage specs, and that with the low barrier configuration, we do not fit the reverse current specifications. Ideally, we should then implement a Schottky contact with a barrier height between the two values. Unfortunately, this is not obvious as the usual metals used in Schottky contacts metallization in microelectronics have either lower barrier than (like Ti, W, Mo) or higher barrier (like Ni,, Al, Au). Even if we could find such metal, with non-usual Schottky metals, we have the risk to get a lot of stability and reliability problems. We have also tested different JBS Ln/Lp dimensions, showing some different distribution of the forward voltage drop versus the reverse leakage current.

Finally, compared with existing commercial technologies, we propose the following possible innovation and trade-off:

- JBS diodes for higher reliability and current peak capability.
- Use of W based Schottky metallization instead of Ti.
- Optimisation of the top metallization layer to increase current peak capability.
- Optimisation of the design of the Schottky metallization edge to withstand a maximum of transitory peak current
- Optimisation of the termination using the novel JTE + rings design to improve reliability
- Use of optimal starting material with last developments of EU supplier

Table 2.- Trade-off conditions

Design/technology parameter	Forward voltage	Reverse current	Switching time (C_{in})
Schottky barrier height increases (Schottky metal choice)	Increases	Decreases	Not sensitive
JBS instead of Schottky	Increases	Decreases	Increases (at high T°)
Diode area increases	Decreases	Increases	Increases

The form factor required to meet the existing system board device footprint was the main driver for the design of the diode package. Using this as the main design rule for the package meant that the design dimensions were constrained to be compatible as a drop in replacement for a D5-B board footprint, similar to a standard 6 Lead Leadless Carrier Chip (6L-LCC) package. In order to accommodate the new die design, the standard 6L-LCC package had to be modified with the lead positions kept constant relative to the device centre line as per figure 2

DOI: 10.1051/e3sconf/20171612005

below:

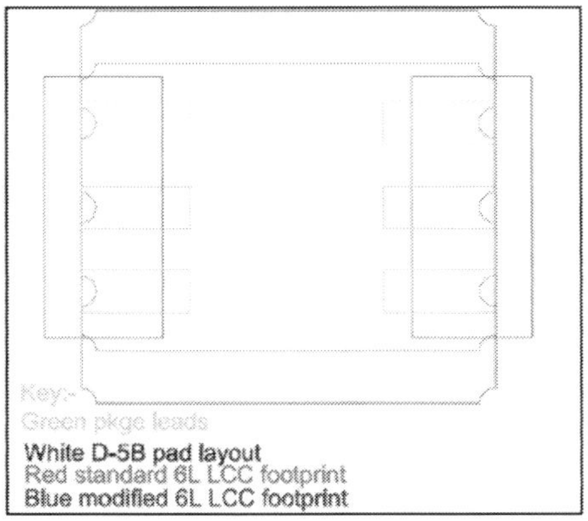

Figure 2: Drawing of the modified footprint compared to the standard D-5B and 6L LCC

A compromise in the package design between its thermal performance and form factor had to be made as the board footprint dictated the dimensions and lead positions for the new package. As a result, a bespoke thermally optimised packaged with a simulated thermal resistance in accordance with JEDEC standard JESD51-14 of under 2 °C / W was designed. Due to the constraints of the design, the compromise of the package lead positions meant that attaching a heat sink to enhance the thermal efficiency of the package when mounted onto the board is not possible. Further works on other designs that incorporate a built-in heat sink are currently being investigated.

3. STATIC ELECTRICAL CHARACTERIZATION

Steady-state electrical characterizations have been performed under a wide temperature range (-55°C, 25°C and 100°C). The evolutions of two parameters, which may vary with temperature, have been monitored: forward voltage and reverse current. Tab. 3 resumes the measures done for 130 devices.

Table 3.- Experimental results of SiC diodes

Temperature	V_f @ 0.1A	I_r @ V_r=1300V
-55°C	0.817 V	5.8 µA
25°C	0.691 V	41.4 µA
100°C	0.590 V	
130°C		1262.6 µA

These parameters have been also measured in two flight existing Silicon diodes (Si-1 and Si-2) an in one commercial SiC diode (SiC) for comparison reasons. Fig. 3 and 4 shown the comparison of these parameters and its evolution with the temperature.

Regarding the voltage forward drop (Vfwd), at 25°C, the best case is the SiC diode developed due to the optimisation process done. At 100°C the worst case is the commercial SiC diode and the best is a silicon diode Si-1 very close to CNM SiC diode developed. The diodes fabricated met the requirements for Vfwd at both room and high temperature.

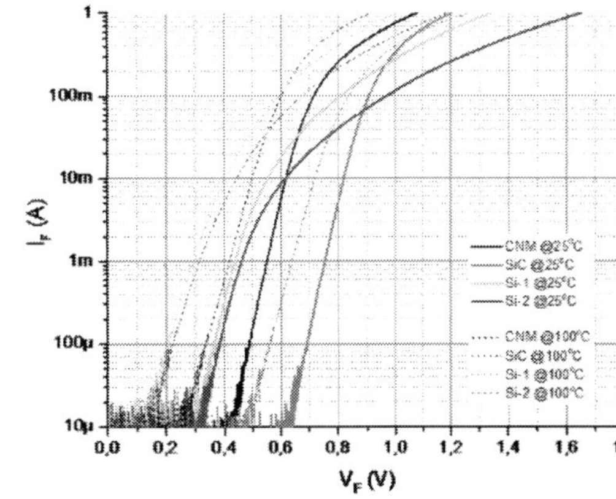

Figure 3.- Forward characteristic I-V

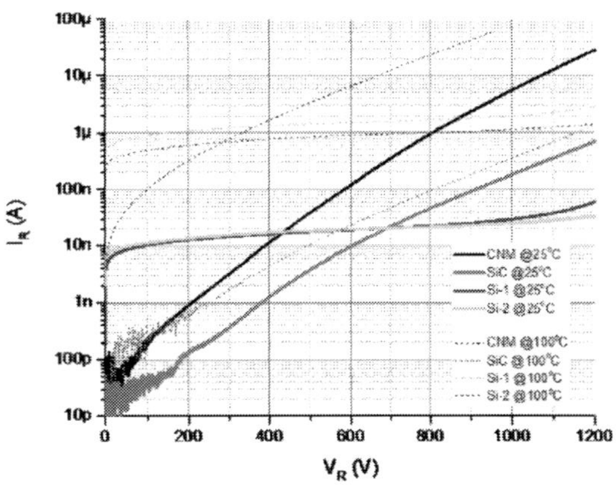

Figure 4.- Reverse characteristic I-V

The good behaviour in forward mode is not conserved to the reverse mode. The silicon diodes present a lower reverse current than the SiC diodes for higher blocking voltage. In Fig. 4 at 100°C up to 300V the SiC diode developed is the best compared with the flight silicon diodes. By sorting the best diodes we can reduce the medium reverse leakage current. In this lot, we did obtained some diodes with reverse leakage current 3 to 4 times lower than the medium values reported in figure 4, with negligible difference in the forward characteristics. As a matter of facts, the starting material has a important effect on the reverse current. In this study e.g. one of the wafers showed 3 times lower leakage currents than the

 DOI: 10.1051/e3sconf/20171612005

other two wafers.
Therefore a trade-off between the optimisation in the forward voltage drop and the reverse current characteristic will be chose.
As future development slight reduction of the diode area to better adjust the forward voltage drop to the 0.7V @0.1A @100ºC specified can be considered. This would also help gain on the input capacitance value to reach specs, together with a decrease of reverse current. On the other hand a smaller die could fit into packages with standard dimension and that offer a more effective heat sink performance.

Taking into account the application, high voltage module for EPC, one of the most important things is the risk for thermal runaway. The rectifiers are placed in a molded module together with the filtering capacitors, and it's mandatory maintain the temperature below a certain value. The thermal runaway is mainly due to the reverse recovery charge (Qrr) during the turn off process, without forgetting the contribution of the leakage current in lesser importance. Therefore we need to assure that the SiC diodes developed are temperature independent, to maintain the Qrr mainly constant.

4. SWITCHING CHARACTERIZATION

To extract switching characteristics of the SiC Diodes, we have designed a double pulse tester, specifically designed with a PCB layout and probes, taking attention to minimize the ringing effects on the measurement.

The following parameters have been measured during the OFF switching process for -55ºC, 25ºC and 175 ºC to characterize as a function of temperature. The turn-off conditions are V_{AK} = 900 V and I_F = 1.5 A:

- Maximum reverse recovery current (I_{RRM})
- Reverse recovery time (t_{rr})
- Reverse recovery charge (Q_{rr})
- Turn-off Energy (E_{off})

The next figures show box charts where has been analysed the temperature dependence of each parameter in function of the switching speed. The boxes contain all devices data for a specific di/dt and temperature. The width and height of the box provide information of key values from summary statistics. The box is determined by the 25th and 75th percentiles and the whiskers are determined by the 5th and 95th percentiles.

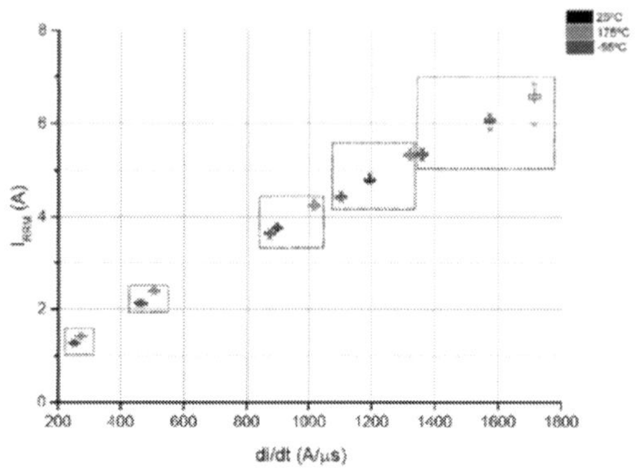

Figure 5.- I_{RRM} temperature and di/dt dependence.

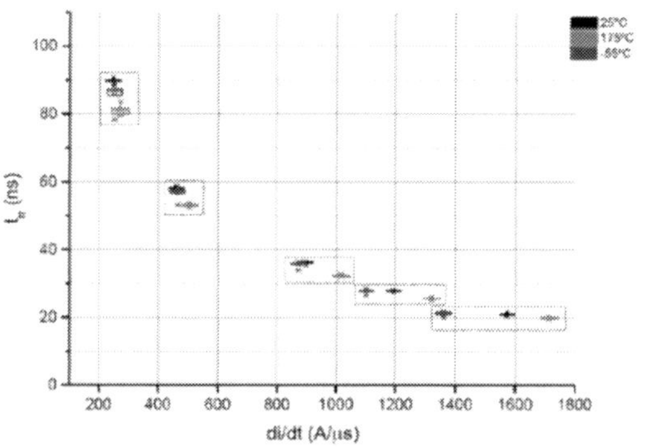

Figure 6.- t_{rr} temperature and di/dt dependence.

For a constant temperature, the reverse recovery time (t_{rr}) decreases with di/dt, representing good switching performances for high frequency applications, although I_{RRM} increases, the energy loss (E_{OFF}) associated with the turn-off is less when it is switched faster.

Regarding the switching behaviour respect to the temperature, when temperature increases the diode switching process is faster, increasing the di/dt, while the t_{rr} and the I_{RRM} are nearly temperature independent. Only a small variation with the temperature is shown to a lower di/dt, where I_{RRM} increase with the temperature and t_{rr} decrease slightly.

At room temperature we have done the switching characterization for the Silicon diodes and for the commercial SiC diode, to compare with the develop SiC diode. Fig. 7 and 8 shows the experimental results.

DOI: 10.1051/e3sconf/20171612005

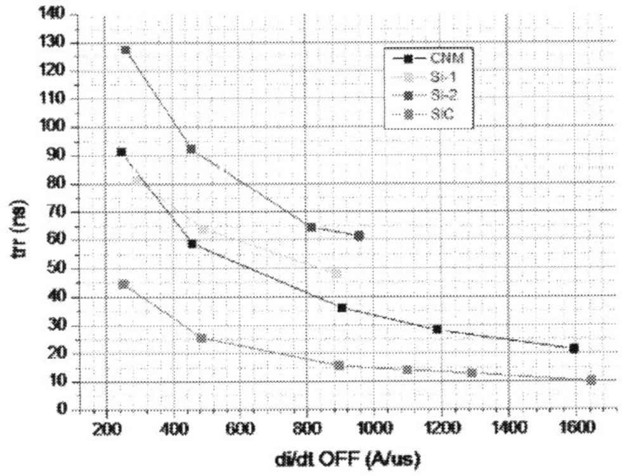

Figure 7. t_{rr} *diodes comparison (RT, V_{AK} = 900 V and I_F = 1.5 A)*

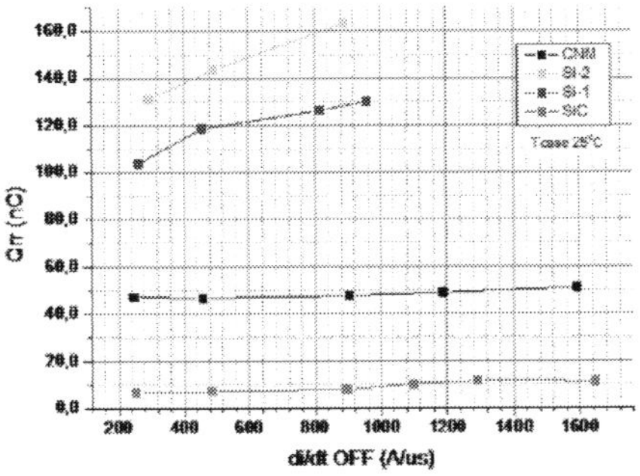

Figure 8.- Q_{RR} diodes comparison (RT, V_{AK} = 900 V and I_F = 1.5 A)

5. DESTRUCTIVE CHARACTERIZATION

Destructive tests were performed according the relevant test procedures. Indeed, specific screening methods and qualification methodologies have to be defined for SiC devices. In reference [3] we proposed and reported a specific characterisation procedure for SiC high temperature diodes. We have done a test campaign for the developed SiC diodes to obtain the absolute maximum currents ratings, repetitive and non-repetitive, and a power cycle test in order to determine the conditions that the diode is able to support more than 2000 power cycles without any detrimental effect.

5.1.- Absolute current ratings.

Four devices have been tested to obtain the maximum absolute current rates. Tab. 4 shows the experimental results.

Table 4.- Experimental results absolute current rate

	Maximum Current (A)	Test conditions
I_{FSM} (A)	415	t_{PULSE}: 10 µs - Tcase: 25 °C
I_{FRM} (A)	215	N° of pulses: 5 - Tcase: 150 °C t_{PULSE}: 50 µs, Duty 0.5%

Due to the higher die size used compared to commercial devices, we obtain a high surge capability. In order to optimise the overall parameters, if the application doesn't need a higher surge current, we can reduce that, reducing the die size and also a leakage currents reduction will be obtained.

5.2 Power Cycle Test.

The Power cycling test is an accelerated lifetime test, to verify the ability of the 1.2kV SiC Diodes under development, to sustain repeated thermo-electrical stresses. An active Power Cycling of constant timing (ton=const. and toff=const.) based in a load current conducted through the SiC diode, which is heated by its internal losses generating a temperature gradient, has been tested. When the desired junction temperature is reached, the load current is switched off. We placed the DUT on a heat sink, without forced cooling, and a natural conducted and radiation-cooling phase concludes the cycle. The heat sink temperature is relatively stable, while the junction of the DUT experiences sharp temperature fluctuations.

The Power Cycle test conditions are:

- $\Delta T_C \simeq$ 40°C, $I_{F\,PULSE}$ = 4 A
- t_{on} = 5 sec , t_{off} = 30 sec.
- $T_{(case\_min)}$= starts at 110°C.
- If $T_{(case)}$=200°C the test shall be stopped.

Considering the robustness, the results of power cycle test shown that the SiC diodes developed can withstand more than 2000 cycles if the case temperature swing is less than 40°C, with a small increase (less than 5%) in the forward voltage at 0.1A. In Fig. 9 the evolution of the forward voltage rising up to 8000 cycles in the worst case for a ΔT_C of 40°C is shown.

DOI: 10.1051/e3sconf/20171612005

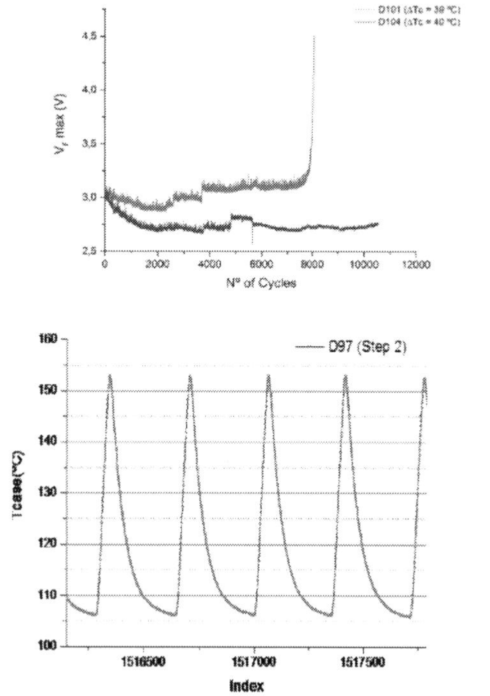

Figure 9.- $V_{F\,(max)}$ vs. n° of cycles (ΔT_C = 40 °C range)

5.3 Specific Screening requirements

Schottky Barrier Stabilization: Drift in the barrier height leading to a Vfwd increase of 1.5% and a decrease of the leakage current, was observed for devices biased with maximum nominal currents (2 amps) for some time. This drift stabilizes after a given number of hours of stress.
Diode full stabilization time depends on the current applied and seems to also depend on the sample. Screening of these diodes shall include a stabilization test in order to ensure a stable final diode. Fig10 shows almost full stabilization after 1 hour with Ifwd = 2 A at 25°C.

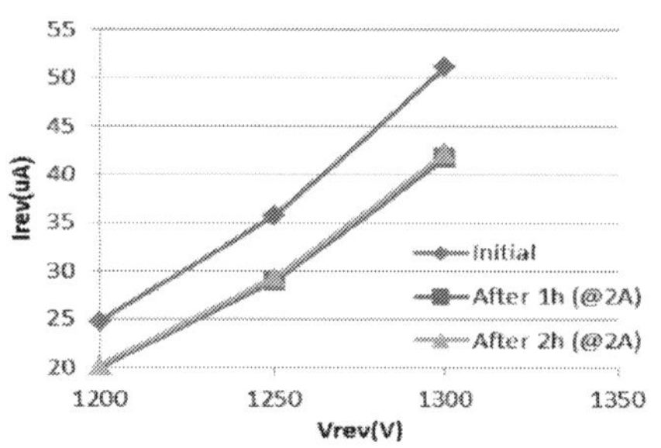

Figure 10.- Drift in Irev at HV after stabilization

Latent defects detection at low temperature: some defects that could lead to early failure of the component arise only while measuring the devices at low temperature like shown in fig 11. Also some failures are observed in a reverse measurement at room temperature after the low temperature characterization. Screening of these devices must always include this test sequence in order to detect and screen out this small population of devices.

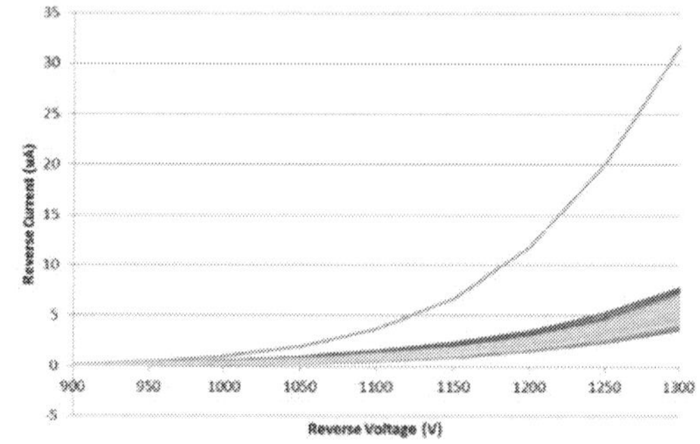

Figure 11.- Out of family behaviour of a diode (-55°C, 128 devices in the graph)

5.4 Endurance test results

Prior to endurance tests, power step stress tests were performed identifying the limits of the device. Temperature step stress arise drift in characteristics for storage temperatures of 200°C and higher because of package degradation. Power step stress test arise different, again package driven, limitations. The SMD solution fitting the final application limited the forward current for all the long lasting tests to 2 amps.

Endurance tests were performed obtaining good results in both reverse (fig. 12) and forward (fig. 13) mode after a thousand hours. In this last figure can be seen that stabilization of the barrier was not achieved in only two hours of sample preparation at 25°C but it is also clear that no degradation occurred since no changes were observed during the last 267 hours of test.

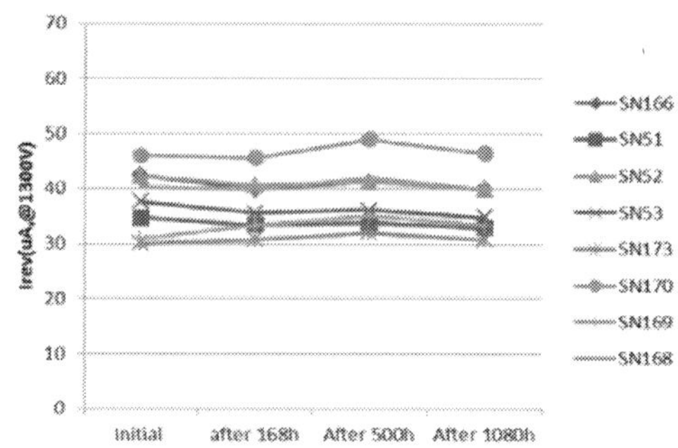

Figure 12.- HTRB (1080h, 100°C, Vrev=1020V) intermediate reverse characterization at 25°C

DOI: 10.1051/e3sconf/20171612005

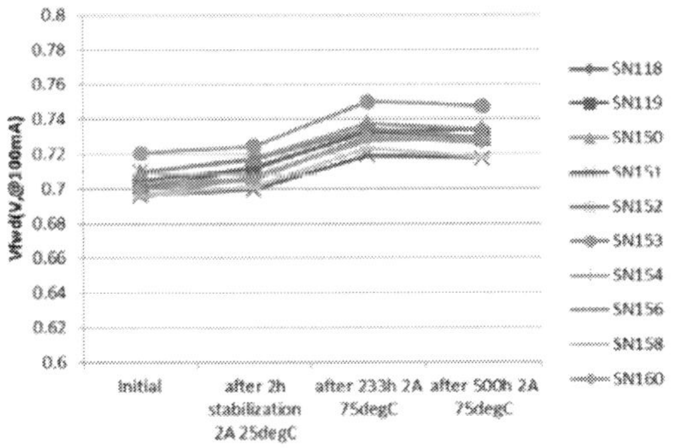

Figure 13.- Accelerated Elec. Endurance (Subset 2, 500h, 75°C, Ifwd=2A) intermediate reverse characterization at 25°C

5.5 Heavy ions test campaign

Based on the experience gathered on previous studies conducted by ESA, an heavy ions test campaign has been performed with the objective to characterise better the already known and shown in many paper and publication SEE sensitivity of SiC Schottky diodes. Fifteen samples were tested in different beam conditions; an analysis of flux dependence has been also performed as well as verification of different annealing methodology.

The damages induced by ions are cumulative and depends on LET of the ions and on the electrical field across the junction. A general very limited Safe Operating Area of Vr ≤ 250V has been confirmed at LET=32.4 MeV cm^2/mg.

A linear degradation of the reverse current (Irev), directly proportional to the fluence of ions heating the junction has been observed. It is not therefore a standard catastrophic failure induced by a single particle but more an accumulation of defects in the active area (metal-SiC interface or in the epi-layer).

The damages are permanent and are affecting only the reverse characteristic. The barrier height is not affected. See fig 14 were an example of the degradation in reverse with a LET=32.4MeV cm^2/mg has been reported.

We have observed a partial recovery of the Irev characteristic after 16 hours of annealing at 200°C. This partial annealing has been considered curing interfaces states induced by ion exposure. However crystal defects in the epilayer seems to play an important role as well and cannot be recovered by the performed high temperature annealing.

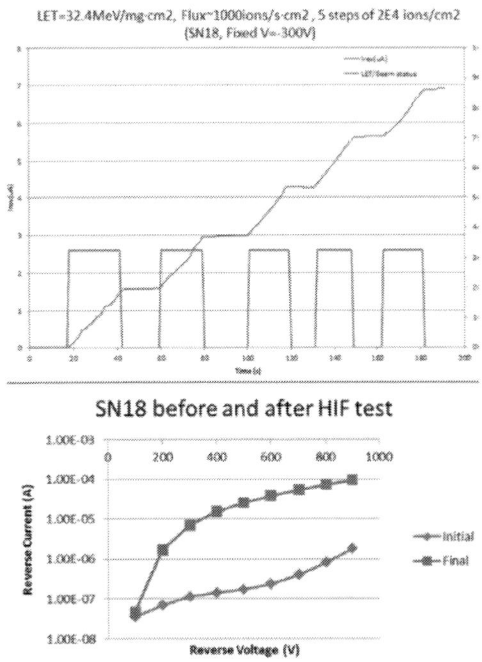

Figure 14.- degradation in situ of the Irev with a LET=32.4MeV cm^2/mg, comparison of Irev characteristic before and after the beam exposure

6. CONCLUSIONS

Different strengths are identified for SiC power devices, among them; high voltage operation seems the most promising in space applications. In the paper the development of an optimized design for a SiC diode able of improving the thermal and forward voltage drop performances of silicon diodes, used as high voltage diodes rectifiers, has been reported.

Samples of the diodes have been sent to potential user for validation in their test bench. They confirmed that the developed diodes meet and exceed the performances they were requiring for prototyping the next generation of EPC. In a second generation, improvement in the reverse current characteristics can be expected by some minor technology adjustments, which would not affect the rest of the characteristics.

Independently of the improvements of SiC semiconductors in the high voltage niche, another challenge is to enable major advances in Electric Propulsion (EP) in-space operations and transportation. Their very good performances in high switching frequency resulted in converter size reduction and improved dynamic response, which are key aspects for increasing system power density.

The Power Processing Unit (PPU) is a technically challenging and expensive part of each EP system (up to 50 % of the cost-indicative). Power key performance beyond state-of-the-art may indeed be more challenging on the PPU side than for the thruster. Also for telecommunication applications dual mode EP systems are required, which can switch between a high thrust and lower thrust mode for station keeping.

The high sensitivity to heavy ions has been confirmed to be the major drawback of SiC Schottky technology for space application. It has been confirmed that heavy ions induce a cumulative degradation of reverse current and no a real sudden catastrophic effect. The concept of derating down to a demonstrated safe operating area seems at the moment the more promising choice to mitigate the heavy ions sensitivity taking advantage of the high blocking capability of SiC.

7. REFERENCES

1. P. Roussel, "SiC Market and Industry Update", International SiC Power Electronics Applications Workshop, ISiCPEAW, Kista, Sweden, 2011.

2. D. Stephani, "Today's and Tomorrow's Industrial Utilization of Silicon Carbide Power Devices", Proc. of the 10th European Conference on Power Electronics and Applications, EPE'2003, paper no. 1199, 2003.

3. P. Godignon, X. Jordà, M. Vellvehí, X. Perpiñà, V. Banu, D. López, J. Barbero, P. Brosselard and S. Massetti, "SiC Schottky diodes for harsh environment space applications", IEEE Trans. Ind. Electron., vol 58, no. 7, pp. 2582-2590, 2011.

4. T. Funaki, J.C. Balda, J. Junghans, A.S. Kashyap, H.A. Mantooth, F. Barlow, T. Kimoto and T. Hikihara, "Power conversion with SiC devices at extremely high ambient temperatures", IEEE Trans. Power Elect., vol. 22, pp. 1321-1329, 2007.

5. R. Singh, D. C. Capell, A. R. Hefner, J. Lai, J. W. Palmour, "High-Power 4H-SiC JBS Rectifiers", IEEE Trans. Electron Dev., vol 49, no. 11, pp. 2054-2063, 2002.

6. Jedec Solid State Technology Association, "Transient Dual Interface Test Method for the Measurement of the Thermal Resistance Junction-to-Case of Semiconductor Devices with Heat Flow Through a Single Path", JESD51-14 Standard, November 2010

E3S Web of Conferences **16**, 12006 (2017) DOI: 10.1051/e3sconf/20171612006
ESPC 2016

LOW SIDE GAN FET DRIVER FOR SPACE APPLICATIONS

O. E. Mansilla [1], J. Broline [1], H. Satterfied [1], L. G. Pearce [1], E. J. Thomson [1]

[1] Intersil Corporation,1650 Robert J. Conlan BLVD, Palm Bay, Florida, USA, 32905, Email:omansill@intersil.com

ABSTRACT

This paper reports the results of preliminary single event effects (SEE) testing of the Intersil ISL70040SEH, a single low side driver specifically designed to drive enhancement mode power GaN FETs.

1. Introduction

Recently, wide band gap semiconductor devices such as GaN and SiC have been gaining interest in power management for space applications. These devices feature higher breakdown voltage, lower $R_{DS(ON)}$, and very low gate charge enabling power management systems to operate at higher switching frequencies while still achieving higher efficiency and a small solution footprint. There is an added benefit, especially from GaN devices, that make these attractive to the space market. Studies have shown that these devices are inherently radiation hard to total ionizing dose. However, GaN power transistors have a limited gate to source voltage range and with that comes the complexities of providing the right grate dive, a low inductive path and the correct timing. In this paper we report results of destructive and non-destructive SEE testing of the ISL70040SEH, a low side GaN FET driver designed for space applications, together with a brief discussion of the part's electrical specifications and how the device solves many of the problems associated with using GaN FETs in power management space applications.

2. The ISL70040SEH Low Side Driver

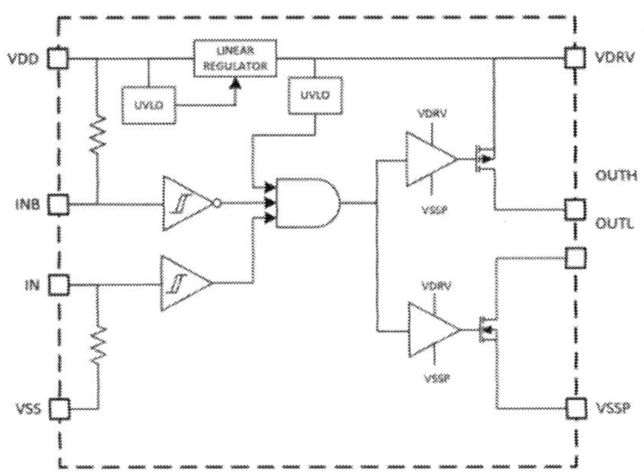

Figure 1. ISL70040SEH block diagram

The ISL70040SEH is a low side driver designed to drive enhancement mode Gallium Nitride (eGaN) FETs in isolated topologies, boost type configurations and synchronous rectification. The ISL70040SEH offers both inverting and non-inverting inputs to satisfy requirements for inverting and non-inverting gate drive in a single device type (refer to Fig. 1).

Inputs of the ISL70040SEH can withstand the input voltages up to 14.7V regardless of the VDD voltage. This allows inputs of the ISL70040SEH to be connected directly to most radiation hardened PWM controllers. The split outputs of the ISL70040SEH offer flexibility to adjust the turn-on and turn-off speed independently by adding additional impedance in either the turn-on path and/or the turn-off path. The nominal operating supply voltage range is 5V to 12V and includes an internal 4.5V series regulator with an under-voltage lockout (UVLO) circuit. When the voltage is below the UVLO threshold, the IN and INB are ignored and an internal pull down device keeps the GaN FET off.

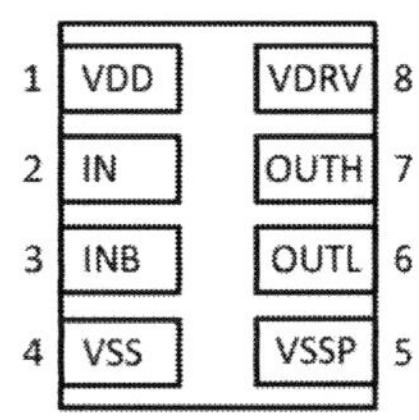

Figure 2. ISL700040SEH pin configuration.

The ISL70040SEH operates over the military temperature range from -55°C to +125°C and is offered in a 8 pin hermetically sealed 6mm x 6mm ceramic leadless chip carrier (CLCC) package. The package was designed to reduce parasitic inductance to mitigate the risk of excessive overshoot when driving the gate of the GaN FET. See Fig. 2 and Tab. 1 for pin configuration and pin description.

The ISL70040SEH is targeting a total dose (TID) rating of 100krad(Si) at a high dose rate (50-300rad(Si)/s) , 75krad(Si) at a low dose rate (< 0.01rad(Si)/s) and will be acceptance tested on a wafer by wafer basis. The ISL70040SEH is also SEE rated for both destructive and non-destructive effects to a linear energy transfer (LET) value of 86MeV·cm^2/mg. Single-event transient (SET) response has emerged as a key issue in power management parts, and this paper will concentrate on ISL70040SEH SEE results.

Table 1. ISL700040SEH pin description.

Pin Name	Description
VDD	This is the supply for the internal linear regulator of the ISL70040SEH.
IN	Non-inverting inputs pin with controls OUTH and OUTL outputs.
INB	Inverting inputs pin with controls OUTH and OUTL outputs.
VSS	Supply ground. Connect this pin to VSSP via the PCB ground plane.
VSSP	Power supply ground, Connect this pin to VSS via the PCB ground plane.
OUTL	The output low pin is the gate driver turn-off output.
OUTH	The output high pin is the gate driver turn-on output.
VDRV	This pin is the output of the internal linear regulator and the gate drive voltage.

3. Performance of the ISL70040SEH

GaN FET devices have special gate drive requirements that must be considered when choosing the correct driver. Typically enhancement mode GaN FETs have a lower rated maximum gate voltage compared to silicon MOSFETs. Therefore, the driver should have a well-regulated gate drive voltage that is high enough to fully turn-on the GaN FET but low enough that it does not damage the gate of the transistor. The driver should also have high source and sink current capability to fully utilize the high switching speed capability of GaN devices. Inversely, high current drive capability also means low pull-up and pull-down resistance output stages. A low resistance turn-off path is essential to mitigate Miller turn-on. Lastly, sometimes the fast turn-on capability can lead to loss of efficiency due to overshoots or increase in EMI. In these cases it is required to slow down the turn-on of the GaN device. This is typically this is achieved with a diode and a parallel resistor series with driver output stage. However, this is not recommended due the low threshold of the GaN device and diode forward voltage drop. A driver with a split output stage allows to tune the turn-on and the turn-off speed without under driving the gate.

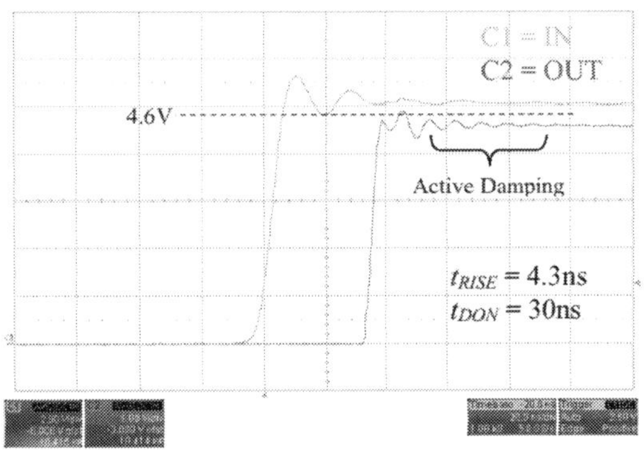

Figure 3. ISL70040SEH typical rising waveform

On the ISL70004SEH the gate drive voltage on the VDRV pin is regulated by the internal linear regulator to 4.5V. This tightly regulated gate voltage is high enough to fully turn-on enhancement mode GaN FETs but low enough to allow some overshoot, alleviating some of the complications of PCB design. The ISL70004SEH also has active damping output stages which changes the output impedance of the pull-up and pull-down paths to dampen overshoot and prevent the gate voltage from exceeding the maximum gate-source rating of enhancement mode GaN FETs.

Fig. 3 shows a typical rising waveform of the ISL70040SEH with a load capacitance of 1000pF and the VDD supply voltage at 12V. The input signal is on channel 1 (C1) and the driver output is on channel 2 (C2). Some of the key performance highlights are:

1. Gate drive voltage steadies to 4.5V with a peak of approximately 4.6V.
2. The rise time for 10% to 90% is 4.3ns.
3. The turn-on delay is 30ns. Some of the delay is due to the internal SET mitigation circuitry.
4. The dampen response of the output due to increased impedance of the output stage.

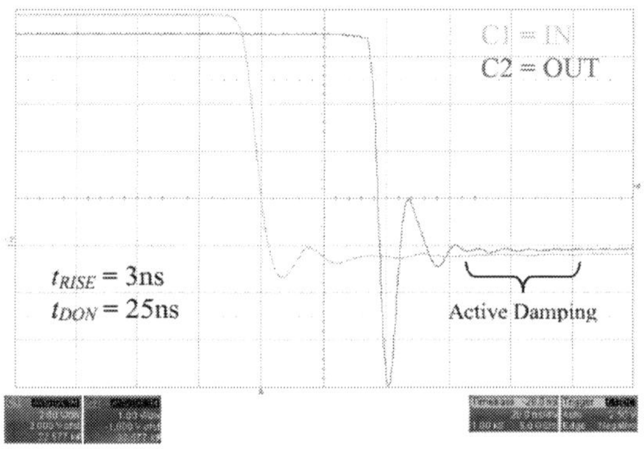

Figure 4. ISL70040SEH typical falling waveform

Fig. 4 shows a typical rising waveform of the ISL70040SEH with a load capacitance of 1000pF and the VDD supply voltage at 12V. The input signal is on channel 1 (C1) and the driver output is on channel 2 (C2). Some of the key performance highlights are:

1. The fall time for 90% to 10% is 3ns.
2. The turn-off delay is 25ns. Some of the delay is due to the internal SET mitigation circuitry.
3. The dampen response of the output due to increased impedance of the output stage.
4. Under shoot peaks at -3V, due to the strong pull-down capability. To minimize undershoot one may add an external resistor on the OUTL pin. The added impedance will also slow down the fall time.

4. SEE Testing of the ISL70040SEH

SEE testing of the ISL700040 took place at the end of August, 2016 at Texas A&M University (TAMU) cyclotron facility. Single event effects characterization was performed to determine its susceptibility to destructive effects including single-event latchup (SEL), single-event burnout (SEB) and single-event gate rupture (SEGR), and to non-destructive effects including single-event transient (SET).

4.1. Destructive SEE testing of the ISL70040SEH

During destructive SEE testing the ISL70040SEH was in an active configuration driving a 10nF capacitance load with a supply voltage on the VDD pin of 14.7V.

The device under test (DUT) was mounted in the beam line and irradiated with heavy ions of the appropriate species. The parts were assembled in 8 lead ceramic leadless packages with the metal lid removed for beam exposure. The beam was directed onto the exposed die and the beam flux, beam fluence, and select voltages and current in the device pins were measured.

The tests were controlled remotely from the control room. All input power was supplied from portable power supplies connected via cable to the DUT. The supply currents were monitored along with the device outputs. All currents were measured with digital ammeters, while all the output waveforms were monitored on a digital oscilloscope for ease of identifying the different types of SEE displayed by the part. Events were captured by triggering on changes in the output. Tab. 2 summarizes the test conditions during heavy ion radiation for the destructive test.

Table 2. ISL700040SEH destructive SEE test conditions

Specification	Condition
VDD (V)	14.7
T_{CASE} (°C)	125
Species & Angle	Gold (Au) ions at normal incidence
LET	86MeV•cm^2/mg
Flux (cm^{-2}s^{-1})	5.00 x 10^4
Fluence (cm^{-2})	1.00 x 10^7
IN(V) & Frequency	0-10V square wave at 500kHz
INB (V)	0
C_{OUT} (nF)	10

The schematic of the test board is given in Fig. 5.

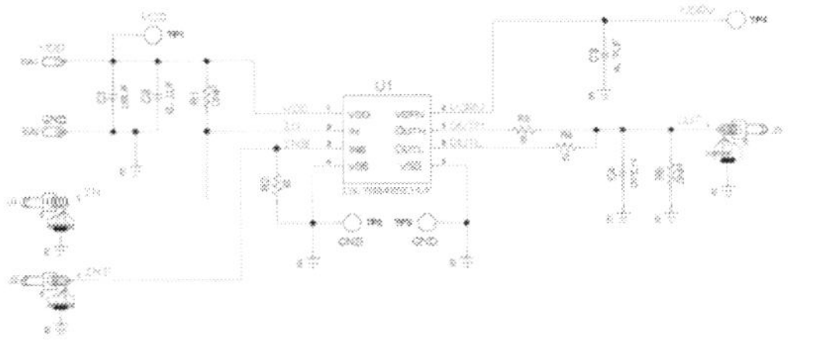

Figure 5. ISL70040SEH SEE test schematic

The supply current, input pin (IN) current and gate output voltage were monitored pre and post heavy ion irradiation. If any of those monitored specifications changed more than the specified criteria the device was considered a failed unit. Tab. 3 summarizes the specifications, conditions and the failure criteria used during destructive SEE testing.

Table 3. Destructive SEE testing failure criteria

Specification	Condition	Failure Criteria
I_{DD} Operating	F_{SW} = 500kHz	5% delta
I_{IN}	V_{IN} = 0V	50% delta
I_{IN}	V_{IN} = 13.2V	50% delta
Static I_{DD}	V_{IN} = 1.35V	5% delta
Static V_{OUT} LO	V_{IN} = 1.35V	50mV delta
Static I_{DD}	V_{IN} = 2.00V	5% delta
Static V_{OUT} HI	V_{IN} = 2.00V	50mV delta

Destructive testing results indicated no failures of the ISL70040SEH for a supply voltage of 14.7V at an LET of 86MeV•cm^2/mg. The final DUT was retested with a VDD supply of 15.1V. That DUT exhibited an increase in operating supply current larger than 5% and was considered a failure. Tab. 4 summarizes the results of destructive SEE testing.

4.2. SET testing of the ISL70040SEH

Single event transient tested utilized the same schematic as shown in Fig. 5. Both static and dynamic tests were conducted to characterize the SET behavior of the ISL70040SEH.

During the static test the driver was supply voltage was 4.5V and 13.2V, the min and max recommended operating range. The outputs were set in a high or low condition and a digital oscilloscope connected to the output was used to trigger on deviations greater than ±500mV. There was no capacitive load connected to output.

For dynamic testing the supply voltage was 4.5V and 13.2V. The input was driven with 0-5V square wave at 500kHz. A digital oscilloscope was connected to the output and use to trigger on a SET event that led to change of pulse width greater than ±50ns. There was no capacitive load connected to output.

Table 4. ISL700040SEH destructive SEE test summary

		IDD @ 10nF & IN = 500kHz, 0-xxV		IIN @ IN =0V		IIN @ IN =13.2V		IDD @ IN =1.35V		VOUT @ IN = 1.35V		IDD @ IN = 2.00V		VOUT @ IN = 2.00V		SEB
Device	Run	Pre (mA)	Post (mA)	Pre (nA)	Post (nA)	Pre (uA)	Post (uA)	Pre (mA)	Post (mA)	Pre (V)	Post (V)	Pre (mA)	Post (mA)	Pre (V)	Post (V)	
1	508	31.34	31.40	25	20	63	63	2.53	2.53	62u	61u	10.69	10.67	4.55	4.55	Pass
2	509	32.12	32.10	20	13	62	62	2.53	2.53	85u	82u	10.67	10.71	4.58	4.56	Pass
3	510	31.20	31.26	7	4	62	61	2.47	2.46	69u	69u	10.68	10.67	4.57	4.57	Pass
4	511	31.87	31.92	170	190	64	64	2.54	2.54	59u	64u	10.66	10.67	4.57	4.57	Pass
4	512	34.15	31.93	630	640	65	63	2.56	2.56	58u	59.00	10.68	10.71	4.57	4.58	Fail

Tab. 5 summarizes the test conditions for the static test in both the low and high states. Four devices were subjected to the test and no triggers were captured with oscilloscope. The ISL70040SEH is SET immune in the static condition. This is a very positive results as in a typical applications the driver will not unintentionally turn-on or turn-off the device and the output state solely relies on the input condition.

Table 5. Static SET test conditions

Test #	VDD	IN	INB	OUT	SET Trigger
1	4.5V	1.35V	0V	LO	0V ± 500mV
2	4.5V	2	0V	HI	4.5V ± 500mV
3	13.2V	1.35V	0V	LO	0V ± 500mV
4	13.2V	2	0V	HI	4.5V ± 500mV

Tab. 6 summarized the test condition for the dynamic test. Four devices were tested utilizing both conditions. There were no triggers captured on oscilloscope. This indicates the pulse width of the controller does not change more than 50ns. For test condition of 500kHz with 50% duty cycle, 50ns translates to 5% of the ON time of the pulse. If the switching frequency is increased to 1MHz for example with a 50% duty cycle a 50ns change means a 10% change of the desired pulse width.

Table 6. Dynamic SET test conditions

Test #	VDD	IN	INB	SET Trigger
1	4.5V	0-5V @ 500kHz	0V	PW ± 50ns
2	13.2V	0-5V @ 500kHz	0V	PW ± 50ns

5. Summary

The ISL70040SEH is radiation hardened low side GaN FET driver which contains all the features necessary to drive enhancement mode GaN FETs in space applications. The single event effect testing demonstrating no SEB or SEL with a supply voltage of upto 14.7V at an LET of 86MeV•cm^2/mg.

Single event characterization demonstrated that in a static condition the output does not change states with regards to the input condition. This means the device is SET immune and will not inadvertently turn-on or turn-off the GaN power transistor due to a heavy ion strike. Dynamic characterization results demonstrate that the output pulse width will not change more than 50ns. For low switching frequency applications 500kHz an below that is less than 5% change of a 50% duty cycle pulse width. As the switching frequency in increased the percentage is much larger and may cause and issue. Follow up SET characterization is planned for next year and the pulse width trigger may be reduced to 20ns or lower detect the minimum change in pulse width due to an ion strike.

6. References

1. A. Ionascut-Nedelcescu, C. Carlone, A. Houdayer, H. J. von Bardeleben, J.-L. Cantin, and S. Raymond, (2002), 'Radiation hardness of Gallium Nitride,' IEEE Trans. Nucl. Sci, vol.49, no.6, pp.2733–2738

2. S. Onoda, A. Hasuike, Y. Nabeshima, H. Sasaki, K. Yajima, S.-i. Sato, and T. Ohshima, (2013), 'Enhanced Charge Collection by Single Ion Strike in AlGaN/GaN HEMTs', IEEE TRANSACTIONS ON NUCLEAR SCIENCE, VOL. 60, NO. 6, pp 4446-4450.

3. A. Lidow, J. Strydom, M. de Rooij, and Y. Ma, (2012). GaN Transistors for Efficient Power Conversion

4. R.D.Harris, L.Z.Scheick, J.P.Hoffman, T. Thrivikraman, M. Jenabi,Y. Gim, and T. Miyahari, (2011), 'Radiation characterization of commercial Gan devices.' in Proc. Radiat. Effects Data Workshop, pp. 1–5.

5. A. Lidow and K. Smalley, "Radiation tolerant enhancement mode gallium nitride (eGaNTM) FET characteristics," (2012), Proc. GOMACTech, pp. 1–3.

ECSS-E-ST&HB-20-20C – A POWER INTERFACE STANDARD & HANDBOOK FOR PRODUCTS DEVELOPMENT

Ferdinando Tonicello[(1)], Bruno Mauret [(2)], Hans Jensen [(3)], Luciano Croci[(4)], Nicolas Deplus [(5)], Nicolas Neugnot[(6)], Peter Rumler [(7)], Tadeusz Kocman[(8)]

[(1)] *ESA-ESTEC, Keplerlaan 1, 2200AG Noordwijk, The Netherlands, Email: ferdinando.tonicello@esa.int*
[(2)] *Airbus Defence and Space SAS, Route de Verneuil, 78130 Les Mureaux, France, Email: bruno.mauret@airbus.com*
[(3)] *Terma A/S, Hovmarken 4, 8520 Lystrup, Denmark, Email: haj@terma.com*
[(4)] *Leonardo S.p.A., Viale Europa snc 20014 Nerviano (MI) Italy, Email: luciano.croci@leonardocompany.com*
[(5)] *Thales Alenia Space, rue Chapelle Beaussart 101, 6032 Mont-Sur-Marchienne, Belgium, Email: nicolas.deplus@thalesaleniaspace.com*
[(6)] *Airbus Defence and Space SAS, 31 rue des Cosmonautes, 31402 Toulouse, France, Email: nicolas.neugnot@airbus.com*
[(7)] *ESA-ESTEC, Keplerlaan 1, 2200AG Noordwijk, The Netherlands, Email: peter.rumler@esa.int*
[(8)] *OHB System AG, Universitätsallee 27-29, 28359 Bremen, Germany, Email: tadeusz.kocman@ohb.de*

ABSTRACT

With progression of space experience, the need of more stable and repeatable interfaces definition becomes clear. Especially for satellite platform needs, some electrical interfaces need a more reproducible application frame, to ensure at the same time the necessary and undisputable quality and reliability needed for institutional and non-institutional missions, and the containment of costs that nowadays is overall requested.
The prerequisite to be able to establish a wide and agreed basis for the definition of the power electrical interfaces on board of a spacecraft is the existence of a consolidated reference architecture.
Such consolidated reference architecture indeed exists for the power subsystem of ESA satellites: due to a number of practical reasons, the concept of using Latching Current Limiters (or LCLs) for the satellite on-board power distribution has consolidated over many decades. This allowed a global European agreement on the relevant interface definition, which culminated in the drafting, reviewing and publication of the ECSS-E-ST-20-20C standard (STD) and of the relevant handbook ECSS-E-HB-20-20A (HB).

1 STD AND HB GENESIS

The idea of preparing a guideline on the subject of power distribution by LCLs started long ago in TEC-EPM section in ESTEC. The idea was to collect and harmonise all the experience around this specific interface, in order to provide a reasoned baseline at the advantage of all the ESA programs.
A first technical note ([1]) was prepared in 2009, and just after the first drafting a workshop was called in ESTEC for exchanging the initially consolidated ideas with industry.
The workshop took place in May 2010, it saw interested and pro-active participation from power electronics European community.
Notably, the requirement set was established after the consolidation of the initial technical note: in fact the technical note contained the description of the LCL power interface, the explanation of the relevant issues, both from source and load side, and the requirements were derived accordingly.
This process allowed understanding some aspects that at that moment were not yet consolidated, and triggered specific research actions on unresolved technical issues (for example, relevant to stability of the LCL source-load interface [2]).

In 2014, an ECSS New Work Item Proposal was approved for the drafting of a standard and of a relevant handbook, with the main objective to give a rationale of the standard requirements.
A dedicated working group (WG) was set-up accordingly, with representatives selected from Satellite Primes, power distribution equipment manufacturers and ESA. The authors of the present paper are the members of the ECSS WG.

The draft versions of the STD and the HB were ready in December 2014, the public review took place mid-2015, and the amended version of both documents have been provided to the ECSS secretariat in November of the same year. The final ECSS publication took place in April 2016.

2 INTRODUCTION

Following the original development chronology, the HB is presented first, after a brief recap of what the power distribution by LCLs is.
Then the STD is explained, especially in relation to its structure, to the relevant key assumptions and to the main (in a certain sense, historical) agreements achieved for what regards the LCL classes.
Some of the most critical power interface requirements are then explained and discussed.

2.1 The Power Distribution by LCLs

The power distribution by Latching Current Limiters, or LCLs, has been widely used in almost all European satellites for some decades as an effective way to achieve a very controlled and reliable load connection and disconnection from the satellite main bus, including power management in case of overload and load short circuit failures.

Additionally, power distribution by LCLs minimises inrush current events due to load filters charging, and for this reason effectively allows the reduction of the loads filters themselves.

A generic architecture for a Latching Current Limiter, or LCL, is shown in Figure 1.

Note that the diagram in Figure 1 is given only as a reference, without losing generality, and some of the features thereby reported can be actually realised differently.

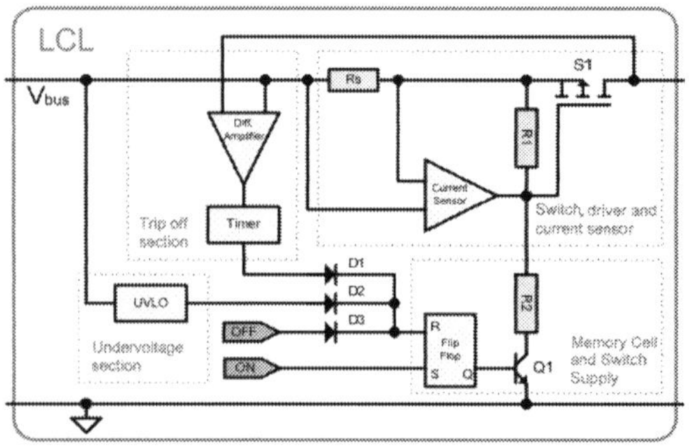

Figure 1, LCL generic block diagram

The Latching Current Limiter, or LCL, is a switch-able, latching, resettable over-current/overload protection placed between a power source and the relevant load.

The LCL can be commanded ON and OFF and a relevant memory cell normally latches its status.

Typically, an LCL presents a minimum residual resistance between power input and power output during nominal operation (i.e. when the switch is commanded closed).

In case of an overload, e.g. when the load current request exceeds a prefixed threshold, the LCL enters current limitation and a time counter is activated.

If the overload condition persists for a given time duration (called trip-OFF time), the time counter commands the LCL OFF.

Normally there should be an external command activation to reset the LCL into its original ON state.

The basic elements of an LCL are the following:

- the section containing **the switch, the driver and the current sensor,**
- the section relevant to the **trip-OFF** timer,
- the section relevant to the **memory cell and switch supply** section,
- the **under-voltage protection** (UVP) section,
- the **auxiliary supply** section (not shown in Figure 1), and
- the **telemetry** section.

For a detailed description of each basic element, refer to chapter 5.2 of the HB [3].

Typical timing diagrams of the reaction of an LCL to an overload are shown in Figure 2 and in Figure 3.

The reaction shown in Figure 3 is normally associated with a (very) abrupt overload and/or with some saturation effects in the current sensor/driver section of the LCL itself.

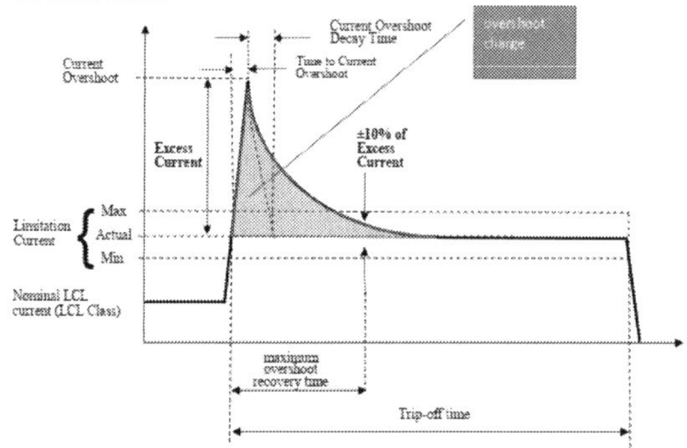

Figure 2, LCL overload timing diagram (case 1)

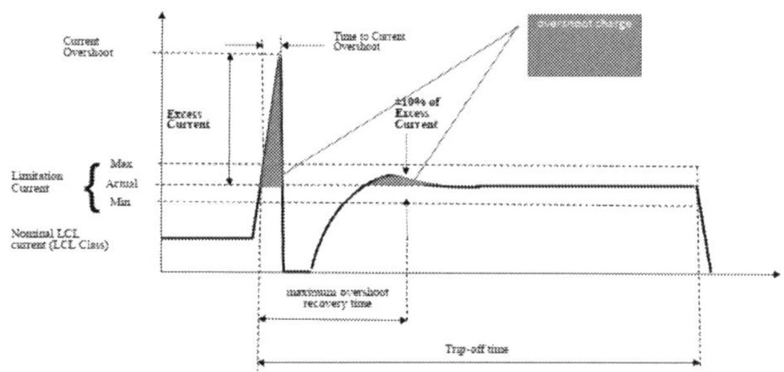

Figure 3, LCL overload timing diagram (case 2)

The LCL behaviour in overload conditions is explained in detail in chapter 5.1.

There are different variants of LCLs: apart the generic LCL described in Figure 1, both the STD and the HB mention the Retriggerable LCL, or RLCL, and the Heater LCL, or HLCL.

The RLCL is an LCL including additional features.

It is basically an LCL not provided with an OFF command, which is set in any case in ON condition during start up, and performing an automatic start up, repeated switch ON sequence after an overload occurred, as long as the overload is present.

In case the overload is removed, the RLCL automatically ends up in ON conditions, e.g. delivering power to the load.

The RLCL is normally used for supplying essential satellite loads, e.g. the ones that are essential for mission success (e.g. receivers and decoders). An example of

RLCL timing diagram is given in Figure 5.

The HLCL is an LCL that is dedicated to the supply and the protection of a group of heaters, as shown in Figure 4.

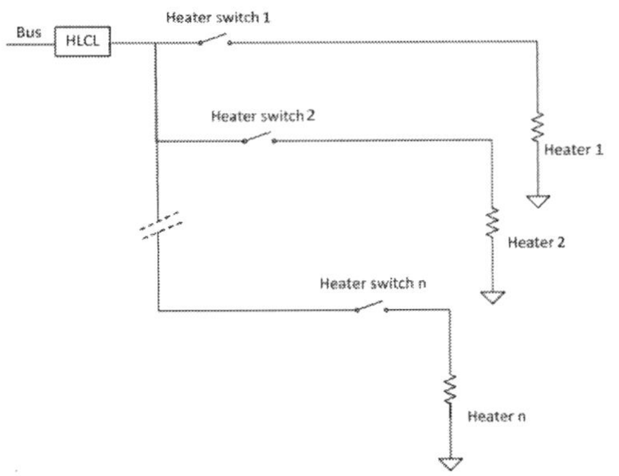

Figure 4, HLCL application

The HLCL has the same functionality of a generic LCL, but its performances are relaxed because the nature of the load is resistive. That is why in the STD the HLCL classes differ from the LCL ones, and not all the requirements valid to the LCLs are also applicable to the HLCLs.

3 THE HANDBOOK

The HB [3] is derived directly from the initial ESA technical note [1] mentioned previously.

The scope of the HB is:

- to explain the principles of operation of power distribution based on LCLs;
- to identify important issues related to LCLs;
- to give some explanations of the requirements set up in the STD for power distribution based on LCLs, for both source and load sides. Key aspects treated in the HB are the interactions between LCL and load (at start up, overload, switch-off conditions, both in time and in frequency domain, considering normal operational conditions but also failure cases).

The HB complements the STD, and it is directed at the same time to power system engineers, who are specifying and procuring units containing LCLs for power distribution and protection, and to power electronics design engineers, who are in charge of designing and verifying power distribution by LCLs.

For the system engineers, this document explains the detailed issues at circuit level and the impacts of the requirements for the design of LCLs.

For design engineers, this document gives insight and understanding on the rationales of the requirements on their designs.

It is important to notice that the best understanding of the topic of Power Distribution based by LCLs is achieved by the contextual reading of both the HB and the STD: the STD requirements are recalled in the HB with the same numbering and titles, so that a document search can quickly bring the reader to the required explanations.

4 THE STANDARD

4.1 Structure

The STD [4] has the typical structure of an ECSS standard (Introduction, Scope, Normative References, Terms, Definitions and Abbreviated Terms, Nomenclature, Principles an Requirements), but it provides in Annex A (Requirements Mapping) an informative table recalling all the requirements, indicating the feature or sub-feature the requirements refer to, the relevant conditions of application – nominal and/or fault case, LCL, RLCL and/or HLCL -, the applicability level (system, subsystem or equipment), and the verification method.

Having a global synoptic has been considered essential for a better understanding of all the LCL distribution specification and its subtleties.

An example is given in
Figure 6.

For verification aspects, apart the usual methods (Inspection – I -, Analysis – A -, Review of Design – RoD -, and Test – T -), a new option was introduced in line with the intention to produce a STD facilitating the production of recurrent solutions: it is the test verification at design qualification level (T*).

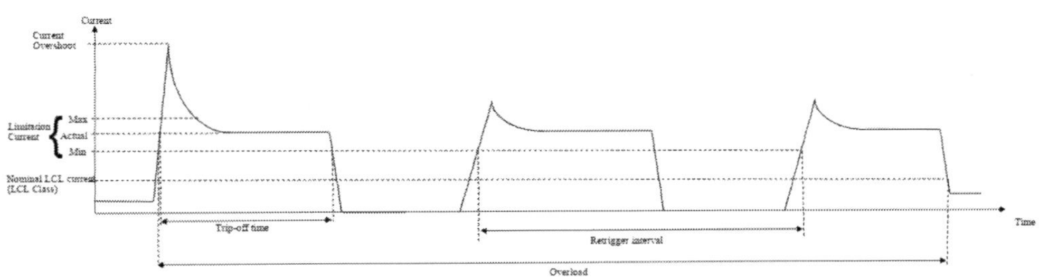

Figure 5, RLCL timing diagram

 DOI: 10.1051/e3sconf/20171613001

Reference	Text of the requirement	Feature	Sub-feature	Conditions	Applicability	Applicability level	Verification
5.2.18.1.1a	The RLCL state shall automatically be recovered to ON conditions after a spurious switch OFF.	Noise immunity	RLCL spurious switch OFF	Nominal	RLCL	Equipment	RoD, A, T*
5.2.18.2.1a	Spurious disable of RLCL retriggering memory cell and of RLCL ON/OFF status memory cell shall not result in the loss of the relevant load.		RLCL spurious effects	Nominal	RLCL	Equipment	RoD, A, T*

Figure 6, Example of requirements mapping (from STD, Annex A)

The test verification at design qualification level (T*) is intended to be performed on an electrical representative version of the hardware, on a set up not necessarily equal to the final flight one, to be established for the LCL distribution product line by the relevant manufacturer.

4.2 Content and coverage

After a chapter identifying the reference power bus specifications within the perimeter of which the LCL requirements are applicable, the LCL requirements are presented.

They have been grouped in the following categories:

- Functional, source (STD section 5.2)
- Functional, load (STD section 5.3)
- Performance, source (STD section 5.4)
- Performance, load (STD section 5.5)

A particular attention has been paid to identify requirements of the LCL source-load interface in order to ensure an optimum matching and anticipate or resolve the need of a late validation by test (specifically for behaviour at switch ON, when the LCL is used to charge the load input filter, and for stability purposes – see later discussion in section 5.2 -).

The target applications covered by the STD are all missions traditionally provided with power distribution and protection by LCLs/RLCLs (science, earth observation, navigation), with exclusion of applications for which the power distribution and protection is provided by fuses (e.g. most of the geostationary telecom satellites).

The STD applies to power distribution by LCLs/RLCLs for power systems, and in general for satellites, required to be Single Point Failure Free.

The STD applies exclusively to the main bus power distribution by LCLs/RLCLs to external satellite loads. Internal power system protections by LCLs/RLCLs are not covered.

Note additionally that paralleling of LCLs to increase power supply line reliability is not covered by the present standard, since this choice does not appreciably change the reliability of the overall function (i.e. LCL plus load).

In fact, a typical reliability figure of the LCL (limited to the loss of its switch ON capability) is 20 FIT or less.

If the load to be connected to the LCL line has a substantially higher failure rate than this, it is not necessary to duplicate the LCL to supply that load.

4.3 Reference Power Bus Specifications and other assumptions

To ensure the development of recurrent power distribution LCL/RLCLs for a number of applications, it is essential to define the envelope of the applicable power bus specifications (see table 1).

The relevant assumptions are derived from state of the art satellite applications, for nominal or abnormal cases.

Table 1 - Reference Power Bus Specifications

Power Bus type :		REGULATED BUS [V]		UNREGULATED BUS [V]	
		28V	50V	28V	50V
Nominal DC Bus Voltage Range at regulation point	Min	28 -1%	50 -1%	22	32
	Max	28 +1%	50 +1%	38	52
Nominal DC Bus Voltage Range at load side	Min	28 -5%	50 -5%	22	38
	Max	28 +1%	50 +1%	38	52
Abnormal DC Bus voltage range	Min	0	0	0	0
	Max (fault tolerance)	N/A	N/A	40	55
	Max (fault emission)			38	52
Nominal Bus ripple voltage	Max	Acc. to ECSS-E-ST-20		Up to ± 500 mVpp in the range of 30 Hz to 50 MHz.	
Nominal Bus voltage transients	Max	Acc. to ECSS-E-ST-20		±1,4V for load steps of 50%, with dI/dt=1A/µs	±2,5V for load steps of 50%, with dI/dt=1A/µs
Abnormal Bus voltage transients	Max	0 to 34 max	0 to 60 max	Within Power Bus abnormal DC limits	

Additional assumptions that have been made refer to the maximum qualification temperature of the unit hosting the power distribution LCLs/RLCLs/HLCLs (70 °C) and to expected main bus voltage time derivative (bus application or removal), considered to be variable between 0 to 0.1 V/ µs.

4.4 Classes

One of the major standardisation outcomes is the definition of LCL classes (see table 2 for LCLs, table 3 for RLCLs and table 4 for HLCLs).

The present LCL/RLCL/HLCL classes definition have been elaborated to ensure an adequate capability to charge load input filters for most of practical applications, to enhance the chances to use single MOSFETs for LCL of lower current classes, and under environmental/application conditions that should be normally respected in all LCL practical design cases.

The LCL classes have been defined with the following purposes:

- to ensure a sufficient number of classes for the users and prime to optimise the class current and associated harness sizing
- to be compatible with the state of the art or with practical design for having only one MOSFET for a given class. This is achievable for low current classes to a given extent
- to ensure for the prime and the user the largest minimum trip-off time to cope with the maximum extent of input filter designs
- to get a consistent input filter energy and sizing when changing the bus voltage level
- to limit and standardise as much as possible the possible numerical values in the standard (e.g. current class, trip-off times).

In particular the following assumptions were made for LCL classes definition:

- Max MOSFET junction temperature T_j in permanent short circuit (transient case) ≤ rated temperature
- Max MOSFET case temperature before limitation ≤ 85°C
- The MOSFET is supposed to be at LCL class current before a hard short circuit is applied
- Use the MOSFET manufacturer Safe Operating Area (SOA) to derive the maximum allowable trip off time especially for trip off time > 10 ms.

The maximum load capacitance per class is also provided, according to the minimum limitation current and trip-off time, and maximum DC bus voltage value (both for regulated and unregulated bus).

An additional factor of 70% is applied, leaving a 10% margin to comply with the maximum allowed input filter charge time of 80% of LCL/RLCL class minimum trip off time.

The 10% margin is the allowance for covering a specific effect affecting LCL performance, the so-called dragging effect, which is explained in Annex D of the HB.

5 CRITICAL REQUIREMENTS

Some critical requirements (indicated by the following subsections) are reported and explained hereafter. In any case, all these requirements are widely covered and explained in the HB.

5.1 LCL dynamic behaviour in overload

In case of a (very) abrupt overload, the LCL presents a transient behaviour as shown in Figure 2 and Figure 3. The so called current limitation response time is a critical performance for an LCL: it has to be fast enough in order to prevent that a dangerous stress level is reached before the LCL enters current limitation.

In order to assess properly this issue, the STD defines:

- the current overshoot -50Amps - : the maximum peak current that can be reached (see req. 5.4.1.1.1a)
- the time to current overshoot – 5µs - : the maximum time from actual current limitation to current overshoot peak (see req. 5.4.1.1.1c)
- the current overshoot decay time: the maximum time constant delay from current overshoot peak to actual current limitation (assumption of exponential law for the decay time) and finally,
- the current overshoot recovery time – 300µs- , the maximum time needed for the current to reduce from its maximum value down to +10% of excess current above actual current limitation (see req. 5.4.1.1.1d and refer to Figure 2 and Figure 3).

These characteristics have to be verified especially in the most critical overload case, when a short-circuit is applied at the output of the LCL (req. 5.4.1.1.1b).

Another reason to properly control the current limitation response time is to respect a maximum energy limit during abrupt overload; this allows ensuring a maximum bus voltage transient (within 5%) due to the LCL overload. The STD introduces therefore the maximum overshoot charge – 1mC - : the charge requested at LCL input for current in excess of the actual limitation current (see req. 5.4.1.1.1e).

One could think that increasing the bandwidth of the current limitation loop is the solution to improve the LCL current limitation time response but this could be detrimental to the LCL stability; a good compromise has to be found between speed of reaction and stability.

The requirements in the STD are applicable also if the LCL presents a characteristic like the one shown in Figure 3 (compared to Figure 2) due to non-linear saturation effects in the control loop.

5.2 Stability of LCL source – load interface

One of the critical areas to cover in the STD is indeed related to stability. The main problem with is that the LCL stability depends on the load, but the load might have a wide variability, being not necessarily known to the system integrator when a generic satellite power

 DOI: 10.1051/e3sconf/20171613001

distribution unit is procured, or to the LCL manufacturer when the design of the LCL needs to be consolidated.
For small-signal stability, the idea was then to identify, and develop, a non conventional approach where the check of the stability is performed in two steps, resorting to a Norton equivalent circuit for the LCL, and addressing separately the stability of the LCL on a voltage sink (e.g. on a zero impedance load), followed by the LCL output impedance - load input impedance matching check, according to [2].
The verification of the LCL control loop stability on a voltage sink is performed for DC voltage of 4 ±1V being this condition a typical worst-case one for the overall MOSFET contribution to the loop stability (in the current MOSFET technology the worst –largest – parasitic capacitance range is expected at low drain-source voltages).
The typical gain and phase margins respectively of 10dB and be 50° are required (req. 5.4.6.1.1a and b).
The condition for the stability on the source and load impedance makes use of the revised Bode approach or other criteria as shown in [2], requires a phase margin of at least 30° for any point in frequency where the source and load impedance magnitudes are equal, and a gain margin of at least 5dB for any point in frequency where the source and load impedance phase difference is equal to -180°±n*360° (see req. 5.5.3.1.1a and 5.5.3.2.1a).
It is clear that the LCL output impedance shall be characterised in current limitation conditions (and for a LCL DC input - output voltage drop of 4±1 V, for the same reasons explained above).
The load impedance instead shall be evaluated considering both nominal load conditions requiring current limitation (e.g. input filter charging) and failure conditions (for example when the load fails in short circuit and the LCL sees an inductive load).
Some additional stability checks are also required in time domain, for a sudden overload with inductive nature (being this the typical worst case stability condition – see req. 5.4.6.3.1a and b) and for start up transients in current limitation mode (for any specified inductive or capacitive load – see req. 5.4.6.4.1a and b).
For time domain transients it is required that the period of the observed oscillation is greater or equal to the envelope decay time (see Figure 7), which is equivalent to say that the number of visible "oscillations" periods in the relevant time domain transient is not higher than 3-4 to be sure to have the requested stability margins.

Much more details and explanations on the LCL stability are provided in the HB chapter 5.7.2.5, where additional hints are also provided for correct requirements verification.

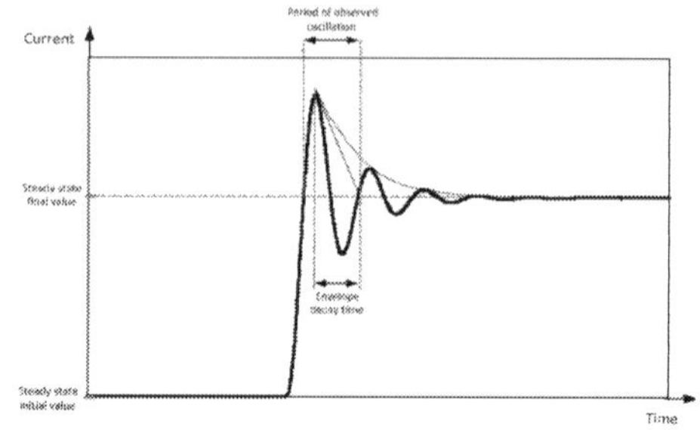

Figure 7, Stability and time domain transients

5.3 Repetitive overload

Especially in case of complex loads, it is virtually impossible to exclude that specific failure modes at load level do not result into a cyclic entry to and exit from LCL current limitation.
This is particularly true when the load converter is provided with local under-voltage protection. If a failure occurs on the secondary side of the converter, causing a local overload, the LCL enters current limitation; the input voltage to the converter decreases and triggers the local under-voltage protection. As a result, the overload disappears, and therefore the input voltage to the converter increases, releasing the under-voltage and enabling the converter operation again. This cyclic entry and exit from current limitation scenario is called hiccup mode.
The main issue of the LCL in hiccup mode is that the internal LCL timer can be periodically reset when the LCL exits current limitation, and for particular hiccup frequencies and duty cycles the timer could never command the LCL OFF.
In this case, the junction temperature of the switch can easily achieve dangerous levels over rating and there is a chance for the switch to fail in short circuit, with catastrophic consequences, possibly main bus short and loss of mission.
Therefore, the standard is requiring that RLCL and LCL ensure a reliable reaction under repetitive overload, i.e. either to switch-off or to maintain the switch junction temperature within derating/rating conditions. This can be resolved rather efficiently by implementing the LCL timer in the trip-off section with different time constants for counting up (during overload duration, and switch current limiting mode) and for counting down (when the switch goes back to ohmic conditions); as an illustration, a ratio of 30 between the count down and count up time constant is an indicative design choice verified on typical LCL design circuit under square wave overload profile and different duty cycles.
The design suitability against repetitive overload requirement is expected to be verified by analysis, in the worst case conditions and for different overload

repetition frequencies and duty cycles, the most critical conditions being verified by test at design qualification level.

5.4 Stress conditions

The overall principle behind a number of requirements addressing allowable stress limits (both for nominal LCL/RLCL/HLCL operation and for overload one) is the following, in accordance with the applicable ECSS derating standard ECSS-Q-ST-30-11C.

For steady state or unlimited repetition of stress conditions, the LCL/HLCL/RLCL components shall respect the applicable voltage, current, power and temperature derating limits.

This is the case for example of the junction temperature of the RLCL switch in overload conditions: there is the possibility to disable the periodical restart of the current limitation, but since this operation is only allowed by operator command, the thermal stress might remain for an uncontrolled duration.

For single transients, or events of limited duration, the LCL/HLCL/RLCL components shall respect the applicable voltage, current, power and temperature rating limits.

This is the case for LCL overloads: the temperature of the LCL switch shall respect the applicable rated limit, in worst-case conditions (including max overload, e.g. short circuit at the output for all expected trip-off time duration).

Note that the stress conditions applicable to the load input filter charge stem directly from the example made above: the temperature reached by the LCL switch during input filter charging will be surely below rated limits since

- the filter has to be fully charged within 80% max of the minimum trip-off time, with the minimum LCL class limitation current (req. 5.4.2.3.1a);
- the output LC voltage is not zero, but increases as long as the filter charging takes place.

5.5 RLCL retrigger disable

The RLCLs are supposed to continuously provide power to the satellite essential loads after start-up. In case of a load malfunction with continuous overload, RLCL would enter current limitation mode for the given trip-off time duration, switch off and attempt to re-start after a given retrigger interval, thereby creating "hiccup" condition on the power bus.

Therefore, it is practical to stop this repeated switch-on sequence after an overload occurred. In order to achieve this, the possibility of disabling the RLCLs retrigger function has to be foreseen.

The disable functionality is initially disabled by default.

Due to potential criticality of a spurious or unwanted disabling of the RLCLs retrigger function, the following measures are required:

- No on-board automatism can disable the retrigger function i.e. this command can only be received as a (hazardous) ground command (req. 5.2.6.4.1a);

- The RLCLs retrigger disable cannot be caused by noise, EMC, ESD, SEE or other unexpected reasons with the exception of hardware failures (req. 5.2.16.1.1 a).

In particular, robustness "by design" of the RLCL against unexpected retrigger disable shall be ensured (req. 5.2.18.2.1a). One potential implementation idea is to "force" the memory cell responsible for the RLCL retrigger status to deliver an enabled signal as long as the number of RLCL retriggering cycles have not reached a specified minimum value. In this way it becomes impossible for a spurious glitch to disable the retrigger function.

Another possibility is to configure the retrigger disable circuit as a mono-stable feature: if an external command is received periodically within maximum time intervals, then the retrigger disable is maintained, on the contrary the retrigger restarts autonomously.

6 FUTURE WORK

The scope of the work started with the STD is not only limited to the power distribution interface by LCLs: the electrical power interfaces standardisation within ECSS continues, with the aim to cover, as far as practical, all power interfaces that are ready and that can benefit from standardisation.

The following effort (2016-2017) will focus on pyro/thermal knife/non explosive actuators interfaces.

7 CONCLUSIONS

An ECSS STD and a HB covering the power distribution interface by LCLs/RLCLs has been presented.

Both the STD and the HB are conceived to facilitate the diffusion of recurrent power distribution products in Europe, and it is expected that this would impact positively both on the quality (reliability, robustness, repeatability) of the different manufacturer's solutions, and on their cost/availability.

Finally, power distribution manufacturers will have the chance to develop their products "irrespective" of a specific mission opportunity, and this will be a global benefit for them, for the primes, and for the Agency.

8 REFERENCES

1. Various authors, *Power Distribution by Latching Current Limiters, Source-Load Specification Rationale*, TEC-EPC/2009/500/FT, v.1, 18 December 2009

2. F. Tonicello, C. Delepaut, M. Martin, M. Triggianese, *Approach to design for stability a system comprising a non-ideal current source and a generic load*, 10th European Space Power Conference, Noordwijkerhout, The Netherlands, 13-17 April 2014

3. European Cooperation for Space Standardisation, ECSS-E-HB-20-20A, *Space engineering* -

 DOI: 10.1051/e3sconf/20171613001

Guidelines for electrical design and interface requirements for power supply, 15 April 2016

4. European Cooperation for Space Standardisation, ECSS-E-ST-20-20C, *Space engineering - Electrical design and interface requirements for power supply*, 15 April 2016
5. Christophe Delepaut, Sara Siconolfi, Olivier Mourra and Ferdinando Tonicello, *MOSFET Gate Open Failure Analysis in Power Electronics*, 28th Applied Power Electronic Conference (APEC), 17-21 March 2013
6. European Cooperation for Space Standardisation, ECSS-E-ST-20C, *Space Engineering -Electrical and Electronic*, 31 July 2

Table 2: LCL classes

Characteristic	LCL class							
	1	2	3	4	5	6	8	10
Regulated Bus voltage [V]	28							
Unregulated Bus voltage [V]	22 to 38							
Class current [A]	1	2	3	4	5	6	8	10
Min limitation current [A]	1,1	2,2	3,3	4,4	5,5	6,6	8,8	11
Max limitation current [A]	1,4	2,8	4,2	5,6	7	8,4	11,2	14
Trip-off min [ms]	10	10	6	6	4	2	2	1,5
Trip-off max [ms]	20	20	12	12	8	4	4	3
Max load capacitance [µF]								
Regulated bus	272	545	490	653	545	327	436	408
Unregulated bus	203	405	365	486	405	243	324	304

Characteristic	LCL class								
	1	2	3	4A	4B	5	6	8	10
Regulated Bus voltage [V]	50								
Unregulated Bus voltage [V]	32 to 52								
Class current [A]	1	2	3	4	4	5	6	8	10
Min limitation current [A]	1,1	2,2	3,3	4,4	4,4	5,5	6,6	8,8	11
Max limitation current [A]	1,4	2,8	4,2	5,6	5,6	7	8,4	11,2	14
Trip-off min [ms]	10	6	4	2	4	2	2	2	1,5
Trip-off max [ms]	20	12	8	4	8	4	4	4	3
Max load capacitance [µF]									
Regulated bus	152	183	183	122	244	152	183	244	229
Unregulated bus	148	178	178	118	237	148	178	237	222

Table 3: RLCL classes

Characteristic	RLCL class				RLCL class			
	0,5	1	2A	2B	0,5	1A	1B	2
Regulated Bus voltage [V]	28				50			
Unregulated Bus voltage [V]	22 to 38				32 to 52			
Class current [A]	0,5	1	2	2	0,5	1	1	2
Min limitation current [A]	0,55	1,1	2,2	2,2	0,55	1,1	1,1	2,2
Max limitation current [A]	0,7	1,4	2,8	2,8	0,7	1,4	1,4	2,8
Trip-off min [ms]	10	10	4	10	10	4	6	4
Trip-off max [ms]	20	20	8	20	20	8	12	8
Max load capacitance [µF]								
Regulated bus	136	272	218	545	76	61	91	122
Unregulated bus	101	203	162	405	74	59	89	118

Table 4: HLCL classes

Characteristic	HLCL class							
	1	2	3	4	5	6	8	10
Regulated Bus voltage [V]	28							
Unregulated Bus voltage [V]	22 to 38							
Class current [A]	1	2	3	4	5	6	8	10
Min limitation current [A]	1,1	2,2	3,3	4,4	5,5	6,6	8,8	11
Max limitation current [A]	1,4	2,8	4,2	5,6	7	8,4	11,2	14
Trip-off min [ms]	0,5	0,5	0,5	0,5	0,5	0,5	0,5	0,5
Trip-off max [ms]	2	2	2	2	2	2	2	2
Max parasitic capacitance [µF]								
Regulated bus	1	1	1	1	1	1	1	1
Unregulated bus	1	1	1	1	1	1	1	1

Characteristic	HLCL class								
	1	2	3	4		5	6	8	10
Regulated Bus voltage [V]	50								
Unregulated Bus voltage [V]	32 to 52								
Class current [A]	1	2	3	4		5	6	8	10
Min limitation current [A]	1,1	2,2	3,3	4,4		5,5	6,6	8,8	11
Max limitation current [A]	1,4	2,8	4,2	5,6		7	8,4	11,2	14
Trip-off min [ms]	0,5	0,5	0,5	0,5		0,5	0,5	0,5	0,5
Trip-off max [ms]	2	2	2	2		2	2	2	2
Max parasitic capacitance [µF]									
Regulated bus	1	1	1	1		1	1	1	1
Unregulated bus	1	1	1	1		1	1	1	1

SPACECRAFT ELECTRICAL PASSIVATION: FROM STUDY TO REALITY

François Bausier [(1)], Maria Nestoridi[(1)], Bruno Samaniego[(2)], Juhani Simola[(3)], Andrew Wolahan[(1)], Julian Austin[(1)], Tiago Soares [(1)]

[(1)] *ESA-ESTEC, Keplerlaan 1, 2200AG Noordwijk, NL, E-mail: francois.bausier@esa.int, maria.nestoridi@esa.int, andrew.wolahan@esa.int,* julian.austin@esa.int, *tiago.soares@esa.int*

[(2)] *Airbus Defence and Space, 31 rue des Cosmonautes 31402 Toulouse France, Email: bruno.samaniego@airbus.com*

[(3)] *RUAG Space Finland, Naulakatu 3, 33100 Tampere, Finland, Email: juhani.simola@ruag.com*

ABSTRACT

Due to the increasing risk posed on satellites from space debris, requirements are now being implemented into satellites to be launched in the future to prevent the release of additional debris. One of the key requirements is the electrical passivation of satellites at end of mission (EoM) in order to prevent the break-up of satellites due to the stored energy in the power subsystem. In order to comply with these new specifications and find suitable solutions for the upcoming missions, a set of workshops and R&D activities have been initiated and conducted in the past few years. In particular it has provided a better understanding of the environmental constraints and the battery safety after EoM, together with the identification of the most appropriate implementations for electrical passivation. The latest available results of these activities and the ongoing work are presented within this paper. In addition, new spacecraft under development are currently implementing electrical passivation solutions taking into account the conclusions of these studies.

1. INTRODUCTION

The increasing number of space debris has become problematic for the sustainment of space activities in earth orbit. Several spacecraft breakups have been observed in the past and some of them are due to a battery breakup. In order to mitigate the risk to generate debris in the future, passivation of the spacecraft after end of mission is now required in some cases. In particular the electrical passivation is intended to deplete and/or make safe the on-board battery. More detailed background information can be found in a previous paper [8].

2. CURRENT SPECIFICATIONS AND GUIDELINES

This section reviews the current space debris mitigation specifications and guidelines related to electrical passivation and applicable to most European spacecraft.

The European standard ECSS-U-AS-10C [1] - adoption notice of ISO 24113:2011 [2] - is applicable to all new ESA missions [3]. In particular the ISO §6.2.2.3 specifies that "*During the disposal phase, a spacecraft or launch vehicle orbit stage shall permanently deplete or make safe all remaining on-board sources of stored energy in a controlled sequence*". The associated ESA handbook ESSB-HB-U-002-Iss1 [4] published in 2015 provides guidelines for interpretation of the requirements, implementation and demonstration of compliance. For example it clarifies that the capability to passivate is not required to be single point failure tolerant or that passivation is not required in case of controlled atmospheric re-entry at EoM. It also provides some possible generic passivation methods.

The French law LOS [5] is applicable to spacecraft operated by French or France-based operators. The associated guidelines [6] provide also clarifications and more detailed specifications. For example it specifies that the battery shall be discharged at EoM and if isolated from the spacecraft the auto-discharge shall be less than 1 year. It also clarifies that the solar generator deactivation can be fulfilled either by solar array short/open circuit, or by stable deactivation of the power conditioning or by opening of all equipment distribution and battery charge lines.

Table 1 compares these main specifications.

	ESA/ECSS/ISO	LOS
Applicability for s/c	ESA procurements and operations of space systems	s/c controlled by French of France-based operator
Applicability date	Published on 16/04/2014	Since 10/12/2010 with transistion period for s/c launched before 31/12/2020
Battery	Depleted or made safe	Depleted or made safe
Power generation	Not specified	Deactivated
LEO duration applicable for passivation	25 years	25 years
MEO/GEO duration applicable for passivation	No limitation specified	No limitation specified

Table 1 – Comparison of specifications

3. R&D ACTIVITIES

In order to comply with the new passivation specifications and find suitable solutions for the upcoming missions, a set of workshops and R&D

activities have been initiated and conducted in the past few years.
First workshops were organised in 2013 (at CNES and at ESTEC) providing the opportunity for experts to share the first investigations and studies on the subject. But it also raised several open points in particular concerning the risk of battery breakup, the environment and behaviour of spacecraft after EoM, the long term behaviour of components and the interpretation of some requirements [7] [8].
Following these events, some R&D studies have been initiated to better understand these open points and identify the most appropriate implementations for electrical passivation. The latest available results of these activities and the work still to be performed are presented below.

3.1 Environmental Impact on Power Systems After End-of-Life [9]

An ESA internal activity was carried out in 2014 with the goal to better understand the environmental effects in terms of thermal and radiative, which can then be used to better understand the type of degradation on the power subsystem of a satellite. An overview of the various environmental conditions which a battery or a PCDU may be susceptible to within 25 years after a 15 year operational life in LEO, MEO and GEO are hence presented.
For the thermal assessment, the following different assumptions were assessed with the goal to cover most of the existing earth orbiting missions and platforms:

- Different orbits, LEO (600 km and 800km SSO), MEO and GEO;
- Alternative configurations where the battery is of various sizes and is located inside or outside;
- Different attitudes of the satellite such as the radiator pointing Nadir or towards the sun, or the radiator rotating in and out of direct sun;
- A range for the thermo-optical properties of the radiators and panels for solar absorptivity and infrared emissivity based on literature and worst case extrapolations.

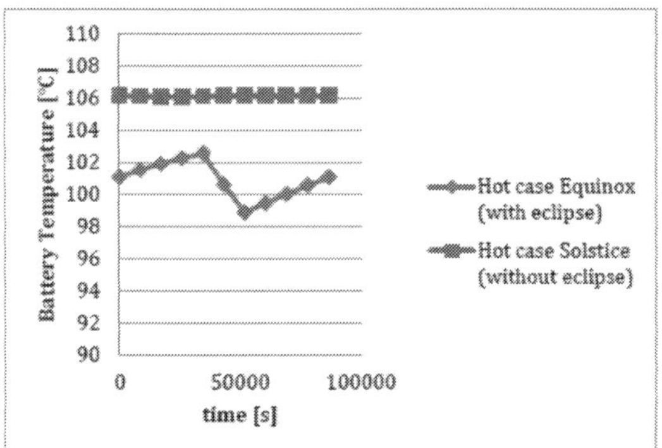

Figure 1: GEO hot case – Effect of Eclipse

In terms of maximum temperature, the hottest case identified was 106°C in GEO, with the radiator pointing towards the sun, and during the solar solstice where there is no eclipse (Figure 1). The minimum temperature in cold case was -167°C.

In LEO case, the maximum temperature was 82°C. The thermal results also indicate that by inducing a spin rate into a satellite the overall temperature range is reduced (see Figure 2). The battery configuration has also an impact.

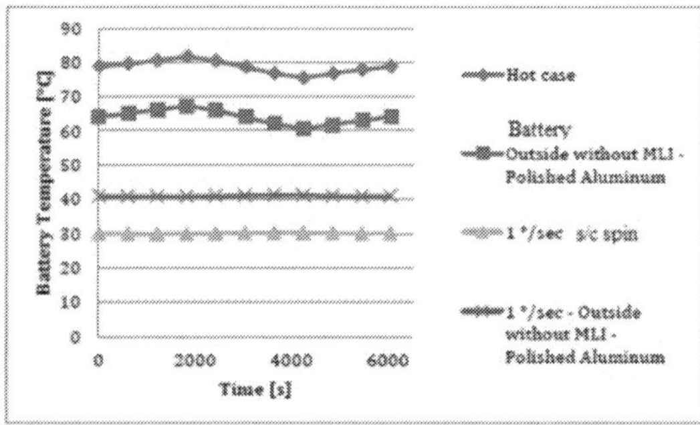

Figure 2: LEO hot case – Effect of battery configuration and spacecraft rotation

For the PCDU the minimum temperature was -162°C and the maximum was 56°C.
For the values provided, generally, worst case assumptions were considered and hence it is expected that the majority of satellites will fall within these temperature boundaries.

A radiation analysis was conducted for three common Lithium-Ion battery types (ABSL 18650HC, Saft VES16 and VL48E) for each of the orbits. A simplified geometric model for the various batteries were identified and then using the Geant4 Radiation Analysis for Space (GRAS) tool [15], it was possible to estimate the radiation dose in each of the battery cores for a total lifetime in-orbit of 40 years, which assumes 15 years of operational lifetime and 25 years during the disposal phase (see Table 2).

Orbit	**Minimum**	**Maximum**
GEO	16.1 kRad	160.1 kRad
MEO GNSS	78.9 kRad	602.0 kRad
LEO SSO	13.4 kRad	26.7 kRad

Table 2 – Radiation total doses in battery, assuming aluminium shielding of 2mm for spacecraft and 1mm for battery case

The doses to be encountered by the bulk battery over the 40 year period are manageable (< 200 kRad for the total battery). However, for the outermost layers of the battery, the doses can be as high as 600 kRad. This

 DOI: 10.1051/e3sconf/20171613002

should not be an issue considering that typically the cells are tolerant to at least 1 MRad. But these values do need to be considered for any potential cell balancing circuit and/or by-pass switch that may be integrated.
For PCDU components, the total dose depends on the shielding and is according to Figure 5.

In addition simulations assessed the meteoroid and space debris flux for the three orbits, and the method to do so has been presented, from which it is possible, for a given design, to identify the minimum size and the number of particles which will penetrate the wall of a satellite or sub-system and possibly damage the battery or PCDU (see Figure 3 below).

Orbit	Fluxes [/m²/year]		
	met	deb	total
LEO (800 km SSO)	1.88 E-3	1.65 E-3	3.53 E-3
LEO (1200 km, SSO)	1.87 E-3	5.11 E-4	2.38 E-3
MEO (23200 km, 55°)	1.32 E-3	1.28 E-6	1.32 E-3
GEO (36000 km)	1.23 E-3	9.08 E-7	1.23 E-3

Figure 3: Example of critical meteoroid and debris fluxes for various orbits

Oxidisation, plasmas and charging were also briefly assessed but their impact was considered less significant.

It has to be noted that these simulations are limited by the assumptions that were taken, and indeed for such a study, a large number of assumptions were required. In order to identify boundary conditions some unlikely extreme case assumptions have also been considered. So the results have to be considered with critical thinking but provided a good basis for the follow on activities described below.

A more in-depth overview of the thermal and radiative analysis are available in reference document [9].

3.2 GSTP - Spacecraft Power System Passivation at End of Mission [10]

The objective of this activity is to study, implement and evaluate the most adequate means to ensure a proper and reliable spacecraft power system passivation at end of mission.
The main tasks performed in Phase 1 are:

- Simulations of environmental conditions (mainly temperature and radiations) for real existing platforms for LEO, MEO and GEO mission after end of mission.
- Study of battery safety, and its dependence on battery conditions (SoC, temperature) after end of mission.
- Study of possible passivation devices and assessment of long term performance of components under harsh environment.
- Review of potential implementations and recommendations.

In order to assess realistic and representative scenarios faced by the satellites at EoM, dedicated thermal and radiation analyses were done on real existing spacecraft platforms for LEO, MEO and GEO missions. The achieved simulation results were compared and combined to those achieved in the previous activity (§3.1 [9]). The main focus of the environmental simulations were to assess thermal and radiative conditions into which the battery and the passivation device (e.g. located in the PCDU) are expected to be exposed after end of mission. The environmental conditions for the battery were simulated in two spacecraft configurations: a) battery mounted inside the spacecraft and, b) outside the spacecraft.
The temperature ranges obtained in the various simulation results are summarized in Figure 4. It has to be highlighted that some simulations considered extreme worst cases such as sun constantly facing the same radiator, heat pipe failure or MLI completely torn. It also considered current existing platforms, knowing that the configuration of future platforms could be improved to reduce the temperature range to a more reasonable level. Thereby the temperature range could be reduced to for example [-100°C to +100°C] or even [-50°C to +80°C], which makes easier to guarantee the battery safety and the passivation device long term performance.

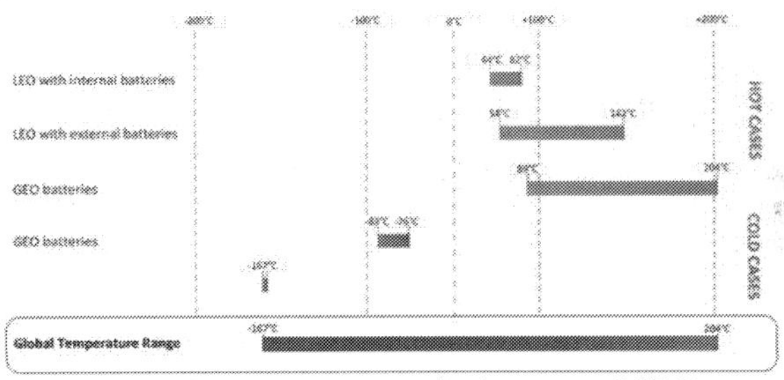

Figure 4 – Temperature range for battery or PCDU

Radiation doses for LEO, MEO and GEO missions after end of life were also simulated. Radiation depth curves for the 3 specified environments after 40 years (15 years operational phase + 25 years disposal phase) are shown in Figure 5. The maximum radiation total dose depends on the orbit, on the shielding and on the duration considered. For the battery cells it has been computed from less than 10 krad (LEO case) to 57 Mrad (MEO case with and equivalent aluminium shielding of 1.5mm from the battery case and spacecraft). For the PCDU components, it has been computed from less than 5 krad to 100 krad.

DOI: 10.1051/e3sconf/20171613002

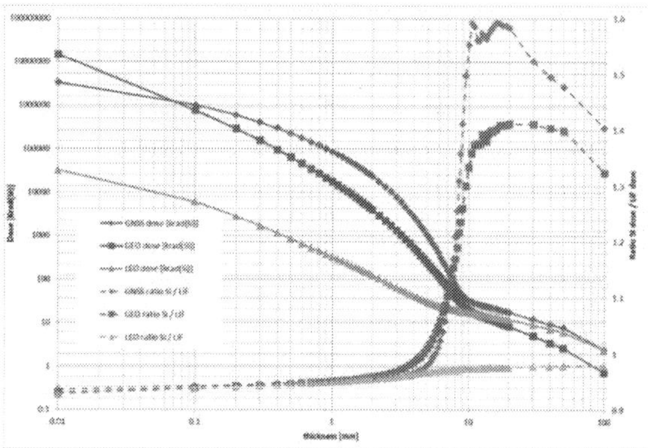

Figure 5 – Radiations for various orbits

The study of the battery safety was focused on various Li-ion batteries from Saft and ABSL.

Battery conditions, in particular State of Charge (SoC) and temperature, were recognised to have a significant impact on the battery safety. The available information about battery performance in various conditions were collected and investigated. The most disastrous phenomenon is the thermal runaway, which may occur in a battery submitted to high temperature (induced by environment or by hazard such as overcharge, short-circuit or damage), and which may lead to battery breakup [12] [13]. Typical behaviour is shown in Figure 6.

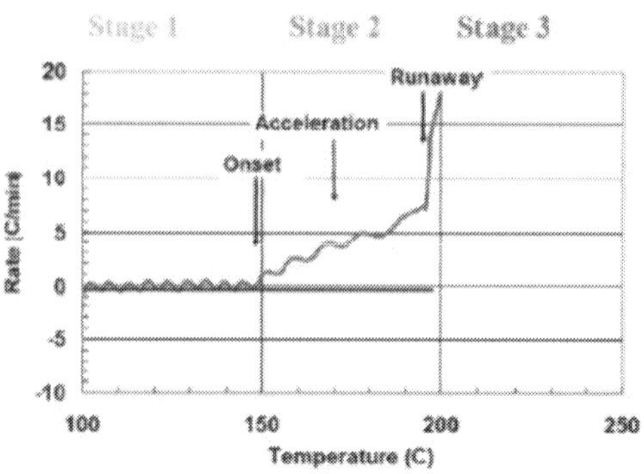

Figure 6 – Battery cell thermal runaway (source [12])

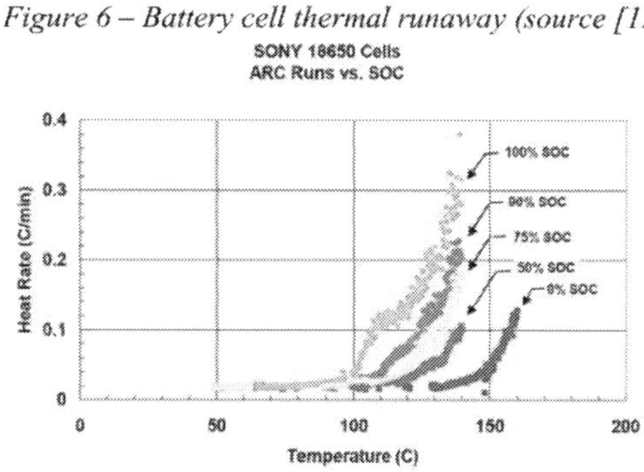

Figure 7 – Thermal runaway vs SoC (source [13])

Figure 7 shows the dependency of the thermal runaway phenomenon on the SoC for new cells. For aged cells thermal runaway starts at slightly higher temperature.

It has been identified that although some battery cells are equipped with various protections which limit the risk, the speed of the thermal runaway reaction may be too high for the protections to react in time in some cases and the risk of breakup cannot be completely excluded.

Recommendations about battery conditions after end of life to reduce the risk of battery breakup were given as follows:

- Discharge the battery as low as possible after EoM (at least 50% SoC but lower is better).
- Avoid temperature higher than 100°C.
- Prevent recharging of the battery after passivation, and in particular avoid overcharging.
- The cumulated radiation doses should not exceed certain limits regarding Li-ion cells themselves but also electronics (if any in the battery).
- The passivation device shall sustain the harsh environment it is exposed to after passivation to guarantee safe conditions for the battery.

Possible components that could be used for passivation device implementation have been studied, focusing on their ability to sustain long lifetime and harsh environment. The following components have been assessed: semiconductors (MOSFET, diode), relay, by-pass switch and cable cutter.

Semiconductors have the advantage to be compact and often easier to accommodate but the main drawback is that their performance is not very well known beyond 20 years, in particular if temperature and radiation conditions become extreme.

Relays are electro-mechanical devices and seem more robust to extreme conditions and to maintain passivation for a long time. The drawback is that they are less compact and their current/voltage ratings can be limited.

Concerning by-pass switches (currently used for battery cell by-pass) it is very robust but bulky, expensive and their voltage capability is not demonstrated above about 5V.

Finally cable cutters are also very robust but bulky, not space qualified for activation after a long time and can produce unwanted short-circuits inside the harness during activation.

A large number (27) of potential implementation solutions were studied:

- solutions at spacecraft level such as adapting the spacecraft configuration or attitude (e.g. spinning) after disposal to protect the battery from the harsh environment, battery containment for small spacecraft, SA mispointing, software patch, decrease of battery charge voltage/current,...
- solutions at battery level such as discharge,

disconnection or short-circuit via various devices, overcharge, overdischarge, use of internal protections.
• solutions at solar array level such as disconnection or short-circuit via various devices.
• solutions at bus level such as short-circuit of the main bus via various devices.
The optimal solution may be a combination of two or more different single solutions.
These solutions were ranked according 25 weighted factors related to performance, compliance with requirements, impact on design and on mission. The most promising solutions according to this ranking are:
• Battery disconnection by relay or by-pass switch and discharge by resistor.
• Disconnection of solar array by relays.
• Short-circuit of solar array by relays via diodes.

In addition the following options were also identified as promising when combined with other solutions or in particular cases:
• adapting the spacecraft configuration or attitude (e.g. spinning) after disposal, decrease of battery charge voltage/current via dedicated levels, over-discharge the battery until it transforms to a pure resistance.
• battery containment for small spacecraft such as cubesat.
• for spacecraft already equipped with batteries with by-pass switches, these switches can be used to disconnect the battery cells (row after row) with minor impact on the design.

A set of recommendations were proposed and are summarized below:
• Perform thermal and radiation analyses after disposal to verify that the battery and passivation device stay inside acceptable ranges in worst case conditions. Avoid high temperature and high radiations.
• For the components used for the passivation device, privilege (electro-)mechanical devices such as relays or by-pass switches which are more robust than semiconductors with respect to long life time and harsh environment. However semiconductors could be used in LEO missions for which the total duration from launch to re-entry is short.
• Regarding battery safety, as detailed above, it is particularly recommended to discharge the battery as low as possible and to avoid high temperature.

Phase 2 of the study is ongoing and is expected to:
1. Provide a "toolbox" document describing in detail the most interesting passivation solutions including the operations.
2. Evaluate and test some components (relay, by-pass switch, Mosfet, diode) that could be used for the passivation device, in particular with respect to extended ratings and/or lifetime.
After a preliminary evaluation with various component manufacturers, it appears that:

• for relays, they are already qualified to be activated until EoM (at the time of passivation). After passivation, they only need to stay in the same stable position (without being activated again), which is a priori not a problem for extended lifetime. Space grade relays are in hermetic package filled with gas (e.g. nitrogen), which can leak with time leading to low gas pressure. This is not a problem provided that the voltage is not above the critical threshold according to the Paschen curve (typically 200-300V for air). Also relays are not sensitive to radiations and their temperature range rating is quite wide (from -65°C to 125°C). If needed the temperature range could probably be extended to <-100°C to 175°C, which is to be confirmed by tests.
• for semiconductors (Mosfets, diodes), as there is no data results for lifetime beyond 20 years, accelerated life tests and cycling tests are necessary to know if they could be used up to 40 years or even longer.
• for by-pass switches (currently used for battery cell by-pass at low voltage below 5V), their capability to be used at higher voltage to short or open circuit battery lines, solar array lines or main bus needs to be evaluated and tested.

3.3 TRP – Battery Passivation [11]

As already explained in the previous section, the battery is considered to be the most dangerous part of the EPS in terms of potential debris generation. A main topic of discussion nowadays is about the battery safety after EoM, following off nominal environmental or failure conditions and subsequent battery abuse in space.
Passivation of the battery is believed to be the solution in mitigating such potential explosion risks. One way to achieve passivation is to deplete the spacecraft battery. However if the latter is to be implemented a very careful assessment of the behaviour of the battery at such conditions should be made.
Such an assessment can be very complex considering that temperature, radiation, SoC and aging of the battery could play a role among other parameters on the risk and possibility of battery explosion. The actual probability of a thermal runaway leading to a cell explosion is impossible to assess without testing.
The ESA TRP objective, performed in collaboration with Airbus DS, SAFT, ABSL and CEA, is to understand and demonstrate through testing the behaviour of a spacecraft lithium-ion battery, under extreme conditions, which the battery can experience after EoM or during/after passivation if any.
Simulation results performed in the GSTP study have been used as an input to the test plan for this TRP. The Li-ion battery safety assessment has been summarized, listing all the issues that can arise during the satellite lifetime. The impact of different parameters such as aging, SoC, temperature, radiation or micrometeoroids will be analysed and tested.

 DOI: 10.1051/e3sconf/20171613002

Both fresh and aged battery cells used in spacecraft batteries will be tested in order to investigate the impact of the aging during LEO and GEO missions under potential off nominal conditions in space that the battery might experience.
The tests foreseen in this TRP are:

- External short-circuit;
- Internal short-circuit;
- Over-charge;
- Over-discharge;
- Accelerating Rate Calorimetry (ARC) test;
- Over-temperature test;
- Micrometeoroids.

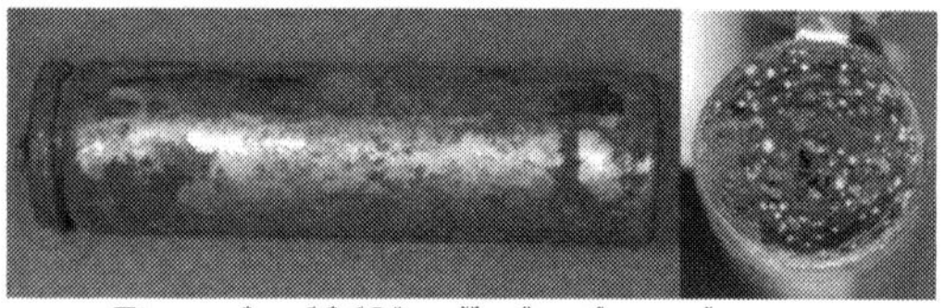

Figure 8 – 18650 cell after thermal runaway

Cells from the two main battery manufacturers in Europe, ABSL and SAFT, will be used for testing including ABSL18650-HC, -HCM, -NL and SAFT VES16, VES140, VES180.

At the time of writing, some tests have already been performed on the VES140 cell. The approach followed in this study is to assess first the impact of different parameters on one cell model and then, depending on the results, identify the most important and refine the test plan for the subsequent models to be tested afterwards.

One of the most interesting results so far are the ARC tests performed on new VES140 cells at different SOC. ARC testing has confirmed that self-heating onset temperature and self-heating rate is a function of the stored electrical energy stored as chemical potential energy within the cell.

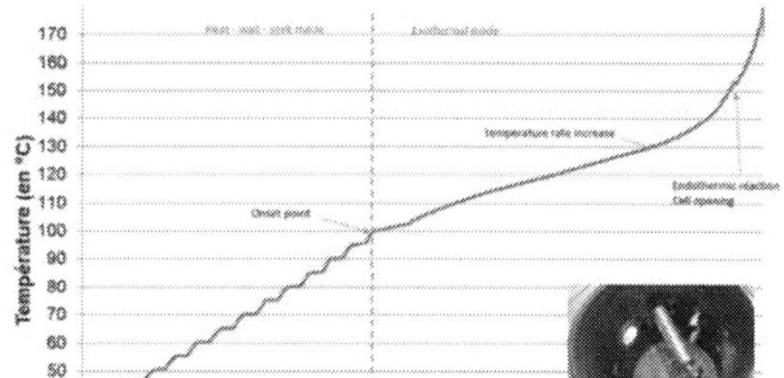

Figure 9 – ARC test result on a VES140 cell at BOL and at 50% SOC

First results show an onset temperature around 90-100°C for new VES140 cells at 50% SOC that is reduced down to 68°C when the cell is fully charged. In both cases the thermal runaway temperature is around 130°C. It is important to clarify that a cell will not automatically go into thermal runaway once it reaches the onset temperature. It means that an exothermal reaction starts, that can potentially lead to a thermal runaway if the cell continues to be heated. Thus, it is of major importance the thermal control design of the spacecraft and its potential degradation after EoM.

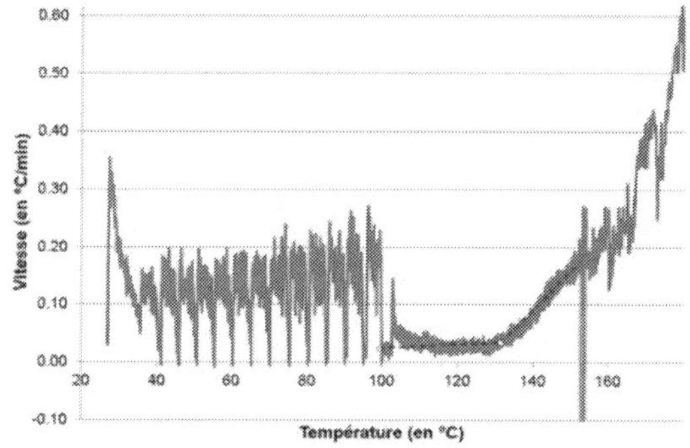

Figure 10 – Heat rate during the ARC test on a VES140 cell at BOL and at 50% SOC

Once the thermal runaway starts, it has been observed that the security mechanism is correctly activated (leak before burst) and the cell opens on the positive side at 150°C, which slightly reduces the temperature. However, if the temperature of the environment continues to increase after that, even if the cell is open, and the fast generation of gases can lead to the expulsion of the windings, generating therefore some debris that could be fatal for the battery and the spacecraft. This has been observed on some samples, that experienced a new reaction at 170°C, expulsing the windings.

Figure 11 –VES140 cell sample after the ARC test with the positive side opened

For now, the main outcome of this study is that environment temperature is key to guarantee the safety of in-flight batteries. Assessments of battery temperatures after the EoM should be performed in order to evaluate the risks of experiencing thermal runaways, especially for the GEO satellites, that will

 DOI: 10.1051/e3sconf/20171613002

last longer than LEO. The study is currently going through all the other tests, the results so far are very promising and will give a broader and clearer perspective of battery safety on spacecraft.

3.4 CleanSat activities

In addition some small activities related to electrical passivation were conducted in the frame of the ESA CleanSat programme, where the goal is to support new developments for space debris mitigation. These activities were jointly managed by the 3 main European large space system integrators (Airbus DS, TAS and OHB) and were occasionally followed by other observers. So it was also a forum to share information among the main European industrial actors. This approach aims at defining harmonised requirements agreed among all the stakeholders (Agency, integrators and suppliers) and a consolidated preliminary design to prepare the technology maturation in GSTP as of 2017.

3.4.1 Battery safety [16]

This study, carried out by ABSL, looked at analysing the safety of Li-ion batteries in LEO, MEO and GEO, with a focus on end of life. Assessments were performed on the risk of thermal runaway occurring, based in differing battery factors, such as temperature and state of charge. The study provided guidelines for the battery conditions to be achieved at the end of life through passivation, as well as highlighting future tests that should be performed to further the understanding on battery safety. Lastly, extra tests relating to battery passivation were proposed for future battery developments.

3.4.2 Solar array disconnection [17]

This study, carried out by Leonardo, investigated different options for isolation or shunting of the solar array in the PCDU. The priority was put in finding a common solution covering the needs of different markets including LEO, MEO and GEO. A trade-off was performed taking also into account the company heritage and a couple of options were identified to perform a preliminary design and a development roadmap. The selected concept was to short-circuit each solar array section by a relay (or 2 for redundancy) and it would be implemented in dedicated modules to be included in the PCDU between solar array interface and SAR modules.

3.4.3 Passivation of solar array in PCDU [18]

Thales Alenia Space Belgium proposed a different approach targeting LEO platforms market which is also the main focus of the CleanSat programme. Therefore the study focused on finding an optimal solution for this market where MPPT architectures are more and more common. The selected concept was to implement an isolated DC/DC converter for the solar array regulator (SAR) and use the galvanic isolation of the transformer to achieve the passivation. This concept has the advantage to offer a real physical and robust separation between solar array and main bus.

4. IMPLEMENTATION ON REAL MISSIONS

New spacecraft under development are currently implementing electrical passivation solutions taking into account the conclusions of the studies. And in many cases there is no much to adapt to implement an adequate electrical passivation.

For example Sentinel-2C and -2D spacecraft (LEO sun-synchronous orbit 786km, 28V unregulated battery bus, prime contractor Airbus DS) will implement a disconnection of the battery by relays and depletion resistors to fully discharge the battery down to 0V (see Figure 12). There was already a quad of relays between battery and main bus on Sentinel-2A and -2B models but the in-flight OFF commands were missing. So the adaptation was to add these OFF commands (direct high level priority commands from ground). In addition, to achieve full battery passivation, depletion resistors in series with arming straps (to avoid battery depletion during AIT) were also added. The depletion resistors dissipate only a few watts maximum and ensure a full depletion of the battery within 1 month (TBC). Once the battery voltage reaches about 4V (0.5V/cell), copper dendrites start to develop inside the cells, which short-circuit the battery irreversibly.

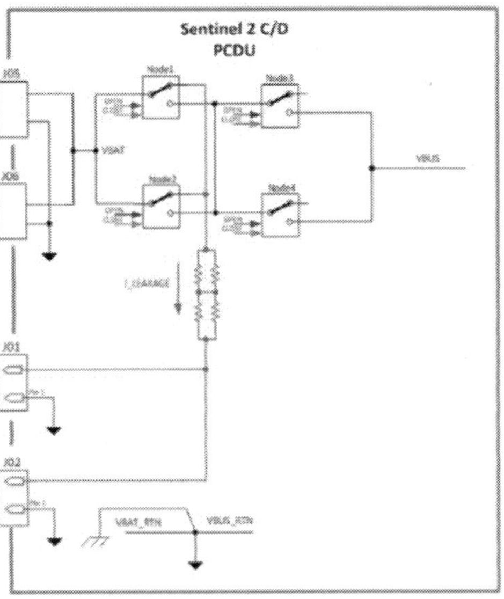

Figure 12 – Sentinel 2 C/D electrical passivation concept

To consolidate the design an environmental analysis during the disposal phase has been performed for the

battery and the PCDU. Also a detailed procedure relates all the operations to be performed at end of mission, including switch off of units, battery discharge, battery isolation by direct commands from ground, confirmation by direct status, etc.
Other earth observation satellites such as Jason-CS/Sentinel-6 and others in preparation will adopt a similar design.
For GEO commercial platforms such as E3000 or Alphabus, the battery by-pass switches can be used at end of mission to disconnect the battery cells after having discharged them as low as possible.
As mentioned in the above paragraphs new developments are also foreseen for future platforms.

5. PERSPECTIVES FOR THE FUTURE

It is expected that more and more earth orbiting spacecraft will implement electrical passivation from now on. As the risk of battery break-up after EoM cannot be completely excluded, a dedicated passivation function is necessary.
In order to optimise the designs, the main activities foreseen to fill the knowledge gaps are:

• tests on batteries (cells and modules) to assess their safety during the disposal phase: via the ongoing TRP activity (§3.3) and/or if necessary future activity (§3.4.1)

• tests on components used for passivation to assess their robustness under extreme conditions and/or extended lifetime: via the ongoing GSTP activity, phase 2 (§3.2)

• development of optimal integrated designs and operations to achieve passivation: via the implementation on ongoing/future projects (§4) or future development activities (§3.4.2, §3.4.3)

6. CONCLUSION

The activities related to spacecraft electrical passivation conducted these last years within ESA and the European space industry have been presented.
The applicable specifications have been clarified by new guidelines.
Various R&D studies led to a better understanding of the environmental impact, the battery safety and the robustness of components during the spacecraft disposal phase. Most promising passivation methods and recommendations have been proposed.
The conclusions of these studies are now being taken into account for the implementation of electrical passivation on new spacecraft.
Finally some studies are still ongoing or in preparation to further characterize the battery safety and optimise the passivation designs.

7. ACKNOWLEDGEMENTS

The authors would like to thank all the people from industry and ESA participating in the above R&D and project activities for the results presented in this paper.

8. REFERENCES

[1] "Space Sustainability, Space debris mitigation requirements", ECSS-U-AS-10C, 10/02/2012

[2] "Space Systems, Space Debris Mitigation Requirements", ISO 24113, Second Edition, 15/05/2011

[3] "Space Debris Mitigation for Agency Projects", ESA/ADMIN/IPOL(2014)2, 28/03/2014

[4] "ESA space debris mitigation compliance verification guidelines", ESSB-HB-U-002-Issue1, 19/02/2015

[5] "Arrêté du 31 mars 2011 relatif à la réglementation technique en application du décret n° 2009-643 du 9 juin 2009 relatif aux autorisations délivrées en application de la loi n° 2008-518 du 3 juin 2008 relative aux opérations spatiales", NOR: ESRR1103737A

[6] « Guide des bonnes pratiques maîtrise d'un object spatial », RNC-LOS-GR-CNF-8-CNES, version 2, 24/03/2014

[7] "Battery Isolation Standardisation & EoL Electrical Passivation Workshop", 10-11 October 2013, ESA/ESTEC, NL, EoL Electrical Passivation MoM TEC-EPM/2013/51/FB, 11/10/2013

[8] "Passivation of spacecraft electrical power system at end of mission", F. Bausier, T. Soares, F. Tonicello, ESPC 2014

[9] GSP "Environmental Impact on Power Systems After End-of-Life", final report, CDF-TN-069 Iss1.0, 04/09/2014

[10] GSTP "Spacecraft power system passivation at end of mission", Phase 1 final report, SPSP.PATRI.RP. 61318 Iss2, 12/10/2015

[11] TRP Battery Passivation, "D1 - Test specification requirements", 2125.NT.SB.16.2143.ASTR

[12] "A General Discussion of Li-Ion Battery Safety", Dan Doughty & E. Peter Roth, The Electrochemical Society Interface, Summer 2012.

[13] "Thermal Abuse Performance of 18650 Li-Ion Cells", E. Peter Roth, Chris C. Crafts, Daniel H. Doughty, James McBreen, Sandia Report SAND2004-0584, Sandia National Laboratories

[14] "Lithium-Ion Cell Passivation In Flight Conditions", G. Bouhours, E. Klein, C. Rebuffel, C. Wallstein, R. Asplanato, Y. Borthomieu, ESPC 2011

[15] "GRAS:A general-purpose 3-D modular simulation tool for space environment effects analysis", G. Santin, V. Ivanchenko, H. Evans, P. Nieminen, E. Daly, IEEE Trans. Nucl. Sci. 52, Issue 6, 2005, pp 2294 - 2299

[16] CleanSat BB01 "Battery safety" Final Report PDEV-ABSL-RP-0098

[17] CleanSat BB22 "PCDU Passivation" Final Report CS-BB22-FNM-RPT-0001

[18] CleanSat BB27 "Passivation of solar array in the PCDU" Final Report CLENSAT-EPS-TASB-TN-0004

PASSIVATION STRATEGIES ON BOARD AIRBUS DS LEO PCDUS

Emilio LAPEÑA [(1)], Daniel RUF[(2)], José Luis HERRANZ [(1)], Fernando GÓMEZ-CARPINTERO [(1)], Miguel RODRÍGUEZ [(1)], José OTERO[(1)]

[(1)]Airbus Defence and Space (CRISA), Torres Quevedo 9, 28760 Tres Cantos, Spain, Email: emilio.lapena@airbus.com – joseluis.herranz@airbus.com – fernando.gomez-carpintero@airbus.com – miguelr.alvarez@airbus.com – jose.otero@airbus.com
[(2)]Airbus Defence and Space (Airbus DS GmbH), Claude-Dornier-Str. 88090 Immenstaad, Germany, Email: Daniel.Ruf@airbus.com

ABSTRACT

Having in mind the increasing amount of spacecraft in orbit, the number of space debris becomes a growing issue. Regarding the space regions around the Earth to be protected, one of them is the Low Earth Orbit (LEO) one, for satellites with altitudes lower than 2000km.

Even if the number of incidents in orbit caused by the spacecraft battery is very low when looking at the total number of satellite breakup events (around 3,6% – none of them were equipped with Li-Ion batteries) the uncontrolled nature of such events and their dramatic potential consequences make necessary a proper treatment, in the form of countermeasures that are fully justified taking into account the limited impact in mass and cost on the PCDU.

This paper deals with the different strategies followed in the Airbus DS LEO PCDUs regarding the implementation of the passivation function in several LEO missions with different architectures (DET and MPPT solar array power conditioning). In the selection of the solution implemented in the frame of every mission, a key driver is the degree of advance in the test performed over flight representative battery modules regarding their safe behavior when deeply depleted after a long period in orbit with the passivation applied over the spacecraft.

1 BACKGROUND

In general terms, "passivation" refers to deplete or make safe the remaining on board sources of energy in a controlled sequence.

Several more or less recent normative are making an effort to regulate this aspect of the spacecraft design for the new missions to come or in course of development, [1]. ESA (through "Space Debris Mitigation for Agency Projects" document, ESA/ADMIN/IPOL (2008)2, dated 1/4/2008), ECSS (through "Space Sustainability, Space Debris Mitigation Requirements" document, ECSS-U.AS-10C, dated 10/2/2012), and ISO (through "Space Systems, Space Debris Mitigation Requirements" document, ISO 24113, dated 15/5/2011) are examples of normative close to our business about this aspect of the design.

Also of relevance is the "Loi Relative aux Operations Spatiales" (LOS), dated 3/6/2008, whose applicability is from 10/12/2010, but establishing a transition period for spacecraft launched before 31/12/2020. This document is applicable for launchers put in orbit from French territory or spacecraft controlled by French or France-based operators. It requires that any satellite launched form French territory will have to provide means for EOL passivation of its energy sources.

Most of these documents require either spacecraft de-orbiting in 25 years or "parking" above 2000 km for more than 100 years. This 25 years period is in line with the requirements asking for 25 years of operative passivation in orbit, even if none of the standards require a fail free passivation electronics behavior. As an alternative to this, it is contemplated the possibility to perform a spacecraft de-orbit with a controlled atmospheric re-entry.

2 PASSIVATION CONSIDERATIONS

Several alternatives are possible to passivate the spacecraft after mission completion, but most of the cases the requirement specifications ask for one of the following strategies:

- Battery disconnection.
- Solar Array Disconnection.
- Solar Array and Battery Disconnection.

In addition, the following considerations are to be taken into account, most of the times also reflected in the PCDU specifications particular of a given mission:

- Passivation activation must be one failure tolerant regarding its unintentional activation, both regarding:
    - Passivation Control electronics, and
    - Power Components involved
- The passivation function is allowed to fail in passivating the spacecraft, i.e. it is not required

DOI: 10.1051/e3sconf/20171613003

to implement a one failure tolerant electronics regarding this aspect of its performance.

- In most cases passivation must be reversible by design, at least down to the moment on which the spacecraft energy sources are no longer available, and/or the satellite On-Board Computer (OBC) is no longer operative.
- And, of course, passivation must be able to operate for the 25 years in orbit, having in mind that most of the time the Satellite Thermal control Subsystem will not be available. The satellite and on-board electronics will experience temperatures in excess of those normally required for this type of equipment, typically lower than the minimum start-up temperature and - many times - lower than the minimum qualification temperatures for some electronic and electromechanical components and/or state of the art associated manufacturing processes.

To complete the scenario, a deep-enough knowledge does not exist about how the batteries will behave in this situation of extreme environmental temperatures, and how safe they will be even if we apply correctly the passivation measures. Activities are ongoing to know more about this aspect of the design.

3 DET AND MPPT

Even if LEO missions share the same Electrical Power Subsystem (EPS) concept, the so called battery follower bus, the passivation detailed implementation strategy is also driven by the mission specific solar array conditioning baseline. The different PCDU designs based on Direct Energy Transfer (DET) or Maximum Power Point Tracking (MPPT) require different implementations of the passivation concept. Recently, the MPPT concept has been upgraded with more powerful performance capabilities, thanks to the development of the Buck Boost Regulator (B2R) concept. It is able to handle a solar array voltage that may be over or under the main bus (battery) voltage depending on the mission phase [4].

In the following this paper presents the Airbus DS practical implementations in a wide range of LEO missions, covering solar array power management based on DET, MPPT and B2R. Separate sections will be devoted to every spacecraft solar array management concept.

4 DET: LEO PCDU EVO CASE

The LEO PCDU EVO is the Airbus DS new generation of Power Conditioning and Distribution Unit designed for Low Earth Orbit applications, for spacecraft powers in the range from some hundreds of Watts up to 3kW. It comes to substitute the previous AS250 PCDU, with more than 12 flight models in the last 10 years, some of them currently in orbit, some of them under FM manufacturing [5].

The PCDU design for this product line implements a Direct Energy Transfer scheme based on S3R cells [3]. These are reliable by design (one failure tolerant) with respect to the loss of the power available from any solar array section. The PCDU implements the M+1 concept (one extra section) with two main functionalities under MEA control:

- To be able to feed the bus with the power coming from a section whose S3R cell failed in open circuit in the power path to the main bus, or
- To handle the power from a solar array section whose S3R cell failed in open circuit in the shunt power path, in case the instantaneous bus consumption is below the power provided by this section.

Nominal S3R cells 1 to 6 are connected to the M+1 section by a pair of power diodes in series as can be seen on Figure 1. The same diodes mentioned before are used to connect the S3R cells to the electronics in charge of passivating the solar array.

Passivation is exercised by a pair of electromechanical power relays in series (to be one failure tolerant w.r.t. a failure leading to the contacts of one relay stuck to on position). Moreover, the activation of the passivation is executed in two steps, Arming + Firing. It is reversible as long as there is sufficient energy available in the spacecraft battery to operate and command the spacecraft avionics.

The passivation relays are physically implemented in the Battery module to optimize the use of the available volume in the unit.

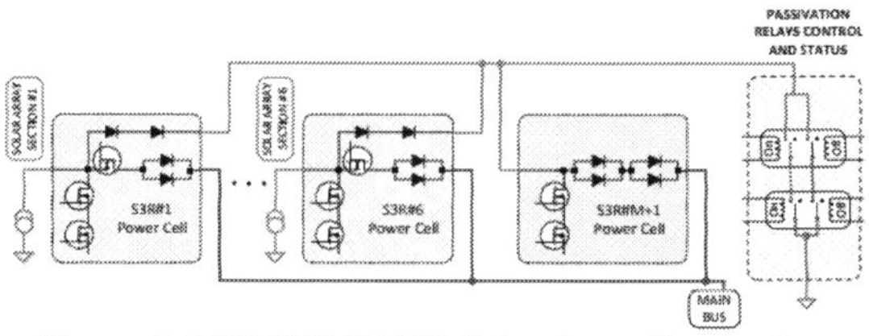

Figure 1. LEO PCDU EVO Solar Array Passivation

The passivation is activated in two steps and through two different command sources:

- First, an ARM command is executed, to enable the function to actuate. This is a High Voltage / High Power Command (HV-HPC) direct

 DOI: 10.1051/e3sconf/20171613003

command sent without intervention of the MIL-1553 serial communication interface. It actuates over a small TL26 relay that keeps its status latched until the moment on which the function is eventually deactivated by the DISARM command, also of HV-HPC type.

- A second command is needed to be executed in the window time opened by the ARM command. This is the so called PASSIVATION CONFIRM, this time received through the MIL-bus 1553 interface.

With this design any single failure either in the HV-HPC generation or in the 1553 interface will not activate the passivation.

Now a sequence of internal processes takes place inside PCDU to go to passivated state. They are depicted in Figure 2.

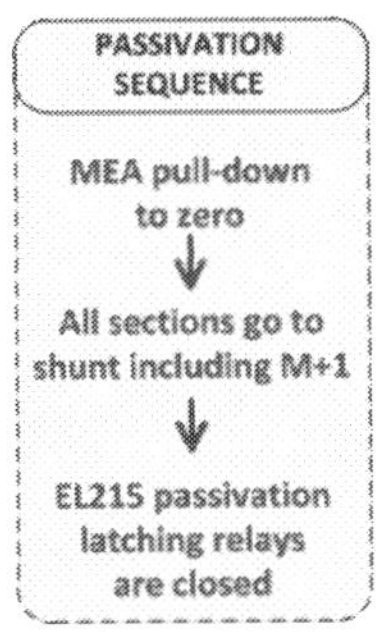

Figure 2. LEO PCDU EVO Passivation steps.

To reverse passivation is required a careful sequence of operations to prevent from stressing the passivation function electronics and the rest of the PCDU hardware. Passivation reverse sequence is detailed in the following diagram.

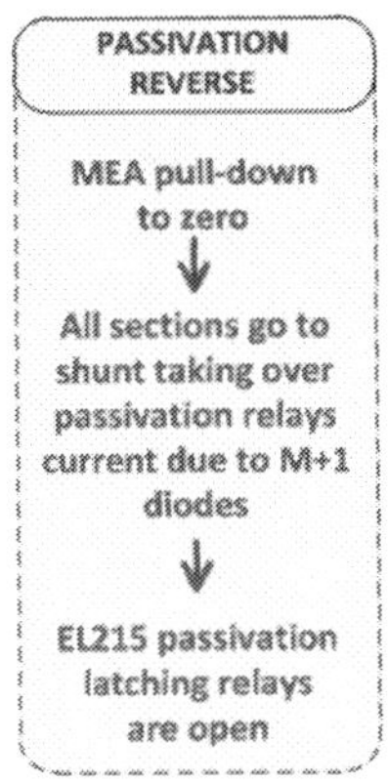

Figure 3. LEO PCDU EVO Passivation Reverse steps.

5 B2R: HP PCDU CASE

The High Power PCDU (HP PCDU) developed by Airbus DS uses solar array regulators in buck/boost (B2R) configuration with Maximum Power Point Tracking. The selected topology allows operating the solar array in a wide voltage range, with input voltages higher or lower than the main bus (battery follower) voltage, whilst providing high conversion efficiency. Thus it removes the limitation of input voltage by the main bus existing in systems with a single buck or a single boost stage. When the solar array voltage is similar to the main bus voltage, the B2R operates intermittently in buck or bust mode, with phases of direct energy transfer in between. To improve the overall system efficiency, the B2R has an internal auxiliary power supply, which is supplying the regulator only during the sunlit part of the orbit, controlled by the solar array voltage.

Redundancy in solar array regulation is achieved by an N+1 concept, i.e. by isolating a failed converter from solar array and bus and by a reconnection of the affected solar array section to the redundant converter [6].

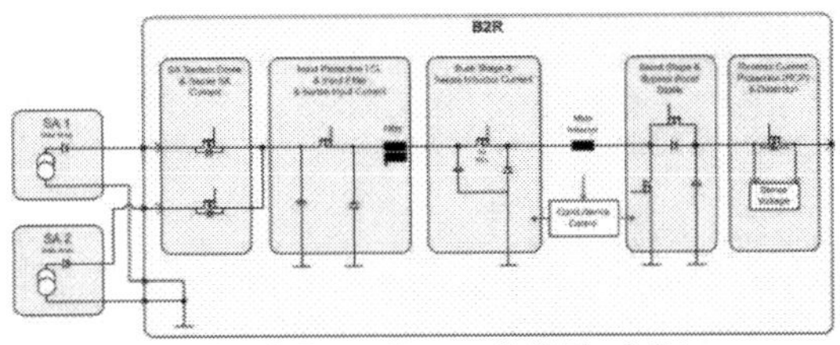

Figure 4. Buck-Boost Power Cell Block Diagram.

In the HP PCDU, the EOL passivation electronics disable the power transfer from the solar array to the main bus, while the battery remains connected to the main bus.

Two independent signals are used to activate the passivation process, each received by a relay:

- Passivation Arming.
- Passivation Confirmation.

The Arming relays are controlled by direct High Power Commands (HPC) while the 'confirmation' switches are activated by MIL-1553 command, so a small driver is needed. All relays are 'latching' as their status needs to be unchanged when the battery is fully discharged after the SA has been passivated.

In order to provide additional operational safety, the Passivation Confirmation is made by two independent MIL-1553 commands, which both have to be received by the PCDU in the right sequence in order to activate the passivation relay.

 DOI: 10.1051/e3sconf/20171613003

The relays are connected in series thus are 'one failure tolerant' w.r.t. unintended activation. In this way, the B2R MPPT modules can be disabled only after receiving an Arm and both Confirmation commands, so no single failure leads to unintended disabling of all SARs.

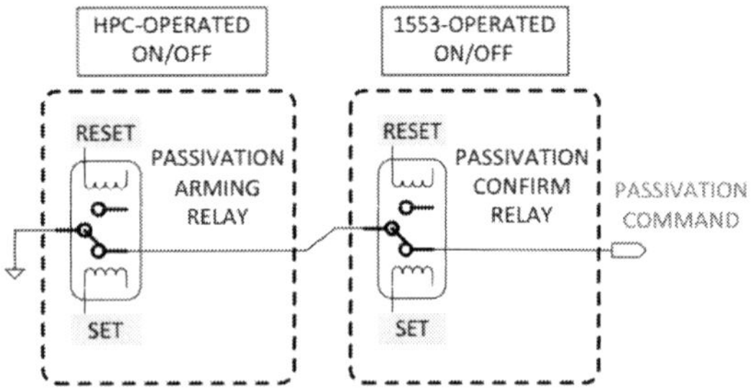

Figure 5. AS400 PCDU Passivation control strategy

The passivation signal generated in this way switches off the B2R auxiliary supplies, thus inhibiting the energy transfer from the solar array to the battery and the main bus. Physically, the passivation relays are located in the Battery module

6 MPPT: SENTINEL-2 C&D AND JASON (SENTINEL-6) PCDU CASES

6.1 Initial Concept

The PCDU design for these two missions Sentinel-2 and JASON is mainly a reuse of the MPPT-based in-orbit Sentinel-2A PCDU. The only difference is the number of MPPT modules required for each spacecraft, as two modules are needed for Sentinel-2C PCDU and four modules for JASON (Sentinel-6) PCDU. While the original Sentinel-2A mission developed before the implementation of the LOS laws did not incorporate any passivation electronics, Sentinel-2C and Sentinel-6 require this function and the related hardware adaptation.

At the beginning the proposed approach was based on pulling down the MEA signal at the end of the satellite operational life. As the MEA signal is a triple redundant majority voted reliable signal, the proposed approach was to act independently on each MEA branch by means of a small latching relay, TL26 type, when required to passivate the spacecraft.

One relay was devoted to each MEA branch, in this way a single failure in one of the relays leading to the loss of one MEA branch is equivalent - from the final effect point of view - to losing one out of three of the MEA branches, which is an already existing failure scenario, self-compensated by the existing design. An arming stage relay, also TL26 type, upstream the MEA branches passivation execution relays, forces the passivation to affect "simultaneously" to the three MEA branches, preventing bus transients that may happen in case the three MEA branches would be passivated (pulled to zero) sequentially. To achieve this, the passivation arming relay must be the last to be closed. Refer to figure 6 for further details.

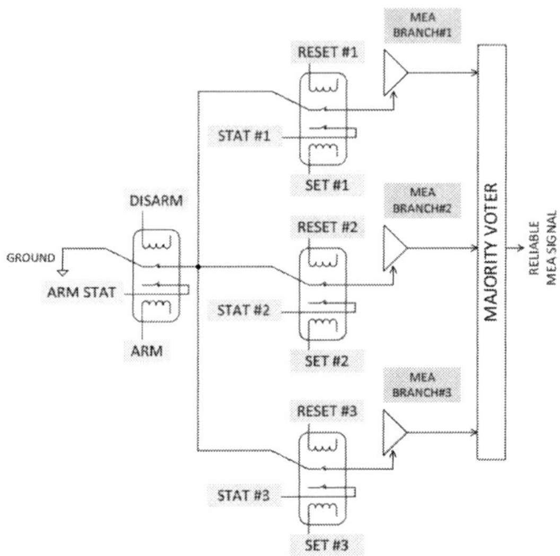

Figure 6. Initial passivation concept for Sentinel-2 MPPT-based missions.

The proposed architecture implements the following failure tolerance concept:

- Passivation will not be executed after a single failure in the passivation relays: two barriers must be closed to exercise the passivation. In particular, the failure in short circuit between the relay contacts and return terminal of the relay coil, would only pull to zero as much as one of the three MEA branches, but not the reliable MEA signal.
- The passivation function can be lost after a single failure (e.g. failure in permanent open condition of the arming relay). This is not against the existing passivation requirements, as was already mentioned in previous paragrahs.

Figure 7 summarizes which would be the PCDU MPPT function behavior in sunlight & eclipse conditions once the passivation function was activated. In the drawing, APR stands for Array Power Regulator, i.e., the power converters in charge of extracting the power from the solar array, either in conductance or MPPT mode as a function of the spacecraft overall energy balance.

DOI: 10.1051/e3sconf/20171613003

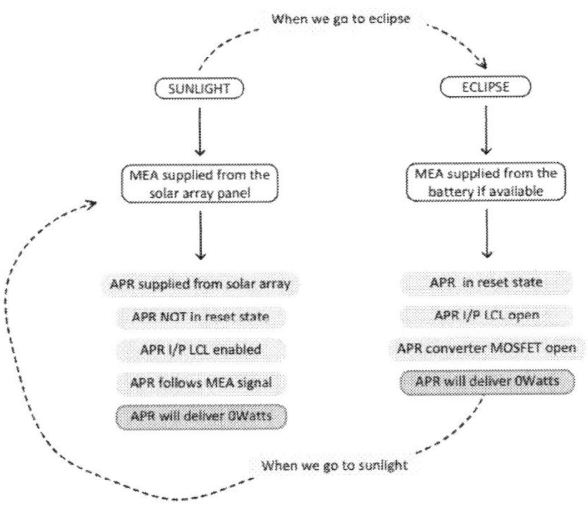

Figure 7. APR behavior in sunlight & eclipse while passivated

6.2 Final Implementation

According to the evolution in time of the EPS customer preferred implementation, for Jason (and Sentinel-2C & D) mission the depletion of the flight batteries is considered a must, but not necessarily the isolation of the solar array power, based on the ISO "Space Systems, Space Debris Mitigation Requirements" document, ISO 24113, whose requirement 6.2.2.3 states that "a spacecraft or launch vehicle orbital stage shall permanently deplete or make safe all remaining on-board sources of stored energy". In view of above considerations, and taking advantage of the existing battery interface electromechanical relays, the final solution adopted for these programs is conceptually depicted in Figure 8.

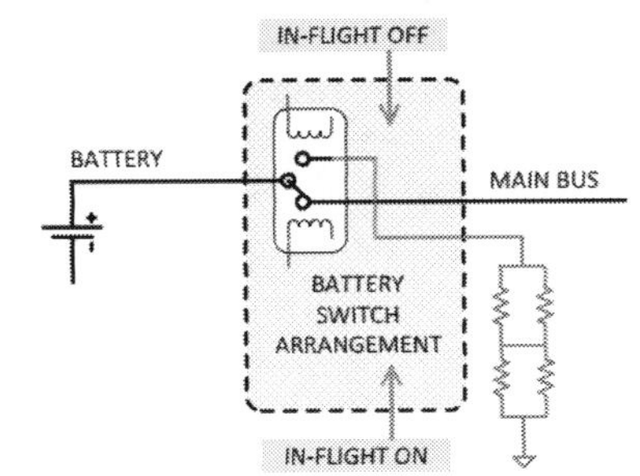

Figure 8. JASON and Sentinel-2 C/D passivation approach

Once the Satellite has reached the end of mission, the passivation process is followed in three steps:

1. Establish a controlled low SOC for the battery (by means programing a special low End-of-Charge (EOC) level).
2. Command Battery switches to open. This enables controlled battery discharge with a fixed built-in leakage (resistors in red in figure 8 above). This is done through flight HPC commands.
3. Entering into passivation phase.

Both Jason and Sentinel-2 C/D share the same architecture, but exhibit some differences regarding battery switch arrangement. A node of the battery switch is always made with two EL-215 relays in parallel, but while the Sentinel-2 C/D battery switch has a 4 nodes quad series-parallel configuration, the Jason one is made with 5 nodes in parallel. Apart from this detail, the performance of the passivation function has no difference.

The quad arrangement of resistors are sized as to provide a slow battery discharge rate leading to a current that can be accommodated in the satellite energy budget in case of unexpected connection. Regarding the Off (battery relay opening) command, an independent command is devoted to each node of two relays in parallel, so switching off (relay opening) is sequential and opportunity exist to reverse the passivation action.

7 THE GRACE FOLLOW-ON PCDU CASE

Being a successor of the GRACE mission launched in 2002 the GRACE FOLLOW-ON (GRACE-FO) PCDU implements an electrically "quiet" bus with negative bus voltage [7]. Solar Array conditioning is provided by MPPT in a two-out-of-three redundancy configuration.
As a major part of the electrical PCDU design was based on previous heritage designs such as e.g. the SWARM PCDUs, the selection of an implementation concept for the passivation function was strongly influenced by boundary conditions of the heritage design.

The concept finally chosen for GRACE-FO is based on pulling the failure tolerant MEA signal downstream the majority voter to a saturation level which sets the power provided by the solar array regulators to zero. Battery discharge is then done by the internal loads of the satellite.

The MEA pulldown is implemented by a series of two latching relays, each switchable by a dedicated discrete HPC-command, i.e. again following two-step logic with:

- Passivation Arming
- Passivation Confirmation

 DOI: 10.1051/e3sconf/20171613003

The status of the relays can be monitored by telemetry and discrete Disarm commands are available for resetting the passivation as long as the satellite can be commanded.

As in the other concepts presented herein, this passivation method is single-failure tolerant w.r.t. unintended activation but may be lost with one single failure.

8 SUMMARY OF STRATEGIES VS MISSION CASES

Table 1 summarizes the approaches selected for passivation electronics in the LEO PCDUs described in this paper.

PCDU Case	LEO PCDU EVO	AS400	S-2 C&D JASON	GRACE-FO
Solar array conditioning architecture	DET (S3R)	MPPT (B2R)	MPPT (Step down)	MPPT (Step down)
Passivation strategy	Passivate solar array	Passivate solar array	Disconnect Battery	Passivate solar array
Passivation ARM command	Direct HV-HPC	Direct HV-HPC	None	Direct HV-HPC
Passivation CONFIRM command	1553	1553	Direct HV-HPC	Direct HV-HPC
Passivation REVERSE capability	YES	YES	YES	YES

Table 1. Summary of passivation implementation strategies

9 CONSIDERATIONS ON EXTENDED PASSIVATION TEMPERATURE RANGE

As commented at the beginning of the paper, the satellite and on-board electronics will experience – once the spacecraft was passivated - temperatures in excess of those normally required for this type of equipment, typically lower than the minimum start-up temperature and - many times - lower than the minimum qualification temperatures for some electronic and electromechanical components and/or state of the art associated manufacturing processes.

In the frame of LEO PCDU EVO program it is under agreement process to perform a set of complementary tests to gain confidence on the behaviour of the critical electronic components for the passivation function in the above mentioned extreme operational conditions.

Disposal phase temperatures are calculated to be in the range [-90°C; +60°C], understood as unit TRP (Temperature Reference Point) levels.

Diodes and relays are very strong and normally qualified for wide temperature ranges. PCB technologies are potentially more fragile. For this activity, it is deemed sufficient to make On/ Off cycles at hot and cold temperatures. The total number of On/Off cycles is very important. Eventually this number could be reduced considering that:

- After some years in disposal phase the solar array and the battery will be to some extent degraded in performances w.r.t. initial ones.
- Also, the amplitude of temperature during an orbital cycling is usually lower than the difference between the minimum and maximum temperature in steady state.

Accelerated On/Off cycling can be made in order to reduce the tests duration: a mixed cycling with steady state temperature combined with On/Off accelerated cycling can be performed, see Figure 9.

A setup will be generated to test the passivation relevant components in the closest possible scenario to the real operational one. The following relevant components are identified:

- Passivation power relays & free-wheeling coils diodes.
- Power Schottky series diodes.
- Shunt power resistors.
- Solar array interface connectors.
- Mechanical connections from the PCBs to main bus.

The setup, see Figure 10, will take into account the different worst case maximum currents circulating through every above mentioned item involved in the passivation, in particular:

- The maximum short circuit current of one solar array section for connector pins, current measurement shunt and Schottky diodes.
- The overall solar array panel current through the passivation relays and mechanical connections from the PCBs to the main bus bar.

Figure 9. Passivation components thermal cycling test.

DOI: 10.1051/e3sconf/20171613003

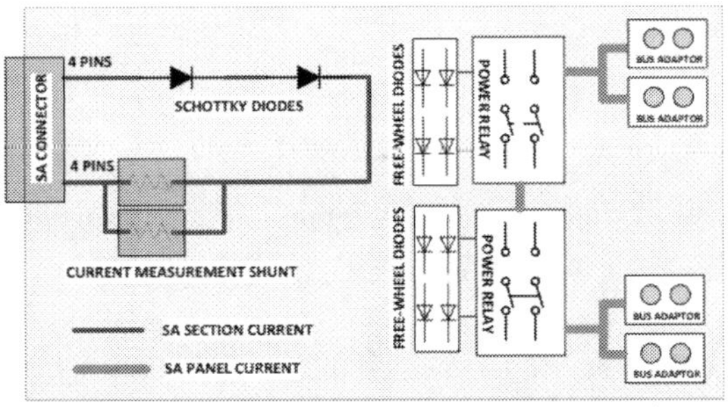

Figure 10. Setup for passivation components cycling.

10 CONCLUSIONS

Several strategies for spacecraft EOL passivation have been presented in this paper, all of them are implemented on board Airbus DS PCDUs developed for LEO applications. Practical cases presented cover a wide power range from a few hundreds of Watts to more than 7kW, and consider all the solar array conditioning techniques available today in the market, namely:

- Maximum Power Point Tracking, either based on conventional step down buck type derived power converters or B2R (buck / boost) power conversion stages.
- Direct Energy Transfer, based in the well-known Sequential Switching Shunt Regulator principle.

The real mission cases considered show a wide variety of implementations either oriented to Battery or Solar Array disconnection. The variety of solutions is driven by two factors:

- Passivation is a "relatively new" functionality that, to some extent, is going in advance with respect to the necessary knowledge on how batteries behave when exposed to the extreme environment of a passivated spacecraft without active thermal control.
- The translation of the passivation requirements (as written in the standards mentioned in section 1 of this paper) into specific requirements in the PCDU specifications is subject to different interpretations that at the end lead to significantly different hardware implementations.

One of the difficulties is the fact that satellite and on-board electronics (both components and processes) will experience temperatures in excess of those normally required for this type of equipment, typically lower than the minimum start-up temperature and - many times - lower than their minimum qualification temperatures.

11 REFERENCES

[1] F. Bausier. (2013). Passivation of spacecraft power system at end of mission. TEC-EPM/2013/33/FB, Technical Note. European Space Agency, Noordwijk, The Netherlands.

[2] Ch. Delepaut. (2013). Battery Isolation Function - Interface Standardisation Rational. TEC-EPM/2013/15CD, Technical Note. European Space Agency, Noordwijk, The Netherlands

[3] E. Lapeña et al. (2016) The LEO PCDU EVO – A Modular and Flexible Concept for Low to Medium Power LEO & Scientific Missions. In Proc. of the 11th European Space Power Conference. Thessaloniki, Greece.

[4] O. Mourra et al. (2010). Buck Boost Regulator (B2R) for Spacecraft Solar Array Power Conversion. In Applied Power Electronics Conference and Exposition (APEC), 2010 Twenty-Fifth Annual IEEE, p. 1313 - 1319

[5] J.Otero et al. (2011). Modular PCDU Product Line For LEO Applications. In Proc. of the 9th European Space Power Conference. Saint-Raphaël. France.

[6] D. Ruf et al. (2016). The AS400 Power Control and Distribution Unit - A Modular And Flexible Unit With B2R Solar Array Regulation For High Power Leo Missions. In Proc. of the 11th European Space Power Conference. Thessaloniki, Greece.

[7] M. Amann et al (2016). The GRACE FOLLOW-ON Quiet Electrical Power System. In Proc. of the 11th European Space Power Conference. Thessaloniki, Greece.

POWER SUBSYSTEM APPROACH FOR THE EUROPA MISSION

Antonio Ulloa-Severino(1), Gregory A. Carr(1), Douglas J. Clark(1), Sonny M. Orellana(1), Roxanne Arellano(1), Marshall C. Smart(1), Ratnakumar V. Bugga(1), Andreea Boca(1), and Stephen F. Dawson(1)

(1)Jet Propulsion Laboratory, California Institute of Technology, 4800 Oak Grove Drive, Pasadena, CA 91109 USA, Email: Antonio.Ulloa-Severino@jpl.nasa.gov

ABSTRACT

NASA is planning to launch a spacecraft on a mission to the Jovian moon Europa, in order to conduct a detailed reconnaissance and investigation of its habitability. The spacecraft would orbit Jupiter and perform a detailed science investigation of Europa, utilizing a number of science instruments including an ice-penetrating radar to determine the icy shell thickness and presence of subsurface oceans. The spacecraft would be exposed to harsh radiation and extreme temperature environments. To meet mission objectives, the spacecraft power subsystem is being architected and designed to operate efficiently, and with a high degree of reliability.

1. EUROPA MISSION OVERVIEW

The planned Europa mission's main focus is to execute an in-depth science investigation of Jupiter's icy moon Europa shown in Fig. 1 [1]. The moon has shown evidence of a liquid water ocean underneath the unknown thickness of icy surface. The ocean could provide an environment suitable for life due its interaction with a volcanic seafloor.

Figure 1: Cutaway diagram of Europa's interior. Artwork credit: Michael Carroll

The combination of harsh radiation, low temperature (30 K) environments, and distance from the sun (5 AU to 5.5 AU) creates challenges on design and implementation of the spacecraft and instrument hardware. The mission design team is expending great efforts to minimize the amount of total dose radiation the flight system would accumulate and maximize science data return.

The NASA-selected instruments intend to characterize Europa by producing high-resolution images and surface composition maps. An ice-penetrating radar intends to determine the crust thickness, and a magnetometer would measure the magnetic field strength and direction to further help scientists determine ocean depth and salinity. A thermal instrument would be used to identify warmer water that could have erupted through the crust, while other instruments would help detect water and particles in the moon's atmosphere.

The nine NASA-selected instruments that would be used for the Europa mission are the following [2]:

1) Plasma Instrument for Magnetic Sounding (PIMS)
2) Interior Characterization of Europa using Magnetometry (ICEMAG)
3) Mapping Imaging Spectrometer for Europa (MISE)
4) Europa Imaging System (EIS)
5) Radar for Europa Assessment and Sounding: Ocean to Near-surface (REASON)
6) Europa Thermal Emission Imaging System (E-THEMIS)
7) Mass Spectrometer for Planetary Exploration/Europa (MASPEX)
8) Ultraviolet Spectrograph/Europa (UVS)
9) Surface Dust Mass Analyser (SUDA)

2. POWER SUBSYSTEM ARCHITECTURE

The power subsystem architecture design approach involves implementing a robust single-fault tolerant design with small fault containment regions, based on a similar approach used for the Cassini mission to Saturn. The design has focused on three main areas: the baseline solar arrays, power management electronics, and batteries as shown in Fig 2.

DOI: 10.1051/e3sconf/20171613004

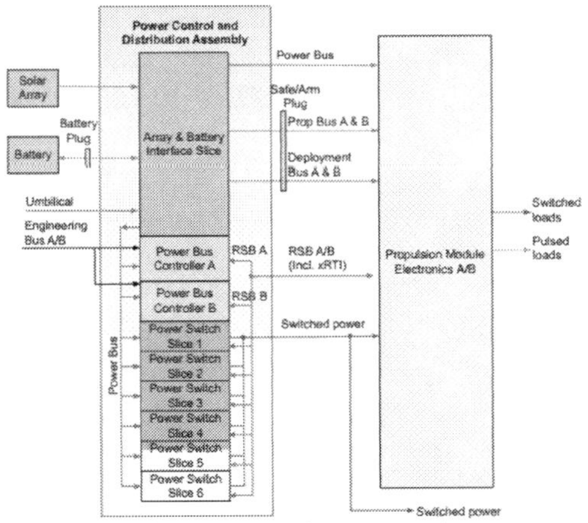

Figure 2. Baseline Power Subsystem Architecture.

2.1 Power Source

The solar array substrates, materials, and components are being tested under the relevant environmental conditions to characterize their performance and reliability in the Jovian environment. The primary challenges for photovoltaic operation at Jupiter are the extremely high radiation from both electrons and protons, extremely low temperatures (30 K), and low-irradiance low-temperature (LILT) solar cell performance. Also, mission design requirements dictate that the solar array mass and size constraints are of critical importance.

The solar array string design and layout has been adjusted to match the capability of the radiation-hard electronics. The number of cells in series was set by the topology of the power management electronics to deliver the most power at the end of the mission in the most severe environment. The challenge was to obtain enough voltage for the power management electronics early in the mission, during the inner solar system cruise, and then still produce efficient power at the end of mission. The power management electronics was designed to deliver close to the maximum power available from the solar array at the end of mission in a robust fault-tolerant architecture. The power management electronics has two primary functions, power control and power distribution.

2.2 Power Electronics

The baseline power control design operates near the maximum power point of the solar array throughout the Jovian tour based on solar cell telemetry and modelling. Power control electronics can charge the battery after a flyby operating near the I-V (current-voltage) curve maximum power point without collapsing the solar array voltage. The design is single fault tolerant. Power bus voltage and solar array voltage set points are controlled via commands. Power control uses an N+K architecture for the number of power stages, where N is number required for the mission and K is the number of additional stages provided for redundancy, as shown in Fig. 3. The command and telemetry interface to the power management electronics is block redundant.

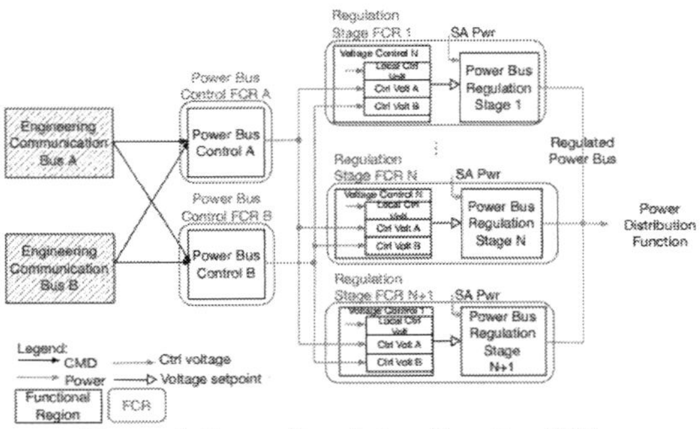

Figure 3. Power Regulation Function FCRs.

The power distribution electronics are designed with internal housekeeping power that provide independent functional slices that can be utilized throughout the spacecraft. The power distribution function incorporates a resettable circuit breaker power switch to separate the fault containment region for each load interface. The power switch is part of the load fault containment region as shown in Fig. 4 and is fail-safe off. Redundant functions on the spacecraft are powered by another power switch in a separate fault containment region. Each power switch has a block redundant command and telemetry interface to maintain the load-level fault containment region.

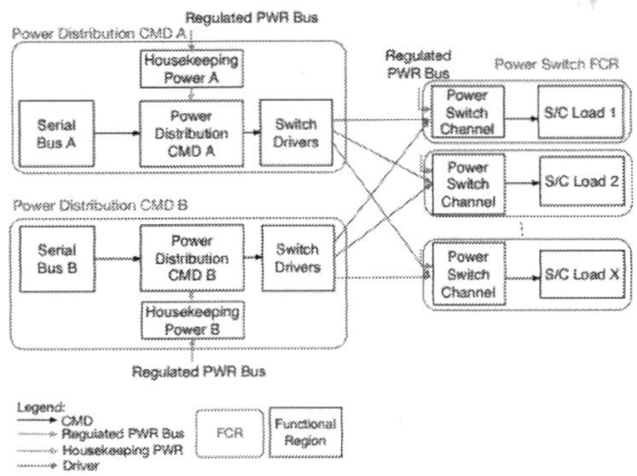

Figure 4. Power Distribution FCRs.

2.3 Energy Storage

The preliminary battery architecture uses high specific energy small 18650-size lithium-ion cells, to capitalize on their internal safety functions, high capacity, and excellent cell-to-cell reproducibility. Utilizing such a format is advantageous, since the approach does not require individual cell monitoring and balancing circuitry. Each string of battery cells is a separate electrical fault containment region. Each battery cell string

3. POWER SUBSYSTEM SYSTEM ENGINEERING

3.1 System Engineering Context, Requirements and Verification and Validation (V&V)

System engineering on the Europa Project is divided into the following domains:

- Project system engineering, which is concerned with everything from securing funding to delivering on the project's stated science goals and requirements.
- Mission system engineering, which is concerned with the launched spacecraft, launch vehicle, flight-to-ground communications networks, and the pre- and post-launch operations of the Europa spacecraft and payload.
- Flight system engineering, which is concerned with the Europa spacecraft and payload.
- Payload system engineering, which is concerned with the science payload instruments.
- Subsystem system engineering for each of the spacecraft's subsystems: power, telecom, guidance, navigation, and control, avionics, mechanical, thermal, and propulsion.

Each of these system engineering domains develops formal requirements for the products (hardware, procedures, algorithms, software, etc.) developed by the domain. Where appropriate, requirements are linked to domains above (parents) and domains below (children). When parent or child link exists, system engineers on both ends of the link must agree and concur on the requirements and links.

Where coordination between subsystems is needed to achieve desired flight system functionality, such as solar array deployment, each of the involved subsystems will have requirement(s) linked to a common parent requirement in the flight system domain. There are no direct links between requirements in different subsystems.

Within a subsystem, system engineers may decompose requirements into finer detail. Within the power subsystem for example, requirements exist at the following levels: subsystem, assembly, slice (board), and field-programmable gate array (FPGA).

During the early phases of the project, system engineering for the Europa power subsystem is primarily concerned with three topics:

- definition of power subsystem functionality,
- determination of required power subsystem capabilities, and
- design of interfaces, both within and external to the power subsystem.

The set of power subsystem requirements is the formal agreement between the power subsystem domain and the flight system domain. Requirements for the power subsystem are being developed and maintained in the Rational Dynamic Object Requirements System (DOORS) Next Generation web-based software tool.

Later, as the Project and subsystems proceed into implementation phases, the attention of power subsystem engineers would shift into V&V activities. Verification involves the formal check that the delivered power subsystem products meet all their requirements. Validation is a check to ensure that the power subsystem would meet the project and mission objectives, when operated in the context of the other subsystems, the flight system, and the mission system.

3.2 Power Subsystem Functionality

The Europa Power Subsystem has the following major functions:

- Power generation,
- Energy storage,
- Power conditioning,
- Power distribution,
- Safety inhibits for radio frequency (RF) transmitters, propulsion, and deployments,
- Detection and response to under-voltage conditions, and
- Response to "dead bus" conditions.

The major internal interfaces within the power subsystem are:

- Solar array power interface to power electronics,
- Battery charge/discharge interface to power electronics,
- Protected bus interfaces to Propulsion Module Electronics (PME),
- Communications bus interface to Power Switch Slice (PSS) and PME.

The major external (outside the subsystem) interfaces are:

- Launch-vehicle break-wire interfaces to power subsystem electronics,
- Switched power interfaces to payload and flight system loads,
- Interface to engineering communication bus,
- Umbilical power interface, and
- Umbilical safety interfaces.

Power generation for the Europa spacecraft would be

provided by a solar array, comprised of approximately ten panels arranged into two five-panel wings. Each wing is mounted on a single-axis gimbal to facilitate sun-pointing the wing.

Energy storage is necessary on the spacecraft to power the flight system and the payload equipment during times where the electrical load power exceeds the power generation capability of the solar array and during eclipse and launch periods. The baseline power subsystem contains lithium-ion batteries to provide energy storage.

Power conditioning is needed to match the variable voltage output of the solar array to approximately the fixed voltage of the power bus and battery.

The power distribution function provides switched power to electrical loads in the flight system and payload. The power distribution function provides controlled voltage ramp to the load (soft-start) and resettable over-current protection.

The power subsystem electronics detects separation of launch vehicle break-wire interfaces. The interfaces convert to "open-circuit" when the spacecraft separates from the launch vehicle. Once an open-circuit condition is detected, the flight software may activate power distribution to RF transmitters. Protected propulsion and deployment busses originating in the power subsystem electronics may also be energized.

Flight software (FSW) running on the Command and Data Handling (C&DH) computer commands the power subsystem via the engineering communication bus. This bus, based on MIL-STD-1553, is also used by the FSW to collect power subsystem telemetry.

The power subsystem electronics also detects the presence of under-voltage (UV) conditions. Upon detection of UV, the power subsystem electronics will remove power to non-critical loads, and make telemetry available to the FSW. The power subsystem is designed to recover from a "dead bus" condition, where the battery is completely discharged. This could be due to a fault that points the solar array off-sun for an extended period of time.

The power subsystem electronics also contain umbilical interfaces that pass-through launch vehicle wiring and terminate in ground support equipment (GSE) at the launch site. These interfaces provide a means to power the flight system and payloads and charge the battery. Certain telemetry functions related to battery safety are available to GSE even when the spacecraft is powered off.

The power generation and energy storage functions of the power subsystem need to be sized adequately to perform the mission, without being oversized and providing a mass penalty. Power subsystem engineers use JPL-developed modelling software and work closely with flight system engineers to predict the performance of the solar array and batteries over the mission timeline.

4. KEY AND DRIVING REQUIREMENTS

Key requirements have a significant impact or high sensitivity on the mission's ability to meet negotiated success criteria; these requirements are mission critical and essential for the overall science objectives.

Driving requirements are those deemed particularly difficult and the likelihood of needing additional resources to meet the requirement is significant, as assessed by the requirement implementer. This set defines the system architecture and has the most impact on the design and implementation.

A summary of the key and driving requirements are summarized below.

4.1 Lifetime and Performance in the Jovian Environment

The spacecraft would be exposed to the high-energy radiation and extreme cold thermal cycles in the Jovian environment. The power subsystem would be required to perform its critical functions during and after the environmental exposure that significantly impact its performance and lifetime. The spacecraft provides a radiation-shielded enclosure that houses sensitive electronics reducing the ionization radiation levels.

Although the power electronics are located inside the enclosure, parts qualified for the environment, including transient effects, are not always available, driving evaluation and testing of electronic devices. The accumulated radiation environment inside the enclosure is 150 krad (Si); however, the power electronics are required to meet a radiation design factor of 2, or 300 krad (Si). The solar array and battery are outside the enclosure, where the radiation environment is predicted to be 3.0 Mrad (Si), and are required to operate reliably at distances of greater than 5 AU. Similar to the electronic devices, the solar cell technologies will undergo a qualification process.

4.2 Fault Tolerant Mission Critical Functions

In order to meet the science objectives, all mission critical functions need to be single fault tolerant to provide a robust and resilient design.

The power subsystem consists of the following mission

 DOI: 10.1051/e3sconf/20171613004

critical functions, shown in Fig. 5:

- Power source,
- Energy storage,
- Power bus regulation,
- Power distribution, and
- Command and telemetry.

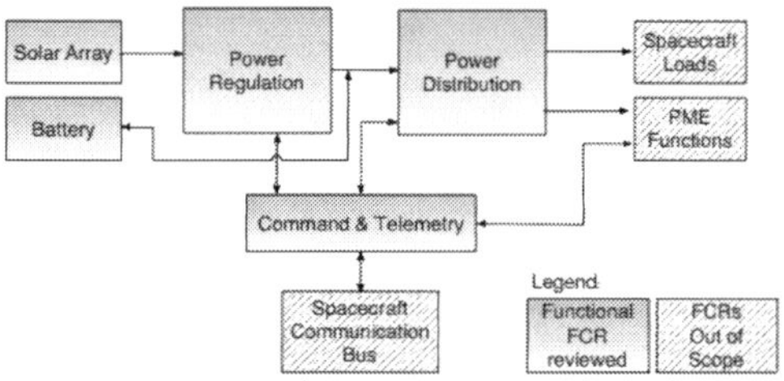

Figure 5. Power Subsystem Functional Block Diagram.

These will be designed such that a single failure can be tolerated and contained within the defined fault containment boundaries preventing fault propagation. Lower-level fault containment with N+1 redundancy minimizes power consumption while providing resiliency and robustness in the design. This primarily drives design implementation complexity.

4.3 Power Subsystem Sizing (Consumption and Load Profile)

Power and energy demands from the spacecraft loads are the key drivers for the solar array and battery sizing. As solar array and battery power demands increase, the power electronics needs to be capable of processing and managing it. As the payload science instrument designs mature, their power demand typically increases and drives mass and cost higher for all hardware elements. At distances greater than 5 AU, the limited sunlight received by the solar array would be more than 25 times less than that observed at 1 AU. Due to this limitation in power availability, the low power consumption requirements imposed on the power electronics is very challenging to meet.

4.4 Planetary Protection Category III Rating [3]

A Category III classification rating ensures rigorous planetary protection and contamination control requirements are implemented, to extensively lower the probability of contaminating the icy moon with microbes carried from Earth. The bioburden reduction level required prior to launch is stringent and challenging to meet. All power hardware materials and components will be subjected to approved sterilization methods that are currently being developed by the project and the subsystem leads. The approved methods include dry heat microbial reduction and vapor hydrogen peroxide treatment for surfaces only; radiation dosage of ~10 Mrad is under evaluation by the project as an alternative approach for other components. These requirements drive cost impacts taking into account fabrication phases, sterilization approaches for each hardware assembly, and vendor conformance to these requirements.

5. POWER SUBSYSTEM ELECTRONICS

5.1 Europa Power Electronics

The primary focus of the power electronics is to provide a robust, highly reliable and efficient power management and distribution system within the expected harsh conditions. In order to meet these criteria, the power electronics have been designed to minimize both power consumption and physical size while providing more visibility and control to the power bus distribution than the systems implemented on previous Class A missions.

The power subsystem is designed to be capable of managing the spacecraft bus while powered by the solar array and battery simultaneously. The Europa Power Electronics Assembly (PCDA) is comprised of three slices, shown in Fig. 2:

1. Array and Battery Interface Slice (ABIS) – one slice per assembly,
2. Power Bus Controller (PBC) Slice – two slices per assembly, and
3. Power Switch Slice (PSS) – up to six slices per assembly.

5.2 ABIS and PBC Overview

The primary function of the ABIS is to provide the interface between the solar array and the battery power to the spacecraft golden node power bus. A step down power converter is used as the interface between the solar array and the spacecraft power bus. The power converter output voltage is an adjustable voltage controlled by the redundant PBC. The battery interface is a direct connection to the spacecraft bus. The power converter majority votes the control signals from the redundant PBCs, and its local control to adjust the output voltage to regulate the spacecraft power bus and limit the battery charge current to ensure battery health. The ABIS also provides the spacecraft power bus interface for the other slices in the power electronics assemblies. The ABIS provides one of three safety inhibit functions to meet safety requirements. The safety inhibit function interrupts the current path to the propulsion and deployment buses which are inhibited by launch vehicle break-wires. For each of the power interfaces, the ABIS provides current telemetry sensors that are sent to the PBC for measurement and digital telemetry.

The PBC slice primary function is to send the control signals to the ABIS power converter to regulate the spacecraft power bus and maintain the charge rate of the battery. A PBC slice contains a FPGA, SRAM memory, analog to digital converter to report analog telemetry as digital telemetry, multiple multiplexers to cycle through all the telemetry channels, independent housekeeping power converters, and a current source circuit which provides a test current to measure Platinum Resistance Thermometer (PRTs) exist throughout the power electronic assemblies. Each PBC continuously monitor the power bus and battery charge current and determines the proper voltage set point of the power converter to either charge or discharge the battery based on a commanded voltage and current set point. Due to the expected harsh environment, two identical PBC slices are provided for redundancy, which result in two independent bus regulation measurements and two control signals routed to the ABIS. The ABIS majority votes these two signals along with a local measurement to determine the correct bus voltage regulation point.

The PBC also measures multiple telemetry inputs (voltages, currents, and temperatures) and reports the information to the flight system through a 1553 digital telemetry interface. The PBC provides the detection of the spacecraft break-wire interfaces and sends the commanded control signals for the inhibited power busses (propulsion and deployment busses). The PBC also measures the bus voltage for the purpose of determining the under-voltage threshold. If the bus voltage falls below a hardwired threshold, the PBC will send a discrete signal to all PSS slices to initiate their predetermined under voltage protocol to recover the bus voltage.

5.3 PSS Overview

The primary function of the PSS slice is to provide the redundant, fail safe off, load switches for spacecraft loads. A single PSS slice contains two FPGAs for command redundancy, two independent housekeeping power supplies, high and low side switches of which individual current telemetry is reported. The design of the PSS ensures that the switches are individually redundantly commanded, which is single fault tolerant. Two FPGAs on the PSS allow switches to have specific behaviors under certain conditions, such as under voltage conditions or critical paired loads. The PSS communicates with the PBCs on a dedicated serial communication bus separate from the spacecraft engineering communication bus to ensure that a single failure will not stop the execution of a switch command. Switch status and telemetry are reported by both FPGAs through digital telemetry to the PBCs and to the flight system via engineering communication bus when requested.

6. BATTERIES

The preliminary architecture for the Europa mission is to use a battery design consisting of high specific energy, small 18650-size lithium-ion cells, to capitalize on their internal safety functions, high capacity, and excellent cell-to-cell reproducibility. Utilizing such a format is advantageous since the approach does not require individual cell monitoring and balancing circuitry. The power subsystem is currently evaluating potential cell chemistries which can provide the desired performance to meet the challenging mission requirements, including (i) tolerance to high levels of radiation, (ii) efficient operation over a wide temperature range, and (iii) excellent storage life characteristics to survive the long cruise period (over 6.5 years).

Due to the strong magnetic field of Jupiter, energetic particles are captured from the solar wind, primarily electrons and protons, creating intense radiation belts that surround the planet. There is a concern that lithium-ion chemistries might degrade under such harsh radiation environments, since they contain organic electrolyte solutions, polymer binders and separators, which may lead to performance loss and potentially premature cell failure. To address this concern, studies subjected the following commercial lithium-ion cells to high levels of γ-radiation using a ^{60}Co source:

- large format $LiNiCoO_2$-based,
- large format $LiNiCoAlO_2$-based [4], and
- $LiCoO_2$-based [5].

With regard to the large format cells, both 7 Ah prismatic cells and 9 Ah cylindrical cells were demonstrated to be tolerant to cumulative levels as high as 25 Mrad, with less than 10% permanent capacity loss observed, some of which may be attributed to the cycling and storage associated with the characterization testing. In a similar study, the Sony 18650-size lithium-ion cells were demonstrated to be tolerant to radiation levels as high as 18 Mrad, displaying 4% to 6% capacity loss upon exposure. A comparable study on recently developed commercially available high specific energy 18650-size li-ion cells was conducted on Panasonic NCR-A and NCR-B cell [6]. Minimal capacity fade and impedance growth was observed with the cells after being subjected to cumulative γ-radiation levels of 18 Mrad from a ^{60}Co source. After completing the radiation testing, these cells also displayed good performance when subsequently cycled at full depth of discharge (DOD) at ambient temperatures using C/2 charge and discharge rates over the voltage range of 3.0 V to 4.10 V (over > 700 cycles completed for each cell type). More recent studies have been devoted to evaluating the radiation tolerance of other candidate commercially available 18650-size high specific energy cells.

Another key requirement is that the spacecraft battery delivers a long mission life (> 11 years) while meeting the power and energy demands over a wide temperature range. Given that there would be a long cruise period (over 6.5 years) prior to reaching the Jovian orbital insertion (JOI) event, it is essential to properly maintain the battery during this time to preserve the health to meet end of life requirements. This can be achieved by operating the battery during cruise at a partial state-of-charge (SOC) and at moderate to low temperatures. For the Europa mission, various candidate high specific energy 18650-size lithium-ion cells are being characterized for storage characteristics over a range of temperatures [7]. These tests involve storing the cells on the spacecraft power bus at a partial state of charge at a fixed potential to mimic the spacecraft environment. To evaluate their performance, capacity and impedance are measured periodically to determine the degradation rates of various cell chemistries. In addition to cell level testing, complementary 8-cell string (connected in series) level tests have been implemented, in which individual cell voltages are monitored throughout the testing. There is some concern that the cell voltage dispersion amongst individual cells within a string can diverge with time, leading to poor performance and the possibility of overcharging the cells. Although only six months of storage time have been completed on the 8-cell modules to-date, minimal increase in the cell voltage dispersion has been observed amongst all cell chemistries evaluated. Furthermore, minimal cell voltage dispersion was also observed with time when modules were subjected to 100% depth-of-discharge (DOD) cycle life testing at 20 °C. As illustrated in Fig. 6, when the open circuit potentials of the eight cells comprising one of the modules evaluated are measured after completing the charge period, very little change in voltage dispersion is observed as a function of cycle life.

In addition to long-term storage at moderate temperature, evaluations were conducted for short duration and high temperature storage testing on cells maintained at a high state-of-charge to simulate the conditions on the launch pad prior to launch. In summary, these tests are essential to enable proper end of life performance projections and appropriate battery sizing.

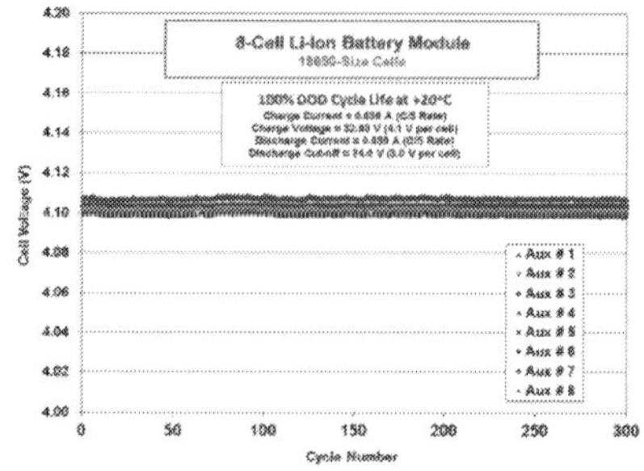

Figure 6. End of charge open circuit cell potentials of an 8-cell Li-ion cell pack during 100% DOD cycle life testing at +20 °C.

7. SOLAR ARRAY

The Europa solar array baseline design is comprised of two deployable solar array wings, one on each side of the spacecraft x-axis. The wings are single-axis gimballed to allow the sun incidence angle to be adjusted independently of the spacecraft pitch axis. Each wing is comprised of five solar panels, where each panel has a total area of approximately 9 m^2. The active solar array area, i.e. the total surface area available to be populated with coverglass interconnected cells (CICs), is 87 m^2. The array delivers power to the spacecraft throughout all mission phases when sunlight is available to the solar array. The environmental conditions driving the size of the array are 5.5 AU, -140 °C at the end of mission, after including the effect of radiation and all other degradation effects. The CICs are based on SolAero ZTJ solar cells with Qioptiq CMG coverglass, DC93-500 transparent adhesive, and silver-clad molybdenum interconnects. All CICs are screened under low-irradiance room temperature (LIRT) conditions of 5.5 AU, 28 °C which eliminates almost all the CICs that would experience performance degradation under LILT flight conditions. All series strings on the array have the same length, and each string is outfitted with a blocking diode. The minimum acceptable string length is driven by Venus fly-by conditions at 0.6 AU, whereas the maximum string length is driven by Jupiter end-of-eclipse conditions at 5.5 AU. Each panel is outfitted with two PRT temperature sensors and two CIC sensors for in-flight I-V sweep measurements.

The Europa project is currently testing ZTJ bare cells and CICs under mission-relevant laboratory conditions of irradiance, temperature, and radiation dose. Fig. 7 shows a typical example of LILT illuminated current-voltage (LIV) sweep data taken on bare ZTJ cells at beginning of life (BOL) prior to irradiation, as well as after exposure to a 2e15 1 MeV e^-/cm^2 radiation dose. The LILT LIV sweeps were obtained at a cell temperature of -140 °C, and at a 5.5 AU irradiance, using an AM0-calibrated 2-zone solar simulator. This type of measurement will be performed over a test matrix of multiple irradiances, temperatures, radiation

DOI: 10.1051/e3sconf/20171613004

doses, and results will be used in making data-driven predictions of the array performance in the flight environment. As a second example, Fig. 8 shows the measured LILT average maximum power (Pmp) remaining after radiation, as a fraction of the pre-radiation power, for various 1MeV electron irradiation doses. The solar simulator test conditions were once again 5.5 AU, -140 °C; however, in this case the test articles were 12 ZTJ CICs with 0.012 in. thick coverglass rather than bare cells. The error bars represent the standard deviation in the measurements at each fluence. In addition to measured data, also shown in Fig. 7 is an exponential least-squares fit to the maximum power (Pmp) remaining fraction, per Eq. 1:

$$\frac{P_{MP}(f)}{P_{MP}(0)} = \alpha + (1-\alpha)e^{-\kappa f} \quad (1)$$

where α and κ are fit parameters, and f is the applied 1MeV electron exposure fluence.

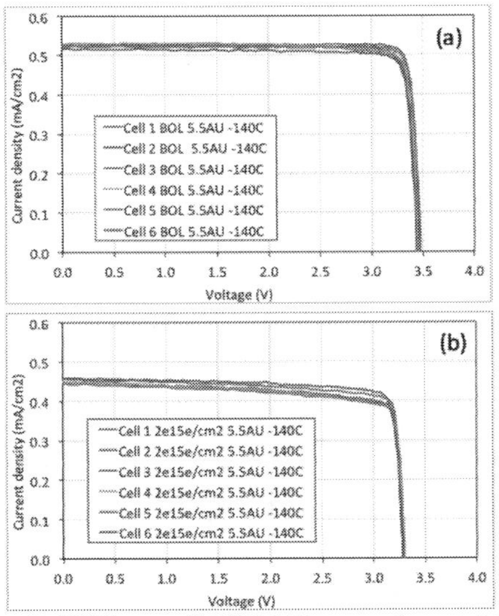

Figure 7. ZTJ bare cell LILT LIV sweep data at 5.5 AU, -140 °C: (a) pre-radiation; (b) after 2e15 1 MeV e⁻/cm².

8. SUMMARY

As described, the Europa mission baseline power subsystem architecture design approach has focused on three areas: solar arrays, power management electronics, and batteries. The power subsystem architecture has been designed to meet a number of challenging requirements, including high levels of ionizing radiation, extreme temperatures, and low solar irradiance. Furthermore, the architecture is a robust, single fault tolerant design with high reliability. The power subsystem architecture provides efficient delivery of power to loads with small fault containment regions. The fault containment regions of the solar array and battery are at the string level. The fault containment region of the power control is N+K. The fault containment of power distribution is at the load level with a block redundant command and telemetry path.

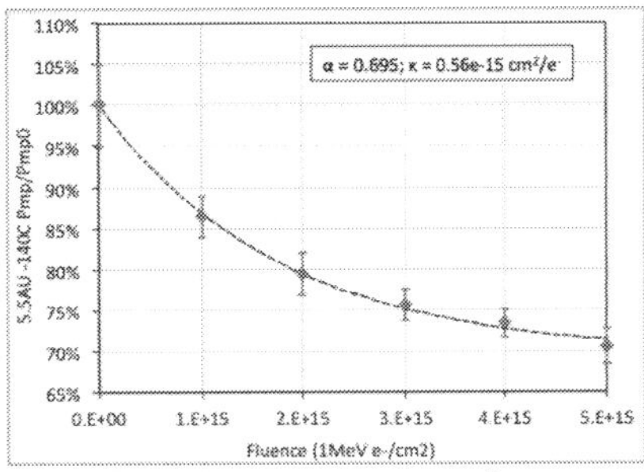

Figure 8. ZTJ CIC remaining power fraction at LILT (5.5AU, -140 °C) as a function of applied radiation dose.

9. ACKNOWLEDGEMENT

The work described here was carried out at the Jet Propulsion Laboratory, California Institute of Technology, under contract with the National Aeronautics and Space Administration (NASA).

10. REFERENCES

1. Europa Public Website, http://solarsystem.nasa.gov/europa/overview.cfm

2. Northon, K, (2015, July 30). NASA's Europa Mission Begins with Selection of Science Instruments.

3. Office of Planetary Protection, https://planetaryprotection.nasa.gov/categories

4. Ratnakumar, B. V., Smart, M. C., Whitcanack, L. D., Davies, E. D., Chin, K. B., Deligiannis, F., & Surampudi, S. (2004). Behavior of Li-Ion Cells in High-Intensity Radiation Environments. Journal of the Electrochemical Society, 151 (4), A652-A659.

5. Ratnakumar, B. V., Smart, M. C., Whitcanack, L. D., Davies, E. D., Chin, K. B., Deligiannis, F., & Surampudi, S. (2005). Behaviour of Li-Ion Cells in High-Intensity Radiation Environments: II. Sony/AEA/ComDEV Cells. Journal of the Electrochemical Society, 152 (2), A357-A363.

6. Krause, F. C., Lawrence, A., Smart, M. C., Dawson, S. F., Ulloa-Severino, A., & Ratnakumar, B. V. (2015). Evaluation of Commercial High Energy Lithium-Ion Cells for Aerospace Applications. 227th Meeting of the Electrochemical Society, Abstract #47580.

7. Ratnakumar, B. V., Smart, M. C., & Whitcanack, L. D., (2010). Storage Characteristics of Lithium-Ion Cells. ECS Transactions, 25 (36), 297-306.

IN-ORBIT TREND ANALYSIS OF GALILEO SATELLITES FOR POWER SOURCES DEGRADATION ESTIMATION

Frederic Bard [(1)], Aurore Carré [(1)], Pablo Fernandez[(1)], Maria Nestoridi[(1)], Sergio Alia[(2)], Aakesh Datta[(2)], Olabisi Durodola[(3)]

(1) European Space Agency – European Space Research and Technology Centre (ESTEC), The Netherlands, Frederic.Bard@esa.int
(2) OHB System AG, Universitätalle 27, D-28359 Bremen, Germany, Sergio.Alia@ohb.de
(3) ADS GmbH, Robert-Koch Str.1, 82024 Taufkirchen, Germany, Olabisi.Durodola2@airbus.com

ABSTRACT

This paper presents the IOV and FOC Battery in orbit trend analysis and degradation modelling based on the Galileo experience in the frame of the ESPC 2016. Galileo provides a unique opportunity to study a constellation of satellites in the Medium Earth Orbit (MEO) to make a statistical analysis of power sources degradations due to its environment.

From a project point of view the objective of the study is to establish a process to obtain the in-orbit battery degradations. Accurate battery degradation knowledge will allow precise battery management from operation teams (battery fade strategy and combined Earth Lunar eclipse power estimations). Another key feature of this study is to be able to assess different battery performance (Saft and ABSL) under very similar mission requirements over a long period (electrical profile, temperature, and environment).

This paper reports on the first four years in orbit of IOV PFM satellite (10 eclipse seasons and still running) and on-going FOC satellite telemetry (TM) analysis (starting with two years of data from FOC GSAT201 and GSAT202).
The results are in all cases better than the predictions, which is expected due to the usage of conservatives assumptions in the design to cover (for both IOV and FOC) worst case scenario for the entire constellation. It should be noted that the FOC GSAT201 and GSAT202 batteries are degrading slightly faster than the 6 others FOC batteries identified GSAT203, GSAT204, GSAT205, GSAT206, GSAT208 and GSAT209, but still below predictions due to their peculiar unexpected orbit reached after launch (higher DoD up to 42% measured due to longer eclipses). These 2 satellites will require specific degradation monitoring.

1. INTRODUCTION

The Galileo satellites compose a constellation in the MEO Orbit. It is composed of 24 reference satellites on 3 orbital planes separated by 120° right ascension of the ascending node (Plane A, B and C). Galileo is on a circular orbit of 29 600 km, at an inclination of 56°. The satellites design lifetime is 12 years.

The battery model correlation consist of tuning the ageing factor that lead to the cell degradations. Applying this analysis on all the batteries of the constellation, the trend analysis is obtained as well as a comparison in performance of the different battery types used on-board Galileo Satellites.

2. GALILEO POWER SUBSYSTEM

The satellite specification remaining unchanged for both projects, both IOV and FOC power subsystem are built on a 50V fully regulated bus. The nominal spacecraft power is in the order of 1550 W based on conservative power budget with a very stable payload consumption. Both spacecraft designs use the same solar array provided by Airbus NL capable of producing more than 1904 W at EOL (prediction assuming worst case degradations and 1 section failed). Each spacecraft are using Sequential Switching Shunt Regulators (S3R) to connect to the SA sections. The Solar array design, based on GaAs triple junction 28% cells, is made of 29 cells per strings for a total of 88 strings. Thanks to a very stable solar array pointing law (solar array always normal to the sun) the operating temperature remains between 60 and 63°C (outside of eclipse) and the power produced is affected only by the seasonal variations and radiation degradation.

The batteries however are coming from two different manufacturers: IOV is equipped with Saft VES180 and VES140 cells and FOC is equipped with ABSL Sony 18650HC cells and 18650HCM cells.

This MEO orbit has a period of 14.08 hours with two Earth eclipse seasons per year. The eclipses are presented in Fig. 2 with a maximum duration of 60min.

DOI: 10.1051/e3sconf/20171613005

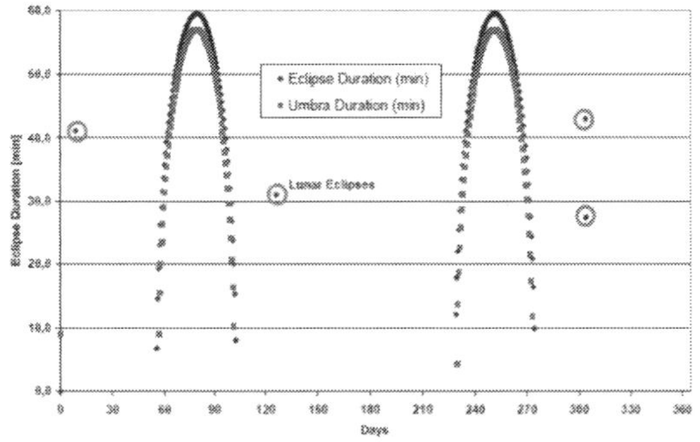

Figure 1. Galileo orbit Earth eclipse profile

During its lifetime, Galileo batteries will endure a maximum of 2064 eclipses. This is significantly more than for typical GEO satellites (1080 cycles over 12 years). However it should be noted that due to the specificities of each orbital planes (which has a small RAAN drift of about -10 degrees per year), the eclipses seasons last between 25 and 47 days. This means that depending on the Sun, Earth and orbital plane alignment the number of eclipses per season could vary between 45 cycles and 86 cycles.

It is worth mentioning that, due to a wrong orbit injection from Soyuz-Fregat in August 2014, the FOC GSAT201 and GSAT202 have reached an elliptical orbit (resonant orbit) with Earth eclipses lasting up to 68min.
This study will establish the actual degradation impact due to higher DoD on these two satellites.

3. IOV BATTERY MODELLING

The IOV batteries have been provided by Saft. Based on a P-S topology, they are composed of 9 cell-packs serial connected. Each cell-pack consists in 3 cells parallel connected, using the VES180 cell on GSAT101 and GSAT102 and the VES140 cell on GSAT103 and GSAT104.

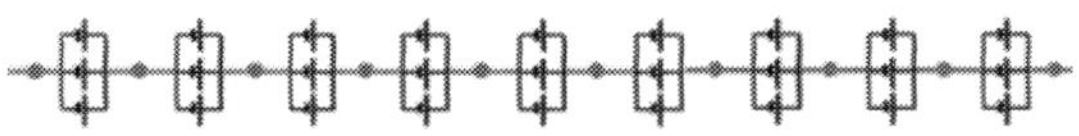

Figure 2. 3P9S IOV battery configuration

3P9S VES180 batteries have a nameplate energy of 4900 Wh, and the VES140 have a nameplate energy of 3585 Wh. The Saft batteries include a passive balancing system optimizing the cell voltages over lifetime and a by-pass system allowing to isolate a failed cell.
The batteries are controlled in temperature to a value of 18.5°C in average for the 12 years of lifetime. IOV predictions are compared to measured data, considering a Used Energy of 45% in average with respect to the EOL energy.

IOV batteries telemetry have been acquired and post-processed. Using the current and temperature profiles during the longest Earth Eclipses and applying this profile into on-ground simulation tools, the battery degradation parameters could be obtained. Fig. 3 shows a view of the IOV battery module

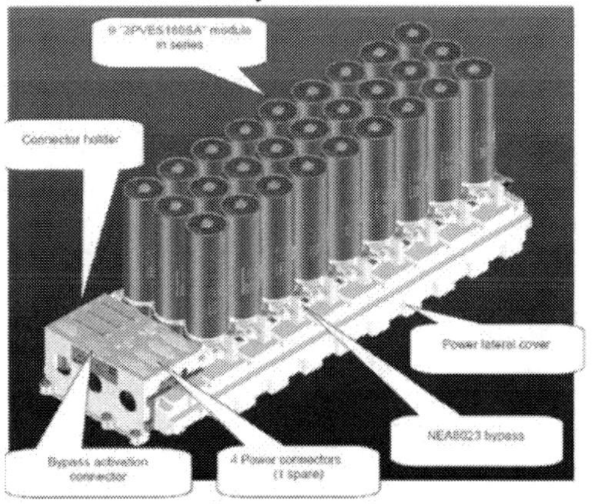

Figure 3. Galileo 3P9S IOV battery configuration

3.1 Inputs and assumptions

The simulations were run for satellite IOV PFM (FM02 battery based on VES180 cell). Table 1 presents the summary of the IOV batteries for both IOV PFM and FM02 satellites. The latest eclipses season was used to establish the latest degradations figures of the batteries.

Table 1. Galileo IOV batteries summary table

S/C	IOV PFM	IOV FM02
LEOP Name	3A5	3A6
Sat Name	GSAT101	GSAT102
battery type	VES180	VES180
Battery name	**FM02**	**EQM**
Plane	B	B
Launch date	21/10/2011	21/10/2011
Eclipse Season 1	Jan-12	Jan-12
Eclipse Season 2	Jul-12	Jul-12
Eclipse Season 3	Jan-13	Jan-13
Eclipse Season 4	Jul-13	Jul-13
Eclipse Season 5	Jan-14	Jan-14
Eclipse Season 6	Jul-14	Jul-14
Eclipse Season 7	Jan-15	Jan-15
Eclipse Season 8	Jul-15	Jul-15
Eclipse Season 9	05/01/2016	26/12/2016
Eclipse Season 10	27/06/2016	29/06/2016

The following hypotheses were considered for the IOV GSAT101 (FM02 battery):

- Battery characteristics provided by Saft during final on-ground testing (energy, internal resistance)
- 581 days on-ground storage at 20°C including 10 cycles at 100% DOD
- LEOP power consumption of 201Wh, at 18°C,

with a total duration about 5.8 hours (to be compared to the 4900 Wh used at worst case dimensioning for the battery sizing)

- The MEO mission profile considering a S/C power consumption of 1286 W in average (to be compared to the 1562W used as worst case dimensioning for the design)
- The battery temperature at 18°C during both eclipse and solstice periods
- The number of eclipses cycles identified on Plane B during the latest eclipse season is only 46 cycles instead of 86

The following plot provides an example of an eclipse season profile on the IOV FM02 battery (on-board GSAT101).

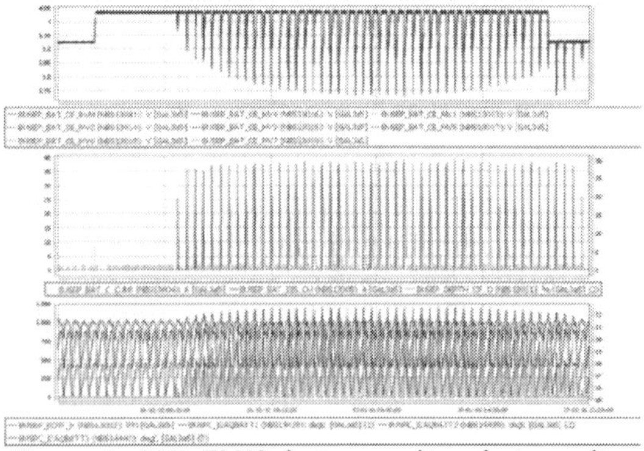

Figure 4. IOV FM02 battery cycling during eclipse season 9

The battery management strategy is presented in Fig. 4. At the end of the eclipse phase, the cell packs voltages are reduced to a voltage of 3.92 V ensuring a lower SoC during sun phase to increase the lifetime.

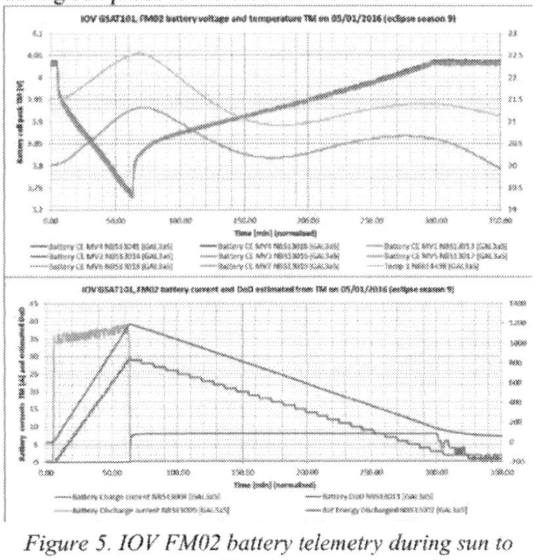

Figure 5. IOV FM02 battery telemetry during sun to eclipse transition in eclipse season 9

Figure 5. shows the transition from sun to eclipse. It could be seen that the dispersion of the cell packs are very limited at this stage.

3.2 Simulation results

For IOV batteries, the correlation is based on the SLIM simulation tool provided by Saft. SLIM (Saft Space Li-ion Model) is a macroscopic electrochemical model developed since 2005. The purpose of this model is to calculate the EOL performances of Saft VES batteries taking into account the battery characteristics (acceptance energy and internal resistance), the specification or the mission profile and the in orbit battery management.

This model is a battery design tool vs satellite mission profile and for the IOV satellites, simulations have been done by using a MEO specific environment (orbital period, number of cycles, charge rate…).

Fig. 6 presents the simulation results (SLIM outputs) compared to the battery voltage telemetries (at battery level). It is worth noting that available voltage telemetries were at cells pack level. Thus, the battery voltage telemetry has been simply calculated by adding them and do not take into account any potential voltage drop due to ohmic losses link to cables. Consequently, the current difference between the telemetries and the predicted voltages are considered conservative and further investigations in this study will be done to fine tune this aspect and lead to more realistic outputs.

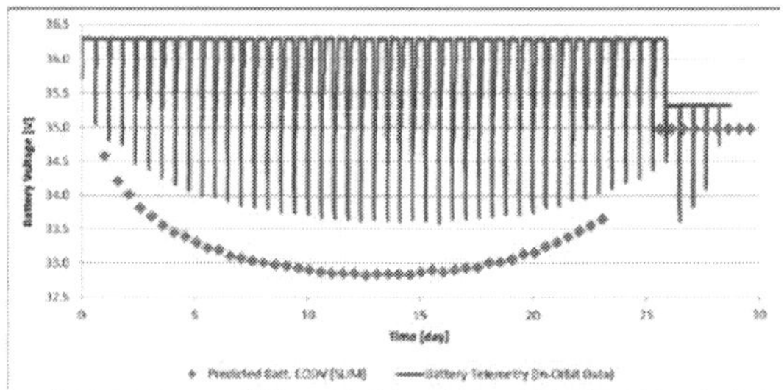

Figure 6. IOV FM02 battery EoD voltage comparison between SLIM and telemetry during eclipse season 9

Considering a single worst case eclipse of 60min, the SLIM output is compared with the battery voltage at battery level in Fig. 7.

 DOI: 10.1051/e3sconf/20171613005

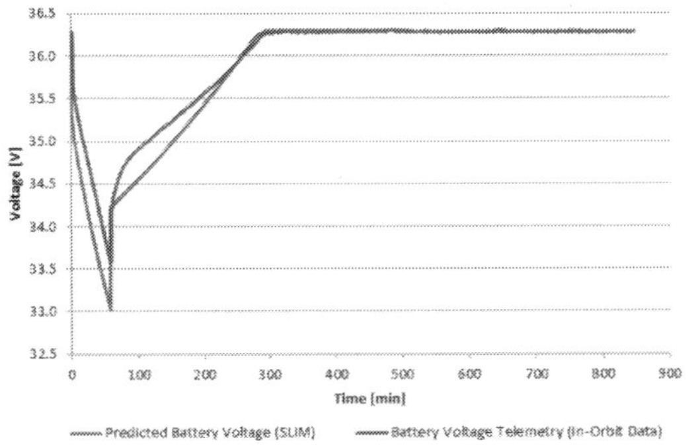

Figure 7. IOV GSAT101 battery voltage comparison during the longest eclipse (cycle 28) of season 9

The correlation between the battery voltage TM from IOV FM02 battery compared to the SLIM simulation results is confirming the fact that the IOV FM02 battery is behaving better than predicted. This is in line with the real in-orbit conditions (lower on-ground storage period, lower power consumption, lower number of eclipse cycles).

The predicted degradation over the complete mission was expected at 1.5 %. With the revised analysis performed after 9 eclipse seasons and extrapolated for the lifetime, the degradation figure is expected at 0.6 %. This is an additional margin which could be used for potential mission extension over the 12 years guaranteed of the IOV satellites.

3.3 Conclusions and remaining activities

The results obtained from the SLIM correlation exercise on IOV FM02 battery confirm that the battery is behaving well. All battery operations are in line with expected value such as :

- The charge and discharge management
- The temperatures range
- The balancing operations

Comparable analysis have already been performed on in-orbit behaviour of VES batteries from Saft (for instance [1] and [2]). This study is in line with these. As already stated, the study will be further processed. Indeed, investigations will be done to fine tune the hypotheses regarding the battery voltage acquisition.

4. FOC BATTERY MODELLING AND TELEMETRY ANALYSIS

Each FOC battery is comprised of one module, which contains 64 strings in parallel, each consisting of 11 cells in series. The design difference between IOV, i.e.Saft batteries and FOC, i.e. ABSL batteries is that the ABSL batteries do not contain a balancing system. The latter is feasible due to the highly uniform COTS cells and ABSL's proprietary screening and matching routine. [3]

The following figure presents the FOC battery provided by ABSL. It is based on the typical Sony 18650HC cell assembled into 4 blocks in a double deck battery module.

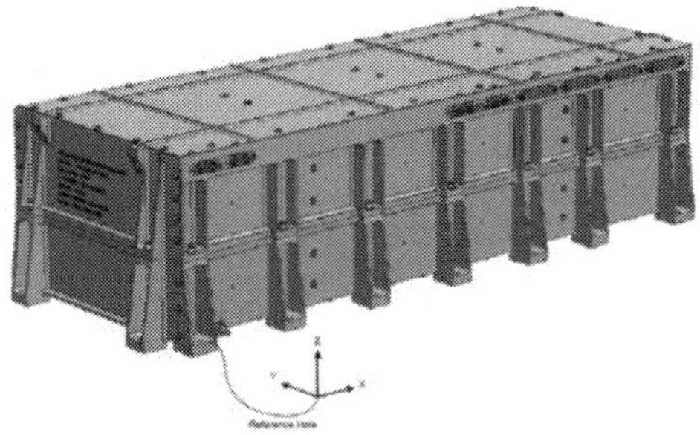

Figure 8. Galileo FOC battery from ABSL

4.1 Inputs and assumptions

The 11S64P nameplate energy is 3800 Wh. In Fig. 10, the predicted degradations over life time considering 4 year of storage and AIT activities followed by 12 years in orbit is presented for different solstice conditions.

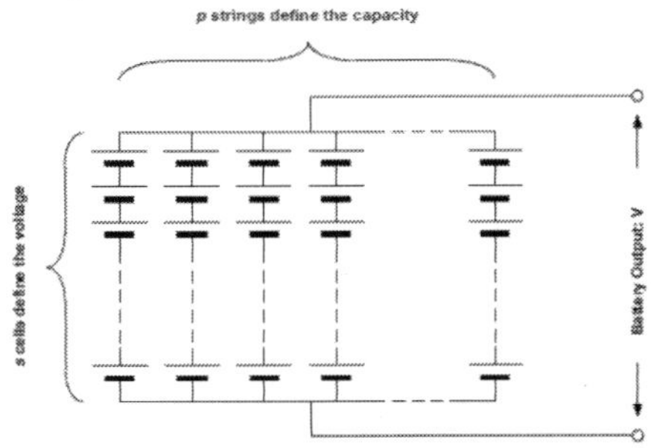

Figure 9. 11S64P FOC battery configuration

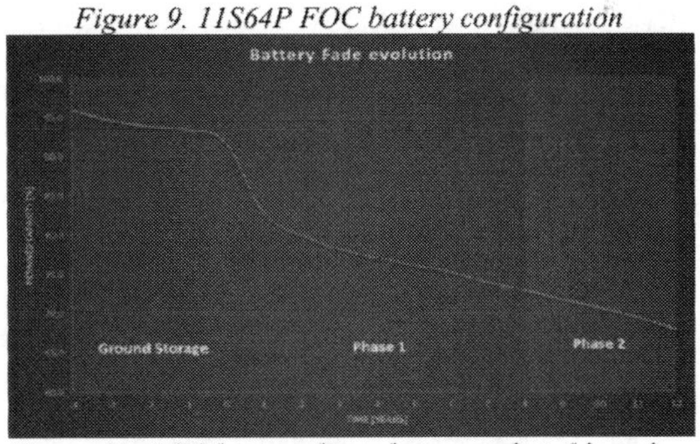

Figure 10. FOC battery degradation predicted based on LIFE model provided by ABSL considering a battery fade strategy implementation

Considering the nominal mission with the above constraints, the degradation are estimated at 32%. This estimation is considering an average DoD of 35% during eclipse seasons (2 seasons per year) and a SoC of 80% during the solstice season. To be noted in the current predictions, the knee of the degradation curve is expected after 14 years in orbit after which the battery performances could not be guaranteed anymore. However the current results confirms that this inflection is to be expected later (allowing potential mission extension).

Fig. 11 presents the battery voltage evolution over the seasons by superposing 3 eclipses profiles of GSAT203. The first information is a clear decrease of the minimum voltage reached during the typical eclipse (60min) confirming a small increase in delta DoD from 27.2% during the first eclipse season up to 32% after one year in orbit (stabilised during the 2 last eclipse seasons).

The battery management approach is also visible with a reduced EoC voltage at 45.2V applicable for the first 8 years in orbit to increase the battery lifetime.

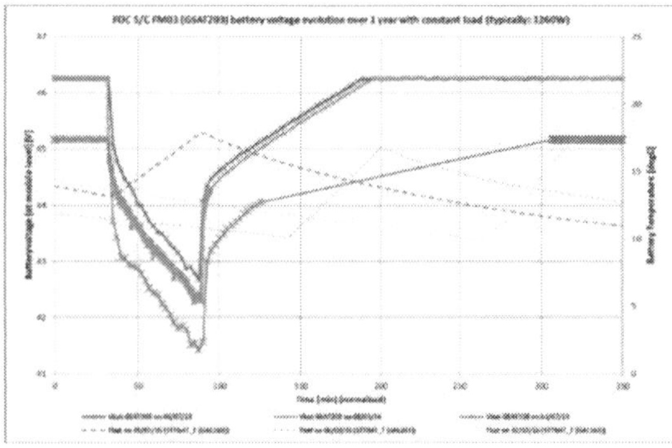

Figure 11. FOC FM03 battery telemetry during sun to eclipse transition over two orbits.

Based on the telemetry data the FOC spacecraft consumptions could be established with an accuracy of less than 3%. The average power consumption for the FOC constellation is:

- In nominal mode: between 1250W and 1300W
- In safe mode: between 900W and 930W

All the telemetries obtained from the in-orbit spacecraft fall into this range (see Table 2).

It is worth to compare these data with the power of 1550W which was used at the time of the design phase for battery lifetime predictions.

4.2 Results and main conclusions

A total of 17 simulations were run for the FOC batteries for a total of 8 spacecraft in nominal operations over the last 2 years.

As a reference case GSAT203 (FOC FM03) will be used to illustrate the methodology of the model correlation. BEAST models are feed with in-orbit telemetry of the battery current and temperatures during the discharge and charge phase, then the Beast simulation voltage is adjusted to the in-orbit battery voltage applying the battery capacity degradation coefficient and the internal resistance increase. This is an iterative process until the simulation voltage match as best as possible the telemetry. This results in estimated degradation figures which could be compared to the predictions made at the design phase. All results are provided in Table 2. Figures 12, 13 and 14 illustrate the correlation of the battery in-orbit voltage to the Beast simulation voltage.

The matching is not perfect due to the EMF jumps at the transitions between charge and discharge.

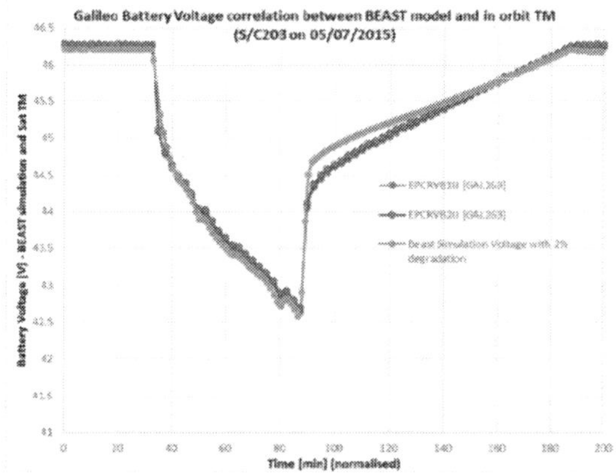

Figure 12. FOC GSAT203 battery voltage correlation for the reference eclipse on 05/07/2015

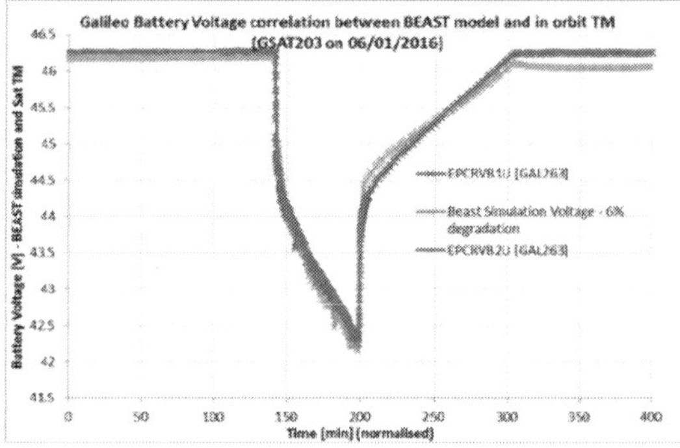

Figure 13. FOC GSAT203 battery voltage correlation for the reference eclipse on 06/01/2016

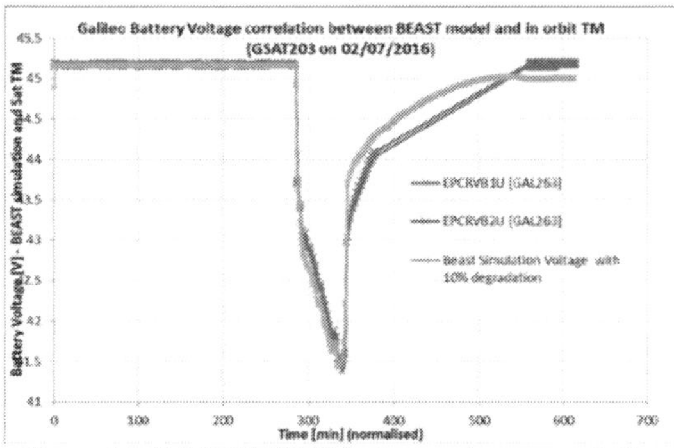

Figure 14. FOC GSAT203 battery voltage correlation for the reference eclipse on 02/07/2016

DOI: 10.1051/e3sconf/20171613005

Table 2. Summary of FOC battery degradations

S/C	Launch Date	Reference Eclipse Date	Difference (months)	Fade Expected [%]	Fade Beast Simulation [%]	Resistance increase Expected[%]	Resistance increase Beast simulation [%]	Power consumption [W]	Delta DoD [%]
GSAT201	22/08/2014	24/06/2015	10	18.8	7	21.8	9	1298	36.20%
	22/08/2014	16/06/2016	22	21.4	15	29.9	20	1249	39.50%
GSAT202	22/08/2014	28/06/2015	10	18.8	7	21.8	9	1265	36.70%
	22/08/2014	12/06/2016	22	21.4	18	29.9	25	1284	42.00%
GSAT203	27/03/2015	05/07/2015	4	13.45	2	17.75	5	1255	27.20%
	27/03/2015	06/01/2016	10	18.8	6	21.8	10	1260	31.70%
	27/03/2015	02/07/2016	16	20.1	10	25.85	15	1264	32.10%
GSAT204	27/03/2015	03/07/2015	4	13.45	3	17.75	4	932	N/A
	27/03/2015	01/01/2016	10	18.8	8	21.8	10	1283	31.40%
	27/03/2015	29/06/2016	15	20.1	10	25.85	10	1305	34.10%
GSAT205	11/09/2015	01/02/2016	5	13.45	4	17.75	5	1271	31.50%
	11/09/2015	18/07/2016	10	18.8	7	21.8	5	1259	27.50%
GSAT206	11/09/2015	01/02/2016	5	13.45	3	17.75	4	1268	29.75%
	11/09/2015	23/07/2016	10	18.8	8	21.8	4	1294.8	32.10%
GSAT208	17/12/2015	07/04/2016	4	13.45	2	17.75	3	1318	31%
GSAT209	17/12/2015	07/04/2016	4	13.45	2	17.75	3	1303	31.20%

Based on Table 2, the following conclusions could be drawn:

- Table 2 presents the capacity fade obtained estimated based on TM analysis compared to the predictions (fade expected in reference to figure 10).
- The average DoD for all S/Cs is in the order of 27% to 32%, with the noticeable exception of GSAT201 and GSAT202 where the maximum average DoD reached 42%
- For the S/C in the nominal orbit, the correlation exercise highlighted a significant margin compared to the life predications. This is due to the conservative assumptions made in the design phase in terms of S/C power consumption (in average the S/C consumed 250W less than predicted) and to a shorter on-ground AIT phase
- The correlation of the battery TM from GSAT201 and GSAT202 highlight a faster degradation (still within prediction levels), which is in-line with a longer eclipse of up to 68min on the resonant orbit.
- At this stage of the analysis, no results were out of family indicating that none of the battery had suffered damages

The results confirmed the good behaviour of Galileo FOC batteries fleet, but this analysis is only the beginning and should be followed over longer time.
GSAT201 and GSAT202 batteries will require a specific monitoring to follow the degradation evolution. At this stage the end of life of the FOC batteries could not be assessed accurately, this will be done at a later stage following the progression over several years.

5. CONCLUSIONS AND FUTURE WORK

This study provides a quantitative data on the in-orbit performance of the Galileo satellite batteries which could be used for in-orbit trend analysis. All telemetries and correlation results acquired at this stage are showing better than predicted performances due to conservative assumptions made in the design phase.

For IOV, only the FM02 battery correlation has been established at this stage, but all in-orbit telemetries acquired on all IOV spacecraft are showing normal behaviour in voltage, current and temperature.

Future work on the Galileo battery in-orbit analysis will involve extension of the correlation exercise to all IOV spacecraft (IOV GSAT102 which include VES180 EQM module, IOV GSAT103 including PFM03 VES140 and IOV GSAT104 including FM04 VES140).

Furthermore the battery model provided by BEAST is sufficient for a first order estimation of the battery degradation but is conservative in terms of the EMF jumps during charge and discharge transitions. To model accurately these effects it is necessary to use a transient model for the battery including capacitive effects (diffusion effect). This activity will be performed using more refined models in the second step of the Galileo in-orbit TM assessment.

6. REFERENCES

1. Borthomieu Y., Gambini D. Saft Li-Ion VES140S battery flight experience return on TAS Space Bus platforms, ESPC 2014

2. Borthomieu Y., Prevot D., Massot J., Tastet P., Simon E. VES100/140 lithium-ion cells LEO life-test results & Proteus flight heritage, ESPC 2011

3. Fredon S., Melac L., Simmons N., Long duration LEO mission experience with ABSL batteries onboard the myriad smallsat platform, ESPC 2011.

E3S Web of Conferences **16**, 13006 (2017) DOI: 10.1051/e3sconf/20171613006
ESPC 2016

THE COSMO-SKYMED SECOND GENERATION SAR ANTENNA ELECTRICAL POWER CHAIN AND PLATFORM POWER DISTRIBUTION

E. Scorzafava[(1)]
G. Gianninoto [(2)], **M. Cantamessa** [(2)], **P. Zanella** [(2)]
D. Carlani [(3)], **E. Scione** [(3)]

[(1)] *Italian Space Agency (ASI) – viale del Politecnico snc Roma (Italy), Email: edmondo.scorzafava@asi.it*
[(2)] *Leonardo S.p.A. – viale Europa Nerviano (Italy), Email:guido.gianninoto@leonardocompany.com, marco.cantamessa@leonardocompany.com, pietro.zanella@leonardocompany.com,*
[(3)] *Thales Alenia Space Italia – via Saccomuro 24 - Roma (Italy), Email: daniele.carlani@thalesaleniaspace.com, emiliano.scione@thalesaleniaspace.com*

ABSTRACT

This paper presents the Electrical power Subsystem (EPS) of the COSMO-SKYMED SECOND GENERATION (CSG) spacecraft, regarding the design and development of the electronics for power generation, conditioning and distribution. It provides a detailed description of the power equipment's and of the Photovoltaic Assembly (PVA).
The design characteristics of the EPS elements are described. A summary of the main performances is given, including a glance to the EMC aspects, which are very important in the definition and characterization of the design of a high power SAR system.
It will be also shown how the design of the power conditioning and distribution electronics has taken advantage of the Leonardo (former Finmeccanica) heritage on Power Management and Distribution (PMD) for scientific platforms [1], [3].

1. COSMO SECOND GENERATION MISSION

COSMO-SKYMED SECOND GENERATION (CSG) is a program founded by the Italian Space Agency (ASI) and Minister of Defense (MoD) aimed to earth observation by means of a Synthetic Aperture Radar (SAR) Antenna. Its mission has been defined to continue the operation of the four spacecrafts fleet operated in the frame of the COSMO SkyMed program first generation.
The present program consists of two satellites. The first satellite launch is foreseen within 2018, the second one within 2019.
Each satellite will carry a Synthetic Aperture Radar Antenna (SAR) of new generation, to perform the observation of the earth. Its operation will require a very big amount of energy, with power peaks up to 19000W. This represent the most challenging requirement for the Electrical Power Subsystem (EPS), for what concerns the handling of such power and the EMC related aspects.
The two satellites will operate in a Dawn/Dusk, sun-synchronous circular orbit.
Based on these operative constraints, the paper presents the way the CSG power generation and conditioning electronics have been designed and developed.
The power system chain to supply the SAR Antenna is made by the following equipment's:

- 10 * 24Ah Lithium-Ion (Li-Ion) battery modules.
- A Solar Array (SA) made by two different wings on each spacecraft.
- One Solar Array Drive Assembly (SADA) per SA wing.
- One Power Conditioning and Distribution Unit (PCDU) internally redounded.
- One SAR Antenna Power Distribution Unit (SPDU) providing an internal redundancy considering the acceptable performances graceful degradation of the SAR Antenna.
- 40 Tile Power Supply Units (TPSU). The SAR Tile is essentially a complex electronic box equipped with radiating panels that represents the elementary brick of the SAR antenna.

The paper provides a detailed description of all these equipment's of which Leonardo is the supplier, with the exception of the SADA and of the Battery modules (not under Leonardo responsibility) and, in particular, of how the extreme operative requirements of the mission are met.
It will be also shown how the design of the power conditioning electronic has taken advantage of the Leonardo (former Finmeccanica) heritage on main bus power conditioning for scientific platforms.

Figure 1. COSMO Second Generation satellite view.

DOI: 10.1051/e3sconf/20171613006

2. EPS OVERVIEW

The architecture of the Electrical Power Subsystem (EPS) of CSG is shown in Fig. 2.

The EPS of CSG provides two different busses:

- an unregulated 49V to 67V bus which supplies power peaks up to 19000W during imaging. It also feeds the payload heaters located on the antenna.
- a regulated (P/F_PWR) 28V bus which supplies up to 1800W, including distribution to Platform (P/F), to P/F Heaters and to NEA and pyrotechnic devices.

The distribution to the Platform is realised via LCL protected lines which are allocated inside the PCDU.

The limitation and switching of the P/F Heaters is performed inside SPDU.

The distribution to the SAR antenna is realised via LCL protected lines which are allocated inside the SPDU.

The power generator consists of a deployable Solar Array (SA) made of two wings of four panels each with a total area of 22 m² and a peak power capability @ BOL of nearly 5kW.

The electrical energy is stored in 10 Lithium-Ion (Li-Ion) battery modules each of them with a capacity of 24Ah.

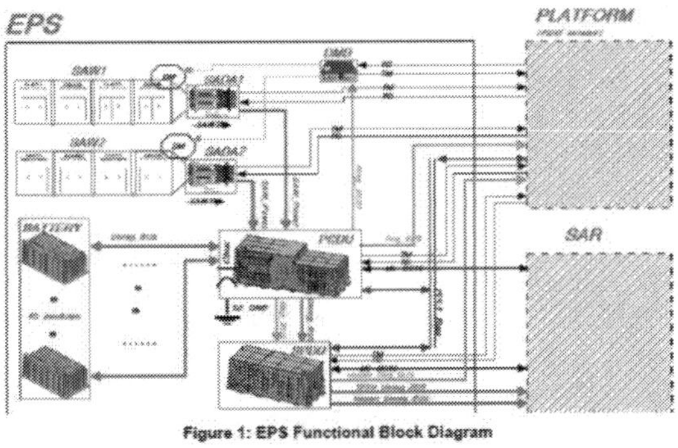

Figure 2: CSG EPS Configuration.

The EPS described in this paper is managed by Thales Alenia Space-Italy (TAS-I), which is leader of an industrial team composed of:

- Leonardo – Finmeccanica S.p.A. as responsible for the PVA, PCDU, SPDU and TPSU manufacturing.
- EnerSys ABSL as manufacturer of the batteries.
- RUAG Space as manufacturer of the SADAs.

TAS-I is also responsible of the whole spacecraft.

3. POWER CONDITIONING AND DISTRBUTION UNIT (PCDU)

The PCDU is part of the CSG EPS.

It houses in a single box the following functions:

- Power conditioning to provide the unregulated bus ranging from 49V to 67V. It is based on 16 Sequential Switching Shunt Regulator (S3R) sections operating under control of a Battery Charge control Module (BCM) that controls the charge current released to the Batteries respect to the set Charge Current and End Of Charge Voltage (EOCV), operating in Constant Current (CC) or Constant Voltage (CV) modes depending on the in-out power balance and performing the full batteries charge in tapered mode.
- Power conditioning to provide the 28V regulated P/F_PWR bus, starting from the 49V-67V unregulated bus. It is based on 6 step-down DC/DC converters, operating under control of a Main Error Amplifier (MEA). The P/F_PWR bus is going to feed all the Platform (P/F) loads, the P/F heaters located inside SPDU, as well as the NEA & PYRO devices to release deployable appendages like Solar arrays and antennas.
- Distribution and protection via Latching Current Limiters (LCLs) and Retriggered Latching Current Limiters (R-LCLs) of the regulated power to the P/F loads (essential ones connected to the R-LCL line).
- Distribution of regulated power to the protection lines for the P/F heaters, located inside SPDU.
- Distribution and limitation of the current needed to actuate the NEA and PYRO devices necessary for the release of the satellite appendages.
- Reception of the Umbilical Supply, to perform nominal PCDU operation and battery maintenance, via external power sources during ground operations.

Fig. 3 shows the block diagram of the PCDU.

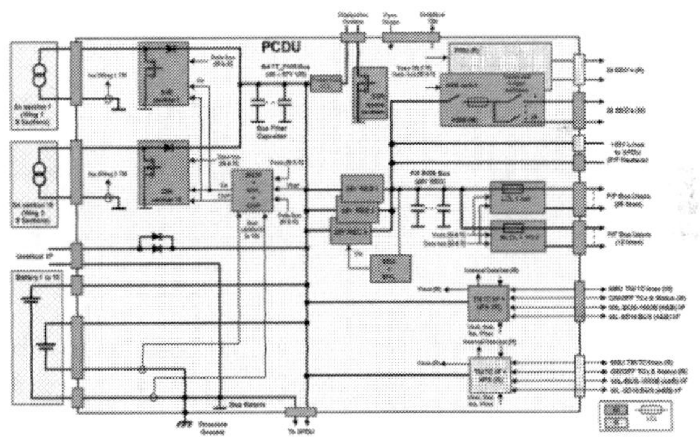

Figure 3. PCDU Block Diagram.

The input power will be provided by the SA during sunlight periods or by 10 Li-Ion battery modules during eclipses and when the SA power is lower than the power demand. The power generation consists of two SA wings made of 16 nominally identical sections.

The PCDU is sized to manage and provide during all mission phases a total peak output power of 19000W towards the SAR antenna on the unregulated bus. This large amount of power will be released by the Solar Arrays according to their capability and to the current lightening conditions, integrated by the battery modules

stored energy.
A suitable internal cabling (AWG8 wires are used) and metal bars system realise such power distribution with a voltage drop from battery power input interfaces to power output interfaces better than 0.4V.

3.1. Unregulated Battery Bus Control

The unregulated Battery Bus power is controlled in order to manage the Battery Charge. A triple redundant electronics (SPFF design) senses the actual Battery Voltage and the Battery Charge Current and compares them to the Battery Charge settings coming from 1553 bus, generating an Error Signal that commands the Sequential Switching Shunt Regulators (S3R) to:

- Open and release all the current available from the related Solar Array section to the Main Bus.
- Close a shunt element to exclude all the current available from the related Solar Array section from the Main Bus.
- Switch between the two above status' with a duty cycle proportional to the amount of the current necessary to reach the equilibrium in the charge control.

In this way the Battery charge is automatically performed in Constant Current (set current charge value) or in Constant Voltage (set battery voltage value). This latter realises a progressive (tapered) charge of the battery up to the final charge Voltage at zero-charge-current.
The S3R sections and the triple redundant control electronic are all fed by two internal power regulators (buck topology), working in hot redundancy, that realise a SPFF supply.
The power conditioning on the unregulated bus is completed by a 825uF capacitor bank made of 25 self-healing PM907 type capacitors. The value has been defined as compromise between the specified maximum voltage ripple and the dynamic performances of the regulating voltage and current loops that perform the battery charge control.
The S3R sections have been designed with a single MOSFET to perform the shunting function and a single diode to connect the SA section to the Bus.
This choice allowed a great reduction in the physical space occupied by the S3R section, which was a critical aspect for the PCDU allocation inside the CSG spacecraft.
The critical failure of the diode in short circuit is controlled via an active electronic current limiter that contains the short current (both peak and limited value) and then opens in latched mode the MOSFET to remove the short circuit on Bus. The failed section is then permanently connected to the Main Bus.
A spare shunt, connected to the unregulated bus sinks exceeding power in case of loss of an S3R section fixed on bus, to avoid overvoltage.
No overvoltage protection has been foreseen inside PCDU, as the SPFF control electronics, the constant presence of the battery and the spare shunt guarantee the nominal operation inside the specified battery voltage limits even in presence of a very low load.

3.2. Regulated Platform Bus Control

The PCDU is also sized to release up to 1800W to the P/F Bus, regulated at 28V with a precision better than ±0.5%, in accordance with ECSS-E-ST-20C. This Bus feeds:

- all the platform loads, via LCL/R-LCL-protected lines.
- the P/F heaters via non protected lines; LCL protection and switch interruption are realised inside SPDU
- the Electro Explosive Devices (EED) accounting the Non Explosive Actuators (NEA) and Pyro devices. PCDU supports two simultaneous NEA and Pyros fires with full power on P/F lines in nominal operation. The EED can be actuated in sequence up to 6 + 6 devices.

The P/F Bus regulation is obtained via 6 DC-DC power regulators commanded in trans-conductance mode by a centralised, triple-redounded (SPFF) electronic.
The voltage control loop controls the current released by power modules outputs on a 3mF capacitor bank sensing the voltage across it. The capacitor bank is made of 30 self-healing PM90 type capacitors, The system meets the impedance requirement of ECSS-E-ST-20C.
The full power regulation is guaranteed by design also when only 5 DC-DC regulators are active as the power regulators have enough power capability to sustain 1/5 of the bus full power.

3.3. PCDU Power Distribution

The power distribution of both the Busses inside PCDU is realised using metal power bars so to reduce at minimum the voltage drops and allow an easy, modular connection of all the modules whether they feed, control or supply the Busses themselves.

The distribution of the P/F bus is performed by a number of protected lines belonging to the following two types:

- LCL for the non-essential loads that are disconnected after a HW-fixed delay when an overload appears at its output. Several sizes of LCL are present, identified by their maximum nominal current value. The intervention delay or Trip Off Time for each LCL type is sized tacking into account thermal stress considerations on the limiting elements (MOSFETs). The higher current LCL are obtained assembling in parallel basic LCL circuits; each LCL has its TM & TC electronics. Each LCL is designed also to survive a multiple, repetitive

 DOI: 10.1051/e3sconf/20171613006

overload (typically due to the presence of an Under Voltage Protection on the user side. LCLs can be externally commanded ON or OFF and give their status and the operating current as telemetries.

- R-LCL for the essential loads have the same principal characteristics as the LCLs, but they cannot be commanded ON or OFF via external commands; in case of over load they automatically re-start after a fixed time bouncing from the conduction and the limitation indefinitely. This bouncing can be excluded or re-activated via telecommands.

Both LCLs and R-LCLs have an input (on MB) UV protection with different thresholds (for the essential and non-essential loads); this protection is centralised at board level and disconnect simultaneously all the LCLs on that board.

The EED actuators perform the driving function of the Pyro or NEA devices feeding them with a fixed current for a fixed time duration. Their combination is sufficient to guarantee the device actuation. The actuator realises a triple barrier against unwanted activation:

- One HW interruption via a latching type relay (line ARM/DISARM).
- Two more HW levels to close the power line after the reception of:
  - ✓ A selection Telecommand that closes the addressed output electronic switch (series redounded to protect against the short circuit of one switch).
  - ✓ A fire Telecommand that closes the power line (common to up to 32 lines) via a non-latching relay, interrupting (commutation at no current) both hot and cold ends and via a series redounded Current Limiter, that sets the actuation current.

3.4. PCDU Command and Monitor Interfaces

The PCDU commanding and monitoring is provided through a Mil-BUS 1553 redounded interface and also through a Memory Load (ML) and Serial Data (SD) bus for some specific functions (Tile Control Unit LCL lines). For both the communication busses each terminal inside PCDU (Main and Redundant) can talk with both the main and redundant control units.
Several telemetries are available, giving a complete snapshot of the unit status during operation.
A number of temperature measurements (unconditioned thermistors) allow the thermal monitoring of the unit during AIT operation.

Fig. 4 shows a picture of the PCDU.

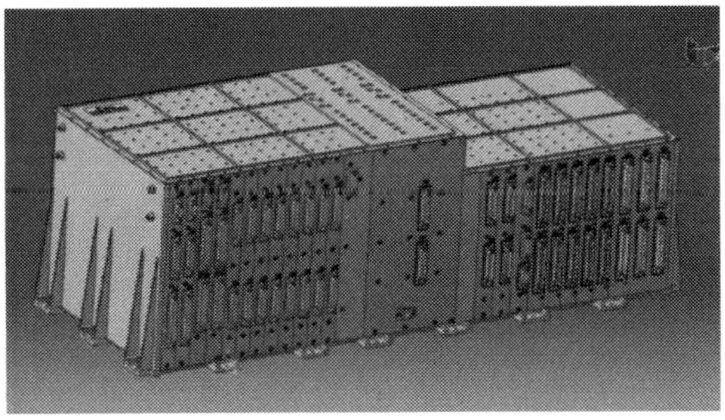

Figure 4: PCDU Box.

3.5. PCDU Main Characteristics and budgets

The main characteristics of the PCDU design are:

- Bus voltage: 49V- 67V unregulated bus voltage.
- Unregulated (Battery) Bus power capability up to 19000W to SAR antenna (both Sun or Eclipse).
- Max Battery Charge Current ripple 5App.
- Battery Charge Voltage ripple depending on Battery and harness impedance. Maximum value is 5A*0.062Ω = 0.31mVpp.
- Battery Charge Current in CC mode can be set via telecommand from 10A to 80A; up to 498 different Current levels setting possible.
- Battery End Of Charge (EOC) Voltage in CV can be set from 62V to 67V; up to 3641 different Current levels setting possible.
- SA current per section: 5A maximum in short circuit.
- 16 SA sections.
- P/F_PWR Bus maximum power: 1800W, including 850W to P/F heaters.
- 28V_REG module output power capability: 420W.
- PCDU power dissipation: < 300W maximum at maximum power on both busses.
- TM/TC interface: MIL-1553 data bus + ML & SD channels + direct lines.
- LCL classes: 1A, 2A, 3A, 5A, 7A, 10A, 12A.
- Total 86 LCL lines available.
- R-LCL class: 1A.
- Total 12 R-LCL lines available.
- Pyro/NEA Current: 5.5A ± 0.3A
- Pyro/NEA Pulse Duration: 40ms.
- Pyro/NEA shot capability: 6 + 6 in 1.25sec.
- 28 + 28 Pyro/NEA lines available.
- Mass: 30.6Kg.
- Dimensions: 740mm x 297mm x 250mm.

4. SAR ANTENNA POWER DISTRIBUTION UNIT

The SPDU is positioned in the EPS of CSG between PCDU and TPSUs, as it represents the distribution unit for the TPSU lines.

4.1. SPDU Architecture

The SPDU box houses the following main functions:

- Distribution of unregulated power, fed by the PCDU, to:
  - ✓ Antenna loads.
  - ✓ SAR Tiles on-board heaters
- Distribution of regulated power (28V bus) to the Platform Heaters.
- Generation of Synchronism (sync) signals and their distribution to:
  - ✓ all the TPSUs. Two operative modes are possible for the sync signals:
    - ▪ All in phase to each other
    - ▪ All shifted with 36° shift between adjacent signals.
  - ✓ SAR Electronics (SES) and SAR Antenna (SAS, including the TPSU units).

The Block Diagram in Fig. 5 summarises the functionality of SPDU.

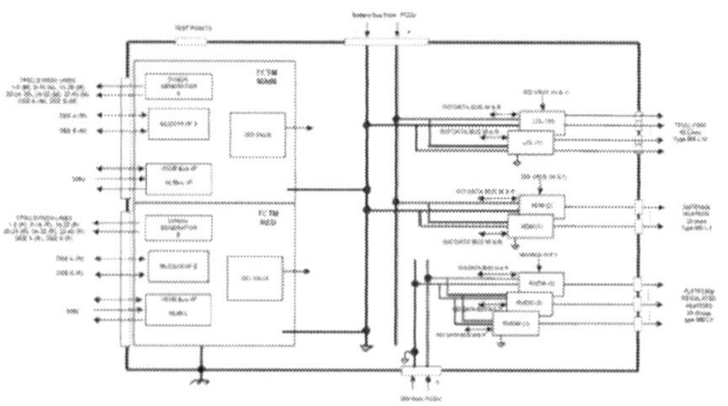

Figure 5. SPDU Block Diagram.

SPDU has the following two main tasks within the EPS:

- To provide Protection for all power outlets (LCL protected lines).
- To perform Filtering action for the Conducted Emission generated by TPSUs towards Batteries.

4.2. Battery Bus Distribution

Each power line towards TPSU inside the Antenna and each Antenna tile heater line are mainly composed of the series of a switch and a current limiter to protect bus against overload and to provide OFF protection to each Tile power line. Both the switch and the LCL are commanded ON/OFF via Memory Load telecommands that drive both the blocks. A delay in the actuation of the LCL ON/OFF with respect to the Switch ON/OFF can be added in the FPGA patterns telecommand.
The limitation current value of the power lines to TPSU has been defined in order to guarantee the operation of the TPSU and its correct switch-on (in-rush) in any specified condition, as well as to cope with thermal environment.
The limitation current value of the heater lines has been defined in order to cope with thermal environment.
Both these power lines implement a protection mechanism, managed by SW, that automatically disconnect the line in case of failure in linear mode of the switch element (MOS), as this dissipative failure cannot be sustained by the mechanical and thermal design. For the TPSU power lines it is possible to inhibit this protection intervention via dedicated 1553 Telecommand sent by on board computer.

On the TPSU power lines boards capacitive filtering are mounted, both on the input, which is common inside the SPDU) and on the output of each power line; the filtering is finalised to improve the EMC features of the chain (CE towards battery).

4.3. Platform Bus Distribution

The distribution of the 28V regulated bus generated inside SPDU to the Platform heaters is organised in groups; each group, made of up to 5 heater lines is protected by a common current limiter.
Two sizes of heaters are fed; the groups have been organised so that the common current limiter, dimensioned to cope with thermal environment, can sustain the whole current required by the heaters belonging to that group.
All the distribution functions are managed in modular way, so that a suitable number of modules (boards) is used to meet the specified budgets.

4.4. SPDU Command and Monitor Interfaces

The SPDU commanding and monitoring is provided through a Mil-BUS 1553 redounded interface and also through a Memory Load (ML) and Serial Data (SD) bus for some specific functions.
SPDU implements a function (EQSOL) that, after the reception of a dedicated telecommand by on board computer, sequentially switches OFF (acting on LCL) all the logic group of TPSU_PWR lines (5 logic groups are foreseen) with a fixed delay (6ms) between two consecutive power OFF.
The EQSOL function is aimed to perform at system level a quick switch-off of the Antenna in case of anomalous battery Under Voltage.

Fig. 6 shows a picture of the SPDU.

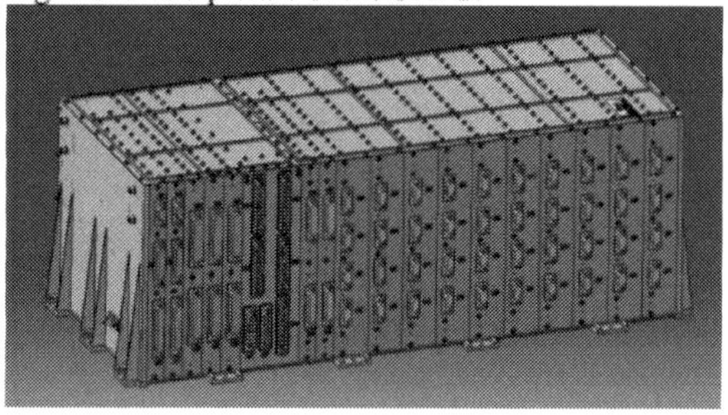

Figure 6: SPDU Box.

DOI: 10.1051/e3sconf/20171613006

4.5. SPDU Main Characteristics

The main characteristics of the PCDU design are:

- Bus voltage: 47V- 67V unregulated bus voltage.
- Unregulated (Battery) Bus power capability up to 18700W towards the SAR antenna.
- Total 40 Power lines to TPSU available.
- Unregulated (Battery) Bus power capability up to 2400W@67V towards the SAR Tile Heaters.
- Total 20 SAR Tiles Heater lines available.
- P/F_PWR Bus maximum power: 800W, towards P/F heaters.
- Total 108 P/F Heater lines available.
  - 80 * 20W lines
  - 28 * 60W lines.
- SPDU power dissipation: < 300W
- 10 sync lines towards TPSU can be operated:
  - All in phase
  - With 36° phase shift between two neighbouring lines.
- 2 sync lines towards Electronics.
- TM/TC interface: MIL-1553 data bus + ML & SD channels + direct lines.
- Two MOS protection SW loops to protect power lines against linear dissipative failures of the switches.
- EQSOL function to perform rapid removal of the power to the antenna in case of contingency.

5. TILE POWER SUPPLY UNIT (TPSU)

In order to properly supply the Antenna, a single Power Supply Unit for each Antenna Tile (TPSU) is foreseen.
TPSU receives the input power directly from SPDU.
Each protected power line in SPDU feeds a single TPSU.
TPSU generates the power lines necessary to supply the following items:

- 2 Electronic Front End (EFE), each composed of two benches of capacitors and 32 single T/R modules.
- 1 True Delay Line (TDL)

TPSU is designed to generate seven different power supply lines, all regulated with an accuracy better than 3% EOL, galvanically isolated versus the primary power voltage.
The V1 and V2 lines supply the most powerful users, TX and RX module, and are generated via a dedicated regulator based on a full-bridge topology which uses the Synchronous Rectification technique.
This regulator manages up to 350W.
All the other Supply Voltages, V3 to V7 used to supply the T/R and EFE digital circuitry, are generated via a second multi-output regulator, based on a Push-Pull topology.
Four of the five output voltages are post regulated to meet the accuracy requirements.
No post regulation is provided for the EFE switch supply line, V7.
Fig. 7 shows a block diagram of TPSU.

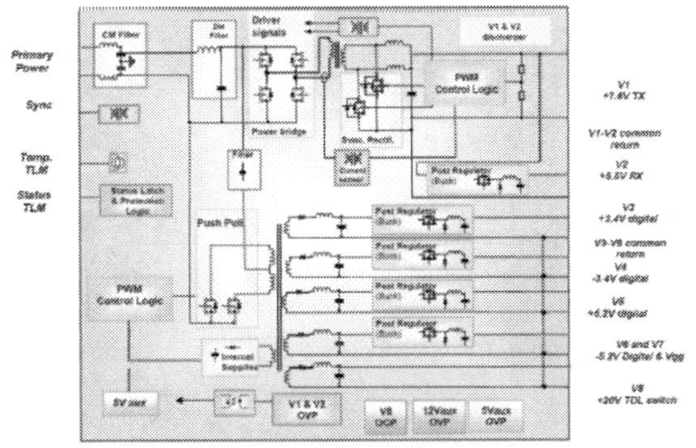

Figure 7. TPSU Block Diagram.

The total output power managed by TPSU is about 400W with an efficiency better than 86%.
The TPSU converters may operate in free running or synchronised by an external signal, TTL standard at 200kHz, generated inside SPDU.
The secondary output rails grounding is provided with two separated return lines here below described:

- Lines V1 and V2 have a common return
- Lines from V3 to V7 have a common return.

The above common returns are referred to TILE chassis at user level, externally to TPSU

The TPSU switches automatically on when the Main Bus voltage is applied, according to SPDU dedicated LCL activation. The TPSU switches off when main bus voltage is removed by SPDU.
Each output line is able to supply the current to allow the fuse blowing in case of any fault at T/R level.

The TPSU mechanical design is based on a baseplate, as shown in Fig. 8, to provide thermal and mechanical interface with the Antenna tile mechanical structure where TPSU will be allocated.

Figure 8. TPSU EM.

6. ELECTRO-MAGNETIC COMPATIBILITY (EMC) DESIGN APPROACH

During imaging very high power changes cause a big perturbation on the power lines within the Pulse Repetition Frequency (PRF) range (from 500Hz to 7000Hz).

The source of such perturbation are the TPSUs that directly feed the SAR antenna RF loads.

As the power of all the TPSUs is modulated synchronously, the resulting Conducted Emission (CE) in the PRF range on the power chain back to the Batteries is the sum of the CE of each TPSU.

Consequently, due to the battery impedance (maximum DC value is 62mΩ), the resulting effect is a main bus voltage ripple that is present on all the Battery Bus users interface.

To validate the behaviour of all the chain both analyses and tests have been performed.

Fig. 9 shows the simulation of the CE injected by each TPSU line on the input Battery line, with maximum battery impedance. The noise at PRF is visible in the range 500Hz÷7.5kHz.

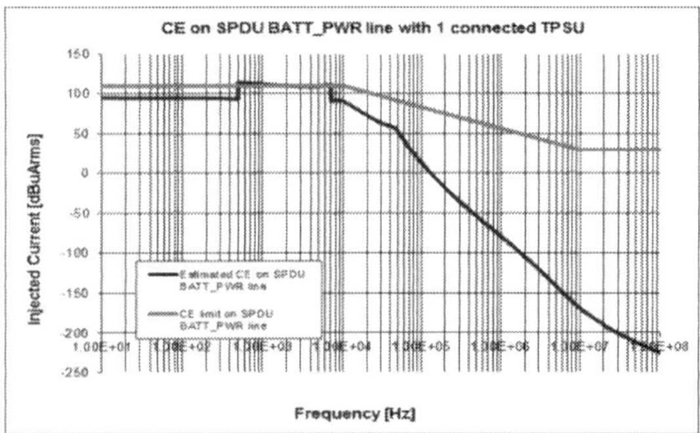

Figure 9. Simulated CE on SPDU Input line, 1 TPSU connected.

An EMC (CE) test has been performed at integrated level between one TPSU EM and one SPDU power module EBB.

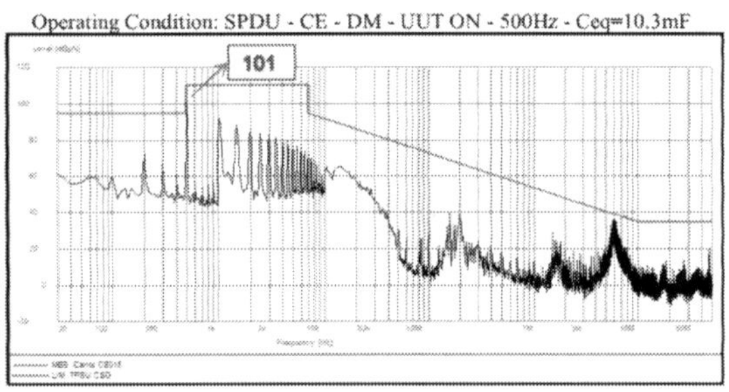

Figure 10. Measured CE on SPDU Input line, 1 TPSU connected (PRF at 500Hz).

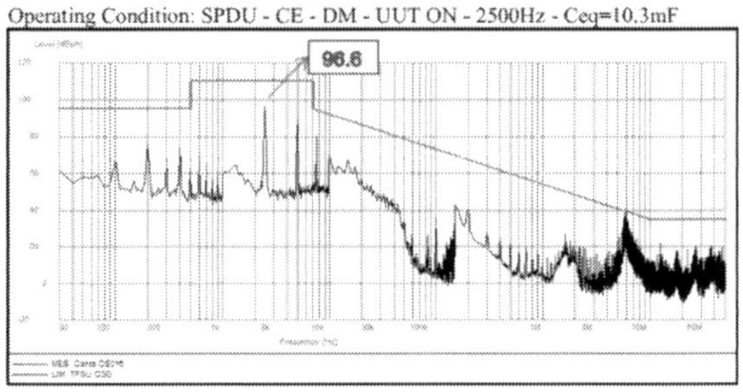

Figure 11. Measured CE on SPDU Input line, 1 TPSU connected (PRF at 2.5kHz).

Fig. 10 and Fig. 11 summarise the test results at two PRF frequencies simulating the filtering effect of all the SPDU lines.

These results substantially confirm the values foreseen from the simulation.

On the basis of the simulation and test results (on one TPSU model) an analysis of the contribution of 40 TPSU simultaneously working has been performed with the Battery impedance as variable parameter. In order to verify complete behaviour of BTA impedance it has been considered the battery impedance variation dependence with:

- Aging
- Orbit
- Temperature
- State of Charge

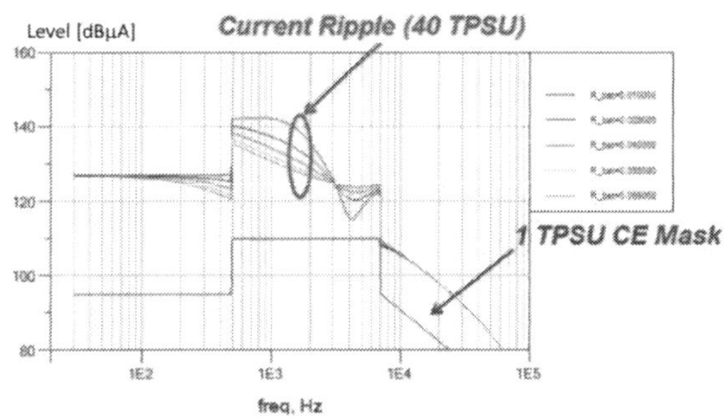

Figure 12. CE simulation on Battery line at varying Battery Impedance.

The S3R function inside PCDU has been sized and to sustain the (worst) specified disturbance at PRF without showing any susceptibility, e.g. S3R sections simultaneous commutation.

A set of test performed on an integrated EBB of the S3R sections of PCDU with a battery impedance of 62mΩ, on the basis of the results summarised in Fig. 12, have also demonstrated the acceptability of such CE levels without susceptibility at power conditioning level. The PCDU SA regulating section EBB, representative of the full S3R section, (including the Battery charge Control

DOI: 10.1051/e3sconf/20171613006

(BCM) module and the capacitor bank) has been submitted to a disturbance (injected on the load side) level of 138dBμA, compatible with the above measured CE levels (at SPDU input) on the PRF range.
Fig. 13, Fig. 14 and Fig. 15, taken at the boundary frequencies of the PRF range and at 2kHz, show the nominal behaviour of the S3R function; one section is switching with the expected proper ripple amplitude and frequency, without being affected by the noisy EMC environment.

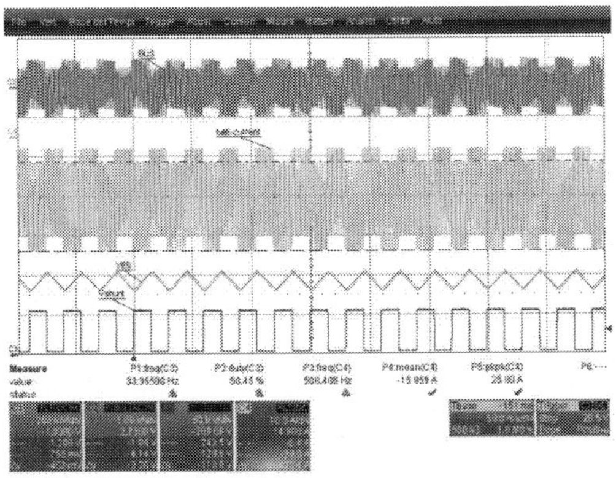

Figure 13. PCDU EMC_CS @ 500Hz.

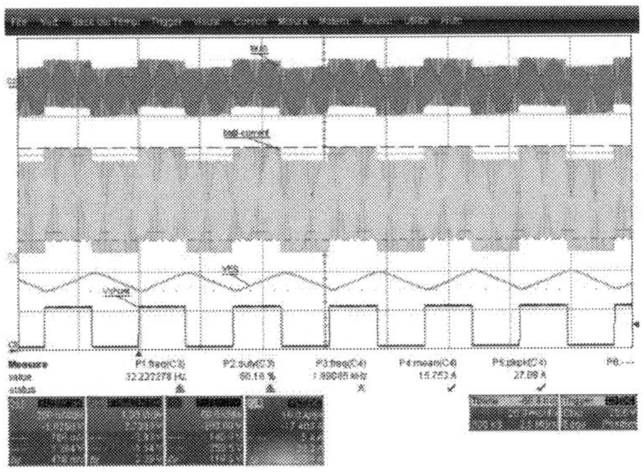

Figure 14. PCDU EMC_CS @ 2KHz

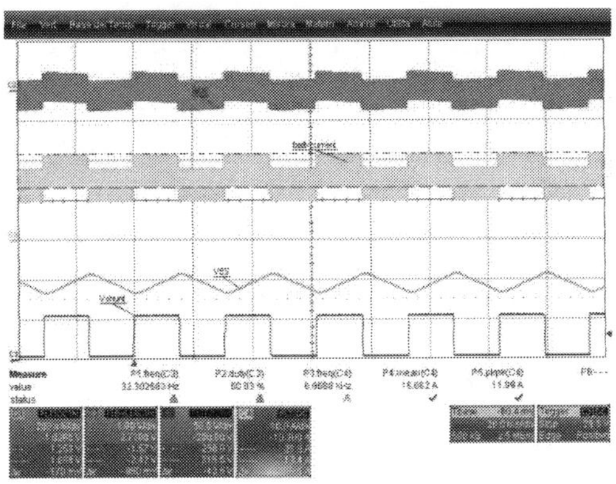

Figure 15. PCDU EMC_CS @ 7KHz

7. PHOTOVOLTAIC ASSEMBLY

The Photovoltaic Assembly is part of the SA, which design is driven by the mission requirements.
The Triple Junction 3G 28% solar cells from AZUR space that represents the European consolidated technology for space-borne solar arrays, have been selected for this project.
The triangular silicon diode (AZUR space) is adopted for solar cell protection.

Figure 16. COSMO SG PVA DVT coupon.

The Solar Array for the Cosmo Second generation satellite is made of 2 identical wings of 4 panel each.
The PVA shall be divided into 2 electrical sections per panel.

The PVA S/C interface characteristics:

- Max open voltage of each section must be lower than 137 volts.
- Vmp shall be higher than 70V
- Voltage drop < 3 V average per wing at 70 V (avg < 3.3 V for Imax BOL)

The Solar Array shall provide the following minimum power @70V:

- 3600W at solar array level
- 450W at panel level
- 500W at panel level (stowed condition)

The PVA network is made of 2 electrical sections with 8 strings each (8p37s configuration).
Each string is protected by 2 blocking diodes grouped at section level into a single diode board.
The remaining factors and the temperature coefficients for Voc, Isc, Vmp and Pmax at the mission dose were calculated using a Matlab® interpolating function, which refer to the piecewise cubic Hermite interpolation.
The add-on mass of the PVA is less than 15 kg.

8. CONCLUSIONS AND FUTURE WORK

CSG is a very demanding program that poses important challenges in terms of cost, mass and performances.

 DOI: 10.1051/e3sconf/20171613006

In order to get the optimum choice between all these sometime conflicting requirements it has been selected a power conditioning architecture that is partly based on already qualified electronic modules.
Some design choices have privileged the simplicity as well as the cost and space saving approach. This is the case of the control of the SA power via an S3R approach, instead of the MPPT regulation approach.
Other design, like the TPSUs have been developed ad hoc for the application, due to the specific requirements given by the upgraded SAR antenna, both in terms of electrical and of mechanical/thermal constraints.

9. REFERENCES

[1] L. Croci, G. Gianninoto, P. Galantini, E. Scorzafava, G. Daprati, M. Manfreda "COSMO – SkyMed. The Electrical Power Subsystem Electromagnetic Compatibility issues", *EDA Workshop on Power Supply and Energy Management for Defense Applications 13-14 June 2007.*

[2] E. Scorzafava, F. Monaci, E. Zampolini et altri " COSMO second generation: spacecraft design and technological challenge - Proceedings of EUSAR 2014; 10th European Conference on Synthetic Aperture Radar;

[3] Lazzeretti, L. Croci, G. Gianninoto, F.G. Re, "Spacecraft power conditioning and distribution: a modular and flexible approach for custom application," *Proceedings of the 9th European Space Power Conference, 2014.*

[4] F. Caltagirone, G. F. De Luca, E. Scorzafava, et altrii "COSMO-SkyMed di Seconda Generazione: Key Technological and Design Evolutions, and Related Potentialities" - Proceedings of 20th Ka and Broadband Communications, Navigation and Earth Observation Conference, Vietri, (2014).

10. ACKNOWLEDGEMENTS

The authors wish to thank all the colleagues that have contributed to the presented activity, with a particular thank to Luca Di Cola for his fundamental contribution.

MPBUS: MODULAR POWER BUS FOR SPACE VEHICLES

Ricardo Martin [(1)], Eduardo Remírez[(2)], Javier Alonso [(3)], Eran Priel [(4)], Alain Carapelle [(5)], Daniel Actis [(6)]

(1) Project Manager, JMP Ingenieros, 26371 Sotés – La Rioja (Spain), ricardo.martin@jmpingenieros.es
(2) CEO, JMP Ingenieros, 26371 Sotés – La Rioja (Spain), Eduardo.remirez@jmpingenieros.es
(3) CTO, Nájera Aerospace, 26371 Sotés – La Rioja (Spain), javier.alonso@nasp.es
(4) HW&SW Engineer, Yamar Electronics, 6249220 Tel Aviv (Isarel), eran@yamar.com
(5) Radiation expert, Centre Spatial de Liège, 4031 Angleur (Belgium), A.Carapelle@ulg.ac.be
(6) HW&SW Engineer, Centre Spatial Liége, 4031 Angleur (Belgium), d.actis@ulg.ac.be

ABSTRACT

Cost reduction in the access to space is leading the space race in the launchers market. Launcher´s companies are looking for a more efficient and economical way to place satellites in orbit.

MPBUS project proposes a decentralized power bus distribution with jettisonable smart batteries combined with the use of PLC technology, data over DC, achieving overall weight reduction and harness simplification, both goals in line with Next Generation Launchers (NGL) targets.

Two different bus interfaces have been developed, Battery Front End (BFE) in charge to connect power sources equipment to the power bus, and Device Front End (DFE) in charge to connect and protect power demanding equipment to the power bus. Same power wire is used for transmitting data for power management and telemetry, reducing overall data harness.

Panel demonstrator prototype integrates these front ends to functionally test the application of MPBUS concepts to the combination of both buses, unregulated 28V Power Bus and Telemetry Subsystem, from VEGA launcher.

Environmental tests have been performed to MPBUS devices successfully.

ABBREVIATIONS AND ACRONYMS

ADC	Analogue Digital Converter
AFE	Analog Front End
AIT	Assembly & Integration & Testing
AVUM	Attitude and Vernier Upper Module
BER	Bit Error Rate
BFE	Battery Front End
CE-I	Conducted Emission Current Transient
DC	Direct Current
DFE	Device Front End
DPA	Decentralized Power Bus Architecture
ECC	Error Correction Codes
EGSE	Electrical Ground Support Equipment
EMC	Electro Magnetic Compability
FPGA	Field Programmable Gate Array
LCL	Latch Current Limiter
LDO	Low Drop Out
MFU	Multi Function Unit
NGL	Next Generation Launcher
OBC	On Board Computer
PLC	Power Line Communication
SPI	Serial Peripheral Interface
TID	Total Ionizing Dose
TRL	Technology Readiness Level
VEGA	Vettore Europeo di Generazione Avanzata

1. INTRODUCTION

While the PLC technology is widely used and accepted in the industrial sector, its application for space is still in development stages, [1], [2]. However the interest for this technology and their application for launchers is raising due to adequate matching to several NGL requirements, such as mass reduction, harness optimisation and AIT cost reduction [3].

The MPBUS project has developed a new decentralized power bus architecture (DPA) applied to launchers, introducing the Smart Battery concept, combining distributed batteries with Battery Front Ends (BFE) connected to the power bus along the launcher, allowing the use of multi-power sources along the power bus and to maintain power management network permanently monitored using PLC over same DC power wires as communication path.

Smart Batteries can be jettisoned following the launcher´s stages separation schedule and MPBUS can adapt to the new power bus layout after every stage separation.

DOI: 10.1051/e3sconf/20171613007

MPBUS can also manage, with DFEs, the protection and connection/disconnection of the different loads to/from de power bus.

The MPBUS PLC maximum rate is 3Mbps, the power network management consumes only 5 Kbps, so there are almost 3 Mbps to be used as data payload coming from other spacecraft subsystem. In MPBUS prototype, Telemetry subsystem data, non-critical subsystem sited along all the launcher´s length, has being used as payload data in PLC. Every Telemetry node is connected to the MPBUS through a DFE to be both powered and data connected using the power wires.

MPBUS combination of DPA and PLC applied to launcher typology achieve several advantages vs traditional power distribution (point to point), as power harness reduction and elimination of data wires for payload subsystem (Telemetry) data transmission.

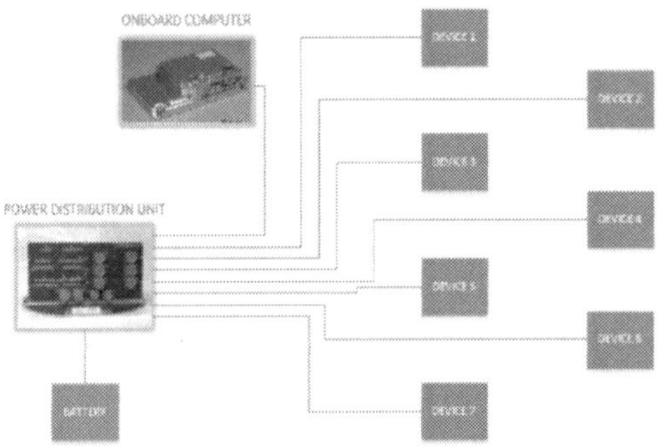

Figure 1. Traditional point to point power distribution

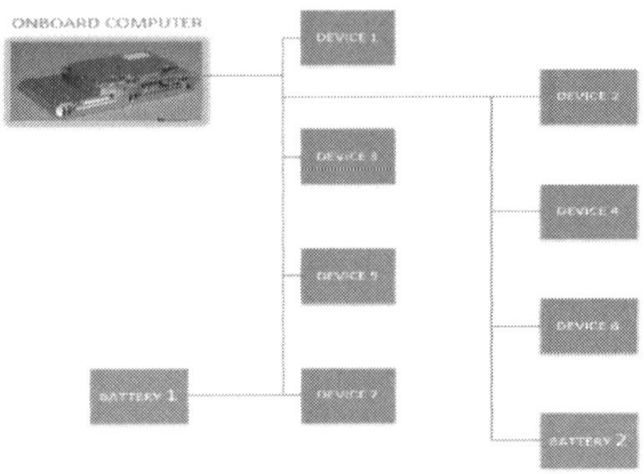

Figure 2. MPBUS decentralized power distribution

These features result in weight, volume and complexity harness reduction and thus in a greater harness design flexibility and an AIT cost reduction.

2. DESIGN

The MPBUS is based in a bus topology with one master node and slave nodes. Communication is based on a Token Pass technique managed by the master node with limited access time to the bus for every slave node, accessing to the bus sequentially according to a unique address. Every message in the bus is broadcast to the rest of nodes and each node process it accordingly.

The current implementation of the MPBUS in the final prototype allows one master and up to 16 slaves, but there is no theoretical limitation for maximum amount of nodes into the bus. The protocol is composed of two different types of frames: master and slave. The master frame is a broadcast message received for all nodes with a command for a single node, telemetry configuration for all nodes and telemetry data from the master node. The slave frame is a broadcast message with information to maintain the token pass, answer to master commands and telemetry data exchange. Figure 3 shows a typical MPBUS communication oscilloscope screenshot, where nodes response can be observed as a carrier frequency over 28 Vdc.

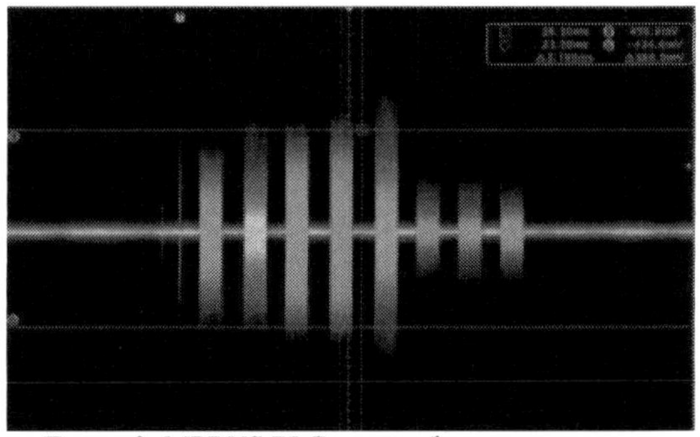

Figure 3. MPBUS PLC carrier frequency response

Distributed power is achieved by connecting /disconnecting the batteries to the two wires power bus through the BFE interfaces. Make-before–break technique is used for switching on/off batteries to the bus, two batteries can be connected to the bus at the same time due to the use of a protective diode in each BFE, ensuring that there is always a battery supplying energy to the bus, eliminating voltage gaps in power source transitions.

Different loads or energy demanding devices can be powered/unpowered when connected/disconnected to the bus through DFE interfaces. Each DFE has its own protections, based in Latch Current Limiter (LCL) circuit, following [4] specification.

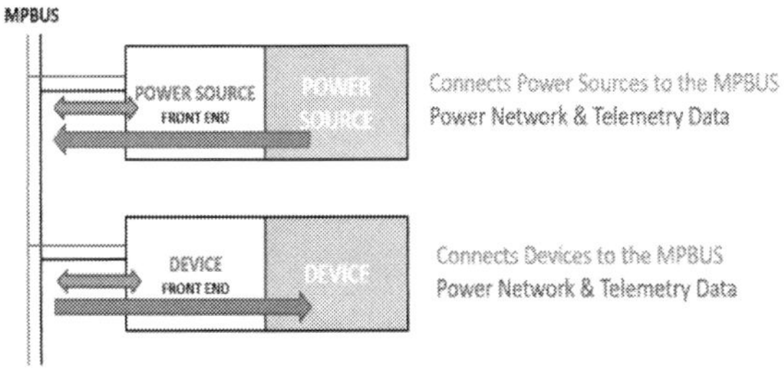

Figure 4. BFE & DFE MPBUS interfaces

For the MPBUS final prototype, both BFEs and DFEs were designed according to VEGA launcher specifications [5].

2.1 Battery Front End

The Battery Front End is the interface that allows the connection/disconnection of a battery to/from MPBUS. The BFE electronics is always powered from its own battery and it is composed of the following main modules: Power Source Management, Power Supply, Control Logic, Communications and Acquisition.

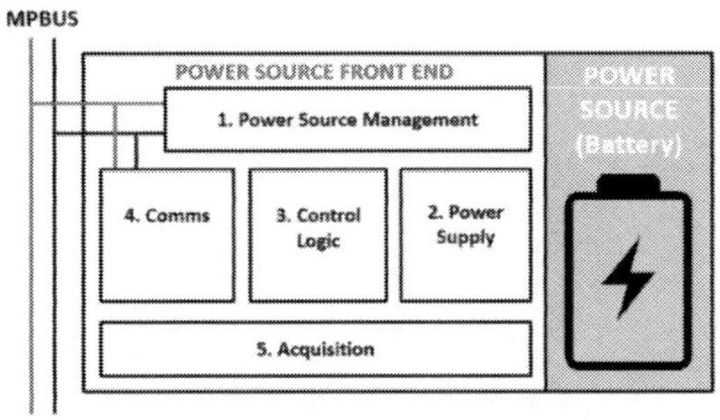

Figure 5. BFE interface

The Power Source Management is the circuit that connects/disconnects the battery to/from MPBUS. It is composed by: switching circuit, on/off sensor and current sensor.

The Power Supply is connected directly to the battery and provides the different voltage levels needed by the front end components: 3,3 Vdc and 1,5Vdc. It is composed by a protection circuit, DC/DC converters and LDOs regulators.

The Control Logic subsystem, based in a Microsemi FPGA ProAsic3 A3P600, is the brain of the interface and manages the different functions of the module (conection/disconection of the battery, communications and acquisition).

The Communications subsystem, the DCB3M modem, is the PLC core and it has been developed using a Microsemi FPGA SmartFusion2 M2S050. It is connected to the MPBUS and powered from the Power Supply module.

The DCB3M modem is based in a phase modulated (QPSK) narrow band carrier that protects the data with several levels of error correction codes. In addition, unique signalling technology allows arbitration over noisy and severely attenuating channels.

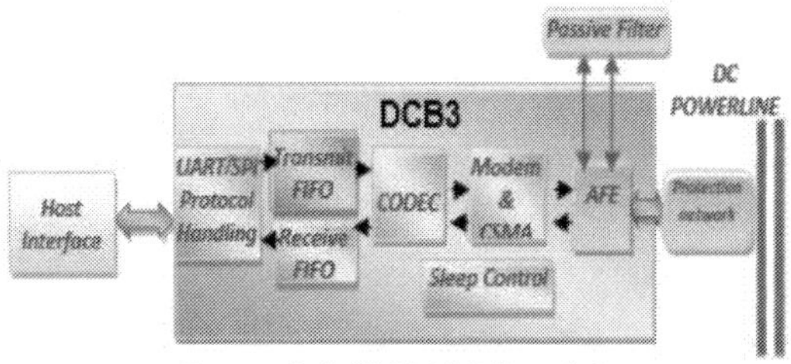

Figure 6. DCB3M PLC module

The modem implementation consists of analogue front end (AFE) that gets the analogue signal from the powerline, filter the carrier signal from the powerline noise and convert it into a digital bit stream. The digital part implemented in the FPGA recovers the digital stream into the phase-modulated carrier and decode the data from the noise. Three levels of Error Correction Codes (ECC) provide the user the ability to select the most effective ECC that mitigates the noise on the powerline.

The Communications module has a selectable carrier frequency channel (13MHz to 30 MHz in steps of 100kHz), a selectable ECC level, resulting in an effective transmission rate from 0.9Mbps up to 3Mbps. This module is managed through a SPI channel by the Control Logic and provides the ability to select the operating carrier frequency. A software tool has been developed to sweep the entire frequency range and map the attenuation between the nodes of the final prototype as function of the carrier frequency in order to find the channels with the lowest signal attenuation.

A study about the integration of the Communication module into an ASIC has been performed in order to assess its feasibility.

The Acquisition system acquires relevant data from the power network (battery voltage, bus voltage, current, battery cells temperature…), subsystems and external sensors. All this data is managed by the Control Logic and sent through the MPBUS by the Communication module.

For the MPBUS final prototype, the BFE has being designed to work under three different operating modes:

 DOI: 10.1051/e3sconf/20171613007

Operating Mode	Battery	Telemetry Data
IDLE	Disconnected	No
TLM	Disconnected	Yes
BATTERY ON	Connected	Yes

Figure 7. BFE operating modes

2.2 Device Front End

The Device Front End (DFE) is the interface that protects the MPBUS from load device anomalies and vice versa. DFE allows the control of the connection/disconnection of a load to/from MPBUS. The electronics of this front end is always powered from the MPBUS and it is composed of the following main modules: Load Management, Power Supply, Control Logic, Communications and Acquisition.

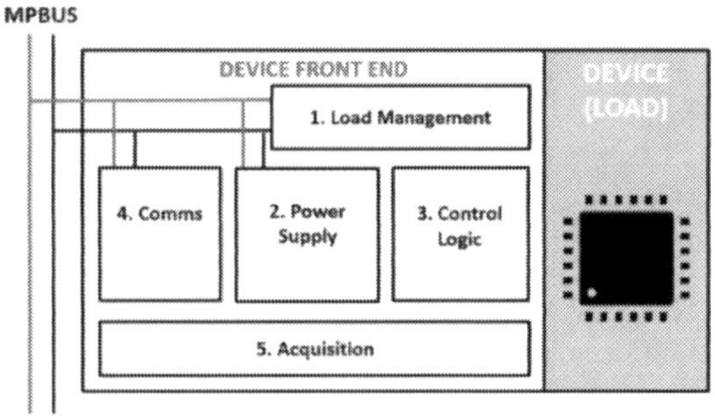

Figure 8. DFE

The Load Management is the circuit that connects/disconnects the device to/from MPBUS and it has a protection built-in LCL. It is composed by: protection circuit, switching circuit, on/off sensor and current sensor.

The Power Supply is connected directly to the MPBUS and provides the different voltage levels needed by the front end: +5V, -5V, 3,3V, and 1,5V. It is composed by a protection circuit, DC/DC converter and LDOs.

The hardware of the Control Logic is similar to the one in the BFE but with different firmware. Rest of DFE are equal than those than BFE.

For the MPBUS final prototype, the DFE has being designed to work under the three different operating modes as for the BFE.

3. MPBUS PROTOTYPE

To assess the MPBUS, two different prototypes have been manufactured. First proto was in charge to demonstrate basic front ends functionalities and served as basis for the final proto.

For final proto MPBUS assessment, unregulated 28V power bus and Telemetry subsystem from VEGA launcher were selected to be implemented according to new MPBUS concepts. This new architecture replicates them in a simplified manner on a 6 meters long panel demonstrator, simulating VEGA launcher different stages and interstages.

Wire length placed at stages has same length than real VEGA stages (total 35 meters). The nodes placed at interstages simulate real VEGA avionics load devices (OBC, UCTM, UCATs, IRS TVCs), demanding from MPBUS power sources up to 23A.

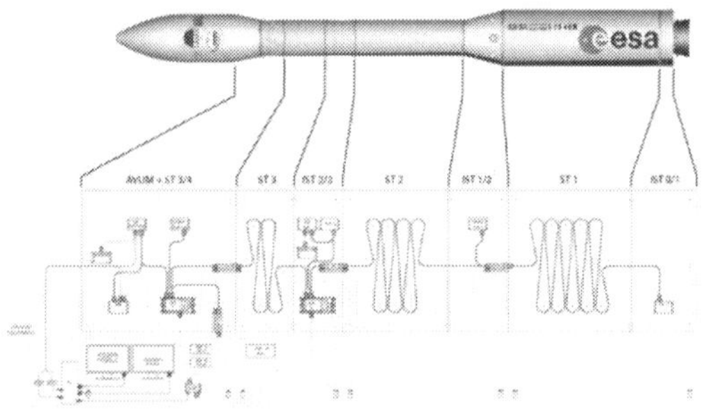

Figure 9. MPBUS Final Prototype layout

Figure 10. MPBUS Final Prototype

Stages are connected by 4 controlled bus separators based on electromechanical spring-loaded solenoid that can simulate stages jettison, allowing MPBUS to be splitted into several sections while launcher´s mission is performed.

Figure 11. Stage's separator device

EGSE subsystem, external to the panel, simulates the VEGA connection to ground equipment allowing to command and control the MPBUS from ground prior to lift off phase. The EGSE part is composed by one special MPBUS node developed as a support equipment that works as a gateway between the PC used as a control unit in the EGSE and the MPBUS panel demonstrator.

There are 3 different power supplies (28V) in the final prototype, one external power supply at the EGSE simulating launcher powering when at ground phase, BAT2 node at interstage 2/3 powered by external power supply and BAT1 node at upperstage powered from Li-ion battery.

There are 7 different nodes simulating VEGA avionics devices, each one is capable to simulate load energy consumption for up to five VEGA devices and it can handle the switching of every of these loads separately.

BAT1 and BAT2 BFE and TVC3 DFE have been designed and manufactured using components with space equivalent. Rest of nodes have been designed and implemented based on industrial components for cost and time constraints.

The MPBUS harness is a two wire shielded twister PTFE insulated AWG 12 cable, selected following [6] and DB15 3W3 connectors with central pin not connected. Auxiliary industrial DB9 connectors and cables, out of the MPBUS core scope, are used as ground-launcher link wires for simulating separator's signal and wired telemetry link between UCTM and EGSE.

3.1 Battery nodes

Real VEGA launcher has two batteries [7], 55V and 28V, placed at the upperstage (AVUM). The 28V battery is made of Li-ion SAFT MP176065 cells in a 8S8P configuration. This battery powers the unregulated 28V bus, which powers several equipment as OBC, UCTM, UCATs, since the EGSE umbilical disconnection at launch pad to the end of mission. When AVUM starts its programmed payload orbit injection maneuver, the 28V battery capacity is drained about 25%. MPBUS proposes to split the original 28V 8S8P battery into two, one with a 28V 8S2P (BAT2) configuration placed at intersatge 2/3 in charge to power the bus from the EGSE umbilical disconnection to just before stage 3 separation and another battery 28V 8S6P (BAT1) in charge to power the AVUM after stage 3 separation. This battery distribution reduces the inert AVUM mass allowing to increase the thrust-to-weight ratio for this last stage.

Figure 12. BAT2 node, upper enclosure removed.

Battery nodes have been designed in a two floor layout aluminum enclosure, upper floor for BAT2 accommodates 28V 8S2P Li-ion battery and BFE board is placed at the bottom floor.

BAT2 has a NTC sensor for temperature monitoring, this analog signal is wired to the BFE where the Acquisition module sample and convert it and the Control module send it to the simulated OBC through MPBUS by the Communication module.

Figure 13. BAT2 BFE with PLC modem removed

Mechanized openings in the aluminum enclosure are available for connectors to distribute power to the several MPBUS branches.

3.2 Device nodes

In a similar manner as with battery nodes, a front end for devices that demands energy from the MPBUS has been developed.

For MPBUS prototype 7 simulated avionics devices have been implemented onto the panel demonstrator: OBC, TLM Controller, TLM, SIMx5, SIMx2, TVC3 and TVC1. The different DFEs have loads to simulate the real consumption of the devices in the VEGA launcher. The consumptions simulated for the TVCs are the one related to the electronic control, not to the TVCs electrical actuators. The SIMx5 and SIMx2 simulated the loads of several devices of VEGA.

The system is managed by the EGSE node at the beginning of the mission simulation. Once the EGSE is removed, the OBC becomes the master of the MPBUS.

The TLM Controller is the node that collects all data sent in the MPBUS and resend them to the EGSE station for post analysis.

Rest of devices just simulates the energy consumption and send their telemetry data.

TVC3 DFE has been designed and manufactured using components with space equivalent. A LCL class 3 circuit has been implemented. Rest of DFEs are developed with industrial components.

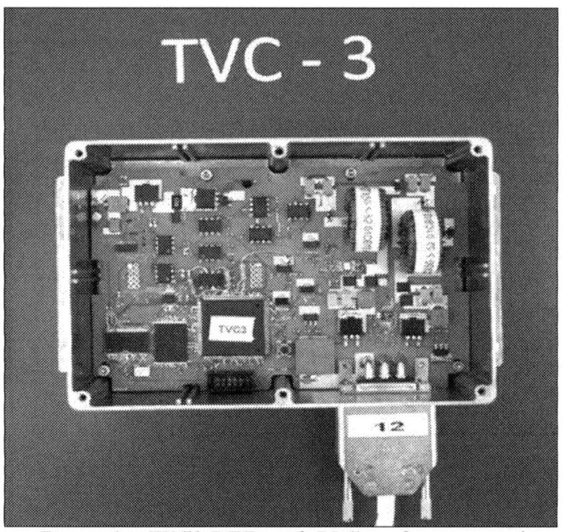

Figure 14. TVC3 DFE with PLC modem removed

4. TESTS RESULTS

The BFE and DFE have been functionally tested into the panel demonstrator at NASP facilities and environmental tests over single nodes have been performed at CSL facilities to assess their compatibilities with VEGA launcher specifications.

4.1 Functional Tests

Three main functional tests have been performed: (i) mission test, (ii) transfer rate and (iii) word error rate. A specific testing software has been developed to manage different missions, collect and analyse data to extract test results.

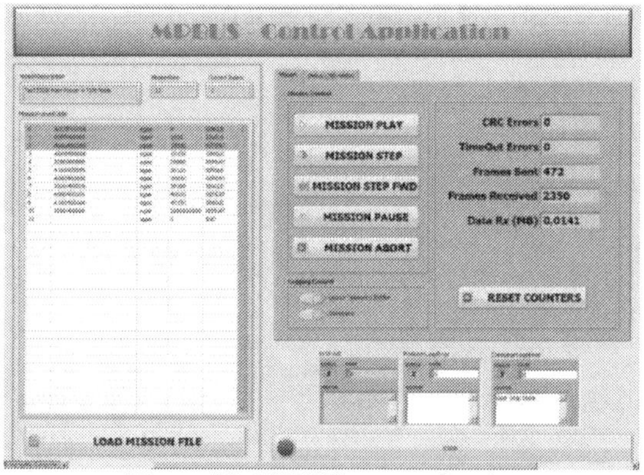

Figure 15. MPBUS test application

The MPBUS final prototype was functionally tested following a simulated typical VEGA mission, covering from previous lift off moments to end of mission phase. During the mission, device and battery nodes were communicating through the MPBUS and executed their respective connection/disconnection in a correct sequence to perform the mission. Launcher stages were sequentially jettisoned and the nodes were connected and disconnected until the last stage remained powered from its own BFE. MPBUS communications, operation modes of every node and different power distribution scenarios, up to 23A, were achieved successfully.

Transfer rate tests was performed with all MPBUS nodes connected configured with Telemetry mode on, maximum sample rate and hard ECC. Average transfer rate obtained was 0,90 Mbps, almost matching with real VEGA Telemetry maximum rate 1 Mbps [8].

Word error rate tests has been performed over all possible MPBUS layout configurations after every stage separation. TLM Controller node role was to sniffer all MPBUS communications frames and send them to EGSE equipment for data post processing. Two different errors, CRC and timeout, have been recorded. MPBUS fails VEGA word rate restrictive specifications [9]. The MPBUS measured byte error rate is 2.35E-7, showing MPBUS potential feasibility at this early development stage.

4.2 Environmental Tests

These tests were performed at CSL facilities to check BFE and DFE behaviour in VEGA/Ariane space environment in terms of vacuum, temperature, radiation and EMC.

4.2.1 Radiation test

The most critical radiation environment of a launcher is a flux of proton and neutron during 30 minutes of mission [10]. It has been established that in nominal solar conditions, the heavy ion aggression does not affect the success of a mission [11].

Electronic components from front ends selected for protons and neutrons testing were both FPGA, the ProAsic3 A3P600 and the SmartFusion2 M2S050 from Microsemi. 16 specimens each model has been tested.

There are mainly two types of effects induced by the radiation in electronic components: Total Ionizing Dose (TID) Effects and Single Event Effects (SEE). As SEE have to be managed at design level, tests focused on TID effects. Neutron dose can be converted to total ionizing dose equivalent value [12].

TID influence can be assessed by measuring the change in gate propagation delay [13]. FPGA design involving large number of serial inverter gates was developed and propagation delay was measured on test specimens before and after radiation.

5.2.1.1 Protons

The estimated fluence at 50 MeV p+ for Ariane5 is around $1.E^7$ p+/cm^2 [14] for a 30 minutes mission. Four FPGA specimens for both models were tested with this fluence and energy (30 minutes irradiation). There were another specimen batch with four specimens each model tested with 3 times the fluence (three times the flux with same 30 minutes irradiation time).

Results manifest that there is no significant change in the propagation delay.

5.2.1.2 Neutrons

The test was done at a flux of 1 n/cm^2*s (corresponding to the mean atmospheric environment [14]) and 3 n/cm^2*s. The irradiation time was 30 min as the typical Ariane5 launcher mission. The selected source (Am-Be: ISO 8529-2 Reference neutron radiations) has a (theoretical) spectrum between 1-10 MeV with a peak around 5 MeV. The flux of 1 n/cm^2*s is obtained by adapting the distance between the source and the target, thanks to the $1/r^2$ effect. The total fluence after 30 minutes is then 1800 n/cm^2.

As in protons test, four FPGA specimens from both models were tested with nominal dose and another four specimens from both models were tested with 3 times the dose during 30 minutes.

Results manifest that there is no significant change in the propagation delay.

4.2.2 Thermal Vacuum test

The MPBUS BFE has been tested into a vacuum chamber. Four thermal cycles between -10°C and +60°C at a pressure of less than 10^{-5} mbar were performed while BFE performed power connection/disconnection procedures and communication with another MPBUS nodes placing outside vacuum chamber.

No error occurs in the range of the test.

4.2.3 EMC test

BFE and DFE have been tested at the University of Liége Laboratory in a semi-anechoic chamber for EMC, following [15] and [16] procedures, to test the devices in radiative/conductive emission and susceptibility.

Except the transient current that exceeded the limit at the CE-I test, the MPBUS passed all tests.

5. CONCLUSION

The MPBUS solution has demonstrated the benefits of combining decentralized power distribution and PLC, achieving mass reduction, harness optimisation, complexity, risk and AIT cost reduction.

Communication over power wires has been stablished by MPBUS nodes despite to the complex environment prototype: stages separation, connectors and separator´s interfaces, bus branches and wire length.

Most of the functional and environmental tests based on VEGA´s specifications were successful, even at current MPBUS TRL 4, showing great potential for next development phases.

Lack of specific standards for spacecrafts using PLC technology has been detected. MPBUS developments can help to settle future standardization about this topic.

6. ACKNOWLEDGMENTS

The research leading to these results has received funding from the European Community's Seventh Framework Programme (FP7/2007-2013) under grant agreement n° 606818 for the MPBUS project.

7. REFERENCES

1. Powerlink: Two-Wire Power and Data Link. Statement of Work – Appendix 1 to ESTEC/ITT AO/1-6774/11/NL/AT, Issue 1, Rev 5.
2. Johannes Wolf. Power Line Communications (PLC) in Space- Current Status and Outlook. Proceedings 2012 ESA Workshop on Aerospace EMC.
3. http://www.esa.int/Our_Activities/Launchers/Technology
4. ECSS-E-ST-20-20C DIR1. Space engineering – Electrical design and interface requirements for power supply. ESA-ESTEC 20 May 2015.
5. VG-SG-1-C-35-SYS Issue 5 Rev.1. VEGA Program – General Design Norms and Rules. ELV S.p.A.
6. ECSS Detail Specification No.3901/024
7. R. Albano, P. Brochard and F. De Coster: "Lithiumion batteries for the VEGA launcher" Proceedings of the conference on Changes In Aeronautical And Space Systems – Challenges For On Board Energy, Avignon, June 2006, France
8. VG-ST-1-C-002-SYS Issue 6 Rev. 1, VEGA Programme–Avionics System Technical Specification, page 74.
9. VG-SG-1-C-037-SYS Issue 3 Rev.1, VEGA Program – Communication System General Specification, pp 28,30.
10. Ariane 5 User's Manual Issue 5 Rev. 1, July 2011
11. Calvet, M.C., Mazon, R., Simon, Ph., Galindo, D., Poivey, C., Garnier, P., Bourrieau, J., Cadot, B., Ecoffet, R., Acknowledgement of the Natural Radiation Environment upon the ARIANE 5 Launcher, Third European Conference on Radiation and its Effects on Components and Systems, RADECS 1995, pp. 170-174 France.
12. A.Holmes, L.Adams, Handbook of radiations effects, 2nd edition, Oxford University Press (2007)
13. Kastensmidt F. L., Pereira Fonseca, E. C., Vaz, R. G., Gonçalez, O. L., Chipana, R. and Wirth G. I., TID in Flash-Based FPGA: Power Supply-Current Rise and Logic Function Mapping Effects in Propagation-Delay Degradation, IEEE Trans. Nucl. Sci., Vol.58, Issue 4, pp 1927-1934, 21 April 2011
14. Colas, G., Neutrons atmosphériques en avionique, Presentation to RIS, Thalès Avionics, CAL/DT/I, 20 Nov. 2002
15. VG–SG–1-C-30-SYS Is. 4 Rev.1
16. MIL-STD-461F standard

DOI: 10.1051/e3sconf/20171613008

FUNCTIONAL STUDY OF A DISTRIBUTED MPPT POWER MANAGEMENT SYSTEM

S. Bifaretti (1)**, S. Pipolo** (1)**, T. F. Catalano** (2)**, G. Daprati** (2)**, V. Iacovone** (2)**, E. Scione** (2)

(1) Department of Industrial Engineering, University of Rome Tor Vergata, Via del Politecnico 1 Rome, Italy
bifaretti@ing.uniroma2.it
(2) Thales Alenia Space Italia S.p.A., Via Saccomuro 24, Rome, Italy, { toni.catalano, giorgio.daprati, vincenzo.iacovone, emiliano.scione, }@thalesaleniaspace.com

ABSTRACT

This paper presents the concept of a control strategy for Solar Array (SA) power regulation using an independent Maximum Power Point Tracking system for each Solar Array section in order to maximize the power extracted from every SA sections. Moreover, it allows to distribute the battery charge current between the power sources in order to evenly divide the switching losses on the power semiconductors of the converters and, thus, extending their life time and to reduce the dissipation power. The proposed strategy can be applied to the Power Control Unit designed for satellites with unregulated power bus architecture. Significant simulation results, obtained using a Matlab/Simulink model, demonstrates the validity of the proposed approach.

1. INTRODUCTION

Solar Array (SA) regulation systems based on Maximum Power Point Tracking (MPPT) have been largely developed and implemented for spacecraft applications in the last years. The architectures related to MPPT systems can be classified, as illustrated in [1], in parallel [2]-[4] or sectional [5]-[6]. In the first solution only one MPPT is used to achieve the voltage reference for the entire SA control. Such approach is valid in case the SA panels or wings are exposed to the same environmental conditions (e.g. temperature, illumination) and aging. However, these are ideal cases because both operational constraints (loss of sun pointing due to payload imaging manoeuvres) and standard entry/exit eclipse situations, imply different thermal and illumination conditions for the SA panels. On the contrary, sectional architectures, which employ separate MPPTs able to perform an individual tracking for each SA section, are preferred due to their better performances in extracting power from the sources. For this reason, the latter is accounted and discussed in this paper. Usually the references for the SA regulators are provided alternatively by the MPPT or by the output of the battery charge control loop on the base of the required load and battery power. The paper proposes a control scheme that employs, at the same time, a combination of both references in order to generate a proper control signal to the buck DC-DC converters chosen for SA regulation. The control method is applied to each converter using two different control loops in order to regulate at the same time the SA and the output currents.

2. SINGLE SECTION CONTROL

The control system, referred to a single section of the SA, combines a classical Perturb & Observe (P&O) MPPT algorithm with a battery charging regulation system, as shown in Fig. 1.

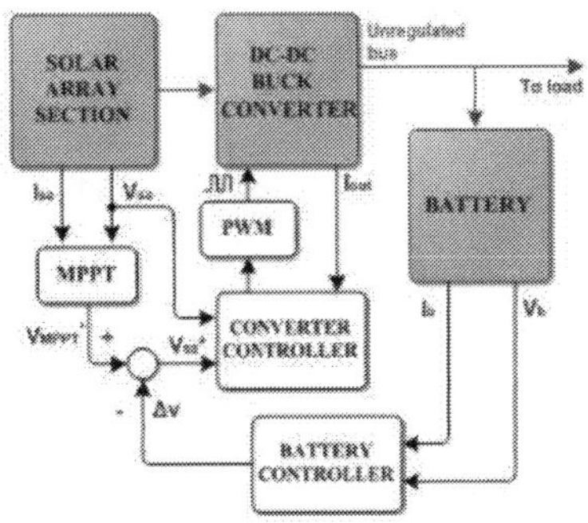

Figure 1: Block diagram of a single SA section.

The voltage reference V_{sa}^* for the buck converter is determined by the difference of V_{MPPT}^* and the output of the battery controller $\Delta v \leq 0$ so that the power balance between the SA, the load and the battery is satisfied.

The charging control of the battery is performed according to the Constant Current - Constant Voltage (CC-CV) algorithm. Based on SA power generation, battery State Of Charge (SOC) and load condition, it is possible to distinguish two different operation modes:

a) Battery discharging mode ($P_{load} > P_{sa}$)

In this case the input voltage reference of the buck converter V_{sa}^* is completely determined by the control signal coming from the MPPT algorithm (e.g. Δv=0), allowing the SA to operate at maximum power.

b) Battery charging mode ($P_{load} < P_{sa}$)

In this case, the system allows the charge of the battery according to the CC-CV algorithm.

As the battery progressively reduces its power demand, Δv becomes more and more negative, so that the voltage operating point of the SA decreases until the condition $P_{load} = P_{sa}$ is satisfied.

2.1. DC-DC Converter Control Loop

The double control loop shown in Fig. 2 has been used to regulate the SA voltage V_{sa} and on the output current of the converter I_{out}.

The conduction interval of the Power MOSFET M_n is determined according to the difference between the voltage reference V_{out}^*, provided by the MPPT, and the output current V_{out}. A feed-forward term *r*, achieved by the battery and SA voltage ratio, is also added to obtain a faster transient response. The reference current I_{out}^* is the sum of two terms: the first one is the difference between the desired voltage V_{sa}^* and the SA voltage measure, whilst the second is obtained by the feed-forward term FF, which takes into account the current supplied by the SA and the voltage ratio r.

These feed-forward terms have been accounted in order to fasten the control and permit to the PI regulators to work in a reduced operating range.

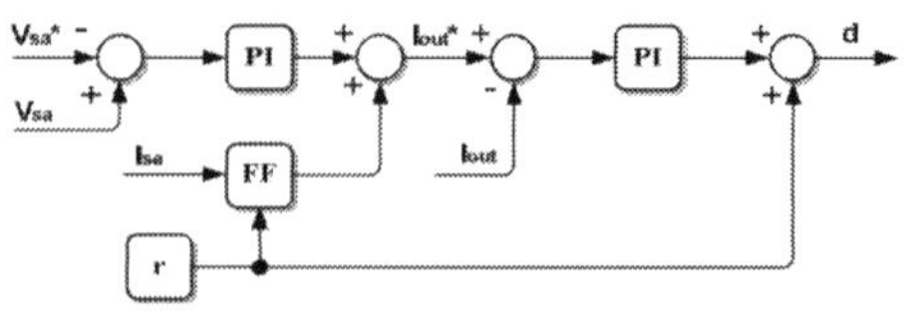

Figure 2: Single converter controller block diagram.

2.2. Battery Charge Control Loop

Fig. 3 shows the detailed block diagram of the battery controller, which is composed of two different control loops: the internal loop used for the battery constant current control, while the external loop employed for the constant voltage control.

The voltage loop sets the current reference at which the battery will recharge. The battery current limiter is used in order to set a maximum charging current level as long as the battery has not yet reached its maximum charged voltage command V_b^*, representing the End Of Charge Voltage (EOCV) value. In this case the system operates in the Constant Current (CC) charge mode. As the battery voltage reaches the EOCV, the controller switches in the Constant Voltage (CV) mode, so the charging current command I_b* progressively reduces to zero as soon as the battery is completely charged.

The battery current controller receives the battery charge current error at input and generates a positive limited control signal, namely Δv, that subtracted from the maximum power point reference V_{MPPT}^* determines the converter input voltage V_{SA}^*.

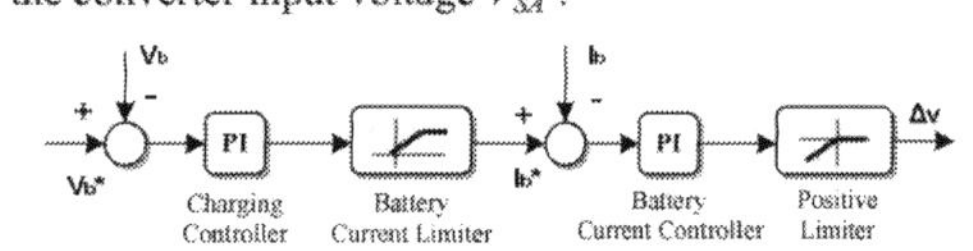

Figure 3. Battery controller block diagram.

3. POWER SYSTEM CONTROL

A SA composed of 4 sections has been considered. Each section is managed by independent MPPT and voltage controller and receives a control signal coming from the battery controller in order to select its appropriate operating point. The SA source can provide about 250W at Maximum Power Point (MPP) and End Of Life (EOL).

The power management proposed in this paper allows maximizing the power extracted from every SA sections. Moreover, it allows to distribute the battery charge current between the power sources in order to evenly divide the switching losses on the power semiconductors of the converters and, thus, extending their life time and to reduce the converter power dissipations. To this aim, as illustrated in Fig. 4, the battery controller provides 4 current error signals, each one proportional to the current supplied by the corresponding SA section. Then, each Δv_i (i=1..4) is calculated on the basis of the battery current error signal multiplied by the term $1\text{-}I_{sai}/I_{tot}$, being I_{sai} the current supplied by *i*-th section and I_{tot} the total current supplied by all the SAs. Using such approach, each section provides a percentage of the battery current proportional to its own current, maximizing the current extraction from the sections providing the lowest power. On the other hand, this control logic resulted in a high robustness towards potential SA strings failures and variations of the operative conditions (both irradiance and temperature), allowing the system to operate correctly under different operative conditions, leading to an higher flexibility and to increase the system efficiency.

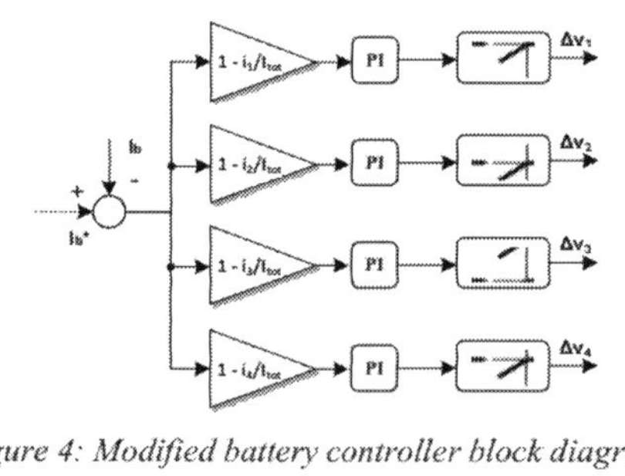

Figure 4: Modified battery controller block diagram.

4. SIMULATION MODELS

Simulations have been performed in Matlab/Simulink/ SimPowerSystem accounting an average model of the buck converter circuit, in order to highlight the performances of the controller independently from the power switching devices. Simulations consider an ideal behaviour of the electronic components.

4.1 Solar Array

Fig. 5 shows the single-diode equivalent circuit of a

Photovoltaic (PV) module whose physical parameters cannot be achieved directly from manufacturers datasheets. To overcome such a problem, in this paper the model of a PV module developed in [7] has been considered. Such a model, described later in detail, doesn't require the use of any complex numerical methods and employs only parameters provided by manufacturers' datasheets.

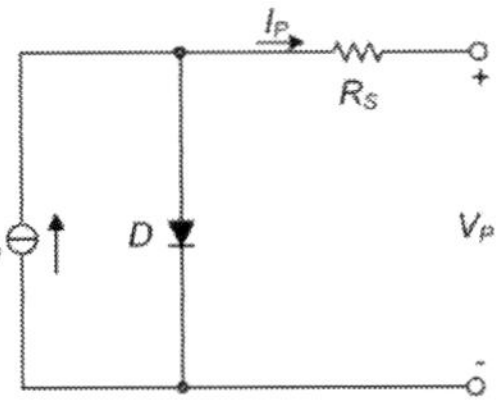

Figure 5: Equivalent circuit of a PV module.

The current of a PV module can be expressed, as function of voltage, by the simplified expression derived from [8]:

$$I_P = I_{SC} \cdot \left[1 - C_1 \cdot \left(e^{\left(\frac{V_P}{C_2 \cdot V_{OC}}\right)} - 1\right)\right] \quad (1)$$

where

$$C_1 = \left(1 - \frac{I_{MPP}}{I_{SC}}\right) \cdot e^{\left(\frac{-V_{MPP}}{C_2 \cdot V_{OC}}\right)} \quad (2)$$

$$C_2 = \frac{\left(\frac{V_{MPP}}{V_{OC}} - 1\right)}{\ln\left(1 - \frac{I_{MPP}}{I_{SC}}\right)} \quad (3)$$

Coefficients C_1 and C_2 depends on the following module parameters:

- short circuit current I_{SC};
- open circuit voltage V_{OC};
- maximum power point voltage V_{MPP};
- maximum power point current I_{MPP}.

Such parameters can be expressed as follow [7]:

$$I_{SC}(G,T) = I_{SCS} \frac{G}{G_S}[1 + \alpha(T - T_S)] \quad (4)$$

$$V_{OC}(G,T) = V_{OCS} + \beta(T - T_S) - \Delta V(G) \quad (5)$$

$$I_{MPP}(G,T) = I_{MPPS} \frac{G}{G_S}[1 + \alpha(T - T_S)] \quad (6)$$

$$V_{MPP}(G,T) = V_{MPPS} + \beta(T - T_S) - \Delta V(G) \quad (7)$$

where parameters I_{SCS}, V_{OCS}, I_{mpps} and V_{mpps} are defined at standard conditions, STC (G_s=1000W/m^2 and T_s=25°C) and α and β are respectively the current and the voltage temperature coefficient; all the above parameters are provided by manufacturers in the cell datasheet. It is possible to note that the parameters referred to currents depends on module solar irradiance G and temperature T, while the voltage ones depends only on temperature.

The correction term $\Delta V(G)$ in (5) and (7) is necessary to account the voltage variation as a function also of solar irradiance and it is calculated as:

$$\Delta V(G) = V_{OCS} - V_{OCm}$$

where voltage V_{OCm} represents the open circuit voltage of the I-V curve translated from STC to the considered irradiance G and it defined as:

$$V_{OCm} = C_2 \cdot V_{OCS} \cdot \ln\left[1 + \frac{\left(1 - \frac{I_t}{I_{SCS}}\right)}{C_1}\right].$$

I_t is the short circuit current at irradiance G and can be written as:

$$I_t(G) = I_{SCS}(1 - \frac{G}{G_S}).$$

Azur Space triple-junction cells have been considered for the Solar Array simulations.

The SA characteristics are considered at EOL conditions to avoid degradation computation, and they are listed in Table 1.

Parameter (@ AM0)	Value
# Sections	4
P_{MPP_TOT}	250 W
V_{MPP}	42 V
I_{SC_MAX} (BOL)	2 A
V_{OC_MAX}	58 V

Table 1 : Solar Array characteristics.

4.2. Battery

The battery used in the simulations is series-parallel configured with a total capacity of 9 Ah (fading at EOL already included). EOCV and nominal current charge parameters have been set to 37.85V and 3A respectively.

The battery model already available in Simulink library has been used in the simulations.

5. SIMULATION RESULTS

In order to demonstrate the effectiveness of the parallel power management strategy, two simulations have been performed. In the first one, a power load transient shown in Fig. 6 has been accounted, together with ecl-sun & sun-ecl transitions considering the relevant variations in illumination and temperatures in addition to string failures accounted in two sections, summarized in Table 2. In particular, for the first half of the simulation the SA is supplying both the load and the battery, the latter with the current imposed by the CC-

DOI: 10.1051/e3sconf/20171613008

CV algorithm. At 250s the load transient happens and the SA sections begins to operate at MPP while the battery discharges.

Parameter	Value
SA section-1 Irradiance	0 → 1000 W/m^2
SA section-2 Irradiance	0 → 1100 W/m^2
SA section-3 Irradiance	0 → 1200 W/m^2
SA section-4 Irradiance	0 → 1400 W/m^2
Temp. variations	1 °C/s
V_{batt_EOC}	37.85 V
V_{batt}	35 V (@ t=0)
I_{batt_charge}	3 A
I_{load}	2 A (flat); 30 A (pulse 60s)

Table 2: Operative conditions of the simulation.

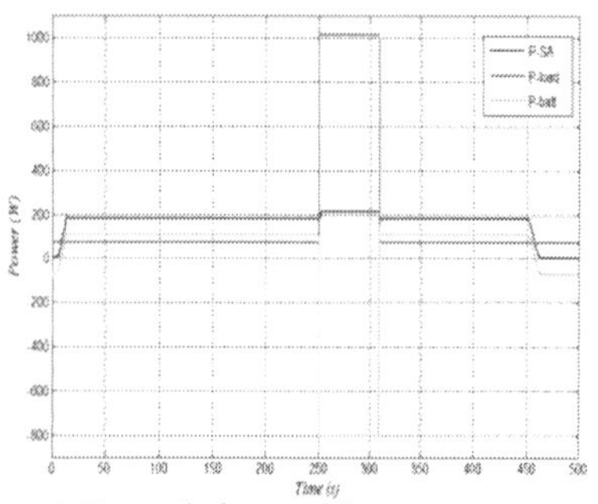

Figure 6: Power balance of the power management.

This situation lasts for 60s, after that the system returns to the previous state, so that the operating point of the SA sections moves from the MPP towards higher voltage in order to balance the power request of both the load and the battery. Figs 7 show the output SA current waveforms during the transition from discharging mode to charging mode. During charging mode, the proportional partition of the output current is being applied, maximizing the current extraction from the sections that can provide the lowest powers. In order to complete the functional verification of this system in a wide range of operative conditions. During the eclipse-sunlight transitions, the system reaches equilibrium as soon as the battery current reaches the value set as the maximum charging current, regardless of the proportional distribution. This means that the SA sections operating points ensure the energy balance between the sources, battery and load and also that the battery charge current never exceeds the imposed value. Only after a load power request that move the operating points of the SA sections to the MPP the battery current is being distributed proportionally.

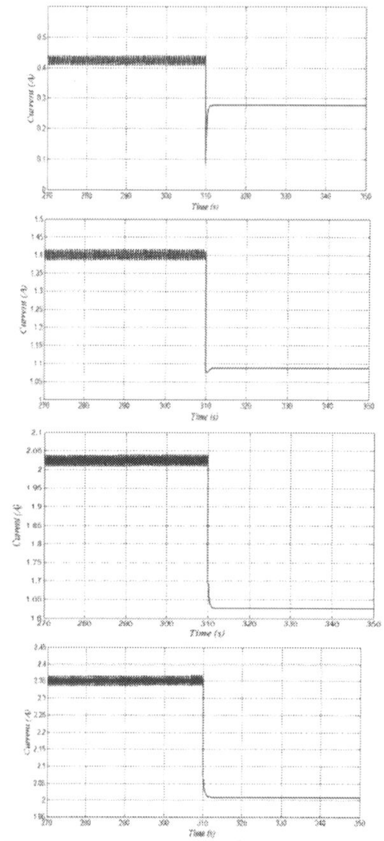

Figure 7: Output current of SA sections 1-4.

In the second simulation, the power load transient shown in Fig. 6 and the transient due to a different maximum battery charge current, starting from 5A and arriving at 1A, as illustrated in Fig. 8, have been simulated. Even in this simulation both the eclipse-sunlight and sunlight-eclipse transitions have been taken into account. The temperature and irradiance variations on the SA sections end after 180s due to the eclipse-sunlight transition. At first the maximum battery charge current is set at 5A and the total power from the SA sections allows to meet this requirement: in fact there's no need to operate at MPP at the beginning, because the maximum power is higher than the one needed for the battery charge at 5A.

This situation lasts till 85s, when the system starts to fully operate at MPP, while not satisfying anymore the 5A battery charge current condition, as the total SA power is lower than the one needed. This is because the temperature transient is not yet finished and it decreases the total SA power. Consequently, the battery current begins to be characterized by a current ripple introduced by the MPPT algorithm. At 180s even the temperature

transient ends and all SA sections reaches a steady state condition, still operating at MPP but supplying the battery with only a 3.9A charge current.
At 250s the maximum battery charge current transient happens, setting the limit at 1A. This instant represent the transition from discharging to charging operating mode, as the system exits from the MPPT phase. All the operating points of the SA sections move from the MPP to higher voltage in order to supply the battery with a lower power.
This new operative condition persists until the sunlight-eclipse transition happens and the total power becomes zero, disabling the control.
Finally, Fig. 9 shows the output currents waveforms of the DC-DC converters.

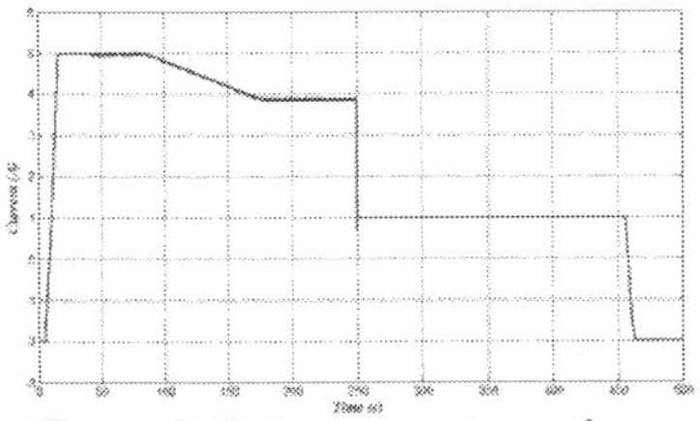

Figures 8: Battery current waveform.

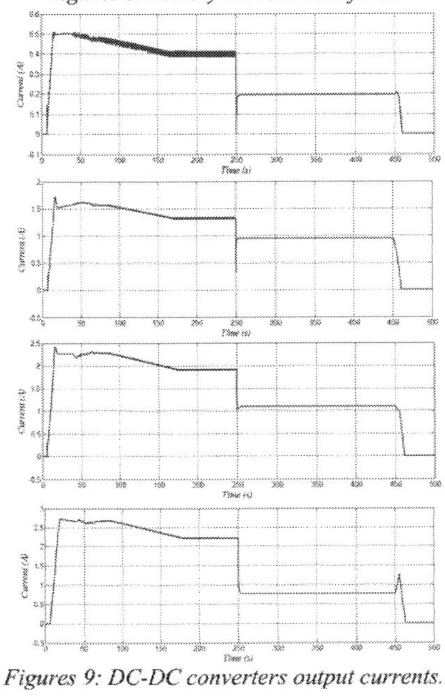

Figures 9: DC-DC converters output currents.

6. CONCLUSIONS

The proposed control strategy allows the maximization of the power extracted from every SA sections, as each one is controlled by an independent P&O MPPT algorithm. Furthermore, the developed parallel power management system allows to equally divide the stress on the power electronic components, in particular the switching devices, extending their life time and decreasing their power dissipations.

In order to validate the proposed approach, several simulation cases have been considered, accounting different operating conditions for the SA sections. The accounted simulation cases envelope the SA worst case profiles, therefore an additional improvement of the system performances and efficiency is expected. A prototype of the power module will be tested in order to validate the simulations and to verify the efficiency increasing.

7. ACKNOWLEDGMENT

This work was finalized thanks to the collaboration of C. Citriniti. This paper has been based on the result achieved in the REGIONE LAZIO Italian Project "Progettazione di un Microsatellite e Sviluppo Tecnologie Abilitanti (Cod. RL 023301).

8. REFERENCES

[1] F. Tonicello, The control problem of maximum point power tracking in power systems, Proceedings of the 7th European Space Power Conference, Stresa, Italy, 9-13 May 2005, ESA SP-589.

[2] F. Tonicello, S.Vazquez del Real, Maximum Point Power Tracker approach to a regulated bus, Proceedings of the 5th European Space Power Conference, Tarragona, Spain, 21-25 September 1998, ESA SP-416.

[3] W. Denzinger, Electrical power subsystem of Globalstar, Proceedings of the 4th European Space Power Conference, Poitiers, France, 4-8 September 1995, ESA-SP-369.

[4] L. Croci, P. Della Putta, B. Vavassori, The RADARSAT-2 Power Control Unit, Proceedings of the 6th European Space Power Conference, Porto, Portugal, 6-10 May 2002, ESA-SP-502.

[5] J. Simola, K. Savela, J. Stenberg, F.Tonicello, Maximum Power Point Regulator System, Proceedings of the 9th European Space Power Conference, Saint Raphael, France, 6-10 June 2011, ESA SP-698.

[6] Z. Varadi, V. Gorocz, J. Szabo, Power Subsystem for Small Satellites with unified operational mode controller, IEEE Conference Publishing 2011 2nd International Conference on Space Technology, Athens, Greece, 15-17 September 2011.

[7] A. Bellini, S. Bifaretti and V. Iacovone, A Simplified Model of A Photovoltaic Module, in Proc. of Applied Electronics 2009, Pilsen (Czech Republic), September 2009.

[8] W. Xiao, W.G. Dunford, and A. Capel, A novel modeling method for photovoltaic cells, IEEE Power Electronics Specialists Conference, Aachen, Germany, 2004.

 DOI: 10.1051/e3sconf/20171613009

Modeling of the Section Capacitance to Interface with a S3R

A. Fernández, J.R. González

European Space Agency, Keplerlaan 1, 2201AZ Noordwijk, The Netherlands,
Email: Arturo.Fernandez@esa.int

ABSTRACT

Cell capacitance is a well-known parameter in silicon solar cells. Its effects on the power units, specially on the S3R, is also well known since it has to be taken into account to design them. The capacitor energy is discharged into the power switch and the dissipation increases. Moreover, it introduces a delay in the energy transfer from the array to the main power bus. In triple junction solar cells, this parasitic element becomes far more complex and is not so well characterized. This paper addresses this issue and proposes a large signal model to be used in combination with a S3R.

1. INTRODUCTION

Junction capacitance is a well-known issue in silicon semiconductors. There are two main contributors to this parasitic effect:

-The depletion layer (or transition) capacitance

-The diffusion capacitance

The first one is very simple to explain because it has the same concept as a conventional capacitor: two metal plates separated by a dielectric material. In the case of a semiconductor junction (e.g in a silicon diode), the depletion layer between the n- and p-sides of a p–n-diode serves as an insulating region that separates the two diode contacts. Thus, the diode in reverse bias exhibits a depletion-layer capacitance, also known as junction capacitance. This concept is shown in Fig. 1.

In reverse bias the width of the depletion layer depends on the bias voltage. In fact, the layer is widened with increasing reverse voltage bias v_R, and the capacitance is accordingly decreased. In a simplified model, the junction capacitance is:

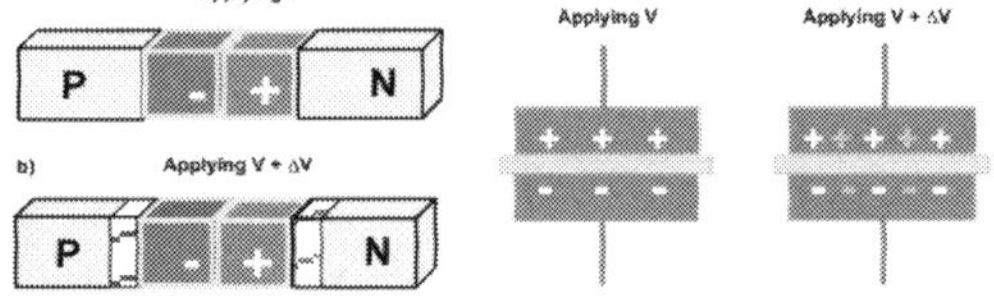

Fig. 1: a) diode with a depletion layer between two conductive blocks; b) The effect of the bias voltage on the depletion layer is shown; c) equivalent effect on a capacitor

$$C_j = k\varepsilon_0 \frac{A}{w(v_R)} \qquad (1)$$

with A being the device area, κ the relative semiconductor dielectric permittivity, ε_0 the electric constant, and w the depletion width.

In forward bias, besides the above depletion-layer capacitance, minority carrier charge injection and diffusion occurs. A diffusion capacitance exists expressing the change in minority carrier charge that occurs with a change in forward bias as shown in Fig. 2. In terms of the stored minority carrier charge, the diode current i_D is:

$$i_D = \frac{Q_D}{\tau_T} \qquad (2)$$

where Q_D is the charge associated with diffusion of minority carriers, and τ_T is the transit time, the time taken for the minority charge to transit the injection region. Typical values for transit time are 0.1–100 ns. On this basis, the diffusion capacitance is calculated to be:

$$C_D = \frac{dQ_D}{dv_D} = \tau_T \frac{di_D}{dv_D} = \frac{i_D \cdot \tau_T}{V_{th}} \qquad (3)$$

Generally speaking, for usual current levels in forward bias, this capacitance far exceeds the depletion-layer capacitance.

2. STATIC TRIPLE JUNCTION CELL CAPACITANCE

Starting from the previous basis, the authors in [3] presented a multi junction solar cell capacitance model. The concept relies on modelling the capacitance of the individual cells one by one and then building the full model connecting all of them in series.

DOI: 10.1051/e3sconf/20171613009

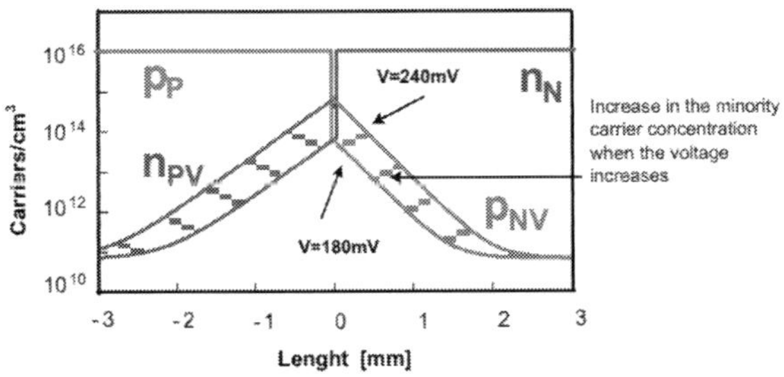

Fig. 2: diffusion capacitance effect due to the increase in the minority carrier concentration when the bias voltage increases

The three junctions have different electrical characteristics (I_F, V_F) and different capacitance values. Moreover, this junction capacitance is voltage dependent as was shown before. Thus, the full model is quite complex and the prediction of the behavior of the full stack is not obvious.

Fig. 3a shows a simplified model of the three junctions in series, each one with its own voltage dependent capacitor in parallel. Fig. 3b shows the capacitance value of the three junctions as a function of voltage. The top cell is the GaInP cell, the middle one is the GaInAs cell and the bottom one is the Ge cell. The basic characteristics are shown in Table 1. The values used in the model are based on the values obtained in [3].

When the power system topology is based on a S3R, the solar array sections will be subject to a large signal excursion since there are two possible states:

- shunted or
- biased to the bus voltage level.

As a consequence, each cell will be subject to an equivalent excursion from zero to a corresponding voltage level (typically close to the Maximum Power Point (MPP) value). The effects of this transient at cell level are not as simple as one would have thought as

TABLE1.: BASIC CELL PARAMETERS OF THE AZUR TRIPLE JUNCTION SOLAR CELL

Cell	V_F [V]	I_F [A]	C_0 [μF]
GaInP	1.41	0.502	2.47
GaInAs	1.021	0.535	1.28
Ge	0.253	0.881	3.92

will be explained in the following paragraphs. To simplify the explanation and to make the test generic, we will assume that the cell excursion will go from zero to open circuit voltage V_{OC}.

3. CELL DYNAMIC CHARGE

When the S3R is shunting the section and, as a consequence, the solar cells, the first assumption could be that the three junctions are short-circuited and that the three cell capacitors have no charge. However, the real situation is totally different. Due to the fact that each subcell (junction) has a different IV characteristic, and the fact that the three subcells are connected in a monolithic series association, the total stack current is limited by the one with the smallest current value. In our case, the GaInP subcell. The main outcome of this very particular configuration is that the electrical equilibrium of the stack is such that the top subcell is reverse biased but the other two are forward biased to its saturation levels (1.0 V and 0.2 V respectively). Thus, the top subcell voltage level will be -1.2 V, so that the total voltage of the series association is zero, consistent with the short circuit condition .

The main consequence of this situation is that, when the short circuit is released, the middle and the bottom subcells are already fully charged. Hence, they will not undergo any sort of transient. Only the top subcell

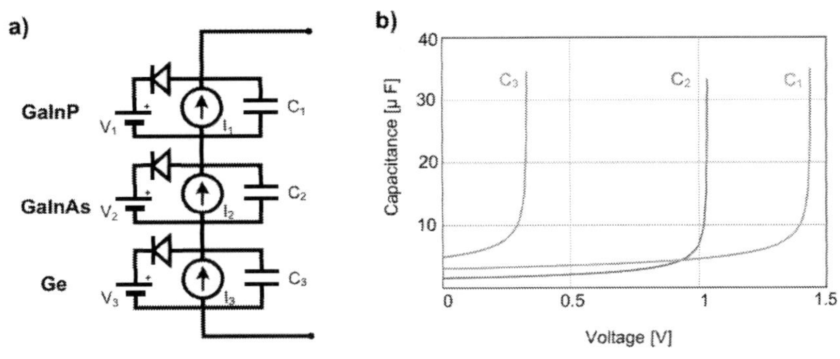

Fig. 3: a) simplified model of a triple junction solar cell; b) capacitance value as a function of voltage for the three sub-cells.

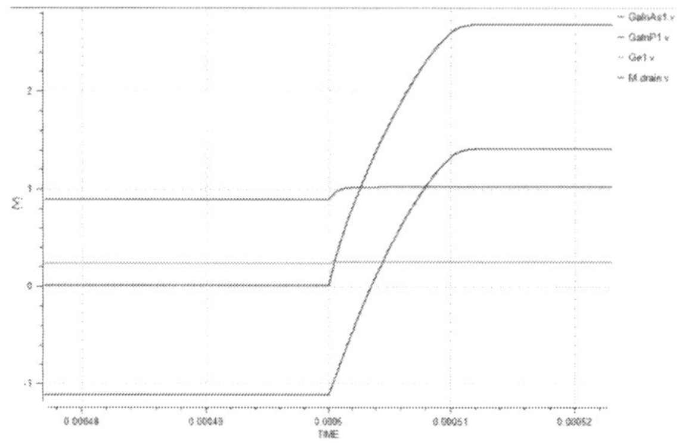

Fig. 4: release transient in a triple junction solar cell. Red: GaInAs, Orange: Ge, Blue: GaInP, Purple: Total cell voltage

(GaInP) will be subject to a voltage excursion from the reverse biased situation (-1.2 V) to the final open circuit situation (+1.4 V). Fig. 4 shows the voltage values on the three cells when the short-circuit is released. The very important conclusion of this is that, during this transient, the dynamic behavior is only influenced by the GaInP subcell.

To obtain a large signal model of this dynamic behavior, we need to model an equivalent capacitor that will be placed in parallel with the full three subcell stack, not only across one sub-cell. Moreover, the voltage excursion to which it will be subject is also going to be different. In the case of the GaInP subcell, it goes from -1.2 V to +1.4 V while in the full cell model it goes from 0 V to + 2.6V. Moreover, as was explained above, the capacitance is voltage dependent.

As was explained in [4], to model this case we need to consider the equivalent electrical charge that is loaded in the capacitor. Hence, the charge equivalent linear capacitance is:

$$C_{eq,Q} = \frac{1}{V_A}\int_0^{V_A} C_x(v_C)dv_C \qquad (4)$$

which will have the same amount of stored charge at V_A as the nonlinear capacitor C_x. As seen from (4), this value is simply the average of the C_x curve with respect to voltage.

In the case of the GaInP subcell, this average value is 2.25 μF. Fig. 5 shows a simulation where both the nonlinear model and the charge average capacitor model are released from short-circuit. As can be seen, they both reach the open circuit status at the same time. The transient shape is not identical since the simplified model is a conventional capacitor and the transient shape is a simple straight line.

However, note that in fact, the transient time is the actual figure that is important for the S3R control loop design. This value has to be taken into account to avoid double section switching. Neither the shape nor the energy are relevant in this case.

4. Cell Dynamic Discharge

The transient going from V_{OC} to zero is completely different though. In this case, the three junctions are in an open circuit state and hence, their capacitors are fully charged. When the cell is short-circuited, all this energy is extracted from the 3-subcell stack and transferred to the switching element which is the S3R MOSFET.

During the transient, the three junctions are actually discharged and undergo a transient from V_{OC} to zero as shown in Fig. 6a. At this point, the transient is finished for an external viewer (as seen in the purple trace). This means that no more energy is extracted and that the voltage stack will not move any more once it has reached zero volts.

However, internally in the cell the action has not finished. The internally generated photocurrent will continue charging the elements and the transient at individual subcell level continues. The final status will be as explained in the paragraph above: the top subcell will be reversed biased while the middle and the bottom subcells will be forward biased again. Fig, 6b shows this full transient.

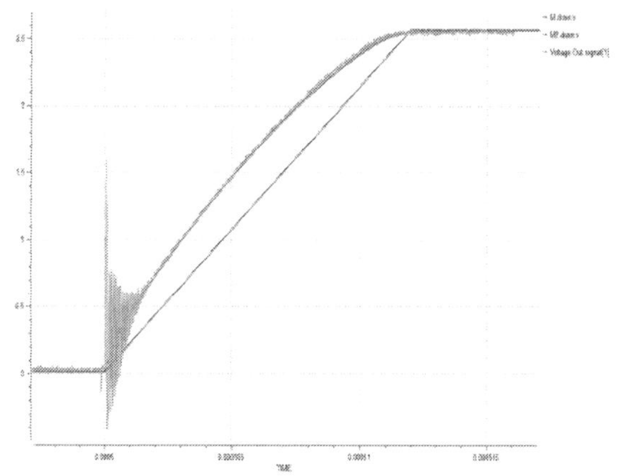

Fig. 5: simulation of the release transient in a triple junction solar cell. Blue: simplified average model, Red: non-linear model, Orange: test results.

DOI: 10.1051/e3sconf/20171613009

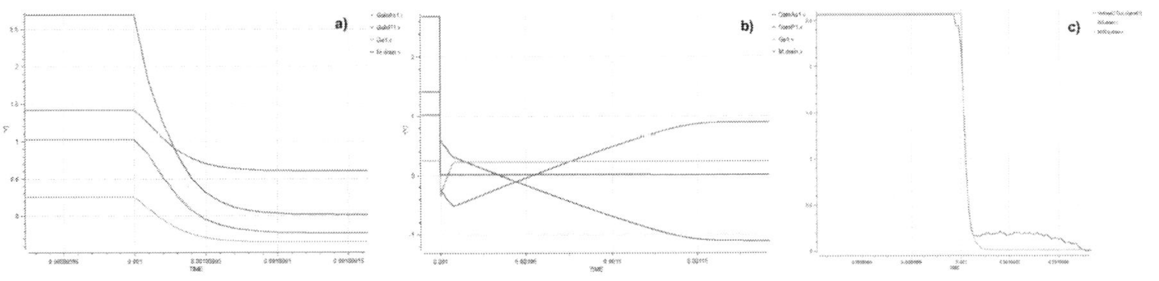

Fig. 6: a) first part of the energy discharge transient; b) full transient of the energy discharge in a triple junction solar cell. Red: GaInAs, Orange: Ge, Blue: GaInP, Purple: Total cell voltage; c) simplified model (orange) compared to the test result (red).

In terms of large signal modeling at cell level, the key point is to model the energy that has been transferred to the switch. In this case, the calculation has to be done in a different way.

The total energy in a nonlinear capacitor is given by integrating the product of C_x and v_c across the voltage range e.g. from 0 to V_{OC} [5].

$$E_C = \int_0^{V_A} v_C C_x(v_C) dv_C \quad (5)$$

This energy can then be used to find a linear capacitance value that contains the same amount of energy at V_A by considering the solution to the energy stored in some equivalent capacitor C_{eqE}, which is linear, constant-valued:

$$E_C = \frac{1}{2} C_{eq,E} V_A^2 \quad (6)$$

By combining (3) and (4) an expression for the energy -equivalent linear capacitor can be obtained:

$$C_{eq,E} = \frac{2}{V_A^2} \int_0^{V_A} v_C C_x(v_C) dv_C \quad (7)$$

where C_{eqE} and C_x store the same amount of energy at $v_c = V_A$.

In the case of the triple junction solar cell, this calculation has to be performed individually for the three subcells. Then, the equivalent capacitor has to be calculated in such a way that its stored energy is the same as the sum of the three individual energies. In the case of the AZUR 3G28 cell, this value is 840 nF. Fig. 6c shows the comparison between a simplified model with a linear capacitor (840 nF) and the test results with that cell on the continuous light sun simulator.

As can be seen, this value is around 2.5 times smaller than the one calculated for the release transient.

It has to be noted that this value is needed for a completely different calculation in the S3R. The section that is switching will dissipate this energy and hence, the thermal interface has to be designed accordingly. The snubber used to soften the peak current on the MOSFET will be also affected by this value.

As can be seen, the model is asymmetric since the equivalent capacitance is different depending on the transition direction. Moreover, both values are needed for two very different design aspects of the S3R: the capacitor modeling the release transition is needed for the control loop design while the capacitor modeling the shunt transition is needed to design the thermal interface and the current snubber.

In terms of circuit modeling, it is now evident that the large signal behavior of a triple-junction solar cell cannot be replicated with a simple linear capacitor. The modeling is not complex though since it can be done with a capacitor which capacitance value changes depending on the current sense. When the capacitor is being charged, the value should be 2.25 µF while, when the capacitor is being discharged, the capacitance should be 840 nF.

5. String Modeling

Once the model of one individual cell has been obtained, the modeling of the full string can be tackled. One could be tempted to think that the string model is a simple series connection of single cells. However, this is not the case. In fact, what actually happens is very similar to what have been described for the triple junction subcell stack.
Fig. 7a shows a simplified model of a cell string. As can be seen, 4 cells are connected in series and each one of them has a bypass diode and a capacitor in parallel.

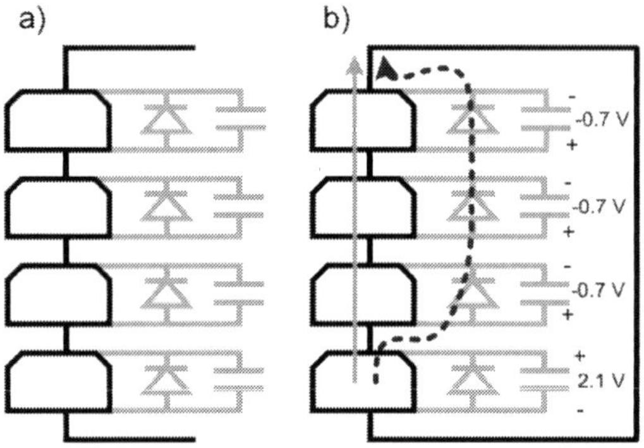

Fig. 7: a) 4 cells stack configuring a string. Each cell has a bypass diode and a capacitor in parallel; b) shortcircuited string with the excess current flowing through the bypass diodes.

In an ideal situation, all 4 cells would have exactly the same shortcircuit current. However, in reality each cell will have a slightly different current value. As a consequence, when the string is shunted it is electrically impossible that all four cells have zero volts and different current values flowing across them.

In the triple junction stack, the current value is limited by the subcell with the lowest current. In the string the situation is different because of the bypass diodes. The equilibrium condition will be that the cell with the highest current value will drive the string current and the excess with respect to other cells current will flow through the diodes.

The fact that the bypass diodes conduct the excess current has another important consequence: the cells in parallel with those diodes will be reverse biased.

Let's assume that the forward voltage drop of the bypass diode is $V_D = 0.7$ V and that the open circuit voltage of the solar cells is $V_{OC} = 2.7$ V. Under these constraints, the equilibrium condition to solve the shortcircuit will be as shown in Fig. 7b.

One cell is forward biased (+2.1 V) and three cells are reverse biased (-0.7 V). The cell in forward will adapt its voltage so that the overall stack has zero volts across it. In this case: 0.7 V x 3 = 2.1 V.

Coming back now to the capacitance, when the short circuit is released, the cell that is already forward biased will contribute very lightly to the transient. Its capacitance is already charged to 2.1 V and will have a negligible effect in the string transient as can be seen in Fig. 8a (blue trace).

At string level, the consequence is interesting: since one capacitance is already charged and does not contribute to the transient, the overall capacitance value will be equivalent to 3 capacitors in series instead of 4. Hence, the string capacitance will be C_{cell} / 3 and will be higher than in an ideal case where the 4 cells contribute to the transient. This can be clearly seen in Fig. 8b. The red trace shows the overall string voltage assuming that the four cells are shortcircuited while the blue trace shows a real 4 cell stack with 4 slightly different current values. As can be seen, the real case is slower than the ideal one: the equivalent string capacitance is C_{cell} / 3 instead of C_{cell} / 4.

Hence, to model a generic string we need to take into account the voltage drop across the bypass diodes (V_D) and the cell V_{OC} voltage. Depending on the ratio of those two voltages V_{OC} / V_D more or less cells' capacitances will need to be removed from the transient. In this example, the ratio is 2.7 / 0.7 = 3.85 so one out of every 4 cells will not contribute to the transient and the equivalent string capacitance will be increased accordingly.

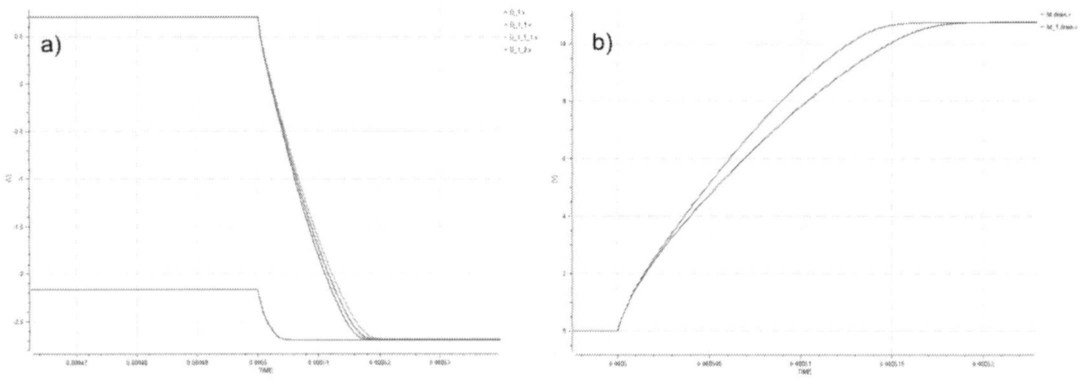

Fig. 8: a) voltage across the 4 cells of the string; b) overall string voltage; red trace: ideal string with 4 cells shortcircuited; blue trace: real string with 3 cells in reverse and 1 in forward.

 DOI: 10.1051/e3sconf/20171613009

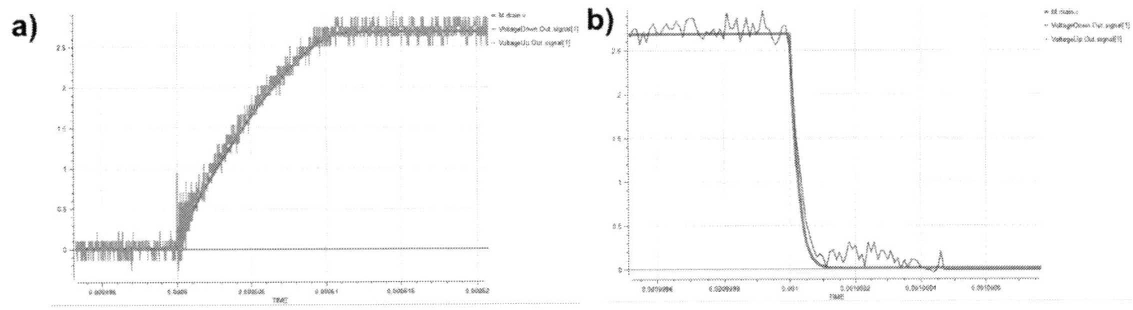

Fig. 9: a) Azur 3G28 release transient. Green: experimental result, Red: simulation result; b) shunt transient, Blue: experimental result, Red: simulation result.

6. Test Results

6.1 Continuous Light Sun Simulator

Several tests have been performed in order to verify this concept. A single cell has been placed in a temperature controlled holder and illuminated using continuous light under a triple source solar simulator. The initial calibration has been adjusted in such a way that the top sub-cell is the one limiting, as it is the case in real BOL conditions. A low r_{DSon} MOSFET has been placed in parallel with the cell and switched continuously at a frequency of 1kHz. The circuit arrangement has been carefully done to minimise the parasitic impedance and to have the cleanest possible switching.
The voltage between drain and source on the MOSFET was measured and compared to the simulation results. The model used in this case is the triple junction model with three separated nonlinear capacitors. Fig. 9a shows the release transient and Fig. 9b shows the shunt transient. In both cases, the model is able to mock the behavior of the real cell fairly well. As can be seen, the time scale is completely different.

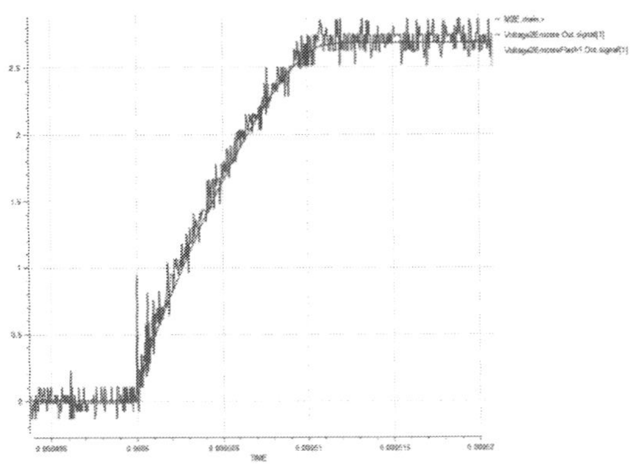

Fig. 10: Solaero cell release transient. Blue: experimental result; Red: simulation result with 3 non-linear capacitances model.

It takes a few µs to go from zero to V_{OC} while it only takes 100 ns to shunt the cell from V_{OC} to zero. Of course, this is due to the fact that both processes are completely different as has been explained in the previous paragraphs.

The same test was repeated for the Solaero ZTJ cell. The model was slightly tuned by adjusting the capacitance values. Solaero cells turned out to have around 15% lower capacitance. The result of the test compared to the simulation is shown in Fig. 10. As can be seen, both test and simulation results fit very well.

6.2 VLASS Continuous Flasher

Another set of testing has been done at the Airbus NL premises using a Vortek Large Area Solar Simulator (VLASS) [6] on an EM panel populated with Solaero ZTJ cells. In this case, the tests have been done at section level. In this case, the section had 46 cells in series and only 1 in parallel (46s1p).
The cell parameters have been adjusted accordingly with the outcome of the continuous light sun simulator and the results are shown in Fig. 11. The orange trace shows the release transient assuming that all 46 cells in series are initially shortcircuited. As can be seen, the experimental result is slower than the model. The red trace shows the model after adjusting the capacitance to take into account the effect of the string. In this case, the Solaero external bypass diode has a voltage drop of 0.8 V and the V_{OC} of the cell is 2.7 V. The ratio V_{OC} / V_D = 3.37, which means that out of every 4 cells, one will not contribute to the transient and 3 will undergo the full voltage swing from –0.8 V to 2.7 V. The adjustment implies an increase of the virtual capacitance of 33%. As can be seen, the simulated result adjust much better to the actual release test.

The shunt transient at panel level is more complicated than the release transient because of the transfer har-

 DOI: 10.1051/e3sconf/20171613009

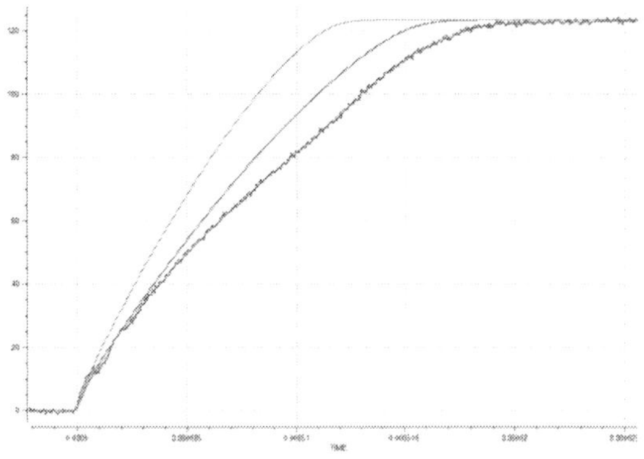

Fig. 11: section level (46s1p) release test with Solaero cells compared to the simulated results. Orange trace: ideal string; Red trace: model compensated with the string effect.

ness. The impedance between the section and the switch is much higher and has an important inductive component. Hence, unless it is very well characterized it is very difficult to match the simulation and the test results. Still, once the impedance value was adjusted, the result is also quite accurate.

6.3 ESTEC Pasan Flasher

To continue investigating in the cell level and the section level modeling, the ESTEC Pasan flasher was also used to perform several tests. Moreover, in the Power Systems laboratory a model of the Herschel solar panel is available and hence, transient tests on a real SA section can be performed. This is very interesting since we could perform test both at cell and section level with the same light source. In this case, the flasher can provide a 14 ms light plateau that allows to visualize 14 full switching transients at 1 kHz.

However, instead of allowing the confirmation of the model, these test results have introduced some uncertainties that so far we have not been able to explain.

The results from these tests are significantly faster than the ones obtained from the other tests. Fig. 12a shows the comparison of a switching transient obtained on the same Azur 3G28 cell with the continuous light sun simulator (blue trace) and the Pasan flasher (orange trace). As can be seen, the latter is significantly faster than the first one. In fact, the equivalent capacitance is 3 times smaller.
Fig. 12b shows the same comparison but this time with the Solaero ZTJ cell. The situation is repeated here since the result from the Pasan Flasher (orange trace) is also faster than the continuous light sun simulator (blue trace).

The tests performed on the Herschel panel also show a transient result that is much faster than the one obtained with the model. Fig. 13 shows this comparison and, as can be seen, the flasher result is considerably faster than the model. In fact, to fit the model the capacitance should have been reduced by a factor of 3.

Furthermore, the test at cell level offer different relative speeds on the Azur and Solaero cells as shown in Fig. 14. When illuminated with the continuous light simulator, the Solaero cell (blue trace) is faster than the Azur (red trace) as shown in Fig. 14a. However, when illuminated with the Pasan flasher the results reverse as shown in Fig. 14b. The Azur cell (red trace) is faster than the Solaero cell (blue trace).

Note also that the results from the VLASS flasher shown in Fig. 11 are also slightly different from the results from the model calibrated with the continuous light even though, in this case, the test results are slower. Overall, it has not been possible to correlate results obtained with different light sources, even though all

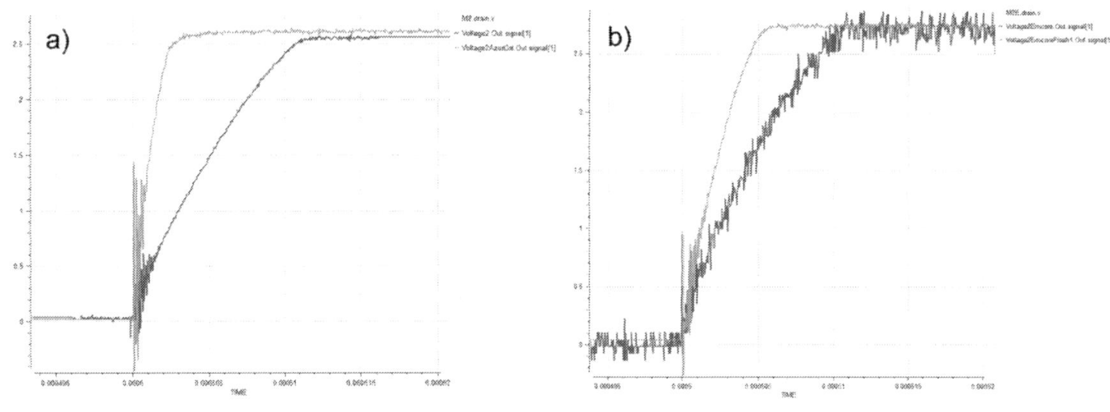

Fig. 12: a) Azur 3G28 release transient at cell level. Orange trace: Pasan flasher; Blue trace: continuous light simulator; b) Solaero release transient at cell level. Orange trace: Pasan flasher; Blue trace: continuous light simulator

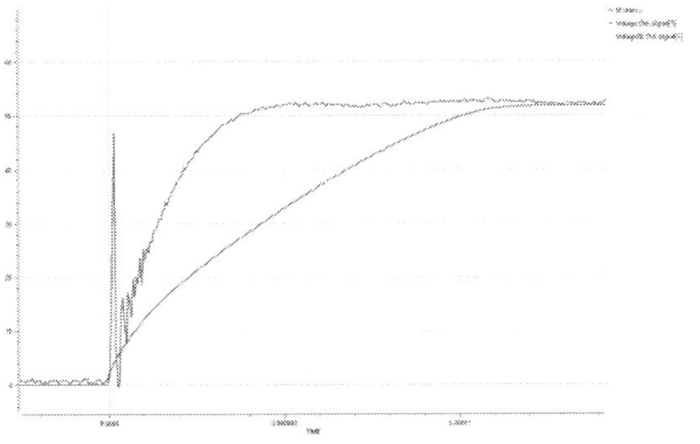

Fig. 13: Release test at section level (20s5p) on the Herschel panel. Blue trace: test result with the Pasan Flasher; Red trace: simulated result with the cell calibrated with the continuous light simulator.

of them are supposed to be calibrated to the AM0 spectrum. Or at least to provide results equivalent to those from AM0. A plausible explanation for these phenomena has not been found yet.

7. Conclusions

The cell capacitance on a triple junction solar cell is a complex parasitic effect to describe. Each one of the subcells has a nonlinear capacitance in parallel that is voltage dependent. The fact that each subcell has different electrical parameters implies that the release and the shunt transient are completely different. As a consequence, a simplified model has to account for this asymmetry and use different capacitance values for either transient. The main conclusion is that the equivalent release capacitor is around 2.5 times larger than the equivalent shunt capacitor. The paper shows how to calculate these simplified values and has demonstrated their validity by comparing test results from continuous light simulation with the model simulated results and showing a very good agreement. The slowing effect of the string arrangement has also been explained and validated in one test set-up.

However, the experiments with different light sources have given different results So far, it has not been possible to provide an explanation for those differences. Hence, there is still some work to be done in order to understand any possible influence of the test light source.

8. References

[1] *John Sparkes (1994).* Semiconductor Devices *(2nd ed.). CRC Press.*

[2] Robert B. Darling (1995). A Full Dynamic Model for pn-Junction Diode Switching Transient. *IEEE Trans. Electron Devices*. vol. 42, no. 5, pp. 971-976

[3] Dynamic Modelling Of Multi Junction Gallium Arsenide Solar Arrays, N. Neugnot, X. Roy, R. Chouffot, M.E. Gueunier-Farret, J.P. Kleider, H. Barde, C. Baur, A. De Luca. ESPC??

[4] Circuit-Oriented Modeling of Nonlinear Device Capacitances in Switched Mode Power Converters, Daniel Costinett, Regan Zane, Dragan Maksimovic. APEC??

[5] R. W. Erickson and D. Maksimovi´c, Fundamentals of Power Electronics, Second Edition. New York, NY: Springer Science+ Business Media, LLc, 2001.

[6] T. Thrum, D. Camm, D. Parleniuk and D. Slootweg, "Characterizing state-of-the-art solar panels a new approach for large area testing," Photovoltaic Specialists Conference, 2000. Conference Record of the Twenty-Eighth IEEE, Anchorage, AK, 2000, pp. 1320 -1323. doi: 10.1109/PVSC.2000.916134

[7] Capacitance Measurements on Multi-Junction Solar cells, P. Rueda, E. Fernandez Lisbona, M. Diez Herrero, World Conference on Photovoltaic Energy Conversion, Poster 817.

[8] R. Blok, E. van den Berg, & D. Slootweg; Solar Cell Capacitance Measurement, ESPC 2002, p.597

[9] R. Anil Kumar, M. S. Suresh, and J. Nagaraju, Effect of Solar Array Capacitance on the Performance of Switching Shunt Voltage Regulator; IEEE Transactions on Power Electronics, Vol. 21, No. 2, March 2006 p.543

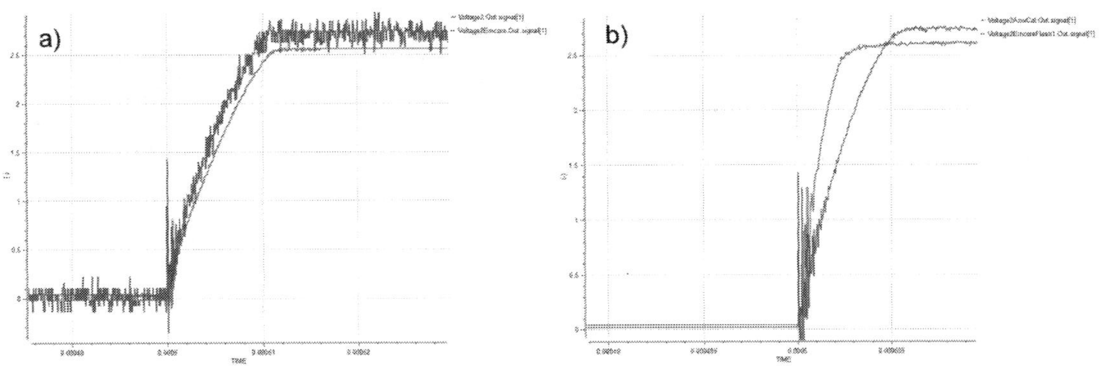

Fig. 14: comparison of the release transient between the Azur 3G28 (red trace) and the Solaero cell (blue trace). a) with the continuous light sun simulator; b) with the Pasan flasher.

 DOI: 10.1051/e3sconf/20171613010

POWER SYSTEM FOR THE EU:CROPIS SATELLITE - RESULTS FROM DESIGN TRADE-OFFS, ANALYSIS, SIMULATION AND TESTING

Jakob Fromm Pedersen

German Aerospace Center, Robert-Hooke-Str 7, 28359 Bremen, Germany, Email: jakob.pedersen@dlr.de

ABSTRACT

The Eu:CROPIS satellite is the first satellite to be launched as part of the DLR Compact Satellite Program. The mission is currently in Phase D and is scheduled to be launched in 2017. The Eu:CROPIS satellite is a small satellite of about $1m^3$, and a mass of 250kg. The nominal power consumption in science mode is approximately 200W from which ~50% is provided to the payloads. The design of the Eu:CROPIS power system design is driven by the primary payload which is a biological payload that has a very narrow temperature range, and the attitude control system which utilizes only magnetic torquers as actuators. The power system has been optimized in order to provide as much power as possible during the Launch and Early Operations Phase (LEOP) of the mission resulting in a design which for the nominal part of the mission is over-dimensioned.
This paper captures the trade-offs affecting the power system carried out on system level, which were made in the design phase of the mission, and will explain in more detail the analysis, simulation and testing related to the LEOP of the mission. Based on the simulations and resulting design iterations performed for the LEOP the probability of power balance in this phase has been increased significantly compared to the power system which was optimized for the nominal mission.

1. INTRODUCTION

The German Aerospace Center (Deutsches Zentrum für Luft- und Raumfahrt, DLR) is conducting research in practically all areas of satellite design. In the Institute of Space Systems, Department of Avionics Systems (RY-AVS) the focus is on the avionics suite; Onboard Data Handling, Communication and Power System.

In the domain of the Electrical Power System (EPS) the majority of resources in the last years have been allocated to supporting the Compact Satellite Program. The Compact Satellite Program is taken on by DLR in order to provide a platform for experiments and research in space. While the Compact Satellite Program strives to provide a standard platform, each mission is tailored specifically to the selected payloads. Since the program focuses on scientific payloads the power requirements and configuration (e.g. available area for solar panels, or the attitude profile) of the satellite will be different for each mission.
In providing support for the Compact Satellite Program, a design methodology has been developed, which allows to quickly assess the requirements for the power system based on requirements for the mission, as well as deriving requirements for the individual units which form the power system. Furthermore a test and verification setup has been established, allowing to simulate the various mission phases. For the Eu:CROPIS mission all power system units have been procured within Europe from established vendors of power system components.

In order to assess and evaluate the performance of the power system throughout all mission phases, the procured units are initially modeled in Matlab Simulink based on the data available from the supplier data sheets, and the mission phases are simulated in order to ensure that the power system is performing as assessed in the initial phases. Later the simulation model is updated based on more detailed values received from the suppliers and, if needed, the model is extended with data measured during unit testing.

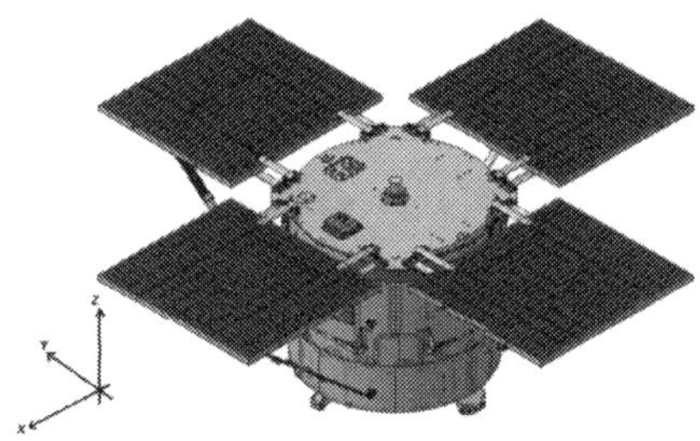

Figure 1. Eu:CROPIS Satellite in deployed configuration

2. EU:CROPIS MISSION AND DESIGN TRADE-OFFS

The first mission from the DLR Compact Satellite program to be launched is the Eu:CROPIS mission, which will be launched in 2017. The Eu:CROPIS mission consists of one satellite of approximately 250 kg, and dimensions of 1x1x1.2 m^3. The primary payload for the Eu:CROPIS mission is a biological experiment which will assess the impact of various levels of gravity on a biological regenerative life support system[1]. The Eu:CROPIS satellite in the deployed configuration can be seen in Figure 1. The final configuration of the Eu:CROPIS satellite consists of a 'barrel' which contains all bus components and payloads, and four deployable solar panels. The payloads which are sensitive to varying gravity are located close to the edge of the 'barrel', and 'artificial gravity' is generated by spinning the Eu:CROPIS satellite around the Z-axis (see Figure 1).
Since the design of the Eu:CROPIS has undergone several iterations, the major trade-offs, design decisions and constraints for the power system will be explained in the following sections.

2.1. Satellite Configuration

One major design driver for the Eu:CROPIS satellite was the ability to be launched as a piggy-back satellite. In order to stay compatible with most launch opportunities the satellite, and thus also the power system, has been designed for low earth orbits between 550 and 650km. For the power system this leads to the worst case assumption of an sun synchronous orbit with an LTAN of 1200, which results in an orbit of which approximately 38% is in eclipse.
In addition to the orbital constraints, the dimensions and mass of the satellite is limited by the availability of piggy-back launches. In the early phases of the project a market survey was performed and the maximum envelope of 1x1x1.2m^3 was defined with a maximum mass of 250kg. A piggy-back launch opportunity is dependant on the available separation systems supported by the launch vehicle. Most opportunities similar in size and mass of the Eu:CROPIS envelope utilize a miniature separation ring, comparable to the separation rings used for bigger payloads. Since the launch profile is determined by the primary payload, no specific attitude can be ensured at the moment of separation. Due to the miniaturization of the separation rings and the comparably low mass of the satellite, the satellite will be separated with a non-zero angular momentum, resulting in a tumble. For Eu:CROPIS a maximum value of $10^o/s$ was fixed during the early design phase. This value was later reduced to $5^o/s$.
Due to the uncertainties in attitude and angular rates after separation, coupled with the limited actuation capability of the attitude control system, it was a requirement to have solar cells on minimum 4 faces of the satellite in order to ensure a minimal power generation. Based on the design of spin-stabilized satellites in the past, body

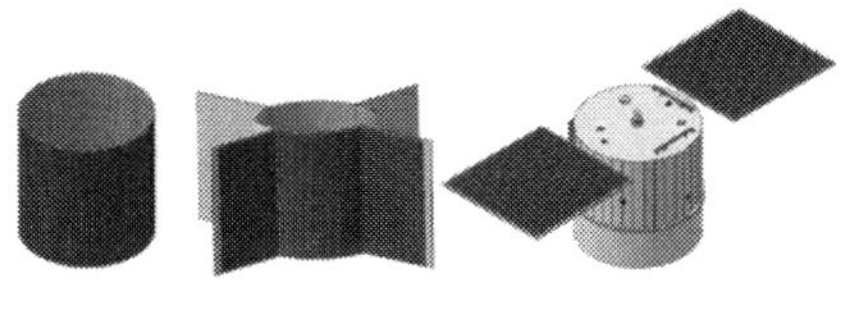

Figure 2. Early design considerations for the solar panels

mounted panels on the 'barrel' were investigated (see Figure 2).
Due to thermal requirements from the payloads and a rising power demand the concept was changed to deployable solar panels already in Phase 0. Based on the power requirements in Phase 0 and A, only two solar panels were required in order to ensure a positive energy balance, however in order to ensure that a minimum of four faces of the satellite was able to generate power, the required solar array area was split to four panels, which are stowed during launch (see Figure 3).

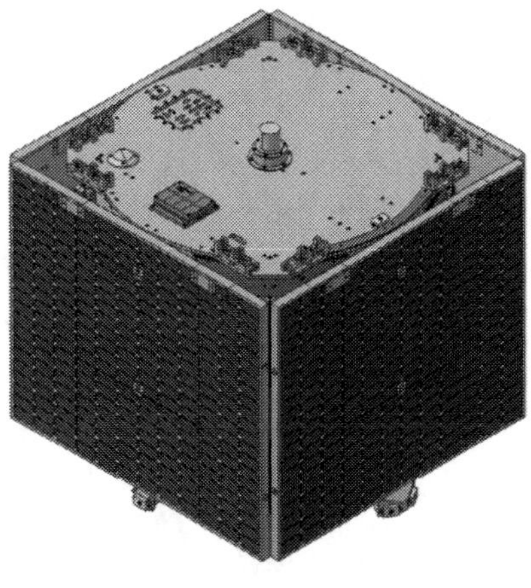

Figure 3. Eu:CROPIS Satellite in stowed configuration

2.2. Power System Architecture

When deciding on a power system architecture for a satellite in low earth orbit, the main decision is whether to use maximum power point tracking (MPPT) for the solar panel conversion, or to use direct energy transfer. The decision depends on the mission profile on the operating conditions for the solar panels, and is a trade-off between between simplicity, efficiency and cost [2][3][4][5]. Since the Eu:CROPIS satellite is relying on the power generation of the solar panels both in stowed and deployed configuration, for which the operating temperature of the solar panels is significantly different, a power regulator implementing MPPT was chosen. Following the trend of small satellites comparable to Eu:CROPIS in power demand and size, an unregulated

(or battery-regulated) power bus with a bus voltage of 22-34V was chosen as the back-bone of the power system. The unregulated power bus simplifies the design of the power system by discarding the battery charge and discharge regulator, however increases the input voltage range for the satellite units compared to a regulated bus. Due to the choice of using an MPPT regulator the maximum power available from the solar panels can be utilized independent of the battery voltage. The Eu:CROPIS Power system architecture can be seen in Figure 4.
As a result from a trade-off between redundancy, mass, volume and cost the PCDU was procured with only two Array Power Regulators (APRs). The APRs are not coupled in parallel, and search the maximum power point for each input independently. When the solar panels are deployed, all panels will be sun-pointing (see Figure 1). Since all four panels are identical, the temperature and thus the panel I-V profile will be identical for all panels. However during the Launch and Early Operations Phase (LEOP) of the mission, Eu:CROPIS is required to generate power in the stowed configuration (see Figure 3). In order to ensure that each APR only sees one panel generating power at any given time, two opposite panels are connected in parallel to one APR. This prevents the overlapping of two different I-V curves, which could trick the MPPT into locking on the wrong maximum power point. Additionally this design allows to reduce the required maximum power handling capability for each APR.

In order to simplify the power system design, it was decided at an early point to only distribute the main power bus to the satellite units. All units which are not compatible with the main bus voltage (e.g. 5V inputs), are required to implement DC/DC converters prior to the unit input. Since most units with a 5V input do not provide galvanic isolation between the primary and secondary power, the galvanic isolation is provided by the DC/DC converter. The drawback of this implementation is an increased power consumption, in particular for units with low input current. Due to the galvanic isolation in most cases the DC/DC dissipates as much power as the unit powered. For future DLR compact satellite missions, a more detailed trade-off will be made regarding this decision, in particular if the power budget is narrow.

2.3. Launch and Early Operational Phase (LEOP)

Seen from the power system, the LEOP is the most critical part of the Eu:CROPIS mission. Due to the narrow temperature requirements of the primary payload, the approach of defining a satellite mode, where all non-essential units are powered off cannot be utilized on Eu:CROPIS. If the temperature of the primary payload gets too low, the biological cells and organisms contained in the payload will start to die, which will lead to mission failure. In order to keep the biology within the required temperature range heaters are mounted to the critical areas of the payload.
The attitude control system for the Eu:CROPIS satellite consists of various sensors for determining the attitude and relies solely on magnetic torquers for control [6]. Since the size and angle of the angular torque which can be generated varies with the Earth magnetic field, initial worst case analysis showed that approximately 4-5 hours were required in order to achieve a sun-pointing attitude with the +Z axis.
During Phase 0 - B the power concept for the LEOP was to rely completely on the spacecraft battery as a power source for worst case analysis, and the requirement specification for the battery was based on this assumption. Due to power creep of satellite units early in the detailed design phase it was decided to perform a larger design iteration on the satellite modes to re-assess the solar panel sizing and redefine the autonomy of the satellite during the LEOP. This exercise forced all subsystems to rethink the power demand during the LEOP and reduce the power demand to "need to have" rather than "nice to have". As an outcome of the design iteration, the modes of the satellite were re-defined (see Table 1).

Satellite Mode	Power Consumption
Low Power	36 W
ACS Mode	72 W
Safe Mode	102 W
Science Mode	190 W

Table 1. Eu:CROPIS Satellite Modes and Power Consumption

The Eu:CROPIS satellite automatically enters the "Low Power" mode, where only the Communication Receivers and Onboard Datahandling is active, upon detecting the separation from the launcher. If the battery state of charge is 'good', the satellite will change to "ACS" mode, where the attitude control system is active, and the satellite will attempt to achieve sun-pointing attitude with the stowed solar arrays. If at any point the temperature in the payload is critically low, the payload heaters are activated. If the battery state of charge is detected as 'low', the satellite will switch to the "Low Power" mode, and wait for the battery to charge until the state of charge is 'good'. If the battery state of charge is detected as 'critical', the satellite will deactivate the payload heater loops regardless of the payload temperature. The autonomy of the Eu:CROPIS satellite is shown in Figure 5.

In addition to iterating the satellite modes and re-assessing the autonomy of the satellite, the solar array sizing was re-analyzed. During Phase 0-B the design driver for the solar panels was the nominal part of the mission, where a solar array output of $\approx 380W$ was required (equal to 95.53W per panel). In order to increase the probability that a positive power balance can be maintained during the LEOP, a trade-off between cost, mass and probability of power balance, in case attitude control can not be obtained, was performed. Based on

 DOI: 10.1051/e3sconf/20171613010

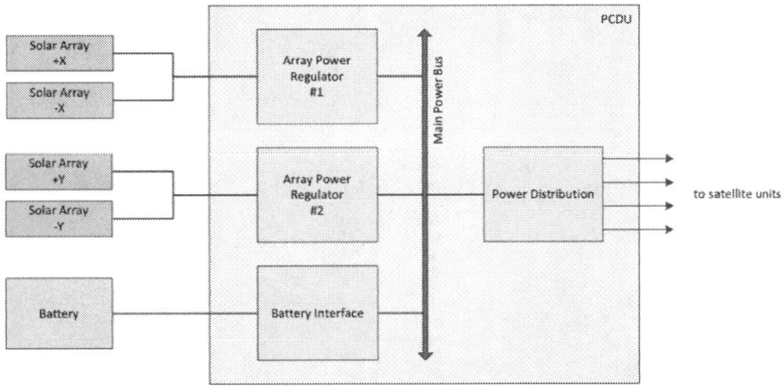

Figure 4. Eu:CROPIS Power System Architecture

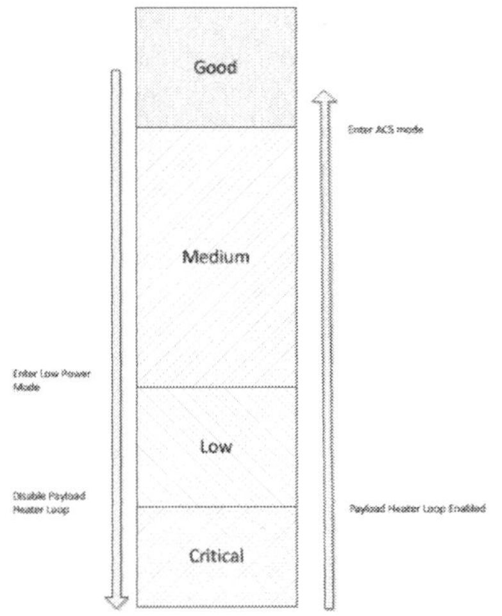

Figure 5. Power System State of Charge and Autonomy

the analysis it was decided that the panel output was to be maximized, resulting in a solar panel capability of $189W$ worst case. The combined power capability of two solar panels is more than what a single APR can convert. However, due to the connection scheme of the solar panels to the APRs, a single APR will never see more than one illuminated solar panel during the LEOP phase.

3. POWER SIMULATION AND ANALYSIS

In order to analyse the power generation and power balance during the LEOP a power system model has been constructed in Matlab Simulink. The model used for the probability analysis includes the kelly cosine relation between solar panel incidence angle and power output [7], and a minimum power deadband for the APR. For the simulation model fixed efficiency parameters are assumed for the PCDU, Battery and Solar Array. The assumptions for the model are either based on worst case values provided by the supplier, or based on tests performed on the delivered units.

In order to assess the probability of a power balance in case of a malfunction onboard the satellite during the LEOP, an analysis of the power generation during a random tumble has been performed. An amount of 10.000 'tumble cases' have been constructed for the satellite, where it is assumed that the satellite is rotating around a fixed random vector. In addition, the location of the sun vector is randomized. The power output for a single revolution has been simulated using the Simulink model. Under the assumption that the satellite has a minimum angular rate of $0.25^o/s$ (equal to approximately 4 revolutions per orbit), the power during one revolution has been assumed as the average power output during the entire orbit, excluding the eclipse period.
Based on the power generation output, the available power for the satellite has been calculated. A histogram of the results can be seen in Figure 6, and the cumulative distribution function can be seen in Figure 7. Based on the cumulative distribution function, the probability of positive power balance has been calculated for "Low Power" and "ACS" mode respectively, with varying payload heater power. The results can be seen in Table 2.

Satellite Mode	Heater Power	Panel Power 189.00 W	95.53 W
Low Power	0	99.60 %	95.32 %
Low Power	10 W	99.17 %	64.04 %
Low Power	20 W	98.53 %	16.12 %
ACS	0	95.19 %	0.01 %
ACS	10 W	87.35 %	0.00%
ACS	20 W	60.78 %	0.00 %

Table 2. Probability For Positive Power Balance

As can be seen in the table, the probability of power balance in all cases is increased significantly with the fully

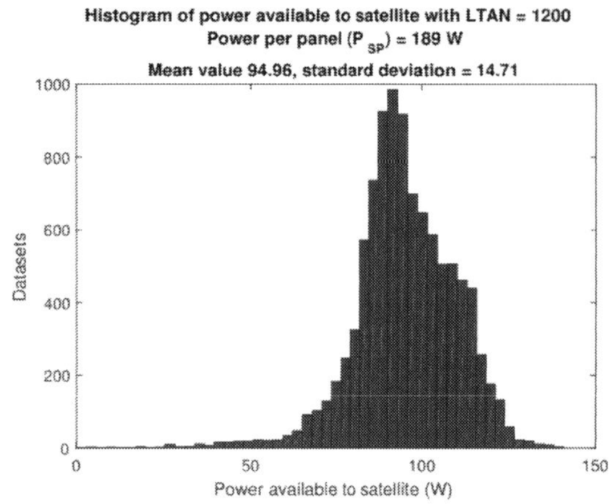

Figure 6. Histogram of simulation results with panel power of 189 W

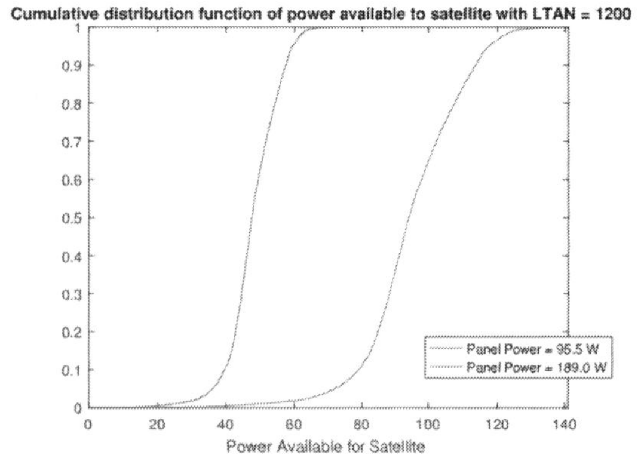

Figure 7. Cumulative distribution function of simulation results

utilized solar panel (189W per panel), compared to the solar panels designed based on power demand during the nominal mission (95.53 W per panel).

4. POWER SYSTEM TESTING

In order to verify that the Eu:CROPIS power system is performing within specifications and according to the design, several tests have been performed on both unit and system level. For the power system a strong reliance on unit testing by the suppliers has been chosen, due to limited resources and facilities available to the project. Instead the focus has been on the system tests, both on power system and satellite level.

On satellite level the interaction between the power system and the onboard data handling system has been verified, by testing the system behavior in various test-cases. The failure handing has been verified by inducing a short-circuit on an LCL output. Based on the configuration of the given LCL it is verified, by telemetry and by measuring the PCDU output, that the system reacts as commanded, e.g. that the output is re-activated by software, or that the satellite automatically enters safe mode. Also the mode-transitioning has been verified by varying the battery voltage using a battery simulator. It has been verified that the satellite is changing the mode autonomously according to the battery state of charge, which is determined based on the battery voltage (refer to section 2.3).
The Eu:CROPIS Engineering Model is not a full representation of the flight model, e.g. most redundant units are not included in the EM. In addition to the missing units, several units utilize COTS components in order to reduce cost and lead time, which results in a different power consumption than for the FM units. As a consequence it has not been possible to verify the satellite power budget and the energy balance during the nominal mission. Due to the large power margin for the nominal mission phase, this is not seen as critical. Rather than the nominal mission phase the majority of the testing performed related to the power system has been focused on the LEOP of the mission. Two extensive test campaigns have been performed related to the LEOP; 'MPPT Efficiency' and 'Verification of Tumble Case', and both will be described in the following sections.

4.1. MPPT Efficiency

In order to maintain a stable attitude, the Eu:CROPIS satellite is required to spin during all mission modes [6]. During the LEOP Mode, the attitude control system will maintain a spin rate of 5 revolutions per minute (RPM) in a barbeque (BBQ) mode around then Z-axis, with the spin axis being perpendicular to the sun, and thereby optimizing the power output. During the nominal mission, the satellite will have a spin rate of 20 or 30 RPM, depending on the experiment phase.

During the LEOP mode the solar panels are stowed. Due to the BBQ mode of the satellite, it is assumed that the panel temperature is relatively constant and therefore it can be assumed that the V_{oc} and V_{mp} parameters of the solar panels are constant during the LEOP phase. However due to the rotation of the satellite, the panel current will vary according to the kelly cosine function of the incidence angle. The satellite is required to rotate with a minimum of approximately 2 RPM in order to remain stabilized during the BBQ mode, however a spin rate of 5RPM is preferred in order to provide a higher margin against external disturbances. A spin rate of 5 RPM, corresponds to an angular velocity of 30 degrees per second, which is also the angular rate of the solar incidence angle on the solar panels. Due to the relative large angular rate of the incidence angle, the current parameters of the solar panel will change with a high slew rate, in particular when the panels are coming 'into sunlight'. In order to verify that the MPPT can follow the change in panel parameters, while still maintaining

 DOI: 10.1051/e3sconf/20171613010

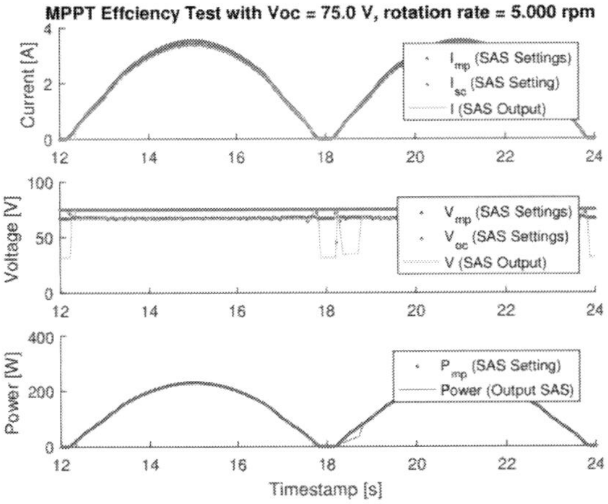

Figure 8. V_{oc} = 75V, 5 RPM, P_{mp} 230W

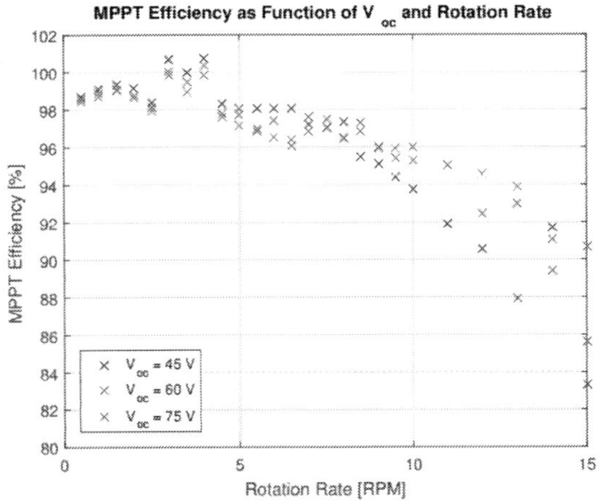

Figure 9. MPPT Efficiency as function of V_{oc} and Rotation Rate with P_{mp} = 230W

the MPP, a series of tests has been performed on the Eu:CROPIS PCDU EM, in order to verify the MPPT efficiency for a rotating satellite.
The MPPT efficiency has been measured for panel parameters with V_{oc} between 45-75V, which is the expected range of the Eu:CROPIS solar arrays, and a P_{mp} of 115W and 230W. The rotational rate was varied between 0.5 and 15 revolutions per minute.

All MPPT efficiency tests have been performed using the Keysight E4360 Solar Array Simulator (SAS), which allows to generate a list of I-V profiles with a minimum dwell time of 30ms. Assuming a fixed V_{oc} and V_{mp} the I_{mp} and I_{sc} parameters have been calculated and transferred to the SAS using the Matlab Instrument Control Toolbox. In order to determine the efficiency of the MPPT the current and voltage on the output of the SAS is logged during 10 rotations. The MPPT efficiency has been determined based on the power being drawn from the SAS, compared to the power available according to the SAS settings during 8 rotations (the first and last rotation are not being used for the efficiency calculations).
In Fig. 8 the measurements from a single rotation can be seen. During the illumination of the first solar panel, the PCDU is tracking the power point during the entire half-sine, however during the illumination of the second panel, the MPP is not tracked during the initial part of the half-sine. Once the MPP tracking has been achieved, the MPP is tracked for the remaining part of the panel illumination.
Due to the technical limitations in general for a SAS in terms of accurately simulating a solar panel [8], and the minimum dwell time between two adjacent sampling points, the efficiency measurements are subject to uncertainties. Due to the dwell time limitations the solar panel parameters are varying with an frequency of 33.3 Hz, while the PCDU is tracking the MPP with a higher frequency. This allows the PCDU to find the MPP and 'stabilize' before the solar panel parameters are changed, whereas the solar panel parameters on orbit will change as a continues function when the satellite is rotating. The SAS output voltage and current is sampled with the maximum frequency of 50Hz, so it is not possible to identify the MPP tracking, and inspect how the PCDU reacts to a sudden change in solar panel parameters. It is however observed that once the PCDU has found the MPP during the illuminated phase, the MPP is maintained. It is estimated that the uncertainty for the efficiency measurements is within $\pm 2\%$, based on the measured values, and the user guide for the SAS [9]. This uncertainty also accounts for measurements with an efficiency higher than 100%. The results from the efficiency tests can be seen in Fig. 9 and Fig. 10 for P_{mp} = 230 W and P_{mp} = 115 W respectively. As can be seen in the figures, the efficiency is constant until a rotational rate of approximately of 5 RPM, after which the efficiency begins to decline. Based on the measurements, no correlation between the V_{oc} parameter and the efficiency can be found. Only a small decline in MPPT efficiency was seen with a P_{mp} = 115W compared to P_{mp} = 230W, however it can not be excluded that the difference is due to measurement inaccuracies. In addition to the results shown in Fig. 9 and Fig. 10 measurements were performed up until 30 RPM, in order to asses whether the primary mission could be performed with stowed solar panels, however the results were inconclusive, and a large variation in the efficiency was seen between the various runs.
Based on the measurements performed it is concluded that the efficiency of the MPPT is sufficient to allow the Eu:CROPIS satellite to rotate with 5 RPM during the LEOP phase.

In addition to verifying the MPPT efficiency for the rotating satellite, the efficiency was also measured for a solar panel with fixed parameters. This test confirmed that the MPPT efficiency is > 99%.

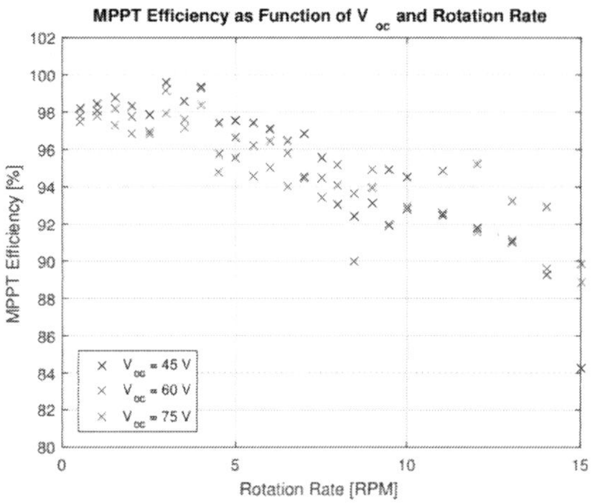

Figure 10. MPPT Efficiency as function of V_{oc} and Rotation Rate with P_{mp} = 115W

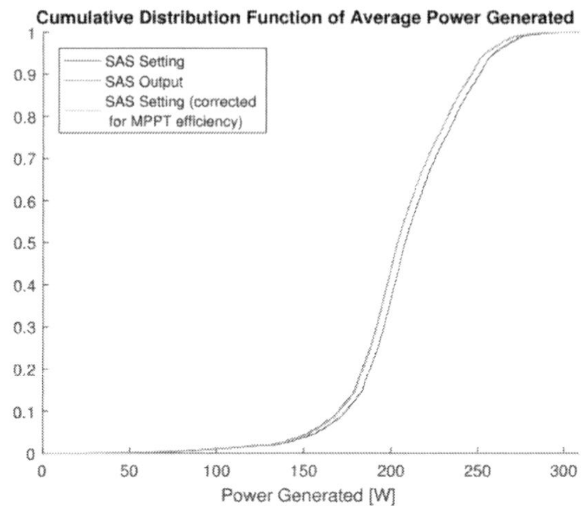

Figure 11. Cumulative distribution function of 4000 random tumble cases

4.2. Verification of Power Generation during random Tumble

As explained in section 3 an extensive simulation and analysis has been performed regarding the probability of power balance in the case that no attitude control is achievable. In order to verify the assumptions used for the analysis, a test campaign has been performed to determine the correlation between the simulation and the measured values.
For the simulation and test the solar array temperature has been assumed to be 50°C, which is the expected maximum temperature for the solar panels during the LEOP BBQ mode, and the satellite is assumed to be rotating with a rate of 5 RPM. The MPPT efficiency has been assumed to be 98.0% based on the results from the MPPT efficiency measurements for a rotating satellite, explained in the previous section. A total of 4.000 tumble cases have been simulated using the same SAS as used for the MPPT efficiency measurements. In Fig. 11 the cumulative distribution function can be seen for the SAS power output, the SAS power settings and the SAS power setting corrected with the MPPT efficiency parameter. As can be seen in the figure the SAS settings corrected for MPPT efficiency correlate very well with the SAS output values, and it is therefore concluded that the MPPT efficiency assumption is valid for the random tumble simulation.

5. CONCLUSION

The Eu:CROPIS power system, like any other subsystem on-board the satellite, has undergone several minor and major changes throughout the design phases. Many changes were triggered due to changes in power demand from the satellite subsystems. However, the biggest trade-offs, some of which are explained in this paper, were due to risk mitigation or requirements imposed due to external factors. The largest design driver in the later design phases of the Eu:CROPIS mission has been the LEOP, and several changes have been carried out due to the requirements and assumptions for this phase.
The Eu:CROPIS satellite has a power system which is over-dimensioned for the nominal science phase, however the robustness against failures in the early phases of the mission has been increased significantly. The increase in robustness has been achieved by applying good engineering when re-assessing the spacecraft modes and autonomy, and by increasing the performance of the solar array to the maximum extent possible within the boundaries of the satellite envelope.

During the Eu:CROPIS development phase, the power system has been designed, simulated and tested. A simulation framework has been established in Matlab Simulink, and is continuously expanded to include more parameters and reflect test results obtained by tests on the real hardware.
A significant effort has been invested in testing the power system, in particular for the solar array input. Utilizing solar array simulators, the solar array of a spinning satellite has been simulated with rotational rates of up to 15 RPM, as well as a significant number of random tumble cases. Based on the tests performed, full confidence has been gained that the power system for the Eu:CROPIS mission will perform according to the expectations.

5.1. Future Work

Based on the Eu:CROPIS project several topics have been identified which need further analysis in order to improve the accuracy of the simulation and testing. In addition several tasks which must be performed in order to sup-

port the flight phase of the Eu:CROPIS mission are still open.
As explained in section 4.1, the MPPT efficiency measurements for a rotating satellite are associated with uncertainties due to the dwell time limitations in the solar array simulator. While the results obtained for the Eu:CROPIS mission provides confidence that the system will perform as expected, more detailed analysis and measurements will have to be performed in order to assess the impact on future missions, and to what extent the measurements are valid.
In order to model the behavior of the flight model the model of the power system will have to be updated to reflect the measured performance during the verification campaign. This includes updating the Matlab Simulink model to reflect dynamic efficiencies for the PCDU. The parameters for the model will initially be based on measurements performed by the supplier on the FM, and will later be updated with measurements performed on the Eu:CROPIS FM, if applicable.
Finally work must be invested in the preparation of the EGSE, in particular for the control software, allowing to perform mission simulations without the need of Matlab Simulink. The EGSE and associated software will be used to perform all functional tests on the Eu:CROPIS FM during the verification campaign on system level, and during the flight campaign. The EGSE must be able to simulate the solar panel input for the nominal and worst case orbits for the nominal mission phase, as well as for the LEOP. Once all tests on the Eu:CROPIS FM are completed, the EGSE will be used to operate the Eu:CROPIS ground reference model.

6. REFERENCES

[1] Jens Hauslage, Michael Lebert, and Hartmut Müller. Eu:CROPIS - euglena and combined regenerative organic-food production in space. In *Life in Space for Life on Earth (Joint Life Sciences Meeting of ISGP, ESA and CSA)*, Juni 2014.

[2] W. Denzinger. Electrical power subsystem of globalstar. In *4th European Space Power Conference*, September 1995.

[3] Antonio De Luca and Giovanni Chirulli. Solar array power conditioning for a spinning satellite. In *8th European Space Power Conference*, September 2008.

[4] Toni Fabio Catalano, Stefano Costantini, and Giorgio Daprati. Sentinel-1 eps architecture and power conversion trade-off. In *9th European Space Power Conference*, June 2011.

[5] Emanniel Colas. Electrical power subsystem of globalstar 2nd generation satellite. In *9th European Space Power Conference*, June 2011.

[6] Ansgar Heidecker, Takahiro Kato, Olaf Maibaum, and Matthew Hölzel. Attitude control system of the Eu:CROPIS mission. In *65th International Astronautical Congress*. International Astronautical Federation, Oktober 2014.

[7] Mukund R. Patel. *Spacecraft Power System*. CRC Press, 2005.

[8] A. Fernandez, C. Baur, and F. Gomez-Carpintero. Solar array hysteresis and its interaction with the mppt system. In *Proceeedings of the European Space Power Conference 2014*, April 2014.

[9] *Keysight Technologies Series E4360 Modular Solar Array Simulator*.

THE GRACE FOLLOW-ON QUIET ELECTRICAL POWER SYSTEM

Manfred Amann [(1)], Mike Gross [(2)], Hauke Thamm [(3)]

[(1) (3)] Airbus Defence and Space GmbH, 88039 Friedrichshafen, Germany, Email: manfred.amann@airbus.com/ hauke.thamm@airbus.com
[(2)] Jet Propulsion Laboratory, California Institute of Technology, 4800 Oak Grove Drive, Pasadena, CA 91109-8099, USA, Email: michael.a.gross@jpl.nasa.gov

ABSTRACT

This paper provides and overview of the GRACE-FO Electrical System and an in depth discussion of the Electrical Power System (EPS). The EPS elements are presented and their features described by their key parameters. These components are inclusive of the body-mounted Solar Array Panels for Power Generation, the Power Control Unit (PCU) and Power Distribution Unit (PDU), a Li-Ion Battery and the Thermal Control System's Heater Control Electronics (HCE's) acting as active current control front-ends. The latter is explained more in detail with respect to its purpose and implementation. The GRACE-FO EPS has been given the designation of "quiet" EPS due to the function of the HCE within the thermal control Subsystem.
The final portion of the paper provides information about the power and energy budgeting and operational aspects of the EPS such as undervoltage and overcurrent H/W features of the PCU/PDU design, and the EPS related On-board Software and On-Board Control Procedures. Finally, the system level power analysis is provided to provide the reader with a better understanding of the expected behaviour of the electrical power parameters.

1. SCOPE

Expected to launch in 2017, the new GRACE-FO Mission* satellite pair will continue and expand upon the measurements initiated by the pathfinder GRACE Mission, launched in 2002. The new mission's primary objective is to estimate the global high-resolution models of the Earth's gravity field for an additional period of five years at a precision and temporal sampling equivalent to that achieved with the original GRACE satellites by means of a microwave interferometer instrument. A second objective is to demonstrate low Earth orbit interferometry by an inter-satellite laser ranging interferometer instrument [1][2].
In addition to the accommodation of the instruments described above, the satellites also carry science Accelerometer Instruments, which are used to determine the non-gravitational forces experienced by the satellites. This instrument suite requires a spacecraft that provides a "quiet" environment for the system to meet its performance requirements. This "quiet" environment includes extreme temperature stability, high mechanical stability, precise alignment, the elimination of platform generated micro-seismic activity and excellent EMC performance.
See an artwork of the GRACE-FO satellites below in figure 1.

Figure 1: GRACE-FO Satellite constellation artwork Courtesy of JPL

DOI: 10.1051/e3sconf/20171613011

2. SATELLITE ELEKTRICAL SYSTEM OVERVIEW

The GRACE-FO Electrical Block Diagram is depicted hereafter.

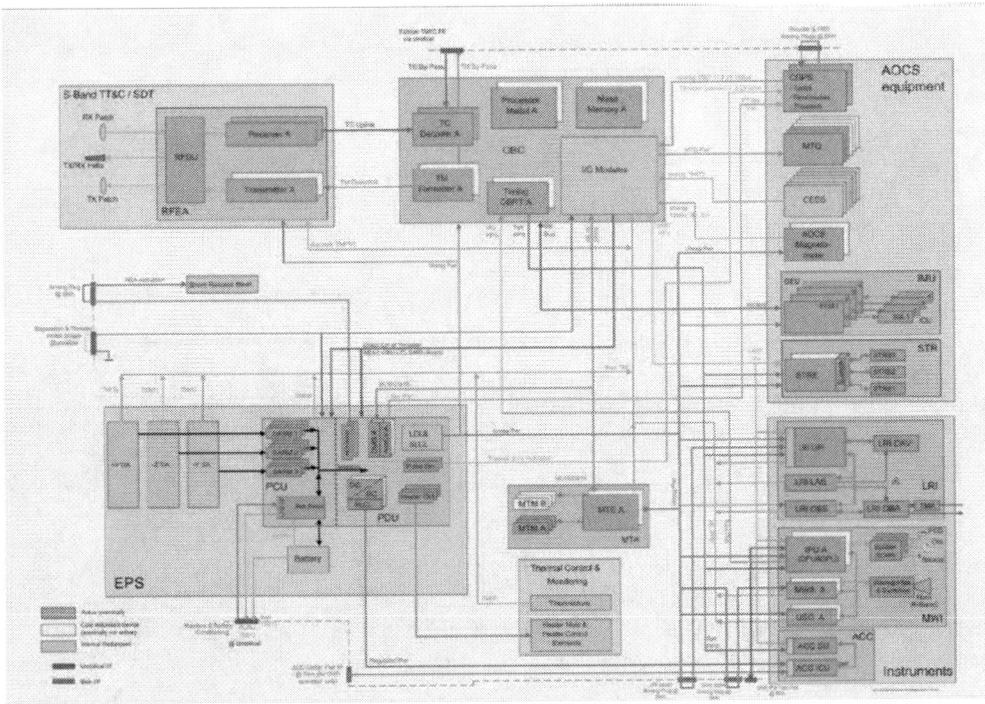

Figure 2: GRACE-FO Electrical Block Diagram

The main blocks are:

- Electrical Power Subsystem (EPS), described hereafter.
- On Board Computer (OBC), the platform data handling system (DHS).
- Telemetry and Telecommand (TT&C) and Science Data Transmission (SDT) subsystem as interface between the ground station and the OBC for reception of telecommands and telemetry
- The Attitude and Orbit Control System (AOCS) units ensure the S/C attitude and orbit control
- Thermal Control Subsystem
- The GRACE-FO instruments, namely the Laser Ranging Instrument (LRI), the Microwave Instrument (MWI) instrument and the Accelerometer (ACC)

3. EPS DESIGN

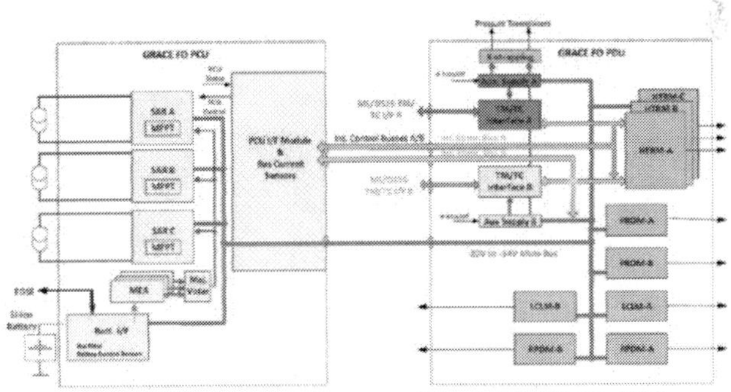

A more detailed block diagram of the GRACE-FO EPS is shown in figure 3.

Figure 3: EPS Blockdiagram

It shows the 3 main blocks,

- The Solar Generator, consisting of 3 body-mounted Solar Panels

- The Battery
- The Power Control and Distribution Unit (PCDU) as central power conversion and power control unit

The PCDU is split into two boxes for accommodation reasons, the Power Control Unit (PCU) and the PDU (Power Distribution Unit).

A GRACE-FO specific feature is the positive grounding of the primary power system. Hence, the power bus is negative with respect to the S/C ground. This implies a dedicated design for the primary power conversion from the Solar Array to the PCDU and to the power users.
The reason for this solution is the advantage of reduced S/C charging [3] and also heritage from previous missions (original GRACE, CHAMP, SWARM).

3.1 Solar Array Design

The Solar Array consists of three body mounted panels, each side panel with 29 strings/combined in 5 sections and the top/zenith panel with 20 strings combined in 4 sections. Each string is equipped with 24 SCAs (Solar Cell Assemblies from AZUR SPACE) of a GaAs triple junction technology with integrated shunt diode. The S/C panels' orientation is optimised for the best energy harvest one can get with the given orbit.
A specific characteristic of the GRACE-FO structure is the solar (and rear/front) panel lay-up. For the mechanical stiffness a conventional CFRP honeycomb panel is used, however, on top a highly thermally isolating foam layer of 40 mm thickness is implemented. The outermost panel layer is a CFRP face sheet. This setup provides excellent thermal isolation for the inner S/C compartments. On the plain solar panels this lay-up supports the SCAs' temperatures to be almost equal across all of the individual panels. This is important for the proper operation of the MPPT control scheme of the PCU.

Figure 4 GRACE-FO Side Panel

3.2 Battery Design

The GRACE-FO Batteries are equipped with Li-Ion cells with a total nameplate capacity of 78 Ah. The Battery configuration is a 8s52p (8 cells per string and 52 strings in parallel) in a flat single module. See figure 5 below. A specific characteristic of the Battery thermal design is that it is mainly based on radiative coupling to the external environment. It is also equipped with internal heaters and HCEs on the top cover. There are two (plus two redundant) independent heater control loops available to stabilize the Battery cell temperature. (See detailed HCE approach in section 4)

Figure 5 GRACE-FO Flight Battery

3.3 PCDU Design

This section lists the PCDU Modules (as shown in figure 3) and provides details regarding the number of channels, the types of outlets and special function incorporation such as battery passivation and regulated outputs. The GRACE-FO PCDU provides a non-regulated power bus (BNR), hence the users have to be tolerant to power bus voltage variations of about +/- 30 %. The PCDU is split in two physical housings, for accommodation reasons.
Find below the PCU/PDU assembly during EMC Testing.

 DOI: 10.1051/e3sconf/20171613011

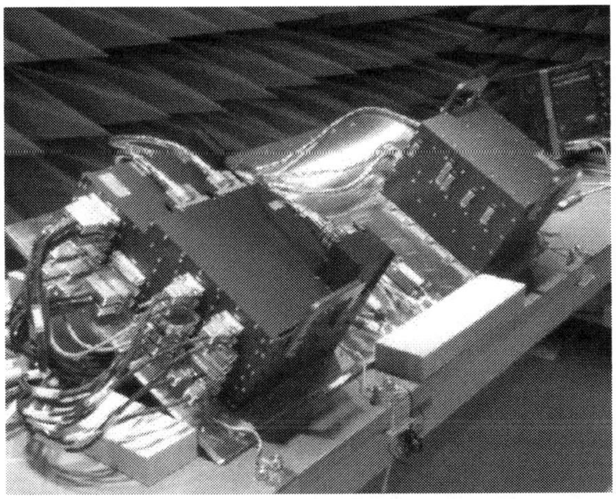

Figure 6 PCU and PDU in EMC Test Chamber

3.3.1 PCU Design

The PCU consists of 6 modules, which are

- 3 SARM (Solar Array Regulator Module), one for each Solar Array Panel; each SARM consists of 3 independent SARs for redundancy reason; 2 out of the 3 are able to convert the full power required
- Main Error Amplifier Module, which controls the SARM power to adjust the battery charging current and hosts the power bus voltage under- and overvoltage protection
- Battery I/F Module with a capacitor bank to stabilize the power bus, Battery relays and power bus current sensors
- PCU Interface Module, which provides the power and signal interfaces to the PDU and contains the Battery passivation circuitry. The passivation function works such that it disables the SARM, which leads to an overdischarge of the battery and the depletion of the electrical energy of the system.

3.3.2 PDU Design

The PDU part of the PCDU contains the modules:

- Redundant Latching Current Limiter Modules (LCLM), each with 11 outlets, which supply most of the platform units from the unregulated power bus
- Redundant Regulated Power Distribution Modules (RPDM). Each RPDM carries 7 outlets to supply the payload units (one with regulated power) and an actuator channel
- Redundant Propulsion Modules (PROM), each carrying 7 thruster, 6 Latch Valve and one Solenoid Valve drive channel
- Redundant Data Management System Modules (DMSM), which serve for interfacing to the OBC and for the generation of auxiliary voltages for the other PDU and PCU modules
- PDU Interface Module to manage the interfaces to the PCU.

4. THERMAL CONTROL SYSTEM (TCS) OF GRACE-FO

Due to the very demanding temperature stability requirements ("quiet" S/C) on the instrument temperatures and mechanical stability requirements, very high attention has been paid to the GRACE-FO TCS and mechanical design. Passive measures are implemented such as

- Structure lay-up with foam layers
- Low CTE selection of structure material
- strong thermal isolation where temperature changing environment is in the view of a thermally sensitive unit or a conductive path is available to a thermally varying element
- Nadir side thermal radiators

In addition complex active temperature stabilisation has been established

4.1 Thermal Control Software and Algorithms

All heater loops are S/W controlled and are individually configurable by parameters to PI control algorithms. The algorithms calculated power demand is realised by 7 bit resolution heater current adjustment of the dedicated channels. To avoid high power dissipation in the PDU and to capture waste heat from the current regulation, a byproduct of the chosen linear regulation scheme, all heater power stages are located next to their designated heaters. The implemented algorithm also allows for some adjustments and settings which are

- adjustable assignment of any on-board thermistor reading for the control loop's actual temperature
- single point failure free triple majority voting of thermistor readings with adjustable criteria

- for more accurate power control, the mainbus voltage can be considered in the algorithm
- adjustable maximal allowed current step size to limit the rate of current change and reduce bus load variations
- total current monitoring of heater groups to safeguard the EPS in case of a single channel malfunction and to avoid excessive power consumption

4.2 Heater Control Element Approach

As mentioned above, each heater channel is a combination of three elements; the On Board Computer (TCS algorithm and telemetry monitoring), the PDU (distribution of primary power and channel current fault protection, and the HCE itself. The core of each HCE is the "linearly" controlled power stage (MOSFET), which allows the TCS to recapture waste heat of the current regulation operation due to its integration alongside the TCS heaters that it regulates current to. Find a picture of an HCE unit in figure 6. The PDU provides 72 (3 modules by 24) channels. This high number of heater channels allows placing heaters and the related HCE's at many locations instead of only a few, and by design, keeps thermal control zones stable, thus minimizing the thermal disturbances in the system. This combination of HCE application with its linear control across the spacecraft system provides for the "quiet' nature of the spacecraft, which has been discovered to be necessary on both the Champ and GRACE Missions. The STAR and SuperSTAR Accelerometers have, in flight, measured micro-seismic events originating from the typical thermostatically controlled TCS. The historic "bang-bang" control associated with thermostats generates mass displacement and recoveries in the spacecraft structure that theorized and analysed to be due to both thermal distortion and Lenz law interactions with the structure.

Figure 7 HCE unit

5. POWER AND ENERGY BUDGET

5.1 On-board Power Demand

The power budget has been set up to service the power demand of the S/C power users in the different orbit scenarios and different mission phases. The expected worst case power has been calculated to be approx. 280 W.
A considerable amount is assigned to heater power, which is by nature the biggest uncertainty in most satellites.

5.2 Power Generation

The EPS relevant GRACE-FO Mission and orbit parameters are:

- LEO (Low Earth Orbit) between 490km (BoL) and 410 km (EoL).
- High inclination angle (close to polar orbit), which imposes a slow drifting orbital plane and a high percentage of orbits with eclipses.
- Satellites shall always be nadir oriented.
- Each satellite may fly forward or backward.
- Body mounted Solar Arrays

These parameters imply a complex analysis of different cases for power generation and thermal conditions.

The analysis has been performed by the Airbus DS satellite power tool Power Systema [4] with the model as shown below in figure.

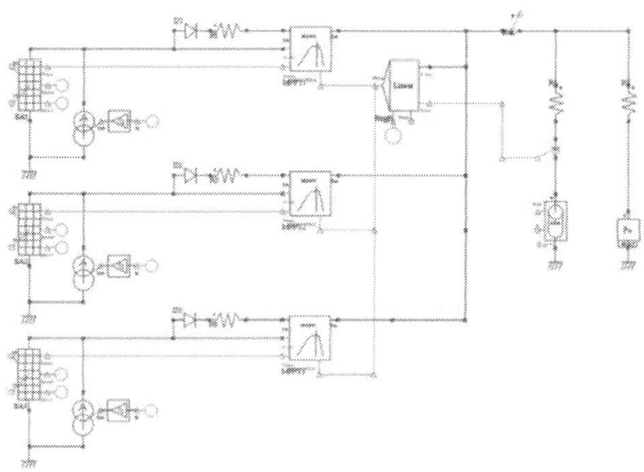

Figure 8 Power Systema model for GRACE-FO

For the above mentioned conditions the possible average power per orbit under worst case conditions has been analysed for each of the 24 LTAN hours. The

DOI: 10.1051/e3sconf/20171613011

results are shown in the following profile chart (average values per orbit):

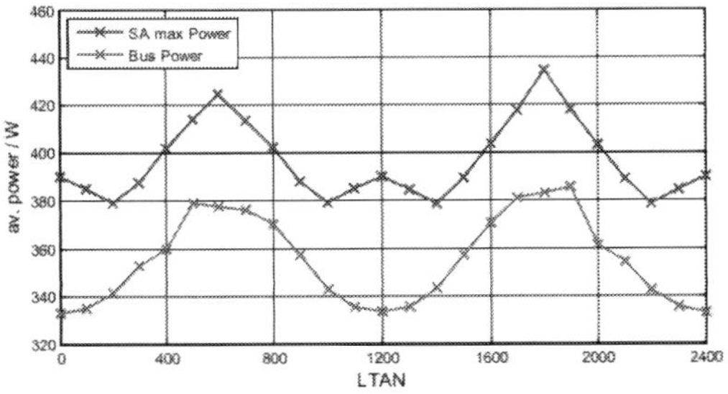

Figure 9 GRACE-FO energy gneration dependency on LTAN

This was the first step before other cases were studied with modified conditions and configurations such as load variations, mission phases, equipment ageing, etc.
These cases were analysed mainly by time domain simulations with 5 to 10 orbits to be able to interpret the results correctly.
An example of a plot set at EoL is shown below in figure 10.

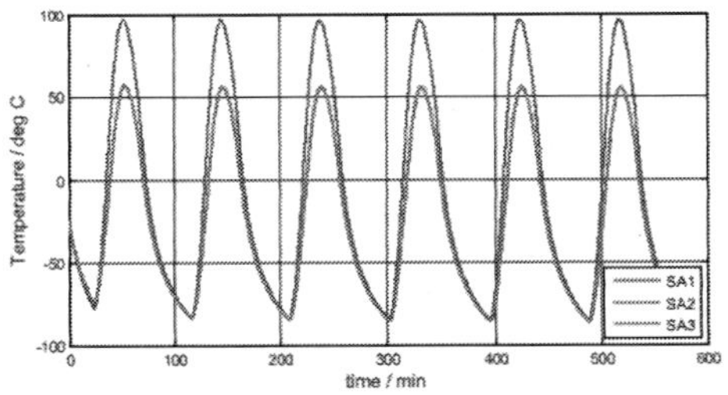

Solar Array temperatures (SA2 and SA3 are identical)

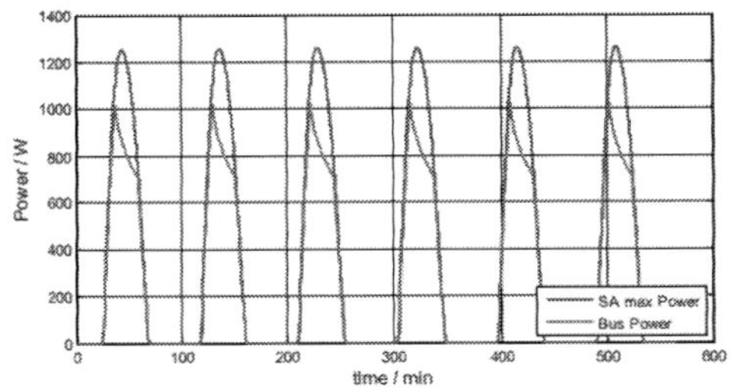

Solar Array max power capability and the used power for the power bus

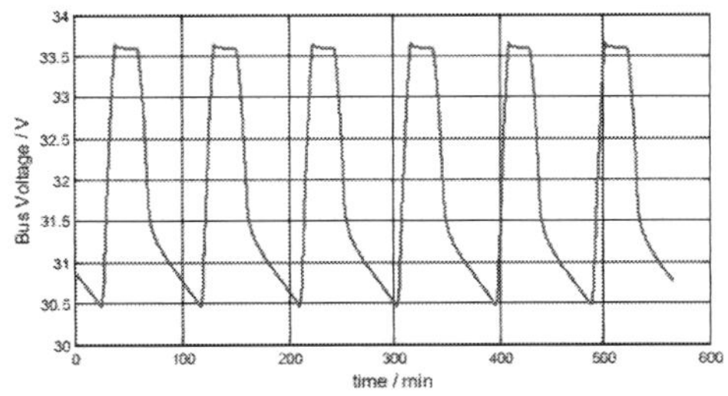

Power Bus voltage

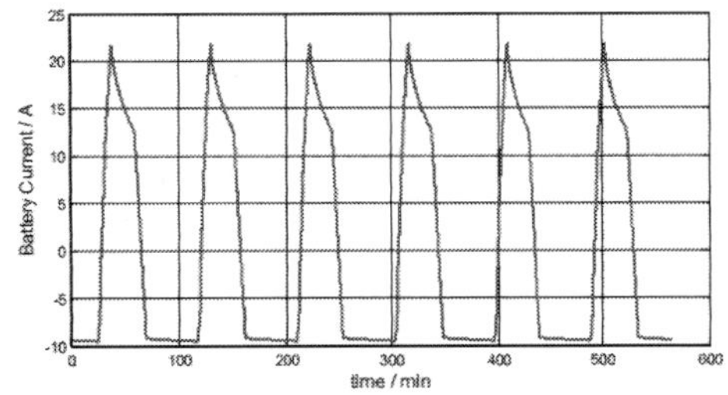

Battery current

Figure 10 Power Systema simulation for an LTAN1200 orbit

5.3 Power System Contingency Design

This section describes features of the GRACE-FO H/W measures in case of contingency cases, in addition to the current limiters at the PDU outputs. It includes also the On-Board Control Procedures (OBCP) implementation to ensure a protected power system operation, on the S/C.

The PCU provides as H/W protection features:

- Battery relay hold current capability to ensure closed battery relays during launcher ascent
- Battery charge and discharge current limitation
- Solar Array Regulator switch-off and interlocking

The PDU provides as H/W protection features

- Disable Non-Essential Loads (DNEL) logic in case of unexpected low mainbus voltage
- Aux Converter switch-off interlocking
- Current limiter output stage protection in case of load oscillation

The Failure Detection and Isolation Recovery (FDIR) concept of the GRACE-FO Platform supports the PCDU H/W detection and serves for additional detection measures by monitoring the on-board telemetry (TM).

EPS relevant FDIR cases are:

- Mainbus voltage monitoring leading to a S/C mode transition in case the mainbus drops under a threshold value
- Supply current monitoring of outlet currents and off commanding when upper and/or lower current values are exceeded.
- Temperature monitoring can be configured to lead to switch the heater loop to the redundant channel
- Any other TM parameter can be used to trigger an on-board control procedure

6. CONCLUSION

The GRACE-FO Electrical Power System has been designed and optimized to fulfil the challenging mission requirements on the S/C platform. The EPS units/equipment with their key figures have been presented. By means of the HCE approach, the EPS provides the functionality for the complex, configurable, efficient and "quiet" Thermal Control System. The TCS Software allows the system to benefit from the multiple adjustment features.
Finally the EPS power analysis has been exemplarily presented and the H/W and S/W supported contingency functionality has been explained.

7. ACKNOWLEDGEMENTS

The work described in this paper was carried out at Airbus Defence and Space, GmbH, under contract to the Jet Propulsion Laboratory, California Institute of Technology. The Jet Propulsion Laboratory, California Institute of Technology is managing the GRACE-FO Project development under a contract with the National Aeronautics and Space Administration. US Government sponsorship acknowledged.

8. REFERENCES

[1] http://gracefo.jpl.nasa.gov/mission/
[2] http://www.gfz-potsdam.de/en/section/global-geomonitoring-and-gravity-field/topics/development-operation-and-analysis-of-gravity-field-satellite-missions/grace-fo/
[3] Solar Array Arcing Mitigation for Polar Low-Earth Orbit Spacecraft,
Grant Bonin et al., UTIAS Space Flight Laboratory
[4] Investigation results on solar array thermal and electrical imbalance phenomenon on power systems equipped with MPPT,
F. Mercier et al., ESPC2014 paper

E3S Web of Conferences **16**, 13012 (2017) DOI: 10.1051/e3sconf/20171613012
ESPC 2016

EXOMARS 2016 MISSION ELECTRICAL POWER SYSTEM

Ezio Ciancetta [(1)], **Giuseppe Cuzzocrea** [(1)], **Jean-Jacques Digoin**[(2)], **Federico Chiusano** [(3)]

[(1)] *Thales Alenia Space, Strada Antica di Collegno 253, 10146 Torino (Italy), Ezio.Ciancetta@thalesaleniaspace.com, Giuseppe.Cuzzocrea@thalesaleniaspace.com*
[(2)] *Thales Alenia Space, 100 Bd du Midi, 06322 Cannes (France), Jean-Jacques.Digoin@thalesaleniaspace.com*
[(3)]*European Space Agency, Keplerlaan 1, 2201 AZ Noordwijk (The Netherlands), Federico.Chiusano@esa.int*

ABSTRACT

ExoMars is the first step of the European Space Agency's Aurora Exploration Programme. Comprising two missions [1], the first one launched in 2016 and the second one to be launched in 2020, ExoMars is a joint ESA-Roscosmos program that will address the scientific question of whether life ever existed on Mars and demonstrate key technologies for entry, descent, landing, drilling and roving on the Martian surface [2].
The Spacecraft Composite (SCC), consisting of a Trace Gas Orbiter (TGO) and an EDL (Entry Descend and Landing) Demonstrator Module (EDM) named Schiaparelli, has been launched on 14 March 2016 from the Baikonur Cosmodrome by a Proton Launcher.
The two modules will separate on 16 October 2016 after a seventh-months cruise. The TGO will search for evidence of methane and other atmospheric gases that could be signatures of active biological or geological processes on Mars and will provide communications relay for the 2020 mission surface assets. The Schiaparelli module will prove the technologies required to safely land a payload on the surface of Mars, with a package of sensors aimed to support the reconstruction of the flown trajectory and the assessment of the performance of the EDL subsystems.
Following separation, the EDM will undergo a three-days coasting phase to reach the Mars atmospheric Entry Interface Point, complete the Entry Descent and Landing phases in six minutes and perform surface science operations for a duration of up to four Martian sols, with data transmission to TGO and to other ESA and NASA orbiters.

This paper outlines the Exomars 2016 Electrical Power System (EPS) design, providing a description of the major design drivers and resulting configuration, with a view to highlight aspects that could be considered for future designs.

1. SPACECRAFT AND MISSION DESCRIPTION

Thales Alenia Space Italia lead the ExoMars industrial team and they are in charge of EDM design, including the electrical power supply chain. Thales Alenia Space France are responsible for the TGO.
The picture in Fig. 1 shows the SCC in launch configuration.

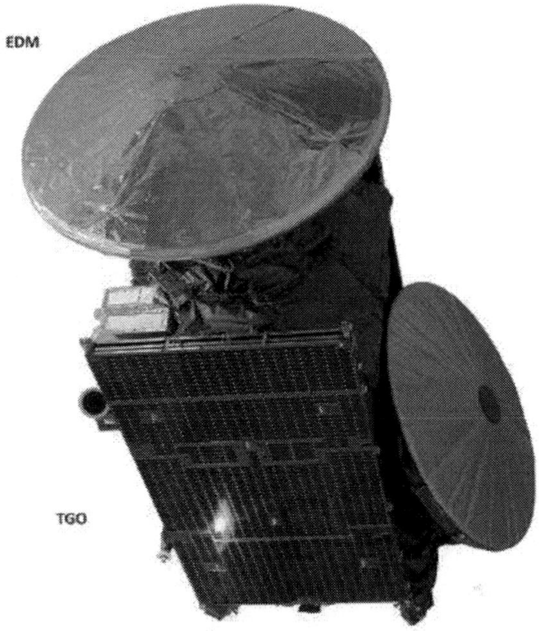

Figure 1. Exomars Spacecraft Composite (SCC) in launch configuration.

The EDM mission will be fully autonomous, with mission phases management performed by On Board Software (OBSW) and the Guidance, Navigation and Control (GNC) in charge of handling mode transitions. GNC functions will be performed in autonomy, without any ground intervention. During the Coasting and EDL phases the EDM power is provided by a set of batteries and GNC functions shall allow the execution of the EDL phase, including the management of two hibernation phases required for energy saving, considering that the entire mission is performed relying on batteries only.

The Exomars 2016 mission phases are shown in Fig. 2.

 DOI: 10.1051/e3sconf/20171613012

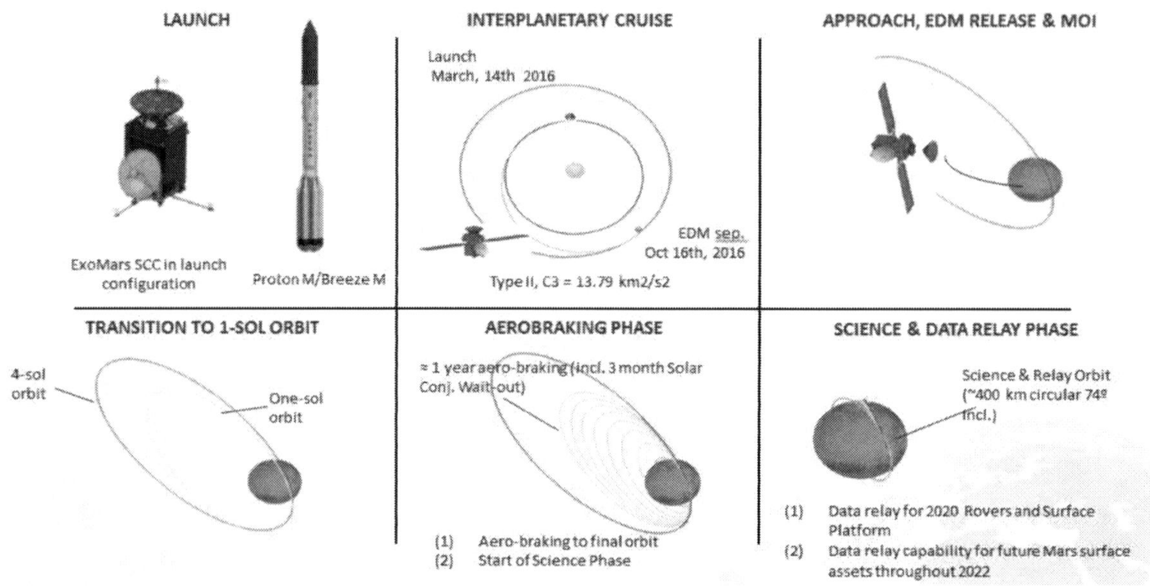

Figure 2: Exomars 2016 Mission

The Schiaparelli module will encounter four distinct mission phases, in sequence:

- seven months of cruise (including short check-outs) attached to the TGO;
- hibernation during the seventy-two hours of coasting to Mars after separation from TGO;
- six minutes of EDL;
- at least four sols on Mars Surface (including hibernation to save energy).

The EDM data will be sent to TGO, to Mars Express orbiter and to the NASA orbiters Mars Reconnaissance Orbiter (MRO), Mars Atmosphere and Volatile Evolution Mission (MAVEN) and Mars Odyssey.

The objectives of the Exomars mission have been modified several times along the history of the program, together with the changes of the international partners. As a consequence, the spacecraft architecture was reconsidered accordingly, with changes affecting also the power supply chain.

2. EDM ELECTRICAL POWER SUBSYSTEM DESIGN DRIVERS AND ARCHITECTURE

The Electrical Power System architecture was built around some major design drivers; in particular, the following aspects have been considered for defining the design of the EDM Power system.

- Reception of power from the TGO. During the cruise phase, the power system has to be capable of receiving power from the Orbiter.
- Unavailability of an autonomous system for power generation. The landing will be done during the global dust storm season, when there is a high probability that no direct sunlight will reach the surface. This condition makes ineffective a Solar array. Other power sources like Radioisotope Thermoelectric Generators (RTG) were discarded for immaturity of the design and risks associated to their usage
- Use of primary batteries. Missing a Solar array allowing the battery recharge, only primary power sources can be used and therefore the maximization of power density (maximum ratio between power and mass) was necessary. The first selection was using primary cells for the batteries, but the energy stored inside any single battery cell was too high and risky for the mission, in particular for the mechanical loads. In fact, during a vibration test a primary cell exploded as a consequence of internal short circuit.
- Optimization of the power consumption. Having a reduced power source, a strong optimization in the power consumption was required. All units were specified to be low power demanding. A hibernation mode was designed for the periods in which the capsule does not need a complete functionality, e.g. in the three days approaching Mars and during the surface operations in all the time frames when the communications with orbiters are not possible. In this mode, only timers are active to wake up the EDM when required.
- Thermal insulation to minimize the power

 DOI: 10.1051/e3sconf/20171613012

consumption for heating the vital units. All the vital units needed on the surface are located in a specific Warm Compartment highly isolated from the external environment. The Battery temperatures have to be kept sufficiently warm to optimize their energy production.

The EDM EPS is designed to be powered by the TGO unregulated power bus or by its internal batteries. The EPS power conditioning and distribution functionalities are implemented into two different units: Central Terminal and Power Unit (CTPU) and Remote Terminal and Power Unit (RTPU). The CTPU implements the part of the EDM EPS that is requested to be operative also on Mars surface and for this reason it is located in the EDM Warm Compartment. The RTPU indeed is located on the crushable side of the lander and it is no longer used after touch-down on Mars surface.

The EPS block diagram is shown in Fig. 3.

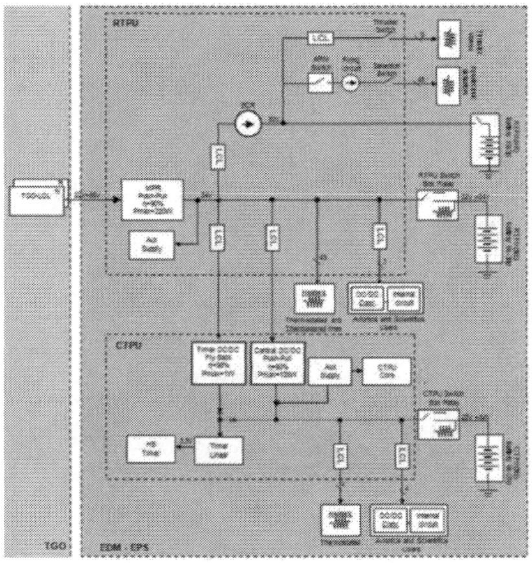

Figure 3. EDM EPS Block Diagram.

The RTPU Main Power Regulator (MPR) is fed by the TGO unregulated power bus and regulates it to a stable voltage of 34V±0.5%, with a power capability of 220W and an efficiency greater than 90%.
The CTPU Central DC/DC converter regulates the power coming from RTPU Main Bus to 34V±2%, with an efficiency greater than 90% at 135W. These regulations are needed to guarantee a bus voltage slightly higher than End of Charge (EoC) battery voltage, i.e. 33.6V.
The CTPU and RTPU regulated bus are distributed via Latching Current Limiters (LCL) to the EDM equipment.
The EPS regulated bus stability has been designed in accordance with the Middlebrook criteria [3] and it has been successfully tested on the EDM Avionic Test Bench (ATB) and Proto-Flight Model (PFM).

An exploded view of the EDM is reported in Fig. 4. The location of EPS equipment is shown in Fig. 5.

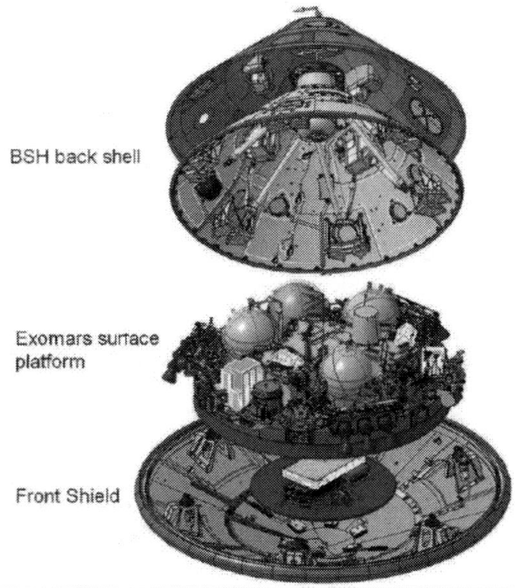

Figure 4. EDM exploded view.

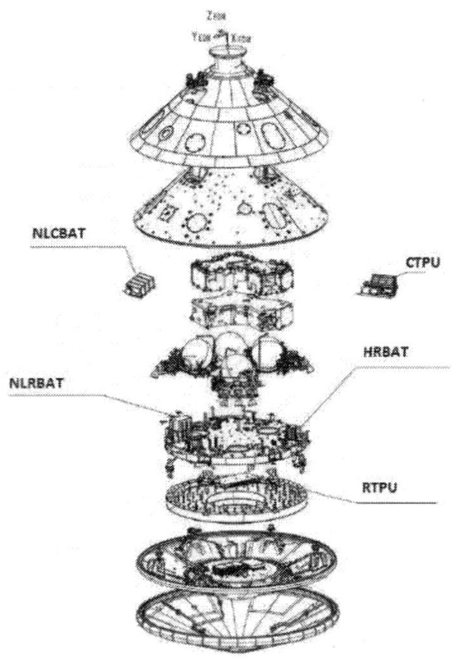

Figure 5. EDM EPS equipment configuration.

DOI: 10.1051/e3sconf/20171613012

During the cruise the RTPU is nominally switched ON (EDM service mode), while the CTPU is operative only during check out phases and during the EDM Batteries recharge phase. When the EDM is in service mode the TGO is committed to manage the EDM Thermal Control System (TCS), executing in its Satellite Management Unit (SMU) the EDM TCS algorithm. This algorithm, especially in coldest condition, is aimed to tailor the TCS consumption as complement of avionic units power demand, keeping the total power below the RTPU MPR capability.
The EDM batteries recharging phase is scheduled at the end of the cruise, in order to compensate the slight self-discharge of the lithium cells.

Three different battery modules, based on lithium cells, have been designed to supply the EDM when it switches to internal power mode:

- RTPU NL Battery (NLRBAT - 87Ah), used in Coasting and EDL phase;
- CTPU NL Battery (NLCBAT - 62Ah), used on Mars surface;
- RTPU High Rate (HR) Battery (HRBAT - 3.3Ah), used to supply during the EDL the 60V unregulated power bus of the RTPU, used for pyro firing and RCS actuation with a peak current capability of 45A as required during pyro actuation.

The HRBAT is connected to the 60V unregulated bus via latching relays embedded into the battery. The charging is managed via a dedicated Battery charge Regulator (BCR) with three selectable levels of EoC voltage (up to 61.5V) and charge current rate (up to 0.3A). The converter has a global efficiency greater than 80% and it implements the taper mode functionality.

The NLRBAT connection to Main Bus (MB) is managed via relay boxes. The charge functionalities are implemented by power resistors, bypassed when the EDM is in internal power mode. The simplicity of the NL battery charge system is paid in terms of charge time. Fig. 6 shows the test results of NLRBAT charged via 4.99 Ω resistor using a bus voltage of 34.5V. The charge resistor is bypassed at 34V and the battery is directly connected to the MB. This design solution has been chosen because the need of recharging this battery was specified when the design was already frozen.

At the separation from the TGO the EDM is switched in internal power, then the EDM goes in coasting hibernation mode. In this phase in order to save energy the RTPU and CTPU Main and Aux converter are switched OFF and the RTPU main bus is handled by NLRBAT. This unregulated bus is distributed via LCLs to heater lines and CTPU. The RTPU distribution losses in this phase are lower than 1W. The exit from hibernation is scheduled by CTPU.

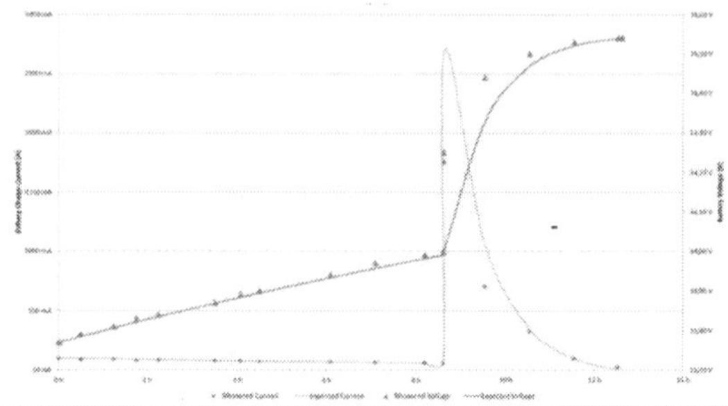

Triangle markers → Battery voltage measurement;
Green curve →Calculated battery voltage;
Square markers → Battery current measurement;
Yellow curve→ Calculated battery charge current.

Figure 6. NLRBAT Charge model vs. measurements.

The CTPU Timer logic is powered by dedicated low power converter. The Timer DC/DC converter takes power from the RTPU unregulated bus and regulates it to 34.5V±5% with a power capability of 1W and an efficiency greater than 90%. During coasting hibernation, the overall CTPU power consumption is less than 1W and the most part of the energy consumed by EDM is spent by the thermal control.

The coasting discharge power profile of NLRBAT has been simulated using EDM correlated thermal model with an overall system margin of 5%, as shown in Fig. 7. The NLRBAT is predicted to complete the mission with a final State of Charge of 16%.

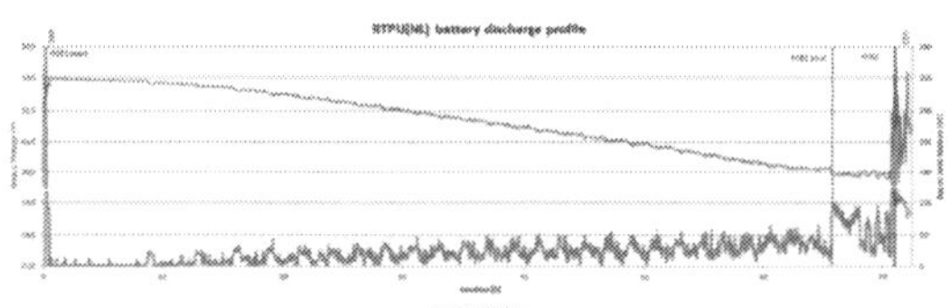

Figure 7. NLRBAT Simulated Discharge Profile.

During Entry, Descent and Landing the HRBAT is in charge to provide high power peaks, with duration of 25 to 50 ms, demanded by pyrotechnic actuators used in propellant circuit valve (up to 360W), Parachute Deployment Device (up to 720W), front shield and back shell separation mechanism (up to 1800W). These performances are met by the HRBAT even with a string in failure.
The main part of the HRBAT energy is spent to supply the Reaction Control System (RCS) flow control valve during breaking manoeuvre at the end of the parachute phase (up to 800W for 60s).
Fig. 8 shows how the under-voltage condition (48V) is never triggered on the RTPU 60V unregulated bus. The TCS is designed to maintain the battery temperature always above 10°C in order to avoid battery Internal Resistance (IR) degradation.

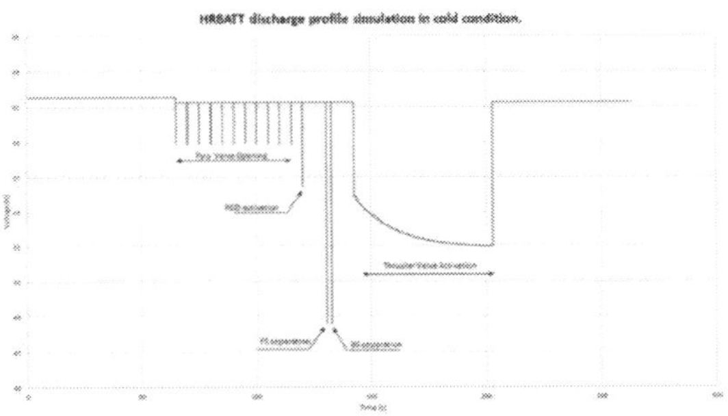

Figure 8. HRBAT Discharge Profile.

On Mars surface the lander is fed by NLCBAT which is accommodated in the Warm Compartment together with Transceiver (TVCR), CTPU and scientific equipment. The relay box connecting the NLCBAT to the main bus is mounted outside the Warm Compartment, being able to withstand Mars coldest temperature.

The DREAM (Dust characterisation, Risk assessment and Environment Analyzer on Martian Surface) Instrument units are powered autonomously by a dedicated battery.

During Mars surface operation the lander is always in Hibernation except during orbiter passes (30 min), where CTPU and the TVCR are fully operative in order to upload the scientific data. The most part of NLCBAT is used by thermostated heater lines, being the CTPU power dissipation lower than 1W: this is because only the 3.3V linear regulator is working.

The discharge power profile of NLCBAT during Mars surface operations has been simulated using EDM correlated thermal model with an overall system margin of 5%, as shown in Fig. 9. The greater contributor in terms of power request is the TCS, that depends of course from the external environment not totally predictable at priori. Therefore in case of very benign external environmental conditions, where the temperature is warmer than expected, the life of Mars can be increased up to ten sols.

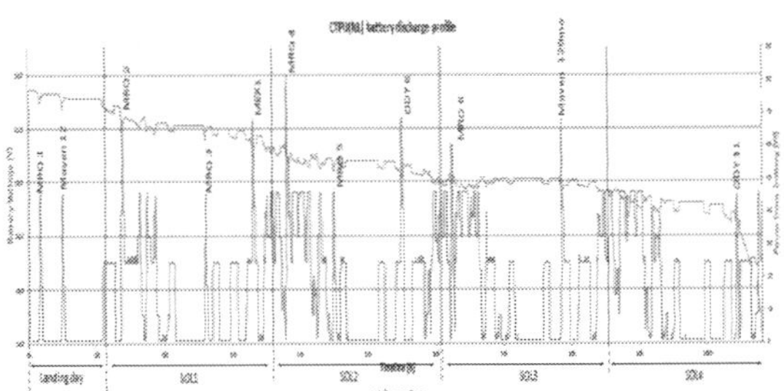

Figure 9.NLCBAT simulated discharge profile.

The NLCBAT is predicted to supply the lander operation through the eleven orbiter passes scheduled in the 98h (landing day + 4 sols) following the touch down on Mars surface with the standard thermal environment.

3. TGO ELECTRICAL POWER SUBSYSTEM DESIGN DRIVERS AND ARCHITECTURE

The TGO EPS is in charge of electrical power generation and energy storage for the Exomars spacecraft composite during launch, cruise and Mars orbiting phases.

EPS design drivers are the variable distance to Sun, the numerous phases associated to different demands in power (with and without telemetry transmission or with and without UHF communication for examples), the occurrence of solar eclipses with varying durations (from few minutes up to 160 minutes).

The Power Subsystem is sized taking into account nominal and contingency situations, against the two most critical scenarios which are the following:

- during the Aerobraking phase, at the end of the sequence where orbit durations are the shortest and the off-sun durations (eclipses + atmospheric passes) the longest;
- in Science phase, taking into account the optimized communication scenarios (with the associated Telemetry Tracking and Command subsystem consumption in Tx mode), as the eclipses may span over 36% of the orbit.

The Power Subsystem shall also withstand the launch phase. In March 2016 scenario the duration of this phase, from launcher lift-off up to solar array deployment, had a duration around 11 hours. During this phase the TGO was spinned and the stowed solar arrays (see Fig. 1) were alternately Sun pointed (one panel faced to Sun) during a period of 600 seconds. This sequence is called "Flip-Flap" sequence. Even if the Battery is sized to supply the TGO during all this sequence with the predicted thermal environment, the recharging during the Sun pointed periods were appreciated to reduce the Depth of Discharge (DoD).

The TGO EPS provides the following power via an unregulated bus to the whole SCC:

- Power needed by TGO, including
  - 190W to the Science Payloads
  - 55W to Science Payloads survival heaters (Payloads Off).
- 60W to EDM during Cruise,
- 120W to EDM during EDM commissioning and health check,
- 240W to EDM during thermal boost before separation,
- 120W to the Main Separation Assembly (MSA)

DOI: 10.1051/e3sconf/20171613012

TCS heaters up to EDM separation.

The overall TGO power loads, not including battery recharge, is variable along the mission phase and configuration and could reach up to 1200W in worst case.

During all the mission phases, in order to reduce its recharge duration, the battery will be recharged with the maximum available current. Charge current limit is 60A, allowing a power recharge up to 2000W (C/3).

Exomars TGO EPS is composed of the following equipment:

- Li-Ion battery (BTY), composed in baseline of two identical modules, having an overall name-plate capacity of 180Ah;
- Power Conditioning Unit (PCU), including the solar array power conditioning performed by Maximum Power Point Tracking (MPPT) Buck regulators and the battery management;
- Propulsion Control and Distribution Unit (PCDU), including the protection and distribution functions, the pyro commands and the RCS interfaces;
- Solar Array (SA), composed of two deployable wings, each wing equipped of two Ga-As panels, one yoke panel (not cells populated) and one SADM mechanism;
- Harness, providing the interconnections between all the electrical and electronic equipment installed on the TGO, down to the EDM, the Payloads electrical interfaces and umbilical interface.

The TGO EPS architecture is depicted in Fig. 10.

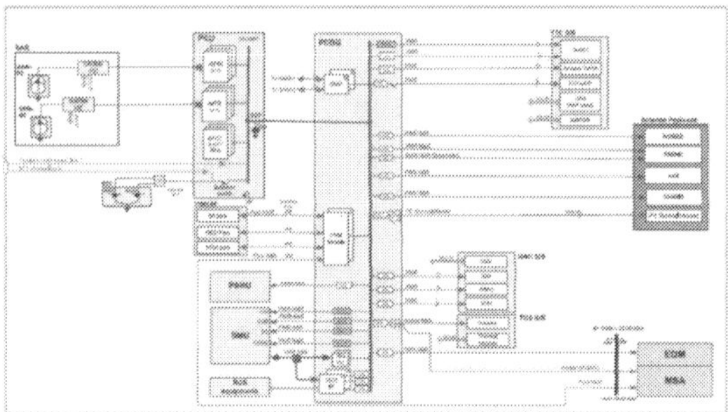

Figure 10. TGO EPS Block Diagram.

The power conditioning functions are totally autonomous and no reconfiguration is expected from the Satellite Management Unit (SMU) in case of single failure. In particular, hot-redundant electronics have been implemented for the power conditioning and battery management functions.

The PCDU distributes an unregulated and protected bus voltage (by means of R-LCL/LCL) to the TGO avionic units, the EDM and the four payloads.
The nominal bus voltage range is 22V÷ 36V defined as follows (measured at PCU output level):

- Maximum operating voltage: 36V (maximum voltage when the battery is in recharge at its maximum EOCV);
- Minimum operating voltage: 22V (lower limit in a safe mode)

The maximum non-operating voltage is 40V, maximum voltage to be withstood by user units.

The Solar Array is a major equipment of the orbiter module bus: its sizing is of particular importance and shall consider the complexity of the mission, in particular the aerobraking sequence.
The aerobraking is a manoeuvre performed near Mars (about 120km of pericenter height) to reduce the orbital period. It consists of several manoeuvres (or passes, about 800 during the overall sequence) through the upper level of the Mars atmosphere. The heat generated by the friction of the spacecraft against the atmosphere dissipates the kinetic energy of the spacecraft, thus imparting a change in velocity or delta-V.
The Solar Array rear face will be exposed to the dynamic flux during each aerobraking pass (up to 2800W/m^2). The thermal analysis verified that sensitive elements (e.g. cells, substrate, blocking diodes,...) do not exceed their qualified temperature. The analysis results show that the maximum temperatures are reached during the walk-out (end of aerobraking, lasting two weeks).
Fig. 11 shows the deployment of TGO solar array in TAS Cannes facilities.

Figure 11. Panoramic photo of the deployed solar array integrated on TGO in Cannes facilities.

The location of TGO power equipment is shown in Fig. 12.

DOI: 10.1051/e3sconf/20171613012

Figure 12. TGO configuration, showing the EPS units location.

Additional data relevant to TGO EPS were reported in [4] and [5].

In order to simulate the different ExoMars phases and verify the correct EPS behaviour, a specific tool has been developed by TAS. Some simulation results are depicted in the figures below.

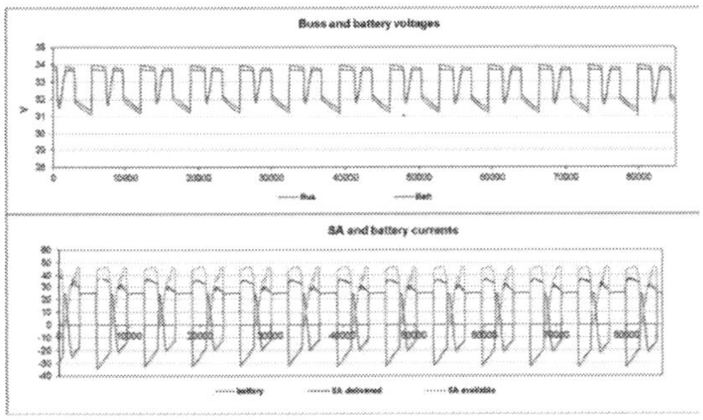

Figure 13. Results of walk-out simulations.

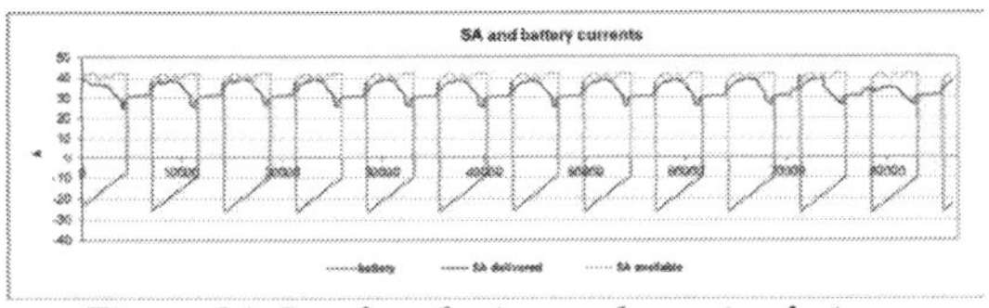

Figure 14. Results of science phase simulations.

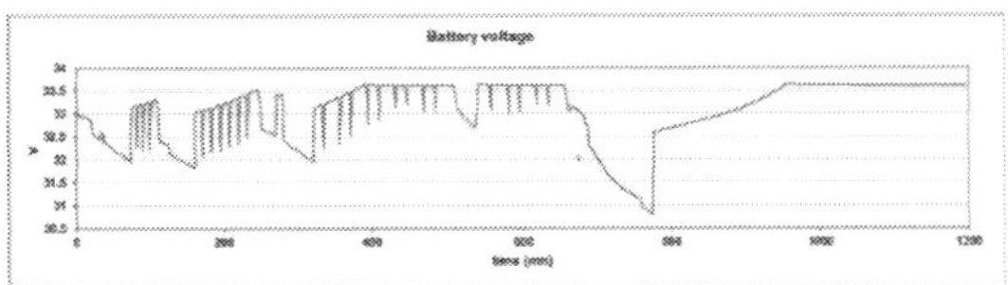

Figure 15. Result of battery voltage simulation during Flip-Flap.

These simulation results have also been checked with flight telemetries and showed a good correlation. An example of battery voltage is reported in the following figure.

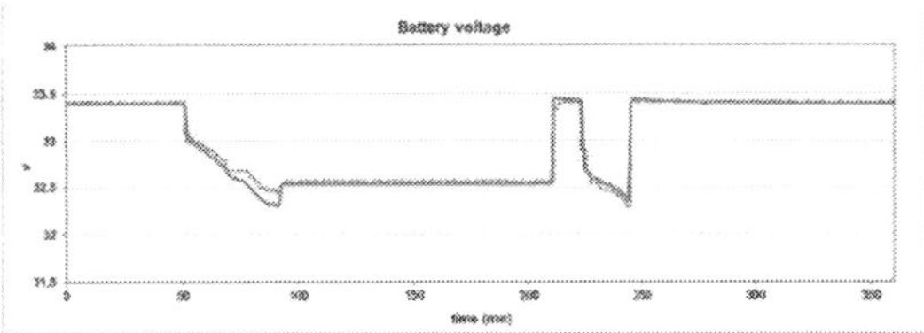

Figure 16. Battery voltage (red bold simulation, red doted TM).

4. CONCLUSIONS

This paper presented the Exomars 2016 mission Electrical Power System, describing the major design drivers, the architecture and the results of the preliminary measurements performed in flight.
At the time of the issue of this paper, the Launch and Early Orbit Phase has been completed with a successful In Orbit Commissioning Review. The Deep Space Manoeuvres (DSM 1 and 2) have been successfully performed at the end of July and mid-August and have put the Spacecraft in the correct orbit for Mars. The landing on Mars surface of the Schiaparelli module is foreseen on 19 October 2016 at about 14:47 GMT. At that time the TGO will perform the critical Mass Orbit Insertion (MOI) manoeuvre.

5. REFERENCES

1. Cassi, C., Vinai, B., Gianfiglio, G. & Baglioni, P. (2012). ExoMars: One Project Two Missions. In Proc. 63rd International Astronautical Congress, IAC-12.A3.3A.3.
2. Capuano, M., Portigliotti, S., Fiumanò, A., Blancquaert, T., Bayle, O. & Lorenzoni, L.. Exomars Mission 2016: EDM Science Opportunities. In Proc. 63rd International Astronautical Congress, IAC-12.A3.3A.4.
3. Middlebrook, R.D. (1976). Input filter considerations in design and application of switching regulators. In Proc. IEEE IAS annual meeting, 1976.
4. Digoin, JJ. & Boutelet, E. (2011). Exomars Mission 2016, Orbiter Module Power System Architecture. In Proc. 9th European Space Power Conference, ESA SP-690 (CD-ROM), European Space Agency, Noordwijk, The Netherlands.
5. Croci, L., Zanella, P. & Riva, S. (2014). The Exomars Trace Gas Orbiter Electrical Power Subsystem. In Proc. 10th European Space Power Conference, ESA SP-719 (CD-ROM), European Space Agency, Noordwijk, The Netherlands.

MULTI-OUTPUT POWER CONVERTER, OPERATED FROM A REGULATED INPUT BUS, FOR THE SIREUS RATE SENSOR

Marcos Compadre Torrecilla [(1)(2)], **Alan H Weinberg** [(1)], **Patrick Hutton** [(3)], **Craig Clark** [(1)]

[(1)] *Clyde Space Ltd, 5B Skypark 5, 45 Finnieston Street, G3 8JU, Glasgow, United Kingdom, Email: marcos.compadre@clyde.space, alan.weinberg@clyde.space, craig.clark@clyde.space*
[(2)] *School of Engineering, University of Glasgow, Rankine Building, Oakfield Avenue, G12 8LT, United Kingdom Email: m.compadre.1@research.gla.ac.uk*

[(3)] *Leonardo Airborne & Space Systems (Selex ES Ltd). Crewe Toll, Edinburgh, EH5 2XS, United Kingdom Email: patrick.hutton@leonardocompany.com*

ABSTRACT

This paper describes a DC to DC converter designed to meet the power supply requirements of the SiREUS Coarse Rate Sensor (CRS) which is a 3-axis MEMS Rate Sensor (MRS) that uses a resonating ring gyro and will be used in different ESA missions. The converter supplies +5V, -5V, 3.3V, 1.8V and 40V and it has been designed and prototyped by Clyde Space Ltd with the EQM and FM units being manufactured by Selex ES. The first model was designed for a 28V un-regulated bus and the second model presented here has been designed for a 50V regulated bus. PWM voltage regulation was not used because of the noise requirements and the regulated input bus allowed an unregulated power stage approach. There are also stringent volume and interface constraints, which also affected the design. For such reasons, a fixed duty-cycle, quasi-resonant single-ended topology with output linear regulators has been implemented; having the advantages of providing low switching losses, low radiated and conducted noise and no over-voltage failure mode. This paper highlights the techniques used to satisfy stringent noise and protection requirements of the load.

1. INTRODUCTION

The CRS is a 3.6W (maximum) 3-axis rate sensor that is used continuously from just after launch to de-spin the spacecraft, until full ADCS stabilization is achieved. Thereafter it is used in standby (active or redundant) for fault detection isolation and recovery. The operational life is the same as the spacecraft mission and the radiation requirements are commensurate with that. On the ESA Sentinel-3 the gyros will be used for identifying satellite motion and also to place it into a pre-set altitude in association with optical sensors after its separation from the launcher, for Sun and Earth acquisition.

Many types of gyros have traditionally been two unit solutions with different (mechanical, thermal and electrical) interfacing characteristics for each unit which resulted in long interconnects and EMC/EMI issues. The CRS presents a more compact single unit, which is more attractive in traditional satellite applications as well as within the increasing range of small-micro satellites [1]. The PSU (Power Supply Unit) for the CRS must therefore follow the low mass and low volume requirements imposed by the design of the sensor. The PSU shall fit in a 75x75mm PCB.
Inside the MRS, a drive signal is applied to the detector drive plates at a nominal 14 kHz, which excites the silicon ring into oscillation [2]. Because the MRS is very sensitive to 14 kHz signals, a design requirement to minimize noise at that frequency, and also at the third harmonic, has been imposed. Filtering such low frequencies requires relatively large input filter components which has a major impact on the physical implementation of the PSU. The topology selected uses quasi-resonant switching to reduce the amount of high frequency noise generated in the system.
The use of Rad-Hard linear regulator integrated circuits has been considered in order to provide a compact solution to meet the regulation requirements of three of the five converter outputs within the volume constraints. The proposed IC is the ITAR free RHFL4913 from STMicroelectronics. This, however, for the 1.8V output, presented a problem in a previous implementation of the PSU for the CRS. The load for the 1.8V output is 800mW and the minimum input voltage of the RHFL4913 is 3V. This produced a 530mW hotspot on the PCB, which in turn, produced thermal issues in the CRS unit. A low-dropout linear regulator has been designed using discrete components to reduce the power dissipation of this output regulator. A very fast current protection has been implemented on this 1.8V series regulator to protect the load, which is an FPGA, from Single Event Effects (SEE).

2. ARCHITECTURE

Because of the limited volume allocated, the only PWM converter at first considered possible to fit the size of PCB was the simple Flyback converter because the inductor and transformer action required for PWM operation are all self-contained in a single Flyback magnetic component. But this was discarded for the

 DOI: 10.1051/e3sconf/20171614001

following reasons:

- The Flyback produces relatively high radiated and conducted noise due to its high peak switching current and voltages.
- PWM operation also produces variation in the magnitude of the harmonics and therefore of the magnitude of the noise.
- In the event of an SEE that keeps the Power MOSFET on for several cycles, the stored energy in the Flyback inductor will be high. When at the end of the SEE event the MOSFET turns -off, this energy is difficult to control in order to avoid an output-over voltage. Detection of an over-voltage fault is also difficult because during the time the MOSFET is on it is not possible for the over-voltage protection at the output to detect the fault. It can only detect over-voltage when the output begins to rise but at that instant the energy in the inductor has been already released. The only solution is to use a high power crow-bar across one or more of the output voltages. This solution is not favoured because it can lead to more complex failure modes.

The architecture selected is shown in Fig. 1. There is a common-mode filter followed by a damped PI filter. These two filters have been designed to provide 30dB of attenuation of conducted noise at the 14 kHz critical frequency. An LCL is used immediately after the filter to protect the converter against over-current due to an SEE produced in the converter circuits. This allows the converter to recover without damage from an SEE. Because the load of the LCL is inductive (i.e. the primary winding of the main transformer), feedback compensation of the LCL was required [3], [4], [5].

The requirement for the outputs to be isolated is satisfied by the single-ended transformer based topology. Each of the converter outputs has linear regulators to provide fine regulation and load current limitation.

2.1 Single-Ended Transformer Based Topology

Since the converter operates from satellite main bus, which is regulated from 46V to 51V, PWM regulation is not required. In addition, a single- ended topology using a single power MOSFET and voltage transformer has been selected. The MOSFET operates at a fixed pulse-width to give output voltages that follow the regulated input voltage. The fine regulation of the outputs is implemented by linear regulators, which also have the advantage of output noise suppression. This set-up gives a simple low component count solution to satisfy the size constraints of the PSU. Without PWM control, over-voltage failure mode is not possible. For example, if, the pulse–width changes due to a SEE, the output voltages will not change,

This topology does not require an inductor for its power stage. This frees up room for the relatively large ($1mH$) input filter inductor needed to provide the required level of attenuation of 30db at the 14 kHz critical frequency.

There is a start-up circuit that regulates the voltage of the control electronics to 14V thus allowing the PSU to start-up upon power-on. The main converter has an auxiliary winding in the main transformer that provides 15V on the auxiliary supply voltage and "backs-off" the operation of the more inefficient start-up regulator.

An under-voltage detection circuit switches the converter off when the voltage on the main satellite bus goes below 46V, and on again at 47V.

2.2 Quasi-Resonant Conversion

With normal square-wave operation the voltage at the drain of the main switching MOSFET, M_{SW}, will increase to values higher than twice the input voltage as it is explained in the following paragraphs. With a 50V input voltage, the voltage at the end of the "off" time will be in the region of 130 to 150 volts. At turn-on, M_{SW} switches from this high voltage to zero in few nano-seconds. This produces a very high $dv(t)/dt$ which, in turn, causes high frequency radiated noise. This noise is capacitively coupled to the input and outputs causing EMC problems. In order to minimize the amount of radiated noise generated in the power converter, quasi-resonant switching conversion has been implemented. The magnetising inductance resonates with the output

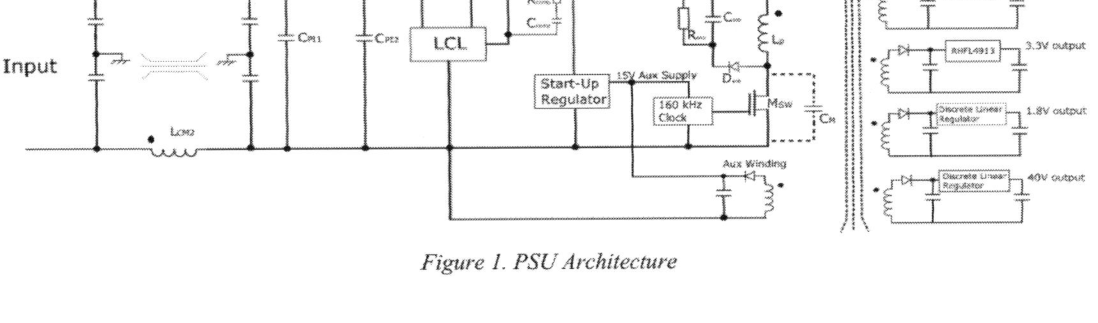

Figure 1. PSU Architecture

DOI: 10.1051/e3sconf/20171614001

capacitance, C_M, of M_{SW}. Assuming a fix value of C_M, the voltage waveform on the drain during the "off" time is dictated by Eq. 1.

$$v(t) = V_{IN} + I_{L_p} \cdot \sqrt{\frac{L_p}{C_M}} \cdot \sin(\omega_o t) - V_{IN} \cdot \cos(\omega_o t) \quad (1)$$

With

$$\omega_o = \frac{1}{\sqrt{L_p \cdot C_M}} \quad (2)$$

It can be seen that, with a value of L_p = *560uH* and C_M = *1nF*, the previous equation is dominated by the sine term. The output capacitance C_M is not a fixed value or even a linear capacitance, but the error produced by assuming a fixed value capacitance is negligible and only noticeable at the beginning of the "off" time, when the drain voltage is low and therefore C_M peaks in capacitance. Fig. 2 shows the *Spice* simulation of the quasi-resonant waveform. In this figure it can be seen that the drain voltage when M_{SW} turns on has drop to less than 30V. Fig. 3 shows the case where the magnetising inductance L_p has not been tuned to resonate with C_M at the switching frequency (L_p = *2mH*) and therefore the drain voltage at turn-on is around 150V, producing a very high *dv(t)/dt* and therefore high radiated noise.

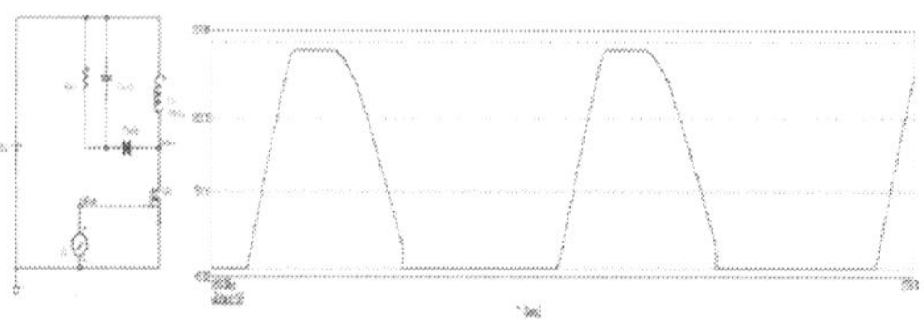

Figure 2. Quasi-Resonant waveform at the drain of the switching MOSFET

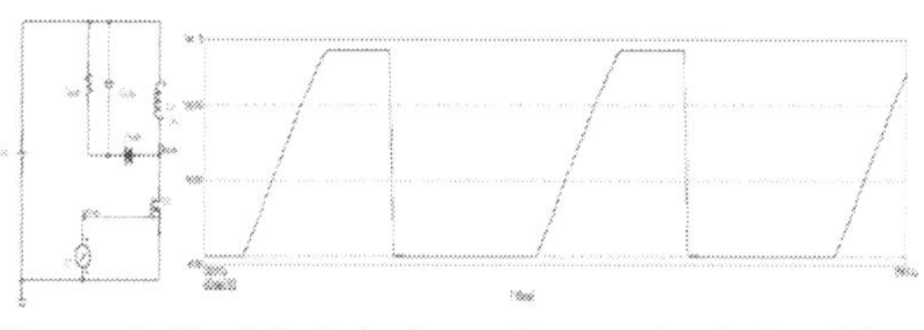

Figure 3. Hard-Switched waveform at the drain of the switching MOSFET (without quasi-resonance)

The main drawback of this topology is the high voltage reached at the drain of the switching MOSFET during the "off" time. This voltage occurs due to the volt/second reset of the magnetising inductance L_p when M_{SW} turns off. Ideally, and for a 50% duty-cycle clock signal, this voltage becomes twice the input voltage. In reality, due to switching delays of M_{SW}, the voltage reached at its drain is higher than twice the input voltage reaching values as high as three times the input voltage. This effect is further exacerbated due to the slow transitions introduced by the quasi-resonant switching. Because of this, an R-C-D snubber has been introduced to keep the drain voltage below its de-rated maximum value. This snubber circuit can be seen in Fig. 2 and Fig. 3.

2.3 Output Series Regulators

As with most multi output converters the outputs are series regulated as follows:

- The +5V, -5V and 3.3V outputs are regulated using an RHFL4913 integrated circuit from STMicroelectronics. It is a Rad-Hard linear regulator IC.
- The 40V series regulator was designed using very basic discrete components.
- For the 1.8V regulator a novel regulation technique was used to achieve very fast feedback current control to limit the over-voltage spike when an SET (Single Event Transient) occurs.

2.4 The 1.8V Output Linear Regulator

The problem with conventional linear regulators, like the RHFL4913, is that if a SET occurs that switches on the main power transistor, even for a short period of time of 1µsec, then the result is a voltage spike that can destroy a sensitive load. One such load is the FPGA being supplied by the 1.8V regulator. A SET may have a duration of 1µsec but can have this period expanded due to capacitive storage or saturation. Since the linear regulator is an analogue device this expansion is most significant when the SET occurs in the voltage feedback loop. We have therefore performed a *Spice* simulation on the discrete 1.8V linear regulation to indicate what improvement can be made for a SET. The SET has been simulated by two pulse generators, one of 1µsec period and another one of 10µsec period. The 1µsec pulse is injected directly on the gate-source of the power MOSFET. The period of 1µsec is used because there is virtually no possibility of pulse expansion at this point. The 10µsec pulse is injected at the output of the operational amplifier that controls the voltage loop.

Fig.4 shows the resulting waveforms for both the above cases and, for comparison, with and without the added second current loop. It is clear that the new design has a major impact on the ability to withstand SET's. SET tests were performed on the breadboard of the converter to confirm the results of the *Spice* simulation. These results are shown in Section 3.

2.4.1 Noise Attenuation

Fig. 5 shows the simulated attenuation achieved. This is around 55dB at 168 kHz which is the frequency of interest because of the ripple voltage coming from the

DOI: 10.1051/e3sconf/20171614001

converter supply. In Fig. 6, the input and output waveforms were simulated with a 200mV peak-peak ripple at the input. The result shows that the resulting ripple voltage at the output level is about 325μV. This confirms the 55dB or rejection.

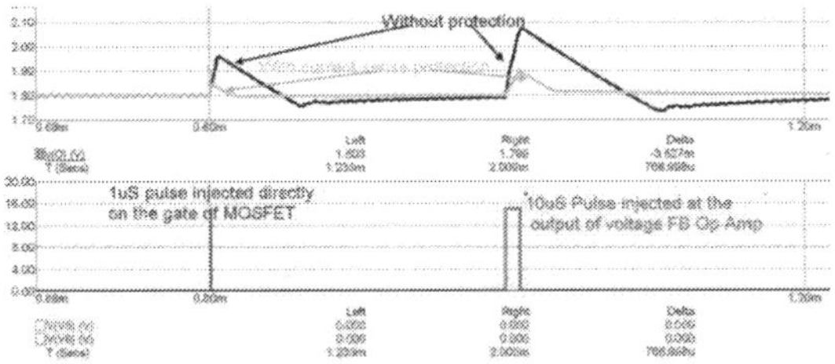

Figure 4. Response of the 1.8V linear regulator to a simulated SET

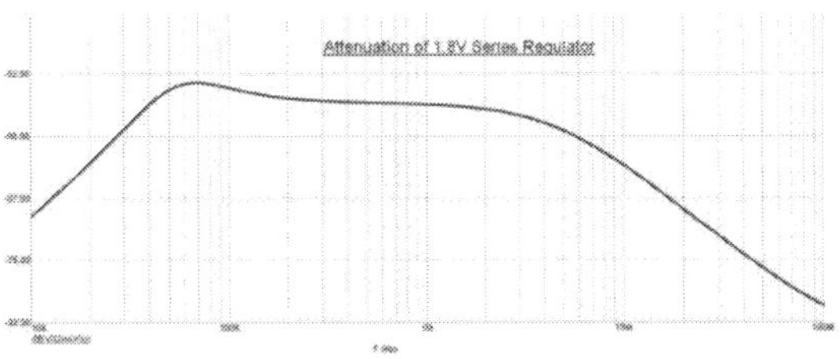

Figure 5. Voltage ripple attenuation of the regulator

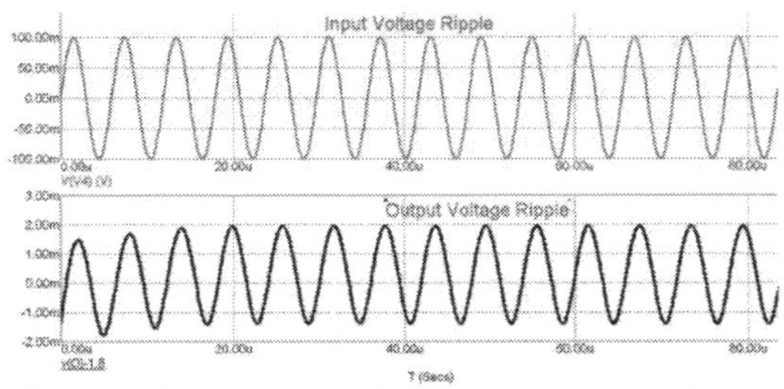

Figure 6. Comparison of the input and output voltages of the 1.8V linear regulator

2.4.2 Headroom

The headroom was tested on the *Spice* circuit by variation of the input supply voltage and monitoring the output voltage. Fig. 7 shows this response, where it can be seen that the headroom required is about 80mV.

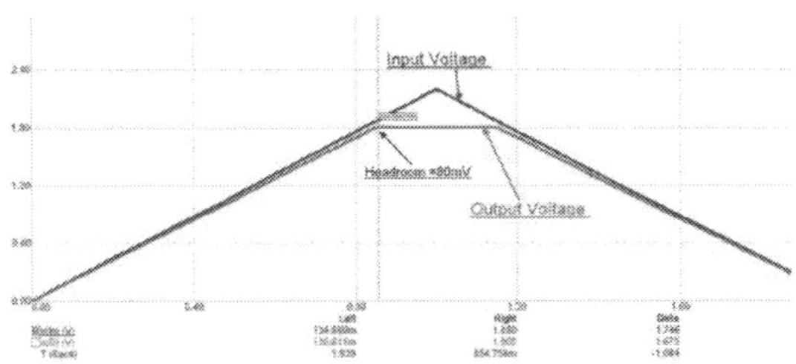

Figure 7. Headroom required for the 1.8V regulator

2.4.3 Load Response

The simulation of load change is shown in Fig. 8. The output voltage response shows a well damped proportional-plus-integral feedback with a voltage overshoot of about 30mV.

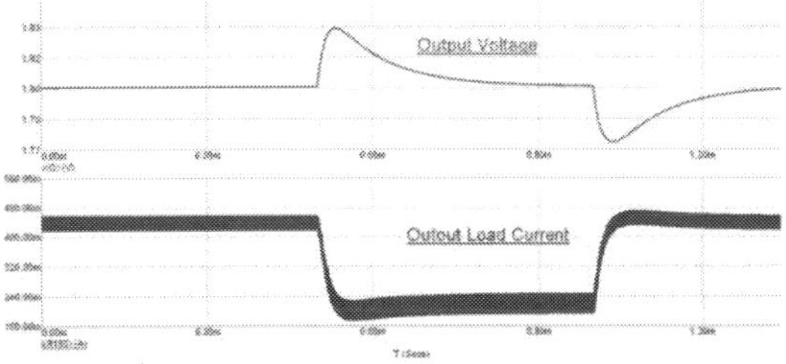

Figure 8. Step load response on the 1.8V output with a 50% load change

2.4.4 Feedback Stability

The voltage feedback open-loop response is shown in Fig. 9. This results in a phase margin of 75° at 119 kHz. Also the transient response with load confirms this stable response.

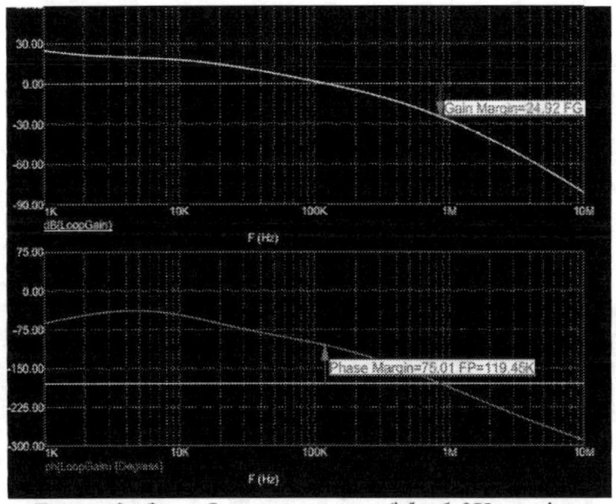

Figure 9. Open-Loop response of the 1.8V regulator

2.5 Input Filter

The input filter was designed to give a 30dB attenuation of conducted ripple current. To satisfy the conducted noise requirements a PI filter was required. This design has been simulated with two values of input harness inductance ($L_{Harness}$), 1μH and 5μH, and is shown in Fig. 10. The resonant peak of the curves is due to the resonance of the harness inductance with the first 500nF input capacitance (C_{PI1}, see Fig. 1) as per Eq. 3.

$$\omega_{peak} = \frac{1}{\sqrt{L_{Harness} \cdot C_{PI1}}} \tag{3}$$

DOI: 10.1051/e3sconf/20171614001

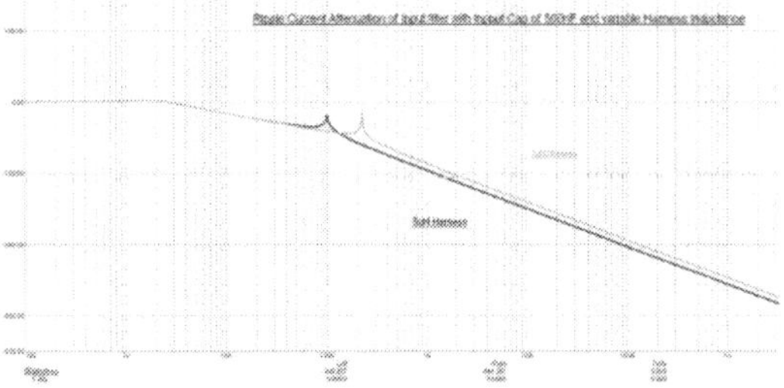

Figure 10. Input filter response

3. RESULTS

The prototype was tested by Clyde Space Ltd. Fig. 11 shows the quasi-resonant waveform on the drain of the main switching MOSFET. In this figure it can be seen the sloping edges of the quasi-resonant waveform and the voltage clamping action provided by the R-C-D snubber shown in Fig. 2 and Fig. 3. It can be appreciated that the drain voltage drops to only 10V before the MOSFET turns-on. The voltage waveform shown in Fig. 11 resembles the waveform produced by the *Spice* simulation shown in Fig. 2. Fig. 2 shows the case where the MOSFET is modelled with a fixed value output capacitance whereas Fig. 11 shows the voltage signature with the actual EM hardware. Any differences are thought to be due to the non-linear capacitance of the switching MOSFET.

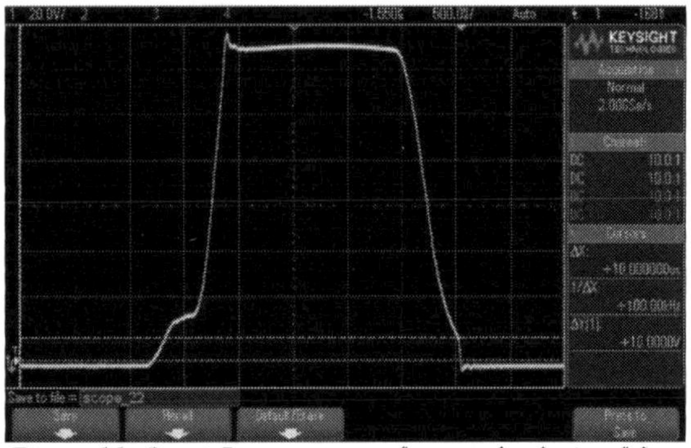

Figure 11. Quasi-Resonant waveform at the drain of the switching MOSFET

The converter passed all the tests for Conducted Emissions and Conducted Susceptibility. These tests were conducted by Selex ES.

The SET test for the OpAmp of the 1.8V regulator was tested injecting two different pulses to the gate of the MOSFET of the linear regulator. The pulses have an amplitude of 10V and a duration of 1μsec and 10μsec, in accordance with the simulations performed and shown in Fig. 4. Fig. 12 and Fig. 13 show the responses of the 1.8V linear regulator to the 1μsec and 10μsec pulses, respectively.

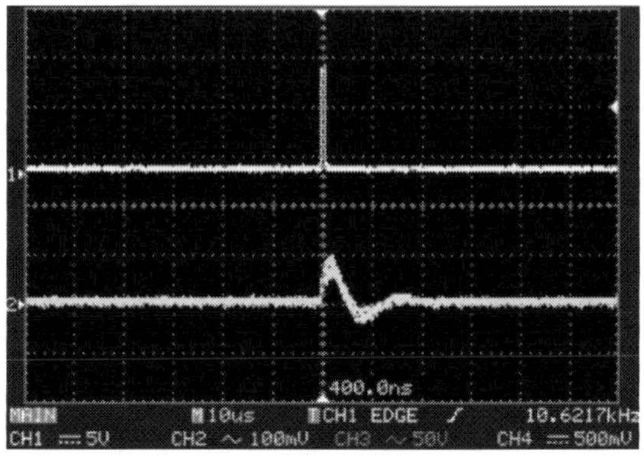

Figure 12. Response of the 1.8V linear regulator to a simulated 1μsec SET

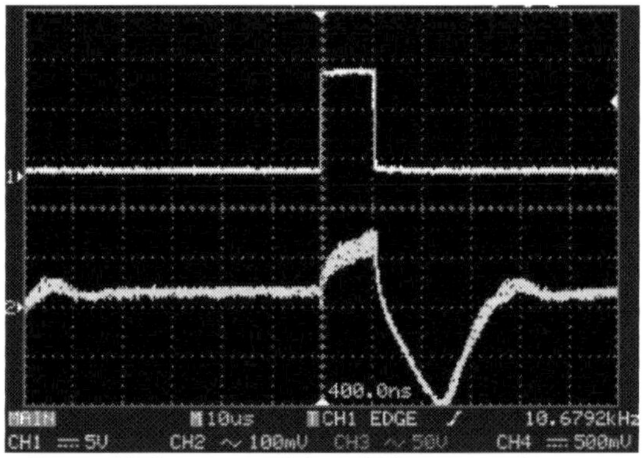

Figure 13. Response of the 1.8V linear regulator to a simulated 10μsec SET

Noise transparency from the input of the PSU to the rectified outputs, at the input of the linear regulators, has been measured in order to check the noise attenuation provided by the converter's input stage. A sinusoidal signal with an amplitude of 1V peak-to-peak and a frequency equal to, approximately, the critical frequency (14.7 kHz) has been injected in series with the converter DC input voltage and the AC coupled noise has been measured at the input of each one of the output linear regulators. Fig. 14 shows the measurements taken at the input of the 1.8V linear regulator, which is the most critical one. It can be observed that the noise transmitted to the input of the linear regulator is 40mV, approximately, for an input AC signal of 1V peak-to-peak.

The load step response test was also performed in all the outputs. The outputs were varied by steps between 100% and 50% of the nominal load currents. Fig. 15 shows the performance of the 1.8V output when the load current changes between 440mA and 220mA. Very small overshoots of less than 50mV have been measured.

The outputs provided by the RHFL4913 linear regulators have step load responses with overshoots of less than

 DOI: 10.1051/e3sconf/20171614001

100mV. The 40V output has overshoots of 550mV.

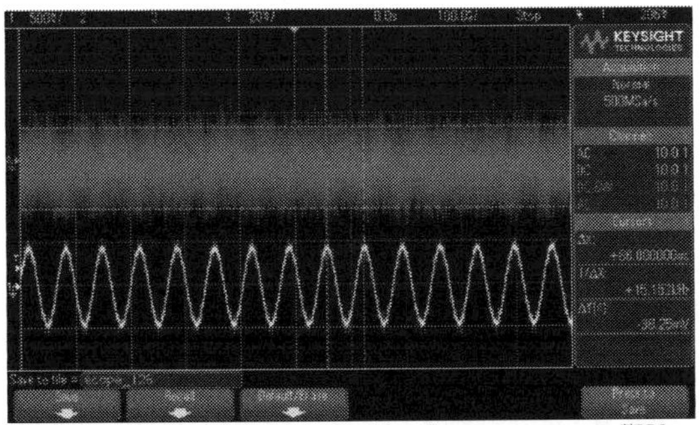

Figure 14. Noise transmitted by the input stage. CH1: Input injected at the input of the converter. CH2: AC signal measured at the input of the 1.8V linear regulator

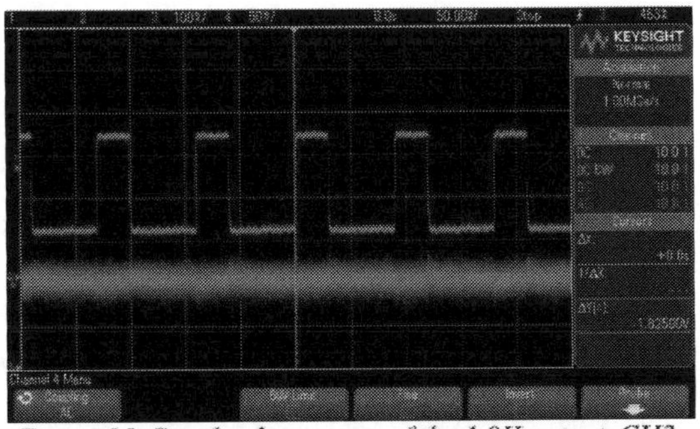

Figure 15. Step load response of the 1.8V output. CH3: Current steps between 100% and 50% nominal output current. CH4: AC coupled voltage at the output of the regulator

Fig. 16 shows the Engineering Model of the PSU assembled inside the SiREUS sensor avionics box. The final solution comprises a mother board and a daughter board with the filter components and the LCL on just one side. The need for a daughter board comes from the small PCB surface area available for the design.

Figure 16. EM of the PSU assembled in the SiREUS CRS avionics box

4. CONCLUSIONS

This paper presents a power converter that has been custom-designed for very stringent low conducted and radiated noise requirements with the restraint of a small PCB volume. The operation from a regulated bus input voltage without PWM control allows a quasi–resonant power stage to be used that has considerably reduced radiated noise. The topology used has proved to be the right choice since the EM, when integrated with its CRS load passed all its noise tests. Also, without PWM control the converter has less failure modes and therefore requires less protection circuits, than a more conventional PWM converter.

The 1.8V supply for the FPGA load has overcome the major disadvantages of the RHFL4913 IC with regard to headroom and SET sensitivity.

The main disadvantage of using an unregulated power stage is low efficiency but, because of the low power required, this is not a concern.

5. REFERENCES

1. D. Durrant, S. Dussy, B. Shackleton, A. Malvern. MEMS Rate Sensors in Space Becomes a Reality. AAIA Guidance, Navigation and Control Conference and Exhibit. 20-23 August 2007, Hilton Head, South Carolina.

2. D. Durrant, H. Crowle, J. Robertson, S. Dussy. SIREUS – Status of the European MEMS Rate Sensor. AIAA Guidance, Navigation and Control Conference and Exhibit. 19-21 August 2008, Honolulu, Hawaii.

3. A. Soto, L. Jimenez, E. Lapeña, C. Delepaut. Stability Analysis for the LCL of GEO-PCDU Product. ESPC 2014, Noordwijk, The Netherlands.

4. C. Delepaut, T. Kuremyr, M. Martin, F. Tonicello. LCL Current Control Loop Stability Design. ESPC 2014, Noordwijk, The Netherlands.

5. F. Tonicello, C. Delepaut, M. Martin Alfonso, T. Perez Soriano, T. Kuremyr, M. Triggianese. Approach to Design for Stability a System Comprising a Non-Ideal Current Source and a Generic Load. ESPC 2014, Noordwijk, The Netherlands.

DOI: 10.1051/e3sconf/20171614002

A NEW EUROPEAN HIGH FIDELITY SOLAR ARRAY SIMULATOR FOR NEAR EARTH AND DEEP SPACE APPLICATIONS

Hjalti Pall Thorvardarson [(1)], Flemming Gøttsche [(2)], Ferdinando Tonicello [(3)]

[(1)] Rovsing, Dyregaardsvej 2 DK-2740 Skovlunde Denmark, Email:hpt@rovsing.dk
[(2)] Rovsing, Dyregaardsvej 2 DK-2740 Skovlunde Denmark, Email:fgo@rovsing.dk
[(3)] ESA- ESTEC, Keplerlaan 1, 2200AG Noordwijk, The Netherlands, Email: Ferdinando.tonicello@esa.int

ABSTRACT

Following an intensive design, development, and testing effort of almost 3 years, Rovsing with ESA assistance succeeded in the development of a new European high fidelity Solar Array Simulator (SAS) for near Earth and deep space applications. ESA now has a versatile, highly modular and efficient SAS at its disposition that serves at simulating modern high power solar arrays for Earth observation, science or telecom satellites as well as for future deep space missions.

The special features compared to other SAS are:

- Ultra-fast dynamic response performance, the key to high fidelity to SA simulation within the space domain.
- I-V curves can be simulated with up to 4096 point resolution.
- Store 600 I/V curves in memory and create full life cycle simulations.
- Up to 685W of Power from a single module (137V / 5A & 68.5V / 10A).
- Class leading energy efficiency from a hybrid design, translating into less heat dissipation and therefore considerable less noise from the SAS rack as a whole.
- The smallest form factor available, 2740W from a single 3U high unit, with no need for additional forced cooling devices or empty space sections between stacked sub-racks.
- Programmable first level protection (OVP, OCP).

Since the beginning of 2016 ESA is able to prepare its demanding missions with a flexible and leading edge SAS solution. The SAS is already foreseen to support missions like JUICE, MetOp-SG, ExoMars and the MPCV European Service Module.
The paper describes the development logic, challenges and lessons learned as well as the successfully concluded qualification testing phase of the new SAS and gives an outlook on future applications and features that are soon to be added.

1 ROVSING PROFILE AND ORGANIZATION

Rovsing was established in 1992 and has been listed on the OMX Nordic Stock exchange since December 2006. Rovsing is located in Skovlunde on the outskirts of Copenhagen, Denmark. Website: www.rovsing.dk.

Rovsing is recognised for the delivery of ground testing equipment and systems, software solutions and consultancy services to the European Space Agency (ESA), European space prime contractors and U.S. defence primes and has achievements within the following business areas:

- Test systems for electrical and functional testing of complex and hybrid electronic and software systems, in particular for satellite subsystems, several of them being best in class.
- Critical and non-critical software systems.
- Independent software verification and validation (ISVV) services.
- Software process assessment and improvement services.
- Engineering and logistics support.

1.1 History of the Development

Initial work started within Rovsing in 2009 with the idea about creating a European Solar Array Simulator based on in-house differentiating topology. Rovsing approached ESA and Danish High Technology Fond for seeking backing in the development.

Rovsing and ESA kicked off phase 1 of the SAS development in December 2009. Phase 1 covered the definition of the requirements, proof of concept, and was concluded in October 2011:

- Task 1: SAS requirements and specification
- Task 2: SAS preliminary design
- Task 3: SAS power module board – breadboard implementation
- Task 4: SAS power module board – breadboard testing
- Task 5: Commercial evaluation

Phase 2 covered the maturing of the requirements and the design into a SAS product:

- Task 1: Consolidation of system requirements and design
- Task 2: Prototyping
- Task 3: Prototypes Control Implementation and Integration

- Task 4: SAS prototypes integrated test
- Task 5: manufacturing and individual test of 10 SAS series-0 modules
- Task 6: Final System Test, Validation and Documentation

2 WHAT IS A SOLAR ARRAY SIMULATOR

2.1 Solar Cell Characteristics

A SAS must emulate the electrical characteristics of an array of solar cells as precisely as possible. A common multi junction solar cell model is shown in Fig 1.

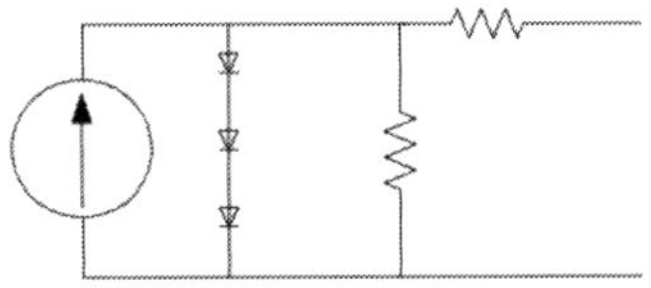

Figure 1: Triple junction solar cell equivalent.

On spacecrafts, solar cells connected in series form a string; multiple strings connected in parallel are denominated a "section", see Figure 2.

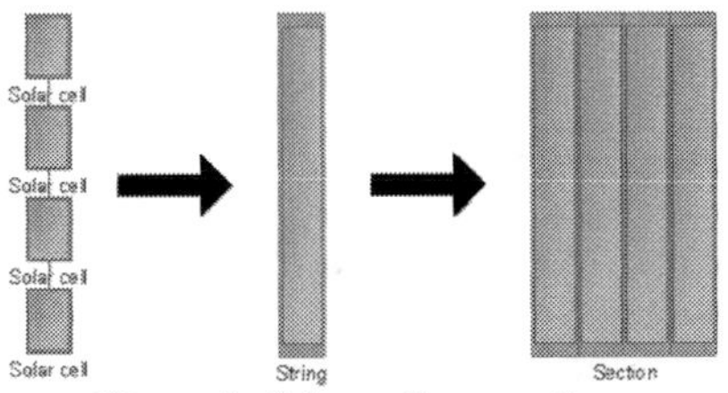

Figure 2: Solar cell connections

One SAS Module output must emulate such a section. A spacecraft solar array (SA) consists of multiple sections; hence multiple SAS Modules are required to emulate the array.

The power from a solar array is expressed as follows:

$$P = I*V$$

The I/V curve of a given SA section is a function of the:

- Illumination
- Temperature
- SA technology used

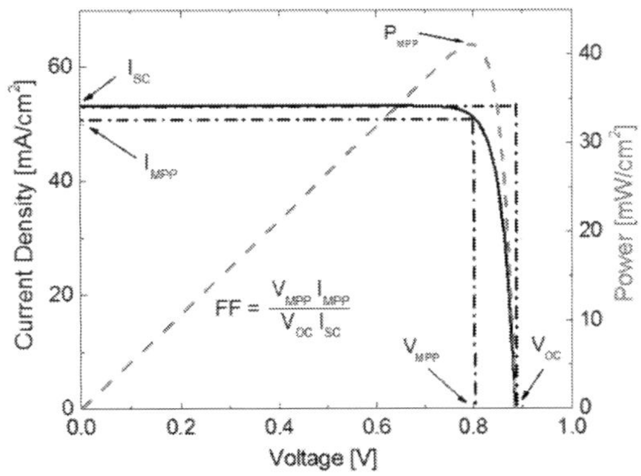

Figure 3: Solar cell characteristic (single junction)

An I/V curve is characterized by the following parameters (see ***Figure 3***): Open circuit voltage (Voc), Short circuit current (Isc), Voltage level at the maximum power point (Vmp), Current level at the maximum power point (Imp), Maximum power point (Pmpp), Fill Factor (FF)

These I/V characteristics are what the SAS must emulate. Hence, the SAS Module must always control the current to achieve an operation point on the curve. Dynamic behaviour of the solar cell section must furthermore also be emulated as close as possible.

[Britton] presents an equation that gives a good representation of the relation between voltage and current based on the above parameters of the string:

$$V = \frac{\dfrac{V_{OC}\ln\left(2-\left(\dfrac{I}{I_{SC}}\right)^{N}\right)}{\ln(2)} - R_S(I-I_{SC})}{1+\dfrac{R_S I_{SC}}{V_{OC}}} \qquad (1)$$

where

$$N = \frac{\ln(2-2^{a})}{\ln\left(\dfrac{I_{MP}}{I_{SC}}\right)} \qquad (2)$$

$$a = \frac{V_{MP}\left(1+\dfrac{R_S I_{SC}}{V_{OC}}\right)+R_S(I_{MP}-I_{SC})}{V_{OC}} \qquad (3)$$

$$R_S = \frac{V_{OC}-V_{MP}}{I_{MP}} \qquad (4)$$

For now, the [Britton] equations are considered to be the most accurate way to simulate the behaviour of triple junction GaAs cells and are used in the remaining part of this document to supply I/V behaviour.

2.2 Dynamic Performance and Response Time

An overall goal for a SAS is to simulate the

performance of real solar cells as close as possible, both statically (voltage and current levels) and also the dynamic behaviour with respect to output capacitance, response time, etc. Therefore the response time of the SAS Module output has been identified as one of the most important design drivers.

It can be useful to look at where the SAS Module is operating on an I/V curve when specifying the dynamic behaviour. Figure 4 defines three areas of the I/V curve; different types of power conditioning units normally work within only one of the three areas when steady state is reached. Systems based on shunt regulators can behave in a way that covers "Case 2" and "Case 3" when the operating point of one section jumps from Isc to MPP.

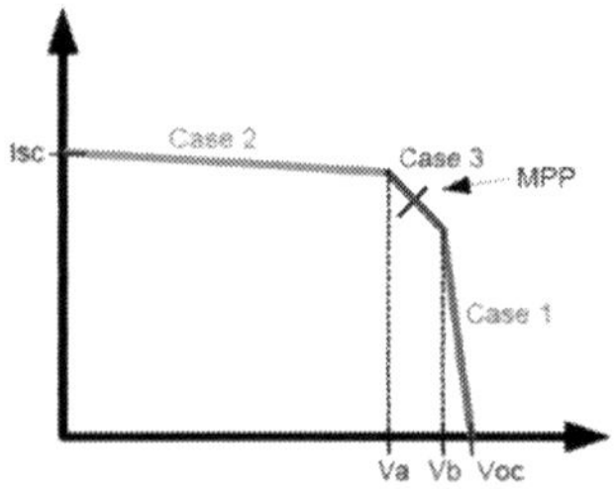

Figure 4: Working areas of the I/V curve.

Examples of regulators working in the 3 areas are:

- Series switching devices or DC-DC converters operating in the area of "Case 1".
- Shunt regulators in the area of "Case 2". Also in "Case 3" when operating point of a section jumps from Isc to MPP.
- Maximum power point trackers in the area of "Case 3".

It is in this perspective the requirements of recovery from short- and open-circuit and requirements of response time below 10μs strain the designs of both the overall digital control and the power stage of the SAS Module. Research and detailed design has been the major results of the proof-of-concept activity.

3 RO-5100 SAS MODULE OVERVIEW

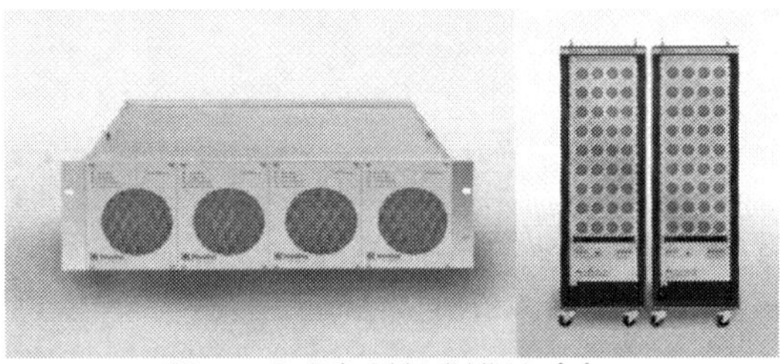

Figure 5: RO-5100 SAS Modules

The Rovsing RO-5000 SAS product range provides the following primary advantages to space customers:

- Ultra-fast dynamic response performance, the key to high fidelity to SA simulation within the space domain.
- Up to 685W of Power from a single module (137V / 5A & 68.5V / 10A).
- Class leading energy efficiency from a smart linear + switch-mode power hybrid design, translating into less heat dissipation and therefore considerable less noise from the SAS rack as a whole.
- Excellent static and dynamic performance
  - Low ripple and noise
  - Fast load switch recovery time
- Up to 600 I/V tables with each 4096 point resolution
  - Voc, Isc, Imp and Vmp
- Highly configurable Curve switching
  - 100ms to 9 hours
- High accuracy for measurement, programming and read-back
- Programmable First Level Protection
  - Over Voltage Protection (OVP), Over Current Protection (OCP) and Over Temperature Protection (OTP)
- Excellent thermal management
  - Low heat dissipation
  - Low noise
- High MTBF
- FMECA - Functional Level
- Control and monitoring through Ethernet
  - SCPI interface
- The smallest form factor available, 2740W from a single 3U high unit, with no need for additional forced cooling devices or empty space sections between stacked sub-racks.

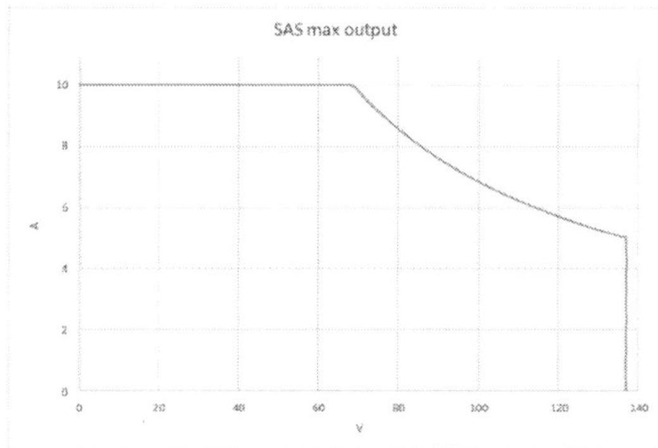

Figure 6: RO-5100 SAS Module Power Profile

The standard SAS Module (RO-5100) provides output voltage up to 137V covering the full range of spacecraft needs, with custom RO-510x variants available on request for higher voltage levels. The RO-5100 provides high programming accuracy; curves can be simulated with up to 4096 point resolution with up to 600 I/V

 DOI: 10.1051/e3sconf/20171614002

curves stored in memory, with switching between curves programmed from 100ms up to approximately 9 hours. The unit provides excellent ripple and noise performance, load switch recovery time of less than 10us and outstanding measurement accuracy. A high fidelity simulation of the actual solar array is provided both statically (voltage and current levels) and also in terms of dynamic behavior with respect to output capacitance and response time. The RO-5100 provides excellent static and dynamic performance across the principle regulation types (series switching, shunt regulation and MPPT), providing the user with the high fidelity required to accurately simulate the behavior of the actual solar throughout its lifecycle.

Each RO-5100 SAS Module is fully self-contained with regards to inputs, outputs and its chassis thus easing configurability and management as units can be slid in and out of the rack without the need for adjusting the rest of the system.

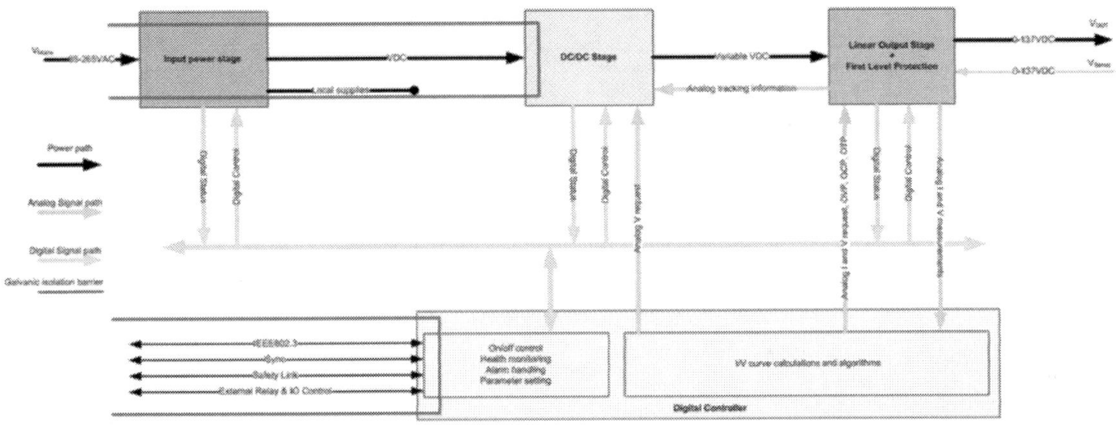

Figure 7: SAS Block Diagram

The unit is designed for high density rack integration achieved via its small size and low power dissipation. Due to this design, the unit possesses no input buttons or keypads at the front, only basic LED indicators. The saved space is utilized by the cooling fan partly responsible for achieving the high density of a 3U high sub rack, housing up to 4x SAS Module. Since the unit does not present a keypad input or an LCD output to the end user, all of its commanding and monitoring is performed remotely through the SCPI interface via standard Ethernet. Hence all SAS Modules are delivered with the standalone "SCPI Client", a simple lightweight JAVA based tool with the following features:

- Connect and configure
- Commanding
- Monitoring
- Full API overview
- Simple scripting

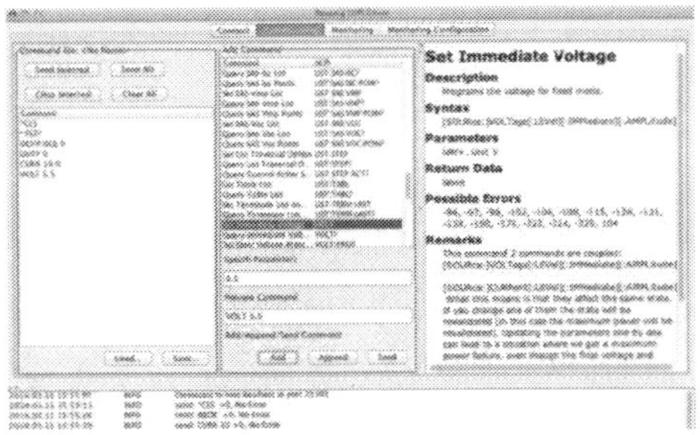

3.1 SAS System Overview

A Rovsing Solar Array Simulator System could consist of:

- SAS Modules that supply power to the spacecraft (S/C) simulating the electrical characteristics of the S/C solar array (or sections thereof).
- 19" Rack, SAS Subracks, isolation transformer for mains input, harness, Ethernet switches and other support equipment such as Second Level Protection (SLP) Modules.
- Controller PC hosting the SAS Module/s Client Software and Control Software, where the operator through a GUI can:
  - Use the Scenario Editor to create scenarios simulating events such as course of a day, spinning of the S/C, and eclipse events. During execution of a scenario, the SAS Module/s is continuously re-configured according to a time-line.
  - Manually configure the SAS Module/s to simulate specific solar array configurations (the configuration reflects a given type of solar cell, insolation, and temperature).
  - Receive status from the SAS Module/s.

The figure below shows a possible SAS System and its contents. Depending on number of SAS Modules, there

DOI: 10.1051/e3sconf/20171614002

can be several racks for the individual project.

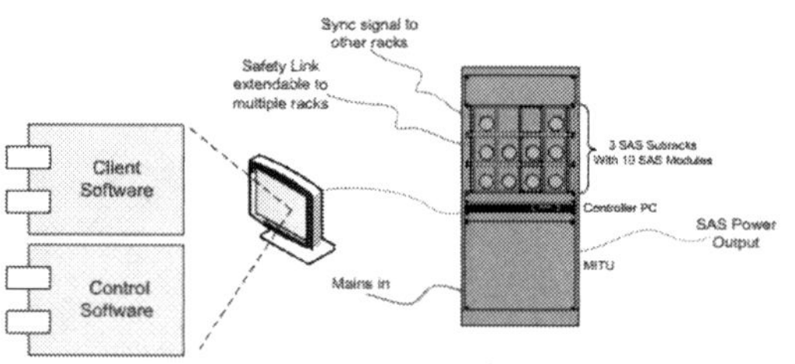

Figure 8: SAS System and its contents.

The following design decisions were made during initial studies and are therefore requirements for the SAS Product Development:

- Each SAS Subrack must hold up to 4 SAS Modules (plugged-in from the front), all cabling being handled on the back. The SAS Modules operate in a stand-alone fashion, i.e. no resources are shared between the SAS Modules. All control and TM communication is handled via Ethernet interfaces.
- The Sync signal shown in *Figure 8* is used to synchronise all SAS Modules connected together when they are programmed to switch between two or more different output characteristics (to simulate the changes in solar cell output characteristics when the illumination or temperature changes, or in case of a defective solar cell section or similar). If multiple racks are present in the SAS, the sync signal must be distributed to these also. The Sync signal master is determined by cable select.
- Safety Link is supported by the SAS Module/s. Safety Link is a concept developed by Rovsing where a safety event can trigger the link causing all connected equipment to power down the output. When the SAS Modules are used in combination with Rovsing's Second Level Protection (or SLP) then a triggering of the Safety Link will instantly galvanically isolate the output towards the S/C. The Safety Link can be extended between multiple SAS Modules and when combined with the Rovsing Measurement Acquisition Simulation Commanding (MASC) it can also be connected to various types of COTS equipment.

3.2 SAS Client Software

The objective of the Scenario Editor is to provide the operator with the ability to create complex scenarios with relative ease, and in the process make sure the user retains a full overview. *Figure 9* shows the major components of Scenario App Software.

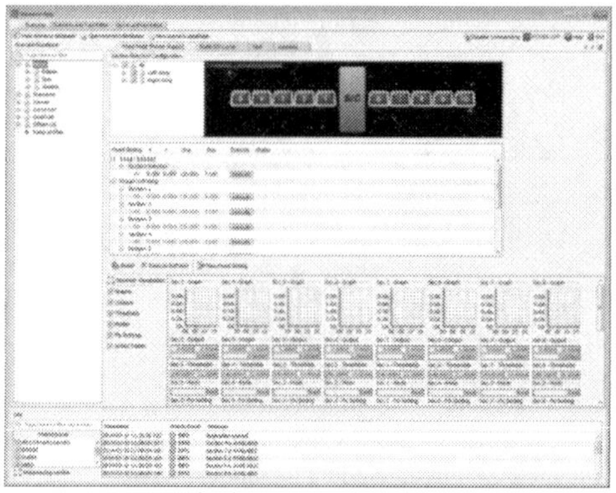

Figure 9: Scenario App Software.

- **Scenario Editor**
  - The Scenario Editor can create, load, modify, rename, save and delete objects in the Scenario Database. Eclipse events and other custom curve progressions can be programmed in the editor.
- **Scenario Executor**
  - The Scenario Executor can send Scenario Databases for preparation in the Scenario Backend, and it can issue execution start, hold, resume and stop commands to the Scenario Backend.
- **Curve Editor**
  - The Curve Editor can be used to create, load, modify, rename, save and delete IV Curve Formulas and IV Curve Tables.
- **Main**
  - The Main Object is the Scenario App framework which ties components of the Scenario App together.
- **Scenario Database**
  - A Scenario Database is a mapping of identifiers to Scenario Database Objects and the Scenario Database Objects. The Scenario Database objects are Scenarios, Tests, Groups, Sections, {I,V} Curve Formulas, {I,V} Curve Formula sequences, {I,V} Curve Tables Sequences, {I,V} Curve Tables, Offset Graphs and SAS Modules.
- **Communication**
  - The Communication Object handles communication between the Scenario App and the Control App.

DOI: 10.1051/e3sconf/20171614002

4 KEY TEST RESULTS

For the key performance parameters and use cases the following sections show the applied test setups and the test results achieved with the RO-5100 SAS Module. In all screen-shots, yellow trace is Voltage and red is Current.

Shunt Switching Testing:

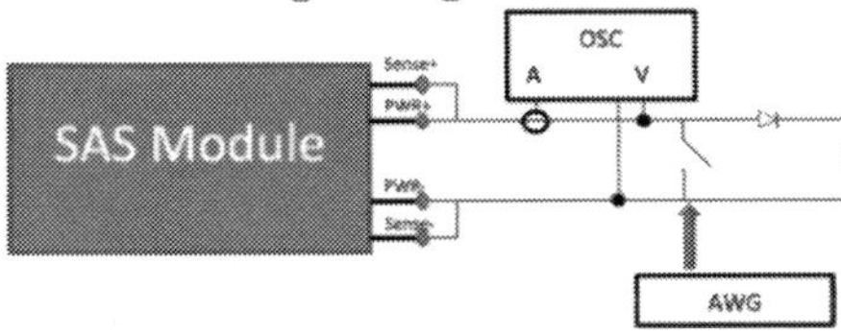

Battery emulator is configured to the desired open circuit battery voltage (by means of power supply) and current sinking capability (by means of fixed resistive load).
S3R emulator includes a heavy duty MOSFET controlled by an arbitrary waveform generator (AWG).

Example of actual performance in shunt switching, configuration shown in title:

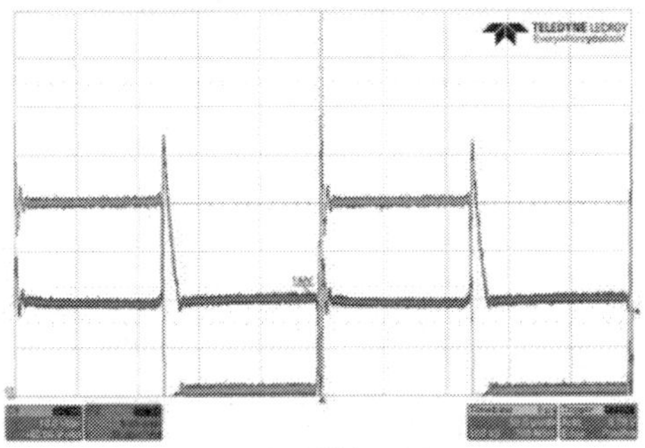

S3R response @ 4kHz 50% duty cycle for I/V curve VOC=65V, ISC=10A, VMP=46V, IMP=9.2A with 40V battery

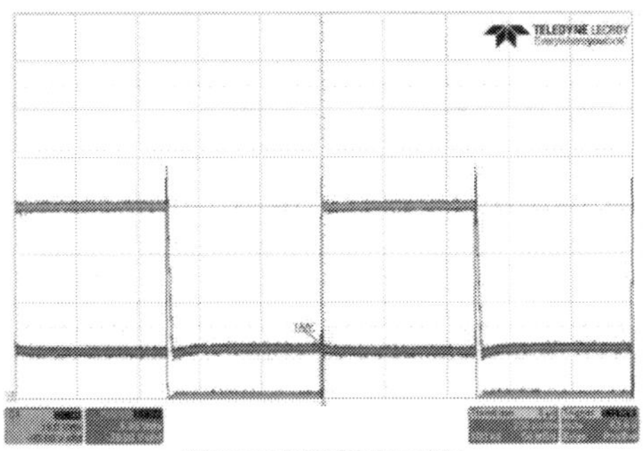

S3R response @ 1kHz 50% duty cycle for I/V curve VOC=130V, ISC=5A, VMP=91V, IMP=4.6Awith 80V battery

Series Switching Testing:

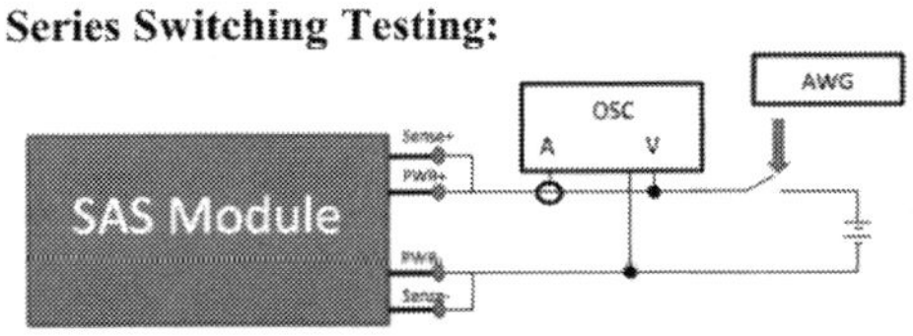

The output is switched by a P-channel MOSFET driven by an OP-Amp with 200V common mode range, connected to the positive output of the SAS Module.

Example of actual performance in series switching, configuration shown in title:

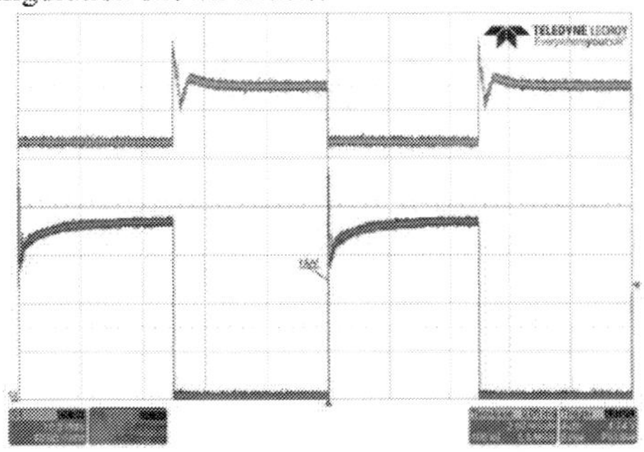

Series switching response @ 100Hz 50% duty cycle for I/V curve VOC=65V, ISC=10A, VMP=46V, IMP=9.2A with 53V battery.

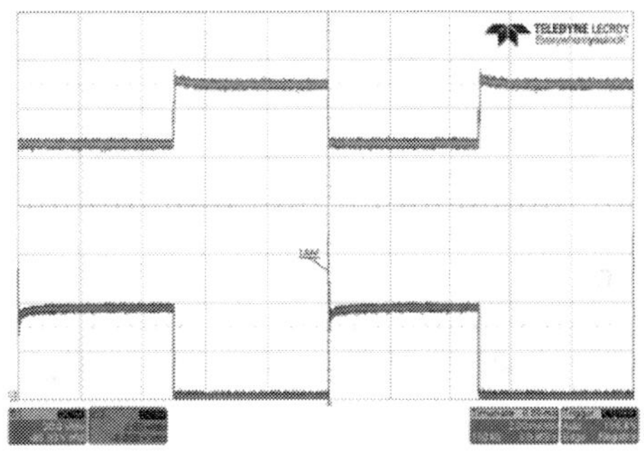

Series switching response @ 100Hz 50% duty cycle for I/V curve VOC=130V, ISC=5A, VMP=91V, IMP=4.6A with 106V battery.

MPPT Testing:

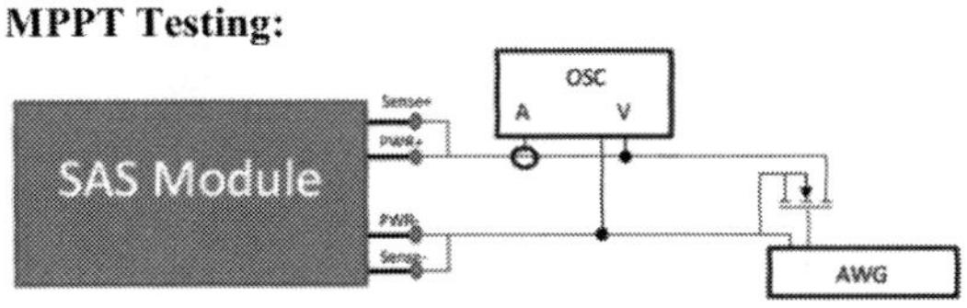

Electronic load is sweeping repeatedly between Vmin and Vmax triangular wave shape 50% duty cycle at noted frequency. (I.e. dV/dt = (Vmax-Vmin)*2*freq).

Example of actual performance during MPPT, configuration shown in title:

 DOI: 10.1051/e3sconf/20171614002

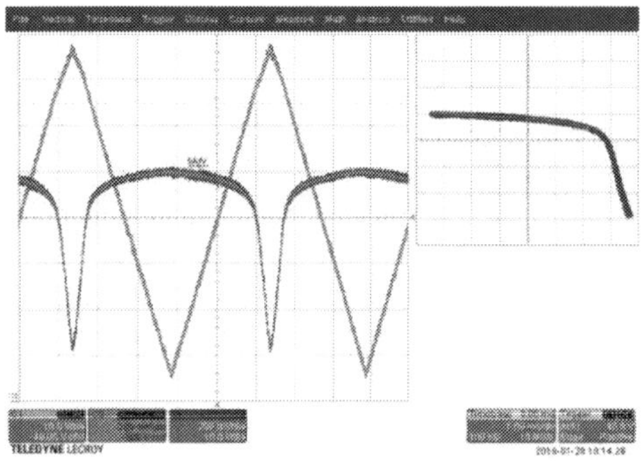

I/V curve 80V 5A 285W FF=71% wide range sweep
Vmin=5V Vmax=79V freq=200 Hz

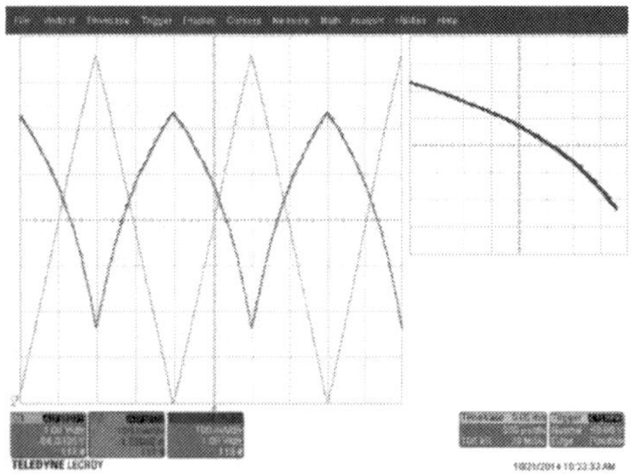

I/V curve 80V 5A 285W FF=71% narrow range sweep
Vmin=62V Vmax=70V freq=500 Hz

Output Noise Testing:
Example of actual noise performance, with output ON and external loading:

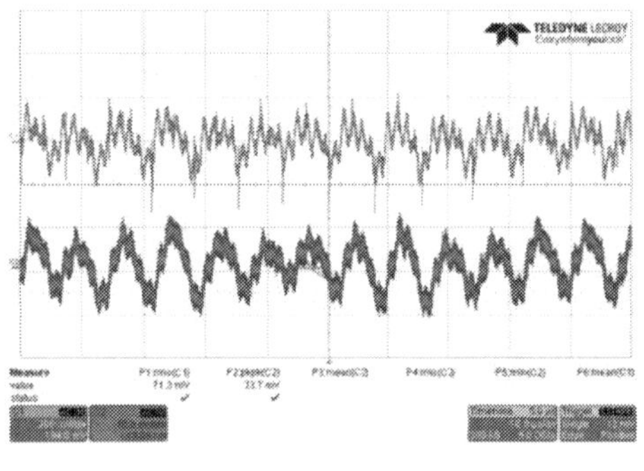

Output Impedance Testing:

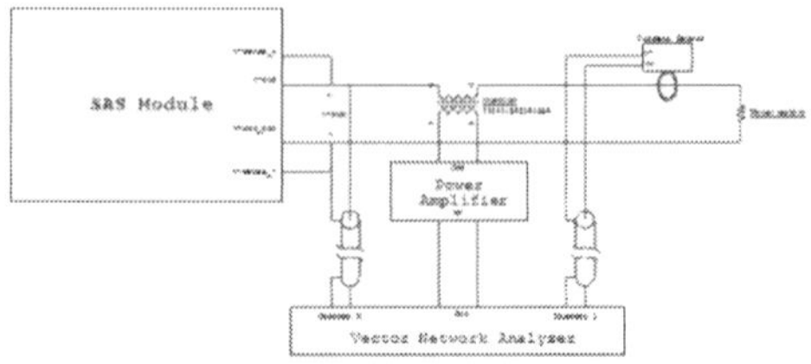

I/V curve 44V, 2A, 75.5W is loaded and different power resistors are applied to vary the SAS Module dc output. The Injection Transformer is connected in series with the power resistor in order to inject a sinusoidal perturbed voltage signal.

Example of actual impedance performance:

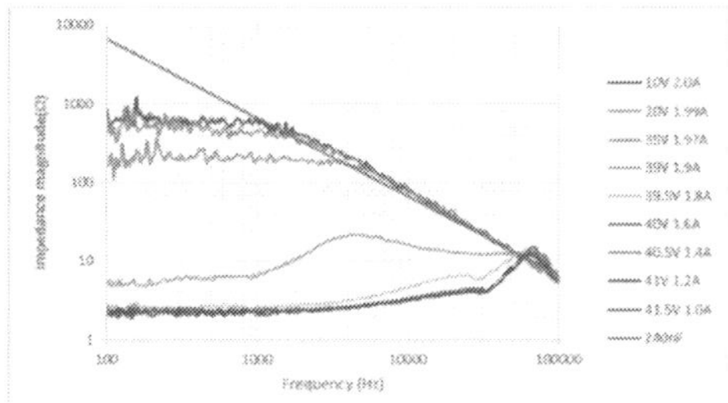

5 Commercial Evaluation

Given the early start of the marketing and sales activities for systems based on the RO-5100 SAS Module and stand-alone SAS Modules, a number of contracts, both commercial and institutional in Europe and the USA, have already been captured since 2013 (numbers w/o spares modules):

- Power SCOE with 72x SAS Modules for MPCV-ESM SAW FEE (+option 36x additional units)
- Power SCOE with 36x SAS Modules for MPCV-ESM PCDU EGSE;
- Power SCOE with 48x SAS Modules for Boeing PSTE (+ option 8x 48 additional units);
- Power SCOE with 8x SAS Modules for Lockheed Martin;
- 13x SAS Modules for Solar Orbiter;
- 17x SAS Modules for ExoMars CM;
- 42x SAS Modules for MetOp-SG;

On the basis of the above 236 (plus optional 420) SAS Modules for the period 2015 – 2018 (options until 2020), the commercial "success" is already there and confirms the order of magnitude derived from the above market assessment.

VERY HIGH FREQUENCY POWER SWITCHING: A ROAD MAP TO ENVELOPE TRACKING

Christophe Delepaut [(1)], **Natanael Ayllon** [(2)]

(1) ESA/ESTEC, Keplerlaan 1, 2201 AZ Noordwijk, The Netherlands, Email:Christophe.delepaut@esa.int
(2) ESA/ESTEC, Keplerlaan 1, 2201 AZ Noordwijk, The Netherlands, Email:Natanael.Ayllon@esa.int

ABSTRACT

RF (Radio Frequency) GaN transistors may be used for power switching at frequencies of 30 MHz and above, thereby reaching voltage control bandwidth of critical interest for the so-called ET (Envelope Tracking) technique. ET is meant at optimising the efficiency of RF amplifiers by supplying them with a voltage adapted to their power level at any time, and is the subject of many R&D works worldwide. The present paper provides an overview on the different activities run at the European Space Agency so far in this context, and present the possible way-forward to embark ET on future spacecraft.

1. INTRODUCTION

Power switching for space applications typically features Si MOSFET (Silicon Metal Oxide Semiconductor Field Effect Transistor) run at frequency of some hundreds of kHz. The applicable power levels stretch from a few W up to tens of kW, and the voltage levels from a few V to hundreds of V [1][2]. Such characteristics are driven by the ON serial resistance time parasitic capacitance FoM (Figure of Merit) of the concerned semiconductor technology which relates to the losses in conduction and in switching.

The emergence of the GaN HEMT (Gallium Nitride High Electron Mobility Transistor) technology for switching has recently resulted in significant improvement of both switching and conductive losses, allowing power switching to be operated in the low MHz region at fair conversion efficiency for similar voltage levels [3][4].

The above-mentioned state-of-the-art switching FET (Field Effect Transistor) are so-called "normally OFF", i.e. open at zero gate to source voltage, and have a positive commutation threshold voltage with respect to drain polarity. To this extent, they oppose to GaN HEMT meant for RF amplification which are "normally ON" and have a negative commutation threshold.

RF GaN FET have even more reduced parasitic capacitance with respect to its counterpart for switching, and may be used for switching, however at a price of appropriate gate biasing (negative voltage for N-FET). Such technology opens the way to switching at VHF (Very High Frequency), namely 30 MHz and above, for targeted voltage levels of up to 100 V [5]. This in turn enables voltage control bandwidth in the 10 MHz region.

On the other hand, payloads of power switching units within spacecraft encompass RF amplifiers, called transmitters, for communication or broadcasting purpose. Such amplifiers dissipate all the more power that their RF output power is low. To lower the dissipated power, an effective mean consists in supplying them with a voltage modulated proportionally to the signal envelope amplitude. Such technique is called ET (Envelope Tracking). Relevant amplitude modulation bandwidths are application dependent, with a low end starting at 10 MHz [6].

Accordingly, VHF power switching may match the requirements for ET. In this context, ESA started its own investigation and placed a number of Contracts with University, Scientific Institute and Industry. The present paper reports the results achieved in the different Contracts so far, and how they fit within an overall ET road map. It supports and highlight the way-forward towards a full scale Industrial application of ET on board of spacecraft.

The paper is organised as follows. Chapter 2 deals with background information on ET and describes the main building blocks. Chapter 3 addresses the most significant past and ongoing Contracts placed by ESA and synthetises their outputs. In Chapter 4, the road map is exposed. The TRL (Technology Readiness Level) of the different technologies involved is identified, and an overview is provided with a GANTT chart. Conclusions are delivered in Chapter 5.

2. BACKGROUND INFORMATION

2.1 Overview

Nowadays, there is a raising interest in using complex and high order digital modulation schemes with high spectral efficiency in order to achieve high amounts of data rate needed for high demanding services. The counterpart of using these high order modulation schemes is the inherent high PAPR (Peak to Average Power Ratio) of the telecommunication signal, which means that there is a large difference between the average amplitude and the peak amplitude of the signal.

DOI: 10.1051/e3sconf/20171614003

Common space-borne transmitters based on SSPA (Solid State Power Amplifiers) use static supply voltages to the RF transistors. In order to have a high efficient transmitter (required for space-borne applications where the DC power available is limited), the high power amplifier needs to be operated close to saturation. However, if the signal to be transmitted has large PAPRs, this will mean that the signal is clipped and heavily distorted. The consequence is a transmitter able to provide the required amount of power, but with a very distorted shape that is not often possible to correctly demodulate in Ground. Therefore, the link is not usable.

The solution to this point is indeed to reduce the operating point to a more linear region so that the amplification is done in a more linear manner. The consequences are lower output power transmitted and a low efficient transmitter. If the DC power supply is static, the power consumption will keep high while the RF output power is decreased. This translates into higher power to be dissipated, directly impacting the reliability of the components on-board the SSPA.

Generally speaking, linearisers are able to pre-distort the signal to compensate for the distortion introduced by the SSPA. Therefore, the operation of the transmitter can become closer to the saturation area, making the transmitter more efficient with a lower signal distortion. Nevertheless, the solution is not flexible and optimum enough against dynamic changes of the modulation scheme or changes in the RF output power, as it is required today in many flexible payloads. In this context, ET can enable a step forward to achieve high level of RF output power flexibility together with high level of efficiencies.

ET is a high efficiency amplifier architecture that consists of extracting the envelope information from the RF carrier, amplify it and supply this envelope to the drain or collector of the high power RF transistor in the RF path, as shown on Fig. 1. This is often called a dynamic supply architecture.

The main benefit of this architecture is that the transmitter is able to keep high levels of efficiency at levels of input power back off of several dBs (up to 10dB). Therefore, in presence of complex and high order digital modulation schemes with significant PAPR, the transmitter operates at high level of efficiency almost all the time.

With the advent of Gallium Nitride technology for the RF devices, the interest for this high efficiency SSPA architecture has even raised.

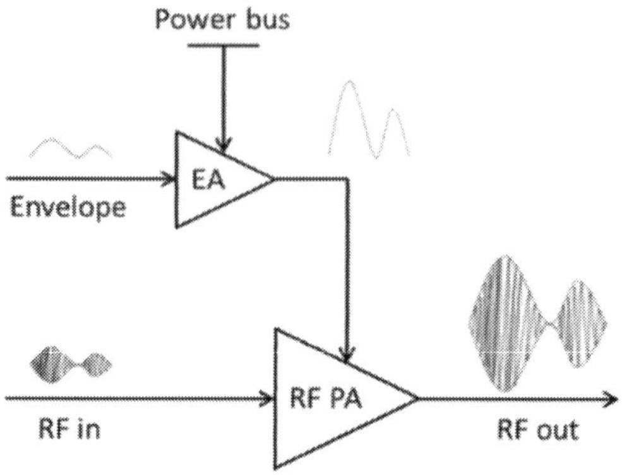

Figure 1. ET-SSPA architecture [5]

The reason is that RF GaN amplification transistors are potentially able to operate at different supply voltage levels without significantly changing the optimum output impedance. This means that a dynamic supply modulation scheme can be applied, guaranteeing optimum operation of the RF amplifier at all bias supply levels. This is particularly important for guaranteeing a good stability of the RF power amplifier, preventing it from undesired oscillations during the supply change.

2.2 Building Blocks

As depicted in Fig. 1, the ET-SSPA has two main building blocks. The first and most important one is the EA (Envelope Amplifier), that is in charge of acquiring the envelope of the RF signal and amplify it to the right level of voltage required by the RF Amplifier.

For space-borne SSPAs targeting L/-S- applications the EA should be able to be fed by 100V bus voltage and should be able to provide voltage variations in the order of 20 to 50 V with current values of 2 to 4 A. All this should be achieved with DC power conversion efficiency in the order of 90 %. At the same time, the EA should be able to smoothly amplify envelope signals with large bandwidth, that is, with very fast time domain response.

Additional requirements for the EA encompass high gain of typically 40 dB, high linearity, low noise, including low switching ripple, and high slew rate capability, in the order of 1 V/ns, as the bandwidth performance must be met not only for small signals but also for large signal variations.

The second key building block is the RF power amplifier. What changes in an ET-SSPA from a standard class-AB amplifier is that now the RF amplifier operates with dynamic drain voltage supply. In addition, the drain bias network typically designed for the RF amplifier now interfaces with the output filter

 DOI: 10.1051/e3sconf/20171614003

of the DC/DC converter of the EA. This is an interesting feature that brings a difficult trade-off: bandwidth of the output filter versus output ripple of the overall ET-SSPA.

3. R&D CONTRACTS AT ESA

3.1 Envelope Amplifier

In 2011, a 3 year long NPI (Networking/Partnering Initiative) Contract has been placed with the University of Mons and the support of TAS (Thales Alenia Space), in Belgium, to demonstrate the feasibility of VHF power switching meant as EA (Envelope Amplifier) for ET applications.

The targeted converter was based on a step down Buck topology. The converter is provided with the envelope reference extracted from the input RF signal and must deliver the amplified envelope with the appropriate gain. A step down topology is appropriate for spacecraft power bus regulated at 50 V or 100 V.

The converter output voltage was to be controlled in closed-loop. There are several reasons why such feedback loop is implemented. First, it ensures a better input to output linearity with respect to component drift and non-linearities. Secondly, it also allows the converter output filter to have a cut-off frequency lower than the control bandwidth. This is desirable as the output voltage ripple must be kept low without compromising the tracking bandwidth. Finally, placing the output filter cut-off frequency within the control bandwidth makes possible active damping of the filter with no dissipative resistor.

From a practical viewpoint, the Buck converter has been implemented in a reverse configuration, meaning that the switching FET is grounded and that the rectifying diode is connected to the supply voltage. Accordingly, the load is to be connected between the power supply and the converter output. The asset of the reverse buck is that its switching operation is much easier as the FET gate source dipole is no longer floating as in a normal FET, which feature may prove critical at VHF. On the other hand, the accommodation of the RF amplifier with the reverse Buck is not straightforward and is addressed below (see § 3.3).

The manufactured breadboard is shown on Fig. 2. It is constituted by power dice directly mounted on the chassis and low level dice assembled on RF substrate. Input to output power conversion efficiencies well above 90 % have been reached at 50 MHz switching frequency, 50 V input voltage and power level up to 45 W. The efficiency measurements are provided on Fig. 3.

Figure 2. VHF power switching breadboard

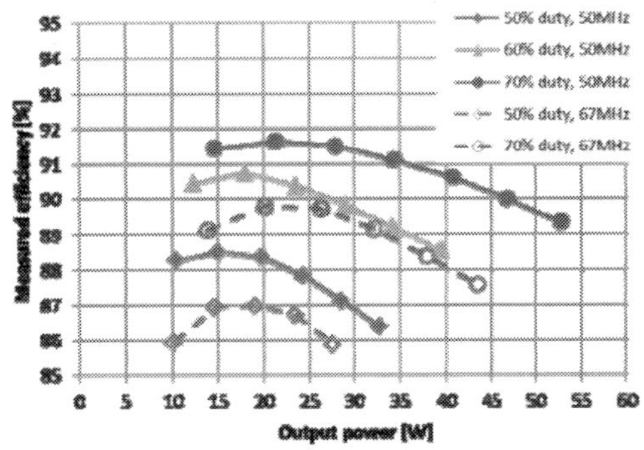

Figure 3. Efficiency at 50 V input voltage

From feedback control viewpoint, the tracking of a 1 MHz envelope reference has been demonstrated, albeit at a price of lowering the switching frequency of the self-oscillating switching controller down to 10 MHz. The tracking measurement is reported on Fig. 4. The large peak to peak switching ripple is a consequence of the lowering of the switching frequency.

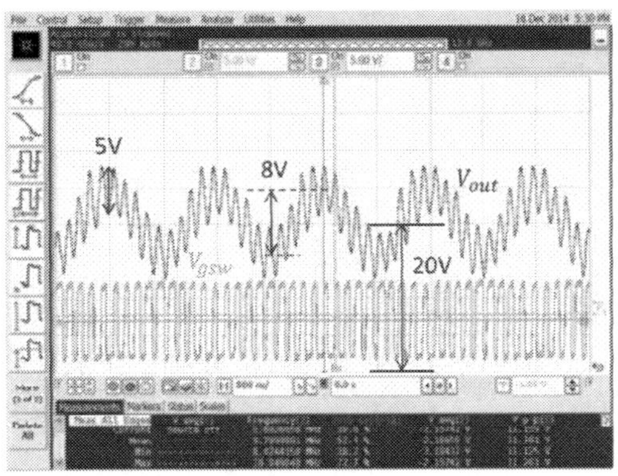

Figure 4. Sine reference envelope tracking

DOI: 10.1051/e3sconf/20171614003

The reason for lowering the switching frequency was the larger than expected open-loop phase shift in function of frequency. This is driven by the choice of a AlN (Aluminium Nitride) substrate selected for its enhanced thermal conductivity desirable for the driver output stage. The drawback of this substrate is its larger dielectric permittivity. Accordingly, the propagation speed of the signal along the tracks was lowered, and the parasitic track capacitance combined with design resistance resulted in stronger parasitic time constant. In both case, the size of the control circuit was critical, which indicates that the integration of the driver circuit is of crucial importance.

3.2 Modulator & Driver Integrated Circuit

In 2015, a 2 year long ARTES 5.1 (Advanced Research in Telecommunications Systems) activity was started with IMST, in Germany. The objective of this Contract is to develop an IC (integrated Circuit) able to drive the switching GaN FET within the EA. The IC is to include not only the FET driver itself, but also modulator which is the circuit in charge of determining the duty cycle to be applied to the switching FET. The modulator & driver function is identified within the ET system diagram on Fig. 5. The closed-loop operation of the EA is not shown on this diagram, but consists in a feedback from the DC/DC converter output towards the modulator input.

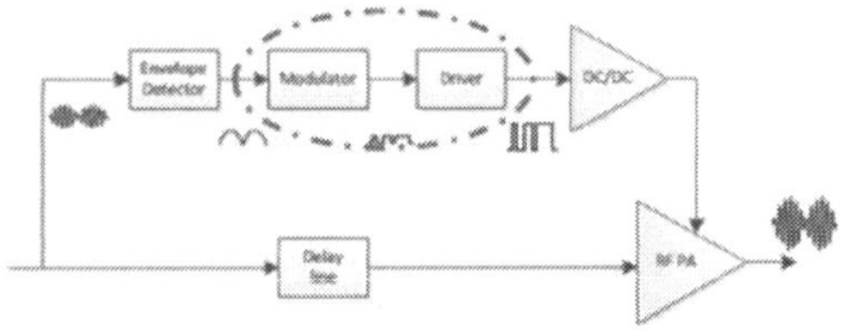

Figure 5. Modulator and driver within ET system

The IC design significantly benefits form the return of experience accumulated during the above-mentioned NPI activity where a discrete modulator and driver design was implemented. As a specific challenge, it has to operate switching up to 100 MHz with peak to peak output voltage of 6.5 V and slew rate of more than 10 V/ns loaded by a switching FET gate capacitance of up to 30 pF.

Two microelectronic technologies have been compared for the design of the circuit, namely the UMS25GH GaN technology and the IHP's SGB25VGD SiGe (Silicone Germanium) one. The trade-off has shown that the SiGe process, combining the advantage of CMOS devices, bipolar transistors and complementary RF LDMOS devices, was superior to the GaN process as to reach the requested output slew rate performance. The technology has been selected accordingly.

The Contract encompasses two successive wafer runs. The intention is to inject into the second wafer run the lessons learned during testing of the first sample. The first wafer of the modulator & driver circuit was designed as two separate dice, as shown on Fig. 6, giving more flexibility from testing and signal accessibility viewpoint. The second and final wafer run will consist in a single die.

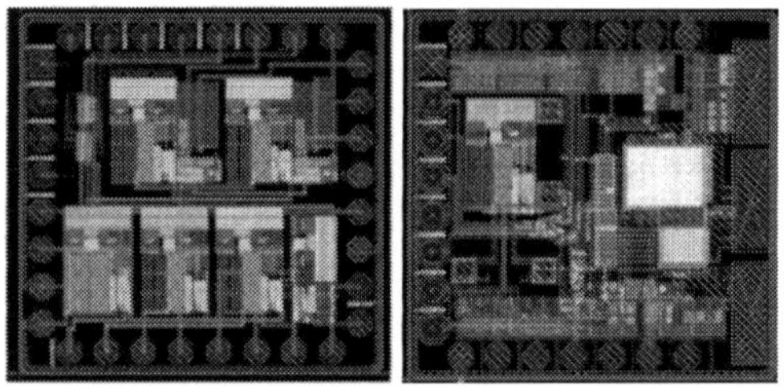

Figure 6. Driver & modulator IC prototype

To demonstrate that the developed IC successfully fits within an EA, the Contractor undertook to design with ESA support a VHF switching converter as test bench to be used to validate the behaviour of the circuit. Relevant simulations of the switching FET gate and drain voltages are shown om Fig. 7.

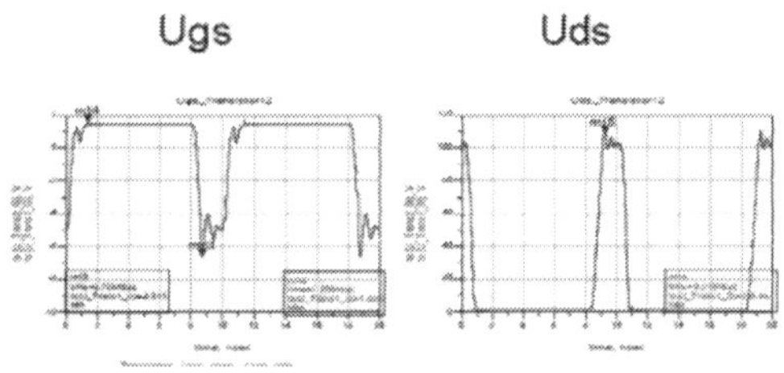

Figure 7. VHF power switching simulation

The targeted power level is 240 W. Simulations indicate an efficiency figure reaching up to 91.7 % at 100 MHz switching frequency. The breadboarding activity is ongoing at the time of writing this paper.

3.3 Envelope Tracking System

By end of 2015, a 2 year long continuation NPI Contract has been set up with FBH (Ferdinand Braun Institute) in Berlin, Germany, and with the support of TAS in Belgium. The focus of this activity is about interfacing the EA with the RF amplifier being the main constituents of the ET system.

DOI: 10.1051/e3sconf/20171614003

The ET system is displayed on Fig. 8, with the EA on the left side, incorporating the reverse buck topology, and the RF amplifier on the right side [7]. The figure highlights that the RF ground is separated from the system ground due to the choice of the reverse buck converter.

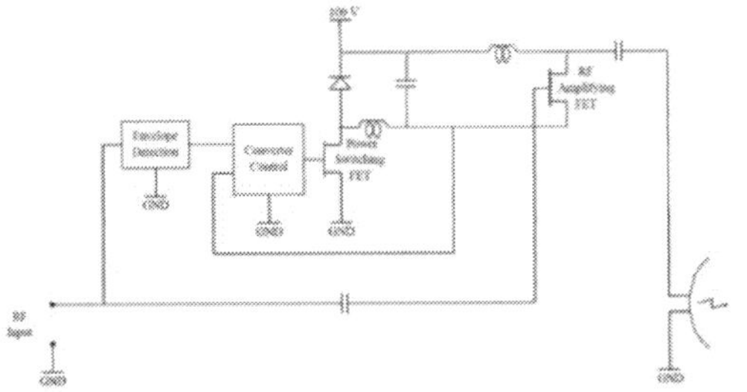

Figure 8. ET system architecture (patent pending)

The benefit of the reverse Buck has been explained earlier (see §3.1). From RF load interface viewpoint, the dedicated RF ground is not an issue due to the capacitor coupling any RF amplifier with its load. Such capacitor is meant to eliminate the envelope voltage before delivering the amplified RF signal.

It is underlined that reversing the Buck has shifted the dV/dt issue from the switching FET source to the RF FET source. However the order of magnitude has been mitigated by a factor three, from about 1 kV/ns on the switching node of the Buck at 100 MHZ switching frequency down to typically 1 V/ns at RF amplifier reference voltage for a 10 MHz tracking bandwidth.

In the frame of the ongoing Contract, an RF amplifier compatible with the above-mentioned architecture has been designed and breadboarded, with convenient attention paid to the in-band and out-band stability. The breadboard is shown on Fig. 9.

The left and right connectors are the RF input and output. The three upper connectors connect the RF FET drain, gate and source. The testing of the amplifier is successful so far [8]. Large signal testing is ongoing.

The RF amplifier is subsequently to be connected to power switching EA. Such configuration will deliver very valuable information on the ET system, including not only overall efficiency performance but also sensibility to switching ripple amplitude, to tracking bandwidth and envelope delay, to residual non-linearities, etc.

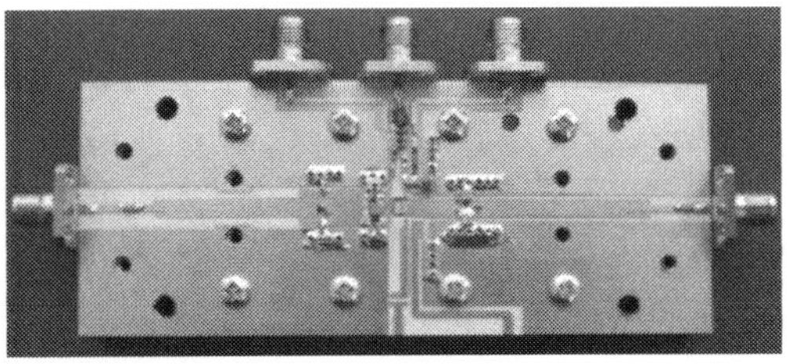

Figure 9. Floating RF amplifier breadboard

Finally, the Contract is going to concentrate on the coupled design of the EA with an RF amplifier, based on the outputs of the previous steps. This last step includes preliminary experimental validation.

4. ROAD MAP

The ESA Road Map is based on activities initiated one decade ago. It concerns the ET system, which encompasses the EA and the RF amplifier. As a critical component, the EA includes the modulator & driver circuit meant for driving the VHF switching operation.

The NPI activity performed with the University of Mons has brought the EA concept, based on a single VHF switching converter, up to TRL3 (Technology Readiness Level).

The ARTES activity with IMST is currently bringing the modulator & driver circuit up to TRL5 by delivering a workable IC chip. The associated validation works at EA level are also bringing the EA concept up to TRL5.

The ongoing NPI activity with FBH is going to bring the whole ET system up to TRL3.

Accordingly, the next activity shall target to increase the maturity level of the ET system from TRL3 to TRL5. All critical functions are to be implemented within an EM (Engineering Model), to be designed, manufactured and tested in a relevant environment.

The Road Map GANTT chart is displayed on Fig. 10. It focuses on year 2016 and beyond. The ongoing ARTES and NPI Contracts are identified, pointing at the future activity for EM development.

DOI: 10.1051/e3sconf/20171614003

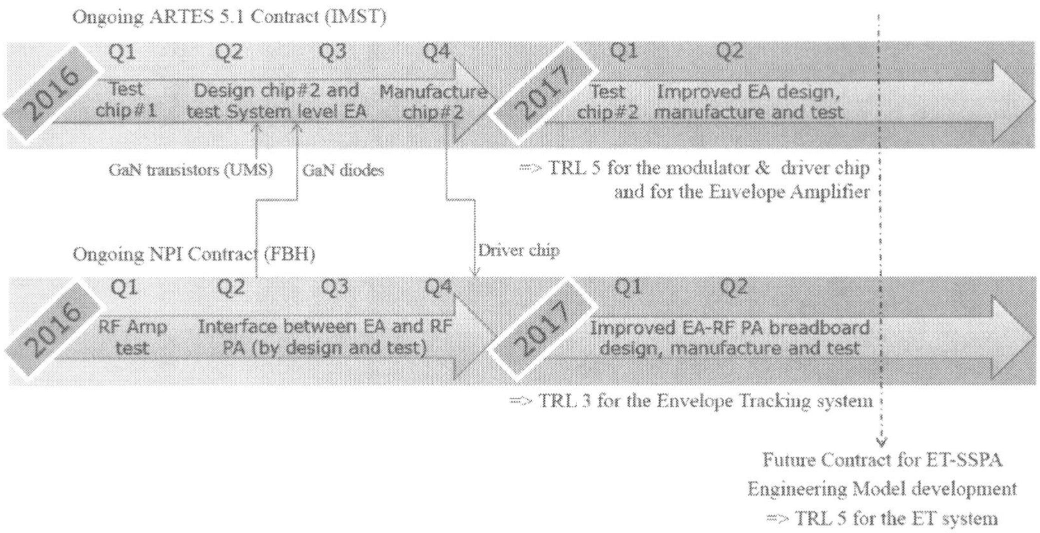

Figure 10. ET road map GANTT chart

5. CONCLUSION

ET is a promising technique to reduce electrical power consumption and thermal dissipation for RF amplification on board of spacecraft. Intense R&D efforts are ongoing worldwide, strengthened by the GaN semiconductor technology, to ever increasing switching frequencies, voltage levels and power levels, and the technique is soon going to be within the reach of space applications. ESA has started its own R&D investigation on the subject many years ago, with TRL 3 for the whole ET system to be reached by 2017. By then, Industry is expected to be ready to bring the maturity level from TRL3 up to TRL5 by developing, manufacturing and testing an EM of the ET system.

6. ACKNOWLEDGEMENT

TAS Belgium is greatly acknowledged for its contribution to the funding of the NPI activities covered by the present paper.

7. REFERENCES

1. "Performance and Simplicity in Power Conversion Functions Made Possible by New European Components", F. Tonicello, O. El Korashy, A. Pesce, 9th European Space Power Conference, Saint-Raphael, France, 6-10 June 2011

2. "High Efficient Universal Buck Boost Solar Array Regulator SAR Module", S. Kimmelmann, W. Knorr, 10th European Space Power Conference, Noordwijkerhout, The Netherlands, 13-17 April 2014

3. "Generation 4 eGan FETs Widening the Performance Gap with the Aging MOSFET", A. Widow, Efficient Power Corporation, EEWeb, 28 October 2014

4. "GaN FET module performance advantage over silicon", N. Mehta, Texas Instrument, March 2015

5. "A 90% Efficient 50 MHz Buck Converter for Space Telecom Systems", F. Leroy, O. Deblecker, F. Dualibe, N. Le Gallou, C. Delepaut, 10th European Space Power Conference, Noordwijkerhout, The Netherlands, 13-17 April 2014

6. "Over 10 MHz Bandwidth Envelope-Tracking DC/DC Converter for Flexible High Power GaN Amplifiers", N. Le Gallou, D. Sardin, C. Delepaut, M. Campovecchio and S. Rochette, IEEE MTT-S International Microwave Symposium, Baltimore, USA, 5-10 June 2011

7. "Radio-Frequency High Power Amplifier with Broadband Envelope Tracking by Means of Reversed Buck Converter", ESA patent, publication numbers: WO2014170710, US 2016/0056770

8. "A 15 W L-Band Floating-Supply GaN RF PA for Envelope Tracking Systems in Space Applications", T. Kuremyr, C. Delepaut, N. Ayllon, Wolfgang Heinrich, O. Bengtsson, to be submitted for publication to IEEE Transactions on Microwave Theory and Techniques

 DOI: 10.1051/e3sconf/20171614004

WATKINS-JOHNSON TOPOLOGY INTEGRATED IN A FULL-BRIDGE CONVERTER

Giulio Simonelli [1], Oliver El Korashy [2], Hadrien Carbonnier [3]

[1] *ESA-ESTEC, Keplerlaan 1, 2200AG Noordwijk, The Netherlands, Email:giulio.simonelli@esa.int*
[2] *ESA-ESTEC, Keplerlaan 1, 2200AG Noordwijk, The Netherlands, Email:oliver.el.korashy@esa.int*
[3] *ESA-ESTEC, Keplerlaan 1, 2200AG Noordwijk, The Netherlands, Email:hadrien.carbonnier@esa.int*

ABSTRACT

Electrical thrusters require high power and High Voltage (HV) of 300V to 2000V depending on the thruster technology. A converter suitable for this application should have a high power handling capability (full bridge converter) and a simple output stage (fly-back converter). Other desirable characteristics would be: simplicity, able to supply a wide range of loads, low stress on switches, high stability and galvanic isolation between input and the output power.

The Watkins-Johnson (W-J) topology is used mainly in HV applications due to the fact that its magnetic elements are on the input side. However the power output of this converter is constrained by the "single switch topology". Further downsides of the basic W-J topology are the relatively high voltage stress on the switching elements and the fact that is not galvanically isolated .

This paper will present a new topology of a Full Bridge Isolated Converter which is based on the W-J topology. Steady state behaviour in Continuous Current Mode (CCM) and in Discontinuous Current Mode (DCM) will be investigated. Dynamic performances of the converter will be investigated with the analysis of its transfer function. An equivalent circuit suitable for PSPICE simulations will be presented.

The topology presented potentially offers the following advantages which are suited to providing power conditioning to sub-systems which require high voltage and high power:

- Simple output stage suitable for HV.
- Buck type transfer function, hence simple control scheme.
- Low stress on components.
- Single stage converter
- Suitable for modular high power converter due to the simplicity of the output stage

1. WATKINS-JOHNSON TOPOLOGY

In figures 1 and 2 the W-J topology and a practical realization of it is presented.

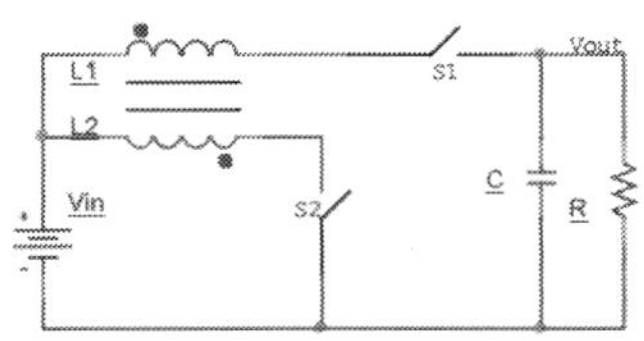

Figure 1. Watkins-Johnson basic topology

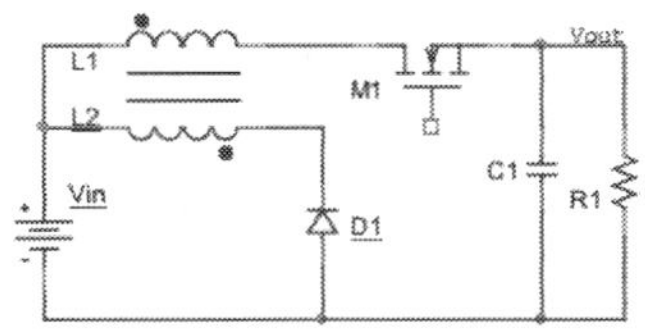

Figure 2. Watkins-Johnson topology practical realization

Its peculiarity is that energy is transferred from input to output during the ON time of M1, while during the OFF time the energy stored in the core of the mutual inductor L1 L2 is given back to the input source.

It can be shown that in CCM the conversion ratio is of the basic topology is (δ refers to duty cycle):

$$\frac{V_{out}}{V_{in}} = \frac{2\delta - 1}{\delta} \quad (1)$$

Equation (1) is also valid for the practical realization of fig. 2 but only for CCM and δ greater than 0.5.

For the practical realization of fig. 2 the max static voltage stress on the switching elements is 2*Vin. Considering that the input voltage for high power satellites is already 100V and could rise higher in the, this level of voltage stress can seriously limit the choice of power components.

2. ISOLATED VERSION OF THE WATKINS-JOHNSON TOPOLOGY AND ITS INTEGRATION IN A FULL BRIDGE

The isolated version of the topology under study can be implemented as a push pull or full bridge that follows a Watkins-Jonson stage, as depicted in figure 3 and 4.

DOI: 10.1051/e3sconf/20171614004

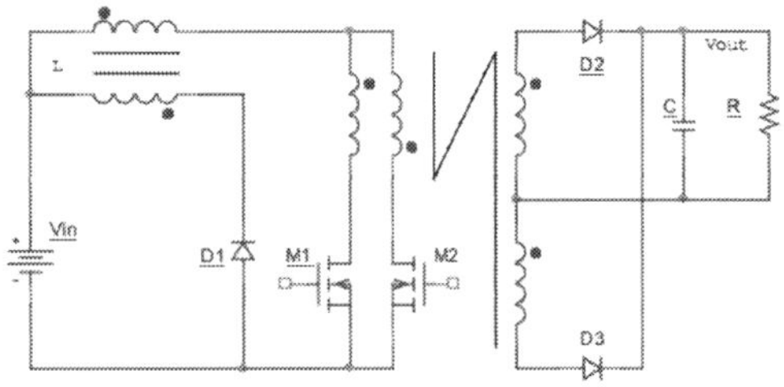

Figure 3. Watkins-Johnson followed by Push-Pull stage

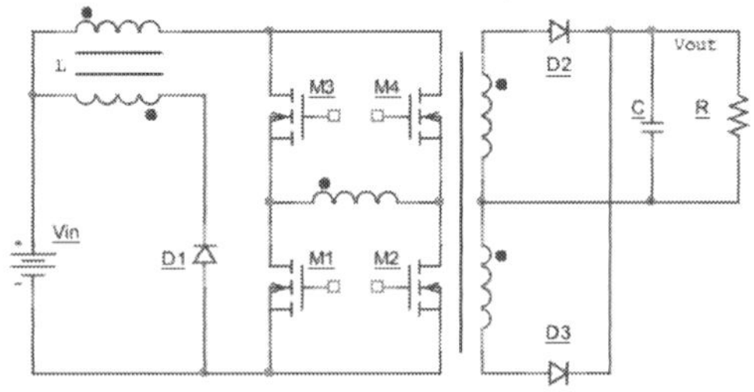

Figure 4. Watkins-Johnson followed by Full-Bridge stage

Despite the introduction of the isolation and the higher power handling capability of the converters in fig. 3 and 4, the conversion ratio is the same as for the basic topology except for a small modification introduced by the transformer ratio $n=N_p/N_s$ where N_p and N_s are respectively the number of the primary and secondary turns of the transformer:

$$\frac{V_{out}}{V_{in}} = \frac{2 \cdot \delta - 1}{n \cdot \delta} \quad (2)$$

Also the maximum voltage stress on the switching elements is the same as the basic topology, i.e. 2*Vin.

Instead of simply cascading the Watkins-Johnson and a full bridge stage, we can "integrate" the two topologies, as depicted in figure 5. It can be seen that, at the cost of one extra diode on the primary side, the maximum voltage stress on the Mosfets is reduced from 2*Vin to Vin, making the converter suitable for systems based on power busses of 100V or higher.

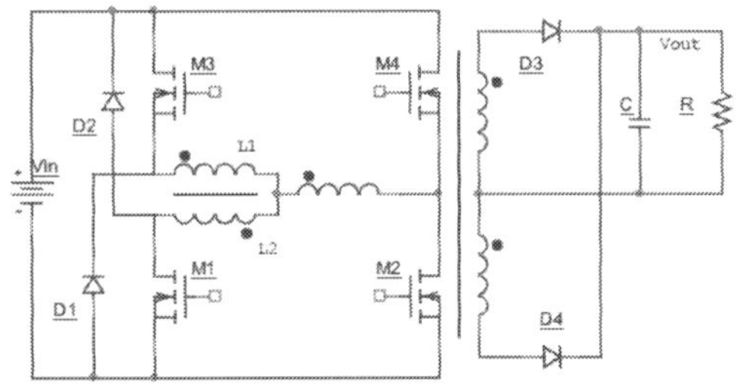

Figure 5. Watkins-Johnson integrated in a Full-Bridge stage

The strong points of this new converter are:

- galvanic isolation
- high power handling capability
- low voltage stress on switches
- simplified output stage

The last point, a simplified output stage, is extremely valuable in case of HV applications and makes this converter a promising candidate for applications like electrical propulsion where high power and high voltage are requested.

It can be also shown that the switches M2 and M4 could operate easily in Zero Voltage Switching (ZVS) mode. The weakest point of this converter is the high input AC current, which will require an adequate filter.

3. CONTINUOUS CURRENT MODE

In continuous current mode, there is always current circulating in the inductor. In this condition the converter operates in 4 phases here described by the names of the switches that are conducting. Referring to fig. 5:

- Phase 1: M3, M2, D3 are ON (and primary current flows in upper inductance branch)
- Phase 2: D1, D2 are ON (and inductor currents recirculates on both inductor branches to the input source)
- Phase 3: M4, M1, D4 are ON (and primary current flows in lower inductance branch)
- Phase 4: D1, D2 are ON (and inductor currents recirculates on both inductor branches to the input source)

Even if the switches are different, phase 1 and 3 are essentially the same, so we can study the CVT considering only phase 1 and 2.

The circuits in phase 1 are depicted in fig. 6 where only the conducting switches have been included. To derive the conversion ratio we will apply the method of the inductor volt seconds balance and the capacitor charge balance under the small ripple approximation [1].

We will use the following convention to distinguish between instantaneous, DC and AC values:

$x = x(t) = X + \underline{x}(t)$

$x = x(t)$: instantaneous value

X : DC value

$\underline{x}(t)$: AC value

In Phase 1 (see fig.6) the governing equations are:

$$v_L = v_{in} - R_L \cdot i_{on} - n \cdot (v_{out} + V_{ds}) \quad (3)$$

$$i_c = n \cdot i_{on} - \frac{v_{out}}{R} \quad (4)$$

DOI: 10.1051/e3sconf/20171614004

Where i_{on} is the inductor current, V_{ds} is the voltage drop on the diode on the secondary side, R_L the parasitic resistance of one inductor winding.

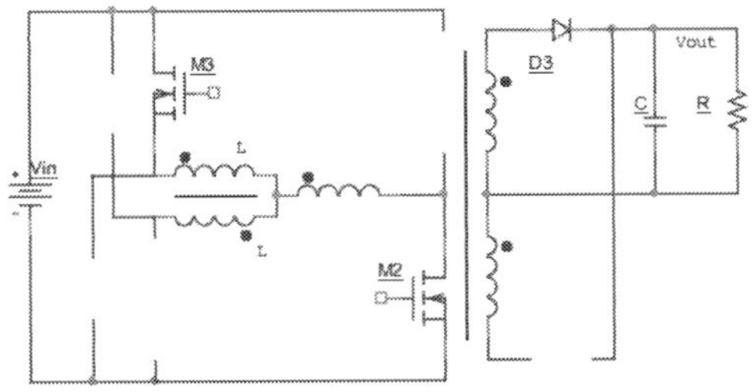

Figure 6. Phase 1 circuit

In Phase 2 (see fig.7) the governing equations are:

$$v_{L_{12}} = -2 \cdot R_L \cdot i_{off} - 2\,V_{dp} - v_{in} \tag{5}$$

$$i_c = \frac{-v_{out}}{R} \tag{6}$$

Where i_{off} is the inductor current, L_{12} is the inductance of the two windings in series of the mutual inductances, v_{L12} the voltage across L_{12}, V_{dp} is the voltage drop on the diode on the primary side.

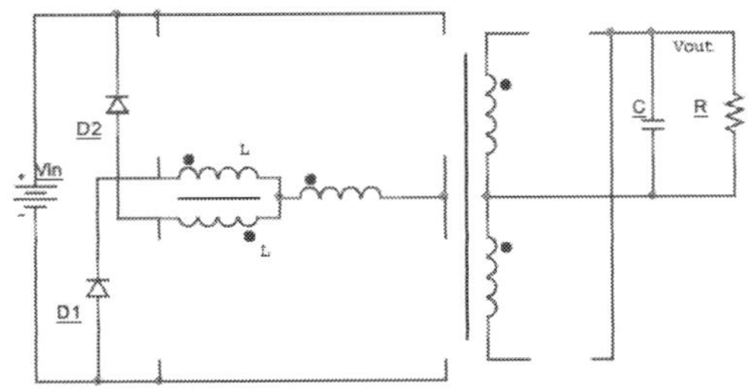

Figure 7. Phase 2 circuit

We now enter in the tricky part of the analysis. In fact the value of the inductance changes between Phase 1 and 2 and in order to apply the principle of VoltSecond balance we have to express V_{L12} in terms of V_L. Considering that:

$$L_{12} = 4 \cdot L \tag{7}$$

(because doubling the number of turn on a core will increase the inductance by a factor of 2^2), and that at the transition times between Phase 1 and 2:

$$i_{off} = \frac{i_{on}}{2} \tag{8}$$

(because the energy in the inductance can't change instantaneously), we can write:

$$v_{L_{12}} = 2 \cdot v_L \tag{9}$$

Substituting (8) and (9) into (5) we have the following governing equations for phase 2:

$$v_L = -R_L \cdot \frac{i_{on}}{2} - V_{dp} - \frac{v_{in}}{2} \tag{10}$$

$$i_c = \frac{-v_{out}}{R} \tag{6}$$

We can now proceed with the Volt Second balance using (3) and (10) and charge balance using (4) and (6) only for the DC values of the concerned variables:

$$0 = \delta \cdot \left[V_{in} - R_L \cdot I_{on} - n \cdot \left(V_{out} + V_{ds}\right)\right] + (1-\delta) \cdot \left(-R_L \cdot \frac{I_{on}}{2} - V_{dp} - \frac{V_{in}}{2}\right) \tag{11}$$

$$0 = \delta \cdot \left(n \cdot I_{on} - \frac{V_{out}}{R}\right) + (1-\delta) \cdot \frac{-V_{out}}{R} \tag{12}$$

From (11) and (12) we can obtain the conversion ratio of the full bridge with the Watkins-Johnson topology, that, in the case the parasitic values R_L, V_{dp} and V_{ds} are zero is:

$$\frac{V_{out}}{V_{in}} = \frac{3 \cdot \delta - 1}{n \cdot 2 \cdot \delta} \tag{13}$$

The conversion ratio modelled by (13) has been plotted in fig.8

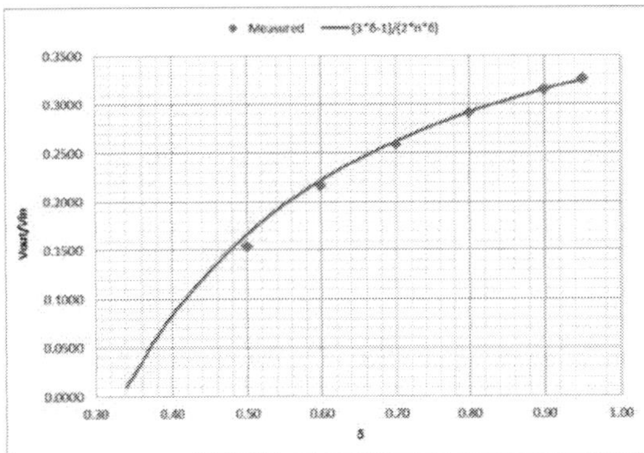

Figure 8. Conversion ratio Vout/Vin(δ) for n=3

4. CONVERTER MAIN WAVEFORMS AND CHARACTERISTICS IN CONTINUOUS MODE

In fig. 9 the major waveforms of the converter in CCM are depicted with reference to the electrical scheme in figure 5.

The major characteristics of the converter in CCM are reported here after.

DOI: 10.1051/e3sconf/20171614004

Input output relationship:

- Vout/Vin=(1/n)*(3*δ-1)/(2*δ)

Switch ratings:

- Drain voltage: Vin
- Peak current: I_Lpk

Primary diodes ratings:

- Voltage: Vin
- Peak current: I_Lpk/2

Secondary diodes:

- Voltage: 2*Vout
- Peak current: (n)*I_Lpk

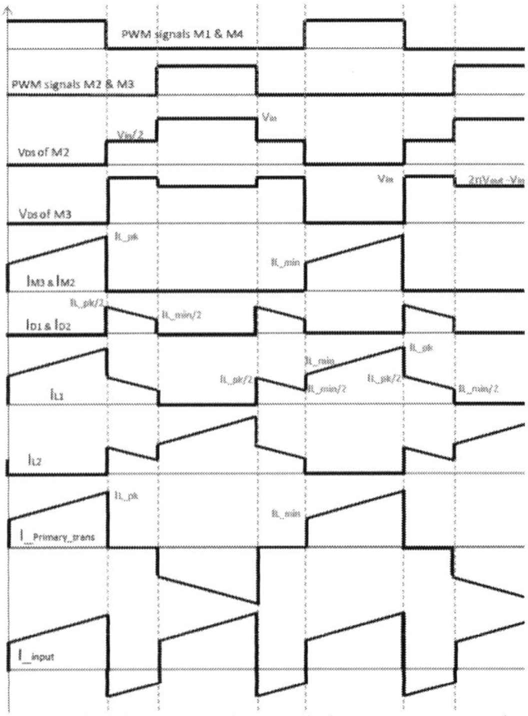

Figure 9. Major waveforms of the converter in the switching period (ideal case)

5. CONTINUOUS AND DISCONTINUOUS MODE BOUNDARY AND K_{crt} and R_{crt}

We know that this converter transfer the current in the inductor to the output only during phase 1, in phase 2 the inductor current goes back to the input source. Hence the average current of the inductor during phase 1 must be the same as the average output current corrected by the transformer ratio. We can then write:

$$I_{L\_avg\_on} = \frac{1}{n} \cdot \frac{V_{out}}{R} \qquad (14)$$

Using (13) in (14) and rearranging we have:

$$I_{L\_avg\_on} = \frac{1}{n^2} \cdot \frac{3 \cdot \delta - 1}{2 \cdot \delta} \cdot \frac{V_{in}}{R} \qquad (15)$$

On the other hand we know that at the boundary condition between CCM and DCM must be:

$$I_{L\_avg\_on} = \frac{\Delta I_{on}}{2} \cdot \delta \qquad (16)$$

From the fig. 6 we can state that

$$\Delta I_{on} = \frac{V_{in} - n \cdot V_{out}}{L} \cdot \delta \cdot T \qquad (17)$$

Where ΔI_{on} is the rise of current in the inductor during phase 1 (t on). Using (13) in (17) and rearranging we have:

$$\Delta I_{on} = \frac{1 - \delta}{2 \cdot L} \cdot V_{in} \cdot T \qquad (18)$$

Equation (16) also allow us to set the boundary condition between the two modes, for CCM must be:

$$I_{L\_avg\_on} > \frac{\Delta I_{on}}{2} \cdot \delta \qquad (19)$$

For DCM must be:

$$I_{L\_avg\_on} < \frac{\Delta I_{on}}{2} \cdot \delta \qquad (20)$$

Substituting in equation (19) equation (15) on the left side and (18) on the right side, we get:

$$\frac{1}{n^2} \cdot \frac{3 \cdot \delta - 1}{2 \cdot \delta} \cdot \frac{V_{in}}{R} > \frac{1}{2} \cdot \frac{1 - \delta}{2 \cdot L} \cdot V_{in} \cdot T \cdot \delta \qquad (21)$$

Rearranging (21):

$$\frac{2 \cdot L}{n^2 \cdot R \cdot T} > \frac{(1 - \delta) \cdot \delta^2}{3 \cdot \delta - 1} \qquad (22)$$

Defining:

$$K = \frac{2 \cdot L}{n^2 \cdot R \cdot T} \qquad (23)$$

$$K_{crt}(\delta) = \frac{(1 - \delta) \cdot \delta^2}{3 \cdot \delta - 1} \qquad (24)$$

We have that we are in CCM if:

DOI: 10.1051/e3sconf/20171614004

$$K > K_{crt}(\delta) \qquad (25)$$

Equation (24) predicts that for δ=1/3 K_{crt} will be infinite so the converter will be always in DCM, which is correct because from (13) Vout is zero for δ=1/3 so no current in the inductor at all since the load is zero Watts.

For δ<1/3 K_{crt} equation (25) loses its meaning and the converter will be always in DCM. Rearranging (22) we can define $R_{crt}=R_{crt}(\delta)$.

$$R < \frac{2 \cdot L}{n^2 \cdot T} \cdot \frac{3 \cdot \delta - 1}{(1-\delta) \cdot \delta^2} \qquad (26)$$

$$R_{crt}(\delta) = \frac{2 \cdot L}{n^2 \cdot T} \cdot \frac{3 \cdot \delta - 1}{(1-\delta) \cdot \delta^2} \qquad (27)$$

Equation (27) has been plotted in fig. 10, under the assumption of L=70uH, T=5us and n=3. The Y axis is R_{crt}. In order to be in CCM, R of the load must be lower than R_{crt} plotted.

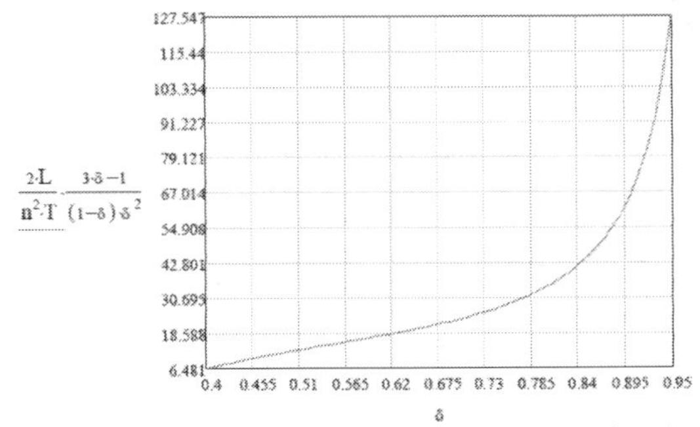

Figure 10. Plot of $R_{crt}(\delta)$ for L=70uH, T=5us and n=3

6. CONVERTER MAIN WAVEFORMS AND CHARACTERISTICS IN DISCONTINUOUS CURRENT MODE

In the following figure the main waveforms of the converter in DCM are reported with reference to the electrical scheme in figure 5.

The major characteristics of the converter (Voltage and Current stress levels) in DCM are essentially the same as in CCM, as can be seen in figure 11.

The converter gain in DCM is given by the following formula in the ideal case:

$$\frac{V_{out}}{V_{in}} = \frac{\delta^2}{n \cdot (K + \delta^2)} \qquad (28)$$

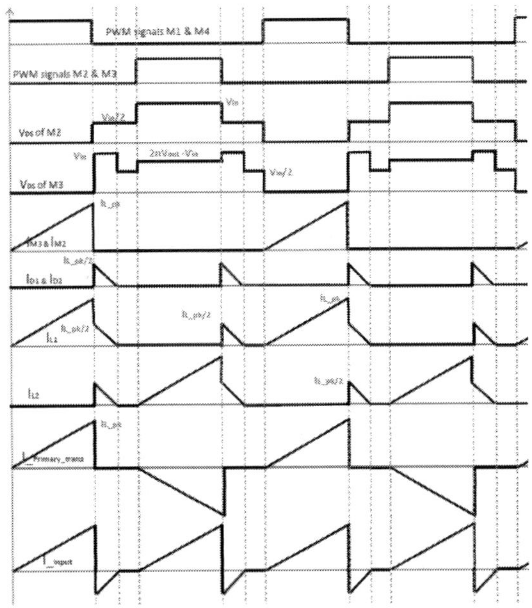

Figure 11. Major waveforms of the converter in the switching period when in CCM (ideal case).

7. MEASUREMENTS

Figure 8 shows measurements of Vout/Vin compared to duty cycle, measured at Vin = 100V, n=3, Pout=50W. It is observed that the measured results follow the derived formula, allowing for a slight error due to voltage drop in the output diodes.

A breadboard of the converter in fig. 5 was configured in closed loop using a UC1825 PWM controller which utilises peak current control. The breadboard was designed for Vin=100V, Vout=30V, fsw=100kHz and max Pout=500W.

Although this topology has been identified as suitable for HV, for the first iteration a low 30V output voltage was selected in order to simplify the testing and debugging.

Figure 12 shows some waveforms measured on the converter working at 113W. It is observed that the general shape of the measured waveforms corresponds to those found in Figure 9. The discrepancies are due to parasitic effects, in particular, the ~1.8MHz damped oscillations (seen on the inductor current and $V_{DS(M3)}$) are due resonance of the transformer leakage inductance with the parasitic capacitance of the Mosfets. These oscillations can be reduced or removed using an RC snubber across each Mosfet.

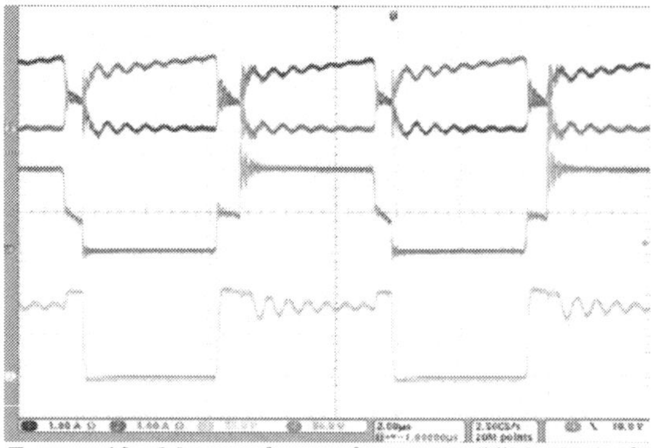

Figure 12. Measured waveforms in CCM: I_{L1}(red), I_{L2}(blue), $V_{D(M2)}$ (pink), $V_{DS(M3)}$ (green) @ Vin=100V, Vout=30V, Pout=113W, fsw=100kHz,

Figure 13 shows the efficiency curve of the breadboard, plotted from a selection of measurements from 15-500W. Above 50W the efficiency is higher than 91% and from 110W-190W it is >94.5%. Above about 380W the converter was operating at maximum duty cycle (96.5%), so without regulation and the output voltage gradually drops with increasing power demand. Note that the breadboard has not been optimised for its highest efficiency, improvements could be made by reducing leakage inductance and resistance in the magnetic components and by selection of different switching components.

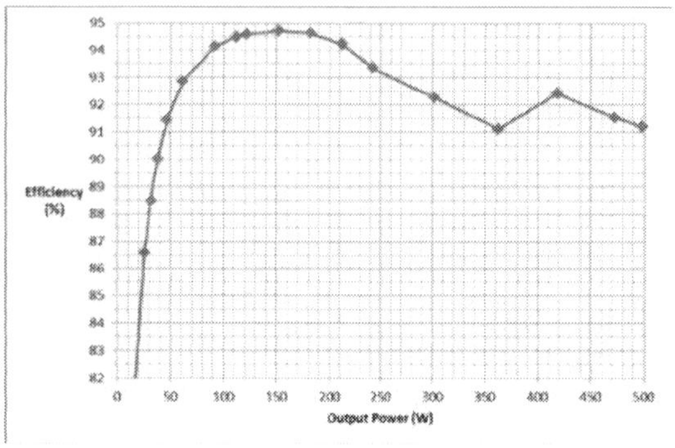

Figure 13. Efficiency curve of the converter @ Vin=100V, Vout=30V, fsw=100kHz

The stability of the converter topology is demonstrated using spectrum analysis which measured phase margin above 60deg and gain margin above 10dB throughout the output power range, including at low power (10W) in DCM. Two bode plots are given to show stability performance at 10W (fig.14) and 200W (fig.15).

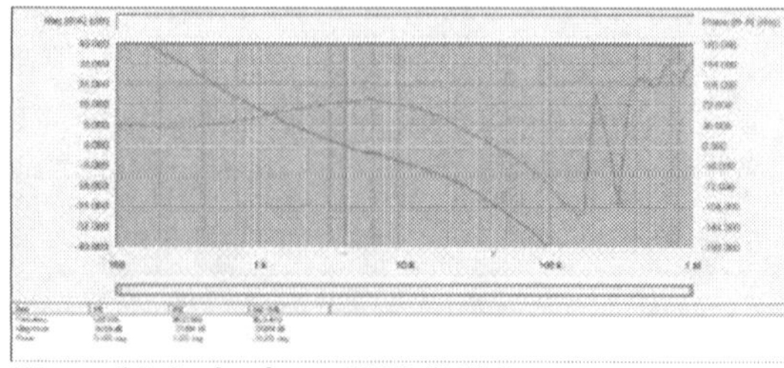

Figure 14. Bode plot at 10W (DCM)

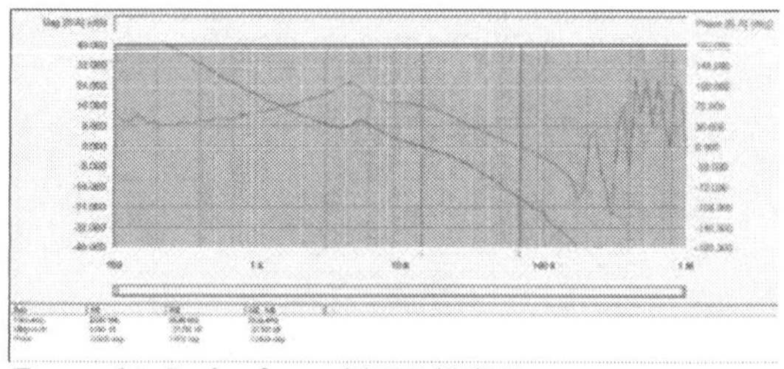

Figure 15. Bode plot at 200W (CCM)

8. CIRCUIT EQUIVALENT MODELLING

We will discuss now a practical and valuable way to model our converter for closed loop simulations without passing through space state averaging and linearization.

Equation (3), (4) for the ON time and (10), (6) for the OFF time, allow us to define an "averaged" equivalent circuit suitable to be simulated in PSPICE.

Averaging equations (3), (4), and (10), (6) and the expressions of the input voltage get:

$$L\cdot\left(\frac{d}{dt}i_{on}\right) = \left(\frac{3\cdot\delta}{2}-\frac{1}{2}\right)\cdot v_{in} - \frac{(\delta+1)\cdot R_L}{2}\cdot i_{on} - (1-\delta)\cdot V_{dp} - \delta\cdot n\cdot V_{ds} - \delta\cdot n\cdot v_{out}$$

(28)

$$C\cdot\frac{d}{dt}v_{out} = \delta\cdot\left(n\cdot i_{on} - \frac{v_{out}}{R}\right) + (1-\delta)\cdot\frac{-v_{out}}{R} \quad (29)$$

$$i_{in} = \delta\cdot i_{on} + (1-\delta)\cdot -\frac{i_{on}}{2} \quad (30)$$

Rearranged and simplified they become

$$L\cdot\left(\frac{d}{dt}i_{on}\right) = \left(\frac{3\cdot\delta}{2}-\frac{1}{2}\right)\cdot v_{in} - \frac{(\delta+1)\cdot R_L}{2}\cdot i_{on} - (1-\delta)\cdot V_{dp} - \delta\cdot n\cdot V_{ds} - \delta\cdot n\cdot v_{out}$$

(31)

$$C\cdot\left(\frac{d}{dt}v_{out}\right) = \delta\cdot n\cdot i_{on} - \frac{v_{out}}{R} \quad (32)$$

$$i_{in} = \left(\frac{3\cdot\delta}{2}-\frac{1}{2}\right)\cdot i_{on} \quad (33)$$

The equivalent circuit described by (31), (32) and (33) is depicted in fig. 16.

DOI: 10.1051/e3sconf/20171614004

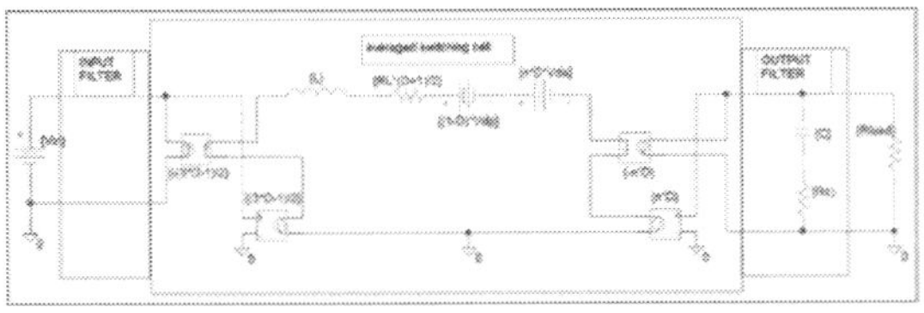

Figure 16. equivalent averaged circuit of the Full Bridge Watkin-Johnson converter.

In the PSPICE model we see three areas: "input filter", "averaged switching cell" and "output filter". In the first and latter areas components can be added as wished without considering "averaging" their value.

When adding extra components in the averaged switching cell, we have to "average" them. For example if we want to add the parasitic capacitance of the output diodes, then the actual value to be used in the circuit will be n*D*Cds, where Cds is the capacitance of the secondary diodes.

Using the circuit of figure 16 it is easy simulate a closed loop converter and plot the transfer functions, enabling the study the transient response while modelling the input and output filters. It is worth noting that the circuit in fig. 16 is a non-linear circuit, and it describes both DC and AC behaviour.

9. CONCLUSIONS

- A Full Bridge DC/DC converter based on the Watkins-Johnson topology has been presented and its static characteristics and waveforms in CCM and DCM have been analytically derived.
- A breadboard have been built to verify the analytical work, to study its limitations and to verify the stability of the converter in voltage closed loop.
- An average PSPICE model of the converter in CCM has been also presented.
- The authors believe that this work has confirmed that the Full Bridge W-J converter is a strong candidate for converters for Electrical Propulsion units.

10. SUGGESTIONS FOR FUTURE WORK

Further work should be in the area of:

- loop stability (with current loop)
- study the effect of the "parasitics"
- efficiency optimization (conditions for partial ZVS operations)
- HV output stage
- develop a modular approach to increase output power

11. REFERENCES

1. Robert W. Erickson (1997), *Fundamentals of Power Electronics*. International Thomson Publishing.
2. F. Tonicello, O. Mourra, *B2R (Buck & Boost Regulator) Control Aspects and Small Signal Analysis*, Space Power Workshop 2011, Los Angeles, CA.
3. V. Vorperian, *Simplified Analysis of PWM Converters Using the Model of the PWM switch: Parts I and II*. IEEE Transaction on Aerospace and Electronic Systems. Vol. AES-26, pp 490-505, May 1990.
4. E. van Dijk, J.B. Klaassens, H.J.N. Spruijt, D.M. O'Sullivan, *PWM Switch Modelling of DC-DC Converter Topologies*, Proceedings of the European Space Power Conference, Graz August 1993, pp 301-306.098

TWO-DOMAINS CONTROL OF A BUCK CONVERTER

Mariel Triggianese (1), Hadrien Carbonnier (1), Ferdinando Tonicello (1), Maria Rodriguez Rogina (2)

(1) *ESA/ESTEC, Keplerlaan 1, 2200AG Noordwijk, The Netherlands, Emails: name.surname@esa.int*
(2) *Campus de Viesques, Gijon-Asturias- Spain GP33204, Email: Rodriguezrmaria@uniovi.es*

ABSTRACT

In the design of a power converter, the maximum output voltage transient under load change is one of the main drivers. The ECSS-E-ST-20C imposes specific requirements on the power quality of a regulated bus, which result most of the time in the use of a large capacitor bank to cope with the linear loop dynamics. Moreover, this sometimes results in slow dynamic response.

In this paper a combined linear and non-linear control applied to a buck converter is proposed. This control technique enhances the transient performance with a faster recovery time and allows smaller output filter capacitors to be used. The proposed control is a combination of a linear control based on a conductance control principle, and a non-linear control, which intervenes only during transients. The latter is based on the detection of the output voltage variation and on the immediate application of maximum or minimum inductor current, as consequence of voltage undershoot or overshoot. Once the controlled output voltage is within the allowed range, the linear control takes back the voltage regulation control.

The buck converter is chosen as a reference topology since most of the current solar array regulators (SARs) use a buck or super buck topology. Most Battery Discharge Regulators (BDRs) are also based on buck-derived topologies.

This concept can find application in the three-domain regulated bus control.

1. INTRODUCTION

The proposed control technique is called "two-domains" because it is given by the combination of a linear control, called conductance control, well described in [5] and widely used in space, and a non linear part (called NL-control in the present paper), which is based on the detection of the output voltage variation and comparison of this variation with a pre-determined threshold. During a load transient, in case the output voltage crosses a threshold, the NL-control is activated and forces maximum or minimum inductor current. As soon as the linear control recovers its dynamic, the linear part of the modulating signal takes back control of the converter. In this way the converter works with fixed frequency in steady-state operation and for small signal variation the behaviour is linear.

In case of large load-steps, the operation of the converter is similar to a hysteresis control around a well-defined reference.

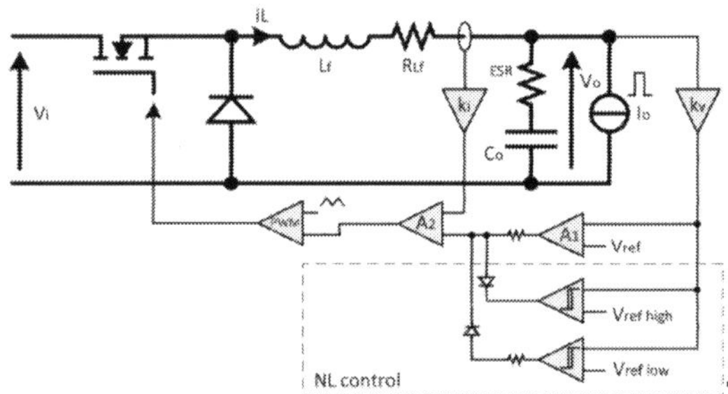

Figure 1: Generic block diagram for the two domain control scheme.

The objective of the proposed work is to design a system satisfying the following criteria:

- Enhanced step load response
- Lower output capacitance when compared to converters with conventional control
- Power quality, transient, ripple and stability requirements as per ECSS-E-ST-20C

1.1 Example of application

We can take as reference a 300 W buck converter providing a regulated 28 V output from an input voltage of 36 to 50 V. For such a system , ECSS-E-ST-20C imposes 1 % bus voltage variation for 50 % load change and 5 % variation under any other load change and inter-domain application. The maximum impedance that the converter can have is given by:

$$Z_0 = \frac{0.02V_0^2}{P_0} \quad (1)$$

For our example application the result is 52 mΩ. For a well-designed conductance control, the maximum output impedance is given by:

 DOI: 10.1051/e3sconf/20171614005

$$Z_0 = \frac{1}{k_v A_1 G} \tag{2}$$

Where k_v is the output voltage sensing gain and A_1 is the voltage loop controller proportional gain, as shown in *Figure 1*. G is the conductance of the power stage under closed loop conductance control. Assuming a high-bandwidth current loop, it can be approximated with the following equation:

$$G = \frac{1}{k_i} \tag{3}$$

Where k_i is the gain of the current sensor. The same well-known design criteria for a conductance control also implies the following:

$$C_0 = \frac{k_v A_1 G}{2\pi f_{BW}} \tag{4}$$

where f_{BW} is the desired bandwidth of the converter, which might be initially set to about 1/10 of the switching frequency. For a 300 W converter, we can suppose that a switching frequency of around 100 to 150 kHz is adopted in order to keep the losses low. This results in a bandwidth in the order of 10 to 15 kHz.

$$C_0 = \frac{1}{Z_0 2\pi f_{BW}} \tag{5}$$

which implies a capacitance of more than 300 µF. In a control system that features a non-linear part, which acts during transients and ideally has an infinite gain, such as the one of a comparator used in a hysteresis control, the output impedance is in theory close to zero, if Eq. (2) is taken as reference.

The transient performance is therefore mainly dominated by the output filter dynamics, i.e. the voltage variation across the capacitor due to the current ripple.

It is worth noticing that the NL-control can be designed to act on the set point of the current loop and not directly on the duty cycle. It is also possible to inject the NL signal at several places in the loop, as shown in *Figure 1*. In this case it is combined with the voltage loop, to change the reference of the current loop as fast as possible. This approach also allows to limit the current in case of overload.

In the following paragraph the design criteria for the combined conductance and NL-control are presented.

2. DESIGN CRITERIA

The converter topology that is considered is a buck converter. The design criteria of a buck converter are well-known. In particular, the design of the output capacitor, based on the maximum ripple requirement in steady state and on the maximum voltage variation during step load. The driving requirement is generally the maximum variation under step-load.

The proposed NL-control part, shown in Figure 1, can be explained looking at *Figure 2*. It consists of two comparators with hysteresis that detect if the output voltage is higher (or lower) than a pre-determined threshold V_{tr_high} (or V_{tr_low}). When this happens, a command called V_{cmd_high} (or V_{cmd_low}) is sent and combined with the voltage control loop signal. This command will therefore impose full (or minimum) current.

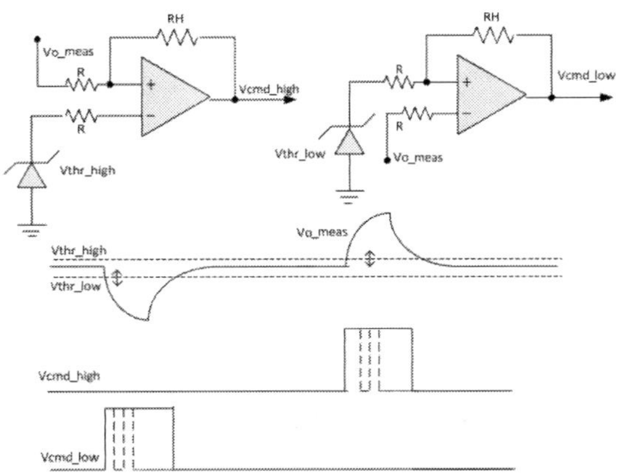

Figure 2: NL-control comparators

This results in a form of hysteresis control applied on the output voltage for a certain period of time, until the linear control dynamic takes back control of the voltage regulation.

2.1 Output capacitor design criteria

The main requirement that is considered for the design of the buck output capacitor is the allowed transient during 50 % step load, which is 1 %. During this operation mode, the NL-control part takes the lead and imposes a sudden change in the current reference, i.e. maximum (or minimum) current. This will result in full (or eventually the one limited by the control circuitry) (or zero) duty cycle. The inductor current will ramp up (or down) imposing the same AC current variation in the output filter capacitor.

As described in [2], the minimum achievable output voltage transient appearing at the onset of an overload is:

DOI: 10.1051/e3sconf/20171614005

$$\Delta V_o = \frac{\Delta I_o^2 L_f}{2(V_{in} - V_o) \cdot C_o} \tag{6}$$

for undershoot and:

$$\Delta V_o = \frac{\Delta I_o^2 L_f}{2 V_o C_o} \tag{7}$$

for overshoot, with ΔI_o being the load (=inductor current) step and L_f the buck inductor value.
In a buck converter, the worst-case transient is the undershoot case. The minimum value of output capacitance necessary for achieving a given ΔV_o is therefore calculated considering the Eq. (6).
Targeting ΔV_o to be less than $0.01 \text{x} V_o$:

$$C_0 \geq \frac{\Delta I_o^2 L_f}{0.02 V_o (V_{in} - V_o)}, \tag{8}$$

which becomes:

$$C_0 \geq \frac{\Delta I_o^2 L_f}{0.02 V_o (V_{in,\min} - V_o)}, \tag{9}$$

The actual ΔI_o varies with the exact moment the transient occurs. The worst case occurs when the inductor current is at the end of the switching period and the MOSFET is OFF. For example, if the load changes form 50% to 100% load and the MOSFET is OFF, without taking into account other delays, the overall ΔI_o in the Eq.(7) becomes: $\frac{I_0}{2} + I_{0,ripple,pk-pk}$.
The bigger the ripple, the deeper the transient and a bigger capacitor needs to be used.

Moreover, to get the minimum voltage variation it is desirable to have the minimum hysteresis in the NL-control as well as minimum delays in the feedback path (for example due to the comparators and to the driver). In fact all these factors result in additional contributions to the output voltage variation. For example the hysteresis band contribution ΔV_{hyst} will result in an increase of the voltage variation of $\frac{\Delta V_{hyst}}{k_v}$.
In principle, the faster the voltage detection and the entire chain of commands, the smaller the resulting voltage transient during application of the NL control.

3. SIMULATION RESULTS

Before presenting experimental results, let us show some simulation results. The case study that has been bread-boarded (and simulated) is a converter of 300 W maximum power, 36 - 50 V input voltage range and 28V output voltage. The control system consists of a conventional conductance control, with conductance gain $G = 0.36$, $k_v = 0.118$ and sized in order to have 60° phase margin. The voltage loop control signal is also combined with the output of the two comparators described in paragraph 2.

The PWM switching frequency is 300 kHz and the inductor is 13 μH. According to Eq.(8), the output capacitor shall be greater than 80 μF. A capacitor 85 μF has been used in the simulation and in the breadboard.

Figure 3 and *Figure 4* show the transient performance when 50 % load change is applied (100%full load to 50% in *Figure 3* and viceversa in *Figure 4*), without NL-control (figure (a)) and with NL-control (figure (b) and (c)). The blue track is the output error signal V_o-V_{ref} and the green one is the output of the NL-comparators. In Figures (b) and (c) the NL-comparator thresholds V_{tr_high} and V_{tr_low} shown in *Figure 2* have been changed in order to verify the effect.

The effect of the NL-control on the overall peak of the voltage variation can clearly be seen.

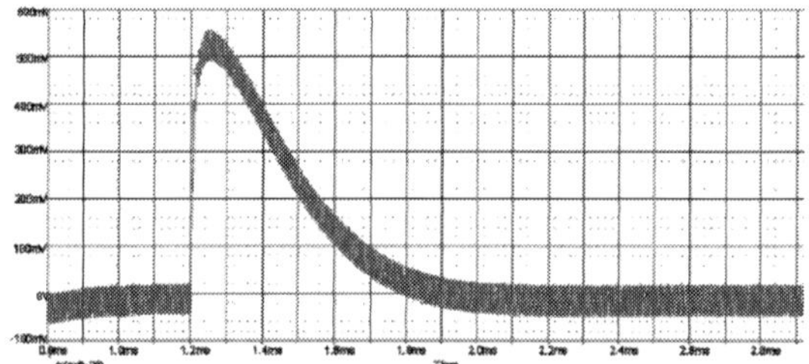

(a)

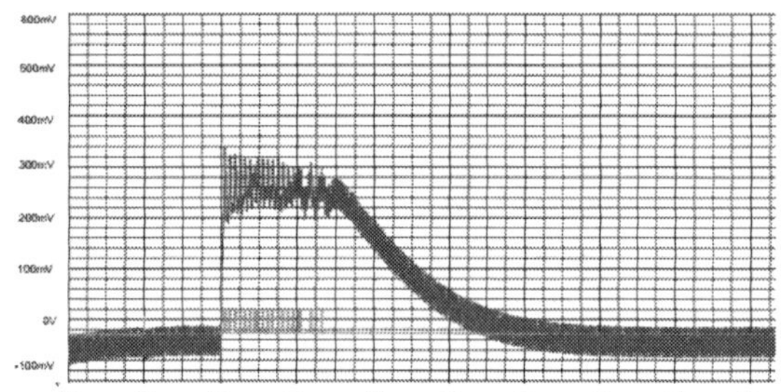

(b)

DOI: 10.1051/e3sconf/20171614005

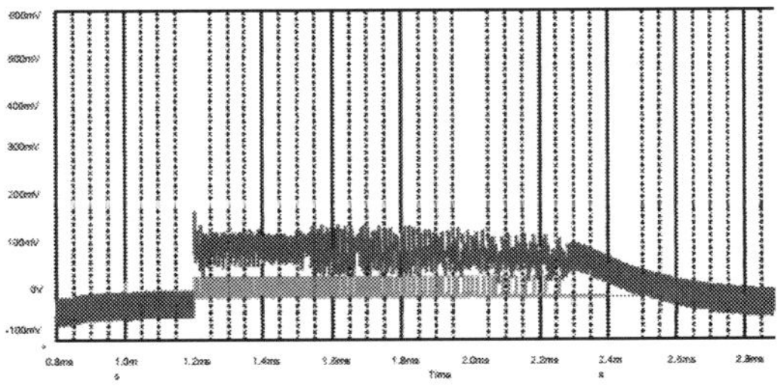

(c)

Figure 3: 100% full load to 50% load transient performances without NL-comparators (figure (a)) and with NL-control at different thresholds ((b) and (c)). Blue: V_o-V_{ref} ; green: NL-comparator output.

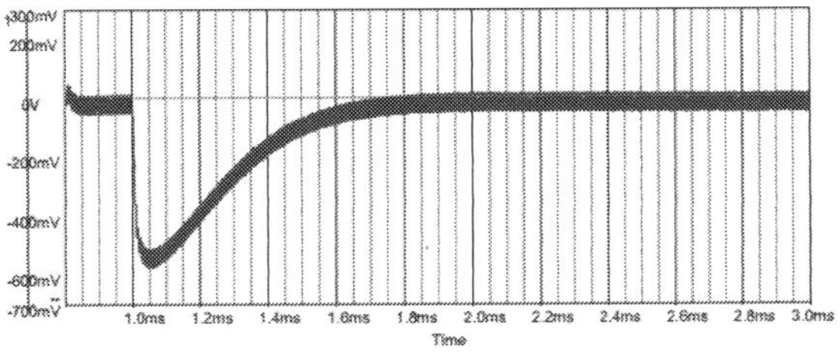

(a)

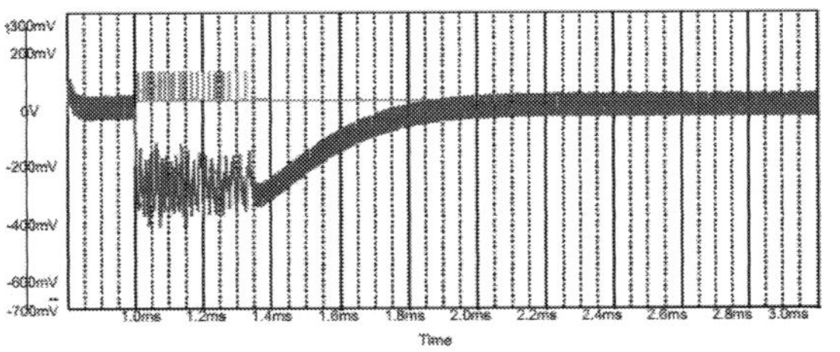

(b)

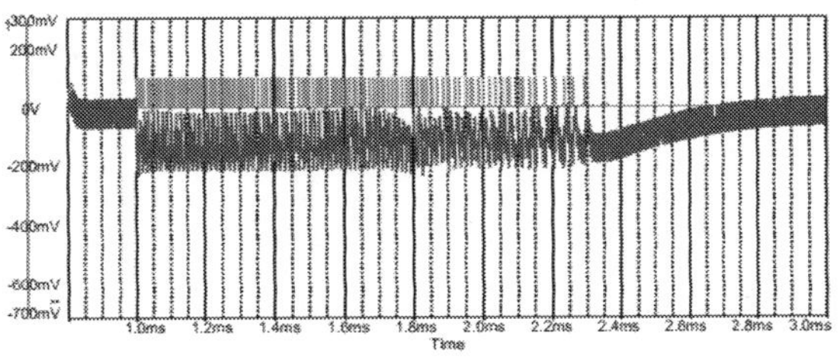

(c)

Figure 4: 50% load to 100% load transient performances without NL-comparators (figure (a)) and with NL-control at different thresholds ((b) and (c)). Blue: V_o-V_{ref} ; green: NL-comparator output.

4. EXPERIMENTAL RESULTS

The converter described in the previous part was built on a PCB in the ESTEC power electronics laboratory. The level of the switching noise needs to be kept very low due to the sensitive nature of the NL control circuits (high speed comparators). This was achieved through the use of ground planes, careful layout of the power cell for low parasitics, controlled switching times for the MFETs and implementation of snubbers. The PWM controller is implemented entirely using discrete components mainly to be able to interface it at several levels with the NL-control circuit part. It is also capable of commanding the power MFETs from 0 % to 100 % duty cycle. The converter also features an output overload protection and a soft start, and reaches an efficiency of 95 % at full load. The prototype is shown in *Figure 5*.

Figure 5: Two-domains buck converter prototype

The voltage loop of the linear control circuit has been designed with a bandwidth of 16kHz and the current loop with a bandwidth of 95kHz. The voltage loop measurements are shown in *Figure 6*.

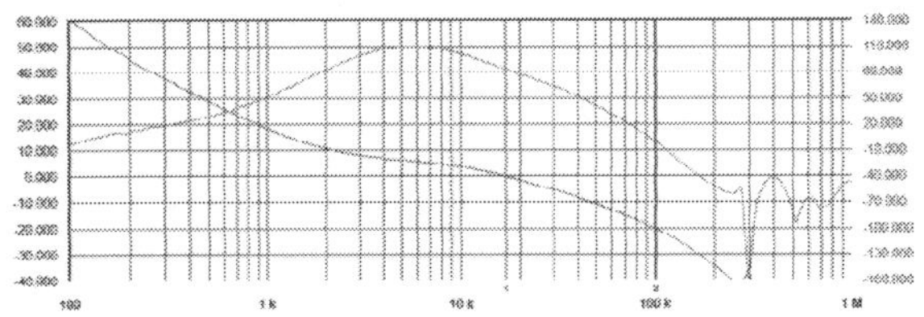

Figure 6: Voltage loop measurements

For the NL-control side, two different methods for detecting the output voltage variation have been used. The first is based on a DC detection, as in a normal conductance control (shown in *Figure 2*), while the other one is based on AC detection, which senses the high-frequency components of the output voltage with a much higher gain, resulting in better noise immunity (as shown in *Figure 7* with the addition of a capacitor across the high side sensing resistor).

The experimental results of output voltage transient at 50% load change are summarised below for a DC coupled detection and AC coupled one. The magenta trace is the AC output voltage component and the green one is the output of the NL-comparator.

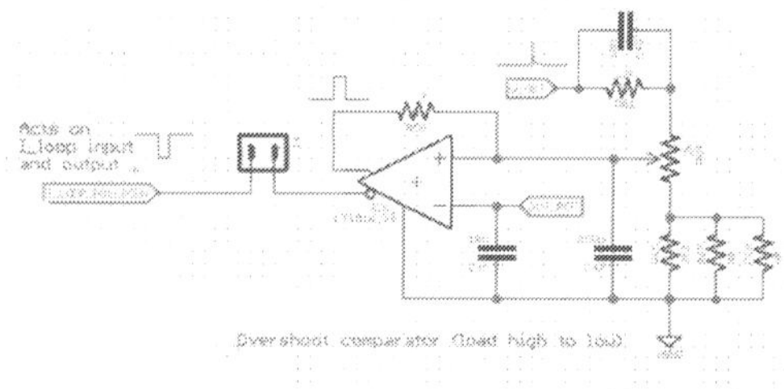

Figure 7: Prototype NL-comparator circuit schematic wired for AC detection.

4.1 DC coupled: Low to High step load tests (50% to 100% load power)

The experimental results show the output voltage transient performance (AC variation) without connecting the NL-comparator (*Figure 8)* and with NL-comparators at three different DC detection thresholds (*Figure 9* to *Figure 11*).

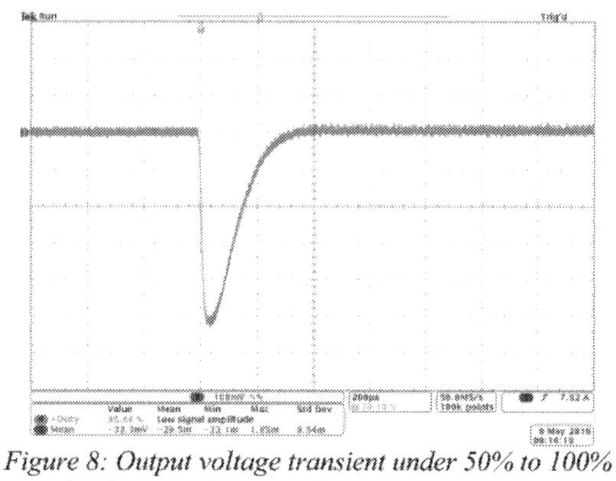

Figure 8: Output voltage transient under 50% to 100% load power change without NL-comparators

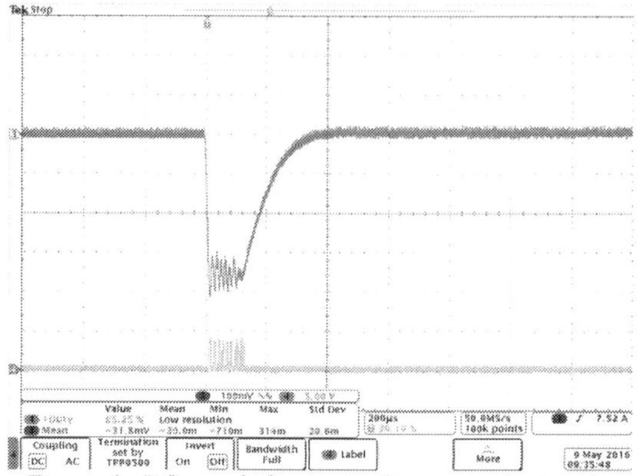

Figure 9: DC coupled output voltage transient under 50% to 100% load power change with NL-comparators

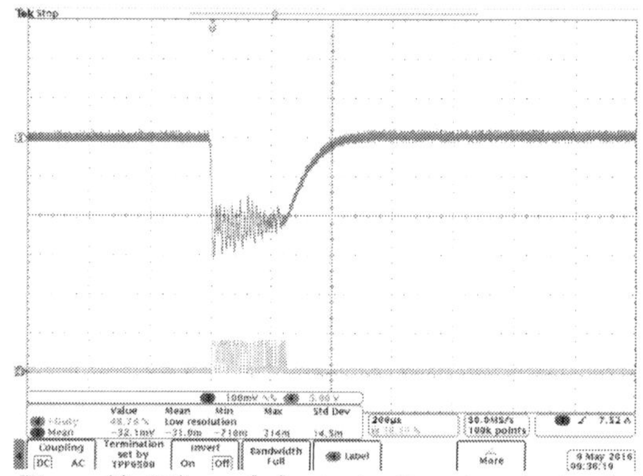

Figure 10: DC coupled output voltage transient under 50% to 100% load power change with NL-comparators

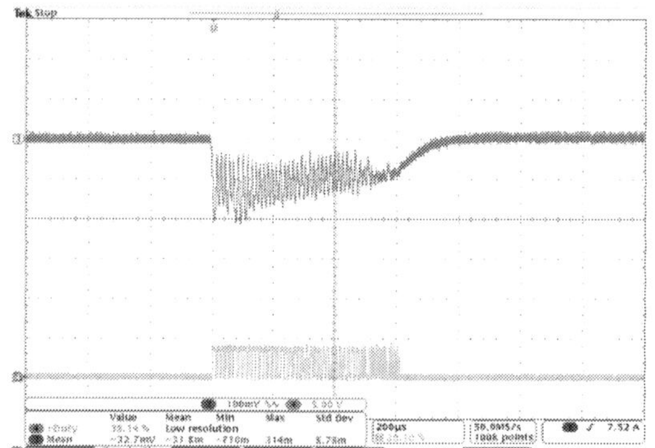

Figure 11: DC coupled output voltage transient under 50% to 100% load power change with NL-comparators

DOI: 10.1051/e3sconf/20171614005

4.2 AC coupled: Low to High step load tests (50% to 100% load power)

The experimental results show the output voltage transient performance (AC variation) with NL-comparators at two different AC variation thresholds (*Figure 12* and *Figure 13*)

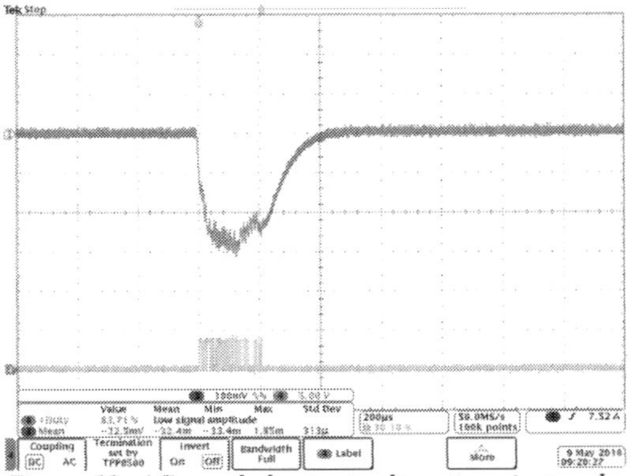

Figure 12: AC coupled output voltage transient under 50% to 100% load power change with NL-comparators

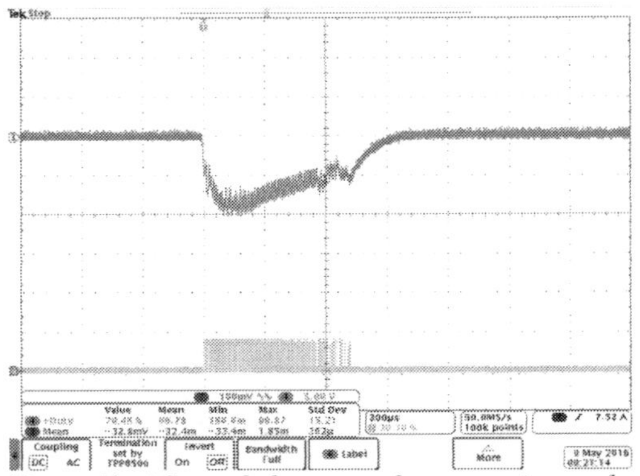

Figure 13: AC coupled output voltage transient under 50% to 100% load power change with NL-comparators

4.3 DC coupled: High to Low step load tests (100% to 50% load power)

The experimental results show the output voltage transient performance (AC variation) without connecting the NL-comparator (*Figure 14*) and with NL-comparators at two different DC variation thresholds (*Figure 15* and *Figure 16*)

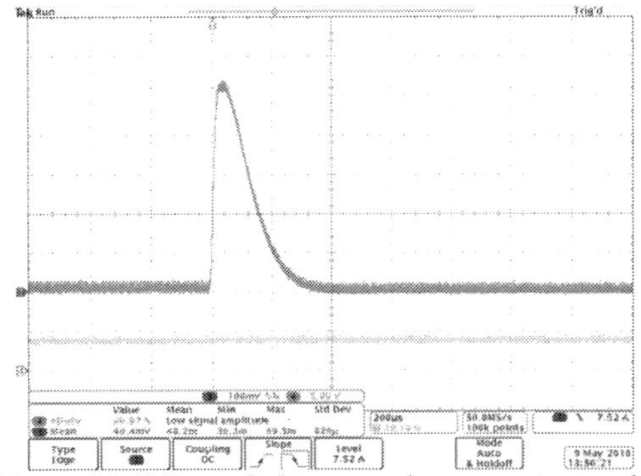

Figure 14: DC coupled output voltage transient under 100% to 50% load power change without NL-comparators

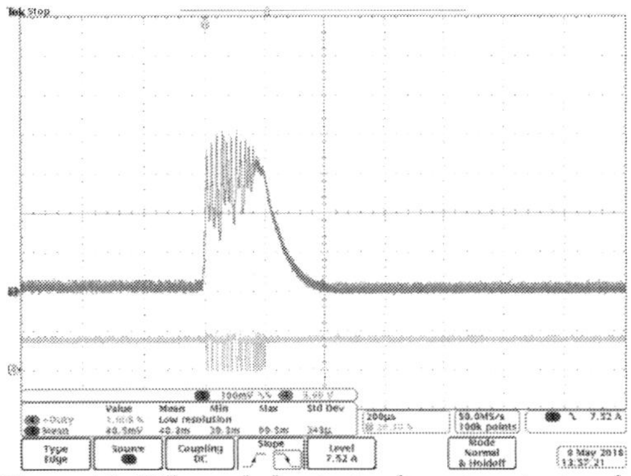

Figure 15: DC coupled output voltage transient under 100% to 50% load power change with NL-comparators

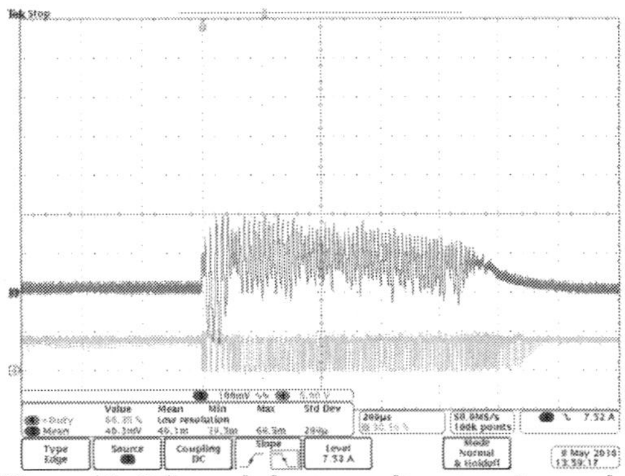

Figure 16: DC coupled output voltage transient under 100% to 50% load power change with NL-comparators

4.4 AC coupled: High to Low step load tests (100% to 50% load power)

The experimental results show the output voltage transient performance (AC variation) with NL-comparator at the smallest reference threshold (*Figure 17*)

 DOI: 10.1051/e3sconf/20171614005

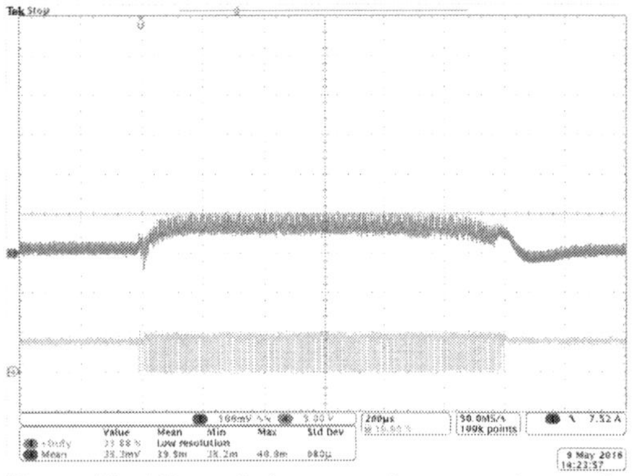

Figure 17: AC coupled output voltage transient under 100% to 50% load power change with NL-comparators

Tests show that the AC coupling gives better results and it will be taken as reference for this type of control.
They also show that this control used 85µF against the min 300µF of a conventional conductance control.

4.5 NL-control ripple characterization

The voltage ripple during the transient NL-control depends on many factors.

In a system where the ESR of the output capacitor is relatively big [3], the ripple reflects the hysteresis of the comparators. Knowing the ESR, the load current and the applied hysteresis window and internal delays it is possible to calculate the frequency.

In our case-study we aim for the best performance in terms of output voltage variation during transition and minimum ripple. This means: the ESR is very small, the hysteresis window is set to be as small as possible and minimum internal delays are aimed. In this case the evaluation of the hysteresis switching frequency becomes very difficult. In a system with zero delays and hysteresis the frequency is in theory infinite. Testing is the best solution for this type of control, where each single delay contribution plays an important role in the performance.

From the testing done on the prototype, it seems that the frequency remains more or less constant with the input voltage, while the ripple of the inductor current slightly changes, as shown in the *Figure 19* and *Figure 18* It can be also seen that the current ripple is given by:

$$\Delta I_{ripple} \approx \frac{\left(V_{in} - V_{o}\right) V_{o} T_{hyst}}{L_{f} V_{in}} \qquad (10)$$

where T_{hyst} is the period of oscillations. The current ripple is indeed around 6.5A in one case and the half of that in the other one.

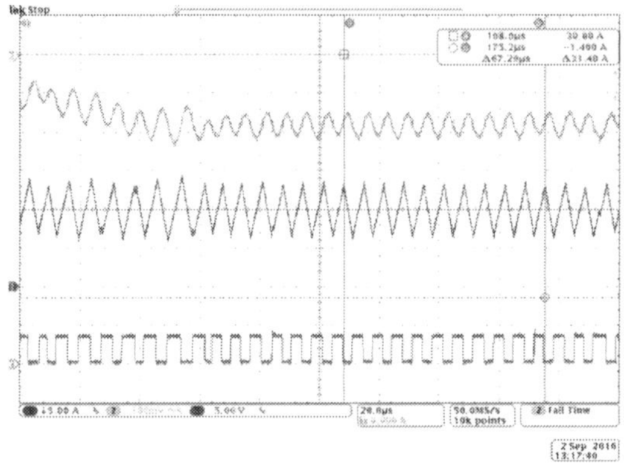

Figure 18: 50% to 100% load transient: output voltage error (light blue), inductor current (blue), NL-comparator signal output (magenta)@50V input

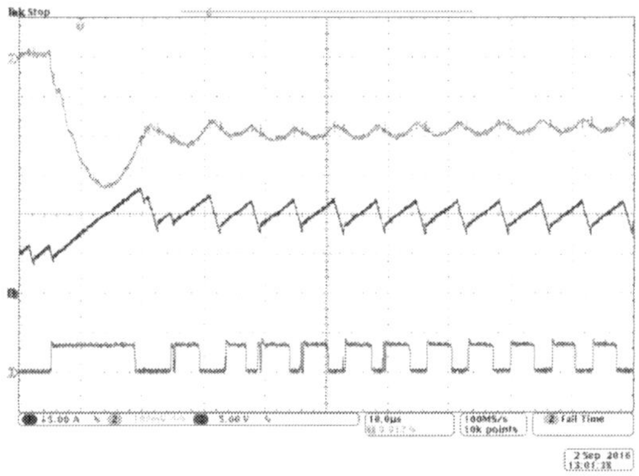

Figure 19: 50% to 100% load transient: output voltage error (light blue), inductor current (blue), NL-comparator signal output (magenta)@36V input

The output capacitor ripple during the NL-control can also be characterized with the following formula:

$$\Delta V_0 = \frac{\Delta I_{ripple}}{8C_0} T_{hyst} \qquad (11)$$

It is in fact linked to the current ripple, as can be seen in *Figure 20*, where the $\Delta I_{ripple,}$ is 6A and T_{hyst} is 6.8µs. The ΔV_0 is indeed 57mV.

DOI: 10.1051/e3sconf/20171614005

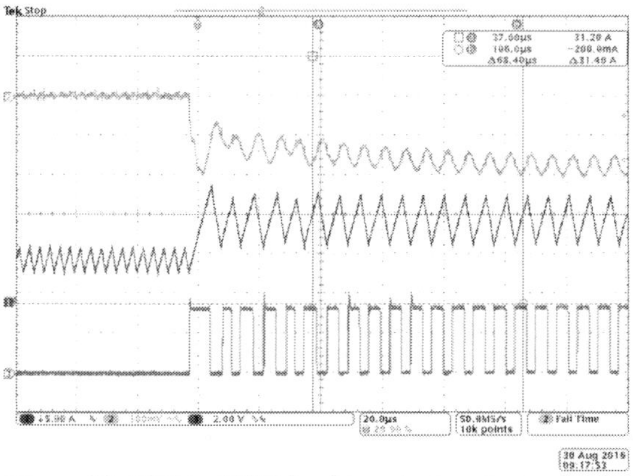

Figure 20: 50% to 100% load transient: output voltage error (light blue), inductor current (blue), NL-comparator signal output (magenta)

5. CONCLUSIONS

In this paper a combined linear and non-linear control applied to a buck converter has been presented. The proposed control is a combination of a linear control based on a conductance control principle, and a non-linear control, which intervenes only during transients. The latter is based on the detection of the output voltage variation and on the immediate application of maximum or minimum inductor current, as consequence of voltage undershoot or overshoot. Once the controlled output voltage is within the allowed range, the linear control takes back the lead of the voltage regulation.

This control technique has been applied to a buck converter, which has been bread-boarded in TEC-EP Power Laboratory. The design criteria have also been presented together with the ripple characterization.

Two different detection methods have also been designed.

Experimental results demonstrated that the proposed control technique is able to satisfy the requirements reducing the output capacitor size w.r.t. a conventional conductance control.

The effects on component stress will actually depend on the specific application, i.e. on the frequency of large load transients occurrence. If large load transients are expected to occur with high rate, then the inductance size might be reviewed w.r.t. the one used in a system with a conventional control, as it is done in the present paper.

6. REFERENCES

1. A. Barrado, R. Vizquez, A. Lizaro, J. Pleite, E. Olias. Fast Transient Response with Combined Linear-Non-Linear Control Applied to Buck Converters. PESC 2002.
2. Richard Red, Brian P. Erisman and Zoltan Zansky. Optimizing the Load Transient Response of the Buck Converter. APEC 1998.
3. L.K. Wong and T.K. Man. Steady State Analysis of Hysteretic Control Buck Converters. Power Electronics and Motion Control Conference, 2008.
4. Miguel Castilla, Luis García de Vicuña, Josep M. Guerrero, José Matas, and Jaume Miret,Designing VRM Hysteretic Controllers for Optimal Transient Response, IEEE Transaction on Industrial Electronics, Vol. 54, No. 3, June 2007.
5. O'Sullivan, H. Spruyt, A. Crausaz, PWM conductance Control, PESC '88.

DOI: 10.1051/e3sconf/20171614006

BENEFITS AND DRAWBACKS OF A HIGH FREQUENCY GAN ZVZCPS CONVERTER

J. M. Blanes [1], **A. Garrigós** [1], **R. Gutierrez** [1], **J. A. Carrasco** [1], **E. Maset** [2], **J.B Ejea** [2], **E. Sanchis-Kilders** [2], **A. Ferreres** [2], **J. L. Lizán** [2]

[1] *Industrial Electronics Group, Miguel Hernández University of Elche (Spain), Email: jmblanes@umh.es*
[2] *Laboratory of Industrial Electronics and Instrumentation, University of Valencia (Spain), Email: emaset@uv.es*

ABSTRACT

This paper presents the benefits and drawbacks of replacing the traditional Si Mosfets transistors with enhancement mode GaN transistors in a Half-Bridge Zero Voltage and Zero Current Switching Power Switching (ZVZCPS) converter. This type of converters is usually used as Electronic Power Converters (EPC) for telecommunication satellites travelling-wave tube amplifiers (TWTAs). In this study, firstly the converter is theoretically analysed, obtaining its operation, losses and efficiency equations. From these equations, optimizations maps based on the main system parameters are obtained. These optimization maps are the key to quantify the potential benefits of GaN transistors in this type of converters. Theoretical results show that using GaN transistors, the frequency of the converter can be pushed from 125kHz to 830kHz without sacrificing the converter efficiency. This frequency increase is directly related to reduction on the EPC size and weight.

To validate the theoretical results, a 150W prototype was built and its performance measured. Experimental results show an unexpected operation at high frequencies, mainly due to the influence of the transformer parasitic capacitance and the rectifiers diodes reverse recovery effect. Finally, it can be concluded that the parasitic capacitance of the converter and the diodes recovery time are, in practice, critical parameters to increase the switching frequency keeping the ZVZCPS.

1. INTRODUCTION

In telecommunications satellites, Travelling Wave Tube Amplifiers (TWTAs) represent about 35% of the payload mass. Therefore, any improvement in its Electronic Power Converter (EPC) size and weight is critical to reduce the overall mass of the satellite. In European satellites, the EPC is usually a two-stage converter; a Buck regulator connected in series with a Zero Voltage Zero Current Power Switching (ZVZCPS) isolated high voltage converter. The ZVZCPS technique was proposed in [1] and its main advantage is that all the converter switches always operate under both Zero Voltage Switching and Zero Current Switching when Switching ON and OFF. This advantage allows achieving vey high efficiencies (better than 98%) and operating at a fixed frequency with very low EMI noise generation.

The emergence of Gallium Nitride (GaN) Transistors, with lower on resistance and lower switching losses than the conventional Silicon Mosfets, and its demonstrated robustness against radiation effects, has promote many research works focused on evaluating the benefits of this technology on power converters for aerospace applications [2-4]

This paper is focused on the potential reduction of the size and the weight of the ZVZCPS converter using GaN Transistors, increasing the switching frequency but without affecting the overall losses of the converter. Section 2 briefly describes the mathematical model of the converter, obtaining its main waveforms and the efficiency equations. In section 3 optimization maps have been developed to optimize the converter design. To validate the theoretical experimental results of a 150W prototype are presented in section 4. Finally, conclusions are presented in section 5.

2. ZVZCPS HALF-BRIDGE OPERATION

For this study, a current-fed half bridge ZVZCPS topology has been chosen. The classical current-fed push-pull topology has been discarded because, in high power European telecommunication satellites, the main bus voltage of the satellite is regulated at 100V. The EPC is supplied from the main bus, so if a push-pull topology is used, following the European Space Agency (ESA) derating standards, 250 V_{ds} transistors are needed. At the beginning of this study no commercial GaN transitors were available with this characteristic so the comparison could not be done.

In Figure 1 it is shown the ZVZCPS Half-Bridge Topology, with only two outputs for simplicity, and in Figure 2 are depicted the main waveforms of the converter.

During a period, there are four different states, named, T_{on1}, T_{gap1}, T_{on2} and T_{gap2}.

State T_{on1}: This interval starts by turning ON switch M_1 when its voltage is zero. The current through M_1 increases in a resonant manner because C_{TUN1} resonates with the leakage inductance of the main transformer, producing a sinusoidal current through M_1. The parasitic capacitance of M_2 (C_{ds2}) is charged to $V_{Ctun1}+V_{Ctun2}$

 DOI: 10.1051/e3sconf/20171614006

voltage and the main transformer magnetizing current (i_{Lm}) grows linearly. This interval ends when the current through the transistor M_1 is almost zero (equal to the magnetizing current).

State T_{gap1}: When the current through M_1 is almost zero (equal to the magnetizing current), switch M_1 is turned OFF. During this interval the energy stored in L_m forces the discharge of the M_2 parasitic capacitance (C_{ds2}), charging the M_1 parasitic capacitance (C_{ds1}) linearly at a constant current. During this interval there is no transfer of energy from the input to the output and all the diodes are OFF. The tuning capacitors C_{tun1} and C_{tun2} are linear charged with the input current of the converter I_{in}.

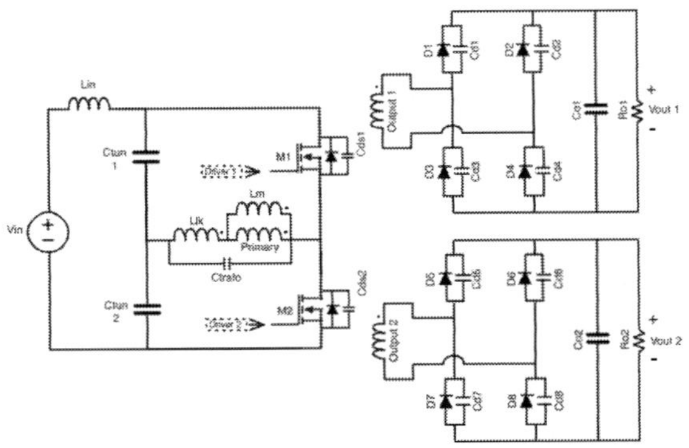

Figure 1. Two output Half-Bridge ZVZCPS Converter

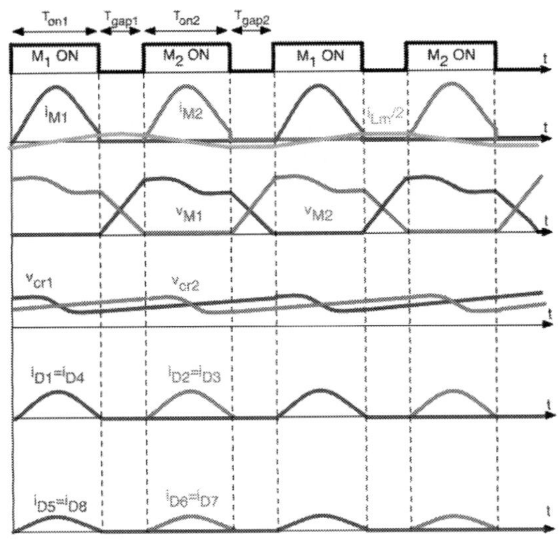

Figure 2. Two output Half-Bridge ZVZCPS Converter main waveforms

Intervals T_{on2} and T_{gap2} are analogous to interval T_{on1} and T_{gap1} respectively. In state T_{on2}, switch M_2 is turned ON and the parasitic capacitance of M_1 (C_{ds1}) is charged to $V_{Ctun1}+V_{Ctun2}$. During interval T_{gap2} The energy stored in L_m forces to discharge the M_1 parasitic capacitance (C_{ds1}), charging the M_2 parasitic capacitance (C_{ds2}).

In order to optimize efficiency of the designed converter, all the converter power losses equations have been calculated.

2.1 Transistor Losses

Conduction Losses: The transistor conduction losses are directly related with the *RMS* current flowing through them and their ON resistance. To be more accurate the magnetizing current is also used in this calculus. The magnetizing current during the ON time has a quasi-triangular shape so its RMS value can be easily calculated:

$$P_{Mos} = 2I_{Mos(RMS)}{}^2 R_{ds} + I_{Lm(RMS)}{}^2 R_{ds} \quad (1)$$

Switching Losses: Although the converter has been described as it always commutates at Zero Voltage and Zero Current and there are not switching losses, this is not really true. When the transistor is turned off, the magnetizing current is flowing through it. To calculate these losses, the voltage of the transistor during the OFF commutation (V_{off}) has been simplified to the voltage of the parasitic capacitance being charged at a constant current I_{Lm} during the time that the switch last to be turned off (T_{off}).

$$V_{off} = \frac{T_{off}\, I_{Lm}}{C_{eff}} \quad (2)$$

$$P_{off} = 2\,(0.5\, V_{off}\, I_{Lm}\, T_{off}\, F_{sw}) \quad (3)$$

Driver Losses: Driver losses have to be considered in both transistors and they are calculated multiplying the energy provided to the transistor gate by the switching frequency.

$$P_{gate} = 2(V_{drv}\, Q_{gate}\, F_{sw}) \quad (4)$$

2.2 Diode Losses

The current flowing through the diodes is proportional to the transistors multiplied by the transformer turn relation ratio.

$$P_D = 4I_{Dsec1(AVG)} \cdot V_{FDsec1} + 4I_{Dsec2(AVG)} \cdot V_{FDsec2} \quad (5)$$

2.3 Tuning Capacitors Losses

The losses are the ones associated to the equivalent serial resistance (ESR) of the capacitors:

$$P_{CTun(RMS)} = 2I_{CTun(RMS)}{}^2 \cdot ESR_{CTun} \quad (6)$$

 DOI: 10.1051/e3sconf/20171614006

2.4 Output Capacitors Losses

As in the previous case, losses are associated to the equivalent serial resistance (ESR) of the output capacitors:

$$P_{Co} = {I_{C_{o1}(RMS)}}^2 \cdot ESR_{Co1} + {I_{C_{o2}(RMS)}}^2 \cdot ESR_{Co2} \qquad (7)$$

2.5 Transformer losses

Core Losses: For the calculations, it is used the equation provided by Ferroxcube in the application note "Planar Transformed Design" [5]. Core losses are calculated as function of frequency (f in Hz), peak flux density (B_{peak} in T) and the temperature (T in °C)

$$P_{core} = C_m f^x {B_{peak}}^y (ct_0 - ct_1 T + ct_2 T^2) \quad (8)$$

In this formula, C_m, x, y. ct_0. ct_1 and ct_2 are parameters which have been found by curve fitting of the measured power loss data. These parameters are specific for a ferrite material and can be found in the Ferroxcube application note.

Copper Losses: Copper losses are the ones produced by the current in the conductors of the transformer windings due to their associated resistance.

$$P_{cu} = \frac{\rho\,(MLT)\,{n_1}^2\,{I_{tot(RMS)}}^2}{W_A\,K_u} \qquad (9)$$

Where: ρ is the wire effective resistivity, MLT is the mean length of the transformer turns, n_1 is the number of turns in the primary, $I_{tot(RMS)}$ is the sum of the rms winding currents referred to the primary, W_A is the window area of the core and K_u is the window utilization factor, usually 0.1878 for planar transformers.

The minimum number of turns (n_1) is defined by the maximum peak flux density allowed to limit the core losses of the transformer.

$$n_1 \geq \frac{\frac{V_{in}}{2} T_{ON}}{2 \cdot B_{peak} \cdot A_e} \qquad (10)$$

3. ZVZCPS HALF-BRIDGE OPTIMIZATION MAPS

From the converter losses equations derived in previous section, a graphic representation of the converter losses and efficiency can be obtained. These graphic representations have been called "*Optimization Maps*" where all the losses and the efficiency are represented as function of the transistors ON time (T_{ON}) and the magnetizing inductance of the transformer (Lm). These parameters have been chosen because the whole converter design depends only on the value of these two parameters. These optimization maps are the key to quantify the potential benefits of GaN transistors in this type of converters.

The comparison has been done for a 150W 50V converter with two 75W outputs (250V and 500V). Two converters have been designed from these specifications, a Si converter operating at a frequency of about 125kHz and a GaN converter operating at a frequency of about 1MHz. For the Si converter the transistors chosen are RJK1056DPB from Renesas Technology and the diodes are MUR130 for the 250V output and the MUR160 for the 500V output. For the GaN converter the transistors are EPC2001 from EPC and the diodes are the same as in the Si converter. It is important to emphasize that, although the GaN transistor have better characteristics than the Si ones, their drain-source parasitic capacitance is of the same order or even higher than in Si transistors. This is one of the major drawback for pushing up the frequency using GaN transistors in this type of resonant converters, because during the GAP time this capacitance have to be charged/discharged as it was explained in section 2.

The transformer has been designed with the restriction that its maximum core losses should be less than the 0.25% of the converter power (325mW). From this restriction the maximum peak flux density is calculated, and once the maximum B_{peak} is fixed, the core material, dimensions and the number of turns are calculated. For this study the cores needed are: E43+PLT43 (material 3F3) with 5 turns in the primary side for the Si design and E32+PLT32 (material 3F3) with 3 turns in the primary side for the GaN design.

The final size and weight of the designed EPC converters are: 225 cm^2 / 397g for the Si one and 121 cm^2 / 198g for the GaN one.

The optimization maps obtained for the Si and GaN designs are depicted in Figure 3. In each map the centre of the overlapped circle shows the chosen operating point for the comparison (T_{on}=4µs, Lm=50µH, Freq=125kHz in the Si Converter and T_{on}=0.5µs, Lm=3.1µH, Freq=830kHz in the GaN Converter). In both designs the converter theoretical efficiency is the same, 98.25%.

From the optimization maps can be deduced that transistor losses are lower as the switching frequency decreases (higher T_{on}). This is due to the fact that both, the switching losses and the driver losses are directly proportional to the switching frequency. Besides, it can be observed the trend to increase losses with the reduction of L_m. This is because a reduction of L_m involves a higher magnetizing current and then also higher switching losses.

In the transformer losses maps, it can be observed that the trends in both designs are different. This is mainly produced by the frequency limits of the core material (*3F3*). In the the GaN design, working at 830kHz, this material is very near to its frequency limit.

DOI: 10.1051/e3sconf/20171614006

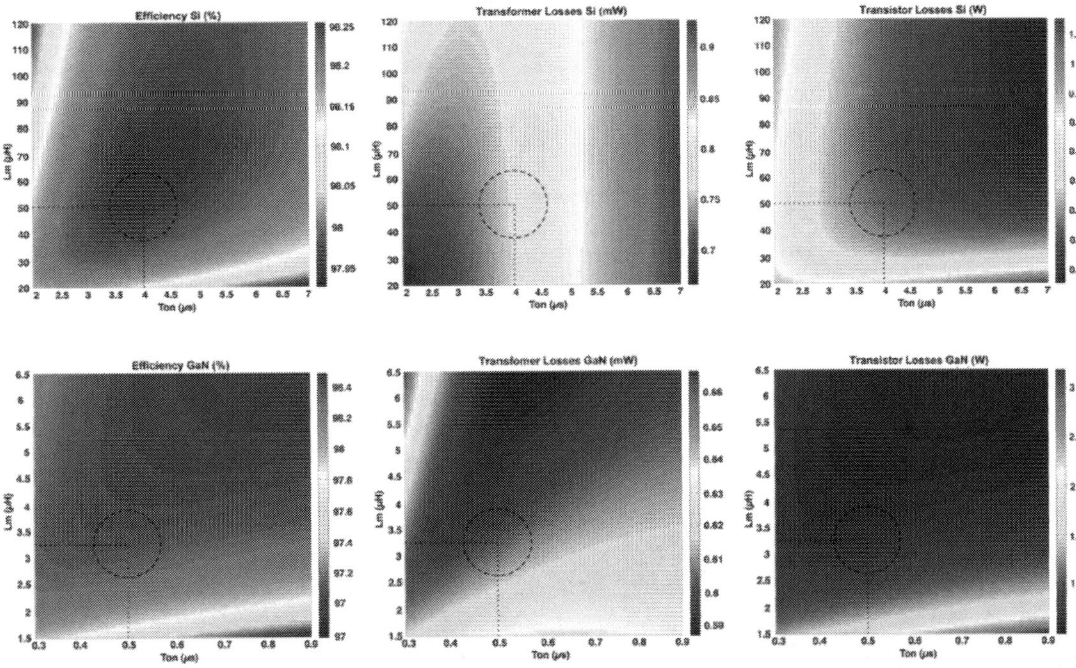

Figure 3. Optimization Maps.

x-axis variable: Transistors ON time, y-axis variable: Transformer Magnetizing Inductance.

Finally, in the efficiency maps it can be observed that none of the converters are working at the maximum efficiency point, maximum efficiency can be achieved decreasing the switching frequency but increasing the size and the weight of the converter. Efficiency higher than 98% has been fixed as target for the comparison. As can be seen, this efficiency can be achieved using GaN transistors switching at a frequency of 830kHz, about seven times higher than the frequency used with the traditional Si transistors (125 kHz).

4. EXPERIMENTAL PROTOTYPE

A 150W experimental prototype was implemented to validate the theoretical results. The desired control signals were generated with an astable 4017 and an LM5113 GaN driver from Texas Instruments was used to switch the GaN transistors.

Initially, a planar transformer was designed and implemented. The core used was E32+PLT32 (material 3F3) with 3 turns in the primary side, 30 turns in one secondary and 60 turns in the other one. The secondaries were implemented in a 16-layer PCB and the primary was implemented using three turns of copper foil. Once implemented, the transformer was measured and characterized, dealing in a parasitic capacitance of 34nF. This parasitic capacitance is too high for a ZVZCPs converter working at 830kHz. It is worth noting that the total parasitic capacitance of the converter must be charged/discharged during the GAP time, using the energy stored in the magnetizing inductors, so if the parasitic capacitance is too high it is needed an increment of the GAP time (reducing the frequency) or an increment of the magnetizing current (incrementing the losses). In figure 4 it is shown the experimental prototype with the planar transformer.

A second transformer was implemented on a RM14 core (material N49) with 3 turns in the primary and 30 and 60 turns in the secondaries. Its magnetizing inductance was adjusted to 3μH and the measured parasitic capacitance was1.32nF. This value is more appropriate for the converter so this transformer was the one used for the experimental tests.

In figure 5 it is shown an oscilloscope screenshot with the main converter waveforms. As can be seen, the experimental waveforms do not match with the theoretical ones during the GAP time. The current through the primary of the transformer is not null, it has a sinusoidal shape, even changing its polarity. This is due to rectifiers diodes reverse recovery effect, when the diodes are switched off, the reverse current is reflected in the primary side, even cancelling the magnetizing current and requiring the body-diode of the transistor to conduct. This effect has been magnified at high frequencies due to

DOI: 10.1051/e3sconf/20171614006

the higher rate of change of the diode current during the ON time, dealing in a higher reverse current. This effect has been neglected in the theoretical studies because at low frequencies is almost null. In further researches it will be interesting to study the influence of the recovery time in the overall behaviour of the converter.

Although the converter is not working as it is desired, the efficiency have been measured with a High Performance Power Analyzer WT1800 from Yokowaga. The measured efficiency is presented in figure 6, this efficiency includes the auxiliary power supply used to power the control part of the converter. At full load (150W) the efficiency is 90.7%, quite lower than the one theoretical calculated. This is due to the fact that the final transformer core is not the same that the one used for the calculations, the auxiliary power supply consumption and the most important reason the behaviour of the prototype during the GAP time.

Figure 4. 150W Half-Bridge ZVZCPS Prototype

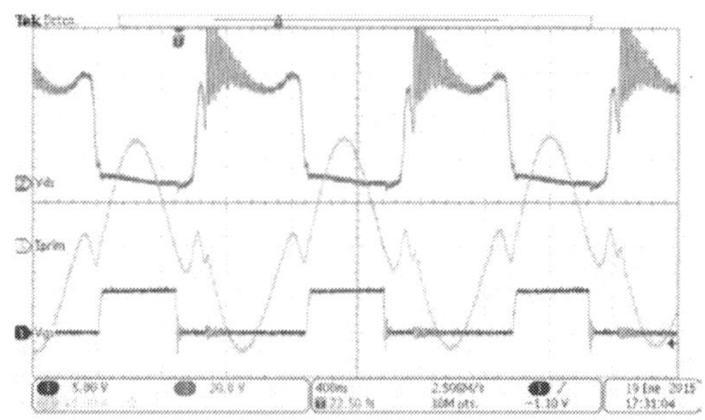

Figure 5. 150W ZVZCPS Prototype waveform Red – Drain-Source Voltage, Green – Primary Current, Blue – Transistor driving signal.

5. CONCLUSIONS

GaN transistors have lot of advantages compared with their Silicon counterparts. They have lower conduction and switching losses, so it is possible to design more efficiency converters or to increase the frequency of the actual converters reducing it size and weight.
In the case of ZVZCPS converters, the GaN transistors benefits is limited because this type of converters has very high efficiency with very low commutation losses.

Besides, increasing the switching frequency is limited by all the parasitic capacitance of the converter. Planar transformers are also a promising technology to reduce the size of the converters but they have a very high parasitic capacitance so its benefits in this type of converters is also reduced. Besides, when the switching frequency it is increased, the diodes reverse recovery effect is magnified, affecting the original converter behaviour.

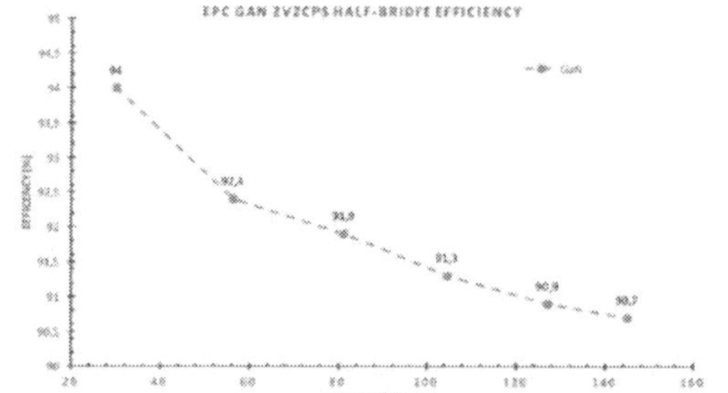

Figure 6. 150W ZVZCPS prototype measured efficiency

6. ACKNOWLEDGMENTS

The authors would like to thank the Spanish Ministry of Economy and Competitiveness which has supported these research activities under projects AYA2012-37444-C02-2 and ESP2015-68117-C2-2-R.

7. REFERENCES

1. A. H. Weinberg and L. Ghislanzoni, "A new zero voltage and zero current power switching technique," IEEE Trans. Power Electron., vol. 7, no. 4, pp. 655–665, Oct. 1992.

2. G. Delamare, P. Maynadier and H. Schneider, "GaN HFET-Based DC/DC Converters for low power RF Satellite Equipment" Proceedings of the 10th European Space Power Conference. Noordwijk, The Netherlands, 2014.

3. F. Leroy et al, "A 90% Efficient, 50MHz Buck Converter for Space Telecom Systems" Proceedings of the 10th European Space Power Conference. Noordwijk, The Netherlands, 2014.

4. W. Serrano, O. Limaux and JM. Mollard "High Temperature EPC for GaN SSPA Applications" Proceedings of the 10th European Space Power Conference. Noordwijk, The Netherlands, 2014.

5. "Design of Planar Power Transformers", Application Note, Ferroxcube.

 DOI: 10.1051/e3sconf/20171614007

DYNAMIC MODELLING OF THE SERIES RESONANT CONVERTER OPERATING IN DISCONTINOUS CONDUCTION MODE AND ITS APPLICATION IN SPACE

A. Soto[(1)], J. Cortes[(1)], F. Pascual[(1)]

[(1)]Airbus Defence and Space, (CRISA), Torres Quevedo 9, 28760 Tres Cantos (Spain), Email:Andres.Soto@airbus.com

ABSTRACT

The Series Resonant Full Bridge operating in Discontinuous Conduction Mode (DCM) at fixed frequency, SRFB-DCM, is an interesting topology to provide isolation in the interface to main buses while having high efficiency thanks to Zero Voltage Switching (ZVS) and Zero Current Switching (ZCS). As this converter cannot be regulated, usually, it is more interesting for regulated buses or in case there is post-regulation. This converter finds application in Power Processing Units for Electrical Propulsion and in any secondary high power front-end converter.

Despite its simple design without regulation, in general, the dynamic behaviour of resonant converters is difficult to predict. A model is necessary to predict compatibility with input and output interfaces (stability, output impedance, start-up in-rush,...), avoiding late discovery of oscillations, for example.

This paper presents a method to obtain the theoretical dynamic model of the SRFB-DCM with fixed frequency and duty cycle which seems easier to understand but surprisingly not treated in literature. Results are contrasted with simulations.

1. INTRODUCTION

A lot of literature can be found about modelling the DC characteristics of the Series Resonant Full Bridge, SRFB, and other resonant converters operating both in Continuous Conduction Mode (CCM) and Discontinuous Conduction Mode (DCM), [1], [2], with frequency/duty as controlling variables, but little about dynamics [3]. This is one of the most difficult topics in power electronics.

The SRFB converter is shown in Fig. 1. All analysis presented in the paper are also valid for half-bridge configuration.

The resonant inductance can be implemented entirely or partially with the leakage inductance of the transformer, thus, the usual problems associated to this inductance are avoided with this topology. This is quite interesting for applications requiring very good isolation between primary and secondary where it is difficult to have a good transformer coupling.

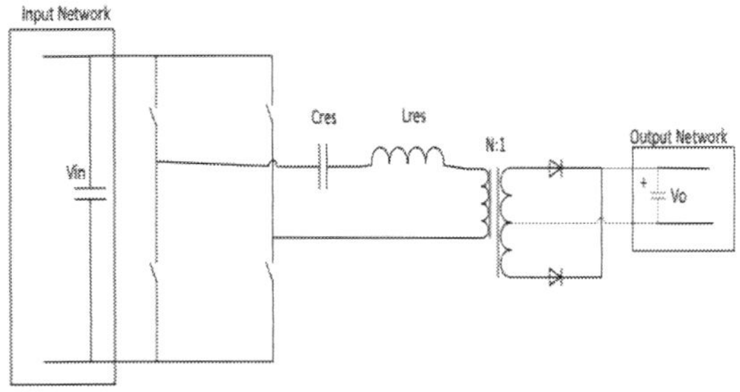

Figure 1. Simplified schematics of SRFB converter.

The SRFB converter working in CCM can only step-down the input voltage affected by the turns ratio seen in secondary side. The output voltage is regulated commanding the frequency as a function of input voltage and load current. However, regulation becomes difficult or impossible at light loads.

In DCM, the DC transfer function is unity, being output voltage (affected with turns ratio) equal to input voltage independently of the load (assuming no converter losses) and switching frequency as long as DCM operation is maintained. Thus, this converter cannot be regulated.

DCM operation takes place when there is a time before the end of one semi-period, dead time, in which the inductor current keeps constant to zero (or equal to the magnetizing current). As a consequence, the capacitor voltage keeps constant during this dead time.

Two conditions are necessary to obtain DCM behaviour. First, the switching frequency should be lower than the resonant frequency and, second, the maximum absolute resonant capacitor voltage in steady state should be lower than $2\cdot\hat{V}_o$, $\hat{V}_o$ is the output voltage affected by the turns ratio.

Thanks to the inductor current dead time, the switches can be turned-off with zero current (ZCS) and the rectifier diodes will not have reverse recovery. It is possible to improve further the efficiency by enabling Zero Voltage Switching (ZVS) having an inductance in parallel with the tank or using the magnetizing current to have a small current in the switches during the dead time.

 DOI: 10.1051/e3sconf/20171614007

2. FUNDAMENTAL EQUATIONS

The following assumptions are done to start the analysis of the voltage and current waveforms of the resonant tank operating in DCM:

- Magnetizing current of converter transformer is assumed zero.
- Even in transient conditions, the inductor current at the beginning of the switching period is zero. It could be seen as a strong constraint, but it will be shown it is not the case.
- Input and output voltages, v_{in} and v_o are assumed to be constant during a switching semi-period. In other words, input and output switching voltage ripples are small. But the analysis allows for slow changes in these voltages from one period to the other.
- Resistive losses are neglected.

Three first conditions are found easily in practical converters. Resistive losses will be treated later. If magnetizing current is significant with respect to resonant inductor current, then the converter is an LLC resonant type [4]. Here, the model obtained will be valid for converters where the magnetizing current is small compared to main inductor.

Fig. 2 shows a conceptual model of the converter with current and voltage reference which helps to get insight in its operation.

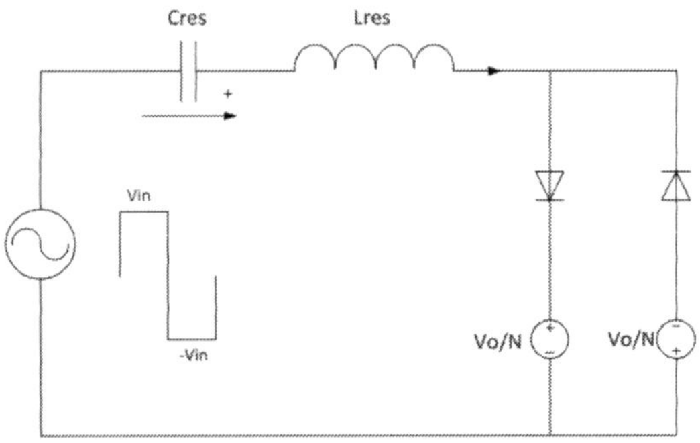

Figure 2. Equivalent simplified circuit and references for voltage and current.

Fig. 3 shows the equivalent circuit for one semi-period. Initial condition for current is zero based on previous assumptions. Initial capacitor voltage $uc0$ will be obtained as a function of the operating point of the converter.

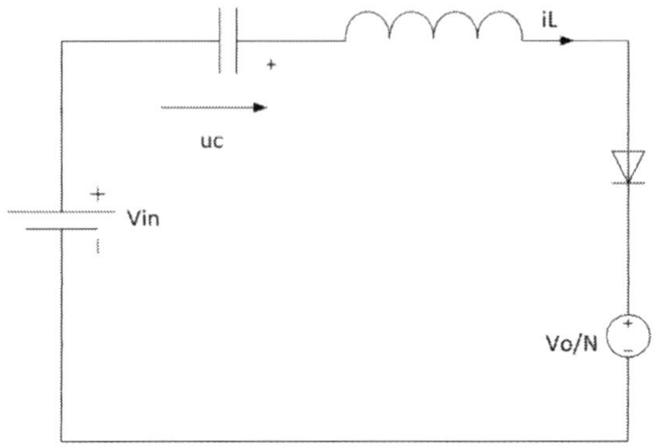

Figure 3. Equivalent circuit for first semi-period (SP1) and references for voltage and current.

The inductor current and capacitor voltage during a switching semi-period, both for transient and steady state operation, have the general forms given by Eq. 1.

$$\begin{cases} iL(t) = A \cdot \sin(\omega_{res} \cdot t), & 0 < t \leq \frac{T_{res}}{2} \\ iL(t) = 0, & \frac{T_{res}}{2} < t \leq \frac{T_{sw}}{2} \end{cases}$$

$$\begin{cases} uc(t) = A \cdot \cos(\omega_{res} \cdot t) \cdot \sqrt{\frac{L_{res}}{C_{res}}} + B, & 0 < t \leq \frac{T_{res}}{2} \\ uc(t) = -A \cdot \sqrt{\frac{L_{res}}{C_{res}}} + B, & \frac{T_{res}}{2} < t \leq \frac{T_{sw}}{2} \end{cases} \quad (1)$$

Constants A and B can be calculated as a function of the initial capacitor voltage $uc0$ by calculating the current slope at the beginning of the semi-period, as shown in Eq.2. The voltage $\widehat{v_o}$ is the output voltage affected by the turns ratio during the switching period. Similarly, v_{in} is the input voltage during the switching period. For the analysis, these two voltages can change slowly from one period to the other, for example during transients.

$$\begin{cases} \left.\frac{diL(t)}{dt}\right|_{t=0} = A \cdot \omega_{res} = \frac{v_{in} - \widehat{v_o} + uc0}{L_{res}} \\ uc0 = uc(0) = A \cdot \sqrt{\frac{L_{res}}{C_{res}}} + B \end{cases} \quad (2)$$

$$\begin{cases} A = (v_{in} - \widehat{v_o} + uc0) \cdot \sqrt{\frac{C_{res}}{L_{res}}} \\ B = -(v_{in} - \widehat{v_o}) \end{cases}$$

 DOI: 10.1051/e3sconf/20171614007

Merging Eq. 1 and Eq. 2, then Eqs. 3-4 are obtained for one semi-period.

$$\begin{cases} 0 < t \leq \frac{T_{res}}{2} \\ iL(t) = (v_{in} - \widehat{v_o} + uc0) \cdot \sqrt{\frac{C_{res}}{L_{res}}} \cdot \sin(\omega_{res} \cdot t) \\ uc(t) = (v_{in} - \widehat{v_o}) \cdot (\cos(\omega_{res} \cdot t) - 1) + uc0 \cdot \cos(\omega_{res} \cdot t) \end{cases} \quad (3)$$

$$\begin{cases} \frac{T_{res}}{2} < t \leq \frac{T_{sw}}{2} \\ iL(t) = 0 \\ uc(t) = -2 \cdot (v_{in} - \widehat{v_o}) - uc0 \end{cases} \quad (4)$$

From Eq. 4, it is interesting to notice that during the dead time and at the end of one switching semi-period, the capacitor voltage is the negative of its initial value minus twice the difference between input and output voltage.

The inductor current will not change direction and remain zero leading to a dead time if the voltage applied to the inductor when the other rectifier diodes try to conduct is in the opposite direction. Later in sub-chapter 2.2, this condition for DCM is explored.

The same set of equations Eqs.3-4 are valid in the next semi-period (SP2) if the opposite resonant tank current and voltage are considered, as shown in Fig.4.

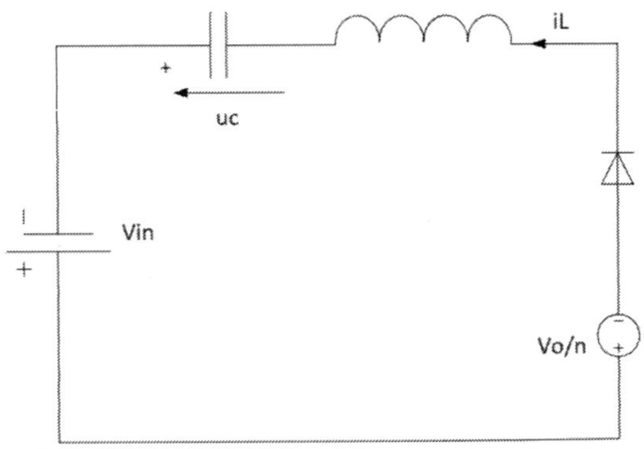

Figure 4. Equivalent circuit for second semi-period (SP2) and opposite references for voltage and current.

The initial condition of the capacitor voltage at each semi-period is the negative of the voltage at the end of the previous semi-period (current is always zero initially), as Eqs. 5-6 show.

$$uc_{SP2}(0) = -uc_{SP1}\left(\frac{T_{sw}}{2}\right) \quad (5)$$

$$uc_{SP3}(0) = -uc_{SP2}\left(\frac{T_{sw}}{2}\right) \quad (6)$$

Using the same set of equations given by Eqs. 3-4 for each semi-period with the initial conditions given by Eqs. 5-6, it is possible to create only one model for the different semi-periods where the evolution with time of capacitor voltage and inductor current are given. It is also possible to remove the rectification diodes and the full bridge so the interaction with the rest of the converter is modelled as Fig. 5 shows.

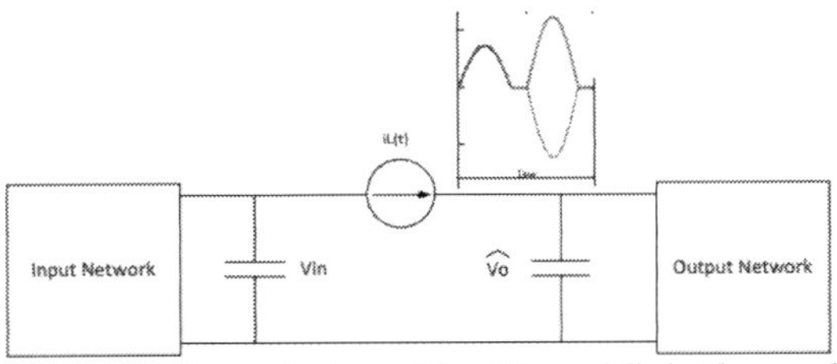

Figure 5. Equivalent model without full bridge and diode rectifiers and definitions of il(t), $v_{in}(t)$ and $\widehat{v_o}(t)$.

It is important to remark that Eqs. 3-6 are valid for both DC and AC analysis. Given an initial condition for the capacitor voltage and the evolution with time of the input and output voltages, it is possible to calculate the evolution of the resonant inductor current, and hence, the dynamics of the converter.

2.1 Steady state equilibrium

Steady state in the capacitor can be found only when there is a periodic voltage. So it is necessary that the voltage at the end of the switching period is equal to the value at the beginning.

In the second semi-period (SP2), the initial condition for Eqs. 3-4 is the opposite of $uc_{SP1}\left(\frac{T_{sw}}{2}\right)$, so at the end of this second semi-period, the end of the switching period, and applying Eq.4, the voltage is given by Eq. 7.

$$uc_{SP2}\left(\frac{T_{sw}}{2}\right) = -4 \cdot (V_{in} - \widehat{V_o}) - uc0_{SP1} \quad (7)$$

Steady state in the capacitor can be found only when there is a periodic voltage so, given Eq. 6, Eq. 8 should be met.

$$uc0_{SP3} = uc0_{SP1} = -uc_{SP2}\left(\frac{T_{sw}}{2}\right) \quad (8)$$

Clearly, from Eq. 7 and Eq. 8, it is necessary for periodic voltage that $V_{in} = \widehat{V_o}$. This is the reason this converter has unity transfer function and why it cannot be controlled with switching frequency or duty if it is operating in DCM. Hereafter, sub-index SP can be removed since both semi-periods have the same waveforms.

On the other hand, converter will be in steady state when the rectified average inductor current is equal to average load current $\widehat{I_o}$, referred to primary, as seen in Fig. 5. Then, considering that $V_{in}=\widehat{V_o}$ and that the current waveform is the same for the two semi-periods, it follows Eq. 9.

$$< iL(t) \geq \frac{1}{T_{sw}} \cdot \int_0^{T_{sw}} iL(t) \cdot dt =$$
$$= \frac{1}{\frac{T_{sw}}{2}} \cdot \int_0^{\frac{T_{res}}{2}} uc0 \cdot \sqrt{\frac{C_{res}}{L_{res}}} \cdot \sin(\omega_{res} \cdot t) \cdot dt \quad (9)$$

Eq.9 leads to the equality given in Eq. 10.

$$\widehat{I_o} =< iL(t) >= \frac{2}{\pi} \cdot \frac{fsw}{fres} \cdot uc0 \cdot \sqrt{\frac{C_{res}}{L_{res}}} \quad (10)$$

The initial resonant capacitor voltage as a function of load current at each semi-period in steady state is then given by Eq. 11.

$$uc0 = \widehat{I_o} \cdot \sqrt{\frac{L_{res}}{C_{res}}} \cdot \frac{\pi}{2} \cdot \frac{fres}{fsw} \quad (11)$$

2.2 Conditions for DCM

It is interesting to have a way to determine when the converter could lose DCM, and then soft-switching. It is desirable this criteria is valid not only for steady state but also for transient operation.

The inductor current will not change direction during a semi-period and remain zero if the resulting voltage applied to the inductor when the current tries to change direction is opposing to this change, as given in Eq. 12.

$$-(-2 \cdot (v_{in} - \widehat{v_o}) - uc0) - (v_{in} + \widehat{v_o}) < 0$$
$$uc0 < 3 \cdot \widehat{v_o} - v_{in} \quad (12)$$

If the capacitor voltage at the beginning of a semi-period is meeting Eq.12, then the converter stays in DCM. This condition is valid for steady state and transient operation.

In steady state, using Eqs. 11-12 and $V_{in}=\widehat{V_o}$, then it is obtained Eq. 13.

$$\sqrt{\frac{L_{res}}{C_{res}}} \cdot \frac{\pi \cdot fres}{4 \cdot fsw} \cdot \widehat{I_o} < \widehat{V_o} \quad (13)$$

Clearly, there is a maximum current allowed to stay in DCM.

Only in especial transients Eq.12 could not be met. For example, during starting-up or short-circuit events.

3. TRANSIENT MODELLING

Based on Eq. 3-4, it is possible to derive a time-discrete model for the averaged model where the switching is removed.

Being k an index to refer to each semi-period, the time evolution of current and voltage during that semi-period is given by Eq. 14, where t=0 is the beginning of that semi-period.

$$\left[\begin{array}{l} 0 < t \leq \frac{T_{res}}{2} \\ iL(t)_k = (v_{in}(k) - \widehat{v_o}(k) + uc0\,(k)\,) \cdot \sqrt{\frac{C_{res}}{L_{res}}} \cdot \sin(\omega_{res} \cdot t) \\ uc(t)_k = (v_{in}(k) - \widehat{v_o}(k) + uc0(k)) \cdot (\cos(\omega_{res} \cdot t) - 1) + uc0(k) \end{array}\right.$$
$$\left[\begin{array}{l} \frac{T_{res}}{2} < t \leq \frac{T_{sw}}{2} \\ iL(t)_k = 0 \\ uc(t)_k = -2 \cdot (v_{in}(k) - \widehat{v_o}(k)) - uc0(k) \end{array}\right. \quad (14)$$

The capacitor voltage evolution at the beginning of each semi-period can be described, based on Eqs. 3-6, as the sequence given in Eq. 15.

$$uc0(k+1) = 2 \cdot (v_{in}(k) - \widehat{v_o}(k)) + uc0(k) \quad (15)$$

The investigated dynamic model should replace the sinusoidal current of the inductor by an averaged current value depending on k, iL(k), that provides the same dynamic response. Particularly, iL(k) evolution should provide the same charge transfer from input to output at each semi-period.

Looking at Fig. 5, the charge transferred from input to output during semi-period k is given by Eq. 16 using Eq. 14.

$$Q(k) = \int_{\frac{T_{sw}}{2} \cdot k}^{\frac{T_{sw}}{2} \cdot (k+1)} iL(t) \cdot dt = \int_0^{\frac{T_{sw}}{2}} iL(t)_k \cdot dt =$$
$$= \frac{1}{\pi} \cdot \frac{1}{fres} \cdot \sqrt{\frac{C_{res}}{L_{res}}} \cdot (v_{in}(k) - \widehat{v_o}(k) + uc0(k)) \quad (16)$$

This charge is a function of $v_{in}(k)$, $\widehat{v_o}(k)$ and uc0(k), which is also a function of input and output voltages as per Eq. 15. Then, Eq. 15 and Eq.16 form a system of equation that should explain the dynamics of the whole converter.

Current iL(k) could be understood as a constant current during the semi-period that gives the charge calculated in Eq. 16, so Eq.17 is obtained.

$$Q(k) = iL(k) \cdot \frac{Tsw}{2} \quad (17)$$

It would mean that from semi-period to semi-period, the current that is being modelled is transiently evolving in steps, which has no physical meaning. Actually, if going this way, a resistor is found in the averaged model which introduces losses. But the analysed converter is ideal with no losses, so the model cannot be fully correct.

Instead, if a linear variation of averaged current is considered during a semi-period and if iL(k) is the averaged current at the beginning of the semi-period k, then Eq. 18 is obtained.

$$Q(k) = \left(iL(k) + \frac{iL(k+1)-iL(k)}{2}\right) \cdot \frac{Tsw}{2} \quad (18)$$

Then, merging Eqs. 15-16 and Eq. 18, the system of equations to solve is given by Eq. 19.

$$\begin{cases} iL(k) + \frac{iL(k+1)-iL(k)}{2} = \frac{2}{\pi} \cdot \frac{fsw}{fres} \cdot \left(v_{in}(k) - \widehat{v_o}(k) + uc0(k)\right) \cdot \sqrt{\frac{C_{res}}{L_{res}}} \\ uc0(k+1) = 2 \cdot \left(v_{in}(k) - \widehat{v_o}(k)\right) + uc0(k) \end{cases} \quad (19)$$

Doing some maths to remove uc0(k) from the system of equations and solve for iL(k), Eq. 20 is obtained.

$$\frac{iL(k+2) - iL(k)}{2} = $$
$$= \frac{2}{\pi} \cdot \frac{fsw}{fres} \cdot \sqrt{\frac{C_{res}}{L_{res}}} \cdot \left(v_{in}(k+1) - \widehat{v_o}(k+1) + v_{in}(k) - \widehat{v_o}(k)\right) \quad (20)$$

Eq. 20 gives the evolution of the current only as a function of the input and output voltages. This equation could be solved numerically with Tsw/2 as time step together with input and output networks to obtain the dynamics of the converter.

This time-discrete equation does not resemble an equation with a known solution. Doing more maths with operators, being the operator Δ defined as $\Delta \cdot x(k) = = \frac{x(k+1)-x(k)}{Ts}$ with $Ts = \frac{1}{2 \cdot f_{sw}}$, then Eqs. 21-23 are obtained.

$$\frac{T_s^{\,2}}{2} \cdot \Delta^2 \cdot iL(k) + T_s \cdot \Delta \cdot iL(k) =$$
$$= \frac{4}{\pi} \cdot \frac{fsw}{fres} \cdot \sqrt{\frac{C_{res}}{L_{res}}} \cdot \left[\frac{T_s}{2} \cdot \Delta + 1\right] \cdot \left(v_{in}(k) - \widehat{v_o}(k)\right) \quad (21)$$

$$T_s \cdot \Delta \cdot \left(\frac{T_s}{2} \cdot \Delta + 1\right) \cdot iL(k) =$$
$$= \frac{4}{\pi} \cdot \frac{fsw}{fres} \cdot \sqrt{\frac{C_{res}}{L_{res}}} \cdot \left[\frac{T_s}{2} \cdot \Delta + 1\right] \cdot \left(v_{in}(k) - \widehat{v_o}(k)\right) \quad (22)$$

$$\Delta \cdot iL(k) = \frac{8}{\pi} \cdot \frac{fsw^2}{fres} \cdot \sqrt{\frac{C_{res}}{L_{res}}} \cdot \left(v_{in}(k) - \widehat{v_o}(k)\right) \quad (23)$$

Now it is easy to recognise in Eq. 23 the time discrete form of a derivative. Assuming slow change of input and output voltages during one semi-period, assumption already done for the derivation, then this expression can be put in continuous time as in Eq. 24.

$$\frac{diL(t)}{dt} = \frac{8}{\pi} \cdot \frac{fsw^2}{fres} \cdot \sqrt{\frac{C_{res}}{L_{res}}} \cdot \left(v_{in}(t) - \widehat{v_o}(t)\right) \quad (24)$$

Eq. 24 is the equation of an inductor with inductance value given by Eq. 25.

$$L_{eq} = \frac{\pi}{8} \cdot \frac{fres}{fsw^2} \cdot \sqrt{\frac{L_{res}}{C_{res}}} = \frac{1}{16} \cdot \frac{1}{fsw^2} \cdot \frac{1}{C_{res}} \quad (25)$$

Remarkably, the equivalent inductance value depends only on the selection of the resonant capacitance once the switching frequency is fixed.

Fig. 6 shows the averaged model obtained so far replacing in Fig. 5 the resonant current source by the inductor just obtained.

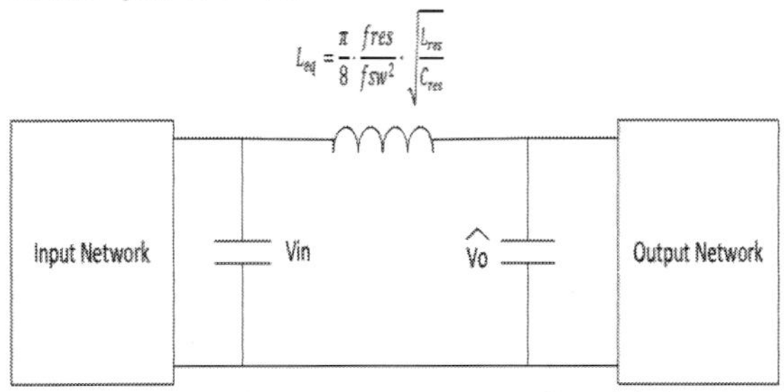

Figure 6. Equivalent dynamic averaged model for the SRFB converter in DCM.

The evolution of the resonant capacitor voltage, although not necessary to know for the dynamics of the converter, can be also calculated easily now. It is important to have it to know the stress of the capacitor and to know if the hypothesis of DCM is kept.

Using Δ operator defined before, Eq.15 becomes Eq.26.

$$\Delta \cdot uc0(k) = 4 \cdot fsw \cdot \left(v_{in}(k) - \widehat{v_o}(k)\right) \quad (26)$$

From Eq.23, it is possible to obtain Eq.27.

$$\begin{gathered} \Delta \cdot uc0(k) = 4 \cdot fsw \cdot Leq \cdot \Delta \cdot iL(k) \\ uc0(k) = 4 \cdot fsw \cdot Leq \cdot iL(k) + C \end{gathered} \quad (27)$$

Constant C in Eq.27 is determined forcing that in steady state Eq. 11 is met, so C becomes 0. Then, Eq. 27 is showing that the initial value of the capacitor voltage at each semi-period is proportional to averaged inductor current.

 DOI: 10.1051/e3sconf/20171614007

3.1 Modelling of losses and damping

So far, it has been assumed no resistive voltage drop in the resonant tank. The approximation is very good to calculate the waveforms of current and capacitor. However, since this assumption implies no resistive behavior in the power path (no Rdson, ESR, inductor resistance,...), it fails to predict the losses of the converter and its damping, as comparison with simulations and test shows.

The approach that is followed to calculate the equivalent resistance is more heuristic but quite effective, as seen later in the simulation verification and also seen in tests verification.

The real total resistance of the topology cannot predict correctly the losses in an average model since the actual currents are much higher due to its sinusoidal shape than the current predicted by the average model.

One method to obtain the equivalent resistance is to make the losses of the averaged model equal to the losses in the switched converter. It means Eq. 28 should hold.

$$< iL^2(t) > \cdot\, Rreal = < iL(t) >^2 \cdot R_{eq} \qquad (28)$$

Doing some calculations, the equivalent resistance given by Eq.29 can be found.

$$R_{eq} = \frac{\pi^2}{8} \cdot \frac{fres}{fsw} \cdot Rreal \qquad (29)$$

The final model is shown in Fig. 7. No small signal assumption has been done during the analysis so this model should be valid also for large transients, as following chapter will show.

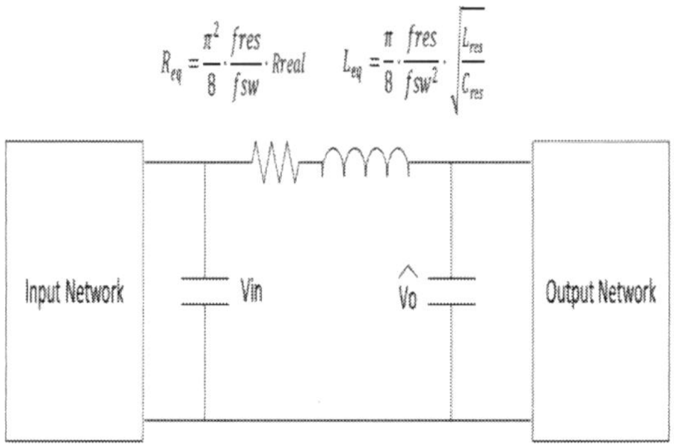

Figure 7. Equivalent averaged model with dynamics and losses.

4. SIMULATIONS RESULTS

A simulator tool able to do frequency analysis of switched converters is needed. Simplis simulator is used since it is possible to obtain the frequency response from a switched time simulation, as if measuring the loop gain of a converter in the lab.

The simulated converter, Fig.8, does not have input filter neither output filter for simplification. The main parameters are R_{real}=60 mΩ, L_{res}=1 μH, C_{res}=1.5 μF, f_{sw}=100 kHz, V_{in}=100 V, N=1.5 and C_o=500 μF. Diodes and MOSFETs are modelled as ideal elements to simplify the comparison.

The output impedance comparison in frequency domain, Fig. 9, shows a perfect matching of the model with simulations results. The output impedance comparison in time domain for large signal (5 A load step) is shown in Fig.10. A perfect matching both in DC and AC can be seen.

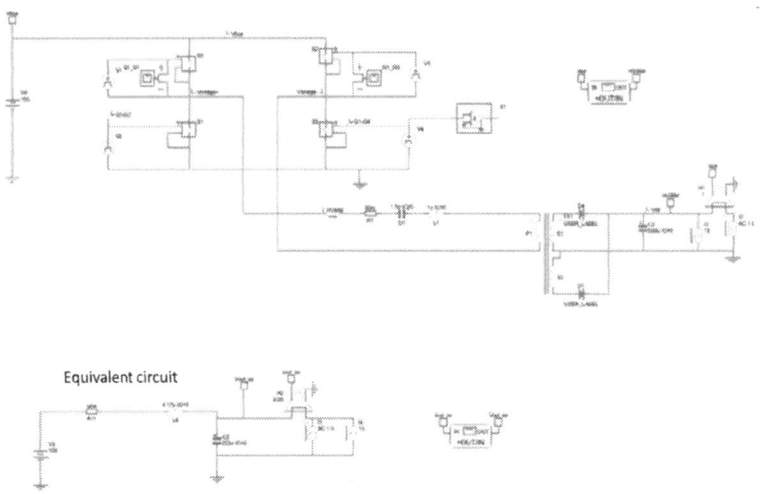

Figure 8. Switched model and equivalent averaged model used in Simplis simulator for model verification.

 DOI: 10.1051/e3sconf/20171614007

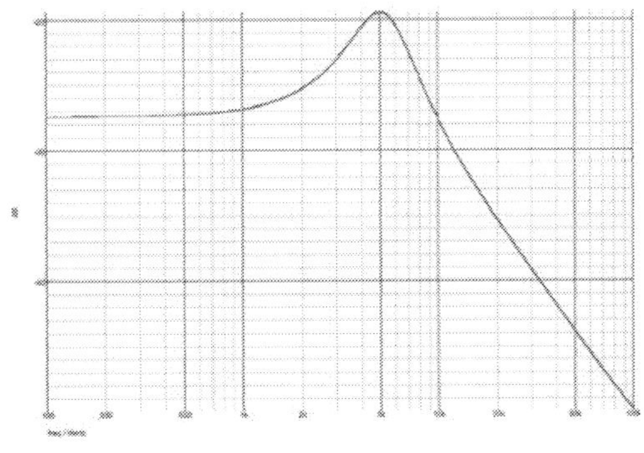

Figure 9. Output impedance frequency response comparison: green switched converter with Simplis, red averaged model.

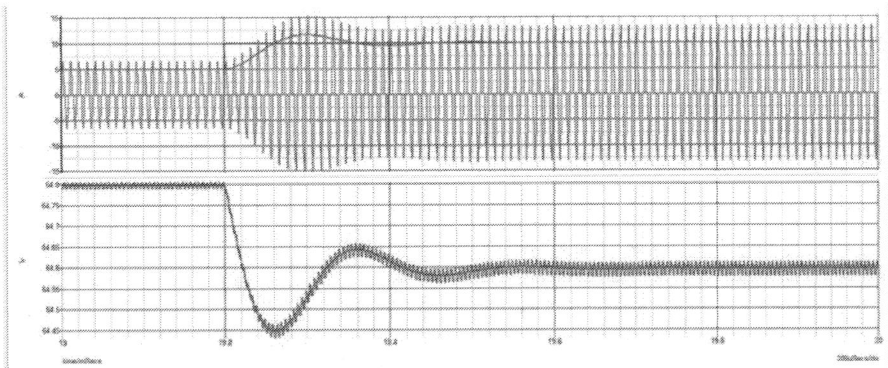

Figure 10. Load step large signal transient. Lower plot: output voltage response, green switched model, red averaged model. Upper plot: inductor current and load step, yellow switched model, cyan averaged model.

5. CONCLUSIONS

The paper derives a dynamic theoretical model for the Series Resonant Converter Full Bridge operating in DCM. This converter has many applications due to its simplicity and high efficiency, especially as front-end converter providing isolation and interfacing regulated buses or un-regulated buses with post-regulations. Despite its simplicity and not being a regulated converter, it is necessary to know its dynamics to design and justify properly the converter interfacing the input and output networks.

This paper obtains a simple model with only one resistor and inductance. Theoretical expressions are given for these elements which enable design-oriented analysis. Paradoxically, being the DCM series resonant converter operated below resonance, capacitive behaviour, the equivalent model is an inductor of value depending exclusively on the resonant capacitance and switching frequency.

The model has been validated with simulation results, showing a perfect prediction both for small-signal analysis and also for large transients.

The model has been used to design and justify properly the input and output filters of a flight converter and also the stability with its load.

6. REFERENCES

1. Erickson, R., Maksimovic, D. *Fundamental of Power Electronics, Second Edition*, 2001 University of Colorado, Boulder, Colorado. pp 706–759.
2. Vorperian, V., Cuk, S., *A Complete DC Analysis of the Series Resonant Converter*, IEEE Power Electronics Specialist Conference 1982, pp. 85-100
3. Witulski, F., Hernándes, A. and Erickson, R. *Small Signal Equivalent Circuit Modeling of Resonant Converters*, IEEE Transactions on Power Electronics, Vol. 6, No. 1 January 1991.
4. Lazar, J.F., Martinelli, R. *Steady-state analysis of the LLC series resonant converter*, IEEE Applied Power Electronics Conference and Exposition 2001

 DOI: 10.1051/e3sconf/20171614008

MC²: A POWER CONDITIONNING AND DISTRIBUTION UNIT FOR STRATOSPHERICS BALLOONS

Bonnet François[(1)], Vivian Bernard[(1)], Stephanie Venel[(1)], Thomas Baldran[(2)]

[(1)]CNES – French Space Agency 18, avenue Edouard Belin 31401 TOULOUSE Cedex 9 – FRANCE
Email: francois.bonnet@cnes.fr; vivian.bernard@cnes.fr; stephanie.venel@cnes.fr
[(2)]EREMS, 13, Chemin de la Madeleine – ZI 31130 - FLOURENS – FRANCE, Email: thomas.baldran@erems.fr

ABSTRACT

For long duration scientific missions with stratospheric balloons (objective of 3 month duration), renewable energy is used. Solar panels with mono crystalline silicon solar cells are mounted on both scientific and avionic gondola. A power conditioning board with Maximum Power Point Tracking (MPPT) is designed and currently tested. This board is called MC²: Communicant Conditioning Module. It allows controlling a Li Ion battery charge through PWM regulators. Moreover, outlets ON/OFF commutations associated to overcurrent's protections are implemented in this board. The battery active thermal control is made by MC² autonomously. The main design drivers are mass, costs and efficiency. A CAN Bus between MC² and On Board Computer allows to have a commandability and observability of MC² through OBC. The overall avionic gondola is designed to be Single Points Failure free by using two segregated chains in order to be compatible with safety rules. The nominal chain is the main chain and use MC² with renewable energy. The secondary chain uses a primary electrochemical cell which feeds loads in case of undervoltage of the main chain. This overall architecture allows both chains to be designed without SPF free constrains. This paper describes the overall requirements and the design of MC². The main innovation described in this paper is the way to implement MPPT: the MPPT algorithm is performed at the output of the power converter. This MPPT extracts maximum power of both solar panel characteristics and power converter. The main advantage is that this MPPT uses only one existing sensor (output current of boost converter) instead of using current and voltage sensor of each solar panel.

1. INTRODUCTION

1.1. Missions Objectives

The scientific mission called STRATEOLE 2 [1] is a campaign dedicated to advance the knowledge of coupling processes between the troposphere and the stratosphere in the deep tropics, and foreseen in 2020-2021 and 2023-2024. The goal is to study the stratospheric dynamics and ozone chemistry in the southern hemisphere by a set of instrumented drifting balloons (objective is 30 balloons) during 3 month between 18km to 20 km of altitude.

Avionics gondola is in charge of balloon flight control (control, localisation, communication…). CNES (French Space Agency) is responsible of the development and control of this gondola. This gondola is a kind of platform in spacecraft vocabulary.

Likewise, scientific gondola corresponds to the payload gondola. In order to save developments costs both scientific and avionic gondolas shall use the same MC² and solar panel cells and packaging technologies.

1.2. Previous mission and experience: Concordiasi

A previous mission was successfully achieved in 2010 to validate infrared measurements made by spacecraft and to improve meteorologic and climate models. Moreover, additional aims were to improve the forecast of chemical transient on stratosphere (upper Antarctic) and to understand the behaviour of ozone evolution in stratosphere during winter and spring.

This mission was composed of 20 balloons which evolved near south-pole during a mean time of 70 days. The solar panels mounted on the avionic gondola of this mission use polycrystalline silicon cells, which were previously cut from standard size cells. A picture of a scientific gondola used during Concordiasi campaign is presented on *Figure 1* .

Figure 1: Concordiasi avionic gondola with cuts solar cells

 DOI: 10.1051/e3sconf/20171614008

The main advantage of using cut cells is to have a serial connection of cells for a small solar panel surface with an open circuit voltage always upper than battery voltage. A simple circuit which connects to or disconnects from solar panels (8 in total) battery was used. This kind of control is usually called DET: Direct Energy Transfer.

Nevertheless, the main drawback is the low efficiency of the power conditioning. Indeed, the number of cells in series is determinate so as to have the maximum power point always upper than the battery voltage for a temperature range inside [-90°C; +110°C]. *Figure 2* shows an example of the working point of the solar panel in case of DET. The battery imposes the polarization voltage of the solar panel and the working point imposed by Vbat could be far from the MPP (depending on solar panel temperature).

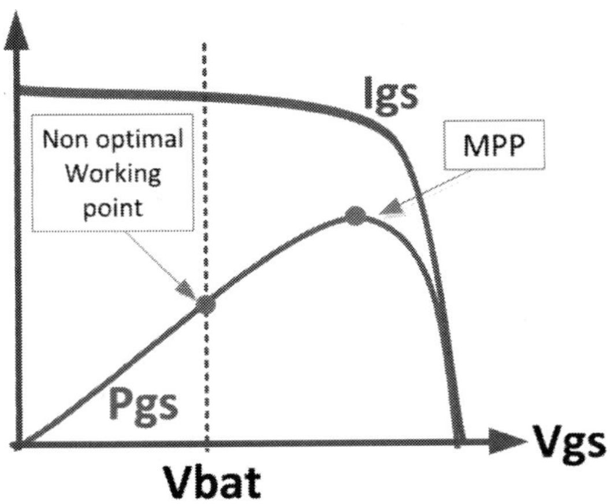

Figure 2 : Concordiasi power conditioning

2. MC² REQUIREMENTS

The first and the main technical requirement is that MC² shall be used for both scientific and avionic gondola.

2.1. Avionic gondola requirements

The preliminary design of avionic gondola is shown in *Figure 3*.

An energetic simulation allowed to determine that the minimum number of solar cells is 9 per solar panel for an overall power consumption of 10W (mean value). These solar cells are associated in serial. The loads to supply by the MC² in this gondola are Inmarsat and Iridium modems, radar transponder, flashing light (for aerial security), pressure and temperature sensors, on board computer and MC². Moreover, outlets uses for thermal control (heaters) are necessary. The overall need is to have 12 outlets ON/OFF controlled with overcurrent protections.

Figure 3 : Avionic gondola preliminary design

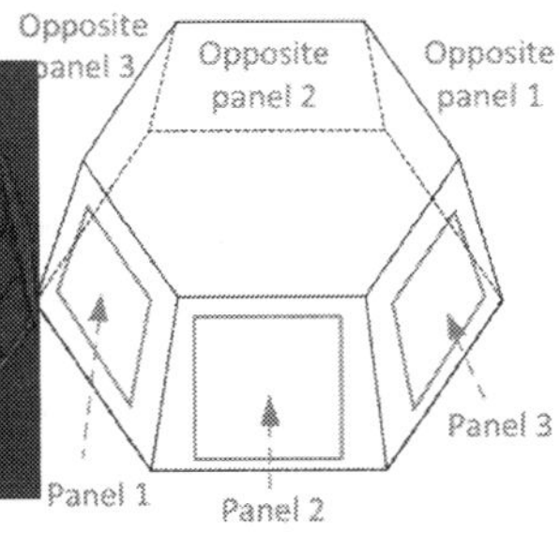

Figure 4 : scientific gondola preliminary design

2.2. Scientific gondola requirements

The preliminary design of scientific gondola is shown in *Figure 4*.

This gondola needs six solar panels with 12 solar cells associated in serial for each solar panel. This gondola needs 12 outlets ON/OFF as the avionic gondola.

2.3. Solar cells characteristics

The solar cells used in both scientific and avionic gondola are monocrystalline silicon cells C60Jp manufactured by SUNPOWER [2].

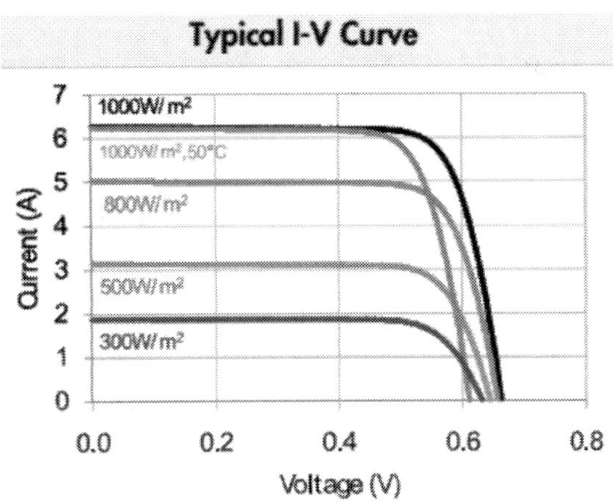

Figure 5 : solar cells electrical characteristics

These cells are usually used at 1000W/m². A dedicated experimental gondola launched in July 2016 from Aire sur l'Adour in France allowed recording the electrical characteristics of this solar panel in real conditions (flux and temperature). *Figure 6* shows the electrical characteristics Solar panel current IGS and voltage VGS on the top graph and solar panel power: PGS vs VGS on the bottom one. The solar panel temperature for a normal solar flux of 1368W/m² is 90°C. The main electrical data for 90°C are:

- Voc=5.1V
- Icc=7.8A
- Vmp=3.75V
- Imp=7A
- Pmp=26.5W

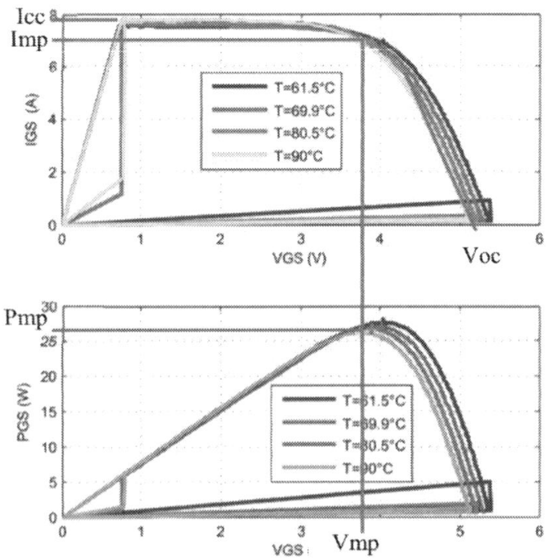

Figure 6: Experimental data of solar pannel 9S versus temperature

The conditioning circuit shall be designed in order to work with a good efficiency for low voltages (3.75V) and high currents (7A).

2.4. Battery interface cells characteristics

Usual loads are already designed and work properly in a supply voltage range [10V - 20V]. A battery with 4 LiIOn cells [3] associated in serial is used. The voltage of this battery (depending on the state of charge) is in the range [10V – 16.8V] which is compliant to the supply voltage range of loads. Consequently, an electrical architecture with an unregulated bus (battery directly connected to the distribution bus) is used.

2.5. Mass and volume constrains

The complete board (PCB, components, connectors) mass shall not exceed 700g for the avionics gondola and 900g for the scientific gondola. Volume restrictions are an area lower than 183*183 mm with a thickness lower than 25mm.

3. PRELIMINARY DESIGN OF PROTOTYPE

3.1. Boost regulator

According to the electrical architecture defined in §2.4 and the solar panel electrical characteristics §2.3, a voltage elevator power converter shall be used between solar panels and battery.

3.2. Power multiplexor

Moreover, the energy analysis already presented in §2.1 and §2.2 assumes a worst case illumination. This worst case considers only a direct solar flux of one panel and neglects the albedo flux which can provide some energy on other panels. This assumption is especially verified during flight over ocean without cloud cover. This case is a realistic case and solar cells number is determinate by considering this case.
According to this assumption, it is not necessary to have two opposite panels simultaneously connected to MC². An innovative solution which consists in using two (or three depending on the gondola) power multiplexers between solar panels and powers converters is chosen. It allows a mass optimization by combining input filters and power converters for two opposite panels. It means that scientific gondola uses 3 inputs filters and power converters, and avionic gondola uses only 2 inputs filters and converters.

3.3. PCB

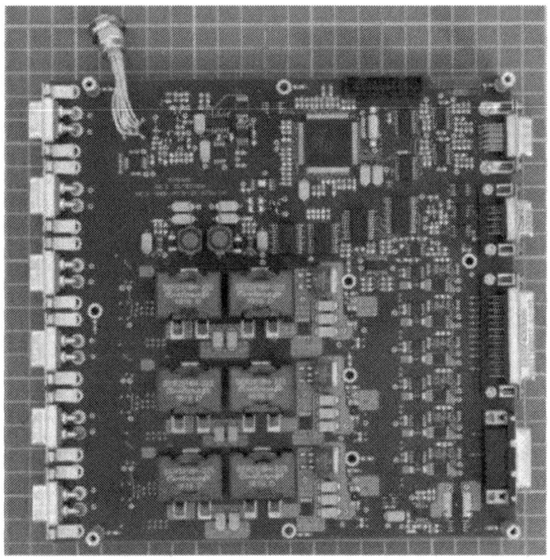

Figure 7 : Printed Circuit Board prototype designed by EREMS. Top face

Figure 8 : Printed Circuit Board prototype designed by EREMS. Bottom face

This PCB is currently tested in EREMS facility's for electrical functioning purpose. CNES has no experimental results at this time, which is why this paper is focussed on simulation results. On left side of *Figure 7*, are the 2W2 connectors which will interface with the 6 solar panels. On the left side of the bottom face in *Figure 8*, the power multiplexors are visible. The differential filters (sized for few volts and 8 amps) are presented in the middle of the top face.

4. OVERALL ARCHITECTURE OF MC²

The overall architecture of MC² is shown *Figure 9*.

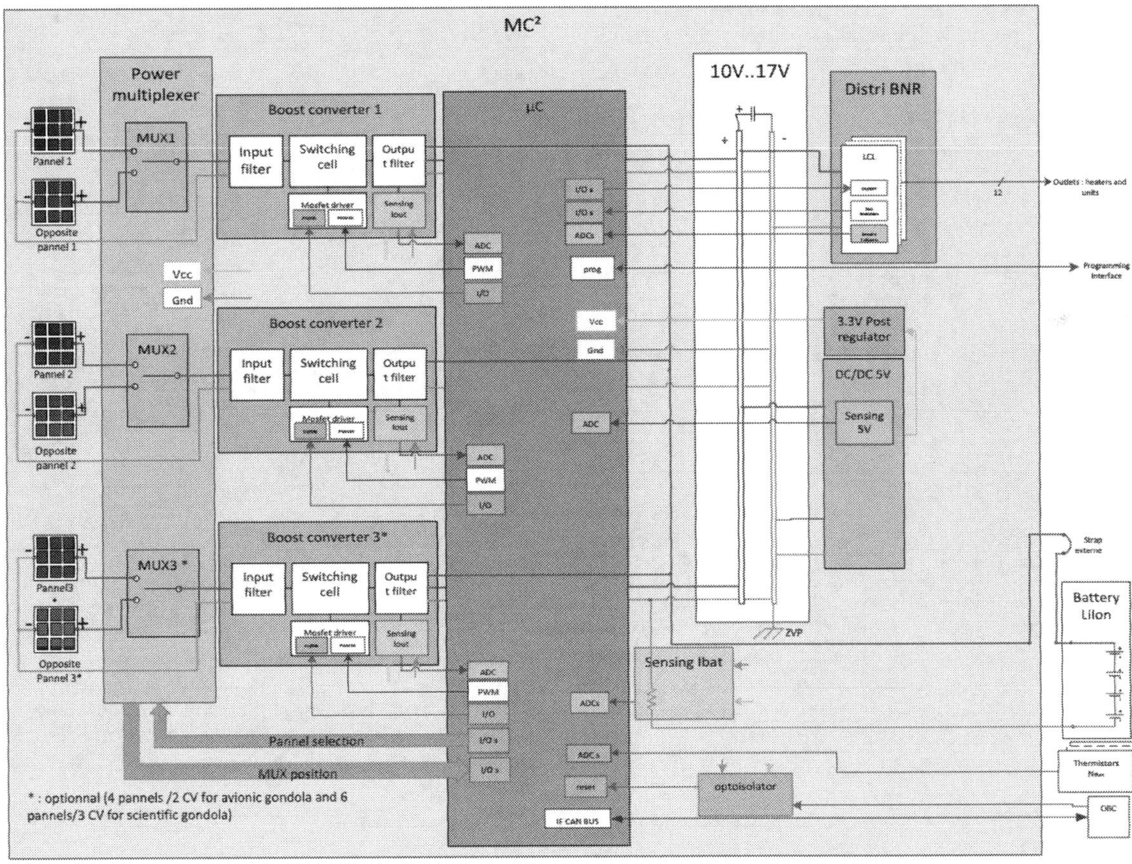

Figure 9 : Overall architecture of MC²

DOI: 10.1051/e3sconf/20171614008

5. POWER CONDITIONNING REGULATION

5.1. MPPT algorithm

The usual way to implement a MPPT controller is a perturb-and-observe method. The perturbation comes from a working point change, and the observation comes from measurement of solar panel current and voltage.

The main idea presented in this paper is to observe the boost converter output current as presented *in Figure 10.*

A duty cycle =1 means that the MOSFET is permanently ON and short circuit the solar panel. A duty cycle =0, means that the solar panel is connected to the battery if the rectifying diode is correctly polarized.

As the output voltage of boost converter is imposed by the battery, the maximum output power is reached for maximum output current of the converter.

Figure 11 shows the solar panel electrical characteristics: IGS vs VGS on the first graph. The next graph shows the output current of the boost converter vs the duty cycle. Clearly, this graph shows a maximum current value which is reached at the MPP working point of the solar panel. The two graphs in the bottom are the solar panel current vs duty cycle and the solar array voltage vs duty cycle.

So, to explain the MPPT operation in this application, the MPPT disturbance comes from an increment of boost duty cycle provided by the PWM output of the microcontroller. This duty cycle increment is directly linked to the clock frequency and PWM bit scaling of microcontroller. The output current evolution of the boost converter is analysed and a tracker acts on PWM duty cycle in order to find the maximal output current.

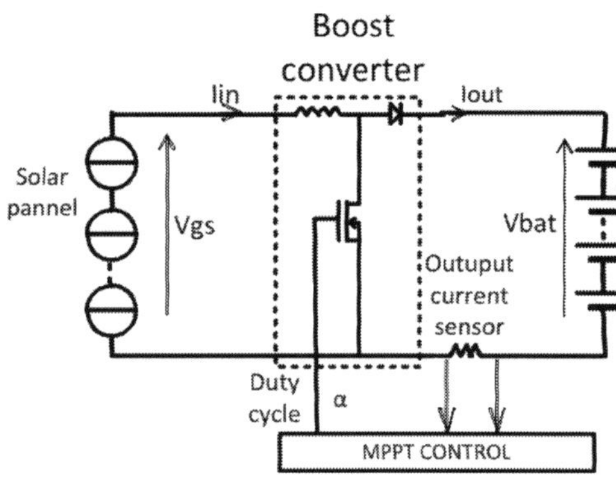

Figure 10 : Sensor and boost converter

It means that only one current sensor at the output of converter is sufficient to implement a MPPT. The disturbance comes from a positive or negative increment on boost converter duty cycle and the observation is the output current of the converter.

The output current sensor is already necessary for boost converter regulation. So, using this existing sensor for MPPT allows to supress (2) 3 solar panel current sensors and (2) 3 voltage sensors (2 for avionic gondola).

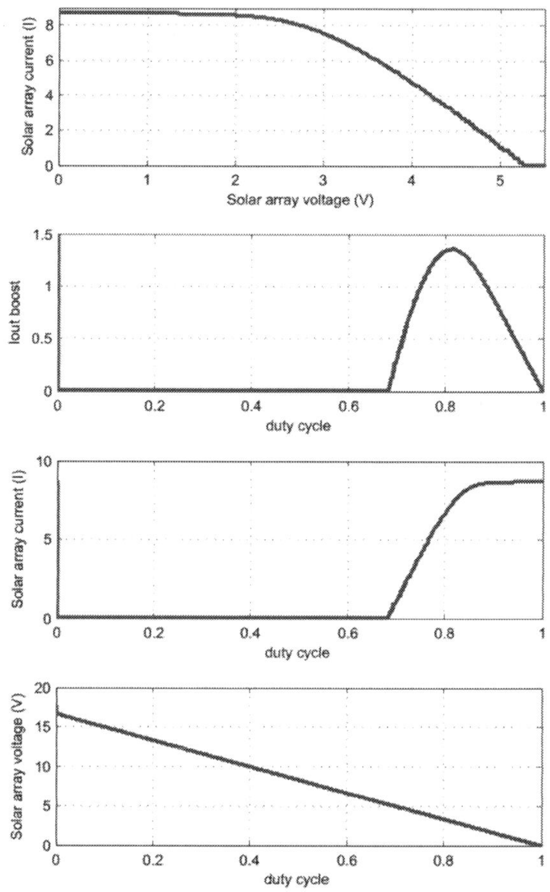

Figure 11: boost converter output current and voltage vs duty cycle.

5.2. Power Multiplexor control

If there is enough power at converters output to supply loads and to charge the battery according the profile shows in *Figure 12*, the power multiplexor keeps the present state.

DOI: 10.1051/e3sconf/20171614008

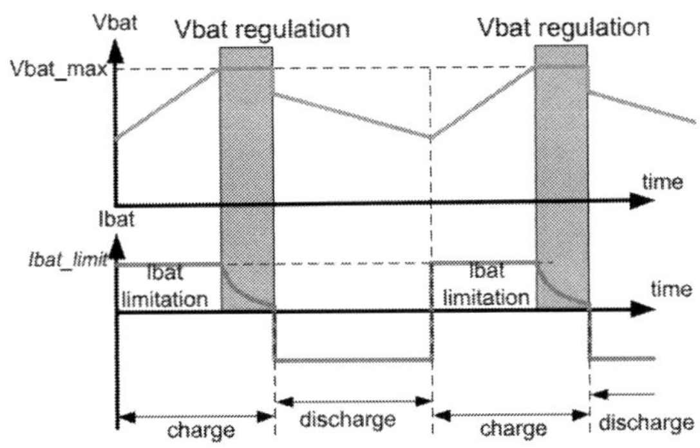

Figure 12 : charge/discharge profile of battery

Nevertheless, if the battery current or voltage are lower than programmed references (Vbat_max or Ibat_limit), it means that the power needed by the gondola is upper than the solar array power. In this case, the power multiplexors are switched one after the other until power at converters output becomes sufficient.

So it means that the control scans sequentially all the pairs of opposite solar panels, and selects the panels which give the more power. This power multiplexor algorithm is based on periodic timers with 120° phase shifts between each other's. In this case, the timers trigged the research of the most powered panel between one panel and its opposite panel, as explained in *Figure 13.*

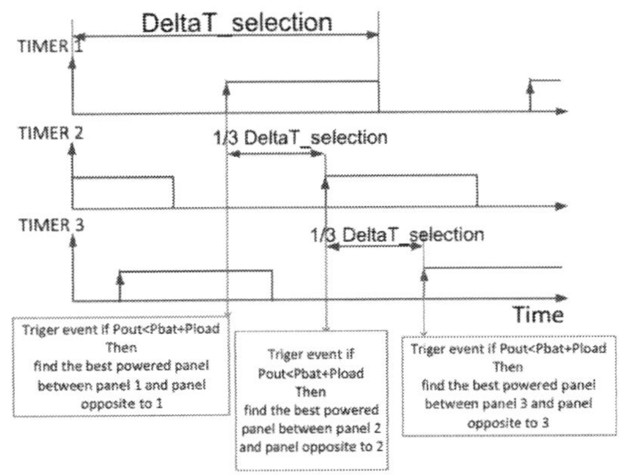

Figure 13 : Power multiplexor timers

5.3. Overall regulation algorithm

The overall regulation is shown in the block scheme. First, the battery reference voltage is compared to the battery voltage at sensor output.

The voltage error goes to a Proportional Integral corrector to define the battery current reference.

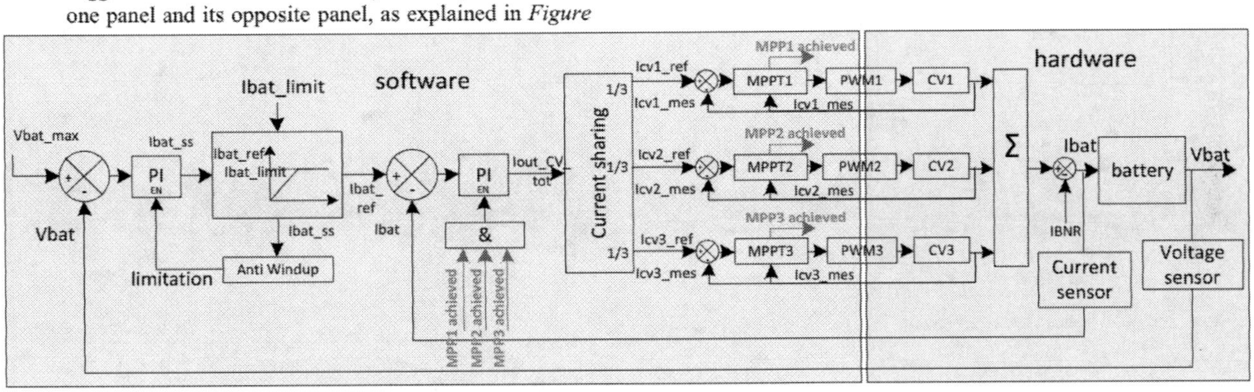

Figure 14: Block scheme of the power conditionning algorithms

Then this battery current reference is limited to the maximum charge rate of the battery. An anti-windup is implemented in this loop to reduce time response in case of saturation. Then, the battery current is regulated by a PI loop. The Integral action of the current could be disabled in case all MPP (1 to 3) are achieved. Indeed, it is not necessary to try to increase the total output currents of converters in case of power deficit between solar array power and power consumption of the gondola. The current sharing is balanced between the 3 converters. Then an independent MPPT algorithm is implemented on each converter.

6. MODELING AND SIMULATION

6.1. Introduction

A model in Matlab/Simulink was developed. This model allows simulating MPPT algorithms behaviour and power conditioning control made by MC². The hardware models are the solar arrays and their solar flux, the power multiplexors, the boost converters, the battery and loads. The software model is a

 DOI: 10.1051/e3sconf/20171614008

representative sampled function. The power conditioning algorithm is representative of the software implemented inside the microcontroller for the real power conditioning. In order to test the algorithms robustness, noises have been added to converter outputs currents measurements. This simulation is focused on avionic gondola architecture (4 solar panels). All the simulation results presented hereafter are for a MC2 configured for avionic gondola architecture.

6.2. Simulation results

Figure 15 shows the MPPT response. The algorithm starts at 0.1s and the convergence is achieved in few ms.
An internal variable shows the achievement of MPP. This variable is used for the software as explained in 5.3.

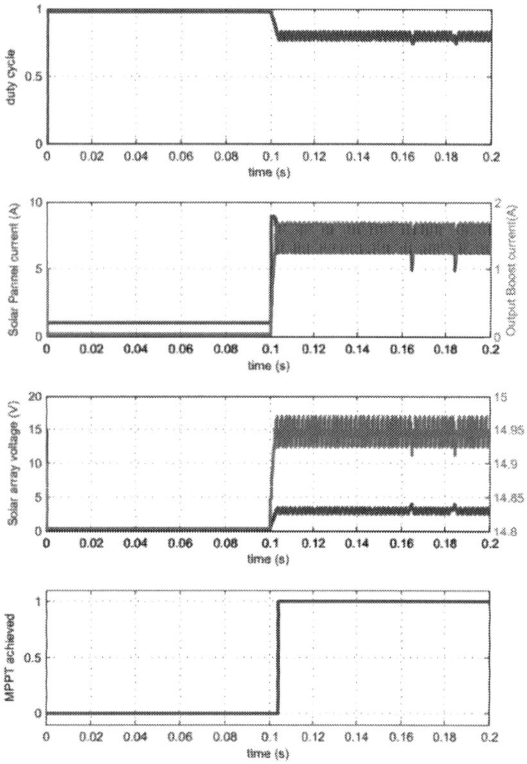

Figure 15 : MPPT response

Duty cycle format is 10 bits and the increment of duty cycle around the MPP is +/-30 Low Significant Bits, which is equal to +/-3% of duty cycle increment. This increment value depends on the capability of the microcontroller (depending on its clock frequency, and PWM frequency). +/-3% of duty cycle ripple leads to a solar panel voltage ripple of 1Vpp and a current ripple of 2Amps. On battery interface, the current ripple is 0.5A. This value of +/-3% of duty cycle ripple is a worst case obtained by a PWM with 5.1 equivalent useful bits, which is a very low granularity value.

The power multiplexor test is presented in *Figure 16.* Remember that avionic gondola architecture with only 4 solar panels is presented. The graph (a) simulates the solar flux view by solar panels (according to *Figure 9*). When the power multiplexor connects the panel to the converter, relative flux become a bold line. The graph (b) shows the current at multiplexor output. Depending on the value of the variable sampling mux (shows in (d) and defined in *Figure 13*), the currents have some transients. Indeed, during this transients, the other panel is temporary connected to the boost converter and the MPP of this panel is found. If the MPP of this panel is lower than the previous one, then the power multiplexor goes back to the previous position. During the temporary connection of another panel, if the MPP becomes upper than the previous MPP, then the power multiplexor keeps this position.

In (c), the variable MPP achieved is showed. Each time the variable sampling MUX is equal to 1, another panel is connected and the MPP is found.

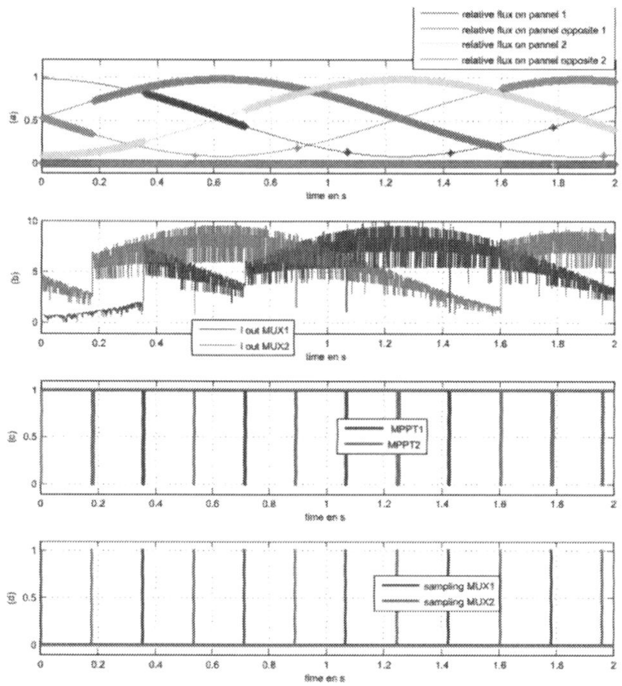

Figure 16: Power multiplexor test

DOI: 10.1051/e3sconf/20171614008

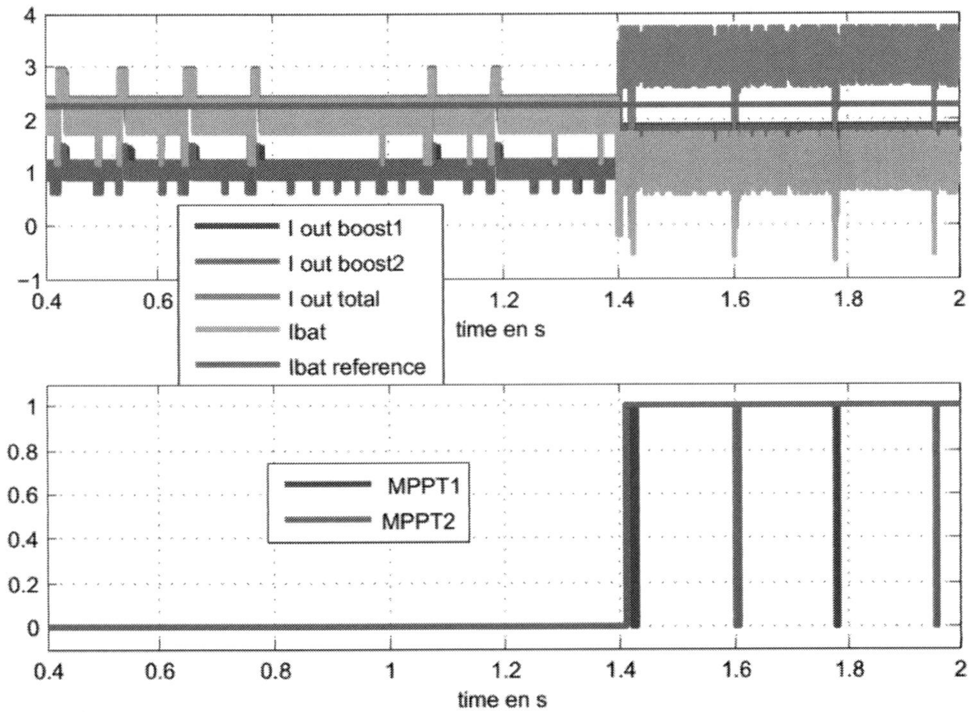

Figure 17: Battery current regulation loop test

Regarding the battery current loop, *Figure 17* shows the behaviour of the regulation. The reference current is equal to the battery capacity divided by 3. At 1.4s, the load current goes from 0.05A to 2A, in order to see the transient response of the loop. With a load of 2A, the solar panel power is not sufficient to maintain the battery current equals to the reference. The battery current goes to 1A, and both MPPT variables 1 and 2 go to 1. It means that the power conditioning is in MPPT mode, and all the available power at solar panels is converted by the boost converters.

7. CONCLUSION

This paper presents a power conditioning and distribution board for stratospheric balloons. Renewable energy is used and power conditioning algorithms using a MPPT are implemented and simulated.

A dedicated architecture based on power multiplexor allows to select 2 among 4 (or 3 among 6 depending on gondola) solar panels. The multiplexor state depends on the maximal extracted power of the opposite solar panel. Priority connection is given to the solar panel which gives the more power.

A new Maximum Power Point Tracker is implemented and associated to a boost converter. This MPPT uses the output current of the converters which is proportional to the power (the converter output is connected to a battery constant DC voltage). Simulation results are presented and the behaviour of the algorithms is showed. The board is currently in test in EREMS facilities. Experimental results could be presented during the oral session.

8. REFERENCES

1. Stratéole-2, a long-duration balloon campaign at the equator, 22nd ESA symposium on European rocket and balloon programmes and related research, june 2015, Tromso.

2. SUNPOWER C60Jp monocrystalline silicon solar cells : datasheet available at http://eshop.terms.eu/_data/s_3386/files/1379942540-sunpower_c60_bin_ghi.pdf

3. SAFT MP 176065 Integration cell. Datasheet Available at : http://www.saftbatteries.com/force_download/MP_176065.pdf

A STUDY OF A B³R CONVERTER BASED PCU, TO CONFIRM THE FLEXIBILITY AND PERFORMANCE OF THE CONVERTER TOPOLOGY

Erik Mache [(1)], Michael Koch [(2)]

[(1)] *ASP Equipment GmbH, Am Wasserstall 2, 88682 Salem, Germany, Email: e.mache@asp-equipment.de*
[(2)] *ASP Equipment GmbH, Am Wasserstall 2, 88682 Salem, Germany, Email: m.koch@asp-equipment.de*

ABSTRACT

The main objective of this activity was to study the B³R converter [1] and to confirm its flexibility and performance (mainly efficiency and mass reduction) versus solutions that is implemented in a satellite PCU. For details see [2] ESTEC SOW.
An architecture with centralised regulated bus, single battery and centralised solar array was selected for this study.
Project Plan:
- Study, design and development of an evaluation prototype, based on the B³R topology (including protections and control loops).
- Study, design and development of an Electrical Elegant Breadboard including three power stages, based on the B³R topology (including protections and control loops).
- Laboratory tests of these converters (power stages, control loops and protections).
- Conclude the study by comparing the achieved performances (including efficiency and mass) versus current solutions implemented on a satellite PCU. Summary of the pros and cons of the use of the B³R concept.

1. INTRODUCTION

In a satellite, the main function of a Power Conditioning Unit (PCU) is to manage the energy coming from several power sources (usually solar arrays and battery) and to deliver it continuously to the users in an appropriate form during the overall mission.
The ESA Power Conditioning Department proposed new solutions based on Buck Buck Boost Regulators (B³R), which appeared to be a promising alternative [1]. They predicted on one hand more efficient regulators and or on the other hand a better utilisation of the source, since they are capable of operating on a wider range of source characteristics. This new B³R topology is a multiple port DC DC converter, which can interface a solar array, a battery and generate a regulated DC bus voltage.
The B³R topology is a flexible topology that can be used in different architectures for different applications.

For this study different architectures of the solar array (centralised or distributed), the regulated bus (centralised or multiple) and the battery (single or segregated) were discovered. After a comparison of properties, the architecture "Centralised Regulated Bus and Single Battery and Centralised Solar Array" was selected.

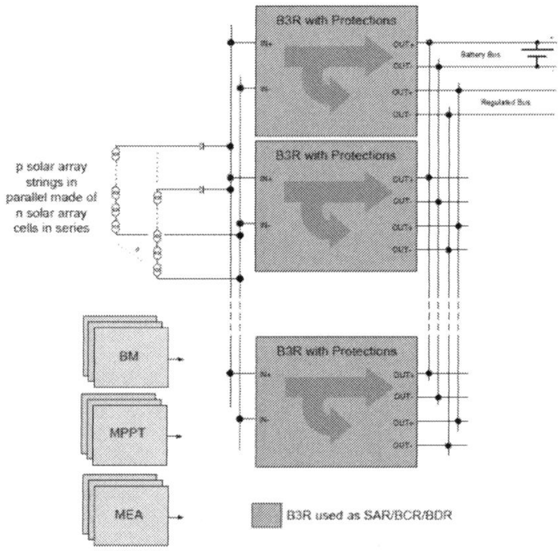

Figure 1-1: Centralised Regulated Bus and Single Battery and Centralised Solar Array

2. STUDY LOGIC

The objective of this task was to design and build an Electrical Elegant Breadboard to study the adequacy of the protections and evaluate the stage performances at B³R level and at architecture level.
According to the selected architecture an Electrical Elegant Breadboard (EEBB) of the single point failure free architecture was designed. Since the design was suitable the Agency decided to approve it.
The power architecture contains three B³R power stages with their dedicated current control loops and protections. The higher level control loops are located on a separate printed circuit board.
For the purpose of the study it was found as adequate to manufacture the Electrical Elegant Breadboard with commercial EEE parts, having same or equivalent functionalities and performances as Hi-Rel EEE-parts..

DOI: 10.1051/e3sconf/20171614009

3. DESIGN DESCRIPTION

The EEBB is built up on two circuit boards and one mechanical frame. One board is the Power Board (blue in Fig. 3-1). It includes three B^{3}R power stages. The Controller Board (green in Fig 3-1) contains the high level controller which is designed to be single point failure free. The following figure shows the internal signal routing of the EEBB and the external connections.

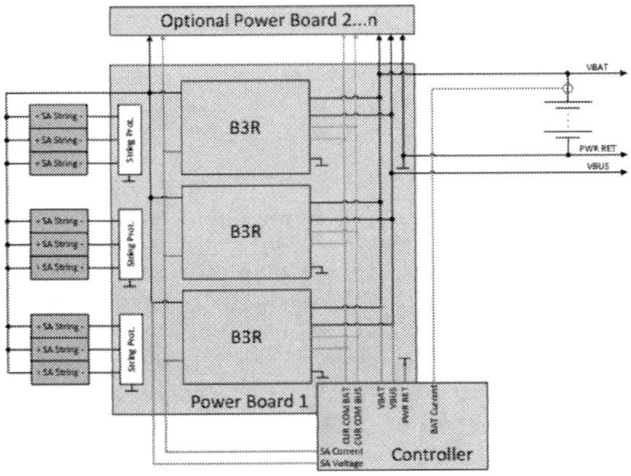

Figure 3-1: EEBB Block Diagram

3.1. B^{3}R Power Stage

The B^{3}R power stage is a three port DC/DC converter. Solar Array (SA) input is passed via a buck/boost stage to charge the battery and supply the bus. Both SA input switches (buck/boost) are controlled by a battery port output current controller. The regulated bus output is supplied by a buck stage, with the buck switch controlled by an bus port output current controller. The power stage can operate in three different operating modes:

- BUCK (SA Voltage > Battery Voltage)
- Direct Energy Transfer (DET) (SA Voltage = Battery Voltage)
- Boost (SA Voltage < Battery Voltage)

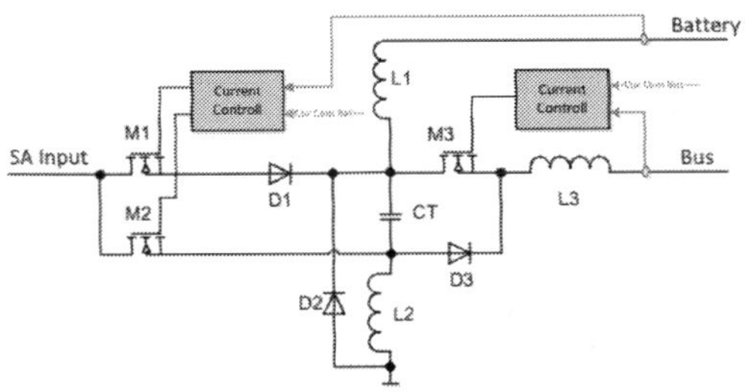

Figure 3-2: B^{3}R Power Stage

3.2. Protection

Every B^{3}R power stage protection has an independent circuit which monitors several signals. Three MOSFET switches are controlled by the protection to isolate the power stage in case of a detected failure (battery protection switch, solar array protection switch and protection switch against short circuit to ground).

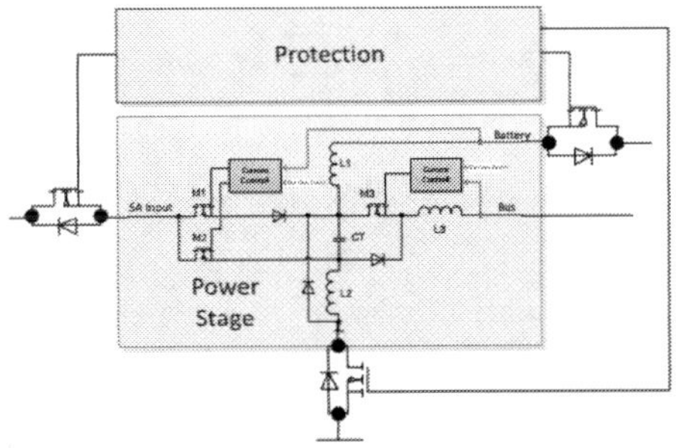

Figure 3-3: Protection Scheme

The switches are controlled by a central latch. When an over load condition is detected, the latch is set, the power stage is disabled and the protection switches opens. The latch stays active until a power cycle occurs (SA and Battery disconnect and reconnect) or an external RESET command is issued. Fig. 3-1 shows the possible sources to set and reset the latch, and the state of the protection switches and the power stage.

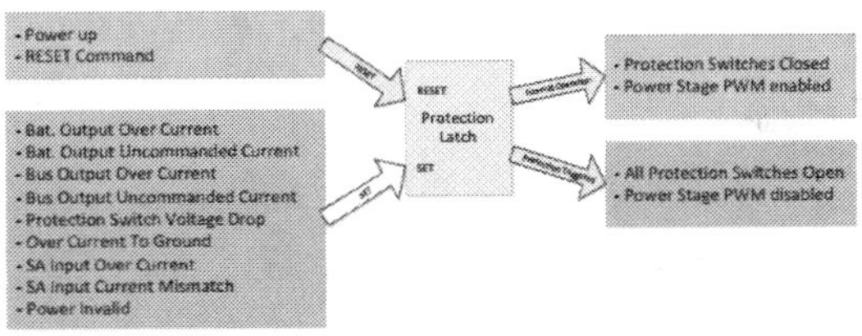

Figure 3-4: Protection Latch

3.2.1. Battery Output Over Current

A segregated current sense amplifier is connected to the shunt resistor in the battery port output. If the charge current to the battery exceeds the limit, the protection will be triggered. This threshold is set around to 14A, which is well above the maximum current that the control circuit in normal operation will permit (10A). Buck and boost PWM signals are disabled when the protections trigger.

3.2.2. Battery Output Uncommanded Current

The current sensed by the battery over current protection is used to detect uncommanded current to the battery. The current command signal from the controller

DOI: 10.1051/e3sconf/20171614009

board is used to detect a zero current command. If a zero current command is detected and there is still current flowing on the battery port, the protection will be triggered. The threshold to detect uncommanded current is 0,5A.

3.2.3. Bus Output Over Current

A segregated current sense amplifier is connected to the shunt resistor in the battery port output. If the charge current to the battery exceeds the limit, the protection will be triggered. This threshold is set around to 9A, which is well above the maximum current that the control circuit in normal operation will permit (7,6A). Buck PWM signals are disabled when the protection trigger.

3.2.4. Bus Output Uncommanded Current

The current sensed by the bus over current protection is used to detect uncommanded current to the bus. The current command signal from the controller board is used to detect a zero current command. If a zero current command is detected and there is still current flowing on the bus port, the protection will be triggered. The threshold to detect uncommanded current is 0,5A.

3.2.5. Protection Switch Voltage Drop

In normal operation, the voltage drop on all three protection switches is monitored. The voltage drop is caused by the RDSon and the current of the protection switch. If a protection switch fails in a way that increases RDSon, the voltage drop will also increase. A voltage drop over 0,7V will trigger the protection latch.

3.2.6. Over Current to Ground

In the normal operating the current through Ground can flow in both directions. This depends on the operating condition. In a failure case, the current flows to ground. To protect from a short circuit to ground in any point of the power stage, the current to ground is sensed. If the current increases over the threshold, the protection will trigger.

3.2.7. SA Input Over-Current

The SA input current will be measured using the voltage drop over the input filter inductor. If the current increases the max input current threshold, the protection will trigger.

3.2.8. SA Input Current Mismatch

Fig. 3-5 shows the connection principle between the different power stages. The SA input current will be measured using the voltage drop over the input filter inductor. The current is compared to the input current of another power stage. In normal operation all B3R power stages get a common current output command. Therefore the output currents are identical. Then also the input currents are identical. If one power stage suffer a failure which increase the power dissipation of that converter, the input current will increase to maintain the commanded output current. This will be detected by the SA input current mismatch monitor and triggers the protection latch.

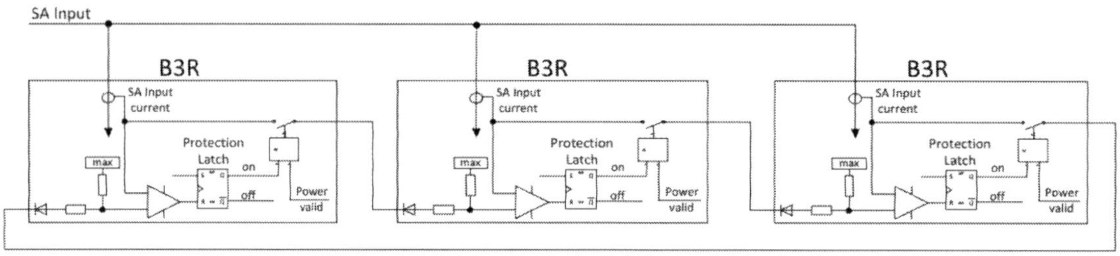

Figure 3-5: SA Input Current Mismatch Monitor

This monitor is combined with the input over current protection. The threshold to activate the protection is either the max input current, or a reduced level transmitted from another power stage. The power stage itself transmits its input current level to the next power stage for comparison. To avoid unintended trigger of that monitor, a current level will only be transmitted if the power stage operates in normal conditions. An activation of the protection latch or an internal supply failure disables the transmission of the input current level to the next stage.

3.2.9. Solar Array Short Circuit Protection

In addition, there are several input protection switches connected in the return line of the solar array that will isolate shorted strings of the solar array avoiding losing the whole section or solar array. As a baseline, 3 switches (MFETs) per B3R are implemented.

If a failure in the solar array side, for example in the driving mechanism, leads to a short circuit of it, and all the sections are directly connected together, as happens to be in a centralised solar array, the whole solar array

DOI: 10.1051/e3sconf/20171614009

will be lost. However, if the solar array is divided in sections, each one connected through a protection switch, in case one fails, only that section will be lost, minimising in this way the risk.

3.3. Control Scheme and MPP Tracking

The control of the PCU is implemented with different control loops.

- Basic current control loops on the power stage
- High level control loops on the controller board

3.3.1. Basic Current Control Loops

The basic current control loops are located on the B^{3}R power board and regulates the output current of the battery port and the bus port. Fig. 3-6 shows the two basic current control loops which get the current command from the high level controller fitted on the control board. These basic current control loops are implemented on each B^{3}R power stage. This arrangement ensures equivalent current sharing over the different power stages. The output of the current controllers is feed to a voltage comparator. It is compared to a saw tooth signal to generate a PWM signal which drives the power stage. To insure proper start up the controller is initialised by the protection latch. With an active protection signal, the integrator of the controller is held in reset state. When the converter gets enabled (power up or enable command) the controller does a soft start.

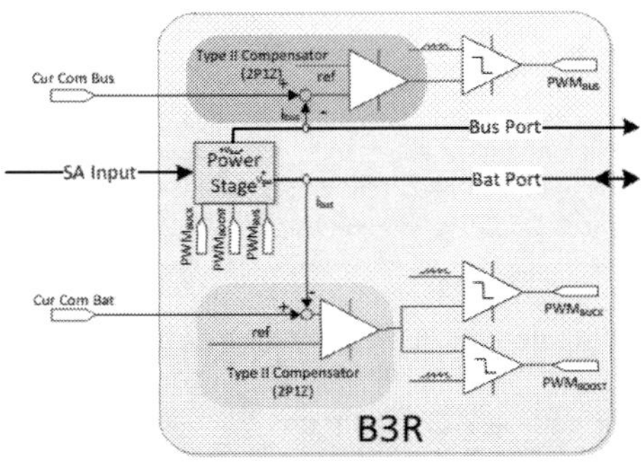

Figure 3-6: Basic Current Control Loops

The power stage for the battery output supports buck and boost operation. Two separate comparators are used to generate the PWM signals for the boost and buck switch. The offset of the two saw tooth signals for buck and boost mode is arranged at different voltage levels. This ensures segregated operation modes dependant on the SA and battery voltages. Boost mode for SA voltage lower battery voltage. Direct energy transfer (DET) for SA voltage equal battery voltage. Buck mode for SA voltage higher battery voltage. The output voltage of the current controller determines the operating mode.

3.3.2. High Level control loops and MPP Tracking

The high level controller generates the current command for the basic control loops. These current command signals are global signals for all B^{3}R power stages. To reach the required system availability they have to be single point failure free. Therefore these signals are generated by majority voted outputs of the triplex controllers.

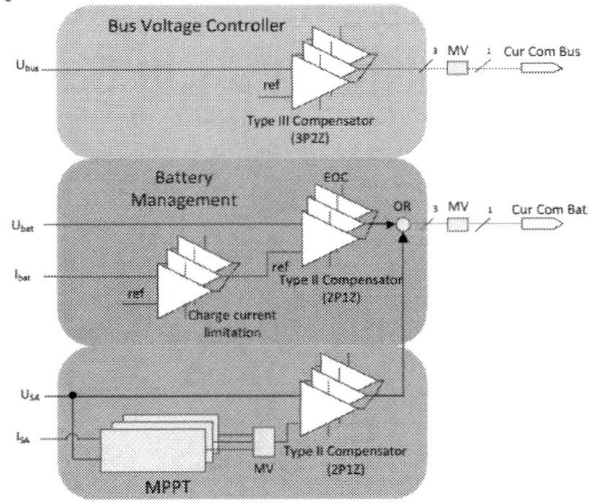

Figure 3-7: High Level Control

3.3.3. Bus Output Voltage Controller

The bus voltage controller is a Type III compensator. It provides three poles and two zeros, which are selected to reach the desired gain and phase margin. The compensator is implemented three times with a following majority voter. The voted output is the current command for the bus output and is distributed to all converters. See Fig. 3-7.

3.3.4. End of charge control Loop

This loop control limits the maximum battery charge voltage and current. It contains two compensators. One is a type II compensator to control the battery charge voltage, the other is a type I compensator which controls the max charge current. The schema is shown in Fig. 3-7, Battery Management. The battery voltage controller has a max voltage reference which can be reduced by the max charge current controller. This arrangement ensures compliance with the battery operating limitations. The output of the battery management controller is linked together with the output of the MPPT. The lowest current command will generate the current command for the battery output.

DOI: 10.1051/e3sconf/20171614009

3.3.5. MPP Tracking

The maximum power point tracker uses the "perturb and observe" method to find and track the maximum power point. Three independent trackers generate a voted SA voltage reference. A SA voltage controller uses this reference to generate a battery output current command. See Fig. 3-7 for the schema. The SA voltage controller is implemented with a type II compensator which provides two poles and one zero. They are positioned in a way to reach the desired phase and gain margin.

4. TEST RESULTS

The B3R module was tested in all operating modes (Buck / DET / Boost / Charge / Discharge / MPPT / Battery management) and behaves as expected. This includes stability measurements (small signals and large signal transient), efficiency measurements (including control consumption and protections dissipation), power stages waveforms measurements, transient measurements from MPPT mode to Charge mode (and the opposite), transient measurements between Buck/Boost modes, and DET mode.

4.1. Efficiency

Efficiency was measured on the connectors of the EEBB and based on one power stage. It includes all internal B3R losses like the cooper tracks, filter inductors, protection switches and current sense shunts, auxiliary power supply and the power stage. The following figures show the efficiency for different operating modes with different load and charge current. The X-axis shows the battery port current. Negative current means battery discharging. The Y-axis shows the bus current.

5	94,5	92,8	91,2	89,5	88,9	88,4	87,8
4		93,6	91,9	89,9	89,5	89,1	88,7
3			92,7	90,3	90	89,8	89,5
2				87,5	88,4	89,3	90,3
1				84,7	86,8	88,9	91
	-3	-2	-1	0	1	2	3

Figure 4-1: Boost Mode (VSA=25V, VBAT=42V, VBUS=28V)

5	94,8	93,7	92,7	91,6	91,7	91,8	91,9	91,8	91,6	91,5	91,3
4		94,2	92,8	91,5	91,7	92	92,2	92	91,9	91,7	91,6
3		94,6	93	91,4	91,8	92,1	92,5	92,3	92,2	92	91,8
2				89,4	90,5	91,6	92,8	92,6	92,5	92,3	92,2
1				87,4	89,3	91,1	93	92,9	92,8	92,6	92,5
	-3	-2	-1	0	1	2	3	4	5	6	7

Figure 4-2:Buck Mode (VSA=42V, VBAT=31V, VBUS=28V)

5	95	94,4	93,9	93,3	93,5	93,6	93,8	93,6	93,4	93,1	92,9
4		94,3	93,9	93,5	93,7	94	94,3	94	93,8	93,5	93,3
3		94,2	93,9	93,6	94	94,3	94,7	94,4	94,2	93,9	93,6
2				91,8	92,8	93,9	94,9	94,6	94,4	94,1	93,9
1				90	91,7	93,4	95,1	94,9	94,6	94,4	94,1
	-3	-2	-1	0	1	2	3	4	5	6	7

Figure 4-3: DET Mode (VSA=VBAT=40V, VBUS=28V)

4.2. Stability Of The Control Loops

In battery charging mode, the MPPT ensures always operating in MPP. If the charging voltage of the battery is reached, the power, delivered by the SA, has to be reduced. There are two different regions to operate the SA, to reduce the power of the SA: Current region and voltage region.

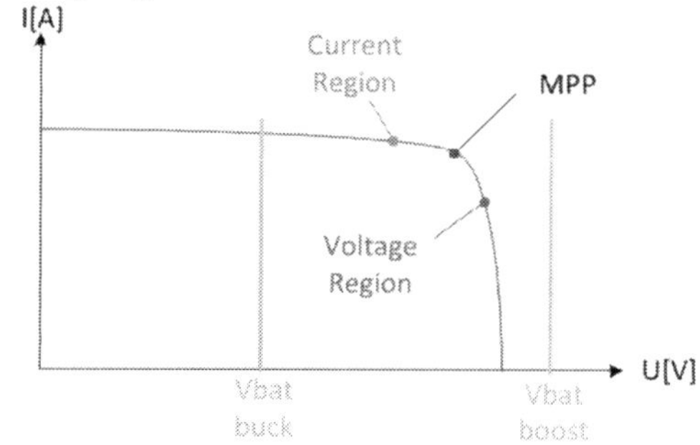

Figure 4-4: SA EOC Operating Modes

4.2.1. Stability EOC In Voltage Region

To operate the EOC in the voltage region of the SA, the current control loop is configured to operate in voltage region. Therefore the pulse wide of the power switches will be reduced, when a lower output current is demanded. Zero current demand cause the buck and boost switches to open, leading to the maximum no load voltage of the SA. All loops were measured and analysed.

The current loops were measured in boost and buck operation and reach all a phase margin > 60° and gain margin > 10 dB with a bandwidth > 17 kHz.

The input voltage loops result where measured in buck and boost operation and reach all a phase margin > 56° and gain margin > 10 dB with a bandwidth > 1,6 kHz.

4.2.2. Stability EOC In Current Region

To operate the EOC in the current region of the SA, the current control loop is configured to operate in current region. Therefore the pulse wide of the power switches will be increased, when a lower output current is demanded. Zero current demand cause the buck and boost switches to close, leading to the maximum short current of the SA. All loops were measured and analysed.

The current loops were measured in boost and buck operation and reach all a phase margin > 60° and gain margin > 10 dB with a bandwidth > 400 Hz.

The input voltage loops result where measured in buck and boost operation and reach all a phase margin > 49° and gain margin > 11 dB with a bandwidth > 1 kHz.

The result shows stable operation in every condition. The bandwidth is much lower compared to operating in

DOI: 10.1051/e3sconf/20171614009

the voltage mode. Further investigation showed the limiting factor is the capacitance on the SA input. In the current region, the SA behaves as a current source for low frequencies (high impedance). As frequency increase the SA behaviour turns to a voltage source (low impedance).

4.3. Internal Failures Stimulation

To verify the detection of power component failures all cases, which are identified to be critical, were stimulated by HW failure injection. The critical failure cases where identified by analysis and simulation and a result of the FMECA.

All tests where done in the following way:

- Applying the failure during normal operation and record the reaction.
- With the failure still applied, enable the converter by an enable command and verify the reaction of the protection.
- Remove the failure stimulation and enable the converter by an enable command.

M2 SD short boost/buck

- OVER CURRENT TO GROUND

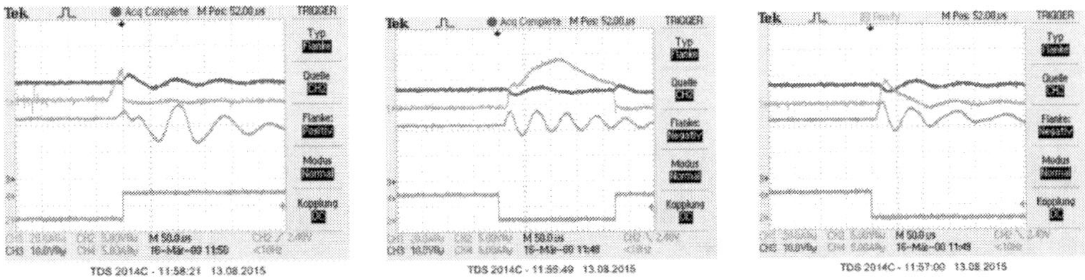

Figure 4-5: M2 SD Short Stimulation

CH1: Ground Cur, CH2: prot. Latch, CH3: Bat Voltage, CH4: SA Current

M2 DG short boost/buck

- SA INPUT CURRENT MISMATCH

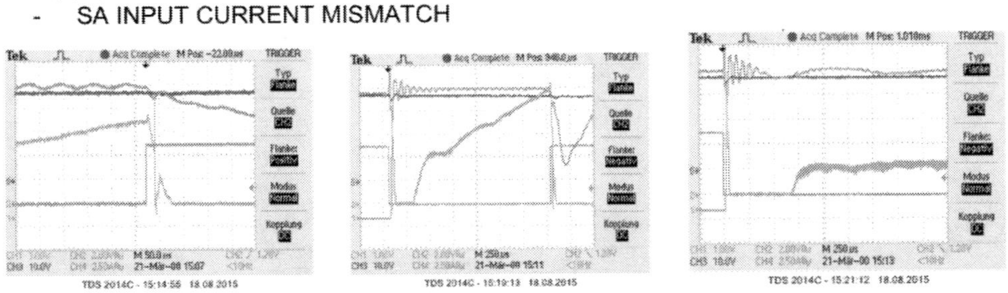

Figure 4-6: M2 DG Short Stimulation

CH1: Converter input current, CH2: prot. Latch, CH3: Bat Voltage, CH4: SA Current

M1 SD short buck

- SA INPUT CURRENT MISMATCH

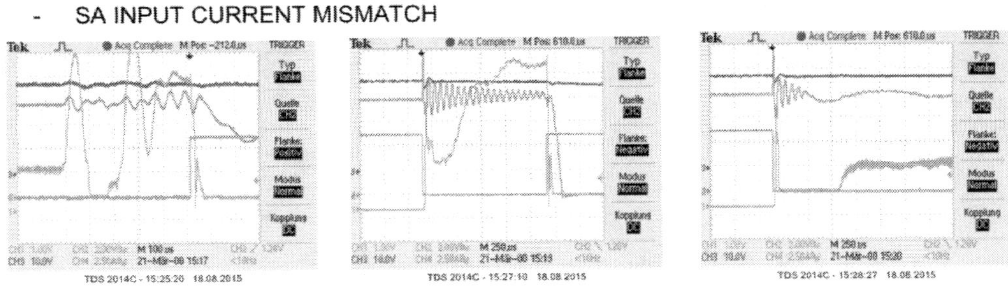

Figure 4-7: T3 M1 Short Stimulation

CH1: Converter input current, CH2: prot. Latch, CH3: Bat Voltage, CH4: SA Current

 DOI: 10.1051/e3sconf/20171614009

D2 open buck

- SA INPUT CURRENT MISMATCH

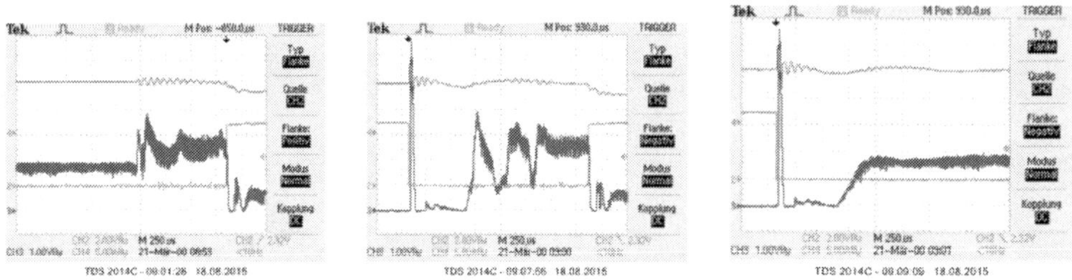

Figure 4-11: D2 open Stimulation
CH2: prot. Latch, CH3: Converter input current, CH4: SA Current

D2 short buck/boost/eclipse

- OVER CURRENT TO GROUND

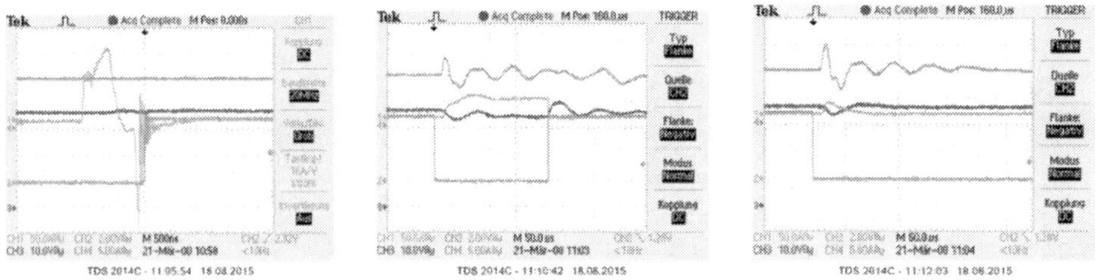

Figure 4-8: D2 short Stimulation
CH1: Ground Cur, CH2: prot. Latch, CH3: Bat Voltage, CH4: SA Current

D3 open buck/boost/eclipse

- SA INPUT CURRENT MISMATCH

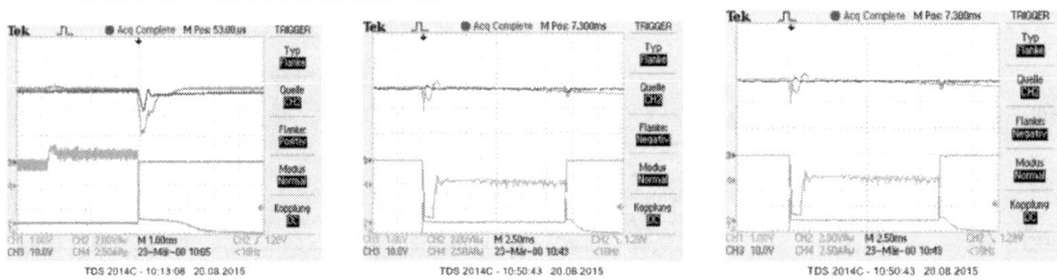

Figure 4-9: D2 open Stimulation
CH1: Converter input current, CH2: prot. Latch CH3: Bus Voltage, CH4: SA Current

D1 short boost

- OVER CURRENT TO GROUND

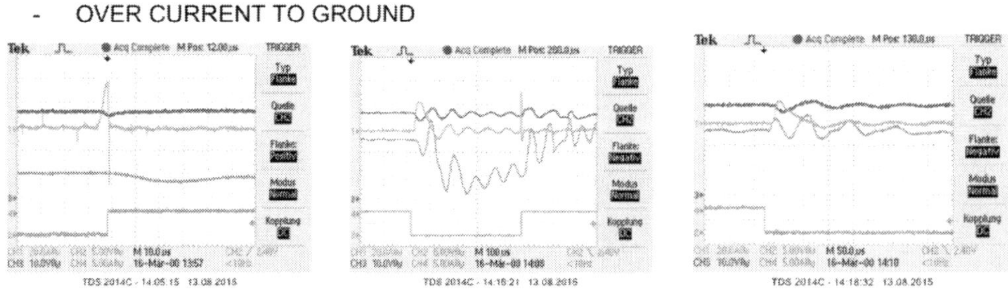

Figure 4-10: D1 A1 short Stimulation
CH1: Ground Cur, CH2: prot. Latch, CH3: Bat Voltage, CH4: SA Current

DOI: 10.1051/e3sconf/20171614009

5. EEBB HARDWARE

Picture 3 shows the EEBB hardware. The three power stages can be identified on the power board mounted in the frame within three vertical sections. The Controller Board is assembled above the Power Board.

Figure 5-1: EEBB Hardware Top View

6. ACTIVIES DURING THE STUDY

- Study, design and develop of an evaluation prototype, based on the B³R topology (including protections and control loops).
- Study, design and develop of an Electrical Elegant Breadboard including three power stages, based on the B³R topology (including protections and control loops).
- Test in laboratory environment (power stages, control loops and protections).
- Concluding the study by comparing the achieved performances (including efficiency and mass) versus current solutions implemented on a satellite PCU. Summary of pros and cons of the use of the B³R concept.

The B3R module was tested in all operating modes (Buck / DET / Boost / Charge / Discharge / MPPT / Battery management) and behaves as expected.
This includes:

- Stability measurements (small signals and large signal transient).
- Efficiency measurements (including control consumption and protections dissipation).
- Power stages waveforms measurements.
- Transient measurements from MPPT mode to Charge mode (and the opposite).
- Transient measurements between Buck/Boost modes, and DET mode.

7. SUMMARY OF THE STUDY RESULTS

As a result of this activity, it was shown that a flexible PCU with several B³R power stages can be implemented. For the comparison with current implementations the following key parameters were determined:

- SA input voltage: 0 – 60V
- Battery Voltage: 32 – 50V
- Regulated Bus Voltage: 28V
- Maximum solar array input power of one power stage: 600W
- Efficiency in buck mode, including supply and protections: 93% max.
- Efficiency in boost mode, including supply and protections: 91% max.
- Weight-Power ration in a two out of three operating condition: 1.725g/W (worst case)
- Weight-Power ration in a five out of six operating condition: 1.38g/W

A PCU implementation of a single point failure tolerant B³R topology was successfully demonstrated. In a primary PCU, single point failure tolerance is essential for the operation in a spacecraft. It was found as important to detect and isolate failures in the power stage. This ensures that after a failure case neither the Solar Array, the Battery and the Bus ports are shorted, nor significant power dissipation will propagate heat and damage redundant power stages. Detailed analysis and tests with stimulated power components failures where done and showed a reliable detection and isolation of the affected power stage.
Operating during end of charge (EOC) in the current or voltage region of the Solar Array was tested. It was shown that both operation modes are feasible with different characteristics.
Also it was shown, that the B³R is able to handle abnormal conditions, like start up from depleted battery.

8. REFERENCES

1. Buck-Buck-Boost Regulator (B³R) for Spacecraft Power Conditioning Unit, O. Mourra, A. Fernández, S. Landstroem & F. Tonicello, Proceedings of the 9th European Space Power Conference, 2011, Saint Raphael, France.

2. ARTES 5.1 Statement of Work PCU module based on Buck Buck Boost Regulator (B³R) Reference nr.: 12.1TT.56 Technical Officer: Olivier Mourra, ESTEC, Noordwijk

DOI: 10.1051/e3sconf/20171614010

INTERLEAVED, MULTI-SWITCH, MULTI-PHASE BOOST CONVERTER FOR BATTERY DISCHARGE REGULATORS

Fernando Sobrino-Manzanares, Ausias Garrigós

Industrial Electronics Group, UMH, Avda. De la Universidad s/n (Spain), Email: augarsir@umh.es

ABSTRACT

Non-isolated step-up voltage converters are commonly required as Battery Discharge Regulators (BDR's). The Weinberg and boost derived topologies are traditionally employed in modules up to one or two kilowatts. The Interleaved Boost Converter (IBC) is a distributed power processing approach that provides some advantages: reduced power processing (current) per phase, distributed losses (power semiconductors and magnetics), input and output current ripple cancellation, improved dynamic response and better modularity, among others. The IBC converter as BDR has been previously treated and advantages and drawbacks discussed. In this paper a new approach is proposed, a multi-switch, multi-phase interleaved boost converter (named Multi Interleaved Boost Converter, MIBC) that provides inductor and switching frequency decoupling, which in turn affects in different aspects, such as, magnetics, input and output current ripples or part count. This paper presents and describes the MIBC, the potential benefits compared to the IBC and its application as BDR.

1. INTRODUCTION

This paper is meant to be a general description of a PWM interleaved method for DC/DC converters. This modulation scheme is aimed to increase converter power density by combining number of modules (phases) in parallel and number of switches per module, please refer to figure 1.

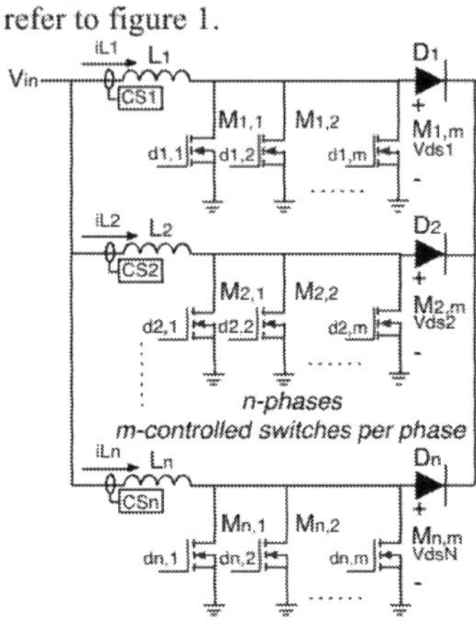

Figure 1. Multi Interleaved Boost Converter

2. MIBC DESCRIPTION

The MIBC has been devised using several multi-switch boost modules, which are out in phase [1]. Traditional multi-switch approach considers the same driving signal for all switches in order to divide the current into several power semiconductors being the reasons for that insufficient current handling capability, excessive power dissipation, complex thermal management or switch redundancy. In the MIBC, the driving signal of each power switch keeps the same switching frequency but the on-time is reduced by a factor *m* and they are displaced each other in phase over *2pi/m* radians. This driving scheme decouples switching frequency and inductor frequency, which actually increases by a factor *m*. Also the common point, mosfet drain, has an effective switching frequency *m* times higher than switching frequency. Another interesting feature is that only one switch conducts at a given time, so average current sharing could be adjusted by increasing the number of switches and decreasing the on time.

On the other hand, phase shifted paralleled modules, so-called interleaved converters, also offer reduction of input and output current ripple, improved dynamic response or power losses distribution, among other advantages. For the MIBC, modules will be displaced *2pi/(nm)* radians.

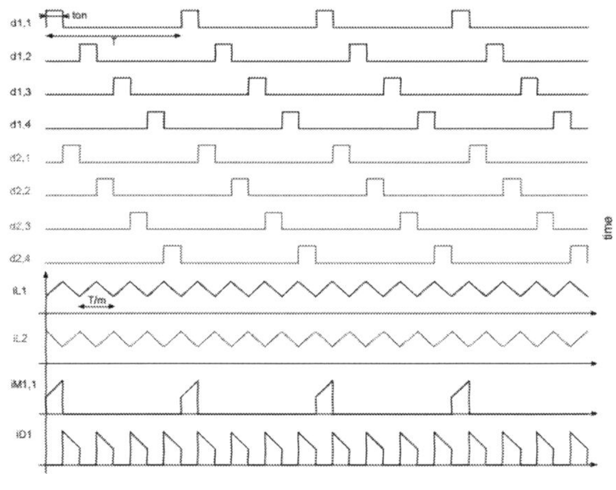

Figure 2. Sketch of the main waveforms of a 2-4-MIBC (2 phase, 4 switches per phase)

DOI: 10.1051/e3sconf/20171614010

2.1 DC equations

The MIBC DC analysis has been performed for the general (n x m) configuration. Main equations are listed in table 1.

Descrip.	Equation	nb
Duty cycle	$D = \dfrac{t_{on}}{T}$	1
Voltage conversion ratio	$M = \dfrac{V_o}{V_{in}} = \dfrac{1}{1-mD}$	2
Inverse voltage conversion ratio	$M' = 1/M$	3
Inductor current ripple (pk-pk)	$\Delta i_L = \dfrac{1}{m}\dfrac{V_o M'(1-M')}{Lf}$	4
Average inductor current	$\langle i_L \rangle = \langle i_{in} \rangle / n$	5
Max. inductor current	$i_{L\max} = \langle i_L \rangle + (\Delta i_L/2)$	6
Min. inductor current	$i_{L\min} = \langle i_L \rangle - (\Delta i_L/2)$	7
RMS inductor current	$I_{RMS\ L} = \sqrt{\langle i_L \rangle^2 + (\Delta i_L^2/12)}$	8
Inductor frequency	$f_L = mf$	9
Minimum input current for CCM	$\left\langle i_{\substack{in\min \\ CCM}} \right\rangle = n\dfrac{V_{in} D}{2Lf}$	10
MOSFET average current	$\langle i_M \rangle = D \langle i_L \rangle$	11
MOSFET RMS current	$I_{RMS\ M} = \sqrt{D} I_{RMS\ L}$	12
Diode average current	$\langle i_D \rangle = (1-mD)\langle i_L \rangle$	13
Diode RMS current	$I_{RMS\ D} = \sqrt{1-mD}\, I_{RMS\ L}$	14
Ripple cancellation index	$R = M' \bmod n^{-1}$	15
Input current ripple (pk-pk)	$\Delta i_{in} = \dfrac{n}{m}\dfrac{V_o}{Lf} R\left[\dfrac{1}{n} - R\right]$	16
Input (output) frequency	$f_{in(out)} = nmf$	17
Output RMS current	$I_{\substack{out \\ RMS}} = \langle i_{in} \rangle \sqrt{\dfrac{\left[\left(D'_e - \dfrac{X}{n}\right)\left(\dfrac{X+1}{n} - D'_e\right)\right] + \dfrac{n}{12D'_e}\left(\dfrac{V_{in} D}{L\langle i_{in} \rangle f}\right)^2}{\left[(X+1)^2\left(D'_e - \dfrac{X}{n}\right)^3 + X^2\left(\dfrac{X+1}{n} - D'_e\right)^3\right]}}$ $D'_e = 1 - mD$; $X = \text{floor}(nD'_e)$	18

Table 1. DC equations of the general (n x m) MIBC converter

2.2 Small-signal transfer functions

Neglecting all parasitic elements, the small-signal transfer functions of the general (n x m) MIBC converter have also been obtained and listed in table 2.

Descrip.	Equation	nb
duty cycle-to-output voltage	$G_{vd} = \dfrac{\tilde{v}_o}{\tilde{d}} = K_{vold}\dfrac{(1 - s/\omega_z)}{(s/\omega_o)^2 + s/Q\omega_o + 1}$ $K_{vold} = mV_{in}/D'^2_e$; $\omega_z = nD'_e R/L$; $\omega_o = D'_e/\sqrt{(L/n)C}$; $Q = D'_e R\sqrt{(nC)/L}$	19
Duty cycle-to-inductor current	$G_{id} = \dfrac{\tilde{i}_L}{\tilde{d}} = K_{iLld}\dfrac{(s/\omega_{z1} + 1)}{(s/\omega_o)^2 + s/Q\omega_o + 1}$ $K_{iLld} = 2m\langle i_{in} \rangle / nD'_e$; $\omega_{z1} = 2/RC$	20
Inductor current-to-output voltage	$G_{vi} = \dfrac{\tilde{v}_o}{\tilde{i}_L} = \dfrac{\tilde{v}_o}{\tilde{d}}\left(\dfrac{\tilde{i}_L}{\tilde{d}}\right)^{-1} = \dfrac{K_{vold}}{K_{iLld}}\dfrac{(1 - s/\omega_z)}{(s/\omega_{z1} + 1)}$	21

Table 2. Small-signal equations of the general (n x m) MIBC converter

3. COMPARATIVE STUDY: 8-IBC vs 4-2-MIBC vs 2-4-MIBC

In order to evaluate and compare the MIBC and the IBC converters used as BDRs, a theoretical study has been carried out using [2] as the baseline specifications, please refer to table 3.

Comparative study: BDR main parameters	
Po	1.6kW
Vin min	56V (mD)max=0.44
Vin max	94V (mD)min=0.06
Vo	100V
L	50uH
Co	88uF
fs	125kHz

Table 3. BDR power module specifications for converter comparison

The study assumes the following, three configurations are evaluated: a) n=8, m=1; b) n=4, m=2; c) n=2, m=4. All have the same number of MOSFETs and consider the same parameters, listed in table 3.

3.1 Input & output current ripple

Substituting values in the equations (16) and (17) result in the input and output current ripples represented in figures 3 and 4.

 DOI: 10.1051/e3sconf/20171614010

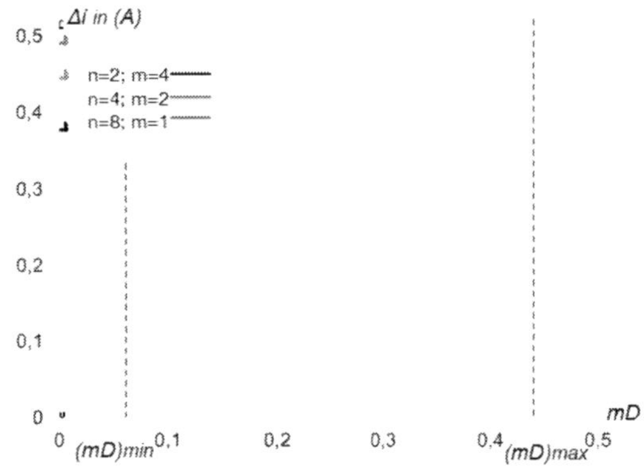

Figure 3. Input current ripple

As observed from figure 3, the 2-4-MIBC configuration exhibits the lowest input current ripple of all three. For certains *mD*, 4-2-MIBC and 8-IBC configurations have less input ripple, but at the extreme values of the duty cycle 2-4-MIBC performs better.

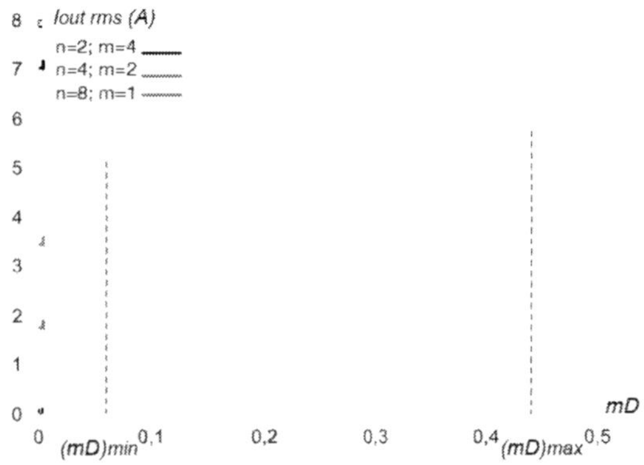

Figure 4. RMS output current (@Vin=56V, Po=1.6kW)

From the point of view of the output current, the 2-4-MIBC has the highest RMS value in the whole range, which is a key parameter for output capacitor losses and voltage ripple. Thus, low ESR capacitors are mandatory to not impair MIBC performance.

3.2 Inductor design

Inductor design, using Magnetics toroid MPP cores, has been considered to evaluate size and volume reduction using MIBC. The inductor design considers Dmax and Po max, the same inductance value and switching frequency for all three configurations. The design procedure has been tuned to achieve similar inductor losses (global losses considering the sum of all inductors) and temperature increment. The most representative inductor parameters have been gathered in table 4.

Comparative study: inductor design			
	n=2; m=4	n=4; m=2	n=8; m=1
Nb of ind.	2	4	8
Iavg	14.2850 A	7.1425 A	3.5713 A
Δi	0.986 A	1.97 A	3.94 A
f	500 kHz	250 kHz	125 kHz
Core ref.	55254	55929	55351
Turns	22	21	34
L[a]	0.051 mH	0.051 mH	0.051 mH
Ploss [b]	8.15 W	4.31 W	2.83W
ΔT	60.6°C	62.1°C	56.2°C
AWG	12	15	18
Win. fact.	18.3%	24.8%	21.5%
DC res.	6.51 mΩ	9.58 mΩ	24.51 mΩ
Finish.OD[c]	45.5 mm	31.1 mm	26.8 mm
Finish.HT[d]	20.1 mm	15.4 mm	12.1 mm
Area [e]	20.70 cm^2	9.67 cm^2	7.18 cm^2
Mass	128.8 g	50.5 g	27.6 g
Total loss.	16.3 W	17.24 W	22.64W
Total area	41.4 cm^2	38.7 cm^2	57.4 cm^2
Total mass	257.6 g	202 g	220.8 g

a) L at full load; b) Ploss per inductor; c) Finished Output Diameter; d) Finished Height; e) Area as OD^2

Table 4. Inductor comparison

From table 4, 4-2-MIBC exhibits the best outputs in terms of total area and mass, 2-4-MIBC has the lowest losses and 8-IBC gives the poorest results in terms of losses and required size, bringing to light that MIBC offers less input ripple, less losses and less area and mass if compared to the traditional IBC.

3.3 CCM/DCM boundary

Another interesting feature concerns to CCM and DCM limits when MIBC is considered. As the number of phases and inductor frequency changes, the minimum inductor current also varies. Since boost transfer functions depend on conduction mode, the wider the range in a particular mode, the better from the control point of view. Analysing the three configurations, please refer to table 5, one realises that 2-4-MIBC requires the lowest output power to remain in CCM.

Comparative study: CCM/DCM boundary			
	n=2; m=4	n=4; m=2	n=8; m=1
Iin min avg CCM	0.986A	3.94 A	15.8 A
Pin min CCM	55 W	221 W	883 W

Table 5. CCM/DCM limit comparison

Table 5 reveals that 8-IBC requires more than 880W to work in CCM. In other words, 8-IBC will work in both, DCM and CCM, in real conditions. On the contrary, 2-4-MIBC only requires 55.2W to work in CCM; thus, problems related to DCM operation, like parasitic ringing or control loop design are virtually eliminated.

 DOI: 10.1051/e3sconf/20171614010

3.4 Power semiconductor stress and impact on losses

An important concern regards to part electrical rating, subsequent part selection and converter losses. One of the most significant differences between the three configurations relates to conduction time and maximum current per switch. Regarding the power mosfet, reduction of both, phases and conduction time per mosfet, implies identical average current value, similar RMS value but higher peak current. Two important considerations arise from this fact, higher peak current capacity will be required and switching losses could increment substantially. Since all configurations will exhibit similar conduction losses, soft-switching techniques can be explored to keep similar performance. Regarding power diodes, one diode per phase is only required, but working at higher current level and *m* times switching frequency. Parallel diodes could be also considered to split power losses.

Comparative study: electrical rating (Vin=56V, Vo=100V, Pin=1600W)			
	n=2; m=4	n=4; m=2	n=8; m=1
Conduction time	0.88us	1.76us	3.52us
MOS avg current	1.57A	1.57A	1.57A
MOS RMS current	4.74A	3.35A	2.34A
MOS max. current	14.78A	8.13A	5.54A
D·(Irms mosfet)^2	2.47A^2	2.47A^2	2.49A^2
DIODE avg current	8A	4A	2A

Table 6. MIBC electrical rating

3.5 Small-signal transfer functions

The MIBC small-signal transfer functions depend on both, *n* and *m* parameters. The number of phases, *n*, determines the equivalent inductance whereas the number of devices, *m*, has impact on the duty cycle.

Figures 5 and 6 represent the Bode plot of the small-signal transfer functions Vo/d and iL/d obtained by the PSIM simulator for the configurations specified in the table 7.

Comparative study: small-signal transfer functions (Vin=56V, Vo=80V, Pin=1024W)			
	n=2; m=4	n=4; m=2	n=8; m=1
Kvo/d	53.2dB	47.2dB	41.2dB
fz	19.5kHz	39kHz	78kHz
fo	2.37kHz	3.36kHz	4.75kHz
Q	8.21	11.6	16.4
Kil/d	40.4dB	28.3dB	16.3dB
fz1	579Hz	579Hz	579Hz

Table 7. MIBC small-signal transfer functions

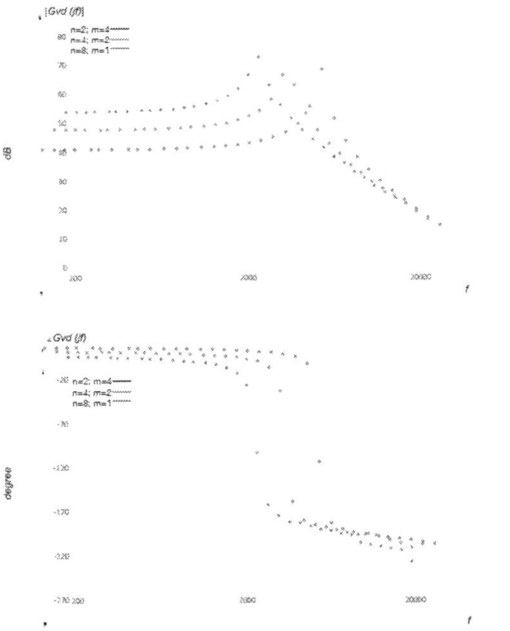

Figure 5. Duty cycle-to-output voltage transfer function. PSIM non-averaged model (Vin=56V, Vo=80V, Po=1024W)

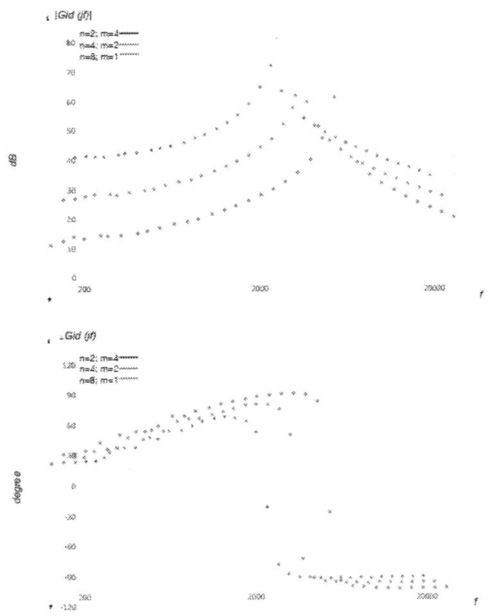

Figure 6. Duty cycle-to-inductor current transfer function. PSIM non-averaged model (Vin=56V, Vo=80V, Po=1024W)

DOI: 10.1051/e3sconf/20171614010

4. MIBC SIMULATION

Computer simulations using PSIM (Powersimtech) have been carried out to compare different configurations: a) n=2; m=4, b) n=4; m=2, c) n=8; m=1, using the values of Table 3.

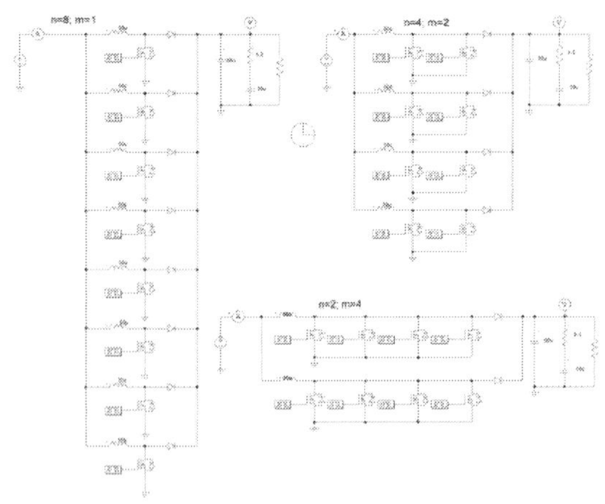

Figure 7. MIBC different simulation schemes.

Input current ripple at the minimum input voltage and maximum power for each configuration is represented in figure 8. The lowest input current ripple corresponds to the n=2; m=4, according to (16) and figure 3.

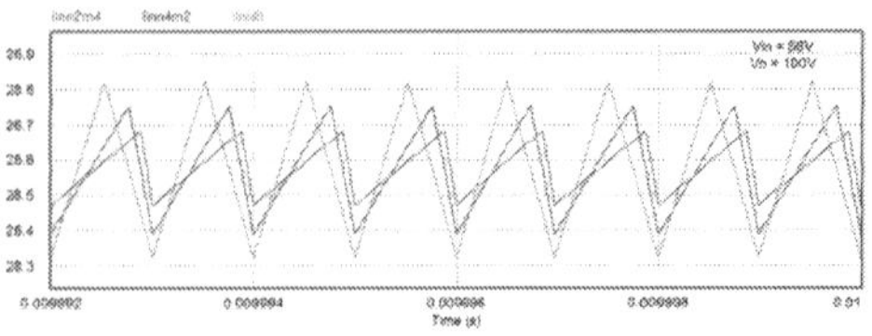

Figure 8. Input current ripple.

Output capacitor current at the minimum input voltage and maximum power for each configuration is represented in figure 9. The lowest RMS value corresponds to the n=8; m=1, according to (18) and figure 4.

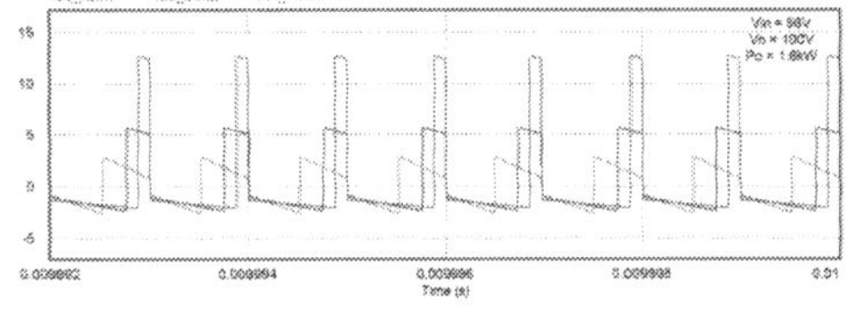

Figure 9. Output capacitor current.

CCM/DCM boundary is represented in figure 10. Considering a load resistor of 181.82 ohms (55W@100Vo), only the n=2; m=4 configuration keeps in CCM operation due to higher inductor frequency. The other two configurations, in open loop mode, increase the output voltage because of DCM operation.

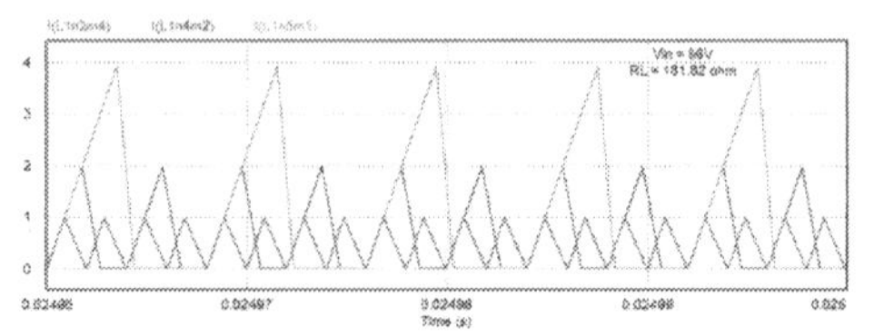

Figure 10. Input current ripple.

Mosfet current is depicted in figure 11. As it can be observed, the average value is the same for all configurations but the peak value and RMS value is higher as *n* decreases.

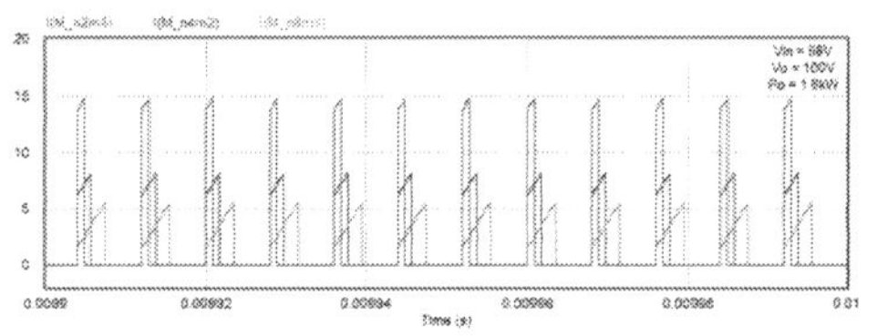

Figure 11. Mosfet current.

5. MIBC: ANALOG PWM GENERATOR

Two methods have been devised to obtain desired PWM signals for the MIBC. First method uses *n* sawtooth signals and the control signal (Vc) splits into *m* signals (Vci) with different offset (Voi), (22).

$$V_{Oi} = (i-1)\frac{V_p}{m} \quad ;i = 1..m \qquad (22)$$

$$V_{ci} = V_c + V_{Oi}$$

Simplified block diagram of the PWM generator for the 2-4-MIBC and the sketch of the main signals are depicted in figure 12.

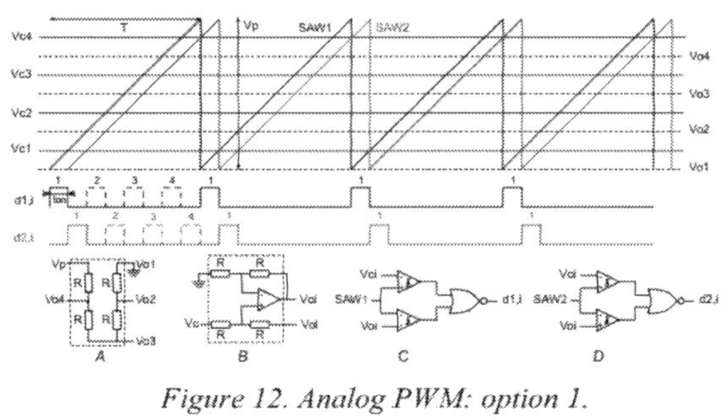

Figure 12. Analog PWM: option 1.

Another option is to use *n·m* sawtooth signals and only one control signal, which is limited to *Vp/m*. Figure 13, shows the sketch of this analog PWM generator. Only one comparator is required for each driving signal.

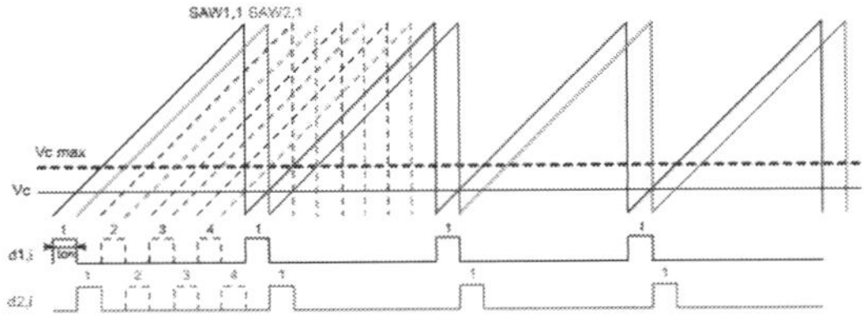

Figure 13. Analog PWM: option 2.

Figure 14 represents the PWM signal (d1,1) and the sawtooth (SAW1,1) of one power MOSFET using the second method with the commercial multiphase integrated oscillator LTC6909 and a constant current source that charges a capacitor.

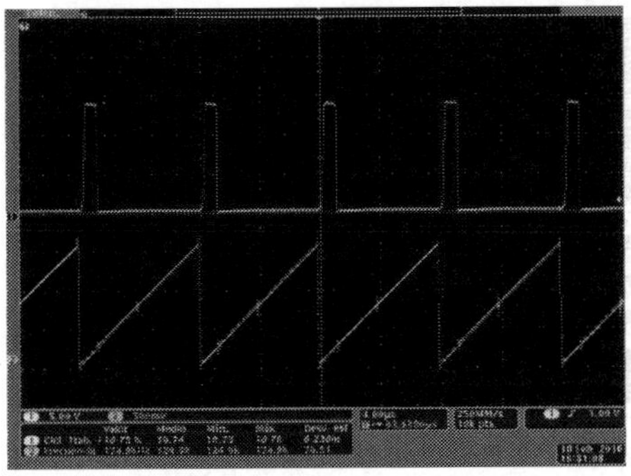

Figure 14. Sawtooth and PWM signal: option 2.

6. MIBC: DIGITAL (FPGA) PWM GENERATOR

Analog implementation of the PWM generator offers simplicity as the main advantage. On the contrary, digital approach, FPGA is considered here, is more complex but has some benefits that should be considered. To highlight some, high time resolution to achieve very low duty cycles, synchronous signals with adjustable dead time for bidirectional conversion or soft-switching, hot reconfiguration for redundancy, protection or efficiency purposes.

In this paper a digital PWM adapted from analog method option 1 is presented. Figure 15 shows the four gate signals of one phase switching at 96kHz.

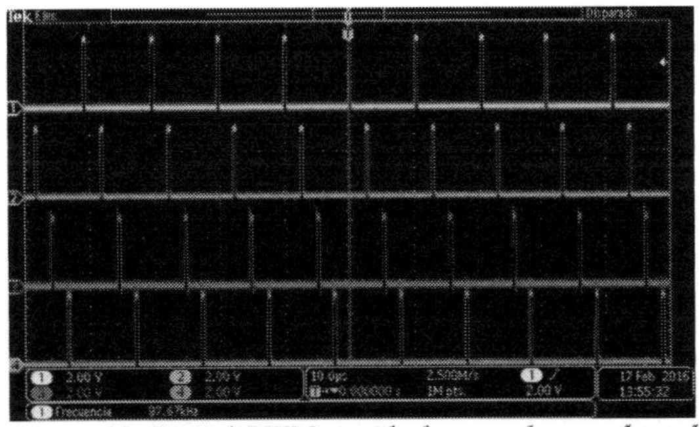

Figure 15. Digital PWM signals for one phase: adapted from circuit option 1.

7. CONCLUSIONS

This paper presents an interleaving technique that is suitable for DC/DC converters, BDR's is one possible application, as studied here, but other power conditioning functions could be explored. Design equations, comparative study and analog and digital (FPGA) practical implementation of the PWM generator system is presented.

8. REFERENCES

1. Hegazy, O., Van Mierlo, J. & Lataire, P. (2012). Analysis, modeling and implementation of a multidevice interleaved DC/DC converter for fuel cell hybrid electric vehicles. IEEE Trans. Power Electron. 27(11), 4445–4458.
2. Carbonnier, H, Fernandez, A., Triggianese, M. & Tonicello, F. (2014). Interleaved boost converter used as a battery discharge regulator for space applications, 10th ESPC, Noordwijkerhout, The Netherlands.

OUTPUT IMPEDANCE IMPROVEMENT USING COUPLED INDUCTORS

E. Sanchis Kilders, A. Ferreres, J.L. Gasent Blesa, D. Osorno, D. Gilabert,
E. Maset, J. Jordán, V. Esteve, J.B. Ejea

Dpt. Ingeniería Electrónica / ETSE, Universidad de Valencia, Avd. de la Universitat s/n, 46100 Burjassot, Spain,
Email: esteban.sanchis@uv.es

ABSTRACT

When using a single DC/DC converter with multiple outputs and having a buck topology, which has one filter inductor per output, the designer can choose to couple these outputs together. This paper demonstrates additional benefits of coupling output inductors together. Apart from saving mass and volume, and due to an improved small signal behaviour it also reduces the output impedance of the regulated output. The paper will analyse a seven output push-pull converter used as a space power converter module and verify the theoretical results with experimental measurements.

Key words: small signal analysis; output impedance; coupled inductors; peak current control.

1. INTRODUCTION

Coupled inductors applied to interleaved topologies have already been studied in [1] [2] [3] [4]. Their use in multiple output converters, usually applying weighted control, is studied in [1] [5] [6] [7] and an improvement of cross regulation has been demonstrated in [5]. Coupled inductors also save volume and mass, which for some applications, like space-borne systems, is definitely an important advantage. A small signal analysis of a coupled inductor forward converter is presented in [6] and the transfer functions for a two outputs case are studied. A study including peak current mode control is presented in [7], providing information not only on the transfer function but also on output impedance and even cross impedance. Other uses of coupled inductors like in photovoltaic systems and its advantages are described in [8], where zero ripple in the source current is achieved by design. In [9], the authors propose to use coupled inductors to improve the transient response of high power output feeding load voltage / high current loads. It is a way of taking advantage of the lower output impedance of coupled inductor converters.

This paper presents a detailed study of the output impedance of seven coupled inductors used together with primary peak current control in a push-pull converter. The model has been simulated and experimentally validated with a converter that is the Power Converter Module (PCM) of a real space application. A detailed description and more practical information of the PCM of the Polarimetric and Helioseismic Imager (PHI), instrument to fly on board of the Solar Orbiter satellite (SolO) was already published in [10].

2. POWER SUPPLY OF THE PHI INSTRUMENT

Although already mentioned in [10] and for the sake of completeness, a short description of the PCM designed for PHI to fly onboard SolO will be given. The PCM is based on a push-pull topology with seven outputs using a single transformer and a single output coupled inductor (see Fig. 1), mainly to save volume, mass and to improve small signal behaviour and output impedance, as this paper will demonstrate.

The PCM is a custom design system providing seven output voltages (+3.4 V, +7 V, −7 V, +14 V, −14 V, +65 V and primary auxiliary +14 V) with galvanic isolation. Two +28 V output voltages are also delivered from the PCM primary side (28 V bus voltage), but are not considered in the study, as they do not pass through the DC/DC converter. The PCM manages a maximum power of around 35 W. The PCM not only includes the DC/DC converter, but also protections, inrush current control, dedicated input filter and a whole power distribution unit on its output to properly supply all required voltages to each user.

The engineering model (EM) of the PCM was already built and is shown in Fig. 2. The EM of the PCM was used to measure the experimental results presented at the end of the paper. Therefore, not just a simple breadboard was used to validate the theoretical results but a real space application subsystem. This has allowed to verify the effect of all additional circuitry on the small signal analysis and output impedance and how to model this effect in a SPICE simulation.

DOI: 10.1051/e3sconf/20171614011

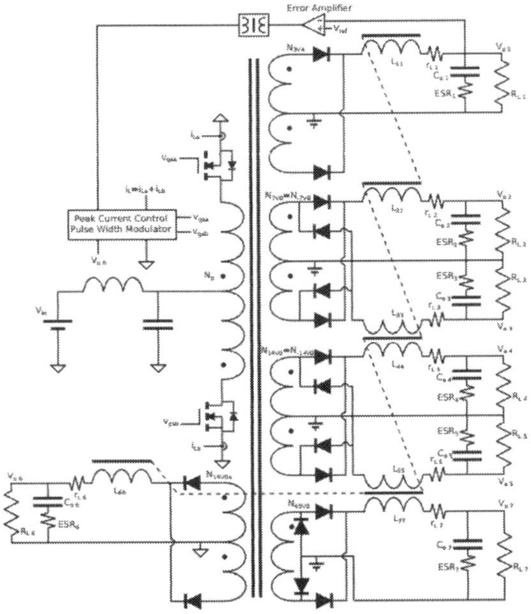

Figure 1. Simplified schematic of the PCM converter.

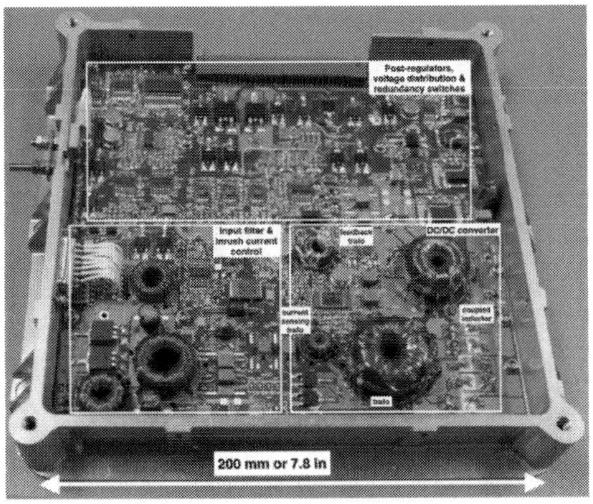

Figure 2. Engineering Model of the PCM. On the right side of the picture, the coupled inductor can be seen and below it the main transformer. The three ring cores on the left belong to the input filter and the two small ring cores in the middle are the magnetic feedback and the current sensing transformer.

3. SMALL SIGNAL ANALYSIS

The theoretical study has been focused on the coupled inductor, which therefore has been considered real. The model used for the coupled inductor is the inductance matrix, which mathematically suits best for the whole equations deduction and is the real and exact description of a coupled inductor (but neglecting parasitic elements, like capacitances). Measuring the inductance matrix has to be done very carefully [1] [11] and the matrix will be physically feasible if, and only if, its eigenvalues are real and positive (demonstrated in [12]).

3.1. Procedure

The listed steps have been followed:

1. The small signal analysis of a simplified BUCK-type converter with coupled inductors has been done.
2. The linearised model has been studied including the experimentally measured coupling inductance matrix. The resulting equations can be particularised for no coupling (changing all mutual inductances of the inductance matrix to zero).
3. A linearised model has been simulated with SPICE to validate the theoretical model (see Fig. 8) and the transfer functions have been compared graphically to the previous obtained equations for verification purposes.
4. A second and more complete SPICE circuit (based on the previous one and shown in Fig. 9) has been built incorporating many additional parasitic and real components, like input filter, duty cycle loss due to the transformer's leakage, parasitic resistance of tracks of the PCB and of the current sensing shunts used in the PCM among others.
5. The frequency response of the complete SPICE model and the real circuit measurement has been compared.

3.2. Analysis

The small signal analysis is based on well-known averaging techniques. Taking into account that the Push-Pull converter is a BUCK type topology and that in our case all output inductors are coupled together, only the secondary side of the j-th output, shown in Fig. 3, is used for the analysis.

In Fig. 3, rectifying diodes have already been linearised and the input transformer has also been converted into a DC transformer, that includes not only the turns ratio but also the effect of the duty cycle. Using now matrix algebra to analyse the circuit, we can take into account that all outputs are coupled together and therefore interact with each other. The sub-indices of the matrices are j and k and both vary from 1 to 7. Of course, the dimensions of the matrices that appear in these formulae are equal to the number of coupled inductors of the converter and are equal to seven in our case. Once the inductance matrix, $\mathbf{L_o}$, is known (including mutual inductances, L_{jk} and self

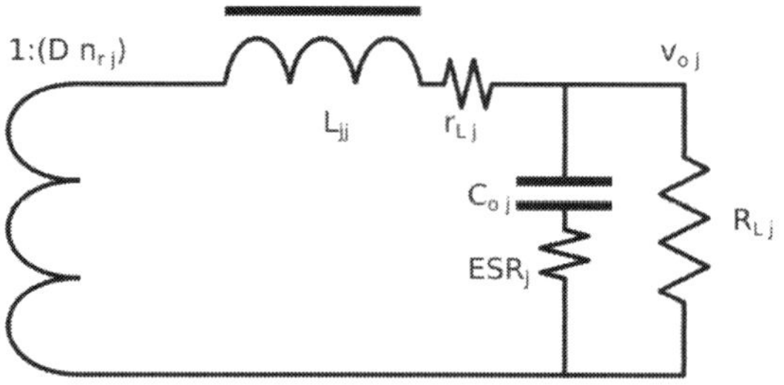

Figure 3. Small signal model of the j-th output of the push-pull converter. D is the duty cycle, $n_{r\,j}$ is the turns ratio, $v_{o\,j}$ is the output voltage, L_{jj} is the self inductance, $C_{o\,j}$ is the output capacitor, $R_{L\,j}$ is the load resistance, $r_{L\,j}$ is the parasitic series resistance of the inductance, ESR_j is the equivalent series resistance of the output capacitor.

inductances, L_{jj}), it has to be checked if the matrix is physically feasible, i.e. it is a semidefinite positive matrix [12].

The block diagram that describes the small signal behaviour is shown in Fig. 4.

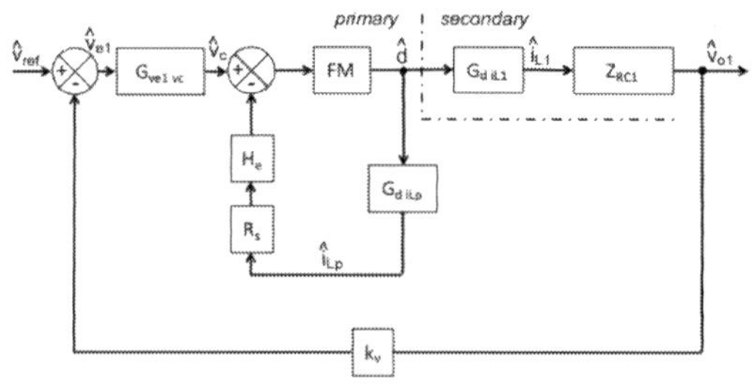

Figure 4. Small signal control block diagram of the PCM. The peak current control is closed measuring the primary current, i_L, which sums up all secondary currents with their corresponding transformer ratio.

The expressions for the different blocks shown in Fig. 4 are given hereafter.

H_e represents the sampling effect of peak current control and its expression (Eq. 1) is demonstrated in literature. FM (Eq. 2) corresponds to the modulators transfer function found in a peak current control loop, where f_s is the switching frequency, R_s is the current sensing resistor, V_{in} is the input voltage and L_p is the equivalent primary inductance.

$$H_e(s) = \frac{\frac{s}{f_s}}{e^{\frac{s}{f_s}} - 1} \approx 1 - \frac{\frac{s}{f_s}}{2} + \left(\frac{\frac{s}{f_s}}{\pi}\right)^2 \qquad (1)$$

$$FM = \frac{f_s}{R_s \frac{V_{in}}{L_p}} \qquad (2)$$

$G_{d\,i_L}$ is the transfer ratio related to primary current i_L and duty cycle d.

$$G_{d\,i_L}(s) = \sum_j n_{r\,j}\, G_{d\,i_L\,j}(s) \qquad (3)$$

where the transfer function of the j-th output related to the primary current is,

$$G_{d\,i_L\,j}(s) = \frac{V_{in} Y_{LRC\,p\,j}(s)}{n_{r\,j}} \qquad (4)$$

The transformer transfer ratio is given by $n_{r\,j}$, where the general expression of the transposed vector, $\mathbf{n_r^T}$, is given by

$$\mathbf{n_r^T} = \frac{1}{N_p}\left(N_{3V4}\ N_{7V0}\ N_{-7V0}\ N_{14V0}\ N_{-14V0}\ N_{14V0a}\ N_{65V0}\right) \qquad (5)$$

The admittance $Y_{LRC\,p\,j}$ of each output seen from the primary side, corresponds to

$$Y_{LRC\,p\,j}(s) = n_{r\,j} \sum_k n_{r\,k}\, [\mathbf{Z_{LRC}}(s)^{-1}]_{jk} \qquad (6)$$

Where the impedance $Z_{LRC\,jk}$ is given by,

$$Z_{LRC\,jk}(s) = \begin{cases} L_{jj}s + r_{L\,j} + Z_{RC\,j}(s) & \text{if} \quad j = k \\ L_{jk}s & \text{if} \quad j \neq k \end{cases} \qquad (7)$$

and $r_{L\,j}$ is the series resistance of each inductance and $Z_{RC\,j}$ is,

$$Z_{RC\,j}(s) = \left(\frac{1}{C_{o\,j}s} + ESR_j\right) \parallel R_{L\,j} \qquad (8)$$

$C_{o\,j}$ is the output capacitance, ESR_j is the equivalent series resistor of the capacitance and $R_{L\,j}$ is the output load of the j-th output of the converter.

The function $G_{v_{e1}\,v_c}$ corresponds to the error amplifier transfer function. This block is fully controlled by the designer and usually built with an opamp that allows to compensate and stabilise the overall control loop.

4. OUTPUT IMPEDANCE

To obtain the output impedance, a current perturbation, i_g must be injected at the output. The output impedance results from measuring the output voltage perturbation

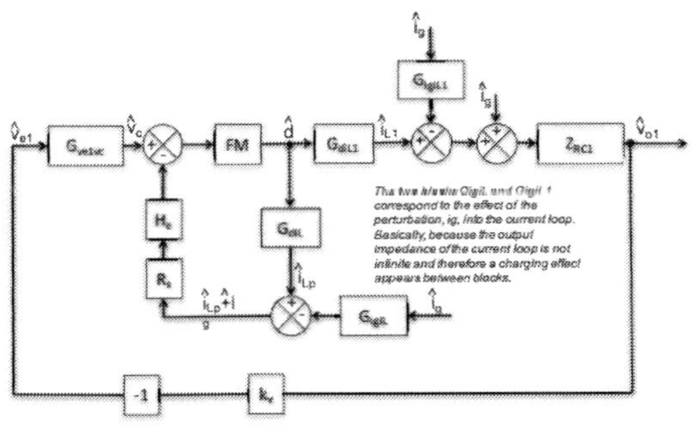

Figure 5. Small signal control block diagram of the PCM showing the current i_g perturbation injection. This perturbation affects also the current loop with two additional blocks, $G_{i_g\,i_{L1}}$ and $G_{i_g\,i_L}$ that weigh this influence.

divided by the injected current perturbation and keeping any other perturbation equal to zero.

But an additional effect must be taken into account based on the block diagram shown in Fig. 5. Although ideally the input impedance of all these blocks is infinite and their output impedance is zero, this is not true in real application. The resulting loading effect counts for additional effects of this injected perturbation at other points of the diagram, as shown in Fig. 5. A more physical explanation would be that the injected current affects the current loop, both at primary side and at secondary side. In fact the perturbation injected in output number one is coupled to all other outputs through the coupled inductor and therefore affects all other outputs. Of course, also the overall current loop is affected. This influence is described by $G_{i_g\,i_{L1}}$ and $G_{i_g\,i_L}$.

Applying matrix algebra and taking into account the small signal analysis of the converter, the expression for both transfer functions can be calculated,

$$G_{i_g\,i_L}(s)\big|_{j=1} = Z_{RC\,j} \sum_{k} n_{r\,k}\, [\mathbf{Z_{LRC}}(s)^{-1}]_{jk} \quad (9)$$

And for the given output, the expression is,

$$G_{i_g\,i_{Lj}}(s)\big|_{j=1} = [\mathbf{Z_{LRC}}(s)^{-1}]_{jj} Z_{RC\,j} \quad (10)$$

The index $j = 1$ stands for injecting the signal at the first output. Both equations, Eq. 9 and Eq. 10, show that the influence of the injection into the whole system depends on the whole output impedance matrix, $\mathbf{Z_{LRC}}(s)$, which, due to the inductance matrix, couples all outputs together (because it is not a diagonal matrix).

The output impedance is given by the following expression (where the dependence on s of the open and closed loop transfer functions G and T and H_e have been omitted in Eq. 11 for simplicity).

$$Z_o(s) = Z_{RC\,1} \frac{\frac{R_s\, H_e\, FM\, G_{d\,i_{L1}}\, G_{i_g\,i_L}}{T_i+1} - G_{i_g\,i_{L1}} + 1}{T_v + 1} \quad (11)$$

The current loop gain T_i is given by

$$T_i(s) = R_s\, H_e(s)\, FM\, G_{d\,i_L}(s) \quad (12)$$

and the voltage loop gain T_v is given by

$$T_v(s) = k_v\, G_{v_{e1}\,v_c}(s)\, Z_{RC\,1}(s)\, G_{v_c\,i_{L1}}(s) \quad (13)$$

where $G_{v_c\,i_{L1}}$ is

$$G_{v_c\,i_{L1}}(s) = \frac{FM}{1+T_i(s)} G_{d\,i_{L1}}(s) \quad (14)$$

The magnitude of the output impedance of both, SPICE simulated circuit (using circuit shown in Fig. 8) and theoretical equations are compared in Fig. 6. It can be clearly seen that both are almost equal, what assures that the equations are right and fit the model and SPICE simulation, as both reach the same results. The expression of the output impedance when uncoupling the output inductors has also been simulated and calculated from the deduced equations, which are basically the same.

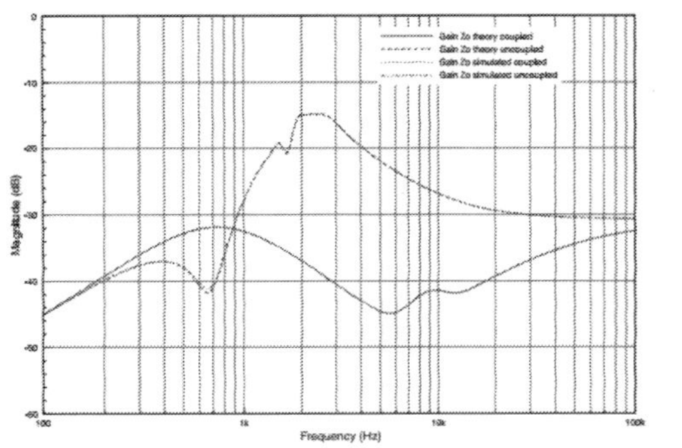

Figure 6. Magnitude of output impedance obtained from the described equations and the SPICE circuit shown in Fig. 8. Both, simulated and calculated curves are identical. The output impedance when uncoupling all outputs is also shown to demonstrate that it is worse.

It is clearly seen in Fig. 6 that the output impedance of the circuit using coupled inductance is lower than the one using uncoupled inductances. In addition, when uncoupling the inductances, the curve shows an uneven trace because

DOI: 10.1051/e3sconf/20171614011

output impedance of all different outputs add together in the primary and their different resonances are then seen by the primary peak current loop. It has to be remembered that coupling inductors together results in a single equivalent primary inductance and therefore a single resonance (see [10]), resulting in a much more even curve. A 15 dB improvement of the impedance comparing the peaks of both curves is reached (see Fig. 6).

5. EXPERIMENTAL RESULTS

Using the EM of the PCM, the output impedance has been measured. As the PCM is much more complex than the model used to study the coupling effect on multiple output converters, the results cannot be directly compared with the simulated and theoretical results shown before. An indirect procedure has been used to validate the measurements. Based on the SPICE model used in the theoretical verification, a new more detailed SPICE circuit (see Fig. 9) of the PCM has been built. This SPICE circuit includes many of the existing parasitic elements, the input filter, internal blocks of the PWM modulator, rectifying output diodes and resistance of tracks and switching devices (these switching devices are responsible of switching ON or OFF the different output voltages).

This complex SPICE circuit has been compared to the measured output impedance (see Fig. 7). It can be seen that both curves are very similar, what confirms that the base model, that demonstrates that output impedance improves when coupling output inductors together, is correct.

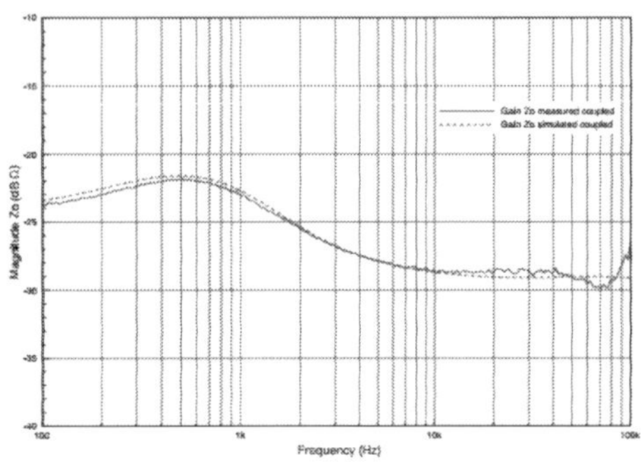

Figure 7. Output impedance measured in the real circuit and compared with simulated output impedance using a more detailed SPICE circuit (see Fig. 9) based on the already verified SPICE circuit (see Fig. 8).

The shape of the real impedance changes compared to the simulated ideal impedance shape. The main resonance peak stays at the same frequency range between, 500 kHz and 1000 kHz. The change of this frequency is because more output capacitors have been added in the real circuit, after the ON/OFF switches. Therefore a kind of RC network (the R's come from the ESR of the capacitors and the resistances of the tracks) is placed after the output of the converter and the injection is done at this point. This is not shown in Fig. 1 but in Fig. 9. This fact increases the output impedance (about 10 dB which corresponds to a factor of 3.2) and also changes its shape. Therefore, the theoretical output impedance measurement is different from the real circuit, because the circuits are different. Real circuits have more parts after the output and this changes its behaviour compared to the ideal circuit.

6. CONCLUSIONS

This paper demonstrates that coupling output inductances in BUCK-type multiple-output converters and regulating a single output, brings as additional benefit a lower output impedance at the regulated output compared to an equivalent uncoupled converter. Other, already known benefits of coupling output inductors are mass and especially volume saving, even if the magnetic design and manufacturing of converter can be more complex.

Anyhow, it must also be stated that the improved output impedance can be slightly masked by parasitic effects found in the real circuit of the whole power converter module. The addition of the power distribution circuitry and distributed output filtering addition of output capacitors at the user side, which adds RC elements to the output path, results in a slightly different topology with a different output impedance (usually larger). As a result, the experimental measurement of the output impedance of the real system is different to the theoretical one, both in magnitude and in shape. But adding, mainly all parasitic resistances to the SPICE circuit allows to model this behaviour very well.

7. ACKNOWLEDGMENTS

This research has been supported with funds of the Ministerio de Economía y Competitividad of Gobierno de España and FEDER, under projects with reference:

- AYA2012-39636-C06-03
- ESP2013-47349-C6-5-R
- ESP2014-56169-C6-4-R

and partially by the Generalitat Valenciana, under project reference PROMETEO/2012/044.

8. REFERENCES

[1] R.W. Erickson and D. Maksimovic. A multiple-winding magnetics model having directly measurable parameters. In *Power Electronics Specialists Conference, 1998. PESC 98 Record. 29th Annual IEEE*, volume 2, pages 1472–1478, 1998.

 DOI: 10.1051/e3sconf/20171614011

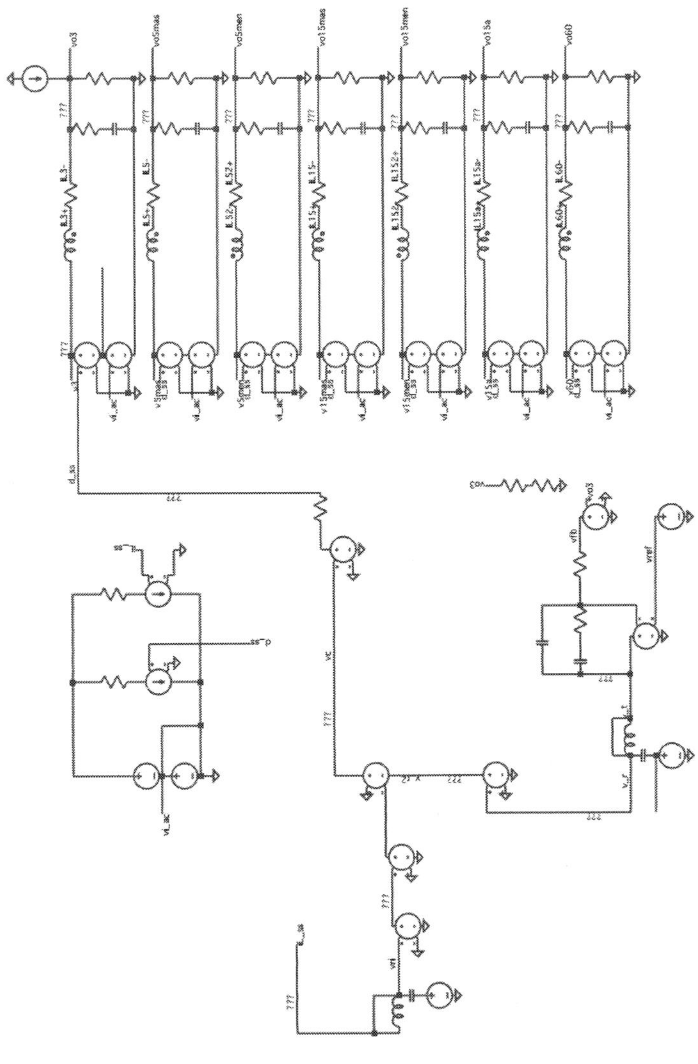

Figure 8. "Ideal" SPICE circuit used to simulate the ideal behaviour of the PCM. The frequency response is shown in Fig. 6.

[2] H. B Shin, J. G Park, S-K Chung, H. W Lee, and T.A. Lipo. Generalised steady-state analysis of multiphase interleaved boost converter with coupled inductors. *Electric Power Applications, IEE Proceedings -*, 152(3):584–594, 2005.

[3] H. B Shin, E.-S. Jang, J. G Park, H. W Lee, and T.A. Lipo. Small-signal analysis of multiphase interleaved boost converter with coupled inductors. *Electric Power Applications, IEE Proceedings -*, 152(5):1161–1170, 2005.

[4] V. Vorperian. The effect of the magnetizing inductance on the small-signal dynamics of the isolated cuk converter. *Aerospace and Electronic Systems, IEEE Transactions on*, 32(3):967–983, July 1996.

[5] D. Maksimovic, R.W. Erickson, and C. Griesbach. Modeling of cross-regulation in converters containing coupled inductors. *Power Electronics, IEEE Transactions on*, 15(4):607–615, 2000.

[6] Qing Chen, F.C. Lee, and M.M. Jovanovic. Small-signal analysis and design of weighted voltage control for a multiple-output forward converter. *Power Electronics, IEEE Transactions on*, 10(5):589–596, 1995.

[7] Qing Chen, F.C. Lee, and M.M. Jovanovic. Current-mode control for multiple-output converters. *Aerospace and Electronic Systems, IEEE Transactions on*, 32(4):1412–1420, 1996.

[8] M. Veerachary. Power tracking for nonlinear

DOI: 10.1051/e3sconf/20171614011

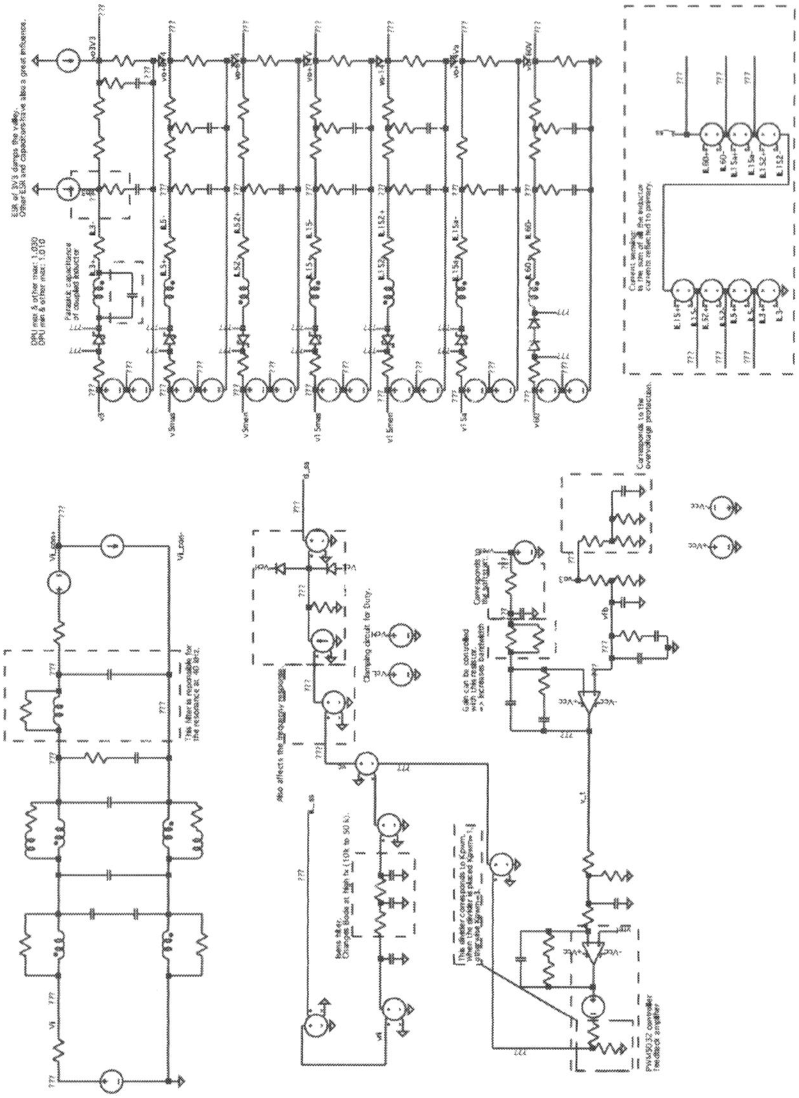

Figure 9. "Real" SPICE circuit used to simulate the real behaviour of the PCM. The frequency response is shown in Fig. 7 and compared to the measured curve. The values of the different parameters are not shown in this picture.

pv sources with coupled inductor sepic converter. *Aerospace and Electronic Systems, IEEE Transactions on*, 41(3):1019–1029, July 2005.

[9] P. Alou, J.A. Cobos, O. Garcia, R. Prieto, and J. Uceda. Coupling inductors for supplying pulsating loads with dc/dc converters. In *Power Electronics Specialists Conference, 1998. PESC 98 Record. 29th Annual IEEE*, volume 2, pages 1790–1795 vol.2, May 1998.

[10] E. Sanchis-Kilders, A. Ferreres, E. Maset, J.B. Ejea, J.L. Gasent-Blesa, V. Esteve, J. Jordan, A. Garrigos, and J.M. Blanes. On the design of a multiple-output dc/dc converter for the phi experiment on-board of solar orbiter. In *Applied Power Electronics Conference and Exposition (APEC), 2013 Twenty-Eighth Annual IEEE*, pages 3305–3310, March 2013.

[11] B. Hesterman. Analysis and modelling of magnetic coupling. Technical report, Advanced Energy Industries, www.denverpels.org, 2007.

[12] Yilmaz Tokad and Myril B. Reed. Criteria and tests for readability of the inductance matrix. *American Institute of Electrical Engineers, Part I: Communication and Electronics, Transactions of the*, 78(6):924–926, 1960.

EFFICIENT POINT OF LOAD CONVERSION FROM THE SATELLITE MAIN BUS

Principal Patrick Franks, Microsemi Corp, 11861Western Avenue, Garden Grove CA92841, USA;
patrick.franks@microsemi.com

Co-Authors Tom Gati; Jean Louis DeBauche, Microsemi Corp, 11861Western Avenue, Garden Grove CA92841, USA;
tom.gati@microsemi.com; JeanLouis.DeBauche@microsemi.com

ABSTRACT

This paper examines some of the issues involved in the distribution of power from the Main Satellite bus to advanced Mixed Signal Electronics payloads. Trade-offs are described in the design of Bulk voltage conversion vs Point of Load conversion. These trade-offs should support mission objectives of high efficiency, fault tolerance and control of crosstalk between subsystems.

1. INTRODUCTION

Satellite digital processing systems present demanding requirements for power distribution from the vehicle's main power bus down to regulated power rails on FPGA devices such as RTG4 FPGAs, Microsemi's fourth-generation flash-based FPGA. Heritage systems consist of bulk sub-bus power converters, typically 5V outputs, driving a number of parallel load assemblies with low impedance distribution and good local decoupling to maintain adequate Point of Load (POL) regulation. The lower voltage rails and higher dI/dt transient requirements of advanced processor systems drive a requirement to integrate bulk power conversion into processor load module printed wiring assemblies (PWAs) for improved regulation and fault tolerance. This paper discusses these regulation issues, broader issues of isolation and fault tolerance and the solutions offered by Microsemi's SB30 bulk POL DC-DC power converter, which in this paper is termed a "Point of Module" converter.

2. FPGA POWER REQUIREMENTS

A generalized concept for PWA power distribution to the FPGA is illustrated in Figure 2.1 [1]

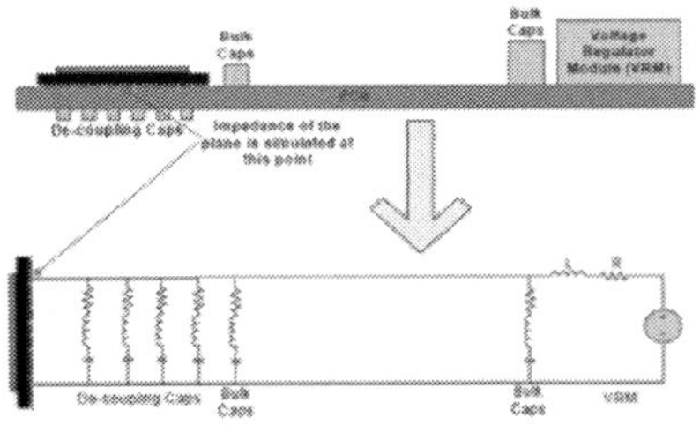

Figure 2.1 FPGA General Power Distribution

Impedance to the power rails must be sufficiently low to maintain static and dynamic voltage regulation limits for the main FPGA functions.

Taking the example of Microsemi's new RTG4 FPGA; power supply requirements are classified as:

- Core (1.2V; 1.5A, up to 8A)
- Input /Output (3.3V; 500mA)
- Serializer/Deserializer (SerDes) (1.2V; 250mA)
- Double data rate (DDR) (1.5V; 250mA)
- Phase-locked Loop (PLL) (3.3V; 10mA)

3. BUS IMPEDANCE CALCULATIONS

Complex FPGA designs have increasing amounts of current transients switching across the power bus. Simultaneously switching outputs (SSO) contribute to a major share of instantaneous current issues. Decoupling with low effective series inductance capacitors provides localized high frequency energy to decouple noise from the switching currents of the device power bus. This is most effective when capacitors are in close proximity to the device. Some of these high frequency capacitors are required to be placed directly by the FPGA.

Target impedance is calculated based on Eq 1:

 DOI: 10.1051/e3sconf/20171614012

$$Z_{Min} = \%\ Ripple \times \frac{V_{supply}}{I_{trans}} \quad (1)$$

Vsupply:	Supply voltage of the power plane
% Ripple:	Percentage of ripple allowed on the power plane
Itrans:	Transient current drawn on the power plane. Generally, transient current is half of the maximum current.
Zmin:	Target impedance of the plane

4. POINT OF MODULE POWER DISTRIBUTION CONCEPT

4.1. Point of Module Power Distribution Considerations

What is desirable for a Module Level power converter? The following features should be considered:

- Bus facing EMI performance
- Bus isolation
- Fault containment
- Fault isolation
- System efficiency
- Number of Bulk supply rails
- Tradeoff between bulk power conversion and POL power conversion, driven by serviced load regulation requirements

4.2. EMI Performance

For general EMI compatibility MIL-STD 461 establishes emission requirements and complementary susceptibility performance that guarantees compatibility between power converters on a bus distribution system. In heritage sub-bus distribution systems several load modules would connect on a sub-bus with no intermediate filter. In a number of applications, notably communications satellites, it is important to isolate noise between channels. A module level regulator with galvanic isolation and a MIL STD 461 EMI compliant filter provides excellent channel to channel isolation for these applications

4.3. Bus Isolation

The important benefit of using an isolating bus converter to connect the Module to the satellite main bus is the ability to establish ground potential on the load side independent of the Main Power bus common potential. Now the potential of the Module's local ground is insulated from steady state and transient voltage variations that may occur on the Main Bus. Signal interfaces between modules are therefore maintained with optimum signal to noise margins.

4.4. Fault Containment

Satellite systems are often arranged in N+M redundant configurations to provide for graceful degradation and prolonged mission life. The Point of Module converter already provides good fault isolation performance through its galvanic isolation and load current limit function.

4.5. Fault Isolation

Further security from Failure Mode Effects may be provided by a fuse at the converter input that opens in the event of an internal high current fault. A further refinement to this is a Solid State current limiter / resettable fuse that will limit the load current during the fault event and help preserve Main Bus voltage.

4.6. System Efficiency

Larger Satellites with higher power use higher Main Bus voltages to transmit power to the payload systems. 100 volt bus is now common The International Space Station (ISS) has 120V operating up to 165V transient. The ideal for size and weight is a single stage conversion from main bus to user voltage. However with FPGA core voltages running as low as 1 volt with potential still to move south, a 100:1 voltage conversion ratio becomes unwieldy.

4.7. Number of Bulk Power Supply Rails

The number of power converter output rails is determined by the module load configuration. Generally a 5V bulk rail is provided for distribution to downstream POL regulators. 3.3V down to 2.5V bulk rails can also be provided for Input / Output and other "non-core" functions that don't have large load transient requirements and have relatively wider line regulation tolerance.

4.8. Load Power Allocations to Bulk and POL

The specific choice between on module bulk supply rail usage and POL usage must be determined by the Mixed Signal module designer based upon steady state and transient regulation requirements achieved considering on-PWA bus distribution impedances. In a general case an efficient POL converter transforms the source impedance of the bulk bus by the square of its voltage transformation ratio over its effective closed loop control bandwidth. This enables the energy in a local decoupling capacitance to be stored more efficiently at a higher voltage level.

 DOI: 10.1051/e3sconf/20171614012

5. SB30 ARCHITECTURE CONCEPT

In traditional power distribution a bulk sub-bus converter supplies power to parallel connected mixed signal processor modules (PWA's).

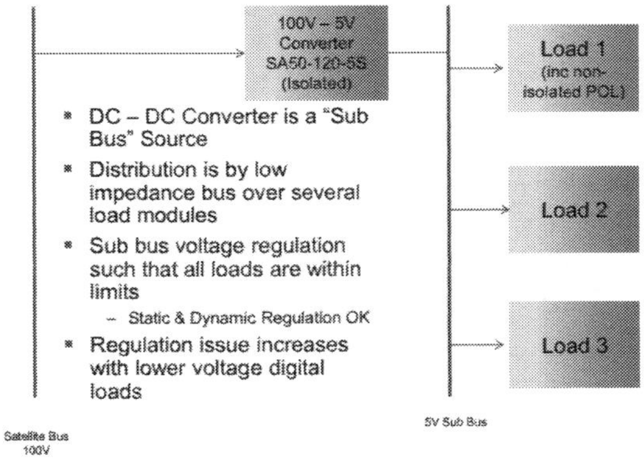

Fig 5.1 Traditional Heritage Sub Bus Power Distribution

As device supply voltages drop and current levels increase the regulation performance of the traditional approach becomes problematic and "Point of Module" power regulation becomes preferable.

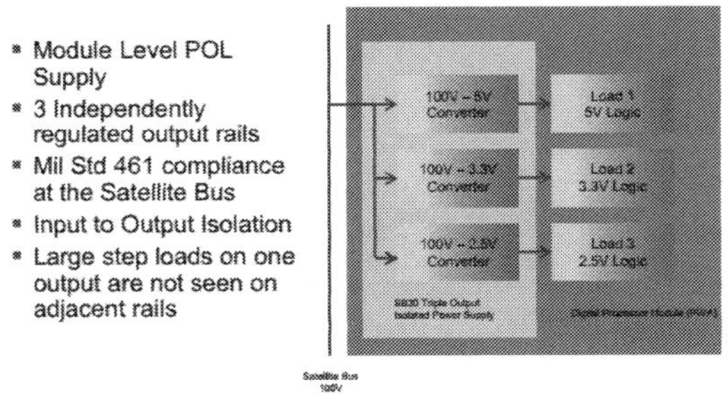

Fig 5.2 Main Bus Distribution to Point of Module Concept

In the Point of Module concept, main satellite bus power is brought to the Payload module input and an on board power converter directly generates necessary bulk voltages for internal distribution on the module.

The SB30 DC-DC Converter and its optional associated Self-Resettable Fuse (SRF) connect directly to the satellite main bus and provide the following services:-

- Efficient, direct conversion from the Main Bus to on- module power distribution rails
- 3 Independently Adjustable Regulated Outputs,
- Power Rail voltages are programmable between 5V and 2.5V.
- Embedded EMI filtering with a high rejection of noise coupling to other systems
- Flexible bus input voltage, 28V to 100V
- Hiccup overcurrent protection
- Fault disconnection from the Satellite power bus
- Enhanced PWA distribution impedance model, transient and static

The SB30 Power Conversion topology is shown in Figure 2. [2]

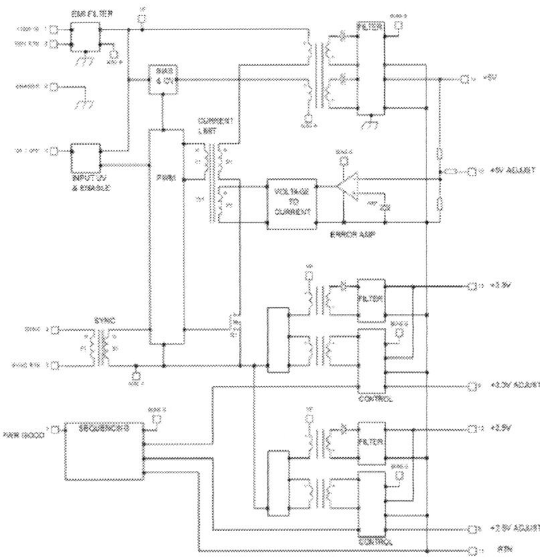

Figure 2: SB30 Power Conversion Topology

The primary power conversion channel converts the 100V Main Bus input to 5V bulk output using a primary side PWM controlled flyback converter. Output voltage is sensed and compared with a secondary side reference to produce an output voltage error signal. This signal is converted to a proportionally pulsed current signal that is combined with sensed primary current to deliver pulse by pulse feedback to the PWM controller. This "V + I" feedback control delivers excellent static and dynamic output voltage regulation performance.

The 5V output channel then powers the secondary side controllers for the other 2 power conversion channels. These auxiliary power converters are also flyback converters. This topology produces 3 independently regulated output rails all with independent current limit protection. Each channel has single stage conversion from the main bus to each output rail.

6. FAULT ISOLATION THROUGH ACTIVE POWER DISCONNECT

The main bus input to the Point of Module power

 DOI: 10.1051/e3sconf/20171614012

converter can be provided with additional protection against downstream load faults through the series connection of a Self-Resettable Electronic fuse. This concept which is currently in product development provides a soft turn on switch into the power converter and limits then turns off the current according to a preset I^2t overload profile. The device is configured to retry the load at pre-determined intervals to check for fault clearance and can be permanently turned off if this does not occur.

The concept for this electronic fuse is shown below:

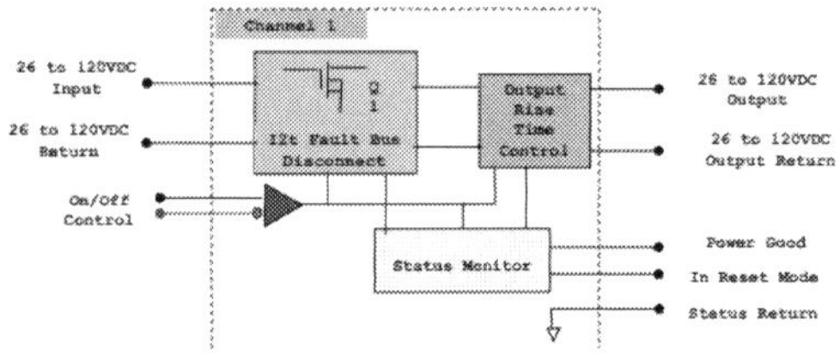

Fig 6.1 Electronic Fuse Concept

In applications that require it, the Electronic fuse can parallel connect 2 redundant power sources into the Point of Module power converter.

7. THE FULL SOLUTION

The concept for the full Point of Module Power Distribution system is shown below:

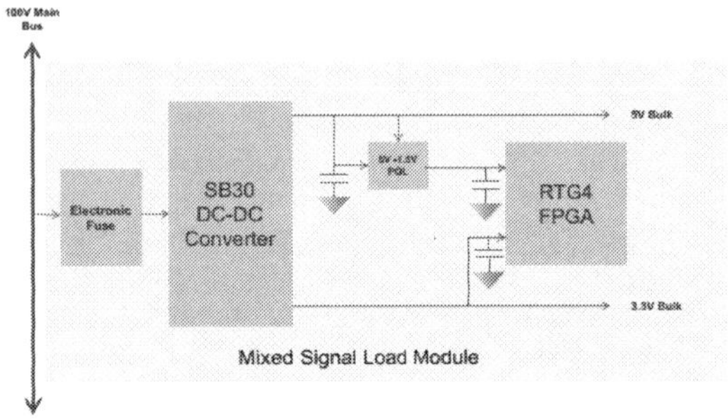

Fig 7.1 Full Point of Module Distribution Concept

Note the local decoupling capacitors at the FPGA which decouple the fast logic transition edges and the local decoupling capacitor at the POL converter which limits the transient voltage from a step load change in the FPGA Core power demand.

8. CONCLUSION

The SB30 Point of Module design concept combined with electronic fusing enables an optimal power conversion and distribution scheme for modern space digital FPGA and processor systems where conversion efficiency, fault tolerance and well regulated power rails are of paramount importance for robust and reliable solutions.

9. REFERENCES

1. Microsemi application note AC453; Layout Guidelines for RTG4-Based Board Design
2. Microsemi Datasheet: SB30-100 Triple Series; Radiation Hardened Isolated DC/DC Converters

A NEW, SINGLE DIODE, SEQUENTIAL SWITCHING SHUNT REGULATOR

G. Gianninoto [(1)], E. Scorzafava [(2)], D. Carlani [(3)]

[(1)] *Leonardo S.p.A. – viale Europa Nerviano (Italy), Email:guido.gianninoto@leonardo company.com,*
[(2)] *Italian Space Agency (ASI) – viale del Politecnico snc Roma (Italy), Email:edmondo.scorzafava@asi.it*
[(3)] *Thales Alenia Space Italia – via Saccomuro 24 - Roma (Italy), Email: daniele.carlani@thalesaleniaspace.com*

ABSTRACT

This paper describes the design improvement of a Sequential Shunt Switching Regulator (S3R) based on a power cell composed of one switching FET and one Diode connecting the Solar Array (SA) section to the Main Bus.

This topology allows to minimize the S3R cells dimensions, to optimize the efficiency figures and complies with the electrical Single Point Failure Free (SPFF) requirement on the Main Bus.

The single switching FET can cause the loss of one Solar array section if it fails in short circuit, while its failure in open can be compensated by the presence of one additional (spare) shunt section positioned on Main Bus to drain the power eventually in excess.

If the series redundancy of the Diode is removed, its failure in short circuit, causes overload on the Main Bus when the switching FET turns ON. To avoid the possible subsequent failure in short circuit of the switching FET due to over-heating, with permanent loss of Main Bus, the maximum current drawn by the FET in failure condition and the duration of this operation mode must be strictly controlled in order to avoid thermal overstress and consequently thermal failure propagation.

The adopted solution foresees a circuitry that controls the current in the FET, both the peak in the first transient and the subsequent limited value, and forces the FET in permanent open status after a short period, compatible with the overheat of the FET junction.

No spurious cause leads to the permanent loss of the S3R section due to permanent activation of the protection, causing the switching element (FET) to be continuously open.

In fact a dedicated circuit, activated by sensing of the Error Signal, unlatches the section to restore the power control.

The presented design has found its first application in the PCDU for the COSMO-SkyMed Second Generation (CSG) satellite power system.

A dedicated Elegant Bread Board of the S3R, integrated with the board designed to control the Battery Current and Voltage has been tested in the design and development plan of CSG PCDU.

1. S3R DEVELOPMENT AND SYSTEM OVERVIEW

The Sequential Shunt Switching Regulator (S3R) has been developed to control the power delivered from a Solar Array (SA) made of 16 sections to condition an unregulated Bus where the Battery Charge Current and Voltage are regulated.

The application has been developed to control a SA and a Battery with the following characteristics:

- 10 Li_ion battery modules each up to 24Ah.
- Battery Voltage from 41V to 68V.
- 16 Solar Array sections.
- Isc = 5A max.
- Voc = 140V max.
- SAS capacitance = 1uF max.

The S3R solution was adopted for its simplicity and cheapness, also because it was not required the MPPT control.

Furthermore, the need for a reduction in space occupation of the S3R cells as well as of their dissipation due to the stringent space available in the Power Conditioning and Distribution Unit (PCDU) in the frame of the COSMO Second Generation program, suggested the single Diode S3R solution.

2. S3R OPERATION

Each S3R cell can operate in three different ways:

- Conduction mode: the cell delivers all power from the SA section to the battery bus;
- Shunt mode: the cell absorbs all the current from the SA section;
- Switching mode: the cell switches between conduction and shunt modes, thus modulating the power delivered to the bus.

Depending on the power demand, in order to control the charge parameters of the Battery (voltage and charge current), the system determines via an Error Signal, the number of S3R cells in conduction and in shunt mode and which S3R section is regulating (switching) its current.

The system is designed so that only one cell at once may operate in switching mode.

3. S3R FUNCTIONALITY AND FAILURE PROTECTION

Fig. 1 depicts the block diagram of the S3R Cell.

Details of the blocks are given in the following paragraphs.

The shunt MOSFET is turned on whenever the current

 DOI: 10.1051/e3sconf/20171614013

from that SA section is not required by the Bus loads. This case occurs when the total power that the SA can deliver exceeds the demand of the system and the control logic responds by 'shunting' one or more sections.

On the contrary when the current from the SA section needs to be delivered to the bus, the MOSFET is turned off, allowing the diode to conduct.

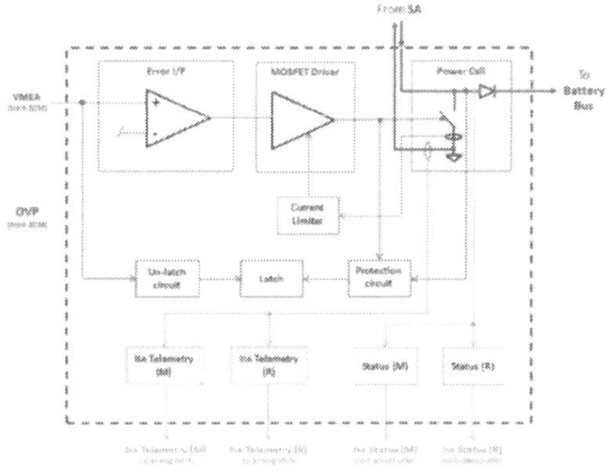

Figure 1. S3R Cell block diagram.

The topology of the power cell is visible in Fig. 2. It guarantees the shunt regulation with minimal component count, thus allowing space and cost saving.

Besides an efficiency above 98% has been estimated for power transfer to Main Bus, with bus voltage above 50V and section current up to 5A.

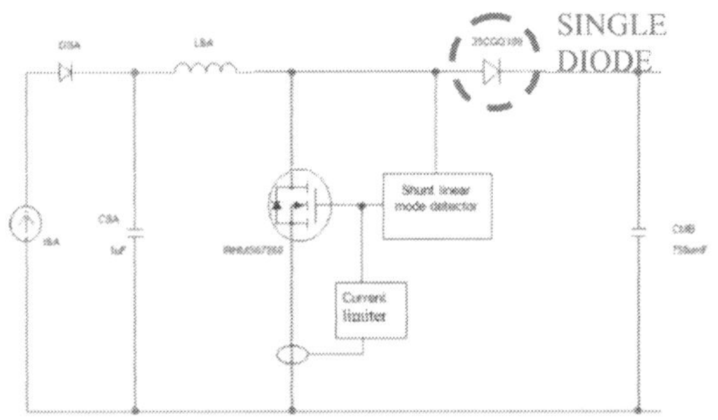

Figure 2. One-series-diode S3R Cell Configuration.

One failure inside PCDU can cause, as worst condition:

- Short circuit of the shunting element (FET). It causes the short circuit to ground of one Solar Array section only, without propagation to other 15 sections and to the system.
- Short circuit of the series diode. It causes one Solar Array section to be permanently connected to the BATT_PWR_Bus. In case of intervention of the shunt FET this condition is very dangerous for the system causing a permanent short circuit on Main Bus.

Normally the series redundancy is applied to control this second failure mode.

The improvement of the presented design is based on the electronic control of such a condition using the shunting element of the power cell to sustain and remove the dangerous operation.

To get this aim a Latching Current Limiter, positioned on the switching leg of the S3R section will trip and force the shunting FET to be permanently opened (latched), to prevent Main Bus overload, in the case the shunt element is commanded closed in presence of the short circuit of the diode.

4. S3R CURRENT LIMITATION

Current Limitation is necessary to prevent overload on Main Bus. Besides the current limiter must force and latch the shunt MOSFET in open.

The definition of the limitation value is limited downwards to optimise the power losses during switching in normal operation, and upwards to guarantee the survival of the FET itself to the failure event.

Due to the criticality of the application the limitation design, to guarantee the reliability of the MOSFET after the intervention of the protection, must guarantee that:

- The maximum peak current do not exceed, by design, the maximum acceptable value specified (180A is the rated one for the device used), derated at 75% (as per ECSS-Q-ST-30-11C-Rev1 relevant to the maximum I_{DRAIN}).
- The maximum temperature of the junction of the switching device (FET) must remain below 110°C, according to ECSS-Q-ST-30-11C-Rev1, during the whole limitation period. The junction temperature rise is caused by two main contributions, during the limitation phase before the trip-off intervention:
  - Initial (peak) transient.
  - Actual Limitation period.

It must be noted that the entity of the maximum peak current is relevant as regards both the MOSFET junction temperature rise and the current capability of the bonding inside the device (fuse effect for too-high currents). The former aspect has been evaluated in the dedicated computations regarding the MOSFET junction temperature rise. The latter has been clarified by the MOSFET manufacturer that provided the plot with the "Fuse Current" of the bonding as function of the overcurrent duration. The plot is reported in Fig. 3.

With the peak current duration observed both in para. §5 and §5.2, in the order of 300nsec, a very high current can be sustained by bonding without any risk.

So, the chosen approach to use the maximum derated value as acceptable value for the peak current of the MOSFET is a very conservative approach.

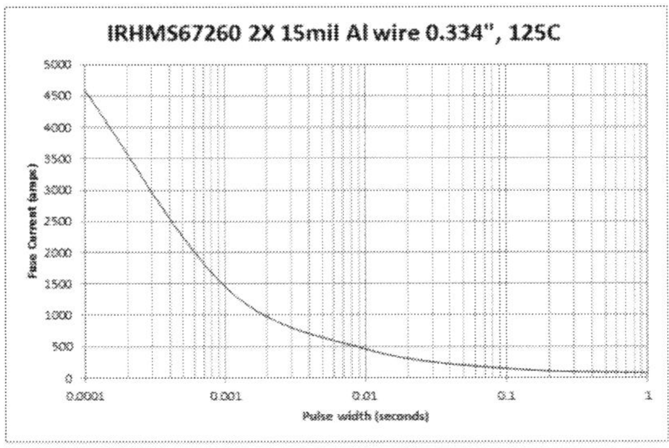

Figure 3. Sensitivity of MOSFET bonding to Pulsed Over Current.

The protection is designed to be active also in case of unit powering in presence of a (latent) failure of one diode in short circuit.

The definition of the limitation value moves from the control of the switching losses and of the current peak during turn-on and turn-off transients.

The switching losses are mainly related to turn-on transient, when the SA capacitance C_{SA} is discharged. In order to avoid large spikes (which can lead to EMC problems or possibly to reliability issues), the discharge current must be limited to a value I_lim (higher than the S/A maximum current I_SA), resulting in higher turn on time. It can be demonstrated that the maximum value of the switching losses is:

$$P_{sw,max} = \frac{1}{2} C_{SA} V_{bus,max}^2 \frac{I_{lim}}{I_{lim} - I_{SA}} f_{sw,max} \tag{1}$$

where $V_{BUS,max}$ and $f_{sw,max}$ are the maximum values of the bus voltage and the switching frequency. The equation shows that the higher I_{lim}, the lower $P_{sw,max}$.

As I_{lim} cannot be arbitrarily high an appropriate compromise between peak current and switching losses must be identified.

Fig. 4 shows the normalized power losses in the shunt MOSFET as a function of the ratio I_{lim}/I_{SA}. The power is normalized to

$$P_{\infty} = \frac{1}{2} C_{SA} V_{bus,max}^2 \frac{I_{lim}}{I_{SA}} f_{sw,max}, \tag{2}$$

which is the asymptotic value for $I_{lim} \rightarrow \infty$.

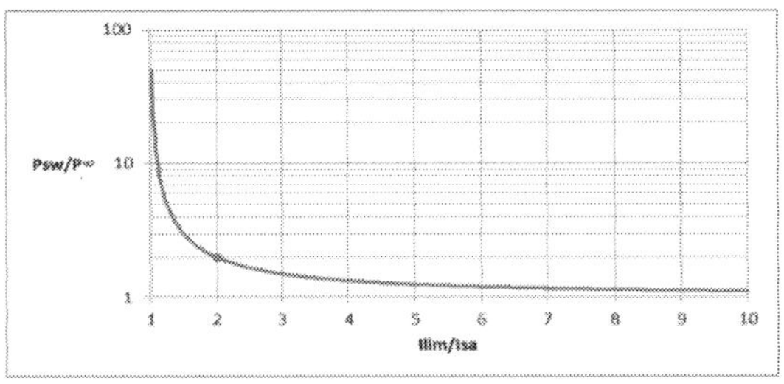

Figure 4. MOSFET switching losses (normalized) as a function of the ratio Ilim/Isa.

Moving from this considerations, a good trade-off in the definition of I_{lim} can be found with a value of I_{lim} equal to about 2 * I_{SAmax}, as further increases in the value of I_{lim} lead to small variations of P_{sw}.

Fig. 5 shows the schematic of the current limitation circuit, where it is shown that an active current limiter and a passive current limiter concur to the function.

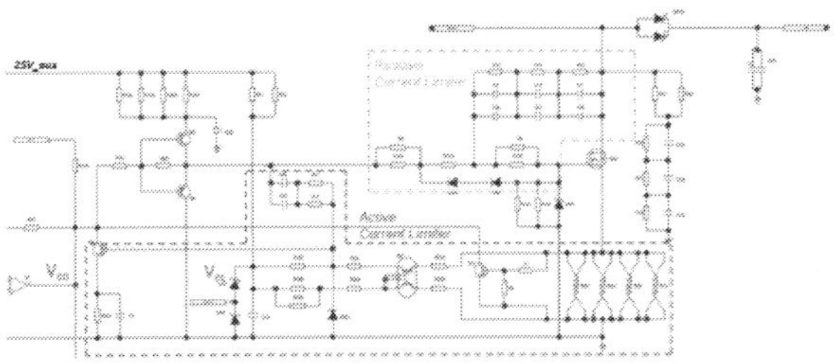

Figure 5. Current Limiter schematic.

4.1. Active Current Limiter

The active current limiter is based on a current mirror sensing circuit surely active during diode short and possibly during normal turn-on transients.

The limitation value is defined by the equation (3), where R_s is the parallel of the four sensing resistors R_{12}, R_{23}, R_{24}, R_{25}, R_c=R_{224}+R_{190}//R_{225}, V_{cc} is the local supply voltage.

$$I_{lim} = \frac{R_{222} - R_{223}}{R_s} \frac{V_{cc} - V_{be9}}{R_{222} + R_c} \tag{3}$$

The circuit has been dimensioned to obtain I_{lim} = 12A.

As the current limiter has an intrinsic delay in intervention during which the FET current can rise uncontrolled, up to very high values that can easily exceed the rated maximum current, a further "fast acting" current limiter, that opens the Gate-Source of the FET, is implemented (Q_{11}) to speed up the circuit and to control the value of the current peak in case of short circuit of the series diode. This additional limiter is silent in normal operation, as its threshold is defined by:

$$I_{lim,facl} = \frac{V_{be11}}{R_s}, \tag{4}$$

Which is significantly greater than I_{lim}, in the order of 30A.

4.2. Passive Current Limiter

The passive current limiter works (in conjunction with the active circuit) to limit current peaks during turn-on transients.

It limits the discharge rate of the parasitic SA capacitance C_{SA}, by controlling the MOSFET Gate to Drain capacitance charge.

The resulting effect is beneficial for all the turn on

transients, during normal operation or after a diode short circuit event, but it doesn't work if the diode short circuit occurs while the mosfet is shunting, which is the heaviest failure condition.

4.3. Fault Detector and trip-off circuit

Fig. 6 shows the fault detection circuit and LCL trip-off timer.

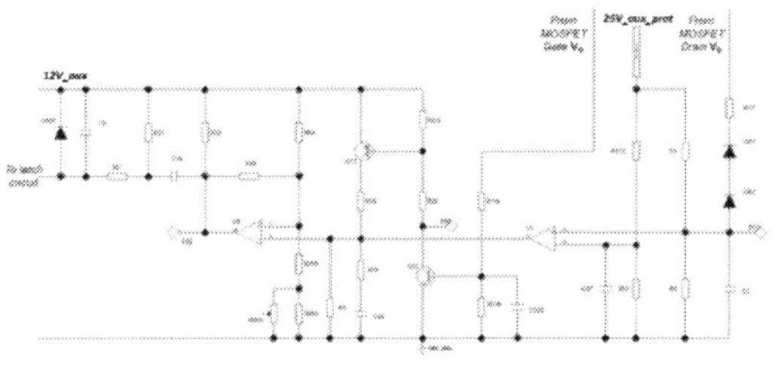

Figure 6. Fault detection circuit.

When a fault is present and the current limiter is activated both the shunt MOSFET gate and drain voltages rise (Gate initially to 12V and successively to V_{gs_th}, during limitation, while Drain to Main Bus voltage).

This condition triggers the fault detection circuit. After a fixed period, defined by the charge of a capacitor (C_{48}) up to meeting a threshold that activates a comparator (U_3), the protection chain acts on the reset of a Flip Flop which latches the status and switches the MOSFET to open.

Of course the protection must not to be triggered during normal operation.

When the S3R MOSFET is off its gate voltage is low and the fault detection circuit is not triggered, so that C_{48} cannot be charged.

When the S3R MOSFET is ON its drain voltage is low and C_{48} charge is inhibited by the port of U_3 activated by the conduction of CR_1 and CR_2.

During the turns on transient of the S3R MOSFET the fault condition is triggered but rapidly removed at the MOSFET closure.

In this case the partial charge of C48 must be sufficiently low to avoid latch-off of the module with adequate margin.

To meet this aim the condition (5) must be satisfied.

$$t_{trip\text{-}off} > t_{turn\text{-}on} \tag{5}$$

Equation (5) sets a lower limit to the Trip-off time.

The maximum value for $t_{turn\text{-}on}$, assuming V_{bus} = 70V, I_{lim_min} = 10A, C_{SA} = 1µF is computed in equation (6).

$$t_{turn-on} = \frac{C_{SA} \times V_{bus}}{I_{\lim} - I_{SA}} = \frac{1\mu\mu \times 70V}{10A - 5A} = 14\mu s \tag{6}$$

On the other side, the upper limit for the trip-off time is set by thermal considerations on the S3R MOSFET when working as limiting element.

Assuming a maximum limitation current of 14A and a maximum drain voltage of 70V, the shunt MOSFET is claimed to dissipate 14A * 70V = 980W.

Considering a mounting temperature for the MOSFET of 70°C the resulting maximum allowable $Z_{\Theta j\text{-}c}$ is

$$Z_{\Theta j-c} = \frac{110-70}{980} = 0.41°C/W, \tag{7}$$

which corresponds to a pulse width of about 170µs (from datasheet curves for $Z_{\Theta j\text{-}c}$).

Equations (5) and (6) show that the trip-off time can be set to a value between 14µs and 170µs.

The nominal trip-off time has been set to 100µs.

5. DESIGN VERIFICATION

The behaviour of both the current limitation and the fault protection intervention have been validated both via simulation and by test. The tests have been performed on an EBB representative of the electrical circuit and of the PCB layout.

5.1. Simulation Results

Computer aided simulations with PSpice , using the actual component models, have been performed.

The failure simulation of the series diode of the S3R section has been evaluated in both the following conditions:

1. Short circuit present at MOSFET turn on.
2. Short circuit applied when MOSFET is already on. This condition is heavier than the previous both for the peak current and for the thermal stress of the MOSFET, as the intervention of the switch-on circuit, including the passive limitation, is by-passed, being the MOSFET already on.

The simulations have been performed at 70V Bus Voltage and 5A SA Current, that round in excess the maximum values defined for the reference CSG program.

Fig. 7 shows the simulation result for condition 1.

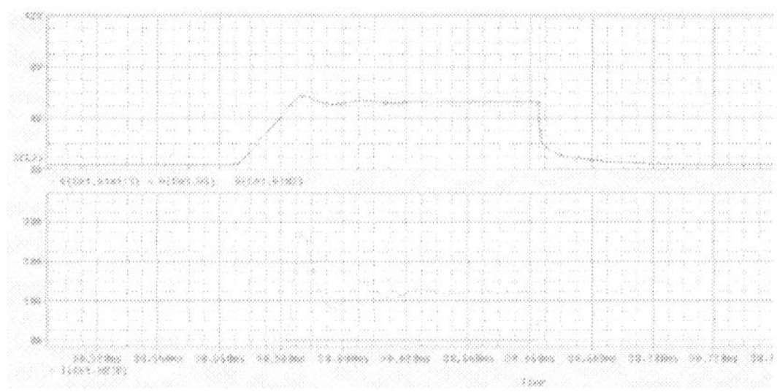

Figure 7. MOSFET turn on with diode in short circuit.

In the upper box the MOSFET V_{GS} (red trace), together with internal voltages related to the timer are shown, precisely the latch output (green trace) and the voltage of the inverting input of the comparator (U3, pin8)

 DOI: 10.1051/e3sconf/20171614013

implementing the protection (yellow trace). The lower box shows the MOSFET I_{DRAIN}.
The limitation transient can be observed, lasting 100μs and limiting current at 12A for a (trip-off). The current peak is very contained (less than 30A). This value is well acceptable as it causes reduced stresses on the limiting device.
Fig. 8 shows the simulation result for condition 2. The V_{GS} (orange trace) and V_{DS} (blue trace) of the MOSFET, together with the latch output (green trace) and the voltage of the inverting input of the comparator (U3, pin8) implementing the protection (red trace) are visible in the window above, while I_{DRAIN} is visible in window below.

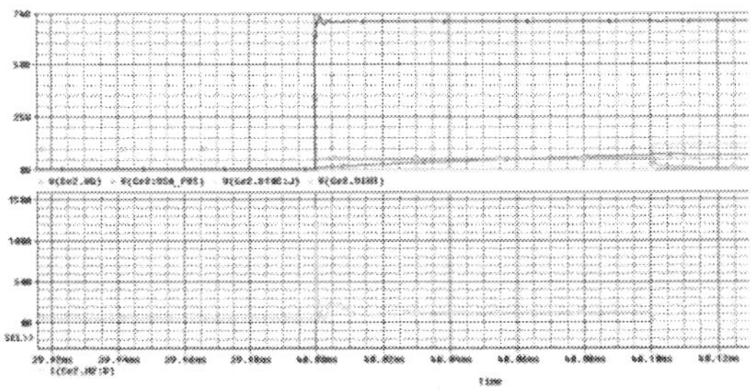

Figure 8. Limitation circuit simulation. Worst case.

The limitation transients is shown in Fig. 8. The limitation Current (12A) and its duration (100μs) are the same as in Fig. 7.
Fig.9***Figure 9*** shows the action of the fast limitation circuit that keeps the current peak below 130A as maximum value. Its shape is triangular, about 250ns in duration at half value.

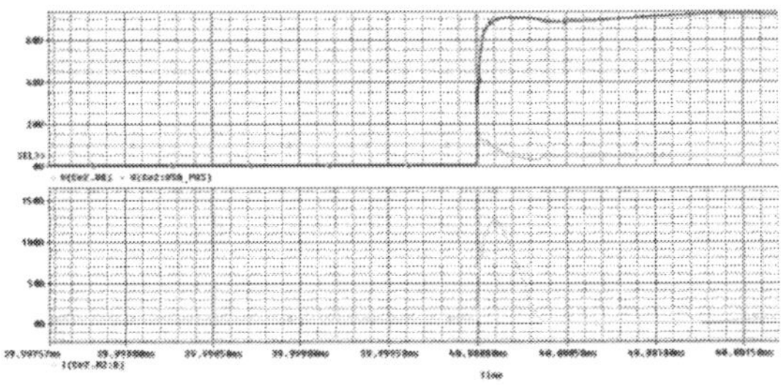

Figure 9. Limitation circuit simulation. Peak control.

5.2. Test Results

The design and the simulation results have been verified on an Elegant Bread Board (EBB) representative of the final design both electrically and for the layout, but not mechanically.
Two S3R boards have been manufactured, tested and integrated with the Battery Control electronics Module (BCM) and with a capacitor bank to reproduce the whole battery charge control function as designed in the frame of the CSG program.
Current Limitation after the short circuit of the series diode has been tested in the same conditions applied in the simulations:

1. Short circuit present at MOSFET turn on.
2. Short circuit applied when MOSFET is already on.

For both the test conditions the failure in short circuit of the series diode has been applied placing across it a Short Circuit Generator.
The current flowing into the FET has been measured in the leg sense resistor (nominal value 15.5mΩ) where the limitation is electronically sensed.

Test condition #1 has been performed on all the S3R sections and at various Bus voltage values, in order to characterise the limitation function. The analysis of the results shows a very contained spread in values and waveforms among the various circuits.
Fig. 10 shows V_{MB} (high end value), MOSFET V_{GS} and I_{DRAIN} current.

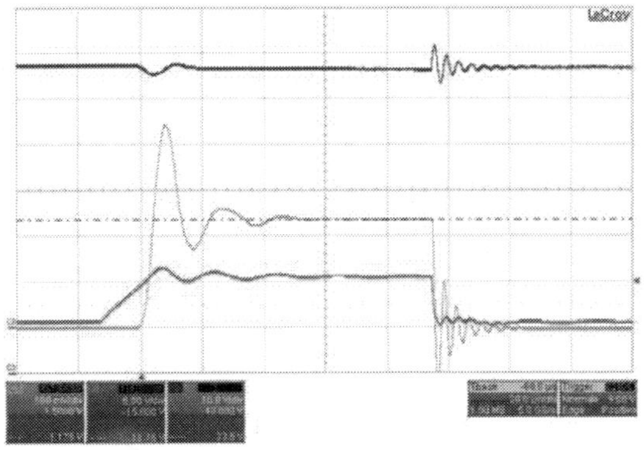

Figure 10. Limitation measurement at Mosfet turn- on.

Test condition #2 has been characterised at both low and high ends of the Bus voltage range, when the FET was conducting, as made for the simulation. This represents the most stressing operating condition for the MOSFET.
The waveforms showing the intervention of the protection are visible in Fig. 11 and in Fig. 12, where the traces have the following meaning:

TR1: Voltage across Rsense. (MOSFET Current).
TR2: V_{DS} of the FET.
TR3: V_{GS} of the FET.
TR4: Internal latch signal.

The maximum measured current peak is about 85A, well inside the simulation results and the acceptability level stated in para. §4.

 DOI: 10.1051/e3sconf/20171614013

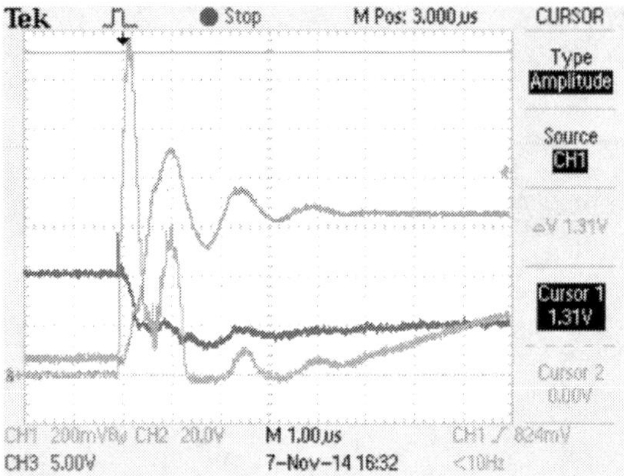

Figure 11. Limitation measurement. Peak control.

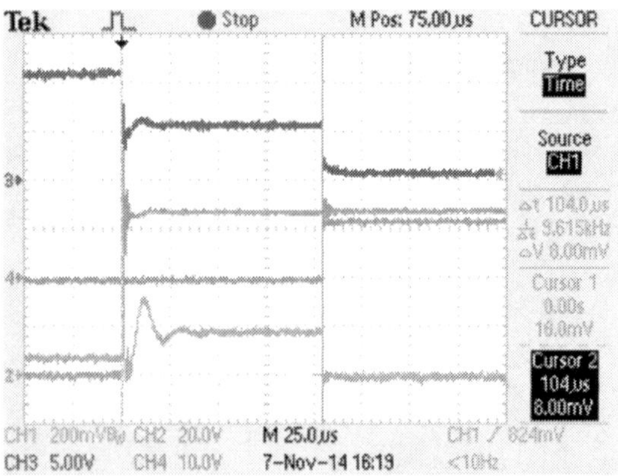

Figure 12. Limitation measurement.

To evaluate the benefit of the adopted solution, dissipation has been also measured during the tests of the full S3R integrated function.
The efficiency of the S3R section has been measured well above 98% at maximum SA power on a temperature range -25°C ÷ +60°C.

5.3. MOSFET Junction Temperature

From the results shown in the paragraph §5.2, the maximum expected temperature rise in the switching elements (FET) can be computed.
Moving from waveforms in Fig. 11and Fig. 12, the power dissipated on the MOSFET has been evaluated by means of the plots of Fig. 13 and Fig. 14, where the time evolution of the dissipated power has been approximated with equivalent P*t discrete envelopes, to simplify the computation.

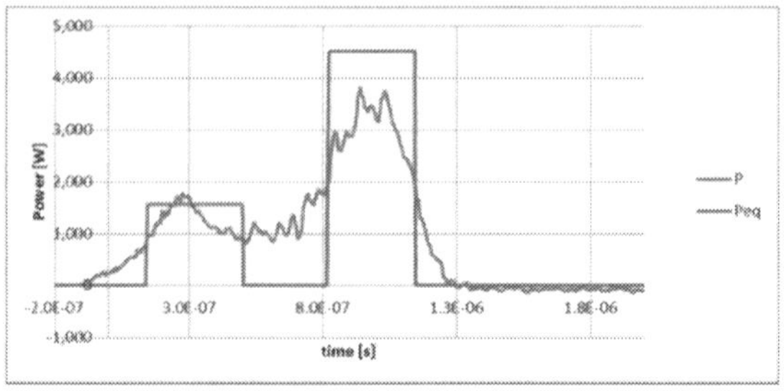

Figure 13. MOSFET power dissipation. Peak current.

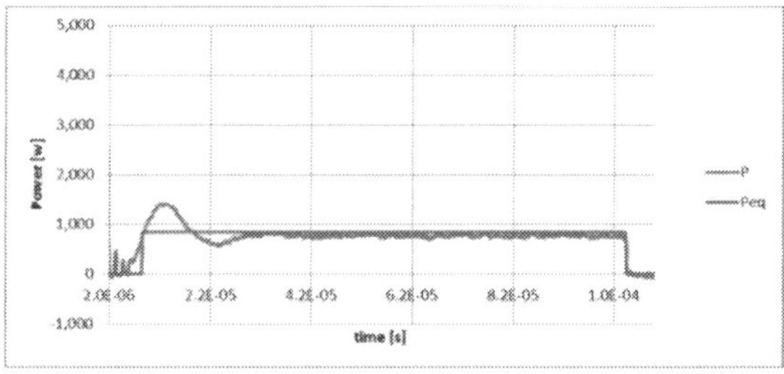

Figure 14. MOSFET power dissipation. Limitation.

Fig. 15 shows the thermal junction-to-case resistance of the FET has been extrapolated from the datasheet.

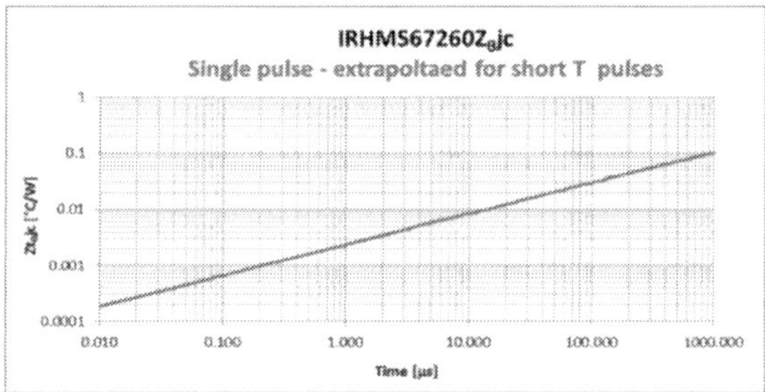

Figure 15. Single pulse thermal impedance versus pulse duration. Short pulses extrapolation.

A temperature rise of about 8°C results in the first (fast) transient (current peak of Fig. 13, about 1.2µs long).
During the nominal limitation (Fig. 14) a temperature rise of about 25°C resulted from the computation.
An overall temperature rise of about 33°C has been estimated at worst from both the simulations and measurements.
Considering a reasonable maximum mounting temperature that does not exceed 70°C, as it is in the specific application, the overall maximum junction temperature has been estimated not exceeding 103°C. This figure is well inside the derated value of 110°C, that is the acceptability level stated in para.§ 4.
It must be further considered that the intervention of the protection occurs only once in the operating life of the component, so that no accumulation of stress is foreseen.

 DOI: 10.1051/e3sconf/20171614013

6. UNLATCH FUNCTION

The limitation circuit output is latched into a Flip-Flop that disables (opens) the S3R MOSFET in case of activation. In normal operation the Flip-Flop is unlatched, leaving free to work (open, shunt, switch) the MOSFET.
An automatic un-latch mechanism has been designed to allow the recovery of unwanted spurious intervention of the latch protection.
This mechanism is driven by the error signal that controls the S3R function.
The reset is activated when the error signal exits its operating range upwards and meets a predefined threshold. When this happens too much power is delivered to the bus; the unlatch function triggers a command that resets the Flip-Flop and restores the normal operation of the MOSFET.
As every section has its own unlatch circuit, the full operation of all the S3R section is restored to regain the nominal operation.
In case of a section actually latched after a diode short circuit event, it is reactivated and soon re-latched due to a further intervention of the protection.
A test on EBB has been performed to verify the transients related to this event. Fig. 16 shows the recording of the two consecutive events: failure of diode in short and unlatch after 10ms. The delay time is representative of the reaction time of the error signal after the diode failure in the specific tested configuration.

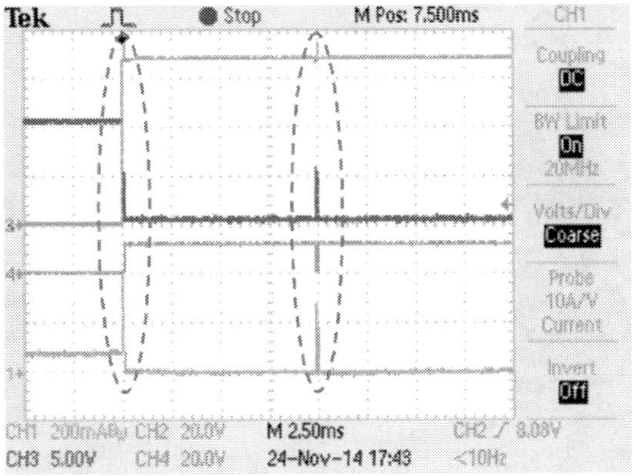

Figure 16. Un-latch after short of diode.

Fig. 17 and Fig. 18 show the zoom of the two transients of Fig. 16.
As expected the unlatch condition is less stressing, both in terms of peak current and of thermal stress, as the MOSFET is initially in "open" condition, so the first current peak results significantly lower.

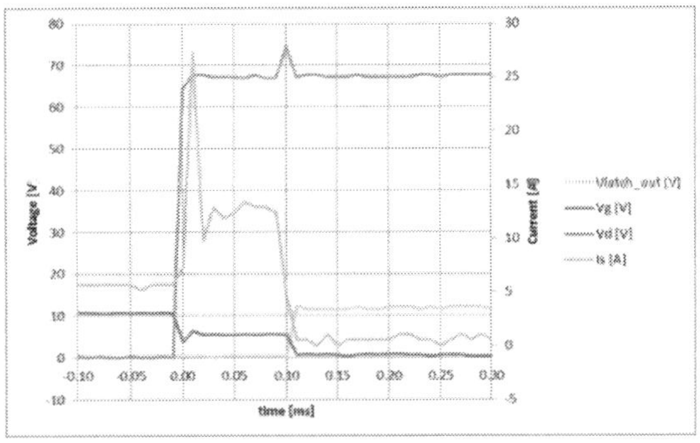

Figure 17. Diode short failure. 1st transient.

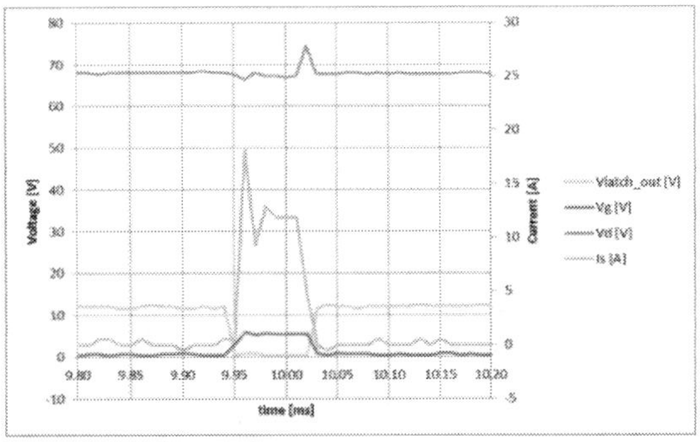

Figure 18. Un-latch event (and subsequent latch). 2nd transient.

The whole latch-unlatch electronics of the S3R section has been designed and verified to guarantee a correct initialisation at first power application. The correct initialisation of all the electrical parameters that concur to set the latch status is guaranteed by means of a delayed application of power supply to the electronics; the delay is obtained conditioning the release of the power supply to part of electronics and to FET driving circuit, by a threshold on the main internal supply.

7. S3R MODULE PHYSICAL REALISATION

The S3R concept described in this paper has been implemented for the first time in the frame of the earth observation program COSMO Second Generation.
The whole S3R function, made of 16 sections, has been allocated on two boards (8 per board) of reduced dimensions (about 240mm * 185mm), with a pace of 25mm between boards.
Fig. 19 and Fig. 20 show the top and bottom view of the S3R built for CSG PCDU.

DOI: 10.1051/e3sconf/20171614013

Figure 19. CSG S3R Top Side.

Figure 20. CSG S3R Bottom Side.

8. CONCLUSIONS

The feasibility of the proposed configuration for the S3R function has been demonstrated, meeting the electrical, thermal and reliability objectives.
The solution, deeply analysed, has been tested and validated on an EBB (complete for all the function) even at high and low temperature (in ambient pressure only).
The design has demonstrated, above all, its great advantages in terms of space occupation, as the electronics necessary to manage the main design aspects has been contained into acceptable dimensions, as the physical implementation clearly show.
As expected also the efficiency of the whole S3R has shown a really high figure (above 98%).

9. REFERENCES

[1] F. Tonicello, L. Bust Llavona, S. Keller "Synchronous Rectified, Single Diode Sequential Switching Regulator (S3R)", Proceedings of the 9th, European Space Power Conference, 2011.

10. ACKNOWLEDGEMENTS

A particular thank goes to Matteo Scrigna and Giorgio Ebale for their fundamental contribution in the definition and verification of the design of the solution presented in this paper.

PPU MK3 FOR 5 KW HALL EFFECT THRUSTERS

11TH EUROPEAN SPACE POWER CONFERENCE

Eric Bourguignon, Stéphane Fraselle

Thales Alenia Space Belgium, B-6032, Mont-sur-Marchienne, Belgium,
eric.bourguignon@thalesaleniaspace.com
stephane.fraselle@thalesaleniaspace.com

ABSTRACT

Thales Alenia Space Belgium (ETCA) has developed and qualified the PPU Mk3, dedicated to 5kW Hall Effect Thrusters, to reply to the market demand of full electric satellites performing the Orbit Raising with Electrical Propulsion.
The PPU Mk3 is based on the PPU Mk2 heritage (itself derived from the PPU Mk1) but features additional optimizations to reduce cost and take advantage of the single cathode configuration and independent magnet of most 5 kW class thrusters.
The PPU Mk3 anode supply was first consolidated by a successful coupling test of the 5kW Anode supply and Filter Unit breadboards in October 2014.
Then, a Demonstration Model (DM) of the PPU Mk3, fully representative of the EQM, has been built and has been successfully validated by coupling test with the SPT140-D thruster from Fakel in May 2015, with the PPS-5000 thruster from Safran in October 2015 and with the XR-5 thruster from Aerojet in December 2015.
The PPU Mk3 EQM has undergone its qualification tests which have included thermal vacuum tests, pressure increase test (corona) and a full EMC test campaign. After the Qualification Review successfully held in March 2016, the PPU Mk3 EQM has been already coupled with a SPT140-D thruster, in May 2016. Next coupling test is planned with the PPS-5000 thruster in October 2016.
In August 2016, the three first PPU Mk3 flight models have been delivered from a total of fifteen PPU Mk3 flight models already ordered by two customers.

After the presentation of TAS-B experience in Electric Propulsion, the PPU Mk3 product is described, with its development, coupling tests and qualification.

1. TAS-B EXPERIENCE IN ELECTRIC PROPULSION

Since 1996, Thales Alenia Space Belgium (ETCA) designs, develops and produces Power Processing Unit (PPU) to supply Hall Effect Thrusters.
The first PPU Mk1 qualification model (Fig. 1), developed for the 50V bus Stentor program, has supplied during 8900 hours an SPT-100 thruster in a vacuum chamber simulating space environment.

Figure 1. Stentor PPU MkI EQM

Qualified for the 100V Spacebus 4000 platform, the 100V PPU Mk1 and Filter Unit (FU) EQM have cumulated 6300 hours ground operation with a PPS1350-G thruster.

Thirty three PPU Mk1 flight models were delivered to Airbus DS, ESA, IAI, OHB, Safran, TAS-F.
In October 2005, the Smart-1 spacecraft reached the Moon after 4958 hours of cumulated operation of the PPU and its PPS1350-G thruster.
Eighteen PPU Mk1, currently in flight for North South Station Keeping on nine telecom satellites, have cumulated more than 32 400 hours flight operation in March 2016.
For the SmallGEO platform, TAS-B has developed and qualified an External Thruster Selection Unit (ETSU) to select one out of four thrusters. Two ETSU flight models were delivered, associated with the PPU flight models.

In the frame of AlphaBus extension program and in partnership with ADS and TAS-F, TAS-B has developed the PPU Mk2 product (Fig. 2). The PPU Mk2 addresses PPS1350-G, PPS1350-E, SPT-100 and Hall Effect Thrusters up to 2.5kW for 100V platforms. Taking benefit of the flight experience of the PPU Mk1, the PPU Mk2 provides 1.6 times more output power (1.5kW → 2.5kW) and more flexibility to thrusters and platforms, with reduced manufacturing cost. Qualified in July 2014, the PPU Mk2 EQM was successfully

 DOI: 10.1051/e3sconf/20171615001

coupled with the PPS1350-G at 1.5kW, the PPS1350-E at 2.5kW and the SPT-100. The first PPU Mk2 flight model is delivered from a total of ten PPU Mk2 flight models ordered by two customers.

Figure 2. PPU Mk2 EQM

2. PPU MK3 OBJECTIVES

The objective of the PPU Mk3 development was to capitalize on the PPU Mk2 product to propose a cost-optimized solution with a reduced time to market to drive 5 kW-class thrusters. These high power thrusters will enable an electrical orbit raising of telecom satellites.

The PPU Mk3 objectives were:

- Competitive product,
- Dedicated to PPS-5000, SPT140-D, XR-5 thrusters.
- Dedicated to SpaceBus Neo, Eurostar 3000, NeoSat and Electra platforms.
    - Bus voltage: 100V regulated
    - MIL-STD-1553B interface
- Qualification and first flight models in 2016

3. PPU MK3 DESCRIPTION

The PPU Mk3 features all the supplies required to operate a Hall Effect Thruster which features a single cathode and a magnet coil independent from the discharge. It also features a switching function which enables to operate one out of two thrusters. It communicates with the satellite platform through a 1553 bus and receives its power from a 100V bus.

Figure 3 presents the functional diagram of the PPU Mk3.

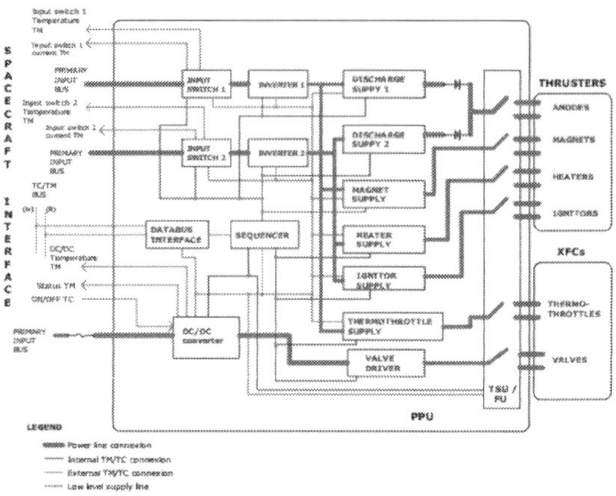

Figure 3. PPU Mk3 Functional Diagram

The main functions of the PPU are:

- 1553 data bus interface to communicate with the satellite.
- DC/DC converter which enables to supply the low-level circuits of the PPU. This DC/DC must be fused protected.
- Sequencer, which is implemented by a FPGA and which controls and monitors all the PPU supplies. The sequencer features an automatic mode where the sequencing of all the supplies is managed to start-up the thruster and operate it in steady mode. Telemetry based protections are also implemented in the FPGA. The sequencer includes a PROM which contains default values and valid ranges to operate different types of thrusters.
  Standard start-up or soft-start start-up sequence where the cathode is ignited and kept in sustain before applying the anode voltage to reduce inrush current may be selected. The sequencer also implements the regulation loop of the discharge current by controlling the setting of the thermothrottle supply.
- Input switch protections (one for each discharge supply) which enables to avoid any failure propagation to the satellite 100V bus in case of an internal failure.
- Two inverters with their transformer provide the insulated voltages required for the thruster. The inverter is a resonant topology in order to optimize the efficiency.
- Two discharge supplies operate in parallel with their outputs summed by diodes. They provide the anode voltage which is commandable from 100V up to 400V. The discharge supplies implement a power limitation: once the knee-current is reached, the voltage drops linearly as the current increases as presented in figure 4 which shows the anode supply characteristic. The short-circuit current is commandable up to 22A. The anode supply, based on two modules of 2.5kW connected in parallel,

 DOI: 10.1051/e3sconf/20171615001

delivers up to 4.74 kW. The power limitation is required because when the plasma forms at thruster ignition, it drains a lot of current. The linear increase of the current as the voltage decreases is required in order to avoid a lock-up with the thruster characteristic.

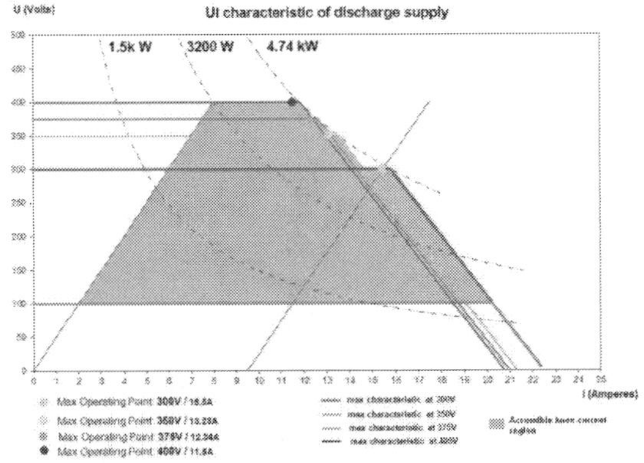

Figure 4. U-I characteristics of the Anode Supply

- Magnet supply drives the thruster magnet coils independent from the discharge, with current up to 7A.
- Cathode heater supply current capacity is 18A.
- Thermothrottle supply providing a current which enable to regulate the Xenon flow.
- The valve driver enables to control the XFC valves.
- The switching unit is relay-based and enables to supply two different thrusters, one at time.
- Two Filter Units are implemented inside PPU Mk3, downstream the TSU. This filter enables to limit the voltage ripple at PPU output due to the thruster noise. Indeed, when the thruster is fired, it generates significant noise. A telemetry providing an image of the thruster noise RMS current value is implemented.

The PPU Mk3 is robust to abnormal pressure increased inside satellite up to 1Pa, by mechanical design.

The mass of the PPU Mk3 that includes TSU and two FUs is 18.6 kg. Its dimensions are 390 x 315 x 263 mm^3 (LxWxH).

The PPU Mk3 most dissipative functions are located as close as possible to the baseplate for cooling purpose. The two anode modules include the input switch protections and the main transformer and inverter. The HI module features the heater and ignitor supplies and the MT module contains the magnet and thermothrottle supplies. These four modules are plugged into the Backplane module which features the sequencer, including the communication to the platform and the DCDC. The TSU/FU module comes on top and includes the switching relays necessary for thruster selection and the filter units.

4. PPU MK3 DEVELOPMENT

The PPU Mk3 development started in 2013, with a Study Phase, to issue and review PPU Mk3 specification with the thruster manufacturers and the primes. The PPU Mk3 architecture was optimized and new packaging was selected to reduce the number of modules and sub-assemblies in order to propose more competitive product. This phase was concluded in January 2014 with the issue of the Technical Requirement Document which is the input for the following phase, the PPU Mk3 Development Phase.
During this phase, a 5kW anode and FU breadboard were coupled with a SPT140-D thruster, in Fakel facilities, at Kaliningrad, Russia, in October 2014, in order to secure the interfaces with the thruster.
By powering the thruster up to 400V and 4.7kW, the anode supply output characteristics and output impedance were validated. The implementation of the FU inside the PPU Mk3 was also validated by testing different harness lengths between PPU, FU and thruster. With these results, the PDR was successfully closed in November 2014.

After the PDR, the PPU Mk3 Demonstration Model (Fig. 5), representative of future flight model (fit, form and function) was developed, manufactured and tested with representative load simulating the thruster and XFC.

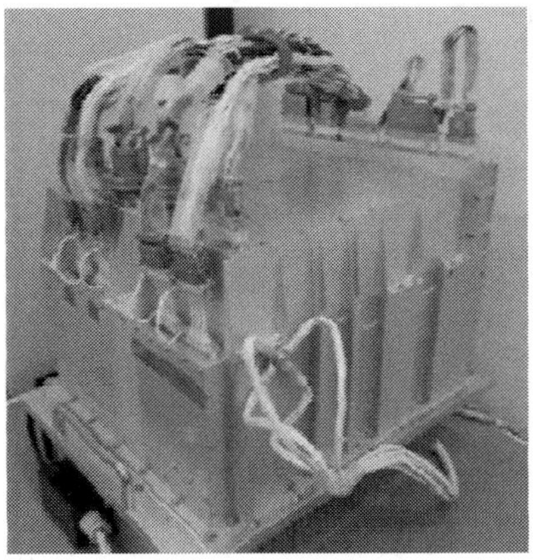

Figure 5: PPU Mk3 DM

The modules of the DM were first tested independently and then assembled. The objective of the DM was to validate the design, including the PCB routing, in a mechanical assembly representative of the flight models. This was important to validate as early as possible that there were no internal EMC issues.

DOI: 10.1051/e3sconf/20171615001

The testing of the DM also enabled to validate the FPGA which controls and monitors all the PPU supplies.
The DM was tested in hot and cold conditions at ambient pressure to validate the good operation of all functions within a wide temperature range as well as to validate the thermal management of the unit. The dissipations in different representative conditions were also measured to check that they were inferior to the worst case computations. The efficiency was measured typically above 95% (Fig. 6). The anode supplies of the PPU Mk3 DM were tested up to 5.2 kW to demonstrate margin on the maximum output power

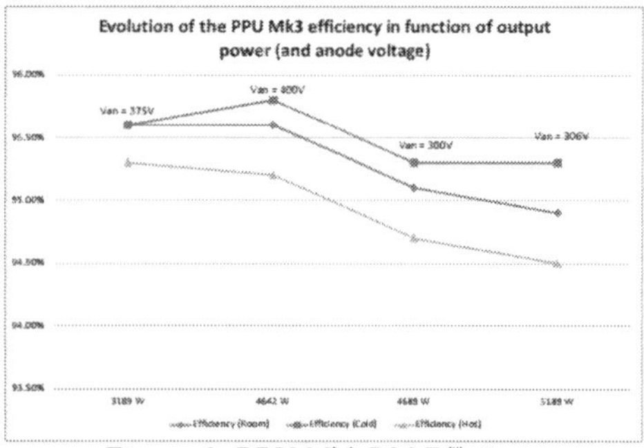

Figure 6: PPU Mk3 DM Efficiency

Another objective of the DM was to provide a representative equipment to validate the Automatic Test Equipment and its test sequence code. The development phase indeed included the implementation of automatic test sequences to implement production tests.

In May 2015, the PPU Mk3 DM has been successfully coupled with a SPT140-D thruster at Aerospazio facilities, in Italy, in partnership with ADS. During that test, the PPU also operated the XFC and its FPGA efficiently regulated the discharge current by controlling the thermothrottle current. This validation test launched the definition and manufacturing of the PPU Mk3 EQM which was anticipated before the CDR, successfully closed in September 2015, in order to reduce the development planning.

In October 2015, the PPU Mk3 DM was also successfully coupled with a PPS-5000 thruster at Pivoine facilities in CNRS Orléans, France.

In December 2015, the PPU Mk3 DM was also successfully coupled to an XR-5 thruster (without XFC) in the frame of an ESA contract led by European Space Propulsion (ESP) as well, at QinetiQ facilities in Farnborough, UK. This test involved experts from ESP, Aerojet Rocketdyne, Mars-Space and TAS-B.

5. PPU MK3 QUALIFICATION AND FM's

The PPU Mk3 EQM (Fig.7) was submitted to a full qualification campaign.

Figure 7 : PPU Mk3 EQM

The qualification tests first covered the mechanical environment: sine and random vibrations were applied and pyro shocks were performed along all three axis.

The thermal and pressure conditions were validated during the thermal-vacuum test campaign as the EQM was submitted to three cold starts and 9 thermal cycles in vacuum conditions. At the beginning and at the end of the TVAC tests, a pressure increase test up to 2 Pa was performed to demonstrate that the PPU Mk3 is robust to an external pressure increase which can cause Paschen discharges in high voltage equipments.

The PPU Mk3 EQM was then submitted to a full EMC campaign including different LISN configurations to cover different platforms. The equipment conducted emissions were fully characterized both in differential and common modes. Conducted susceptibility tests were also performed to check the good behaviour of the unit in case of bus transient and with injections simulating the thruster worst-case noise. The radiated emissions were measured and radiated susceptibility tests were implemented. Direct ESD tests were performed on the PPU thruster outputs to check that they are robust to an electrical discharge occurring on the electrodes of the thruster. Bundle and ground plane ESD tests were also performed with success.

After the Qualification Review successfully held in March 2016, the PPU Mk3 EQM has been coupled with a SPT140-D thruster at Aerospazio facilities, in May 2016. Next coupling test is planned with the PPS-5000

 DOI: 10.1051/e3sconf/20171615001

thruster in October 2016.

In August 2016, the three first PPU Mk3 flight models have been delivered from a total of fifteen PPU Mk3 flight models already ordered by two customers.

6. PPU MK3 VARIANT FOR XR-5

Similarly to PPU Mk1 and Mk2, the PPU Mk3 supplies the thermothrottle of the Xenon Flow Controller to adjust the thruster xenon flow. The thruster discharge is regulated by the PPU sequencer through control of the thermothrottle supply. As the XR-5 thruster is currently qualified with proportional valves (PFCV) instead of thermothrottle to adjust the xenon flow, TAS-B has performed, with ESP, a Neosat pre development activity to validate TAS-B PPU Mk3 capacity to supply and regulate a Proportional Valve. TAS-B has developed and manufactured a breadboard model, that was coupled to a PFCV in November 2015 at ESP/TAS-UK facilities in Belfast, UK. The test setup included a Thruster simulator reproducing the Discharge current, allowing to test the Discharge current regulation control through the PFCV. The coupling test was successfully carried out. The digital model used for regulation study and optimization was thereby validated. The Discharge current stability, accuracy, response time and overshoot were within required performances.

Thanks to this pre-development and the coupling test of PPU Mk3 DM with XR-5 thruster, the development of a PPU Mk3 variant for XR-5 thruster is secured.

7. CONCLUSION

TAS-B has designed, developed and qualified a new PPU Mk3 product dedicated to 5kW-class Hall Effect Thrusters and 100V satellite platforms. The PPU Mk3 took advantage of a strong heritage with the already qualified PPU Mk2 to ensure a short time to market. Cost reduction measures were implemented in the design, mainly in terms of mechanical packaging.

The development started in 2013 and the equipment was successfully qualified in March 2016. Coupling tests were successfully performed with the PPS-5000, the SPT140-D and the XR-5 thrusters.

Up to now, a total of 37 PPU (Mk1, Mk2, Mk3) flight models have been delivered by TAS-B from a total of 60 PPU flight models ordered by our customers: Airbus DS, ESA, IAI, ISS, OHB, Safran, TAS-F.

These orders demonstrate the confidence our customers in TAS-B experience in Electric Propulsion, based on:

- 8 900 hrs ground coupling test with EQM Stentor
- 6 400 hrs ground coupling test with EQM SB4000
- 4 958 hrs flight operation of SMART-1
- 32 400 hrs in orbit on telecom satellites

HALL EFFECT THRUSTER DIRECT DRIVE PPUs, EXPERIMENTAL INVESTIGATION OF THE CATHODE POTENTIAL GROUNDING PROBLEM

Luca Ghislanzoni [1], Luca Benetti [1], Tommaso Misuri [1], Giovanni Cesaretti [1], Lorenzo Fontani [1].

[1] SITAEL S.p.A., via Livornese 1019, 56122 San Piero a Grado, Italy. Email: luca.ghislanzoni@sitael.com

ABSTRACT

In order to maximize system mass and cost savings, a Hall Effect Thruster Direct Drive Approach shall ideally get away with the need to provide galvanic isolation for the Anode voltage. In this respect, we present some experimental data aiming at characterising the implications of a direct connection between the HET Cathode Return Potential and the Spacecraft Ground.

1. INTRODUCTION

Modern High Power GEO Telecom Platforms offer sufficient power capability (20 kW and even more in the near future) to afford the possibility to complete, within reasonable times, the Orbit Raising phase by means of Electric Propulsion only. Considering that for such platforms a solar array and electrical power bus voltage range of 200-300V represents the target of various technology developments currently being researched, significant savings could be achieved by supplying a Hall Effect Thruster (HET) Anode directly from said voltage, the so called *Direct Drive* approach, thus avoiding the need for the High Power Converter otherwise required for adapting the typical 100-120V bus voltage to the about 250-500V Anode voltage range required by high power (5kW and upward) Hall Effect Thrusters. At system level, such an approach would hence allow significant mass and cost savings. As an example for a 20kW application, such a flight-standard converter would probably weight about 30 kg, dissipate > 1kW, and cost in the range of several 100s of k€. However, maximum savings can only be achieved by renouncing to the galvanic isolation usually provided by said High Power Converter.

Concerning the subject of Hall Effect Thruster Grounding Concepts, a review of the presently available literature (see for example [1] for an introduction to the various approaches) suggests a lack of experimental data. At SITAEL we have thus decided to start an in-house test campaign for the collection of useful experimental data on this subject.

2. CONVENTIONAL HET GROUNDING

On most spacecrafts utilizing electric propulsion the typical grounding approach (let us call it ***Floating CRP***) implements galvanic isolation between the Cathode Return Potential (CRP) and the spacecraft Electrical Power System (EPS) Ground. In practice, the otherwise floating CRP is then referenced to the EPS Ground via an RC network composed of about 200 nF (max) in parallel with about 100 kΩ resistor (suitable values typically range from about 50 kΩ to about 200 kΩ). Any beam stray current, I_{BSC}, which may "leak" to the EPS Ground, through a plurality of complex diffusion and interaction mechanisms involving some of the spacecraft conductive surfaces exposed to the plasma environment, would then result in a voltage drop across R. Such voltage drop would thus move the CRP in the direction of reducing I_{BSC}, until a point of equilibrium is finally achieved. Typically, at equilibrium, the CRP can reach few tens of volts negative with respect to the EPS Ground, but also the case with CRP few volts positive cannot be entirely ruled out (as shown by measurements performed on-

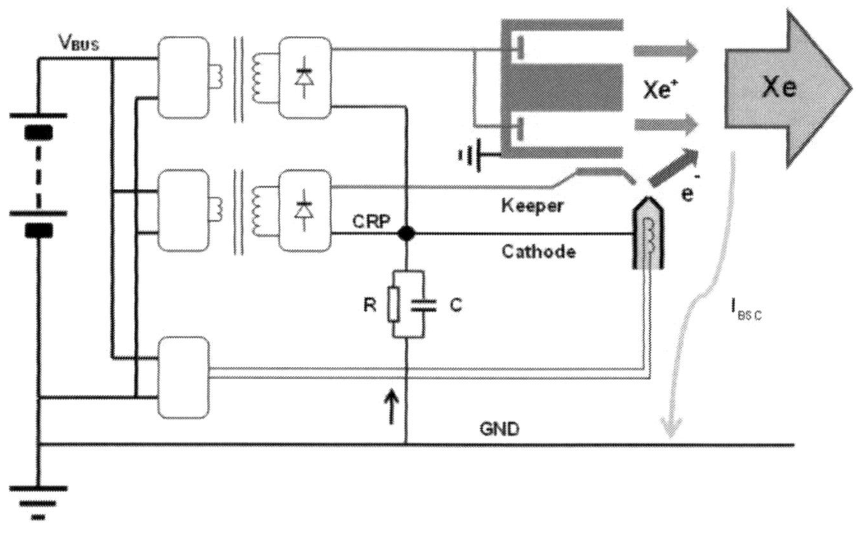

DOI: 10.1051/e3sconf/20171615002

board during the SMART-1 mission [2]). In fact, given a certain value of R, the actual value of I_{BSC} will depend on the value of relevant environmental variables:

- plasma environment
- orbit altitude
- sunlight or eclipse
- solar activity
- season

It seems obvious that a test bed environment might significantly differ from the actual orbital environment, at least for what concerns:

- the level of vacuum achievable
- the much larger collecting surface for I_{BSC} (the internal walls of the chamber, which is grounded), with respect to the exposed conductive surfaces of a spacecraft.

By definition, such a ***Floating CRP*** approach requires the full anode power to undergo conversion (i.e. dc → ac → transformer → rectifier → anode voltage) in order to provide galvanic isolation, and hence incurring in the mass and cost penalties recalled above.

3. THE GROUNDED CRP APPROACH

Ultimately, a Direct Drive application would imply a ***Grounded CRP*** approach, consisting in forcing the CRP potential to the spacecraft Ground potential. In such a condition the current flowing between the CRP and the Test Bed Ground, coincident with I_{BSC}, will need to be monitored in order to study its dependence from the relevant environmental variables. The functional diagram below helps in understanding the ***Grounded CRP*** experimental setup.

The ***Grounded CRP*** configuration, corresponding to what a typical Direct Drive application would require, is obtained when SW1 is closed. In such a situation one shall consider the possibility that after the cathode has reached its nominal operating temperature, and thus thermally emitted electrons are already present, and the flow of Xenon is started, the application of the discharge-igniting pulse at the cathode Keeper electrode might instead strike a discharge to some ground referenced conductive parts others than the intended Cathode electrode. Such potential catastrophic impossibility to ignite the Cathode discharge can be prevented by adopting galvanic isolation (with respect to the EPS Ground) of the Keeper voltage, while also adding diode ***D*** as to reverse block any current discharge path not striking the Cathode. In other words, by design (because of the galvanic isolation) the igniting discharge current is in this way forced to flow between the positive and return outlets of the Keeper supply. Hence, connecting said Keeper return directly to the Cathode, while decoupling it from the Main Ground by means of diode ***D***, will ensure that any Keeper current will automatically be forced to return through the Cathode. Once the Anode discharge is also ignited, diode ***D*** will then allow the bulk of I_{AN} to flow through. For diagnostic and characterization purposes, a current sensor (***A***) will monitor the value of I_{BSC}.

With both SW1 and SW2 closed it will then be possible to verify under what conditions the risk of not being able to correctly ignite the cathode discharge might become a concern. The following $I_{BSC}(V_{CRP} - V_{GND})$ characterization was performed:

- ***SW1 & SW2 open***: while the thruster is firing, the output voltage of a two quadrant (i.e. capable of both sourcing and sinking current) laboratory supply (***V*** in the figure above), can be adjusted from few 10s of Volts negative to few 10s of Volts positive, and plots of $I_{BSC}(V_{CRP} - V_{GND})$ obtained.
- ***SW1 & SW2 closed***: ***V*** is then set to 0V and SW1 closed, as to obtain a ***Grounded CRP*** configuration. Once the thruster is stably firing, SW2 can also be closed in order to avoid unnecessary power dissipation through diode ***D***. The measured values of I_{BSC} are then recorded while sweeping the thruster through various operating points.

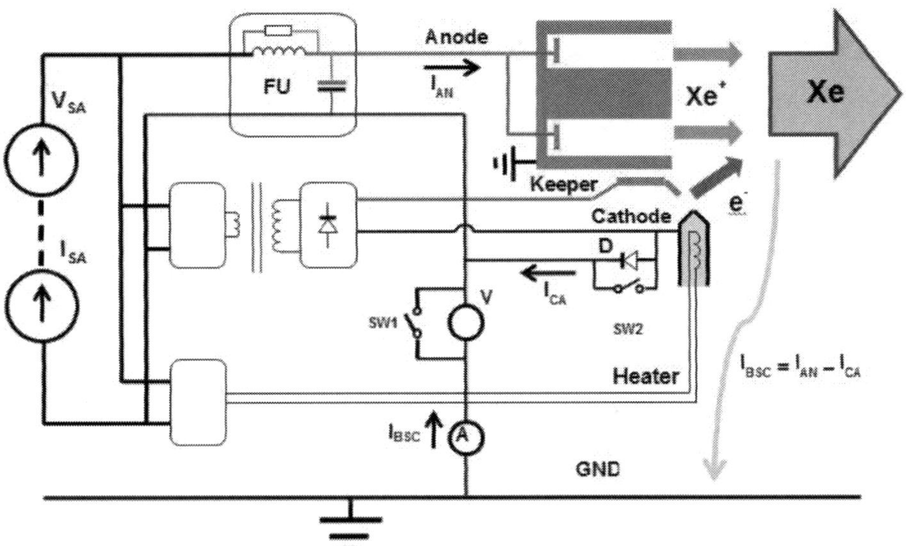

 DOI: 10.1051/e3sconf/20171615002

Although said I_{BSC} $(V_{CRP} - V_{GND})$ characterization, performed in a vacuum chamber test setup, might not be fully representative of what may be observed in an actual orbital environment, we believe that its comparison with similar experimental data that will be later collected during the μHET In Orbit Validation Mission of SITAEL's low power HT-100 thruster may nevertheless provide useful indications. Such set of experimental data will aid in assessing the actual differences between the two environments, possibly suggesting with respect to which phenomenon one can be considered worst case with respect to the other. Furthermore, such set of experimental data could later be used for helping in the validation of models developed for the prediction of such effects.
In this respect a recent paper from Katz et al. [3] suggests that:
"The most important parameters that go into predicting cathode common voltages are solar array electron collection and the thruster body I-V curve. Both of these can be measured in laboratory tests. This data combined with a Hall thruster plume model such as HALLPLUME2D and interactions codes, e.g. NASPAS can be used to choose the value of the resistor that both isolate the spacecraft chassis from thruster power system noise and reduce cathode common voltages to prevent undue sputter damage."
By following the suggested methodology it appears therefore possible, starting from laboratory measurements, to predict the above mentioned I_{BSC} $(V_{CRP} - V_{GND})$ Volt – Ampere characteristics.

4. TEST SETUP

We have thus proceeded with a first measurement of said I_{BSC} $(V_{CRP} - V_{GND})$ Volt – Ampere characteristics for the following example:

- HET was SITAEL's HT 5K operated at 2.5 kW.
- The firing chamber was SITAEL's IV4 facility.
- The voltage between CRP and GND was forced by means of a two-quadrants power supply.

The IV4 steel chamber has a diameter of 2m and a total volume of about 8 m^3. The facility is equipped with a series of cryogenic heads, a cryogenic pump and two turbomolecular pumps. IV4 can reach an ultimate vacuum level in the 10^{-8} mbar range.
This vacuum chamber is equipped with a complete array of diagnostics, including plasma probes, as well as non-intrusive capabilities such as infrared thermography, ultra-high speed imaging and spectroscopy.
In order to damp the plasma beam energy and to reduce the contamination of the thruster due to back-sputtering from the facility's walls during long duration tests, a bi-conical beam target lined with high purity graphite, has been designed and accommodated in front of the thruster inside the opening cap of the facility.

SITAEL's IV4 facility (open during servicing).

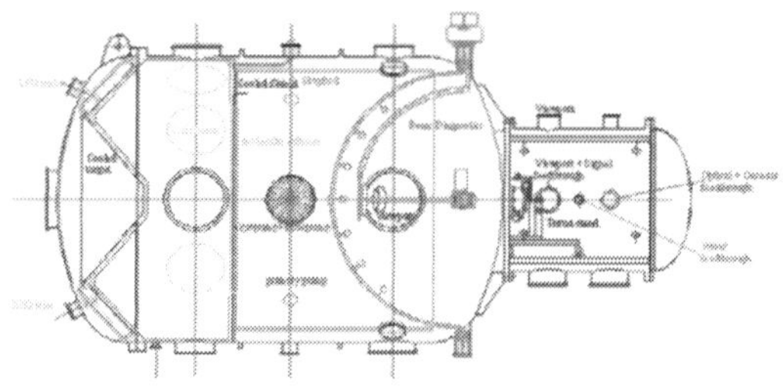

IV4 facility schematics.

5. MEASUREMENT RESULTS

The measured I_{BSC} $(V_{CRP} - V_{GND})$ Volt – Ampere characteristic is illustrated in the following graph:

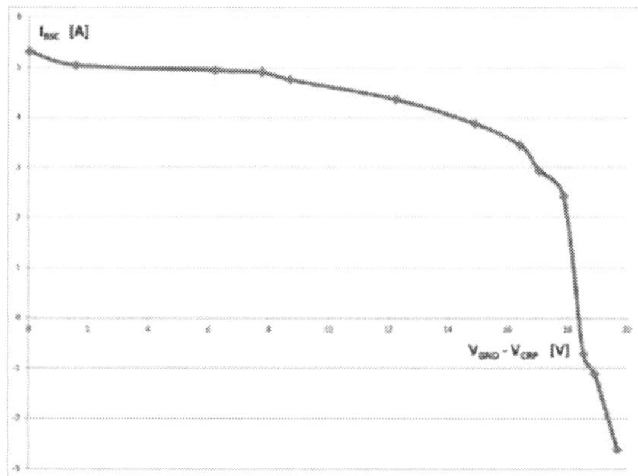

I_{BSC} $(V_{CRP} - V_{GND})$

about which we can make the following observations:

- I_{BSC} changes sign at about the same V_{CRP} value which the cathode would reach at equilibrium with the conventional ***Floating CRP*** approach.
- Slowly varying the voltage of the two quadrants supply (i.e. slowly varying V_{GND} - V_{CRP}) around said equilibrium voltage results in a large variation of I_{BSC}.

We are not aware of any such I_{BSC} $(V_{GND} - V_{CRP})$ measurements performed on flying spacecrafts.

6. CRP BIAS POWER SUPPLY

At first sight it may appear that a Direct Drive approach would allow a considerable simplification of the Power Processing Unit, thanks to the removal of the Anode Voltage Supply. However, if a direct connection to GND of the Cathode Potential may result in a large I_{BSC} current also when in an actual orbital environment, then galvanic isolation of the anode voltage would anyway appear to become a practical necessity. Thus falling back again to the need for a massive and costly Anode Voltage Supply. Although a laboratory firing chamber environment might significantly differ from the actual environment experienced in orbit, we nevertheless believe that flying, on a large GEO Telecom platform, a Direct Drive HET propulsion system with its CRP directly connected to GND might expose the spacecraft to the risk of potential damages resulting from a too large I_{BSC} current.

Of course, the very same test setup illustrated above suggests an interesting alternative: insert a series power supply between the CRP and GND (see simplified schematics at the bottom of this page). Said "CRP Bias Supply" would then typically need to handle solely about 1/20th to 1/10th the power of the otherwise required Anode Supply, considerably reducing cost, mass, volume, and burden on the spacecraft thermal control system.

The negative output voltage of said CRP Bias Supply could then be controlled in order to achieve the condition $I_{BSC} = 0$. Such a feedback loop would need to measure I_{BSC} as the difference between the positive anode current and the cathode return current. A differential current measurement of this kind can optimally be obtained by means of a Hall Effect Current Sensor coupled to both currents by means of two sensing windings in oppositions.

Because the adjustments to the output voltage of said CRP Bias Supply would need to follow only the very slow variations of I_{BSC} (as determined by the slowly varying orbital environment), the actual controller for said feedback loop could then be implemented by the FPGA hosting also the overall PPU Sequencer and Controller.

7. CONCLUSIONS

A Direct Drive approach to the supply of the anode voltage for a Hall Effect Thruster would reap major benefits (mass, cost, dissipation) if it were possible to directly connect the Cathode Potential to the spacecraft's Ground. However, this may result in damaging I_{BSC} values. It has been shown how an interesting solution might consist in introducing, between CRP and GND, a CRP Bias Supply controlled by means of a feedback loop aimed at minimizing I_{BSC}.

Said CRP Bias Supply would then typically need to process only about 1/20th to 1/10th of the actual HET anode power, and thus resulting in significant savings.

8. REFERENCES

1. Gollor, M. (2011). "Accommodation Aspects of Electric Propulsion with Power Processing Units and the Spacecraft" 9th Annual International Energy Conversion Engineering Conference, AIAA 2011-5651.

2. Koppel, C. R., Marchandise, F., Estublier, D. & Jolivet, L. (2004). The SMART-1 Electric Propulsion Subsystem In Flight Experience. AIAA-2004-3435.

3. Ira Katz et al. "Effect of Solar Array Plume Interactions on Hall Thruster Common Potentials". Spacecraft Charging Technology Conference 04 - 08 April 2016, ESA/ESTEC.

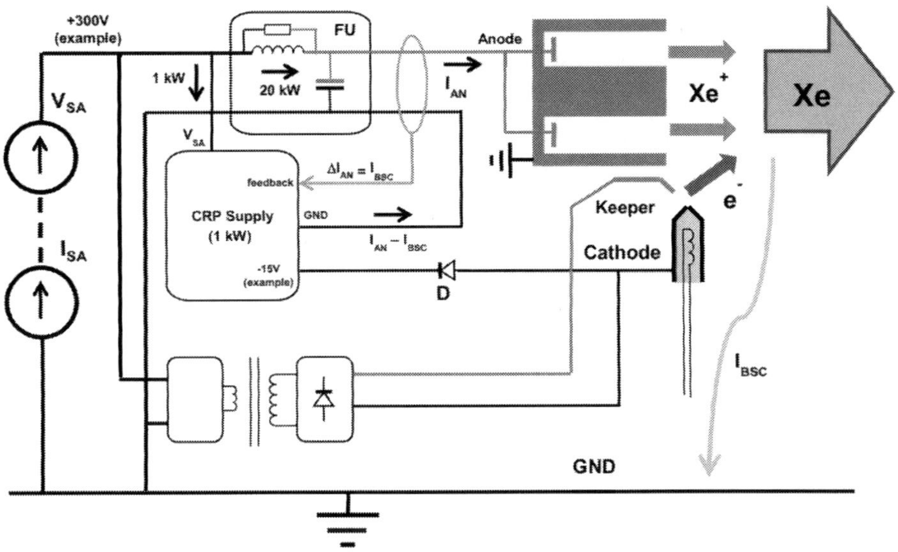

DOI: 10.1051/e3sconf/20171615003

DEVELOPMENT OF A LOW COST PPU FOR FEEP ELECTRIC PROPULSION USING COTS COMPONENTS

Bernhard Seifert, Alexander Reissner, David Jelem, Thomas Hörbe

FOTEC Forschungs- und Technologietransfer GmbH, seifert@fotec.at

ABSTRACT

In the frame of the development of an electric-propulsion thruster for Micro- and Nanosatellites, FOTEC has put forward the in-house development of a low cost PPU, capable of driving the thruster with up to 10 kV emitter and -10 kV extractor voltage. After 5 generations of development, the PPU currently used in a thruster testing and qualification campaign has achieved 85% total efficiency including high voltage multiplier stages and protection circuitry. The PPU includes the functionality of the step-up conversion for emitter and extractor, propellant heating and neutralizer-operation within a volume of 90 x 94 x 78 mm, at a weight of less than 230 g and a total cost in production of less than € 1,000. This low cost approach using COTS components is targeting constellations and mega-constellations and is certainly the other extreme compared to the conventional design approach for spacecraft PPUs. In the light of the ongoing discussions of a change in the philosophy of designing spacecraft, going from fail-safe components to a "mass production" strategy which aims for fail-safety by redundancy. This paper also gives a starting point for the discussion on how a possible trade-off between cost and reliability can be achieved.

1. Introduction

For several years, FOTEC has developed the mN-FEEP thruster to be used in future ESA missions [1]. This thruster can provide highly accurate thrust ranging from 1 µN to above 1 mN [2]. It allows large satellites to control their position with an unprecedented accuracy enabling e.g. formation flight of scientific spacecraft [3]. In order to emit ions at an I_{sp} of up to 5,000 s at a dynamic thrust range from 1 µN to 1 mN, a so called crown emitter is being used, that has been developed by FOTEC for the ESA NGGM mission [4, 5, 6, 7]. This emitter was initially developed to enable fixed formation flight missions and therefore combines a large thrust range with outstanding accuracy (see Figure 1).

As the need of a propulsion system for micro- and nano-satellites became an urgent matter, it has been realized, that the thrust level provided by this mN-FEEP thruster already allows significant increase of the mission range of such satellites. The high I_{sp} on the other hand allows for very high Δv maneuvers at high propellant utilization efficiency.

Keeping the emitter in the configuration for the NGGM mission, the housing of the thruster has entirely been re-designed in order to fit into a standard CubeSat structure [8]. In parallel, a PPU has been developed and been integrated into the thruster module (see Figure 2). The result of this redesign is the new IFM nano thruster module with the particular advantage that no miniaturization of the emitter itself was necessary. The thruster module, therefore, can build on an extensive development history that has started long before the present application has emerged. The result is a highly performant compact thruster module, that includes all subsystem from the thruster and PPU up to the propellant and neutralizer in less than 1 U at less than 1 kg.

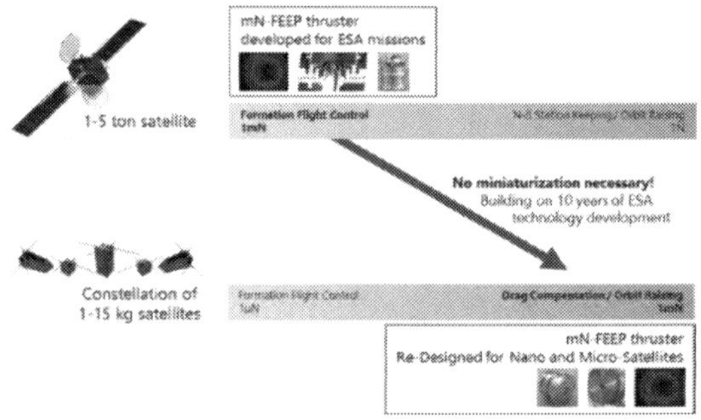

***Figure 1** Evolution of the IFM nano thruster from the mN-FEEP thruster developed for ESA.*

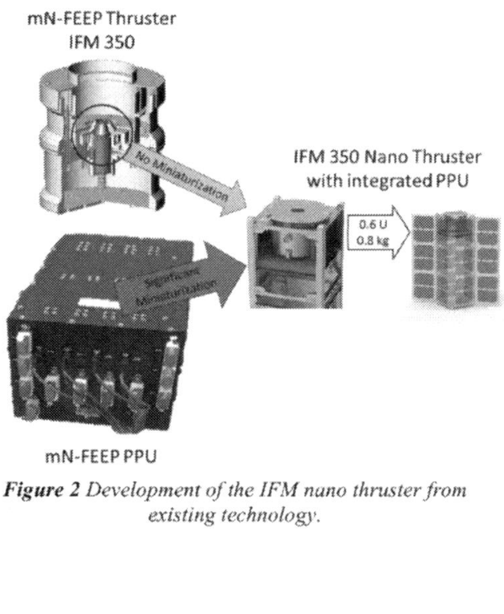

***Figure 2** Development of the IFM nano thruster from existing technology.*

2. The Indium FEEP Technology – State of the art and achievements from previous developments

Field emission is an effect which is closely tied to the presence of strong electric fields. In practice, this means that the fundamental structure on which field emission takes place is shaped like a needle, due to the field-enhancing effect at its tip. An important application of this effect is the so-called *Liquid Metal Ion Source* (LMIS), because it uses the process of field emission to ionize a thin film of liquid metal covering the needle which is biased towards several kV with respect to the counter electrode (extractor). The thusly created ions are then accelerated by the strong electric fields and can be used for ion implantation in semiconductor industry or for micromachining with a focused ion beam (FIB).

The *Department of Aerospace Engineering* at FOTEC (the former department *Space Propulsion and Advanced Concepts* at the Austrian Institute of Technology, AIT) has a long and successful heritage in manufacturing LMIS for space. They have been used for mass spectrometer (for example on ROSETTA) or for spacecraft potential control (for example on the NASA mission MMS). Up to date, FOTEC is the only supplier worldwide for LMIS with flight heritage (see Table 1).

Experiment	Function	Spacecraft / No. of LMIS	Operation Time
LOGION	Test of LMIS in µ-Gravity	MIR / 1	24 h (1991)
MIGMAS/A	Mass Spectrometer	MIR / 1	120 h (1991-94)
EFE-IE	S/C Potential Control	GEOTAIL / 8	600 h (1992 -)
PCD	S/C Potential Control	EQUATOR-S / 8	250 h (1998)
ASPOC	S/C Potential Control	CLUSTER / 32	Ariane 5 launch failure 1996 Still operational after crash
ASPOC-II	S/C Potential Control	CLUSTER-II / 32	6516 (since 2000)
COSIMA	Mass Spectrometer	ROSETTA / 2	Fully Operational, Experiment Ongoing
ASPOC/DSP	S/C Potential Control	DoubleStar / 4	8979 h (2004 - 2007)
MMS ASPOC	S/C Potential Control	MMS / 32	Commissioned successfully in 2015

Table 1 *Flight heritage of FOTEC LMIS.*

This principle of generating positive ions and accelerating them by the very same field can also be used to generate thrust. When a liquid metal ion source is used in this way, it is termed *Field Emission Electric Propulsion* (FEEP).

Since the number of ions and their energy can conveniently be controlled by the applied voltage levels, outstanding thrust accuracy can be achieved. The main advantage of using FEEP thrusters lies in their capability to produce thrust from the sub-µN level up to several tens of µN per needle.

For more than 15 years, research has been carried out to use this technology for generation of ultra-precise thrust in the µN-range suitable for formation flights of multiple spacecraft. The FEEP technology has been put forward in ESA's *Basic Technology Research Program* and in the frame of early preparatory studies for the *Next Generation Gravity Mission* (NGGM).

In this environment, the so-called porous Tungsten crown emitter has been developed, which employs 28 needles for field emission. Apart from the multiple emission sites, the most important new feature was the porous Tungsten matrix which enables internal flow of the liquid metal (similar to a capillary type emitter) to a very sharp tip (similar to a solid needle). It therefore combines the advantages of both, the capillary emitter and the solid needle.

Another key feature of the porous Tungsten crown emitters is the rapid manufacturing method from using micro powder injection molding. In this process, a mixture of Tungsten powder and binding polymer is injected into the mold and subsequently sintered to achieve final hardness and purity. In the course of this procedure, the structure becomes porous and facilitates flow of propellant through its volume.

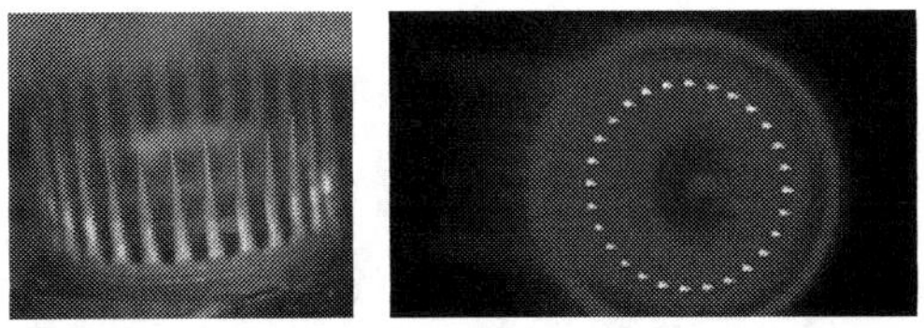

Figure 3 *Porous tungsten crown emitter after wetting with Indium (left), high-thrust operation (right).*

Depending on the voltage applied to the crown emitter, a certain number of needles is actually emitting ions. As indicated in the picture above at high thrust levels, almost all needles are contributing to the total current whereas at low thrust levels only a couple of needles, whose electrical impedance is lowest, is emitting ions. This behavior allows operation at a thrust level ranging from few µN up to several mN.

In Figure 4 the typical performance of a 28-needle porous crown emitter, operated at 3 mA ion current, is shown. One recognizes the ability to shift the extractor voltage which allows a trade-off between high specific impulse (low extractor voltage) and high thrust-to-power operation (high extractor voltage). The I-V characteristic of this emitter is shown in Figure 5.

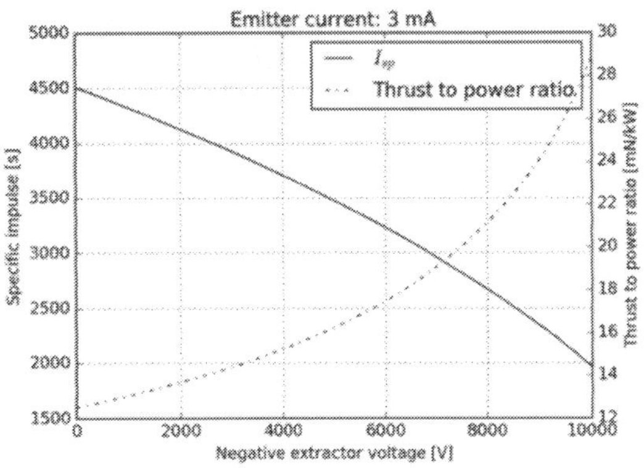

Figure 4 *Typical performance of a 28-needle porous crown emitter.*

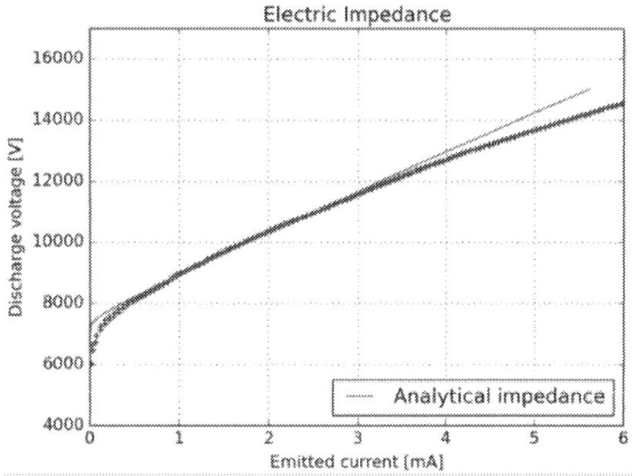

Figure 5 *Measured I-V characteristic of porous Tungsten crown emitter with theoretical model [9].*

Extensive testing of these emitters includes voltage/current characteristic measurements, direct thrust measurements, characterization of the start-up behavior, determination of the mass efficiency and a lifetime test.

At the time of finalization of this paper the ongoing lifetime test has demonstrated more than 12,000 h of operation without detectable performance degradation, targeting 20.000 h and more [10].

3. Thruster Design Concept

The following thruster design has been established in order to accommodate the present emitter technology on a micro- or nano-satellite (see Figure 6).

Figure 6 *Simplified design of the IFM nano thruster module. It consists of the top plate with crown emitter and extractor ring, the housing, the high voltage PPU (blue) and the low voltage PPU (bottom).*

Each module includes the emitter itself, the propellant with heater, the power processing unit, the high voltage multipliers and a two redundant neutralizers. The thruster uses either pure Indium or an alloy of Indium and Gallium as propellant which are both solid at room temperature. When the thruster shall be operated in orbit, the propellant is heated above its melting point and becomes liquid. Avoiding any liquid and reactive propellants as well as pressurized tanks during integration and launch significantly simplifies corresponding procedures and regulations.

Electrical interfaces between the IFM nano thruster module and the spacecraft are reduced to a low voltage power bus (12 V) and a digital communication interface. Auxiliary voltages are generated internally from the supply voltage.

Table 2 shows the key parameters of the IFM nano thruster module.

Parameter	Value
Dynamic thrust range	1 μN to 1 mN
Nominal thrust	350 μN
Specific impulse	2,000 to 4,500 s
Propellant mass	30 to 250 g
Total impulse	more than 5,000 Ns
Power at nominal thrust	31 W incl. neutralizer
Outside dimensions	94 x 90 x 78 mm
Mass (dry / wet)	640 / 870 g

Table 2 *Key performance figures of the IFM nano thruster module.*

4. System Overview

For proper operation of the FEEP porous crown emitter in the IFM nano thruster module, the integrated PPU has to provide several controllable outputs:

- Positive DC high voltage generation for the FEEP emitter: 0 to 10 kV, 0 to 3 mA
- Negative DC high voltage generation for the extractor electrode: 0 to -10 kV, 0 to -300 μA
- Positive DC low voltage generation for the propellant heater: 0 to 20 W
- AC low voltage generation for each of the two neutralizers: 0 to 3 W
- Negative high voltage biasing of both neutralizers: fixed to -200 V

The PPU is operated at the nominal voltage of 12 V which has to be provided by the spacecraft power sourcing unit (PSU). The given voltage and current figures are maximum values which are only reached at maximum thrust levels. For lower thrust requirements or at lower specific impulse levels, these figures can be significantly lower.

The conversion efficiency is crucial for long-term operation since no active cooling means like active heat pipes or levers can be implemented in the small size of the module. A prototype PPU has been developed with a maximum total deliverable power of 40 W, at the final average efficiency of 85%, the total heat losses are therefore below 6 W. During the development of the PPU, the focus is laid on efficiency optimizations, reliability and controllability.

For proper operation, the PPU continuously monitors several temperatures across the module, voltages and currents. Therefore, remote diagnostics enable the detection of any malfunctioning during operation.

5. High Voltage Generation

In order to keep the output-to-input ratio of both high voltage sections for the emitter and extractor as low as possible, a highly efficient pulse width modulation (PWM) controlled, fixed frequency, full-bridge topology was chosen. The bridge voltage corresponds to the nominal supply voltage and the optimum switching frequency was designed to be 60 kHz. Commercially available cold cathode fluorescent lamp (CCFL) small-size transformers are being utilized to generate output voltages of up to 1200 V (ratio of 1:100).

A Villard type diode-capacitor multiplier ladder circuit is connected to generate the desired output voltage of up to 10 kV for the emitter and up to -10 kV for the extractor respectively. The multiplier circuitry is located on a separate board (high voltage PPU) to avoid any spark-overs or creeping currents which might occur due to miniaturization of the design. In addition, the high voltage PPU is potted with STYCAST FT-2850 low outgassing epoxy resin for additional protection against coronal discharges. In Figure 7 the most current design of the low voltage PPU is shown.

Figure 7 *Prototype of low voltage PPU suitable for standard CubeSat bus.*

Both, the emitter and the extractor can be individually operated in current-, voltage- or power-control mode. Since the power of the emitter section is up to 25 W, four CCFL transformers are used in parallel to share the load. For the extractor circuitry a single transformer is sufficient.

For nominal operation the extractor is voltage controlled and set to a fixed voltage. The emitter is current controlled to control the required thrust. The theoretical thrust is internally estimated and a feed-back loop controls the power supplies to maintain the commanded thrust. In case of erroneous voltage or current feedback measurements, both full bridges can be duty-cycle controlled.

6. Heater and Temperature Control

The heater power section, used to melt the propellant, uses a half-bridge topology with attached LC-filtering to maintain highest efficiency at low AC ripple. The heater itself is comprises of high-temperature resistance wire encapsulated in Aluminum oxide ceramic. During hot-standby the heater has to keep the propellant liquid, which allows instantaneous thruster operation. For Indium propellant the reservoir temperature is being controlled to 180 °C and for Indium-Gallium alloy 100 °C is sufficient. If the heater is disabled, the propellant reservoir cools down to ambient temperature and it takes up to 60 minutes to heat up the reservoir again before the thruster itself can be re-enabled.

Since the emitter and the attached propellant reservoir is kept at up to 10 kV during operation, accurate temperature sensing is challenging. The PPU therefore uses a contact-less so called thermopile which uses the pyroelectric effect in conjunction with absorption of heat radiation from the reservoir. Alternatively, electrically insulated temperature sensors, such as thermocouples, can be used. However, tests have shown significant measurement errors since the insulator is acting as a thermal barrier as well.

In Figure 8 the temperature profile during heat-up is shown. The melting point of pure Indium at 156.6 °C can be observed which is used for additional verification of the thermal calibration.

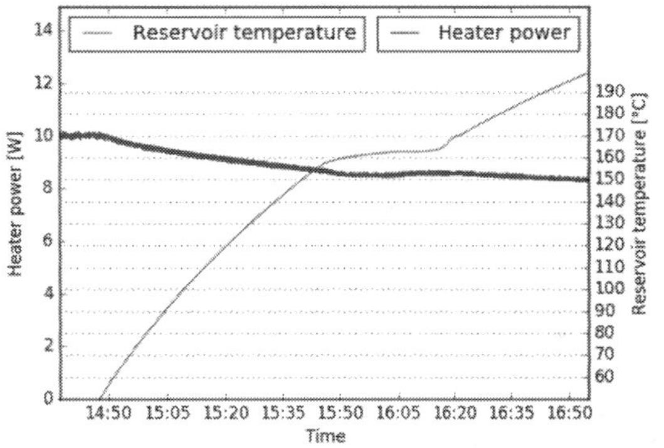

Figure 8 *Temperature profile of the reservoir during controlled heat-up.*

7. Neutralizer and Spacecraft Charging

Since ions are expelled at a total current of up to 3 mA, the FEEP thruster module needs means to balance spacecraft charging. This is achieved by the use of two cold-redundant electron sources acting as neutralizers. Such an electron source consists of a Tantalum disc which is heated up to 2,200 K and biased to -200 V. The electrons are expelled when the spacecraft is negatively charged whereas the active biasing is required to overcome the negative potential generated by the extractor ring electrode. Once electrons have left the neutralizer, they will be pulled towards the positive potential of the ion plume. (see Figure 9). The PPU is able to measure this charge balancing current and it can automatically control the heating power of the neutralizers.

The neutralizer discs are connected to the PWM output of a half bridge via an inductor and decoupling / pumping capacitors. This allows biasing the neutralizer discs and allows measurement of the charge balancing current.

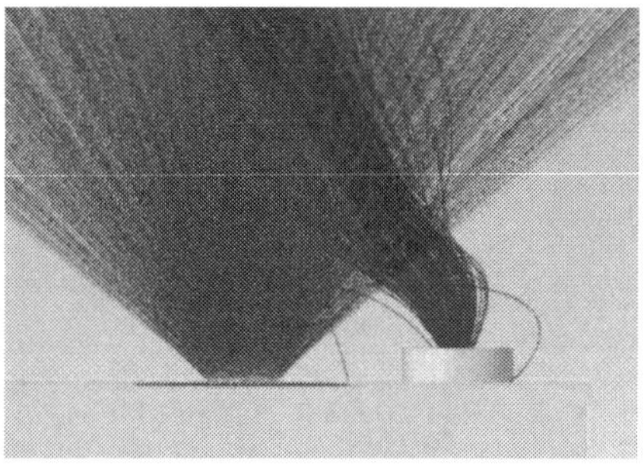

Figure 9 *Electrons (blue) being pulled into the emitted ion plume (red).*

8. Communication and Software

The IFM nano thruster module is based on an 8-bit Atmel AVR8 type microcontroller and uses a 3.3 V digital communication interface (UART, I²C or RS-422). It provides a memory map being read and written by the on-board computer (OBC) to read the status or control the thruster respectively. Each message uses an 8-bit cyclic redundancy check (CRC8) checksum which protects against bit flips on the communication line.

Firmware updates are supported which allows in-orbit bug-fixes or adaptation of the software, if required. Since the microcontroller provides in-system programming capability, no additional interface to the OBC is required.

Crucial non-volatile values, such as the calibration factor of the thermopile, are stored in electrically erasable programmable read-only memory (EEPROM) section of the microcontroller multiple times and are auto-corrected during read-out with voters.

Software-configurable fuses are implemented to detect over-voltage, over-current or over-temperature conditions. In case of sensor malfunction or analog-digital converter (ADC) failure, they can individually be disabled during operation.

9. System Efficiency

Any power losses of the PPU need to dissipated by the spacecraft structure in order to prevent overheating of the PCBs and populated components. Since the PCB mostly consists of FR4 glass-fiber enhanced epoxy resin, the thermal conductivity is rather low compared to Aluminum or Copper. Preliminary tests have revealed that the heat gradient through the PCB from the transformers to the interface connections at the corners is too large to keep the PPU continuously below 85 °C at full-power operation (see Figure 10).

Thermal management is crucial to ensure operation under varying thermal environments and interface temperatures. Therefore, a solid Aluminum heat spreader was developed to guide the heat flux from the transformers (which are the major source of power losses) through cut-outs in the PCB directly to the rear side of the IFM nano thruster module and further to the interface of the spacecraft. Threads allow tight connection and thus good thermal contact to the spacecraft bus (see Figure 10). Additionally, in order to increase the effective contact surface area, thin Indium foil washers were used at all connection points along the heat path to prevent temperature drops.

Figure 10 *Solid Aluminum heat spreader at the rear side of the low voltage PPU.*

First FEM simulations and experimental tests have shown significant improvements and confirm the design. The PPU could continuously be operated at nominal output power at an interface temperature of 40 °C. The following thermal image was taken from a setup where only the two right transformers were in thermal contact with the heat spreader (see Figure 11, white circle). One can recognize the significantly lower temperature of the ferrite cores compared to the two left transformers.

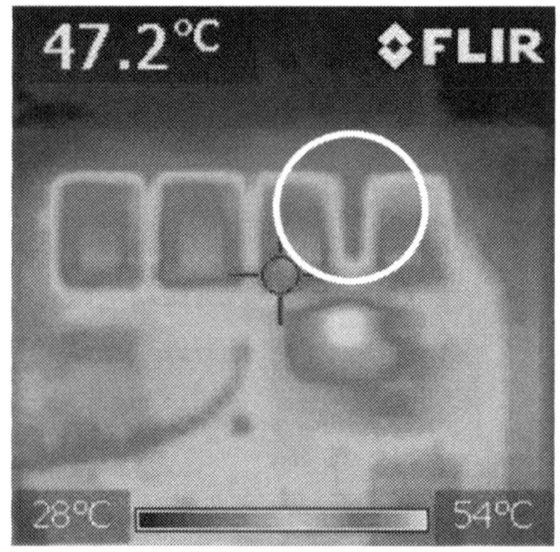

Figure 11 *Temperature distribution of the low voltage PPU at high power operation in air.*

The PPU was tested when fully integrated into an engineering model and subjected to typical electrical load

DOI: 10.1051/e3sconf/20171615003

conditions gathered from known I-V FEEP emitter characteristic. In the following test, the temperature of reservoir was being held constant at 170 °C and the emitter was controlled from 0 to 20 W output power. A 4 MΩ high voltage power resistor was being used to load its output (see Figure 12).

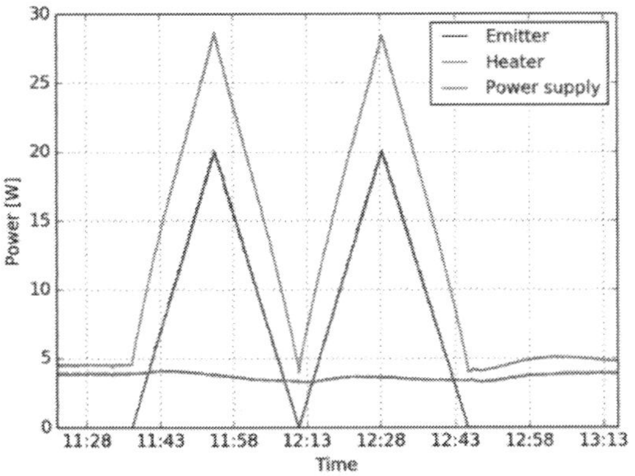

Figure 12 *High output power test with temperature controlled heater (green), emitter power (blue) and total input power (red).*

At emitter output power of 20 W and heater output power of 4 W the total input power was measured to be 28 W which corresponds to ca. 85% overall system efficiency. One recognizes that less heater power is required when the emitter is operated at high power.

As the PPU is optimized for nominal output power and therefore, the relative losses increase at lower emitter power levels. The load-independent constant power losses were found to be 0.18 W and the output power dependent losses increase almost linearly with the output power (see Figure 13).

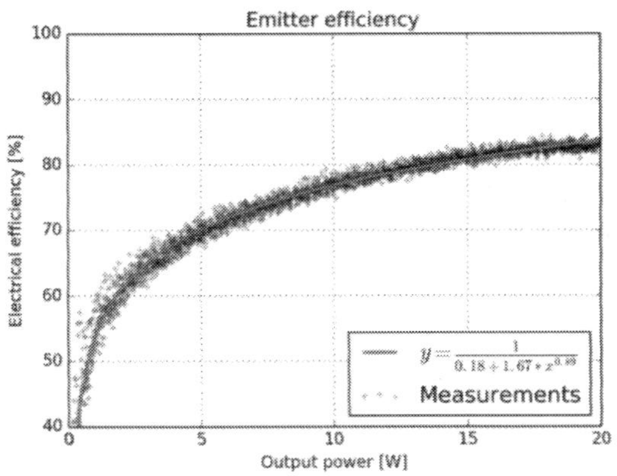

Figure 13 *Electrical emitter efficiency vs. emitter output power. The constant losses reduce the relative efficiency at low emitter power levels.*

The heater efficiency shows similar behavior and was found to be 87% for a typical output power of 4 W which is required to keep the reservoir at 175 °C (see Figure 14).

At higher output power levels, the efficiency reaches more than 97%.

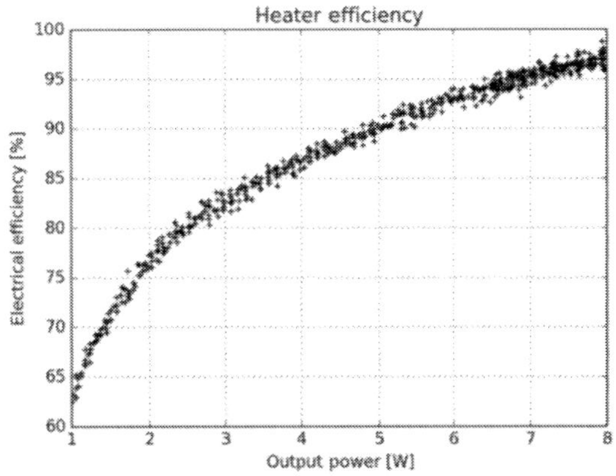

Figure 14 *Electrical heater efficiency vs. heater output power. The constant losses reduce the relative efficiency at low heater power levels.*

For system engineers, being mainly interested in the generated thrust for a given input power, the following chart is important. As already summarized in Table 2 the nominal thrust level of 300 µN can be achieved at an input power of 31 W. This figure includes all electrical efficiency losses, the required heater power to maintain the reservoir temperature and the power to operate the neutralizer to compensate for spacecraft charging (see Figure 15).

The constant power use of the PPU is 7 W even when no thrust is generated. This results from the heater power (4 W) and the neutralizer power (3 W).

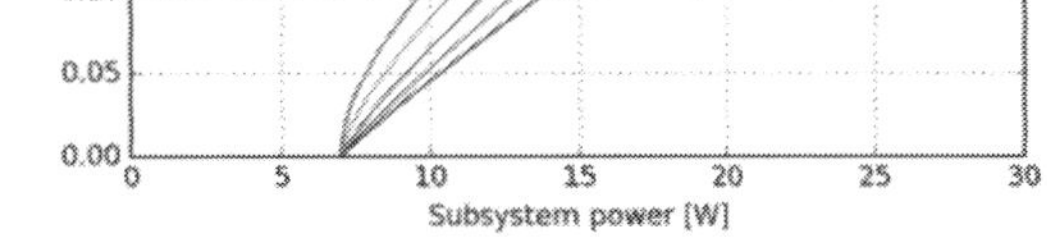

Figure 15 *System power requirements for different thrust levels at various extractor voltages.*

10. In Orbit Demonstration

FOTEC is currently building flight modules of the IFM Nano thruster to be launched in Q3 2017 on a commercial platform. The spacecraft will accommodate the IFM Nano thruster module in a seven module configuration. A common housing as seen in Figure 16 will be placed inside a 15-inch separation ring being used to detach the spacecraft from the launcher. Multiple units of the IFM nano thruster modules can be combined and be operated in parallel to provide higher thrust levels without degradation of system efficiency. By the use of seven modules an overall thrust of more than 2 mN can be achieved at an input power below 200 W. This takes into account the reduced heater power and the reduced number of neutralizers to perform charge balancing.

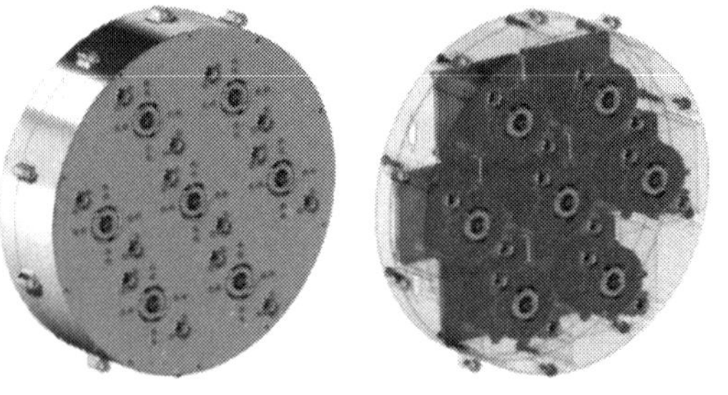

Figure 16 *Propulsion module being placed inside the separation ring. It consists of seven independent IFM Nano thruster modules.*

11. Conclusions

The IFM nano thruster module offers a unique continuously scalable I_{sp} range combined with high electrical efficiency, compact dimensions and ultra-high propellant density. The herein presented PPU solely consists of COTS components and is capable of powering and controlling all necessary subsystems and supports a large portion of the operational envelope a FEEP emitter can offer with regard to I_{sp} and thrust.

It was shown that the PPU sustained a test campaign under realistic load conditions such as thermal vacuum or vibrational tests. The PPU showed outstanding electric efficiency compared to competitors' high voltage generators. Ongoing developments of the PPU will extend the accessible thrust and I_{sp} levels even further which, in turn, increases the range of missions profiles for nano- and micro-satellites.

As the trend towards electric propulsion is strongly emerging, FOTEC is well positioned as the only supplier of integrated FEEP propulsion systems.

12. References

[1] I. Vasiljevich, M. Tajmar, W. Grienauer, F. Plesescu, N. Buldrini, J. Gonzalez del Amo, B. Carnicero Domunguez, and M. Betto, *Development of an Indium mN-FEEP Thruster*, 2008

[2] A. Reissner, N. Buldrini, B. Seifert, F. Plesescu, C. Scharlemann, and J. G. del Amo, *mN-FEEP Thruster Module Design and Preliminary Performance Testing*, presented at the 33rd International Electric Propulsion Conference, Washington, D.C., USA, 2013

[3] A. Reissner, N. Buldrini, B. Seifert, and F. Plesescu, *Detailed Performance Characterization of the mN-FEEP Thruster*, presented at the Space Propulsion 2014, Cologne, Germany, 2014

[4] I. Vasiljevich, N. Buldrini, F. Plesescu, M. Tajmar, M. Betto, and J. Gonzalez del Amo, *Porous tungsten crown multi-emitter testing programme using three different grain sizes and sintering procedures*, in Proceeding of the International Electric Propulsion Conference, Wiesbaden, Germany, 2011

[5] I. Vasiljevich, N. Buldrini, F. Plesescu, M. Tajmar, M. Betto, and J. del Amo, *Consolidation of milli-Newton FEEP Thruster Technology based on Porous Tungsten Multiemitters*, 2011

[6] M. Tajmar, I. Vasiljevich, and W. Grienauer, *High current liquid metal ion source using porous tungsten multiemitters*, Ultramicroscopy, vol. 111, no. 1, pp. 1–4, Dec. 2010

[7] M. Tajmar, I. Vasiljevich, F. Plesescu, W. Grienauer, N. Buldrini, M. Betto, and J. G. del Amo, *Development of a porous tungsten mN FEEP thruster*, in Proceedings of the Space Propulsion Conference, San Sebastian, 2010.

[8] California Polytechnic State University, *CubeSat Design Specification*, Rev. 12

[9] G. L. R. Mair, *Theoretical determination of current-voltage curves for liquid metal ion sources*, J. Phys. D: Appl. Phys., 1984

[10] A. Reissner, N. Buldrini, B. Seifert, F. Plesescu, and T. Hörbe, *Lifetime Testing of the mN-FEEP Thruster*, presented at the Space Propulsion 2016, Rome, Italy, 2016

E3S Web of Conferences **16**, 15004 (2017) DOI: 10.1051/e3sconf/20171615004
ESPC 2016

POWER PROCESSING UNIT FOR MICRO SATELLITE ELECTRIC PROPULSION SYSTEM

Spiridon Savvas, Dimitrios Malamas, Alexandros Manoudis, Pavlos Ramnalis [(1)], Luca Benetti, Silvia d'Argliano, Luca Onida, Davide Caniglia, Lorenzo Fontani, Tommaso Misuri [(2)]

[(1)] *Sitael Hellas S.LTD, Dionisiou Solomou 4, Veria, 59132 Imathia, Greece, Email: spiridon.savvas@sitael-hellas.com*
[(2)] *Sitael, Via Livornese, 1019 S. Piero a Grado 56122 Pisa, Italy, Email:luca.benetti@sitael.com*

ABSTRACT

The Micro Satellite Electric Propulsion System (MEPS) program has been originated by the increasing need to provide a low-cost and low-power Electric Propulsion System (EPS) for small satellites (<300Kg) and finds applications in orbit insertion after launcher separation, orbit maintenance, orbit transfers and deorbiting.
The MEPS design was conceived as a "plug and play" subsystem that can be easily installed in a number of small spacecraft platforms, consisting of two Thruster Units (TU), Power Processing Unit (PPU), Xenon Propellant Tank Assembly (PTA) and Propellant Management Assembly (PMA). Being the "heart" of the system, the PPU supplies and monitors all the EPS voltages and currents.
The objective of this paper is to present the design of the PPU whose concept is focused on high efficiency (> 92%), small size and weight and high reliability. Its functional modules and preliminary results obtained at breadboard level are also presented.

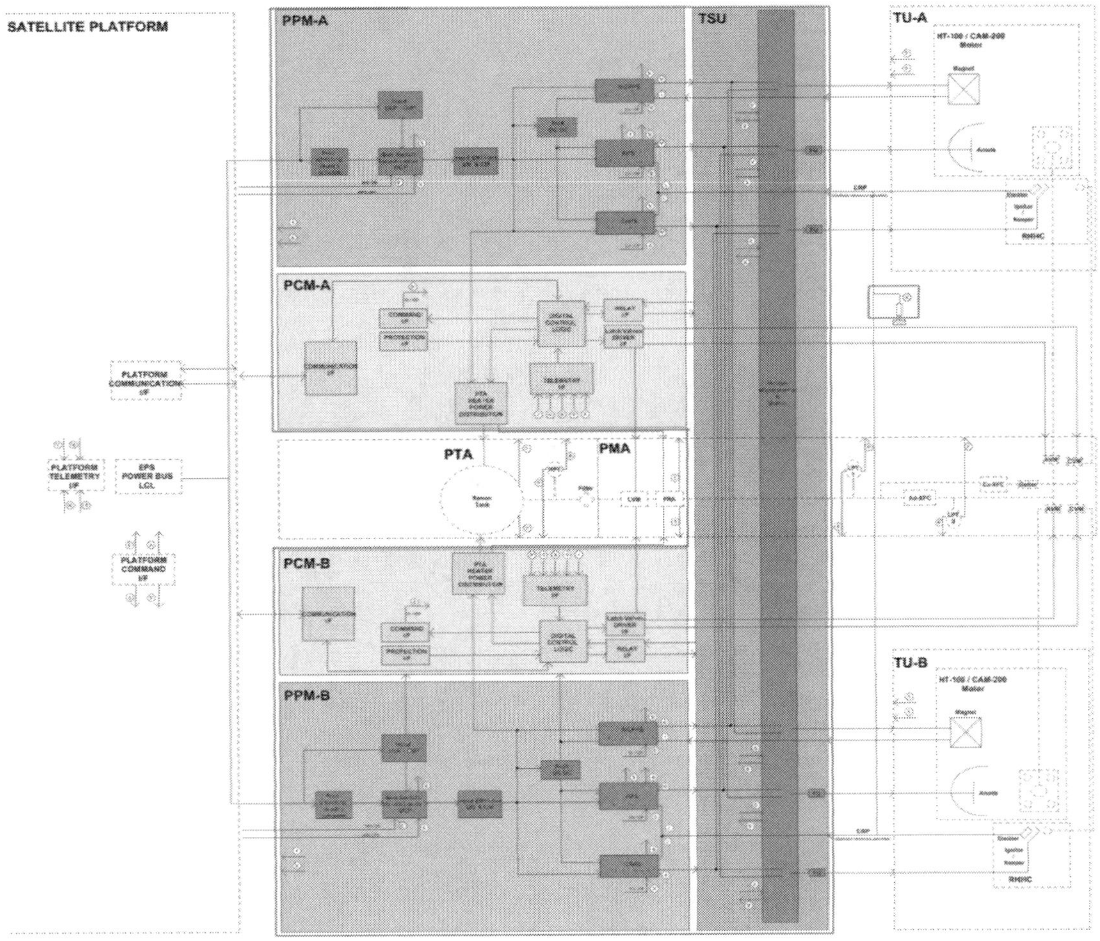

Figure 1. PPU block diagram w.r.t. other EPS elements and Platform.

DOI: 10.1051/e3sconf/20171615004

1. INTRODUCTION

The MEPS program aims at the qualification of an EPS based on low power Hall Effect Thruster (HET).
The EPS consists of two TU each composed of a Thruster/Cathode Assembly, a single PPU and a single Xenon PTA/PMA.
The PPU, whose functional blocks are shown in Fig. 1, is the central part of the EPS. It is supplied by the satellite's Power Distribution Unit (PDU) and provides power conditioning and control for two types of HET (Sitael's HT100 and Rafael's CAM-200), for Rafael's Heater-less Hollow Cathode (RHHC) and for Rafael's PMA/PTA. The PPU also manages bidirectional communication with the satellite on board computer (OBC).
The PPU has the following main functions:

- power conditioning and monitoring of all voltages and currents required by the EPS in all of its operating modes
- to enable the control of the EPS in order to achieve the best performance according to mission requirements
- to manage bidirectional communication with satellite computer including TM and TC in order to command EPS functions and monitor all EPS relevant parameters
- to ensure isolation between the high voltage electronics that supply the thrusters and the low voltage electronics supplied by the input power bus
- to protect the main bus from the TU EMI

2. POWER PROCESSING UNIT

2.1 General Description

The PPU enables the single operation of any one of the thrusters in their complete thrust range, ensures full performance for an unregulated power bus voltage from 22V to 34V and is composed of the following functional blocks:

- PPU Power Module A and B (PPM-A and PPM-B, for redundancy) which implement the power electronic circuits necessary to supply any of the two Thruster Units (TU-A and TU-B)

- Thruster Switching Unit (TSU) which contains the electromechanical circuits enabling the selection of any one of the TUs

- Filter Unit (FU) which has the purpose to filter the high frequency current noise typically associated with the TUs

- PPU Control Module A and B (PCM-A and PCM-B, for redundancy), which provide the communication with the satellite and the control of the PPU and EPS valves.

In order to test the PPU functionalities and to permit the coupling with the TUs, a PPU BB (see Fig. 2) has been assembled and tested.
The breadboard does not implement the redundancy of the final unit; consequently, the TSU is not implemented on the breadboard as well.
In particular the breadboard implements the modules necessary to drive each type of TU:

- Anode Power Supply suitable for both the CAM-200 and HT100
- Cathode Power Supply for the RHHC
- Magnetic Field Power Supply only for the CAM-200 (HT100 has permanent magnets)

Figure 2. PPU Breadboard detail.

The functions of the PCM necessary to drive the two TUs are implemented using the Labview GUI which is displayed in Fig. 3. The GUI performs the start-up sequence necessary to ignite the cathode and to establish the anode discharge current in automatic mode and is flexible enough to permit the use of each converter independently from the others in manual mode.

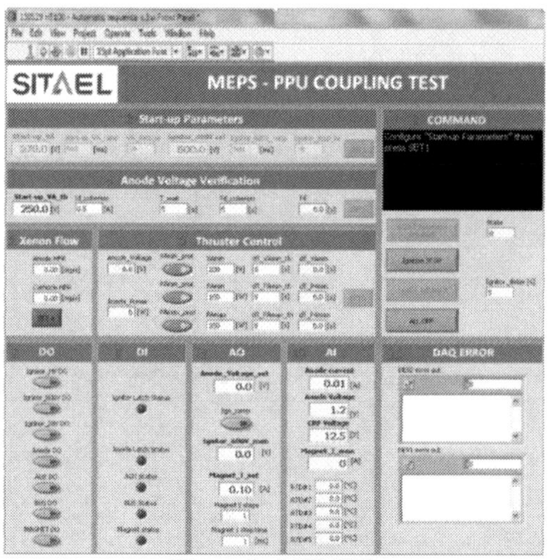

Figure 3. PPU GUI.

DOI: 10.1051/e3sconf/20171615004

2.2 PPU Power Module (PPM)

Each PPM is composed of:

Anode Power Supply (APS)

The APS supports the plasma discharge current and sets the anode voltage during operation. It is an isolated DC/DC power supply whose output is connected between the thruster anode and the Cathode Reference Potential (CRP). It is supplied by the primary input power Bus downstream the EMI input filter and the inrush current control circuit. It provides:

- Power up to 250W to drive the thruster anode
- Acquisition of anode temperature, driving voltage and discharge current
- Regulation of anode voltage according to the thrust required by the satellite OBC. The maximum regulation voltage is 300V
- Galvanic isolation between input power bus and the thrusters. The isolation voltage is 2500Vdc

The topology selected for the APS DC/DC converter is derived from the Boost-Integrated Half Bridge (BHB) topology which is particularly suitable to provide isolation between a low input voltage source and a high output voltage load thanks to its boosted link voltage. Moreover the continuous input current of the BHB has the advantage to allow the realization of a simple average input current mode control loop and to reduce the EMI. The PPU implementation of the BHB converter presents a modified output stage with respect to the original BHB topology: the output inductor is absent and a voltage doubler replaces the full-wave rectifier. Therefore a magnetic component is avoided, yielding the reduction of weight and power losses, while the transformer realization is simpler because the voltage doubler does not need the central tap of the transformer secondary windings. The simplifications of the magnetics are particularly important as they are a critical part of the design, especially when dealing with high voltages and insulations. Finally, the modified BHB topology allows both Zero Voltage Switching transitions of the primary mosfets and Zero Current Switching transitions of the secondary diodes, permitting high-efficiency even with a relative low power. The APS converter has been initially validated using a dynamic load simulator with an output operating voltage of 300V and currents between 0.3A and 1A, which are encountered with the various flow rates. Fig. 4 presents the test results at three input voltages (22, 28 and 34 Volts) with dummy resistive loads.

An important detail of the PPU EM design regards the implementation of the output current and voltage telemetries. Since high accuracy is required on all TU measurements, it was preferred to perform the analogue to digital conversion directly on the secondary and to realize the isolation using a digital isolator based on magnetic coupling, implemented using discrete components.

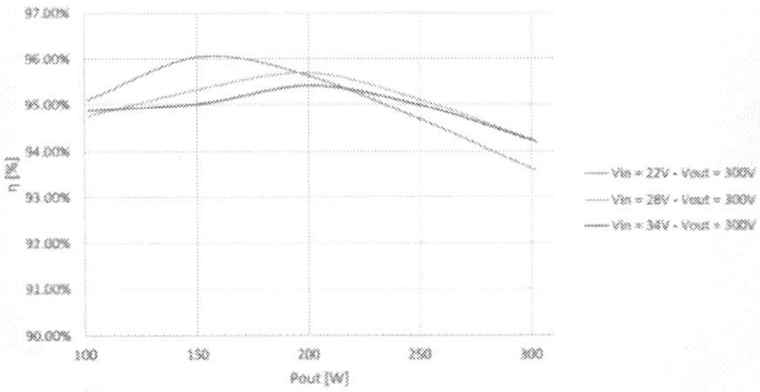

Figure 4. Efficiency for Vout 300V and different Input Voltages.

Magnetic Field Power Supply (MGFPS)

MGFPS is an isolated DC/DC settable current source whose purpose is to supply the magnetic field coils of the CAM-200 thruster with a current up to 3.5A.
The MGFPS is connected to the unregulated bus coming from the PDU downstream the EMI input filter and inrush current control circuit
In order to improve the efficiency and reliability of the PPU EM, the EM implementation of the MGFPS will be with only one isolated conversion stage with forward topology.

Cathode Power Supply (CAPS)

CAPS is an isolated DC/DC Power Supply providing the following features:

- High voltage driving of thruster Cathode during the breakdown phase which is expected to be lower than 1200V
- Current driving during keeping phase
- Acquisition of cathode driving voltage and current
- Filtering of the noise generated by the cathode during ignition
- Galvanic isolation between input power bus and the thruster and the chassis. The isolation voltage is 2500Vdc.

Fig. 5 depicts the block schematics of the Cathode power supply implemented for the PPU BB.

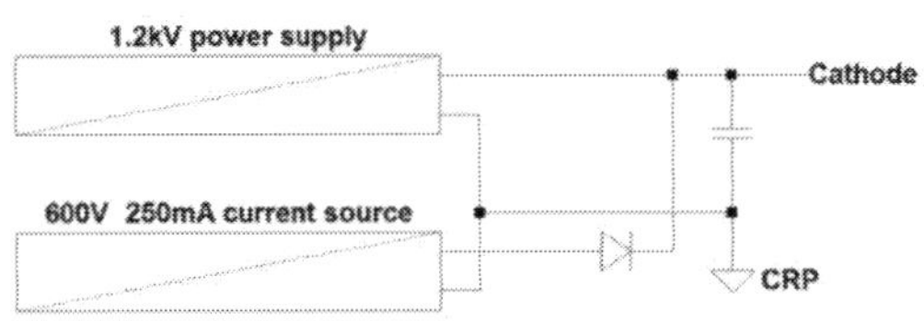

Figure 5. CAPS simplified diagram.

DOI: 10.1051/e3sconf/20171615004

In order to achieve the flexibility necessary to support the testing of the RHCC, the CAPS module is implemented using two different and separately tunable power supplies, each taking care of one of the two phases of the cathode power up sequence which are represented in Fig. 6 and described hereafter:

- The Breakdown phase is the first stage of the cathode ignition sequence. In this phase cathode supply serves as discharge initiator to the cathode. The breakdown pulse is applied in order to ionize a low Xenon flow rate, passing through the cathode and creates an arc discharge between the emitter and keeper. At the breakdown phase the cathode will be driven by a discharge capacitor, whose value can be changed between 220nF and 1uF on the PPU BB in order to tune the energy delivered to the ignitor during breakdown. The ignition phase is obtained by means of a 1200V power supply driven by a current source in order to simulate the behavior of a current generator charging the output capacitor with a constant current. The design of the isolation stage is based on the same architecture used for the APS.
- The Keeping phase is the phase which aims to raise cathode temperature to favor the thermionic emission. During this stage the current within the cathode is controlled by an isolated DC/DC converter.
  During this phase the current is limited to 250mA and the maximum voltage is clamped to 600V. The maximum voltage limit is settable in order to allow some flexibility during the test of the cathode.
  The design of the isolation stage is based on the same architecture used for the APS.

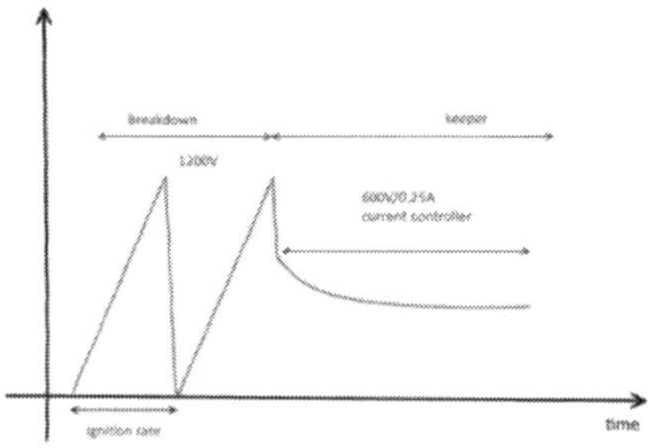

Figure 6. CAPS sequence.

Auxiliary Power Supply (AUX)

The galvanically insulated Auxiliary DC/DC converter provides all the housekeeping supplies necessary to the PPU modules and the auxiliary supplies to control the PMA/PTA fluidic subsystem. The Auxiliary converter design for the PPU EM consists in a multiple output Forward converter with peak current mode control.

2.3 Filter Unit (FU)

FU is a passive low pass LC filter with resistive damping which is aimed at reducing the high frequency noise current generated by the TU during its operation.
The capability of FU will be evaluated during the coupling test activities and its operation shall be optimized in order to be compatible with the EMC/ESD requirements.

2.4 PPU Control Module (PCM)

The EPS is externally controllable by the spacecraft's OBC while its internal control is performed by the PCM. In particular the PCM controls the APS, CAPS and MGFPS, the TSU high voltage relays and the PMA/PTA valves so as to enable the driving and control of one thruster at a time.
Specifically, the PCM provides the following features:

- Temperature monitor via thermal sensors located in the thrusters and PMA
- Anode, cathode and magnetic field driving voltages and currents monitors
- Heaters and valves driving currents monitors
- Output signals voltages and currents monitor
- Voltage settings for APS and current settings for MGFPS
- Sequencing and management of thruster's operation in various modes (i.e. start-up, shutdown, power setting, thrust regulation and failure recovery)
- Management of MEPS redundancy policy, in particular commanding the TSU
- Management of Telecommand / Telemetry communication

The PCM will be an FPGA-based digital board including non-volatile memory with EDAC for parameters storage, DAC control for Anode and Cathode voltages, ADCs and MUXs (with relative signals conditioning circuits) for external/internal sensors/voltages readout.
The FPGA approach will make the programmed sequences control more flexible with the possibility of adjusting the internal parameters.
The PCM is not considered critical part. Thus, it is being directly developed at EM level.

2.5 Thruster Switching Unit (TSU)

The TSU provides the means to enable the active selection between the two TUs, disconnects the possibly failed thruster from PPMs and connects the isolated TU to a resistive network in order to avoid its complete floating. The switching function is performed by high voltage relays which are operating while PPM supplies are OFF in order to avoid electric arcs.

3. COUPLING TESTS

Coupling tests have been performed in order to verify the capability of the PPU BB to ignite the two TUs (Rafael's CAM-200 with RHHC and Sitael's HT100 with RHHC) and keep the discharge stably on in steady state condition

 DOI: 10.1051/e3sconf/20171615004

for different power levels. The test setup used is presented in Fig. 7.

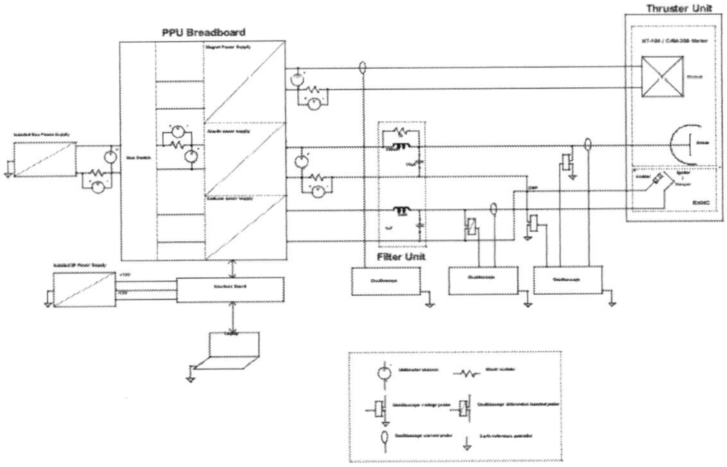

Figure 7. Coupling test setup.

For the needs of the tests, Sitael's IV4 vacuum chamber has been used.

The operating points tested are indicated in Tab. 1 and 2 respectively for both TUs in the preliminary phase (each TU was coupled with a laboratory PPU) and in the coupling phase with the PPU BB.

Table 1. HT100 Operating points verified during the coupling test with PPU BB.

Operating Point	Power, [W]	$\dot{m}_A$, [mg/s]	$\dot{m}_C$, [mg/s]	V_{dc}, [V]
1	130	0.6	0.25	300
2	160	0.75	0.25	275
3	190	0.75	0.25	325
4	220	1	0.25	275
5	250	1	0.25	300

Table 2. CAM-200 Operating points verified during the coupling test with PPU BB.

Operating Point	Power, [W]	$\dot{m}_A$, [mg/s]	$\dot{m}_C$, [mg/s]	V_{dc}, [V]	I_{mag}, [A]
1	110	0.62	0.25	250	2
2	130	0.69	0.25	250	2.5
3	160	0.75	0.25	275	2.5
4	180	0.83	0.25	275	2.5
5	200	0.90	0.25	275	2.5

3.1 Coupling test results for CAM-200

The coupling test with the PPU BB succeeded the ignition of the CAM-200 thruster in all operating points.

Tab. 3 and Fig. 8 present the efficiency of the APS recorded for the various operating points of the CAM-200 thruster for three different input voltages 22, 28 and 34 volts.

Table 3. CAM-200 Anode efficiency.

A_{MFR} [mg/s]	C_{MFR} [mg/s]	V_{in} [V]	V_{OUT_AN} [V]	I_{OUT_AN} [A]	P_{OUT_AN} [W]	n_{AN} [%]	T_{An_MFET} [°C]
0.62	0.25	22	250	0.458	114	93.9	33.60
0.69	0.25	22	250	0.513	128	93.7	31.30
0.75	0.25	22	275	0.578	159	94.3	33.10
0.83	0.25	22	275	0.648	178	93.9	35.00
0.9	0.25	22	274	0.710	195	93.8	35.60
0.62	0.25	28	251	0.458	115	93.1	36.10
0.69	0.25	28	251	0.512	129	93.5	32.10
0.75	0.25	28	275	0.573	157	94.1	34.30
0.83	0.25	28	276	0.648	179	94.0	34.30
0.9	0.25	28	275	0.711	196	93.9	36.20
0.62	0.25	34	251	0.457	114	92.2	36.10
0.69	0.25	34	251	0.513	129	92.9	34.80
0.75	0.25	34	275	0.576	158	93.4	36.10
0.83	0.25	34	275	0.648	178	93.6	34.30
0.9	0.25	34	276	0.711	196	93.7	36.60

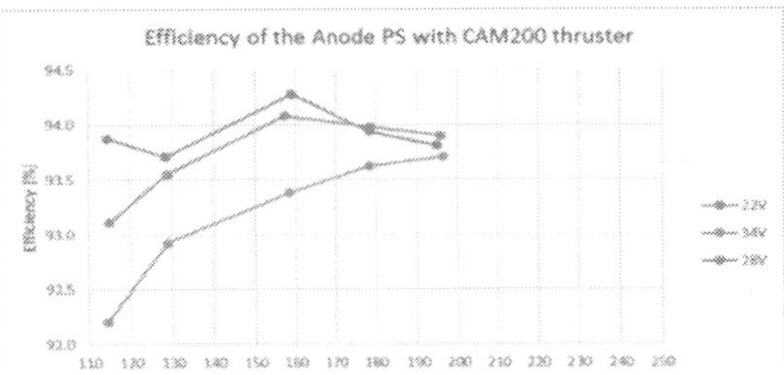

Figure 8. Anode efficiency with CAM-200.

Start-up and steady state plots were recorded for the various operating points and are presented in Fig. 9 and 10 respectively. The waveforms presented in the plots are the following:

DOI: 10.1051/e3sconf/20171615004

Ch1	V_{an} vs GND
Ch2	V_{crp} vs GND
Ch4	I_{an}
Math	V_{an}-V_{crp}

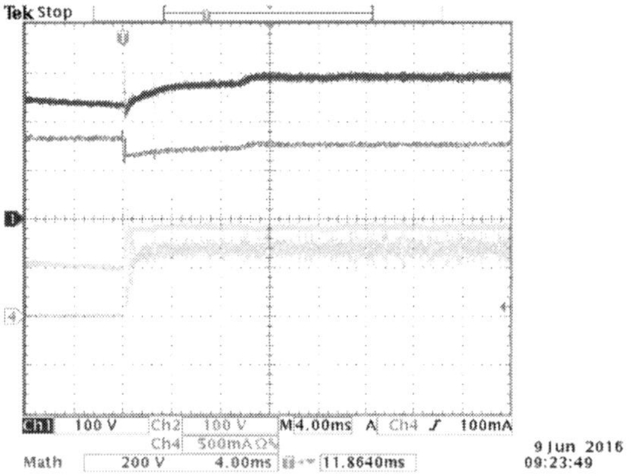

Figure 9. CAM-200 Anode start-up current and voltage with Xe_MFR = 0.83 mg/s and output voltage set to 300V. Magnet current is 2.5A.

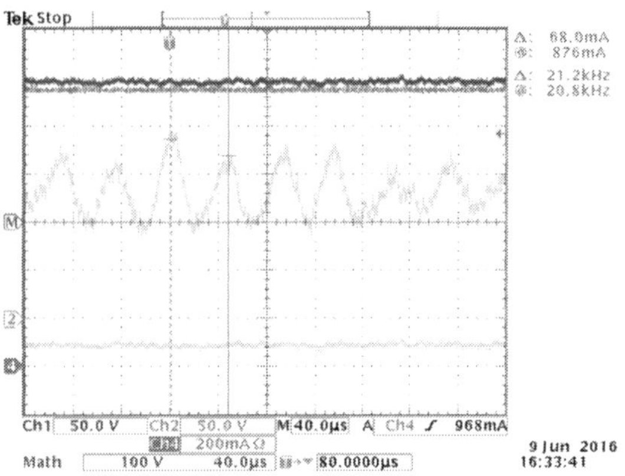

Figure 10. CAM-200 Anode steady state current and voltage with Xe_MFR = 0.75 mg/s and output voltage set to 275V. The characteristic frequency of the anode current is about 20kHz for the CAM-200.

3.2 Coupling test results for HT100

The coupling test with the PPU BB succeeded the ignition of the HT-100 thruster in all operating points. Tab. 4 and Fig. 11 present the efficiency of the APS for three different Input Voltages 22, 28 and 34 volts.

Table 4: Anode efficiency with HT100.

A_{MFR} [mg/s]	C_{MFR} [mg/s]	V_{in} [V]	V_{OUT_AN} [V]	I_{OUT_AN} [A]	P_{OUT_AN} [W]	n_{AN} [%]	T_{An_MFET} [^{0}C]
0.6	0.25	22	300	0.460	138	94.0	34.30
0.75	0.25	22	275	0.583	160	94.2	35.70
0.75	0.25	22	325	0.569	185	94.6	36.60
1	0.25	22	275	0.824	227	93.2	41.30
1	0.25	22	300	0.822	247	93.6	41.80
0.6	0.25	28	301	0.460	138	93.7	35.10
0.75	0.25	28	275	0.582	160	94.0	36.10
0.75	0.25	28	326	0.570	186	94.6	37.60
1	0.25	28	275	0.824	227	93.6	41.30
1	0.25	28	301	0.823	247	94.0	42.30
0.6	0.25	34	301	0.460	139	93.1	35.90
0.75	0.25	34	275	0.582	160	93.5	38.30
0.75	0.25	34	328	0.570	187	93.8	38.70
1	0.25	34	275	0.824	227	93.3	41.80
1	0.25	34	301	0.822	247	93.8	41.40

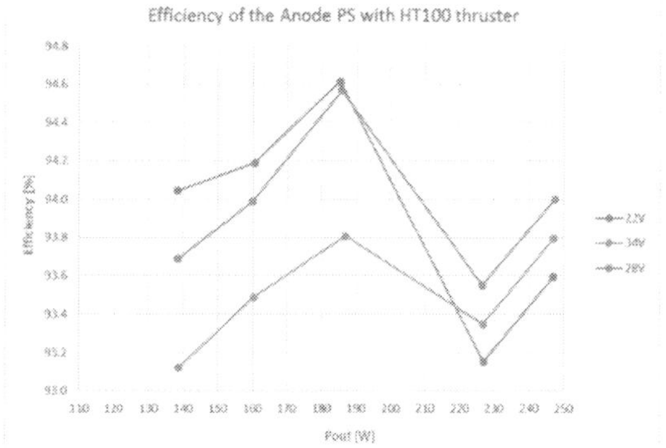

Figure 11. HT100 Anode efficiency.

Start-up and steady state plots were recorded for the various operating points and images of the anode and cathode waveforms are presented in Fig 12. The waveforms presented are the following:

Ch3	V_{cath} vs V_{crp}
Ch4	I_{an}
Math	V_{an} - V_{crp}

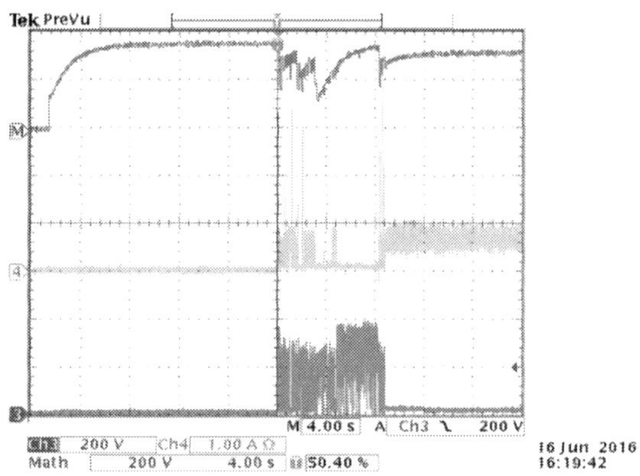

Figure 12. HT100 power-up sequence and switch-on transient with AN_Xe_MFR = 0.75 mg/s and CAT_Xe_MFR = 0.25 mg/s. Output voltage set to 325V. The cathode switches on at maximum voltage (1200V).

DOI: 10.1051/e3sconf/20171615004

The plot shows the sequence used for the start-up:

- Initially the anode voltage is applied to the APS.
- The HV module and cathode power supply are then switched on.
- Cathode breakdown occurs at maximum voltage (1200V) and a current limited phase (keeping phase) follows. During this phase the current is limited to 250mA.
- The anode's discharge starts occurring and once the steady state phase is reached the keeping phase is terminated.

Fig. 13 and 14 represent a detail of the cathode voltage and current during breakdown transient and the inception of keeping phase respectively.

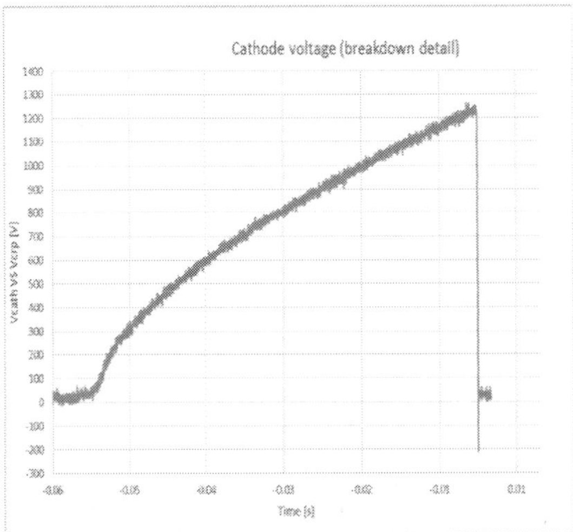

Figure 13. Cathode voltage during breakdown.

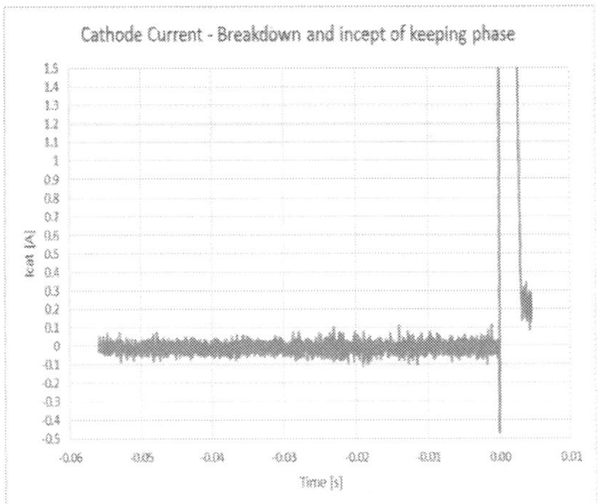

Figure 14. Cathode current transition from breakdown to keeping phase.

Finally, the steady state waveforms are presented in Fig. 15. The waveforms presented are:

Ch1	V_{cath} vs V_{crp}
Ch3	I_{cath}
Ch4	I_{an}

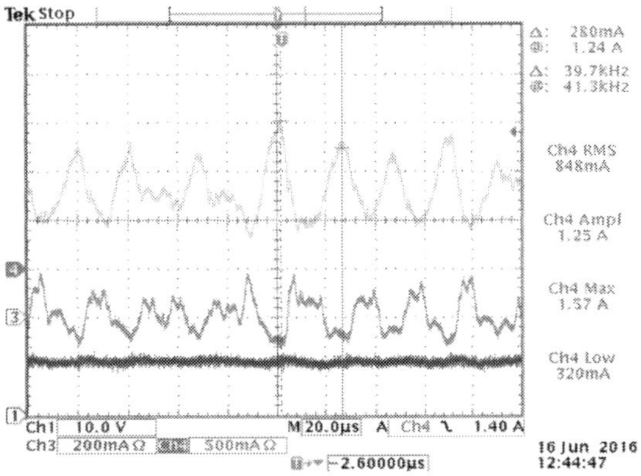

Figure 15. HT100 Anode steady state current and Cathode current with AN_Xe_MFR = 1 mg/s and CAT_Xe_MFR= 0.25 mg/s. Output voltage set to 300V. The characteristic frequency of the anode current is about 40kHz for the HT100.

Fig. 16 presents an image of the HT100 thruster and RHHC firing in the IV4 Sitael vacuum chamber facility after their successful ignition during the coupling tests with the PPU.

Figure 16. HT100 and RHHC firing during the coupling test with MEPS PPU.

4. FUTURE ACTIVITIES

The most critical parts of the PPU design have been evaluated by the means of the modular PPU BB and are currently being updated for the EM phase. Moreover, some of the PPU functional blocks need to be designed from the beginning as they were not considered critical. After the EM implementation and validation, which will involve a complete coupling with the TUs and the PMA/PTA subsystems, the QM will be realized and qualified.

5. CONCLUSIONS

In the framework of MEPS program an EPS based on low power HET is going to be qualified. Sitael Hellas, supported by Sitael is responsible for the design of the PPU. This PPU is suitable for both Sitael HT100 and Rafael CAM-200 HET. The operational design principles of the proposed DC/DC converters and of all the power modules have been discussed and their advantages have been described.
MEPS PPU BB implementation has been briefly described and the coupling test results have been presented. The PPU BB contains critical parts of the design and has been successfully coupled to HT100 and CAM-200 thrusters, in particular the CAPS and the APS have been able to switch on both thrusters in any operating point and provide efficiency higher than 93% in a power range between 130W to 250W.
PPU's final development will display considerable technological aspects in the existing space systems:

- High Efficiency: PPU is designed to exceed 92% efficiency.
- High Reliability: PPU is designed to have reliability higher than 0.99. Redundancy is assessed on the critical parts of the PPU, at module level.
- Small Budget: PPU will include both TSU and FU with an the envelope of 300mm x 240mm x 230mm (L x W x H) volume

Optimization of the PPU behaviour is foreseen during the next stages of the project (EM and QM). On the road towards the EM that is currently being developed, a complete boards' layout as well as, the housing and mechanical design have been implemented and two representative images are presented in Fig. 17 and 18.

6. REFERENCES

1. T. Misuri, R. Albertoni, C. Ducci, B. Waldvogel, L. Appel, R. Eytan, D. Lev, K. Dannenmayer, D. Di Cara, "MEPS: A Low Power Electric Propulsion System for Small Satellites", 10th IAA Symposium on Small Satellites for Earth Observation, Berlin, Germany, 20-23 April 2015

2. T. Misuri, C. Ducci, R. Albertoni, M. Andrenucci, D. Pedrini, "Sitael Low Power Hall Effect Thrusters for Small Satellites", IEPC-2015-102 ISTS-2015-b-102, Kobe, Japan, 1-10 July 2015

3. Amoz Davidson, Pavel Gontmacher, Jacob Herscovitz, and Igal Tidhar. "Development, Qualification and Integration of the VENμS PPU", 50th AIAA/ASME/SAE/ASEE Joint Propulsion Conference, Propulsion and Energy Forum, (AIAA 2014-3608).

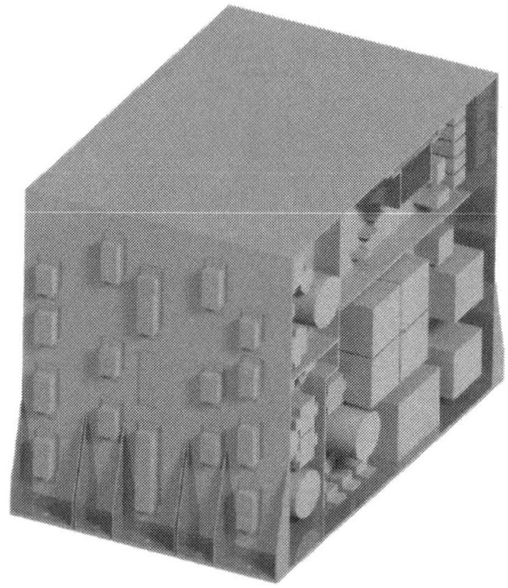

Figure 17. Mechanical design (Courtesy of INASCO Hellas).

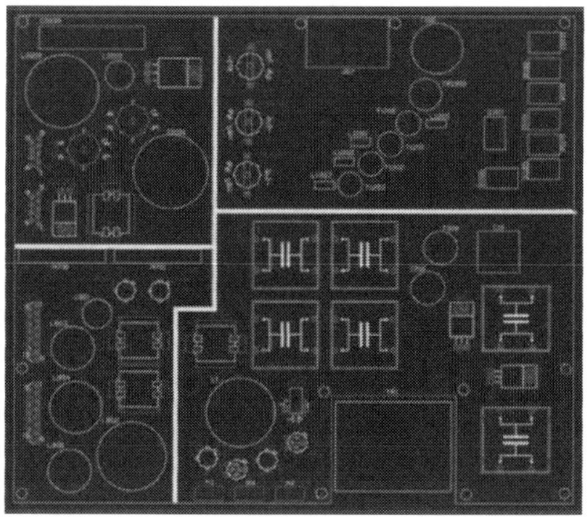

Figure 18. APS Board layout.

THE HIGH-VOLTAGE SYSTEM OF CALET APPARATUS

Francesco Petroni [(1)], Lorenzo Fontani [(1)], Luca Onida[(1)], Pierpaolo Pergola[(1)],

Guido Castellini [(2)], Andrea Viciani [(1)]

[(1)] SITAEL, Via Livornese 1019, Pisa, Italy, Email: francesco.petroni@sitael.com
[(2)] IFAC CNR, Via Madonna del Piano 10, Firenze, Italy, Email: g.castellini@ifac.cnr.it

ABSTRACT

CALET (CALorimetric Electron Telescope) is installed on the International Space Station since August 2015, is a mission devoted to measure the intensity of cosmic ray electrons and protons accelerated to near the speed of light. The apparatus is also designed to observe high energy gamma rays, nearby source of high energy radiation and it may even detect signatures of the elusive dark matter.
One of the core components of the whole apparatus is the high voltage power supply system. It is based on hundred of independent DC/DC converters providing precise voltage regulation to two sections of CALET detectors.
This paper presents the experience gained in the implementation of the CALET HV system: the system design requirements and the technical solutions adopted, the two system sections block schemes and the main characteristics of the DC/DC converters that guarantee the CALET HV system technical performance.

1. The CALET experiment

CALET is a sophisticated scientific apparatus for measurements of high-energy (GeV - TeV) fluxes of electrons and photons, and also protons and atomic nuclei, originating from outside the solar system. It has been developed within an international Japanese led cooperation involving nearly 50 researchers from Japan, Italy and the USA [1].
CALET aims at addressing several high-energy astrophysics questions: the origin of cosmic rays; how these accelerate and travel across our galaxy; evidence of the existence of dark matter; the origin of nearby cosmic-ray; etc. Furthermore, CALET expects also to provide an important insight of rare interactions between matter and dark matter due to its long permanence in space [2].
CALET will complete the observations of CREAM, PAMELA and AMS-02 on a possible deviation from a pure power-law of proton and He spectra from the few hundred GeV to multi-TeV region [2].
The module has been launched in orbit aboard the HTV-5 vehicle and installed on the International Space Station (ISS) in 2014. CALET is currently attached to the Japanese Experiment Module (JEM), Kibo [3] .
The main CALET telescope consists of an array of scintillation detectors to determine the electric charge of the incoming cosmic ray particles, an imaging calorimeter of scintillating fibers to determine particle trajectory, and a tungstate calorimeter to measure particle and gamma ray energies up to 20 TeV [4], see Fig. 1.

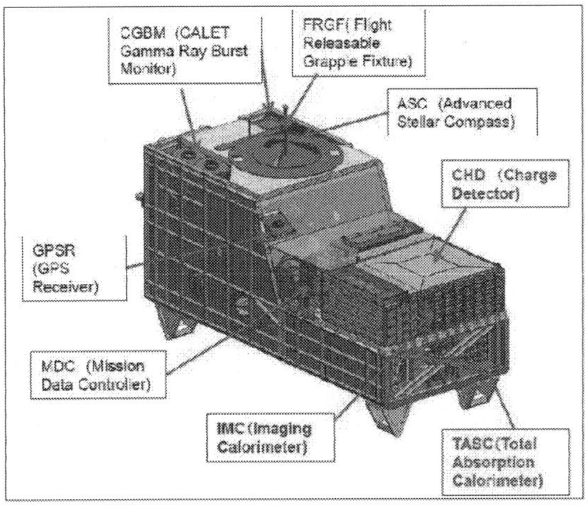

Figure 1. Overview of the main CALET telescope

The electromagnetic calorimeter has an approximately cubic size, 50 cm side length, and it is horizontally segmented in a set of layers alternating high-density and scintillating materials [5].
The working principle of the calorimeter is that a single high-energy charged particle or photon, traversing the high-density layers, produces a shower of secondary charged particles. These particles interact in the scintillating layers (scintillating fibres, lead tungstate PWO) generating scintillation light that is collected by different types of detectors:

- photomultiplier tubes (PMT);
- photodiodes (PD);
- avalanche photodiodes (APD).

The charge detector is composed of several layers of plastic scintillator, to be crossed by the particles before getting into the calorimeter, that provide information on intensity and on distribution of the scintillation signal.

 DOI: 10.1051/e3sconf/20171615005

These data are thus collected to determine species and associated energy of the primary incident particle [5].

2. The CALET HV system design criteria

The CALET high-voltage (HV) system (surnamed brick) fulfils the task of generating the high voltages (from 500 V to 1000 V) required for the operation of the PMT and APD photo-detectors.
The HV system has been designed and implemented by SITAEL in coordination with CNR-IFAC utilizing COTS components.
The HV system has been conceived, manufactured and tested to cope with the stringent requirements deriving from long time operations in space environment. Among these requirements, it is worth to mention:

- The brick structure was required to be stiff enough to support the mechanical stresses encountered during the launch phase;
- The materials used for each HV module was required to be compatible with the foreseen temperature and pressure conditions during orbital flight. For the qualification of the system temperatures between -20 °C and +65 °C and pressures below 10^{-5} mbar have been used;
- The overall system designed was required to be compatible with the high level of ionizing radiation present in low-Earth orbit. Qualification up to at least 10 krad of total absorbed ionizing dose, and immunity for electrical events caused by single ionizing particle (Single Event Effect, SEE) for the whole duration of the mission (>5 years) were the two main drivers;
- The electromagnetic interference and compatibility (EMI/EMC) was required to cope with the stringent requirements of payloads installed on the ISS space station.

A first specimen of the HV system, the "HV Box 001" was delivered to JAXA in 2012, with the purpose of performing characterization and qualification. The "HV Box 002", to be integrated with the CALET apparatus and launched for flight operation, was completed and delivered to JAXA on April 2013.

3. The CALET HV system configuration

The CALET HV system is composed of an aluminium support structure 290x240x240 mm^3, housing:

- 80 modules in the "PMT section" (derived from the design of AMS02 HV System) delivering a corresponding number of negative high-voltage lines (up to -1000 V) for the PMT photo-detectors of CALET system;
- 22 modules in the "APD section", delivering a corresponding number of negative lines, which voltages up to -500 V, for the APD photo-detectors;
- the related electronic boards (PHCB and AHCB) for control and interface with the Mission Data Controller (MDC) unit managing the operation of the CALET apparatus.

The HV brick features a total of 102 redundant independent HV lines. The system weights in total about 14 kg and its power consumption is approximately 22 W in the nominal operating configuration, see Fig. 2

Figure 2. The CALET HV system front (up) and back (bottom)

On the front side of the HV unit, 102 miniaturized HV

E3S Web of Conferences **16**, 15005 (2017) DOI: 10.1051/e3sconf/20171615005
ESPC 2016

coaxial connectors are installed for the connection of each line. On the rear side a total of 4 power connectors are installed on the mechanical structure for input power lines and digital signals.

The PMT and APD sections are physically divided into two electrically identical halves (Fig. 3). Each half is composed of a hot and a cold (identical) part for redundancy: each DC/DC and Linear Regulator (LR) is doubled, while each control board is formed by two parts (hot and cold part).

The doubled parts are identified as hot (normally powered-on) and cold (normally powered-off and used for substitution in the case of unrecoverable damage). When the hot HV output of a given channel (LR or DC/DC) is on, the corresponding cold HV output has been turned off, or vice versa. The hot and cold part of each control board can be powered independently (with commands addressing each part independently).

Figure 3. The CALET HV system (PMT and APD sections) open in its two halves

The PMT HV section of each brick contains two (hot/cold) DC/DCs and 5 (hot/cold) groups (boxes) each composed of 8 LRs, and correspondingly a total of 40 HV output channels (one HV connector per channel). Each (hot/cold) LR generates a typical HV (varying with channel).

The APD HV section of each brick is formed by a total of 11 (hot/cold) DC/DC converters, driving 11 FEC channels for a total of 88 APD's.

The converters are grouped into 4 mechanical boxes: two boxes contain 6 (hot/cold) converters each, while the other two boxes contain 5 (hot/cold) converters each. The boxes are equally divided between the two bricks, in such a way that each brick contains one box with 6 hot/cold converters and one box with 5 hot/cold converters.

The HV outputs of each hot/cold pair of LR's (for PMT section) or DC/DC's (for APD section) are ORed through a diode protection circuit, thus producing a single HV output channel.

The block schemes for half PMT and APD section is illustrated in Fig. 4. The HV system is designed to be interfaced with the Mission Data Controller (MDC) unit managing the sequence of operation of the whole CALET apparatus.

Up to three different power voltages are required to be provided by each section of the CALET HV system. The digital interface, for data and commands is implemented with LVDS bidirectional serial links, with a typical transmission rate of 1 Mbps. The HV lines have been designed and characterized for a linearity better than 3% and a peak-to-peak noise (ripple) less than 150 mV.

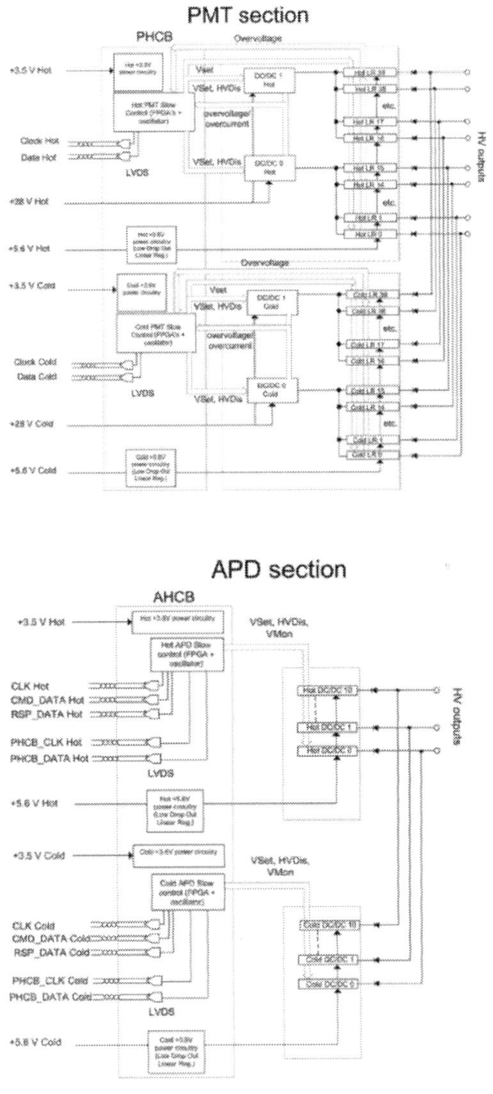

Figure 4. Block schemes for half PMT and APD section of the CALET HV system

The logic structure of the HV system has been

implemented by means of FPGA programmable logic devices on the PHCB and AHCB (see fig.4) boards.
The FPGA family used in the PHCB is the Actel A54SX-A. This "antifuse" FPGA family guarantees correct and reliable operations in space under the foreseen radiation environment, without cumulative effects from absorbed Total Ionization Dose (TID) and a negligible rate of Single Event Effects (SEE), specifically SEU and SEL.
For the AHCB the considered FPGA family is the Actel ProASIC3E. The ProASIC3E family features the "flash programmable" technology, implementing the configuration memory of the FPGA logic structure on non-volatile flash EEPROM bits. This technology allowed for fast and zero-cost reprogramming on the same chip during the development phase. The FPGA features similar performances (and it is produced by the same vendor) to those of the A54SX-A family for what concerns radiation effects.
In general the logic architecture allows to:

- power on/off a single HV line or groups of HV lines;
- set the output voltage for each HV line, with 0.1% resolution over the whole dynamic range;
- real-time monitor the output voltages for the APD section;
- save and read the configuration of single lines;
- real-time detection of anomalous conditions, (overvoltage or overcurrent in the high-voltage modules, or SEE);
- automatically configure (at system power-on) the different HV lines, without needing to receive any external command, according to a scheme defined on the basis of the characteristics of the PMT or APD channels.

4. The CALET HV DC/DC converters

The power lines for PMT HV system are: +28 V DC, +5.6 V DC and +3.5 V DC (nominal values). The power lines for APD HV system are: +5.6 V DC and +3.5 V DC (nominal values).
The DC/DC converter and linear regulators adopted for the PMT HV system are the same utilized in the past for AMS02 experiment.
The DC/DC converter adopted for each line of the APD HV system is a variation of the SITAEL S9097/8 family (more than 1000 flight module delivered mostly for the Chinese market in the last decade), modified (S9106) in such a way to have a reduced output voltage range lower than the 3000 V range as per specifications of these modules. Fig. 5 presents a string of 6 modules before assembly into the CALET brick.

Figure 5. String of S9106 modules assembled in the CALET configuration

The S9106 modules ensure a minimum radiation tolerance of 30 Krad. The converter offers compact size, low power consumption, wide operating temperature range (-40 to +70 °C), low ripple and high temperature stability. S9106 modules operate with a +5 V DC input and deliver a remotely programmable output voltage ranging from 0 to 500 V; the DC/DC converter is provided with an output voltage monitor and an HV disable feature. Fig.6 presents the S9106 block diagram.

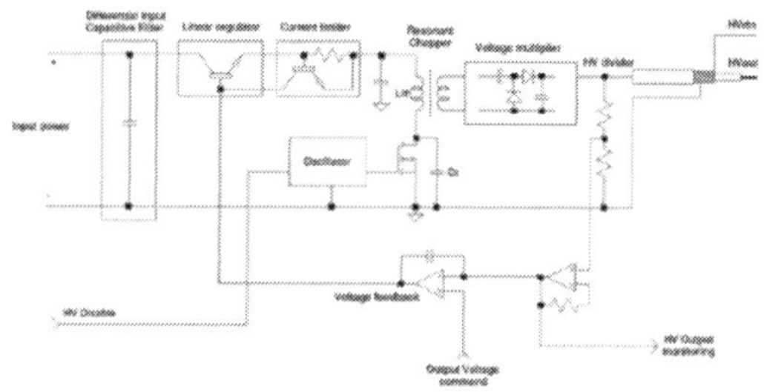

Figure 6. S9106 block diagram in the CALET configuration

More in detail the S9106 converters feature:

- capacitive input filter, which reduce the current ripple and noise injected into input bus;
- primary current limiter which intervenes in case of anomalous absorbed current;
- linear regulator which modulates the chopper voltage amplitude according to feedback signal;
- resonant chopper which is used to drive the step-up transformer with low switching losses
- voltage multiplier which raises the transformer voltage up to the desired output level;
- voltage monitor which senses the output voltage for both internal voltage feedback circuit and external output voltage monitor;
- voltage feedback which comparing the output voltage monitor and the set voltage, drives the linear regulator;
- voltage set which programs the output high

voltage;

- HV Disable, which can disables the output voltage generation.

The switching converter works at fixed frequency, avoiding any unpredictable noise due to frequency beat which may increase noise level.
The resonant chopper has been implemented by adopting a ZVS technique (switching turn-on takes place at Zero Voltage) and ZCS technique (turn-off takes place at Zero Current) to drastically reduce switching losses, increment efficiency and lower conducted and radiated noise. The resonant condition will be maintained over the entire load range.
A 3-cell Cockroft-Walton voltage multiplier is used to lift the secondary voltage of the transformer up to the required output voltage. The usage of components with lower rated voltage than the output voltage yields mass and volume saving and high efficiency.
The use of a linear regulator to control the output voltage not only simplifies the circuit with respect to classic PWM solution but allows to employ a fixed frequency resonant chopper and to greatly ease feedback compensation network design since the converter behaves like a first order system.
S9106 configured for the CALET experiment is able to operate at any pressure, including the 0.1mbar to 10mbar range (minimum of Paschen curve), without any HV discharge or corona effect.

5. Conclusions

The HV power supply system for the Calorimetric Electron Telescope is one of the biggest ever realized for a space mission with more than 200 independently regulated HV channels designed and tested to behave in any vacuum condition. COTS components selected and utilized for PAMELA (in orbit since 10 years), AMS02 (in orbit since 5 years) and CALET (in orbit since August 2015) have shown outstanding characteristics and reliability as well as the compromise solutions adopted for the assembly. These successful criteria will be proposed as a baseline for SITAEL low cost line of products dedicated to micro and mini satellites for LEO missions.

6. REFERENCES

1. Torii, S., (2011), Overview of the CALET Mission to the ISS, 32nd International Cosmic Ray Conference, Beijing, China

2. Torii, S., et al. (2005),The CALET Instrument for Experiment on the ISS, 29th International Cosmic Ray Conference, Pune, India

3. Torii, S., (2015) The CALorimetric Electron Telescope (CALET) on the ISS for High Energy Astroparticle Physics, ISPC ISS Research Investigations and Experiments, IKI RAS/Moscow, Russia

4. Marocchesi, P. S., (2015) CALET: a high energy astroparticle physics experiment on the ISS, World Scientific Publishing Co., arXiv:1512.08059 [astro-ph.IM]

5. Bigongiari, G., (2015) CALET perspectives for calorimetric measurements of high energy electrons based on beam test results, Proceedings of Science, 34th International Cosmic Ray Conference, The Hague, The Netherlands

E3S Web of Conferences **16**, 15006 (2017) DOI: 10.1051/e3sconf/20171615006
ESPC 2016

HIGH VOLTAGE SMART POWER MODULE FOR FAULT-TOLERANT LAUNCHER APPLICATIONS

Debrouwere Richard[(1)], **Guillaume Michel**[(1)]

[(1)] *Thales Alenia Space Belgium, Rue Chapelle Beaussart 101, B-6032 Mont-sur-Marchienne, Email : firstname.name@thalesaleniaspace.com*

ABSTRACT

Generalization of electromechanical actuators use in thrust vector control launchers applications has increased electronics power level needs by a factor of ten, in a global context which puts costs under pressure.

The need for a modular and affordable power solution being versatile to fulfil different launcher mission profiles has become thereby stronger.

In order to target different power and functional needs, an attractive design should be modular and multipurpose, will work at higher voltages and will be able to deliver powers up to a dozen of kilowatts.

Moreover, an efficient design to cost approach will consider alternative technologies, widely used by other high-reliability industries, including Components Off-The-Shelf (COTS) integration.

This paper presents the design of a low cost and highly integrated smart power module (SPM) intended to be used into launchers applications, embedding technologies and components from automotive world and being mainly producible by large-scale industry.

1. SPM SPECIFICATION

1.1. Introduction

The presented Smart Power Module (SPM) targets to drive motors but architecture and technologies are intended to be a future base element of a wider range of applications (DC/DC, DC/AC, satellite,…), its versatility being within others provided by embedded intelligence.

The design must be compliant to specific launcher constraints: radiative environment (mainly robustness to SEE, cumulated radiations being negligible), missions profiles (from several minutes into launchers lower stages and up to 30 hours into upper stages), thermal interfaces (from adiabatic conditions during several minutes to conductive interfaces during up to 30 hours) and launchers applications specific fault-tolerance requirements.

1.2. Functional architecture

A full H power bridge has been integrated together with power electronics drive and isolation.

The choice of a H bridge has been made as it delivers a voltage level that is twice the high power supply one, delivering there through a higher power to the load. Moreover, H topology will be to manage star shaped

An accurate current measure, several protections and embedded intelligence are associated to power electronics in order to satisfy to applications where control can be highly distributed :

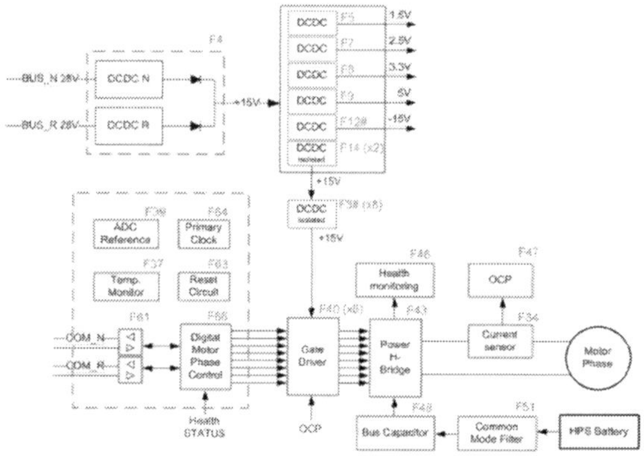

Figure 1. SPM functional architecture

Over-current protection is implemented in order to detect any malfunction into switching signals generation electronics. Both OCP detection and module switch-off are fully independent from the other functions.

A major concern before launcher take-off is to ensure that not any non-detectable failure has occurred during electronics integration. The result would be a start of mission with an electronics that is not single failure proof anymore and must be avoided. IGBT health monitoring has been therefore integrated.

Finally, a high capacity digital component is embedded in order to provide versatility when integrating the SPM into various applications.

1.3. Fault tolerance

A typical thrust vector control launcher application uses poly-phased motors, each phase driven by a SPM. The whole system being sized in such a way that loss of one phase keeps ensuring the full required performance.

The consequence is that loss of a SPM is allowed when facing a single component failure.

However, all SPM's being fed by a unique high power source (HPS), a single failure into a SPM may not cause any short-circuit on the HPS. This is especially applicable to power components composing the H-Bridge where an IGBT failure could lead to mission loss.

A SPM has to be manageable by two nominal / redundant controllers and will thereby offer two separated control interfaces. There is no need for a single failure proof communication interface as communication links are made of a point to point network topology. In this case, loss of both communication channels leads to power module loss but has no impact on other modules used in parallel.

In order to avoid a counter-productive behaviour (e.g. a constant controlled current into a motor phase) of one module when losing control link, communication channel health has to be monitored and a time-out has to trigger power switch-off autonomously.

Finally, as a SPM is intended to be powered by two nominal/ redundant low power supplies, no single failure causing a short-circuit on both LPS is allowed as it would lead to loss of both N/R controllers.

1.4. Main performance characteristics

- **Electrical characteristics**

SPM average power output target is 5kW and peak power is up to 10kW. It will thereby work at voltages up to 500V and be able to deliver a peak current of 200A.

In order to provide the needed control bandwidth, its switching frequency will be up to 10kHz, which is compliant with use of IGBT's.

- **Thermal Interfaces**

SPM design is foreseen to respect a temperature derating of 40°C wrt to maximum rating.

Adiabatic mode:

The module must be designed to work several minutes without transferring power to launcher structure.

It will thus make use of a high thermal capacity that will be sized to store dissipation caused by a delivery of 75 A_{RMS} during 3 minutes with an initial temperature of 50°C.

Conductive mode:

The SPM must also support longer missions, composed of high power delivery periods which will increase electronics temperature and quiet periods where low power is delivered allowing the temperature to decay.

SPM target is to limit thermal resistance between module baseplate and each power die to 0.5 K/W, a large analysis showing that this thermal resistance associated with the huge thermal capacitance will cover most of long mission needs.

- **Paschen effect**

Launcher applications present the specificity that powered-on electronic will see a depressurization from 1 bar to vacuum within a couple of minutes.

Working with voltages over 200V into air implies to consider Paschen effect.

Thales Alenia Space has chosen not to put the Paschen constraint onto power modules but at equipment level by integrating all the electronics into airtight cases.

- **Environment**

Vibration and shock constraints are not presented here as they must be combined with equipment transfer function. Radiations and more particularly single event effect (SEE) have to be considered.

2. SMART POWER MODULE DESIGN

2.1. Fault-tolerant power architecture

A major concern of the power hybrid conception was the requirement forbidding power supply short-circuit in case of a single power component failure.

Protections using IGBT desaturation detection were ruled out as they have to be triggered within a dozen of microseconds and were thereby too sensitive to perturbations.

Moreover, use of desaturation detection based short-circuit protection increases design power complexity as very high currents (up to several thousands of Amperes) have to be interrupted, causing higher overvoltages; this, in turn, demands active components with higher voltage ratings at the cost of switching and conduction losses increasing.

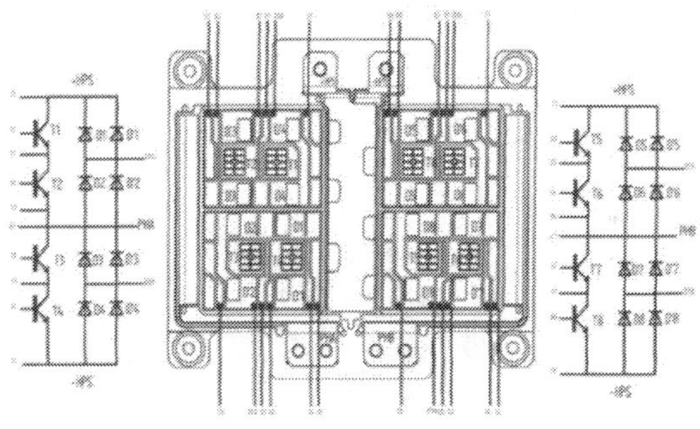

Figure 2 Fault tolerant power architecture

Considered alternative proposed in this design is the doubling of IGBT's in series (see Fig. 3). This solution has the advantage of preventing short-circuit instead of trying to isolate it, allowing the use of lower loss components.

Moreover, doubling the IGBT's in series increases robustness against SEE as both IGBT's of a same half-branch must see a single event at the same time in order to provoke an effect.

2.2. Physical segregation of power dies

A dissipative power die failure may induce a very high and concentrated power emission and thus cause a high local temperature elevation. The impact could be failure propagation to other power dices.

DOI: 10.1051/e3sconf/20171615006

A first level of protection is provided by silicon gel which is poured within the hybrid.

A second level of protection against failure propagation is the use of separated substrates for each part of the Left/Right and Low/High half branches of the H bridge

Figure 3: 4 identical substrates to compose H-bridge

However, laboratory experience has shown that these protections are not sufficient to border a high power default on a faulty die area.

As IGBT have been doubled in series, an additional physical segregation (walls) between each power dies of a substrate has been added.

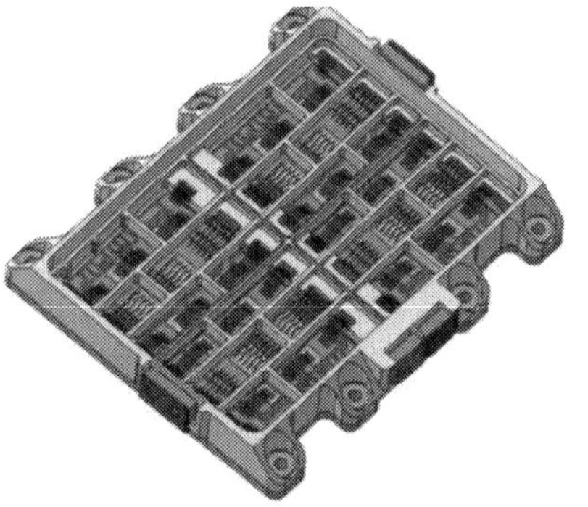

Figure 4: Physical segregation between dies

Thermal propagation in case of failure has been simulated and does not require any extra-protection.

2.3. Optimized power dies implementation.

When designing power hybrid substrates, a specific care has been taken to dies implementation.

An original way of implementing IGBT's and diodes dies to build-up a H-bridge, which lowers loop inductances, limits strongly overvoltages and EM emissions, especially at switching frequency, has been realized and patented by Thales Alenia Space.

The idea is to optimize currents paths and loops by placing diodes of the H-bridge lower side close to IGBT's of the higher side and vice-versa.

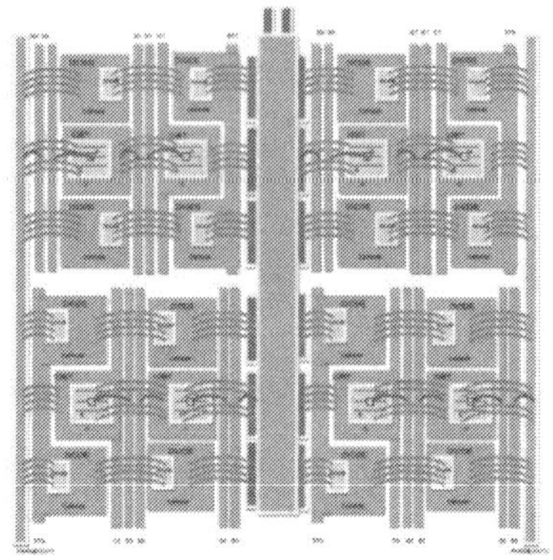

Figure 5: H-bridge diodes and IGBT dies patented implementation

A potential drawback is the vicinity of dies from different half bridge sides, leading to less segregation. However, this point has been mitigated by physical segregation of each die.

Tests results presented later show that this inventive topology reduces overvoltage in a very efficient way.

2.4. COTS power module

The first idea was the use of a COTS power module from industrial world. However, Fault-tolerance constraints did not allow the use of existing power modules.

The chosen alternative strategy was thus to design a custom hybrid, respecting the specific fault tolerance constraints and being producible by automotive large series production means, in order to keep cost under control.

2.5. Use of automotive-proven technologies

Fortunately, electrical and hybrid vehicles make a large use of power hybrids which are currently produced at a large scale while providing high-reliability and affordable costs.

The first phase of power design was the identification of power hybrid technologies already used into automotive industry and establish a design that could be produced by mature and highly automated production means.

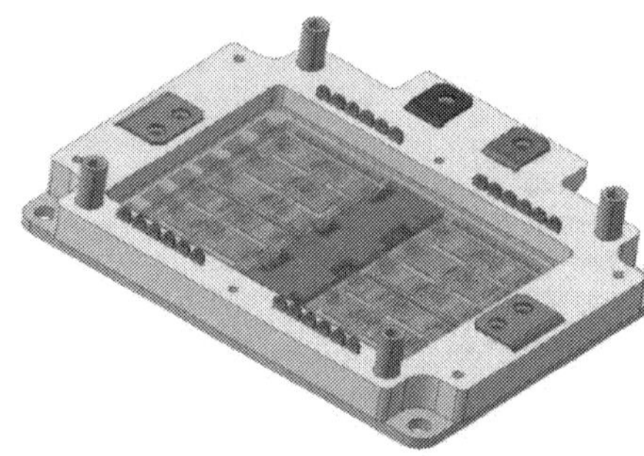

Figure 6: Power hybrid presentation

 DOI: 10.1051/e3sconf/20171615006

The baseplate, made of copper, has been designed in such a way that it is able to store a high power dissipation during a typical launcher lower stage mission, that is, a few minutes.

Substrate technology is a DBC (Direct Bonded Copper) which is directly soldered on the copper baseplate. Dies are soldered on the DBC. Soldering of substrate and dies is aimed at optimizing thermal resistances.

A special focus has been set on the soldering process which has to limit voids under chip and substrate areas, voids increasing linearly thermal resistance path to the baseplate.

Interconnection of dies with substrates and substrate with package and bus-bars is carried out by means of an automated ultrasonic wire bonding process, using Al wires. Internal bus-bar lines are made of thick copper with aluminium cladding, central bus-bar being a superposed copper strip structure, required to be double-isolated as it transfers energy from the unique HPS.

Housing is based on injected polymer resin around a copper structure in order to obtain an insulated package frame which will assure several functions:

- Protection of the electronic content from bad handling, or mechanical shocks, or vibration,...
- Good stiffness and stability
- Ability to be rigidly fixed to the copper base plate

Finally, an internal coating and protective gel has to be poured inside the device as to protect all internal conductors from being in contact with ambiance.

Different gels have been tested in order to ensure that they are compliant to vacuum environment and thermo-mechanical environment.

2.6. Thermal aspects

Before limiting thermal impact of dissipation on dice temperature by acting on thermal impedance, power dissipation has been limited as far as possible by using appropriate IGBT and diodes dies:

- V_{CE} typ < 1.6V @125°C / 150A
- E typ : E_{ON} < 1.75mJ
- V_F < 1.50V typ @150A, 125°C

Detailed simulations of mission profiles show that worst case power dissipation of the whole power hybrid is more than 200W in average and can reach up to 1 kW during a few milliseconds.

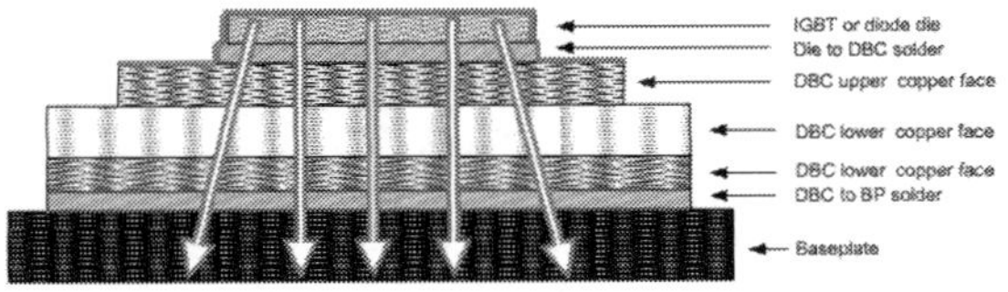

Figure 7: Thermal resistance between power die and baseplate is made of different layers.

At IGBT level, power dissipation is limited to around 40W in average but could reach more than 300W during a few milliseconds.

Conclusion of thermal analysis is that thermal design must achieve a thermal resistance of less than 0.5 K/W per power component, from the die to the copper baseplate.

2.7. Power hybrid integration with SPM structure

A SPM has to be produced in large series (up to 1000 pieces per year) and has therefore to be simple to assemble.

The number of elements and the number of different articles has been kept as low as possible in order to be able to produce it in the simplest way.

Power drive electronics being partially at high power potential, they have been put on a dedicated PCB in order not to interfere and limit high voltage design constraints on low level electronics.

The result is an easy to produce compact module, which is simple to integrate into more complex applications.

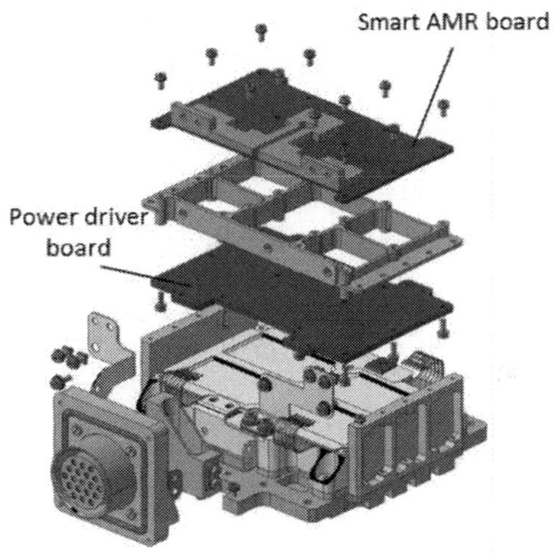

Figure 8: SPM mechanics

3. TESTS

3.1. Test goals

The test campaign pursued the following objectives:

- To demonstrate that a modular concept is adapted to multi-phases motor drive applications.
- To measure overvoltage spikes due to electronics switching and confirm thereby voltage margins.
- To confirm thermal studies estimating the worst case end of mission power dies temperature around 110°C for a launcher low stage typical mission.

 DOI: 10.1051/e3sconf/20171615006

3.2. Mock-up

An electrical and thermal representative mock-up, whose base element is a high-reliability automotive quality power hybrid prototype, has been built-up, made of six SPM's and driven through their communication bus by one FPGA based controller.

Figure 9: SPM prototype integration into motor drive application.

The controller sends appropriate duty cycles to SPM's in order to generate 6 sinusoidal currents in each of six air-inductances, each current being shifted by 60°.

This test is representative of a six-phased synchronous motor drive application, which is typical of a 60 kW launcher electromechanical actuator application.

3.3. Tests results

- **Multiphase motor application:**

Conclusion is that both communication link bandwidth and embedded electronics allow to generate currents with amplitudes and bandwidth which are compatible to targeted applications.

- **Switching overvoltage:**

Overvoltage generated is strongly limited to a few dozens of Volts which is a very interesting improvement factor with regards to common IGBT power modules.

- **End of mission temperature:**

Typical lower stage launcher RMS Current levels have been injected for a period of 3 minutes, temperatures are in line with expected results :

Figure 10: Power dies temperature at end of mission.

4. CONCLUSIONS

This paper has presented the design of a multipurpose smart power module intended to answer to a large number of needs into launcher applications.

Specific launcher RAMS constraints have been considered and the way the module fits these requirements demonstrated.

Technologies from automotive industry have been used as far as possible in order to reduce production costs and have been detailed .

Remarkable fact are that overvoltage level related to power switching is kept to a very low level, thanks to an innovative dies implementation and that thermal performances are very good, comparable to COTS power module.

This automotive quality level concept could be extended to satellite application into a next future.

5. REFERENCES

[1]. C. Lihua, F. Z. Peng, and C. Dong (2008), "A smart gate drive with self-diagnosis for power MOSFETs and IGBTs", in *Proc. 23rd Annu. IEEE Appl. Power Electron. Conf. Expo., pp. 1602–1607.*

[2]. V. Choudhary, E. Ledezma, R. Ayyanar, and R. M. Button (2008), "Fault tolerant circuit topology and control method for input-series and output-parallel modular DC-DC converters", *IEEE Trans. Power Electron., vol. 23, no. 1,pp. 402–411.*

[3]. B. Francois and J. P. Hautier (2002), "Design of a fault tolerant control system for a N.P.C. multilevel inverter", in *Proc. IEEE Int. Symp. Ind. Electron. L'Aquila, Italy, vol. 4, pp. 1075–1080.*

[4]. B.A. Welchko, TA. Lipo, ,T.M Jahns, S.E. Schulz (2004) : "Fault tolerant three-phase AC motor drive topologies, a comparison of features, cost and limitations", *IEEE Transactions on Power Electronics, Vol.19, Issue4, pp.1108-1116.*

DOI: 10.1051/e3sconf/20171616001

Si-DOPED InAs/GaAs QUANTUM DOT SOLAR CELL WITH AlAs CAP LAYERS

Dongyoung Kim [(1), *], Mingchu Tang [(1)], Jiang Wu [(1)], Sabina Hatch [(1)], Yurii Maidaniuk [(2)], Vitaliy Dorogan [(2)], Yuriy I. Mazur [(2)], Gregory J. Salamo [(2)], Huiyun Liu [(1)]

[(1)] Department of Electronic and Electrical Engineering, University College London, Torrington Place, London WC1E 7JE, United Kingdom,
[(2)] Institute for Nanoscience and Engineering, University of Arkansas, Fayetteville, Arkansas 72701, United States of America
[*] Email: d.kim@ucl.ac.uk

ABSTRACT

In this work, the effect of Si doping on InAs/GaAs quantum dot solar cells with AlAs cap layers is studied. The AlAs cap layers suppress the formation of the wetting layer during quantum dot growth. This helps achieve quantum dot state filling, which is one of the requirements for strong sub-bandgap photon absorption in the quantum dot intermediate band solar cell, at lower Si doping density. Furthermore, the passivation of defect states in the quantum dots with moderate Si doping is demonstrated, which leads to an enhancement of the carrier lifetime in the quantum dots, and hence the open-circuit voltage.

1. INTRODUCTION

Since the concept of the intermediate band solar cell (IBSC) was proposed in 1997, significant efforts have been made to realise IBSCs with efficiencies that exceed the Shockley-Queisser limit of 31 % [1]. The IBSC utilises a collection of intermediate levels within the bandgap, called the intermediate band (IB), to absorb sub-bandgap energy photons. Its potential to exceed the Shockley-Queisser limit and reach the theoretical limit of 63.2 % [2] arises from the additional photocurrent generated by the sub-bandgap photon absorption. One of the candidates for the high-efficiency IBSC is the quantum dot intermediate band solar cell (QD-IBSC), which utilises the discrete nature of the carrier density of states and the bandgap tunability of quantum dots (QDs). In a QD-IBSC, the IB is formed by the confined electron states in the QD array.

However, there have been several challenges in implementing high-efficiency QD-IBSCs. Firstly, thermal decoupling needs to be achieved between the IB and the conduction band (CB) of wetting layer (WL). The thermal coupling between the IB and CB enables the carriers to be thermally excited or relaxed between the IB and the CB [3]–[5]. This thermal process leads to suppression of the second-photon absorption, and hence a lower open-circuit voltage (V_{OC}) [6], [7]. Tutu et al. have demonstrated suppression of thermal escape of electrons in QDs by removing the WL [8]. Lam et al. have also reported reduction of thermal coupling of QD states from the WL by introducing a potential barrier between the QDs and WL via Si doping [9]. Secondly, the accumulated strain from the QDs leads to the formation of threading dislocations [10]–[12]. The strain-induced dislocations of the QDs result in short minority carrier lifetime, and cause difficulties in increasing the number of the QD layers that are needed to maximise the photon absorption by QDs [11]–[14]. Furthermore, the performance of IBSCs relies on an IB that is partially filled with electrons [15], [16]. For strong sub-bandgap photon absorption, the IB needs to have empty states to receive the electrons pumped from the VB, and states filled with electrons to pump electrons to the CB. Luque et al. proposed the use of doping as a method to achieve partial filling of the confined states in the IB [17].

In this work, we report the saturation of strain-induced dislocations and QD state filling by Si-doping QDs in InAs/GaAs QDSCs with AlAs CLs. Previously, we demonstrated that the deposition of AlAs CLs on InAs QDs could suppress the formation of the WL [8]. Consequently, the effective bandgap of the QDSC was increased, which in turn led to the increase in the thermal activation energy and the V_{OC}. In this study, in addition to the enhancement of the V_{OC} by applying AlAs CLs, a further increase in the V_{OC} by ~ 44 mV is observed by passivating the defect states with moderate Si doping (6 e/dot). In addition, QD state filling, which is essential for strong two-photon absorption, is observed with significantly lower Si doping densities when compared with that of our previous study [9].

2. GROWTH DETAILS

The InAs/GaAs QDSC samples with AlAs CLs were grown by a solid-source molecular beam epitaxy (MBE) on n^+ GaAs (100) substrates. All SCs were grown with the same p-i-n structure that consists of a 200 nm n^+ GaAs buffer layer, 30 nm n^+ $Al_{0.35}Ga_{0.65}As$ back surface field, 1000 nm n^- GaAs base, 250 nm p GaAs emitter, 30 nm p $Al_{0.75}Ga_{0.25}As$ window layer, and p^+ 50 nm GaAs contact layer. The intrinsic region of the SCs consists of 20 stacks of InAs QDs with 2 ML AlAs CLs

 DOI: 10.1051/e3sconf/20171616001

separated by a 20 nm GaAs spacer. The QDs were grown by the Stranski-Krastanov mode at ~ 500 °C. The QDSCs were Si-doped with four different doping densities of 0, 6, 12, and 18 e/dot.

Atomic-force microscopy (AFM) was used to analyse the morphology of the InAs QDs grown on GaAs. The average diameter of the QDs measured from the AFM image (shown in Figure 1) was ~ 30 nm, with an average height of ~ 5 nm. No large defective clusters were observed, which indicates a high structural quality [18]. The in-plane density of the InAs QDs was estimated to be ~ 2.3×10^{10} cm^{-2}.

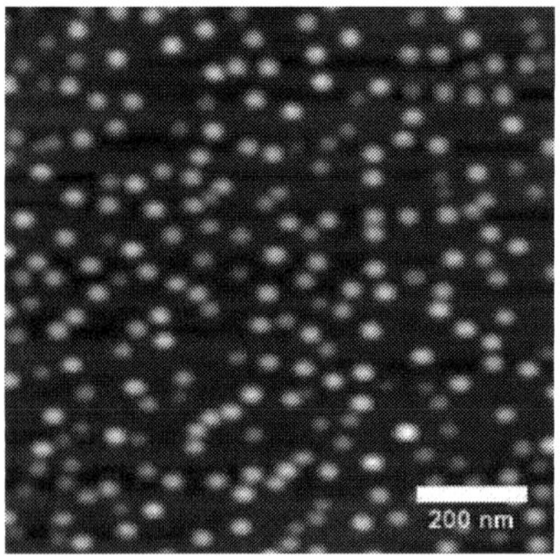

Figure 1 AFM images of InAs QDs grown on GaAs in 2D (top) and 3D (bottom).

3. OPTICAL CHARACTERISTICS

The optical properties of the Si-doped InAs/GaAs QDs with AlAs CLs were compared. Figure 2 (a) shows the normalised PL spectra for the QDSCs at 10 K. The spectra displays additional QD emission peaks (950 – 1100 nm) for the samples with higher Si doping. This can be explained by the emissions related to excited state transitions observed at shorter wavelengths alongside the ground state emission [19]. In contrast to our previous study [9], the formation of WLs by applying AlAs CLs on QDs during the MBE growth [8]. In this manner, most Si dopants are concentrated inside the QDs rather than the WL, and contributes to state filling. Figure 2 (b) shows the integrated PL intensity as a function of Si doping density, at 10 K and 300 K. It can be seen that the QDSC with doping density of 6 e/dot exhibits the strongest PL peak intensity. The enhancement of the PL intensity can be attributed to the passivation of the defect states by Si doping, which leads to suppression of the non-radiative process [20], [21]. Figure 2 (b) shows that any further increase in the Si doping level results in a dramatic decrease in the PL intensity. This can be due the presence of too many Si atoms, which can cause destruction of the crystal lattice of the InAs QDs and formation of non-radiative recombination centres [13].

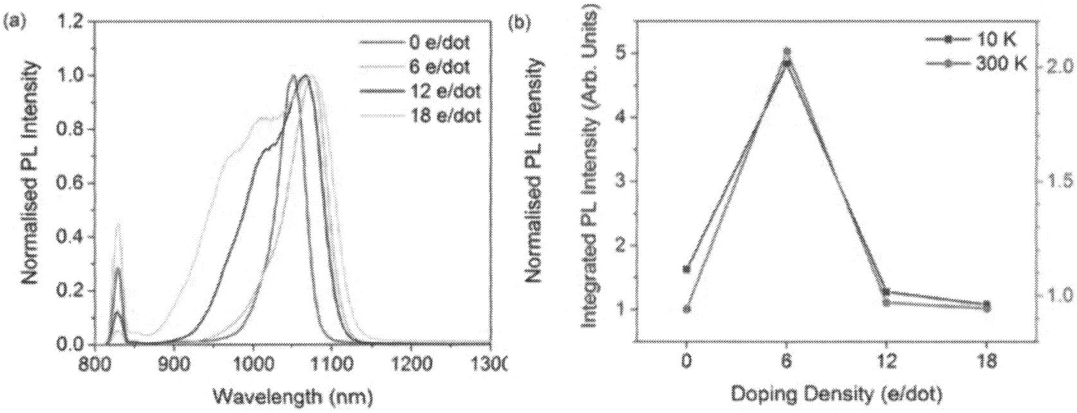

Figure 2 (a) Normalised PL spectra of Si-doped QDSCs with AlAs cap layers measured at 10 K (P_{ex} = 37 mW), and (b) Integrated PL intensity vs. Si doping density at 10 K (I_{ex} = 386 W/cm^2, λ_{ex} = 532 nm) and 300 K (I_{ex} = 459 W/cm^2, λ_{ex} = 635 nm).

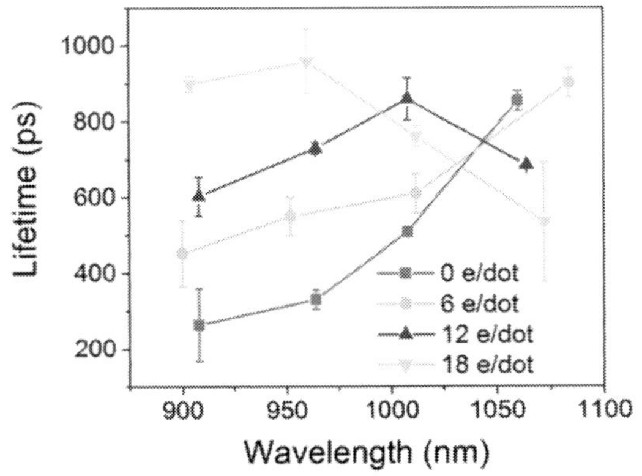

Figure 3 Carrier lifetime versus wavelength obtained from the transient photoluminescence spectra of the Si-doped QDSCs with AlAs cap layers at 10 K (λ_{ex} = 750 nm)

In order to provide further evidence of the QD state filling and the passivation of defect states achieved by Si doping, a time-resolved PL study was performed for the QDCSs at 10 K. Figure 3 shows that the carrier lifetime in the region of 875 – 925 nm increases with increasing Si doping density. This can be linked to the state filling effect observed from the steady-state PL measurements in Figure 2 (a). The appearance of the higher-energy QD emission peaks with higher Si doping densities in Figure 2 (a) indicates that the additional electrons supplied by the Si dopants are used to fill the QD states. With more QD states filled with electrons, fewer available decay channels exist for the carriers, which in turn results in longer carrier lifetimes. In contrast, the carrier lifetime of the QD ground state (1050 – 1100 nm) in Figure 3 shows that increasing the Si doping density leads to an initial increase with moderate doping density (6 e/dot), followed by a significant drop in the carrier lifetime with higher doping densities (12, 18 e/dot). This trend is similar to that observed in the integrated PL intensity measurements, as shown in Figure 2 (b). Therefore, it can be inferred that the initial increase in the carrier lifetime with doping density of 6 e/dot originates from the passivated defect states. However, once all defect states are filled, further addition of Si atoms leads to the formation of non-radiative recombination centres, and/or increase Auger scattering rate, which decreases the carrier lifetime.

4. SOLAR CELL PERFORMANCES

The EQE spectra of the Si-doped QDSCs with AlAs CLs are presented in Figure 4. For all doped samples, there is a drop in EQE at ~ 870 nm which corresponds to the bandgap of GaAs. At the higher wavelengths, flat spectral responses without a WL peak (~ 915 nm) are observed. It can be also noted from Figure 4 that the supra-bandgap (400 – 870 nm) absorption of all QDSCs with Si doping is significantly reduced when compared with that of the undoped QDSC. This can be attributed to the decrease in depletion region after Si doping, which reduces the effective absorption area.

The J-V characteristics of the QDSCs in Figure 5 show a clear dependence on the Si-doping density. Firstly, all Si-doped samples display a significant drop in the current density when compared with the undoped sample. This can be related to the reduction in the EQE contribution from the supra-bandgap region observed in Figure 4. Secondly, a further gradual decrease in the current density with increasing doping is observed amongst the doped QDSCs. This is likely to be due to the point defects formed by Si dopants substituting Ga and As or existing as interstitials [22], [23]. Although Si doping has a negative impact in the current density, an enhancement of the V_{OC} is observed with moderate doping density, which is likely to be due to the moderate Si doping passivating the defect states.

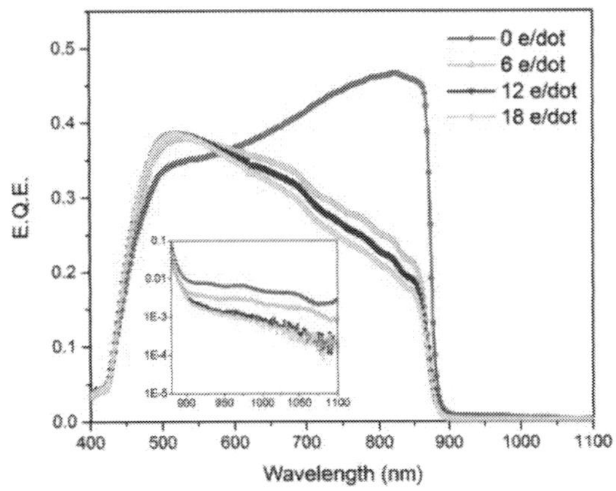

Figure 4 Linear-scale external quantum efficiency spectra of Si-doped QDSCs with AlAs cap layers. Inset shows semi-log scale of sub-bandgap external quantum efficiency.

DOI: 10.1051/e3sconf/20171616001

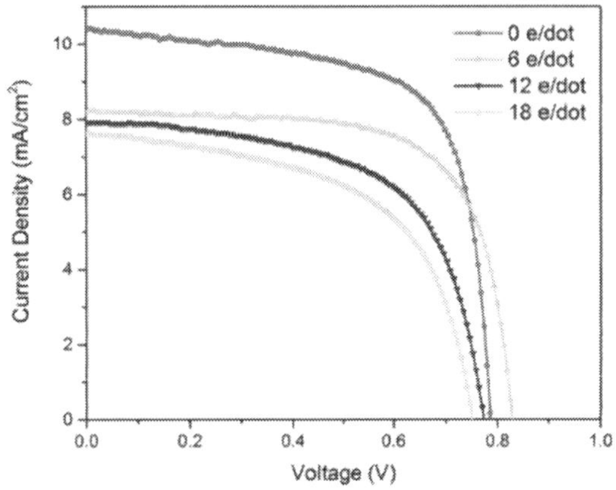

Figure 5 Current density versus voltage behaviour of Si-doped QDSCs with AlAs cap layers under one-sun (AM .5G) illumination

5. CONCLUSION

In conclusion, an improvement in the V_{OC} (~ 44 mV) has been achieved by introducing direct Si doping to QDs with AlAs CLs. The increase in the V_{OC} is attributed to the passivation of the defect states with moderate Si doping (6 e/dot), alongside the reduced CB-IB thermal coupling with AlAs CLs. Also, with the AlAs CLs, the QD state filling effect is observed with low Si doping densities. The factors that contributed to the decrease in the supra-bandgap EQE and the photocurrent after Si doping, such as the decrease in depletion region and the formation of point defects, should be taken into account when designing future Si-doped QDSCs. Nonetheless, the results presented in this article hold some promise for overcoming some of the main challenges in implementing QD-IBSCs with the efficiencies close to that of the theoretical model of the IBSC.

6. REFERENCES

[1] W. Shockley and H. J. Queisser, 'Detailed Balance Limit of Efficiency of p-n Junction Solar Cells', *J. Appl. Phys.*, vol. 32, no. 3, pp. 510–519, Mar. 1961.

[2] A. Luque, A. Martí, and C. Stanley, 'Understanding intermediate-band solar cells', *Nat. Photonics*, vol. 6, no. 3, pp. 146–152, Mar. 2012.

[3] S. Sanguinetti, M. Henini, M. Grassi Alessi, M. Capizzi, P. Frigeri, and S. Franchi, 'Carrier thermal escape and retrapping in self-assembled quantum dots', *Phys. Rev. B*, vol. 60, no. 11, pp. 8276–8283, Sep. 1999.

[4] E. Antolín, A. Martí, C. D. Farmer, P. G. Linares, E. Hernández, A. M. Sánchez, T. Ben, S. I. Molina, C. R. Stanley, and A. Luque, 'Reducing carrier escape in the InAs/GaAs quantum dot intermediate band solar cell', *J. Appl. Phys.*, vol. 108, no. 6, p. 64513, Sep. 2010.

[5] A. Luque and A. Martí, 'The Intermediate Band Solar Cell: Progress Toward the Realization of an Attractive Concept', *Adv. Mater.*, vol. 22, no. 2, pp. 160–174, Nov. 2010.

[6] C. G. Bailey, D. V. Forbes, R. P. Raffaelle, and S. M. Hubbard, 'Near 1 V open circuit voltage InAs/GaAs quantum dot solar cells', *Appl. Phys. Lett.*, vol. 98, no. 16, p. 163105, Apr. 2011.

[7] A. Mellor, A. Luque, I. Tobías, and A. Martí, 'Realistic Detailed Balance Study of the Quantum Efficiency of Quantum Dot Solar Cells', *Adv. Funct. Mater.*, vol. 24, no. 3, pp. 339–345, Jan. 2014.

[8] F. K. Tutu, P. Lam, J. Wu, N. Miyashita, Y. Okada, K.-H. Lee, N. J. Ekins-Daukes, J. Wilson, and H. Liu, 'InAs/GaAs quantum dot solar cell with an AlAs cap layer', *Appl. Phys. Lett.*, vol. 102, no. 16, p. 163907, Apr. 2013.

[9] P. Lam, S. Hatch, J. Wu, M. Tang, V. G. Dorogan, Y. I. Mazur, G. J. Salamo, I. Ramiro, A. Seeds, and H. Liu, 'Voltage recovery in charged InAs/GaAs quantum dot solar cells', *Nano Energy*, vol. 6, pp. 159–166, May 2014.

[10] H. Y. Liu, I. R. Sellers, T. J. Badcock, D. J. Mowbray, M. S. Skolnick, K. M. Groom, M. Gutiérrez, M. Hopkinson, J. S. Ng, J. P. R. David, and R. Beanland, 'Improved performance of 1.3μm multilayer InAs quantum-dot lasers using a high-growth-temperature GaAs spacer layer', *Appl. Phys. Lett.*, vol. 85, no. 5, pp. 704–706, Aug. 2004.

[11] F. K. Tutu, I. R. Sellers, M. G. Peinado, C. E. Pastore, S. M. Willis, A. R. Watt, T. Wang, and H. Y. Liu, 'Improved performance of multilayer InAs/GaAs quantum-dot solar cells using a high-growth-temperature GaAs spacer layer', *J. Appl. Phys.*, vol. 111, no. 4, p. 46101, Feb. 2012.

[12] A. Martí, N. López, E. Antolín, E. Cánovas, A. Luque, C. R. Stanley, C. D. Farmer, and P. Díaz,

'Emitter degradation in quantum dot intermediate band solar cells', *Appl. Phys. Lett.*, vol. 90, no. 23, p. 233510, Jun. 2007.

[13] X. Yang, K. Wang, Y. Gu, H. Ni, X. Wang, T. Yang, and Z. Wang, 'Improved efficiency of InAs/GaAs quantum dots solar cells by Si-doping', *Sol. Energy Mater. Sol. Cells*, vol. 113, pp. 144–147, Jun. 2013.

[14] T. Sugaya, Y. Kamikawa, S. Furue, T. Amano, M. Mori, and S. Niki, 'Multi-stacked quantum dot solar cells fabricated by intermittent deposition of InGaAs', *Sol. Energy Mater. Sol. Cells*, vol. 95, no. 1, pp. 163–166, Jan. 2011.

[15] A. Martí, L. Cuadra, and A. Luque, 'Partial filling of a quantum dot intermediate band for solar cells', *IEEE Trans. Electron Devices*, vol. 48, no. 10, pp. 2394–2399, Oct. 2001.

[16] A. Martí, N. López, E. Antolín, E. Cánovas, C. Stanley, C. Farmer, L. Cuadra, and A. Luque, 'Novel semiconductor solar cell structures: The quantum dot intermediate band solar cell', *Thin Solid Films*, vol. 511–512, pp. 638–644, Jul. 2006.

[17] A. Luque and A. Martí, 'A metallic intermediate band high efficiency solar cell', *Prog. Photovolt. Res. Appl.*, vol. 9, no. 2, pp. 73–86, Mar. 2001.

[18] H. Y. Liu, I. R. Sellers, M. Gutiérrez, K. M. Groom, W. M. Soong, M. Hopkinson, J. P. R. David, R. Beanland, T. J. Badcock, D. J. Mowbray, and M. S. Skolnick, 'Influences of the spacer layer growth temperature on multilayer InAs/GaAs quantum dot structures', *J. Appl. Phys.*, vol. 96, no. 4, pp. 1988–1992, Aug. 2004.

[19] M. V. Marquezini, M. J. S. P. Brasil, J. A. Brum, P. Poole, S. Charbonneau, and M. C. Tamargo, 'Study of temperature-dependent exciton dynamics in a single quantum well with self-assembled islands', *Surf. Sci.*, vol. 361–362, pp. 810–813, Jul. 1996.

[20] K. A. Sablon, J. W. Little, V. Mitin, A. Sergeev, N. Vagidov, and K. Reinhardt, 'Strong Enhancement of Solar Cell Efficiency Due to Quantum Dots with Built-In Charge', *Nano Lett.*, vol. 11, no. 6, pp. 2311–2317, 2011.

[21] T. Inoue, S. Kido, K. Sasayama, T. Kita, and O. Wada, 'Impurity doping in self-assembled InAs/GaAs quantum dots by selection of growth steps', *J. Appl. Phys.*, vol. 108, no. 6, p. 63524, Sep. 2010.

[22] D. T. J. Hurle, 'A comprehensive thermodynamic analysis of native point defect and dopant solubilities in gallium arsenide', *J. Appl. Phys.*, vol. 85, no. 10, pp. 6957–7022, May 1999.

[23] D. T. J. Hurle, 'A thermodynamic analysis of native point defect and dopant solubilities in zinc-blende III–V semiconductors', *J. Appl. Phys.*, vol. 107, no. 12, p. 121301, Jun. 2010.

DOI: 10.1051/e3sconf/20171616002

SUGGESTED TEST SEQUENCE AMENDMENTS OF THE CURRENT ECSS E-ST-20-08C STANDARD FOR BARE SOLAR CELLS QUALIFICATION

Blanco, G. [(1)], **Jüngst, G.** [(2)], **Grás, A.** [(1)]

[(1)] *INTA-SPASOLAB, Careterra de Ajalvir km. 4, 28850 Torrejón de Ardoz, Spain, Email: blancomg@inta.es, grassa@inta.es*

[(2)] *ISDEFE Consulting, C/ Beatriz de Bobadilla 3, 28040 Madrid, Spain, Email: jungstg.pers_externo@inta.es*

ABSTRACT

INTA-SPASOLAB as independent testing facility for more than 25 years has an extensive expertise in the certification and qualification of photovoltaic devices for space applications. Recently INTA-SPASOLAB has participated in the qualification of the latest technology upgrade of lattice-matched III-V GaInP/GaInAs/Ge triple-junction (TJ) bare solar cells manufactured by AZUR SPACE Solar Power GmbH (Germany). This participation has consisted of carrying out test sequences of subgroup B, Beginning of Life (BOL) performance, C1 (Electron Irradiation) and P (Proton Irradiation) according to the current ECSS E-ST-20¬08C, Rev.1 reference standard.

At these subgroup qualifications, some performance tests, in particular spectral response and electric performance tests, have been agreed to be either modified in the test sequence order or added, in order to understand better the evolution of the electrical cell parameter results.

In this work, test sequence variations and obtained results on those subgroups B, C1 and P, are presented. The gained experience at this qualification and other previous work about solar cell characterization is used to suggest amendments.

1. INTRODUCTION

Solar cells have been used for electrical power generation since the Vanguard 1 satellite mission. One of the main problems encountered on their mission operation is their vulnerability to the harmful environmental space conditions, and in consequence a lost in power generation.

For this and other reasons of space product assurance, the European Space Agency (ESA) and European industry associations established so called European Cooperation for Space Standardisation (ECSS) qualification standards. In particular, for photovoltaic solar cell assemblies the document ECSS-E-ST-20-08C, Rev 1 is the reference standard **[1]**.

2. TEST SECUENCE DESCRIPTION

All test measurements of this qualification have been carried out by following accordingly the test sequence and requirements of the ECSS-E-ST-20-08C standard.

In addition, following test deviations have been agreed, as highlighted in ***Table 1***:

Subgroup B (BOL Performance)	Subgroup C1 (e- Irradiation)	Subgroup P (p+ Irradiation)
01VI	01VI	01VI
02DW	02DW	02DW
04FT	03EP	03EP
05EP	04DC	04DC
06DC	05SR[**]	05SR[**]
07PH	06EI	06PI
07EP[*]	07VI	07VI
07TA	08EP[*]	08EP[*]
08EP	08PH	08PH
09DC	08EP[*]	08EP[*]
10TC&TB	08TA	08TA
11OP	08EP[*]	08EP[*]
12RB	09EP	09EP
13EP	10DC	10DC
14VI	11TC&TB	11TC&TB
15SR[****]	12RB	12SR[***]
	13EP	13VI
	14SR[***]	14SR[****]
	15VI	
	16SR[****]	

[*] Additional EP step included
[**] SR step moved before 03EP
[***] SR step moved before 09EP
[****] SR step at extreme temperatures added

***Table 1**. Subgroups B, C1 and P Qualification Test Plan for Bare Solar Cells, plus agreed modifications (marked in colour) according to ECSS-E-ST-20-08C, Rev 1.*

(1) One additional electric performance (EP) test between photon irradiation (PH) and thermal annealing (TA) and three additional electric performance (EP) tests, first EP before PH and second EP after PH and before TA, were included in subgroups B, C1 and P, respectively, in order to determine separately possible effects of photon irradiation and thermal annealing on electrical cell performance.

(2) Spectral response measurements (SR) planned at step 05 have been anticipated before test step 03 EP in both subgroups, C1 and P, respectively, so that electrical cell performance of beginning of life (BOL) cells could be determined more accurately applying spectral mismatch factor correction before the next electron or proton irradiation test step.

(3) SR measurements at lowest (1175°C) and highest (+140°C) cell temperature have been added as final test in subgroup B, C1 and P, respectively with two

main objectives. Firstly, to determine possible temperature dependent spectral response degradation of EOL cells compared to those results at 28°C (this is not reported in this paper). Secondly, to compute spectral mismatch correction factors at these two temperatures and to determine potentially the impact on temperature coefficient (TC) results of electrical cell parameters.
In this work the results of these agreed deviations (1) to (3) are presented. Eventually, it is suggested that at test sequence of subgroup B, C1 and P some EP and SR testing steps might be added.

3. RESULTS

3.1 Electrical cell performance stability after photon irradiation and thermal annealing

According to the applied ECSS standard, the stability of the electric solar cell performance under equivalent light and temperature of 1 solar constant (s.c.) AM0 conditions shall be verified by the two-step test, photon irradiation (PH) and subsequent thermal annealing (TA), which is to be performed at BOL, after the Electron Irradiation and after the Proton Irradiation test step, respectively. For a better analysis of the relative changes of electrical cell parameters before and after the PH or TA test, additional electrical performance measurements have been included in between the PH and TA test step. The obtained changes are highlighted in Figure 1 for the selected parameters short circuit current (Isc), maximum power (Pmax) and open circuit voltage (Voc). For non-irradiated BOL cells solar cell performance has been within determined measurement uncertainty. By doing so, this performance has been considered to be stable. A pass-fail criterion of this verification item is not written explicitly in the ECSS standard. Concerning the irradiated cells, short circuit current retained its value within the 1% measurement uncertainty whose contribution is made up of, while the after EI or PI decreased Pmax and Voc results recovered. At the 1 MeV electrons per cm2 irradiated cells this recovery was by tendency increasing continuously from lowest to highest fluence. In particular, from the 5E+14 to the 1E+15 fluence, the recovery contribution of Pmax might be addressed to thermal annealing, but, since this increase stays within a 1% standard measurement uncertainty, this is not sufficient evidence. Hence, taking into account the applied fluences at EI and PI testing, it is considered that there is no need to add any EP tests related to this PH plus TA test, respectively.

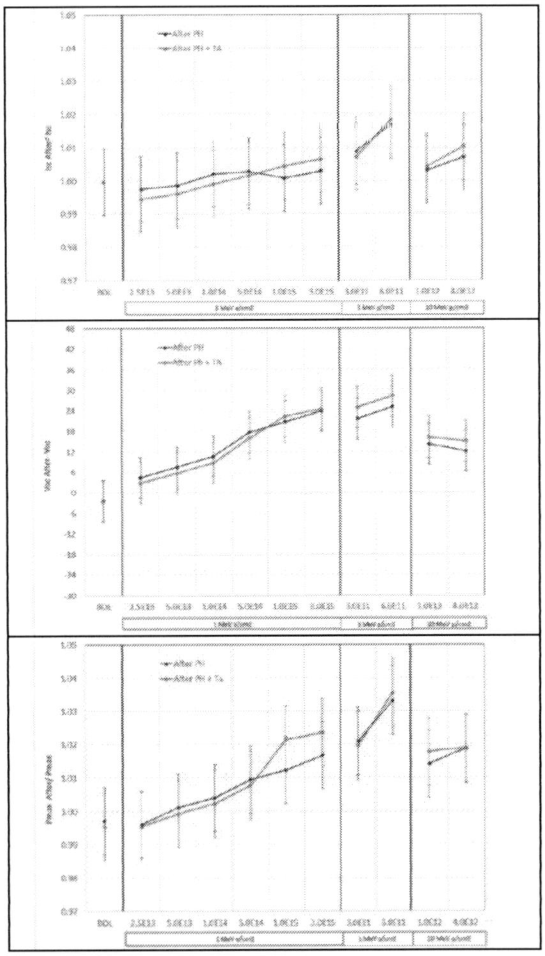

Figure 1: Obtained relative changes of Isc, Voc and Pmax of BOL, electron or proton irradiated cells after photon irradiation (PH) and thermal annealing (TA) test steps, respectively.

3.2 Accuracy benefits from Spectral Response Measurements before Electrical Performance Characterization

At the electric performance (EP) tests first 1 s.c. AM0 effective irradiance conditions on the solar simulator are adjusted, as described in [2], by using single component reference (SCRC) cells calibrated in Isc with regards to the WMO reference spectrum, spatial irradiance uniformity corrections, if applicable, and spectral mismatch correction factors Mj at each subcell j of the triple junction (TJ) test cell. By applying these Mj factors, this adjustment can be set more accurately with the benefit of more accurate and reliable electrical cell parameter results.
At the computation of Mj according to equation (1), the

 DOI: 10.1051/e3sconf/20171616002

spectral response SR_{test} of each TJ cell's subcell j need to be determined previously before the EP test:

$$M_j(E_k) = \frac{\int SR^j_{test} E_{AM0} d\lambda \cdot \int SR^j_{ref} E_k d\lambda}{\int SR^j_{test} E_k d\lambda \cdot \int SR^j_{ref} E_{AM0} d\lambda} \quad (1)$$

where SR^j is the spectral response of the SCRC cell, E_{AM0} the spectral distribution of the AM0 reference spectrum WMO, while E_{SIM} corresponds to the spectral irradiance of the solar simulator's light sources measured by a calibrated spectroradiometer.

Due to the fact that in the current ECSS the SR test was either not included in the test sequence of subgroup B, or in subgroup C1 and P was chronologically situated after the 03-EP step and as well after the 09-EP step, which is the step after electron or proton irradiation, SR tests have been performed before the aforementioned EP tests in this qualification test.

In this section spectral mismatch factor results Mj at respective testing steps are presented except those Mj of the Germanium (Ge) bottom subcell. Due to the excessive short circuit current generation of the Ge subcell of the TJ test cells compared to the GaInP and InGaAs subcell currents at EP tests, Mj correction of this subcell was not relevant.

It is to be pointed out that the employed SCRC cells were 3G30 isotype component cells of the triple junction test cells to be qualified with same cell dimension and geometry.

Regarding to the BOL EP tests (05-EP of subgroup B, and 03-EP of C1 and P), Mj results were 0.994 +/-0.001 for the GaInP (top) subcells, while 1.001 +/-0.001 for the InGaAs (middle) subcells, respectively.

This means, that, without applying spectral mismatch correction at these EP tests, the BOL TJ testing cells, whose current limiting subcell is precisely the GaInP subcell, are illuminated by a 0.6% less light intensity than required. As consequence, the measurement uncertainty u(Ycell) of electrical cell parameter Ycell increases, especially in Isc, Impp and Pmax. Apart from the 1% calibration standard uncertainty of the SCRC, the u(Ycell) increases by this additional error. The amount of increase depends on the sensitivity coefficient δYcell/δMj, which depends on the cell type and can be determined at Spasolab [3] by spectrometric characterization [4].

Regarding to the EP tests at 28°C after electron or proton irradiation (09-EP and 13-EP in subgroup C1 and 09-EP in subgroup P), the spectral mismatch factor changes with respect to its BOL value have been determined for the respective GaInP and InGaAs subcell of the End-Of-Life (EOL) testing cells after electron or proton irradiation step at different fluences.

As highlighted in *Figure 2*, spectral mismatch factors of the BOL results changed for the InGaAs subcell slightly increasing, but by tendency towards the maximum fluences, being 0.3% the maximum deviation at 3E+15 1MeV electron fluence per cm^2 in this qualification test.

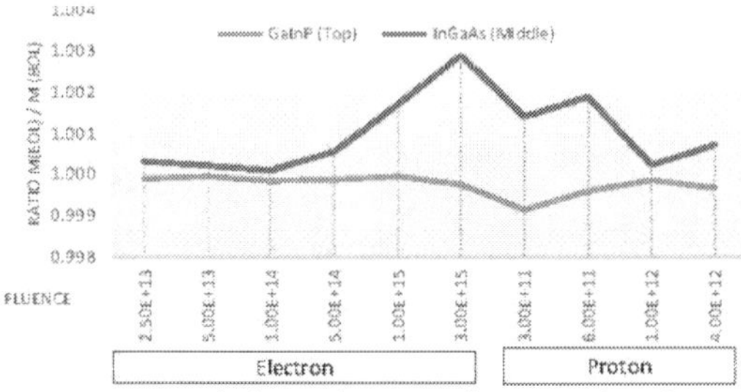

Figure 2: Relative change of spectral mismatch correction factor of EOL testing cells after electron or proton irradiation at different fluences with respect to its BOL value.

3.3 Spectral Response Measurements at extreme temperatures for temperature coefficient analysis

Spectral Response measurements of the EOL TJ testing cells of subgroup C1 and P at the extreme cell temperatures -175°C and +140°C have been included as additional step, 16-SR, as depicted in Table 1, at the test sequence end. The objectives of these SR tests have been twofold:

(i) Firstly, to determine the change of spectral mismatch correction factors M_j (T) at these extreme temperatures with regards to 28°C, and

(ii) Secondly, at test step 11-TC, to assess the impact of the produced error on electrical cell parameters Ycell at extreme temperatures when not using at the temperature coefficient (TC) measurements of Ycell, the temperature corresponding correction factor M_j (T), but Mj (T = 28° C) at 28°C instead.

The results of the first objective are highlighted in next two figures, *Figure 3* at -175°C and *Figure 4* at +140°C.

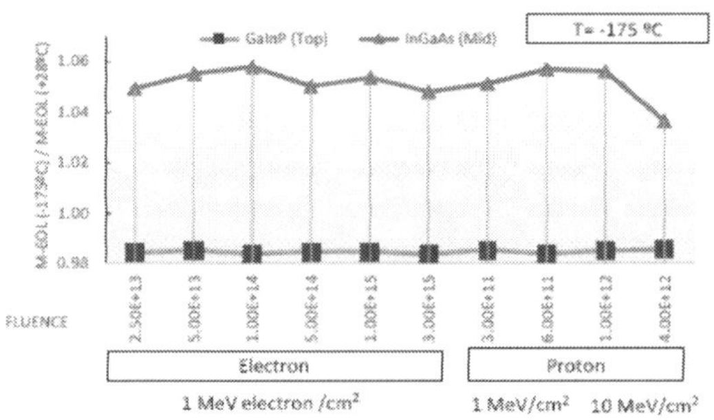

Figure 3: Relative change of spectral mismatch correction factor M at cell temperature of -175°C with respect value M at +28°C for EOL testing cells after electron or proton irradiation at different fluences.

 DOI: 10.1051/e3sconf/20171616002

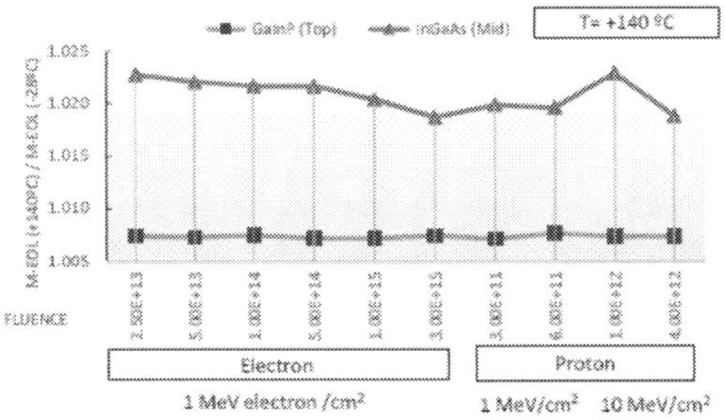

Figure 4: Relative change of spectral mismatch correction factor M at cell temperature of +140°C with respect value M at +28°C for EOL testing cells after electron or proton irradiation at different fluences.

While the relative changes of Mj (T) of the GaInP subcell at -175°C were constantly at 0.984 +/-0.001, these relative changes turned to 1.007 +/-0.001 at +140°C. In contrast, the relative changes Mj (T) of the InGaAs subcell have been superior to unity at both extreme temperatures, 1.05 at -175°C and 1.02 at +140°C, respectively, and varied in a broader value range +/-0.01. It is to be pointed out that these obtained mismatch results cannot be taken as general results, since these are not independent on the utilised solar simulator.

Regarding to the second objective, the error of applied M_j (T) affect to Ycell results depending on which the current limiting subcell at a certain temperature is, either the GaInP or to the InGaAs subcell in the tested TJ cells. While this effect can be determined directly in case of Isc, at other cell parameters, e.g. Pmax, the spectrometric characterization tool, described in [4], needs to be used.

In ***Figure 5*** the Isc case is illustrated on the example of an irradiated TJ testing cell, whose current limiting subcell has been the InGaAs subcell. Due to the relative change of the mismatch correction factor at +140°C, by 1.02, the resulting short circuit current at +140°C has been overestimated by 2%. In change, at lowest temperature of -175°C the overestimation of shortcircuit current has been major, around 5% according to the error change of 1.05.

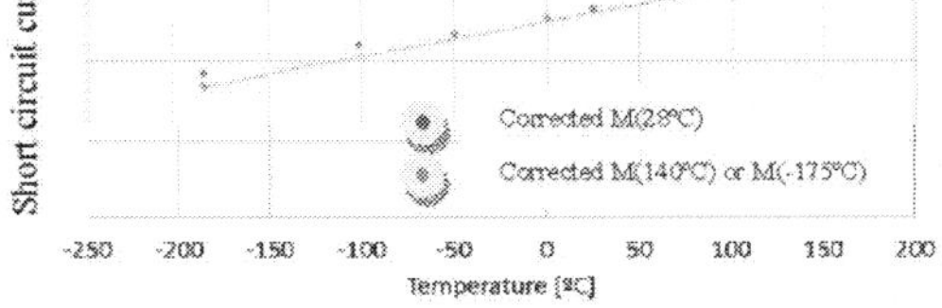

Figure 5: Isc results as effect of applied or not applied spectral mismatch correction at an irradiated TJ test cell of this qualification.

4. CONCLUSIONS

In the qualification of subgroup B, C1 and P of bare TJ solar cells, the agreed test variations on test sequence according to the ECSS-E-ST-20-08C, Rev 1 standard [1] and their obtained results on electrical cell performance have been presented in this work.

Related to the test of photon irradiation (PH) and thermal annealing (TA) to verify the solar cell performance stability, it is shown that electric cell parameter results after respective PH and TA step have not varied more than determined measurement uncertainty for the particular TJ test cells. As stability criterion the obtained measurement uncertainty has been utilised, although it is suggested to have an explicit criterion available in this ECSS standard. Concerning the electrical parameters (EP) characterization, before and after each irradiation, I-V curves were recorded for all cells to follow the evolution of the tested parameters. This method requires the characterization and accurate adjustment of the solar simulator. For doing this, it is suggested to perform spectral response measurements of the tested cells at different testing temperatures previous to the EP characterization at the same temperatures. This method allows achieving a more accurate match to the AM0 reference spectrum and controlling the uncertainty of the electrical cell parameters. It is to be pointed out that spectral mismatch corrections results are typical of each solar simulator. Hence, these corrections could not be used as a general case for all the used set up.

5. REFERENCES

[1] ECSS-E-ST-20-08C, Rev 1 (2012). Space Engineering Photovoltaic assemblies and components. ECSS Secretariat, ESA-ESTEC Requirements & Standards Division, Noordwijk, The Netherlands.

[2] A. Grás, et al., Analysis for Multi-Junction Solar Cell Measurements at Spasolab, Proc. '6th European Space Power Conference', Porto, Portugal, 577-580, 6-10 May 2003.

[3] G. Jüngst, A. Grás, et. al. "Measurement uncertainty for electric performance calibration of triple junction solar cells at Spasolab", 29th PVSEC conference, Paris (France), 2013.

[4] C. Baur, A. Bett, "Measurement uncertainties of the calibration of multi-junction solar cells", IEEE proceedings of 31st IEEE PVSC, Florida, 2005.

DEDICATED TOOL FOR IRRADIATION AND ELECTRICAL MEASUREMENT OF LARGE SURFACE SAMPLES ON THE BEAMLINE OF A 2.5 MEV PELLETRON ELECTRON ACCELERATOR: APPLICATION TO SOLAR CELLS

Jérémie Lefèvre*, Patrice Le Houedec, Jérôme Losco, Olivier Cavani, and Bruno Boizot

Laboratoire des Solides Irradiés, CNRS-UMR 7642, CEA-DRF-IRAMIS, Ecole Polytechnique, Université Paris-Saclay, 91128 Palaiseau Cedex, France

ABSTRACT

We designed a tool allowing irradiation of large samples over a surface of A5 size dimension by means of a 2.5 MeV Pelletron electron accelerator. *in situ* electrical measurements (I-V, conductivity, etc.) can also be performed, in the dark or under illumination, to study radiation effects in materials. Irradiations and electrical measurements are achievable over a temperature range from 100 K to 300 K.

The setup was initially developed to test real-size triple junction solar cells at low temperature within the framework of the ESA's JUICE mission. It will be made available to users at the beginning of 2017.

After a brief description of the SIRIUS irradiation facility hosted at Laboratoire des Solides Irradiés (LSI), this paper gives detailed information about the Large Surface Irradiated-Cell (LSIC) device. Preliminary results obtained during the ongoing qualification phase of the setup are also discussed.

Key words: SIRIUS; electron irradiation; triple jonction solar cell; JUICE mission.

1. INTRODUCTION

1.1. Laboratoire des Solides Irradiés

The LSI laboratory at Ecole Polytechnique (Palaiseau, France) is a mixed unit affiliated to CEA, CNRS and Ecole Polytechnique. It studies fundamental properties of the solid state and its diverse interactions with radiation (photons, electrons, ions).

The laboratory has a 2.5 MeV electron accelerator with the unique feature of having a large variety of *in situ* characterization techniques. This installation, opened in 2013, is called SIRIUS for *Système d'IRradiation pour l'Innovation et les Utilisations Scientifiques*.

*Email: jeremie.lefevre@polytechnique.edu

1.2. SIRIUS facility

The SIRIUS facility is equipped with a new generation Pelletron accelerator that delivers monoenergetic electrons beams between 150 keV and 2.5 MeV with a current beam ranging from 150 nA to 200 μA. It is equipped with two beamlines, and a host of equipment for online experiments, including two cryostats to study the low temperature electrical transport properties of irradiated materials. Room Temperature (RT) experiments include visible and near-infrared PhotoLuminescence (PL), cathodoluminescence, and Raman spectroscopy. Low temperature *in situ* electron paramagnetic resonance and PL will be installed on the SIRIUS beamline in 2017. Such a large variety of online equipments makes the SIRIUS facility a unique tool in the world.

The accelerator is planned to work more than 200 days per year. It is available to a large community of scientists and industrials from France and abroad. It has also been developed to be complementary with other irradiation facilities like ion beams for example. SIRIUS is actually part of the french national accelerator network for the study of materials under irradiation (EMIR). [1]

The scientific topics currently studied around the accelerator are :

- Effects of electron irradiation on material properties (mechanical, optical, magnetic and electrical transport in high TC superconductors);
- Ageing under irradiation;
- Characterization of induced defects (type of defects, displacement thresholds);
- Radio-grafting of polymers (in particular applications to biomaterials);
- Corrosion under irradiation.

Electron irradiation can also be used to simulate the radiative environment in space. For this purpose, the LSIC device has been developed to study the degradation of

DOI: 10.1051/e3sconf/20171616003

electrical properties in space solar cells. This setup will be first used to undertake analysis of real-size solar cells within the framework of the ESA's JUICE mission.

2. THE LARGE SURFACE IRRADIATED-CELL

2.1. Setup description

The LSIC device allows irradiation of a wide variety of samples up to a A5 size paper dimension. It includes the possibility to perform *in situ* 4-wire electrical measurements, ① in the dark, or under illumination through a window made of radiation-resistant quartz glass ② (Fig. 1). The setup can be coupled to the SIRIUS beamline by using a metal piece ③ including a water-cooled removable stopper actionable with a hydraulic piston to stop incident electrons.

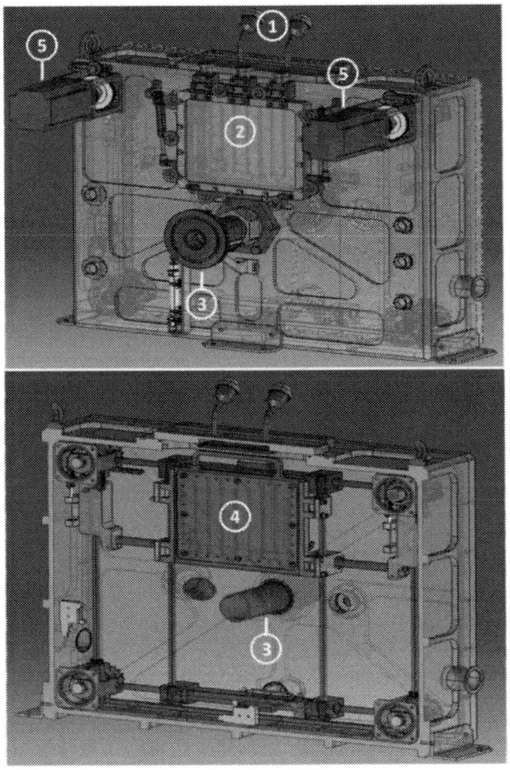

Figure 1. 3D views of the large surface cell device. (a) Back-side and (b) front-side without the hermetic lid.

The principle of the experiment consists of scanning the sample holder ④ on the electron beamline. Two servomotors ⑤ control the movement of pulleys and chains allowing a fine and smooth XY displacement of the sample holder.

A LabVIEW program drives the motion of the sample holder and manages a complete irradiation cycle. The surface to be irradiated, with the appropriate parameters (electron beam energy, fluence, current and temperature of irradiation) can be set easily and precisely. The displacement velocity of the sample holder is enslaved to a digital charge integrator to ensure a good homogeneity in terms of electron fluence all over the irradiated area.

The sample holder, shown in Fig. 2, is equipped with an adjustable system ⑥ intended to ensure both the positioning and maintaining in place of several samples simultaneously ⑦. This assembly made of conductive and insulating pieces includes grooves to guide wires in order to perform up to 4 4-wire *in situ* electrical measurements in a same configuration of experiments.

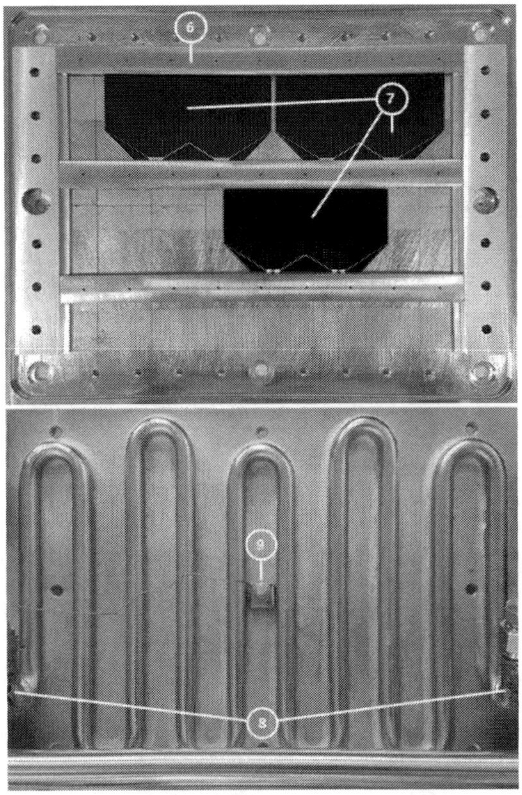

Figure 2. Pictures of the sample holder : (a) front-side and (b) back-side with the welded liquid nitrogen heat exchanger.

Liquid nitrogen circulating in a coil welded on the opposite side of the sample holder allows to work in a wide range of temperatures ranging from 100 K to 300 K (Fig. 2). A specific automatic proportional-integral-derivative algorithm is used to regulate the temperature by means of a coaxial thermocouple located the entrance of the heat exchanger. Temperatures values at the entrance and exit of the coil are obtained with 2 Cernox sensors ⑧. A third one is screwed at the middle of the sample holder ⑨ to get the temperature of samples during experiments.

2.2. Project Schedule Development

The qualification phase of the LSIC device has started in July 2016 and some results are discussed in the next part of this paper. Irradiation experiments with *in situ* electrical measurements will be performed by the end of this year. The public availability of the setup is planned for early 2017.

3. PRELIMINARY RESULTS

3.1. Sample holder displacement

Due to the harsh radiation environment during experiments at the SIRIUS facility, electronic components of the LSIC setup have to be deported a few meters from the beamline. The use of two-conductor shielded twisted pair cables reduces drastically electromagnetic interferences that is usually a major source of problems for systems using servomotors. Online tests have shown that the sample holder can be driven instantly and precisely with the LabVIEW control program (Fig. 3). This allows to choose the motion speed and velocity, as well as the scanning frequency of the sample holder.

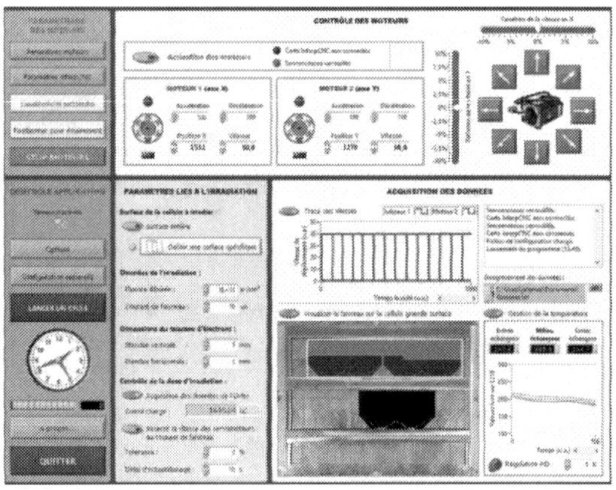

Figure 3. LabVIEW control program of the LSIC device.

The relative position of the sample holder on the electron beamline during irradiation experiments is represented in the LabVIEW interface by a red spot (Fig. 3). It includes an alpha component that allows to visualize the displacement history of the sample holder. This ensure an homogeneous irradiation all over the surface to be irradiated (Fig. 3).

3.2. Temperature control

Preliminary tests performed with the heat exchanger have shown a very satisfactory behavior in terms of cooling power capability of the LSIC device, since the sample holder can be cooled from RT to 80 K in around 30 mn (Fig.3).

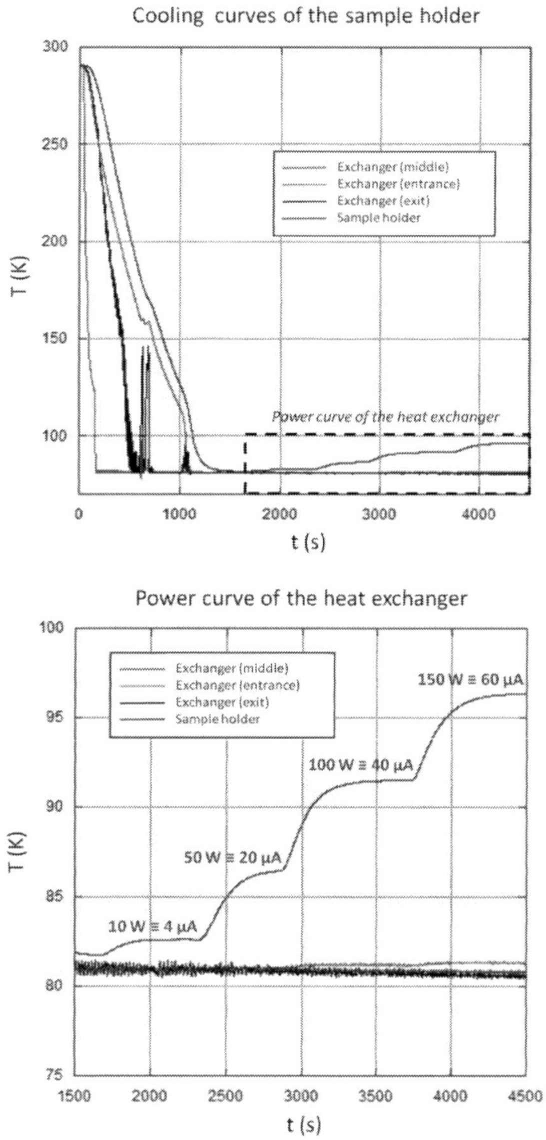

Figure 4. (up) Cooling down curves of the sample holder and (down) power curve of the heat exchanger.

High flux irradiation with a high beam current is expected to cause significant increase in sample temperature because of vast beam heating. This effect is due to the lost since electrons lose their energy mostly by electronic interaction in materials. The heating power transferred to a sample during irradiation (P_{th}) can be written as follows:

$$P_{th} = P_{el} - P_R$$

with P_{el} as the power of the incident electrons, and P_R as the power lost through X-ray radiation.

In order to simulate the power of the electron beam, a heating resistance was screwed of the sample holder. The P_{el} parameter can be expressed as follows:

$$P_{el} = U.I$$

with U as the energy of incident electrons in MeV, and I as the current beam in μA. In a conservative approach, P_R is not taken into account, so that the current beam can be directly deduced from the power supplied to the resistance.

Fig. 5 shows that the sample holder temperature is below 100 K for current beam up to 60 μA that is a significantly high value.

4. REFERENCES

[1] http://emir.in2p3.fr/-Le-reseau-EMIR-

DOI: 10.1051/e3sconf/20171616004

CHEOPS SOLAR CELL ASSEMBLIES LIFE TEST

Emilio Fernández Lisbona[(1)], Gianfelice D'Accolti[(1)], Joel Asquier [(1)], Robert Holloway [(1)], Bob Witteveen [(1)]

Pietro Zanella [(2)], Luigi Ferrante[(2)]

[(1)] *ESA-ESTEC - Postbus 299 – NL 2200 AG Noordwijk – The Netherlands, emilio.fernandez.lisbona@esa.int, Gianfelice.Daccolti@esa.int, Joel.Asquier@esa.int, Robert.Holloway@esa.int, Bob.Witteveen@esa.int*
[(2)]*Leonardo Finmeccanica, Viale Europa, 20014 Nerviano (MI) – Italy, Pietro.Zanella@leonardocompany.com, Luigi.Ferrante@leonardocompany.com*

ABSTRACT

Life tests are performed to determine the stability of solar cell assemblies under worst case operation conditions for long duration. This test is part of the qualification programme to which solar cell assemblies are submitted according to ECSS standards. This paper presents the test set-up configuration (recently implemented in the European Space Power Laboratory), test conditions and test results obtained of the solar cell assembly life test performed under the ESA CHEOPS satellite programme.

1. INTRODUCTION

1.1 The CHEOPS Mission

The CHEOPS (CHaracterizing ExoPlanet Satellite) mission is dedicated to searching for planet transits by means of ultrahigh precision photometry on bright stars already known to host planets. The mission was selected by the Science Programme Committee in November 2012 as the first small mission in the ESA Science Programme. The project passed the system CDR in May 2016 and is targeting a shared launch opportunity in 2018.

To cope with the very challenging schedule and budget, the design and implementation of CHEOPS is articulated around the re-use of an existing, "off-the-shelf" platform, the AS250 product line from Airbus DS with some minor adaptations. The aim of the modifications is to meet the power requirements of the mission (battery and solar array resizing) and to provide a cost efficient solution where a downsizing of elements is possible while satisfying the performances. The manufacture of the platform and the integration & testing of the satellite (including the integration of the scientific instrumentation) is carried out by AIRBUS DS Spain.

The satellite (Fig. 1) is composed of a prismatic body platform, an optical instrument (a photometer developed by a consortium led by the University of Berne) mounted on top of the platform, and three body-mounted solar array panels. The three body-mounted solar array panels have a 67 degree deviation between them, in order to achieve the best illumination in any mission attitude. See [1] for more information about CHEOPS mission.

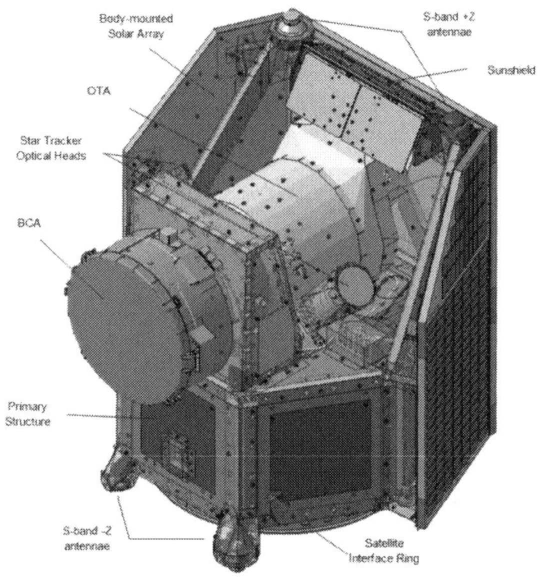

Figure 1. CHEOPS Satellite.

To account for the specific requirements of the CHEOPS mission in terms of sky visibility and stray-light rejection (e.g. minimizing Earth stray light) and the mission constraints (e.g. shared launch scenario), a dawn-dusk circular SSO (LTAN 6 a.m. or 6 p.m., and altitude from 650 to 800 km) was selected as the nominal orbit. Due to the attitude range and characteristics (e.g. roll around the line-of-sight to maintain the instrument's radiators pointed to cold space), the solar cells are subject to a large number of thermal cycles and elevated cell temperatures, specifically for the central panel, which has resulted in the need for delta-qualification tests of the solar cell assemblies.

 DOI: 10.1051/e3sconf/20171616004

1.2 Delta-Qualification Tests

The solar cell assembly delta-Qualification test for CHEOPS mission consisted of the performance of a life test in accordance with ECSS [2] and the specific mission requirements.

2. LIFE TEST DEFINITION

The solar cell assembly life test is specified to be performed in a vacuum (less than $2x10^{-5}$ mBar) and under 1 solar constant AM0 illuminated simulated conditions.

The duration of the life test is calculated to be considerably less than the real duration of the mission and follows the Arrhenius law, taking into account the total duration of the CHEOPS mission (3.5 years), solar cell activation energy (0.7eV, being a rather conservative value and in accordance with [2]) and the worst case solar array operational temperature during the mission (i.e. in the range between 120°C and 130°C). Therefore, CHEOPS mission conditions require a life test exposure much larger than for GEO and LEO missions and it has been calculated to be 6990 hours at 160°C.

The life test is split in 3 phases, where the solar cells are exposed at three different temperatures. Therefore, the duration of each phase is calculated to be 2000 hours at 140°C, 2000 hours at 160°C and 2078 hours at 175°C. The solar cells are loaded close to the maximum power point for 93% of the time of each phase. During the remaining 7%, the solar cells are biased in reverse voltage to simulate the worst case condition of solar cells current matching in the string with a 0.05V margin.

3. CHEOPS LIFE TEST SAMPLE

Leonardo-Finmeccanica provided the test sample plate (see Fig.2) consisting of 12 flight representative solar cell assemblies, that were previously qualified for GEO missions under ESA's MTG programme.

The substrate is an stainless steel plate of 300 x 300 mm². The use of stainless steel material helps to avoid cell cracking during vacuum and temperature exposure and provides a reasonable thermal conductivity.

The Solar cell assembly is comprised of 12 AzurSpace 3G30 Triple Junction solar cells with CMG 100 µm coverglass with AR coating and Ag plated Invar interconnectors. These are high grade solar cells. The solar cell shunt diode is an AzurSpace SIBD type, but they are not connected in parallel to the cell.

The bus bars are InvarAg single strips and the solar cell laydown adhesive and bus bar potting is RTV-S691.

Figure 2. CHEOPS life test sample.

The bus bar bonding patterns and adhesive is CV-2-1142 and the wire type is 3901-020 ESCC specification.

The plate is equipped with 6 thermocouples of type T bonded below 6 different solar cells.

The 4xPolymide heaters for active thermal control and attached with RTV-S691used for the MTG life test, were removed for the CHEOPS life test.

4. ESTEC LIFE TEST FACILITY

The test facility is located in the Solar Generator Laboratory of the Power Systems Division (TEC-EP) at ESTEC and is comprised of:

4.1 Thermal vacuum facility

The Vacuum Solar Cell Illumination Facility (VSCIF) is equipped with a quartz glass window that allows AM0 illumination of the solar cells samples. The VSCIF facility is equipped with a XDS10 scroll pump and a BOC Edwards STP-301/451 Series turbomolecular pump and the vacuum level is typically lower than 10^{-6} mbar. The VSCIF facility is equipped with a hot plate that uses flat Mica based heaters and allows exposure at constant temperature of the solar cell samples between 120°C and 400°C. No cold trap is used for the CHEOPS life test as the VSCIF facility walls are at much lower temperature (40-60°C) than the solar cells.

Calcium fluoride windows, placed in the chamber cold locations and in close proximity to the test samples, are used to monitor the deposition on these surfaces of organic contamination during the test.

4.2 Solar Simulator

The solar simulator (SESO SSMD 1x2,5-0-3,5) is a single source system with a Xenon arc lamp of 2500 W. The light beam is composed by reflection on 8 internally mounted mirrors. The SESO solar simulator is placed outside the vacuum facility. The vertical light beam illuminates the solar cell sample through the VSCIF facility window. The illuminated plane has a diameter of 350mm at a distance of 650mm from the exit window of the simulator.

The solar simulator performance is as follows:

- Intensity: 0.7 solar constant (SC) to 1.7 SC.
- Uniformity: 2.0% at 1 SC in the 350mm beam target diameter.
- Spectrum: AM0 at 1 SC, Class A [2].
- Long term stability: 0.2% over 24 h at 1 solar constant (SC).

4.3 Solar cells biasing electrical device

Solar cells under test are electrically connected to an electric switch box to allow operation in loaded mode and in reverse mode. Each cell is looped with two other solar cells of the same row by means of the open positive terminal. Two groups of 6 cells in parallel are connected to two different power supplies for working in reverse mode. Each solar cells are loaded with a resistor (2Ω) when working in loaded mode.

4.4 Data Acquisition Unit

The current equivalent and voltage are measured for each solar cell independently during the complete test. The DACQ also records the solar cell temperatures and the equivalent current of two silicon solar cells (one filtered to be sensitive to UV spectral range) in order to control the solar simulator light intensity that degrades due to lamp ageing.

5. TEST SET-UP VERIFICATION

5.1. Test set-up spike verification

The solar cell electrical circuit was verified to be free of electrical transients for planned (transitions between safe mode and forward/reverse modes) and unexpected events that may occur during the life test execution. A dummy cell was used to verify the transients in the electrical circuit. The dummy cell terminals were directly connected to the measurement oscilloscope. The verification helped to refine the test operational procedure and to prevent transients during nominal test operations. Furthermore an oscilloscope has been connected to one of the cells, in order to monitor unexpected transients, during the complete life test.

5.2. Solar Simulator Calibration

The uniformity was measured with the CHEOPS test solar cells that were previously electrically characterized with a calibrated pulsed solar simulator. The overall uniformity is verified to be in the range of ± 2.0%.

The solar simulator intensity at the solar cell sample test plane is measured with a set of representative component solar cells in the centre of the beam and in the same plane of the CHEOPS test solar cells. In order to characterize the lamp spectral ageing, the light beam intensity is measured at the beginning and at the end of each phase or lamp change. The nominal solar simulator lamp life is around 2000 hours.

6. LIFE TEST EXECUTION

The following figures show the temperature, current and voltage of the solar cells monitored during part of phase 2, as an example of the execution of the CHEOPS life test.

Fig. 3 shows the solar cells temperature in loaded mode and in reverse mode. The temperature increase corresponds to the reverse mode step.

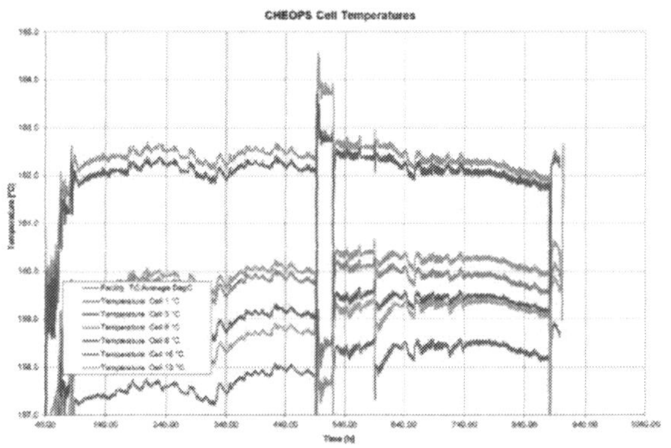

Figure 3. Solar cells temperatures.

Fig. 4 shows the solar cells currents in loaded mode and for a short period in reverse mode. The solar simulator light intensity was increased at 600h to compensate for the aging of the lamp.

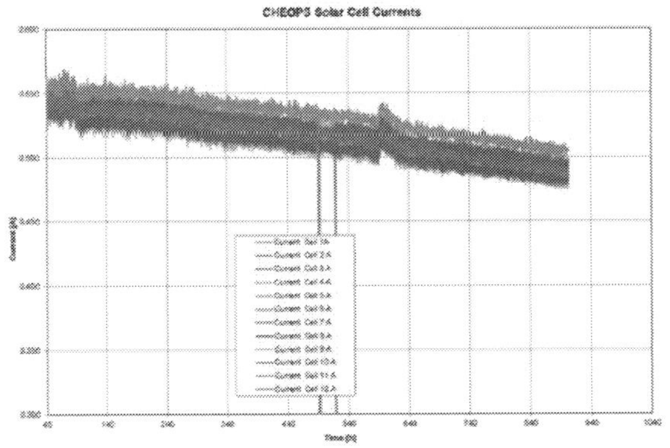

Figure 4. Solar cells currents.

Fig. 5 shows the solar cells voltage in loaded mode and few hours in reverse mode.

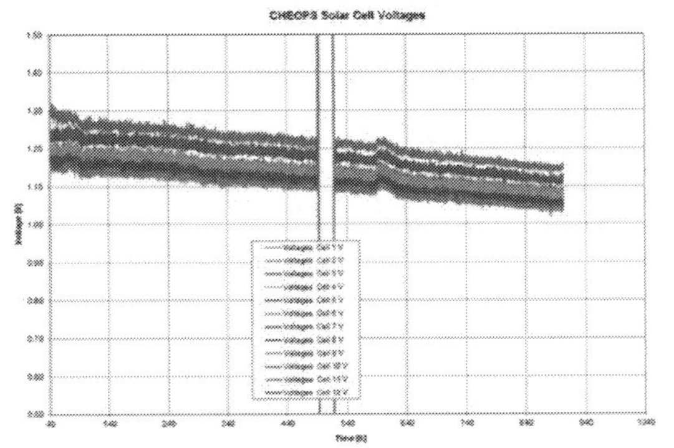

Figure 5. Solar cells voltages.

7. LIFE TEST RESULTS

The solar cells are inspected and electrically characterised before the exposure test starts and at the end of each phase. The electrical performance of the solar cell assemblies is measured, at 1 solar constant conditions and at 25°C. Visual inspection, including electroluminescence inspection of the solar cells, is also carried out. The contamination content of the calcium fluoride windows is measured using IR spectrometry analysis.

After the completion of phases 1 and 2 and more than 1800 hours of exposure at 175°C, no degradation of solar cell assembly's electrical performance has been recorded and no defects have been identified. The test is planned to be completed towards the end of September 2016.

8. LESSONS LEARNED

The Solar simulator optics are sensitive to contamination that becomes fixed to their surfaces as a result of the exposure time, this is particularly due to Ultra Violet (UV) radiation. This contamination decreases the performance of the solar simulator in the UV region and requires the cleaning of the optic surfaces in order to restore it. The cleaning is performed with solvent agents that do not damage the optic coatings.

The process to set-up the reverse bias condition on the solar cells is time consuming because of inter-cell interaction due to the electric switch box design. The solution to implement individual power supplies to bias each solar cell, is seen as a much efficient, even if more expensive.

The solar simulators lamp life may be linked to specific lamp manufacturing batches. Therefore, the implementation of an individual lamp life control has been stablished.

The use of reference component solar cells and monitor, silicon and silicon VI & IR filtered, solar cells to calibrate and monitor the solar simulator intensity, helped to verify the CHEOPS solar cells performance, together with their measured current at test temperature, after unforeseen test interruptions.

9. ACKNOWLEDGEMENTS

The authors are grateful to Airbus DS Spain, in particular to Natalia Perez and Ignacio Diaz, for their work in the definition of the test duration in accordance to the CHEOPS mission requirements. Furthermore the authors are grateful to Natascha Klein and Narimane Brahimi from ESTEC TEC-EPB section for their support in the test set-up preparation and the life test execution.

10. REFERENCES

1. Corral van Damme C., Rando N., Asquier N., Ratti F., Southworth R. & Isaak R. (2016). CHEOPS: ESA First small science mission – From mission concept to CDR in 3.5 years. Small Satellites Systems and Services Symposium, 30 May – 3 June 2016, Valletta, Malta.
2. European Cooperation for Space Standardization, ECSS-E-ST-20-08C Rev. 1 "Photovoltaic assemblies and components".

DOI: 10.1051/e3sconf/20171616005

COMMERCIAL PHOTODIODE SUITABILITY FOR SOLAR SIMULATOR LIGHT MONITORING

C. Barber [(1)], **P. López** [(2)], **G. Martínez** [(1)]

[(1)] *INTA-SPASOLAB, Carretera de Ajalvir km. 4, 28850 Torrejón de Ardoz, Spain*

Email: barberdcl@inta.es, martinezfg@inta.es

[(2)] *ISDEFE Consulting, C/ Beatriz de Bobadilla 3, 28040 Madrid, Spain, Email: lopezmp.pers_externo@inta.es*

ABSTRACT

The use of commercially available photodiodes for developing diverse solar simulator light monitoring applications is a very convenient approach due to availability, quality, performance and cost effectiveness.

Photodiodes as light monitors are meant to be operated in constant current zone, commonly referred to as photoconductive mode. In this operation mode, they offer linear response to impinging light over a few orders of magnitude. However, while presenting desirable characteristics in the manufacturer's specification sheets, there is a general lack of information regarding current response saturation for higher irradiances as it is the case of placing the photodiode directly under AM0 like conditions leading to possible misuse.

This paper will provide a guided approach on how to determine photodiode eligibility for solar simulator light monitoring applications. Needing to be linear around AM0 irradiances, suitability will rely on empirical IV tests at different irradiation conditions of three commercial silicon photodiodes performed in one of Spasolab's Xe based solar simulators. Further testing data including spectral response will also be provided from a specially manufactured off the production line photodiode from Hamamatsu, which has never been published, proving it to be an excellent option. Test data will serve to give guidelines on suitability for the application focusing on series resistance, having enough reverse bias headroom and containing power consumption to avoid excessive self-heating, yielding an appropriate bias voltage to assure linearity. Finally, a discussion on the photodiode's upper saturation limit determination will be introduced.

1 INTRODUCTION

Photodiodes are an interesting option in solar simulator behaviour characterization in applications such as short and long term stability metering and irradiance uniformity mapping. To do so, photodiodes have to be operated in constant current operation zone, or what is normally referred to photoconductive operation(see fig. 1), normally presenting near constant current due to high shunt resistances in Silicon.

Higher shunt resistance is better as in the case of solar cell performance, since it implies higher output current to bias voltage rejection. However, photodiodes are prone to saturation under AM0 irradiance conditions. This mainly has to do with the fact that photodiodes are manufactured to be high sensibility light sensors for a wide variety of applications and not optimized for carrier extraction under high irradiances (see fig. 1). This is a major difference with solar cells, which have much more complex and dense metallization grids, providing lower series resistances for the carrier paths.

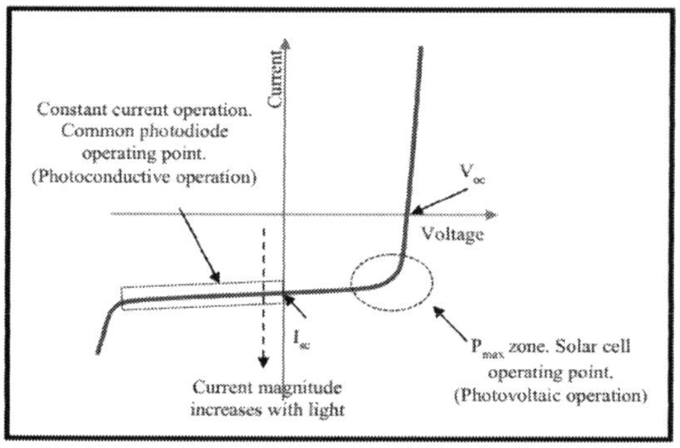

Figure 1 Comparing Solar Cell and Photodiode operating points on an IV characteristic

In practice, current saturation under a certain voltage operating point is originated by the lower density metallization grid found in the photodiodes which increases the series resistance found at the terminals of the device, as can be seen in fig. 2. Having the photodiode self-biased by driving its photocurrent through a low resistance is not enough in most applications. The photodiode's upper linear limit can be expanded by increasing reverse bias, however, doing so will make the photodiode consume power translating in self-heating of the device, which under higher irradiations and without active temperature control could quickly surpass temperature operational limits.

DOI: 10.1051/e3sconf/20171616005

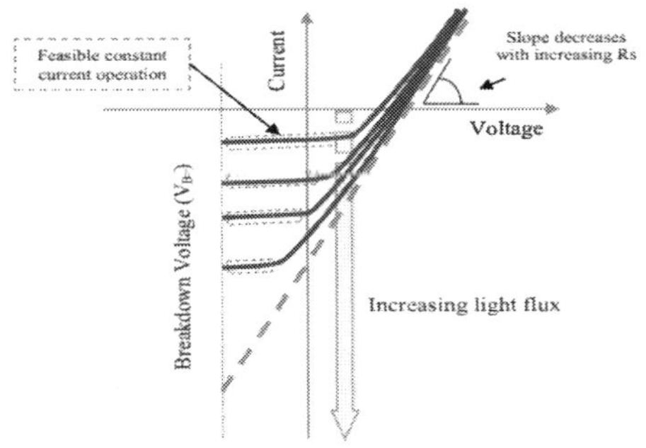

Figure2 Photoconductive zone reduction with increasing light flux

However, employing commercial photodiodes for such purposes presents good benefits for solar cell testing facilities highlighting their availability, cost, variety and overall robustness.

According to the previous, the work presented will explore the suitability of photodiodes for AM0 like irradiance conditions, determining appropriate working operation points and assuring linearity in a desired irradiance range. Furthermore, it is interesting to establish a way to determine an upper linear limit by extracting the photodiode's series resistance and shunt resistance from a measured IV curve under darkness. For the latter, parameter extraction and curve generation will be performed with IVFIT, comparing it to the measured IV curves. Although focusing on silicon photodiodes, which do have a broad spectral response range from 300 to 1100nm (being sensible to the simulator zones that address top and various centre junctions), the approach is analogous for infrared InGaAs or Ge based photodiodes.

2 DUT AND TEST SETUP DESCRIPTION

2.1 Device under test

Three silicon photodiodes are compared in this work.

- Centronics OSD 100-6 (RS303-674 alternate coding). Circular 1cm^2 effective area, metal encapsulation with unspecified window material.
- Hamamatsu s1387-1010BQ. 1cm^2 effective area, enhanced UV response photodiode. Quartz window with ceramic encapsulation
- Hamamatsu s2387-1010R. Square 1cm^2 effective area. Resin cover with ceramic encapsulation.
- Modified off the production line Hamamatsu s2387-1010x. 1cm^2 effective area. Special request substituting epoxy resin by quartz cover with ceramic encapsulation.

2.2 Test setup

For this test, a triple source solar simulator at nominal conditions, employing only Xenon lamp and Spasolab's FEP [1] system will be employed. The data acquisition system (FEP) employs a calibrated Keithley 2602 sourcemeter to undertake electrical performance tests that are self triggered once light is shed upon the device, performing a fast IV sweep, setting IV sweep limits between -3.5V and the photodiode's Voc. This is the main characteristic of the FEP system and allows for lack of thermal control. The devices will suffer little temperature variation during the test and will initially be around laboratory conditions. The solar simulator will only be operated with the xenon source working at nominal power conditions. Irradiance steps will be made by using several neutral density filters (NDF).

For the spectral response data, SPASOLAB's double monochromator spectral response facility will be employed. Spectral response tests will only be performed to Hamamatsu's s2387-1010x.

3 TEST RESULTS

3.1 Irradiance linearity data:

Linearity tests have been separated in two due to photodiode availability at the time of the tests. The first group is comprised by S2387-1010R, S1337-1010BQ and Centronics OSD100-6 (with alternate code RS303-674). Three mesh NDF filters of 3%, 11% and 43% have been introduced in the solar simulator which is set to near AM0 conditions, aiming to observe behaviour at operational injection levels. Second tests is performed over the tailored s2387-1010X, further increasing irradiance up to the maximum achievable without concentrating optics for the solar simulator employed, which sheds a total irradiance close to 1.5·AM0. This is compared to Silicon primary reference standard. A better approach to determine linearity could have been performed by applying superposition of light sources [2]; unfortunately there was no option to do so. Despite not having adopted an orthodox linearity test approach, the main interest is to observe if substantial IV deformations appear at irradiance in the environment of AM0 to make the constant current zone disappear when swept in the defined voltage range.

Data for s1337-1010BQ is shown with an additional 10% transmission optical NDF due to the total loss of linear operation zone at the previous injections. The latter could have been tested up to higher reverse voltages to observe the appearance of constant current linear operation zone, however, this blue enriched silicon photodiode, due to its construction, has a rated reverse voltage breakdown of -5V.

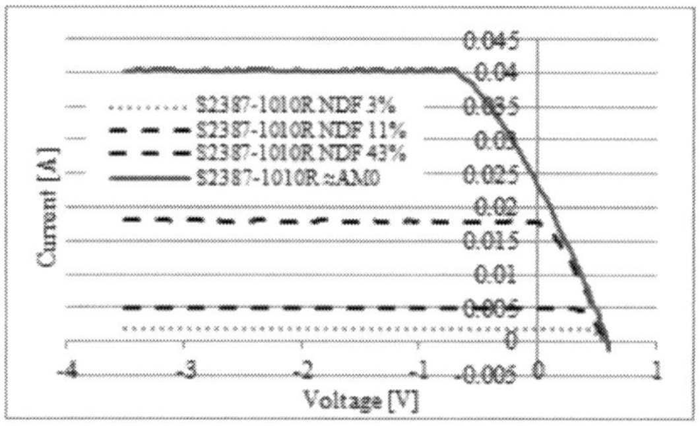

Figure3 S2387-1010R. Linearity Test IV Curves

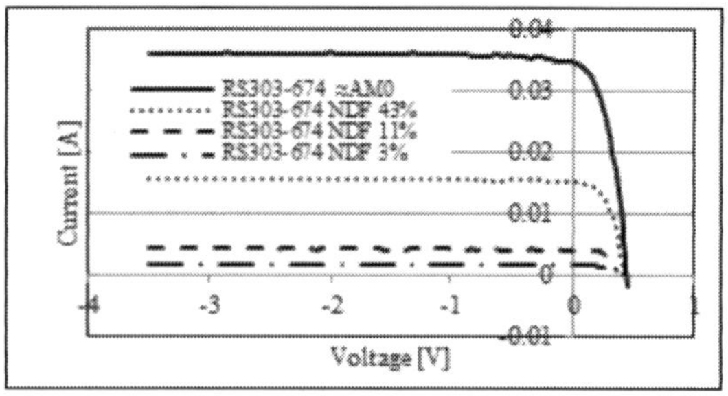

Figure4 RS303-674. Linearity Test IV Curves

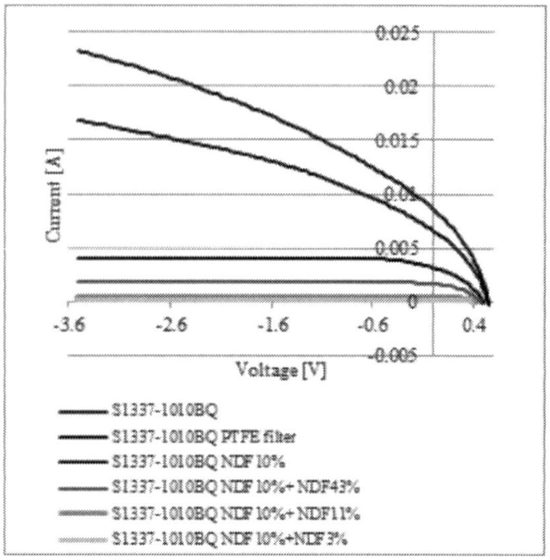

Figure5 s1337-1010BQ. Linearity Test IV curves

IV curve results in fig. 3 and fig. 4 show good form factor, with constant current zones entering the reverse bias region but with enough headroom to bias at low negative voltage levels.

For the blue enriched photodiode (see fig. 5), however, the IV curves tested prove it impossible to operate in linear operation zone as no constant current operation is seen and its breakdown voltage is near as aforementioned (this is due to the shallow junction required to trap higher energy photons).

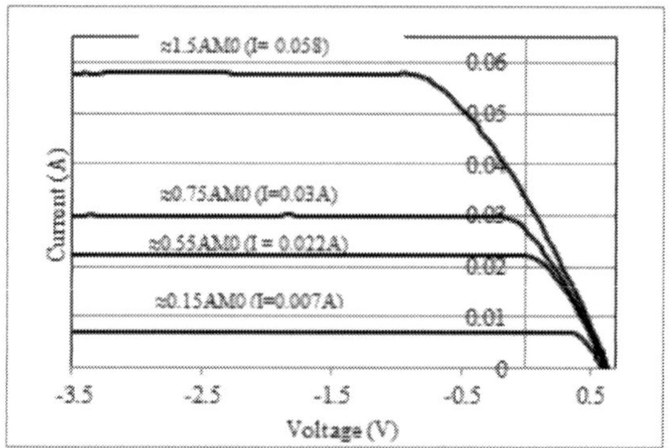

Figure6. s2387-1010X. Linearity Test IV curves

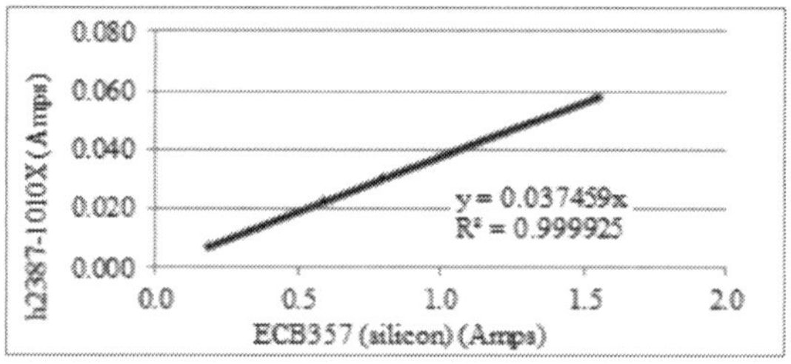

Figure7 s2387-1010x vs Silicon Primary linearity data correlation

For the case of the specially fitted s2387 (see fig. 6), IV curves show constant current operation with enough headroom to increase irradiance up to 2AM0 if the photodiode is biased at -3.5V, which is well under the breakdown voltage of the device. The linear correlation with the primary standard has an excellent match, noted by the R2 parameter as seen in fig. 7. Therefore, this photodiode makes a great irradiance monitoring device, despite the fact that it has to be operated in reverse voltage bias to do so.

3.2 Spectral response:

Spectral response for 2387-1010x has been tested between the range 280-1100nm. The starting point has been selected since it was feasible that the photodiode could provide output at such wavelengths having a quartz cover instead of the epoxy resin. For the standard commercial s2387-1010R, the spectral response is only provided down to 340nm. The measured spectral response is compared to the rest of the photodiodes, as seen in fig. 7, however the latter data is obtained from their respective datasheets by digitizing the provided curve. For better representation, all values have been interpolated from 190 to 1100nm at 1nm steps.

 DOI: 10.1051/e3sconf/20171616005

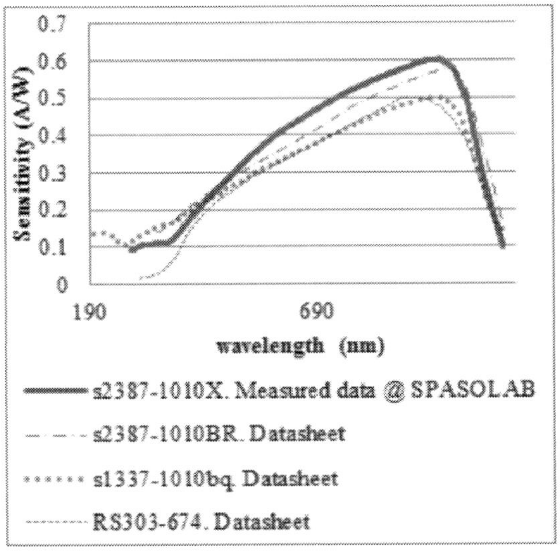

Figure 8. SR for 2387-1010X compared to the rest of photodiodes

Comparing the spectral response to that of s2387-1010R, a general increase of spectral gain is obtained just by replacing epoxy resin by quartz. This however, does not affect the linear behaviour of the device, which has been proven linear even up to 1.5AM0 irradiances.

Having a quartz cover on silicon photodiodes for solar simulator monitoring applications is key to attain response under 300nm, as well as to avoid cover aging due to higher energy photons that affect epoxy resins. Further testing to increase the SR range could not be done, however, the interest of exploring the lower SR limit has not been overseen.

A total of four 2387-1010x photodiodes have been tested for SR in order to check consistency, the external quantum efficiency of all is shown in fig. 9. Curves show a close match with 40% EQE under 300nm.

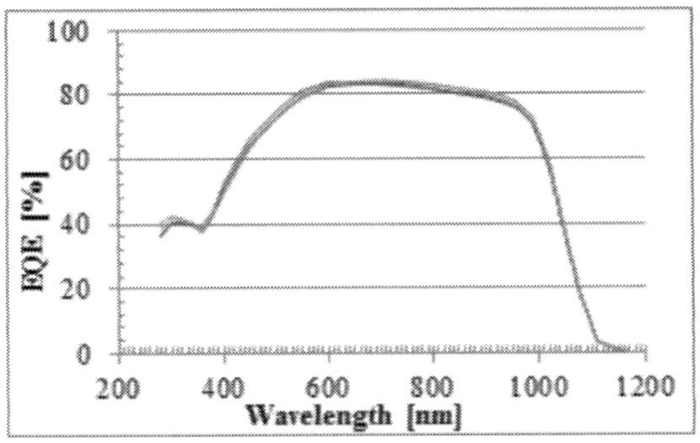

Figure 9. EQE of 4 different s2387-1010x photodiodes

4 UPPER LINEARITY DISCUSSION

When determining if a desired photodiode is suitable for an application that involves irradiances over $1000W \cdot m^{-2}$, linear operation at such irradiance is the only concern.

From the manufacturer's literature and datasheets, it is not possible to extract an analytical conclusion, furthermore, they normally don't risk to include a figure that expresses suitability for high irradiances expressing high linearity but only up 10mW over the photodiode area [3], as something applicable to silicon photodiodes. However, as seen from the IV data presented herein, this irradiance level can be heavily increased maintaining linear operation, even without requiring very high reverse voltage biasing.

The ultimate limiting parameter referred in the manufacturer's literature is series resistance. The effect of this parameter is quite straight forward, as it decreases the slope, shifting constant current operation to the 2nd quadrant (when taking current flow as a generator).

A preliminary approach on this work regarded the extraction of the series resistance from a dark IV curve with IVFIT software [4], forecasting the operation with the corresponding current injection at AM0, employing a two diode lumped model. The latter was obtained by digitizing the datasheet SR graph and interpolating in nm steps (as mentioned for fig. 8) and AM0 spectrum by employing the current density expression Eq.°1. The values obtained for the different photodiodes are gathered in tab. 1.

$$J_{@AM0} = \int_{200}^{1100} SR_{PD}(\lambda) \cdot E_{AM0}(\lambda) \cdot d\lambda \qquad (1)$$

Table 1 Calculated current densities for each photodiode at AM0

Photodiode	I @ AM0 (A)
s2387-1010X	**0.03797**
S2387-1010BR	**0.03581**
s1337-1010BQ	**0.03284**
rs303-674	**0.03085**

However, when generating a curve with the series resistance extracted from the dark IV curve, the modelled behaviour fails completely to recreate the tested data. By comparing the tested data series resistance in dark condition with that near AM0, the latter has an increased value of 1 decade, behaving in the same fashion for all photodiodes.

As series resistance has been observed to vary with the illumination level, there is no way to employ the general model to predict suitability with only employing a dark IV curve and parameters from the datasheet.

4.1 Series Resistance variation with higher

current injection.

The variation of the series resistance with the irradiance is an effect that has been described multiple times within the literature, and was first well described in 1963 [5]. This phenomenon was observed specifically when comparing IV curves of solar cells with low and high density metallization grid at increasing irradiances. In their work, the tests showed a heavy increase in the series resistance with illumination for the low density cell and not such a difference in the case of higher density, appearing rounding effects in the IV curves. This effect can clearly be seen in fig. 5, where the shape of the IV curve of s1337-1010BQ, ends up changing drastically.

Recent works have progressed in obtaining numerically such varying series resistances as found in [6]. However, it requires the previous acquisition of various IV curves up to high injection levels.

4.2 Manufacturer's saturation limit notes

The most commonly expressed definition for saturation in the manufacturer's literature is shown in Eq. °2, which is taken from [3]. This is affected by the uncontrolled parameter Rs. This again, requires to be obtained prior to determining saturation, as aforementioned.

Also, *Psat* is suggested to be under 10mW impinging the effective area of the device, which limits the use to 10%AM0 equivalent total irradiance. However, they also recommend increasing linearity by applying reverse voltages. This is a general definition; however, taking the s2387-1010 for example, it has been proven to operate in linear operation up to irradiances 15 times the limit.

$$Psat = \frac{V_{BI} + V_R}{(R_S + R_L)\cdot S_\lambda} \cdots \qquad (2)$$

$Psat \leq 10mW$
V_{BI}--> Built in voltage
V_R= Applied Reverse Voltage
S_λ--> Sensitivity at wavelength λ

Therefore, series resistance is a key parameter that affects the saturation limit of the device, however it still requires to be obtained previously.

5 PHOTODIODE ELIGIBILITY

Taking into account the reverse breakdown voltage and maximum operation temperature ratings supplied by the manufacturer, linear operation will be defined as having enough reverse voltage breakdown overhead to bias the photodiode in constant current zone at the defined irradiance. To keep within safe operation conditions, bias voltage will have to be within 80% of the reverse bias voltage and special considerations will be made taking into consideration the dissipated power within the junction. However, upper irradiance level has not been possible to determine from the information provided in the datasheets, since series resistance needs to be measured before. This, however, can be performed without sophisticated light sources, just enough to provide high current injection levels. The current injection level at AM0 can be determined with the digitized SR response from the datasheet and the standard AM0, and light, from a simple source, can be increased until those current levels are met. This inconvenience wouldn't exist if manufacturers included series resistance as a function of light intensity or their maximum light intensity operation points at different reverse bias levels.

6 CONCLUSIONS AND FUTURE WORKS

In this paper we place focus on determining commercial photodiode suitability for Multiple Zone solar simulator monitoring by introducing a series of irradiance linearity verification and operation point guidelines. Linearity tests by performing IV curves at different irradiances have been performed on four silicon photodiodes, comparing linearity to that of a silicon reference standard. Intending to forecast irradiance linearity upper limit without the need of performing previous tests has been deemed not possible with the current information provided by the manufacturers, since the series resistance varies heavily with the irradiance on devices that have low metallization density. This effect has been contrasted with the literature, finding that previous IV curve knowledge is required. However, only current injection is required to obtain the desired series resistance, therefore, linearity can be tested with simple halogen lamps that provide enough power to recreate desired operation points.

Additionally, results for a tailored photodiode have been presented and its suitability for solar simulator monitoring when biased actively under -1V has been shown.

7 ACKOWLEDGEMENTS

The research was co-supported by the European Space Agency (ESA) under contract No. 4000107040/12/NL/GLC and the European Metrology Research Programme under the project ENG 55 (Photoclass). The EMRP is jointly funded by the EMRP participating countries within EURAMET and the European Union.

8 REFERENCES

1. C. Barber and A. Gras (2014), "SPASOLAB's Low Temperature Solar Array Coupon Electrical Performance under Continuous Light Test Setup," in *10th European Space Power Conference (ESPC)*, Noordwijkerhout, Netherlands.
2. K. Emery, S. Winter, S. Pinegar and D. Nalley (2006), "Linearity Testing of Photovoltaic Cells," in *4th World Conference on Photovoltaic Energy Conversion (WCPEC-4) IEEE*, Waikoloa.
3. Hamamatsu Corporation (2014), Opto-Semiconductor Handbook. Chapter 2 Si Photodiode, Hamamatsu Photonics Solid State Division.
4. Energy Research Centre of the Netherlands (ECN), "IVFIT curve fitting program," ECN, 2015. [Online]. Available: https://www.ecn.nl/units/zon/products-services/i-v-curve-fitting-program-ivfit. [Accessed february 2016].
5. H. R. M. Wolf (1961), "Series resistance effects on solar cell measurements," in *Pacific General Meeting of the AIEE*, Salt Lake City, Utah.
6. T. Luka, S. Eiternick and M. Turek (2014), "Determination of the Current and illuminatino dependent series resistance using LED Solar Simulators," in *29th European Photovoltaic Solar Energy Conference and Exhibition*, Amsterdam, NL.

DEVELOPMENT OF A NEW TIME RESOLVED IRRADIANCE UNIFORMITY MAPPER AT SPASOLAB

C. Barber [1], A Martínez de Olcoz [2], J.M. Fernández[2], P. López[2], G. Martínez[1]

[1] *INTA-SPASOLAB, Carretera de Ajalvir km. 4, 28850 Torrejón de Ardoz, Spain*

Email: barberdcl@inta.es, martinezfg@inta.es

[2] *ISDEFE Consulting, C/ Beatriz de Bobadilla 3, 28040 Madrid, Spain, Email: martinezoam.pers_externo@inta.es, fernandezdjm.pers_externo@inta.es, lopezmp.pers_externo@inta.es,*

ABSTRACT

Irradiance uniformity mapping of solar simulators is a key characterization required for performing accurate solar cell electrical performance tests, alongside spectrum matching and stability. Within photovoltaic space standards such as [1] and [2], uniformity mapping is specified to verify the solar simulator non-conformance class. However, uniformity mappers are a relevant tool for solar simulator manufacturers as well as for space solar cell testing facilities, aiding to adjust test plane irradiance uniformity after maintenance, lamp replacement or on a more analytic approach, applying correction factors to the test plane. Thus, it is desirable to have a tool that allows for easier and faster acquisition of irradiance uniformity maps.

This paper presents a time-resolved uniformity mapper developed at Spasolab based on two main premises: easy adaptability to different solar simulators and versatility in terms of irradiance operation range, extended up to 2 AM0. A modular two-stage design features independent bias PCB's per each photodiode that allow for easy sensor swapping or different biasing for future sensor replacements. A new type of photodiode active biasing is also presented, with the main feature of being capable of switching on the bias voltage only during the acquisition process, hence reducing self-heating. Ad-hoc software developed in LabVIEW and PXI instrumentation enable the system to have different acquisition modes further expanding the irradiance range to be tested without risking hazard to the system due to overheating. Finally, outdoor calibration method and results for the system are also presented.

1 INTRODUCTION

The development of photovoltaics as an important power source in space missions during the 1960s brought a need to establish synthetic methods to test photovoltaic devices at array and cell level under AM0. This required a better understanding of illumination conditions that were to be recreated in order to pursue repeatability in photovoltaic characterizations throughout testing laboratories. This implied not only recreating light to match a certain spectrum and irradiance level but also to ensure that the light was shed upon the test plane in a uniform collimated fashion. As irradiance uniformity is tied to solar simulator optics and light source displacement, measurements were soon to be established as a feedback loop while developing solar simulators. In these early stages, JPL developed a test apparatus for measuring the irradiance uniformity of 15-foot Space Simulator upgrading, consisting on a single sensor swiped in polar coordinates with a reference sensor to correct stability [3], as depicted in fig. 1. This type of system is still in use and while benefiting from taking data from only one sensor, corrections with the second sensor must be applied and the automated sweep is normally time-consuming.

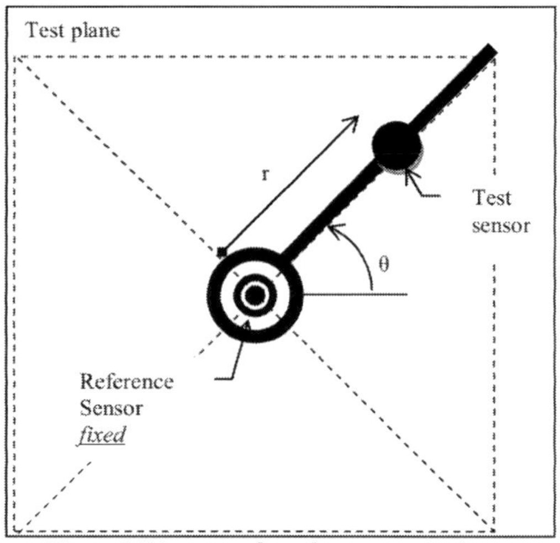

Figure 1 Diagram of a polar sweep single sensor uniformity mapper

DOI: 10.1051/e3sconf/20171616006

With advances in instrumentation and data acquisition, more affordable approaches for a multiple sensor simultaneously acquired systems have been made available. In the past years, the development of these time resolved multiple sensor uniformity mapping systems have been published both for large area flash simulators and for steady state such as in [4]. Such systems present an array of sensors that are synchronized and simultaneusly acquired, hence, obtaining irradiance maps quickly and rejecting light source instabilities. Such system is depicted in fig. 2.

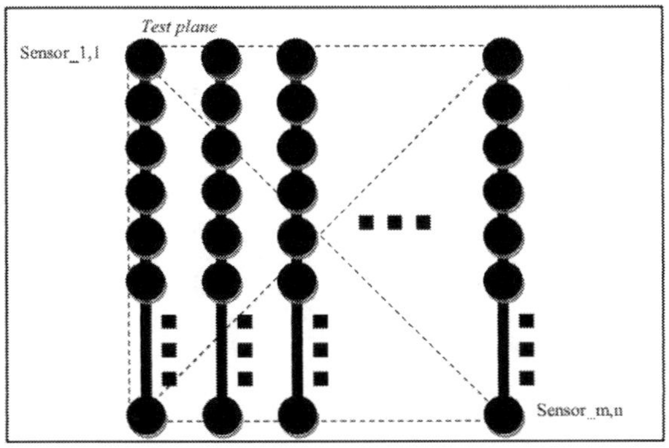

Figure 2 Two dimension sensor array uniformity mapper diagram

However, these types of systems have plenty of room for different strategies and developments yet. In this paper we will present the developed time resolved uniformity mapper at Spasolab, focused on steady state solar simulators but with potential capabilities for performing under flash systems. Special attention has been placed on having a wide irradiance linearity response from the sensors, surpassing 2 AM0. Also, effective sensor area and spectral response makes the system suitable to comply with [1] as well as other terrestrial standards.

2 HARDWARE DEVELOPMENTS

2.1 Overview

A two stage approach has been taken for developing the printed circuit board (PCB) harness. Up to 40 photodiodes, of 1cm^2 effective area, can be housed in the main printed circuit board. The design currently includes high quality silicon photodiodes with quartz cover. Sensor deployment covers an area of 146 by 146 mm, placed in a checkered pattern (subsequent rows of 4 and 5 sensors). This pattern presents a uniform distribution, without weighing specific areas. Sensors are mounted on their individual pcb's which harbour the bias circuit and are connected to the main PCB through polarised flat 9 pin connectors. This allows for quick replacement, and for various configurations.

The hardware is completed with a custom manufactured aluminium chassis that also houses the connectors and led indicators. A window lid has also been developed to prevent irradiation from hitting the main PCB and from undesired reflexions from the coating. This lid will also enforce flatness throughout the sensors, as interconnectors can wobble. The chassis and the system's compact form enables easy mounting on various tests platforms in vertical or horizontal orientation, even being capable of placement within a testing chamber.

Figure 3 A complete view of the system, including bias PCB with sensors mounted on the main PCB. Custom chassis and top window lid also shown.

2.2 Biasing Circuit

A simple active bias circuit that hasn't been found within the literature has been developed. Requiring negative voltage bias to further increase the photodiode's linear operation, a simple option employing a single small profile integrated circuit has been used, which is a low voltage dropout linear regulator (LDO). These small footprint voltage regulators present a controlled voltage output, presenting low voltage drop with load current. A common feature nowadays is the addition of enable pins, that allow fast transitions from on (regulated voltage out) and off (high impedance) states. For this application, a CMOS based 3.3V regulator from Microchip, the TC1185, has been employed. The developed topology can be seen in fig. 4. Having the capability of disconnecting the biasing voltage is of great value, avoiding self-heating of the devices due to power consumption.

DOI: 10.1051/e3sconf/20171616006

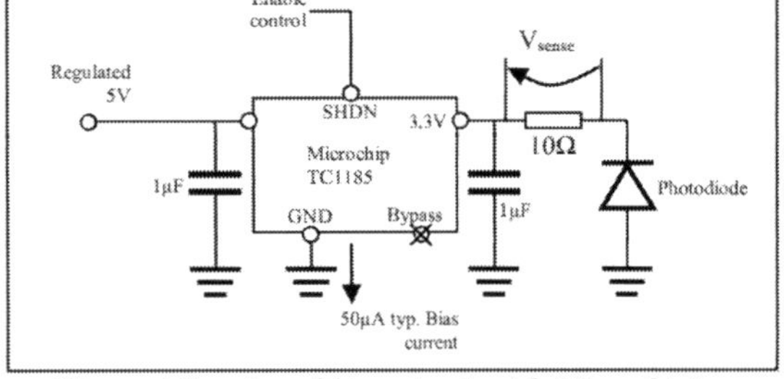

Figure 4 Developed bias circuit with Microchip's TC1185

The current loading capabilities of the LDO are 150mA, which presents enough headroom for the expected photodiode current, rated in the vicinity of 40mA for AM0. This topology presents some inconveniences when compared to transimpedance topologies, such as not being able to include a gain stage or having less bandwidth. For the latter, bandwidth is not an issue in steady state simulators where xenon flicker components fall under 1KHz and it is yet to be tested under flash simulators. Despite not having a gain stage, a greater resistor for the voltage readout can be introduced, in the present case being a 10Ω precision resistor.

Taking the photodiodes response and the output behaviour of the LDO as shown in the manufacturer's datasheet, and also taking into account the voltage drop in the test resistor, an operation value of 2 AM0 has been calculated to be within safe operation boundaries, both of the photodiode presenting linear response as for the LDO.

The LDO has a typical rise time of under 200µs when activating through the enable pin. This has been monitored in 0 load, however, the software enables the circuit 50ms prior to the data acquisition. Therefore, no transients have been observed at the tested irradiance levels, neither after increasing acquisition rates up to 100Ks/s.

Circuit layout with the sensor mounted can be seen in fig. 5 and fig. 6.

Figure 5 Front view of the silicon photodiode mounted on the bias PCB.

Figure 6 Rear view of the bias PCB showing the interconnector and the low dropout voltage regulator.

2.3 Data Acquisition and control software

The system has been based in a PXI unit with five simultaneous acquisition 8 analogue channel boards from National Instrument's S series. High data throughput is available, and acquisition rates can be set to 100Kilosamples/second simultaneously and synchronized throughout all 40 channels. A bespoke software application has been developed in LabVIEW. The software controls data acquisition, on state timing and raw data management. As mentioned earlier, software only activates bias voltage to the photodiodes during acquisition, avoiding uncontrolled temperature increase due to the photodiode's power consumption. In idle state the sensor is biased at 0V.

Taking advantage of the versatility offered by the PXI unit, three operation modes have been developed:

- Normal operation: Measurement is triggered on demand, activating LDO only during acquisition time.
- Continuous operation: In this mode the software triggers measurements until stopped by the user. With the 3D graphic view of the irradiance map, this mode enables for a near real time view of the light uniformity.
- Light Triggered: This mode triggers the acquisition when light is shed upon the device. This mode allows measuring higher irradiances without incurring in an uncontrolled temperature increase to hazardous levels as well as maximizing temperature uniformity. This allows for stabilization across sensors at laboratory conditions prior to the test.

3 CALIBRATION

3.1 Linearity Test

Prior to outdoor calibration, an indoor linearity test has been performed employing the light superposition method [5]. To do so, a halogen lamp ring placed on one of Spasolab's solar simulator facilities has been employed. The halogen ring is comprised by 6 lamps,

 DOI: 10.1051/e3sconf/20171616006

all unfiltered, and at 75% nominal power. These have been separated into three groups (LA,LB,LC), conforming a total of 4 irradiance steps.

Table 1 Superposition analysis lamp step configuration

Measured	Summed
LALB	LA+LB
LALC	LA+LC
LBLC	LB+LC
LALBLC	LA+LB+LC

Raw data has been plotted for each step, and is shown in separate 3D maps for a qualitative perspective on the contributions of each lamp configuration, which is shown in fig. 7. Vertical unit is taken directly from the raw data, and is expressed as the voltage measured across the precision resistance.

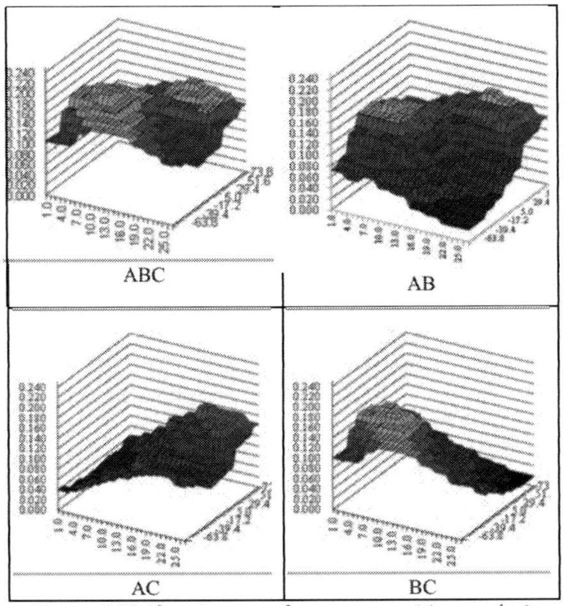

Figure 7 Uniformity maps from superposition analysis

Data later has been analyzed, evaluating the sum of individual sources and comparing it to the measured value. Data fitting has been evaluated using least squares method. Slope and coefficient of determination (R2) factor have been condensed in fig. 8 and in fig. 9. Sensors positions 12 and 14 have been omitted due to an issue with the acquisition channel and failure in the cable harness respectively.

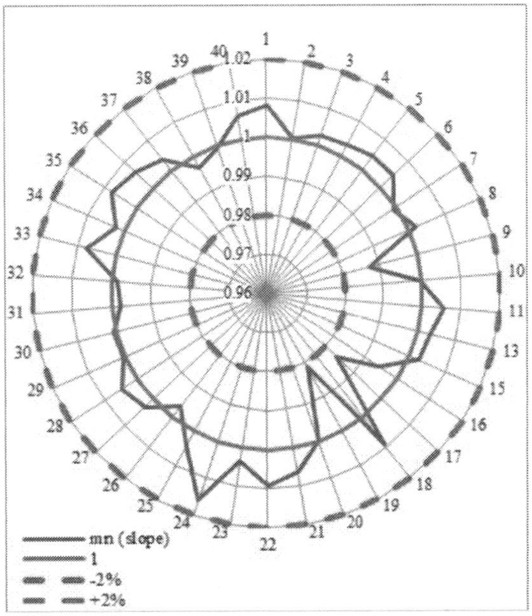

Figure 8 Radar chart with linear estimation of m(slope) employing superposition analysis

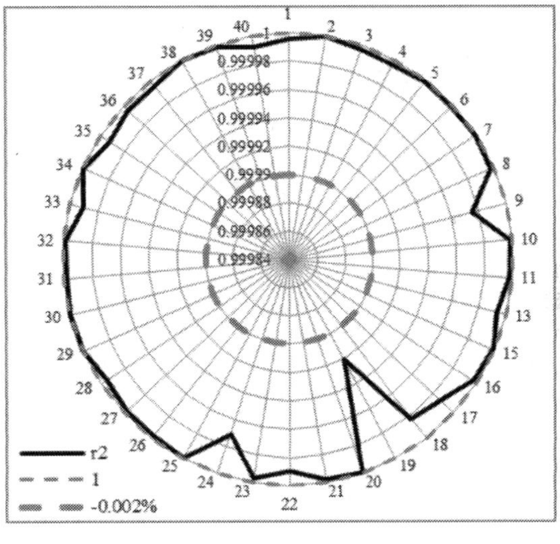

Figure 9 Radar chart with R2 from the linear estimation of the superposition analysis

Data fitting has shown excellent results for the linearity of the complete system. It is interesting to observe the high match for the R2 of each position.

Summing up what this test has yielded; linearity of the complete system has been evaluated and proved to have a linear response throughout the signal chain. Unfortunately, this analysis will have to be repeated

 DOI: 10.1051/e3sconf/20171616006

with other light configuration capable of injecting a response of up to 1.5AM0 equivalent irradiance. The current setup has been chosen due to the high stability of halogen lamps power by precision power sources. However, this paves the way to establish an easier calibration map for the device.

3.2 Outdoor calibration

Outdoor calibration has been performed with the premise that irradiance throughout the device is ideally uniform. The configuration for the test is comprised by the uniformity mapper, a calibrated pyranometer and a primary working standard solar cell. At the time of the test, the elevation angle was adjusted by the use of the primary standard, fixing the structure at its maximum short circuit current Isc. The aim is to achieve maximum uniformity of light. At the time of the tests, sky was clear, GNI was greater than $1000 w \cdot m^{-2}$ with a variation of less than 2% in any 5 minute interval. No visible clouds or hazy conditions in 45° view angle around the sun. The wind speed was also under 2m/s, hence, temperature is considered uniform throughout the device.

Figure 10 Outdoor calibration test equipment displacement

Two tests are compared. The first will include data that is employed to create a calibration matrix by normalizing all the sensors with respect to sensor 16, which has been chosen due to its centred position as well as for its good coefficient of determination and slope in the linearity analysis. Software has been configured to perform tests with 1KHz sampling rate and a total of 1000 samples. Triggering method is set to normal (triggers when launched).

The irradiance value measured with the piranometer at the time of each test is shown in tab. 2.

Table 2 Irradiance levels at the time of the tests

Test 1. Calibration	Test 2. Compensating with calibration matrix
1004.873 W·m-2	1005.994W·m-2

As mentioned the calibration matrix is obtained by normalizing with respect to sensor position 16. This matrix is stored and new uniformity values are compensated using the relation in Eq. 1.

$$S_i^* = \frac{S_i}{cm_i} \rightarrow value\ in\ mV \qquad (1)$$

Where S* is the compensated value at position i, S the raw data input at position i and cm the compensation matrix. For data treatment and representation, sensor positions 12 and 14 have been removed as aforementioned. For representation purposes, gaps between sensors have been filled by averaging the surrounding sensor values, which doesn't modify non conformance calculation, since maximums and minimums are left untouched. The uniformity map of "Test 2" without compensation and after compensating with the normalized matrix from test 1, is shown in fig. 11 and fig. 12 respectively. The scale shown is centimetres for the plane and volts for the sensor output level.

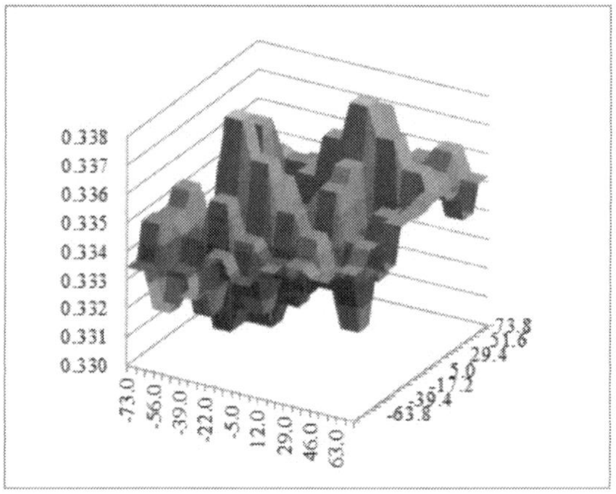

Figure 11 Test data without compensating values

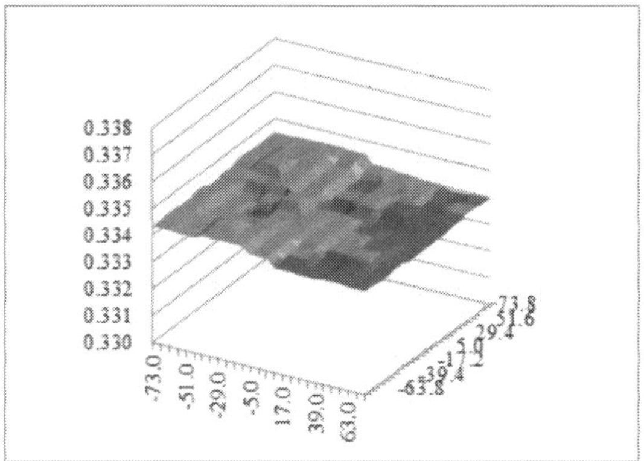

Figure 12 Test Data after compensation

The non conformance values for each non compensated and compensated uniformity maps are shown in tab. 3.

Table 3 Non conformance values with and without compensation of Test 2 throughout the test plane

NC(%) w/o compensation	NC(%) with compensation
3.551	0.134

This shows clearly the data dispersion straight out of the raw data file must be compensated. A valid compensation matrix should be averaged over a higher number of test samples and for different uniform intensities. However, at the time of edition of this paper, test data was not available.

4 CONCLUSIONS AND FUTURE WORK

This paper presents a versatile and high linear range photodiode based time resolved uniformity mapper developed at SPASOLAB. Novel features include a new and simple active bias topology that allows disabling bias voltage to prevent the photodiode from consuming power while in idle, enabling bias voltage only during acquisition. The biasing circuit along with the photodiodes have been developed to test uniformity maps up to 2AM0, although such tests are not included within this work. A modular approach allowing quick sensor swapping and a bespoke compact aluminium chassis enhances versatility for performing irradiance uniformity measurements to a wide variety of solar simulators. Custom developed LabVIEW software, along with PXI instrumentation, provides different modes of operation that can trigger tests automatically with light.

Future works comprehend a tighter characterization of the device, exposing it to different light sources and configurations, and performing an adequate uncertainty analysis. Swapping silicon for infrared photodiodes for longer wavelength uniformity mapping would also be of interest. The verification of the system's capabilities to test irradiance uniformity maps of ultra violet simulators is also in sight.

5 ACKOWLEDGEMENTS

The research was co-supported by the European Space Agency (ESA) under contract 4000107040/12/NL/GLC and the European Metrology Research Programme under the project ENG 55 (Photoclass). The EMRP is jointly funded by the EMRP participating countries within EURAMET and the European Union.

6 REFERENCES

1. European Cooperation for Space Standarization (2012), "ECSS-E-ST-08 "Space Enineering. Photovoltaic assemblies and components"," ESA, Nordwijk ZH.
2. ISO (2005), "ISO15387-05. Space Systems - Single-junction solar cells - Measurement and calibration procedures," ISO.
3. NASA (1968), "Design and Construction of the 15-ft-Beam Solar Simulator SS15B," NASA, Pasadena, CA.
4. D. Rivola, S. Dittman, M. Pravettoni, G. Friesen and D. Chianese (2014), "High-Speed Multi-Channel System for Solar Simulator Irradiance Non-Uniformity Measurment," *Photovoltaic Specialist Conference (PVSC), IEEE 40th.*
5. K. Emery, S. Winter, S. Pinegar and D. Nalley (2006), "Linearity Testing of Photovoltaic Cells," in *4th World Conference on Photovoltaic Energy Conversion (WCPEC-4) IEEE*, Waikoloa.

ENHANCEMENT OF SOLAR CELL EFFICIENCY FOR SPACE APPLICATIONS USING TWO-DIMENSIONAL PHOTONIC CRYSTALS

P.A. Postigo, I. Prieto, E. Baquedano

IMM-Instituto de Microelectrónica de Madrid (CNM-CSIC), Isaac Newton 8, PTM, E-28760 Tres Cantos, Madrid, Spain, Email: pabloaitor.postigo@imm.cnm.csic.es

ABSTRACT

The effects of having a nanopatterned photonic crystal (PC) structure in the surface of a solar cell can be usefully employed to increase the energy conversion efficiency, which may be critical for space applications. In this work, we have measured the reflectance (R) and transmittance (T) of thin InP layers (270 nm thick) bonded to a glass substrate and nanopatterned with holes down to the glass in a triangular symmetry lattice separated by a lattice parameter a=450nm and maintaining a value of r/a=0.32. The optical spectra were measured with angular resolution in the range from 0.55 to 2.0 eV. There are noticeable changes in the spectra of the PC sample, with minima and maxima of the R and T clearly shifted with respect to the unpatterned sample, and new features that alter significantly the overall lineshape of each spectrum. Those features correspond in a first approximation to the well-known Fano-like resonances of the discrete photonic modes of the PC lattice and they have been used before to determine experimentally the position of the PC bands. The observed features can be translated to the optical absorption (A) defined as A=1-R-T provided there are low or negligible scattering effects. The generated absorption spectra show enhancements above and below the electronic band edge of the InP that can be correlated with the photonic band structure. Even using a thicker semiconductor layer, the above-mentioned effects can justify to use a photonic crystal front surface with sub-wavelength motifs. In this way, we have fabricated and characterized a complete Ge/InGaP solar cell with a 2D-PC on its front surface. An increase in the photocurrent up to a 8% was achieved on a solar cell with a 40% of its surface covered with a PC pattern. Enhancements of the external quantum efficiency (EQE) of 22% for a wide range of wavelengths and up to a 46% for specific wavelengths have been measured, without use of any anti-reflection coating (ARC). A correlation with the area of photonic crystal patterning has been clearly observed. Finally, a low-cost nanofabrication procedure to obtain high quality two-dimensional photonic crystals in large areas (up to square cm) is described.

1. INTRODUCTION

During the last years, much attention has been paid to photonic crystals (PC) for different applications, including solar cells (SC) [1] and in very promising designs that take advantage of slow-light modes in photonic crystal structures [2]. Moreover, there has been an important amount of work on light extraction out of LED devices by two-dimensional photonic crystals [3]. In a similar way, a 2D-PC layer can enhance transmission of the incident light inside a device and there are a number of theoretical [4] and experimental [5] works related to this. Efficiency of solar radiation conversion can be increased not only by increasing the number of p–n junctions but also by reducing optical losses induced by solar radiation reflection from the SC front surface. Optical losses can be reduced via optimization of the structure using some technological operations, including the deposition of one- or two-layer antireflection coatings [6,7]. Usually, one-layer ZnS or two-layer MgF_2–ZnS antireflection coatings (ARC) are used in the cascade solar cells. ZnS at temperatures above 970°C forms porous films, and MgF_2 films added in the deposition of these coatings, i.e., mixtures of film forming materials are used [8,9]. To improve light absorption in thin-film solar cells, it is mandatory to minimize the Fresnel reflection at the air/silicon interface in the range of the entire solar spectrum. Moreover, to achieve high absorption efficiency for the entire day, an angle independent antireflection property is required. Although multilayer antireflection coating is commonly used, it has problems related to material selection, thermal mismatch and sensitivity to the variation in thickness and instability of the thin-film stacks [10]. Over the past decade, submicrometer patterned structures with a tapered feature have been focused on as a more practical method for ultra broadband and omnidirectional antireflections [11] where excellent antireflection properties have been demonstrated via reflectivity characterizations [12]. These structures can be called PCs in the sense of that there is a periodic arrangement of the index of refraction on the surface, being the motifs columns or holes with any shape. Depending on the thickness of the layer in which they are patterned and the index contrast with the surrounding layers or materials, the properties of these PCs may vary in a significant way. In this work we will

DOI: 10.1051/e3sconf/20171616007

show experimental measurements of the reflectivity of PCs patterned on InP material with infinite thickness (InP bulk) and on very thin InP layers (270 nm) where the PC effects are more prominent. In the first case, most of the expected effect is just due to diffraction and it can act as a good antireflection coating (ARC). We have found an optimized PC-ARC that can combine good optical properties (low reflectivity over a wide range) with a good mechanical stability based in the use of a thin layer of dielectric perforated with the PC pattern. A slight patterning of the PC inside the InP layer is necessary for an optimum performance. Thus, a mixed thin dielectric layer-bulk patterning can be used for best optical and mechanical results over other alternatives like moth-eyelike fabrication [13], nanorod [14], and nanostructures employing oblique-angle deposition methods [15] which can be seriously damaged by even a smooth contact with other surface or during handling.

2. PHOTONIC CRYSTAL ANTI-REFLECTION COATINGS ON InP

InP substrates were nanopatterned with two different photonic crystal lattices with triangular and square symmetry. Lattice parameters *a*=400, 470, 530 and 600 nm were used for both types of lattices. The *r*/*a* ratio for the triangular symmetry lattices was, respectively, 0.25, 0.26, 0.23 and 0.2 where *r* is the radius of the holes. After processing, measurement of the reflectivity was performed on normal incidence and varying the angle of incidence for the wavelength range from 450 to 810 nm. The results show that the nanoholes-PC structures fabricated can be used as effective ARCs. Fig.1 shows the reflectivity measured at normal incidence on the two best coatings, one corresponding to a triangular symmetry and one to a square lattice with *a*=470 and 400 nm respectively. For comparison, the reflectivity of an InP bulk sample is included. The obtained values are comparable to other results obtained in tapered subwavelength etched structures although the mechanical resistance of our nanopatterned film could be better due to the use of holes instead of columns. In both of our samples, the total etching depths were 140 nm through the ARC and 160 nm inside the InP. Deeper etchings can greatly enhance the anti-reflection effect, since the refractive index value changes smoothly and continuously but further work is needed to obtain an optimum value. The influence of the incident angle of light on the efficiency of a solar cell is crucial for the device performance. In general, the cell efficiency of solar cells with a flat surface drops rapidly as the incident angle increases due to the increased reflection loss, which can decrease the efficiency more than a 50% at incidences of 70°. In the PC ARC, however, the reflectivity is sustained at an incident angle in excess of 60°, and it is degraded by only 20% at 75°. This angle-independent optical property is required for the device to operate efficiently during the whole daylight.

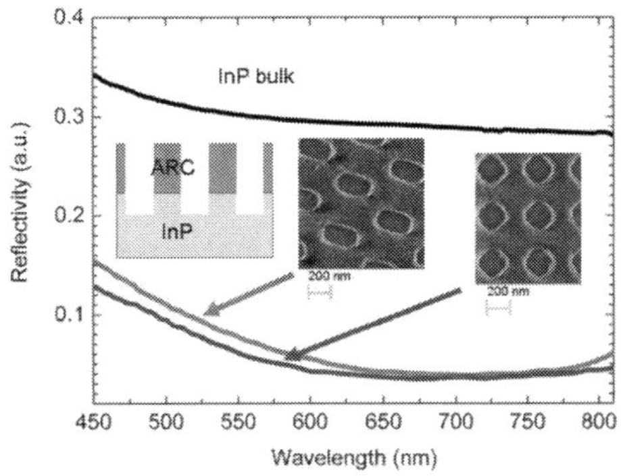

Figure 1. Reflectivity measured at normal incidence on two photonic crystal ARCs with triangular and square lattices with a=470 and 400 nm respectively. For comparison, the reflectivity of an InP bulk sample is included. The insets show a layout of the cross section and SEM pictures of the surfaces. [16]

Soft nanolithography allows for the simple and low-cost fabrication of nanopatterns with different shapes and sizes over large areas. This simple, fast, wide-area (cm^2) and extra low-cost nanopatterning method has been used to fabricate 2D nanostructures (nanoholes and nanopillars) with aspect ratios ~5.5 and diameters of ~200 nm. Figure 2 shows a AFM image of a 2D nanopattern of nanoholes obtained by this method.

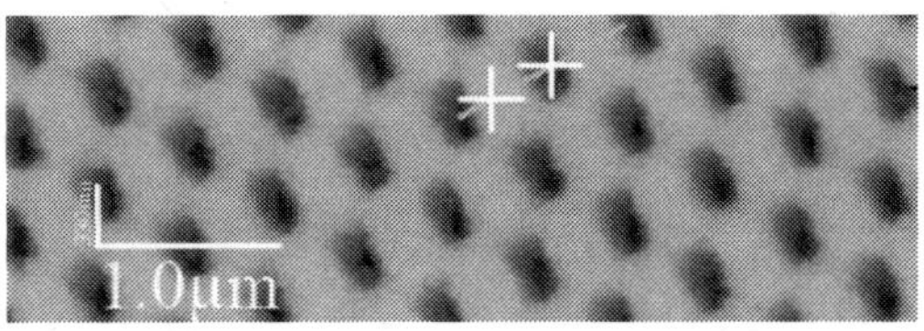

Figure 2. AFM image of a 2D nanopattern of nanoholes obtained by low cost, wide area soft-nanolithography.

3. PHOTONIC CRYSTAL ANTI-REFLECTION COATINGS ON A III-V SOLAR CELL

In a similar way than for the InP bulk case, light can be enhanced to enter in a device by a 2D-PC layer and there are a number of theoretical works related to this [17]. Nevertheless, little work has been done in the actual manufacture and characterization of a solar cell with a two-dimensional photonic crystal (2D-PC) on its front surface, conceived as a periodic distribution of the dielectric constant in the plane (the surface of the solar cell) and involving sub-wavelength motifs. All of the reasons mentioned above can be applied for a photonic crystal front surface as has previously done with sub-wavelength gratings of pyramids [18]. With the aim to prove the validity of PC as an additional strategy to improve solar cells performance, we have used a PC on a single p-n junction solar cell based on Ge with a photonic crystal nanopatterned InGaP window layer. [19] In order to quantify the improvement introduced by the PC, the external quantum efficiency (EQE) of the solar cells manufactured in the same batch with and without the photonic crystal layer has been measured. We have found an increase in EQE around 22% in average for the solar cells with the photonic crystal structure, which in terms of photocurrent implies an increase between 16% and 25% for direct terrestrial spectrum (AM1.5D). Despite the intrinsic performance of the used solar cells is not optimum (filling factors around 52%) due to their experimental nature (these cells were obtained with heteroepitaxial techniques) there is an improvement of the efficiency of about 11% for the cell with PC pattern. In this prototype, only a 40% of the surface was patterned with the PC. By linear extrapolation, an enhancement of 25-26% can be obtained for a solar cell with the entire surface patterned by a PC. The measured increment in the photocurrent is 8%, so this gives an enhancement in the photocurrent of 22% for a fully covered surface. The open-circuit voltage (V_{OC}) was not degraded. Also, the dark I-V curves of both devices, with and without PC are observed to be exactly the same, so none degradation has been introduced to the device due to the PC pattern fabrication.

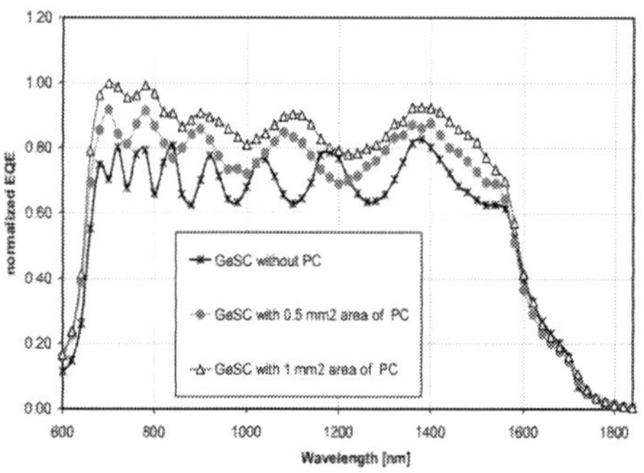

Figure 3. Relative EQE for two PC-patterned solar cells with two different areas patterned, 1 mm^2 and 0.5 mm^2.

Figure 3 shows the relative EQE for two PC-patterned solar cells with two different areas patterned, 1 mm^2 and 0.5 mm^2. The experimental EQE data have been normalized to the highest EQE value. The EQE in both the PC-patterned solar cells are always higher than in the non-patterned cell. For the cell with PC-pattern covering 1 mm^2 the enhancement of current using the spectrum AM1.5D is 25%, whereas for the cell with half of the area patterned the increase is 16%. For the cell with pattern covering 1 mm^2 the enhancement is up to a 46% -i.e. 0.46 times more EQE than the non patterned device for the same wavelength- for some specific wavelengths (1150 nm) and a 22% enhancement in average for the entire spectrum, whereas for the cell with half of the area patterned ranges up to a 32% (also for 1150 nm) with an average of 11% for the whole spectrum. In this case there is a decrease in the EQE of around 10% at 1200nm compared to the cell without PC. This decrease is counterbalanced in the case of the first cell with a positive increase of only a 2%, but above zero. This shows clearly the validity of our approach to enhance the performance of solar cells, which could be also applied to other photon-sensing devices like photodetectors and thermo-photovoltaic devices.

In summary, we have used a photonic crystal ARC made on a single-junction Ge/InGaP solar cell. An increase in the photocurrent up to a 8% has been achieved with a 40% of its surface covered with a PC pattern. Enhancements of the external quantum efficiency (EQE) of 22% for a wide range of wavelengths and up to a 46% for specific wavelengths have been measured. A correlation with the area of photonic crystal patterned has been observed. We believe that PC ARCs have several advantages over single layer or bilayer ARCs that can suffer much more easily failure by mechanical or thermal cracking.

4. Acknowledgements

We acknowledge funding from MINECO TEC2014-54449-C3-3-R and PCIN-2013-179.

5. REFERENCES

1. L. Zeng, Y. Yi, C. Hong, J. Liu, N. Feng, X. Duan, L. C. Kimmerling, and B. A. Alamariu, "Efficiency enhancement in Si solar cells by textured photonic crystal back reflector," Appl. Phys. Lett. 89, 111111-111113 (2006)
2. C. Seassal, Y. Park, A. Fave, E. Drouard, E. Fourmond, A.Kaminski, M. Lemiti, X. Letartre, P. Viktorovitch, "Photonic crystal assisted ultra-thin silicon photovoltaic solar cell" Photonics for Solar Energy Systems II, edited by Andreas Gombert, Proc. of SPIE Vol. 7002, 700207, (2008)
3. B. S. Cheng, C. H. Chiu, K. J. Huang, C. F. Lai, H. C. Kuo, C. H. Lin, T. C. Lu, S. C. Wang and C. C. Yu, "Enhanced light extraction of InGaN-based green LEDs by nano-imprinted 2D photonic crystal pattern," Semicond. Sci. Technol. 23, 055002-055006, (2008)
4. K. R. Catchpole and M. A. Green, "A conceptual model for light-coupling by pillar diffraction gratings", Journal of Appl. Phys. **101**, 063105-063105-8 (2007)

5. Y. Kanamori, M. Sasaki, and K. Hane, "Broadband antireflection gratings fabricated upon silicon substrates," Op. Lett. 24, 1422-1424 (1999)
6. Goetzberger, A., Knobloch, J., and Voss, B., *Crystalline Solar Cells*, John Wiley & Sons, 1998
7. V. G. Dyskin, Z. S. Settarova, and M. N. Tursunov, "Antireflecting Coatings for Cascade Solar Cells", *Applied Solar Energy, 2009, Vol. 45, No. 3, pp. 151–153.*
8. Validov, M.A., Ivanov, V.A., and Akhmadiev, M.Kh., USSR Inventor's Certificate no. 246890, *Byull. Izobret.*, 1969, no. 21.
9. Tret'yakova, V.I., Gavrilova, G.A., and Anufriev, R.L., USSR Inventor's Certificate no. 793956, *Byull. Izobret.*, 1981, no. 1.
10. Alferov, Zh.I., Andreev, V.M., and Rumyantsev, V.D., Trends of Solar Photoenergetics, *Fizika i tekhnika poluprovodnikov,* , vol. 38, 8, pp. 937–948. (2004)
11. Born, M. and Wolf, E., Principles of Optics, 1969.
12. J.Y. Chen, K.W.Sun "Enhancement of the light conversion efficiency of silicon solar cells by using nanoimprint anti-reflection layer", Solar Energy Materials & Solar Cells, 94, 629–633,(2010).
13. C.-H. Sun, W.-L. Min, N.C. Linn, P. Jiang, B. Jiang, "Templated fabrication of large area subwavelength antireflection gratings on silicon", Appl. Phys. Lett. 91 (2007) 231105–231107.
14. Y.-J. Lee, D.S. Ruby, D.W. Peters, B.B. McKenzie, J.W.P. Hsu, "ZnO nano- structures as efficient antireflection layers in solar cells", Nano. Lett. 8, 1501–1505, (2008).
15. M.-L. Kuo, D.J. Poxson, Y.S. Kim, F.W. Mont, J.K. Kim, E.F. Schubert, S.-Y. Lin, "Realization of a near-perfect antireflection coating for silicon solar energy utilization", Opt. Lett. 33, 2527–2529 , (2008).
16. P. A. Postigo; M. Kaldirim; I. Prieto; L. J. Martínez; M. L. Dotor; M. Galli; L. C. Andreani, Proc. SPIE 7713, Photonic Crystal Materials and Devices IX, 771307 (2010)
17. Dutta, P.S., 2007 International Workshop on Physics of Semiconductor Devices (IWPSD '07), p 828 (2007)
18. Y. Kanamori, M. Sasaki, and K. Hane, Op. Lett. 24, 1422 (1999).
19. I. Prieto, et al. "Enhanced quantum efficiency of Ge solar cells by a two-dimensional photonic crystal nanostructured surface" Appl. Phys. Lett. **94**, 191102 (2009)

PROTON EXCHANGE MEMBRANE FUEL CELL WITH ENHANCED DURABILITY USING FLUORINATED CARBON AS ELECTROCATALYST

Yasser Ahmad(1)(2), **Katia Guérin**(1)(2), **Laetitia Dubau**(3) (4), **Marian Chatenet**(3) (4), **Sandrine Berthon-Fabry** (5)

(1) *Université Clermont Auvergne, Université Blaise Pascal, Institut de Chimie de Clermont-Ferrand, BP 10448, F-63000 CLERMONT-FERRAND, France, Diane.Delbegue@univ-bpclermont.fr*
(2) *CNRS, UMR 6296, ICCF, F-63171 Aubière, France*
(3) *Univ. Grenoble Alpes, LEPMI, F-38000 Grenoble, France*
(4) *CNRS, LEPMI, F-38000 Grenoble, France*
(5) *MINES ParisTech, PSL Research University PERSEE - Centre procédés, énergies renouvelables et systèmes énergétiques, CS 10207 rue Claude Daunesse F-06904 Sophia Antipolis Cedex, France*

ABSTRACT

This study evaluates the fluorination of a carbon aerogel and its effects on the durability of the resulting electrocatalyst for Proton Exchange Membrane Fuel Cell (PEMFC). Fluorine has been introduced before or after platinum deposition. The different electrocatalysts are physico-chemically and electrochemically characterized, and the results discussed by comparison with commercial Pt/XC72 from E-Tek. The results demonstrate that the level of fluorination of the carbon aerogel can be controlled. The fluorination modifies the texture of the carbons by increasing the pore size and decreasing the specific surface area, but the textures remain appropriate for PEMFC applications. Two fluorination sites are observed, leading to both high covalent C-F bond and weakened ones, the quantity of which depends on whether the treatment is done before or after platinum deposition. The order of the different treatments is very important. The presence of platinum contributes to the fluorination mechanism, but leads to amorphous platinum rather inactive towards the Oxygen Reduction Reaction. Finally, a better durability was demonstrated for the fluorinated then platinized catalyst compared both to the same but not fluorinated catalyst and to the reference commercial material (based on the loss of the electrochemical real surface area after accelerated stress tests).

RESULTS AND DISCUSSION

To minimize the corrosion of the catalyst support, several approaches have been tested. Graphitizing carbons improves their electronic conductivity and corrosion resistance [1-3]. Coatings with conductive polymers have also been studied [4], because of their high surface area, low ohmic resistance and stability, but the apparent electronic conductivity of the composite material often decreases during the catalyst deposition because it becomes disconnected electronically. Metal oxides, such as tin and indium oxide (ITO), TiO_x, SnO_2, WO_x, IrO_2, and carbides, such as WC and W_2C, have also been proposed as promising durable catalyst supports [5,6], and to overcome the unsufficient electronic conductivity of the metal oxide, doping has been used [7,8]. Some researchers investigated TiO_2/carbon composites [9,10] and showed that mixing carbon and TiO_2 is beneficial to the electrocatalyst durability. Nevertheless, none of these solutions are completely satisfactory on the field.

In this research, the surface chemistry of the carbon aerogel CA was modified using a controlled fluorination. In order to limit the corrosion due to the oxidation of the carbon by its own oxygen content at its surface and the effect of the water environment, some of the dangling bonds were saturated with fluorine. So far, few fluorine-doped carbons (of the carbon blacks or template mesoporous carbon families) have been used as catalysts for their activity in alkaline medium [11,12] or in acidic media [13]. Because a high specific surface area of the carbon support is important for an optimized dispersion of platinum nanoparticle, and an adapted porosity is necessary to limit the mass transport losses, the study has been done on a carbon aerogel substrate, which possesses these 2 advantages [14]. In this work fluorinated and platinized carbon aerogels were synthesized and characterized as oxygen reduction reaction (ORR) electrocatalyst in acidic media, and their durability in PEMFC conditions was evaluated by accelerated stress tests performed in four-electrode cell. The results are discussed and compared to a benchmark electrocatalyst material (commercial Pt/XC72, from E-Tek).

The bulk density of the monolithic pristine CA, determined is ρbulk = 0.145 ± 0.002 g.cm-3. The texture of the CA materials, analyzed by nitrogen adsorption desorption, shows in any case a type IV curve with a H1 hysteresis loops in the IUPAC classification suggesting that the pristine and the modified materials are mainly mesoporous. However, the various treatments (insertion of platinum nanoparticles or fluorination) modify the texture within this "mesoporous range" (Table 1). The modification of the texture is different according to the treatment.

DOI: 10.1051/e3sconf/20171617001

Concerning the fluorination, the tendencies observed are the same on the CA or on the Pt-CA: the specific surface area decreases in both cases, all types of pore being modified. The total porous volume (determined by single point at $P / P° = 0.98$) increases owing to the increase of the macroporous volume, because the micropore and mesopore volumes both decrease. The limitation of the smallest pores leads to a decrease of the specific surface area of the material. This modification of the carbon texture during the fluorination (even if the level is low) could be compared to a strong activation phenomenon with the probable formation of gaseous CF_4 or C_2F_6.

Sample	S_{BET}[a] (m² g⁻¹) ± 5	V_{mic} (cm³ g⁻¹) ± 0.004	V_{meso} (2-7.5nm) (cm³ g⁻¹) ± 0.05	V_{meso} (2-50nm) (cm³ g⁻¹) ± 0.05	Vt, (cm³ g⁻¹) ± 0.1
CA	610	0.257	0.07	0.28	4.0
F-CA	393	0.142	0.02	0.17	4.7
Pt-CA	414	0.190	0.03	0.23	2.0
Pt/F-CA	333	0.132	0.02	0.20	4.0
F/Pt-CA	294	0.093	0.02	0.18	3.4

Table 1. Textural characterizations for the various materials owing to N2 isotherms. SBET : specific surface area, Vmic : micropore volume (< 2 nm), Vmeso : mesopore volume (width between 2 and 7.5 nm), Vmeso : mesopore volume (width between 2 and 50 nm), Vt : Total pore volume.

On the platinized material, the fluorination produces PtF_6 gas [35] which can be able to open small porosity blocked by the platinum nanoparticles. On the contrary, during the platinum deposition (on CA or on F-CA), the accessible volume decreases; this effect probably results from the fact that platinum nanoparticles sit (are stabilized) preferentially at the mouth of the small pores, thereby closing the corresponding porosity. The decreasing of the small pore volume naturally leads to a decrease of the specific surface area. This effect is more important in the non-fluorinated sample; indeed, the fluorination destroys the small pores, which are the main promoter of the specific surface area. Now, whatever the treatment, the specific surface area of the sample remains interesting in terms of value ($S_{BET} > 294$ m² g⁻¹) and isotherm shape for use as PEMFC electrocatalyst and the pore size distribution is in accordance with the necessity of the water management: few micropores, few small mesopores (<5 nm) and a balance of the porosity between mesopores and

TGA experiments show that the pristine CA material is stable in air up to 450°C. At higher temperature, the carbon burns with the oxygen into CO_2 (and CO). All the treatments modify the thermal stability of the resultant material and the decomposition temperature of the oxidation of the non-fluorinated carbon parts (T_C value) deceases with the number of treatment, whatever the order of the treatments. From ambient temperature to 150°C, the mass lost is attributed to the elimination of the atmospheric water absorbed on the high specific surface areas carbons. F/Pt-CA is the less sensitive material to moisture. Above 220°C, the platinum nanoparticles catalyze the degradation of the carbon in the air, as it can be seen on Pt-CA curve. On F-CA, the C-F bonds formed are not removed until 170°C, which confirms their high stability. For higher temperatures of thermal analysis, the C-F bonds are broken, and finally total defluorination is reached at 325°C. The weight loss between 170°C and 347°C confirms the initial fluorination level (25 wt.%) resulting of the fluorination by XeF2 of CA observed by weighting the sample before and after fluorination. For the platinized and fluorinated samples, the decomposition mechanism is more complex (and different according to the order of the treatment) due to the interaction between platinum (see T_{C-F}), fluorine, carbon and the different compounds produced during the degradation. A weak bond leads to the liberation of atomic fluorine and a strong one to the formation of C_2F_4 and CF_2 [36]. On the Pt/F-CA, defluorination begins at lower temperature and the weight loss is faster. On the F/Pt-CA, the behaviour is quite different. The differences could be linked to the type of bond rupture and to the reactivity of the platinum with the atomic fluorine formed during the decomposition, leading to the creation of volatile PtF_6. The lack of crystallinity of the platinum in this sample also impacts the reactivity. The lower remaining weight at 600°C on the Pt/F-CA and F/Pt-CA samples compared to Pt-CA one is consistent to this hypothesis.

The IR reveals two fluorination sites and explains the different behaviours of the samples. For all the fluorinated samples, the IR spectra show both highly covalent C-F bonds for wavenumber close to 1200 cm⁻¹, together with weakened C-F bonds (1100 cm¹). With platinum, new peaks appear between 600 and 750 cm⁻¹. The F/Pt-CA leads to the same behaviour between the weakly and the highly covalent bonds, with higher quantity of weak bonds; a new peak appears at 600 cm⁻¹, which could be attributed to a link between the platinum and the fluorine. This link could explain the higher degradation temperature observed in TGA compared to F-CA. The presence of platinum contributes to the fluorination mechanism (higher fluorine loading) and limits the formation of highly covalent bonds, which is *a priori* a drawback; this will be discussed later. The Pt/F-CA spectrum is slightly different from the F-CA spectrum. The process leading to the platinum insertion (insertion of the sample in a water-in-oil solution, $NaBH_4$ reduction and washing with acetone, ethanol and MilliQ water) modifies the proportion between the highly and the weakly covalent C-F bonds. A part of the

 DOI: 10.1051/e3sconf/20171617001

weak bonds could be involved with platinum. A peak at 750 cm^{-1} appears on Pt/F-CA, while the one at 700 cm^{-1} is not observed anymore. This difference can be also attributed to a weak link between the platinum and the fluorine and also explains the different thermal behaviour of Pt/F-CA. Similar spectra are observed for CA and Pt-CA samples, confirming that no covalent bond exists between the carbon and the platinum.

By XRD it is possible to see the impact or the various treatments on the platinum particles size and their crystallinity. The mean crystallite size ($\overline{d}_{XRD}$) was calculated from the Debye-Scherrer equation. The platinum particle sizes are comparable after the deposition on fluorinated or not CA sample. When the fluorination is done on the platinized sample, the platinum particles become essentially amorphous, as evidenced by the absence of diffraction peak.
The TEM micrographs confirm these values and permits to see the repartition of the particles at the CA or F-CA surface. For Pt-CA (Figure), the platinum nanoparticles are quite well dispersed on the surface of the CA (around 5 nm and 10 nm) and agglomerates can be occasionally observed (less than 20 nm). On the contrary, for Pt/F-CA, nearly all the Pt deposit is constituted of agglomerates of *ca.* 3-8 nm-sized crystallites (Figure 2). It appears that the Pt deposit is not interacting enough with the F-CA substrate to enable sufficient dispersion. On the F/Pt-CA electrocatalyst (Figure 3), the platinum nanoparticles are neither well defined nor well distributed on the carbon surface, on the contrary to the Pt-CA. The fluorination of the Pt-CA sample induces large modifications of the platinum texture and chemistry. Platinum nanoparticles of about 3 nm in size are observed and well distributed over the carbon substrate.
These different materials (with various texture, chemistry and quality of the platinum nanoparticles) have been characterized electrochemically in order to investigate the impact of fluorination treatment on the Oxygen Reduction Reaction (ORR) activity and stability.

The base voltammograms (Figure 10 A) exhibit the classical H_{UPD} behavior observed on platinum catalyst between 0.05 and 0.4 V *vs.* RHE. Interestingly, all the catalyst supported on carbon aerogel show a very important double layer charge contribution, whether fluorinated or not. This typical behavior is linked to the wettability properties of the CA as well as the chemistry of its surface. It is also important to take in mind that the carbon support of our reference catalyst (Vulcan XC 72) has a lower specific surface area (around 250 m^2 g^{-1}) than the aerogel carbon studied here.

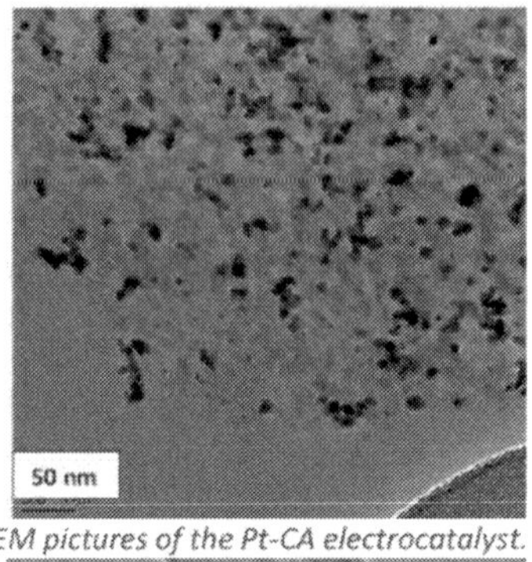

Figure 1. TEM pictures of the Pt-CA electrocatalyst.

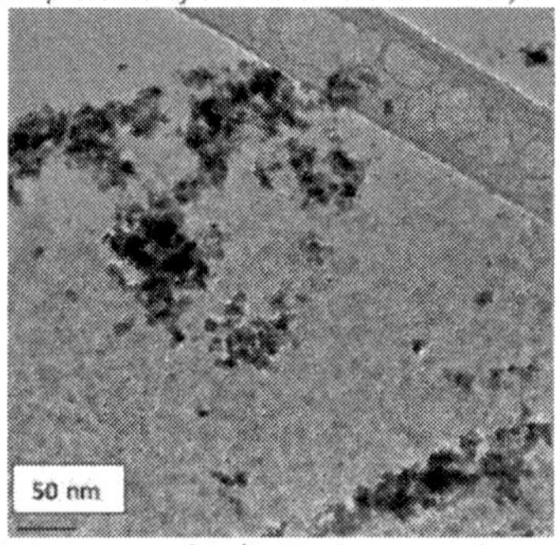

Figure 2. TEM pictures of Pt/F-CA electrocatalyst.

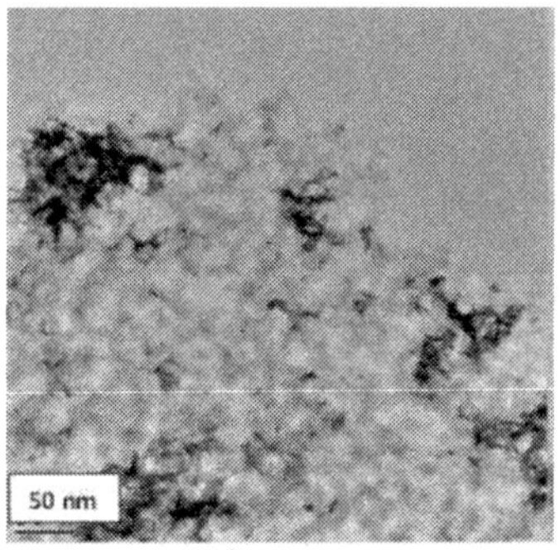

Figure 3. TEM pictures of F/Pt-CA electrocatalyst

CO stripping voltammograms are in agreement with the TEM observations of the different catalyst. The reference material shows a single peak which is the fingerprint of a very well-dispersed catalyst, as evidenced in the representative TEM images. In contrast, the CO stripping peaks performed on the electrocatalysts supported on the CA samples are larger and clearly present several contributions. In addition to the contribution corresponding to isolated particles, others contributions at lower potential sign the presence of agglomerated particles, since CO is more easily oxidized on the grain boundaries present between two aggregated particles .

The Tafel plots of the mass-transport-corrected kinetic current for Pt/CA, Pt/F-CA, F/Pt-CA electrodes and the reference Pt/C 20% (E-Tek) material show that the specific activity measured at 0.85 and 0.90 V vs. RHE are summarized in Table 5. Pt/CA and Pt/F-CA are the most active electrocatalysts for the ORR and their specific activities are three times superior to the activity of the reference material at 0.90 V vs. RHE. It is well known that ORR is a structure-sensitive reaction and in that respect, platinum particle size plays a major role. The lower particle size presents a larger fraction of low-coordinated sites on their surface which strongly adsorb ORR intermediates thus impacting negatively the activity [38,40]. It is therefore not surprising to obtain a higher activity in the case of the CA-supported electrocatalysts, owing to their larger Pt particles size [40]. According to the results derived from the XRD analysis (Table 2), the particle size of the reference material (2.3 nm) is well below the values obtained for the Pt/CA and Pt/F-CA catalyst (around 4 nm). It is also interesting to notice that the fluorination treatment made initially (before the platinum deposit) does not modify significantly the ORR activity.

In contrast, when the fluorination has been made after the Pt deposition, the ORR activity is very low; this result is probably due to the fact that the post-fluorination treatment destroys the crystallinity of the Pt particles as shown in the XRD pattern in Figure 5.

Electrode	SA / µA $cm^{-2}{}_{Pt}$	
	@ 0.85 V	@ 0.9 V
Pt-CA	250	59
Pt/F-CA	212	55
F/Pt-CA	71	16
Pt/XC72 E-Tek	103	25

Table 2. ORR kinetic parameters for Pt/CA, Pt/F-CA and F/Pt-CA electrodes and the reference Pt/C 20% (E-Tek) determined in O2-saturated 0.1 M H2SO4; sweep rate 0.002 V s-1; rotation rate 1600 rpm, positive-going potential sweep from 0.4 to 1 V vs. RHE; T = 25 ± 1°C.

Accelerated stress tests (AST) were performed to investigate the robustness of the fluorinated Pt electrocatalyst and to get some insights into the role played by carbon–fluoride functionalization on the resistance towards carbon corrosion and consequent particle detachment. The aging procedure consisted of 5000 potential cycles between 0.60 V *vs.* RHE and 1.05 V *vs.* RHE at a sweep rate v = 0.05 V s^{-1} in 0.1 M H_2SO_4 and T = 298 K (liquid electrolyte). Such experimental conditions simulate the typical potential range experienced by a cathode material during PEMFC operation (*i.e.* at high and low current density, during idling and open circuit conditions). Different organizations such as the American Department of Energy (DoE) and the Fuel Cell Commercialization Conference of Japan (FCCJ) have their own AST protocols. In the "Electrocatalyst Cycle and Metrics" test from the American DOE, a triangle sweep cycle between 0.6 and 1.0 V *vs.* RHE is used to investigate the stability of the metal nanoparticles. In the Japanese FCCJ procedure, start/stop cycles are mimicked by a triangular wave potential cycle between 1.0 and 1.5 V *vs.* RHE, and load cycles by a rectangular wave potential cycle between 0.6 and 1.0 V *vs.* RHE. Since carbon is thermodynamically unstable at a PEMFC cathode ($E^0_{CO2/C}$ = 0.207 V *vs.* NHE at 298 K) and its corrosion is further catalyzed by Pt, the choice of the potential window 0.6 – 1.05 V *vs.* RHE appears adequate to evaluate the materials durability in PEMFC conditions, and notably to study the robustness of the metal nanoparticles, which is directly correlated to the stability of the carbon support, as previously characterized in this potential window by IL-TEM or indirectly by the detachment/migration of the platinum crystallites. Furthermore, even in PEMFC, the corrosion is visible when the potential remains in this range. It is even more true that the carbon is amorphous and is supporting platinum.

The changes in the electrochemically active surface area (ECSA) were monitored *via* CO_{ad} stripping coulometry. The results, shown in Figure 4 demonstrate the better durability of the fluorinated catalyst compared to non-fluorinated catalyst, but also compared to the reference material. Here again, the particle size must be taken into account. The smallest particles are more prone to dissolve and redeposit (*e.g.* via Ostwald ripening), leading to larger crystallites and thus inducing larger ECSA losses upon ageing. The promising result of this study comes from the comparison of Pt/CA and Pt/F-CA, which directly allows checking the impact of fluorination on the durability (the particle sizes of these samples are essentially identical). Fluorination induces a lower ECSA decrease upon cycling (10.5% *vs.* 17 % ECSA loss) when comparing the fluorinated with non-fluorinated electrocatalyst. Unfortunately, it was not possible to quantify the degree of fluorination after the AST due to the very low amount of catalyst available on the working electrode (30 µg). At the light of this study, it is not possible to conclude on the origin of this spectacular result. Two hypotheses can be considered, a better resistance to corrosion and/or an optimization of the anchoring of Pt nanoparticles on the fluorinated carbon support. Concerning this last point, we have to keep in mind the obvious interaction between the carbon, the fluorine and the platinum. The presence of the platinum on fluorinated carbon aerogel modifies the two different bonds between the carbon and the fluorine [14] and the fluorination on the platinized carbon, not only modifies the both C-F interactions but leads to amorphous platinum.

 DOI: 10.1051/e3sconf/20171617001

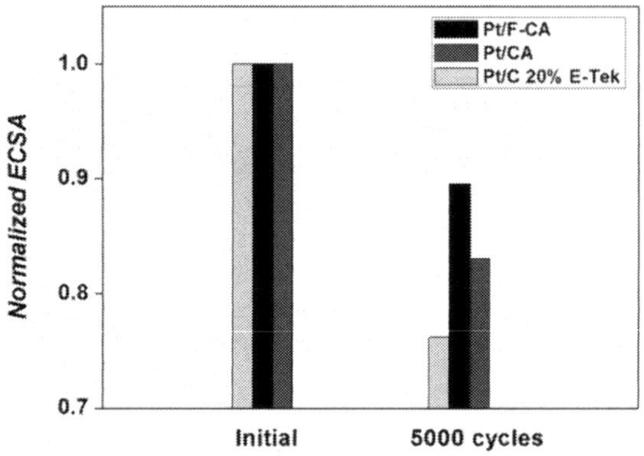

Figure 4. Comparison of the electrochemical surface area (ECSA) estimated by CO stripping coulometry before and after an AST consisting of 5000 potential cycles between 0.60 V vs. RHE and 1.05 V vs. RHE at a sweep rate v = 0.05 V s-1 in 0.1 M H2SO4 and T = 298 K.

References

[1] D. Zhai, H. Du, B. Li, Y. Zhu, F. Kang. Porous graphitic carbons prepared by combining chemical activation with catalytic graphitization Carbon 49 (2011) 725–729

[2] H. Jin, H. Zhang, Y. Ma, T. Xua, H. Zhong, M.J. Wang. Stable support based on highly graphitic carbon xerogel for proton exchange membrane fuel cells. Power Sources 195 (2010) 6323–6328

[3] F. Alcaide, G. Álvarez, O. Miguel, M.J. Lázaro. Stability of carbon nanofibers as catalyst supports for PEMFC. Book of Abstracts. 2nd CARISMA International Conference on Progress in MEA 2010, La Grande Motte (France), 2010 p. 151

[4] G. Inzelt, M. Pineri, J.W. Schultze, M.A. Vorotyntsev. Electron and proton conducting polymers: recent developments and prospects Electrochim. Acta 45 (2000) 2403–2421.

[5] E. Antolini, E. Gonzalez. Solid State Ionics. Ceramic materials as supports for low-temperature fuel cell catalysts. 180 (2009) 746-763

[6] S.Y. Huang, P. Ganesan, B.N. Popov. Titania supported platinum catalyst with high electrocatalytic activity and stability for polymer electrolyte membrane fuel cell. Applied Catalysis B: Environmental 102 (2011) 71–77

[7] I. Cerri, T. Nagami, B.E. Hayden, J.C. Davies. Durable electrocatalysts for PEMFC. Proceeding of the Fundamentals and Developments of Fuel Cell Conference, Grenoble (France), 2011 p 282-283

[8] K. Sasaki, F. Takasaki, Y. Shiratori, Z. Noda. Alternative electrocatalyst support materials for polymer electrolyte fuel cells : semiconducting oxides and carbon nanofibers. Proceedings of the 18th World Hydrogen Energy Conference (2010)

[9] L. Xiong, A. Manthiram. Synthesis and characterization of methanol tolerant Pt/TiOx/C nanocomposites for oxygen reduction in direct methanol fuel Cells, Electrochim. Acta 49 (2004) 4163-4170

[10] A. Bauer, C. Songa, A. Ignaszaka, R. Huia, J. Zhanga, L. Chevallier et al. Improved stability of mesoporous carbon fuel cell catalyst support through incorporation of TiO2. Electrochim. Acta 55 (2010) 8365-8370

[11] H. Wang, A. Kong, Mesoporous fluorine-doped carbon as efficient cathode material for oxygen reduction reaction, materials letters 136 (2014) 384-387

[12] X. Sun, Y. Zhang, P. Song, J. Pan, L. Zhuang, W. Xu, W. Xing, Fluorine-Doped Carbon Blacks: Highly Efficient Metal-Free Electrocatalysts for Oxygen Reduction Reaction ACS Catalysis 2013, 3, 1726 1729

[13] X. Sun, P. Song, T. Chen, J. Liu, W. Xu, Fluorine-doped BP 2000: highly efficient metal-free electrocatalysts for acidic oxygen reduction reaction with superlow H2O2 yield, **Chem. Commun.**, 2013,**49**, 10296-10298

[14] M. Ouattara-Brigaudet, S. Berthon-Fabry, C. Beauger, M. Chatenet, N. Job, M. Sennour, P. Achard, Influence of the carbon texture of platinum/carbon aerogel electrocatalysts on their behavior in Proton Exchange Membrane Fuel Cell cathode, International Journal of Hydrogen Energy 37 (2012) 9742-9757

DOI: 10.1051/e3sconf/20171617002

ADVANCED CARBON FLUORIDES FOR PRIMARY LITHIUM BATTERIES

K. Guérin (1)(2), **M. Dubois** (1)(2), **A. Hamwi**(1)(2)

(1) *Université Clermont Auvergne, Université Blaise Pascal, Institut de Chimie de Clermont-Ferrand, BP 10448, F-63000 CLERMONT-FERRAND, France, Diane.Delbegue@univ-bpclermont.fr*
(2) *CNRS, UMR 6296, ICCF, F-63171 Aubière, France*

ABSTRACT

Li-CFx battery using a specific fluorinated nanocarbon as cathode material exhibits a capacity exceeding the expected theoretical value when used as an electrode material in primary lithium battery. Carbon nanodiscs were partially fluorinated by atomic fluorine released by thermal decomposition of TbF_4, and the capacity of this material was up to 1180 mAh.g^{-1}, whereas a theoretical value of 847 mAh.g^{-1} for the $CF_{0.95}$ sample was calculated. The obtained value is also higher than the maximum one of 865 mAh.g^{-1} expected for CF_1 carbon fluorides. The discharge mechanism was investigated using mainly SEM and solid state NMR in order to understand this "extracapacity". Both the unfluorinated carbon and the LiF covering, which is formed outside the carbon lattice during the discharge mechanism, play a key role for the achievement of the extracapacity by the consumption of Li^+ to form Li_2F^+ species stabilized by the carbon host structure formed after the electrochemical defluorination.

RESULTS AND DISCUSSION

Carbon nanodiscs CND were fluorinated by atomic fluorine release by TbF_4 decomposition at different temperatures. The fluorination progressively turns the structure as a function of the reaction temperature into a (CF)***n*** one. SEM images underlined the maintaining of the geometry for fluorination, whatever the reaction temperature. Nevertheless, as revealed by atomic force microscopy, the swelling due to the accommodation of the fluorine atoms differs according to the fluorination route as underlined by AFM, which was performed in tapping mode. The disc surface is smooth and cracks are observed on the whole surface. The accommodation of fluorine atoms in the sample is then homogeneous across discs thus leading to a homogeneous non-localized swelling. This therefore generates cracks on the material surface to accommodate the volume expansion associated with this process. Fluorinated CND were used as cathode material in primary lithium batteries. The two samples obtained by those fluorination way (C-500 and C-550 with atomic ratio F/C = 0.72 and 0.95, obtained at fluorination temperatures of 500 and 550°C, respectively) were used as cathode in primary lithium batteries. Samples having ***F/C*** of about 0.7-0.8 were investigated because of its known good performances. [1, 2, 3].

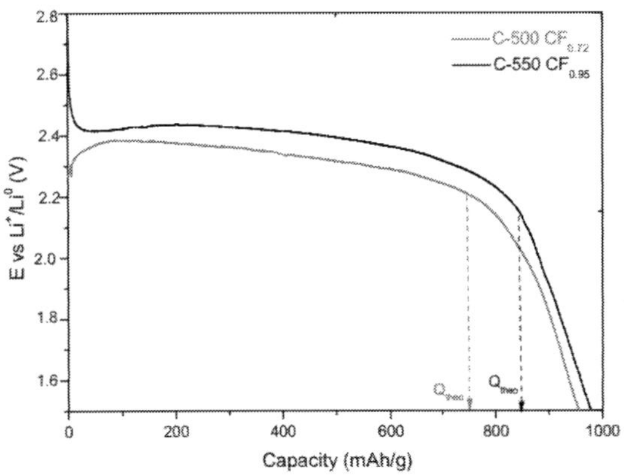

Figure 1. Galvanostatic discharge curves in EC/PC/3DMC LiPF6 1 M electrolyte at a current density of 10 mA.g-1 of fluorinated carbon nanodiscs (C-500 and C-550)

Figure 1 shows the galvanostatic discharge curves of C-500 and C-550 samples with EC/PC/3DMC LiPF6 electrolyte. Both samples exhibit the behavior usually expected for carbon fluorides with a well-defined plateau and initial overvoltage because of its insulating behavior. The voltage initially drops and then increases steadily to reach the main voltage plateau after few minutes; such a behavior has been already observed for graphite fluorides. [4, 5] The average potential is stable during the discharge in agreement with the unique C-F bonding (determined by ^{13}C and ^{19}F solid state NMR), which is still covalent whatever the fluorination conditions. The most unexpected observation is that the experimental capacity (Qexp) significantly exceeds the theoretical value (Qtheo), for those two materials. Indeed, for C-550, the theoretical capacity is 847 mAh.g-1 while an experimental capacity of 978 mAh.g^{-1} is measured. This is also the case for C-500 sample, where 955 mAh.g^{-1} as experimental capacity is measured instead of 751 mAh.g^{-1}.

DOI: 10.1051/e3sconf/20171617002

Different depths of discharge DoD, 0%, 25%, 50%, 100% (relative to the theoretical capacity extracted from the fluorine content) and 127% (corresponding to the extracapacity), have been applied for a primary lithium battery with C-500 as cathode in EC/PC/3DMC $LiPF_6$ 1 M. After the discharge, the cathode material was dried under dynamic primary vacuum, transferred and ground in the argon-filled glove box and put into a rotor that allows NMR spectra to be recorded (without exposure to air) in order to investigate the discharge using ^{19}F nucleus. After the recording of the NMR spectra, the electrode was investigated by EPR and SEM also without exposition to air.

SEM images of the C-500 composite electrode before and after discharges are shown in Figure 2. The good dispersion of PVDF binder is well highlighted for the raw electrode composite. The disc surface is covered with a film made with binder fibril (Figure 2a). In accordance with the well-known mechanism of formation of LiF with the reduction of CFx [1, 6, 7], these particles can be attributed to LiF. The high concentration on the surface and edges suggests that LiF is formed outside the carbon lattice during the discharge. By analogy with lithium-graphite intercalation compounds, such as LiC_6, F^- ions, which are formed during the fluorocarbon reduction, diffuse within the interlayer and combine with solvated Li+ on the sheet edges. [8, 9] A recent study on the discharge mechanism of fluorinated carbon nanofibres showed that such particles made of LiF, are formed outside the carbonaceous matrix. [1] After the full theoretical discharge (100% of faradic yield), the fluorocarbon matrix is totally converted into carbon. The peak related to the (CF)n phase disappeared. No diffraction peak of ordered carbon lattice is observed meaning that amorphous carbon is formed. LiF peaks are observed at 39, 45 and 65° in 2θ values. The formation of crystalline LiF particles is underlined by the sharpness of the peaks. It is to note that the discharge mechanism is nearly similar regardless the nanocarbon used, discs of fibres. [1] The LiF covering becomes thicker with increasing the discharge rate (DoD) up to 100%.

Figure 2. SEM images of C-500 at 0 (a), 25 (b), 50 (c) and 100% discharge (d and e). Figure (e) shows the sample with extracapacity (DoD of 127 %)

After half of the discharge, discs and cones are still observed and are covered by particles with 20-30 nm size. The amount of these particles is higher on the cone and discs edges.

Finally, when an extracapacity (DoD of 127%) is measured, the surface of the discs are covered by cracks that recall the disruptions observed on pristine fluorinated discs, well underlined by AFM images. Moreover, the discs seem to be split into two parts starting from their center.

Such data are in complete agreement with 19F MAS NMR, recorded at high spinning rate (34 kHz) (Figure 3). The PVDF used as a binder is characterized by the presence of a line at -94 ppm assigned to CF_2 groups. The amount of CF_2 groups in the fluorinated materials underlined by a chemical shift of -120 ppm/$CFCl_3$ on the ^{19}F NMR spectra does not change upon discharge. These groups are electrochemically inactive. The lines near -80 ppm are related to phosphorous species. The spectrum of solid $LiPF_6$ is added for comparison and exhibits a line at -74 ppm.

The pristine fluorinated CND sample contains small amount of CF_2 groups, the lines for these groups being present with a very low intensity. As for the carbon nanofibres, the decrease of the intensity of the C-F line at - 190 ppm as a function of the depth of discharge, is in accordance with the electrochemical breaking of the C-F bonds during the discharge and it is accompanied by the formation of LiF. [1] For DoD of 100%, this line is no more present underlining the total conversion of the fluorocarbon matrix into carbon lattice. The line at -204 ppm is unambiguously assigned to the fluorine nuclei present in the particles of LiF. The two lines with a chemical shift at -153 and -147 ppm are observed for discharged fluorinated CND with DoD of 127% unlike fluorinated CND discharged at DoD of 0 and 100% (Figure 3). A fluorinated specie, which is Li2F+is obtained. Thus, our hypothesis to explain unprecedented extracapacity involves an insertion phenomenon of lithiated species Li2F+ into the carbon newly formed by electrochemical defluorination. This additional process could be the origin of the obtained extracapacities.

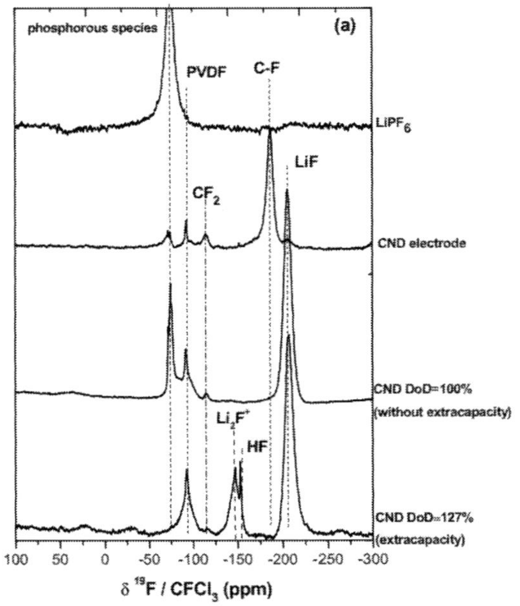

Figure 3. (a)19F MAS NMR spectra (34 kHz) of C-500/PVDF/CB (80/10/10 w/w) composite electrode before discharge (composite with simple contact with the electrolyte), and after discharge with and without extracapacity (depths of discharge DoD = 100 and 127% respectively). Lines at -74 ppm are related to phosphorous species as shown by the comparison with $LiPF_6$ powder

Taking into account structural (NMR and EPR) and morphological (SEM and AFM) characterizations, the discharge mechanism, quasi-independent of the electrolyte may be described as follows. The discharge occurs successively according to the following steps: i) the growth of LiF particles because the breaking of C-F bonds and the recombination with Li+ ions to form a continuous shell on the surface of the discs at 100% of discharge ii) carbon lattice is rebuilt onto the central disc, which serves as a reinforcement, iii) to exceed the theoretical capacity, an insertion process into the reformed carbon lattice may occur; the access of the ions into the carbon host takes place through the edges of discs but also via cracks on the initial fluorinated discs highlighted by AFM. The shell of LiF formed seems to have a major role in extracapacity; it seems to allow the diffusion of the lithium ions, either Li+ or Li_2F^+ through the edges or the surface after the full discharge. As a matter of fact, the LiF shell seems to be locally consumed on the access areas for insertion. This mechanism is similar to that of the lithium-thionyl chloride batteries ($Li/SOCl_2$) where the LiCl layer reacts with lithium ions to give Li_2Cl^+ cations according to the reaction:

$$Li+ + LiCl \rightarrow Li2Cl+ \qquad (1)$$

which improves the performance of the cell. The splitting in two parts of the discs exactly on their center may be explained by two phenomena: either consumption of LiF as in the surface cracks or a volumic expansion because of the insertion.

The extracapacity may be obtained because of the maintaining of the geometry of the carbon lattice. This point must be explained taking into account the intrinsic properties of CND. Using freezing fracture, Naess et al [10] observed for the same nanodiscs from N-Tec an in-plane cleavage that was usually parallel to the disc surface. The particles were fractured in a plane close to the geometrical centre of the disc. To establish this conclusion, a large number of TEM images (185) were analyzed. The histogram of the thickness ratio x1/x2 (x1 is the height of the lower part of the disc and x2 the height of the entire disc) shows that a significant number of discs rupture in an area close to the core. The splitting occurs between the graphitic layers, which are only bonded by weak van der Waals forces, and not in the amorphous carbon surrounding the disc. The authors concluded then the as-synthesized discs can be described as graphitic core within an envelope of non-crystalline carbon. The graphitic core is first grown to its full size, whereupon layers of amorphous carbon are deposited, the thickness of the graphitic core being estimated to only 2-5 nm. Moreover, if the material is

DOI: 10.1051/e3sconf/20171617002

heat-treated at 2700°C, the graphitization takes place simultaneously in different parts of the amorphous carbon. Thus the temperature-induced conversion from amorphous carbon to layers of graphite is not fully coupled geometrically to the original graphite core. Central discs are more organized than the outer discs after graphitization. The higher the graphitization, the lower the reactivity with fluorine. So the fluorination is higher in the outer parts than in the central discs, not or less fluorinated.

This history of the nanodiscs has important consequence during the fluorination/defluorination. An analogy between thermal and electrochemical defluorination is interesting. Ahmad et al. described the preparation of multilayer carbon nanomaterials obtained by thermal treatement (defluorination) of these sub-fluorinated nanocarbons obtained by the controlled fluorination process. [11] The defluorination occurs with a progressive thinning of the discs because of evolution of gaseous CF_4 and C_2F_6 species. Some carbons reorder onto the non or less fluorinated central discs resulting in very thin discs (thickness in the 5-20 nm range) which fold up on themselves. Homogeneous dispersion of fluorine atoms into the outer discs thanks to the controlled fluorination and lowering of the structural defects result in equivalent thinning in the whole discs. Any thinning occurs during the electrochemical defluorination but reduced carbon rebuilt onto the central discs, which act as a reinforcement to preserve the discotic geometry.

An insertion of lithiated ions may takes place into the complex structure of defluorinated carbon discs covered with LiF. The LiF covering may acts both as lithium ions source and as diffusive layer for those ions. LiF is then probably consummed during the additional electrochemical process, assimilated as an insertion of lithiated ions Li+ or Li2F+ (in accordance with the additionnal line observed in 19F and 7Li NMR spectra and charge transfer seen by EPR). LiF layer is then thinner at any zones of insertion, i.e. onto the sheet edges and onto the disruptions of carbon sheets, initially present in the raw fluorinated materials (well underlined by AFM).

Scheme 1 summarizes the electrochemical discharge mechanism and the insertion phenomenon which leads to extracapacity.

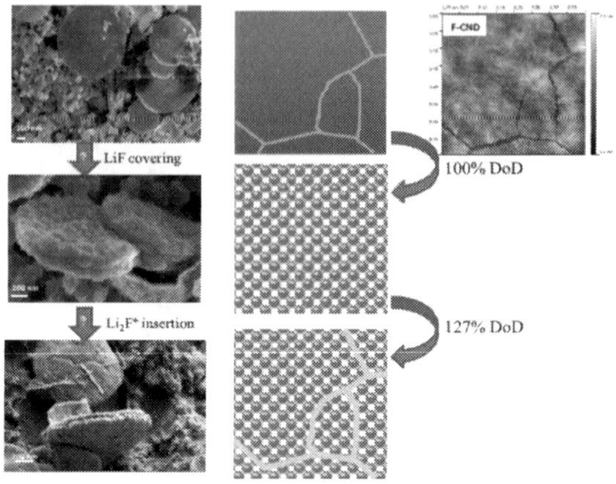

Scheme 1. Mechanisms explaining the extracapacity obtained with fluorinated CND

Since the LiF layer and the formation Li_2F^+ species play a major role in extracapacity, the thicker the LiF layer is, the greater the insertion of lithiated species and the extracapacity.

Taking into account that the ternary mixture of alkylcarbonates used as solvent for EC/PC/3DMC $LiPF_6$ 1M electrolyte is able to dissolve higher quantity of LiF (SLiF =9.08 $g.L^{-1}$) compared to PC solvent (where SLiF=0.13 $g.L^{-1}$), extracapacity should be higher with PC $LiClO_4$ 1M electrolyte rather than EC/PC/3DMC $LiPF_6$ 1M electrolyte.
Thus, galvanostatic discharges at low current density (10 mA/g) were performed in 1M LiClO4 PC electrolyte. A capacity up to 1180 mAh/g was observed, higher than 847 mAh/g expected for CF0.95 composition and enhanced in comparison with the one obtained in EC/PC/3DMC $LiPF_6$ 1M electrolyte. Such experimental facts clearly underline that the extracapacity is related to the LiF covering and it can be improved via the choice of the electrolyte to reach 140% faradic yield.

DOI: 10.1051/e3sconf/20171617002

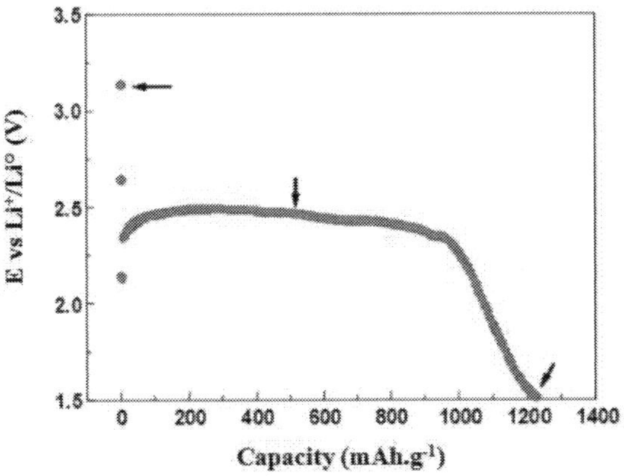

Figure 4. Galvanostatic discharge curve in PC LiClO4 1M electrolyte at a current density of 10 mA.g-1 of CND fluorinated by controlled fluorination (C-550).

The extracapacity obtained there is the higher ever known in CFx primary lithium battery.

References

[1] Y. Ahmad, K. Guérin, M. Dubois, W. Zhang, A. Hamwi, Electrochim. Acta 2013, 114, 142.
[2] K. Guérin, M. Dubois, A. Houdayer, A. Hamwi, J. Fluorine Chem. 2012, 134, 11.
[3] R. Yazami, A. Hamwi, K. Guérin, Y. Ozawa, M. Dubois, J. Giraudet, F. Masin, Electrochem. Commun. 2007, 9, 1850.
[4] K. Guérin, R. Yazami, A. Hamwi, Electrochem. Solid-State Lett. 2004, 7, 159.
[5] M.J. Root, R. Dumas, R. Yazami, A. Hamwi, J. Electrochem. Soc. 2001, 148, 339.
[6] S. S. Zhang, D. Foster, J. Wolfenstine, J. Read, J. Power Sources 2009, 187, 233.
[7] E. Rangasamy, J. Li, G. Sahu, N. Dudney, C. Liang, J. Am. Chem. Soc. 2014, 136, 6874.
[8] J. Giraudet, M. Dubois, J. Inacio, A. Hamwi, Carbon 2003, 41, 453.
[9] A. Naji, P. Willmann, D. Billaud, Carbon 1998, 36, 1347.
[10] N. Stine Nalum, E. Arnljot, H. Geir, D. K. Kenneth, Sci. Technol. Adv. Mat. 2009, 10, 065002.
[11] Y. Ahmad, E. Disa, M. Dubois, K. Guérin, V. Dubois, W. Zhang, P. Bonnet, F. Masin, L. Vidal, D. A. Ivanov, A. Hamwi, Carbon 2012, 50, 3897.

TOWARDS SUPERCAPACITORS IN SPACE APPLICATIONS

Brandon Buergler [1], Bertrand Faure [2], David Latif [3], Lukas Diblik [3], Petr Vasina [3], Valéry Gineste [2],Marek Simcak [4]

[1] ESTEC, Keplerlaan 1, PO Box 299, 2200 AG Noordwijk, The Netherlands, Email: brandon.buergler@esa.int
[2] AIRBUS DEFENCE & SPACE - Space Systems, 31 rue des Cosmonautes - Z.I. du Palays,31402 Toulouse Cedex 4, France, Email: bertrand.faure@astrium.eads.net
[3] EGGO Space, EGGO Space s.r.o., Dvořákova 328, 563 01 Lanškroun, Czech Republic, Email: vasinap@eggo.cz
[4] CSRC, spol. s.r.o., Janska 12, 602 00 Brno, Czech Republic, Email: marek.simcak@csrc.cz

ABSTRACT

Supercapacitors offer a wide range of applications for space flight. The aim of this activity was to pursue life tests on commercial off the shelf (COTS) supercapacitors from different manufacturers, to evaluate their performance after long term vacuum exposure and to investigate balancing designs for the use of these cells in banks of supercapacitors (BOSC). This study enabled to select the most suitable part for space applications and to confirm the design rules at unit level and deratings at component level, which need to be applied. All those complementary results have paved the way to the on-going activities related to Nesscap 10F qualification and associated modular Bank Of Supercapacitors development for space applications.

1. INTRODUCTION

Supercapacitors also called electrochemical double layer capacitors are energy storage devices that store electric energy in the double layer between a high surface area electrode and an electrolyte [1]. As compared to batteries no chemical reactions should occur at the electrodes. Thus, with exception of parasitic electrochemical reactions, the storage of energy in supercapacitors is a purely physical phenomenon [2]. The use of supercapacitors on spacecrafts and launchers can be manifold. Currently, applications as power system stabilisation and energy storage for pyrotechnic mechanisms and electro-mechanical thrust vector actuation are envisaged [3]. The supercapacitor cells must be capable of withstanding the harsh environments present for space applications. The application of supercapacitors in micro-satellites is also an option, which has been studied by Shimizu and Underwood [4]. The advantage of supercapacitors is their extremely high power density of several tens of kW/kg and their capability of operating at temperatures down to -40°C. The main challenge of this technology is the limited energy density, which is in the range of 5 Wh/kg [5]. Due to the purely physical operation principle supercapacitors are capable of performing a large amount of cycling. Furthermore, the performance of supercapacitors with decreasing temperature (e.g. -40°C) does not deteriorate as strongly as compared to the currently used Li-ion chemistries.

The ageing of carbon based supercapacitors becomes evident by the increase of the Equivalent Series Resistance (ESR) and in parallel a decrease of capacity with lifetime. Two types of electric aging tests are usually done. On the one hand the supercapacitor is subjected to a long duration at constant voltage and somewhat increased (compared to room temperature) temperature, on the other hand the supercapacitor is electrically cycled at e.g. room temperature. In the calendaric test the aging rate has been shown to increase with increasing temperature and voltage and in the cycling tests the capacity fade and ESR increase become more pronounced with increasing cycle number [6]. The results shown in this paper have been elaborated within an ESA led activity [7]. The aim of this work was to evaluate the application of supercapacitors in space applications and has partially been presented in [8]. This paper is a continuation of the work in terms of life test duration and an extension to the balancing system, which is of major importance for the lifetime of supercapacitor banks. Environmental tests, electrical characterisation and life tests were done both on single cells and on module level and these results shall be reported here. The Nesscap 10 F cell was selected owing to its good performance in the life tests. This paper focuses on this cell type. Not included are the modelling of supercapacitor cells electric performance and ageing, which are reported in [9]

2. EXPERIMENTAL

2.1 Single Cell Tests

The tests on three COTS supercapacitors reported in [7] and [8] were continued:

 DOI: 10.1051/e3sconf/20171617003

Figure 1: The four tested commercial off the shelf supercapacitor types from left to right Cap XX HS130 (2.4 F), Maxwell PC10 (10 F), Nesscap EHSR 0010C0-002R7 (10 F), Maxwell BCAP0010 P270.

Cap-XX HS130 (2.4 F), Maxwell PC10 (10 F) and Nesscap 10F (10 F). In addition, a new COTS part from Maxwell, BCAP0010 P270 was submitted for evaluation testing. The different supercapacitor types are shown in Figure 1.

The cells were submitted to calendaric aging test at different conditions with temperatures of -35° C, 22° C, 45° C and 65° C and voltages of 0.6, 0.8, 1.0 and 1.2 times the rated voltage (Vop). As compared to Nesscap cells, a reduced testplan with shorter durations was carried out on Maxwell cells (6000 hours). Different cycling tests with different electric profiles were also performed on the individual supercapacitors at room temperature. The results obtained with the new Maxwell BCAP0010 P270 during the 75% discontinuous energy cycling as shown in Figure 2 shall be shown here. The results are compared to the Nesscap EHSR 0010C0-002R7 cell.

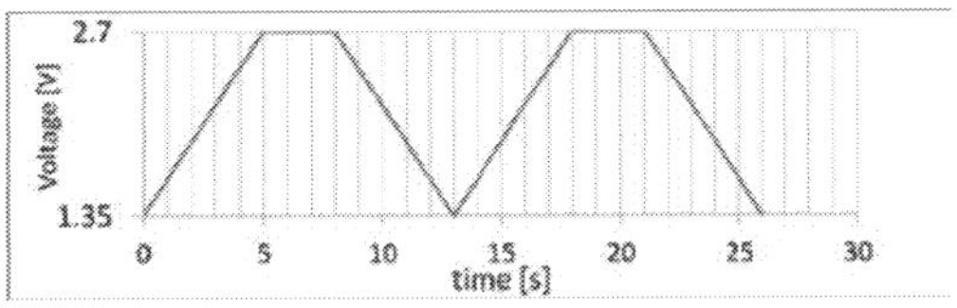

Figure 2 Cycling profile for the 75% discontinuous energy cycling test.

Finally, 10 Nesscap parts were submitted to life test under vacuum conditions (<10mbar) at 0.8·Vop at 45° C and under ambient pressure also at 0.8·Vop in a temperature range from 35 to 45° C. This test was done to check the cells long term compatibility with space environment.

2.2 Bank of Supercapacitor Tests

As previously described the Nesscap cell was chosen as the baseline owing to the good performance in the tested conditions. In [7] the design and manufacturing and testing of a 15s2p bank of supercapacitors was described. The unit had in internal resistance of less than 160 mΩ (DC) and a capacity of 1.26 F at beginning of life (BOL). In this paper the extension of the cycling test to more than 4 Million cycles is reported.

A further topic was to assess the dependence of the lifetime of the BOSCs on the balancing resistor. For the 15s2p BOSC each cell pair was fitted with a parallel resistor of 532 Ω. This allows for having balanced voltages for all the cells in the BOSC, however, it also induces a loss of efficiency in the unit. In order to study the influence of the balancing resistor, 4s1p BOSCs were manufactured by CSRC and tested at EGGO. An overview of the manufactured BOSCs is given in Table *1*. Three BOSCs with different balancing factor were based on the Nesscap cell, one on the Maxwell cell and one on the Cap-XX cell. The balancing factor is defined as the ratio between balancing current at 0.80·Vop and the leakage current of a cell at BOL. The manufactured BOSCs (Figure 3) were tested without housing.

BOSC No	Components manufacturer and reference	ESR [Ω]	Capacity [F]	Balancing Factor	Balancing resistor [Ω]
1	Nesscap EHSR 0010C0-002R7	0.078	2.24	50	2150
2	Nesscap EHSR 0010C0-002R7	0.078	2.22	100	1000
3	Nesscap EHSR 0010C0-002R7	0.080	2.25	No balancing	No balancing
6	Maxwell BCAP0010 P270	0.228	2.48	100	681
7	Cap-XX HS130F	0.098	0.56	100	3160

Table 1: Overview of the BOSCs tested in the activity.

The PCB design was common for all the cells and is shown in Figure 4. The total balancing resistor per cell could be tuned by choosing the right resistor values in the two strings of two series connected resistors.

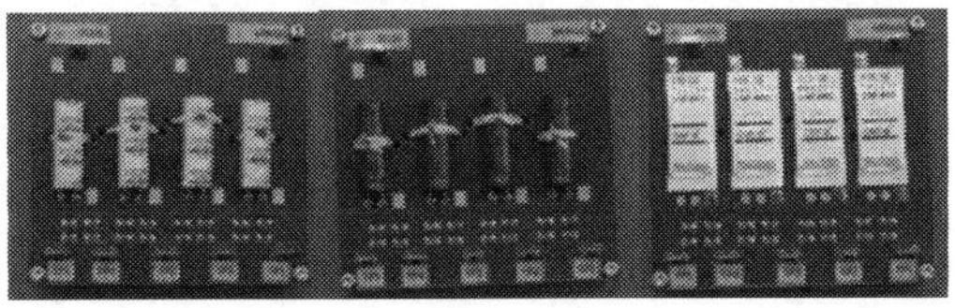

Figure 3: The manufactured banks of supercapacitors based on Nesscap (only BOSC Nr.1 shown), Maxwell and Cap-XX cells respectively.

DOI: 10.1051/e3sconf/20171617003

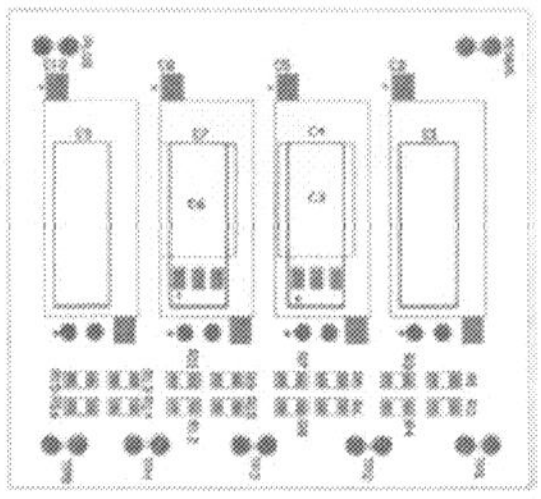

Figure 4: PCB of the 4s1p BOSCs allowing to use a single PCB design for all the studied cells and allowing flexibility in terms of balancing resistors.

The 4s1p BOSCs were initially characterised and then subjected to a cycling profile as shown in Figure 5 at 45°C. It should be noted that no initial "balancing" of the BOSCs was done to adjust to the same level the individual cell voltages before the tests began. The capacity, ESR and leakage current were determined in regular intervals. The individual cell voltages were also measured in order to be able to check the balancing state of each BOSC.

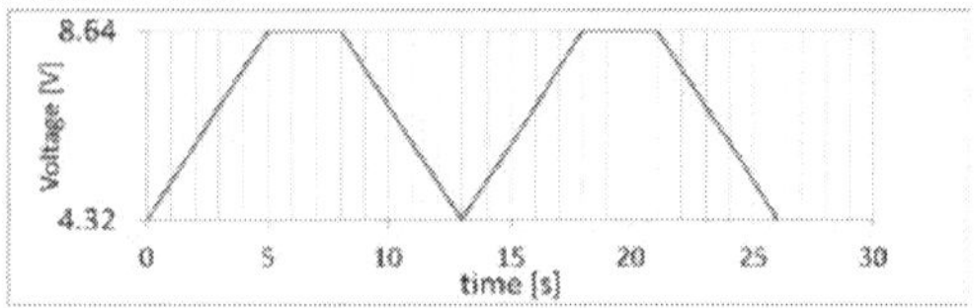

Figure 5 Cycling profile for the 4s1p BOSCs.

In the electric test profile for the BOSCs as shown in Figure 5, the maximum operation voltage was chosen to be 8.64 V, which corresponds to 2.16 V (= 0.8·Vop) per cell. The cell rated voltage Vop for the Nesscap cell is 2.7 V. Therefore using 2.16 V as the maximum voltage during cycling corresponds to a derating of 20%, which was shown to be adequate to allow for long enough lifetimes (see cell life tests). The test profile consisted of charging the BOSC from 4.32 V to 8.64 V within 5 seconds. A current of around 2 A was needed at BOL for doing so. Then there was a short rest time of 3 s followed by a discharge to 4.32 V. The BOSC was immediately recharged after it reached 4.32 V.

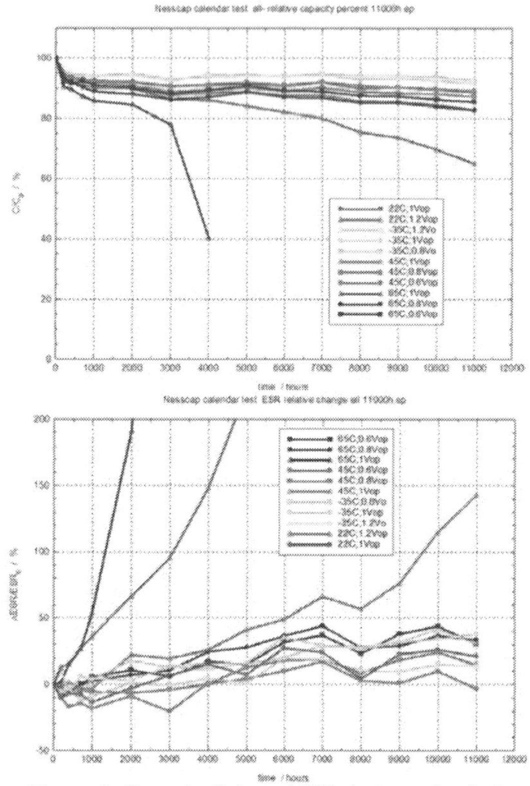

Figure 6: Capacity fade and ESR during calendaric aging of the Nesscap supercapacitors at different operating voltages and temperatures.

Thereafter the cycle was repeated. Before each intermediate check-up every 100'000 cycles of the BOSCs the individual cell voltage measurement was done. This was done by discharging the BOSC and acclimatising to 23°C ± 2°C for two hours. Thereafter, the BOSCs were charged to 8.64 V with a current of 0.8 A. Then the individual cell voltages were measured while the BOSCs were kept at constant voltage of 8.64 V. The capacity of the BOSCs was determined by cycling the BOSCs at 0.45 A (0.11 A for BOSC07) for 10 times between 0 and 10,8 V (11 V for BOSC07). The capacity was determined from the last discharge to 0 V.

3. RESULTS AND DISCUSSION

3.1 Single Cell Tests

The calendaric ageing of the Nesscap supercapacitors is shown in Figure 6, those of Maxwell in Figure 7. The average of five individual specimens is shown per test condition. Vop corresponds to the values given in the data sheet of the cells (for Nesscap, Cap-XX and Maxwell: 2.7 V). At 45°C and 0.8·Vop, which will be the expected maximum temperature for a cell in a conventional spacecraft, the Nesscap and Maxwell specimens showed capacity fade of around 10% after

 DOI: 10.1051/e3sconf/20171617003

6000 hours (Figure *6*). In the same conditions, the ESR increase is around 20% for Nesscap and for Maxwell.

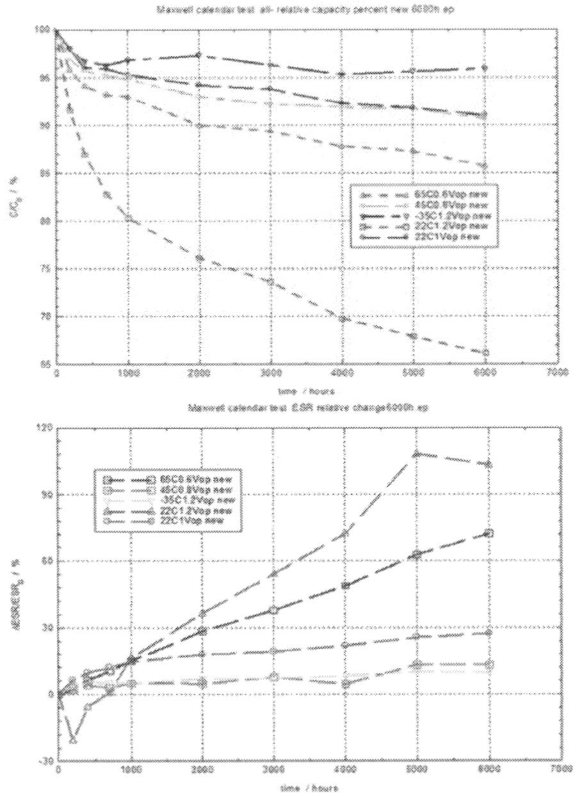

Figure 7: Capacity fade and ESR during calendaric aging of the Maxwell BCAP0010 P270 supercapacitors at different operating voltages and temperatures.

A derating of 0.85·Vop and 45°C appears to be a good compromise between sufficient durability for space applications and simplified thermal management.

Cycle life test up to 800.000 cycles in continuous 100%, discontinuous 75% and continuous 75% conditions were also performed on Maxwell BCAP0010 P270. The results of the 75% discontinuous cycling is shown in Figure 8. The relative capacity losses of both cell types are similar. The evolution of ESR with cycles is more pronounced for Maxwell cells, however, after 600'000 cycles the cells have a similar relative ESR increase. It has to be noted that the DC ESR at the BOL of the Maxwell cell is about twice the value of the Nesscap cell (55 mΩ vs 27 mΩ).

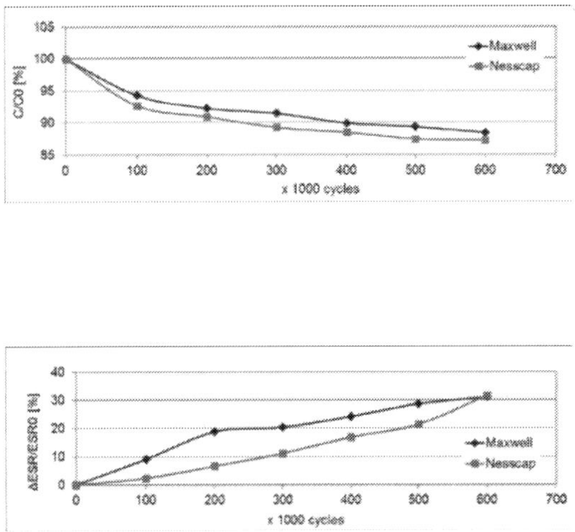

Figure 8: Capacity fade and ESR evolution of Maxwell BCAP0010 P270 and Nesscap EHSR 0010C0-002R7 during 75% discontinuous cycling.

The temperature measured on the cells during the cycling is shown in Figure 9. The average temperature is around 30° C with some minor variations. Due to the higher ESR of the Maxwell cells, they operate at slightly higher temperature than the Nesscap cells. Both cells show a similarly good performance during the cycling with a capacity fade of around 12% and an increase of ESR of 30%. After 600'000 cycles.

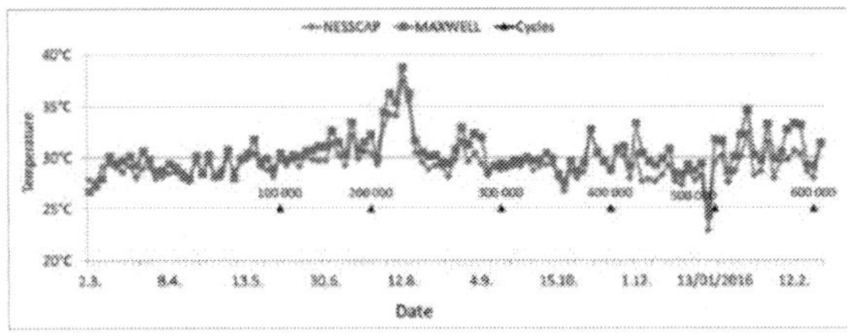

Figure 9: Temperature measured on the cells during the 75% discontinuous cycling test.

Finally, a life test under vacuum conditions was performed on the Nesscap cells.

The capacitance evolution is plotted on Figure 10 and compared with results under ambient pressure.

DOI: 10.1051/e3sconf/20171617003

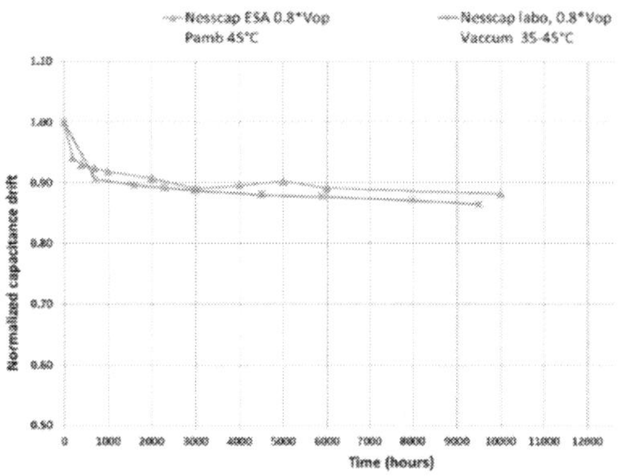

Figure 10: Capacity fade of Nesscap 10F in ambient pressure (green) and vacuum conditions.

The observed drifts are very similar to those observed in ambient pressure under the same conditions of voltage and temperature. This test campaign showed that the long term vacuum exposure does not deteriorate the performances of the Nesscap 10F cells more strongly than in ambient pressure.

3.2 Bank of Supercapacitor Tests

15s2p BOSC lifetest update

BOSC2 (of initial activity as described in [7,8]), continued to cycle after it had completed 1.5 million cycles. After around 1.5 Million cycles corresponding to 290 full days of cycling the capacity fade was around 11% and the ESR increase around 16%. The continued test results are shown in Figure 11. The capacity and the impedance at 1 kHz are shown. There is a relatively pronounced capacity fade within the first 600'000 cycles, of which the rate decreases. The slope of the ESR curve decreases with increasing cycle number. After 4.5 million cycles the capacity has decreased by about 13% and the ESR has increased by around 36%.

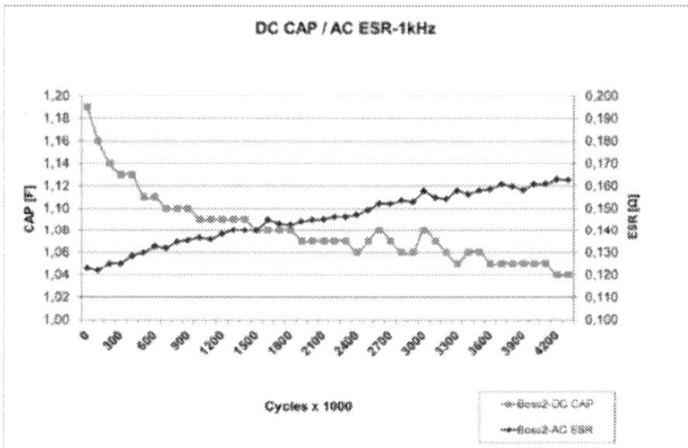

Figure 11: Capacity and ESR evolution during the life cycle test with the 15s2p BOSC at 45° C with a discontinuous 75% energy cycling profile.

4s1p BOSC life test results

The results of the 4s1p BOSCs with different balancing resistors are shown in Figure 12.

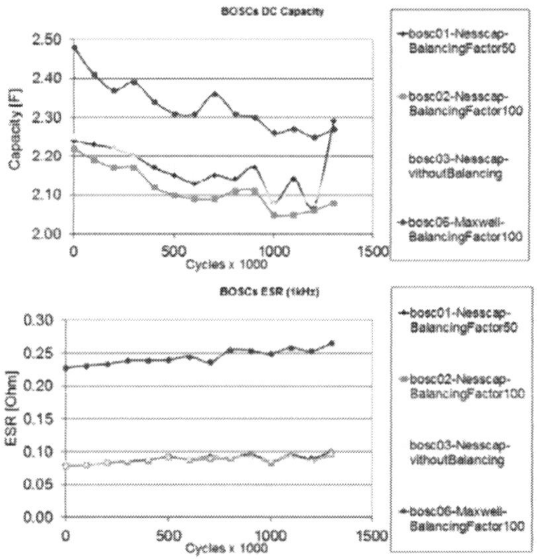

Figure 12: Overview of the Capacity and ESR evolution during the life cycle test of the 4s1p Nesscap and the Maxwell BOSCs with different balancing factors.

The capacity fade after around 1.2 million cycles is around 8% for the Nesscap BOSCs (average of all BOSCs), which is in excellent agreement with the results of the 15s2p BOSC in Figure 11. Around 9% were found for the Maxwell BOSC, which had a slightly higher capacity but also higher ESR as compared to the Nesscap BOSCs.

In Figure 13 the performance of the Cap-XX BOSC is shown. After around 700'000 cycles this BOSC failed due to the failure of cell 2. It seems the negative contact of the cell became highly resistive by disconnection due to a venting event. The reason for this failure is not fully clear.

 DOI: 10.1051/e3sconf/20171617003

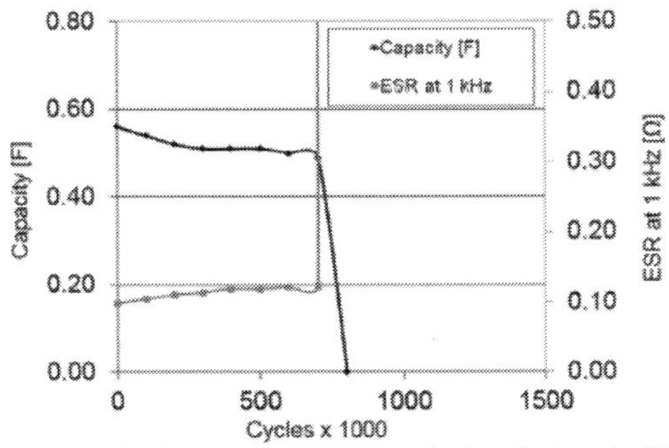

Figure 13: Capacity and ESR evolution during the life cycle test the 4s1p Cap-XX BOSC with a balancing factor of 100.

Balancing performance

The individual cell voltage measurements of BOSC01 are shown in Figure 14. At the beginning of the test the cells are unbalanced but they quickly equilibrate. The difference between the maximum and minimum cell voltage (Delta Vmax) becomes quite constant and remains at a value of around 10 mV. The balancing therefore works as it should with a balancing factor of around 50.

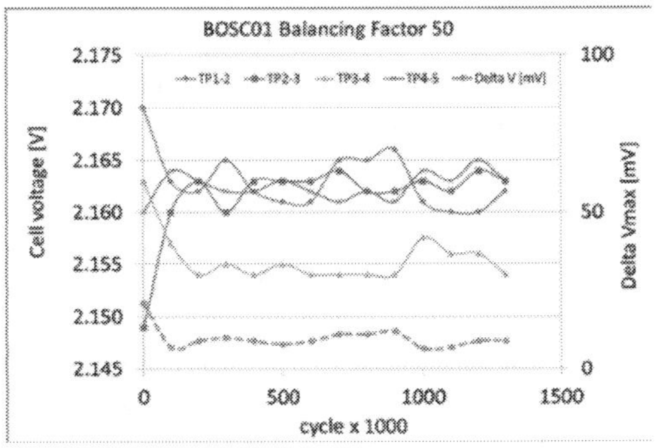

Figure 14: Individual cell voltages and Delta Vmax (right axis) of BOSC01 with balancing factor of 50.

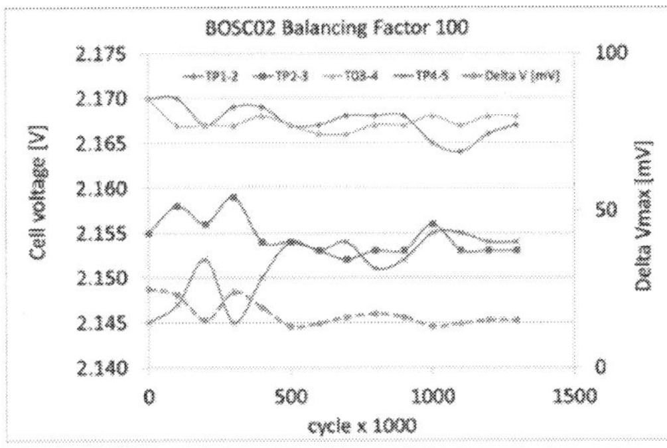

Figure 15: Individual cell voltages and Delta Vmax (right axis) of BOSC02 with balancing factor of 100.

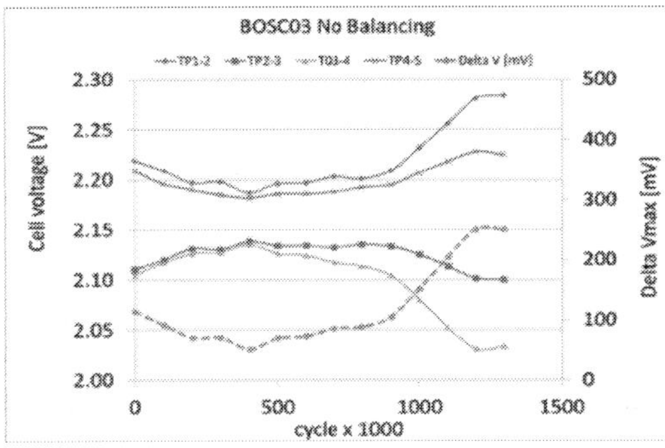

Figure 16: Individual cell voltages and Delta Vmax (right axis) of BOSC03 without cell balancing.

With balancing factor of 100 the BOSC performed as shown in Figure 15. The difference of maximum and minimum cell voltages was around 15 mV and stayed constant.

In contrast the BOSC without balancing shows diverging cell voltages as shown in Figure 16. A suboptimal life time can be expected with this pronounced imbalance of cell voltages.

After the tests the BOSCs were disassembled and each cell characterised. The measured capacities and ESRs were compared to the initial values. It seems that the presence of a balancing system allowed for a more homogeneous capacity fading of each cell in the BOSC as can be seen in Figure 17. Without balancing the average capacity fade was higher and the spread of the cells individual capacity fade was much larger at the end of the cycling test. No clear influence of the balancing on the ESR evolution could be seen.

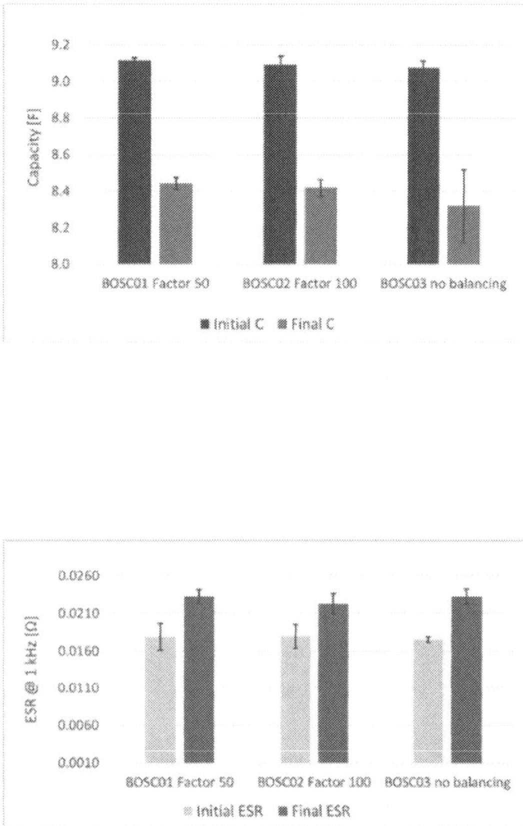

Figure 17: Average cell Capacities and ESRs before and after the cycling tests for BOSC01, BOSC02 and BOSC03 without cell balancing.

The cell voltages of BOSC06, which was based on Maxwell cells is shown in Figure 18. The voltages of cell 1 and cell 3 showed a slightly increasing trend whereas the voltage of cell 4 decreased during the cycling. Cell 2 showed a more or less stable voltage. As compared to BOSC01 and BOSC02 Delta Vmax is about three times higher (40 mV), possibly due to the higher internal resistance of the Maxwell cells.

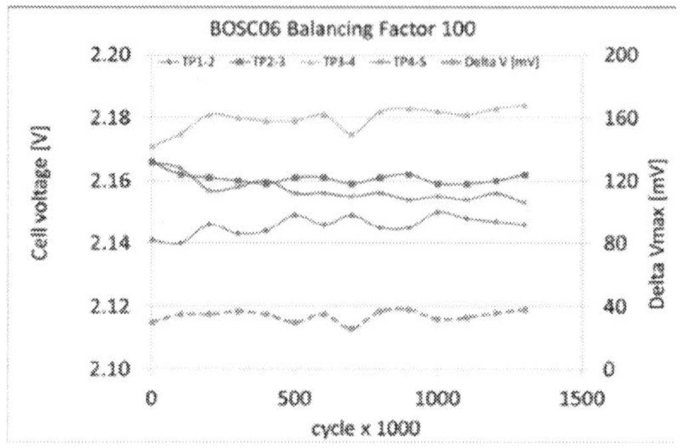

Figure 18: Individual cell voltages and Delta Vmax (right axis) of BOSC06 with Maxwell cells with a balancing factor of 100.

In Figure 19 the individual cell voltages of BOSC07 based on the Cap-XX cells are shown. There were pronounced changes in the cell voltages during the cycling. The cell that failed after around 700'000 cycles was the one with the lowest cell voltage (cell 2).

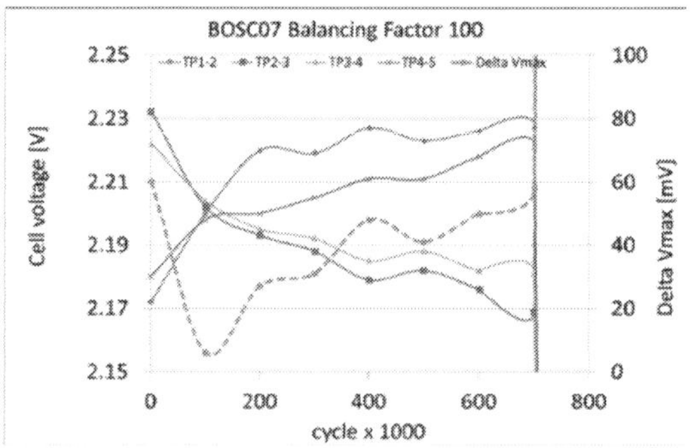

Figure 19: Individual cell voltages and Delta Vmax (right axis) of BOSC07 with Cap-XX cells with a balancing factor of 100.

4. CONCLUSIONS AND FUTURE WORK

The Nesscap and Maxwell supercapacitor cells are suitable for space applications if they are derated to 0.85·Vop and operated at a maximum temperature of 45°C. On BOSC level, more than 4 Million cycles could be demonstrated at these conditions. Based on the results of the 4s1p BOSCs a balancing system seems to be mandatory to guarantee a sufficiently long lifetime. The balancing system allows to equally distribute the aging of a BOSC to the individual cells. It is not trivial to find a correlation between the individual cell characteristics and the balancing status during the cycling tests. This would be an interesting topic for future investigations.
The benefits of the activity are numerous: confirmation of ageing models to extrapolate BOSC End-of-Life performances to 10 or 15 years, a first indication of the balancing factor and design to be applied and demonstration of vacuum performance.
From those results, the Nesscap cell is preferred when compared to Maxwell because of the low ESR, superior cycling performances, vacuum test demonstration and the manufacturer agreement for 5 years production continuity and traceability. Thanks to those activities, Maxwell could be identified as a credible back-up solution for the Nesscap cells.

All those complementary results have paved the way to a new project entitled "Generic Space Qualification of 10F Nesscap Supercapacitors" aiming firstly to qualify Nesscap 10F supercapacitors for space applications and secondly to develop and qualify the associated BOSC (both shown in Figure 20). A generic and modular unit including components parallel/series arrangements compatible with several space systems is envisaged.

Figure 20: Nesscap ESHSR-0010C0-002R7UC supercapacitor under qualification and modular BOSC under development.

5. ACKNOWLEDGEMENTS

This research activity was conducted under ARTES 5.1 program element funded by the European Space Agency (Evaluation of Supercapacitors and Impact at System Level (4000105661/12/NL/NR). The view expressed herein can in no way be taken to reflect the official opinion of the European Space Agency.

6. REFERENCES

1. B.E. Conway, *Electrochemical Super capacitors Scientific Fundamentals and Technological Applications*, Kluwer and Plenum, New York, USA, 1999.
2. J.R. Miller, "CAPACITORS, Overview", *Encyclopedia of Electrochemical Power Sources*, Elsevier, Amsterdam, 2009, 587-599.
3. V. Gineste, D. Loup, P. Mattesco, N. Neugnot, "Preliminary Evaluation of Commercial Supercapacitors for Space Applications", *9th European Space Power Conference*, Saint Raphael, France, 2011.
4. T. Shimizu, C. Underwood, "Super-capacitor energy storage for micro-satellites: Feasibility and potential mission applications", *Acta Astronautica*, 85, 138-154, 2013.
5. P. Kurzweil, "CAPACITORS, Electrochemical Double-Layer Capacitors", *Encyclopedia of Electrochemical Power Sources*, Elsevier, Amsterdam, 2009, 607-633.
6. H. Gualous, R. Gallay, M. Al Sakka, A. Oukaour, B. Tala-Ighil, B. Boudart, "Calendar and cycling ageing of activated carbon supercapacitor for automotive application", *Microelectronics Reliability*, 52, 2477-2481, 2012.
7. "ESA contract n°1-6859/11/NL/NR "Evaluation of supercapacitors and impacts at system level" including CCN1 and CCN2",
8. B. Buergler, E. Simon, P. Vasina, D. Latif, L. Diblik, V. Gineste, M. Simcak, "Design, Construction and Test of a Supecapacitor Bank for Space Applications" 10th European Space Power Conference, ESA, Noordwijkerhout, The Netherlands, 2014.
9. V. Sedlakova, J. Sikula, J. Majzner, P. Sedlak, T. Kuparowitz, B. Buergler, P. Vasina, "Supercapacitor equivalent electrical circuit model based on charges redistribution by diffusion", *Journal of Power Sources*, 286, 58-65, 2015.

DOI: 10.1051/e3sconf/20171617004

USE OF FUEL CELLS AND ELECTROLYZERS IN SPACE APPLICATIONS: FROM ENERGY STORAGE TO PROPULSION/DEORBITATION

Javier Brey [(1)], Delia Muñoz [(1)], Verónica Mesa [(1)], Tamara Guerrero [(2)]

(1) H2B2 Electrolysis Technologies S.L., c/Torre de los Herberos, 21, 41703 Dos Hermanas, Seville (Spain), Email: javier.brey@h2b2.es
(2) Solar-Mems Technologies, nave 1, Calle Earle Ovington, 24, 41309 Seville (Spain)

ABSTRACT

In this paper, two main groups of technologies and applications that can be distinguished regarding hydrogen field: gas production (hydrogen and oxygen), and power generation, are presented.

Firstly, for gases generation, reforming of bioethanol and PEM electrolysis are presented as possible choices for hydrogen production for terrestrial plants related to space activities. In the case of on board gases production, electrolysis is presented as the most suitable option.

Secondly, regarding power generation with fuel cells, different applications are presented (landers, aircrafts, stationary back-ups, APUs, UAVs, etc.).

Thirdly, two options of combination of power generation with fuel cells and gases production are shown: reversible fuel cell systems and regenerative fuel cell systems, and a comparison between the current batteries used and these regenerative fuel cell systems.

In the conclusion, two applications scales are presented: one for power generation depending of the level of power generated, and another one for gases generation depending of the gases production rate.

1. INTRODUCTION

The use of hydrogen as an energy carrier in space applications is growing day by day because of its advantages in comparison with other technologies and its environmentally friendly nature.

First applications of fuel cells were developed for aerospace sector. In the 1960s, General Electric developed the first Proton Exchange Membrane (PEM) fuel cell for Gemini missions (NASA), and, on the other hand, hydrogen is the main fuel for spacecrafts and oxygen is necessary for all manned missions, so the development of these technologies has an important impact in the aerospace sector.

Figure 1. PEM fuel cell used in Gemini missions.

2. GASES GENERATION

Regarding gases generation there are many technologies to be used not only for spacecrafts but also for terrestrial plants related to space activities.

Reforming of any fuel or biofuel (in order of being more environmentally friendly) is one of the available technologies for hydrogen production for terrestrial plants to support spatial needs.

In case of, in example, bioethanol steam reforming, the overall reaction that occurs is:

$$CH_3CH_2OH\ (g) + 3H_2O\ (g) \rightarrow 6\ H_2\ (g) + 2CO_2\ (g) \quad (1)$$

In this reaction, one mole of bioethanol reacts with three moles of steam water, producing hydrogen and CO_2. The aim of this process consists of maximizing the yield of hydrogen, limiting the formation of CO which is a catalyst poison. To achieve this goal, it is necessary to correctly choose operating conditions: temperature optimization, pressure and mole ratio of water/bioethanol and optimum choice of the catalyst.

Other possibility for hydrogen production is electrolysis process.

Proton exchange water (PEM) electrolysis is an electrochemical process in which pure water split up into hydrogen and oxygen by means of electricity using a polymer electrolyte membrane as a medium of transfer.

DOI: 10.1051/e3sconf/20171617004

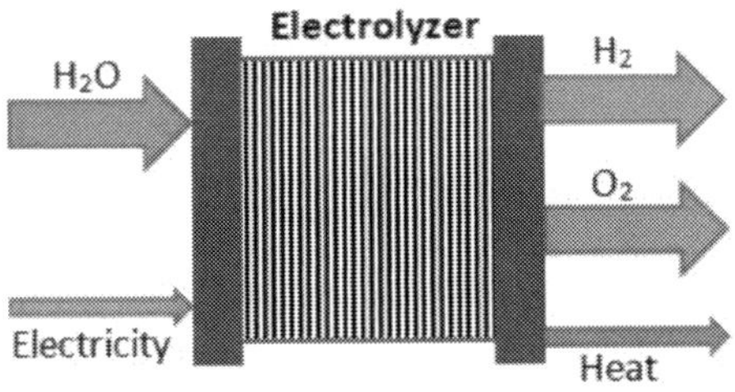

Figure 2. PEM Electrolyzer schema.

The reaction that occurs in the cell could be summarized as in Eq. 2:

$$\begin{aligned} &\text{Anode reaction:} \quad H_2O \rightarrow \tfrac{1}{2} O_2 + 2H^+ + 2e^- \\ &\text{Cathode reaction:} \quad 2H^+ + 2e^- \rightarrow H_2 \qquad (2) \\ &\text{Global reaction:} \quad H_2O + \text{electricity} \rightarrow H_2 + \tfrac{1}{2} O_2 \end{aligned}$$

The authors have experience on the design of this kind of plants because of their work in the "Environmentally Friendly Hydrogen Production" contract with ESA, where the identification of an environmentally friendly technology for hydrogen production and a techno-economical study for the implementation of a plant in Kourou to support Arianne 5 needs were carried out.

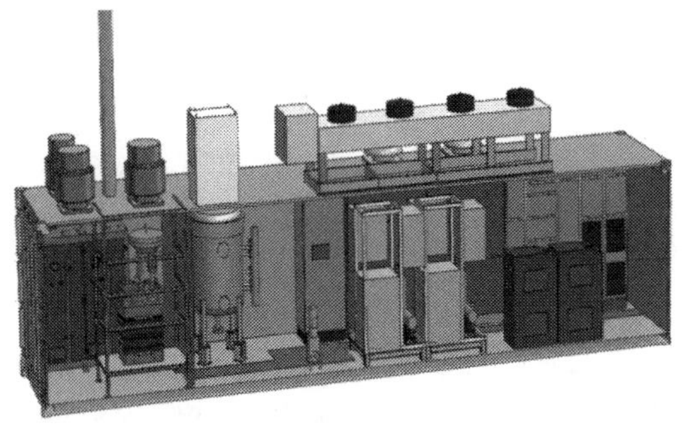

Figure 3. Example of PEM electrolyser.

Gases production is also relevant for space applications, where the applications of generation of hydrogen and oxygen can be distinguished.

Electrolysis process is the most suitable technology for gases production on board. Even if the overall electrochemical reaction is the same as for a terrestrial plant, some operational and design details have to be modified in order to fulfil with on orbit conditions.

Hydrogen produced by electrolysis is essential for space missions because it is the main fuel used for propulsion of spacecrafts during launch and it can also be used for on orbit propulsion and deorbiting at the end of the operative life of the spacecraft.

On the other hand, the other product gas of the electrolysis process, oxygen, is necessary for life support in manned missions, like in the International Space Station.

3. POWER GENERATION

Since the first applications of fuel cell were developed for aerospace sector, this technology has substantially improved.

From the alkaline fuel cell used in missions of Apollo program to the PEM (high and low temperature) fuel cell specially designed for space applications in our days, much knowledge and improvements in the technology and balance of plant for the operation has been achieved.

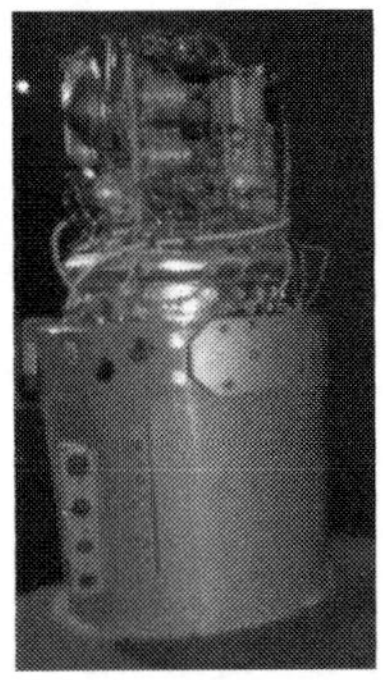

Figure 4. Alkaline fuel cell in Apollo missions.

The reaction (Eq. 3) for any of these fuel cell technologies consists of power generation by the electrochemical reaction of hydrogen and oxygen, becoming water as only product and avoiding any noxious product.

$$H_2 + \tfrac{1}{2} O_2 \rightarrow H_2O + \text{power} \qquad (3)$$

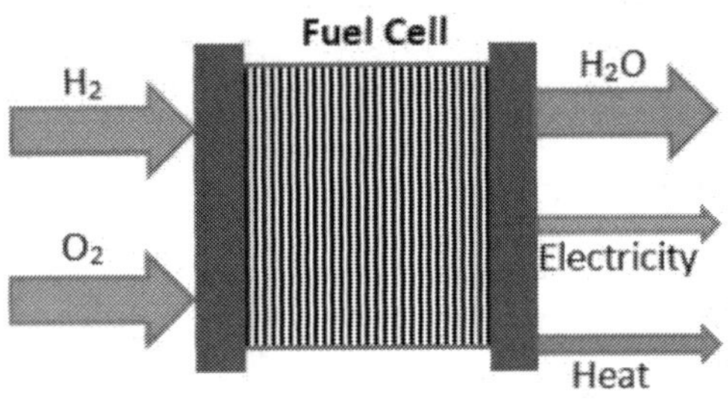

Figure 5. Fuel cell schema.

Power generated by fuel cell systems can be used for many aerospace applications:

DOI: 10.1051/e3sconf/20171617004

- Landers,
- Aircrafts,
- Stationary back –ups,
- Military/ Aeronautic auxiliary power units (APUs),
- Unmanned aerial vehicles (UAVs).

4. COMBINATION OF POWER GENERATION AND GASES PRODUCTION

The combination of the operation of a fuel cell and an electrolyzer working in a close loop can be used as an energy storage system [1].

3.1 Reversible fuel cell system

A reversible fuel cell system consists of a stack that operates in both ways:

- Fuel cell mode: the stack consumes hydrogen and oxygen and produces water and power.
- Electrolyzer mode: when power is supplied to the stack, it consumes water and produces hydrogen and oxygen.

The authors [2] have worked in the development, tests and integration of a reversible fuel cell system for a picosatellite.

This reversible fuel cell system is composed of a stack where reactant gases and product water are contained in. There are no storage independent parts. Hydrogen, oxygen and water are adsorbed in the membranes of the stack and circulate in close loop during its operation. Water produced by the fuel cell is then used as reactant in electrolyzer mode; and gases produced by the electrolyzer are used as reactants in fuel cell mode.

This system has successfully passed vibration, thermal and vacuum tests and has overcome 50 operation cycles. Its reliability has been demonstrated and it presents several advantages as:

- Possible increase in available power,
- Reduction of the energy storage system mass (possibility to increase payload),
- Compact energy generation and storage systems on board.

3.2 Regenerative fuel cell systems (RFCS)

On the other hand, a regenerative fuel cell systems consist of two different stacks (one fuel cell and one electrolyzer) connected to three storage tanks (for hydrogen, oxygen and water) that operate in a closed loop [3].

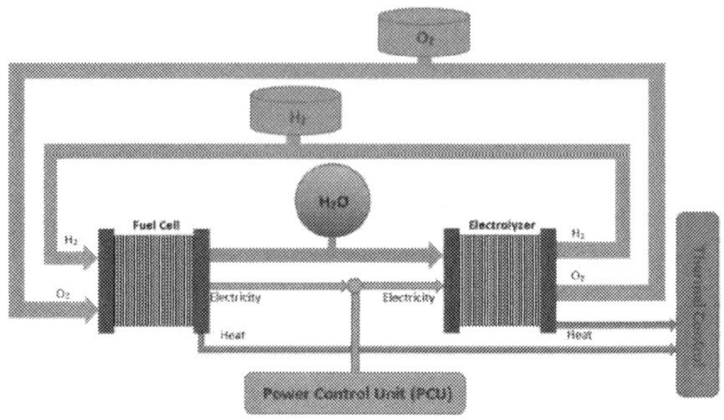

Figure 6. Regenerative fuel cell system schema.

This kind of systems can be used as auxiliary power supply for spacecrafts (satellites, rovers, etc.). There are three possible operational modes of the system:

- Discharge mode: when the spacecraft needs power (i.e. shadow periods), the fuel cell operates using hydrogen and oxygen stored in the tanks (discharge process of gases storage tanks) and producing the demanded power and water to be stored in the water tank.
- Charge mode: when there is available power from the spacecraft (light periods), the electrolyzer operates using water from the water tank and recharging the gases tanks with the oxygen and hydrogen generated.
- Idle mode: no stack is operating but fuel cell is ready to supply power when the spacecraft requires it and tanks are plenty of gases. The fuel cell and the electrolyzer cannot operate at the same time.

The authors have evaluated the feasibility of this system collaborating with the main suppliers of fuel cells and electrolyzers for space, identifying issues and solutions:

- Thermal control solutions identified to reduce mass,
- Water management in space conditions,
- Fast response for power supply,
- Passive system design to avoid maintenance and failures.

The feasibility of this regenerative fuel cell system has been demonstrated. It could be a possible substitute for current batteries presenting a mass saving and higher energy density that improves with the increase of the power level.

DOI: 10.1051/e3sconf/20171617004

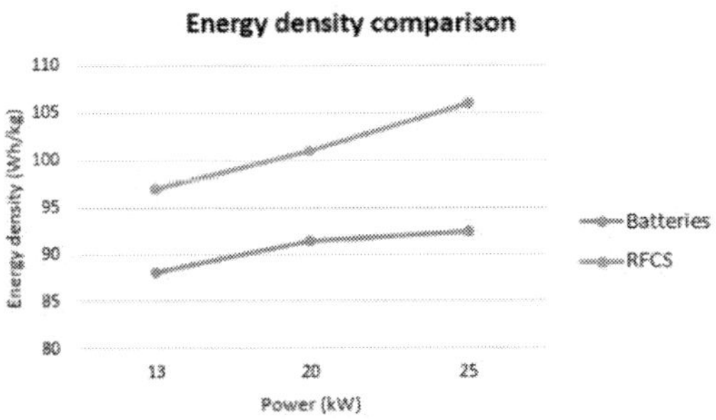

Figure 7. Energy density comparison. RFCS vs batteries.

5. CONCLUSIONS

The use of fuel cells and hydrogen production technologies in the aerospace sector is becoming more important day by day.

Its advantages in comparison with other technologies have great impact for different spatial activities and its development for different applications is a fact.

6. REFERENCES

1. "Fuel cell development for space applications: fuel cell system in a closed environment", Journal of Power Sources, Volume 137, Issue 2, 29 October 2004, Pages 269–276. Yoshitsugu Sone, Mitsushi Ueno, Saburo Kuwajima.

2. IAC-14-B4.2.6 "Cepheus, A multi-project satellite for technology qualification", Multiple authors. Published on 65th International Astronautical Congress, Toronto, Canada (2014).

3. "Regenerative fuel cell systems", Energy Fuels, 1998, 12 (1), pp 56-71. Fred Mitlitsky, Blake Myers, Andrew H. Weisberg

Saft SOLUTIONS FOR FULL GEO ELECTRIC PROPULSION

D. Prevot [1], Y. Borthomieu [1], F. Malet [2], C. Rebuffel [3], A. Carre [4], C. Cenac-Morthe [5]

[1] *Saft Space and Defence Division BP 1039, 86060 Poitiers, France,*
Yannick.Borthomieu@saftbatteries.com, Didier.Prevot@saftbatteries.com

[2] *AIRBUS D&S, 31 avenue des Cosmonautes, 31402 Toulouse, France, Fabien.Malet@airbus.com*

[3] *Thales Alenia Space, 5 allée des Gabians, 06156 Cannes la Bocca cedex, France,*
Christophe.Rebuffel@thalesaleniaspace.com

[4] *European Space Agency – European Space Research and Technology Centre (ESTEC), The Netherlands,*
Aurore.Carre@esa.int

[5] *CNES, 18 avenue Edouard Belin, 31401 Toulouse Cedex9, France, Celine.Cenac-Morthe@cnes.fr*

ABSTRACT

Today, the full electric propulsion is becoming more commonly used on board geostationary satellites for the launch up to the Electrical Orbit Raising (EOR) and all on-orbit manoeuvres.
This electrical propulsion induces additional cycles for the battery that may result in higher ageing.
The Electrical Orbit Raising (EOR) is currently simulated by two additional GEO seasons up to 80%DoD while the Plasmic Propulsion System (PPS) cycles are carried out, at least 162 cycles up to 24%DoD or 198 cycles up to 20%DoD per solstice season and at least 52 cycles up to 15%DoD or 64 cycles up to 10%DoD per eclipse season for 15 years.
The need for an appropriate demonstration by life-test was anticipated and addressed in collaboration with the primes and the agencies, in the frame of ARTES 3-4 contract with ESA for the VES180SA cell & CNES contract for the VES16 cell.
The article will present the whole cycling results available, and will provide afterwards the correlation status of Saft Li-ion Model (SLIM) with the experimental data acquired.

Keywords: **DOD**: Depth of Discharge; **OCV**: Open Circuit Voltage; **EOCV**: End Of Charge Voltage; **EODV**: End Of Discharge Voltage; **EOR**: Electrical Orbit Raising; **GEO**: Geosynchronous Orbit; **IR**: Internal Resistance; **PPS**: Plasmic Propulsion System; **SG**: Strain Gauge; **SLIM**: Saft Li-Ion Model; **SoC**: State of Charge.

1. VES180SA LI-ION CELL WITH & WITHOUT PPS GEO LIFE-TEST RESULTS

This chapter focuses on the analysis of the life-test results performed on VES180SA lithium-ion cells with and without PPS in semi-accelerated mode covering different mission profiles proposed by primes, as described in Tab. 1. The life time test performed without PPS is used as reference test in order to evaluate the impact of the Electrical Orbit Raising (EOR) and all on-orbit manoeuvres (PPS) on the cell ageing.

Test Ref.	Test conditions	Nb Cells Tested	Achieved Seasons	Test Status
NDTF17 Semi-Acceler.	Semi-Accelerated GEO cycling at 75%DoD @15°C EOCV = 4.075V without PPS (Reference Test)	1	16	On-going
	Semi-Accelerated GEO cycling at 75%DoD @15°C Two initial seasons without PPS EOCV = 4.075V with PPS (10%DoD in equinox & 20%DoD in solstice)	2	16	On-going
NDTF18 Semi-Acceler.	Semi-Accelerated GEO cycling at 75%DoD @20°C EOCV = 4.075V without PPS (Reference Test)	1	18	On-going
	Semi-Accelerated GEO cycling at 75%DoD @20°C Two initial seasons without PPS EOCV = 4.075V with PPS (15%DoD in equinox & from 19% to 24%DoD in solstice)	2	18	On-going

Table 1: VES180SA Li-ion cells On-Going GEO with & without PPS Life-Tests Program

DOI: 10.1051/e3sconf/20171617005

1.1 NDTF17 & 18 Test Set-Ups

The test set-ups applied to the NDTF 17 and 18 lifetime tests are similar, except for the base plate regulation temperature which varies respectively from 15 to 20 °C.

For each lifetime test, three VES180SA cells were encapsulated "head down" in 3P cell-package and instrumented with two circumferential strain-gages, one thermocouple and one voltage telemetry in order to measure the PPS effect on the evolution of the temperature, the EOD voltage and the cell swelling. Then, they have undergone the specified mechanical and vacuum environmental tests without presenting leakage mark and abnormal degradations of their electrical performances.

Before starting the lifetime tests, the two cell-packages were assembled with a CHO- THERM layer on thermally regulated baseplate and insulated from ambient temperature by insulating box, as presented in the Fig. 1, 2 and 3.

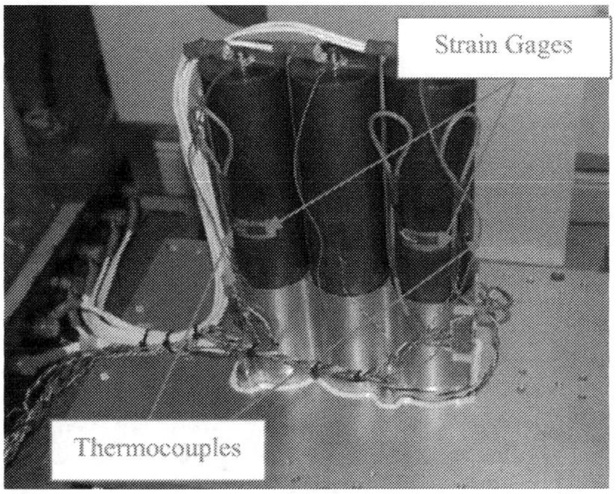

Figure 1. Cell-Package instrumented and mounted on baseplate

Figure 2. Regulated Baseplate with thermocouple and CHO-THERM under Cell-Package

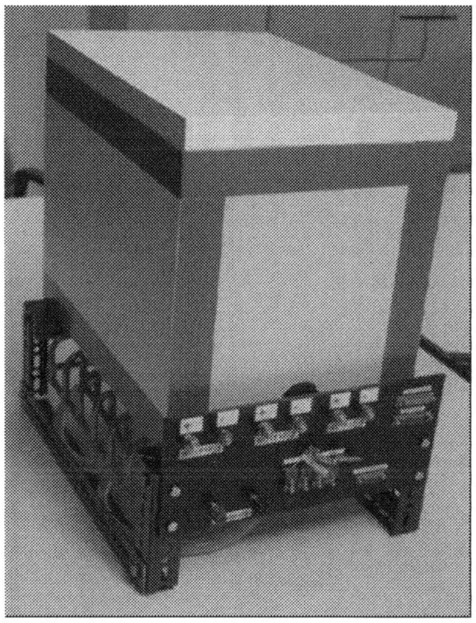

Figure 3. Regulated Baseplate with Insulating box

1.2 NDTF17 Lifetime Test Results

- In cycling, we have not observed significant impact of PPS on EoD voltage at the longest eclipse (cycle 23), as presented in Fig. 4 and 5.

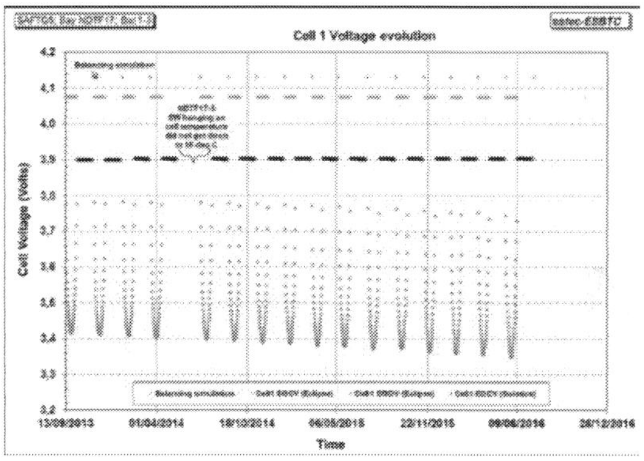

Figure 4. EoC/EoD Voltage variation in GEO Cycling without PPS

DOI: 10.1051/e3sconf/20171617005

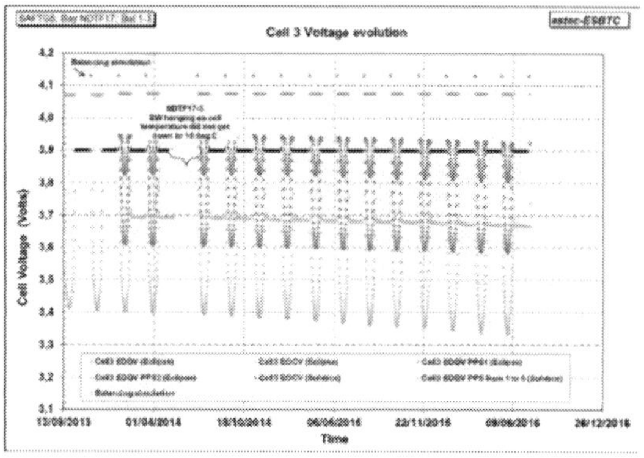

Figure 5. EoC/EoD Voltage variation in GEO Cycling with PPS

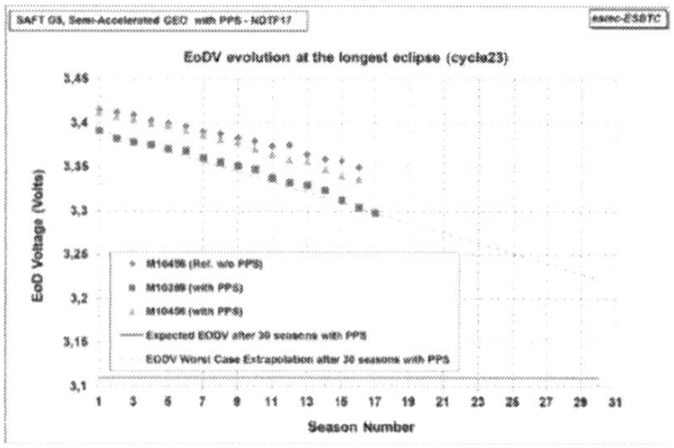

Figure 6. EoD Voltage at cycle 23 evolution vs. season

EOD voltage variation slopes are respectively between -5.8 & -5.1 mV per season for M10399 & M10458 cells cycled with PPS compared to -4.4 mV per season for M10456 cell cycled without PPS. These slopes should permit to maintain EOD Cell Voltage over the expected value after 30 seasons with PPS (i.e. ≥ 3.11Volts), as presented in Fig. 6.

-Also during cycling, no difference was identified on the evolution of the cell swelling at the end of charging (similar circumferential strains measured around 1000 ppm at the longest eclipse with or without PPS) and on the cell temperature increase at the end of discharge (similar temperature measured around 22°C at the longest eclipse with or without PPS), refer respectively to Fig.7 and 8, then Fig.9 and 10.

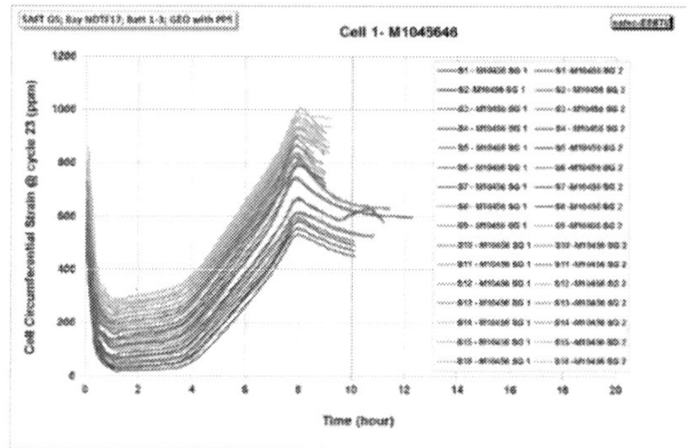

Figure 7. Cell 1 Circumferential strains evolution at cycle 23 without PPS vs. season

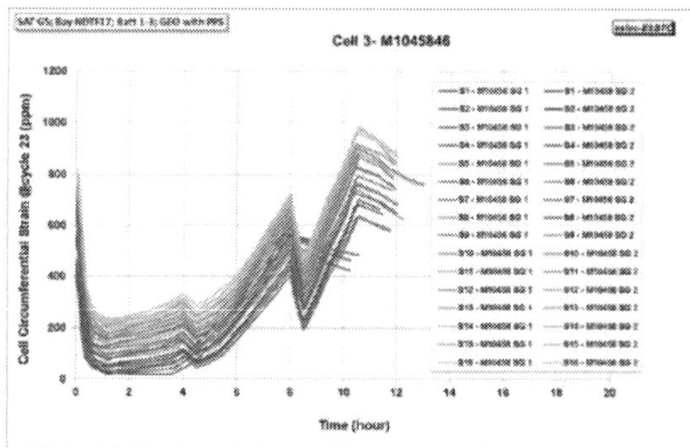

Figure 8. Cell 3 Circumferential strains evolution at cycle 23 with PPS vs. season

DOI: 10.1051/e3sconf/20171617005

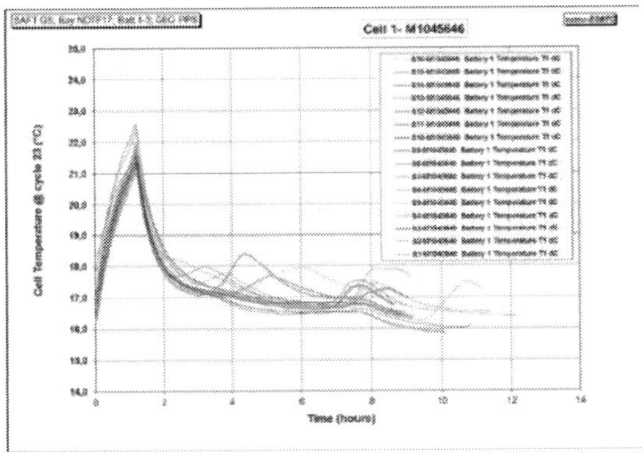

Figure 9. Cell 1 Temperature evolution at cycle 23 without PPS vs. season

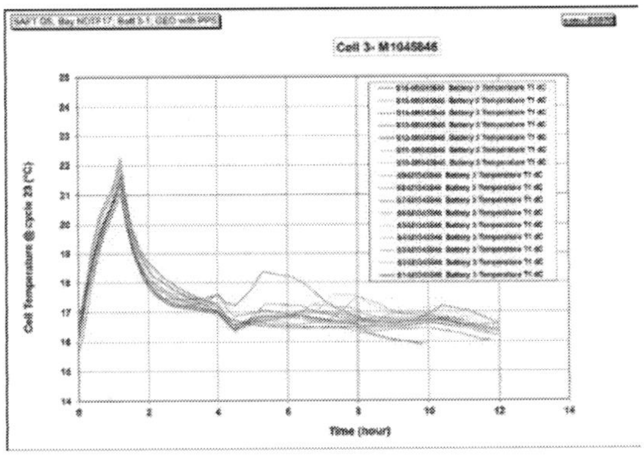

Figure 10. Cell 3 Temperature evolution at cycle 23 with PPS vs. season

- The evolution of the cell characteristics, such energy, IR20%DoD and IR60%DoD is measured through a check-up sequence, involving a residual discharge and a reference test performed at EOCV=4.1V @15°C every two seasons. This reference test is based on charge up to 4.1V, then discharge down to 2.7V under C/2 discharge rate @15°C with pulses 1.5C-30seconds at 20% and 60%DoD.

After 16 GEO seasons, no significant impact of PPS is visible on the energy variation slopes.

That should permit to maintain energy loss under the expected value after 30 seasons with PPS (i.e. ≤10.5%), as presented in Fig.12. The changes noticeable on the energy evolution are mainly induced by EoD temperature variation, as presented in Fig.11.

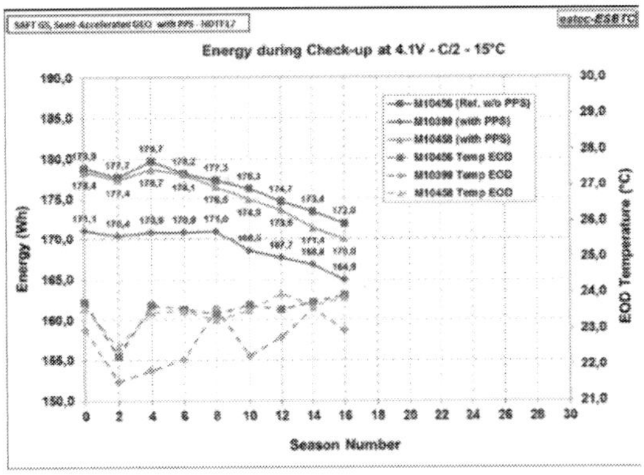

Figure 11. Energy & Temperature evolution in GEO cycling with & without PPS

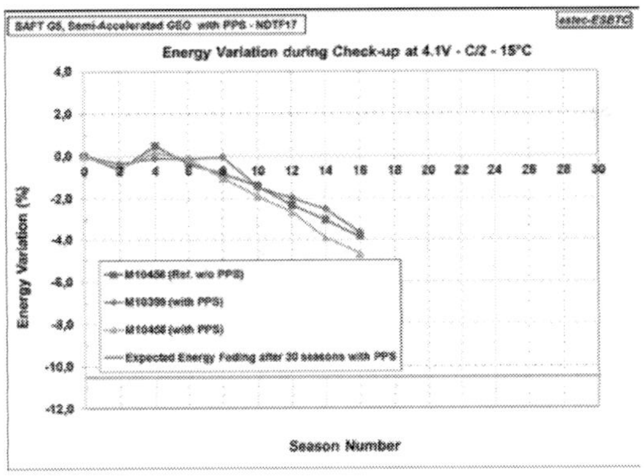

Figure 12. Energy variation in GEO cycling with & without PPS

Also, no significant difference was identified on the evolution of Internal Resistance at 20% & 60%DoD

DOI: 10.1051/e3sconf/20171617005

after 16 GEO seasons with or without PPS, as presented respectively in Fig.13 and 14.

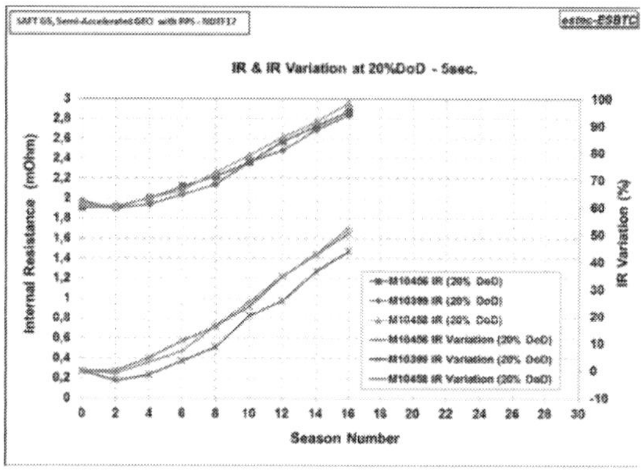

Figure 13. IR20%DoD variation with & without PPS

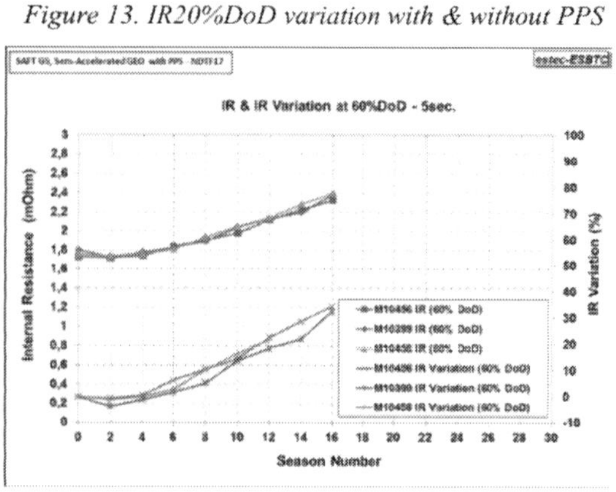

Figure 14. IR60%DoD variation with & without PPS

1.3 NDTF18 Lifetime Test Results

- In cycling, we have not observed significant impact of PPS on EoD voltage at the longest eclipse (cycle 23), as presented in Fig. 15 and 16.

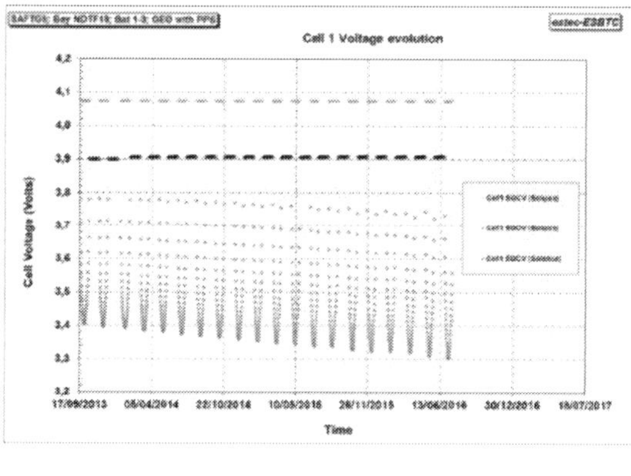

Figure 15. EoC/EoD Voltage variation in GEO Cycling without PPS

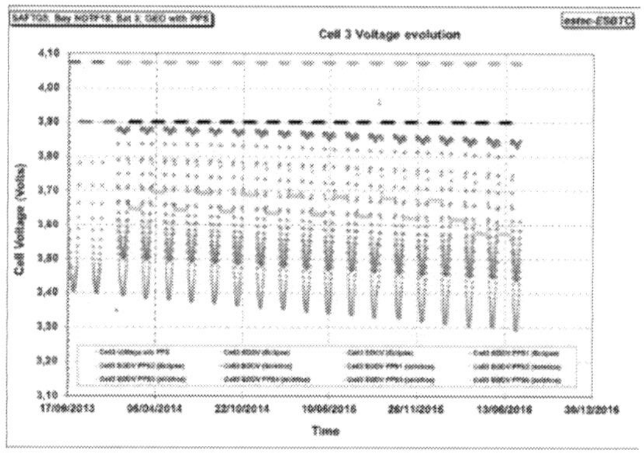

Figure 16. EoC/EoD Voltage variation in GEO Cycling with PPS

EOD voltage variation slopes are respectively between -5.2 & -5.9 mV per season for M10459 & M10377 cells cycled with PPS compared to -5.3 mV per season for M10376 cell cycled without PPS.

These slopes should permit to maintain EOD Cell Voltage over the expected value after 30 seasons with PPS (i.e. $\geq$ 3.14Volts), as presented in Fig. 17.

DOI: 10.1051/e3sconf/20171617005

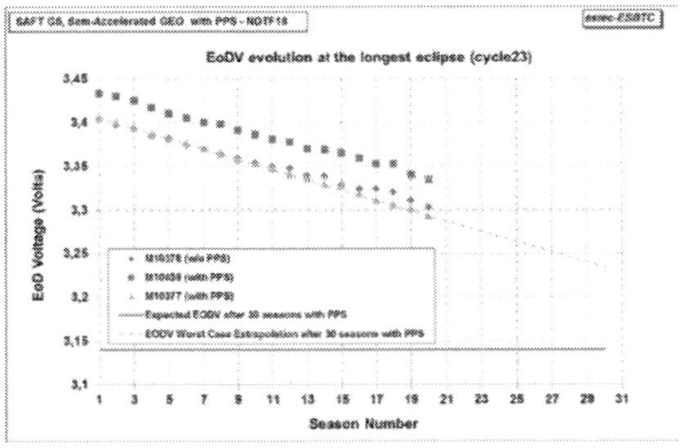

Figure 17. EoD Voltage at cycle 23 evolution vs. season

-Also during cycling, no significant difference was identified on the evolution of the cell swelling at the end of charging and on the cell temperature at the end of discharge, respectively about 100 ppm increase and 1°C decrease were measured in GEO cycling with PPS at the longest eclipse, refer respectively to Fig.18 and 19, then Fig.20 and 21.

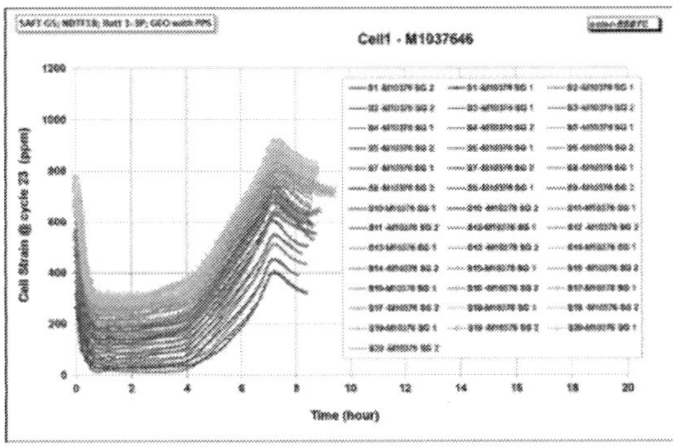

Figure 18. Cell 1 Circumferential strains evolution at cycle 23 without PPS vs. season

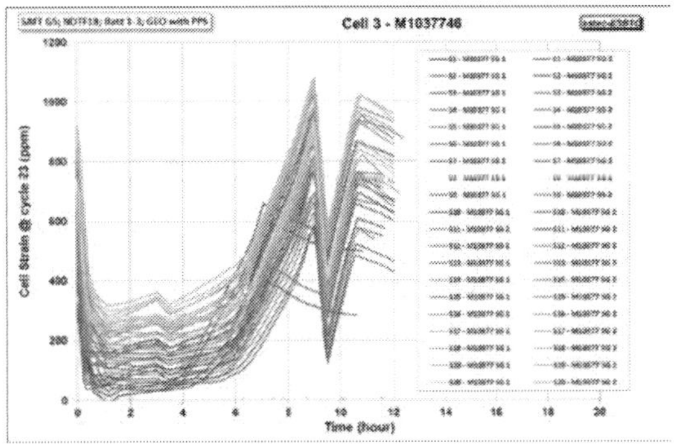

Figure19. Cell 3 Circumferential strains evolution at cycle 23 with PPS vs. season

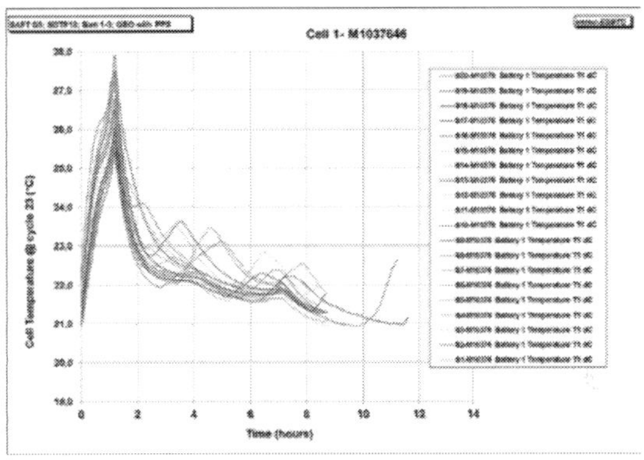

Figure 20. Cell 1 Temperature evolution at cycle 23 without PPS vs. season

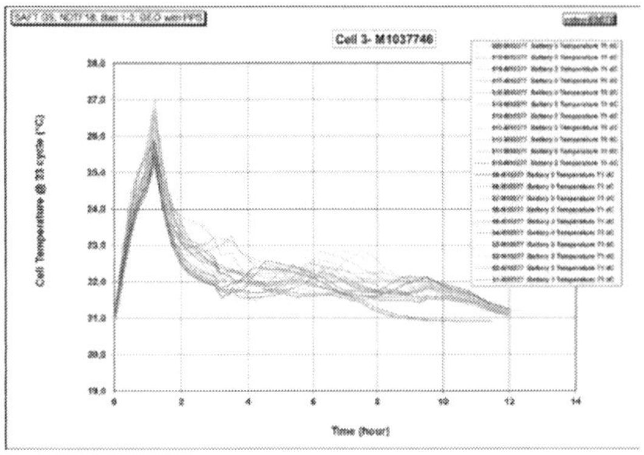

Figure 21. Cell 3 Temperature evolution at cycle 23 with PPS vs. season

- The evolution of the cell characteristics, such energy, IR20%DoD and IR60%DoD is measured every two seasons through a check-up sequence, involving a residual discharge and a reference test performed at EOCV=4.1V @20°C.

This reference test is based on charge up to 4.1V, then discharge down to 2.7V under C/2 discharge rate @20°C with pulses 1.5C-30seconds at 20% and 60%DoD.

After 18 GEO seasons, no significant impact of PPS is visible on the energy variation slopes. That should permit to maintain energy loss under the expected value after 30 seasons with PPS (i.e. ≤11.2%), as presented in Fig. 23.

The changes noticeable on the energy evolution are mainly induced by EoD temperature variation, as presented in Fig.22.

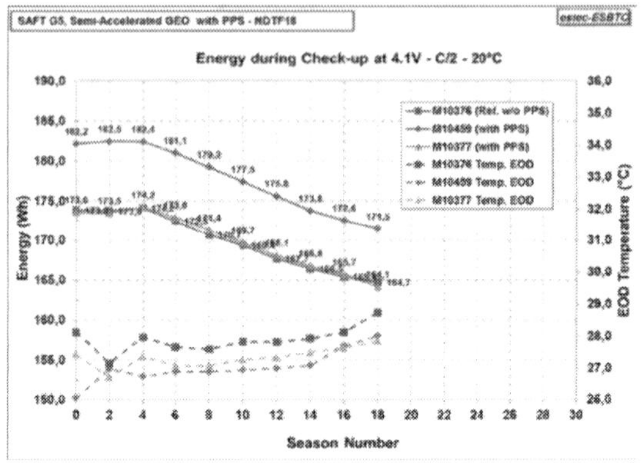

Figure 22. Energy & Temperature evolution in GEO cycling with & without PPS

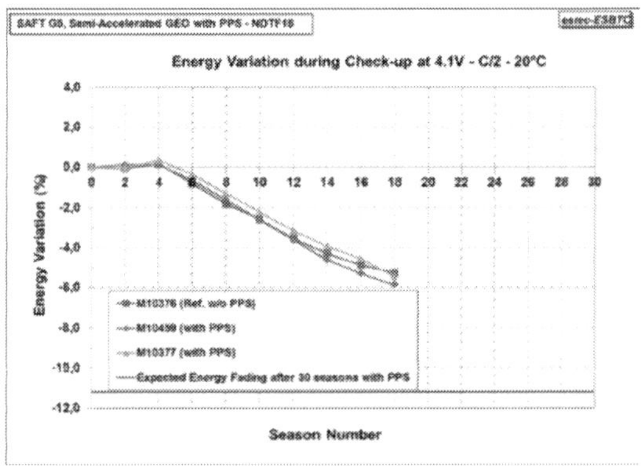

Figure 23. Energy variation in GEO cycling with & without PPS

Also, no significant difference was identified on the evolution of Internal Resistance at 20% & 60%DoD after 18 GEO seasons with or without PPS, as presented respectively in Fig.24 and 25.

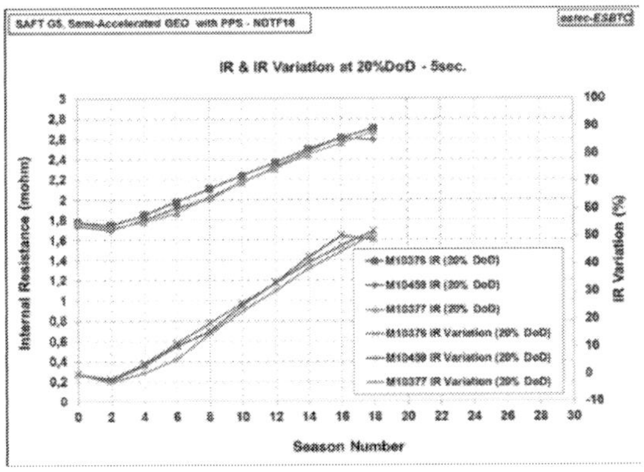

Figure 24. IR20%DoD variation in GEO cycling with & without PPS

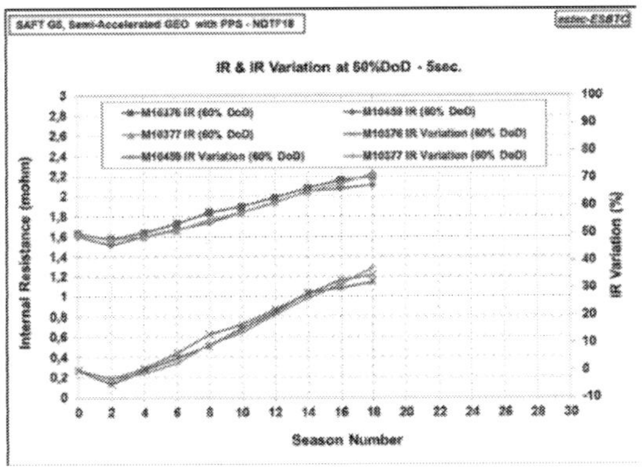

Figure 25. IR60%DoD variation in GEO cycling with & without PPS

1.4 Conclusion

The effect of PPS discharge on those GEO cycling tests @ 4.075V are negligible in term of energy loss, EoDV drop and internal resistance increase, which confirm the excellent ability of VES180 cell to sustain GEO PPS missions.

As of today after respectively 16 and 18 seasons, the expended EoL PPS trends are in line with SLIM predictions on the two experimented PPS profiles, including some margin as the results are better than prediction so far.

2. VES16 LI-ION CELL WITH & WITHOUT PPS GEO LIFE-TEST RESULTS

This chapter focuses on the analysis of the life-test results performed on VES16 lithium-ion cells with and without PPS in semi-accelerated mode, as described in Tab. 2. The life time test performed without PPS is used as reference test in order to evaluate the impact of the Electrical Orbit Raising (EOR) and all on-orbit manoeuvres (PPS) on the cell ageing.

Test Ref.	Tests Conditions	Nb Cells Tested	Achieved Seasons	Test Status
G10a Semi-Acceler.	80%DoD per cell duration per GEO Pofile @20° EOCV = 4.05 V No PPS	1	14 seasons	On-going
G10b Semi-Acceler.	Equinox periods: 2 GEO cycles per day at 80%DoD with 2 PPS cycles at 10%DoD @20° EOCV = 4.05 V Solstice periods: 6 PPS cycles per day at 20%DoD	2	14 seasons	On-going

Table 2: VES16 Li-ion cells On-Going GEO with & without PPS Life-Tests Program

2.1 G10a and G10b Test Set-Ups

The test set-ups applied to the G10a and G10b lifetime tests are similar.

All VES16 cells were instrumented with one thermocouple and one voltage telemetry in order to measure the PPS effect on the evolution of the temperature and the EOD voltage.

Before starting the lifetime tests, all VES16 cells were located in a ventilated thermal chamber set at 20°C.

2.2 G10a and G10b Lifetime Test Results

- In cycling, we have not observed significant impact of PPS on EoD voltage at the longest eclipse (cycle 23), as presented in Fig. 26 and 27.

DOI: 10.1051/e3sconf/20171617005

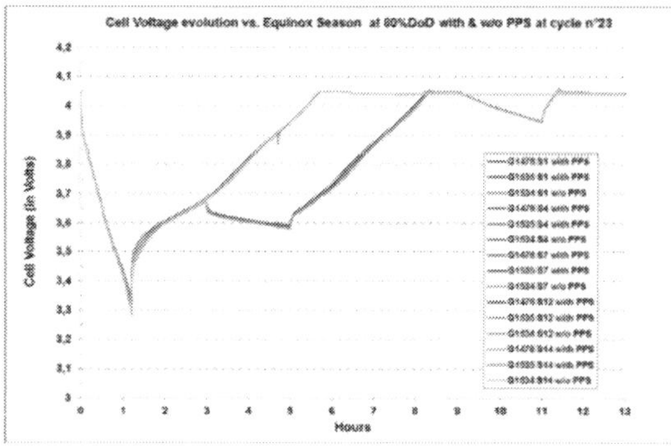

Figure 26. EoC/EoD Voltage variation in GEO Cycling with & without PPS (Equinox period)

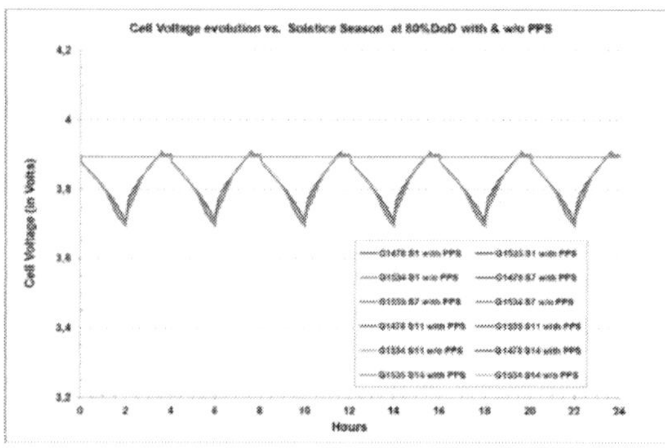

Figure 27. EoC/EoD Voltage variation in GEO Cycling with & without PPS (Solstice period)

EOD voltage variation slopes are all same at -3.0mV per season.

These slopes should permit to maintain EOD Cell Voltage over the expected value after 30 seasons with PPS (i.e. ≥ 3.22Volts), as presented in Fig. 28.

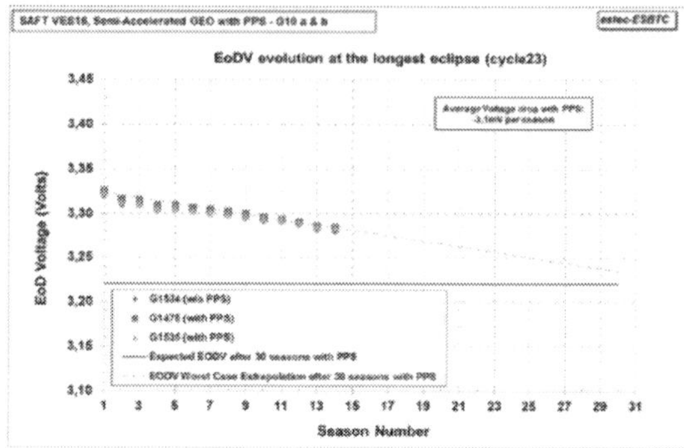

Figure 28. EoD Voltage at cycle 23 evolution vs. season

-Also during cycling, no difference was identified on the evolution of the cell temperature increase at the end of discharge (similar temperature measured around 24°C at the longest eclipse with or without PPS), refer to Fig.29.

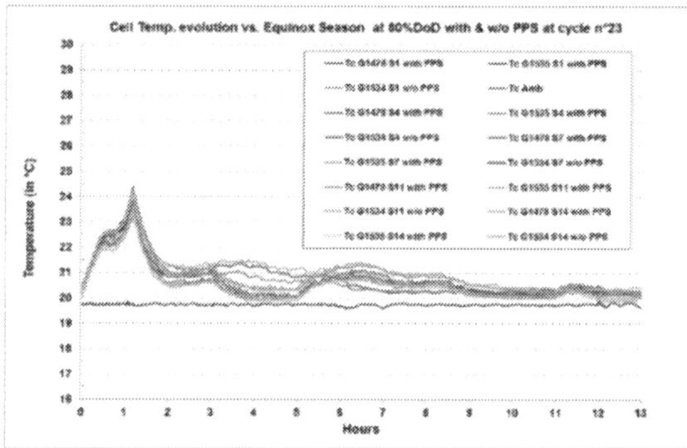

Figure 29. Cell Temperature evolution at cycle 23 with & without PPS vs. season

- The evolution of the cell characteristics, such energy, IR20%DoD and IR60%DoD is measured every two seasons through a check-up sequence, involving a residual discharge and a reference test performed at EOCV=4.1V @20°C.

This reference test is based on charge up to 4.1V, then discharge down to 2.7V under C/2 discharge rate @20°C with pulses 1.5C-30seconds at 20% and

 DOI: 10.1051/e3sconf/20171617005

60%DoD.

After 14 GEO seasons, no significant impact of PPS is visible on the energy variation slopes. That should permit to maintain energy loss under the expected value after 30 seasons with PPS (i.e. ≤ 9.9%), as presented in Fig. 30.

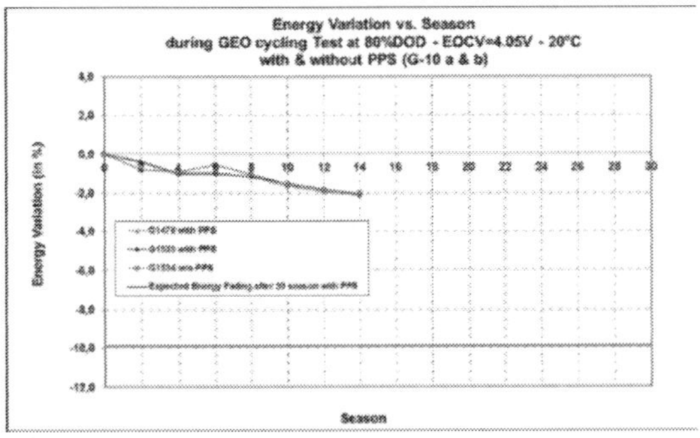

Figure 30. Energy variation in GEO Cycling with & without PPS

Also, no significant difference was identified on the evolution of Internal Resistance at 20% & 60%DoD after 14 GEO seasons with or without PPS, as presented respectively in Fig.31 and 32.

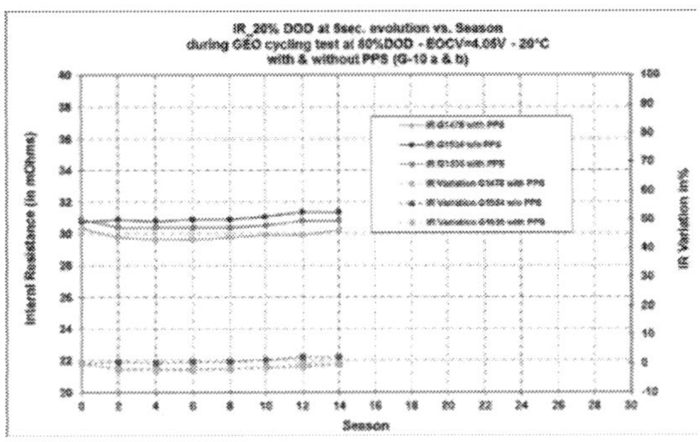

Figure 31. IR20%DoD variation in GEO cycling with & without PPS

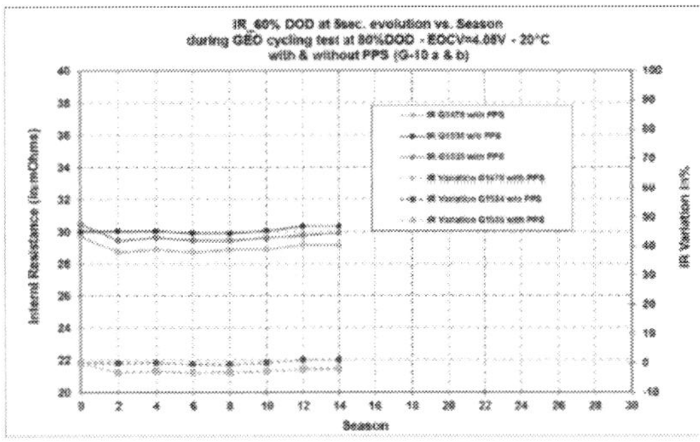

Figure 32. IR60%DoD variation in GEO cycling with & without PPS

2.3 Conclusion

The effect of PPS discharge on those GEO cycling tests @ 4.05V are negligible in term of energy loss, EoDV drop and internal resistance increase, which confirm the excellent ability of VES16 cell to sustain GEO PPS missions.

As of today after respectively 14 seasons, the expended EoL PPS trends are in perfectly line with SLIM predictions.

3. REFERENCES

1. Rémy, S., Prevot, D., Reulier, D. & Dr. Vigier, F. VES16 Li ion cell for satellite, 9th ESPC 2011 - Article 2103465 / Session 3.
2. Borthomieu, Y. & Prevot, D. SLIM Battery Modelling Features, 9th ESPC 2011 - Article 2103465 / Session 23.
3. Defer, M., Du Peyrat, D., Prevot, D. & Borthomieu, Y. Qualification of High Specific Energy Li-Ion Cell VES180SA, 8th ESPC 2008 - Article S13-2 / Session 13.

REVIEW OF COMMERCIAL CELLS FOR SPACE APPLICATIONS

Rachel Buckle [(1)], Sam Roberts [(1)]

[(1)] *ABSL Space Products, F4 Culham Science Centre, Abingdon, UK. Email: rachel.buckle@absl.enersys.com*

ABSTRACT

Advances in Li-ion technology mean that manufacturers currently produce cells that have higher energy densities, higher rate capabilities, and longer quoted cycle lives than the currently qualified ABSL space cells. It is therefore possible that cells are available that could enable a lower mass/volume solution for a given application. ABSL carry out regular reviews of commercial off the shelf (COTS) cells, to determine whether any available cells are suitable for space applications, and offer superior performance to those currently qualified.

This paper describes the process that is followed to review and test commercial cells.

1 INTRODUCTION

ABSL has used the 18650HC(M) for space batteries since 1998; it was employed on the first ever Li-ion space battery for the PROBA-1 mission. Although the detailed performance and life-time characteristics of these cells are very well understood, advances in Li-ion technology mean that manufacturers currently produce cells that have double the energy density of this heritage cell. In addition, cells are available with the same energy density as the HC(M) but with considerably higher rate capabilities. Cells with longer cycle life have also been developed, although there is normally a trade-off in energy density such that a simple comparison cannot be made for end-of-life (EoL) performance.

To date, ABSL have qualified three Li-ion cells for space applications, these are:

- **18650HC(M)**. The heritage ABSL cell has a nominal capacity of 1.5Ah (133 Wh/kg) and has been extensively characterised. The design was modified in 2009 to include a central mandrel; this was to ensure a clear vent path in the case of overpressure. The cell was completely re-qualified following this change.
- **18650HR**. This cell was qualified specifically for a launch vehicle program and later adopted for MARES on the ISS. The cell has lower capacity than the HC(M) cell, but can deliver significantly higher currents.
- **18650NL**. This cell provides a considerable increase in energy density compared to the HC(M) cell: 189 Wh/kg compared to 133 Wh/kg.

COTS cells have been recently assessed to identify those that may provide a performance benefit compared to the currently qualified cells. Due to the long durations for full qualification, and the quantity of life test data required to predict the long term performance, these reviews are carried out regularly.

2 REVIEW PROCESS

The initial stage of any COTS assessment is to determine what cells are available on the market. During the last market survey, cells from over 40 companies were assessed. For this review, a commercially available database was used for the initial search [1]. The database was used to filter the available cells by chemistry and form factor; these were then ranked by performance requirements, e.g. energy density and cycle life.

Once the initial list of cells was acquired from the database, the cell specification sheets were obtained. The data from these sheets were then assessed in detail against the cell performance requirements described below.

3 CELL REQUIREMENTS

The cell requirements are based on three factors: energy density, cycle life, and rate capability. There are obviously other desirable qualities for a space cell, for example launch vibration tolerance and survival under vacuum, however data on these traits is not available for the majority of COTS cells.

3.1 Energy Density

Since the introduction of Li-ion space batteries in the late 90's, the technology has developed and become more accepted. This has led to the steady reduction in mass budgets and an increase in the energy and rate requirements.

Some projects have requirements that necessitate the EoL energy density to be around 150 Wh/kg, with development projects demanding more. Obviously, the beginning-of-life (BoL) energy density needs to be greater than this; the actual value will depend on the fade.

The other energy criterion that was used to assess cells is from an ESA project to develop a Li-ion cell, with a specific energy of 250 Wh/kg or greater [2].

3.2 Cycle Life / Fade

LEO missions are now commonly in excess of 11 years comprising ~60,000 cycles(without electric propulsion and significantly more with), with GEO missions commonly specified for ≥15 years operation. The fade of any selected cells should allow the EoL energy density requirement (described above) to be achieved.

However for COTS cells, the only data commonly available is the retained capacity after 200-300 cycles between maximum and minimum voltage, rather than for more realistic lower depth of discharge (DoD) regimes. Therefore, this COTS cell assessment bases the cycle performance on a comparison with the ABSL 18650HC(M) cell. Measured ABSL data show that the HCM cells complete at least 1500 cycles at 100% DoD to 80% retained capacity.

3.3 Rate Capability

For missions that require SAR operation, the maximum discharge currents and system minimum voltage requirements for typical high energy COTS cells can be a limiting factor. This may be due to high cell resistance, the operation of safety devices, or the temperature rise.

The ability to discharge at rates >2C with minimal impact to fade would therefore be considered an advantage.

4 SURVEY OF MANUFACTURERS AND CELLS

4.1 Manufacturers

As part of the market review, an assessment of the manufacturer, as well as that of the cell type, is required. In this phase of the project, the cell manufacturer was assessed based on historical relationships. For the manufacturer of each selected cell type, the following areas have been considered.

- **Quality**. Issues can include variability of cell capacity, fade, corrosion, seal leakage.
- **Relationship**. Responses to queries, open discussion on the end use of the cells, and the battery design.
- **Flexibility/customisation**. Possibility of custom cells or specific delivery conditions.
- **Experience/heritage**. Particularly important for Space applications, where customers are trying to minimise risk of battery causing spacecraft failure.

4.2 Cells

Data from the manufacturers' specification sheets were assessed against the cell requirements detailed in Section 3. In order to summarise the cell performance and allow comparison between manufacturers, the cell performance was scored according to the criteria in Table 1. The energy density and cycle life were considered the most important requirements, whereas the rate capability is less significant.

Table 1. Cell scoring system

Requirement	Scoring
Energy Density (Wh/kg)	0 = <130 1 = 130-160 2 = 160 – 190 3 = 190 – 210 4 = > 210
Cycle Life (no. of cycles to 80% capacity)	0 = <299 1 = 300-399 2 = 400 - 599 3 = 600 - 999 4 = >1000
Rate Capability (C)	0 = max discharge rate ≤ 2 1 = max discharge rate 2 - 5 2 = max discharge rate > 5

Following the review (Table 2), the cells could be described in three groups:

- Very high energy density: up to 260 Wh/kg (Types 5, 7, 8).
- High cycle life (can complete >1000 100% cycles with less than 20% fade). These cells typically have lower energy densities (Types 1, 2, 4).
- In between these two extremes (Types 3, 6).

Table 2. Cell performance scores

Cell	Energy Density	Cycle Life	Rate Capability	Score
Type 1	3	4	2	**9**
Type 2	1	4	2	**7**
Type 3	3	2	1	**6**
Type 4	0	4	0	**4**
Type 5	4	2	1	**7**
Type 6	2	3	2	**7**
Type 7	4	1	1	**6**
Type 8	4	2	1	**7**

5 TEST DATA

Following the review of COTS cells, testing must be carried out on the short-listed cells in order to ascertain

 DOI: 10.1051/e3sconf/20171617006

whether they are suitable for typical space conditions.

The initial stage is restricted to those tests that provide go/no-go information, i.e. if the cell cannot survive vacuum, it will be unsuitable for space. In addition, long term tests will be started, so that useful data can be obtained as early as possible. The selected tests are shown in Figure 1.
To date, testing has been started for four of the eight cell types (1-4).

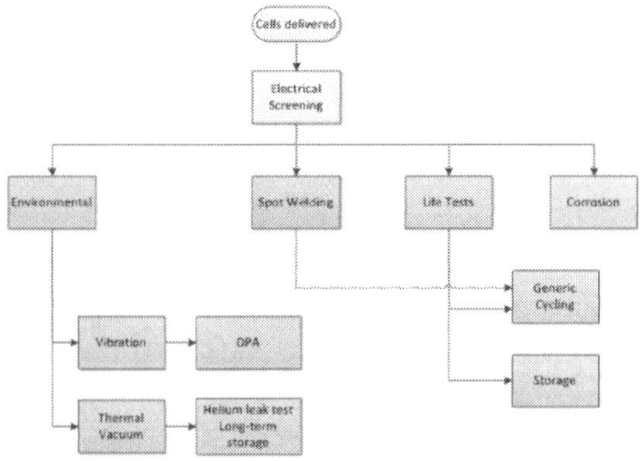

Figure 1. Technology watch test overview

5.1 Electrical Screening

The first test any cells undergo is an electrical cycle to determine the capacity and resistance. As ABSL manufacture batteries without balancing and bypass electronics, the cells are closely matched to ensure they stay appropriately balanced throughout the life of the battery. Therefore, cell types with a wide capacity or resistance spread may be more difficult to use. The level of cell rejection is also an important parameter, as it gives an indication of the manufacturing quality.

The normalised capacity results from four cell types are shown in Figure 2. The screening data have been referenced to the manufacturers' 'nominal' capacity from the cell datasheets; the conditions for these values are different between manufacturers and cell types. The screening at ABSL has been carried out using the same equivalent regime (i.e. same C rates) for all cell types, to provide a fair comparison. The data show that for three of the four cell types, the returned capacity is in line with the manufacturers' nominal value. However, Cell Type 3 returned more capacity than expected.

Aside from the actual capacity, the spread of values was reviewed. Cell Type 3 can be seen to have a narrower distribution than the other cells, which may mean a higher utilisation could be achieved. However, these results are from batches of less than 300 cells, so care should be taken when considering scaling to a full batch.

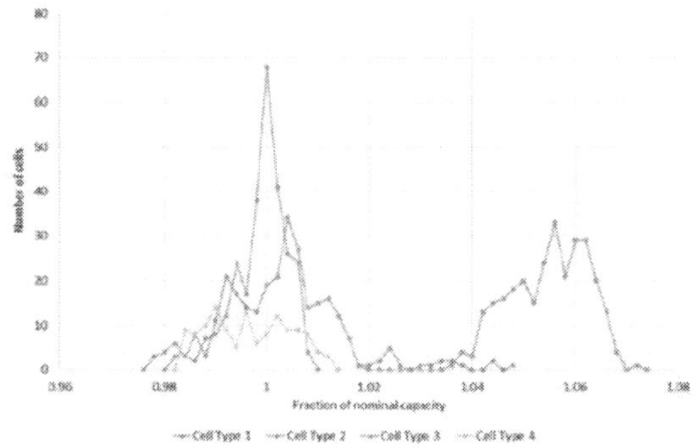

Figure 2. Normalised screening capacity

Another important factor is the cell resistance. Data at end of charge (EoC) is shown in Figure 3. The resistance is dependent on the design purpose of the cells; those designed for high rate applications have lower resistance (for example Cell Types 1 and 2).

As for the capacity, the spread of EoC resistance is important, due to the effect on the level of balancing that can be achieved. Three of the four cell types show low levels of variation in the resistance, however Cell Type 4 shows a much higher level of 40 mOhm out of an average of 180 mOhm (22%).

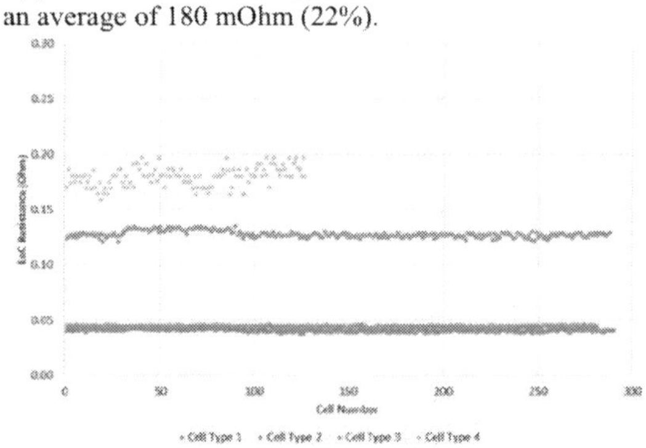

Figure 3. Cell EoC resistance (screening)

The reject rates for the four screened cell types are shown in Table 3. These cells have only been rejected due to electrical parameters being out of family; rejections due to dents, scratches are not taken into account. Three of the cell types have very low reject rates. However, one cell type had a high rate; most of these cells were rejected due to low delivery voltage, which may indicate a high level of self-discharge.

 DOI: 10.1051/e3sconf/20171617006

Table 3. Screening electrical rejection rate

Cell Type	Reject Rate
Type 1	1 / 290 (0.3%)
Type 2	2 / 281 (0.7%)
Type 3	2 / 290 (0.6%)
Type 4	15 / 127 (12%)

5.2 Environmental

To survival of the cells in the space environment is extremely important. Therefore, the performance of the cells will be assessed under typical vibration conditions encountered when an ABSL battery design is subjected to its specified random and sine vibration. The cells must also be able to survive sustained periods at hard vacuum without failure of hermeticity.

In addition to the space environment, batteries and cells are stored on ground prior to launch. Under these circumstances, the corrosion resilience of the cells needs to be evaluated.

To date, the robustness of two of the cell types to vibration has been tested. Figure 4 shows the vibration environment the cells were subjected to. Following the testing, the cells underwent a destructive parts analysis (DPA). The cells were inspected to ensure no signs of damage, for example cracks, tears, or delamination. No differences were found between non-vibrated and vibrated cells, demonstrating their ability to withstand severe mechanical testing.

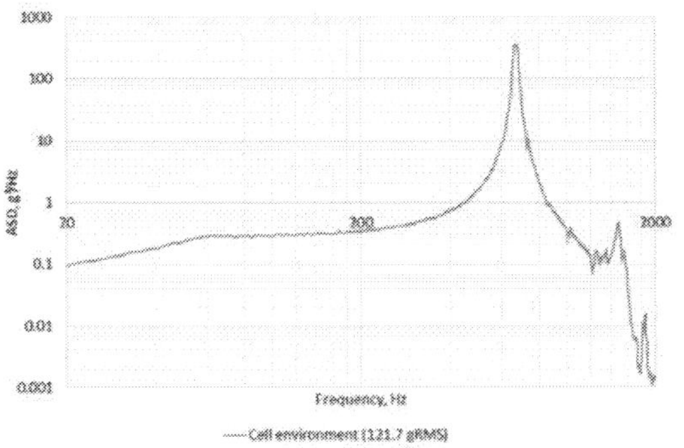

Figure 4. Random vibration test condition

With respect to vacuum and corrosion testing, no data are currently available.

5.3 Life Testing

Lifetime performance was identified as an important factor in selecting another cell for space qualification. These tests need to be started as early as possible, due to the long timescales needed to acquire sufficient data for comparison and battery sizing.

For general applications, it is desirable for the cells to show:

- Good consistency within a batch to allow cells to remain balanced through life.
- Low variability within batches to ensure fade/resistance prediction tools do not have to be batch specific.
- Lower fade than the current ABSL cells.

The cell performance will be characterised under regimes typical for space applications. Therefore, the life tests will be conducted at typical LEO conditions (20% DoD), as well as higher DoD (40%), which is similar to the average DoD during GEO for ABSL batteries. At present, only the 40% DoD tests have been started. Cell tests to determine the fade under storage regimes are also being started.

5.3.1 Cycling

The end of discharge (EoD) voltages for three cell types cycled at 40% DoD are shown in Figure 5. To date, all cell types show a lower rate of decrease in their EoD voltage compared to the heritage HCM cell, as well as the newer NL cell. Therefore, these three cell types currently look promising in terms of long term cycling performance. It should be noted that the HCM cell was cycled at a higher current (2C/3), which explains why the EoD voltage is seen to be lower than for the tests cycled at C/2.

From previous test experience, the relative behaviour of cells at 40% DoD has corresponded well with their performance at lower DoDs. It is therefore expected that the cell performance at lower DoDs will also show an improvement over the current ABSL cells. This will be subject to test.

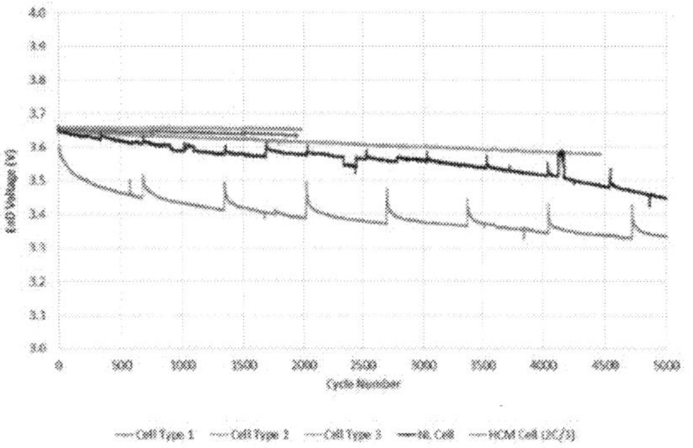

Figure 5. Cycling data (40% DoD, 4.1V, 20°C, C/2 except where stated)

As well as the relative performance of the selected cell types against the HCM and NL cells, the variation within a batch and batch-to-batch needs to be

DOI: 10.1051/e3sconf/20171617006

determined. If the variation between batches is large, a generalised fade tool is not feasible, and any battery sizing should be conducted using specific batch data (or the worst case). Whereas, variation within a batch may mean the cells would not remain balanced. For example, data for three tests at 40% DoD on Cell Type 2 are shown in Figure 6. The testing on Batch 1 and Batch 2 were carried out at separate ABSL sites using different batches. These early data show that the results are extremely consistent within a batch, and between batches. These tests will continue to be monitored to determine whether the trends are maintained in the long term.

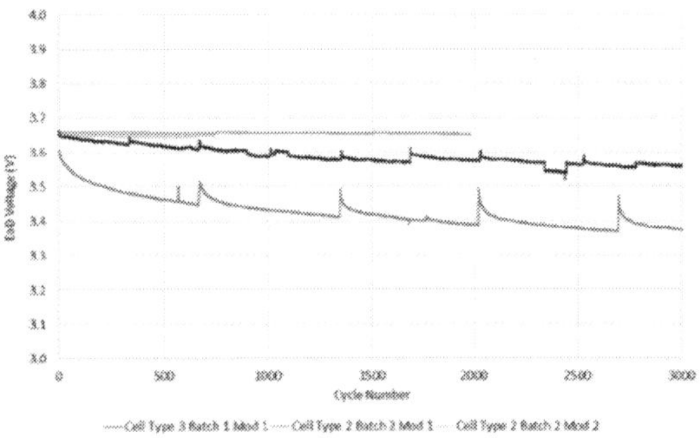

Figure 6. Cycling data showing in batch, and batch-to-batch variation (40% DoD, 4.1V, 20°C, C/2)

5.3.2 Storage

Most applications will require cells to be stored for a certain period, for example ground storage prior to integration, during a cruise phase, or on a launch pad. 100% state of charge (SoC) is the worst case and used prior to launch or during satellite test campaigns, whereas long term storage is likely to be between 50% and 5% SoC depending on the need for emergency power.

At present, data for storage are only available for one cell type: Cell Type 2. A comparison of the retained capacity against ABSL's current cells, as well as the effect of SoC, are shown in Figure 7. As expected, the higher the SoC, the higher the rate of fade. In terms of the comparison to the HCM and NL, this cell type is showing performance that is at least as good. Therefore, the very early data is promising.

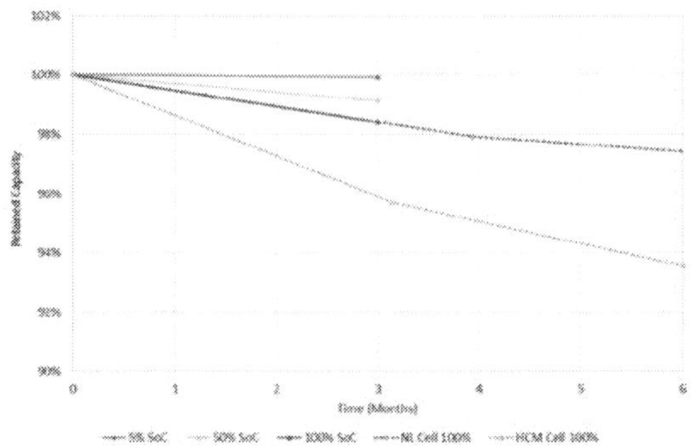

Figure 7. Retained capacity of Cell Type 2 during storage at 20°C

6 CONCLUSION

A review of COTS Li-ion cells has shown that multiple types are available that have higher energy density, longer cycle life, and/or better rate capability than the current ABSL range.

Testing has been started on four or the eight selected cell types. Initial data show that there are some promising candidates for further investigation. Results from cell types with less fade than the HCM and NL cells, as well as low variability, have been presented.

The COTS cell review will continue to amass data on available cells. Four further cell types are about to undergo electrical screening, and additional life tests are due to be started to demonstrate the performance at lower DoD. The environmental testing will continue once all cell types have been screened.

7 REFERENCES

1. Power Sources Database, Schmuel De-Leon Energy Ltd.
2. Ligneel, E., Defer, M., Fischer, F., Dumont, E., Fusalba, F., Chami, M., and Gineste, V.; High specific Li-ion cells based on high voltage materials; Executive Summary; ESA Contract 4000107101/12/NL/LvH; 2015.

DOI: 10.1051/e3sconf/20171617007

HIGH PERFORMANCES OF OXYFLUORIDE ELECTRODE USED IN LITHIUM ION BATTERY

K. GUERIN [1], N. LOUVAIN [1], M. EL-GHOZZI [1], C. CENAC-MORTHE[2]

[1] Clermont Université, Université Blaise Pascal, Institut de Chimie de Clermont-Ferrand, BP 10448, F-63000 Clermont-Ferrand, France. Katia.araujo_da_silva@univ-bpclermont.Fr
[2] Centre National d'Etudes Spatiales, Toulouse, France, Email: Celine.Cenac-Morthe@cnes.fr

ABSTRACT

Reactivity of pure molecular fluorine F_2 allows the creation of new materials with unique electrochemical properties. We demonstrate that titanium oxyfluoride $TiOF_2$ can be obtained under molecular fluorine from anatase titanium oxide TiO_2, while the fluorination of rutile TiO_2 leads only to pure fluoride form TiF_4. Contrary to most fluorides, $TiOF_2$ is air-stable and hydrolyses poorly in humid conditions. Such stability makes it possible for $TiOF_2$ to be studied as an electrode material in Li-ion secondary batteries systems. It shows capacities as high as 220 mAh g^{-1} and good cyclability at high current rates at an average potential of 2.3 V vs Li^+/Li. At such a potential, only Li^+ insertion occurs, as proven by in operando XRD/electrochemistry experiments.

RESULTS AND DISCUSSION

According to thermodynamics, the reaction of molecular fluorine F_2 with rutile titanium dioxide TiO_2 at 25 °C should lead to an equimolar mixture of TiF_4 and O_2:

$TiO_2(s) + 2\ F_2(g) = TiF_4(s) + O_2(g)$ (eq. 1)

From room temperature, the reaction is supposedly quantitative, with solid TiF_4 formed. When the temperature increases, TiF_4 starts to sublimate from approximately 200 °C, and the presence of gaseous TiF_4 becomes non-negligible above 250 °C, which is lower than expected (Tsub = 284 °C).[1] The major difficulty is that only TiF_4 is the only stable phase reported and no other, hypothetical or not, metastable fluoride or oxyfluoride phases can be included in the calculation. Moreover the difference of reactivity between anatase and rutile under molecular fluorine has never been analysed. In order to investigate the reactivity of titanium dioxide, anatase and rutile TiO_2 powders have been fluorinated with a mixture of molecular fluorine and nitrogen gases at 250 °C for 18 hours in dedicated fluorination equipment.

Figure 1 displays the results of the in situ pressure data collected during the static fluorination process. First, let us consider the case of micrometre-sized powders of rutile and anatase TiO_2. When rutile is heated up to 250°C in one atm. of a F_2/N_2 mixture, the pressure reaches about 1350 mbar. Then, when temperature is kept at 250 °C, the pressure keeps on slowly increasing. This indicates the formation of a gaseous phase of TiF_4 because of its high tensile strength at such temperature, in good accordance with thermodynamics. On the contrary, when anatase is fluorinated in similar conditions (Figure 1), the pressure profile obtained is clearly dissimilar, suggesting a different reactivity with pure fluorine between both polymorphs, contrary to what was expected. At 250 °C, the pressure drastically decreases from 1350 to 1200 mbar, bespeaking the formation of a new solid phase with a net consumption of F_2 gas. All parameters being kept identical, the only reason for such difference is polymorphism.[2] The first interface encountered by F_2 gas is the crystal surface, thus crystal structure and morphology may affect the reactivity during a gas-solid process.[3] Indeed, such a difference between rutile and anatase is often discussed in adsorption of active species in photocatalysis.

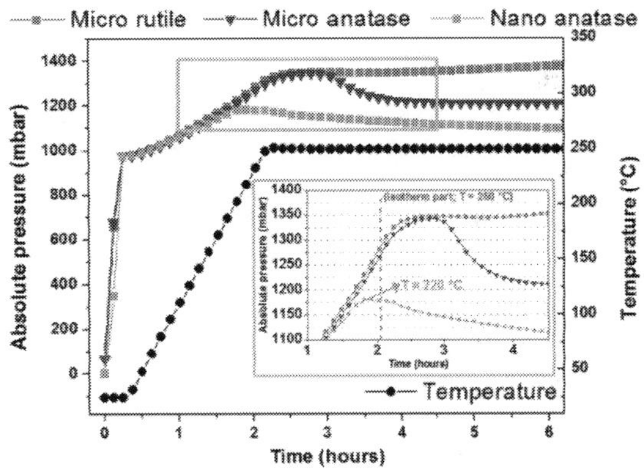

Figure 1. In situ total pressure data collected during static fluorination of rutile and anatase TiO_2 with 1 atm. of a mixture of N_2 and F_2 gases at 250 °C; the inset shows a zoom on the isotherm part of the curves.

Based on results from in situ pressure data, XRD and TG analyses, 30-nm nanocrystals of pure $TiOF_2$ were

DOI: 10.1051/e3sconf/20171617007

obtained from a nanopowder of anatase TiO_2.[4]

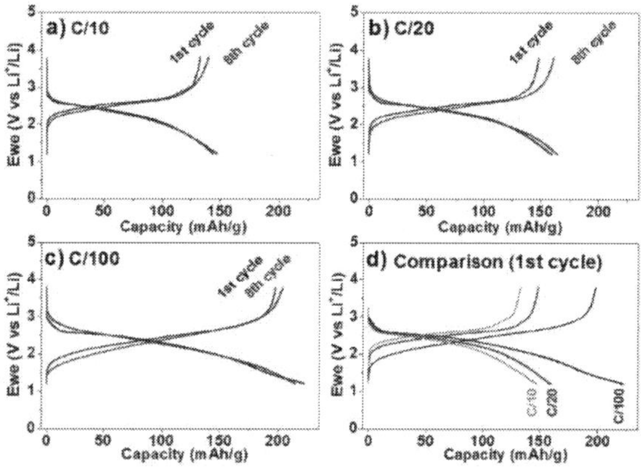

Figure 2. Galvanostatic voltage profiles for $TiOF_2$/Li cells cycled at C/10 (a), C/20 (b), and C/100 (c) between 3.8 and 1.2 V, and a comparison of their first cycle (d).

The as-prepared nanocrystals of $TiOF_2$ were tested as electrode of Li-ion battery in Swagelok-type half-cells. The galvanostatic charge-discharge voltage profiles at three different current rates (C/10, C/20 and C/100, where C is set at 263.15 mAh g^{-1} for a 1-e- process) of the first and eighth cycles for $TiOF_2$ versus Li between 3.8 and 1.2 V are shown in Figure 2. Such potential window as been chosen according to the studies by Chowdari et al. [5] showing that under 1.0 V no more potential plateau occurs in reduction. For each current density, the discharge curves show a relatively flat plateau around 2.25 V vs Li^+/Li. The $TiOF_2$ nanocrystals electrode delivers a reversible specific capacity of 145 mAh g^{-1} at C/10 (Figure 2a) over the first eight cycles. The differences in shape and potential between the first cycle and the following ones are thus relatively small, indicating that the reaction mechanism is most probably the same. At slower rates (Figure 2b-d), the reversible specific capacity is increased up to 155 mAh g^{-1} (C/20) and an impressive 220 mAh g^{-1} (C/100). The fact that the specific capacity increases when reducing the current density would indicate a kinetically limited ionic or electrical diffusion mechanism, which could be explained either by the insulating nature of $TiOF_2$, a common characteristic of fluorinated materials, or a none convenient volume/area ratio of the $TiOF_2$ grains. Noteworthy, the potential of $TiOF_2$ is 0.6 V higher than the one of anatase TiO_2, as expected from the high electronegativity of the fluoride ion. We endeavoured to study the Li-insertion mechanism into $TiOF_2$ in order to discriminate between a typical insertion, yielding a Li_xTiOF_2 phase, and a conversion reaction, yielding a composite mixture of LiF and LiyTiO2 (the formation of Ti0 in our working potential window has been ruled out). In a first step, an ex situ XRD study has been carried out on electrodes that were submitted to a C/10 rate and stopped at 2.5, 2.0, and 1.2 V during their first discharge. The impact of Li-insertion is relatively small on the XRD profile and no lithium fluoride can be detected, thus a priori indicating that no conversion reaction may occur in the chosen potential window.

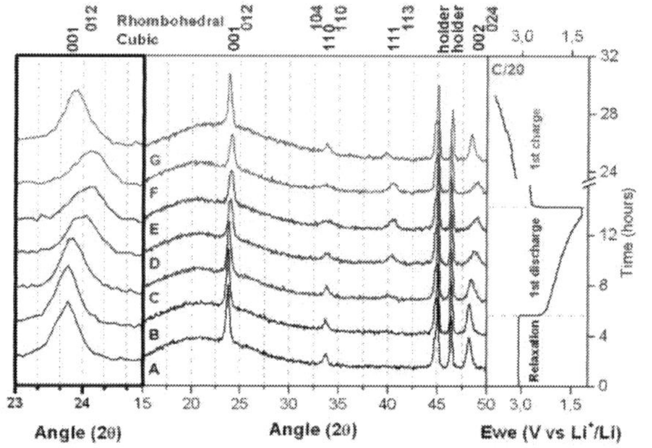

Figure 3. In operando evolution of the XRD pattern recorded at C/20 with a close view on the 001 reflection (left) and its corresponding voltage profile (right). Miller indices for the cubic (black) and rhombohedral (blue) cells are shown. Peaks between 45° and 47.5° are originating from the measurement cell.

In order to go further into details in the reaction mechanism of $TiOF_2$ with Li, and to probe cell parameters modulation, the first charge-discharge cycle was investigated by in operando XRD and the results are shown in Figure 3. During the open-circuit, only the well-defined peaks of crystalline cubic $TiOF_2$ are visible in the diffraction pattern (Figure 3, patterns A and B) at 2θ = 23.4°, 33.3°, and 47.8°, corresponding to (001), (011), and (002) hkl planes, respectively (ICSD #160661), and no evolution, either in intensity or in position, can be detected. At the beginning of the reaction (Figure 3, pattern C), the intensities of these peaks slightly decrease without observing any new peak in the diffraction pattern until half discharge, where Bragg peaks broaden and shift to higher angles (Figure 3, pattern D). This is clearly noticeable on the close view on the 001 reflection (Figure 3). Please note that at this point of discharge a new peak is unambiguously appearing at 2θ = 40.1°, and that such XRD behaviour is characteristic of Li-intercalation materials.[6] At the end of discharge (Figure 3, pattern E), the set of reflections of the initial cubic $TiOF_2$ phase are broadened and

DOI: 10.1051/e3sconf/20171617007

shifted to higher angles, and their intensity is slightly lower. The angle shift may point to the cubic-to-rhombohedral phase transition described for chemically lithiated $TiOF_2$, [7] and the hypothesis has been verified by a profile matching refinement of the fully discharged pattern using a rhombohedral model (R-3c). All the diffraction features are well modelled by this phase with cell parameters a = 5.206 Å and c = 13.334 Å, and this allows us to conclude that the phase at the end of the discharge is a lithiated $TiOF_2$ of general formula Li_xTiOF_2; thus, the present peaks at 2θ = 23.8°, 33.4°, 34.4°, 40.1°, and 48.7° would correspond to (012), (104), (110), (113), and (024) hkl planes, respectively. While at the early oxidation process the diffraction pattern can still be indexed by the rhombohedral cell (Figure 3, pattern F), the cubic non-lithiated $TiOF_2$ phase is fully recovered at the end of charge (Figure 3, pattern G), therefore corroborating our hypothesis for Li-insertion and confirming the reversibility of the electrochemical performances obtained. With a concrete picture of the reactions of lithium with titanium oxyfluoride, it is now possible to deduce the lithium content, i.e. x in Li_xTiOF_2 formula, from our electrochemical measurements. At a C/10 rate, a specific capacity of 145 mAh g^{-1} was obtained (Figure 2a) and this leads to a lithiated Li_xTiOF_2 with x = 0.55 (with a theoretical capacity of 263.15 mAh g^{-1} for 1 lithium); the lithium content can be as high as x = 0.87 at C/100, for a specific capacity equal to 220 mAh g^{-1} (Figure 2c).

The Li-insertion mechanism being validated, the cycling performances of our $TiOF_2$ nanocrystals were investigated (Figure 4). The galvanostatic charge-discharge composition-voltage profiles of key cycles of a $TiOF_2$/Li cell consecutively cycled at C/20, C/10, C/20, C, and finally C/20 between 3.8 and 1.2 V are displayed in Figure 4a. The most striking feature of this material is that the specific capacity obtained at C/20 is systematically resumed after being cycled at higher C/10 and C rates. In addition to its excellent cyclic capacity retention at sequential current rates, $TiOF_2$/Li cells demonstrate a stable specific capacity at the high current density of C for more than 30 charge/discharge cycles between 3.0 and 1.2 V (Figure 4b). The discharge capacity slowly increases during the first thirty cycles and reaches 60 mAh g^{-1}. This relatively high capacity retention at C may stem from the presence of O^{2-} ion in $TiOF_2$, which makes it a semiconductor with a band-gap of approximately 2.5 eV,[8] whereas metal fluorides are commonly referred to as insulator with band-gaps larger than 4 eV that cannot be cycled at such high density current without nanocarbon grafting.[9-11] An impressive capacity of 200 mAh g^{-1}, close to the theoretical one, can even be obtained when the half-cell is cycled at C/100 (Figure 4b). The system retains a high capacity of 180 mAh g^{-1} after tenth of cycles, a value that can surely be improved by optimizing the battery engineering.

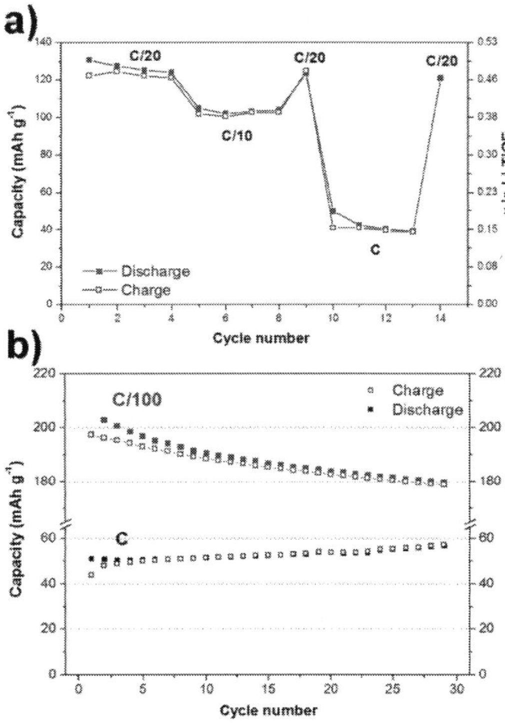

Figure 4. Composition-voltage profiles for one $TiOF_2$/Li cell cycled at C/20, C/10, C/20, C and C/20 consecutively between 3.8 and 1.2 V (a); cycling performance curves of $TiOF_2$ nanocrystals at current

REFERENCES

1. G. Brauer and G. Brauer, Handbook of Preparative Inorganic Chemistry, Academic Press, Inc., 1963.
2. D. S. Torkhov, P. E. Meskin, Y. V. Kolen'ko, V. A. Ketsko, A. A. Burukhin, B. R. Churagulov and N. N. Oleinikov, Dokl. Chem., 2004, 394, 36-38.
3. S. C. Li and U. Diebold, J. Am. Chem. Soc., 2010, 132, 64-66.
4. N. Louvain, Z. Karkar, M. El-Ghozzi, P. Bonnet, K. Guérin, P. Willman, J Mater. Chem A, 2014, 2, 15308-15315
5. M. V. Reddy, S. Madhavi, G. V. Subba Rao and B. V. R. Chowdari, J. Power Sources, 2006, 162, 1312-1321.
6. J. N. Reimers and J. R. Dahn, J. Electrochem. Soc., 1992, 139, 2091-2097.
7. D. W. Murphy, M. Greenblatt, R. J. Cava and S. M. Zahurak, Solid State Ionics, 1981, 5, 327-330.

DOI: 10.1051/e3sconf/20171617007

8. C. Xue, T. Narushima and T. Yonezawa, J. Inorg. Organomet. Polym., 2012, 23, 239-242.
9. R. Ma, Y. Dong, L. Xi, S. Yang, Z. Lu and C. Chung, ACS Appl. Mater. Interfaces, 2013, 5, 892-897.
10. C. Li, L. Gu, J. Tong and J. Maier, ACS Nano, 2011, 5, 2930-2938.
11. S. W. Kim, D. H. Seo, H. Gwon, J. Kim and K. Kang, Adv. Mater., 2010, 22, 5260-5264.

E3S Web of Conferences **16**, 17008 (2017) DOI: 10.1051/e3sconf/20171617008
ESPC 2016

TOWARD STATE ESTIMATION OF SATELLITE-BORNE LITHIUM-ION BATTERY BASED ON LOW FREQUENCY IMPEDANCE DATA

Kohei Tanaka [(1)], **Omar S. Mendoza-Hernandez** [(2)], **Yoshitsugu Sone** [(1) (2)],

Seisuke Fukuda [(1) (2)], **Masayuki Itagaki** [(3)]

[(1)] *SOKENDAI, School of Physical Sciences, 3-1-1 Yoshinodai, Chuo-ku, Sagamihara, Kanagawa, 252-5210 Japan, tanaka.kohei@ac.jaxa.jp:*
[(2)] *Japan Aerospace Exploration Agency, 3-1-1 Yoshinodai, Chuo-ku, Sagamihara, Kanagawa, 252-5210 Japan*
[(3)] *Tokyo University of Science, Faculty of Science and Technology, 2641 Yamazaki, Noda city, Chiba, 278-8510 Japan*

ABSTRACT

The satellite borne batteries should be composed by safe materials if we don't want to have a risk of explosion caused by batteries. Therefore, we focused on two safe batteries. One is a lithium-ion battery with an ionic liquid electrolyte, and the other is a $LiFePO_4/C$ type lithium-ion battery. To check whether the batteries are suit for space applications or not, we demonstrate the ionic liquid type batteries and $LiFePO_4/C$ type battery in orbit by mounting on "Hodoyoshi-3" microsatellite, and test $LiFePO_4/C$ type cell on the ground at various conditions for a better understanding.

On the ground tests, AC impedance and capacity of the cells were initially measured, and charge/discharge cycling was constantly repeated at 10, 23 and 45℃. The cells were discharged by constant current (CC) protocol to DOD 50% with 1.0 C for 30 minutes. They were then charged by a constant-current/constant voltage (CC-CV) protocol to 3.6 V for 65 minutes with 0.5 C. For capacity check, the cells were charged at 1.0 C in CC-CV mode until their charge current becomes 60 mA, and discharged at 1.0 C in CC mode to 2.0 V at 23℃. The AC impedance was measured by applying 100 mA of AC oscillation over the frequency range from 0.01 Hz to 10 kHz at SOC 50%.

As a result, the decrease in the impedance for the charge transfer through the cycles was observed at each test condition. Furthermore, especially in over recommended charge condition at 10℃, cells that were charged and discharged at 1.1 A/1.1 A were led to internal short circuit. The results suggested that the negative electrode performed as a "lithium-ion excess" by cycles. We define "lithium-ion excess" that lithium-ion happens to stay inside the negative electrode without desorption after cells discharge.

1. INTRODUCTION

In traditional satellite operations, batteries are controlled only by their voltage, current and temperature data, so there are issues in long-term operations because deterioration is not considered. However, it is equally important to take into account deterioration to detect true conditions as operation terms have become longer by mission requirements.

There are two main issues that make it difficult to continue satellite operations on orbit. The first is difficulty for the operators to obtain the true battery capacity since satellites do not charge or discharge for capacity checks despite the deterioration of battery capacity. The second one is that the battery internal chemical condition changes by continuing charge and discharge cycles. Therefore, battery deterioration has been a major issue for long-term operation.

Impedance helps operators to understand battery deterioration. Impedance is defined as the frequency domain ratio of the voltage to the current, as in [1]. From impedance data, operators can determine the internal chemical conditions of a battery directly. Even though there are dedicated devices to measure the impedance data on the ground, it has been difficult for satellites to add the devices because it has limitations to its size and weight. Therefore, we have to calculate impedance by using House Keeping data.

On the ground, researchers are trying to calculate impedance data from the House Keeping data. Dinh Vinh [2] monitored battery voltage and current with a higher sampling rate of over 1 kHz, and tried to calculate impedance data using the "*Kalman filter*" technique in his research. He obtained high frequency impedance in his research and the result was sufficient, however the sampling time was too fast to apply on satellites. T. Momma [3] tried to obtain impedance data by decreasing the temperature, for example minus 20 ℃, because the impedance data increases when the temperature decreases. The results showed that the impedance data could be obtained analytically by monitoring battery voltage and current. However, they were not suitable for practical use since the condition was not suitable for the utilization of the operation of satellites. In orbit, T. Yairi [4] proposed a data-driven anomaly detection framework and checked with the framework whether satellites were having trouble or not. Through those researches, the deterioration of satellite-borne battery may be able to be detected accurately. However, even if the deterioration or risk of troubles can detect while they are operating, the risk of propagation continues after the satellites end the operation because the lithium-ion batteries are a fire

 DOI: 10.1051/e3sconf/20171617008

hazard due to containing oxygen. Therefore, the satellite borne batteries should be composed by safe materials if we don't want to have a risk of explosion caused by batteries.

For these reasons, we focused on two safe batteries. One is a lithium-ion battery with an ionic liquid electrolyte [5], and the other is a $LiFePO_4/C$ type lithium-ion battery [6]. The ionic liquid type batteries are safe because they have lower flammability and non-volatility. $LiFePO_4/C$ lithium-ion battery is reputable for being safe since there is no process to outgas combusted oxygen. Therefore, the batteries are suitable for space applications. Furthermore, to demonstrate that these batteries are suited for space applications, we mounted both of the batteries on the "Hodoyoshi-3" microsatellite and continue to test them in orbit.

Furthermore, to understand the $LiFePO_4/C$ type cells, we tested the cells on the ground. As a related work, Dubarry et.al, tested $LiFePO_4/C$ type cells with 0.04 C, 0.2C, 0.5C, 1C, and 2C [7]. As their results, when the charge and discharge rate is over 0.5C, the capacity increase in fist 150 cycles. Furthermore, the capacity increase when the rate increase. Therefore, we assume that $LiFePO_4/C$ type cells deteriorate in different process with the other types of lithium-ion cells [8].

In this paper, we report the overview of demonstration results about ionic liquid type batteries and $LifePO_4/C$ type batteries in the orbit, and the test results of $LiFePO_4/C$ type lithium-ion secondary cell on the ground.

2. DEMONSTRATION IN ORBIT

2.1 Overview of the "Hodoyoshi-3" Microsatellite

The "Hodoyoshi-3" microsatellite is a 60 kg-class satellite developed by the university of Tokyo and NESTRA as shown in Fig.1. Hodoyoshi-3 was successfully launched on a Dnepr rocket at the Yasny Launch Base in Russia (7:11 pm, 19 June, 2014 U.T.).

Hodoyoshi-3 satellite is an earth observation satellite, and it has mission space, called "rental space", for demonstration and we mounted a demonstration module to test the ionic liquid type battery and the $LiFePO_4/C$ type battery as shown in Fig.2.

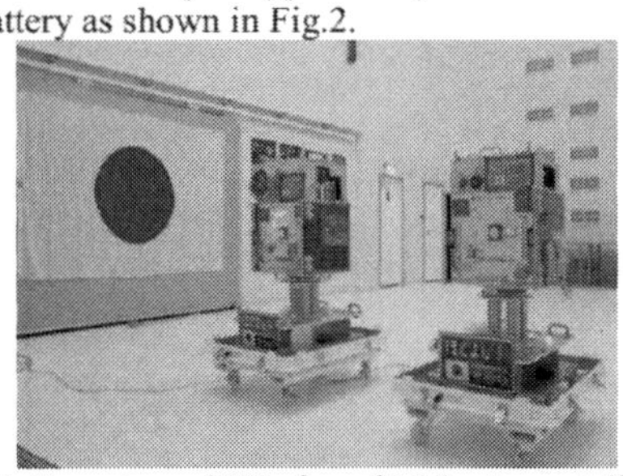

Figure1 *Overview of "Hodoyoshi-3" microsatellite. The right side is "Hodoyoshi-3" and the left side is "Hodoyoshi-4". These satellites were developed by the university of Tokyo and NESTRA.*

Figure2 *Overview of demonstration module.*

2.2 Test results in the orbit

The module charges with constant power (10 W) and discharges with constant resistance (100 Ω). The $LiFePO_4/C$ type batteries are composed of three series cells and the ionic liquid type batteries are composed of two series cells. Fig.3 shows the test results of the $LiFePO_4/C$ type batteries and Fig.4 shows the test results of the ionic liquid type batteries in the orbit [9]. The test condition is set as in Tab.1. Both of the batteries successfully charged and discharged, and they are still working well in orbit even after 2 years.

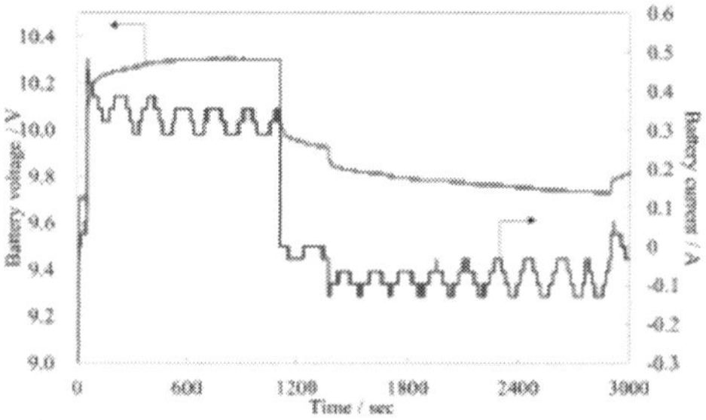

Figure3 *The charge and discharge results of the $LifePO_4$ type batteries.*

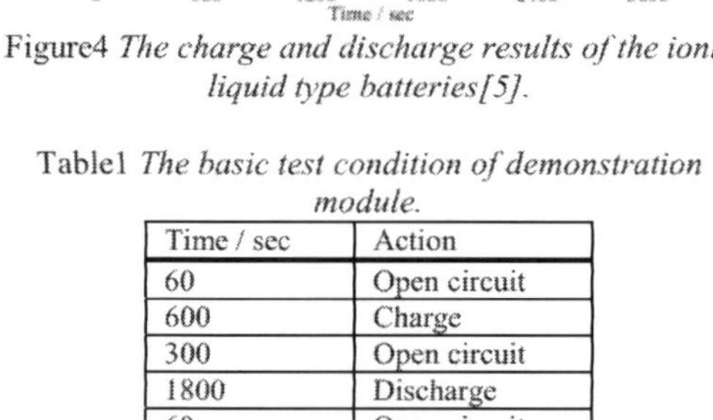

Figure4 *The charge and discharge results of the ionic liquid type batteries[5].*

Table1 *The basic test condition of demonstration module.*

Time / sec	Action
60	Open circuit
600	Charge
300	Open circuit
1800	Discharge
60	Open circuit

DOI: 10.1051/e3sconf/20171617008

3. EXPERIMENTS ON THE GROUND

3.1 Test conditions

To understand battery characteristics on the ground, the cell impedance is measured with an impedance analyser (NF corporation: FRA5014/As-510-ECA) and we get impedance transition data through charge and discharge cycles. The charge and discharge cycling tests are repeated with a charge and discharge test system (Kikusui electronics corporation: PFX2011). Tab.2 shows the specifications about the $LiFePO_4/C$ type cells and Fig.5 shows appearance of the cycling tests.

Table2 *Specifications about olivine-type lithium-ion secondary cells.*

Size	18650
Initial capacity	1050 mAh
Operative temp.	0 ~ 40℃
Voltage range	2.0 - 3.6 V

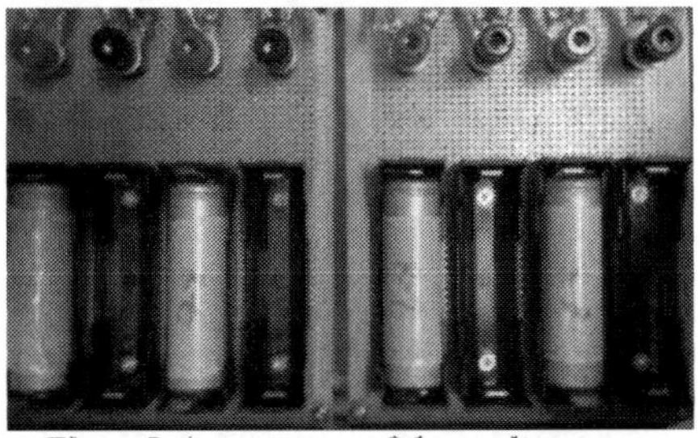

Figure5 *Appearance of the cycling tests.*

The effect of impedance is taken into account from temperature parameters and deterioration parameters. Therefore, to understand the cell characteristics changing, the test conditions of capacity check and cycle test were set as shown in Tab.3 and Tab.4.

AC impedance and capacity of the cells were initially measured, and charge/discharge cycling was constantly repeated at 10, 23 and 45℃. The cells were discharged by constant current (CC) protocol to DOD 50% with 1.0 C for 30 minutes. They were then charged by a constant-current/constant voltage (CC-CV) protocol to 3.6 V for 65 minutes with 0.5 C. For capacity check, the cells were charged at 1.0 C in CC-CV mode until their charge current becomes 60 mA, and discharged at 1.0 C in CC mode to 2.0 V at 23℃. The AC impedance was measured by applying 100 mA of AC oscillation over the frequency range from 0.01 Hz to 10 kHz at SOC 50%.

Table3 *Capacity check condition.*

Charge	CC-CV, 1 C/3.6V,Time-cut (2.5 h), 23℃ Wait time 30 min.
Discharge	CC, 1 C , 2.0 V-cut, 23℃, Wait time 30min.

Table4 *Charge and discharge cycle conditions.*

Charge	CC-CV, 0.5 C/3.6V, It-cut(60 mA: max 65min.), 10, 23, 45℃
Discharge	CC, 1 C, Time-cut (30 min.), 10, 23, 45℃, DOD50%

In this case, the results about capacity check are shown in Fig.6, and the test results about impedance are shown in Fig.7, Fig.8 and Fig.9 [10]. Fig.7, Fig.8 and Fig.9 show the changing of impedance through cycling tests. The cell was tested at 10, 23 and 45℃. Fig.10 shows the discharge curve trend at capacity check.

As per the results, the decrease in the impedance for the charge transfer resistance through the cycles was observed at each test condition as shown in Fig.7, Fig8 and Fig.9. Furthermore, the capacity of cells increased at 23℃ by cycles as shown in Fig.6. Despite of the increase of capacity, the open voltage after discharge go decrease as shown in Fig.10.

Fig.7 shows the Nyquist plot of tested cells at 10℃. There are over two semi-circles in all cases. The point of intersection at high frequency indicates the electrolysis solution resistance (R_{sol}) and the radius by semi-circles indicate the charge transfer resistance (R_{ct}). R_{ct} becomes smaller at each test condition through charge and discharge cycles. Furthermore, especially for cells tested at 10℃, the R_{sol} increased by cycles. However, the total impedance ($R_{sol} + R_{ct}$) consequently became smaller. We assume that the reason why the resistance becomes smaller is the lithium-ion excess to the negative electrode.

The test samples we used were $LiFePO_4/C$ type cells and the cells are said to be able to charge and discharge at a high rate without degradation. Furthermore, they have a plateau voltage at wide SOC. The negative electrode is a graphitized carbon. If there is lithium-ion excess inside the negative electrode through charge and discharge cycles, the SOC at negative electrode increases. This situation will occur if the desorption at negative electrode becomes difficult. When the SOC at negative electrode increases, the impedance gets smaller through charge and discharge cycles.

From these results, if only the charge curve trend was checked, it appears that DC resistance got bigger, so the timing of CV charge became earlier. However, if the impedance trend was checked also, the impedance got smaller, so we can assume that the results from Fig.6 to Fig.10 suggested that the negative electrode performed as a lithium-ion excess by cycles. Therefore, it is important to check the impedance. Furthermore, the impedance of the cells that were charged and discharged at 1.1 A/1.1 A in 10℃, which is in over limits of recommended charge condition, significantly decreased, and finally, the cells led to internal short as shown in Fig.11.

DOI: 10.1051/e3sconf/20171617008

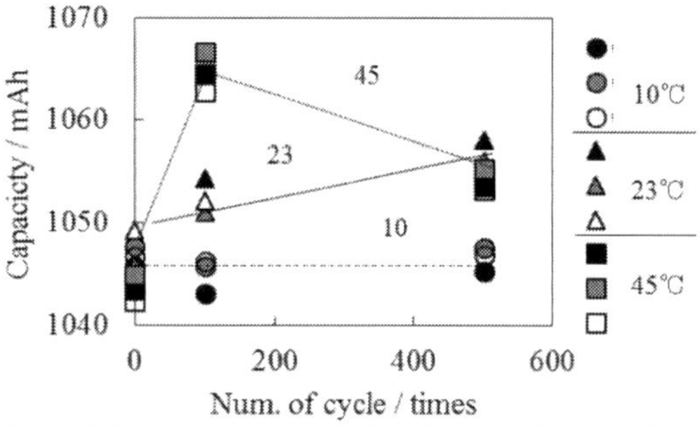

Figure6 *Capacity changing of each cells. The circled data show the tested results at 10 ℃. The triangled data show the tested results at 10 ℃[10].*

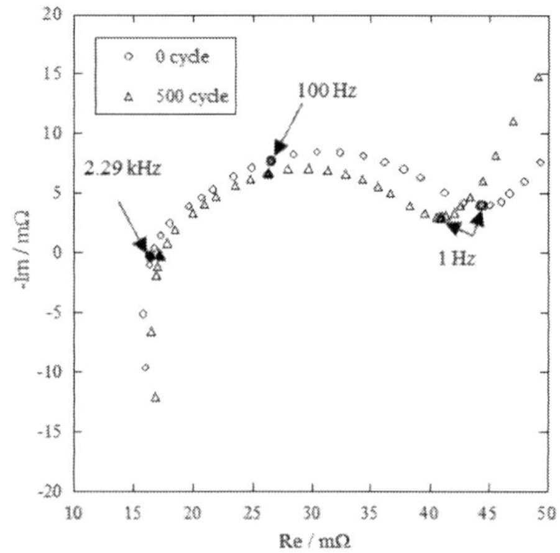

Figure7 *Impedance changing through the cycling test. The impedance was measured at SOC50% with 23 ℃. The cell was tested at 10 ℃[10].*

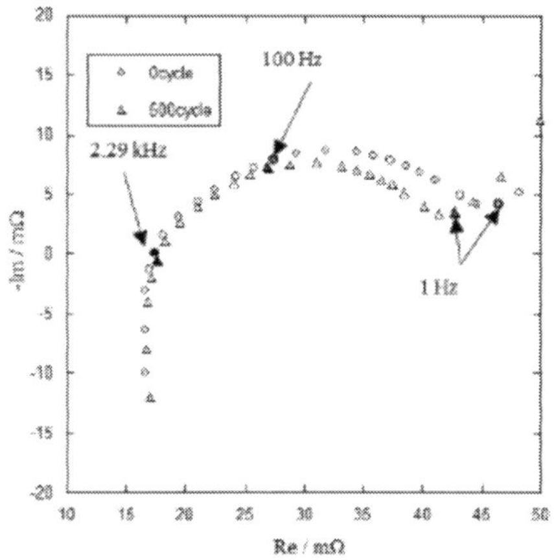

Figure8 *Impedance changing through the cycling test. The impedance was measured at SOC50% with 23 ℃. The cell was tested at 23 ℃[10].*

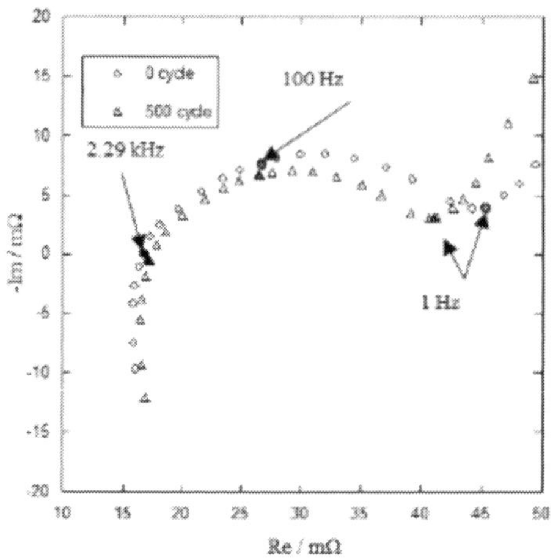

Figure9 *Impedance changing through the cycling test. The impedance was measured at SOC50% with 23 ℃. The cell was tested at 45 ℃[10].*

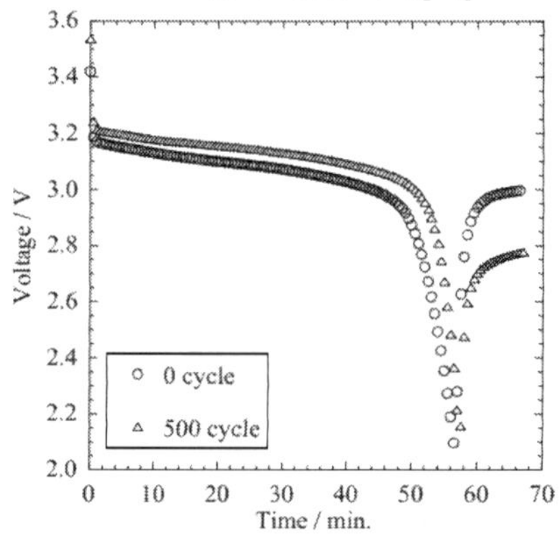

Figure10 *Discharge curve trend with 1C/23degC after the cycling at 23 ℃[10].*

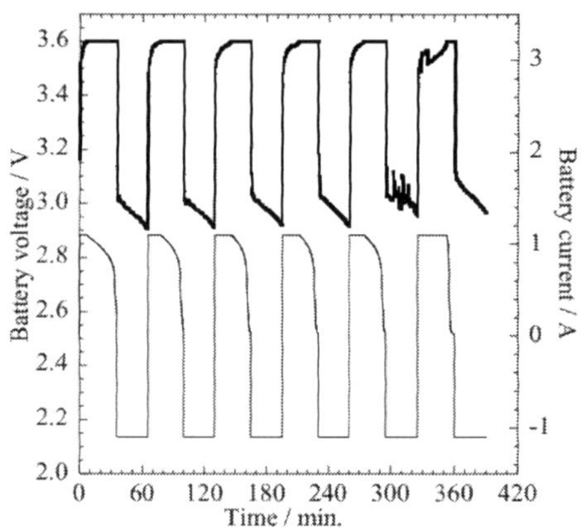

Figure10 *Charge and discharge data when internal short happened [11].*

4. DISCUSSION

After two years, ionic liquid type batteries and $LiFePO_4/C$ type batteries are working well in orbit. The comparative testing between ground and orbit was already conducted in the case of ionic liquid type batteries [9]. There results indicated that the cells are working well. We will test $LiFePO_4/C$ type batteries on the ground, and compare with the data obtained in orbit.
On the ground, $LiFePO_4/C$ type cells are tested at 10, 23, and 45℃. The impedance at charge transfer got smaller and the impedance at electrolyte got bigger by cycles. Furthermore, the capacity at 23℃ continued to increase and the open voltage of discharge curve trend go decrease. As compared with Dubarry, the point that the capacity go increased was the same at 23℃.
As a results, we assume that the balance between positive electrode and negative electrode is shifting because the negative electrode continues to stock the lithium-ion due to fail of desorption from negative electrode.

5. CONCLUSION

For the future space applications, we tested ionic liquid lithium ion battery and $LiFePO_4/C$ in orbit by mounting on Hodoyoshi-3 microsatellite. Both of the batteries seem to be working well after two years.
To understand the characteristics of $LiFePO_4/C$ type cells, we also tested the cells on the ground at several conditions.
In the case of over limits of recommended charge condition at 10 ℃, the impedance significantly decreased at low frequency, and finally, the cells led to internal short. From the results, we assume that the lithium-ion excess were observed by checking the impedance data. Therefore, it is important to check the impedance even if it is a satellite-borne battery. Furthermore, changes in impedance occurred in low frequency, so the changes will be able to detect by HK data of satellite.

6. ACKNOWLEDGEMENT

The cells were provided by Sony Energy Devices Corporation.
This research is partially supported by 'International Joint Research Program for Innovative Energy Technology' of Ministry of Economy, Trade and Industry (METI).

7. REFERENCES

1. Huet,F., (1998). A review of impedance measurements for determination of the state-of-charge or state-of-health of secondary batteries, Journal of Power Sources, p 59-69

2. Dinh, V. D., Christophe F., Khadija E. K. B., & Guy F.,(2009). Impedance observer for a Li-ion battery using Kalman filter, IEEE TRANSACTIONS ON VEHICULAR TECHNOLOGY, VOL. 58, NO. 8, p3930-3937

3. Momma, T., Matsunaga, M., Mukoyama, D., & Osaka, T., (2012). Ac impedance analysis of lithium ion battery under temperature control, J. Power Sources, 216, 304-307

4. Yairi, T., Inui, M., Kawahara, Y., & Takata, N., (2011). Spacecraft Telemetry Monitoring Method Based on Dimensionality Reduction and Clustering, JSASS journal, Vol.59 No.691, p. 197-205 (written in Japanese)

5. Yamagata, M., Tanaka, K., Tsuruda, Y., Sone, Y., Fukuda, S., Nakasuka, S., Kono, M., and Ishikawa, M., (2015). The First Lithium-ion Battery with Ionic Liquid Electrolyte Demonstrated in Extreme Environment of Space, Electrochemistry, p 918.

6. Kang, B., and Ceder, G., (2009). Battery materials for ultrafast charging and discharging, Nature, p 190.

7. Dubarry, M., Truchor, C., and Liaw, B.Y., (2014). Cell degradation in commercial $LiFePO_4$ cells with high-power and high-energy designs, Journal of Power Sources, p 408.

8. Vetter, J., Novák, P., Wagner, M. R., Veit, C., Möller, K. C., Besenhard, J. O., Hammouche, (2005). Ageing mechanisms in lithium-ion batteries. Journal of Power Sources, p 269–281.

9. Tanaka, K., Yamagata, M., Tsuruda, Y., Sone, Y., Fukuda, S., Kono, M., Ishikawa, M.,Shirasaka, S., and Nakasuka, S., (2015). Charge and Discharge Characteristics on Orbit of High Safety Lithium-ion Battery, Proceeding of Conference on Space Science and Technology (in Japanese).

10. Tanaka, K., Sone, Y., Mendoza-Hernandez, O.S., Fukuda, S., & Itagaki, M., (2016). Typical Trend of the Impedance Spectrum Obtained from the Life Cycle Testing of the LiFePO4 type Lithium-Ion Secondary Cells, Electrochemistry, Vol. 84 No. 8 p. 601-604.

11. Tanaka, K., Sone, Y., Mendoza-Hernandez, O.S., Fukuda, S., & Itagaki, M., (2016). The AC Impedance Changing of the Olivine-type Lithium-Ion Secondary Cells through Charge/Discharge Cycling, 67th Annual Meeting of the International Society of Electrochemistry, 21-26, August.

E3S Web of Conferences 16, 17009 (2017) DOI: 10.1051/e3sconf/20171617009
ESPC 2016

QUALIFICATION AND LIFE TESTING OF LI-ION VES16 BATTERIES

Stéphane Remy [(1)], Serge Lawson [(1)], Stéphane Lefeuvre [(1)], Edouard Mosset [(1)], Maria Nestoridi [(2)]

[(1)] *Saft, Defense & Space Division, BP 1039, 86060 Poitiers - France,*
[(2)] *ESA/ESTEC, Keplerlaan 1, Postbox 299, 2200 AG Noordwijk - The Netherlands,*

stephane.remy@saftbatteries.com ; serge.lawson@saftbatteries.com ; stephane.lefeuvre@saftbatteries.com; edouard.mosset@saftbatteries.com; Maria.Nestoridi@esa.int

ABSTRACT

In the frame of an ESA GSTP 5.2 activity (contract 4000105105), the qualification and life testing of a Saft range of Li-ion batteries based on VES16 cells and theirs autonomous simplified balancing system (SBS) has been carried out. In this abstract the development, qualification plan and successful results from all various tests conducted on the VES16 qualification battery modules are synthetized.

Up to the present time, Saft batteries have been mainly utilizing for space applications high capacity cells, like the 45 Ah VES180 and the 35 Ah VES140 cells, targeting predominantly space missions in Geostationary Earth Orbit (GEO). However following the qualification and commercialization of the Saft 4.5 Ah VES16 cell in October 2011 [1] & [2], Saft has been developing and qualifying in the frame of this ESA GSTP 5.2 contract, VES 16 batteries for space missions, targeting both GEO and Low Earth Orbit (LEO) satellite missions. The electrochemistry of the VES16 cells used for the battery modules under the ESA qualification program is not novel. For VES16 cells, the Saft knowhow from large capacity space cells used for space applications, since SMART 1 mission in 2003, has been tailored for a cell with smaller capacity in order to facilitate the modular philosophy that has been deployed in this battery range.

1 BATTERY MODULES DESIGN

The VES16 battery design is based on standard building packs of different voltages, capacities and mechanical arrangements, in order to address most of the present and future space missions' needs.
The GSTP-5 tested VES16 battery designs are characterized by modularity, thus enhancing customization capabilities as per mission requirements and minimizing the number of relevant delta-qualification activities required for individual space missions.

VES16 battery range, proposed by Saft, adopts a S-P configuration, since cells are connected in series (S) to form a string and strings were connected in parallel (P) to form a battery pack. The latter can individually form a battery module or a number of packs can be connected in parallel in order to form battery modules depending on capacity requirements.

The Saft VES16 battery line offers batteries ranging from 6 to 10 S and 4 to 8 P configurations (Figure 1). This range includes a single deck also called "flat" (Figure 2) and a double deck (Figure 3) design aiming to fit any satellite accommodation. The aim of this qualification is not to test all the relevant modules but to test a number of representative ones for space applications, which have been called **QM1** (2x10S8P), **QM2** (10S5P) and **QM3** (8S4P) detailed later in this document.

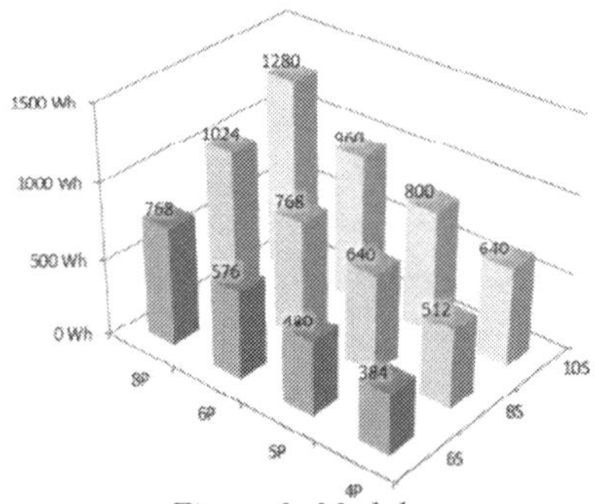

Figure 1: Modules energy range from 384 Wh to 1280 Wh (EOCV = 4.1 V per cell)

The range can be extended to higher voltages configurations by simply increasing the number of cells in series configuration. Indicatively, an 11S or 12S case can be derived from the 10S case by a simple homothetic arrangement.

Figure 2: Single deck modules family

DOI: 10.1051/e3sconf/20171617009

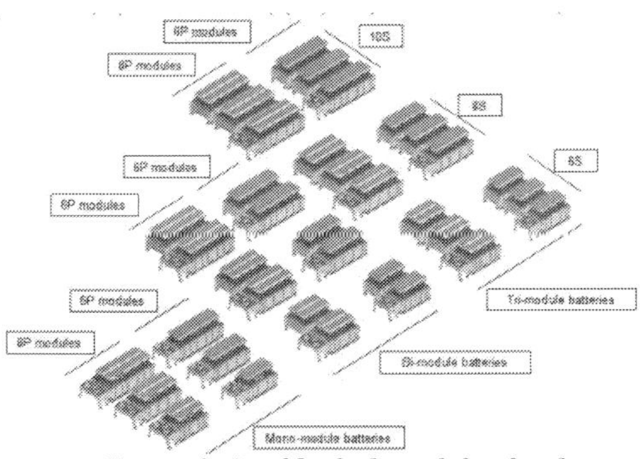

Figure 3: Double deck modules family.

The single deck battery module is an arrangement of cells, glued by resin, between a baseplate and a top plate. Connection to the cell can is made with tabs to the cell case, for the cathode, and to the cell top cap, for the anode. All strings of cells are parallel connected by bus bars. On the terminals of each individual cell a balancing system is implemented, called simplified balancing system (SBS). The principle of this balancing electronics system is based on cell current consumption above a given voltage threshold U0 (Figure 4). The system is fully autonomous and requires neither command nor telemetry. To achieve this balancing function, the SBS architecture is based on a simple circuitry which core component is a micro shunt. This micro shunt component quality level, selected by Saft in order to optimize SBS function, is QML-Q (Class 2 component with specific additional upgrade tests and characterization) as baseline. But the same component is qualified in class 1 (QML-V) and fully interchangeable.

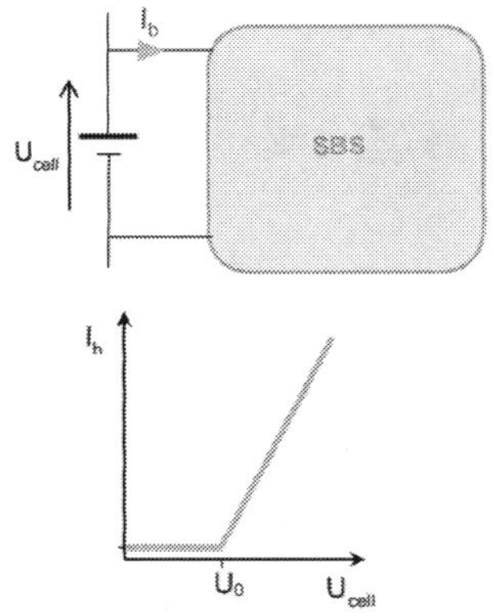

Figure 4: SBS principle

The battery module can be customised to include the required type and number of thermal sensors, heaters and monitoring connectors, even thermostats if needed. The connectors can be rectangular or circular, redundant or not, with separated or mixed polarities, as specified by Battery General Design and Interface Requirements. It is noticeable that the individual cell voltages are accessible on ground for any maintenance or health status check, by dedicated ground connectors, until the last days prior to launch.

The double-deck battery design is exactly the same as the single deck. The main principle is that the baseplate is shared between the two decks, so that the cells are glued on both sides of the baseplate. Therefore, the heart of the modules remains unchanged for all the VES16 batteries range.

2 DEVELOPMENT OF VES16 BATTERIES

This section describes the development and the design & analysis results at cell, SBS and battery level. The development plan is based on three main steps; the VES16 qualification plan, the SBS electronics qualification plan and the battery range design.
The design of the modules was justified mainly by the electrical, electronical, thermal and mechanical analyses.
In the frame of the GSTP-5.2 qualification Tests campaign, Saft completed these analyses, by detailed qualification tests results.

2.1 Design main justifications

Detailed analyses have been conducted for single and double deck modules, for LEO and GEO cases.
These analyses have demonstrated that the battery can sustain the various stresses generated by space environment, especially, the space missions' specific electrical, electronical, thermal and mechanical environmental requirements.
Hereafter, the main findings from the thermal, mechanical, reliability and radiation analyses are presented.

As a summary of thermal analysis, all the battery parts (cells, electronic components and wiring) remain within their derated values, (rated values in failed cases) with margins, for all 9 cases studied for single and double deck batteries.

The gradients within a string or within a module remain lower than 4°C, in all cases, as can be seen in Figure 7.

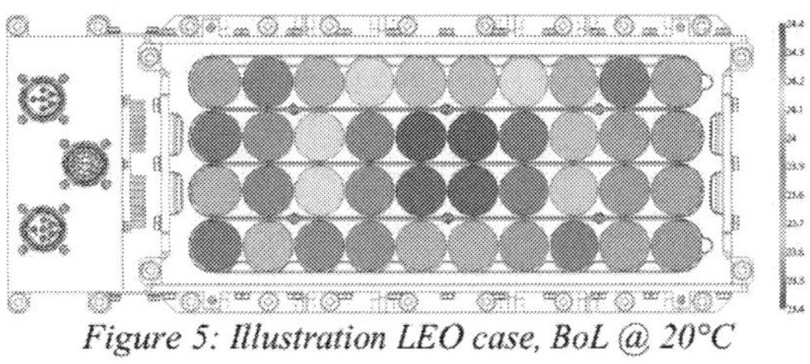

Figure 5: Illustration LEO case, BoL @ 20°C

Battery modules mechanical behaviour:

The modal analyses provided the natural frequencies of the structures and cells, which were the following:

- Above **400 Hz** for the 10S8P module (QM1),
- Above **1000 Hz** for the 10S5P module (QM2),
- Above **1090 Hz** for the 8S4P module (QM3),

The mechanical analysis showed that for random loads the structure and cells MOSyield are all positive and the minimum is locally for the cells' structure when the battery is loaded along the vertical-axis.

Reliability

The reliability analysis concluded that for a 12.5 years LEO mission, taking into account the loss of one string, a probability of success is 99,86 % and 99,99 % for the QM1 battery (2x10S8P) and the QM2 battery (10S5P) respectively.

Radiations:

The VES16 cell can withstand radiation levels up to 5 MRads. The only battery item that can be sensitive to radiations, is the SBS core component.

The component has been qualified, by ELDRS test, with 36 rad.s-1 rate, at 140 krads total dose.

In order to verify that this SBS core component will not be a limitation for the VES16 battery in orbit life, radiation and shielding analyses of the VES16 battery design studies have been carried out for the QM1 module.

Four mission cases have been considered:

- 12 years LEO mission at 700 km;
- 12 years LEO mission at 1200 km;
- 10 years MEO mission at 23616 km;
- 15 years GEO mission at 35870 km.

Electronics:

The electronics SBS system is fully autonomous and requires neither command, nor telemetries. This system was developed with CNES support, to address all relevant space design requirements and issues.

The battery system has been modeled, using P-spice, following ESA experts approach guidelines, to assess its behaviour in worst case conditions @ EOL. [7]

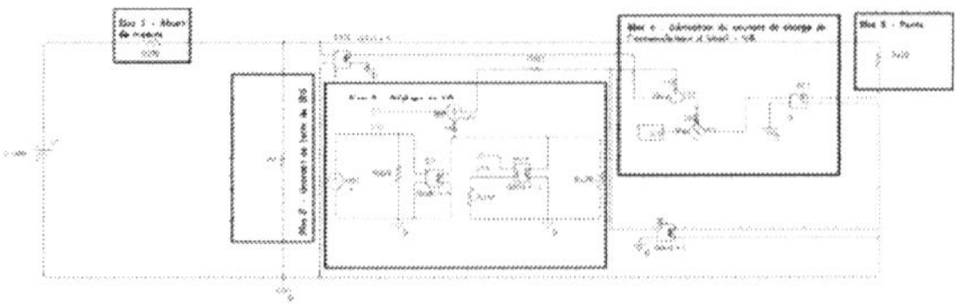

Figure 6: SBS model diagram

The model has been correlated with real battery balancing test, achieved on the QM3 battery.

3. GSTP-5 QUALIFICATION CAMPAIGN

In the frame of the ESA GSTP 5.2 Qualification activity, three different batteries have been selected to be submitted to the qualification campaigns.

Those include two different single deck "flat" modules; a 8S4P (called QM3, shown in Figure 7) a 10S5P module (called QM2, shown in Figure 8) and also a 2x10S8P double deck module (called QM1, shown in Figure 9).

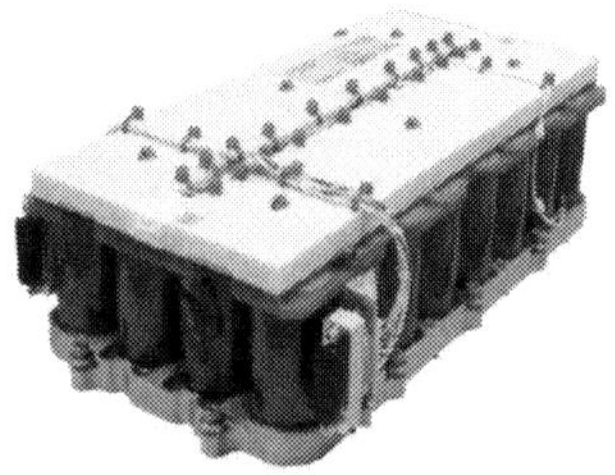

Figure 7: QM3 – 8S4P VES16 module

Figure 8: QM2 – 10S5P VES16 module

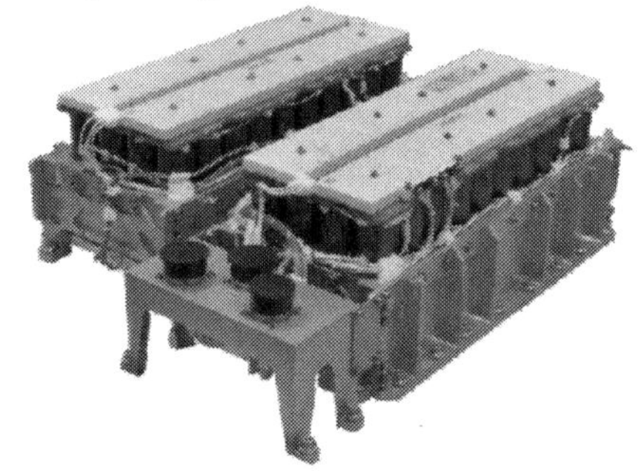

Figure 9: QM1 – 10S16P VES16 battery

DOI: 10.1051/e3sconf/20171617009

In Table 1, the Qualification test matrix for the three modules is presented.

GSTP-5.2 QUALIFICATION PROGRAM	QM1 2x10S8P VES16	QM2 10S5P VES16	QM3 8S4P VES16
CHARACTERIZATION TESTS			
Conformity check	X	X	X
Functional tests and inspections:	X	X	X
Initial charge retention	X	X	X
Stored energy measurements at various discharge temperatures and discharge currents		X	
Self-discharge measurement in several conditions (SOC, Temp.)			X
Internal Resistance measurement at 0%DOD, 20% DOD and 60% DOD		X	
Charge retention (72 hours, @20°C)	X	X	X
Impedance at 0%DOD, 50%DOD		X	X
Balancing efficiency Tests	X	X	X
ENVIRONMENTAL TESTS			
Sine vibrations (3 axes) including resonance frequencies search before and after.	X	X	X
Random vibrations (3 axes) including resonance frequencies search before and after.	X	X	X
Charge retention (72 hours, @20°C)	X	X	X
Stored energy measurements @20°C with IR measured at 20% & 60% DOD	X	X	X
Shocks tests (3 impacts per axes)	X	X	
Charge retention (72 hours, @20°C)	X	X	
Stored energy measurements @20°C with IR measured at 20% & 60% DOD	X	X	
CORONA TEST	X	X	
Thermal cycling in vacuum conditions	X	X	
Leak Test	X	X	
Magnetic Moment Measurements		X	
EMC test		X	
Impedance Measurements	X	X	X
Charge retention (72 hours, @20°C)	X	X	
Stored energy measurements @20°C with IR measured at 20% & 60% DOD	X	X	
Conformity check	X	X	X
LIFE TESTS			
LEO Life tests 30% DOD, @20°C, in real time			X

Table 1: Qualification Test Matrix

As planned and foreseen, all three modules were submitted to characterization and environmental tests. QM3 battery is presently submitted to LEO life tests.

3.1 Characterisation Tests

Characterisation tests have been successfully completed for the three batteries QM1 (10S16P VES16), QM2 (10S5P VES16) and QM3 (8S4P VES16). Main results are presented hereafter.

3.1.1 Characterisation Tests on QM1

In Figure 10, the effect of discharge current while keeping a fixed charge rate C/10 is shown.

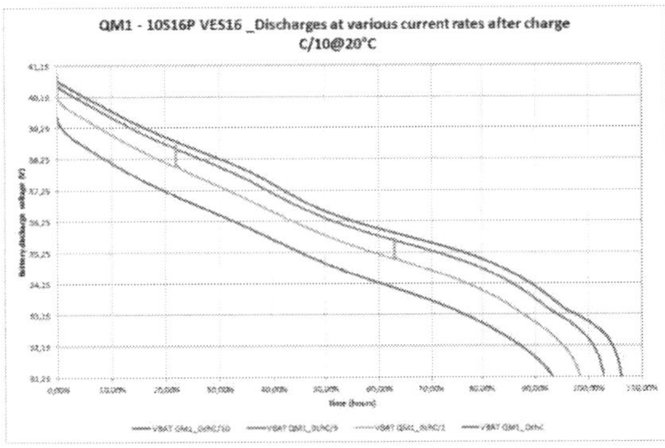

Figure 10: Discharge voltages vs Discharge Currents after charge C/10 @20°C.

In Figure 11, the battery charge and discharge voltage profiles of each cell are shown, allowing assessing the cells balancing inside the battery, thanks to AIT connectors, set on balancing (SBS) boards.

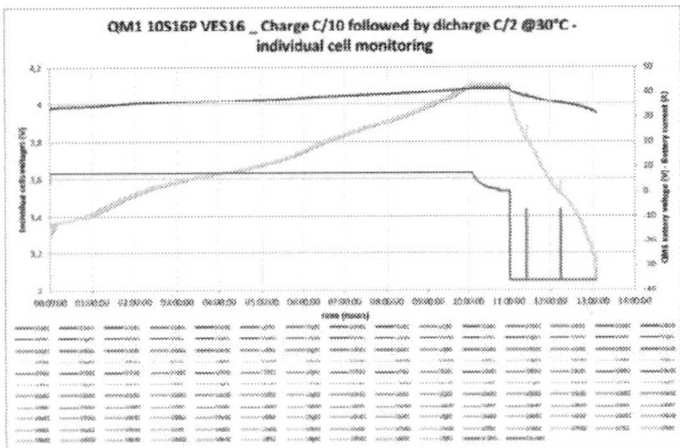

Figure 11: Battery Charge & Discharge @ 30°C – continuous monitoring of individual cells

All results obtained during electrical characterization of the QM1 battery were in line with expectations.

3.1.2 Characterization Tests on QM2 – 10S5P VES16

The main objective of the QM2 battery characterization test campaign was to characterise the battery at different charge / discharge rates and temperatures. The aim was to ease construction of a battery performance detailed meshing, for future batteries sizing, including not only cell, but also battery architecture contribution (parallel and series connections in particular).

 DOI: 10.1051/e3sconf/20171617009

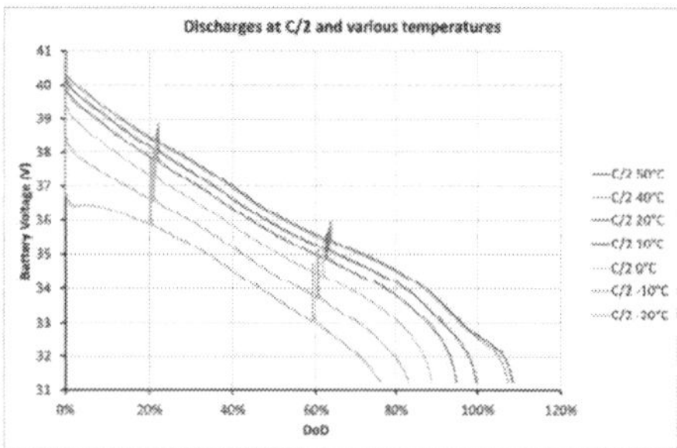

Figure 12: Battery Discharge @ C/2 vs Temperature

The characterization campaign has been completed successfully, for QM2 battery, and the battery has been authorised to pass through the environmental tests campaign as described in the following paragraphs.

3.1.3 Characterisation Tests on QM3 – 8S4P VES16

Balancing system capability demonstration was performed, by introducing a 50 mV voltage deviation/unbalance between a single cell and the rest of the cells, in the module. The modules were then submitted to LEO (20% DoD), and GEO (70% DoD max), cycling in order to check the duration or number of charge / discharge cycles needed to re-equilibrate / balance the string voltages.

The 50 mV value, between average cell and unbalanced cell, was chosen since this value corresponds typically to the relevant unbalance obtained after 6 months of storage period. In Figure 13 and Figure 14, it is shown that the initial unbalance is recovered after less than 30 days in LEO, and less than 15 eclipse days in GEO cycling.

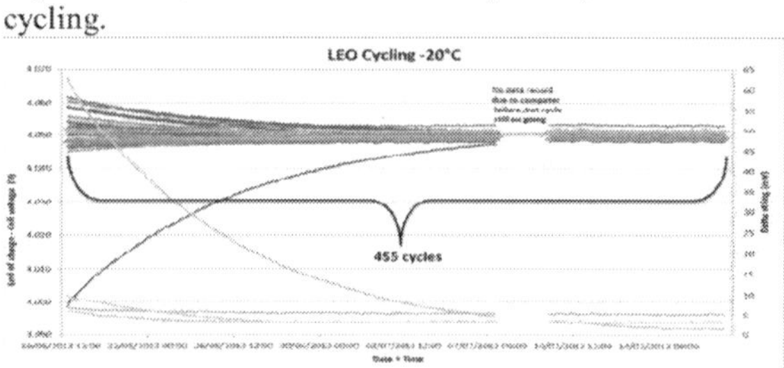

Figure 13: QM3 Battery voltage spread recovery after unbalance in LEO

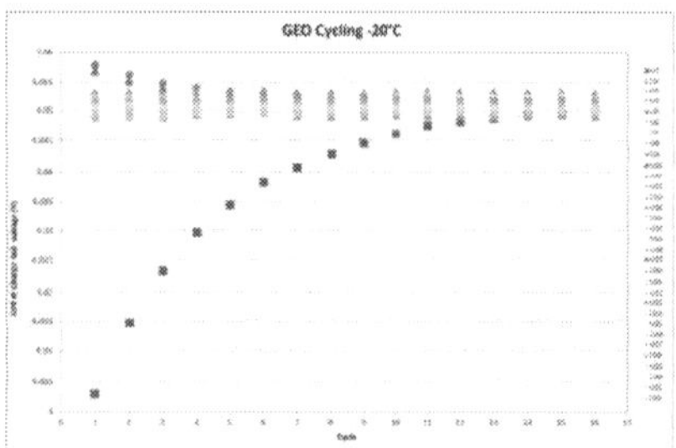

Figure 14: QM3 Battery cells' voltages spread recovery after unbalance during GEO cycles.

Once electrical behaviour of the three batteries QM1, QM2 and QM3 was assessed, by the characterization tests completion, authorization was granted to continue the qualification tests sequences with thermal, mechanical and other Space missions' environments.

3.2 environmental Test Results

3.2.1 Environmental Tests on QM1 module

The QM1 module was submitted to environmental tests, consisting in:

- Mechanical vibrations;
- Shocks tests and;
- Thermal vacuum cycles

3.2.1.1 QM1 - Mechanical vibrations

QM1 battery was tested in following mechanical vibrations:

- Sine vibrations (24g)
- Random vibrations (12.6 Grms).

Figure 15: QM1 during shock tests.

3.2.1.2 QM1 10S16P VES16 - Mechanical Shocks

QM1 battery was tested in mechanical shocks (hammer test):

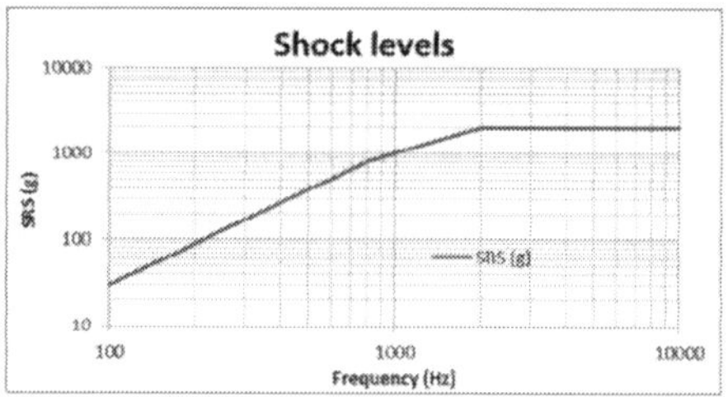

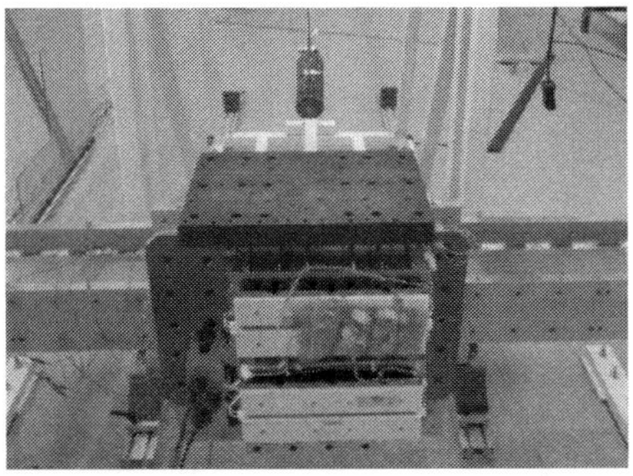

Figure 16: QM1 during shock tests.

Three impacts were realized per axis, with battery discharges and battery voltage noise measurements, during shocks.

In conclusion, the QM1 environmental mechanical sequence has been performed successfully.

3.2.1.3 QM1 - Thermal Vacuum Test

The QM1 battery was submitted to thermal environmental tests, in thermal vacuum chamber, in order to be as close as possible to VES16 batteries mission operating conditions.

PHASE	BATTERY POWER CHARGE/DISCHARGE	MOUNTING PLANE T°C
CORONA TEST	OPEN CIRCUIT	20°C
CONDUCTANCE MEASUREMENT AT STEADY STATE	OPEN CIRCUIT	0°C 10°C 40°C
HEATERS FUNCTIONNAL TESTS	OPEN CIRCUIT	0°C
CAPACITY TEST 1	I charge = C/10, until EocV = 41 V. EOCV = 4.1V/cell Discharge at I = C/2.	20 °C
CAPACITY TEST 2	I charge = C/10, until EocV = 41 V. EOCV = 4.1 V/cell Discharge at I = C/2.	40 °C
CAPACITY TEST 3	I charge = C/30, until EocV = 41 V. EOCV = 4.1 V/cell, Discharge at I = C/2.	0 °C
CAPACITY TEST 4	I charge = C/10, until EocV = 41 V. EOCV= 4.1 V/cell, Discharge at I = C/2.	10 °C
CAPACITY TEST 5	I charge = C/10, until EocV = 41 V. EOCV = 4.1 V/cell, Discharge at I = C/2.	30 °C
LEO CYCLE (2 cycles per temp.)	I charge=C/3 during 65 min, until EocV = 40.5V. I discharge = C/1.95 during 35 mn	+10 °C 30 °C
GEO CYCLE (2 cycles per temp.)	I charge=C/10 during 1368 min, until EocV = 40.5V. I discharge = C/1.5 during 72 mn	+10 °C +30 °C
PULSES TEST (repeated 3 times)	1 pulse of 1.9 C for 5 minutes 4 pulses of 1.9 C for 75 seconds	+10 °C
Thermal vacuum Cycling (8 cycles)	OPEN CIRCUIT	- 5 °C 45 °C

Table 2: Thermal vacuum tests sequence

Figure 17: QM1 prepared for thermal vacuum tests.

3.2.1.4 QM1 10S16P VES16 – Leak Tests

Battery leak rate was measured to 2.8 10-7 mbar.l/s, compliant to the leak rate criterion.

QM1 battery successfully passed all the Qualification tests Sequence.

3.2.2 QM2 10S5P VES16 - Environmental Tests.

The QM2 module was submitted to environmental tests, consisting in:

- Mechanical vibrations;
- Shocks tests and;
- Thermal vacuum cycles

3.2.2.1 QM2 10S5P VES16 - Mechanical vibrations

QM2 battery was tested in vibrations:

- Sine vibrations (24 g)
- Random vibrations (14.8 Grms)

Figure 18: QM2 on vibration shaker.

The first resonance frequencies measured, in module width axis, were above 1000 Hz, which correlates well with the relevant simulations.

 DOI: 10.1051/e3sconf/20171617009

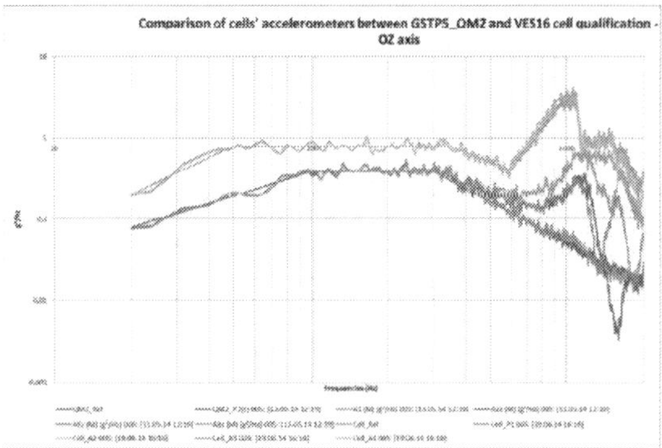

Figure 19: QM2 cells' accelerometers comparison with VES16 cell qualification.

3.2.2.2 QM2 10S5P VES16 - Mechanical Shocks

QM2 battery was tested in mechanical shocks (hammer test) according to following levels:

Shock tolerances -3dB/ +6dB

Frequency (Hz)	SRS (g)
100	30
800	800
2000	2000
10000	2000

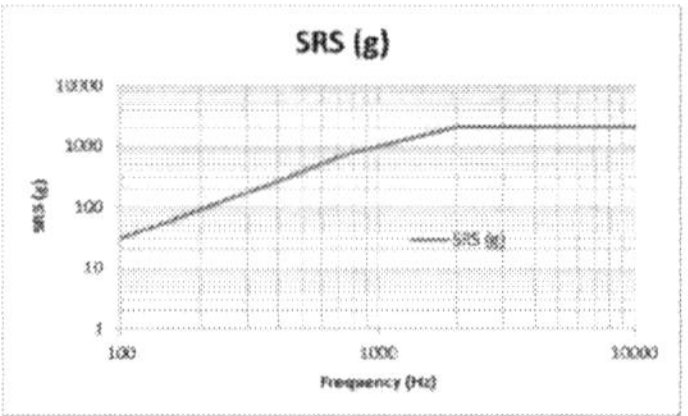

Figure 20: QM2 shocks tests levels.

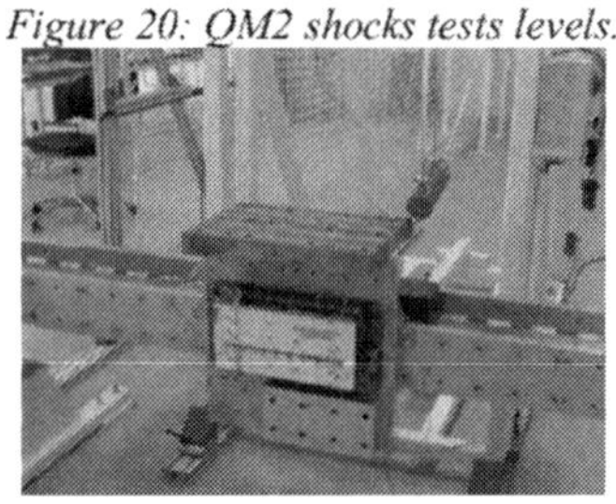

Figure 21: QM2 during shocks tests.

Three impacts were realized, per axis, with battery discharges and battery voltage noise measurements during shocks.

In conclusion, the QM2 environmental mechanical sequence has been passed successfully.

3.2.2.3 QM2 10S5P VES16 - Thermal Vacuum Test

The QM2 battery was submitted to thermal environmental tests in thermal vacuum chamber to be as close as possible to VES16 battery mission operating conditions.

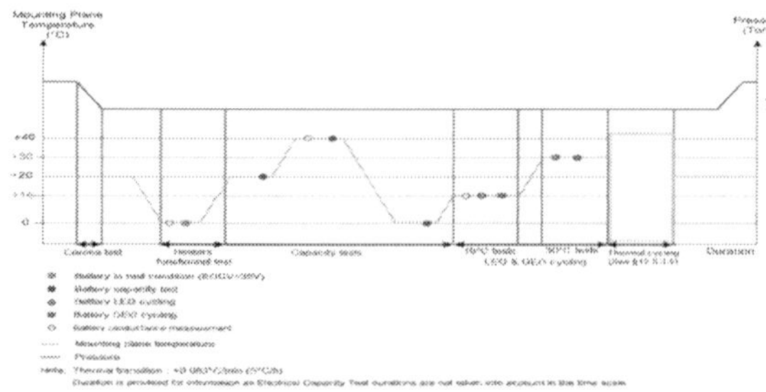

Figure 22: QM2 Thermal vacuum tests sequence

3.2.2.4 QM2 10S5P VES16 – EMC Tests

Following mechanical and thermal environments tests, on this QM2 10S5P VES16 battery, we seize the opportunity of this QM2 battery availability and qualification sequence to characterize battery magnetic moment and battery Electro-Magnetic Compatibility through tests and measurements, in various configurations:

- Magnetic Moment measurement sequence was in line with ECSS-E-ST-20-07C and battery operating conditions described
  Stray field: battery operating, measurement on the six semi-axes, at the reference distances. Characteristics of the battery charge/ discharge rates are:
  - Ich=C/10, Idch=C/1.5 (GEO case)
  - Ich=C/3, Idch=C/2.5 (LEO case)
  - Ich=C/3, Idch=C (radar peak case)
- EMC test sequence was in line with ECSS-E-ST-20-07C and cover cases in table:

Topic	Title
Characterization	System characterization
CS	CS, power leads, differential mode, 30 Hz to 50 MHz
	CS, power & signal leads, common mode, 50 kHz to 100 MHz
	CS, power leads, short spike transients
RS	RS, electric field, 30 MHz to 40 GHz
ESD	Susceptibility to electrostatic discharges - Conducted
	Susceptibility to electrostatic discharges- Radiated
Characterization	System characterization

Table 3: QM2- 10S5P VES16 EMC Test sequence

3.2.2.5 QM2 10S5P VES16 – Final Tests

The leak rate and final functional tests and conformity inspections results allow concluding that the QM2 battery successfully passed all the environmental tests Sequence.

3.2.3 Environmental Tests on QM3 module

As part of the environmental tests, the QM3 module has been submitted to sine vibration (24 g) and random vibrations (11 Grms in lateral axis, and 17 Grms on vertical axis) at qualification levels.

The resonance frequencies were above 1100 Hz, on all axes, which correlate well with the relevant simulations.

In short, the QM3 environmental mechanical sequence has been performed successfully, as shown by battery impedance and final functional tests results.

3.3 LEO Life Testing on QM3 module

LEO life tests have been initiated with the QM3 battery, at the following conditions:

- Real Time LEO, 20°C
- DoD = 30%
- Charge C/3 during 65 minutes up to 4.05 V per cell (or I < 80 mA)
- Discharge C/2 during 35 minutes.

The above conditions were selected to be stringent enough to fit with mostly needed life test qualification conditions; but also to be comparable with existing Saft cell level running life tests.
Relevant Saft Life test, at cell level, is referenced E1-07: 20000 cycles have been achieved on 3 cells and 18000 cycles achieved on 3 other cells.

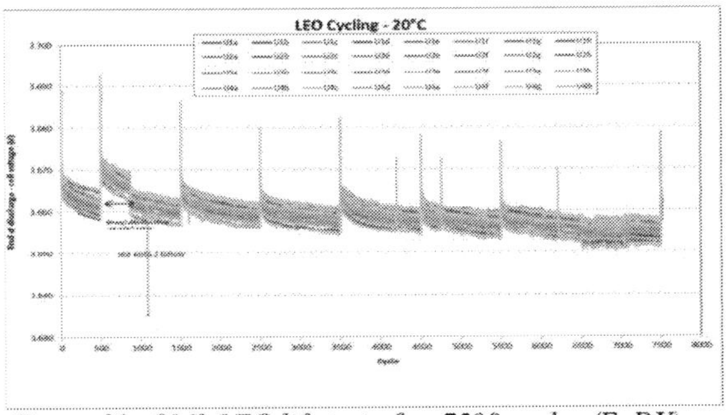

Figure 23: QM3 LEO life test after 7500 cycles (EoDV)

In Figure 23, the EoDVs for the 24 cells of QM3 battery during its LEO life test are shown. Despite the fact that this is only the beginning of the life test, the results obtained to this moment are satisfactory. As it is shown in Figure 30, there is only a 7.5 mV deviation between all cells at 3.65 V, at the end of each discharge.

The current life test is of great interest for potential future LEO applications since the QM3 battery is fully representative of a complete flight system.

4 CONCLUSION

All characterisation and environmental tests carried out on those 3 qualification batteries have been successfully carried out, which allow providing a complete battery designs ranges, qualified to stringent environments and available for the space community.
This activity, supervised by ESA, has greatly helped Saft to reinforce its battery design performances justification.
Today, more than 15 contracts, embedding VES16 batteries, have been awarded to Saft, by all European major primes and agencies both in LEO and GEO projects.

VES16 cells and batteries are cutting edge designed, multi-purpose missions and extended demanding needs, robust, modular, made for long life cycle with space mission required reliability.

5 REFERENCES

[1] D. Prévot, F. Vigier. "Saft VES16 Li-ion qualification synthesis report", Ref. S0115-12, May 2015.

[2] S. Remy, D. Prévot, D. Reulier, F. Vigier. "VES16 Li-ion cell for satellites", 9th ESPC Proceedings, Article 2103465, June. 2011.

[3] E. Mosset. "Active & Autonomous Balancing System" 9th ESPC Proceedings, Article 2103452, June. 2011.

[4] S. Remy, E. Mosset, S; Lefeuvre, M. Nestoridi. "VES16 li-ion batteries range qualification" 10th ESPC Proceedings, April. 2014.

[5] M. Nestoridi, O. Mourra, E. Mosset, S. Lefeuvre, S. Remy " Effect of an induced cell soft-short at a VES16 battery" 10th ESPC Proceedings, April. 2014.

[6] VES16 battery range, GSTP-5 design, Final Data Package, ref. S3561-15

[7] S. Lefeuvre, A. Benkhalfia, E; Mosset, O. Mourra, M. Nestoridi "Battery simulation Tool" 11th ESPC Proceedings, Oct. 2016.

DOI: 10.1051/e3sconf/20171618001

The Global Precipitation Measurement (GPM) Spacecraft Power System Design and Orbital Performance

George Dakermanji [(1)], **Michael Burns** [(2)], **Leonine Lee** [(3)], **John Lyons** [(4)], **David Kim** [(5)], **Thomas Spitzer** [(6)], **Bradford Kercheval** [(7)]

[(1,2,7)] AS&D, Inc. Seabrook, MD 20706 USA
[(3,4,5,6)] NASA Goddard Space Flight Center, Code 563, Greenbelt Road, Greenbelt, Maryland 20771 USA
[(1)]Email:george.dakermanji-1@nasa.gov [(2)], michael.g.burns@nasa.gov [(3)], leonine.s.lee@nasa.gov,
[(4)]john.w.lyons@nasa.gov, [(5)]david.kim@nasa.gov, [(6)] thomas.j.spitzer@nasa.gov, [(7)] bradford.p.kercheval@nasa.gov

ABSTRACT

The Global Precipitation Measurement (GPM) spacecraft was jointly developed by National Aeronautics and Space Administration (NASA) and Japan Aerospace Exploration Agency (JAXA). It is a Low Earth Orbit (LEO) spacecraft launched on February 27, 2014. The spacecraft is in a circular 400 Km altitude, 65 degrees inclination nadir pointing orbit with a three year basic mission life. The solar array consists of two sun tracking wings with cable wraps. The panels are populated with triple junction cells of nominal 29.5% efficiency. One axis is canted by 52^0 to provide power to the spacecraft at high beta angles. The power system is a Direct Energy Transfer (DET) system designed to support 1950 Watts orbit average power. The batteries use SONY 18650HC cells and consist of three 8s x 84p batteries operated in parallel as a single battery.

The paper describes the power system design details, its performance to date and the lithium ion battery model that was developed for use in the energy balance analysis and is being used to predict the on-orbit health of the battery.

INTRODUCTION

The Global Precipitation Measurement (GPM) spacecraft, shown in Fig. 1 is a Low Earth Orbit (LEO) spacecraft launched on February 27, 2014 from Tanegashima, Japan for a three year basic life and five years extended mission life. GPM succeeds the Tropical Rainfall Measuring Mission (TRMM) spacecraft, which was also a joint mission between National Aeronautics and Space Administration (NASA) and Japan Aerospace Exploration Agency (JAXA) to study rainfall for weather and climate research that operated in a 400Km 65 ° inclination orbit. The TRMM satellite mission ended on April 15, 2015 after 17 years of operation.

Since launch, GPM has been providing accurate, frequent global detailed measurements of precipitation to assist researchers in studying global climate, and improving the forecasting of extreme weather events. The spacecraft has two instruments; the Dual-frequency Precipitation Radars (DPR) and the GPM Microwave Imager (GMI). The spacecraft development was funded jointly by NASA and JAXA. NASA/Goddard Space Flight Center (GSFC) was responsible for the spacecraft, the GMI instrument, performed the Integration and Test (I&T) and is responsible for the Mission Operations from GSFC. JAXA built the DPR instrument, provided the launch vehicle and performed the launch services.

The spacecraft is in a circular 400 Km altitude, 65 degree inclination orbit. The GPM orbit and inclination were selected to provide full global coverage updated in a 24 hour period. The beta angle varies during the year approximately -90 and +90 degrees. The orbit period is 93 minutes. The maximum eclipse duration is about 37 minutes. Fig. 2 shows the beta angle and eclipse duration variation during the year. The spacecraft is nadir pointing in the +Z direction and flies in the X direction. To maintain the DPR instrument pointed away from the sun with orbit precession during the year, 180 degrees yaw manoeuvres are performed at around beta zero. The solar array consists of two sun tracking wings with cable wraps. Each wing consists of four panels. One axis is canted by 52^0 to provide power to the spacecraft at high beta angles.

Figure 1. GPM Spacecraft On-Orbit Configuration

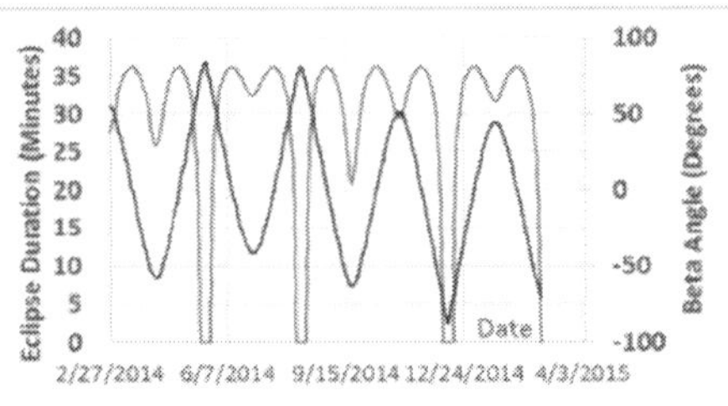

Figure 2. Eclipse time (red) and beta angle (black) variations during a year

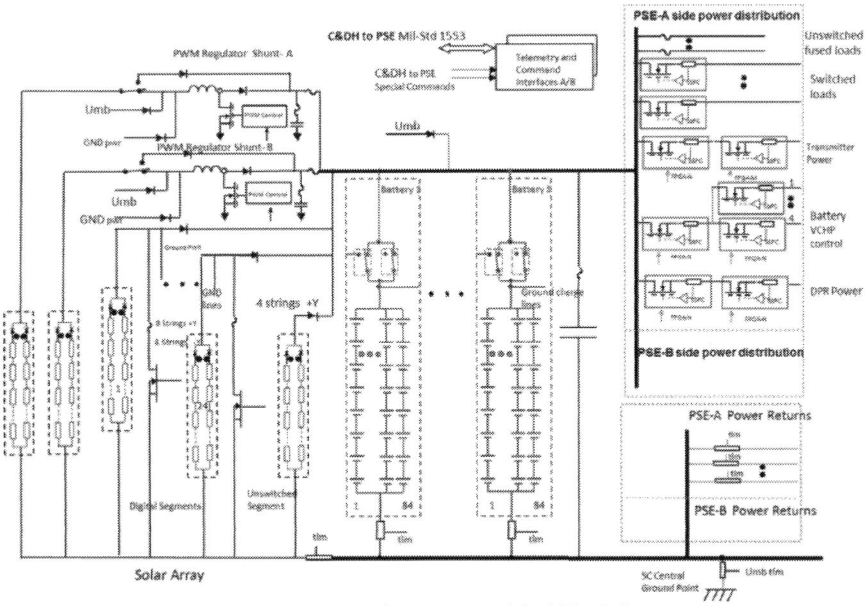

Figure 3. Power subsystem simplified block diagram

Power System Description

The GPM power system is a Direct Energy Transfer (DET) System with solar array and batteries connected directly to the spacecraft power bus as shown in Fig. 3. The battery dominated bus voltage range for essential loads is 21 to 35 V depending on the state of charge of the batteries.

The Electrical Power System (EPS) consists of: The Power System Electronics (PSE), three Sony 18650HC based lithium ion batteries and two solar array wings. The power system topology and PSE hardware elements have heritage to Solar Dynamics Observatory (SDO), Lunar Reconnaissance Orbiter (LRO) and other recent NASA/GSFC spacecraft. The power system is designed to support 1950 watts orbit average power. The solar array wings are divided into segments. Segments from each wing are connected to the power bus in the PSE through isolation diodes in the "Segment Module". A transistor connected from the diode anode to power return is used to shunt the circuit from the power bus when the solar array segment power is not needed to perform "coarse" power control. There are 24 switched segments. The fine solar array control is performed by redundant boost topology Pulse Width Modulated (PWM) converter in the Solar Array Module (SAM). The solar array strings connected to each of the redundant SAM PWM modules are from strings of each wing. The PWM module strings of the flat +Y wing experience extensive shadowing caused by the spacecraft body as well as the spinning GMI instrument antenna dish. The PWM solar cell strings from the +Y wing are placed on the outermost panel that is the last to be shadowed with increased beta angle. The PWM circuit solar array strings from the canted –Y wing, which is always in the Sun, are located on the second panel. The PWM converters operate at a fixed 125 KHz. Only one PWM controller is in control at a time.

1. Power System Electronics (PSE)

The PSE contains the circuits needed to control the solar array power, charge the batteries as well as the electronics for power distribution and current monitoring of spacecraft loads. Fig. 4 shows the PSE simplified block diagram indicating the power services and their current ratings. The PSE is in one box that is internally redundant. Both primary and redundant sides A&B are powered simultaneously during normal operations. The PSE contains the electronics to perform regulation of power bus, battery charging and solar array power control with the PWM SAM and digital segments control modules, monitoring battery voltages, currents and temperatures. It contains the circuitry for spacecraft power distribution using over-current protected switched services using hybrid solid state resettable circuit breakers. The PSE also contains fuse protected un-switched power services. The fuses are placed in accessible Fuse Plugs on the PSE. The PSE interfaces to the spacecraft Communication and Data Handling (C&DH) system on Primary and Redundant Mil-STD-1553 interfaces. In addition, there are Special Hardline interface commands and telemetry lines to the C&DH for recovery from Mil-STD-1553 bus and internal PSE failures. The PSE contains the Umbilical and ground Direct Power Conditioner (DPC) interfaces. The spacecraft Central Ground Point is in the PSE.

The PSE controls the battery charging with current limited selectable levels until the battery voltage reaches a ground selectable voltage limit. The voltage is then held constant while the charge current tapers toward zero.

DOI: 10.1051/e3sconf/20171618001

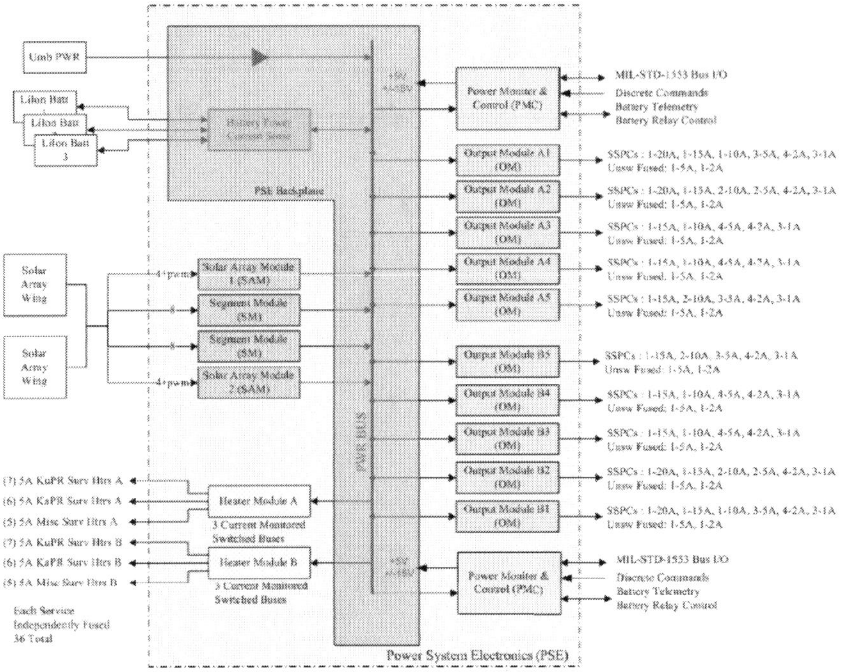

Figure 4. PSE simplified block diagram

The current and voltage limits are set with 12 bit Digital to Analog Converters (DAC)s. The battery current control loop also allows controlled constant current discharge at selectable levels. This feature was used during ground testing to simplify and accelerate battery discharge. In flight, it is used to perform special tests such as battery internal resistance characterization during full sun periods. In addition to the battery current and voltage control loops, a separate power bus overvoltage control loop is incorporated in the design to limit the power bus voltage to safe levels. This protects the spacecraft from inadvertent removal of the batteries from the power bus (particularity during ground testing). This feature also protects the bus from over-voltage conditions in the event of a failure in the voltage or current control loops.

The overvoltage control is performed by a hysteresis controller that switches ON / OFF groups of eight solar array (SA) segment shunting transistors. During the overvoltage control loop operation, the bus voltage oscillates very slowly as the battery charges and discharges with removal or addition of the group of segments depending on loads and Solar Array Simulator (SAS) power settings. Four overvoltage limits are available. The PSE is housed in a single machined aluminum alloy box. The overall enclosure dimensions are 245 mm x 785 mm x 319 mm (HxLxW), the mass is 47.3 Kg. The PSE uses parts selected to a radiation dosage of 10 KRad total ionizing dose (TID) with 2.54 mm Al shielding, but many parts were 100 KRad parts in order to meet the single event effects requirements.

2. Batteries

The GPM batteries were built by ABSL, U.S.A. in Longmont, Colorado, Using SONY 18650HC cells. The battery set consists of three 8s x 84p batteries operated in parallel as a single battery. The nameplate End–of-Life (EOL) battery capacity is 80 Ampere-Hour (Ahr) when tested to an end-of-discharge voltage of 24 V (3 V/cell). I&T batteries, identical to flight units but with cells of a different lot, were used during spacecraft I&T. The flight batteries were installed at the launch site, as is typical in GSFC battery management plans for GPM class spacecraft.

Each battery consists of seven 8s x 12p blocks arranged in two decks; three blocks at lower deck and 4 blocks on upper deck. One block volume at the lower deck, houses the two parallel battery disconnect 100 A Hartman relays, the connectors wiring and power bus pc boards with voltage sense resistors and relay drive decoupling diodes. Each battery has six 2 KΩ Platinum Resistor Thermometers (PRT) mounted on selected cells at the predicted hottest and coldest locations on six of the seven blocks to monitor battery temperature during flight. In addition there are three 2 KΩ thermistors mounted at predicted hottest and coldest cells for ground I&T monitoring. The three batteries are mounted on a single baseplate with heat pipes to maintain low temperature gradients between the three battery cells. Figure 5 shows the photo of one GPM battery. A single battery mass and dimensions is 41.2 Kg and 676 mm x 306 mm x 166 mm.

Figure 5. GPM battery

3. Solar Array

The two articulated solar array wings rotate independently around the Y axis. Cable wraps are used to transfer the power from the rotating wings to the spacecraft. The cable wraps are rewound during the eclipse period. Fig. 6 shows the two deployed wings. The solar cell laydown was performed by EMCORE in Albuquerque, NM, USA using their ZTJ cell with nominal 29.5% efficiency.

The GSFC provided panel substrates were aluminium honeycomb core with composite facesheets. The cell side was co-cured with 50 µm Kapton insulator sheets. The backside of the panels was painted with A-276 white thermal paint. Protection of the Kapton insulator and the silver plating on cell interconnect from Atomic Oxygen (AO) was required.

Figure 6. Solar array deployed wings. -Y top, +Y bottom

AO is prevalent at the 400 km altitude orbit and was predicted to be 1.464 x 10^{22} atoms/cm^2 during the mission. AO erodes the silver converting it into silver oxide and destroys the Kapton insulator exposing the conductive composite facesheet. Prior to cell laydown, GSFC thermal coating personnel sprayed a few microns of low viscosity room temperature vulcanization (RTV) silicone AO protective overcoat CV1144 on all Kapton side panel areas that are expected to be exposed after cell laydown using cell patterned masks. AO converts a very thin layer of the silicone RTV into SiO_2 which protects the Kapton from further erosion. Test on a coupon that was sprayed with CV1144 and then populated with solar cells and exposed to atomic oxygen environment verified and qualified the concept's use on the flight substrates.

EMCORE ZTJ, 140 µm thick, 76.1 x 37.16 mm, InGaP/InGaAs/Ge triple junctions, n/p implementation cells were used. A discrete silicon bypass diode on each cell is placed in one of the corners. The cover glass is 76.25 x 37.31 x 0.1 mm, 0214 glass (CMG equivalent) with anti-reflective (AR) coating from OCLI. The cells were interconnected with silver plated Kovar 2-toe interconnects.

The silicon bypass diode protects the cell from reverse voltages expected during the mission and during fabrication and handling of the panels and I&T due to shadowing. Strings of 20 cells in series were used on the PWM strings, the switched segment strings had 19 or 20 series cells per string depending on the predicted temperature at their location on the panels.

The expected charged particle radiation dosage during the five-year extended mission was estimated to be 8.6E+11 equivalent 1-MeV electrons/ cm^2.

There are two 2 KΩ PRTs on each –Y wing panels and one on each +Y wing panel. Each PRT is bonded to the cell side composite facesheet through a hole made from the panel back side.

Two reference cells for V_{oc} and I_{sc} are placed on the third –Y wing panel counting inward to spacecraft. They are used to evaluate cell degradation during the mission. The two PRTs on that panel are located near the reference cells. There are 218 and 228 strings on the +Y and –Y wing panels respectively.

Spacecraft Instrument Operation Transients Effect on Batteries Investigation

At initial DPR instrument integration with the spacecraft at GSFC, unexpected high current transients were noticed on the DPR power lines and the batteries when the DPR was operated. A typical scope current wave shape in one the three batteries is shown in Fig. 7.

Investigation revealed that the DPR phased array radar generates cyclical high rate large current transients on the spacecraft power bus. The interaction of the DPR high current transients with the limited bandwidth PSE operation lead to the battery transient current profile observed. The pattern repeated at 700 milliseconds with the low amplitude component frequency of the transients being around 75 Hz.

Small discharge/charge battery cycling will occur during full sun periods when the batteries are fully charged with zero charge current and during short eclipse orbits where the charge current tapers to very low values.

The small discharge/charge cycling impact on the battery was a concern and was investigated [1].

An accelerated life test was conducted using two GPM I&T batteries. One battery was used as the test battery, the second as the control.
A worst case piecewise linear profile simulating the observed wave shape consisted of four ±3 A square wave pulses, followed by ±0.5 A square wave for about 30 cycles. The pattern was repeated at 10 Hz rate. The two batteries were maintained at room temperature during the life test. Periodic capacity tests were conducted. Tests over a 15 month period did not indicate measurable difference between the reference and test batteries. The test profile was changed to a ±2.5 A 100 Hz test. Capacity test after six month with the new profile did not indicate noticeable difference in degradation due to the micro discharge/charge cycling either indicating that there should be negligible impact on GPM batteries life due to the DPR transients.

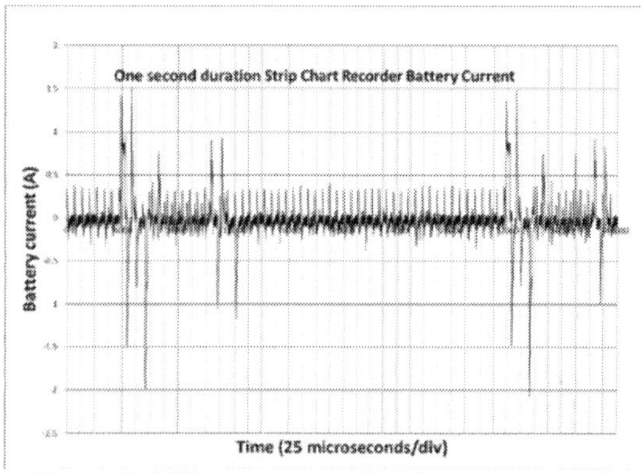

Figure 7. DPR instrument induced currents in each GPM battery

Power System Orbital Operation

The actual orbital load power of the spacecraft components particularly the instruments and spacecraft heater power was considerably lower than the 1950 W orbit average design specification. Orbital data indicated a power demand of around 1250 W at low beta and around 1350 W at high beta angles with a corresponding maximum eclipse battery Depth-of-Discharge (DoD) of around 10%.
To take advantage of the low spacecraft power with its resultant low battery DoD, the operational charge voltage was lowered to reduce the stresses in the battery; the battery operation was at 33.6 V at launch and during the very early period after launch, It was lowered to 33.1V (around 90% State-of Charge (SOC)). It was further reduced after about one year to 32.9 V (around 85% SOC). Since September 2015, it is operating at a voltage limit of 32.7 V corresponding to about 4.1 V/cell (around 80% SOC). The batteries still store ample energy for emergencies at this lower SOC.
On board software for solar panel sun-pointing control was modified to reduce the drag caused by the large solar panels during flight by feathering the two wings alternatively depending on beta angle variations; the +Y wing is made to track between beta angles 0 to ± 20 while –Y wing is feathered with its panel edges in the velocity vector direction, while for beta angles between ±20 to ±55 degrees, the +Y is feathered with its edge in the velocity vector direction while +Y wing is tracking the sun, then for beta angles above ±55 degrees, both wings are feathered. Feathering the wings minimize the projected area in the velocity vector directions and lowers substantially the drag on the spacecraft which reduces the number of propulsion burns required to maintain the spacecraft in the desired altitude. Feathering also reduces the power generated by the solar array that must be processed by the PSE leading to several degrees reduction in PWM controller transistor and diode temperatures. In addition, feathering require fewer solar array drive movements thus less stress and wear on solar array drive system. The reduction in propulsion maneuvers for altitude adjustment has resulted in considerable savings in propulsion fuel to date. It will allow further mission extensions.

Solar Array orbital Performance

Fig. 8 show the observed variation of V_{oc} and I_{sc} to date adjusted for temperature, solar distance and angle. The 28℃, one Sun AM0, normal sun incidence measured values by the cell vendor of I_{sc} and V_{oc} were 0.4762 A and 2.71 V respectively.
Minimal degradation is indicated. The small reduction in I_{sc} value observed is likely due to UV radiation degradation of the cover glass adhesive. The V_{oc} results indicate also minimal degradation. The low degradations observed are expected due to the low predicted charged particle radiation environment of the spacecraft orbit and the low activity of present solar cycle.
Solar array power analysis from flight data indicate that there have been no string losses to date. Telemetry resolution is not adequate to estimate individual cell short circuit failures.
Fig. 9 and Fig. 10 show the minimum and maximum temperatures during an orbit of the +Y and -Y wing panels observed to date. The temperatures extremes are less than the thermal predictions.

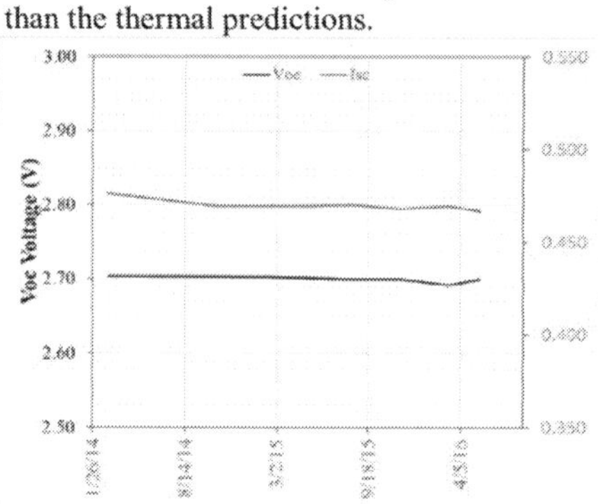

Figure 8. V_{oc} *and Isc variations during the mission*

Battery Orbital Performance

The end of charge/discharge battery voltages are shown in Fig. 11 along with eclipse duration variations. The plot shows the charge voltage limit changes implemented. The voltage limit is occasionally raised to 33.6 V during full sun periods to perform special characterization tests and for special spacecraft operational scenarios that require full battery SOC. The battery baseplate temperature is controlled by the C&DH software. Fig. 12 shows data from nine PRT sensors on the three batteries during two orbits duration. The low DoD and good thermal control maintains temperature excursions during an orbit to around 1 ℃. The temperature gradient between the nine temperature sensors on the three batteries at any time during the orbit is around 0.5 ℃.

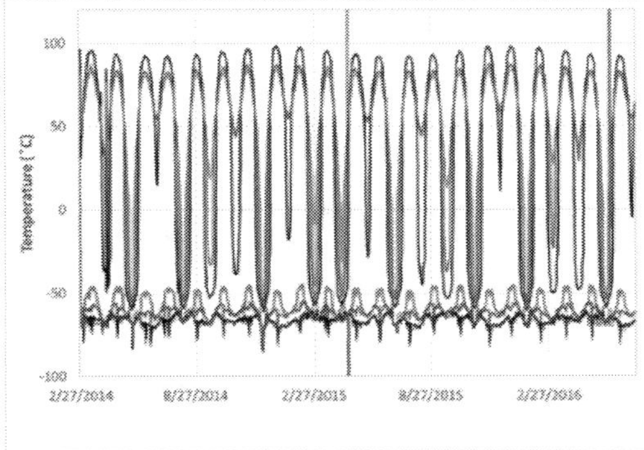

Figure 9. Plus Y panels minimum and maximum temperatures to date

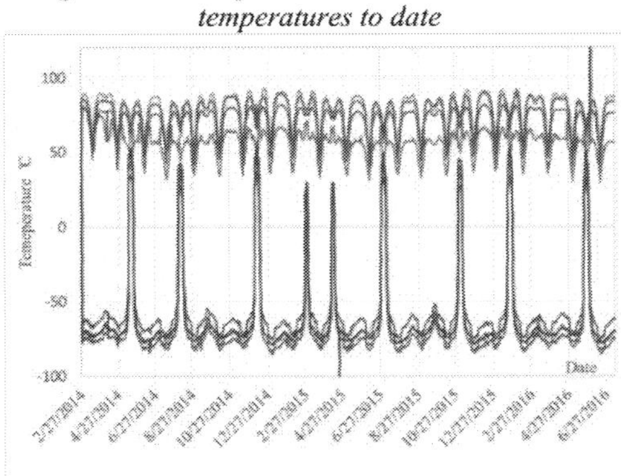

Figure 10. Minus Y panels minimum and maximum temperatures to date

The three GPM battery cell temperatures vary between 12^0 and 14.5^0 during the year depending on the beta angle and sun distance variations. Colder temperatures occur during full sun periods when there are no battery discharges.

Figure 13 shows the orbit average temperature variations of battery-1, the plots for the other two batteries are identical.

Battery Modelling and Power System Energy Balance

A detailed GPM energy balance program was developed which included the lithium ion battery model, the solar array model, Sun to panels orbital angle variations, solar array temperature, solar array shadowing effects, power system solar array/battery control electronics and load models.

The energy balance program was used as part of the solar array sizing analysis, verification for EOL worst case conditions, predicting the battery and power system performance for various mission operation scenarios and the impact on the available spacecraft power of various fault scenarios. It is presently used for battery health

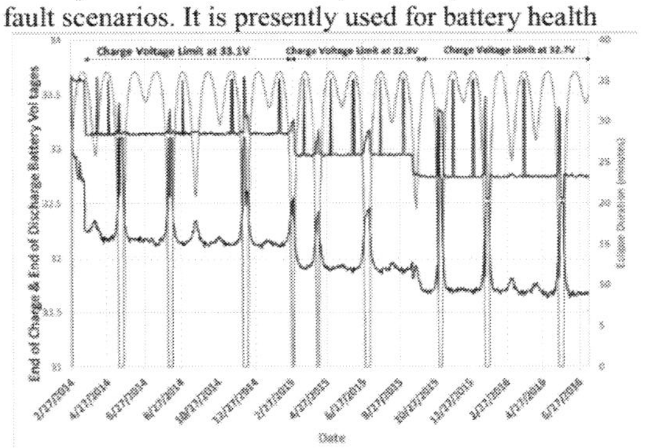

Figure 11. Battery End of Charge & Discharge voltages and eclipse durations to date

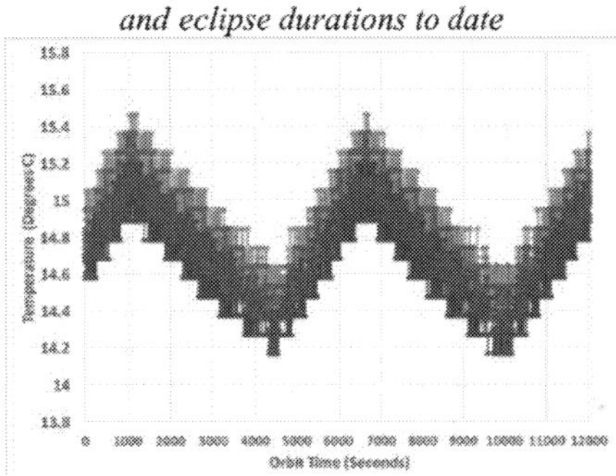

Figure 12. Typical temperature gradients between the cells of the three batteries during an orbit

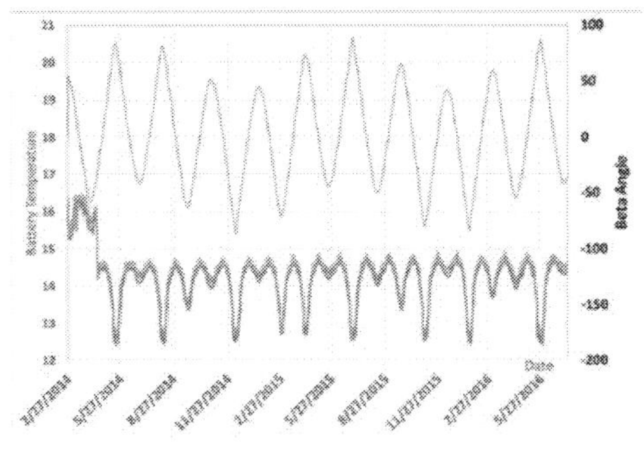

Figure 13. Orbit average battery-1 temperatures with beta angle variations

evaluation, orbital power operation scenarios, and solar/lunar eclipse effects.

The energy balance program is written in Excel.

The model development used for GPM follows ideas from [2,3,4] (G. Dudley & all) but instead of using individual cell positive and negative electrodes test & modeling data for cell capacity-Ahr. The GPM model uses measured battery voltage and normalized state-of-charge data, characterized at very low charge/discharge rates along with measured internal resistance values derived from flight or test data.

Figure 14 shows a measured GPM battery voltage versus normalized SOC plot for a 33.6 V (4.2 V/cell) battery case showing the hysteresis. The tests were conducted on GPM Test and Flight Spare batteries both 8s x 84p SONY 18650HC batteries maintained at 20 °C. The charge and discharge rate used was 0.8 A. The measured Beginnin-of-Life (BoL) capacity at 20^0C was around 109 Ahr considering 24V (3 V/cell) as the 0% SOC. SONY HC cells exhibit minor loops [2, 3]. Battery minor loop characterization over GPM operating range was performed and the normalized minor loop voltage-SOC are used in the model.

Experimental tests on SONY 18650HC cell blocks and batteries indicated that with aging and over typical spacecraft operational battery temperature range the relationship of voltage to normalized SOC remain approximately the same especially in the linear range below above 55% SOC.

Figures 15 and 16 show typical GPM orbital flight data and model results of battery voltage and total battery assembly current of the three batteries.

For energy balance model purposes, the battery internal resistance R_{in} consists mainly of R_{dc} and R_p. R_{dc} is the resistance in the battery due to wiring, cell current collectors and electrolyte resistances. This resistance can be obtained from measurements or orbital data of voltage/current at step charge/discharge transitions after taking into account the hysteresis voltage at the corresponding SOC point.

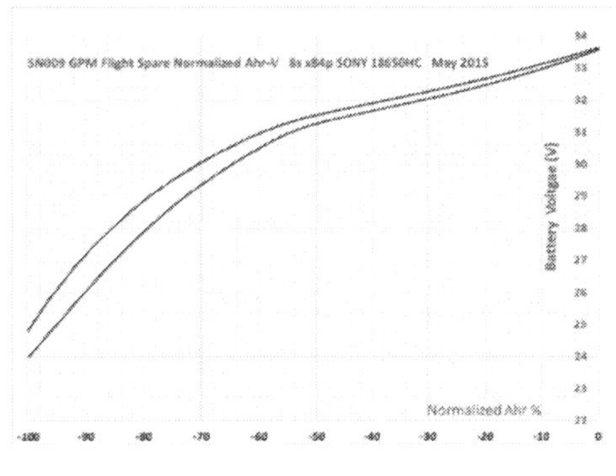

Figure 14. Battery Voltage v.s. Normalized SOC

R_p is the chemical ionic polarization/diffusion resistance due to lithium diffusion in the electrodes. R_p is time varying that typically takes 10 to 15 minutes to settle. The total internal resistance $R_{in} = R_{dc} + R_p(t)$

R_p is assumed in the GPM model, based on experimental data, to vary to a first approximation as $R_p(1-\exp(-\frac{t}{\tau}))$, where τ is the time constant and R_p is the steady value of R_p. It is derived from the measured/derived R_{in} and R_{dc}.

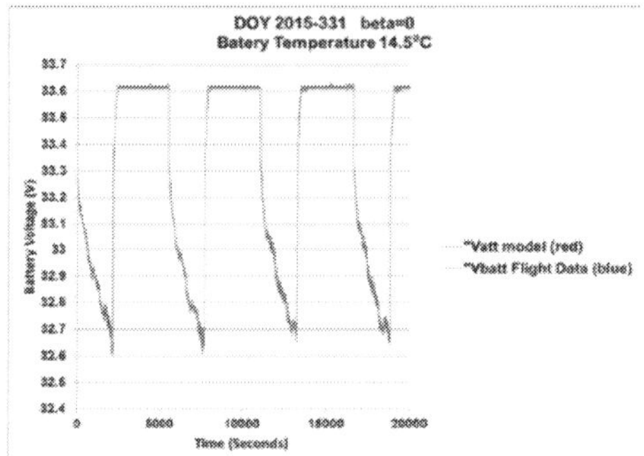

Figure 15. Flight battery voltage and model results

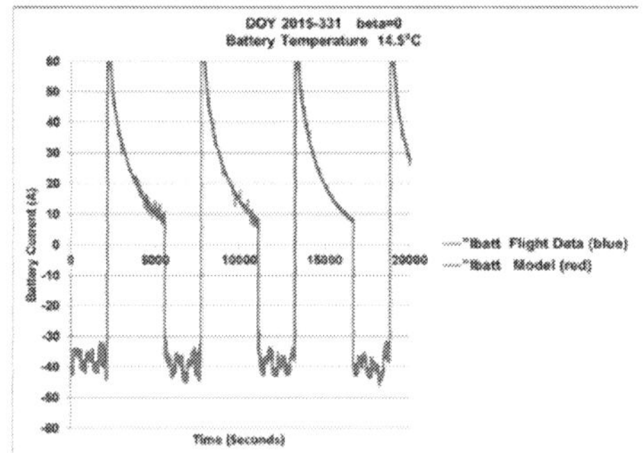

Figure 16. Flight battery assembly current and model results

The battery internal resistance depends on temperature, aging and SOC [5]. In the region above 50% SOC the variations are almost linear. For GPM analysis with its low DoD and very small battery temperature variations during an orbit, the resistance can be assumed constant in its operation range. The model can be made to accommodate varying internal resistance with deeper SOC and temperature if needed. The total internal resistance is derived from cycling data looking at the region after the transient settling period for R_p, taking into account the hysteresis effects or by discharging the battery at different constant current rates and calculating the resistance from the voltage difference data.

The lumped total resistance derivation, the resistance measurements and methodology accuracy verification were conducted in the GSFC Battery Lab by performing charge /discharge cycling and deriving the R_{in} as done with flight data. The resistance measurements were also done by discharging the battery at different rates and calculating the lumped R_{in} as delta V/ delta I at the same discharged Ahr values.

The battery internal total resistance value in the model are derived from flight data. The capacity in the model is adjusted to best fit the model results with the flight

DOI: 10.1051/e3sconf/20171618001

battery voltage and current data.

The internal resistances of the GPM batteries are very small and the flight current data are noisy making accurate measurement difficult. The resistance values used are adjusted for model-flight data best fit. The technique predicts good results when done using ground life cycling data where there is minimal noise on the measured data used.

Figure 17 shows the variations to date of R_{in} derived from flight data along with model predictions of the percent capacity remaining in the batteries. The data used in the plots were from orbits with beta angles near zero corresponding to longest eclipses in order to compare data from orbits with similar load and thermal environment. The first data point was for a battery temperature of about 17.5˚C. The rest were at 15˚C

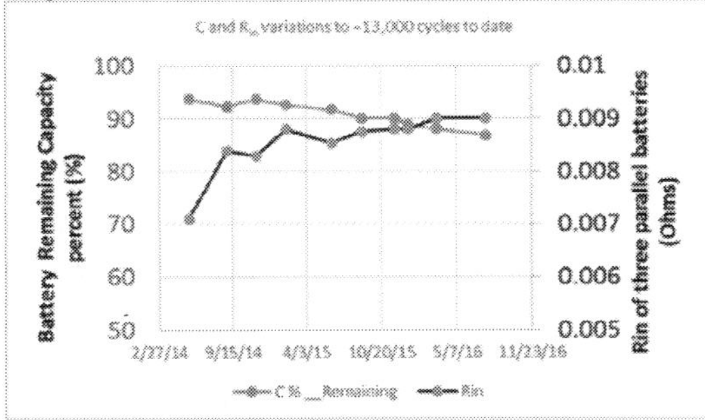

Figure 17 Model battery capacity degradation predictions and Rin (three batteries) variations to date

The model is also being used to predict effects of special operations/manoeuvres including effects of eclipses by the moon. Figure 18 is an example of battery voltage model results and orbital data of a recent lunar partial shadowing during three orbits and the corresponding model predictions. Lunar shadowing percentage is also shown.

The system level information required for the model are: battery charge voltage & current limits, orbital beta angle, battery & solar array temperatures and day of year for sun-earth distance. Only the instantaneous orbital spacecraft power is needed for the model to calculate the battery voltage, current and SOC. The start of eclipse discharge is used in model calculation as the starting point with initial values of voltage and SOC.

Battery on-orbit internal resistance characterization is also being performed during full sun periods by utilizing the PSE capability for setting the charge current to a negative value. After stabilizing the taper current to zero at a selected operating voltage, the batteries are discharged at constant rates starting from same SOC. The internal resistance is derived as delta V/delta I.

The R_{in} characterization were conducted during recent two full sun periods about 3 month apart. The differences were within measurement uncertainties during the period.

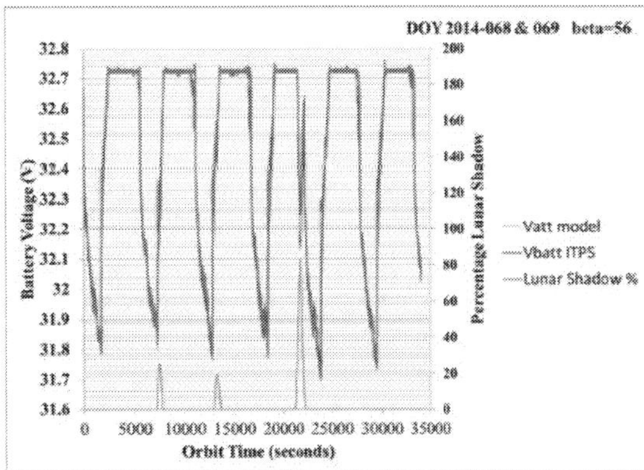

Figure 18. Battery voltage flight and model results with lunar shadowing profile

Summary and Conclusions

The GPM spacecraft has been providing important science data for global weather monitoring and predictions. The power system performance has been excellent. There is ample solar array power and the battery DoD is low. Operational changes implemented reduced the propellant use and the stresses in the battery and electronic components. To date there have been two –Y wing panels PRT temperature sensors that have been intermittent. These two PRTs provide data as telemetry information that are not critical for power system operation. The power degradation in the battery and solar array have been less than predictions. The power system is expected to support the mission for substantially longer period than the baseline extended mission plan.

REFERENCES

1. Dakermanji, G. & all (2016). Global Precipitation Measurement (GPM) Spacecraft Lithium Ion Battery Micro-Cycling Investigation. NASA Aerospace Battery Workshop (planned in November 2016)
2. Dudley, G.J. (2003). A Simple Electrochemical Model of a Lithium-Ion Cell. NASA Aerospace Battery Workshop
3. Dudley, G.J. & all. (2005). Electrical/Thermal Models of a SONY 18650HC Li-Ion Cell. Proceedings of the Seventh European Space Power Conference. , Stresa, Italy. ESA SP-589, ESA, Noordwijk, Netherlands.
4. Dudley G. Twaite C. Ferretti R. Walpole A. (2006) Use of Electrical/Thermal Cell Model for Mars Express Battery Health Assessment , 2006 NASA Aerospace Batter Workshop.
5. Twaite C. LIon-AEA-0009_05 Modelling Li-ion Battery Internal Resistance.AEAEngineering Note.

E3S Web of Conferences 16, 18002 (2017) DOI: 10.1051/e3sconf/20171618002
ESPC 2016

MODULAR AND DECENTRALIZED PCU

F. Bonnet [(1)], L. Rivière [(1)], C.Elisabelar [(1)], J. Chirat [(1)] ,S. Brun, H. Gras, J. Domingo-Salvany [(2)]

[(1)] CNES, francois.bonnet@cnes.fr, laurent.riviere@cnes.fr, christian.elisabelar@cnes.fr
[(2)] NEXTER Electronic, h.gras@nexter-group.fr, j.domingo@nexter-group.fr

ABSTRACT

This paper presents the studies made on the definition and design of a decentralized and modular electrical architecture of Power Conditioning Unit (PCU).
The modular PCU is able to cover a large range of mission demands by adapting the number of power modules (PM) while the electrical interfaces remain the same.
A decentralized architecture is proposed where each module is able to control the solar arrays and to manage the batteries. It appears that this kind of architecture becomes feasible thanks to digital circuits and using a communication bus [1].
Breadboards are being tested in order to validate the concept.
Reliability and robustness aspects are studied and a redundant architecture is tested.

1. CONTEXT

Studies made by CNES and TAS from 2008 to 2014, focused on a modular PCU based on BUCK solar array regulator with MPPT control [2]. This architecture is composed by N+1 power modules which control the solar array, manage the battery and use a reliable digital bus for communication. This solution is compatible with unregulated or regulated busses, and the MPPT can address very large needs in spacecraft power supplies.
CNES decided to go on a second way by considering DET (Direct Energy Transfer) topologies for solar array regulator, unregulated bus and the possibility to implement remote power modules. For these reasons, this architecture is called Ultra Modular PCU. The trade-offs made for the architecture and the electronic design are more adapted for low cost applications or micro-satellites.

2. ARCHITECTURE

2.1 Architecture general considerations

The architecture is based on remote power modules connected on the same primary Unregulated Power Bus (UPB). Each module controls one or several Solar Array sections and is connected to one or several battery modules (Figure 1 and 2). A digital bus interconnects all the modules via control circuits, and is directly connected to the On-board Computer (OBC).

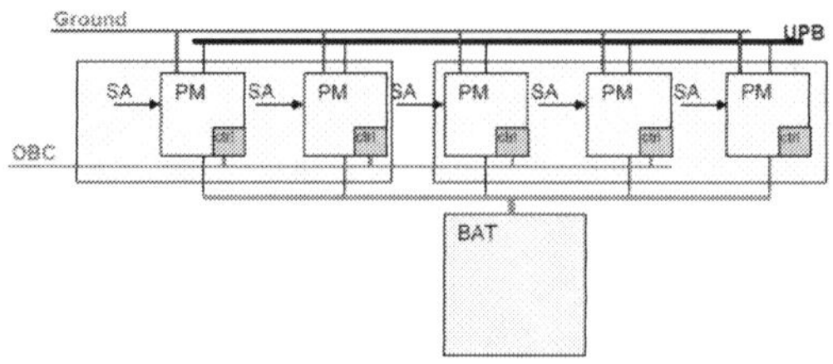

Figure 1. PCU Architecture with a single battery.

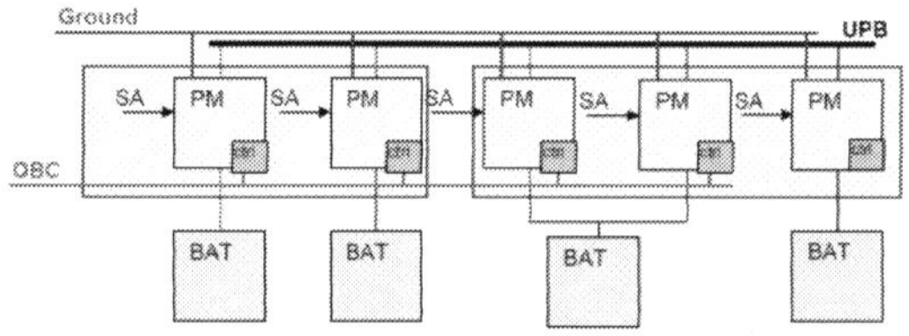

Figure 2. PCU Architecture with battery modules.

Lots of possibilities exist for the PM topology. In our objective to build a modular PCU, the elementary PM is sized for 8A to 10A which allows supplying a microsatellite and with 10 PM in parallel a big LEO spacecraft can be supplied.
For example the figure 3 shows different topologies studied.

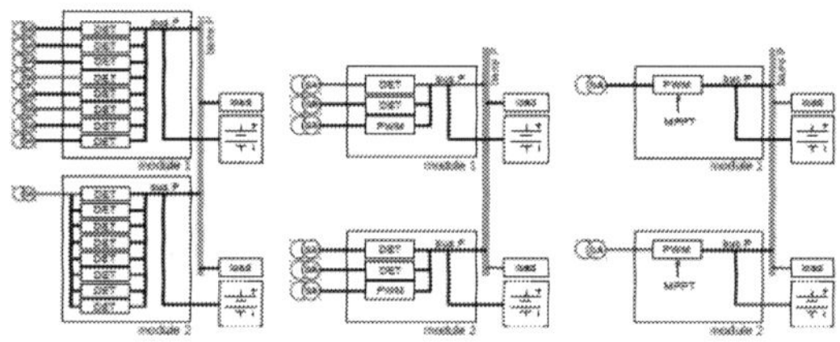

Figure 3. Some DET topologies assessed.

The first one is composed of 8 SA sections with the possibilities to connect in parallel some or all of them for modularity and adaptability aspects. Each section is sized for the total PM current divided by 8.
Second topology uses 3 SA sections, 2 DETs and one PWM.
The third one is a DCDC converter which can operate at MPPT.
We want to keep the possibility to have remote PMs and then solutions with PWM DCDC converters (second and third examples) were not chosen because of EMC

DOI: 10.1051/e3sconf/20171618002

perturbations induced associated with the difficulty of multiple groundings of PMs.
The choice of DET with low frequency control is preferred and much more compatible with PM implemented on different places in the spacecraft.
DET topologies suppose that the SA is decomposed in several sections. This assumption is well verified because the electrical network of a SA is always made of strings with a maximum current of 2.5A (with big size solar cells) and the elementary PM that we want to size is with a minimum of 10A and then 4 SA sections are needed.
The option with 8 DET is complex because the accommodation of the SA sections (parallelization) needs to configure connectors. This solution gives some constraints for the synchronization of the DET commands.

2.2 SA regulator topologies

The main requirements of the PM are:

- Control the operating point of the SA,
- Insure the power transfer to the battery and the spacecraft,
- Battery current and voltage regulation,
- Electrical passivation at the end of service by disconnecting the power source,
- Connect the battery to the power bus,
- Insure the full performance operation after one single failure,
- Acquisition of voltages and currents of the PM,
- Digital communication with the On Board Computer,
- ON or OFF launch.

The power of each module is also considered. The ratios of the different average powers from SA, Battery and output delivered power are summarized on table 1.

P_{GS}/P_{sl}	2.5 à 3
E_{batt}/P_{sl} [s]	4 à 5
E_{batt}/P_{GS} [s]	2
E_{batt} [Wh]	$4 \cdot V_{batt} \cdot I_{sl}$

Table 1. Power ratios

An average output delivered power (Psl) of 100W is chosen, which give a SA power (Pgs) from 250W to 300W. The global SA current is then from 7A to 13A on a Non-Regulated Bus standard of (24V – 38V).

Several topologies were assessed and finally 3 of them are short-listed for the trade-off:

- 6 configurable SA sections (Figure 5),
- 6 configurable SA sections with an additional one for redundancy (Figure 6),
- 3 reliable SA sections (Figure 7).

Shunt DET is preferred instead of serial DET (figure 4), considering transistor command (floating on serial DET) and failure cases are more advantageous for shunt DET.

Figure 4 : Shunt DET and Serial DET

The current sharing in case of a configurable DET presents some drawbacks on the MOSFETs sizing and on their commands.

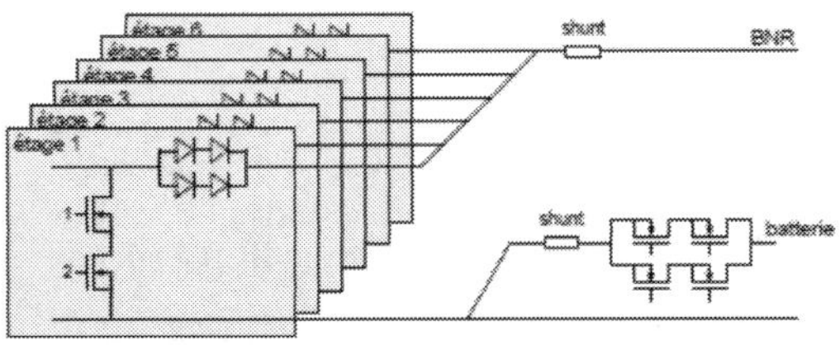

Figure 5 : 6 configurable SA sections

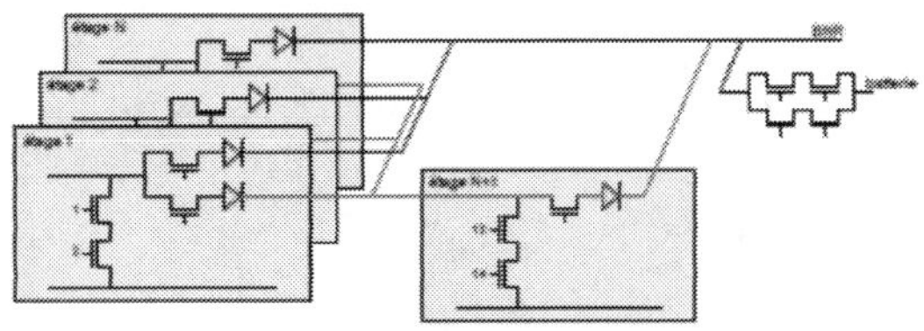

Figure 6 : 6+1 configurable SA sections

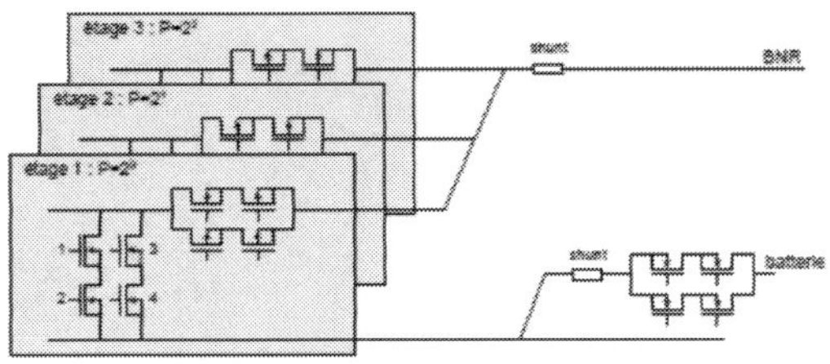

Figure 7 : 3 Reliable SA sections

On the topologies of figures 5 and 6 each SA section is sized by the current of the section, while in topology of figure 7, each SA section is sized for the global SA current delivered to the PM. 6 elementary DETs are compared with 3 oversized DETs.

2.3 Control

The key circuit for the control of the PM is a microcontroller. We take the option where the microcontroller is not fully rad hard and has some sensibilities regarding radiation environment. It is the

 DOI: 10.1051/e3sconf/20171618002

case of some components which were evaluated at CNES and remain tolerant for cumulative dose but have some internal parts of the circuit sensitive to SEE. In order to cope with this situation and thanks to our experience on microcontrollers used in flight the following precautions are implemented:

- Corruption of the memory: the firmware rewrites sequentially the program memory, the program file is written in an EEPROM not sensitive, the microcontroller restarts in case of external watchdog triggering.
- Latch-up protection circuit is added with current protection and restart of the microcontroller.
- For the cumulated radiation dose, the components shall be rad tolerant.

If a single event effect occurs, even if the design can recover the microcontroller, the operation is interrupted during a few milliseconds. The global command of the PMs switches shall not be disturbed during these recoveries and then specific circuit shall maintain the correct statuses of the power parts.

Reliability and safety shall be also taken into account.

To cope with these considerations some hardware topologies are proposed:

- 3 voted microcontrollers (Figure 8),
- 5 voted microcontrollers,
- 2 microcontrollers with reliable multiplexed DET commands with or without memorization (Figure 9),
- One microcontroller per DET switch,
- Full rad hard microcontroller.

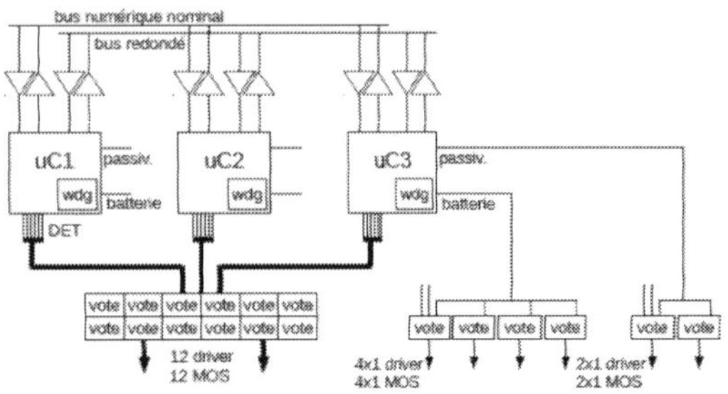

Figure 8 : 3 voted microcontrollers

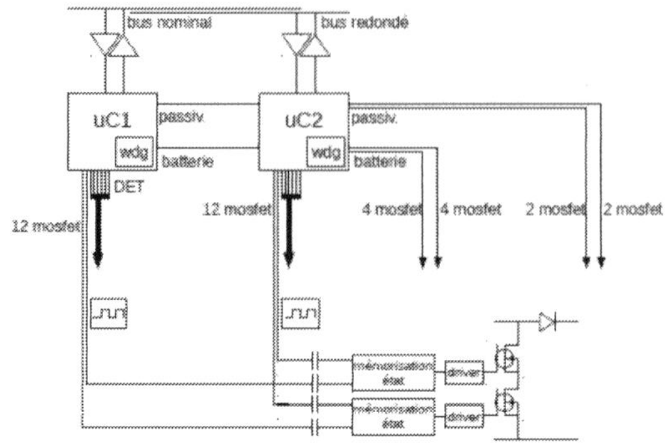

Figure 9 : 2 microcontrollers with DET commands memorization

Erroneous command (bit-flip) is also considered. This case is particularly a problem when several SA sections are connected together in parallel (case of configurable SA sections), the full SA current of the PM can run through one section. This failure case induces to oversize each SA section.

2.4 PM topologies trade-off

The following criteria are used for the trade-off:

- Volume,
- Number of connectors,
- Maximum delivered output power of the module,
- Efficiency (or dissipation),
- Number of power components mounted on module structure,
- Cost,
- Management of bit-flip, latch-up, CC drain-source of the MOSFETs, CC drain-gate of the MOSFETs, CC gate-source of the MOSFETs.

For the sizing aspects, it is considered that the maximum dissipated power per power component (MOSFET, diode) mounted on PCB is 500mW and when mounted on module structure 10W.

The PM topology selected is the 3 reliable SA sections in shunt configuration, with 2 redundant microcontrollers and command memorization.

3. BREADBOARD DESCRIPTION

3.1 Global breadboard description

The breadboard is composed by 3 PM modules connected together with 9 SA sections and one battery as shown on figure 10.

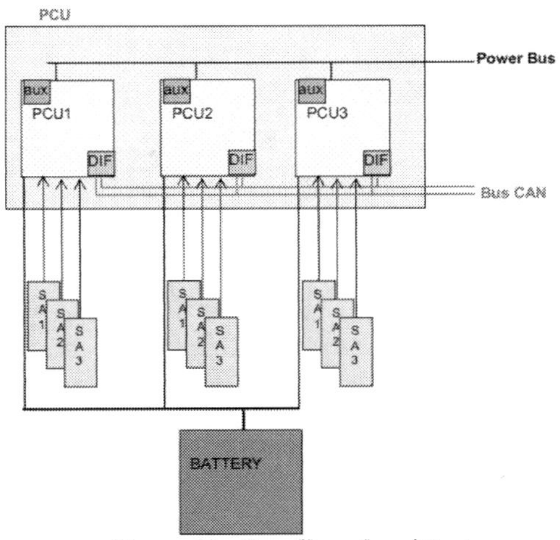

Figure 10 : Breadboard architecture

3.2 PM hardware

The synoptic of the PM breadboard is given in figure 11 and compliant with the topology selected at the trade-off.

Each shunt switch is composed by a quad of MOSFET where each branch is controlled by the same microcontroller (branch A (B) controlled by microcontroller A (B).

On the breadboard the serial switch is composed of a quad of diodes. Other solutions using MOSFETs are studied and could be preferably implemented but not for this breadboard. The main objective here remains to test functional aspects instead of efficiency performances on real power condition.

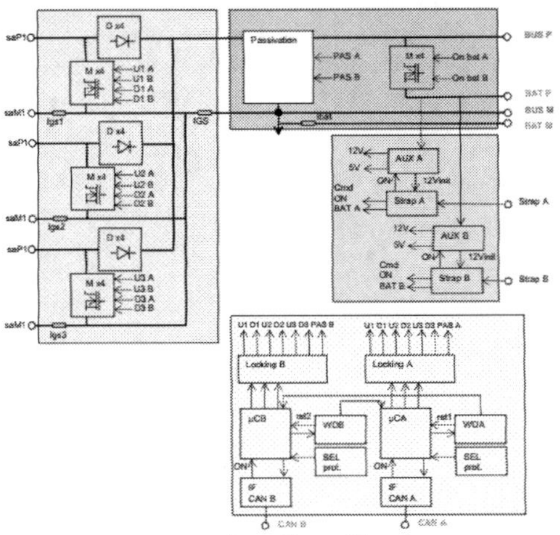

Figure 11 : PM breadboard hardware synoptic

The connection with the battery or battery module is performed by a quad of MOSFETs.

A specific passivation circuit allows disconnecting the SA regulator from the other parts of the PM.

Redundant auxiliary supplies give the secondary voltages to supply all the module circuits.

A strap detection circuit maintains each auxiliary supply OFF during launch (in launch OFF configuration) and at the same time gives polarisations to shunt all the DETs. When the separation is detected, the auxiliary supplies start nominally, and switch ON the battery then the PCU starts.

The global control of the power module is done by 2 redundant digital circuits. Each one is composed of a microcontroller, a rad-hard watchdog, a latch-up protection, a rad-hard locking device and a CAN transceiver.

The breadboard is reduced power model with a bus voltage of 30V and a SA current per section of 100mA.

3.3 PM Software

The system operates in hot redundancy where the two chains execute the same firmware.

Some of the parameters used by the PCU software are stored in EEPROM memory: ID of the PCU module, battery voltage setting for the regulation algorithm, maximum battery current allowed, allocation table used by the sections allocation algorithm, array with the address of the sections grouped in the allocation table.

The initialization process selects the clock frequency, the inputs / outputs, initializes the operational mode, loads the EEPROM parameters, initializes the CAN bus functions, the ADC, the timer, the regulation algorithm, and the commands.

The PCU can be in different working modes. A working mode corresponds to a sequence of actions which are executed each 1s.

- Operational mode is the main mode of the PM. Acquisitions are shared between the PMs, the regulation is working and OBC commands are processed. This working mode is automatically operational after starting or after a reset.
- Passivation mode is used to receive and start arming/passivation commands. In this mode, all the SA sections are commanded shunted.
- Failure mode is used when an error occurs mainly during EEPROM parameters reading. In this mode, all the DET shunt switches are opened to let the redundant line operate.

DOI: 10.1051/e3sconf/20171618002

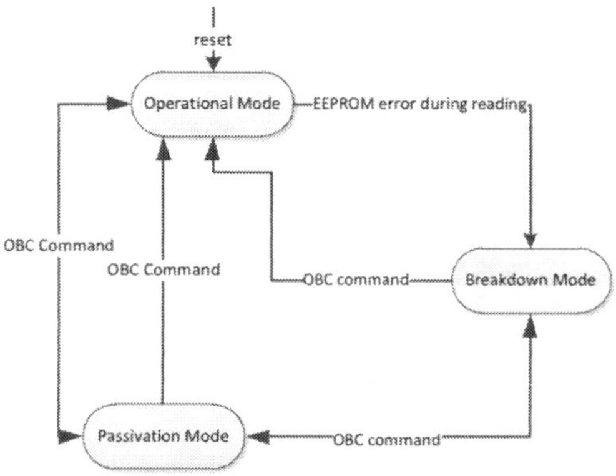

Figure 12 : Software modes

Operational mode is composed of 2 parts:

• The regulation process: acquisitions are done and shared between all the PMs. The regulation algorithm is executed and the sections are closed or opened using the allocation table. Service messages are sent (PM statuses, sections currents and voltage values).

• The commands process: Commands from the OBC are monitored and processed.

Failure mode is composed of 2 parts:

• The acquisition process: acquisitions are done and shared between all the PMs. All DET shunts are opened to let the redundant line performing its actions. Then, service messages are sent.

• The commands process: Commands from the OBC are monitored and processed.

Each PM executes its sequence of actions every 1s. This period of 1s is defined by a timer which is used by the PM as a counter. The timer is based on the internal oscillator of the microcontroller which works at a frequency of 8MHz. Because the internal oscillators of the PMs are not perfectly equal, a way to synchronize the PCU is needed. A specific algorithm combined with the CAN bus protocol is implemented to insure the synchronization of all the PMs.

The regulation algorithm computes the current setting to be provided by the solar generators. It also implements the allocation algorithm in charge of determining the set of SA sections to be activated to reach the current setting computed by the regulation algorithm.

The regulation algorithm is composed of 2 loops:

• The internal regulation loop is in charge of the regulation of the battery current delivered by the all PMs. This regulation is done by an integral corrector. Anti-windup scheme is used to prevent integration wind-up: limitation of the error during eclipse period for example.

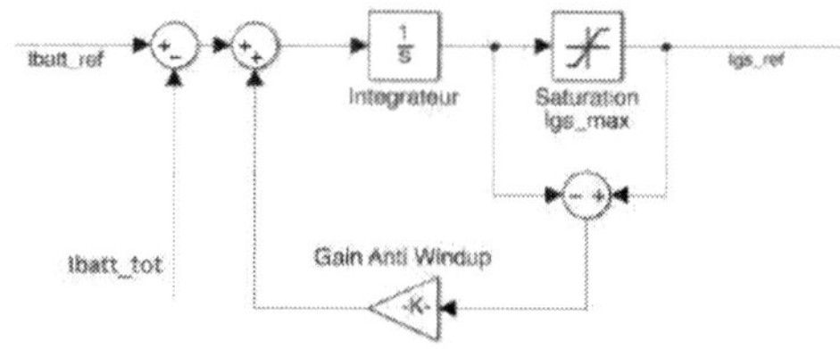

Figure 13 : Internal regulation loop

• The external regulation loop is in charge of the regulation of the battery voltage. This regulation is done by a proportional-integral corrector. The following figure presents the regulation used by each PM:

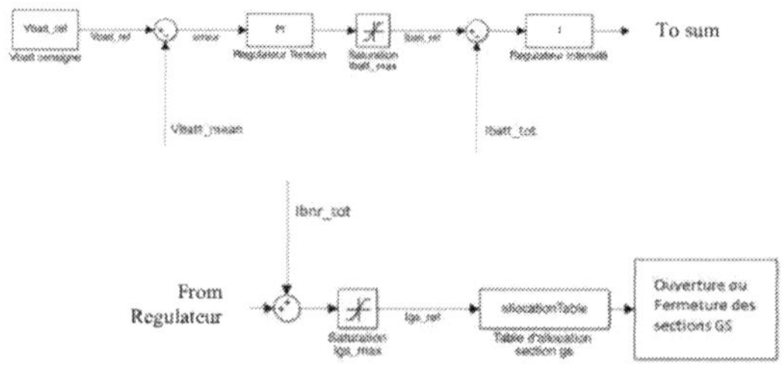

Figure 14 : External regulation loop

The average bus voltage measured by all the PMs is computed before regulation loop and for the currents it is the sum of output currents and the sum of battery currents which are processed.

The allocation table is used to choose which combinations of SA sections have to be opened or closed to reach the current setting computed by the regulation algorithm.

The table groups the short-circuit current of all SA sections of the system.

allocationTable[0]	150	→	sectionIDvector[0]	PCU3 Section2
allocationTable[1]	150	→	sectionIDvector[1]	PCU3 Section3
allocationTable[2]	75	→	sectionIDvector[2]	PCU1 Section3
allocationTable[3]	75	→	sectionIDvector[3]	PCU3 Section1
allocationTable[4]	50	→	sectionIDvector[4]	PCU2 Section1
allocationTable[5]	50	→	sectionIDvector[5]	PCU2 Section2
allocationTable[6]	50	→	sectionIDvector[6]	PCU2 Section3
allocationTable[7]	37.5	→	sectionIDvector[7]	PCU1 Section1
allocationTable[8]	37.5	→	sectionIDvector[8]	PCU1 Section2

In order to reach the current setting, the short-circuit currents are added successively starting with the highest current (beginning of the allocation table). A section is activated if it permits to get closer to the setting.

The allocation table is very important for modularity or mission adaptation. It is in the table that it is configured the total number of SA sections with their current capabilities.
Others allocation table can be programmed in order to optimize some particular performance of the PCU. For example priority could be given for selected SA sections to switch at working frequency while others sections are only commanded from time to time.

The OBC interface provides a way to monitor acquisitions values: SA section currents, battery voltages and currents, power bus voltages, switches statuses and active working mode of each PM.

Once the acquisitions have been shared between all the PMs, each PM computes the following values in order to execute the same regulation algorithm:

- mean value of the bus voltage acquisitions,
- sum of the battery current acquisitions,
- sum of the SA sections current acquisitions,
- computed output current.

The OBC can send commands to start Arming/Passivation process (only in Passivation working mode), modify the working mode of the PM, modify regulation parameters and command SA section measurements.

3.4 Robustness considerations

The reliability and robustness of the system are studied. First of all power circuits (SA sections, battery interface), digital control circuits, and communication bus were redounded. With a focus on low cost the use of a COTS microcontroller can be decided, then protection circuits against SEL are implemented. Logic circuits insensitive to radiation allow monitoring of the microcontrollers, recovery and voting the SA sections orders. The two CAN buses work in hot redundancies.
Because of the topology where each branch of the shunt is controlled by one command way, the only case which can cause the loss of the power bus generation is where the 2 MOSFETs of the same branch are commanded ON permanently. This situation can occur in the event of a failure on voltage or current measurement that increases the value too high and then the regulation forces the shunting. This situation can be detected by the firmware by implementing a specific monitoring algorithm.
In the end, the OBC is in charge of monitoring the PCU thanks to the telemetries and can detect any failures or drifts which could occur and then is able to reconfigure the PMs.

The PCU can operate only with one way (A or B), but in this failure case some short interruptions no longer than 1s can occur if one microcontroller is restarted. This situation is not a problem because the non-regulated bus concept can insure a permanent supply of the spacecraft, and also because the commands of the DETs are locked at the previous states during recovery.

4. EXPERIMENTAL RESULTS

The figure 15 shows the general experimental setup. The 3 PM are stacked together and connected to 9 SA simulators and one Li-Ion battery.

The breadboard is a reduced power model with a capability of 200mA per SA section. A SA simulator composed of 24 sections of 100mA has been developed for these activities. The battery is composed by one string of 8 Li-Ion cells. The power bus is loaded by a resistance and an active load.
The digital CAN busses are connected to a PC simulating the OBC and allowing the display on an MI of all the date exchanges on the 2 redundant CAN buses.
The system operate at 1Hz : all the SA sections status are refreshed each second.

Figure 15. The 3 PMs during test.

Figure 16 shows the recharge of the battery, at the beginning the battery current is limited and then when the bus voltage reaches the limitation reference at 30.4V, the current in the battery decrease. The ripple on the voltage is directly due to the DC impedance of the battery multiplied by the current switching steps.

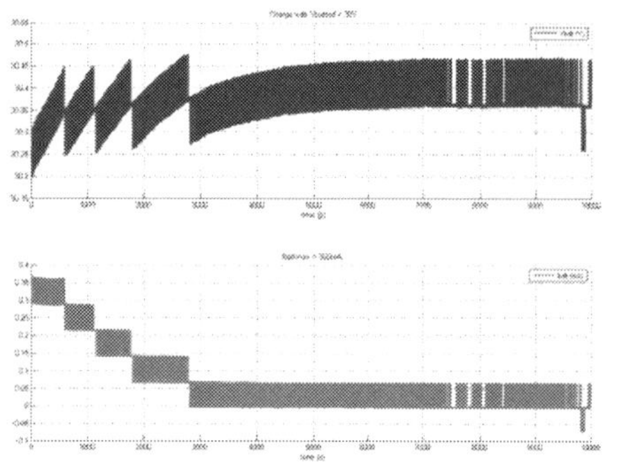

Figure 16. Experiment results, bus voltage (top) and battery current (down).

5. CONCLUSION AND FUTURE WORKS

The architecture of a decentralized and modular PCU gives performances as expected. Thanks to the digital control, lots of possibilities are offered to control the SA regulation or the battery management, and the capability to customize the PCU operation. Architecture and circuits design implementation are compliant with reliability and robustness needs and the system will be single failure free.

Future works will focus on achieving a real power model including all the redundancies and a spatial design compatible. Moreover, control laws improvements such as FDIR, robustness improvements, are still in progress.

The hardware architecture will remain decentralized, and the software architecture too. But with the same hardware, centralized software architecture is feasible where all the commands and control are made by the On Board Computer. In this case, each PM firmware keeps only the execution of the OBC commands, the TM acquisitions and the dialog with the OBC. It is very similar to PROTEUS or PLEÏADES PCU, but here there is modularity and the algorithms developed have lots of possibilities: adaptation with the number of PMs, allocation table for better regulation performances.

6. REFERENCES

1. C.Elisabelar, L. Lebaratoux : "Decentralized and Modular electrical architecture" ESPC2014.

2. M.Fossion, V. Lempereur, S. Fredon, C. Elisabelar, J.M. Cresp : « A modular and digitally controlled power conditioning unit" ESPC 2011.

 DOI: 10.1051/e3sconf/20171618003

LOSS MODELLING AND EXPERIMENTAL VERIFICATION OF A 98.8% EFFICIENCY BIDIRECTIONAL ISOLATED DC-DC CONVERTER

Rakesh Ramachandran, Morten Nymand

University of Southern Denmark, Odense, Denmark, Email:rar@mmmi.sdu.dk

ABSTRACT

High efficiency in power converters means less power wasted, which implies reduction in heat sink requirement, thereby small in size and less in weight. This power loss reductions in space application leads to smaller solar panels and less fuel on-board, hence reduces the spacecraft size, weight and launch cost. One of the major challenges for power converters in space application is to boost their efficiency, to reduce thermal dissipation problems.

In this paper, design and implementation of an ultra-high efficiency isolated bi-directional dc-dc converter utilizing GaN devices is presented. Loss modelling of the GaN converter is also included in this paper. The converter has achieved a maximum measured efficiency of 98.8% in both directions of power flow, using the same power components. Hardware prototype of the converter along with the measured efficiency curve is also presented in this paper.

1. INTRODUCTION

Solar energy is a major source of energy for the satellite in orbit. A bidirectional dc-dc converter is highly desirable to charge the battery with solar energy and in eclipse period to power the satellite equipment.

Some of the mandatory requirement for a converter used in satellite includes high conversion efficiency to reduce thermal dissipation problems, electrical isolation between input and output stages and low mass [1], [2]. Since, the size, reliability and efficiency of the overall system mainly depend on the power losses of the converter; efficiency is a key parameter in designing such a dc-dc converter.

GaN FETs have lower switching loss and conduction loss compared to Si MOSFET. Hence the efficiency of the whole converter can be increased to reduce the thermal dissipation and size, thereby making it more compact [3], [4].

Various bidirectional converters to achieve high efficiency have been discussed in the literature [5]-[7]. In [5], a 2 kW, 20 kHz with current doubler topology is presented. The maximum efficiency achieved is 96% at 600W in boost mode and 96% at 1200W in buck mode. A 1 kW resonant bidirectional converter is presented in [6] with an efficiency of 97.5% in forward and 97% in backward direction.

This paper presents an ultra-high efficiency isolated dc-dc converter. GaN devices are used as the switching device to realize the hardware prototype of the converter. This paper also discusses the design of high efficiency magnetics along with design considerations of using GaN FETs as the switching device. The loss modelling of the GaN converter is also presented in the paper.

The proposed converter is a bidirectional converter which allows the power flow in both directions using the same power components. The hardware prototype of a 1.7 kW GaN converter is presented in this paper. The measured maximum efficiency of the converter is 98.8% in both the directions of power flow. The converter has also achieved an efficiency of above 98.5% over a wide range of output power.

2. ISOLATED BI-DIRECTIONAL DC-DC CONVERTER

The circuit diagram of an isolated full bridge bidirectional dc-dc converter is shown in Figure 1.

The converter works as an isolated buck converter in forward direction and isolated boost converter in backward direction. Figure 2 show the operational waveforms of the bidirectional converter in both directions of power flow.

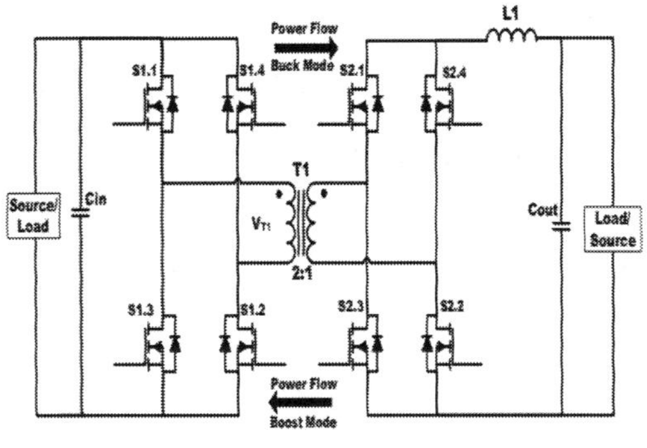

Figure 1. Isolated full-bridge bidirectional dc-dc converter

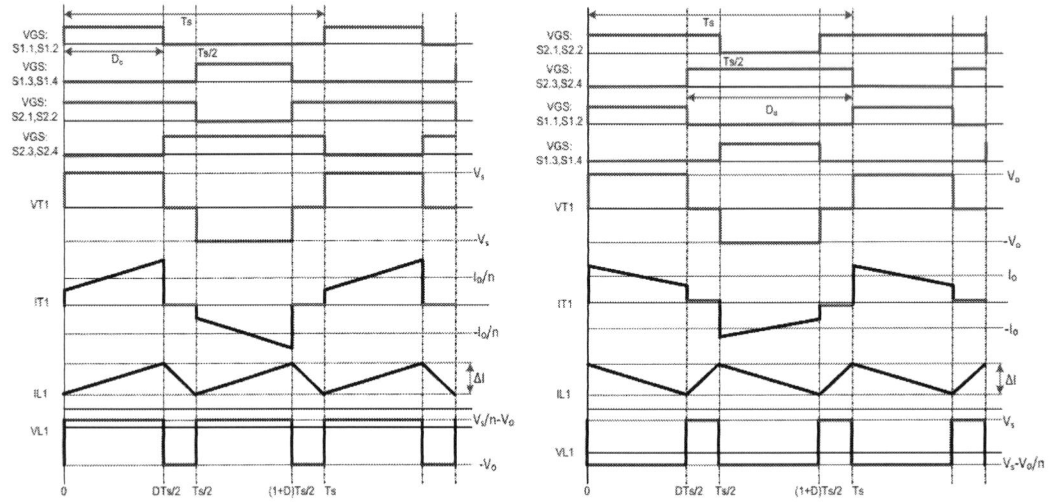

(a). *(b).*

Figure 2. Operational waveforms of an isolated bidirectional dc-dc converter, (a) buck mode and (b) boost mode

The operation principle of the converter in buck mode and boost mode are explained in [8] and [9], respectively.

3. CONVERTER DESIGN

In buck mode of operation, the designed isolated dc-dc converter has an input voltage of 130V and output voltage of 52V. The 130V side is referred as the high voltage side and corresponding secondary is referred as the low voltage side. The design specifications of the converter are shown in Table.1.

Output Power	1.7 kW
High Voltage Side	130 V
Low Voltage Side	52 V
Switching Frequency	50 kHz
Transformer Core	Ferroxcube EE55/28
Inductor Core	Kool Mμ E40
HV side GaN FETs	EPC2010C
LV side GaN FETs	EPC2001C

Table 1. Bidirectional converter design considerations

3.1. Transformer

High frequency transformer galvanic-ally isolates the source and the load. In order to have high efficiency in transformer, both core loss and copper loss has to be reduced. Ferrites E-core has been selected as the core material to have lower core loss. The major challenge to achieve high efficiency in power transformer is to have very low ac-resistance. Extensive interleaving of primary and secondary reduces the proximity effect and hence dramatically reduces the ac-resistance of the transformer [7]. Extensive interleaving with fewer turns also reduces the leakage inductance of the transformer. Hence, low ac-resistance and low leakage inductance increases the efficiency of the transformer even at higher power levels.

3.2. Inductor

In an isolated dc-dc converter, storage inductor is designed to operate with a dc bias current with a small ac ripple current. Core materials with high saturation flux will help in minimizing the inductor size and increasing the efficiency. Thus, Magnetics Kool Mμ is selected for the inductor core. For high dc bias current application, copper cross section has to be higher; hence copper foils are used for the winding. In order to reduce the ac-winding loss, two winding technique, is used for inductor design, explained in [11]. A thin ac winding is used to reduce the ac resistance and a thick dc winding to reduce the dc resistance. Thereby reducing the losses and hence high efficiency in inductor can be realized.

3.3. GaN FETs

GaN FETs are used as the switching devices. 200V and 100V FETs are used at the high voltage and low voltage sides of the transformer respectively. In order to reduce the conduction losses, four devices are used in parallel.

One of the main reasons for selecting GaN device is that they have zero reverse recovery losses compared to Si MOSFET. This is highly advantageous in low voltage high current synchronous rectification applications. Synchronous rectification helps in improving the efficiency of the overall converter.

The major challenges in using GaN FETs in high power

DOI: 10.1051/e3sconf/20171618003

converters are the optimum PCB layout design and the design of gate drive circuit. They have a very small threshold voltage and low gate-source voltage rating. A small noise injected into the gate of these devices, due to any parasitic inductance, can cause the device to turn-on or turn-off inadvertently and can cause damage to the whole converter.

LM5114 from Texas Instruments is chosen as the driver for the GaN devices. These drivers are capable of driving multiple GaN FETs due to their large sink current capability. Each driver drives four parallel devices. They also have a split gate output, which provides flexibility to adjust the turn-on and turn-off independently. The driver is compactable with TTL/CMOS signals and can withstand voltages up to 14V. This allows the driver to provide input directly from any PWM controller. LM5114 has a fast switching speed and minimized propagation delay of 12nS. An LCR tank circuit is usually formed with the parasitic source impedance, gate capacitance and the drive pull down path. This LCR tank circuit causes ringing in the gate of the GaN devices, an optimized resistor is used to damp this ringing [10].

3.4. Filter capacitors

Ceramic capacitors are used as the filter capacitors at both high voltage and low voltage sides of the converter. Compared to electrolitic capacitors, ceramic capacitors occupies very small volume. This helps in improving the power density of the converter. Even though the capacitance value in ceramic capacitor reduces with the dc-bias voltage, twice the number of capacitors has been used, which provides enough capacitance at both high voltage and low voltage side of the converter. These capacitors also have a very low ESR value, hence contribute to very low losses.

4. ANALYTICAL LOSS MODELLING

Loss modelling of individual power components is performed to predict the performance characteristics of the converter. The losses in the converter are dependent on many functions, mainly on switching frequency, input voltage and load current. The major source of power losses in an isolated converter are distributed among magnetics and the power semiconductor devices.

The power loss in individual components, transformer, inductor and semiconductor devices, can be expressed by a second order quadratic equation as a function of output current, I_o.

$$P_{loss} = k_o + k_2 I_o^2 \tag{1}$$

where, k_o is the idle losses and k_2 is the loss factor associated with the resistive losses in the respective power component.

The loss modelling of the individual power components are discussed below.

4.1. Transformer

The idle loss in the transformer is mainly the core loss. The core loss can be calculated using the Steinmetz formula [12],

$$P_v = k\, f^{\alpha} \hat{B}^{\beta} \tag{2}$$

where, P_v is the power loss per unit volume, B is the flux density, f is the frequency, and k, α, β are the material parameters.

The ac-resistance of the transformer can be expressed by the equation [9]

$$R_{ac} = R_{dc}\, \varphi \frac{\sinh 2\varphi + \sin 2\varphi}{\cosh 2\varphi - \cos 2\varphi} + \frac{2(m^2-1)}{3} \varphi \frac{\sinh \varphi - \sin \varphi}{\cosh \varphi + \cos \varphi} \tag{3}$$

where, $\varphi = {}^{h}/_{\delta}$, h is the height of the conductor, δ is the penetration depth, R_{ac} is the ac resistance of the winding, R_{dc} is the winding dc resistance and m is the number of layers in the winding.

The rms current through the primary winding can be written as

$$I_{RMS,tfr} = \frac{I_o}{n}\sqrt{D_c} \tag{4}$$

where, n is the turns ratio and D_c is the duty cycle.

The resistive losses in the transformer can be calculated from the ac-resistance and the RMS current through the winding.

The transformer power loss equation can be expressed as a function of output current, I_o as

$$P_{tfr} = 2.18 + 0.00262\, I_o^2 \tag{5}$$

4.2. Inductor

Similar to transformer, inductor losses also includes both core loss and winding loss. Core loss can be estimated from the Steinmetz equation as given in (2).

Since the inductor is designed using a two-winding technique, the winding losses includes both ac and dc-winding losses. The losses are calculated assuming ⅔ of the ac ripple current flows through the ac winding and the remaining ⅓ flows through the dc-winding [11].

 DOI: 10.1051/e3sconf/20171618003

The power loss equation for the inductor is given as

$$P_{Idr} = 0.83 + 0.001175 I_o^2 \qquad (6)$$

4.3. GaN FETs

The losses in GaN FETs include both switching loss and resistive loss.

Gate drive losses and output capacitive losses are included in the switching losses and can be estimated from the equation given below.

Gate drive loss is given by,

$$P_{drive} = V_d Q_G f \qquad (7)$$

where, V_d is the drive voltage and Q_G is the total gate charge and f is the switching frequency.

The analysis of capacitive switching loss in an isolated dc-dc converter is explained in [13]. From [13], the capacitive switching loss in the GaN FETs can be written as,

$$P_{cap(loss)} = \{2Q_{oss,p(V_{in})} - [\tfrac{5}{8}Q_{oss,p(\frac{3V_{in}}{8})} + \tfrac{3}{8}Q_{oss,p(\frac{5V_{in}}{8})}] + \tfrac{2}{n}Q_{oss,s(\frac{V_{in}}{n})}\}\, V_{in}\, f \qquad (8)$$

where, $Q_{oss,p(V_{in})}$ is the output charge of four parallel primary switches at V_{in}, $Q_{oss,p(\frac{5V_{in}}{8})}$ is the output charge of four parallel primary switches at $\frac{5V_{in}}{8}$, $Q_{oss,p(\frac{3V_{in}}{8})}$ is the output charge of four parallel primary switches at $\frac{3V_{in}}{8}$, $Q_{oss,s(\frac{V_{in}}{n})}$ is the output charge at $\frac{V_{in}}{n}$ of four parallel secondary switches.

Resistive losses in GaN FETs can be estimated from the on-resistance of the respective FETs and the RMS current through them.

Losses in the high voltage FET can be calculated by,

$$P_{HLOSS} = R_{DS(ON)H} \frac{P_o\sqrt{D_C}}{n_C V_o} \qquad (9)$$

where, $R_{DS(ON)H}$ is the on-resistance of high voltage FETs, P_o is the output power, and V_o is the output voltage in buck mode of operation.

Similarly, the conduction losses in low voltage FET can be expressed as,

$$P_{LLOSS} = R_{DS(ON)L} \frac{P_o}{V_o} \qquad (10)$$

where, $R_{DS(ON)L}$ is the on-resistance of low voltage FETs.

The power loss equation for the GaN FETs can be written as

$$P_{GaN} = 3.79 + 0.00652 I_o^2 \qquad (11)$$

The diectric loss in the filter capacitors can be calculated by,

$$P_{di\_cap} = 2\pi f C V^2 \tan\delta \qquad (12)$$

where, C is the filter capacitance value , V is the voltage across the capacitor and *tan* δ is the loss tangent or dissipation factor.

So the total loss in the converter can be expressed as a second order quadratic equation as

$$P_{tot} = 6.8 + 0.0117 I_o^2 \qquad (13)$$

From the above loss modelling, efficiency of individual power components are calculated and plotted in Figure 3. Figure also shows the calculated efficiency curve for the converter.

5. EXPERIMENTAL RESULTS

The hardware prototype of a 1.7 kW Isolated bidirectional dc-dc converter is shown in Figure 4. The proposed converter is realized on a PCB having size of 14.3cm in length and 7.7cm in width.

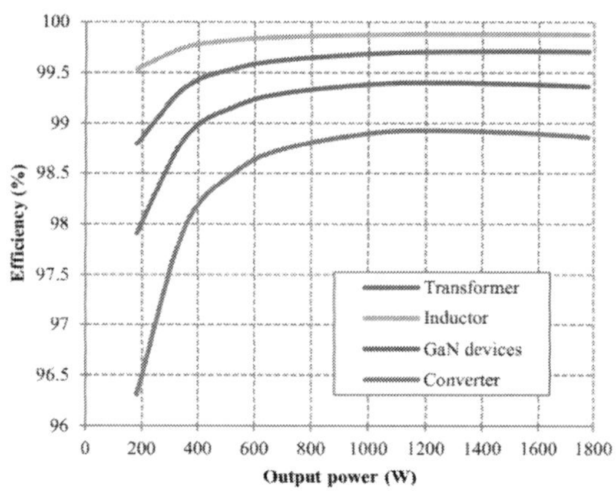

Figure 3. Calculated efficiency curve of the converter along with the efficiency curve of power components

DOI: 10.1051/e3sconf/20171618003

Figure 4. Hardware prototype of the bidirectional isolated dc-dc converter

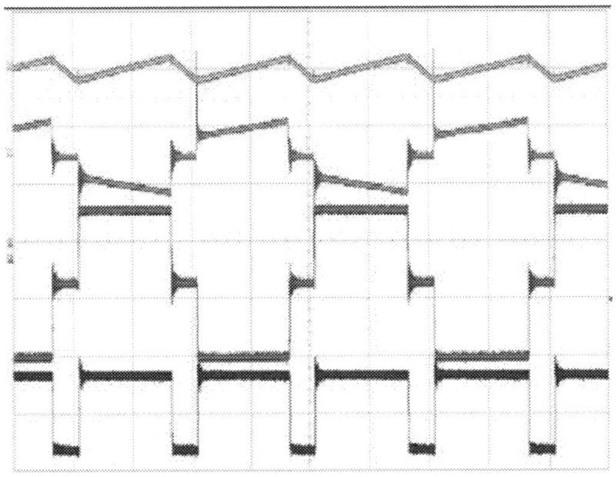

Figure 5. Measured waveforms of the bidirectional dc-dc converter in buck mode of operation (Green: Inductor current 20A/div, Yellow: Transformer primary current 20A/div, Red: Transformer primary voltage 100V/div, Blue: Rectifier voltage 50V/div, Time scale: 5μS/div)

When the efficiency becomes higher, for e.g. above 95%, extensive care has to be taken to measure the efficiency precisely and accurately. Current measurement is performed using a highly stable and precise sense resistors and voltage is measured using Keysight's high precision digital multi-meters.

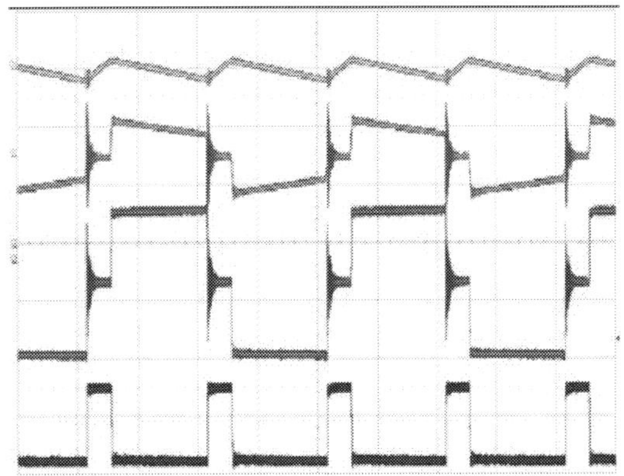

Figure 6. Measured waveforms of the bidirectional dc-dc converter in boost mode of operation (Green: Inductor current 20A/div, Yellow: Transformer secondary current 20A/div, Red: Transformer secondary voltage 100V/div, Blue: Rectifier voltage 50V/div, Time scale: 5μS/div)

The bidirectional operation of the converter is verified by operating the converter in buck mode and then in boost mode. In buck mode of operation, high voltage side is connected with a voltage source of 130V and an electronic load is connected at the low voltage side.

In boost mode of operation, the low voltage side is connected with a voltage source of 52V and the high voltage side is connected with an electronic load. The operational waveforms of the converter in both buck and boost mode are measured and shown in Figure 5 and Figure 6, respectively.

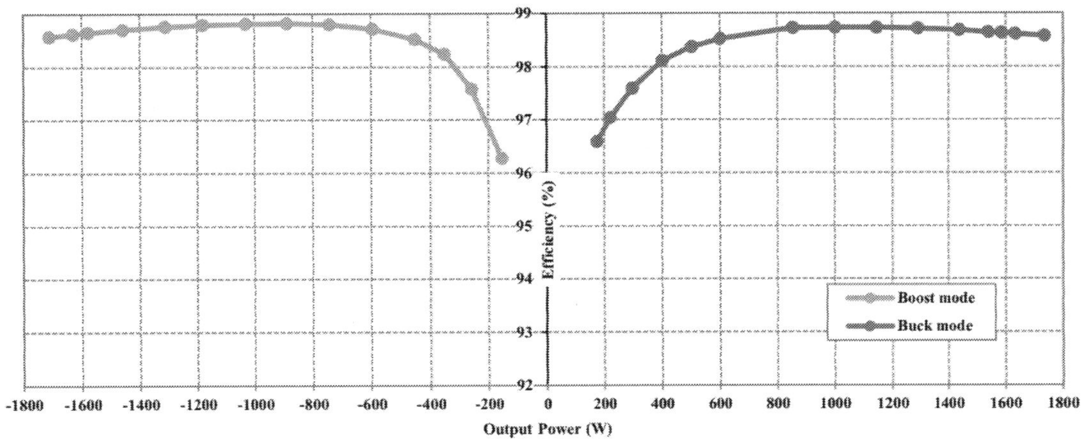

Figure 7. Measured efficiency curve of the bidirectional isolated dc-dc converter

The efficiency of the converter, both in buck mode and boost mode of operation are also measured and plotted in Figure 7. The isolated bidirectional dc-dc converter has achieved a measured maximum efficiency of 98.8% in buck mode as well as in boost mode of operation. The converter also exhibited a measured efficiency of above 98.5% over a wide operating power range, in both directions of power flow.

6. CONCLUSION

This paper presents the design and implementation of an isolated bidirectional dc-dc converter. Due to zero reverse recovery losses in GaN FETs, they are used as switching devices to improve the efficiency of the bidirectional dc-dc converter with synchronous rectification.

The design consideration for high efficiency magnetics as well as GaN devices are discussed in this paper. The analytical loss modelling of the individual power components along with calculated efficiency curve for the converter is presented in the paper. The measured efficiency curve of the converter is also shown in the paper. The converter has achieved a measured maximum efficiency of 98.8% in both buck and boost mode of operation. The efficiency of the isolated bidirectional converter is above 98.5% over a wide range of output power.

7. ACKNOWLEDGMENT

The project is sponsored by the Danish National Advanced Technology Foundation under Intelligent Efficient Power Electronics (IEPE), strategic research center between the industries and universities in Denmark.

8. REFERENCES

1. Kankam, M. D. & Elbuluk, M. E. (2001). A Survey of Power Electronics Applications in Aerospace Technologies. In Proc. 36th Intersociety Energy Conversion Engineering Conference, Savannah, Georgia.

2. Weinberg, A. H. & Schreuders, J. (1986). A High-power High-voltage dc-dc Converter for Space Applications. In Proc. IEEE Transactions on Power Electronics, pp.148-160.

3. Kaminski N. (2009). State of the art and the future of wide band-gap devices. In Proc. 13th European Conference on Power Electronics and Applications, pp. 1-9.

4. Millan J. et al. (2014) A survey of wide bandgap power semiconductor devices. In Proc. IEEE Trans. on Power Electronics, pp. 2155– 2163.

5. Chiu, H. J. & Lin, L. W. (2006). A Bidirectional dc-dc Converter for Fuel Cell Electric Vehicle Driving System. In Proc. IEEE Trans. Power Electron., pp. 950-958.

6. Jiang, T. et al. (2015). A Bidirectional LLC Resonant Converter with Automatic Forward and Backward Mode Transition. In Proc. IEEE Transactions on Power Electronics, pp. 757-770

7. Yu, X. & Yeaman, P. (2014). A New High Efficiency Isolated Bi-directional dc-dc Converter for DC-bus and Battery-bank Interface. In Proc. 29th Applied Power Electronics Conference and Exposition (APEC), pp. 879-883.

8. Ramachandran, R. et al. (2014). Design of a Compact, ultra-high Efficient Isolated dc-dc Converter utilizing GaN Devices. In Proc. 40th Annual Conference of the IEEE Industrial Electronics Society, IECON, pp. 4256-4261.

9. Nymand, M. & Andersen, M. A. E. (2010). High-efficiency Isolated Boost dc-dc Converter for High-power Low-voltage Fuel Cell Applications. In Proc. IEEE Trans. Ind. Electron., pp. 505-514.

10. LM5114 datasheet, Peak current low-side gate driver, (www.ti.com)

11. Nymand, M. et al. (2009). Reducing ac-winding losses in High-current High-power Inductors. In Proc. 35th IEEE IECON Conf., Portugal, pp. 774-778.

12. Steinmetz, C. P. (1984). On the Law of Hysteresis. In Proc. IEEE, pp. 197 -221.

13. Ramachandran R. & Nymand M. (2015). Analysis of Capacitive Losses in GaN Devices for an Isolated Full Bridge DC-DC Converter In Proc. International Conf. on Power Electronics and Drive Systems, pp. 467-472.

DOI: 10.1051/e3sconf/20171618004

ON-ORBIT OPERATIONS OF A POWER SYSTEM FOR JAPAN'S VENUS EXPLORER AKATSUKI

Hiroyuki Toyota [(1)], Nobuaki Ishii [(1)], Masato Nakamura [(1)], Hiroki Ooto [(2)], Kazuya Koide [(2)], Tomohiko Sakamoto [(2)], Hidetoshi Abe [(2)], Hideki Yoshino [(3)], Hiroyuki Takamura [(3)], Takayuki Ose [(3)], Yukishige Nozaki [(3)]

[(1)] ISAS/JAXA, 3-1-1 Yoshinodai, Chuo-ku, Sagamihara, Kanagawa 252-5210, Japan, Email: htoyota@isas.jaxa.jp
[(2)] The Furukawa Battery Co., Ltd., 23-6 Kuidesaku, Joban-shimofunaomachi, Iwaki, Fukushima 972-8501, Japan, Email: h-ooto@furukawadenchi.co.jp
[(3)] NEC Space Technologies, Ltd., 1-10 Nissin-cho, Fuchu, Tokyo 183-8551, Japan, Email: h-yoshino@ex.jp.nec.com

ABSTRACT

This paper describes on-orbit operations of a power system for Japan's Venus explorer Akatsuki. It was launched on May 20, 2010, and approached Venus on December 7, 2010, but orbit insertion failed because of a propulsion system malfunction. After the failure, the spacecraft stayed on an orbit nearer to the sun than Venus for 5 years before successfully entering an orbit around Venus in 2015. Telemetry data that show radiation degradation of the solar arrays, and here analytical results by the relative damage coefficients method are presented. The capacity loss of the batteries was successfully suppressed by lowering the state of charge and temperature. We plan to extend the mission period from 4.5 years to longer than 9 years to obtain more fruitful scientific results.

1. INTRODUCTION

Akatsuki is Japan's first Venus orbiter. It observes the mysterious atmospheric circulation on Venus using five cameras covering from ultraviolet to near-infrared light. The spacecraft approached Venus as planned on December 7, 2010, but orbit insertion failed because of a propulsion system malfunction. On December 7, 2015, Akatsuki approached Venus again and was successfully inserted into an elliptic orbit around Venus. After the failed first orbit insertion, the spacecraft survived on an orbit nearer to the sun than Venus for 5 years. Now it is aiming at scientific observations on an orbit around Venus for over 2000 days, despite a design lifetime of 4.5 years in total. This paper describes the on-orbit operation of the power system for Akatsuki.

2. OVERVIEW

Figure 1 depicts the Akatsuki spacecraft in its flight configuration. The spacecraft mass is 502 kg, including 189 kg of fuel and oxidizing agent and 33 kg of observation cameras. The dimensions of the main bus are 1.45 × 1.04 × 1.4 m. It is equipped with two solar array paddles, which are mounted on the north and the south body surfaces and rotate to track the sun.

Figure 2 shows a block diagram of the power system for the Akatsuki spacecraft. Electrical output of the solar arrays is delivered through solar array drive assemblies (SADAs) and solar array blocking diodes (SBDs) into a series-switching regulator (SSR) to be regulated to 50 V.

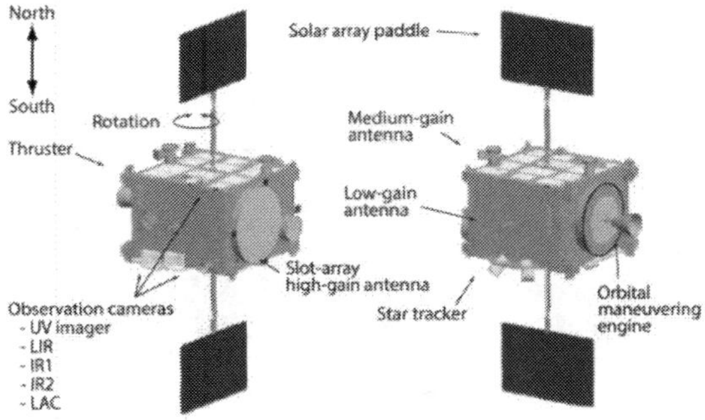

Figure 1. Akatsuki spacecraft

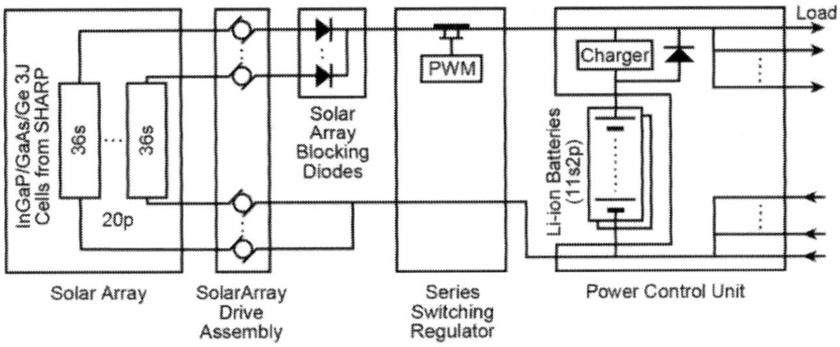

Figure 2. Block diagram of the Akatsuki power system

DOI: 10.1051/e3sconf/20171618004

A power control unit (PCU) charges and discharges lithium ion batteries, and simultaneously distributes electricity to power buses at several voltage levels.
Akatsuki was launched aboard an H2A rocket from the Tanegashima Space Center and directly injected into an orbit toward Venus on May 20, 2010. Figure 3 shows Akatsuki's trajectory. Akatsuki approached Venus on December 7, 2010 after a half year of cruise, and started burning its 500-N-class orbital maneuvering engine (OME) during Venus orbit insertion (VOI). The planned burn period was for 715.5 s, but the OME stopped burning at 158 s because of a malfunction in its fuel supply line. The OME subsequently lost its nozzle, rendering it unusable. The spacecraft thus failed to enter Venus orbit, instead entering an orbit around the sun with a period of 203 days, a perihelion distance of 0.6 AU, and an aphelion distance of 0.73 AU. After orbiting the sun nine times over five years, Akatsuki again approached Venus and conducted VOI-R1 by burning its four 23-N-class reaction control system (RCS) thrusters for 1228 s on December 7, 2015. The spacecraft successfully entered an elliptic orbit around Venus with a period of 13.6 days, an inclination of 3°, and an apoapsis altitude of 440,000 km. To decrease the apoapsis altitude and avoid long eclipses, a trim maneuver was performed at the first periapsis, resulting in an apoapsis altitude of 360,000 km and a period of 10.5 days. The orbit will be continuously adjusted to extend the mission life from the originally planned 800 days to 2000 days on Venus orbit.

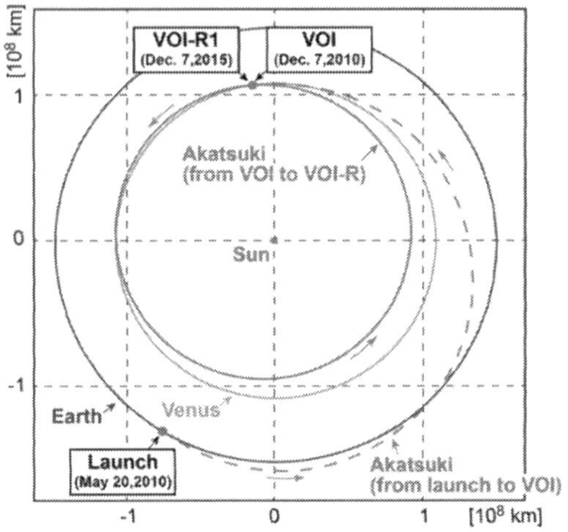

Figure 3. Trajectory of Akatsuki from launch to Venus orbit insertions VOI and VOI-R1.

3. SOLAR ARRAY PANELS

Figure 4 shows a solar array panel of ESA's Venus Express, which was operated until about a year before Akatsuki entered an orbit around Venus. On the solar array panels of Venus Express, solar cell strings and optical solar reflector strings are mounted alternately to lower the panel temperature.
Akatsuki, on the other hand, was originally planned to be launched aboard a small solid rocket, M-V, which was far smaller than the H2A rocket. The solar array panels of Akatsuki were designed to fit in the M-V payload fairing, which has a diameter of 1.8 m. As a result, the front side of Akatsuki's solar array panel, which was 1.43 m wide and 1.036 m long, was covered by InGaP/GaAs/Ge 3J solar cells from Sharp Corp. to obtain the required electrical power, as shown in Figure 5. The solar cells are covered by 100-µm-thick CMG cover glasses with AR coats. The predicted generated power is more than 480 W at 1.0781 AU and more than 660 W on the Venus orbit. The rear side was covered by optical solar reflectors (OSRs) to lower albedo input from Venus. The designed temperature range of the solar array panels was from −170 °C to +184 °C [3].
Figure 6 shows telemetry data of the solar array output voltage and temperature. Between VOI and VOI-R1, the solar array temperature varied from +100 °C to +140 °C, depending on the distance to the sun. The output voltage varied according to the temperature, and gradually decreased over time due to radiation damage caused by solar protons. The output voltage trend also shows four sudden drops, which are assumed to have been caused by solar flares [4].
We analyzed radiation damage to the solar array by the relative damage coefficients (RDCs) method [5]. Figure 7 shows the RDCs of the solar cells used in the analysis. Figure 8 shows the constant solar proton environment that caused the gradual voltage drop estimated by the SPENVIS [6].

Figure 4. Solar array panel on Venus Express [2]

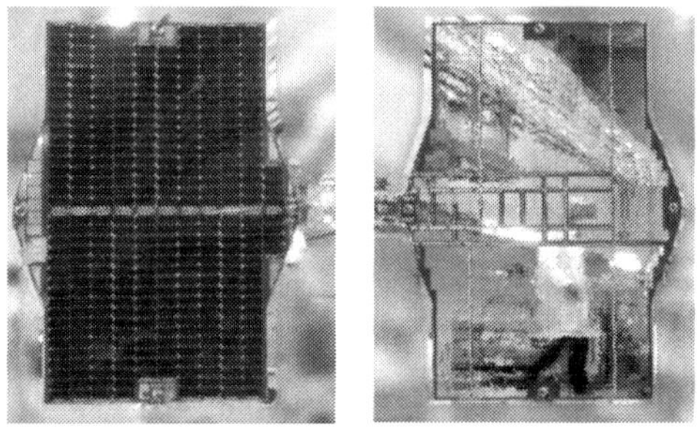

(a) Front side *(b) Rear side*

Figure 5. Solar array panels on Akatsuki

DOI: 10.1051/e3sconf/20171618004

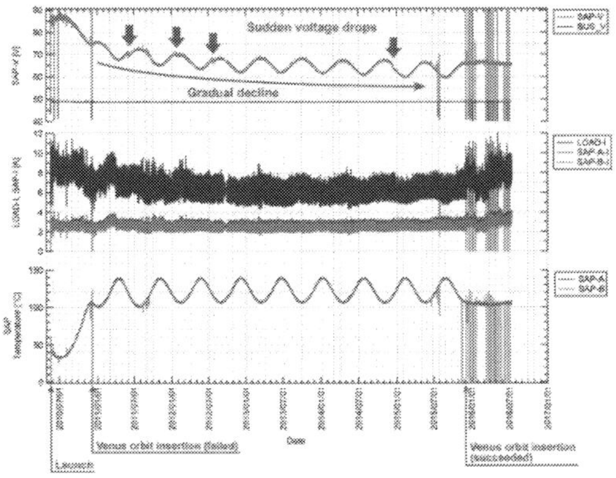

Figure 6. Telemetry data of solar array output voltage and temperature

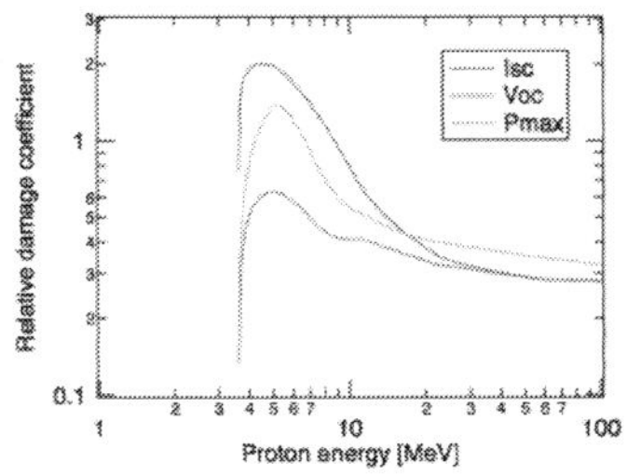

Figure 7. Relative damage coefficients of solar cells used in analysis

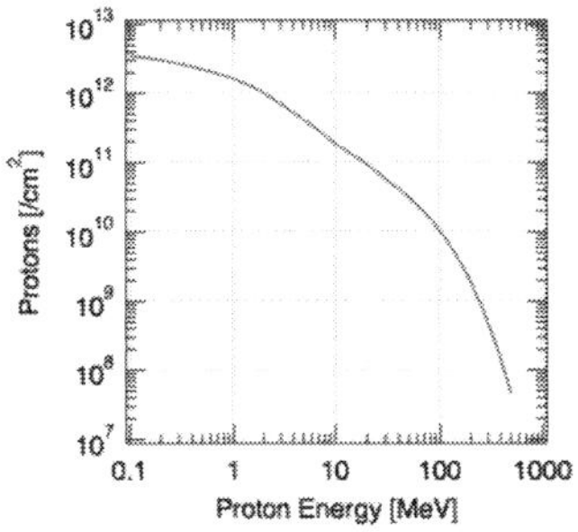

(a) Solar proton spectrum accumulated from May 21, 2010 to December 7, 2017

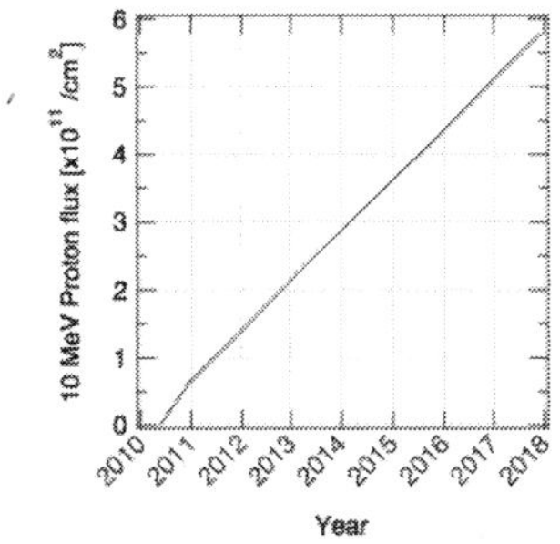

(b) Time evolution of equivalent 10-MeV proton fluence for P_{max}

Figure 8. Solar proton environment

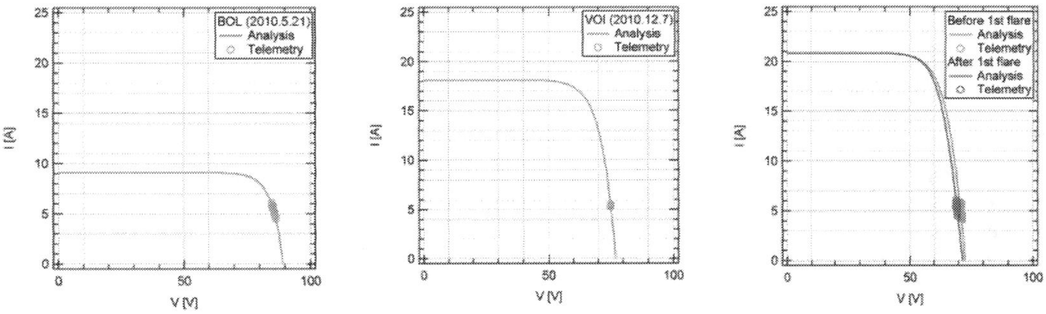

(a) BOL (May 20, 2010) *(b) VOI (Dec.7, 2010)* *(c) 1st flare (June 5, 2011)*

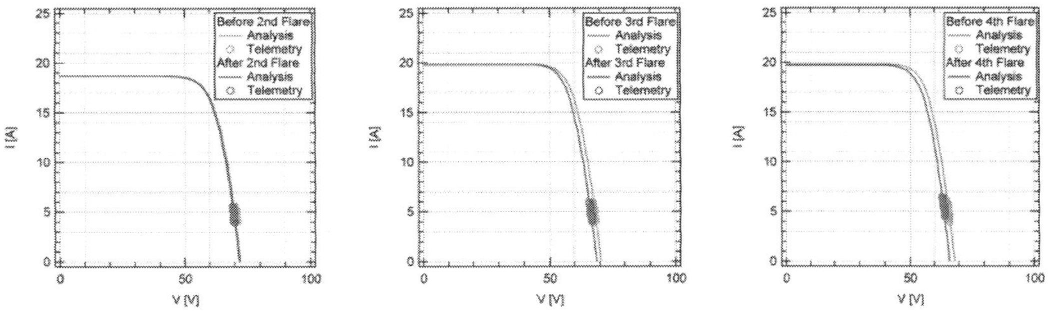

(d) 2nd flare (Jan. 23, 2012) *(e) 3rd flare (July 23, 2012)* *(f) 4th flare (Dec. 13, 2014)*

Figure 9. Current-voltage characteristics of the solar array as calculated by the RDC method and telemetry data

For each sudden voltage drop event, we estimated a proton dose that matches the telemetry data, because there are no observation data available.

Figure 9 shows the analytical results and telemetry data. The calculated current-voltage (IV) characteristics by the RDC method agree well with the telemetry data when the radiation damage margin (RDM) is 0.5. This is plausible, because the solar activity was lower than expected over this period [7].

Figure 10 shows IV and power-voltage (PV) characteristics at BOL, VOI, VOI-R1, and 2000 days from VOI-R1 calculated using the RDC method. The power required by Akatsuki, which is less than 500 W at 56 V, can be generated even 2000 days from VOI-R1.

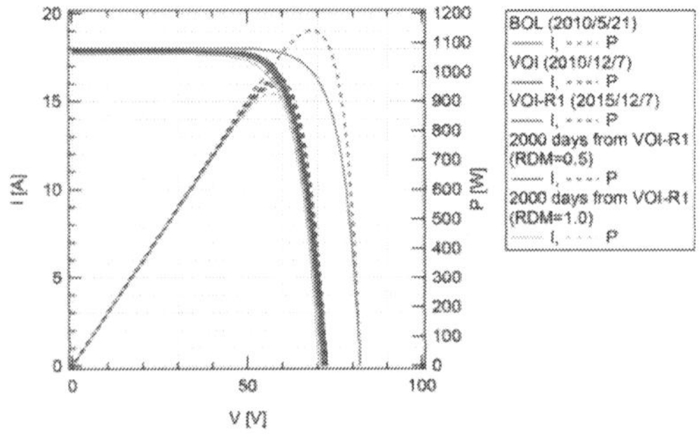

Figure 10. Current-voltage and power-voltage characteristics calculated using RDCs

4. BATTERIES

Akatsuki carries two 23.5-Ah lithium-ion batteries connected in parallel. Each battery consists of eleven cells connected in series and protection circuits (Figure 11). The capacity of lithium ion batteries degrades faster at higher temperatures and at a higher state of charge (SOC). We therefore planned to keep the temperature and the SOC as low as possible to minimize capacity degradation.

Figure 12 and Table 1 show the original battery operation plan. When not in use in ground tests, the batteries were kept at 10% SOC in open circuit at room temperature for 1.5 years prior to launch to prevent over-discharge. After the spacecraft was launched, the batteries were kept at 41% SOC at 10 °C in preparation for recovery from an unexpected attitude loss of the spacecraft until VOI (except during reset operations, in which all cells were charged to 4.1 V then discharged to balance cell voltages). This period was 0.5 years, while the longest trajectory was supposed to take 2.5 years. After VOI, the batteries were used as a power supply during eclipses for two years. During this period, the batteries are repeatedly charged and discharged 550 times by changing the SOC from 41% to 85% depending on the eclipse duration, which could be up to 90 minutes. [8]

After Akatsuki failed in the VOI on December 7, 2010, we revised the operation plan as shown in Figure 13 and Table 2 to reduce the capacity loss and support an extended mission. The SOC was reduced from 41% to 30% and the temperature was lowered from 10 °C to 0 °C, based on the capacity actually required when the spacecraft recovered from its attitude loss during the failed VOI.

After the successful VOI-R1, Akatsuki entered an elliptic orbit around Venus with an apoapsis altitude

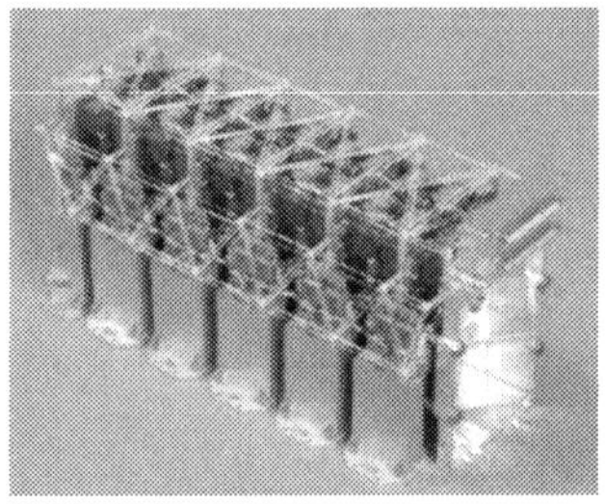

Figure 11. Lithium ion battery for Akatsuki

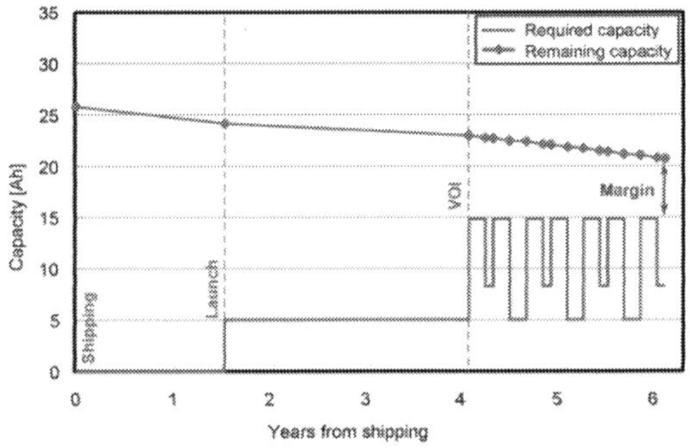

Figure 12. Originally planned required capacity and predicted trend of battery capacity

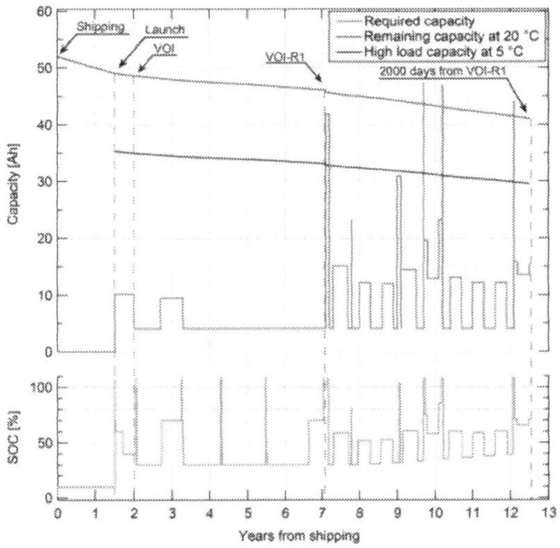

Figure 13. Required capacity and predicted trend of battery capacity revised after VOI and VOI-R1

Table 1. Original battery operation plan

Mission phase		Time [Years]	SOC [%]	Temperature [°C]	Battery state
Shipping to launch		1.5	10	25	Open circuit
Launch to VOI	Nominal	0.5	41	10	Float charge
	Backup	2.5			
Venus orbit		2	41-85	10	Charge and discharge cycle

Table 2. Battery operation plan revised after failed VOI

Mission phase	Time [Years]	SOC [%]	Temperature [°C]	Battery state
Shipping to launch	1.5	10	25	Open circuit
Launch to VOI	0.5	41	10	Float charge
VOI to VOI-R1	5	30	0	
Venus orbit	6	30-108	5-20	Charge and discharge cycle

much greater than originally planned. As a result, the longest eclipse duration was extended from 90 minutes to 7.9 hours and the greatest required energy during an eclipse increased from 750 Wh to 1600 Wh, while the number of the eclipses was reduced from 550 times to 110 times over 2000 days on Venus orbit. As shown in Figure 13, eclipses at 9.7, 10.2, and 12.1 years after shipping require capacities larger than the remaining capacities. During these eclipses, the batteries will meet the requirements by being charged to 108% SOC and putting the spacecraft into survival mode, which requires less power than nominal mode.

5. SUMMARY

This paper described the operation of Japan's Venus explorer Akatsuki. The spacecraft's orbit insertion failed because of a malfunction in its OME, but successfully entered an orbit around Venus using its thrusters for attitude control after orbiting around the sun for 5 years, despite a designed spacecraft life of 4.5 years. Radiation degradation of the solar array was analyzed by the RDC method, and the results agreed well with telemetry data. The battery life was extended by lowering the SOC and the temperature to meet the demands of a mission extended to 12.5 years. Akatsuki has already smoothly started scientific observations and a mission life of 2000 days on Venus orbit is expected.

6. REFERENCES

1. Masato Nakamura et al. (2016). AKATSUKI return to Venus. Earth, Planets and Space. 68:75.
2. http://www.esa.int/spaceinimages/Images/2005/08/Venus_Express_ready_for_shipping_to_Baikonur2
3. Hiroyuki Toyota et al. (2008). Venus Climate Orbiter Planet-C Solar Array Development, in Proc. 8th European Space Power Conference, European Space Agency, Constantz, Germany.
4. Yu Takahashi et al. (2014). Analysis of Radiation Damage in On-orbit Solar Array of Venus Explorer Akatsuki, in Pro. 9th European Space Power Conference, European Space Agency, Noordwijkerhout, The Netherlands.
5. B. E. Anspaugh (1996). GaAs Solar Cell Radiation Handbook, Jet Propulsion Laboratory Publication, 96-9.
6. SPENVIS web site. https://www.spenvis.oma.be/
7. Ian G. Richardson et al. (2016). 25 MeV Solar Proton Events in Cycle 24 and Previous Cycles, Advances in Space Research.
8. Hiroki Ooto et al. (2008). Development of Lithium-Ion Battery for the Venus Climate Orbiter "PLANET-C", in Proc. 8th European Space Power Conference, European Space Agency, Constantz, Germany.

ELECTRICAL POWER SUBSYSTEM FOR THE EUCLID SPACECRAFT

Ezio Ciancetta [(1)], **Marco Cimino** [(1)], **Giuseppe Cuzzocrea** [(1)], **Giuseppe Gervasio** [(1)], **Elena Maiorano** [(2)], **Ignacio Martinez** [(3)], **Luc Sanchez** [(4)]

[(1)] *Thales Alenia Space, Strada Antica di Collegno 253, 10146 Torino (Italy), Ezio.Ciancetta@thalesaleniaspace.com, Marco.Cimino@thalesaleniaspace.com, Giuseppe.Cuzzocrea@thalesaleniaspace.com, Giuseppe.Gervasio@thalesaleniaspace.com*
[(2)] *European Space Agency, Keplerlaan 1, 2201 AZ Noordwijk (The Netherlands), Elena.Maiorano@esa.int*
[(3)] *SpaceTech GmbH, Seelbachstraße 13, D-88090 Immenstaad (Germany), Ignacio.MartinezAlonso@spacetech-i.com*
[(4)] *Saft, Rue Georges Leclanché, BP1039, 86060 Poitiers (France), Luc.Sanchez@saftbatteries.com*

ABSTRACT

Euclid is the second medium class mission of the European Space Agency in the frame of Cosmic Vision 2015-2025 program [1]. It is a cosmology mission whose prime objective is to study the geometry and the nature of the dark matter and the dark energy with unprecedented accuracy.
The spacecraft will be launched in 2020 by a Soyuz launcher, to perform a six-year survey of the extragalactic sky from a large-amplitude orbit around Lagrange point L2 of the Sun-Earth system.

This paper outlines the Euclid Electrical Power Subsystem (EPS) design, providing a description of the major design drivers and resulting configuration, with a view to highlight aspects that could be considered for future designs.

1. SYSTEM DESCRIPTION

The spacecraft is shown in Fig. 1. It is composed by:

- a Payload Module (PLM) including the Telescope, the baffling system, the Optical Bench supporting the instrument detectors and front ends;
- a Service Module (SVM) comprising the spacecraft platform with its subsystems and the Sunshield protecting the PLM from solar radiation and supporting the PhotoVoltaic Assembly (PVA).

Euclid will produce a massive survey of the extragalactic sky (above 30 deg galactic latitude), simultaneously in the visual band and in the near-infrared band, and a deep survey of two 20 deg² fields observed on a monthly basis.

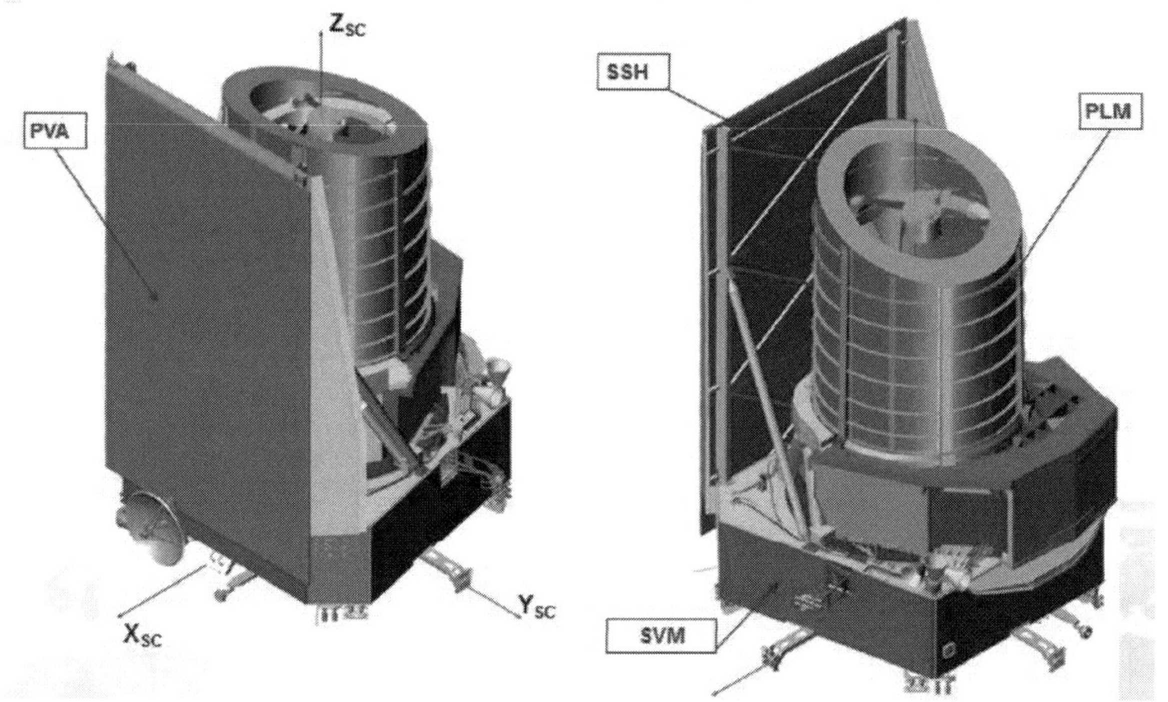

Figure 1. Euclid configuration

DOI: 10.1051/e3sconf/20171618005

The spacecraft will carry two instruments for spectroscopy, imaging and photometry: NISP (Near Infrared Spectro Photometer) and VIS (Visible Instrument); a 1.2 m mirror telescope will focus the incoming images on the focal plane units of these instruments.

The performance of this survey requires an extreme pointing stability associated to the ability to move from one target to the next in few tens of seconds, several thousand of times during the mission without degradation. Stability is the key word of Euclid: stability of the telescope, of the experiment detectors, of the temperatures, of the pointing. A prerequisite for image stability is a near-constant temperature environment, maintained by scanning the sky along circles of near-constant Solar Aspect Angle (SAA). High stability is requested to the structural materials and the temperatures have to be controlled carefully. This constraints had to be taken into account also in the EPS architecture, driving the selection of the Sunshield and solar array characteristics and imposing a high stability of the heater power distribution.

The flight-proven platform architecture of Herschel [2] is taken as reference for the SVM, with six panels and a central cone in which the propellant tanks are mounted. The size of the SVM is scaled to fit the Soyuz launcher (Herschel launcher was Ariane 5 and the SVM had eight panels), but it is also driven by the PLM shape and by the overall dimensions of the Spacecraft. Additional modifications have been introduced to take into account partial obsolescence of the equipment. This architecture allows the modular accommodation of the platform equipment and provides room for the warm payload electronics. Fig. 2 shows the Euclid SVM in closed configuration, while the its exploded view is depicted in Fig. 3.

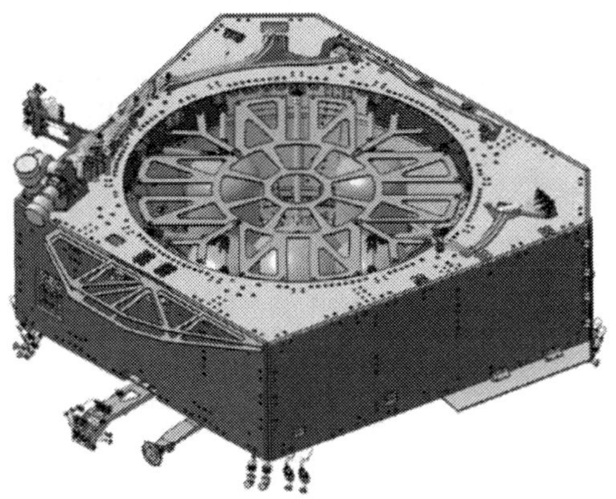

Figure 2. Euclid SVM.

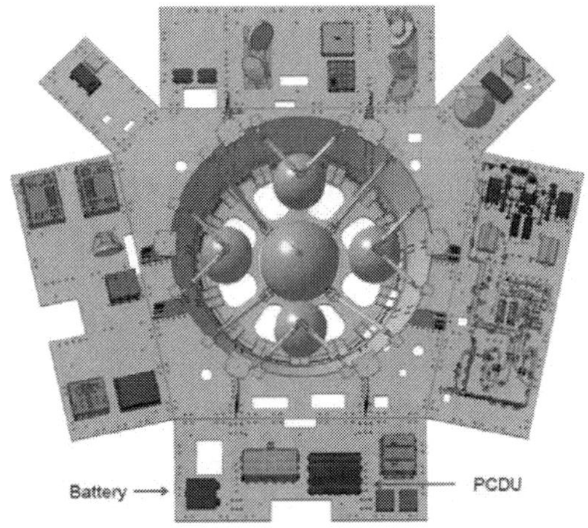

Figure 3. SVM configuration – Exploded view.

The Lagrange point L2 lies approximately on the Earth-Sun line at 1.5×10^6 km from the Earth in the anti-Sun direction. In Euclid nominal attitudes the Sun is in $+X_{SC}$ direction, with elevation angle up to 31° and azimuth angle up to 8°. The lifetime will be 6.25 years, including one month for the transfer phase to the operational orbit around L2.

Thales Alenia Space Italia leads the Euclid industrial team and it is in charge of SVM design, including the electrical power supply chain.

2. ELECTRICAL POWER SUBSYSTEM DESIGN DRIVERS

The spacecraft power demand is up to 1700 W. The EPS is in charge of generating, conditioning, controlling, storing the electrical power and distributing it to instruments and spacecraft equipment. The mission is nominally eclipse free, hence the Battery is designed to be used only during launch phase and in case of attitude anomaly failures, while the Solar Array (SA) is required to generate the electrical power for the whole lifetime. A Power Conditioning and Distribution Unit (PCDU) provides the core of EPS functions and interfaces with the on-board computer for telemetry and telecommands.

The Euclid EPS block diagram is shown in Fig. 4.
The power link Battery–PCDU is interrupted at spacecraft skin level for allowing the replacement of the Battery with the battery simulator during the Assembly, Integration and Test (AIT) activities.

The Euclid Power architecture has been built around some major design drivers. In particular, the following aspects have been considered for defining the design.

- Mass saving. In order to avoid a launch mass exceeding the launcher capability, all the trade-offs relevant to the EPS had to take into account the contribution of mass.

- Solar Array configuration. The possibility of using a deployable SA was discarded since phase A, due to the requirement of minimising the use of mechanisms: the requirement came from the need of avoiding perturbations. In addition, that option would have caused an impact on mass due to the use of hinges, hold down mechanisms and additional screens between PLM and SA backside. For these reasons, the solution with fixed solar arrays was selected.
- SA power conditioning. At the beginning of phase B2 a trade-off was conducted in order to select the best regulation concept between Maximum Power Point Tracking (MPPT) and Direct Energy Transfer (DET). The characteristics of the mission, with limited variation of illumination (maximum Sun depointing of 33°) and stable array temperature conditions, led to select the DET regulation, in a Sequential Switching Shunt Regulator (S^3R) topology.
- Number of shunt sections. Different options were taken into account for defining the number of SA sections most suitable for the EPS. The number of fifteen sections was considered as the optimum for allowing the split of PVA surface in three or five identical panels.
- Regulated vs. Unregulated Power Bus. The eclipse-free characteristics of the Euclid mission would allow avoiding the use of a Battery Charge and Discharge Regulator (BCDR) and the bus maintained regulated by the Battery, under the condition that spacecraft peak power consumption is always lower than the SA power. However, in order to maximize the reuse of equipment from similar missions as Herschel and Planck, a fully regulated $28V_{DC}$ bus was selected for Euclid as baseline since the definition phase.

3. SOLAR ARRAY

SpaceTech GmbH (DE) is responsible for the Sunshield and Solar Array (SSH) subsystem, with the Photo-Voltaic Assembly (PVA) made by Leonardo (IT), the Solar Panels and Support Structure manufactured by RUAG (CH) and the Thermal Control Hardware manufactured by HPS Ltd. (PT).

The PVA area is close to 11 m^2. It is located on the central part of the SSH, which size is driven by the PLM shape and limited by the launcher fairing. The environment and the worst case illumination conditions impose additional requirements for the SA sizing:

- the minimum solar flux at summer solstice in L2 is 1293 W/m^2;

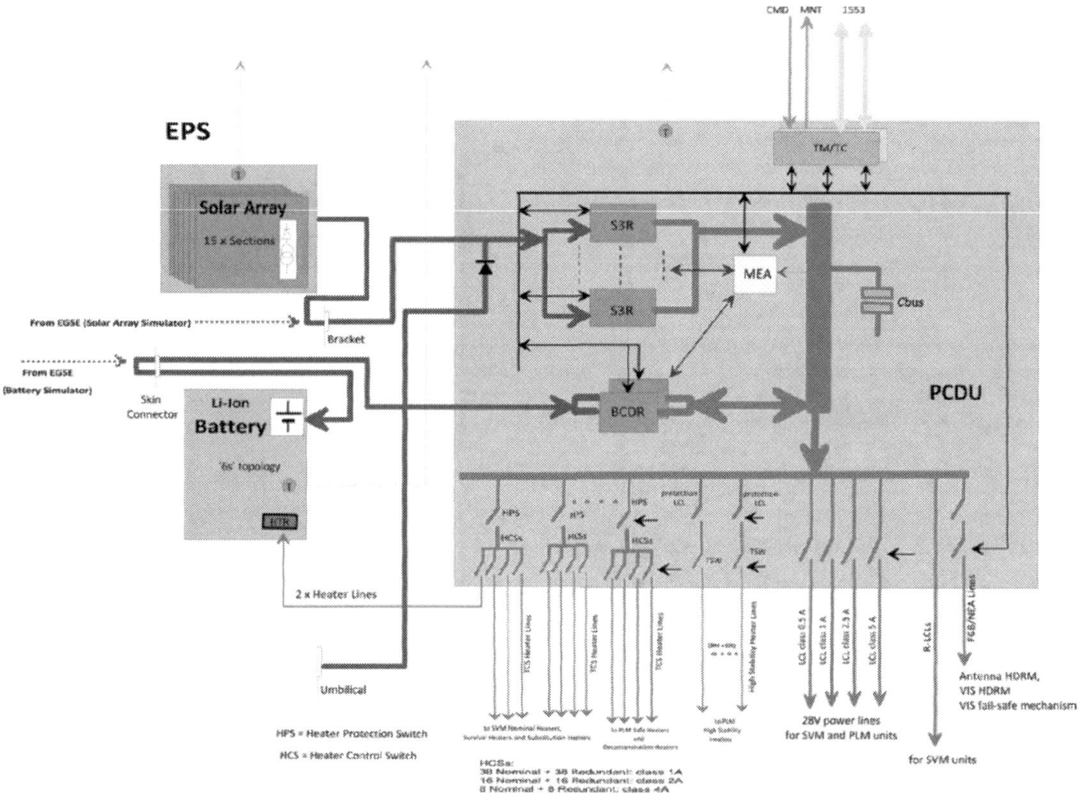

Figure 4: EPS Block Diagram.

 DOI: 10.1051/e3sconf/20171618005

- the degradation of the solar cells after six years and three months in L2;
- the maximum sun angle of 121° with respect to the telescope axis and the maximum alpha angle of 8° around X_{SC} axis; adding 1° of tolerance, the resulting incidence on SA will be therefore between 0° and 33°.

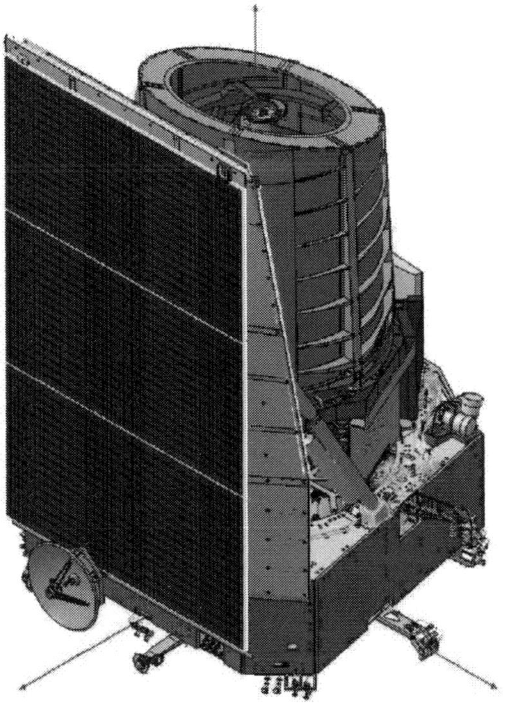

Figure 5. PVA implementation on the spacecraft.

As shown in Fig. 5, the PVA is composed of three panels of the same dimensions and the same cell/string layout. Each panel consists of 51 strings distributed in four sections of ten strings and one section of eleven strings. In total there are fifteen sections. All strings are composed by nineteen solar cells connected in series.

The solar cells used in the PVA are triple junction Gallium Arsenide 3G30C by Azur Space, covered by a 100μm CMG coverglass as a protection against solar ultraviolet radiation.
The 3G30C cell has a size of 40 x 80 mm and two cropped corners. All solar cells shall include a bypass protection diode to prevent the adjacent solar cell from unacceptable high reverse voltages in case of shadowing or cell damage. One option is the internal bypass diode, which is integrated into the cell itself. This diode technology is qualified to a temperature of 140°C and assumed to withstand temperatures up to 150°C for a limited time. As mentioned in the previous section, the telescope and the instruments are extremely sensitive to temperature and temperature variations: to protect them against thermal radiation, the SSH rear side is covered by Multi-Layer Insulation (MLI) sheets, resulting in high PVA operation temperatures, close to the diode qualification limits. For this reason, the current baseline is to use an external Si bypass diode mounted in one cropped corner of the solar cell: this is the same choice that was made for Herschel and Planck. In that case, the data available from the temperatures measured in flight showed a maximum temperature of 130°C, reached at Winter Solstice with 0° SAA and non-operating solar cells [3]. This bypass diode is performing a specific life test at 195°C in the frame of Euclid PVA qualification. With the mentioned flight and qualification heritage, the selected design is expected to withstand temperatures up to 150°C without any performance degradation.

Each string is terminated by a blocking diode to prevent current flow back and reverse voltages in case of (unexpected) shadowing or failure.

All the harness connecting the PVA parts is redundant, ensuring a robust power transfer concept.
The SA sections are connected to the PCDU via power harness. In order to reach a good compromise between mass and voltage losses, an AWG 22 has been selected for power harness.

A set of Sun sensors (FSS and SAS) is accommodated on top of the SSH. These sensors are part of Attitude and Orbit Control Subsystem and connected to the on-board computer.

In order to guarantee a robust design in terms of bonding, isolation and grounding, different technical solutions have been implemented.
The solar array panel cores are connected to the spacecraft structure through bleed resistors integrated into the wiring on the back side of the panels. In order to comply with the redundancy concept, two bleed resistors per panel are implemented.
The complete SSH structure and the sensor brackets are referenced to the spacecraft potential. The insulation between the interface points of the structure and the solar array panels is achieved by means of Glass Fibre Reinforced Plastic (GFRP) brackets.
A double insulation concept has been applied for all the front and rear side components in order that, even after a single failure, the elements are still insulated each other. Apart from the standard measures, on the panel rear side the electrical components have been covered with Kapton foil. This configuration has been adopted to avoid the risk of electrical contact between the rear side metallic parts (e.g. bleed resistors, blocking diode terminals) and the MLI internal layers, which are grounded to the spacecraft.

The MLI external layer (facing the panel substrate) is

 DOI: 10.1051/e3sconf/20171618005

made by uncoated Kapton foils, in order to guarantee the double insulation concept described above. The other MLI layers are electrically interconnected and bonded to the grounding rail by means of bonding straps. This grounding rail is connected to the spacecraft voltage reference. Fig. 6 provides an overall scheme of the grounding and isolation concept.

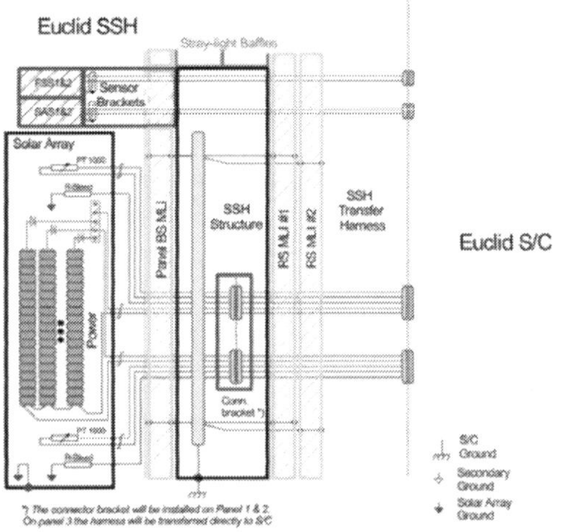

Figure 6. SSH bonding and grounding philosophy

The implemented grounding concept avoids the build-up of electric charges on solar panels. Furthermore, it ensures that the involved equipment (structure and Thermal Control Hardware) is connected to the spacecraft reference with low resistive paths.

As characteristic of the sequential shunt switch regulation, the number of SA sections which are connected by the PCDU to the power bus depends on the spacecraft power demand. As shown in Fig. 7, the sections have been arranged in order that the ones facing the most sensible PLM area will be always connected during the science operations: this configuration ensures a lower and stable temperature in the area with higher radiative coupling between SSH and PLM.

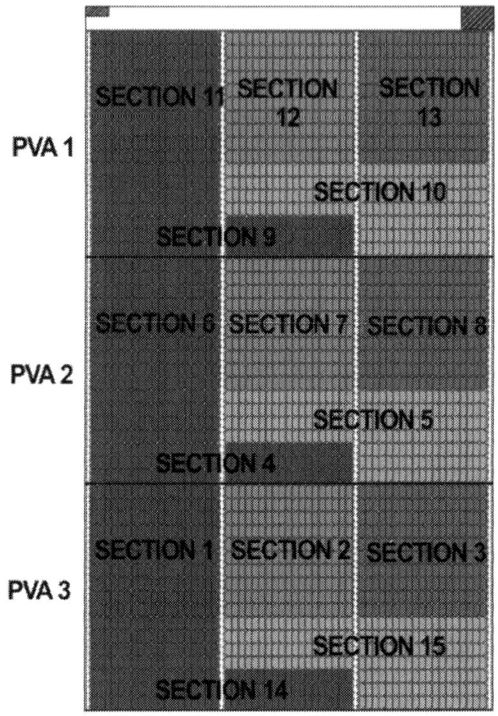

Figure 7. SA sections configuration

In view of maximum decoupling between high temperature solar panels and instruments, further features have been implemented in the SSH thermal design. As shown in Fig. 8, a multiple MLI-stack installed on the back side of the SSH central part allows a minimized heat flux from the (PVA) panel rear side towards the PLM, as the three MLI stacks have defined intermediate spaces to radiate also laterally.

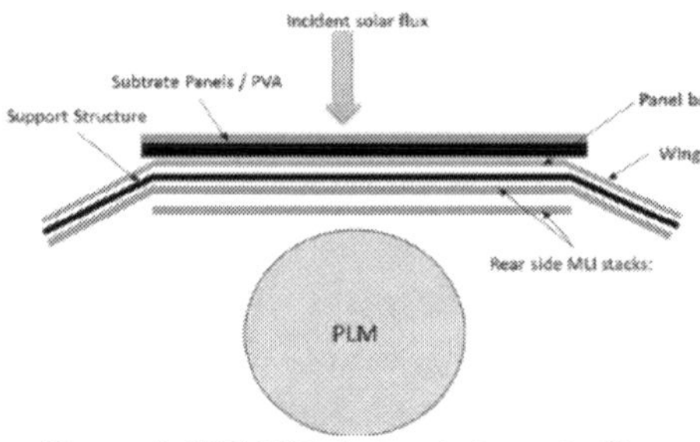

Figure 8. SSH MLI design (schematic illustration)

The selected configuration with nineteen 3G30C cells in series guarantees a stable working condition under different temperature cases, as shown in Fig. 9.

 DOI: 10.1051/e3sconf/20171618005

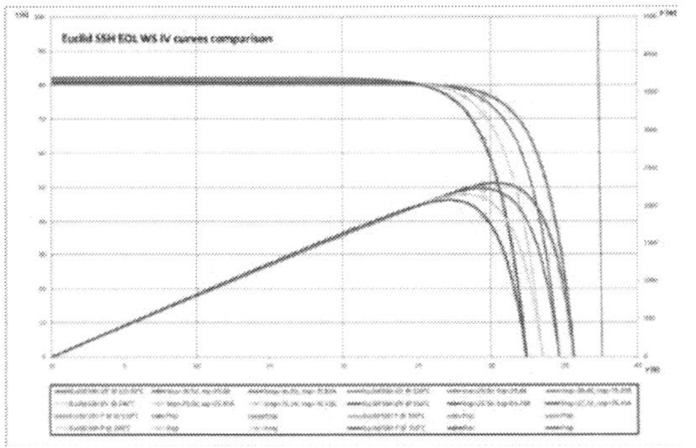

Figure 9. SA curves vs. temperature

According to the thermal analysis, the expected maximum temperature on shunted SA sections is 137°C at L2 and, as described previously, the maximum temperature measured on Herschel and Planck was lower than 130°C. Under these extreme conditions, the SA operating voltage (Vop) might exceed the maximum power point voltage (Vmpp) but, once the SA sections are connected to the power bus, the temperature decreases and a self-recovery is ensured, with Vop on the left side in the SA curve (Vop < Vmpp).

Two temperature sensors (PT1000) are installed on each panel in order to monitor the substrate rear side temperature. These sensors are connected to the on-board computer.

The Power generated by the SA during the mission is between 1750 W and 2450 W, depending on SAA, solar flux and ageing.

4. BATTERY

The Battery is made by Saft (FR). It is composed of VES16 Li-Ion cells developed by Saft, arranged in 6s-6p topology.

VES16 cell has been qualified in 2011 in the frame of a contract with CNES [4]. It is a D-size Li-Ion cell with 4.5Ah capacity, initially dedicated to both Geostationary Earth Orbit (GEO) and Low Earth Orbit (LEO) programs, which utilization can be extended to other space missions. Its electrochemistry is based on Saft heritage on high capacity cells for space application, tailored for a cell with smaller capacity. The main cell characteristics are depicted in Fig. 10.

The arrangement of the cells in Euclid battery is done in a single deck version: this is one of the configurations qualified in the frame of ESA's General Support Technology Programme #5.2 (GSTP-5.2) activity [5]. It consists of a parallel wiring of six strings, each one composed of six cells in series. The cells are glued by resin between the baseplate and the top plate.

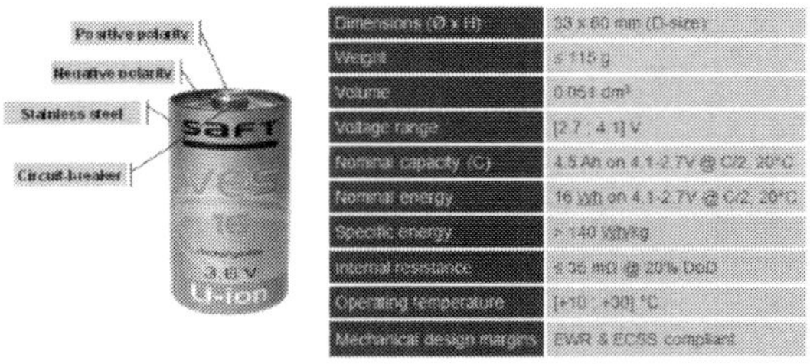

Figure 10. VES16 Battery cell

A Simplified Balancing System (SBS) is implemented on the terminals of each individual cell. The principle of this balancing electronics system is based on cell current consumption above a given voltage threshold. The system is fully autonomous, i.e. it does not need to be commanded/monitored by the PCDU. The SBS architecture is based on a simple circuitry with a core micro shunt associated to a set of resistors, in order to adjust the threshold at which the balancing function is activated
The serial connection of the cells in a string and the SBS is implemented with an equipped printed circuit board (PCB). This PCB provides also the access for cell voltage monitoring.

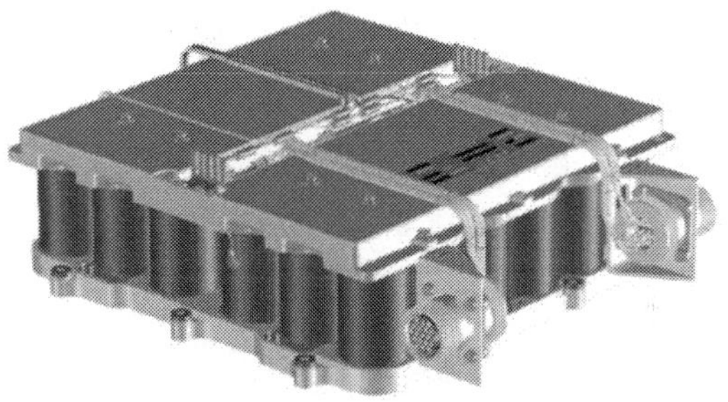

Figure 11. Euclid battery general view

Euclid is nominally an eclipse-free mission and therefore the Battery is nominally used during Launch and Early Orbit Phase (LEOP) only. Contingency cases might imply the use of the Battery in case of attitude loss or to cover any peak power demand from the spacecraft. The sizing case is represented by the energy required by the spacecraft from lift-off until few minutes after the separation.
The Battery has been specified to sustain one hundred cycles for all the mission duration, considering an average power discharge of 300W for seventy minutes, including a peak power discharge of 850W lasting ten minutes at each cycle, in the temperature range 10°C to 20°C.
In order to evaluate the impact of micro-cycles on electrical performances, three VES16 cells are subjected to an accelerated life-test (discharge at 4W for five minutes at 10°C, for several thousand cycles). At the time of the issue of this paper the test is still running in

the European Space Battery Test Centre at Estec.

The Euclid battery has a nameplate energy of 576 Wh: this size allows to cover the worst case duration of launch with a Depth Of Discharge (DoD) lower than 50%.
Even if Euclid battery presents a design coming from GSTP-5 battery range, the size difference regarding GSTP-5 Qualification Models (QM) and the specific electrical design to insure high discharge current (52.5A during pulse) justify its classification in Category C.
Then, in addition of electrical, thermal and mechanical analysis performed, Saft is manufacturing two battery models:

- one Structural and Thermal Model (STM) dedicated to simulate thermally the battery dissipation during different phases, by the use of specific heater bands. This battery is mechanically representative of Flight Model (FM);
- one Engineering Qualification Model (EQM) which aims to qualify the battery to Euclid requirements.

The qualification and space mission of this battery will extend VES16 battery range and will give more heritage to Saft for cosmology missions.

5. POWER CONTROL AND DISTRIBUTION UNIT

Terma (DK) manufactures the Power Control and Distribution Unit (PCDU).

The main functions required to the PCDU are to generate and distribute a single, fully regulated bus of $28V_{DC}$ with a 1700 W (average) and 2200 W (peak) power capability in sunlight mode and 400W power capability during eclipse.

To meet its functional requirements, the PCDU is required to include the following elements:

- a three-domain regulation working in majority voting control;
- a bus capacitor bank;
- fifteen Single Point Failure Free Sequential Switching Shunt Regulator (SS3R) modules;
- two Battery Charge and Discharge Regulator (BCDR) modules;
- sixty-eight Latching Current Limiters (LCL) and six Re-triggerable Latching Current Limiters (R-LCL), with different current limitation capabilities;
- 144 heater lines
- three Non-Explosive device Actuators (NEA) and three Frangibolt actuators
- Two Command and Monitoring modules, including Mil-Std-1553 Remote Terminal interfaces.

The PCDU block diagram is shown in Fig. 4.
The PCDU manages the SA power and the battery power to provide a regulated 28V ± 0.5% Main Bus.
A Main Error Amplifier (MEA) function manages the power regulation such that the 28V bus supply will always be provided when sufficient energy is available, either from SA, from the Battery or from both. Three independent error amplifiers are voted into one reliable MEA line that regulates the shunts, the battery charging and battery discharging in a classical three-domain regulation scheme. Each error amplifier is voted together by means of standard 2/3 majority voting circuitry forming the reliable MEA control line. The MEA line is distributed reliably to all power regulators.
The PCDU will always use SA power for Main Bus regulation as long as sufficient power can be derived for the applied main bus load. Otherwise, the remaining power required will be provided from the battery.
The PCDU does not include software for any power management function. The transitions between the three domains are automatic and lead to negligible transients to the main bus voltage.

The SA power conditioning is performed by fifteen SS3Rs. Each shunt is connected to a SA section and it is required to regulate up to 7A. The architecture is such that no single failure can lead to the loss of a SA section. The excess SA power, not needed to satisfy the main bus and the battery charge, will be dissipated on the SA (i.e. shunted) and only a minor fraction is dissipated in the PCDU. Three SS3Rs include also an EGSE interface in order to power the spacecraft via umbilical lines during the ground testing.

The battery charge regulators (BCR) and the battery discharge regulators (BDR) manage the battery charging and discharging according to available power and battery power.
BCR and BDR functionalities are implemented on the same BCDR module. For ensuring a redundant configuration the PCDU is equipped with two BCDR modules operating in parallel. Each BDR can sustain the power needed to supply the spacecraft at launch.

A capacitor bank is foreseen at main bus level in order to:

- stabilise the bus voltage regulation loop
- ensure a low output impedance
- filter the switching noises due to the regulation.

The PCDU provides seventy-four LCL outlets to protect the users/loads. The LCLs are divided in five different classes, with different limitation capabilities. All the LCLs can be switched-on/off by 1553 commands, with the exception of six R-LCLs, devoted to critical loads, which cannot be switched OFF and automatically attempt to switch ON when powered or after a retrigger interval when a trip off event occurred. Part of the LCLs are directly commandable via High Power Command. Some LCLs are required to be configured as "ON-

Protected", i.e. no single failure can lead to have an LCL which cannot be switched OFF. All LCLs have an under-voltage lockout function that ensures switch-off in case of main bus under-voltage.

The on-board computer is in charge of the SVM and PLM thermal control, but the heaters actuations pass through the PCDU. Each heater line can be individually actuated. To avoid permanent supply of an heater due to switch failure or heater-structure contact, a second serial switch is implemented. Heaters are grouped up to a maximum of eight to reduce the number of these switches.
The status of the switches and the current of each heater group is monitored via telemetry. In addition to these, sixteen high stability heaters are requested from the PLM thermal control. They are implemented with high output voltage stability.

6. CONCLUSIONS

This paper presented the Euclid EPS, describing the major design drivers, the architecture and the characteristics of the equipment.
The Euclid System Preliminary Review (PDR) has been successfully completed in 2015 and the program is in its phase C. At the present time, all the power equipment PDR's have been completed, while the relevant Critical Design Reviews are under preparation. The outcomes of the detailed design phases have been presented: analyses and test results up to the issue of this paper are promising.

7. REFERENCES

1. Bladon, G. (2015). ESA's Core Technology Programme - Paving the way for ESA's Cosmic Vision Plan. SRE-A-COEG-2015-001, European Space Agency, Noordwijk, The Netherlands. Online at http://sci.esa.int/core-technology-programme-brochure/ (as of 17 March 2016).

2. Jackson, B. & Rueda, P. (2005). Power System for the Herschel & Planck Spacecraft. In Proc. 7th European Space Power Conference, ESA SP-589 (CD-ROM), European Space Agency, Noordwijk, The Netherlands.

3. Ciancetta, E., Deplus, N., Zanella, P., Neto, A. & Fernadez, E. (2014). Herschel and Planck Power System Flight Return Experience. In Proc. 10th European Space Power Conference, ESA SP-719 (CD-ROM), European Space Agency, Noordwijk, The Netherlands.

4. Prévot, D. & Vigier, F (2012). Saft VES16 Li-Ion qualification synthesis report. S0115-12, January. 2012.

5. Remy S., Mosset E., Lefeuvre S., Nestrodi M. (2014). VES16 Li-Ion Batteries Range Qualification. In Proc. 10th European Space Power Conference, ESA SP-719 (CD-ROM), European Space Agency, Noordwijk, The Netherlands.

 DOI: 10.1051/e3sconf/20171618006

INTRODUCING THE RADIATION-HARDENED VOLTAGE REGULATORS RHFL4913 AND THE RHFL6000

G. Bonna [(1)], **C. Ribellino** [(2)], **S. Pappalardo** [(3)]

[(1)] *STMicroelectronics S.p.A., Stradale Primosole n.50, 95121 – Catania, Italy. Email: giovanni.bonna@st.com*
[(2)] *STMicroelectronics S.p.A., Stradale Primosole n.50, 95121 – Catania, Italy. Email: lillo.ribellino@st.com*
[(3)] *STMicroelectronics S.p.A., Stradale Primosole n.50, 95121 – Catania, Italy. Email: salvo.pappalardo@st.com*

ABSTRACT

With the progressive scaling of the modern CMOS processes arising during the recent years, for the devices operating in space the susceptivity to the over-voltage events has increased. They have become more stringent the requirements in terms of accuracy of the supply voltage worth their reliability. For the case of CMOS memories operating at 1.5V it is essential to guarantee a variation of the nominal V_{OUT} to within 5% including those arising on the supply voltage during the transients.

Finally, for a voltage regulator operating in space one of the most salient effects are the voltage transients induced on the output by the single events (SET) that may exceed the aforementioned 5% threshold. This can be cause of damage into the circuitry fed downstream.

Among the devices in ST portfolio designed to cope with the mentioned effects there are two voltage regulators, the RHFL4913 and the RHFL6000.

The first one was initially developed for the harsh environment of the Large Hadron Collider. It was tested by the CERN and it was proven that it is able to operate without problems at up to 100Mrad TID at High Dose Rate without problems (1). Subsequent tests performed at low dose rates (expected in space environment), showed that this device is also ELDRS free up to 1Mrad (2). Further tests performed by NASA highlighted also a certain susceptivity to the SET (3) but this issue can be circumvented by the adoption of a properly sized up RC low pass filter put between the regulator output and the load as shown in fig.1 (4).

With the above provision, this regulator is suitable for operating in space. Currently, the RHFL4913 is QMLV qualified.

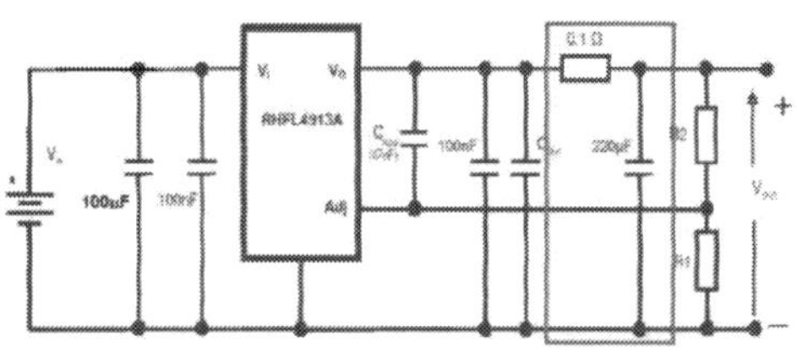

Figure 1 - Typical configuration recommended in space applications using the RHFL4913 with the RC filter (marked in red) for mitigating the Single Event Transients on the output

1. INTRODUCING THE RHFL6000

Nevertheless, a drawback of the filter mentioned above is the increase of the minimum V_{IN}-V_{OUT} dropout needed for the regulation due to the presence of the 0.1Ω resistor in series to the output. In some applications, it would be preferable to gain efficiency by using a regulator capable of performing this action with the only capacitors used for the compensation at both input/output terminals. In this respect, the RHFL6000 can be considered an evolution of the RHFL4913. It was intended for circumventing the SET mitigation issue without need of any filter. By specific tests, it was seen that, with a combination of capacitors of proper characteristics put at both I/O terminals, a PCB configured as shown in fig.2 and with proper applicative rules it is possible to achieve SET of amplitude limited to within 3.3% of absolute V_{OUT} for up to 120MeV LET. Finally, this device is also QMLV qualified.

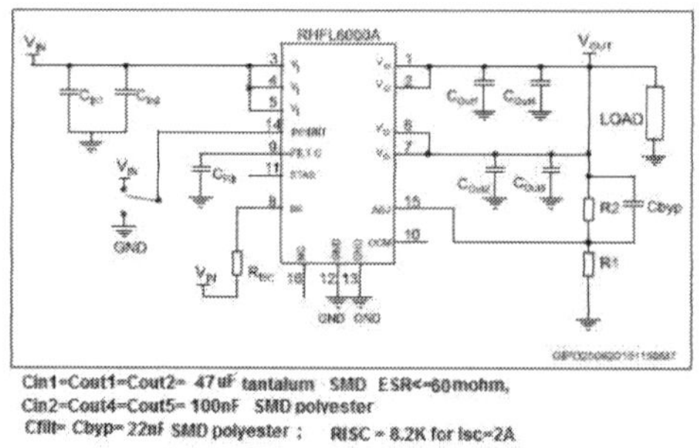

Figure 2 - Pcb proposed for RHFL6000 with the recommended values for the components

For both regulators, pictures of the dices are given in figg.3 and 4 for RHFL6000 and RHFL4913 respectively.

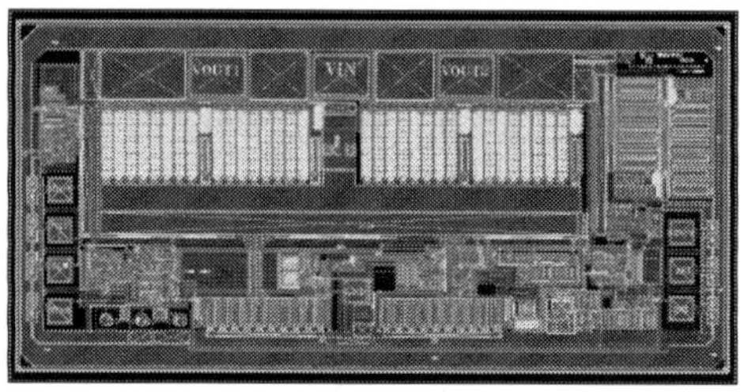

Figure 3 - RHFL6000 die

DOI: 10.1051/e3sconf/20171618006

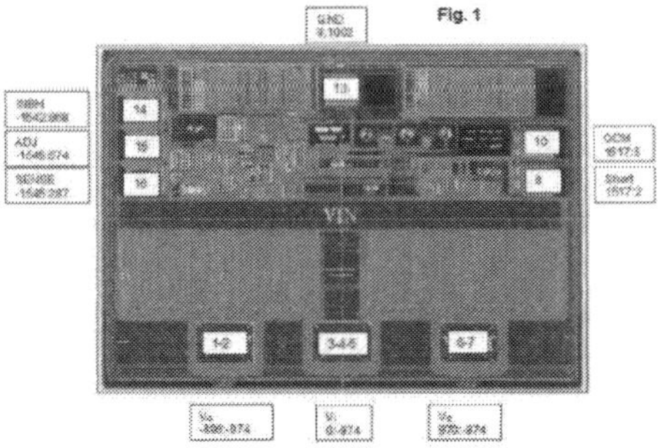

Figure 4 - RHFL4913 die

The immunity to radiation is achieved through a precise strategy involving the process selection and the definition of proper rules for both the design and the layout.

For what concern the first point, the ST process features high-speed bipolar transistors (Ft = 6Ghz) intrinsically robust to the TID: tests performed at CERN at HDR showed a degradation of the β of 40% at 500Krad TID for both PNP and NPN. These bipolar transistors have an also relatively thin base (and this is, very likely, what makes these cells sensitive to the SEE).

As mentioned, the radiation immunity is achieved also through layout rules common to the rad-hard practice of design. These are intended to act against the parasitic effects induced into the single cells (one of these being the parasitic N-channels activated by the TID) by the extensive use of n+ ring into the NPN and PNP cells and the resistor pockets and p+ all over the dice.

A similar approach applies to cope with the Single Event Effects. On this subject, the layout of the pass element was also drawn by applying proper rules aiming at limiting the short transient current arising during the single events and causing peaks on the output proportional to the VIN-Vout dropout.

The basic layouts of the cells implemented are shown in figg. 5, 6 and 7 respectively for NPN, PNP and resistor pockets.

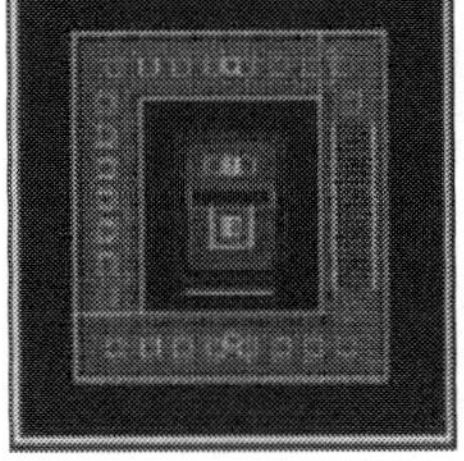

Figure 5 – NPN basic layout cell

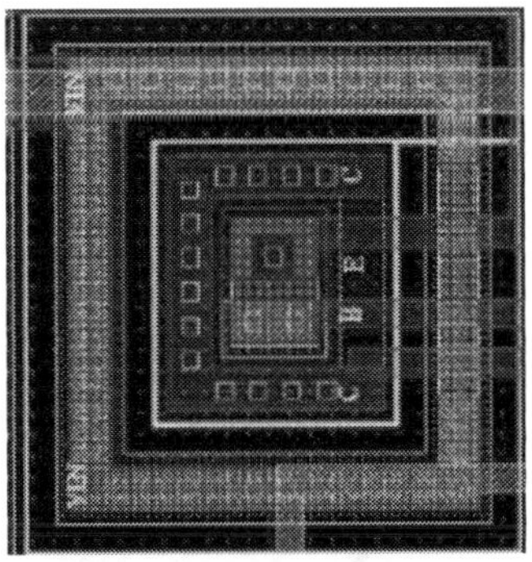

Figure 6 – PNP basic layout cell

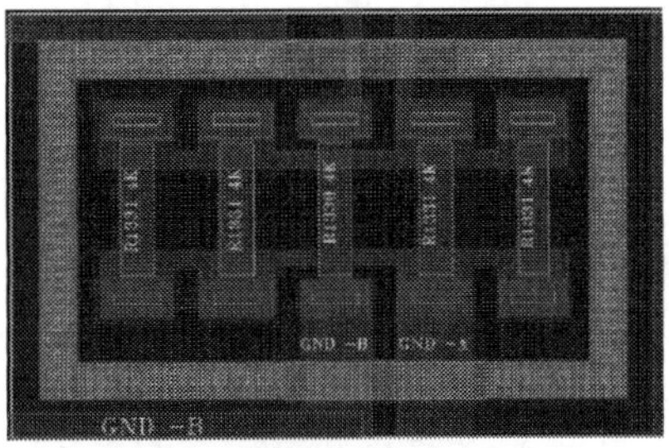

Figure 7 – Layout cell for Resistor pockets

Still on the TID provisions, at design level care also has been taken to set a minimum level of bias current to minimize degradation effects on the β and potentially bad operation of the regulator caused by the increased leakage.

In addition, both regulators are realized using only bipolar transistors which are known to be intrinsically more robust compared to their MOS counterpart.

Finally, for the SET immunity of the RHF6000, a proper feedforward scheme embedded (patent proposal is pendign) switches ON during positive variations of the output voltage and provides a sinked current on the output (see Isinked in fig.8) contributing to the discharge of the bulk capacitor Cout.

This helps limiting the duration of the positive output transients and allows a fast recovery to the stationary regime of the regulator in a time shorter compared to a device with single ended output stage (fig.8). It turn helpful in case of long-duration SET induced.

E3S Web of Conferences **16**, 18006 (2017)
ESPC 2016 DOI: 10.1051/e3sconf/20171618006

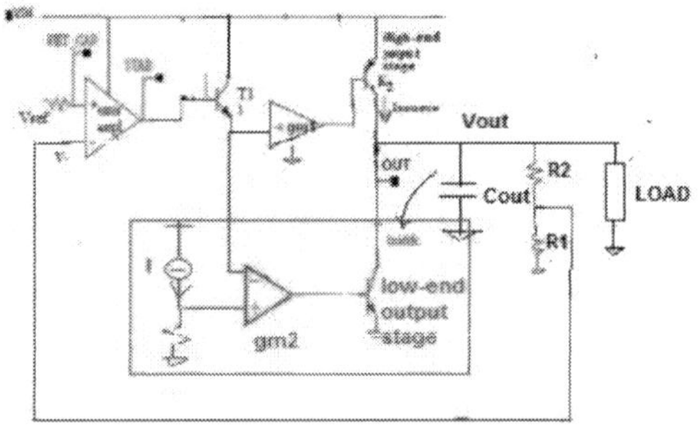

Figure 8 - RHFL6000: basic control loop architecture view

2. TECHNICAL SPECIFICATIONS AND EXPERIMENTAL RESULTS

As shown in TABLE1 the RHFL6000 is tested in accordance to the MIL-STD-883J Method 1019.9 in ELDRS conditions. For details regarding the TID tests, please refer to the table 1 [5]:

Table 1 - ELDRS Condition for tests on RHFL6000

TABLE 1

Type	Conditions	Value	Unit
TID	50 rad(Si)/s high dose rate up to	300	krad
	10 mrad(Si)/s low dose rate up to (1)	100	
	ELDRS free up to (1)	100	
Output voltage radiation drift	From 0 krad to 300 krad at 50 rad/s, MIL-STD-883J method 1019.9	<1.1	ppm/krad
Quiescent current (ON state)	From 0 krad to 300 krad at 50 rad/s, MIL-STD-883J method 1019.9 V_I = 2.5 V to 12 V, I_O = 5 to 30 mA, T_J = -55 to +125 °C	<15	mA

For what concern the SET, the test were performed with the following bias configurations and trigger thresholds [5]:

Table 2 - SET: bias configurations and trigger thresholds

For each configuration the parts were tested with the following bias conditions:

- Bias 1: VIN = 3V, VOUT=1.5V, Vinhibit=0V, IOUT = 1A, VINHIBIT =0V,
- Bias 2: VIN = 9V, VOUT = 0V, Vinhibit=9V, IOUT = 0A, VINHIBIT = 9V
- Bias 3: VIN = 4V, VOUT = 2.5V, Vinhibit=0V, IOUT = 1A, VINHIBIT = 0V
- Bias 4: VIN = 7V, VOUT = 5V, Vinhibit=0V, IOUT = 300mA, VINHIBIT = 0V
- Bias 5: VIN = 9V, VOUT = 1.5V, Vinhibit=0V, IOUT = 100mA, VINHIBIT = 0V
- Bias 6: VIN = 12V, VOUT = 9V, Vinhibit=0V, IOUT = 300mA, VINHIBIT = 0V

Trigger thresholds for SET test are defined, according to the bias conditions, in the following table:

	Bias 1	Bias 2	Bias 3	Bias 4	Bias 5	Bias 6
Vcc	3V	9V	4V	7V	9V	12V
V_{out}	1.5V	0V	2.5V	5V	1.5V	9V
Upper trigger threshold	1.55V	0.1V	2.58V	5.15V	1.55V	9.27V
Lower trigger threshold	1.45V	-	2.43V	4.85V	1.45V	8.73V

Results and conclusions are summarized in tables 3, 4, figgs. 9 and 10 [5].

Table 3 – RHFL6000: SEL/SET Perfromances

Feature	Conditions	Value	Unit
SEL/B performance	LET = 120 MeV*cm²/mg V_I = 12 V	No Latchup/burnout	-
SET performance during events	LET = 32 MeV*cm²/mg Saturated cross-section = 6.18*10⁻⁵ cm² V_{IN} up to 9 V V_I - V_O ≤ 7.5 V I_{OUT} < 300 mA	± 15% max over less than 300 ns	% of V_O
	LET = 120 MeV*cm²/mg V_{IN} up to 12 V V_I - V_O < 3.0 V I_{OUT} < 300 mA	No SET above ± 3%	
	LET = 120 MeV*cm²/mg V_{IN} up to 4 V V_I - V_O < 1.5 V I_{OUT} < 1 A	No SET above ± 3.3%	

Table 4 - RHFL6000: SET Cross Section

7.3.5.1. SET Cross sections

LET Eff (MeV.cm².mg⁻¹)	SET Cross Section (cm²)			
	N° 9	N° 10	N° 11	N° 12
120	3.00E-06	4.20E-05	6.20E-05	5.40E-05
102.8	1.00E-06	1.30E-05	3.25E-05	3.05E-05
86.37	2.00E-06	8.00E-06	2.70E-05	2.40E-05
70	<1.00E-06	8.00E-06	2.30E-05	2.05E-05
60	<1.00E-06	9.33E-06	3.40E-05	3.35E-05
46.5	<1.00E-06	<1.00E-06	1.30E-05	1.50E-05
32.3	<1.00E-06	<1.00E-06	<1.00E-06	<1.00E-06

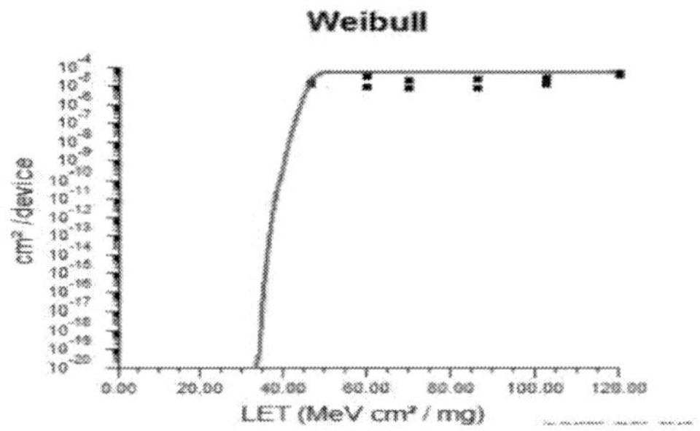

Figure 9

The main objective of this test was to evaluate the sensitivity of the RHFL6000, a Rad-Hard Voltage Regulator to Single Event Latch up (SEL) and Single Event Transient (SET).
The irradiation was performed at RADEF with a maximum LET at **120 MeV.cm²/mg**.
The mains conclusions are the following :

- **No SEL** was observed up to LET of 120 MeVcm²/mg and up to temperature of 125 °C.
- **No SET** was observed on bias 1 to 4 and 6
- **SET** was observed in **bias 5** (VIN= 9V, VOUT= 1.5V, IOUT= 100mA, VINHIBIT= 0V).
  The LET threshold was 32.3 **MeV.cm²/mg** with a saturation cross-section of **6.18E-5** cm²

Figure 10

During these experiments, the trigger threshold for SET detection was set a 3.3% of the absolute value of V_{OUT}.

It is worth noticing that, of all the configurations bias under examination, SET exceeding the mentioned target were detected only on the one labelled as bias 5. The corresponding LET threshold was 32.3MeV*cm²/mg and the saturation cross section 6.18 e-05 cm².

The worst case SET observed is shown in fig.11

DOI: 10.1051/e3sconf/20171618006

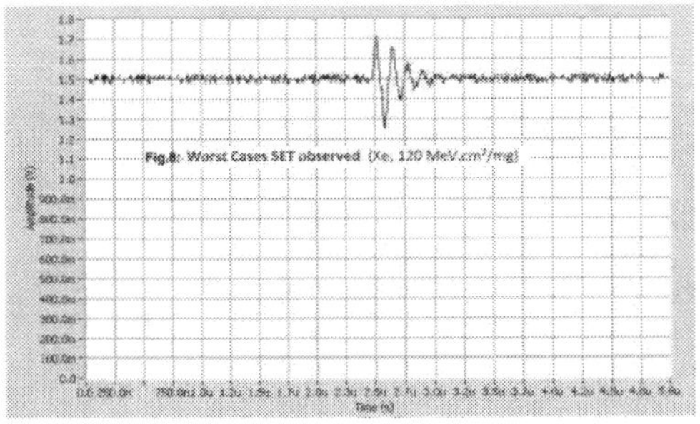

Figure 11 - RHFL6000: SET Worst Case

The standard performances can be summarized into the table below:

- Input voltage range from 2.5V to 12V
- 2A guaranteed output current
- Low dropout: 0.3V typ. @ 0.4A
- Embedded over-temperature and overcurrent protection
- Adjustable overcurrent limitation
- Output overload monitoring/signalling
- Adjustable output voltage
- Internal control loop accessible via an external pin
- Inhibit (ON/OFF) TTL compatible control
- Programmable output short-circuit current
- Remote sensing operation
- Rad-hard: guaranteed up to 300krad
  MIL-STD-883J Method 1019.9 high dose rate and 0.01 rad/s in ELDRS conditions
- Radiation environment: SET/SEL/SEB:
  - SEL free @ LET=120MeV*cm^2/mg
  - SET: less than 3.3% of V_{OUT} @ 120MeV
- Heavy Ions SET dedicated internal circuitry implemented for absorbing output transient
- Operating junction temperature range:
  -55°C to +125°C

3. ADDITIONAL FEATURES OF THE RHFL6000 STABILITY

In order to comply with the requirement of the 5% absolute max target of Vout during a transient, it was seen that it is essential that the regulator operates with capacitors, at both I/O, having the following constraints: Cout>=100µF; ESR <=30mΩ. This means that the regulator must be stable with these capacitors. In some applications, the mentioned condition can lead to stability margins not compliant with the targets in space. One of these cases is when a load supplied at the output is too low. In this case, it is possible that the zero brought by the output capacitor is not sufficient to increase the phase near the cutoff frequency Ft because located too far from it. To overcome this issue, in the RHFL6000 design it was considered the possibility of adding a further external zero Z_L, close to Ft, using the classical lag compensation approach. By inserting an RC group between the terminals CSTAB and GND, if such zero falls within one decade left to Ft, it can provide a positive phase shift to the Bode diagram and the phase margins are thus increased. To achieve this effect, one must consider that an external RC group in that location adds also a pole P_L left to the zero. If the P_L/Z_L ratio is not adequately calibrated, the negative shift of P_L can compensate that of Z_L. Therefore, it is fundamental to determine a convenient ratio for the doublet (P_L,Z_L) to avoid this effect. In this respect, a model compatible with Pspice is made available by ST with which it is possible, starting from the application data, to determine Ft and optimize the mentioned external impedance.

For a generic series RC group with an impedance Ro in parallel we have

$$P_L = 1/[2\pi(R+Ro)] \quad (1)$$

and

$$Z_L = 1/(2\pi RC) \quad (2)$$

For the RFL6000, Ro is the dc impedance seen at the output of the error amplifier and is ≅ 1MΩ @T=25V

Therefore,

$$P_L/Z_L = R/(R+Ro) \quad (3)$$

Concerning the main poles and zeroes determinant for the stability, with reference to the general LDO scheme in fig.12, we have

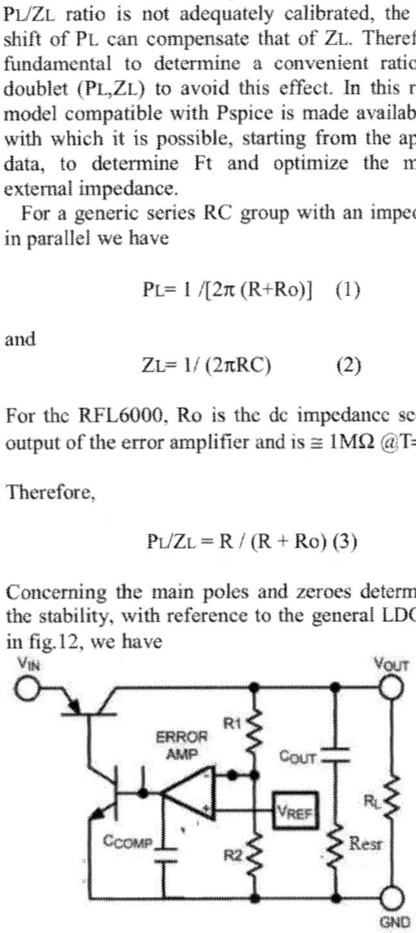

Fig. 12: Basic scheme of an LDO

$$Po = \frac{1}{2\pi RoutCout} \quad (4)$$

and

 DOI: 10.1051/e3sconf/20171618006

$$Zo = \frac{1}{2\pi ResrCout} \quad (5)$$

where:
Rout = impedance seen between Vout and GND, given by the parallel combination of the load RL with the resistor divider couple R1+R2;
Resr= series resistance associate to the output capacitor Cout.
There is also an internal pole, Pint, located at the output of the error amplifier. For the RHFL6000, Pint ≅ 800 Hz.

An applicative example of the achievable result is shown in fig. 13. Application conditions are:

Vin=5V Vout=2.5V; RLOAD=250ohm, Cout=220µF, ESR=30mΩ, R1=R2=10KΩ.

In this example, Ft=24Khz, ZL was set at 10khz and PL at 9Khz, that is away on the right from Pint.and Po. This was achieved with an RC group Cp=300pF, Rp=160Kohm. Notice the increased phase near to Ft

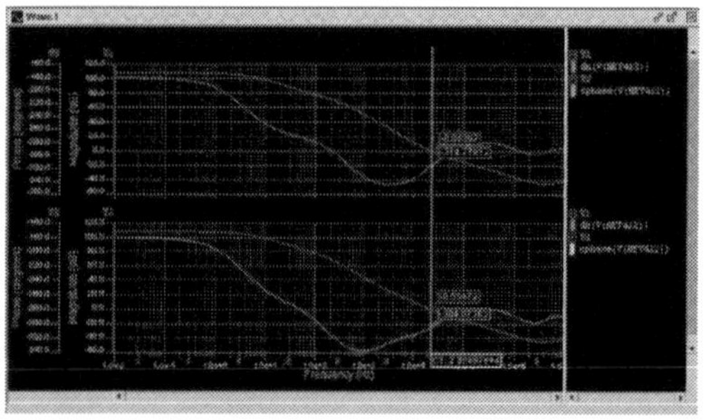

Figure 13: Effect, on the Bode diagram. of an RC series group (R=200KΩ, C=300pF) put between CSTAB and GND.

About the impact on the transient response accuracy, relevant plots are shown in fig.14.

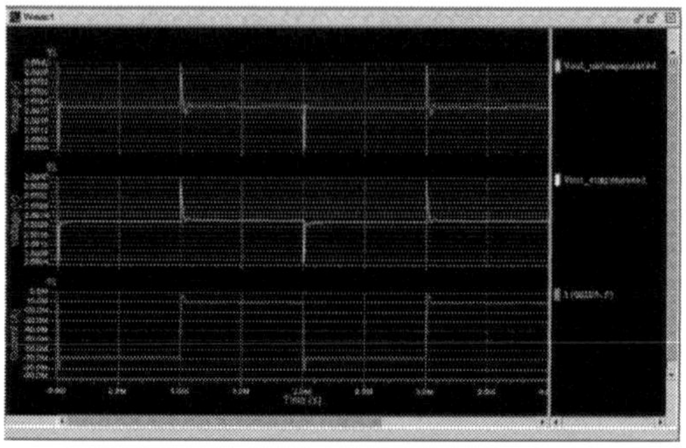

Figure 14: Effect of the added RC group on the load transient accuracy of the RHFL6000 with the RC group between CSTAB and GND inserted (yellow) and the same group unconnected (green).

4. CONCLUSIONS

The RHFL6000 is a regulator showing high performances in terms of immunity to the TID absorbed and to the Single Event Effect.

Tests showed that, with a proper configuration PCB, part selection and applicative guidelines this regulator is suitable for sourcing the supply voltage to devices requiring a tight control of the supply voltage amplitude, such as the 1.5V FPGA memories, microprocessors and other devices realized into the modern submicron CMOS process.

5. REFERENCES

(1) Radiation performances of the L4913 voltage regulator, N. Boetti, P.Jarron, B. Kisielewsky, F. Faccio , Radiation Effects Data Workshop, 2002 IEEE Conference Publications

(2) Evaluation of STMicroelectronics RH-L4913 Positive Hardened low drop voltage regulator. P. BRIAND, STMicroelecrronics, Marielle BELASIC,CNES, Jean-Francois VADROT, STMICROELECRONICS, 2002

(3) SEE test Report V.1.0 Heavy ions SEE test of RHFL4913 from STMicroelectronics, Christian Poivey, Hak Kim, Anthony Sanders, NASA Jan. 23, 2006 - T121805_RHFL4913

(4) SEE test Report V 3.0 Heavy Ino test of RHFL4913 from STMicroelectronics with SET mitigation circuitry, Christian Poivey, John Pandolf, Anthony Sanders, Hak Kim, Sept 19, 2007

(5) TRAD - HEAVY IONS TEST REPORT
Single Event Effects RHFL6000
Sept. 23, 2014

 DOI: 10.1051/e3sconf/20171618006

(6) Relashionship between frequency response and settling time of operation amplifiers - by Yeshwant KAMATH, Robert MEYER, Paul GRAY, IEEE JSSC, dec 1974

DOI: 10.1051/e3sconf/20171618007

MICROSATELLITE POWER CONTROL AND DISTRIBUTION UNIT FOR THE INNOSAT PLATFORM

Moysis Tsamsakizoglou [(1)], **Henrik Löfgren** [(1)], **Marcus Gunnarsson** [(1)]

[(1)] ÅAC Microtec AB, Uppsala Science Park, SE-75183 Uppsala, Sweden,
Email:moysis.tsamsakizoglou@aacmicrotec.com

ABSTRACT

The Power Control and Distribution Unit (PCDU) described in this paper is a custom design for the InnoSat satellite platform. Particular attention is given to the architecture, design techniques and general failure mitigation approach that has been adopted to meet the low cost constraints of the platform while maintaining reliability for its entire mission lifetime. A point of importance is the scalability of the design in order to meet different payload power demands from mission to mission while minimizing NRE work. The very important areas of component selection, testing and qualification are also described. In order to quantify some of the topics discussed, the description is mostly focused on the PCDU of the first InnoSat based satellite named MATS and which is expected for launch at 2018.

1. INTRODUCTION

InnoSat is a micro-satellite platform (40 – 100 kg) that targets scientific and earth observation missions [1]. The platform in its low power configuration, on which the first satellite MATS is based, has an average power consumption of 80W (in its high power mode can this can reach up to 180W). Therefore two of the design requirements have been scalability and modularity so that changes in power requirements can be easily absorbed by the design with minimal NRE effort. Another constraint coming from the system has been the cost which has had a large impact on the component selection. However, the plan for commercializing the platform places a strong requirement for reliability and being able to support mission lifetimes up to 5 years is the platform baseline. Therefore producing a PCDU that provides the best trade-off between reliability and cost was necessary.

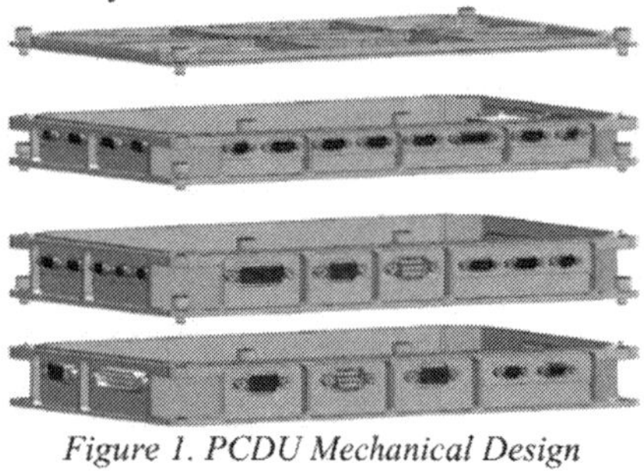

Figure 1. PCDU Mechanical Design

2. PCDU ARCHITECTURE

The MATS satellite architecture is based on a 28V main power bus while it provides isolation towards the low-power 5V bus used primarily by the data handling system. An 8s5p LiPo battery configuration is used for energy storage and a 19s9p solar panel for energy generation. The PCDU has been split in three separate units which are placed in a stack as shown in figure 1. The bottom unit is the battery management unit which handles the input power from the solar panel, performs battery charge regulation and under voltage detection on the 28V bus. The middle unit is the power management unit which receives the 28V battery regulated bus and through a number of latching current limiters performs power distribution to all the 28V consumers of the satellite. An isolated DC-DC converter generates the 5V low-voltage bus. The third unit on top is the control unit. Its core functionality of this unit is to receive the 5V power and distribute it through latching current limiters to the 5V consumers on the satellite. This unit also constitutes the brain of the MPDU since it contains the FPGA logic that collects housekeeping from all three units and controls all latching current limiters. The above functionality partitioning enables a flexible design since, within the InnoSat platform concept, it allows quick adaptation to requirements that can differ from mission to mission.

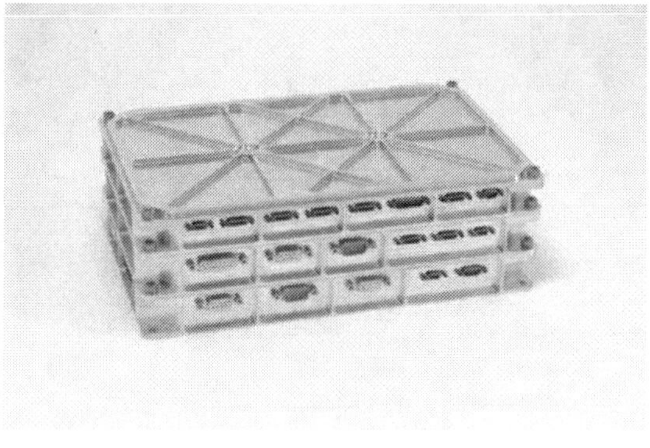

Figure 2. PCDU Engineering Model

3. DESIGN PHILOSOPHY

3.1. Component Selection

In order to keep the cost low while maintaining a high-enough reliability, a mix of Commercial-Of-The-Shelf

DOI: 10.1051/e3sconf/20171618007

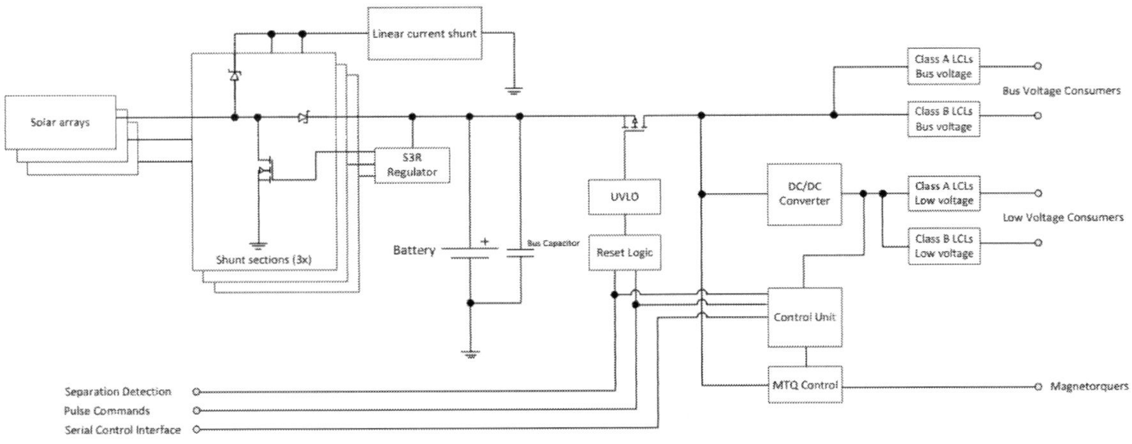

Figure 3. Simplified Power System Block Diagram

(COTS) and space-qualified components have been used in the design.

When selecting the COTS components, parts for which radiation test results are available has been used where possible. Parts which have previously been used successfully in actual flight missions have also been selected. In cases where no such data has been available, other factors have been taken into account such as similarity to components with known radiation performance, e.g. if it is produced with the same process at the same fab. Regarding general reliability, automotive parts have been preferred throughout the selection. Derating according to ECSS is also applied.
For critical parts where single-event failure modes exist which may be mission critical, such as the switch MOSFET in the flyback converter, space qualified parts have been used as no other form of mitigation has been identified. For some passive components, specifically tantalum capacitors and power inductors, military or space grade components have been used to ensure reliable functionality in the space environment.

Single Event Effects (SEE) are mitigated throughout the design with a number of different approaches:

- Internal supply voltages are current limited with automatic restart
- Digital signals are filtered to mitigate transients
- Discrete bipolar design used for many functions rather than integrated circuits.
- Triple-Voting is used on critical external signals such as separation detection.
- Triple-Modular-Redundancy (TMR) is used in the FPGA.

Overall, the component selection has been done as a trade-off where cost, reliability and power consumption has been taken account. The resulting approach maintains a higher level of reliability than a pure COTS solution, while the price and power consumption is kept significantly lower compared to a fully qualified solution. The resulting quality level is also well matched with the relatively short life-time requirements of the mission.

4. BATTERY MANAGEMENT

The battery charging is done through Direct Energy Transfer (DET), where the solar panels are connected directly to the battery through diodes. The solar panel strings are arranged into three groups, each group being connected through a separate diode. The system bus is unregulated and no battery discharge regulator is used.

The battery charge regulation is performed using Sequential Switching Shunt Regulation (S^3R) [2] with one shunt section per solar array string group. A maximum of one group is switching at any given time, with the other two groups being either fully shunted or fully on, depending on the battery voltage. The battery charge current is limited by a separate charge current regulator with linear shunts. The linear shunt MOSFETs are chassis mounted in order to provide low thermal resistance and maintain an acceptable temperature during shunting. With the specific configuration of battery and solar arrays that is used in the InnoSat platform, the linear current shunt will only operate in very specific cases (and for a limited period of time) where the battery voltage is low (so that no switching shunts are activated) and the satellite is in safe mode so that the bus current is low.

The battery is protected from deep discharge by an Under Voltage Lock Out (UVLO) switch which will turn off all parts of the system except the battery charging if the bus voltage reaches a critically low level. This switch is also used to keep the system powered down until the separation detection indicates that the satellite has separated from the launcher. The same

 DOI: 10.1051/e3sconf/20171618007

switch is also controlled by a hardware pulse command decoding circuit so that a full system reset can be performed through a pulse command. The pulse command decoding is edge triggered and requires both a rising and falling edge to trigger. It also uses timer circuits to only trigger on pulses with lengths within a specific interval. Together, these functions protect the system reset function from both transient and permanent errors on the pulse command inputs.

4.1. Scalability

As the Battery Management module is located in its own separate mechanical housing, and the interfaces are separated so that it can operate independently, the power capability of the platform can easily be extended by adding a second module providing the required interfaces and shunt sections for additional solar panel strings. The limiting factor for this scalability is mainly the current rating of the connectors and other upstream components such as the system UVLO switch. This modularity also enables the possibility to change from a DET system with an unregulated bus to a system which uses for instance Maximum Power Point Tracking (MPPT) or which has a regulated bus, without modifying any other modules in the unit.

5. POWER DISTRIBUTION

The PCDU handles power distribution for both the 28V and 5V power channels. Generation of 5V takes place at the power management unit while the distribution of it happens at the control unit.

5.1. DC/DC Conversion

The DC/DC converter is a 25W flyback type operating in always-on mode. It provides a 5V main output as well as a few auxiliary output voltages. The design is made with special attention to protection against single-event upset conditions. The converter has fault sensing which will detect:

- primary side controller errors
- output overload conditions or S/C
- output overvoltage conditions

A detected error condition triggers a complete shutoff of the converter and a timer circuit is then used for automatic restart after 2 seconds.

Components that are vital for the system performance have been selected for high reliability; e.g. the main MOSFET switch in the flyback circuit is a space-qualified type, and a military-grade two-channel optocoupler with known good radiation tolerance is used for isolation of the control loop and error signals from the secondary side.

5.2. Distribution channels

All 28V or 5V power outputs have latching current limiter protection and they are designed using discrete semiconductors and passive components. Following a discrete design approach not only provides improved reliability but makes the design more analysable. The current limit and trip time for each LCL is selected by choosing the values of particular programming resistors at the time of board assembly.

There are three types of distribution circuitry. Class A type LCLs are by default in the ON state after power-up and they recover automatically after a latch condition with a delay of 5 seconds. Such LCLs are used for systems that need to be continuously ON such as the data handling and the receiver section of the radio. Class B LCLs are by default in the OFF state and they require a command in order to turn ON. When tripped, they remain in the LATCHED state until they are reset explicitly with a command. The third type is a current limited switch which is used for release mechanism deployment of QWKNUT type. For release mechanisms the Arm and Fire strategy is adopted. A main arm switch is followed by current limited switches and that prevents from an accidental deployment.

Each LCL has ON/OFF control and trip status signals that connect to the Control Unit. The status of these signals is treated as telemetry and allows knowing the state of each LCL.

In addition, each LCL has a measurement amplifier to monitor the output current. Multi-channel ADCs are used to provide readout of the output current values from all LCLs through the SPI bus. The purpose of these measurements are to allow detecting anomalies that result in increased power consumption but are low enough to not trip the LCL.

5.3. Design Scalability

Adding more LCLs in the present design does not require a complete redesign of the PCDU thanks to the modularity of the design. The serial control interface, described later, allows for a different number of LCLs without any hardware changes in the OBC interface.

6. MAGNETORQUER CONTROL

For the MATS mission, the satellite platform is equipped with three magnetorquers for attitude control. A digitally controlled three-channel bipolar current drive circuit is implemented on the Power Unit for this purpose.

6.1 Current drive principle

Each of the three channels consists of a voltage-

DOI: 10.1051/e3sconf/20171618007

controlled precision DC current generator, in series with an H-bridge circuit for polarity selection.

A 12-bit DAC is used to generate the control voltage for the current generator. The digital control signals needed for complete control of all three axes can be summarized as:

- SPI-bus communication for setting the individual DC current values
- On/off control (one per channel)
- Polarity control (one per channel)

The on/off function and current control form a redundant means of disabling the current drive in case of a single fault condition in a drive channel.

7. PCDU HOUSEKEEPING AND CONTROL

The top unit of the PCDU stack, the *control unit*, is responsible for collecting all housekeeping and controlling the LCLs. Therefore usage of FPGA logic on the control unit allows the satellite OBC to interface directly the PCDU through a point to point RS485 interface. This approach has the following advantages:

- No matter what number of LCLs or amount of housekeeping, the interface of the PCDU towards the system remains simple. This simplifies harnessing significantly and reduces cost. The approach also relaxes the OBC requirements regarding input/output interfaces.
- Using a balanced point to point interface towards the OBC prevents troublesome ground loops since the FPGA operates on the 28V ground domain while the OBC is on the 5V ground domain. This also demonstrates the cost reduction contribution of this solution since in a different case isolation would be necessary for every input/output interface from the OBC towards the PCDU.
- The FPGA logic gives the capability to the PCDU to very precisely and deterministically control the duty cycle of the magnetorquers. Due to the point to point interface and the entirely hardware based FPGA logic (no soft processor used) the PCDU has the capability to duty cycle the magnetorquers even down to a duration of 1ms. That relaxes the OBC which just passes the duration information to the PCDU and lets it handle the ON/OFF-timing of the magnetorquers.

The RS485 interface towards the OBC is UART based and thus asynchronous. This feature relaxes the OBC design since UARTs are widely used and commonly found in on-board computers. This also simplifies testing of the PCDU since the unit now can be directly connected through a low cost USB-to-RS485 converter to a standard PC. In figure 4 is shown a snapshot of the test software application tool that has been developed for this purpose. This tool in combination with a USB-to-RS485 converter can replace an OBC in unit level tests by controlling the PCDU and visualizing received telemetry. This not only assists unit testing and verification but only contributes to cost reduction since the engineer who runs the tests can interpret and log data directly thus saving a lot of time. The same tool is also planned to be used during the qualification process of the FM unit.

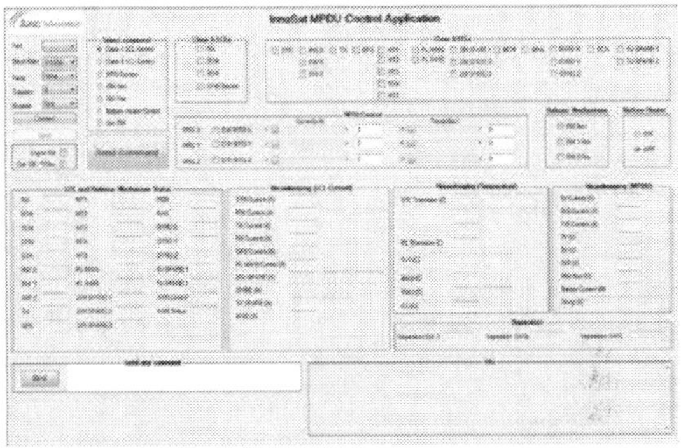

Figure 4. PCDU PC Test Software Screenshot

8. TESTING AND RESULTS

The engineering model of the PCDU has passed all electrical tests and is delivered for system integration with the rest of the spacecraft subsystems. To ease the integration process and make the EPS subsystem transparent to the AIT personnel, a power electrical ground support equipment (EGSE) had to be built to perform the following:

- Allow the user to easily switch between different power modes without any need to mount/dismount connectors or prepare any harnesses.
- Allow the user to have control of the mating/separation process.
- Allow complete power OFF of the S/C through a single key switch which can bring the S/C in a known safe state in which it can stay during long periods of inoperability.
- The design shall support not only the power requirements of the MATS mission but also the high-power configuration of the platform. This enables using the same power EGSE equipment for all missions based on the InnoSat platform.

The power EGSE allows the user to choose between three power modes through a rotary selector switch. Those power modes are:

DOI: 10.1051/e3sconf/20171618007

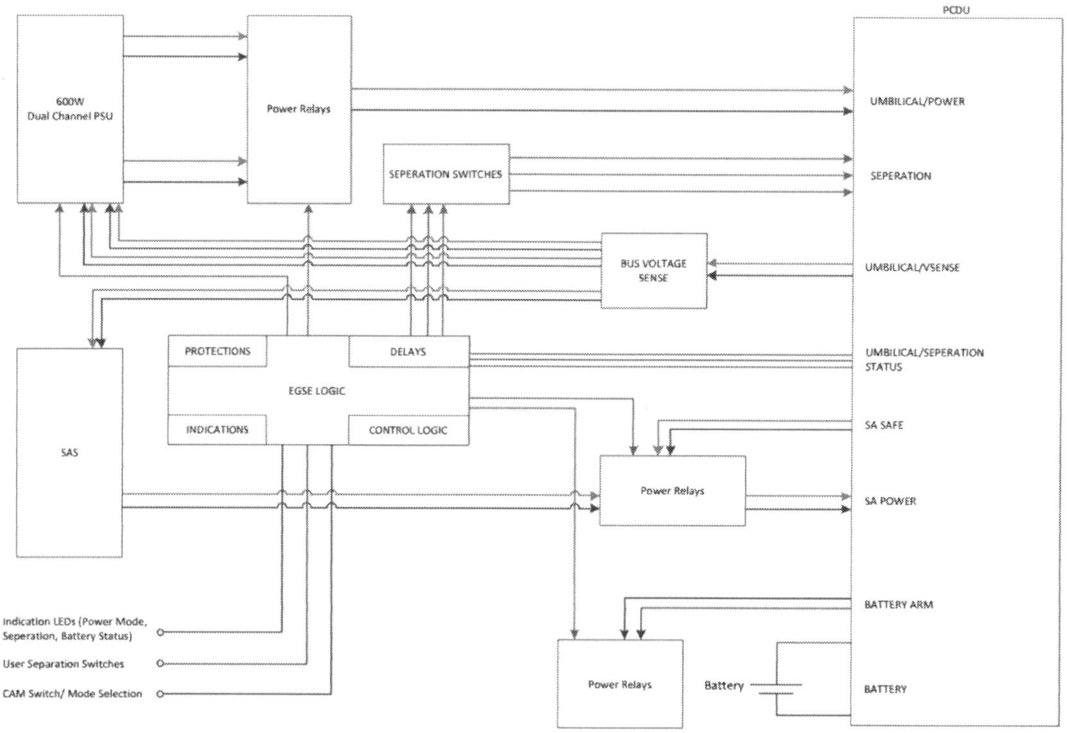

Figure 5. Simplified Power EGSE Block Diagram

- *External Power Mode.* In this mode the S/C is powered externally from a power supply. During the long phase of AIT, this mode is preferred since it reduces the usage of the battery and as a result extends its life. The battery is disconnected and the Solar Array Simulator or the Solar Panel is inhibited (shunted).
- *SAS Power Mode.* In this mode the battery and the Solar Array Simulator are connected thus having an orbit like set-up.
- *External Battery Charge Mode.* In this mode the power EGSE mates the spacecraft and charges the battery from an external power supply.

For the design of the power EGSE, apart from the main functional requirements, some critical usability and reliability requirements have been also taken into account:

- The user interface/panel shall be designed such that there is no possible action or sequence of actions from the user that can lead to S/C failure or even put it to a risk. Isolating the EPS subsystem reduces the risk for accidents.
- There shall be no single failure in the EGSE that can lead to S/C failure. There is no need for a single point failure free EGSE as long as it does not propagate the failure to the S/C.
- For every potential failure there shall be an indication or a combination of indications that will inform the user to take actions about it.

8.1. EGSE Implementation

To realize the power EGSE in a cost effective way, an external dual channel power supply has been chosen. One power supply channel is used for the external power mode and therefore its output is overcurrent and overvoltage protected. The second channel used to charge the battery is current limited and overvoltage protected. By using the external power supply with all necessary built in protections and given that an external Solar Array Simulator (SAS) is available, the main functionality of the built power EGSE is to handle the power routing and implement the user interface. A brief block diagram of the power EGSE is shown in figure 5.

A voltage sense distribution in the EGSE allows the Solar Array Simulator and both power supply channels to sense accurately the battery bus voltage in the PCDU.

 DOI: 10.1051/e3sconf/20171618007

For the power supply this is critical in order to compensate for the long harness to the umbilical. The SAS on the other hand uses the voltage sense only for protection purposes (e.g. overvoltage). The sense distribution is implemented as three separate voltage followers thus providing high impedance towards the battery bus. Power switching between the power supply channels takes place through high power industrial relays which are controlled from the EGSE logic and the user mode selection switch. Power from the Solar Array Simulator is applied only during the SAS Power Mode. The PCDU has two interfaces; the SA power from which it receives power and the SA safe which can be used to shunt externally the power from the SA power interface. Power relay logic ensures that SAS power is applied only when is needed. Another important functionality of the EGSE is the inserted delays between user actions. On-delay timers ensure that we avoid race conditions and that power is applied safely after 3 seconds. The battery arm interface is a critical one since it connects the battery electrically to the system. The battery has its own physical interface. Since InnoSat is based on a battery regulated bus, it is very critical to control reliably the battery connection. The only power mode in which the battery is not needed is the external power mode. At this mode the battery is disconnected and the S/C is powered entirely from an external power supply. Also when the user turns the EGSE OFF through the key switch, the EGSE ensures that nothing in the S/C is powered by disconnecting the battery and the rest of the power interfaces.

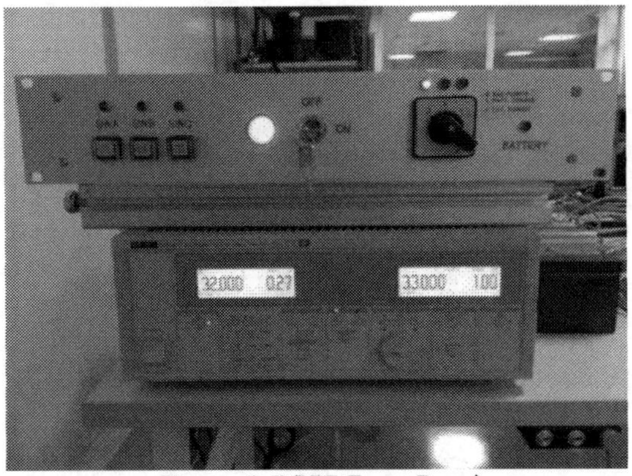

Figure 6. EGSE Front Panel

The EGSE has been already tested and integrated in the avionics test bench. In figure 6 is shown the external power supply and the user front panel. On-going work now is to assemble that in movable rack installation so that it is more compact and easy to move. The power supply has been selected and the EGSE designed already having the rack installation in mind from the start. In figure 7 the PCDU is installed in the avionics test bench.

Figure 7. The PCDU integrated in the system

9. CONCLUSIONS AND FUTURE WORK

A low cost PCDU, meeting the reliability requirements of a small satellite platform has been developed. Through the use of both COTS and space grade components, together with design mitigation of failure cases, good performance and reliability is achieved while keeping the cost low. The modular design and the use of a serial control interface allows for the system to be easily extended. An external software test tool and a power EGSE have been built that ease integration and aim to support the platform in its entire lifetime.

The next step in the development is the manufacturing of the qualification model and the qualification testing will follow. An extensive qualification/environmental campaign is foreseen according to the flow in figure 8.

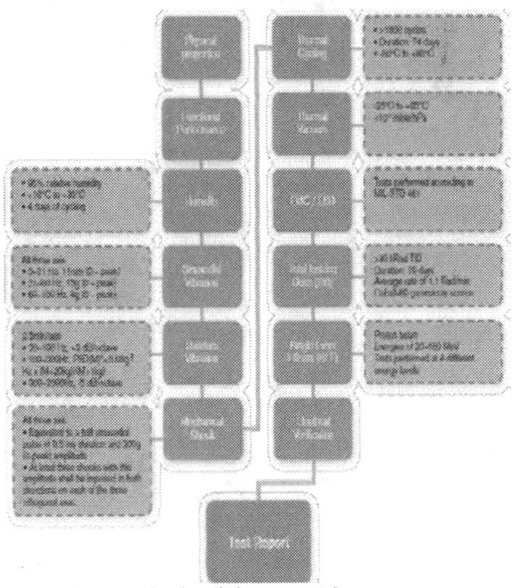

Figure 8. Qualification Flow

A potential future upgrade of this unit is a higher power configuration, which can be performed with minimal changes to most parts of the unit thanks to the modularity of the design. To change the battery charging principle from DET to MPPT is another possible future addition.

10. REFERENCES

1. N. Larsson, R. Lilja, M. Örth, S. Söderholm, J. Köhler, R. Lindberg, J. Gumbel. (2015). InnoSat and MATS – An Ingenious Spacecraft Platform applied to Mesospheric Tomography and Spectroscopy. 10th IAA Symposium for Earth Observation.

2. O'Sullivan, D., Weinberg, G. (1977). The sequential switching shunt regulator S3R. ESA SP-126, pp123-131

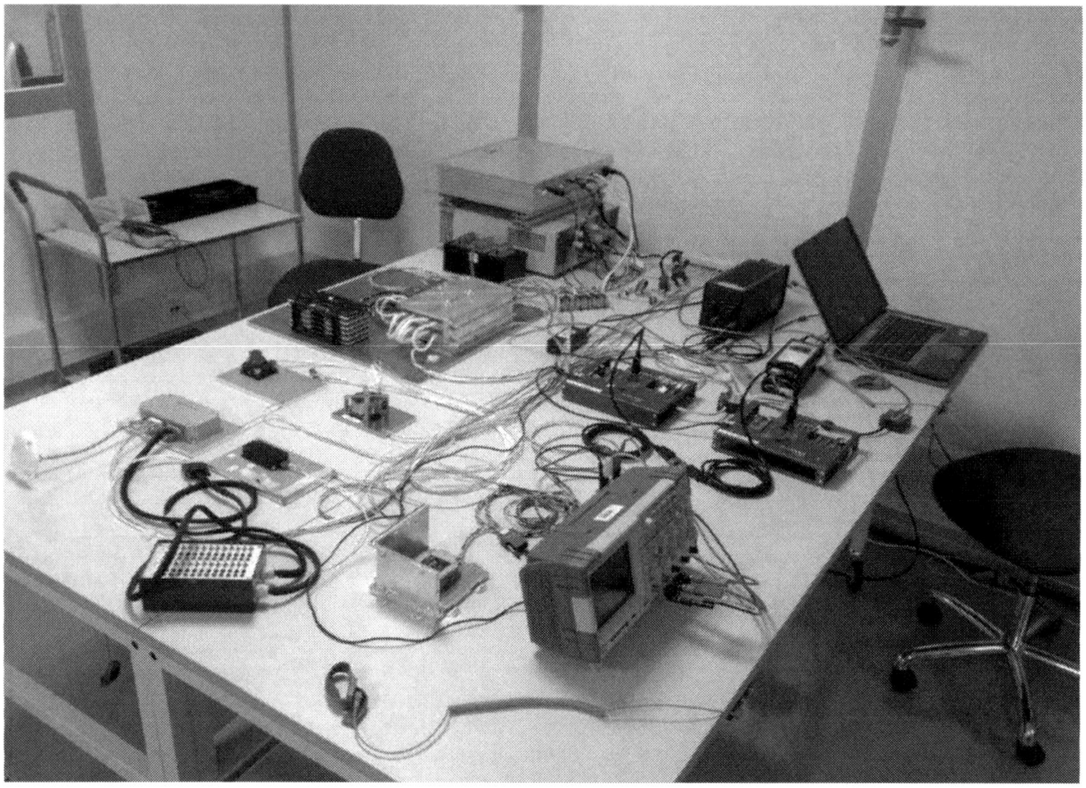

Figure 9. Avionics test bench at an early stage. The functionality and performance of the PCDU are verified in the system.

THE AS400 POWER CONTROL AND DISTRIBUTION UNIT - A MODULAR AND FLEXIBLE UNIT WITH B2R SOLAR ARRAY REGULATION FOR HIGH POWER LEO MISSIONS

Daniel RUF [(1)], Erwin HUG [(1)], Axel VOGT [(1)], Ignacio BARBERO [(2)]

[(1)] Airbus Defence and Space (Airbus DS GmbH), Claude-Dornier-Str.88090 Immenstaad, Germany, Email: Daniel.Ruf@airbus.com, Erwin.Hug@airbus.com, Axel.Vogt@airbus.com
[(2)] Airbus Defence and Space (CRISA), Torres Quevedo 9, 28760 Tres Cantos, Spain, Email: Ignacio.Barbero@airbus.com

ABSTRACT

This paper presents the 'High Power' Power Control and Distribution Unit (HP-PCDU) which has been designed as modular and generic equipment in order to cope with the growing power demand of Low Earth Orbit (LEO) missions. Next to the AS400 satellite platform of Airbus Defence and Space it will be used for the MetOp-SG satellites.

The HP-PCDU supports unregulated bus voltages of 50V and 100V and provides a 30V secondary regulated power bus.

Solar Array Regulation with Maximum Power Point Tracking (MPPT) is performed by Buck-Boost-Regulators (B2R) operated in a cold-redundant 'N+1' topology. A new FDIR strategy based on discrete hardware protections and FPGA-logic for autonomous failure handling is presented.

Next to an outline of other power control and distribution functions this paper also includes a summary of the novel High Rate Diagnostics Mode that allows monitoring telemetry signals with a time resolution up to 200µs.

1 MOTIVATION

In a system study on New Modular and Flexible Power Systems for LEO Missions performed by ESA and carried out by Astrium (now Airbus Defence and Space), architectures and sizing of various existing and upcoming LEO missions have been analysed. As a conclusion of this study, Solar Array (SA) control by Buck-Boost Regulation (B2R) with MPPT has been identified as the most promising recurrent power architecture suitable for all LEO missions [1].

Following the results of this study, Airbus Defence and Space has defined the AS400 'High Power' LEO platform for which the AS400 PCDU has been developed as the key equipment of the Electrical Power System (EPS).

The High Power PCDU extends the range of generic LEO PCDUs available from Airbus Defence and Space beyond the power class of the LEO PCDU EVO [5] for a Solar Array Power of more than 7kW, bus voltages of 50V or 100V and Maximum Power Point Tracking.

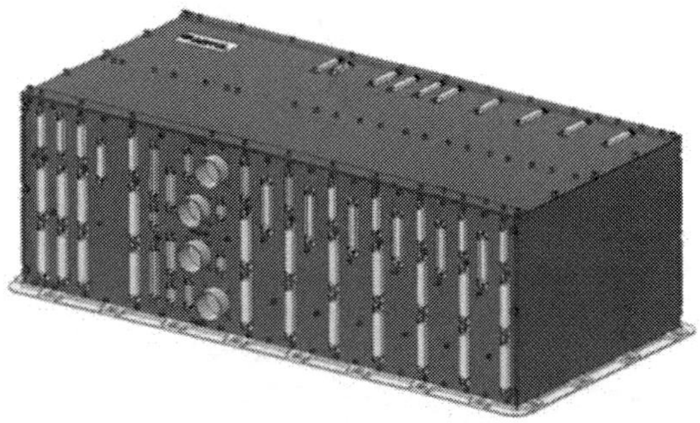

Figure 1. High Power PCDU in the configuration for MetOp-SG Satellite-A

During the design phase special attention was put on creating a modular and scalable product which can be used in a wide range of satellite and EPS configurations enveloping the following:

- Unregulated Main Bus voltage adaptable to 50V or 100V (with an option down to 28V)
- Solar Array power capability from 1000W to 7200W
- Solar Array V_{OC} up to 150V
- Solar Array configuration with up to 11 individual MPPT groups
- Battery Charge/Discharge current up to 110A
- Provision of a secondary regulated power bus with 30V, providing up to 2500W with single failure tolerance
- Provision of up to 124 Latching Current Limiters (LCL) with current capabilities ranging from 0.5A to 20A nominal trip-off current
- Provision of up to 288 (nominal and redundant) switchable heater outlets
- Provision of up to 72 (nominal and redundant) release initiator lines, configurable to drive either electro-explosive or heating device actuators.

2 PCDU ARCHITECTURE

In order to meet the required level of flexibility and scalability, a PCDU design based on individual modules has been chosen, allowing stacking the modules together in a very flexible way.
In total six different module types have been defined, each covering a wide performance and functional range.

While the minimum configuration of the PCDU is defined by one Data Handling Module, one Battery Interface Module and two Solar Array Regulator Modules, the maximum configuration can reach up to 22 modules in total.
A digital internal communication bus based on a Serial Peripheral Interface (SPI) type structure, allows for easily changing the number of modules without a need to adapt the internal communication system or even the FPGAs.

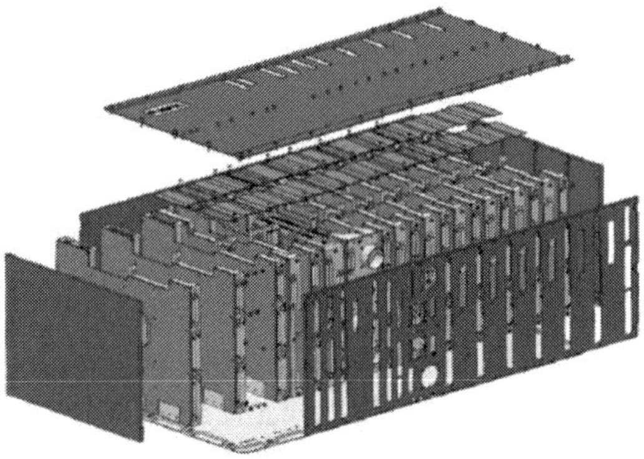

Figure 2. Exploded View of the AS400 PCDU in the configuration for the MetOp-SG Satellite A, showing the modularity of the design

The following table (*Figure 3*) shows the different functions available for the generic High Power PCDU product and their performance range.

Function	HP-PCDU Performance Parameter	
Power Control Section		
Solar Array Regulation	Available Regulator Types	B2R / Buck / Boost
	Regulators per Module	2
	Max. SA total I_{MPP}	165A (in 50V Version) 126A (in 100V Version
	SA Voltage Range	40V - 150V
Battery Interface Function	Main Bus Voltage Range	0V - 100.8V
	Max. Charge / Discharge Current	110A
	Number of Battery Connectors	Up to 4
	Launch Power Configuration	ON or OFF
	EOL Passivation Function	✓
	Number of EOCV Levels	8
	Number of Charge Current Lim. Levels	8
Secondary Regulated Bus	Regulated Voltage	30V
	Maximum Power	2500W
	Single Failure Tolerance	✓
Power Distribution Section		
Power Distribution Functions	Max. total Number of CL Outlets	124
	Thereof max. Number of LCLs	124
	Thereof max. Number of PO-CLs	18
	Thereof max. Number of SO-CLs	27
	Max. number of Heater Groups	36
	Max. number of Heater Outlets per Group	8
	Max. number of Voltage Groups (UMB/RSB)	18
	Supported Actuator Types	EED (NSI/NEA), HDA
	Total Number of Release Actuator Outlets	72
Communication Section		
	Number of Mil-1553 Bus Interfaces	2 (each N+R)
	Equipment Swicth-Off Outputs	✓
	System-UV Outputs	✓
	Switch-Off Line Inputs	✓
	DNEL-Trigger Inputs	✓
	Direct-On Command Inputs	✓

Figure 3. HP-PCDU Functional and Performance Envelope (details are described in the following chapters)

 DOI: 10.1051/e3sconf/20171618008

3 BUCK-BOOST SOLAR ARRAY REGULATOR B2R

The Solar Array Regulator (SAR) based on B2R and a cold-redundant N+1 architecture is a key novelty of the High Power PCDU concept.
By using a Buck power stage followed by a Boost stage, the converter input voltage (i.e. Solar Array Voltage) can be above or below the main bus voltage level [2]. The MPPT function is available in both domains.

Due to the wide main bus voltage range for which the HP-PCDU is designed, Solar Array regulation by B2R is the most flexible solution creating a high level of independence between the operational SA voltage range and the main bus voltage range.

Each Solar Array Regulator Module (SARM) contains two B2R regulators, each with individual input sections and individual MPPT-control.

3.1 Regulator Concept

The basic block diagram of a single B2R regulator is shown in *Figure 5* below.
Each regulator provides two inputs which can be used for independent SA sections. Each input line is protected by an electronic diode against short circuits at SA or harness level.
An input protection formed by a latching current limiter (LCL) is used for SAR input filter charging during switch-on and separates the regulator in case of failure from its SA sections.
The buck stage is formed by a standard buck converter using parallel N-channel MFETs for loss reduction and buck diodes. A dedicated hardware-based protection circuit detects any failure causing excessive losses (e.g. Linear Mode failure in FETs) and can isolate the regulator if necessary. It is followed by a standard boost converter.

To protect the main bus against any short circuit failure in the regulator a dedicated Reverse Current Protection (RCP) monitors the output current of each SAR.
While in a pure buck or boost regulator the relations between voltages and currents at input and output are clear, the location of the highest current in a B2R configuration is not obvious. Therefore both input and output currents are monitored and limited if necessary by an internal control loop in order to control the maximum dissipation of the regulator.
This leads to a maximum power envelop as shown in *Figure 4* below.

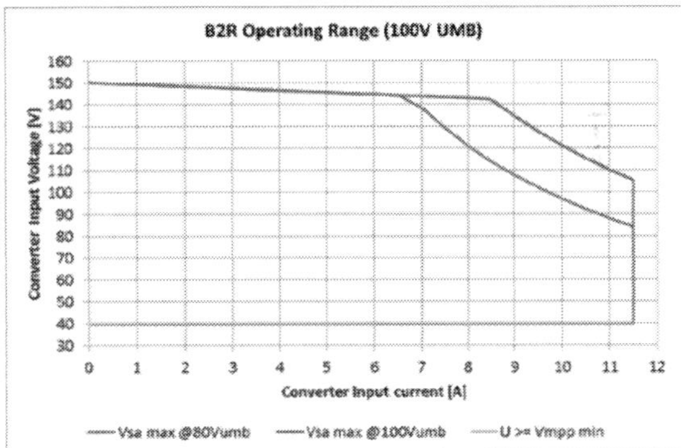

Figure 4. B2R Regulator Operating Range incl. thermal limitations for Main Bus Voltages of 80V and 100V

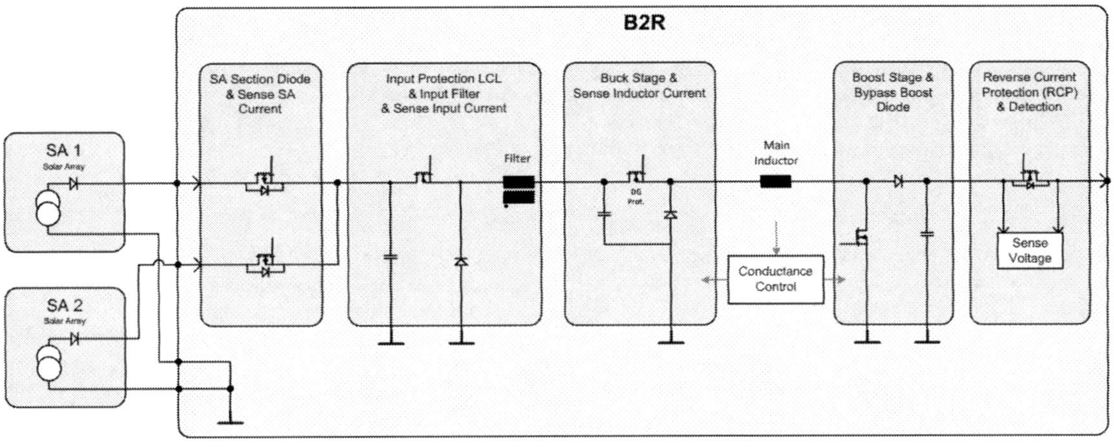

Figure 5. Simplified Block Diagram of a single Buck-Boost Regulator

 DOI: 10.1051/e3sconf/20171618008

3.2 The N+1 Redundancy Scheme

3.2.1 Motivation and Trade-Off

During the definition phase of the High Power PCDU extensive trade-offs have been performed to identify the optimum redundancy concept for the SA regulation function.
While in the past operation with classical two-out-of-three redundant regulators has proven to be very robust due to its simplicity, the key drawback of this approach is the approximately 50% overhead in terms of EEE-parts, mass and volume on top of the nominal demand.

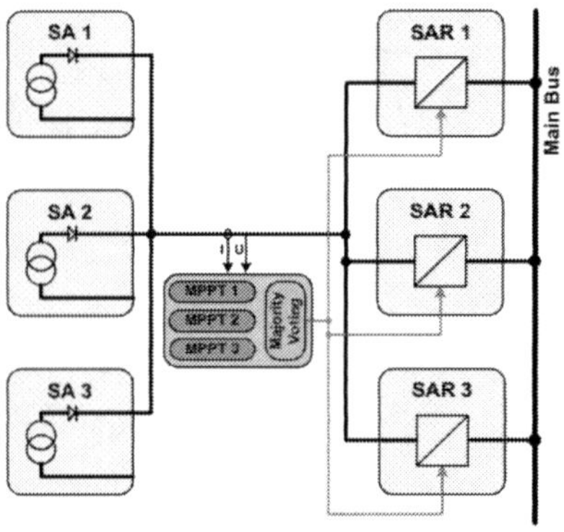

Figure 6. Two-out-of-Three Redundancy Concept with common failure tolerant MPPT and failure tolerant power capability of two regulators

As a more mass and volume efficient concept, the N+1 redundancy scheme with segregated regulators operated in cold redundancy has been identified.
It connects each group of SA sections to one single regulator and provides one single, common regulator for redundancy. Together with an individual MPPT function per regulator, this concept additionally allows better tracking granularity compared to the two-out-of-three redundancy concept which needs to share a failure tolerant MPPT control for the group of three regulators.

In case of a failure in any of the nominal regulators, the failed one is disconnected from SA and Main Bus and the SA power is connected to the redundant regulator. With an overhead of approximately *(N+1)/N* plus some extra hardware for failure detection and activation of the redundancy, it typically becomes more mass and volume efficient for $N \geq 3$ regulators. The obvious drawback of this architecture is the higher technical complexity necessary for detection and isolation of a failed regulator. Nevertheless, as the High Power PCDU is especially optimized to cope with a SA power of above 2kW and typically up to more than 7kW, the mass saving and higher system flexibility was found to be the dominating argument for the selection of this redundancy scheme.

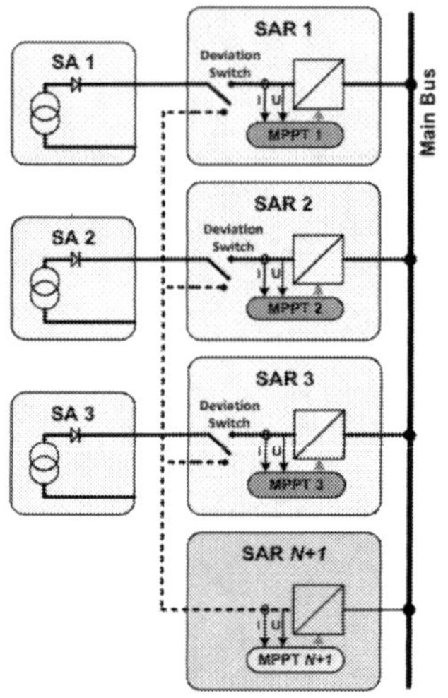

Figure 7. N+1 Redundancy Concept with segregated regulators, here shown with failure tolerant capability of three regulators

3.2.2 Deviation Switch

One of the most critical functions in the N+1 redundancy scheme is the switch-over between any nominal to the redundant regulator.
Each B2R SAR contains a so-called 'Deviation Switch' downstream its input diodes which is able to connect the SA sections to a common power rail feeding the input of the redundant SAR.
This failure tolerant Deviation Switch is formed by an antiparallel quad of MFETs. The FET drivers are supplied from the second regulator on the same module, therefore are able to activate the Deviation Switch fully independent of the state of its own SAR.
In case of failure of the regulator, the Deviation Switch is activated and the failed regulator is disconnected from SA and power bus.
The Deviation Switch can be activated by the internal hardware protection of the regulator, by an FPGA based health check or by external command. Its activation is fully reversible.

3.3 Failure Protection Strategy

3.3.1 FDIR Approach

As already mentioned in the previous sections, the N+1 architecture in combination with Buck-Boost Regulation requires a higher complexity in PCDU-based Failure Detection, Isolation and Recovery (FDIR) compared to the two-out-of-three redundancy schemes.
As a key requirement, and in order to achieve full compliance to chapter 5.7.2 of ECSS-E-ST-20C, the switching between any nominal SAR and the redundant one needs to be autonomously done by the PCDU itself.

A detailed FMECA has been established to identify all credible failure modes at regulator and unit level and their effects on the overall system. Two major classes of failure criticalities are distinguished:

a) Failures which require immediate reaction (typ. within microseconds) to avoid e.g. significant loss of power, overstress to parts or electrical or thermal failure propagation inside or outside the PCDU

b) Failures which cause a long term degradation of the overall system through slow loss of electrical energy, which can be for example degradation of a MPPT function, input to output short circuit, input or output open circuit on one regulator etc.

For all failures of type a) dedicated hardware-based protection functions have been implemented which, in case of failure, lead to an immediate activation of the Deviation Switch.

Failures of type b) are covered by a Health Check algorithm implemented in the FPGAs of the Data-Handling Module of the PCDU which can activate the redundant regulator. This function is described in more detail in the following.

3.3.2 Health Check

The Health Check algorithm is implemented in both FPGAs of the DHSM. It case of hot redundant operation only one of them executes the function.
The principle of the Health Check is to compare the behaviour of a pair of Solar Array regulators to each other and – in case of significant differences – activates the Deviation Switch of the regulator with the lower power output. This monitoring is continuously repeated for any definable pairs of regulators.

Before starting the comparison of regulators, two main conditions must be fulfilled:

- The SAR must be in MPPT mode, i.e. taper charge with reduced SA power demand has not been reached
- The total SAR input and output current must be above a minimum threshold, which means that the satellite is not in eclipse or abnormal illumination conditions

Using High Data Rate Telemetry (see section 8) for input voltage and current, the FPGA calculates average values for the input power (without the typical MPPT tracking ripple). The input power figures of the pair of regulators are compared to each other. If the difference in power input exceeds a configurable delta for a predefined number of times, a failure in the regulator with the lower power is assumed and its Deviation Switch is activated by the FPGA.

In order to confirm a failure of the regulator and to exclude failures at Solar Array level, the same check is repeated between the redundant regulator and the healthy regulator of the original pair as well as between the redundant and the failed one.

If a failure in the regulator is confirmed, i.e. significantly better electrical performance is achieved by the redundant one, this configuration is kept and the health check is deactivated, otherwise the original regulator is reactivated.

The configuration of the PCDU (e.g. number of Solar Array Regulator Modules, regulator pairs for comparison etc.) and the health check threshold parameters are fully configurable.

3.4 B2R Module Options

In the baseline design, the SARM is built to cope with the maximum performance envelope defined for the High Power PCDU.

In order to optimize the module for missions with lower requirements, a number of options have been foreseen in the design.

These include for example:

- For systems requiring only a buck-based regulation, the boost stage can be removed and bridged to reduce mass and improve the transfer efficiency.
- The modules can be populated with EEE-parts of different voltage ratings in order to optimize cost and efficiency for each main bus voltage range.

Figure 8. EM Module of B2R Solar Array Regulator Module with two Regulators

DOI: 10.1051/e3sconf/20171618008

3.5 B2R Typical Performances and Test Results

For performance testing and hardware characterisation, three EM modules of the B2R have been built, using flight-quality representative EEE-parts for the power cells. One module has been built in the 50V UMB version, another two in a 100V UMB version.

The figures below show the measured efficiency of one single regulator in the 100V UMB version over temperature, one with the output voltage at 100V and the other at 80V.
As it can be seen, the BOL efficiency in typical operating areas is far above 96% and above 98% in the DET mode, when the input voltage is similar to the output voltage.

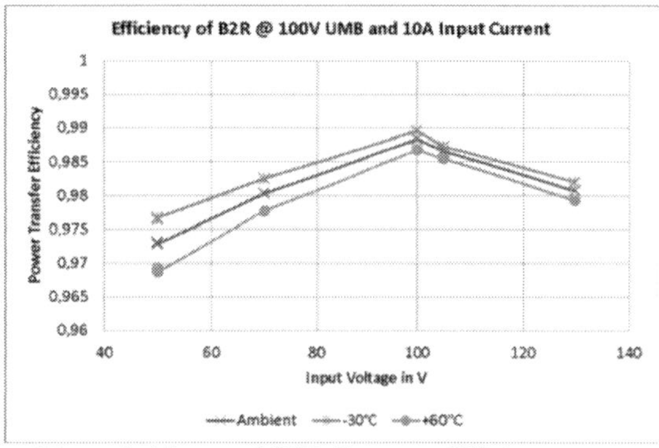

Figure 9. Measured BOL Converter Efficiency over Input Voltage and Temperature for 10A Input Current and Output to a 100V Bus

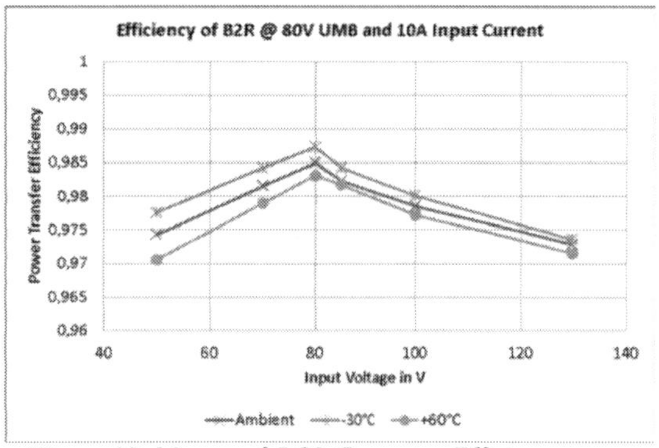

Figure 10. Measured BOL Converter Efficiency over Input Voltage and Temperature for 10A Input Current and Output to a 80V Bus

4 BATTERY INTERFACE MODULE

The Battery Interface Module (BATM) provides all key functions related to battery connection and basic power control in the system.
Next to the failure tolerant Main Error Amplifier (MEA) with eight configurable limits for *End Of Charge Voltage* (EOCV) and eight *Charge Current Limitation Levels* it contains a centralized *Disconnect of Non-Essential Loads* (DNEL) logic, able to switch off of non-essential power outlets in case of a critical undervoltage scenario.

A solid state battery separation switch made from 10 parallel N-channel MFETs is able to connect and disconnect the battery from the main bus during on-ground and in-orbit operations. It contains a current-limited pre-charging function for connection of the battery to the main bus filter and an end-of-life (EOL) passivation function.

In order to activate the EOL passivation, two commands (*Passivation Arming* and *Passivation Confirmation*) must be sent through discrete communication lines. Once activated, the function inhibits Solar Array Regulation and interrupts the power transfer from Solar Array to Main Bus. The battery remains connected to the Main Bus and will consequently be fully discharged by the power demand of the satellite after the passivation has been initiated. Each of the two passivation commands acts on one series connected latching relay, which generate the internal passivation signal.
This kind of passivation commanding is single-failure tolerant against unintentional activation but can be lost with one single failure (see ref. [6] for details).

The High Power PCDU can be launched in powered or unpowered configuration. In case of an unpowered launch, the battery switch is kept open for launch and the Solar Array regulation is inhibited. Once two of the three available separation straps have been pulled off after successful separation from the upper stage of the launcher, the battery switch automatically connects the battery to the main bus and activates the B2R modules.

5 30V REGULATOR MODULE

The 30V DC/DC Regulator Modules (DCDCM) are supplied by the unregulated main bus and generate a secondary regulated power bus distributed via a power rail to all other modules inside the PCDU.
Each DCDCM contains two parallel buck-type converters operated in hot redundancy.
The power capability of each buck converter is up to 750W peak in case one single DCDCM is implemented, but has to be limited for thermal reasons in case two or three modules are mounted. A maximum failure tolerant

DOI: 10.1051/e3sconf/20171618008

power capability of 2500W is available in the full configuration of the PCDU.
The DCDCM is able to cope with the full voltage range of the unregulated main bus with one single electrical design thus provides a maximum flexibility for a wide range of applications.

6 POWER DISTRIBUTION MODULE

Next to the SA regulation the power distribution is typically the function of a PCDU which requires the highest flexibility and the one with the highest number of individual elements needed. Therefore special care was applied for definition and design of the Power Distribution Module (DISTM) in order to achieve a maximum in flexibility and configurability.

The DISTM of the HP-PCDU contains Latching Current Limiter (LCL) outlets and Heater Group outlets.

Depending on the required PCDU configuration, the power outlets can be allocated in three independent voltage groups with separated connectors, for which each can be either connected to the unregulated main bus (50V or 100V) or the regulated secondary bus (30V).

Four types of different LCLs are available on each module and can be configured as needed:

- **Standard LCLs** with one single switching barrier with configurable trip-off currents between 0.5A and 12A and trip-off times up to 20ms
- **Safe-Off (SO-) LCLs** with two independent switching barriers allowing to switch off the outlet even in case of one failure inside the LCL. Trip-off currents up to 5A are available.
- **Permanent-On (PO-) LCLs** which are intended to supply the vital loads of a spacecraft with up to 2A. They are automatically re-activated after trip-off until reactivation is inhibited by command.
- **Heater Group (HG-) LCLs** supply and protect a group of heater outlet switches. They are standard LCLs available in classes up to 10A trip-off current and can be used as Standard LCLs if not needed for heater groups.

Additionally up to 16 LCLs with a trip-off current up to 20A can be implemented for specific mission needs.

Further on each DISTM includes individually switchable heater line outlets (1.5A / 2A max), protected by the upstream HG-LCLs.

7 DEPLOYMENT MODULE

In the maximum configuration the redundant Deployment Modules (DEPLM) can provide up to 72 (nominal and redundant) deployment actuator lines supporting both Electro-Explosive Devices (EED, driven by a current source) or Heating Device Actuators (HDA, e.g. thermal knifes driven by a voltage source).

To fire an actuator three independent barriers need to be activated:

- A latching relay to select one output line
- A latching Arming Relay, common for all outlets on the module
- A common LCL protecting the main bus against short circuit failures in the deployment chain

The firing pulse itself will be generated by a DC/DC converter which can be configured to act either as current or voltage source.

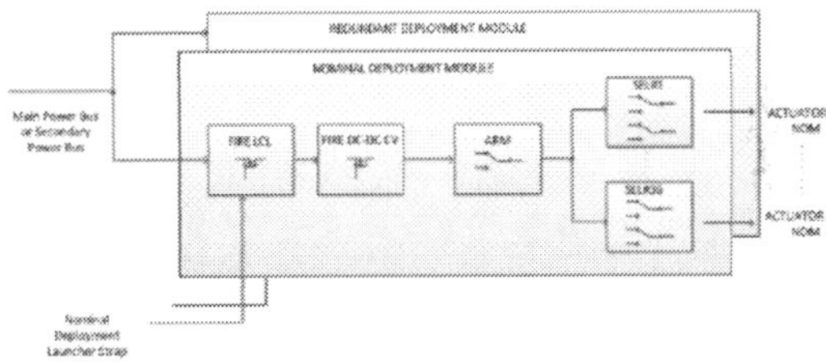

Figure 11. Deployment Actuation Chain with nominal and redundant DEPLM

8 DATA HANDLING SYSTEM INTERFACE MODULE

8.1 Overview

The high flexibility and configurability of the HP-PCDU product is a major driver also for the design of the PCDU-internal command and control elements and the related external communication interfaces. These functions are provided by the generic Data Handling System Interface Module (DHSM).
The DHSM communicates with all other internal modules through a redundant serial bus and SECOIA ASICS on each module.
It provides discrete interfaces for special functions, such as Equipment Switch-off Line (EQSOL, pre-warning for upcoming LCL switch-off) Outputs, System-Undervoltage Outputs, Switch-Off Line (SOL, triggering of LCL deactivation) Inputs, Disconnect of Non-Essential Loads (DNEL)-Trigger Inputs and Direct –On (DOC, activation of predefined LCLs) Lines

8.2 High Rate Diagnostics Mode

Very often the observability of the EPS is limited by the maximum time resolution available through the spacecraft internal communication buses. Interfacing the PCDU as non-packet terminal through the MIL-1553 bus the cyclic acquisition typically limits the resolution of analogue EPS data to 10Hz.
The High Power PCDU contains a novel High Rate Diagnostic Mode (HRDM) to overcome this limitation.

The HRDM allows to log up to 10 different analogue or digital telemetry values available on the PCDU internal bus system with a time resolution down to 200µs. The maximum length of the data log is only limited by the storage volume available (500kB). Data can be downloaded through the MIL-1553 bus for further processing outside the PCDU.

Using the HRDM the HP-PCDU can monitor critical analogue TM data with up to 5kHz and therewith creates plenty of new possibilities for spacecraft testing and operation, especially for failure investigation in orbit. As very interesting scenarios one could imagine e.g.:

- Observe the inrush current behaviour of equipment supplied by the PCDU when switching LCLs or a function changing the internal consumption of the equipment
- Monitor the Solar Array MPP tracking behaviour
- Monitor pyro initiator current profiles during firing
- Observe the voltage ripple or spikes on the main bus or detect current variations on a power outlet

9 CONCLUSION

This paper illustrates the functionality and the wide performance range of the High Power PCDU product developed by Airbus Defence and Space. The unit is implemented in the AS400 satellite platform and other LEO missions (e.g. MetOp-SG). Through its generic design approach with a very high scalability in terms of bus voltage, power and functionality it can be easily adapted to the need of many other missions, not only limited to LEO. It provides the following key advantages:

- High scalability and flexibility, therefore simple adaptation without significant non-recurring development for various applications possible
- Reduced risk and cost for the user due to highly recurring design
- Application with a wide range of Solar Array designs and battery voltages possible, thus allows an efficient standardization also for these components
- Functions unused for a specific mission can be efficiently removed or bypassed to fly only necessary hardware and avoid unnecessary risks

This paper also outlines the novel concept of using B2R Solar Array Regulators in a N+1 cold redundancy scheme and explains the necessary protection features including a digital health check implemented in the central control FPGA of the unit.

10 ACKNOWLEDGEMENT

The Buck-Boost Solar Array Regulator described in this paper was developed in the frame of a GSTP-program under contract 4000112434/14/NL/LvH.
The authors of this paper wish to thank Mr. Olivier Mourra and Mr. Ferdinando Tonicello of ESA ESTEC for their support in the execution of this project and the fruitful technical discussions.

11 REFERENCES

[1] N. Neugnot et al. (2011). A New Modular and Flexible Power System For LEO Missions. In Proc. of the 9th European Space Power Conference. Saint-Raphaël. France.

[2] O. Mourra et al. (2010). Buck Boost Regulator (B2R) for Spacecraft Solar Array Power Conversion. In Applied Power Electronics Conference and Exposition (APEC), 2010 Twenty-Fifth Annual IEEE, p. 1313 - 1319

[3] F. Bausier. (2013). Passivation of spacecraft power system at end of mission. TEC-EPM/2013/33/FB, Technical Note. European Space Agency, Noordwijk, The Netherlands

[4] Ch. Delepaut. (2013). Battery Isolation Function - Interface Standardisation Rational. TEC-EPM/2013/15CD, Technical Note. European Space Agency, Noordwijk, The Netherlands

[5] E. Lapeña et al. (2016) The LEO PCDU EVO – A Modular and Flexible Concept for Low to Medium Power LEO & Scientific Missions. In Proc. of the 11th European Space Power Conference, Thessaloniki, Greece.

[6] E. Lapeña et al. (2016) PASSIVATION STRATEGIES ON BOARD AIRBUS DS LEO PCDUS. In Proc. of the 11th European Space Power Conference, Thessaloniki, Greece.

THE LEO PCDU EVO - A MODULAR AND FLEXIBLE CONCEPT FOR LOW TO MEDIUM POWER LEO & SCIENTIFIC MISSIONS

Emilio LAPEÑA [(1)], **José Luis HERRANZ** [(1)], **Fernando GÓMEZ-CARPINTERO** [(1)], **Miguel RODRÍGUEZ** [(1)]

[(1)] *Airbus Defence and Space (CRISA), Torres Quevedo 9, 28760 Tres Cantos, Spain, Email: emilio.lapena@airbus.com – joseluis.herranz@airbus.com – fernando.gomez-carpintero@airbus.com – miguelr.alvarez@airbus.com*

ABSTRACT

After a successful AS250 LEO PCDU product life [1], with more than 12 FMs manufactured (4 of them currently in orbit), and 10 years in the market, an evolution is required in the product to make available a new PCDU for Low Earth Orbit, scientific, interplanetary and export missions in line with evolving customer needs.

Market trend towards volume & mass reduction, plus new required capabilities (in particular End of Life Passivation and Launch Off to support multiple launch) justify the need for a new LEO PCDU EVO generation. This paper will deal with a simple question with no immediate answer: how to save 20% PCDU mass and volume while keeping or improving existing performances.

1 MOTIVATION

The targeted unit is intended to satisfy the needs of LEO and scientific spacecraft with a design and construction optimized for small missions, but covering also medium power range units up to 3kW. The proposed concept is also applicable to the distribution part of the 28V regulated buses classical of interplanetary missions like the PCDU for JUICE. A novel approach has been developed for that, on which the interface with the solar array (DET function) and the battery is scalable to the total solar array and battery energy to be managed.

2 PCDU ARCHITECTURE

Figure 1 is an image of the LEO PCDU EVO EM as built for the Merlyn flight mission.

A brief summary will be provided on unit capabilities, in order to have an overall view in advance to implementation aspects, see table 1 hereafter.

	Configuration capabilities	
Module Type	Minimum	Maximum
DHS I/F	1	1
DET	1	2
BATT/MEA	1	2
DISTRIBUTION	0	6
DEPLOYMENT	0	2
TOTAL	3	13

Table 1. LEO PCDU EVO modularity vs. configuration

Figure 1. LEO PCDU EVO EM during integration in the configuration for Merlyn mission

A single Data Handling Interface module is required regardless the unit configuration selected, since the master FPGA (either nominal or redundant) contained on it already encloses all the required VHDL code to govern any PCDU configuration from maximum to minimum one.

In general terms the PCDU is built with a number of modules of catalogue of existing ones. No modifications are required on the building blocks to configure the PCDU for a given mission. Automatic software tools have been developed to configure the distribution modules to the particular needs for LCL classes and types of a specific application case.

Table 2 below shows the different functionalities available for the LEO PCDU EVO and their most relevant performances.

3 HOW TO DO THE SAME (OR EVEN MORE) IN LESS VOLUME

Mass & volume reduction while adding new performances was identified as the design target for this PCDU evolution. Packaging optimization, proper selection of the modularity and functionality integration are the key drives to achieve these targets. This must be

DOI: 10.1051/e3sconf/20171618009

Function	LEO PCDU EVO Performance Parameter	
DHS I/F Data Handling Subsystem Interface Module	Number of Mil-1553 Bus Interfaces	2 (each N+R)
	Number of APS	1xN + 1xR
	Number of SOL Inputs	1xN + 1xR
	DOC (Direct On Command) Lines available	YES, N+R
	TM/TC Control	Parallel for Conditioning Serial for Distribution
DET Direct Energy Transfer (S3R)	Reliable sections	YES M+1 concept
	Maximum section current	9.5A
	Efficiency	>98,5%
BATT/MEA Battery Interface & Main Error Amplifier	Max. Charge / Discharge Current	50A/65A per BAT Module
	Number of EOCV Levels	Up to 16
	Number of Charge Current Lim. Levels	Up to 16
	EoL Passivation	YES
	Launch Off Capability	YES
Distribution Capabilities	Max. Number of CL Outlets	Up to 104
	Max. Number of PO-LCL	Up to 16
	Max. Number of Heater Outlets	Up to 192
	Types of Limiters available	SO-LCL, PO-CL, LCL
	Classes of Current Limiters available	0.5A, 1A, 2A, 3, 4A, 5A, 6A, 8A, 10A
	Classes of PO-CL available	0,5A, 1A, 2A, 3A, 7A
Deployment Devices	Supported Actuator Types	EED & LSRU
	Safety Deployment Strap	YES
	Max. Number of Deployment Outlets	24+24

Table 2. LEO PCDU EVO Module and functionality

achieved in both Power Conditioning and Power Distribution parts of the whole PCDU. These two parts of the PCDU will be discussed now separately.

3.1 Power Conditioning

When the same implementation of a concept is used for a wide power range, normally the lower power application cases are penalised in mass, and/or the higher power application cases are penalised in efficiency. Implementations for high current paths do not use to be light in terms of weight.

The innovative idea used in this LEO PCDU EVO is based on making totally modular the PCDU Power Conditioning part. The higher the power, the higher the number of "power slides" to be used. This modular approach is extended to the EPS implementation. This modularity applies to the Battery module and to the DET module.

The Power Conditioning modularity concept is reflected in Figure 2. Several design aspects have been also optimized in order to achieve not only a more optimum use of the available volume, but also an easier manufacturing:

- Sub-D connectors for battery I/F instead of bulky circular ones,
- Solid state-based battery switch,
- Flex connections for power devices, and
- Exhaustive use of commercial space grade inductors SMT type.

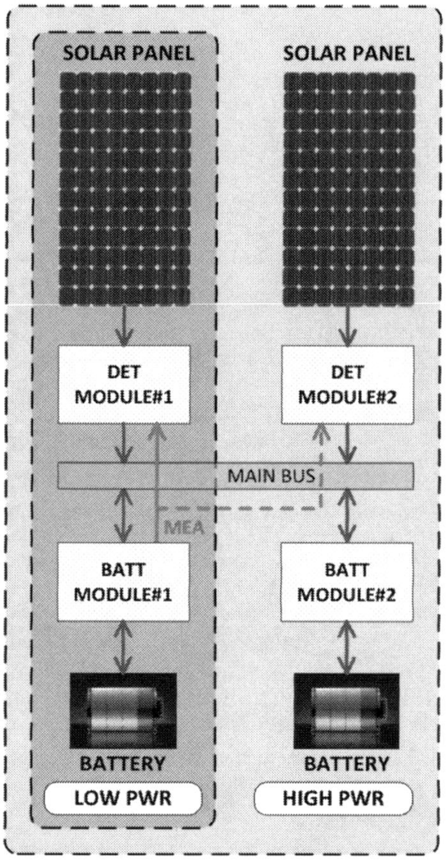

Figure 2. LEO PCDU EVO Conditioning modularity

 DOI: 10.1051/e3sconf/20171618009

New packaging solutions are also considered, enabling pitch reduction on all modules, e.g. using SMT capacitors of better volumetric efficiency for main bus capacitor. Deep analysis on component type minimization was also run. Figure 3 shows the EM module of DET function.

Figure 3. DET Conditioning Module EM

It is to be highlighted that LEO PCDU EVO mass and volume reduction was not made thanks to performances reduction. All functionalities of previous product were maintained if not improved.

In particular it is to be highlighted that DET S3R cells are reliable by design (one failure tolerant) with respect to the loss of the power available from any solar array section.

DET implements the M+1 concept (one extra section) with two main functionalities under MEA control, see figure 4 for reference:

- To feed the bus with the power coming from a section whose S3R cell failed in open circuit in the power path to the main bus, or
- To handle the power from a solar array section whose S3R cell failed in open circuit in the shunt power path, in case the instantaneous bus consumption is below the power provided by this section.

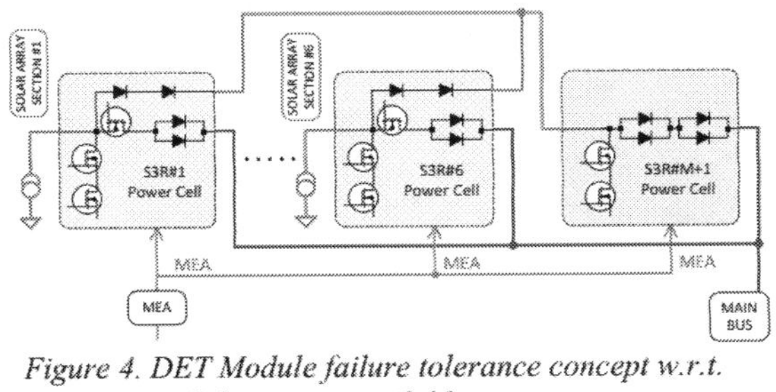

Figure 4. DET Module failure tolerance concept w.r.t. Solar Array available energy

Nominal sections are connected to M+1 section by a couple of power diodes in series as can be seen on figure 4.

In addition, there are new functionalities in line with market trends and customer requirements regarding green space regulations. In this sense, the LEO PCDU EVO implements electronics to passivate the solar array once the end of mission was reached [4].

The same diodes mentioned before that are connecting every S3R cell to the M+1 cell, are used to connect the S3R cells to the electronics in charge of passivating the solar array, refer to Figure 5.

Passivation is exercised by a pair of electromechanical power relays in series (to be one failure tolerant w.r.t. a failure leading to the contacts of one relay stuck to on position).

Moreover, passivation command is executed in two steps, Arming + Firing, and passivation action is reversible as long as there are energy enough available in the spacecraft battery.

Power Conditioning modularity as shown on Figure 4 is also applied to passivation function: the passivation relays are physically implemented in the BAT/MEA module to optimize the use of the available volume in the unit and to provide the same solution regardless the power range.

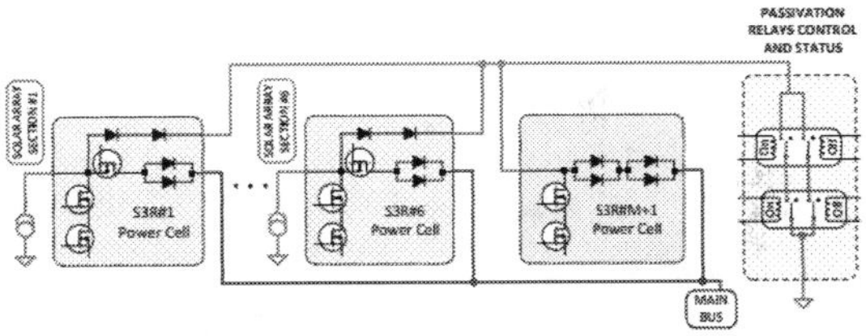

Figure 5. EOL Solar Array Passivation

Another performance added to PCDU design is the launch off capability for multiple launch scenarios. It is based on a couple of series redundant separation straps.

Opening one of the straps is enough to start up the PCDU in a controlled manner. A parallel control of the PCDU status through an interface with EGSE is also possible while on ground via two switches in series. This feature – in conjunction with the solid state-based battery switch - easies the handling of the PCDU on ground while spacecraft integration is taking place.

The scheme for the launch off interface (and EGSE interface) implementation is shown on Figure 6 hereafter.

 DOI: 10.1051/e3sconf/20171618009

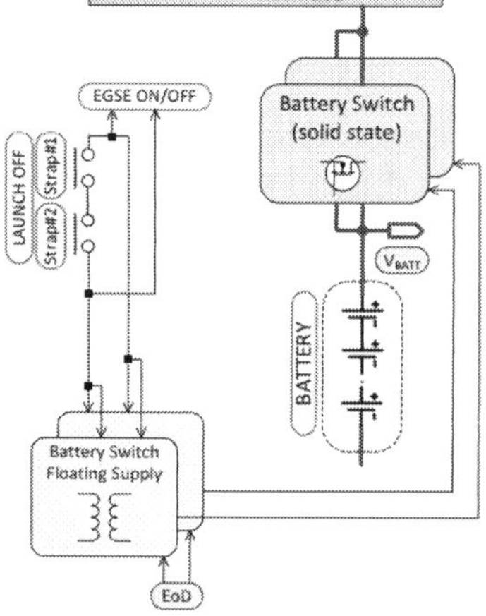

Figure 6. Launch off Interface implementation

3.2 Power Distribution

Power Distribution Modules are normally the most numerous ones in a PCDU. Any effort oriented to their optimum implementation has a multiplicative effect and – at the end - may drive the physical characteristics (mass and dimensions) of the PCDU. A relevant integration effort has been conducted to optimize the implementation of the Power Distribution Modules, using the heritage coming from previous GEO and interplanetary missions. Design drivers can be summarised in the key aspects that follow.

<u>Internal Communications</u>

Distribution Modules implement as much outlets as possible to reach the maximum achievable degree of integration. As consequence, the number of individual telecommands and telemetries to be handled is proportionally high. Any module size and mass optimization passes through local TM/TC control function integration.

A Digital ASIC (**SECOIA** – **SE**rial **CO**mmunication **I**nterface **A**SIC) performs this task operating as slave of the master FPGA in the DHS Module, see Figure 7.

Figure 7. ***SECOIA – SErial COmmunication Interface*** *ASIC*

This product is the evolution of many FPGA based serial link designs used in the past, in particular the PCDUs for BEPI Colombo MPO and Solar Orbiter mission.

<u>Distribution Modularity</u>

The use of SECOIA opens the door to an increase in the number of distribution items that can be implemented in a single board. In fact the limitation per module now comes from the number of pins available in the module output connectors that can be implemented in the unit's front panel to be compatible with the targeted box height.

Second step of optimization was to select the adequate number of Distribution items to configure the standard multi-mission Distribution board. Several different configurations were challenged in the scenarios provided by previous real flight mission cases. From that analysis, an optimum configuration was selected, while still feasible in terms of electronics packaging surface budget. In fact, the use of this modularity meant the reduction of one board less in all mission cases studied. Again, the proposed modularity is possible thanks to the savings in Distribution Module local TM/TC control part surface budget due to SECOIA use. Figure 8 shows the EM of the Distribution Module implementation.

Figure 8. Distribution Module EM

The distribution concept flexibility is such that:

- All heater group LCLs can be used as "pure" LCLs if needed.
- 2 Current Limiters per module are configurable as Re-armable protection devices.
- 4 Current Limiters per module are configurable as high current outlets (up to 10A).
- 4 Current Limiters per module are configurable as "Safe Off" device, i.e. with a second switch in series.

DOI: 10.1051/e3sconf/20171618009

Current Limiter Control

The implementation of the control elements of the function is also responsible for the high degree of integration achieved. In particular:

- Current Limiters undervoltage protection is centralised. Three levels, for essential, non-essential and permanent loads, are selectable for every Current Limiter. Every protection threshold detector is triple redundant and majority voted to be one failure tolerant.
- LCL control electronics is supplied from the OR-ing of one secondary voltage from every APS, N & R, plus a linear regulated power source located in MEA module. In this way the status and control of every LCL is kept whatever the status of the TM/TC I/Fs and their potential switching in time from nominal to redundant or vice versa.
- Current Limiters control loop and current telemetry is based on dual bipolar transistors, easily replaceable in case of future eventual obsolescence.

Deployment module architecture

Regarding the deployment function, number of safety barriers and architecture are the classical ones with Firing + Arming + Selection stages. A total of 24 deployment channels are available.

Every deployment channel is able to handle either EED or LSRU devices, thanks to a firing device based on a step-down buck converter able to work in two operational modes:

- Current source, with a current level intended for classical pyrotechnic devices, and
- Voltage source (current limited) for non-explosive actuators.

FMEA and dedicated thermal analyses in failure case have been run to demonstrate that no electrical and/or thermal failure propagation will take place from the distribution function to the deployment function. See figure 9.

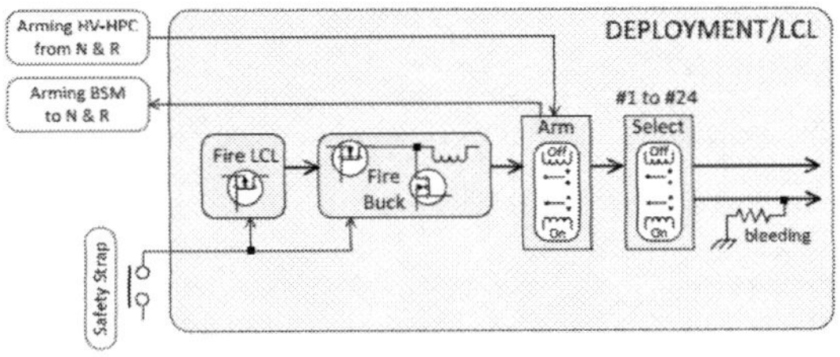

Figure 9. Deployment architecture

Arming function is nominally controlled by SHP commands, while the rest of the function is governed by 1553 I/F. Safety strap is also provided to inhibit firing function on ground. Figure 10 shows the EM of Deployment module for Merlyn flight mission.

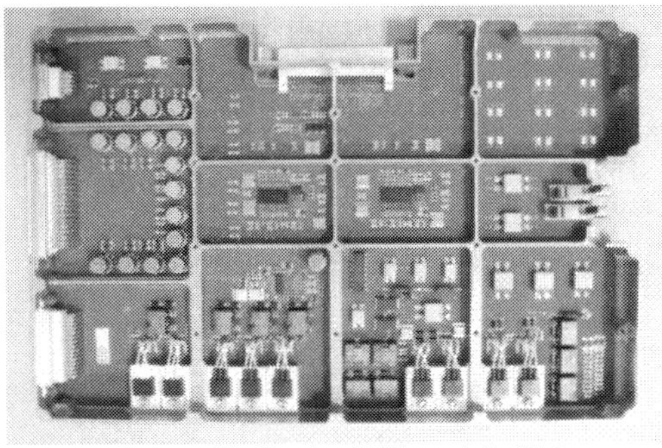

Figure 10. Deploy/LCL module EM

4 ACHIEVEMENTS

It was said at the beginning of the paper that main targets were to save 20% mass and volume while keeping or improving existing performances, and adding at the same time new functionalities in line with new requirements.

Previous chapters highlighted the most relevant electrical and functional aspects of the new design. Now attention will be focused on PCDU physical properties, mass & dimensions.

4.1 Dimensions

Unit dimensions are: 292mm x 302mm x 210mm w.o. mounting feet & connectors protrusions, lenght x depth x height. Depth is 350mm with protrusions.

This represents a 23% saving in volume for the same modularity and some added functionalities w.r.t. previous AS250 PCDU concept. It is remarkable the reduction in height of 12,5% for improved compatibility with the existing platforms from different primes.

4.2 Mass

PCDU mass is <16,5kg for a configuration as to satisfy the needs of the previous typical AS250 mission. This represents a 15% saving in mass for the same modularity and some added functionalities.

5 CONCLUSIONS

Staring from an existing LEO PCDU product, with more than 12 FMs manufactured, strategies have been presented for product optimization and mass & volume reduction, while adding new features in line with emerging customer requirements.

The activity carried out is intended for unregulated bus LEO missions, but is directly applicable to science missions and to the PCDU distribution part of the 28V regulated buses classical of interplanetary missions, as is the case of the PCDU for JUICE. Merlyn spacecraft will be the first LEO flight application case of this new concept, being FM expected by beginning 2017.

Moreover, the use of this concept for JUICE will bring to the product new building blocks developed in the frame of this interplanetary mission, developed to be compatible in size and performances with the LEO PCDU. This will complement the product portfolio of available functionalities for 28V buses (either belonging to the category of regulated or battery follower).

Main achievements are:

- ✓ Multi-mission configurable PCDU intended to cover the user power range from ≤500W up to 3kW.
- ✓ Distributions module flexibility to be used in the distribution part of 28V regulated bus PCDUs for interplanetary missions.
- ✓ Optimized height and width for full compatibility with most platforms currently on the market.
- ✓ Relevant mass reduction w.r.t. previous LEO PCDU product.
- ✓ Built-in EOL passivation in line with new 'green space' regulations.
- ✓ Built-in Launch-off capability for multiple launch.
- ✓ Compatible to drive the deployment devices existing in the market with the same design.
- ✓ Reliable S3R cell design:
  - ✓ No power loss after a single failure, and
  - ✓ Power in the bus from solar array interface can be regulated down to 0 Watts after a single failure.
- ✓ Solid state based battery switch.
- ✓ Very high LCL configuration capabilities:
  - ✓ Wide range of classes available from 0.5A to 10A.
  - ✓ Three different types of LCL available.
    - ✓ PO-LCL: Limiter that is permanently On, also called R-LCL (R standing for "Re-armable")
    - ✓ SO-LCL: Safe Off Limiter, with a second switch in series with independent control, and
    - ✓ Normal LCL.

6 REFERENCES

[1] J.Otero et al. (2011). Modular PCDU Product Line For LEO Applications. In Proc. of the 9th European Space Power Conference. Saint-Raphaël. France.

[2] F. Bausier. (2013). Passivation of spacecraft power system at end of mission. TEC-EPM/2013/33/FB, Technical Note. European Space Agency, Noordwijk, The Netherlands

[3] Ch. Delepaut. (2013). Battery Isolation Function - Interface Standardisation Rational. TEC-EPM/2013/15CD, Technical Note. European Space Agency, Noordwijk, The Netherlands

[4] E. Lapeña, (2016) Passivation strategies on board Airbus DS LEO PCDUs. In Proc. of the 11th European Space Power Conference. Thessaloniki, Greece

DIGITAL CONTROL FOR POWER MANAGEMENT SYSTEMS

Bekemans Marc [(1)], Acconci Terence [(1)], Van Humbeeck Thierry [(1)], Van Esbeen Alain [(1)]

[(1)] *Thales Alenia Space Belgium, Rue Chapelle Beaussart 101, B-6032 Mont-sur-Marchienne, Email: firstname.name@thalesaleniaspace.com*

ABSTRACT

The present document aims to show the advantages of a digital implementation of the regulation(s) used for the control of DC/DC converters. In a first step, a general overview of the capabilities provided by digital solution, w.r.t. analogue, will be introduced. An example of a pure digital algorithm (the predictive programmed current control) will also be presented. In a second step, a practical application of a digital control will be considered. In this example, the output voltage of two Buck DC/DCs, connected in parallel, is controlled via a digital regulation. This application also integrates different possible features (soft start, current sharing, embedded scope...), easy to implement in digital, and whose analogue equivalent would be hard or impossible to obtain. Finally, a quick overview of the radiation-hardened ASIC developed by Thales Alenia Space will be given. This component, called "Digital Programmable Controller" (DPC) is totally suitable to implement digital regulations for space applications.

1. DIGITAL REGULATION CONCEPT

1.1. Introduction

The introduction of digital regulation in power converters is a good opportunity to imagine new concepts of regulation. We will first expose several way to "digitalize" classical control principle like : Peak Control, PWM Control, and Hysteresis Control. And finally present a pure predictive controller which is only realizable by digital processing and not be limited by the classical stability criterion liked to the phase lack due to calculation time.

1.2. Digital "equivalent" to analog regulations

Current average mode (CAM)

Current average mode (CAM) regulation algorithm is one of the most common inner-and-outer-loops regulation systems. The principle here is to use an inner current loop which will regulate the average current in the inductance of the converter. The reference average current will be provided by the outer voltage loop, which will be in charge of regulating the output voltage. As illustrated in Figure 1, the output measured voltage is compared to a reference voltage, and the error goes into a first regulator. This regulator provides an average current reference which will be compared to a measured average current, giving a current error which will be processed into a second regulator to give a duty cycle to apply to the Buck system.

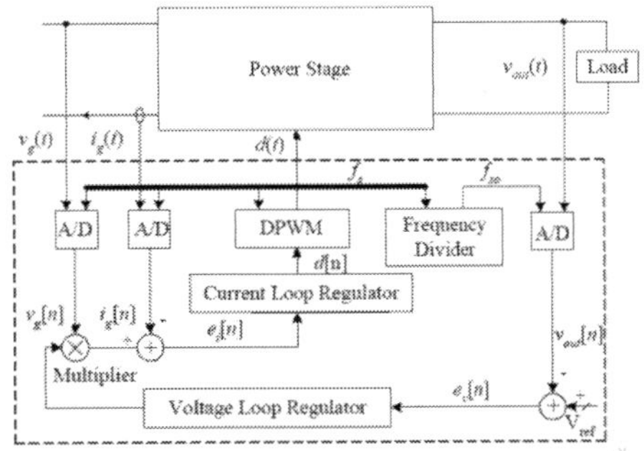

Figure 1 : Current average mode regulation

Current peak mode (CPM)

Current peak mode principle is almost the same than for the current average mode, except that here, only one regulator will be used. The output measured voltage is compared to a reference voltage, and this error goes into the regulator. This one provides the maximum current that the inductance current of the Buck has not to exceed. This maximum current reference is compared to the measured current through an analog comparator, and the result drives the "reset" input of a RS flip-flop, which cuts the transistor of the switching power system, making the current into the inductance decrease.

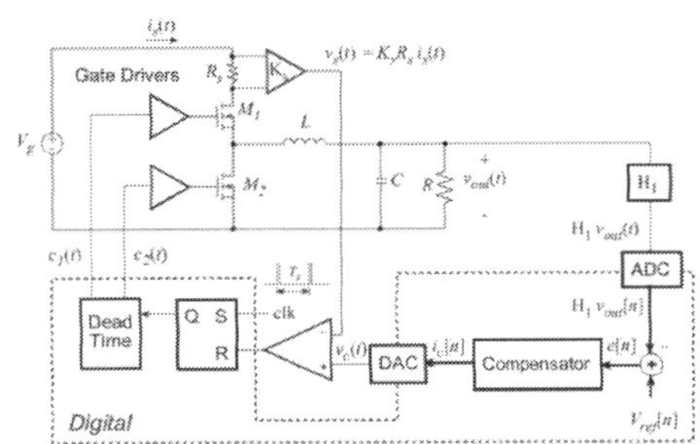

Figure 2 : Digital implementation of current peak mode

It is important to note that the current amplifier, comparator & flip-flop are all components of the analog chip and that the analog was foreseen such as to be the "analog components" of a current peak mode regulator.

 DOI: 10.1051/e3sconf/20171618010

1.3. Novelty : Digital Predictive Controller

These two methods (digital current average mode & digital current peak mode) have been successfully tested. However further thoughts lead us to conclude that this approach was a bad copy of analog implementation.
Everyone agrees that the digital technology brings more flexibility, more versatility, more stability of the parameters…but everyone also contends that the price to pay is a reduction of dynamic performance due to calculation time and sampling delay. It is, indeed, generally the case but in switching converters a predictive controller could take into account the sampling time due to switching and reach dynamic performance very close to that of peak control even if we sample the output signal only once per switching period. The leitmotif in digital design is: don't transpose to digital what is being done in classic analogue circuits. The search for improvements brings us to distinguish two different axes on which we can work:

- **Improvement of dynamic response** of algorithm in order to be as good as analogue controller in small-signal dynamic but better in large signal or in non-linear operation.
- **Improvement of observability**, for two reasons: the first is the improvement of the "knowledge" necessary to achieve a more effective control, the second is that the digital concept could help us to reduce the number or the performance of sensors by numerical reconstruction of state variables.

The dynamic response is linked to the bandwidth capability of the system under the constraint of conserving the stability margin. A fully digital predictive peak controller could avoid the problem of stability but transpose peak control to digital is difficult because the "off" decision is made on a cross value criterion. That means we need an a comparator to define the ON-OFF transition. Another way is to implement the recurrent equation of "peak control" in a numerical form and let it work on a sampled value. We cannot speak any more in that case of peak current control; we shall rather call it Programmed Current Control (PCC) but the dynamic performance is kept. These alternatives are summarized below:

Digital PID Controller

- carries over the drawbacks of average current control
- has an additional delay due to sampling and the computation time. This reduces the dynamic of the control
- seems difficult to perform as well as the analog solution
- exploits the advantages of digital for adaptive parameterization of the corrector

Predictive Programmed Current Control:

- carries over the advantages of Peak Current Control
- Doesn't use an external analog comparator
- Reserves one switching period for computation
- Only uses one current sample per switching period.

In analog it's possible to control only the valley or the peak because the "measurement" instants (crossing of the "off" threshold) are simultaneous to the "action" instants. With a digital system, any point of the current sequence can be sampled. This can be done by synchronizing the sampling time with the PWM sequence. The consequence is that we could regulate any intermediate value between peak and valley.

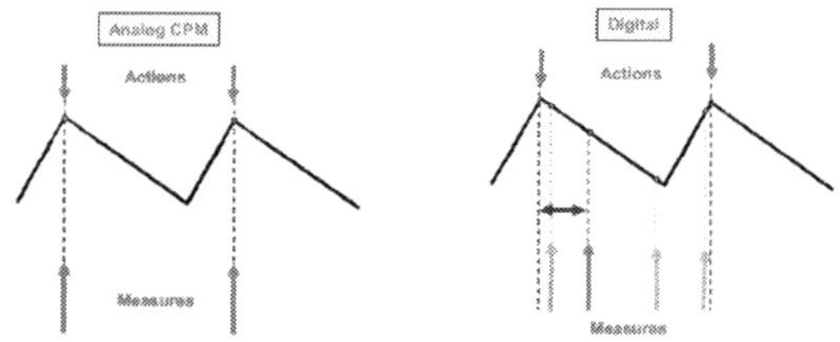

Figure 3: Digital controller could regulate any point between peak and valley

With the assumption of reacting within one switching period to any perturbation, the mathematical analysis of the recurrent system yields the following recurrent equations which links the duty cycle and the current. The variables m_1 and m_2 are respectively the rising and falling slope of the current; me_1 and me_2 are the estimated rising and falling slope set into the corrector. Let us notice that this form is usable for any topology: only the definition of m_1 and m_2 changes. Example for à Buck:

$$m_1 = \frac{V_{in} - V_{out}}{L} \qquad m_2 = \frac{-V_{out}}{L}$$

The Recurrent Equation of the Predictive Programmed Current Control are:

$$\mathbf{X}_{n+1} = \mathbf{A} \cdot \mathbf{X}_n + \mathbf{B} \cdot \mathbf{U}_n \tag{1}$$

With:

$$\mathbf{X}_n = \begin{bmatrix} I_n \\ \theta_n \end{bmatrix} \qquad \mathbf{U}_n = \begin{bmatrix} 1 \\ I_{ref} \end{bmatrix} \tag{2}$$

$$\mathbf{A} = \begin{bmatrix} 1 & (m_1 + m_2) \cdot Ts \\ \dfrac{-1}{(me_1 + me_2) \cdot Ts} & -1 \end{bmatrix} \tag{3}$$

 DOI: 10.1051/e3sconf/20171618010

$$\mathbf{B} = \begin{bmatrix} -m_2 \cdot Ts & 0 \\ \dfrac{2 \cdot me_2}{(me_1 + me_2)} & \dfrac{-1}{(me_1 + me_2) \cdot Ts} \end{bmatrix} \quad (4)$$

This equation materializes into the following block diagram where the structure is independent of the topology: only the internal gain has to be adjusted to the power topology used. It's interesting to see that we recover all feed forward effect naturally present in Analog Peak Current Control.

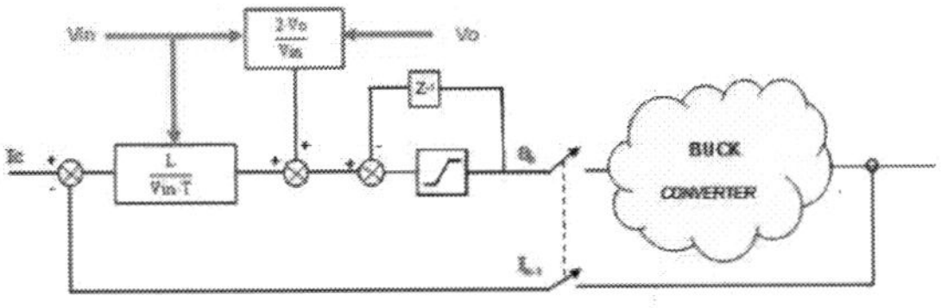

Figure 4: Control block diagram of predictive Programmed Current Control for a Buck converter

This approach is completely different from that of classic control. To convince ourselves, let us look at the loop gain: it is equal to one for all frequencies. This is not a feedback system! Indeed, in feedback systems we replace the lack of knowledge by the gain. Here, the loop gain is 1 for all frequencies! This is a pure predictive controller without analog equivalent.

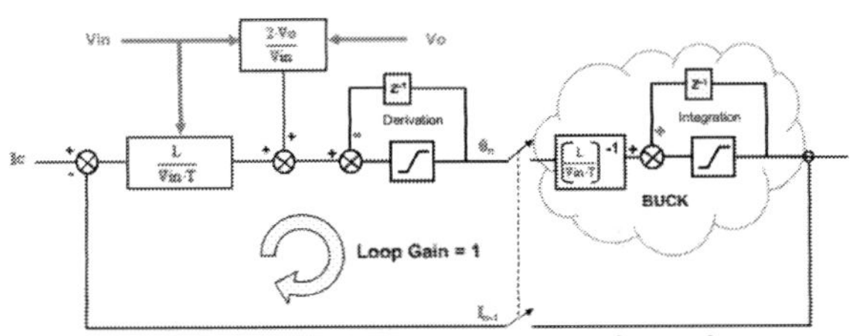

Figure 5: Control block diagram of predictive Programmed Current Control

2. APPLICATION OF DIGITAL PROCESSING : DIGITAL CONTROL OF POLYPHASE PARALLEL BUCKS

Plenty of applications for DC/DC regulation can be imagined using a digital regulation. Indeed, once the different measurement are digitalized using ADCs, they can be combined in many different ways, in order to achieve the expected regulation functionality and performance. This makes a strong difference w.r.t. an analogue regulation, and this is what the application developed in the current paragraph aims to show.

The hardware of this application is the following : two Bucks DC/DC connected in parallel to a single external load, and whose output voltage is regulated thanks to a digital regulation. In practical, the DPC component (see paragraph 3) was used and placed on its DPC Reference Kit.

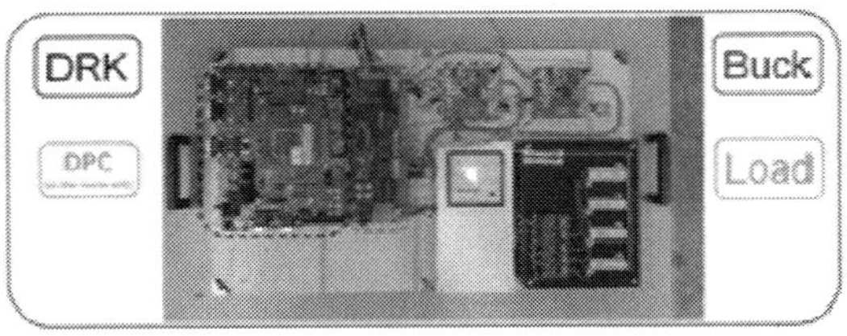

Figure 6 : Digital control of polyphase parallel Bucks : hardware view.

The next paragraphs aim to show the different functionalities that were implemented here.

2.1 Embedded scope feature

Plenty of different measurement were implemented, like input and output currents/voltages of both DC/DCs. In order to be able to easily follow the evolution of all these measurements, an embedded scope feature has been implemented. All the measurements are stored and can be displayed, on demand, on several graphs. Here, not less than 6 measurements were retrieved in the same time. This kind of embedded scope is a strong advantage of the digital. With an analogue regulation, several oscilloscopes, with synchronisation between them, would have to be used to achieve the same result. In addition, it is important to note that not only measurement can be retrieved via this feature, but also any digital variable used in the algorithm (for instance the calculated duty cycle, or any required variable useful for debug purpose). *Figure 7* to *9* illustrate the results obtained via the embedded scope.

2.2 Soft start

During the start-up phase, the output capacitance is charged, leading to a potential high in-rush current at the input of the DC/DC converter. This can be handled by using a soft-start control algorithm, smoothing the initial duty cycle evolution in order to control the in-rush input current. This is illustrated in *Figure 7*, where we can see that the input current ("Iin") and the output current of the Bucks ("IoutA", "IoutB") are controlled. The soft start parameters (duration, slope of the current) can easily be changed on-line, by only modifying some digital parameters in the regulation code.

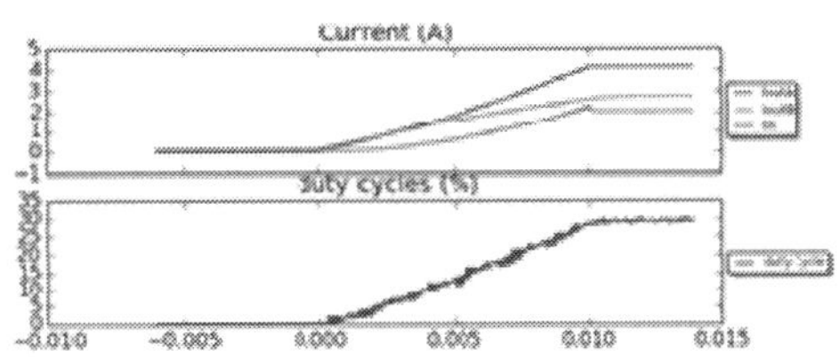

Figure 7 : Soft start behaviour.

2.3 Output current sharing

The output current of both Bucks are measured. This information can be used in order to tune the regulation to achieve a perfect current sharing at the output of both Bucks. This is illustrated in *Figure 8*, showing what happens when the current sharing algorithm is activated, starting from an initial condition where both Bucks are strongly unbalanced.

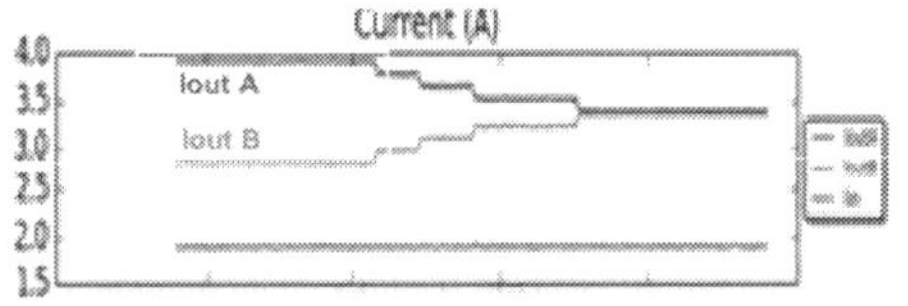

Figure 8 : Activation of the output current sharing algorithm.

2.4 Activation/deactivation of the second Buck

Using two Bucks in parallel has several advantages : on one hand, the maximum power provided to the output load can be higher, on the other hand the output ripple can be smaller, if both Bucks are used in a polyphase way (i.e. with the PWM signal shifted by one half of the switching period from one Buck to the other). In addition, the load transient performance will be better (because each Buck only "sees" one half of the current transient). However, using two Buck in parallel can lead to dramatic loss of DC/DC efficiency if the output current is too small. A simple way to avoid this phenomenon is too deactivate the second Buck if the output current becomes too small, and to reactivate it when the output current is increasing. This kind of algorithm is quite simple to implement in digital. It can also easily be re-configured (e.g. change of output current threshold, implementation of hysteresis), or adapted. For instance, we could imagine another criterion based on the DC/DC efficiency, computed from the voltage and current measurements performed at the input and the output of the DC/DCs. Figure 9 illustrates what happens when one Buck is deactivated.

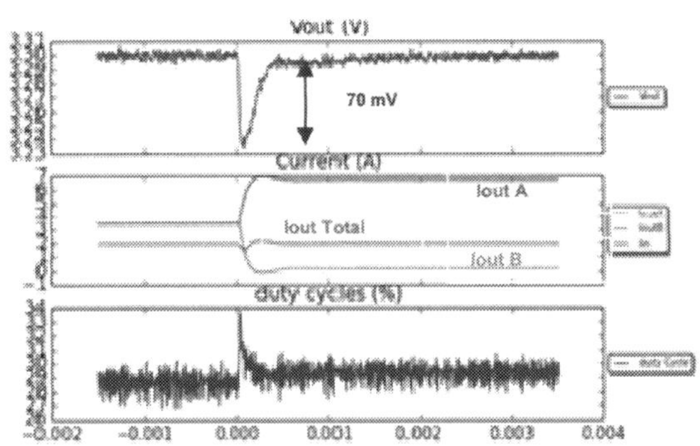

Figure 9 : Deactivation of one Buck

2.5 Regulation algorithm

The output voltage regulator implemented here is a PID. Since it is a digital regulation, the parameters of the regulator can easily be changed, in order to tune the regulation bandwidth, the recovery time,... Additional features can also be added, like saturations, slew rates, or measurement pre-processing (e.g. median on successive samples). All these elements are easily implementable in digital. In addition, more advanced algorithm can be considered, like the predictive algorithm presented in paragraph 1.3 (and in [1]), involving three different measurements, and including the processing time in its definition. Other algorithms, like sliding mode control ([2]), capacitor charge balance ([3]) or additional feed-forward ([4]) could also be considered. In the end, the easy tuning of the algorithm parameters as well as the possibility of using more evolved algorithms are strong assets of the "digital world".

3. DIGITAL PROCESSING IMPLEMENTATION

3.1 Implementation means needs

Digital power control tasks require signal digital acquisition (ADCs), signal digital processing, and PWM generation functions. Some auxiliary functions are often required, like protection, soft start, telemetries & telecommands (TM/TC) handling via a data communication interface, signal observation capabilities, ...

From performance point of view, the following parameters are critical: the ADC conversion and digital processing latencies, and the ADC & voltage reference accuracies. The first will determine the regulation bandwidth (a few µs maximum), the second will determine the residual error (typically a few tenth of percent maximum).

3.2 Implementation targets

Given the needs introduced in previous section 0, digital functions can be targeted in an FPGA, with peripheral ADCs, reference voltage source, and oscillator. For space applications, those components shall be radiation hardened, leading to high components bill of material (BOM).
To overcome this cost problem, Thales Alenia Space takes benefit of the Digital Programmable Controller ("DPC") ASIC, which embeds all required functions in the same radiation hardened component, at a much cheaper BOM.

3.3 The Digital Programmable Controller ("DPC")

The DPC ASIC (see *Figure 10*) is a radiation hardened mixed-mode and multi-core micro controller, developed in the frame of an ESA ARTES program. It targets DC/DC converter control applications, together with

DOI: 10.1051/e3sconf/20171618010

other applications like motor control, general purpose analog & digital TM/TC interface on CAN or MIL-1553B buses, intelligent sensor management, monitoring and protection.
The DPC architecture is shown on *Figure 11*. For power management functions, a regulation arithmetic sequencer running at 40 MHz handles digital power control tasks. It includes 4 low latency (1µs) 13 bits accurate ADCs, multiplexed to 16 inputs. Those inputs are processed by a dedicated 16 bits micro controller core with multiplication & division functions (in red on *Figure 11*). The control feedback action is applied via up to 6 complementary PWM outputs clocked at 120MHz, and up to 3 linear analog outputs (DACs).

The ASIC embeds its internal power supply LDO, an accurate voltage reference, and a frequency reference. Besides the regulation core, three other micro-controller cores are assigned to respectively TM/TC communication tasks (in green on *Figure 11*), supervision tasks, and activity and signal observation tasks.

The circuit is radiation hardened: >60krad(Si), SEL LET_{th} >68 MeV.cm²/mg, SEU LET_{th} >40 MeV.cm²/mg. The performances are guaranteed in the -55°C to +125°C temperature range.

Figure 10: the DPC ASIC

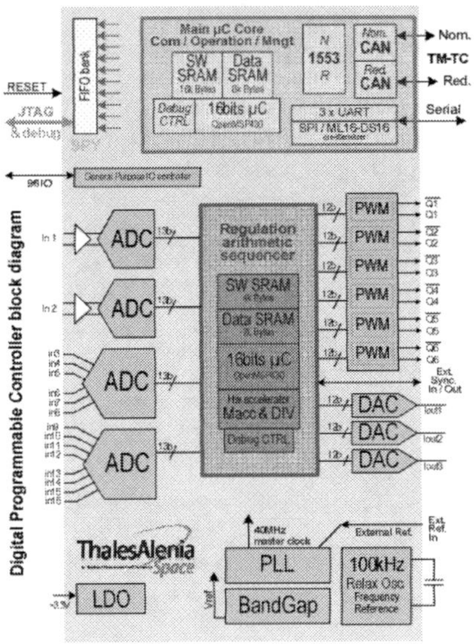

Figure 11: DPC block diagram

The circuit is fully validated since 2015, and is already used in Thales Alenia Space observation and telecom equipments.

3.4 Rapid prototyping and FM implementation

Besides the DPC component, Thales Alenia Space has developed software and hardware toolkits that allow fast prototyping any power controller.

For instance, the DPC Reference Kit (DRK) is a hardware board that implements the DPC and all its signal interfaces. Once connected to a development laptop via USB, and to power components, it allows application engineers focusing on the design of the controller algorithm.

The DPC Plugin Module ("DPM") is an FM grade mezzanine board that contains the DPC chip and some auxiliary circuits like CAN transceivers, APS and LCL. Once plugged on the application power hardware (FM), it is ready to be programmed with the power controller (or any other application). This approach drastically reduces implementation and integration costs.

4. CONCLUSION

Today, all the required elements are available to handle the digital control for power converters. Indeed, suitable digital algorithms as well as a dedicated component compatible with space environment exist. These elements have been included in a DC/DC demonstrator and are now ready to be part of future space power products.

5. REFERENCES

1. J. Chen, A. Prodic, R.W. Erickson, D. Maksimovic (2003). Predictive Digital Current Programmed Control. IEEE transactions on Power Electronics, vol. 18, no. 1, 411-419

2. E. Vidal-Idiarte, C.E. Carrejo, J. Calvente, L. Martinez-Salamero (2011). Two-Loop Digital Sliding Mode Control of DC-DC Power Converters Based on Predictive Interpolation. IEEE transactions on Power Electronics, vol. 58, no. 6, 2491-2501

3. G. Feng, W. Eberle, Y-F. Liu (2007). A New Digital Control Algorithm to Achieve Optimal Dynamic Performance in DC-to-DC Converters. IEEE transactions on Power Electronics, vol. 22, no. 4, 1489-1498

4. S. Chae, B. Hyun,W. Kim, and B. Cho (2008). Digital load current feed-forward control method for a dc-dc converter. APEC 2008 - Twenty-Third Annual IEEE Applied Power Electronics Conference and Exposition, 498-502.

E3S Web of Conferences **16**, 18011 (2017)
ESPC 2016
DOI: 10.1051/e3sconf/20171618011

POWER PROCESSING UNIT ACTIVITIES IN AIRBUS DS SPACE EQUIPMENTS

Author: Fernando PINTÓ [(1)]

Co-authors: Javier PALENCIA [(2)], **Nicoletta WAGNER**[(3)], **Guillaume GLORIEUX** [(4)]

[(1)] *Airbus Defence and Space (Airbus DS Crisa), Torres Quevedo 9, 28760 Tres Cantos, Spain, Email: Fernando.Pinto@airbus.com*

[(2)] *Airbus Defence and Space (Airbus DS Crisa), Torres Quevedo 9, 28760 Tres Cantos, Spain, Email: Javier.Palencia@airbus.com*

[(3)] *Airbus Defence and Space (Airbus DS GmbH), Claude-Dornier-Str.88090 Immenstaad, Germany, Email: Nicoletta.Wagner@airbus.com*

[(4)] *Airbus Defence and Space (Airbus DS SAS),Rocade Jean Moulin ,Elancourt France, Email: Guillaume.Glorieux@airbus.com*

ABSTRACT

This paper presents the Power Processing Units in Airbus DS Space Equipment.

Electric propulsion has (EP) become a reality in the commercial market, and therefore is not anymore a niche market for earth observation and/or deep space missions. Major players worldwide for Commercial satellite manufacturers are targeting in their current programs, electric satellites solutions for orbit raising (OR) and station keeping (SK) maneuvers. Several EP technologies are competing against each other, in particular Hall Effect Thruster (HET), Gridded Ion Engine (GIE) technology (including Kaufmann and Radio Frequency technologies) and High Efficiency Multistage Plasma (HEMP). The main two parameters used to trade-off the main EP technologies are Thrust/Power ratio and Specific Impulse (ISP). Currently the market is mainly thrust driven, and therefore the dominant technology is HET. HET is the most efficient technology for the Thrust/Power ratio parameter.

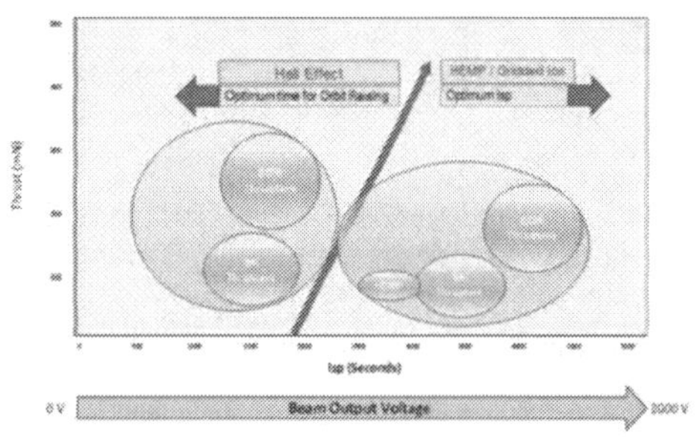

Figure 1. Electric Propulsion Technologies landscape

Airbus DS Space Equipment include in our portfolio PPU solutions for high Thrust ratio and high ISP solutions.

1 MOTIVATION

EP business case is driven by current technologies, operator's mindset, and future trends. Several questions with difficult answer can be raised about the future trends for EP. Would the major operators accept OR transfer period above 5 months? How launcher direct injection would impact on the current technologies?

Currently the answers to these questions are not obvious. Airbus DS Space Equipment believes the future is not clear and will most likely not be based on just one single technology, to answer the mid/long term market needs.

In this sense Airbus DS Space Equipment supported by the National Agencies and the European Space Agencies is evolving their current PPU products for high voltage technologies (PPU HV) and developing a new HET PPU solution (PPU Elektro)-able to answer mid-long term market needs.

Airbus DS contains the following PPUs in their portfolio:

- ✓ PPU "Elektro" for HET technology
- ✓ PPU HV for high voltage applications products: GIE and HEMP technologies
- ✓ Generic building blocks to maximize communalities between the different technologies

Figure 2 illustrates the PPU mapping with respect the main EP technologies in the market:

DOI: 10.1051/e3sconf/20171618011

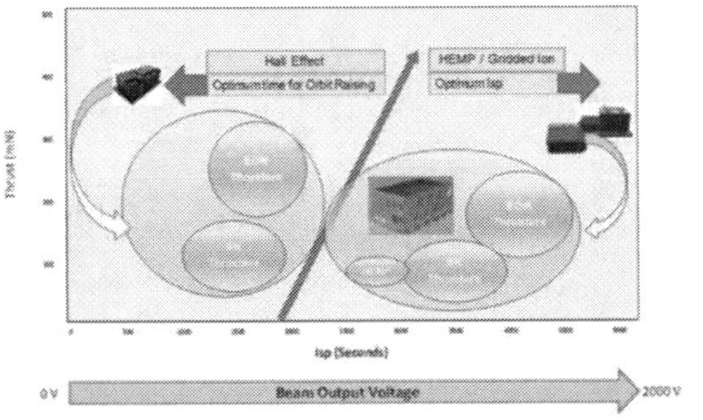

Figure 2 Airbus DS PPU activities to answer to the different Electric Propulsion Technologies

- ✓ PPU "Elektro" product is a joint development between Airbus DS Elancourt (France) and Airbus DS Crisa (Spain). It is a flexible and modular equipment with a new innovative concept of PPU, capable of adapting to a wide range of EP subsystem configurations and compatible to the main HET thrusters in the market. Recently, EM PPU coupling test has been successfully accomplished and the EQM is currently under manufacturing.

- ✓ PPUs for HV applications product. Currently there are already several PPU products available already qualified as follows:

  - o PPU for T5 is a joint development between Airbus DS Crisa (Spain) and Airbus DS Germany. PPU is a flight proven design in the GOCE mission
  - o PPU for T6: is a joint development between Airbus DS Crisa (Spain) and Airbus DS Germany. PPU is a qualified design. First mission will be ESA Bepi-Colombo mission. Next step currently under development, is to industrialise existing product in the frame of EGEP program
  - o PPU for HEMP developed by Airbus DS Germany. PPU is a qualified design and FM already delivered to the customer.
  - o Next step is to have available the maximum communality between the different technologies. For this-purpose Airbus DS Space Equipment is running a development of generic building blocks to easily customize the PPU to any EP thruster independently of its technology. The maximum communalization will significantly reduce the development efforts, but also the recurrent efforts by applying new technologies and innovative solutions.

- ✓ Generic building blocks are the basis for the PPU configuration for the different thrusters: PPU for T5 & T6, PPU for RIT10 and RIT22 and PPU for HEMP. The core element is the generic High Voltage Module, which can be described as a functional block that is designed to be customized to specific thruster needs by exchanging a sub-module without need for a delta-qualification. This modular approach provides the necessary flexibility and maximum of recurring elements. Airbus DS has managed to harmonize these products based on the generic High Voltage Power Supply (HVPS).

2 PPU ELEKTRO FOR HET

The PPU Elektro motivation is to respond to Satcom market towards full Electrical Propulsion solutions, capable not only to perform Station Keeping maneuvers, but also Orbit Raising and transfer. Airbus Defence and Space initiated PPU development in order to answer to these needs targeting, but not limited to, 5 kW class Hall-Effect Thrusters.
This joint development between Airbus DS Elancourt (France) and Airbus DS Crisa (Spain) offers a new innovative concept of PPU, capable to adapt to a wide range of EP subsystem configurations.

Figure 3 Airbus DS PPU Elektro for HET. Mock-up in 15 kW configurations

The PPU Elektro offers:

- ✓ An optimised cost solution thanks to an innovative architecture and Design For Manufacturing (DFM) approach based on extensive experience on Telecom Platforms
- ✓ A flexible and modular equipment → Same solution able to deliver from 1.5 kW up to 20 kW. PPU Elektro is thus able to drive up to a max. power of 20 kW (4 thrusters).
- ✓ The capacity to answer OR and SK needs, with in-flight flexibility to configure the unit into a wide range of operating points between 300 to 400V (optimize ISP vs thrust)
- ✓ A flexible design compatible to all main HET thrusters in the market.
- ✓ Compatible with a 100V power bus (standard for the Telecom Satellites), it could be easily adapted to other power bus voltages.

 DOI: 10.1051/e3sconf/20171618011

Elektro PPU is physically implemented in three main building blocks: Heater-Keeper-Ignitor module (HKI), anode module and Filter thruster switching unit (FTSU) as follows:

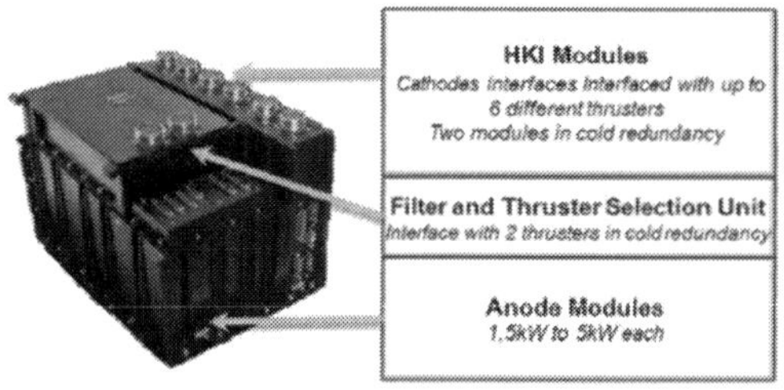

Figure 4 Airbus DS PPU "Elektro" Building Blocks (5 kW configuration)

The electrical architecture is presented in Figure 5 and 6 The main principle of this architecture is that all functions which are not used permanently in the subsystem (Cathodes Power Supply: Heater, Keeper and Ignitor and sequencing of the overall PPU) are gathered in a module called HKISeq. This module is able to address up to 6 cathodes. All other functions (Anode 5kW power supply, Magnet and Xenon Flow Control supplies) are gathered in the module called Anode module. The Filter/Thruster Switching Unit (FTSU) modules are used to filter and switch the Anode power supplies towards 2 thrusters.

Therefore, Elektro PPU architecture and design enables two different configurations of the PPU:

- ✓ Bi-thruster configuration: Able to interface with up to two thrusters
- ✓ Multi-thruster configuration: Up to six cathodes, offering redundancy at PPU level

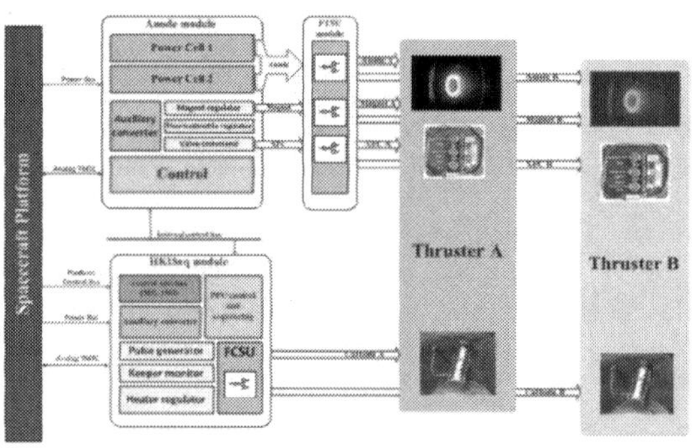

Figure 5 Airbus DS "Elektro" Bi-thruster configuration

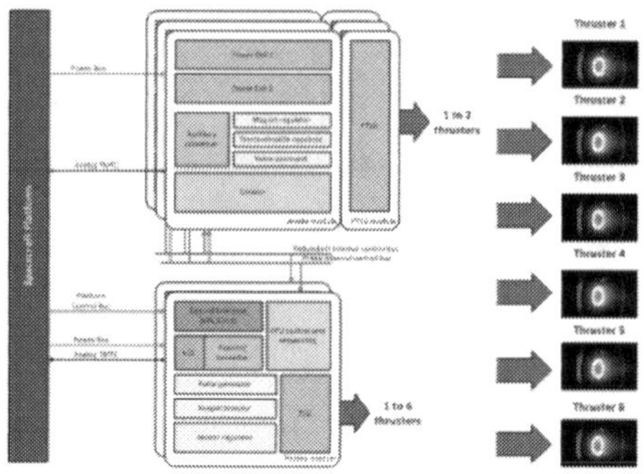

Figure 6 Airbus DS "Elektro" Multi-thruster configuration

All modules are physically independent and the communication between HKISeq and Anode modules is made by a CAN bus through a harness. Thus, the modules can be assembled altogether with standard fixing points (no specific tie-rods) to form a single unit, or they can be implemented separately.
Examples of the possible configurations are presented in the Figures 3 and 4. This modular approach aims at facilitating accommodation of the PPUs in the spacecraft.

The PPU Elektro development is on-going. The EM PPU already delivered and CDR and EQM manufacturing are currently on going and close to finalisation. The BB and EM model has demonstrated the functionality and performances. The efficiency was measured around 96% and a characterization of the power limitation has been performed

Figure 7 Airbus DS EM PPU Elektro under validation tests in 5 kW configurations

 DOI: 10.1051/e3sconf/20171618011

3 HIGH VOLTAGE TECHNOLOGIES PPUS

3.1 PPU for T5

The PPU for QinetiQ T5 thruster, the so called IPCU has been developed by Airbus DS Crisa (Spain) and Airbus DS Germany for the beam output supplies. The IPCU has successfully flown in the ESA GOCE mission.
This unit is meant to be used for low power missions, mainly for earth observation purposes. The main asset of this unit is the high level of controllability from 0.6 mN to 20 mN with a resolution of 12 μN, with a beam output voltage supply of 1200V. Refer to Figure 8.

Figure 8 Airbus DS PFM IPCU for T5

3.2 PPU for T6

The Power Supply and Conditioning Unit, PSCU, has been developed as a parallel contract to AlphaBus (ESA under ARTES Program). The PSCU has been developed by Airbus DS Crisa and Airbus DS Germany for 1850 Vdc Beam Supply Modules

This unit is the first 5 kW PPU ever developed in Europe, able to answer to the new full electric satellites needs for the Telecom market and the most demanding deep space missions.

The first flight mission for the PSCU will be ESA BepiColombo mission to Mercury.

The PSCU (see Figure 9) is composed of two Accelerator-Neutralizer Supplies (DANS), housed in single box together with the Beam Supply (BS).

The PSCU T6 design has been upgraded for BepiColombo in order to provide full redundancy and flexibility at PPU level:

- ✓ PSCU T6 enables the system to interface with up to four thrusters, operating one at a time with a total power of 5kW
- ✓ PSCU T6 is able to offer full redundancy for each of the power supplies

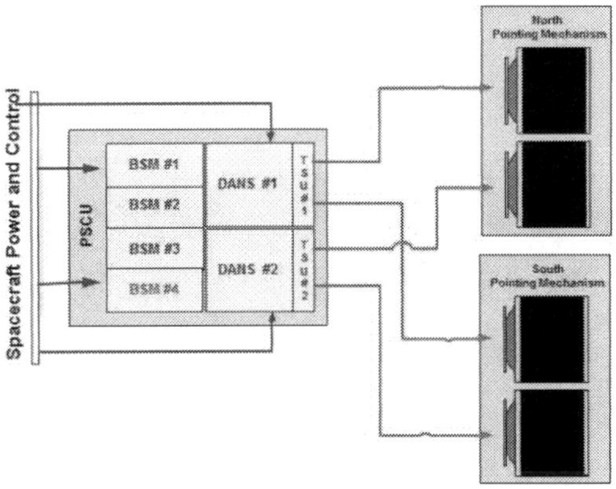

Figure 9 Airbus DS EQM PSCU Architecture

The BEPI PPU EQM has been successfully coupled with the thruster and fully qualified. The flight models (2 FMs) are manufactured and delivered to end customer to support the first flight mission.

Figure 10. Airbus DS EQM PSCU for T6

The next steps for the PSCU T6 product are the product industrialisation. In the frame of the European GNSS Evolution Program (EGEP), Airbus DS Crisa together with Airbus DS Germany is performing the industrialization of the BEPI PPU Unit. This activity will provide to the market a competitive EP system based on the T6 engine targeting Galileo Second Generation platform as the first customer.

The main goals of the industrialization of the BEPI PPU leading to the EGEP PPU design are therefore:

- ✓ Minimization of recurrent cost: It is the most important goal for a recurrent production as the required by Galileo mission.
- ✓ Keep the Technology Readiness Level (TRL)
- ✓ System flexibility: As part of the industrialization it has been also considered a goal to get a unit design allowing a greater flexibility at system level, so that with minor (or none) unit adaptations it would be possible to configure different system architectures.

DOI: 10.1051/e3sconf/20171618011

The EGEP PPU is built of two sub-units:

- ✓ The DANS built off four modules:
  - o Power Conditioning Module
  - o LV Control Module
  - o LV Power Module
  - o HV Electronics Module
- ✓ The Beam Supply (BS)

One of the major differences is that the unit will not include any type of internal redundancy

The DANS together with the BS and the harness connecting both sub-units built off the complete PPU.

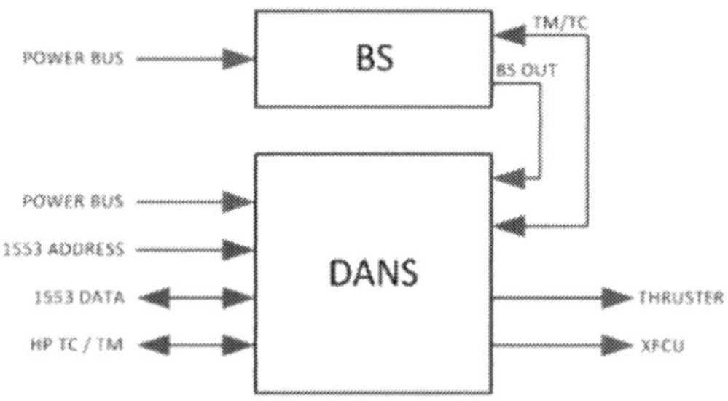

Figure 11 EGEP PPU

This configuration allows expanding the unit to connect two or three thrusters increasing the number of DANS but not BS modules as far as only one of the thrusters is powered at full power or two of them are powered simultaneously with at half power each.

Figure 12 shows the dual thruster configuration.

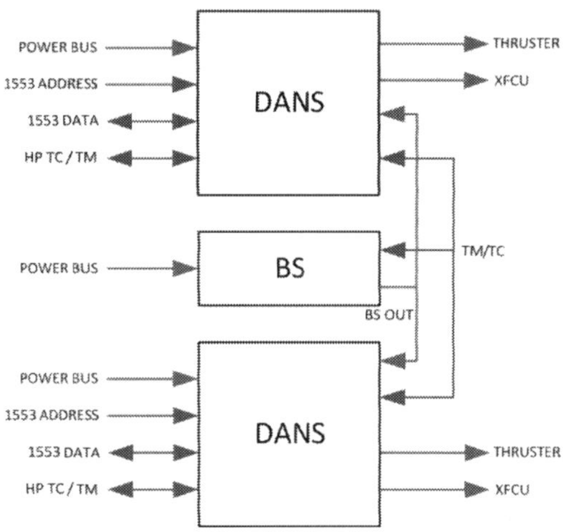

Figure 12 EGEP PPU Two thrusters configuration

The proposed PPU concept also helps to distribute the power dissipated by placing DANS and BS properly separated and allocated to different panels.

After industrialization of the unit the electrical performances expected are the same (or better) than those obtained for BEPI PPU. In particular the expected improvements/design goals expected with respect to former design are:

- ✓ Mass reduction: 29%
- ✓ Volume reduction: 40%
- ✓ EEE parts reduction: 30%
- ✓ Power Dissipation reduction: 22%

EGEP PPU development includes a design phase and EM model manufacturing and testing. The design phase is finished and EM delivery of the complete PPU for coupling with the thruster is expected in Q2 2017

3.3 HEMP PPU

Airbus DS in Germany has developed the PPU for HEMP thruster, the so called PSCU. The PSCU has been developed in the frame of an On-Orbit Demonstration opportunity (SmallGeo). The already qualified hardware is now foreseen for a first flight within the German Heinrich-Hertz (H2) mission.
The Airbus DS PSCU is designed as a modular product (see Figure 13), providing two high voltage power supplies (Anode Supplies) in parallel and operating as main supply for the thrusters, each delivering 1000V / 1400W. Redundancy concept approach includes redundancy at PPU level for most of the different power supplies and the capability to drive several thrusters with one single PPU.

Figure 13 Airbus DS PFM PSCU for HEMP

3.4 Generic Building Blocks for the PPUs

The new PPU generation for these technologies will be focused on two main axis:

- ✓ To develop new modular, flexible and cost effective solutions
- ✓ To maximize the communalization or building blocks between the different PPUs for the several technologies.

For this purpose, Airbus DS Germany is running in the frame of Artes 5.1 a development of generic building blocks to easily customize the PPU to any EP thruster independently of its technology. The maximum communalization will significantly reduce the development efforts, but also the recurrent efforts by applying new technologies and innovative solutions. This activity excludes the High Voltage Power Supply as this already is the main common building block baselined for the different High Voltage PPU designs (see next paragraph).

4 GENERIC HIGH VOLTAGE POWER SUPPLIES FOR HIGH VOLTAGE THRUSTERS

Generic High Voltage Power Supply has been developed in Airbus DS Germany and is currently commonly used in all high voltage PPU applications in particular GOCE, AlphaBus, BepiColombo and SmallGEO. The actual generation of modules is used for the Anode Supply Modules for HEMP3050 plasma thruster and the Beam Supply Module (BSM) of the QinetiQ T5 and T6 Ion thruster. Refer to figure 14

Figure 14 Airbus DS High Voltage Power Supply

The development of a new generation of the generic High Voltage Power Supply has been started in order to meet the needs of Orbit Raising applications. The main technology developments and improvements are:

- ✓ Increase of power from 1,4 kW up to 5 kW
- ✓ Modular approach (1,4/2,5/5 kW)
- ✓ Configurable output voltage to optimise the efficiency for orbit raising and station keeping
- ✓ Design for manufacturing / cost
- ✓ Planar transformer & AMB / direct bonded copper

Generic HVPS qualification is expected to be concluded in 2017 and widely used for the new High Voltage generation PPUs, in particular EGEP PPU as described in the previous chapter

Figure 15 Airbus DS New Generation Generic High Voltage Power Supply

5 CONCLUSIONS AND OUTLOOK

This paper has summarized the PPU activities in Airbus DS Space Equipment. The complete portfolio contains solutions for all the Electric Propulsion HET and High Voltage technologies. Airbus DS Space Equipment is continuously improving the designs in order to be able to answer not only to the short term needs, but also the mid and long term needs.

6 ACKNOWLEDGEMENT

The authors of this paper wish to thank ESA ESTEC and delegations for their support, in all the relevant activities, summarised as follows:

- ✓ The HKI module as part of the PPU Elektro for HET, described in this paper was developed in the frame of Artes 3/4 program under ESA contract E/0534-01B
- ✓ The PPU industrialisation for T6 has been run under the contract EGEP ID 76.02.02 – Industrialization of Power Processing Unit for T6 Electric Propulsion Subsystem
- ✓ DLR Zuwendung "Generic High Voltage Power Supply", Ref. No. FKZ.: 50 RN 1302

7 REFERENCES

[1] IEPC-2015-225/ISTS-2015-b-225 Power Processing Units - Activities in Europe 2015

[2] AIAA 2016-5032 (Propulsion and Energy Forum). Electric propulsion Electronics Activities in Europe 2016

[3] SP2016_ 3125086 (Space propulsion conference 2016). Optimized 5 kW PPU for Gridded Ion Engine

[4] SP2016 (Space propulsion conference 2016) Power Processing Units activities in Airbus DS Space Equipment.

AUTHOR INDEX

AUTHOR INDEX

AUTHOR INDEX

AUTHOR INDEX

AUTHOR INDEX

AUTHOR INDEX

EDP Sciences – Web of Conferences
17, Avenue du Hoggar
Parc d'Activité de Courtabouf
BP 112
F-91944 Les Ulis Cedex A
France

ISBN 978-1-5108-4119-2

9 781510 841192

CURRAN ASSOCIATES INC.
proceedings.com